Brief contents

Contents

LEWIN'S
CELLS

Jones and Bartlett Titles in Biological Science

LEWIN'S
CELLS

SECOND EDITION

LEAD EDITORS

Lynne Cassimeris
Lehigh University

Vishwanath R. Lingappa
University of California, San Francisco
Prosetta Bioconformatics, Inc.

George Plopper
Renssealaer Polytechnic Institute

JONES AND BARTLETT PUBLISHERS
Sudbury, Massachusetts
BOSTON   TORONTO   LONDON   SINGAPORE

World Headquarters

Jones and Bartlett Publishers
40 Tall Pine Drive
Sudbury, MA 01776
978-443-5000
info@jbpub.com
www.jbpub.com

Jones and Bartlett Publishers
Canada
6339 Ormindale Way
Mississauga, Ontario L5V 1J2
Canada

Jones and Bartlett Publishers
International
Barb House, Barb Mews
London W6 7PA
United Kingdom

Jones and Bartlett's books and products are available through most bookstores and online booksellers. To contact Jones and Bartlett Publishers directly, call 800-832-0034, fax 978-443-8000, or visit our website, www.jbpub.com. Substantial discounts on bulk quantities of Jones and Bartlett's publications are available to corporations, professional associations, and other qualified organizations. For details and specific discount information, contact the special sales department at Jones and Bartlett via the above contact information or send an email to specialsales@jbpub.com.

Production Credits

Chief Executive Officer: Clayton Jones
Chief Operating Officer: Don W. Jones, Jr.
President, Higher Education and Professional Publishing: Robert W. Holland, Jr.
V.P., Sales: William J. Kane
V.P., Design and Production: Anne Spencer
V.P., Manufacturing and Inventory Control: Therese Connell
Publisher, Higher Education: Cathleen Sether
Acquisitions Editor: Molly Steinbach
Senior Editorial Assistant: Jessica S. Acox
Editorial Assistant: Caroline Perry
Production Manager: Louis C. Bruno, Jr.
Senior Marketing Manager: Andrea DeFronzo
Text Design: Anne Spencer
Cover Design: Kristin E. Parker
Illustrations: Imagineering Media Services, Inc., Newgen Imaging Systems
Photo Research and Permissions Manager: Kimberly Potvin
Composition and Production Services: Newgen Imaging Systems
Cover Image: © Paul D. Andrews, College of Life Sciences, University of Dundee
Printing and Binding: Courier Kendallville
Cover Printing: Courier Kendallville

Library of Congress Cataloging-in-Publication Data

Cassimeris, Lynne.
  Lewin's cells. — 2nd ed. / Lynne Cassimeris, Vishwanath R. Lingappa,
George Plopper.
    p. cm.
  ISBN-13: 978-0-7637-6664-1
  ISBN-10: 0-7637-6664-X
  Rev. ed. of: Cells / lead editors, Benjamin Lewin ... [et al.]. c2007.
  1. Cells. I. Lewin, Benjamin. II. Lingappa, Vishwanath R. III. Plopper,
George. IV. Title. V. Title: Cells.
  QH581.2.C445 2011
  571.6—dc22
                        2010000314
6048

Printed in the United States of America
14 13 12 11 10   10 9 8 7 6 5 4 3 2 1

About the cover: *Xenopus* (frog) XLK2 cells. For each cell, a portion of the cell's internal skeleton is shown in green and the nucleus in red.

3 DNA replication, repair, and recombination 63
Jocelyn E. Krebs

4 Gene expression and regulation 105
David G. Bear

8 Protein trafficking between membranes 345

Vivek Malhotra, Graham Warren, and Ira Mellman

Part 3 The nucleus 391

9 Nuclear structure and transport 393

Charles N. Cole and Pamela A. Silver

10 Chromatin and chromosomes 439

Benjamin Lewin and Jocelyn E. Krebs

Part 4 The cytoskeleton 501

11 Microtubules 503
Lynne Cassimeris

12 Actin 557
Enrique M. De La Cruz and E. Michael Ostap

16 Apoptosis 713

Douglas R. Green

17 Cancer—Principles and overview 739

Robert A. Weinberg

Part 6 Cell communication 767

18 Principles of cell signaling. . 769

Elliott M. Ross and Melanie H. Cobb

19 The extracellular matrix and cell adhesion 821

George Plopper

Feature Boxes

Preface

Eighty years ago, the cellular world opened up. The electron microscope granted us, for the first time, a detailed perspective of basic cellular structures, and the ultracentrifuge allowed us to biochemically isolate and characterize fractions of cytoplasmic and nuclear material. Geneticists could investigate the relationship between the ever-shifting chromosomal structure and the molecular mechanisms of genetic inheritance—an effort that culminated with the triumphant revelation of the structures of DNA and RNA and a translation of the genetic code.

But we have come a long way from there. We have perfected our understanding of genes themselves, adjusting our definition from "determinants of a genetic phenotype," to "protein-encoding segments of DNA," and now, more precisely, "units of genomic information required for the transcription of functional messenger RNA or noncoding RNA." And we are still learning about the proteins these mRNAs produce. The RSCB Protein Data Bank (PDB) was established in 1971 as an international repository for structural data, but it did not truly begin to grow until the early 1990s. Now, in 2010, it holds more than 60,000 structures and is expanding at the rate of about 7,000 structures per year. For now, X-ray crystallography and nuclear magnetic resonance are the only techniques available for the determination of macromolecular structures at high resolution. Important advances in other methods, however—including visualization of fluorescently tagged proteins in living cells and new types of electron microscopy—are describing cellular structures and processes in ever-increasing detail.

What this all means is that the scope of biological questions that can be asked has been fundamentally changed. The new field of structural genomics has enabled us to relate increased structural resolution to functional changes, providing powerful insights at deeper levels of understanding. With our growing ability to process huge data sets, complete characterizations of cellular structures such as the nuclear pore complex and the centrosome, which are constructed from hundreds of proteins, may soon be attainable.

Perhaps most exciting is the combination of structural and mechanistic information with developments in genetics, biochemistry, and physiology—the primary vision of the emerging field of systems biology. Most cell biologists today recognize that only a comprehensive approach to research, from the nuclear pore complex to the extracellular matrix, will begin to lift the veil from the cellular processes underlying cystic fibrosis, epilepsy, and cancer.

Any cell biology textbook must provide a current perspective of the structure, function, and regulation of biological systems, but in today's world it is imperative that we also present the subject in the context of biochemistry and molecular biology, genomics, histology and pathology, and physiology. Thoroughly revised and updated, *Lewin's CELLS, Second Edition*, turns a new and sharper lens on the fundamental units of life.

Audience

This second edition, expanded and updated from Benjamin Lewin's *CELLS*, is geared for advanced undergraduate and graduate students taking a first course in cell biology. A key objective in developing this book was to present the concepts and mechanisms underlying cell structure and function, gleaned from decades of research, in a format that provides students with the information necessary for a solid foundation in cell biology, without overwhelming them with too much detail. The major goal of the team of lead editors and 26 expert authors has been to incorporate the current research in the field, thoroughly cover each topic, and provide ample illustrations of cellular processes at the molecular level—but without being unwieldy.

New and Key Features

Lewin's CELLS, Second Edition, covers the structure, organization, growth, regulation, movement, and interactions of cells, with an emphasis on those in the eukaryotic domain. These topics are presented in 21 chapters grouped into seven parts, beginning with the definition of a cell, providing background on basic cellular processes, continuing on to the components of cells and the regulation of cell functions, and ending with cell diversity. Plant cells and prokaryotic cells are covered in separate chapters to emphasize their diversity while highlighting the properties shared by all cells.

Areas of New Coverage

Chapters from the first edition were thoroughly updated and revised by their original authors, 26 experts in diverse areas of cell and molecular biology and biochemistry.

This second edition also includes several entirely new chapters:

Chapter 2, Bioenergetics and Cellular Metabolism
Chapter 3, DNA Replication, Repair, and Recombination
Chapter 4, Gene Expression and Regulation
Chapter 5, Protein Structure and Function

The following list highlights some other areas of key content updates:

- Chapter 9, Nuclear Structure and Transport, discusses the dramatic increase in our understanding of nuclear pore complex structure, organization, and biogenesis, and the nature of the molecular environment found in the central channel of the NPC, which ensures selectivity in transport. Also updated substantially is the discussion of RNA export, focusing on recent advances in our understanding of export of mRNA, tRNA, ribosomal subunits, and microRNAs.
- Chapter 10, Chromatin and Chromosomes, now contains an extensive discussion of histone variants and the roles they play in chromosome segregation, transcription, and DNA repair.
- Chapter 13, Intermediate Filaments, shows how mutations in keratin genes have been linked to skin blistering diseases.
- Chapter 14, Mitosis, explains how errors in chromosome attachment to the mitotic spindle are detected and corrected. It also discusses mitosis as a pharmacological target for development of anticancer drugs.
- Chapter 15, Cell Cycle Regulation, explains the mechanisms responsible for cell proliferation and the way these mechanisms are controlled to prevent chromosome damage.
- Chapter 16, Apoptosis, includes an expanded discussion of the inflammasome, a structure that senses danger signals and responds to them.
- Chapter 18, Principles of Cell Signaling, features a discussion of Abl and the development of inhibitors and resistance in the treatment of chronic myelogenous leukemia. The authors have also added improved protein structures that illustrate important regulatory principles.
- Chapter 19, The Extracellular Matrix and Cell Adhesion, discusses the role of the extracellular matrix during the evolution of multicellularity. It also contains an expanded discussion of various integrin-based complexes in vivo.
- Chapter 21, Plant Cell Biology, covers newly discovered proteins that predict the plane of cell division. It

also includes advances showing that microtubules provide tracks for the movement of cellulose-synthesizing enzymes.

Design

The design of *Lewin's CELLS, Second Edition*, is specifically intended to enhance pedagogy. Chapters are divided into sections with declarative titles that emphasize the main points. Each section begins with a set of **Key Concepts** that enable readers to grasp the important ideas at the outset. To stimulate students' interest in future work, chapters include a section called **What's Next**? that describes some of the interesting questions that researchers are tackling. Key review articles have been listed for students interested in the experiments that led to the current understanding of each topic, and additional references to original research papers and reviews are available on this book's Student Companion Web Site. Each chapter in *Lewin's CELLS, Second Edition*, now includes several **Concept and Reasoning Checks**, allowing students to test their understanding of the material just presented. Pedagogy has also been enhanced by adding special feature boxes to highlight **Medical Applications**, **Historical Perspectives**, and **Methods and Techniques** used to study cell processes (for a list of these features, see page xvii).

The artists, in collaboration with the authors and editors, have developed the illustrations to be as self-explanatory as possible, with such features as text boxes that lead the reader through a figure. Liberal use of well-labeled micrographs and molecular structures helps students to recognize cellular components and understand relationships between structure and function. Whenever possible, the schematic figures take into account the relative sizes of molecules. Colors and molecular shapes, the latter based on atomic structures where known, are used in a consistent manner throughout the book.

Supplements to the text

Jones and Bartlett Publishers offers an impressive array of ancillaries to assist instructors and students in teaching and mastering the concepts in this text. Additional information and review copies of any of the following items are available through your Jones and Bartlett Publishers sales representative or by going to www.jbpub.com/biology.

For the student

The **Student Companion Web Site** we developed exclusively for the second edition of this text, http://biology.jbpub.com/lewin/cells, offers a variety of resources to enhance understanding of cell biology. Students will find chapter summaries and study quizzes that help them to review the key concepts, as well as an interactive glossary, flashcards, and crosswords to aid with memorization of key terms. The site also contains a selection of interactive figures, animations, and videos, visual aids that are essential to understanding the dynamic nature of cells. These online images are indicated by the symbol ⁝ to the left of figure legends in this book. The interactive figures include diagrams and micrographs with labels that can be turned on and off as well as short videos with labels showing the progression of key processes. For those students who wish to explore topics in cell biology in greater depth, a list of important research papers and reviews is also provided for every chapter in the book, along with links to related sites on the Web.

For the instructor

Compatible with Windows® and Macintosh® platforms, the *Instructor's Media CD-ROM* provides instructors with the following traditional ancillaries:

- The **PowerPoint® Image Bank** provides the illustrations, photographs, and tables (to which Jones and Bartlett Publishers holds the copyright or has permission to reproduce digitally) inserted into PowerPoint slides. Instructors can quickly and easily copy individual images or tables into their existing lecture slides.
- The **PowerPoint Lecture Outline Slides** presentation package provides lecture notes and images for each chapter of *Lewin's CELLS, Second Edition*. Instructors with the Microsoft® PowerPoint software can customize the outlines, art, and order of presentation.
- The Instructor's Media CD also contains more than 350 interactive **figures, animations, and videos**.

A *Test Bank* (prepared by Esther Siegfried at Pennsylvania State University, Altoona) is also available through your Jones and Bartlett Publisher's representative. The questions are presented in straight text files that are compatible with most course management software.

Acknowledgments

We thank the many scientists who provided advice informally throughout the development of this book and the following scientific advisors, who read portions of the text and made many valuable suggestions:

Stephen Adam	Northwestern University Feinberg School of Medicine, Chicago, IL	Vivek Malhotra	University of California, San Diego, CA
Tobias Baskin	University of Massachusetts, Amherst, MA	Frank McCormick	University of California, San Francisco, CA
Harris Bernstein	National Institutes of Health, Bethesda, MD	Akira Nagafuchi	Kumamoto University, Kumamoto City, Japan
Fred Chang	Columbia University, New York, NY	Roel Nusse	Stanford University, Palo Alto, CA
Louis DeFelice	Vanderbilt University, Nashville, TN	Andrew Osborne	Harvard Medical School, Boston, MA
Paola Deprez	Institute of Microbiology–ETH, Zurich, Switzerland	Erin O'Shea	Harvard University, Cambridge, MA
Arshad Desai	University of California, San Diego, CA	Marcus Peter	University of Chicago, Chicago, IL
Paul De Weer	University of Pennsylvania, Philadelphia, PA	Suzanne Pfeffer	Stanford University, Stanford, CA
Biff Forbush	Yale University, New Haven, CT	Tom Rapoport	Harvard Medical School, Boston, MA
Joseph Gall	Carnegie Institution, Baltimore, MD	Ulrich Rodeck	Thomas Jefferson University, Philadelphia, PA
Emily Gillett	Harvard Medical School, Boston, MA	Michael Roth	University of Texas Southwestern Medical Center, Dallas, TX
Rebecca Heald	University of California, Berkeley, CA	Lucy Shapiro	Stanford University, Palo Alto, CA
Alistair Hetherington	Bristol University, Bristol, United Kingdom	Thomas Shea	University of Massachusetts, Lowell, MA
Harald Herrmann	German Cancer Research Center, Heidelberg, Germany	David Siderovski	University of North Carolina, Chapel Hill, NC
Philip Hinds	Tufts–New England Medical Center, Boston, MA	Mark Solomon	Yale University, New Haven, CT
Jer-Yuan Hsu	University of California, San Diego, CA	Chris Staiger	Purdue University, West Lafayette, IN
Martin Humphries	University of Manchester, Manchester, United Kingdom	Margaret A. Titus	University of Minnesota, Minneapolis, MN
		Livingston Van De Water	Albany Medical Center, NY
James Kadonaga	University of California, San Diego, CA	Miguel Vicente-Manzanares	University of Virginia, Charlottesville, VA
Randall King	Harvard Medical School, Boston, MA	Patrick Viollier	Case Western Reserve University, Cleveland, OH
Roberto Kolter	Harvard Medical School, Boston, MA	Claire Walczak	Indiana University, Bloomington, IN
Susan LaFlamme	Albany Medical Center, NY	Junying Yuan	Harvard Medical School, Boston, MA
Rudolf Leube	Johannes Gutenberg University, Mainz, Germany	Sally Zigmond	University of Pennsylvania, Philadelphia, PA

We are also grateful to all the scientists who made this book possible by providing essential micrographs and other images, as well as to the journal and book publishers for permission to reproduce them. The credits are listed in the figure legends.

We welcome suggestions for revisions or corrections, which can be sent to us at info@jbpub.com.

Contributors

Lead Editors

Benjamin Lewin founded the journal *Cell* in 1974 and was Editor until 1999. He also founded the Cell Press journals *Neuron, Immunity,* and *Molecular Cell.* In 2000, he founded Virtual Text, which was acquired by Jones and Bartlett Publishers in 2005. He is the original author of *Genes* and *Essential Genes.*

Lynne Cassimeris is a Professor in the Department of Biological Sciences at Lehigh University in Bethlehem, PA. She studies microtubule assembly dynamics and mitosis.

Vishwanath R. Lingappa is Senior Scientist at Bioconformatics Laboratory, CPMC Research Institute; Chief Technology Officer at Prosetta Corporation; and Emeritus Professor of Physiology at the University of California, San Francisco. His research is in protein biogenesis. He practices internal medicine as a volunteer physician at San Francisco General Hospital and has coauthored textbooks of physiology and pathophysiology.

George Plopper is an Associate Professor at Rensselaer Polytechnic Institute. He studies signal transduction and cellular behavior induced by extracellular matrix binding.

Authors

Raymond Ochs is a Professor of Pharmaceutical Sciences at St. John's University in New York. His research concerns regulation of intermediary metabolism and control by intracellular calcium ion.

Jocelyn E. Krebs is an Associate Professor of Biological Sciences at the University of Alaska, Anchorage. Her lab studies DNA repair and transcription in the context of chromatin structure in the yeast *Saccharomyces cerevisiae* and the role of chromatin remodeling during vertebrate development in the frog *Xenopus laevis.*

David G. Bear is Professor of Cell Biology and Physiology and Assistant Dean for Admissions at the University of New Mexico School of Medicine and is Professor and Chair of Chemistry and Chemical Biology at the University of New Mexico. His research interests focus on the assembly and intracellular trafficking of messenger RNA ribonucleoprotein complexes in striated muscle cells, muscular dystrophies, and muscle cell cancers.

Stephen J. Smerdon is joint head of the Division of Molecular Structure at the MRC National Institute for Medical Research, UK, where he works on the structural biology of a variety of cell signaling pathways, particularly the regulation of DNA-damage complex assembly by phosphorylation-dependent mechanisms.

Stephan E. Lehnart is a Professor of Translational Cardiology at the Center of Molecular Cardiology of the University of Goettingen Medical Center and an Adjunct Associate Professor at the University of Maryland Biotechnology Institute. His major research interests are membrane transport mechanisms that control intracellular calcium cycling in the heart and other organs that contribute significantly to disease processes.

Andrew R. Marks is the Clyde and Helen Wu Professor of Medicine and Chair and Professor of the Department of Physiology and Cellular Biophysics at Columbia University. He works on how macromolecular signaling complexes regulate ion channel function in muscle and nonmuscle systems. He is a member of the Institute of Medicine of the National Academy of Sciences, American Academy of Arts and Sciences, and the National Academy of Sciences.

D. Thomas Rutkowski is an Assistant Professor in the Department of Anatomy and Cell Biology at the University of Iowa Carver College of Medicine. His lab studies how cells adapt to chronic protein misfolding stress in normal physiology and disease.

Vivek Malhotra is the Chair of Cell and Developmental Biology at CRG in Barcelona, Spain. His laboratory has long been involved in the mechanism of protein secretion and Golgi organization.

Graham Warren is the Scientific Director of the Max F. Perutz Laboratories, Vienna, Austria, and his laboratory studies the structure, function, and biogenesis of the Golgi apparatus.

Ira Mellman is Vice President of Research Oncology at Genentech, Inc., in San Francisco, California. His research focuses on the cell biology of the immune response (specifically the role of dendritic cells in T-cell stimulation) and on the signals that control the formation and function of epithelial cells.

Charles N. Cole is a Professor of Biochemistry and of Genetics at Dartmouth Medical School. His interests have included nuclear transport, regulation of cellular transformation and immortalization, RNA metabolism, microRNAs, and breast cancer.

Pamela A. Silver is a Professor of Systems Biology at Harvard Medical School. Her interests have included nuclear transport, organization of the genome, RNA dynamics, and synthetic biology.

Enrique M. De La Cruz is an Associate Professor of Molecular Biophysics and Biochemistry at Yale University. His laboratory uses biochemical and biophysical techniques to investigate the mechanisms of actin- and myosin-based motility. His work also focuses on motor properties of RNA helicases.

E. Michael Ostap is a Professor of Physiology at the University of Pennsylvania School of Medicine and is a member of the Pennsylvania Muscle Institute. His laboratory uses cell biological, biochemical, and biophysical techniques to investigate the mechanisms of cell motility. His work is currently focused on the study of unconventional myosins.

Birgit Lane is Executive Director of the Institute of Medical Biology in Singapore. She studies intermediate filaments, particularly keratins, and the part they play in normal tissue resilience and in human diseases.

Conly L. Rieder is a Senior Research Scientist and Chief of the Wadsworth Center's Laboratory of Cell Regulation. The Wadsworth Center is a research arm of the New York State Department of Health. He is also a Professor in the Department of Biomedical Sciences, State University of New York at Albany. Dr. Rieder has spent over 30 years researching how cells divide.

Kathleen L. Gould is Professor and Vice-Chair of Cell and Developmental Biology at Vanderbilt University School of Medicine and an Investigator of the Howard Hughes Medical Institute. Her laboratory studies the mechanism and regulation of cytokinesis in the fission yeast *Schizosaccharomyces pombe*.

Susan L. Forsburg is a Professor in Molecular and Computational Biology at the University of Southern California. Her laboratory studies DNA replication and genome dynamics in the fission yeast *Schizosaccharomyces pombe*.

Douglas R. Green is the Chair of Immunology at St. Jude Children's Research Hospital in Memphis, TN. His laboratory studies the process of apoptosis and related forms of cell death.

Robert A. Weinberg is the Daniel K. Ludwig and American Cancer Society Professor for Cancer Research at the Massachusetts Institute of Technology and a founding member of the Whitehead Institute for Biomedical Research. His research is focused on the molecular mechanisms that control cell proliferation and the formation of tumors.

Elliott M. Ross is a Professor in the Graduate Programs in Molecular Biophysics and Cell Regulation and the Department of Pharmacology at the University of Texas Southwestern Medical Center in Dallas. His group studies information processing in G-protein signaling networks.

Melanie H. Cobb is a Professor in Pharmacology and the Graduate Programs in Cell Regulation and Cancer Biology at the University of Texas Southwestern Medical Center at Dallas. Her group studies regulation and function of protein kinases with an emphasis on MAPK pathways and WNKs.

Matthew Chapman is an Associate Professor at the University of Michigan and the 2009 recipient of the university's Class of 1923 Teaching Award. His lab studies the function and biogenesis of bacterial amyloid fibers.

Jeff Errington is Director of the Institute for Cell and Molecular Biosciences at Newcastle University. He works on the cell cycle and cell morphogenesis of bacteria.

Clive Lloyd is a Project Leader at The John Innes Centre, Norwich, UK. He studies the role of the cytoskeleton in plant growth and development.

Abbreviations

A	Adenine or adenosine
ADP	Adenosine diphosphate
AMP	Adenosine monophosphate
cAMP	Cyclic AMP
ATP	Adenosine triphosphate
ATPase	Adenosine triphosphatase
bp	Base pair(s)
C	Cytidine or cytosine
cDNA	Complementary DNA
CDP	Cytidine diphosphate
CMP	Cytidine monophosphate
CTP	Cytidine triphosphate
DNA	Deoxyribonucleic acid
DNAase	Deoxyribonuclease
G	Guanine or guanosine
GDP	Guanosine diphosphate
GlcNAc	N-Acetyl-D-glucosamine
GMP	Guanosine monophosphate
GTP	Guanosine triphosphate
ΔG	Free energy change
kb	Kilobases or kilobase pairs
Mb	Megabases or megabase pairs
mRNA	Messenger RNA
MW	Molecular weight
Pi	Inorganic phosphate
PPi	Inorganic pyrophosphate
RNA	Ribonucleic acid
RNAase	Ribonuclease
rRNA	Ribosomal RNA
tRNA	Transfer RNA
T	Thymine or thymidine
U	Uracil
UDP	Uridine diphosphate
UMP	Uridine monophosphate
UTP	Uridine triphosphate

Units

Å	Angstrom
D or Da	Dalton
g	Gram
h or hr	Hour
M	Molar concentration
m	Meter
m or min	Minute
N	Newton
S	Svedberg unit
s or sec	Second
v	Volt

Amino acids

A	Ala	Alanine
C	Cys	Cysteine
D	Asp	Aspartic acid
E	Glu	Glutamic acid
F	Phe	Phenylalanine
G	Gly	Glycine
H	His	Histidine
I	Ile	Isoleucine
K	Lys	Lysine
L	Leu	Leucine
M	Met	Methionine
N	Asn	Asparagine
P	Pro	Proline
Q	Gln	Glutamine
R	Arg	Arginine
S	Ser	Serine
T	Thr	Threonine
V	Val	Valine
W	Trp	Tryptophan
Y	Tyr	Tyrosine

Prefix (Abbreviation)	Multiple
mega (M)	10^6
kilo (k)	10^3
deci (d)	10^{-1}
centi (c)	10^{-2}
milli (m)	10^{-3}
micro (μ)	10^{-6}
nano (n)	10^{-9}
pico (p)	10^{-12}

PART

1

Introduction

Photo © Mopic/ShutterStock, Inc.

What is a cell?

1

Vishwanath R. Lingappa
Prosetta Bioconformatics, San Francisco, CA

Benjamin Lewin
Founding Publisher/Editor, Cell Press and
Virtual Text

A HUMAN OOCYTE (EGG), as visualized by scanning electron microscopy. All of the cell types in an organism develop from the fusion of an egg and a sperm. Photo © Dennis Kunkel Microscopy, Inc./Phototake/Alamy Images.

CHAPTER OUTLINE

1.1 Introduction

Key concepts

- Cells arise only from preexisting cells.
- Every cell has genetic information whose expression enables it to produce all its components.
- The plasma membrane consists of a lipid bilayer that separates the cell from its environment.
- Functions within cells are mainly carried out by proteins, alone or complexed with other proteins, RNA, or lipids.

A number of scientific disciplines have distinctive perspectives on this question. The focus of this textbook is a cell biologic perspective on the current understanding of the structure, function, and regulation of biologic systems. But of necessity that requires reaching the interface with the disciplines of biochemistry and molecular biology, histology and pathology, and physiology, at the very least. Likewise, to know what cells are, we must consider what we know of how they came to be and how they differ from one another in structure, function and regulation.

The vast diversity of living organisms is based on a single basic building block: the cell. The basic principle of biology, as recognized in the cell theory of the nineteenth century, is that cells can be produced only by the division of preexisting cells.

The simplest organisms are unicellular: the cell itself is an organism, existing as a single entity that reproduces to make more copies of itself. Unicellular organisms may be adapted to survive in many different types of environments, extending between extremes of cold or heat, living in aerobic or anaerobic conditions, or even surrounded by methane gas. Some live within other organisms.

Cells also may form multicellular organisms, in which case different cells are specialized to have different functions. The cells within a multicellular organism communicate with one another in order to enable the organism to function as a whole. The organism has the ability to reproduce, although individual cells within it may or may not reproduce. Cells within an organism that gain the ability to reproduce in an unrestrained manner result in cancer.

Cells vary enormously in size and shape, as shown in **FIGURE 1.1**. The smallest cells are unicellular organisms that exist as spheres with a diameter no more than 0.2 μm. One of the

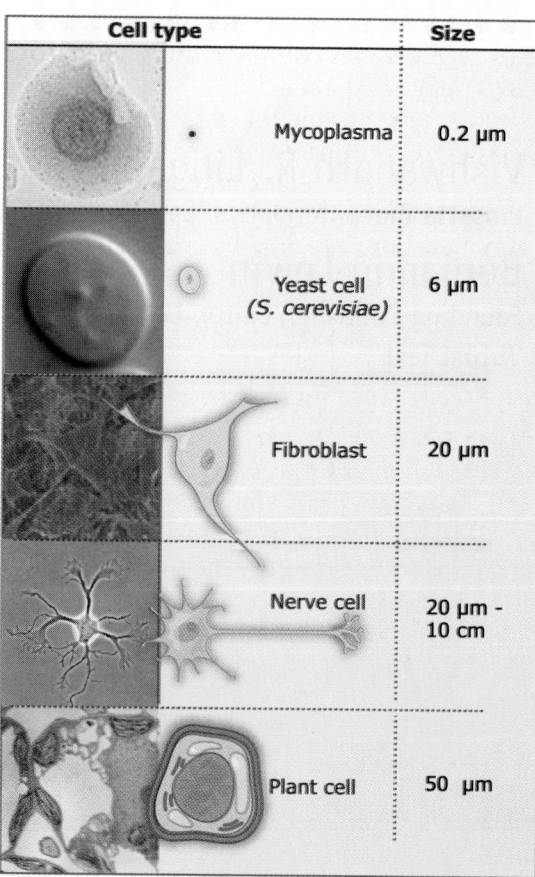

Cell type	Size
Mycoplasma	0.2 μm
Yeast cell (*S. cerevisiae*)	6 μm
Fibroblast	20 μm
Nerve cell	20 μm – 10 cm
Plant cell	50 μm

FIGURE 1.1 Cells vary greatly in size and shape. Some cells are spherical, whereas others have long extensions, with many variations in between. Mycoplasma photo courtesy of Tim Pietzcker, Universität Ulm; yeast cell photo courtesy of Fred Winston, Harvard Medical School; fibroblast photo courtesy of Junzo Desaki, Ehime University School of Medicine; nerve cell photo courtesy of Gerald J. Obermair and Bernhard E. Flucher, Innsbruck Medical University; plant cell photo courtesy of Ming H. Chen, University of Alberta.

largest cells is a neuron (a nerve cell) of the giant squid, which is 5000 times larger at a diameter of 1 mm. Extending from the body of the neuron are processes (axons) that are 20 μm in diameter (100 times wider than the smallest cell) and that may be as long as 10 cm! Between these extremes, human (and other mammalian) cells often have diameters in the range of 3–20 μm.

A cell may have little variation in its shape, forming a simple sphere that lives bathed in fluid, or it may have a well-defined structure, such as the neuron with its extended processes, or epithelial cells that have distinct apical and basolateral surfaces carrying out distinct functions. A cell may live free in solution, may be attached to a surface, or may be attached to other cells. Cells may communicate with other cells or may attack them.

But underlying the abilities of cells to take such different forms, the construction of all cells is based on several common properties as stated below:

- A membrane called the **plasma membrane** segregates the interior of the cell from the external environment.
- The plasma membrane contains systems that control import into and export out of the cell.
- Cellular components are constructed from food sources using internal systems for energy conversion.
- Genetic material contains the information needed to renew all the cell's components.
- Gene expression enables the cell to put to practical use the information encoded in genes.
- Individual protein products are encoded by genes and, upon synthesis, can assemble into larger structures.

A cell is circumscribed by a membrane consisting of a lipid bilayer. **FIGURE 1.2** summarizes the unique properties of a lipid bilayer. It is a macromolecular structure built from lipids. The key property of the lipid is that it is amphipathic—one end is a hydrophilic "head" and the other end is a hydrophobic "tail." Each "leaf" of the lipid bilayer has one side consisting of an array of the hydrophilic heads, while the other side consists of the hydrophobic tails. An aqueous environment causes the hydrophobic tails to aggregate, so that the hydrophobic sides of each leaf can come together to form a nonionic center, much like an oil drop in water. The hydrophilic heads of the two leaves face into the ionic milieu on either side of the lipid bilayer. The lipid bilayer has the important property of fluidity, which allows it to fuse with other membranes, generate new membranes by fission, and provide a solvent for proteins that can reside within the bilayer and move around within it.

The lipid bilayer is somewhat permeable to water but impermeable to ions, small charged molecules, and all large molecules. Differences in the ionic environment on the two sides of the membrane can create osmotic pressure—a situation in which water moves across the membrane to dilute the concentration of ions on whichever side has the more concentrated ionic environment.

The plasma membrane separates the contents of the cell from the external environment. For a unicellular organism, the "external environment" is the exterior world; for a multicellular organism, it is both the exterior world outside the organism as well as the interior world created by other cells (e.g., by cells that form blood vessels) within the organism. The plasma membrane has no structural strength; in fact, it is rather fragile and easily broken. As a result, the integrity of the cell usually requires the plasma membrane to be supported by some underlying structure with more tensile strength.

Most processes in living cells are catalyzed by enzymes whose binding constants and other set points dictate the tolerance for variation in concentrations of small and large molecules in the internal and external environments that are compatible with life. However, organisms are adjusted to a wide range of environments, and those that live in extreme conditions have enzymes that can function under conditions that would be lethal for more "normal" organisms.

A cell needs to regulate its interior environment to enable its systems to work properly. In particular, ionic levels and pH must be controlled. The impermeability of the membrane creates a need for special systems within the membrane to move ions in and out of the cell.

A cell must import material from the outside. In particular, it must import sources of energy (which are substrates for metabolism) and small molecules that are precursors for the components that are assembled into larger molecules and structures. Fatty acids are used to make lipids, amino acids are used to make

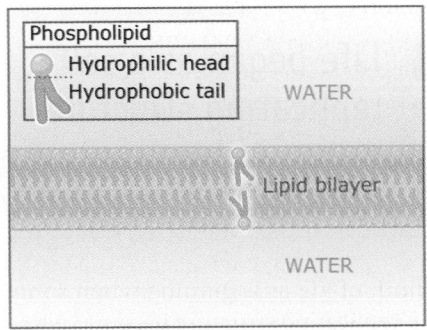

FIGURE 1.2 The lipid bilayer of membranes primarily contains amphipathic phospholipids.

proteins, and nucleotides are used to make RNA and DNA.

Just as all cells must import material from the environment, so they must be able to expel material. Cells export a variety of ions, small molecules, and even proteins. Export and, to a significant extent, import are extremely specific processes: they allow what is supposed to go out (or in) to do so in a highly selective manner.

In order to survive and reproduce itself, a cell must be able to obtain energy from its environment and use that energy to synthesize its own components. The source of energy can be material that is ingested from the environment, most typically a mix of simple and complex carbon compounds. Light can also be used as an energy source. The way energy is utilized is specific for each cell type.

Because the production of new cells requires the division of a preexisting cell, a cell must carry within it the information for reproducing all of its components. The form of this information is a single type of genetic material, DNA, which codes for all the proteins of the cell. The proteins in turn can assemble into large structures or take responsibility for catalyzing the internal reactions of the cell. The apparatus that is used to interpret the genetic code has the same types of components in all cells. Because a cell is under a barrage of insults from the environment, a means of repairing damage to its genetic information is essential for long-term survival.

Cells perpetuate by dividing themselves. A special apparatus confers the capability to divide to generate two progeny cells, each identical to the parent cell in its content of genetic material, and also containing approximately half of the other structures of the cell (with some exceptions as noted in *1.23 Differentiation creates specialized cell types, including terminally differentiated cells*).

FIGURE 1.3 summarizes the minimal features needed to make a cell. In summary, a membrane segregates the cell from its environment, and many basic requirements for the way all living cells interact with their environment follow from the intrinsic properties of membranes. Utilizing energy in order to build more complex components from smaller building blocks is required to construct the cell. Genetic material carries the information necessary to produce all the features of the cell, and all cells have a system for utilizing this information.

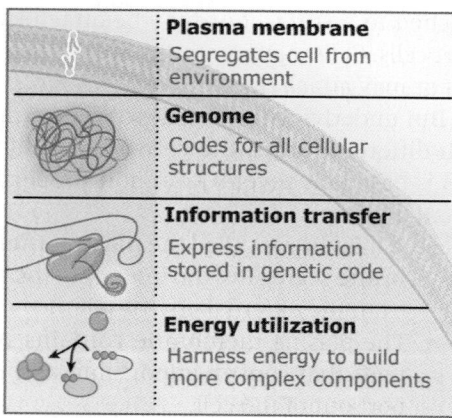

FIGURE 1.3 A cell has a genome that codes for all its structures, a means of expressing the genetic information, a system for utilizing energy, and a plasma membrane that controls communication with the outside world.

Cells carry out various functions through specific proteins expressed from genes. Proteins participate in cellular functions in many different ways. For example, by binding other proteins they can stabilize or silence other proteins; they can serve as structural components, as channels, and as enzymes. A challenge for our understanding of how cells work is to be better able to grasp how the individual actions of proteins are coordinated to make a coherent and smoothly functioning organism. Critical dimensions that make such coordination possible include signal transduction mechanisms, acting on enzymes organized into pathways, as part of the overall program of homeostasis: maintenance of the constancy of the internal environment in a multicellular organism.

Concept and Reasoning Check

1. Describe some physical properties of membranes that are crucial for their biologic functions.

1.2 Life began as a self-replicating structure

Key concept

- The first living cell was a self-replicating entity surrounded by a membrane.

We think of life as beginning when some sort of self-replicating structure became segregated from its environment. This would be necessary both to protect it from damage by adventitious events and to prevent dilution of the primeval soup. Discoveries in recent years have suggested that the first self-replicating structure may have been based on ribonucleotides, so we may think of some primitive RNA becoming surrounded by a protective membrane to generate the first living object. Let us call this object the protoplasmic blob.

The RNA would have required the capacity to replicate itself by assembling nucleotide precursors into a copy of itself. We do not know what other catalytic activities it may have had, but these would most likely have been concerned with the metabolic activities required to provide the precursors for replication.

Some sort of coupling between the self-replicating entity and a surrounding membrane must have occurred at an early stage. One activity that could propel the formation of a cell would be for the RNA to be able to assist in extension of the membrane. Some recent work provides a possible answer on the nature of such a connection. When RNA is enclosed by a lipid membrane, the RNA's charged groups generate an osmotic pressure. This creates a tension in the membrane that can be relieved by increasing its area, that is, by incorporating more lipids into the membrane. This type of physical interaction between nucleic acid and membrane could have been the initial force that held the protoplasmic blob together.

Impermeability to aqueous solutes is a fundamental biologic characteristic of membranes. However, there must have been some movement of material in and out of the blob, because otherwise the resources of the interior would soon become exhausted. Perhaps some exchange of materials with the exterior was made possible by sporadic breaks in the membrane, which made the membrane temporarily unstable. The role of the first "membrane" would, therefore, be to segregate the self-replicating structure from the exterior as suggested in **FIGURE 1.4**, but it probably would not otherwise distinguish the interior environment from the exterior.

Replacing the initially uncontrolled communication between interior and exterior of such systems with the capacity to maintain different conditions in the interior from the exterior would have been an important development in converting the protoplasmic blob into something that we could call a primitive cell, as summarized in **FIGURE 1.5**. This would have been accomplished by development of the first systems within the membrane that allowed

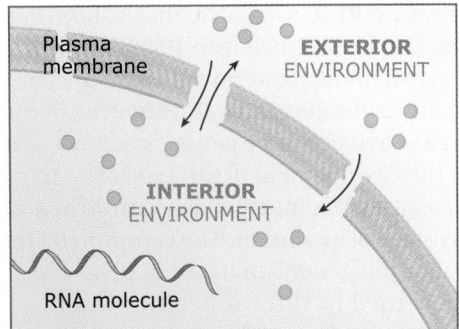

FIGURE 1.4 The very first self-reproducing cellular entity had a self-replicating RNA surrounded by a membrane.

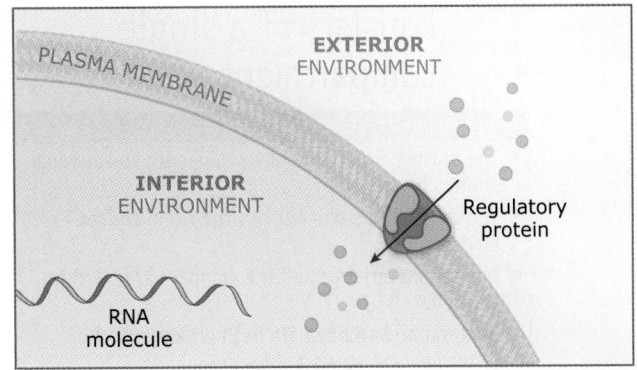

FIGURE 1.5 A primitive cell had a self-replicating genome and a plasma membrane that controlled import and export of materials.

precursors for cell components to be imported and unwanted products to be exported.

What other features would be necessary to call this a cell? The self-replicating entity would need not merely to be self-replicating but also be able to determine, directly or indirectly, the properties of its environment, including the surrounding membrane. There would need to be metabolic systems for converting precursors imported from the environment into the molecules that the cell needs and for assembling them into structures. A system would be needed for utilizing supplies of energy and for storing them in some form that would enable the cell to use the energy store to drive energetically unfavorable reactions.

The hereditary material of a modern cell is not self-replicating but has been restricted to a coding function. It now requires a specific apparatus for its replication and expression. The advantage of using proteins rather than RNA is their much greater catalytic activity and wider diversity of forms. We can see fairly easily how the genetic material could evolve from a self-replicating RNA to a DNA that is both repli-cated and transcribed into RNA, but it is not so easy to understand how it came to code for protein via the genetic code. However, because all cells have the same genetic code, we know that the development of this type of system for gene expression must have occurred at a very early stage of evolution. The common features are that the genetic material is DNA and that RNA is used in three different roles in gene expression: as messenger RNA, transfer RNA, and a structural component of the ribosome, the protein-synthesizing machine.

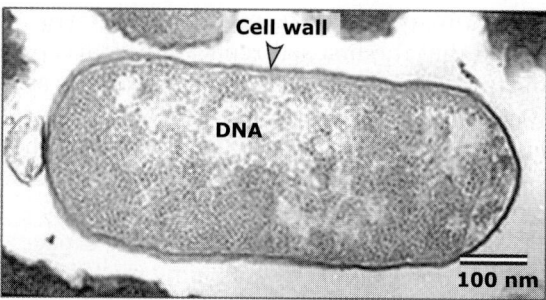

FIGURE 1.6 A bacterium has a single compartment, although different internal regions may be distinguished. Photo courtesy of Jonathan King, Massachusetts Institute of Technology.

1.3 A prokaryotic cell consists of a single compartment

Key concepts

- The plasma membrane of a prokaryote surrounds a single compartment.
- The entire compartment has the same aqueous environment.
- Genetic material occupies a compact area within the cell.
- Bacteria and archaea are both prokaryotes but differ in some structural features.

All living cells can be classified into one of two general types according to how they are divided internally into compartments (a "compartment" is defined for this purpose simply as a volume that is enclosed by a membrane):

- A **prokaryote** has a single compartment that contains its genetic material, the apparatus for gene expression, and the products of gene expression. The compartment is bounded by a membrane, and there are no other compartments within it.
- A **eukaryote** has at least two compartments; the entire cell is contained within a surrounding plasma membrane; but, the interior has a second compartment that contains the genetic material.

Prokaryotes fall into two classes. Originally, all prokaryotes were thought to be **bacteria**, but now we consider some of them to be **archaea**. Both bacteria and archaea exist only as unicellular organisms (although some bacteria may show aggregation properties in populations). The area contained by the plasma membrane is called the **cytoplasm**. The prokaryotic plasma membrane is surrounded by a **cell wall** whose rigid structure gives physical protection against the environment.

FIGURE 1.6 shows that the genetic material occupies a compact area within the single compartment of a bacterium but is not segregated by membranes from the rest of the contents. The simplest forms of bacteria are mycoplasma, which are not free living; because they cannot produce many of the basic compounds required for life, they have to live within other organisms that provide those molecules. They have small genomes, containing only ~500 genes that code for the bare minimum of structural features required to make a cell. Free-living bacteria have genomes with more than 1500 genes and code for the metabolic enzymes needed to metabolize small molecules, as well as for a more complex apparatus for regulating gene expression.

Bacteria are divided into two groups, which probably diverged about 2 billion years ago. The groups are called Gram-negative and Gram-positive, depending on whether they react with the Gram stain. *Escherichia coli* is the best characterized Gram-negative, and *Bacillus subtilis* is the best characterized Gram-positive. Gram sensitivity is determined by the interaction of the stain with the cell wall. Gram-positive bacteria have a cell wall surrounding the plasma membrane, and the stain reacts directly with components of the wall. Gram-negative bacteria have a second membrane surrounding the cell wall. The presence of this membrane, and differences in the constitution of the wall, prevent the stain from reacting. The region between the outer and inner membranes is called the periplasmic space, and it has its own characteristic set of proteins and other components. By the criterion that a compartment is a membrane-enclosed space, we might say that Gram-negative bacteria have two compartments, but the periplasmic space is concerned only with the interaction between the bacterium and its environment and does not change the basic fact that the synthetic activities of the bacterium occur in the same compartment that contains the genetic material.

The cytoplasm of the bacterium is a single aqueous environment, meaning that all enzymes work under similar ionic conditions. However, bacteria are far from being "bags of enzymes," and it is now clear that many proteins are targeted to specific subcellular locations or structures. Some bacteria can undergo developmental changes involving the formation of specialized cell types, reminiscent of development in higher organisms.

Bacteria comprise a wide variety of forms and have become highly divergent in evolution. Defining their phylogenetic relationships is difficult; there is no counterpart to the fossil record that has been used for higher eukaryotes. However, molecular methods, initially based on sequencing of ribosomal RNAs, and more recently based on complete genome sequencing, have revolutionized prokaryotic phylogenetics and have led to the identification of archaea as a separate class of prokaryotic organisms, as **FIGURE 1.7** shows.

Archaea resemble bacteria in their appearance and construction: they are small, single-cell organisms. They tend to live under extreme environmental conditions (such as very high temperatures) and were originally mistaken for bacteria that had adapted to these conditions. Like bacteria, archaea have only a single cellular compartment, with no internal membranes. They may have many of the same morphologic features, such as a rigid wall or capsule surrounding the plasma membrane, and flagella that protrude into the environment. The major

differences come at the molecular level, where their components are dissimilar from those of bacteria. The archaeal apparatus for gene expression is more closely related to that of eukaryotes than that of bacteria, and the cell walls are made from different subunits than those of bacteria or plants. They contain a different set of lipids in their membranes. Their genetic complexity is comparable to free-living bacteria. (For further details on prokaryotes, see *20 Prokaryotic cell biology*.)

Concept and Reasoning Check
1. How do archaea differ from bacteria?

1.4 Prokaryotes are adapted for growth under many diverse conditions

Key concept
- Prokaryotes have adapted to many extreme environmental conditions and highlight the variations that are possible in constructing living cells.

Prokaryotes have a wide diversity of life styles, and many have adapted to specialized niches, including some environmental extremes that extend the conditions under which life can survive. Prokaryotes can be classified into three groups according to their abilities to grow at different temperatures as follows:

1. The majority of known prokaryotic species, called mesophiles, grow best between 25°C and 40°C. This group of bacteria has been extensively studied because it includes bacteria that are human pathogens.
2. Psychrophiles grow best at temperatures between 15°C and 20°C, but there are some species able to live at 0°C. Their habitat is cold water and soil.
3. Thermophiles grow at temperatures between 50°C and 60°C, but some can tolerate temperatures as high as 110°C.

Prokaryotes can also be classified by their ability to grow in acidic and alkaline environments. Many microbes can withstand extreme variations in pH. Acidophiles grow best at a pH below 5.4. Some soil prokaryotes are among the most alkaline-tolerant organisms known, sometimes growing in pH 12. These prokaryotes maintain an internal pH around 7 and are

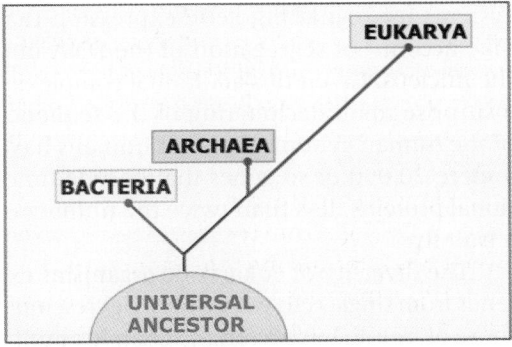

FIGURE 1.7 Phylogenetic analysis by molecular methods shows that organisms can be classified into three domains.

protected by their cell walls against the external extremes.

You might think that oxygen would be necessary in the environment for all cells. The basic role of oxygen is to be used as an electron acceptor during respiration. However, not all organisms require oxygen to carry out respiration, and some produce their energy through fermentation instead of respiration, enabling them to function under anaerobic conditions (in the absence of oxygen). Some single-celled eukaryotes, such as the yeast *Saccharomyces cerevisiae*, can exist under either aerobic or anaerobic conditions.

The abilities to function under these extreme conditions are properties of the individual types of cells and, thus, greatly extend the limits of the definition of a cell in terms of its interaction with the environment. But all cells have the same basic molecular constitution, with adaptations of individual components, such as proteins and lipids, to enable survival in specific environments. For example, enzymes made by psychrophiles are adapted to work best at low temperatures, while enzymes of thermophiles are stable at high temperatures (making them useful for certain applications in biotechnology).

1.5 A eukaryotic cell contains many membrane-delimited compartments

Key concepts
- The plasma membrane of a eukaryotic cell surrounds the cytoplasm.
- Within the cytoplasm there are individual compartments, each surrounded by a membrane.
- The nucleus is often the largest compartment within the cytoplasm and contains the genetic material.

A major increase in complexity of organization comes with the eukaryotes. A eukaryotic cell does not have a homogeneous internal environment but is divided into compartments, each of which is surrounded by a membrane. **FIGURE 1.8** shows that the interior of a eukaryotic cell is divided into two major compartments: the cytoplasm and the **nucleus**. The membrane-bounded compartments within the cell are often called **organelles**. The nucleus is often the most prominent organelle. The cyto-

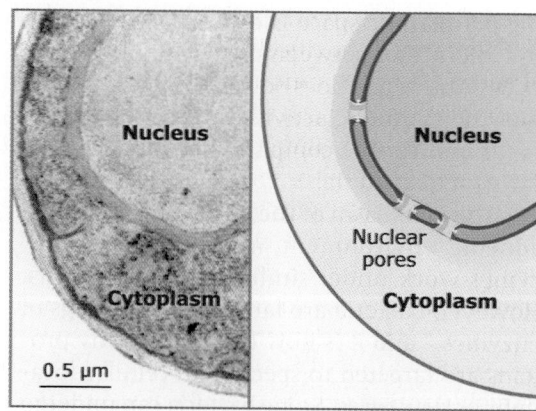

FIGURE 1.8 A eukaryotic cell is divided into a nucleus and cytoplasm. Pores in the nuclear envelope are used to transport molecules between these two compartments. Photo © Wright, et al., 1988. Originally published in **The Journal of Cell Biology**, 107: 101–114. Used with permission of Rockefeller University Press. Photo courtesy of Jasper Rine, University of California, Berkeley.

plasm is usually defined as the rest of the cell, meaning the volume between the nucleus and the plasma membrane (i.e., everything excluding the nucleus).

Macromolecules are transported into and out of the nuclei via nuclear pores, which are proteinaceous channels through the nuclear membranes. The pores are large enough to be completely permeable to smaller molecules, so there is no difference in the aqueous environment of the nucleus and the cytoplasm.

The genetic material is contained within the nucleus. The genetic complexities of eukaryotic cells vary extensively. The simplest unicellular organisms have genomes containing ~5000 genes. In addition to the functions coded by prokaryotic genomes, a eukaryotic cell needs to specify all of the additional structural elements, systems for localizing proteins to its individual compartments, and more complex systems for regulating gene expression that take account of segregation of the DNA into the nucleus. Given this additional complexity a surprise upon deciphering of the sequence of the human genome was that humans have a mere 20,000 or so genes that encode functional proteins, less than twice the number of a fruit fly.

The diversity of eukaryotic organisms extends from single cells whose lifestyles resemble those of free-living bacteria to complex multicellular organisms with many different types of constituent cells. Eukaryotes originated as unicellular organisms, and most of their characteristic features were established before the development of multicellular organisms, explaining the conservation of basic cellular features in fungi, plants, and animal cells.

An important point to remember is the very high concentration of macromolecules in all cell compartments. The modern cell has greatly intensified the characteristic that defined the first primitive cells and does not merely segregate macromolecules from their environment but also concentrates them. In the nucleus, the concentration of DNA is equivalent to a gel of high viscosity. In the other compartments, proteins are concentrated at a high density. One consequence of this organization is that localization becomes very important.

1.6 Membranes allow the cytoplasm to maintain compartments with distinct environments

Key concept
- Organelles that are surrounded by membranes can maintain internal milieus that are different from the surrounding cytosol.

Eukaryotic cells usually have other membrane-bounded organelles as well as the nucleus within the cytoplasm. Sometimes the term **cytosol** is used to describe the aqueous environment of the cytoplasm excluding all the membrane-bounded compartments within it. The cytosol can be thought of as a single compartment that is circumscribed by the plasma membrane and is in contact with the outer surfaces of all internal organelles. We might regard it as a specialized compartment, one of whose major functions is to synthesize proteins both for its own use and for import into the organelles.

Membranes within a cell have the same general structure—a lipid bilayer—as the plasma membrane that surrounds the cell. The exact nature of the individual lipids can be different for each membrane, but the general properties remain the same. Just as the impermeability of the plasma membrane distinguishes the interior and exterior of the cell, the impermeability of an organelle membrane distinguishes the interior of the organelle from the surrounding cytosol. There is no free exchange of ionic solutes across the membrane. This is the basic feature that allows organelles

to provide specialized environments within the eukaryotic cell. (The nucleus is an exception due to the presence of nuclear pores.)

Transport of both small molecules and macromolecules into and out of membrane-bounded compartments is controlled by proteins embedded in the membrane (analogous to the use of protein complexes in the plasma membrane to control import and export for the cell). The interior of an independent membrane-delimited compartment is called the **lumen**. Its aqueous environment can differ from that of the surrounding cytoplasm.

Specialized functions occur within the lumen of each organelle. This requires specific proteins to be localized within the organelle. With the exception of mitochondria and chloroplasts (which synthesize some of their own proteins), organelles do not synthesize proteins, so the proteins must be imported from the cytosol, where they are made.

FIGURE 1.9 shows the location of the most important membranous organelles found within the eukaryotic cytoplasm. The milieu of the lumen in each organelle is adjusted for its functions.

The **endoplasmic reticulum (ER)** is a series of convoluted membrane sheets contiguous with the outer membrane of the nuclear envelope. The lumen of the ER provides an oxidizing environment (like the exterior of the cell) that is important for one of its functions: folding proteins and assembling multisubunit oligomers.

A series of the membrane-delimited compartments in a typical eukaryotic cell are

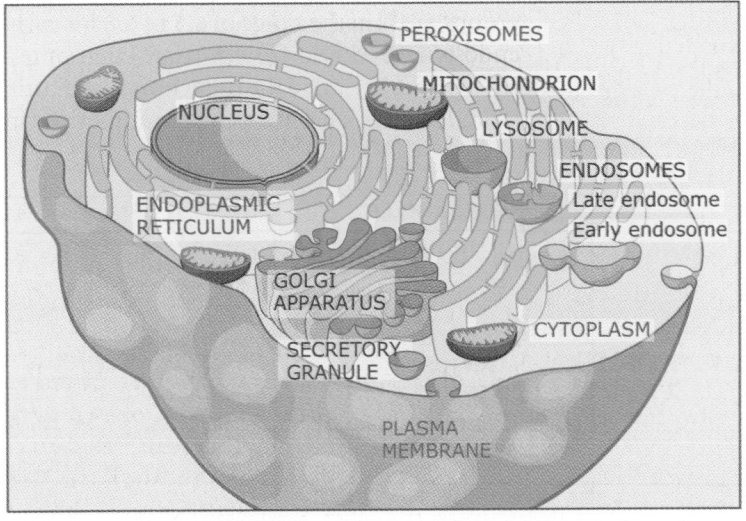

FIGURE 1.9 The cytoplasm of a eukaryotic cell contains several membrane-bounded compartments.

related and interact with one another by fission and fusion of their membranes. The endoplasmic reticulum, the **Golgi apparatus** (a series of "stacks" of flat disks each surrounded by membrane), and the *trans*-Golgi network constitute the major components of the **secretory pathway**: their membranes pinch off and fuse, moving content and membrane proteins from compartment to compartment. Secretory vesicles leave the *trans*-Golgi compartment and fuse with the plasma membrane. Covalent modifications of proteins involving the addition of small sugar molecules occur in the ER and Golgi apparatus. Other organelles that are part of this network are the endosomes and lysosomes, where degradation of proteins occurs.

Mitochondria (found in all eukaryotes) and chloroplasts (found in plants) are concerned with energy handling. Mitochondria undertake the basic reactions that provide the cell with its supply of the energy-rich intermediate, adenosine triphosphate (ATP). Chloroplasts undertake photosynthesis of carbon dioxide, allowing green plants to generate small carbon molecules to use as part of the food supply. The presence of chloroplasts is one of the major features that distinguish cells of the plant kingdom from cells of the animal (and other) kingdoms.

The concentrations of ions or small molecules typically differ within each cytoplasmic compartment, as shown in **FIGURE 1.10**. One striking difference is that the endoplasmic reticulum has a very high concentration of calcium ions. Another is that the pH of endosomes and lysosomes is much lower than in the cytosol. Endosomes are divided broadly into two groups: pH is in the range of 6.5 to 6.8 for early endosomes and as low as 4.5 in late endosomes. By contrast, the pH of the mitochondrial matrix is higher than that of cytosol, as summarized in

ORGANELLE	pH
Early endosome	6.5 - 6.8
Late endosome	5.0 - 6.0
Lysosome	4.5
trans-Golgi network	6.5 - 6.7
Mitochondria Matrix	8
Intermembrane space	7
Cytosol	7.4
Nucleus / ER	7.4

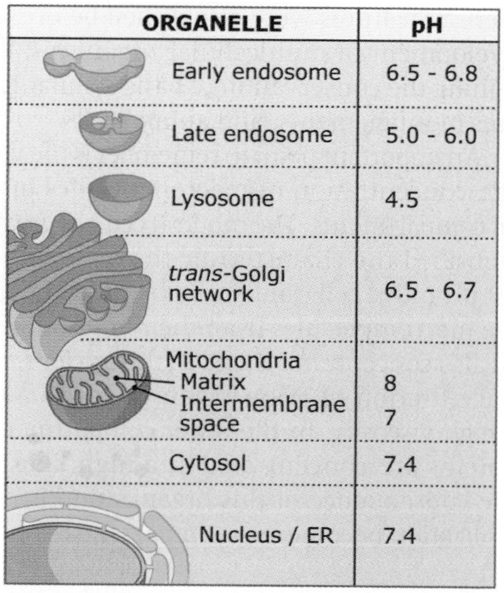

FIGURE 1.11 A pH map of the cell shows that different organelles have different pH values.

FIGURE 1.11. (For further details on organelles and the secretory pathway, see 7 *Membrane targeting of proteins* and 8 *Protein trafficking between membranes*.)

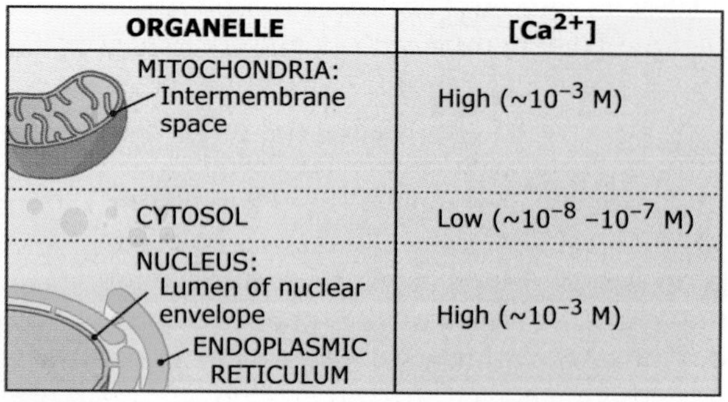

Concept and Reasoning Check

1. Name some of the component membranes of the secretory pathway.

1.7 The nucleus contains the genetic material and is surrounded by an envelope

Key concepts

- The nucleus is the largest organelle in the cell and is bounded by an envelope consisting of a double membrane.
- Genetic material is concentrated in one part of the nucleus.
- Nuclear pores provide the means for transport across the envelope for large molecules to enter or leave the nucleus.

The nucleus is often the largest visible compartment in a eukaryotic cell, as seen in **FIGURE 1.12** and contains almost all of the genetic material (in fact, all except for the limited number of genes carried by the mitochondrion or chloroplast). The size of the nucleus is related to the amount of DNA it contains, so the proportion of the cell volume that it occupies varies widely; typically it is 1%–2% in yeast cells

ORGANELLE	$[Ca^{2+}]$
MITOCHONDRIA: Intermembrane space	High (~10^{-3} M)
CYTOSOL	Low (~10^{-8} –10^{-7} M)
NUCLEUS: Lumen of nuclear envelope — ENDOPLASMIC RETICULUM	High (~10^{-3} M)

FIGURE 1.10 Compartments in a eukaryotic cell may have different ionic milieus.

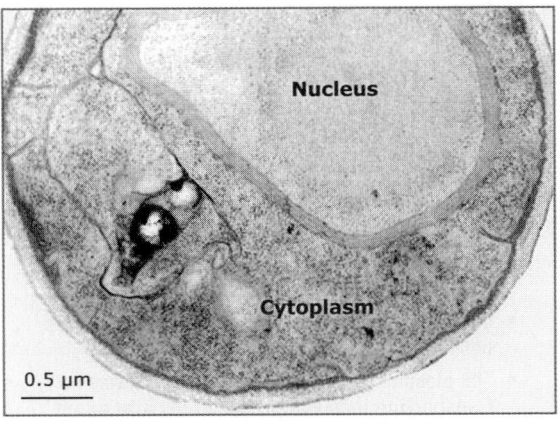

Nucleus

Cytoplasm

0.5 µm

FIGURE 1.12 Although the proportion of the cell that is taken up by the nucleus varies according to cell type, the nucleus is usually the largest and most noticeable compartment in eukaryotic cells. Photo © Wright, et al., 1988. Originally published in **The Journal of Cell Biology**, 107: 101–114. Used with permission of Rockefeller University Press. Photo courtesy of Jasper Rine, University of California, Berkeley.

and ~10% in most somatic animal cells (see *9.2 Nuclei vary in appearance according to cell type and organism*). The genetic material forms a mass called **chromatin** that is concentrated in one part of the nucleus.

The nucleus is surrounded by an **envelope** that consists of two concentric membranes, the outer nuclear membrane and the inner nuclear membrane, as shown in **FIGURE 1.13** (see *9.6 The nucleus is bounded by the nuclear envelope*). The two membranes are separated by a lumen. The outer membrane of the nuclear envelope is continuous with the ER membrane, and the lumen of the nuclear envelope is continuous with the lumen of the ER. The inner nuclear membrane is usually supported by a network of filaments called the nuclear lamina, located in the nucleus and anchored to the inner membrane.

Because small molecules move freely between the cytosol and the nucleus, the aqueous environment of the compartments is the same. However, material with a greater mass than ~40,000 daltons (corresponding to a small protein) can only enter or leave the nucleus if it is transported through **nuclear pore complexes** that are embedded in the envelope. The nuclear pores are the most conspicuous features of the nuclear envelope at the level of electron microscopy (see *9.9 Nuclear pore complexes are symmetrical channels*). Each nuclear pore complex has a central channel that is used for both import and export of material above the size limit for free diffusion. This enables the nucleus to have a different composition from the cytosol with regard to proteins and other large molecules.

The nucleus contains subcompartments that have specialized functions, although they

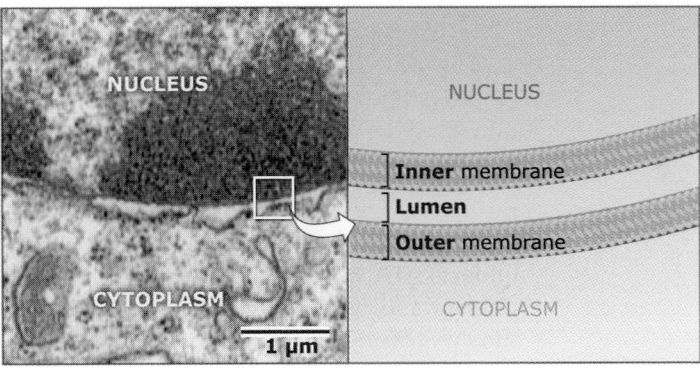

NUCLEUS

CYTOPLASM

1 µm

NUCLEUS

Inner membrane
Lumen
Outer membrane

CYTOPLASM

FIGURE 1.13 The nucleus is surrounded by the nuclear envelope, which consists of an inner membrane and an outer membrane. The membranes are separated by a lumen that is contiguous with the lumen of the endoplasmic reticulum. Photo courtesy of Terry Allen, University of Manchester.

are not delineated by membranes (see *9.4 The nucleus contains subcompartments that are not membrane-bounded*). The major subcompartment in the nucleus is the **nucleolus**, which can be seen by light microscopy and is where ribosomal RNAs are synthesized and ribosomal subunits are assembled.

What is the advantage to a eukaryotic cell of having a nucleus? The nucleus protects the DNA and helps to concentrate the regulatory proteins and repair enzymes. The human genome is 750× larger than the *E. coli* genome, so any particular sequence is a correspondingly smaller part of the genome. This requires regulatory factors to achieve higher concentrations in order to find their targets. It helps to confine the target (i.e., the genome) and the regulatory proteins within a smaller part of the cell (i.e., the nucleus). This also enables the genome to be better protected against adventitious damage.

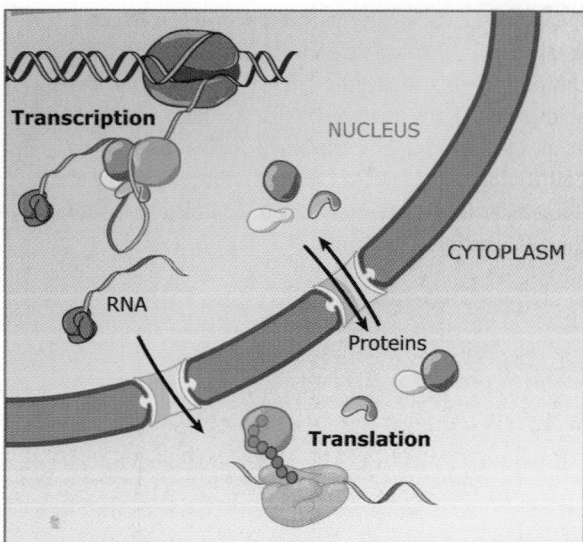

FIGURE 1.14 RNA is transported from the nucleus to the cytoplasm, and proteins are transported into the nucleus (and sometimes out again).

Having a nucleus has important consequences. **FIGURE 1.14** shows that the transport of macromolecules between the nucleus and the cytoplasm is a two-way process. All proteins that are needed in the nucleus (including those required for replication and transcription) must be imported from the cytoplasm. By contrast, mRNAs are transcribed in the nucleus but must be exported to the cytoplasm, where the machinery for protein synthesis is located. This is quite different from the situation in prokaryotes, where transcription and translation are coupled, occurring in the same time and place. Movement of molecules into and out of the nucleus is an important target for regulation.

A typical eukaryotic cell has one nucleus. However, there are some exceptional circumstances that create cells having many nuclei. This occurs notably in early insect development, as in *Drosophila*, when many nuclear divisions occur without cell division, producing a **syncytium** that contains hundreds of nuclei within a common cytoplasm. Another circumstance in which a syncytium forms is when animal muscle cells fuse together. At the other extreme, a few differentiated cell types, such as mammalian mature red blood cells, lack nuclei. (Begging the issue of whether these can be considered "cells," note that they are end-differentiated products that are descended from cells.)

1.8 The plasma membrane allows a cell to maintain homeostasis

Key concepts

- Hydrophilic molecules cannot pass across a lipid bilayer.
- The plasma membrane is more permeable to water than to ions.
- Osmotic pressure is created by ionic differences between the two sides of a membrane.
- The plasma membrane has specific systems for transporting ions and other solutes into or out of the cell.
- The transport systems allow the cell to maintain a constant internal environment that is different from the external milieu.
- Ion channels are proteinaceous structures embedded in membranes that allow ions to cross the membrane while remaining in an aqueous environment.

The plasma membrane fulfills several functions required for the integrity of the cell as given below:

- It contains the contents of the cell within a limited volume.
- It allows the aqueous conditions within the cell to be different from those outside it.
- It contains proteinaceous complexes that control the import and export of molecules.
- It has systems for signaling between the interior of the cell and the external environment.

The intrinsic properties of membranes create a need for cells to regulate movement of water and ions. **FIGURE 1.15** shows that membranes are permeable to hydrophobic compounds. The difference in permeability to water and to ionic solutes has an important

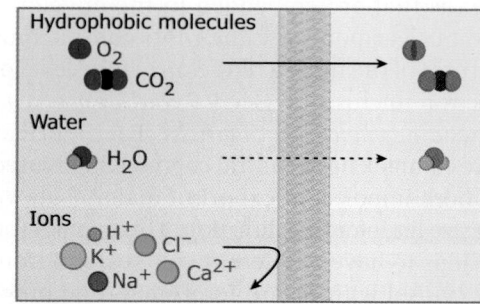

FIGURE 1.15 In contrast to hydrophobic molecules and water molecules, ions cannot rapidly traverse the lipid bilayer itself.

consequence in allowing **osmotic pressure** to develop across the membrane in response to differences in concentrations of dissolved substances on either side.

The rate at which sodium or potassium ions (e.g.) can diffuse across a lipid bilayer is 10^{10} times the rate of water diffusion. As a result, when there is a difference in ion concentrations on either side of the membrane, water moves across to equalize the concentration of solutes on either side, as shown in **FIGURE 1.16**.

If a cell had no mechanism to control solute levels, it would shrink or expand in response to osmotic pressure, whenever the concentration of solutes outside was greater than inside or vice versa. Its size would, therefore, fluctuate depending on external conditions. In extreme conditions, this would be lethal in either condensing the cell to a nonfunctional mass or in bursting it.

The cell responds to this situation by controlling the movement of ions and water across the plasma membrane. Its ability to maintain a constant internal environment is called **homeostasis**. This is an important function of all cells, whether they are part of unicellular or multicellular organisms. One major role of homeostasis in animal cells is to cope with osmotic pressure by balancing the ionic composition so as to avoid the accumulation of water. To maintain homeostasis, ions and water may need to be moved into or out of a cell in a regulated manner.

For unicellular organisms, homeostasis is necessary because the exterior environment

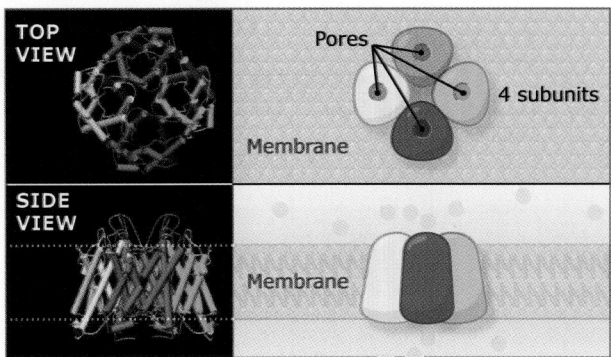

FIGURE 1.17 Protein complexes embedded in a membrane provide aqueous channels (pores) for the transport of ions and small molecules from one side of the membrane to the other. Structures created from RCSB Protein Data Bank.

may be subject to significant fluctuations. For multicellular organisms, it enables individual cells to maintain internal environments that are distinct from that of the extracellular fluid. Typically, the interior of the cell has higher potassium, but lower sodium and calcium, concentrations than in the exterior medium.

FIGURE 1.17 shows that passage across the membrane is made possible by the presence in the lipid bilayer of protein complexes that provide channels through the membrane. The outer surface of the protein complex contacts the lipid bilayer, but the inner surface surrounds an aqueous environment. An ionic solute, or a hydrophilic protein, passes through the aqueous channel without ever contacting the lipid bilayer. Channels are specialized for the passage of particular substrates.

The mechanism used to move an ion across a membrane depends on whether it is moving from high to low concentration or in the reverse direction. The difference in an ion concentration between the interior and exterior of the cell creates a gradient across the membrane. When an ion is moving with the gradient—that is, from the side of the membrane where its concentration is high to the side where its concentration is low—an **ion channel** allows the concentration gradient to drive it across. When it is necessary to move an ion against the concentration gradient, a **carrier protein** that uses an energy supply is required (see *6.2 Channels and carriers are the main types of membrane transport proteins*).

If a channel simply created an aqueous passage across the membrane, ionic conditions on either side would rapidly equalize. To maintain the integrity of the ionic milieu on either side, a channel is closed unless material is being trans-

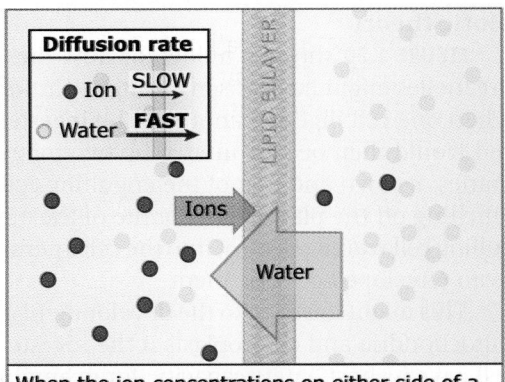

When the ion concentrations on either side of a membrane differ, water moves across to equalize the concentration of solutes on both sides.

FIGURE 1.16 The movement of water across a membrane occurs in response to osmotic pressure. The direction of movement depends on the relative concentrations of solutes on either side of the membrane.

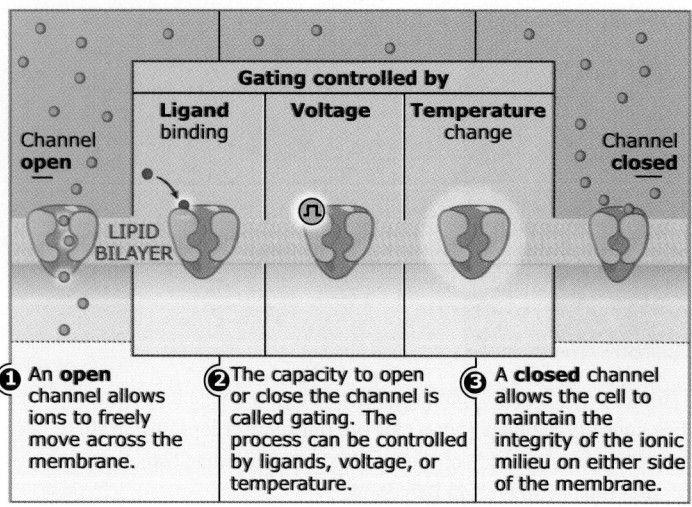

Gating controlled by

Channel open	**Ligand** binding	**Voltage**	**Temperature** change	**Channel closed**

LIPID BILAYER

❶ An **open** channel allows ions to freely move across the membrane.

❷ The capacity to open or close the channel is called gating. The process can be controlled by ligands, voltage, or temperature.

❸ A **closed** channel allows the cell to maintain the integrity of the ionic milieu on either side of the membrane.

FIGURE 1.18 The process of gating controls the opening and closing of an ion channel.

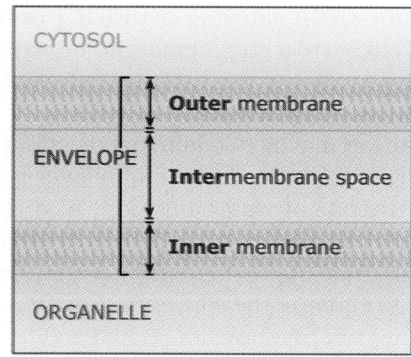

CYTOSOL

ENVELOPE
- **Outer** membrane
- **Inter**membrane space
- **Inner** membrane

ORGANELLE

FIGURE 1.19 An envelope consists of an outer membrane separated from the inner membrane by an intermembrane space. Each membrane is a lipid bilayer.

ported across. The capacity to open or close the channel is called **gating**. **FIGURE 1.18** shows that a gate accomplishes a conformational change that allows ions to move through the channel or that blocks it. Gates can be controlled in various ways such as by small molecule ligands, voltage, or temperature.

In addition to controlling the movement of solutes, all cells have **aquaporins** that can transport water across the plasma membrane (see *6.11 Selective water transport occurs through aquaporin channels*). The aquaporins react to osmotic pressure by moving water through a special channel.

Plant cells deal with osmotic pressure in a different way. In plant cells, water is accumulated inside special compartments called vacuoles, the internal pressure is contained by the rigid cell wall, and this is, in fact, used to drive the expansion of the cell (see *21.10 Cell expansion is driven by swelling of the vacuole*).

1.9 Cells within cells: Organelles bounded by envelopes may have resulted from endosymbiosis

Key concept
- Organelles bounded by envelopes probably originated by endosymbiosis of prokaryotic cells.

Three organelles of the eukaryotic cell—the nucleus, mitochondrion, and chloroplast—

are each bounded by an envelope. **FIGURE 1.19** shows that the envelope is a double membrane in which the outer membrane faces the exterior and is separated by an intermembrane space from the inner membrane, which circumscribes the interior compartment. All of the organelles surrounded by envelopes contain genetic material. (The other membrane-bounded organelles are surrounded by a single lipid bilayer and do not contain genetic material.)

The structure of an envelope surrounding a compartment that contains genetic material immediately reminds us of the prokaryotic cell. This similarity provided the basis for the idea that these compartments might have evolved from prokaryotic cells that were engulfed by a host cell. This resembles the situation of **endosymbiosis**, in which certain types of bacteria enter and then live within the cytoplasm of a eukaryotic host cell. Accordingly, this model for organelle evolution is called the **endosymbiotic theory**.

FIGURE 1.20 shows a model for how these organelles might have arisen during evolution when one cell ingested another. The ingested cell would then be surrounded by two membranes: its own and that of the engulfing cell. Pinching off the plasma membrane of the engulfing cell would release it into the cell interior as an enveloped compartment.

This might have led to the development of mitochondria and chloroplasts if the ingested cell gave its host a new capacity; for example, the ability to perform photosynthesis. In due course, the ingested cell would lose functions that it no longer required, because they were provided by the surrounding cytoplasm, and it would come to specialize in providing the functions needed for its host.

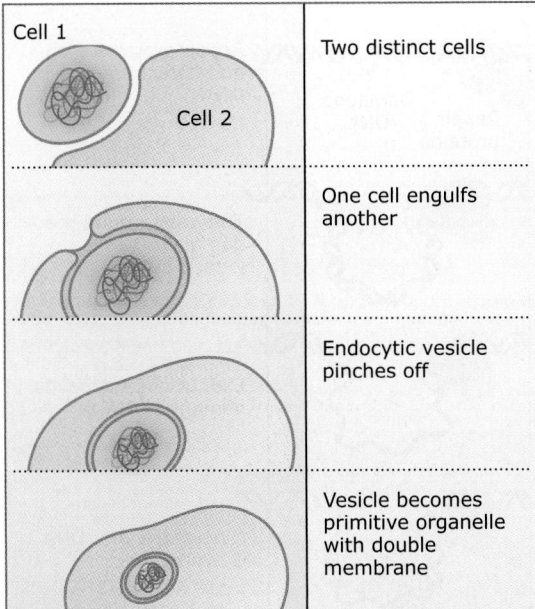

FIGURE 1.20 Enveloped organelles could have evolved when one cell ingested another.

Cell 1
Cell 2

Two distinct cells

One cell engulfs another

Endocytic vesicle pinches off

Vesicle becomes primitive organelle with double membrane

Mitochondria and chloroplasts have far fewer genes than an independent bacterium and have lost many of the genetic functions that are necessary for independent life (such as those coding for metabolic pathways). The majority of genes coding for organelle functions in fact are now located in the nucleus. (The proteins are synthesized in the cytosol and then imported into the organelle.) These genes must have been transferred into the nucleus from the organelle at some time after the endosymbiotic event.

We can trace the exchange of material between the nucleus and an organelle by comparing the locations of the corresponding genes in different species. Since the original engulfment, most of the exchange of genetic information between each organelle genome and the nuclear genome has involved the transfer of functions to the nuclear genome, although there can also be transfer in the other direction.

Both organelles have retained the capacity to synthesize some proteins by expressing their own genetic information. In fact, the best evidence for the endosymbiotic origin of chloroplasts and mitochondria lies with their apparatus for gene expression. In each type of organelle, this is closely related to the apparatus of modern prokaryotes and is much more distantly related to the apparatus of eukaryotic cells. Sequence homologies suggest that mitochondria and chloroplasts evolved separately. Mitochondria share an origin with

α-purple bacteria, and the chloroplast appears most closely related to cyanobacteria.

A relationship with bacteria can also be seen in the way that organelles propagate. Mitochondria and chloroplasts divide much like bacteria, by invaginating the surrounding membrane to make a division across the interior. The components used for division are related to those used in bacteria.

The origin of the nucleus is less obvious. Perhaps at some early stage in eukaryotic evolution, one prokaryotic cell engulfed another, and the engulfed cell took over the genetic functions for the combined unit. Once again, the nature of the apparatus for gene expression is suggestive, with eukaryotes and archaea better related to one another than to bacteria. However, eukaryotes also have genes related to those of bacteria, most closely where metabolic functions are concerned. One possibility is that eukaryotes arose by a fusion event involving both bacteria and archaea, with genes from both ending up in the nucleus formed from the engulfed cell.

Concept and Reasoning Check

1. What is the endosymbiotic theory of origin of cellular organelles?

1.10 DNA is the cellular hereditary material, but there are other forms of hereditary information

Key concepts

- DNA carries the genetic information that codes for the sequences of all the proteins of the cell.
- Information can also be carried in cellular structures that are inherited.

The double helix of DNA carries the basic hereditary information for every living cell. For bacteria and archaea, all of the sequence information is usually carried in a single chromosome. For eukaryotes, all but a handful of genes are carried in the chromosomes of the nucleus, and small amounts of sequence information are carried in mitochondria and (for plants) in chloroplasts.

DNA can also be the genetic material of viruses, but some viruses use RNA. All viruses surround the genetic material with a protein coat called the capsid. Viruses are not living organisms, of course, but function genetically during an infection of a cell in the same way as the apparatus of the host cell.

Cells are also able to perpetuate information that is not carried in the sequence of DNA. This is called **epigenetic** inheritance. Formally, this describes a situation in which two cells have different phenotypes although their DNA sequences are identical at the locus responsible for the phenotype.

We do not know whether cells *require* information in some form other than the sequence of DNA. If we could read out the sequence of DNA to form all of the proteins, would they be able to interact to form all of the cell structures and functions? If not, what is the nature of the missing information that enables cells to arise only from preexisting cells? Is it necessary to have preexisting structures to form templates for the assembly of cell structures (see *1.19 Localization of cell structures is important*)?

Concept and Reasoning Check

1. How is inherited information conveyed by cells?

1.11 Cells require mechanisms to repair damage to DNA

Key concepts

- The genetic material is continually damaged by environmental forces or by errors made by cellular systems.
- Repair systems to minimize damage to DNA are essential for the survival of all living cells.

Maintaining the integrity of the genetic information is of equal importance to replicating it accurately. In fact, there are more genes in the human genome specifically devoted to repairing damage to DNA than there are for providing the basic replication apparatus.

Individual errors in the DNA sequence occur in two ways. First, mistakes may be made during replication, with the wrong base being inserted into the new chain. Replication systems have proofreading mechanisms to guard against such errors, which reduce such corrections to a very low rate. Second, DNA may be damaged by environmental effects, including radiation or chemicals that modify the bases. The cell contains many **repair** systems that act to restore a damaged DNA sequence to its original state. **FIGURE 1.21** illustrates the action of a repair system that functions by recognizing damage to DNA, removing it, and replacing the excised material.

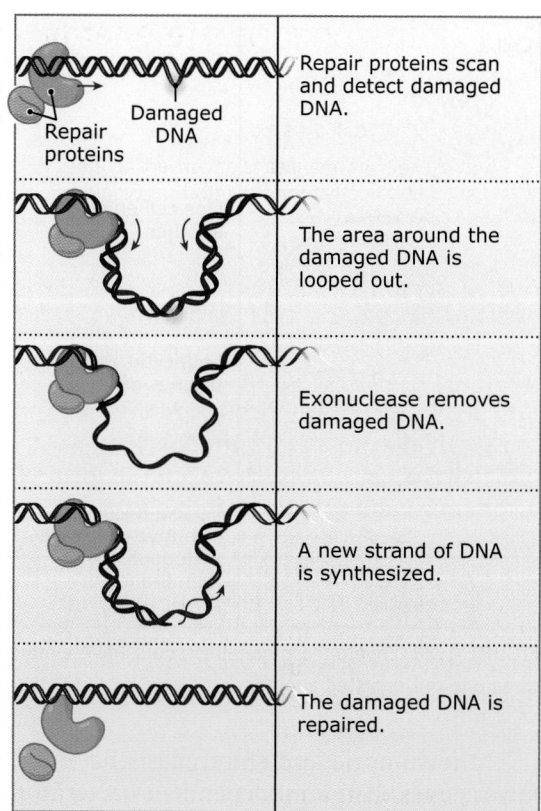

	Repair proteins scan and detect damaged DNA.
	The area around the damaged DNA is looped out.
	Exonuclease removes damaged DNA.
	A new strand of DNA is synthesized.
	The damaged DNA is repaired.

FIGURE 1.21 A repair system can recognize a damaged site in DNA, excise a stretch of nucleotides containing the damage, and resynthesize a replacement strand.

Some mutations occur in spite of the operation of these systems, but at a rate that can be tolerated—indeed, some rate of mutation is essential to provide the variations required for evolution. Mutations occur at a rate of $\sim 10^{-6}$ events per gene per generation or at an average rate of change per base pair of 10^{-9} to 10^{-10} per generation over a range of organisms from bacteria to higher eukaryotes. Even organisms that live under extreme conditions have a similar rate. These similarities suggest that the overall rate of mutation has been subject to selective forces that have balanced the deleterious effects of most mutations against the advantageous effects of some mutations.

No cell can survive without its repair systems. When all repair systems are eliminated from the bacterium *E. coli*, for example, a single irradiation event with ultraviolet light can be lethal, whereas a normal bacterium can repair quite large numbers of damaged bases.

Concept and Reasoning Check

1. How are errors introduced into DNA sequences?

1.12 Mitochondria are energy factories

Key concept
- All living cells have a means of converting energy supplied by the environment into the common intermediate of ATP.

A cell obtains energy from the food supplied by its environment. This energy then has to be converted into some form that can be distributed throughout the cell. The common solution (not just for mitochondria but also for the equivalent energy-handling systems in prokaryotes) is to store energy in the form of a common molecule that can be used whenever and wherever it is needed in the cell. The details of energy handling vary with different types of cells, but the feature common to all living cells is the ability to convert energy supplied by the environment into ATP, which provides the common molecule that can drive individual chemical reactions as required.

ATP can be generated in two ways: in the cytosol, and in the mitochondrion. The first pathway exists in the cytosol of a eukaryotic cell, or within a bacterium, where the process of glycolysis degrades glucose to pyruvate, and during this reaction two molecules of ATP are produced. This reaction can proceed anaerobically (in the absence of oxygen).

The second pathway is the main source of energy production and occurs in the mitochondrion in the eukaryotic cell. The process that makes ATP in the mitochondrion is called oxidative phosphorylation and involves the electron transport chain. Pyruvate generated from glycolysis enters the matrix (lumen) of the mitochondrion, where it is degraded and combined with coenzyme A to form acetyl CoA. The acetyl part of the acetyl CoA is then degraded to carbon dioxide by the citric acid cycle, releasing hydrogen atoms. The hydrogen atoms are used to reduce the carrier NAD+ to NADH, and then oxidation of NADH releases a proton and an electron.

Hydrogen ions (protons) are translocated across the membrane, from the matrix into the intermembrane space, while electrons are transported along the membrane through a series of protein carriers. **FIGURE 1.22** illustrates this process. The result is to create a proton gradient across the membrane. The gradient drives the protons back across the membrane through the large protein complex of ATP synthase, and the current drives the synthesis of

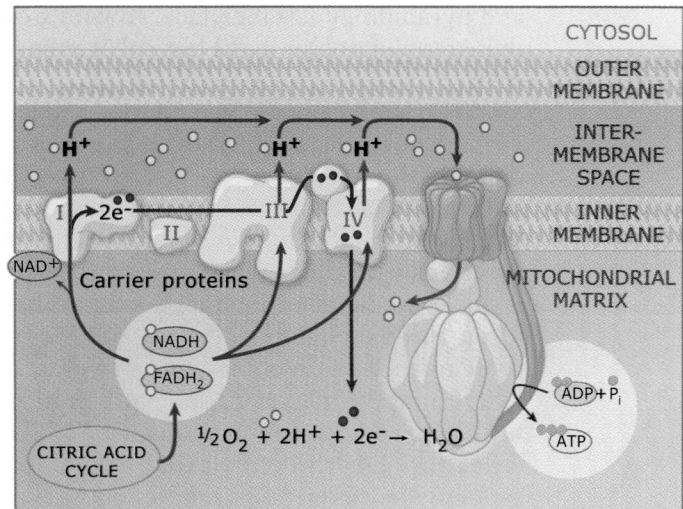

FIGURE 1.22 Chemiosmosis uses proton transport to drive synthesis of ATP.

ATP from ADP and inorganic phosphate. This process is called **chemiosmosis**.

Mitochondria are often called the power centers of the (eukaryotic) cell; they provide the energy that is required for its metabolic processes and for effecting structural changes. To be more precise, they convert energy provided by the environment into forms that can be used by the cell. The conservation of mitochondrial structure and function across all eukaryotes argues that the endosymbiotic event that created them must have occurred at the very start of eukaryotic evolution.

Concept and Reasoning Check
1. How and where is ATP generated?

1.13 Chloroplasts power plant cells

Key concept
- Plastids are membrane-bounded organelles in plant cells and can develop into chloroplasts and other specialized forms.

Plastids are membrane-bounded organelles that are present only in plant cells (see *21.20 Plants contain unique organelles called plastids*). Many of the basic metabolic reactions of plant cells occur within plastids. Plastids are highly specialized into several different types, but they all perform some common reactions. They are the organelles in which synthesis of fatty acids, many amino acids, and purines and pyrimidines all take place. In contrast, these reactions occur in the cytosol of animal cells.

Two closely apposed membranes, the inner and outer membranes, enclose plastids, and an intermembrane space separates the membranes, as is the case for mitochondria. The **stroma** is the interior of plastids that is bounded by the inner membrane. The stroma is similar to the mitochondrial matrix in that it contains the plastid DNA, RNA, and many proteins such as ribosomes and enzymes. It differs from the mitochondrial matrix in that the stroma contains membrane-enclosed disks called thylakoids; the thylakoid membrane contains the energy-producing systems.

All of the types of plastids develop from a common precursor organelle, called a proplastid. Proplastids are smaller than differentiated plastids, lack internal membranes, and are not specialized. When a plant cell develops into a particular differentiated type, its proplastids differentiate and acquire functions suited to that cell type. Thus, the type of plastid that develops depends on the cell type.

Chloroplasts are plastids that enable a plant to make ATP from a system in which the electrons are provided by chlorophyll molecules that have been activated by light, instead of by the chemical breakdown of glucose. In the presence of light, chloroplasts develop in the parts of a plant, such as leaves, in which light gathering and photosynthesis will occur. Plants that are grown in the dark do not develop chloroplasts but instead develop a different type of plastid in their leaves. Seeds and tubers have yet another type of plastid, the amyloplast, which synthesizes starch that is stored as granules in the stroma. Some types of plastids contain enzymes for the synthesis of certain small compounds. Chromoplasts synthesize and store pigments called carotenoids, which are red, orange, or yellow molecules that give some flowers and fruits their color.

Differentiated plastids can interconvert between different types according to environmental or developmental signals. For example, chloroplasts develop into chromoplasts when tomatoes ripen from green to red and when green leaves of deciduous trees turn red, orange, or yellow. During this interconversion, which is regulated through expression of nuclear genes, the plastids lose chlorophyll and thylakoid membrane and synthesize carotenoids.

1.14 Organelles require mechanisms for specific localization of proteins

Key concept
- All organelles import proteins from the cytosol.

The specialization of organelles for different functions means that each requires a unique composition of small molecules and macromolecules. However, synthesis of these components does not necessarily occur in the organelle where they function. Many components are imported from the cytosol into the organelle, including the proteins that form its basic structure. (Mitochondria and chloroplasts synthesize a small number of their proteins internally.)

How do organelle components get to the site where they function? As summarized in **FIGURE 1.23**, there are at least eight major types of organelle. Import and export of small molecules is controlled by proteins embedded in the organelle membrane(s). Import of proteins during or following their synthesis in the cytosol requires special mechanisms.

Organelle membranes define the types of locations to which proteins must be imported, as summarized in **FIGURE 1.24**. When a compartment, such as the endoplasmic reticulum or the Golgi stack, is bounded by a single membrane, a protein may be directed to the interior or may be incorporated in the membrane. When a compartment has an envelope, such as the nucleus, mitochondrion, or chloroplast, there are more possibilities: specific proteins are found in the outer membrane, intermembrane space, inner membrane, and the interior.

The basic principle of protein localization is that any protein whose destination is a specific organelle carries within it a short amino acid sequence that constitutes a **sorting signal** (or targeting signal). Each type of organelle has one or more distinct classes of signals. Sorting signals are recognized by a specific cellular machinery at one or more steps along the route to the protein's final destination. Typically, these signals work by recognizing protein receptors that drive correct localization.

ORGANELLE	Function
Nucleus	Gene expression proteins exported and imported; RNA exported
Endoplasmic reticulum	Protein modification; proteins imported cotranslationally
Golgi apparatus	Protein modification; proteins arrive by trafficking from ER
Endosome Early endosome Late endosome	**Sorting of internalized proteins for transport to other compartments; proteins that function in endosomes are targeted from the secretory pathway**
Lysosome	**Degradation of internalized proteins; degradation of cytosolic proteins in stressed cells; proteins that function in lysosomes are targeted from the *trans*-Golgi network**
Mitochondrion	Energy handling; proteins imported from cytosol; some proteins synthesized in organelle
Peroxisome	**Oxidative processes; proteins imported from cytosol**

FIGURE 1.23 Each organelle has a distinct composition and structure that relate to its function.

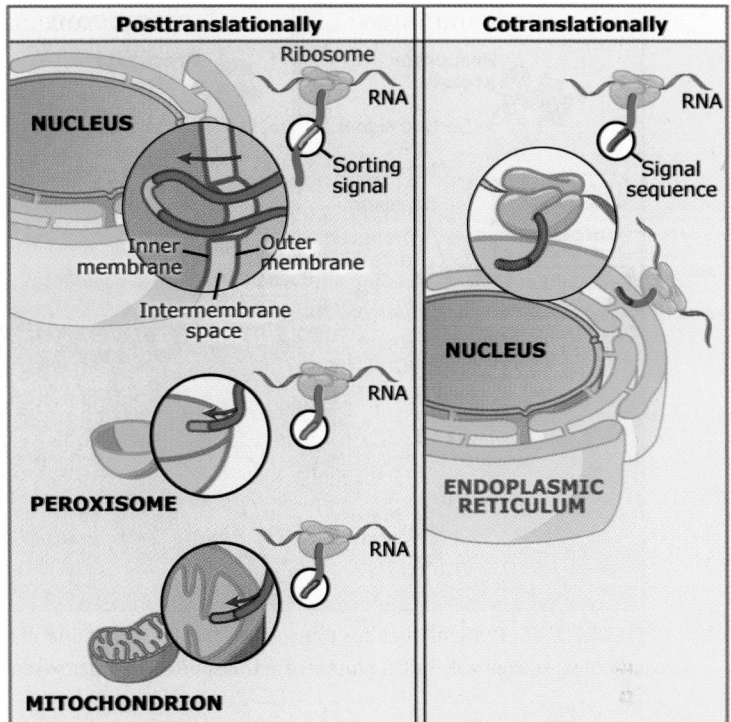

FIGURE 1.24 Proteins may be localized within the membrane or the lumen of an organelle; in the case of organelles with envelopes, proteins may also be localized within either membrane or the intermembrane space.

1.15 Proteins are transported to and through membranes

Key concepts

- Proteins are transported into organelles through receptor complexes that are embedded in the organelle's membrane.
- For the nucleus, and for organelles such as mitochondria or chloroplasts, proteins are released into the cytosol after synthesis and then associate with the organelle.
- For the endoplasmic reticulum, proteins are transferred into the receptor complex on the ER membrane during synthesis.

All membrane-bounded organelles have the same general problem: how to get a protein that is made in the cytosol to pass into or through the surrounding membrane. There is a large energetic obstacle to passing a hydrophilic protein through the hydrophobic membrane. Each organelle has developed its own variation on the same general solution. Its membrane contains a proteinaceous hydrophilic pore that is used to transport proteins so that they do not need to interact with the hydrophobic membrane. The nature of the pore, and the nature of its interaction with the transported protein, depends on the organelle.

Nuclear pores are massive structures that use a complex transport apparatus to identify proteins that must be transported into or out of the nucleus. The proteins are collected on one side of the envelope, escorted through the pore, and released on the other side (see *9.11 Proteins are selectively transported into the nucleus through nuclear pores*). The pore extends across both membranes of the nuclear envelope, and the protein is transported in its mature form from one side to the other.

Organelles such as mitochondria and chloroplasts have proteins within both their outer and inner membranes whose function is to transport target proteins into the organelle. (There is no export.) The imported proteins are synthesized by cytoplasmic ribosomes and released into the cytosol. They have specific sequences that interact with the receptors in the organelle membrane (see *7.27 Import into mitochondria begins with signal sequence recognition at the outer membrane*). FIGURE 1.25 shows that the channel through the membrane is very restricted, and the transported protein must be unfolded in order to enter and then refolded into its mature conformation on the

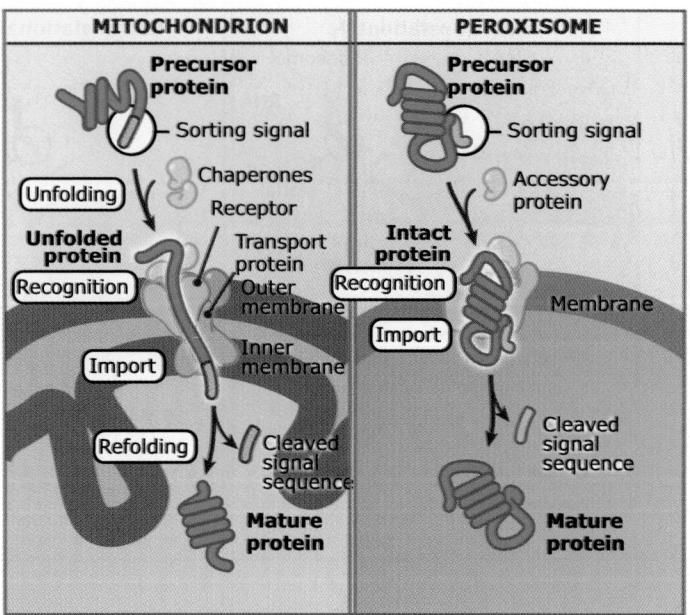

MITOCHONDRION	PEROXISOME

MITOCHONDRION

Precursor protein — Sorting signal

Chaperones

Unfolding

Receptor

Unfolded protein

Recognition — Transport protein / Outer membrane

Inner membrane

Import

Refolding — Cleaved signal sequence

Mature protein

PEROXISOME

Precursor protein — Sorting signal

Accessory protein

Intact protein

Recognition — Membrane

Import

Cleaved signal sequence

Mature protein

FIGURE 1.25 Proteins that are transported into mitochondria must be unfolded; in contrast, intact proteins are transported into peroxisomes.

other side. This requires extensive involvement of accessory proteins—chaperones—that control protein folding (see *1.17 Protein folding and unfolding is an essential feature of all cells*). One interesting exception is the peroxisome, which has developed a system for importing proteins in their mature, folded conformation (see *7.30 Proteins fold before they are imported into peroxisomes*).

The endoplasmic reticulum, Golgi apparatus, endosomes, and plasma membrane are separate organelles but use the same system for transporting and localizing proteins. The process starts when a nascent protein, while still in the process of being synthesized by a ribosome in the cytosol, associates by means of a special "signal sequence" with receptors at the surface of the endoplasmic reticulum (see *7.2 Proteins enter the secretory pathway by translocation across the endoplastic reticulum membrane (an overview)*). This interaction leads to the formation of a ribosome-membrane junction followed by the insertion of the protein into the channel. The ribosome synthesizing the protein remains associated with the membrane of the endoplasmic reticulum while synthesis continues.

From this starting point, the protein either may pass through the membrane into the lumen of the ER or may become incorporated in the membrane. If its ultimate destination is beyond the endoplasmic reticulum, in one of

the stacks of the Golgi apparatus, or in an endosome or the plasma membrane (or if it is to be secreted from the plasma membrane into the environment), it is recognized by means of specific amino acid sequences and is transported by the process known as protein trafficking (see *1.16 Protein trafficking moves proteins through the endoplasmic reticulum and Golgi apparatus*).

There is a more difficult problem for proteins that are localized within the membrane. The same pores are used to associate the protein with the membrane, but instead of passing through, the protein is transferred laterally to reside within the membrane. Just as signal sequences *initiate* translocation across the ER membrane through a protein-lined channel, likewise specific "stop transfer" sequences appear to terminate translocation and set in motion events leading newly synthesized membrane proteins to a final location within or spanning the lipid bilayer. This mechanism is much less well understood but is thought to involve some transient disruption of the translocation channel to allow hydrophobic regions of the protein to associate with the surrounding lipids.

1.16 Protein trafficking moves proteins through the endoplasmic reticulum and Golgi apparatus

Key concepts

- All proteins that are localized in the ER, Golgi apparatus, or plasma membrane initially associate with the ER during synthesis.
- Proteins are transported from one compartment to another by membranous vesicles that bud from one membrane surface and fuse with the next.
- Proteins are transported into the cell from the exterior by vesicular transport in the reverse direction.

Entry into, or passage through, a membrane is a unique event that a protein undertakes on a single occasion. When a protein is required to move through a series of membranes, as in the case of a protein that initially associates with the ER but ultimately is released from the plasma membrane, its association with the membrane occurs at the beginning of the process. After that, it remains in a membranous environment and is transported from one membrane to another in a **vesicle**. Similar systems are used to transport proteins both out of and into the cell.

Exocytosis is the process by which proteins are directed to either the plasma membrane or the extracellular space (see *8.2 Overview of the exocytic pathway*). Some proteins are secreted constitutively, that is, they are always transported out of the cell after they have been synthesized. Other proteins, especially those made by some specialized cells, such as those producing digestive enzymes, are released only when the cell receives an appropriate stimulus.

Once a protein enters the ER, it typically remains within the membrane or the lumen unless and until it is transported to another part of the system by a vesicle or returned to the cytosol to be degraded. Transport vesicles are small spheres (typically 100–200 nm) of membrane that form by "budding" from a membrane, as illustrated in **FIGURE 1.26** (see *8.4 Concepts in vesicle-mediated protein transport*). A transport vesicle buds from one membrane surface and then moves to another membrane surface with which it fuses. The vesicles are called **coated vesicles** because the membrane is surrounded by a protein coat. Different types of vesicles can be distinguished by their coats, which are responsible for targeting and for selecting the proteins that are carried by the vesicle. A soluble protein is carried from a donor compartment to a recipient compartment within the lumen of the vesicle; a membranous protein is carried within the vesicle's membrane. By this means, a protein may be transported from the ER to the Golgi apparatus, across the Golgi stacks, and to the plasma membrane.

The import of proteins into the cell also uses coated vesicles. Vesicles form at the plasma membrane and move into the cell; this is called **endocytosis** (see *8.3 Overview of the endocytic pathway*). Endocytosis works on the same principle as exocytosis, but the pathway proceeds in the opposite direction. Vesicles budding from the plasma membrane are used to internalize material from the extracellular medium and to retrieve material from the plasma membrane. The vesicles that are used for endocytosis are different from those used in exocytosis, as defined by their protein coats. Some pathogens use endocytosis to enter their host cells; indeed, an endocytic event may have been responsible for the origins of mitochondria and chloroplasts (see *1.9 Cells within cells: Organelles bounded by envelopes may have resulted from endosymbiosis*).

FIGURE 1.27 summarizes the similar mechanisms that are used to transport newly synthesized proteins along the exocytic pathway and to import proteins into the cell via the endocytic pathway. In each case, a protein may be transported by a series of budding and fusion events as it moves from one membrane surface to another.

One consequence of this process is that there is continuous movement of membrane components from one membrane to the next as successive budding and fusion events occur. Quantitatively, exocytosis is greater than endocytosis, so forward (anterograde) transport results in a net flow of lipids from the ER to the plasma membrane. Empty vesicles may return the lipids to earlier parts of the pathway by retrograde transport. In effect, this creates a continuity of membranes throughout the system.

Transport of proteins in vesicles is a highly selective process. The basic principle is that a protein is selected by its possession of a special

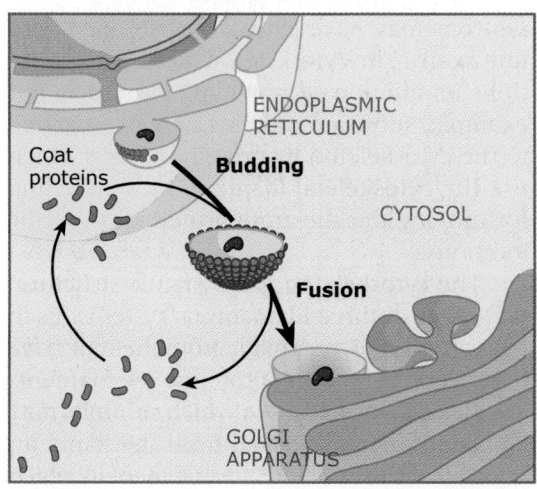

FIGURE 1.26 Transport of proteins between membrane-bounded compartments occur when vesicles containing the proteins bud from one compartment and subsequently fuse with another compartment.

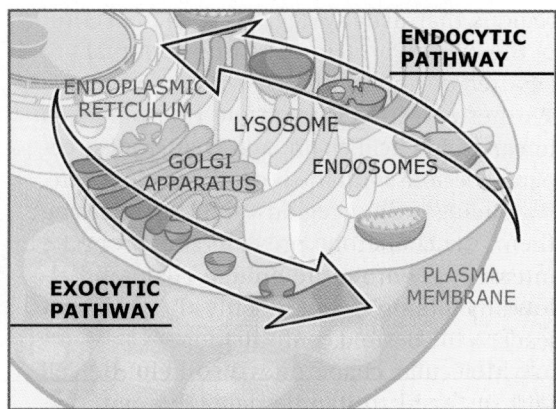

FIGURE 1.27 Transport vesicles move proteins through the ER, Golgi apparatus, and plasma membrane.

signal (most often a short sequence of amino acids) for inclusion in a vesicle moving to the next compartment. If it does not have this signal, it remains in the compartment it has reached or moves much more slowly. Budding vesicles can selectively incorporate the proteins to be transported, leaving resident proteins of the organelle behind.

1.17 Protein folding and unfolding is an essential feature of all cells

Key concept

• Protein conformation is a consequence of primary sequence but often cannot be achieved by spontaneous folding and requires assistance from molecular chaperones.

The activities of a protein are a consequence of its tertiary structure. An enzyme typically has an active site at which the catalytic activities occur. A structural protein that is incorporated into an oligomeric assembly has sites that interact with other protein subunits and/or a general structure designed for a specific purpose. In either case, the ability of the protein to fold into the correct structure is crucial. This is rarely accomplished spontaneously; more often, it requires the protein to interact with other proteins while it is being synthesized. Even when folding to the correct structure can be accomplished spontaneously, the speed of reaction is usually too slow to be useful in the living cell and requires assistance. Proteins that assist the folding of other proteins are called molecular **chaperones**.

Molecular chaperones work by detecting features in proteins that are not exposed in the mature conformation—typically hydrophobic regions that usually aggregate in the center of the mature protein structure—but that are exposed during synthesis. When a protein is synthesized, it exits from the ribosome as a linear amino acid chain, and chaperones recognize regions with the potential to aggregate. By binding to these regions, and later releasing them, the chaperone prevents inappropriate interactions within the protein chain and allows its folding to follow only the path that leads to the desired conformation.

Molecular chaperones also help the cell with its need to handle damaged material. When a protein is damaged, for example, because of excess heat, its conformation changes, exposing the same sorts of regions (typically hydrophobic) that were available to the chaperones while it was being synthesized. These regions can now be used to distinguish the damaged protein and to mark it for degradation.

Molecular chaperones are found in all cell compartments. In the cytosol, they assist folding of nascent proteins as they are synthesized on ribosomes. In the lumens of the organelles, they assist folding of proteins as they are translocated through the membrane. Many of the components of the complex machinery for protein translocation across the ER membrane are molecular chaperones.

Concept and Reasoning Check

1. How are molecular chaperones believed to work? (For further details on protein chaperones, see *7 Membrane targeting of proteins*.)

1.18 The shape of a eukaryotic cell is determined by its cytoskeleton

Key concepts

• The eukaryotic cell cytoskeleton is an internal framework of filaments, including microtubules, actin filaments, and intermediate filaments.
• It provides an organizing template for many activities, including anchoring organelles in place.

The term cytoskeleton describes frameworks of filaments that are found in most eukaryotic cells. They create a more rigid internal structure that determines the shape of the cell. For example, epithelial cells may be cuboidal, while neurons may have extremely extended very fine axons. The cytoskeleton has several functions in addition to providing structure. For example, substrate proteins may be attached to the cytoskeleton by protein motors, which use the cytoskeletal filaments as a guideline for moving the substrate proteins to specific locations.

The cytoskeleton is a dynamic structure. It consists of three filamentous systems. Each consists of a polymer made from the repetitive interactions of subunit proteins. The filaments are dynamic structures in which subunits may be added to or subtracted from the filament, often by a "treadmilling" arrangement in which they are added to one end and subtracted from the other end. The three systems are the **microtubules**, **actin filaments**, and **intermediate filaments**.

Microtubules are polymers of tubulin, a dimer of two closely related proteins, α-tubulin, and β-tubulin. They form a hollow tube ~25 nm in diameter. Microtubules are intrinsically unstable and are stabilized by interactions with other proteins. They are important for maintaining cell structure; drugs that dissociate microtubules cause most cells to lose their shape and to collapse into spheres. Dissolution of the microtubules causes the endoplasmic reticulum to collapse around the nucleus and the Golgi apparatus to fragment into pieces, showing that microtubules have an important role in maintaining the structure of these organelles.

The variety of cell structures that can be determined by microtubules is seen in the extremes of fibroblasts and neurons. Fibroblasts are motile cells that move within the organism; the microtubules radiate in a star-like pattern from a single point near the nucleus, as shown in **FIGURE 1.28**. By contrast, the long processes (axons and dendrites) that project from the body of a neuron are extended by means of parallel bundles of microtubules that run for the whole length, as shown in **FIGURE 1.29**. They are both structural elements, conferring strength on the extension, and are used as the tracks that motors use to transport proteins along the extension.

Microtubules undergo a dramatic change in organization, with a complete reorganization occurring every time the cell divides. **FIGURE 1.30** shows the reorganization that occurs at mitosis, when the microtubule network dissociates completely and is replaced by the spindle.

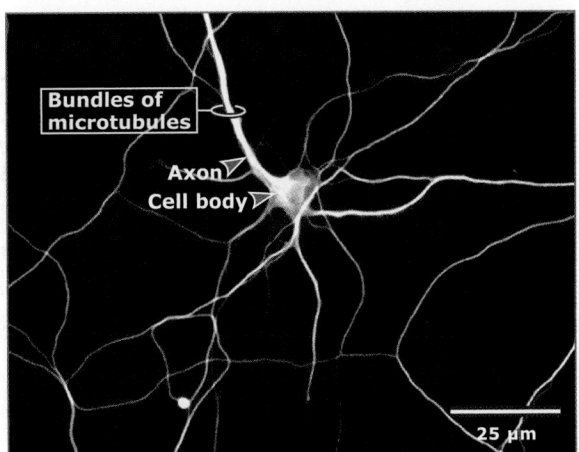

FIGURE 1.29 The extended processes of neurons have very long microtubules. Photo courtesy of Ginger Withers, Whitman College.

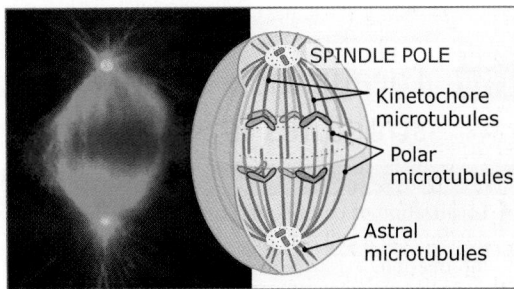

FIGURE 1.30 A dividing cell has a spindle formed from microtubules. In the fluorescence micrograph, microtubules, chromosomes, and the centriolar areas are stained green, blue, and yellow, respectively. Photo courtesy of Christian Roghi, University of Cambridge.

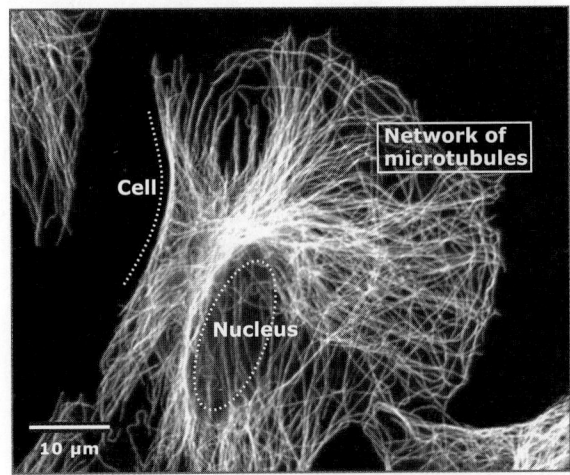

FIGURE 1.28 Fluorescence micrograph of fibroblast cells stained for microtubules. The position of the nucleus and part of the cell membrane are indicated. Photo courtesy of Lynne Cassimeris, Lehigh University.

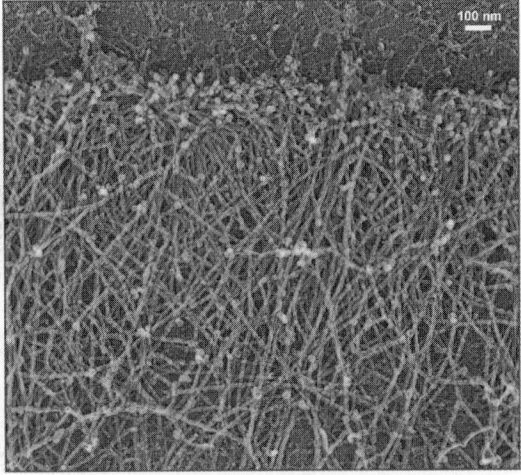

FIGURE 1.31 Electron micrograph showing the network of actin filaments at the edge of a fibroblast cell. Photo courtesy of Tatyana M. Svitkina, University of Pennsylvania.

Actin filaments are made from subunits of an actin protein. Actin is one of the most abundant proteins in the eukaryotic cell and has been highly conserved during evolution. All actin subunits have the same polarity in a filament, with an ATP-binding site at one end contacting the next subunit. An actin filament is a dimeric polymer of subunits arranged like two strings of beads twisted together to form a thread of ~8 nm in diameter.

In addition to crossing the cell, actin filaments extend into specialized structures that protrude from the surface and that allow a cell to move. FIGURE 1.31 shows the actin network in a fibroblast. The movement represents mechanical work, powered by the hydrolysis of ATP. It is the polymerization of the filament that drives motility, which, of course, is an essential property for cells in a multicellular organism as well as for unicellular organisms.

1.19 Localization of cell structures is important

Key concepts
- Localization of certain structures at specific positions in a cell may be part of its hereditary information.
- Positional effects are important in early development.

Cells have **positional information**: certain structures are localized in specific places. This is especially evident in some differentiated cells. For example, in **polarized cells**, one side of the cell is distinct from the other, and material may be selectively transported across the cell in a specific direction.

The organization of new structures can depend on the orientation of preexisting structures. One of the early experiments to show this relationship was performed in the 1960s with the protozoan *Paramecium*. FIGURE 1.32 shows that the organism is a single cell with an oval shape that has rows of cilia oriented asymmetrically on its surface. It is possible to invert the orientation experimentally by a technique in which two attached cells give rise to a single cell with inverted rows of cilia. Then the inverted pattern is inherited by daughter cells resulting from division. The underlying responsibility for the pattern is a property of the way that microtubules assemble on basal bodies to form the cilia, but there has been no change in the structures of the protein subunits.

Positional effects are often associated with microtubules. Another structure made from microtubules is the centriole, which is found in most eukaryotic cells. Centrioles are small proteinaceous structures that lie close to the microtubule organizing centers (MTOCs) that form the poles at either end of the cell where

FIGURE 1.32 Scanning electron micrograph of a *Paramecium* shows the rows of cilia. Photo © Tamm, 1972. Originally published in **The Journal of Cell Biology**, 55: 250–255. Used with permission of Rockefeller University Press. Photo courtesy of Sidney L. Tamm, Boston University.

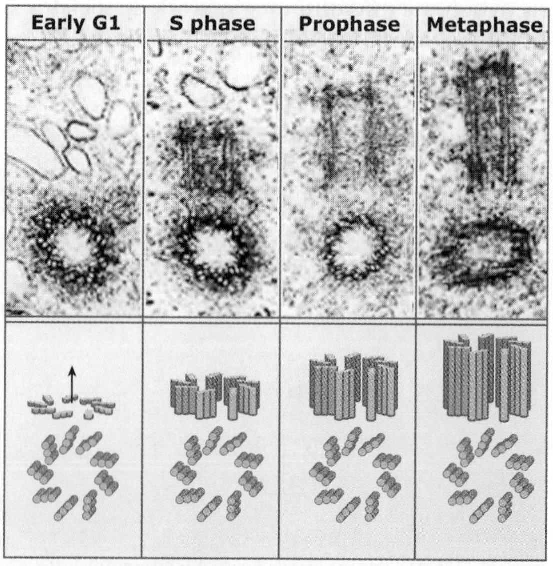

| Early G1 | S phase | Prophase | Metaphase |

FIGURE 1.33 Centrioles arise by templating on existing centrioles. Photos © Rattner and Phillips, 1973. Originally published in **The Journal of Cell Biology**, 57: 359–372. Used with permission of Rockefeller University Press. Photo courtesy of Jerome B. Rattner, University of Calgary.

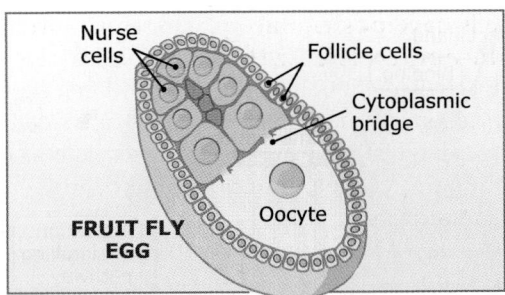

FIGURE 1.34 Nurse cells surround the oocyte and are responsible for transferring material to it that creates an asymmetrical pattern.

the microtubules of the spindle terminate. **FIGURE 1.33** shows that a new centriole always grows perpendicular to an existing centriole, possibly by some sort of templating mechanism. This suggests that centrioles cannot be formed *de novo* but can only arise from existing centrioles. This would imply that the centriole could be regarded as carrying information essential for heredity that is not directly carried by DNA (see *1.10 DNA is the cellular hereditary material, but there are other forms of hereditary information*).

Positional distinction may start in the egg. The egg of the fruit fly, for example, has gradients of proteins running along both the anterior-posterior (front to back) and dorsal-ventral (top to bottom) axes. **FIGURE 1.34** shows that these gradients are imposed on the egg by the nurse cells that surround it, and they are essential for controlling the early stages of development. Subsequent development of the egg into a female adult regenerates the system, in which each oocyte is surrounded by nurse cells that impose positional asymmetries on it. So the positional information in each generation is needed to recreate the positional information in the next generation and should be regarded as an important component of the hereditary information of the organism.

1.20 Cellular functions: Enzymes, pathways, and feedback

Key concepts

- Critical roles for enzymes in cellular function.
- Dimensions of homeostasis.

Thus far, we have focused on the structures that comprise a cell and discussed their larger roles.

Let us consider how those functions are actually carried out: through action of enzymes.

Enzymes are proteins that serve as catalysts. That is, they greatly speed up chemical reactions, typically by factors of at least a million, without being consumed or altered by the reaction. They do this by lowering the activation energy of the catalyzed reaction. A number of fundamental principles and distinguishing characteristics of enzymes are crucial for understanding how they carry out the functions associated with cells. They are as follows:

- Enzymes decrease the time it takes for a reaction to reach equilibrium, but do not alter the reaction equilibrium itself—they speed up the forward and reverse reactions by the same factor.
- Enzymes are highly specific in their choice of substrates (the reactants in the chemical reaction to be catalyzed).
- Enzymes are typically organized in pathways that allow cells to achieve a concerted goal (e.g., synthesis of fatty acids) in stepwise fashion.
- A number of feedback mechanisms occur to regulate enzymes and their pathways and can be thought of as contributing a fine granularity to homeostasis.

At the cellular level, homeostasis is brought about by each cell being able to sense critical goings on in its environment, by virtue of receptors, and then responding to what is sensed. The molecular details include signal transduction triggered by receptor activation, which alters the activity of a wide range of enzymes that govern pathways involved in the functions of the cell.

Hormones and nerves represent two extremes of receptor-mediated communication in multicellular organisms. Hormones act at a distance, via the bloodstream, and are typically lower in concentration, but act for a longer time due to their higher affinity for receptors. Neurotransmitters of the nervous system are delivered rapidly and precisely and are typically low affinity short action high concentration signaling molecules. Regardless of the nature of receptor-mediated communication, its proper function depends on signal transduction.

Changes in cell organization also contribute to homeostasis. For example, a high level of a hormone can result in endocytosis and clearance of receptors from the cell surface. In the absence of receptors, no amount of hormone can have an effect.

Finally, it must be remembered that the cell is a very crowded place where the interactions between proteins may influence many parameters including protein folding, biologic activity, and regulation.

1.21 Signal transduction pathways execute predefined responses

Key concepts

- Events on the outside of the cell can trigger actions inside the cell by using receptor proteins embedded in the membrane.
- A receptor spans the membrane and has domains on both the exterior and interior.
- The receptor is activated when a ligand binds to the exterior domain.
- Ligand binding causes a change in the structure or function of the interior domain.

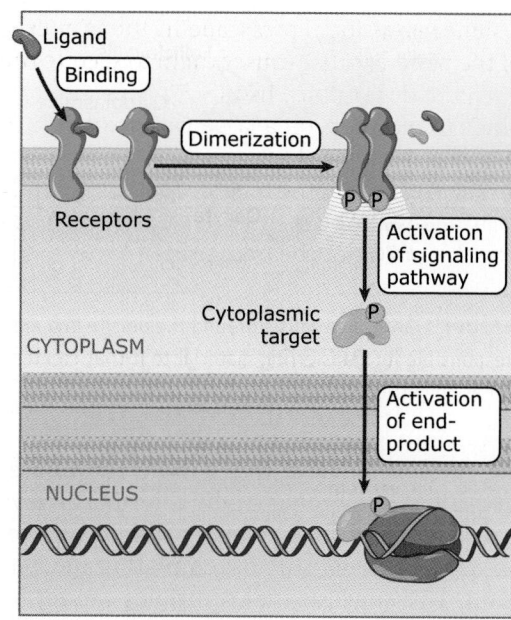

FIGURE 1.35 Ligand-receptor binding at the cell surface triggers a signal transduction pathway that generates a specific response.

The ability to respond to its environment is an essential property for every cell. For a unicellular organism, the environment is the milieu of the outside world. The response may be as simple as importing a nutrient that becomes available or as complex as migrating toward a food supply. For multicellular organisms, the environment may be created by other cells, and the ability to communicate between cells becomes essential, especially when cells acquire specialized functions that depend on their interactions with other cells.

Physical movement of molecules across the plasma membrane is not the only way a cell can respond to its environment. **Signal transduction** allows an event on the external side of the membrane to trigger a response within the cell. The mechanism also allows a signal to be amplified and to trigger a predefined response from the cell, as summarized in **FIGURE 1.35**.

The common feature in all signal transduction pathways is that a component in the environment is recognized, typically by a protein in the plasma membrane. The environmental trigger is called the **ligand**, and the plasma membrane protein is called the **receptor**. The receptor usually spans the membrane, and binding to the ligand on the extracellular side triggers a change that activates its function on the intracellular side. One common mechanism is for ligand binding to cause receptor monomers to dimerize;

formation of the dimer creates an enzymatic activity. This enzyme acts on some other component, which in turn acts on its substrate, and a whole chain of interactions ensues. Finally, the end product of the pathway is activated. The reaction can be amplified at some or all of its stages, so that a single ligand-receptor event produces many copies of the final product(s). The end product can cause a change in gene expression or a structural change in the cell.

The nature of the response is tailored to the situation. When a bacterium recognizes a nutrient in its vicinity, a signal transduction cascade resulting from the interaction of the nutrient with a receptor causes a change in the utilization of flagella so that the bacterium is propelled toward the food supply. When a yeast of one mating type recognizes the small polypeptide pheromone secreted by a cell of the other mating type, signal transduction triggers a response in which the two cells grow toward one another in a polarized manner and fuse together.

In multicellular organisms, signal transduction pathways may be connected to produce a coordinated physiologic response. When a mammal has a meal, the sugar content increases in the bloodstream. The sugar triggers a signal transduction cascade in the islet cells of the pancreas that causes them to secrete the polypeptide hormone insulin. The insulin then binds to receptors on a large variety of cells to trigger a pathway that causes the cells to take up the sugar from the blood stream. The sugar is stored as glycogen, providing a longer-term energy source for the organism.

1.22 All organisms have cells that can grow and divide

Key concepts

- The simplest form of division is shown by some organelles where the membrane is pinched inward.
- Bacteria often divide by growing a rigid septum across the cell as an extension of the cell wall.
- During mitosis, eukaryotic cells are extensively reorganized and form the specialized structure of the spindle that partitions the chromosomes to daughter cells.

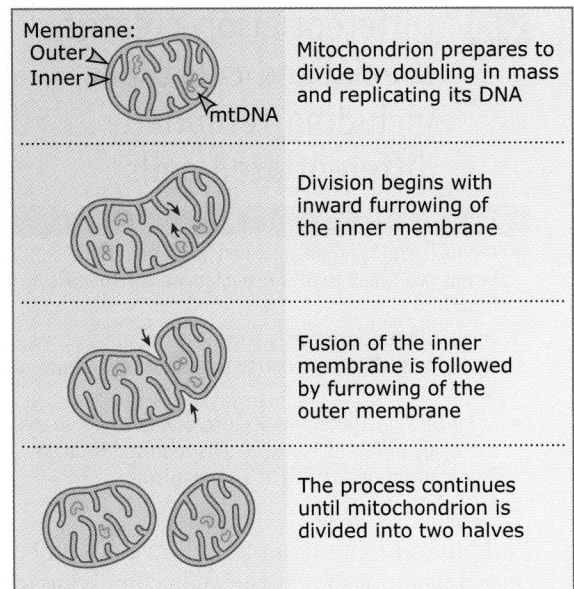

Membrane:
Outer
Inner
mtDNA

Mitochondrion prepares to divide by doubling in mass and replicating its DNA

Division begins with inward furrowing of the inner membrane

Fusion of the inner membrane is followed by furrowing of the outer membrane

The process continues until mitochondrion is divided into two halves

FIGURE 1.36 A mitochondrial membrane is pinched to separate the parental compartment into two daughter compartments.

A cell's basic requirement for ensuring that it can reproduce is that, after it has doubled its size, it can divide into two. The two copies of its genetic material are segregated by a specific mechanism that ensures each daughter cell gets one copy. Other components are divided stochastically by virtue of their distribution throughout the cytosol. For a eukaryotic cell, it is necessary to reproduce the internal compartments so that they can be apportioned to the daughter cells. This can involve either dissolution of a compartment into components that are subsequently reassembled into two daughter compartments, or a physical division of a parental compartment whose needs in principle resemble those of the cell itself.

If we include the division of organelles together with bacterial division and eukaryotic cell division, we can classify division mechanisms into three general groups.

The simplest form of division is shown by mitochondria in some cells, in which the membrane is pinched inward in a process that is akin to budding a vesicle from a membrane. **FIGURE 1.36** shows how this results in dividing the parental mitochondrion into two. This may be the oldest form of a division mechanism, reflecting the mechanism for division in the ancient bacteria from which mitochondria are descended.

Bacteria and archaea often rely on mechanisms related to the construction of their cell

walls. First, they ensure that one copy of each daughter chromosome finds itself at opposite ends of the cell, and then the cell is divided by growing the membrane and cell wall across the center. The common feature in most cases is the use of the division protein FtsZ, but there are examples where some other molecular apparatus must be used.

Eukaryotic cells undergo a complex reorganization in which the nucleus dissolves, the chromosomes are partitioned in equal sets to the opposite ends of the cell, a separation occurs into two daughter cells, and the cell structure reforms. The most common form of this **mitosis** is for the dissolution of the nucleus to allow the compartmentalized structure of the cell to be replaced by a single structure called the spindle; the individual chromosomes become compacted, and then they are attached to an apparatus that moves them so that one member of each daughter pair is localized on either side of the equator of the spindle. The critical stage in the evolution of this process may have been the acquisition of the capacity to bind the chromosomes to the structures of the division apparatus. This apparatus is responsible for the mechanics of cell division, but in addition a eukaryotic cell has pathways for controlling the cell cycle, that is, for determining if and when it activates a division.

1.23 Differentiation creates specialized cell types, including terminally differentiated cells

Key concepts

- A multicellular organism consists of many different cell types that are specialized for specific functions.
- Many differentiated cells have lost the ability to divide and/or to give rise to cells of different types.
- Stem cells have the potential to divide to generate the many different types of cells required to make an organism or a tissue of an organism.

Early in our definition of a cell, we described one of its properties as the ability to divide to generate copies of itself. While this is true of the cells of all unicellular organisms, the specialization achieved by cells in multicellular organisms requires a variation on this theme. **Differentiation** describes the process by which a cell acquires a new phenotype or gives rise to progeny cells whose phenotype differs from the parent.

Development of an organism is the process by which an initial single cell (a fertilized oocyte in the case of mammals) gives rise by sequential cell divisions to an organism consisting of cells of many different types. The fertilized zygote is **totipotent**: it has the capacity to give rise to all of the cells of the organism. Development might be regarded as the process during which there are successive restrictions on the capacity of progeny cells, so that they can give rise only to certain cells of the body.

The existence of multicellular organisms with specialized tissues makes it necessary to draw a distinction between two types of cells. **Somatic cells** are all the cells of the organism except the **germ cells**. Somatic cells are specialized for specific purposes in that organism and make no contribution to future generations. In a mammal, they are diploid (having one paternal and one maternal set of genetic information). Production of the next generation is accomplished by the formation of specialized germ cells. The germ cells are specific for each sex and are haploid; sperm and oocytes each contain one set of genetic information.

During development, and in the adult organism, some cells retain the capacity to generate all the cells of specific tissues, while others can generate only subsets of the cells or may in fact be unable to generate any other type of cell. Cells with broad activities in regeneration are called **stem cells**. For example, the cells of the immune system are descended from immune stem cells. One major question in the biology of organisms is whether the stem cells for any particular tissue exist only during development or are retained by the adult.

The development of a tissue requires a series of differentiation events, in which progeny cells continue to change phenotype until finally all of the necessary cell types have been produced. At the end of the process, some cells may have lost the capacity to divide; they are said to be **terminally differentiated**. Cells can also senesce and die. Many somatic cells are terminally differentiated; so are germ cells, although they have the unique quality that the union of two germ cells generates a zygote that is able to divide.

Terminally differentiated cells can have properties so extreme as to call into question whether they really fit our definition of a cell. A mammalian red blood cell differentiates to the point at which it loses its nucleus and consists of nothing more than a membrane that is supported internally by a network of actin and spectrin and encloses a solution of hemoglobin. Because the red blood cell is descended from fully functional cells, we regard it as a cell, although it has lost almost all of the usual defining features.

There used to be a debate as to whether cellular differentiation involved any permanent change in genetic capacity of the cell or whether the restrictions on phenotypic development were solely epigenetic. The development of the ability to clone organisms by placing a somatic cell nucleus into an oocyte decisively answers this question: almost every cell of the organism retains the genetic information needed to support development. (There are some exceptions for cells in which changes have occurred to the genetic material, such as those producing antibodies, or where genetic information has been lost, such as red blood cells.)

Concept and Reasoning Check

1. What are some consequences of terminal differentiation?

http://biology.jbpub.com/lewin/cells

To explore these topics in more detail, visit this book's Interactive Student Study Guide.

References

Brown, J. R., and Doolittle, W. F. 1997. Archaea and the prokaryote-to-eukaryote transition. *Microbiol. Mol. Biol. Rev.* v. 61 p. 456–502.

Chen, I. A., Roberts, R. W., and Szostak, J. W. 2004. The emergence of competition between model protocells. *Science* v. 305 p. 1474–1476.

Fulton, A. B. 1982. How crowded is the cytoplasm? *Cell* v. 30 p. 345–347.

Funder, J. W. 1988. Receptors, hummingbirds, and refrigerators. News in Physiological Sciences vol 2.

Lang, B. F., Gray, M. W., and Burger, G. 1999. Mitochondrial genome evolution and the origin of eukaryotes. *Annu. Rev. Genet.* v. 33 p. 351–397.

Sonneborn, T. M. 1970. Gene action in development. *Proc. R. Soc. Lond. B Biol. Sci.* v. 176 p. 347–366.

Bioenergetics and cellular metabolism

2

Raymond Ochs
School of Pharmacy, St. John's University, Queens, NY

George Plopper
Rensselaer Polytechnic Institute, Troy, NY

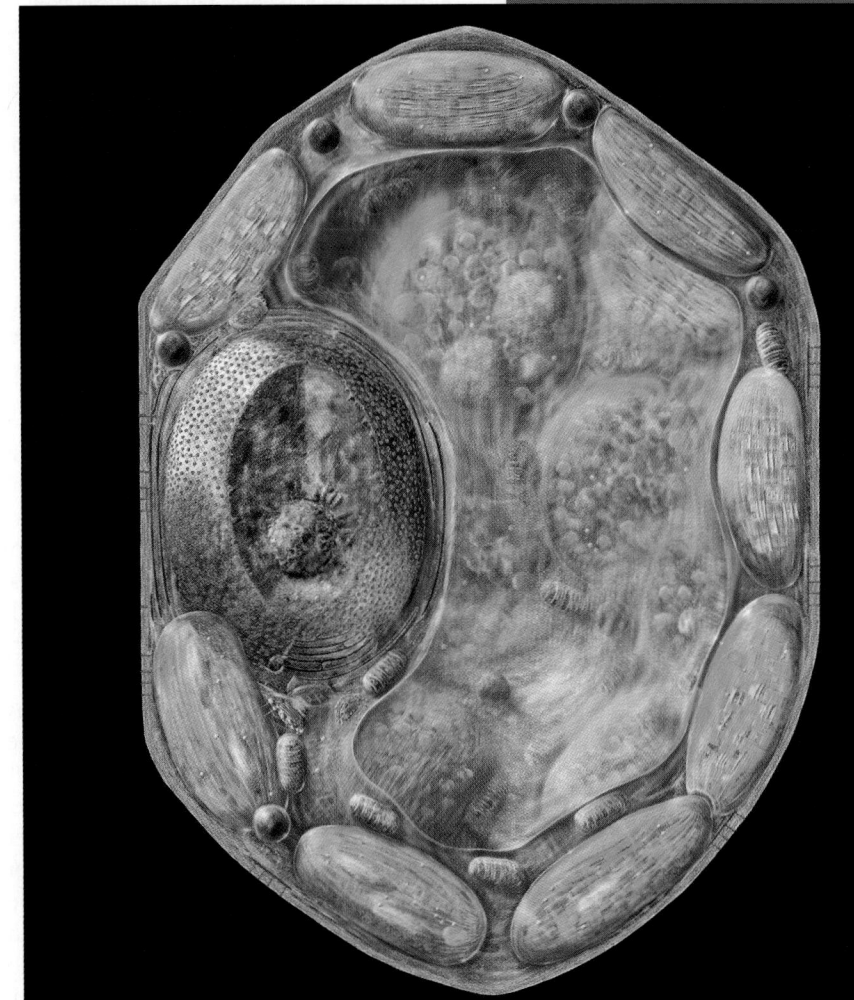

A TYPICAL PLANT CELL contains chloroplasts, which capture light energy to convert carbon dioxide and water into sugars, and mitochondria, which release this energy by converting sugars and other biomolecules back into carbon dioxide and water. Some of the released energy provides the power necessary to maintain life. Photo © Russell Kightley/Photo Researchers, Inc.

CHAPTER OUTLINE

2.1 Introduction

Key concepts

- The terms metabolism and catabolism refer to the breakdown and assembly of biologic molecules, respectively.
- Metabolism and catabolism are linked to energy changes in these biologic molecules.
- The basic principles of chemistry form the foundation of metabolism. In particular, collisions between molecules determine the rate of chemical reactions in cells and thereby influence how, where, and when biomolecules change their structure.
- Metabolism involves pathways composed of several steps, each involving a small change in the energy of the intermediate molecules.

One of the first principles in cell biology is that, to remain alive, all organisms must maintain a chemical disequilibrium with their environment. This means that the types and concentration of molecules in cells must always differ from that in the extracellular spaces. This requires a constant expenditure of energy, so cells devote a considerable amount of effort toward capturing energy from their environment. In many cases, this energy is in the form of complex biomolecules (e.g., food) that must be broken down to extract the useful energy from them. The process of capturing energy by breaking complex molecules into simpler molecules is called **catabolism**. For example, cells convert glucose to CO_2 and H_2O, and capture much of the energy released during this conversion. In many cases, this captured energy is used to build complex molecules that cells need, and this is called **anabolism**. Anabolic reactions are synthetic, building simple precursors into more complex molecules. For example, cells use dozens of chemical reactions to assemble multiple copies of a two-carbon compound to form cholesterol, which contains 27 carbons.

Thermodynamics is the study of energy changes. An examination of its principles provides insight into how reactions in cells proceed, and leads to an understanding of why reactions proceed as they do. It is possible to categorize reactions into two groups: those that achieve **near-equilibrium** in cells—which describes most of them—and those that are so far displaced from equilibrium that they never are reversed under cellular conditions; these are **metabolically irreversible**.

Much of this captured energy is stored in the form of potential energy (e.g., within chemical bonds, or as chemical gradients across membranes) that can later be used to power cellular activities. Muscle contraction, cell division, and cell migration are examples of how potential energy is converted into kinetic energy.

The purpose of this chapter is to examine how cells convert energy from one form to another. From *2.2 Chemical equilibrium and reaction kinetics are linked* to *2.5 Standard free energy, the mass action ratio, and the equilibrium constant characterize reaction rates in metabolic pathways*, we will explore the principles of thermodynamics to understand how and why chemical reactions are organized in cells. We will address two critical questions about cellular metabolism, and see how the answers to these questions evolve as we dig deeper into the principles of thermodynamics. In *2.5 Standard free energy, the mass action ratio, and the equilibrium constant characterize reaction rates in metabolic pathways* and *2.6 Glycolysis is the best understood metabolic pathway*, we will focus our attention on the metabolic pathways used to convert food molecules into potential energy, because they form the foundation for all life. In so doing, we will review fundamentals of chemistry, including atomic structure, chemical bonds, reduction and oxidation (aka redox), and acid-base behavior. We will use metabolism of the simple sugar molecule glucose as our primary example.

2.2 Chemical equilibrium and reaction kinetics are linked

Key concepts

- The field of kinetics uses equations to represent collision frequencies between molecules.
- Chemical equilibrium is achieved when a reaction's forward rate equals its reverse rate.

The first key question we will address is, **what is the definition of chemical equilibrium, and how is it measured?** To answer this question, we need to define some terms. Let us begin by examining a generic chemical reaction, written in the form:

$$A + B \Leftrightarrow C + D$$

Here, A and B are called *reactants* or *substrates*; C and D are *products*. Each of the terms A, B, C, and D is a different molecule, and in metabolic pathways they may also be called *compounds* or *metabolites*. In order to visualize the reaction, consider the process:

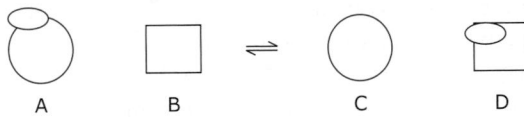

A   B   C   D

In this "mechanistic" view, a piece of molecule A moves onto B, creating C and D. To characterize the reaction, we need to know the **rate** (number of reactions per unit of time) at which A + B generates C + D (the **forward** reaction, written as A + B → C + D) and the rate of the **reverse** reaction (C + D → A + B). If three molecules of A and five molecules of B participate in the forward reaction, each A can collide with any of the Bs, so the total possible collisions are 3 × 5 = 15. This type of analysis is an example of **collision theory**, which states that the rate of a reaction is proportional to the number of collisions between reactants. Hence, the rate of this reaction is proportional to 3 × 5. In the general case, the rate is proportional to [A] × [B], often represented as [A]*[B] or simply [A] [B], where the square brackets indicate the concentration of the molecule.

Another means of representing a reaction includes both the concentration of reactants and the rate of the reaction:

$$\text{rate}_f = k_f[A][B] \qquad \text{(reaction 2.1)}$$

where:

k_f is the forward **rate constant**, a term that incorporates all factors in the reaction except for concentration of the substrates (e.g., temperature and ionic strength)

rate_f is the rate of the forward reaction as written (note the subscript f). Note that the rate refers to the number of events (reactions) that occur over time as a function of concentration.

For the reverse reaction, the same reasoning leads to:

$$\text{rate}_r = k_r[C][D] \qquad \text{(reaction 2.2)}$$

The rate of a reaction is therefore composed of two terms: the rate constant and the concentrations of the reactants. This development can provide insight into the concept of **reaction equilibrium**. A chemical equilibrium exists when *the forward and reverse rates of a reaction are equal*. This is represented by the equation:

$$k_f[A][B] = k_r[C][D] \qquad \text{(reaction 2.3)}$$

A simple rearrangement of the equation yields

$$k_f/k_r = [C][D]/[A][B] \qquad \text{(reaction 2.4)}$$

The ratio k_f/k_r is called the **equilibrium constant**, $\mathbf{K}_{eq}$.

It is possible to extend equilibria from single reactions to multiple reactions connected in series, such as

$$A \Leftrightarrow B \Leftrightarrow C \Leftrightarrow D \Leftrightarrow E$$

This is useful in situations where several related reactions are typically considered together, such as the binding of oxygen to hemoglobin, or the dissociation of polyprotic acids.

Thus, the answer to our first question is that chemical equilibrium is defined as the condition wherein the forward and reverse rates of a chemical reaction are equal, where the rates equal the number of reactions that take place in each direction per unit of time. Furthermore, we see that rate, and hence equilibrium, are directly influenced by the relative concentrations of each reactant.

2.3   The steady state model is essential for understanding the net flow of reactants in linked reactions

Key concepts

- Pathways that are in steady state generate a net flow of reactants in one direction.
- Intermediates in a steady state are at a fixed concentration over time.
- Metabolic pathways are best represented by a steady state model.

Because cells are never in equilibrium with their environment, chemical reactions in cells always "move," meaning reactants are always consumed to produce products. In many cases, this involves several reactions linked together to form a pathway. The next fundamental question we need to answer is **how is the overall rate of an entire metabolic pathway determined?** We will develop three different answers to this question, each reflecting a different level of sophistication.

While every single chemical reaction has a rate constant, when we want to understand the rate of several chemical reactions linked into a cellular pathway, we have to introduce a new term, the *steady state*. This means that if the rate of each reaction is equal (i.e., steady), there will be a net conversion of the starting reactant into the final product. Consider the following three-molecule pathway as an example:

$$S \rightarrow I \rightarrow P$$

Two reactions are occurring in the forward direction.

$$S \rightarrow \mathbf{I} \text{ and}$$

$$\mathbf{I} \rightarrow P$$

At steady state, the rate of the first reaction equals the rate of the second. Coupling the two reactions results in a net flow of material from S to P. This can be visualized as $\mathbf{I}$ being converted into P, such that the concentration of $\mathbf{I}$ does not drop, since S is converted into $\mathbf{I}$ to replace it. When these two reactions happen at the same rate, the amount of S decreases and the amount of P increases, but the amount of $\mathbf{I}$ remains steady. Extending the steady state model to three or more reactions in a linear sequence, the concentration of each of the intermediates between the initial substrate and final product remains steady, the reaction rates linking them are identical, and only the concentrations of the first and last molecules change. Thus, a single rate applies to all of the reactions, *when they are in steady state*. In the example

$$S \rightarrow I \rightarrow J \rightarrow K \rightarrow L \rightarrow P$$

all of the intermediate rates (e.g., $S \rightarrow I$, $I \rightarrow J$, etc.) are equal, as is the overall rate, $S \rightarrow P$. The result is that the concentrations of the intermediates (I, J, K, and L) do not vary with time; they are the molecules forming the *steady state*. The outer metabolites (S and P in this case) change with time. Thus, unlike equilibrium, there is no corresponding back reaction. For example, the rate of $J \rightarrow I$—as well as any other back reaction—is zero.

In order to clearly picture the situation, consider a line of people forced to enter a subway station from a street. Suppose that, as diagrammed in **FIGURE 2.1**, people are admitted at a rate that enables them to flow into the station through an entry gate without backup and then exit the station by boarding a train. The number of people entering is analogous to the [S]; the number of those leaving is analogous to the [P]. The behavior of the people in the building represents the steady state: while they are constantly moving, the net flow in and out of the building remains constant. By comparison, equilibrium would occur when the flow of people entering the building via the gate is equal to the number of people moving out of the building through the same gate (i.e., in the opposite direction).

Metabolic pathways are readily modeled by a steady state. While the steady state is clearly distinct from equilibrium, equating the two is a common error. It is tempting to use the term *equilibrium* for all situations where there is an element of balance, not recognizing that this balance is not appropriate when a

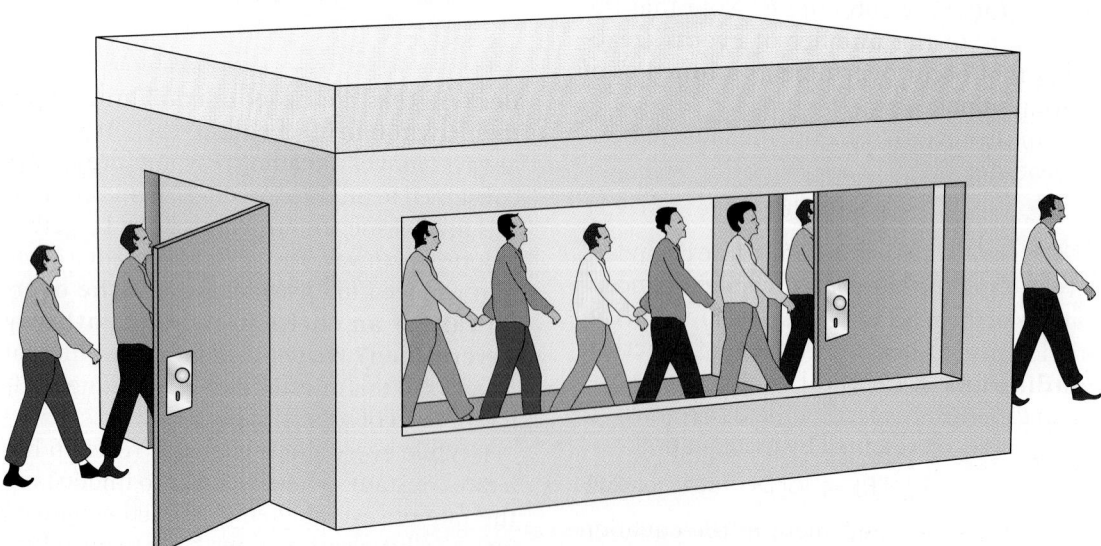

FIGURE 2.1 The steady state window. The notion of steady state is visualized as the constant number of people in a picture window of a building despite constant entry and exit. The maintenance of this steady state requires that the same number enter as leave. The identity of the people changes with time; their number does not unless the rate of entry and exit change.

series of chemical reactions produces *net flow*. In that condition, only the steady state model is appropriate.

Finally, it is important to stress that both equilibrium and steady state are *models*. While they are often appropriate to the situation, they are always approximations. In the time period prior to an enzymatic reaction or series of reactions reaching constant intermediate concentrations, neither model applies. This is called a *presteady state* condition, and it is analyzed by a different model.

Thus, the first answer to our second question is, under steady state conditions, the rate of a metabolic pathway is equal to the rate of conversion of initial reactants into final products. No intermediate rates need be considered.

2.4 Thermodynamics is the systematic treatment of energy changes

Key concepts

- Energy conservation, the constancy of the sum of heat and work, is the first law of thermodynamics.
- A need for some energy dissipation, entropy, is the second law of thermodynamics.
- The free energy, combining the first and second laws, is the driving force for reaction direction.
- Reactions catalyzed close to equilibrium may be reversed in cells and are not regulatory.
- Reactions far displaced from equilibrium are not reversed metabolically and may be regulatory.

To reach the next level of answer to our second question, we now have to consider that all molecules possess energy, and this energy has an important impact on how molecules change form, and pathway regulation. Chemical thermodynamics is the study of energy changes in reactions. We can begin introducing thermodynamics into our sample pathways by simply adding a single term to represent the energy change that accompanies the reaction:

$$A + B + energy \Leftrightarrow C + D$$

To understand the nature of how this energy changes in cells, we will use the laws of thermodynamics, first developed in the 19th century to describe how steam engines convert heat energy (fire) into kinetic energy (motion). It may be surprising that the same laws of energy conversion developed for steam engines also apply to cells, but this gives us a strong foundation for considering metabolic path-

ways. Application of these laws to cells requires us to review the definition of some commonly used words that are strictly defined in the fields of thermodynamics and physics.

The **universe** in thermodynamics is everything we know to exist. The portion of the universe that we are interested in studying, denoted by a border of some kind, is called the **system**. Everything apart from the system is called the **surroundings** (**FIGURE 2.2**). Usually the region near the boundary between a system and its surroundings is the most important part of the surroundings, since most of the universe that is affected by the system tends to be very close to the system. The **state** of a system is most commonly defined by the values set for a small number of variables such as temperature, pressure, volume, and number of molecules of a substance. Once in a particular state, the variables are called **state variables**. Functions that define a state, or define the shift from one state to another, are called **state functions**.

Interactions that are biologically important take place across the boundary between the system and the surroundings. Both material and energy may flow across the boundary, subject to restrictions imposed by the boundary. **Open systems** permit complete exchange, and isolated systems allow no exchange at all. In cell biology, the most interesting interactions between a system and its surroundings are those

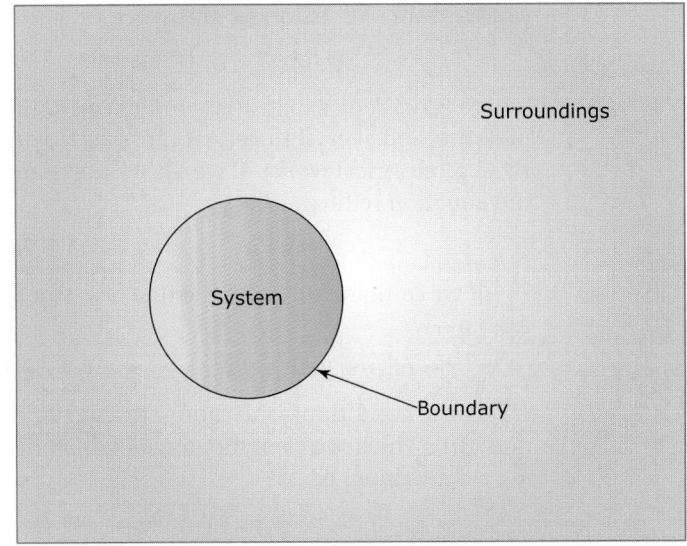

FIGURE 2.2 System and surroundings defined as a Venn diagram. Thermodynamics requires a definition of a system as the province of interest for study; the remainder of the universe is the surroundings. In reality, it is the portion of the surroundings close to the boundary between system and surroundings that is important for the analysis.

that permit only partial exchange. **Energy** is a combination of heat and work exchanges. To keep track of the direction of exchange, we call heat exchange positive if it passes from the surroundings to the system (i.e., heat *added to* the system); in contrast, work is positive if it is transferred from the system to the surroundings (i.e., work *done by* the system).

The **first law of thermodynamics** holds that the energy in any reaction is conserved. Formally, if we define the change in energy from one state to another (or a change in net internal energy) of a system as ΔE, the conservation equation is:

$$\Delta E = q - w \qquad \text{(equation 2.5)}$$

where q is the **heat** absorbed by the system and w is **work** done by the system.

Heat and work are not state functions, as they have no defined state variables associated with them. They are "birds of passage"—vagrants of the thermodynamic world. A system can *contain* no heat or work. Most of us have an innate sense that this is true of work; kick a ball and the imparted motion is obviously not an entity that was transferred from your foot. Likewise, heat is energy transferred only as a result of a difference in temperature between two bodies.

If we set certain limits, it is possible to arrive at a heat change that is a state function. Suppose we allow a system to change only by the work of expansion (increased volume) and that the change occurs at constant pressure. We can write an expression for this work as:

$$w = P\Delta V + w' \qquad \text{(equation 2.6)}$$

where $P\Delta V$ is our expansion work at constant pressure, and everything else is represented as w' (e.g., electrical work). If we set w' = 0, then the equation reduces to:

$$w = P\Delta V \qquad \text{(equation 2.7)}$$

If we combine this with the first law, since $\Delta E = q - w$

$$q = \Delta E + w = \Delta E + P\Delta V \qquad \text{(equation 2.8)}$$

These conditions only apply at constant pressure, which we can emphasize by including a subscript in q:

$$q_p = \Delta E + P\Delta V \qquad \text{(equation 2.9)}$$

We define q_p as **enthalpy**, which is "heat content," under conditions of constant pressure. Usually, we substitute the term **ΔH** for q_p; this emphasizes that the new entity is a state function. With this notational change, we have:

$$\Delta H = \Delta E + P\Delta V \qquad \text{(equation 2.10)}$$

If $\Delta H < 0$, heat is evolved (lost) in a reaction; equivalently, we say the reaction is exothermic, meaning it releases heat. In biochemical systems that interest us, ΔV is also near zero. Thus, in most biologic circumstances, we can often ignore the difference between ΔH and ΔE. However, because reference works have tabulated numerous values of reaction enthalpies, by convention cell biologists still use ΔH.

Heat change means three possible types of alterations in the potential movements of molecules: translation (atoms or whole molecules moving throughout space), vibration (displacement of the two nuclei across a covalent bond), and rotation (circular motion of the two nuclei across a covalent bond). An important property of ΔH values, as with state functions generally, is that they are additive. Thus, we can sum enthalpy changes for reactions, or sets of reactions, and arrive at heats of reactions, which cannot be determined directly.

Often an exothermic reaction ($\Delta H < 0$) is favored; that is, it reacts from left to right as conventionally written. Accordingly, this is the direction many reactions take. But consider what happens when different salts are dissolved in water. The reactions causing separation of some salts into their constituent ions are exothermic, while others are endothermic. We might expect, consistent with our everyday experience, exothermic and endothermic reactions (i.e., those containing negative and positive ΔH values) to move in the forward and reverse direction, respectively, while a reaction with a ΔH value of zero might move in either direction with equal probability. But, as **FIGURE 2.3** shows, dissolution of many salts is favorable, regardless of the change in enthalpy.

Dissolution	ΔH
$H_2SO_4 + H_2O \rightarrow 2H^+ + SO_4^{2-}$	negative
$NaCl + H_2O \rightarrow Na^+ + Cl^-$	zero
$KCl + H_2O \rightarrow K^+ + Cl^-$	positive

FIGURE 2.3 Examples of enthalpy changes when substances are dissolved in water. Note that dissolution of H_2SO_4 releases heat, but dissolving KCl consumes heat—the solution becomes colder as the KCl separates into K^+ and Cl^- ions. Dissolving NaCl causes no change in enthalpy. Because all three of these reactions are favorable, reaction direction is unrelated to enthalpy change.

To help explain why this is true, let us introduce the concept of **entropy**. While often casually equated with "disorder," this word does not entirely capture the sense in which this principle applies to chemistry. The classical view of entropy was developed by several investigators in the 19th century whose primary interest was the improvement of the efficiency of steam engines. A steam engine converts heat energy into work. Since all engines must run in an "energy input-work output" cycle, sufficient heat must be introduced in each cycle of a steam engine to replace that lost during the creation of work. A major problem with early steam engines was their poor efficiency. Minimizing heat loss was therefore a major focus of those attempting to improve the efficiency of these engines. For example, the Watt engine used a condenser in the circulating steam to minimize cooling of the steam in the hot reservoir.

However, further modifications were sometimes considerably *less* efficient. To help explain why, Nicolas Léonard Sadi Carnot developed the first theoretical engine and made two important discoveries. Surprisingly, he learned that while 100% efficiency is theoretically possible in an engine cycle, no work could be done by the cycle unless the efficiency is less than 100%. Second, the ratio of heat transferred to the system (q) and the system temperature (T) is constant during a perfectly efficient **Carnot cycle**. This yielded the ratio:

$$q/T$$

that was later called the **entropy**, an invented word deliberately reminiscent of energy yet distinct from it. The symbol assigned to this was ΔS, the Δ required because it is a system change from one state to another. In the Carnot engine cycle, $\Delta S = 0$, and this is why it generates no work.

Ludwig Boltzmann developed a separate, statistical view of entropy for systems composed of a very large number of molecules. The Boltzmann equation for entropy is:

$$S = k \ln W \qquad \text{(equation 2.11)}$$

where S is the entropy, k is the molecular gas constant (the universal gas constant R divided by Avogadro's Number), and W is the number of ways of arranging a system (or the number of possible states). What makes this equation important is that it provides a mechanistic explanation for entropy in terms of the number of most probable states in a system. To see how the numbers of states can predict reaction direction, consider a splitting reaction:

$$A–B \rightarrow A + B$$

wherein the atoms A and B are attached in substrate, but separate in the products. Clearly, the products can be arranged in more ways (i.e., number of states) than the substrate. Therefore, from an entropy standpoint, the forward reaction is more favorable than the reverse. For the overall process,

$$A–B \Leftrightarrow A + B$$

we would write:

$$\Delta S > 0$$

Positive changes in entropy indicate the forward direction is favored. However, this is still just a *factor* in determining the favorable direction, that is, the direction in which a reaction will occur.

Neither ΔH nor ΔS alone can determine the direction of a reaction. However, Josiah Willard Gibbs devised a means of using both terms that can. The **free energy**, symbolized as **G**, was derived using a combination of the first and second laws of thermodynamics. As with enthalpy and entropy, we typically focus on the change in free energy (**ΔG**) during a reaction rather than on the absolute free energy values of each reactant. The relationship, defined for constant temperature (T), is:

$$\Delta G = \Delta H - T\Delta S \qquad \text{(equation 2.12)}$$

Now we can determine how likely a reaction will proceed in the forward direction from the sign of ΔG for that reaction, as shown in **FIGURE 2.4**. If it is negative, then the forward reaction is possible; if positive, the reverse reaction is possible. Note that this does not mean that it *must* occur; whether a reaction takes place is ultimately determined by kinetics. If $\Delta G = 0$, the reaction is at equilibrium and will not take place. A negative ΔG value simply means a reaction is energetically possible. By convention, a reaction with negative ΔG value is called

ΔG IS	REACTION PROCEEDS	TERMINOLOGY
negative	forward	exergonic
positive	backwards	endergonic
zero	unchanged	equilibrium

FIGURE 2.4 Free energy and reaction direction.

exergonic, and a reaction with a positive ΔG is called **endergonic**.

In order to understand the distinct forces driving chemical reactions, we focus on the individual terms in the free energy equation. We consider ΔH the heat term, that is, typically the energy of bond formation or breakage. ΔS is the entropy term, proportional to probabilities of arrangement (number of states). When one of these terms is dominant, we say that a reaction is *driven* by that component; hence, dissolving KCl in Figure 2.3 is *entropy driven*.

Thus, the second answer to our second fundamental question is **the direction and control features of a metabolic pathway are determined by the change in free energy that occurs between the reactants and the products**. The change in free energy is determined by the sum of two independent variables (change in entropy and change in enthalpy); as long as the sum of these two variables is negative, the reaction will proceed in the positive direction.

Concept and Reasoning Checks

1. In physics, conservation laws require that some measurable quantities, such as mass, remain constant in a closed system. Explain why entropy is *not* conserved in a closed system.
2. Use the free energy equation to explain how a chemical reaction can proceed even if it consumes heat energy.

2.5 Standard free energy, the mass action ratio, and the equilibrium constant characterize reaction rates in metabolic pathways

Key concepts

- Chemical potential is a measure of the free energy in every molecule.
- Standard free energy is the change in free energy for a reaction that occurs under standard conditions of 1 atmosphere of pressure, 25°C, and 1 Molar concentration of all components of the reaction.
- Standard free energy is related to the equilibrium constant of reactions.

To advance to the next level of understanding of chemical reaction rates in cells, we need to examine an additional concept of energy exchange during these reactions. The amount of energy consumed or released varies for different reactions, and can also vary for the same reaction if the environmental conditions change. The term **chemical potential** is used to express the free energy of individual reactants and products in a reaction, and is represented by the term G_i, where the subscript i is one component of the reaction. To simplify matters, we usually compare the energy changes in different chemical reactions by comparing the reactions under the same environmental conditions. For example, the relationship between the chemical potential of a reactant or product and its concentration is expressed by the equation:

$$G_i = G_i^o + RT \ln [i] \qquad \text{(equation 2.13)}$$

Note this equation contains the term G_i^o. By convention, any variable containing the superscript zero is called the **standard** value, and this reflects a very specific set of environmental conditions: temperature of 25°C, concentration of 1.0 moles/liter, and 1.0 atmospheres of pressure. The G_i^o of this equation defines the chemical potential of a reactant under standard conditions.

Conversion of free energy for reactions under cellular conditions can thus be viewed as a displacement from standard conditions depending upon the prevailing concentration. For example, for the reaction $A + B \Leftrightarrow C + D$, we can express the change in free energy as $\Delta G = (G_C + G_D) - (G_A + G_B)$. By substituting $G_i^o + RT \ln [i]$ for the free energy G_i of each reactant, the resulting equation for free energy change becomes:

$$\Delta G = \Delta G^0 + RT \ln [C][D]/[A][B] \qquad \text{(equation 2.14)}$$

Note that at equilibrium, $\Delta G = 0 = \Delta G^0 + RT \ln [C][D]/[A][B]$

By substituting K_{eq} for the term $[C][D]/[A][B]$ (see equation 2.4), we can simplify this to

$$\Delta G^0 = -RT \ln K_{eq}. \qquad \text{(equation 2.15)}$$

This is one of the most important equations cell biologists use to understand the chemical reactions that drive metabolism. By inserting 8.3 J/K/mol (the gas constant) for R and 298 K (standard temperature) for T, and simply keeping track of the signs on both sides of the equation, we see that a reaction that has a very high (positive) K_{eq} will have a very negative ΔG^0.

In *2.2 Chemical equilibrium and reaction kinetics are linked*, we discussed the fact that although living cells cannot be at equilibrium with their environment, biochemical reactions that make up the pathways of these cells are in many cases very close to equilibrium. This is why K_{eq} is so useful for understanding the free energy changes that accompany chemical reactions in cells. But the enzymes that ultimately control the rates of cellular metabolic pathways can be greatly displaced from equilibrium. To determine the extent of displacement of a reaction from equilibrium, we define a new ratio, called the **mass action ratio**, or **Q**. Replacing RT ln K_{eq} with RT ln Q results in:

$$\Delta G = \Delta G^0 + RT \ln Q \qquad \text{(equation 2.16)}$$

If we substitute for −RTlnK_{eq} for ΔG^0, the relationship between K_{eq} and Q is defined as:

$$\Delta G = -RT \ln K_{eq} + RT \ln Q \qquad \text{(equation 2.17)}$$

It is always the case that $Q < K_{eq}$ for reactions, but the difference between Q and K_{eq} can vary over a wide range. We therefore classify enzymes operating in cellular pathways into two groups, depending on the magnitude of the difference. The majority of enzymes mediate reactions with values of Q and K_{eq} that are very close in value, within about an order of magnitude, and these are called **near-equilibrium** reactions. These reactions are sensitive to change in substrates and products, and may proceed in the reverse direction depending upon circumstances. In contrast, in **metabolically irreversible** reactions, Q is two or more orders of magnitude less than K_{eq}. Because these reactions are greatly displaced from equilibrium, they are not reversed in cells, and are not a part of pathways that require the reverse direction reaction. Irreversible reactions are less sensitive to changes in substrate concentration than those near equilibrium are. The enzymes catalyzing metabolically irreversible reactions play a dominant role in regulating the overall behavior of a pathway, and are sometimes called **control points** for this reason.

Concept and Reasoning Checks

1. Why must the free energy of each reaction in a cell be negative?
2. For many diseases, drugs are being developed that target enzymes. Discuss the reasoning for not developing drugs targeted to near-equilibrium enzymes.

2.6 Glycolysis is the best understood metabolic pathway

Key concepts

- Every step in a pathway must have a negative ΔG.
- The mobile cofactor nicotinamide adenine dinucleotide (abbreviated NAD$^+$ or NADH, depending on its oxidation state) is recycled back to NAD for glycolysis by the lactate dehydrogenase reaction.
- The mobile cofactor, adenosine triphosphate (ATP) is produced by glycolysis and converted to adenosine diphosphate (ADP) by energy-consuming reactions such as the sodium pump.

Ten reactions of glucose metabolism, collectively called **glycolysis** (**FIGURE 2.5**), form the most studied metabolic pathway in all cells. Each reaction in the pathway (often referred to as a *step* of glycolysis) is catalyzed by a different enzyme. Before we examine them in detail, let us take a global look at how the 10 reactions are organized.

- First, all the reactions take place in the cytosol, and are virtually identical in all cells. This strongly suggests that glycolysis arose in the earliest cells, before they diverged into archaea, bacteria, and eukaryotes.
- Second, cells can use glycolysis to extract energy from a number of different compounds in addition to glucose. For example, other sugars such as galactose and fructose "merge" into the glycolytic pathway when they are converted into glycolytic intermediates. Amino acids and the sugar portion of nucleotides can also merge via similar mechanisms.
- Third, the 10 steps of glycolysis collectively convert one six-carbon sugar, glucose, into two three-carbon compounds called pyruvate. This means that only one carbon-carbon bond in glucose is broken in these reactions, so relatively little energy is extracted from the glucose molecule as it is processed. When cells must rely mainly on glycolysis to capture energy from their environment, this severely limits the amount of work they can perform. For example, very active cells such as fast skeletal muscle can remain active for a period of only several minutes.
- Fourth, most of the glycolytic pathway is linear, meaning the product of one reaction is the substrate of the next.

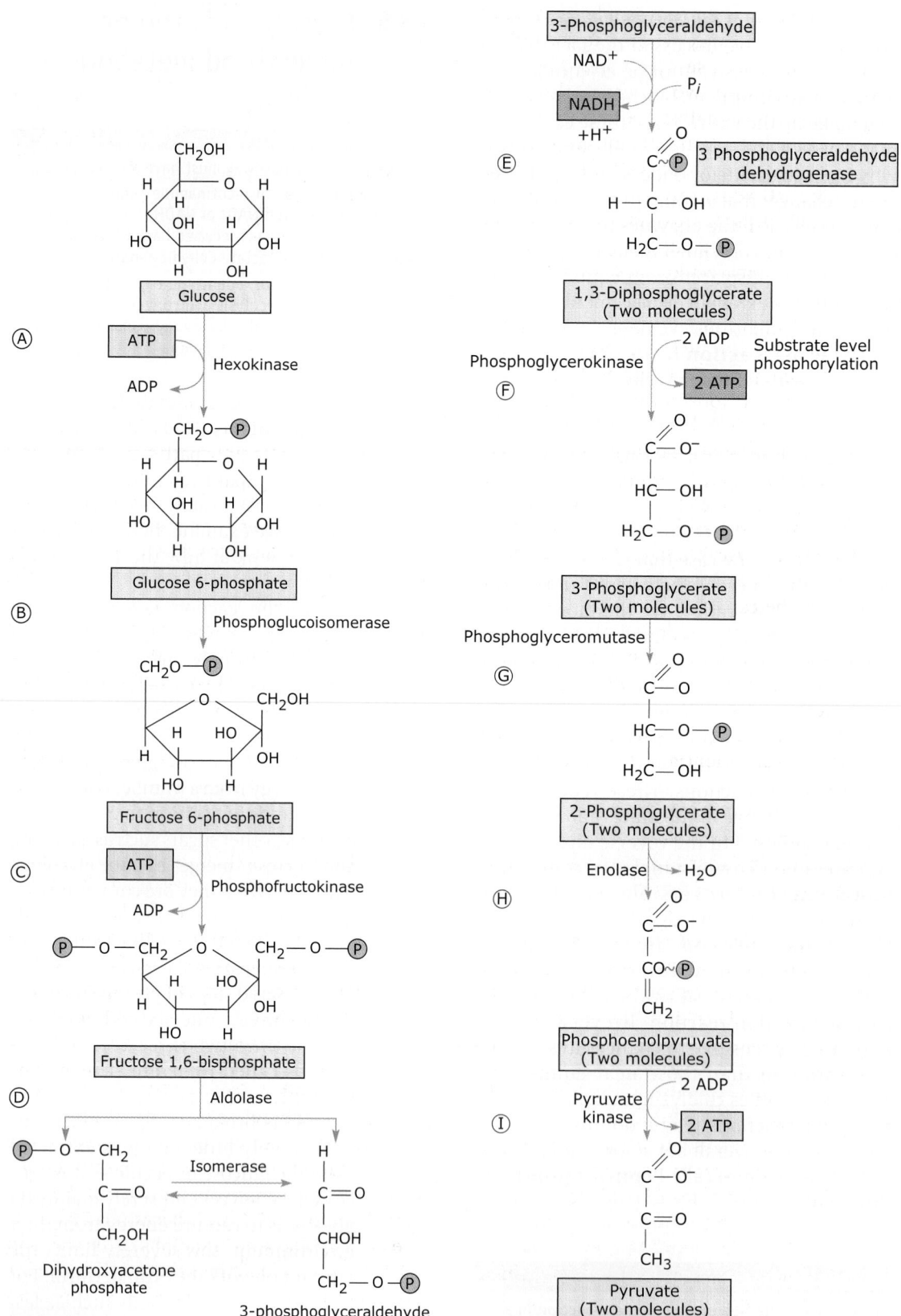

FIGURE 2.5 Glycolysis. The glycolytic pathway as it occurs in most species is indicated as beginning with glucose and ending with lactate. The majority of the reactions are near-equilibrium ones; they are indicated by forward and reverse arrows. The forward arrows are larger indicating the flux direction in the operation of glycolysis. Three steps are metabolically irreversible: hexokinase, phosphofructokinase, and pyruvate kinase. Note that in glycolysis the yield is two ATPs. Four ATPs are produced and two ATPs are consumed in the process.

However, step four is a branch point in the pathway where the six-carbon sugar fructose-1,6-bisphosphate is cleaved into two different, three-carbon sugars called dihydroxyacetone phosphate (DHAP) and glyceraldehyde 3-phosphate (G3P). In step five, the DHAP is converted into a second G3P. Steps six through ten resume the linear pattern, but take place twice per glucose molecule, because two copies of G3P are generated per glucose.

- Fifth, glycolysis serves several purposes in cells. The first and most obvious is to generate ATP, which is used to drive other reactions in the cell. A second is to generate pyruvate, which is used by other metabolic pathways to extract additional energy from the original glucose molecule that entered the cytosol. In addition, since every intermediate of glycolysis is also the intersection of another pathway in the cell, glycolysis can be considered as a portion of numerous other metabolic routes.

While values for ΔG^0 can be derived directly from equilibrium constants, those of ΔG must be made by analysis of concentrations of the intermediates of glycolysis at steady state. Both ΔG^0 and ΔG values for each reaction in glycolysis are shown in **FIGURE 2.6**. Notice that the values for ΔG^0 are distinct from those for ΔG; some of the former are positive, but all of the latter are negative. In order to have pathway flow, each step must have a negative free energy, reflected in the ΔG values. The few steep slopes (large negative ΔG) represent the *metabolically irreversible* reactions: steps catalyzed by the enzymes hexokinase, phosphofructokinase, and pyruvate kinase. All of the other reactions have nearly flat slopes (ΔG close to zero), and are *near equilibrium*. The principal control point in the pathway is step three, catalyzed by the enzyme phosphofructokinase.

Like all pathways, glycolysis employs several enzyme **cofactors** in addition to its intermediates. A cofactor extends the catalytic possibilities of an enzyme beyond the side chain amino acids of a protein. As they are needed in amounts comparable to the enzymes themselves, only a small absolution concentration of cofactors is required for an enzymatic reaction. Cofactors are grouped into two classes, depending on how they interact with proteins. The first is called **mobile cofactors**, because they dissociate from the enzyme during each

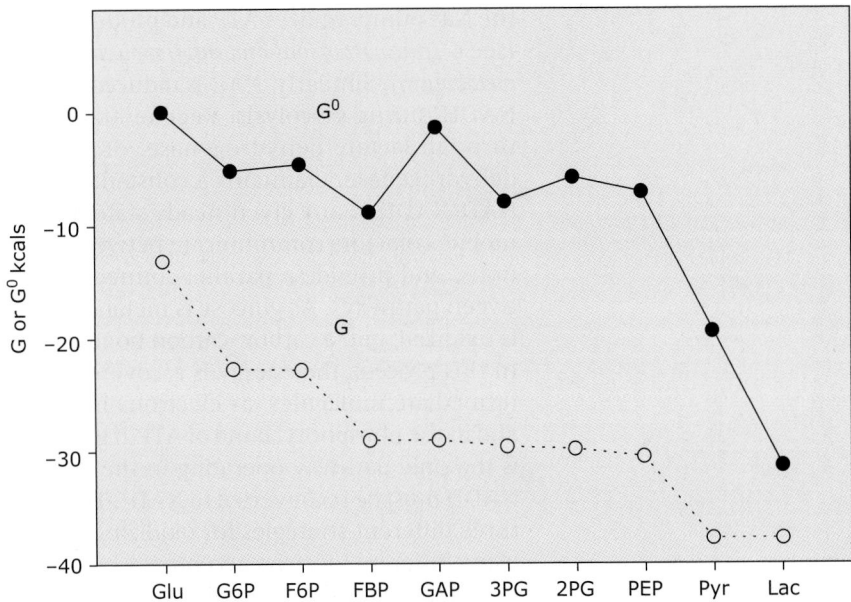

FIGURE 2.6 Standard and actual free energies of glycolysis in the red blood cell. The plot of G^0 and G values against the intermediates of glycolysis is calculated from K_{eq} and measured values of red cell glycolysis. The G^0 value for glucose (Glu) was assigned to zero; the corresponding value for G was fixed at -13 kcal (which represents the difference between 1 M and 5 mM glucose). Thus, the slopes represent the direction and magnitude of the corresponding changes in free energies. Note that standard free energy can even be positive; it is not possible for any change in actual free energy to be positive in an operating metabolic pathway. The figure graphically demonstrates that ΔG and not ΔG^0 values must be used to assess energy changes in pathways. The x-axis represents the pathway flow; each pathway intermediate is plotted in an equally spaced sequence from glucose to lactate. The y-axis represents values of G or G^0, such that the slope of the line provides a visual impression of each reaction's free energy change.

catalytic cycle. As such, they travel from one protein to another as part of their function. The second are the **prosthetic groups**, which remain attached to enzymes during the catalytic cycle. These are unable to participate in any other cellular reaction. While common, not all enzymes require cofactors.

In glycolysis, half of the enzymatic steps require the mobile cofactors ATP, ADP, inorganic phosphate (abbreviated P_i), and nicotinamide adenine dinucleotide. During glycolysis, the concentrations of ATP and P_i remain almost constant, while changes take place in the concentration of ADP. This is because ATP and P_i are in far greater concentrations in the cell than ADP. These considerations apply only to the *free* concentrations of these molecules. This is an important consideration in particular for ADP, as 95% of that molecule is bound to proteins within cells, leaving a very small amount free to react. For the continued operation of glycolysis, an energy utilizing reaction, such as the Na+ pump, utilizes ATP and produces ADP (see *6 Transport of ions and small molecules across metabolism*). Similarly, NAD is reduced to form NADH during glycolysis. Regenerating NAD through lactate dehydrogenase, or another dehydrogenase, maintains a constant ratio of NAD/NADH at any given steady state. Hence, mobile cofactors communicate between reactions, and provide a parallel connection between pathways. As glucose is metabolized, it is oxidized, and a carbon-carbon bond is split. In this process, the energy is recovered in intermediate molecules, as electrons in NADH and in the phosphoryl bond of ATP. If glycolysis is the only pathway operating in the cell, the NADH must be reconverted to NAD+. There are three different strategies for oxidizing NADH, using different electron acceptors.

Modern prokaryotes, the closest living descendents of the first cells, use a wide range of electron acceptors to oxidize NADH. Many of these molecules are created by the cells themselves, including pyruvate, the end product of glycolysis. When cells oxidize NADH by reducing pyruvate, the product is **lactate**. Transport proteins in the plasma membrane permit lactate to diffuse into the extracellular space, preventing lactate buildup. Alternatively, pyruvate can also be converted into a compound called **acetaldehyde**, which then acts as an electron receptor for NADH and is subsequently converted into the alcohol **ethanol**. Ethanol is capable of crossing the plasma membrane by free diffusion, and thus can leave the cell without a transport protein. The pathway of glycolysis or fermentation is performed by all cells, reflecting its ancient origins during evolution.

The other energy intermediate of glycolysis is ATP, which is converted to ADP in the processes of cellular energy utilization. Examples of energy utilizing processes include the motor proteins that exert force on the cytoskeleton (see *11 Microtubules* and *12 Actin*), proteins in cellular membranes involved in active transport (see *6 Transport of ions and small molecules across membranes*), and many signaling proteins (see *18 Principles of cell signaling*). All of these proteins cleave ATP to generate ADP.

Despite the fact that glycolysis is a largely linear pathway, in most cells each intermediate of glycolysis intersects with at least one other pathway, so that glycolysis also represents a metabolic crossroad of the cytosol. For example, glucose-6-phosphate, the product of the first step, can also be used to form the ribose portion of nucleotides.

2.7 Pyruvate metabolism by the pyruvate dehydrogenase complex leads to oxidative respiration

Key concepts

- Pyruvate dehydrogenase complex is composed of enzymes bound together that directly exchange intermediates.
- Three enzymes and five cofactors—NAD+/NADH, CoA, thiamine pyrophosphate, lipoic acid, and flavine adenine dinucleotide—are involved in the pyruvate dehydrogenase complex reactions.

A major step in evolution occurred ~2.5 billion years ago when sufficient oxygen gas accumulated in the atmosphere to serve as an electron acceptor in metabolism. Molecular oxygen (O_2) has a very high, positive **redox potential** (+0.82 V), meaning it has a strong affinity for electrons and is easily reduced. This makes O_2 a very attractive electron acceptor (oxidizer) in cells, and many organisms, including all metazoans, now rely on it to fully metabolize their food sources. These organisms are called **aerobes** for this reason. The elaborate strategy these organisms developed to fully oxidize food is called **oxidative respiration**, and we will examine it closely later in *2.7 Pyruvate metabolism by the pyruvate dehydrogenase complex leads to oxidative respiration*.

Under *aerobic* conditions, pyruvate formed by glycolysis does not have to undergo fermentation. Instead, it can be attached to a cofactor called **CoA** and converted into a compound called **acetyl-CoA**. In prokaryotes, this occurs in the cytosol. In eukaryotes, it takes place in the matrix (interior) of the mitochondrion. The mitochondrial pyruvate transporter allows entry of pyruvate into the mitochondrial matrix and access to the **pyruvate dehydrogenase complex** composed of three distinct enzymes. It is called a **complex** rather than a pathway,

	Nicotinamide Nucleotide (NAD^+)	Flavin Nucleotide (FAD)
Reduced form	NADH	$FADH_2$
Electrons transferred	Two at a time only	One or two at a time
Type of cofactor	Mobile	Prosthetic group

FIGURE 2.7 A comparison of nicotinamide and flavin nucleotides.

because the proteins are physically associated with each other. Consequently, intermediates are not connected to any other reactions in the cell.

The overall reaction for the pyruvate dehydrogenase complex is:

$$\text{pyruvate} + NAD^+ + H^+ + CoA \rightarrow$$
$$\text{acetyl-CoA} + NADH + CO_2$$

NAD^+, NADH, and CoA are mobile cofactors. Three prosthetic groups are required as well: **lipoic acid**, **thiamine pyrophosphate**, and the redox pair called flavin adenine nucleotide (**FAD/FADH$_2$**). A comparison of $FADH_2$ and NADH is provided in **FIGURE 2.7**.

Note that this complex completely oxidizes one of the carbons in pyruvate, yielding CO_2. The electrons thus extracted from pyruvate are initially stored in the FAD prosthetic group as the reduced form, $FADH_2$. To permit the dehydrogenase complex to continue functioning, these electrons must be transferred elsewhere to regenerate FAD. For this enzyme complex, the acceptor is the mobile cofactor NAD^+. Thus, while FAD is converted into $FADH_2$ by the pyruvate dehydrogenase complex, it is also regenerated by the same complex and therefore does not appear in the overall reaction.

Concept and Reasoning Check

1. How is the pyruvate dehydrogenase complex distinct from a pathway like glycolysis?

2.8 Fatty acid oxidation is the major pathway of aerobic energy production

Key concepts

- Fatty acids are carried as thio esters to CoA.
- Cytosolic steps lead to formation and transport of a carnitine ester.
- CAT-1 is rate limiting for fatty acid oxidation and regulated by malonyl CoA.
- Mitochondrial oxidation of fatty acids produces reducing equivalents and CO_2.

The oxidation of fatty acids yields far greater energy per molecule than the oxidation of glucose. In addition to serving as important components of membrane phospholipids, fatty acids are used by virtually all organisms as a form of energy storage. Fatty acids are long chain hydrocarbons that contain a carboxylic acid group at one end. In many animals, fatty acids are stored as triplets attached to a "core" molecule called glycerol. These **triglycerides (fats or oils)** are essential components of our diet. When attached to another molecule, fatty acids are called **acyl groups**, so another name for a triglyceride is **triacylglycerol**. Fatty acids vary in length (between 16–22 carbons are present in the most common fatty acids) and saturation (the extent of double, and occasionally triple, bonds between carbons in the chain). In humans, fatty acid oxidation is especially important in cells that consume large amounts of energy, such as skeletal and cardiac muscle.

FIGURE 2.8 shows the pathway for the oxidation of palmitate, a 16-carbon saturated fatty acid. Similar to glycolysis, the first stage of the pathway requires energy investment, using the high-energy bonds of ATP. The enzyme involved, acyl-CoA synthetase, catalyzes the attachment of a cofactor called **CoA** to the fatty acid, yielding a (**fatty) acyl-CoA**. For example, palmitic acid activated in this manner is called palmitoyl CoA. The formation of the CoA ester in place of the carboxyl group at the end of the fatty acid provides enzyme-binding sites for subsequent steps, and allows easy transfer of the acyl group. Creation of the thioester is called the activation step because this molecule is more reactive than the carboxyl group and can be further metabolized for energy extraction.

In step two, palmitoyl CoA is converted to the ester palmitoyl-carnitine, catalyzed by the enzyme CAT-I (carnitine acyl transferase I). This enzyme, which is located in the outer mitochondrial membrane, is the key regulatory step of the overall process of fatty acid oxidation. The palmitoyl-carnitine created in the

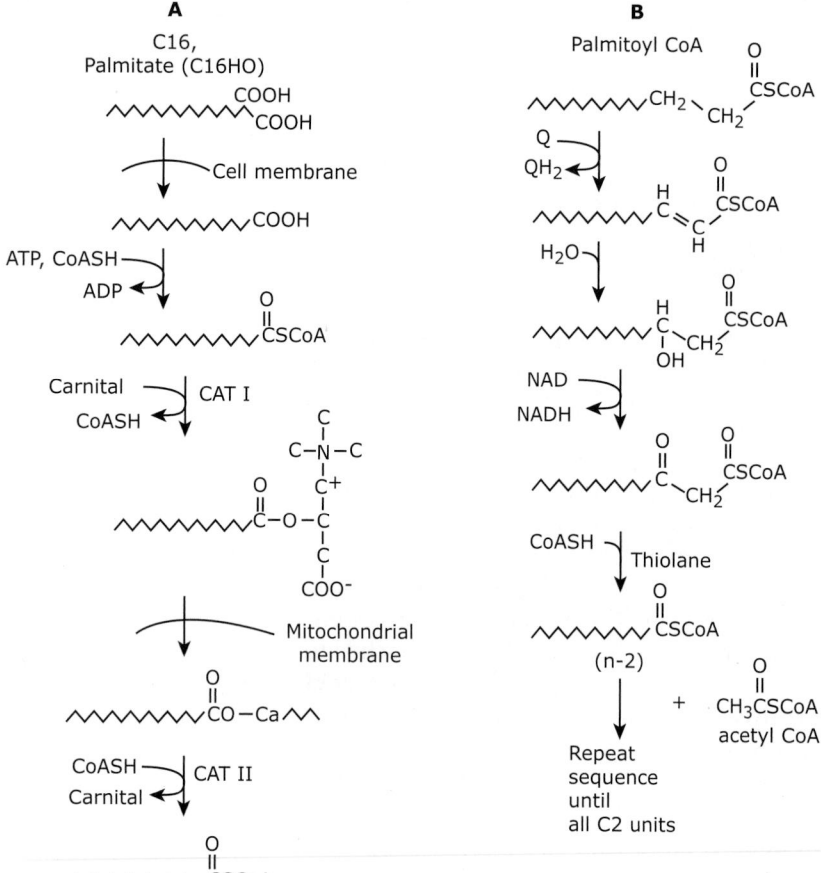

A

C16,
Palmitate (C16HO)

B

Palmitoyl CoA

FIGURE 2.8 Palmitate oxidation. Outline of the steps in the oxidation of palmitate, a representative fatty acid. Note that the first step is the entry of palmitate into the cell, and the third, entry of the acyl-CoA into the mitochondria where most of the reactions take place. The pathway is sometimes referred to as a spiral since the last four reactions repeat with (n-2) shortened versions of the CoA ester until the entire fatty acid is converted to acetyl CoA.

This proceeds until the entire fatty acid chain has been converted to acetyl-CoA molecules and the electrons are trapped in NADH and $coQH_2$. A shorthand name for this series of reactions is **beta oxidation**.

In *2.9 The Krebs cycle oxidizes acetyl-CoA and is a metabolic hub* we will move to the next stage of metabolism, which focuses on the complete oxidation of the acetyl-CoA molecules generated by the pyruvate dehydrogenase complex and the acyl-CoA dehydrogenase enzymes.

2.9 The Krebs cycle oxidizes acetyl-CoA and is a metabolic hub

Key concepts

- The Krebs cycle is specialized to oxidize only one substrate: acetyl CoA.
- Intermediates of the Krebs cycle are components of other pathways.

Thus far in this chapter, we have progressed through two of the four phases of cellular metabolism: glycolysis and generation of acetyl-CoA. Note that despite the variety of energy sources (foods) that we eat, cellular metabolism converts most of them into a few intermediates: acetyl-CoA, NADH, $coQH_2$, and ATP. While each of these molecules can be used to build new molecules (via catabolism), in most cells the goal of metabolism is the complete oxidation of food to generate ATP. Thus, the last two phases are dedicated to converting acetyl-CoA to reduced forms of cofactors (thereby releasing CO_2) and then converting the energy stored in these cofactors into ATP. The topic of *2.9 The Krebs cycle oxidizes acetyl-CoA and is a metabolic hub* is phase three, conversion of acetyl-CoA to reduced cofactors.

As with glycolysis and beta oxidation, metabolism of acetyl-CoA first evolved in prokaryotes. In modern prokaryotes and in mitochondria, nine chemical reactions are required to fully oxidize acetyl-CoA to CO_2 + CoA, and the captured electrons are stored

intermembrane space is then delivered to the mitochondrial matrix via a transport protein. Once inside the mitochondria, another CAT isoform (CAT-II) catalyzes the reformation of palmitoyl CoA from palmitoyl-carnitine. Thus, these two closely related enzymes (isozymes), located in different cellular spaces, shuttle the fatty acyl-CoA into the mitochondrial matrix.

Mitochondrial palmitoyl CoA is then processed through the series of reactions shown in Figure 2.8. The pathway extracts electrons (and hydrogen atoms) from the molecule in a stepwise fashion to reduce two one NAD+ and one coQ per cycle. A series of reactions produces the molecule **acetyl-CoA,** one NADH, one $coQH_2$, and a fatty acyl-CoA shortened by two carbons (C14-CoA). The latter undergoes the same series of reactions, producing another acetyl-CoA, NADH, $coQH_2$, and a C12-CoA.

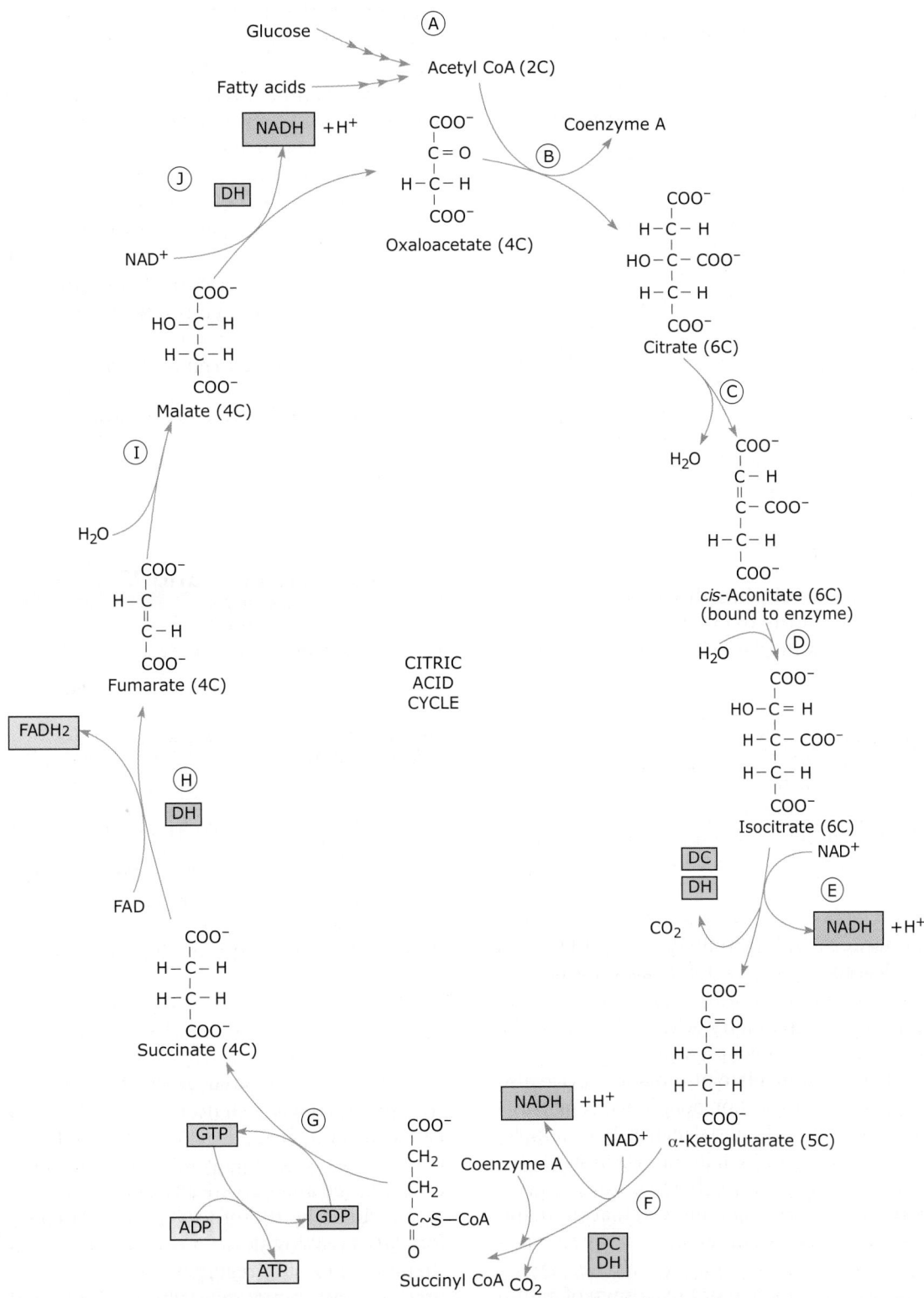

FIGURE 2.9 The citric acid cycle, or Krebs cycle. The pathway exclusively oxidizes acetyl-CoA to two molecules of CO_2, and the energy molecules NADH, QH_2, and GTP. The steps are: (1) citrate synthase, (2) aconitase, (3) isocitrate DH, (4) 2-ketoglutarate DH, (5) succinyl CoA synthetase, (6) succinate DH, (7) fumarase, and (8) malate DH. Most of the intermediates are also intersection points of other pathways. For example, oxaloacetate and 2-ketoglutarate can be interconverted with aspartate and glutamate, respectively.

in cofactors NADH and a reduced coenzyme called **quinone** (abbreviated **coQH$_2$**). These nine reactions form a cycle, shown in **FIGURE 2.9**. Sir Hans Adolph Krebs established the sequence as a cycle so they are commonly called the Krebs cycle in his honor. The first step is the condensation of the acetyl group of acetyl-CoA with the four-carbon compound oxaloacetate to form the six-carbon compound, citric acid. (The cycle is also called *citric acid cycle* and *tricarboxylic acid cycle* for this reason.) CO$_2$ is released in two subsequent steps, thereby reducing the number of carbons in the intermediates of the cycle to five, and then four. As the bonds holding these carbons to the intermediates are broken, the electrons are moved to NADH and coQH$_2$, and one high-energy phosphate bond is directly formed in guanosine triphosphate (GTP) (energetically equivalent to ATP). These molecules—all mobile cofactors—represent the extractable energy from acetyl CoA and the bulk of all energy extraction in the cell; by comparison, glycolysis produces little net ATP. The GTP formation step is sometimes called **substrate level phosphorylation** to distinguish it from the oxidative phosphorylation processes taking place in the inner membrane of mitochondria and the plasma membrane of aerobic prokaryotes, which we will discuss in *2.10 Coupling of chemical reactions is a key feature of living organisms*.

As the reaction sequence is cyclic, all intermediates are regenerated, so that only small, catalytic concentrations are necessary to maintain a steady state. As a corollary, no intermediates of the Krebs cycle enter the pathway for oxidation. The only input to the cycle is acetyl-CoA, and the only output is CO$_2$. CoA is released in the first step (citrate formation) and reincorporated into CoA esters for other pathways, such as fatty acid oxidation, or the pyruvate dehydrogenase complex.

Each intermediate of the Krebs cycle also serves as an intersection point for several pathways in the mitochondrial matrix. For example, succinyl CoA is utilized in the synthesis of porphyrins, which become heme groups of cytochromes and hemoglobin. Oxaloacetate and 2-ketoglutarate are interconverted with their amino forms, aspartate and glutamate, playing roles in the formation or breakdown of amino acids. As such, it is possible that the flow of carbon in the Krebs cycle is different from that viewed exclusively for its role in energy formation. Some reactions are dedicated to restoring intermediates lost to other such processes and

are called **anaplerotic** (refilling) reactions. One such enzymatic route is that catalyzed by pyruvate carboxylase, which exists in many cells:

$$\text{pyruvate} + CO_2 + ATP \rightarrow \text{oxaloacetate} + ADP + P_i$$

This reaction was a curiosity when first discovered in animal cells, as it appears to allow CO$_2$ incorporation, which is not known to exist outside of photosynthetic organisms. However, there is actually no net incorporation of CO$_2$. When pyruvate carboxylase occurs in a pathway of significant flux, such as synthesis of glucose (gluconeogenesis) and lipid biosynthesis, the CO$_2$ incorporated is released in a subsequent step. For example, in gluconeogenesis, pyruvate is converted to phosphoenolpyruvate (PEP) by first forming oxaloacetate, and net flux through pyruvate carboxylase is required:

$$\text{pyruvate} + CO_2 \rightarrow \text{oxaloacetate} \rightarrow PEP + CO_2$$

No net CO$_2$ incorporation (also called "CO$_2$ fixation") occurs in animal cells.

Concept and Reasoning Check

1. Explain why an intermediate (e.g., malate) cannot be oxidized by the Krebs cycle.

2.10 Coupling of chemical reactions is a key feature of living organisms

Key concepts

- Creatine phosphate provides a supplemental energy molecule.
- Adenylate kinase serves to phosphorylate adenosine monophosphate (AMP).
- Nucleoside diphosphate kinase rephosphorylates non-ADPs.

Living organisms have many examples of coupled chemical reactions. The concept of **reaction coupling** is a subtle one: cells use reactants of higher energy to drive what would otherwise be an energetically unfavorable reaction. Usually, the process is illustrated by formally breaking down a reaction into parts that do not exist in reality but allow us to appreciate how energy redistribution is effected by molecular coupling. An example is the change in free energy that accompanies the reactions catalyzed by the enzyme glutamine synthetase. This enzyme couples the generation of the amino acid glutamine from glutamate and ammonia to the cleavage of ATP. We can use ΔG^o for each reaction to illustrate how this coupling works. The ΔG^o for the glutamine generation reaction is positive (+ 14.2 kJ/mol) while that for the conversion of ATP to ATP + P_i is negative (−30.5 kJ/mol). Because ΔG^o is a state function, we can sum the two values, yielding a net negative ΔG^o (−16.3 kJ/mol). This means that, *under standard conditions*, the coupled reaction is favorable. We must be careful about overinterpreting this kind of analysis, however, because these reactions do not occur under standard conditions in living cells. Still, the use of standard conditions allows us to examine the reactions in isolation and examine strictly energy redistributions between the molecules. Since ΔG^o is directly related to K_{eq} we can also use the value of ΔG^o as an index of the degree of displacement from equilibrium that the reaction components need to achieve in cells to allow the actual ΔG to become negative.

ATP serves as the energy source for most energy-linked reactions in cells. It is considered a "high-energy" molecule for a number of reasons, including the increased entropy resulting from the splitting of the molecule (i.e., ADP and P_i have more degrees of freedom than ATP). In addition, ADP and P_i have more resonance states than ATP, and are accordingly more stable. Finally, the three phosphoryl groups in ATP exhibit more electrostatic repulsion, due to proximity of their negative charges, than the two phosphoryl groups in ADP.

In addition to providing energy to enzymatic reactions, ATP and other nucleotides serve many other functions. For example, formation of RNA and DNA requires four different nucleotide triphosphates (ATP, CTP, GTP, TTP) and deoxyribose nucleotide triphosphates (dATP, dCTP, dGTP, dUTP), respectively. Also, ATP and GTP help drive the assembly of actin filaments and microtubules (see *11 Microtubules* and *12 Actin*). ATP serves as the initial energy source for creating all of these different nucleotides.

One important enzyme that helps create them is **nucleoside diphosphokinase** (or **NDP kinase**), which catalyzes the reaction

$$ATP + NDP \rightarrow ADP + NTP$$

where, NDP and NTP represent diphosphate and triphosphate forms of a different nucleotide, respectively. These nucleotides include cytosine, guanosine, thymidine, and uridine. The energy contents of product and substrate molecules of the NDP kinase reaction are approximately equal. ΔG^o for this reaction is ~0, so K_{eq} is ~1.0. But because the mass action ratio Q of the reaction in the cell varies considerably, the actual ΔG for this reaction varies as well.

2.11 Oxidative phosphorylation is the final common pathway converting electron energy to adenosine triphosphate

Key concepts

- Reduction potentials allow energy analysis of electron transporters.
- Protein complexes exist within the mitochondrial membrane as a route for electron and proton movement.
- The isolated complexes are connected by the mobile carriers NADH, $coQH_2$, and cytochrome c.
- The chemiosmotic hypothesis of a proton gradient as the energy intermediate explains the action of uncouplers.

The oxidation of the reduced cofactors NADH and $coQH_2$ coupled to the conversion of ADP and P_i into ATP is known as **oxidative phosphorylation,** and represents an excellent application of reaction coupling. This is the last phase of metabolism, the site of virtually all oxygen utilization by cells, and, correspondingly, virtually all ATP formation. We have already seen that GTP (converted to ATP by NDP kinase) is directly formed in a single reaction through bond rearrangement (e.g., the *substrate level phosphorylation* performed by succinyl CoA synthetase in the Krebs cycle). However, oxidative phosphorylation is far more complex, for several reasons as given below:

- First, it requires an additional form of temporary energy storage, an ion gradient across a membrane. Like high-energy electrons, this ion gradient is a form of potential energy that is tapped when the ions are allowed to flow back across the membrane.

- Second, the proteins required to complete oxidative phosphorylation must be either embedded in, or closely associated with, the membrane that maintains the ion gradient. Whereas glycolysis, fermentation, beta oxidation, and the Krebs cycle take place in the fluid phase of a cellular compartment, oxidative phosphorylation is restricted to specific regions in a membrane.
- Third, the spatial orientation of oxidative phosphorylation proteins is very specific. Because they create an ion gradient across a membrane, they must move ions in the right direction to be useful.

In prokaryotes, oxidative phosphorylation occurs in the plasma membrane. This complicates matters even further, because the same membrane is also responsible for maintaining all of the other chemical gradients these cells require to live. Eukaryotes enjoy the benefit of performing oxidative phosphorylation in the inner mitochondrial membrane, which is a derivative of a prokaryotic plasma membrane (**FIGURE 2.10**). This simplifies matters considerably, because the mitochondrion performs only a subset of tasks for the cell, and is therefore specialized for performing oxidative phosphorylation.

Like all previous phases of metabolism, oxidative phosphorylation takes place in several steps, each mediated by proteins. What makes this phase of metabolism distinct from the previous phases is that the intermediates in this pathway are not all chemical modifications of a metabolic substrate. Instead, this pathway is characterized by a series of oxidation and reduction reactions taking place in proteins and cofactors, resulting in the sequential transport of high-energy electrons from one protein to the next. The pathway is often called an **electron transport chain** (**ETC**) for this reason. All of the components of the ETC exist as physically isolatable complexes, numbered I-IV. Notice that this pathway is branched: NADH generated by glycolysis, the pyruvate dehydrogenase complex, and the Krebs cycle enter via complex I, while the electrons donated to coQ during the Krebs cycle enter via complex II. All electrons then pass through complexes III and IV before reducing O_2 to form H_2O. This is why the presence of atmospheric oxygen is so important for aerobic organisms: without O_2 as the final electron acceptor, electron flow through the ETC would come to an immediate halt.

FIGURE 2.11 illustrates how oxidative phosphorylation converts high-energy electron energy into a proton gradient. After electrons are transferred from NADH or QH_2, reduction and subsequent oxidation of cofactors within proteins in complexes are linked with the generation of a proton gradient in the case of mitochondria. In some cases, redox activity alters the conformation of proteins that leads to movement of the proton from the interior to

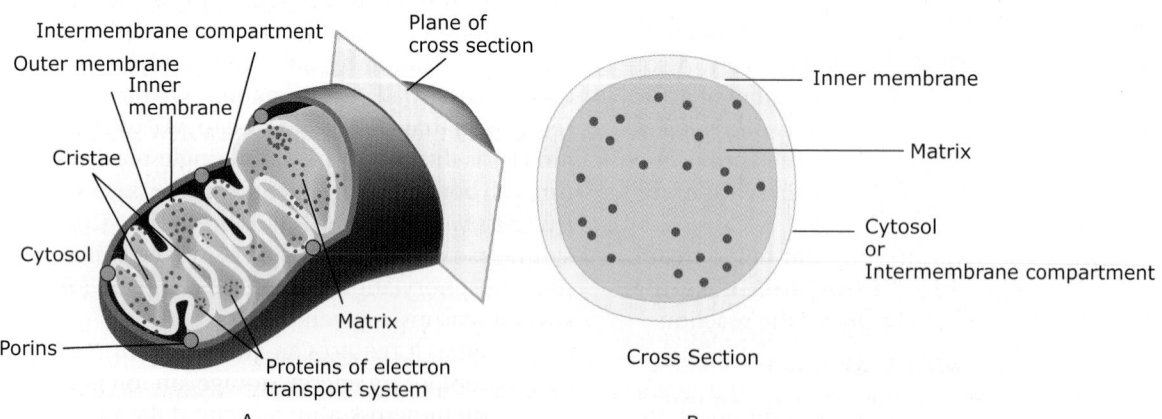

FIGURE 2.10 The mitochondrial essentials for oxidative phosphorylation. A, indicates both membranes of the mitochondria; the presence of porin in the outer membrane means that relatively low molecular weight molecules (up to 10,000) are permeable to this membrane. As all of the enzymes of the process of oxidative phosphorylation are associated with the inner membrane, the representation in B, is sufficient for its understanding.

Complex	Name	Substrate	Product	Cofactors	Inhibitor
I	NADH oxidase	NADH	QH_2	Fe/S protein FMN	Rotenone
II	Succinate dehydrogenase	Succinate	QH_2	Fe/S protein FMN	Malonate
III	Q-c reductase	QH_2	Cyt. c_{red}	Fe/S protein cyt. 560/566	Antimycin A
IV	cyt c oxidase	Cyt. c_{red}, O_2	H_2O	cyt. a/a_3 Cu protein	Cyanide
V	F_1F_0ATPase	ADP, Pi, H^+ in matrix	ATP, H^+ in cytosol	None	Oligomycin

FIGURE 2.11 Properties of the five complexes of oxidative phosphorylation. The ordering of the complex corresponds to the ranking of the reduction potentials (in decreasing order) of the mobile cofactors for the respiratory chain: NADH, QH_2, and Cyt. c_{red}.

the exterior of the mitochondrion, or from the cytosol to the extracellular space for a prokaryote. As the pathway moves electrons *through* the membrane, it simultaneously moves protons *across* the membrane. Note that complexes I, III, and IV move protons across the membrane. The complete path of electron transfer is typically summarized as NADH → complex I → coQ → complex III → cytochrome c → complex IV → O_2. Because electrons donated by QH_2 bypass complex I, they power only two of the three proton pumping complexes, and therefore contribute proportionally less to the formation of the gradient.

To better understand why the electrons are moved through the ETC in a specific order, we need to understand how the change in energy of these reactions varies. As we saw in the glycolysis reactions, the net change in free energy for the pathway is negative, and this change is distributed to several reactions, each with a small free energy change. To understand the relationship between electron transfer and free energy, let us look more closely at redox reactions. For analysis, they are usually written as half-reactions, which is a conceptual division of a reaction into electron acceptor and donor pairs. Under standard conditions, a half-reaction is conventionally written in the direction of reduction, and the value for **standard reduction potential** symbolized as $\Delta\epsilon^o$; the units are volts. The relationship between this value and the standard free energy is:

$$\Delta G^0 = -nF\Delta\epsilon^o \qquad \text{(equation 2.18)}$$

For example, zinc has a tendency to oxidize copper. The reaction can be set up in an electrochemical cell, with an overall $\Delta\epsilon^o$ of 0.42 V. Calculated separately,

Zn → Zn^{++} + $2e^-$	$\Delta\epsilon^o$ = + 0.76
Cu^{++} + $2e^-$ → Cu	$\Delta\epsilon^o$ = −0.34
Zn + Cu^{++} → Zn^{++} + Cu^o	$\Delta\epsilon^o$ = 0.42

A positive $\Delta\epsilon^o$ corresponds to a negative ΔG^0, which provides the reaction direction under standard conditions.

One feature of the ETC simplifies energetic analysis considerably: virtually the entire sequence passes electrons directly from one carrier to the next in a roughly 1:1 ratio, without side reactions. Because most of the reactions occur in complexes, the concentrations of intermediates are roughly equal, and thereby cancel out, with only the concentration of H^+ remaining as the disparate value of 10^{-7}. We can account for this by defining a new ΔG^0, in which all the properties are the same as ΔG^0, except that pH = 7. Thus, the values used in calculating energy changes for biochemical reactions are a modification of this standard state with this single proviso, often termed a biochemical standard state. This is symbolized by adding a superscripted dash (called a prime) to the conventional free energy change in standard state: $\Delta G^{0'}$.

FIGURE 2.12 shows the relationship between redox potential and reaction progress for the ETC. For example, NAD$^+$/NADH starts with a lower redox potential than coQ/coQH$_2$, and

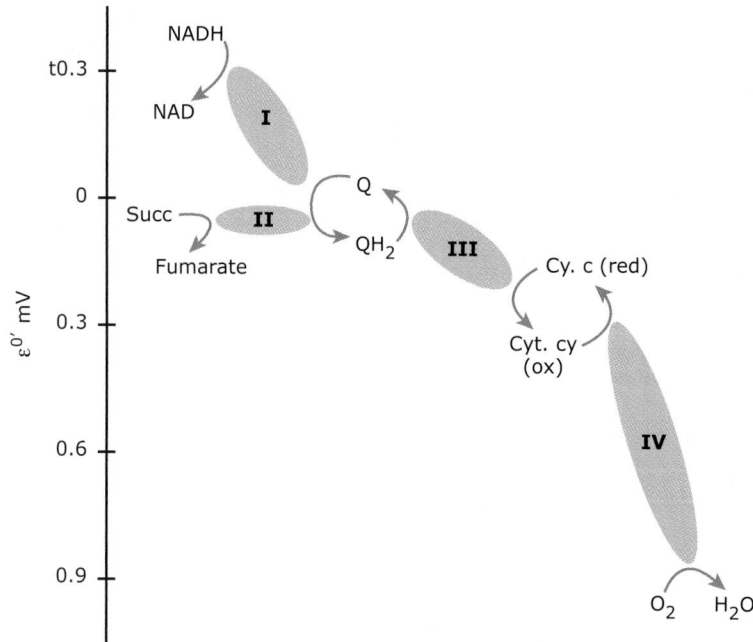

FIGURE 2.12 Energy diagram of mitochondrial electron transport. The four complexes are drawn sequentially on a scale of reduction potentials. The steepness of the drop in redox potential indicates the potential amounts of energy capture at each step. There are relatively large changes in complexes I, III, and especially IV, for which proton translocation occurs. The very slight drop at complex II is consistent with the fact that electron transfer at this step does not produce any proton gradient. Adapted from W. M. Becker, et al. *The World of the Cell, Seventh edition.* Benjamin Cummings, 2008.

this corresponds to a greater amount of free energy. By contrast, electrons entering the ETC from $coQH_2$ bypass complex I. They do not carry enough energy to make the transfer to complex I favorable, and instead pass them to coQ. This step is unable to contribute to the proton gradient, a point that is clear from the small energy change.

The mechanism of H^+ pumping in complexes I and IV is similar to the mechanism of glucose transport described in *6 Transport of ions and small molecules across membranes*. In complexes I and IV, H^+ can be donated by an acidic group on a protein component facing the inner side of the membrane. As a redox reaction takes place involving this donor, the energetic change forces reorientation of this protonated group within the protein such that it now faces the outer side of the membrane. This resembles the glucose transporter in that, once bound, these proteins can reorient to face the opposite side of the membrane. Complex III uses a different loop mechanism to move protons wherein the diffusible cofactor coQ binds H^+ ions at the inner face, flows to the outer face, and releases them. The driving force for the binding and release of protons in the loop is simply the redox state of coQ, and

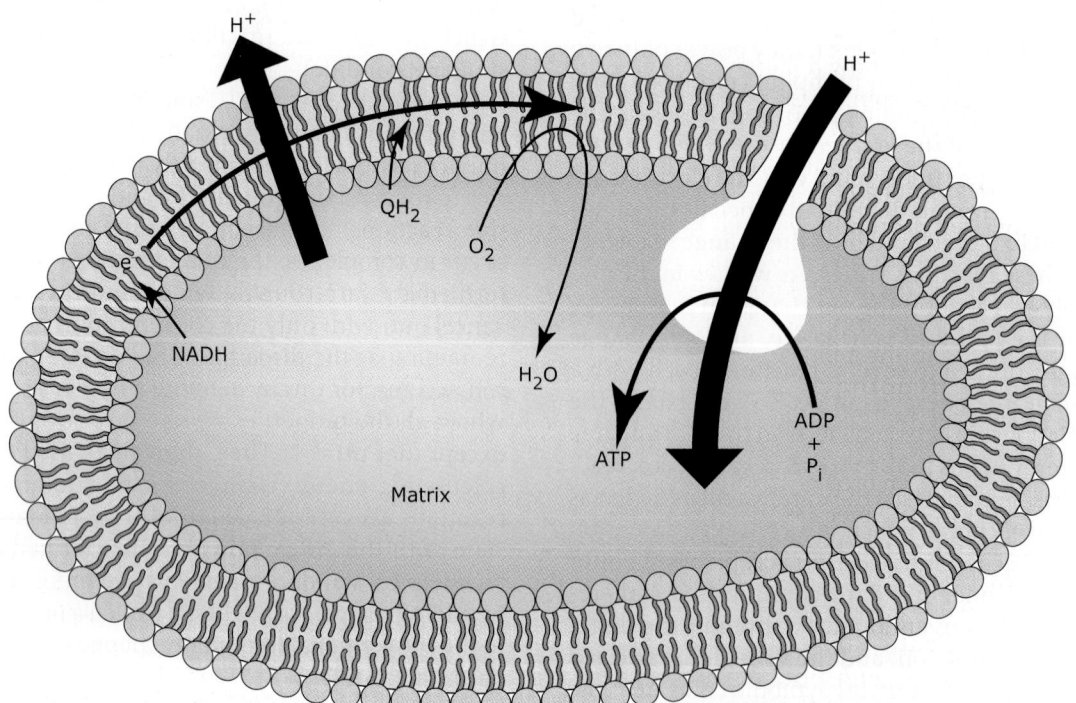

FIGURE 2.13 Oxidative phosphorylation. Using the simplified representation of the mitochondria from Figure 2.10B, the electron pathway through complexes embedded in the inner membrane is indicated, coupled to the flow of protons from the matrix to the cytosol. The protons reenter only through a special channel linked with ATP synthesis.

this is controlled by the reactions immediately up- and downstream of coQ. Overall, in each of the three complexes, H+ transport across the membrane is strictly coupled to electrons moving through it.

The resulting proton gradient is a form of potential energy. As protons move across the membrane from "outside" (intermembrane space/extracellular space) to "inside" (matrix space/cytosol), they create an electrical and chemical gradient. As they are positively charged, the inside is relatively negative; this separation of charge is called a **membrane potential**, symbolized as $\Delta\Psi$. The chemical gradient of protons created by this movement is measured as a ΔpH across the membrane. The $\Delta\Psi$ and ΔpH together comprise a quantity known as the **proton motive force**. This means that two driving forces move protons back across the membrane: the negative charge *and* the lower proton concentration.

FIGURE 2.13 shows how the proton motive force is converted to ATP. The key is the coupling of proton flow, a form of kinetic energy, to the phosphorylation of ADP, an energetically unfavorable reaction. This takes place in a large protein complex called the F_1F_o ATP synthase. It consists of a channel that selectively transports H+ through the membrane and back to the inner space, and an **ATP synthetase** that adds P_i to ADP to generate ATP. The flow of protons literally causes a portion of the synthetase to spin, much like how a flowing river causes a waterwheel to spin, and this kinetic energy is used to create the ATP.

A key concept that helps explain the mechanism of ATP generation is that electron flow and proton flow are coupled in the ETC. Peter Mitchell, who received the Nobel Prize in Chemistry in 1978 for elucidating the mechanisms of energy transfer in cells, proposed the coupling relationship in his **chemiosmotic hypothesis**. Strong support for his ideas came when he demonstrated that several molecules are capable of uncoupling electron flow from proton flow, and also block ATP synthesis. The mechanism of uncoupling is illustrated for the classical uncoupler dinitrophenol in **FIGURE 2.14**. All uncouplers are lipophilic weak acids capable of dissipating a protein gradient by binding protons on one side of the membrane and releasing them on the other, accelerating the flow of electrons through the ETC. This simple explanation of uncouplers was a triumph for the chemiosmotic hypothesis, and helped explain

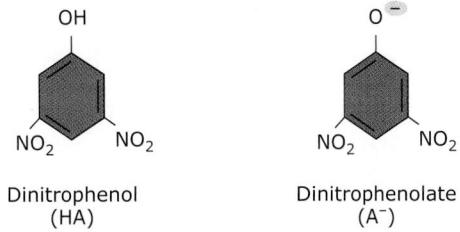

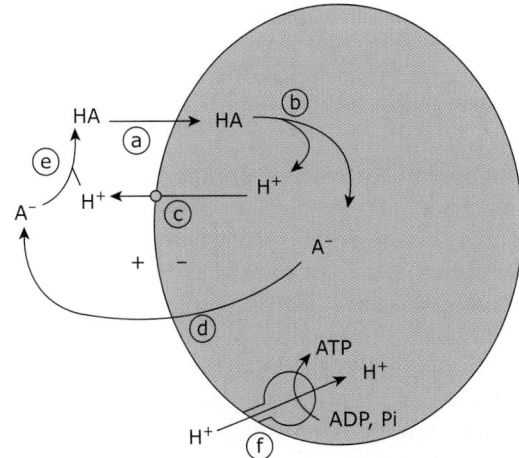

FIGURE 2.14 The protonated form of uncoupler is represented as HA, and its dissociated form as A−. A, The HA form crosses the mitochondrial membrane. B, Inside the matrix, HA dissociates to H+ and A−. C, Protons are pumped out by the respiratory chain; D, the A− is expelled due to its delocalized charge and the electrical gradient favoring its export. E, In the cytosol, H+ and A− combine according to their equilibrium, forming HA, establishing a cycle. This cycle provides a route of entry of H+ that is an alternative to the F_1F_oATPase.

uncoupling by a large variety of chemical structures. It also explained why these compounds trigger large increases in respiration (oxygen consumption) but a steep decline in ATP synthesis.

Concept and Reasoning Checks

1. Why do inhibitors of electron flow (respiratory inhibitors) also block proton movement across the mitochondrial membrane?
2. How are free energy and redox changes related?
3. How is electron flow normally linked to the proton gradient and ATP formation?
4. Explain why uncouplers increase electron flow, but decrease ATP synthesis.

2.12 Photosynthesis completes the carbon cycle by converting CO_2 to sugar

Key concepts

- The light reactions of photosynthesis are similar to a reverse mitochondrial electron transport chain.
- Energy from light absorbed by chlorophyll is the key driving force for photosynthesis.
- The proton gradient across the thylakoid membrane of chloroplasts drives ATP synthesis.

The overall pathway of photosynthesis is the conversion of CO_2 and H_2O to sugar, producing O_2. In photosynthetic prokaryotes, photosynthesis occurs in deep folds of the plasma membrane called **thylakoids**. In plants and other eukaryotes, the reactions of photosynthesis occur in organelles called **chloroplasts**, which are descendents of an internalized single-celled photosynthetic organism. As such, they contain two membranes like mitochondria, surrounded by a third membrane analogous to the plasma membrane, as **FIGURE 2.15** shows. These membranes define two important spaces: the thylakoid lumen, surrounded by thylakoid membrane, and the chloroplast stroma, enclosed by the inner chloroplast membrane.

The reactions of photosynthesis are traditionally divided into the **light reactions** and the **dark reactions**. Light reactions use the energy of photons to strip electrons from water, raise their energy level, and use the excited electrons to assemble two essential ingredients for the dark reactions: the reduced form of the cofactor **nicotinamide adenine dinucleotide phosphate** (**NADPH**), and ATP. Overall, the light reactions resemble the steps of oxidative phosphorylation run in *reverse*, while the dark reactions are effectively a *reversal* of glycolysis and the Krebs cycle. If the reactions of photosynthesis are written together, they become:

light reactions: $ADP + P_i + H_2O + NADP + light \rightarrow O_2 + ATP + NADPH$

dark reactions: $CO_2 + ATP + NADPH \rightarrow (CH_2O) + ADP + P_i + NADP$

net: $CO_2 + H_2O + light \rightarrow (CH_2O) + O_2$

The light reactions (illustrated in **FIGURE 2.16**) are driven by sunlight, yet the similarities to mitochondrial respiration are striking as given below:

- The reactions take place in a membrane (the thylakoid membrane), and are focused on the progression of high-energy electrons through protein complexes connected by mobile electron carriers.
- As the electrons pass through these complexes, protons are ferried across the membrane and into the thylakoid lumen, building a proton gradient.
- The mobile electron carriers are similar to those in the mitochondrion. Plastoquinone (PQ) is structurally simi-

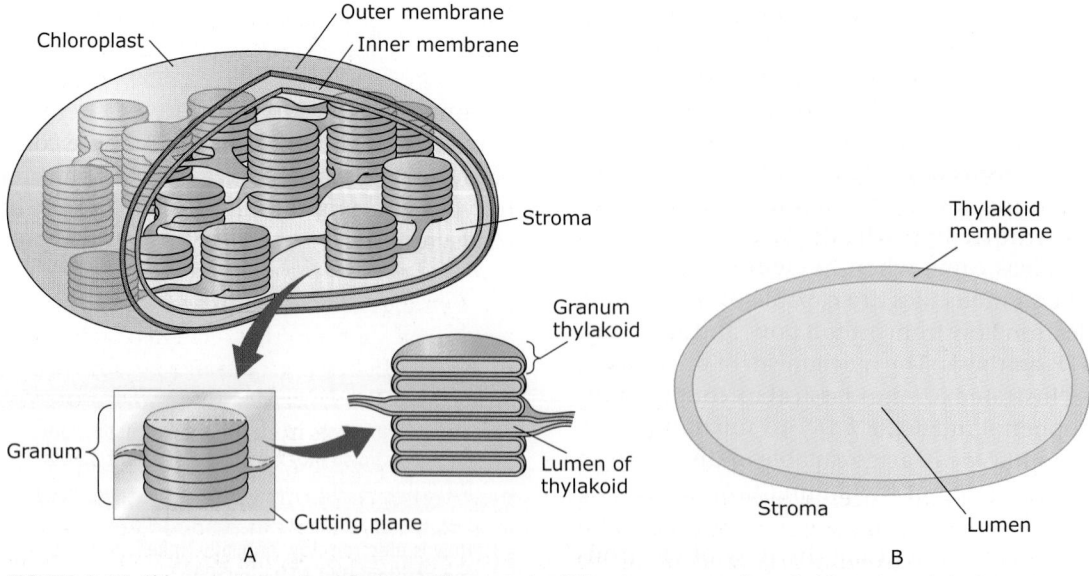

FIGURE 2.15 Chloroplast structural features. The more traditional representation of chloroplasts is indicated in A, which suggests the stacking of the thylakoid membrane and the presence of inner and outer membranes. The essentials required for the operation of chemiosmosis are displayed in B.

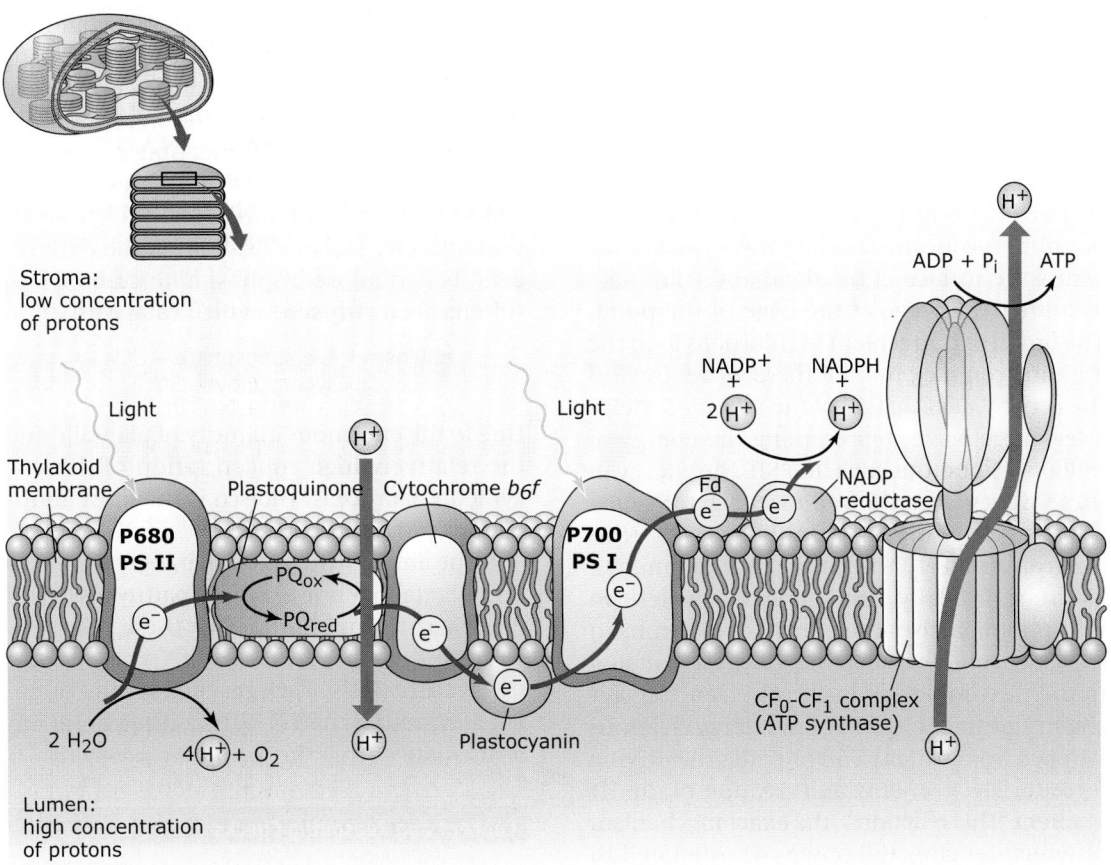

Stroma:
low concentration
of protons

Light

Thylakoid
membrane

**P680
PS II**

PQ_{ox}

PQ_{red}

e^-

Plastoquinone

Cytochrome $b6f$

H^+

e^-

Plastocyanin

Light

e^-

**P700
PS I**

e^-

Fd
e^-

e^-

$NADP^+$
+
2H^+

NADPH
+
H^+

NADP
reductase

H^+

$ADP + P_i$ ATP

CF_0-CF_1 complex
(ATP synthase)

H^+

2 H_2O

4H^++ O_2

H^+

Lumen:
high concentration
of protons

FIGURE 2.16 Chemiosmosis and the chloroplast. Similar to the case of mitochondria and bacterial chemiosmosis, the chloroplast uses electron transfer through membrane complexes and parallel movement of protons across it. Two major differences are (1) the requirement for light to drive the energetically uphill electron movement at two photosystems (PS) and (2) the reversal of the direction of the proton movement (from outside to inside the vesicle during the energy generation step).

lar to ubiquinone. Plastocyanin (PC), like cytochrome c in mitochondria, is a membrane surface protein that shuttles electrons between the last two complexes of the electron transport chain.

- The proton gradient generated by the ETC is used to synthesize ATP. The CF_oCF_1 ATP synthase is analogous to the mitochondrial F_oF_1 ATP synthase, and harnesses the flow of protons from the thylakoid lumen to the stroma. Both the CF_o subunit in chloroplasts and the F_o subunit in mitochondria are sensitive to the inhibitor oligomycin, as indicated in Figure 2-10.

The key differences between the light reactions and mitochondrial respiration are represented by the structure and function of the protein complexes in their respective electron transport chains. Whereas mitochondrial respiration begins with *high-energy* electrons do-

nated by NADH, the electrons entering the light reactions are *low energy* electrons donated by water. Low energy electrons can do very little work, so the first major difference between the two processes is that *the light reactions energize the electrons they receive*. This is achieved in two of the protein complexes, called **photosystems** (**PSI** and **PSII**). The proteins in each complex contain porphyrin rings chemically similar to the heme groups in mitochondrial proteins, with a Mg^{2+} as the central chelated ion instead of the Fe^{2+} in heme groups. Porphyrin rings in **chlorophyll molecules** serve to capture the light energy in photons of sunlight.

Acting together, several hundred chlorophylls in each photosystem complex (called an **antenna complex**) act as a net to capture 90% of the photons that strike them, exciting electrons in the chlorophyll molecules. Because the chlorophylls have overlapping electron orbitals, energy is propagated through a resonance energy transfer process, rapidly reaching a specialized pair of chlorophyll molecules called a **reaction center** at the center of the photosystem. These are named **P700** in PSI, and **P680** in PSII. One way to visualize how this works is to imagine hailstones striking the surface of a pond; the vibration caused by each hailstone is concentrated into waves that move across the surface of the pond and culminate as a breaking wave at the edge of the pond. The orderly arrangement of chlorophylls in the antenna complex funnels these waves toward the center, such that all of the vibration yields a few large waves representing the energy of millions of photons each. In PSII, the energy in these waves is transferred to a cofactor called the **oxygen-evolving complex**, which strips electrons from H_2O, releasing protons and O_2. (More specifically, the energy equivalent to four photons raises the energy of electrons in a chelate complex, such that they are donated to the next molecule in the electron transfer chain, and are replaced by low energy electrons stripped from water.) This provides the driving force for electron flow and creation of the H^+ gradient. This reaction—the exact mechanism of which remains unknown—is estimated to generate ~260 Gigatons of O_2 per year.

As electrons pass from PSII to PSI through the chain, generating the proton gradient across the thylakoid membrane, they lose most of the energy they received in PSII. PSI contains a reaction center that reexcites these electrons via a mechanism very similar to that used by PSII, but the high energy in these restored electrons are not used to pump more protons. Instead, they are used to reduce $NADP^+$ to NADPH. These electron carriers, plus the ATP generated from the proton gradient, fuel the dark reactions.

It is remarkable that Mitchell's chemiosmotic hypothesis explains both energy extraction by mitochondria and energy capture by chloroplasts. Historically, the hypothesis was first accepted for chloroplasts, possibly because the process was experimentally simpler to demonstrate. Unlike mitochondria, most of the H^+ gradient in chloroplasts exists as a chemical (i.e., pH) gradient, rather than an electrical gradient, which is the predominant part of the mitochondrial proton motive force. A rapid shift in the pH of the media of a chloroplast preparation—called an "acid bath" experiment—can drive the synthesis of ATP. This dramatic result was early convincing evidence in favor of the chemiosmotic hypothesis.

The name *dark reactions* is somewhat misleading, in that the reactions can actually occur in the light; they simply have no requirement for light energy. These are the steps involved in **carbon assimilation**, meaning conversion of CO_2 into simple organic molecules that cells use to build their membranes, nucleic acids, proteins, etc. The key enzyme in the dark reactions is ribulose bisphosphate carboxylase (often called **rubisco**), which catalyzes:

$$\text{ribulose -1, 5-bisphosphate} + CO_2 \rightarrow \\ \text{(2) Glyceraldehyde-3-P}$$

Due to the enormous quantity of plant life and the relatively high concentration of this enzyme in plant cells, rubisco is the most abundant enzyme on earth.

The dark reactions are strikingly similar to other metabolic and catabolic pathways, again emphasizing how important these reactions were in the early forms of life. One distinctive feature of the dark reactions is the use of NADPH, rather than NADH, as the energy intermediate during the formation of sugars.

Concept and Reasoning Checks

1. What are the functions of the photosystems in chloroplasts?
2. How is the proton motive force in photosynthesis different from the same force in mitochondria?

2.13 Nitrogen metabolism encompasses amino acid, protein, and nucleic acid pathways

Key concepts

- The nitrogen cycle connects the atmosphere, bacteria, plants, and animals.
- Amino acid metabolism integrates metabolic pathways, protein metabolism, and nucleic acid pathways.

All organisms require nitrogen-containing compounds, which are essential components of amino acids, nucleic acids, and energy carriers such as NADH and ATP. Plants and bacteria synthesize nitrogen-containing compounds called **alkaloids** as products of minor metabolic routes called **secondary metabolism**. These serve as defense mechanisms for those organisms, and are the source of a large number of pharmaceuticals. Here we consider in more detail two aspects of nitrogen metabolism: the **nitrogen cycle**, and amino acid metabolism.

Nitrogen compounds are cycled between the abundant N_2 in the atmosphere and the earth's organisms. The ability to incorporate N_2 into organic molecules, however, is limited to a very few organisms that can convert it to NH_3. One is a species of bacteria that forms nodules on the root of legumes. These bacteria, of the genus *Rhizobium*, contain the enzyme **nitrogenase,** which catalyzes the six-electron reduction of N_2 to NH_3. This complex and energy-consuming reaction (at least 12 ATP molecules are utilized to drive the reaction) is catalyzed by another electron transport chain involving enzyme complexes. While NH_3 can be incorporated into a variety of metabolic and catabolic pathways in most organisms, other more oxidized forms of nitrogen, such as NO_2^- and NO_3^- can be produced by soil bacteria. Additionally, a substantial quantity of nitrogen oxides are produced in the atmosphere from N_2 and O_2, catalyzed by metals (called the Haber process) and lightning sparks. The global conversion of nitrogen species between N_2 and its other forms can be returned to N_2 by certain bacteria; collectively, this iterative conversion between forms of nitrogen is called the **nitrogen cycle**.

Because cells invest most of their captured nitrogen in the amino group of amino acids, amino acid metabolism plays a central role in the nitrogen cycle. The amino acids alanine, aspartate, and glutamate can be converted into ketoacid intermediates of glycolysis and the Krebs cycle (pyruvate, oxalate, and 2-ketolutarate, respectively) by a single chemical reaction, the transfer of an amine group. This reaction, called **transamination,** is catalyzed by **transaminases**. For example, the enzyme aspartate amino transferase catalyzes the reaction:

oxaloacetate + glutamate ⇔
aspartate + 2-ketoglutarate

The enzyme catalyzes a near-equilibrium reaction in cells. Thus it serves as a means to form aspartate from oxaloacetate, or in the reverse direction, as a means to form glutamate from 2-ketoglutarate. Similarly, the enzyme alanine aminotransferase, which catalyzes

pyruvate + glutamate ⇔
alanine + 2-ketoglutarate

can produce alanine from pyruvate, or metabolize alanine; accordingly, it is also near equilibrium.

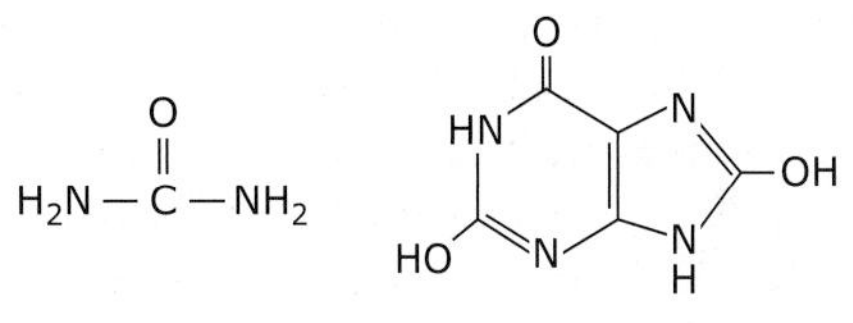

FIGURE 2.17 Structures of urea and uric acid.

Cells of higher animals synthesize only about half of the amino acids needed to form proteins; the other half must be supplied by the diet. Thus, amino acids are classified as **essential** (dietary) or **nonessential** (biosynthesized). Some amino acids are covalently modified after they are incorporated into proteins (e.g., phosphorylation, glycosylation), while others can also be modified as individual molecules (e.g., sulfation).

Amino acids catabolism follows a strict pattern. The nitrogen portion is removed (by transamination or deamination), and the carbon skeleton is degraded, either to CO_2 (via the Krebs cycle) or to glucose (by gluconeogenesis). Vertebrates convert the nitrogen to urea, a molecule that has a simple structure (**FIGURE 2.17**) through the urea cycle in the liver. Birds excrete nitrogen in the form of uric acid.

Concept and Reasoning Checks

1. What is the role of animals, plants, and bacteria in the nitrogen cycle?
2. Explain how transaminase reactions, which have a reactant in the Krebs cycle, can be considered anaplerotic (filling reactions).

2.14 The Cori cycle and the purine nucleotide cycle are specialized pathways

Key concepts

- The Cori cycle is a pathway between liver and muscle; the liver converts lactate to glucose, while the muscle converts glucose to lactate.
- The purine nucleotide cycle converts muscle aspartate to fumarate, used to replenish Krebs cycle intermediates.

While metabolic pathways are traditionally considered to be confined within single cells, some pathways span multiple cells. For example, the Cori cycle is a pathway performed by teams of muscle and liver cells. Due to their high ATP demand, many muscle cells

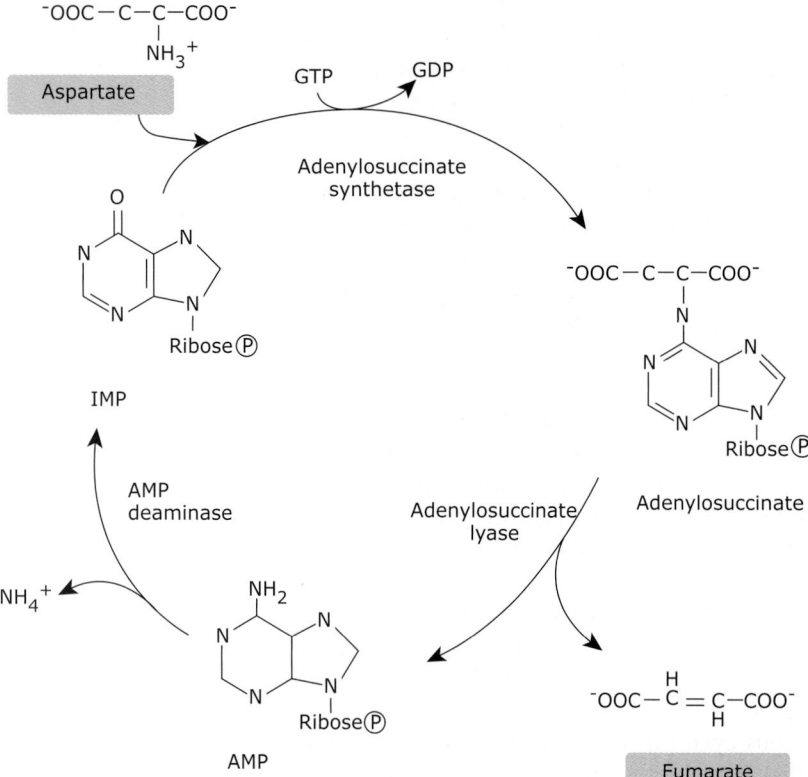

FIGURE 2.18 The purine nucleotide cycle. Conversion of aspartate to fumarate is catalyzed by the cycle illustrated involving enzymes of nucleotide metabolism. The purpose of the transformation is to provide an input of carbon into the Krebs cycle under conditions in which they may be depleted. This is important in skeletal muscle, which has little pyruvate carboxylase, the enzymatic step used in other tissues to convert pyruvate to oxaloacetate for the same purpose.

cycle, the conversion of the ketoacid pyruvate to alanine—and to other ketoacids such as oxaloacetate and 2-ketoglutarate—removes intermediates of the Krebs cycle. Replenishment of Krebs cycle intermediates is commonly achieved by the operation of the enzyme pyruvate carboxylase, which generates oxaloacetate from pyruvate.

An alternative method to replenishing Krebs cycle intermediates is the **purine nucleotide cycle**. As depicted in **FIGURE 2.18**, the cycle relies upon the deamination and reamination of AMP and the conversion of the carbon of aspartate to fumarate. Although formed in the cytosol, the fumarate can be converted to malate due to the presence of a cytosolic fumarase, and the malate transported into the mitochondria. Because the amino acid input may be derived from other tissues, the purine nucleotide cycle is another example of a multicellular metabolic pathway.

Concept and Reasoning Checks

1. How are cyclic pathways that span different cells distinct from those that occur in a single cellular compartment?
2. Muscle cells have little pyruvate carboxylase. How does the purine nucleotide cycle fill this need?

2.15 Metabolic viewpoints provide insight into cellular regulation—only metabolically reversible reactions are possible regulatory sites

Key concepts

- ATP is not a cellular regulator because its concentration is constant.
- Long-term regulation requires protein synthesis and is distinct from rapid alterations such as increases in allosteric modulators of enzymes.

In the previous sections, we have examined specific issues of metabolism and catabolism, such as reaction equilibrium constants, enzyme mechanisms, and regulation. If we adopt a more global view of metabolism, new questions emerge:

- Can ATP, ADP, or AMP serve as regulators of metabolic pathways?
- Is enzyme induction a separate mechanism from rapid enzyme control?

frequently secrete lactate, the product of anaerobic glucose metabolism. This is especially common during periods of intense exercise when oxygen is limiting, or during anaerobic exercise such as weight lifting. Cells in the liver can capture some of this lactate, and use it to synthesize glucose via **gluconeogenesis**. Gluconeogenesis utilizes most of the *glycolytic* enzymes that run in the reverse direction, plus *gluconeogenic* enzymes that catalyze the metabolically irreversible steps of glycolysis. Because the newly made glucose can be released back into the bloodstream for use by the muscle cells, the pathways of glycolysis in muscle plus gluconeogenesis in liver constitute a cycle.

Several ketoacids may be converted to amino acids, which can be released into the extracellular space. For example, alanine created by transamination of pyruvate (by alanine amino transferase) can be transported to liver cells, which can convert it to glucose and urea. While this seems similar to the Cori

Many investigators have posited that ATP itself or perhaps ADP or AMP might be regulators. Usually, the concentration of cellular ATP is invariant, and thus the molecule itself is not a regulator. There are instances where, pathologically, or during apoptosis, ATP content drops. However, there are hundreds of kinases that could respond kinetically to changes in ATP concentration; its variation would prove disastrous to normal cellular homeostasis. Cells have been shown to achieve remarkably constant levels of ATP. For example, muscle, which has enormous ATP turnover, has a separate high-energy phosphate compound and attendant enzyme, creatine and creatine phosphokinase, to ensure constancy of ATP levels. ADP is more complex; it is likely that it does change, but the actual free concentration of this nucleotide is likely to be very small, and changes in its concentration control mitochondrial respiration rather than participate in regulation of kinases. AMP has recently been shown to be important in the activation of **AMP-dependent kinase** (**AMP-kinase**). However, control of this regulatory enzyme is more complex; a group of kinase kinases exists that target and activate the AMP-kinase. AMP is known to alter a few other enzymes *in vitro*, but the extent of its actual influence on these enzymes and clear involvement in the AMP-kinase remains unknown.

Finally, we consider if enzyme induction uses a separate mechanism from rapid enzyme control. Most studies suggest that the same cell signal molecules involved in short-term regulation also regulate the increase in transcription and consequently affect the amount of total enzyme protein. The latter is affected in different ways and clearly has further means of control, but it now appears that similar mechanisms, such as tyrosine kinases, cyclic AMP, and Ca^{2+}, can trigger rapid alterations in enzyme activity as well as a longer-term regulation.

Concept and Reasoning Checks

1. How does recognition of a steady state alter thinking about pathways?
2. What is the distinction between a primary and secondary regulatory site?
3. How is rapid regulation distinct from enzyme induction?

2.16 What's next?

A new analytic approach for metabolism has developed recently, termed **metabolomics**. Following the overwhelming success of nucleotide sequencing of the entire genome for several organisms, including the human, many researchers have extended the approach to proteins—**proteomics**—and to metabolism. The suffix *-omics* is now widely used to indicate an application of simultaneous large-scale data collection and analysis. Genomics approaches have succeeded in large part because the information being sought is one-dimensional: despite conformational changes of the DNA molecule, only its linear sequence is involved in information transfer.

To date this approach for studying protein and metabolic pathway structure and function areas has yielded limited success. In the case of metabolism, several databases of protein sequence and interaction data currently exist, though they have yet to meet the level of detail in the Gene Bank collection of DNA sequences. One of the attractive features of more comprehensive protein and metabolism databases is the possibility that they will reveal patterns characteristic of a disease state. However, the concentrations of intermediates and flux of pathways change routinely in healthy cells, so finding a characteristic signature of a disease will be difficult to tease out.

A separate technologic advance resulting from genetic analysis is the ability to target and ablate (remove) the expression of a specific protein enzyme in single cells. This is based on the introduction of short, interfering RNAs (siRNAs) that trigger cleavage of specific mRNAs. siRNA technology is currently in wide use to study the role of specific proteins in cellular function, and could yield important insights into the function and regulation of metabolic pathways. But this genetic removal of an enzyme is different from short-term application of an inhibitor, and will likely yield different outcomes over time as cells process the siRNAs and then recover. In addition, genetic silencing can lead to alterations of other, unintended, protein targets. Use of gene and protein array technologies may help keep track of such global changes, though these too are subject to many complications. In the future, as more extensive arrays of proteins and multiple measurements of metabolites are available, the combination of array methodology—"-omics"—with selective ablation of enzymatic steps may be extended to systematically examine cellular systems in the absence of distinct enzymes

The dramatic rise in obesity and type II diabetes in the United States in recent years

Type II diabetes has become an epidemic. The disease is characterized by a relative lack of insulin, distinguishing it from Type I diabetes, which is usually precipitated by a loss of pancreatic beta cells. The older names for these two forms are Type I (juvenile) and Type II (adult-onset), reflecting the common patterns of occurrence. However, in large part to changes in nutrition, we now have a sharp increase in childhood obesity, often concurrent with instances of Type II diabetes. Obesity, even in the absence of diabetes, causes insulin resistance, a bellwether of the disease.

It has long been known that normalizing blood glucose can improve insulin sensitivity. Hence, strategies other than increasing insulin intake have been part of the therapy for Type II diabetes for some time. Stimulation of insulin release from the beta cells of the pancreas is an established approach. A newer one is **incretin**-based therapy, using agonists of the peptide GLP-1, or inhibitors of its proteolysis. GLP-1 is the peptide hormone secreted by cells in the duodenum that stimulates insulin release in response to sugar in the diet.

A drug more than fifty years old, metformin (marketed as glucophage), has recently been shown to act as an activator of a central energy modulator, AMP-activated protein kinase (AMP-kinase). The subsequent actions, to decrease hepatic gluconeogenesis, increase muscle glucose transport, and increase fatty acid oxidation in both tissues, cause an increased energy consumption and lowering of blood glucose concentrations. Because AMP-kinase itself is a central regulator for a variety of cells and modulates essential energy pathways, a new interest in cellular energy and metabolism has arisen. Diabetes itself is essentially a metabolic disorder, akin to fasting, albeit with the added complication of an elevated glucose level in the blood.

It is also well established that a key clinical concern of Type II diabetes is related to the immune system; the disease can be considered to confer on the body a state of chronic low-level infection. Additionally, with the discovery of many new hormones produced by the adipocytes and involved in homeostatic regulation of other tissues, including cells of the liver, muscle, and brain, a wide range of cells are involved. There are other disease states that have a similar metabolic character and respond to the same drugs. For example, polycystic ovary disease shares several metabolic features with diabetes. In addition, side effects of certain drug treatments cause a similar set of pathologies, such as a new class of antipsychotic drugs (atypical) and protease treatment for AIDS patients. The set of similar symptoms produced by these disparate situations is referred to as **metabolic syndrome**.

We can achieve a specific understanding of mechanisms by understanding what occurs at the level of the cell. Yet we must consider a wider scope of events due to the connections between various tissues and different conditions. A unifying aspect of our current understanding of diabetes does allow us to conclude that there is a fundamental unity of cell behavior and that understanding the most basic metabolic level reaps rewards well beyond what we may be able to picture in studying our own particular area of interest.

suggests a greater need to understand metabolism. A better fundamental understanding of metabolism and its control, coupled with new tools to analyze metabolic events, is needed to address these problems. It is ironic that, in the face of world hunger, a modern medical crisis is the result of overfeeding. It harkens back to the early description of diabetes, "starvation in the midst of plenty" (see *Historical Perspectives* box on page 60).

2.17 Summary

Metabolism is a process of energy extraction and energy utilization that proceeds at a steady state. The intracellular concentrations of most of the intermediate molecules of metabolic pathways are constant at any given flux rate, yet materials are utilized to yield waste products. The transfer of energy during metabolism can be analyzed using the formalism of thermodynamics, which separates the energy forms as enthalpy, effectively the bond energy, and entropy, effectively the energy due to position and arrangement. Together, the two forms can be combined into an equation and represented by the free energy (ΔG), which accounts for reaction changes at a given temperature in terms of the two thermodynamic variables. Reactions that are close to equilibrium (ΔG close to 0) are called near equilibrium, and not likely sites of metabolic control. Those displaced from equilibrium ($\Delta G \ll 0$) are candidates for regulatory sites in pathways. Cells have several pathways for energy exchange. All cells use glycolysis, which produces lactate from glucose and yields a small amount of ATP. Most cells use oxidative pathways, which are fed by either carbohydrates or lipids, and employ the Krebs cycle and ETC to produce prodigious amounts of ATP. Some of the extracted energy is used in catabolic (biosynthetic) pathways. In addition, plants and some single-celled organisms use light energy to drive photosynthesis, which provides the energy to incorporate CO_2 into carbohydrates. The metabolism of nitrogen, an atom essential in proteins and nucleic acids, is typically separate from direct energy-providing molecules. Finally, pathways can span separate cells in multicellular organisms, such as the Cori cycle, in which glucose is cycled between muscle and liver.

http://biology.jbpub.com/lewin/cells

To explore these topics in more detail, visit this book's Interactive Student Study Guide.

References

Griffin, J. L., and Nicholls, A. W. 2006. Metabolomics as a functional genomic tool for understanding lipid dysfunction in diabetes, obesity and related disorders. *Pharmacogenomics* v. 7 p. 1095–1107.

Kaufman, F. R. 2005. *Diabesity: The Obesity-Diabetes Epidemic That Threatens America—and What We Must Do to Stop It.* New York: Bantam Books.

Klotz, I. M., and Rosenberg, R. M. 2000. *Chemical Thermodynamics.* New York: Wiley.

Morowitz, H. J. 1970. *Entropy for Biologists.* New York: Academic Press.

Newsholme, E. A., and Leech, A. R. 1983. *Biochemistry for the Medical Sciences.* New York: Wiley.

DNA replication, repair, and recombination

3

Jocelyn E. Krebs
University of Alaska, Anchorage, AK

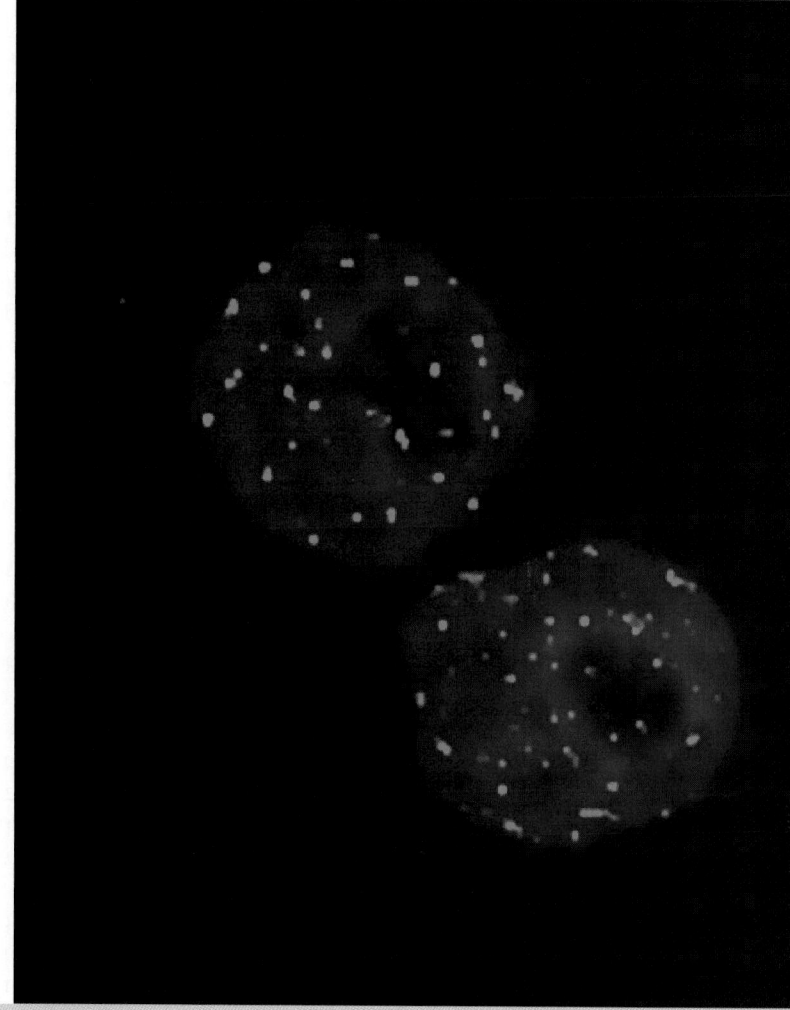

PHOSPHORYLATED H2AX FOCI (green) appear at sites of double strand breaks after cells are irradiated with X-rays. Photo courtesy of Jack A. Taylor, National Institutes of Health.

CHAPTER OUTLINE

3.1 Introduction

The hereditary basis of every living organism is its **genome**, a long sequence of DNA that provides the complete set of genetic information carried by the organism. The genome includes chromosomal DNA as well as DNA in plasmids and (in eukaryotes) organellar DNA as found in mitochondria and chloroplasts. Physically, the genome may be divided into a number of different DNA molecules, or **chromosomes**. The ultimate definition of a genome is the sequence of the DNA of each chromosome. Functionally, the genome is divided into genes. Each gene is a sequence of DNA that codes for at least one RNA or polypeptide as its final product.

In this chapter, we describe the experiments that led to the discovery of DNA as the genetic material of most organisms and the physical structure of DNA and its constituent parts. The description of the complementary base-paired structure of DNA in 1953 by Francis Crick and James Watson led to an immediate (and correct) prediction of the basic mechanism of replication, the process by which DNA is duplicated so that accurate copies of the genetic material can be passed on to progeny. Replication of duplex DNA is a complicated endeavor involving multiple enzyme complexes. We will discuss the different enzymatic activities that are involved in the stages of initiation, elongation, and termination.

DNA replication is a highly accurate process, but it is still subject to error. Errors introduced by replication, or by any of a large number of sources of physical/chemical DNA damage, must be corrected if they are not to lead to mutation. Injury to DNA is minimized by systems that recognize and correct errors or damage. Two systems ensure that DNA sequences are accurately maintained. First, the replication machinery includes proofreading steps to ensure accurate replication. Second, repair systems monitor the genome for DNA damage and fix the damage to restore the original sequence. Repair systems are as complex as the replication apparatus itself and are specialized to recognize a wide array of errors. When a repair system reverses a change to DNA, there is no consequence. A mutation may result, though, when it fails to do so. The measured rate of mutation reflects a balance between the number of damaging events occurring in DNA and the number that have been corrected (or miscorrected). While estimates of the daily rate of damage to the human genome range from 1000 to 1,000,000 lesions per cell day, the overall mutation rate in humans appears to be approximately one mutation per 30 million bp per generation. Adequate repair systems are required for genome integrity, and loss of function in a DNA repair pathway has serious repercussions, as illustrated by human diseases associated with such losses. Diseases resulting from repair defects are frequently associated with increased cancer risk, due to the increased mutation rate caused by these defects.

It is important to remember, however, that some mutation rate is essential for the process of evolution to occur. In addition to sequence changes caused by unrepaired errors or damage, cells also have mechanisms that ensure that the genetic material is subject to some variation before it is passed to offspring. The process of meiotic recombination, in which cells exchange material between homologous chromosomes, allows favorable and unfavorable mutations to be separated and tested as individual units in new assortments. It provides a means of escape and spreading for favorable alleles, and a means to eliminate an unfavorable allele without changing allele frequencies for all the other genes with which this allele is linked. This is the basis for natural selection.

3.2 DNA is the genetic material

Key concepts
- Bacterial transformation provided the first evidence that DNA is the genetic material of bacteria.
- Bacteriophage infection showed that DNA is the genetic material of viruses.

Today, it seems as though we have always known that DNA is the hereditary material that carries the blueprint of the cell. In fact, this was a surprisingly recent discovery that ushered in the modern era of molecular genetics. During the first half of the 20th century, before the structure of DNA was known, there were several key experiments that demonstrated that DNA, and in some cases RNA, was in fact the genetic material. While the conclusions drawn from these experiments were debated, and many scientists did not feel that DNA was sufficiently complicated to contain the instructions required to build an organism, these key experiments provided the scientific evidence required to prove that DNA is the hereditary material.

Frederick Griffith made a key series of observations in 1928 that set the stage for demonstrating that DNA is the hereditary material. Griffith studied a bacterium called *Pneumococcus* (now *Streptococcus*) *pneumonia*, which causes pneumonia in humans and is lethal in mice. Griffith took advantage of the availability of two different strains of *S. pneumonia*, a pathogenic strain that kills infected mice and nonpathogenic strain that does not kill infected mice. The two strains differ primarily in their outer polysaccharide coat, one having a rough-looking outer coat ("R" strain) and the other a smooth-looking outer coat ("S" strain). When the pathogenic S bacteria are destroyed by exposure to heat, they are unable to kill infected mice as expected. However, Griffith discovered that if the heat-killed S bacteria are mixed with the nonpathogenic R strain and the mice were infected with this mixture, the mice died. When bacteria were isolated from the dead mice they were found to have the characteristics of the lethal R strain of bacteria. Somehow, the nonpathogenic bacteria had been "transformed" by the heat-killed bacteria into a lethal strain. This process is called transformation and involves the change of the genetic background of one strain into that of another strain. Griffith's experiment is depicted in **FIGURE 3.1**.

What caused this transformation of one strain of bacteria into a different strain of bacteria? Whatever the substance, it had changed the hereditary properties of the nonlethal strain and thus was a candidate for the genetic material. This question was answered in a series of experiments conducted by Oswald Avery, Colin MacLeod, and Maclyn McCarty in 1944. These scientists theorized that a specific biochemical component of the heat-killed bacteria must survive the heat treatment and be transferred to the nonpathogenic recipients. Since the polysaccharide coat of the two strains of bacteria was different, this was an obvious candidate. However, purified polysaccharides were not capable of transformation, indicating that polysaccharides were not the transforming chemical. The researcher then took the approach of destroying each of the major biochemical components of an S strain extract, one at a time, to determine which component was responsible for the transformation, as shown in **FIGURE 3.2**. Destruction of proteins, fats and RNA in extracts from S strain bacteria did not prevent transformation. Only when the DNA was degraded was the ability to

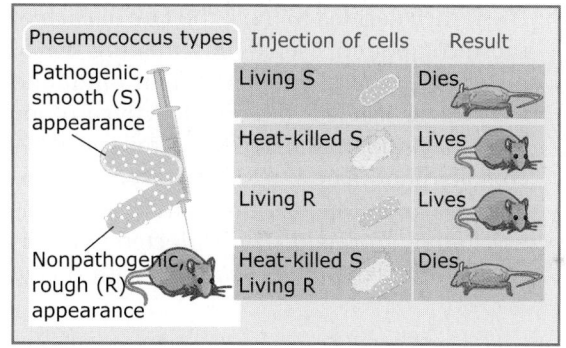

FIGURE 3.1 Neither heat-killed S-type nor live R-type bacteria can kill mice, but simultaneous injection of both can kill mice just as effectively as the live S-type.

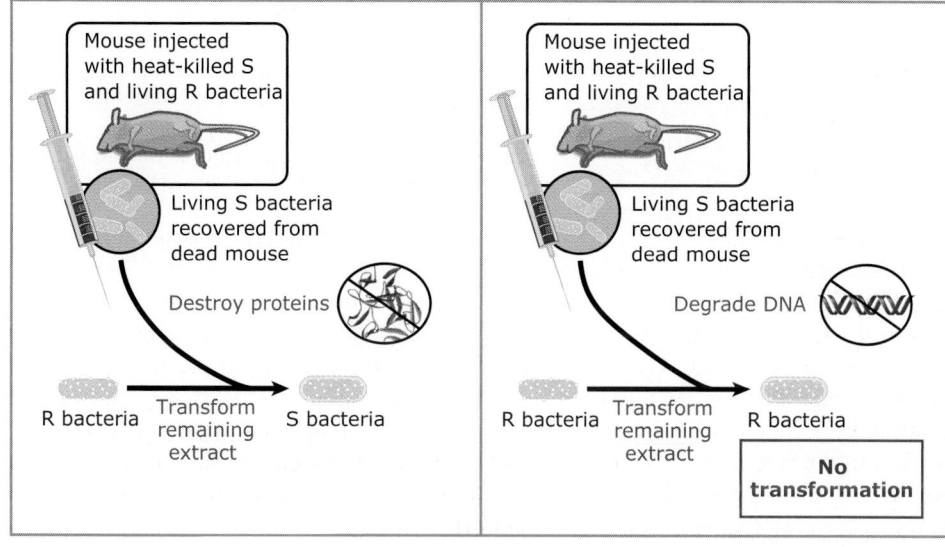

FIGURE 3.2 Extracts of the S strain bacteria lacking protein can still transform R strain bacteria into S strain, but extracts lacking DNA cannot. Structure from Protein Data Bank 3HIA. Z. -Y. Zhang, et al., *Crystal structure of the choline binding domain of Spr1274 in Streptococcus pneumoniae.*

transform the nonlethal strain of bacteria into a lethal strain lost. These were the first results to strongly suggest that DNA is the transforming chemical and thus that DNA is, in fact, the hereditary material.

However, there were still scientists who remained unconvinced. In 1952, Alfred Hershey and Martha Chase provided yet another demonstration that DNA was the hereditary material. Their experiments made use of a bacteriophage, a virus that infects bacterial cells. The bacteriophage they chose for their work was phage T2, a simple phage composed of nothing but DNA and protein (**FIGURE 3.3**). The DNA is contained within a protein shell called the phage head. Hershey and Chase reasoned that the hereditary material must be injected into the bacteria after the phage attaches to the outside of the bacterial cell, so that it can direct the production of progeny phage that ultimately kill the bacterial host. Since the phage is composed of protein and DNA, they decided to track the fate of each of these components separately making use of radioactive labels. Proteins contain sulfur and they can be labeled with a radioactive isotope (^{35}S) that is not found in DNA. Similarly, DNA contains phosphate and can be traced by labeling with an isotope of phosphate (^{32}P) that is not found in protein. They infected two cultures of the bacteria *E. coli*, one with phage labeled in their protein coat with ^{35}S and one with phage labeled in their DNA with ^{32}P, and then followed the fate of the radioactive labels. After a brief period of infection, they put each mixture into a blender to separate the bacterial cells from the empty phage "ghosts," and then centrifuged to recover the infected cells. The radioactivity associated with DNA (i.e., the ^{32}P) was found associated with the cells. The radioactivity associated with the protein (i.e., the ^{35}S) was found in the supernatant with the phage ghosts. Importantly, the infected cells produced progeny phage as expected. Thus, the material that enters the cell to direct the production of new phage is the DNA and not the protein. The results of this experiment served to convince the remaining skeptics that DNA was, in fact, the hereditary material. (We now know that in some viruses RNA serves as the genetic material, but DNA is the hereditary material in the vast majority of organisms.)

Concept and Reasoning Check

1. If Hershey and Chase had observed that nearly none of the DNA but a substantial fraction of the protein of phage T2 enters *E. coli* cells during infection, what would they have concluded?

3.3 The structure of DNA

Key concepts

- Nucleotides consist of a purine or pyrimidine base linked to the 1' carbon of a pentose sugar and a phosphate group on either the 5' or 3' carbon of the sugar.
- The pentose sugar of DNA is deoxyribose; for RNA it is ribose.
- Polynucleotide chains are joined by phosphate groups linked to the 5' carbon of one sugar and the 3' carbon of the next sugar, resulting in a continuous sugar-phosphate backbone.
- DNA contains the bases adenine, thymine, guanine, and cytosine; RNA contains uracil instead of thymine.
- DNA is a double helix consisting of two antiparallel strands held together by complementary base pairing.
- The DNA helix contains parallel major and minor grooves.

When it became clear that DNA (RNA in some viruses) was the carrier of heredity, it became crucial to understand the structure of this important biomolecule. After all, this relatively simple molecule had to be capable of directing three crucial functions: (1) it had to be capable of faithful duplication at every cell division in

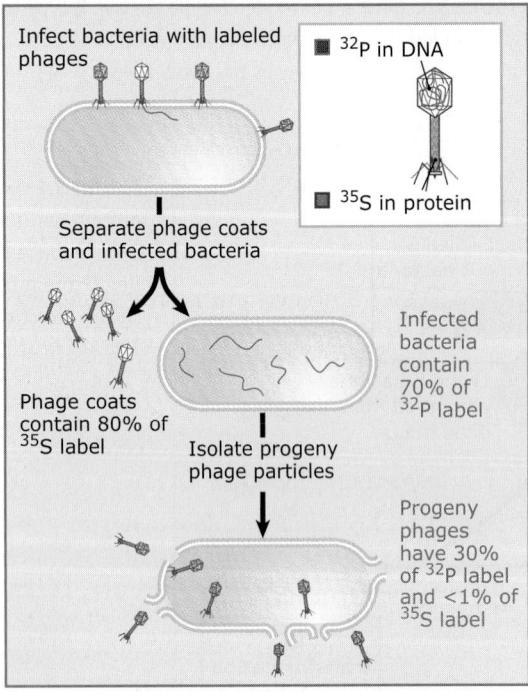

Infect bacteria with labeled phages

^{32}P in DNA

^{35}S in protein

Separate phage coats and infected bacteria

Phage coats contain 80% of ^{35}S label

Isolate progeny phage particles

Infected bacteria contain 70% of ^{32}P label

Progeny phages have 30% of ^{32}P label and <1% of ^{35}S label

FIGURE 3.3 The genetic material of phage T2 is DNA.

order to pass hereditary information on to the next generation, (2) it had to somehow provide the information required to direct the production of the vast array of proteins and RNA molecules that are expressed in the organism (i.e., it had to have information content), and (3) it had to be capable of change (i.e., mutation) at a low rate consistent with the principles of evolution, while remaining a (reasonably) stable repository of the genetic blueprint of the cell. Yet chemical studies indicated that DNA, at least from the chemical point of view, was a fairly uncomplicated molecule composed of only four building blocks. So how could this comparatively simple molecule provide all the information required to produce a new organism? This was a major question driving molecular biology and genetics, and an understanding of the structure of DNA was expected to provide key insights into how these functions might be accomplished. Indeed, once the structure of DNA was proposed by Watson and Crick in 1953, it revealed a simple and elegant means for accomplishing all of these functions. We will start by describing the building blocks of the DNA molecule and then move to a discussion of the complete three-dimensional structure of double-stranded DNA.

The overall structure of the building blocks of the DNA molecule is quite straightforward from a chemical point of view. Each of the four building blocks consists of three chemical components: a phosphate, one of four **purine** or **pyrimidine** bases, and the sugar deoxyribose. The structure of each of the four **nucleotide** building blocks is shown in **FIGURE 3.4**. Note that each nucleotide is assembled from its components in the same way. The significant difference is the nitrogenous base that is attached to the sugar. It should also be noted that the chemical composition of RNA is similar with two exceptions—the sugar in RNA is ribose instead of deoxyribose and RNA contains the base uracil instead of thymine. In his studies of the DNA molecule, Erwin Chargaff provided two important empirical rules regarding the nitrogenous base content of most DNA molecules. First, he noted that the total amount of pyrimidine nucleotides (thymine and cytosine) was always equal to the total amount of purine nucleotides (adenine and guanine). In other words, the amount of T+C is always equal to the amount of A+G in a double-stranded DNA molecule. Second, the amount of A always equals the amount of T and the amount of G always equals the amount of C. These facts

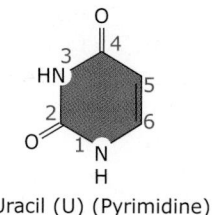

Uracil (U) (Pyrimidine)

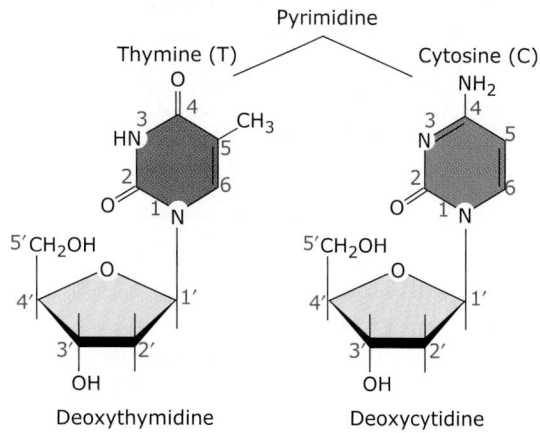

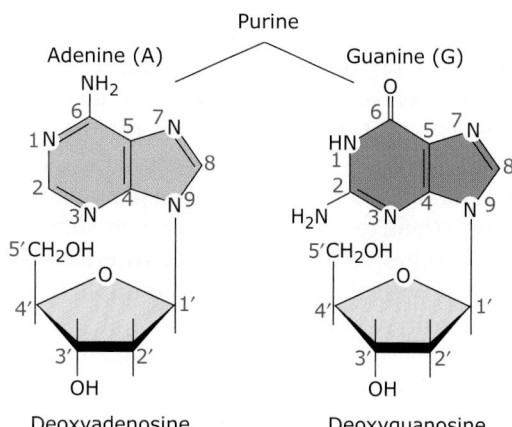

FIGURE 3.4 The four deoxyribonucleotides that make up DNA, and the ribonucleotide containing the RNA-specific base uracil.

must be taken into account in any structure for duplex DNA and were quite important to Watson and Crick as they worked to establish the three-dimensional structure of the DNA molecule.

The final piece of experimental evidence that led to a solution of the structure of DNA came from X-ray diffraction data that had been gathered by Rosalind Franklin and Maurice Wilkins. The X-ray data suggested that DNA had a helical structure. Using all of this information as well as model building Watson and Crick were able to correctly deduce the double-helical structure of the DNA molecule.

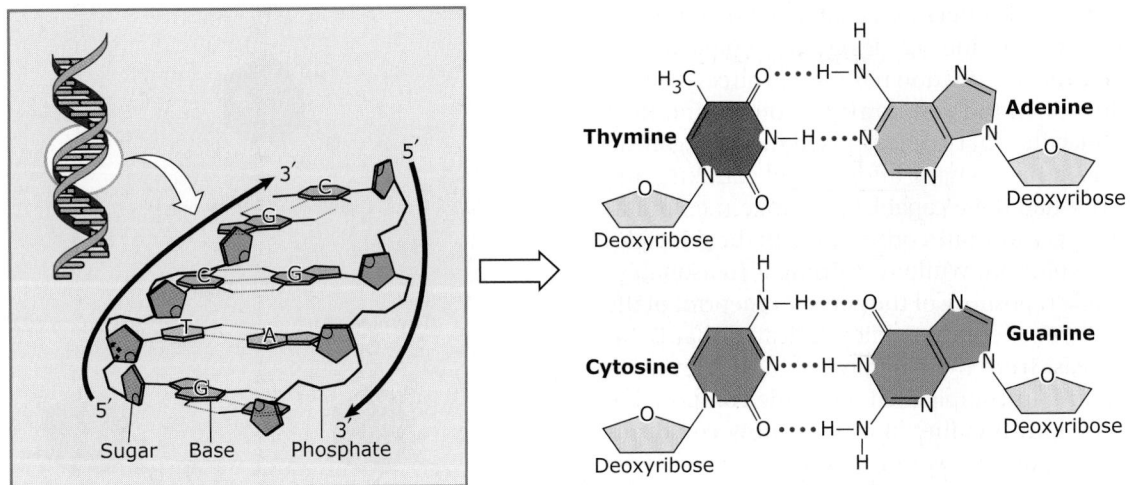

FIGURE 3.5 Base pairs in DNA.

In their paper describing this structure in 1953 they noted that this structure provided an immediate answer to the question of how the DNA molecule could be accurately duplicated at each cell division. It was clear that A on one strand of the double helix always pairs with T on the opposite strand and that C on one strand always pairs with G on the other strand. Therefore, one DNA strand carries all the information required for the synthesis of the second strand. If the two strands are separated, then each strand can be used as a template to direct the synthesis of the second, **complementary** strand with the result that the duplex DNA molecule would be faithfully replicated. Of course, we now also understand that the order of the bases on a DNA strand is responsible for determining the RNAs and proteins encoded by DNA, which provides the information content contained in DNA. In addition, we know it is possible for bases to change due to mutation, providing the molecular basis for evolution.

The three-dimensional structure of the DNA molecule is essential for the role this molecule plays in heredity. DNA is a double-stranded helix, 20 Å in diameter, composed of two single strands of DNA held together by hydrogen bonds between the bases of each strand as seen in **FIGURE 3.5**. The DNA helix makes a complete turn every 34 Å, with 10 bp per turn. The A-T base pair has two hydrogen bonds and the G-C base pair has three hydrogen bonds. These relatively weak chemical interactions allow the two strands to be separated in the process of DNA replication while also maintaining the stability of the DNA molecule due to their enormous numbers over the length

of a double helix. Note that a single chromosome in a human (or any other organism) is composed of a single DNA double helix and can be several million base pairs in length. Thus, there are millions of hydrogen bonds holding the two strands together to give DNA the stability required to serve as the genetic material.

An important aspect of the structure of the DNA double helix centers on the fact that the two single-stranded chains of the duplex molecule run antiparallel to each other as viewed from the backbone of the DNA molecule. The negatively charged backbone of the DNA molecule is composed of repeating sugar and phosphate groups, while the nitrogen-containing bases lie on the inside of the molecule where they can minimize their exposure to water. A careful look at the sugar-phosphate backbone reveals a chemical polarity, shown in **FIGURE 3.6**. On one strand, the backbone runs in a 5′ (the 5′ C on the deoxyribose) to 3′ (the 3′ C on the deoxyribose) direction while on the opposite strand the polarity of the backbone is 3′ to 5′. Thus, either end of a linear double-stranded DNA molecule has one free 5′ phosphate and one 3′ hydroxyl. This polarity of the two strands is important in the process of DNA replication (see *3.4 DNA replication is semiconservative and bidirectional*) and in the process of gene expression. In addition, many of the enzymes that interact with and process the DNA molecule do so with specific attention to chemical polarity of the DNA chain. For example, both DNA and RNA polymerases are only able to synthesize nucleotides in the 5′ to 3′ direction. Thus, the polarity of each strand is an important aspect of the DNA structure.

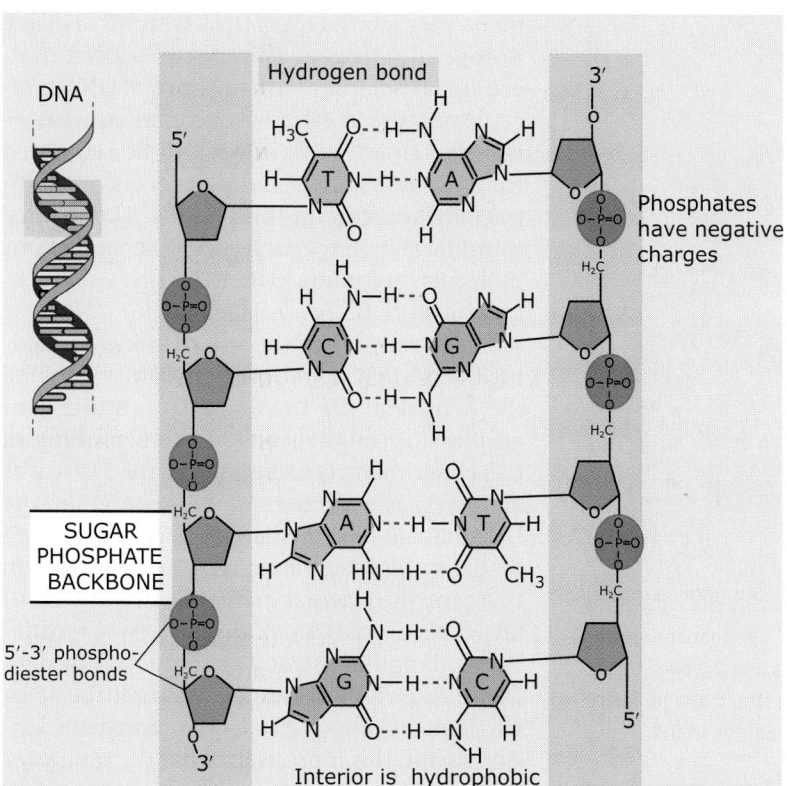

FIGURE 3.6 DNA strands consist of a series of 5'-3' links that form a negatively charged backbone from which the bases protrude. Two strands are held together in an antiparallel orientation via base pairing.

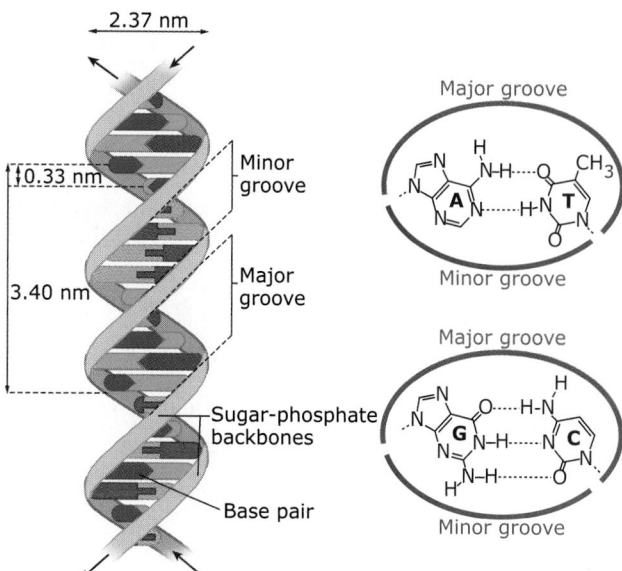

FIGURE 3.7 The DNA helix consists of continuous major and minor grooves created by the angle of the bonds between base and sugar.

In addition, the three-dimensional structure of DNA reveals the presence of two grooves that run in a spiral along the length of the DNA molecule (**FIGURE 3.7**). The larger groove (22 Å across) is called the **major groove** and the narrower groove (12 Å) is called the **minor groove**. These grooves are created by the angle of the bonds between the bases and the deoxyribose sugar, which results in a larger "face" of each base pair presented in the major groove. The major groove provides sufficient information to allow proteins that recognize specific DNA sequences to bind to those sequences. The major groove is also wide enough to accommodate a protein α-helix, thus many sequence-specific DNA-binding proteins use α-helices to read DNA sequences. This is especially important in the regulation of gene expression (see *4.6 Activators and repressors regulate transcription initiation*).

Concept and Reasoning Check

1. If the G-C content of a DNA duplex is 0.44, what are the proportions of the four bases?

3.4 DNA replication is semiconservative and bidirectional

Key concepts

- Heavy isotope labeling revealed that DNA replication is semiconservative.
- DNA replication initiates at a single site known as the origin.
- Replication forks move bidirectionally from the origin, creating a replication bubble.

Accurate duplication of the DNA molecule during each cell division cycle is essential for the maintenance of genetic information through generations. One of the first questions asked about the replication of DNA molecules was how the template-directed synthesis of a new DNA strand, as suggested by Watson and Crick, might take place. A straightforward model, called **semiconservative replication**, envisions the two strands of the double helix separating and each strand being used as a template to direct the synthesis of a new or daughter strand of DNA, as shown in **FIGURE 3.8**. This

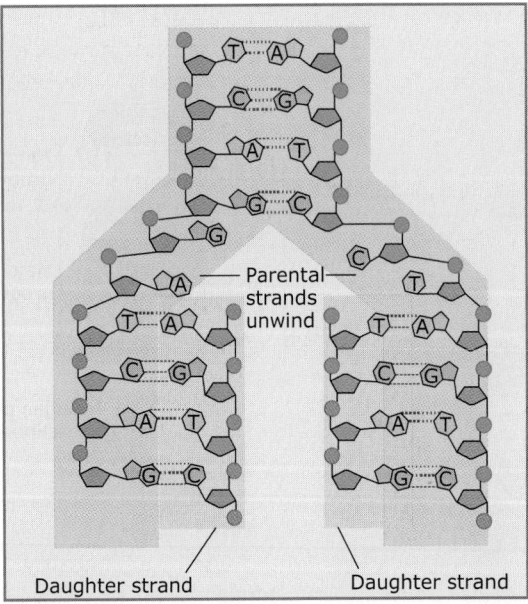

FIGURE 3.8 DNA replication is semiconservative. Replication is accomplished by separating the two strands of a parental duplex, then using each strand as a template for synthesis of a complementary daughter strand.

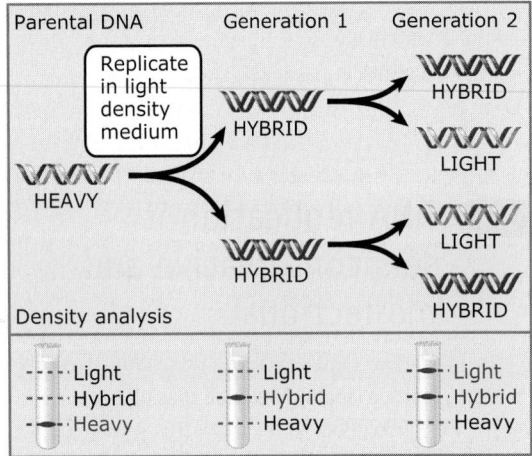

FIGURE 3.9 The Meselson and Stahl experiment showed that DNA replication was semiconservative. *E. coli* grown in medium containing the "heavy" nitrogen isotope ^{15}N have heavy DNA. When these cells are switched to "light" ^{14}N medium, after a single round of replication all DNA is of hybrid density (½ heavy, ½ light). The second round of replication results in the half hybrid and half light DNA, which can only be explained by a semiconservative mode of replication.

model posits that each new double helix consists of one original (also called parental) strand of DNA and one newly synthesized daughter strand of DNA.

A definitive test of this model for DNA replication was provided by Meselson and Stahl in 1958, as depicted in FIGURE 3.9. In their experi-

ment, they labeled *E. coli* DNA with ^{15}N, a heavy isotope of nitrogen. This creates a DNA molecule that is denser than the normal DNA molecule because its DNA bases contain the heavy isotope. This "heavy" DNA can be separated from normal, light DNA by a process called ultracentrifugation. In this process, a centrifuge spinning at high speeds is used to separate DNA molecules according to their density. Using cells in which the DNA was labeled with ^{15}N, these scientists transferred the cells to media containing the normal, light isotope ^{14}N and evaluated the density of the DNA after 0, 1, and 2 generations of cell division. At the beginning of the experiment (generation 0), the DNA was all heavy, as expected. After 1 generation, the DNA had an intermediate density as predicted by the semiconservative replication model. In this case, the original (parental) strand is composed of heavy DNA while the newly synthesized (daughter) strand is composed of light DNA. After 2 generations, one-half the DNA was light and one-half was of intermediate density. Again, this is precisely what the model for semiconservative replication would predict. DNA replication was subsequently shown to occur by a semiconservative mechanism in a number of different organisms, including eukaryotes. This is the only mode of replication that has been observed in any species to date. Even organisms that have single-stranded genomes, like many viruses, must go through a double-stranded intermediate during replication using a semiconservative mechanism.

The basic structure of a chromosome undergoing replication was first visualized by John Cairns in 1963 using *E. coli* cells and a technique called autoradiography, shown in FIGURE 3.10. In these experiments, Cairns grew *E. coli* cells in a medium containing a radioactive form of one of the bases found in DNA ([^{3}H] thymidine) for various amounts of time, and then very gently lysed the cells and collected the chromosomes on filters, which were exposed to a photographic emulsion that detects the radioactive disintegrations. This provided a visual image of a replicating chromosome. The original interpretation of this image suggested that DNA synthesis began at a unique site and proceeded in a single direction around the circular molecule. This is referred to as *unidirectional* synthesis and the site at which DNA synthesis is taking place is called the **replication fork**. We now know that DNA replication does initiate at a unique site (the **origin** of DNA replication) and proceeds in both directions around the circular

(a)

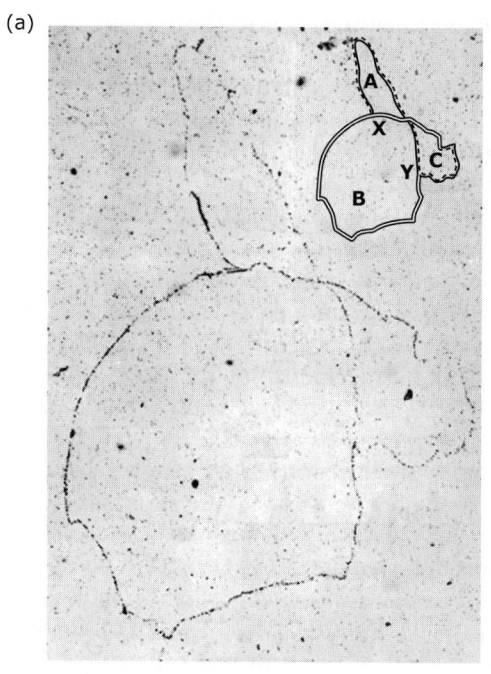

(b)

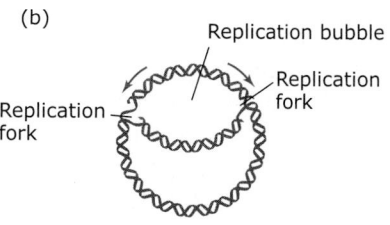

Replication bubble

Replication
fork

Replication
fork

FIGURE 3.10 Replication of a circular chromosome initiates at a single point (the origin) and two replication forks proceed in opposite directions. The image in part (a) shows an experiment and which *E. coli* cells were cultured in the presence of [³H]thymidine for slightly longer than one generation. Incorporated [³H]thymidine was detected by exposing extracted DNA to a photographic gel. The diagram in the upper right is an interpretation of the structure showing the presence of a replication bubble. Note that this early experiment did not distinguish between unidirectional and bidirectional replication. Photo reproduced from J. Cairns, *Cold. Spring Harb. Symp. Quant. Biol.* 28 (1964): p. 44. Copyright 1964, Cold Spring Harbor Laboratory Press. Photo courtesy of John Cairns, University of Oxford.

chromosome. Thus, there are two replication forks emanating from a single origin. This is referred to as **bidirectional** replication. On a circular chromosome, the two forks will eventually meet and merge. Replication of linear eukaryotic chromosomes is similar, with replication beginning at (multiple) origin sites and then extending in both directions away from each origin. The impact of DNA ends in eukaryotic replication as discussed in *3.11 Replicating the ends of a linear chromosome.*

3.5 DNA polymerases replicate DNA

The replication of a DNA molecule is both fast and highly accurate. The rate of replication is ~30,000 nucleotides added per minute in bacteria and about 3000 nucleotides added per minute in eukaryotic cells. To attain this incredible speed without sacrificing accuracy, there must be exceedingly well-regulated machinery in the cell. A DNA polymerase enzyme capable of synthesizing new DNA strands lies at the core of this replication machinery. The first DNA polymerase was discovered and characterized in 1958 by Arthur Kornberg. This enzyme, called DNA polymerase I, was isolated from *E. coli* cells and shown to be capable of synthesizing a new complementary DNA chain by adding nucleotides sequentially onto the 3' end of an existing DNA chain. In the subsequent years, a vast array of DNA polymerases have been discovered and shown to function in myriad processes in the cell requiring the synthesis of new DNA, particularly in replication and in DNA repair processes. The enzyme responsible for synthesis of the chromosome in *E. coli* is DNA polymerase III. The enzyme originally discovered by Kornberg is involved in DNA repair reactions (discussed later in this chapter) and in the maturation of Okazaki fragments. In eukaryotes, there are two major replicative polymerases in the nucleus, DNA polymerases δ and ε (see *3.9 Leading and lagging strand synthesis is coordinated*).

All DNA polymerases isolated to date exhibit two properties that must be considered in the context of replicating a duplex DNA molecule. First, they all require an existing 3' end on which nucleotides can be added. Therefore, they require a **primer**, a short strand of RNA (or RNA + DNA in eukaryotes) to begin DNA synthesis. Second, all DNA polymerases add new nucleotides onto the 3′ end of the primer and thus synthesize DNA in an overall 5′ to 3′ direction, as shown in FIGURE 3.11.

Since the two strands of the DNA double helix are antiparallel, in order to synthesize both complementary strands in the 5′ to 3′ direction, the two new DNA chains must be made in opposite directions, as seen in FIGURE 3.12. On one strand, known as the **leading strand**, the daughter DNA is made continuously in the 5′ to 3′ direction. However, the other daughter DNA must be made discontinuously, one segment at a time, and then the segments are stitched together to form the complete DNA chain. This discontinuous daughter strand is called the **lagging strand**. Each segment on the lagging strand, called an **Okazaki fragment**, is synthesized in the 5′ to 3′ direction. This method of making one DNA chain continuously and the other DNA chain discontinuously satisfies the need for all new DNA chains to be synthesized in a 5′ to 3′ direction.

In order to provide accurate copies of the DNA for the next generation, it is critical for cells to have mechanisms to ensure the fidelity of DNA replication. Replication depends upon the accuracy of base pairing; however, DNA polymerases have intrinsic error rates that vary between polymerases, but are generally in the range of ~1 misincorporated base per 10^5 bp replicated. That may not sound like a high error rate, but in a human genome of 3 billion bp, that error rate would result in 30,000 errors in every cell division! To avoid such disastrous results, many DNA polymerases have mechanisms that allow **proofreading** of each newly incorporated base.

Proofreading depends on a 3′ to 5′ **exonuclease** activity that selectively excises mismatched base pairs present in the active site of the polymerase, depicted in FIGURE 3.13. All bacterial DNA polymerases possess this proofreading activity, as do the major elongating polymerases in eukaryotic cells (see *3.9 Leading and lagging strand synthesis is coordinated*). In some cases, the 3′ to 5′ exonuclease activity is part of the DNA polymerase enzyme itself, most commonly in single subunit enzymes such as those used in DNA repair (e.g., DNA polymerase I). However, the major replicative polymerases in both prokaryotes and eukaryotes are multisubunit enzymes in which the 3′ to 5′ exonuclease activity is provided by a separate subunit from that containing the 5′ to 3′ polymerization activity.

Proofreading results in a 100-fold reduction of the replication error rate from ~10^{-5} to ~10^{-7}, which would still result in ~300 errors per cell division in the human genome. Replication fidelity is further increased by repair mechanisms that correct errors missed by the proofreading activity, which reduces the error rate to less than one mismatch in 10^9 bp replicated. These repair mechanisms are discussed further in *3.14 Mismatch repair corrects replication errors*.

While the central player in DNA replication is the DNA polymerase (DNA polymerase III in the case of replication in *E. coli* and DNA

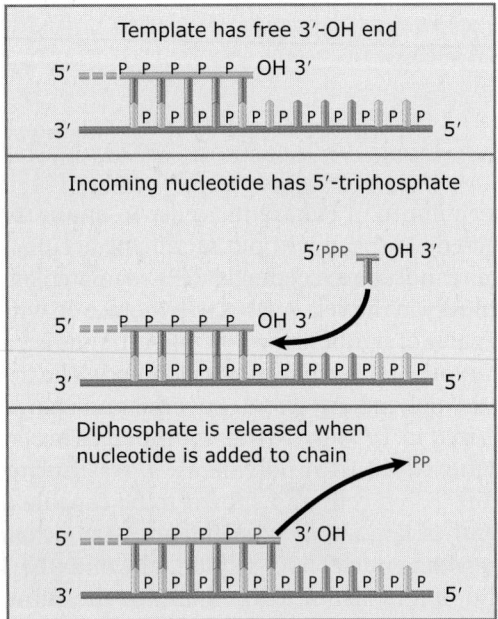

FIGURE 3.11 DNA is synthesized by adding nucleotides to the 3′-OH end of the growing chain, so that the new chain grows in the 5′ to 3′ direction. The precursor for DNA synthesis is a nucleoside triphosphate, which loses the terminal two phosphate groups in the reaction.

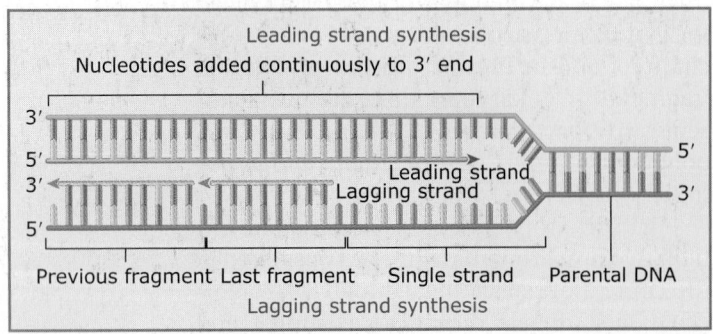

FIGURE 3.12 The leading strand is synthesized continuously 5′ to 3′, while the lagging strand is synthesized discontinuously in short 5′ to 3′ fragments, known as Okazaki fragments.

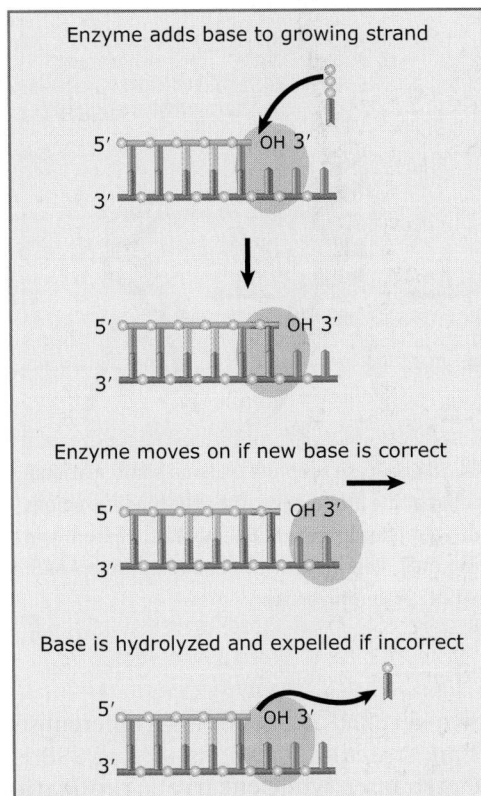

Enzyme adds base to growing strand

5' OH 3'
3'

5' OH 3'
3'

Enzyme moves on if new base is correct

5' OH 3'
3'

Base is hydrolyzed and expelled if incorrect

5' OH 3'
3'

FIGURE 3.13 Proofreading consists of 3' to 5' exonuclease activity that excises newly added mismatched bases.

polymerases δ and ε in eukaryotes), the polymerase is by no means the only enzyme involved in this complex process. To accomplish the replication of a DNA molecule the two strands of the double helix must be separated, the DNA chain must be initiated, the nascent chain must be extended according to the directions provided by the template strand, the initiating RNA primer must be removed, and in the case of the lagging strand, the Okazaki fragments must be ligated to form a complete DNA chain. In the following sections, we will discuss these other essential activities. Like the processes of transcription and translation that will be discussed in *4 Gene expression and regulation*, the overall process of DNA replication is conceptually divided into three phases: initiation, elongation, and termination. We will consider each of these phases beginning with elongation since this phase occupies most of the replication cycle.

Concept and Reasoning Check

1. How does the DNA polymerase respond to a misincorporated base?

Helicases, single-strand binding proteins, and topoisomerases are required for replication fork progression

Key concepts

- Helicases use the energy of ATP hydrolysis to unwind the parental DNA strands during replication.
- Single-strand binding proteins are required to maintain a single-stranded template for replication.
- Topoisomerases relieve torsional strain caused by DNA unwinding and disengage interlocked DNA circles or loops after replication.

DNA replication requires the presence of single-stranded DNA to serve as the template for polymerization. Single-stranded DNA must be generated during initiation when a replication bubble is first formed (discussed in *3.10 Replication initiates at origins and is regulated by the cell cycle*), and then must be provided throughout the elongation phase of replication, when the replication fork is continually advancing. However, the base pairing that holds together the double helix provides a significant energetic obstacle to the unwinding of DNA. Therefore, energy must be expended to open the two strands.

The enzyme responsible for separating the two strands of the helix ahead of the advancing replication fork is called a DNA **helicase**. Helicase enzymes utilize energy provided by ATP hydrolysis to translocate along and separate the two strands of the double helix. In *E. coli* the DNA helicase involved in DNA replication is called the DnaB protein; in eukaryotes, the replicative helicase is composed of MCM proteins. Helicases are commonly multimeric—DnaB and MCM are both hexameric. Six identical molecules of DnaB associate to form a donut-shaped hexamer with a central channel, shown in **FIGURE 3.14**. The lagging strand template occupies the central channel and the protein translocates along this DNA strand in the 5' to 3' direction and unwinds the DNA duplex as the other DNA strand (the leading strand template) is excluded from the central channel. This allows DnaB to actively unwind the duplex ahead of the advancing DNA polymerase. Interestingly, the eukaryotic MCM helicase differs from all known viral or prokaryotic helicases in that it is composed of six nonidentical subunits (MCM2–7).

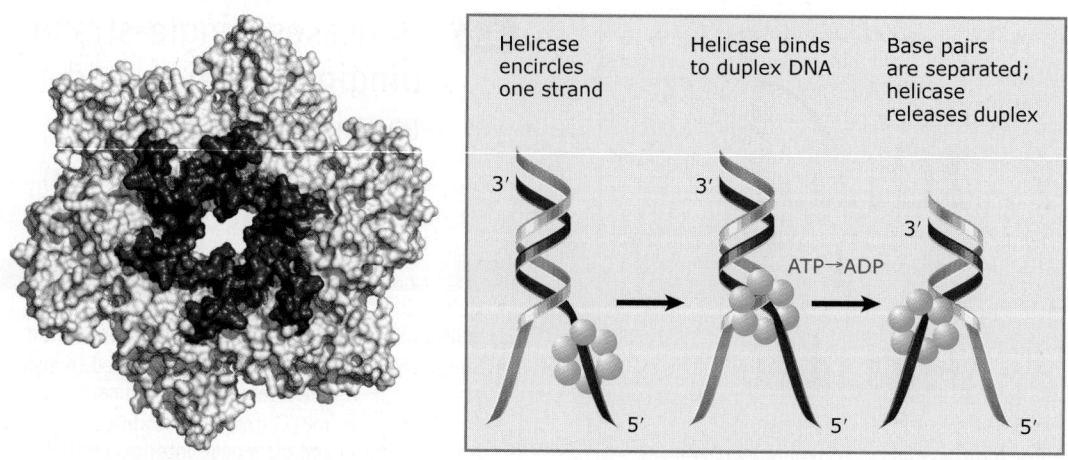

FIGURE 3.14 A hexameric helicase moves along one strand of DNA. The helicase subunits undergo conformational changes when bound to ATP or ADP; ATP hydrolysis results in unwinding of the DNA duplex. The DNA-binding regions of the helicase are shown in blue in the structure on the left. Photo reproduced from S. Bailey, W. K. Eliason, and T. A. Steitz, *Science* 318 (2007): 459–463 [http://www.sciencemag.org]. Reprinted with permission from AAAS. Photo courtesy of Scott Bailey, William K. Eliason and Thomas A. Steitz, Yale University.

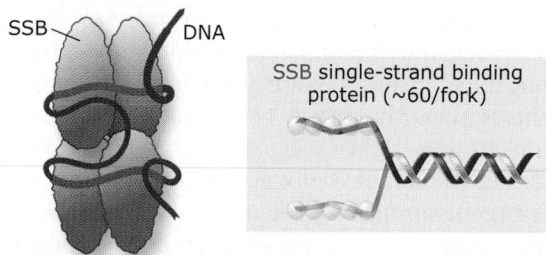

FIGURE 3.15 Single-strand binding proteins bind to single-stranded DNA and prevent it from reverting to the duplex state. Left illustration adapted from T. M. Lohman and M. E. Ferrari, *Annu. Rev. Biochem.* 63 (1994): 527–570.

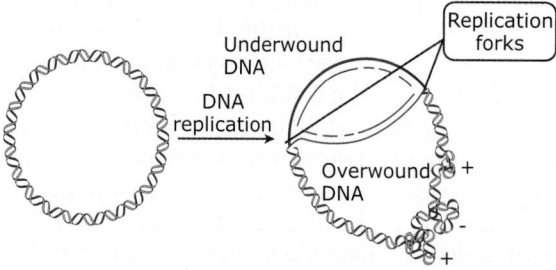

FIGURE 3.16 Replication forks generate overwound DNA (positive supercoiling) ahead of the forks.

Under the normal conditions of the cell, single-stranded DNA reanneals rapidly to regenerate duplex DNA, either to re-form the original parental duplex or even to form intramolecular double-stranded regions where possible. To prevent re-formation of the duplex during replication, single-stranded DNA generated by helicase action is promptly bound by **single-strand binding proteins (SSBs)**, as shown in **FIGURE 3.15**. SSBs from different species are structurally diverse—*E. coli* SSB is a homotetramer, while eukaryotic **replication protein A (RPA)** is a heterotrimer—but all perform the same basic function of maintaining single-stranded regions as templates for the replication machinery.

While SSBs ensure that the replication bubble remains open behind an advancing helicase, another class of enzymes, the **topoisomerases**, play a critical role in allowing helicases to continue to advance. When replication forks advance, the process of unwinding the DNA creates torsional strain ahead of the fork in the form of overwinding, or positive **supercoiling**, as seen in **FIGURE 3.16**. In a short linear DNA molecule, this torsional strain would be resolved by rotation of the DNA ends to relieve the overwinding. However, in a circular molecule or a long linear molecule (which is not free to rotate due to protein binding or frictional drag), this overwinding will continue to accumulate. Eventually, the tension would become so great that the helicase would no longer be able to advance. Cells avoid this outcome through the action of topoisomerases, enzymes that change the topology of DNA by transiently breaking one or both strands of the DNA and passing the strand(s) through the gap. Topoisomerases form a transient covalent attachment to the DNA during the reaction (a phosphodiester bond to a tyrosine in the active site of the enzyme) to prevent loss of the free end, which would result in DNA damage.

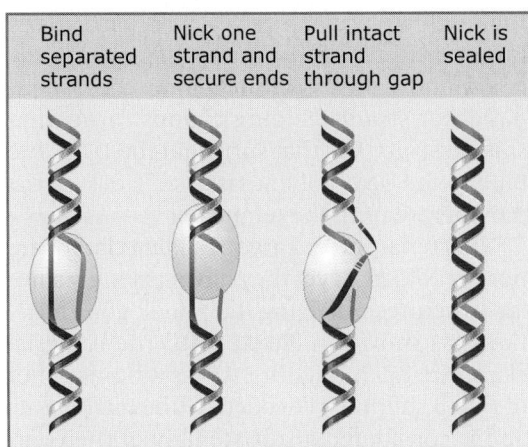

Bind separated strands	Nick one strand and secure ends	Pull intact strand through gap	Nick is sealed

FIGURE 3.17 Topoisomerases change the topology of DNA by transiently making DNA breaks and passing strands through the breaks. Type I topoisomerases (shown) break one strand of DNA, while Type II topoisomerases make double-strand breaks.

There are two classes of topoisomerases: Type I enzymes break only one strand of the duplex (**FIGURE 3.17**), while Type II enzymes create double-strand breaks. In either case, the net result is a change in the level of supercoiling in the DNA substrate. During replication, continuous action of topoisomerases ahead of the replication fork ensures that the overwinding due to fork movement does not accumulate. Either Type I or II enzymes can relieve this tension; however, Type II enzymes have a second critical role in separating the interlocked circular DNAs (known as *catenanes*) that result from replication of circular chromosomes. These enzymes can also resolve tangled or interlocked regions in linear eukaryotic chromosomes.

Concept and Reasoning Check

1. Since SSB does not have an enzymatic function, why is it essential for replication?

3.7 Priming is required to start DNA synthesis

Key concepts

• All DNA polymerases require a 3′-OH end to initiate DNA synthesis.
• During replication, the 3′-OH end is provided by a special RNA polymerase known as a primase.
• DnaG is the primase required for replication in *E. coli*.
• The polα/primase complex provides primers during eukaryotic replication.

As described in *3.5 DNA polymerases replicate DNA*, all DNA polymerases require a free 3′-OH on which to add subsequent nucleotides. In replication, a different type of polymerase must be used to provide a primer for the DNA polymerase. In fact, RNA polymerases can synthesize RNA strands *de novo*, in the absence of any primer (discussed in *4.4 RNA polymerases are large multisubunit protein complexes*). Therefore, replication incorporates specialized RNA polymerases to create primers for DNA polymerase, as shown in **FIGURE 3.18**. These replication-specific RNA polymerases are known as **primases**. In *E. coli*, primase is encoded by the *dnaG* gene, and the DnaG protein associates directly with the DnaB helicase. This ensures its presence at the right place to synthesize a short RNA primer to initiate both the leading strand and each Okazaki fragment on the lagging strand. The DnaG primase synthesizes a short 10–12 nucleotide RNA primer that provides the 3′-OH that DNA polymerase requires to begin DNA synthesis. The primer is synthesized directed by the sequence on the template strand and there are preferred sites for primer synthesis. In *E. coli*, Okazaki fragments are generally 1000 to 2000 bases in length.

In eukaryotes, primase is found in a four-subunit complex known as **polα/primase**. Two small subunits in this complex provide the RNA polymerase (primase) activity, and

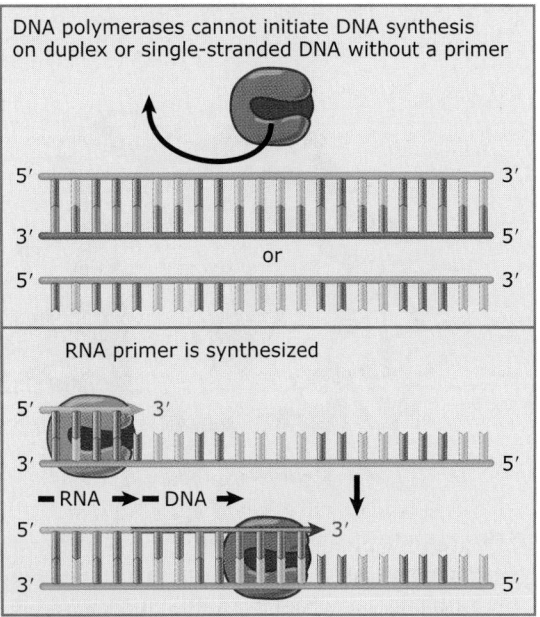

FIGURE 3.18 DNA polymerases require a 3′-OH to initiate synthesis. During replication, this 3′-OH is provided by an RNA primer synthesized by a primase, a specialized RNA polymerase.

another subunit (B) is required for complex assembly. The largest subunit in this complex is actually a DNA polymerase (DNA polymerase α), which serves to elongate the short RNA primer. Thus, the eukaryotic primer consists of ~10 bases of RNA synthesized by primase, followed by ~20–30 bases of DNA synthesized by pol α. This longer primer is then the substrate for the replicative polymerases (described further, in *3.9 Leading and lagging strand synthesis is coordinated*). Eukaryotic Okazaki fragments are surprisingly short, only ~100–150 bases in length. The processing of Okazaki fragments is also described in *3.9 Leading and lagging strand synthesis is coordinated*.

3.8 A sliding clamp ensures processive DNA replication

Key concepts

- Ring-shaped sliding clamps ensure that DNA polymerases replicate processively.
- The *E. coli* sliding clamp is the dimeric β ring.
- The eukaryotic sliding clamp is the trimeric PCNA protein.
- Clamp loaders use the energy of ATP hydrolysis for assembly of sliding clamps.

During replication, **sliding clamps** serve to tether the DNA polymerases firmly to the DNA templates. This ensures that polymerases elongate *processively*; that is, that they do not fall off the template prematurely. This is particularly important on the leading strand, where the polymerase must replicate a long, continuous daughter strand. Sliding clamps form ring-shaped structures that surround the DNA and bind to the DNA polymerase itself, locking the polymerase onto the template.

Bacterial and eukaryotic sliding clamps are not homologous, yet they have strikingly similar structures reflecting their analogous functions, as shown in **FIGURE 3.19**. The bacterial clamp (*E. coli* in Figure 3.19) is a homodimer of two β subunits, encoded by the *dnaN* gene, and is considered a subassembly of the complete DNA polymerase III **holoenzyme** (see *3.9 Leading and lagging strand synthesis is coordinated*). The β dimer forms a doughnut shape with a central cavity large enough to encircle double-stranded DNA. The eukaryotic sliding clamp also forms a DNA-encompassing ring, but the eukaryotic clamps are homotrimers (the human clamp is shown in Figure 3.19). The eukaryotic clamp is also called the **proliferating cell nuclear antigen** (**PCNA**), a name reflecting its identification as a prevalent protein in dividing cells long before its function was elucidated. Note that while these structures have two- or threefold radial symmetry, the two faces of the sliding clamps are not identical. In other words, a clamp has a distinguishable front and back that dictates which side interacts with DNA polymerase and which side is available for other interactions.

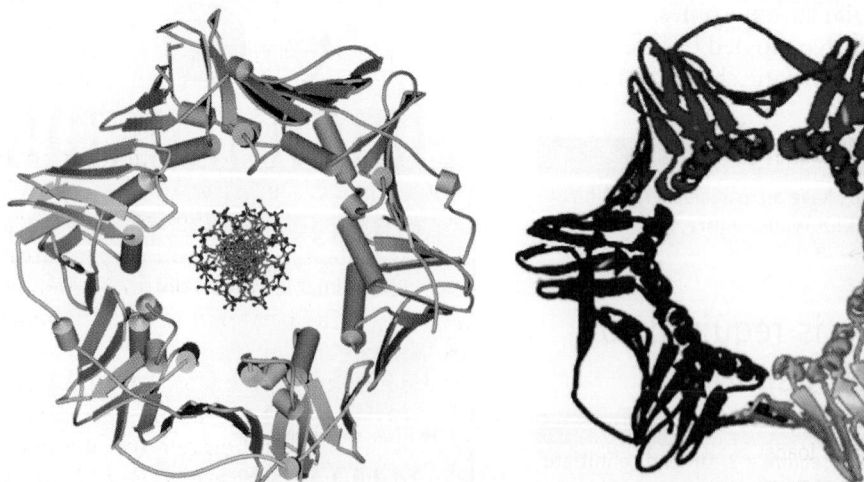

FIGURE 3.19 Sliding clamps in bacteria and eukaryotes are unrelated yet structurally similar. The *E. coli* β ring is composed of two identical β subunits (left; shown with DNA modeled into the cavity), while the human PCNA is a trimer. Different colors indicate individual monomers in each structure. Left structure from Protein Data Bank 1MMI. A. J. Oakley, et al., *Acta Crystallogr. D Biol. Crystallogr.* 59 (2003): 1192–1199. Right structure reproduced from I. Bruck and M. O'Donnell, *Genome Biol.* 2 (2001): reviews3001.1–3001.3. Photo courtesy of Michael O'Donnell, Rockefeller University.

Logically, to place a ring-shaped structure on an unbroken DNA template, the ring must be opened in order to insert the DNA in the central cavity and then closed around the DNA. These steps are accomplished by the **clamp loader**, an ATP-dependent complex that recognizes primer-template junctions as a site for clamp assembly. The clamp loader in *E. coli* is a 5-subunit complex ($\tau_2\delta\delta'\chi\psi$), sometimes called the "tau complex," that comprises another subassembly within the DNA polymerase III holoenzyme (when not present in the holoenzyme, the clamp loader has the slightly different composition $\gamma_2\delta\delta'\chi\psi$). In eukaryotes, the clamp loader is also a heteropentameric complex, known as replication factor C (RFC). The clamp loading cycle for the *E. coli* proteins is depicted in **FIGURE 3.20**; the cycle is comparable in eukaryotes. In the absence of ATP, the clamp loader is in a conformation in which it is unable to interact with the clamp. Upon ATP binding, the clamp assumes a new conformation in which it binds tightly to the clamp and opens the ring. The subunit that opens the ring, δ, is referred to as the "wrench." This complex now has a high affinity for a DNA template-primer complex. DNA entry into the central cavity of the sliding clamp triggers ATP hydrolysis, resulting in release of the clamp loader and closing of the ring around the DNA.

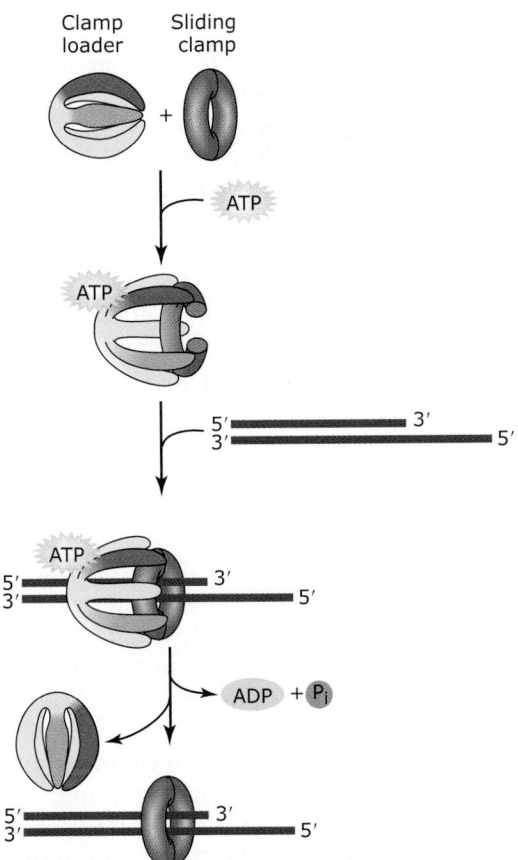

FIGURE 3.20 The clamp loader binds to ATP, which allows it to form a tight complex with the clamp, simultaneously opening the clamp. This complex recognizes primer-template junctions; binding to these junctions stimulates ATP hydrolysis and releases both the clamp and DNA. The clamp now encircles DNA with the correct orientation for use by DNA polymerase. Adapted from G. Bowman, et al., *FEBS Lett.* 579 (2005): 863–867.

Concept and Reasoning Check

1. Why is energy required to load clamps onto DNA?

3.9 Leading and lagging strand synthesis is coordinated

Key concepts

- Replicative DNA polymerases are large holoenzyme complexes that act as dimers during replication.
- The DNA polymerase on the leading strand requires only a clamp loading and priming event.
- The DNA polymerase on the lagging strand dissociates at the end of each Okazaki fragment and reassembles for the next, requiring priming and clamp loading for each cycle.
- A eukaryotic replication fork has one complex of pol α/primase and two complexes of DNA pol δ and/or ε.
- A polymerase "switch" must occur once on the leading strand and at every Okazaki fragment on the lagging strand.
- Okazaki fragment maturation requires removal of the RNA primer and ligation of the fragments.

The replicative DNA polymerases in both prokaryotes and eukaryotes function as large holoenzymes, containing many factors in addition to the catalytic core of the polymerases necessary for replication. In addition, other factors not directly associated with the holoenzyme are also needed for replication, such as single-strand binding proteins or topoisomerases, as discussed above. These factors are summarized in **FIGURE 3.21**, and their relative sites of action at a replication fork are shown. *E. coli* DNA polymerase III is a complex enzyme consisting of more than 12 different polypeptides, divided into three subassemblies. Two of these, the β_2 sliding clamp and the $\tau_2\delta\delta'\chi\psi$ clamp loader, were discussed in *3.8 A sliding clamp ensures processive DNA replication*. The core enzyme

(a)

Function	Prokaryotes	Eukaryotes
Helicase	DnaB	MCM2-7
Single-strand binding protein	SSB	RPA
Primase	DnaG	Primase subunit of polα-primase
Sliding clamp	β ring	PCNA
Clamp loader	$\gamma_2/\tau_2\delta\delta'\chi\psi$ ("tau complex")	RFC
Relaxation of supercoils, decatenation	Topoisomerase I, IV, Gyrase	Topoisomerase I, Topoisomerase II
Polymerase	α subunit of DNA polymerase III holoenzyme	DNA polymerase ε and DNA polymerase δ
Proofreading exonuclease	ε subunit of DNA polymerase III holoenzyme	Part of polymerase subunit (δ or ε)

(b)

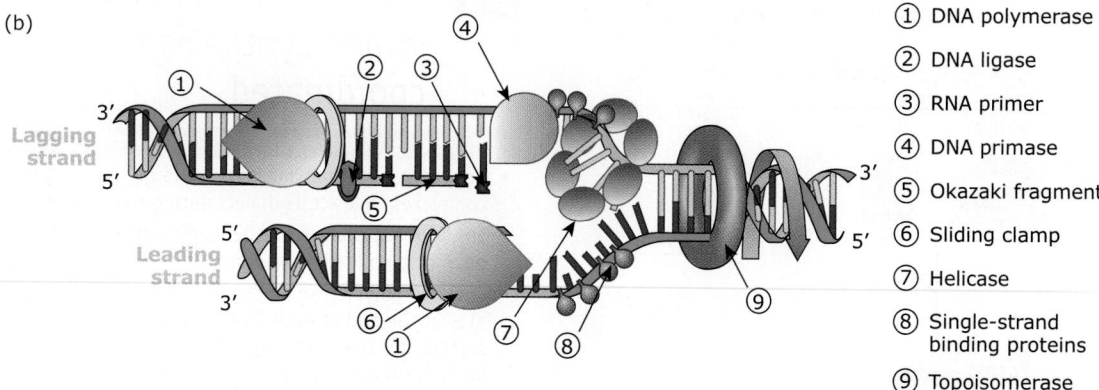

① DNA polymerase
② DNA ligase
③ RNA primer
④ DNA primase
⑤ Okazaki fragment
⑥ Sliding clamp
⑦ Helicase
⑧ Single-strand binding proteins
⑨ Topoisomerase

FIGURE 3.21 Summary of factors acting at a replication fork. The table (a) summarizes the functions required for replication elongation in both prokaryotes and eukaryotes. The diagram (b) shows the relative sites of action of these factors at a replication fork.

subassembly consists of the α subunit (which contains the DNA polymerase activity), the ε subunit (3′ to 5′ proofreading exonuclease), and the θ subunit, which stimulates the activity of ε. The eukaryotic replicases are similarly complex, and eukaryotes additionally use two different replicative polymerases, DNA polymerases δ and ε.

Even though polymerases at the replication fork are moving in different directions relative to the fork movement, the replicative polymerases actually act as dimers to allow for the coordinated synthesis of both the leading and lagging strands. Replication of the lagging strand poses a unique challenge since it is replicated in segments, and the direction of polymerization is away from the direction of fork movement. Physical association of the two polymerases therefore constrains the movement of the lagging strand polymerase, and

instead results in a looping out of the lagging strand. While the leading strand polymerase moves continuously, the lagging strand polymerase must repeatedly release and reengage the template. This dynamic looping cycle, sometimes referred to as the "trombone" model, is shown for a prokaryotic replication fork in **FIGURE 3.22** (the organization is similar for a eukaryotic fork). As one Okazaki fragment is completed on the lagging strand, one molecule of DNA polymerase III transiently dissociates (from both the DNA and the sliding clamp) and then reassociates with the RNA primer and a new sliding clamp to start synthesis on the next Okazaki fragment. In the case of DNA polymerase III, the τ subunits of the clamp loader tether the two polymerases, to ensure that the lagging strand polymerase does not diffuse away and is in place to rapidly engage the next RNA primer.

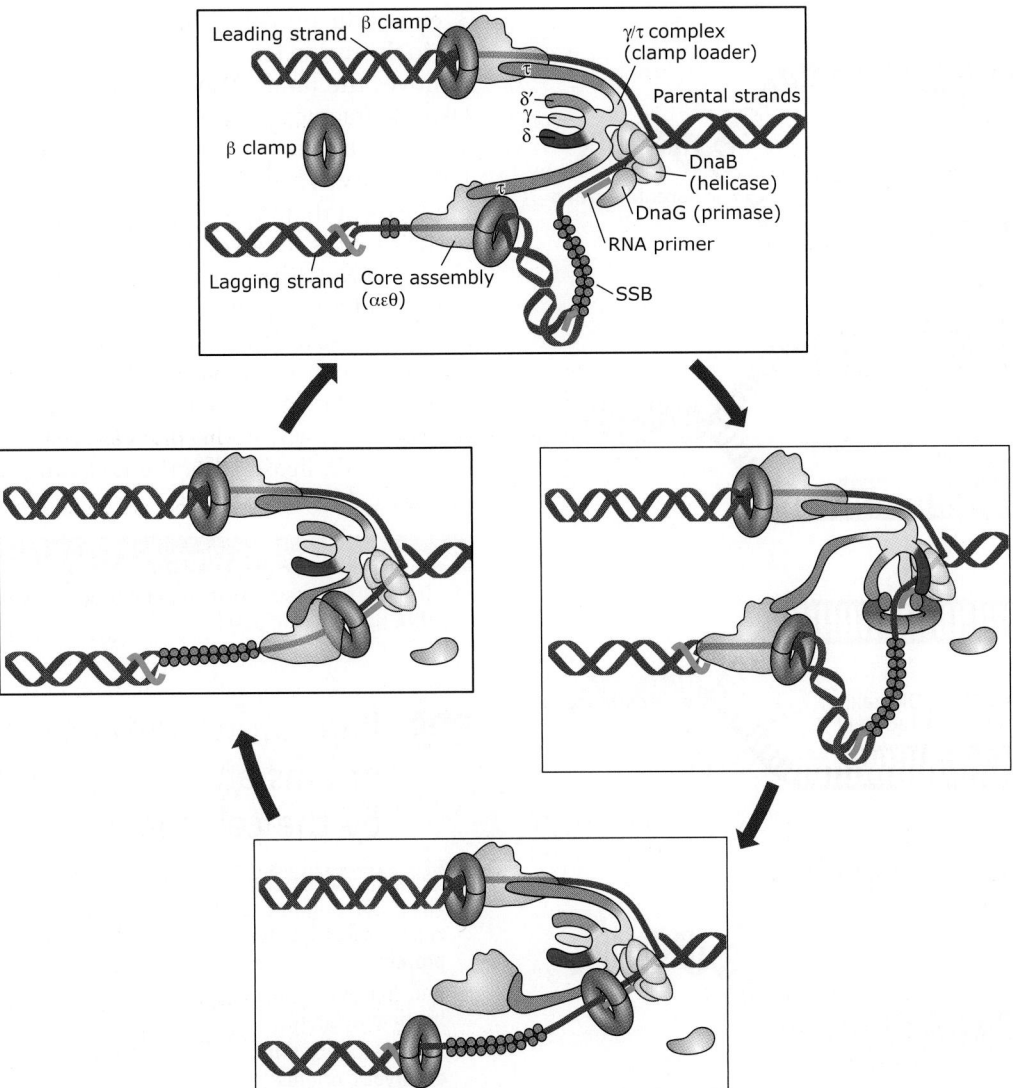

FIGURE 3.22 Polymerase cycling at the replication fork. As DNA polymerase III advances, the clamp loader loads a sliding clamp on the RNA primer. When the lagging strand polymerase reaches the 5' end of the previous primer, it dissociates from the DNA and sliding clamp and moves to the newly loaded clamp. Prokaryotic proteins are shown, but an analogous process occurs at eukaryotic forks. Adapted from A. Johnson and M. O'Donnell, *Annu. Rev. Biochem.* 74 (2005): 283–315.

In eukaryotes, there are actually three DNA polymerases present at the replication fork: pol α, which adds deoxynucleotides to the RNA primer created by primase (which exist in the same complex; see *3.7 Priming is required to start DNA synthesis*) and the two replicative polymerases pol δ and pol ε. Pol α must be replaced by pol δ or pol ε after each primer is synthesized; this event is referred to as the "polymerase switch." Some studies have suggested that the two eukaryotic replicative polymerases may be specialized such that pol δ replicates the lagging strand and pol ε replicates the leading strand, but this remains controversial.

The cycles of polymerase movement ensure that the lagging strand is completely replicated, but the net result is the production of Okazaki fragments, which contain RNA (or, in eukaryotes, RNA/DNA) primers and are disconnected from each other. The completion of lagging strand synthesis therefore requires maturation of the Okazaki fragments, depicted in **FIGURE 3.23**. In prokaryotes, replacement of RNA with DNA is accomplished by the combined action of a ribonuclease (RNase) and a polymerase. RNase H removes all but the last ribonucleotide from the 5' end of the Okazaki fragment. DNA polymerase I both extends the

(a)

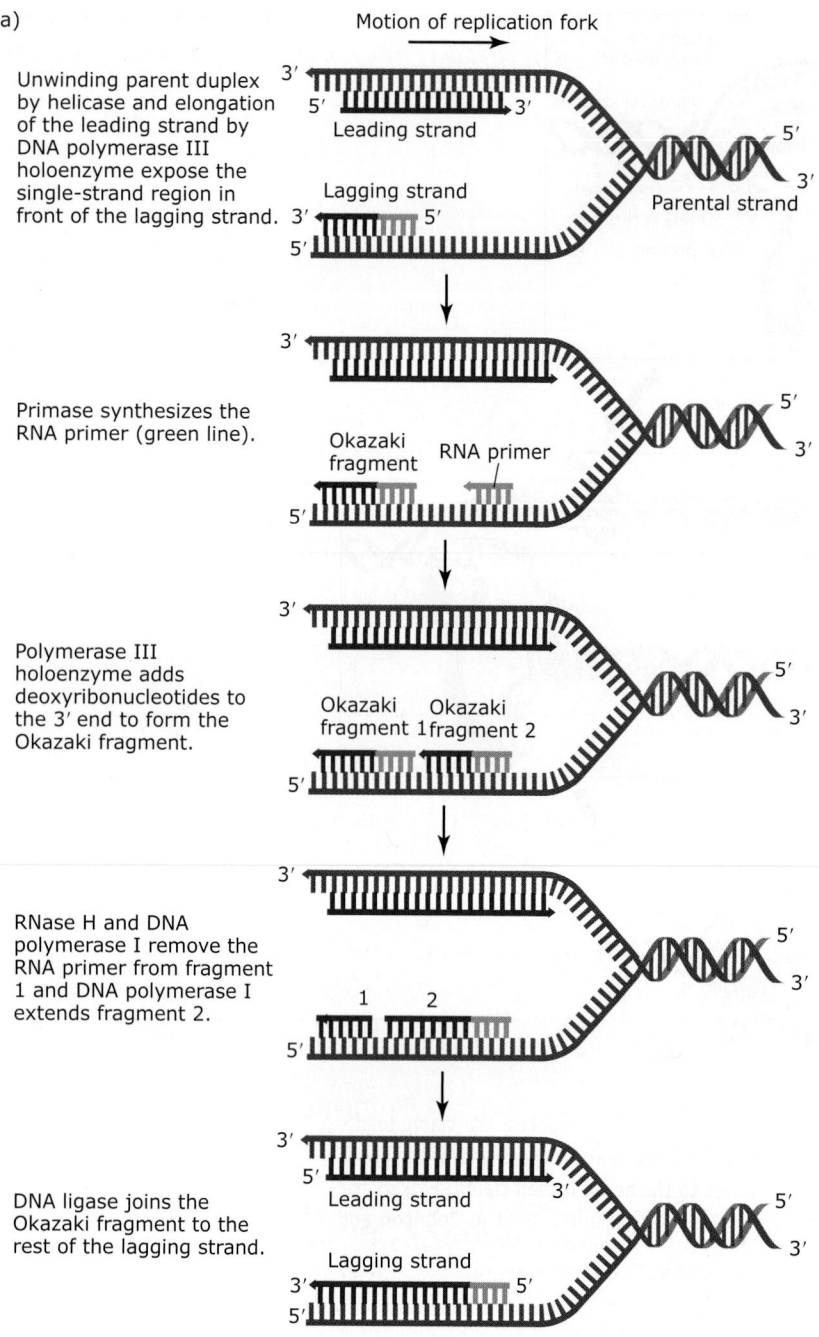

Motion of replication fork →

Unwinding parent duplex by helicase and elongation of the leading strand by DNA polymerase III holoenzyme expose the single-strand region in front of the lagging strand.

3'
5'
Leading strand
3'
5'
Parental strand

Lagging strand
3'
5'

Primase synthesizes the RNA primer (green line).

3'

Okazaki fragment   RNA primer

5'

Polymerase III holoenzyme adds deoxyribonucleotides to the 3' end to form the Okazaki fragment.

3'

Okazaki fragment 1 Okazaki fragment 2

5'

RNase H and DNA polymerase I remove the RNA primer from fragment 1 and DNA polymerase I extends fragment 2.

3'

1   2

5'

DNA ligase joins the Okazaki fragment to the rest of the lagging strand.

3'
5'
Leading strand
3'
5'

Lagging strand

3'
5'

(b)

Function	Prokaryotes	Eukaryotes
Removal of RNA primer	RNase H	RNase H1, FEN-1
Replacement of RNA with DNA	DNA polymerase I	DNA polymerase δ
Ligation of fragments	DNA ligase	DNA ligase

FIGURE 3.23 (a) Processing of Okazaki fragments. (b) Summary of factors required for maturation of Okazaki fragments.

3' end of the previous Okazaki fragment (filling in the region left by RNase H digestion) and removes the final 5' ribonucleotide from the Okazaki fragment through a 5'-3' exonuclease activity. The nicks that remain are sealed by DNA **ligase**. The process is similar in eukaryotes, except that the single ribonucleotide left by RNase H is removed by the "flap endonuclease" FEN-1. The gap is filled in by one of the major replicative polymerases, most likely pol δ. Pol δ is also able to displace the 5' end of an Okazaki fragment, including the DNA synthesized by pol α. Any single-stranded DNA displaced in this way is also processed by FEN-1. Finally, DNA ligase seals the nicks to form a complete daughter strand.

Concept and Reasoning Check

1. In what processes other than replication would DNA ligase participate?

3.10 Replication initiates at origins and is regulated by the cell cycle

Key concepts

- The *E. coli* origin, *oriC*, is ~250 bp in length and contains multiple binding sites for the DnaA protein.
- *oriC* does not support initiation when it is hemimethylated.
- The origin recognition complex (ORC) binds eukaryotic origins.
- ORC recruits Cdc6, which recruits Cdt1 and the MCM helicase.
- Cdc6 is degraded or exported after origin firing, which prevents reinitiation.

We have discussed the processes by which the activities of helicases and primases ultimately lead to replication of both the leading and lagging strands of a replication fork. However, a discussion of the initiation of DNA replication must distinguish between the event that initiates the *overall* process of DNA replication and the initiation of each Okazaki fragment on the lagging strand. Initiation of the overall process of replication occurs at a single specific site in *E. coli* and at multiple sites on the eukaryotic chromosome. This process requires a different constellation of proteins dedicated to the task of initiating DNA replication.

Replication in *E. coli* begins at a unique location on the chromosome called *oriC*, depicted in **FIGURE 3.24**. This region of ~250 bp contains

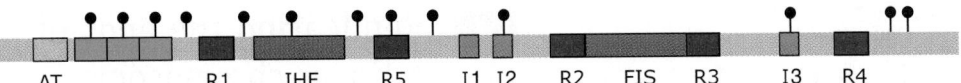

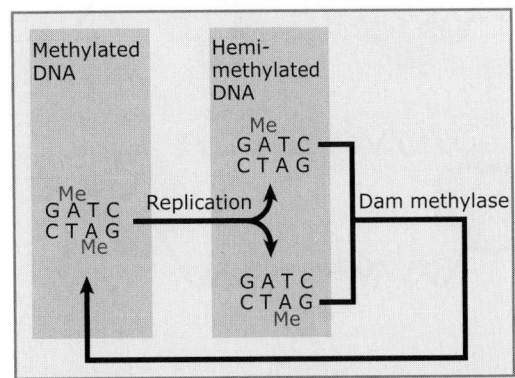

FIGURE 3.24 The minimal *oriC* region. Adapted from S. Dasgupta and A. Lobner-Olesen, *Plasmid* 52 (2004): 151–168.

- ■ Dna A box (R1–R5)
- ▨ 13 mer
- ❘ GATC Dam-methylation sites
- ▢ AT cluster

several binding sites (DnaA boxes and I boxes) for a protein called DnaA. DnaA binds ATP and alters the topology of this region, allowing a cluster of AT-rich sequences at the origin to open. This permits binding of the DnaB helicase (with the assistance of a protein called DnaC). Subsequent synthesis of an RNA primer by DnaG allows initiation of leading strand synthesis on both strands catalyzed by DNA polymerase II. As the leading strand is elongated this reveals sites for priming by DnaG on the lagging strand templates. Thus, two forks proceeding bidirectionally from the origin are established. Termination of DNA replication occurs when the two forks meet on the opposite side of the chromosome.

Replication initiation is a tightly controlled event, and a number of factors exist to promote or inhibit initiation. One important regulatory mechanism in *E. coli* is the methylation of GATC sequences in *oriC*. The adenines in GATC sequences are methylated by the DNA adenine methyltransferase (Dam) throughout the genome. During replication, newly synthesized strands are unmethylated (which is important for DNA repair; see *3.14 Mismatch repair corrects replication errors*), and then gradually become methylated after the replication fork passes, as shown in **FIGURE 3.25**. *oriC* is particularly enriched in GATC sites, and after replication initiates, a protein called SeqA binds and sequesters the hemimethylated origin to prevent reinitiation within a single round of the cell cycle.

Replication in eukaryotic cells is similar but more complex. In all eukaryotes, there are multiple origins of replication, unlike the single origin found on most bacterial chromosomes. This makes sense since there are multiple chromosomes and each must contain at least one origin of replication. In addition, replication is slower in eukaryotes than in prokaryotes and thus requires multiple origins of replication on each chromosome to complete the task of replication during the S phase of the cell

FIGURE 3.25 Replication of methylated DNA gives hemimethylated DNA, which maintains its state at GATC sites until the Dam methylase restores the fully methylated condition.

cycle. Eukaryotic replication origins are also more complex, ranging from the relatively short and recognizable **autonomously replicating sequence (ARS)** elements in budding yeast to large initiation domains of diverse sequence in multicellular eukaryotes. Similar to the situation observed in prokaryotes, the eukaryotic origin of replication is recognized by the specific protein complex called the **origin recognition complex (ORC)**. ORC was originally identified in yeast, but has since been found in a number of other species, including human. This complex of six proteins remains bound to origins throughout the cell cycle, where it recruits several key proteins required for cell-cycle dependent initiation of replication, as shown in **FIGURE 3.26**. These factors include Cdc6 and Cdt1, which are needed to recruit the eukaryotic helicase composed of six MCM proteins (introduced in *3.6 Helicases, single-strand binding proteins, and topoisomerases are required for replication fork progression*). Cell cycle-dependent phosphorylation events result in the release of Cdc6 and Cdt1 and the activation of the MCM helicase to initiate replication. Replication from multiple origins terminates as replication forks meet between origins.

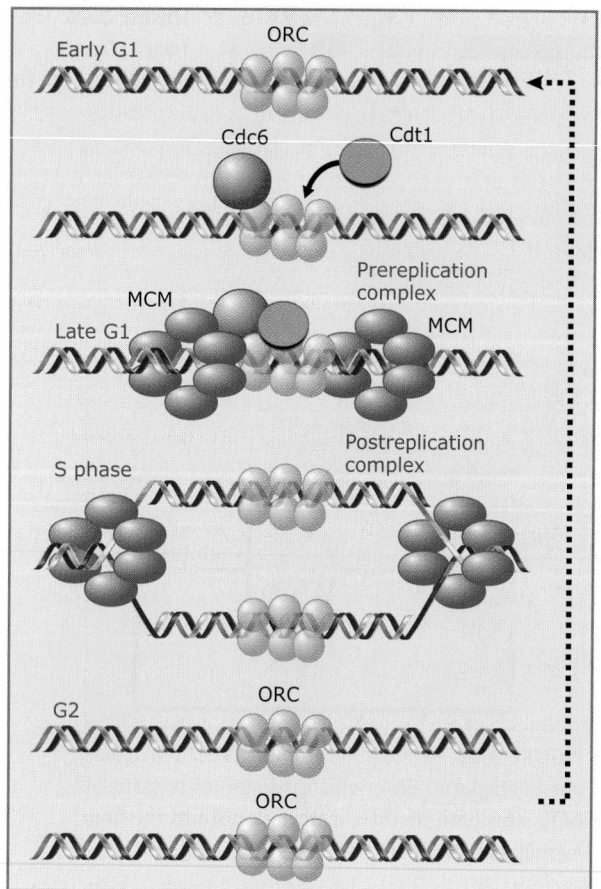

FIGURE 3.26 Initiation at a eukaryotic origin.

3.11 Replicating the ends of a linear chromosome

Key concepts

- Replication of a linear chromosome results in sequence loss from the chromosome ends.
- Telomeres are structures at the ends of chromosomes composed of multiple repeats of a short sequence.
- The G-T-rich strand of the telomere extends in a short 3′ overhang called a G tail.
- The telomerase enzyme uses the 3′-OH of the G tail to prime synthesis of tandem telomeric repeats.
- Telomerase contains an RNA component that anneals to the G tail and provides the template for synthesis of telomeric repeats.

Replicating the end of a linear chromosome provides special challenges that do not exist for the replication of a circular DNA molecule. Since DNA synthesis requires a primer,

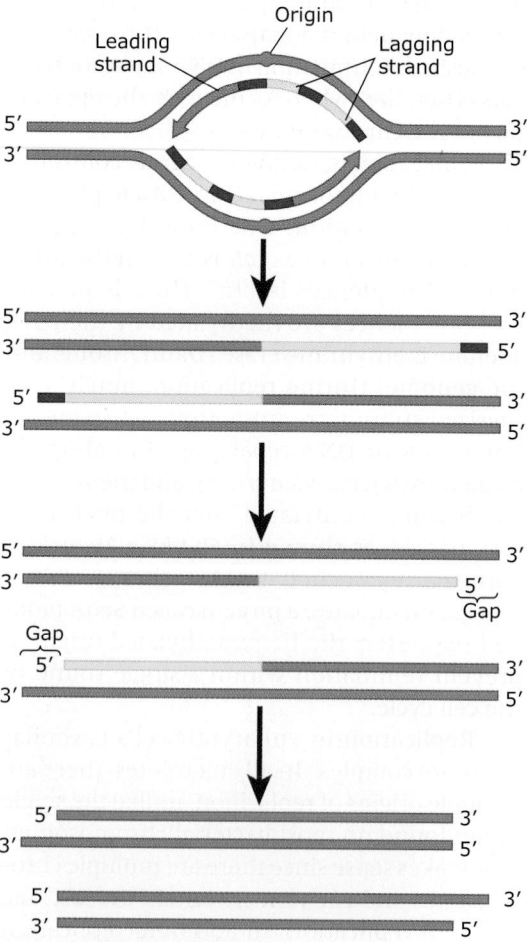

FIGURE 3.27 The end replication problem results in shortening of linear chromosomes after each round of replication.

Replication initiation is highly regulated in eukaryotic cells. Not only is the timing of origin firing controlled, but mechanisms also exist to prevent reinitiation from any origin that has fired within a single cell cycle. This is known as **replication licensing**, and it is controlled by a number of factors. One important licensing factor is Cdc6, the protein required to recruit Cdt1 and the MCM helicase. Cdc6 is only present at origins during mid-late G1 in the cell cycle. After it is evicted from the origin at the time of MCM helicase activation, Cdc6 is either degraded (in yeast) or exported from the nucleus (in mammals). It is not re-synthesized or allowed enter into the nucleus until G1 of the next cell cycle, ensuring that it cannot rebind to origins that have already fired during S phase.

Concept and Reasoning Check

1. Why is it important for the cell to know whether or not a replication origin has fired?

typically supplied by a small RNA, it is not possible to completely replicate the end of the lagging strand on a linear DNA molecule, as shown in **FIGURE 3.27**. Even if the primer were placed at the extreme end of the DNA molecule, its removal after replication is complete would render the very end of the DNA single stranded. This single-stranded DNA would be subject to attack by nucleases and would be lost with the potential loss of genetic information.

Eukaryotic cells solve this problem by using special structures at the end of each chromosome called **telomeres**. Telomeres are composed of an extensive set of repeats of a short DNA sequence that is G-rich on one strand, as shown in **FIGURE 3.28**. Note that the G-rich strand extends beyond the complementary strand (due to limited degradation of ~15 bases from the C-A-rich strand), providing a 3'-single-stranded region of DNA at the end of the chromosome. This single-stranded region of DNA, sometimes referred to as a "G tail," provides two important functions for the telomere. First, this single-stranded DNA can invade a duplex region at the telomere to form what is known as a T-loop at the end of the chromosome (shown in the lower panel

of Figure 3.28). The T-loop is believed to stabilize the end of the chromosome to ensure that chromosome ends are not mistaken for double-strand breaks and used as recombination intermediates, which would cause chromosomes to fuse. Second, this single-stranded DNA serves as a site for binding of the enzyme **telomerase**, to allow extension of the 3'-end of the chromosome.

Telomerase is a specialized DNA polymerase that is directed by an RNA template that is part of the telomerase enzyme complex. Telomerase binds to the single-stranded G-rich strand at the end of the chromosome, and the telomerase RNA associates with the DNA strand by complementary base pairing, as shown in **FIGURE 3.29**. The RNA then serves as a template, which the RNA-directed DNA polymerase component of the telomerase utilizes to add telomeric repeats to the 3' end of the chromosome. Iterative binding and extension by the telomerase causes the 3' end of the chromosome to be extended. Subsequently, this single-stranded DNA provides a priming site for DNA polymerase to allow extension of the complementary strand. Thus, the ends of the chromosomes are protected from loss of genetic information by having a special struc-

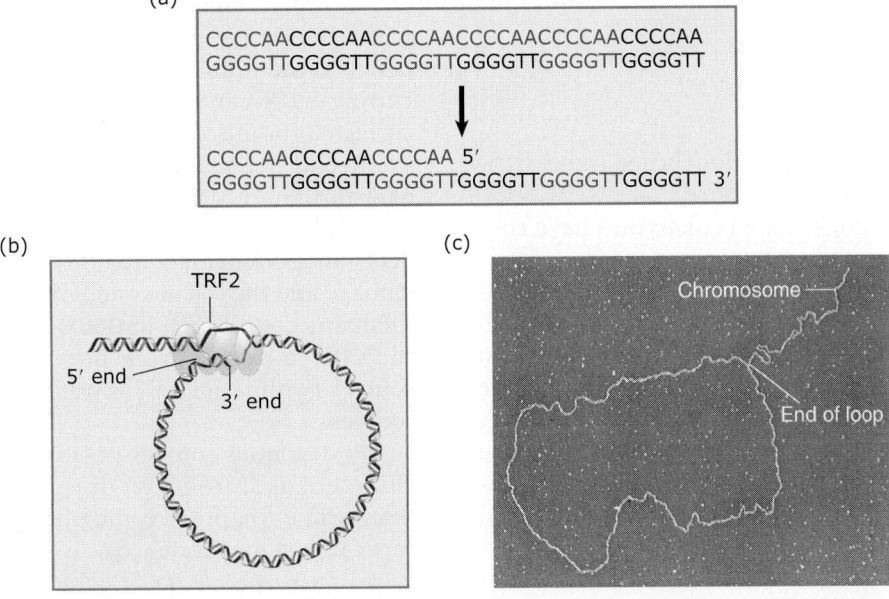

FIGURE 3.28 A typical telomere has a simple repeating structure with a G-T-rich strand that extends beyond the C-A-rich strand. The G-tail, generated by limited degradation of the C-A-rich strand, can invade the double-stranded region of the telomere to form a loop known as a T-loop. The protein Trf2 catalyzes T-loop formation. Photo courtesy of Jack Griffith, University of North Carolina at Chapel Hill.

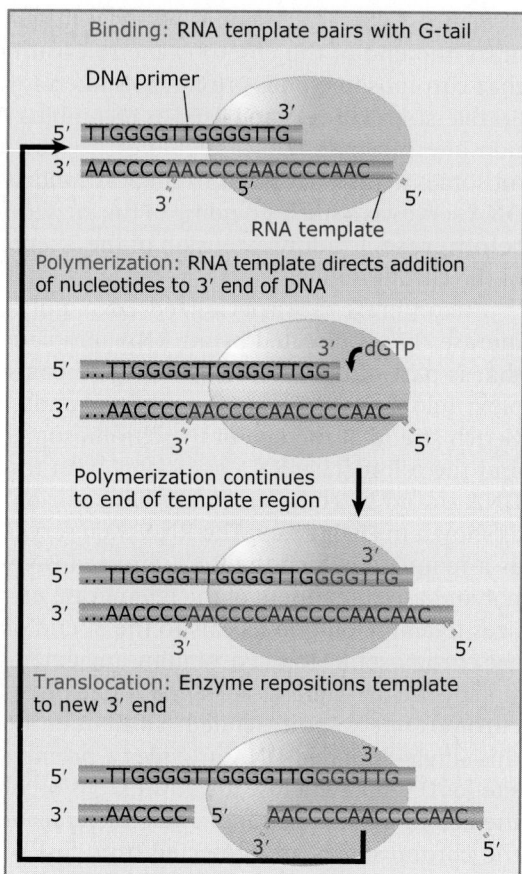

Binding: RNA template pairs with G-tail

DNA primer

5′ TTGGGGTTGGGGTTG 3′

3′ AACCCCAACCCCAACCCCAAC 5′

RNA template

Polymerization: RNA template directs addition of nucleotides to 3′ end of DNA

5′ ...TTGGGGTTGGGGTTGG 3′ dGTP

3′ ...AACCCCAACCCCAACCCCAAC 5′

Polymerization continues to end of template region

5′ ...TTGGGGTTGGGGTTGGGGTTG 3′

3′ ...AACCCCAACCCCAACCCCAACAAC 5′

Translocation: Enzyme repositions template to new 3′ end

5′ ...TTGGGGTTGGGGTTGGGGTTG 3′

3′ ...AACCCC 5′ AACCCCAACCCCAAC 5′

FIGURE 3.29 Telomerase positions itself by base pairing between the RNA template and the protruding single-stranded G-tail. It adds G and T nucleotides as directed by the template, then translocates to reposition the RNA template on the newly synthesized region.

ture and an enzyme uniquely dedicated to the replication of this structure.

While single-celled eukaryotes have constitutively active telomerase in order to protect the integrity of their chromosomes, in multicellular eukaryotes only certain cell types (such as stem cells) contain active telomerase. Most somatic cells do not have an active telomerase enzyme, thus as the cells divide the telomeres shorten until they reach a "crisis" length and the cell enters senescence and eventually undergoes apoptosis (programmed cell death). This appears to be one mechanism used to regulate the lifespan of an individual cell. The majority of cancer cells, on the other hand, have an active (reactivated) telomerase enzyme and there is great interest in understanding if inhibition of the telomerase in cancer cells might provide a suitable therapy to stop the growth of cancer.

3.12 DNA is subject to damage

Key concepts

- DNA is subject to continual damage by both endogenous and exogenous sources.
- Different sources of DNA damage result in specific lesions.
- Replication errors result in mispairs and insertion/deletions.
- Spontaneous reactions result in deamination of bases as well as formation of abasic sites.
- DNA is subject to oxidative damage.
- Alkylating agents create numerous DNA lesions, including interstrand cross-links.
- Ultraviolet (UV) radiation results in dimerization of adjacent pyrimidines.
- Ionizing radiation causes severe DNA damage.

In the normal cellular environment DNA molecules are not absolutely stable. Each base pair in the DNA double helix has a certain probability of changing or mutating. Any event that introduces a deviation from the usual double-helical structure of DNA is a threat to the genetic constitution of the cell. While a certain mutation rate is required for the process of evolution, the typical amount of DNA damage sustained daily would be rapidly lethal if not repaired with high accuracy. Injury to DNA is thus minimized by systems that recognize and correct the damage. When a repair system reverses a change to DNA, there is no consequence. A mutation may result, though, when it fails to do so. The measured rate of mutation reflects a balance between the number of damaging events occurring in DNA and the number that have been corrected (or miscorrected).

Repair systems often can recognize a range of distortions in DNA as signals for action, and a cell is likely to have several systems able to deal with DNA damage. Major sources of DNA damage and the repair systems that cope with them are summarized in **FIGURE 3.30**. Damage to DNA can come from both endogenous sources (within the cell) as well as exogenous sources.

Endogenous sources of DNA damage include errors introduced during replication that escape the proofreading function of the replicative polymerases (discussed in *3.5 DNA polymerases replicate DNA*), which range from single-base mismatches to insertions or deletions of stretches of DNA. The chemistry of the cellular environment itself is a significant source of damage. DNA is subject to spontaneous water-mediated reactions that include deamination of cytosine, guanine, and adenine,

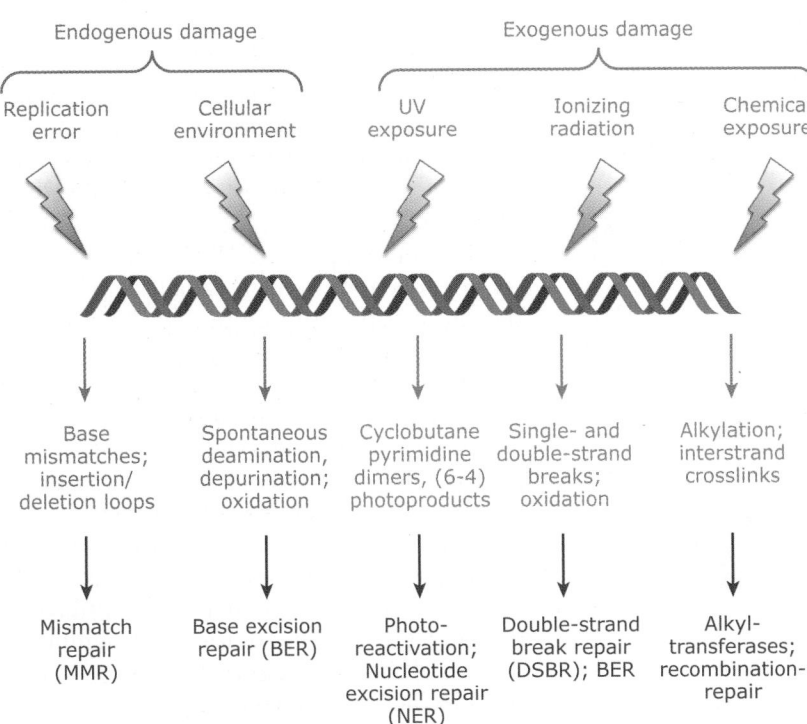

Endogenous damage

Replication error | Cellular environment

Exogenous damage

UV exposure | Ionizing radiation | Chemical exposure

Base mismatches; insertion/ deletion loops

Spontaneous deamination, depurination; oxidation

Cyclobutane pyrimidine dimers, (6-4) photoproducts

Single- and double-strand breaks; oxidation

Alkylation; interstrand crosslinks

Mismatch repair (MMR)

Base excision repair (BER)

Photo-reactivation; Nucleotide excision repair (NER)

Double-strand break repair (DSBR); BER

Alkyl-transferases; recombination-repair

FIGURE 3.30 DNA is subject to specific types of damage from a variety of sources, and is repaired by specialized repair systems. Some repair systems recognize multiple types of damage; primary repair pathways are indicated in this figure.

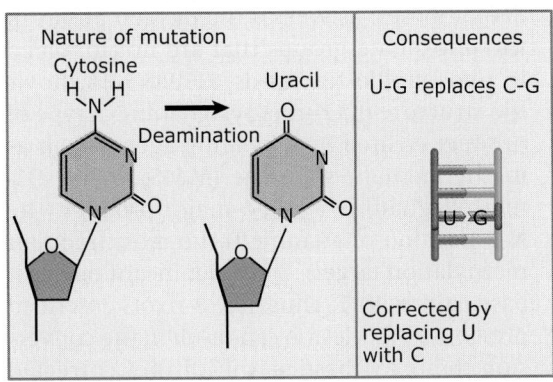

Nature of mutation

Cytosine → Deamination → Uracil

Consequences

U-G replaces C-G

U × G

Corrected by replacing U with C

FIGURE 3.31 Spontaneous deamination of cytosine creates a U:G base pair.

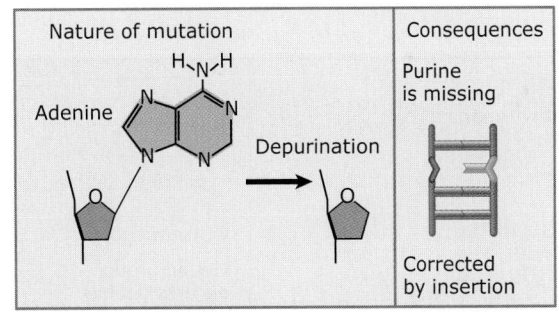

Nature of mutation

Adenine → Depurination

Consequences

Purine is missing

Corrected by insertion

FIGURE 3.32 Spontaneous depurination creates an abasic site. Spontaneous depyrimidation also occurs at a lower rate.

which converts these to uracil, xanthine, and hypoxanthine respectively. Cytosine deamination, shown in **FIGURE 3.31**, is the most frequent, estimated at ~100–500 events/cell/day in a mammalian cell. Since uracil, xanthine, and hypoxanthine are normally not present in DNA, repair machinery preferentially removes these bases.

Hydrolysis of the N-glycosyl bond (which links the base to the sugar) also occurs spontaneously in the cellular environment, most frequently when the base is a purine (~10,000 events/cell/day), shown in **FIGURE 3.32**. Loss

of a pyrimidine is less common (~500 events/cell/day). Cleavage of the N-glycosyl bond thus results in an **abasic site** (a site lacking a base); also known as an **AP** (for apurinic/apyrimidinic) **site**.

Another significant source of damage results from the reactive oxygen species (ROS), which are generated both by normal cellular metabolism (endogenous source), as well as resulting from ionizing radiation (exogenous source). Cells have a number of mechanisms to prevent damage from ROS production, including enzymes that convert reactive species

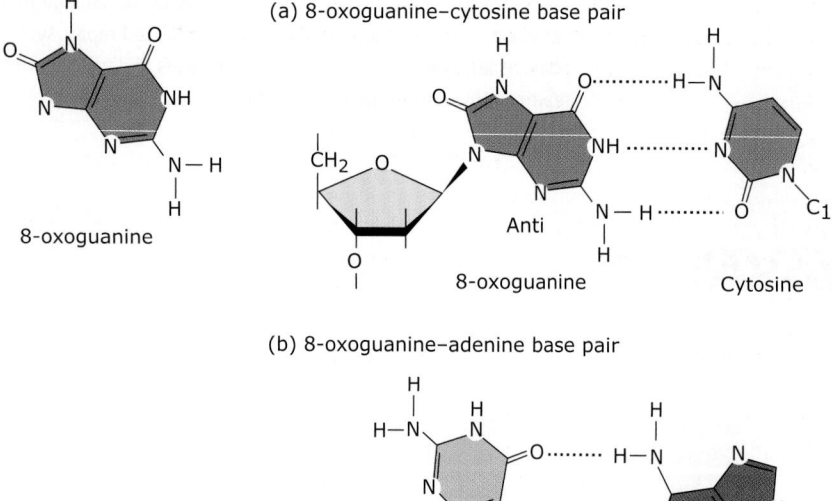

(a) 8-oxoguanine–cytosine base pair

8-oxoguanine

Anti

8-oxoguanine   Cytosine

(b) 8-oxoguanine–adenine base pair

Syn

8-oxoguanine   Adenine

FIGURE 3.33 Oxidative damage results in a variety of DNA lesions; some (such as 8-oxoguanine, shown) cause base mispairing or structural distortions of the DNA.

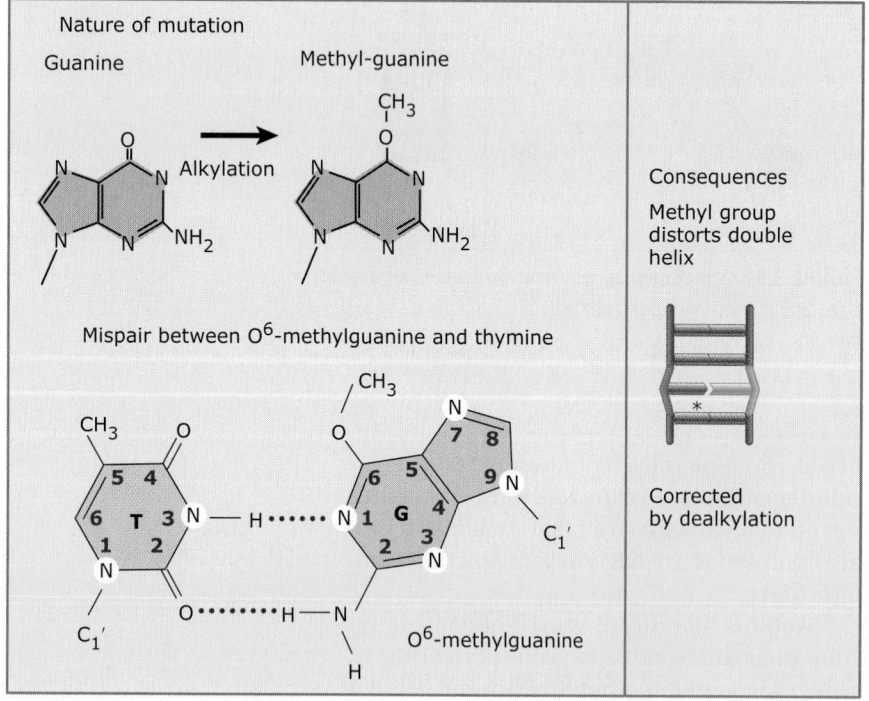

Nature of mutation

Guanine   Methyl-guanine

Alkylation

Mispair between O^6-methylguanine and thymine

Consequences

Methyl group distorts double helix

Corrected by dealkylation

O^6-methylguanine

FIGURE 3.34 Base alkylation can cause distortions of the double helix. In addition, some alkylation products, such as the O^6-methylguanine shown here, can also cause base mispairing. Right illustration adapted from A. Memisoglu and L. D. Samson. *Encyclopedia of Life Sciences.* John Wiley & Sons, Ltd., April 2001. [doi: 10. 1038/npg.els.0000579].

such as superoxide and hydroxyl radicals to less reactive products. It has been difficult to estimate how much damage from endogenous ROS production really occurs in a healthy cell, but endogenous ROS production typically increases during cellular aging as a result of decreased efficiency of the electron transport chain. Whether produced by internal chemistry or exogenous ionizing radiation, hydroxyl radicals can result in more than 80 different kinds of specific base damage. One of the most common, 8-oxoguanine (8-oxoG), is shown in **FIGURE 3.33**. 8-oxoG is capable of base pairing with both cytosine and adenine; 8-oxoG:A base pairs can be converted to T:A base pairs during replication if the damage is uncorrected.

Exogenous sources of damage are far more varied, including numerous sources of radiation and a vast array of chemical mutagens that attack DNA. One of the most significant sources of chemical damage to DNA is a family of molecules called **alkylating agents**, which transfer methyl, ethyl, or occasionally larger alkyl groups to the DNA. Alkylation can occur at most of the nitrogen and oxygen atoms in the bases (except the nitrogens linked to deoxyribose), as well as the oxygen atoms in the phosphate groups that are not involved in phosphodiester bonds. **FIGURE 3.34** shows the structure of O^6-methylaguanine, a type of product created by alkylating agents such as methylmethane sulfonate (MMS). While O^6-methylaguanine is not a major product (the N-7 position of guanine is the most frequent methylation target), it is a significant one as it base pairs with T. Thus, like 8-oxoG described previously, this lesion can result in the conversion to an A:T base pair if it is not corrected prior to replication.

While alkylating drugs like MMS are used frequently in the laboratory, they are (fortunately!) not frequently encountered in the environment. There are some environmental agents that are DNA alkylators, though in many cases they are harmless until modified by the cellular metabolism. Polycyclic aromatic hydrocarbons (PAHs) are a diverse family of molecules containing two or more fused aromatic rings; some members of this family are inadvertently converted into alkylating epoxides by the action of cytochrome P450 enzymes. Cytochrome P450 also converts *aflatoxins* into reactive epoxides. Aflatoxins are a class of carcinogens produced by fungi in the *Aspergillus* genus, which grow on grains such as peanuts, rice, and corn and constitute a serious public health threat in the US. **FIGURE 3.35** shows the result of an aflatoxin epoxide attack on guanine, which creates a lesion that results in replication errors.

Some alkylating agents have two reactive sites, allowing them to create two covalent links to the DNA. These can be intrastrand (within the same strand) or interstrand (across the double helix) cross-links. One of the first agents recognized to create interstrand cross-links was

Aflatoxin attached to the N7 position of the guanine base, which is shown in red

FIGURE 3.35 Cytochrome P450 converts some environmental agents into potent alkylating agents. Shown is aflatoxin attached to the N7 position of guanine.

bis[2-chloroethyl]methylamine, more infamously known as the chemical warfare agent nitrogen mustard gas. This molecule and the cross-link it forms are shown in **FIGURE 3.36**. Despite its inauspicious beginnings, nitrogen mustard gas was the first of a number of cross-linking compounds that have proven to be useful chemotherapeutic drugs, including the simple molecule cisplatin (also shown in Figure 3.36), though these drugs do have serious side effects due to their general cytotoxicity.

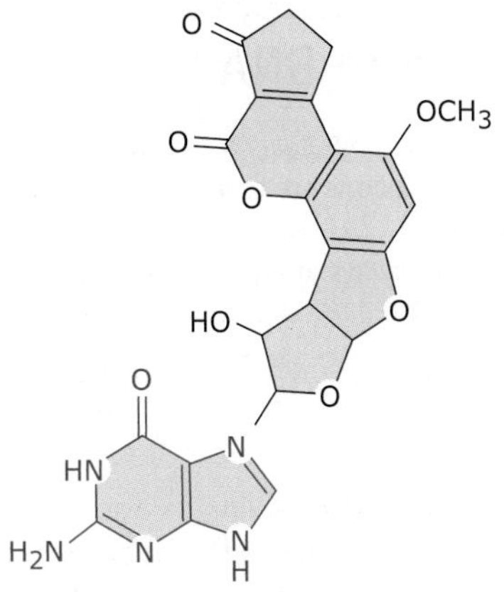

A major external source of DNA damage is radiation damage. There are two types of radiation that cause serious DNA lesions: UV radiation and ionizing radiation (which includes X-rays and gamma rays). Most UV damage is caused by UV-B, which is a band of UV light with wavelengths in the 296–320 nm range, and primarily causes DNA lesions in the skin. (UV-C, 100–295 nm, is the most dangerous UV band, but most UV-C is screened by the ozone layer. Ozone depletion would lead to a much greater fraction of UV-C reaching the earth's surface and a concomitant increase in skin cancer in humans.)

Nearly all UV exposure results in one of two major photoproducts: **cyclobutane pyrimidine dimers (CPDs)** and **(6–4) photoproducts (6–4PPs)**, shown in **FIGURE 3.37**. Both occur when adjacent pyrimidines are present on the same strand of DNA; CPDs account for 75% of the lesions and are most commonly formed between adjacent thymines.

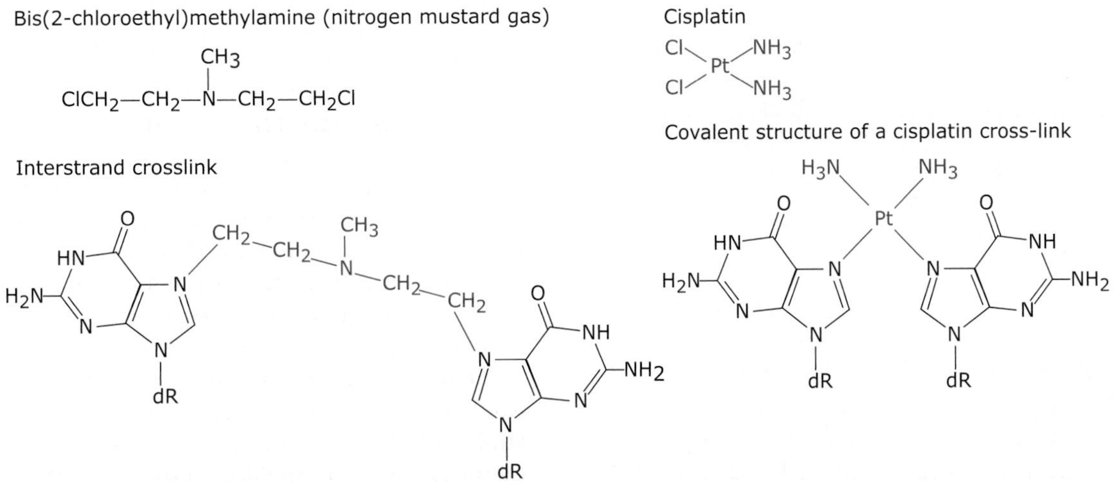

FIGURE 3.36 Nitrogen mustard gas and cisplatin are cross-linking agents used as chemotherapeutic drugs. Left illustration adapted from E. C. Friedberg, et al. *DNA Repair and Mutagenesis, Second edition*. ASM Press, 2005; middle illustration adapted from H. Huang, et al., *Science* 270 (1995): 1842–1845; right illustration adapted from E. C. Friedberg, et al. *DNA Repair and Mutagenesis, Second edition*. ASM Press, 2005.

(a) Cyclobutane pyrimidine dimer

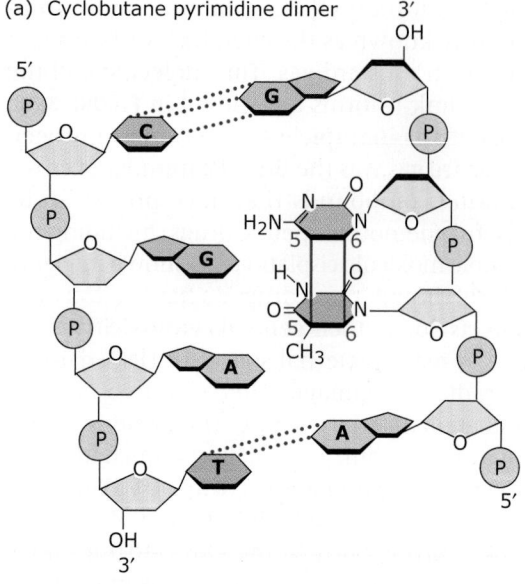

(b) (6–4) photoproduct

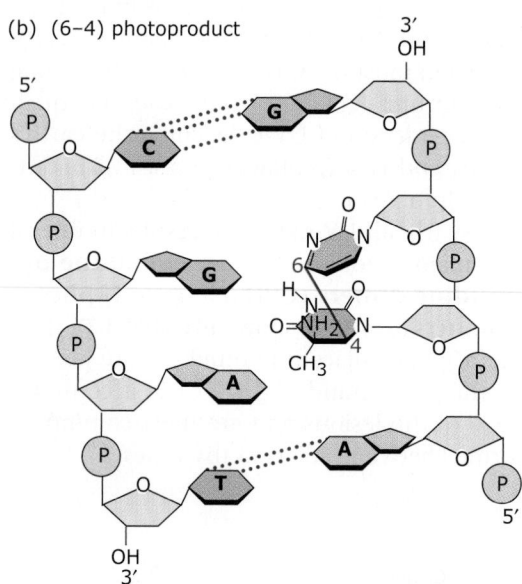

FIGURE 3.37 UV light results in dimerization of adjacent pyrimidines to form either cyclobutane pyrimidine dimers (CPDs) or (6–4) photoproducts (6–4PPs). Adapted from E. C. Friedberg, et al. *DNA Repair and Mutagenesis, Second edition.* ASM Press, 2005.

X-rays and gamma rays, on the other hand, create a large variety of DNA lesions, which can be caused by direct damage to the DNA by the radiation, as well as indirect damage when the radiation creates other reactive species (primarily ROS) in the vicinity of the DNA. The most serious type of DNA damage, the double-strand break, is a frequent outcome of ionizing radiation. Unrepaired (or inaccurately repaired) double-strand breaks result in serious chromosomal aberrations, including fusions, deletions, inversions, and translocations. It is not surprising that ionizing radiation is extremely lethal to cells.

Concept and Reasoning Check

1. If a replication error that creates an A-G mismatch is not repaired before the next round of replication, what will be the sequences of the two daughter DNAs after replication?

3.13 Direct repair can reverse some DNA damage

Key concepts

- Different lesions are repaired via specific repair pathways.
- Repair pathways are divided into those that effect direct repair, and those that replace damaged regions.
- Photoreactivation reverses UV damage.
- Alkyltransferases remove alkyl groups from DNA.

As described in *3.12 DNA is subject to damage*, the numerous sources of DNA damage result in a wide spectrum of DNA lesions that must be recognized and accurately repaired by the cell. Cells have evolved a number of repair systems that are specialized for different classes of damage, each characterized by factors that recognize the initial lesion. Most repair pathways function by removing the lesion (either the damaged base alone or a larger surrounding region) and using the information on an intact complementary strand to replace the damaged area. These types of pathways, which involve multiple enzymes acting at numerous steps, will be discussed in subsequent sections.

In certain cases, however, DNA damage can be directly reversed through the action of a single enzyme. The first known example of **direct repair** came from the discovery of **photoreactivation**, a phenomenon in which bacterial cells exposed to UV irradiation exhibited better survival rates if they were kept in the light after UV exposure, rather than placed in the dark. It was eventually determined that a single enzyme, **photolyase**, uses the energy of an absorbed photon to directly cleave the cyclobutane ring of a CPD, restoring the original pyrimidines, as shown in **FIGURE 3.38**. Photolyase catalyzes a similar repair of (6–4)PPs. Photolyases are

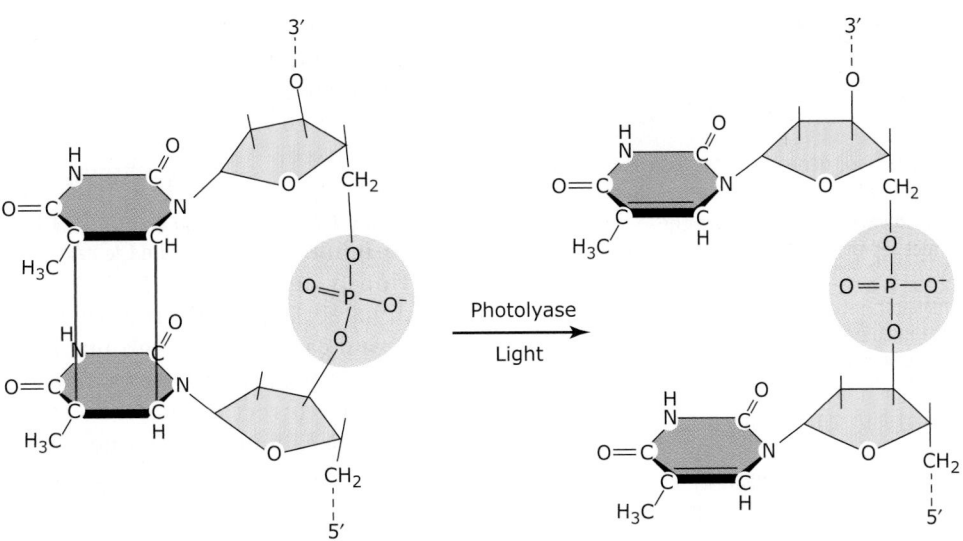

FIGURE 3.38 Photolyase reverses UV-induced pyrimidine dimers, using a light-driven reaction to disrupt the cyclobutane ring.

nearly ubiquitous in nature, but interestingly appear to have been lost in placental mammals (such as human) some time after the split between marsupial and placental mammals. Placental mammals utilize a more complex system to repair UV damage (see *3.16 Nucleotide excision repair removes bulky DNA lesions*), although humans have begun to use bacterial photolyases in clinical applications (see the *Medical Applications* box on page 96). Thus, modern medicine may provide us with a very useful enzyme that we have inconveniently lost during evolution.

The second type of direct repair is the removal of alkyl groups by a group of enzymes known generically as **alkyltransferases**. A large number of these enzymes exist, with specificities for different lesions. A human enzyme, alkyladenine DNA glycosylase (AAG), recognizes and removes a variety of alkylated substrates, including 3-methyladenine, 7-methylguanine, and hypoxanthine. **FIGURE 3.39** shows the structure of AAG bound to a methylated adenine, in which the adenine is flipped out and bound in the glycosylase's active site. This base flipping mechanism is used by numerous repair enzymes (including photolyase) to both recognize and repair DNA damage. A surprising feature of the alkyltransferases is that they are **suicide enzymes** that are inactivated by their own catalysis. The alkyltransferases act by transferring the alkyl group from the damaged base to a cysteine in the

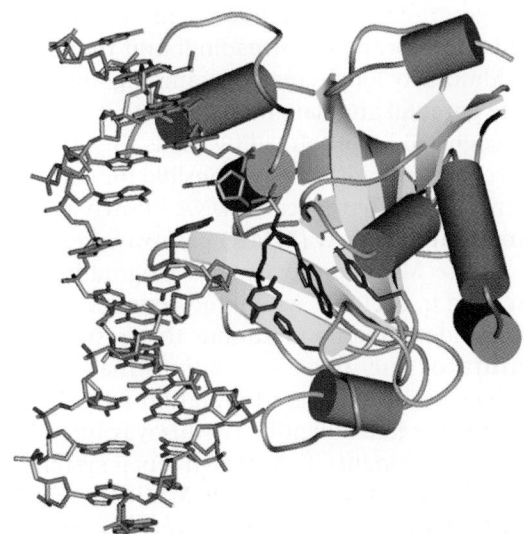

FIGURE 3.39 Crystal structure of the DNA repair enzyme alkyladenine DNA glycosylase (AAG) bound to a damaged base (3-methyladenine). The base (black) is flipped out of the DNA double helix (blue) and into AAG's active site (orange and green). Reproduced from A. Y. Lau, et al., *Proc. Natl. Acad. Sci. USA* 97 (2000): 13573–13578. Copyright © 2000 National Academy of Sciences, U.S.A. Photo courtesy of Tom Ellenberger, Washington University School of Medicine.

alkyltransferase itself. This restores the base, but the alkyl group is now irreversibly attached to the alkyltransferase, which is rendered permanently inactive.

3.14 Mismatch repair corrects replication errors

Key concepts

- The bacterial *mut* genes code for a mismatch repair system that corrects mismatched base pairs.
- Repair pathways are divided into those that effect direct repair, and those that replace damaged regions.
- The mismatch repair system preferentially corrects the base in the daughter strand following replication fork passage.
- The direction of mismatch repair in prokaryotes is dictated by DNA methylation.
- Eukaryotic mismatch repair targets mismatches and insertion/deletion loops.
- Eukaryotic mismatch repair is tightly linked to replication.

As discussed in *3.5 DNA polymerases replicate DNA*, DNA replication is a highly accurate process, in part due to proofreading activities contained in replicative DNA polymerase complexes. However, a significant number of base mispairs escape proofreading, and there are additionally loops created by misalignment of the parental and daughter stands that are not recognized by proofreading. Cells thus utilize a repair system that is responsible for surveillance of newly replicated DNA known as the **mismatch repair (MMR)** pathway.

The MMR pathway was originally described in bacteria, where the genes involved were identified by mutations that result in a **mutator** phenotype—mutations that themselves lead to increased mutation frequency. Many *mut* genes identified in this way turn out to be components of mismatch repair systems. The key genes in prokaryotic MMR are *mutH*, *mutS*, and *mutL*. The mismatch repair system is a type of **excision repair**, which (in contrast to direct repair) entails the removal of a region of DNA containing the error, and replacement of the correct sequence using the complementary strand as a template.

The prokaryotic MMR pathway is shown in **FIGURE 3.40**. In this pathway, MutS binds to the mismatch (as either a dimer or a tetramer) and is joined by MutL. The challenge of repairing mismatched bases is that, unlike forms of DNA damage in which a base is noticeably aberrant, both bases in a typical mismatch are normal, undamaged bases. Thus, the mismatch machinery has no intrinsic means of determining which base in a mismatch actu-

ally represents an error. If the mutated base is excised, the wild-type sequence is restored. If it happens to be the original (wild-type) base that is excised, the new (mutant) sequence becomes fixed.

When mismatch errors occur during replication in *E. coli*, it is possible to distinguish the original strand of DNA. As described previously, immediately after replication of methylated DNA, only the original parental strand carries methyl groups (see Figure 3.25). This provides the basis for a system to accurately correct replication errors. MutS has two DNA-binding sites, one of which specifically recognizes mismatches. The second is not specific for sequence or structure, and it has been suggested it may be used to translocate along DNA until a GATC sequence is encountered, driven by the ATPase activity of MutS.

The MutS · MutL complex activates the MutH endonuclease, which then cleaves the unmethylated GATC. This unmethylated strand is then excised from the GATC site to the mismatch site. The excision can occur in either the 5' to 3' direction (using RecJ or exonuclease VII) or in the 3' to 5' direction (using exonuclease I), and is assisted by helicase II (also known as UvrD). A new DNA strand is then synthesized by DNA polymerase III. The excision can be quite extensive; mismatches can be repaired preferentially for >1 kb around a GATC site. The result is that the newly synthesized strand is corrected to the sequence of the parental strand.

Eukaryotic cells have systems homologous to the *E. coli mut* system. In addition to recognizing mismatches, eukaryotic MMR also repairs insertion/deletion loops caused by **replication slippage**. In a region such as a microsatellite, where a very short sequence is repeated several times, realignment between the newly synthesized daughter strand and its template can lead to a stuttering in which the DNA polymerase slips backward and synthesizes extra repeating units. These units in the daughter strand are extruded as a single-stranded loop from the double helix, as shown in **FIGURE 3.41**. Alternatively, the template strand can form the loop, resulting in a net decrease in the number of repeats synthesized in the daughter strand.

Three MutS homologs, MSH2, -3, and -6, and two MutL homologs, MLH1 and PMS2, are involved in mismatch recognition and repair

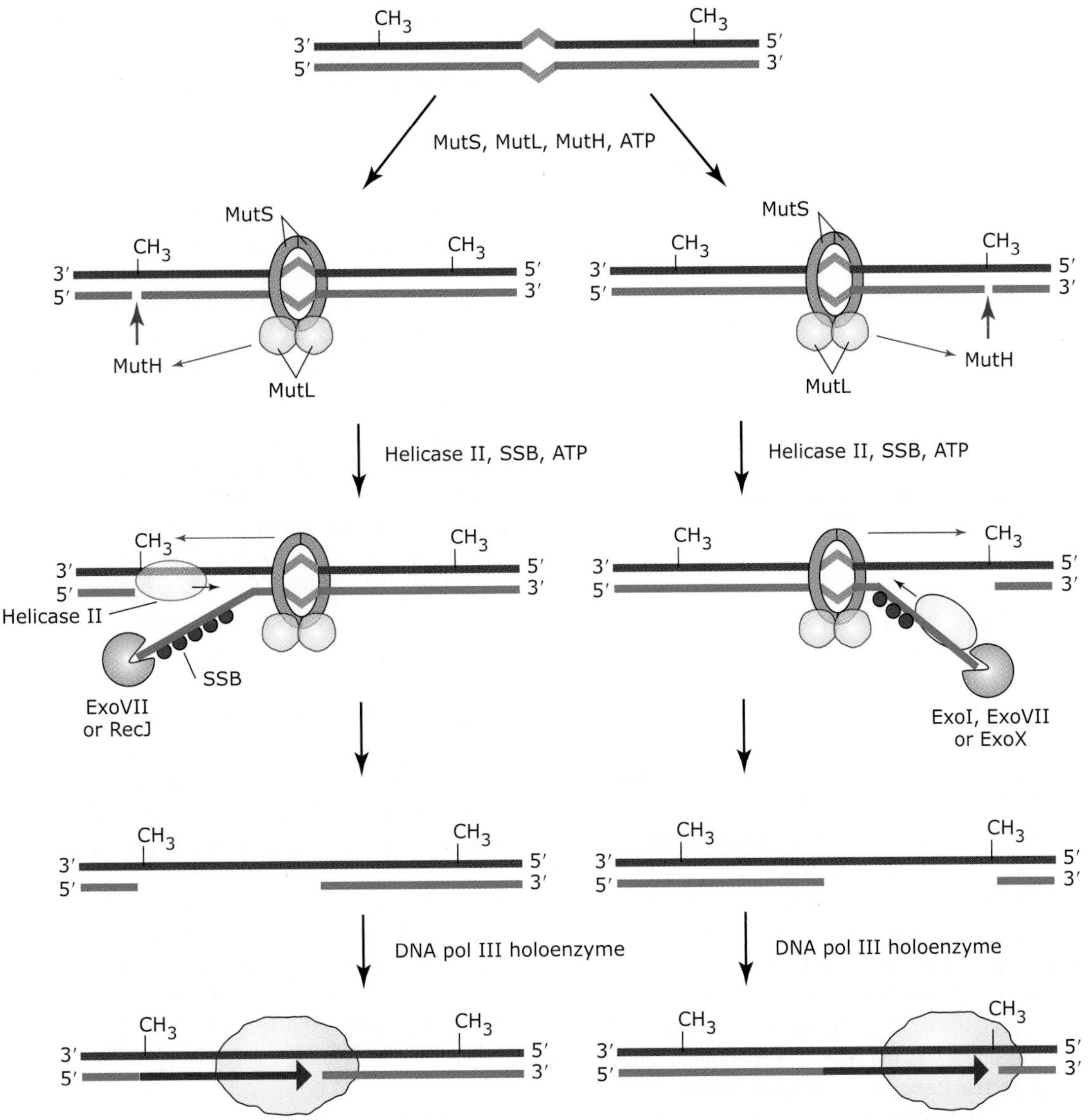

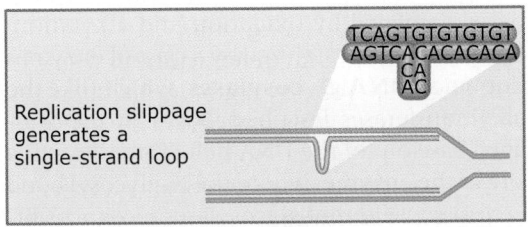

FIGURE 3.40 Mismatch repair in *E. coli* involves recognition of a mismatch by the MutS and MutL proteins. The MutH endonuclease nicks the DNA at the nearest unmethylated (daughter strand) GATC site, which then provides an entry point for a helicase to unwind the DNA, up to and including the mismatch. The unwound end is degraded by exonucleases, and the resulting gap is filled in by polymerase III. DNA ligase seals the nick to complete the repair. Adapted from R. R. Lyer, et al., *Chem. Rev.* 106 (2006): 302–323.

FIGURE 3.41 Replication slippage results in insertion/deletion loops in regions of repetitive sequence. If uncorrected, increases or decreases in repeat numbers result.

in eukaryotes, as shown in **FIGURE 3.42**. MutSα, an MSH2-MSH6 heterodimer, is responsible for recognizing mismatches and small insertion/deletion loops, while MutSβ, an MSH2-MSH3 heterodimer, only recognizes insertion/deletion loops in a range of sizes. MutSα or -β not only recognize the mismatches/loops but also interact directly with components of the replication machinery, including PCNA and the clamp loader RFC. These interactions are

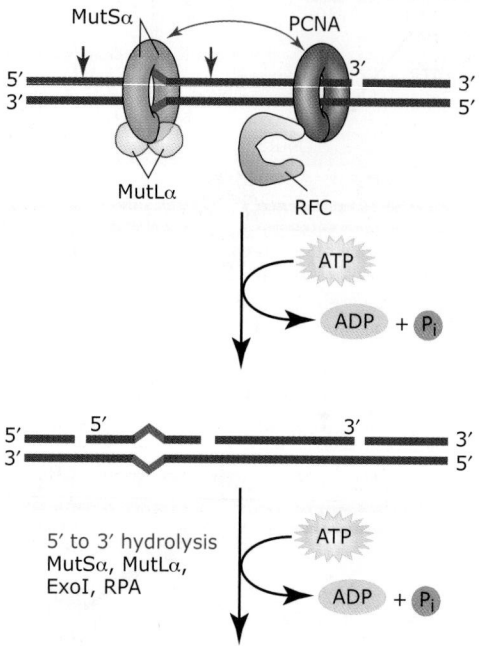

3'-directed

MutSα   PCNA

5'
3'
    MutLα   RFC

ATP

ADP + P$_i$

5'   5'   3'   3'
3'   5'

5' to 3' hydrolysis
MutSα, MutLα,
ExoI, RPA

ATP

ADP + P$_i$

5'   3'
3'   5'

FIGURE 3.42 Human mismatch repair is performed by homologs of the bacterial MMR system. MutSα (or β for insertion/deletion loops) binds the mismatch and interacts with PCNA and RFC. MutLα's endonuclease activity is then activated and results in nicking of whichever strand already contains a nick (i.e., the lagging strand). MutSα activates Exo I, which removes the mismatch-containing region. The reaction is shown for a template with a nick 3' to the mismatch; a similar process occurs when the nick is 5' to the mismatch. Adapted from F. A. Kadyrov, *Cell* 126 (2006): 297–308.

important for ensuring the association of the MMR system with sites of replication, and also appear to facilitate the specific repair of the daughter strand. Surprisingly, even though most eukaryotes possess DNA methylation, eukaryotic mismatch repair systems do not use DNA methylation to select the daughter strand for repair. Instead, MutSα (in the presence of PCNA, RFC, and ATP) activates an endonuclease activity in MutLα (a MLH1-PMS2 heterodimer), which then nicks the DNA with a strong preference for a DNA strand that already contains a nick. In other words, MutLα cleaves the lagging strand efficiently, thus targeting the correct strand for excision by the MutSα-activated exonuclease ExoI. It is not

known how specificity for the daughter strand is achieved on the leading strand.

The importance of the MutS/L system for mismatch repair is indicated by the high rate at which it is found to be defective in human cancers. Loss of this system leads to an increased mutation rate, and mutations in MutS/L components can lead to hereditary nonpolyposis colorectal cancer (HNPCC); patients with HNPCC are also prone to cancers in a variety of other tissues, such as skin, ovary, stomach, and kidney. Defective mismatch repair is associated with a significant fraction of sporadic tumors as well.

Concept and Reasoning Check

1. Part of the diagnosis of HNPCC includes comparison of microsatellite repeats in the healthy and cancerous cells of a patient. What would be the expected difference and why would this be diagnostic for HNPCC?

3.15 Base excision repair replaces damaged bases

Key concepts

- Base excision repair entails direct removal of a damaged base from DNA.
- Base removal triggers the excision and replacement of a stretch of DNA.
- Base excision repair removes bases damaged by oxidation, alkylation, and deamination.
- Apurinic/apyrimidinic (AP) sites enter the base excision pathway directly.

Base excision repair (BER) is an excision repair pathway that is initiated by the precise removal of a damaged base from the DNA, without perturbing the sugar-phosphate backbone, to produce an abasic site. Abasic, or AP (apurinic/apyrimidinic), sites generated by spontaneous depurination or depyrimidation are self-made substrates for the BER pathway.

The BER pathway is used to repair a large array of DNA single-base lesions, including bases damaged by oxidation and alkylation. Base removal is catalyzed by a class of enzymes known as **DNA glycosylases**, which (like the alkyltransferases described above) flip the damaged base out of the DNA helix into the active site of the enzyme, where the N-glycosyl bond is cleaved. Different glycosylases recognize different substrates; the reaction of a glycosylase that recognizes uracil (generated by cytosine deamination) is shown in **FIGURE 3.43**. Some enzymes, in addition to their glycosylase activ-

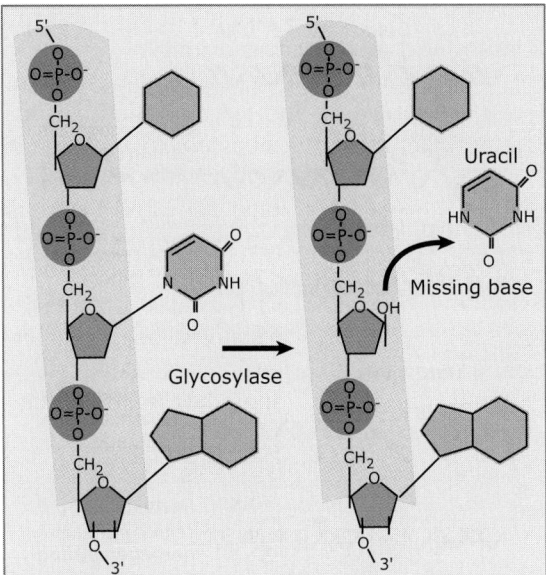

FIGURE 3.43 A glycosylase removes a base from DNA by cleaving the bond to the deoxyribose.

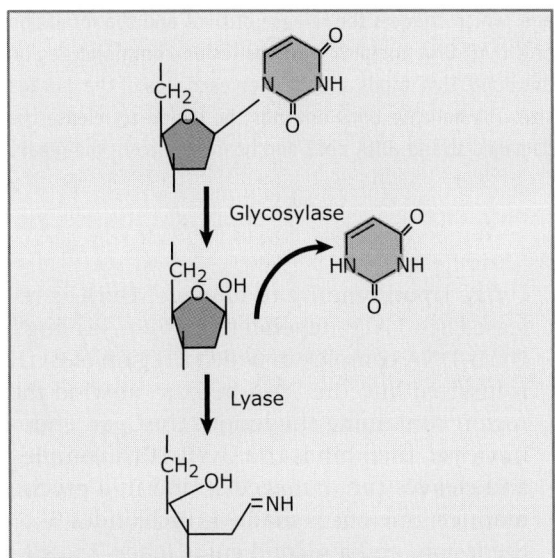

FIGURE 3.44 A glycosylase hydrolyzes the bond between base and deoxyribose, but a lyase takes the reaction further by opening the sugar ring.

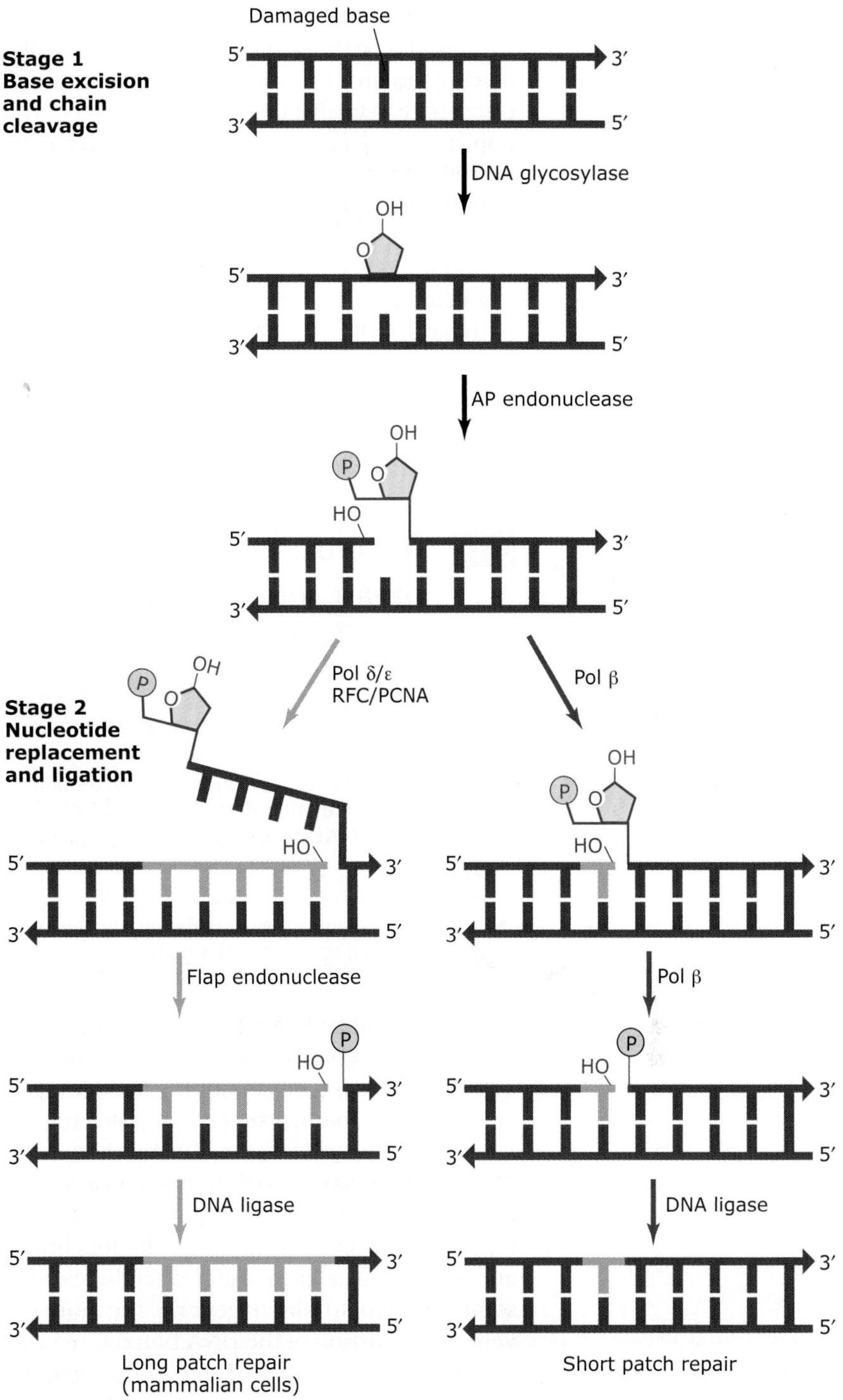

ity, also contain **AP lyase** activity that cleaves the bond between the sugar and phosphate 3′ to the AP site, as shown in **FIGURE 3.44**. These enzymes are called DNA glycosylases/lyases.

No enzymes exist to replace a base in an abasic site, so the next stage of BER is to cleave the phosphodiester backbone so that the DNA containing the abasic site can be excised and replaced. Spontaneous AP sites enter the pathway at this stage. AP sites are recognized by **AP endonuclease**, which cleaves the DNA backbone 5′ to the AP site. This results in a gap

**Stage 1
Base excision
and chain
cleavage**

**Stage 2
Nucleotide
replacement
and ligation**

Long patch repair
(mammalian cells)

Short patch repair

FIGURE 3.45 The gap resulting from glycosylase/AP endonuclease action can be repaired via one of the two pathways in eukaryotes. Long patch repair entails polymerization by DNA pol δ or ε and displacement and removal or 2–8 nt. Short patch repair uses DNA pol β to add a single nucleotide and to remove the 5′-deoxyribose phosphate. Ligase seals the nick in both pathways. Adapted from O. D. Schaerer, *Angew. Chem. Int. Ed. Engl.* 34 (2003): 2946–2974.

with a 5′-deoxyribose phosphate (the abasic sugar) and a 3′-OH. This gap can be repaired in one of two different ways, shown in **FIGURE 3.45**. In *long patch* repair, DNA pol δ or ε (in conjunction with PCNA and RFC) adds 2–8 nucleotides to the 3′ end, simultaneously displacing the DNA containing the 5′-deoxyribose phosphate to form a flap, which is cleaved by a flap endonuclease. In *short patch* repair, a conserved repair polymerase, DNA pol β, inserts a single nucleotide. DNA pol β contains an intrinsic 5′-deoxyribose phosphate lyase activity that removes the 5′-deoxyribose phosphate. DNA ligase seals the remaining nick in both pathways. When BER is initiated by glycosylase/lyase action, AP endonuclease actually removes the sugar left at the 3′ end, so that when DNA pol β inserts a single nucleotide, the product is a suitable substrate for ligase.

3.16 Nucleotide excision repair removes bulky DNA lesions

Key concepts

- The prokaryotic *uvr* excision repair system excises a ~12 base region containing the lesion.
- Xeroderma pigmentosum (XP) is a human disease resulting from mutations in human nucleotide excision repair (NER) genes.
- Global genome repair recognizes lesions anywhere in the genome.
- Transcription-coupled repair preferentially repairs the transcribed strand of active genes.

Like MMR and BER, **NER** is an excision pathway that removes damaged bases—in this case, a sizeable region containing the damaged base—and fills in the resultant gap using the intact strand as a template for a DNA polymerase. The NER pathway is used to repair a variety of bulky DNA lesions, including the dimeric photoproducts resulting from UV irradiation, bulky adducts such as aflatoxin-damaged bases, and some DNA cross-links. In general, the more severe the distortion of the DNA helix, the more readily the lesion is recognized and repaired by NER.

In *E. coli*, genes responsible for NER were identified by mutants that were unable to repair UV-induced damage in the dark (i.e., when photolyase was unable to repair the damage) and thus were named *uvr* for UV resistance. Four *uvr* genes—*uvrA, uvrB, uvrC,* and *uvrD*—encode the core proteins required for NER. They act essentially in the order of their names, as seen in **FIGURE 3.46**. First, a heterotrimer consisting of two UvrA and one UvrB subunits scans the DNA using the helicase activity of

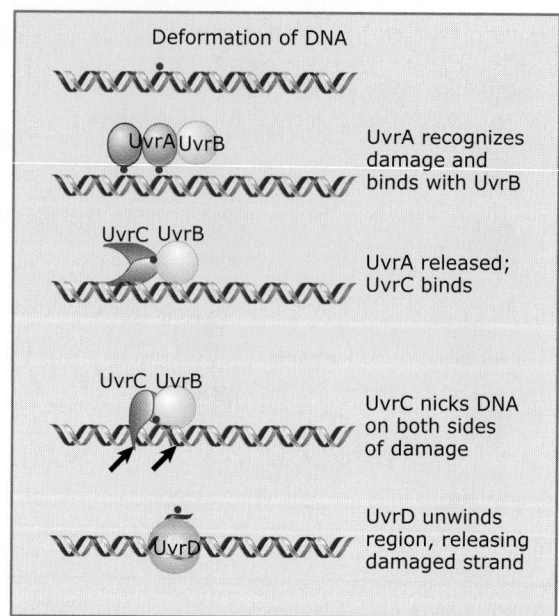

FIGURE 3.46 In the Uvr repair pathway, and UvrA₂UvrB heterotrimer translocates along DNA and recognizes damage, which triggers the release of UvrA and the formation of a UvrB-DNA complex with the lesion-containing region unwound. UvrC binds and nicks on each side of the damage site. The helicase UvrD unwinds the region to release the damaged strand. DNA pol I and ligase complete the repair.

UvrB. Upon binding to damage, UvrA is released in an ATP-dependent manner, leaving a UvrB-DNA complex in which a region of UvrB is inserted into the DNA helix to unwind the region containing the lesion. UvrC, an endonuclease, then binds the UvrB-DNA complex and cleaves the damaged strand in a precise manner: one cut is made 4 nucleotides 3′ to the lesion, and a second cut is made 7 nucleotides 5′ to the lesion. The UvrD helicase then removes the excised fragment, leaving a gap to be filled in by DNA pol I. As always, ligase seals the final nick.

A conserved NER system repairs bulky adducts in eukaryotes—in mammalian cells, this is the only system that can repair UV-induced photoproducts. The importance of NER for repair of UV damage in mammals is revealed by **XP**, a disease caused by loss of NER in humans and which results in up to a 1000 times increase in the incidence of skin cancer. XP is discussed in greater detail in the accompanying *Medical Applications* box on page 96. Most of the proteins required for NER have XP in their names, reflecting their initial genetic identification in patients with XP.

NER in eukaryotes is actually divided into two subpathways, illustrated in **FIGURE 3.47**. The major difference between the two pathways is how the damage is initially recognized. In **global genome repair (GG-NER)**, a protein called XPC, complexed with a protein called hHR23b, detects the damage and initiates the repair pathway. XPC can recognize damage

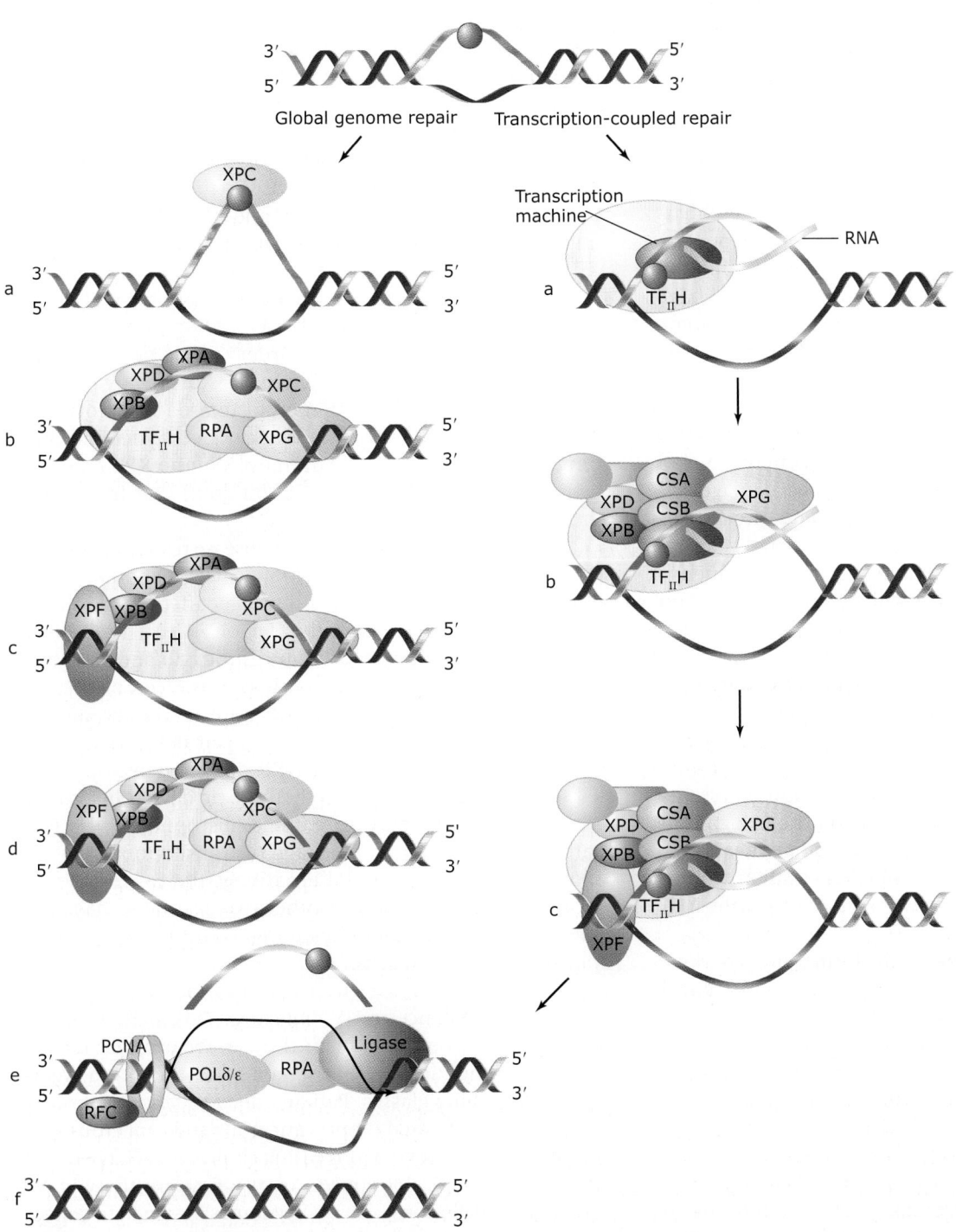

FIGURE 3.47 Nucleotide excision repair occurs via two major pathways: global genome repair, in which XPC recognizes damage anywhere in the genome, and transcription-coupled repair, in which the transcribed strand of active genes is preferentially repaired, and the damage is recognized by an elongating RNA polymerase. Adapted from E. C. Friedberg, *Nature Revs. Cancer* 1 (2001): 22–23.

XP is an autosomal recessive disorder associated with extreme UV light sensitivity, increased risk of skin cancer, and neurologic degeneration. The name *xeroderma pigmentosum* literally means "dry pigmented skin." XP, and two other related disorders, Cockayne syndrome and trichothiodystrophy, are associated with defects in DNA nucleotide excision repair (NER). XP patients exhibit sun-induced freckles and other skin pigmentation as infants, which is rare in the general population, and this is often the basis of the clinical diagnosis. About half the patients with XP are extremely light sensitive and many exhibit sunburn and blistering with minimal sun exposure; others tan normally but still exhibit pigmentation on sun-exposed skin. With continued sun exposure, the skin becomes dry and looks prematurely aged. Children with XP frequently develop precancerous lesions, and have a 1000-fold increased risk of developing either melanoma or nonmelanoma skin cancer before age 20. The average age of XP patients that have skin cancer is under 10 years old, which is 50 years earlier than the average age of the appearance of skin cancers in the general population. With continued exposure to sunlight, XP patients may develop chronic UV-induced eye inflammation, corneal damage and the loss of eyelashes. About 30% of XP patients also exhibit neurological defects including progressive mental retardation, loss of deep tendon reflexes, and loss of high-frequency hearing. Patients may also develop difficulties in walking and swallowing over time.

XP is a relatively rare genetic disorder; the overall prevalence in the United States is estimated to be 1/1,000,000 but is somewhat more common in some populations. In Japan, the prevalence is 1/22,000. There are seven distinct genes (*XPA, XPB, XPC, XPD, XPE, XPF,* and *XPG*) that are associated with this disease. Each complementation group represents mutations in different proteins involved in the recognition and repair of DNA damage induced by UV radiation, and the proteins have been named for the complementation groups. Approximately 50% of all XP is due to mutations in the *XPA* and *XPC* genes.

UV radiation damages DNA by the formation of pyrimidine dimers, including cyclobutane pyrimidine dimers and (6,4)-photoproducts. These photoproducts are normally recognized, excised, and repaired by the NER pathway (see Figure 3.47). The NER pathway also removes other bulky DNA modifications induced by other carcinogens. As discussed in this chapter, DNA repair by the NER pathway differs depending on whether or not the damaged DNA is in a region that is actively transcribed. Actively transcribed regions of the genome are repaired more rapidly and are recognized through a distinct complex of proteins of the transcription-coupled repair (TCR) pathway. Damaged DNA is detected by the presence of stalled RNA polymerase, which, in turn,

recruits DNA-binding proteins CSA and CSB. In contrast, DNA damage in the nontranscribed regions of the genome is detected by the binding of proteins of the global genome repair (GGR) pathway, which include XPE and XPC proteins. After the recognition of the DNA lesions by these two distinct mechanisms, the pathways converge and the damaged DNA is unwound, excised, and repaired by common proteins.

There is a very interesting genotype/phenotype correlation between mutations in the NER pathway and the sun sensitivity, cancer risk, and neurologic manifestations of the XP disease. Patients with mutations in the proteins that are unique to the GGR pathway (XPC and XPE) exhibit XP without neurologic defects. Patients with mutations in the other XP proteins that are common to GGR and TCR exhibit XP with the associated neurologic defects. Interestingly, mutations in CSA and CSB, the DNA-binding proteins in the TCR pathway, result in the disorder Cockayne syndrome. Cockayne syndrome is characterized by sun sensitivity, mental retardation, and developmental retardation; however, there is no increased risk of skin cancer. Cockayne syndrome can also result from specific mutations in the XPB and XPD helicases. These mutations are at different sites from those that result in XP.

Trichothiodystrophy is another disease resulting from defects in proteins involved in NER; specifically, another subset of mutations in XPB and XPD that may affect their roles as components of the transcription factor $TF_{II}H$. This disorder is associated with photosensitivity, brittle hair and nails, and growth retardation, but not with sun-induced pigmentation or cancer, and these patients appear to have specific transcriptional defects rather than repair deficiency. These observations of XP and other NER defective disorders has led to the suggestion that the neurologic defects associated with these diseases may arise due to disruption of the TCR pathway, resulting in a blockage of transcription and the apoptosis of neural cells. Alternatively, the cancer risk associated with XP, but not the other two disorders, may arise due to genomic instability resulting from the disruption of the GGR pathway in cells.

There is no cure or any treatment for XP, other than aggressive avoidance of UV exposure. This includes limiting time spent outdoors or in the car, wearing protective clothing, using high SPF sunscreen at all times and wearing UV absorbing glasses. Patients need to be continuously monitored and treated for precancerous and cancerous skin lesions. Recently, several promising clinical trials have tested the effects of topical lotions to treat photosensitivity in XP patients. These lotions contain liposomes designed to deliver DNA repair enzymes (such as prokaryotic photolyases) to the site of application, replacing the absent repair functions in the skin cells and reducing the risk of skin cancer.

Source: Esther Siegfried, Pennsylvania State University, Altoona

anywhere in the genome. On the other hand, **transcription-coupled repair (TC-NER)**, as the name suggests, is responsible for repairing lesions that occur in the transcribed strand of active genes. In this case, the damage is recognized by RNA polymerase II itself.

Both pathways then use an overlapping set of proteins to effect the repair. The strands of DNA are unwound for ~20 bp around the damaged site. This action is performed using the helicase activity of the transcription factor TFIIH (discussed in *4.5 Promoters direct the initiation of transcription*), which includes the products of two XP genes (XPB and XPD). Then cleavages are made on either side of the lesion by endonucleases encoded by the XPF and XPG genes. The single-stranded stretch including the damaged bases can then be replaced by new synthesis using pol δ or ε.

3.17 Double-strand breaks are repaired by two major pathways

Key concepts

- DNA double-strand breaks can be repaired by homologous recombination.
- In the absence of a homologous donor, double-strand breaks are repaired via nonhomologous end joining (NHEJ).

DNA double-strand breaks—broken chromosomes—are the most severe form of DNA damage that can occur, particularly in eukaryotes. Unrepaired breaks in a linear chromosome result in loss of any chromosome fragments lacking centromeres, and aberrant repair can result in chromosome–chromosome fusions, translocations, inversions, and deletions. Even if the two correct broken ends are matched up and rejoined, loss of sequence from the broken ends is a serious risk. Cells have evolved two very different methods to repair double-strand breaks. The first method of **double-strand break repair (DSBR)** is **homologous recombination (HR)**, which is actually a collection of related pathways. HR depends on the presence of a homologous donor sequence (a sister chromatid or homologous chromosome) that can be used to accurately replace

any sequences that may have been lost from a broken DNA end. In the absence of an available homologous donor, cells instead use a pathway called **nonhomologous end joining (NHEJ)**, which essentially entails the direct ligation of DNA ends, at some risk of loss of sequence prior to repair. Interestingly, both DSBR pathways share most of their components with two normally occurring pathways: HR is at the heart of meiotic recombination during gamete formation (discussed in *3.18 Homologous recombination is used for both repair and meiotic recombination*), and NHEJ is the pathway used during somatic recombination in the immune system.

The basic steps of double-strand break repair via HR are shown in **FIGURE 3.48**. Many of the factors required for HR are encoded by *RAD* genes, so named because they were originally identified in yeast mutants sensitive to ionizing radiation (X-rays). The key steps of this repair pathway following recognition of the break are as follows:

1. The 5′-ends of the break are exonucleolytically degraded to produce single-stranded overhanging ends with 3′-OH termini. This **strand resection** requires a complex of proteins known as MRN (Mre11, Rad50, Nbs1) in mammals or MRX (Mre11, Rad50, Xrs2) in yeast. This complex is required for resection, but does not have all the enzymatic activities to perform the resection itself and must therefore recruit other (unknown) factors. The single-stranded DNA is immediately coated with RPA.
2. With the assistance of Rad52 and other factors, Rad51 replaces RPA to form a Rad51 filament.
3. The 3′-end of the single-strand DNA invades a homologous duplex to form a D-loop. Rad51 catalyzes this **strand invasion** step.
4. DNA polymerase extends from the invading 3′-end, enlarging the D-loop. The second 3′-end anneals to the D-loop and also primes synthesis by DNA polymerase.
5. Ligase joins the free ends.
6. The resulting crossover structures (Holliday junctions) are resolved. **Resolution** entails either the joint action of helicases and topoisomerases to *dissolve* the junctions, or the action of resolvases that cleave and religate the junctions.

The net result of HR is that the sequence surrounding a DSB is accurately repaired using intact information from another chromosome. If a sister chromatid is used, identical information is used for repair. If a homologous chromosome is the source of information, any sequences lost from the site of the DSB will be replaced with the sequences present on the homolog—which may or may not be identical to the sequences lost during damage. If these sequences differ, the result can be that a previously heterozygous region becomes homozygous during repair, a type of *gene conversion*.

In the absence of homologous sequences to guide accurate repair, cells must still repair DSBs as efficiently as possible in order to avoid chromosomal aberrations. This is often necessary during G1, when no sister chromatids are readily available, and homologous chromosomes may be difficult to access (particularly in large mammalian genomes), or absent in typically haploid eukaryotes such as yeast. In this case, repair is carried out using the NHEJ pathway, depicted in **FIGURE 3.49**.

The central player in NHEJ is the **Ku** heterodimer, a complex of the Ku70 and Ku80 proteins. Ku was first identified as an autoantibody target in patients with the autoimmune disease scleroderma and is highly conserved

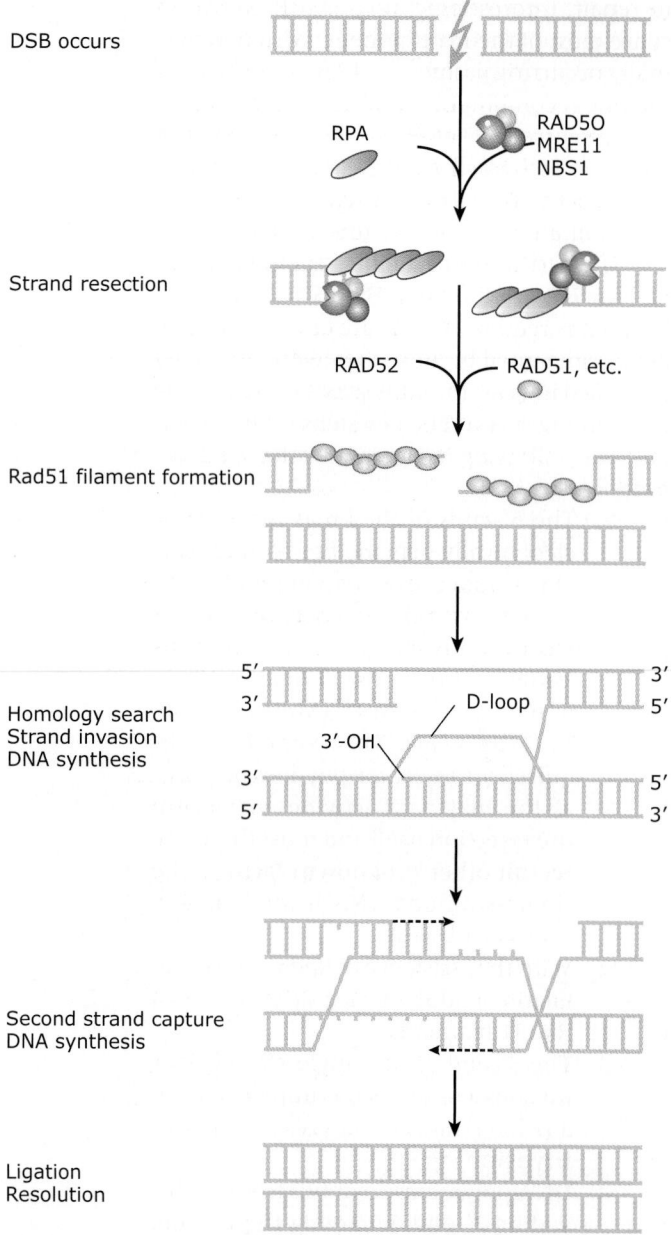

FIGURE 3.48 Double-strand break-repair by homologous recombination. Recruitment of the MRN/MRX complex leads to nuclease degradation of the ends, called DNA resection, leaving single-strand tails with 3'-OH ends. These ends are coated with RPA, which is then replaced by Rad51 (with the assistance of Rad52). Strand invasion by one end into homologous sequences forms a D-loop. Extension of the 3'-OH end by DNA synthesis enlarges the D-loop. Once the displaced loop can pair with the other side of the break, the second double-strand break end is captured. DNA synthesis to complete the break repair, followed by ligation results in the formation of two Holliday junctions. Resolution of the Holliday junctions produces separated duplexes.

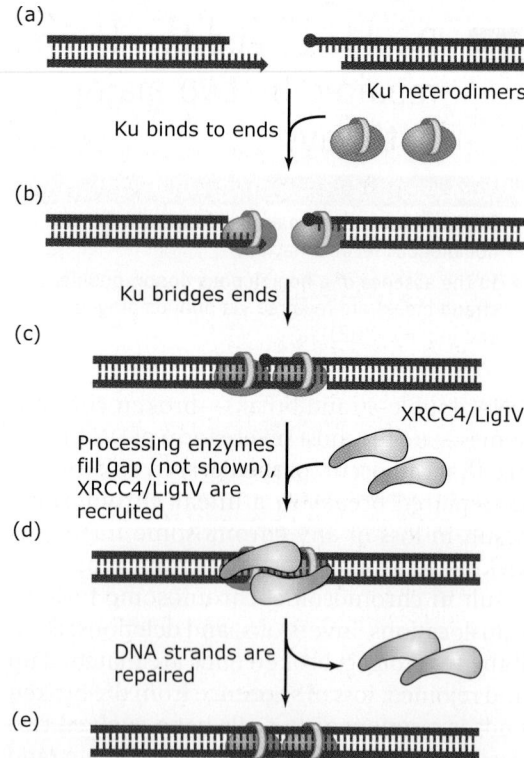

FIGURE 3.49 In nonhomologous end joining, double-strand ends are bound by the Ku heterodimer. Ku-DNA complexes come together to bridge the ends, and the processing enzymes fill in gaps, if present. Ends are ligated by the NHEJ-specific ligase LigIV and its partner XRCC4. Adapted from J. M. Jones, M. Gellert, and W. Yang, *Structure* 9 (2001): 881–884.

throughout eukaryotic species. The Ku heterodimer recognizes DNAs ends, and two heterodimers come together to form a scaffold that holds the ends together and allows other enzymes to act on them. The MRN/MRX complexes that function in HR also appear to assist Ku in bringing the ends together during NHEJ. DNA-dependent protein kinase (DNA-PK), which is activated by DNA to phosphorylate protein targets, also participates in end juxtaposition. One of DNA-PK's phosphorylation targets is the protein Artemis, which in its activated form has both exonuclease and endonuclease activities, and can trim overhanging ends to provide blunt ends compatible with ligation (Artemis also has an key role in cleaving the hairpins generated during recombination of immunoglobulin genes, and thus is essential in immune system function.) Specialized DNA polymerases (PolL and PolM in mammals) fill in any remaining single-stranded gaps. These processing steps can result in loss or alteration of sequence near the breakpoint, but are essential to create ligatable ends if the initial DSB does not have blunt ends that terminate properly with 5' phosphates and 3'-OH. The actual joining of the double-stranded ends is performed by DNA ligase IV, which functions in conjunction with the protein XRCC4 (Lif1 in yeast). Mutations in any of these components render cells more sensitive to radiation.

3.18 Homologous recombination is used for both repair and meiotic recombination

Key concepts

- Chromosomes must synapse (pair) for meiotic recombination to occur.
- Recombination is initiated by the formation of a double-strand break in a recipient DNA.
- Resolution of Holliday junctions can result in recombinant genomes or patch recombinants.

We have introduced homologous recombination as a mechanism for repair of DNA, but HR also occurs as a normal (and essential) process during meiosis, where it serves to provide a reshuffling of genetic information between maternal and paternal homologous chromosomes. The frequency of recombination is not constant throughout the genome, but is influenced by both global and local effects. The overall frequency may be different in oocytes and in sperm: recombination occurs twice as frequently in female humans as in male humans. Within the genome, its frequency depends upon chromosome structure; for example, crossing over is suppressed in the vicinity of the condensed and inactive regions of heterochromatin, and there are recombination "hotspots" where recombination frequency is very high.

As in the case of double-strand break repair, meiotic recombination is initiated by a double-strand break in DNA. In this case, however, the break is deliberately induced by the action of a meiotic endonuclease, Spo11. Spo11 is related to the type II topoisomerases, and it undergoes a similar reaction cycle in which it induces double-strand breaks and becomes covalently attached to the generated ends, as shown in **FIGURE 3.50**. Yeast Spo11 makes ~150–200 breaks throughout the genome in each meiosis, which results in ~1 crossover per chromosome arm. The MRX/N complex, assisted by other factors (Sae2 in yeast), then releases Spo11 and goes on the resect the DNA ends as occurs during double-stand break repair. The subsequent steps are very similar to those of repair (see Figure 3.48), except that in some steps

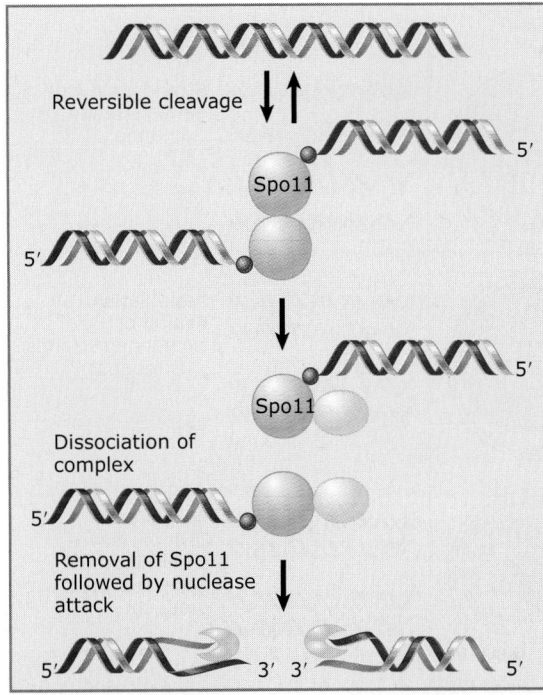

FIGURE 3.50 The Spo11 endonuclease initiates meiotic recombination by creating a double-strand break in a recipient chromosome. Spo11 is covalently attached to the DNA ends. The MRN/X complex, in combination with other factors, releases Spo11 and resects the ends.

meiosis-specific proteins perform the needed functions. For example, while Rad51 is active during meiosis, a Rad51-related recombination enzyme, Dmc1, is expressed exclusively during meiosis, and both proteins are needed to complete meiosis.

The different stages of meiotic recombination correlate with the visible progress of chromosomes through the five stages of meiotic prophase, as depicted in **FIGURE 3.51**. The beginning of meiosis is marked by the point at which individual chromosomes become visible. Each of these chromosomes has completed

replication and consists of two sister chromatids, each of which contains a duplex DNA. The homologous chromosomes approach one another and begin to pair in one or more regions, and pairing extends until the entire length of each chromosome is apposed with its homolog. The process is called **synapsis** or chromosome pairing. It is thought that initiation of sites of crossovers may be what leads to the initial pairing (as opposed to pairing preceding crossover formation).

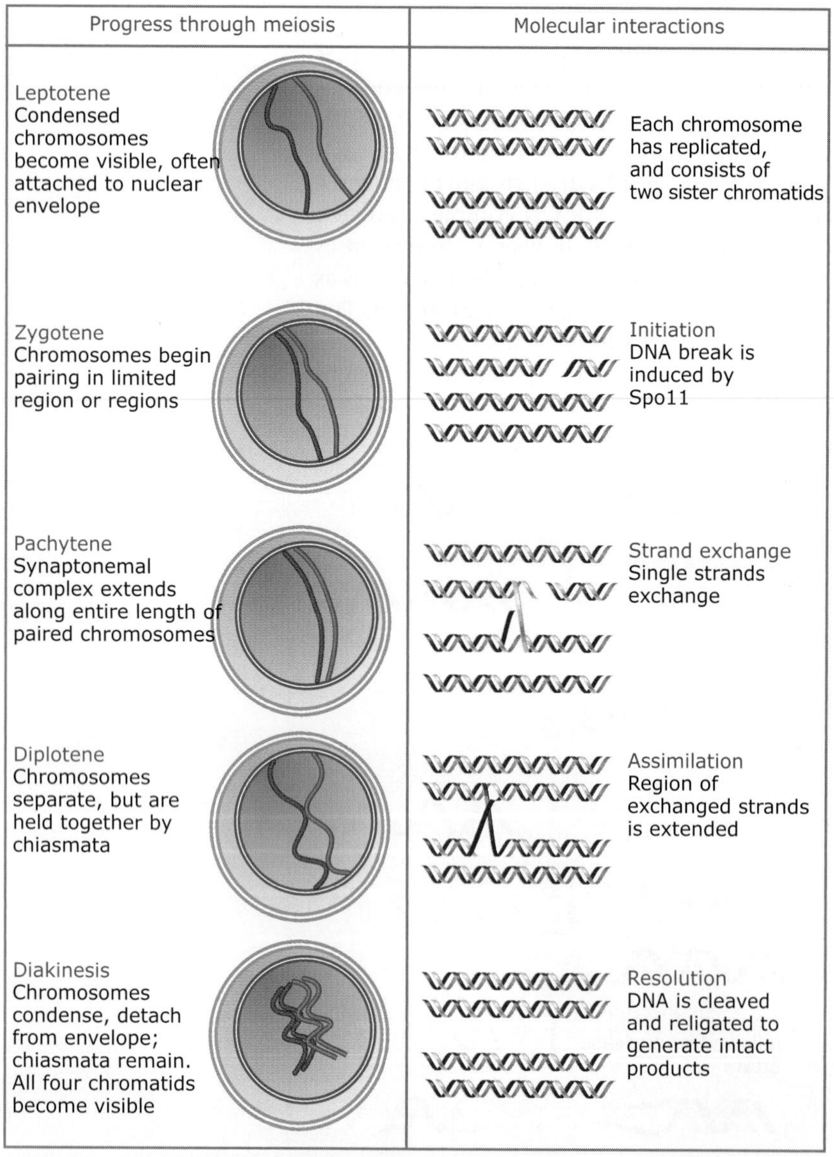

When the synapsis process is completed, the chromosomes are laterally associated in the form of a **synaptonemal complex**, which has a characteristic structure in each species, although there is wide variation in the details between species. **FIGURE 3.52** shows an example. During synaptonemal complex formation, each chromosome (sister chromatid pair) condenses around a proteinaceous structure called the **axial element**. The axial elements of corresponding chromosomes then become aligned, and the synaptonemal complex forms as a tripartite structure, in which the axial elements, now called **lateral elements**, are separated from each other by a central element. Each chromosome at this stage appears as a mass of chromatin bounded by a lateral element. The two lateral elements are separated from each other by a fine, but dense, central element. The triplet of parallel dense strands lies in a single plane that curves and twists along its axis. Two groups of proteins play central roles in formation of these structures. First, the cohesins form a single linear axis for each pair of sister chromatids from which loops of chromatin extend, forming the lateral elements. (Cohesins belong to a general group of proteins involved in connecting sister chromatids so that they segregate properly at mitosis or meiosis.) Second, the Zip proteins form the central element, creating transverse filaments that connect the lateral elements.

The distance between the synapsed homologous chromosomes is greater than 200 nm, which is considerable in molecular terms (the diameter of DNA is 2 nm). Thus, a major problem in understanding the role of the synaptonemal complex is that although it aligns homologous chromosomes, it is far from bringing homologous DNA molecules into contact.

As described for repair via homologous recombination, an intermediate joint molecule is formed in which there is a connection between the two DNA duplexes. The point at which an

Progress through meiosis	Molecular interactions
Leptotene Condensed chromosomes become visible, often attached to nuclear envelope	Each chromosome has replicated, and consists of two sister chromatids
Zygotene Chromosomes begin pairing in limited region or regions	Initiation DNA break is induced by Spo11
Pachytene Synaptonemal complex extends along entire length of paired chromosomes	Strand exchange Single strands exchange
Diplotene Chromosomes separate, but are held together by chiasmata	Assimilation Region of exchanged strands is extended
Diakinesis Chromosomes condense, detach from envelope; chiasmata remain. All four chromatids become visible	Resolution DNA is cleaved and religated to generate intact products

FIGURE 3.51 Recombination occurs during the first meiotic prophase. The stages of prophase are defined by the appearance of the chromosomes, each of which consists of two replicas (sister chromatids), although the duplicated state becomes visible only at the end. The molecular interactions of any individual crossing-over event involve two of the four duplex DNAs.

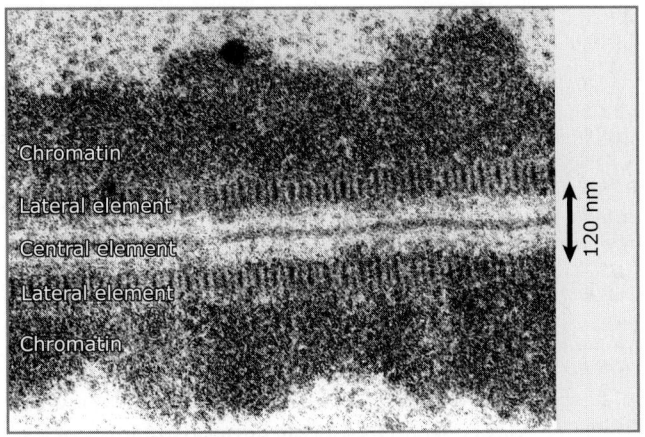

Chromatin
Lateral element
Central element
Lateral element
Chromatin

120 nm

FIGURE 3.52 The synaptonemal complex brings chromosomes into juxtaposition. Each pair of sister chromatids has an axis made of cohesins. Loops of chromatin project from the axis. The synaptonemal complex is formed by linking together the axes via zip proteins. Photo reproduced from D. von Wettstein, *Proc. Natl. Acad. Sci. USA* 68 (1971): 851–855. Photo courtesy of Diter von Wettstein, Washington State University.

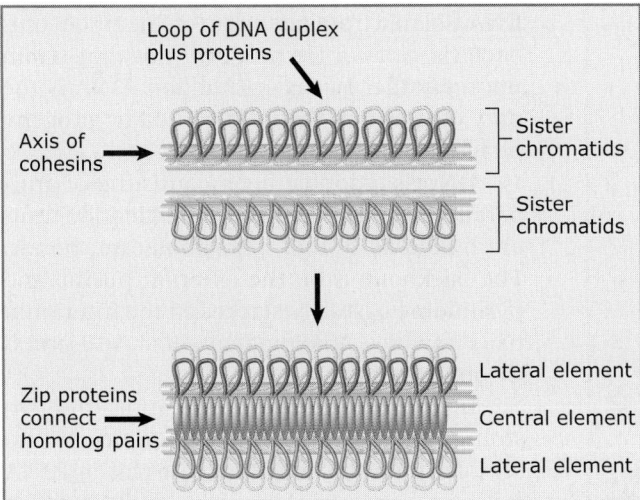

Loop of DNA duplex plus proteins

Axis of cohesins

Sister chromatids

Sister chromatids

Zip proteins connect homolog pairs

Lateral element

Central element

Lateral element

individual strand of DNA crosses from one duplex to the other is called the recombinant joint. An important feature of a recombinant joint is its ability to move along the duplex. Such mobility is called **branch migration**. **FIGURE 3.53** illustrates the migration of a single strand in a duplex. The branching point can migrate in either direction as one strand is displaced by the other. Strand invasion and branch migration thus result in the structure in Figure 3.48, a molecule with two recombinant joints, which can be at varying distances from the original breakpoint.

These crossovers formed during recombination must be *resolved* to restore two separate duplex molecules. Resolution requires nicking of the DNA backbones. We can most easily visualize this process by viewing the joint molecule in one plane as a Holliday junction. This is illustrated in **FIGURE 3.54**, which depicts a molecule with a single recombinant joint (for simplicity) with one duplex rotated relative to

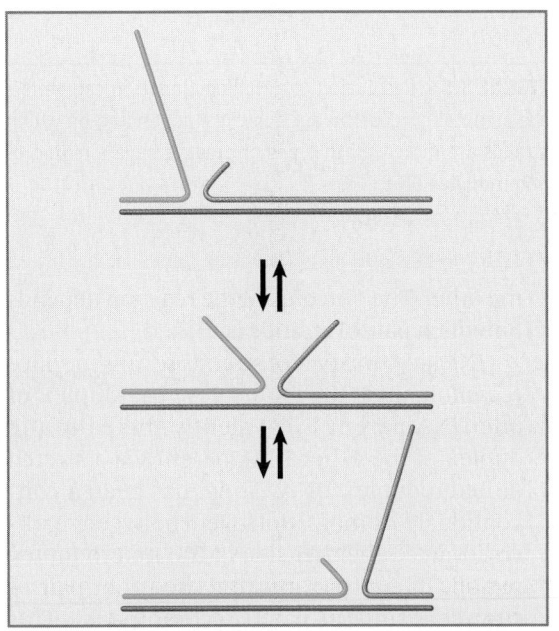

FIGURE 3.53 Branch migration can occur in either direction when an unpaired single strand displaces a paired strand.

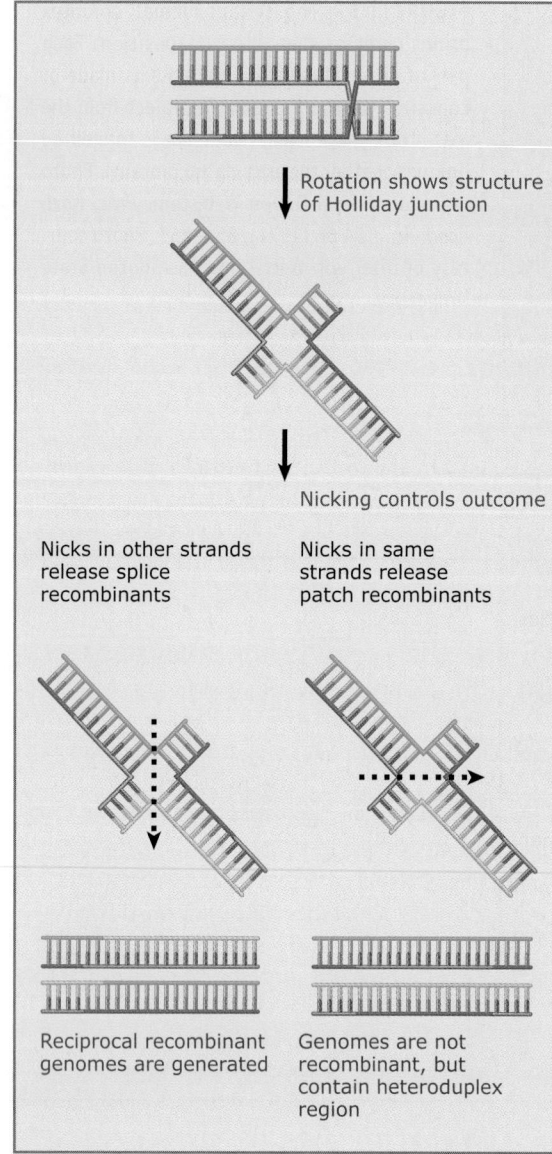

Rotation shows structure of Holliday junction

Nicking controls outcome

Nicks in other strands release splice recombinants

Nicks in same strands release patch recombinants

Reciprocal recombinant genomes are generated

Genomes are not recombinant, but contain heteroduplex region

FIGURE 3.54 Resolution of a Holliday junction can generate parental or recombinant duplexes, depending on which strands are nicked. Both types of product have a region of heteroduplex DNA.

the other. The outcome of the reaction depends on which pair of strands is nicked.

Nicking one pair of strands results in splice recombinant DNA molecules. The duplex of one DNA parent is covalently linked to the duplex of the other DNA parent via a stretch of heteroduplex DNA. There has been a conventional recombination event between markers located on either side of the heteroduplex region. In contrast, nicking the other pair of strands results in patch recombinants. This nicking releases the original parental duplexes, which remain intact with the exception that

each contains a length of heteroduplex DNA. For a molecule with two recombinant joints, if both joints are resolved in the same way, the original noncrossover molecules will be released, each with a region of altered genetic information that is a footprint of the exchange event. If the two joints are resolved in opposite ways, a genetic crossover is produced.

3.19 Summary

Two classic experiments provided strong evidence that DNA is the genetic material of bacteria and eukaryotic cells (and many viruses). DNA isolated from one strain of Pneumococcus bacteria can confer properties of that strain upon another strain. In addition, DNA is the only component that is inherited by progeny phages from parental phages.

DNA is a double helix consisting of antiparallel strands in which the nucleotide units are linked by 5′ to 3′ phosphodiester bonds. The backbone is on the exterior; purine and pyrimidine bases are stacked in the interior in pairs in which A is complementary to T and G is complementary to C.

In semiconservative replication, the two strands separate and daughter strands are assembled by complementary base pairing. DNA synthesis occurs semidiscontinuously, in which the leading strand of DNA growing 5′–3′ is extended continuously, but the lagging strand that grows overall in the opposite 3′–5′ direction is made as short Okazaki fragments, each synthesized 5′–3′. The leading strand and each Okazaki fragment of the lagging strand initiate with an RNA primer, synthesized by primase, which is extended by DNA polymerase. Prokaryotes and eukaryotes each possess more than one DNA polymerase activity. DNA polymerase III synthesizes both lagging and leading strands in *E. coli*, while DNA polymerases δ and ε are the eukaryotic replicative enzymes.

Replication is performed by dimers of DNA polymerase complexes; each new DNA strand is synthesized by a different polymerase complex containing a catalytic subunit. Processivity of the polymerase is maintained by a sliding clamp (β ring in prokaryotes, PCNA in eukaryotes), which forms a ring around DNA. The clamp is loaded onto DNA by the clamp loader complex.

The looping model for the replication fork proposes that, as one-half of the dimer advances to synthesize the leading strand, the other half of the dimer pulls DNA through as a single loop that provides the template for the lagging strand. The transition from completion of one Okazaki fragment to the start of the next requires the lagging strand catalytic subunit to dissociate from DNA and then reattach to a sliding clamp at the priming site for the next Okazaki fragment.

Replication fork movement requires the action of a helicase (DnaB in prokaryotes, MCM helicase in eukaryotes) to unwind the DNA, and single-strand binding proteins (SSB in prokaryotes, RFA in eukaryotes) to maintain the unwound region. Topoisomerases are necessary to relieve positive supercoiling (overwinding) ahead of the replication fork.

Origin activation involves an initial limited melting of the double helix, followed by more general unwinding to create single strands. Several proteins act sequentially at the *E. coli* origin. Replication is initiated at *oriC* in *E. coli* when DnaA binds to a series of repeats, where it uses hydrolysis of ATP to generate the energy to separate the DNA strands. The prepriming complex of DnaC–DnaB displaces DnaA. DnaC is released in a reaction that depends on ATP hydrolysis; DnaB is joined by the replicase enzyme, and replication is initiated by two forks that set out in opposite directions. Several sites that are methylated by the Dam methylase are present in the *E. coli* origin, including some binding sites for DnaA. The origin becomes hemimethylated and is in a sequestered state for ~10 min following initiation of a replication cycle. During this period, reinitiation of replication is repressed.

Eukaryotic origins are recognized by the ORC, which binds to the origin and recruits Cdc6, Cdt1, and the MCM helicase. Cell cycle-dependent signals result in phosphorylation of many of these proteins, leading to Cdc6 displacement from the origin and activation of the helicase to initiation replication fork unwinding. Cdc6 is either degraded or exported from the nucleus, so that it is not available to rebind the origin until G1 of the next cell cycle. This is a component of replication licensing, which prevents origins from firing more than once in a given cell cycle.

Telomeres protect chromosome ends. Almost all known telomeres consist of multiple short repeats in which one strand (which is G-rich) has a protruding end that provides a template for addition of individual bases in defined order. The enzyme telomerase is a ribonucleoprotein whose RNA component provides the template for synthesizing the G-rich strand. This overcomes the problem of the inability to replicate at the very end of a duplex. The telomere stabilizes the chromosome end because the overhanging single strand displaces sequences in earlier repeating units in the telomere to form a loop, so there are no free ends.

All cells contain systems that maintain the integrity of their DNA in the face of damage or errors of replication. DNA can be damaged in many ways, including by radiation (UV light or ionizing radiation such as X-rays), or chemical assaults that result in oxidation or alkylation. DNA is also spontaneously damaged in the normal cellular environment. Repair systems can recognize mispaired, altered, or missing bases in DNA, as well as other structural distortions of the double helix. Excision repair systems cleave DNA near a site of damage, remove one strand, and synthesize a new sequence to replace the excised material. Direct repair systems (such as photolyase for UV damage, or alkyltransferases for alkylation damage) directly reverse DNA damage to restore the original state.

The prokaryotic *mut* and *dam* systems are involved in correcting mismatches generated by incorporation of incorrect bases during replication and function by preferentially removing the base on the strand of DNA that is not methylated at a dam target sequence. Eukaryotic homologs of the *E. coli mut* system are involved in repairing mismatches as well as insertion deletion loops that result from replication slippage; mutations in this pathway are common in certain types of cancer.

Excision repair systems cleave DNA near a site of damage, remove one strand, and synthesize a new sequence to replace the excised material. Base excision repair recognizes damaged bases and abasic sites, and removes damaged bases via glycosylase (or glycosylase/lyase) action. AP endonuclease then cleaves the phosphate backbone to create a substrate for repair synthesis and ligation.

NER recognizes bulky adducts in DNA. The *Uvr* system provides the main excision-repair pathway in *E. coli*. In eukaryotes, NER is divided into two pathways, global genome repair and transcription-coupled repair. Both pathways overlap significantly and depend on products of the XP genes; in GG-NER XPC and other proteins recognize DNA lesions, while in TC-NER the damage is recognized by an elongating RNA polymerase. Mutations in XP genes in humans lead to the disease XP.

Double-strand DNA breaks are a serious type of DNA damage that is repaired by two general systems: homologous recombination and NHEJ. HR uses a homologous sequence (such as a sister chromatid) to accurately repair the damaged sequence. HR involves MRX/N-dependent resection of one strand at each broken end to create a 3'-OH overhang, which is assembled into a Rad51 filament and invades a homologous duplex. The 3'-end then provides a primer for repair synthesis. NHEJ entails direct ligation of DNA ends and depends on the Ku heterodimer.

Meiotic recombination uses the same machinery as that used for double-strand break repair via HR. Meiotic recombination is initiated by a Spo11-dependent double-strand break. Recombination occurs in the synaptonemal complex, in which homologous chromosomes are paired (synapsed) in a structure that requires cohesin and zip proteins. Resolution of Holliday structures during meiotic recombination determines whether the products are reciprocal splice recombinants, or patch recombinants in which the original parental duplexes are intact except for a heteroduplex region.

References

Bergoglio, V., and Magnaldo, T. 2006. Nucleotide excision repair and related human diseases. *Genome Dyn.* v. 1 p. 35–52.

Cromie, G. A., and Smith, G. R. 2007. Branching out: Meiotic recombination and its regulation. *Trends Cell Biol.* v. 17 p. 448–455.

Hsieh, P., and Yamane, K. 2008. DNA mismatch repair: Molecular mechanism, cancer, and ageing. *Mech. Ageing Dev.* v. 129 p. 391–407.

Johnson, A., and O'Donnell, M. 2005. Cellular DNA replicases: Components and dynamics at the replication fork. *Annu. Rev. Biochem.* v. 74 p. 283–315.

Li, X., and Heyer, W. D. 2008. Homologous recombination in DNA repair and DNA damage tolerance. *Cell Res.* v. 18 p. 99–113.

Patel, S. S., and Picha, K. M. 2000. Structure and function of hexameric helicases. *Annu. Rev. Biochem.* v. 69 p. 651–697.

San Filippo, J., Sung, P., and Klein, H. 2008. Mechanism of eukaryotic homologous recombination. *Annu. Rev. Biochem.* v. 77 p. 229–257.

http://biology.jbpub.com/lewin/cells

To explore these topics in more detail, visit this book's Interactive Student Study Guide.

Gene expression and regulation

4

David G. Bear

Department of Cell Biology and Physiology,
University of New Mexico School of Medicine,
Albuquerque, NM

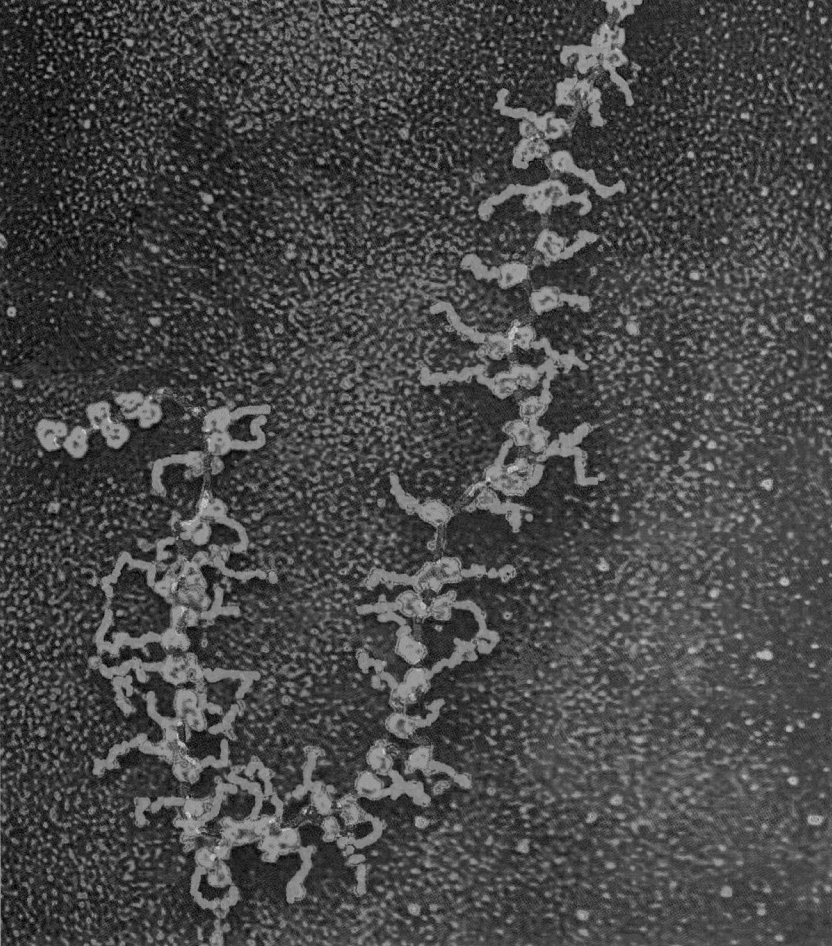

TRANSLATION OF A MESSENGER RNA (mRNA) isolated from a cell in the salivary gland of an insect *Chironomous tentanis* in this digitally colored transmission electron micrograph. Translation is the process by which ribosomes (blue particles) translocate along the mRNA (pink strand) and read its sequence in three nucleotide units called codons. As the ribosome moves to each codon on the mRNA, it inserts the appropriate amino acid into the growing protein chain (green filaments). Photo © Dr. Elena Kiseleva/SPL/ Photo Researchers, Inc.

CHAPTER OUTLINE

4.1 Introduction

Key concepts

- DNA functions as the template for the synthesis of RNA by RNA polymerase (transcription). Messenger RNA (mRNA) directs the synthesis of proteins by the ribosome (translation).
- The genetic code refers to the set of 64 base triplets (codons) that are read by the ribosome during translation.
- Each of the 61 codons codes for one of the 20 common amino acids while three of the codons cause the ribosome to stop translation
- In prokaryotes, transcription and translation of mRNA occur concomitantly in the same region of the cell
- In eukaryotes, transcription and mRNA processing occur concomitantly in the nucleus and translation occurs subsequent to the export of the mRNA to the cytoplasm.

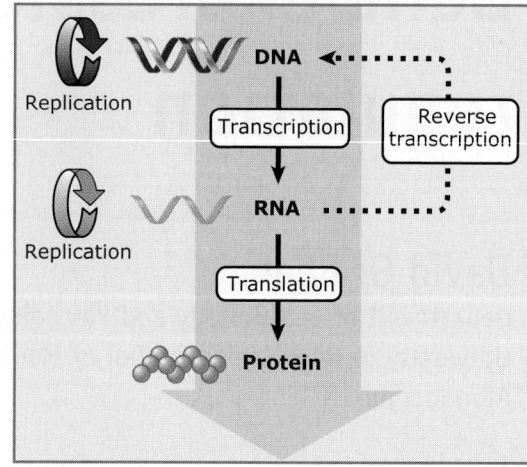

FIGURE 4.1 The Central Dogma of molecular biology. The Central Dogma states that the DNA serves as the template for the synthesis of RNA by the enzyme RNA polymerase (transcription), and the RNA transcript is decoded by the ribosome to synthesize a polypeptide (translation). The enzyme reverse transcriptase can synthesize DNA from RNA. Reverse transcription is a rare event in normal cells. Many RNA viruses contain a reverse-transcriptase gene in their genome.

Modern biology began in the early part of the 20th century with the direct examination of the structure and function of cells, first by improved microscopy techniques that permitted the detailed study of cell structure, and then by high-speed centrifugation methods that facilitated biochemical isolation and characterization of cytoplasmic and nuclear subcellular fractions. Geneticists focused on the relationship between the dynamic structure of chromosomes within the cell nucleus and mechanisms of genetic inheritance; this effort culminated with the elucidation of the structures of DNA and RNA and an understanding of the genetic code. The scientific revolution of molecular biology that took place in the mid-20th century provided a framework known as the **Central Dogma of Molecular Biology**: DNA functions as the template for the synthesis of RNA (**transcription**) and RNA directs the synthesis of protein by ribosomes (**translation**) as depicted in **FIGURE 4.1**.

Initially, molecular biology focused on reconciling the classical concept of the **gene**—the unit of information controlling inherited traits—with the more modern tenets of the Central Dogma. These studies led to the **One-Gene-One-Protein-Hypothesis** that posited a gene to be the unit of genetic information that directs the synthesis of a specific **messenger RNA (mRNA)**, which in turn encodes an individual protein. Much of the early work on the Central Dogma was carried out using a nonpathogenic strain of the bacterium *Escherichia coli* (*E. coli*) because of the ease of genetic manipulation and the ability to easily fraction-

ate the cellular components for biochemical analysis. Studies with *E. coli* provided our first understanding of the mechanisms of transcription and translation, with details of the structural and functional properties of **RNA polymerases** that transcribe the genome and **ribosomes** that translate mRNA to make protein. This work also revealed that transcription and translation are coupled in bacteria—the mRNA is translated concomitantly as the newly synthesized RNA emerges from the active site of RNA polymerase (**FIGURE 4.2**).

In the latter half of the 20th century, investigations of gene expression in several eukaryotic model systems, including the yeast *Saccharomyces cerevisiae*, the insect *Drosophila melanogaster*, and the mammalian HeLa cultured cell line, showed that many of the fundamental mechanisms of gene expression are conserved among all organisms; however, there are some basic differences between prokaryotes and eukaryotes, particularly with regard to the posttranscriptional pathways of mRNA biosynthesis (**FIGURE 4.3**).

One of the most surprising discoveries was that most protein-coding genes in higher eukaryotes are interrupted with DNA segments termed **introns** that are transcribed

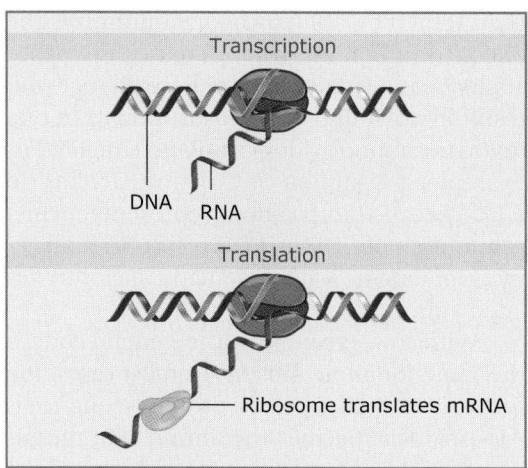

FIGURE 4.2 Gene expression in prokaryotes. Prokaryotes do not have a cell nucleus. Thus, transcription and translation take place in the same cellular compartment and occur concomitantly.

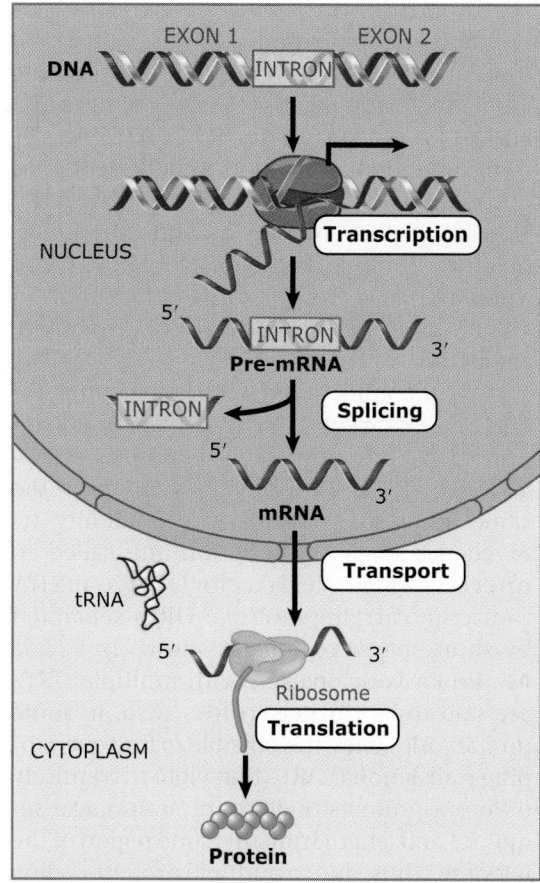

FIGURE 4.3 Gene expression in eukaryotes. Eukaryotes contain a cell nucleus where transcription and RNA processing occur. The mRNA transcript is exported from the nucleus to the cytoplasm where translation takes place.

into RNA, but are removed prior to translation. **Ribonucleoprotein (RNP)** complexes referred to as **spliceosomes** were found to excise the introns from the original mRNA, or **pre-mRNA**, and to ligate the remaining expressed mRNA sequences, termed **exons**, together to form the **mature mRNA**. In addition to splicing, eukaryotic mRNAs undergo other posttranscription modifications including the addition of a **7-methylguanosine (7-MeG)** cap to the 5' end and a **poly (A) tail** comprising 100–300 adenosine ribonucleotides on the 3' end. Selected mRNAs may also have additional bases added posttranscriptionally in a process referred to as **mRNA editing**. Analogous to transcription-translation coupling in prokaryotes, almost all of the eukaryotic mRNA processing reactions are coupled to transcription.

Some of the greatest similarities, as well as some of the biggest differences, between prokaryotic and eukaryotic cells are at the final step of gene expression—translation. Prokaryotes lack a cell nucleus and the translation complex (ribosomes and associated factors) is in direct contact with the transcription apparatus on the chromosomal DNA, while in eukaryotes, the transcription and RNA processing reactions occur in the nucleus and translation occurs in the cytoplasm. This separation of RNA biosynthesis from translation necessitates the packaging of mRNA into an mRNA RNP (mRNP) complex that then must be exported through the pores of the nuclear envelope into the cytoplasm prior to translation. The details of this nucleocytoplasmic export process are presented in *9 Nuclear structure and transport*. In spite of the extra steps between transcription and translation in eukaryotes, the prokaryotic and eukaryotic translational machineries are very similar. In fact, some of the highest conservation of macromolecular structure and function among species is found in the ribosomes and associated translational factors.

The information contained in a mature mRNA for making a protein is decoded by the ribosome. The ribosome reads the mRNA sequence in three-nucleotide units that are referred to as **codons**. Because there are four different nucleotide bases in DNA and RNA—A, G, C, and T (U in RNA)—there are 64 possible codons (**FIGURE 4.4**). Each codon specifies the insertion by the ribosome of a specific amino acid into the protein. There are 20 amino acids that are commonly found in proteins; thus,

First Position (5'-end)	Second Position				Third Position (3'-end)
	U	C	A	G	
U	UUC Phe	UCU Ser	UAU Tyr	UGU Cys	U
	UCC Phe	UCC Ser	UAC Tyr	UGC Cys	C
	UUA Leu	UCA Ser	UAA Stop	UGA Stop	A
	UUG Leu	UCG Ser	UAG Stop	UGG Trp	G
C	CUU Leu	CCU Pro	CAU His	CGU Arg	U
	CUC Leu	CCC Pro	CAC His	CGC Arg	C
	CUA Leu	CCA Pro	CAA Gln	CGA Arg	A
	CUG Leu	CCG Pro	CAG Gln	CGG Arg	G
A	AUU IIe	ACU Thr	AAU Asn	AGU Ser	U
	AUC IIe	ACC Thr	AAC Asn	AGC Ser	C
	AUA IIe	ACA Thr	AAA Lys	AGA Arg	A
	AUG Met Start	ACG Thr	AAG Lys	AGG Arg	G
G	GUU Val	GCU Ala	GAU Asp	GGU Gly	U
	GUC Val	GCC Ala	GAC Asp	GGC Gly	C
	GUA Val	GCA Ala	GAA Glu	GGA Gly	A
	GUG Val	GCG Ala	GAG Glu	GGG Gly	G
Start Codon					
Stop Codon					
Nonpolar Side Chain					
Uncharged Polar Side Chain					
Charged Polar Side Chain					

FIGURE 4.4 The genetic code. The genetic code is nearly universal among all species. The table shows the 64 codons, 61 of which specify one of the 20 naturally occurring amino acids commonly found in proteins, while three codons do not code for an amino acid, but serve as signals for the ribosome to end translation (stop codons). The codon AUG codes for methionine and also serves as the initiation codon.

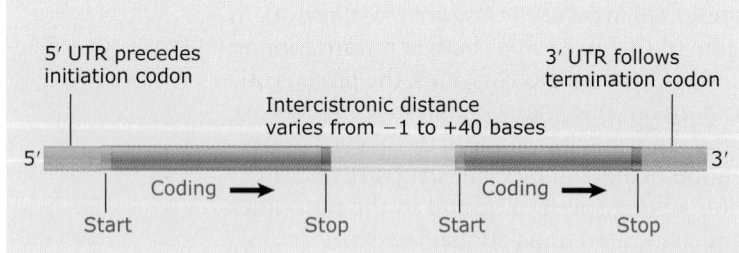

FIGURE 4.5 mRNA structure in prokaryotes. Most prokaryotic mRNAs contain several tandem open reading frames coding for proteins and are said to be "polycistronic." The distance between the stop codon of one coding region and the start codon of the next coding region can vary from a one-base overlap (−1) up to about 40 bases.

some amino acids are specified by more than one codon. In general, codons that encode the same amino acid differ by only one base. The tendency for similar amino acids to be represented by related codons minimizes the effects of mutations and increases the probability that a single random base change will result in no amino acid substitution or in one involving amino acids of similar character. For example, a mutation of CUC to CUG has no effect, because both codons represent leucine. A mutation of CUU to AUU results in replacement of leucine with isoleucine, a closely related amino acid.

With few exceptions, the genetic code is the same for all organisms. In most cases, the first codon that specifies a protein sequence is AUG, which specifies the amino acid methionine, and is referred to as the **start codon**. Three of the codons (UAA, UAG, and UGA) do not code for any amino acid, but instead cause the ribosome to terminate translation; each of these is referred to as a **stop codon**. The ribosome translocates along the mRNA in a 5' to 3' direction reading the codons in frame. The region between the start codon, AUG, and one of the stop codons is referred to as the **open reading frame** or **ORF**, and is generally synonymous with a peptide-coding region. The translation process carried out by the ribosome entails facilitating the base pairing of the **anticodon loop** of a **transfer RNA (tRNA)** carrying an amino acid with the codon sequence in the mRNA, followed by the transfer of the growing peptide chain located on the previous tRNA that docked with the mRNA onto the amino acid from the incoming tRNA, in what is known as the **peptidyl transferase reaction**.

An important difference between prokaryotic and eukaryotic mRNAs is the organization of the protein-coding sequences. Many genes in bacteria that code for proteins in the same functional pathway are tandemly arrayed as a single transcription unit called an **operon**, which yields a single large mRNA transcript carrying multiple ORFs separated by short spacer regions as shown in **FIGURE 4.5**. Prokaryotic operons with multiple ORFs are said to be **polycistronic**. Also, in some prokaryotic genes, it is possible to find overlapping translational ORFs that yield two entirely different proteins of different amino acid sequence and length from the same region of the genome. Thus, due to multiple protein-coding sequences residing in operons and overlapping ORFs, the number of proteins encoded by relatively compact prokaryotic genomes can be quite large. By contrast, most eukaryotic mRNAs do not contain more than one primary start and stop codon, and are said to be **mono-cistronic** (**FIGURE 4.6**). However, eukaryotic mRNAs are often alternatively spliced so that there can be more than one protein made from the same transcript.

In addition to the protein-coding region, most mRNAs in both prokaryotes and eukaryotes contain sequences located before (upstream) and after (downstream) the protein-coding regions, referred to as the **5′ untranslated** and **3′ untranslated regions** (**UTR**s), respectively. 5′ UTRs often encode signals that regulate the translation initiation efficiency of the start codon and 3′ UTRs often contain sequence elements that regulate translational initiation efficiency as well as mRNA stability. The ability of 3′ UTR sequences to affect translation initiation suggests that proteins bound at the 5′ and 3′ ends of the mRNA mediate the formation of circularized mRNA. Also within the 5′ and 3′ UTRs are specific sites that are required for translational repression and RNA stabilization; the sites function as binding sites for specific RNA binding proteins and for small noncoding RNAs called **microRNAs** (**miRNAs**), which will be discussed later on.

The goal of this chapter is to present an overview of the basic molecular mechanisms of how genes are expressed and to provide a brief description of the strategies that the cell uses for regulation of gene expression. Other chapters in the book will describe how cellular structures organize and compartmentalize gene expression (*7 Membrane targeting of proteins* focuses on cytoplasmic structure and organization in protein synthesis; *9 Nuclear structure and transport* discusses the spatiotemporal organization of genes and their expression inside the nucleus, and the nucleocytoplasmic trafficking of proteins and RNAs; *10 Chromatin and chromosomes* provides an extensive overview of the importance of chromosome structural dynamics in the control of transcription; and *18 Principles of cell signaling* explores the response of gene expression to cellular signaling). These chapters underscore the concept that gene expression behaves as an integrated system, which is self-organized into cellular membrane-bound compartments and nonmembranous subcompartments. This dynamic organization facilitates the regulation of gene expression during cell growth, proliferation, differentiation, and development, and in response to nutritional and physical stress or changes in the extracellular environment.

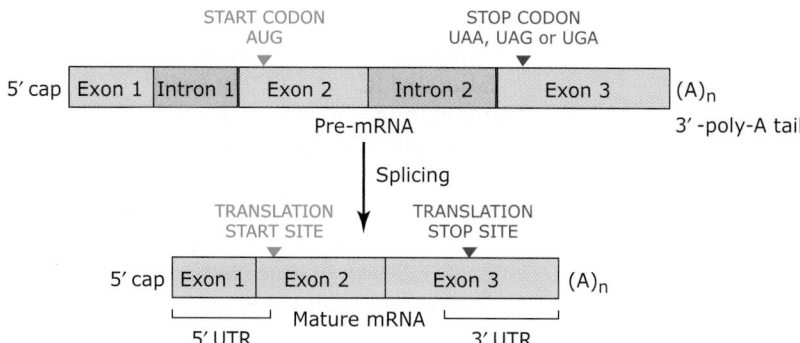

FIGURE 4.6 mRNA structure in eukaryotes. Eukaryotic mRNAs must be processed from the primary transcript (pre-mRNA) to form the mature mRNA. Processing includes removal of the introns (splicing) as well as the addition of the 7-methylguanosine (7-MeG) cap and the poly(A) tail. In general, once a mature mRNA is formed, only a single protein is produced. The AUG start codon and the stop codon can be located in any of the exons.

Concept and Reasoning Check

1. List the steps in the expression of a protein-encoding gene in a eukaryotic cell and indicate the location within the cell that each of the steps takes place. What could be the advantage of performing these steps in separate cellular locations?

4.2 Genes are transcription units

Key concepts

- The number of genes within a genome is generally a function of the complexity of both the structure and life cycle of an organism.
- Eukaryotic cells have large nuclear genomes as well as smaller genomes associated with mitochondria and chloroplast cytoplasmic organelles.
- Most RNA transcription within eukaryotic cells generates RNA that does not encode proteins but that has other important functions. Thus, the most general definition of a gene is a segment of the genome that encodes a functional transcript.

Over the past two decades, the sequencing and functional analysis of prokaryotic and eukaryotic genomes, known as **genomics**, has led to an evolving definition of the term "gene." In this section, we will focus on both the structural and functional definitions of a gene and how genes are organized within the genome.

A surprising finding from genomic studies is that the exact number of genes in various organisms is often difficult to determine from the DNA sequence alone in the absence of information about the RNA transcripts and protein products. In prokaryotic organisms such as mycoplasma and bacteria, the genomes are quite compact, ranging in size from 0.6 to 7 Mb with ~500 to ~8,000 genes. By comparison, the number of genes in eukaryotic genomes is estimated to be ~6,000 in yeasts, ~25,000 in humans, and ~30,000 in rice (FIGURE 4.7). The exact number of proteins in higher eukaryotes is still unknown, but is estimated to be 3 to 10 times more than the number of genes. In higher eukaryotes, most protein-coding genes encode mRNAs that are alternatively spliced to give rise to multiple protein species from a single gene. In some cases, the proteins derived from alternative splicing are related in structure and function, while in other cases, the proteins have significantly different sequences and functions. In addition, eukaryotic genes can give rise to more than one protein due to alternative transcriptional start sites or alternative translational start and stop sites. Thus, the size of a genome and the number of proteins encoded by the genome are not always correlated.

In eukaryotic organisms the majority of the genome is within the chromosomes located inside the nucleus. However, the cytoplasmic mitochondria and chloroplasts contain circular double-stranded DNAs, which are referred to as the **mitochondrial and chloroplast genomes**. Mitochondrial genomes vary in size from 80 to 100 kb in fungi and plants to less than 20 kb in mammals. The human mitochondrial genome is 16.6 kb and encodes 37 genes (FIGURE 4.8). Some mitochondrial genes in fungi and plants contain introns, while the much more compact mammalian mitochondrial genes do no have any introns and can overlap, similar to the organizations of genes in bacteria. Not all of the proteins required for mitochondrial function are contained in the mitochondrial genome; in fact, most mitochondrial proteins are encoded in the nuclear genome and synthesized in the cytoplasm, and then imported into the mitochondria. By contrast, chloroplast genomes are relatively large compared to mitochondrial genomes, with sizes of about 140 to 200 kb, and may be present in 20 to 40 copies with over 100 to 120 genes. The organelle genomes provide an additional number of proteins and RNA that are related to their compartmentalization within the cytoplasm.

Our concept of the gene has changed significantly with the recent discovery that most cellular RNAs do not encode proteins; in fact, much of the transcription that takes place in the cell is geared toward the production of **functional noncoding RNAs (ncRNAs)**.

Species	Genomes (Mb)	Genes	Lethal loci
Mycoplasma genitalium	0.58	470	~300
Rickettsia prowazekii	1.11	834	
Haemophilus influenzae	1.83	1,743	
Methanococcus jannaschi	1.66	1,738	
B. subtilis	4.2	4,100	
E. coli	4.6	4,288	1,800
S. cerevisiae	13.5	6,034	1,090
S. pombe	12.5	4,929	
A. thaliana	119	25,498	
O. sativa (rice)	466	~30,000	
D. melanogaster	165	13,601	3,100
C. elegans	97	18,424	
H. sapiens	3,300	~25,000	

FIGURE 4.7 Genomes and gene numbers. The genome sizes and gene numbers have been determined for many common organisms. A number of "lethal loci," which correspond to essential genes, have been determined for several genetically tractable organisms.

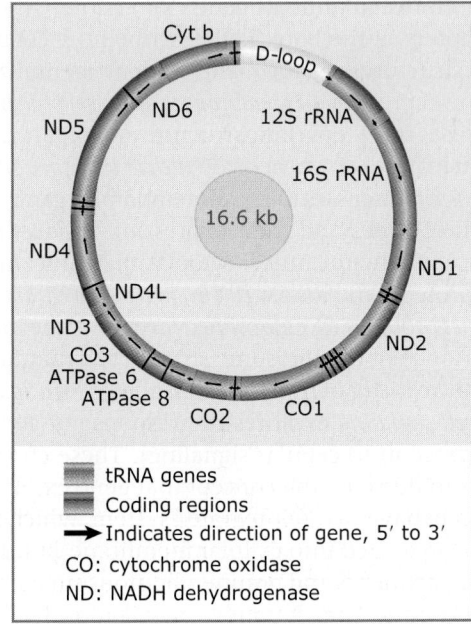

tRNA genes
Coding regions
→ Indicates direction of gene, 5′ to 3′
CO: cytochrome oxidase
ND: NADH dehydrogenase

FIGURE 4.8 The mitochondrial genome. The human mitochondrial genome encodes 37 genes. However, the vast majority of mitochondrial proteins are encoded by nuclear genes and imported into the mitochondria.

ncRNA Class	Function(s)
ncRNAs Involved in Translation	
Ribosomal RNA (rRNA)	Structural and functional components of the ribosome; some rRNAs have catalytic activities
Transfer RNA (tRNA)	Covalent attachment with amino acids for interaction with ribosome and mRNA during protein synthesis
7SL RNA	Co-translational translocation of secreted proteins into the endoplasmic reticulum
ncRNAs Involved in Nuclear RNA Processing	
Small nuclear RNA (snRNA)	RNA Processing—mRNA splicing and rRNA processing
Small nucleolar RNA (snoRNAs)	RNA Processing—Processing of ribosomal RNA; guidance of covalent modification of rRNA, tRNA, mRNA, and snRNA
ncRNAs Involved in Regulation of Gene Expression	
Large noncoding RNA	Suppression of transcription of specific gene loci on chromosomes
Small interfering RNA (siRNA)	Gene-specific mRNA degradation & transcriptional regulation
MicroRNA (miRNA)	Gene-specific mRNA degradation & translational repression/translational activation
Short hairpin RNA (shRNA)	mRNA degradation
Piwi-Interacting RNA (piRNA)	Transposon silencing
Other ncRNAs	
Vault RNA	Structural component of the cytoplasmic vault particle, which has been implicated in intracellular detoxification processes and intracellular transport

FIGURE 4.9 Noncoding RNAs. Most cellular RNAs do not code for proteins and are referred to as noncoding RNAs (ncRNAs).

Surprisingly, more than 70% of the human genome is transcribed, and much of this transcription results in ncRNAs that perform a wide range of functions (**FIGURE 4.9**). These include **ribosomal RNAs** (**rRNAs**) that become part of the ribosome, **tRNAs** that are involved in translation, **small nuclear RNAs** (**snRNAs**), **small nucleolar RNAs** (**snoRNAs**) that are part of the rRNA processing reactions, and various other RNAs that are involved in cellular structure and function. More recently, a large class of ncRNAs referred to as **miRNAs** has been identified. miRNAs regulate gene expression by base pairing to specific mRNAs, and inhibiting translation or stimulating degradation of mRNA. It is highly likely that discoveries of many new ncRNA functions will be made in the future.

In summary, the concept of the "gene" has evolved over the past century from the classical definition as a determinant of a genetic phenotype, to a protein-encoding segment of DNA, and finally to the more modern definition as the unit of information contained in the genome required for the transcription of a functional mRNA or ncRNA. In the next several sections, we will explore the process of transcription and its regulation in detail.

Concept and Reasoning Check

1. Why are the sizes of eukaryotic genomes significantly larger than the amount of DNA required to encode proteins and functional RNAs?

4.3 Transcription is a multistep process directed by DNA-dependent RNA polymerase

Key concepts

- Specific sequences referred to as "promoters" and "terminators" direct the location on the DNA template where RNA polymerase begins and ends transcription.
- The four phases of transcription are promoter recognition, initiation, elongation, and termination.
- During transcription, RNA polymerase opens duplex DNA and uses the template strand to direct the polymerization of ribonucleotides into a complementary RNA strand.

The first step in gene expression is the transcription of DNA to make RNA, which is carried out by the enzyme DNA-dependent RNA polymerase. RNA polymerases are found in all

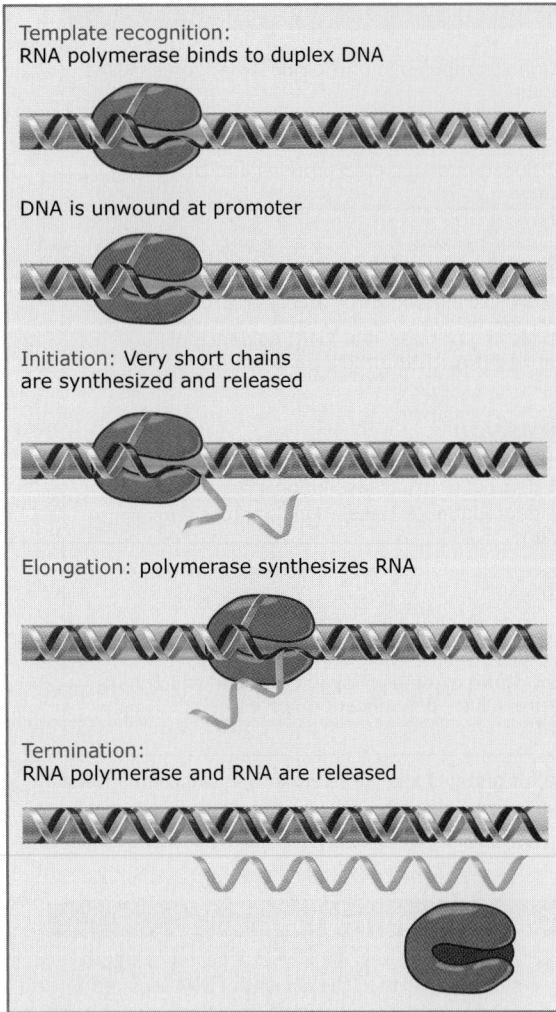

Template recognition:
RNA polymerase binds to duplex DNA

DNA is unwound at promoter

Initiation: Very short chains
are synthesized and released

Elongation: polymerase synthesizes RNA

Termination:
RNA polymerase and RNA are released

FIGURE 4.10 The transcription cycle. The transcription cycle has four phases.

cellular organisms and in mitochondria and chloroplasts. The transcription cycle can be separated into four different phases: promoter location, initiation, elongation, and termination as illustrated in **FIGURE 4.10**. They are as follows:

1. Promoter recognition—RNA polymerase initiates transcription at locations on the genome referred to as **promoters**. In prokaryotes, the promoter region is relatively small (~40 bp) and RNA polymerase can locate the promoter through direct recognition of the DNA sequence. In eukaryotes, the promoter regions are much larger (often more than 100 bp). A **preinitiation complex** (**PIC**), consisting of a large number of initiation factors, assembles onto the promoter. The eukaryotic nuclear RNA poly-merases recognize and bind to promoter regions through recognition of the PIC, rather than through direct recognition of a DNA sequence.

2. Initiation—To initiate RNA synthesis, RNA polymerase binds to the promoter and unwinds the DNA to produce a "bubble" of separated strands, exposing one of the strands to serve as the template for the RNA polymerization reaction. Unlike DNA polymerases, which initiate DNA synthesis using a short primer synthesized by a primase enzyme (see *3 DNA replication, repair, and recombination*), RNA polymerases do not require a primer to initiate polymerization of RNA from **ribonucleoside triphosphate** (**NTP**) precursors. Thus, because the first nucleotide in RNA synthesis is an NTP, there are three phosphate groups at the 5′ end of an RNA, while each phosphoester linkage in the internal sequence of the RNA contains a monophosphate and a ribose sugar. The template DNA strand directs the synthesis of a complementary strand of RNA in a 5′ to 3′ direction using NTPs as substrates. The 3′-OH group on the ribose at the growing 3′ end of the RNA is a nucleophile that attacks the α-phosphate group on the incoming NTP liberating the β-γ pyrophosphate group (**FIGURE 4.11**). The initiation phase of transcription can be dissected into the following steps: (1) binding of RNA polymerase to the core promoter region; (2) separation of the DNA strands by RNA polymerase to form the open complex—a process termed **isomerization**; and (3) RNA polymerase moving away from the promoter and down the DNA template—a process termed **promoter escape**. Initiation formally ends only when RNA polymerase succeeds in leaving the promoter region by extending the transcript to a length that more than fills its active site and moving the transcription bubble.

3. Elongation—The 12 to 14 bp transcription bubble of open DNA within the RNA polymerase is propagated inside RNA polymerase as it translocates along the DNA template (**FIGURE 4.12**). The transcription bubble remains es-

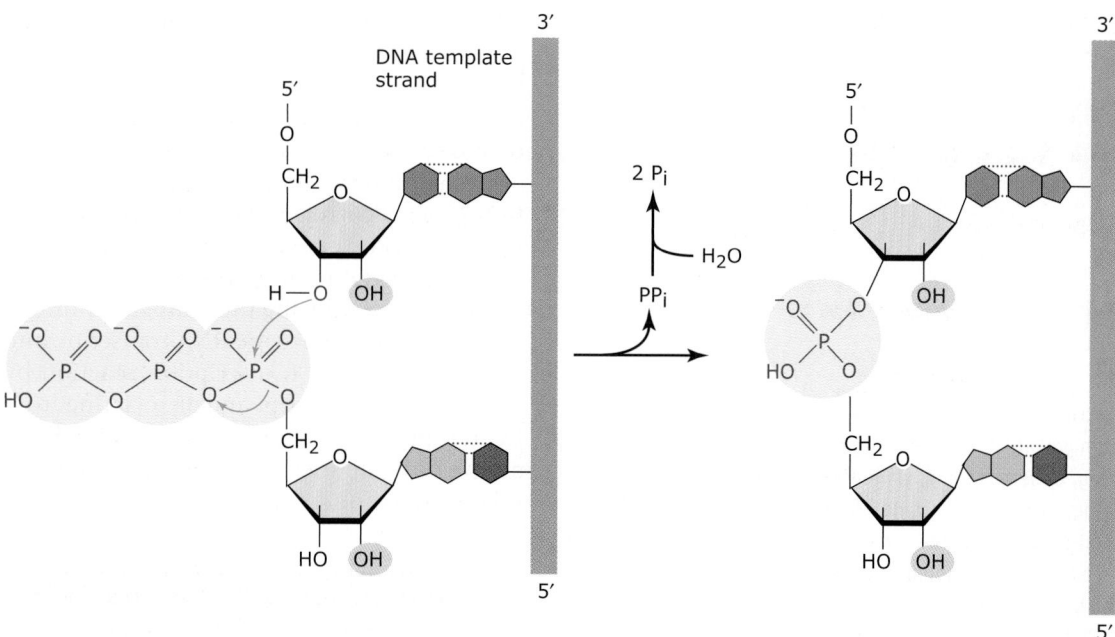

FIGURE 4.11 The mechanism of RNA synthesis. RNA polymerase catalyzes the template-directed polymerization of RNA in the 5′ to 3′ direction. The 3′ hydroxyl group on the 3′ end of the RNA carries out a nucleophilic attack of the α-phosphate on the nucleoside triphosphate monomer, liberating pyrophosphate (PP_i). Adapted from J. M. Berg, et al. *Biochemistry, Fifth edition*. W. H. Freeman and Company, 2002.

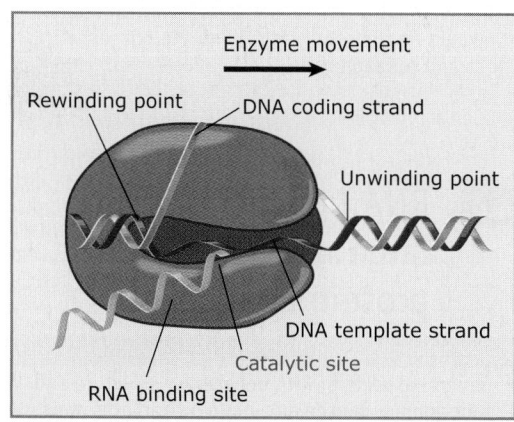

FIGURE 4.12 The active site of bacterial RNA polymerase. The catalytic pocket of RNA polymerase transiently unwinds a small region of the duplex DNA template creating a "bubble" that permits the template strand to direct transcription. As the polymerase moves along the DNA template, the region in front of the polymerase is unwound and the region behind the polymerase is rewound, thereby maintaining the bubble at a constant size of 10–14 base pairs.

sentially the same size throughout the elongation reaction: RNA polymerase unwinds the DNA in front of the bubble and rewinds the DNA at the back of the bubble, which creates a major change in the DNA topology dispersed over about one turn of the DNA helix. Within the bubble, 8 to 10 bases of the template strand are transiently paired with the transcript. During elongation, the enzyme moves along the DNA and extends the growing RNA chain at a rate that varies between 10 and 100 nucleotides per second. Transcription elongation does not occur at a constant rate; RNA polymerase can pause at specific sites for up to several minutes before resuming elongation. The nature of these pause sites is still under active investigation, but the purpose of the pausing is to facilitate the coordination of cotranslational processes such as translation (in prokaryotes) and RNA splicing (in eukaryotes).

4. Termination—Termination of transcription involves three steps: (1) the cessation of RNA polymerase movement; (2) the release of the transcript; and (3) the release of RNA polymerase. Eukaryotic transcription termination can occur hundreds or thousands of nucleotides from the translational stop codon. The signals and proteins

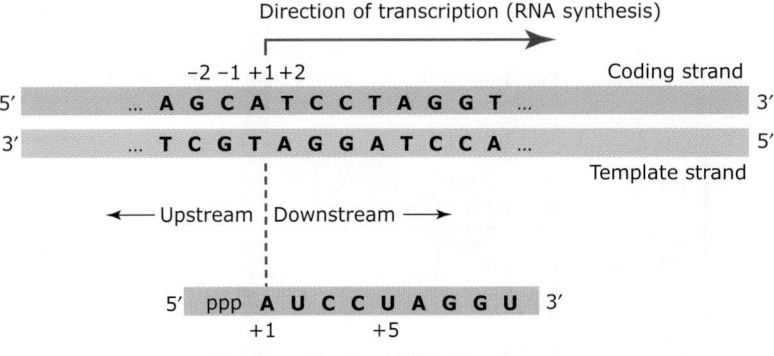

Direction of transcription (RNA synthesis)

−2 −1 +1 +2

Coding strand

5′ … A G C A T C C T A G G T … 3′

3′ … T C G T A G G A T C C A … 5′

Template strand

← Upstream ┆ Downstream →

5′ ppp A U C C U A G G U 3′

+1   +5

Newly synthesized RNA strand

FIGURE 4.13 The DNA transcription template. The conventions for describing transcription are the same for both prokaryotes and eukaryotes. The top strand is referred to as the nontemplate, coding, or sense strand, while the bottom strand is referred to as the template, noncoding, or nonsense strand. Because the RNA transcript is complementary to the template strand, it is the same sequence as the nontemplate or coding strand. In the RNA transcript, all monomer units are ribonucleoside monophosphates. In addition, the base thymine found in the DNA nontemplate strand is replaced with uracil in the RNA transcript. The direction in which the polymerase moves and transcribes the DNA template is referred to as downstream, while the opposite direction is referred to as upstream. The first base pair to be used as a template for the synthesis of the RNA is referred to as the transcription start site and is labeled as position +1. The next nucleotide is labeled as position +2, etc. The position immediately upstream of +1 is labeled as −1.

required for transcription termination in mammals are still being identified and characterized.

By convention, the DNA sequence of a DNA transcription unit is written such that the **coding strand** or **sense strand** is in the 5′ to 3′ direction and is the **top strand** in the sequence (**FIGURE 4.13**). Thus, the coding strand shares the same sequence as the RNA transcript—the only difference is that in RNA the base thymine (T) is replaced with the base uracil (U). The **bottom strand** is the **template** that directs the nucleotide specificity of polymerization by RNA polymerase and is written in the 3′ to 5′ direction. In most cases, when a DNA sequence is published, only the sense strand is written out—the template strand is easily inferred from the complementary base pairing rules. The **transcription start site**—the location on the template that directs the insertion of the first base in the RNA—is designated by convention as +1. The base on 5′ side of the start site (+1) is designated as −1 and the base on 3′ side of the start site is designated as +2 (notice that there is no "0" in this system). Sequences on the DNA template behind the transcribing polymerase (in the 5′ direction

on the sense strand of the template) are said to be **upstream**, while sequences located in the front of the transcribing polymerase (in the 3′ direction on the sense strand) are in the **downstream** direction.

In summary, the phases of the transcription process correlate with RNA polymerase recognizing the promoter, opening the duplex DNA to access the template strand, leaving the promoter and utilizing the template strand as a guide to polymerize the RNA transcript, and finally ending the transcription reaction by releasing the transcript and dissociating from the template. In the next section, we will examine how the structure of RNA polymerase relates to its function.

Concept and Reasoning Check

1. A number of prokaryotic and eukaryotic genomes contain genes that are overlapped, and which are transcribed by RNA polymerases using the opposite DNA strands as transcription templates. The promoters, which are located in front of each gene at opposite ends of the DNA sequence, are said to be "converging." What might be the biological consequences of convergent transcription? Draw a diagram with overlapping genes and convergent promoters, indicating which strands would be the sense strand and the nonsense strand for each gene.

4.4 RNA polymerases are large multisubunit protein complexes

Key concepts
- General aspects of RNA polymerase structure and function are conserved among prokaryotes and eukaryotes.
- Cellular RNA polymerases are comprised of multiple protein subunits that carry out structural, enzymatic, or regulatory functions.

The simplest RNA polymerases are those encoded by bacterial viruses, also called **bacteriophages**. Bacteriophage DNA-dependent RNA polymerases usually consist of a single subunit that is able to carry out all of the enzymatic functions of the much larger and more complicated cellular RNA polymerases discussed below, including being able to recognize the promoter and terminator sequences. However, the extra subunits of the cellular RNA polymerases facilitate the ability of transcription

Subunit	Number of Subunits in Holoenzyme	Gene	Map Position	Molecular Mass (Da)	Function
alpha (α)	2	rpoA	74.10	36,511	Required for enzyme assembly; interacts with some regulatory proteins
beta (β)	1	rpoB	90.08	150,616	Forms a pincer and is the site of rifampicin action
beta' (β')	1	rpoC	90.16	155,159	Forms a pincer and provides an absolutely conserved -NADFDGD- motif that is essential for catalysis
omega (ω)	1	rpoZ	82.34	10,105	Helps in enzyme assembly but is not required for enzyme activity
sigma (σ)	1	rpoD	69.21	70,263	Directs enzyme to promoters but is not required for phospho-diester bond formation.

FIGURE 4.14 The *E. coli* RNA polymerase subunits. The subunits of *E. coli* RNA polymerase each have a functional role in transcription.

to respond to the environment and to be regulated during cell proliferation, growth, development, and differentiation.

Prokaryotes usually contain a single RNA polymerase that consists of two large subunits with molecular weights of 150 to 160 kDa named β' and β, as well as two copies of the smaller ~30 to 40 kDa α subunit. The $β'βα_2$ complex is referred to as the **core enzyme**. A fourth subunit type called ω is sometimes associated with the core enzyme, but is not present in stoichiometric ratios. ω is important in the transcription of specific genes when the bacterial cell is under nutritional or other forms of environmental stress. The molecular characteristics and function of each of these subunits are shown in **FIGURE 4.14**.

The bacterial core enzyme is capable of carrying out the RNA polymerization reaction, but requires an exchangeable subunit called **σ factor** to recognize the promoter. The core enzyme plus the σ factor are collectively referred to as the **holoenzyme**. Prokaryotes generally have a number of different exchangeable σ factors that permit the holoenzyme to recognize different classes of promoters. Transcription initiation of the majority of genes in the bacterium *E. coli* is carried out by RNA polymerase complexed to a σ factor with a molecular weight of 70 kDa referred to as $σ^{70}$. The crystal structure of a bacterial RNA polymerase enzyme (**FIGURE 4.15**) shows that the channel for DNA lies at the interface of the β' and β subunits

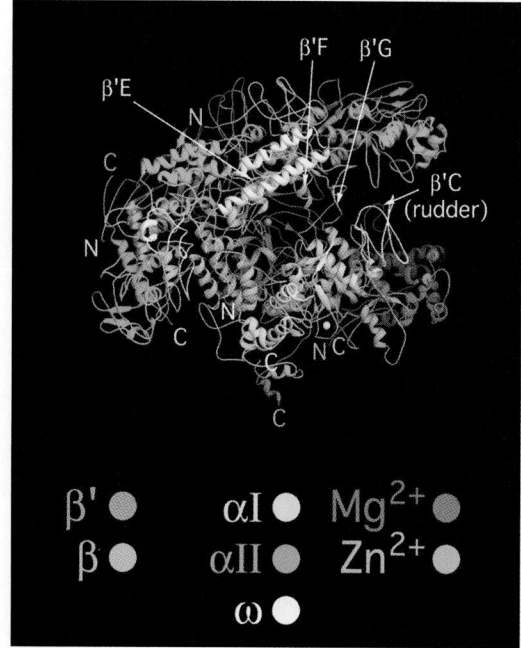

FIGURE 4.15 The structure of a bacterial RNA polymerase. The bacterial core RNA polymerase has a claw-shaped structure. In this X-ray crystal structure of the *T. aquaticus* RNA polymerase the individual subunits are shown in separate colors as delineated in the color code below the structure. One α subunit is on one side of the structure and interacts with the β' subunit, while the other α subunit interacts with the β subunit on the other side. There are also Mg^{2+} and Zn^{2+} ions tightly associated with the RNA polymerase. The various structural domains of the largest subunit β' are delineated as C, E, F, and G. The ω subunit is hidden in this view of the structure. Modified from *Cell*, vol. 98, G. Zhang, et al., Crystal Structure of Thermus Aquaticus..., pp. 811–824, Copyright (1999) with permission from Elsevier [http://www.sciencedirect.com/science/journal/00928674]. Photo courtesy of Seth Darst, Rockefeller University.

Enzyme	Location	RNA Products	Sensitivity to α-amanitin	Sensitivity to actinomycin D
RNA polymerase I	Nucleolus	Pre-rRNA (leading to 5.8S, 18S, and 28S rRNA	Resistant	Very sensitive
RNA polymerase II	Nucleoplasm	Pre-mRNA and some snRNAs	50% inhibition at 0.02 μg/mL	Slightly sensitive
RNA polymerase III	Nucleoplasm	tRNA, 5S rRNA, U6 snRNA (spliceosome), and 7SL RNA (signal recognition particle)	50% inhibition at 20 μg/mL	Slightly sensitive

FIGURE 4.16 The eukaryotic nuclear RNA polymerases. The three nuclear RNA polymerases each transcribe a different class of genes. RNA polymerase II is the target of α-amanitin, a toxin produced in the poisonous mushroom *Amanita pholloides*; the other forms of RNA polymerase are less sensitive to α-amanitin. Actinomycin D is an antibiotic that binds to DNA and blocks RNA polymerases. RNA polymerase I shows the greatest sensitivity to this drug.

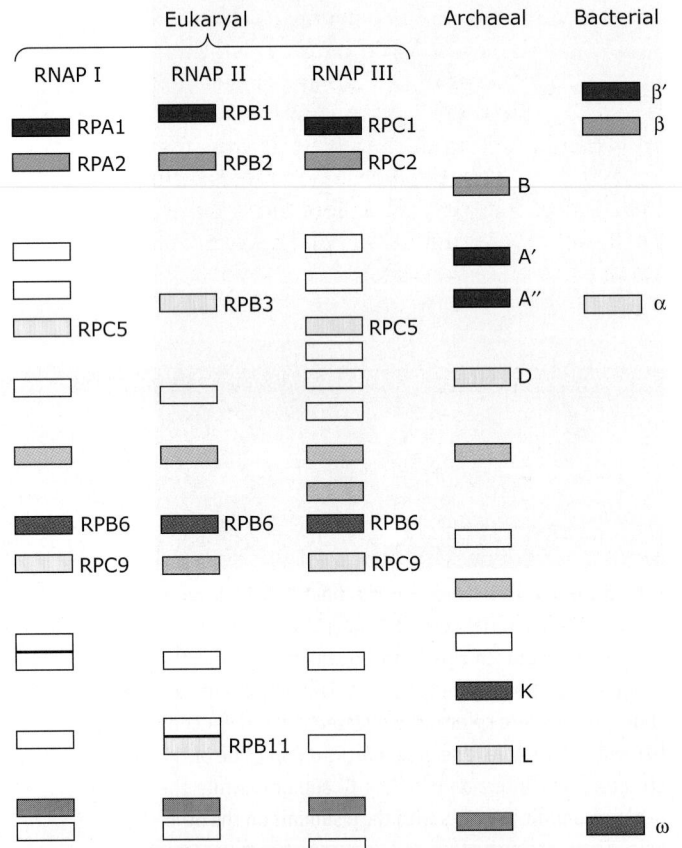

FIGURE 4.17 Comparison of the subunits of RNA polymerases across kingdoms. Nuclear eukaryal, archeal, and bacterial RNA polymerases display some homologies as well as significant differences in subunit size and function. The subunits of the RNA polymerases subunits are shown as they would appear on an SDS-polyacrylamide gel (the higher the band on the gel, the larger the molecular weight). Bands with the same color represent structural (and perhaps functional) homologs. Adapted from S. D. Bell and S. P. Jackson, *Nat. Struct. Mol. Biol.* 7 (2000): 703–705.

(the o subunits are not visible in this view). The DNA is unwound at the active site where the RNA transcript is being synthesized. The β' and β subunits contact DNA at many points downstream of the active site, as well as on the coding strand in the region of the transcription bubble, thereby stabilizing the separation of the strands. The α subunit is important for interacting with proteins that modulate RNA polymerase elongation. RNA is contacted largely in the region of the transcription bubble.

In contrast to prokaryotes, eukaryotes contain three RNA polymerases in the nucleus, designated RNA polymerases I, II, and III (**FIGURE 4.16**). Some plant species have two additional nuclear RNA polymerases that transcribe specific genes encoding small ncRNAs. RNA polymerase I is localized to the nucleolar compartment within the nucleus and transcribes the ribosomal RNA genes; RNA polymerase II transcribes genes encoding mRNAs and genes encoding many small RNAs involved in mRNA processing; RNA polymerase III transcribes the 5S ribosomal RNA genes, tRNA genes, and genes encoding many small RNAs with different functions. These polymerases generally contain 10 to 12 subunits with a total molecular weight of >500 kDa. Each of the polymerases has subunits that are homologous in both structure and function to the core subunits in prokaryotes; in addition, the nuclear polymerases also have a large number of other small subunits with functions that are still under investigation (**FIGURE 4.17**). Some of the small core subunits are common to all three nuclear RNA polymerases, while other subunits are unique to only one polymerase. The general structure of the eukaryotic RNA polymerase II enzyme, as typified in the yeast *S. cerevisiae*, is shown in **FIGURE 4.18**. The overall claw-like shape of the yeast RNA polymerase II is quite similar to bacterial RNA polymerase, but the dimensions of the yeast polymerase are larger due to the larger mass and number of the subunits. The largest subunit in RNA polymerase II has a carboxy-terminal domain (CTD) that consists of multiple repeats of the consensus sequence YSPTSPS. The sequence is unique to RNA polymerase II. There are ~25 repeats in lower eukaryotes ~50 repeats in mammals. The serine residues in the repeats are substrates for phosphorylation.

The mitochondrial and chloroplast cytoplasmic organelles each carry out transcription of their own genomes and utilize RNA polymerases that are distinct from the nuclear polymerases. The mitochondrial RNA polymerase core enzyme (also known as POLRMT) is a single subunit enzyme that displays significant structural similarities to bacteriophage RNA polymerases. However, unlike most bacteriophage-encoded RNA polymerases, POLRMT cannot recognize promoters by itself, and requires a transcription initiation factor, similar to the requirement of bacterial RNA polymerase for a σ factor. In most metazoan organisms, POLRMT is encoded by the nuclear genome and must be imported from the cytoplasm into the mitochondria. Chloroplasts contain two RNA polymerases: one called plastid-encoded RNA polymerase or PEP that is encoded by the chloroplast genome, and another called nuclear-encoded RNA polymerase or NEP that is encoded by the nuclear genome. PEP shares structural homologies to bacteria RNA polymerases and has several subunits homologous to the bacterial subunits β′, β, and α. By contrast, NEP is monomeric and has structural similarities to mitochondrial RNA polymerase.

In summary, cellular RNA polymerases are large oligomeric proteins with two large subunits and a variable number of smaller subunits. There is only one form of the prokaryotic core RNA polymerase, but a number of different exchangeable σ factors can provide specificity for the initiation of the enzyme at different classes of gene promoters. By contrast, there are multiple forms of the eukaryotic nuclear core RNA polymerases. Each polymerase has a set of unique subunits, as well as subunits that are common to all the nuclear polymerases. In the next section, we will explore the structure and function of promoters in prokaryotes and eukaryotes in greater detail.

(b)

(a)

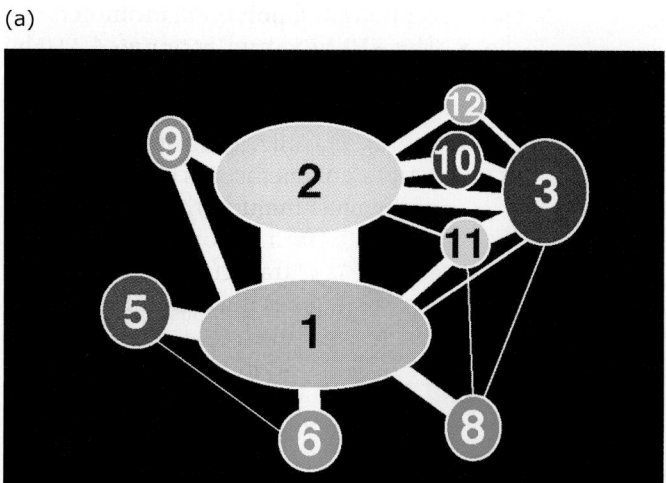

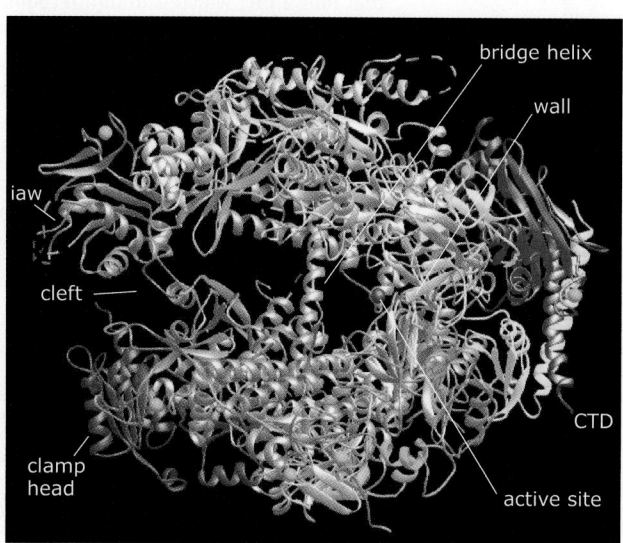

FIGURE 4.18 The structure of RNA polymerase II. The structure of eukaryotic RNA polymerase II is similar in overall shape to bacterial RNA polymerase. (a) A diagram of the interactions between the 12 yeast RNA polymerase II subunits. Note that the colors and numbers correspond to each of the subunits shown in (b). (b) This X-ray crystal structure of the *S. cerevisiae* enzyme (lacking 2 of the 12 subunits) shows the jaw, clamp, and cleft regions of the polymerase, all of which have counterparts in the bacterial enzyme. The carboxy-terminal domain (CTD) of the largest subunit is found on the right-hand side of the polymerase. Reproduced from P. Cramer, D. A. Bushnell, and R. D. Kornberg, *Science* 292 (2001): 1863–1876 [http://www.sciencemag.org]. Reprinted with permission from AAAS. Photo courtesy of Roger D. Kornberg, Stanford University School of Medicine.

4.5 Promoters direct the initiation of transcription

Key concepts

- Promoters are DNA sequences that direct the initiation of transcription by binding RNA polymerase and by promoting the melting of the duplex DNA strands to facilitate the exposure of the template strand to direct RNA polymerization.
- Promoters have conserved sequences that usually differ from the consensus at one or more positions.
- Prokaryotic promoters contain several short conserved sequence elements dispersed over 40–50 bp that are recognized directly by the RNA polymerase holoenzyme.
- Each of the three nuclear RNA polymerases has distinct promoter structures that consist of multiple sequence modules dispersed over a large region near the transcription start site.
- Unlike prokaryotic promoters, eukaryotic promoters are not recognized directly by RNA polymerase, but instead, assemble with transcription factors that form a PIC, which is in turn recognized by RNA polymerase holoenzyme.
- The transition from transcription initiation to elongation at RNA polymerase II promoters involves the phosphorylation of the CTD of the largest polymerase subunit.

Promoters are sequences that function to bind RNA polymerase and facilitate the initial RNA polymerase-dependent strand separation or isomerization process. The most common strategy to identify promoter sequences has been to search promoter regions for common sequences that are present in most, if not all, promoters of the same classes of genes. Such a sequence is said to be **conserved**. However, a conserved sequence is not necessarily conserved at every single position; some variation usually occurs. Putative promoters are defined in terms of an idealized sequence that represents the base most often present at each position. A consensus sequence is defined by aligning all known examples so as to maximize their homology. For a sequence to be accepted as a consensus, each particular base must be reasonably predominant at its position, and most of the actual examples must be related to the consensus by only one or two substitutions. Once conserved nucleotides have been identified by sequence comparison, their importance in transcription initiation can be evaluated by mutagenesis or deletion. This approach was originally applied to bacterial promoters. There are several conserved features of prokaryotic promoters, typified by those found in the σ^{70} promoters of *E. coli*, as shown in **FIGURE 4.19**. These sequences are the following:

- The **start site** is usually located at an adenosine or guanosine nucleotide (e.g., a common start site is C$\underline{A}$T). However, there are a number of exceptions in many *E. coli* promoters.
- The **−10 hexamer** is centered ~10 bp upstream of the start site. The consensus sequence is summarized in the form $T_{80} A_{95} T_{45} A_{60} A_{50} T_{96}$, where the subscript denotes the percent occurrence of the most frequently found base.

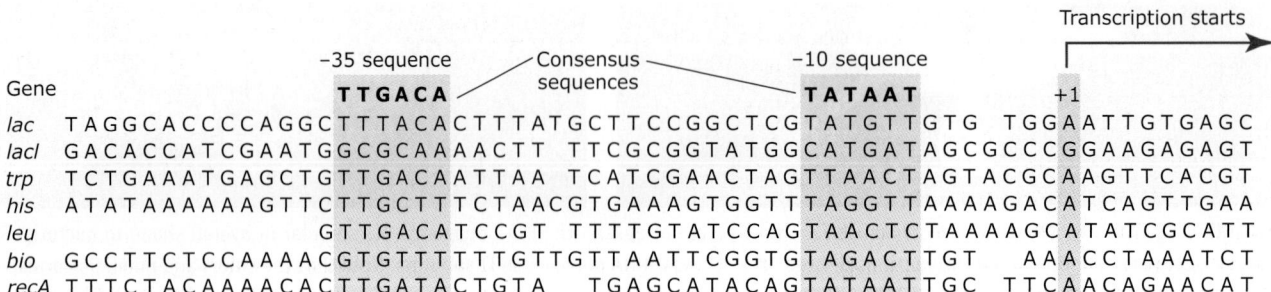

FIGURE 4.19 The *E. coli* σ^{70} promoters contain conserved sequences elements. Sequence elements at −10 and −35 from the start sites of several σ^{70} RNA polymerase promoters are conserved. Much of the variation in promoter strength results from the differences between the consensus sequence and the actual sequence of the promoter element. In addition, the distance variations between the −10 and −35 regions also affect promoter strength.

- The **–35 hexamer** is centered ~35 bp upstream of the start site. The consensus is summarized as $T_{82} T_{84} G_{78} A_{65} C_{54} A_{45}$.

- The **spacer region** between the –35 and –10 hexamers is 17 ± 1 bp, although it can vary between 15 and 20 bp. The actual sequence in the intervening region appears to be unimportant, but the distance is important for optimal promoter function and deviations from the 17 ± 1 bp spacing decrease the strength of the promoter.

- Upstream elements can influence promoter strength. For example, the **UP element** is an A-T-rich sequence found within 50 bp upstream of the –35 hexamer in some promoters that are transcribed at high levels, such as the promoters for rRNA genes.

Based on these observations, the "consensus" *E. coli* promoter consists of the –35 hexamer and –10 hexamer separated by 17 ± 1 bp. It is important to emphasize that these consensus sequences represent the average of hundreds of different gene promoters, and no naturally occurring *E. coli* promoter actually has this exact sequence. However, in general, those promoters that are closest to the consensus sequence are highly efficient in stimulating transcription initiation; and, conversely, promoters that deviate significantly from the consensus sequence tend to be less efficient. Promoter strength (defined as the affinity of the promoter for RNA polymerase and the rate at which the promoter undergoes isomerization to the open complex) for a specific gene is "optimized" rather than "maximized" to provide the correct basal level of transcription for cell growth, proliferation, and homeostasis.

In contrast to prokaryotic promoters, eukaryotic promoters are much more complex; conserved sequences can be very difficult to identify by comparison alone. Part of the reason for the lack of sequence conservation is that nuclear RNA polymerases do not recognize the DNA sequence of the promoter region directly, but instead recognize an oligomeric protein complex called the **PIC** that assembles at the promoter. For RNA polymerases I and III, the PIC is relatively simple, being comprised of just a few proteins, while the PIC formed on RNA polymerase II promoters is extremely large, containing a number of transcription factors with multiple subunits.

In the case of the RNA polymerase I promoter, which controls the transcription of the ribosomal RNA precursor, the protein known as UBF binds to the promoter region as a dimer and recruits a second factor known as SL1/TIF1B to form the PIC (**FIGURE 4.20**). The PIC recruits RNA polymerase I bound to TIF1A. Once RNA polymerase I escapes the promoter and moves into the elongation phase, TIF1A is released.

The PIC formed at RNA polymerase III promoters, which control the synthesis of a number of small RNAs of diverse functions, is slightly more complex than for the polymerase I PIC (**FIGURE 4.21**). The biggest difference is that the promoter elements are usually located, with some exceptions, within the transcribed region, downstream of the transcription start site. There are three types of RNA polymerase III promoters that are classified on the basis of the elements found in the internal or external control regions: Type 1 promoters are found in genes encoding 5S ribosomal RNA, and consist of three internal elements located within the gene beginning about 50-bp downstream from the transcription start site. Type 2 promoters are found in genes encoding tRNAs and

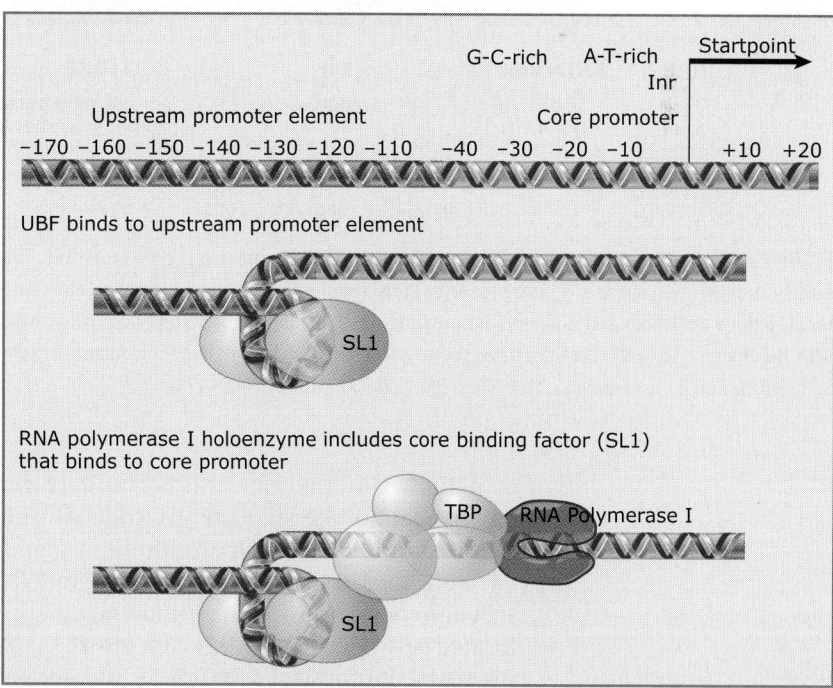

FIGURE 4.20 The eukaryotic RNA polymerase I promoter. The RNA polymerase I promoter consists of a core element located just upstream of the transcriptional start site and the upstream promoter element (UPE) located approximately at –100 bp, which binds to the transcription factor UBF (homodimer). The transcription factors TBP and SL1 are also required for RNA polymerase bind the core promoter.

Internal promoters

(a) Type 1 promoter (5S rRNA genes)

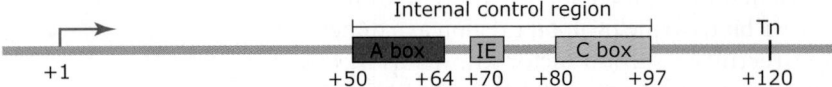

(b) Type 2 promoter (transfer RNA genes)

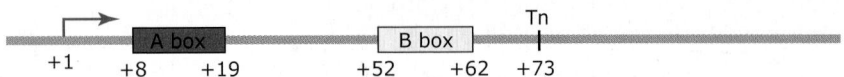

External promoters

(c) Type 3 promoter (mammalian U6 snRNA gene)

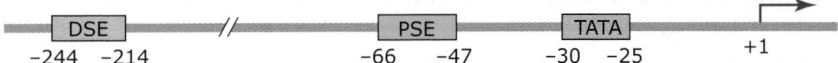

FIGURE 4.21 The three classes of RNA polymerase III promoters. The type 1 promoter directs transcription of 5S rRNA genes and contains internal consensus sequences. The type 2 promoter directs transcription of tRNA genes and also contains internal consensus sequences. The A box (red) is found in both type 1 and type 2 promoters, while the B box (yellow), C box (green), and IE (blue) are sequences specific to the promoter type. The type 3 promoter directs transcription of the genes encoding U6 snRNA, which is part of the spliceosome, and contains external consensus sequences located proximal and distal to the transcription start site. Adapted from M. R. Paule and R. J. White, *Nucleic Acids Res.* 28 (2000): 1283–1298.

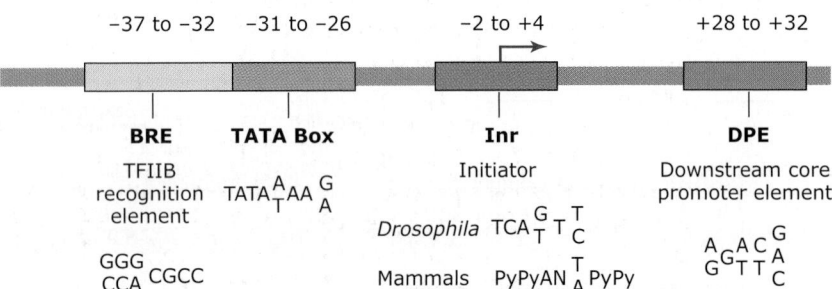

FIGURE 4.22 The modular structure of an eukaryotic RNA polymerase II core promoter. The eukaryotic RNA polymerase II core promoter consists of multiple short sequence elements located both upstream and downstream proximal to the transcription start site. Any given RNA polymerase II promoter may have some, all or none of these elements. Adapted from S. T. Smale and J. T. Kadonaga, *Ann. Rev. Biochem.* 72 (2003): 449–479.

contain two internal elements beginning 8-bp downstream of the start site. By contrast, the type 3 promoters, which are found in genes encoding the U6 small RNA involved in mRNA splicing, contain upstream promoter elements located between −25 and −244 bp upstream of the transcription start site. Type 1 and 2 RNA polymerase III promoters use one or more of the transcription factors called TFIIIA, B, and C, while type 3 RNA polymerase promoters may use other transcription factors specific to each promoter.

Promoters utilized by RNA polymerase II show considerably more variation in sequence and location than those of RNA polymerases I and III, but have a similar modular organization. Short-sequence elements that function as transcription factor-binding sites are located upstream of the start site. There are four short **core promoter elements** located −40 to +40 relative to the transcriptional start site (**FIGURE 4.22**) that are critical for basal level transcription of the gene. These core promoter elements are rarely all found in a single promoter, and are more likely to be found in various combinations, depending on the level of the gene's expression. The **basal transcription factors**, also referred to as the **general transcription factors**, bind to the core promoter elements and/or to RNA polymerase. They are labeled with the prefix TFII to designate that they are general transcription factors for RNA polymerase II. Collectively, the promoter, the general transcription factors, and RNA polymerase are referred to as the **basal transcription apparatus**. The core promoter elements are the following:

- The **TATA box**, named for the consensus sequence (T)ATA, is usually found 25 to 30 bp upstream of the transcription start site. While the TATA box is often cited as a main feature of RNA polymerase II promoters, genes that are not highly expressed generally lack the TATA box. In fact, the TATA box is present in less than 35% of all RNA polymerase II promoters that have been characterized. The transcription factor involved in almost all RNA polymerase PICs is the multisubunit protein complex called TFIID. One of the components of TFIID is the TATA box-binding protein (TBP). However, TFIID also binds to promoters that do not contain a TATA box by interacting with one or more of the other core promoter elements. These interactions may involve the other subunits of TFIID called TATA-box binding protein associated factors (TAFs).

- The **BRE (TFIIB recognition element)** is a GC-rich heptanucleotide sequence that binds to TFIIB, the other basal transcription factor involved in PIC formation. The BRE is often located immediately adjacent to the TATA box (when there is one present) and helps to stabilize the binding of TFIID.

- Two other elements commonly found in core promoters, especially those lacking a TATA box, are the **INR (initiator region)**, which overlaps the transcription start site, and the **DPE (downstream promoter element)**, which is located ~30-bp downstream from the transcription start site. The combination of the INR and the DPE appears to compensate for the lack of a TATA box by interacting with the TAFs of the TFIID complex.

Once TFIID and TFIIB bind to the core promoter, the remaining basal transcription factors, TFIIA, TFIIF, TFIIE, as well as RNA polymerase II, assemble with a large complex of proteins (**FIGURE 4.23**). The proposed function of each of the basal transcription factors is shown in **FIGURE 4.24**. Transcription initiation commences upon complete assembly of the PIC and the polymerization of the first phosphodiester linkages of the RNA transcript.

During the transition between the initiation and elongation phases of transcription, many of the basal transcription factors dissociate from the initiation complex, but TFIIH plays an essential role in the transition. TFIIH possesses both kinase and helicase activities that are activities required for the isomerization process of melting the DNA and the phosphorylation of the CTD of the large subunit of RNA polymerase II. Phosphorylation converts the CTD into a target for the binding of RNA-processing proteins that are carried along with RNA polymerase II (**FIGURE 4.25**). The mRNA splicing and 3′ end processing factors bind to the CTD at various locations along the DNA template and transfer from the polymerase to the mRNA as the factor-binding sequences on the transcript emerge from the active site of the polymerase.

Once RNA polymerase II has left the promoter and entered the elongation phase, there are two important transcription factors that increase the fidelity and elongation rate of the polymerase. TFIIS (also known as SII) is a fidelity factor that causes RNA polymerase II to backtrack and remove any noncomplementary nucleotide that has been erroneously inserted. P-TEFb is an elongation factor that is a cyclin-dependent protein kinase (see *15 Cell cycle regulation* for a complete discussion

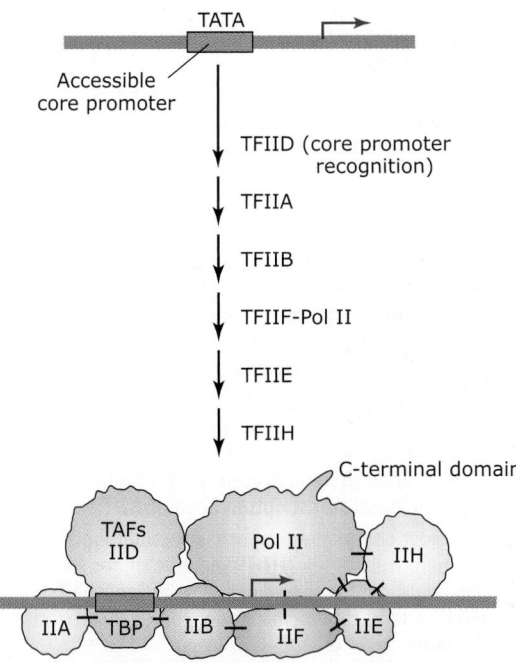

FIGURE 4.23 The assembly of the general transcription factors on an RNA polymerase II core promoter. The general transcription factors (TFIID, TFIIA, TFIIB, TFIIE, and TFIIH) assemble with TFIIF-RNA polymerase II to form the initiation complex at the core promoter. Adapted from R. G. Roeder, *Nature Med*. 9 (2003): 1239–1244.

Factor	No. of Subunits	Functions
TFIIA	2	Stabilizes TBP and TFIID binding. Blocks inhibition by an inhibitory TAF subunit.
TFIIB	1	Binds TBP, RNA polymerase II, and promoter DNA. Its N-terminal region has a finger structure that inserts into the RNA polymerase II active site cleft. It helps fix the transcription initiation site.
TFIID (TBP and TAFs)	1 14	Binds TATA element and deforms promoter DNA. Platform for the assembly of TFIIB, TFIIA, and TAFs. Binds Inr and DPE promoter elements.
TFIIE	2	Binds promoter near transcription initiation site. May help to stabilize the transcription bubble in the open complex.
TFIIF	3	Prevents nonspecific DNA binding. Binds RNA polymerase II and is involved in recruiting RNA polymerase II to the preinitiation complex. It has regions that are homologous to the bacterial σ factor.
TFIIH	10	Functions in transcription and DNA repair. It has kinase and helicase activities and is essential for open complex formation.

Adapted from Hahn, S., *Nat. Struct. Mol. Biol.* 11 (2004): 394–403.

FIGURE 4.24 The function of the RNA polymerase II general transcription factors in yeast. The general transcription factors in yeast (*S. cerevisiae*) have specific functions. Adapted from S. Hahn, *Nat. Struct. Mol. Biol.* 11 (2004): 394–403.

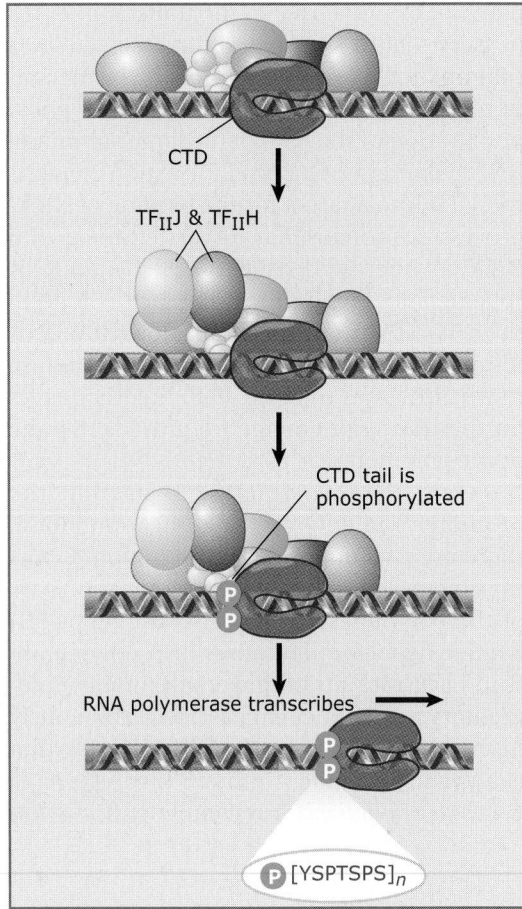

FIGURE 4.25 Phosphorylation of the RNA polymerase II carboxy-terminal domain (CTD). The CTD of RNA polymerase II is phosphorylated by the kinase activity of TFIIH prior to the transition of the polymerase from the initiation phase to the elongation phase of transcription. The human CTD contains 52 repeats of the amino acid sequence YSPTSPS.

of CDKs). P-TEFb phosphorylates the CTD on the large subunit of RNA polymerase II, resulting in the inhibition of transcription inhibitory proteins, thereby stimulating RNA polymerase elongation.

In summary, the interaction of RNA polymerase with the promoter region can occur directly (in prokaryotes), or indirectly (in eukaryotes) through the formation of PICs using the general transcription factors. Both prokaryotic and eukaryotic promoters are constructed out of conserved short modular sequences that contact the individual subunits of RNA polymerase or the basal transcription factors. The variation in promoter structure is correlated with variation in promoter strength and can be regarded as the first level of gene-specific transcriptional control. Once RNA polymerase

leaves the promoter region, other transcription factors help the polymerase to avoid errors and inactivation. In the next section, we will explore the structure and function of the gene-specific transcription factors that activate or repress the basal level of transcription in response to the cellular environment and during cell growth, proliferation, differentiation, and development.

Concept and Reasoning Check

1. The following 80-nucleotide DNA sequence (sense strand) contains a classical RNA polymerase II core promoter. Identify the conserved sequence elements and nucleotide that most likely serves as the transcription start site (+1).

5′TCATACCTTCCTATCTCCGATCCTAGGCTATTTATAACCA TGGTATTTCATGCATTACCTTCATTCCTGTGGCCTACGCA 3′

4.6 Activators and repressors regulate transcription initiation

Key concepts

- Transcription factors modulate RNA biosynthesis at the initiation phase by interacting with the basal transcription apparatus.
- Transcription factors bind to DNA sequences proximal or distal to the basal initiation complex at the core promoter region.
- Enhancers are transcription factor-binding sites located distal to the promoter that function to modulate transcription initiation through DNA looping.
- The *E. coli* lactose (*lac*) operon serves as a paradigm for understanding the general principles of regulation of transcription initiation. The *lac* repressor is a negative regulatory factor that represses transcription initiation at the *lac* promoter by blocking RNA polymerase, while the catabolite activator protein activates transcription by recruiting RNA polymerase to the *lac* promoter.
- Transcription factors are comprised of structural motifs that function in DNA recognition, dimerization of subunits, and protein-protein interactions with the basal transcription apparatus.
- The most common eukaryotic structural motifs are the helix-turn-helix (HTH) and zinc finger (ZF) DNA-binding domains and the helix-loop-helix (HLH) and basic leucine zipper (bZIP) dimerization domains.

Virtually every aspect of cell structure and function is directly or indirectly controlled by the coordinated regulation of gene expression. Gene regulation also directs the differentiation and development of all multicellular organisms through its interface with signal transduction. It is now clear that regulation can occur at every level of gene expression including transcription, RNA processing, nucleocytoplasmic transport, translation, and RNA and protein degradation/recycling. However, we currently know the most about transcriptional regulation, and in particular, control of transcription factors that function as activators and repressors of the initiation step.

In addition to the core promoter and the basal factors, almost all protein-coding genes contain binding sites for transcription factors that direct the increase (upregulation) or decrease (downregulation) of transcription initiation. Transcription factor-binding sites may be located proximal (less than 10 bp) to the core promoter, facilitating short-range interactions between the transcription factor and the basal transcription apparatus; or the binding sites may be distally located (greater than 100 bp) from the core promoter. In general, those transcription factor-binding sites that function distal to the core promoter require that the intervening DNA sequence form a loop that brings the factor in proximity to the promoter. An important type of transcription factor-binding site that functions at such large distances from the core promoter is called an **enhancer**. Enhancers can function hundreds or even thousands of bp from the core promoter, and may be located either upstream or downstream. Due to the extreme flexibility of the large intervening DNA loop, enhancers can also function in either a forward or backward orientation (**FIGURE 4.26**). For enhancer-binding proteins to exert their effects on the basal transcription apparatus, other proteins called **coactivators** may facilitate the conformational rearrangement or the condensation of the DNA loop, thereby bringing the factor in close proximity to the core promoter. Some transcription factors, including enhancer-binding proteins, connect to the basal transcription apparatus through a large oligomeric protein complex called **Mediator**, which has a molecular weight of 1.2 MDa and is essential for transcription activation of almost all protein-coding genes (**FIGURE 4.27**). Because enhancers can function at such large distances and in both directions from the core promoter, there are also sequences called **silencers** that set boundaries providing limitations in both distance and directionality to the effects that enhancers exert on their promoter targets.

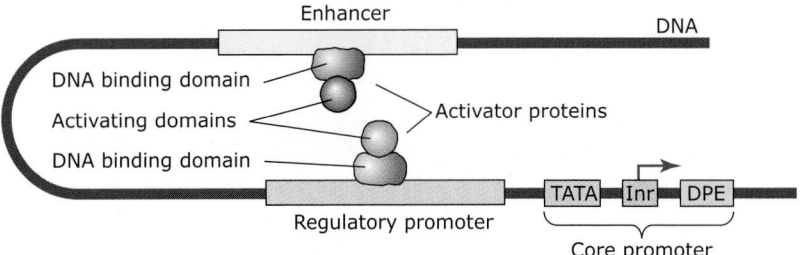

FIGURE 4.26 Enhancer function through DNA looping. Enhancer-binding transcription factors contact the other transcription factors located proximal to the core promoter region via DNA looping.

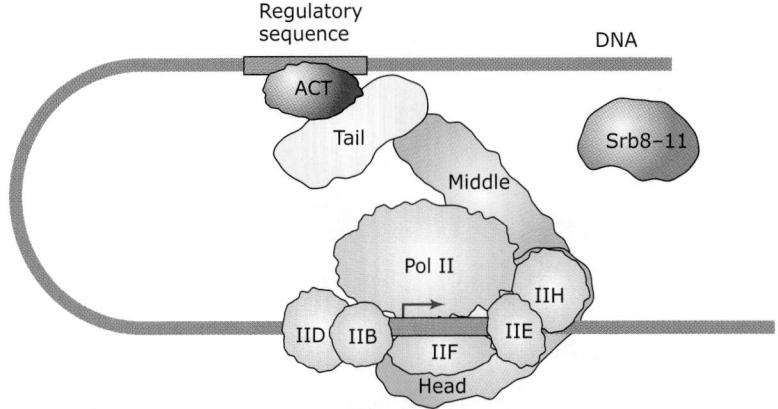

FIGURE 4.27 RNA polymerase II mediator. Mediator functions as a bridge between gene activator (ACT) transcription factors bound at distal sites and the basal transcription apparatus bound at the core promoter. Mediator has three oligomeric structural domains (head, middle, and tail). Each domain consists of several protein subunits. The entire mediator complex consists of more than two dozen subunits. The protein Srb8-11 (also known as Cdk8) can inhibit the interaction of mediator with RNA polymerase. Adapted from S. Björkland and C. M. Gustafson, *Cell* 30 (2005): 240–244.

The first biological regulatory circuit to be characterized genetically and biochemically was the **lactose (lac) operon** in *E. coli* (**FIGURE 4.28**), which has served as a paradigm for all other transcriptional regulatory systems. *E. coli* is a bacterium that inhabits the intestine of most mammals and utilizes the host's dietary intake of glucose as its primary source for energy metabolism. When glucose is scarce, the *E. coli* cell will induce the synthesis of several proteins for the utilization of lactose as an energy source—lactose is a disaccharide composed of glucose and galactose. The enzyme β-galactosidase cleaves the bond between glucose and galactose, while subsequent metabolic steps convert galactose to glucose, thereby yielding two molecules of glucose from a single lactose molecule. The *lac* operon contains three lactose utilization

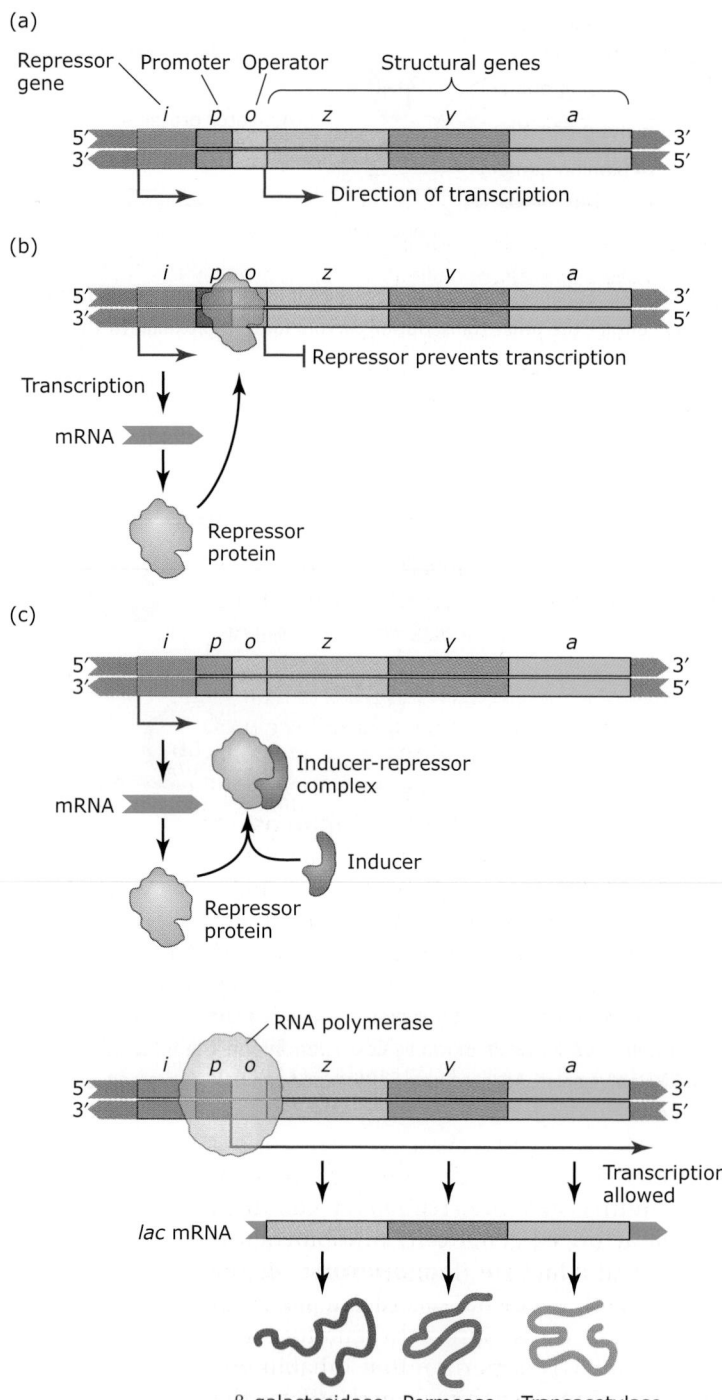

FIGURE 4.28 The *E. coli lac* operon. The *E. coli lac* operon (not drawn to scale) consists of three genes encoding proteins necessary for the utilization of lactose as an energy source (z = β-galactosidase; y = galactose permease; a = transacetylase). (a) The promoter region contains a negative regulatory site called the operator (o), which overlaps the *lac* promoter (p). (b) In the absence of lactose, the lac repressor, encoded by the repressor gene (i), binds to the operator and inhibits transcription. (c) In the presence of lactose, the lactose is converted to allolactose, which binds and stimulates the release of the repressor from the operator, permitting RNA polymerase to transcribe the operon.

genes arranged in tandem: the gene encoding β-galactosidase, which breaks down lactose; a gene encoding a lactose permease, which allows lactose to enter the cell; and a non-essential gene encoding a galactoside acetyl transferase, which acetylates a number of proteins, but whose biological function is still unclear. All three genes are encoded in a single polycistronic mRNA that is transcribed from the *lac* operon promoter called P_{lac}.

P_{lac} is under both negative and positive regulation. Normally, when glucose levels are high and the lactose-utilizing enzymes are not needed, P_{lac} is shut down by a protein called the ***lac* repressor**, which binds to the **operator region** adjacent to the promoter, and blocks binding and/or and promoter clearance by RNA polymerase. If glucose levels fall and lactose is detected in the surrounding medium, lactose is taken up by the *E. coli* cell. A small amount of β-galactosidase is constitutively present in the cell, and one of its minor catalytic activities is to convert some of the lactose to the compound allolactose. Allolactose acts as an **inducer** of the *lac* operon by binding to and eliciting a conformational change in the repressor, which causes the repressor to dissociate from the operator, thereby allowing RNA polymerase to transcribe the *lac* mRNA and produce the enzymes required to break down lactose.

A positive regulatory pathway for induction of the *lac* operon also functions in *E. coli*. When cellular glucose levels are low, the bacterium synthesizes the messenger molecule **cyclic AMP** (cAMP). cAMP binds to a transcription activator protein called the cAMP receptor protein (CRP), whose DNA-binding site is located adjacent to P_{lac}. The cAMP-CRP complex binds to the promoter region and significantly increases the affinity of RNA polymerase for P_{lac}, thereby enhancing transcription of the *lac* operon and resulting in the upregulation of the synthesis of the *lac* utilization enzymes (**FIGURE 4.29**). Thus, full induction of the *lac* operon genes requires both the absence of glucose and the presence of lactose.

The *lac* operon illustrates several important points that are applicable to virtually all other examples of transcriptional regulation in both prokaryotes and eukaryotes as given below:

- *There is a strong link between cellular signaling responses and transcription.* The low level of constitutive synthesis of lactose permease and β-galactosidase allows the cell to import lactose into the cell and for β-galactosidase to con-

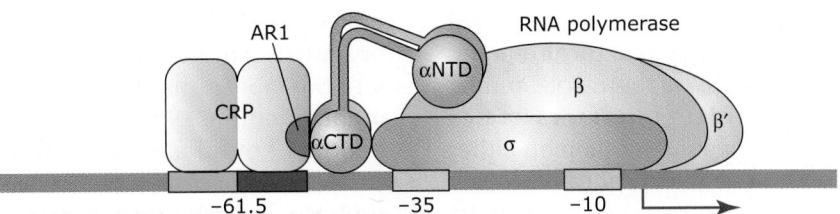

FIGURE 4.29 Positive control of the *lac* operon. The *lac* operon is under positive regulatory control by the cAMP activator protein, also known as catabolite repressor protein (CRP). CRP binds to the site just upstream from the −35 hexamer in the lac promoter and binds to a carboxy-terminal domain of the α subunit of RNA polymerase (α-CTD); this interaction causes an allosteric conformational change that induces the interaction between the amino-terminal domain of the α subunit with large subunits, β′ and β. These protein-protein interactions strengthen the affinity of RNA polymerase for the *lac* promoter. Adapted from S. Busby and R. H. Ebright, *J. Mol. Biol.* 293 (1999): 199–213.

vert lactose to allolactose, which then induces the operon by causing *lac* repressor to dissociate from the promoter. The cAMP formed in response to a drop in intracellular levels of glucose induces the operon by increasing the affinity of RNA polymerase for the promoter. The relationship between cell signaling and gene expression will be discussed in greater detail in *18 Principles of cell signaling*.

- *Key protein-protein interactions between transcription factors and the basal transcription apparatus are mediated by DNA-protein interactions.* Both *lac* repressor and CRP interact with RNA polymerase. These protein-protein interactions are illustrative of three important mechanisms for transcription factor modulation of transcription: The first mechanism is referred to as **physical interference**. In the case of the *lac* operon, the *lac* operator lies just downstream of P$_{lac}$ and binding of the *lac* repressor to the operator physically blocks polymerase from binding to the promoter, although some studies have suggested that polymerase can still bind, but is unable to escape the promoter after initiation. The second mechanism, known as **recruitment**, is illustrated by the interaction of CRP with the CRP-binding site, located just upstream of P$_{lac}$. In this case, the transcription factor increases the affinity of RNA polymerase for the promoter. Eukaryotic transcription factors can also increase the local concentration of RNA polymerase II and the basal transcription apparatus at the core promoter. A

third key mechanism for transcription regulation is **allosteric modification**. Transcription factors can interact with polymerase or other components of the basal transcription apparatus and cause conformation changes in the core promoter complex, leading to a modulation of transcription initiation.

- *Transcriptional regulation often involves both positive and negative regulatory mechanisms.* The *lac* operon is an example of a gene system that is normally off due to constitutive repression by the *lac* repressor, and that requires that the repressor be released for transcription to occur. However, activation of RNA polymerase also occurs by cAMP-bound CRP. Transcriptional regulation of eukaryotic genes also involves both depression and activation. Most genes in eukaryotes are constitutively repressed by highly condensed chromatin near and at the core promoter region. Before basal transcription can occur, the chromatin in the promoter region must be rearranged and remodeled to allow access of the basal transcription apparatus to the DNA template. Thus, the transcription activation factors include a number of proteins that covalently modify histones to loosen their interaction with DNA within the chromatin. In addition, other activator proteins act to physically rearrange or move the histones away from the DNA. This subject will be discussed in greater detail in *10 Chromatin and chromosomes*.

- *Transcription activation may occur at proximal or distal sites.* While the CRP-binding site and primary *lac* operator are located immediately adjacent to the promoter, there are other *lac* repressor-binding sites or **secondary operators** located about −100 bp and +400 bp on either side of the promoter that can facilitate the interaction of *lac* repressor with the primary operator. Thus, action at distance appears to be a universal strategy for transcription factor function in both prokaryotes and eukaryotes.

The regulation of mRNA transcription by RNA polymerase in eukaryotes is remarkably complex compared to the *lac* operon. There are thousands of transcription factors encoded in higher eukaryotic genomes that activate or repress transcription. A comprehensive discussion of eukaryotic transcription factors is beyond the scope of this chapter. However, a few general principles are important for understanding how transcription factors activate or repress transcription in response to the cellular environment and during cell growth, proliferation, development, and differentiation.

The range of different types of eukaryotic transcription factors is vast, but these proteins often share a number of structural features that have specific functions in activation or repression. Transcription factors commonly function as monomers or as dimers. For those transcription factors that interact directly with DNA, a distinct region of the three-dimensional structure of the transcription factor, referred to as a **domain**, interacts with a specific DNA sequence element, while another domain of the

protein interacts directly with RNA polymerase or with some component of the basal transcription apparatus to increase the efficiency of initiation by RNA polymerase through recruitment or an allosteric effect. Thus, such transcription factors contain a **DNA-binding domain**, an **activation domain**, and in some cases a **dimerization domain**. Other transcription factors function as internal subunits of a multisubunit complex, and contain only domains that are involved in protein-protein interactions.

Transcription factors are often classified into families according to the specific combinations of secondary and tertiary structures of their functional domains that are the result of distinctive primary structures, that is, amino acid sequences, resulting in a structural fold or **structural motif**. Motifs are sometimes referred to as **super-secondary structures** to indicate that they contain a specific order of secondary structures. Although a large number of structural motifs have been identified in the crystal structures of transcription factors, four families of motifs are especially common: the helix-turn-helix, the zinc finger, the helix-loop-helix, and the leucine zipper.

The helix-turn-helix or **HTH motif** is found in many prokaryotic transcription factors, including the *lac* repressor and the bacteriophage λ cro and cI repressors. In the HTH motif, one α-helix interacts with the major groove of DNA and the other α-helix, which is separated from the first helix by a β turn, lies at an angle across the DNA. The amino acids within the α-helices contain functional groups that form hydrogen bonds and ionic interactions with functional groups on the nucleotides of the DNA.

In eukaryotes, a motif very similar to the HTH, called the **homeodomain**, is found in a number of transcription factors (**FIGURE 4.30**). The term homeodomain comes from a highly conserved gene sequence known as the **homeobox**, which codes for a 60 amino acid domain commonly found in transcription factors that regulate the development of the body plan of many animals. Examples of eukaryotic transcription factors with homeodomains include the POU family members Pit-1, Oct-1, and unc-86, as well as the HOX proteins involved in pattern formation during development.

The zinc finger or **ZF** motif is found in the DNA-binding domains of various eukaryotic transcription factors, including several nuclear hormone receptors. In addition, other tran-

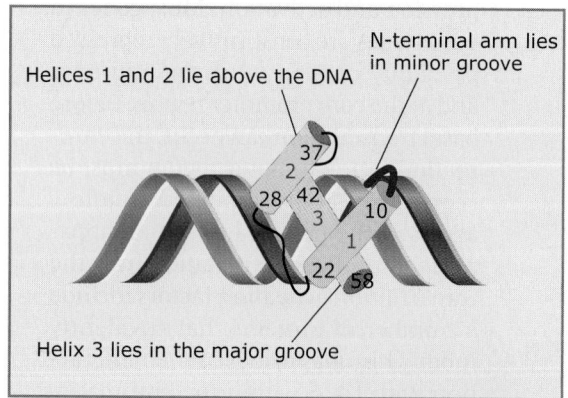

FIGURE 4.30 The homeodomain motif. The homeodomain found in a number of eukaryotic transcription factors important in development contains three α-helices. Helices 1 and 2 lie above the DNA, while helix 3 (the recognition helix) lies in the major groove and contains amino acids that make specific hydrogen bond contacts with bases in the DNA-binding site.

scription factors including **Sp1** and the **Wilms tumor protein** (WT-1) also contain ZF motifs. The ZF motif derives its name from the stable loop of amino acids that forms through the coordination of a zinc ion by cysteine and histidine residues (**FIGURE 4.31**). The finger itself comprises ~20–25 amino acids, and the linker between adjacent fingers is usually 7–8 amino acids. The finger sequences are usually organized into tandem arrays ranging from 2 to 9 finger repeats. Sp1 is an important transcription factor that binds to GC-rich segments located adjacent to the core promoter sequences of many highly expressed housekeeping genes. Sp1 contains three ZF motifs.

Many transcription factors function as homodimers or heterodimers. The subunits of these dimers are often the products of related gene families. Different combinations of the family members form heterodimers can create activating or inhibitory complexes with varying capacities to modulate transcription. In some cases, a family includes inhibitory members, whose participation in dimer formation prevents the partner from activating transcription.

The **basic helix-loop-helix or bHLH motif** is a common motif found in the dimerization domains of many transcription factors that regulate development in eukaryotes. Each amphipathic helix contains hydrophobic residues on one side and charged residues on the other side that enable proteins to dimerize, while a short segment of basic amino acids adjacent to the bHLH region contacts DNA by ionic interactions with the phosphodiester backbone (**FIGURE 4.32**). The two helices are separated by a relatively unstructured loop that allows the two helices to act independently of each other. Two classes of the many bHLH transcription factors are the MyoD family of proteins, which are involved in muscle cell differentiation, and the E-box proteins, which play important roles in the regulation of genes involved in cell growth and proliferation.

The basic leucine zipper or **bZIP motif** is another example of a commonly encountered motif found in the dimerization domains of transcription factors. Each side of the "zipper" consists of a stretch of amino acids with a leucine residue in every seventh position on one monomer subunit that interacts with the corresponding region on the other monomer subunit to form the dimer. Adjacent to each zipper is a stretch of positively charged residues that is involved in binding to DNA, similar to the bHLH motif. The two sides of the leucine zipper in effect form a Y-shaped structure, in which the zipper motifs comprise the stem and the two basic regions protrude out to form the

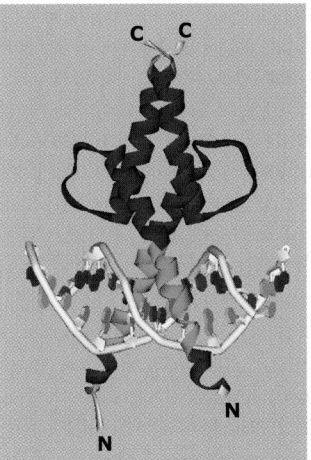

FIGURE 4.32 The helix-loop-helix motif. The helix-loop-helix motif is used by transcription factors to form homo- and heterodimers. In the structure shown above, the transcription factor MyoD is a homodimer with two recognition helices containing highly basic regions that contact specific bases in the DNA. Structure from Protein Data Bank 1MDY. P. C. Ma, et al., *Cell* 77 (1994): 451–459.

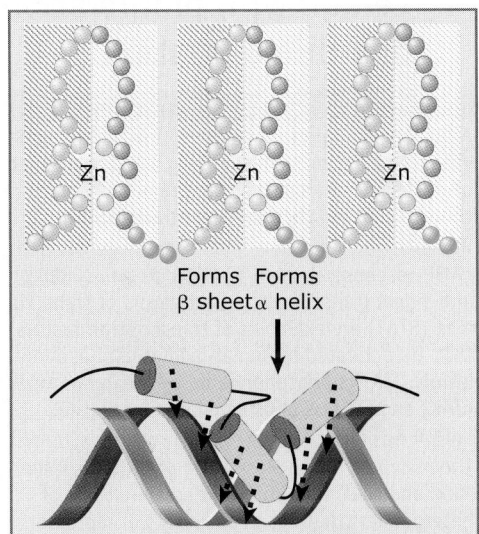

Forms Forms
β sheet α helix

FIGURE 4.31 The zinc-finger motif. Zinc fingers form through the chelation of Zn^{2+} ion by histidine and/or cysteine amino acids. Often multiple zinc fingers are found within adjacent α-helices that insert into the major groove of the DNA.

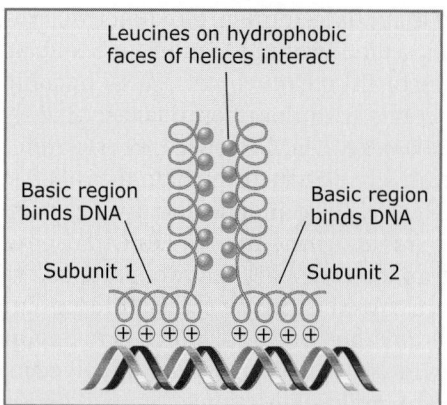

FIGURE 4.33 The leucine zipper motif. The leucine zipper motif is used by many transcription factors to form homo- and heterodimers. The zippers are held together by hydrophobic interactions between the α-helical stretches of leucine residues on each monomer. The basic regions located near the leucine zipper motifs interact with specific bases within the transcription factor-binding site.

segments that interact with the DNA (FIGURE 4.33). As with the bHLH motif, zippers can be involved in either homodimer or heterodimer formation. An important example of a homodimer transcription factor is C/EBP, which binds to a sequence known as the CAAT box, usually located 50–100 bp from the core promoters of highly expressed genes. An example of a heterodimer that forms due to the bZIP motif is the Jun-Fos transcription factor known as AP1, which activates the transcription of a number of genes important in cell growth and proliferation. Mutations in the Jun and Fos subunits of AP1, which are oncoproteins, are associated with certain forms of cancer as discussed in greater detail in *17 Cancer—Principles and overview*.

Aside from the use of structural motifs in the taxonomic classification of transcription factors, knowledge of the location of motif structures can be very useful in the assessment of potential effects of specific gene mutations, particularly disease-causing mutations. Molecular biology techniques are becoming increasingly important in diagnosis and prognosis of diseases, and are changing their entire classification systems. Diseases that were formerly classified on the basis of signs and symptoms are now being associated with specific gene mutations. Diagnosis and prognosis of various diseases based on specific mutations and levels of expression of specific genes (referred to as molecular biomarkers) are becoming increas-

ingly commonplace in medical practice. In many cases, some of the most deleterious disease-causing mutations associated with genes encoding transcription factors are found in the key structural motifs of the transcription factor functional domains.

In summary, the genomes of both prokaryotes and eukaryotes encode a large number of transcription factors that act to increase or decrease the basal level of transcription of specific genes, often in response to signals from the cellular environment or other cells. The mechanisms of transcriptional activation include recruitment of the basal transcription machinery, alteration of chromatin structure to clear the core promoter region, and allosteric activation of RNA polymerase. In the next section, we will discuss some specific examples of signal-mediated control of transcription initiation in mammalian cells.

Concept and Reasoning Check

1. Enhancers can function at very large distances to activate initiation at core promoters. There are often several different core promoters within the maximum distance for enhancer-mediated activation. Propose some possible mechanisms for selective activation of one promoter over another.

4.7 Transcriptional regulatory circuits control eukaryotic cell growth, proliferation, and differentiation

Key concepts

- A number of important transcription factors activate transcription by facilitating the displacement of nucleosomes from the core promoter region, thereby clearing the region for transcription initiation.
- cAMP-response element binding protein (CREB) and signal transducers and activators of transcription (STAT) are examples of transcription factors that are activated by signal-mediated phosphorylation. STAT is phosphorylated by the Janus kinase (JAK) and CREB is phosphorylated by protein kinase A (PKA).
- The members of the Myc family of proteins (Myc, Max, and Mad) form different combinations of heterodimers that act as activators or repressors depending on the dimer composition.
- The steroid hormone receptors are transcription factors that when bound to hormone activate transcription by binding to DNA sequences known as hormone-response elements within the promoters of hormone-responsive genes.

We previously discussed how transcription activation occurs in both prokaryotes and eukaryotes through interactions between the transcription factors and the basal transcription apparatus. Activation plays an especially important role in eukaryotes where the genome is constitutively repressed due to the assembly of the DNA with histone proteins to form **chromatin**. An important function of many transcription factors is to participate in the displacement of the **nucleosomes** from the core promoter region. The details of this process are discussed in detail in *10 Chromatin and chromosomes*. However, for our purposes here, it suffices to say that the chromatin remodeling process occurs in two steps: (1) the loosening of nucleosomes through covalent modification of specific histones including acetylation, phosphorylation, and methylation and (2) the physical displacement of the loosened histones from the core promoter region by ATP-dependent structural modification proteins. Some of the gene-specific transcription factors that we will discuss in this section carry out these activities.

In any given cell type within a tissue, a select set of transcription factors becomes activated in response to extracellular signals. Various combinations of transcription factors produce different levels of transcription of their target genes, thereby modulating the amount of mRNA and protein that is produced. We will discuss several examples to illustrate the linkage between cell signaling and transcription as a prelude to a detailed discussion of cancer and signal transduction in *17 Cancer—Principles and overview* and *18 Principles of cell signaling* respectively.

A very important signal transduction-mediated, gene-specific transcription factor is the **CREB**. We saw previously in our discussion of the *E. coli lac* operon that cAMP is a messenger molecule that increases when bacterial cell glucose levels are low. Binding of cAMP to the catabolite activator protein CAP promotes the binding of CAP to the *lac* promoter, which activates transcription through the recruitment of RNA polymerase to the promoter. However, cAMP also plays an important role in eukaryotic cells by acting as a messenger molecule in response to different types of extracellular signals (see *18 Principles of cell signaling* for more details). When cAMP levels rise, cAMP binds and activates PKA in the cytoplasm. The cAMP-PKA complex enters the nucleus where it phosphorylates CREB. Phosphorylated CREB binds to the cAMP response element (CRE) within the promoters of a number of cAMP-response genes and activates their transcription (**FIGURE 4.34**). Activation of transcription by CREB does not occur through a direct interaction with the basal transcription apparatus; instead, CREB recruits a complex containing the CREB-binding protein complex, **CBP/p300** and the protein **PCAF**. The CREB-CBP/p300-PCAF complex acetylates the histones in the region of the core promoter, causing their displacement and thereby facilitating the assembly of the basal transcription apparatus at the core

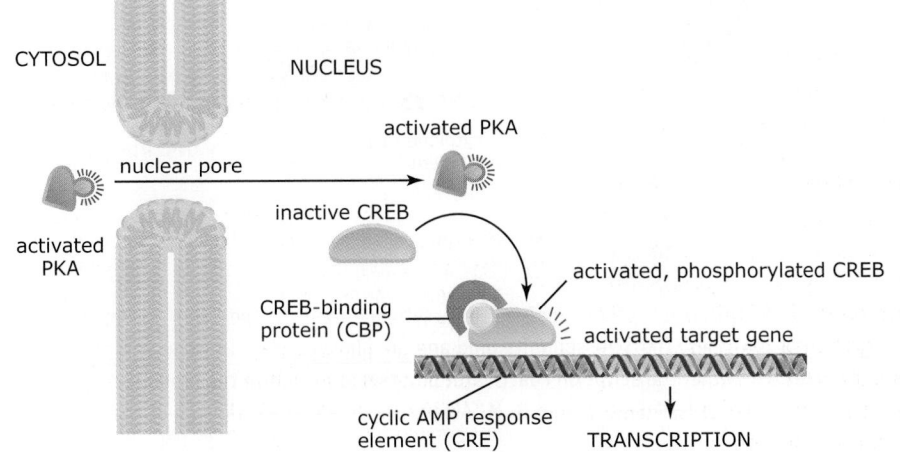

FIGURE 4.34 Regulation of PKA-activated genes by CREB. The cyclic-AMP-dependent protein kinase (PKA) phosphorylates and activates the cyclic-AMP response element binding (CREB) protein. Phosphorylated CREB recruits the coactivator CBP, which as a complex activates a number of genes that are important in hormonal control of cellular metabolism and in brain function. Adapted from B. M. Alberts, et al. *Molecular Biology of the Cell, Fifth edition*. Garland Science, 2008.

promoter. Histone displacement occurs because acetylation of the amino groups on the side chains of lysine residues neutralizes their positive charges and reduces the electrostatic interactions between histones and the negatively charged backbone of DNA.

Another example of the signal-mediated control of transcription factor phosphorylation is the **JAK-STAT pathway**, which is stimulated by the binding of **cytokines** to their membrane-bound cytokine receptors (see *18 Principles of cell signaling* for a detailed discussion of the JAK-STAT signaling pathway). Cytokines are a class of small protein or peptide growth factors that bind to receptors in the plasma membrane of various cells in the immune system and stimulate cell proliferation and growth. Upon binding of the cytokine to the receptor, the receptor phosphorylates and activates a tyrosine kinase called **JAK** (a member of the Janus kinase family). JAK binds, phosphorylates, and activates a transcription factor known as **STAT**. Once phosphorylated, STAT forms a homodimer and translocates into the nucleus where it recognizes sequences known as **cytokine response elements** located within the promoter regions of a number of different genes involved in cell division (**FIGURE 4.35**). As in the case of CREB, STATs activate transcription indirectly by recruiting histone acetylases to clear nucleosomes from the promoter region.

The **c-Myc family** of proteins constitute another important group of transcription factors and include the proteins Myc, Max, and Mad. In highly differentiated nonproliferating

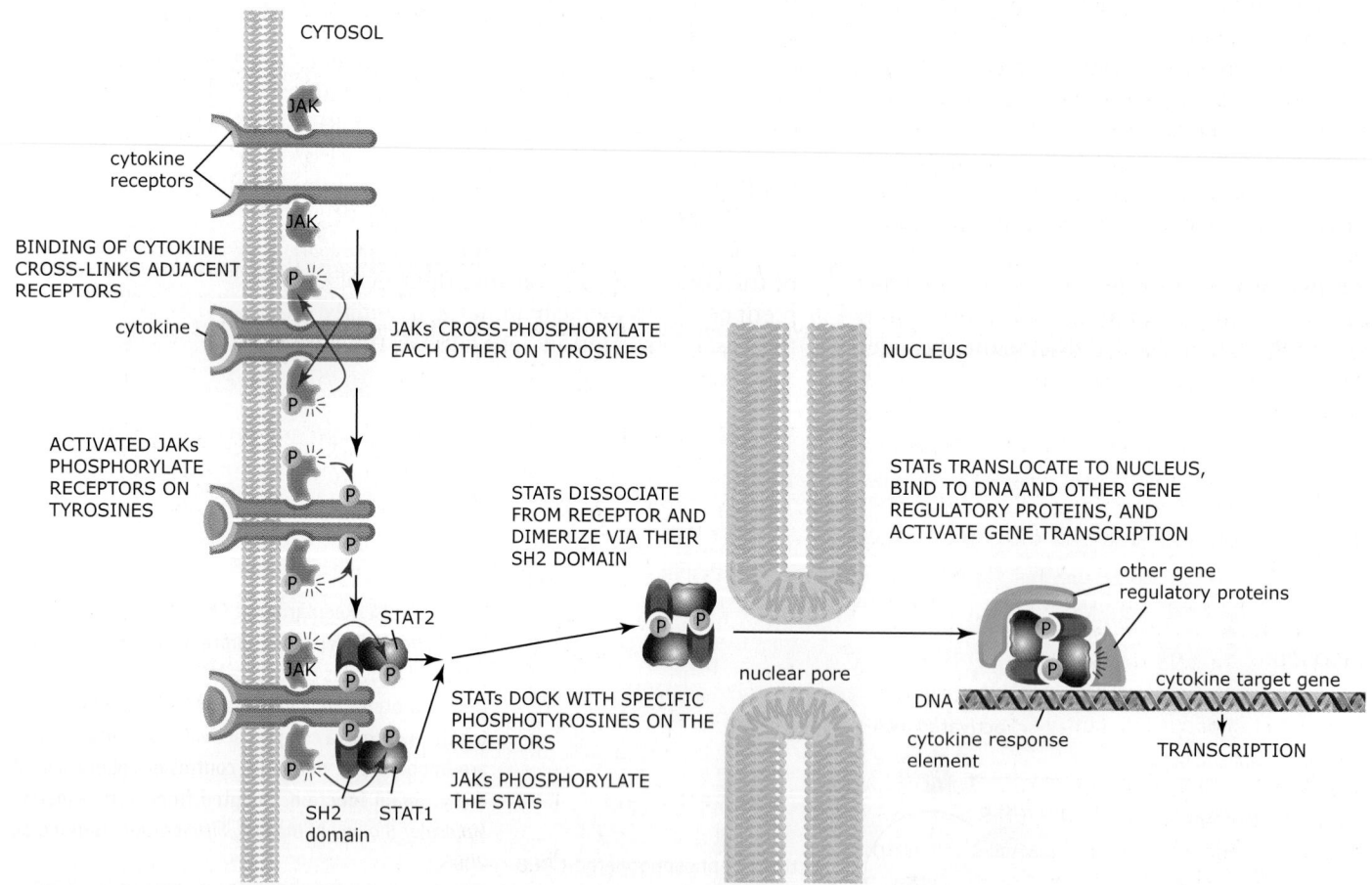

FIGURE 4.35 JAK-STAT signaling. Cytokine signaling works via JAK-STAT transcriptional activation. A cytokine binds to the Janus receptor kinase (JAK) and stimulates the dimerization and autophosphorylation of tyrosines. The STAT subunits bind and are phosphorylated by the JAK dimer. The phosphorylated STAT dimer enters the nucleus and forms a complex with other transcription coactivator proteins to recognize the promoters for genes encoding the differentiation, growth, and proliferation of various blood cells. Adapted from B. M. Alberts, et al. *Molecular Biology of the Cell, Fifth edition*. Garland Science, 2008.

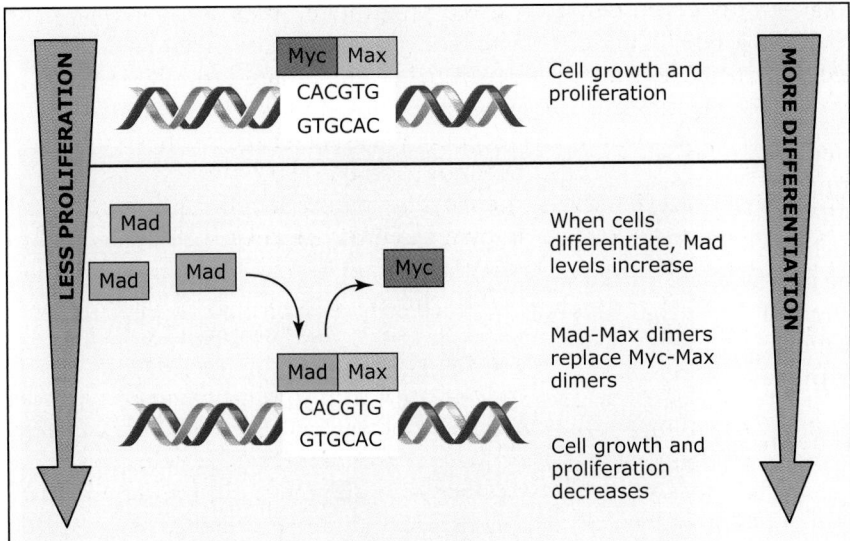

FIGURE 4.36 Transcription activation by Myc family proteins. Myc is the founding member of a bHLH (basic helix-loop-helix) family of proteins that can form homo- and heterodimers. The Myc-Max dimer activates the transcription of genes involved in cell growth and proliferation. The Mad-Max dimer represses the transcription of these same genes. When cells differentiate, Mad levels increase, thereby increasing the formation of Mad-Max dimers and inhibiting the formation of Myc-Max dimers. The end result of the switch from Myc-Max to Mad-Max dimers is a decrease in cell growth and proliferation. Adapted from R. A. Weinberg. *The Biology of Cancer.* Garland Science, 2007.

cells, the expression of the Myc family genes is low. Upon stimulation of cells with growth factors, the Myc and Max genes are activated. All of the members of the Myc family contain a leucine zipper dimerization motif adjacent to a bHLH dimerization-DNA binding motif (**FIGURE 4.36**). Once synthesized, Myc and Max form a heterodimer transcription activation factor that binds to an E-box sequence in the promoter regions of specific genes involved in cellular growth and proliferation including the cyclin genes, thereby activating DNA synthesis and cell division. When growth factors are not present, Mad protein is synthesized and forms a heterodimer with Max. The Mad-Max dimer inhibits Myc-Max formation and also competes with Myc-Max for binding to the E-box sites, thereby acting as a transcriptional repressor to turn off cell proliferation and keep the cells in a quiescent, differentiated state. Thus, control of cell proliferation and differentiation is due to the maintenance of the correct ratio of Myc-Max to Max-Mad complexes. In certain forms of cancer, Myc is overexpressed leading to uncontrolled cell proliferation. In this regard, Myc is referred to as an **oncoprotein** (discussed in more detail in *17 Cancer—Principles and overview.*

Our final example of gene-specific transcription factors is the family of nuclear receptors that includes the steroid hormone receptors for estrogen, progesterone, and testosterone, as well as related nuclear receptors for vitamin D, retinoic acid, and cortisol. These hormones are small molecules that are partially soluble in water, but which freely diffuse across the cellular plasma membrane (**FIGURE 4.37**). Each

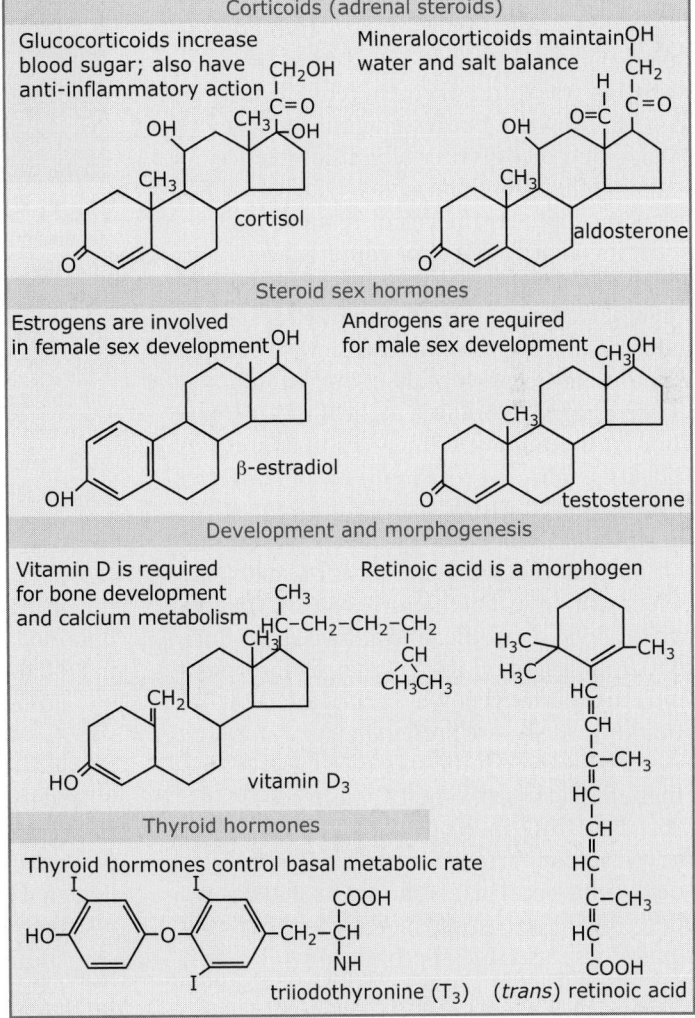

FIGURE 4.37 Steroid and related compounds that activate transcription. Steroids and related small hydrophobic ligands bind to transcription factors that activate specific promoters.

For many years, the expression of genes was investigated one at a time. However, the study of gene expression during cell growth and proliferation and in development and different diseases has required a more global approach in which the temporal expression of a large number of genes can be measured simultaneously. These needs led to the development of two very powerful approaches for gene expression profiling: **microarrays** and the **quantitative reverse-transcriptase polymerase chain reaction (qRT-PCR)**.

Microarrays are arrays of small pieces of DNA from known segments of individual genes that are printed as individual spots on a solid support such as a silicon chip or a glass slide. A single microarray may have hundreds to tens of thousands of different spots representing all or a subset of the known genes in a genome (Figure 4.B1 in the *Methods and Techniques* box). The entire population of mRNAs isolated from cells is labeled with a fluorescent dye and placed in contact with the array. Each of the labeled mRNAs base pairs (**hybridizes**) with those arrayed DNA sequences derived from the gene from which the mRNA was transcribed, thereby causing the hybridized DNAs to fluoresce. A fluorescence-imaging device scans the microarray and records the color, fluorescence intensity, and position of the fluorescent spots, which provides an mRNA expression profile for the cell or tissue.

To compare the levels of mRNAs from different cell types or disease states, two different approaches can be taken. In the dual-color microarray experiment, the mRNAs are isolated from two different cell types, and each mRNA sample is labeled with a different fluorescent dye. The two samples are then mixed together and hybridized to the same microarray. An estimation of the relative differences in expression of specific genes between the two cell types is obtained by measuring the color ratio of the hybridized fluorescent RNAs. For example, the mRNA is isolated from cell type 1 is labeled with a dye that fluoresces red, while mRNA is isolated from cell type 2 is labeled with a dye that fluoresces green. The two samples of mRNA are mixed together and allowed to hybridize to the microarray. If a spot on the microarray fluoresces yellow (green and red produce yellow), one can interpret the result to mean that the mRNAs are present in equal amounts in the two cell types. However, if the spot is red, the gene is expressed at a higher level in cell type 1 than in cell type 2. A green spot would indicate the converse. A fluorescence micrograph of a typical microarray chip with mRNA hybridized using the dual-color approach is shown in **FIGURE 4.B1**. The significant decrease in the cost and increase in the quality of microarrays has led to a second approach in which each individual mRNA sample is hybridized to a separate array. The fluorescence intensities from corresponding spots on the different arrays can then be quantitatively compared by computer-assisted image analysis. This approach eliminates

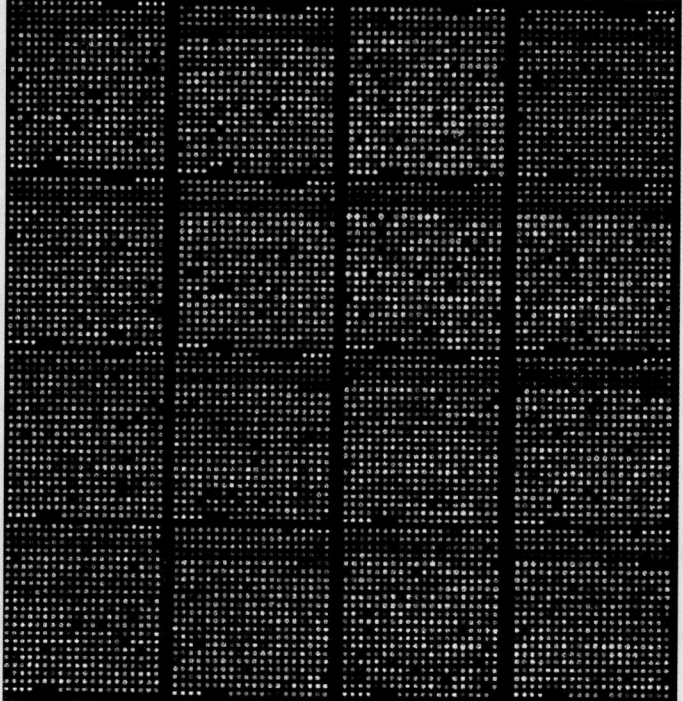

FIGURE 4.B1 A fluorescence microscopy image of a yeast (*S. cerevisiae*) genome microarray spotted with fragments of 6,000 different yeast genes. The microarray was hybridized to a mRNA sample isolated from yeast cells that were grown under two different metabolic conditions. The transcripts from each population were labeled with different fluorescent dyes (Cy3 and Cy5), mixed together and hybridized to the same array. The colors of the spots indicate that the level of expression of a particular mRNA was increased (red) or decreased (decreased) under one condition versus the other. Yellow indicates that there was no difference in expression between the two conditions. Photo courtesy of the University of New Mexico Health Sciences Keck UNM Genomic Resource (KUGR).

the necessity to have dual-color labeling. A display of the data obtained from a comparative microarray analysis is shown in **FIGURE 4.B2**.

Protocols have also been developed to increase the sensitivity of microarrays for detection of mRNAs that are expressed at very low levels. The enzyme reverse transcriptase is used to convert the isolated mRNA to complementary DNA (cDNA) that can be amplified by a millionfold by the PCR. The cDNA is used as a template to prepare RNA that can be detected by direct fluorescent labeling or by secondary fluorescent antibody or streptavidin labeling.

The use of DNA microarrays has been perfected over the past decade and is providing a wealth of information on the differential regulation of gene expression during cellular differentiation, proliferation, and development. Microarray analysis is also yielding clues to the changes in gene expres-

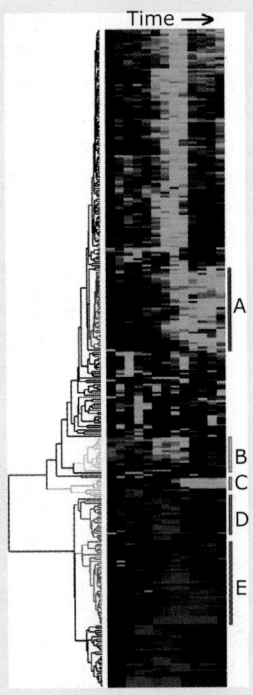

FIGURE 4.B2 An example of an analysis of a microarray experiment. In this example, human fibroblast cells were profiled over time after stimulation with serum. Those mRNAs whose expression increased relative to unstimulated cells are shown in red, while mRNAs whose expression was decreased relative to unstimulated cells are shown in green. Black indicates no significant change in expression. Groups A–E delineate sets of genes that are grouped together by related functions and expression patterns. Reproduced from M. Eisen, et al., *Proc. Natl. Acad. Sci. USA* 95 (1998): 14863–14868. Copyright © 1998 National Academy of Sciences, U.S.A. Photo courtesy of Michael Eisen, University of California, Berkeley.

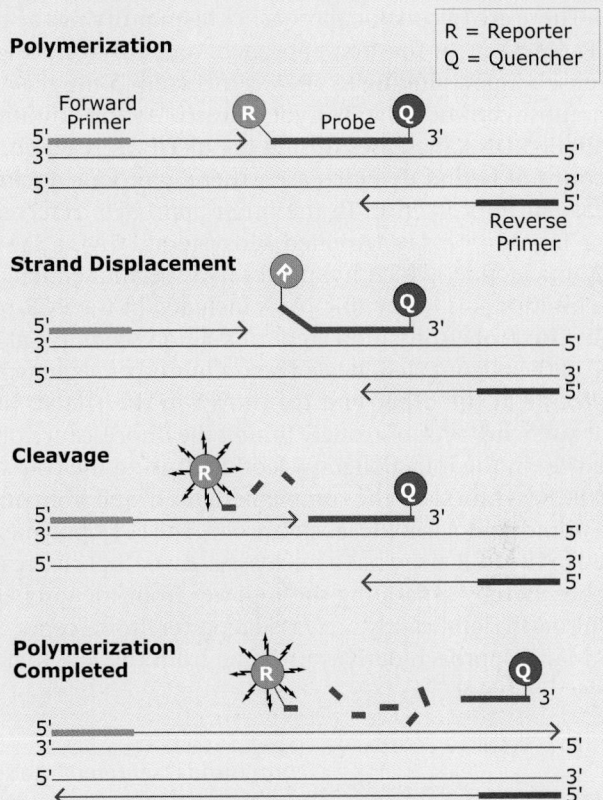

FIGURE 4.B3 The TaqMan® polymerase chain reaction (PCR) assay (Applied Biosystems) is used to quantify the amount of a specific mRNA in an RNA sample. A labeled oligonucleotide probe, which hybridizes to a specific sequence in the interior part of the mRNA, is included in the PCR reaction, which contains a forward and a reverse primer. The probe contains a "reporter" fluorescent dye at one end of the oligonucleotide probe and a non-fluorescent "quencher" molecule at the other end. The quencher acts to strongly inhibit the fluorescence of the reporter in the intact oligonucleotide. During the PCR reaction, synthesis of the complementary strand displaces the oligonucleotide from the strand that is undergoing amplification. The DNA polymerase used in the PCR assay has a nuclease activity that cleaves the reporter from the quencher, resulting in significantly increased reporter fluorescence. The TaqMan® approach derives its name from the fact that the thermostable DNA polymerase used in the assay, Taq DNA polymerase, was originally isolated from the thermophilic bacterium *T. aquaticus*. Courtesy of Life Technologies, Carlsbad, CA.

sion that occur in various diseases including cancer, diabetes, and cardiovascular disease.

A companion technique to microarrays is the **real-time, quantitative reverse transcriptase PCR or qRT-PCR**. Standard **PCR** uses specific DNA oligonucleotide primers to initiate DNA replication across a specific section of the genome. By repeating this process for 20–40 cycles, one can amplify a very small amount of DNA from a specific segment of the genome. A variation of PCR is to use fluorescent dyes to monitor the PCR reaction and to quantify the amount of PCR product produced as a function of time (real-time), which is referred to as a **quantitative PCR** or **qPCR**. Under optimal conditions, the amount of PCR product is directly proportional to the amount of initial starting concentration of DNA at the beginning of the PCR reaction. The qPCR technique can determine the level of a specific mRNA in a specific cell type by using the enzyme reverse transcriptase, which copies mRNA into DNA. In qRT-PCR, total RNA is isolated

from a cell or tissue and is used to produce a **cDNA** copy of each mRNA transcript. The qRT-PCR reaction is carried out on the cDNA using primers to a specific mRNA of interest. Several housekeeping genes, whose mRNA expression levels are known to be relatively constant, are used as standards to obtain quantitative results.

There are two major approaches to quantifying the qRT-PCR reaction. In the first approach, a dye that fluoresces intensely upon binding to DNA is included in the reaction. The most common dye is Cyber Green. As the amount of amplified DNA increases during the qRT-PCR reaction, the amount of bound dye increases, thereby producing an increase in fluorescence. In the other approach, referred to as a TaqMan® assay (Applied Biosystems, CA), a labeled oligonucleotide, which hybridizes to a specific sequence in the interior part of the mRNA, is included in the PCR reaction. The probe contains a "reporter" fluorescent dye at one end of the oligonucleotide and a nonfluorescent "quencher" molecule at the other end (as shown in the **FIGURE 4.B3**). The quencher acts to strongly inhibit the fluorescence of the reporter in the intact oligonucleotide. During the course of the PCR, synthesis of the complementary strand from one of the primers displaces the labeled probe. The DNA polymerase used in the PCR assay has a nuclease activity that cleaves the probe, thereby separating the reporter from the quencher, resulting in significantly increased reporter fluorescence. The TaqMan® approach derives its name from the fact that the thermostable DNA polymerase used in the assay, **Taq DNA polymerase**, was originally isolated from the thermophilic bacterium *Thermus aquaticus*, which lives in hot springs and hydrothermal vents. Taq DNA polymerase can withstand the constant heating-cooling cycles of the PCR process.

The qRT-PCR method has become a popular and widespread approach to examine changes in the expression of specific genes in relationship to cell proliferation, differentiation, and tissue development. qRT-PCR is also frequently used to verify the changes in gene expression detected in large microarray studies. Used together, microarrays and qRT-PCR provide us with a quantitative profile of mRNA synthesis that facilitates the understanding of changes in gene expression. These technologies are revolutionizing our approach to the study of human disease. Many diseases that were previously classified based on physical signs and symptoms are now being classified according to molecular signatures based on differences in gene expression. These signatures are becoming invaluable in disease prognosis and evaluating treatment options.

compound is secreted by a specific cell type and plays a role in tissue-specific transcription activation. In contrast to the membrane-associated receptor signaling discussed in the previous examples, steroid hormone receptors usually bind their ligands in either the cytoplasm or the nucleus, depending upon the receptor. The resulting hormone-receptor complex is an active transcription factor that can bind to its specific activation sequence, known as a **hormone-response element**, located in the enhancer regions of target genes (**FIGURE 4.38**). Each steroid hormone receptor contains a ligand-binding domain near its C-terminus, a sequence-specific DNA-binding domain in the middle of the protein, and an activation domain at the N-terminus. The DNA-binding domains of most steroid hormone receptors contain a ZF motif located adjacent to a basic sequence that recognizes the specific hormone-response element DNA sequence. The steroid hormone-response elements consist of two short half sites separated by 1 to 5 bp; thus, the hormone receptors function only as dimers—some steroid receptors function as homodimers while others are heterodimers. In a manner very similar to CREB, the hormone-bound steroid receptor binds to its response element and recruits the

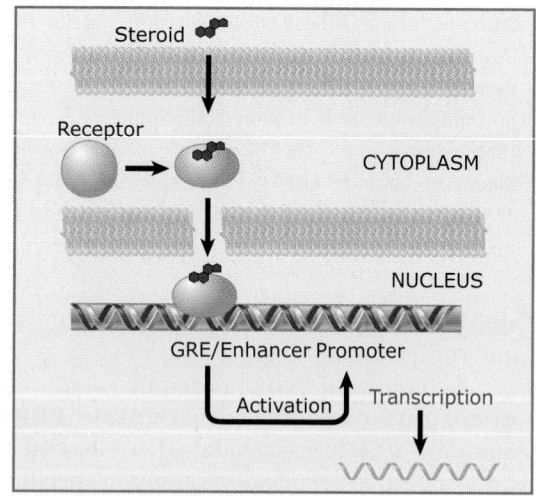

FIGURE 4.38 Steroid-receptor transcription factors. The glucocorticoid receptor is an example of a steroid hormone-receptor transcription factor, which binds to cortisol and interacts with the glucocorticoid response element (GRE) located within the enhancers of promoters for specific genes.

coactivators PCAF and dimeric protein CBP/p300 to the promoter, causing acetylation of histones and displacement of nucleosomes within the core promoter regions of steroid hormone-responsive genes (**FIGURE 4.39**).

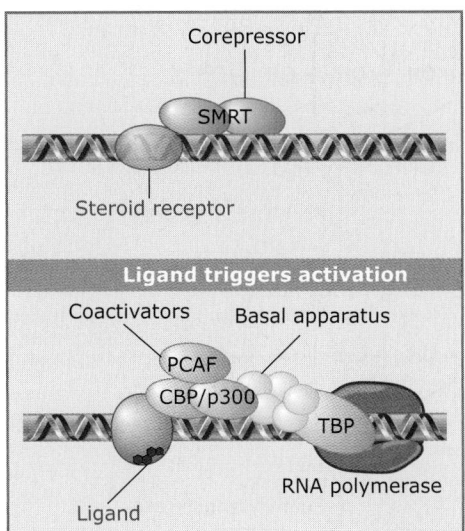

FIGURE 4.39 Steroid receptors use coactivators and corepressors to control transcription activation. The retinoic acid receptor (RAR) binds the SMRT corepressor in the absence of retinoic acid, thereby repressing transcription. When retinoic acid binds to the RAR, SMRT is replaced by the co-activators complex containing the proteins PCAF, CBP, and p300, which leads to activation of the general transcription apparatus bound at the core promoter of retinoic acid-responsive genes.

In summary, we have seen examples of three different mechanisms for signal transduction-mediated regulation of transcription factor activation: phosphorylation in the cases of CREB and STAT, heterodimerization in the case of Myc-Max, and allosteric-induced conformational changes by hormone binding in the case of the steroid receptors. Although there are cases where transcription factors may interact directly to recruit or stabilize the assembly of the basal transcription apparatus, many activator proteins can function indirectly by promoting the displacement of nucleosomes and clearing the core promoter region for the assembly of the basal transcription apparatus and other transcription factors. Thus, transcription activation in metazoan cells is a complex process that is based on combinations and permutations of factors whose assembly and function are governed by signals from the cellular environment.

Concept and Reasoning Check

1. Some breast cancer cells express estrogen receptors that increase the growth rate of the tumor. A number of anticancer drugs are estrogen analogs that inhibit the binding of estrogen to the estrogen receptor. Over time, these drugs often lose their efficacy. Propose at least two different hypotheses for how the resistance to the drugs could develop.

Key concepts

- The 5' end of a eukaryotic mRNA is "capped" with a 7-MeG ribonucleotide that protects the 5' end of the transcript from enzymatic degradation, and functions as a binding site for proteins involved in mRNA export from the nucleus to the cytoplasm and for translation initiation factors in the cytoplasm.
- The 3' end of the mRNA is generated by an endonucleolytic cleavage reaction and the polymerization of a large number of adenosines to form the poly(A) tail. The poly(A) tail protects the 3' end of the transcript from degradation and modulates the efficiency of translation initiation.

Most mRNA transcripts in eukaryotic cells are not functional when first synthesized and must go through one or more steps to convert them into active forms. The initial transcript (also called the primary transcript or pre-mRNA) is converted to the active or mature transcript through a series of covalent modifications. The first processing reaction is the addition of the **7-MeG** cap; the second processing reaction is the addition of 50–300 adenine ribonucleotides onto the 3' end to form the poly(A) tail. In addition, most eukaryotic mRNAs are also spliced. In this section, the capping and polyadenylation processing reactions will be described, and in the following section, RNA splicing will be presented.

Capping of the RNA transcript with 7-MeG is unique to RNA polymerase II transcripts. The cap is important for mRNA stability and translation, and also plays a role in mRNA nucleocytoplasmic transport. The cap is added to the 5' end of the transcript as soon as the RNA emerges from the active site of RNA polymerase II. The structure of the 7-MeG cap is shown in **FIGURE 4.40** and the steps in the capping process are shown in **FIGURE 4.41**. In the first step, an RNA triphosphatase removes the γ-phosphate of the triphosphate group on the first base at the 5' end (which is usually a purine, A or G). Then a guanyltransferase puts a GMP on the 5' end. This reaction is not the conventional 5' to 3' linkage found in all other types of nucleotidyl transfer reactions; instead, the linkage is 5' to 5'. In the final step, the enzyme N^7G-methyltransferase transfers a methyl group from *S*-adenosylmethionine to the N7 position

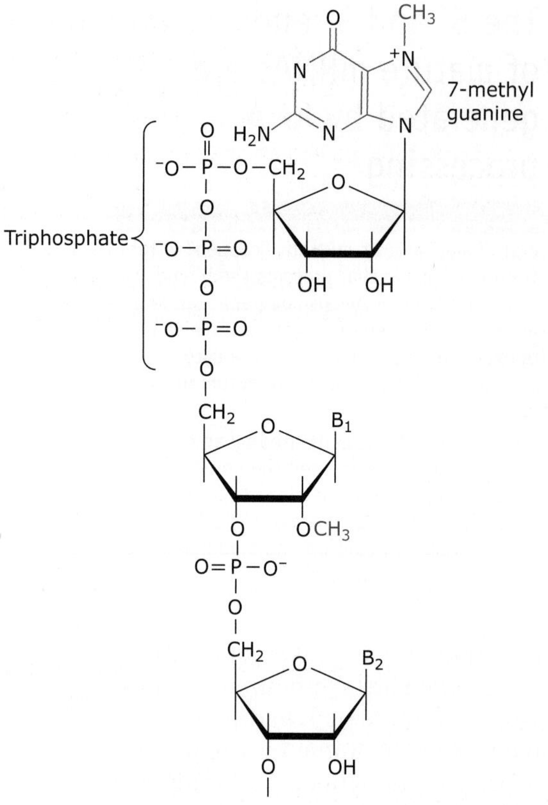

FIGURE 4.40 Structure of the 7-methylguanosine cap. The 7-methylguanosine (7-MeG) cap contains a methyl group on the 7-position of guanine and on the 2′ position of the ribose on the base of the original 5′ end of the mRNA. The 7-MeG cap is attached to the 5′ end via an unusual 5′→5′ triphosphate linkage.

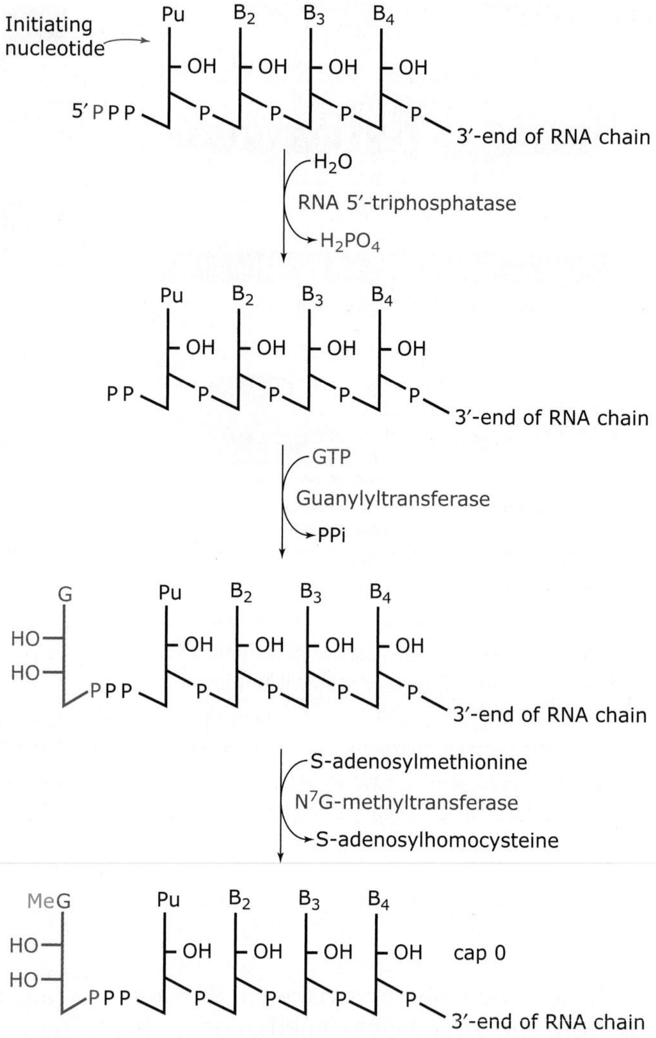

FIGURE 4.41 Pathway for the synthesis of the 7-MeG cap. The γ-phosphate on the 5′ end of the pre-mRNA is removed by a RNA triphosphatase. Guanalyltransferase catalyzes the addition of guanosine monophosphate to the 5′ end of the pre-mRNA. N^7G-methyltransferase catalyzes the methylation of the terminal guanosine.

of the added guanine at the 5′ end. Subsequent to the addition of the G, the 5′ end becomes the substrate for several methylation events. While every mRNA in the cell is capped, the proportions of the different methylated types of cap are characteristic for a particular organism.

In addition to the 7-MeG cap, almost all mRNAs undergo an additional posttranslational modification in which a poly(A) tail is added to the 3′ end of the transcript (histone mRNAs are an important exception). The poly(A) tail is not encoded by the gene sequence, but is added by the enzyme poly(A) polymerase in concert with a number of polyadenylation factors. The polyadenylation reaction occurs in the following two steps:

1. In the first step, a large complex consisting of more than a dozen proteins recognizes the **polyadenylation signal** AAUAAA on the pre-mRNA and directs the cleavage of the transcript just after the first CA dinucleotide located within 10 to 30 nucleotides downstream of the beginning of the AAUAAA (FIGURE 4.42).

2. In the second step, poly(A) polymerase adds the poly(A) tail. The initial length of the poly(A) tail ranges from 50 to 80 adenines in lower eukaryotes such as yeasts to 200 to 300 adenines in mammals. Elements located upstream and downstream from the polyadenylation

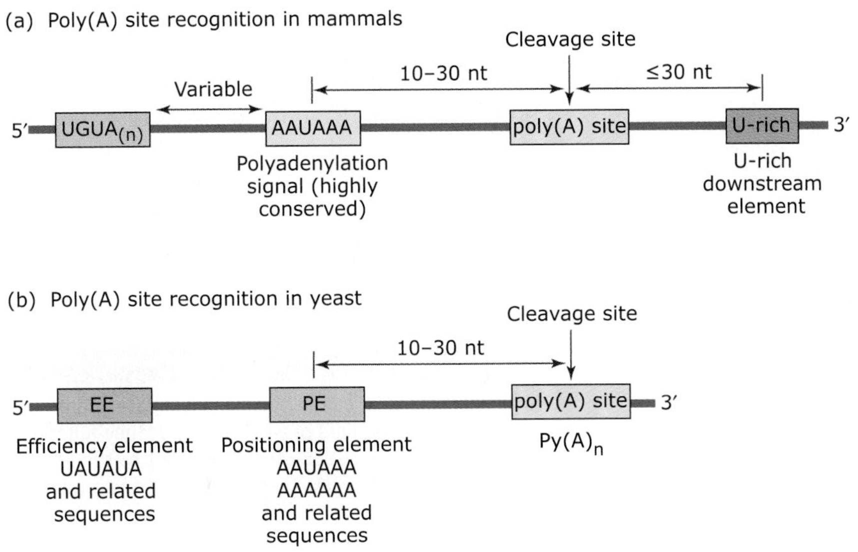

(a) Poly(A) site recognition in mammals

(b) Poly(A) site recognition in yeast

FIGURE 4.42 Polyadenylation signals in mammals and yeast. (a) The poly(A) signal in mammals is comprised of the highly conserved sequence AAUAAA located 10–30 nucleotides upstream of the cleavage site. In most cases, additional polyadenylation enhancer elements are located upstream and downstream of the cleavage site. (b) The poly(A) signal in yeast is comprised of elements that are more variable than mammals. Part A adapted from G. M. Gilmartin, *Genes Dev.* 19 (2005): 2517–2521. Part B adapted from J. Zhao, L. Hyman, and C. Moore, *Mol. Biol. Rev.* 63 (1999): 405–445.

Factor	Processing Step	Function
CPSF cleavage/polyadenylation specificity factor	Cleavage and poly(A) addition	Contains four subunits. CPSF-73 probably cleaves pre-mRNA at the poly(A) site. CPSF-160 binds to AAUAAA and CPSF-30 and CPSF-100 subunits increase specificity.
CstF cleavage stimulation factor	Cleavage	Contains three subunits (CstF-77, CstF-64, and CstF-50). CstF-64 binds to the U-rich sequence; CstF-77 bridges CstF-64 and CstF-50 and contacts CPSF-160.
CFI cleavage factor I	Cleavage	Recognizes sequence elements in poly(A) site.
CFII cleavage factor II	Cleavage	Unknown
PAP poly(A) polymerase	Cleavage and poly(A) addition	Catalyzes poly(A) formation.
PAB II poly(A) binding protein	Poly(A) elongation	Binds poly(A) and CPSF-30; responsible for processive poly(A) elongation and for the tail length
CTD Carboxyl terminal domain of large subunit in RNA polymerase II	Cleavage	Binds CPSF and CstF.
Ssu72 and PC4	Cleavage	Ssu72 and PC4 proteins interact with the general transcription factor TFIIB and with components of cleavage/polyadenylation machinery.
Symplekin	Cleavage and poly(A) addition	Symplekin helps to assemble or stabilize the CstF complex and there by helps to hold the complete cleavage/polyadenylation machinery together.

FIGURE 4.43 Components of the mammalian cleavage/polyadenylation machinery.

signal can enhance the efficiency of the polyadenylation reaction.

Several additional proteins are also required for cleavage and polyadenylation as depicted in **FIGURE 4.43**. The key protein involved in mammalian polyadenylation is the **cleavage/polyadenylation specificity factor** (**CPSF**), which binds to the AAUAAA sequence and directs both cleavage and polyadenylation (**FIGURE 4.44**). Homologs of CPSF are found in

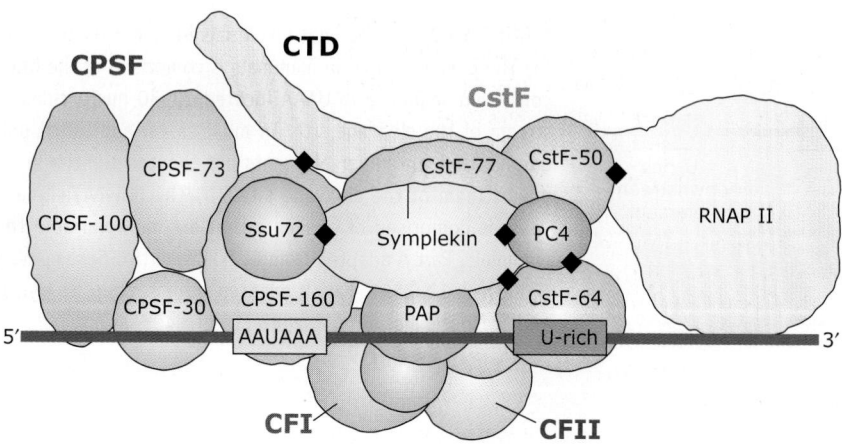

FIGURE 4.44 The mammalian polyadenylation machinery. The black diamonds indicate experimentally confirmed protein-protein interactions between specific components of the complex. CPSF-160 is bound to the AAUAAA signal and CstF-64 is bound to U-rich downstream element. The complex is assembled with the aid of the carboxyl-terminal domain (CTD) of the largest subunit of RNA polymerase II. Adapted from O. Calvo and J. L. Manley, *Genes and Dev.* 17 (2003): 1321–1327.

ing or shortly after the transcript is exported to the nucleus.

In summary, eukaryotic mRNAs are post-transcriptionally modified to add a 7-MeG cap to the 5′ end and a poly(A) tail to the 3′ end. The role of the cap and poly(A) tail in mRNA export will be discussed in 9 *Nuclear structure and transport*. In the next section, we will explore how RNA polymerase terminates the transcription reaction.

Concept and Reasoning Check

1. All RNAs synthesized by RNA polymerase II are capped, regardless of whether or not they function as mRNAs or are ncRNAs. What might be the function of cap for RNAs that do not encode proteins?

4.9 Terminators direct the end of transcription elongation

Key concepts

- Transcription terminators are sequences that signal RNA polymerase to pause, release the RNA transcript, and dissociate from the polymerase.
- There are two classes of prokaryotic transcription terminators: intrinsic terminators that are RNA structures that do not require protein factors for function, and factor-dependent terminators. Intrinsic terminators consist of a GC-rich stem-loop structure followed by 4–8 uridine residues. The most common type of factor-dependent terminator requires the ATP-dependent Rho helicase for transcript release.
- Each of the three eukaryotic nuclear RNA polymerases uses a different transcription termination strategy. RNA polymerase I uses a combination of DNA sequences and DNA-binding proteins. RNA polymerase III uses RNA sequences. RNA polymerase II couples transcription termination to cleavage and polyadenylation of the transcript.

all eukaryotes. Mammalian CPSF consists of four subunits, CPSF-160, CPSF-100, CPSF-73, and CPSF-30. CPSF-73 is thought to be the actual cleavage enzyme that cuts the mRNA just after the CA dinucleotide to form the 3′ end for the addition of the poly(A) tail. The cleavage stimulatory factor CStF binds to a GU-rich sequence downstream of the AAUAAA site and participates only in the cleavage reaction. Two other components, **cleavage factors I and II (CFI and CFII)** are required to stimulate the cleavage reaction. The second function of CPSF is to hold the poly(A) polymerase on the end of the poly(A) tail so that the polymerization proceeds rapidly without the polymerase falling off the end. The first ~10 adenines are added slowly, but once these are polymerized, the reaction proceeds rapidly to produce the full 200 to 300-residue length tail.

Another protein important for the regulation of poly(A) tail length is the nuclear poly(A) binding protein, PABPN1. PABPN1 binds to the poly(A) tail as a linear array of monomers as the tail is synthesized and helps CPSF to maintain the speed of the reaction. However, once the tail has reached the 200 to 300 adenine length, the poly(A)-bound PABPN1 condenses into an oligomeric particle that functions to dissociate poly(A) polymerase from CPSF and the 3′ end of the poly(A) tail, thereby terminating the polyadenylation reaction. PABPN1 remains bound to the poly(A) tail for the duration of the time that the mRNA is in the nucleus, but dissociates from the mRNA poly(A) tail dur-

Although transcription initiation is of central importance in gene regulation, transcription termination is an essential process for ensuring that adjacent genes are transcribed in an independent and orderly manner and that the transcription apparatus is recycled once synthesis of the transcript has been completed. The termination reaction is comprised of three steps: (1) the cessation of RNA polymerase elongation, (2) the dissociation of the transcript from the DNA template, and (3) the dissociation of RNA polymerase from the template. The sequence that directs this process is called the **transcription termination signal**, and the **transcription stop site** is the last nucleotide

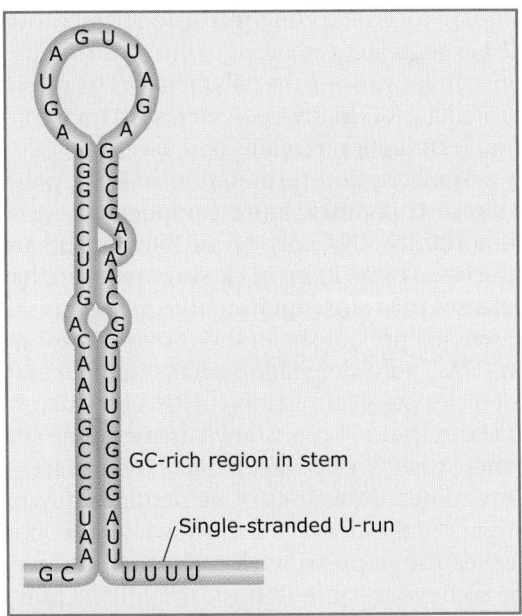

GC-rich region in stem

Single-stranded U-run

FIGURE 4.45 The structure of a bacterial intrinsic terminator. Intrinsic terminators in bacteria have an RNA structure that includes a thermodynamically stable GC-rich stem-loop structure followed by several uridine nucleotides.

in the template that directs the synthesis of the transcript. In prokaryotes, termination generates the 3' termini of most transcripts, although some transcripts may be processed by endo- and exonucleases. However, in eukaryotes, the 3' termini of nearly all species of RNAs are generated by posttranscriptional processing, which may include endonucleolytic or exonucleolytic cleavages, and in the case of mRNAs addition of a poly(A) tail. In this section, we will focus on the signals and proteins responsible for transcription termination.

Terminators are classified in prokaryotes according to whether or not RNA polymerase requires additional protein factors to recognize the termination signal. If RNA polymerase efficiently terminates transcription at specific locations on the DNA in the absence of any other factors, these sites are called **intrinsic terminators**. Intrinsic prokaryotic terminators have two structural features (**FIGURE 4.45**): there is a stable GC-rich stem-loop structure followed by a stretch of U residues; both features are required for termination. The

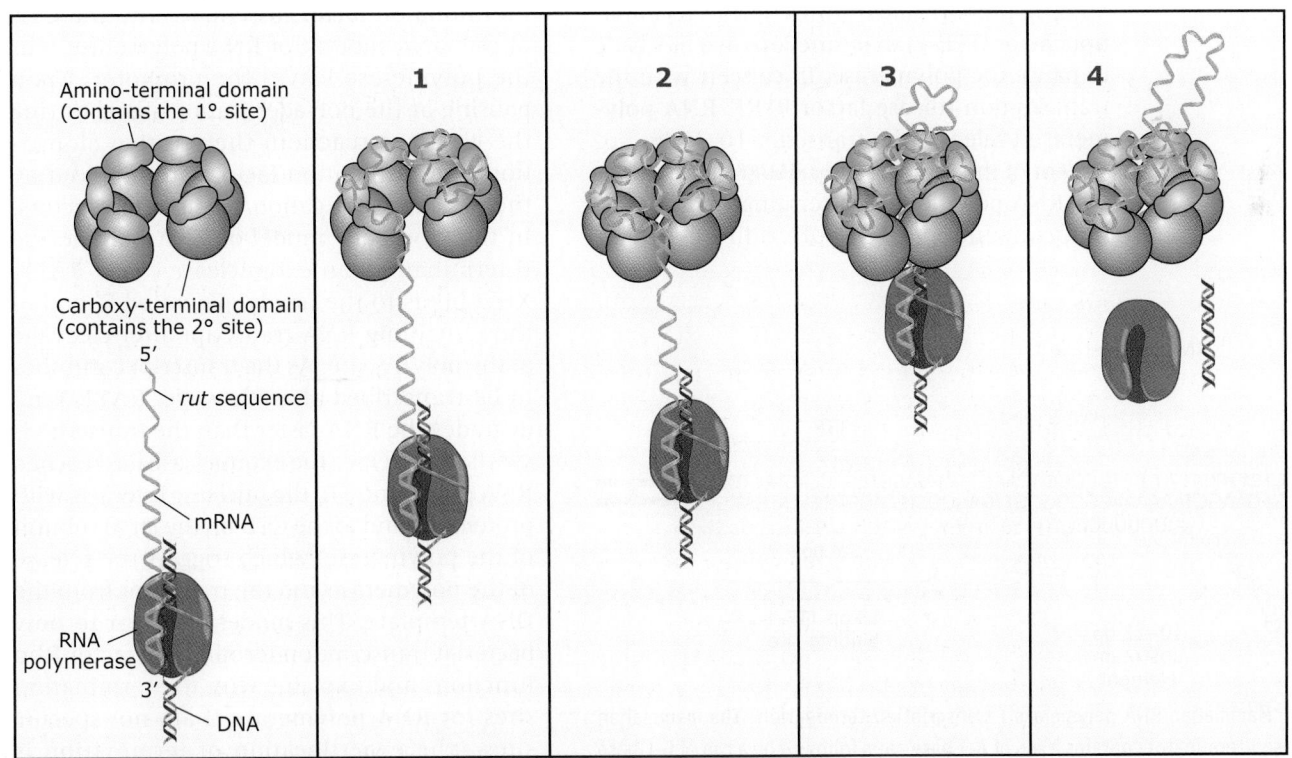

FIGURE 4.46 A model for Rho-dependent transcription termination. The Rho-dependent transcription termination reaction occurs in four steps. Step 1: The amino-terminal domain of Rho in the open-ring conformation binds the Rho-utilization (rut) sequence on mRNA. Step 2: The ring closes and rho binds a second site on the mRNA, thereby activating the rho ATPase. Step 3: Rho reels the mRNA through its hexamer ring by coupling its helicase activity to the hydrolysis of ATP. Step 4: the mRNA-Rho complex is dissociated from RNA polymerase, which disengages RNA polymerase from the DNA. Adapted from D. L. Kaplan and M. J. O'Donnell, *Curr. Biol.* 13 (2003): R714–R716.

stem-loop structure signals RNA polymerase to pause while the U-rich region increases the instability of the DNA-RNA hybrid inside the active site of the polymerase, promoting the dissociation of the transcript from the ternary complex. Often, the actual termination event takes place at any one of several positions toward or at the end of the U-rich region.

Rho is the most common bacterial transcription termination factor, and is an ATP-dependent helicase comprised of six identical subunits. **FIGURE 4.46** depicts the current model for how Rho functions. First Rho binds to the **rho-utilization (*rut*) sequence** within the the RNA transcript, which stimulates the Rho ATP-dependent helicase activity. Rho remains tethered to the *rut* site while translocating along RNA transcript in a process referred to as **tethered tracking**. RNA polymerase continues to transcribe the template until it stops at a strong pause site, allowing Rho to catch up and release the transcript and RNA polymerase from the DNA template.

Our understanding of transcription termination in eukaryotes is less detailed than in prokaryotes. RNA polymerase I transcription termination involves a sequence-specific DNA-binding protein called transcription termination factor (TTF-1) that functions as a blockade to pause the polymerase. In concert with the transcription release factor PTRF, RNA polymerase I releases the transcript 10–12 bp upstream of the binding site (**FIGURE 4.47**).

RNA polymerase III terminates at a location containing a series of uridine residues,

similar to prokaryotic intrinsic terminators. Other sequences adjacent to the U run are important for pausing the polymerase. The role of protein factors in RNA polymerase III transcription termination remains obscure.

Transcription termination of RNA polymerase II is much more complicated. Most importantly, RNA polymerase II transcripts are processed by an internal cleavage reaction that releases the transcript from the polymerase as discussed previously. In the case of almost all mRNAs, polyadenylation occurs immediately after cleavage. Termination of RNA polymerase II transcription occurs downstream of the site where the RNA cleavage reaction took place—sometimes thousands of nucleotides downstream of the cleavage site. At first glance, this leaves the impression that termination may be somewhat superfluous during RNA polymerase II transcription. However, experimental evidence demonstrates that the cleavage and polyadenylation reactions are tightly coupled to transcription termination.

FIGURE 4.48 shows to primary models of how coupling of transcription termination to polyadenylation might occur. In the "antitermination" model, positive elongation/antitermination factors assemble onto the CTD of the large subunit of RNA polymerase II as the polymerase leaves the promoter. Upon pausing at the polyadenylation signal during the cleavage reaction, the positive elongation/antitermination factors are replaced by the negative elongation/termination factors. In the "torpedo" model of RNA polymerase II termination, an exonuclease called RAT1/Xrn2 binds to the newly generated 5′ end of the remaining RNA transcript after cleavage at the poly(A) site. As the transcript continues to be transcribed after cleavage, RAT1/Xrn2 degrades the RNA faster than the transcript is synthesized. Once the exonuclease has reached RNA polymerase II, the nuclease interacts with proteins bound to the CTD on the large subunit of the polymerase, which triggers the release of the polymerase and the transcript from the DNA template. This model is similar to how bacterial transcription termination factor Rho functions and explains why the termination sites for RNA polymerase II are not specific sites—the exact location of termination is determined by the rate of elongation of RNA polymerase on the specific gene sequence and the speed at which the exonuclease degrades the specific transcript. In both models, the

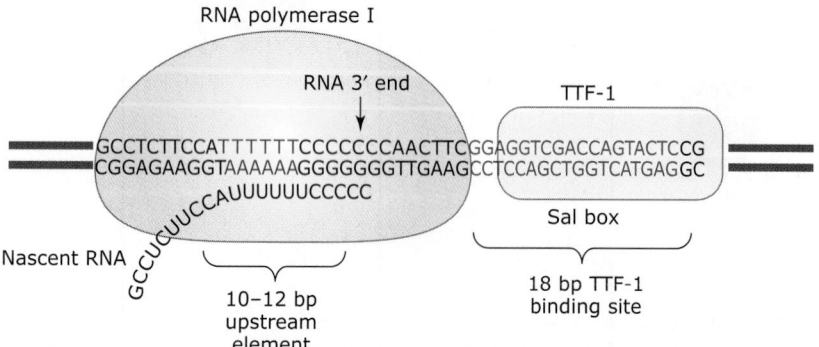

FIGURE 4.47 Mammalian RNA polymerase I transcription termination. The mammalian RNA polymerase I terminator contains a run of A-T base pairs followed by a run of G-C base pairs, which causes the RNA polymerase I to pause on the DNA template. These sequence elements are located upstream of the binding site for the transcription termination factor TTF-1, which helps to release RNA polymerase I from the DNA template. Adapted from R. H. Reeder and W. H. Lang, *Trends Biochem. Sci.* 22 (1997): 473–477.

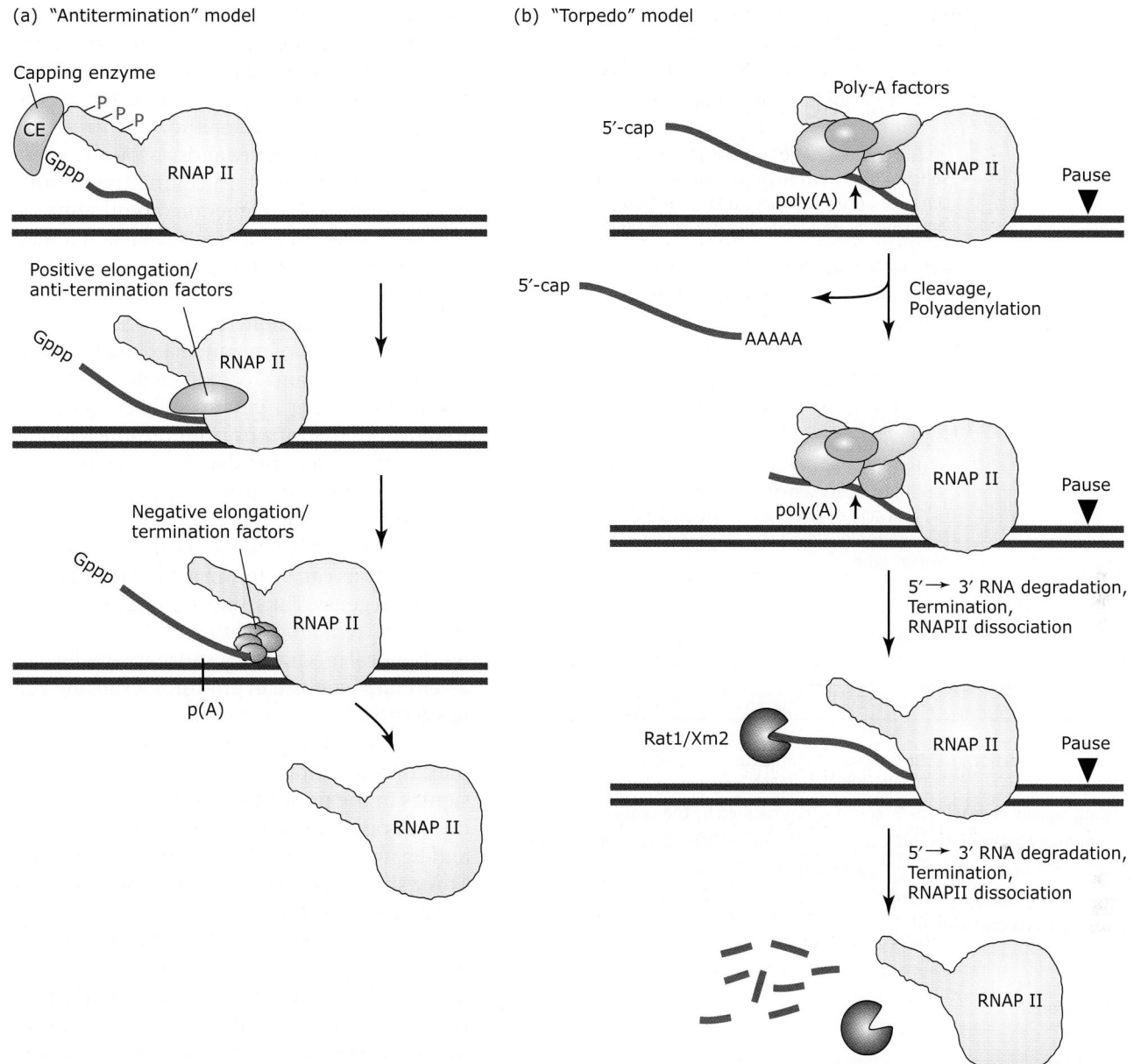

(a) "Antitermination" model

Capping enzyme

CE
Gppp
P P P
RNAP II

Positive elongation/
anti-termination factors

Gppp
RNAP II

Negative elongation/
termination factors

Gppp
RNAP II
p(A)

RNAP II

(b) "Torpedo" model

Poly-A factors

5'-cap
poly(A) ↑
RNAP II
Pause

5'-cap
Cleavage,
Polyadenylation
AAAAA

poly(A) ↑
RNAP II
Pause

5' → 3' RNA degradation,
Termination,
RNAPII dissociation

Rat1/Xrn2
RNAP II
Pause

5' → 3' RNA degradation,
Termination,
RNAPII dissociation

RNAP II

FIGURE 4.48 Two models for termination of RNA polymerase II transcription. (a) The "antitermination model" proposes that elongation factors are loaded onto the phosphorylated CTD of RNA polymerase II, thereby preventing premature termination during transcription elongation. Once RNA polymerase II passes the poly(A) site, termination factors replace the elongation factors and stimulate termination. (b) The "torpedo" model proposes that when cleavage of the transcript occurs at the poly(A) site, the exonuclease RAT1/Xrn2 digests the nascent transcript from the uncapped 5' end in a 5'→3' direction, leading to the dissociation of RNA polymerase from the DNA template. Adapted from S. Buratowski, *Curr. Opin. Cell Biol.* 17 (2005): 257–261.

cleavage event provides an indirect trigger for termination of transcription at a distal site from the polyadenylation signal. The intricacy of the coupling between polyadenylation and termination suggests that transcription termination is extremely important to the cell, most likely for the elimination of interference by nonterminated RNA polymerase II complexes in the process of transcribing adjacent genes, and for the recycling of RNA polymerase II molecules for new rounds of transcription.

Concept and Reasoning Check

1. Why is it necessary to terminate transcription of RNA polymerase II if the transcript is already released by the cleavage step in polyadenylation?

4.10 Introns in eukaryotic pre-mRNAs are removed by the spliceosome

Key concepts

- Splicing of introns from pre-mRNAs is a highly conserved process found in all eukaryotic organisms. Almost all mRNAs in eukaryotic organisms contain a number of introns ranging from 1 to 2 per gene in yeast to an average of 8 per gene in mammals. However, some genes have hundreds of introns.
- Most exon-intron junctions have the sequence AG/GU at the 5' boundary and AG/G at the 3' boundary.
- The major steps of splicing are: (1) cleavage of the 5' exon-intron junction, (2) covalent attachment of the 5' end of the intron to the branch point located just upstream of the 3' exon-intron junction, (3) cleavage at the 3' exon-intron junction, and (4) ligation of the upstream exon to the downstream exon.

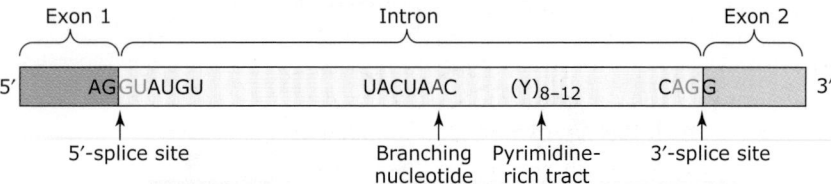

FIGURE 4.49 Splicing signals in yeast. The 5' splice site is defined by the sequence AG/GU (the / denotes the boundary between the exon and intron). The branching nucleotide is the A denoted in red inside the branch site. The 3' splice site is defined by the sequence AG/G (the / denotes the boundary between the intron and exon). The sequence $(Y)_{8-12}$ denotes a tract of pyrimidines (C or U) that enhances splicing.

(a) Human GU-AG introns

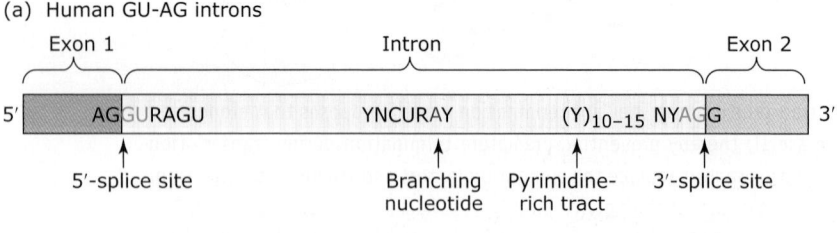

(b) Human AU-AC introns

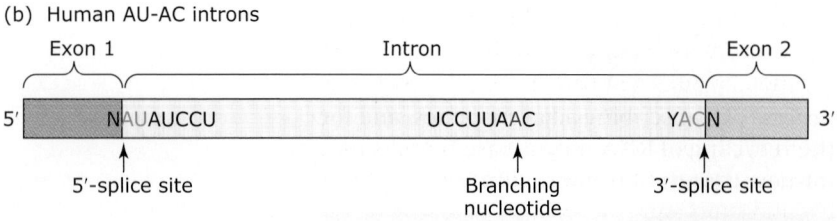

FIGURE 4.50 Splicing signals in mammals. There are two classes of introns in mammals named on the basis of the 5' intron boundary sequences: (a) the major class designated GU-AG introns and (b) a minor class designated AU-AC introns. The splicing signals in mammals are less conserved than in yeast. Y designates a pyrimidine base (U or C); R designates a purine base (G or A).

The removal of introns from pre-mRNA transcripts is a critical intermediate step in the expression of protein-coding genes. All organisms or the viruses that infect them have interrupted genes in their genomes, although introns are very rare in prokaryotes and are found in only a small proportion of unicellular eukaryotic genes. However, in higher eukaryotes introns in protein-coding genes are the rule rather than the exception, with the average human gene containing ~8 introns. At least a dozen genes in the human genome have more than 100 introns. The sizes of mammalian introns range between ~1,000 and 700,000 nucleotides, with the average being 1,000 to 2,000 nucleotides. Removal of the introns from pre-mRNA produces mature mRNAs that range in size from ~1,000 to 100,000 nucleotides. Although introns were originally thought to correspond to specific structural or functional domains in proteins, introns can be found in the UTRs as well as ncRNA genes. Thus, the origins and purposes of introns remain unclear. In this section, the biochemical mechanisms and macromolecular assemblies involved in pre-mRNA splicing will be discussed.

The spliceosome is a very large RNP complex that functions to recognize the splicing signals in the pre-mRNA that delineate exon-intron boundaries and to catalyze the cleavage and ligation reactions in the splicing pathway. Three short conserved RNA sequences known as **splicing signals** are required for splicing. The splicing signals in yeast are shown in FIGURE 4.49 while the analogous signals in mammals are shown in FIGURE 4.50: the **5' splice junction** that determines the 5' end of the intron; the **3' splice junction** that determines the 3' end of the intron; and the **branch point**, which is a sequence where the 5' end forms a covalent linkage at an internal segment of the intron to form a lariat-like RNA intermediate. The mechanism of splicing involves **transesterification** reactions catalyzed by RNA-protein components of the spliceosomes as shown in FIGURE 4.51. While only mRNA splicing in higher eukaryotes will be discussed in this chapter, there are ncRNAs that can autocatalyze the removal of their own introns by transesterification reaction mechanisms very similar to those in pre-mRNA splicing. Thus, it appears that RNA splicing is a highly conserved process that has evolved in complexity throughout biological evolution.

Exon-intron boundaries are identified by the splicing apparatus through recognition

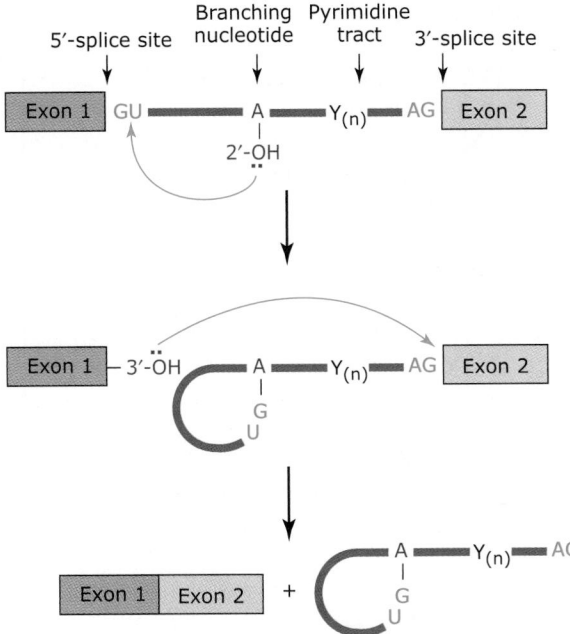

Branching Pyrimidine
5'-splice site nucleotide tract 3'-splice site

FIGURE 4.51 The transesterification steps in mRNA splicing. The mRNA splicing reaction consists of two transesterification steps. In the first step, the 2′ hydroxyl group on the branching nucleotide (and adenosine) carries out a nucleophilic attack on the first nucleotide on the intronic side of the 5′ splice junction (usually a uridine). In the second step, the hydroxyl group on the cleaved 3′ end of the exon carries out a nucleophilic attack on the last nucleotide of the intronic side of the 3′ splice junction, thereby joining the two exons and releasing a covalently closed lariat intermediate.

of short 5′ and 3′ splice site and branch point consensus sequences. Figure 4.50 shows the two major classes of mammalian exon-intron boundary sequences. In the most commonly found splice sites, the 5′ exon-intron boundary is AG/GU, where the last two nucleotides at the 3′ end of the exon are AG and the first two nucleotides of the 5′ end of the intron are GU. Likewise, the 3′ exon-intron boundary is AG/G, where the last two nucleotides of the 3′ end of the intron are AG and the first nucleotide of the 5′ end of the next exon is G. Thus, the intron defined by these signals begins with the dinucleotide GU and ends with the dinucleotide AG. It is obvious that these sequences alone cannot be the entire story for exon-intron recognition—the random chance of encountering the sequence AGGU is on average once per 256 nucleotides, and even more frequent for the sequence AGG. However, other regions of the transcript that are not easily identifiable by sequence alone play an important part in exon-intron recognition. The capacity of RNA to form stem-loop secondary structure and higher-order conformation guide the splicing reaction so that the correct splice sites are recognized and the appropriate introns are removed and the exons are spliced together in the correct order.

The steps in the mRNA splicing reaction pathway are shown in FIGURE 4.52. It is important to keep in mind that the fragments generated in the splicing reaction are held together during splicing by the spliceosome, and

not released until after splicing has been completed. The splicing apparatus contains both proteins and small RNAs (in addition to the pre-mRNA). The five small nuclear RNAs are referred to as **snRNAs**, and include the U1, U2, U4, U5, and U6 snRNAs. There are 10^5 to 10^6 copies of each of these snRNAs per cell. Each of the snRNAs exists as a complex with a set of specific proteins; these **small nuclear RNP complexes (snRNPs)** are sometimes referred to as **snurps**. The snRNPs assemble with a large number of other proteins into a complex with the pre-mRNA to form the spliceosome, which has a size greater than 50 to 60S (see *4.13 Translation is catalyzed by the ribosome* for an explanation of the S value). By point of reference, ribosomes range in size from 50 to 80S. The snRNP particles are involved in the recognition of the splice junctions and the branch point and interact to guide the assembly of the other protein components to form the complete spliceosome. There is significant evidence to suggest that the snRNAs play a more direct role in the catalysis of the splicing reaction, while the proteins help the snRNAs and pre-mRNA adopt active conformations and promote the formation of critical base-pairing interactions necessary for spliceosome assembly and function.

In the first step of the splicing reaction, cleavage occurs at the 5′ end of the exon-intron splice junction, resulting in the separation of the 5′ end of the intron from the 3′ end of the left exon. Next, the 5′ end of the intron forms

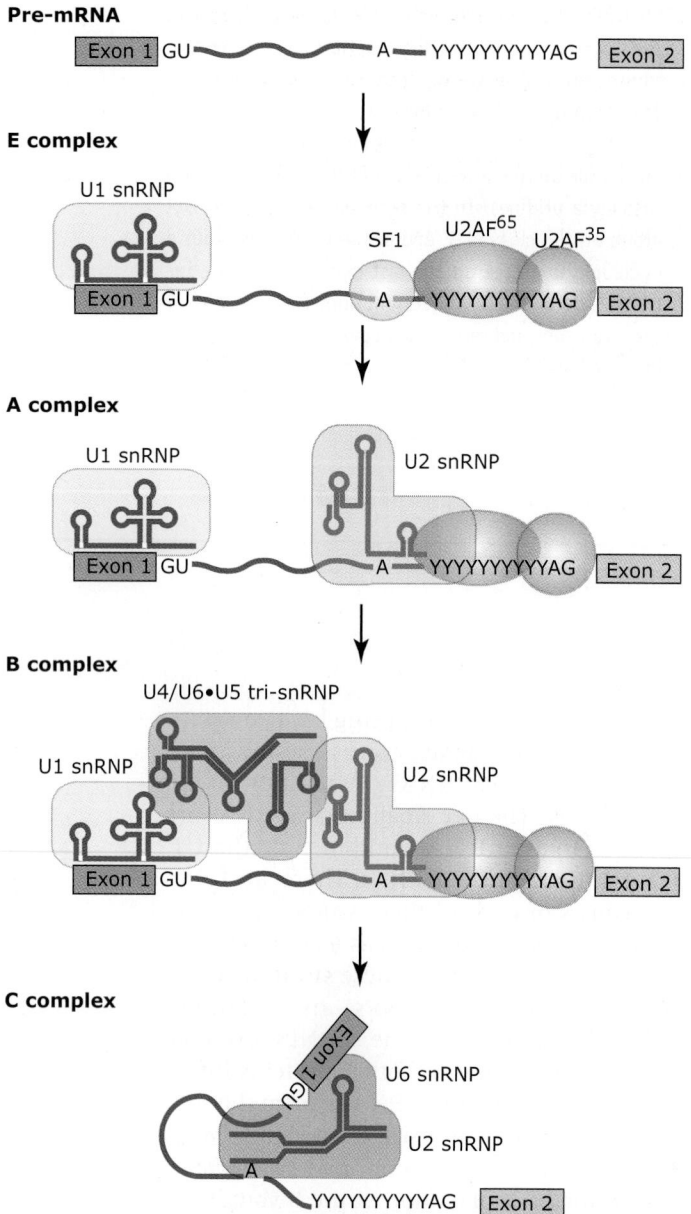

FIGURE 4.52 Spliceosome assembly. The spliceosome is comprised of five snRNA ribonucleoprotein complexes (snRNPs), as well as a large number of auxiliary proteins, including SF1 and U2AF, which help guide the splicing reaction. The steps of splicing are designated according to the spliceosome intermediates, which can be isolated by biochemical techniques. In the formation of the E complex, the U1 snRNP binds to the 5' splice site. SF1 binds to the branch point guided by U2AF, which binds to the polypyrimidine tract near the 3' splice junction. The U2 snRNP displaces SF1 to form the A complex. The U4/U6 • U5 snRNP complex joins the spliceosome to form the B complex. Upon release of the U4 and U1 snRNPs, the C complex containing U2, U5 (not shown), and U6 completes the splicing reaction to yield the mature mRNA. Adapted from K. J. Hertel and B. R. Graveley, *Trends Biochem. Sci.* 30 (2005): 115–118.

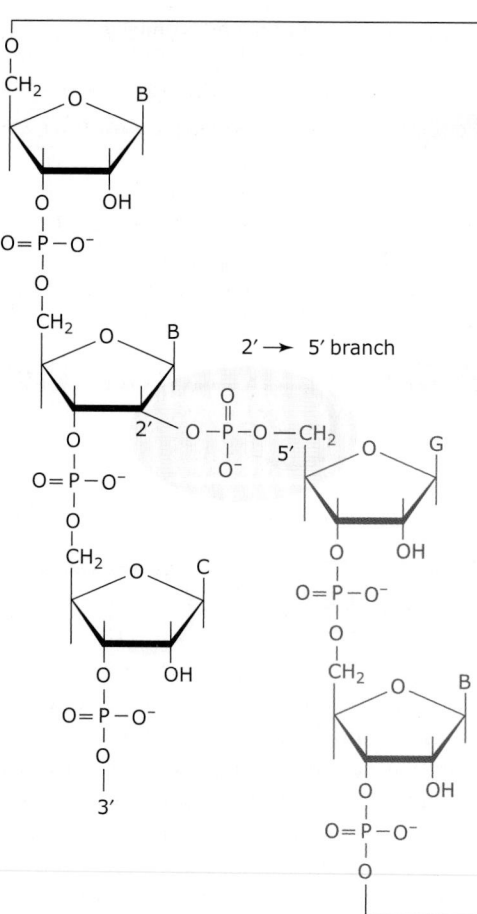

FIGURE 4.53 The structure of the lariat branch formed during mRNA splicing. The transesterification reaction described in Figure 17.23 yields a 2'→5' branch that joins the 2' ribose of the branching nucleotide to the nucleotide on the intronic site of the 5' splice junction.

a 5'–2' linkage with a specific adenine within the interior of the intron in a region known as the branch site, forming the lariat intron intermediate (**FIGURE 4.53**). Next, a cleavage takes place at the 3' end of the exon-intron splice junction, resulting in the separation of the lariat intron intermediate from the downstream exon. Finally, the left exon and the right exon are ligated together. The lariat intermediate is "debranched" and usually degraded by specific nucleases within the cell. Although these reactions are described for illustrative purposes as occurring sequentially, they actually occur very rapidly in a highly coordinated fashion. The branch site plays a critical role in determining the 3' splice junction and has a highly conserved sequence in yeast, which is UACUAAC; however, in higher eukaryotes, the branch site sequence is less conserved, as shown in Figure 4.49 and Figure 4.50. In general, the branch site in most introns is located 20 to 40 nucleotides upstream of the 3' splice junction, and in higher eukaryotes this proximity is important in defining which sequence serves as the branch site.

It is evident from our discussion that the spliceosome is a very large and dynamic complex of proteins and RNAs, which assembles sequentially on the pre-mRNA. In fact, several assembly intermediates can be isolated by biochemical analysis. Splicing occurs only after all of the components have assembled. The known intermediates in this assembly process are shown in Figure 4.52.

One of the most intriguing questions is how 5′ and 3′ end processing (capping and polyadenylation) and splicing occur in an orderly and progressive manner, especially considering that some genes must splice together many dozens or even hundreds of exons. As alluded to previously, the phosphorylated form of the CTD of the large subunit of RNA polymerase II becomes a platform for the attachment of RNA processing factors, including splicing proteins, to RNA polymerase. Once bound to the CTD, some of the splicing factors can transfer to the nascent RNA transcript as the splicing regulatory sites become exposed after being synthesized in the active site of the enzyme (**FIGURE 4.54**).

In summary, splicing occurs through the recognition of the 5′ and 3′ splice sites and the branch site by the dynamic assembly of the components of the spliceosome. However, splicing is a much more complicated than merely removing all the introns and splicing together all the exons. In genes with multiple exons and introns, splicing can be regulated to produce a diverse set of transcripts where different combinations of exons and introns are retained in the mature mRNAs. In the next section, we will explore this process of alternative splicing in greater detail.

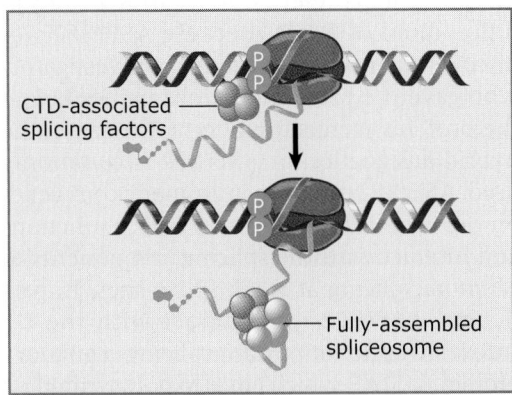

FIGURE 4.54 The carboxy-terminal domain (CTD) helps to load splicing factors onto the pre-mRNA from the CTD. The CTD of the largest subunit of RNA polymerase II serves as a staging platform for specific splicing factors that help to assemble the spliceosome onto the pre-mRNA.

Concept and Reasoning Check

1. The following RNA transcript sequence contains a 5′ and 3′ splice junction and a branch point sequence. Identify these sequences and write the sequence of the exon-exon fusion that occurs after splicing.

5′UCAGGUUUACAUAAUGACCCUGAACGCGCGCGUGCU AACCGCAUGCCCCCAUGCCCUUCCAGGCAACAG3′

4.11 Alternative splicing generates protein diversity

Key concepts

- More than one protein can be generated from a single transcription unit through alternative splicing, resulting in the inclusion or elimination of specific introns or exons, and the modification of the amino acid sequence in the resulting protein product.
- Alternative splicing is controlled by splicing activator and repressor proteins that bind to enhancer and silencer sequences near exon-intron boundaries.
- Splicing activator and repressor proteins are expressed in a cell-type-specific manner that leads to tissue-specific expression of mRNA species encoding alternative proteins.

Although most mammalian genomes contain ~25,000 genes, hundreds of thousands of proteins can be derived from these genes because of alternative splicing. Tissue-specific alternative splicing results in the expression of different protein products in different cell types with the same gene sequences. The alternative inclusion of some introns as exons, or the skipping of some exons, may result in multiple mRNA species with different ORFs. In some cases, different amino acids encoded by the included intron may be inserted into the interior segment of the ORF, and translation of the new mRNA will result in a protein with a different amino acid sequence and a longer or shorter length. In other cases, the alternative inclusion of the intronic sequences may result in either the generation or the loss of a start or a stop codon, resulting in an mRNA with either a completely new reading frame or a longer reading frame. **FIGURE 4.55** shows the diversity of mRNA products that can be gener-

(a) An optional exon (blue) can be skipped or included

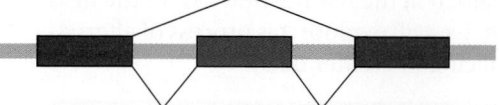

(b) Only one exon in an array of optional exons (blue and green) may be included in mature mRNA

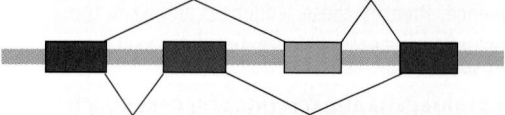

(c) An intron (yellow) may be retained or excluded

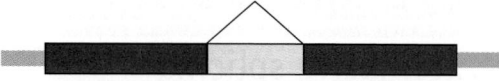

(d) An exon may have alternative splice sites at its 3′-end

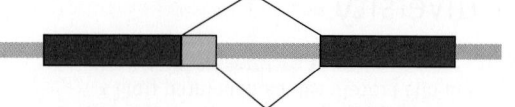

(e) An exon may have alternative splice sites at its 5′-end

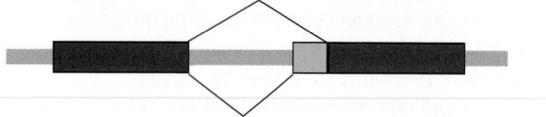

FIGURE 4.55 Alternative splicing. Alternative splicing generates protein diversity by including or excluding selected introns as exons. The red boxes correspond to constitutive exons that are always included in the mature mRNA, while the blue and green boxes correspond to optional exons that may or may not be retained. The yellow boxes correspond to introns that may be either retained or excluded in the mature mRNA. The cyan boxes designate regions within the constitutive exons that have alternative 3′ or 5′ splice sites, and which can lead to larger or smaller constitutive exons. The gray regions correspond to introns. Adapted from B. R. Graveley, *Trends Genet.* 17 (2001): 100–107.

ated through alternative splicing from a single gene. The myriad of different RNA and protein products that may be produced through alternative splicing can blur the distinction between introns and exons. The term "exon" refers to those RNA segments that are found in the final mature mRNA transcript, and the exons will be different for each of the different mRNA species that is generated from a transcription unit.

More than 70% of all human genes encode alternatively spliced mRNAs, underscoring the importance of splicing in the regulation of gene expression. While the mechanisms that control alternative splicing remain under intensive study, it is clear that the decision to splice at any given potential intron-exon boundary in a pre-mRNA is not simple. Similar to the regulation of transcription, there are specific sequences and binding proteins that activate or repress splicing of introns. As one might expect, these protein-binding sites are strategically located to enhance or inhibit the use of specific splice sites. Subtle shifts in the binding of specific proteins to these sites can influence the likelihood that the splice event will occur. For example, if a splice site junction signal or a branch point deviates from the canonical sequence or is located in an RNA stem-loop structure, proteins that increase the affinity of the spliceosome components for the sequence or that denature the stem-loop structure can increase the likelihood that the intron will be spliced out. Such sequences are sometimes referred to as **splicing enhancers**, which can be located within the exon (exonic) or within the intron (intronic). Similarly, proteins can shield the splicing junction and branch point signals from the spliceosome, resulting in repression or silencing of splicing of the intron and the inclusion of the intron within the mRNA—these sites are sometimes referred to as **splicing silencers**.

There are several examples of splicing regulatory proteins that play important roles in exon/intron definition. One example is the mRNA-binding protein **ASF/SF2**—a member of the **SR family** of proteins with amino acid sequences that contain segments rich in the amino acids serine (S) and arginine (R), known as **RS domains**. More than 10 different SR proteins have been found in human cells. In addition to the serine/arginine-rich domain, the SR proteins also contain one or more segments referred to as an RNA recognition motif or **RRM**. Because ASF/SF2 was discovered by different groups of investigators who gave the protein different names before the proteins were shown to be the same, the hybrid designation of ASF/SF2 is commonly used. ASF/SF2 can form multimeric complexes on the exon sides of 5′ and 3′ splice junctions and interact with the splicing components to promote splicing at weak splice sites. In particular, ASF/SF2 can interact with the U1 snRNP. Another important splicing regulatory protein is U2AF, which binds to polypyrimidine tracts often located within introns between the branch site and the 3′ splice junction. U2AF is a heterodimer consisting of 65 and 35 kDa subunits and can interact with the U2 snRNP as well as with the SR proteins to help recruit spliceosome components to the pre-mRNA.

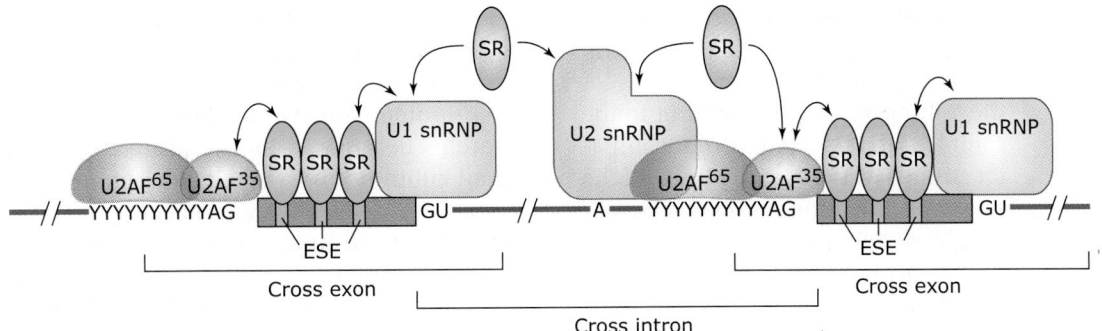

FIGURE 4.56 Regulation of splice site recognition. The SR proteins bind to exonic splicing enhancers (ESEs) and intronic splicing enhancers (ISEs) to increase the affinity of the snRNPs for the splice junctions and branch site. The splicing factor U2AF, which binds to the polypyrimidine tract acts in conjunction with the SR proteins to enhance lariat formation during the splicing reaction. Adapted from T. Maniatis and B. Tasic, *Nature* 418 (2002): 236–243.

FIGURE 4.56 shows a model of how ASF/SF2 works in conjunction with the SR and U2AF proteins to regulate splicing.

In summary, the regulation of splicing to produce mRNA diversity, in turn leading to multiple proteins from the same transcription unit, is a key strategy in generating protein diversity. In the next several sections, we will explore the translation of the mature mRNA transcript.

4.12 Translation is a three-stage process that decodes an mRNA to synthesize a protein

The final step in the expression of protein-coding genes is the translation of the mRNA to synthesize a polypeptide chain. Translation is catalyzed by ribosomes, which are large RNP complexes comprised of several rRNA species and 50 to 80 proteins. There are three phases

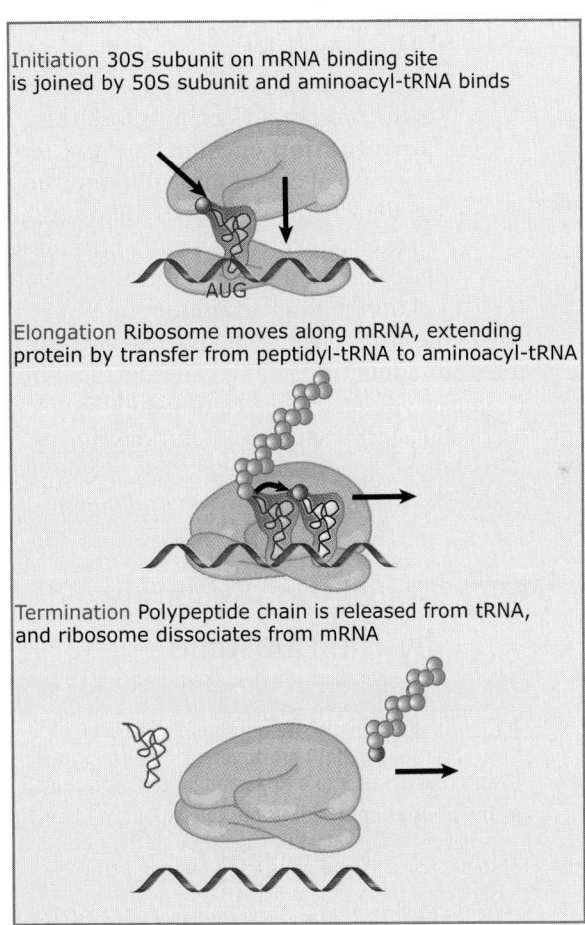

FIGURE 4.57 The three phases of translation. Translation can be divided into three phases: initiation, elongation, and translation.

to the translation process: initiation, elongation, and termination as discussed below (**FIGURE 4.57**):

1. **Initiation** involves those reactions that lead up to the polymerization of the first two amino acids. The steps include the dissociation of the small and

large subunits from the ribosome, the binding of the small ribosomal subunit to the initiation codon on the mRNA, the joining of the large subunit to the complex, and the binding of the initiator aminoacyl tRNA to the P-site of the ribosome. Initiation is a relatively slow step in protein synthesis and usually determines the rate at which an mRNA is translated.

2. **Elongation** includes those reactions from synthesis of the first peptide bond to addition of the last amino acid. The peptidyl transferase step transfers the peptide from the previous step in elongation to the incoming aminoacyl tRNA. During elongation, the mRNA moves through the ribosome as each of the codons is translated. Elongation is the most rapid step in translation.

3. **Termination** encompasses the recognition of one of three different stop codons by the ribosome, the cleavage of the complete polypeptide from the last peptidyl tRNA, and the dissociation of the ribosome from the mRNA.

In the next several sections, we will explore the translation process in greater detail. Other chapters will detail the relationship between intracellular cellular protein trafficking and protein synthesis (see 7 *Membrane targeting of proteins* and 8 *Protein trafficking between membranes*).

4.13 Translation is catalyzed by the ribosome

Key concepts

- The ribosome is the site of protein synthesis (translation) in both prokaryotes and eukaryotes, and consists of 3 to 4 RNAs and 50 to 80 proteins.
- The translation reactions are carried out in large part by the ribosomal RNAs, although the ribosomal proteins and translation factors are necessary cofactors in the process.
- Translation is an RNA-guided process, involving interactions among tRNAs, rRNAs, and the mRNA. The peptidyl transferase reaction results in the polymerization of the polypeptide chain and is catalyzed by ribosomal RNA.
- tRNAs are 70 to 95 nucleotides in length and contain a number of covalently modified nucleotides. All tRNAs fold into similar secondary and tertiary structures.
- tRNAs function as adapter molecules that covalently attach to amino acids and allow the ribosome to decode the mRNA through tRNA anticodon base pairing with the mRNA codon.

The ribosome is perhaps the most complex and interesting of all biological supramolecular assemblies. All ribosomes are composed of two multiprotein subunits—a large subunit with proteins designated by the prefix "L" and a small subunit with proteins designated "S." *E. coli* ribosomes have a molecular weight of ~2.5 MDa. The small ribosomal subunit consists of a single RNA (16S rRNA) and 21 (S1–S21) proteins; the large subunit contains two RNAs (5S and 23S rRNAs) and 34 (L1–L34) proteins (**FIGURE 4.58**). Eukaryotic ribosomes are somewhat larger; mammalian ribosomes have a molecular weight of more than 3 MDa. In mammals, the small subunit consists of a single RNA (18S rRNA) and more than 30 proteins, while the large subunit has three RNAs (5S, 5.8S, and 28S rRNAs) and more than 45 proteins. Historically, ribosomes and their components have been named according to their rate of sedimentation in an ultracentrifuge, using the unit of the sedimentation coefficient called the Svedberg coefficient or "S value." The S value is based on both the molecular mass of the particle as well as its shape. Bacterial ribosomal subunits sediment as 30S and 50S particles and combine to form 70S particles. Ribosomal subunits in higher eukaryotes sediment as 40S and 60S particles and combine to form 80S particles. Because the S value is determined by both mass and shape, the final size of the ribosome is not simply the sum of the S values for the individual subunits, but reflects the change in overall shape that occurs when the small and large subunits interact. While a few of the roles of the individual ribosomal proteins have been defined in specific steps of the translation reactions, as we saw in the case of

Ribosomes			rRNAs	r-proteins
Bacterial (70S) mass: 2.5 MDa 66% RNA		50S	23S = 2904 bases 5S = 120 bases	31
		30S	16S = 1542 bases	21
Mammalian (80S) mass: 4.2 MDa 60% RNA		60S	28S = 4718 bases 5.8S = 160 bases 5S = 120 bases	49
		40S	18S = 1874 bases	33

FIGURE 4.58 Ribosome structure and composition. Both prokaryotic and eukaryotic ribosomes contain a large and a small subunit comprised of a number of proteins and 1–3 rRNAs. The "S value" refers to the sedimentation coefficient and reflects both molecular weight and shape.

the spliceosome, most of the proteins appear to be involved in promoting the conformational rearrangements and changes in base pairing between various RNA molecules.

Over the past two decades, it has become apparent that the steps of translation are guided by a combination of RNA-RNA and protein-RNA interactions. Most importantly, rRNA catalyzes the central reaction of translation, the peptidyl transferase step, which catalyzes the polymerization of the polypeptide chain. RNA is thought to have preceded proteins in biomolecular evolution; thus, in hindsight, it is not surprising that RNA catalysis is necessary for protein synthesis; however, it took many decades to prove this experimentally. Through a series of elegant studies carried out in a number of laboratories, translation has been shown to consist of a set of dynamic RNA-mediated reactions among rRNAs, tRNAs, and the mRNAs that are modulated by ribosomal proteins and protein factors. During the steps of translation, the rRNAs, tRNAs, and mRNAs change conformations to facilitate the formation and disruption of specific base-pairing interactions between the RNAs. Remarkably, in spite of some differences in the size and organization of the ribosomes and some of the regulatory protein factors, translation is the most highly conserved step in gene expression. Almost all of the RNAs and proteins involved in translation have both structural and functional homologs that can be readily identified among all prokaryotes, eukaryotes, and archaea.

The tRNAs function as adapters to bring the correct amino acids to the reactive center of the ribosome as directed by the mRNA codon sequence. The tRNAs are 70–95 nucleotides in length and have three levels of structure. The primary structure of the tRNA is the linear sequence of nucleotides, which in addition to the conventional bases A, G, C, and U contains a number of bases that are modified posttranscriptionally by a series of tRNA modification enzymes. The secondary structure of a tRNA is determined by the intramolecular base pairing, which results in the formation of adjacent stem-loop structures. All tRNAs can be formed into a similar secondary structure referred to as the **cloverleaf structure** (FIGURE 4.59). The tertiary structure of the tRNA refers to the three-dimensional folding of the entire molecule, which is stabilized by specific base pairs between nonadjacent nucleotides and by base-stacking interactions. The tRNA cloverleaf secondary structure folds into a three-

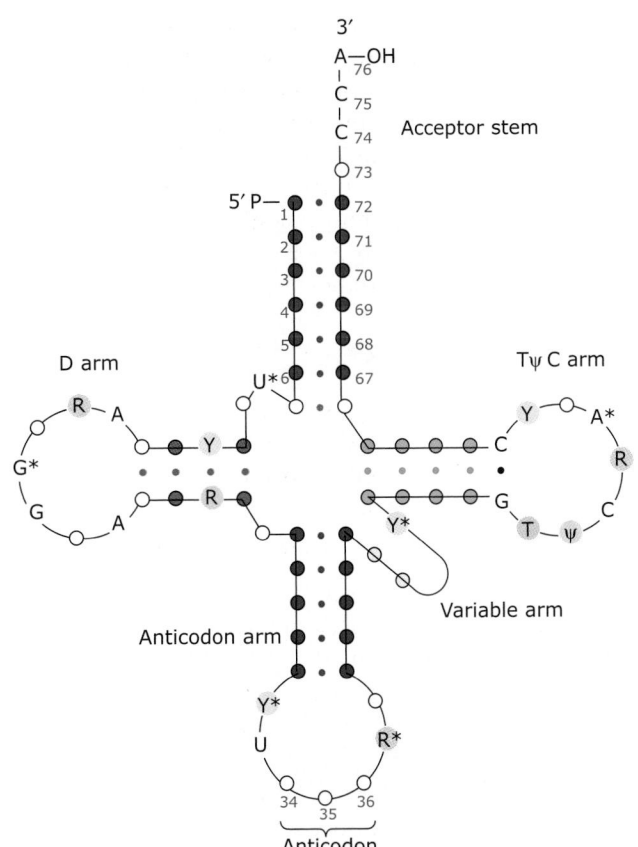

FIGURE 4.59 The tRNA cloverleaf secondary structure. All tRNAs can be folded into a conserved secondary structure that contains short base-paired helical regions (dots). Some positions contain invariant pyrimidine (Y) or purine (R) bases, or contain modified bases such as pseudouridine (Ψ) or methylated bases (*), while other positions contain nucleotides that vary between specific tRNAs. Adapted from D. Voet, J. G. Voet, and C. W. Pratt. *Fundamentals of Biochemistry, First edition*. John Wiley & Sons, Ltd., 1999.

dimensional (tertiary) structure that has an overall general L-shape, as shown in **FIGURE 4.60**. Each tRNA has a slightly different tertiary structure, but the anticodon loop is always located at the top of the L, while the aminoacyl acceptor stem is located at the foot of the L. The L-shape maximizes stacking of the bases by the formation of a helix containing the anticodon arm and the D arm, as well as the formation of a perpendicular helix containing the TΨC arm and the acceptor arm (see Figure 4.60 and **FIGURE 4.61**).

Thus, tRNAs have the following two functional ends:

1. The **3' CCA end**: The amino acids to be polymerized into the polypeptide chain come to the ribosome as covalent adducts to the 3' ends of tRNAs. The OH group at the end of the CCA sequence

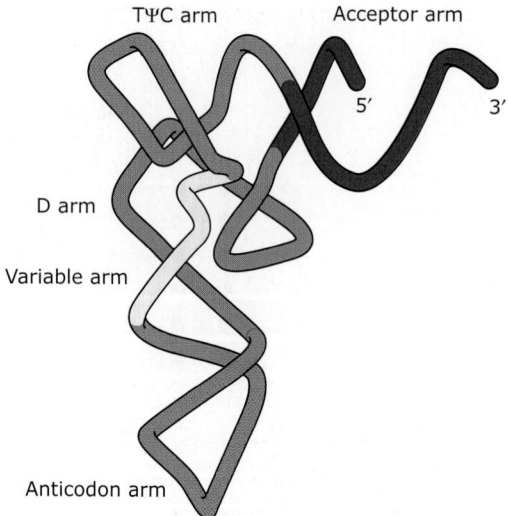

FIGURE 4.60 The tRNA L-shaped tertiary structure. All tRNAs can be folded into a conserved L-shaped tertiary structure that stacks the helices of the secondary structure as shown. The tertiary structure is stabilized by base pairing between distal nucleotides on regions that are brought into proximity by the folding of the tRNA. Adapted from J. G. Arnez and D. Moras, *Trends Biochem. Sci.* 22 (1997): 211–216.

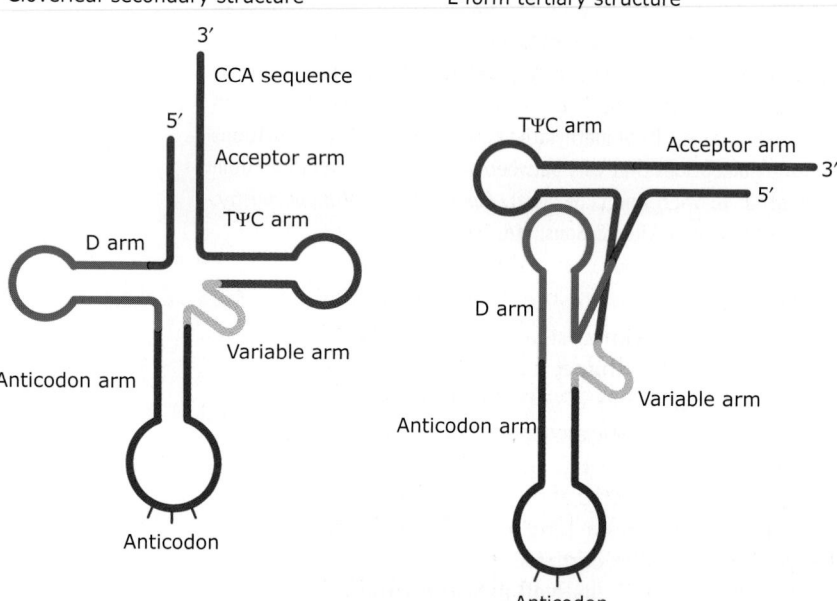

FIGURE 4.61 The folding of tRNA secondary structure into L-shaped tertiary structure. The helical regions of secondary structure stack with one another to form extended helices, which are perpendicular to each other.

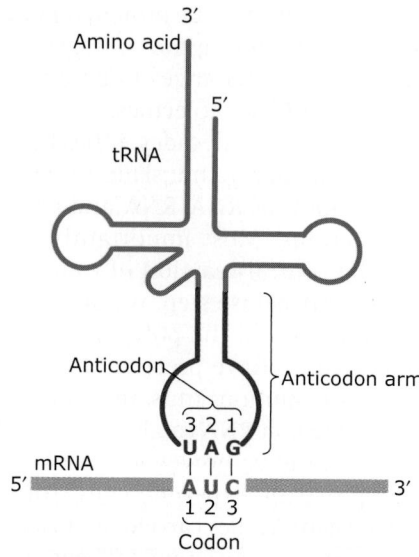

FIGURE 4.62 The interaction between the tRNA anticodon and the mRNA codon. The base pairing between the anticodon loop of the tRNA (magenta) and the codon of the mRNA (green) is antiparallel. Thus, the anticodon sequence is designated as 5'-GAU-3' in keeping with the convention that all nucleic acids are written in the 5'→3' direction. Adapted from D. C. Nelson and M. M. Cox. *Lehninger Principles of Biochemistry, Third edition*. W. H. Freeman & Company, 2000.

forms an acyl linkage with the amino group on the amino acid. The covalent conjugate between an amino acid and its cognate tRNA is called an **aminoacyl tRNA**. The aminoacylation of the tRNA is catalyzed by an **aminoacyl-tRNA synthetase**, which has to match the specific tRNA with the correct amino acid before the aminoacyl reaction can take place. Each tRNA has its own tRNA synthetase to catalyze its aminoacylation with the correct amino acid.

2. The **anticodon loop:** The anticodon loop base pairs with the complementary mRNA codon, thereby "reading" the genetic code embedded in the mRNA sequence, as shown in **FIGURE 4.62**. As discussed previously, there are 64 different codons and 61 of the codons specify one of the 20 amino acids commonly found in proteins; thus, there should be 61 different anticodons and thus, in theory there should be 61 different tRNAs. However, there are actually many more than 61 tRNAs; each has a different sequence in the remainder of the molecule but has the same anticodon loop. The three termination codons do not have cognate tRNAs.

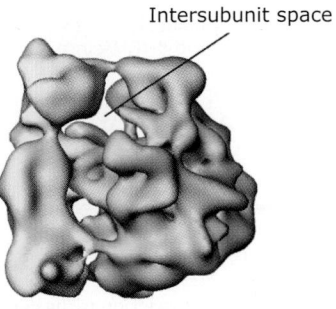

Intersubunit space

(a) Intact ribosome

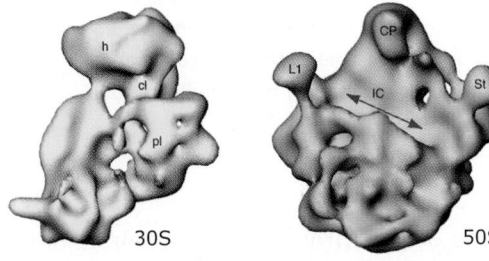

30S   50S

(b) 30S subunit   (c) 50S subunit

FIGURE 4.63 Cryo-electron microscopy reconstruction of the *E. coli* 70S ribosome and its constituent subunits. The image shows the reconstruction of the *E. coli* ribosome from electron micrographs at a resolution of 25 Å. Reproduced from J. Frank, R. K. Agarwal, and A. Verschoor, Ribosome structure and shape, Encyclopedia of Life Sciences, December 12, 2001 [doi: 10.1038/npg. els.0000534]. Copyright © 2001 John Wiley & Sons, Ltd. Reproduced with permission of Blackwell Publishing Ltd. Photos courtesy of Joachim Frank, HHML/HRI, Wadsworth Center.

The initiation codon in both prokaryotes and eukaryotes is generally AUG, although a few genes use another triplet. AUG specifies the amino acid methionine. However, in prokaryotes, there are two tRNAs with an anticodon for AUG, which is CAU (remember that codon-anticodon base pairing conforms to the same rules as all other nucleic acid base pairing in that one sequence is 5′ to 3′ and the other is 3′ to 5′, but all sequences are always written 5′ to 3′). One of the prokaryotic tRNA^met species is called the initiator tRNA, and carries a methionine that has been covalently modified with a formyl group to yield *N*-formylmethionine; it is designated tRNA^f-met. For the insertion of all other methionines in the polypeptide chain, a noninitiator (elongator) tRNA^met is utilized. By contrast, in eukaryotes, there is only one tRNA^met that serves as both the initiator and elongator tRNA for methionine.

The details of the structure of prokaryotic and eukaryotic ribosomes have emerged from several high-resolution electron microscopy and X-ray diffraction studies. An example of the structure of a bacterial ribosome derived

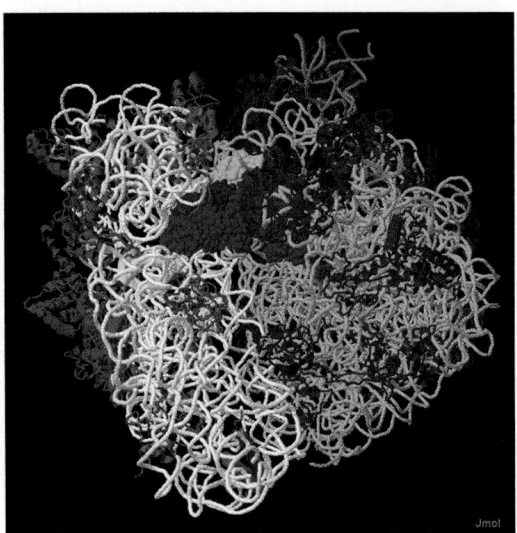

FIGURE 4.64 The tRNA binding sites of the bacterial ribosome. The structure of the ribosome form the thermophilic bacterium *Thermus thermophilus* was determined to 5.5 Å resolution by X-ray crystallography. The rendering shows the relative locations of the A-site (red), P-site and peptidyl transferase site (purple), and E-site (cyan). Structure from Protein Data Bank IJGO. G. Z. Yusupova, et al., *Cell* 106 (2001): 233–241. Protein Data Bank 1GIY. M. M. Yusupova, et al., *Science* 292 (2001): 883–896.

from high-resolution electron microscopy is shown in **FIGURE 4.63**. There are three binding sites for the aminoacyl tRNAs on the prokaryotic ribosome are depicted in **FIGURE 4.64**: The A-site binds the incoming aminoacyl tRNA that now carries the next amino acid to be added to the growing polypeptide chain, the P-site or peptidyl tRNA site contains the tRNA that was previously in the A-site and which carries the polypeptide chain, and E-site is the exit site where the tRNA that previously carried the polypeptide chain leaves the ribosome. The P- and E-sites do not appear to be distinct separate sites in eukaryotic ribosomes.

In summary, the key reaction in protein synthesis is the transfer of the peptidyl tRNA in the P-site to the new aminoacyl tRNA in the A-site, catalyzed by the **peptidyl transferase** activity of the ribosomal RNAs, followed by the movement of the peptidyl tRNA in the A-site to the P-site, and the dissociation of the tRNA from the ribosome (**FIGURE 4.65**). Different sets of accessory factors assist the ribosome at each stage. Energy is provided at various stages by the hydrolysis of guanine triphosphate (GTP). In the next section, we will explore each step in translation in greater detail and describe the roles of a myriad of accessory protein factors that guide the process.

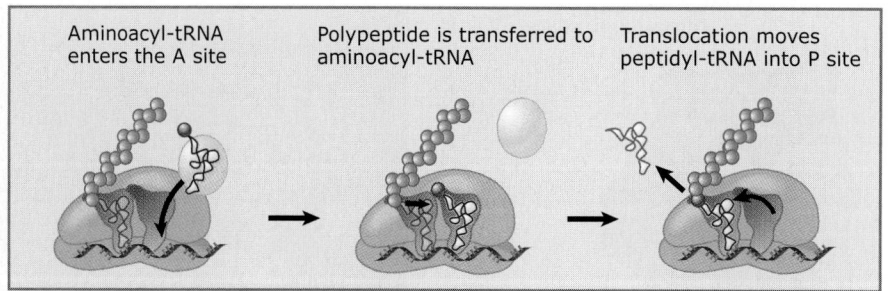

Aminoacyl-tRNA enters the A site

Polypeptide is transferred to aminoacyl-tRNA

Translocation moves peptidyl-tRNA into P site

FIGURE 4.65 The peptidyl transferase reaction. The peptidyl transferase reaction is the central biochemical reaction that takes place during protein synthesis. The aminoacyl-tRNA enters the A-site on the ribosome. The growing polypeptide chain on the tRNA in the P-site is transferred to the aminoacyl-tRNA in the A-site and translocation of the ribosome to the next codon on the mRNA moves the peptidyl-tRNA to the P-site and the deacylated tRNA from the P-site to the E-site.

Concept and Reasoning Check

1. Trace the steps in the occupation of the aminoacylated tRNA-binding sites required to synthesize the tripeptide, met-val-lys.

4.14 Translation is guided by a large number of protein factors that regulate the interaction of aminoacylated tRNAs with the ribosome

Key concepts

- The energy source for translation is the hydrolysis of GTP by G-protein translation factors.
- The translation initiation factors control the assembly of the small ribosomal subunit with the mRNA and the location of translation initiation site.
- The translation elongation factors control the assembly of the aminoacylated tRNAs with the ribosome.
- The translation termination factors control the cessation of the elongation reaction and the release of the completed polypeptide chain from the ribosome.

Here, we will briefly discuss the roles of the translation factors in the various steps of protein synthesis, and describe the important role of GTP in the translation process.

During initiation, the small ribosomal subunit binds to mRNA (the 30S ribosomal subunit in prokaryotes and the 40S subunit in eukaryotes). In prokaryotes, recognition of the start codon occurs through the base pairing of the 16S rRNA within the 30S subunit to a short sequence on the mRNA immediately upstream of the start codon, commonly referred to as the **ribosome-binding site** or the **Shine-Dalgarno sequence**, and named after the investigators who discovered the site (**FIGURE 4.66**). The initiation factors IF1, IF2, and IF3 (**FIGURE 4.67**) are required for this recognition process to occur. IF3 keeps the 30S subunit from reassociating with the 50S subunit, an event that inhibits initiation, and facilitates the 16S RNA-mRNA interaction. IF2 binds the initiator tRNA, which carries a formylmethionine, and controls the binding of the formylmethionine tRNA to the empty P-site on the 30S subunit. IF2 also promotes the association of the 50S subunit with the 30S-mRNA-tRNA$^{f\text{-met}}$ complex, and the subsequent release of all of the initiation factors. This process requires energy, and IF2 is a ribosome-dependent GTPase. This initiation step is the only time when an entering aminoacylated tRNA binds to the P-site; all other aminoacyl tRNAs that subsequently bind to the ribosome enter at the A-site. To make sure the A-site is not available for this initial tRNA interaction, the initiation factor IF1 blocks the A-site.

The translation initiation reaction in higher eukaryotes is very similar in most respects to the prokaryotic reaction, although there are more than 12 eukaryotic initiation factors compared with just three in bacteria. As mentioned earlier, another major difference is that in eukaryotes the first amino acid to be incorporated in response to the AUG start codon is methionine, as opposed to formylmethionine in prokaryotes.

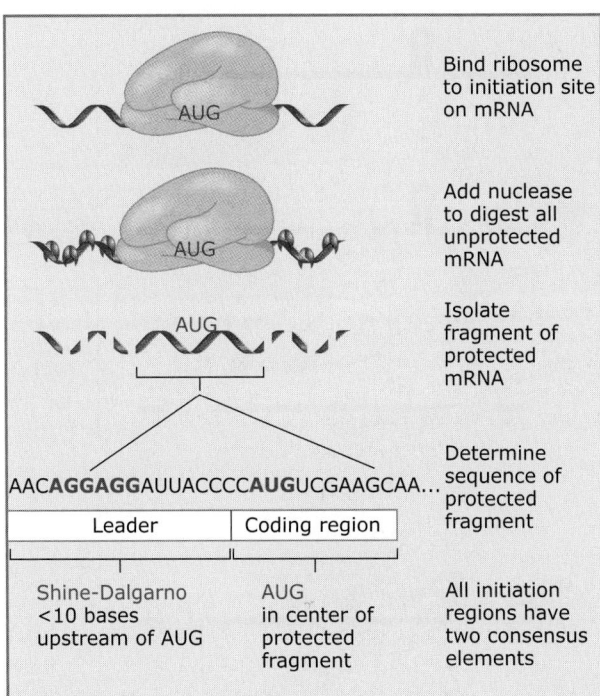

Bind ribosome to initiation site on mRNA

Add nuclease to digest all unprotected mRNA

Isolate fragment of protected mRNA

Determine sequence of protected fragment

AACAGGAGGAUUACCCCAUGUCGAAGCAA...

| Leader | Coding region |

Shine-Dalgarno <10 bases upstream of AUG

AUG in center of protected fragment

All initiation regions have two consensus elements

FIGURE 4.66 The prokaryotic ribosome-binding site on mRNA. During the initiation phase of translation, the 30S ribosomal subunit binds to a sequence that includes the AUG start codon as well as an adjacent upstream region called the Shine-Dalgarno sequence, named after its discoverers.

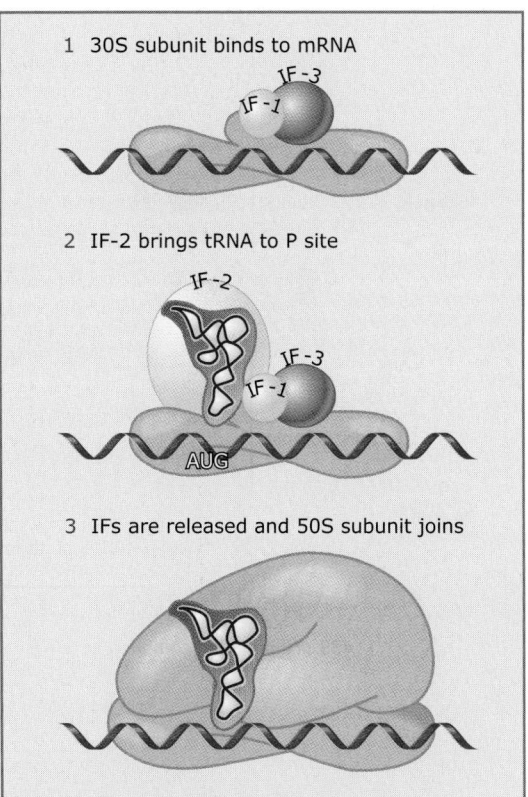

1 30S subunit binds to mRNA

2 IF-2 brings tRNA to P site

3 IFs are released and 50S subunit joins

FIGURE 4.67 The prokaryotic translation initiation reaction. The initiation factors IF1 and IF3 stabilize the interaction of the 30S ribosomal subunit with the ribosome-binding site on the mRNA, and IF2 stimulates the binding of the tRNA to the 30S-mRNA complex.

FIGURE 4.68 shows that the eukaryotic initiation pathway can be broken down into five stages as follows:

1. Stage 1 consists of the dissociation of the 80S ribosome by the initiation factors eIF1A and eIF3. eIF2 binds to the aminoacylated tRNA^{met} along with GTP, analogous to prokaryotic IF2, and interacts with eIF5 and eIF1 to form the 43S preinitiation complex.
2. Stage 2 is when the 43S preinitiation complex assembles with the mRNA. In contrast to the ability of the prokaryotic 30S ribosome to recognize internal ribosome-binding sites, in most cases, the 43S preinitiation complex must first bind to a complex of initiation factors located at the 5′ 7-MeG cap on the mRNA.
3. Stage 3 occurs independently of stage 1, and encompasses the formation of the cap-binding complex, which consists of initiation factor eIF4E bound directly to the cap and also bound to eIF4G and EIF4A. The cap-binding complex requires energy in the form of ATP hydrolysis to unwind any stable stem-loop structures in the mRNA that might interfere with initiation codon recognition.
4. Stage 4 is when the 43S preinitiation complex joins the cap-binding complex and scans the mRNA in the 5′ to 3′ direction searching for the first AUG that is in the proper sequence context. In contrast to prokaryotes, there is no easily recognized ribosome-binding sequence analogous to the Shine-Dalgarno sequence in eukaryotes. The nucleotides immediately adjacent on the 5′ and 3′ sides of the AUG appear to be critical to whether or not the initiation complex chooses to use the sequence to initiate translation. The consensus sequence for an AUG in the proper context to be used as an initiation codon is NNPuNNAUGG, where N is any one of the four nucleotides and Pu is a purine (A or G).
5. Stage 5 is when eIF5 and eIF5B each catalyze the hydrolysis of GTP to stimu-

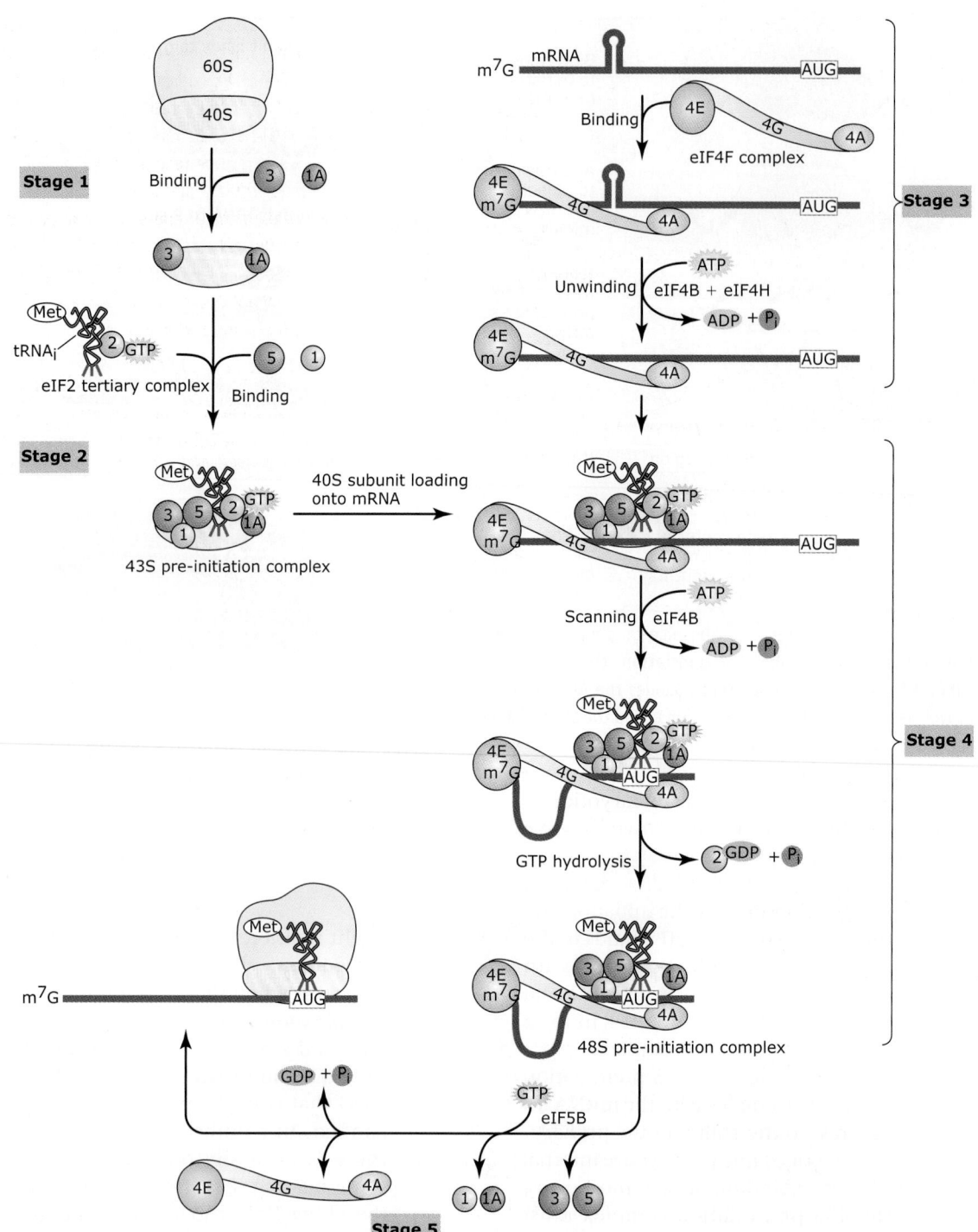

FIGURE 4.68 The eukaryotic translation initiation pathway. The eukaryotic translation initiation reaction can be subdivided into five stages. The stages are guided by the eukaryotic initiation factors (eIFs). In stage 1, the 80S is dissociated into the 40S and 60S subunits by eIF1A and eIF3A. In stage 2, eIF1 and eIF5 stimulate the eIF2-GTP-tRNA^{met} complex to join the eIF1A-eIF3-40S complex to form the 43S preinitiation complex. In stage 3, the eIF4E-eIF4G-eIF4A complex interacts with the 5′ end of the mRNA and eIF4B and eIF4H catalyze the ATP hydrolysis-dependent unwinding of any mRNA secondary structure located between the mRNA 5′ end and the initiation site. In stage 4, the 43S preinitiation complex joins the eIF4E-eIF4G-eIF4A-mRNA complex. The entire assembly scans in the 5′→3′ direction until it locates the initiation site, at which time eIF2 hydrolyzes its bound GTP to GDP and is released resulting in the formation of the 48S preinitiation complex. In stage 5, additional hydrolysis of GTP results in the dissociation of all remaining initiation factors from the 40S-mRNA complex and the assembly of the 60S subunit onto the 40S subunit to form the initiated 80S-mRNA complex. Adapted from R. J. Jackson, *Biochem. Soc. Trans.* 33 (2005): 1231–1241.

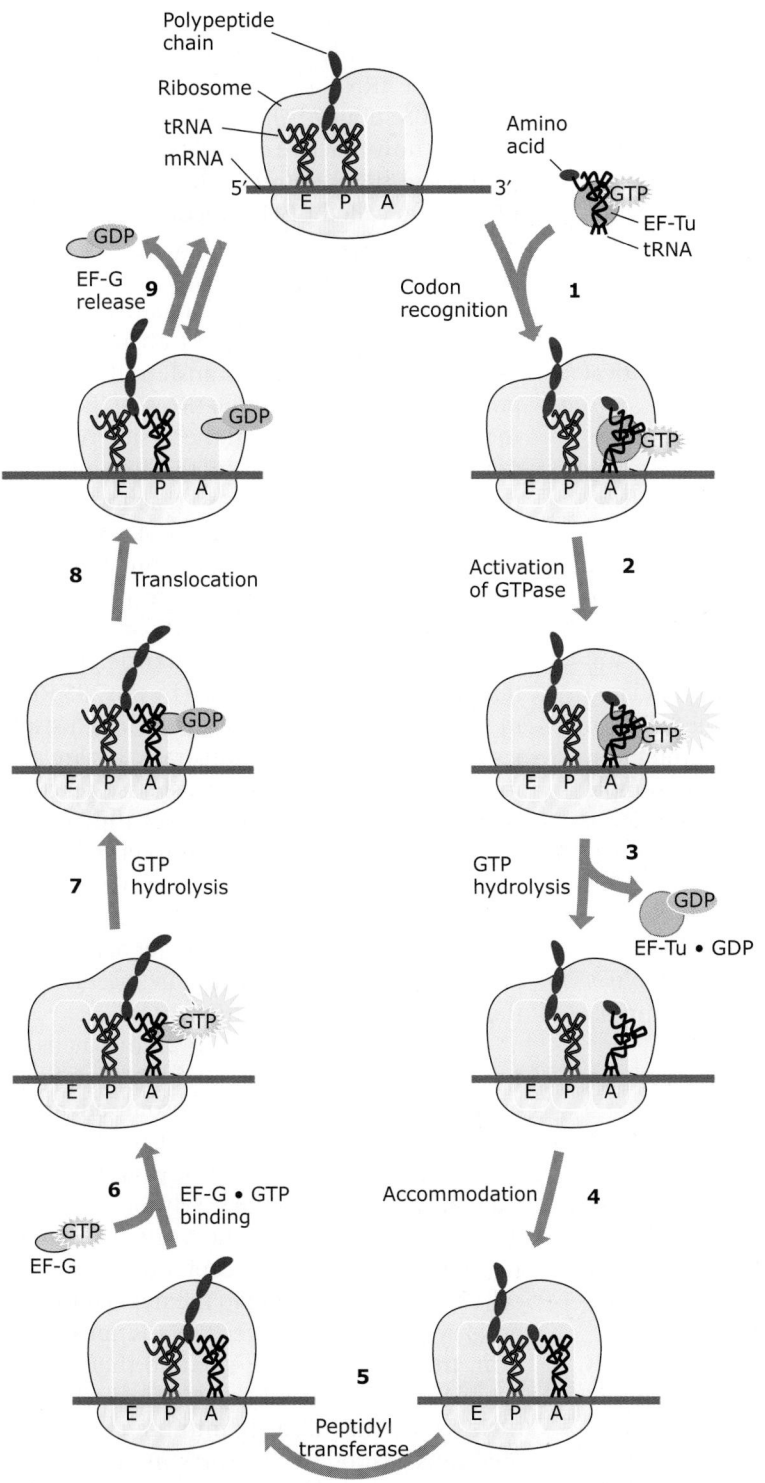

Polypeptide chain
Ribosome
tRNA
mRNA

Amino acid

GTP
EF-Tu
tRNA

GDP

EF-G release **9**

GDP

E P A

Codon recognition **1**

GTP

E P A

8 Translocation

Activation of GTPase **2**

GDP

E P A

GTP

E P A

7 GTP hydrolysis

GTP hydrolysis **3**

GDP

EF-Tu • GDP

GTP

E P A

E P A

6 EF-G • GTP binding

Accommodation **4**

GTP
EF-G

E P A

E P A

5

Peptidyl transferase

FIGURE 4.69 The translation elongation pathway. The translation elongation pathway in bacteria can be divided into nine steps. Step 1: EF-Tu · GTP-tRNA complex binds to the A-site on the ribosome. Step 2: The EF-Tu GTPase activity is activated. Step 3: The ribosome catalyzes the hydrolysis of GTP and the release of EF-Tu · GDP. Another factor, EF-Ts (not shown) mediates the regeneration of EF-Tu · GDP back to EF-Tu · GTP. Step 4: The acylated amino acid on the 3' end of the tRNA in the A-site moves next to the 3' end of the polypeptidyl-tRNA sitting in the P-site. Step 5: The polypeptide is transferred to the aminoacyl tRNA in the A-site. Step 6: EF-G · GTP binds to the peptidyl tRNA in the A-site. Step 7: GTP is hydrolyzed to form the EF-G · GDP complex. Step 8: The polypeptidyl tRNA translocates from the A-site to the P-site and the deacylated tRNA to the E-site. Step 9: EF-G dissociates from the A-site to yield an empty A-site for the next aminoacylated tRNA. The steps in eukaryotic translation elongation are very similar; however, the eukaryotic homologs of EF-Tu and EF-G are named EF1A and EF2, respectively. Adapted from V. Ramakrishnan, *Cell* 108 (2002): 557–572.

late the assembly of the 60S ribosomal subunit onto the initiation complex and the dissociation of the initiation factors.

Eukaryotic initiation usually occurs by the scanning mechanism as previously outlined; however, there is an alternative means of initiation, used especially by certain viral RNAs, in which a 40S subunit associates directly with an internal site called an IRES, bypassing any upstream AUG codons. There are few sequence homologies between known IRES elements, but their use is especially important in specific types of viral infections where the virus inhibits host protein synthesis by destroying 7-MeG cap structures on the 5′ end of the mRNAs and inhibiting the initiation factors that bind them.

Once the complete ribosome has assembled at the initiation codon, the transition to the elongation phase occurs. As in the case of initiation, the prokaryotic and eukaryotic steps in elongation are very similar. There are three elongation factors that function to modulate the elongation reactions. The steps in the prokaryotic elongation reaction are illustrated in **FIGURE 4.69**. These steps are nearly identical in eukaryotes.

At the beginning of the elongation phase, the A-site of the ribosome is vacant due to the loss of the initiation factor that blocked it from occupancy by the initiator tRNA. The aminoacyl tRNA with the amino acid to be inserted at the next position of the polypeptide binds to the A-site of a ribosome, as the P-site is occupied by initiator peptidyl-tRNA. In prokaryotes, entry of the aminoacyl tRNA into the A-site is mediated by elongation factor EF-Tu (also called EF1A). EF-Tu is a highly conserved protein throughout prokaryotes and eukaryotes and is an example of a monomeric GTP-binding protein called a G-protein. We will encounter various examples of monomeric G-proteins that are involved in nucleocytoplasmic trafficking (see *9 Nuclear structure and transport*) and involved in receptor-mediated signal transduction (see *18 Principles of cell signaling*). All G-proteins work by a common set of principles as follows:

- When GTP is present, the protein is in its active state.
- When the GTP is hydrolyzed to GDP, the factor becomes inactive.
- Activity is restored when the GDP is replaced by GTP, which is an exchange reaction—this is not the result of phosphorylation of the GDP by the G-protein.

The binary complex of EF-Tu · GTP binds aminoacyl-tRNA to form a ternary complex of aminoacyl-tRNA-EF-Tu · GTP. The ternary complex binds only to the A-site of the ribosome when the P-site is already occupied by peptidyl-tRNA. This is the critical reaction in ensuring that the aminoacyl-tRNA and peptidyl-tRNA are correctly positioned for peptide bond formation during the peptidyl transferase reaction. The aminoacyl-tRNA is loaded into the A-site in two steps: First, the anticodon end binds to the A-site of the small subunit. Then, codon-anticodon recognition triggers a change in the conformation of the ribosome,

which stabilizes the tRNA binding and causes EF-Tu to hydrolyze its bound GTP. The CCA end of the tRNA now moves into the A-site on the 50S subunit. The EF-Tu · GDP complex, which is now inactive due to its low affinity for aminoacyl-tRNA, is released. Another factor, EF-Ts (also called EF1B), mediates the regeneration of EF-Tu · GDP back to the active EF-Tu · GTP complex.

In the next step of the elongation reaction, which is the actual peptidyl transferase step, the polypeptide chain is transferred from the polypeptide attached to the tRNA in the P-site to the aminoacyl-tRNA in the A-site. As mentioned earlier, a large body of research indicates that the peptidyl transferase reaction is catalyzed by the 23S RNA in the prokaryotic 50S subunit, but supported by ribosomal proteins that hold the 23S RNA in the correct conformation for its catalytic function. By analogy, it is hypothesized that the peptidyl transferase reaction in eukaryotes is catalyzed by the 28S RNA in the 60S ribosomal subunit, although studies to unequivocally demonstrate this are still in progress. The peptidyl transferase reaction is triggered when EF-Tu releases the aminoacyl end of its tRNA. The aminoacyl end of the tRNA in the A-site then pivots into proximity with the end of the peptidyl-tRNA in the P-site, whereupon there is a rapid transfer of the polypeptide chain from the tRNA in the P-site to the aminoacyl-tRNA in the A-site.

The cycle of addition of amino acids to the growing polypeptide chain is completed by translocation, when the ribosome advances three nucleotides along the mRNA with the concomitant expulsion of the uncharged tRNA from the P-site, so that the newly formed peptidyl-tRNA can enter from the A-site. The empty A-site is then ready to accept a new aminoacyl-tRNA corresponding to the next codon. In prokaryotes, the discharged tRNA is transferred from the P-site to the E-site and then to the cytoplasm, while in eukaryotes the discharged tRNA is expelled directly into the cytoplasm. This translocation step requires another monomeric G-protein called EF-G (also called EF2). EF-G · GTP stimulates the transition of the movements of both the deacylated tRNA and the peptidyl tRNA. The mechanism by which EF-G drives this transition reaction is still under investigation; there is evidence supporting the movement through energy derived from the hydrolysis of the GTP, as well as other evidence that supports a more indirect role in which the EF-G · GTP stabilizes the pretranslocation state, while EF-G · GDP stabilizes the posttranslocation state. EF-G is a major constituent of the cell, present at ~1 copy per ribosome.

The final step in translation is termination, which is stimulated when the ribosome encounters one of the three stop codons, UAG, UAA, or UGA. The strength of these codons in terminating translation is dependent on the base on either side.

Translation termination is a two-step process that requires release factors (**FIGURE 4.70**). Bacteria have two classes of release factors. Class 1 release factors include RF1 and RF2, while RF3 is a class 2 release factor. RF1 specifically recognizes termination codon UAG, while RF2 recognizes termination codon UGA. Both recognize termination codon UAA. RF1 and RF2 function in the first step of translation termination by interacting directly with the termination codons on the mRNA to block tRNAs from entering the A-site. Specific domains of the class 1 release factors have structural homologies to tRNAs, allowing these domains to fit into the A-site on the ribosome. Additionally, the RFs stimulate the intrinsic ribosomal pep-

tidyl hydrolase activity that cleaves the polypeptide from peptidyl tRNA and releases the polypeptide from the ribosome. The class 2 release factor RF3 stimulates the release of RF1 or RF2 from the ribosome. Although the general mechanism of termination is similar in prokaryotes and eukaryotes, there are some differences in the details. The eukaryotic class 1 release factor, eRF1, is a protein that recognizes all three termination codons. Its sequence is unrelated to the bacterial factors. The class 2 eukaryotic release factor is called eRF2 and it appears to have the same activity as prokaryotic RF3.

Once translation termination has occurred, the mRNA and deacylated tRNA remain associated with the ribosome. A ribosome-recycling factor acts in conjunction with EF2 bound to GTP to dissociate the tRNA, mRNA, and the small and large subunits, using energy from the hydrolysis of the GTP.

In summary, translation is a complex process guided by a large number of translation factors. The translation reaction mechanism and the structure and function of the transla-

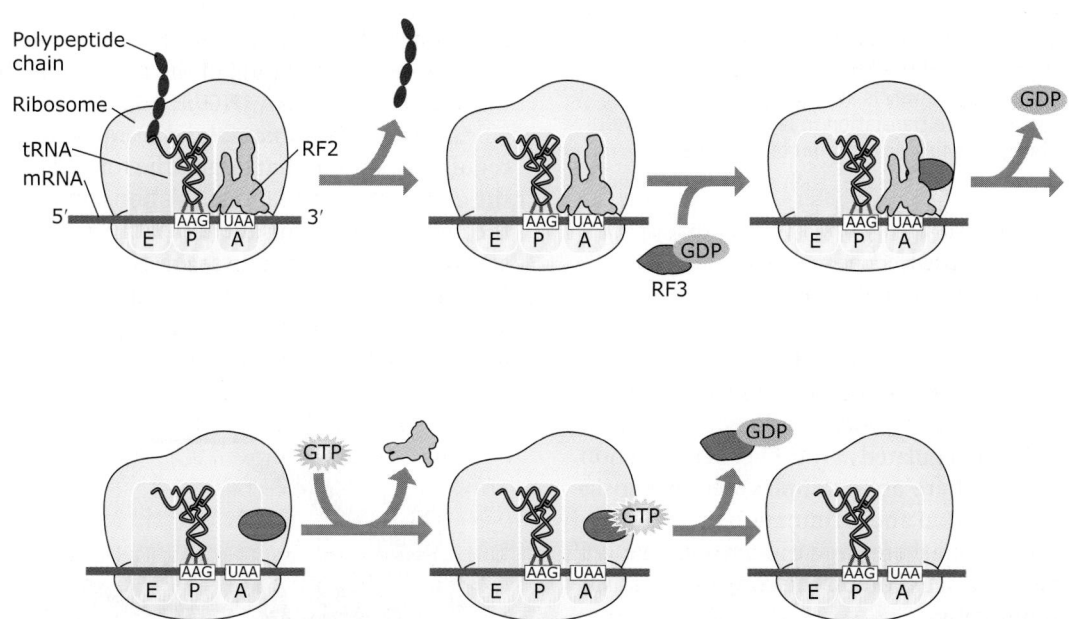

FIGURE 4.70 The translation termination pathway. The translation termination reaction in bacteria is regulated by one of several release factors (in this case, RF2). When the ribosome encounters a stop codon (UGA, UAA, UAG), the release factor binds to the stop codon in the A-site. The ribosome hydrolyzes the aminoacyl linkage between the tRNA in the P-site and the polypeptide, thereby releasing the polypeptide from the ribosome. The RF3-GDP complex binds to RF2, causing the dissociation of the GDP. RF2 binds to a GTP molecule and dissociates from the ribosome. RF3 binds a GTP, and the subsequent hydrolysis of GTP results in the dissociation of RF3. The deacylated tRNA remains bound to the P-site until the complex is dissociated by the binding of ribosome release factor (RRF) to the A-site. Adapted from L. Kisselev, et al. *EMBO J.* 22 (2003): 175–182.

tional components are conserved across species. The different roles of mRNA, tRNA, and rRNA in the mechanisms of the various steps of translation provide the best evidence that cells have evolved from a primordial RNA world. In the next section, we will explore some examples of the regulation of gene expression at the level of translation.

4.15 Translation is controlled by the interaction of the 5′ and 3′ ends of the mRNA and by translational repressor proteins

Key concepts

- The cytoplasmic poly(A) binding protein, PABPC1, functions as a translation initiation factor by binding to the poly(A) tail at the 3′ end of the transcript and interacting with the cap-binding proteins on the 7-MeG cap at the 5′ end of the transcript, via looping of the mRNA back on itself to form a circular structure.
- The mTOR pathway is an example of how the cell can modulate translation in response to the availability of amino acids for protein synthesis.
- Translational repressors, such as the iron-response element binding protein, can regulate translation initiation by binding to specific RNA sites within the 5′ UTR and blocking the 40S ribosome from locating the initiation codon.

While most of our previous discussion of gene regulation has focused on control of mRNA biosynthesis and processing, gene expression can also be regulated at the level of translation. Although there are examples where translation elongation and termination are regulated in both prokaryotes and eukaryotes, we will focus our discussion on the control of translation initiation.

In prokaryotes, the most important factors that control translational initiation are the intrinsic strength and physical accessibility of the Shine-Dalgarno sequence within the ribosome-binding site, which are in turn influenced by both the sequence and secondary structure of the mRNA. Those initiation sites whose Shine-Dalgarno sequences deviate greatly from the consensus sequence, or whose Shine-Dalgarno sequences are located in regions of stable RNA secondary structure, require translational activators to increase the affinity of the 30S ribosomal subunit for the ribosome-binding site. These activators interact directly with the 30S ribosomal subunit or function to denature the RNA secondary structure around the initiation site.

In contrast to the regulation of prokaryotic translation initiation, in eukaryotes the principle strategies are either to modulate the formation of the initiation complex at the 7-MeG cap or to inhibit or enhance the ability of the 40S ribosome to locate and properly initiate translation at the initiation codon. Control signals for these strategies can be found at both the 5′ and 3′ ends of the eukaryotic mRNA. As discussed in the previous section, the eukaryotic initiation factors eIF4E, eIF4G, eIF4A, and eIF4B form the 7-MeG cap-binding complex that prepares the mRNA for assembly with the 43S complex consisting of eIF2, eIF3, tRNA^{met}, and the 40S ribosomal subunit. The activity of the cap-binding complex can be greatly enhanced by the cytoplasmic poly(A) tail binding protein, PABPC1. It is proposed that this stimulation occurs by the interaction of the PABPC1-bound poly(A) tail at the 3′ end of the mRNA with the initiation factor complex on the cap at the 5′ end through the looping of the poly(A) tail to form a circular mRNA (**FIGURE 4.71**).

An important example of a translational regulatory system that targets the formation of the eukaryotic translation initiation complex is the mTOR signal transduction pathway. mTOR stands for mammalian target of rapamycin. Rapamycin is an antibiotic that binds to mTOR and inhibits its function. mTOR is a protein that forms a complex with two other proteins, GβL

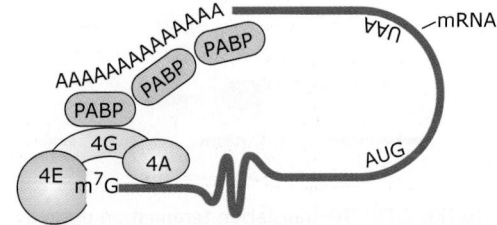

FIGURE 4.71 The looping of mRNA during translation. During translation initiation, mRNA circularizes through the interaction between the poly(A) binding protein on the poly(A) tail and the eIF4E-eIF4G-eIF4A complex on the 5′ 7-MeG cap. Adapted from T. Preiss and M. W. Hentze, *Bioessays* 25 (2003): 1201–1211.

and Raptor (**FIGURE 4.72**). Together this complex is an active protein kinase and the node for the intersection of a number of different receptor-mediated signaling pathways that are involved in the regulation of cell growth and stress responses to nutrient deprivation. When cells are deprived of nutrients, including amino acids, the cell shuts down translation. Two important repressors of protein synthesis are the eIF4E-binding protein called 4E-BP1, which binds to eIF4E and blocks its assembly with the 7-MeG cap, and the S6 kinase, which phosphorylates ribosomal protein S6 causing inhibition of the 43S ribosomal preinitiation complex. Under normal conditions when nutrients are plentiful and there is active translation, various cellular signaling pathways result in an increase in the kinase activity of the mTOR-Raptor-GβL complex. The two primary targets of the kinase are 4E-BP1 and the S6 kinase. Thus, the mTOR-Raptor-GβL complex acts to block the translational inhibitory effects of 4E-BP1 and S6 kinase.

Another important example of translational regulation in eukaryotes is the control of iron levels in mammals by the proteins **ferritin** and **transferrin**. Because iron is a toxic metal that can cause oxidative damage to cellular components, it is important to keep excess iron stored safely as a complex with ferritin in the blood stream until needed for synthesis of hemoglobin and other iron-containing enzymes. Transferrin helps iron enter cells through transferrin receptor-mediated endocytosis. When iron levels are low, it is advantageous to the cell to inhibit the levels of ferritin so that any remaining free iron can be available for endocytic transport by transferrin via the transferrin receptor.

To reduce the levels of ferritin and increase the levels of transferrin receptor, there is a family of iron regulatory proteins called IRPs, which bind to a site located on both the ferritin and the transferrin receptor mRNAs called the **iron-response element** or **IRE**. The IRE is comprised of consensus sequence elements and a distinct stable RNA secondary structure, and is located upstream of the translation initiation site on ferritin mRNA (**FIGURE 4.73**). When the IRE is bound by the IRP, the complex blocks the scanning 40S ribosome from reaching the start codon.

In the case of the transferrin receptor, there are multiple IREs located in the 3' UTR. Binding of an IRP to the IREs in the transferrin receptor 3' UTR greatly stabilizes the receptor's normally

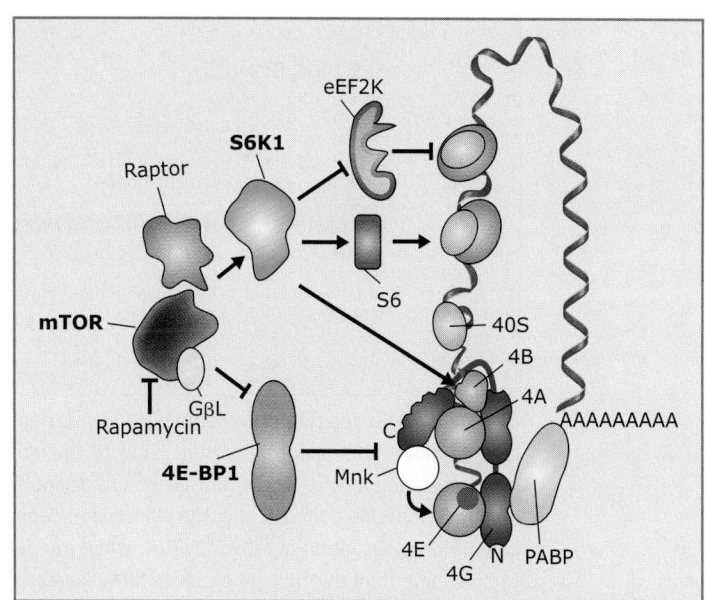

FIGURE 4.72 The mTOR pathway regulates translation in eukaryotes. The mammalian target of rapamycin (mTOR) signal transduction pathway regulates translation in mammalian cells. Rapamycin is an antibiotic that binds to mTOR and inhibits its function. mTOR, GβL, and Raptor assemble into a complex to form an active protein kinase that specifically inactivates two inhibitors of protein synthesis, 4E-BP1 and S6 kinase. 4E-BP1 binds to eIF4E and blocks its assembly with the 7-MeG cap, and the S6 kinase phosphorylates ribosomal protein S6 causing inhibition of the 43S ribosomal preinitiation complex. mTOR-Raptor-GβL complex acts to block the translational inhibitory effects of 4E-BP1 and S6 kinase. When cellular nutrients are plentiful and there is active translation, various cellular signaling pathways result in an increase in the kinase activity of the mTOR-Raptor-GβL complex. Under conditions of starvation, mTOR activity is downregulated, and 4e-BP1 and S6 kinase act to inhibit protein synthesis. Adapted from M. B. Mathews, N. Sonenberg, and J. W. B. Hershey (eds.). *Transitional Control in Biology and Medicine, Third edition.* Cold Spring Harbor Laboratory Press, 2007.

unstable mRNA. Thus, when iron levels are low in the blood, the cell acts to decrease the level of ferritin translation initiation and to increase the stability of the transferrin receptor mRNA. The IRP has an iron-binding site; when the iron-binding site on the IRP is filled with iron, the affinity of IRP for the IRE is significantly reduced. Consequently, when iron levels are high in the blood, ferritin synthesis is increased to enhance iron storage. High iron levels also release the binding of IRPs to the transferrin receptor IREs and thereby stimulate the degradation of IRP.

In summary, translational control plays an important role in the regulation of specific genes, using a set of activator and repressor proteins and specific sequences to modulate

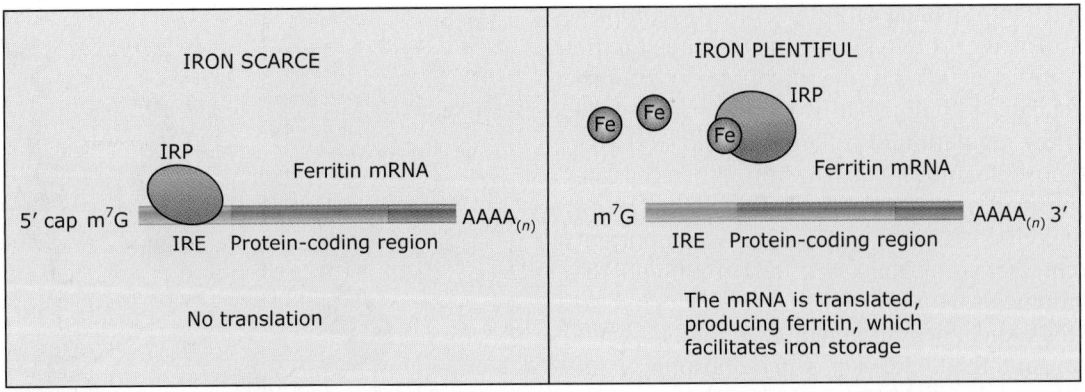

FIGURE 4.73 Regulation of ferritin mRNA translation initiation by the iron-response element-binding protein. The binding of an iron regulatory protein (IRP) to the iron-response element (IRE) in the 5′ untranslated region of the mRNA encoding ferritin blocks translation initiation. When the IRE is bound by the IRP, the complex blocks the scanning 40S ribosome from reaching the start codon. When iron levels are low in the blood, IRP binds tightly to the IRE and inhibits translation of ferritin mRNA. When iron levels are high in the blood, iron binds to the IRP and causes it to dissociate from the IRE site on the mRNA, thereby increasing ferritin synthesis to enhance iron storage. Adapted from J. M. Cooper and R. E. Hausman. *The Cell: A Molecular Approach, Fourth edition.* Sinauer Associates, Inc., 2006.

ribosome assembly with mRNA and the recognition of the translation initiation site. Another very common strategy for regulating translation initiation is to modulate the interaction of the 43S complex or the poly(A) tail with the cap-binding complex. The levels of translation are directly related to the amount of mRNA available. In the next section, we will explore another type of cellular control of protein synthesis—the localization of mRNAs to specific regions of the cell for the purpose of site-specific translation.

Concept and Reasoning Check

1. How might translational control be achieved at the other steps in translation subsequent to initiation?

4.16 Some mRNAs are translated at specific locations within the cytoplasm

Key concepts

- Translation of selected mRNAs in specific regions of cells generates structural and functional cell polarity.
- Zipcode-binding proteins (ZBPs) bind to zipcodes in the mRNA and help to deliver some proteins to specific sites in the cytoplasm for translation using the microfilament and microtubule cytoskeletal systems.

In the last section, we saw how mRNA-binding proteins could regulate translation. However, not only can translation of mRNA be controlled temporally, it can also be regulated spatially; that is, some mRNAs are localized to specific regions in the cytoplasm before translation occurs. mRNA localization leads to spatially restricted translation of an mRNA, allowing for multiple rounds of translation of the same mRNA and providing a large localized concentration of newly synthesized protein at a specific cellular location and time. The localization of protein synthesis to the site where the proteins will be used is much more efficient than transporting the protein from a large number of spatially unrestricted translation sites to a specific cellular location.

One of the best examples of mRNA localization is β-actin mRNA. Fibroblast cells can move along substrates by extending out processes called lamellipodia. Lamellipodia form at the leading edge of a cell, extend forward and adhere to the surface in front of the cell, and then pull the cell toward the adhesion point. This movement is governed by the local synthesis and polymerization of β-actin into filaments at the site of lamellipodium formation. β-actin mRNA is localized to the leading edge of the cell and translated where the lamellipodium is to be formed. The following mechanism based on studies of β-actin mRNA

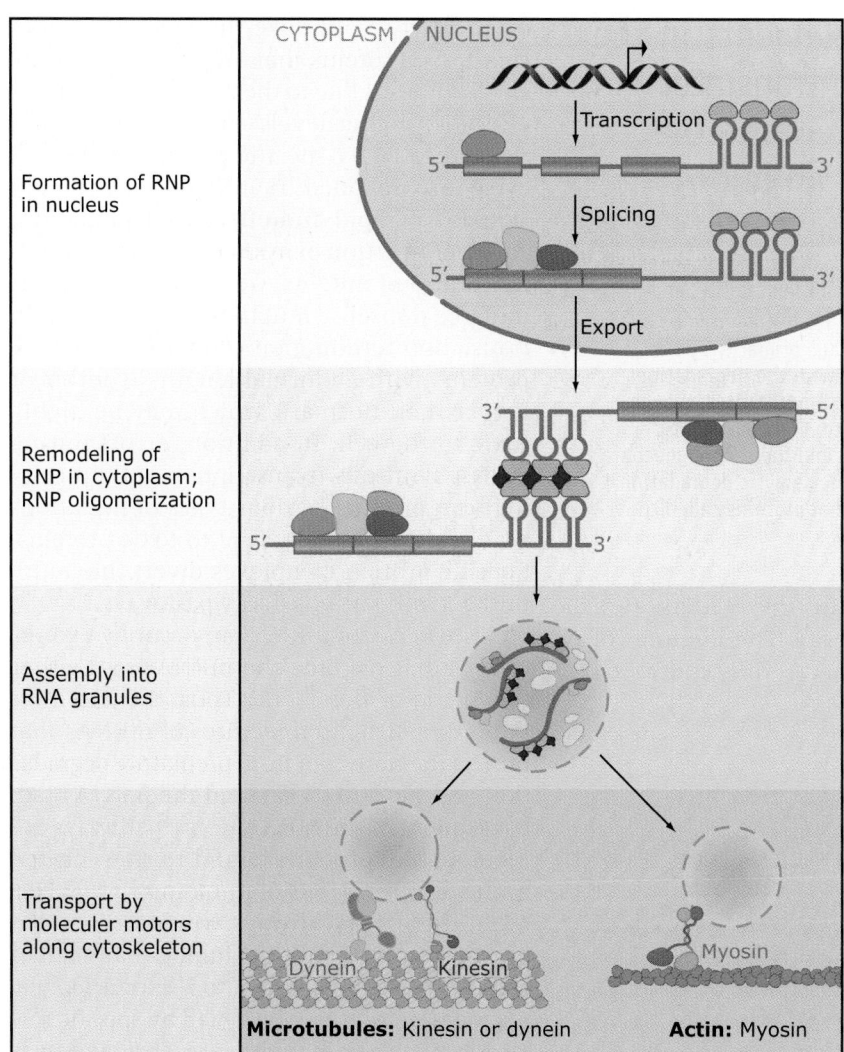

On the left side of the figure, the following stage labels appear:

- Formation of RNP in nucleus
- Remodeling of RNP in cytoplasm; RNP oligomerization
- Assembly into RNA granules
- Transport by moleculer motors along cytoskeleton

Figure labels: CYTOPLASM · NUCLEUS · Transcription · Splicing · Export · 5′ · 3′ · Dynein · Kinesin · Myosin · **Microtubules:** Kinesin or dynein · **Actin:** Myosin

FIGURE 4.74 Cytoplasmic localization and site-specific translation of mRNA. mRNAs that are to be localized to specific regions of the cytoplasm contain a sequence in the 3′ UTR called a zipcode, which binds to proteins called zipcode-binding proteins. After export to the nucleus, the mRNP complex assembles into granular aggregates that are transported to specific locations in the cytoplasm by dynein or kinesin along microtubules or by myosin along actin filaments. Reprinted from *Cell*, vol. 136, K. C. Martin and A. Ephruss, mRNA Localization..., pp. 719–730, Copyright (2009) with permission from Elsevier [http://www.sciencedirect.com/science/journal/00928674].

localization (**FIGURE 4.74**) may also be applicable to many localized mRNAs:

- Located within the 3′ UTR of the localized mRNA is a sequence element known as a **zipcode**, which binds to a protein known as a **zipcode-binding protein**.
- The mRNA together with the ZBP-zipcode complex is exported to the cytoplasm where the mRNP is remodeled and several mRNPs assemble into an **RNP granule**.
- The ZBP-containing RNP granule binds motor proteins **dynein** or **kinesin**, which move molecular cargo along the **microtubule** cytoskeleton (see *11 Microtubules*). Alternatively, the RNP granule can bind to the motor protein **myosin**, which moves cargo along the

actin microfilament cytoskeleton (see *12 Actin*). Both types of motor proteins utilize ATP as an energy source for the mechanochemical transport of the RNP granule.

Other examples of mRNA localization include the localization of specific mRNAs to the axons and dendrites of neurons and the movement of specific mRNAs to distinct locations in *Drosophila* embryos during the early stages of development.

Concept and Reasoning Check

1. During the localization of mRNAs to specific sites in the cytoplasm, translation of the mRNA is inhibited. Suggest a mechanism for this inhibition and for activation of translation once the mRNA reaches its final destination.

4.17 Sequence elements in the 5′ and 3′ untranslated regions determine the stability of an mRNA

Key concepts

- The 5′ and 3′ UTRs of the mRNA contain sequences that regulate translation, mRNA stability, and mRNA localization in the cytoplasm.
- Degradation of mRNAs occurs primarily by two pathways that both begin with the degradation of the poly(A) tail: The first mechanism involves decapping followed by 5′ to 3′ exonucleolytic degradation of the mRNA, and the second mechanism involves 3′ to 5′ exonucleolytic degradation of the mRNA by a complex of proteins called the exosome.

An additional strategy for the regulation of translation is mRNA decay. The lifetimes of specific mRNAs are often carefully controlled by the cell to optimize the level of protein synthesis. Proteins that are required in only small amounts due to their potential cytotoxic effects at higher levels, or proteins that are needed for only specific periods in the cell cycle, are encoded by mRNAs that contain signals for rapid and effective degradation. Another function of mRNA decay is the quality control of mRNAs. Genes that have accumulated nonsense mutations or mutations in translation termination codons can produce proteins with abnormal lengths (shorter or longer than normal), which may be highly toxic to the cell. In addition, errors during mRNA synthesis (transcription and processing) can produce various types of mutations that could potentially lead to toxic proteins. Specific protein complexes divert the faulty mRNAs into various decay pathways.

The decay of mRNA can occur by exonucleases that degrade the mRNA from either the 5′ end or the 3′ end. Thus, the two most important structural features of mRNAs that protect the transcript from premature degradation are the 7-MeG cap and the poly(A) tail; consequently, the mRNA decay pathways are regulated by proteins bound to sites located in the 5′ and 3′ UTRs of the transcript (**FIGURE 4.75**). These two pathways, which are found in most eukaryotic cells, including mammalian cells, both begin with the 3′ to 5′ exonucleolytic degradation of the poly(A) tail by specific nucleases that are still under intensive investigation. Once the poly(A) tail has been degraded to less than 10 adenine residues, the remainder of the mRNA is degraded by one of the two following sets of events:

1. Removal of the 5′ 7-MeG cap followed by 5′ to 3′ exonucleolytic digestion of the remaining body of the mRNA.
2. 3′ to 5′ exonucleolytic digestion of the rest of the poly(A) tail followed by the remaining body of the mRNA.

While there are a number of specific protein-binding sites involved in mRNA stability, the best-characterized and most common protein-binding site is the **AU-rich element** or **ARE** (**FIGURE 4.76**). AREs range between 50 and 150 nucleotides in length and represent a diverse set of nucleotide sequences. Many AREs contain numerous copies of the short sequence element AUUUA or the longer sequence UUAUUUAUU along with one or more stretches of uridine residues. More than 15 ARE-binding proteins have been identified in mammalian cells. ARE elements are most of-

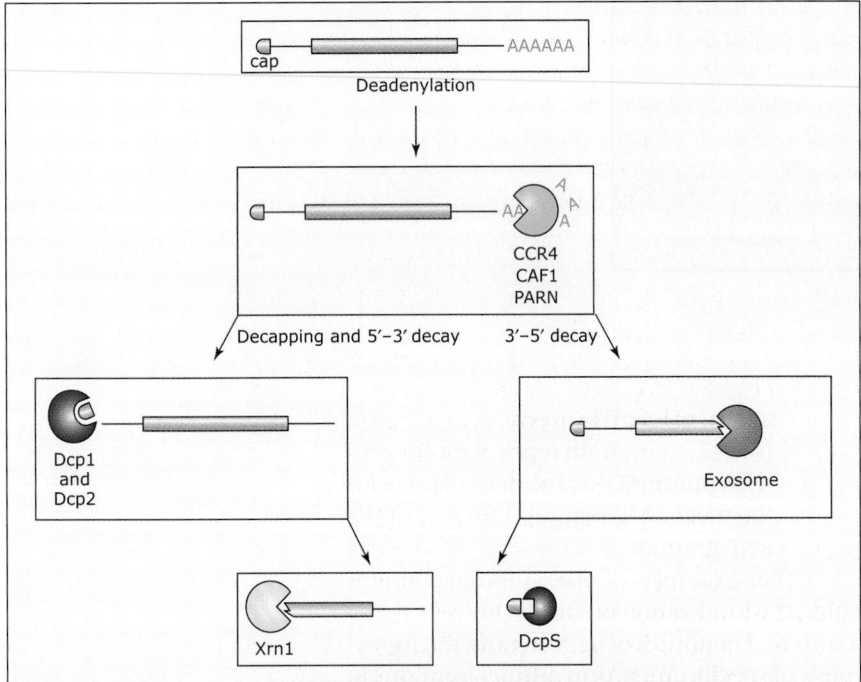

FIGURE 4.75 Two pathways for eukaryotic mRNA decay. The decay of mRNA can occur by exonucleases that degrade the mRNA from either the 5′ end or the 3′ end. The 7-MeG cap and the poly(A) tail protect the mRNA from degradation while the mRNA is translated. Decay of the mRNA in both pathways begins with the 3′→5′ exonucleolytic degradation of the poly(A) tail by specific nucleases that are still under intensive investigation. Once the poly(A) tail has been degraded to less than 10 adenine residues, the remainder of the mRNA is degraded by either removal of the 7-MeG cap followed by 5′→3′ exonucleolytic digestion of the remaining body of the mRNA by the Xrn1 nuclease, or by 3′→5′ exonucleolytic digestion of the rest of the poly(A) tail followed by degradation remaining body of the mRNA by the exosome. Adapted from C. J. Wilusz and J. Wilusz, *Trends Genet.* 20 (2004): 491–497.

Sequences involved in mRNA post-transcriptional regulation

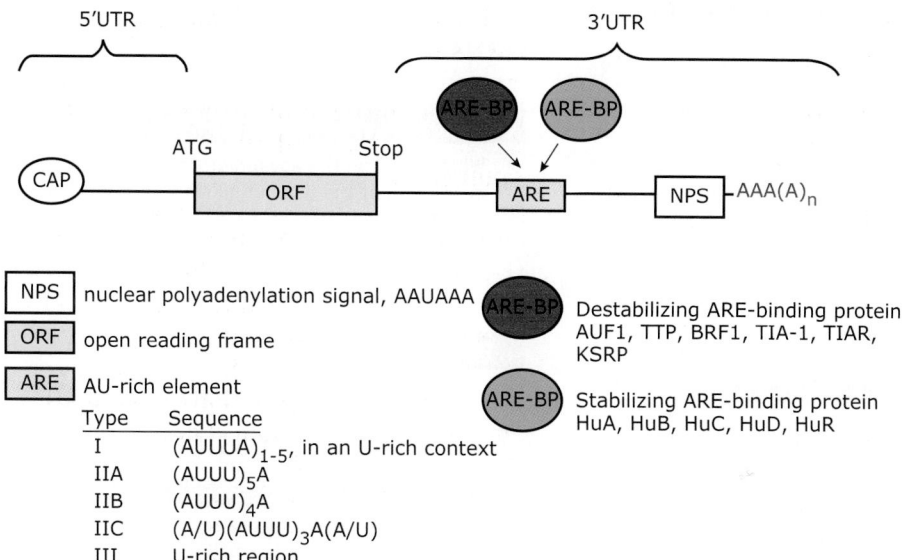

FIGURE 4.76 AU-rich elements control mRNA stability. AU-rich elements (AREs) are found in the 3' UTRs of mRNAs that have short half-lives. The ARE binds to one of several different ARE-binding proteins, which stimulates the degradation of the mRNA. Courtesy of Rebecca Hartley, University of New Mexico School of Medicine. Reproduced from an illustration by Rebecca Hartley, University of New Mexico School of Medicine.

ten found in mRNAs encoding proteins whose levels must be carefully controlled and whose expression is only desirable for a specific short period during the cell cycle. Some ARE-binding proteins are expressed in all tissues, while others are tissue-specific. In addition, some ARE-binding proteins are present at only specific stages of the cell cycle. ARE-binding proteins can function to either stabilize or destabilize the mRNA transcript, depending upon the ARE-binding protein and the presence of other mRNA stability proteins associated with the 3' UTR. A complex called the **exosome**, which contains multiple exonucleases, has been found in most eukaryotic cells and has been shown to be involved in the 3' to 5' nucleolytic degradation of mRNAs. The exosome has a preference for mRNAs that contain AREs.

In addition to regulating the levels and temporal expression of specific proteins, mRNA decay is an importance surveillance mechanism to identify and remove mRNAs containing mutations that could lead to the production of proteins that are deleterious to the cell. Such mutations include premature stop codons caused by frameshift mutations that occur due to mRNA splicing errors. The normal ORFs of protein-coding regions are normally protected through evolution from acquiring inherited stop codons that would cause premature translation termination and truncated proteins. However, if an aberrant splicing event occurs that generates a frame

shift mutation, the chances of encountering a stop codon becomes 3/64 or about 1 in 20 for every codon that is read by the ribosome. Thus, it is almost inevitable that a truncated protein will result from a splicing error.

To avoid the generation of potentially deleterious truncated proteins, the cell uses a pathway called **nonsense-mediated decay** or **NMD** (FIGURE 4.77). The NMD complex assembles with a ribosome to carry out a single round of nonproductive translation that scans the mRNA for premature stop codons. It is clear that not all mRNAs go through this "pioneer round" of translation; the conditions that govern which mRNAs go through this surveillance are still unclear.

A pathway similar to NMD detects the lack of a translation termination signal, called **nonstop-mRNA decay**. Nonstop-mRNA decay protects the cell against the production of potentially deleterious proteins that arise from a lack of a stop codon, and ensures that ribosomes and other translational components are not tied up in unproductive protein synthesis and are properly recycled.

In summary, mRNA decay is an important mechanism for controlling the amount of mRNA available for translation. The signals that control mRNA decay are located in the 5' and 3' UTRs of the transcript and interact with a variety of proteins that in turn interact with various mRNA binding proteins. In addition, mRNA decay plays a critical role in mRNA

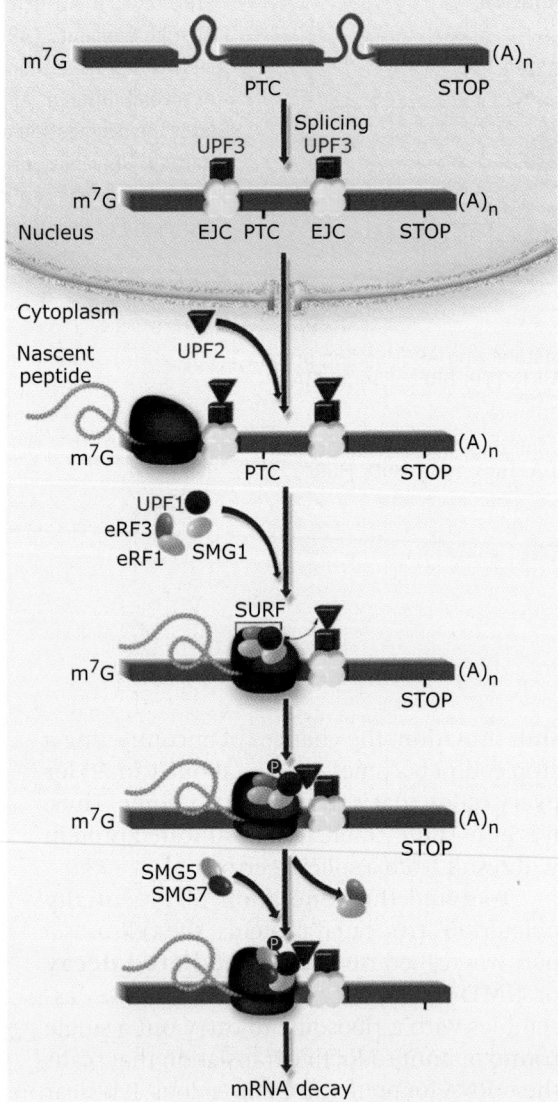

FIGURE 4.77 A model for nonsense-mediated mRNA decay. The nonsense-mediated decay (NMD) pathway is an mRNA quality control process to degrade mRNAs that contain a premature termination codon (PTC). PTCs usually result from splicing defects caused by mutations in splicing signals, and can lead to the synthesis of deleterious proteins. After mRNA splicing, the NMD core protein UPF3 binds as part of the exon junction complex (EJC) marking each junction between adjacent exons. After cytoplasmic export, UPF2 joins UPF3. If a ribosome encounters a premature termination codon, four other proteins, UPF1, SMG1, eRF1, and eREF3 form a complex at the stalled ribosome called SURF. UPF1 binds to UPF2 at the nearest EJC and causes SMG1 to phosphorylate UPF1, leading to the assembly of SMG5 and SMG7, and to the subsequent degradation of the mRNA. Reprinted from *Trends Genet.*, vol. 24, L. Linde and B. Kerem, Introducing sense into nonsense..., pp. 552–563, Copyright (2008) with permission from Elsevier [http://www.sciencedirect.com/science/journal/01689525].

quality control to prevent the production of deleterious proteins. In the next section, we will explore another mechanism for control of mRNA utilization.

Concept and Reasoning Check

1. What would be the advantages of regulating gene expression at the level of mRNA stability versus transcription?

4.18 Noncoding RNAs are important regulators of gene expression

Key concepts

- RNA interference (RNAi) is a recently discovered set of pathways that use small ncRNAs to inhibit expression of mRNAs.
- Small interfering RNAs (siRNAs) are double-stranded RNAs (dsRNAs) found in plant cells and some animal cells; they contain sequences complementary to specific mRNAs. siRNAs as part of the protein containing RNA-induced silencing complex (RISC) base pair to the mRNA and induce mRNA decay. Synthetic siRNAs can be used to knockdown gene expression in mammalian cells.
- miRNAs are originally generated as precursors from segments of introns or from their own transcription units and are processed in the cytoplasm to mature 23 nucleotides RNAs that can function as part of the RISC to promote mRNA decay or can inhibit translation.

Perhaps the single most important event in molecular cell biology during the past decade has been the discovery of two classes of ncRNAs that cause sequence-specific gene silencing by a process known as RNA interference or **RNAi**. The first class of small regulatory ncRNAs to be identified was the **siRNAs**. Originally identified in virus-infected plants and in nematodes, siRNAs are derived from larger dsRNA precursors that are transcribed and processed in the nucleus into shorter dsRNAs; these short dsRNAs are in turn exported to the cytoplasm for further trimming by a complex, including the nuclease **Dicer**, to short 21–23 bp siRNAs (**FIGURE 4.78**). The duplex siRNA is denatured and one strand, known as the **guide strand**, assembles with one of several members of the **Argonaute (Ago)** protein family to form the **RNAi complex** or **RISC**. The RISC uses the guide strand to recognize and cleave specific mRNAs in the cytoplasm. Only a few naturally occurring mammalian siRNAs have been reported thus far, but synthetic siRNAs have emerged as a powerful tool for the study of mammalian gene expression. Synthetic siRNAs can be designed to target specific mRNAs and can be transfected directly into mammalian cells or synthesized inside the cell in situ on transfected expression plasmid vectors. Once inside the cell, an siRNA can assemble in the RISC and degrade the mRNA target, often resulting in a near elimination or at least a significant decrease (knockdown) in a specific protein.

The second class of ncRNAs, known as **miRNAs**, has been discovered with very similar properties to siRNAs, and has had a profound effect on our understanding of gene regulation. It is estimated that there may be over 1,000 different mammalian miRNAs. As in the case of siRNAs, miRNAs bind to specific mRNAs through complementary base pairing. Although the miRNA target sequences are most often found in the 3' UTR, targets may be located in any region of the mRNA, including the 5' UTR and the coding region. The identification of target sequences for miRNAs is under very active investigation, and more than 500 different target sequences have been identified. In vertebrates, individual miRNAs have been found to play important roles in differentiation and development. miRNAs can function to either destabilize mRNAs through the RISC complex or can repress translation.

The original precursor of miRNAs can come from a primary transcript (**pri-miRNA**), synthesized from its own promoter or from an excised mRNA intron. Pri-miRNAs and intronic miRNA precursors are processed into **pre-miRNAs** within the nucleus by an enzymatic complex including the nuclease **Drosha** and the **DCGR8 protein**. The pre-miRNA is exported to the cytoplasm, where the same Dicer complex that was used for siRNA generation processes the pre-miRNA into the 18–23 nucleotide final miRNA product (**FIGURE 4.79**). The miRNA assembles along with an Ago protein into the RISC, similar to the siRNA pathway. Depending on the miRNA and its target sequence, the attachment of the miRNA-RISC leads to mRNA degradation or to translational interference. Interaction between the miRNA and the target mRNA is mediated by base pairing between the target site on the mRNA and an 8-nt segment on the miRNA referred to as the **guide sequence** (**FIGURE 4.80**). Computer algorithms have been developed to identify

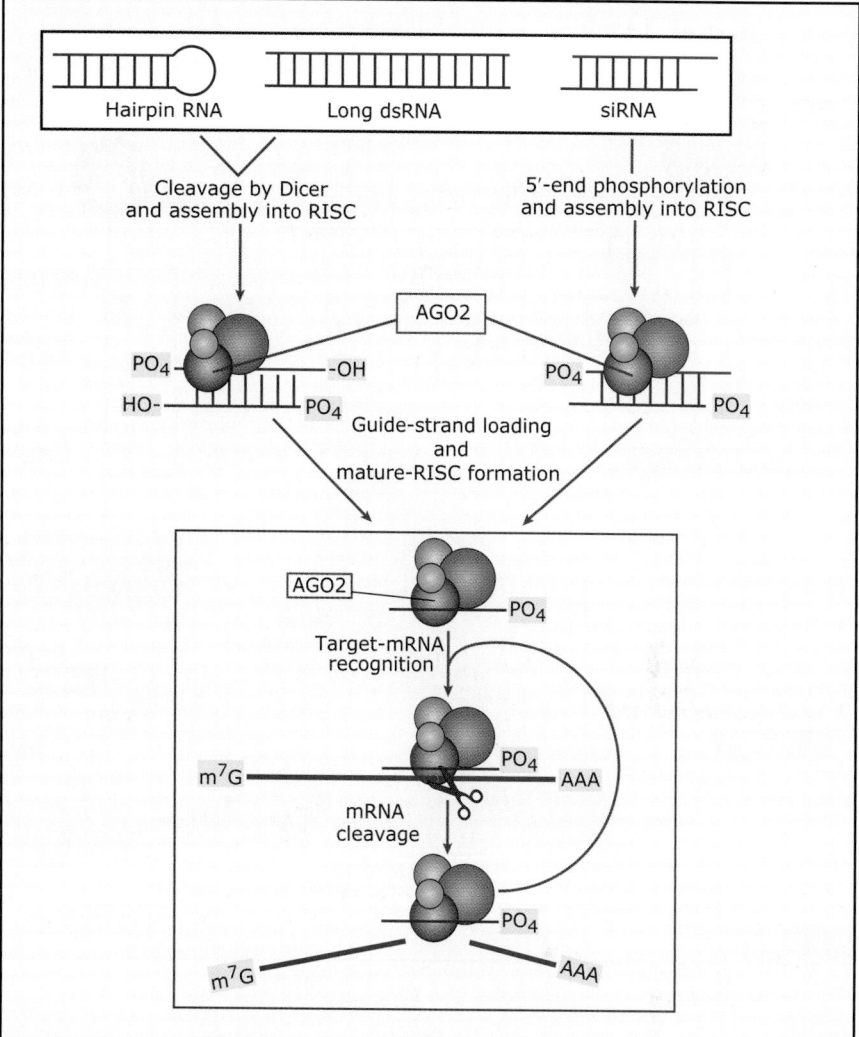

FIGURE 4.78 The siRNA pathway. Small interfering RNAs (siRNAs) target and degrade specific mRNAs. Naturally occurring siRNAs are derived from larger double-stranded RNA (dsRNA) precursors, made in the nucleus and processed in the cytoplasm by the nuclease Dicer, to a final 21–23 bp siRNA product. siRNAs can also be synthesized in the laboratory as short double-stranded products and transfected into cells. The siRNA is unwound and one strand, known as the guide strand, along with one of the Argonaute (Ago) family proteins, assembles into what is known as the RNA interference complex or RISC. The RISC uses the guide strand to recognize and cleave specific mRNAs in the cytoplasm. Reprinted by permission from Macmillan Publishers Ltd: *Nat. Rev. Mol. Cell Biol.*, T. M. Rana, Illuminating the silence..., 2007, vol. 8 (1): 23–36.

potential miRNA target sites in mammalian genomes, and it is common to encounter at least a half-dozen different miRNA targets within any given mRNA 3' UTR or 5' UTR; confirmation of the target site is carried out experimentally.

The study of the role of miRNAs in cell differentiation and embryonic development is one of the most active areas of investigation in cell biology. It is now clear that miRNAs are also

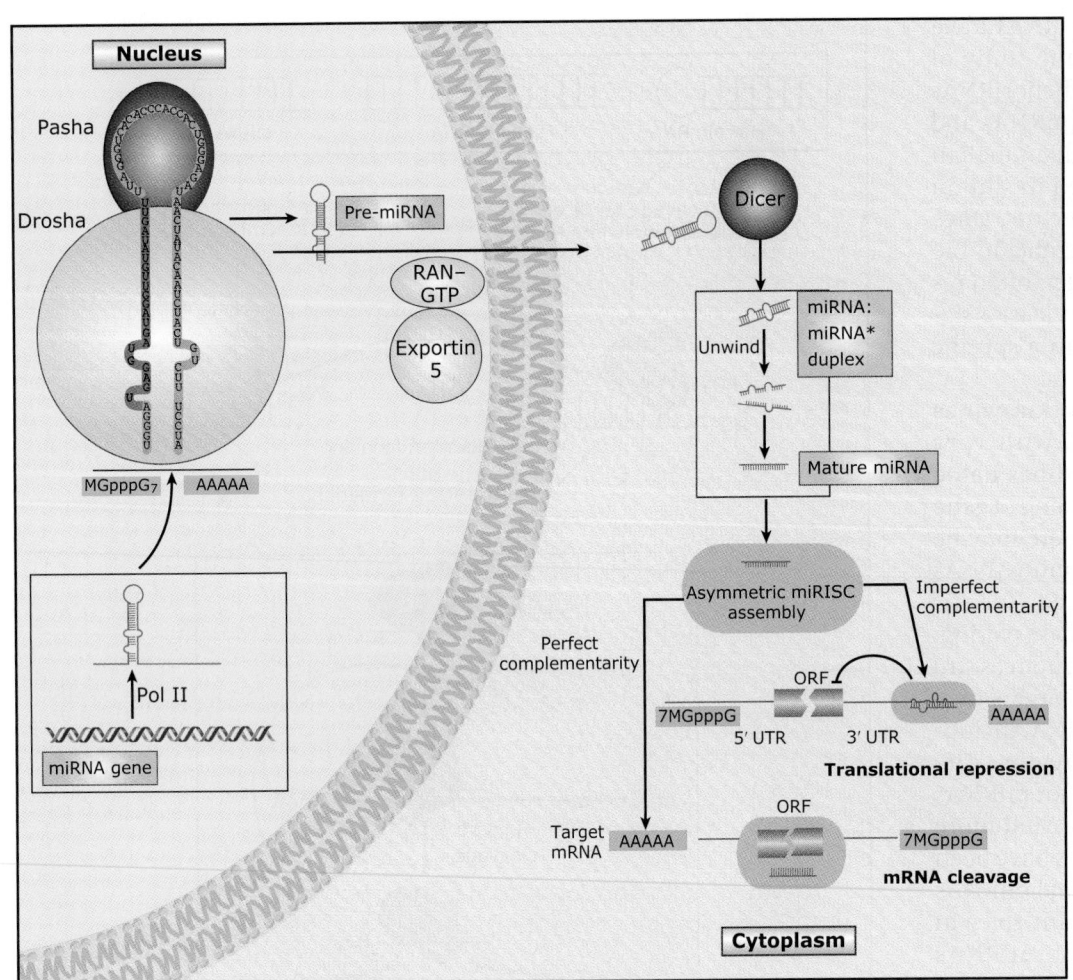

FIGURE 4.79 The miRNA pathway. MicroRNAs (miRNAs) can be derived from intronic regions of mRNA transcripts or can be transcribed from their own promoters. In this figure, the primary transcript, called a pri-mRNA, is transcribed from its own promoter and processed into a pre-miRNA within the nucleus by an enzymatic complex including the nuclease Drosha and the DCGR8 protein. The pre-miRNA is exported to the cytoplasm, where the Dicer complex processes the pri-miRNA into an 18–23 nucleotide final miRNA product. The miRNA assembles along with an Argonaute protein into the RISC in a manner similar to the siRNA pathway. Depending on the degree of complementarity between the miRNA and its target mRNA sequence, the attachment of the miRNA-RISC leads to mRNA degradation or to translational interference. Adapted from A. Esquela-Kerscher and F. J. Slack, *Nat. Rev. Cancer* 6 (2006): 259–269.

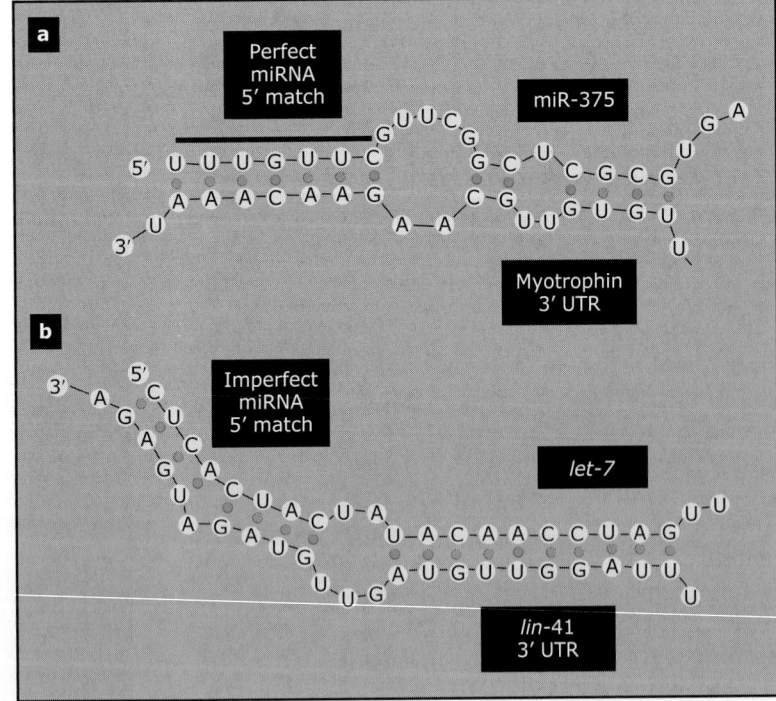

FIGURE 4.80 The miRNA "seed" sequence. Interaction between a miRNA and its target mRNA is mediated by base pairing between the target site on the mRNA and an 7–8 nucleotide segment on the miRNA guide strand referred to as the seed sequence. (a) Perfect complementarity between the miRNA miR-375 seed sequence and the myotrophin mRNA target, but imperfect complementarity outside the seed sequence. (b) Imperfect complementarity between miRNA let-7 and lin-41 mRNA, but more extensive complementarity outside of the seed sequence. Adapted from N. Rajewsky, *Nat. Genet.* 38 (2006): S8–S13.

involved in many disease processes including cancer and cardiovascular disease.

4.19 What's next?

In this chapter, we have reviewed the individual events in gene expression and considered how each step is regulated. However, these events do not occur as individual biochemical processes, but function as an **integrated, complex dynamic system**, linked together by regulatory factors that act in multiple steps to keep the reactions coupled to each other, facilitating feedback and autoregulatory behavior. It is also important to emphasize that gene expression is not monotonic; in other words, genes are not simply on or off, but have a dynamic range of expression. By using multiple layers of regulation at different stages of expression, it is possible to control RNA and protein levels temporally with a much larger dynamic range. In addition to temporal regulation, gene expression is controlled spatially within the context of the structural organization of the cell. Trafficking of RNAs and proteins between the various membrane-bound organelles as well as the nonmembranous subcompartments will be more fully explored elsewhere in the book, but it suffices to say that the interplay between temporal and spatial regulation adds an increasing level of complexity to control of gene expression.

While there are a plethora of applications and implications for studies of gene regulatory systems, some of the most compelling are in medicine. There are more than 210 distinct cell types in the adult human body. Each of the cells in the body contains the same linear sequence of DNA—the same set of genes (the same genome). However, it is the gene expression pattern of the genome that determines the differentiation state and structural and metabolic homeostasis of each of these cell types. We will see in other chapters of the book that the failure to control the growth, proliferation, and differentiation of just a single cell can have enormous ramifications for the health of the entire organism. The complete understanding of the pathways that lead to cellular growth, proliferation, and differentiation will provide us with great capabilities for understanding the nature of health and disease at the levels of the organism and population.

Ultimately, by understanding gene expression at the integrative systems level, it may be possible to modify and correct aberrant gene expression to treat a variety of human diseases. **Stem cells** are **pluripotent** cells that can differentiate into a wide variety of different cell types. Programming stem cells to differentiate into functional tissues and organs will pave the way for effective disease intervention strategies. It may even be possible to partially or completely reprogram differentiated cells to change cell types. However, even the ability to repair an inherited mutation in a single faulty master regulatory gene would have enormous benefits.

What would be necessary to successfully intervene in the control of cellular gene expression? First, we will need to know the expression states at the multicellular level—not just at the single cell level. We are also becoming increasingly aware that cells do not exist as isolated entities. Even prokaryotes can function as communities of cells, communicating their functional states between one another through intercellular signaling pathways (see *20 Prokaryotic cell biology* for a further discussion). For eukaryotic metazoans, this communication includes not only interacting as individual cells, but also interacting physically to build the next levels of biological structure within the organism—tissues and organs. Thus, the expression of genes in one cell must be coordinated with the expression of genes in other cells in its immediate neighborhood as well as in tissues and organs located at a significant distance. These **intercellular gene control** pathways are almost a complete mystery at the moment, but understanding them at this higher level will be critical if we are to be able to exploit our knowledge of gene regulation to the fullest extent in the promotion of health in organisms and populations.

To address the future challenges of understanding the coordinated function of cellular genomes at the organism and population levels, we will have to combine experimental biology with mathematical and computational approaches. We are just beginning to see the application of sophisticated computer algorithms and network theory to analyze global gene regulatory systems. The first attempts using these approaches are being applied to the understanding of networks of transcription activators and repressors involved in cell differentiation, development, and cancer. However, as we have seen in this chapter, it is now clear that alternative splicing, mRNA stability, and ncRNAs are also very important and common pathways in gene regulation and will have to be integrated into truly comprehensive modeling of gene expression systems. The iterative approach of testing theoretical models experimentally and using the results to modify the models will increase our level of understanding of how gene expression systems provide the direction for complex dynamic structural and functional integrity of the cell—in the context of its physical environment and in the context of a tissue and organ within the organism. Thus, the biggest challenge for the future of the study of gene expression and regulation at the systems level will require effective collaborative efforts between experimental and theoretical scientists and mathematicians in which members of the research teams will have to learn new vocabularies, approaches, instrumentation, and tools.

4.20 Summary

In this chapter, we have focused on the basic mechanisms of gene expression and regulation. The modern definition of the gene is a chromosomal or organellar DNA sequence that encodes the information to transcribe a functional RNA. The RNA transcript may code for a protein or be an ncRNA with structural, regulatory, or catalytic functions. In the first step of gene expression, the gene is transcribed to produce the primary RNA transcript; however, the vast majority of primary RNA transcripts, especially in eukaryotes, must be processed into mature transcripts. In the case of eukaryotic protein-coding genes, RNA processing includes 5′ capping, 3′ polyadenylation, and the removal of intronic sequences. Transcription and RNA processing occur in the nucleus, and are directly coupled to mRNA export to the cytoplasm where translation, localization, storage, and degradation occur. The export of mRNAs from the nucleus to the cytoplasm is regulated to ensure that only functional mRNAs leave the nucleus and are translated at the correct time and location within the cytoplasm. Translation, mRNA stability, and mRNA cytoplasmic localization are regulated by proteins and small ncRNAs. Many of the gene regulatory systems in eukaryotic cells are targets of signaling mechanisms that connect the extracellular environment, cell growth, proliferation, and differentiation to gene expression. It is clear that aberrant regulation of gene expression is a major mechanism of inherited and acquired diseases. Throughout the remainder of this book, we will encounter numerous examples of how cell cellular structure and function are inextricably linked to gene expression and regulation.

http://biology.jbpub.com/lewin/cells

To explore these topics in more detail, visit this book's Interactive Student Study Guide.

References

Carthew, R. W., and Sontheimer, E. J. 2009. Origins and mechanisms of miRNAs and siRNAs. *Cell* v. 136 p. 642–655.

Danckwardt, S., Hentze, M. W., and Kulozik, A. E. 2008. 3′ end mRNA processing: molecular mechanisms and implications for health and disease. *EMBO J.* v. 27 p. 482–498.

Davidson, E. H. 2006. *The Regulatory Genome: Gene Regulatory Networks in Development and Evolution.* Burlington, MA: Academic Press.

Kornberg, R. D. 2007. The molecular basis of eukaryotic transcription. *Cell Death Different* v. 14 p. 1989–1997.

Krebs, J. E., Goldstein, E. S., and Kilpatrick, S. T. 2010. *Lewin's Essential Genes.* 2nd Ed. Sudbury, MA: Jones and Bartlett.

Martin, K. C., and Ephrussi, A. 2009. mRNA localization: gene expression in spatial dimension. *Cell* v. 136 p. 719–730.

Scherer, S. 2008. *A Short Guide to the Human Genome.* Cold Spring Harbor, NY: Cold Spring Harbor Press.

Sonenberg, N., and Hinnenbusch, A. G. 2009. Regulation of translation initiation in eukaryotes: mechanisms and biological targets. *Cell* v. 136 p. 731–745.

Thomas, M. C., and Chiang, C. M. 2006. The general transcription machinery and general cofactors. *Crit. Rev. Biochem. Mol. Biol.* v. 41 p. 105–178.

Tropp, B. 2008. *Molecular Biology: Genes to Proteins.* 3rd Ed. Sudbury, MA: Jones and Bartlett

Wahl, M. C., Will, C. L., and Lührman, R. 2009. The spliceosome: design principles of a dynamic RNP machine. *Cell* v. 136 p. 701–718.

Protein structure and function

5

Stephen J. Smerdon

MRC National Institute for Medical
Research, London, UK

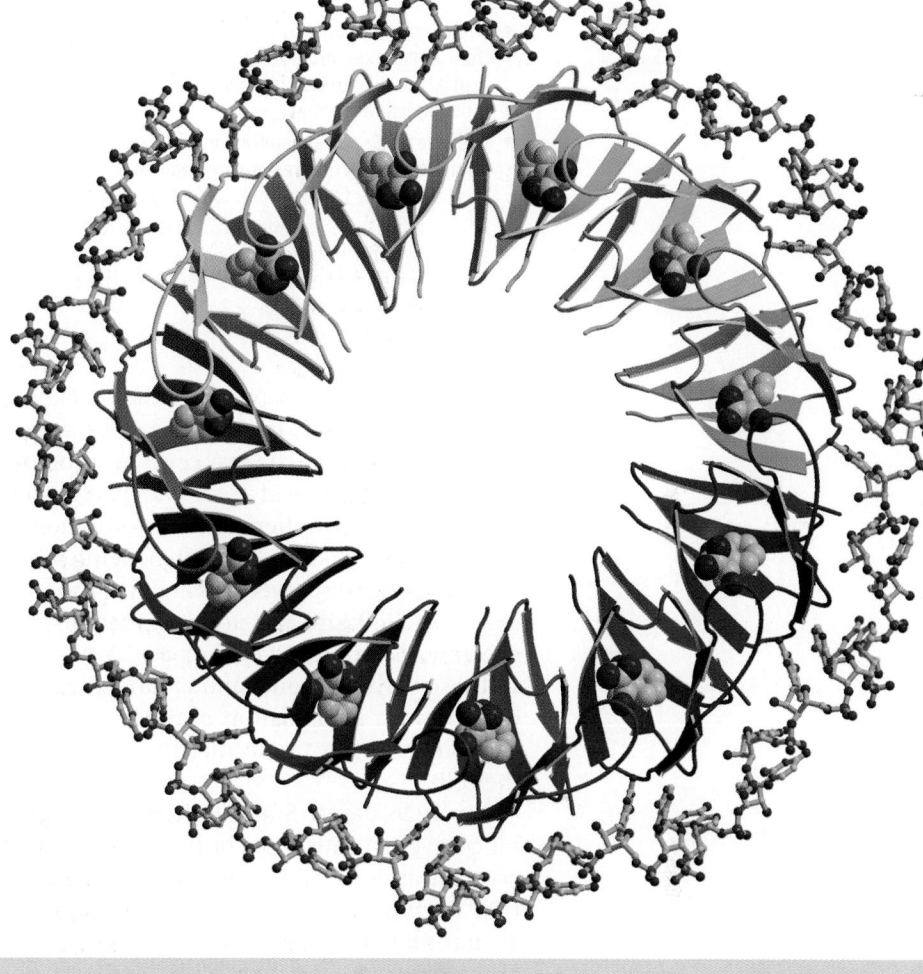

CRYSTAL STRUCTURE of the trp RNA-binding attenuation
protein (TRAP) bound to RNA. Structure courtesy of Paul
Gollnick, State University of New York at Buffalo, USA,
and Alfred Antson, University of York, UK.

CHAPTER OUTLINE

5.1 Introduction

Key concepts

- Proteins are large, complex polymers. Their three-dimensional structures dictate their biologic function.
- The three-dimensional structure of proteins and their complexes provides a framework that is essential for a full comprehension of their myriad biochemical activities.
- The size and spatial separation of atoms that make up molecular structures is too small to be directly observable by, for example, light microscopy.
- At present, three methods are available for protein structure determination: X-ray crystallography, nuclear magnetic resonance (NMR) spectroscopy, and electron microscopy.

The word "protein" derives from the Greek *proteos*, meaning "first" or "of first importance." In the early part of the last century, proteins rather than nucleic acids were widely regarded as the repository of hereditary information—the "genetic material." Classic biochemical experiments that disproved this are now the stuff of biologic folklore and are described on this site and elsewhere. Over the past 75 years or so these early misconceptions have been replaced by an appreciation of the importance of proteins as the "molecular workhorses" of the cell.

In 1926, James Sumner showed that urease from jack beans could be highly purified by crystallization, enabling him to demonstrate that enzymes were proteins. Nevertheless, at that time proteins were still thought of as heterogeneous substances with random structure. This dogma was challenged in 1934 when John Desmond Bernal and a graduate student, Dorothy Crowfoot, demonstrated that a crystal of the proteolytic enzyme pepsin produced a pattern of discrete diffraction spots on a film when exposed to a beam of X-rays. This experiment showed unequivocally that proteins possess an ordered, well-defined arrangement of atoms, and the field of structural biology was born.

Proteins are a diverse class of biologic polymers that play an extraordinary variety of functional roles. In the form of enzymes, proteins catalyze most of the chemical reactions that take place in the cell. Protein function is not, however, limited to chemical catalysis. For example, interactions between protein hormones and receptors are responsible for the transmission of many developmental and physiologic signals and represent just one of many activities that are mediated through highly specific protein-binding events.

A chemist can utilize an almost limitless variety of conditions to increase the efficiency of chemical reactions in the laboratory. In contrast, synthetic (anabolic) and degradative (catabolic) processes, along with the host of interactions necessary for life, must occur in a largely aqueous environment and within a rather narrow range of temperatures. To a great extent, these constraints have driven the evolution of the large and complex protein molecules that we observe in living organisms. Clearly, biologic activity derives directly from the relative spatial arrangement of the atoms and chemical groups from which proteins are constructed. For this reason, biologic mechanisms can only be truly understood in the light of the three-dimensional atomic structure of the macromolecules involved. This chapter focuses on how our understanding of these fundamental molecular processes has evolved through the elucidation and analysis of the three-dimensional structure of proteins.

The primary goal of all structural techniques is the determination of the precise spatial relationship between each and every atom in the molecule of interest. In this respect it is important to recall that the chemical bonds between atoms within a protein are of the order of 10^{-10} m. Optical theory shows that in order to "resolve" two objects, we must illuminate them with radiation of a wavelength that is no longer than about twice the distance between them. Given that the wavelengths of the visible electromagnetic spectrum are between ~400 and 800 nm, it is clear that light microscopy is not useful when investigating objects as small as proteins, and thus other methods must be employed.

At present, X-ray crystallography and NMR are the only available techniques for the determination of macromolecular structures at high resolution. Significant advances in other areas, however—particularly electron microscopy—are providing important structural information in ever-increasing detail. A thorough treatment of the theoretical background to these methods is beyond the scope of this chapter and the interested reader is provided with references to a number of excellent textbooks, review articles, and online information. In *5.2 X-ray crystallography and structural biology; 5.3 Nuclear magnetic resonance;* and *5.4 Electron microscopy of biomolecules and their complexes,* the aim is instead to provide a brief historical background along with suffi-

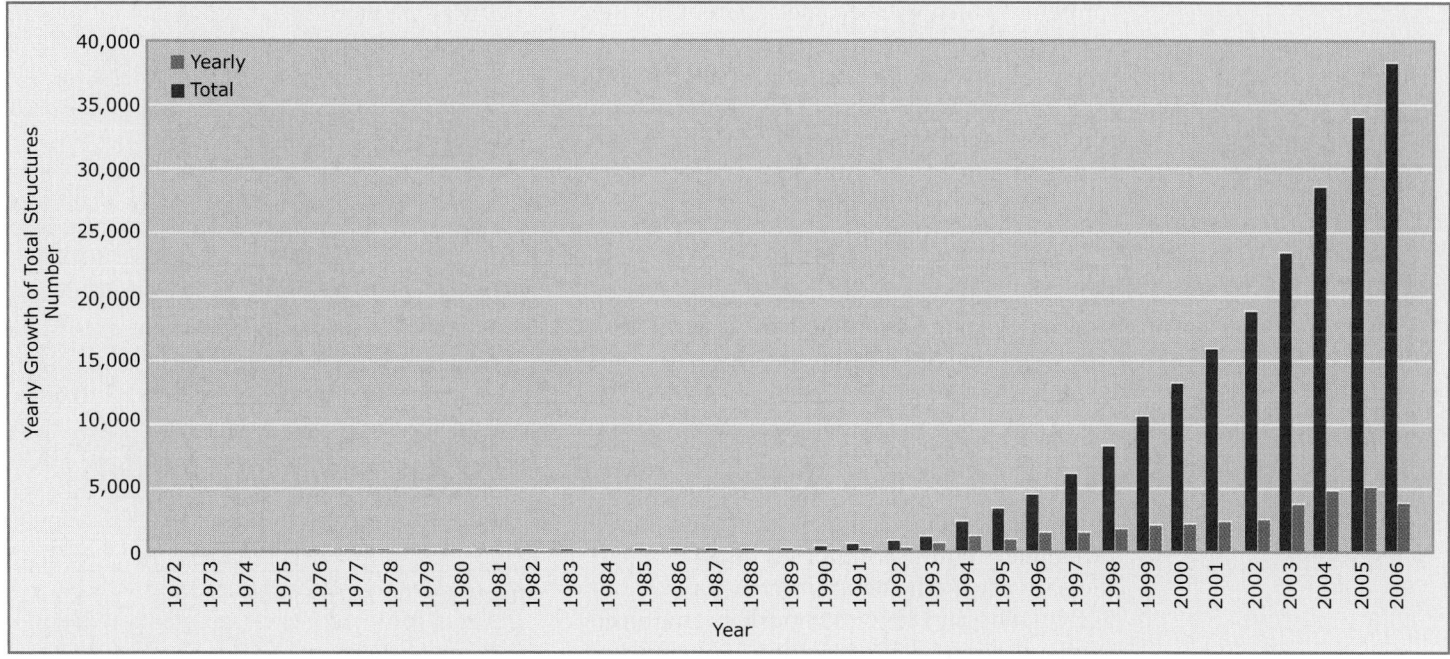

FIGURE 5.1 The rate of growth of the Protein Data Bank.

cient technical information to guide the reader through the various examples provided in the following sections, and to describe the technical advances in currently popular structural techniques that have resulted in the recent explosion of structural information.

The first protein structure, that of myoglobin, was reported by John Kendrew and coworkers in the late 1950s. Since then, the number of structures determined each year has increased exponentially. This expansion of structural information has occurred in parallel with, and as a result of, advances in the fields of molecular biology and physics (see *5.2 X-ray crystallography and structural biology*). In 1971, the Protein Data Bank (PDB) was established as an international repository for structural data. At present, a total of ~60,000 structures have been deposited and structures are currently being determined at the rate of ~7000 per year (**FIGURE 5.1**).

Structural methods are increasingly being incorporated into the pantheon of routine but powerful methodologies that can be brought to bear on an experimental system. Furthermore, the scope of biologic questions that can be asked has been fundamentally changed. The new field of *structural genomics* (see *5.17 What's next? Structural biology in the postgenomic era*) has emerged with the goal of determining the

structures of all proteins from a number of target organisms ranging from simple prokaryotes to humans. Given that the human genome encodes upward of 30,000 proteins, this undertaking is ambitious. If successful, though, the benefits to basic biologic science and to medicine (see *5.16 Structure and medicine*) could be considerable.

5.2 X-ray crystallography and structural biology

Key concepts

- At present, X-ray crystallography is the primary method for investigating macromolecular structure at atomic resolution.
- Diffraction from a crystal produces a diffraction pattern that can be related to the electron densities of each atom in the molecule by a Fourier transform.
- Phase information crucial for reconstructing an image of the molecule within the crystal is lost in the diffraction experiment but can be recovered by techniques of isomorphous replacement, molecular replacement, and anomalous scattering.

X-ray crystallography is, undoubtedly, the most effective and widely employed method for high-resolution structure determination. In light of the size, flexibility, and complexity of proteins, which will become apparent in later sections, it is perhaps amazing that these molecules can be enticed to form highly ordered three-dimensional crystalline arrays that are the first basic requirement of the method. In fact, this phenomenon was first documented in 1847 by the embryologist Karl Reichart, who observed crystallization of the oxygen transporter hemoglobin, a molecule that was to play a central role in the development of modern X-ray crystallographic methods. On December 28, 1895, Wilhelm Conrad Roentgen gave his preliminary report, entitled *Über eine neue Art von Strahlen (On a New Kind of Rays)*, to the president of the Wurzburg Physical-Medical Society, accompanied by experimental radiographs of his wife's hand. Not everyone shared Roentgen's enthusiasm for his discovery, particularly the mathematician and engineer Lord Kelvin who, in 1897, infamously pronounced X-rays to be an "elaborate hoax"—certainly not the highlight of an illustrious career!

X-rays have extremely short wavelengths of the order of one Ångstrom (Å), and in 1912, the physicist Max von Laue suggested that they might be used to investigate the atomic structure within crystals of small molecules. Lawrence Bragg was able to demonstrate X-ray diffraction by crystals of sodium chloride and solve its crystal structure. Diffraction of X-rays occurs as a result of the scattering of X-rays by the electrons that orbit each atom. Bragg formalized the diffraction of X-rays in terms of the reflection of incident radiation from imaginary planes of atoms that result from their periodic arrangement in a crystal. For this reason, diffraction "spots" are now known as "reflections" and the famous equation he derived ($n\lambda = 2d\mathrm{Sin}\theta$) is known as Bragg's Law (**FIGURE 5.2**).

Crystals are periodically repeating arrays (see the *Methods and Techniques* box on page 173), and as a result the pattern and the relative intensities of diffracted spots are related to the underlying arrangement of atoms by a mathematical summation known as the Fourier transform (**FIGURE 5.3**). The intensities of the reflections are simply the Fourier transform of the electron density around each atom; the pattern that they form on the detector (photographic film in the case of Bragg) is the transform of the crystalline lattice within which the atoms are arranged. The observed diffraction pattern is a product (strictly a convolution) of the two. Importantly, this dictates that each atom in the crystal contributes, to a greater or lesser extent, to every reflection. In

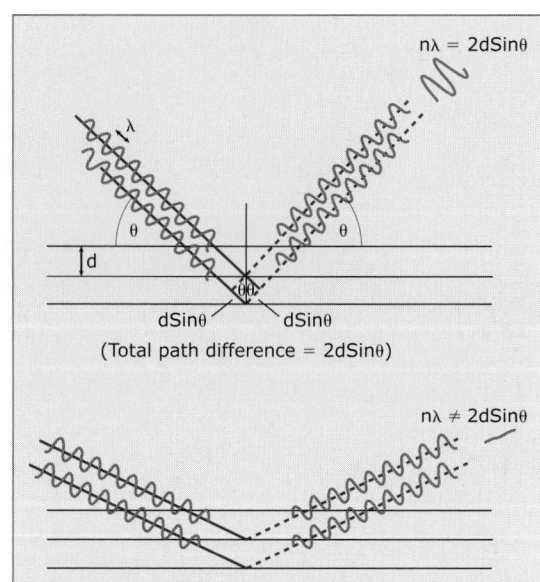

FIGURE 5.2 The Bragg construction.

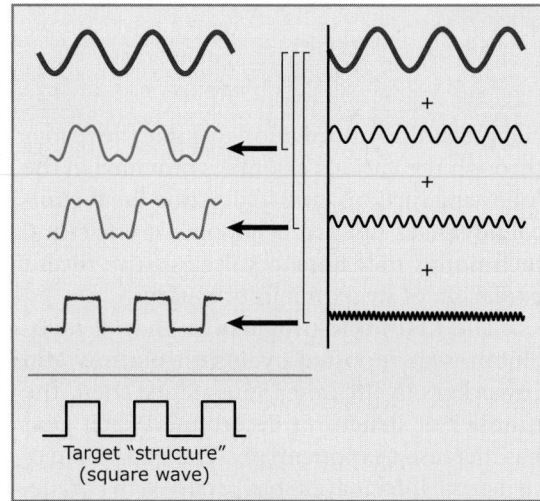

FIGURE 5.3 The Fourier transform. The addition of waves of different phase, frequency, and amplitude results in improved approximation of the square wave.

order to calculate the electron density within the crystal, and thus the atomic positions, the relative phase angles of each reflection resulting from the constructive interference between scattered X-rays must be known. These are, however, completely lost in the experiment. This is a fundamental difference between diffraction and microscopy, where phase information is preserved through the use of lenses that are not available for very short wavelength X-rays. This information loss is known as the "phase problem."

Crystals are formed from basic repeating motifs or unit cells that are related to each other by translation only (**FIGURE 5.B1**). Within a unit cell, the individual molecules (or groups thereof) that constitute the crystal's asymmetric unit are arranged according to a total of 230 possible symmetries or space-groups that are described in terms of different rotational (two-, three-, four-, or sixfold rotations), translational, and mirror operations. In fact, as we will see, proteins are chiral and for this reason can only take a subset of these, representing a total of 65 available space-group symmetries.

Only the size of the unit cell decrees the number of X-ray reflections that can be observed at any given resolution. X-ray data from protein crystals may involve the measurement of tens of thousands of reflections, even for a small protein crystallized in a low symmetry space-group (**FIGURE 5.B2**). The resolution of a structure is directly related to the level of accuracy at which atomic positions are known. From Bragg's Law, we know that the more finely the unit cell is sampled, the closer together the Bragg planes become. At smaller d-spacings, the Bragg requirement that, for a spot to be observed, the total path difference must be an integral number of wavelengths (see Figure 5.2) will only be fulfilled at progressively higher values of θ. Thus the resolution is defined by the minimum value of d (in Å units) for which reflections are represented in the final set of diffraction data. In terms of the Fourier transform, the higher-resolution reflections are those that contribute the highest-frequency terms in the summation, and therefore contribute the most detailed structural information. **FIGURE 5.B3** shows how the final calculated electron density varies with data resolution. Initial estimates of the phase angle for each reflection are generally poor. Remember, however, that the diffraction pattern is a Fourier transform of the contents of the asymmetric unit, and therefore we can calculate a theoretical diffraction pattern once we know the locations of the atoms in the crystal. This is the basis of crystallographic refinement, where the atomic model is adjusted using molecular graphics and computational procedures (see *5.5 Protein structure representations—a primer*) so as to maximize the agreement of the calculated pattern with the experimentally observed diffraction data.

FIGURE 5.B1 A crystal lattice. Structures from Protein Data Bank 1NSF. R. C. Yu, et al., *Nat. Struct. Biol.* 5 (1998): 803–811.

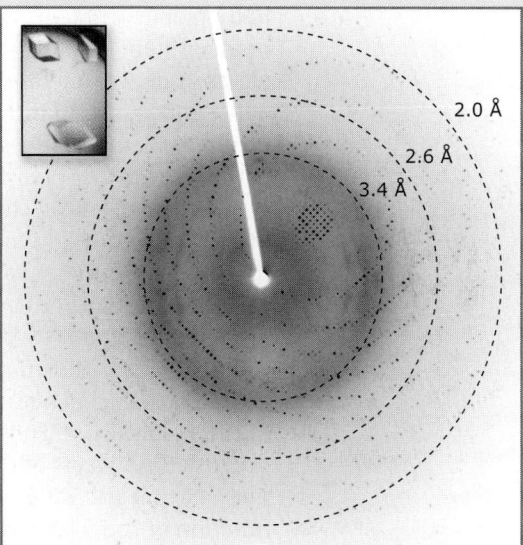

FIGURE 5.B2 A part of the diffraction pattern obtained from crystals of hen egg-white lysozyme (inset), the first enzyme structure to be solved by X-ray crystallography. Circles show different limits of resolution.

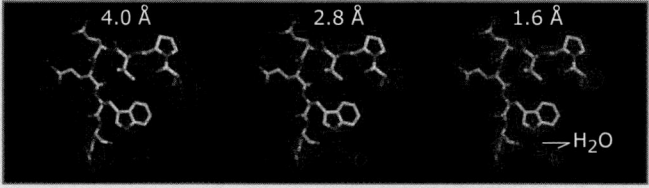

FIGURE 5.B3 Electron density maps at increasingly high resolution. Data from S. Smerdon. Structures created from RCSB Protein Data Bank.

As mentioned earlier, in 1934 Bernal and Crowfoot had shown that protein crystals, like small inorganic compounds, had sufficient internal order to diffract X-rays. Nevertheless, these and subsequent experiments dramatically illustrated the problems of investigating molecules of the size of proteins by diffraction methods. Such were the technical difficulties that the first structure determination of a protein by X-ray crystallography did not happen for more than 20 years. The structure of myoglobin, a small molecule of 153 amino acids that acts as a store of molecular oxygen, was truly a revelation, showing for the first time many of the fundamental architectural principles of protein structure that we now take for granted. The crucial technical advances that enabled this extraordinary achievement, however, were developed on a related but much larger molecule, hemoglobin, which had been first crystallized in the early part of the nineteenth century. Vernon Ingram (working with Max Perutz, who had already been laboring for many years on the hemoglobin problem) was able to soak crystals of the protein in dilute solutions of heavy (i.e., electron dense) metal salts and collect X-ray data from them. In most cases this treatment resulted in large changes in the pattern of spots or a complete loss of diffraction. Occasionally, though, the data collected were similar (or isomorphous) enough with those derived from unsoaked, *native* crystals that further analysis was possible. Perutz used a mathematical procedure related to the Fourier transform called a Patterson function to determine the positions of bound heavy atoms in the derivatized crystal. Isomorphism is important here because the success of the procedure relies on the fact that the differences in the Patterson function of the underivatized and derivatized crystal data derive solely from the addition of the heavy atom. From this knowledge of the heavy atom positions, an approximate phase angle for each reflection could be calculated, enabling a map of the electron density of the hemoglobin molecule to be produced at 5.5 Å resolution. The smaller size of myoglobin (~17 kDa, compared with hemoglobin at ~65 kDa) meant that John Kendrew, working in the same department as Ingram, was able to solve its structure well before the first structure of hemoglobin was published in the early 1960s. The contributions of both Perutz and Kendrew were acknowledged in 1962 by the award of the Nobel Prize for Chemistry.

In total, the structure of hemoglobin had taken some 30 years to solve. How is it that structures are now being determined at a rate of more than 5000 per year? By and large, this phenomenal increase in speed can be attributed to four major developments that have gradually come into common use over the last 10 to 15 years: recombinant DNA technology; cryo-crystallography; multiwavelength anomalous scattering methods; and the availability of high-brightness, tunable synchrotron X-ray sources.

The extent of the improvements in the X-ray crystallographic method has meant that the rate-limiting steps in structure determination are now the ability to grow well-diffracting crystals, and therefore, the availability of *crystallizable* samples. The considerable difficulties associated with the need to purify potentially scarce proteins from the cells/tissues within which they naturally reside have been largely removed by recombinant DNA technology. It is now possible to produce tens or hundreds of milligrams of highly pure protein by expressing its cloned gene in a variety of host cells, most commonly the bacterium *Escherichia coli*, but also in cultured yeast, insect, and mammalian cells. Crystallization can now be performed automatically with a variety of commercially available crystallization "screens" and robotic liquid-handling devices.

Having produced well-diffracting single crystals, data can now be collected at a number of high-intensity synchrotron radiation sources around the world (**FIGURE 5.4**). These large installations produce hard (i.e., short wavelength) X-rays as a by-product of accelerating packets of electrons at velocities approaching the speed of light, in a circular orbit with a diameter measured in hundreds of meters. As the electrons are forced to follow a circular path under the

FIGURE 5.4 A modern synchrotron source. Courtesy of Diamond Light Source, UK.

influence of a high magnetic field, energy is lost as electromagnetic radiation at wavelengths ranging from γ-rays into the infrared region. Using sophisticated optics, a beam of X-rays, with an intensity that may exceed that available from a laboratory source by several orders of magnitude, can be focused onto a protein crystal with great precision. This, in combination with modern electronic charge-coupled device (CCD) detectors in place of X-ray film, results in a considerable reduction in the time required to collect diffraction data. Complete data sets that would otherwise require days to collect can now be measured in a matter of minutes. Unfortunately, the use of radiation of such intensity would destroy many protein crystals in a few seconds due to effects of localized heating and the production of chemically reactive free radicals. Fortunately, radiation damage can be largely eliminated by preserving crystals at liquid nitrogen temperatures (100°K or −173°C) during exposure to the X-ray beam.

In cases where the structure of a homologous protein is available, a structure solution can be achieved using the technique of molecular replacement. This method attempts to place a known structure (the search model) into the crystal of the unknown protein by comparison of the Patterson functions calculated from the search structure and the target diffraction data. Here, the rotational orientation and the translation of the search model that best fit the observed diffraction data are determined and applied, providing an approximate starting structure for model building and crystallographic refinement. Obviously, in many cases, homologous structures may not be available. However, a combination of the use of recombinant DNA technology, cryo-crystallography, and synchrotron radiation sources has enabled the phase problem to be directly solved rather trivially using a technique known as multiwavelength anomalous diffraction (MAD). This method derives from the fact that, at characteristic wavelengths, chemical elements interact with X-rays in such a way that the resultant scattered wave gains a shift in its phase.

Although laboratory X-ray sources are limited to X-rays of a single fixed wavelength, synchrotron radiation can be "tuned" to supply X-rays at different, but well-defined, wavelengths over a useful range of 0.5 Å to 2.5 Å. In 1990, Wayne Hendrickson and colleagues showed that phases could be determined directly from a single crystal by exploiting the anomalous scattering of selenium atoms in-troduced by expressing a recombinant protein in bacteria grown on broth containing selenomethionine as the sole source of methionine. Anomalous scattering from sulfur atoms found in the amino acid cysteine had been used previously to determine the structure of a small protein crambin. Until recently, this was not considered to be a generally applicable approach because the anomalous scattering effect is rather small for sulfur. Nevertheless, the experiment was successful largely due to the high degree of order for the crambin crystals that enabled extremely accurate data to be collected at very high resolution. Selenium is a much more effective anomalous scatterer, and the ability to produce derivatized protein straightforwardly considerably facilitated the determination of phases and thus structure determination. In practice, the experiment involves the collection of data sets at and around the wavelength that corresponds to the peak of the anomalous differences. Although the theoretical details of the method are beyond the scope of this chapter, it can be conceptualized as a kind of isomorphous replacement experiment in which the necessary intensity differences are produced by physics (i.e., variable wavelength X-rays) rather than chemistry (i.e., the addition of heavy atoms). All data are collected from one crystal, and as a result the problems of lack of isomorphism that frustrated crystallographers for so long are effectively removed to the point that in a favorable case, protein structures can now be determined in a matter of hours.

Concept and Reasoning Checks

1. What is the "phase problem" and how can it be circumvented?
2. How is the resolution of a structure related to the number of X-ray reflections measured for a given crystal, and why?
3. Summarize the technical advances that have fueled the extraordinary increase in speed of macromolecular structure determination by X-ray crystallography.

5.3 Nuclear magnetic resonance

Key concepts

- NMR is a powerful method for investigating structures of proteins and their complexes in solution.
- NMR methodologies can provide detailed information about macromolecular dynamics that are difficult or impossible to extract from X-ray crystallographic data.

The phenomenon of NMR was predicted by quantum mechanics before its first experimental observation by Isidor Isaac Rabir in the 1930s, and subsequently in solution by Felix Bloch and Edward Mill Purcell in the 1940s. The physics underlying the technique involves the realization that many atomic nuclei (those with spin quantum number > 0) possess a magnetic moment as a consequence of possessing both spin and charge. When placed in an external magnetic field, the magnetic moment adopts one of a fixed number of orientations, as its behavior is quantized. In biomolecular studies, we are concerned almost exclusively with nuclei (^1H, ^{13}C, ^{15}N) that are spin-½, and adopt two possible orientations that correspond to low and high-energy states. Irradiation with electromagnetic radiation of appropriate wavelength leads to transitions between these states, giving an absorption spectrum. NMR transitions lie in the radiofrequency region of the spectrum, with wavelengths in the MHz range. The experimental realization of NMR was recognized by the award of Nobel Prizes for Physics to Rabir (1944) and to Bloch and Purcell (1952).

First viewed as a physicist's tool for extracting the magnitudes of the magnetic moments for atomic nuclei, NMR soon became an indispensable technique for chemists, following the observation that the exact resonance frequency of a nucleus was exquisitely sensitive to its local chemical environment. For example, when the NMR spectrum of ethanol (CH_3CH_2OH) is recorded at sufficiently high resolution (**FIGURE 5.B4**), separate signals can be seen for each of the three chemically distinct types of proton (hydrogen nucleus) present. Each compound therefore gives a characteristic NMR "fingerprint," making NMR an invaluable analytical tool in chemical investigations. In addition to the different resonance positions ("chemical shifts"), NMR signals contain fine structure, arising from through-bond communication between the magnetic moments ("J-coupling"). Analysis of these two effects allows the different signals to be attributed, or *assigned*, to the hydrogen type (i.e., the C**H**$_3$, C**H**$_2$, or O**H**

moiety). The utility of NMR spectroscopy in protein structure investigations was not immediately obvious. The low energies of the NMR transitions render it an insensitive technique requiring large quantities of sample, and the complexity of the NMR spectrum of even a small protein, with say 500 hydrogen types, was considered intractable. The 800 MHz NMR spectrum of the 14 kDa protein lysozyme (Figure 5.B4) contains several hundred peaks.

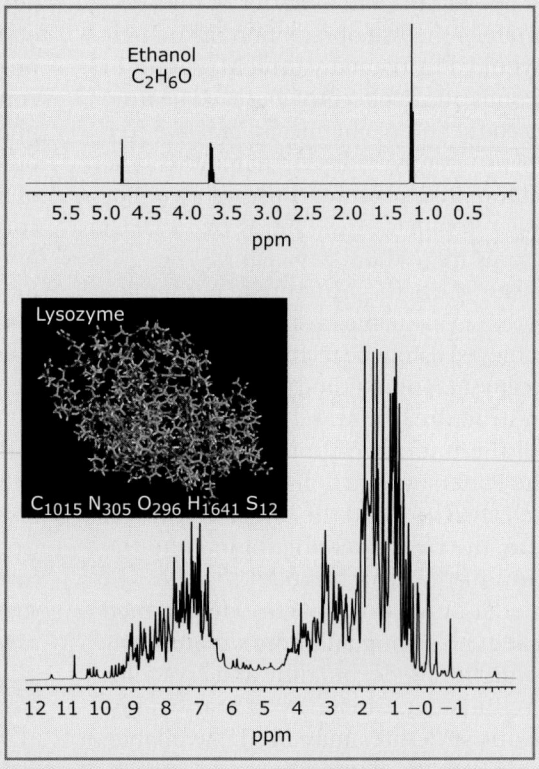

FIGURE 5.B4 One-dimensional ^1H NMR spectra of ethanol (top) and a small protein lysozyme (bottom). Structure from Protein Data Bank 1GXV. M. Refaee, et al., *J. Mol. Biol.* 327 (2003): 857–865.

As mentioned earlier in this chapter, X-ray crystallography is not the only means of determining the structures of proteins, and over the last 20 years or so enormous progress has been made in the use of NMR to examine biomolecular structures *in solution*, and at resolutions comparable to those derived for the more "traditional" X-ray diffraction techniques.

Several technical and methodological advances were central to the development of NMR as a tool to investigate the structures of

biomolecules. Richard Ernst introduced Fourier transform NMR, which increased the sensitivity of the technique by orders of magnitude. Ernst also addressed the issue of complexity in his introduction of multidimensional NMR methods, which spread the signals out into a second (or higher) frequency dimension. An example of such a two-dimensional spectrum can be seen in **FIGURE 5.5**. The conventional one-dimensional spectrum lies along the diagonal of the two-dimensional spectrum. The off-diagonal

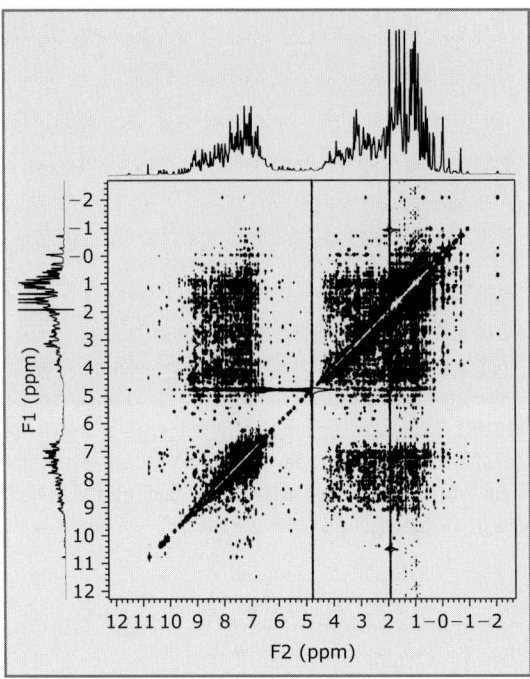

FIGURE 5.5 Two-dimensional NOESY spectrum of lysozyme collected at high-field strength (800 MHz), with the one-dimensional spectrum for each separate dimension shown alongside and above the x and y axes.

FIGURE 5.6 A modern 900 MHz magnet. Photo courtesy of National High Magnetic Field Laboratory at Florida State University.

peaks represent correlations between different hydrogen types. In this instance the correlations are a result of the nuclear Overhauser effect (NOE), and they signify that the protons sharing the correlation lie within ~5 Å of each other in the tertiary structure. In addition, the sensitivity and resolution problems have been lessened somewhat by the availability of ever-increasing external magnetic field strengths, from a few tenths of a Tesla (T) produced by a permanent magnet in the early days of NMR up to the 20T and larger fields accessible using superconducting magnets today (**FIGURE 5.6**). Ernst's contributions to the development of NMR spectroscopy were recognized with the award of the Nobel Prize for Chemistry in 1991.

The use of these methods in protein structure determination was pioneered by Kurt Wuthrich and coworkers, who used a combination of the J-coupling (through-bond) information with "through-space" information from the NOE to assign the NMR spectrum of proteinase inhibitor IIa. They then went on to use the NOE information to calculate its three-dimensional structure. Initial skepticism was allayed by a blind trial in which the struc-

ture of the α-amylase inhibitor tendamistat was solved independently using X-ray crystallography and NMR spectroscopy. Wuthrich's realization of the potential of NMR to solve the three-dimensional structures of proteins, together with his development of methodologies toward this goal, was also acknowledged with the award of the Nobel Prize for Chemistry in 2002.

The use of ^{1}H NMR spectroscopy for the assignment of a protein's spectrum, and the elucidation of its three-dimensional structure, remained a daunting undertaking. This situation was transformed in the early 1990s by employing molecular biology techniques for heterologous protein expression in bacterial hosts. This enabled the production of isotopically labeled protein samples using expression in *E. coli* cultured on a minimal growth medium supplemented with ^{13}C-labeled glucose and ^{15}N ammonium chloride as the sole carbon and nitrogen sources. (The common isotopes of these nuclei, ^{12}C and ^{14}N, are not amenable to study by high-resolution NMR techniques.) This facilitated the development of a huge arsenal of "triple resonance" (^{1}H/^{13}C/^{15}N) NMR methods, notably by Ad Bax and coworkers. These

NMR methods permitted much more efficient through-bond communication between nuclei, as one-bond ^1H-^{15}N, ^{15}N-^{13}C, and ^{13}C-^{13}C J-couplings could now be used instead of three-bond ^1H-^1H J-coupling. In addition, they allowed the extension of multidimensional NMR to include ^{13}C and/or ^{15}N frequencies. Hitherto the assignment procedure was predicated on the observation of NOEs between sequential residues—a painstaking process fraught with ambiguity. The use of triple-resonance techniques allows an unambiguous step-by-step journey along the polypeptide backbone.

Armed with a complete or near-complete assignment of the ^1H, ^{13}C, and ^{15}N nuclei in the protein, it is possible to extract a huge amount of structural information from various NMR parameters. The two most fruitful sources traditionally have been NOEs (the observation of an NOE between two ^1H nuclei, and its magnitude, are constraints on the maximum distance between the nuclei) and coupling constants—the values of three-bond coupling constants, for example, $^3J(H_N–H_\alpha)$—which are functions of the intervening dihedral angle. These structural constraints, if sufficient in quantity, can be included as additional energy terms, along with known covalent bond lengths and angles, in restrained molecular dynamics protocols with simulated annealing schedules to calculate the structure. The progress of such a calculation is depicted in **FIGURE 5.7**. Owing to the nonexact and possibly incomplete nature of the experimental constraints, the calculation is performed many times. The results are superimposed to give a family of structures (**FIGURE 5.8**), all of which are compatible with the experimental data;

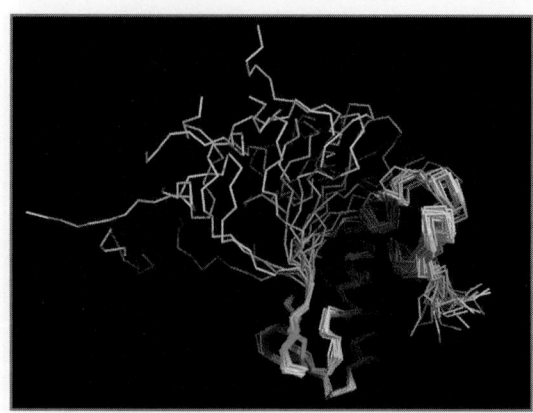

FIGURE 5.8 An ensemble of structures, all of which are consistent with the experimental NMR data. Structure from Protein Data Bank 1G2K. W. Schaal, et al., *J. Med. Chem.* 44 (2001): 155–169.

this gives some impression of the precision of the structure determination.

Although the many examples of structures determined by both NMR and crystallography show that proteins in solution and in the hydrated crystalline state are, by and large, very similar, NMR has a real and important advantage over X-ray methods in its ability to access the dynamics of biomolecules in the solution state. It is possible to infer dynamics from crystal structures, but the return to equilibrium of NMR signals contains direct information about atomic motions. The relaxation of ^{15}N nuclei has been most widely exploited in this regard. The relaxation of the ^{15}N nucleus is dominated by its attached amide hydrogen nucleus, and the efficiency of this interaction is governed by the rate of reorientation of the ^1H–^{15}N bond vector with respect to the external magnetic field. The resulting analysis gives exquisite, residue-specific information on protein dynamics over a wide range of time scales, from picoseconds (fast internal motion of a residue with respect to the protein overall) to nanoseconds (overall tumbling) to micro/milliseconds (conformation exchange processes).

A powerful use of NMR spectroscopy is the rapid analysis of protein–protein or protein–ligand interactions. These experiments are most commonly performed using heteronuclear single quantum coherence (HSQC) spectra acquired from ^{15}N-labeled protein. Peaks in the HSQC spectrum are derived from protons attached to nitrogen atoms. Thus there is at least one peak for almost every residue in a protein (from the amide hydro-

FIGURE 5.7 Computational "folding" of a protein into a structure consistent with restraints derived from a variety of NMR experiments. Data from G. Kelly and M. Conte. Structures created from RCSB Protein Data Bank.

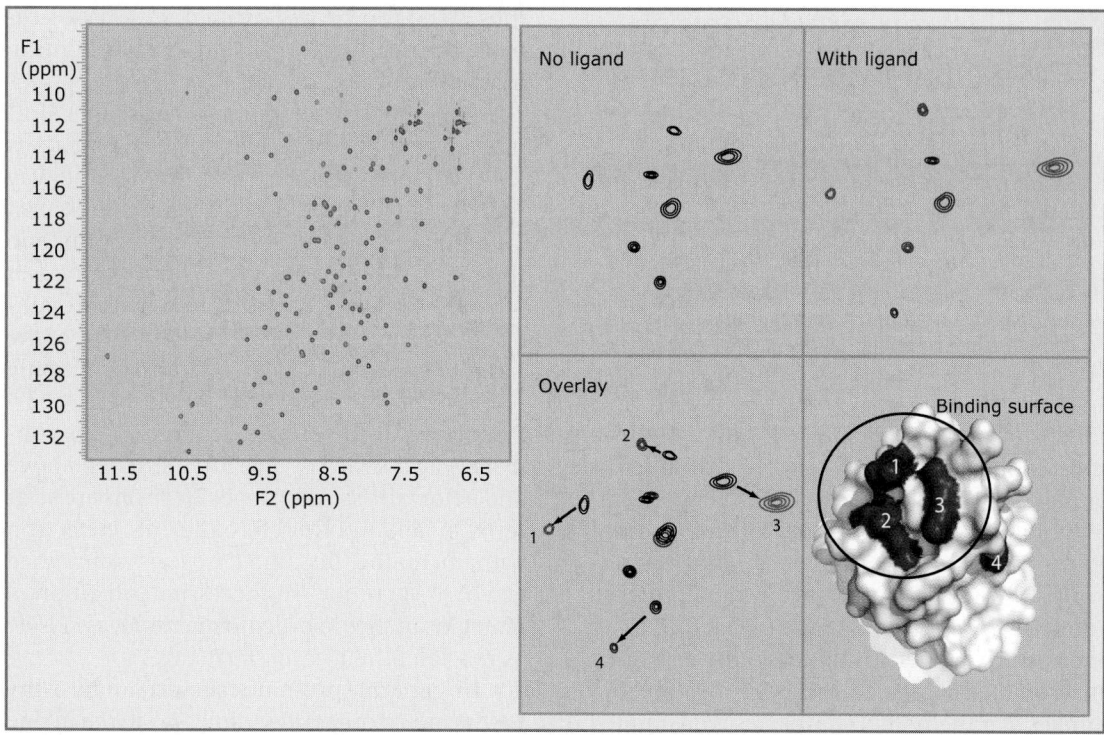

FIGURE 5.9 Chemical shift mapping experiment. The HSQC spectrum of the unliganded protein is shown on the right. The changes in position and intensity of assigned chemical shifts that occur as a ligand is added can be "mapped" onto the molecular surface to reveal the potential binding site.

gen of the peptide bond, except proline) and additional peaks for any side-chain NHs. The ^{15}N-HSQC spectrum of a 10 kDa protein is shown in **FIGURE 5.9**. Addition of an unlabeled binding partner contributes no new peaks but will affect the peaks corresponding to protein residues at the binding site. These peaks will either shift gradually as the binding partner is added, or gradually disappear and reappear elsewhere in the spectrum, according to the kinetics of the complex in question. Mapping of the perturbed residue positions on the overall structure can then provide an excellent picture of the interaction surface, without the need to laboriously determine the structure of the complex. Quantitative analysis of the pattern of shifts may allow determination of dissociation constants. This approach is also used extensively to screen for binding in a drug discovery context. Analogously, conducting a pH titration of the protein sample allows residue-specific pKa values to be determined—information difficult to obtain by other methods.

When compared with X-ray crystallography, NMR spectroscopy of proteins remains an immature discipline, in which significant meth-odological and technical advances are still common. Recent years have seen the introduction of new types of structural constraints (residual dipolar couplings) to augment those conventionally employed. Transverse relaxation optimized spectroscopy (TROSY) has permitted the extension of conventional methods to the study of proteins and their complexes up to molecular weights of ~100 kDa, and the introduction of cryogenic detector electronics has yielded a gain in sensitivity of approximately threefold. Structural genomics initiatives have prompted new, faster data acquisition schemes and ever-increasing automation of the data analysis and structure calculation procedures. Most recently, advances in NMR methodology (CRINEPT, CRIPT) have provided some insight into structural aspects of systems as large as the GroEL/GroES system with a molecular weight of ~1 MDa.

Concept and Reasoning Checks

1. List two advantages of NMR spectroscopy over X-ray diffraction methods.
2. Why is isotopic labeling of protein samples for NMR necessary?

5.4 Electron microscopy of biomolecules and their complexes

Key concepts

- Cryo-electron microscopy is capable of imaging macromolecular complexes that may be too large or flexible for X-ray diffraction or NMR approaches.
- In favorable cases, electron microscopy (EM) methods can produce structural information at or near atomic resolution.

Electron micrographs are a familiar sight in most biology textbooks, and electron microscopy has been used for over 50 years to image biologic samples at a resolution that far exceeds anything possible with light microscopy. Beyond its obvious use in imaging cell ultrastructure, EM has come to the fore as an increasingly powerful method for investigating macromolecular structure. For a long time, EM studies of proteins and complexes were limited to resolutions of the order of 20 Å to 30 Å, but more recent developments in hardware and in experimental approach are providing information at much higher resolution, in some special circumstances approaching that of X-ray crystal structures.

Remarkably, modern **transmission EM**s are still built following the original concepts laid out in the first electron microscope designed in the 1930s by Ernst Ruska. The electron source is generally a tungsten or lanthanum hexaboride filament from which electrons "boil" off at very high temperatures in a process called thermionic emission. Higher-resolution studies employ a field-emission gun (FEG) that provides a bright beam of approximately coherent (i.e., parallel and in-phase) electrons that have a very narrow distribution of energies. The high energies, and therefore short wavelengths, of these electron beams explain why EM is capable of much higher resolution than optical microscopy.

For imaging, electron microscopes employ a system of high-field magnets that act as lenses to focus the electrons, in the same way that glass lenses are used in light microscopy. The obvious advantage here is that unlike in the X-ray diffraction experiment, the phases of the electron waves are maintained throughout the process, allowing direct imaging of the sample on a suitable detector, most commonly photographic film or CCDs. Lastly, beam characteristics such as size and coherence are controlled by a series of apertures situated above and below the sample stage. Electrons are scattered by air, and therefore the entire electron path, including the sample stage, is maintained at a high vacuum. This necessitates fixation of biologic samples in order to maintain, as closely as possible, their native structure.

In general, protein structure analysis by EM is carried out with samples prepared in one of two ways. Negative staining involves the deposition of heavy metal salts, usually uranyl acetate or phosphotungstic acid, on and around the molecules of interest spread onto carbon support. These heavy elements interact with the electron beam very strongly, and as a result this method provides extremely high-contrast images. In addition, the presence of the metal largely protects the protein complex from the damaging effects of electron bombardment and allows high electron doses to be employed. The sample is essentially coated in heavy salts, though, and because of this details of internal structure are lost (**FIGURE 5.10**). This leaves an image of the outline of the biomolecule and limits the attainable resolution to around 10 Å to 30 Å. In addition, the effects of heavy metal binding and dehydration may result in some distortion of the native structure.

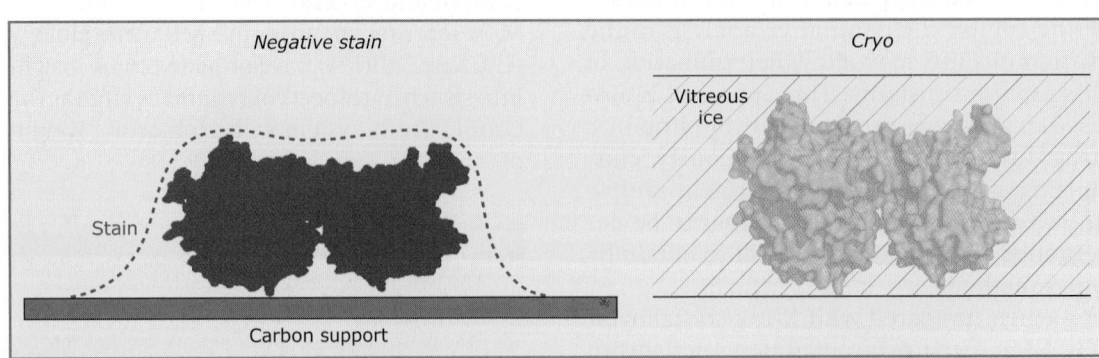

FIGURE 5.10 Negative stain versus cryo-EM.

Clearly, it is desirable to maintain the sample in as close to physiological conditions as possible, and to be able to investigate its overall architecture, both external and internal. This has been made possible by the development of cryo-EM methods. Here, the sample molecules are spread, under largely native solution conditions, onto a carbon grid and flash frozen by being plunged into a reservoir of liquid ethane. The rapidity of the freezing process prevents formation of ice and instead results in vitrification, where the water molecules adopt an amorphous or "glass-like" noncrystalline state (Figure 5.10). In order to maintain this frozen state, the sample stage in a cryo-EM is maintained at low temperatures with liquid nitrogen (~80 K) or helium (~5 K). This limits the damaging effects of the electron beam, but the effects are not eliminated and cryo-EM studies must be carried out using electron doses that are much lower than are possible in negative stain experiments. This limitation is exacerbated by the fact that images of unstained samples have a much lower contrast due to the poor interaction of electrons in carbon, nitrogen, and oxygen from which organic molecules are made. The resulting low signal–noise ratio complicates interpretation and makes assignment of orientation difficult, placing a lower limit on the molecular weight of around 200 kDa to 300 kDa in the single-particle approaches described below.

The highest-resolution structures determined thus far have been derived from two-dimensional arrays (or crystals) of identically oriented protein complexes. This, then, allows both electron diffraction and imaging experiments to be carried out. Here, the advantage over X-ray diffraction is that the phases of the measured amplitudes can still be directly determined and the image reconstructed by Fourier transform. Electron crystallography is not, however, without disadvantages and problems of crystal quality, sample preparation, and other technical issues make electron crystallography a rather challenging endeavor. Most notably, a single diffraction image collected at a single orientation of the two-dimensional crystal will only allow a two-dimensional image, or projection, of the sample to be generated. In order to extend the information to three dimensions, a number of images must be collected where the sample is tilted with respect to the electron source. In practice, physical limitations prevent tilt angles of more than about 60°, which results in a cone of missing data. Inevitably, the

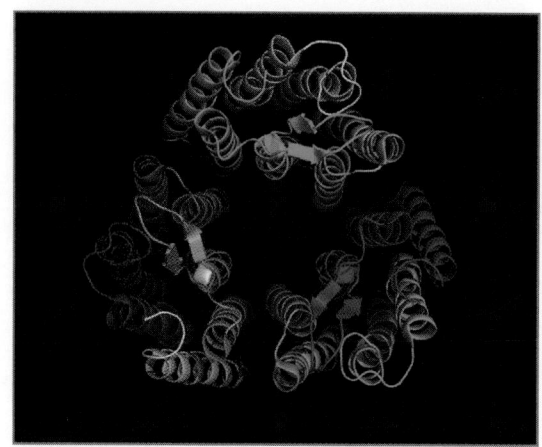

FIGURE 5.11 Bacteriorhodopsin structure determined by electron crystallography. Structure from Protein Data Bank IFBB. S. Subramaniam and R. Henderson, *Nature* 406 (2000): 569–570.

resulting three-dimensional structure will be less well defined (i.e., at a lower resolution) along the direction of the electron beam than in the plane of the sample stage. In spite of the experimental difficulties, this method has been used with considerable success to obtain the structures of several membrane proteins, most notably bacteriorhodopsin and other helical assemblies at effective resolutions of 3 Å to 4 Å (**FIGURE 5.11**).

It has been suggested that the preparation of highly ordered two-dimensional protein crystals of high-molecular-weight protein complexes presents as many technical challenges as growing three-dimensional crystals for X-ray analysis. However, the requirement for ordered arrays may now be circumvented using single-particle methods that were originally developed for negative stain imaging, but which are now being applied to great effect in cryo-EM. In this approach, individual proteins are scattered onto a carbon surface, where they sit in a range of orientations. Thus different "faces" of the molecule complex are imaged as a series of two-dimensional projections, and from knowledge of the relative orientations of each of these a three-dimensional view of the molecule can be built (**FIGURE 5.12**). Although conceptually simple, in practice this method is technically challenging and somewhat laborious. In addition to, and partly as a result of, the low contrast, deriving the relative orientation of each particle is demanding. The simplest assumption is that all molecules have identical structures and are related by a set of simple rotations. In practice, this may not

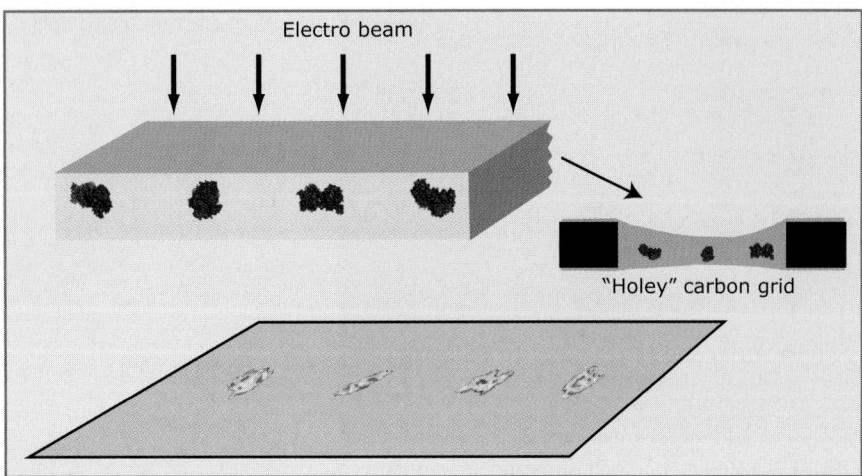

FIGURE 5.12 Transmission cryo-EM.

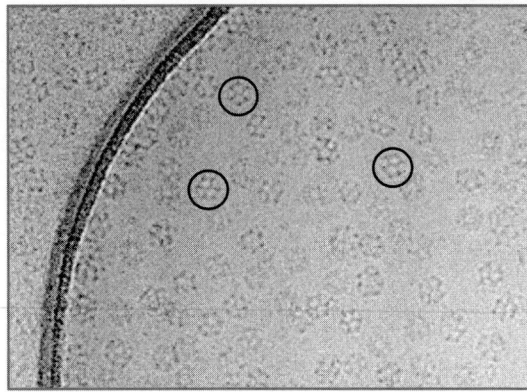

FIGURE 5.13 Cryo-EM data. Three similarly oriented particles are highlighted. Photo courtesy of Peter B. Rosenthal, MRC National Institute for Medical Research, London.

be the case and conformational heterogeneity arising from intrinsic disorder/flexibility or from the existence of different liganded states of molecules within the sample may be difficult to account for, and the problem becomes increasingly acute as higher resolutions are sought. Also, a complete high-resolution image requires a complete sampling of possible orientations, a situation that is rarely achieved because any asymmetric objects randomly scattered onto a surface will inevitably settle most often in the most stable orientations. This problem can be partially circumvented by use of "holey" carbon grids, which have pores that contain particles in suspension such that they can adopt essentially random orientations (Figure 5.12). Nonetheless, the technical difficulties mean that the most complete and highest-resolution cryo-EM reconstructions may require analysis of tens or hundreds of thousands of individual particles. An example of a field of single particles in a cryo-EM image is shown in **FIGURE 5.13**.

Probably the most common use of single-particle cryo-EM is in the investigation of the structures of large, multiprotein complexes that may be difficult or impossible to crystallize for X-ray crystallographic analysis, or that are too large for NMR studies. In many cases, individual proteins within such complexes may be amenable to X-ray or NMR methods, and high-resolution structures may be available. In such cases, it may be possible to orient or "dock" the structure of the isolated subunit into the lower-resolution EM envelope, potentially providing valuable information about the structural and functional roles of individual components in the context of the biologically relevant complex. Although these docking procedures may be carried out manually, a variety of computational fitting procedures have now been developed to guide and accelerate the process. This combined approach has had a number of notable successes, none more impressive than the reconstruction of an atomic model for the entire T4 colivirus by Michael Rossmann and coworkers.

Finally, a recently developed and exciting use of cryo-EM called cryo-electron tomography is beginning to reveal the structures and distribution of large protein complexes in cells. In this method, individual cells are frozen on a support, as for single-particle studies, and a set of EM snapshots is then collected as the frozen specimen is rotated by small angular increments. Each image therefore represents a two-dimensional projection of the sample "viewed" from different directions, and from this series the original image can be reconstructed in three dimensions by back-projection. The basic principle is shown schematically in **FIGURE 5.14** for an isolated protein complex as a simple example. Clearly, the amount of information potentially contained in the reconstructed image is truly immense, representing the three-dimensional arrangement of every molecule in the cell! Extracting information about specific complexes, however, represents a formidable challenge. In theory, specific labeling of particular complexes would be helpful, but is extremely difficult to achieve for complexes within a cell. An alternative approach is to use structural "templates" derived from complexes of known structure to search the cytoplasmic milieu. This method has already been used to locate large symmetric complexes such as the 26S proteasome in cryo-preserved cells.

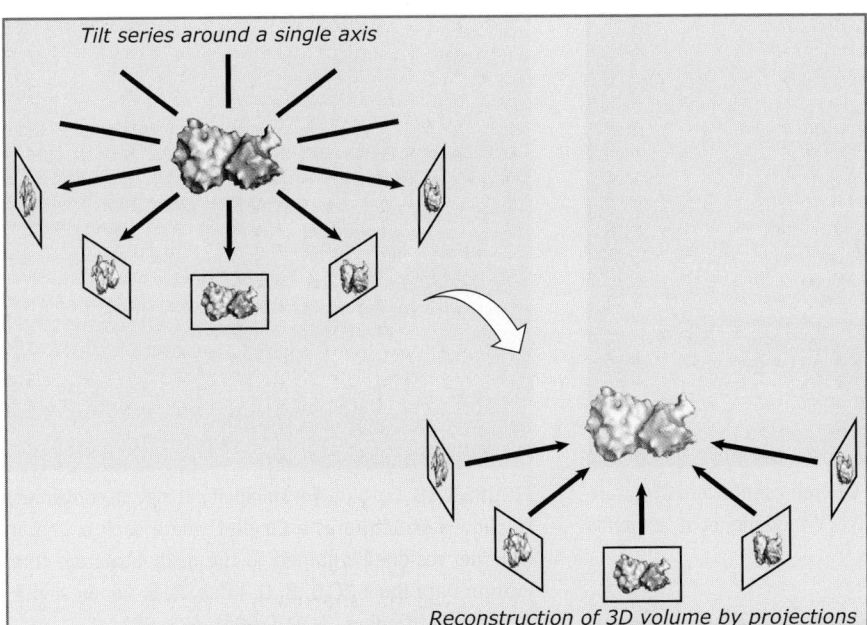

FIGURE 5.14 The principle of cryo-electron tomography.

Tilt series around a single axis

Reconstruction of 3D volume by projections

Concept and Reasoning Checks

1. List some similarities and differences between electron and light microscopy.
2. How is a three-dimensional image of a molecule derived from two-dimensional EM images?

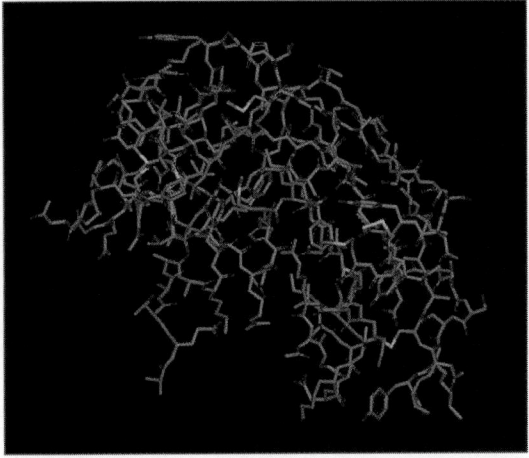

FIGURE 5.15 Ball and stick representation. The colors are in standard CPK (Corey, Pauling, Kultin) color scheme. Carbon is gray, hydrogen white, nitrogen blue, and sulfur yellow. Structure from Protein Data Bank 1JVT. L. Vitagliano, et al., *Proteins* 46 (2002): 97–104.

5.5 Protein structure representations—a primer

Key concepts

- Proteins are three-dimensional objects, and even a relatively small example of ~10 kDa molecular weight will contain upward of 1000 atoms. This causes considerable difficulties in presenting structural data in a clear and understandable way.
- Displaying each atom and chemical bond certainly conveys the degree of complexity in proteins, but little other useful information is discernable. For this reason, a number of different schematic representations have been devised in order to illustrate different features of protein structure.

Arguably, the most obvious way in which to represent a molecular structure is to draw each atom along with the chemical bonds between atoms. This kind of representation is known as a "ball-and-stick" type, a name that admirably describes it. Here, each atom is shown as a sphere, generally of arbitrary radius, and each bond as a cylinder (or sometimes a cone), as shown in **FIGURE 5.15**. These kinds of representation are most used to convey stereochemical details of, for example, active-site regions of enzymes for which details of relative atomic positions are of most interest. A related method is the "space-filling" or CPK (for Corey, Pauling, and Kultin) representation (**FIGURE 5.16**). In this case, atoms are shown as larger spheres scaled according to their atomic radii. For large molecules with thousands of atoms and bonds, the ball-and-stick representation contains far too much information and the overall architecture of a protein is largely obscured. Similarly, the space-filling representation suffers from the inevitable property that atoms "inside" the protein are not visible at all, and only surface atoms are discernable.

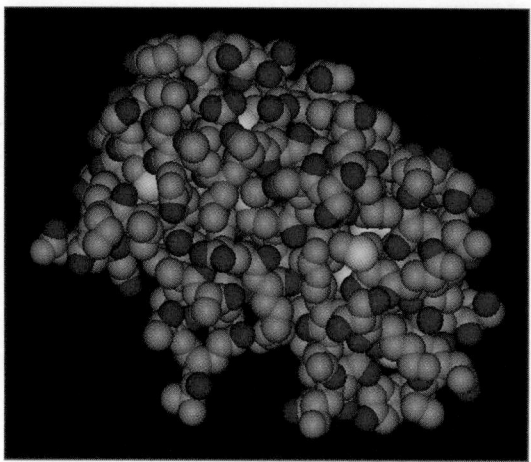

FIGURE 5.16 Space-filling CPK representation. Structure from Protein Data Bank 1JVT. L. Vitagliano, et al., *Proteins* 46 (2002): 97–104.

FIGURE 5.17 Ribbon representation with α-helices in red and β-conformations in yellow. Structure from Protein Data Bank 1JVT. L. Vitagliano, et al., *Proteins* 46 (2002): 97–104.

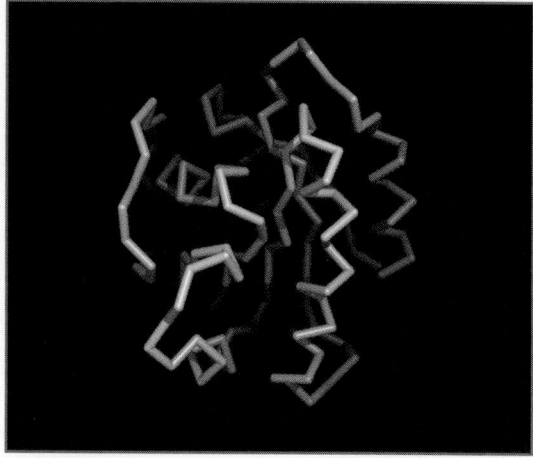

FIGURE 5.18 Cα-plot. For simplicity, it may be convenient to show a structure as a Cα plot where each α-carbon (one per residue) is joined to the next. Structure from Protein Data Bank 2CLO. B. U. Klink, R. S. Goody, and A. J. Scheidig, Biophys. J. 91 (2006): 981–992.

By far the most popular and effective means of conveying the overall structure of a protein is the "ribbons" representation (**FIGURE 5.17**), in which β-strands are shown as arrows to indicate directionality (N-terminal to C-terminal) and helices are represented as ribbon-like coils or as tubes. As will be seen later in this chapter, the overall path in three dimensions of a protein chain is quite accurately described by linking the Cα atoms of each consecutive amino acid—thereby creating a Cα plot (**FIGURE 5.18**). This gives a much less cluttered view of the shape and architecture, but to the inexperienced eye it contains limited information about secondary structural content. In the "ribbons" diagram, each secondary structural element is distinguished by a different shape or motif. Thus β-strands are shown as arrows and α-helices as

tubes or helical ribbons. Segments of random-coil structures that link strands and helices together are represented as a "worm" that may either rigorously follow the positions of the Cα atoms or, more commonly, trace an approximate and much smoother path determined by mathematical interpolation procedures.

Armed with an overall picture of the structure of a protein, together with the possibility of being able to illustrate more detailed aspects at the level of individual atomic arrangements, we may still wish to examine topographical and physico-chemical properties of the protein surface, that is, the regions of the molecule that are directly in contact with bulk solvent (water, ions, and so forth), small-molecule ligands (cofactors, substrates, and so forth), and other proteins with which biologically interesting interactions are made. Surfaces are generated by computationally "rolling" a sphere or probe with a given radius (usually 1.4 Å, which approximates that of a water molecule) over a hard-sphere model of the entire protein structure. If the path is taken as that defined by the center of the probe, the result is a "solvent-accessible" surface. A commonly used variation of this method generates surface points from the volume boundary of the probe, which tends to smooth out sharp grooves and crevices. This kind of representation is often called a "molecular" or "Connolly" surface (**FIGURE 5.19**).

Having generated the surface diagram, it is possible to "map" various features of the molecule onto it. One of the most popular such uses is to display the electrostatic properties of the protein surface—that is, the regions of

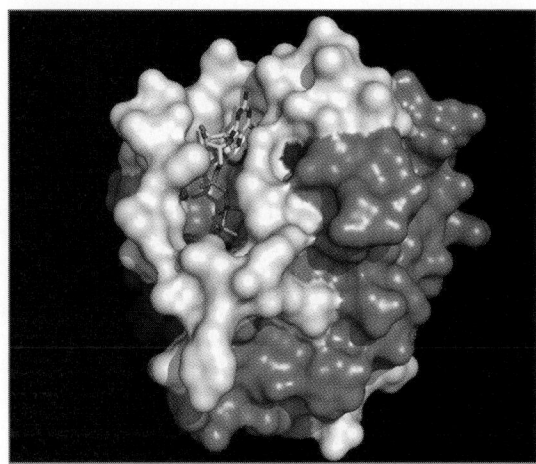

FIGURE 5.19 Surface representation. External, solvent-accessible features such as ligand-binding grooves and clefts may be revealed, and various properties (secondary structure in this case) of the underlying atoms can be mapped onto the surface. Structure from Protein Data Bank 2CL0. B. U. Klink, R. S. Goody, and A. J. Scheidig, *Biophys. J.* 91 (2006): 981–992.

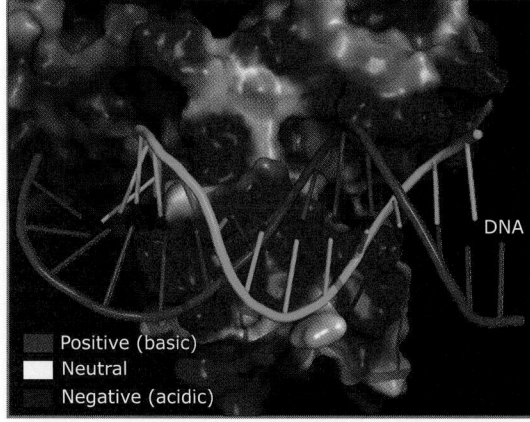

FIGURE 5.20 Electrostatic potential of a DNA-binding protein mapped onto the molecular surface. The highly negatively charged DNA molecule interacts predominantly with regions of positive charge (blue) on the protein. Structure from Protein Data Bank 1CGP. S. C. Schultz, G. C. Shields, and T. A. Steitz, *Science* 253 (1991): 1001–1007.

positive and negative charge that are most often associated with clusters of basic (Arg, Lys) or acidic (Asp, Glu) amino acids. This information can reveal likely binding sites for ligands, such as DNA (FIGURE 5.20). Similarly, algorithms have been developed to calculate the relative hydrophobicity of different parts of the molecular surface. Nonpolar interactions generally contribute greatly to the formation of stable protein–protein interactions; as a result, this approach can also reveal potential binding sites and may even be used to estimate binding affinities. Additional indicators of functionally important regions can often be derived from

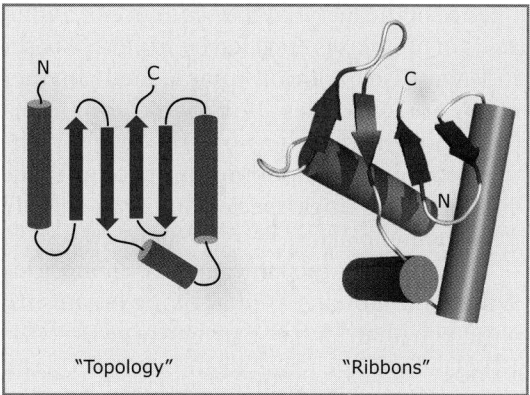

FIGURE 5.21 A topology diagram (left) can be used to simplify a three-dimensional structure (right) for comparison or other purposes.

mapping primary sequence homology within a family of related proteins onto the molecular surface of one of its members, given that the most highly conserved amino acids that are accessible to solvent are likely to be involved in an evolutionarily conserved function.

Finally, it is often convenient to be able to schematically represent the secondary structural elements of proteins in terms of their relative position and the order in which they occur in the polypeptide chain. This is often called a topology diagram, and such diagrams are used extensively in the classification and comparison of different families of protein folds (see *5.8 Tertiary structure and the universe of protein folds*). Unfortunately, there are almost as many variations on this theme as protein structures themselves; an example of one that will be used in this chapter is shown in FIGURE 5.21.

5.6 Proteins are linear chains of amino acids—primary structure

Key concepts
- Proteins are composed of linear chains formed by condensation reactions between amino and carboxyl groups of amino acids.
- Only 20 amino acids are commonly found in proteins; all are L-enantiomers with α-configuration.
- Differences in physical and chemical properties of amino acids are fundamental to the diversity in protein structure and function.

Proteins are linear and unbranched polymers of amino acid building blocks. (There are a small number of exceptions, such as cyclic peptides, which are formed posttranslationally.) This basic structural property is a consequence of the fact that the sequence of amino acids in proteins is encoded by triplets of bases in DNA—itself a linear, unbranched polymer chain of nucleotides. Unlike in common organic polymers such as polythene (polyethylene), the basic monomer units of proteins comprise not a single species, but 20 chemically distinct amino acids.

Remarkably, the same set of 20 amino acids is found in proteins from all living organisms. Nineteen share a basic structure, NH_2-(CH-R)-COOH, and each is distinguished by the composition of the R-group (**FIGURE 5.22**). The exception is the imino-acid proline, in which the nitrogen atom of the backbone is locked into a five-membered pyrrolidine ring. The conformational rigidity imposed by this arrangement plays a number of important roles in protein folding. Amino acids are generally referred to by their three-letter or single-letter abbreviation (Figure 5.22). For the most part, the three-letter form will be used here.

A wide variety of "nonstandard" amino acids, such as selenocysteine, hydroxyproline, hydroxy-lysine, ornithine, and γ-carboxy-glu, occur as by-products of metabolic reactions,

through posttranslational modification or by the activity of specific biosynthetic pathways. These amino acids play highly specialized roles, some of which will be mentioned in forthcoming sections.

Amino acids generally found in proteins all have the α-configuration (**FIGURE 5.23**), meaning that the amino group is attached to the α-carbon. Thus α-amino acids contain only a single carbon atom (excluding the carbonyl carbon) in the backbone. Alternative forms, such as β-amino acids that have two backbone carbon atoms, are not found in proteins, but β-alanine is found in some naturally occurring peptides such as carnosine. Presumably, β-amino acids were selected against early in evolution due to the increased degree of rotational freedom around the C–C bond that would prevent higher-order folding, although some oligopeptides made from β-amino acids are known to adopt novel "secondary structures" in solution.

With the exception of glycine, which has a single hydrogen atom as its "side-chain" and is thus symmetric, amino acids all possess a chiral center at the Cα atom; that is, they can adopt a right- (D) or left-handed (L) form, so named because of the way in which these forms rotate polarized light (**FIGURE 5.24**). While D-amino acids do occur in some specialized

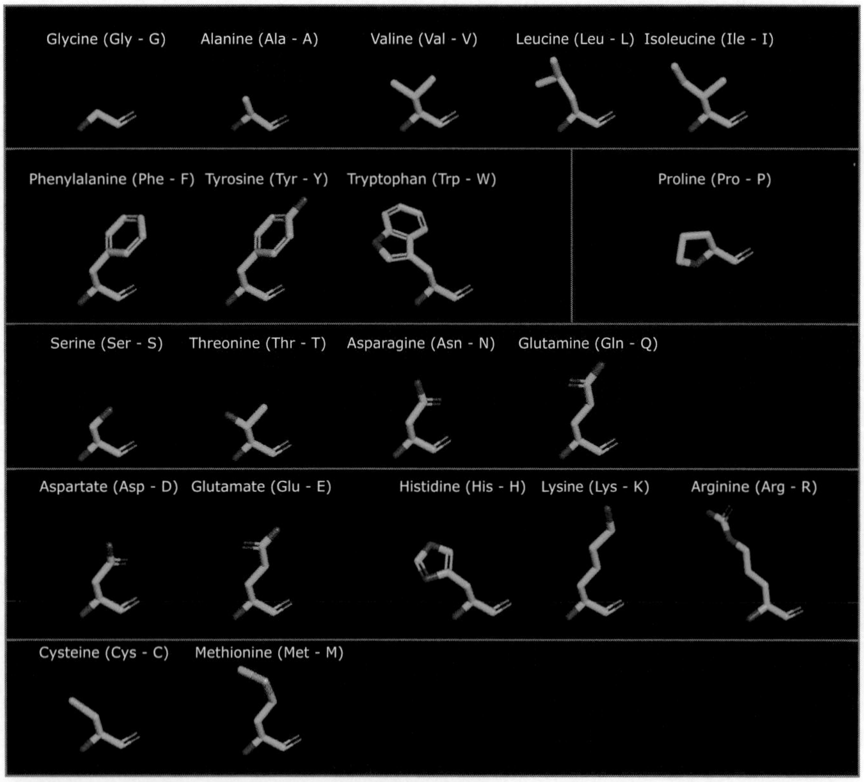

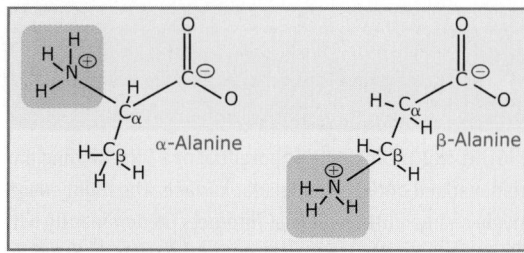

FIGURE 5.23 α (left) and β (right) amino acids shown in their zwitterionic form.

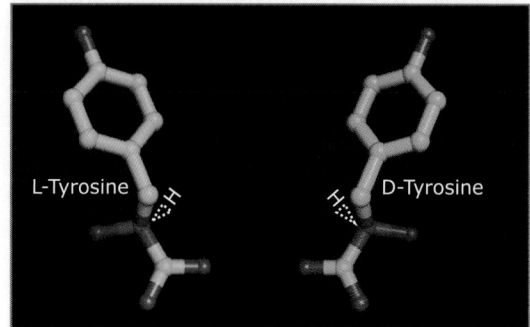

FIGURE 5.24 D- and L-tyrosine.

FIGURE 5.22 The 20 naturally occurring amino acids grouped roughly according to chemical properties, with three-letter and single-letter codes indicated. Carbon = green, oxygen = red, nitrogen = blue, and sulfur = yellow.

circumstances, all proteins are built from L-amino acids, although there appears to be no fundamental reason why evolution could not have selected the D-form.

In general, amino acid residues in proteins adopt the *trans* configuration (**FIGURE 5.25**). In a *cis* configuration, the Cα atoms (and thus the side-chains) of adjacent residues are brought into close proximity; this configuration is, therefore, sterically disfavored. The major exception is the amino acid proline, for which the energy barrier between the *cis* and *trans* forms is much lower, and which has been observed in the *cis* configuration more often than any other amino acid. Indeed, the utility of *cis-trans* proline isomerization is exploited in a number of signaling systems, and can be "catalyzed" by a family of proteins called *cis-trans* prolyl isomerases that have evolved specifically for this purpose.

An extremely important chemical feature of amino acids is that they are amphoteric. At physiological pH the α-amino and α-carboxylic acid groups are essentially completely ionized, and amino acids (at least those with nonionizable R-groups) are therefore also zwitterions with no net charge. As we will see, some amino acids have side-chains that contain additional ionizable groups. Together, these properties are important for the solubility of amino acids, the ability to form polypeptide chains, and the overall charge characteristics of the folded protein molecule that may be important for biologic function in many contexts.

The amino acid sequence encoded in a DNA sequence is translated into a protein polymer by the "decoding" of the mRNA template nucleotide sequence by the ribosome. These large, protein–RNA complexes catalyze the formation of "peptide" or amide bonds between the amino (NH_2) and carboxyl (COOH) terminal groups in a **condensation** reaction, proceeding in the N- to C-terminal direction. The chemist Linus Pauling used available crystal structures of small molecules to show that the peptide bond is, in fact, a resonance hybrid of two forms. This results in a partially double-bonded character and confers rigidity (**FIGURE 5.26**). If this were not the case, the additional degree of rotational freedom would prevent amino acid chains from folding into the stable protein structures we see in biologic systems. Thus the backbone of a polypeptide chain has only two degrees of rotational freedom: one around the N–Cα bond (φ), and the other around the Cα–C = O bond (ψ). Steric effects

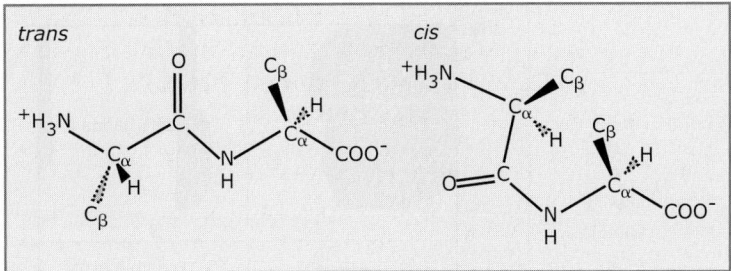

FIGURE 5.25 *cis* and *trans* peptide bonds. Most amino acids adopt the *trans* configuration.

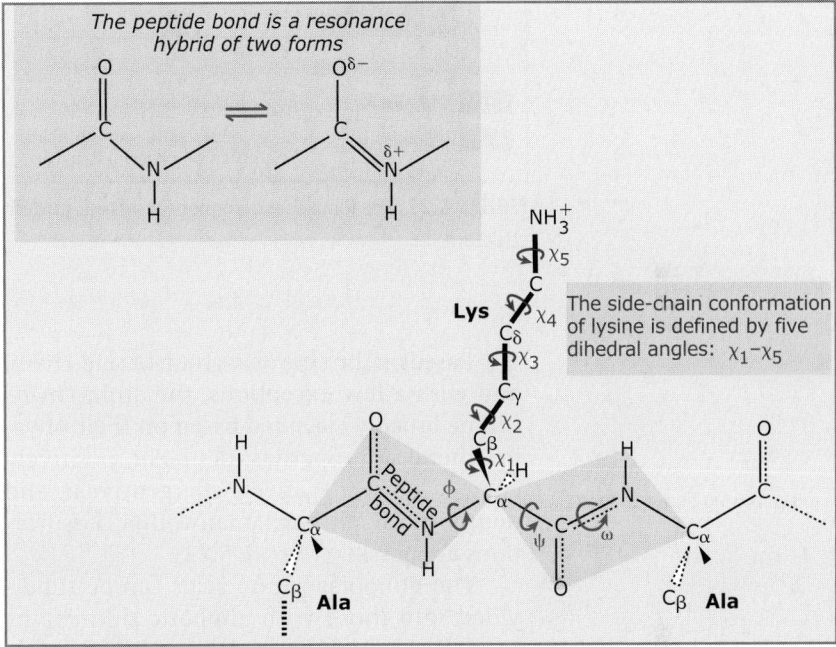

FIGURE 5.26 Planarity of the peptide bond. In addition, the five dihedral angles for lysine about which essentially free rotation can occur are shown.

result in a rather limited range of φ-ψ dihedral angles that are energetically favorable in a polypeptide, and this accessible conformational space is classically represented in graphical form as the familiar φ-ψ or Ramachandran plot (**FIGURE 5.27**). Glycine is an exception, however, and the absence of a side-chain results in a much more extensive range of accessible φ-ψ combinations. The Ramachandran plot was originally derived from empirical considerations based on van der Waals contact distances (see below), observed in the limited database of small-molecule structures available at the time. Nevertheless, it has been largely confirmed by the protein structures determined since its introduction, and is one of the commonly used means of assessing the reliability of newly determined structures.

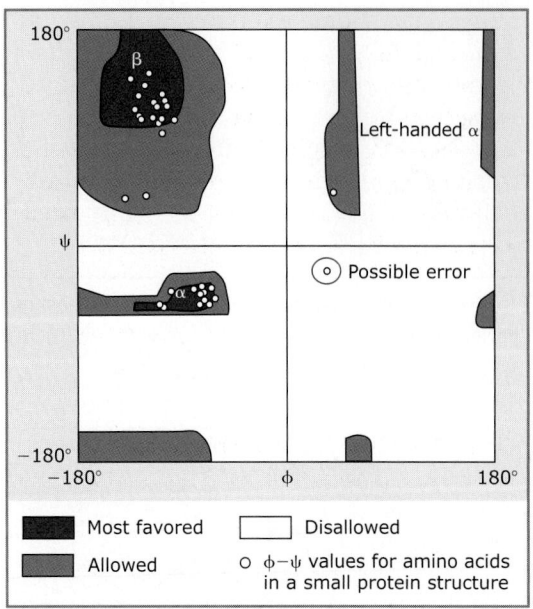

180°

Left-handed α

○ Possible error

−180°
−180°   φ   180°

■ Most favored   □ Disallowed
■ Allowed   ○ φ–ψ values for amino acids
                          in a small protein structure

FIGURE 5.27 The Ramachandran plot for a small protein structure.

Based on the type of R-group or side-chain, and with a few exceptions, the amino acids can be broadly classified based on their physico-chemical properties: nonpolar, positively charged-polar, negatively charged-polar, and neutral-polar, although many other classifications are possible (Figure 5.22).

The nonpolar amino acids can be subdivided into those with aliphatic side-chains (Ala, Val, Leu, and Ile) and those with aromatic side-chains (Phe, Tyr, and Trp). It is the presence of the aromatic residues that confers the characteristic absorption spectra of polypeptides in the UV-visible range of wavelengths. In general, they are rather **hydrophobic** and do not favor interactions with polar solvents such as water. Thus, they are most often, but not exclusively, located in the core of globular protein molecules. Indeed, the hydroxyl group of Tyr and the pyrrole nitrogen atom within the large indole side-chain of Trp have significant polar character. In addition, the pyrrolidone side-chain of proline contains three carbon atoms in a ring, and therefore has a significant hydrophobic nature.

Of the more **hydrophilic** amino acids, the four neutral-polar residues are distinguishable by the presence of either hydroxyl (Ser, Thr) or amide (Asn, Gln) groups within their side-chains, and these are uncharged at physiological pH. Carboxylic acid groups distinguish the two negatively charged (acidic)-polar amino

acids, aspartate and glutamate. Lys, Arg, and His are the basic or positively charged amino acids, and are characterized by primary amine, guanidine, and imidazole groups, respectively. The imidazole group of His is highly versatile and is often employed in enzyme active sites due to its basicity at physiological pH and its nucleophilicity. These important properties will be revisited later.

Two amino acids contain sulfur in their side-chains: The sulfhydryl of Cys is extremely reactive and is able to form disulfide bridges within or, less often, between polypeptide chains. This property is important for the stability and folding of some proteins, particularly secreted proteins or those that are otherwise exposed to the harsh conditions of the extracellular milieu. Methionine is rather nonpolar, and also contains a sulfur atom in its side-chain. As we have seen from the use of its selenium-substituted cousin in modern methods of crystallographic phase determination, it occupies a rather special place in the hearts of X-ray crystallographers!

As should be clear from the structures of the amino acids, many of their side-chains have rotational freedom around single bonds (Figure 5.26). These dihedral angles are referred to by the Greek letter χ such that χ_1 describes rotation around the Cα–Cβ bond, χ_2 around Cβ–Cγ, and so forth. Clearly, not all χ rotations are possible. For example, the χ rotations of Pro are restricted to very small angular increments associated with different puckers of the pyrrolidine ring. In addition, χ_5 rotations around the Arg Nε–Cζ bond are highly restricted at physiological pH because, when protonated, the N–C bond of the guanidinium has significant double-bonded character due to resonance. Examination of the structural database has revealed that additional restrictions imposed by steric and other effects result in favored conformations or "rotamers" for many side-chains. This information has been usefully incorporated into a number of commonly used computer graphics programs to aid in the interpretation of electron density maps during the structure determination process.

The characteristics of the amino acids and the aqueous environment in which proteins exist determine the type of interactions that are observed within and between proteins and their ligands. Most of these are relatively weak and, with the exception of the disulfide linkage, do not involve the formation of chemical bonds.

By far the strongest and the major driving force in protein folding and the interactions between proteins is the hydrophobic effect. As mentioned earlier, "hydrophobic" literally means "water hating" and describes the property of certain atoms, such as carbon, that prevents them from interacting with water in a thermodynamically favorable way. Protein folding itself can be thought of as arising from the requirement that nonpolar atoms are buried in the interior or "hydrophobic core" or a protein in its folded state. In a similar way, the hydrophobic effect contributes greatly to the intermolecular interactions between proteins and other proteins or ligands (see *5.14 Protein–protein and protein–nucleic acid interactions*).

The hydrogen bond occurs most commonly between a hydrogen atom covalently bonded to an electronegative atom and possessing a partial positive charge, and a second electronegative acceptor atom. In proteins, the most common hydrogen bond (or H-bond) donors are NH groups (main-chain peptide NH), NH_2 groups (amino termini; asparagine/glutamine side-chains), and OH groups (serine and threonine side-chains). Many potential H-bond acceptors exist in proteins, including C = O (main-chain carbonyl oxygen; asparagine/glutamine side-chains), -N = (histidine side-chain), -O- (serine/threonine side-chains), and occasionally, the S-H groups (cysteine side-chains). The strength of hydrogen bonds varies greatly depending on the identity of the donor/acceptor groups and geometry, but is generally in the range of 2 to 3 kcal/mol. A number of observations from high-resolution crystal structures also suggest that CH groups can act as H-bond donors, although such interactions are estimated to be much weaker. Some examples of H-bond interactions are shown in **FIGURE 5.28**.

Interactions between positively and negatively charged atoms in proteins most often occur between the side-chains of the basic amino acids lysine and arginine (and in some circumstances histidine) and those of glutamic/aspartic acids. These interactions are often classified as salt bridges and are most usefully thought of as H-bonds that also involve a significant contribution from electrostatic effects. Electrostatic interactions are complex and depend on a variety of factors, such as local **dielectric constant** and the ionization state of the atoms involved. The latter is described by the pK_a value and is dependent on environmental and chemical factors such as the pH, solvent polarity, and local electrostatic effects. Clustering of similarly charged groups on the surface of a protein can create regions or patches of positive or negative electrostatic potential that often define interacting surfaces for cationic or anionic ligands. This is best exemplified by the interactions between positively charged surfaces on nucleic acid-binding proteins with the highly negatively charged polyanions, RNA and DNA.

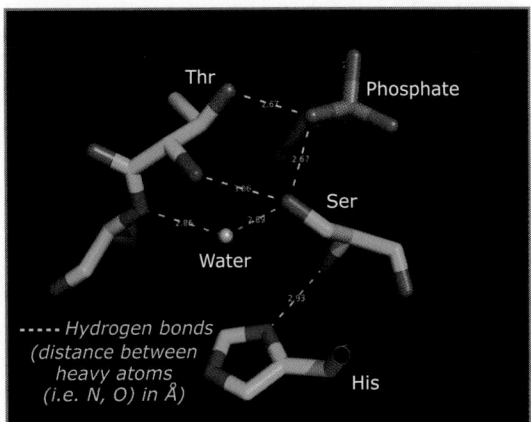

FIGURE 5.28 Hydrogen-bonding interactions. Data from S. Smerdon. Structures created from RCSB Protein Data Bank.

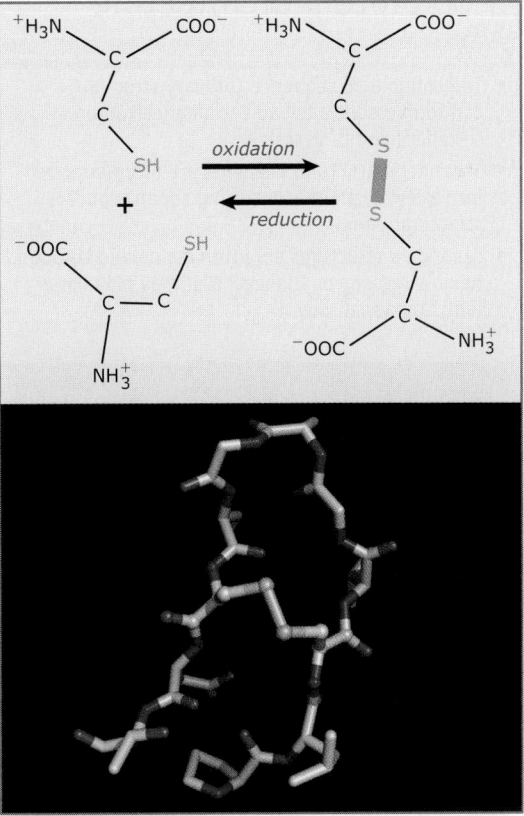

FIGURE 5.29 The disulfide bond. Structure from Protein Data Bank 1GC1. P. D. Kwong, et al., *Nature* 393 (1998): 648–659.

Disulfide bonds or linkages are, except for some unusual examples described later (such as green fluorescent protein [GFP]), the only covalent interaction that takes place between protein side-chains. They are formed by oxidation of the sulfydryl (S-H) side-chains of cysteine residues to form a sulfur–sulfur (disulfide) bond. Although uncommon in cellular proteins due to the highly reducing conditions of the cytoplasm, they are often found in secreted proteins (such as hormones) or proteins anchored to the plasma membrane but exposed to the extracellular milieu, where the additional structural stabilization provides rigidity and resistance to proteolytic degradation (**FIGURE 5.29**).

5.7 Secondary structure— the fundamental unit of protein architecture

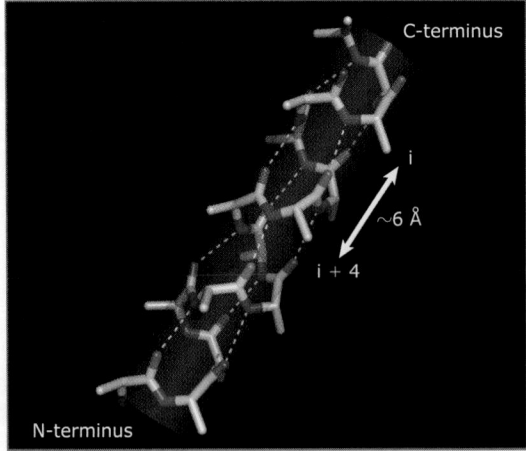

FIGURE 5.30 The α-helix. Structure from Protein Data Bank 1RRQ. J. C. Fromme, et al., *Nature* 427 (2004): 652–656.

Protein structures are not usually composed of extended chains of amino acids, but rather are formed from the association of one or more segments of a rather limited number of regular structural elements. These elements adopt characteristic backbone φ-ψ angles (see *5.6 Proteins are linear chains of amino acids—primary structure*), and the majority of proteins utilize two major **secondary structures**, the α-helix and the β-strand. This is exemplified by the fact that α and β main-chain configurations are the most highly populated regions of the Ramachandran plot. In the absence of crystal or NMR structures, the secondary structural content of a protein molecule can be estimated using spectroscopic methods such as circular dichroism.

The structure of the α-helix was first proposed by Pauling based largely on model building and profound chemical intuition. In fact, α-helices were the first secondary structural element revealed by crystallography. Initial 5 Å electron density maps of sperm whale myoglobin showed eight rod-like structures, and subsequent higher-resolution studies essentially confirmed the most important aspects of the Pauling model. The α-helix is formed and stabilized by a characteristic hydrogen-bonding pattern formed between the main-chain NH group of residue **i** with the main-chain carbonyl oxygen of residue **i+4** (**FIGURE 5.30**). The peptide bonds have a small dipole. As a result, and because they all point in the same direction in an α-helix, the α-helix behaves as a "macro-dipole" with small positive and negative charges at the N- and C-terminal ends, respectively. This charge characteristic may, in turn, be exploited in protein–protein interactions or ligand binding.

Amino acid chirality dictates that the helical twist thus formed is always right-handed, and no examples of left-handed α-helices have been observed in any biologic structure yet determined, nor are they likely to be. In a geometrically ideal helix, the direction of the interresidue hydrogen bonds is exactly parallel to the helix axis. This situation is only rarely observed in protein structures, and some non-linearity of the hydrogen-bond geometry is most common.

A number of variants of the α-helix are, of course, possible. Of these, the so-called 3_{10} helix, in which residue **i** is hydrogen-bonded to **i+3**, is regularly seen to form short turns in an otherwise extended structure. A helix in which residue **i** is hydrogen-bonded to **i+5** is known as the π helix, but is predicted to be rather unstable and, presumably for this reason, does not occur in any known protein structure.

The β-strand constitutes the second commonly observed secondary structural element in proteins. β-strands were first observed in early crystal structures of lysozyme, and consist of an essentially fully extended polypeptide chain. Alone, the β-strand is unstable and is almost always found in combination with others to form β-sheet structures.

The conformation of a single strand is such that the peptide NH and C = O groups project in opposite directions, and these directions are reversed in successive residues within the strand. Due to the fundamental structure of α-amino acids described earlier, the side-chains of each adjacent residue then project in opposite directions, but in a plane that is essentially orthogonal to that defined by the main-chain groups (**FIGURE 5.31**). Thus, free of potential steric clashes between side-chain atoms, strands can associate through NH—-[C1] O = C hydrogen bonds (**FIGURE 5.32**).

β-sheets are not flat, but show a left-handed twist that may be more or less pronounced in different structural contexts (**FIGURE 5.33**). As may be already clear, association of β-strands can occur in either a parallel or an antiparallel orientation, and β-sheets are often composed of both (**FIGURE 5.34**).

In considering the association of individual secondary structural elements to form a folded tertiary structure (see *5.8 Tertiary structure and the universe of protein folds*), it is clear that some motifs are highly represented in the structural

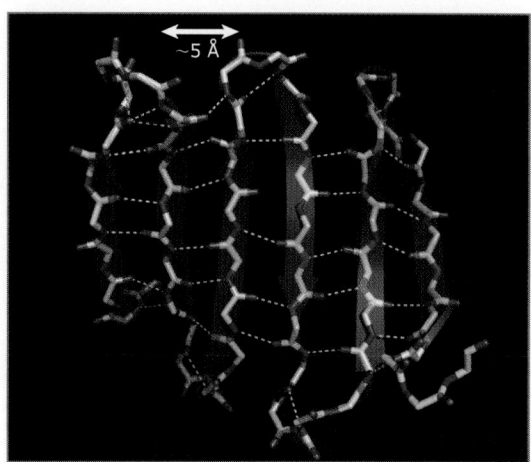

FIGURE 5.32 An antiparallel β-sheet showing the characteristic interstrand hydrogen-bonding pattern. Structure from Protein Data Bank 1RRQ. J. C. Fromme, et al., *Nature* 427 (2004): 652–656.

FIGURE 5.33 The β-barrel. Structure from Protein Data Bank 1BXW. A. Pautsch and G. E. Schulz, *Nat. Struct. Biol.* 5 (1998): 1013–1017.

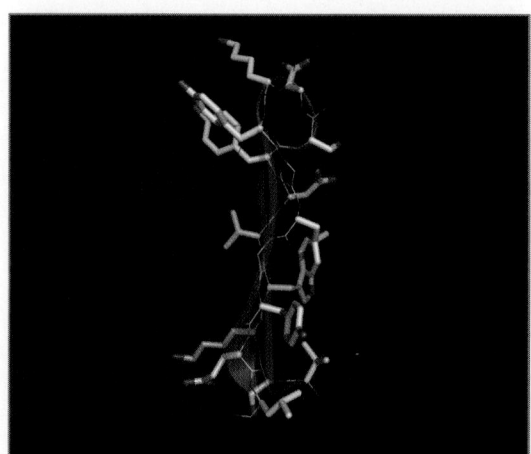

FIGURE 5.31 A β-sheet viewed from the side. Amino acid side-chains project away from the plane of the sheet. Structure from Protein Data Bank 1RRQ. J. C. Fromme, et al., *Nature* 427 (2004): 652–656.

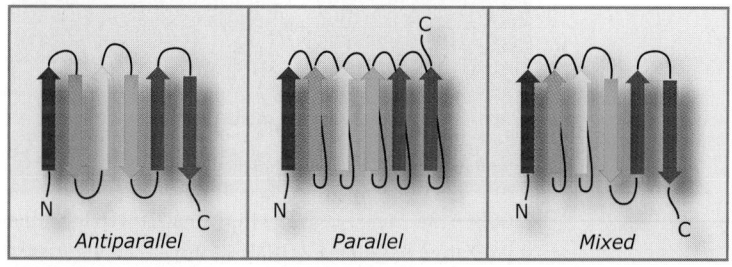

FIGURE 5.34 Different β-sheet topologies: parallel, antiparallel, and mixed.

database. These are often referred to as elements of supersecondary structure, and examples include certain types of β-hairpins that connect consecutive β-strands in a sheet, certain combination of β-strands (e.g., β-meander, Greek key motif), and turns that link together α-helices to form the well-known helix-loop-helix and helix-turn-helix motifs (**FIGURE 5.35**).

Among the most common supersecondary structures is the coiled coil that arises from the interaction of two, three, four, or even more bundles of α-helices with a characteristic pattern of hydrophobic and charged amino acids that repeats every seven residues—the so-called heptad repeat (**FIGURE 5.36**). Coiled-coil motifs were first predicted by Francis Crick in a theoretical analysis of how α-helices pack together. We now know of many thousands of examples of proteins with a diversity of functions that are observed or predicted to contain coiled-coil regions. The heptad repeat results in a pronounced amphipathic nature where each helix has polar and nonpolar faces. The nonpo-

lar side-chains pack together at the interface, leaving the more hydrophilic side-chains exposed to solvent. Often, basic and acidic residues are found juxtaposed in the coiled coil, where they form salt-bridging interactions and thus provide additional structural stabilization. Although parallel coiled coils with a left-handed superhelical twist are most common, antiparallel configurations, and some with a right-handed twist, are also known. Coiled coils in which leucine dominates the "a" position of the heptad repeat (-abcdefg-abcdefgetc) are often referred to as "leucine zippers."

Although not generally considered to be secondary structures as such, most proteins contain regions of extended conformations that serve to join together the α-helices and β-strands. These regions are often referred to as "linkers" for this reason. In their most extreme form they can be rather long, comprising tens or even hundreds of amino acids. Alternatively, they may consist of only two or three residues forming tight turns that, themselves, have been observed to fall into a number of common supersecondary structural classes described earlier. In many cases, linker regions tend to be poorly defined in X-ray and NMR structures but, as will be seen later, they may become ordered, or even adopt a canonical secondary structure, upon interaction with specific ligands or partner proteins.

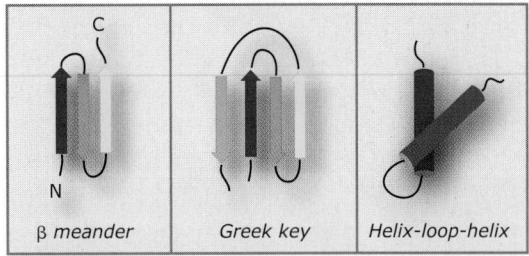

FIGURE 5.35 Examples of supersecondary structures.

β meander | Greek key | Helix-loop-helix

Concept and Reasoning Checks

1. What would be the effect of building an α-helix from amino acids with "D" configuration?
2. How many different topologies of a 3-stranded β-sheet are possible?

5.8 Tertiary structure and the universe of protein folds

Key concepts

- In spite of the complexity of protein sequences, it appears that the number of ways in which polypeptides fold into their final tertiary structures is limited; structure and function are more highly conserved than in an amino acid sequence.
- Some protein folds are seen to carry out numerous biologic functions, whereas others appear to have evolved to perform specialized activities.

The folded state results from the formation and association of secondary structural elements described earlier, together with intervening linker regions and turns to form the tertiary structure

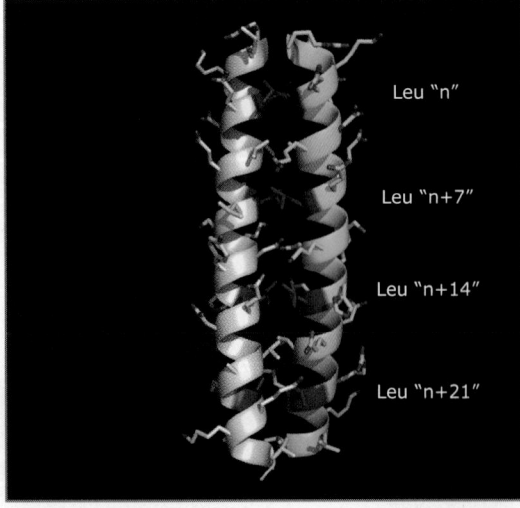

Leu "n"

Leu "n+7"

Leu "n+14"

Leu "n+21"

FIGURE 5.36 Coiled-coils and the heptad repeat. Structure from Protein Data Bank 2ZTA. E. K. O'Shea, et al., *Science* 254 (1991): 539–544.

The magnitude of the "folding problem" was noted by Levinthal in the 1960s. Levinthal presented the problem, now known as "Levinthal's paradox," in the following way: If we consider a protein of a 100 residues and assume that the main-chain Φ-φ angles can take one of three possible values (side-chain rotamers are completely ignored), then each peptide can adopt nine (3 × 3) possible conformations.

Therefore, our 100-residue protein can adopt 9^{100} possible three-dimensional structures. Making a further assumption that changes in peptide conformation can occur on the femtosecond (10^{-15} s) time scale, finding the correct folded conformation would take, on average, ~10^{70} years, rather than the few seconds or minutes that might be actually observed in the test tube.

of the molecule. On the basis of his studies of the reversible folding–unfolding of ribonuclease, Anfinsen suggested that the information that determined the final folded state of a protein was encoded entirely within its amino acid sequence. Since then, there have been considerable efforts to unravel the complex processes involved in the formation of a folded structure from a string of amino acids (see the *Historical Perspectives* box on page 193). Although we still do not fully understand how this occurs, progress has been made through a combination of experimental and theoretical approaches.

So, the number of possible three-dimensional structures for even a small protein is astronomical. Nonetheless, it is clear that proteins do only seem to adopt a relatively small number of tertiary structures. Broadly speaking, two classes of tertiary structure are discernable: fibrous and globular. Fibrous proteins, as their name suggests, are characterized by rather elongated architectures and are exemplified by "structural" proteins such as keratin and collagen. In contrast, most tertiary structures fall into the second broad class and are generally referred to as "globular," reflecting their more spherical shape.

A question that fascinates many structural biologists is, how many different structures of globular proteins are there? The structural database, at present, contains some 40,000 structures (some of which may be, e.g., mutant forms of the same protein). But how many of these represent different protein folds, and how many such folds remain to be determined? The importance of these questions has resulted in an increasing number of "structural genomics" programs (see *5.17 What's next? Structural biology in the postgenomic era*) whose aims are, in part, to determine the structures of all possible protein folds in biology. This is not merely a stamp-

collecting exercise but, if successful, will play an important role in our efforts to understand how a protein sequence determines its final three-dimensional structure.

The overall structure of large (>50 kDa) proteins can almost always be subdivided into combinations of smaller, compact entities that are generally referred to as domains, and are often thought of as segments that are capable of independent folding. Indeed, this notion has been extensively used to experimentally define domain boundaries within larger molecules by limited proteolytic digestion. In this method, small quantities of proteases are mixed with varying molar excesses of the protein of interest with the hope that flexible, exposed linker regions, or unfolded N- and/or C-terminal sequences, will be more easily cleaved than the compact globular domains (**FIGURE 5.37**).

If we assume that all globular proteins are formed from one or more "domains," then our

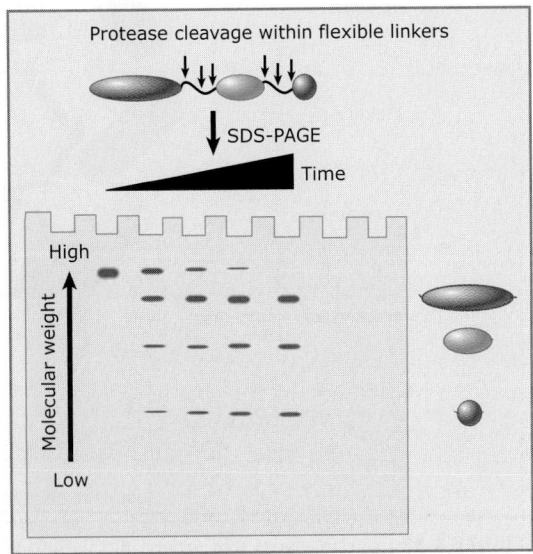

FIGURE 5.37 A "limited proteolysis" experiment.

"fold" problem becomes one of deciding how many topologically discrete combinations of α-helices and β-strands are likely to be represented in the structural proteome. A number of approaches have been taken to this problem, including the reduction of "fold-space" into a periodic table-like classification of likely topological arrangements of beta and alpha secondary structural elements. Despite the exponential increase in structure determinations, and the fact that large-scale structural genomics efforts often choose target proteins on the basis that they may contain a novel fold, the rate at which new folds are being revealed does not follow the same exponential behavior. This is, in part, related to an overarching question of how relationships between different structures are measured, defined, and classified. Regardless, it seems likely that as more and more variations in topology are revealed, a picture may emerge that describes "fold-space" as a continuum of structures, each subtly different from the next, rather than the somewhat "quantized" view that has been favored for so long.

To date, around 800 protein folds have been revealed and classified. It is clear, however, that a subset of these must have arisen early in evolution and occur in many hundreds of known protein structures. In general, these "common" folds fall into two distinct classes: those that seem to possess functional versatility and appear in proteins with extraordinarily diverse biologic activities, and those that appear especially well suited for a specific role. For example, the "triosephosphate isomerase (TIM) barrel" that is made up of eight copies of one β-strand and one α-helix (**FIGURE 5.38**) was first observed in an early structure of the metabolic enzyme triose phosphate isomerase, but has subsequently been seen in over 100 different structures of molecules with a diversity of largely enzymatic functions. Similarly, the all-beta immunoglobulin or "Ig" fold, first seen in structures of the Fab fragment IgG, is now known to occur in molecules with diverse functions, such as chaperones and transcription factors. In contrast, the α-helical globin fold is exquisitely tailored to bind macrocyclic cofactors called metalloporphyrins (see Figure 5.61). More than 100 structures of globin fold proteins have been determined yet, remarkably, this appears to be their major function.

Given that <1000 distinct protein folds have been seen in the structures determined to date, and that considerably greater diversity is seen in the vast database of protein sequence

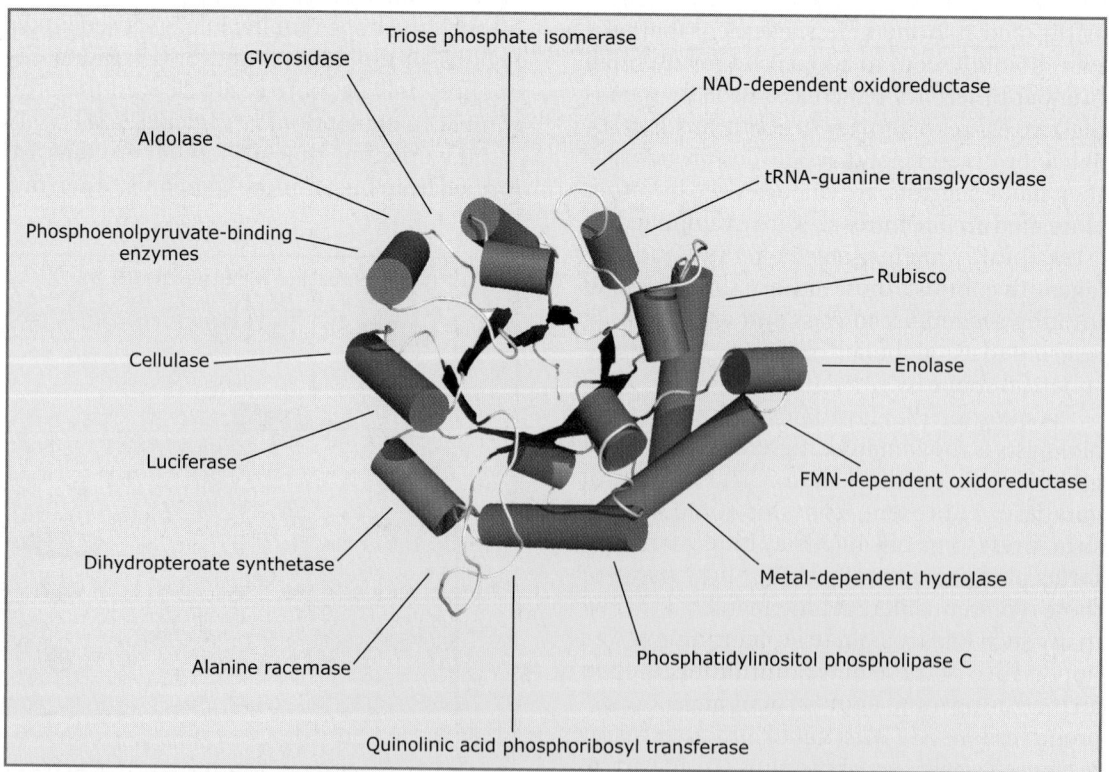

FIGURE 5.38 The TIM-barrel fold occurs in a variety of functional contexts. Structure from Protein Data Bank 8TIM. P. J. Artymiuk, W. R. Taylor, and D. C. Phillips, *Triose Phosphate Isomerase*.

data, it is clear that their structure and function are much more strictly conserved than in an amino acid sequence. This situation arises through two distinct mechanisms of convergent and divergent evolution.

Convergent evolution was noted early in the history of structural biology. The X-ray structures of two proteolytic enzymes, trypsin and chymotrypsin, revealed a close relationship in both sequence and overall structure. In particular, the precise three-dimensional arrangement of a histidine-aspartate-serine triad of catalytic residues was seen to be almost identical at the active sites of each protein. It was with some surprise that the later structure of a completely unrelated protease, subtilisin, showed a completely different fold but retained the identical stereochemical arrangement of catalytic triad residues that, nonetheless, occur in a different relative order in the sequences of the two enzymes (**FIGURE 5.39**). Thus, in these enzymes, the same enzymatic function appears to have been "invented" more than once in evolutionary history.

Although this represents convergent evolution of the structural features of an active site, it would appear that convergence of overall protein fold might also have taken place. Proteins have been shown to have near-identical tertiary structures, yet share no detectable homology other than that expected for alignment of a pair of random sequences. This is something of a gray area, where it may be difficult or impossible to decide whether two structures have arisen by convergent or divergent processes.

Divergent evolution is most commonly observed and is manifested by conservation in sequence and overall structure, or the structure of a core segment, along with the position of functionally important residues (where function is also maintained). In its most generally accepted sense, divergent evolution implies that multiple, related protein sequences share a common evolutionary origin. Observed differences then arise from selective pressure to evolve, for example, differences in substrate specificity between members of a family of enzymes. Again, the serine proteases represent a good example, where trypsin and chymotrypsin share close structural homology, but differ in their preference for different classes of amino acids, C-terminal to the substrate's scissile bond.

In the pantheon of structural folds/motifs observed to date, a number of interesting "outliers" have been observed. We started this section

with the tentative assertion that related primary structures (i.e., amino acid sequences) must give rise to similar tertiary structures in the folded state. Although this is generally true, it has occasionally been seen that two sequences that are apparently closely related can fold in similar but topologically distinct ways, often with important functional consequences. For example, members of the KH family of RNA-binding domains have similar overall structures that, nevertheless, fall into several classes differing in the order of their alpha and beta secondary structural units. Nowhere is this phenomenon more obvious than in prion-related diseases that are caused by the aggregation of a small prion protein (PrP), following a dramatic conversion from a predominantly helical form to a form containing a preponderance of β-structure. Prions are discussed in more detail in *5.16 Structure and medicine*.

A related phenomenon called cyclic permutation occurs when one or more secondary structural elements at one end of the molecule

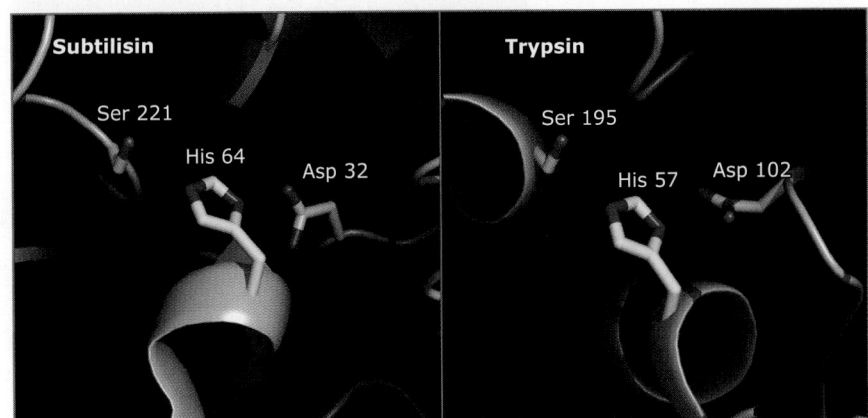

FIGURE 5.39 Convergent evolution of an enzyme active site. Left structure from Protein Data Bank 1N65. B. Jiménez, L. Poggi, and M. Piccioli, *Biochemistry* 42 (2003): 13066–13073. Right structure from Protein Data Bank 1GNS. O. Almog, et al., *J. Biol. Chem.* 277 (2002): 27553–27558.

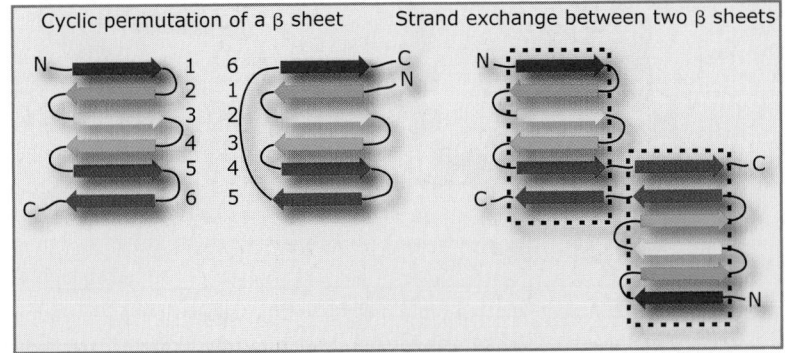

FIGURE 5.40 Schematic representation of cyclic permutation and strand exchange/domain swapping.

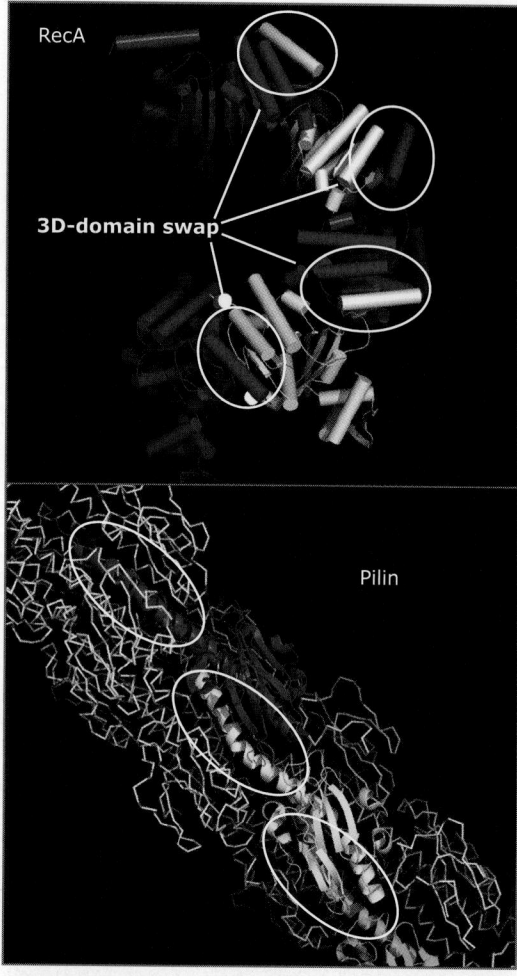

RecA

3D-domain swap

Pilin

FIGURE 5.41 Domain swapping in two example structures, RecA and bacterial pilins. Top structure from Protein Data Bank 2REB. R. M. Story, I. T. Weber, and T. A. Steitz, *Nature* 355 (1992): 318–325. Bottom structure from Protein Data Bank 2HIL. L. Craig, et al., *Mol. Cell* 23 (2006): 651–662.

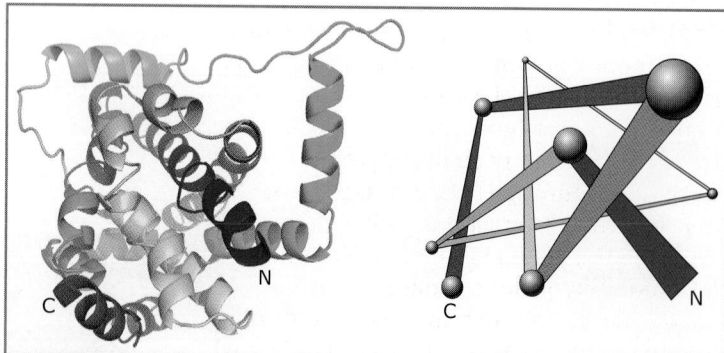

FIGURE 5.42 A deep-knotted protein (left) with a schematic representation (right) to show how the knot is formed. Structure from Protein Data Bank 1YEV. J. S. Kavanaugh, P. H. Rogers, and A. Arnone, *Biochemistry* 44 (2005): 6101–6121.

are transposed to the other end, where they form identical or near-identical interactions with remaining secondary structural elements (**FIGURE 5.40**). This is only possible because of the fact that, in many structures of diverse proteins, the N- and C-termini are observed to be located close together in space, despite being distant at the level of the linear amino acid sequence.

A number of examples of structures have now been reported that show exchange of one or more secondary structural elements between two, or rarely three or four, identical domains, an effect that is variously referred to as strand exchange, segment swap, or domain swap (**FIGURE 5.41**). This arrangement obviously implies that self-association (dimer-, trimer-, tetramerization, or even polymerization) may be functionally important. The biologic relevance of segment swapping is not always clear, however, and because of its predominance in crystal rather than NMR structures, it has sparked several arguments about the relative merits of the two approaches! In many cases, this phenomenon has been shown to be artifactual, but nonetheless it does play well characterized roles in biologic processes such as RecA filament formation and the assembly of bacterial pili (Figure 5.41).

As should be clear from the structures presented so far, if one conceptually held the N- and C-termini of most proteins in either hand and pulled, the structure would unravel back to the original linear chain of amino acids. That is, unless different parts of a protein chain are covalently cross-linked—for example, by disulfide bond formation, as observed in so-called cysteine-knot structures—proteins do not form knots! Until relatively recently, this was the widely held view and for good reason: a knotted structure would have considerable implications for folding pathways. In 2000, however, bioinformatics analysis detected a disulfide-independent knotted structure in a plant protein, acetohydroxy acid isomeroreductase (**FIGURE 5.42**). Although a few additional examples have since been discovered, these "deepknotted" structures remain uncommon and a real structural curiosity!

Concept and Reasoning Checks

1. How do prion proteins provide an exception to Anfinsen's hypothesis.
2. How would the inclusion of proline in a Levinthal-style calculation affect the outcome?
3. Why are knotted structures so rare?

5.9 Modular architecture and repeating motifs

Key concepts

- Many eukaryotic proteins do not consist of a single globular domain with a single biologic activity. Instead, they consist of multiple domains or modules, each with a specific function, connected together like beads on a string.
- Modular architectures provide evolutionary flexibility, where different functions may be added or removed from a protein by the insertion or deletion of modules with specific activities.

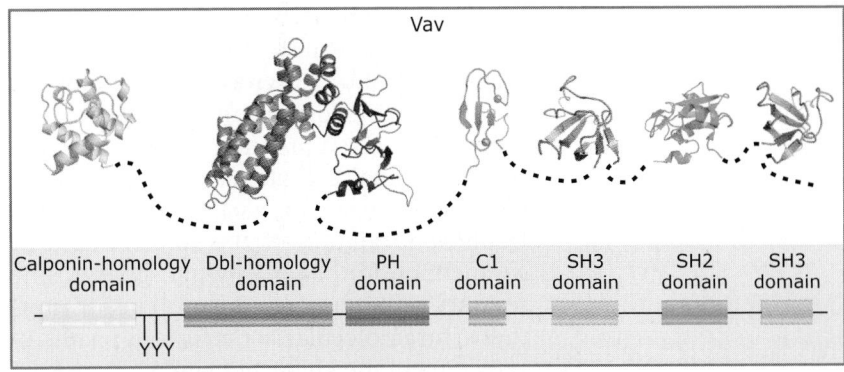

FIGURE 5.43 Modular architecture of the Vav oncogene product. This representation is deceptive and a number of physiologically important intramolecular interactions between domains are known to occur. Structures from Protein Data Bank 1AAZ. H. Eklund, et al., *J. Mol. Biol.* 228 (1992): 596–618. Protein Data Bank 1KBE. M. Zou, et al., *J. Mol. Biol.* 315 (2002): 435–446. Protein Data Bank 1GCP and 1DBH. M. Nishida, et al., *EMBO J.* 20 (2001): 2995–3007.

Any complex biologic process, such as the biosynthesis of an amino acid, may require the activity of many proteins. In prokaryotes, genes are most often organized in operons, whereby they are transcribed as a single, polycistronic mRNA. This elegant mechanism ensures that the expression of a particular set of genes can easily be temporally coordinated in response to a single biologic stimulus. In contrast, eukaryotic genes are generally isolated, and expression of each protein occurs from an individual monocistronic mRNA.

Genome sequencing has revealed that eukaryotic proteins are often constructed as a linear array of protein "modules" joined together, much like beads threaded onto a string. Here, each module has a specific function that contributes to the overall activity of the "polyprotein" within which it resides. Individual modules may play specific roles that are otherwise structurally or functionally independent of those of their associated partner domains. In contrast, the activities of individual domains within a modular protein may be interdependent and may be closely associated in space through intramolecular domain–domain interactions.

Analysis of available genome sequence data has revealed the existence of several hundred protein modules with diverse biologic functions. These modules have been classified and curated in several databases (e.g., SMART, PFAM, and ProDom). Although many have been characterized with respect to structure and biologic activity, many more remain to be investigated, and yet more to be discovered.

The degree of modularity is, in some cases, breathtaking. For example, the Vav protooncogene is a regulator of Rho family small GTPases and contains a total of seven modules that variously mediate binding to lipids, phosphotyrosine, proline-rich motifs, and actin, in addition to a domain that mediates guanine-nucleotide exchange on small GTPases of the Rho family (**FIGURE 5.43**). This architecture is evolutionarily very flexible, affording the possibility of facile generation of proteins with novel functions though domain/exon shuffling. Structures of many of these isolated domains are now available and have generally shown that the N- and C-termini are often located close together in space. Thus additional domains may be added by insertion into the linking regions between preexisting modules without any structural disruption.

Perhaps the most extreme example of the use of modular architecture is seen in the sarcomeric protein titin. The human protein contains around 37,000 amino acid residues and has a molecular mass of ~4 MDa. Sequence and structural analysis has identified ~300 immunoglobulin and fibronectin domains, along with a kinase domain at the C-terminal end and stretches of sequence unique to titin.

In many cases, the precise biochemical roles of a given domain can only be inferred from known activities of well characterized homologues. In a few examples, one or more domains within a protein may exhibit dual functionality through both intramolecular and intermolecular interactions. This is true of Vav (Figure 5.43), but is perhaps best illustrated by the Src-family protein kinases (**FIGURE 5.44**). These molecules all contain three distinct functional domains: an N-terminal SH3 is followed by a short proline-rich linker, which connects to an SH2 domain and a tyrosine kinase domain. Regulation of the activity of Src kinases critically depends on the phosphorylation status of a highly conserved tyrosine residue at the extreme C-terminus. When phosphorylated, the phosphotyrosine (pTyr) binds to the central SH2 domain, and additional interactions between a proline-rich sequence and the N-terminal SH3 domain serve to maintain the kinase in an inactive conformation. Dephosphorylation of the tyrosine and/or competition for binding by phosphotyrosine residues on Src-interacting proteins disrupts these intramolecular contacts and results in activation of the kinase domain.

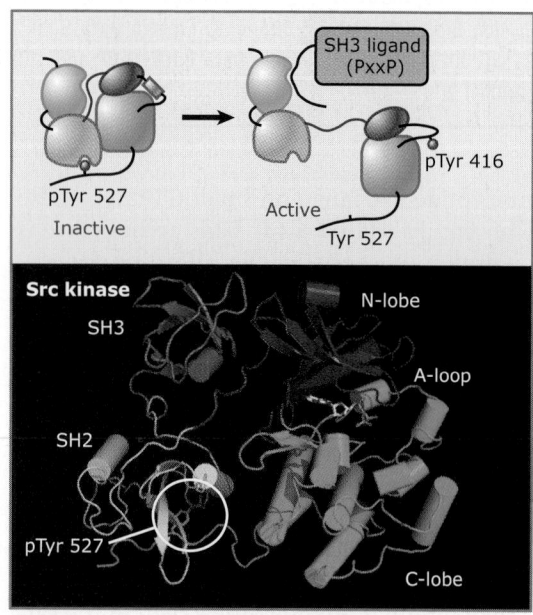

FIGURE 5.44 Intra- and intermolecular interactions regulate Src-family kinase activation. Structures from Protein Data Bank 2SRC. W. Xu, et al., *Mol. Cell* 3 (1999): 629–638.

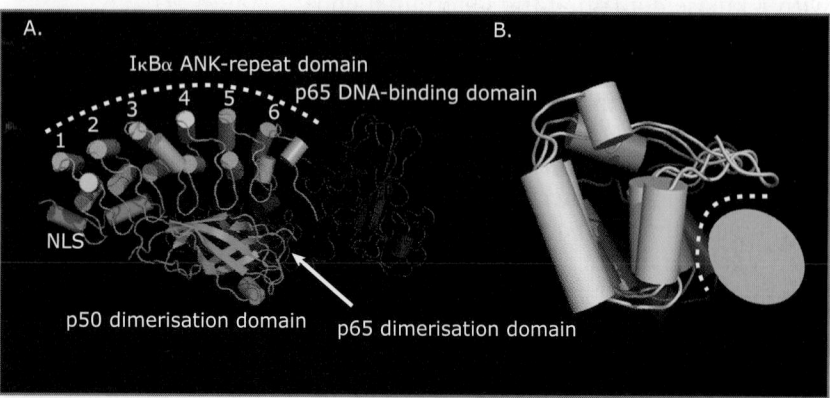

FIGURE 5.45 Ankyrin-repeat structure and binding in the IκBα-NFκB complex. The NLS binds into the ankyrin groove, but binding occurs in other regions of the ankyrin repeat stack. Structures from Protein Data Bank 1NFI. M. D. Jacobs and S. C. Harrison, *Cell* 95 (1998): 749–758.

An additional level of modularity can be seen in protein domains that in turn consist of many tandem copies of short repeating sequences (ankyrin [ANK], leucine-rich repeats [LRRs], tetratricopeptide repeats (TPRs), WD40, HEAT/ARM, pumilio, and so forth). These molecules are functionally diverse; they may play purely structural roles, mediate protein–protein or protein–ligand interactions, or a combination of all of these. As is true of protein structures in general, modular-repeat proteins can be loosely classified on the basis of their secondary structural content. For simplicity of presentation we will consider two classes: those that have predominantly α-helical structure (although some β-content may be evident) and those in which β-structure dominates.

A major characteristic of the helical-repeat proteins is that the number of repeats within any one molecule is highly variable, a feature presumably generated through recursive gene duplication. Evolutionarily, this type of tertiary architecture is attractive, given that it results in molecules with extended interaction surfaces that can be tailored in size by the simple addition or subtraction of repeating units.

Helical repeating motifs were first noted in comparisons of budding and fission yeast transcription factors, and were initially known as Swi6/Cdc10 repeats after the molecules in question. Subsequently, they were noted in the ankyrin cytoskeletal proteins, which may contain up to 50 repeats of the characteristic 30- to 35- residue motif and are now most commonly known as ANK repeats. At present, over 3500 ANK repeat proteins containing more than 15,000 individual motifs are known. Structural studies have revealed the ANK repeat to consist of a short, tight β-hairpin-like turn followed by a pair of antiparallel α-helices and a partially conserved linker region. The helices of each repeat pack against those of the neighboring motif through conserved hydrophobic residues. This generates an elongated structure that is both curved and twisted around the long axis of the stack (**FIGURE 5.45**). The hairpin loops extend away from the helical bundle at an angle of approximately 90°, forming a groove that may be used for interaction with a variety of protein targets and other ligands (Figure 5.45). Although this is the most common mode of ANK domain binding, it is clear that other regions of the surface may also be utilized for inter- and/or intramolecular interactions in specific contexts.

Other repetitive helical-repeat motifs include LRRs, TPRs, Pumilio repeats that are in turn related to the HEAT/ARM family of motifs, and others. Association of multiple copies of these motifs can give rise to a variety of tertiary structures, and in the majority of cases where the activities have been well characterized, helical-repeat architectures assemble to produce extended surfaces for interaction with protein partners or other ligands. LRRs form a characteristic horseshoe shape (**FIGURE 5.46**), which provides a highly concave binding surface. TPR and HEAT/ARM repeats associate to form a superhelical array of helical motifs, which results in a binding groove that spirals along the length of the molecule. In the case of TPR repeats, the twist on the helical stack is rather small. In contrast, HEAT/ARM proteins tend to show a superhelical twist that can be quite spectacular and of a size that can wrap around entire proteins (Figure 5.46).

In contrast to the diversity seen in helical-repeat architectures, β-repeat motifs are less common and essentially fall into two classes known as β-propellors and β-helix structures. The WD40 domain is named after a conserved Trp-Asp dipeptide that is embedded in a repeating sequence of ~40 residues. Although the number of repeats is quite variable, seven are most commonly observed. The first structure of a WD40-repeat protein was that of the β-subunit of heterotrimeric G-proteins, which revealed that each motif folds into a four-stranded β-sheet. These sheets then pack together to form an extended barrel-like structure that resembles a propeller in which each repeat forms an individual "blade." Functionally, WD40 domains play a variety of roles in mediating protein–protein interactions. Most recently, it has become clear that they constitute a family of modules that may bind to peptide motifs that are specifically modified posttranslationally (see *5.12 Posttranslational modifications and cofactors*). **FIGURE 5.47** shows an example of this phenomenon where the WD40-repeat domain of WDR5 recruits lysine methytransferases to dimethylated lysine 4 of histone H3 by binding the modified lysine side-chain through a recognition pocket formed at the center of the β-propeller domain.

RCC1 (regulator of chromatin condensation 1) is a multifunctional signaling molecule that localizes to the nuclear compartment. Its major role appears to be as a nucleotide exchange factor for a small GTPase Ran that regulates the trafficking of cargo into and out of the nucleus. Structurally, RCC1 resembles WD40-repeat domains in that it adopts a seven-bladed

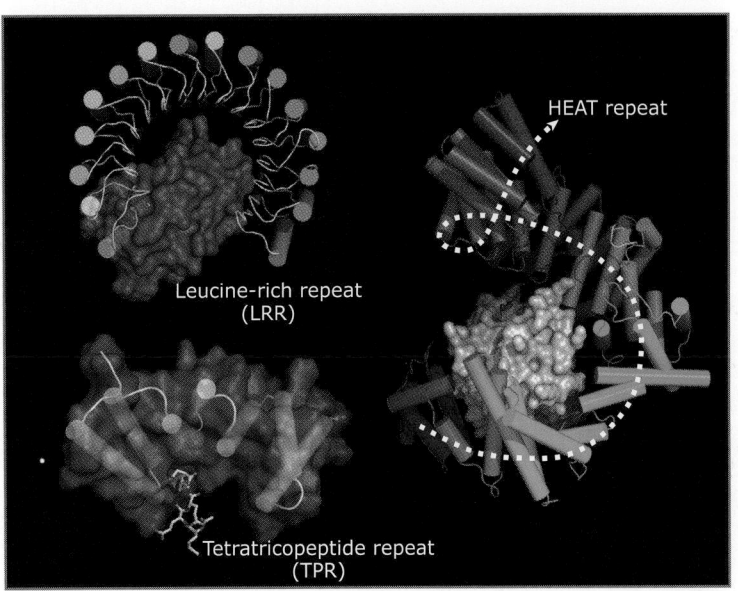

FIGURE 5.46 LRR, TPR, and HEAT helical-repeat architectures. Structures from Protein Data Bank 1DFJ. B. Kobe and J. Deisenhofer, *Nature* 374 (1995): 183–186. Protein Data Bank 1WA5. Y. Matsuura and M. Stewart, *Nature* 432 (2004): 872–877. Protein Data Bank 2BUG. M. J. Cliff, et al., *Structure* 14 (2006): 415–426.

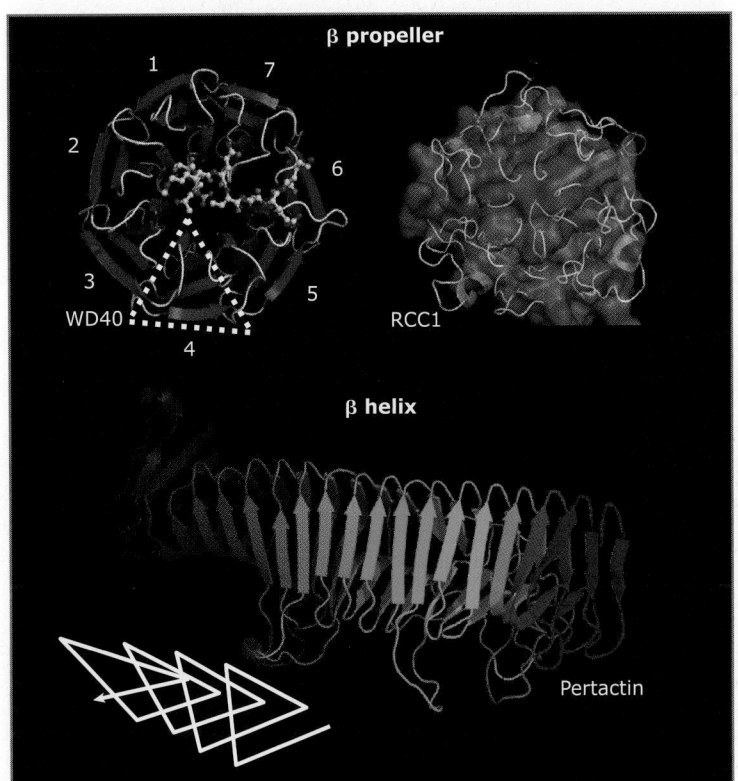

FIGURE 5.47 WD40, RCC1, and β-helix architectures. Structures from Protein Data Bank 2HK6. M. D. Hansson, et al., *Biochemistry* 46 (2007): 87–94. Protein Data Bank 1DAB. P. Emsley, et al., *Nature* 381 (1996): 90–92. Protein Data Bank 1I2M. L. Renault, et al., *Cell* 105 (2001): 245–255.

β-propeller-like fold. There is, nonetheless, no detectable sequence homology between WD40 and RCC1 repeat motifs and, as can be seen from the complex of RCC1 with nucleotide-free Ran (represented as a transparent surface in Figure 5.47), they can be distinguished by differences in the topological arrangement of β-strands, and the inclusion of a short region of α-helix that packs at the outer edge of each of the blades.

As we have seen, WD40 and RCC1 repeats tend to form closed, circular arrays. β-repeat structures, however, are also able to form the kinds of highly extended structures that are generated in helical-repeat molecules such as karyopherins (Figure 5.46). Such an example is the β-helix architecture where repeating pairs or triplets of β-strands associate to form "helical" arrays that can be highly elongated. The β-helix fold has been observed in many different proteins with a plethora of functions. For example, pertactin, a virulence factor that mediates adhesion of the pathogenic *Bordetella pertussis* bacterium with host cells (Figure 5.47), is constructed from consecutive strands that, conceptually, form triplets and pack around a right-handed spiral (Figure 5.47 inset). Conversely, other examples such as insect antifreeze proteins have similar structures, but with a left-handed twist.

Concept and Reasoning Checks

1. What are the evolutionary advantages of modular protein architectures?
2. What are the functional benefits of extended protein structures, such as those formed from repeating units, as protein-interaction scaffolds?

5.10 Quaternary structure and higher-order assemblies

Key concepts

- The quaternary association of individual proteins is extremely common.
- The simplest quaternary structures are dimers of two identical monomers, but considerably complexity is observed and assemblies such as viral capsids may contain hundreds of protein chains.
- Protein assemblies facilitate allosteric and cooperative effects that most often arise from structural changes within the assembly and provide for exquisitely precise regulation of activity.
- Changes in quaternary structure may involve the reversible binding of regulatory subunits to a stable core assembly.

Quaternary structure describes the fact that many proteins do not exist or function as monomers, but associate to form oligomers. The simplest form of oligomer arises from the homotypic association of two identical subunits to form a dimer (more precisely a homodimer). Alternatively, two different proteins may bind to form a heterodimer (a heterotypic association). Obviously, any number of combinations is possible and, as we will see, many variations are observed to occur in biologic systems.

Most dimers arise from the interaction between identical (or near-identical) surfaces on each protomer. Such interactions are known as isologous and are the most commonly observed in structures of protein oligomers. The interaction surfaces are buried in the dimer, and as a result they are not available for interaction with another subunit. As shown in **FIGURE 5.48**, isologous interactions are not restricted to dimers, but can and do occur in higher-order homo-oligomers (trimers, tetramers, pentamers, and so forth). In contrast, heterologous interactions between protomers involve distinct surfaces that do not overlap. In its simplest form, this situation is less commonly observed because it implies that the protomer will become infinitely polymerized. Additional rotational symmetry, however, can result in "closing" of the system, and this closed heterologous oligomerization is much more prevalent. Nonetheless, "nonclosed" heterologous oligomerization (effectively polymerization) of proteins such as actin and tubulin is extremely important in muscle contraction and formation of the actin cytoskeleton and microtubules.

Why be oligomeric? In some cases this is a difficult question to answer, because a functional relevance may not be obvious from the structure or from known biologic function. It is clear, however, that oligomerization can confer a degree of structural or chemical stability. From an evolutionary point of view, formation of oligomers, and particularly hetero-oligomers, affords a great deal of functional and architectural flexibility, in a similar manner to the modular architectures that were discussed in *5.9 Modular architecture and repeating motifs*.

The combination of subunits that constitute a particular protein complex may provide different structural characteristics that are tailored to specific biologic activities. In addition, activity and specificity can be modulated through regulated association and dissociation of accessory subunits. An excellent example of such a system is provided by RNA polymerases,

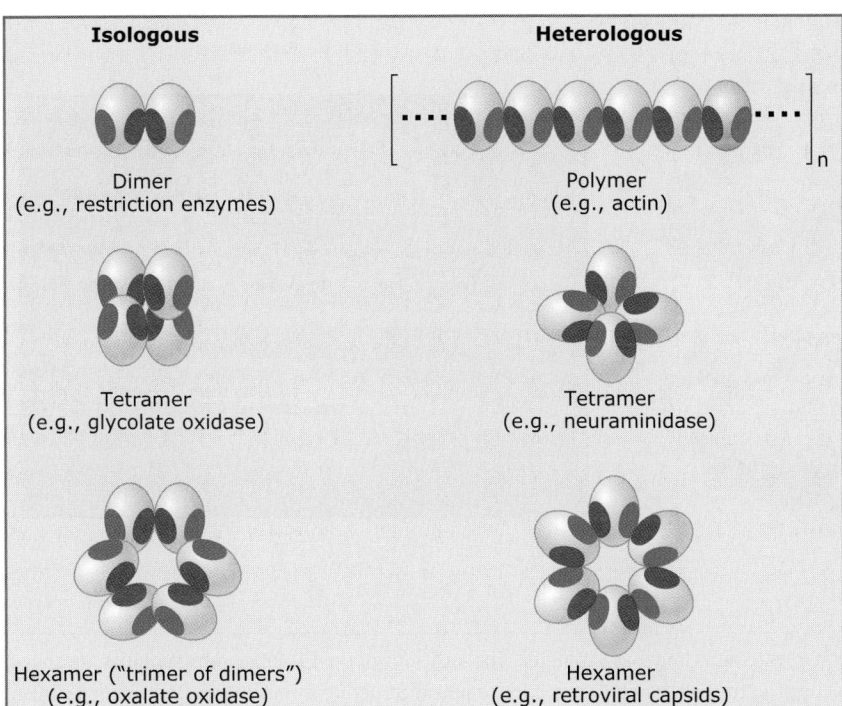

Isologous	Heterologous

Dimer
(e.g., restriction enzymes)

Polymer
(e.g., actin)

Tetramer
(e.g., glycolate oxidase)

Tetramer
(e.g., neuraminidase)

Hexamer ("trimer of dimers")
(e.g., oxalate oxidase)

Hexamer
(e.g., retroviral capsids)

FIGURE 5.48 Isologous and heterologous tertiary structures.

high-molecular-weight complexes that carry out the fundamental process of gene transcription into mRNA. Prokaryotic RNA polymerases have a basic structure composed of five subunits ($\alpha_2\beta\beta'\omega$; the subscript denotes that the α-subunit is, itself, a dimer) that carry out the core activities of DNA binding, RNA synthesis, and translocation along the DNA template. The stable, core assembly binds nonspecifically to DNA. Binding of an accessory subunit, however (α, which directly interacts with DNA in a sequence-specific manner), is required to allow the RNA polymerase holoenzyme to be able to recognize sites on chromosomal DNA at which transcription should begin (called promoter regions). Through the binding of different σ subunits, specificity for different classes of promoters (with different DNA sequences) can occur, allowing precise temporal regulation of transcription of particular classes of genes (**FIGURE 5.49**). RNA polymerase is just one example of an asymmetric quaternary arrangement of multiple protein subunits that shows both hetero- and homotypic character. As we will now see, however, symmetrical quaternary arrangement has been exploited by evolution in a number of ways.

Pyruvate dehydrogenase is a three-enzyme complex comprising E1 (pyruvate dehydrogenase), E2 (dihydrolipoyl transacetylase), and E3

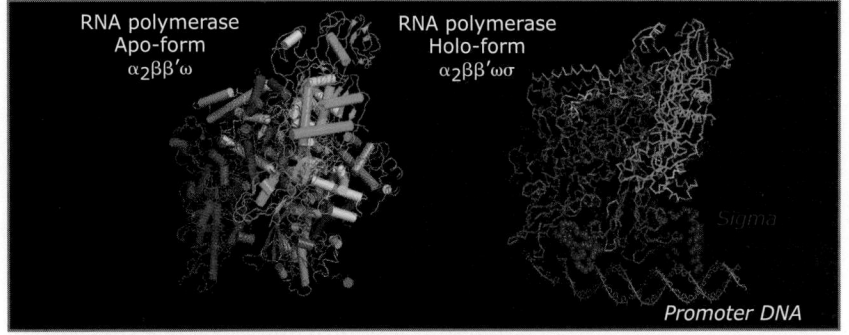

RNA polymerase
Apo-form
$\alpha_2\beta\beta'\omega$

RNA polymerase
Holo-form
$\alpha_2\beta\beta'\omega\sigma$

Sigma

Promoter DNA

FIGURE 5.49 Promoter specificity of RNA polymerase may be regulated by binding of sigma factors that supply additional, sequence-specific DNA interactions. Left structure from Protein Data Bank 1HQM. L. Minakhin, et al., *Proc. Natl. Acad. Sci. USA* 98 (2001): 892–897. Right structure from Protein Data Bank 1L9Z. K. S. Murakami, et al., *Science* 296 (2002): 1285–1290.

(dihydrolipoyl dehydrogenase), which catalyze five distinct reactions in a pathway that leads to oxidative decarboxylation of pyruvate to acetyl CoA. The bacterial enzyme consists of a core of eight E2 trimers that interact to form the corners of a cube of approximate dimensions $80 \times 80 \times 80 \text{ Å}^3$ (**FIGURE 5.50**). The cube is further decorated with twelve E1 dimers and six E3 dimers to form a complex of around 4.5 MDa. The size and complexity are truly staggering, but considerable advantages are achieved in these large multienzyme aggre-

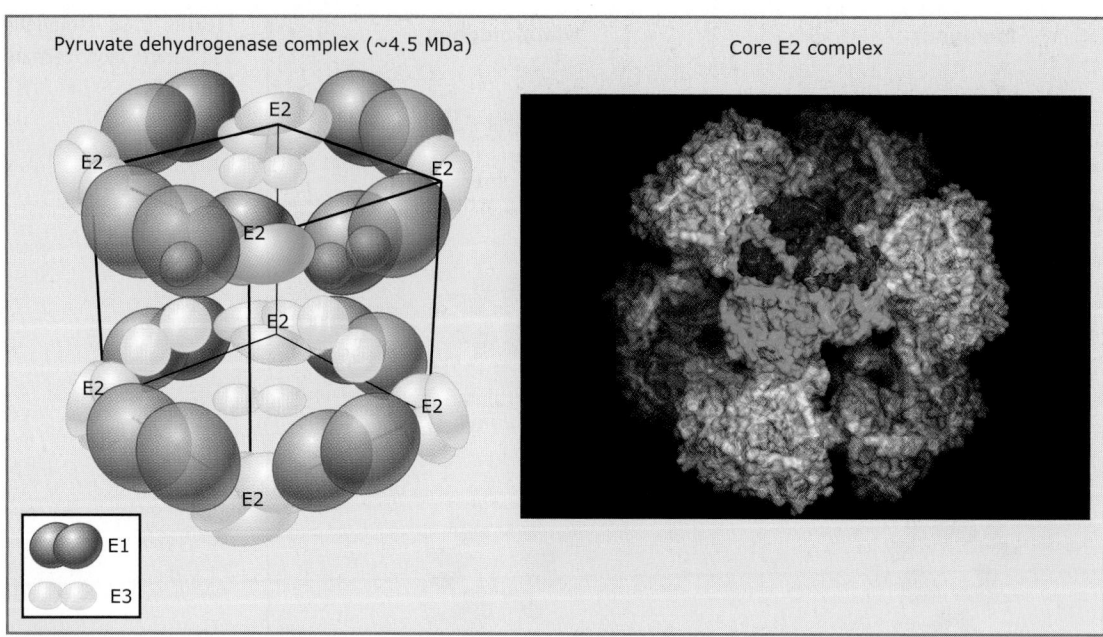

Pyruvate dehydrogenase complex (~4.5 MDa)

E2
E2
E2
E2
E2
E2
E2
E2

E1
E3

Core E2 complex

FIGURE 5.50 The prokaryotic pyruvate dehydrogenase complex. The locations of the E1, E2, and E3 subunits are shown on the left, along with the X-ray structure of the core, cubic E2 complex. Structure from Protein Data Bank 1EAA. A. Mattevi, et al., *Biochemistry* 32 (1993): 3887–3901.

gates. Structural intimacy of different catalytic domains provides opportunities for tight and coordinated regulation through, for example, binding of allosteric effector molecules, or by posttranslational modifications such as phosphorylation. Furthermore, the multiprotein "lattice" generates a cage-like environment that protects the products of one reaction from the unwanted attention of other enzymes and cytoplasmic components, and maintains substrates in close proximity to active sites. Alone or in combination, these effects can provide for considerable enhancement of catalytic rate/efficiency. As mentioned earlier, the oxygen-carrying molecule hemoglobin was among the first protein structures to be determined. Hemoglobin forms an $\alpha_2\beta_2$ heterotetramer of subunits, each of which carries a single heme (Fe-protoporphyrin IX) prosthetic group as the site of binding of diatomic ligands such as oxygen and carbon monoxide. As we will see in *5.13 Dynamics, flexibility, and conformational changes*, the α- and β-subunits are not functionally isolated, but are, in contrast, highly coupled. Loading of oxygen onto successive subunits progressively increases the affinity of the unfilled sites such that the oxygen affinity for the fourth and final subunit is increased several hundred times compared to the uncharged (deoxy) hemoglobin tetramer molecule.

In some cases, the symmetry of a protein in a homomeric complex is correlated with structural characteristics of its binding partner. Restriction endonucleases are the workhorses of molecular biologic techniques due to their absolute specificity for particular sequence motifs in DNA. The major (and arguably the most useful) class of restriction enzymes includes those that recognize "palindromic" motifs for which the sequence of bases of one strand of double-stranded DNA read in the standard 5' to 3' direction is exactly the same when read on the opposite strand (**FIGURE 5.51**). The astute observer will note that this arrangement generates a twofold axis of symmetry at the center of the double-stranded sequence motif. The remarkable ability of restriction endonucleases to cut the phosphodiester backbones of both strands at identical positions in the sequence is achieved straightforwardly by dimerization of the catalytic domains of the enzyme. This generates a twofold axis of symmetry that coincides with that of the DNA substrate in the specific and catalytically competent protein–DNA complex.

An impressive use of symmetry in large protein assemblies is seen in the structures of viral capsids. In all cases, many copies of one or a small number of protein chains are used to build a viral shell that may be 1000 Å in

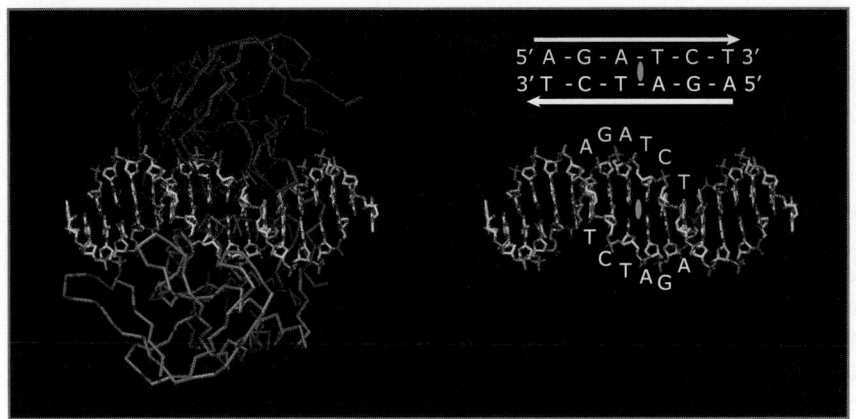

FIGURE 5.51 A twofold symmetric restriction enzyme binds to a twofold pseudosymmetric DNA sequence. Structures from Protein Data Bank 1DFM. C. M. Lukacs, et al., *Nat. Struct. Biol.* 7 (2000): 134–140.

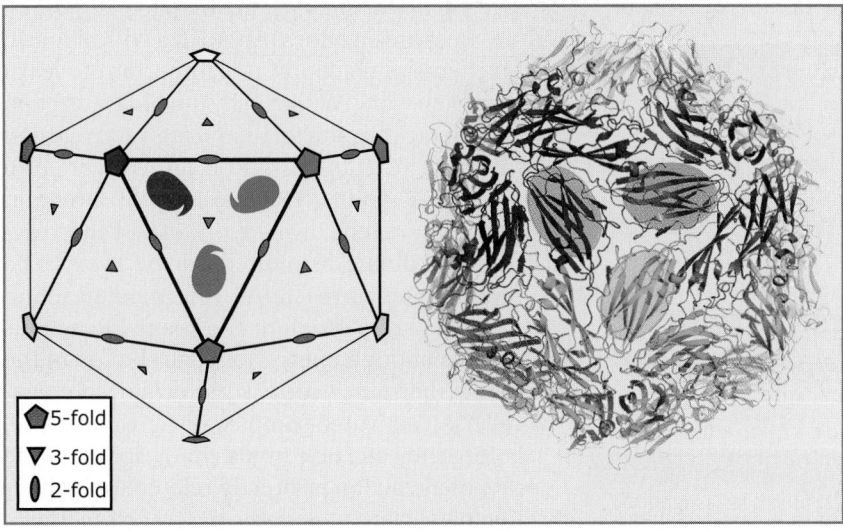

FIGURE 5.52 Icosahedral architecture in a simple (T = 1) satellite virus. Structure from Protein Data Bank 1STM. N. Ban and A. McPherson, *Nat. Struct. Biol.* 2 (1995): 882–890.

diameter or more, which is necessitated by the obvious problem that viral genomes must be relatively small in order to be accommodated within the capsids that they ultimately encode.

In broad terms, viruses can be classified in terms of the architecture of their shells. Enveloped viruses, such as human immunodeficiency virus (HIV) and influenza, are coated with a lipid bilayer, ultimately derived from the plasma membrane of the infected cell within which the virus was replicated and shed. Viral proteins involved in binding to receptors displayed on the surface of target cells are embedded within this bilayer. These viruses will not be considered further.

Of the nonenveloped viruses, two major architectures are seen, which are generally referred to as helical, and icosahedral or spherical. Tobacco mosaic virus is the archetypal helical virus, and its capsid is formed from ~2100 copies of a single protein subunit of 154 amino acids. These are arranged to form a helical rod with 16 copies per turn and a total length of ~3000 Å.

Icosahedral viruses have been by far the most intensively studied by X-ray crystallography and cryo-EM methods. The icosahedron is one of the few ways of symmetrically assembling identical subunits into a roughly spherical shape. It has 20 triangular faces and contains fivefold, threefold, and twofold axes of symmetry (532 symmetry). Each face has threefold symmetry and can thus contain a minimum of three identical subunits (**FIGURE 5.52**). Therefore the simplest icosahedral virus (such as plant satellite viruses) must contain 60 (20 × 3) subunits with each located in an identical environment to the other 59. In order to construct larger shells, the number of subunits must increase and, clearly, this number must be a multiple of 60. In fact, it was suggested by Donald Caspar and Aaron Klug in the 1960s that only certain multiples of 60 are consistent with formation of closed spherical shells (1, 3, 4, 7, 13, …). Caspar and Klug termed these multiples triangulation or "T" numbers. Thus a T = 3 virus (e.g., tomato bushy stunt virus) contains 180 subunits, whereas a T = 13 virus such as reovirus contains 780! Caspar and Klug also invoked the notion of "quasi-equivalence" to explain how the same protein must be able to form more than one type of contact in order to form an icosahedral shell in spherical viruses with T > 1. It was not, however, until the structure of a small T = 3 plant virus— tomato bushy stunt virus or TBSV—was solved that the structural basis of quasi-equivalent packing was revealed.

Concept and Reasoning Checks

1. How does oligomerization favor functional diversity?
2. Why do viruses use such highly symmetric capsid shells made of many copies of only one or just a few different subunits?

5.11 Enzymes are proteins that catalyze chemical reactions

Key concepts

- Enzymes catalyze such vital processes as replication, transcription, and translation.
- Enzymes act on substrates.
- A reaction's energy of activation is the amount of energy needed to overcome the energy barrier to form a transition state.
- Enzymes increase reaction rate by lowering the energy of activation.
- Since the reaction rate is directly proportional to the enzyme-substrate concentration, the theoretical maximum rate of reaction occurs when all enzyme molecules are in complex with substrate.
- The turnover number is the number of substrate molecules converted to product by an enzyme in a given amount of time.
- The selectivity of enzyme catalysis is due to specificity in the active site; this specificity can be conferred by the lock-and-key mechanism (the shape of the active site is complementary to the shape of the substrate), or by induced fit (the shape of the active-site changes to fit the substrate subsequent to binding).
- The catalytic properties of enzymes are intimately associated with the three-dimensional structure.

Virtually all chemical reactions that take place in the cell are catalyzed by enzymes or RNA catalysts called ribozymes. Enzymes catalyze the vast majority of the reactions involved in DNA, RNA, and protein metabolism. A large number of different enzymes are required to catalyze a complex process such as replication, transcription, or translation.

The molecule on which an enzyme acts is its substrate. Only a small number of substrate molecules, sometimes only one, participate in a single catalyzed reaction. Enzymatic activity is measured by following the rate of substrate breakdown or product formation. Several different physical techniques can be used to monitor an enzyme-catalyzed reaction. The two most common techniques used are colorimetric and radiotracer techniques (see the *Methods and Techniques* box on page 206).

Because enzymes play such a critical role in biochemical reactions, it is important to learn how they work. We begin by drawing a reaction coordinate diagram for an uncatalyzed reaction in which reactants A and B are converted to products C and D (**FIGURE 5.53**; plot shown in red). The extent of reaction, called the reaction coordinate, is plotted on the x-axis and the energies of reactants, intermediates, and products are plotted on the y-axis. Reactants require enough energy to reach the top of the energy barrier to form a molecular complex, called an activated complex or transition state, before they can be converted to products. The rate of a reaction is directly related to the fraction of reactant molecules that reach the transition state in a given period of time and depends on the energy of activation (E_a), that is, the energy needed to reach the transition state. A catalyst increases the reaction rate by providing an alternative reaction path with a lower energy of activation (Figure 5.53, plot shown in blue). Catalysts do not change the equilibrium position for a reaction, only the rate at which the reaction takes place.

The catalytic power of enzymes exceeds all man-made catalysts. A typical enzyme accelerates a reaction 10^8- to 10^{10}-fold, though there are enzymes that increase the reaction rate by a factor of 10^{15}. Enzymes are also highly specific in that each catalyzes only a single reaction or set of closely related reactions. Several different but nonexclusive hypotheses have been proposed to explain the catalytic properties of enzymes (see the *Historical Perspectives* box on page 207).

In any enzyme-catalyzed reaction, the enzyme, E, always combines with its substrate, S, to form an enzyme-substrate complex, ES, which can then either dissociate to reform sub-

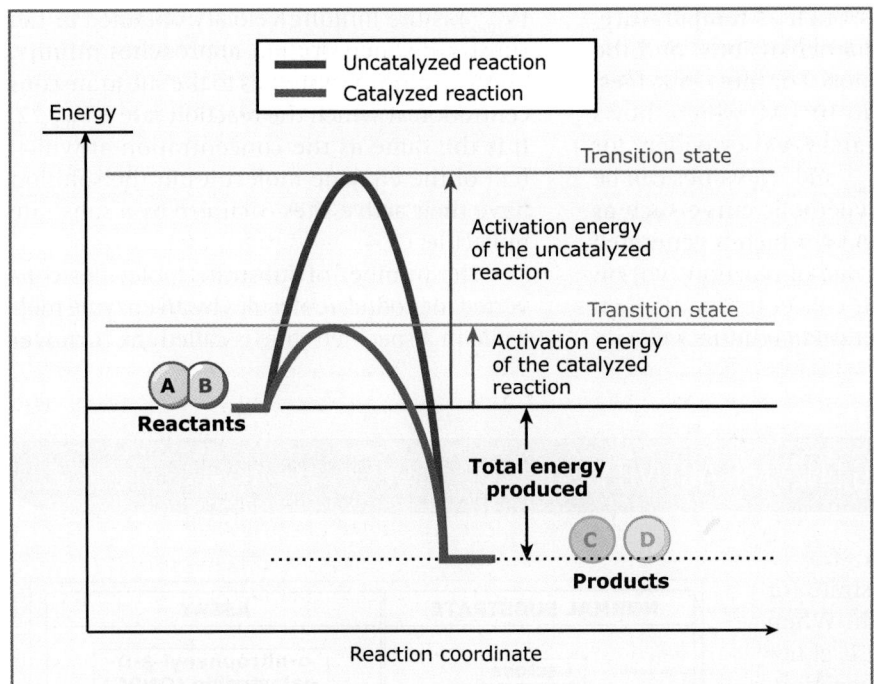

Legend:
— Uncatalyzed reaction
— Catalyzed reaction

Energy

Transition state

Activation energy of the uncatalyzed reaction

Transition state

Activation energy of the catalyzed reaction

A B
Reactants

Total energy produced

C D
Products

Reaction coordinate

FIGURE 5.53 Reaction coordinate diagram for an uncatalyzed reaction (red) and an enzyme catalyzed reaction (blue). The enzyme lowers the activation energy but does not affect the energy released by the reaction. The enzyme therefore increases the reaction rate but does not affect the equilibrium.

strate or go forward to product, P, and enzyme, E. After the ES-complex forms, the substrate is usually altered in some way that facilitates further reaction. The process can be summarized by the following equation:

$$E+S \underset{k_{-1}}{\overset{k_{+1}}{\rightleftarrows}} ES \xrightarrow{k_{+2}} E+P$$

where k_{+1}, k_{-1}, and k_{+2} are rate constants for the reaction. (By convention, kinetic constants for forward reactions k_{+1} and k_{+2} have a (+) symbol in their subscript, and kinetic constants for reverse reactions (k_{-1}) have a (−) symbol in their subscript.)

The rate of reaction is directly proportional to the ES concentration. Hence, the theoretical maximum rate of reaction (V_{max}) is observed when all of the enzyme molecules are present in enzyme-substrate complexes. The ratio (k_{-1} + k_{+2})/k_{+1} is called the Michaelis constant (K_M). For most enzymatic reactions, formation of ES is reversible in the sense that ES can dissociate, yielding free E and free S.

Usually, dissociation of the ES-complex is more rapid than conversion of the complex to enzyme and product. When this is the case, the value of K_M is a measure of the strength of the ES binding. That is, when $k_{-1} \ggg k_{+2}$, K_M approaches k_{-1}/k_{+1}, the dissociation constant

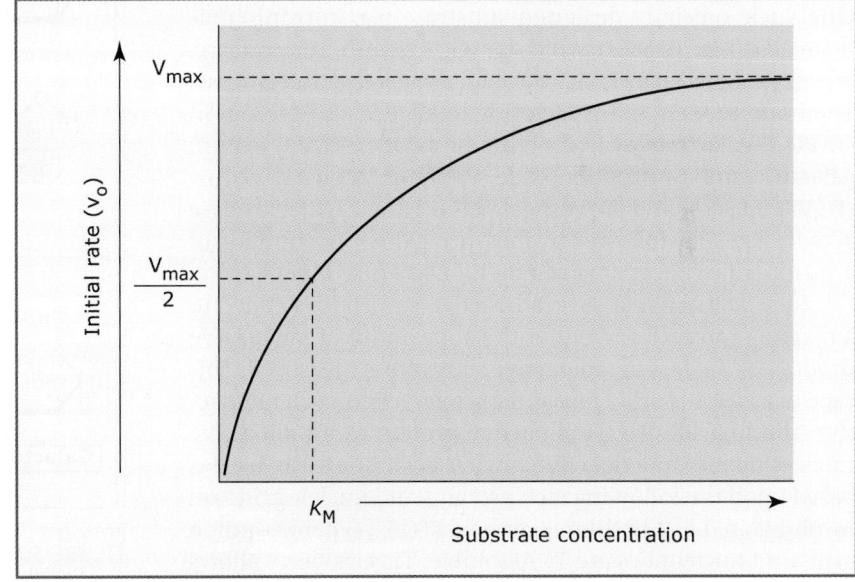

FIGURE 5.54 Calculation of V_{max} and K_M. V_{max} and K_M values can be determined from a hyperbolic curve generated by plotting the initial rates of reaction, v_o, on the y-axis, and substrate concentration on the x-axis. The theoretical maximum velocity, V_{max}, is the limiting velocity obtained as the substrate concentration approaches infinity. The K_M value corresponds to the substrate concentration at which the reaction rate is $V_{max}/2$.

for enzyme-substrate and K_M is a measure of an enzyme's affinity for its substrate. A high value of K_M indicates low affinity (and weak binding), and a low value of K_M means high affinity (and strong binding). Strength of binding depends on several conditions, such as temperature, pH, the presence of particular ions, and the overall ion concentration. For most enzymes, K_M ranges from 10^{-6} to 10^{-1} M, which shows that the affinity of E and S varies widely for different enzymes. V_{max} and K_M values can be determined from a hyperbolic curve such as that shown in **FIGURE 5.54**, which is generated by plotting the initial rate of reaction (vo) on the y-axis and substrate concentration ([S]) on the x-axis. The theoretical maximum velocity (V_{max}) is the limiting velocity obtained as the substrate concentration approaches infinity. The K_M value corresponds to the substrate concentration at which the reaction rate is $V_{max}/2$. It is the same as the concentration at which half of the enzyme molecules in the solution have their active sites occupied by a substrate molecule.

The number of substrate molecules converted to product molecules by an enzyme molecule in a specified time is called the turnover

METHODS AND TECHNIQUES: MEASURING ENZYME ACTIVITY

Colorimetric assays are based on the fact that a substrate (or product) absorbs light of a particular wavelength. When a substance that absorbs visible or ultraviolet light is either a substrate or product of an enzyme-catalyzed reaction, then substrate breakdown or product formation can be followed using a spectrophotometer. For this reason, investigators often work with specially designed substrates that generate products with unique light absorption properties. One such specially designed substrate is α-nitrophenyl-β-D-galactoside (ONPG), which is used to assay β-galactosidase, an enzyme that cleaves lactose to form galactose and glucose. The bond broken, a β-galactoside linkage is also present in ONPG. β-galactosidase hydrolyzes the colorless ONPG to form galactose and *o*-nitrophenoxide, which is intensely yellow (**FIGURE 5.B5** [A] and [B] respectively). Thus, β-galactosidase activity is readily followed by measuring the concentration of *o*-nitrophenoxide at a wavelength of 420 nm (blue light).

In a radioactive assay, radioactive substrate is added to a reaction mixture and either the appearance of radioactive product or the loss of radioactive substrate is measured. This type of assay is used to measure the conversion of a radioactive amino acid into a radioactive protein or a radioactive nucleotide into a radioactive nucleic acid. The assay used is based upon the following fact: proteins and nucleic acids are insoluble in 0.5 *M* trichloracetic acid (TCA), whereas amino acids and nucleotides are TCA-soluble. This property allows protein synthesis to be measured by adding a radioactive [^{14}C] or [^{3}H] amino acid to a reaction mixture that contains the other 19 nonradioactive amino acids and the appropriate enzymes and factors. After a period of time, TCA is added and the mixture is filtered and washed with TCA. The [^{14}C] amino acid is soluble and passes through the filter; any [^{14}C] protein that is made will precipitate and be retained on the filter. Counting the filter-bound radioactivity is then a measure of the extent of reaction.

Synthesis of DNA can also be measured in this way by using a mixture containing the four deoxynucleotide precursors

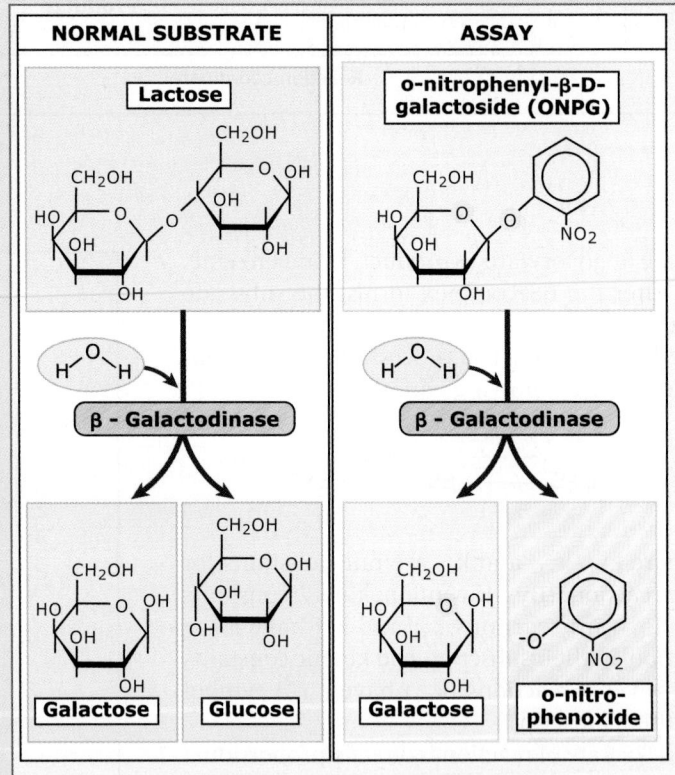

FIGURE 5.B5 β-galactosidase assay. β-galactosidase converts (A) its natural substrate lactose to galactose and glucose and (B) a synthetic substrate o-nitrophenyl β-D-galactoside to galactose and nitrophenoxide, which has an intense yellow color.

of DNA, one of which is radioactive, and other appropriate components of the mixture. For example, when creating a mixture using [^{14}C] thymidine triphosphate (TTP), after the reaction, TCA is added and the mixture is filtered. The [^{14}C] TTP passes through the filter but any DNA that has been synthesized is retained on the filter.

number. The turnover number is determined by dividing V_{max} by the molar concentration of the total enzyme present in the reaction mixture. The site on the enzyme at which the substrate binds and the chemical reaction takes place is called the active site. The extraordinary selectivity in enzyme catalysis is almost entirely a result of specificity of enzyme-substrate binding. The lock-and-key and induced fit mechanisms have been proposed to explain how enzymes bind their substrates (**FIGURE 5.55**).

In the lock-and-key mode, the shape of the active site of the enzyme is complementary to the shape of the substrate (Figure 5.55 [A]). In the induced fit mode, the enzyme changes shape upon binding the substrate and the active site has a shape that is complementary to that of the substrate only *after* the substrate is

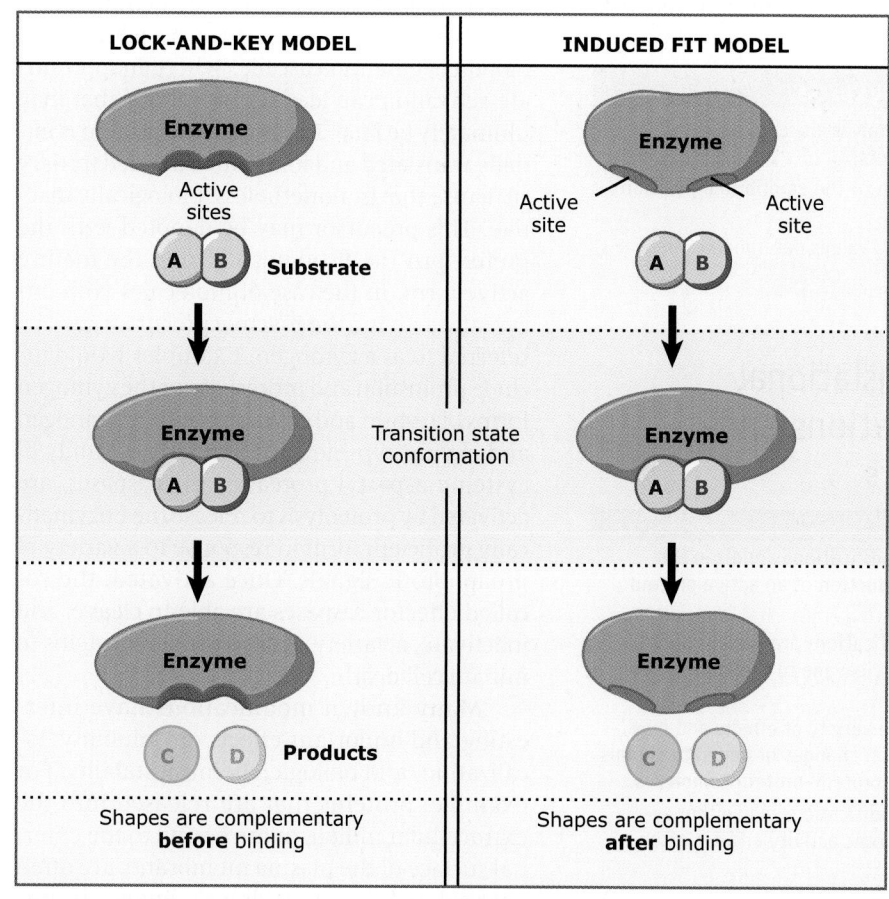

FIGURE 5.55 Enzyme-substrate interaction. There are two major mechanisms of enzyme binding, (a) lock-and-key and (b) induced fit. In the lock-and-key model, the shape of the active site of the enzyme is complementary to the shape of the substrate. In the induced fit mode, the enzyme changes shape upon binding the substrate and the active site has a shape that is complementary to that of the substrate only after the substrate is bound.

HISTORICAL PERSPECTIVES: THE ORIGINS OF ENZYME CATALYSIS

- Enzymes lower the energy of activation by stabilizing the transition state. This hypothesis is supported by studies that show IgG molecules, which are formed when animals are injected with a stable analog of the transition state, act as catalysts.
- Enzymes lower the energy of activation by putting a strain on a susceptible bond.
- Distortion of the susceptible bond makes it easier to break the bond.

- Enzymes lower the energy of activation for reactions involving two or more substrates by holding the substrates near to each other and in the proper orientation.
- Enzymes lower the energy of activation by forming a covalent bond with a reactant molecule that destabilizes some other bond.
- Enzymes lower the energy of activation by acting as proton donors and acceptors.

bound. Although the induced fit model is primarily concerned with the change in enzyme shape, it is important to note that substrates often change shapes when they bind to enzymes (Figure 5.55 [B]). For every enzyme-substrate interaction examined to date, one of these two mechanisms applies. It is often the case, though, that the substrate itself undergoes a small change in shape. In fact, the strain to which the substrate is subjected is often the principal mechanism of catalysis; that is, the substrate is held in an enormously reactive conformation.

Concept and Reasoning Checks

1. How do enzymes catalyze the conversion of substrates to products?
2. What do enzymes do to the equilibrium position of a reaction?
3. How are K_M and V_{max} values determined experimentally?

5.12 Posttranslational modifications and cofactors

Key concepts

- Posttranslational modifications are often the final step in the production of an active protein or enzyme.
- Many different modifications are known, ranging from proteolytic cleavage to the addition of chemical groups.
- Modifications exert a variety of effects and may induce conformational changes or generate signals for the formation of protein–protein complexes.
- Posttranslational modifications may either activate or inhibit biologic activity and are often reversible.

So far, we have seen how proteins are produced as amino acid chains that fold into a myriad of different tertiary and quaternary structures. In some cases, further processing of the polypeptide chain must occur and this can happen in several ways. For most proteins, posttranslational chemical modifications are the rule rather than the exception and can have a variety of different effects on the structure of proteins and many aspects of their behavior. Indeed, a single modification may exert all of these effects on certain proteins, or a number of modifications of a single protein may occur. Furthermore, these need not occur at the same time and may take place at different stages of a protein's lifetime. An exhaustive description of all modifications known at present is unrealistic here. Instead, this section will focus on some specific examples in order to give a flavor of the biologic and structural versatility and flexibility that they provide.

Proteolysis is one of the most dramatic posttranslational modifications, and is seen in the activation of some precursors of proteases, hormones, viral polyproteins, and other molecules (FIGURE 5.56). Functionally, the necessity for proteolytic activation most often is related to a need for precise and timely regulation. This is no better demonstrated than by proteases of the blood coagulation cascade, where inappropriate activation can lead to thrombosis that may ultimately be fatal. In general, the protein is initially translated and folds into a defined tertiary structure that is, nonetheless, biologically inactive. This precursor may be denoted with the prefix "pro" to distinguish it from the mature active form. In the case of molecules with enzymatic activity, the inactive precursor may be referred to as a zymogen. Examples would include proinsulin and procaspase or the zymogen forms of trypsin and chymotrypsin (trypsinogen and chymotrypsinogen). Similarly, a family of cysteine-aspartyl proteases, the caspases, are activated by proteolysis to release the enzymatically proficient form in response to a variety of proapoptotic signals. Once activated, the so-called effector caspases are able to cleave, and inactivate, a variety of downstream proteins to initiate cell death.

Many known modifications have interesting and important effects on solubility, localization, and biologic/chemical stability. For example, proteins that are released into the extracellular milieu, or are bound to the external surface of the plasma membrane, are often glycosylated on asparagine (N-linked) or serine (O-linked) residues (FIGURE 5.57). This may directly aid in protein folding, protect against proteolytic attack, provide immune surveillance, and generate binding sites for interacting partners. This diversity of effects arises, in large part, from the complexity of glycosylation patterns that may involve a number of different sugars and glycosidic linkages. The extent of the modification may be so great as to represent 40% or more of the overall mass of the glycoprotein. This, plus the heterogeneity of the attached sugar chains, provides an enormous barrier to high-resolution structural studies. In most cases, crystallization of highly glycosylated molecules is only possible after extensive enzymatic deglycosylation, removal of known

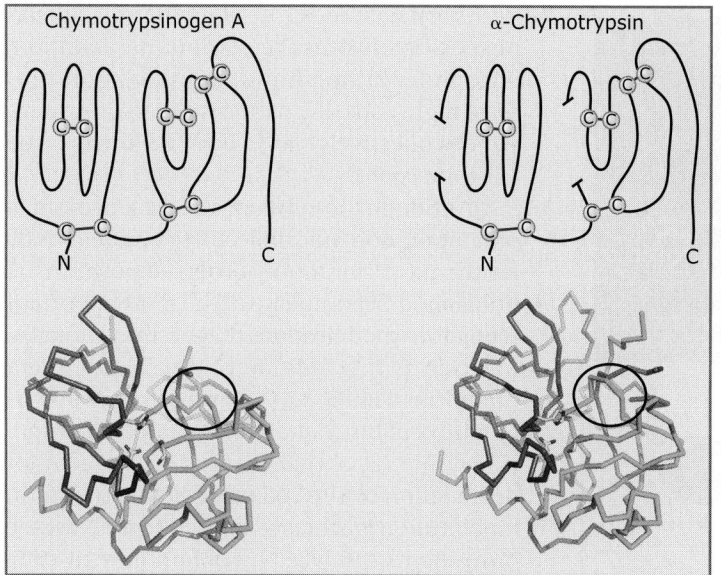

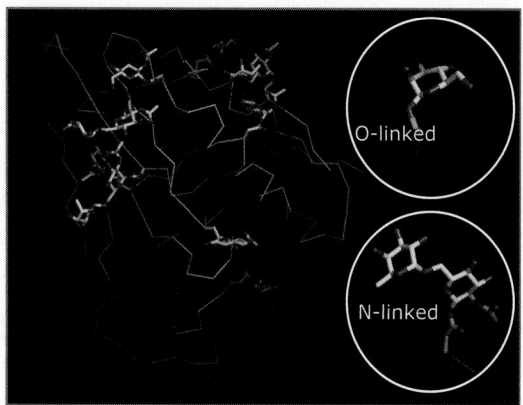

FIGURE 5.56 Proteolytic processing of an inactive zymogen (chymotrypsinogen) to form the active enzyme, α-chymotrypsin. Left structure from Protein Data Bank 1CHG. S. T. Freer, et al., *Biochemistry* 9 (1970): 1997–2009. Right structure from Protein Data Bank 4CHA. H. Tsukada and D. M. Blow, *J. Mol. Biol.* 184 (1985): 703–711.

FIGURE 5.57 O- and N-linked glycosylation. Structures from Protein Data Bank 1GC1. P. D. Kwong, et al., *Nature* 393 (1998): 648–659.

glycosylation sites by mutagenesis, inhibition of glycosylation during recombinant protein expression, or a combination of all three.

Of greatest interest here are the variety of modifications that directly or indirectly influence protein structure and activity. Phosphorylation most often occurs on the hydroxylated amino acids, tyrosine, threonine, and serine. It can, however, occur on histidine and aspartate residues, but in these cases is highly unstable. It is the most prevalent posttranslational modification that occurs in human cells, and it has been estimated that ~30% of all proteins in the human proteome are phosphorylated at some stage during their lifetime! The role of phosphorylation in driving conformational change is best exemplified by protein

kinases themselves, and we have already seen an example of how this occurs in the Src kinase family (Figure 5.44). This example also highlights the fact that phosphorylation can create binding sites for interaction with other proteins/domains capable of specifically recognizing phosphorylated serine, threonine, or tyrosine (see *5.14 Protein–protein and protein–nucleic acid interactions*). The paradigm for phosphorylation-driven complex formation is undoubtedly the SH2 domain (Src-homology-2), which features prominently in receptor tyrosine kinase signaling pathways through its ability to specifically bind to phosphotyrosine motifs. It is now clear, however, that a diversity of proteins and domains function in all aspects of protein kinase signaling. This is best exemplified by the proliferation of protein modules now known to function as phospho-dependent binding domains in serine/threonine kinase signaling pathways. Additionally remarkable is the diversity in architecture seen in these molecules, which ranges from the all-helical 14-3-3 family through to the all-beta Forkhead-associated domains.

Phosphorylation is not unique in its ability to mediate protein–protein interactions and several other modifications, notably acetylation and methylation of the basic amino acids lysine and arginine, and its ability to stimulate interactions with a number of domains such as Tudor, PHD, and Bromo domains, most notably in the context of modification of histone tails in epigenetic regulation of chromatin structure (see *5.14 Protein–protein and protein–nucleic acid*

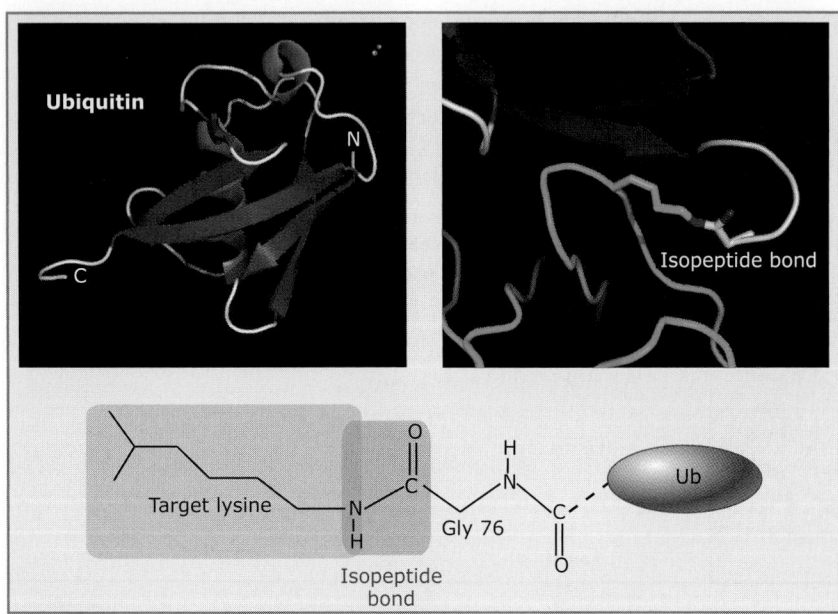

FIGURE 5.58 Ubiquitinylation and the isopeptide bond. Structures from Protein Data Bank 1AAR. W. J. Cook, et al., *J. Biol. Chem.* 267 (1992): 16467–16471.

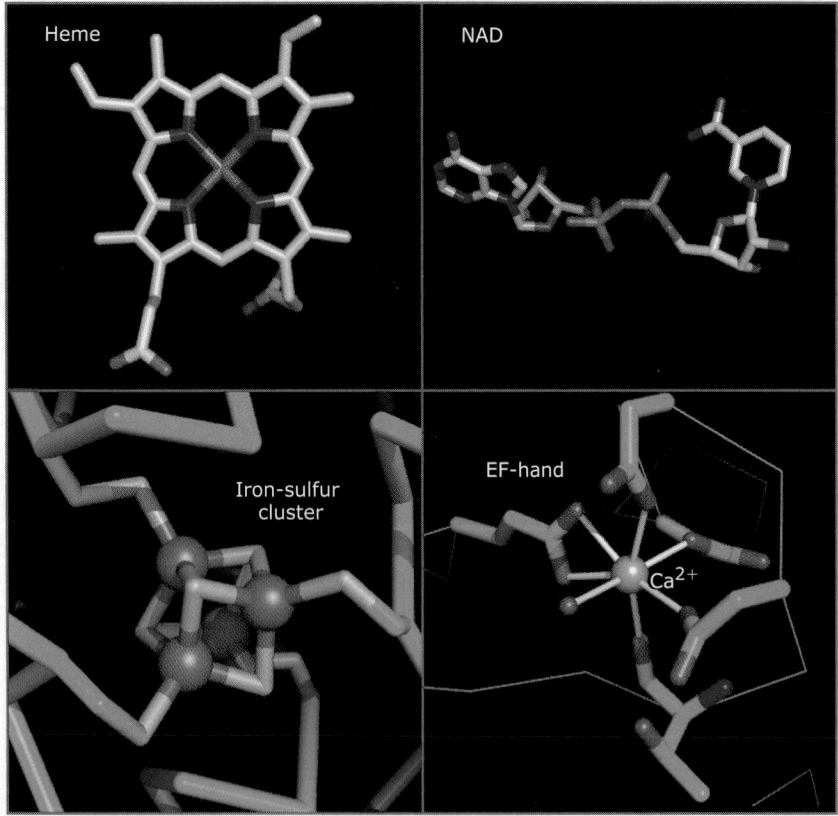

FIGURE 5.59 Examples of four prosthetic groups: heme (iron-protophorphyrin IX), NAD, an iron-sulfur cluster, and a calcium-binding "EF" hand. Structures from Protein Data Bank 1MBN. H. C. Watson, *The stereochemistry of the protein myoglobin.* Protein Data Bank 1OG3. H. Ritter, et al., *Biochemistry* 42 (2003): 10155–10162. Protein Data Bank 1CP2. J. L. Schlessman, et al., *J. Mol. Biol.* 280 (1998): 669–685. Protein Data Bank 1CLL. R. Chattopadhyaya, et al., *J. Mol. Biol.* 228 (1992): 1177–1192.

interactions). Indeed, the combinatorial effects of specific histone acetylation, methylation, phosphorylation, and ubiquitinylation produce highly specific patterns of modifications, which collectively have become known as the "histone code."

Although the covalent attachment of small organic or inorganic molecules to proteins is by far the most commonly observed posttranslational modification, one of the most important regulatory modifications that occurs in eukaryotic cells is the addition of the small protein ubiquitin (**FIGURE 5.58**). Formation of polyubiquitin chains that are conjugated via a specific lysine (Lys48) flags the target protein for degradation by the 26S proteasome. Ubiquitin modification is achieved in a sequential cascade of reactions catalyzed by ubiquitin-activating enzyme (E1). Modification of ubiquitin-conjugating enzyme (E2) and ubiquitin ligase (E3) results in the formation of an isopeptide bond between the activated C-terminal carboxylate group of ubiquitin and the terminal amino group of one or more lysines in the target protein. More recently, it has become apparent that monoubiquitination, or polyubiquitination through Lys63, may play a role in cell signaling. This notion is supported by the fact that ubiquitin-binding proteins exist and the discovery of deubiquitinating enzymes that ensure that this modification is reversible in a manner akin to protein phosphorylation.

Among the most common modifications encountered is the binding of cofactors or **prosthetic groups** that play central roles in biologic function (**FIGURE 5.59**). The diversity of prosthetic groups is considerable, ranging from single metal ions to large organic macrocycles. In the case of the hemoglobins and myoglobins, the heme prosthetic group itself contains a centrally coordinated iron atom that constitutes the binding site for oxygen. Indeed, the binding of various metal ions is observed to occur in a variety of different contexts, where they may play direct catalytic roles, as in phosphoryl transfer (such as Mg^{2+} and Mn^{2+}) and electron transfer processes (metalloporphyrins/iron-sulfur clusters), or may contribute to structural integrity as observed in Zinc-finger (Zn^{2+}) and EF-hand motifs (Ca^{2+}). Other common prosthetic groups include nicotinamide adenine dinucleotide and its reduced form (NAD/NADH), nicotinamide adenine dinucleotide phosphate and its reduced form (NADP/NADPH), flavin adenine dinucleotide (FAD), and flavin mononucleotide (FMN), which are present in many enzymes involved in intermediary metabolism and other pathways.

Finally, one of the most remarkable post-translational modifications occurs in a family of fluorescent proteins well known to modern cell biologists. These are single-domain molecules that form a β-barrel tertiary fold. GFP is the archetypal member of this family, and was originally purified from a species of jellyfish, *Aequoria victoria*. Its fluorescent property derives from the formation of a fluorophore within the hydrophobic core, by means of a series of reactions involving a triplet of conserved Ser-Tyr-Gly amino acids that result in rapid cyclization of the main-chain between the serine and glycine residues, along with a slow oxidation of the tyrosine side-chain (**FIGURE 5.60**). The structural stability of GFP has allowed extensive modification of its fluorescent properties by site-directed mutagenesis through alteration of the local environment of the fluorophore

and even changing its structure. These mutational variants, along with fluorescent proteins identified and cloned from other organisms, have provided a veritable arsenal of molecules with different excitation and emission spectra. These, in turn, constitute powerful tools for investigation of the subcellular localization and interactions of proteins to which GFP and its multicolored offspring have been fused.

Concept and Reasoning Checks

1. List two posttranslational modifications that can sponsor protein–protein interactions.
2. What are the physiological advantages of post-translational proteolytic processing?

5.13 Dynamics, flexibility, and conformational changes

Key concepts

- Proteins are often mistakenly construed to be rather static and rigid structures. Structural information derived from NMR, other solution spectroscopy, and even X-ray crystallography, however, has shown that proteins are highly dynamic.
- A wide variety of atomic motions have been observed in proteins, and may occur on a broad range of time scales.
- The dynamics of a protein can be, and most often are, intimately linked to their biologic function.
- Conformational changes within a single protein or a multiprotein complex can involve fluctuations in chemical bonds, amino acid side-chain motions, or large movements of domains or entire proteins within a complex.

The small oxygen-storage protein, myoglobin, has been called the "hydrogen atom" of molecular biology, reflecting its prominence in some of the major developments over the last 50 years. Mammalian myoglobins are extremely highly conserved proteins, containing 153 amino acids together with a macrocyclic iron-binding cofactor, heme. Myoglobin's importance in our understanding of the role of protein dynamics in biologic function arises from the early observation that the site of oxygen binding on the distal side of the heme (**FIGURE 5.61**) is inaccessible to bulk solvent. Nevertheless, oxygen binding to myoglobin in solution occurs with association kinetics only marginally more slowly than the diffusion limit. This implies that significant conformational displacements of atoms from their positions observed in crystal structures of myoglobin must take place in order that even small, diatomic ligands such as oxygen can

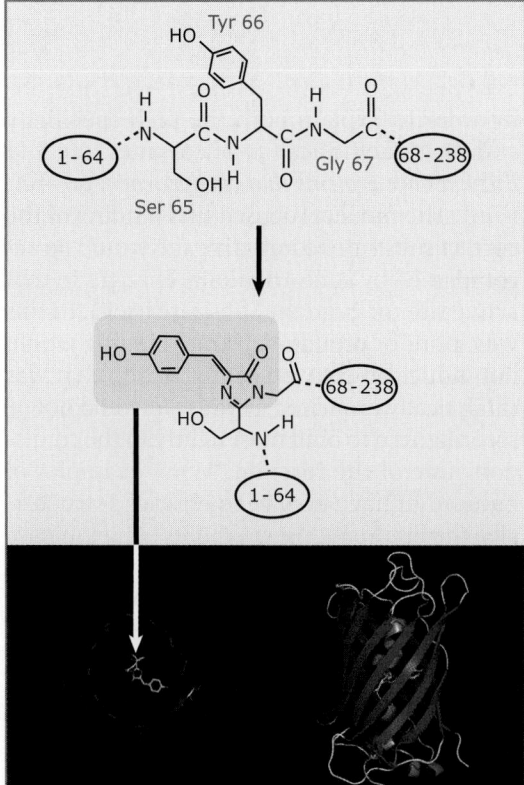

FIGURE 5.60 The GFP fluorophore forms upon folding of the protein itself and remains buried in the hydrophobic core of the β-barrel. Structures from Protein Data Bank 1GFL. F. Yang, L. G. Moss, and G. H. Phillips, Jr., *Nat. Biotechnol.* 14 (1996): 1246–1251.

The structural motions that occur in proteins can be crudely classified as "local" or "global." Local changes involve, for the most part, extremely small movements resulting from thermal fluctuations in covalent bonds that take place on the femto (10^{-15}) to pico-second (10^{-12}) time scale and reflect positional displacements of less than 1 Å. Conformational changes in amino acid side-chains (such as flipping of the aromatic rings of phenylalanine and tyrosine) or the aromatic rings of phenylalanine and tyrosine occur on the millisecond time scale, whereas more extensive "global" motions of secondary structural elements or whole domains can take place on time scales ranging from pico to milliseconds, and sometimes considerably longer.

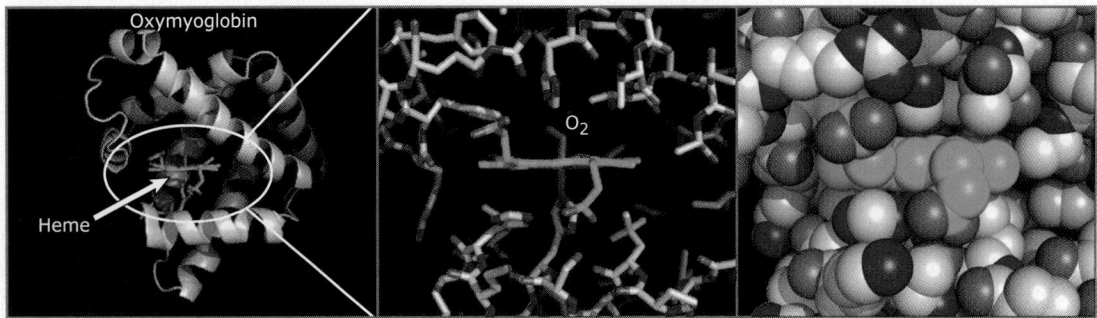

FIGURE 5.61 A conformational barrier to oxygen binding in myoglobin. Structures from Protein Data Bank 1MBO. S. E. Phillips, *J. Mol. Biol.* 142 (1980): 531–554.

bind to the heme iron. Through a variety of computational, biochemical, biophysical, and structural studies on myoglobin and a host of other systems, we now view proteins as existing not as a single rigid structure, but rather as an ensemble of rapidly interconverting conformations (see the *Methods and Techniques* box on page 212). In this way, the motions experienced by atoms within myoglobin are sufficient to open up channels in the structure, allowing access of oxygen and carbon monoxide to the protein interior.

The idea that conformational change and flexibility are important in protein function has been around for many years. The effects of mutational disruption in hemoglobin as a cause of sickle-cell anemia were recognized as a conformational defect long before hemoglobin's structure was finally determined. Indeed, we now recognize that mutation is a major cause of structural change in proteins, which we consider in more detail later (see *5.16 Structure and medicine*).

Some of the major insights into the biologic significance of conformational changes have been from studies of enzyme catalysis. In 1958, Koshland suggested the notion of induced-fit in order to explain both the high specificity and the catalytic activity of enzymes (see *5.11 Enzymes are proteins that catalyze chemical reactions*). The model proposed that binding of the correct substrate to an active site would be accompanied by conformational changes in that active site (or even the substrate itself). In this way, non- or pseudo-substrate binding would not induce an enzyme conformation that was catalytically proficient, because it would not be reconfigured to bind most tightly to the transition state of the reaction. Many examples of induced fit have now been observed structurally, the first being the structures of hexokinase when free and bound to its substrate glucose, which show large changes in the relative orientation of two domains as the enzyme closes around the substrate (**FIGURE 5.62**).

We have already addressed the concept of allosteric conformational changes in the context of the oxygen carrier hemoglobin. Here binding of oxygen to the ferrous heme-iron atom of one subunit results in a shortening of the coordination bond that connects the heme group to the so-called proximal histidine residue. This change in bond length exerts a pull on the F-helix within which the proximal his-

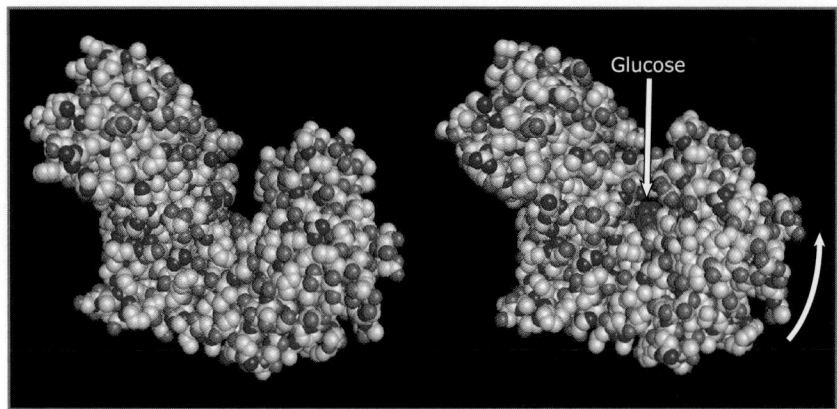

Glucose

FIGURE 5.62 Induced fit in enzyme regulation. Left structure from Protein Data Bank 2YHX. C. M. Anderson, R. E. Stenkamp, and T. A. Steitz, *J. Mol. Biol.* 123 (1978): 15–33. Right structure from Protein Data Bank 1HKG. T. A. Steitz, M. Shoham, and W. S. Bennett, Jr., *Philos. Trans. R. Soc. Lond. B. Biol. Sci.* 293 (1981): 43–52.

tidine resides, and the ensuing conformational change is transmitted through the breaking of salt bridges to the other subunits, raising their oxygen affinity. As such, hemoglobin is a model of allostery, where binding of a ligand or "allosteric effector" causes a conformational change that either positively or negatively affects interactions at a second site. The effect of allostery on the affinity of successive binding sites is known as cooperativity.

Allostery is one of the most common regulatory mechanisms found in biologic systems and is facilitated by quaternary association. The concept of allostery was originated by Monod, Wyman, and Changeux in a classic paper published in the *Journal of Molecular Biology*. The "MWC" or "concerted" model considers allostery as acting on only two symmetric and rigid quaternary structures called the "tense" or "T"-state, which has low ligand affinity, and the "relaxed" or "R"-state, which has higher affinity. These two states are in equilibrium, and binding of a ligand/effector molecule to successive subunits "pulls" the conformational equilibrium toward the high-affinity R-state. Alternatively, Koshland and coworkers described a "sequential" model, which allows for binding of a ligand to one subunit to directly influence the binding site conformation of other subunits in the oligomer. In fact, both models may be necessary to explain the overall behavior of hemoglobin in response to pH (the "Bohr effect"), binding of diatomic ligands (oxygen, carbon monoxide), and allosteric effectors such as diphosphoglycerate.

Allosteric effectors need not always be small molecules. They can be whole proteins, and there are many instances of conformational changes that result from binding of regulatory subunits to activate or even inhibit activity of the resulting complex. As an example, we will

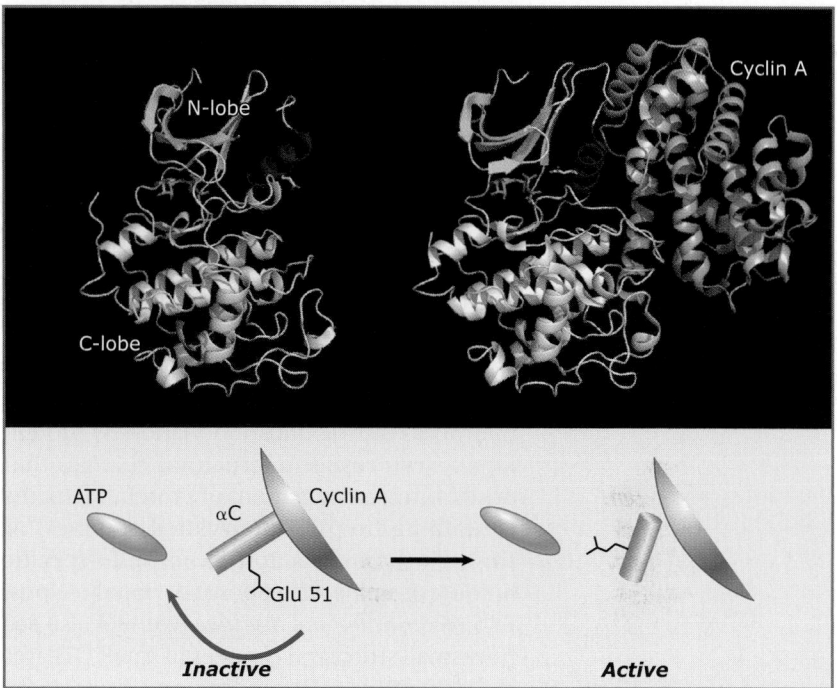

N-lobe

Cyclin A

C-lobe

ATP αC Cyclin A

Glu 51

Inactive *Active*

FIGURE 5.63 Protein–protein and phospho-dependent conformational changes in the activation of cyclin-dependent kinases. Left structure from Protein Data Bank 1HCL. U. Schulze-Gahmen, H. L. De Bondt, and S. H. Kim, *J. Med. Chem.* 39 (1996): 4540–4546. Right structure from Protein Data Bank 1QMZ. N. R. Brown, et al., *Nat. Cell Biol.* 1 (1999): 438–443.

consider the cyclin-dependent kinases (CDKs) that function as master regulators of cell-cycle progression in eukaryotes. As such, their kinase activity must be highly controlled, and this occurs at a number of levels. Each CDK associates with different but specific activating subunits called cyclins at precise times in the cell cycle. The major effect of cyclin binding is a conformational change in the kinase subunit that pushes an α-helix containing a catalytically important glutamate residue into the active site (**FIGURE 5.63**). This is not the whole story, however, because additional events are required for full activation. First, an inhibitory tyrosine phosphorylation must be removed by protein phosphatase. Second, an activating phosphorylation on a specific threonine residue located in the "activation" or "T"-loop is required for it to adopt an ordered conformation required for substrate binding and catalysis. Indeed, the central role of phosphorylation-dependent conformational changes in CDK activation, and kinase activation in general, exemplifies how posttranslational chemical modifications can result in biologically important structural changes.

Possibly the most common driver of conformational changes in proteins is the hydrolysis of nucleoside triphosphates (NTPs), particularly ATP. NTPs are remarkable molecules because they are relatively stable in isolation in spite of the fact that the phosphate–phosphate bonds are referred to as "high-energy": hydrolysis of the bond between the β- and γ-phosphates yields around 12 kcal mol^{-1} of free energy. This occurs efficiently only when hydrolysis of ATP is catalyzed by enzymes called ATPases, an essential characteristic given that spontaneous hydrolysis by water would otherwise render ATP too unstable to be useful!

In earlier sections we have seen how ATP hydrolysis during phosphorylation by protein kinases can result in structural changes that arise from rearrangement of protein segments containing the phosphorylated residues, or through the interaction of phospho-specific binding proteins with phosphorylated regions. In *5.16 Structure and medicine*, we will also see how small structural changes in small GTPases are driven by GTP binding and hydrolysis. ATP hydrolysis can, however, be coupled to large conformational changes, a phenomenon perhaps best exemplified by so-called motor proteins such as myosin, kinesin, and dynein. These molecules most generally produce mechanical movement through interaction with fibrous cellular substructures such as actin filaments and microtubules formed by polymerization of actin and tubulin, respectively.

The most complete structural picture of how the chemical energy of ATP hydrolysis is transformed into mechanical force has emerged from studies of a proteolytic fragment of myosin called S1, which contains the globular "head" or "motor" domain that binds to actin filaments and ATP itself, and a regulatory domain, which The myosin power stroke is often referred to as the lever arm. In full-length myosins, a third "tail" domain is present that is

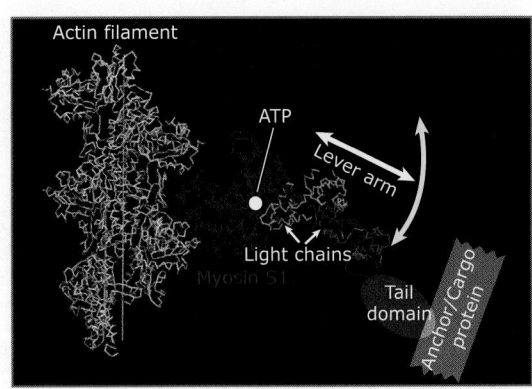

FIGURE 5.64 A model of the myosin power stroke. Structures created from RCSB Protein Data Bank.

responsible for interacting with other cellular proteins. Movement is produced by changes in affinity of the regulatory domain for the actin filament, which is, in turn, coupled to a cycle of ATP binding, hydrolysis to ADP plus inorganic phosphate (P$_i$), and finally, ADP and phosphate release (**FIGURE 5.64**). These changes in affinity result in the successive binding and dissociation of the myosin/actin complex during the cycle. The small conformational changes that occur during the ATP cycle are then transduced to and magnified by the lever arm, an extended coiled-coil structure that is stabilized through calcium-binding EF-hand proteins known as light-chains. The lever arms of different classes of myosins (of which some 20 are currently known) may differ substantially in length. This provides for different lengths of power stroke that are adapted for specific biologic functions.

Concept and Reasoning Checks

1. What was the major observation suggesting that oxygen binding by myoglobin must involve a dynamic process and conformational change?
2. What are the major ways in which conformational changes in proteins are induced?

5.14 Protein–protein and protein–nucleic acid interactions

Key concepts

- The interactions that proteins make define their biologic function and are dictated by their structure, dynamics, and physico-chemical properties.
- Protein interfaces are formed by combinations of polar, nonpolar, and electrostatic interactions whose relative contributions define binding specificity and affinity.

The biologic functions of proteins are exercised through the interactions that they make with other molecules, and the diversity of these interactions can be seen in specific contexts elsewhere in this chapter. This section will address the molecular basis of how protein interactions occur, the nature of the interfaces that have been observed, and the ways that these characteristics regulate binding affinity, stability, and specificity. Although the discussion will focus on protein–protein and protein–nucleic acid complexes, the principles described are equally applicable to interactions of proteins with other ligands, such as sugars, lipids, cofactors, and substrates.

The structure and physico-chemical properties of protein–ligand interfaces are necessarily dictated by the distribution and conformations of the amino acids at that interface and, importantly, by their interactions with surrounding solvent. For this reason, interfaces can be roughly described in terms of the relative contributions of hydrophobic interactions, hydrogen bonding, and electrostatic effects. Quantitatively, the overall size of the interface is often reported in terms of the solvent-accessible surface that is buried or rendered inaccessible to solvent upon complex

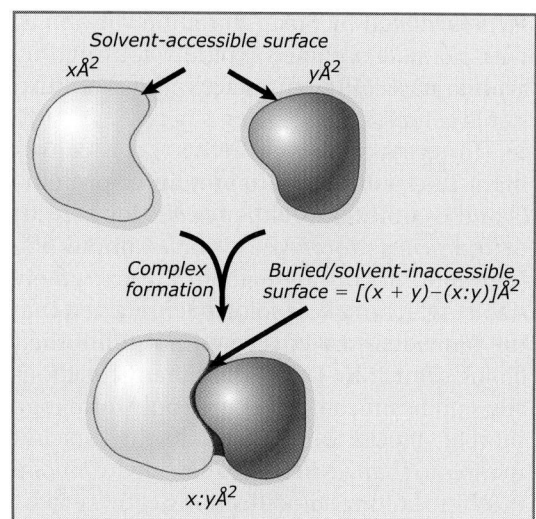

FIGURE 5.65 Solvent-accessible surface is buried at protein interfaces.

formation (**FIGURE 5.65**). Surfaces actually observed in crystal structures show buried surface areas in the range of around 800 $Å^2$ at the lower end up to or exceeding 5000 $Å^2$, with a mean value around 1500 $Å^2$. A more detailed analysis of the binding surface can supply information about the relative extents of interfacial con-

METHODS AND TECHNIQUES: BINDING EQUILIBRIA

Reversible binding of two molecules to form a bimolecular complex is most often described by the affinity of the interaction. The term affinity refers to the equilibrium constant that is defined by the relative concentrations of the bound and free species in a mixture at equilibrium. In thermodynamic terms, the equilibrium constant is related to the Gibbs free energy (ΔG) of binding (**FIGURE 5.B6**), and this must always be negative for a spontaneous interaction. The association equilibrium constant, K_a, has units mol^{-1}, so we will refer to affinity in terms of dissociation constant K_d, which has the more intuitive units of mol. Thus tighter binding complexes have lower K_d values such that an affinity of 1 mM is weaker than 1 μM, and so on. The free energy, ΔG, can also be described as a sum of enthalpic (ΔH) and entropic (ΔS) terms. The relative enthalpic and entropic components of protein interactions can be measured in several ways, but the structural basis for the contributions of each of the two components is often difficult or impossible to delineate. Nonetheless, some general trends are discernable and the ability to predict ab initio how proteins form complexes is, along with the

protein-folding problem, a major goal of computational and experimental biochemists.

For binding of two molecules A and B at equilibrium

$$A + B \rightleftharpoons AB$$

$$K_a = \frac{[AB]}{[A][B]} \; M^{-1}$$

$$K_d = \frac{1}{K_a} = \frac{[A][B]}{[AB]} \; M$$

where [ ] denotes concentration

The association equilibrium constant K_a is related to the Gibbs free energy ΔG by

$$\Delta G = -RT \ln K_a$$

where T is the temperature (in K) and R is the gas constant

ΔG can be expressed as the sum of enthalpic and entropic components:

$$\Delta G = \Delta H - T\Delta S$$

FIGURE 5.B6 Binding, equilibria, and free energy.

tacts mediated by polar and nonpolar atoms, with nonpolar contacts typically contributing around 60% of the interface in high-affinity, stable complexes.

In terms of biologic activity, it is convenient to classify protein–protein or protein–ligand complexes as either *stable* (long-lived) or *transient* (short-lived). For example, the four subunits of hemoglobin are extremely stable, which makes biologic sense given that the individual proteins could not fulfill their biologic function. Conversely, many functionally significant complexes may only form for a short but precise time period, allowing rapid responses to changes in the cellular environment. Posttranslational modifications can "switch" a weak and transient interaction into a tight and stable one. In the absence of such modifications, though, interfaces formed between weak, transiently interacting proteins tend to be small and somewhat less hydrophobic in nature.

Although the numbers of protein–protein and protein–ligand complexes represented in the protein databank continues to increase, it remains difficult or even impossible to accurately predict the affinities of interactions from structure alone. This is because interaction energies that occur between large and complex interfaces consist of enthalpic contributions from van der Waals forces, hydrogen bonding, and electrostatic interactions, along with entropic effects derived from the formation of nonpolar interactions that displace surface water molecules (and therefore increase entropy). Often, a loss of entropy arises from a reduced conformational flexibility upon complex formation. Broadly speaking, the overall size of the interface is loosely correlated with affinity, which has the inevitable consequence that for the interaction of two globular proteins, an increase in affinity can only be achieved by an increase in the overall size of one or both partners. As we will see in *5.15 Function without structure?* some proteins have circumvented this evolutionary "limitation" by utilizing tracts of unstructured sequence to maximize interaction surfaces.

Specificity is one of the most critical features of any interaction and can be usefully defined as the relative affinities of a protein for different binding partners. For this reason, high affinities are not necessarily an indicator of high specificity, and vice versa. As a general rule, it is important to realize that affinities and specificities of interactions in biologic

systems are optimized rather than maximized. Nonpolar interactions are thought to contribute most to overall affinity, whereas specificity is mostly derived from shape complementarity (which may have a nonpolar component) and hydrogen bonding. In this respect, the pioneering work of Alan Fersht in the early 1980s is particularly significant. Using the enzyme tyrosyl tRNA synthetase as a model system, Fersht and coworkers employed the newly developed technique of site-directed mutagenesis to introduce individual amino acid substitutions into the protein and examine the effects on catalysis and specificity. In particular, these studies showed that single uncharged hydrogen bonds contribute relatively little to binding and specificity, whereas those involving a charged donor or acceptor are much more significant.

As mentioned in the preceding discussion of nonpolar interactions, water molecules play an important, albeit indirect, role in the formation of protein-binding interfaces. Ordered water molecules are commonplace in protein–protein and protein–ligand interfaces, where they may mediate linking hydrogen-bonding contacts between side-chains across the binding surface (**FIGURE 5.66**), or fill "holes" and thus increase surface complementarity. How and if they contribute to interaction specificity has been a controversial question that initially gained prominence in the context of protein–nucleic acid interactions. Most structures of complexes solved at high resolution contain interfacial waters, many of which make their full complement of four hydrogen bonds (two H-bond donors and two acceptors). These

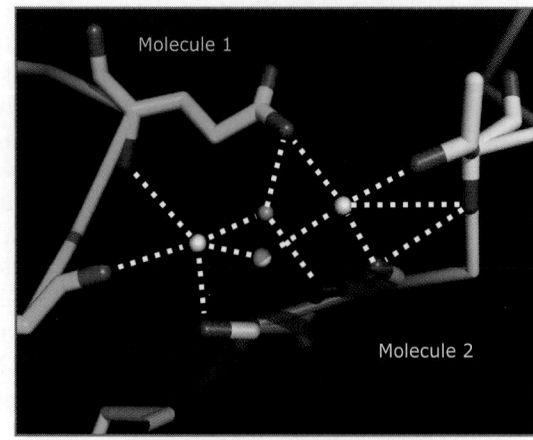

FIGURE 5.66 Interfacial, water-mediated hydrogen bonds. Structure from Protein Data Bank 1TX4. K. Rittinger, et al., *Nature* 389 (1997): 758–762.

water molecules clearly play an important structural role, and in some cases appear to have been conserved through evolution.

Nucleic acids constitute a major class of binding partners for proteins, and these interactions are central to the regulation of the processes of transcription, translation, and DNA replication. The first protein–DNA complexes characterized by X-ray crystallography were those of prokaryotic transcriptional regulators, and these structures showed, for the first time, how proteins are able to exploit the characteristics of B-form DNA in generating specific and high-affinity interactions. The structure of double-stranded DNA, as suggested by Watson and Crick in the 1950s, has become something of an icon. To recap, the two strands of connected nucleotide bases intertwine in an antiparallel arrangement to form a right-handed double-helical structure with the phosphate groups of each nucleobase located on the periphery of the double helix, linking to the 3′ ribose hydroxyl group of the next through a phosphodiester bond. The strands are held together through specific patterns of hydrogen bonds (the familiar base-pairing interactions) of the nucleobases (adenine with thymine, cytosine with guanine) at the center.

From the point of view of a DNA-binding protein, the major structural features of classical B-form double-stranded DNA (dsDNA) are two "grooves" that differ in width and depth (FIGURE 5.67). The minor groove is rather narrow, but the major groove is much wider, allowing access of binding proteins to nonpolar and hydrogen-bonding groups on the edges of the base pairs. In fact, the width of the major groove is ideally suited to accommodate a single α-helix, and many X-ray and NMR structures have shown how helices in the context of helix-turn-helix (HTH), helix-loop-helix (HLH), basic leucine zipper (bZIP), Zn²⁺, and other motifs interact with the major groove (FIGURE 5.68). Major groove recognition, however, is not limited to α-helices, and the structure of another bacterial repressor, MetJ, showed that a pair of antiparallel β-strands can function in a very similar and equally effective way (Figure 5.68). Indeed, it has subsequently become clear that combinations of α, β, and extended structure may be used in combination in protein–DNA recognition to provide specificity through interaction with the base edges, along with increased affinity through largely nonspecific electrostatic interactions with the negatively charged phosphate backbone. It has also emerged that proteins do not only bind to linear DNA duplexes, and major distortions in bound DNA have been observed that range from significant bends in the DNA helical axis to dramatic deformations in the phosphodiester backbone, allowing access to the minor groove itself.

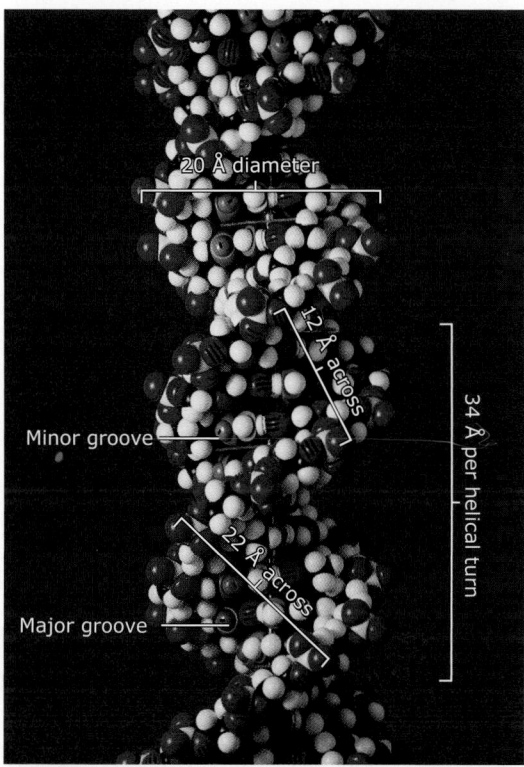

FIGURE 5.67 Structure of double-stranded DNA. © Photodisc.

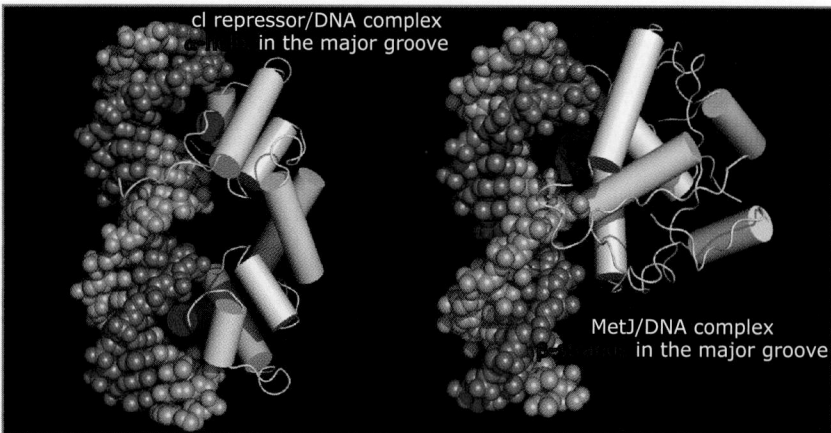

FIGURE 5.68 Interactions with the major groove of DNA by α-helices and β-strands. Left structure from Protein Data Bank 1LMB. L. J. Beamer and C. O. Pabo, *J. Mol. Biol.* 227 (1992): 177–196. Right structure from Protein Data Bank 1MJ2. C. W. Garvie and S. E. Phillips, *Structure* 8 (2000): 905–914.

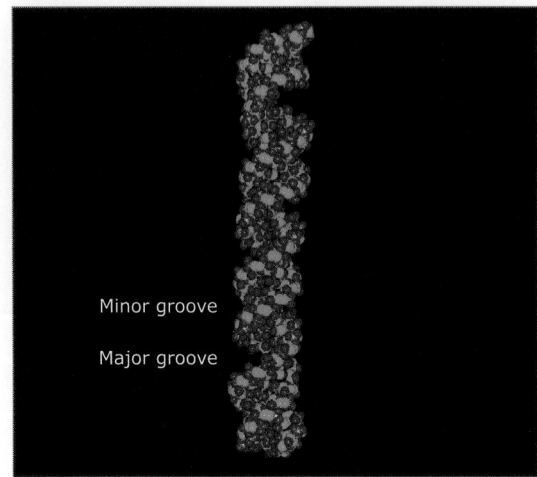

FIGURE 5.69 Structure of double-stranded RNA. Structure from Protein Data Bank 3CIY. L. Liu, et al., *Science* 320 (2008): 379–381.

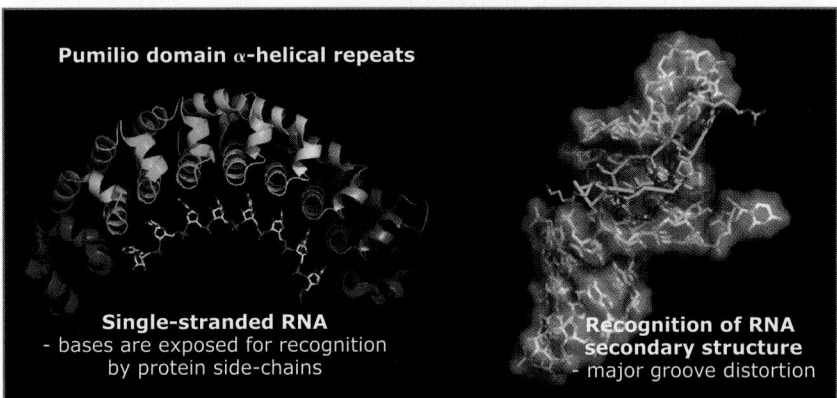

FIGURE 5.70 Sequence- and structure-dependent protein–RNA interactions. Left structure from Protein Data Bank 1M8Y. X. Wang, et al., *Cell* 110 (2002): 501–512. Right structure from Protein Data Bank 1ZBN. V. Calabro, M. D. Daugherty, and A. D. Frankel, *Proc. Natl. Acad. Sci. USA* 102 (2005): 6849–6854.

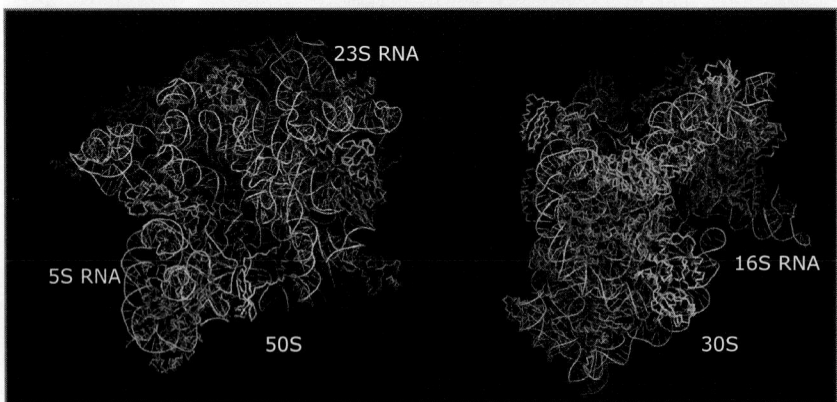

FIGURE 5.71 RNA structure and RNA–protein interactions in the bacterial ribosome. Left structure from Protein Data Bank 1KC8. J. L. Hansen, P. B. Moore, and T. A. Steitz, *J. Mol. Biol.* 330 (2003): 1061–1075. Rights structure from Protein Data Bank 1HNX. D. E. Brodersen, et al., *Cell* 103 (2000): 1143–1154.

In comparison to DNA, RNA presents a much more varied and complex spectrum of problems in molecular recognition by proteins. This arises from the fact that RNA is produced as a single-stranded molecule by transcription from a DNA template. As such, RNA can and does form a bewildering array of secondary and even tertiary structures that are, nevertheless, technically difficult to investigate by either X-ray or NMR methods. Double-stranded RNA regions form a so-called A-form structure in which the minor groove is wider than in B-form DNA, whereas the major grove is deeper but narrower, presenting different structural features to potential binding proteins (**FIGURE 5.69**).

The diversity in RNA structure appears to be matched by the diversity of ways in which proteins interact with it. Single-stranded regions in RNA may allow direct "reading" of the base sequence through interactions with the base-pairing hydrogen-bond donor/acceptors that are otherwise inaccessible in double-stranded RNA or DNA. In addition, proteins may recognize three-dimensional surfaces created by tertiary interactions within any given RNA molecule. Examples of many or all of these strategies are available (**FIGURE 5.70**), but the generalizations that enable us to broadly classify protein–DNA interaction mechanisms are less obvious for protein–RNA binding systems. Suffice it to say that the high-resolution structures of the 50S and 30S subunits of prokaryotic ribosomes revealed a host of new RNA structural motifs within the 23S, 16S, and 5S RNAs. At the same time, these remarkable structures have revealed a host of novel interactions of the ribosomal RNAs with the complement of ribosomal proteins (**FIGURE 5.71**). Nevertheless, even including the information from the structures of ribosomes and their components, protein–DNA complexes dominate the complement of nucleic acid structures available at present and there is much more to be learned about RNA structure and the ways in which proteins and RNA interact.

Concept and Reasoning Checks

1. What kinds of interactions generate specificity and affinity in protein binding to other proteins or ligand molecules?

2. How are nucleobase-specific interactions with B-form DNA achieved?

3. Why are RNA molecules so much more structurally diverse than DNA?

5.15 Function without structure?

Key concepts

- Many protein functions may be carried out by, or depend on, unstructured regions of amino acids.
- In the formation of complexes involving unstructured regions, complexes may remain completely or partially unfolded, or may adopt secondary or tertiary structures upon binding.

To this point we have seen many examples of how the three-dimensional structure of a variety of proteins and protein complexes is exquisitely related to function. It has, however, become increasingly apparent that some proteins with clearly defined and important biologic activities appear to lack any obvious tertiary or even secondary structure. These have become known as "natively unfolded" or "intrinsically unstructured" proteins. These molecules are not as uncommon as one might expect. Although only a few examples have been found or predicted in prokaryotes, it appears that upward of 30% of proteins encoded in eukaryotic genomes may fall into this class. Although some proteins may consist entirely of unstructured regions, a hierarchy of organization exists, with some proteins containing some secondary structure and others containing unstructured regions within which globular domains may be embedded.

Unfolded proteins or regions within proteins are generally characterized by amino acid sequences that are of "low complexity." Such sequences may contain extended tracts of polar residues in many combinations, and such primary structures are not able to adopt a globular fold due to the absence of nonpolar groups that could form a hydrophobic core. Given the remarkable functional characteristics that are bestowed upon proteins by virtue of the secondary, tertiary, and quaternary structures, the advantages of unfolded conformations might appear to be somewhat obscure. Indeed, there would appear to be some obvious disadvantages, most notably sensitivity to proteolytic degradation. Several useful features of disordered regions are, however, discernable, and we will consider them in the context of a few of the structurally and biochemically characterized biologic systems in which the function of natively unfolded proteins has been investigated.

One of the first structures of a complex of a natively unfolded protein (NUP) with a binding partner to be described was that of the cyclin/CDK inhibitor p27 and its cognate kinase cyclinA/CDK2. In fact, this is one of a number of examples of the activity of NUPs in the general area of cell-cycle regulation that seems to be something of a focus for this class of molecules. The X-ray structure shows clearly how p27 wraps around the cyclin/CDK complex and inhibits the enzyme by rearranging the position of the kinase N-lobe and intruding into the ATP binding site (**FIGURE 5.72**). In forming the complex, a large solvent-accessible surface is buried by fewer than 70 residues of p27. This exemplifies the fact the extent of binding interfaces generated by NUPs is much greater than would be possible for a globular protein of the same number of amino acids.

The general importance of posttranslational modifications has been discussed (see *5.12 Posttranslational modifications and cofactors*) and the extended, unstructured nature of natively unfolded regions would be expected to facilitate access when modifying enzymes to target residues. This notion correlates well with the observation that NUPs are prevalent in cell-cycle regulatory mechanisms where a good deal of posttranslational modification, particularly phosphorylation, is employed. In addition to cell-cycle proteins, natively unfolded regions are characteristic of many proteins involved in nucleic acid recognition and transcriptional activation, many of which may also be subject to phosphorylation or other regulatory mechanisms. In terms of DNA binding, a globular domain may mediate sequence-specific recognition whereas an

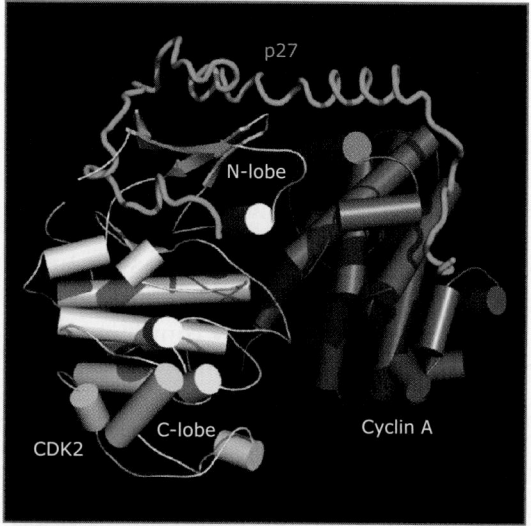

FIGURE 5.72 p27-inhibition of cyclin-CDK. Structure from Protein Data Bank 1JSU. A. A. Russo, et al., *Nature* 382 (1996): 325–331.

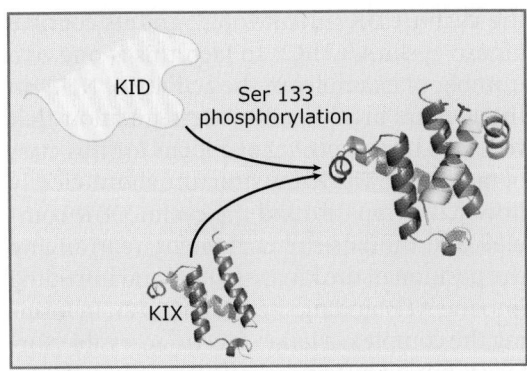

FIGURE 5.73 Phosphorylation-dependent folding-upon-binding in KIX/KID interactions. Structures from Protein Data Bank 1KDX. I. Radhakrishnan, et al., *Cell* 91 (1997): 741–752.

extended, basic "tail" region binds nonspecifically to the phosphate backbone, contributing to the overall interaction affinity. Similarly, transcriptional activation regions such as the classical "acid blob" segment of herpes simplex virus VP16 execute their biologic function in a natively unfolded form. In other transcriptional activator proteins, however, natively unfolded regions may spontaneously fold into globular structures upon interaction with partner proteins or activation targets. For example, the kinase-inducible transcriptional activation domain (KID) of the cyclic AMP-response element binding protein (CREB) binds to the KIX domain of CREB-binding protein (CBP) upon phosphorylation of a specific serine residue, Ser 133. Both the phosphorylated and unphosphorylated forms of KID are natively unfolded in solution but fold into a roughly helix-turn-helix motif upon binding to KIX (**FIGURE 5.73**). As the name implies, KID refolding and binding to CBP is absolutely dependent on phosphorylation, and the extensive hydrogen-bonding interactions between phospho-Ser 133 and KIX residues in the complex presumably provide a favorable enthalpic contribution that allows formation of the folded structure. Similarly, folding-upon-binding has been observed to occur in a number of DNA-binding proteins following interaction with their DNA targets, and this effect can directly contribute to the specificity of the interaction.

Concept and Reasoning Checks

1. How do interactions between folded, globular proteins and proteins that are natively unfolded differ?

2. Why, for a given number of residues, can natively unfolded proteins generate a larger interaction surface than globular molecules?

5.16 Structure and medicine

Key concepts

- The accumulated knowledge of the detailed three-dimensional atomic structures of many thousands of proteins and their complexes has provided unparalleled insights into a great many aspects of the biologic functions of proteins.
- Protein structure analysis has opened up new vistas of opportunity for understanding the molecular basis of disease and development of new therapeutic approaches.
- Availability of high-resolution structures of disease target proteins is now generally considered to be essentially to the drug-design process.

Mutation as a cause of disease has long been a focus for molecular and structural biology. Indeed, the existence of naturally occurring mutations, and the ability to induce mutations chemically, underpinned much of the early research into the nature of the gene and the genetic code itself. As we have seen, the precise sequence of amino acids in any given protein has evolved over millions of years to provide a precise architecture tailored to biologic function. Clearly, not all mutations are necessarily deleterious or the process of natural selection could not work, and we now know of thousands of sequence polymorphisms within the human genome that are functionally (or phenotypically) silent. Equally, however, we also know of many point mutations, frameshifts, deletions, insertions, and genetic rearrangements that have devastating medical consequences.

The concept that disease may arise as a result of genetic aberration has a long history and stems from the observations of Archibald Garrod in the late nineteenth century, who coined the term "inborn errors of metabolism" to describe a variety of congenital metabolic diseases such as phenylketonuria. Garrod insightfully attributed these to defects in enzymes long before it was known that enzymes were proteins! In the 1950s, Max Perutz's studies on sickle-cell hemoglobin were the first to apply structural methods—in this case, X-ray crystallography—to attack a human disease at the molecular level, showing how a single mutation of glutamate to valine in the β-subunits of adult human hemoglobin causes pronounced changes in structure and solubility of the affected $\alpha_2\beta_2$ Hb tetramer, causing it to aggregate as fibrillar structures in red blood cells and induce the sickle shape long known to microscopists (see the *Historical Perspectives* box on pages 222–223).

Sickle-cell anemia is one example of how the disruption of structural integrity can lead to disease. Among these are the "diseases of aggregation" such as Alzheimer's, Huntington's, and prion-related disorders that are often associated with neurodegeneration. Huntington's disease is representative of a class of disorders caused by expansion of cytosine-adenine-guanine nucleotide triplets that encode the amino acid glutamine. The number of consecutive glutamines within the expanded glutamine tract is linked to the onset of disease with a threshold of 36 or more. The physical basis of this observation remains unclear, but as the expansion exceeds this number, affected molecules become deposited in insoluble aggregates known as inclusion bodies. In fact, disorders resulting from conformational disruption may arise not only by mutation, but also through other physico-chemical effects, as appears to be the case with prion-related diseases. The native prion protein (PrPc) is a largely α-helical molecule (**FIGURE 5.74**) that undergoes a dramatic structural transition to form fibrillar aggregates with a characteristic β-sheet structure. Again, the cause of this conformational change is still under debate and has been variously proposed to involve posttranslational modifications, metal binding, and other events such as the three-dimensional domain swapping that we described earlier. It appears, however, that the extended beta structures may be a feature of many, if not all, diseases of aggregation, suggesting a common, but still poorly understood, assembly mechanism.

Among the best-studied diseases of mutation is cancer, and thousands of pro-oncogenic genetic lesions have been identified and mapped within a large array of cell-cycle regulators, signaling molecules/complexes, and others. The first oncogene identified and characterized was derived not from human cells, but from a "transforming" retrovirus, Rous sarcoma virus. The oncogene product was shown to be nearly identical in sequence to cellular *ras*, a small-GTPase. The crucial difference between the virally encoded protein (v-Ras) and the cellular homologue (H-Ras) is a point mutation that results in substitution of a glycine at position 12 to valine (G12V). Ras is the archetypal member of a large superfamily of GTP-binding proteins that function in many different signaling pathways. They generally have a low intrinsic GTP-hydrolysis activity (GTPase) and exist in either GTP-bound or GDP-bound forms that differ in the structural arrangement of two "switch" regions of the protein, switches I and II (**FIGURE 5.75**). They are in turn controlled by two classes of regulatory molecules that inactivate the GTP-bound form through stimulation of the intrinsic GTPase activity (GTPase-activating proteins or GAPs) and guanine-nucleotide exchange factors (GEFs) that catalyze the replacement of GDP with GTP. In the GTP-bound form, small GTPases are able to bind to and regulate the activity of a wide variety of downstream effector molecules such as protein kinases, whereas the GDP-bound state is inactive for effector interaction. The oncogenic G12V mutation is

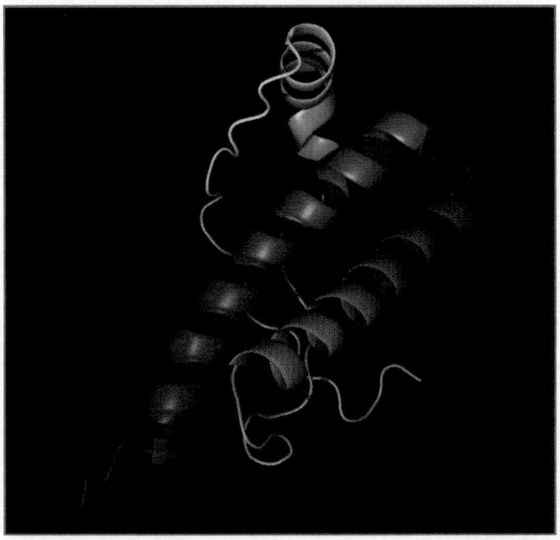

FIGURE 5.74 Sheep prion protein fragment. Structure from Protein Data Bank 1UW3. L. F. Haire, et al., *J. Mol. Biol.* 336 (2004): 1175–1183.

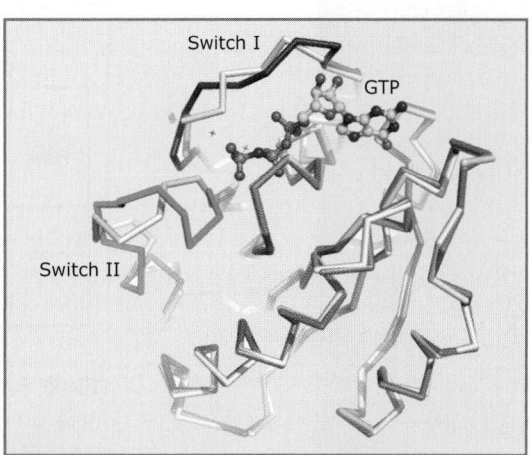

FIGURE 5.75 The GTPase "switch." Structure form Protein Data Bank 2CL0. B. U. Klink, R. S. Goody, and A. J. Scheidig, *Biophys. J.* 91 (2006): 981–982. Protein Data Bank 4Q21. M. V. Milburn, et al., *Science* 247 (1990): 939–945.

In 1904, a West Indian student of African origin who was suffering from anemia, recurrent pains, leg ulcers, jaundice, a low red blood cell count, and an enlarged heart sought the assistance of James Herrick, a Chicago physician. Upon examining the student's blood under the microscope, Herrick observed that many red blood cells had a sickle or crescent shape rather than the biconcave appearance normally observed. Herrick hypothesized that the sickle-shaped red blood cells might be the key to understanding the patient's anemia and other symptoms. The disease suffered by the student, which is now known to be an inborn error of metabolism, was given the name sickle-cell anemia.

Herrick's hypothesis received additional support in 1927, when investigators observed that red blood cells isolated from individuals with sickle-cell anemia change from a biconcave to a sickle shape when deprived of oxygen, whereas red blood cells from normal individuals remain biconcave (FIGURE 5.B7).

Linus Pauling and coworkers tried a new approach to the study of sickle-cell anemia in 1949. They knew that hemoglobin, the major protein in red blood cells, was responsible for binding oxygen and suspected that individuals with sickle-cell anemia have a variant form of hemoglobin. Their hypothesis was tested by comparing the electrophoretic mobilities of normal adult hemoglobin (HbA) and sickle-cell hemoglobin (HbS). HbA migrated as though it were slightly more negative than HbS, suggesting that sickle-cell anemia is caused by one or more amino acid substitutions in hemoglobin. However, the studies by Pauling and coworkers did not reveal whether the amino acid substitution (s) altered the α-globin chain, β-globin chain, or both, nor did the studies reveal which amino acid (s) was (were) altered.

The problem was solved in an ingenious fashion by Vernon Ingram in 1954. Modern techniques for sequencing polypeptide chains were not yet available, so Ingram could not compare the two forms of hemoglobin amino acid by amino acid. Instead, he used trypsin to cut HbA and HbS into well-defined fragments, which he spotted onto filter paper and partially separated by electrophoresis. Then he dried the paper, turned it on a right angle, and developed it in a second direction by chromatography. Because each type of protein treated in this fashion yields a unique two-dimensional pattern, the method is called peptide fingerprinting or peptide mapping. With one important exception, the "fingerprint" of HbS is identical to that of HbA (FIGURE 5.B8).

Amino acid and sequence analyses showed that the fragment obtained from HbA that does not match up with one

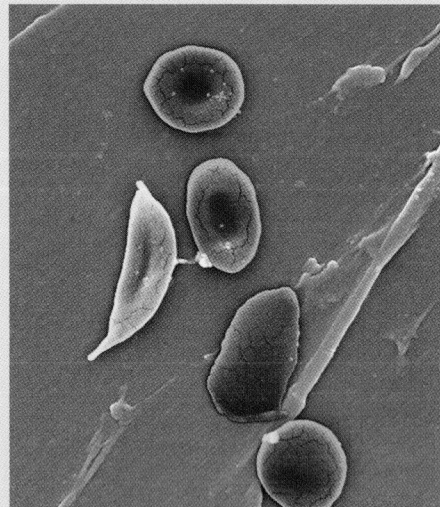

FIGURE 5.B7 Biconcave- and sickle-shaped red blood cells. Photo courtesy of Sickle Cell Foundation of Georgia/Janice Haney Carr/CDC.

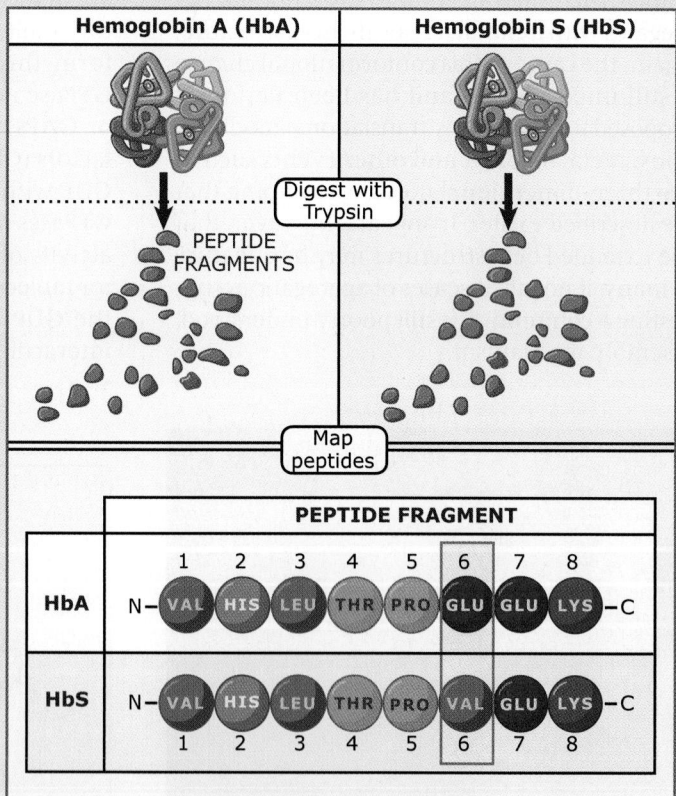

FIGURE 5.B8 Peptide map of hemoglobin A (HbA) and hemoglobin S (HbS). Schematic diagram shows peptide maps created by trypsin digestion of (A) HbA and (B) HbS. All of the peptides but one are identical in HbA and HbS. The one peptide that is different, which is shown in red, has a single amino acid replacement. (C) Glutamate in position 6 of the β-globin chain is replaced by valine.

from HbS is an octapeptide derived from the N-terminal end of the β-chain. The only difference between the two fragments is that HbS contains a valine residue in place of the glutamate residue found in HbA. This substitution was shown to be in the sixth residue from the N-terminal residue of the β-chain (see Figure 5.B8 [C]). The observed substitution is consistent with the observation by Pauling and coworkers that HbA migrates as a slightly more negatively charged protein than does HbS. Ingram's studies showed for the first time that a single amino acid substitution can cause a profound change in protein function. This finding stimulated great interest in the nascent field of molecular biology and reinforced the importance of studying structure–function relationships. Moreover, Ingram's studies introduced peptide mapping, a very powerful technique that is still used today.

Investigators originally thought that capillaries are blocked because the poorly deformable sickle-cells cannot fit through them. More recent studies show that sickle red blood cells, but not normal red blood cells, adhere abnormally to the endothelial cells that line blood vessels. Research activity in the field of sickle-cell anemia is now directed toward trying to correct the genetic error and developing an effective therapy to treat the symptoms.

Why have selective pressures not worked to eliminate sickle-cell anemia? The rather surprising answer is that selective pressure for another factor actually favors the perpetuation of the sickle-cell gene. Individuals who have inherited a nonsickling gene from one parent and a sickle-cell gene from the other parent have the sickle-cell trait associated with a mild form of anemia. When exposed to the protozoan that causes malaria, they are more resistant to malaria than are individuals with two nonsickling genes. As might be expected, the greatest incidence of sickle-cell anemia and trait occurs in those regions of Africa where malaria is most prevalent.

located in the phosphate-binding loop (P-loop) that forms a structural cradle for the β- and γ-phosphates of GTP. The mutation has two major effects. First, it lowers the intrinsic rate of GTP hydrolysis, maintaining the active conformation and thus continuously providing Ras-dependent growth and proliferation signals. Second, the G12V mutation blocks the productive association of GTP-bound Ras with RasGAP, effectively protecting the GTP-bound state even further (**FIGURE 5.76**). We now know that somatic G12V mutation of normal cellular ras occurs in a high proportion of hu-

man cancers, classifying the Ras gene as a protooncogene.

A second class of molecules that are intimately involved in protecting cells against the effects of cancer-promoting mutations are the so-called tumor suppressors. Of these, one of the best characterized is a tetrameric protein called p53 (53 kDa is its apparent molecular weight on SDS-PAGE gels). p53 has been called the "gatekeeper" of the cell cycle. It is a modular protein comprising an N-terminal regulatory region, a central DNA-binding domain, and a C-terminal tetramerization motif, and it functions primarily as an activator of the transcription of an inhibitor of CDKs. Mutations that directly interfere with biochemical activity, or result in reduced expression of p53, are found in the majority of tumors, and occurrence of p53 mutations in the germline result in a familial predisposition to cancer. Although different classes of mutations may be associated with different cancer types, many occur within the central DNA-binding domain of p53. Here, mutations have a variety of structural effects, including overall destabilization of the domain, but the X-ray structure shows clearly that many mutations occur at residues that are intimately involved in nucleic acid binding (**FIGURE 5.77**).

One of the greatest problems facing modern medicine is that of drug resistance, and

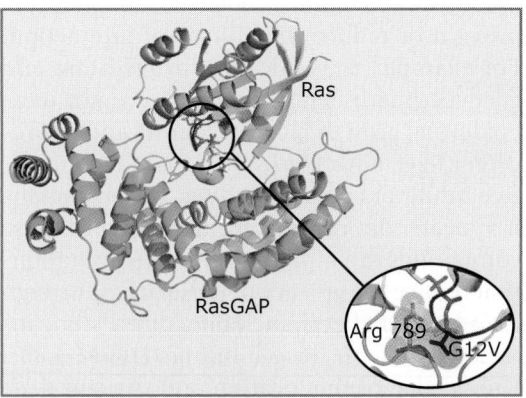

FIGURE 5.76 Steric effects of the G12V oncogenic Ras mutation. Structure from Protein Data Bank 1WQ1. K. Scheffzek, et al., *Science* 277 (1997): 333–338.

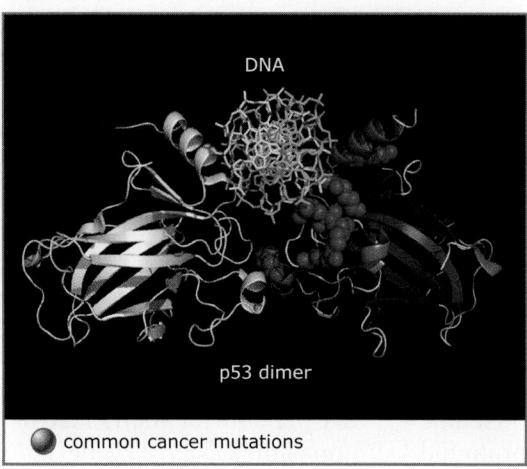

FIGURE 5.77 Cancer-causing mutations in the p53 tumor suppressor. Structure from Protein Data Bank 2GEQ. W. C. Ho, M. X. Fitzgerald, and R. Marmorstein, et al., *J. Biol. Chem.* 281 (2006): 20494–20502.

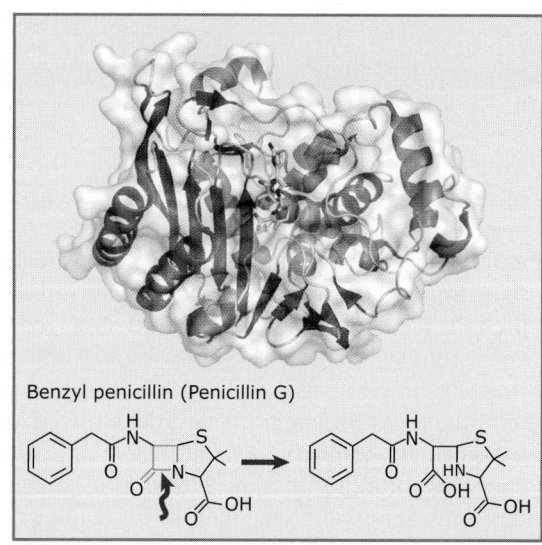

FIGURE 5.78 β-lactam ring cleavage by β-lactamases. Structure from Protein Data Bank 1IEL. R. A. Powers, et al., *Biochemistry* 40 (2001): 9207–9214.

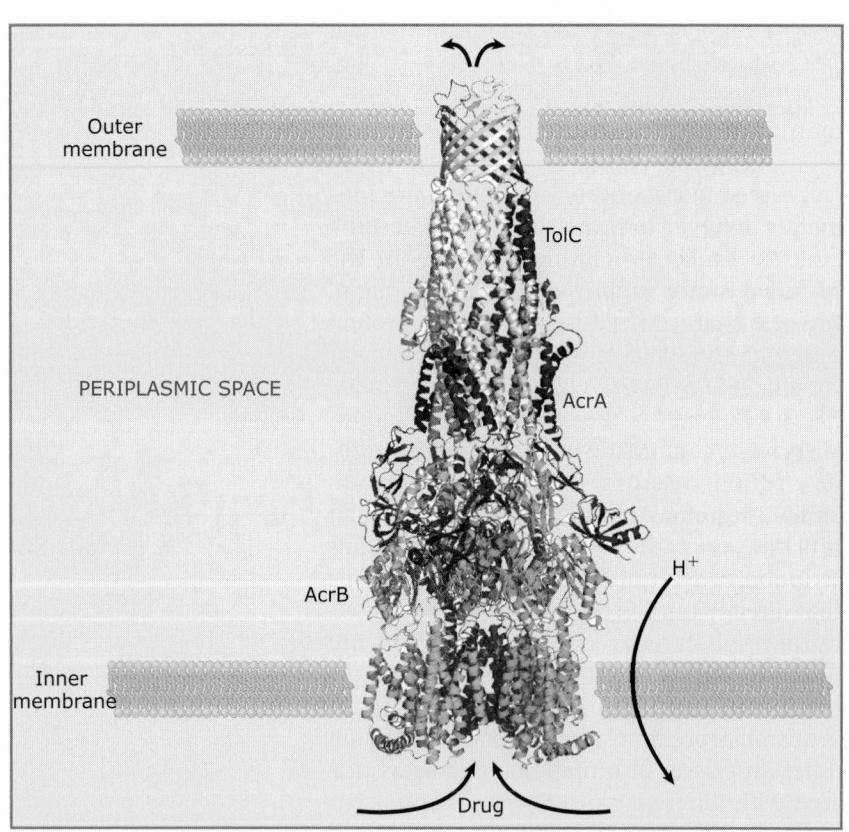

FIGURE 5.79 A model of the TolC/AcrA/AcrB complex model. Reproduced by permission of Ben Luisi, University of Cambridge.

structural biology continues to play an important role in understanding and combating it. The extent of the problem is exemplified by the fact that a number of highly pathogenic micro-

organisms are resistant to virtually every antibiotic in clinical use. Resistance to the action of drug molecules can occur in a number of ways. First, enzymes that are able to use the drug molecules as substrates and chemically inactivate them are common. For example, penicillin, the first antibiotic identified by Alexander Fleming nearly a century ago, is a member of a large family of β-lactam antibiotics in common clinical use. Resistance to β-lactams, however, is common and often mediated by a group of enzymes called β-lactamases that are able to cleave the β-lactam ring (**FIGURE 5.78**). Second, membrane-bound efflux pumps such as the *E. coli* TolC/AcrA/AcrB complex are able to efficiently export a broad spectrum of drug molecules from target cells (**FIGURE 5.79**). Finally, resistance may be mediated by mutations in the protein targets of small-molecule inhibitors that prevent or reduce the affinity of interaction. For example, the anticancer drugs Iressa and Gleevec/Imatinib target several receptor (e.g., epidermal growth factor receptor) and nonreceptor (e.g., c-Abl) tyrosine kinases as competitive inhibitors of ATP binding. Unfortunately, it appears that cancer cells treated with these compounds can rapidly accumulate mutations that reduce the efficacy of these drugs through a number of effects, including direct steric interference with drug binding (**FIGURE 5.80**). Knowledge of the location and structural effects of these mutations, however, can assist in the design of new compounds that circumvent these problems.

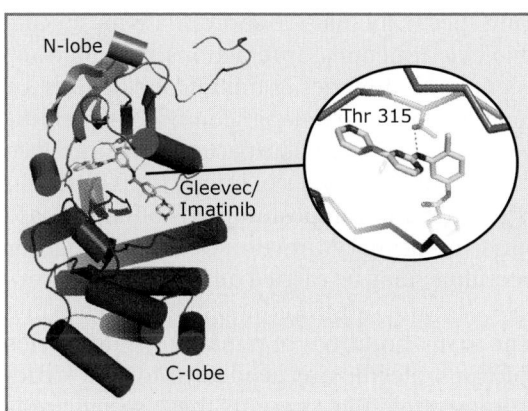

FIGURE 5.80 Drug-resistance mutation in c-Abl kinase. Structure from Protein Data Bank 1IEP. B. Nagar, et al., *Cancer Res.* 62 (2002): 4236–4243.

Concept and Reasoning Checks

1. From a structural and mechanistic viewpoint, how do oncogenic mutations in the small-GTPase Ras result in cancer?
2. What are the major challenges in the design and use of small-molecule drugs against disease?

5.17 What's next? Structural biology in the postgenomic era

The advent and success of large-scale genome sequencing has simultaneously shown how diverse biologic systems are at all levels of organization and complexity. Although the structural database now contains information for around 60,000 proteins and mutational variants, we must still remember that the human genome alone may encode upward of 30,000 different proteins, of which we know the structures of only a relatively small fraction. If we remember that this basic set of proteins may be posttranslationally modified and that numerous variants may be produced by, for example, differential mRNA splicing, it is clear that much remains to be done merely to characterize individual molecules. To this end, a number of structural genomics consortia have been established around the world, with a view toward substantially increasing the available database of protein structures. These large-scale efforts have had a degree of success, although the question of whether they have fulfilled expectations is still open to debate. It remains to be seen whether the current rate of progress is maintained as more difficult problems, be they individual proteins or complexes, move to the top of the list of targets!

Still elusive is a real understanding of the physical processes that drive and regulate protein folding. The exponential increase in structural information, however, is beginning to influence the efforts of mathematical biologists to predict structure from sequence—not necessarily from first principles, but from a set of empirical rules or guidelines derived from the database of known structures. In addition, although we know the basic principles involved in the binding and specificity of proteins with other proteins and ligands, it is still difficult or even impossible to confidently predict the structural, kinetic, and thermodynamic bases of all but the very simplest of interacting systems. This is crucial because as we have seen, proteins, in general, do not function alone, and the most interesting—and therefore most difficult—challenge for structural biologists is the detailed understanding of the biologically relevant complexes that exist in biologic systems.

Many examples of how structural biology has begun to address these outstanding issues have been described in the foregoing sections, and it is clear that technological developments are still being made, allowing more complex and larger structures to be determined at high resolution. In particular, the growth of NMR and cryo-EM into mature methods has considerably added to the arsenal of structural biologists, and the powerful combination of EM with crystal and NMR analyses has already had some impressive successes. Nonetheless, the detailed structural characterization of cellular structures, such as the nuclear pore complex, the centrosome, and other "mega-complexes" that may be constructed from hundreds of proteins, still seems only a distant possibility.

Perhaps most exciting is the comprehensive integration of structural and mechanistic information into the framework of gene expression, biochemistry, cell biology, and physiology—a goal that is central to the emerging field of "systems biology." Clearly, there is much left to do!

5.18 Summary

The three-dimensional structure of protein molecules is intimately associated with their biologic function. Over the last 50 years or so we have seen an explosion in the growth of structural databases. X-ray crystallography has, to date, been by far the most successful method for the high-resolution analysis of protein structures and complexes. Modern high-field heteronuclear NMR approaches, however, and the developments in single-particle cryo-EM are now making substantial contributions.

Proteins are made up from a basic and universal set of 20 α-amino acids with L-configuration. The sequence of the amino acids in a protein chain encodes the final folded or tertiary structure that may contain elements of secondary structure, α-helices, and β-strands. In spite of the diversity of proteins sequences that is observed, the number of distinct structures that exist is likely to be rather small, with perhaps only a few thousand folds covering all globular protein domains.

A specific protein or enzyme may associate with itself or with others to form tight quaternary arrangements that may be crucial for biologic activity, and provide structural stability or functional flexibility. Such higher-order arrangements are necessary for allosteric control of activity, one of the most commonly observed regulatory mechanisms. Additional functional and evolutionary versatility is provided by modular protein architectures where individually folded protein domains with different and complementary activities are encoded within a single protein sequence. Fine-tuning and precise control of protein function may also be achieved through posttranslational modifications that may directly affect activity, stability, localization, and so forth.

Proteins are, in the main, rather dynamic, and motions range from small and rapid atomic "vibrations" through to large-scale conformational changes in different biologic contexts. Such motions may arise and be driven in different ways. They may be linked to the hydrolysis of ATP, they may occur as a result of—or prerequisite for—binding of proteins to themselves and/or other ligands, or they may be a product of posttranslational modification. Unwanted conformational changes may also be brought about by mutation, resulting in many genetic diseases, including cancer.

All proteins function through interacting with other proteins or a variety of ligands such as DNA, RNA, lipids, carbohydrates, and small organic and inorganic molecules. The surfaces that they employ in these interactions are tailored by evolution to have affinity and specificity that are appropriate for specific biologic functions. Interactions may be transient, with lifetimes of milliseconds or less, or may be longer lived, as seen in many multisubunit complexes. The nature of the interface in each of these types of interactions differs in structure and composition. It is often the case that the primary function of an enzyme, for example, may be carried out by only a few active-site and substrate-binding residues, while the many hundreds of remaining amino acids of the molecule are needed to form a structural scaffold that presents these specific residues to the substrate with exquisitely precise stereochemistry.

In many cases, proteins contain both folded domains along with unstructured regions, usually at the N- and C-terminal ends, that may play a variety of regulatory roles. Indeed, regions within modular proteins are usually connected through unstructured "linker" regions that allow sufficient structural flexibility to permit, for example, intramolecular interdomain interactions. It now appears that up to ~30% of all protein sequences within eukaryotic genomes may encode unstructured regions. In some cases, an entire protein may contain no tertiary structure, and these molecules have been classified as "natively unfolded." The functional advantages remain largely unclear, but certainly include properties of flexibility, ease of posttranslational modification, and the ability to engage binding partners through more extensive interacting surfaces than are possible for folded domains.

Knowledge of the structure of proteins and their complexes informs not only the understanding of underlying biologic processes and pathways, but has provided invaluable insights into the nature and cause of many human diseases. For this reason, structural analysis is now firmly established as a major tool in the identification of novel therapeutic compounds and will continue to underpin and drive the developments of new and more powerful computational approaches to the design of new drug molecules and the improvement of existing therapies.

http://biology.jbpub.com/lewin/cells

To explore these topics in more detail, visit this book's Interactive Student Study Guide.

References

Anfinsen, C. B. 1973. Principles that govern the folding of protein chains. *Science* v. 181 p. 223–230.

Fersht, A. ed. 1998. *Structure and Mechanism in Protein Science: A Guide to Enzyme Catalysis and Protein Folding*. New York: W. H. Freeman.

Fink, A. L. 2005. Natively unfolded proteins. *Curr. Opin. Struct. Biol.* v. 15 p. 35–41.

Frank, J. 2002. Single-particle imaging of macromolecules by cryo-electron microscopy. *Annu. Rev. Biophys. Biomol. Struct.* v. 31 p. 303–319.

Goodsell, D. S., and Olson, A. J. 2000. Structural symmetry and protein function. *Annu. Rev. Biophys. Biomol. Struct.* v. 29 p. 105–153.

Kendrew, J. C., Bodo, G., Dintzis, H. M., Parrish, R. G., Wyckoff, H., and Phillips, D. C. 1958. A three-dimensional model of the myoglobin molecule obtained by x-ray analysis. *Nature* v. 181 p. 662–666.

Monod, J., Wyman, J., and Changeux, J. P. 1965. On the nature of allosteric transitions: a plausible model. *J. Mol. Biol.* v. 12 p. 88–118.

Pawson, T. 1995. Protein modules and signalling networks. *Nature* v. 373 p. 573–580.

Perutz, M. F., and Mitchison, J. M. 1950. State of hemoglobin in sickle-cell anaemia. *Nature* v. 166 p. 677–679.

Ramachandran, G. N., Ramakrishnan, C., and Sasisekharan, V. 1963. Stereochemistry of polypeptide chain configurations. *J. Mol. Biol.* v. 7 p. 95–99.

Richardson, J. S. 1981. The anatomy and taxonomy of protein structure. *Adv. Protein Chem.* v. 34 p. 167–339.

Seet, B. T., Dikic, I., Zhou, M. M., and Pawson, T. 2006. Reading protein modifications with interaction domains. *Nat. Rev. Mol. Cell Biol.* v. 7 p. 473–483.

Wuthrich, K. 1990. Protein structure determination in solution by NMR spectroscopy. *J. Biol. Chem.* v. 265 p. 22059–22062.

Membranes and transport mechanisms

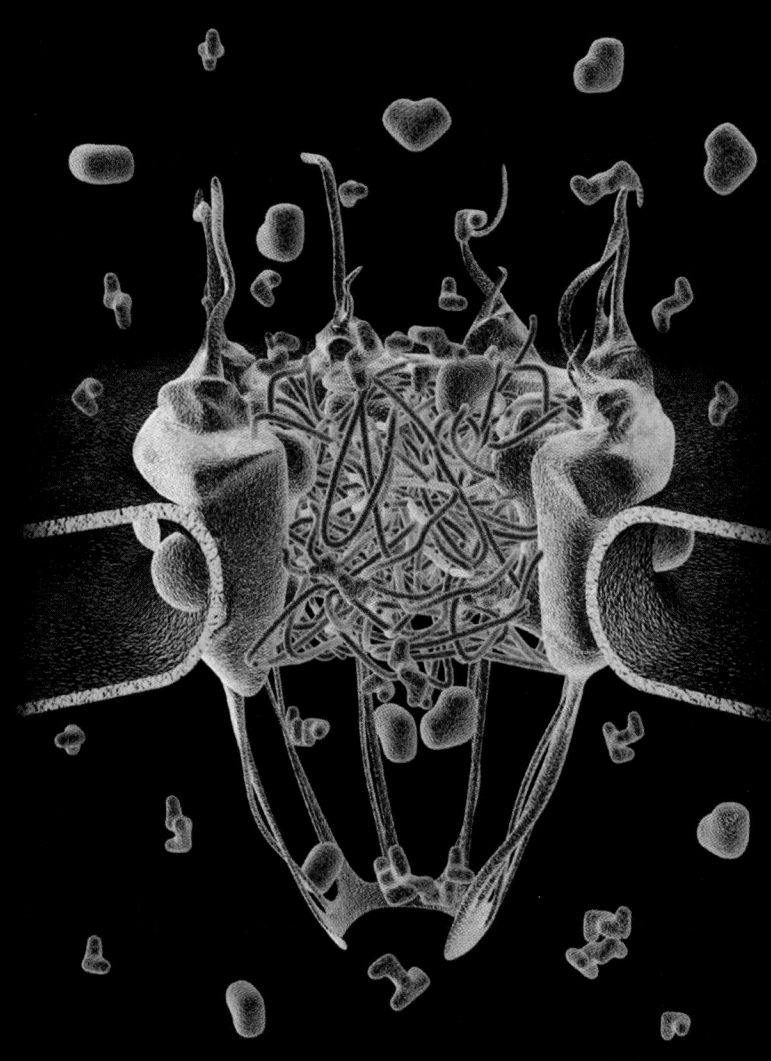

PART

2

Membranes and transport
mechanisms

Transport of ions and small molecules across membranes

6

Stephan E. Lehnart
Georg August University, Germany

Andrew R. Marks
Columbia University, New York, NY

COMPOSITE IMAGE OF CRYSTAL STRUCTURES of mammalian and prokaryotic transport proteins embedded in membranes.

CHAPTER OUTLINE

6.1 Introduction

Key concepts

- Cell membranes define compartments of different compositions.
- The lipid bilayer of biologic membranes has a very low permeability for most biologic molecules and ions.
- Most solutes cross cell membranes through transport proteins.
- The transport of ions and other solutes across cellular membranes controls electrical and metabolic functions.

Biologic membranes are selectively permeable barriers that surround cellular compartments. The plasma membrane separates the inside of the cell from the extracellular environment, and, within eukaryotic cells, additional membranes separate specialized compartments from the cytosol (see Figure 1.9). Cellular compartments differ markedly in the composition of their membranes and internal milieus. Throughout evolution, cells have developed mechanisms to maintain and regulate the composition of each compartment. The maintenance of solute concentrations across membranes is a prerequisite for cellular homeostasis, which is the ability of cells to maintain a relatively constant internal environment for metabolic processes that are vital for survival. The homeostatic regulation of cytosolic ion concentrations also determines the relative osmotic pressures on each side of the cell membrane and thereby regulates cell volume. Moreover, rapid and transient changes of the rate of ion transport across the cell membrane can be employed to adapt the cell to altered metabolic situations, to process information (such as stress signals), and to transport nutrients into, or waste products out of, the cell.

Because the interior of the lipid bilayer is hydrophobic, it is essentially impermeable to polar, hydrophilic, and large biologic molecules, as **FIGURE 6.1** shows. How do inorganic ions and charged and water-soluble molecules move into and out of cells and across intracellular membranes in a selective manner? We now know that membrane-spanning proteins facilitate their movement from one compartment to another. Transport proteins reside in the plasma membrane and in the membranes of intracellular organelles such as the endoplasmic reticulum, Golgi apparatus, endosomes, lysosomes, and mitochondria. Each type of membrane has a distinct complement of transport proteins, as do different cell types. In this chapter, we will discuss the membrane proteins that transport ions and small molecules such as glucose. We will first give an overview of the general classes of membrane transport proteins (see *6.2 Channels and carriers are the main types of membrane transport proteins*) and then discuss the functions and structures of specific proteins in more detail. We will also consider how the different types of transport proteins in a cell function together in the physiological context. The transport of proteins (and other large particles) across membranes, into cells, and within cells is discussed in other chapters (see *7 Membrane targeting of proteins; 8 Protein trafficking between membranes;* and *9 Nuclear structure and transport*).

Much of this chapter is devoted to the transport of ions across membranes. Cells use membrane transport proteins to maintain intracellular ion concentrations that are much different from the extracellular concentrations, as **FIGURE 6.2** shows for the major physiological ion species in an animal cell. The net effect of these concentration differences is that for resting animal cells, the inside of the cell is charged slightly negative relative to the outside. These concentration and charge differences together create **electrochemical gradients**, which are used by cells as a store of potential energy. (Some of the

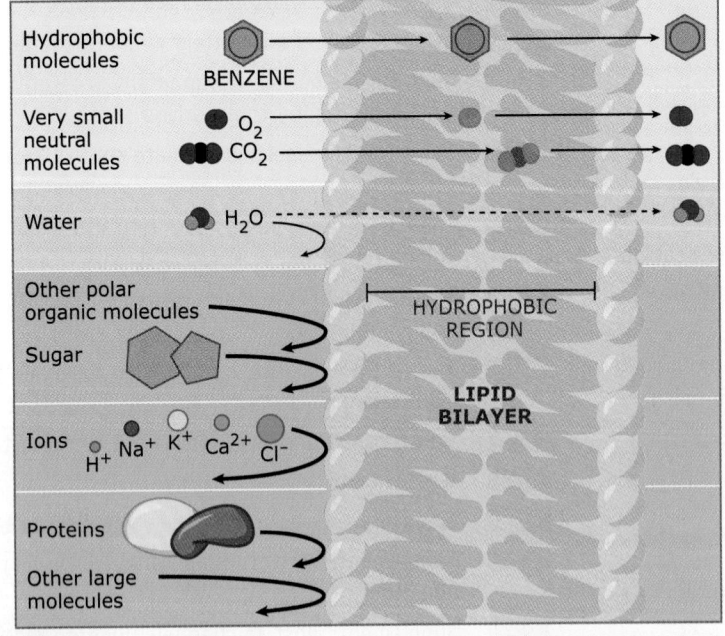

FIGURE 6.1 Permeation of lipid bilayers by biologically important molecules.

	Na⁺	K⁺	Ca²⁺	Cl⁻
Extracellular concentration (mM)	145	4	1.5	123
Intracellular concentration (mM)	12	155	10^{-4}	4.2

FIGURE 6.2 Ion gradients across the plasma membrane of mammalian skeletal muscle cells. The relative atomic radii of the nonhydrated ions are indicated.

physical properties of ions and the ionic basis of membrane potentials are introduced in *6.3 Hydration of ions influences their flux through transmembrane pores* and *6.4 Electrochemical gradients across the cell membrane generate the membrane potential*.) Regulation of electrochemical gradients across cell membranes allows for a large number of basic cellular functions, such as energy production and processing of electrical signals in and out of cells (see *6.19 The Na⁺/K⁺-ATPase maintains the plasma membrane Na⁺ and K⁺ gradients; 6.20 The F_1F_o-ATP synthase couples H⁺ movement to ATP synthesis or hydrolysis*; and *6.12 Action potentials are electrical signals that depend on several types of ion channels*, respectively).

Some of the main tools used to study membrane transport proteins are summarized here and are mentioned throughout the chapter. The flow of charged species (ion currents) across a membrane is detected by electrophysiological techniques applied to whole cells or to membrane fragments. These techniques allow the measurement of the effect of manipulations such as changes in ion composition or addition of inhibitors or activators. Ion channels were first identified and purified using naturally occurring toxins (venoms) that inhibit their function. These toxins have also been used as specific probes for channel function. The relationship between structure and function has been studied using recombinant transport proteins, site-directed mutagenesis, reconstitution of purified proteins into artificial membranes, and expression of transport proteins in heterologous cells. Our understanding of transport proteins has been revolutionized by the ability to solve the structures of a small number of membrane transport proteins at the atomic level. In addition to showing how solutes are transported and permeate membranes, these structural "snapshots" contribute to the proposal of general models for transmembrane transport mechanisms.

6.2 Channels and carriers are the main types of membrane transport proteins

Key concepts

- There are two principal types of membrane transport proteins: channels and carriers.
- Ion channels catalyze the rapid and selective transport of ions down their electrochemical gradients.
- Transporters and pumps are carrier proteins, which use energy to transport solutes against their electrochemical gradients.
- In a given cell, several different membrane transport proteins work as an integrated system.

Some transport proteins are present in the plasma membrane, whereas others are present in the membranes of intracellular organelles. In order to maintain the composition of the cell and its intracellular compartments, it is important that transport proteins are *selective* for a particular solute species over others. Membrane transport proteins can be classified into two groups, **channels** and **carriers**, depending on the mode of transport, as **FIGURE 6.3** shows. Channels, composed of channel proteins, contain a pore region through which solutes pass at high flux rates when the channel is open. Carriers, composed of carrier proteins, bind solutes on one side of the membrane, undergo an allosteric (conformational) change, and release the solutes on the other side of the membrane.

There are several types of proteins that form channels in membranes. Porins, which are present in some prokaryotes and in mitochondria, and gap junctions, which connect the cytoplasms of adjacent cells, allow the passage of solutes mainly based on size (see *19.21 Gap junctions allow direct transfer of molecules between adjacent cells*). Examples of more selective channels are nuclear pore complexes and channels that mediate protein translocation across

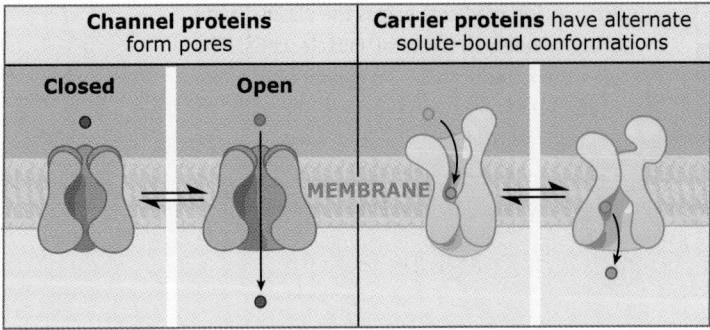

Channel proteins form pores		Carrier proteins have alternate solute-bound conformations
Closed	Open	

MEMBRANE

FIGURE 6.3 Channels and carrier proteins are the two basic types of membrane transport proteins. Solutes diffuse near their maximal diffusion rate through the pore of channel proteins, whereas carrier proteins bind solutes on one side of the membrane, undergo a conformational change, and release it on the other side at a significantly slower rate.

the endoplasmic reticulum membrane (see *9 Nuclear structure and transport* and *7 Membrane targeting of proteins*). The channels that we will consider in this chapter are ion channels and aquaporins, which catalyze the highly selective movement of ions or water molecules across membranes, respectively. Thus far, more than 100 different types of these channel proteins have been described. Channel proteins have the following characteristics:

- solute selectivity,
- a rapid rate of solute permeation, and
- a gating mechanism that regulates solute permeation.

The part of the channel protein through which solutes pass from one side of a membrane to the other is called the **channel pore**. There are several different channel pore configurations. Some channels consist of a single protein whose transmembrane segments form a pore. Others exist as oligomers of identical or different subunits that together form a pore. Still other channels consist of two or more subunits, each of which forms a pore by itself. Often, the oligomeric channel complexes are regulated or targeted to a specific membrane or location by additional subunits.

Most channel proteins are highly selective for a particular solute species, such as sodium (Na^+), potassium (K^+), calcium (Ca^{2+}), chloride (Cl^-), or water. Other channel proteins are nonselective cation or anion channels. As will be apparent in our discussions of specific channel proteins, the pores of distinct channel proteins have a structural feature called the selectivity filter that allows them to discriminate among different solutes.

The electrochemical gradient of the solute dictates the direction of net ionic flux

through the channel. In other words, solutes move through the channel permeation pathway in the energetically favorable direction, which is down their electrochemical gradient. For example, for the resting cell discussed in Figure 6.2, there would be inward currents through Na^+, Ca^{2+}, and Cl^- channels and an outward current through K^+ channels. Since no energy source other than the energy of the electrochemical gradient is involved, this type of transport has been called passive transport. Channel proteins have fast conduction rates; for ion channels, the measured rates are up to 10^8 per second, which is close to the maximal rate of diffusion of ions in water.

To control diverse cellular functions, membrane transport proteins are regulated by **gating**, the process by which these proteins undergo conformational changes to open and close ion transport through the pore in response to specific stimuli. For example, ion channels may be ligand-gated, voltage-gated, stretch-activated, or temperature-activated. Ion channels can be rapidly activated, making them ideal for fast processing of signals. For example, neuronal signaling depends on small electric currents created by ion channels in the cell membrane. Ion channels are also important for regulating cell volume and intracellular pH, transport of salt and water through epithelial cells, acidification of intracellular organelles, and intracellular signaling.

Carrier proteins transduce the free energy stored in electrochemical gradients, ATP, or other energy sources into transport of substrates against a concentration gradient. Because energy is used, this type of transport has been called active transport. Carrier proteins can be divided into two groups, **transporters** and **pumps**, as shown in **FIGURE 6.4**. Transporters couple the energy stored in electrochemical membrane gradients to facilitate the movement of substrates across cell membranes. These transporters can be further divided into **uniporters**, **symporters** (also called cotransporters), and **antiporters** (also called exchangers). On the other hand, membrane pumps use energy, such as the energy released by ATP hydrolysis, directly, to drive energetically less favorable substrate accumulation or efflux pathways. Compared with channel proteins, carrier proteins have a much slower rate of transport, on the order of 1000 solute molecules per second.

There are two types of active transport, **primary active transport** and **secondary**

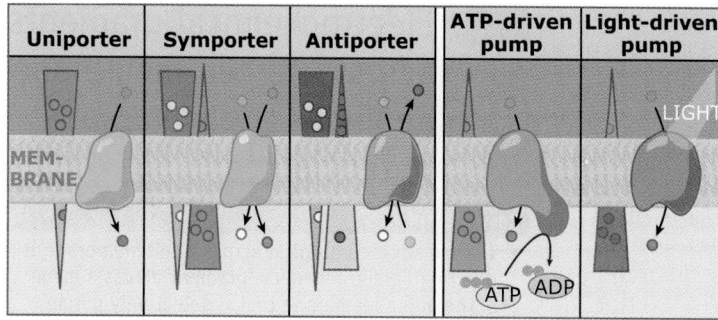

FIGURE 6.4 Transporters and pumps are the two basic types of carrier proteins. Uniporters, symporters, and antiporters are the three types of transporters. The insets depict the solutes' electrochemical gradients across the membrane. Depending on the carrier protein, transport occurs either down the gradient (from high to low) or against it.

PASSIVE	ACTIVE
Down electrochemical gradient (**No** energy required)	Against an electrochemical gradient (**Energy** required)
Diffusion	Membrane pumps or ATPases
Facilitated diffusion (channels, uniport, coupled transport)	Endocytosis
Solvent drag secondary to transepithelial transport	

FIGURE 6.5 Mechanisms of solute transport across membranes can be categorized as passive or active.

active transport. Carrier proteins that mediate primary active transport use ATP as the energy source to drive transport of solutes against their electrochemical gradients. Thus, these proteins help to maintain the concentration gradients of solutes across cellular membranes. The Ca^{2+}-ATPase and Na^+/K^+-ATPase pumps are important examples of transport proteins that mediate primary active transport (see *6.18 The Ca^{2+}-ATPase pumps Ca^{2+} into intracellular storage compartments* and *6.19 The Na^+/K^+-ATPase maintains the plasma membrane Na^+ and K^+ gradients*). Carrier proteins that mediate secondary active transport do not use ATP directly. Instead, they use the free energy stored in the electrochemical gradients, which are generated by primary active transporters, to drive transmembrane solute transport. Accordingly, symporters and antiporters mediate secondary active transport (see *6.15 Symporters and antiporters mediate coupled transport*). The mechanisms of solute transport are summarized in **FIGURE 6.5**.

In a given cell, all types of membrane transport proteins—channels, transporters, and pumps—work in concert such that the function of one type of transport protein depends on the function of others. We will discuss several examples of this interdependence throughout this chapter. For example, the ionic gradients across cell membranes are maintained by the complex interplay of several types of transport proteins. In addition, the proper functioning

of epithelial cells, such as those in the kidney, intestine, and lungs, involves the transport of many types of ions and solutes. We will also describe the role of defective membrane transport proteins in various diseases.

6.3 Hydration of ions influences their flux through transmembrane pores

Key concepts

- Salts dissolved in water form hydrated ions.
- The hydrophobicity of lipid bilayers is a barrier to movement of hydrated ions across cell membranes.
- By catalyzing the partial dehydration of ions, ion channels allow for the rapid and selective transport of ions across membranes.
- Dehydration of ions costs energy, whereas hydration of ions frees energy.

Since the lipid bilayer of cellular membranes is hydrophobic, polar ions are not able to cross the membranes on their own (see Figure 6.1). In order for ions to cross a membrane, they must pass through specialized transmembrane proteins: ion channels and carrier proteins. In this section, we will consider some of the physical properties of ions in solution and their effects on ion transport.

Ions in solution are hydrated; that is, they are surrounded by water molecules. The positive or negative electric charge of an ion attracts dipolar water molecules, which have a partial negative charge from the oxygen atom and partial positive charges from the hydrogen atoms. Ion hydration is the basis for the rapid dissolution of NaCl crystals in water, for example. This process is energetically favorable because water molecules are attracted to free Na^+ and Cl^- ions.

Water molecules form layers called **hydration shells** around an ion, thereby partially neutralizing its charge in solution. Membrane bilayers are highly effective barriers for hydrated ions. Because hydration of an ion is energetically favorable, a relatively large amount of energy would be required for an ion to shed its hydration shell and move into the hydrophobic environment of the lipid bilayer. During the process of transmembrane ion transport, ion channels help to overcome this energy barrier.

The form of the hydration shell depends on the size and charge of the ion. Water dipoles orient toward cations and anions according to their charge and size. Smaller ions have a more localized charge that results in a higher **charge density** compared to larger ions with the same charge. The higher charge density results in a stronger local electric field, which attracts more water molecules and increases the thickness of the hydration shell. Thus, a smaller ion has a larger hydration shell than a larger ion with the same charge, resulting in a larger effective radius when permeating a channel pore.

Why is hydration important in the context of ion transport? Ion channels create an environment that is similar to or mimics water-filled pores and thus facilitates the partial dehydration of ions as they travel through the channel. As the ion traverses the channel, it forms weak electrostatic bonds with charged or partially charged amino acid residues, which substitute for and mimic the specific hydration shell and thereby help to make the transport process energetically favorable and selective at the same time. The selectivity of an ion channel depends on its ability to catalyze partial dehydration, in an energetically favorable way, for a particular ion species but not others, using specific channel dimensions and ion-binding sites. (These concepts will be discussed in more detail in *6.5 K+ channels catalyze selective and rapid ion permeation*.)

6.4 Electrochemical gradients across the cell membrane generate the membrane potential

Key concepts

- The membrane potential across a cell membrane is due to an electrochemical gradient across a membrane and a membrane that is selectively permeable to ions.
- The Nernst equation is used to calculate the membrane potential as a function of ion concentrations.
- Cells maintain a negative resting membrane potential with the inside of the cell slightly more negative than the outside.
- The membrane potential is a prerequisite for electrical signals and for directed ion movement across cellular membranes.

A defining feature of cells is that they maintain concentrations of solutes that are significantly different from the extracellular environment. In the case of ions, their concentration differences across the membrane give rise to a charge difference: the inside of cells is slightly negatively charged compared to the outside. The combination of the charge difference and the concentration difference is called the electrochemical gradient. The electrochemical gradient is maintained by the actions of selective channels and carrier proteins in the plasma membrane.

To understand how an electrochemical gradient is established across a membrane, we will first consider the simple case in which a membrane is permeable only to one ionic species. **FIGURE 6.6** shows two compartments, A and B, which are separated by a thin membrane. Compartments A and B contain solutions with different concentrations of KCl, which dissociates into hydrated K^+ and Cl^- ions. Because there are equimolar concentrations of K^+ and Cl^- ions in both compartments, each compartment is neutral with respect to charge. If the membrane were impermeable to ions, the electrical potential across the membrane, as measured with a voltmeter, would be zero.

Next, consider the situation in which the membrane is permeable only to K^+ ions (e.g., when the membrane has K^+ channels embedded in it) (see Figure 6.6). The diffusion of solutes down their concentration gradients is energetically favorable (expressed as a negative energy difference ΔG). Therefore, K^+ ions would diffuse down their concentration (chemical)

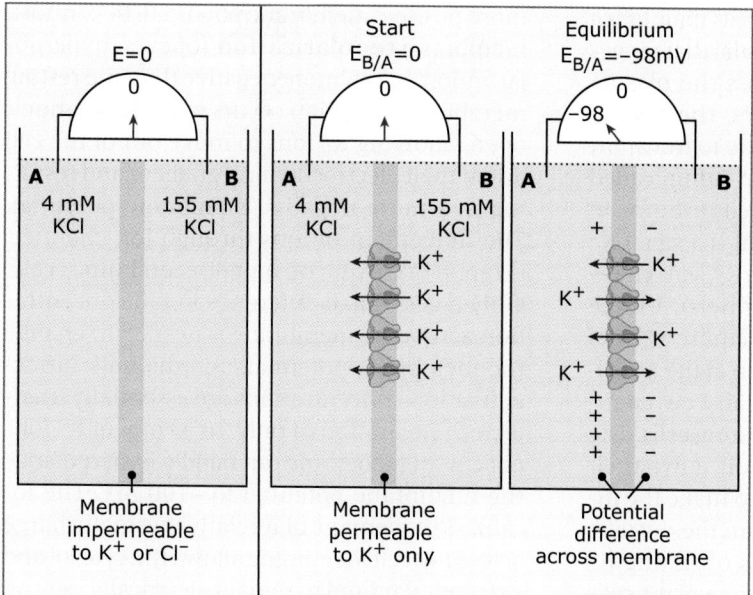

gradient from compartment B to compartment A, thereby altering the charge distribution across the membrane. As positively charged ions accumulate in compartment A and repel each other, these repulsive forces will oppose the movement of K^+ ions from compartment B to A. When the system reaches its electrochemical equilibrium, the forces of the concentration and electrical gradients exactly balance each other, and there is no *net* movement of K^+ ions across the membrane. At this point, movement of K^+ ions from one compartment counters the movement of K^+ ions from the other compartment. However, compartment A will have relatively more positive-charged ions than compartment B. These excess K^+ ions (in compartment A) are attracted to the excess Cl^- ions (in compartment B) across the thin membrane so that the charges line up along either side of the membrane. The difference in charge across the membrane results in a potential difference, also called the **membrane potential**. The equilibrium (membrane) potential of compartment B relative to compartment A is negative. This example shows the two essential conditions for cells to generate a non-zero membrane potential:

- a difference in the concentrations of ionic species on either side of the membrane, resulting in a separation of charge, and
- a membrane that is selectively permeable for at least one of these ionic species.

The membrane potential is, therefore, a function of ion concentrations. At equilibrium, this rela-

tionship can be expressed quantitatively for an ion, X, by using the Nernst equation:

$$E_X = 2.3 \frac{RT}{zF} log_{10} \frac{[X]_B}{[X]_A}$$

- E: equilibrium potential (volts)
- R: the gas constant (2 cal mol^{-1} K^{-1})
- T: absolute temperature (K; 37°C = 307.5 K)
- z: the ion's valence (electric charge)
- F: Faraday's constant (2.3 × 10^4 cal volt^{-1} mol^{-1})
- $[X]_A$: concentration of free ion X in compartment A
- $[X]_B$: concentration of free ion X in compartment B

For animal cells, K^+, Na^+, and Cl^- ions are the major contributors to the membrane potential (see Figure 6.2 for ion concentrations in skeletal muscle cells). Ca^{2+} and Mg^{2+} ions contribute little to the resting membrane potential. The plasma membrane is selectively permeable to any of these ions (i.e., the membrane contains ion channels selective for each ionic species). The Goldman-Hodgkin-Katz voltage equation, which is an expanded version of the Nernst equation, takes into account this complexity and the membrane permeability (P) of each ion. For the major ionic species, the equation expresses the membrane potential as a function of their permeabilities and their concentrations inside (i) and outside (o) the cell:

$$E = 2.3 \frac{RT}{zF} log_{10} \frac{P_K [K^+]_o + P_{Na}[Na^+]_o + P_{Cl}[Cl^-]_i}{P_K [K^+]_i + P_{Na}[Na^+]_i + P_{Cl}[Cl^-]_o}$$

Depending on the cell type, cells maintain a negative resting membrane potential that ranges from −200 mV to −20 mV across the plasma membrane. In mammalian cells, the resting membrane potential is mainly due to transport through K+ channels and an ion pump called the Na+/K+-ATPase. The major contributor to the negative membrane potential is a small K+ flux through plasma membrane K+ leak channels (also called resting K+ channels). Unlike most other K+ channels, which require a signal to open, K+ leak channels are open at the (negative) resting membrane potential. Few channels for other ions are open in resting cells. The movement of K+ ions out of the cell, down their electrochemical gradient, helps to make the inside of the cell more negative than the outside. We do not know all the sources of resting K+ conductance. In some cells, such as plant cells and bacteria, and in organelles such as mitochondria, the resting membrane potential is due to a proton gradient rather than a K+ gradient.

In order for K+ ions to diffuse out of cells through K+ channels, the K+ concentration must be greater inside the cell than outside. This concentration gradient is maintained by the action of the Na+/K+-ATPase, which pumps two K+ ions into the cell for every three Na+ ions that it pumps out. It is, therefore, electrogenic: more *net* charge is pumped out than in. Thus, in addition to the K+ leak channels, Na+/K+-ATPases help to make the inside of the cell more negatively charged than the outside. If a cell's Na+/K+-ATPases were inactivated, the concentrations of Na+ and K+ ions would eventually be the same on both sides of the membrane. This is because the lipid bilayer has some permeability to ions, albeit very low. In other words, without the primary active transport of the Na+/K+-ATPase, the membrane potential would become zero.

The membrane potential of a resting cell is relatively constant. However, changes in the membrane potential occur when ion channels open in response to ligand binding, mechanical stress, or a voltage change, thus increasing their permeability to a specific solute. In the case of voltage-sensitive ion channels, changes in the membrane potential influence the flow of ions through the channels. The control of ion channel opening and closing is called **gating**. The membrane potential is determined by the ions for which there are the most open channels. For example, membrane **depolarization** occurs when Na+ or Ca2+ channels open, allowing the respective ions to flow into the cell down their electrochemical gradients. This results in a more positive membrane potential. In contrast, membrane **repolarization** (or even hyperpolarization, when more negative than the resting membrane potential) occurs when K+ channels open, allowing K+ ions to move out of the cell down their electrochemical gradient and resulting in a more negative membrane potential. The movement of ions through ion channels is rapid, occurring on a millisecond time scale. Only a very small difference in ion concentration across the membrane is needed to change the membrane potential, and the bulk intracellular ion concentrations are essentially unaffected. Separation of only 10^{-12} mol of K+ ions per cm2 of membrane can rapidly hyperpolarize the membrane potential to −100 mV. The localized movement of a relatively small charge across the cell membrane allows the cytosol and extracellular fluid to remain electrically neutral and minimizes the forces of charge repulsion.

From an energetics standpoint, the membrane potential represents a form of energy reservoir that can readily be used to perform work. With negatively charged ions stored on the cytosolic side and positively charged ions stored on the extracellular side of the membrane, the plasma membrane functions analogously to an electrical capacitor or a battery—devices that store electrical energy and are used to provide energy in electrical circuits. The energy is released in the form of ions moving down their respective electrochemical gradients and can be coupled to the uphill transport of other ions or solutes against their concentration gradients. (For examples see *6.16 The transmembrane Na+ gradient is essential for the function of many transporters*.)

(For more on the Nernst and Goldman-Hodgkin-Katz equations see *6.24 Supplement: Derivation and application of the Nernst equation*.)

6.5 K+ channels catalyze selective and rapid ion permeation

Key concepts

- K+ channels function as water-filled pores that catalyze the selective and rapid transport of K+ ions.
- A K+ channel is a complex of four identical subunits, each of which contributes to the pore.
- The selectivity filter of K+ channels is an evolutionarily conserved structure.
- The K+ channel selectivity filter catalyzes dehydration of ions, which confers specificity and speeds up ion permeation.

Potassium (K+) channels are integral proteins of the plasma membrane and mediate the flow of K+ ions between the inside and outside of the cell. K+ channels are conserved throughout evolution from bacteria to humans, and have many biologic functions in different cell types, ranging from stabilization of the resting membrane potential to termination of the action potential and to maintenance of electrolyte balance. (In contrast, the physiological function of K+ channels in prokaryotes is not well understood.) Similar to other ion channels, K+ channels are gated by different mechanisms: that is, they open and close in response to electrical or chemical signals. (For more on gating of K+ channels see *6.6 Different K+ channels use a similar gate coupled to different activating or inactivating mechanisms*.) Since cells maintain a significantly higher intracellular than extracellular K+ concentration (see Figure 6.2), opening of K+ channels results in K+ efflux as the ions move down their electrochemical gradient. (For more on maintenance of K+ concentrations see *6.19 The Na+/K+-ATPase maintains the plasma membrane Na+ and K+ gradients*.)

As with other ion channels, K+ channels form a narrow water-filled pore through which ions pass. K+ channels are highly selective for K+ over other cations. This selectivity is important for maintenance of cellular ion concentrations. Electrophysiological measurements have shown that K+ flux through the channels occurs at rates approaching 10^8 ions sec^{-1}, which is close to the maximal diffusion rate of solutes in water. In this section, we will discuss the molecular basis for this selective yet rapid permeation of K+ ions.

Each K+ channel is a homotetramer, with the four subunits together forming a central pore, as **FIGURE 6.7** shows. Two main types of K+ channels can be defined according to the transmembrane topology of the subunits. In 2TM/1P-type K+ channels, each subunit has two transmembrane (2TM) α helices, M1 and M2, separated by a pore (P) loop. The P-loop contains a short helical segment called the pore helix. These features are common to all K+ channels. The 2TM/1P class includes the inward rectifier, KATP and G protein-coupled K+ channels. In the other main type of K+ channel, called 6TM/1P, each subunit contains six transmembrane (6TM) domains plus one P-loop. In this type of K+ channel, the S5-P-S6 domains form the pore region in a manner analogous to the M1-P-M2 domains of the 2TM/P channels. The four additional transmembrane segments

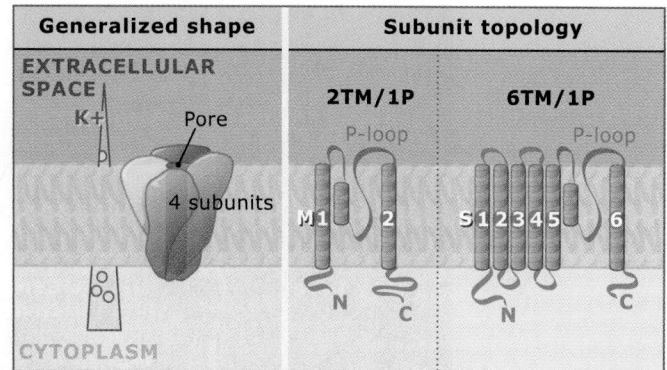

FIGURE 6.7 K+ channels are tetramers of identical subunits, with each subunit contributing to the central pore region. There are two main types of subunits, those with two transmembrane domains and those with six. Both types have a pore loop. The inset at left indicates the resting K+ gradient of animal cells.

(S1–S4) are involved in channel gating. The 6TM/1P channel class includes voltage-gated K+ channels (K$_V$ channels) and ligand-gated channels such as Ca^{2+}-activated K+ channels. (For more on voltage-gated or Ca^{2+}-activated K+ channels see *6.6 Different K+ channels use a similar gate coupled to different activating or inactivating mechanisms*.)

Our understanding of the selectivity of K+ channels comes from electrophysiological and mutagenesis experiments and from X-ray crystallographic analyses. In addition to confirming predictions made from earlier experiments, the crystallographic analyses provide views of the channel architecture at the atomic level, allowing mechanistic insights at the molecular level. The first crystal structure of a K+ channel was solved for the bacterial KcsA channel, a 2TM/1P type channel, in its closed conformation. The pore, through which ions pass, consists of two main parts, the **selectivity filter** and the central cavity, as **FIGURE 6.8** shows. The M2 (inner) helices and the P-loops line the pore. The selectivity filter, which is the narrowest part of the pore, is near the extracellular opening and binds K+ ions. The selectivity filter is 12 Å long and 3 Å wide, and each subunit contributes one P-loop to it. Six ion-containing sites have been visualized in the channel: four sites (P1-P4) within the selectivity filter, and two other sites (P0 and P5), one on the extracellular side of the filter and one in the central cavity. These sites represent a composite image from all the K+ channel protein configurations within the crystal structure.

Each of the four P-loops in the K+ channel selectivity filter contains a highly conserved

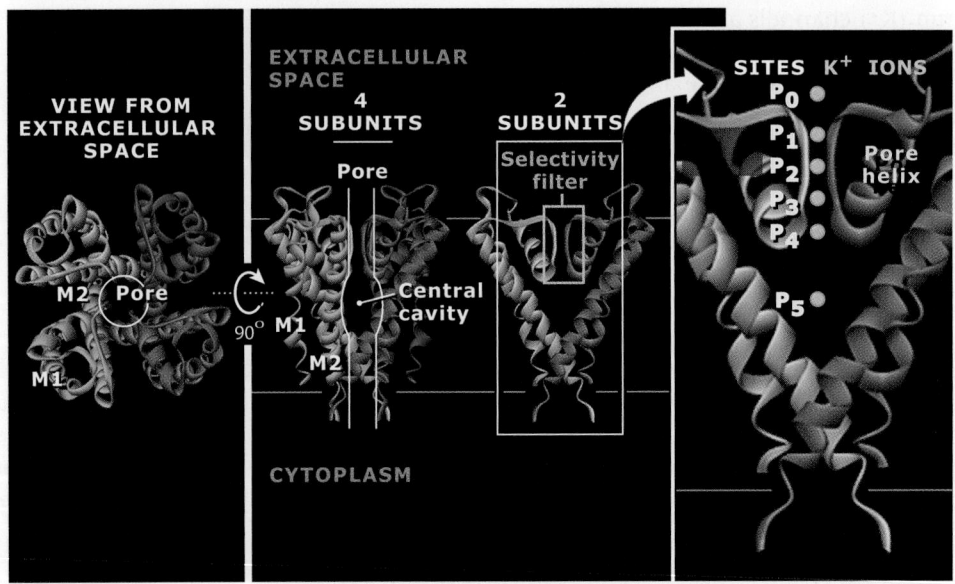

FIGURE 6.8 The crystal structure of the KcsA K+ channel, shown as a ribbon diagram that depicts the protein backbone. The presumed position of the membrane is indicated. For clarity, some views show only two subunits. The crystal structure is of a truncated KcsA channel and lacks some of the N-terminal and C-terminal residues. Structures from Protein Data Bank 1K4C. Y. Zhou, et al., *Nature* 414 (2001): 23–24.

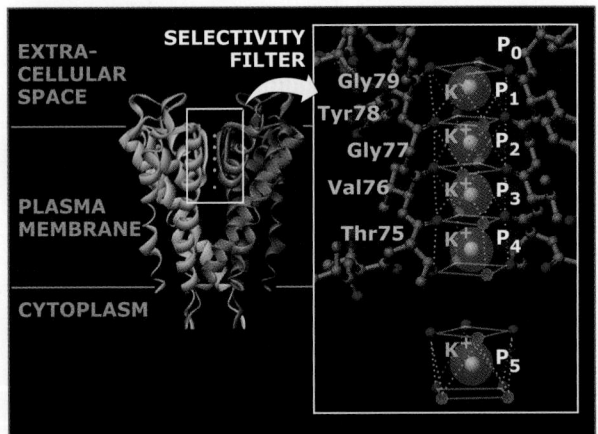

FIGURE 6.9 All four subunits of the bacterial KcsA K+ channel contribute to the selectivity filter. For clarity, the pore loop residues of only two subunits are shown, together with the relevant oxygen atoms from the pore loops of the other two subunits. P0-P5 are the six putative K+-binding sites. The P0 site lies at the extracellular opening of the channel pore (K+ ion not shown). The P1-P4 sites in the selectivity filter are each formed by eight oxygen atoms (red) from the P-loops. The P5 site in the central cavity is occupied by a hydrated K+ ion (the oxygen atoms of eight water molecules are shown in red). The structure represents a composite of all proteins in the crystal, so all sites are occupied by K+ ions. The presumed position of the membrane is indicated. Structures from Protein Data Bank 1K4C. Y. Zhou, et al., *Nature* 414 (2001): 23–24.

sequence called the *signature sequence*: Thr-X-Gly-Tyr-Gly or Thr-X-Gly-Phe-Gly (TXGYG or TXGFG, respectively), where X can be one of several amino acid residues. In the KcsA channel, the signature sequence is Thr-Val-Gly-Tyr-Gly (TVGYG). Each of these residues directs a carbonyl oxygen (either from the backbone or side chain) into the K+ permeation pathway. As **FIGURE 6.9** shows, these carbonyl oxygen atoms, which carry a partial negative charge, form the corners of a series of four "cages" that coordinate the positions of permeating K+ ions. The coordinated positions of K+ ions within the selectivity filter resemble a linear chain.

K+ ions in solution are hydrated, that is, they are surrounded by water molecules. In order for K+ ions to enter the narrow selectivity filter, water molecules must be removed. However, dehydration of ions is energetically costly because it must overcome the strong attractive forces between positively charged K+ ions and the partial negative charges of water dipoles. The K+ channel solves this problem because the partial negative charge of the oxygen atoms in the selectivity filter act as surrogate water molecules, creating a hydrophilic environment that lowers the dehydration energy for the permeating K+ ions and mimics the weak negative charges of water dipoles forming the hydration shell.

This dehydration step is the basis for ion selectivity. Consider another abundant monovalent cation, Na⁺, which has a smaller diameter than K⁺, yet is 10^4 times less likely than K⁺ to permeate K⁺ channels. The reason that K⁺ channels are virtually impermeable to Na⁺ ions is that the size of the K⁺ channel selectivity filter compensates for the energetic cost of dehydration for K⁺ but not for dehydration of Na⁺ ions. The dehydrated K⁺ ions fit well within the selectivity filter, but a dehydrated Na⁺ ion is too small to coordinate efficiently with the carbonyl oxygen atoms in a way that would allow dehydration to be energetically favorable. Thus, the structure of the selectivity filter favors passage of K⁺ but not Na⁺ ions.

In order for a K⁺ channel to catalyze diffusion at a rate of 10^8 ions per second, the process by which an ion is partially dehydrated, passed through the selectivity filter, and rehydrated would have to occur in about 10 nsecs. A model for how high throughput could be achieved has been proposed based on analysis of different crystal structures, each obtained with different concentrations of K⁺ ions, and based on computer simulations. The model is based on the observation that K⁺ ions in the selectivity filter occupy either the P1 and P3 positions (1,3-configuration) or the P2 and P4 positions (2,4-configuration) at any given point in time, as **FIGURE 6.10** shows. These configurations take

into account that K⁺ ions are unlikely to occupy adjacent positions due to repulsion between the positive charges. The two configurations are energetically similar and in equilibrium with each other. As K⁺ ions enter the selectivity filter from the central cavity, electrostatic repulsion causes the ions to move along the filter in either one of the double-occupied configurations. As a result, the ion on the extracellular side of the queue is pushed out. The high conductance is achieved in part because there is essentially no energy barrier for ions to move between the 1,3- and the 2,4-configuration.

In addition, a general principle for facilitating rapid ion permeation is that when the channel opens, the channel pore can widen to allow essentially free access of ions until they reach the selectivity filter. Thus, in K⁺ channels, the effective distance of ion permeation is shortened to the 12 Å length of the selectivity filter, as opposed to being the width of the membrane. (For more on the open conformation of K⁺ channels see *6.6 Different K⁺ channels use a similar gate coupled to different activating or inactivating mechanisms.*)

Another feature of the K⁺ channel structure is the ~10 Å diameter wide central cavity at the intracellular side of the selectivity filter. The central cavity is lined with hydrophobic residues, which helps to minimize the interaction of hydrated ions with the permeation pathway. At the same time, this central cavity also helps K⁺ ions to overcome repulsion from

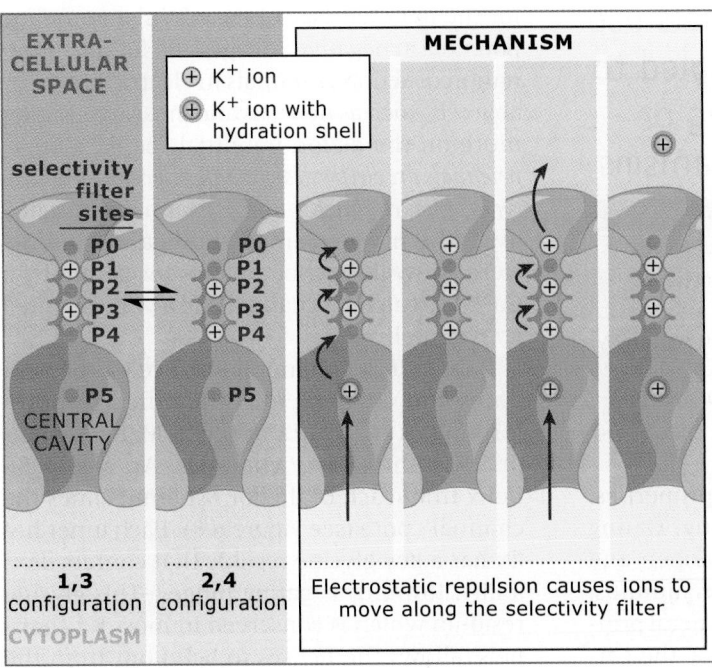

FIGURE 6.10 K⁺ ions move down their concentration gradient, from the intracellular side to the extracellular side of the plasma membrane. The model predicts that two partially dehydrated K⁺ ions occupy the selectivity filter at any given time. These ions move rapidly between two configurations that are equivalent in terms of free energy. Electrostatic repulsion from incoming ions causes ions to move forward through the selectivity filter. Thus, the selectivity filter eliminates energetic barriers and allows high throughput of K⁺ ions.

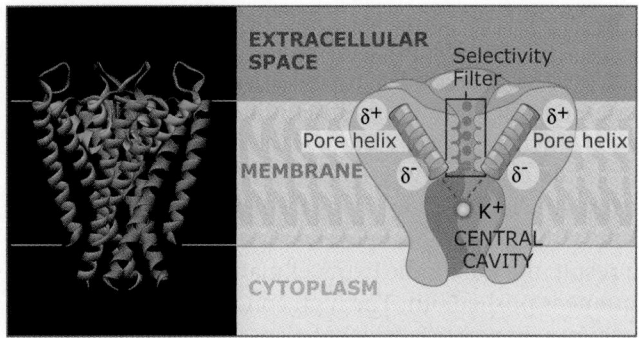

FIGURE 6.11 Four pore α-helices (green) direct their negative dipole charges toward the central cavity and stabilize the positively charged K⁺ ion occupying the P5 site (for clarity, only two pore helices are shown in the right schematic). The partial charges of helix dipoles are indicated. Structure from Protein Data Bank 1K4C. Y. Zhou, et al., *Nature* 414 (2001): 23–24.

the hydrophobic center of the lipid bilayer. Two characteristics of the central cavity help to stabilize K⁺ ions within the channel at the center of the membrane, where the energy barrier for ion permeation is maximal. First, the central cavity keeps the K⁺ ion surrounded by water molecules. Second, the pore helices from each channel subunit direct their partial negative charges toward the center of the central cavity, as **FIGURE 6.11** shows. (The partial negative and partial positive charges of the α helices arise from hydrogen bonds between adjacent turns; as a result, the C-terminus is more negatively charged than the N-terminus.)

6.6 Different K⁺ channels use a similar gate coupled to different activating or inactivating mechanisms

Key concepts

- Gating is an essential property of ion channels.
- Different gating mechanisms define functional classes of K⁺ channels.
- The K⁺ channel gate is distinct from the selectivity filter.
- K⁺ channels are regulated by the membrane potential.

Ion channels have two essential properties, selective ion conduction and gating. Gating is the ability of an ion channel to open and close in response to an appropriate stimulus. In this section, we will discuss the general principles of channel gating and describe the gat-

ing mechanisms of different K⁺ channels. (For details on ion conduction through K⁺ channels see *6.5 K⁺ channels catalyze selective and rapid ion permeation*.)

Because uncontrolled ion flow consumes energy and compromises cell function, the opening of ion channels is tightly controlled. Taking into account cell volume, extracellular and intracellular K⁺ concentrations, and the rate of transport of ions through K⁺ channels, it is estimated that the opening of only 10 channel molecules could potentially drain a cell of K⁺ ions within one second. In fact, K⁺ channels open and close within milliseconds, which prevents ion leak yet is sufficient to maintain the negative resting membrane potential. To achieve open and closed gating mechanisms, K⁺ channels undergo conformational changes in response to specific stimuli.

Different subfamilies of K⁺ channels are gated by different extracellular or intracellular signals. For some K⁺ channels, the gate opens intrinsically by a sensing mechanism that links channel activity to the metabolic state of the cell. In a second type of mechanism, ligand binding to the channel's intracellular domains causes a conformational change that opens the channel gate. Ca²⁺, ATP, trimeric G proteins, and polyamines are examples of such ligands. Lastly, in voltage-gated K⁺ channels, changes in the membrane potential lead to conformational changes in the transmembrane segments that open the channel gate. An example of voltage gating is the voltage-sensing mechanism of voltage-dependent K⁺ channels that allows for the maintenance of the resting membrane potential at negative voltages and for the termination of action potentials in electrically excitable cells such as neurons and muscle cells. (For more on the action potential see *6.12 Action potentials are electrical signals that depend on several types of ion channels*.) Some channels are gated by both voltage change and ligand binding. Here we will describe models for gating by a Ca²⁺-activated K⁺ channel and a voltage-gated K⁺ channel.

How do K⁺ channels achieve gating? An important component is the inner, or pore-forming, α helix, M2 in the 2TM/P channels or S6 in the 6TM/P channels. An M2 or S6 helix from each of the four subunits lines the channel's pore (see Figure 6.8). Each inner helix has a key glycine residue that confers flexibility, forming a "gating hinge." This glycine residue, which is conserved in most K⁺ channels, allows the helices to bend out from the channel pore, as **FIGURE 6.12** shows. Thus, in the straight conformation the four inner helices form a closed pore near the channel's intracellular surface. In a bent configuration, the inner helices bend outward, creating an ~12 Å diameter wide opening.

An archaeal Ca^{2+}-activated K^+ channel, shown in **FIGURE 6.13**, has the tetrameric pore structure typical of K^+ channels. In addition, each subunit has at its C-terminus a large intracellular domain called the regulator of K^+ conductance (RCK) domain, which binds two Ca^{2+} ions. It is thought that these four RCK domains associate with four soluble, cytoplasmic RCK domains. From the crystal structure of this channel in its Ca^{2+}-bound, open conformation, and the structure of unliganded RCK domains from a different K^+ channel, a model has been proposed for how the channel could be gated by Ca^{2+}. In this model, the free energy released upon Ca^{2+} binding is used to move the RCK domains outward, which would cause the inner helices to come apart and widen the pore on the intracellular side.

Voltage-dependent K^+ channels are members of the family of voltage-dependent cation channels that includes Na^+ and Ca^{2+} channels. These channels are highly sensitive, opening in response to small changes in membrane voltage. How does a channel "sense" the membrane voltage and translate it into movement from the closed to open conformation? All of the voltage-dependent cation channels have subunits with six transmembrane segments (see Figure 6.7). The first four transmembrane segments (S1-S4) of each subunit form a voltage-sensing module that controls the opening and closing of the pore. Each S4 segment has several positively charged arginine residues (4–7, depending on the channel), called gating charges, which sense changes of the membrane electric field. The large number of gating charges per channel allows for a steep increase in ion conduction in response to a small change in membrane voltage, as the graph in **FIGURE 6.14** shows for a voltage-gated K^+ channel.

It has been shown that a positive change in the membrane potential, detected by the S4 segments, causes movement of the voltage sensor modules, which exert a force that pulls the inner pore helices away from each other, thereby opening the channel gate. However, it has not been known exactly how movement of the voltage sensors causes the channel to open. There are three models for how the voltage sensors move in response to charges of

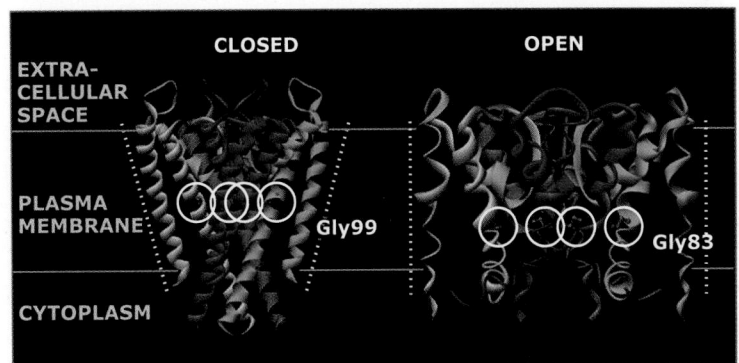

FIGURE 6.12 Crystal structures of the KcsA and MthK K^+ channels in their proposed closed and open configurations, respectively. The glycine residues that form the gating hinges are circled. For the MthK channel, the cytoplasmic domains are not shown. The presumed position of the membrane is indicated. Left structure from Protein Data Bank 1K4C. Y. Zhou, et al., *Nature* 414 (2001): 23–24. Right structure from Protein Data Bank 1LNQ. Y. Jiang, et al., *Nature* 417 (2002): 501–502.

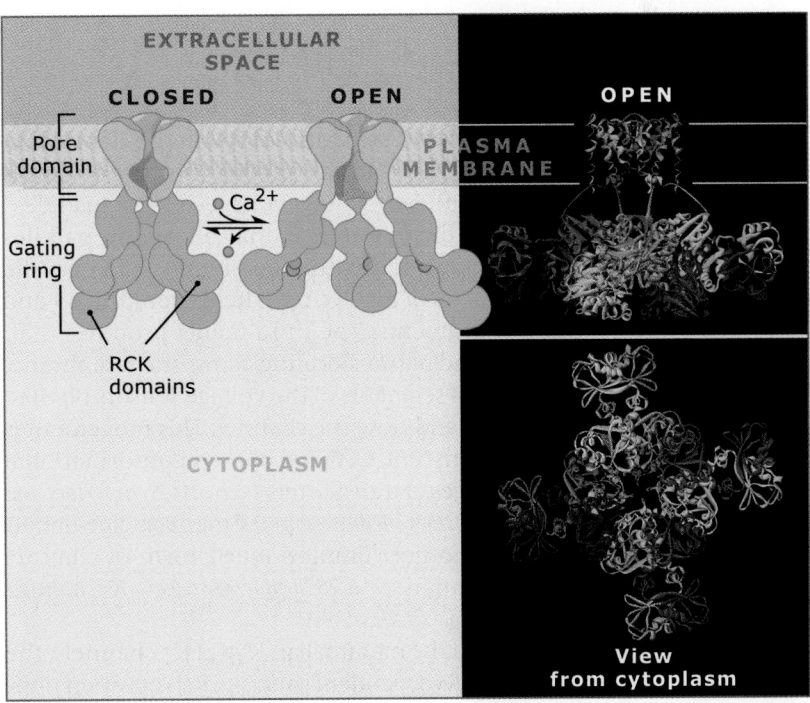

FIGURE 6.13 The schematic shows three of the four K^+ channel subunits. After Ca^{2+} binds to the RCK domains of the gating ring, the diameter of their ring structure increases, exerting force on the pore domain and opening the channel gate. The presumed position of the membrane is indicated. Structures from Protein Data Bank 1LNQ. Y. Jiang, et al., *Nature* 417 (2002): 501–502.

the membrane electric field (see Figure 6.14). One model is that movement of the positively charged S4 segments across the width of the membrane causes the channel to open. A second model has been suggested based on

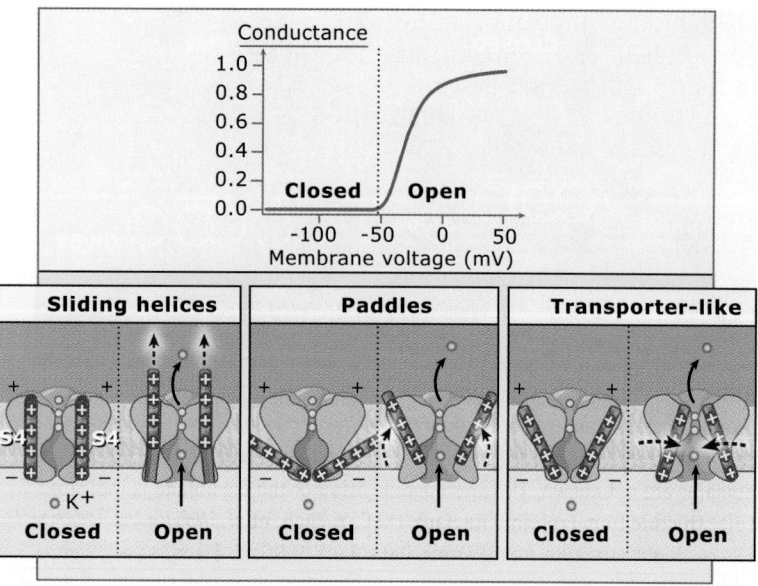

FIGURE 6.14 Three different models for how voltage sensing by K⁺ channels might be translated into channel opening. The S4 transmembrane helices are the positively charged transmembrane segments of the voltage sensor domain (S1-S4). The schematics show cutaway views of the tetrameric channels and only two of four S4 helices per channel.

the crystal structure of a voltage-dependent K⁺ channel. In this model, the pore region is surrounded by four voltage-sensor "paddles" with flexible hinges that allow the paddles to move across the membrane. Each paddle consists of a helix-turn-helix from the S3 and S4 segments. The third model proposes that, instead of translocating across the membrane, the S4 segments of the voltage sensors pivot to open and close the channel. This movement is reminiscent of the proposed conformational changes of transporters (see *6.15 Symporters and antiporters mediate coupled transport*, for details on another common mechanism of channel regulation see *6.25 Supplement: Most K⁺ channels undergo rectification*).

In the mammalian K_V 1.2 K⁺ channels, the voltage-dependent gating switch between nonconductive and conductive states over a small millivolt range is based on movement of independent domains inside the membrane. These voltage sensors (paddles) perform mechanical work on the pore through linker helices analogous to ligand-binding domains of ligand-gated ion channels.

Gene mutations in different K⁺ channels have helped to link K⁺ channel isoforms to their specific physiological functions and to distinct diseases, some of which are described briefly here. Episodic ataxia type-1 is an example of a disease in which a K⁺ channel mutation affects

the nervous system. This inherited disorder is characterized by brief, stress-induced attacks of missing coordination caused by hyperexcitability of motor nerves. This form of ataxia is associated with loss-of-function mutations in the *KCNA1* gene, which encodes a K_V channel. Long QT syndrome is an electric form of heart disease caused by mutations in the *KCNQ1* or other genes, which encode K_V or other cation channels. These mutations result in defective repolarization of the heart that can lead to recurrent arrhythmias and sudden death (see *6.12 Action potentials are electrical signals that depend on several types of ion channels*).

6.7 Voltage-dependent Na⁺ channels are activated by membrane depolarization and translate electrical signals

Key concepts

- The inwardly directed Na⁺ gradient maintained by the Na⁺/K⁺-ATPase is required for the function of Na⁺ channels.
- Electrical signals at the cell membrane activate voltage-dependent Na⁺ channels.
- The pore of voltage-dependent Na⁺ channels is formed by one subunit, but its overall architecture is similar to that of 6TM/1P K⁺ channels.
- Voltage-dependent Na⁺ channels are inactivated by specific hydrophobic residues that block the pore.

Cells maintain an inwardly directed membrane Na⁺ gradient (see Figure 6.2) that is a prerequisite for a large number of Na⁺-dependent membrane transport mechanisms. The electrochemical Na⁺ gradient is generated by the Na⁺/K⁺-ATPase in the plasma membrane (see *6.19 The Na⁺/K⁺-ATPase maintains the plasma membrane Na⁺ and K⁺ gradients*). This ATPase is also called a Na⁺ pump; it uses the energy of ATP hydrolysis to transport Na⁺ and K⁺ ions against their respective electrochemical gradients. Three Na⁺ ions are transported out of the cell for every two K⁺ ions transported in. Thus, the Na⁺ pump is electrogenic, as there is a net movement of electrical charge out of the cell. The net flow of positive charge out of the cell charges the intracellular side more negative relative to the extracellular side. The action of the Na⁺ pump allows the cell to maintain a negative resting membrane potential. The energy required to establish the electrochemical Na^+ gradient comes from ATP hydrolysis and is stored as the electrochemical Na^+ and K^+ gradients at the plasma membrane.

The physiological importance of the electrochemical Na^+ gradient becomes immediately clear upon seeing the large number of Na^+-dependent channels and carrier proteins that depend on this gradient for their functions. These Na^+- or voltage-dependent secondary transport systems, some of which are listed in FIGURE 6.15, use the inwardly directed electrochemical Na^+ gradient and the stored energy to drive solute accumulation against a concentration gradient or to generate electrical signals in the form of action potentials. Two major classes of Na^+-dependent membrane proteins include the voltage-dependent Na^+ channels, which are discussed in this section, and the epithelial Na^+ channels (see *6.8 Epithelial Na+ channels regulate Na+ homeostasis*). A third class is the Na^+/substrate transporters (see *6.16 The transmembrane Na+ gradient is essential for the function of many transporters*).

Excitable cells, such as neurons, muscle cells, and endocrine cells, produce and/or

	Transport protein	Transport stoichiometry	Physiological function
Na^+ EFFLUX	Na^+/K^+–ATPase	3 Na^+ out : 2 K^+ in	Maintenance of transmembrane Na^+ and K^+ gradients
	Na^+/Ca^{2+}-exchanger (*reverse* mode)	3 Na^+ out : 1 Ca^{2+} in	Ca^{2+} entry during phase I of a cardiac action potential
	Na^+/K^+/Ca^{2+}-exchanger (*reverse* mode)	4 Na^+ out : 1 K^+ in : 1 Ca^{2+} in	Light adaptation
	Na^+/HCO_3^--cotransporter (kidney)	1 Na^+ out : 3 HCO_3^- out	Maintenance of the blood and urine pH
	Channels		
	Voltage-dependent Na^+		Rapid Na^+ influx during propagation of action potentials
	Epithelial Na^+		Multiple functions in many tissues, including Na^+ reabsorption in kidney; Maintenance of ionic composition of lung surface liquid, Na^+ absorption in gastrointestinal tract
	Exchangers		
	Na^+/Ca^{2+} (*forward* mode)	3 Na^+ in : 1 Ca^{2+} out	Removal of cytosolic Ca^{2+}
	Na^+/K^+/Ca^{2+} (*forward* mode)	4 Na^+ in : 1 K^+ out : 1 Ca^{2+} out	Light adaptation
	Na^+/H^+	1 Na^+ in : 1 H^+ out	Regulation of intracellular pH and cell volume
	Na^+/Mg^{2+}	2 Na^+ in : 1 Mg^{2+} out	Maintenance of intracellular Mg^{2+} concentration
Na^+ INFLUX	**Cotransporters**		
	Na^+/Cl^-	1 Na^+ in : 1 Cl^- in	NaCl reabsorption in kidney
	Na^+/HCO_3^- (pancreas)	1 Na^+ in : 2 HCO_3^- in	Control of the pH of pancreatic digestive fluids
	Na^+/K^+/Cl^-	1 Na^+ in : 1 K^+ in : 2 Cl^- in	NaCl reabsorption in kidney
	Na^+/glucose	2 Na^+ in : 1 glucose in	Glucose uptake in intestine and reabsorption by kidney
	Na^+/iodide	2 Na^+ in : 1 iodide in	Iodide uptake in thyroid gland and other tissues
	Na^+/proline	1 Na^+ in : 1 proline in	Proline uptake in bacteria; Similar transporters for amino acid reabsorption by kidney

FIGURE 6.15 The Na^+/K^+-ATPase maintains the Na^+ gradient across the plasma membrane by pumping Na^+ ions out of the cell. Na^+ channels mediate the transport of ions down this gradient into cells. Some transporters use the energy released by the flow of Na^+ ions down this gradient to move other solutes up their concentration gradients.

respond to electrical signals. These cells are activated by rapid and transient changes in their resting membrane potential, such as those that occur during the initiation and propagation of an action potential. (For more on action potentials, see *6.12 Action potentials are electrical signals that depend on several types of ion channels*.) During activation, these cells use the transmembrane Na^+-gradient established by the Na^+/K^+-ATPase to translate electrical signals at the cell membrane into intracellular functions. Voltage-gated Na^+ channels are crucial in this process because they are activated by depolarization, during which the membrane potential rapidly becomes more positive. When the membrane potential reaches a critical threshold, Na^+ channels open, selectively conducting Na^+ ions down their electrochemical gradient, into the cell. The Na^+ channels inactivate (close) spontaneously within milliseconds and Na^+ influx stops. Under physiological conditions, the rapid influx of a relatively small number of Na^+ ions changes the electric field of the plasma membrane from negative to positive potentials (referred to as depolarization) but does not affect the global intracellular Na^+ concentration.

Na^+ channels are integral membrane proteins of the plasma membrane that transport Na^+ ions down their electrochemical gradient into cells. They catalyze high rates of ion flux, up to 10^8 Na^+ ions per channel per second. Voltage-gated Na^+ channels have a pore-forming α subunit, shown in **FIGURE 6.16**, and

auxiliary $\beta 1$ and $\beta 2$ subunits. The α subunit consists of four repeat domains, I-IV, which are similar to one another. Each repeat domain contains six predicted α-helical transmembrane segments (1–6), with a pore (P)-loop connecting segments 5 and 6. The voltage-sensors are localized within segments 4, which contain positively charged amino acids. The structure of a voltage-dependent Na^+ channel has yet to be solved at atomic resolution. However, it is proposed that they have a four-fold symmetry around the ion conduction pathway, which is lined by four domains composed of segments 5 and 6 and P-loops. The proposed structure is analogous to the voltage-dependent K^+ channels, except that K^+ channels are assembled from four separate, identical subunits (see *6.5 K^+ channels catalyze selective and rapid ion permeation* and *6.6 Different K^+ channels use a similar gate coupled to different activating or inactivating mechanisms*).

Mutagenesis experiments, in which the effects of amino acid substitutions on ion channel function are assayed, suggest that the P-loops contribute to the selectivity filter of the Na^+ channel. The P-loops contain the signature sequence Trp-Asp-Gly-Leu. Mutations in this sequence affect the channel's selectivity for monovalent cations. The P-loops also bind tetrodotoxin from puffer fish. Tetrodotoxin binding causes paralysis by inactivating voltage-gated Na^+ channels involved in initiation and propagation of action potentials in nerve cells.

It is thought that during membrane depolarization a conformational change and movement of the voltage sensors underlies rapid activation (opening) of voltage-dependent Na^+ channels, similar to the mechanisms proposed for voltage-gated K^+ channels (see Figure 6.14). The β subunits modify the kinetics and voltage-dependence of the gating process and may be important for transport and localization of Na^+ channels to the plasma membrane.

Voltage-dependent inactivation (closing) of Na^+ channels occurs *after* voltage-dependent activation. Inactivation is thought to occur by an initial fast inactivation step followed by a second, slow inactivation step. The cytoplasmic loop between domains III and IV of the Na^+ channel a subunit has been identified by site-directed mutagenesis as an important component of fast channel inactivation. It has been proposed that this cytoplasmic loop occludes the ion pore via a hinged-lid mechanism, with the sequence Ile-Phe-Met (IFM) serving as a

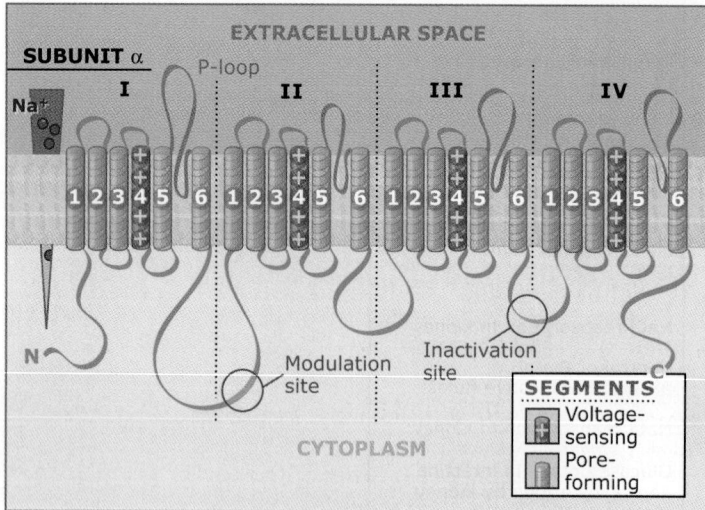

FIGURE 6.16 Proposed topology of the α subunit of voltage-gated Na^+ channels. The four pore forming segments, consisting of transmembrane helices 5 and 6 and the connecting P-loops, are predicted to form the selectivity filter and the channel gate, similar to the architecture of K^+ channels. The inset at left indicates the transmembrane Na^+ gradient.

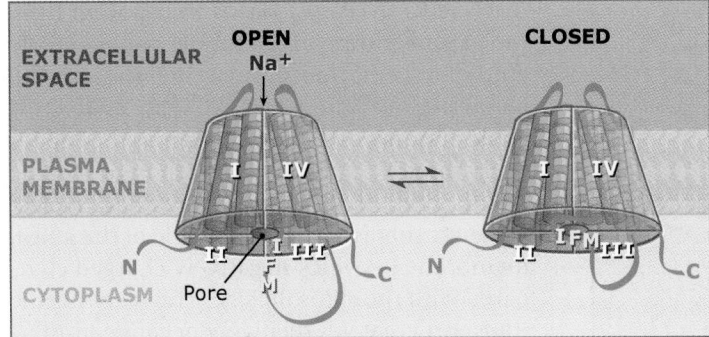

FIGURE 6.17 Na+ channels may use a hinged-lid mechanism at the intracellular side of the pore for fast inactivation. The intracellular loop connecting domains III and IV of the Na+ channel contains the hydrophobic inactivation sequence motif IFM (Ile-Phe-Met), which blocks the pore of the closed channel.

hydrophobic "latch," as **FIGURE 6.17** shows. The glycine and proline residues flanking the IMF motif would provide flexibility to the "hinges" that allow the "lid" to close. In contrast, it is thought that the slow inactivation gate is defined by the P-loops and involves conformational changes in the outer pore. A region near the midpoint of segment 4 of the voltage sensor may also play a role in slow inactivation.

Na+ channels are important targets of anesthetic drugs. Mutagenesis studies have shown that local anesthetics bind with high affinity to domain IV and transmembrane segment 6, which lines the Na+ channel pore. Some drugs used to treat cardiac arrhythmias also act by inhibiting ionic currents through voltage-gated Na+ channels. These drugs, called class I antiarrhythmic drugs, are chemically related to tertiary amine local anesthetics that act on neurons. Cardiac arrhythmias are characterized by abnormal membrane depolarization. Class I antiarrhythmic drugs are thought to inhibit selectively voltage-gated Na+ channels in their inactivated state. The class I antiarrhythmic drugs and related local anesthetics appear to bind on the intracellular side of the Na+ channel and thereby inhibit membrane depolarization leading to potentially deadly arrhythmias.

6.8 Epithelial Na+ channels regulate Na+ homeostasis

Key concepts

- The epithelial Na+ channel (ENaC)/degenerin family of ion channels is diverse.
- The ENaCs and Na+/K+-ATPase function together to direct Na+ transport through epithelial cell layers.
- The ENaC selectivity filter is similar to the K+ channel selectivity filter.

ENaCs are a major class of channels that transport Na+ into cells.

In contrast to the primarily electrical function of the voltage-dependent channels, epithelial Na+ channels mediate bulk flow of Na+ ions and influence water transport across cell layers. As for many other Na+ transport proteins, the function of ENaCs depends on the Na+ gradient established by the Na+/K+-ATPase (see 6.19 The Na+/K+-ATPase maintains the plasma membrane Na+ and K+ gradients; 6.7 Voltage-dependent Na+ channels are activated by membrane depolarization and translate electrical signals; and 6.16 The transmembrane Na+ gradient is essential for the function of many transporters). The ENaCs are not strongly voltage dependent and do not show rapid inactivation but are subject to complex regulation by hormones over a longer timescale. The ENaCs were first discovered in epithelial cells; however, they are also expressed in neuronal and other cell types. ENaCs are related by sequence to the C. elegans degenerin proteins that are involved in touch sensitivity.

ENaCs support a variety of functions, such as blood pressure regulation, reproduction, digestion, and coordination. For example, in the kidney and the gastrointestinal tract, ENaC channels are important for maintaining Na+ and K+ concentrations in the blood plasma and the composition of the urine and stool, respectively. In the lung and salivary glands, ENaC-dependent Na+ transport helps maintain the ion composition of the airway surface liquid and the fluids that are secreted in order to digest food, respectively. In this section, we will describe the role of ENaCs in absorption and transport of Na+ ions across epithelial cells of the kidney.

The major function of the kidneys is to process blood plasma by filtering out metabolic waste products, such as urea, into the urine. In doing so, the kidney maintains homeostasis by regulating the amount of bodily fluids and the concentrations of solutes, which in turn influ-

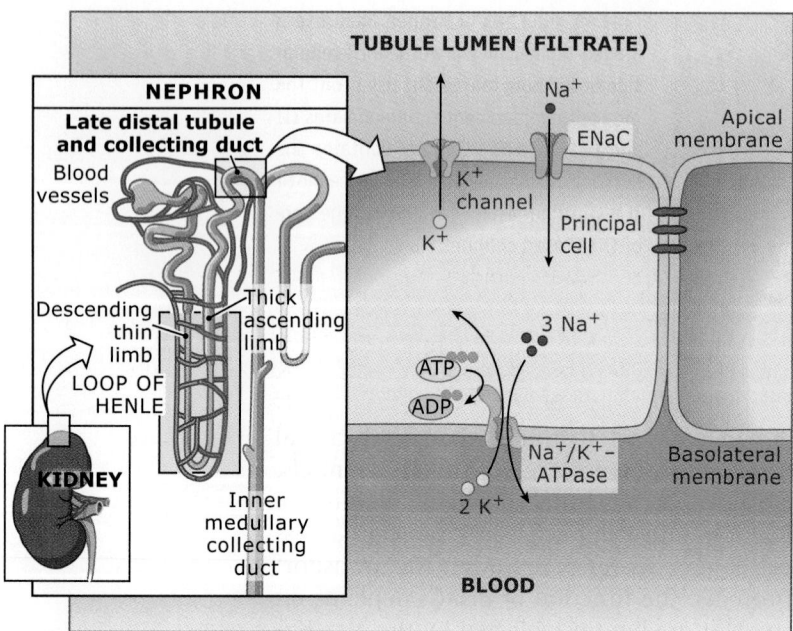

FIGURE 6.18 Epithelial Na$^+$ channels (ENaCs), which are expressed in the apical membrane of cells lining the kidney's collecting tubules, mediate Na$^+$ reabsorption from the kidney plasma filtrate.

ences blood pressure. As part of the complex filtering, reabsorption, and secretion process, ions, sugars, amino acids, and small proteins pass into the renal tubules, which contain the plasma ultrafiltrate. The renal tubule and duct lumen is lined with polarized epithelial cells. The apical membrane faces the filtrate (lumenal) side and the basolateral membrane faces the blood vessel lumen, as shown in **FIGURE 6.18**. These specialized epithelial cells reabsorb many components of the plasma filtrate, such as salts and water, and transport them back into the blood plasma, thereby maintaining the correct concentrations of these components and the blood volume.

ENaCs are located in the apical membrane of specialized epithelial cells in the distal tubule and collecting duct of the kidney. These are among the primary sites for Na$^+$ entry into cells from the plasma filtrate (the precursor of urine) (see Figure 6.18). Na$^+$ ions move through ENaCs down their electrochemical gradient (more Na$^+$ ions outside the cell than inside). Na$^+$ transport by ENaC channels at the apical membrane is coupled to Na$^+$ transport across the basolateral membrane by Na$^+$/K$^+$-ATPases, which results in Na$^+$ transport back into the blood capillary. By transporting Na$^+$ out of the cytosol, the Na$^+$/K$^+$-ATPases create the large electrochemical Na$^+$ gradient necessary for ENaC function. As a result of the combined activities of ENaCs and

the Na$^+$/K$^+$-ATPases, Na$^+$ is transported from the plasma filtrate in the tubule lumen, across epithelial cells, and ultimately back into the blood plasma. The Na$^+$/K$^+$-ATPase is electrogenic: it transports two K$^+$ ions into the cell for every three Na$^+$ ions that it transports out of the cell. In addition, Na$^+$ transport by ENaCs is electrogenic, that is, the lumenal side of the apical membrane becomes negatively charged compared with the cytosolic side. This electrogenic transport creates a relatively negative intralumenal environment, which favors K$^+$ secretion into the filtrate through apical K$^+$ channels. Thus, transepithelial Na$^+$ transport is important to maintain the composition and the volume of the fluid on both the apical and basolateral sides of the epithelial cell layer. ENaC function in the distal tubules and collecting ducts of the kidney contributes to reabsorption of ~7% of filtered Na$^+$ and Cl$^+$ ions, and the secretion of variable amounts of K$^+$. (For details on the Na$^+$/K$^+$-ATPase see *6.19 The Na$^+$/K$^+$-ATPase maintains the plasma membrane Na$^+$ and K$^+$ gradients*.)

Na$^+$ reabsorption from the kidney filtrate through ENaCs is regulated by the hormones aldosterone and vasopressin. These hormones are released from the adrenal glands and pituitary gland, respectively, upon dehydration or salt deprivation and bind to receptors on kidney cells. This leads to expression of ENaCs at the plasma membrane and increased Na$^+$ reabsorption from the filtrate through ENaC and transport of Na$^+$ to the blood plasma. Hormonal regulation allows for Na$^+$ and fluid balance to be maintained according to acute metabolic needs. (For a discussion of the role of other channels in kidney function see *6.10 Cl$^-$ channels serve diverse biologic functions* and *6.11 Selective water transport occurs through aquaporin channels*.)

The ENaC consists of three homologous subunits, α, β, and γ, which form a multimeric channel complex. High-resolution structures of ENaCs are not yet available. However, from *in vitro* reconstitution studies and sequence analysis, it has been predicted that each ENaC complex consists of three subunits (α, β, and γ), with each subunit contributing to form a central channel pore, as **FIGURE 6.19** shows. The predicted topology of each subunit is that of two transmembrane segments, a large extracellular loop, and intracellular N- and C-terminal cytoplasmic domains. Each subunit of the tetrameric channel assembly would contribute to the pore through the second transmembrane segment. The stoichometry of these subunits

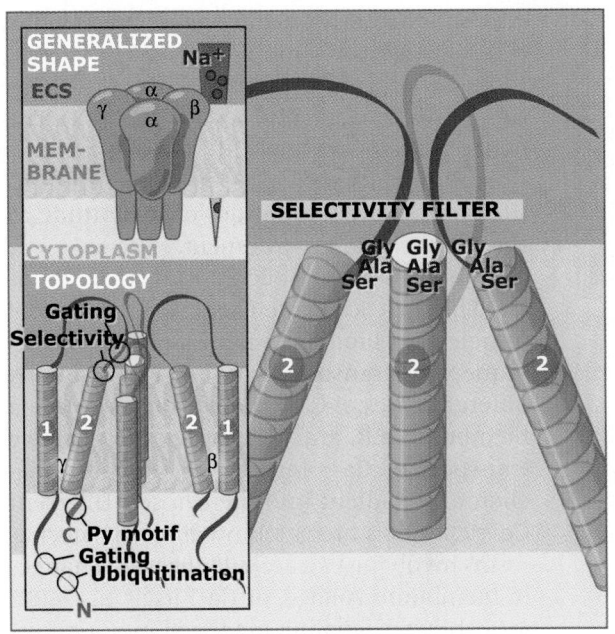

FIGURE 6.19 Epithelial Na+ channels (ENaCs) consist of two α-subunits, one β-subunit, and one γ-subunit. Each subunit is predicted to have two transmembrane segments, an extracellular loop, and intracellular N- and C-termini. Each subunit contributes Gly-Ala-Ser residues to the selectivity filter. Ubiquitination of ENaC is one of the mechanisms that regulates internalization and Na+ transport activity. Additional mechanisms include aldosterone-induced phosphorylation of intracellular residues. Insert shows side view of the acid-sensing ion channel ASIC1, a homotrimeric structure embedded in the plasma membrane which is representative for the ENaC/degenerin family of ion channels.

has to be verified, but ENaC is very likely a heterotrimeric protein like the recently analyzed acid-sensing ion channel 1 (ASIC1), which belongs to the same ENaC/degenerin family, as outlined in Figure 6.19.

The ENaC selectivity filter allows the channel to selectively transport Na+ ions. The conserved selectivity filter of the ENaC/degenerin family contains the signature motif Gly-Ala-Ser adjacent to the second transmembrane segment, at the narrowest part of the pore (see Figure 6.19). Mutation of any of these three residues results in a channel with drastically altered ion conduction properties, as determined by electrophysiological measurements. It is thought that backbone carbonyl oxygens from Gly-Ala-Ser line the selectivity filter and stabilize partially dehydrated ions of a specific diameter. This arrangement is similar to the selectivity filter of certain K+ channels; however, these K+ channels are not related to ENaCs by sequence homology (see Figure 6.9 and *6.5 K+ channels catalyze selective and rapid ion permeation*).

The diuretic drug amiloride blocks Na+ reabsorption by ENaC in the lumenal membrane of the kidney distal tubule and collecting duct (see Figure 6.18). (ENaC function is, therefore, characterized as an amiloride-sensitive Na+ current in electrophysiological measurements.) Amiloride competes with Na+ ions for binding to the extracellular side of ENaC near the selectivity filter. The effect of amiloride on Na+ reabsorption in the distal nephron of the kidney is the basis for its clinical use as a diuretic and in the treatment of hypertension. Because amiloride treatment results in decreased Na+ reabsorption from the filtrate, there is increased elimination of Na+ ions and water in the urine. The decreased Na+ reabsorption into epithelial cells results in a lower Na+ concentration in the blood, which lowers or normalizes blood pressure.

Mutations in the ENaC genes can result in severe abnormalities of blood pressure regulation. In a rare hereditary form of hypertension called Liddle's syndrome, mutations in ENaC cause overactive channels and abnormally high levels of Na+ reabsorption in the distal nephron of the kidney, resulting in increased blood plasma volume, arterial hypertension, and low plasma K+ levels. Mutations that cause Liddle's syndrome are localized to either the β or γ channel subunit genes. In contrast, mutations that reduce ENaC function in the disease pseudohypoaldosteronism type I are associated with hypotension (decreased blood pressure) and decreased Na+ and increased K+ plasma levels. Characterization of these genetic diseases has significantly increased understanding of the complex regulation of epithelial Na+ channels by aldosterone and its contribution to blood pressure regulation and plasma homeostasis.

6.9 Plasma membrane Ca²⁺ channels activate intracellular and intercellular signaling processes

Key concepts

- Cell surface Ca²⁺ channels translate membrane signals into intracellular Ca²⁺ signals.
- Voltage-dependent Ca²⁺ channels are asymmetric protein complexes of up to five different subunits.
- The α_1 subunit of voltage-dependent Ca²⁺ channels contains the conduction pore, the pore loop structures, the voltage sensors, the gating apparatus, and drug binding sites.
- The Ca²⁺ channel selectivity filter forms an electrostatic trap.
- Ca²⁺ channels are stabilized in the closed state by channel blockers.
- Ca²⁺ channel mutations cause neurologic diseases.

Calcium (Ca²⁺) is a universal second messenger that controls many cellular functions, including contraction of cardiac and skeletal muscles, visual processing in the retina, immune responses by T lymphocytes, neuronal excitability and mood behavior, and insulin secretion by pancreatic α cells. Activation of diverse cellular functions is mediated by changes in the cytosolic Ca²⁺ concentration, which is ~10,000 times lower than the extracellular Ca²⁺ concentration for resting cells (see Figure 6.2). The endoplasmic reticulum (ER) and sarcoplasmic reticulum (SR) store Ca²⁺ at concentrations that are on the same order of magnitude as the extracellular environment.

Changes in cytosolic Ca²⁺ levels are regulated by the coordinated action of a variety of soluble Ca²⁺ binding proteins and transmembrane Ca²⁺ transport proteins. For example, different types of Ca²⁺ channels in the plasma membrane, ER, and SR catalyze the selective transport of Ca²⁺ ions down their electrochemical gradient into the cytosol. Different Ca²⁺ channels are gated by different mechanisms involving extracellular ligands, changes in membrane voltage, or Ca²⁺ itself, as **FIGURE 6.20** shows. Signaling mediated by Ca²⁺ ions terminates when plasma membrane Ca²⁺ channels close and Ca²⁺ is extruded from the cytosol through specialized transport proteins (see *6.18 The Ca²⁺-ATPase pumps Ca²⁺ into intracellular storage compartments*). (For more on release of Ca²⁺ from the ER and SR see *6.13 Cardiac and skeletal muscles are activated by excitation-contraction coupling*.) Ca²⁺-dependent inactivation and voltage-dependent inactivation are the two principal mechanisms that control closing of plasma membrane Ca²⁺ channels and prevent cell damage by excessive Ca²⁺ entry. Ca²⁺ binding to the protein calmodulin, which then binds to the intracellular domain of certain Ca²⁺ channels in response to Ca²⁺ influx, allows for negative feedback inactivation of many types of Ca²⁺ channels. In this section, we will discuss the proposed mechanism of ion flux through Ca²⁺ channels, focusing on a class of voltage-gated Ca²⁺ channels, and compare it with the mechanism used by K⁺ channels.

Voltage-gated Ca²⁺ channels allow Ca²⁺ to enter cells when the membrane potential becomes more positive during action potential mediated depolarization (see Figure 6.20). For example, neuronal action potentials activate voltage-gated Ca²⁺ channels and the release of neurotransmitters at the synapse. Thus, these channels translate electrical signals at the plasma membrane into an intracellular signal. The influx of Ca²⁺ increases the intracellular Ca²⁺ concentration to a level that triggers a variety of processes such as muscle contraction, hormone or neurotransmitter release, activation of Ca²⁺-dependent signaling cascades, and gene transcription (for more on muscle con-

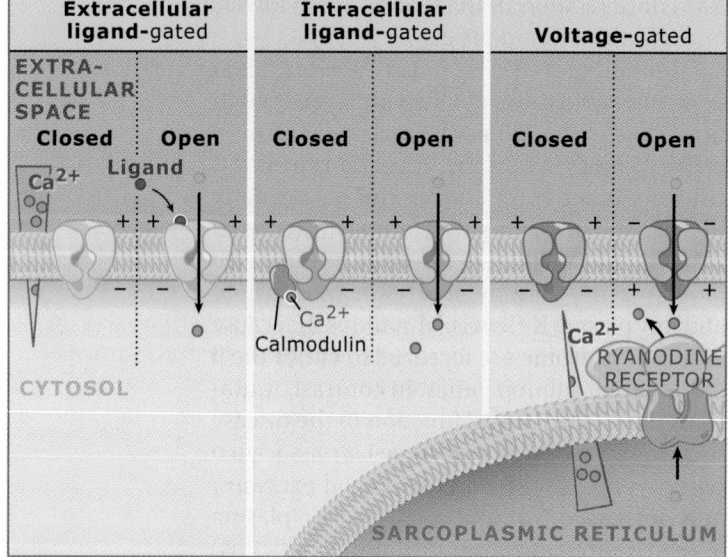

FIGURE 6.20 Three basic types of channels transport Ca²⁺ ions from the outside of the cell into the cytosol. In addition, ion channels transport Ca²⁺ ions from intracellular Ca²⁺ stores to the cytosol. The insets indicate the Ca²⁺ gradients maintained across the plasma membrane and sarcoplasmic reticulum membrane.

traction see *6.13 Cardiac and skeletal muscles are activated by excitation-contraction coupling*). There are different types of voltage-gated Ca^{2+} channels, categorized according to their electrophysiological and pharmacological properties. Here we will describe the L-type Ca^{2+} channels, which were the first to be cloned and are the most studied thus far. L-type Ca^{2+} channels are present in the plasma membranes of skeletal, cardiac, and smooth muscle cells, as well as neurons, and are activated by membrane depolarization.

L-type Ca^{2+} channels are so-called because they exhibit long-lasting openings. These channels include the major $Ca_v1.X$ isoforms of skeletal, cardiac, and smooth muscle cells, as well as neuronal, endocrine, and retinal cells. L-type Ca^{2+} channels are hetero-oligomeric protein complexes composed of five subunits: α_1, α_2, δ, β, and γ, which are depicted in **FIGURE 6.21**. The transmembrane $\alpha_2\Delta$ subunit complex is transcribed from a single gene, posttranscriptionally cleaved, and connected by disulfide bonds.

Ca^{2+} channel modulation and targeting to the cell membrane requires all subunits. Analysis of the largest α_1 subunit sequence for hydrophobic and hydrophilic residues predicts four homologous domains, each consisting of six transmembrane α helices (S1-S6), including the pore region S5-pore loop-S6 contributing to the selectivity filter, and a voltage-sensor module comprised of four transmembrane segments (S1-S4) including the positively charged S4 segment. This type of transmembrane organization is similar to voltage-gated Na^+ channels (see Figure 6.16 and *6.7 Voltage-dependent Na^+ channels are activated by membrane depolarization and translate electrical signals*) and to voltage-gated K^+ channels (see *6.6 Different K^+ channels use a similar gate coupled to different activating or inactivating mechanisms*). The β subunit contributes to cell surface expression of the α_1 subunit and functional regulation of the channel. It engages with the α_1 subunit and may influence gating through an interaction with the pore-forming S6 transmembrane segment in domain I.

A three-dimensional surface structure of an L-type voltage-gated Ca^{2+} channel has been obtained by electron cryomicroscopy and is consistent with the proposed subunit assembly shown in Figure 6.21. The largest portion of the Ca^{2+} channel includes the extracellular structure of the α_2 subunit, and the extracellular loops of the α_1 and γ subunits.

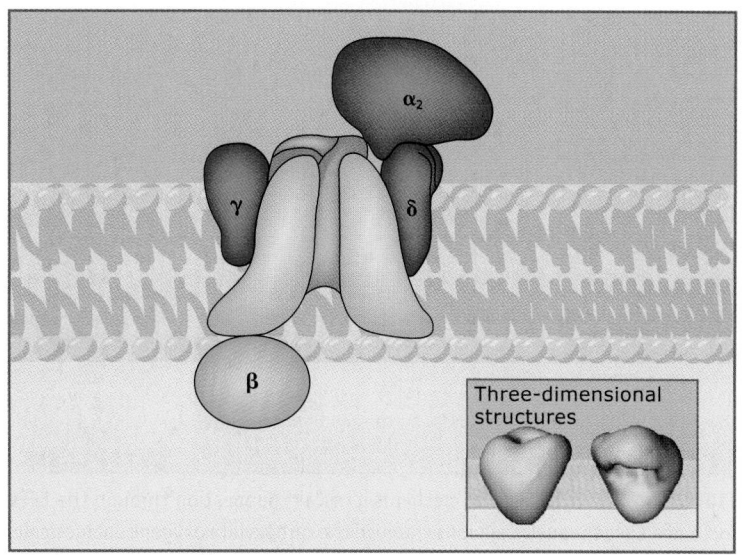

FIGURE 6.21 Predicted membrane topology of L-type Ca^{2+} channel subunits. The three-dimensional surface structure, for which two views are shown, in the inset, was obtained by cryoelectron microscopy. Top adapted from W. A. Catterall and A. P. Few, *Neuron* 59 (2008): 882–901; bottom reproduced from I. I. Serysheva, et al., *Proc. Natl. Acad. Sci. USA* 99 (2002): 10370–10375. Copyright © 2002 National Academy of Sciences, U.S.A. Photo courtesy of Susan L. Hamilton, Baylor College of Medicine.

Ca^{2+} channels are highly selective for Ca^{2+} over Na^+ ions yet allow high rates of ion permeation. Na^+ ions are the most abundant extracellular cations and are ~100-fold more numerous than Ca^{2+} ions (see Figure 6.2). Na^+ ions have approximately the same diameter as Ca^{2+} ions, 2.0 Å. Thus, a simple molecular sieve would not be able to discriminate between Ca^{2+} and Na^+ ions. How do Ca^{2+} channels achieve high selectivity for Ca^{2+} ions and allow rapid Ca^{2+} flux at the same time?

The pores of K^+ channels have selectivity filters that act as surrogate water environments to stabilize the permeating ions, which are dehydrated as they pass through the selectivity filter. For example, the selectivity filters of K^+ channels have four P-loops that form relatively rigid structures lined with backbone carbonyl oxygen atoms that coordinate K^+ ions at specific positions in a linear permeation pathway (see *6.5 K^+ channels catalyze selective and rapid ion permeation*). A model of the selectivity filter of voltage-gated Ca^{2+} channels is most likely also formed by four P-loops. However, it is thought to be a more flexible structure, with each P-loop contributing a glutamate residue such that the carboxylate oxygen atoms, rather than the carbonyl oxygens, line the cation per-

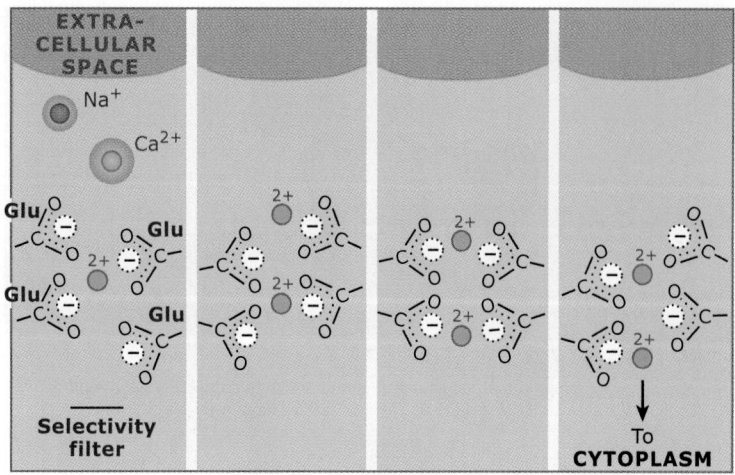

FIGURE 6.22 A hypothetical mechanism of Ca²⁺ permeation through the EEEE locus of a Ca²⁺ channel. Ca²⁺ ions stabilize the carboxylate oxygens energetically and prevent permeation by Na⁺ ions, which do not have sufficient positive charge to fit and partially dehydrate efficiently within the EEEE locus. Electrostatic repulsion of Ca²⁺ ions within the pore may facilitate rapid diffusion.

meation pathway, as **FIGURE 6.22** shows. (This arrangement resembles EF hand Ca²⁺-binding sites, in which Ca²⁺ sits in a pocket containing oxygen atoms, many of which are contributed by carboxylate groups.) The four glutamate residues form the so-called EEEE locus, a structure that is conserved among Ca²⁺ channels.

The EEEE locus is responsible for the selective permeability of Ca²⁺ ions over other physiological cations. Replacement of any of the four glutamate residues by other amino acids results in a channel with compromised selectivity for Ca²⁺ ions. The glutamate side chains are thought to mediate high-affinity binding of Ca²⁺, but not other ions, near the extracellular pore entrance. This idea is based on *in vitro* electrophysiological measurements showing that in the absence of Ca²⁺ ions, Na⁺ ions can permeate Ca²⁺ channels. In addition, the Na⁺ ions permeate the channels more rapidly than Ca²⁺ ions. The slower rate for passage of Ca²⁺ ions suggested that Ca²⁺ ions bind to the channel pore with higher affinity than Na⁺ ions, which would effectively prevent most Na⁺ ions from entering the pore even though Na⁺ ions are more abundant than Ca²⁺ ions in the extracellular space.

The affinity of Ca²⁺ channels for Ca²⁺ is about 10^{-6} M. However, this affinity gives a calculated rate of ion permeation that is 1000 times lower than the measured rate of 10^6 per second. A model to reconcile the affinity and the rate of flux has been proposed. In this model, the EEEE locus accommodates multiple Ca²⁺ ions, and a Ca²⁺ ion entering the pore would cause release of a bound Ca²⁺ ion from the other side due to electrostatic repulsion between the Ca²⁺ ions (see Figure 6.22). Thus, the electrostatic repulsion helps to overcome the affinity of Ca²⁺ binding, which would tend to slow Ca²⁺ permeation. This model is similar to the model proposed for K⁺ channels, which have been shown to have a series of distinct ion binding sites (see *6.5 K⁺ channels catalyze selective and rapid ion permeation*).

Voltage-dependent Ca²⁺ channels are affected by genetic lesions, for example, missense mutations in presynaptic Ca$_v$2.1 channels cause familial hemiplegic migraine and different forms of ataxia whereas mutations in Ca$_v$1.2 channels cause the Timothy syndrome including cardiac arrhythmias and autism spectrum disorder. Voltage-dependent Ca²⁺ channels are major clinical targets for drugs that treat hypertension, cardiac arrhythmias, and other diseases. Commonly used drugs include phenylalkylamines, benzothiazepines, and dihydropyridines, which are often referred to as Ca²⁺ antagonists. Ca²⁺ antagonists were originally introduced in medical practice as vasodilators to treat hypertension. They decrease Ca²⁺ levels in smooth muscle, for example, in blood vessels, resulting in relaxation of vessel tonus and reduction in blood pressure. Mutagenesis and binding studies have identified drug binding sites in segments S5 and S6 of domain III and segment S6 of domain IV, on the cytoplasmic side of the proposed selectivity filter (see Figure 6.21). Phenylalkylamine inhibitors block the Ca²⁺ channels by direct interaction with glutamate residues in the P-loop of the selectivity filter, from the cytoplasmic side, whereas dihydropyridines and benzothiazepines enter the channel pore from the extracellular side.

6.10 Cl⁻ channels serve diverse biologic functions

Key concepts

- Cl⁻ channels are anion channels that serve a variety of physiological functions.
- Cl⁻ channels use an antiparallel subunit architecture to establish selectivity.
- Selective conduction and gating are structurally coupled in Cl⁻ channels.
- K⁺ channels and Cl⁻ channels use different mechanisms of gating and selectivity.

Chloride (Cl⁻) channels are members of a large family of anion channels. As with other ion channels, Cl⁻ channel proteins form pores in biologic membranes. Cl⁻ channels allow for the transport of negatively charged Cl⁻ ions down their electrochemical gradient. In *in vitro* assays, Cl⁻ channels function as nonselective anion channels that in some cases conduct other anions better than Cl⁻. However, Cl⁻ is the most abundant anion in organisms and is, therefore, the predominant ion transported by these channels *in vivo*.

Cl⁻ channels are present in the plasma membrane and in membranes of intracellular organelles. Several important functions of Cl⁻ channels include regulation of cell volume, ionic homeostasis, and transepithelial ion transport. Plasma membrane Cl⁻ transport is also important for the regulation of membrane excitability in muscle and neurons. In addition, during acidification of intracellular compartments such as endosomes, transport of Cl⁻ into organelles through Cl⁻ channels neutralizes the positive charges of protons transported by H⁺-ATPases (see *6.21 H⁺-ATPases transport protons out of the cytosol*).

Cl⁻ channels can be divided into three different gene families. First, the CLC gene family has several members that are targeted either to the plasma membrane or the membranes of intracellular compartments. CLC channels are conserved from bacteria to humans and can be further divided into subclasses based on homology. Second, the cystic fibrosis transmembrane conductance regulator is the only member of the family of ATP binding cassette (ABC) transporters that is known to function as an ion channel (see *6.26 Supplement: Mutations in an anion channel cause cystic fibrosis*). Third, the ligand-gated g-aminobutyric acid receptor and glycine receptor are members of a distinct family of Cl⁻ channels that have specialized functions in the central nervous system.

The CLC Cl⁻ channels are homodimers, with each subunit forming its own ion-conducting pore. No X-ray crystal structures of eukaryotic CLC Cl⁻ channels are available to provide insight at the atomic level. However, crystal structures of bacterial CLC proteins have been characterized. The bacterial CLC proteins function as carrier proteins that transport Cl⁻ ions in exchange for protons, rather than as Cl⁻ ion channels, as was originally proposed. However, based on sequence homology, some features of this bacterial transporter are likely to be similar to eukaryotic CLC Cl⁻ channels.

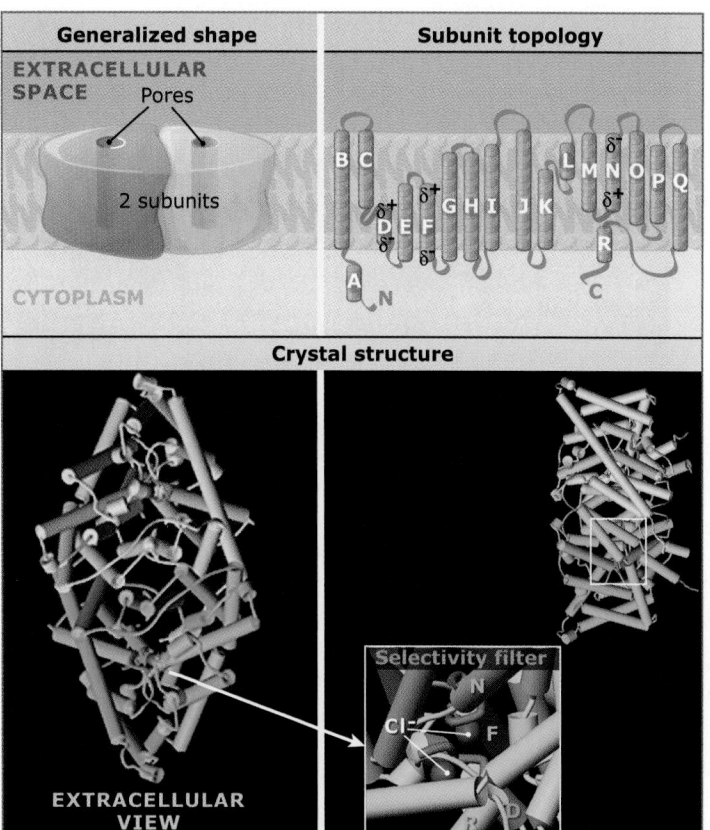

FIGURE 6.23 Schematics and X-ray crystal structure of a bacterial CLC Cl⁻ transporter complex. The Cl⁻ transport proteins of the CLC family form homodimers, with each subunit forming an individual pore. The parts of one protein subunit that contribute to the selectivity filter are highlighted in orange. The extracellular view is shown smaller than the side view. The presumed position of the membrane is indicated. Bottom structures adapted from R. Dutzler, et al., *Nature* 415 (2002): 276–277.

Indeed, the bacterial CLC transporter is also a homodimer, as **FIGURE 6.23** shows. Each subunit contains 18 α helices that are arranged in a complex intramembrane topology. The N-terminal half (helices A to I) of one subunit is structurally related to its C-terminal half (helices J to R). The two halves of a subunit have opposite (antiparallel) membrane orientations, similar to the architecture of the unrelated aquaporin channels (see *6.11 Selective water transport occurs through aquaporin channels*).

An important characteristic of all CLC Cl⁻ channels is that ion conductance and gating are intrinsically coupled. This characteristic is evident from the structure of the prokaryotic CLC selectivity filter. The pore forms an hourglass shape, with the selectivity filter at the narrowest part of the pore in the center of the protein. There are two components that

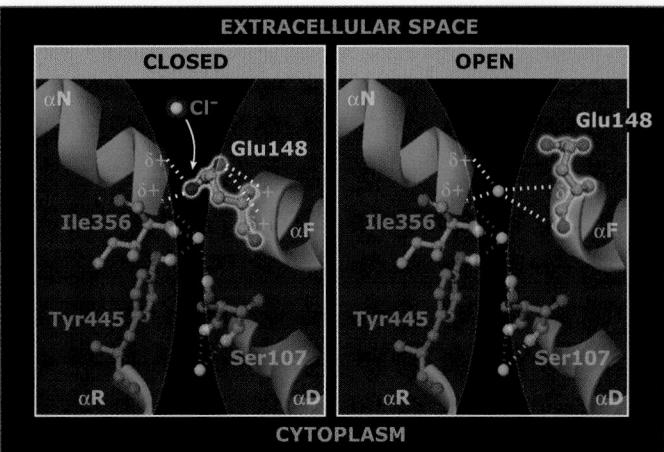

FIGURE 6.24 The selectivity filter of Cl⁻ channels contains three Cl⁻ ion binding sites. In the closed conformation, a glutamic acid side chain (Glu148) occupies the Cl⁻ binding site that is closest to the extracellular vestibule. In the open conformation, this side chain swings out of the permeation pathway, allowing a Cl⁻ ion to bind. Residues involved in coordinating two other Cl⁻ ions are shown. Images are based on the crystal structure of an E. coli CLC channel. Structures from Protein Data Bank 1OTS. R. Dutzler, E. B. Campbell, and R. MacKinnon, *Science* 300 (2003): 108–112.

confer selectivity for anions over cations. First, four a-helical segments (D, F, N, and R) point their partially positive-charged N-termini into the center plane of the membrane (see Figure 6.23). Second, specific amino acids that line the pore form hydrogen bonds with Cl⁻ ions, as **FIGURE 6.24** shows. These components provide a favorable electrostatic environment for Cl⁻ by stabilizing the Cl⁻ ions in the selectivity filter and repelling positively charged cations.

A glutamate side chain on one side of the selectivity filter is thought to act as a gate (see Figure 6.24). The model for how this gate works is based on analysis of X-ray crystal structures of wild-type bacterial CLC transporters and mutant transporters in which glutamate is substituted with a different residue. When the channel is closed, this specific glutamate side chain occupies a Cl⁻ binding site, thereby mimicking occupation by a Cl⁻ ion. The glutamate side chain swings out of the permeation pathway to the extracellular vestibule to open the gate and is replaced by a permeating Cl⁻ ion. This glutamate residue is conserved in almost all CLC channels. It is, therefore, likely that the selectivity filter of the bacterial CLC transporter is similar to that of the eukaryotic CLC Cl⁻ channels.

What causes the glutamate gate to rotate into the open position? Different Cl⁻ channels are gated by ligands, voltage, or changes in in-

tracellular Ca^{2+} concentration. Chloride channels can also be activated by Cl⁻ ion gradients. The model shown in Figure 6.24 predicts that above a certain extracellular Cl⁻ concentration, the Cl⁻ displaces the glutamate gate out of the Cl⁻ binding site and allows for anion permeation. Thus, Cl⁻ ions and negatively charged carboxylate groups of the glutamate side chain undergo electrostatic repulsion and compete with each other for ionic bonds with the partial positive charge of a helix N. This arrangement allows for rapid anion conduction and direct regulation of the gate by the permeating Cl⁻ ions. Since changes in membrane voltage may exacerbate the electrochemical potential of transmembrane Cl⁻ gradients, this model also explains the voltage-dependent opening of most CLC Cl⁻ channels. CLC Cl⁻ channels have no charged transmembrane domains that may act as voltage sensors, which are present in voltage-dependent cation channels (see *6.6 Different K⁺ channels use a similar gate coupled to different activating or inactivating mechanisms*).

Cl⁻ channels and K⁺ channels use fundamentally different mechanisms to establish ion conduction and selectivity. First, in CLC Cl⁻ channels, the selectivity filter and gate form one structural unit. In contrast, in K⁺ channels, the selectivity filter and gate are structurally separated on the extracellular and intracellular sides of the channel, respectively (see Figure 6.8). The separation of these two structural components allows for ligand-binding domains or voltage sensor domains to open and close the pore through conformational changes without affecting the selectivity filter, the structure of which must be maintained in order to discriminate among cations with only small differences in their radius. Second, the Cl⁻ channel gate appears to involve a smaller movement (rotation of the glutamate side chain) than the larger conformational change proposed for K⁺ channels (see *6.6 Different K⁺ channels use a similar gate coupled to different activating or inactivating mechanisms*). Third, the favorable electrostatic environment for Cl⁻ or K⁺ ions, respectively, arises from partial positive or partial negative charges, respectively. The CLC Cl⁻ channels use an antiparallel architecture to direct the partial positive charges of N-terminal helix dipoles toward the selectivity filter, as **FIGURE 6.25** shows. In contrast, K⁺ channels use a parallel architecture that focuses partially negative C-terminal helix dipole charges toward a water-filled cavity as part of their selectivity filter, as Figure 6.11 shows.

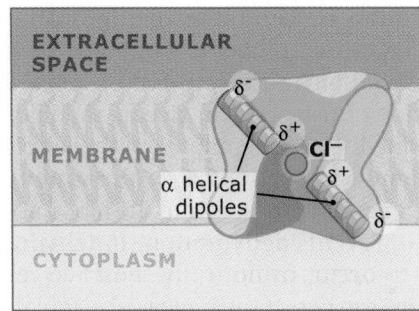

FIGURE 6.25 The two halves of a CLC Cl⁻ channel subunit have an antiparallel orientation such that the partially positive-charged dipoles of certain α helices point toward the selectivity filter and influence Cl⁻ anion binding from opposite sites of the membrane. Only one subunit is shown.

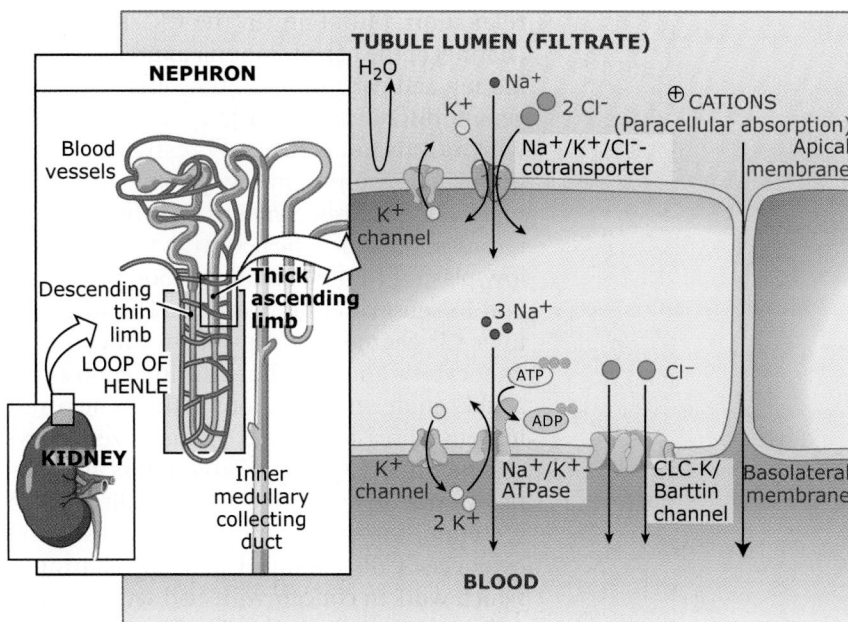

FIGURE 6.26 Transepithelial NaCl transport in the kidney. Epithelial cells in a specialized nephron segment (the thick ascending limb of the loop of Henle) reabsorb Na⁺ and Cl⁻ from the filtrate (urine) through Na⁺/K⁺/Cl⁻-cotransporters in the apical membrane and through CLC channels in the basolateral membrane.

The functions of Cl⁻ channels in transepithelial ion transport are exemplified by the roles of ClC-K channels in the kidney, as **FIGURE 6.26** shows. Expression of these channels at the basolateral membrane in the thick ascending limb of Henle's loop requires association with the b subunit barttin. The ascending limb of Henle's loop functions as the anatomic site of 25% of ion (NaCl and K⁺) reabsorption and is impermeable to water. In particular, the ClC-K channels are required for NaCl reabsorption and indirectly affect maintenance of the K⁺ concentration in blood plasma, which is important because nervous system and heart functions are especially sensitive to altered concentrations of K⁺ ions. The kidney filters out waste products together with ions from the blood. During this process, anions and cations are lost into the plasma filtrate that becomes the urine, and these ions must be reabsorbed in quantities that maintain their correct concentrations in the blood. In the epithelial tubule segment of the thick ascending limb of Henle's loop, Na⁺ and Cl⁻ ions are reabsorbed from the filtrate (for more on reabsorption of Na⁺ ions see *6.8 Epithelial Na⁺ channels regulate Na⁺ homeostasis*). Reabsorption of Cl⁻ ions from the filtrate occurs through Na⁺/K⁺/Cl⁻-cotransporters in the apical membrane. This transport is driven by the basolateral Na⁺/K⁺-ATPase, which establishes a transmembrane Na⁺ gradient by pumping Na⁺ ions out of the cell. The low intracellular Na⁺ concentration provides a favorable electrochemical gradient for the movement of Na⁺-coupled transport processes. This reabsorption of Cl⁻ ions occurs via apical Na⁺/K⁺/Cl⁻-cotransporters and

increases the intracellular Cl⁻ concentration so that it is higher than the extracellular, lumenal Cl⁻ concentration and allows Cl⁻ to leave the cell down its concentration gradient. Cl⁻ ions exit the cell through ClC-K chloride channels in the basolateral membrane, into the blood. K⁺ ions, which enter the cell through the apical Na⁺/K⁺/Cl⁻-cotransporter and the basolateral Na⁺/K⁺-ATPase, exit the cell through K⁺ channels in the apical and basolateral membranes. Without the ClC-K Cl⁻ channels, ion flux through the Na⁺/K⁺/Cl⁻-cotransporter would be insufficient, resulting in decreased NaCl and K⁺ absorption.

In contrast to most mammalian cells, for which the resting membrane potential is dominated by K⁺ conductance, in skeletal muscle the ClC-1 chloride channel isoform provides ~80% of the resting membrane conductance. The important roles of different CLC Cl⁻ channels, in conjunction with other ion transport proteins, have been revealed by gene mutations that result in human diseases and by knockout mice lacking expression of specific Cl⁻ channel isoforms. For example, mutations that result in inactive ClC-1 channels cause different forms of prolonged muscle contraction called myotonias, in which abnormal excitability leads to skeletal muscle stiffness and impaired relaxation. Mutations in the ClC-K chloride channel or the barttin subunit expressed in the kidney and ear result in autosomal-recessive genetic disorders, called Bartter's syndrome, in which Cl⁻ is not sufficiently reabsorbed from the urine, resulting in low plasma K+ levels, metabolic alkalosis, compensatory hyperaldosteronism (in which the body tries to counteract low plasma volume and low blood pressure), and in sensorineural deafness. The intracellular Cl⁻ channel isoform ClC-5 functions in endocytosis and as a shunt during endosomal acidification, and mutations in this gene result in Dent's disease, which is characterized by excessive urinary Ca^{2+} and protein loss and kidney stones. To ensure electroneutrality of acid secretion, the membranes of bone-resorbing osteoclast cells contain ClC-7 Cl⁻ channels, which work in concert with acid secreting H+-ATPases. Mutations in the chloride channel isoform ClC-7 lead to defective bone resorption with compression of the bone marrow space and bone deformations called osteopetrosis.

6.11 Selective water transport occurs through aquaporin channels

Key concepts

- Aquaporins allow rapid and selective water transport across cell membranes.
- Aquaporins are tetramers of four identical subunits, with each subunit forming a pore.
- The aquaporin selectivity filter has three major features (size restriction, electrostatic repulsion, and water dipole orientation) that confer a high degree of selectivity for water.

Water is a major component of our bodies, representing 70% of the mass, and the proper distribution of water is important for maintaining fluid balance. The movement of water across cellular membranes is important for many physiological processes. However, the passive diffusion of water through the lipid bilayer of biologic membranes is unregulated and has a relatively low and finite permeability. The rapid and selective transport of water across cell membranes occurs through aquaporins, which are specialized transmembrane channels. Aquaporins are conserved from bacteria to humans and constitute a superfamily of transport proteins.

In animals, aquaporins are involved in physiological processes that include the thirst mechanism; concentration of urine by the kidneys; digestion; regulation of body temperature; secretion and absorption of spinal fluid; secretion of tears, saliva, sweat, and bile; and reproduction. For example, epithelial cells of the kidney reabsorb 99% of the water from the primary filtrate back into the blood vessels, which prevents dehydration. If dehydration starts to occur, osmotically sensitive cells of the nephron detect the increased osmolality of extracellular fluids, which stimulates release of the hormone vasopressin (also called antidiuretic hormone) from the pituitary gland. High plasma vasopressin levels result in a smaller volume of highly concentrated urine. Binding of vasopressin to its cell membrane receptor signals for the rapid expression of aquaporin-2 at the apical membrane of the kidney's collecting duct. This is mediated by the fusion of intracellular vesicles, which contain aquaporin-2, with the apical membrane of the epithelial cells. Thus, in response to increased osmotic gradients during dehydration, the aquaporins increase water absorption from the urine back into the blood.

Aquaporins are homotetrameric water channels that regulate the movement of water directed by osmotic gradients. As **FIGURE 6.27** shows, each subunit forms an independent pore, in contrast to K+ channels, where four subunits contribute to a single pore (see *6.5 K+ channels catalyze selective and rapid ion permeation*). Each aquaporin subunit contains six transmembrane segments (M1, M2, M4, M5, M6, and M8) that form tandem repeats of three segments each. The loops connecting the second and third transmembrane segments in each repeat contain the signature motif asparagine-proline-alanine (NPA), which is conserved among all aquaporins. These two NPA sequences are juxtaposed in the center of the aqueous pore as part of the selectivity filter. One aquaporin subunit has an extremely high unit water permeability of 3×10^9 water molecules per second, while transport of other solutes or ions is negligible. Remarkably, they have an aqueous path that passes water at high rates and bidirectionally across the cell membrane without being permeable to common ions or even protons in the form of H_3O^+. This highly selective permeability is essential for kidney function, since simultaneous reabsorption of both water and acid would result in life-threatening acidosis.

The aquaporin pore can be divided into three regions: an extracellular vestibule, a

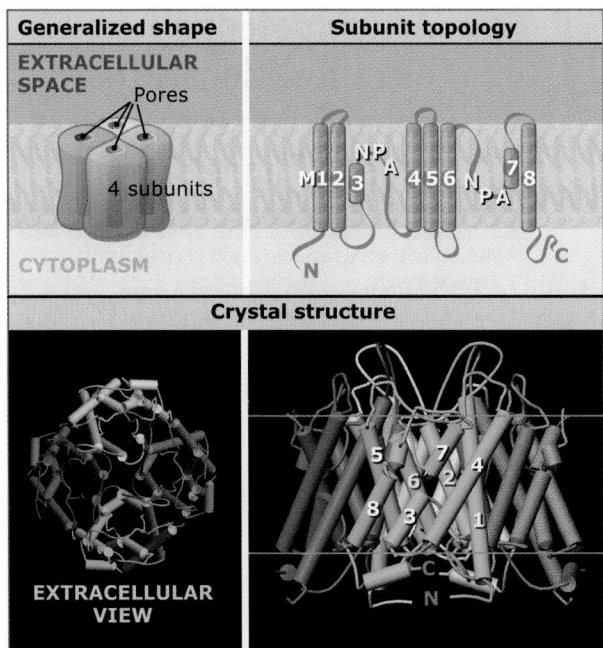

FIGURE 6.27 Schematic and X-ray crystal structure of an aquaporin channel complex. The complex consists of four identical subunits; each subunit forms a pore. The extracellular view is shown smaller than the side view. The presumed position of the membrane is indicated. Bottom structures from Protein Data Bank 1J4N. H. Sui, et al., *Nature* 414 (2001): 872–878.

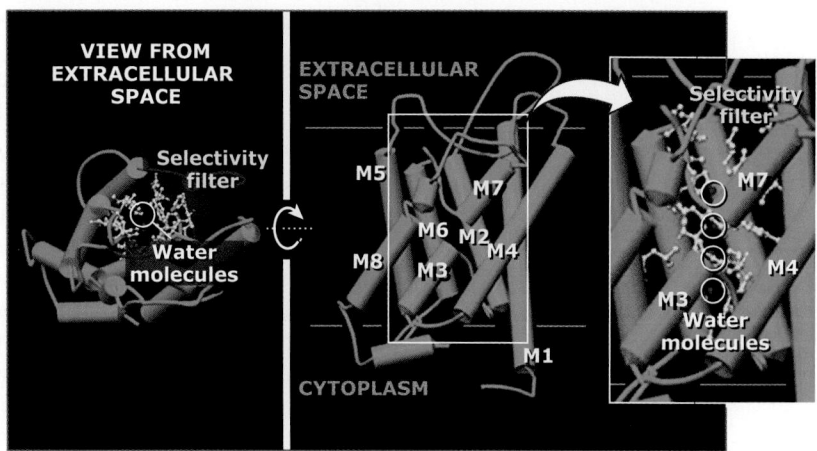

FIGURE 6.28 One subunit from the X-ray crystal structure of the AQP1 aquaporin channel. The presumed position of the membrane is indicated. The residues of the selectivity filter are shown at the atomic level (yellow); for the rest of the protein, the protein backbone is depicted as loops and helices (cylinders). Structures from Protein Data Bank 1J4N. H. Sui, et al., *Nature* 414 (2001): 872–878.

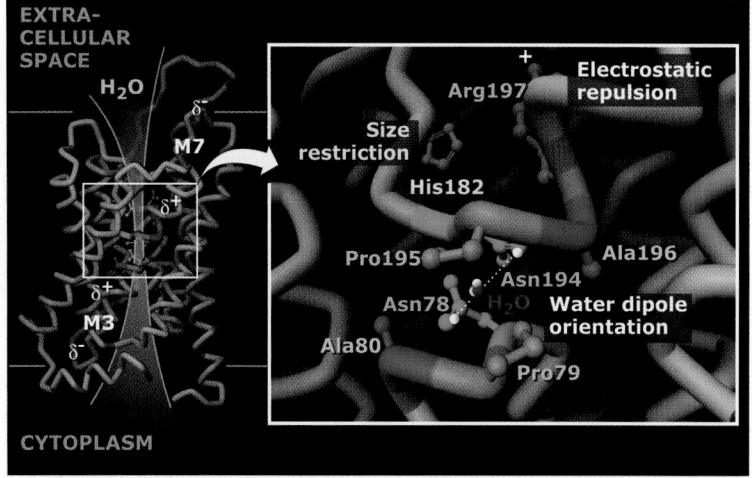

FIGURE 6.29 The pore region of aquaporin channels has three features that confer selective water permeation. Only one of the permeating water molecules in the crystal structure is shown in the inset; its oxygen atom forms hydrogen bonds with the two Asn side chains. Structures from Protein Data Bank 1J4N. H. Sui, et al., *Nature* 414 (2001): 872–878.

narrow pore region containing the selectivity filter, and an intracellular vestibule that covers a distance of 20 Å. The three regions together form an hourglass-shaped water permeation pathway. The residues that contribute to the selectivity filter are highlighted in FIGURE 6.28. Most of the channel wall along the selectivity filter is made up of hydrophobic residues. The hydrophilic residues provide the chemical groups that are essential to establish a pathway for selective water transport.

There are three features of the aquaporin pore region that confer specificity for water permeation, as FIGURE 6.29 shows for human AQP1:

- Size restriction. The extracellular vestibule tapers down to about 2.8 Å in diameter at its narrowest point, which is called the constriction region. Water molecules move through the constriction region in single file, whereas hydrated ions and protons are prevented from entering.
- Electrostatic repulsion. The positive charge of the pore-lining residue Arg197 contributes to electrostatic repulsion for positively charged

molecules and prevents hydronium ions (H_3O^+, which result from protonation of water molecules) from passing through the pore. In addition, the non-membrane spanning a helices M3 and M7 contribute partial positive charges that serve to block proton conduction.

- Water dipole orientation. Reorientation of the water dipole by simultaneous hydrogen bonding with the partial positive charges from the side chains of two asparagine residues in the NPA motifs (Asn78 and Asn194) at the center of the channel contributes to selectivity. These charges interact with water and force it into a particular orientation. This interaction is a second barrier to the entry of H_3O^+. The ability of aquaporin to bind water reduces the energy barrier to transport across a predominantly hydrophobic pathway. However, the number and affinity of the interactions are sufficiently low to allow for rapid transport of water.

Together, these features of the aquaporin selectivity filter allow water to be rapidly transported across membranes, while protons in the form of H_3O^+ and other ions are excluded.

Different aquaporin isoforms are important to maintain fluid homeostasis at the organ and systemic levels. Several aquaporin isoforms are expressed in the kidney and function in water absorption from renal tubules. High water permeability by aquaporin-1 is constitutive in epithelial cells of the proximal convoluted tubules and descending thin limbs of the loop of Henle. In humans, aquaporin-1 proteins help to concentrate 180 liters of blood filtrate per day into a urine volume of 1.5 liters per day by reabsorbing ~178.5 liters of water from the primary plasma filtrate back into the blood vessels via apical cell membranes (water is also reabsorbed via paracellular pathways). Patients with genetic defects in aquaporin-1 are not able to concentrate their urine efficiently. Aquaporin-2 is expressed in kidney epithelial cells different from those that express aquaporin-1. As mentioned earlier in this section, the hormone vasopressin stimulates expression of aquaporin-2 in the collecting ducts, resulting in increased urine concentration. Excessive water intake or inhibition of vasopressin release by alcohol or coffee consumption causes the kidney to excrete large volumes of diluted urine. Patients with nephrogenic diabetes insipidus have genetically defective aquaporin-2 channels and release up to 20 liters of urine per day. In the brain, aquaporin-4 is expressed in cells close to small blood vessels and regulates the movement of water between brain parenchyma and the vascular space. Aquaporin-4 is a potential pharmacological target to aid in the rapid reduction of brain edema, upon which recovery from head trauma or stroke depends. Aquaporin-0 is expressed only in fiber cells of lenses, and missense mutations can cause congenital cataracts in children.

6.12 Action potentials are electrical signals that depend on several types of ion channels

Key concepts

- Action potentials enable rapid communication between cells.
- Na^+, K^+, and Ca^{2+} currents are key elements of action potentials.
- Membrane depolarization is mediated by the flow of Na^+ ions into cells through voltage dependent Na^+ channels.
- Repolarization is shaped by transport of K^+ ions through several different types of K^+ channels.
- The electrical activity of organs can be measured as the sum of action potential vectors.
- Alterations of the action potential can predispose for arrhythmias or epilepsy.

Neurons, muscle cells, and endocrine cells are known as excitable cells because they can produce and/or respond to electrical signals. These cells undergo rapid and transient changes of their membrane potential that are translated into electrical signals, such as the nerve impulse that travels along the axons of neurons or the signals that induce muscle contraction. The electrical signal is called the **action potential**. In the brain, perception involves processing an enormous number of widely distributed action potentials that occur before, during, and after stimulus arrival and that have differing spatiotemporal patterns. In skeletal muscle and heart, action potentials are critical for initiating and coordinating simultaneous contraction of cells. The amplitude and duration of the action potential can be measured by electrophysiological techniques. The action potential lasts a few milliseconds in neurons and hundreds of milliseconds in heart muscle. The longer duration of the cardiac action potential is necessary to allow for coordinated activation of the millions of muscle cells that produce contraction of the heart (see *6.13 Cardiac and skeletal muscles are activated by excitation-contraction coupling*). Action potentials can propagate at a rate of meters per second along cell membranes, thereby allowing for rapid, long-distance communication between cells and forming the basis for complex physiological functions such as those of the brain and heart.

The electrical potential across the cell membrane plays a pivotal role in the generation of action potentials. Cells at rest maintain a negative membrane potential, that is, the inside of the cell is charged slightly negative relative to the outside. This resting membrane potential is maintained mostly by the action of Na^+/K^+-ATPases, which pump three Na^+ ions in for every two K^+ ions that it pumps out of the cell, and by K^+ leak channels (see *6.4 Electrochemical gradients across the cell membrane generate the membrane potential* and *6.19 The Na+/K+-ATPase maintains the plasma membrane Na+ and K+ gradients*).

Based on electrophysiological measurements, a model of the mechanism underlying the action potential was proposed over fifty years ago. The model included two key elements mentioned below, both of which have been shown to contribute to the action potential:

1. The cell membrane undergoes transient and sequential changes in its selective permeability to Na^+ or K^+ ions, and

2. These permeability changes depend on membrane voltage.

Action potentials are mediated by the co-ordinated activation and inactivation of several different types of ion channels. Electrical signals in the cell membrane are made possible by the ability of voltage-gated ion channels to rapidly sense and respond to changes in the membrane potential. The opening and closing of different types of ion channels occur sequentially during the different phases of the action potential, as shown for cardiac cells in **FIGURE 6.30**. The rapid changes of the membrane potential are due to localized transmembrane changes in ion concentrations, with little effect on the total concentrations of intracellular ions.

The action potential starts with a rapid upstroke (phase 0) that is initiated by the opening of voltage-gated Na^+ channels and permits the rapid flow of Na^+ ions down their concentration gradient into the cell (see Figure 6.30). The effect of this movement of Na^+ ions results in membrane depolarization, in which the intracellular environment is charged more positive relative to the outside of the cell. (For details on voltage-gated Na^+ channels see *6.7 Voltage-dependent Na+ channels are activated by membrane depolarization and translate electrical signals*.)

Depolarization stops within milliseconds as Na^+ channels undergo rapid inactivation, and early repolarization begins in phase 1. When voltage-gated Na^+ channels close in the heart, opening of voltage-gated Ca^{2+} channels and transient outward K^+ currents, which are ac-

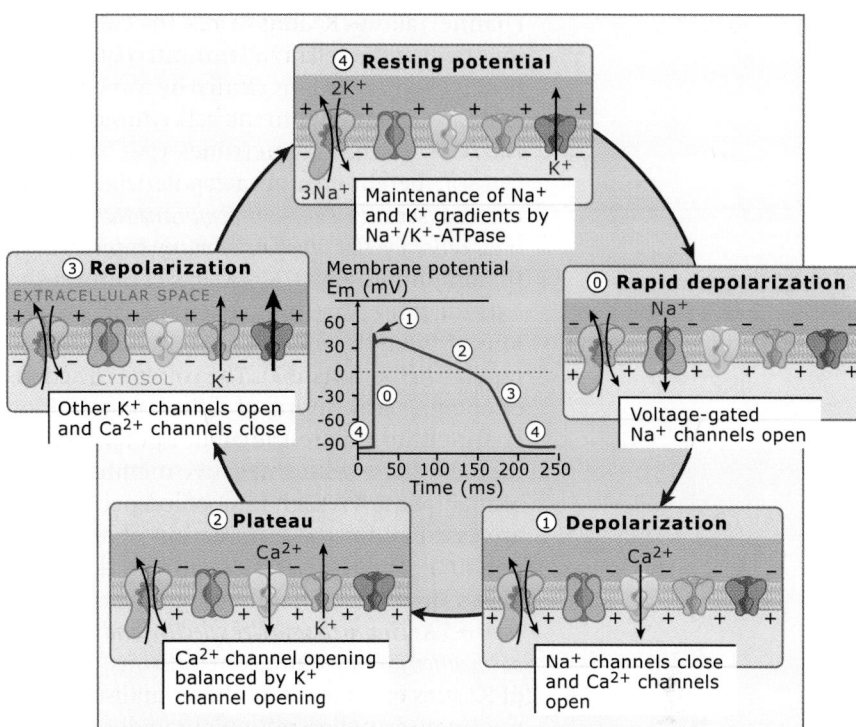

FIGURE 6.30 The action potential is generated and mediated by ionic currents. The cardiac action potential can be separated into five phases. A distinct set of ion channels opens and/or closes during each phase. The different types of K^+ channels that open and close during different phases of the action potential are not specified. The Na^+/K^+-ATPase pumps throughout the action potential, but the rate changes as the membrane potential changes.

tivated at more positive membrane potentials during membrane depolarization, set a new level of the membrane potential. This level is defined by the intricate balance between depolarizing and repolarizing membrane currents at phase 2. Na^+ channels require repolarization of the membrane potential to recover from inactivation before they can become activated again. A small fraction of Na^+ channels fail to enter the inactivated state and create a small but persistent current that, together with a sustained Ca^{2+} inward current, contributes to the prolonged depolarized state of the cardiac action potential. Relative to neuronal action potentials, the cardiac action potential is relatively long. This longer action potential seems necessary to allow for sufficient time to activate intracellular Ca^{2+} release for muscle contraction and to prevent aberrant membrane depolarizations during intracellular Ca^{2+} release (see *6.13 Cardiac and skeletal muscles are activated by excitation-contraction coupling*).

In larger animal species and humans, phases 1 and 2 of the action potential are separated by a notch. In phase 2, the plateau phase, sequential activation of several different types of K^+

channels allows K+ ions to exit the cell. The K+ ions exiting the cell rapidly counterbalance the positively charged ions carried by Na+ channels and Ca^{2+} channels into the cell. Moreover, the Na+/Ca^{2+}-exchanger extrudes Ca^{2+} from the cytosol, thereby creating a depolarizing inward current (see *6.16 The transmembrane Na+ gradient is essential for the function of many transporters*). In addition, the Na+/K+-ATPase continues to extrude three Na+ ions in exchange for two K+ ions, driving the membrane potential toward repolarization (phase 3). The combined actions of opening and closing of different sets of ion channels terminate the action potential and reestablish the resting negative membrane potential (phase 4). High rates of ion permeation are essential for the termination of an action potential, and K+ channels allow for fast rates of ion flow while maintaining ion selectivity (see *6.5 K+ channels catalyze selective and rapid ion permeation*). For example, in neurons, millions of K+ ions exit the cell within a millisecond to terminate an action potential rapidly.

A large variety of voltage-gated K+ channels allow for specialized electrical signaling in different cell types. For example, inward rectifier K+ channels are essential for the stable resting membrane potential and for the long plateau phase of the cardiac action potential. At positive membrane potentials, inward rectifier K+ channels are mostly closed, which allows for a more sustained membrane depolarization (for more on inward rectifier K+ channels see *6.25 Supplement: Most K+ channels undergo rectification*). In cardiac myocytes, only minimal K+ currents flow initially when membrane potentials become more positive than −40 mV. This maintains the influence of the depolarizing inward Na+ and Ca^{2+} currents on the duration of the action potential until delayed rectifier K+ channels become activated after a certain time and drive the membrane potential toward the resting state (see Figure 6.30).

The sum of electrical activity created by the individual action potentials of all neurons in the brain, all muscle cells in one muscle group, or all cardiac cells in the heart can be amplified and visualized as the electroencephalogram (EEG), the electromyogram (EMG), or electrocardiogram (ECG), respectively. These surface recordings of electrical activity are used to monitor abnormalities such as uncontrolled electrical activity in epilepsy, myotonias, or arrhythmias. Such abnormalities can result from mutations that affect the function of specific types of ion channels.

Mutations in voltage-dependent Na+, K+ or Ca^{2+} channels are associated with abnormalities in brain and heart function. For example, mutations in the *SCN5A* gene encoding the cardiac voltage-dependent Na+ channel have been linked to several forms of heart disease. Some gain-of-function mutations in *SCN5A* cause long QT syndromes, in which incomplete inactivation of Na+ channels results in a prolonged action potential. In addition, mutations in the cardiac Ca$_v$1.2 channel, which defines the plateau phase of the action potential, result in long QT syndrome and arrhythmias. These defects result in delayed repolarization of the heart and increase the risk for sudden death due to arrhythmias. Other mutations in voltage-dependent Na+ channels result in different forms of heart disease, paralysis of skeletal muscle or inherited forms of epilepsy. A mutation in the *HERG* K+ channel gene is an example of a voltage-gated K+ channel that is associated with heart disease. This mutation increases the rate of channel deactivation, which reduces the flow of K+ ions out of the cell and slows the repolarization phase of the action potential. Thus, this mutation in a K+ channel gene prolongs the action potential, similar to the effect of some of the mutations in genes encoding voltage-gated Na+ channels.

6.13 Cardiac and skeletal muscles are activated by excitation-contraction coupling

Key concepts

- The process of excitation-contraction coupling, which is initiated by membrane depolarization, controls muscle contraction.
- Ryanodine receptors and inositol 1,4,5-trisphosphate receptors are Ca^{2+} channels through which Ca^{2+} ions are released from intracellular stores into the cytosol.
- Intracellular Ca^{2+} release through ryanodine receptors in the sarcoplasmic reticulum membrane stimulates contraction of the myofilaments.
- Several different types of Ca^{2+} transport proteins, including the Na+/Ca^{2+}-exchanger and Ca^{2+}-ATPase, are important for decreasing the cytosolic Ca^{2+} concentration and controlling muscle relaxation.

Ca^{2+} ions are **second messengers** in numerous signaling pathways in diverse cell types. In higher organisms, intracellular Ca^{2+} mediates processes as diverse as synaptic transmission, muscle contraction, insulin secretion, fertilization, and gene expression. In this section, we will discuss how Ca^{2+} signaling regulates muscle contraction and heart rhythm (for details on the role of the cytoskeleton in muscle contraction see *12 Actin*). The process in which membrane depolarization results in production of force by muscles is called **excitation-contraction coupling**. This is the fundamental mechanism that controls the function of skeletal and cardiac muscles. Resting muscle cells maintain a much lower concentration of free Ca^{2+} in the cytosol ($\sim10^{-7}$ M) relative to the concentrations outside of the cell ($\sim10^{-3}$ M) and in the sarcoplasmic reticulum. Ca^{2+} enters the cytosol at the start of excitation-contraction coupling, then exits the cytosol as the cell returns to its resting state. This temporal increase and decrease of the cytosolic Ca^{2+} concentration is referred to as the intracellular Ca^{2+} transient. Several different types of Ca^{2+} transport proteins are required for this process.

The process of excitation-contraction coupling can be separated into four stages, as shown in **FIGURE 6.31** for a cardiac muscle cell. First,

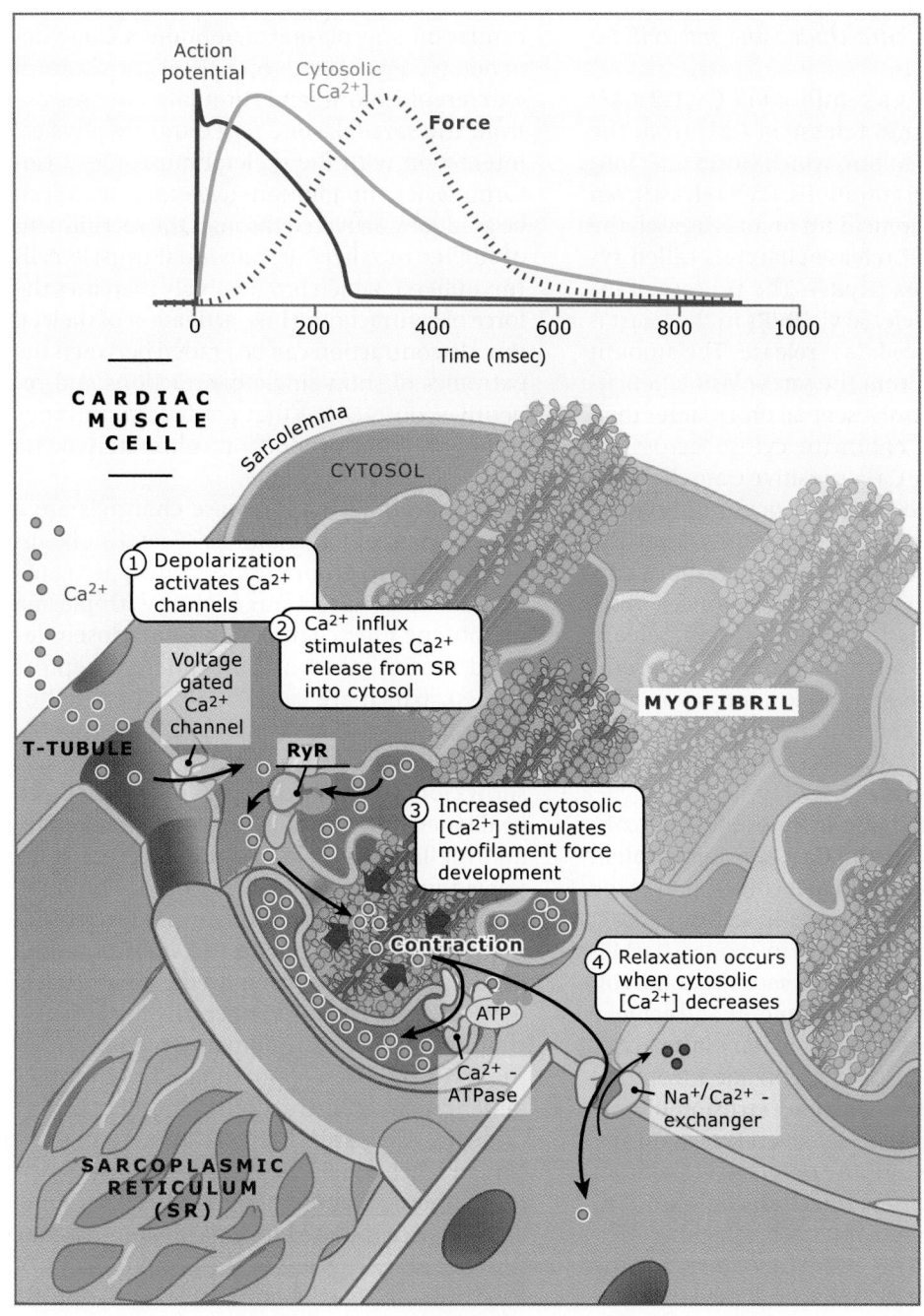

FIGURE 6.31 During an action potential in cardiac muscle cells, several different types of Ca^{2+} transport proteins function to increase and decrease the cytosolic Ca^{2+} concentration.

the signal is initiated at the plasma membrane (sarcolemma) when the membrane undergoes depolarization, in which the membrane potential becomes more positive relative to the resting potential, due to an incoming action potential (see Figure 6.30 and *6.12 Action potentials are electrical signals that depend on several types of ion channels*). Voltage-dependent Ca^{2+} channels (called $Ca_v1.2$ Ca^{2+} channels) sense this change in the membrane potential and open in response to it during phase 2 of the cardiac action potential, allowing a small flux of Ca^{2+} ions to move down their electrochemical gradient into the cell. (For more on voltage-dependent Ca^{2+} channels see *6.9 Plasma membrane Ca^{2+} channels activate intracellular and intercellular signaling processes*.)

Second, the Ca^{2+} influx via $Ca_v1.2$ Ca^{2+} channels stimulates release of Ca^{2+} from the sarcoplasmic reticulum, which stores Ca^{2+} ions at millimolar concentrations. Ca^{2+} release from the sarcoplasmic reticulum occurs through the intracellular Ca^{2+} release channels called ryanodine receptors (RyRs). The process of intracellular Ca^{2+} release via RyRs in the heart is called Ca^{2+}-induced Ca^{2+} release. The amount of Ca^{2+} released from the sarcoplasmic reticulum into the cytosol is several times larger than the amount that enters the cytosol across the sarcolemma. The Ca^{2+}-sensitive Ca^{2+} channels in the cardiac sarcoplasmic reticulum are called ryanodine receptors because they bind the plant alkaloid ryanodine with high specificity, which blocks the channel. Different cell types express different intracellular Ca^{2+} release channels, which open in response to various stimuli that regulate muscle contraction. The major intracellular Ca^{2+} release channel of the sarcoplasmic reticulum in cardiac muscle cells is the RyR2 isoform.

Third, the increase in cytosolic Ca^{2+} concentration activates the Ca^{2+}-sensitive protein troponin C, which stimulates contraction of the myofilaments. An increase in cytosolic Ca^{2+} concentration from 100 nM to about 1.0 µM is necessary to effectively activate the total volume of intracellular myofilaments and achieve simultaneous muscle contraction of the heart.

Fourth, when Ca^{2+} is extruded from the cytosol, the muscle relaxes. Extrusion of Ca^{2+} from the cytosol occurs by several mechanisms. The major pathway is the reuptake of Ca^{2+} ions into the sarcoplasmic reticulum Ca^{2+} store by the action of the sarcoplasmic reticulum Ca^{2+}-ATPase pump. This pump accounts for the reuptake of the bulk of Ca^{2+} released from the sarcoplasmic reticulum Ca^{2+} store by RyRs (for more on the Ca^{2+}-ATPase see *6.18 The Ca^{2+}-ATPase pumps Ca^{2+} into intracellular storage compartments*). In addition, Ca^{2+} transport proteins, such as the Na^+/Ca^{2+}-exchanger in the plasma membrane, remove Ca^{2+} ions from the cytosol. This exchanger accounts for the extrusion of the smaller fraction of Ca^{2+} that entered from outside the cell through voltage-dependent $Ca_v1.2$ Ca^{2+} channels. A small fraction of Ca^{2+} ions is also exchanged between the cytosol and mitochondria.

Overall, the process of excitation-contraction coupling is similar in skeletal and cardiac muscles, with some exceptions. In contrast to cardiac muscle, plasma membrane voltage-dependent Ca^{2+} channels of skeletal muscle are of a different isoform and stimulate Ca^{2+} release from the sarcoplasmic reticulum by physical interaction with the skeletal muscle RyR isoform, RyR1. In addition, skeletal muscle can be gradually activated through the recruitment of higher numbers of individual muscle cells (myofibers), which progressively increases the force of contraction. Thus, activation of skeletal muscle contraction can be graded between the extremes of short single contractions and repetitive contractions that produce continuous or tetanic force production, ultimately being limited by muscle fatigue.

Intracellular Ca^{2+} release channels are a unique class of ion channels. They can be divided into two groups, RyRs, which are gated by Ca^{2+} or by direct interaction with plasma membrane Ca^{2+} channels, and the closely related inositol 1,4,5-trisphosphate receptors (IP_3Rs), which are gated by IP_3. Four RyR or IP_3R subunits assemble in a fourfold symmetrical complex to form a channel, as **FIGURE 6.32** shows for the IP_3R. Both types of channels can be divided into two domains: the pore domain and the large cytoplasmic domain that is involved in gating the channel pore. The IP_3Rs are predicted to have six transmembrane segments and one pore loop per subunit, and a similar topology has been predicted for RyRs.

RyR Ca^{2+} channels are the largest ion channels known. The domain organization of the RyRs has been approximated from three-dimensional reconstructions using electron microscopy. They are ~10 times larger than Ca^{2+}, Na^+, or K^+ channels, as **FIGURE 6.33** shows. Each RyR subunit is about 5000 amino acids long, almost twice the size of the partly homologous IP_3R subunit. The size of the pore domains of RyRs and IP_3Rs is about the same as the K^+

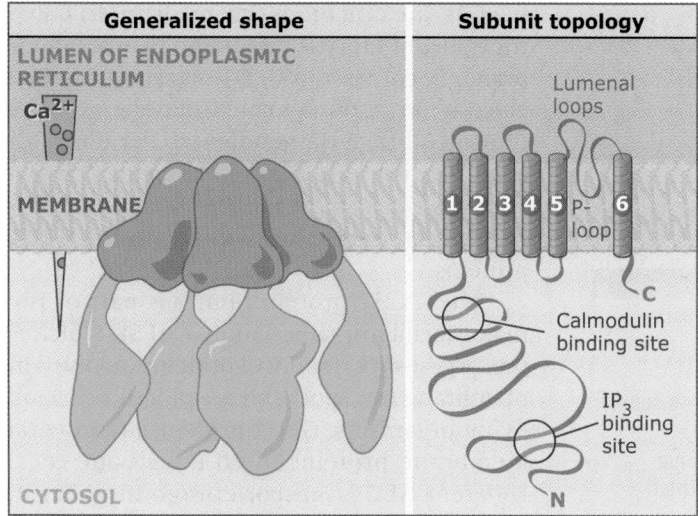

Generalized shape	Subunit topology

LUMEN OF ENDOPLASMIC RETICULUM

Ca²⁺

MEMBRANE

Lumenal loops

1 2 3 4 5 P-loop 6

C

Calmodulin binding site

IP₃ binding site

N

CYTOSOL

FIGURE 6.32 Proposed structure of the inositol 1,4,5-trisphosphate receptor in the endoplasmic reticulum. The inset at left indicates the Ca²⁺ gradient across the membrane of the endoplasmic reticulum in resting animal cells.

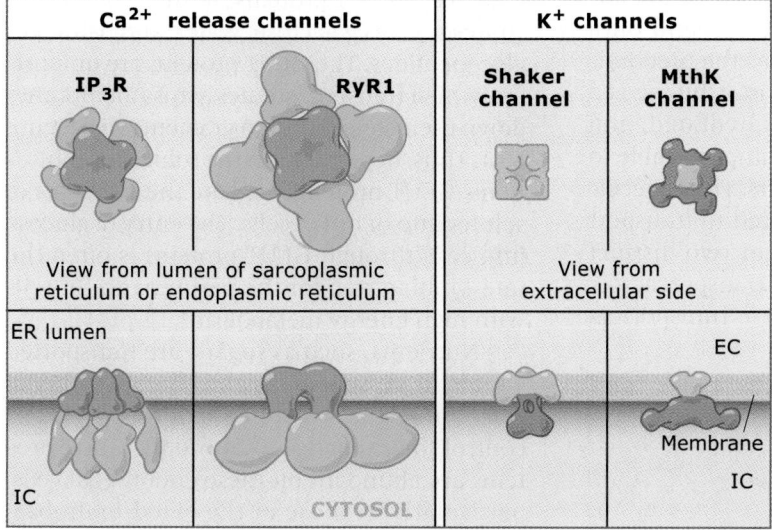

Ca²⁺ release channels	K⁺ channels

IP₃R   RyR1

Shaker channel   MthK channel

View from lumen of sarcoplasmic reticulum or endoplasmic reticulum

View from extracellular side

ER lumen

IC   CYTOSOL

EC

Membrane

IC

FIGURE 6.33 The two intracellular Ca²⁺ release channels, the IP₃ receptor (IP₃R) and the ryanodine receptor (RyR), and two K⁺ channels, the voltage-dependent Shaker channel and the Ca²⁺-gated MthK channel differ in size. The pore domains are indicated in blue or yellow, respectively.

channel pore domain. The large cytoplasmic domains of RyRs and IP₃Rs control gating of the channels by Ca²⁺ or IP₃, respectively, similar to ligand-gated K⁺ channels (see *6.6 Different K⁺ channels use a similar gate coupled to different activating or inactivating mechanisms*).

Mutations that affect the gating of intracellular Ca²⁺ release channels cause distinct forms of disease. For example, missense mutations of the cardiac RyR2 are linked to two genetic forms of arrhythmias and exercise-induced sudden death. The mutant RyR Ca²⁺ channels have a reduced affinity for calstabin2 (also called FKBP12.6). Calstabin2 is a Ca²⁺ channel subunit that stabilizes the closed state of the cardiac RyR and prevents aberrant activation. The result of the RyR mutations is increased Ca²⁺ leak from the sarcoplasmic reticulum during the resting, or diastolic, phase

of the heart. Moreover, dysregulation of RyR2 in heart disease appears to contribute to sudden death and worsening of heart failure. One common mechanism seems to be intracellular Ca²⁺ leak that can potentially trigger fatal arrhythmias by causing aberrant membrane depolarizations.

Mutations in the skeletal muscle RyR isoform also result in aberrant intracellular Ca²⁺ release, which is associated with the disease malignant hyperthermia. Patients are susceptible to uncontrolled intracellular Ca²⁺ release, which causes high temperatures, muscle contractions, and a life-threatening metabolic crisis following exposure to certain inhalation anesthetics and muscle relaxants. Mutations in the gene that encodes the voltage-dependent Ca²⁺ channels of the skeletal muscle plasma membrane that physically interact with and activate RyRs also confer susceptibility to malignant hyperthermia.

6.14 Some glucose transporters are uniporters

Key concepts

- To cross the blood-brain barrier, glucose is transported across endothelial cells of small blood vessels into astrocytes.
- Glucose transporters are uniporters that transport glucose down its concentration gradient.
- Glucose transporters undergo conformational changes that result in a reorientation of their substrate binding sites across membranes.

Glucose is a principal energy source for eukaryotic cells, and many cells depend on a continuous supply of glucose as the predominant source for ATP generation. Glucose is a polar molecule that becomes hydrated, and cell membranes are relatively impermeable to small polar solutes such as sugars. Thus, specific membrane proteins are required to transport glucose into cells. Members of two distinct gene families carry out glucose transport across the plasma membrane. Glucose transporters

(GLUTs) are uniporters that mediate facilitated transport of glucose across the plasma membrane. In contrast to GLUT proteins, the Na^+/glucose cotransporters couple the energy of the transmembrane Na^+ gradient to the transport of glucose (see *6.16 The transmembrane Na^+ gradient is essential for the function of many transporters*). In this section, we will discuss the GLUT proteins.

The GLUT protein family is part of the major facilitator superfamily (MFS), which is the largest superfamily of proteins involved in membrane transport and are ubiquitous in all living organisms. GLUT proteins are integral membrane proteins of all eukaryotic cells. Different GLUT transporter isoforms differ in their kinetic properties, sugar specificity, tissue localization, and regulation. In addition to glucose, some GLUT proteins transport other substrates such as galactose, water, and painkiller glycopeptides. The GLUT proteins are uniporters, which transport solutes across membranes down their concentration gradients (see Figure 6.4). Thus, depending on the solute concentrations, GLUT proteins mediate the transport of solutes into or out of cells. The entry of glucose into cells through GLUT proteins is often the rate-limiting step for the performance of cells with high energy metabolism.

Nutrients, such as sugars, are transported via blood vessels to organs. The endothelial cells lining the walls of small blood vessels control the exchange of nutrients. GLUT proteins are abundant on these endothelial cells, particularly in those of the blood-brain barrier. The brain has a high metabolic demand for glucose utilization, and brain function is especially sensitive to decreased nutrition. The high capacity for glucose transport into neuronal tissues through brain microvessels occurs in several steps involving transport by the GLUT-1 isoform, as **FIGURE 6.34** shows. GLUT-1 is expressed in the endothelial cell membrane at its interfaces with the blood and intercellular space, as well as the plasma membrane of astrocytes, which are cells important for the function of the blood-brain barrier. GLUT-1 proteins at these sites transport glucose from the blood into the endothelial cells, out of the endothelial cells, and into the astrocytes. The astrocytes convert glucose into other energy sources that are transported into neurons.

Other GLUT isoforms are important in different tissues. For example, GLUT-4 mediates glucose uptake by muscle and adipose tissues. During and after a meal, one of the actions

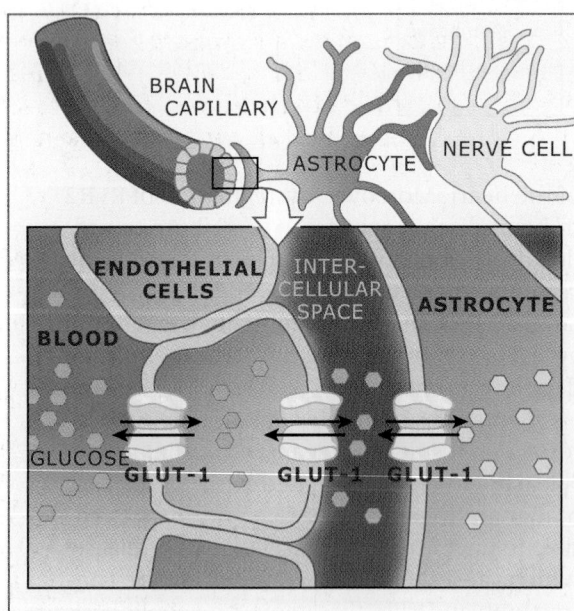

FIGURE 6.34 Glucose is selectively transported across the blood-brain barrier through the glucose transporter isoform 1 (GLUT-1). Transport of glucose from the blood to the brain and central nervous system involves a multistep mechanism through several cell types.

of insulin is to increase glucose uptake into the cells of these tissues. The GLUT-4 isoform, which is also called the insulin-responsive transporter, undergoes regulated transport to the cell surface. The GLUT-4 protein is located in intracellular vesicles that fuse with the plasma membrane, thereby delivering GLUT-4 transporters to the plasma membrane to increase glucose transport capacity. Binding of insulin to its cell surface receptor initiates intracellular signaling cascades that result in the rapid fusion of these vesicles with the cell membrane. This in turn allows for a rapid increase in the transport of glucose through GLUT-4 into the cell. In type II diabetes, the uptake of glucose from blood plasma into muscle and adipose tissue is impaired, apparently due to diminished targeting of GLUT-4 to the plasma membrane. (For more on regulated secretion see *8.19 Some cells store proteins for later secretion*.) Moreover, GLUT-2 *exports* glucose from cells in glucose-synthesizing organs such as the liver.

The predicted topology of GLUT transporters, shown in **FIGURE 6.35**, is similar to that of other MFS members. Based on hydropathy sequence analysis, GLUT-1 transporters are predicted to have 12 transmembrane a helices with intracellular N- and C-termini, and intracellular loops. The intracellular loops contain a substrate binding site and phosphorylation sites.

A model for the structure of GLUT-1 has been proposed, based on site-directed muta-genesis and measurement of glucose transport by the mutant proteins and on the X-ray crystal structure of the bacterial lactose permease, which is a related member of the MFS, in the oligosaccharide/H+ transporter subfamily (see *6.15 Symporters and antiporters mediate coupled transport*). In this model, the orientation of transmembrane helices allows for the formation of a pore cavity for glucose permeation and for hydrogen bonding between the GLUT protein and glucose, as **FIGURE 6.36** shows. Kinetic analysis of glucose transport in red blood cells suggested a mechanism in which GLUT proteins alternate between two major conformations. This mechanism is similar to that proposed for the bacterial lactose permease (see Figure 6.39). In one conformation, the glucose binding site faces the extracellular space, and in the other, it faces the cytosol. Binding of glucose on either side would induce a conformational change, resulting in reorientation of the glucose-binding sites to the opposite side of the membrane and in release of glucose. Thus, although the GLUT transporters are uniporters and the bacterial lactose permease is a symporter, it is thought that they operate by similar mechanisms.

Mutations in the GLUT-1 gene can cause major developmental defects. In children, the brain glucose demand is three to four times higher than in adults and represents up to 80% of the glucose utilization of the body. Mutations in the human GLUT-1 gene are

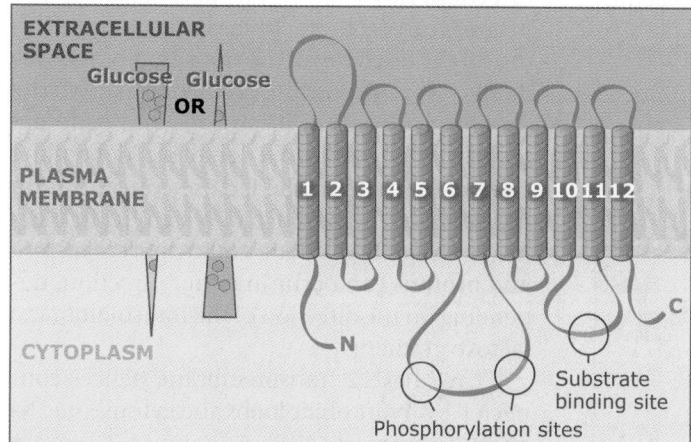

FIGURE 6.35 The predicted topology of GLUT transporters consists of 12 transmembrane segments with intracellular C- and N-termini. The intracellular loops contain phosphorylation and substrate binding sites. The insets indicate that the glucose gradient across the plasma membrane can occur in either direction depending on the cell type and the metabolic state. The gradient dictates the direction of transport.

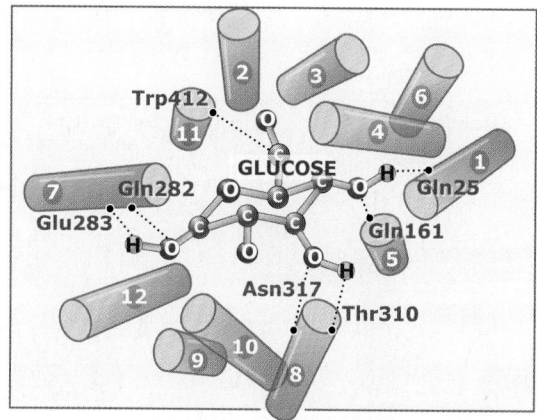

FIGURE 6.36 The predicted orientation of the transmembrane a helices of the glucose transporter GLUT-1. Residues involved in glucose binding are indicated (glucose molecule is not to scale). The helices are viewed from the intracellular side of the plasma membrane. This model is based on homology modeling using the *E. coli* lac permease structure as a template.

linked to GLUT-1 deficiency syndrome, which is a rare disease characterized by seizures and developmental delay that are thought to be caused by impaired glucose transport into the brain. In mice, embryos deficient in GLUT-1 have growth retardation and developmental malformations. Similar malformations are found in mouse embryos that develop under conditions of maternal diabetes, in which excessive blood glucose concentrations suppress GLUT-1 expression in the organs of the embryo.

6.15 Symporters and antiporters mediate coupled transport

Key concepts

- Bacterial lactose permease functions as a symporter that couples lactose and proton transport across the cytoplasmic membrane.
- Lactose permease uses the electrochemical H^+ gradient to drive lactose accumulation inside cells.
- Lactose permease can also use lactose gradients to create proton gradients across the cytoplasmic membrane.
- The mechanism of transport by lactose permease likely involves inward and outward configurations that allow substrates to bind on one side of the membrane and to be released on the other side.
- The bacterial glycerol-3-phosphate transporter is an antiporter that is structurally related to lactose permease.

Transporter proteins are membrane proteins that facilitate the passage of molecules across the membrane bilayer (see Figure 6.4). Uniporters move solutes down their transmembrane concentration gradients (see *6.14 Some glucose transporters are uniporters*). Symporters and antiporters move one solute against its transmembrane concentration gradient; this movement is powered by coupling to the movement of a second solute down its transmembrane concentration gradient. Many transporters are part of the MFS of transport proteins that translocate sugars, sugar-phosphates, drugs, neurotransmitters, nucleosides, amino acids, peptides, and other solutes across membranes. In this section, we will discuss two MFS transporters, the bacterial lactose permease LacY, which functions as a monomeric oligosaccharide/H^+ symporter, and the bacterial glycerol-3-phosphate transporter GlpT, an antiporter.

LacY was the first gene to be isolated that encodes a membrane transport protein. The LacY symporter uses the free energy released from the translocation of H^+ down its electrochemical gradient (usually turning the cytosol alkaline) to drive the accumulation of nutrients such as lactose against its concentration gradient, as **FIGURE 6.37** shows. The H^+ gradient across the cytoplasmic membrane is established by the respiratory chain and by the action of the F_1F_0-ATPase, which couples ATP hydrolysis to the export of protons from the cell (see *6.20 The F_1F_0-ATP synthase couples H^+ movement to ATP synthesis or hydrolysis*). For LacY, the stoichiometry of lactose and H^+ translocation is 1:1, with both substrates moving in the same direction. However, cotransport of lactose and protons can occur in either direction, depending on the direction of the transmembrane gradients. Thus, the lactose gradient can drive the uphill translocation of protons and generate an inward or outward H^+ gradient, depending on the direction of the lactose concentration gradient. In the absence of a significant electrochemical H^+ gradient, cotransport of lactose and protons can occur in either direction, depending on the direction of the transmembrane lactose gradient.

LacY has 12 transmembrane helices connected by hydrophilic loops and cytoplasmic N- and C-termini, as **FIGURE 6.38** shows. There are two domains of six transmembrane segments each, forming a symmetrical, heart-shaped structure, as determined by X-ray crystallographic analysis, as **FIGURE 6.39** shows. The substrate binding site is a hydrophilic cavity that is approximately at the center of the lipid

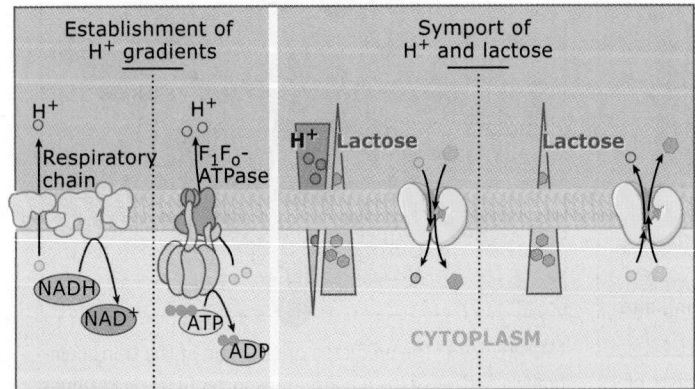

FIGURE 6.37 The bacterial lactose permease LacY catalyzes lactose uptake, driven by the energy of the inward H^+ gradient, which is established by bacterial respiration or ATP hydrolysis that exports protons from the cell. Alternatively, LacY can use the energy of the lactose gradient to drive the uphill translocation of protons out of the cell.

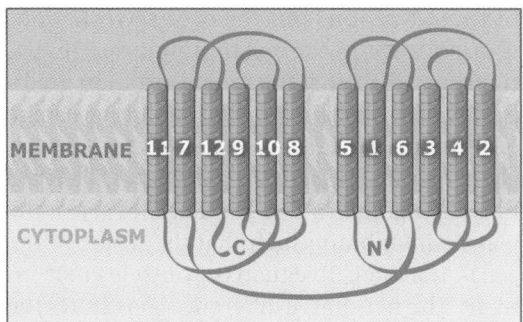

FIGURE 6.38 Topology of the bacterial lactose permease LacY.

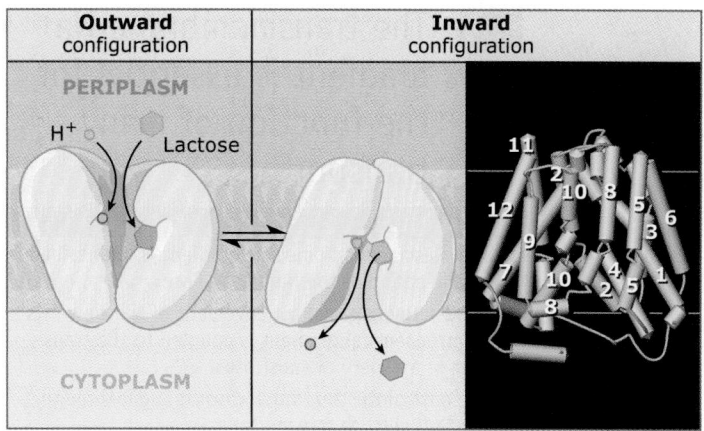

FIGURE 6.39 Model of conformational changes between the inward- and outward-facing configurations of the lactose permease LacY. The X-ray crystal structure is of the inward-facing conformation of the transporter with the substrate-bound cavity open to the cytoplasm. The presumed position of the membrane is indicated. Right structure from Protein Data Bank 1PV6. J. Abramson, et al., *Science* 301 (2003): 610–615.

bilayer. The crystal structure of LacY is shown in an inward configuration, with the open cavity facing the cytoplasm.

An allosteric model for the mechanism of lactose/proton transport by LacY had been proposed previously and is consistent with the crystal structure. In the alternating access model, shown in Figure 6.39, the LacY substrate binding site is accessible from either the intracellular or the extracellular side of the membrane, but never to both sides simultaneously. Protonation and binding of lactose in the outward-facing conformation induces a conformational change, resulting in the inward-facing conformation. This structural arrangement, in which both substrates bind before the conformational change occurs, allows for coupled and simultaneous transport. Release of the lactose and protons into the cell then induces a transition back to the outward-facing conformation. In this way, substrate binding and release lowers the energy barrier between the inward- and outward-facing conformations and facilitates their interconversion.

The bacterial glycerol-3-phosphate transporter GlpT is an antiporter that is related to LacY. GlpT accumulates glycerol-3-phosphate into the cell for energy production and phospholipid synthesis. GlpT is an organic phosphate/inorganic phosphate (P_i) exchanger that is driven by a P_i gradient, as **FIGURE 6.40** shows. Similar to LacY, GlpT has symmetrical N- and C-terminal domains, each consisting of six transmembrane segments surrounding the substrate translocation pathway. Although LacY functions as a symporter and GlpT is an antiporter, both transporters probably use the basic alternating access mechanism of transport. However, for GlpT, glycerol-3-phosphate binds and phosphate is released in the outward conformation, and the opposite occurs in the

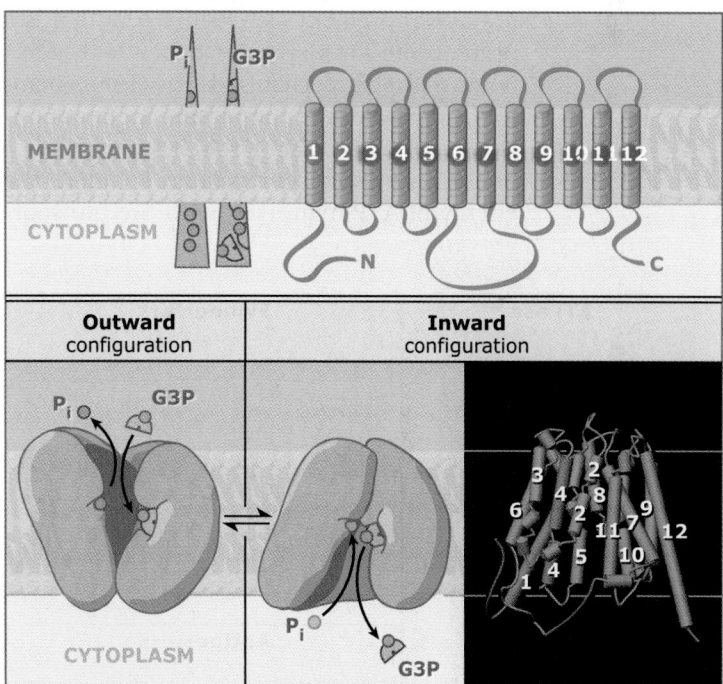

FIGURE 6.40 Schematic and X-ray crystal structure of the bacterial glycerol-3-phosphate transporter and proposed alternating access model for transport. This transporter is similar to the lactose permease LacY. The insets at top left show the phosphate and glycerol-3-phosphate gradients across the bacterial cell membrane. The crystal structure is of the transporter in the inward-facing configuration. The presumed position of the membrane is indicated. Inset structure from Protein Data Bank 1PW4. Y. Huang, et al., *Science* 301 (2003): 603–604.

inward conformation. The alternating access mechanism has also been proposed for glucose uniporters (see *6.14 Some glucose transporters are uniporters*).

6.16 The transmembrane Na⁺ gradient is essential for the function of many transporters

Key concepts

- The plasma membrane Na⁺ gradient is maintained by the action of the Na⁺/K⁺-ATPase.
- The energy released by movement of Na⁺ down its electrochemical gradient is coupled to the transport of a variety of substrates.
- The gastrointestinal tract absorbs sugar through the Na⁺/glucose transporter.
- The Na⁺/Ca²⁺-exchanger is the major transport mechanism for removal of Ca²⁺ from the cytosol of excitable cells.
- The Na⁺/K⁺/Cl⁻-cotransporter regulates intracellular Cl⁻ concentrations.
- Na⁺/Mg²⁺-exchangers transport Mg²⁺ out of cells.

Cells maintain an inwardly directed transmembrane Na⁺ gradient (see Figure 6.2) that is a prerequisite for many Na⁺-dependent membrane transport mechanisms. The electrochemical Na⁺ gradient at the plasma membrane is generated by the Na⁺/K⁺-ATPase, which is a primary active transport protein. It couples the energy of ATP hydrolysis with the transport of three Na⁺ ions out for every two K⁺ ions transported into the cell. Both ion species are transported against their concentration gradients. Since there is a net flow of positive charge out of the cell, the Na⁺ pump is referred to as electrogenic. The action of the Na⁺/K⁺-ATPase helps to maintain a charge difference across the plasma membrane such that the inside of the cell is more negative than the outside environment. The net charge difference constitutes the negative resting membrane potential. In other words, the energy that is required to maintain the electrochemical Na⁺ gradient becomes stored at the plasma membrane. (For more on the membrane potential see *6.4 Electrochemical gradients across the cell membrane generate the membrane potential*; for details on the Na⁺/K⁺-ATPase see *6.19 The Na⁺/K⁺-ATPase maintains the plasma membrane Na⁺ and K⁺ gradients*.)

The electrochemical Na⁺ gradient is important for the maintenance of physiological functions in many tissues. Many Na⁺- or voltage-dependent channels and secondary active transport proteins, some of which are listed in Figure 6.15, use the energy stored in the inwardly directed electrochemical Na⁺ gradient to drive solute accumulation against a concentration gradient or to generate electrical signals in the form of action potentials. Two major classes of Na⁺-dependent membrane proteins include the voltage-dependent Na⁺ channels and the epithelial Na⁺ channels (see *6.7 Voltage-dependent Na⁺ channels are activated by membrane depolarization and translate electrical signals* and *6.8 Epithelial Na⁺ channels regulate Na⁺ homeostasis*). A third class is the Na⁺/substrate transporters, which are grouped into different families based on sequence similarities. In this section, we will discuss some of these transporters. Na⁺-dependent transporters involved in pH regulation are discussed in the next section (see *6.17 Some Na⁺ transporters regulate cytosolic or extracellular pH*).

Some of the Na⁺-dependent substrate transporters are illustrated in **FIGURE 6.41**. Most Na⁺/substrate transporters are not evolutionarily related and, therefore, family members can differ in the substrate or the transport process. Na⁺/substrate transporters may be part of catabolic pathways by providing substrates for a metabolic function of the cell. These transporters use the Na⁺ gradient to drive the uphill transport of substrates such as ions, sugars, amino acids, vitamins, and urea into or out of the cell. Among these are the Na⁺/glucose cotransporter and the Na⁺/iodide cotransporter.

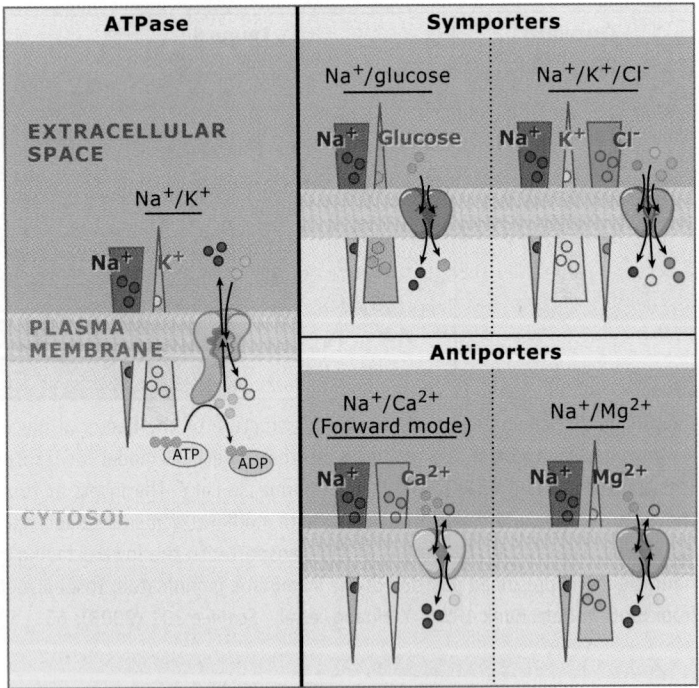

FIGURE 6.41 Some examples of Na⁺-dependent transporters. The energy of the plasma membrane Na⁺ gradient maintained by the Na⁺/K⁺-ATPase is used by these and many other transport systems. The insets indicate ion gradients for a typical animal cell.

In some bacteria, the Na⁺/proline cotransporter is involved in osmotic regulation; for other bacteria, Na⁺/substrate symporters increase their survival during infection of their host.

Some cells, such as those in the intestinal brush border, use the transmembrane Na⁺ gradient to drive the transport of sugars against their concentration gradient during nutrient absorption. The Na⁺/glucose cotransporter (see Figure 6.41) mediates the absorption of two dietary sugars, D-glucose and D-galactose, from the intestinal lumen. This cotransporter is a member of the *SGLT1* gene family, which has a common core structure of 13 transmembrane helices. The sugar binding and translocation pathway of the Na⁺/glucose cotransporter is formed by four transmembrane helices near the C-terminus. It has been proposed that the N-terminal domain of the Na⁺/glucose cotransporter is responsible for Na⁺ binding and that Na⁺/glucose cotransport results from interactions between the N- and C-terminal domains of the protein. A major question is how sugar transport is coupled to the electrochemical Na⁺ gradient. A current model proposes that an extracellular Na⁺ ion binds to an empty binding site in the transporter, inducing a conformational change that allows sugar to bind. Sugar binding would induce a second conformational change that exposes Na⁺ and the sugar to the intracellular side of the membrane. Upon release of Na⁺ and sugar into the cell, the empty Na⁺/glucose cotransporter would change conformation, returning to its starting position. Overall, the conformational changes proposed for this model are similar to those proposed for the uniporter GLUT-1, which functions as a glucose transporter (see *6.14 Some glucose transporters are uniporters*).

Many extracellular signals induce intracellular Ca²⁺ signaling (see *18 Principles of cell signaling*). Unstimulated cells maintain a low concentration of cytosolic Ca²⁺ of about 10–7 M. Ca²⁺ signaling depends on Ca²⁺ activation, in which the cytosolic Ca²⁺ concentration is raised by Ca²⁺ influx from outside the cell and release of Ca²⁺ ions from the sarcoendoplasmic reticulum Ca²⁺ store. Transport of Ca²⁺ out of the cytosol, across the plasma membrane and into the sarcoendoplasmic reticulum, occurs as the signal terminates. The two main Ca²⁺ extrusion proteins in the plasma membrane of most animal cells are the Na⁺/Ca²⁺-exchanger (NCX) (see Figure 6.41) and an ATP-driven Ca²⁺ pump. The Na⁺/Ca²⁺-exchanger has about a 10-fold lower affinity for Ca²⁺ ions but a 10-

to 50-fold higher turnover rate than the plasma membrane Ca²⁺-ATPase. Therefore, in excitable cells, the Na⁺/Ca²⁺-exchanger constitutes the primary Ca²⁺ extrusion system to the extracellular side of the plasma membrane. (For more on Ca²⁺ signaling in excitable cells see *6.13 Cardiac and skeletal muscles are activated by excitation-contraction coupling.*)

The Na⁺/Ca⁶⁺-exchangers form a family of proteins. The cardiac exchanger is predicted to have nine transmembrane segments, a large intracellular loop that is important for exchanger regulation, and antiparallel a-repeat domains important in ion translocation, as **FIGURE 6.42** shows. The two a-repeat regions contain portions of the transmembrane segments 2 and 3, and 8 and 9, and have opposite orientations, similar to the architecture of aquaporin water channels. In addition, it has two P-loops that have Gly-Ile-Gly sequences, reminiscent of the Gly-Tyr-Gly sequences in the selectivity filter of K⁺ channels. (For more on aquaporins see *6.11 Selective water transport occurs through aquaporin channels*. For more on the K⁺ channel selectivity filter see *6.5 K⁺ channels catalyze selective and rapid ion permeation*.) This topological model for the Na⁺/Ca²⁺-exchanger is based on sequence analysis, mutagenesis, and functional experiments. Thus, in contrast to the model for the glucose transporters GLUT-1 and Na⁺/glucose cotransporter, the Na⁺/Ca²⁺-exchanger is thought to have similarities with channel proteins.

The Na⁺/Ca²⁺-exchanger is believed to transport three Na⁺ ions in exchange for one Ca²⁺ ion, which creates a net electrogenic current of one positive charge per transport cycle. The exchanger can move Ca²⁺ ions either into or out of cells, depending on the net electrochemical driving force. For example, in cardiac myocytes at rest, the membrane potential E_m is

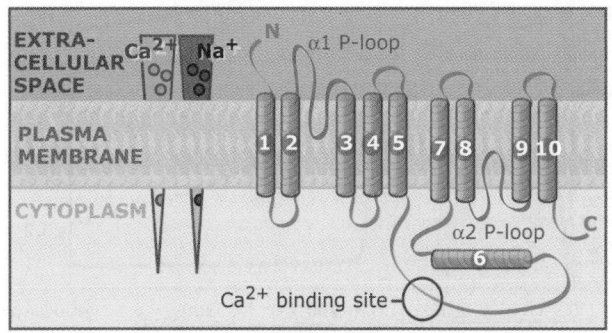

FIGURE 6.42 Predicted topology of the cardiac Na⁺/Ca²⁺ exchanger. The insets indicate the Na⁺ and Ca²⁺ ion gradients for a resting animal cell.

less than the equilibrium potential for E_{NCX}, and Ca^{2+} ions are extruded from the cell. However, during the upstroke of the cardiac action potential in phases 0 and 1, E_m exceeds E_{NCX}, which leads to a brief period of Ca^{2+} entry into the cell through the exchanger. Progression of membrane repolarization again results in a more negative membrane potential, with $E_m < E_{NCX}$, and Ca^{2+} ions are extruded again. Thus, the net Ca^{2+} movement mediated by the exchanger briefly changes direction when the membrane potential becomes more positive relative to the reversal potential of the Na^+/Ca^{2+}-exchanger, as **FIGURE 6.43** shows. Influx of Ca^{2+} ions over several milliseconds occurs physiologically during the beginning of the relatively long cardiac action potential, when the transmembrane Na^+ or Ca^{2+} gradients are altered. Under physiological conditions, the exchanger works mainly in the Ca^{2+} extrusion mode, driven by the Na^+ gradient. However, positive membrane potentials during the action potential plateau phase can limit Ca^{2+} extrusion. (For more on the cardiac action potential see *6.12 Action potentials are electrical signals that depend on several types of ion channels* and *6.13 Cardiac and skeletal muscles are activated by excitation-contraction coupling.*)

Another transporter that uses the energy of the Na^+-gradient to drive transport of ions across the plasma membrane is the $Na^+/K^+/Cl^-$-cotransporter, which mediates electroneutral transport with a stoichiometry of 1 Na^+:1 K^+:2 Cl^- (see Figure 6.41). Under physiological conditions, the cotransporter catalyzes the movement of these ions into cells. The cotransporter maintains the intracellular Cl^- concentration; in specialized epithelial cells, it accumulates intracellular Cl^- concentrations above the electrochemical equilibrium. For example, in the kidney thick ascending limb of Henle's loop, the $Na^+/K^+/Cl^-$-cotransporter is present on the apical membrane of specialized epithelial cells facing the tubule lumen. The cotransporter is crucial for the reabsorption of NaCl from the kidney filtrate, as Figure 6.26 shows.

Mutations in the gene that encodes a $Na^+/K^+/Cl^-$-cotransporter isoform of the kidney result in autosomal recessive disorders called Bartter's syndrome. Patients with Bartter's syndrome have urinary water and salt loss, decreased urinary concentrating ability, and increased urinary Ca^{2+} excretion. These effects of a mutation in one transporter isoform show that transcellular ion transport is mediated by the concerted action of many types of transport proteins and that dysfunction of one protein may affect the function of several other proteins. In fact, mutations in other transport proteins that work in concert with the $Na^+/K^+/Cl^-$-cotransporter to mediate salt absorption in the kidney can also result in Bartter's syndrome. These include mutations in a K^+ channel and in a chloride channel (CLC-K) that is a component of the basolateral chloride conductance (see Figure 6.26).

The Na^+/Mg^{2+}-exchanger is one more example of a Na^+-dependent, secondary transport protein (see Figure 6.41). Because a high concentration of Mg^{2+} interferes with many cellular functions, the intracellular Mg^{2+} concentration is tightly controlled by a buffering mechanism. For example, Mg^{2+} ions compete with Ca^{2+} ions at combined Ca^{2+}/Mg^{2+}-activation/inactivation sites in a variety of proteins. Since Mg^{2+} permeates the plasma membrane only at a low rate, it must be constantly extruded from the cytoplasm to maintain a physiological concentration. The major extrusion mechanism is the Na^+/Mg^{2+}-exchanger. It has been proposed that the Na^+/Mg^{2+}-exchanger is electroneutral under physiological conditions, transporting two Na^+ ions into the cell for each Mg^{2+} ion extruded from the cell. The Na^+/Mg^{2+}-exchanger is also an important drug target for a widely used antidepressant drug.

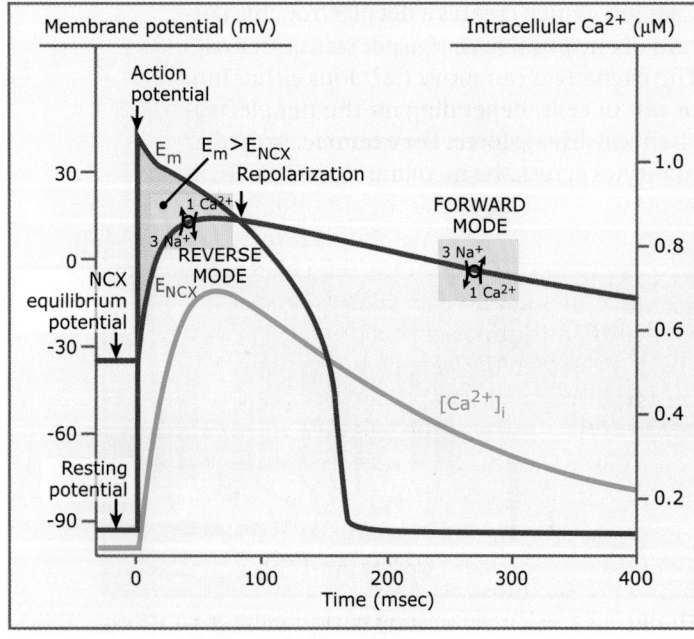

FIGURE 6.43 The Na^+/Ca^{2+}-exchanger (NCX) operates in forward and reverse modes depending on the membrane potential.

6.17 Some Na⁺ transporters regulate cytosolic or extracellular pH

Key concepts

- Na⁺/H⁺ exchange controls intracellular acid and cell volume homeostasis.
- The net effect of Na⁺/HCO₃⁻-cotransporters is to remove acid by directed transport of HCO₃⁻.

As described in the previous section, many transport proteins use the energy of the transmembrane Na⁺ gradient to drive the movement of other ions and solutes across the plasma membrane (see *6.16 The transmembrane Na⁺ gradient is essential for the function of many transporters*). In this section, we will discuss two additional transporters, the Na⁺/H⁺-exchanger and the Na⁺/HCO₃⁻-cotransporter, which require the energy of transmembrane Na⁺ gradient. These transporters are involved in pH regulation, which is important for the function of proteins, many of which are pH-sensitive.

The Na⁺/H⁺-exchangers and Na⁺/HCO₃⁻-cotransporters help to maintain the intracellular and extracellular acid-base balance. The bicarbonate/carbon dioxide exchange reaction facilitated by the enzyme carbonic anhydrase is the most important acid buffer system in the extracellular fluids, such as the plasma. The pH of blood is normally slightly basic (pH 7.4) and is tightly regulated because many cell and protein functions are critically pH-sensitive. The lungs and the kidneys are the organs involved in maintaining the acid-base balance of the plasma by excreting volatile (CO_2) and nonvolatile acids, respectively. The kidneys help control the blood pH by regulated transport of HCO_3^- through epithelial cells lining the early segment of the proximal tubule, as **FIGURE 6.44** shows. There are two sources of the CO_2 in these cells: diffusion across the cell membrane and synthesis by specific enzymes. Carbonic anhydrase in the proximal tubule catalyzes the formation of H_2O and CO_2 from HCO_3^- and H^+. CO_2 can freely diffuse into the epithelial cell of the proximal tubule, where HCO_3^- and H^+ are synthesized by carbonic anhydrase from H_2O and CO_2. Thus, the secretion of protons (acid) into the lumenal filtrate (urine) by the apical membrane Na⁺/H⁺-exchanger (NHE) is coupled to the transport of an equal number of bicarbonate base equivalents into the blood via the Na⁺/HCO₃⁻-cotransporter in the basolateral membrane by the proximal tubule. The coupled

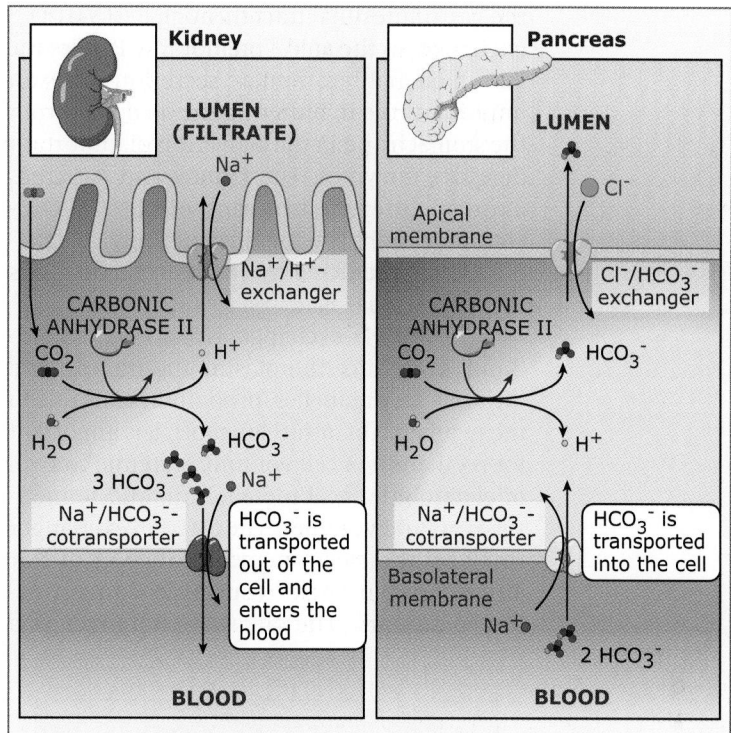

FIGURE 6.44 Models of bicarbonate transport across epithelial cells in the kidney and pancreas.

secretion of H^+ and HCO_3^- from the cell is electroneutral and is an important mechanism for secreting nonvolatile acid equivalents into the urine (contributing to metabolic acid homeostasis). The net effect is to *reabsorb* bicarbonate from the plasma filtrate in the form of CO_2, which is then transported as HCO_3^- across the apical membrane into the plasma. At the same time, NHE contributes to bulk Na⁺ reabsorption by the early proximal tubule from the plasma filtrate; the Na⁺ ions are then transported back into the plasma. The key element in proximal tubule acid regulation and reabsorption is the basolateral Na⁺/K⁺-ATPase. In summary, H^+ excretion and Na⁺ entry into the cell across the apical membrane is mediated by specific symporter and antiporter proteins.

In contrast to the kidney, in the pancreas and other secretory organs, epithelial cells *secrete* bicarbonate across the apical membrane. In secretory organs, the inward Na⁺ gradient drives the cotransport of HCO_3^- into the cell in a stoichiometry of 1 Na⁺ to 1 or more HCO_3^- ions (see Figure 6.44). In the pancreas, the Na⁺/HCO₃⁻-cotransporter in the basolateral membrane mediates bicarbonate uptake from the plasma, HCO_3^- is produced by carbonic anhydrase, and bicarbonate is transported into the gastrointestinal tract through a Cl^-/HCO_3^- exchanger in the apical membrane. Pancreatic and duodenal bicarbonate secretion plays an important role in buffering the acid load from the stomach and in the activation of important digestive enzymes. The kidney and pancreas express different isoforms of Na^+/HCO_3^--cotransporters. The topology of these cotransporters, as predicted by sequence analysis, is shown in **FIGURE 6.45**.

The Na^+/H^+-exchanger is an integral membrane protein of the plasma membrane and intracellular organelles. In addition to its role in the regulation of intracellular pH, it is important for regulation of cell volume, systemic control of electrolytes, acid metabolism, and homeostasis of fluid volumes. The Na^+/H^+-exchanger catalyzes the electroneutral transport of Na^+ and H^+ down their respective concentration gradients. The Na^+/H^+-exchanger is a secondary active transporter. The inward Na^+ gradient established by the plasma membrane Na^+/K^+-ATPase drives the countertransport of protons with a stoichiometry of 1 Na^+ for 1 H^+. The transmembrane Na^+ influx contributes to regulation of the cell volume and to the absorption of salt and water across epithelial cells. Na^+/H^+-exchangers are regulated in response to intracellular pH changes, to a variety of extracellular stimuli such as growth factors and hormones, and to mechanical stimuli, such as osmotic stress and cell spreading. Under physiological conditions, plasma membrane Na^+/H^+ exchange activity extrudes excess acid accumulated by cellular metabolism and by $H^+/$ acid leak pathways. Together with bicarbonate-transporting systems, Na^+/H^+-exchangers play a crucial role in maintaining cytoplasmic acid-base balance.

The NHE isoform 1 (NHE1) is considered the prototypical mammalian exchanger of the plasma membrane; other isoforms differ in tissue expression and membrane localization. All Na^+/H^+-exchanger isoforms share a common structure. Computer modeling of the secondary structure predicts a common membrane topology with 12 conserved membrane-spanning segments at the N-terminus and a more variable cytoplasmic C-terminus that contains numerous phosphorylation sites and binds regulatory proteins and mediators, as **FIGURE 6.46** shows. The N-terminal transmembrane domain is thought to contain the catalytic core structure important for Na^+/H^+ exchange. A large extracellular reentrant loop (R-loop) between transmembrane segments 9 and 10 resembles the pore-loop structure identified in K^+ channels and may form part of the ion conduction pathway. Two Na^+/H^+-exchanger proteins can form a homodimeric structure.

Bacteria have Na^+/H^+-exchange proteins that use the energy of the electrochemical H^+ gradient across the membrane to excrete Na^+ in exchange for H^+ flow into the cell. In *Escherichia coli*, the primary Na^+/H^+-exchanger is NhaA, which regulates the cell's Na^+ content and pH and is especially important for survival under conditions of high salinity and/or alkaline pH. NhaA exists as a dimer in the cytoplasmic membrane; each monomer has 12 transmembrane segments. **FIGURE 6.47** shows the X-ray crystal structure of an NhaA monomer in an inactive conformation, obtained at acidic pH. At both sides of the membrane, the protein has funnel-shaped surfaces that are lined by negatively charged amino acids, which attract

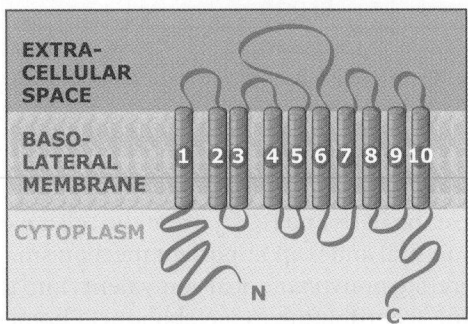

FIGURE 6.45 Predicted membrane topology of Na^+/HCO_3^--cotransporters. The insets indicate the Na^+ and HCO_3^- gradients across the basolateral membrane of epithelial cells in the kidney and pancreas.

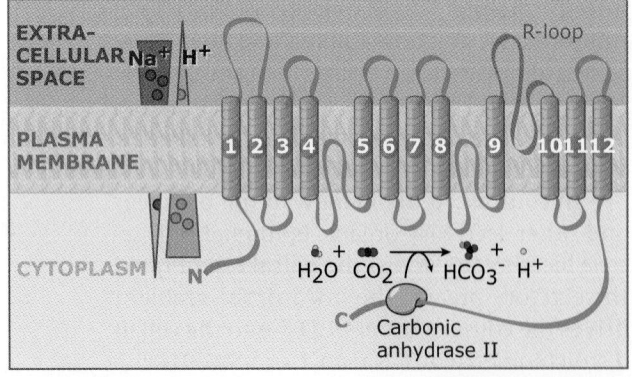

FIGURE 6.46 Predicted membrane topology of the Na^+/H^+-exchanger isoform 1 (NHE1). NHE1 function is regulated in several different ways. Shown is the regulated binding of carbonic anhydrase II, which stimulates NHE1 transport activity. The insets at left indicate the Na^+ and H^+ gradients across the plasma membrane of animal cells.

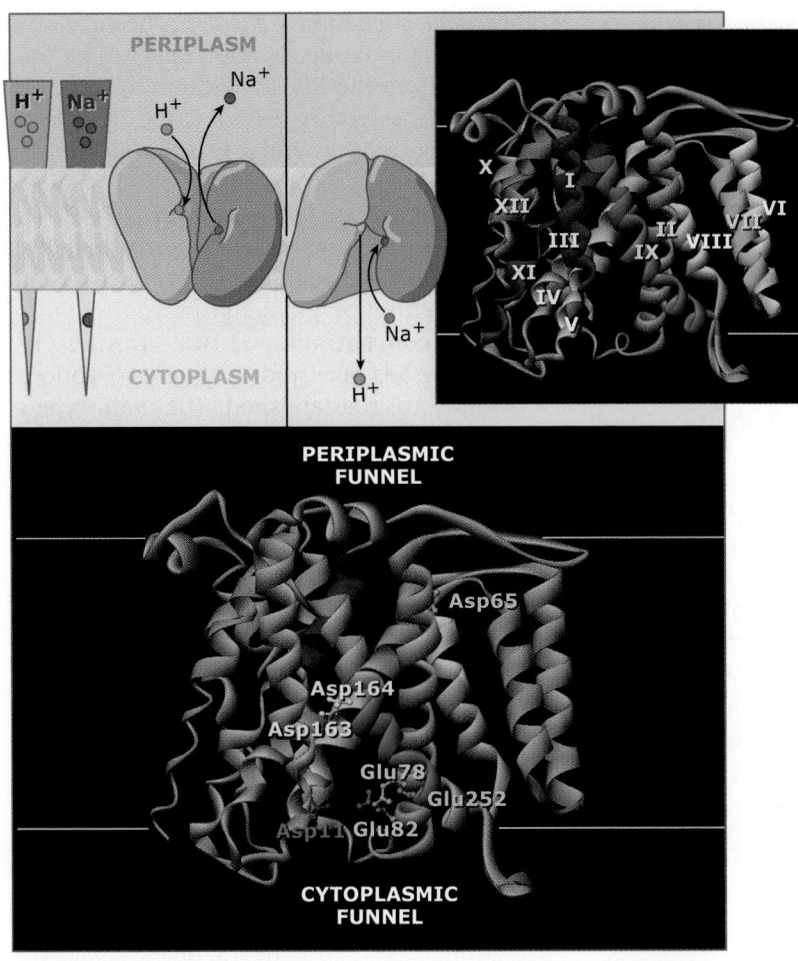

FIGURE 6.47 Schematic and X-ray crystal structure of a bacterial Na⁺/H⁺-exchanger, NhaA, showing the structural basis for Na⁺/H⁺ translocation and pH regulation. The crystal structure is depicted as a ribbon diagram of the protein backbone. Residues whose alterations affect pH regulation and cation transport are shown with side chains. The schematics show the alternating access model of translocation suggested by the structural features. The insets show the Na⁺ and H⁺ gradients across the cell membrane. Structures from Protein Data Bank 1ZCD. C. Hunte, et al., *Nature* 435 (2005): 1197–1202.

cations. At the membrane center, amino acids Asp163 and Asp164 form the Na⁺ binding site. At acidic pH, only Asp164 is exposed to the permeation pathway, and the cytoplasmic passage is blocked by helix XIp.

It is proposed that transport of Na⁺ and H⁺ by NhaA uses the alternating access mechanism (see *6.15 Symporters and antiporters mediate coupled transport*). The model is that a change to alkaline pH at the cytoplasmic side induces a conformational change in helix IX that in turn causes reorientation of helices XIp and IVc, thereby exposing the Na⁺ binding site and removing the cytoplasmic barrier. Binding of Na⁺ on the cytoplasmic side would trigger an additional, small conformation change of helices XIp and IVc that closes the cytoplasmic side and exposes Na⁺ to the periplasm. Upon release of Na⁺ to the periplasmic (extracellular) side, Asp163 and Asp164 would be protonated, which induces a conformational change to expose the Na⁺ binding site to the

cytoplasm, where the aspartate residues would be deprotonated.

Altered Na⁺/H⁺-exchanger activity in human cells has been linked to the pathogenesis of several diseases, including hypertension, diarrhea, diabetes, and tissue damage caused by ischemia (obstructed or slowed blood flow). In cardiac and neural tissues, ischemia leads to increased Na⁺/H⁺ exchange activity. This results in increased intracellular Na⁺ concentrations and a secondary increase in intracellular Ca²⁺ concentrations due to activation of the Na⁺/Ca²⁺-exchanger in reverse mode (see *6.16 The transmembrane Na⁺ gradient is essential for the function of many transporters*). The Ca²⁺ overload causes a cascade of events that lead to cardiac arrhythmias or damage to neuronal tissues following stroke. Pharmacological inhibition of Na⁺/H⁺ exchange activity may have beneficial effects during or following ischemia and is actively investigated as a therapeutic strategy.

6.18 The Ca²⁺-ATPase pumps Ca²⁺ into intracellular storage compartments

Key concepts

- Ca²⁺-ATPases undergo a reaction cycle involving two major conformations, similar to that of Na⁺/K⁺-ATPases.
- Phosphorylation of Ca²⁺-ATPase subunits drives conformational changes and translocation of Ca²⁺ ions across the membrane.

Stimulation of cells by hormones or electrical signals at the plasma membrane is translated into transient increases in the cytosolic Ca²⁺ concentration, which subsequently regulates a wide variety of intracellular responses. Ca²⁺ activates specific cellular functions, which range from gene transcription and protein synthesis, to functions such as regulated secretion of hormones, immunologic activation, cell movement, neuronal excitation, and cell contraction. Almost all cells depend on release of Ca²⁺ from intracellular stores to rapidly increase the cytosolic Ca²⁺ concentration (for an example see *6.13 Cardiac and skeletal muscles are activated*

by excitation-contraction coupling). The increase in cytosolic Ca²⁺ concentration depends on Ca²⁺ release from the ER. In addition to its role as part of the secretory pathway (see *7 Membrane targeting of proteins* and *8 Protein trafficking between membranes*), the ER functions as the intracellular Ca²⁺ storage compartment. The Ca⁶⁺ concentration in the ER is maintained at around 10^{-3} M, creating a steep electrochemical concentration gradient toward the cytosolic Ca²⁺ concentration of 10^{-7} M. After Ca²⁺ signaling has occurred, resting Ca²⁺ concentrations are reestablished. The main type of Ca²⁺ transport protein involved in extrusion of Ca²⁺ from the cytosol is the Ca²⁺-ATPase, different isoforms of which exist in the ER and the plasma membrane.

In this section, we will discuss one of the major Ca²⁺-ATPases present in muscle cells and neurons. Skeletal muscle cells have a specialized structure of large intracellular Ca²⁺ stores called the SR, which effectively controls Ca²⁺ uptake and release throughout the cell volume. The sarcoendoplasmic reticulum Ca²⁺-ATPase (SERCA) in these cells is responsible for the majority of the Ca²⁺ extrusion from the cytosol in muscle cells. Muscle cells are among the largest cells in the body, and for effective activation and relaxation of contractile proteins throughout the cell volume, a network of intracellular Ca²⁺ release and reuptake sites is necessary. (For more on muscle contraction, see *6.13 Cardiac and skeletal muscles are activated by excitation-contraction coupling* and *12 Actin*.)

The SERCA pump is a Ca²⁺-activated ATPase that is part of a large family of ATP-dependent pumps, known as P-type ATPases. P-type ATPases use a structural mechanism by which the energy released by ATP hydrolysis is coupled to the uphill transport of H⁺, Na⁺, or Ca²⁺ ions. A hallmark of P-type ATPases is that in the presence of the appropriate cation, they use ATP to autophosphorylate a conserved aspartic acid residue. Phosphorylation by ATP results in a high energy phosphoprotein intermediate that drives alternating steps of Ca²⁺ binding and phosphate transfer during the reaction cycle of the pump. In this way, the SERCA pump uses the chemical energy of ATP hydrolysis to rapidly decrease the intracellular Ca²⁺ concentration by directing Ca²⁺ reuptake into the SR Ca²⁺ store.

FIGURE 6.48 shows the proposed reaction cycle of the SERCA pump. There are two basic conformations in this cycle:

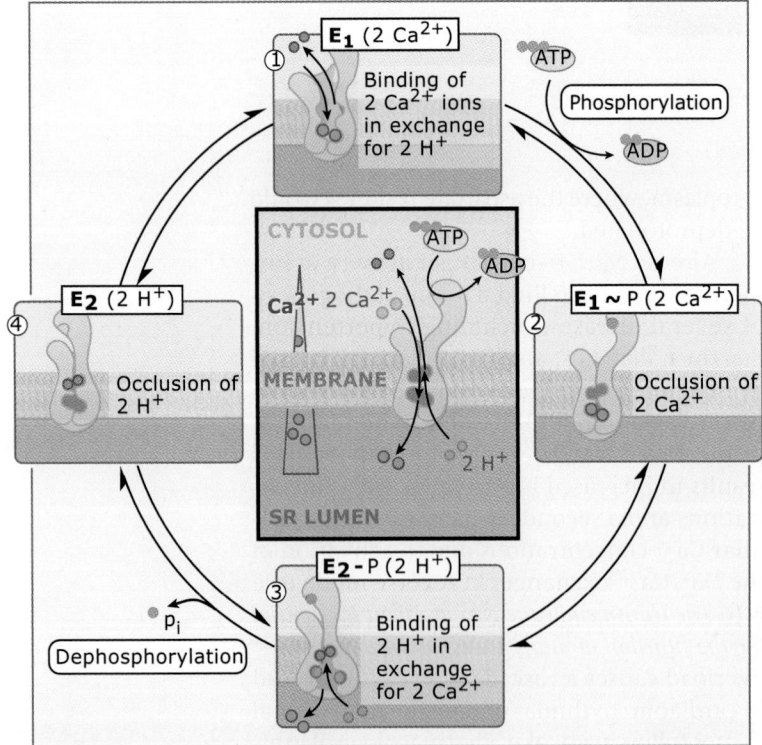

FIGURE 6.48 The reaction cycle for transport of Ca²⁺ into the sarcoplasmic reticulum (SR) by SERCA pumps. The E_1 conformation binds Ca²⁺ with high affinity, whereas the E_2 conformation binds Ca²⁺ with low affinity. The high-energy phosphate bond conformation is indicated as $E_1{\sim}P$.

1. The E_1 conformation, which binds Ca^{2+} on the cytosolic side with high affinity, and

2. The E_2 conformation, which binds Ca^{2+} with a significantly lower affinity and therefore releases Ca^{2+} from the Ca^{2+}-binding sites into the SR lumen.

During each enzymatic cycle, two Ca^{2+} ions are transported into the SR lumen, and it is thought that protons are transported in the opposite direction. High-affinity binding of two Ca^{2+} ions from the cytosol induces a change to the E_1 (2 Ca^{2+}) conformation. In this conformation, the protein undergoes autophosphorylation to form the high-energy $E_1 \sim P(2\ Ca^{2+})$ intermediate. The energy of this phosphoprotein intermediate fuels a conformational change to the E_2-$P(2H^+)$ conformation, which has a lower affinity for Ca^{2+} ions. In this conformation, the ion gate from the cytosol is closed, the affinity of the transport site for Ca^{2+} ions is reduced, and the ion gate toward the lumen of the SR is opened. The Ca^{2+} ions are released into the SR Ca^{2+} store in exchange for H^+ binding. The phosphate group is then hydrolyzed, which completes the cycle. In summary, the reaction cycle of the SERCA pump consists of a sequence of phosphorylation and dephosphorylation events that power the uphill transport of two Ca^{2+} ions into the SR per hydrolyzed ATP, in exchange for two H^+ ions.

The SERCA pump has an asymmetrical arrangement of transmembrane and cytosolic domains that undergo major movements during transport of Ca^{2+} ions. As shown schematically in **FIGURE 6.49**, it consists of 10 transmembrane a helices (M1-M10) and two large cytoplasmic loops between the transmembrane helices, one loop between M2 and M3 and another between M4 and M5. The transmembrane a helices form the Ca^{2+} binding sites, which cooperatively bind two Ca^{2+} ions from the cytosol in the E_1 conformation. The two cytoplasmic loops form three separate domains. The loop between helices M4 and M5 forms the nucleotide binding (N) domain that binds ATP and the P domain that contains the catalytic phosphorylation site. The loop between M2 and M3 forms the actuator (A) domain, which is important for transmission of conformational changes between the cytosolic and transmembrane domains.

Two X-ray crystal structures representing the SERCA pump in the E_2(2 H^+) and E_1(2 Ca^{2+}) conformations are shown in **FIGURE 6.50**. The transmembrane segments contain hydrophilic residues that form a pathway for Ca^{6+} perme-

ation (see view from lumen in Figure 6.50). The opening and closing of the pathway appears to be coupled to conformational rearrangements of the cytoplasmic domains through the S4 and S5 extensions of the M4 and M5 transmembrane segments, respectively (see Figure 6.49). The high-affinity Ca^{2+}-bound structure has a wide gap between the cytosolic domains, which allows ATP to bind to the N-domain. It is thought that after Ca^{2+} ions bind to the transmembrane domains to form the E_1(2Ca^{2+}) conformation, ATP binding allows for phosphorylation of a crucial aspartic acid residue, Asp351, in the P-domain. This phosphorylation ultimately results in a conformational change to the low-affinity E_2 conformation, in which a rearrangement of the transmembrane domains disrupts the Ca^{2+} binding sites, opens the gate on the lumenal side of the SERCA pump, and releases the Ca^{2+} ions into the lumen of the sarcoplasmic reticulum. During the transition from the E_2(2H^+) conformation back to the E_1(2Ca^{2+}) configuration, the cytosolic domains undergo large movements, from a compact structure into an open one. This open structure allows water to enter and catalyze dephosphorylation, which is important to return the SERCA pump to its

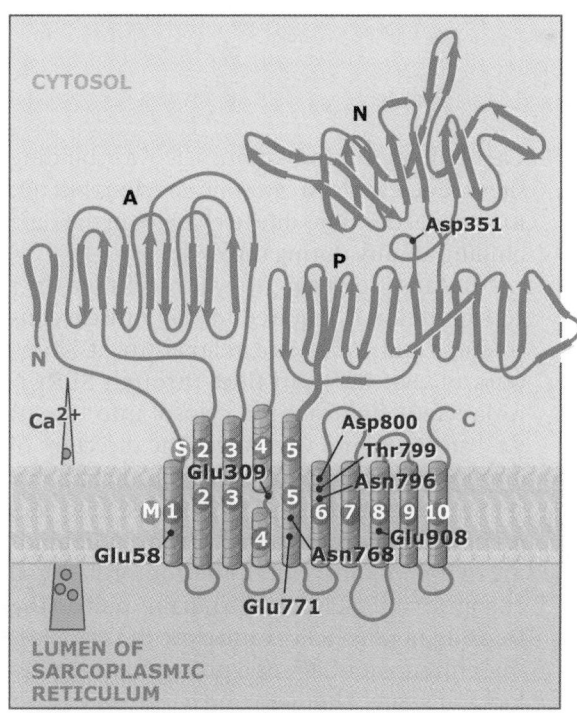

FIGURE 6.49 The location of the phosphorylation site (Asp351) and the key residues involved in Ca^{2+} binding are indicated for the SERCA1 isoform. The inset indicates the Ca^{2+} gradient across the membrane of the sarcoplasmic reticulum.

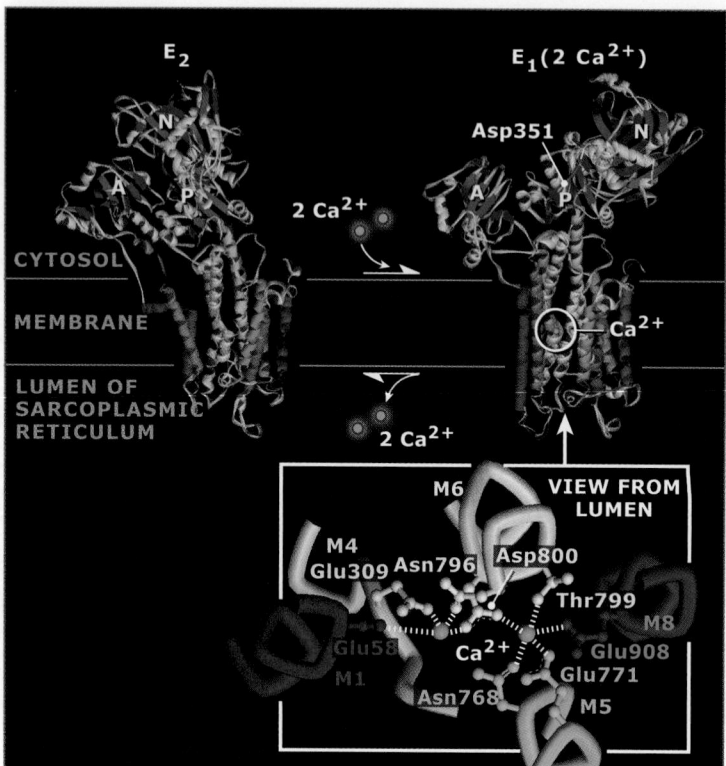

E_2 $E_1(2\ Ca^{2+})$

FIGURE 6.50 The E_2 structure is Ca^{2+}-free, whereas the E_1 structure has two bound Ca^{2+} ions. The transmembrane helices are depicted in a spectrum of colors, from M1 in red to M10 in dark blue. Structures from Protein Data Bank 1EUL and 1IWO. S. Toyoshima, et al., *Nature* 405 (2000): 647–655 and C. Toyoshima and H. Nomura, *Nature* 418 (2002): 605–611.

Ca^{2+}-binding conformation. The Ca^{2+} binding sites are accessible to cytosolic Ca^{2+} ions but not to lumenal Ca^{2+} ions due to changes in the Ca^{2+} binding affinity during the cycle.

Because lowering the intracellular Ca^{2+} concentration is required to stop muscle contraction and to initiate relaxation, it is important that Ca^{2+} ion flow through SERCA is directed only from the cytosol into the SR Ca^{2+} store and not vice versa. The direction of ion flow through Ca^{2+}-ATPases is dictated by sequential coupling of ATP hydrolysis and vectorial Ca^{2+} transport steps. During ATP binding and phosphorylation, the conformational changes in the A, N, and P domains pull on the transmembrane helices and close the cytoplasmic entrance for Ca^{2+}, and phosphoryl transfer locks this gate. Therefore, no backflow of Ca^{2+} ions from the SR to the cytosol occurs under physiological conditions.

The SERCA pump works together with the plasma membrane Ca^{2+}-ATPase (PMCA)

to export Ca^{2+} ions from the cytosol and to set the resting level of the cytosolic Ca^{2+} concentration. Both enzymes have high Ca^{2+} affinities that are above those of other Ca^{2+} binding proteins or buffers in the cytosol. The PCMA and SERCA pumps share similar structures and mechanisms of action, with ten transmembrane sequences and three large cytosolic domains. In addition to having different cellular localizations, the PMCA pump differs from the SERCA pump in transporting one Ca^{2+} instead of two per ATP hydrolyzed. In excitable cells, the higher transport efficiency and the higher density of the SERCA pump in the membranes of the SR Ca^{2+} store explain its quantitatively overwhelming contribution to extrusion of Ca^{2+} from the cytosol.

The Na^+/K^+-ATPase and the H^+-ATPase, which also belong to the P-type family of ATPases, share common structural features with the SERCA pump (see *6.19 The Na^+/K^+-ATPase maintains the plasma membrane Na^+ and K^+ gradients* and *6.21 H^+-ATPases transport protons out of the cytosol*). The homology between Na^+/K^+-ATPase and the Ca^{2+}-ATPases is more extensive than for other ATPases, and comparison of their structural features supports the concept that they share the same structural mechanisms to couple ATP hydrolysis to cation translocation across membranes. All three types of cation pumps use a similar catalytic cycle that involves ATP binding, conformational rearrangements resulting in occlusion of the transported cations, release of the cations on the other side of the membrane, and dephosphorylation to return the enzyme to the original ion-binding conformation.

6.19 The Na^+/K^+-ATPase maintains the plasma membrane Na^+ and K^+ gradients

Key concepts

- The Na^+/K^+-ATPase is a P-type ATPase that is similar to the Ca^{2+}-ATPase and the H^+-ATPase.
- The Na^+/K^+-ATPase maintains the Na^+ and K^+ gradients across the plasma membrane.
- The plasma membrane Na^+/K^+-ATPase is electrogenic: it transports three Na^+ ions out of the cell for every two K^+ ions it transports into the cell.
- The reaction cycle for Na^+/K^+-ATPase is described by the Post-Albers scheme, which proposes that the enzyme cycles between two fundamental conformations.

All cells are negatively charged in comparison with the extracellular environment. This charge difference is due to the presence of slightly more positively charged molecules than negatively charged molecules in the extracellular fluid and the opposite situation in the cytosol. The electrochemical gradient across the plasma membrane is essential for cellular function; it is similar to a battery, which maintains a charge separation that can be applied to perform work. In mammalian cells, the Na^+ and K^+ gradients are two of the major components of the electrochemical gradient across the plasma membrane. Cells maintain a lower intracellular Na^+ concentration and a higher intracellular K^+ concentration than in the extracellular milieu. The generation and maintenance of the electrochemical gradients for Na^+ and K^+ in animal cells require the activity of the Na^+/K^+-ATPase, which is an ion pump that couples ATP hydrolysis to cation transport. It helps to set the negative resting membrane potential, which regulates the osmotic pressure to avoid cell lysis or shrinkage and allows for secondary Na^+-dependent transport of molecules (see *6.7 Voltage-dependent Na^+ channels are activated by membrane depolarization and translate electrical signals; 6.8 Epithelial Na^+ channels regulate Na^+ homeostasis; 6.16 The transmembrane Na^+ gradient is essential for the function of many transporters; and 6.17 Some Na^+ transporters regulate cytosolic or extracellular pH*).

The Na^+/K^+-ATPase belongs to the family of P-type ATPases, which includes the sarcoendoplasmic reticulum Ca^{2+}-ATPase discussed in the previous section (see *6.18 The Ca^{2+}-ATPase pumps Ca^{2+} into intracellular storage compartments*). P-type ATPases are enzymes that form a phosphorylated intermediate by autophosphorylation of an aspartic acid residue during ion transport. During the autophosphorylation process, P-type ATPases transfer the γ-phosphate group from ATP to the active site in the enzyme. For every molecule of ATP hydrolyzed by the Na^+/K^+-ATPase, three Na^+ ions from the cytosol and two K^+ ions from the extracellular fluid are exchanged. The Na^+/K^+-ATPase cycle operates at a rate of 100 times per second. This transport rate is very slow in comparison with ion flux through channel pores, which occurs at a rate of 10^7 to 10^8 ions per second, close to the diffusion rate of ions in water.

The major steps in the enzymatic ion transport cycle of the Na^+/K^+-ATPase have been identified biochemically and are summarized in the Post-Albers reaction scheme shown in

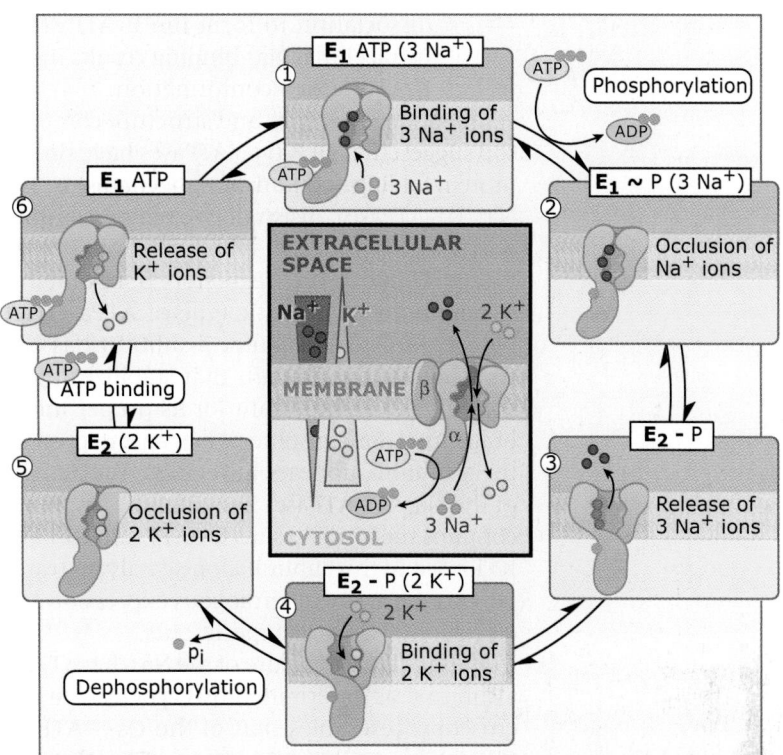

FIGURE 6.51 The Post-Albers reaction cycle of the Na^+/K^+-ATPase. The high-energy phosphate bond is indicated as $E_1 \sim P$. The center panel summarizes the reaction cycle, with insets indicating the Na^+ and K^+ gradients across the plasma membrane of a resting animal cell.

FIGURE 6.51. The Post-Albers scheme was originally proposed for the Na^+/K^+-ATPase and is useful for identifying particular states in all P-type ATPases. In this model, P-type ATPases use two different conformations, called enzyme 1 (E_1) and enzyme 2 (E_2), to bind, occlude, and transport ions. These conformational changes are driven by a phosphorylation-dephosphorylation reaction:

- In the E_1 conformation, intracellular ATP and Na^+ ions bind to the ATPase with high affinity to form the E_1ATP(3 Na^+) state, which immediately induces ATP-phosphorylation at an aspartic acid residue and occlusion of the three Na^+ ions in the $E_1 \sim P(3 Na^+)$ state.
- A subsequent conformational change to the E_2-P state leads to a decreased Na^+ affinity and release of Na^+ ions to the extracellular space, and an increased K^+ ion affinity.
- Binding of extracellular K^+ to the ATPase induces dephosphorylation of E_2-P(2 K^+) and occlusion of two K^+ ions in the E_2(2 K^+) state.
- Intracellular binding of ATP induces a conformational switch followed by K^+

dissociation to form the E_1ATP state. Intracellular Na$^+$ binding results in the E_1ATP(3 Na$^+$) conformation.

Sequence analysis and structure comparison suggest that all P-type ATPases have similar protein folding and transport mechanisms. The Na$^+$/K$^+$-ATPase consists of two major subunits, a catalytic α subunit, which is similar in all P-type ATPases, and a regulatory β subunit that is unique to the specific type of ATPase (see Figure 6.51). The smaller β subunit has one transmembrane domain that stabilizes the α subunit and is important for its proper membrane insertion. In some tissues a third protein, the γ subunit, appears to regulate the activity of the Na$^+$/K$^+$-ATPase. The catalytic α subunit contains the binding sites for ATP and Na$^+$ and K$^+$ ions, and this subunit alone catalyzes transport, as shown by heterologous expression and electrophysiological experiments.

The overall structure of the Na$^+$/K$^+$-ATPase a subunit, as determined by cryoelectron microscopy, resembles that of the Ca^{2+}-ATPase SERCA, as **FIGURE 6.52** shows. The Na$^+$/K$^+$-ATPase a subunit has an arrangement of 10 transmembrane a helices similar to the SERCA pump (see Figure 6.49). The intracellular P-domain between the transmembrane segments 4 and 5 contains the phosphorylation site that is highly conserved among all P-type ATPases. This reversible phosphorylation site is the Asp376 residue in the characteristic Asp-Lys-Thr-Gly-Thr-Leu-Thr sequence motif. Binding of ATP and Na$^+$ ions induces a substantial conformational change in the hinge region connecting the N- and P-domains, such that the ATP-binding site in the N-domain and the phosphorylation site in the P-domain are brought closer together.

The Na$^+$/K$^+$-ATPase is an electrogenic ion pump. Under normal physiological conditions, the free energy of ATP hydrolysis (ΔG_{ATP}) fuels the Na$^+$/K$^+$-ATPase to transport three Na$^+$ ions out of the cell for every two K$^+$ ions pumped into the cell, against their respective concentration gradients. Thus, a net flow of positive charge out of the cell occurs, against the electrical field across the cell membrane. This helps to create a negatively charged cytosol relative to the extracellular fluid. The net outward current creates both an electrical potential difference and an osmotic ion gradient across the cell membrane. (For more on membrane potentials see *6.4 Electrochemical gradients across the cell membrane generate the membrane potential.*)

P-type ATPases are ion pumps that use the energy of ATP hydrolysis to maintain ion gradients across cell membranes. Since each step in the reaction cycle is reversible, P-type ATPases can in principle produce ATP using the energy stored in the membrane potential. Thus, the Na$^+$/K$^+$-ATPase has the potential to run in the reverse mode, such that Na$^+$ ions are transported into the cell and K$^+$ ions are transported out, resulting in a net inward current. Normal transport of Na$^+$ ions out of the cell and K$^+$ ions into the cell proceeds as long as ΔG_{ATP} exceeds the electrochemical energy of the respective ion gradients. When the energy required for the active transport of Na$^+$ and K$^+$ ions equals ΔG_{ATP}, no net ion flux occurs. This would represent the reversal potential of the Na$^+$/K$^+$-ATPase, that is, the membrane potential below which it operates in backward mode. The reversal potential has been estimated to be about −180 mV, which is far more negative than any membrane potential found under physiological conditions. It is, therefore, unlikely that a deleterious inward Na$^+$ current would occur. However, this can change when reduced blood supply during myocardial infarction or intoxication, which can result in cellular ATP deprivation or steeper ionic gradients that can ultimately reverse the movement of ions by the Na$^+$/K$^+$-ATPase and lead to cell death.

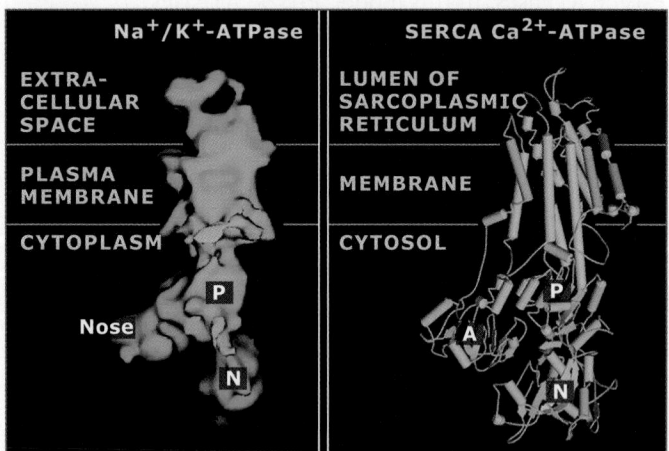

FIGURE 6.52 The Na$^+$/K$^+$-ATPase subunit is similar in structure to the sarcoendoplasmic reticulum Ca^{2+}/ATPase (SERCA). The structure of the Na$^+$/K$^+$-ATPase subunit was determined by cryoelectron microscopy. Left photo reprinted from *Biophys. J.*, vol. 80, W. J. Rice, et al., Structure of Na$^+$, K$^+$-ATPase at 11-Å Resolution..., pp. 2187–2197, Copyright (2001) with permission from Elsevier. [http://www.sciencedirect.com/science/journal/00063495]. Photo courtesy of David L. Stokes, New York University Medical Center; right structure from Protein Data Bank 1IWO. C. Toyoshima and H. Nomura, *Nature* 418 (2002): 605–611.

The Na+/K+-ATPase is the target of a large number of toxins and an important drug target. For example, the naturally occurring plant steroids called cardiac glycosides, such as ouabain and digitalis, specifically inhibit ion transport by the Na+/K+-ATPase. Other toxins like palytoxin from marine corals or sanguinarine from plants are also specific inhibitors. Unlike the cardiac glycosides, which inhibit ion flux through the Na+/K+-ATPase, palytoxin and sanguinarine block the ATPase in an open state, allowing ions to flow down their concentration gradient, which destroys the electrochemical gradients. Cardiac glycosides bind reversibly to the extracellular side of the Na+/K+-ATPase, inhibiting ATP hydrolysis and ion transport. Carefully titrated inhibition of the Na+/K+-ATPase in the heart by cardiac glycosides like digitalis represents a treatment option for patients with heart failure. Partial inhibition of a subpopulation of the Na+/K+-ATPases by cardiac glycosides slightly increases the intracellular Na+ concentration, which results in higher intracellular Ca^{2+} concentrations due to transport through the Na+/Ca2+-antiporter. Slightly increased intracellular Ca^{2+} concentrations are known to increase the contractility of the heart (see *6.13 Cardiac and skeletal muscles are activated by excitation-contraction coupling*).

6.20 The F_1F_0-ATP synthase couples H+ movement to ATP synthesis or hydrolysis

Key concepts

- The F_1F_0-ATP synthase is a key enzyme in oxidative phosphorylation.
- The F_1F_0-ATP synthase is a multisubunit molecular motor that couples the energy released by movement of protons down their electrochemical gradient to ATP synthesis.

Synthesis of ATP during oxidative phosphorylation provides the bulk of cellular energy in eukaryotes and almost all prokaryotes. It is a multistep, membrane-localized process. A 70-kg human with a sedentary lifestyle will generate 2 million kg of ATP during a 75-year life span. In eukaryotic cells, ATP synthesis, which is one of the most frequent enzyme reactions in biology, occurs in specialized organelles called mitochondria. Cellular ATP synthesis is catalyzed by the enzyme F_1F_0-ATP synthase,

which is an extraordinary molecular motor that couples the energy of the electrochemical proton gradient (the **proton motive force**) across the membrane to ATP synthesis. This electrochemical proton gradient is generated by electron transfer complexes during oxidative phosphorylation and consists of two components, the membrane potential and the difference in proton concentrations on either side of the mitochondrial membrane.

The overall structure of the F_1F_0-ATP synthase is similar in all cells, although the subunit composition may differ. The simplest form of the F_1F_0-ATP synthase, in the bacterial cytoplasmic membrane, has eight types of subunits with a stoichiometry of $\alpha_3\beta_3\gamma\Delta\epsilon ab_2c_{10-14}$ and a total molecular mass of ~530 kDa. In mitochondria, 7–9 additional regulatory subunits contribute to a minor fraction of the total mass. The ATP synthase has two major domains, as **FIGURE 6.53** shows:

- the membrane-bound F_0 domain (ab_2c_{10-14} in bacteria), which is involved in translocation of protons down their electrochemical gradient, and
- the globular F_1 domain ($\alpha_3\beta_3\gamma\Delta\epsilon$ in bacteria), which contains catalytic sites responsible for ATP synthesis. (When it is isolated from the F_0 domain, the F_1 domain hydrolyzes ATP on its own.)

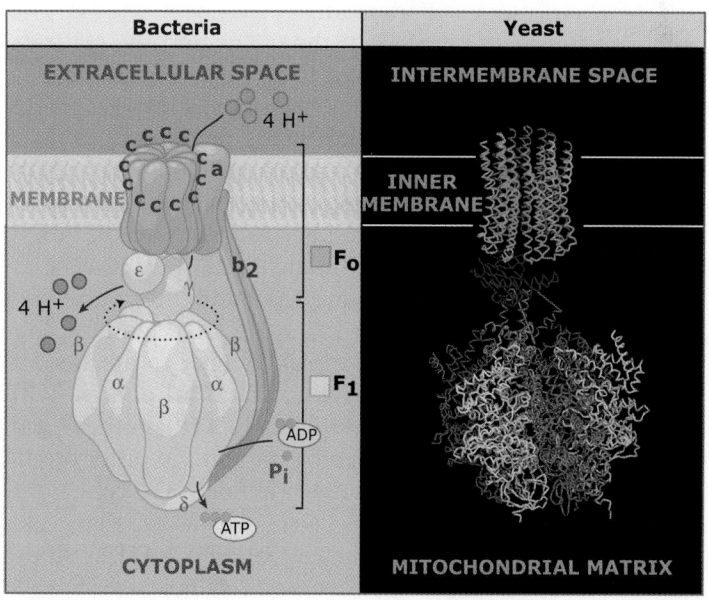

FIGURE 6.53 Schematic of the complete F_1F_0-ATP synthase complex from *E. coli* and X-ray crystal structure of part of the enzyme from yeast. The presumed position of the membrane is indicated. Right structure from Protein Data Bank 1Q01. D. Stock, A. G. W. Leslie, and J. E. Walker, *Science* 286 (1999): 1700.

The c subunits of the F_o domain form a ring that interacts with the subunit a. The γ subunit forms a central rotor stalk that is connected to the c-ring at its base and penetrates into the F_1 catalytic domain. The F_1 domain is an assembly of three α and three β subunits arranged alternately in the form of a hexagonal cylinder around the g subunit. The peripheral stator stalk is composed of $b_2\delta$ subunits, with the δ subunit binding to the F_1 domain and b_2 anchoring the F_o domain in the mitochondrial membrane and interacting with the subunit a.

How does the F_1F_o-ATP synthase use the energy of the transmembrane proton gradient to drive ATP synthesis? The basic parts of the proposed model are as follows:

- subunits a and c control proton transport in such a way that the c-ring rotates relative to subunit a (see Figure 6.53). In other words, the energy of the electrochemical proton gradient is converted into mechanical rotation of the c subunits;
- the γ subunit of the central stalk rotates with the c-ring and couples the transmembrane proton motive force over a distance of 100 Å to the F_1 domain; and
- the mechanical energy of rotation is used to release ATP, whose synthesis is catalyzed by the β subunits, from the F_1 domain.

Rotation of the c-ring and the central γ subunit relative to the a_3b_3 subdomain is, therefore, essential for coupling the proton motive force across the membrane to drive ATP formation and release. Since each c subunit carries one proton, there are 10-14 protons transported per complete revolution of the c-ring (depending on the specific ATPase), and roughly four protons are transported per ATP synthesized. The F_1F_o-ATP synthase converts electrochemical energy to mechanical energy, and back to chemical energy, with nearly 100% efficiency. ATP synthesis can occur at a maximal rate on the order of 100 sec^{-1} and sustains millimolar ATP concentrations in cells.

In some bacteria, the F_1F_o-ATP synthase works in the reverse direction, in which the energy released by ATP hydrolysis drives the translocation of protons out of the cell, generating a proton gradient across the cytoplasmic membrane. The energy of the proton gradient is then used to drive transport of solutes, such as lactose, into the cell (see Figure 6.37 and *6.15 Symporters and antiporters mediate coupled transport*).

6.21 H+-ATPases transport protons out of the cytosol

Key concepts

- Proton concentrations affect many cellular functions.
- Intracellular compartments are acidified by the action of V-ATPases.
- V-ATPases in the plasma membrane serve specialized functions in acidification of extracellular fluids and regulation of cytosolic pH.
- V-ATPases are proton pumps that consist of multiple subunits, with a structure similar to F_1F_o-ATP synthases.

Because protein function is often pH-sensitive, the pH is an important parameter that affects a large number of cellular processes. Metabolic processes in the cytosol constantly produce acidic molecules, and protons must be removed to maintain a constant cytosolic pH. Proteins that transport protons are, therefore, important in pH regulation. For example, the cytosolic pH, which is of general importance for growth control and all metabolic functions, can be regulated by membrane transporters such as the Na^+/H^+ exchanger (see *6.17 Some Na+ transporters regulate cytosolic or extracellular pH*). In addition, mitochondria provide a "sink" for the protons generated in the cytosol, in that the F_1F_o-ATP synthase in the inner mitochondrial membrane uses the proton motive force to drive ATP synthesis (see *6.20 The F_1F_o-ATP synthase couples H+ movement to ATP synthesis or hydrolysis*).

In contrast to the cytosol, which maintains a slightly basic pH, the function of some organelles requires an acidic interior. These compartments include those of the endocytic pathway (clathrin-coated vesicles, endosomes, lysosomes) and of the exocytic pathway (secretory granules). The vacuolar-type proton pumps (V-ATPases) are H^+-ATPases that are essential for maintaining the pH of these organelles, which ranges from 4.5 to 6.8. The V-ATPases constitute a family of ATP-driven proton pumps that transport protons from the cytosol across the membrane into the organelle lumen. They have an important role in endocytosis and intracellular targeting, as **FIGURE 6.54** shows. Acidification of early endosomes is required for the dissociation of internalized ligand-receptor complexes in endosomes and recycling of cell membrane receptors to the plasma membrane. H⁺-ATPases also acidify late endosomes, which is required for the delivery of lysosomal enzymes, such as those containing the mannose 6-phosphate targeting signal, from the *trans*-Golgi network. In secretory granules, such as synaptic vesicles and chromaffin granules, the proton gradient and/or membrane potential generated by H⁺-ATPases provides the driving force for coupled transport and storage of small molecules and ions in the granules. For example, in synaptic vesicles, the uptake of noradrenaline depends on the proton gradient, whereas glutamate uptake depends on the membrane potential. Some enveloped viruses, such as influenza, exploit the acidic pH of endosomal compartments to gain cell entry when the viral protein hemagglutinin becomes activated by low pH and promotes fusion between viral and endosomal membranes. (For more on the endocytic and exocytic pathways see *8.3 Overview of the endocytic pathway* and *8.2 Overview of the exocytic pathway*, respectively.)

Other H⁺-ATPases function at the plasma membrane of certain specialized cells such as renal intercalated cells, neutrophilic immune cells, and osteoclasts, where they play important roles in urine acidification, cytoplasmic pH maintenance, and bone resorption, respectively. The apical membrane of kidney intercalated epithelial cells contains a high density of H⁺-ATPases that actively excrete protons into the urine, as **FIGURE 6.55** shows. Mutations in these H⁺-ATPases cause defective proton extrusion, which can lead to decreased ability of the kidney to regulate blood pH as well as kidney and metabolic acidosis. Macrophages and neutrophilic immune cells rely on H⁺-ATPases in the plasma membrane to maintain a neutral cytosolic pH during heavy production of acids due to metabolic activity. During bone remodeling, osteoclast cells attach to the bone matrix and target H⁺-ATPases to the active contact zone. Acidification of the extracellular contact zone helps to dissolve the bone matrix and to activate hydrolases involved in bone resorption.

The H⁺-ATPases are multisubunit protein complexes that form two functional domains, V_1 and V_0, as **FIGURE 6.56** shows. The cytosolic V_1 domain binds and hydrolyses ATP, and the free energy of ATP hydrolysis provides the energy for proton translocation across the membrane-

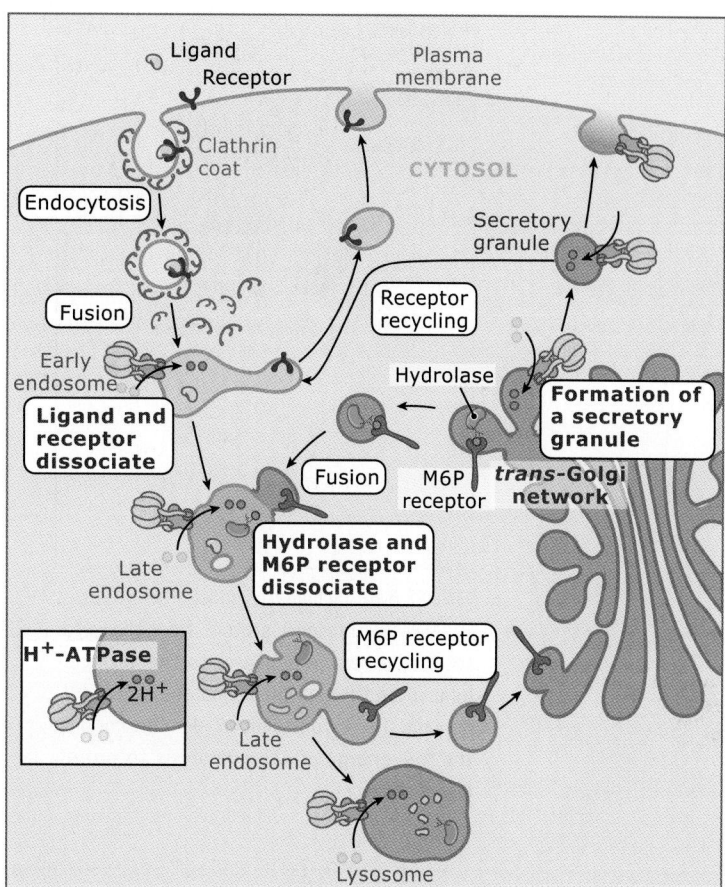

FIGURE 6.54 Some V-ATPases function in vesicle-mediated transport. The acidification of endosomal compartments is important for dissociation of protein-receptor complexes and receptor recycling. In addition, H⁺-ATPases are required for formation of secretory granules.

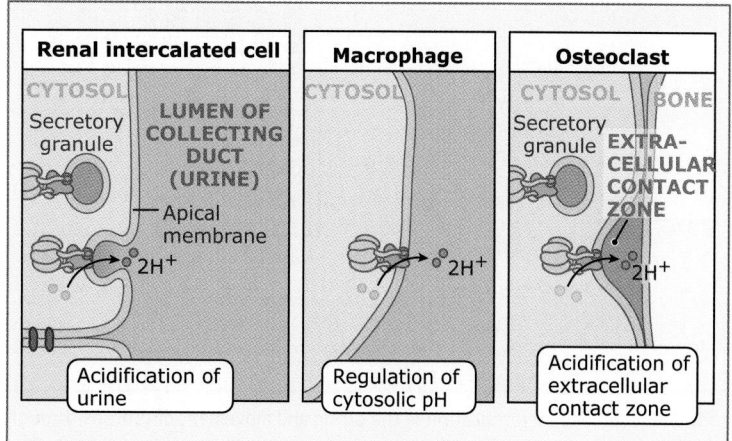

FIGURE 6.55 Plasma membrane H⁺-ATPases serve different functions in specialized cell types.

bound V_0 domain. The proposed structure is based on electron microscopy, protein cross-linking experiments, and site-directed mutagenesis. Although the exact pump components

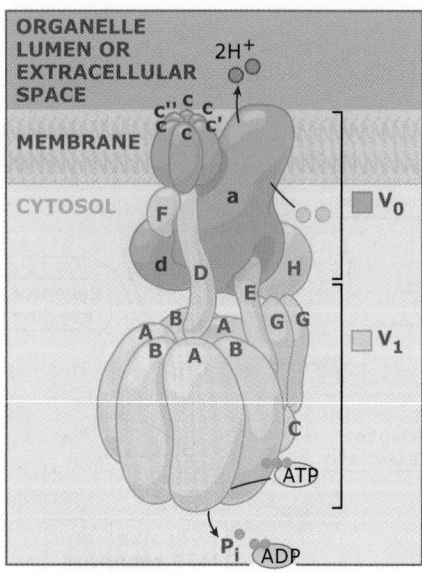

FIGURE 6.56 Schematic model of H⁺-ATPases. H⁺-ATPases transport protons from the cytosol to the vesicle lumen or to the extracellular space through the membrane integral V_0 domain. The cytoplasmic V_1 domain couples the free energy of ATP hydrolysis to proton transport.

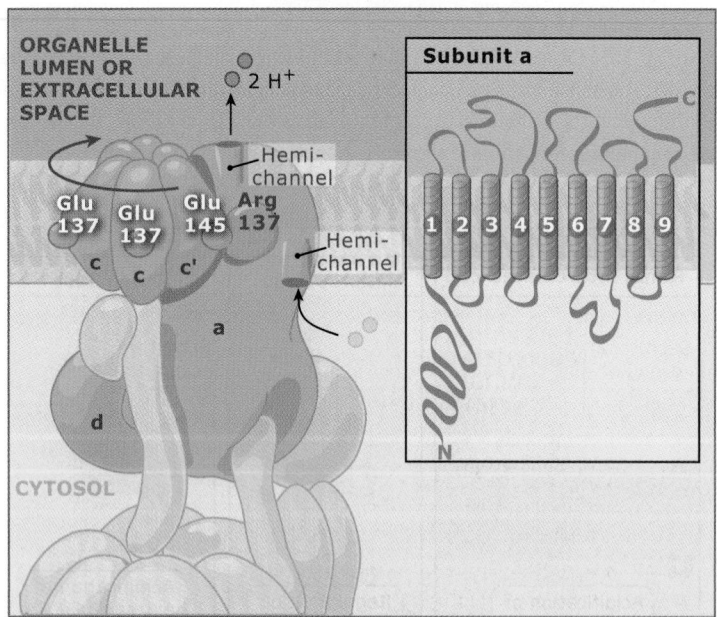

FIGURE 6.57 Model for rotation of the c-ring and movement of protons through the V_0 domain (driven by ATP hydrolysis in the V_1 domain, not shown). The inset shows the predicted topology of subunit a of the V_0 domain based on mutagenesis and chemical modification.

have yet to be identified, biochemical studies suggest that the 640 kDa cytoplasmic ATP-binding V_1 domain comprises subunits A-H in the stoichiometry $A_3B_3C_1D_1E_1F_1G_2H_1$, while the 260 kDa proton pumping V_0 domain contains five types of subunits in a complex of $a_1d_1c''_1c'_1,c_4$.

The H⁺-ATPases are related to the F_1F_0-ATP synthases, which function in the opposite direction to *produce* ATP in mitochondria, chloroplasts, and bacteria (see *6.20 The F_1F_0-ATP synthase couples H⁺ movement to ATP synthesis or hydrolysis*). Both types of enzyme have two analogous domains, F_0 or V_0, and F_1 or V_1 (see Figure 6.53 for F_1F_0-ATP synthase). The V_1 domain contains noncatalytic nucleotide-binding sites on the B subunits and catalytic ATP-binding sites on the A subunits, as determined by site-directed mutagenesis and modification by sulphydryl reagents. Because the structures are similar, a mechanism can be proposed for the H⁺-ATPases based on the greater understanding of the F-ATP synthases.

The F-ATP synthases are thought to function as rotary motors in which a proton gradient drives rotation of the c subunit ring relative to subunit a in the F_0 domain. The energy of this rotation is translated through the central γ and ϵ subunit stalk into the F_1 ring structure and used for ATP synthesis and release by the α and β subunits. The c subunits contain aspartic acid residues that are essential for proton transport, and subunit a is thought to allow access of protons to these sites.

FIGURE 6.57 shows the analogous model for H⁺-ATPases. In this model, the energy of ATP hydrolysis by the V_1 domain is translated into rotation of the c-ring and proton transport. Subunit a of the V_0 domain transports protons via the nine predicted transmembrane segments located in its C-terminal half. A positively charged arginine residue in subunit a would stabilize a negatively charged glutamate residue in one of the c subunits prior to protonation. Subunit a would use two hemichannel structures to transport a proton and reversibly protonate the glutamate residue, which would release it from its interaction with the arginine residue in the c subunit. The electrostatic attraction of the arginine residue with the unprotonated glutamate of the next c subunit would allow rotation of the c-ring, bringing each c subunit in contact with subunit a. This mechanism would allow for ATP hydrolysis to drive unidirectional proton transport.

The activity of H⁺-ATPases can be regulated by several mechanisms. One mechanism involves regulated targeting and fusion of intracellular vesicles containing proton pumps with the plasma membrane. For example, in renal intercalated cells, secretory vesicles containing H+-ATPases reversibly fuse with the apical membrane in response to increased acidity in the cell. Second, formation of disulfide bonds between conserved cysteines near the catalytic site of subunits a leads to reversible inhibition of ATPase activity. Third, under certain conditions, reversible dissociation of the complex into V_1 and V_0 domains may be an important mechanism to control H+-ATPase activity.

6.22 What's next?

A recent breakthrough in the field of membrane transport has been the visualization of ion channels at atomic level resolution. For example, the analysis of X-ray crystal structures has revolutionized our understanding of the underlying principles of selectivity and permeation for a few types of ion channels and for aquaporins. However, X-ray crystal structures are available for a relatively small number of membrane transport proteins, and in many cases, these are derived from bacterial proteins that can be prepared in large quantities. The analysis of crystal structures for other transport proteins will allow more general principles to emerge. In addition, analysis of crystal structures of eukaryotic transport proteins will form the basis for insight into more complex processes such as neuronal function.

The crystal structures have been invaluable in guiding the proposal of models for how membrane transport proteins work. The ability to image transport proteins in real time using nuclear magnetic resonance (NMR) or other spectroscopic methods will complement the work on crystal structures by providing information on the dynamics, as proteins move from one conformation to another, whether it be gating of ion channels from closed to open configurations or the conformational changes that occur as carrier proteins bind and release solutes.

A multitude of potential membrane transporters, ion channels, and pumps have been identified on a genetic or functional level. Genomics and proteomics approaches will allow the identification of even more transport proteins. Moreover, subunits in addition to the pore-forming subunit(s) are known to be essential for the function of some transport proteins. The identification of other such subunits will be important for a more complete understanding how transport protein function is regulated. Microarray and proteomic analysis will also allow comparison of the expression levels of transport proteins and their associated regulators in normal and diseased cells.

As a long-term goal, there is a great need to relate information obtained from structural analysis to a more physiological context, in cells and organisms. We need to be able to study transport proteins in their lipid and cellular environments and to test the proposed mechanisms of transport and regulation. Many diseases have been associated with mutations in membrane transport proteins. However, in order to understand the role of a particular ion channel in disease, it is important to understand the function of that type of ion channel in the context of the whole cell and, ultimately, at the level of the organ and organism. For example, it has been shown that mutations in the cystic fibrosis transmembrane conductance regulator (CFTR) Cl− channels cause abnormal Na+ reabsorption in the lungs and gastrointestinal tract of cystic fibrosis patients, resulting in sticky mucus that obstructs a variety of organs that depend on this type of epithelial transport function. These types of studies will require the development of new biologic assays to test specific functions. A molecular understanding of the function of all the different channels in a cell will be instrumental in elucidating the mechanisms that underlie the higher-order activity of neuronal networks or to understand electrical abnormalities that result in epilepsy, cystic fibrosis, or arrhythmias and sudden cardiac death.

6.23 Summary

Three principal classes of membrane transport proteins, channels, transporters, and pumps, have been characterized. These proteins reside in the plasma membrane and in membranes of intracellular organelles such as the endoplasmic reticulum, endosomes, lysosomes, and mitochondria. They are important for a myriad of cell functions, ranging from uptake of nutrients such as glucose to more complex physiological tasks such as the reabsorption of solutes by the kidney and the propagation of action potentials.

Channels have pores that allow for the rapid, passive diffusion of ions or other solutes across biologic membranes. Channels vary in their degree of selectivity: some are highly selective for K+, Na+, Ca2+, Cl− ions, or H_2O, whereas others are selective for certain anions or cations. Selectivity is imparted by a region of ion channels called the selectivity filter, which makes partial dehydration of the permeating ions energetically favorable over other ions of similar size. Ion channels undergo regulated opening and closing in a process called gating, which is coupled to many different types of regulation. For example, ion channels may be ligand gated, voltage gated, stretch activated, or temperature activated. The direction of net ion transport through channel pores depends on the electrochemical gradient of the ionic species, and transport is associated with the flow of electric current across the membrane. The general architectures for several channel proteins have been characterized, and models have been proposed to account for their selectivity, rapid rates of transport, and gating.

Transporters and pumps are solute-selective carrier proteins, which differ from channels in their mechanism of transport. These transport proteins alternate between two basic conformations: one that binds solute(s) on one side of the membrane and one that releases solute(s) on the other side. Transporters and pumps use different sources of energy to accomplish transport. Transporters couple transport to the energy stored in electrochemical gradients across the membrane. In contrast, pumps use energy from ATP or external sources (light) to drive transport.

Electrochemical gradients across the cell membrane are established by the interplay of the different classes of membrane transport proteins. A large fraction of a cell's metabolic energy is spent to establish transmembrane ion gradients across the cell membrane, as well as across intracellular membranes. For example, the Na^+/K^+-ATPase pump helps to establish the transmembrane Na^+ and K^+ gradients. Channels and carrier proteins use the electrochemical gradients stored at the plasma membrane to perform work. For example, these gradients are used by voltage-gated ion channels to produce electrical signals or by other types of ion channels to activate intracellular signal transduction pathways, to control the cell volume, or to mediate fluid and electrolyte transport. In addition, transporters use the energy of one solute's electrochemical gradient to move another solute against an energetically uphill gradient.

Mutations or dysregulation of ion channel, transporter, or pump proteins can cause a loss- or gain-of-function defect. Many ion channel diseases affect neurons and muscle cells and cause diseases such as epilepsy, ataxia, myotonias, and cardiac arrhythmias. Examples of ion channel diseases that affect other organs are cystic fibrosis, salt-wasting Bartter's syndrome of the kidney, defective insulin secretion of the pancreas, kidney stones, and osteopetrosis. The consequences of mutated ion channels help us to understand ion channel function and their role in organ physiology.

6.24 Supplement: Derivation and application of the Nernst equation

The equilibrium between the opposing chemical and electrical forces defines the resting membrane potential (as discussed in *6.4 Electrochemical gradients across the cell membrane generate the membrane potential*). This equilibrium is reached when the free energy difference between these forces equals zero (in other words, when net flux = 0):

$$\Delta G_{conc} + \Delta G_{elec} = 0$$

The free energy change for movement of solute X across a membrane is:

$$\Delta G_{conc} = -\, RT \ln \left(\frac{[X]_o}{[X]_i} \right)$$

- R: the gas constant (2 cal mol^{-1} K^{-1})
- T: absolute temperature (K; 37°C = 307.5 K)
- $[X]_o$: concentration of X outside of the cell
- $[X]_i$: concentration of X inside of the cell
  and the free energy change due to movement, across the membrane, of charge associated with solute X is

$$\Delta G_{elec} = zFE_m$$

- E_m: equilibrium potential (volts)
- z: the ion's valence (electric charge)
- F: Faraday's constant (2.3 × 10^4 cal $volt^{-1}$ mol^{-1})
  At equilibrium,

$$zFE_m = RT \ln \left(\frac{[X]_o}{[X]_i} \right)$$

and rearranging gives

$$E_m = \left(\frac{RT}{zF}\right) \ln \left(\frac{[X]_o}{[X]_i}\right)$$

Thus, the equilibrium or Nernst potential for a monovalent ion X at 37°C is:

$$E_m = 61.5 \, \log_{10} \left(\frac{[X]_o}{[X]_i}\right) \text{ millivolts (mV)}$$

When ion concentrations have been measured, the Nernst equation can be used to calculate the equilibrium membrane potential for each type of ion. FIGURE 6.58 shows the results of these calculations for the muscle cell plasma membrane. Thus, for $[K^+]_o = 4$ mM and $[K^+]_i = 155$ mM, the membrane potential $E_m = -98$ mV when only K^+ flux is considered, that is, $E_m = E_K$.

For nearly all resting mammalian cells, the plasma membrane is preferentially permeable to K^+ ions. The Na^+/K^+-ATPase maintains the transmembrane K^+ ion gradient, creating a relatively high intracellular K^+ concentration ($[K^+]_i$). Opening of certain types of K^+ channels allows K^+ ions to flow down their concentration gradient, which charges the cell membrane positive on the extracellular side and negative on the intracellular side. This negative membrane potential, E_m, is the electrical driving force that opposes further flow of K^+ out of the cell along the concentration gradient (the chemical driving force). Thus, opening of a specialized subset of K^+ selective ion channels in resting cells determines the level of the negative resting membrane potential, at which no net transmembrane K^+ flux occurs and E_m is stable.

Similar Nernst potentials can be calculated for other ion species. If the membrane becomes more permeable to a given ionic species, the membrane potential E_m will change toward the Nernst potential of that ion (typically becoming more positive during membrane depolarization).

For example, if the plasma membrane becomes more permeable to Na^+ ions, they will flow down their concentration gradient, thereby charging the cell membrane more negative on the extracellular side and more positive on the intracellular side. Under physiological conditions in resting cells, the opening of a few Na^+ channels, together with Na^+ influx through leak currents, will cause an inward Na^+ current and drive the resting membrane potential toward slightly more positive

	● Na⁺	○ K⁺	● Ca²⁺	◐ Cl⁻
Extracellular concentration (mM)	145	4	1.5	123
Equilibrium potential (mV)	+67	-98	+129	-90
Intracellular concentration (mM)	12	155	10^{-4}	4.2

FIGURE 6.58 Free ion concentrations and equilibrium potentials of mammalian skeletal muscle cells. The equilibrium potentials were calculated for 37°C assuming a −90 mV resting potential for the muscle cell membrane. The relative radii of the nonhydrated ions are indicated at the top.

potentials (e.g., −82 mV). In contrast, pure K^+ conductance results in a resting membrane potential of −89 mV, as calculated above. In cell types that have less K^+ permeability, the same background Na^+ leak and other depolarizing ion currents set the resting membrane potential E_m toward more positive potentials (e.g., −50 mV). When a cell becomes electrically excited and Na^+ channels open, the theoretical equilibrium between inward Na^+ currents and outward K^+ currents defines the level of the membrane potential closer toward E_{Na} (rather than E_K). According to the Nernst equation, the equilibrium potential for Na^+ at 37°C is:

$$E_{Na^+} = 61.5 \, \log_{10} \left(\frac{[Na^+]_o}{[Na^+]_i}\right)$$

If $[Na^+]_o = 145$ mM and $[Na^+]_i = 12$ mM (as for the muscle cell in Figure 6.58), the membrane potential $E_m = +67$ mV when Na^+ flux alone is considered. Thus, the net effect of opening Na^+ channels will be an inward current of Na^+, which causes the negative resting E_m to move toward +67 mV.

Similarly, when the plasma membrane becomes more permeable to Ca^{2+}, Ca^{2+} will flow down its concentration gradient, which charges the cell membrane more negative on the extracellular side and more positive on the intracellular side. When a cell is electrically excited and Ca^{2+} channels open, the theoretical equilibrium between inward Ca^{2+} currents and outward K^+ currents defines the level of the membrane potential closer toward E_{Ca}. According to the Nernst equation,

$$E_{Ca^{2+}} = \left(\frac{RT}{2F}\right) \ln \left(\frac{[Ca^{2+}]_o}{[Ca^{2+}]_i}\right) = 30.75 \, \log_{10} \left(\frac{[Ca^{2+}]_o}{[Ca^{2+}]_i}\right)$$

If $[Ca^{2+}]_o = 1.5$ mM and $[Ca^{2+}]_i = 0.1$ μM, as for the muscle cell in Figure 6.58, the membrane potential $E_m = +129$ mV. Thus, the net effect of Ca^{2+} channel opening will be a net Ca^{2+} inward current causing the negative resting E_m to move toward +129 mV. [With respect to action potentials, Ca^{2+} channels open at more positive potentials than Na^+ channels, meaning that they open in a later phase of the action potential (see *6.12 Action potentials are electrical signals that depend on several types of ion channels*).]

For Cl^-, according to the Nernst equation,

$$E_{Cl^-} = -\left(\frac{RT}{F}\right)\ln\left(\frac{[Cl^-]_o}{[Cl^-]_i}\right) = -61.5 \log_{10}\left(\frac{[Cl^-]_o}{[Cl^-]_i}\right)$$

If $[Cl^-]_o = 123$ mM and $[Cl^-]_i = 4.2$ mM, then the membrane potential $E_m = -90$ mV for Cl^- alone. Thus, the net effect of Cl^- channel opening will be a net Cl^- efflux, which stabilizes the negative resting E_m.

6.25 Supplement: Most K⁺ channels undergo rectification

Key concept

- Inward rectification occurs through voltage-dependent blocking of the pore.

How do different K^+ channels allow for transient changes in membrane potential to occur and contribute to termination of membrane depolarization? Most ion channels undergo some form of **rectification**, which is a change of ion conductance in response to a change in membrane voltage. **Inward rectification** means that an ion channel conducts inward current better than outward current. An example is the family of inward rectifier K^+ (Kir) channels, which are evolutionarily highly conserved. Inward rectification refers to a decrease in K^+ conductance during membrane depolarization and an increase of K^+ conductance during repolarization to more negative membrane potentials. Without inward rectification, K^+ conductance would be constant. However, with weak inward rectification, the outward K^+ current is decreased at membrane potentials that are more positive than the K^+ equilibrium potential. With strong inward rectification, the outward K^+ current can be completely blocked at more positive membrane potentials. Two

examples of the function of inward rectification are maintenance of the resting membrane potential at negative voltages and deactivation of outward K^+ currents to conserve the intracellular K^+ concentration during the long plateau phase of the cardiac action potential. Kir channels conduct K^+ ions out of resting cells, but are blocked during depolarization. This makes sense because membrane depolarization is due to transport of Na^+ ions into the cell, and simultaneous efflux of positively charged K^+ ions would oppose or slow membrane depolarization.

Channels can be classified according to their degree of rectification. Strong inward rectifier channels exhibit a sharp cutoff in conductance above a particular membrane potential, whereas weak inward rectifiers continue to transport ions but at a slower rate. Inward rectifier channels are of either the 2TM/1P or 6TM/1P class. Strong inward rectifier channels like Kir belong to the 2TM/1P class of K^+ channels, as **FIGURE 6.59** shows. The various Kir channels are ligand-gated channels that are regulated by different molecules, such as Mg^{2+}, polyamines, ATP, or trimeric G proteins. In Kir channels, the reduced K^+ current at positive potentials results from blockage of the pore from the cytoplasmic side by cations such as Mg^{2+} or by high-affinity ligands such as polyamines. In contrast to inward rectifier channels, the outward or delayed rectifier channels are of the 6TM/1P type and are voltage-dependent or K_V channels. K_V channel conductance increases with membrane depolarization at more positive potentials due to opening of more channels; however, individual channels do not exhibit rectification properties. This is referred to as outward or delayed rectification because channel activation is relatively slow.

An example of the importance of channel rectification is provided by the inward-rectifier ATP-sensitive K^+ channel (K_{ATP} channel), which is an important link between cell metabolism and electrical activity. In pancreatic β cells, insulin secretion can be triggered when rising glucose and ATP concentrations cause consecutive closure of K_{ATP} channels and membrane depolarization. K_{ATP} channels consist of the pore-forming Kir6.2 a subunits and regulatory subunits, which function as a metabolic sensor to control K^+ channel activity. Mutations in either subunit result in loss of K_{ATP} channel function and persistent membrane depolarization with continuous insulin secretion independent of blood glucose levels.

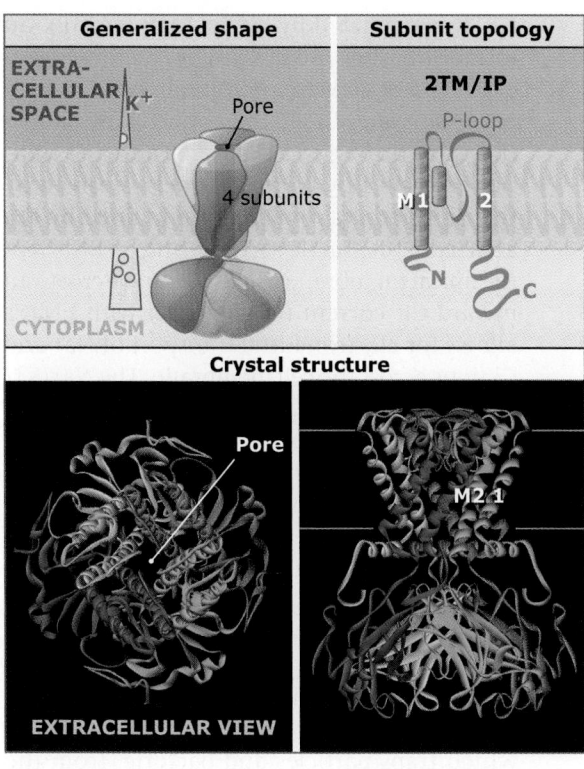

Generalized shape	Subunit topology
EXTRA-CELLULAR SPACE	2TM/IP

Generalized shape: EXTRA-CELLULAR SPACE, K+, Pore, 4 subunits, CYTOPLASM

Subunit topology: 2TM/IP, P-loop, M1, 2, N, C

Crystal structure: Pore, M2 1, EXTRACELLULAR VIEW

FIGURE 6.59 Schematics and X-ray crystal structure of an inward rectifier K+ (Kir) channel. The pore domain is similar to that of other K+ channels. The extracellular view is smaller than the side view. The presumed position of the membrane is indicated. The inset at top left indicates the K+ gradient in resting animal cells. Bottom structures from Protein Data Bank 1P7B. A. Kuo, et al., *Science* 300 (2003): 1922–1926.

This uncontrolled insulin secretion may result in pathologically low blood glucose levels in infancy and neurologic damage.

6.26 Supplement: Mutations in an anion channel cause cystic fibrosis

Key concepts

- Cystic fibrosis is caused by mutations in the gene encoding the CFTR channel.
- CFTR is an anion channel that can transport either Cl⁻ or HCO_3^-.
- Defective secretory function in cystic fibrosis affects numerous organs.

Ion channels, transporters, and pumps serve many important cellular functions, and ion channel dysfunction can cause diseases in many tissues. In the case of ion channels, these diseases are called *channelopathies*. Mutations in ion channel genes may cause either a gain-of-function or a loss-of-function defect that could explain the pathophysiological changes occurring in a specific disease. In previous sections, we have briefly discussed a variety of channelopathies that affect epithelial transport function of the kidney, secretory functions of different organs that can cause bone disease, neurologic or muscle disorders, and cardiac arrhythmias and sudden death (e.g., see *6.8 Epithelial Na+ channels regulate Na+ homeostasis; 6.12 Action potentials are electrical signals that depend on several types of ion channels;* and *6.13 Cardiac and skeletal muscles are activated by excitation-contraction coupling*). In this section, we will discuss **cystic fibrosis**, which is a prominent ion channel disease that is becoming better understood as the alterations in epithelial membrane transport are characterized in more detail.

Cystic fibrosis, one of the most common fatal genetic diseases, is caused by recessive mutations in the gene encoding an anion channel called the (CFTR). CFTR functions as an ATP-gated anion channel that is expressed in the apical membranes of epithelial cells and is regulated by phosphorylation by protein kinase A. CFTR is a member of the large family of transporters that contain an ABC. In general, ABC transporters are ATP-powered pumps that translocate a wide variety of substrates, including amino acids, peptides, ions, sugars, toxins, lipids, and drugs, across cell membranes. In addition to cystic fibrosis, mutations in ABC transporters are implicated in a number of other diseases, including disorders of the immune system, as well as drug resistance to antibiotics and cancer drugs.

CFTR transports Cl⁻ or HCO_3^- across the plasma membrane, as **FIGURE 6.60** shows. The transport of Cl⁻ ions by CFTR was shown by the purification and reconstitution of functional CFTR channels in liposomes. The function of CFTR is regulated by cyclic AMP (cAMP), an intracellular second messenger. Increased levels of cAMP activate protein kinase A, which phosphorylates CFTR, thereby activating anion transport through CFTR. Transepithelial chloride transport by CFTR channels is driven by the Na⁺ gradient, which is maintained by the activity of the Na⁺/K⁺-ATPase and by ENaCs in the apical membrane. CFTR channels create an inward Cl⁻ current that together with ENaCs allows for electroneutral transport of Na⁺ and Cl⁻ ions across the cell membrane. The Na⁺/K⁺/Cl⁻-cotransporter in the basolateral membrane mediates transport of Cl⁻ ions into the cell after stimulation by hormones.

CFTR is expressed in the apical membrane of various epithelial cells, including those of the intestine, lungs, pancreas, and sweat glands. Fluid secretion is important for the function of these organs. For example, in the lungs, a thin layer of airway surface fluid that covers the epithelial cells is necessary for cilia to clear the overlying mucus layer, which traps particles and bacteria, from the airways. CFTR is important for the production and correct ionic composition of this fluid. In patients with cystic fibrosis, the mucus layer has a higher viscosity than normal, which inhibits the clearance of pathogens, and the bronchi are, therefore, less resistant to bacterial infections. Accumulation of sticky mucus will eventually result in respiratory failure. In addition, blockade of the gastrointestinal tract, pancreas, and liver by sticky mucus can cause poor digestion and malnutrition. The function of CFTR during fluid secretion from the apical membrane requires that it work in conjunction with other transport proteins in the apical and basolateral membrane of epithelial cells (see Figure 6.60). CFTR function is important for fluid secretion because the transport of salts (in the form of Na⁺ and Cl⁻ ions) provides the osmotic driving force for the transport of water across membranes.

The CFTR protein contains two nucleotide-binding folds (NBFs) and a unique regulatory (R) domain in the cytoplasmic region and two membrane-spanning domains, as **FIGURE 6.61** shows. Each membrane-spanning domain is composed of six transmembrane segments, and the two domains, together with the nucleotide-binding domains, form a tandem repeat structure. Each of the nucleotide-binding domains contains an ATP-interacting region consisting of a phosphate-binding loop

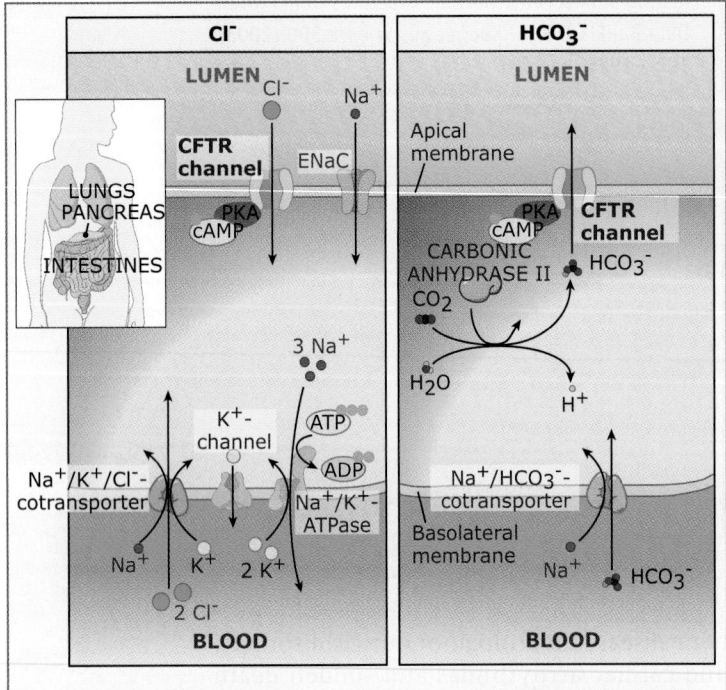

FIGURE 6.60 Transepithelial chloride transport and bicarbonate transport by the CFTR channel. CFTR works in close proximity with and inhibits Na⁺ transport by ENaCs in the apical membrane. Decreased CFTR function results in increased Na⁺ and water reabsorption through epithelial cell layers.

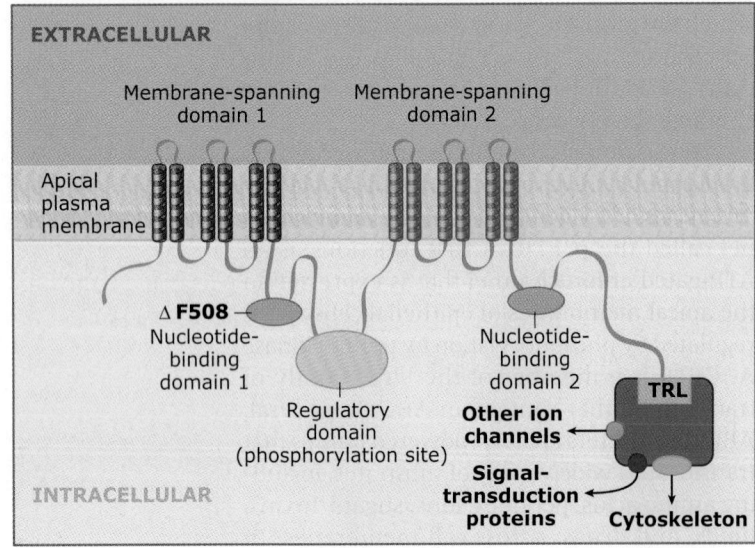

FIGURE 6.61 Hypothetical structure of the cystic fibrosis transmembrane conductance regulator (CFTR) anion channel. Reproduced from S. M. Rowe, S. Miller, E. J. Sorscher, *N. Engl. J. Med.* 352 (2005): 1992–2001. Copyright © 2005 Massachusetts Medical Society. All rights reserved.

formed by a Walker A sequence motif that is conserved in many ATP-binding proteins, including F_1F_0-ATP synthases. Several hundred different mutations in the *CFTR* gene have been linked to cystic fibrosis, but the most common mutation is a deletion that results in a defective CFTR protein lacking Phe508, which normally lies in the N-terminal NBF. The CFTRΔPhe508 protein is not transported to the plasma membrane and remains in the endoplasmic reticulum due to defective protein trafficking.

Since CFTR is a channel, ions are transported down their electrochemical gradient, without an additional energy requirement. Why does CFTR bind ATP? The NBF of some ABC transporters hydrolyzes ATP, and these transporters use the energy of ATP hydrolysis to pump substrates across membranes. In contrast, gating of CFTR appears to be regulated by the intrinsic adenylate kinase activity of its C-terminal NBF, which catalyzes the transfer of phosphate from ATP to AMP to form ADP. A proposed mechanism is that ATP binding may induce the NBFs to dimerize and cause the channel to open, whereas ADP produced by the adenylate kinase activity would cause the NBFs to dissociate and the channel to close. The CFTRΔPhe508 mutation interferes with channel function at a putative NBF interface and disrupts functional channel expression.

CFTR activity can be regulated in other ways. Channel regulation can occur through phosphorylation at the regulatory domain. In addition, the anion selectivity of CFTR can be altered by glutamate activation, such that the channel conducts Cl^- ions but not HCO_3^- ions. Thus, CFTR appears to operate as a ligand-activated anion channel and not as a ATP-dependent pump.

In cystic fibrosis, genetic defects of CFTR channels disrupt transepithelial transport and, therefore, anion and fluid secretion in a variety of organs. The compositional hypothesis posits that CFTR works primarily as a Cl^- channel that transports Cl^- ions into cells to produce a low-salt airway surface liquid and inhibit Na^+ reabsorption by ENaCs. In cystic fibrosis, defective CFTR Cl^- secretion results in Na^+ and water absorption by epithelial cells, which decreases the airway surface liquid volume, increases its viscosity, and impairs mechanical clearance of the fluids. Depending on the mutation site in the *CFTR* gene, anion transport can be disrupted to different extents. In addition to the airway infections described above, patients with cystic fibrosis may have insufficient secretion of digestive enzymes from the pancreas, intestinal obstruction, infertility, and increased salt loss through the sweat glands. The majority of cystic fibrosis patients are born with pancreatic insufficiency that results in maldigestion and diarrhea. The defective bicarbonate and fluid secretion caused by CFTR channel mutations result in loss of pancreas function and progressive destruction of pancreatic tissue. Ongoing studies include pharmacological targeting of defective CFTR or ENaC function to correct epithelial transport in cystic fibrosis patients.

http://biology.jbpub.com/lewin/cells
To explore these topics in more detail, visit this book's Interactive Student Study Guide.

References

Abramson, J., Smirnova, I., Kasho, V., Verner, G., Kaback, H. R., and Iwata, S. 2003. Structure and mechanism of the lactose permease of *Escherichia coli. Science* v. 301 p. 610–615.

Canessa, C. M., Schild, L., Buell, G., Thorens, B., Gautschi, I., Horisberger, J. D., and Rossier, B. C. 1994. Amiloride-sensitive epithelial Na^+ channel is made of three homologous subunits. *Nature* v. 367 p. 463–467.

Doyle, D. A., Morais Cabral, J., Pfuetzner, R. A., Kuo, A., Gulbis, J. M., Cohen, S. L., Chait, B. T., and MacKinnon, R. 1998. The structure of the potassium channel: Molecular basis of K^+ conduction and selectivity. *Science* v. 280 p. 69–77.

Dutzler, R., Campbell, E. B., and MacKinnon, R. 2003. Gating the selectivity filter in ClC chloride channels. *Science* v. 300 p. 108–112.

Ellinor, P. T., Yang, J., Sather, W. A., Zhang, J. F., and Tsien, R. W. 1995. Ca^{2+} channel selectivity at a single locus for high-affinity Ca^{2+} interactions. *Neuron* v. 15 p. 1121–1132.

Hille, B. 2001. *Ion channels of excitable membranes.* Sunderland, MA: Sinauer Associates, Inc.

Hunte, C., Screpanti, E., Venturi, M., Rimon, A., Padan, E., and Michel, H. 2005. Structure of a Na^+/H^+ antiporter and insights into mechanism of action and regulation by pH. *Nature* v. 435 p. 1197–1202.

Jentsch, T. J., Hübner, C. A., and Fuhrmann, J. C. 2004. Ion channels: Function unravelled by dysfunction. *Nat. Cell Biol.* v. 6 p. 1039–1047.

Jiang, Y., Lee, A., Chen, J., Cadene, M., Chait, B. T., and MacKinnon, R. 2002. The open pore conformation of potassium channels. *Nature* v. 417 p. 523–526.

Jung, J. S., Preston, G. M., Smith, B. L., Guggino, W. B., and Agre, P. 1994. Molecular structure of the water channel through aquaporin CHIP. The hourglass model. *J. Biol. Chem.* v. 269 p. 14648–14654.

Stühmer, W., Conti, F., Suzuki, H., Wang, X. D., Noda, M., Yahagi, N., Kubo, H., and Numa, S. 1989. Structural parts involved in activation and inactivation of the sodium channel. *Nature* v. 339 p. 597–603.

Toyoshima, C., Nakasako, M., Nomura, H., and Ogawa, H. 2000. Crystal structure of the calcium pump of sarcoplasmic reticulum at 2.6 Å resolution. *Nature* v. 405 p. 647–655.

Membrane targeting of proteins

7

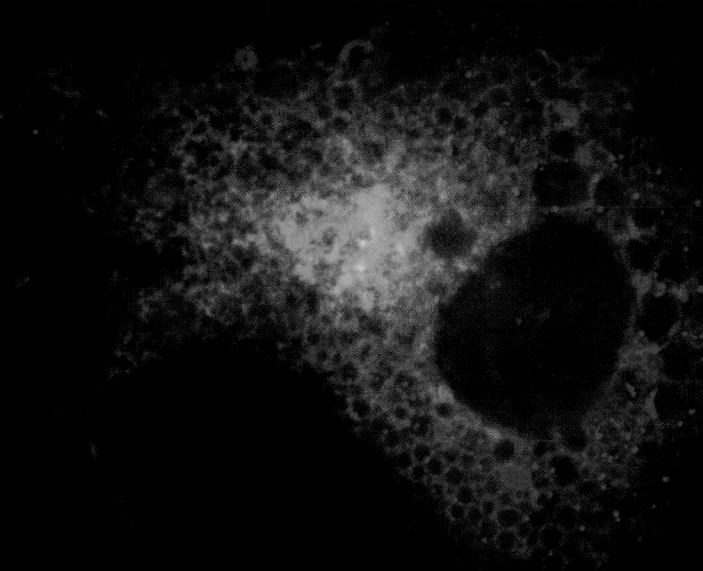

D. Thomas Rutkowski
Department of Anatomy and Cell Biology, University of Iowa Carver College of Medicine, Iowa City, IA

Vishwanath R. Lingappa
Prosetta Bioconformatics, Inc., San Francisco, CA

A LIVE FIBROBLAST. This fluorescent image shows the ER (green), mitochondria (red), and peroxisomes (blue). The cell is simultaneously expressing three similar fluorescent proteins, each with a different targeting signal that directs that protein to a specific organelle. Photo courtesy of Holger Lorenz, Zentrum für Molekulare Biologie der Universität Heidelberg, Germany.

CHAPTER OUTLINE

7.1 Introduction

Key concepts

- Cells must localize proteins to specific organelles and membranes.
- Proteins are imported from the cytosol directly into several types of organelles.
- The endoplasmic reticulum, or ER, is the entry point for proteins into the secretory pathway and is highly specialized for that purpose.
- Several other organelles and the plasma membrane receive their proteins by way of the secretory pathway.

The ability of a cell to interact with and respond to its environment is critical to its survival and function. The extracellular milieu in a multicellular eukaryote is filled with nutrients, growth factors, hormones, and other molecules that can direct a cell to proliferate, differentiate, or undergo programmed cell death. These extracellular cues must be sensed by the cell if it is to respond appropriately. Likewise, cells alter their environments for their own benefit: they can secrete proteins that build up, or disrupt, the extracellular matrix; they can relay signals to adjacent cells by direct contact; and in multicellular organisms, some cells, such as endocrine cells, secrete hormones that influence the activity of other cells at a distance.

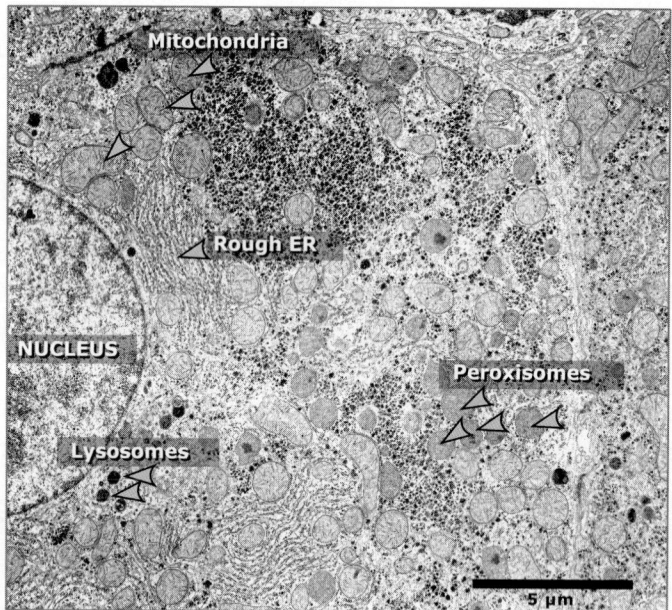

:•: **FIGURE 7.1** An electron micrograph of the interior of a liver cell. The photograph shows the variety and density of membrane-bounded organelles in a eukaryotic cell. Visible are the nucleus, mitochondria, lysosomes, peroxisomes, and the rough endoplasmic reticulum. Photo courtesy of Daniel S. Friend, University of California, San Francisco.

The primary means used by a cell to interact with its environment are secretory and transmembrane proteins. Every cell, from the simplest bacterium to the most highly differentiated and specialized mammalian cell, synthesizes secretory proteins and proteins that span the plasma membrane for this purpose. Secretory proteins are released into the extracellular milieu, whereas proteins spanning the cell's plasma membrane have one portion exposed to the outside environment and one portion remaining inside the cell.

The need to direct proteins across the plasma membrane, either wholly or partially, presents a problem of sorting for the cell. Cellular proteins are synthesized by ribosomes in the cytosol. Therefore, there must be a mechanism for selectively bringing secretory and transmembrane proteins, but not others, to the plasma membrane. In bacteria, this selective transport involves only the cell making a distinction between proteins destined for the plasma membrane and those that are not. In eukaryotes, however, the problem is greatly compounded. While typical prokaryotes have a plasma membrane and no internal organelles, eukaryotic cells contain numerous membrane-enclosed structures, including the nucleus, mitochondria, chloroplasts (in plant cells), peroxisomes, the ER, the Golgi apparatus, and lysosomes, as **FIGURE 7.1** shows. Each of these organelles contains a unique complement of proteins, so in addition to properly localizing secretory and transmembrane proteins, eukaryotic cells must also faithfully sort the proteins of these organelles. In a typical eukaryotic cell, the organelles take up approximately one half of the total cell volume, meaning that a large portion of the proteins being synthesized at any one time must be selectively localized to one of these organelles.

How are proteins localized to organelles? In general, the cell uses targeting signals—discrete stretches of amino acids in a protein's primary structure—to direct proteins to specific organelles. The nature of the targeting signal determines the organelle to which the protein is taken, and that organelle has protein machinery that specifically recognizes the appropriate signal on the proteins brought to it. If a protein is synthesized without a targeting signal, it remains in the cytosol after its synthesis.

For some organelles (mitochondria, chloroplasts, the nucleus, and peroxisomes), proteins are moved from the cytosol once their translation by ribosomes is complete. However, for several organelles (the ER, Golgi apparatus, lysosomes and endosomes, and plasma membrane), the localization process is more complex. Together, these organelles and the transit of proteins through them are referred to as the **secretory pathway**. The Golgi apparatus, lysosomes, and the plasma membrane do not have proteins brought to them directly. Rather, as shown in **FIGURE 7.2**, all proteins destined for any of these membrane-enclosed structures, or proteins to be secreted, are instead first brought to and moved across the membrane of the ER. The ER can thus be considered the gateway to the rest of the secretory pathway. There, proteins are folded into their correct three-dimensional structures, in many cases covalently modified and assembled with other proteins into complexes, and transported first to the Golgi apparatus, and from there on to their final destinations—either back to the ER, or onward to lysosomes or the plasma membrane. Trafficking of proteins between the organelles of the secretory pathway occurs by small vesicles that bud off of the membrane of the source organelle and fuse with the membrane of the destination organelle, releasing the protein contents enclosed within.

An important feature of the secretory pathway is that the **lumen** (interior) of these organelles has an environment that in many ways mimics the environment outside of the cell. (Organelles are believed to have evolved from plasma membrane invaginations specialized for protein secretion that were pinched off into internal vesicles.) Accordingly, proteins that eventually must be secreted will be able to fold within the cell, but in an environment very similar to that in which they will eventually find themselves, as shown in **FIGURE 7.3**.

This chapter focuses on the initial events required to localize proteins among the various organelles and membranes within a cell. Because almost all proteins are synthesized in the cytosol, these events occur at the membranes of the organelles that import proteins directly: the ER, mitochondria, chloroplasts, and peroxisomes. They involve correctly distinguishing the proteins that an organelle must import from all the other proteins in the cytosol and then moving them across or into the membrane of the organelle, a process known as **protein translocation**. Protein translocation is best understood for the ER, and so that process is described in most detail here. This chapter also describes the specialized roles that

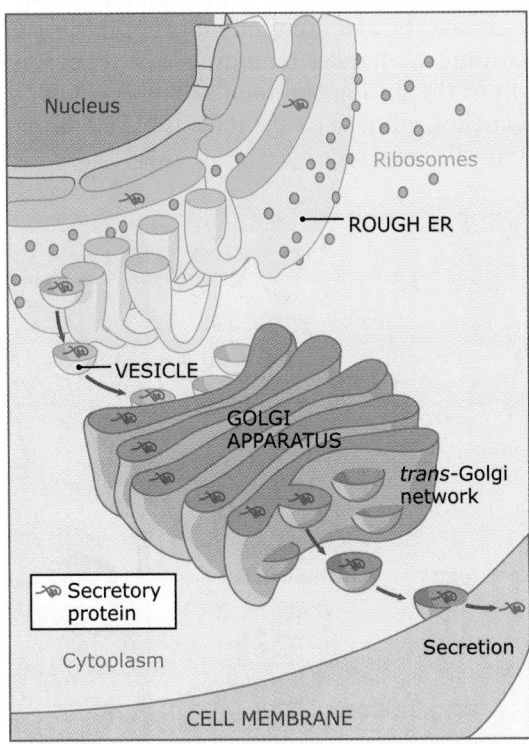

FIGURE 7.2 Proteins enter the secretory pathway by targeting to and translocating across the membrane of the rough endoplasmic reticulum. After folding and modification, they leave the ER in vesicles bound for the Golgi apparatus. Most proteins progress from the Golgi apparatus to the cell surface via secretory vesicles.

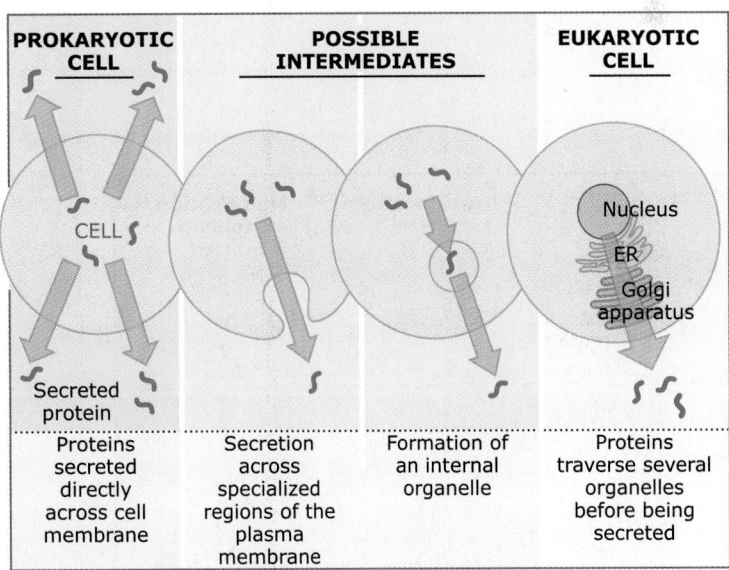

FIGURE 7.3 The organelles of the secretory pathway might have evolved by the internalization of a segment of the plasma membrane specialized for protein secretion. Rather than being secreted directly across the plasma membrane, proteins in eukaryotic cells are transported instead into the lumen of the ER, which is similar in environment to the outside of the cell.

the ER plays in preparing proteins for distribution to other locations accessible via the secretory pathway, as well as a variety of other activities that the ER performs. How proteins move through the secretory pathway and how they are sorted within the Golgi apparatus for delivery to specific locations are discussed in *8 Protein trafficking between membranes.*

7.2 Proteins enter the secretory pathway by translocation across the endoplasmic reticulum membrane (an overview)

Key concepts

- Signal sequences target nascent secretory and membrane proteins to the ER for translocation.
- Proteins cross the ER membrane through an aqueous channel that is gated.
- Secretory proteins translocate completely across the ER membrane; transmembrane proteins are integrated into the membrane.
- Before leaving the ER, proteins are modified and folded by enzymes and chaperones in the lumen.

The cell faces a number of challenges in guiding **nascent proteins** (proteins whose synthesis has been newly initiated) into the secretory pathway, as shown in **FIGURE 7.4**. First, these proteins must be selectively recognized and brought to their sites of translocation at the ER membrane—a process known as **protein targeting**. Next, they must translocate across the ER membrane either completely (in the case of soluble proteins) or partially (in the case of membrane proteins). This must occur without allowing the general exchange of other molecules between the lumen of the ER and the cytosol. Finally, all translocated proteins must be properly folded and in many cases must also be modified or assembled with other proteins in the ER lumen. This section provides an overview of these processes.

The ER is only one of several membrane-enclosed organelles to which targeting can occur. Others include mitochondria, chloroplasts, peroxisomes, and the nucleus. Proteins that are destined for the ER must be distinguished from those that are bound for other organelles or destined to remain in the cytosol. Cells accomplish this segregation through the use of **signal sequences**. These sequences are discrete stretches of amino acids within a protein's primary structure that are recognized by machinery associated with the target organelle. As **FIGURE 7.5** illustrates, in much the same way that postal codes are used to sort packages according to their destinations, signal sequences direct the sorting of proteins to their target organelles, with each organelle using a different type of signal sequence.

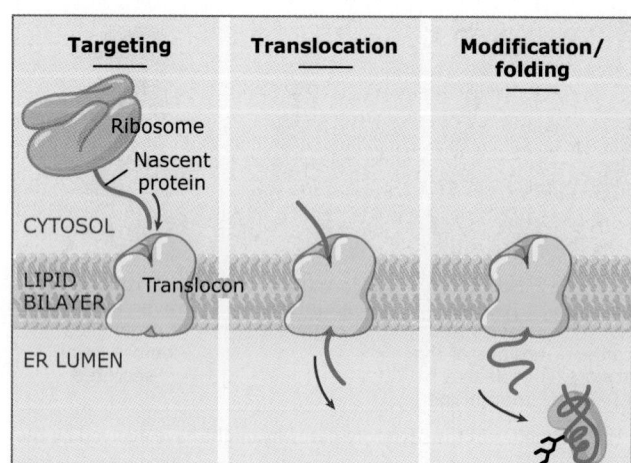

FIGURE 7.4 The three major events that occur for a nascent secretory or membrane protein at the ER are (1) targeting, (2) translocation, and (3) folding and modification.

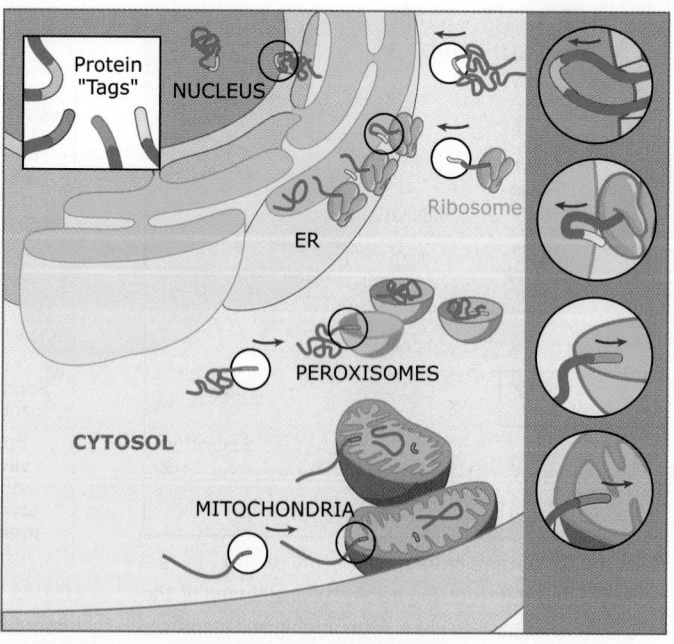

FIGURE 7.5 Organelle-specific signal sequences allow proteins to be accurately distributed within the cell. Proteins that are synthesized without signal sequences remain in the cytoplasm.

The mechanism by which a signal sequence for the ER is recognized influences how the protein is translocated. The most common form of translocation into the ER is **cotranslational translocation**, meaning that it occurs while the protein is being synthesized by a ribosome bound to the membrane. This form of translocation is initiated when the signal sequence is recognized in the cytosol by a complex known as the **signal recognition particle**, or **SRP**. SRP binds to the signal sequence and brings both the protein and the ribosome that is synthesizing it to the ER by interacting with a receptor (a protein that specifically associates with SRP) on the ER membrane. However, some ER signal sequences do not interact with SRP. This causes the proteins to be translocated **posttranslationally**, after their synthesis in the cytosol is complete. Different organisms use the two forms of translocation to different extents. In mammals, almost all translocation is cotranslational, but in simpler eukaryotes, such as the yeast *Saccharomyces cerevisiae*, both forms occur.

When a protein is targeted to the ER, it must cross the lipid bilayer enclosing the organelle. This occurs through a channel that provides an aqueous passageway across the hydrophobic membrane. The proteins that make up and associate with this channel are collectively called the **translocon** to emphasize their function as an integrated unit. The channel is gated, meaning that it opens only during the translocation of a nascent protein. Gating prevents the passage of other material, such as small molecules, ions, and other proteins, and allows the cytosol and the interior of the ER to be maintained as distinct compartments.

How is gating accomplished? The signal sequence of the nascent-chain interacts with the channel, which brings about a conformational change that leads to the channel opening and the chain being inserted into the open pore. As **FIGURE 7.6** shows, this recognition ensures that the channel opens only in response to the targeting of a translocation substrate and not in response to a cytosolic protein that has inappropriately arrived at the ER membrane. Importantly, the channel seems to open just widely enough to allow the chain to pass through it as an unfolded extended polypeptide, without other molecules being able to pass through at the same time. Because the channel is gated by the nascent polypeptide chain, the channel remains closed unless it is occupied by a chain, and so the permeability barrier provided by the ER membrane is maintained.

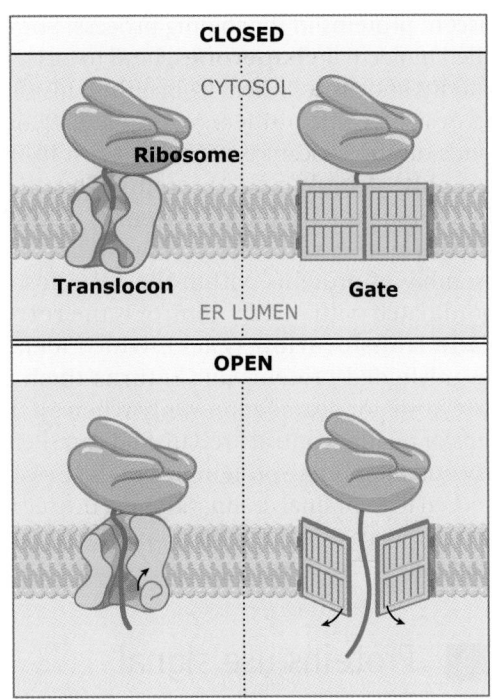

FIGURE 7.6 The translocation channel functions like a gate that is normally closed but opens in the presence of a translocation substrate. The "gate" opens only wide enough to allow the chain to pass through.

When translocation begins, nascent transmembrane proteins must be distinguished from proteins that need to pass through the channel completely. This distinction, like recognition of the signal sequence, is made by the channel. Movement across the lipid bilayer halts when a **transmembrane domain**, a hydrophobic segment destined to span the lipid bilayer, is recognized by the translocon and moved laterally out of the channel into the lipid bilayer. This process can occur multiple times in a single polypeptide, helping create the complex topologies of large multispanning membrane proteins (*topology* refers to the orientation of a protein with respect to the membrane) (see *7.10 Transmembrane proteins move out of the translocation channel and into the lipid bilayer*).

For both secretory and transmembrane proteins, translocation is also coordinated with activities that modify the polypeptide. For instance, most signal sequences are removed early during translocation. Many translocating proteins also have complex carbohydrate structures added, or disulfide bonds formed. Others are cleaved near the end of translocation and covalently attached to a phospholipid.

Finally, each translocated protein must fold. Several types of proteins in the ER assist nascent proteins in the folding process. Some, called molecular **chaperones**, bind to nascent proteins and protect them from either misfolding or aggregating; others rearrange disulfide bonds on the nascent proteins or assist in the assembly of multimeric proteins. Together, all of these proteins form a system of **quality control** that ensures the proper folding and assembly of proteins within the ER. Closely coordinated with quality control is the **retrograde translocation** system, which identifies misfolded proteins and returns them to the cytosol for degradation. Only when all the steps of quality control are satisfied can secretory and membrane proteins leave the ER and proceed to their final destinations via the secretory pathway (see *8 Protein trafficking between membranes*).

7.3 Proteins use signal sequences to target to the endoplasmic reticulum for translocation

Key concepts

- A protein targets to the ER via a signal sequence, a short stretch of amino acids that is usually at its amino terminus.
- The only feature common to all signal sequences is a central, hydrophobic core that is usually sufficient to translocate any associated protein.

All nuclear-encoded proteins begin their biogenesis in the cytosol. The first challenge in translocating proteins across the ER membrane is targeting: bringing secretory and membrane proteins, but not cytosolic ones, to their sites of translocation at the ER. The cell accomplishes this selection by use of a sequence on newly synthesized proteins that directs those proteins to the ER membrane and that is often removed after targeting has occurred.

The idea that a protein could be targeted to the ER by an extension of its amino acid sequence is called the signal hypothesis, which was proposed in the mid 1970s in a classic set of experiments that gave the first indication of how cells direct proteins to specific compartments. The experiments demonstrated that a secretory protein begins its synthesis in the cytosol with an extension of amino acids at the amino terminus. This extension was cleaved from the protein only after it associated with the ER but before its synthesis was complete.

Removal of the extension apparently accompanied movement of the protein across the membrane, because cleaved proteins were only found inside of the ER and not in the cytosol. In contrast, as the experiment in **FIGURE 7.7** illustrates, when a protein that is ordinarily secreted is instead synthesized *in vitro* in the absence of ER, the extension is not cleaved. Thus, the signal hypothesis proposed that the extension allowed nascent proteins to be targeted to the ER, and was removed after the protein began its translocation. The hypothesis predicted that the extension would be present on all secretory and membrane proteins but would not be present on cytosolic proteins. In general, this has proved to be true, and garnered its chief proponent, Günter Blobel, the Nobel Prize in Physiology or Medicine in 1999. This characteristic stretch of amino acids, now called the signal sequence, is indeed a nearly universal mechanism for targeting nascent secretory and membrane proteins. In most cases, these sequences are cleaved from preproteins, yielding **mature proteins** that continue through the secretory pathway.

The most surprising feature of signal sequences is their diversity. Every secretory pro-

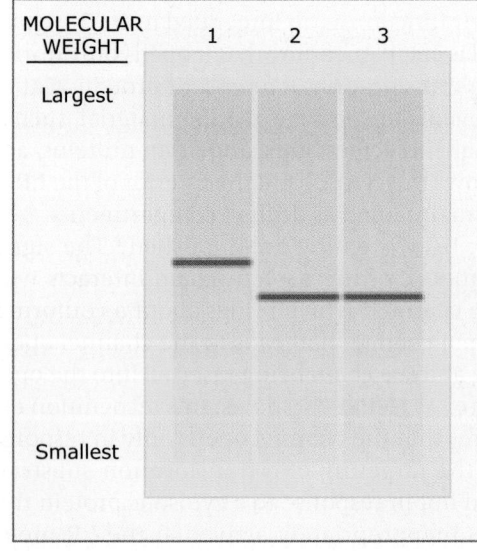

FIGURE 7.7 Signal sequences were discovered when secretory proteins, synthesized *in vitro* in a cell-free system, were found to be larger and, therefore, to migrate more slowly during gel electrophoresis (lane 1) than when those proteins were synthesized by cells (lane 2). If the proteins were synthesized *in vitro* in the presence of purified ER, they were made in the smaller form (lane 3) and were translocated across the membrane of the purified ER.

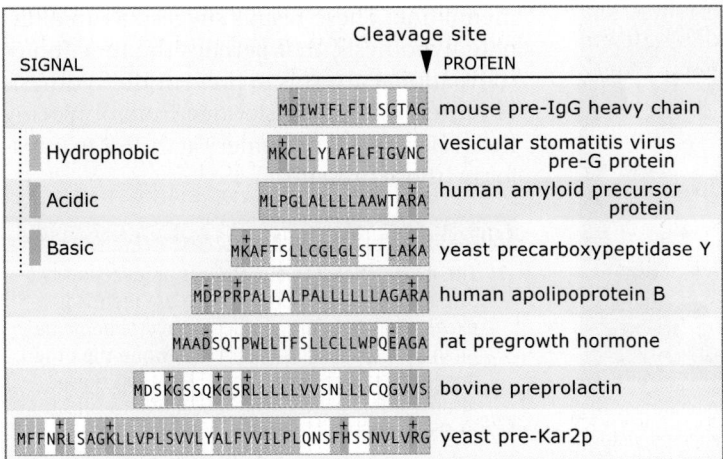

SIGNAL	Cleavage site ▼	PROTEIN
Hydrophobic	MDIWIFLFILSGTAG	mouse pre-IgG heavy chain
	MKCLLYLAFLFIGVNC	vesicular stomatitis virus pre-G protein
Acidic	MLPGLALLLLAAWTARA	human amyloid precursor protein
Basic	MKAFTSLLCGLGLSTTLAKA	yeast precarboxypeptidase Y
	MDPPRPALLALPALLLLLLAGARA	human apolipoprotein B
	MAADSQTPWLLTFSLLCLLWPQEAGA	rat pregrowth hormone
	MDSKGSSQKGSRLLLLLVVSNLLLCQGVVS	bovine preprolactin
	MFFNRLSAGKLLVPLSVVLYALFVVILPLQNSFHSSNVLVRG	yeast pre-Kar2p

FIGURE 7.8 Signal sequences that direct proteins to the ER differ in length and sequence. The common feature that allows them all to act as signals is a long, central region that is very hydrophobic and is often flanked by charged amino acids.

tein has its own unique signal sequence. The only element common to such sequences is a central segment of between 6 and 20 hydrophobic amino acids that has no specific sequence in common from protein to protein. Most signal sequences also contain several polar amino acids at their N-terminus. The hydrophobic domain is usually followed by a C-terminal region of small polar amino acids where signal sequence cleavage occurs. However, neither of these polar domains is strictly required for targeting.

As **FIGURE 7.8** shows, a broad range of sequences can serve as functional signal sequences, based on the characteristics just described. Yet despite their diversity, signal sequences are more or less interchangeable in their ability to target proteins to the ER and to allow for translocation. The signal sequence from one protein can usually be replaced with the signal sequence from another without affecting the ability of the protein to be targeted and translocated. It is amazing that such a seemingly nonspecific sequence can be responsible for the specificity of the targeting process.

7.4 Signal sequences are recognized by the signal recognition particle

Key concepts

- SRP binds to signal sequences.
- Binding of SRP to the signal sequence slows translation so that the nascent protein is delivered to the ER still largely unsynthesized and unfolded.
- The structural flexibility of the M domain of SRP54 allows SRP to recognize diverse signal sequences.

How does a signal sequence direct a protein to the ER for translocation? The discovery of SRP established that signal sequences are recognized by specific protein-protein interactions. SRP is a small cytosolic ribonucleoprotein particle containing six polypeptides and a small RNA molecule. It binds to the signal sequence of a nascent protein emerging from the ribosome and by doing so enables the ribosome-nascent polypeptide complex to interact with the membrane of the ER.

Only one subunit of SRP is required for recognizing signal sequences. This subunit, called SRP54 because of its molecular weight (54 kiloDaltons), is widely conserved among species, demonstrating the general importance of protein-mediated recognition of signal sequences.

SRP54 contains the following three distinct domains:

1. The G domain, which binds guanosine triphosphate (GTP) and hydrolyzes it to guanosine diphosphate (GDP);
2. The N domain, an N-terminal domain that interacts with the G domain; and
3. The M domain, which is a C-terminal domain containing a large number of methionine residues.

The structure of the M domain allows SRP to bind the broad array of signal sequences used in targeting. **FIGURE 7.9** illustrates the M domain, which contains several α helices bundled together to form a cleft in which the signal sequence binds. The methionines of the M domain line this cleft, radiating into it throughout its length. Because methionine side chains are both flexible and hydrophobic, the methionine residues projecting into the cleft act like a

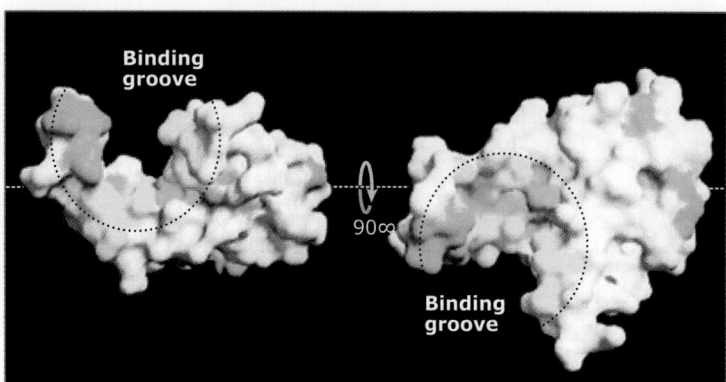

FIGURE 7.9 The structure of the M domain of a bacterial homolog of SRP54. The positions of hydrophobic amino acid residues are depicted in green and yellow. These residues form the surface of a deep groove in which signal sequences are proposed to bind. The groove is seen end-on on the left and from above on the right. Reprinted from *Cell*, vol. 94, R. J. Keenan, et al., Crystal Structure of the Signal Sequence..., pp. 181–191, Copyright (1998) with permission from Elsevier [http://www.sciencedirect.com/science/journal/00928674]. Photos courtesy of Robert Keenan, University of California, San Francisco.

collection of hydrophobic bristles. They allow SRP54 to bind to the wide variety of hydrophobic segments found in signal sequences.

Two other SRP subunits, SRP9 and SRP14, along with the 7S RNA molecule, bind to the ribosome and decrease the rate at which it continues to elongate the protein, probably by physically interfering with the binding of the translational elongation factor. The strength of the slowing varies from substrate to substrate, but in all cases it is relieved only when the ribosome successfully docks at the ER and SRP is released. By delaying or arresting synthesis, SRP might ensure that the nascent polypeptide is delivered to the membrane before a significant length of the chain has emerged from the ribosome. This would make it unlikely that the chain will be able to fold before it arrives at the channel, and it is, therefore, easier to translocate. The importance of this delay is supported by the observation that proteins generally lose their competence for translocation *in vitro* if they are recognized by SRP after significant translation has already occurred. More recently, it was shown that cells with a genetic mutation in SRP, rendering the protein incapable of mediating translational arrest, could not efficiently translocate most proteins. However, translocation efficiency could be restored in these cells by either experimentally slowing the rate of protein translation, or by manipulating the cells to produce more of the protein complex at the ER that is responsible for bringing SRP-bound translocation substrates to the ER

membrane. These results suggest, as an alternate hypothesis, that perhaps the function of translational arrest is to prevent the synthesis of proteins to be translocated from outpacing the availability of machinery at the ER to translocate those proteins.

Concept and Reasoning Checks

1. The mRNA encoding a protein is produced in two alternately spliced forms, which are identical except that one form encodes a hydrophobic domain at the protein's N-terminus while the other encodes charged amino acids. How will these alternate forms affect the subcellular localization of the protein?
2. Why are signal sequences located at the N-termini of polypeptides, rather than the C-termini?

7.5 An interaction between signal recognition particle and its receptor allows proteins to dock at the endoplasmic reticulum membrane

Key concepts

- Docking of SRP with its receptor brings the ribosome and nascent chain into proximity with the translocon.
- Docking requires the GTP binding and hydrolysis activities of SRP and its receptor.

Recognition of a nascent secretory or membrane protein by SRP completes only the first half of the targeting process. Once bound to SRP, the nascent chain must be recruited to the ER membrane and transferred to the translocation channel. A protein complex known as the **SRP receptor (SR)**, localized to the cytosolic face of the ER membrane, serves as the intermediary during this step.

The SR is a dimer of two related subunits. The cytosolically oriented "alpha" subunit (SRα) interacts with SRP, and the membrane-spanning "beta" subunit (SRβ) interacts with SRα and tethers it to the ER membrane, as depicted in **FIGURE 7.10**. Like SRP54, SRα and SRβ each have a domain for GTP binding and hydrolysis, and these three proteins form a related subfamily of GTPases. The GTP binding and hydrolysis activities of these proteins are critical for proper targeting of nascent chains to the ER, as well as for their transfer to the

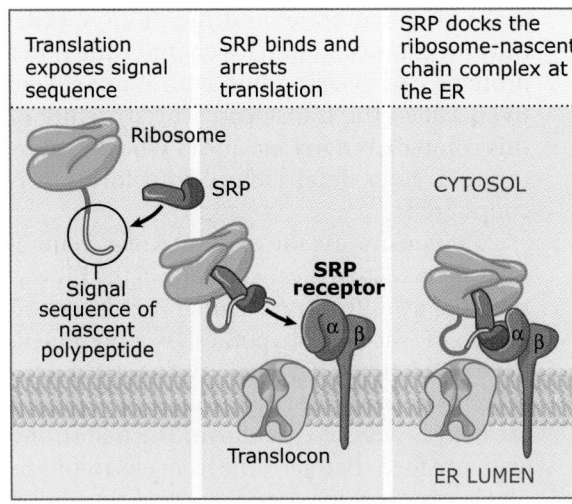

FIGURE 7.10 Shortly after a signal sequence emerges from the ribosome, SRP binds to it, arrests translation, and docks the ribosome-nascent chain complex at the membrane of the ER through an interaction with SRP receptor.

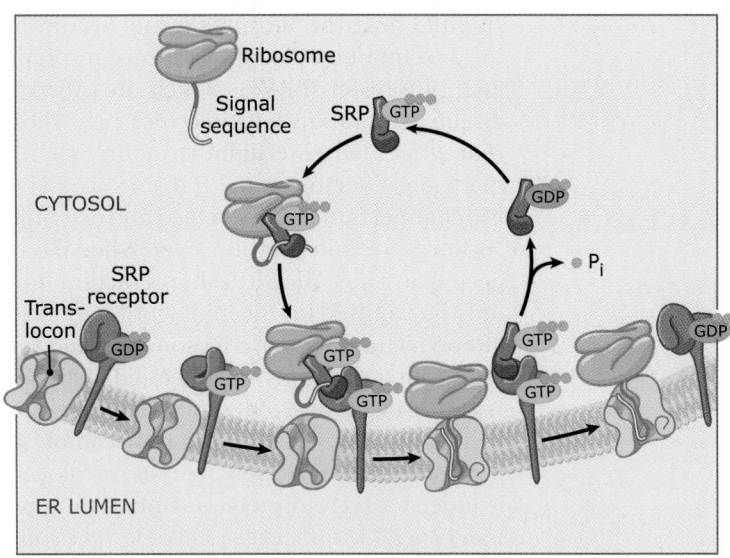

FIGURE 7.11 Both SRP and its receptor bind GTP, releasing the signal sequence and allowing it to be inserted into the translocation channel. GTP hydrolysis stimulates SRP and its receptor to separate afterward.

translocation channel and the recycling of SRP to the cytosol after targeting.

As shown in **FIGURE 7.11**, targeting requires coordinated GTP binding and hydrolysis by SRP and SR. Whether SRP and SRα exist in the GTP-bound state prior to their association with each other, or whether their mutual association leads to GTP binding, is not yet known. In either case, the signal sequence-bound form of SRP associates with SR, as does the translating ribosome; SR associates with the translocon, and binding of GTP by SRβ is probably assisted by a subunit of the translocation channel. Furthermore, the translocon associates with the ribosome-nascent chain complex. Thus, a complex is formed that depends upon the coordinated association of multiple components. In this complex, SRP and both subunits of SR exist in the GTP-bound state. It is likely that only successful complex formation brings about conformational changes necessary to lead to the hydrolysis of GTP by SRP and SR. SRP and SRα seem to facilitate each other's GTP hydrolysis, leading to extensive structural rearrangements of these two components. The ribosome might also accelerate GTP hydrolysis by SRP and SRα. The consequence of these conformational changes is the release of the ribosome-nascent chain complex by SRP and SR, with the interactions between the chain, the ribosome, and the channel keeping the chain in place and ready for translocation.

The multiple interactions and nucleotide hydrolysis steps required for targeting probably are used to ensure the speed and fidelity of the process. If a ribosome translating a cytosolic protein finds itself in the vicinity of the translocon at the ER membrane, the absence of an SRP-SR association will ensure that its presence there is transient and that translocation does not occur. Similarly, if a ribosome targets to an SR complex that is not associated with a channel, the interaction will lack the strengthening influences of the ribosome-translocon and translocon-SR associations, causing GTP hydrolysis by SRP and SR to be disfavored so that the chain will not be released in the absence of a translocon that it can cross.

When released from SRP, the ribosome engages the translocon. This occurs through direct interactions between the ribosome and the proteins that make up the channel. These contacts position the ribosome immediately over the cytosolic end of the channel, allowing nascent polypeptides to move directly between the two. The interaction between the ribosome and the channel is initially weak but strengthens and lasts in one form or another for the duration of translocation.

One aspect of targeting and translocation that remains poorly understood is how the assembly of multiple ribosomes on a single mRNA (called a polysome) affects the targeting process and the assembly of translocons. Once the protein being synthesized by the first ribosome on the mRNA targets to the ER, the other ribosomes following behind it might not require SRP, since they would already be positioned near the ER membrane. Studies using a technique called fluorescence resonance energy transfer (FRET), which measures the distance separating two molecules, have shown that while there are slight structural changes in engaged versus empty translocons, overall the channel remains assembled and essentially "primed" for translocation, even when it is unoccupied. Thus, the SRP-SR interaction might only be required for the initial targeting event, after which subsequent ribosomes would associate with translocons in fairly rapid succession, with the event of signal sequence recognition by the channel serving as a proofreading step to ensure that each polypeptide translocated is indeed a secretory or transmembrane protein (see *7.7 Translation is coupled to translocation for most eukaryotic secretory and transmembrane proteins*).

7.6 The translocon is an aqueous channel that conducts proteins

Key concepts

- Proteins translocate through an aqueous channel composed of the Sec61 complex, located within the ER membrane.
- Numerous accessory proteins that are involved in translocation, folding, and modification associate with the channel.

Proteins cross the ER membrane by passing through a water-filled (aqueous) channel that

No ribosome	Translating ribosome	Nascent chain released
CYTOSOL		
ER LUMEN		
No conductance		Conductance

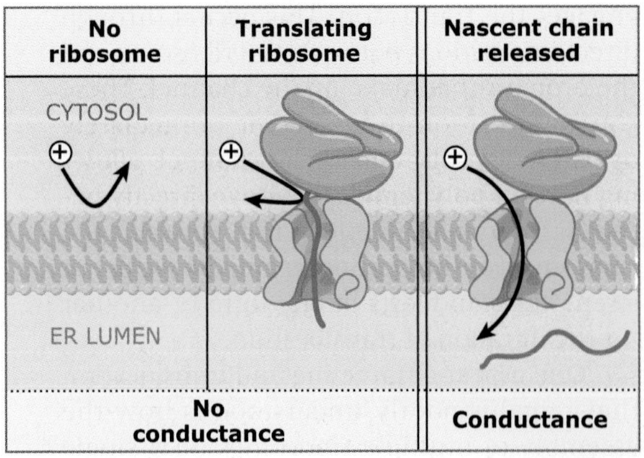

FIGURE 7.12 Ions pass across the membrane of the ER only if translocating nascent chains are released from the bound ribosomes. This suggests that the nascent chain crosses the membrane through an aqueous channel. The channel requires a ribosome to remain open because no conductance occurs if the ribosome is removed after the nascent chain is released.

spans the membrane and functions specifically for translocation. This channel and other proteins closely associated with it are collectively called the translocon. The structure of this complex is dynamic and is worth exploring in greater detail before translocation is considered.

Demonstrating the existence of a channel for proteins across the membrane of the ER was remarkably difficult. A channel was suggested at the time the signal hypothesis was presented, but a number of other ideas involving secretory proteins passing directly through the lipids of the bilayer were also put forth. The first strong evidence for a channel came from electrophysiologic experiments that examined the ability of ions to cross the membranes of vesicles (called microsomes) that were derived from the rough ER. Ions cannot cross a pure lipid bilayer, so any conductance detected would indicate the presence of channels across the membrane. Microsomes in which no translocation was occurring (because of the removal of their associated ribosomes) showed very little electrical conductance. Similarly, ions could not cross a membrane occupied by translocating ribosomes. However, as **FIGURE 7.12** shows, conductance occurred if the nascent proteins were released from the ribosomes without dissociating the ribosomes from the membrane.

This experiment led to the following two conclusions:

1. Ions are able to pass through an aqueous channel that is stabilized by the ribosome; no ions can pass when ribosomes are absent.
2. The nascent chain occupies this channel; ions can only pass through the channel when the nascent chain is released.

Later experiments, using other techniques, confirmed the existence of a channel by demonstrating directly that nascent secretory proteins are in an aqueous environment as they cross the plane of the membrane.

What proteins make up the translocation channel? Candidates were identified by several criteria, including proximity to a translocating nascent chain and interaction with membrane-associated ribosomes. To assign functions to the proteins that fit these criteria, microsomes were dissolved in detergent to solubilize the membrane proteins. **FIGURE 7.13** shows how the proteins were then separated to find the minimum set of components required for translocation. Individual proteins were reconstituted

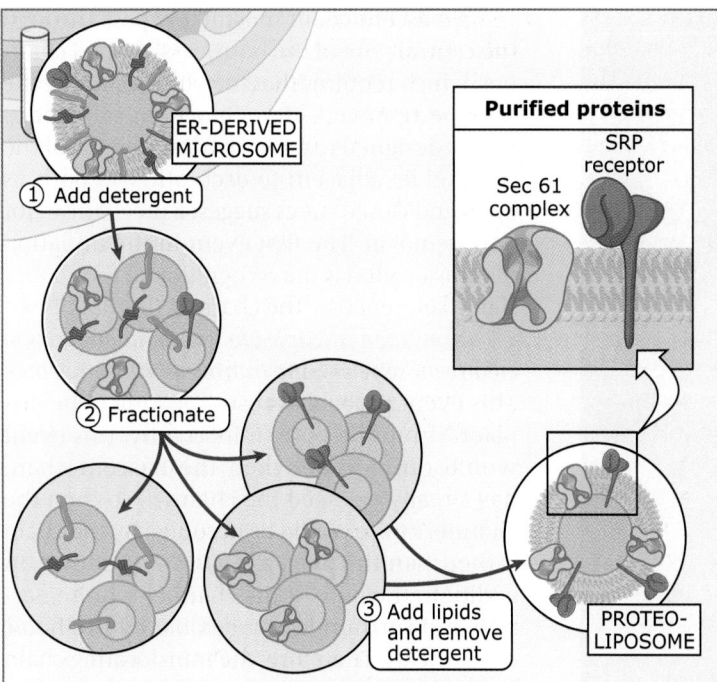

FIGURE 7.13 Detergents can be used to dissolve microsomes derived from the ER into small micelles of protein and lipid. These micelles can then be fractionated to purify the proteins of interest. When the detergent is removed and new lipids are added, vesicles containing only the proteins of interest are formed.

Labels in figure:
ER-DERIVED MICROSOME
① Add detergent
Purified proteins
Sec 61 complex
SRP receptor
② Fractionate
③ Add lipids and remove detergent
PROTEO-LIPOSOME

into lipid-enclosed vesicles called **proteoliposomes**, which substituted for microsomes in translocation reactions *in vitro*. In this manner, the exact composition of this surrogate ER could be controlled, and the importance of each type of protein in translocation could be determined. Conceptually, this approach is like taking apart a car, and then reassembling combinations of components to find only those that are absolutely necessary to make it move. An engine and a few other parts are absolutely required, but regulatory features such as the brakes will be missed.

This approach demonstrated that only SRP, SR, and a complex of three transmembrane proteins known collectively as the Sec61 complex were required for the translocation of certain proteins. Both SRP and SR were known to be involved in targeting, leaving the Sec61 complex as a good candidate to form the channel through which the translocating protein passes.

The conservation of Sec61 across a wide variety of species indicates the unique and critical role this complex plays in translocation. Proteins in the Sec61 complex were originally identified in yeast, using a genetic screen specifically designed to find genes required for the entry of secretory proteins into the ER. This approach yielded several genes, including SEC61. The protein it encodes, Sec61p, is an integral

membrane protein that spans the ER ten times. A homolog, Sec61α, is present in mammals. Biochemical experiments in vitro suggest that Sec61p surrounds translocating proteins and, thus, very likely forms the wall of the channel. Genetic and biochemical analyses have revealed that Sec61p associates tightly with the following two other, smaller proteins, whose exact functions are less clear:

1. Sss1p (Sec61γ in mammals) and
2. Sbh1p (Sec61β in mammals).

Together, these three components comprise the heterotrimeric Sec61 complex.

The detailed structure of the Sec61 complex has been solved using the SecY complex of the archaebacterium *Methanococcus jannaschii*; this complex is structurally and functionally homologous to the Sec61 complex in eukaryotes. The structure immediately suggests several important features of the translocation channel. Most notably, the channel's interior, composed of the transmembrane domains of Sec61α, is shaped essentially like an hourglass, with what is likely the pore occluded by a small plug contributed by a region of Sec61α, as **FIGURE 7.14** shows. From the point of view of the ribosome (i.e., from the cytosol), transmembrane domains 1–5 of Sec61α form half of a clamshell-like arrangement, with transmembrane domains 6–10 forming the other half. Sec61γ is located between the two halves.

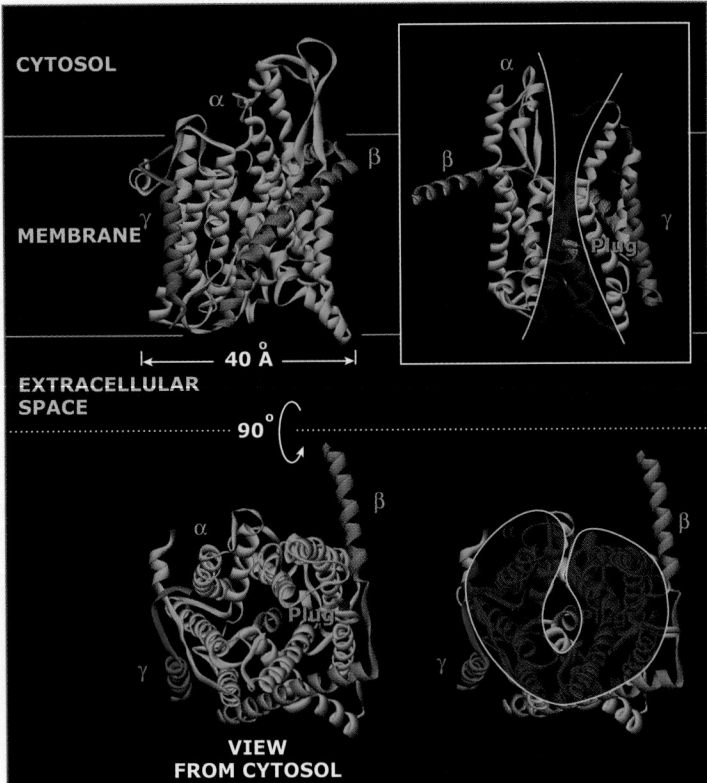

FIGURE 7.14 A model of the sec61 complex in mammals. The channel forms an hourglass-like structure, with the central pore occluded by a small "plug" that is thought to move out of the pore as a consequence of signal sequence binding to the channel. In the inset at top right, some of the transmembrane helices are removed to show the hourglass-shaped pore. The cytosolic view shows the clamshell arrangement of the transmembrane helices. Structures from Protein Data Bank 1RHZ. B. Van den Berg, et al., *Nature* 427 (2004): 24–26.

CYTOSOL

α

β

γ

MEMBRANE

|← 40 Å →|

EXTRACELLULAR SPACE

90°

VIEW FROM CYTOSOL

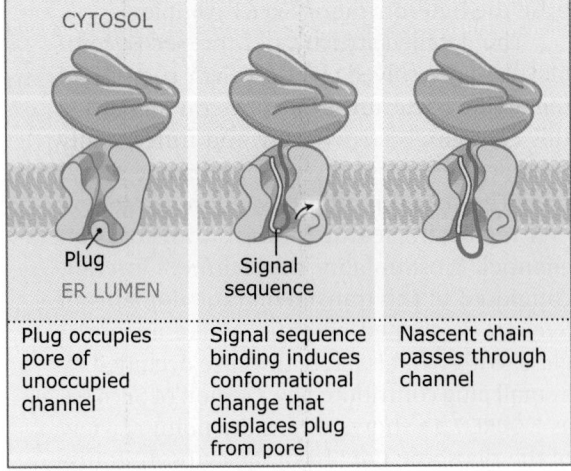

FIGURE 7.15 Signal sequence binding to the channel probably brings about a conformational change that displaces the plug and allows the chain to pass through the narrowly opened channel.

A nascent chain must likely pass through the central pore of the hourglass-shaped channel, which requires that the plug occluding the pore be removed. Here, biochemical experiments designed to address what regions of the channel lie adjacent to each other in both its open and closed states suggest a mechanism for pore removal. The first event in the initiation of translocation is the recognition of the chain's signal sequence by the channel (for details see *7.7 Translation is coupled to translocation for most eukaryotic secretory and transmembrane proteins*). This event probably causes the plug to be displaced from the pore. Importantly, this event would only occur when the nascent chain has already engaged the channel, so then the channel's pore would be occupied by the chain rather than the plug, as **FIGURE 7.15** shows. In addition, the pore of the channel is quite narrow and surrounded by flexible hydrophobic amino acids. Therefore, the translocating chain probably occupies the entire space of the pore, with the movement of ions across the channel strongly disfavored.

It was known prior to the determination of the Sec61 complex structure that the complex associated as trimers or tetramers, and it was previously thought that multiple copies of the complex were needed to form a single channel. However, higher resolution structures of the channel and the ribosome-channel complex now make this idea seem unlikely—a single Sec61 complex (i.e., one molecule each of Sec61α, Sec61β, and Sec61γ) most likely forms the functional channel. Whether additional Sec61 complexes self-associate during translocation in a way that impacts the process of translocation is not clear.

Although Sec61 forms the channel, other proteins that participate in protein translocation or modification are present near the channel, and the Sec61 complex can be thought of as a scaffold around which the proteins involved in targeting, translocation, and protein folding and modification in the ER assemble. For instance, in addition to the assembly of the ribosome and SR complex described above, the signal peptidase complex, which cleaves the signal sequence from translocating proteins, is present at every active translocon. Oligosaccharyltransferase (OST), an enzyme complex that covalently attaches sugars to translocating chains, is also present. A number of proteins whose functions are less well defined also interact with the nascent chain. The translocating chain-associating membrane

(TRAM) protein is often found in close association with the signal sequences and transmembrane domains of translocating proteins. Unlike the enzymes that cut or modify the nascent chain, the TRAM protein is required for the translocation of some proteins and has been implicated at multiple steps of the process. Also present near translocating proteins is the translocon-associated protein (TRAP) complex. This complex is just as abundant as the Sec61α protein and has a role in assisting the recognition of signal sequences by the channel, although the mechanism by which it accomplishes this is not known. Most proteins translocate quite inefficiently into reconstituted ER *in vitro* if only the Sec61 complex and SR are present. The addition of TRAM and the TRAP complex substantially improves the efficiency of translocation for most proteins, as does the addition of lumenal proteins. It is likely that many other proteins and protein complexes influence the efficiency and kinetics of protein translocation in ways that are not yet understood.

Concept and Reasoning Checks

1. Why is an aqueous channel necessary for proteins to cross the ER membrane?
2. What is the function of the plug domain of Sec61α?

7.7 Translation is coupled to translocation for most eukaryotic secretory and transmembrane proteins

Key concepts

- An interaction between the translocon and the signal sequence causes the channel to open and initiates translocation.
- While translocation occurs through the Sec61 channel, the exact steps of translocation and the accessory machinery required are unique to each translocating protein.

Following targeting and docking of a ribosome-associated nascent chain, movement of the polypeptide across the membrane must begin. Much of what is known about the process has come from the study of a relatively small number of model proteins in a cell-free system that reconstitutes translocation. Successful translocation depends on the coordination of a series of interactions among the chain, the channel, and the ribosome. These interactions lead to

changes in the organization of the channel and its association with the ribosome and the nascent chain.

Following release of the signal sequence from SRP at the ER membrane, the sole interaction keeping the ribosome in contact with the membrane is between the ribosome and the channel. This association alone is insufficient to initiate translocation without a nascent-chain recognition step at the ER membrane. Such a step is required because ribosomes have a low but significant intrinsic affinity for the channel, regardless of the substrate being translated. If the ribosome-translocon interaction alone were sufficient to initiate translocation, cytosolic proteins could also be translocated. This is prevented by a step that follows docking, in which the channel must recognize a functional signal sequence.

The changes in the ribosome-translocon interaction brought about by signal sequence recognition are illustrated in **FIGURE 7.16**. After docking and release of SRP, the ribosome is only loosely bound to the translocon. This

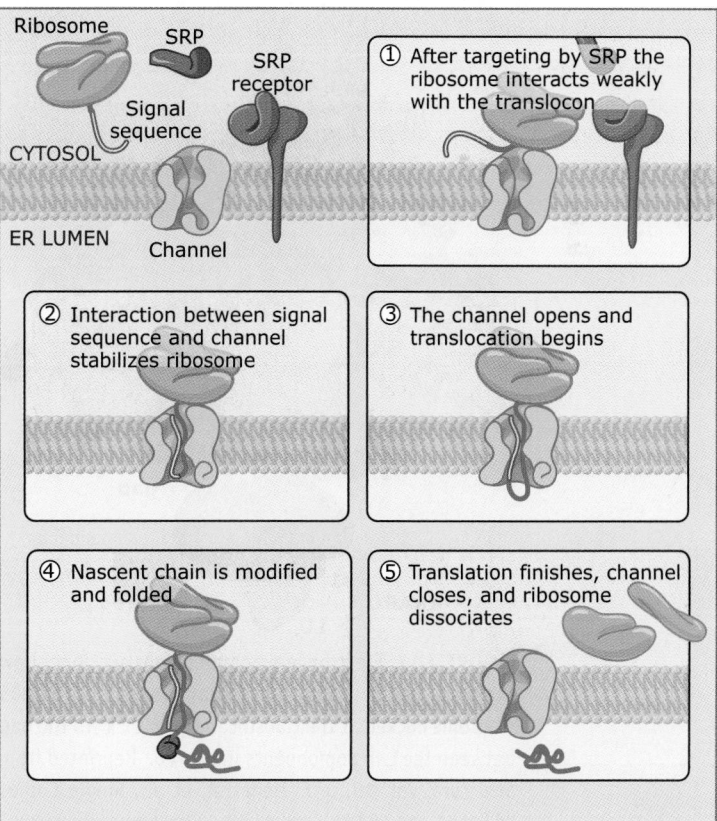

FIGURE 7.16 Elongation of the nascent chain into the lumen of the ER cannot begin until the channel has recognized the signal sequence and bound the ribosome tightly. Folding and modification start very soon after translocation has begun.

leaves the signal sequence and the rest of the nascent chain exposed to the cytosol but immediately adjacent to the end of the channel. Soon after elongation of the nascent chain resumes, the signal sequence is recognized by Sec61α. Recognition is thought to require that the signal sequence insert itself into the channel as a loop with its N-terminus facing the cytosol and its C-terminus oriented toward the ER lumen. This orientation positions the protein so that its preferred pathway of exit from the ribosome is through the channel and into the ER lumen.

Recognition and insertion of the signal sequence initiate translocation and also probably cause the displacement of the central plug occluding the channel and insertion of the mature portion of the chain (i.e., the portion just after the signal sequence) into the pore. As this happens, the ribosome and the channel become much more tightly associated. All of these events, from insertion of the signal sequence to opening of the channel, take place very rapidly following targeting. Little additional elongation of the polypeptide is required, and the events are complete well before most of it is synthesized. In principle, for an ideal secretory protein, only about forty amino acids need to have emerged from the ribosome for the nascent chain to engage the protein translocation machinery.

Once the channel is open, most secretory proteins proceed into the lumen until translation ends. It had previously been widely accepted that the interaction between the ribosome and translocon was so strong that a physical seal formed between these two components, effectively preventing escape of the translocating chain into the cytosol. However, there is a physical gap between these two components as shown in **FIGURE 7.17**, and—at least in certain circumstances—portions of the nascent chain are able to slip through this gap into the cytosol. Because translocating chains then apparently have access to the cytosol, it is not known what biases their movement into the ER lumen so that they do indeed translocate, rather than simply fall into the cytosol. One possibility is that interactions with components in the ER lumen hold the newly emerged chain in the lumen and prevent it from falling out (see *7.9 Adenosine triphosphate hydrolysis drives translocation*). Whatever the mechanism, it is thought that the preferred movement of a translocation substrate is into the lumen. When the ribosome reaches the stop codon, the chain is released into the lumen and the channel closes, although the sequence of these events is still unclear. In particular, whether the channel closure is caused by the chain passing completely through it or by dissociation of the ribosome is not understood.

Although these steps define a basic pathway for translocation, the process varies from one nascent protein (translocation substrate) to another. For instance, signal sequences apparently differ in their interactions with the channel and the manner in which they open it. In turn, the mode of signal sequence recognition can affect subsequent steps in the biogenesis of certain proteins. Some proteins also require ER factors in addition to Sec61 and TRAM to translocate successfully. Others begin translocation along the same basic pathway but later diverge from it well after translocation has begun. For example, some proteins seem to transiently slip out of the gap between the ribosome and translocon, exposing large segments to the cytosol before moving back into the channel. Therefore, while there are certain features of translocation that are essential to the process, it is probably overly simplistic to think of translocation as an invariant mecha-

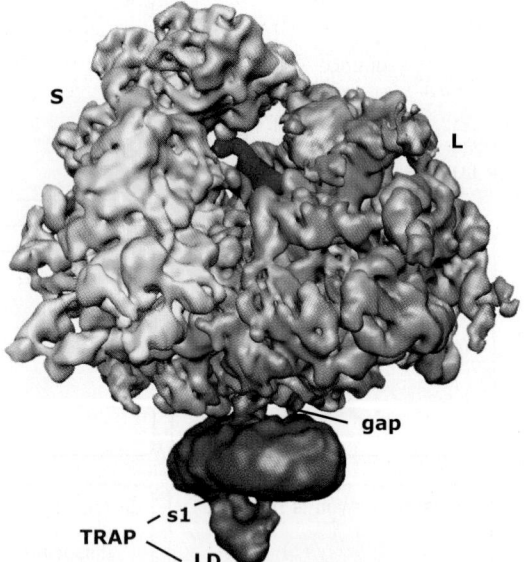

FIGURE 7.17 Electron micrograph reconstructions show the small (yellow) and large (blue) subunits of the 80S ribosome docked at the translocon (purple), with the gap between the two components indicated. Reprinted from *Structure*, vol. 16, J. F. Ménétret, et al., Single Copies of Sec61 and TRAP..., pp. 1126–1137, Copyright (2008) with permission from Elsevier [http://www.sciencedirect.com/science/journal/09692126]. Photo courtesy of Christopher W. Akey, Boston University School of Medicine.

nism for moving proteins across membranes; more likely, the specific pathway taken by a protein across the translocon will have consequences for that protein's biologic activity in ways not yet appreciated.

7.8 Some proteins target and translocate posttranslationally

Key concepts

- Posttranslational translocation proceeds independently of both ribosomes and SRP.
- Posttranslational translocation is used extensively in yeast but is less common in higher eukaryotes.
- The posttranslational translocon is distinct in composition from the cotranslational translocon, but they share the same channel.

During cotranslational translocation, protein targeting and the initiation of translocation occur very early in nascent chain synthesis. This prevents the chain from assuming a folded conformation in the cytosol that would be refractory to translocation. Another pathway of translocation allows proteins to be completely synthesized in the cytosol, yet kept in an unfolded state so that they can be translocated afterward. Such posttranslational translocation is used extensively in unicellular eukaryotes and may occur in higher eukaryotes as well. This form of translocation is independent of both SRP and ribosomes and differs from cotranslational translocation in both machinery and mechanism.

The first evidence for a second pathway for targeting and translocation was the observation that many yeast proteins can translocate *in vitro*, even after being released by the ribosome. Consistent with this idea, *S. cerevisiae* cells lacking SRP remain viable, and many proteins retain their ability to translocate in these cells. It is now clear that under normal conditions, some yeast proteins use only one of the two pathways, but that a large number are capable of using either one efficiently.

There is no cytosolic factor that recognizes signal sequences in the posttranslational pathway. What seems to determine the probability that a protein will use one or the other pathway is whether its signal sequence is sufficiently hydrophobic to interact with SRP. If it cannot, there is nothing to delay translation, and the protein is targeted and translocated after its synthesis is complete.

One consequence of rapid targeting during cotranslational translocation is that translocation substrates are prevented from folding in the cytosol. Before targeting, the length of peptide outside the ribosome is too short to fold, and once within the confined space of the channel, folding cannot occur. In contrast, the folding of posttranslationally translocated substrates in the cytosol is inhibited by association with chaperones of the hsp70 family. These proteins use Adenosine triphosphate (ATP) hydrolysis to continuously bind to and release posttranslational substrates, preventing them from folding or aggregating. Thus, it allows proteins the opportunity to interact with the channel.

Posttranslational translocation does not involve a component like SRP that recognizes signal sequences before they arrive at the membrane. Instead, posttranslational substrates are recruited to the translocon by a multiprotein complex that is part of the posttranslational translocon. Like the cotranslational translocon, the posttranslational translocon includes the trimeric Sec61 complex. However, the posttranslational translocon also contains four other proteins (Sec62p, Sec63p, Sec71p, and Sec72p), as seen in **FIGURE 7.18**. These proteins form a subcomplex that exposes large domains to both the cytosol and the lumen. One or more of them may be involved in the targeting of substrates, although the mechanism is still unclear.

In general, this association of a core translocation channel with alternate groups of accessory proteins gives the cell the flexibility to regulate the translocation of certain substrates or groups of substrates without having to evolve separate channels for each type of

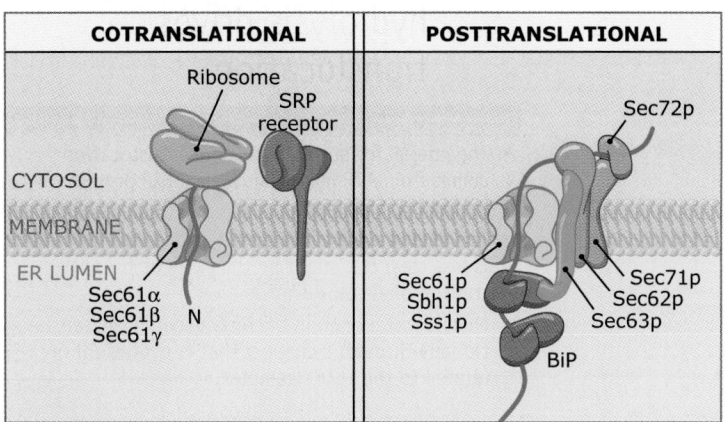

FIGURE 7.18 Different proteins interact with the same channel to promote cotranslational or posttranslational translocation.

substrate to be translocated. It is analogous to the regulation of gene expression by transcription factors that bind to the promoters of only certain subsets of genes, while interacting with the RNA polymerase complex to turn transcription on or off.

After the chain arrives at the translocon, the channel recognizes the signal sequence as it does in cotranslational translocation. Because targeting is independent of SRP, this recognition event is particularly important to ensure that cytosolic proteins are not translocated. Signal sequence recognition presumably brings about the opening of the channel in posttranslational translocation as in the cotranslational mode. The ability of a posttranslationally targeted signal sequence to be recognized by the channel but not by SRP suggests that, while hydrophobicity of the signal sequence is the chief determinant of binding to both components, recognition of the signal sequence by SRP or the channel must be influenced by other, as yet unknown, factors.

While *S. cerevisiae* makes extensive use of posttranslational translocation, evidence for this pathway in higher eukaryotes is less compelling. This form of translocation has so far only been demonstrated *in vitro* for very small substrates. In such cases, folding into stable secondary structures in the cytosol is unlikely. Whether these substrates actually target posttranslationally in vivo has not yet been clearly established. However, Sec62p and Sec63p have homologs in higher eukaryotes. Thus it seems probable either that posttranslational translocation operates in some form, or that this machinery has been co-opted for other uses.

7.9 Adenosine triphosphate hydrolysis drives translocation

Key concepts

- The energy for posttranslational translocation comes from ATP hydrolysis by the BiP protein within the ER lumen.
- The energy source for cotranslational translocation is less clear but might be the same as for posttranslational translocation.
- Most translocation in bacteria occurs posttranslationally through a channel that is evolutionarily related to the Sec61 complex.

What is the driving force for translocation into the ER lumen? Biochemical experiments mea-

suring the passage of fluorescent probes across the membrane of microsomal vesicles suggested that the ribosome docks with the channel so tightly as to prevent the movement of the chain anywhere but across the channel into the lumen. However, cryo-electron microscopy imaging of the ribosome-translocon complex suggests that there may be a gap between the ribosome and channel. In addition, biochemical experiments assessing the exposure of a translocating chain to the cytosol have shown that for many proteins, there are multiple occasions when the chain is exposed to the cytosol, and so its movement into the ER would not necessarily be guaranteed. In addition, such a mechanism would not explain how proteins targeted posttranslationally could be driven across the translocation channel. While there could still be a role for the ribosome in "pushing" nascent chains directly into the ER lumen, it now seems more likely, based on studies of the mechanism of posttranslational translocation, that the movement of chains into the lumen is biased by their association with proteins in the lumen. While the energy source for cotranslational translocation remains an open question, the role of ATP hydrolysis in posttranslational translocation is better understood and is described here.

The energy source driving posttranslational translocation into the ER is ATP hydrolysis by the lumenal hsp70 BiP. BiP is brought close to the posttranslational channel by a transient association with the lumenal domain of Sec63p. **FIGURE 7.19** illustrates a likely model to account for the role of BiP in translocation. In the **Brownian ratchet model**, the function of BiP is to bind to the nascent chain as it emerges from the channel and prevent it from sliding back into the cytosol. (BiP is folded and so cannot pass through the channel.) As each new segment enters the lumen, it is bound by a new BiP molecule, and translocation proceeds in this incremental fashion. BiP thus biases the motion of the chain in one direction. In this model, ATP hydrolysis strengthens the interaction of BiP with the chain, and Adenosine diphosphate (ADP)/ATP nucleotide exchange later causes BiP to release the substrate. This model is supported by experiments showing that BiP depletion from the ER lumen compromises translocation, but translocation efficiency can be restored by adding back to the lumen any fairly large molecule capable of binding the nascent chain. It is even plausible that the folding of the nascent chain in the lumen might pre-

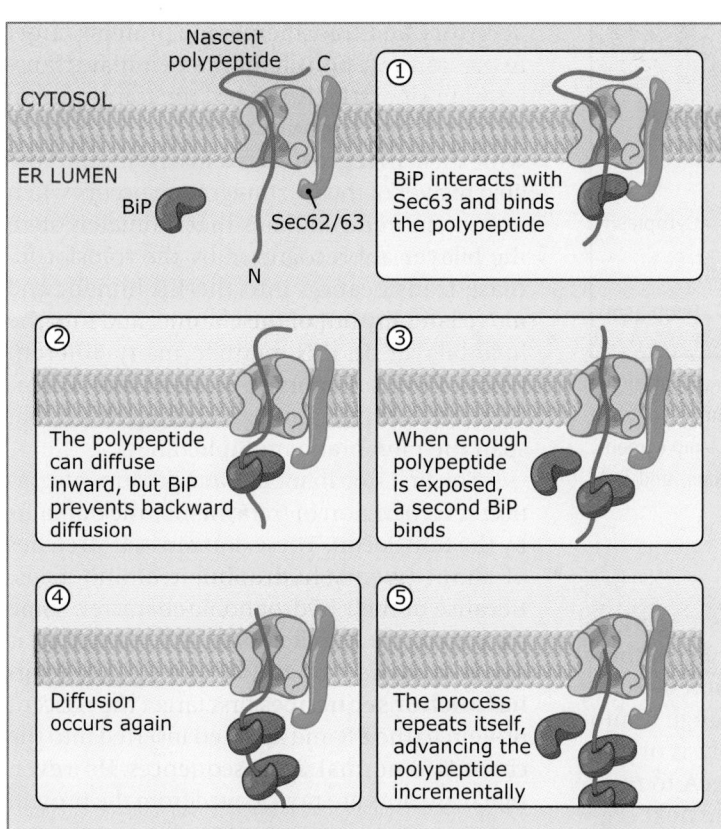

FIGURE 7.19 The Brownian ratchet model of protein translocation. After a brief interaction with Sec62/63, BiP molecules bind to the nascent polypeptide. Because they are too large to fit inside the channel, the bound BiP molecules allow the polypeptide to diffuse inward but limit its diffusion back toward the cytoplasm. Each BiP molecule traps a new segment of the polypeptide in the lumen.

vent the chain from falling back into the cytosol, implying that proteins that fold more readily might translocate more easily than proteins that tend to remain in an unfolded state during a greater portion of their translocation (though this has not been proven). For this model to be true in its most simplistic form, the chain, once it interacts with the channel to begin translocation, must then presumably be freed from its interactions with cytosolic hsp70 and other cytosolic proteins; otherwise the chain would be just as likely to remain outside the ER as to pass through the channel. However, the interactions of the translocating chain with cytosolic proteins during translocation are not yet fully characterized. It is worth noting that while the Sec61 complex and SR are the only proteins absolutely required for translocation (see *7.6 The translocon is an aqueous channel that conducts proteins*), most proteins translocate quite inefficiently under these minimal conditions, perhaps because of the absence of lumenal proteins that would assist their movement.

In contrast to the Brownian ratchet model, the **active pulling model** proposes that ATP hydrolysis causes a conformational change in BiP that results in the bound chain being actively pulled through the channel. Because

pulling (the generation of force) is difficult to demonstrate experimentally, it has not yet been possible to distinguish the contributions of these two possible modes of posttranslational translocation.

Although this chapter focuses on eukaryotes, the problem of protein sorting is shared by prokaryotes, which also produce both secretory and membrane proteins. Translocation is posttranslational in the bacterium *E. coli*. It is similar to posttranslational translocation in yeast, in that substrates are kept unfolded beforehand, and signal sequence recognition occurs at the channel. As in eukaryotes, translocation in bacteria proceeds through an aqueous channel composed principally of three proteins, collectively called the SecYEG complex. The SecY protein is the bacterial homolog of Sec61p and Sec61α. However, because bacteria lack membranous organelles, translocation proceeds directly across the plasma membrane. One consequence of this is that the energy source driving translocation, the SecA protein (SecAp), acts at the membrane's cytosolic face rather than on the receiving side of the membrane as in eukaryotic posttranslational translocation. The crystal structure of SecA, shown in **FIGURE 7.20**, suggests that SecA binds to the

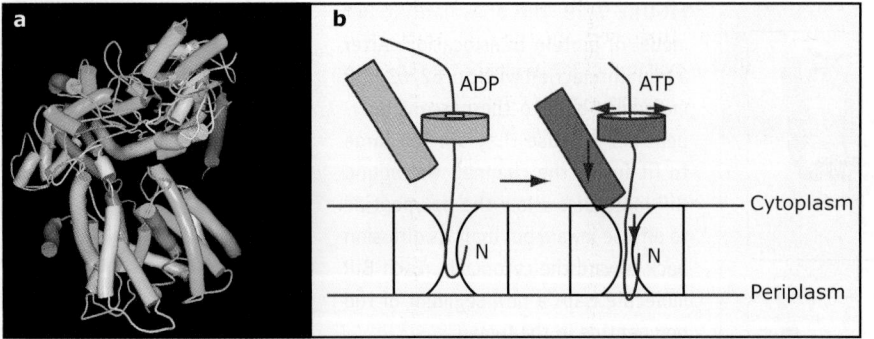

FIGURE 7.20 A model for the action of SecA, based on its crystal structure and that of the SecY channel, proposes that a two-helix finger of SecA (brown) moves the nascent polypeptide (red) forward into the channel, accompanied by the loosening of the ring-like clamp (green). Structure from Protein Data Bank 3DIN. Illustration adapted from Zimmer, Y. Nam, and T. A. Rapoport, *Nature* 455 (2008): 936–943.

translocating polypeptide and forms a ring-like clamp around it. ATP binding by SecA then loosens the ring and allows a portion of SecA to associate with the chain and plunge it into the opening of the channel. Hydrolysis of ATP to ADP then is thought to cause SecA to reset the protein to allow it to rebind the next segment of the translocating chain. In principle, the ring-like clamp of SecA would prevent the chain from slipping back into the cytosol. An additional factor influencing prokaryotic translocation is the presence of an electrochemical potential across the membrane that helps drive translocation, though the mechanism is poorly understood.

The comparison of cotranslational, posttranslational, and bacterial translocation highlights both the conservation of fundamental aspects of passage across membranes and the ways in which the process has been adapted for different situations. In all cases, the fundamental aspect—passage through an aqueous channel—remains the same. However, the substrates and the ways in which they are targeted and moved through the channel differ.

7.10 Transmembrane proteins move out of the translocation channel and into the lipid bilayer

Key concepts

- The synthesis of transmembrane proteins requires that transmembrane domains be recognized and integrated into the lipid bilayer.
- Transmembrane domains exit the translocon by moving laterally through a protein-lipid interface.

Secretory and transmembrane proteins target to the translocation channel and initiate translocation similarly. However, the translocation of membrane proteins must be coordinated with their **integration**, or insertion, into the lipid bilayer of the ER. Integration occurs when transmembrane domains that ultimately span the bilayer are recognized by the translocon, cease translocation into the ER lumen, and move laterally out of the channel and into the lipid bilayer. In this manner, many different types of transmembrane proteins can be synthesized and integrated, including those that span the membrane multiple times.

The first step in membrane protein integration is recognition of transmembrane domains by the translocon. These domains are stretches of about twenty hydrophobic amino acids. Because of their hydrophobic character, some transmembrane domains are also recognized as signal sequences by SRP. These so-called **signal anchor sequences** first target the nascent protein to the ER and are then inserted into the channel as normal signal sequences. However, signal anchors are not cleaved from the protein but are instead integrated into the membrane. In contrast to signal anchor sequences, most transmembrane domains are recognized by the translocon as they emerge from the ribosome, after targeting by a conventional N-terminal signal sequence has already taken place. For either a signal anchor sequence or transmembrane domain, the simplest indication that a transmembrane domain is within the translocon is the hydrophobicity of the domain. Because of its architecture, the translocation channel can detect this. As **FIGURE 7.21** shows, the structure of the translocon suggests that the channel is capable of opening like a clamshell, allowing transmembrane domains to simultaneously contact both the channel and the lipid bilayer. It is likely, in fact, that signal sequences and transmembrane domains bind to the Sec61α protein in the vicinity of the mouth of the clamshell structure, and then this binding probably causes the channel to open laterally. Evidence for this type of interface comes from experiments demonstrating that transmembrane domains within the channel are in contact with both Sec61α and lipid. As a result, even though the translocon provides an aqueous channel across the membrane, sufficiently hydrophobic regions of translocating polypeptides can sample the lipid environment of the membrane as they translocate. Segments containing polar amino acids would be expected to move through the channel without stopping,

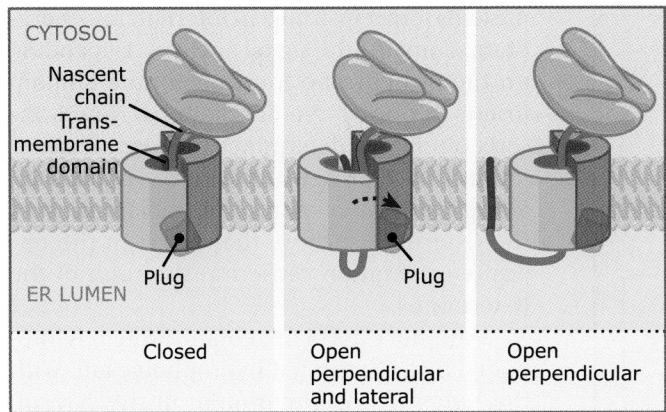

CYTOSOL

Nascent
chain
Trans-
membrane
domain

Plug   Plug

ER LUMEN

Closed   Open   Open
            perpendicular perpendicular
            and lateral

FIGURE 7.21 The translocon is depicted as a cylinder that opens and closes in two ways to allow movement of the nascent chain through the pore and movement of transmembrane domains into the membrane.

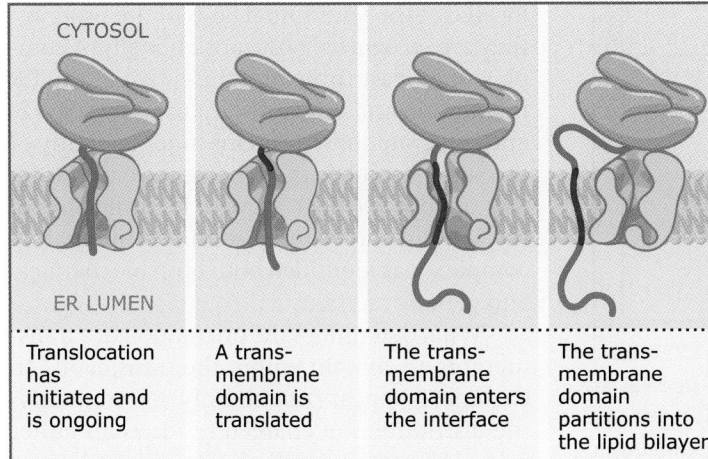

CYTOSOL

ER LUMEN

Translocation   A trans-   The trans-   The trans-
has   membrane   membrane   membrane
initiated and   domain is   domain enters   domain
is ongoing   translated   the interface   partitions into
                                               the lipid bilayer

FIGURE 7.22 A gap in the wall of the translocation channel exposes translocating proteins to the lipid bilayer and allows transmembrane domains to be recognized and integrated. Because of their hydrophobicity, transmembrane domains will favor the lipid environment and move out of the channel into the lipid bilayer.

while hydrophobic domains would interact strongly with the lipids and remain associated with the sides of the channel, thus halting translocation. This is pictured in **FIGURE 7.22**. As with the recognition of signal sequences by the channel, the interaction between the channel and signal anchor or transmembrane sequences is probably also subject to modulation by accessory factors based on features of the transmembrane sequences themselves besides just their hydrophobicity.

7.11 The orientation of transmembrane proteins is determined as they are integrated into the membrane

Key concepts

- Transmembrane domains must be oriented with respect to the membrane.
- The mechanism of transmembrane domain integration may vary considerably from one protein to another, especially for proteins that span the membrane more than once.

Transmembrane domain recognition and integration are complicated by the need to orient each protein in relation to the ER membrane. Some membrane proteins must have their N-terminal domains on the cytosolic side of the membrane, but others require the opposite orientation. Events that occur as a membrane protein is targeted and translocated establish its orientation.

It is simplest to conceptualize the orientation of membrane proteins that span the membrane once and have a cleaved N-terminal signal sequence. These proteins initiate translocation much as secretory proteins do, with the signal sequence directing the targeting and the initiation of translocation. Translocation continues until the transmembrane domain emerges from the ribosome, when it is integrated. As a result, the C-terminal domain does not translocate but remains on the cytosolic side of the membrane, as shown in **FIGURE 7.23**. The Ire1 and PERK proteins (*see 7.20 Communication between the endoplasmic reticulum and nucleus prevents the accumulation of unfolded proteins in the lumen*) are examples of this type of membrane protein.

In contrast to transmembrane proteins with a cleaved signal sequence, signal anchor

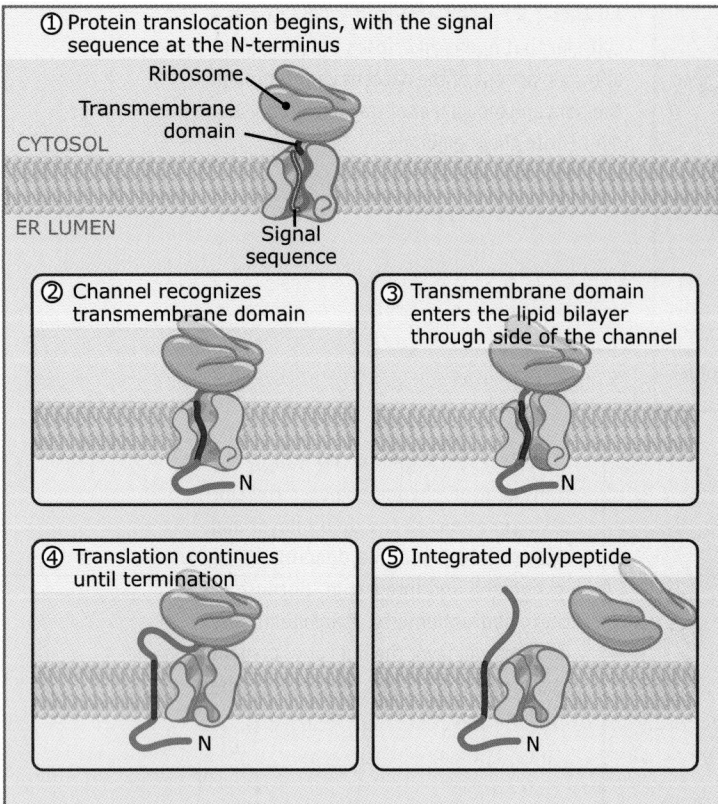

① Protein translocation begins, with the signal
   sequence at the N-terminus

Ribosome

Transmembrane
domain

CYTOSOL

ER LUMEN   Signal
                       sequence

② Channel recognizes
   transmembrane domain

N

③ Transmembrane domain
   enters the lipid bilayer
   through side of the channel

N

④ Translation continues
   until termination

N

⑤ Integrated polypeptide

N

FIGURE 7.23 The signal sequence initiates translocation, which proceeds as it would for a secretory protein until a transmembrane protein is translated and recognized by the channel. This process can only integrate proteins in one orientation.

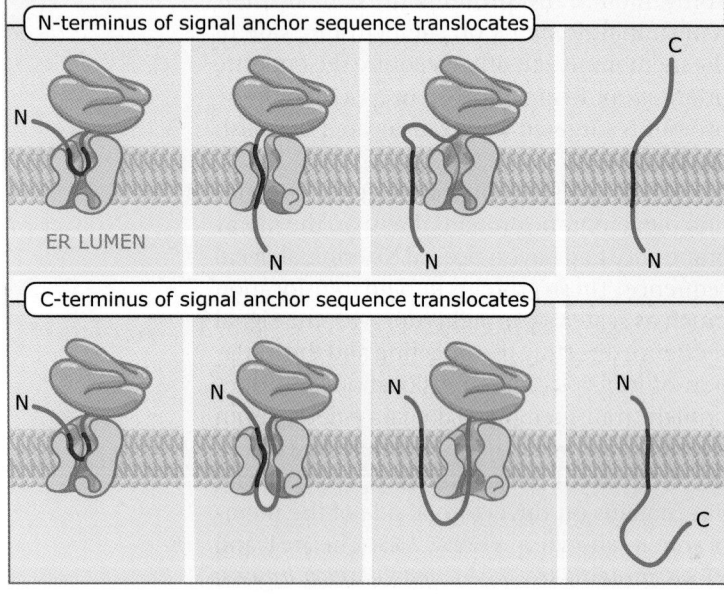

N-terminus of signal anchor sequence translocates

N

ER LUMEN

N

N

C

N

C-terminus of signal anchor sequence translocates

N

N

N

N

C

FIGURE 7.24 Depending on the protein, either the N- or the C-terminus of a signal anchor sequence is translocated after its initial interaction with the channel. Membrane proteins with different orientations result from the two situations.

proteins target by using an internal transmembrane domain (the signal anchor). Depending on the protein, the transmembrane domain inserts in one of two orientations. How it inserts ultimately determines the overall orientation of the protein, as seen in **FIGURE 7.24**. Some domains result in the translocation of the C-terminus; those that insert in the opposite orientation cause translocation of the N-terminus.

Another class of membrane protein is integrated by a C-terminal hydrophobic tail, with the remainder of the protein in the cytosol. Because these transmembrane domains are located at the C-terminal ends of the proteins, they are recognized posttranslationally. A complex of at least three proteins—two at the ER membrane and one at the cytosol—recognize and integrate these proteins independently of the Sec61 complex, though the mechanistic details are not yet clear. As with identification of the Sec61 complex, the C-terminal integrating complex was identified both both biochemical and genetic methods.

What determines the orientation of a transmembrane domain within the translocon? In bacteria, it appears that the key determinant is the distribution of charged residues on either side of the transmembrane domain. Charged lipids are asymmetrically distributed between the two leaflets of the bacterial plasma membrane, providing a physical basis for the mechanism. However, lipids in the ER membrane are not asymmetrically distributed with respect to charge, so this cannot be the explanation in eukaryotes. In addition, some proteins can adopt multiple topological forms, suggesting that their orientation may be influenced by other factors. In general, it is still unclear how transmembrane domains are oriented in eukaryotes.

For multispanning membrane proteins, the simplest model proposes that transmembrane domains are integrated one by one as they appear in the channel, and the orientation of the whole protein would thus be dictated by the properties of its first transmembrane domain. Recognition of a transmembrane domain by the channel probably halts translocation, allowing the subsequent portion of the nascent chain to slip through the gap between the ribosome and channel into the cytosol. The next hydrophobic domain to emerge from the ribosome likely has a high affinity for the signal sequence/transmembrane domain binding site of the channel, and so reinserts the chain into the channel and reinitiates translocation. Cycles of translocation stopping and restarting continue until all the transmembrane domains of the protein are synthesized and recognized. As with the translocation of secretory proteins, however, it is clear that not all transmembrane proteins follow this idealized stepwise pathway of integration. Transmembrane domains do not necessarily integrate immediately after the channel recognizes them, and the translocon is apparently capable of accommodating at least two transmembrane domains at the same time in certain circumstances, by mechanisms that are not clear. In general, the establishment of protein topology with respect to the membrane is studied *in vitro* by translating artificially truncated transmembrane proteins in the presence of purified ER, and using probes that have access to the cytosol but not to the lumen of the ER to infer how the protein is oriented. Such experimental tools, however, might or might not faithfully reflect how transmembrane proteins integrate in living cells.

Concept and Reasoning Checks

1. A protein contains an N-terminal cleavable signal sequence and two further transmembrane domains. How is the protein most likely to be oriented with respect to the membrane?
2. A protein contains an N-terminal cleavable signal sequence and one further transmembrane domain. A genetic mutation changes a leucine in the middle of this transmembrane domain to an arginine. How is this mutation most likely to affect the protein's subcellular localization and topology?

7.12 Signal sequences are removed by signal peptidase

Key concepts

- Nascent chains are often subjected to covalent modification in the ER lumen as they translocate.
- The signal peptidase complex cleaves signal sequences.

As proteins translocate into the ER, they are often covalently modified. Three types of modification are particularly common:

1. Removal of the signal sequence;
2. Attachment of a complex carbohydrate structure (**N-linked glycosylation**); and
3. Attachment of the phospholipid **GPI**.

One or more of these forms of modification occur to almost every protein that enters the ER or is integrated into its membrane.

Of these modifications, the reasons for signal sequence cleavage are the obvious. Uncleaved signal sequences might interfere with the folding of proteins after translocation or be recognized as signal anchors and cause proteins that the cell needed to secrete to be integrated into the ER membrane. Consequently, signal sequences are "disposable." After serving their functions in targeting nascent polypeptides to the ER and guiding them through their initial interactions with the translocation machinery, signal sequences are removed and discarded. Signal sequence cleavage is nearly universal for secretory proteins and for transmembrane proteins in which the signal sequence does not also serve as a transmembrane domain.

Signal sequence cleavage is carried out by a membrane protein complex made up of five subunits, and is called the **signal peptidase** complex (SPC). Only two of the subunits have proteolytic activity; the other three presumably serve in some regulatory capacity. Because the nascent chain is inserted into the translocation channel in a loop orientation, the cleavage site is thought to be positioned just inside the lumenal face of the ER membrane when it is cut by the SPC.

The site of signal sequence cleavage varies from protein to protein and is most directly influenced by the identity of amino acids in the immediate vicinity of the cleavage site. For cleavage to occur, in general, the residue that is located just N-terminal to the cleavage site must have a short side chain and the amino acid residue located three amino acids N-terminal to the cleavage site must be uncharged. For some proteins, several such sites around the true cleavage site exist, and it is not known how the proper site is selected. Poorly defined properties of the signal sequence can also influence the position of cleavage.

Signal sequence cleavage increases the mobility of an *in vitro*-synthesized protein on sodium dodecyl sulfate-polyacrylamide gel electrophoresis (SDS-PAGE), allowing the timing of the event to be studied; it typically takes place after the synthesis of a hundred or more amino acids, though the precise timing varies from one signal sequence to another. In fact, cleavage can occur much later for some proteins. Before it is cleaved, a signal sequence may sometimes influence the protein's interactions with other ER factors, including those responsible for modification or folding of the nascent chain. The features of a signal sequence that influence the timing of its cleavage are not understood.

After signal sequence cleavage, a further proteolytic event sometimes processes the cleaved signal peptide; this cleavage event is carried out by an enzyme complex called signal peptide peptidase (SPP, as opposed to signal peptidase that removes the signal peptide from the nascent chain). SPP processes only a subset of cleaved signal peptides, suggesting that its function is more complex than simply disposing of "used" signal peptides. However, exactly what this function is remains unclear.

7.13 The lipid glycosylphosphatidylinositol is added to some translocated proteins

Key concept

• GPI addition covalently tethers the C-termini of some proteins to the lipid bilayer.

A small but significant number of proteins translocated into the ER are modified by the covalent addition of a phospholipid. Through a covalent linkage to a membrane glycolipid (a membrane phospholipid linked through its head group to a sugar structure) called GPI, a translocated protein can be tethered to the lumenal face of the bilayer. As a consequence, a protein that would otherwise be fully translocated can remain in association with the membrane, while a transmembrane protein may have an additional point of linkage. The reasons that a protein may be linked to the membrane by GPI addition rather than by integration are unclear. There is evidence that a GPI linkage may mark a protein for particular intracellular transport pathways, for example, to the apical surface of polarized cells; or to caveolae or lipid rafts, which are subdomains of the plasma membrane specialized for initiating signal transduction cascades (*see 8.18 Polarized epithelial cells transport proteins to apical and basolateral membranes*). A protein linked to a membrane by a lipid might also have greater mobility in the plane of the membrane than an integrated membrane protein, because lipids diffuse faster in the membrane than do proteins, and because the GPI-anchored protein would not be subjected to interactions with cytoskeletal elements near the plasma membrane that can trap restrict the mobility of transmembrane proteins. In addition, integrated membrane proteins cannot be easily liberated from the membrane, but proteins attached to the membrane by a GPI linkage can be freed by enzymatic removal of the anchor. GPI linkage could, therefore, allow a protein to be released from the membrane in response to signals.

GPI is an elaborate structure that must be synthesized before it can be added to proteins, by the stepwise action of at least 20 proteins. The identities of these genes and their roles in the process were revealed by genetics (i.e., isolating yeast mutants unable to synthesize GPI-linked proteins) and by biochemically identifying the GPI precursors that accumulated in each mutant strain. As shown in **FIGURE 7.25**, GPI synthesis begins on the cytosolic leaflet of the ER membrane with the joining of the membrane phospholipid phosphatidylinositol (PI) to N-acetylglucosamine (GlcNAc). In a series of subsequent reactions, GlcNAc-PI is then deacetylated, the lipid is flipped to the lumenal side of the membrane, and three mannose residues are added. Phosphoethanolamine is added to each mannose residue, yielding the final GPI substrate.

GPI addition requires recognition of the nascent chain that will be the substrate, and

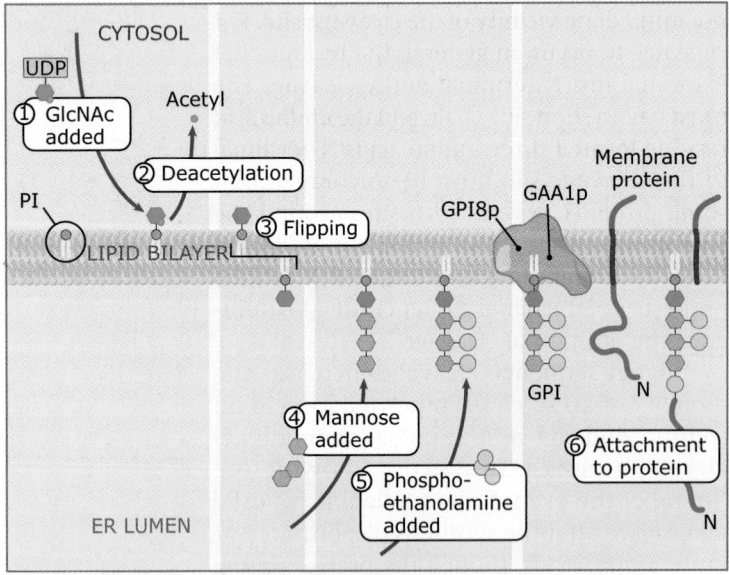

FIGURE 7.25 GPI is synthesized in a series of steps. The first several steps occur on the cytoplasmic face of the ER membrane, and later ones on its lumenal face. When complete, GPI is covalently attached to a protein by a reaction that cleaves the protein near its C-terminus. Modification with GPI tethers a protein to the membrane. GlcNAc stands for N-acetylglucosamine.

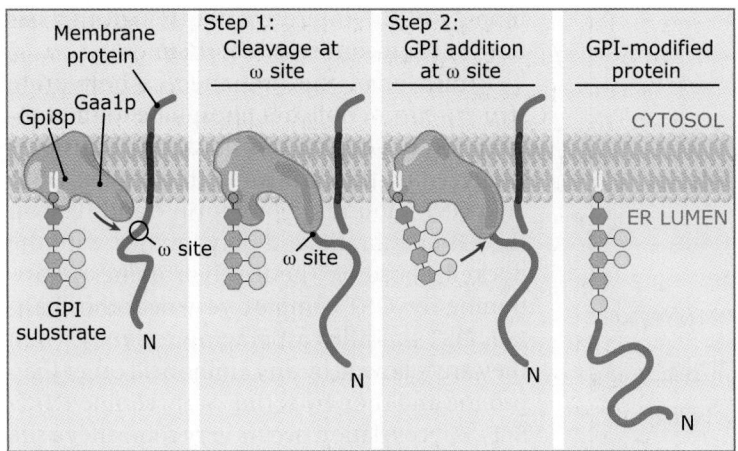

Membrane protein

Gpi8p Gaa1p

Step 1:
Cleavage at
ω site

Step 2:
GPI addition
at ω site

GPI-modified
protein

CYTOSOL

ER LUMEN

ω site

ω site

GPI
substrate

N

N

N

N

FIGURE 7.26 Addition of GPI to a protein occurs in two steps. One of the proteins in the enzyme complex first cleaves the substrate protein at a specific site, forming a covalent bond to it in the process. That bond is then displaced by one of the phosphoethanolamine groups of GPI to produce the GPI-modified protein.

transfer of the GPI moiety to the acceptor site in the protein. The signal for GPI addition is a small C-terminal hydrophobic domain of variable length (typically 10 to 30 residues). Like N-terminal signal sequences, GPI signals vary in sequence among proteins and are interchangeable from one protein to another. When GPI addition occurs, the GPI signal is removed from the protein and the GPI moiety is added to the new C-terminal residue, termed the omega (ω) site. Thus, like a signal sequence, GPI signals are disposable—they are used and then removed.

An integral membrane protein complex, made up of several components, catalyzes GPI addition, with Gpi8p being the catalytically active component. The most likely mechanism for its action involves a two-step transamidation reaction, as shown in **FIGURE 7.26**. In the first step, the enzyme forms a covalent intermediate with the omega site, resulting in cleavage of the C-terminal peptide from the rest of the protein. Then the terminal phosphoethanolamine residue of GPI, brought into close contact with the nascent protein by the enzyme, is added to the omega site. This forms the GPI-linked protein and liberates the enzyme. Whether ATP or GTP is required in this process and whether GPI addition requires that the GPI signal first be integrated into the bilayer are not yet clear.

7.14 Sugars are added to many translocating proteins

Key concept
• OST catalyzes N-linked glycosylation on many proteins as they are translocated into the ER.

As many proteins are translocated into the ER, they are covalently modified with large complexes of sugars. This form of modification is very common: more than half the secretory and membrane proteins in a cell may have sugars added, and many are modified in several different locations within the polypeptide chain. Because this process occurs on asparagine residues (abbreviated "N"), it is called N-linked glycosylation.

A number of roles are proposed for glycosylation. Testing them is difficult because many normally glycosylated proteins show no apparent loss of function when glycosylation is prevented. In several cases, however, a role has been identified. While proteins are still in the ER, modification with sugars assists in the proteins' folding, disulfide bond formation, or degradation (see *7.17 The calnexin/calreticulin chaperoning system recognizes carbohydrate modifications*). Beyond the ER, the functions of these modifications are less clear. Some evidence suggests that changes in glycosylation may serve as a mechanism for altering a protein's function. For instance, the secretion and activity of follicle-stimulating hormone are altered in response to changes in its glycosylation. In other cases, carbohydrates may increase the solubility of proteins or protect them from degradation by extracellular proteases. As **FIGURE 7.27** shows, pathogens can also use glycosylation as a way of evading immune system recognition.

During N-linked glycosylation, a large, preformed intermediate is added as a unit to a substrate in the ER lumen. Sugars are added to dolicholphosphate, a minor membrane phospholipid, first on the cytosolic leaflet of the

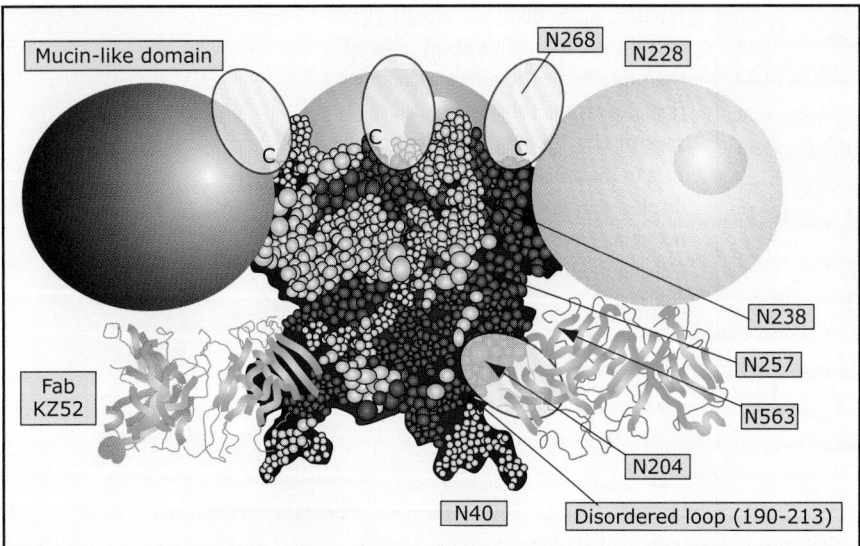

FIGURE 7.27 The Ebola virus surface glycoprotein mediates viral attachment and fusion to host membranes. Modeling of the glycoprotein structure predicts that N-linked glycans (shown in yellow, or also as orange ovals for glycans that could not be precisely modeled) prevent antibodies from binding to the glycoprotein and interfering with its association with its cell surface receptor. Adapted from J. E. Lee, et al., *Nature* 454 (2008): 177–182.

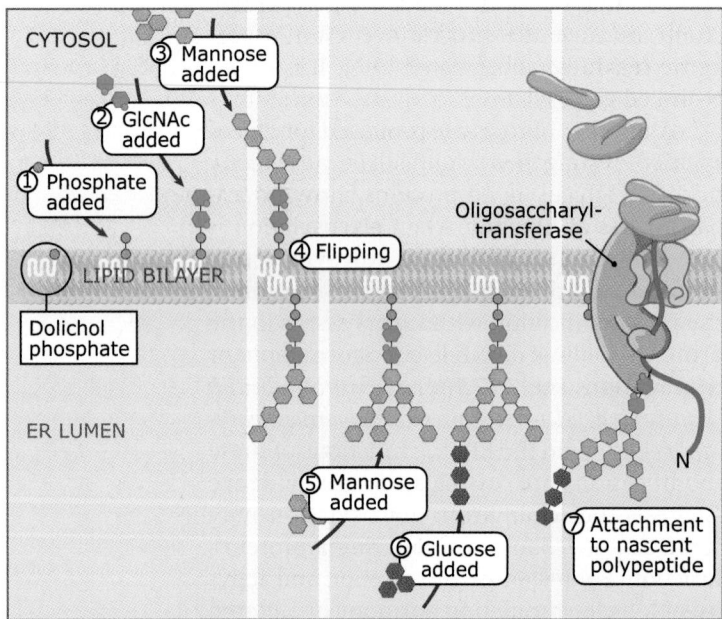

FIGURE 7.28 In a series of steps, an elaborate carbohydrate structure is synthesized on the phospholipid dolichol phosphate. Synthesis begins on the cytoplasmic face of the membrane of the ER but is completed on its lumenal face. The entire oligosaccharide structure that results is transferred onto translocating proteins by the enzyme OST. Modification takes place on asparagine residues that appear in a particular sequence.

ER membrane and then, after the structure is flipped, on the lumenal leaflet. **FIGURE 7.28** illustrates the process. Genetic studies have suggested that a protein called Rft1p might catalyze

flipping of the intermediate, though this has not yet been confirmed biochemically.

Transfer of the branched carbohydrate structure from dolichol phosphate to the substrate occurs during translocation. Transfer is catalyzed by the multisubunit complex **OST**. Two OST subunits, ribophorin I and II, span the ER membrane and may interact with the docked ribosome (hence their names), positioning the OST complex very near the channel. OST modifies asparagine residues when they are followed by any amino acid other than proline and then by serine or threonine (N-X-S/T). Glycosylation occurs very soon after a site emerges from the channel; only ten to twelve amino acids need to have entered the lumen before glycosylation can occur. Recognition of sites is quite efficient: within cells, ~90% of the potential sites are used, but some sites are never used. The extent of a particular protein's glycosylation may change under different conditions. After a protein is initially glycosylated, various modifications of the oligosaccharide structure, including the removal of some sugar residues and the addition of others, take place at different points in the ER and Golgi apparatus.

7.15 Chaperones assist folding of newly translocated proteins

Key concept

- Molecular chaperones associate with proteins in the lumen and assist their folding.

As nascent polypeptides are translocated and modified, they also begin to fold. With large numbers of newly translocated proteins constantly arriving, folding is one of the major activities in the lumen of the ER. Because misfolded proteins can have very harmful effects on the cell, one of the primary responsibilities of the ER is to ensure that only properly folded proteins move on into the rest of the secretory pathway. To allow this, the ER has a very active "quality control" system that recognizes unfolded or misfolded proteins and either gives them the opportunity to fold properly or initiates their destruction.

In general, proteins acquire posttranslational modifications once their folding and maturation in the ER is complete. These often include further alteration of the N-linked glycan structure in the Golgi apparatus and the formation of the correct disulfide-bonded structure. The acquisition of these modifications can typically be detected by SDS-PAGE analysis, and so these are diagnostic for the progression of model proteins through the quality control process, that allow quality control to be studied. However, the small number of substrates studied so far necessarily limits the understanding of the general principles that govern the process.

Proteins that fold in the ER face the same problems as those that fold in the cytosol. Protein folding is driven by hydrophobic interactions: the hydrophobic domains of a polypeptide tend to associate with one another rather than stay exposed to their aqueous surroundings. However, hydrophobic domains may associate incorrectly, resulting in a misfolded protein, or may cause aggregation with other proteins. In vivo, molecular chaperones facilitate the folding process by allowing a protein to fold and refold in a protected environment until the correct form is achieved, and by distinguishing properly folded from misfolded forms. Chaperones are very active within the ER, and form the basis of the quality control system. As long as a protein is associated with chaperones, it cannot progress through the secretory pathway.

Many of the most common chaperones in the ER are related to chaperones that participate in protein folding in the cytosol. The best-characterized of these lumenal chaperones is the hsp70 family member BiP. BiP, also known as Grp78, stands for Binding Protein, as it was originally identified by virtue of its interaction with (i.e., binding to) the immunoglobulin heavy chain during antibody assembly in the ER of B lymphocytes. BiP is typically the most abundant protein in the ER lumen and interacts with many proteins very early during their folding.

Because hydrophobic regions are usually buried within the cores of globular proteins, the presence of an exposed hydrophobic patch on a protein is a strong signal that the protein has not yet finished folding. These segments serve as a cue for BiP to bind to the nascent chain. By successive cycles of binding and release driven by ATP hydrolysis, BiP protects the nascent protein from aggregation and gives it multiple opportunities to achieve its proper conformation. After a protein has folded into a compact structure with its hydrophobic segments buried, it no longer associates with BiP,

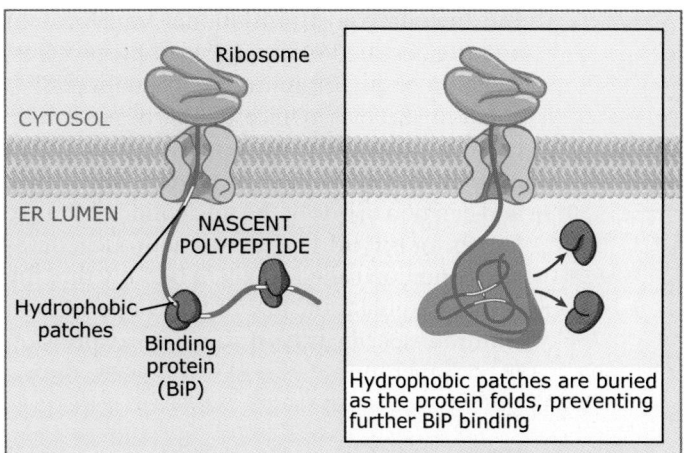

FIGURE 7.29 BiP binds to exposed hydrophobic patches in recently translocated proteins. After a protein has folded, its hydrophobic patches are buried within its structure and are no longer accessible to BiP.

as shown in **FIGURE 7.29**. The extremely high concentration of BiP in the lumen of the ER ensures that it is one of the first chaperones encountered by most nascent proteins, and that as a protein tries to fold, there are always likely to be BiP molecules close at hand. Because it acts from the beginning of the process, BiP plays a very general role in assisting protein folding. Mice homozygous for deletion of the *BiP* gene die very early in embryonic development, attesting to the essential role for BiP in cellular function.

Another common lumenal chaperone with a cytosolic counterpart is Grp94, a member of the hsp90 family. Though Grp94 is also abundant in the ER lumen, unlike BiP it tends to bind proteins that are at least partially folded rather than those that are newly emerged in the lumen and completely unfolded. Grp94 interacts with a narrower group of substrates than BiP, and what property of a protein it recognizes is unclear. Presumably, the function of Grp94 is more to refine the folding of proteins after BiP and other chaperones assist in global folding. This activity of Grp94 illustrates that the quality control machinery acts at multiple levels during the folding of nascent proteins.

Chaperones often function in concert with other proteins called cochaperones, which influence chaperone activity. A class of proteins called J proteins (so named after the bacterial cochaperone DnaJ), for example, interact with HSP70 family members and stimulate their ATPase activities. At least 6 J proteins are active in the ER lumen, and they have been implicated in processes that depend upon

BiP, including posttranslational translocation and the destruction of misfolded ER proteins. Deletion of one ER lumenal J protein, p58[IPK], results in diabetes in mice. Similarly, nucleotide exchange factors can influence the rate of ATP/ADP exchange for BiP or other chaperones, and deletion of one of the nucleotide exchange factors for BiP, SIL1, leads to a neurodegenerative condition called ataxia. It is likely that the cell can regulate the activity of chaperones in substrate-specific ways through the expression of cochaperones.

7.16 Protein disulfide isomerase ensures the formation of the correct disulfide bonds as proteins fold

Key concept
- Protein disulfide isomerases (PDIs) catalyze disulfide bond formation and rearrangement in the ER.

While BiP and Grp94 have cytosolic counterparts, the lumen of the ER requires specialized chaperoning systems as well. It is particularly important that proteins be able to form disulfide bonds when necessary. The formation of disulfide bonds is a consequence of the oxidizing environment that exists both in the lumen of the ER and in the extracellular milieu. Disulfide bond formation occurs between cysteine residues as a protein folds (although disulfide bonds sometimes occur between proteins as well). It is catalyzed by a family of enzymes called **PDIs**. PDI is not restricted to forming correct disulfide bonds and often forms incorrect bonds as proteins attempt to fold. These would trap proteins in incorrect conformations or aggregates if the bonds could not be broken and rearranged. PDI solves this problem by catalyzing disulfide bond rearrangement as well as formation. PDI and other thiol isomerases thus give nascent proteins the flexibility to refold without being constrained by disulfide bonds.

PDI catalyzes disulfide bond formation in other proteins through the use of cysteine residues in its own active site, as **FIGURE 7.30** shows. Disulfide bond formation is an oxidation-reduction reaction, with electrons exchanged between the cysteines of the folding protein and those of PDI. For PDI to participate, its cysteines must start the reaction in their oxidized state (as a disulfide bond), and a system must be present to reoxidize them afterward. It was originally thought that the small molecule glutathione reoxidized PDI. The oxidized form of this small molecule is selectively imported into the ER lumen and can maintain PDI in its oxidized state. However, it was more recently discovered that disulfide bonds can be formed and rearranged in at least some proteins even when glutathione is absent. Although glutathione may play a role, it now seems that an ER protein called Ero1p is primarily responsible for the oxidation of PDI. Ero1p is oxidized by flavin adenine dinucleotide (FAD), which is ultimately oxidized by molecular oxygen. Disulfide bonds in Ero1p that are not part of its catalytic activity are thought to be sensitive to the oxidation/reduction status of the ER lumen, and to bring about different changes in the conformation of Ero1p depending upon whether they are oxidized or reduced. These changes would then influence Ero1p activity to stimulate or inhibit protein oxidation as appropriate to cellular condition.

PDI uses a different mechanism to isomerize the existing disulfide bonds within a misfolded protein. In this reaction, one of the cysteine residues in the active site of PDI forms a transient disulfide bond with the misfolded protein. While the bond exists, the protein can resume folding despite its being bound to PDI.

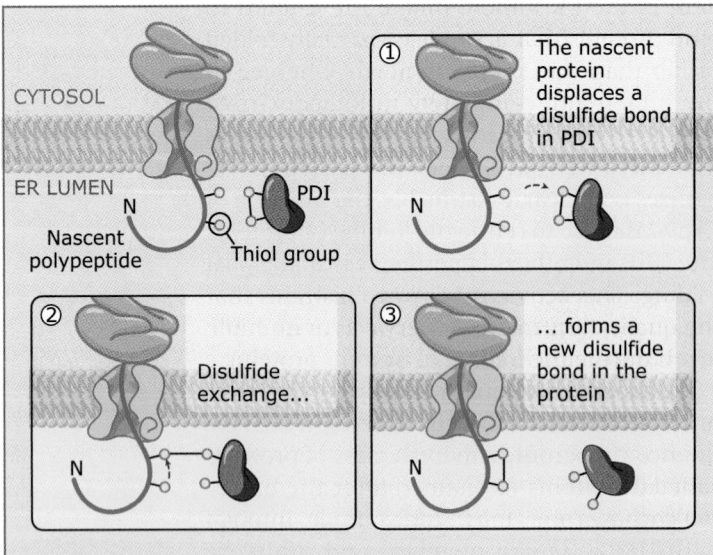

FIGURE 7.30 A disulfide bond in PDI is used to form one in the nascent polypeptide. An enzyme regenerates the disulfide bond in PDI afterward so that it can act multiple times.

When the folding protein finds a conformation in which another disulfide bond is possible, the bond with PDI is broken. How PDI can sense that a protein has the proper disulfide bonds and cease interacting with it is not yet established. It is possible that the tighter packing of the polypeptide chain within a properly folded protein makes its disulfide bonds inaccessible to the enzyme.

While PDI is the best-described thiol isomerase, and is expressed essentially ubiquitously in different cell types, at least twenty other redox-based chaperones of the ER lumen have been identified, and the possibility exists that some or even most of these are cell-type specific or perhaps even substrate-specific.

7.17 The calnexin/calreticulin chaperoning system recognizes carbohydrate modifications

Key concept

- Calnexin and calreticulin escort glycoproteins through repeated cycles of chaperoning, controlled by addition and removal of glucose.

Another form of quality control unique to the ER is lectin-mediated chaperoning. In this pathway, glycoproteins associate with carbohydrate-binding proteins (**lectins**) in the ER lumen. These lectins—most commonly, the integral membrane protein calnexin and its lumenal homolog calreticulin—are probably not themselves chaperones. Instead, they likely bring nascent glycoproteins into contact with the PDI family member ERp57. Thus, glycosylation of translocated proteins provides an additional point of entry for proteins into quality control. Although calnexin and calreticulin recognize slightly different sets of glycoproteins, the two function similarly, and so only calnexin will be considered here.

FIGURE 7.31 shows that entry into the calnexin cycle is controlled by specific modification of the carbohydrate structure that is added to a protein as it translocates (see *7.14 Sugars are added to many translocating proteins*). When the structure is first added, one of its several branches ends with three glucose residues. The two distal glucose residues are quickly removed (by glucosidase I and II, respectively), leaving only the proximal glucose. Calnexin then associates with this form of the

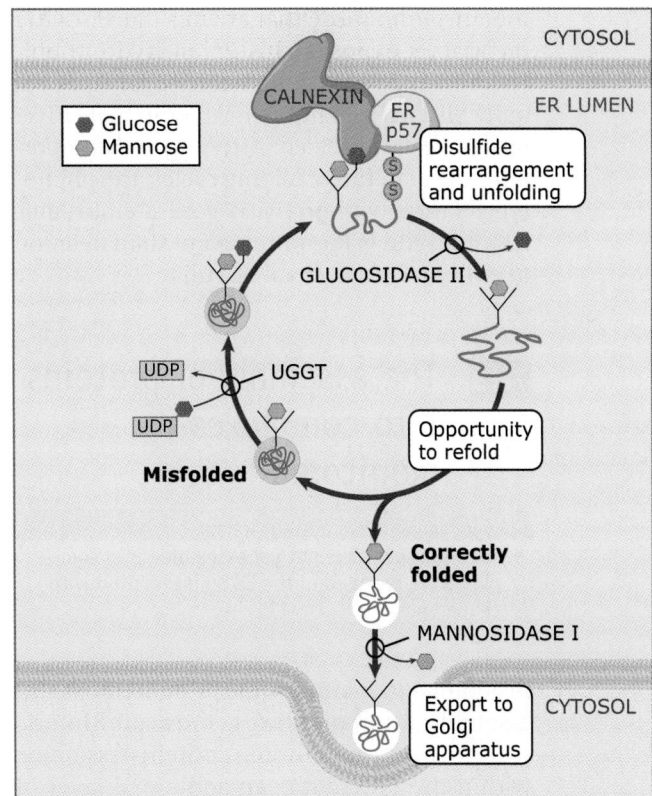

FIGURE 7.31 The presence of a single glucose molecule on an N-linked oligosaccharide allows a misfolded protein another chance to fold by binding to calnexin or calreticulin. The enzyme UGGT adds this glucose residue to misfolded proteins that lack it and must, therefore, be able to discriminate between properly folded and misfolded proteins.

protein, bringing it into contact with ERp57, which catalyzes disulfide bond formation or rearrangement and allows folding or refolding. The requirement for glucose trimming was demonstrated by treating cells with the glucosidase inhibitor castanospermine, which prevented glycoproteins with associating with calnexin or calreticulin.

After ERp57-mediated refolding, the last glucose is removed from the protein by glucosidase II. The unglucosylated protein is no longer a calnexin substrate. However, if the protein has failed to fold properly after a single calnexin cycle, a glucose may be readded by a lumenal enzyme called UGGT (UDP-glucose-glycoprotein glucosyl transferase). This addition allows the misfolded protein to rebind to calnexin and repeat the process. Thus, a protein may have glucose removed and added multiple times, until it achieves its correctly folded form. UGGT plays a critical role in this cycle because it is the sensor that distinguishes correctly folded proteins from those that are misfolded. UGGT recognizes exposed clusters of hydrophobic residues and can reglucosylate sugars that are quite distant from the actual region that UGGT recognizes as unfolded. Attesting to the important role of iterative folding cycles, the phenotype of mice with no UGGT gene is embryonic lethality, and is far more severe than deletion of either calnexin or calreticulin.

7.18 The assembly of proteins into complexes is monitored

Key concept

- Subunits that have not yet assembled into complexes are retained in the ER by interaction with chaperones.

In addition to folding, some proteins must also assemble into complexes. Most multimeric secretory and membrane proteins assemble within the ER, adding an additional layer of complexity to the quality control process. The same types of molecular interactions that influence protein folding also drive oligomerization: hydrophobic associations, disulfide bonding, and ionic interactions. Consequently, the same ER machinery that assists protein folding also monitors assembly. In the absence of assembly, many proteins are retained within the ER in association with specific chaperones. For instance, in antibody-producing B lymphocytes, immunoglobulin heavy chains remain bound to BiP until they become associated with functional light-chain molecules. This ensures that unassembled heavy chains are retained in the ER, and that only fully assembled antibodies exit the ER en route to the Golgi apparatus. **Thiol-mediated retention** is also used to retain unassembled proteins in the ER. Specific cysteine residues that form intermolecular disulfide bonds in an oligomer are somehow monitored as an indication of the assembly state of certain proteins. It is thought that in the absence of oligomerization, these cysteines form intermolecular disulfide bonds with PDI family members, resulting in retention of the incompletely assembled protein in the ER. Finally, some proteins contain unique signals for ER retention in their primary structure. The signals are masked when the proteins are assembled properly, allowing them to leave the ER. For example, the transmembrane domain of the T-cell receptor α chain causes ER retention unless it is part of a fully assembled multimeric receptor.

In addition to the processes already described, other, more substrate-specific forms of quality control operate in the ER lumen. For instance, the proteins prolyl 4-hydroxylase and Hsp47 participate specifically in the folding and assembly of procollagen, which folds into a distinctive triple helical structure unique to members of the collagen family (see *19.2 Collagen provides structural support to tissues*). Substrate-specific quality control proteins often assist in folding and secretion in ways that are not yet understood.

Nascent proteins often interact with more than one chaperoning system in the course of their maturation and assembly. With limited space on the protein, each interaction must be transient. As a consequence, a protein may pass from one chaperoning system to another before leaving the ER. The specific factors with which a protein interacts and the order in which it interacts with them vary from substrate to substrate and are likely to depend on the amino acid sequence. For instance, proteins with glycosylation sites near their N-termini will interact first with calnexin or calreticulin. Proteins with glycosylation sites later in their sequence may instead interact with BiP before progressing to the calnexin/calreticulin pathway. Because several different systems of quality control can guide a protein's folding, blocking its access to a single pathway—for instance, by eliminating glycosylation sites or deleting cysteine residues—often does not prevent the protein from being folded and secreted. Proteins may take up to several hours to progress through these quality control steps.

Although many of the proteins that mediate quality control in the ER are known, their molecular mechanisms of action are still largely unknown. It is clear that they can identify improperly folded proteins but unclear what properties of the proteins they use to make the distinction. Unfolded proteins would seem to be fairly easy to identify, but how are misfolded proteins distinguished from those that are properly folded? In particular, how does UGGT make this distinction? How are properly oligomerized proteins distinguished from improperly aggregated proteins? Answering these and similar questions will be necessary before there is a complete understanding of how quality control in the ER works.

7.19 Terminally misfolded proteins in the endoplasmic reticulum are returned to the cytosol for degradation

Key concepts

- Translocated proteins can be exported to the cytosol, where they are ubiquitinated and degraded by the proteasome—a process known as ER-associated degradation.
- Proteins are returned to the cytosol by the process of retrograde translocation, which is not as well understood as for translocation into the ER.

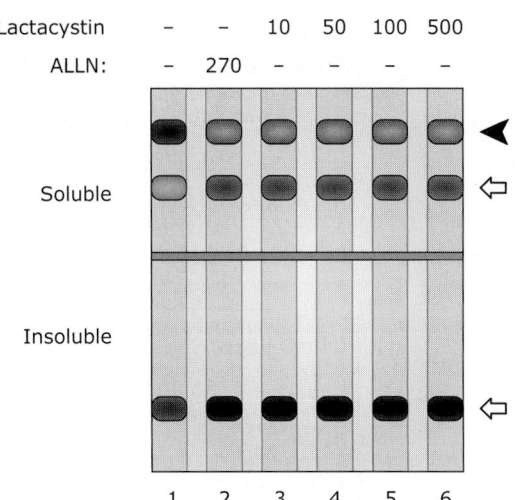

FIGURE 7.32 When cells expressing the CFTR protein were treated with either of two proteasome inhibitors (ALLN or lactacystin), CFTR accumulated in the cell, particularly in immature and insoluble forms that were likely to be misfolded CFTR molecules. Adapted from C. L. Ward, et al., *Cell* 83 (1995): 121–127.

The quality control machinery of the ER ensures that only properly folded secretory and transmembrane proteins are exported, while unfolded proteins are retained by chaperones and given the opportunity to fold correctly. What happens, though, to proteins that never achieve the correct conformation? Somehow, proteins that are unable to fold must be eliminated. As one solution to this problem, the cell uses retrograde translocation (sometimes called *dislocation* or *retrotranslocation*) to export misfolded proteins back into the cytosol. There they are covalently modified with the small protein ubiquitin and degraded by the proteasome, a large protease complex. This degradative pathway is called **ER-associated degradation (ERAD)**.

Discovery of the ERAD pathway began with the understanding that proteins that misfold in the ER are degraded and not allowed to accumulate. Initially, this degradation was presumed to occur within the ER, but despite considerable effort, proteases were never found within the organelle. Instead, evidence began to build that proteins are degraded by the proteasome in the cytosol. First, as shown in **FIGURE 7.32**, chemical inhibitors of the proteasome block the degradation of a newly synthesized integral membrane protein (CFTR, or cystic fibrosis transmembrane conductance regulator). Because the proteasome is a cytosolic protein complex, this observation suggested that the proteins had to be exported from the ER before degradation. Subsequently, it was shown that inhibition of the proteasome could make proteins that had already been completely translocated into the ER accumulate in the cytosol.

How is a protein committed to ERAD? Unlike translocation into the ER, there is no signal sequence in a protein that targets it for ERAD. Rather, the ERAD machinery must recognize physical properties that distinguish terminally misfolded from properly folded or transiently unfolded proteins. For instance, a mutation that destabilizes a protein's final folded form can cause it to be degraded. In fact, the quality control machinery may recognize a mutated protein as a substrate for ERAD even if the mutation has little effect on the function of the protein. Even proteins with no discernible folding defects may be targeted to the ERAD pathway, although there must be cues that the cell recognizes that have yet to be uncovered.

A general pathway for ERAD is shown in **FIGURE 7.33**. The first step in committing a protein to ERAD is recognition of the misfolded form. Although it is not yet known how recognition occurs, chaperones are likely to play a significant role. However, association with a chaperone cannot be the only requirement for ERAD, because chaperones help all nascent proteins to fold. Instead, the signal for retrograde translocation may be related to the length of time that a protein spends associated with chaperones. Chaperones interact with a protein until it folds correctly, so one that cannot fold correctly will remain associated with them much longer than normal. However, it is not generally clear how prolonged association of a protein with chaperones might be

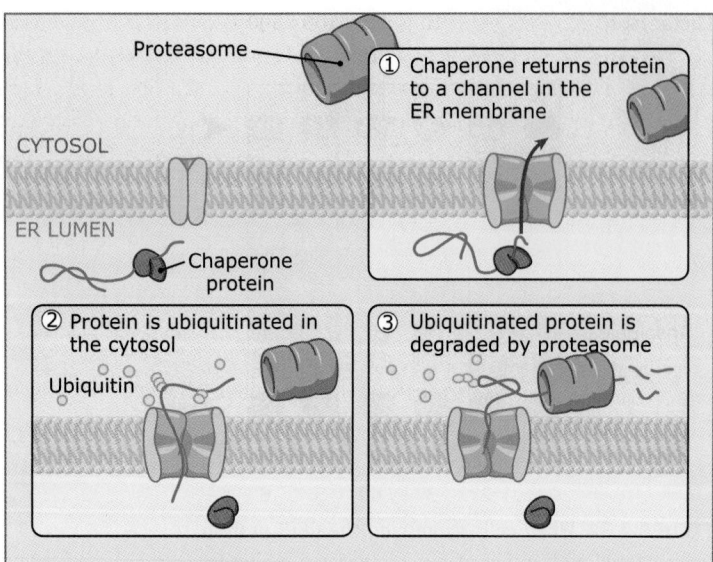

FIGURE 7.33 Terminally misfolded proteins in the lumen of the ER are returned to the ER membrane for retrograde translocation to the cytoplasm. There they are degraded by the proteasome. Lumenal chaperones are likely to play a role in retargeting because misfolded proteins must be associated with them.

Labels in figure:
- Proteasome
- CYTOSOL
- ER LUMEN
- Chaperone protein
- ① Chaperone returns protein to a channel in the ER membrane
- ② Protein is ubiquitinated in the cytosol
- Ubiquitin
- ③ Ubiquitinated protein is degraded by proteasome

sensed. The ER protein α-mannosidase I is a slow-acting enzyme that trims a mannose from the carbohydrates on nascent proteins, and it is proposed that this trimming step is a signal for the cell to either send the protein on to the Golgi if it is properly folded, or to degrade it if it is misfolded. Mannose trimming is thought to make misfolded glycoproteins poorer substrates for calnexin or calreticulin, and better substrates for a protein known as EDEM, that is presumed to be important in shepherding these proteins for ERAD.

After a protein is committed to retrograde translocation, it must cross the ER membrane. The identity of the channel is still somewhat controversial. Some mutations in yeast Sec61p impair ERAD without impacting forward translocation, suggesting that at least some substrates might use the translocon for retrograde transport to the cytosol for ERAD. However, proteins in the process of retrograde translocation can often be found in the proximity of the transmembrane protein derlin-1, which has also been proposed as the retrograde channel. Whatever the identity of the channel, the same challenges facing a translocation substrate also confront an ERAD substrate. For a protein to be exported, it must first be brought to the channel, the channel must open, and the protein must then be moved through it into the cytosol.

Most proteins seem to be targeted for degradation only after translocation is complete. Whether such proteins remain near the translocon until the cell decides whether or not to degrade them is unknown. It is also unknown how misfolded proteins engage the retrograde translocation channel. In fact, the emerging picture seems to suggest that the pathway taken by a substrate out of the ER, and the cytosolic protein required to remove it, depend highly on the type of protein to be degraded and where in the protein the misfolded region is located. It is likely that soluble and transmembrane proteins are degraded by pathways at least partially distinct from each other, at least in part because transmembrane proteins must be extracted from the membrane while soluble proteins must move across it. Even among transmembrane proteins, the pathway taken depends also on whether the mutation is in a lumenal or cytosolic domain, which will influence whether it is recognized as misfolded by ER or cytosolic chaperones.

Once a protein is within the retrograde translocon, it is likely moved through the channel by forces from the cytosol. Ubiquitination plays a role in the export of many (though not all) substrates. Most degradation substrates are polyubiquitinated, and ubiquitination occurs while they are still associated with the membrane. Mutation of components of the ubiquitination machinery leads to the formation of aggregates of misfolded proteins in the ER. This suggests that the channel becomes blocked when the substrate cannot be modified. However, ubiquitination alone is insufficient to extract a protein into the cytosol. A cytosolic ATPase (p97; Cdc48 in yeast) that associates with the ER membrane is also required, and p97 associates with cytosolic cofactors that bind to the ERAD substrate directly. It is not yet clear how this ATPase causes the release of the substrate into the cytosol, but similar ATPases in bacteria and mitochondria bind directly to integral membrane proteins and extract them from the membrane. It is possible that the ATPase required for retrotranslocation acts in a similar manner, binding either directly to the substrate or to its attached ubiquitins, then pulling it from the channel into the cytosol. What role the proteasome might play in the retrograde translocation process is not yet clear.

Regardless of whether or not it is the channel, derlin-1 is essential for the retrograde translocation of some misfolded proteins, and interacts both with the misfolded protein in

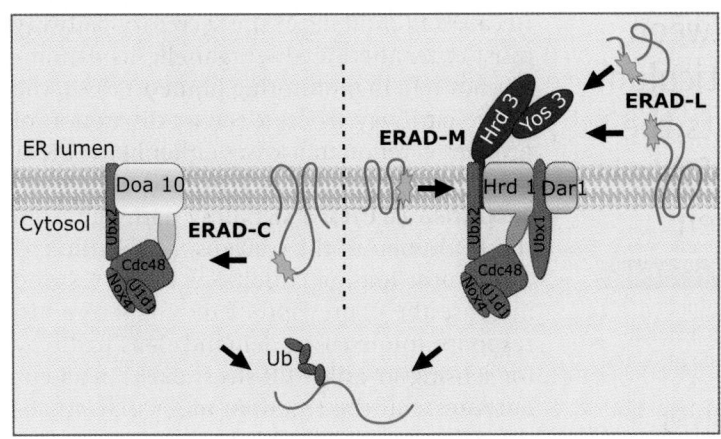

FIGURE 7.34 Whether the problem with protein folding arises in the cytosol (ERAD-C), membrane (ERAD-M) or ER lumen (ERAD-L) determines the specific proteins required for degrading an ERAD substrate. Reprinted from *Cell*, vol. 126, P. Carvalho, V. Goder, and T. A. Rapoport, Distinct ubiquitin-ligase complexes..., pp. 361–373, Copyright (2006) with permission from Elsevier [http://www.sciencedirect.com/science/journal/00928674]. Illustration courtesy of Pedro S. Carvalho, Harvard Medical School.

the ER and with p97 through an association with the VIMP protein. Therefore, derlin is at least one component of the molecular "bridge" between the ER lumen and the degradation machinery in the cytosol, although other components surely exist, as shown in **FIGURE 7.34**. HRD1 is needed for ERAD, and is a type of protein called a ubiquitin ligase that attaches ubiquitin to proteins to be degraded. HRD1 can be found associated with derlin-1 but not Sec61, which is one of the lines of evidence suggesting that derlin-1 might be the retrograde channel.

In a few cases, proteins appear to be identified for degradation even as they are still translocating into the ER. Particularly large substrates might use this form of degradation to avoid the energetic expense of translocating the entire protein if it begins misfolding very early in its synthesis. Designating proteins for retrograde translocation while they are being imported eliminates the need to retarget them. The best-characterized example of this type of degradation is for apolipoprotein B, which is a very large secreted protein that associates in the ER lumen with lipids and fatty acids as the first step in assembly of low-density lipoprotein (see *7.22 Lipids must be moved from the endoplasmic reticulum to the membranes of other organelles*). Failure of an enzyme in the ER lumen to transfer lipids to apolipoprotein B as it translocates causes degradation of the protein to begin before its synthesis and translocation are complete.

When proteins reach the cytosol, they are usually recognized and degraded. In some cases, however, there is evidence that exported proteins instead specifically aggregate in the cytosol. These proteins accumulate in bodies called aggresomes, which might be used to sequester and degrade large amounts of protein. In addition, there is emerging evidence that retrograde translocation with proteasomal degradation is not the sole pathway for disposal of misfolded ER proteins. Evidence exists for degradation pathways both through the lysosome, and through autophagocytic digestion of large regions of ER when there is a pervasive protein folding problem.

The molecular mechanisms involved in ERAD are still poorly understood. Similarly to forward translocation, there is much known about the ERAD pathway taken by a small handful of model ERAD substrates, but the general applicability of these principles is not yet clear. These include such fundamental questions as how misfolded proteins are identified. Recent work has identified the yeast Yos9p protein as an important component of the machinery that targets misfolded proteins to the ERAD machinery, but exactly how it works is not yet clear. Getting a better picture of the process will ultimately require that it be reconstituted *in vitro*.

Concept and Reasoning Checks

1. The CFTR protein is a transmembrane chloride channel located on the plasma membrane. The most common cystic fibrosis mutation results in the deletion of a phenylalanine from the protein. This mutated CFTR protein is recognized by the quality control machinery as misfolded. Why would this cause disease?

2. How do the steps required for a transmembrane protein to be degraded by ERAD differ from those required for a fully translocated protein?

7.20 Communication between the endoplasmic reticulum and nucleus prevents the accumulation of unfolded proteins in the lumen

Key concepts

- The unfolded protein response monitors folding conditions in the ER lumen and initiates a signaling pathway that increases the expression of genes for ER chaperones.
- The protein Ire1p mediates the unfolded protein response in yeast by becoming activated in response to conditions of cellular stress.
- Activated Ire1p splices HAC1 mRNA and results in the production of the Hac1 protein, a transcription factor that localizes to the nucleus and binds to the promoters of genes with a UPR response element.
- The unfolded protein response in higher eukaryotes has evolved more layers of control beyond those seen in yeast.

Even with ER-associated degradation operating at full capacity, under conditions of cellular stress such as elevated temperature, viral infection, or exposure to certain chemicals, misfolded proteins still accumulate in the ER. These threaten to saturate the quality control machinery and interfere with secretion. Or, worse yet, misfolded proteins can form amyloid or other toxic forms of protein aggregates, that are associated with various neurodegenerative diseases such as Alzheimer's Disease.

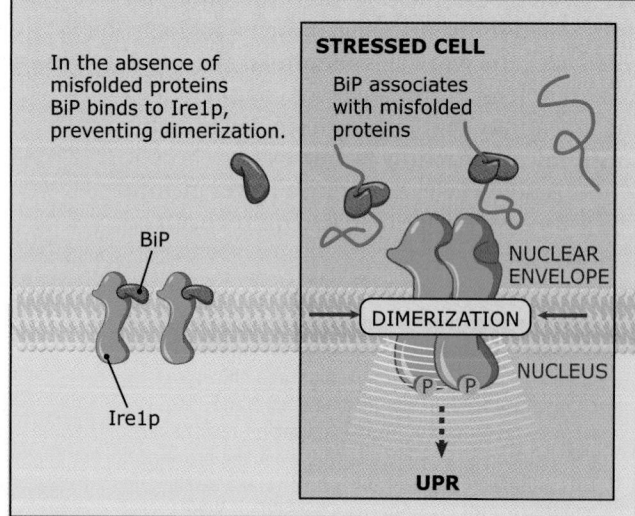

FIGURE 7.35 Under normal conditions, BiP is available to bind to Ire1p and prevent it from dimerizing. In stressed cells, BiP is occupied with unfolded proteins. Ire1p dimerizes to initiate the unfolded protein response.

Because all proteins of the secretory pathway must enter the ER, the organelle has an important role in monitoring homeostasis in the entire pathway. The ER senses disruption of protein secretion that arises either in the ER or at points further along. The **unfolded protein response (UPR)**, a signaling pathway from the ER lumen to the nucleus, allows the cell to monitor folding conditions in the ER and increase the expression of ER chaperones in response to increased demand. This pathway for sensing so-called ER stress exists in all eukaryotes, but the essential molecular details were first described in yeast.

For the yeast UPR, the critical mediator—the protein that senses folding conditions in the lumen of the ER and transmits the information—is a protein in the membrane of the ER called Ire1p. Yeast cells lacking Ire1p are incapable of mounting a UPR in response to ER stress. Ire1p can dimerize by self-association of its lumenal domain, but under normal conditions, BiP binds to that domain and inhibits dimerization. Accumulation of unfolded proteins, however, causes BiP to associate with those proteins instead, and also causes Ire1p to be activated by homo-dimerization, as diagrammed in **FIGURE 7.35**. Signaling occurs through a cytosolic domain of Ire1p, which is a serine-threonine kinase. (It is possible that Ire1p is localized to the subregion of the ER that makes up the nuclear envelope, and that the "cytosolic" domain of Ire1p is therefore actually localized inside the nucleus, although this issue has not yet been definitively addressed.) Dimerization of Ire1p induces autophosphorylation and activation of the cytosolic domain.

As shown in **FIGURE 7.36**, once Ire1p is activated, a second domain in the cytosolic region of the protein catalyzes the removal of an intron from the mRNA of a specific gene, *HAC1*. When Ire1p has removed this intron, tRNA ligase joins the exons to form a new mRNA. Thus, splicing by Ire1p is unrelated to splicing catalyzed by the spliceosome. Splicing of *HAC1* mRNA is critical to the UPR because the intron causes ribosomes to stall on the mRNA and prevents translation of the Hac1p protein. After the intron is removed, however, Hac1p is translated.

The Hac1 protein is a transcription factor that binds to the promoter of the gene encoding BiP and stimulates its transcription. (The yeast homolog of BiP is called Kar2p.) Binding occurs through a specific regulatory sequence called an unfolded protein response element

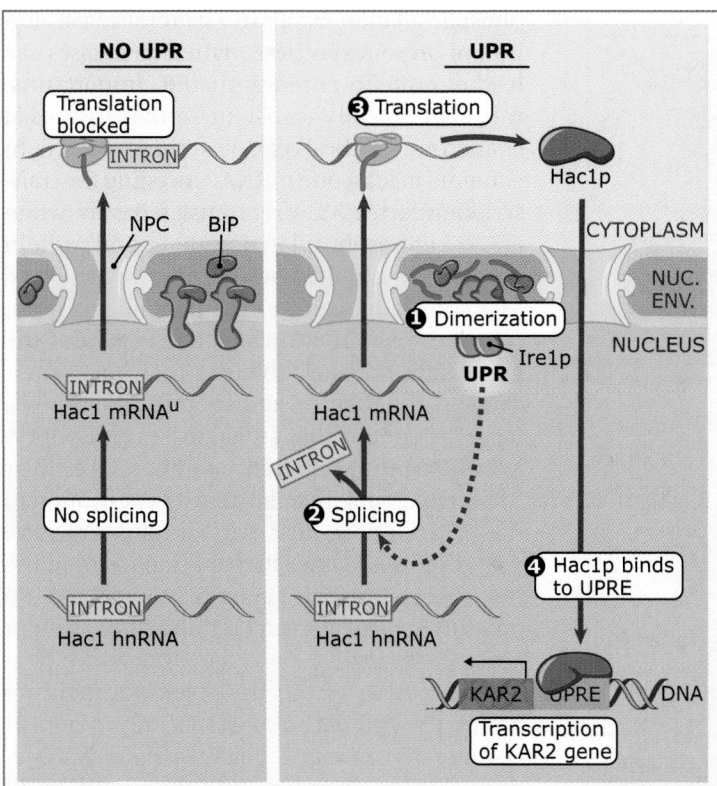

FIGURE 7.36 Dimerization of Ire1p results in the splicing of the Hac1 mRNA, allowing it to be translated. The transcription factor Hac1p is produced, stimulating the expression of genes that encode ER chaperones.

(UPRE). This sequence also appears in the promoter regions of a number of other genes, and Hac1p also stimulates transcription of these genes. Thus, by sensing the presence of misfolded proteins through their association with BiP, Ire1p initiates a cascade of events that leads to the production of additional ER chaperones. These chaperones enhance the ability of the quality control machinery to cope with the burden of unfolded proteins during stressful conditions. The response is much broader than this, however. Although the UPR was originally discovered for its effect on the expression of specific chaperones, the expression of literally hundreds of genes is affected. The nature of the genes induced suggests that elements of translation, translocation, secretion, and ER membrane proliferation are all altered by the cell in order to respond to the problem of unfolded proteins. One such response is the induction of lipid synthesis for ER expansion (see *7.25 The endoplasmic reticulum is a dynamic organelle*).

The UPR in mammals is similar to the UPR in yeast but is more complex. Mammalian Ire1p homologs also catalyze splicing to increase expression of ER chaperones, but the transcription factor whose message is spliced is unrelated to Hac1p. The mammalian factor is called Xbp1. However, mammalian IRE1 appears to be only a minor participant in the mammalian UPR, as

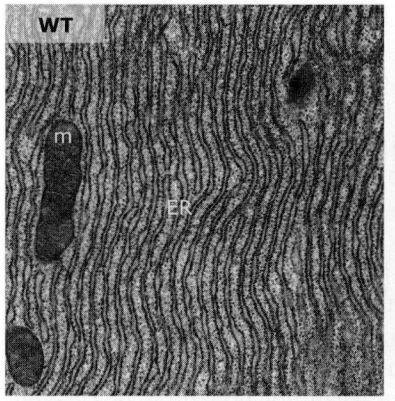
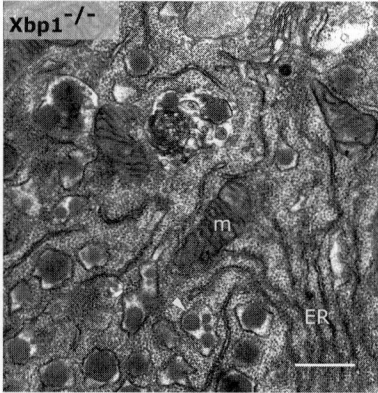

FIGURE 7.37 Compared to a wild-type animal, the pancreatic acinar cells of animals lacking the XBP1 gene are disordered and essentially do not properly secrete digestive enzymes, m = mitochondrion, er = endoplasmic reticulum. Photos courtesy of Ann-Hwee Lee, Harvard University.

cells lacking IRE1 remain capable of activating the UPR during stressful conditions and of upregulating chaperone expression. The current hypothesis is that IRE1 in mammals controls the expression of genes needed to stimulate ERAD. IRE1 and its target XBP1 also seem to have a specific and critical function in the development and differentiation of certain cell types that must dramatically expand their ER in order to accommodate antibody synthesis and secretion (see *7.25 The endoplasmic reticulum is a dynamic organelle*). **FIGURE 7.37** shows how

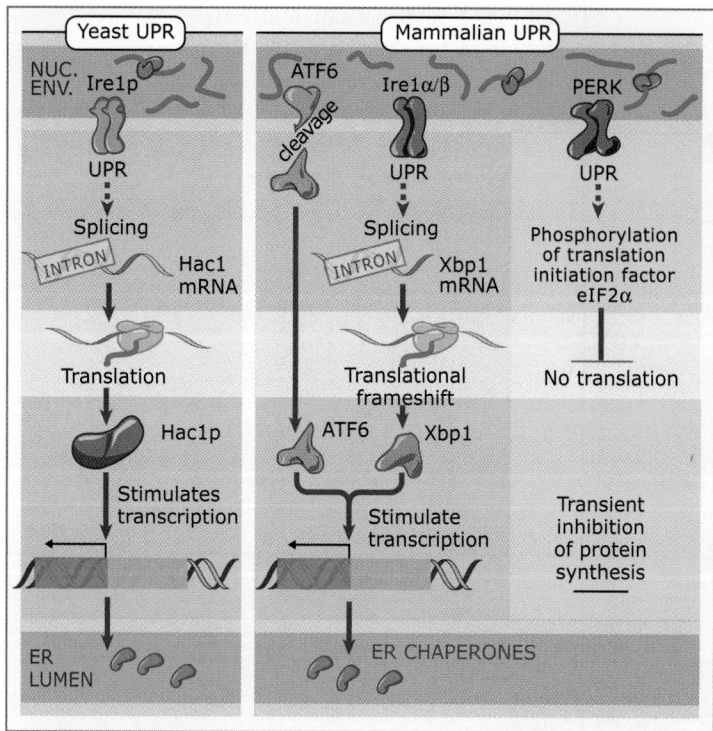

FIGURE 7.38 Several pathways are unique to the UPR of higher eukaryotes. These include inhibition of translation via PERK and cleavage of ATF6 to form a transcription factor.

pancreatic acinar cells, which secrete digestive enzymes, fail to develop appropriately in mice lacking Xbp1.

The mammalian response, depicted in **FIGURE 7.38**, also involves other elements and pathways not present, or not as prominent, in the yeast response. In mammals, ER stress leads to the activation of a second transcription factor, called ATF6. This protein normally spans the ER membrane, but induction of the UPR causes its membrane-spanning domain to be cleaved. The liberated cytosolic domain moves to the nucleus, where it binds to the promoters of genes for ER chaperones. In addition to increasing the capacity of the quality control system, in mammals, ER stress also causes a decrease in protein synthesis, so that fewer proteins are delivered to the ER. This effect is mediated by the PERK protein, a transmembrane ER kinase with a lumenal domain that is homologous to Ire1p but that has an unrelated cytosolic kinase domain. PERK is highly expressed in the pancreas, and its inactivation results in pancreatic dysfunction, indicating the importance of the UPR in organs that secrete large amounts of protein. The activated cytosolic domain of the PERK protein phosphorylates the subunit of the general translation initiation factor eIF2. This

phosphorylation results in a generalized inhibition of protein synthesis, which decreases the load of proteins entering the ER. Importantly, this effect is only transient so that increased chaperone synthesis can be accommodated. In addition, at least one mRNA, encoding the transcription factor ATF4, is translated only when eIF2α is phosphorylated and general protein synthesis is inhibited. ATF4, when synthesized as a consequence of eIF2α phosphorylation, translocates to the nucleus and regulates the expression of genes involved in cellular energetics, redox balance, and amino acid synthesis. It is worth noting that other kinases phosphorylate eIF2α during different conditions of cellular stress (e.g., viral infection leads to activation of the eIF2α kinase PKR), and so the targets of ATF4 are those genes that are important in the response to stresses in general, while the ATF6 and IRE1 pathways can be thought of as more ER stress-specific.

Activation of the UPR is implicated in a broad array of pathologies, including neurodegenerative diseases, diabetes, liver dysfunction, clotting disorders, and many others. In experimental systems, persistent activation of the UPR leads to the activation of cell death pathways. However, many of the diseases in which the UPR is implicated take place over the time course of months or years, suggesting that the UPR, when activated in physiologic situations, must be more of an adaptive response than an apoptotic (i.e., death-promoting) one. Given that the UPR activates both prosurvival and proapoptotic pathways, it is not well understood how survival of a cell can be favored as an outcome under conditions of chronic stress.

7.21 The endoplasmic reticulum synthesizes the major cellular phospholipids

Key concepts

- The major cellular phospholipids are synthesized predominantly on the cytosolic face of the ER membrane.
- The localization of enzymes involved in lipid biosynthesis can be controlled by the cell to regulate the generation of new lipids.
- Cholesterol biosynthesis is regulated by proteolysis of a transcription factor integrated into the ER membrane.

In addition to translocating and preparing proteins for the secretory pathway, the ER is the primary site of synthesis of the cell's phospholipids. These molecules are synthesized in the membrane of the ER and are then distributed among the many separate membranes and membrane-bounded organelles of the cell. These include the plasma membrane, mitochondria, and the organelles of the secretory pathway.

The cell must be capable of expanding its membranes in response to particular needs. Most obvious is the doubling of the plasma membrane and all membrane-bounded organelles that must occur over the course of the cell division cycle. In addition, particular organelles may expand when a high burden is placed on them. For instance, when B-lymphocyte precursors mature into antibody-secreting plasma cells, the ER undergoes a dramatic expansion to accommodate the increased traffic through the secretory pathway (see *7.25 The endoplasmic reticulum is a dynamic organelle*).

Synthesis of phospholipids from soluble precursors takes place primarily on the cytosolic leaflet of the ER membrane, in a process known as the **Kennedy pathway**. As **FIGURE 7.39** shows, the membrane of the ER grows when two fatty acid molecules linked to acetyl CoA are joined with glycerol 3-phosphate to form **diacylglycerol (DAG)**. In contrast to its precursors, DAG is sufficiently hydrophobic to insert into the ER membrane.

Once within the membrane, DAG can be joined to a head group to form a complete phospholipid. The major phospholipids differ principally in the nature of their head groups, as shown in **FIGURE 7.40**. All are synthesized by the attachment of a head group to DAG. The head group is first phosphorylated, then attached to cytidine diphosphate (CDP). The head group, along with one of the phosphates, is then transferred to DAG, completing the synthesis process.

The attachment of the choline head group to CDP to form phosphatidylcholine illustrates how the cell can use subcellular localization of proteins to regulate biosynthetic processes. This reaction is carried out by the enzyme cytidylyl transferase (CT), and is the rate-limiting step in phosphatidylcholine biosynthesis. The cell controls phosphatidylcholine synthesis by maintaining CT in two separate pools: an inactive cytosolic pool and an enzymatically active pool on the cytosolic face of the ER membrane.

The mechanism that maintains these pools is not yet well understood. However, it is clear that conditions in the cell that communicate the need for increased phosphatidylcholine

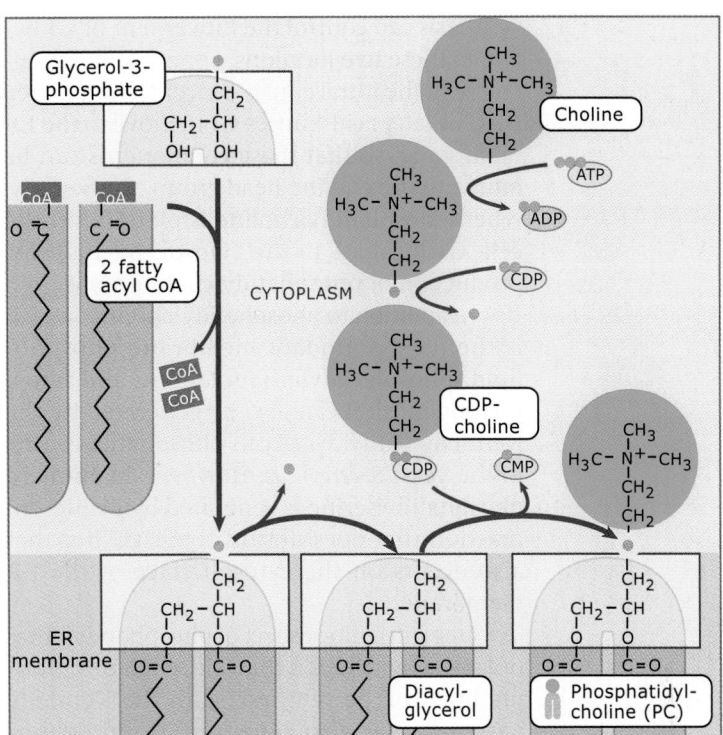

FIGURE 7.39 A phospholipid molecule is formed when water-soluble cytoplasmic components (glycerol-3-phosphate, fatty acyl CoA, and a head group) are combined. The new, lipid-soluble molecule is added to the membrane.

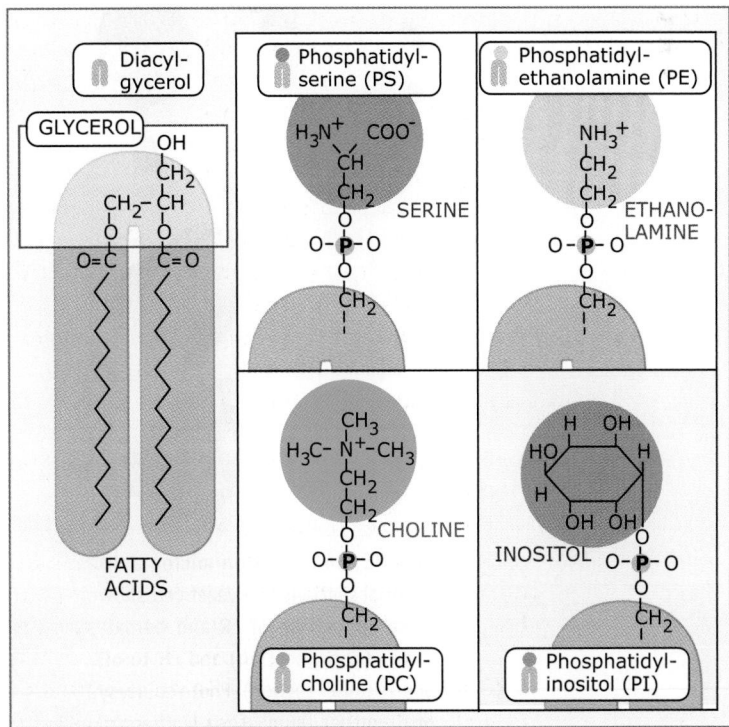

FIGURE 7.40 The major membrane phospholipids differ in the composition of their head groups.

synthesis can control the movement of CT between these two locations. For example, an increase in the intracellular concentration of free DAG or fatty acids causes CT to move to the ER membrane, so that these components can be joined to the choline head group. Conversely, when phosphatidylcholine is plentiful in the cell, CT localizes to the cytosol. In this way, production of phosphatidylcholine stops.

In addition to phosphatidylcholine, which is the most abundant membrane phospholipid, phosphatidylethanolamine and phosphatidylinositol can also be generated by the Kennedy pathway, as can phosphatidylserine in the yeast *S. cerevisiae*. However, in animals, phosphatidylserine is generated by a different reaction that does not involve CDP but that also occurs on the cytosolic face of the ER membrane.

One intriguing aspect of phosphatidylethanolamine synthesis is that, although this phospholipid can be synthesized in the Kennedy pathway, it is also generated in mitochondria. There, it results from the modification of phosphatidylserine synthesized in the ER.

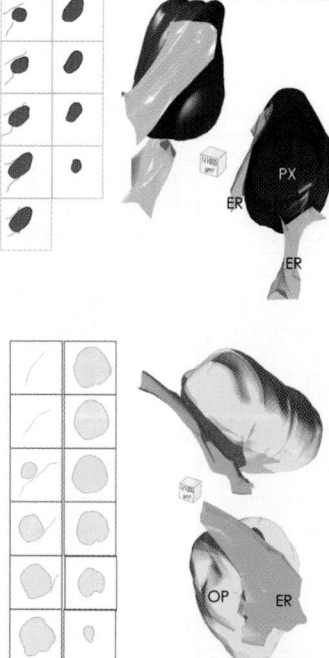

FIGURE 7.41 Electron micrographs of serial sections of a yeast cell show close apposition of ER and peroxisomes (upper section) and ER to oil bodies (lower section). Photo courtesy of Guenther Daum, Graz University of Technology and Guenther Zellnig, University of Graz, Austria.

Production of phosphatidylethanolamine in a different cellular compartment from that of its precursor requires a mechanism for lipid transport between these two locations. Evidence suggests that this transport takes place in a specialized subdomain of the ER called the **mitochondrial-associated membrane (MAM)**. The MAM is a region of the ER that physically associates with a mitochondrion, flattening out over its surface and making very close contact with its membrane. This association can be seen by electron microscopy, as shown in **FIGURE 7.41**, and also demonstrated biochemically by enzymes that occupy the MAM being purified with mitochondria rather than ER. The enzymes of phosphatidylserine synthesis are concentrated in the MAM. Although the mechanism is not yet understood, it seems likely that this region of contact enables phosphatidylserine to be rapidly transferred to mitochondria.

In addition to its role in synthesizing phospholipids, the ER is the site of sterol synthesis. Cholesterol is the major sterol of cell membranes; its biogenesis occurs via multiple steps. The first several steps take place in the cytosol, and the rest within the ER membrane. The ER also houses several elements required by the cell to regulate stimulation and inhibition of cholesterol biogenesis. The key mediator of this regulatory pathway is a protein called sterol regulatory element-binding protein 2 (SREBP2). This protein is ordinarily integrated into the ER membrane, with a small lumenal loop connecting two transmembrane domains, and cytosolically oriented N- and C-termini. Also localized in the ER is a protein called SCAP (for SREBP cleavage-activating protein). SCAP is responsive to the levels of intracellular cholesterol. When increased cholesterol synthesis is needed, SCAP is thought to escort SREBP2 from the ER to the Golgi apparatus. There, proteins cleave SREBP2, liberating the protein's N-terminus into the cytosol. (These are the same proteases that cleave ATF6 during ER stress [see *7.20 Communication between the endoplasmic reticulum and nucleus prevents the accumulation of unfolded proteins in the lumen*].) The N-terminal domain of SREBP2 then moves into the nucleus, where it acts as a transcription factor, activating the expression of genes in the cholesterol biosynthetic pathway. A related protein, SREBP1, is regulated in a similar fashion, and controls the synthesis of triglyceride. Triglyceride is the principle form in which fat is stored and transported in the body, and it is also synthesized by ER-resident enzymes.

7.22 Lipids must be moved from the endoplasmic reticulum to the membranes of other organelles

Key concepts

- Each organelle has a unique composition of lipids, requiring that lipid transport from the ER to each organelle be a specific process.
- The mechanisms of lipid transport between organelles are unclear but might involve direct contact between the ER and other membranes in the cell.
- Transbilayer movement of lipids establishes asymmetry of membrane leaflets.

After lipids are synthesized at the ER, they must be transported to the other membranes of the cell. This process is complicated by the fact that the membranous organelles differ in the lipid composition of their membranes; thus, transfer must be a specific process rather than just a nonspecific exchange of lipids between membranes. Consequently, there must exist mechanisms not only for moving lipids around the cell but also for directing this transfer so that specificity is achieved.

Although the mechanisms of lipid transport are still unclear, several types of movement are proposed. One possible mechanism is through direct contact of membranes, such as occurs between the ER and mitochondria at the MAM. Presumably, a close juxtaposition of the membranes of two organelles could allow phospholipids to move more easily between them. This mechanism of lipid distribution is plausible because the ER is known to make direct contacts with all of the major organelles of the cell. These include the plasma membrane, the *trans*-Golgi network, peroxisomes, vacuoles, and endosomes/lysosomes. Although the functional significance of these contacts is not as well defined as for the MAM, if they are indeed analogous, they could provide a means both for moving lipids from one organelle to another and for ensuring specificity in the lipid transfer process.

Other hypotheses are proposed to account for lipid transfer. The most common, until recently, involved phospholipid transfer proteins. These proteins were identified in *in vitro* experiments on the basis of their ability to exchange lipids between membranes. The proteins extract a lipid from one bilayer and diffuse with it in a hydrophobic binding pocket until they encounter another membrane. Although these proteins could conceivably participate in directed transfer of phospholipids between organelles, they cannot account for expansion of the lipid content of an organelle, because whenever a lipid dissociates from the protein into a membrane, another lipid from that membrane then associates with the empty protein. Evidence of a role for these proteins in lipid transport in vivo is lacking, with the notable exception of ceramide transport protein (CERT). CERT transfers the lipid ceramide from the ER to the Golgi in a directed fashion, for the synthesis of the membrane lipid sphingomyelin (SM). CERT was originally identified in cultured mammalian cells resistant to a cell toxin that requires SM to exert its effect; a clone of resistant cells exhibited a defect in SM biosynthesis. Overexpressing the gene encoding CERT restored the sensitivity of these cells to the toxin. CERT binds to ceramide *in vitro*, and an inhibitor of CERT blocks SM synthesis in cells. CERT might function at spaces of close contact between the ER and Golgi.

It has also been suggested that new lipids are distributed along with proteins as vesicles move through the secretory pathway. This mechanism seems unlikely, however, because drugs that prevent the movement of vesicles through the pathway do not block the movement of lipids from the ER to the plasma membrane. Secretory vesicles could also not deliver lipids to mitochondria and chloroplasts, because the ER is not even connected to these organelles via the secretory pathway.

In some cases, lipids synthesized in the ER are exported to other cells. To achieve this, certain specialized cell types move specific lipids through the secretory pathway in the form of **lipoproteins**. Lipoproteins are large aggregates of protein and lipid that are used to transport water-insoluble material, especially cholesterol and triacylglycerols, through the bloodstream. They contain a core of cholesterol and triacylglycerols surrounded by a monolayer of phospholipid and protein. There are several types of lipoproteins, characterized by their density. Of special interest here are low-density lipoproteins (LDLs), because these are synthesized in the lumen of the ER of liver and intestinal cells and then transported through the secretory pathway out of these cells. The lipids of very low-density lipoproteins (VLDLs), the precursors of LDLs, are synthesized on the cytosolic face of the ER, and then presumably flipped across the ER membrane. In the ER, lipids interact with the ApoB protein as it translocates into the ER lumen. Once assembled, the VLDL particle is moved through the secretory pathway for secretion. If the particle fails to assemble properly in the ER lumen, ApoB is returned to the cytosol by retrograde translocation, where it is degraded (see *7.19 Terminally misfolded proteins in the endoplasmic reticulum are returned to the cytosol for degradation*).

7.23 The two leaflets of a membrane often differ in lipid composition

Key concepts

- Movement of lipid molecules between the leaflets of a bilayer is required to establish asymmetry.
- Enzymes ("flippases") are required for movement of lipids between leaflets.

Lipids are often asymmetrically distributed between the two leaflets of a membrane. For instance, phosphatidylcholine is enriched in the extracellular leaflet of the plasma membrane, but phosphatidylethanolamine and phosphatidylserine predominate in the cytosolic leaflet. Certain cellular signaling pathways and cell-cell interactions depend on this asymmetry. The appearance of phosphatidylserine on the extracellular face of the plasma membrane, for instance, is a hallmark of cells undergoing programmed cell death. Lipid asymmetry is maintained because the polar head groups of lipids cannot pass through the hydrophobic interior of the bilayer. As a result, spontaneous movement of lipids between the two leaflets of a membrane is extremely slow.

The inability of lipids to "flip" spontaneously requires the existence of enzymes called flippases to catalyze the movement. At least some of these enzymes must be present in the membrane of the ER because lipid synthesis is restricted to its cytosolic leaflet. Without catalyzed flipping, only that leaflet would be able to expand as new lipids were synthesized. Although there is ample biochemical evidence for the existence of flippases, purification and characterization of these proteins is difficult. It has not been established how widespread they are within the cell or where they are located.

7.24 The endoplasmic reticulum is morphologically and functionally subdivided

Key concepts

- The ER is morphologically subdivided into specialized compartments, including the rough ER for protein secretion, the smooth ER for steroidogenesis and drug detoxification, and the sarcoplasmic reticulum for calcium storage and release.
- The functions of the smooth ER can be specialized according to the needs of the particular cell type.
- The ER may also be subdivided at the molecular level, in ways not morphologically evident.

Examination of the ER by electron microscopy shows that the structure of the organelle is remarkably heterogeneous. In some regions of the cell, it forms large, flat sheets (*cisternae*), that often run parallel, and in some cases are stacked very closely together, as shown in Figure 7.1. In other regions, it takes the form of long, curving tubules. These two forms of the organelle are interconnected and form a single, continuous structure. The sheets and stacks are generally found in proximity to the nuclear envelope, which is part of the ER. The tubules extend as a network throughout the cell and make contact with the plasma membrane and the membranes of other organelles.

The structural diversity of the ER reflects the spatial separation of its functions into specialized subdomains. This specialization can be viewed as a continuation of the same process that first specialized a part of the plasma membrane for secretion and led to the formation of the ER. Once formed, the organelle acquired new functions and its own membrane subdivided into specialized regions. While it is clear that the ER is highly subdivided on the basis of its functions, the mechanisms used to maintain these divisions are still unclear.

The ER network is divided between **smooth ER** and **rough ER**. Rough ER is uniformly covered with attached ribosomes, which are translating proteins to be secreted or inserted into its membrane. These ribosomes are clearly visible by electron microscopy and give the membrane a studded or rough appearance (hence its name). Rough ER is particularly abundant in cells that secrete large amounts of protein. These include endocrine cells, such as the β cells of the pancreas, which secrete hormones into the bloodstream; and B lymphocytes, which produce and secrete antibodies. As **FIGURE 7.42** illustrates, in these types of cells, the rough ER can expand to occupy much of the cell's interior, whereas the rough ER is less prevalent and less organized in many other cell types.

The smooth ER, in contrast, lacks ribosomes, as shown in **FIGURE 7.43**. In most cells, the smooth ER is largely restricted to the region where protein-containing vesicles bud from the ER for transport to the Golgi apparatus. This region is known as the **transitional ER**. The smooth ER, when broadly defined as any ER without ribosomes, is also responsible for several other functions, including lipid metabolism, steroid synthesis, glycogen metabolism, and drug detoxification. These functions are more specialized than those of the rough ER, and the smooth ER is typically much less abundant than the rough ER. In cells that use one or another of these processes extensively, though, the smooth ER can expand to become as common as the rough ER. For example, Leydig cells of the testes utilize a vast network of smooth ER in the synthesis of testosterone. Similarly, an extensive network of smooth ER in liver hepatocytes houses cytochrome P450s and other classes of enzymes that detoxify chemicals for excretion. In these cells, the amount of smooth ER fluctuates in response to exposure to certain drugs, increasing when they are applied and decreasing again when they are removed. Despite the clear morphologic differences of the rough and smooth ER, there is still no good idea how the divisions between these domains of the ER are maintained, and there may be overlap in their functions.

In skeletal muscle cells, the smooth ER is further differentiated into a subdomain called the **sarcoplasmic reticulum (SR)**. This subdomain is enormous compared to other parts of the ER and forms an extensive network that surrounds the sarcomeres of the muscle, as pictured in **FIGURE 7.44**. Calsequestrin, a protein with multiple low-affinity calcium-binding sites, stores intracellular calcium in the SR. Muscle contraction is stimulated when calcium channels in the membrane of the SR open, releasing calcium into the cytosol. The calcium is then rapidly taken back up by calcium-specific pumps in a different portion of the SR. Thus, the SR can be viewed as a type of smooth ER specialized for calcium uptake, storage, and release. Although it has been best studied in skeletal muscle, the SR has less extensive counterparts in other cell types that also require rapid calcium regulation.

Although the SR has evolved specifically for rapid calcium release, calcium uptake and storage are major functions of the ER in most

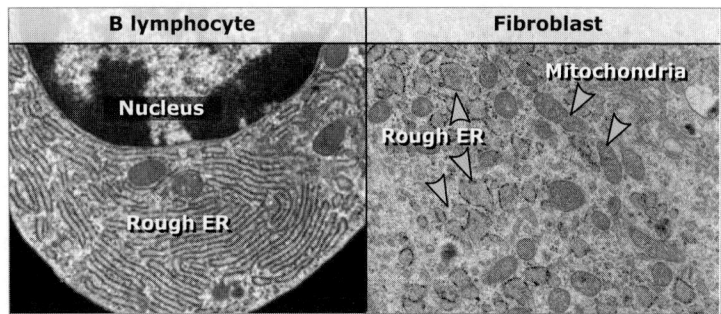

FIGURE 7.42 Transmission electron micrographs showing the morphology of the rough ER. In some cells, such as B lymphocytes, which synthesize large quantities of immunoglobulin, the rough ER occupies a large amount of the cytoplasm and takes the form of flattened stacked sheets. In contrast, the rough ER of an embryonic fibroblast, which secretes collagen and other extracellular matrix proteins, appears more as tubules. Left photo courtesy of Dr. Don W. Fawcett. Reprinted from *The Cell*, 1981; right photo courtesy of Tom Rutkowski, The University of Iowa.

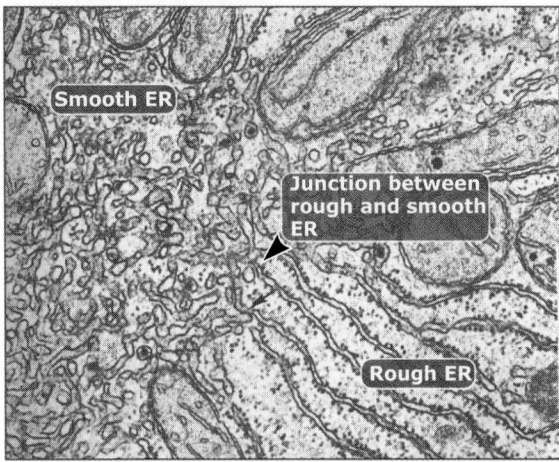

FIGURE 7.43 An electron micrograph of a hepatocyte in the liver. Hepatocytes detoxify drugs in an extensive network of smooth ER, which is visible in the left half of the picture. The smooth ER has the form of curving tubules and often appears as circles or ovals in cross section. Connections between the smooth and rough ER are indicated. Note how clearly the two types of ER are separated into different regions. Photo © Bolender, 1973. Originally published in **The Journal of Cell Biology,** 56: 746–761. Used with permission of Rockefeller University Press. Reprinted courtesy of Dr. Don W. Fawcett. *The Cell*, 1981. Micrograph courtesy of Ewald R. Weibel, University of Berne.

cell types. This role is particularly important, given the use of transient changes in intracellular calcium concentrations in cell signaling pathways (see *18 Principles of cell signaling*). In nonmuscle cells, calcium is bound within the ER by calreticulin, which serves the dual role of calcium-binding protein and glycoprotein chaperone (see *7.17 The calnexin/calreticulin*

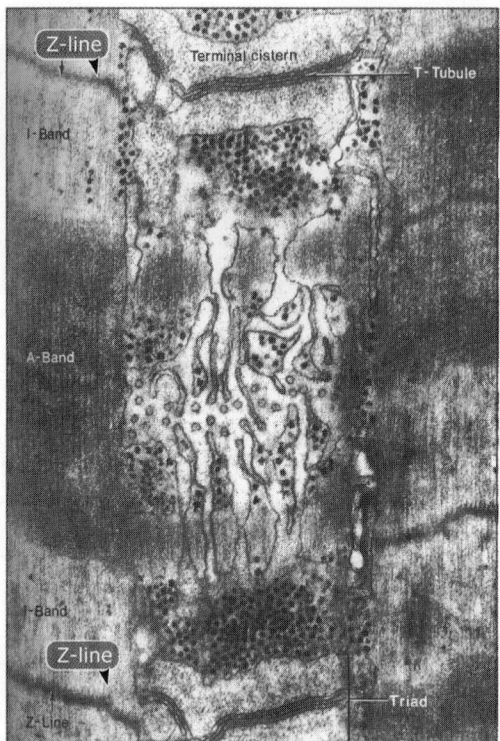

Z-line

Terminal cistern

T - Tubule

I - Band

A-Band

I - Band

Z-line

Z - Line

Triad

FIGURE 7.44 The sarcoplasmic reticulum, evident in this electron micrograph of a skeletal muscle cell, is a version of the smooth ER specialized for calcium storage and release during muscle contraction. Three myofibrils run from top to bottom in the picture. An extensive network of sarcoplasmic reticulum surrounds the myofibril in the middle. Photo © Reese, 1965. Originally published in **The Journal of Cell Biology**, 25: 209–230. Used with permission of Rockefeller University Press. Reprinted courtesy of Dr. Don W. Fawcett. *The Cell,* 1981. Micrograph courtesy of Lee D. Peachey, University of Pennsylvania.

chaperoning system recognizes carbohydrate modifications). Mice lacking calreticulin do not undergo proper heart development. The heart is essentially a specialized form of muscle, and this result attests to the role of calreticulin in maintaining calcium homeostasis. Other abundant ER lumenal proteins, including BiP, Grp94, ERp72, PDI, and calnexin, bind calcium *in vitro* and might also contribute to the organelle's ability to store calcium. Like calsequestrin in the SR, these proteins bind calcium at multiple sites and with low affinity. Because of the role of the ER in calcium storage, the chaperones in its lumen generally require calcium to function optimally. For example, BiP, calnexin, and calreticulin all function less efficiently when ER calcium stores are depleted—so much so that prolonged calcium depletion can induce the unfolded protein response.

Calcium release from the ER is important not only for initiating signaling cascades, but also for triggering apoptosis under certain circumstances. Regions of the ER that closely associate with mitochondria (see *7.21 The endoplasmic reticulum synthesizes the major cellular phospholipids*) seem to allow the mitochondria to rapidly take up calcium released by the ER, which is thought to spur changes in mitochondria that lead to cell death.

Although subdomains of the ER such as the smooth and rough ER are readily apparent morphologically, the organelle also has other, less visible heterogeneity. For instance, despite the continuity of the nuclear envelope and the ER, some proteins are restricted to the inner nuclear membrane and are not found in the outer nuclear membrane or in the rest of the ER. This restriction is apparent from visualization of such proteins in cells. The localization of these proteins is usually based on their physical association with the nuclear lamina or with chromatin. Other functional subdomains of the ER may be present at points where it makes contact with other organelles. Phosphatidylserine synthase activity is enriched in regions of the ER membrane associated with mitochondria. Also, several proteins localize to the points where the ER makes close contact with peroxisomes. The ER is also heterogeneous in its distribution of phospholipids, cholesterol, and even ribosomally associated mRNAs that encode translocated proteins. Any functional significance of the heterogeneity of these components is not yet understood, though. Now that many of the molecular components that play a role in functions within the ER are identified, the mechanisms by which those components are spatially restricted can be addressed.

7.25 The endoplasmic reticulum is a dynamic organelle

Key concepts

- The extent and composition of the ER change in response to cellular need. The ER moves along the cytoskeleton.
- The protein composition of the ER influences its shape, most notably through the action of reticulons that induce ER membrane curvature to form tubules.
- The signaling pathways that control ER composition are not yet understood but may overlap with the unfolded protein response.

Like many other organelles, the ER must adapt to the changing needs of the cell. Subregions of the ER must expand and contract as necessary, and the cell must be able to move and reorganize the organelle, as well as maintain its shape. During mitosis, the ER must be partitioned to the daughter cells. Although relatively little is known about how these processes are achieved at the molecular level, it is clear that the organelle is very dynamic.

The shape of the ER changes constantly. In particular, the tubular elements of the ER are in continual flux, elongating, retracting, branching, and fusing. These movements are somehow related to the cytoskeleton; in mammalian cells, the tubules of the ER often align with microtubules, visibly extending and retracting along them. In yeast, the ER is frequently associated with actin filaments. However, disrupting the cytoskeleton does not cause the ER to break down or make its extensive network of tubules collapse. Neither is the cytoskeleton required for the formation of tubules or networks *in vitro*. Thus, it appears that the formation of tubules and their characteristic organization into a network are an inherent property of the ER and do not depend on the cytoskeleton. Rather, the role of the cytoskeleton seems to be in ensuring that after the ER network forms, it is reliably distributed throughout the cell.

The mechanisms by which the cell is able to form or expand its ER are unknown, but several key features are known. *In vitro*, generating ER tubules requires ATP and GTP, and possibly a protein similar to one (N-Ethylmaleimide-Sensitive fusion [NSF]) that participates in ER-Golgi vesicle fusion. These requirements suggest that proliferation of the ER involves fusion events that convert small ER vesicles into a large, interconnected network. Some form of regulation is necessary to make fusion result in tubules instead of an amorphous mass. The tubular shape of the ER is seen in all eukaryotes and has a conserved cross-sectional diameter of 40–70 nm, suggesting that the formation of the shape is controlled and functionally important. The tubular shape of the ER appears to be formed and maintained as tubules at least in part by the presence of proteins on its cytosolic face that impose curvature on the membrane. Both *in vitro* and in vivo, proteins of the reticulon family promote the formation of tubular ER at the expense of cisternal ER. It is proposed that these proteins form hairpin wedge-like structures that

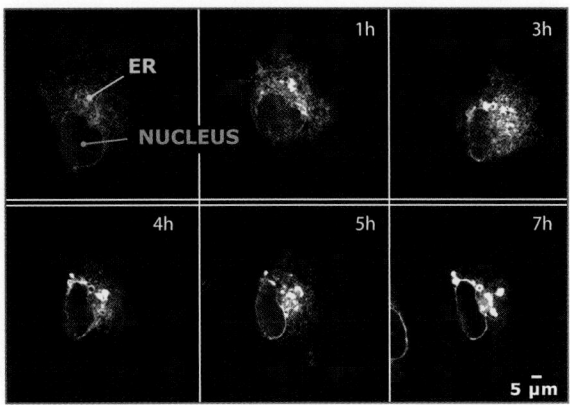

FIGURE 7.45 Overexpression of a fluorescent transmembrane protein leads to alteration of ER structure over a time course of several hours. Photos © Snapp, 2003. Originally published in **The Journal of Cell Biology,** 163: 257–269. Used with permission of Rockefeller University Press. Photos courtesy of Jennifer Lippincott-Schwartz, National Institutes of Health.

give the ER its tubular cross-sectional shape. The roles reticulons play besides shaping the ER, if any, are not clear. Further, it is also not known how the flattened stacks of the cisternal ER remain a stereotypical distance apart from each other. One possibility is that weak interactions between ER-resident transmembrane proteins ensures a relatively constant separation of ER cisterne. Manipulating the protein composition of the ER can be shown to influence its structural organization as shown in **FIGURE 7.45**.

The need for the cell to regulate the size and shape of its ER is underscored by the behavior of the organelle in cells that place extensive demands on their ER network. For example, in comparison to other cell types, those with a high demand on the secretory pathway, such as endocrine cells, undergo a dramatic expansion of the rough ER. This could be achieved by a feedback mechanism that communicates a need for increased synthesis of lipids and ER proteins in response to an increase in protein transiting through the ER. Drugs such as phenobarbital that are detoxified by the smooth ER also induce an expansion of the organelle, and their removal brings about rapid return of the smooth ER to its normal size. Another puzzling manifestation of the capacity of the ER to change size is the formation of **karmellae** during the overexpression of certain membrane proteins. An example of a cell in which karmellae have formed is shown in **FIGURE 7.46**. These structures are stacks of ER that are closely associ-

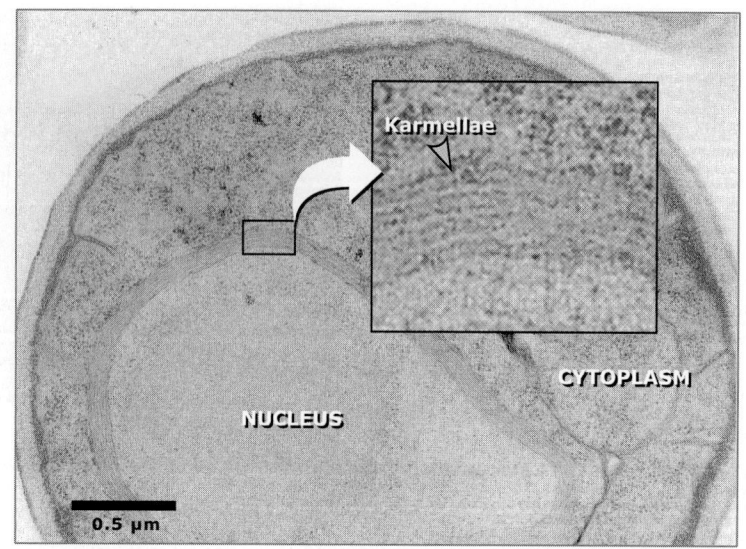

FIGURE 7.46 The overexpression of some transmembrane proteins leads to the formation of stacks of ER called karmellae. This electron micrograph of a yeast cell shows karmellae surrounding the nucleus. Photo © Wright, 1988. Originally published in **The Journal of Cell Biology**, 107: 101–114. Used with permission of Rockefeller University Press. Photo courtesy of Jasper Rine, University of California, Berkeley.

ated with the nuclear envelope and filled with the overexpressed protein. They seem to serve as a reservoir of membrane in which the excess protein can be stored. Again, a pathway of feedback from the ER to gene expression and lipid synthesis seems likely.

Intriguingly, the unfolded protein response, the one signaling pathway already identified between the ER and the nucleus, may play a role in proliferation of the ER. Activation of the UPR in cells not only stimulates the synthesis of chaperones for the ER but also increases the synthesis of inositol. This is the head group of phosphatidylinositol, one of the four major lipids found in membranes. As in upregulation of chaperone synthesis, Ire1p and Hac1p mediate this response in yeast. The synthesis of other phospholipids may be stimulated as well, to allow the cell to increase the total amount of its ER membrane. Thus, the UPR might be responsible for sensing a wide array of changes in the ER and effecting a remodeling of the organelle. Though appealing, this idea is far from proven.

What is the mechanism by which a cell is able to sense the need for ER proliferation without succumbing to ER stress-induced cell death? In the differentiation of B cells into antibody-secreting plasma cells, activation of the UPR and expansion of the ER membrane seems to precede the massive increases in im-

munoglobulin synthesis that occur during this differentiation. This finding suggests that there may be mechanisms by which the UPR can be activated in anticipation of increased ER load rather than strictly in response to it. Likewise, ER-derived compartments in plants are used extensively for storage of various proteins and oils important in seed development and defense against pathogens. These storage proteins are thought to aggregate within the ER-derived compartment, but in a way that leads to their retention rather than their degradation. Upregulation of ER chaperones also seems to precede the accumulation of proteins in such compartments.

Another major challenge in maintaining the ER is partitioning it during mitosis (see *15 Cell cycle regulation*). The organelle cannot be synthesized *de novo* but must originate from preexisting ER, making it essential that it be reliably distributed. It is also advantageous to be able to distribute it equally between the two daughter cells. Were one cell to receive significantly less than half, a delay in the next cell cycle would result, which would be harmful in many situations. In most circumstances, the ER is thought to be maintained as a continuous network throughout mitosis and is apparently just pinched into two roughly equal parts during cytokinesis. In some cases, such as early embryos, the association of the ER with

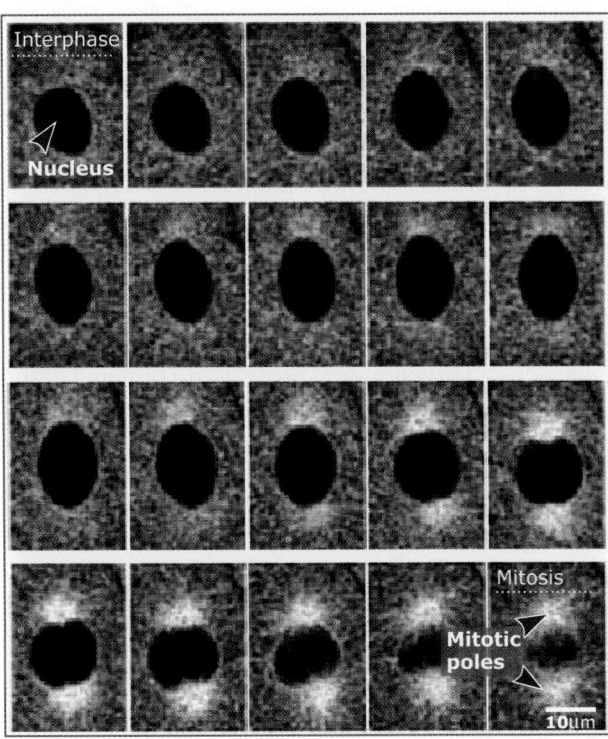

Interphase

Nucleus

Mitosis

Mitotic poles

10μm

FIGURE 7.47 Fluorescence images of the ER in a sea urchin blastomere as it enters mitosis. Interphase is at the upper left and metaphase is at the lower right. The ER remains as a continuous network during cell division and accumulates at the mitotic poles. Expression of a fluorescent protein that is restricted to the ER makes it visible. Photos reproduced from M. Terasaki, *Mol. Biol. Cell* 6 (2000): 897–914. Copyright 2000 by American Society of Cell Biology. Reproduced with permission of the American Society of Cell Biology in the format of Textbook via Copyright Clearance Center. Photos courtesy of Mark Terasaki, University of Connecticut Health Center.

the microtubules of the mitotic spindle may help distribute it evenly. An example of this is shown in **FIGURE 7.47**.

7.26 Signal sequences are also used to target proteins to other organelles

Key concepts

- Signal sequences are used for targeting to and translocation across the membranes of other organelles.
- Mitochondria and chloroplasts are enclosed by a double membrane, with each bilayer containing its own type of translocon.
- Two distinct pathways target matrix proteins to peroxisomes.

The principles underlying protein translocation into the ER also apply to translocation into other organelles. This is true of the nucleus (see *9 Nuclear structure and transport*), mitochondria, chloroplasts, and peroxisomes. Comparing the translocation process into these different organelles reveals that signal sequences and channels are general mechanisms for protein sorting.

Almost all mitochondrial and chloroplast proteins are encoded by nuclear genes and are translated in the cytosol. (~10% to 15% of the

ORGANELLE	SIGNAL	SIGNAL LOCATION
ER	MDPPRPALLALPALLLLLAGARA...	N-terminal
Nucleus	...LAEADRKRRGEFRKE...	Internal
Mitochondria	MLSNLRILLNKAALRKAHTSMVRNFRYGKPVQ...	N-terminal
Chloroplast	MRTRAGAFFGKQRSTSPSGSSTSASRQWLRSSPGRTQRPAAHRVLA...	N-terminal
Peroxisome -PTS1	...VVVGGGTPSRL	C-terminal
Peroxisome -PTS2	MNLTRAGARLQVLLGHLGRP...	N-terminal

Hydrophobic   Acidic   Basic

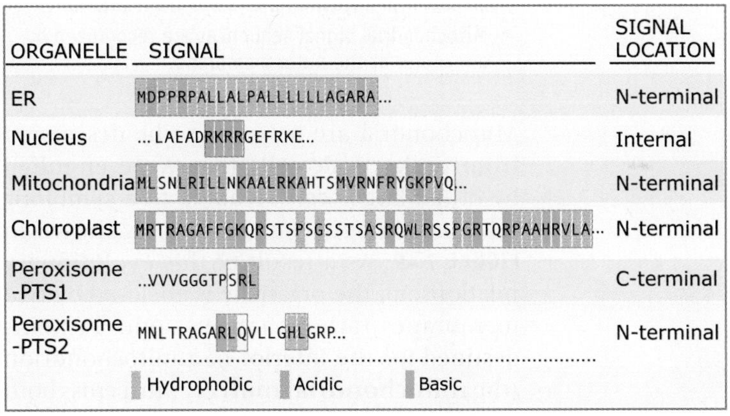

FIGURE 7.48 Several representative signal sequences that are used to target proteins to different organelles. They differ significantly in length and location within the protein, as well as in the distribution of charged and hydrophobic amino acid residues they contain. These differences lead to different physical properties that allow them to be distinguished.

human genome encodes mitochondrial proteins; the genomes of both mitochondria and chloroplasts, in contrast, encode only a small number of proteins that are synthesized within the organelle.) Peroxisomal and nuclear proteins are also synthesized in the cytosol. As with proteins destined for the ER, proteins destined to be localized within the mitochondria and peroxisomes must be targeted to the appropriate organelle and transported across the membrane. **FIGURE 7.48** shows how the specific delivery of proteins to various organelles is possible because proteins target to the ER, mitochondria, chloroplasts, peroxisomes, and the nucleus by using significantly different signal sequences. Each type of signal sequence is recognized by a targeting system and a channel that are unique to the organelle. Most of the components from the several organelles are unrelated to one another, suggesting that protein targeting and import evolved on several different occasions. The discovery that ER proteins target by virtue of a signal sequence led to this hypothesis being rapidly tested and proven for these other organelles as well.

7.27 Import into mitochondria begins with signal sequence recognition at the outer membrane

Key concepts

- Mitochondria have an inner and an outer membrane, each of which has a translocation complex.
- Import into mitochondria is posttranslational.
- Mitochondrial signal sequences are recognized by a receptor at the outer membrane.

Mitochondria are believed to be descended from prokaryotic cells that were engulfed by other, larger cells, leading to a symbiotic relationship between the two, as shown in FIGURE 7.49. As a result of this evolutionary relationship, the organelle is enclosed by two membranes rather than just one. Proteins destined for the interior of a mitochondrion (the **mitochondrial matrix**) must cross both membranes and the space between them (the **intermembrane space**). Each membrane is spanned by a different multiprotein complex that includes a translocation channel. A **TOM** (translocase of the outer membrane) complex spans the outer membrane and a **TIM** (translocase of the inner membrane) complex spans the inner one. These two complexes associate physically at specific points of contact between the two membranes but are capable of acting independently.

Mitochondrial proteins are targeted to the membrane posttranslationally by recognition of an N-terminal signal sequence, usually twenty to fifty amino acids long. Mitochondrial signal sequences are enriched in both basic (i.e., positively charged) and hydrophobic amino acids. The sequences form amphipathic helices, with the charged residues on one side and the hydrophobic residues on the other. The signal sequence of a mitochondrial protein is first recognized by Tom20, an integral membrane protein that is a component of the TOM complex. Tom20 binds the hydrophobic face of the signal sequence in a shallow groove (unlike the deep and flexible groove used by SRP during targeting to the ER). As FIGURE 7.50 illustrates, the interaction between the signal sequence and Tom20 is a weak one that positions the positively charged residues of the signal sequence outside of the binding pocket. The interaction of the signal sequence with the TOM complex is then strengthened by a second protein, Tom22. Tom22 contains an acidic cytosolic domain that interacts with the exposed basic face of the signal sequence. Principally as a result of these two interactions, the protein is brought into contact with the TOM complex.

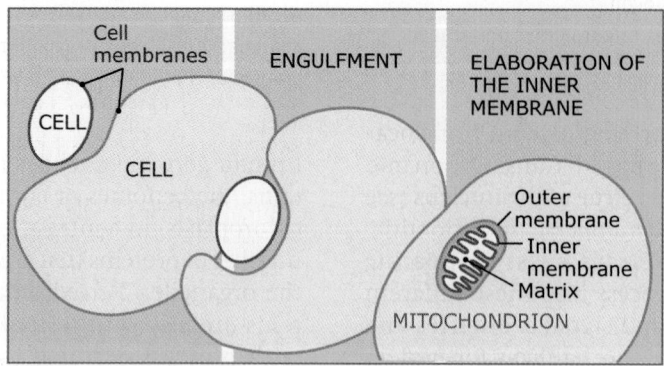

FIGURE 7.49 Mitochondria may have evolved from prokaryotic cells that were engulfed by larger cells and then became specialized to produce energy. This would explain why each is surrounded by two membranes and contains its own genome.

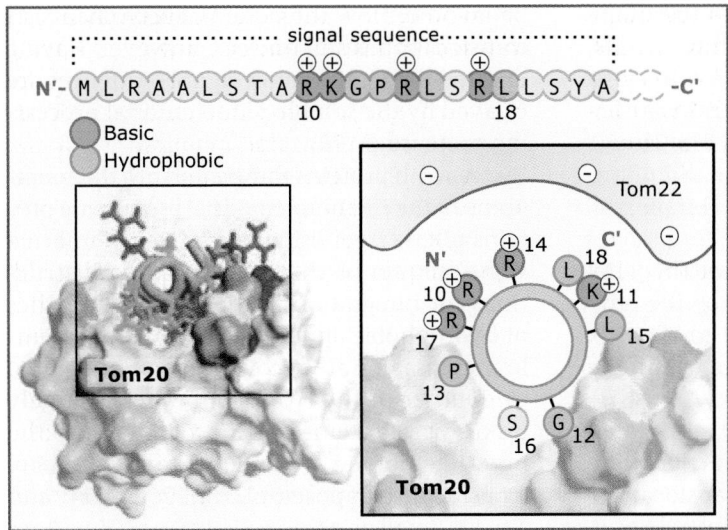

FIGURE 7.50 A mitochondrial signal sequence has basic and hydrophobic surfaces when arranged as a helix. A typical mitochondrial signal sequence is shown at the top, and a view of it end-on, arranged as a helix, is on the right. Its hydrophobic face interacts with a hydrophobic cleft in Tom20; its positive amino acids interact with the acidic domain of Tom22. At left is the structure of Tom20, bound to a signal sequence, with the hydrophobic surfaces of the protein in yellow. Reprinted from *Cell,* vol. 100, Y. Abe, et al., Structural Basis of Presequence Recognition..., pp. 551–560, Copyright (2000) with permission from Elsevier [http://www.sciencedirect.com/science/journal/00928674]. Photo courtesy of Daisuke Kohda, Kyushu University.

During targeting, mitochondrial proteins are kept unfolded and competent for translocation by association with cytosolic hsp70. The Tom70 protein serves as an alternate receptor to Tom20 for mitochondrial proteins, principally those that target by an internal signal sequence (i.e., transmembrane proteins of the outer membrane).

7.28 Complexes in the inner and outer membranes cooperate in mitochondrial protein import

Key concepts

- The TOM and TIM complexes associate physically, and the protein being imported passes directly from one to the other.
- Hsp70 in the mitochondrial matrix and the membrane potential across the inner membrane provides the energy for import.

The passage of a mitochondrial protein through the TOM and TIM complexes is illustrated in FIGURE 7.51. The TOM complex is composed of several proteins in addition to Tom20 and Tom22, with Tom40 likely to form the actual channel. The channel is probably stabilized by several so-called Small TOMs. Currently, neither the recognition of the translocating protein by the channel nor the gating of the channel is well understood. However, after it enters the channel, the basic signal sequence of the substrate might be electrostatically attracted to

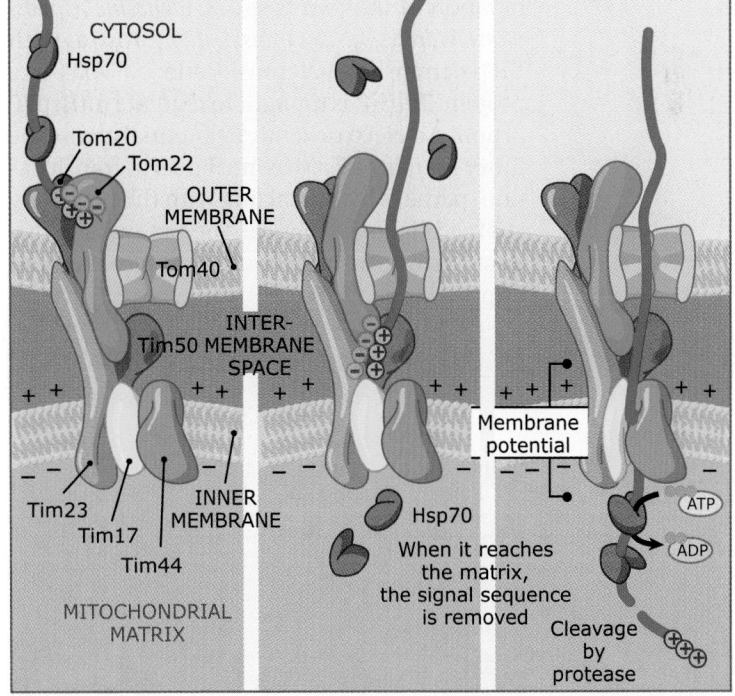

FIGURE 7.51 A protein to be translocated into the mitochondrial matrix is recognized at the outer membrane and then passed directly between channels in the outer and inner membranes. When it reaches the matrix, the signal sequence is removed. Transport is driven both by ATP hydrolysis by chaperones in the matrix and by the membrane potential across the inner membrane. The Tim23 protein is integrated into both membranes and links the channels in them. Adapted from M. J. Baker, et al., *Trends Cell Biol.* 17 (2007): 456–464.

the acidic domains located in the intermembrane space of both Tom22 and the TIM complex protein Tim23. The proximity of Tom22 and Tim23 ensures that the protein is passed between complexes without dissociating into the intermembrane space. There is evidence that Tim23 may be integrated into the outer membrane at its N-terminus, which would bring the TIM and TOM complexes into close association and facilitate the efficient transfer of the chain from one to the other. The Tim50 protein also is thought to play a role in guiding precursor proteins from the TOM complex to the TIM23 channel.

After the chain has docked at the TIM complex (which is formed mainly by the proteins Tim23, Tim17, Tim50, and Tim44, with the first two probably forming the channel), two forces move it into the matrix. An hsp70 homolog in the matrix, mtHsp70, associates with the channel through an interaction with Tim44. As in posttranslational translocation into the ER, mtHsp70 binds the substrate as it emerges from the channel and apparently acts as either a ratchet or motor or possibly a combination of the two (see *7.9 Adenosine triphosphate hydrolysis drives translocation*). Also as with ER posttranslational translocation, a DnaJ-like protein, Tim14, is thought to regulate mtHsp70 action. An electrochemical potential across the inner membrane also contributes to translocation, perhaps by interacting with the positively charged signal sequence. The potential is oriented across the membrane so that movement of positive charges toward the matrix would be favored. How the potential actually assists translocation is still unclear, however. Having entered the matrix, most signal sequences are cleaved by the soluble mitochondrial processing protease (MPP).

As with proteins integrated into the membrane of the ER, mitochondrial membrane proteins often target using an uncleaved internal signal sequence that also contains instructions for integration in the form of stretches of hydrophobic amino acids. How transmembrane domains are recognized and integrated into either of the two membranes is largely unknown. Proteins to be integrated into the inner membrane are targeted to an alternate translocon composed of Tim22, Tim54, and Tim18. Small proteins, called "Small Tims," are thought to help guide hydrophobic nascent inner membrane proteins through the aqueous environment of the intermembrane space. This sequence is illustrated in **FIGURE 7.52**. Little else is known about these pathways.

Inner membrane proteins, whether they are encoded in the nucleus or in mitochondria, can also be integrated by a pathway that involves the inner membrane protein Oxa1p. Nuclear-encoded inner membrane proteins are first translocated across both membranes into the matrix. The signal sequence that directed import is then cleaved off the protein, and a second sequence then directs the protein back to the inner membrane, where it is integrated by Oxa1p. Proteins encoded in the mitochondrial genome are synthesized by ribosomes in the matrix and then integrated directly into the inner membrane by the Oxa1p pathway. The details of the Oxa1p pathway are not yet well established.

Yet another translocon in mitochondria, named the SAM (for sorting and assembly machinery of the outer membrane) complex, has been identified. Although its composition and mechanism of action are yet to be fully elucidated, it seems to favor the translocation and integration of outer membrane proteins that span the membrane with β strands, rather than the far more common α helices. Proteins that use the SAM machinery are thought to first translocate fully into the intermembrane space through the TOM complex, and then use the SAM complex for insertion back into the outer membrane. As with the Tim22 complex, the SAM complex seems to rely on Small Tims to guide substrates from the TOM channel.

One particularly interesting question is how inner membrane proteins are able to by-

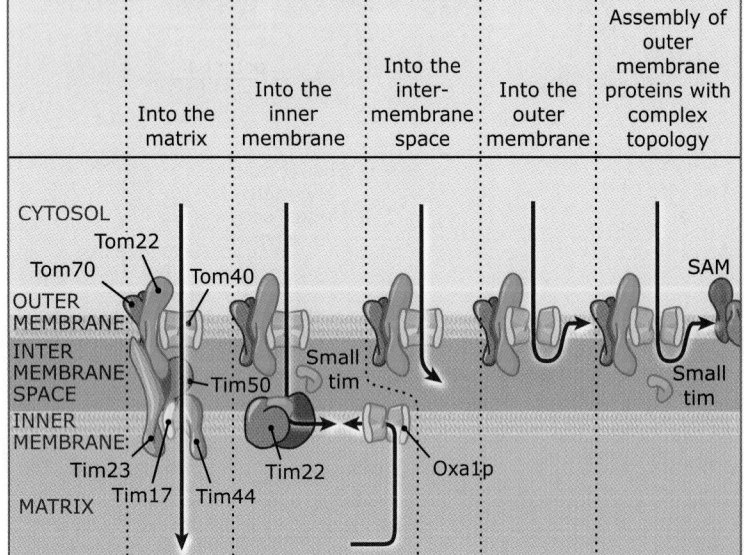

FIGURE 7.52 Different pathways are required to import proteins to different destinations within mitochondria. The TOM complex acts in several different pathways, but complexes in the inner membrane are specialized. There are two pathways by which proteins can reach the inner membrane. Some inner membrane proteins use a combination of pathways: they are first imported into the matrix and are then inserted back into the membrane. Adapted from M. J. Baker, et al., *Trends Cell Biol.* 17 (2007): 456–464.

pass the outer membrane without integrating. This specificity could arise from differences between the transmembrane domains of inner and outer membrane proteins that make inner membrane proteins unrecognizable by the TOM integration machinery. Alternatively, the TIM complex might recognize inner membrane proteins before the TOM complex has a chance to integrate them into the outer membrane. At present, the gating, regulation, and coordination of the TOM and TIM complexes are incompletely understood. Though many of the functions they perform are analogous to those performed by the translocon of the ER, it remains to be seen how similar the mechanisms employed will be.

7.29 Proteins imported into chloroplasts must also cross two membranes

Key concepts

- Import into chloroplasts occurs posttranslationally.
- The inner and outer membranes have separate translocation complexes that cooperate during the import of proteins.

Like mitochondria, chloroplasts contain a double membrane, and the vast majority of their protein products are translated from nuclear-encoded genes. Translocation into the interior of the organelle (the **stroma**) is similar to transport into the mitochondrial matrix, as illustrated in **FIGURE 7.53**, although the core components of the import machinery in the two organelles are not closely related in amino acid sequence. Thus, the machinery appears to have evolved independently for each organelle. Proteins translocate into chloroplasts posttranslationally and require the participation of cytosolic hsp70s to prevent their folding beforehand. A protein destined for the stroma typically contains an N-terminal **transit peptide** that is between twenty and one hundred twenty amino acids long. The transit peptide has a central region rich in serine, threonine, and basic amino acids, and an amphipathic domain at its C-terminus. Information encoding the targeting of proteins to chloroplasts resides within the transit peptide. The preprotein docks at the **TOC** (translocon of the outer envelope of chloroplasts) complex via the GTPase proteins Toc34p and Toc159p. Whether these proteins form simply as a receptor to bind translocation

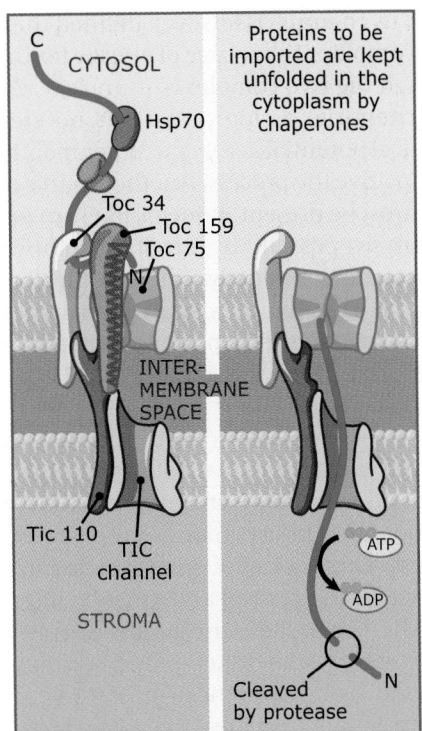

FIGURE 7.53 Chloroplast proteins must cross two membranes to enter the stroma. Proteins to be imported are kept unfolded in the cytoplasm by chaperones. The proteins move through connected channels in the outer and inner membranes. When they have entered the stroma, their transit peptides are removed. A source of energy is required for import but has not yet been identified. Adapted from P. Jarvis, *New Phytol.* 179 (2008): 247–565.

substrates and transfer them to the channel, analogous to SRP receptor, or whether they act to impel substrates into the channel like bacterial SecA, is not yet known.

The channel is formed by the Toc75 protein, which may also recognize the transit peptide. There is evidence that Toc75 might associate with different accessory proteins depending on the substrate being translocated, with highly abundant photosynthetic proteins utilizing one channel and less abundant housekeeping proteins using an alternate channel. Thus, the idea that different substrates require different accessory factors to facilitate their translocation is not restricted just to ER translocation.

The substrate crosses the inner membrane via the **TIC** complex (for translocon of the inner envelope of chloroplasts). The TOC and TIC complexes associate physically and are thought to be engaged simultaneously by the substrate. The TIC complex, including the identity of its channel, is less well defined than the TOC complex. The source of energy for import through the two complexes is unclear. Unlike mitochondria, a chloroplast has no electrochemical potential across the inner membrane to help drive the process. Another source of energy must be present. It seems likely to be ATP hydrolysis, possibly by stromal components. However, a completely different mechanism could be used. When a substrate reaches the stroma, the transit peptide is cleaved by the chloroplast-processing enzyme (CPE).

In addition to their inner and outer membranes, chloroplasts contain internal membranous organelles called thylakoids. The thylakoid membrane encloses a distinct compartment that houses many of the enzymes of photosynthesis. Most of these proteins are synthesized in the cytosol and must be imported into the organelle. One mode of transport across the thylakoid membrane occurs when a second signal sequence—used only after a protein is already in the stroma—is recognized by machinery related to the bacterial SecYEG translocon (see *7.9 Adenosine triphosphate hydrolysis drives translocation*).

Approximately half of all proteins that translocate into the lumen of the thylakoid utilize the Tat (for Twin-Arginine-Translocation) system. While the mechanism of Tat translocation is not known, Tat-dependent signal sequences appear very similar to ER-targeting signals, except that they are somewhat less hydrophobic, and they contain two adjacent arginines from which the name of the system is taken. Translocation across the thylakoid membrane is entirely dependent upon the maintenance of a pH gradient across the membrane, but the reason for this is not known.

Alb3, a homolog of the mitochondrial inner membrane protein Oxa1p, is responsible for integration of proteins into the thylakoid membrane. Proteins are targeted to the Alb3 machinery by interaction with an SRP homolog in the stroma. How the Alb3 protein functions is not yet understood.

Finally, there is circumstantial evidence that at least a few proteins are capable of targeting to chloroplasts via the secretory pathway, as these proteins are modified by N-glycosylation, which is an ER-specific modification. A poison that disrupts ER to Golgi trafficking also prevents these proteins from reaching chloroplasts. However, the mechanism by which the secretory pathway might connect to chloroplasts is not known.

Given the generally amphipathic nature of their signal sequences, how are mitochondrial and chloroplast proteins distinguished in the cytosol of a plant cell? This may be possible because the plant version of Tom22 lacks an acidic cytosolic domain, suggesting that plants have evolved a somewhat different mode of mitochondrial recognition from those of yeast and mammals in order to ensure that proteins are properly targeted.

7.30 Proteins fold before they are imported into peroxisomes

Key concepts

- Peroxisomal signal sequences are recognized in the cytosol and targeted to a translocation channel.
- Peroxisomal proteins are imported after they are folded.
- The proteins that recognize peroxisomal signal sequences remain bound during import and cycle in and out of the organelle.
- Peroxisomal membranes originate by budding from the ER.

As with the ER, mitochondria, and chloroplasts, proteins are also targeted and translocated directly into peroxisomes. These organelles are involved in oxidative processes such as fatty acid oxidation and hydrogen peroxide production and are bounded by a single lipid bilayer. Unlike mitochondria and chloroplasts, peroxisomes have no genome of their own, and so all peroxisomal proteins must be imported from the cytosol. The origin of peroxisomes was a bit controversial until it was convincingly demonstrated that they arise from the ER. The peroxisomal precursors that bud from the ER are composed of only minimal peroxisomal machinery, with most of the peroxisomal protein content imported after budding.

Proteins destined for the peroxisomal interior, or **peroxisomal matrix**, may be targeted to the organelle posttranslationally by one of two pathways. One pathway uses a C-terminal peroxisomal targeting signal 1 (PTS1). The simplest PTS1 sequences are much smaller than the signal sequences for most other organelles, and often consist of only three amino acids. Like other import signals, they vary in sequence. (The consensus sequence is a serine, cysteine, or alanine, followed by a basic amino acid, and then by leucine.) Additional amino acids outside of the PTS1 sequence may strengthen the targeting signal, particularly when the consensus is not strictly followed. PTS1 proteins are targeted to peroxisomes by Pex5p, a cytosolic protein that binds to the targeting sequence. PTS2 proteins, which are much less common than PTS1 proteins, target using a longer signal, which is usually located at the N-terminus of the protein. This targeting sequence is part of a larger peptide that is cleaved after transport into the matrix. PTS2 proteins are recognized and targeted by another cytosolic protein, Pex7p. In mammals, an alternatively spliced form of Pex5p can also mediate targeting of PTS2 proteins. As shown in **FIGURE 7.54**, the PTS1 and PTS2 pathways share common components in the membrane, including a docking complex consisting of at least three proteins, Pex17p, Pex14p, and Pex13p. However, protein factors unique to each pathway exist as well. Other pathways for targeting proteins to the peroxisomal matrix independently of the PTS1 or PTS2 machinery may exist, but these pathways are poorly defined.

Although the mechanism of peroxisomal matrix import is not completely understood, it differs significantly from ER, mitochondrial, and chloroplast translocation in at least one major respect: matrix proteins can be imported after they have folded, or even oligomerized, in the cytosol. This point was dramatically demonstrated, as illustrated in **FIGURE 7.55**, by coating 90-Å gold particles with PTS1 peptides. This caused the particles to be imported despite their large size. In this respect, import of proteins into the peroxisomal matrix may be similar to transport through the nuclear pore complex (see *9 Nuclear structure and transport*). In addition, during both nuclear and peroxisomal transport, the receptor recognizing the targeting sequence (i.e., Pex5p for peroxisomal transport and importins for nuclear transport) is carried with the translocating substrate across the membrane. The receptor is subsequently exported alone, for further use. The internal environment of peroxisomes would be toxic to the cell if the peroxisomal membrane were as permeable to protein and small molecule movement as is the nuclear envelope. Therefore, the peroxisomal translocation channel must be very closely gated to ensure that nothing diffuses out of the organelle. How this is accomplished is not known.

By contrast to their fully translocated counterparts, peroxisomal membrane proteins might follow an entirely different pathway of import. Several proteins, including Pex19p,

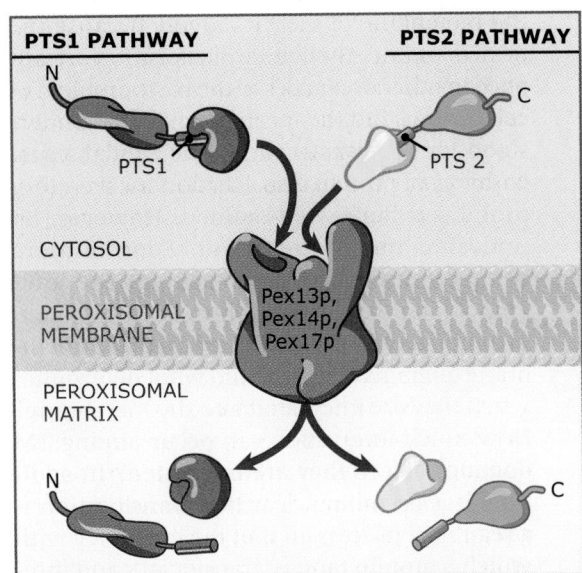

FIGURE 7.54 Peroxisomal proteins are imported after they have folded and are recognized by one of two pathways. Both pathways use some components of the channel, but other components are specific to one. The proteins that recognize peroxisomal signal sequences are imported along with their substrates.

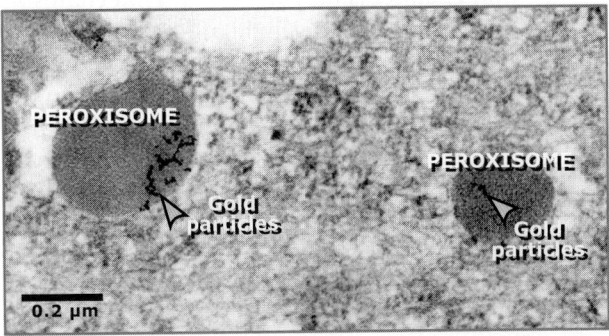

FIGURE 7.55 This electron micrograph shows gold particles coated with the PTS1 peptide that have been imported into peroxisomes. The gold particles have a diameter of 90 Å, significantly larger than most globular proteins. Photo reproduced from P. A. Walton, et al., *Mol. Biol. Cell* 6 (1995): 675–683. Copyright 1995 by American Society of Cell Biology. Reproduced with permission of the American Society of Cell Biology in the format of Textbook via Copyright Clearance Center. Photo courtesy of Suresh Subramani, University of California, San Diego.

Pex3p, and Pex16p, are thought to be involved, with Pex19p possibly serving as the import receptor. However, the roles of these proteins are not understood.

7.31 What's next?

The largest part of this chapter is devoted to explaining how proteins translocate across the membrane of the ER, largely because the principles underlying ER translocation are applied in other organelles to the basic problem of cell sorting, and the mechanism is best understood for ER translocation. The fundamental pathway of protein translocation for secretory proteins is fairly well defined. However, an understanding of the translocation of more complex substrates is still lacking. The integration of membrane proteins, in particular, is incompletely understood. How are transmembrane domains oriented, and what determines when they are integrated into the membrane? How much interaction can occur among TM domains before they are integrated? In addition, it is becoming clear that translocation is a regulated process, in that the efficiency with which a protein targets, translocates and integrates, and the factors it requires to do so, can vary considerably from substrate to substrate, and potentially from one cellular condition to another. An understanding of how translocation is regulated and can be modified by the cell according to need is still lacking.

Even though the structure of the translocation channel is now known, the mechanisms of translocon engagement and gating are still being explored. How does recognition of a signal sequence lead to structural changes in the channel? Is the channel structure sufficiently flexible to allow the translocating chain to be reoriented or for multiple transmembrane domains to accumulate? What happens to the channel structure when translocation terminates? How and when do all of the other proteins acting in the vicinity of the translocon become associated with it?

Less well defined than translocation are the means by which the ER carries out protein folding and recognizes and deals with misfolded proteins. The most fundamental aspect of the process—how the organelle "senses" that a protein is folded properly or improperly—is almost totally unclear. Much more information about how chaperones actually interact with proteins will be needed. To what extent the multiple chaperoning systems in the ER interact, and whether they play distinct or overlapping functions, are issues that also need to be clarified. In addition, it is unclear how the decision is made that a protein is unlikely ever to fold correctly and should be degraded. The existence of a mechanism that times its interaction with chaperones is an attractive idea, but there is no evidence about how it might work. Finally, it is unknown how a protein selected for degradation is retargeted to the channel and moved back into the cytosol, or indeed, even the mechanism by which the channel can be stimulated to open from the inside.

The role of the ER in lipid synthesis and sorting is much less well defined than its role in protein sorting. In part, this is because lipid molecules are much more difficult to study than proteins, because they are much harder to manipulate experimentally. The most fundamental issue of lipid distribution—how lipids are selectively transported from the ER to target membranes—is still the subject of controversy. The discovery of physical associations between the ER and other organelles, such as the mitochondrial-associated membrane, suggests a possible mechanism for transfer. However, at this point, the role of such contacts is not known. It is also unclear how the ER regulates the synthesis of particular phospholipids.

The mechanisms governing the dynamics of the ER as a whole are not understood. How does the organelle maintain its characteristic shape? How are its subdivisions maintained? What is its relationship with the cytoskeleton, and how is it moved about the cell? How is its size determined, and how does it expand when necessary? The discovery of the unfolded protein response points to communication between the ER and the nucleus that may influence some of these behaviors. The contacts that the ER makes with other compartments suggest that it may also communicate with them, but how this might occur is unknown.

Finally, it is now becoming clear that the many functions of the ER—protein folding, lipid synthesis, maintenance of calcium homeostasis, etc.—are very sensitive to perturbation by genetic or environmental factors. ER stress, as these perturbations are broadly defined, is associated with a number of human diseases, ranging from diabetes to cancer to neurodegeneration. What role the ER plays in protecting normal cellular function, and how the unfolded protein response allows cells and organisms to adapt to stress, is a major ongoing area of investigation.

7.32 Summary

The cell contains many types of membrane-bounded organelles, several of which import proteins directly from the cytosol. Mitochondria, chloroplasts, and peroxisomes all import proteins for their own use. The ER also imports proteins from the cytosol, but sends most of them on, either to be secreted or to function in organelles and membranes that cannot import proteins themselves. These include the plasma membrane and the organelles of the secretory or endocytic pathways.

Proteins to be imported into organelles are identified by a signal sequence, a short stretch of amino acids that is usually located at a protein's N-terminus. Signal sequences for various organelles differ in their lengths and chemical features. Signal sequences for the ER are often about twenty amino acids long and have a long, continuous stretch of hydrophobic amino acid residues. Mitochondrial signal sequences are about the same length. However, they alternate hydrophobic and charged amino acids so that when the signal sequence is coiled as an α helix, one side is hydrophobic and the other is charged. Peroxisomal signal sequences are usually only three amino acids long. In all cases, it is the physical properties of a signal sequence rather than its exact sequence that identify it and allow it to specify a protein's destination.

Each type of signal sequence is recognized by binding to a receptor that targets the protein to a specific organelle. When it arrives there, the protein is translocated into the interior of the organelle through a channel in its membrane. Whether a signal sequence is recognized during or after protein synthesis determines which of two types of translocation occurs. Cotranslational translocation occurs when the signal sequence is recognized while the protein is being synthesized. It results in the binding of the ribosome synthesizing the protein to the membrane, where it transfers the nascent protein into the translocation channel. Posttranslational translocation occurs when the signal sequence is recognized after synthesis of the protein is complete.

Most proteins enter the ER by cotranslational translocation. SRP binds the signal sequence soon after it emerges from the ribosome. The ribosome and the nascent protein are targeted to the membrane of the ER through an interaction between SRP and the SRP receptor. The ribosome and the nascent chain then engage the channel through which the protein crosses the membrane. The channel core consists of the Sec61 complex and is the center of a large group of proteins involved in translocation. Together, these are called the translocon. Interaction with the signal sequence opens the channel and allows the chain to be inserted into it in a way that presumably prevents the passage of other molecules across the ER membrane. Translation continues while the ribosome is bound to the channel and the protein moves through it into the lumen.

Integration of a protein into the membrane of the ER begins when a transmembrane domain is translated and enters the channel. Transmembrane domains are recognized by the channel because of their hydrophobicity and are allowed to move through its walls into the lipid bilayer. Recognition of a transmembrane domain causes movement of the nascent protein across the membrane to halt. Translation continues, releasing subsequent segments of the polypeptide into the cytosol. The complete integration of membrane proteins with multiple transmembrane domains is likely to require repeated opening and closing of the channel.

The problem of integrating membrane proteins is complicated by the need to orient them in relation to the membrane. Orientation is apparently determined by properties of a protein's transmembrane domains. It is unclear, however, how they interact with the channel, the lipids of the membrane, or with each other to determine the orientation of the protein.

Some proteins are translocated into the ER posttranslationally. Binding of cytosolic chaperones prevents these proteins from folding and keeps them competent for translocation while they are being targeted. Targeting occurs through binding of their signal sequences to Sec61 in the translocation channel. The channel is the same one used for cotranslational translocation, but contains four additional proteins, including Sec62 and Sec63. Nascent proteins move through this channel as a result of ATP hydrolysis by the BiP protein in the lumen of the ER. BiP uses ATP hydrolysis to bind and release the nascent protein at the lumenal end of the channel. BiP molecules remain bound long enough to act as a ratchet that moves the protein into the lumen.

Many proteins are covalently modified as they are translocated into the ER. The signal sequence is usually removed by signal peptidase soon after a nascent protein is inserted into the channel. Later segments of a translocating protein are often modified soon after they enter the lumen. Sugars may be added by OST, or disulfide bonds may be formed by PDI. A few fully translocated proteins are cleaved near their C-termini and attached to GPI, a phospholipid that anchors them to the membrane.

After they have entered the lumen, proteins begin to fold with the help of a large collection of chaperones. BiP and Grp94 interact directly with unfolded forms of proteins. Calnexin and calreticulin bind to the sugars added to proteins during translocation and participate in a cycle in which the presence or absence of a glucose residue on a protein indicates whether or not it is correctly folded. PDI rearranges disulfide bonds as proteins attempt to fold. When they have folded correctly, proteins no longer interact with chaperones and are allowed to leave the ER for the Golgi apparatus. If a protein fails to fold correctly after repeated attempts or fails to assemble properly with other proteins, it is returned to the translocation channel and moved back to the cytosol by retrograde translocation. When it reaches the cytosol, the protein is degraded by the proteasome.

The accumulation of large amounts of unfolded proteins in the ER causes the unfolded protein response. This is a signaling pathway from the ER to the nucleus that stimulates the production of additional chaperones. The signal is mediated by ER-resident transmembrane proteins that sense the presence of unfolded proteins. Their activation leads to the induction of signaling cascades that change patterns of gene expression to allow the ER to cope with the excess burden of unfolded proteins or, in the case of higher eukaryotes, lead to cell death if severe ER stress persists.

Mitochondria and chloroplasts both import proteins posttranslationally. Both types of organelle have two membranes. Proteins may be localized to either membrane, to the intermembrane space, or to the interior of the organelle. Each membrane has a separate translocon. These are called TOM for the outer membrane and TIM for the inner membrane of mitochondria, and TOC and TIC for the corresponding membranes in chloroplasts. Signal sequences are recognized by the translocon in the outer membrane. The translocons in the inner and outer membranes physically associate so that imported proteins are transferred directly between them. Mitochondrial proteins may be translocated across both membranes and then retargeted to the inner membrane by a separate signal sequence. A separate signal sequence may also continue the translocation of chloroplast proteins across the membrane of an internal organelle called a thylakoid.

In mitochondria, translocation is driven both by the electrochemical gradient across the inner membrane and by interactions between the protein being imported and chaperones in the mitochondrial matrix. How translocation is driven in chloroplasts is unknown. We do not know how proteins are integrated into the membrane of either organelle.

Transport into peroxisomes is posttranslational but is different from translocation into any of the other organelles. Import into peroxisomes crosses only a single membrane and occurs after the protein has folded in the cytosol. Peroxisomal transport sequences are recognized in the cytosol by proteins that remain bound to the substrate as it is translocated. The carrier proteins dissociate after they are inside the organelle, and are returned to the cytosol for reuse. The origin of peroxisomal membrane proteins is unclear.

The ER serves a number of roles in the cell in addition to its primary role in the import, maturation, and distribution of proteins. Its different functions are reflected in its structure. Translocation and maturation of proteins occur in the rough ER, which is covered with bound ribosomes. Branching off the rough ER is the smooth ER. The smooth ER is usually in the form of tubules, which form a constantly rearranging network extending throughout the cytosol. Smooth ER is often associated with cytoskeletal elements, and makes contact with other membranes in the cell. Among the functions of the smooth ER is the synthesis of lipids for all of the cell's membranes. Lipids must somehow be transported from the ER to the other membranes, but how this is achieved is unclear. The contacts that the smooth ER makes with the other membranes suggest a possible route. The ER also acts as a storage site for intracellular calcium. Calcium is released in response to signals from outside the cell, and is pumped back into the organelle afterward. In specialized cells, the smooth ER may also synthesize lipid-soluble hormones or detoxify potentially harmful chemicals. In cells specialized for any of the functions that the ER performs, such as cells that secrete large quantities of protein or that synthesize steroid hormones, the rough or smooth ER can expand to occupy much of the cytosol. In other highly specialized cells, such as the cells of skeletal muscle, the ER can be highly specialized in terms of both its composition and its structure. The sarcoplasmic reticulum, a specialization of smooth ER, is wrapped in sheets around the sarcomeres of skeletal muscle so that it can deliver calcium to stimulate contraction.

The signal hypothesis is now nearing forty years old, and much has been learned about the membrane targeting of proteins in that time.

The availability of new experimental tools will allow the ER to be studied in a global and quantitative way. These approaches, supplemented by classical genetics and biochemistry, will lead to a view of the ER over the next forty years that moves from reductionist to integrative, allowing us to better understand the contribution of the ER to normal cellular physiology and to disease.

http://biology.jbpub.com/lewin/cells

To explore these topics in more detail, visit this book's Interactive Student Study Guide.

References

Anelli, T., and Sitia, R. 2008. Protein quality control in the early secretory pathway. *EMBO J.* v. 27 p. 315–327.

Blobel, G., and Dobberstein, B. 1975. Transfer of proteins across membranes. I. Presence of proteolytically processed and unprocessed nascent immunoglobulin light chains on membrane-bound ribosomes of murine myeloma. *J. Cell Biol.* v. 67 p. 835–851.

Cox, J. S., Shamu, C. E., and Walter, P. 1993. Transcriptional induction of genes encoding endoplasmic reticulum resident proteins requires a transmembrane protein kinase. *Cell* v. 73 p. 1197–1206.

Lakkaraju, A. K., Mary, C., Scherrer, A., Johnson, A. E., and Strub, K. 2008. SRP keeps polypeptides translocation-competent by slowing translation to match limiting ER-targeting sites. *Cell* v. 133 p. 440–451.

Neupert, W., and Herrmann, J. M. 2007. Transport of proteins into mitochondria. *Annu. Rev. Biochem.* v. 76 p. 723–749.Rapoport, T. A. 2007. Protein translocation across the eukaryotic endoplasmic reticulum and bacterial plasma membranes. *Nature* v. 450 p. 663–669.

Rutkowski, D. T., and Kaufman, R. J. 2004. A trip to the ER: Coping with stress. *Trends Cell Biol.* v. 14 p. 20–28.

Tu, B. P., Ho-Schlever, S. C., Travers, K. J., and Weissman, J. S. 2000. Biochemical basis of oxidative protein folding in the endoplasmic reticulum. *Science* v. 290 p. 1571–1574.

Van den Berg, B., Clemons, W. M., Collinson, I., Modis, Y., Hartmann, E., Harrison, S. C., and Rapoport, T. A. 2004. X-ray structure of a protein-conducting channel. *Nature* v. 427 p. 36–44.

Voeltz, G. K., Prinz, W. A., Shibata, Y., Rist, J. M., Rapoport, T. A. 2006. A class of membrane proteins shaping the tubular endoplasmic reticulum. *Cell* v. 124 p. 573–586.

Protein trafficking between membranes

8

Vivek Malhotra
CRG (Center for Genomic Regulation), Barcelona, Spain

Graham Warren
Max F. Perutz Laboratories, Vienna, Austria

Ira Mellman
Genentech, Inc., San Francisco, CA

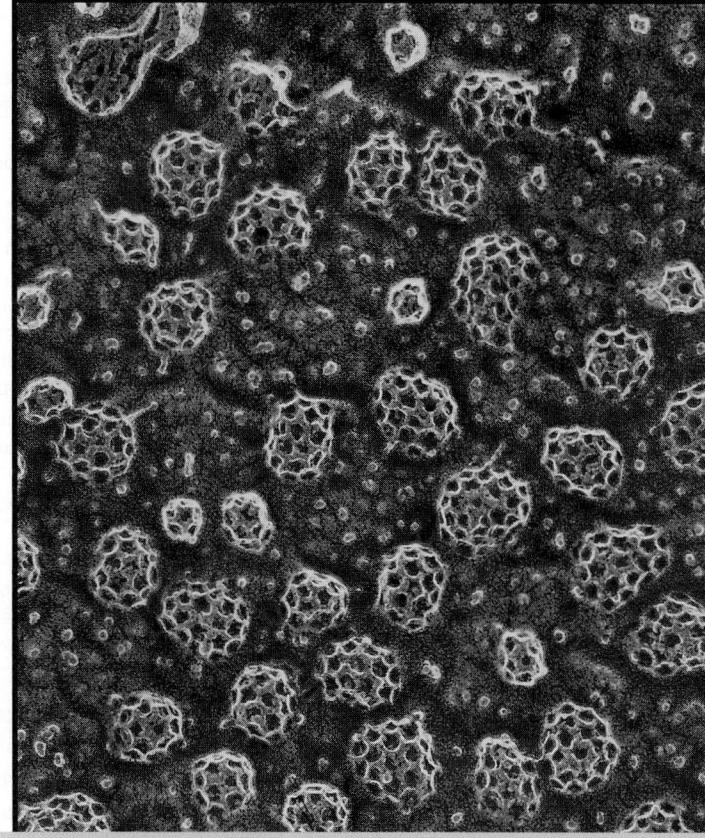

ELECTRON MICROGRAPH OF CLATHRIN-COATED PITS, which are involved in the uptake of extracellular materials by cells. Photo courtesy of John Heuser, Washington University School of Medicine.

CHAPTER OUTLINE

8.1 Introduction

Key concepts

- Eukaryotic cells have an elaborate system of internal membrane-bounded structures called organelles.
- Each organelle has a unique composition of (glyco)proteins and (glyco)lipids and carries out a particular set of functions.
- An organelle comprises one or more membrane-bounded compartments.
- Organelles may act autonomously or in cooperation to accomplish a given function.
- In the endocytic and exocytic pathways, cargo proteins are transferred between compartments by transport vesicles that form by budding from an organelle's surface and subsequently fuse with the target membrane of the acceptor compartment.
- Transport vesicles can selectively include material destined for transfer and exclude material that must remain in the organelle from which they bud.
- Selective inclusion into transport vesicles is ensured by cargo receptors and by signals in cargo's amino acid sequence or carbohydrate structures.
- Transport vesicles contain proteins that target them specifically to their intended destinations with which they dock and fuse.

One of the defining characteristics of eukaryotic cells is the presence of an elaborate system of internal membrane-bounded structures called **organelles**. Whereas all living cells have an outer, or delimiting, membrane bilayer, eukaryotes have membranes that partition the cell interior into functionally distinct, membrane-bounded **compartments**. One advantage of compartmentalization is that it allows cells to have specialized environments for functions that have different chemical requirements.

FIGURE 8.1 illustrates the variety and organization of membrane-bounded organelles typically found in a eukaryotic cell, in this case, an animal cell (see also 7 *Membrane targeting of proteins* and 9 *Nuclear structure and transport*). Each organelle comprises one or more compartments. For example, the **endoplasmic reticulum (ER)** is a single compartment. In contrast, the **Golgi apparatus** consists of a series of membrane-bounded compartments that have distinct biochemical functions. A mitochondrion has two compartments, the matrix and the intermembrane space, which contain distinct sets of macromolecules.

The **cytosol** can be thought of as a single compartment that is delimited by the **plasma membrane** and in contact with the outer surfaces of all internal organelles. The **cytoplasm** consists of the cytosol and organelles. Similarly, the nucleoplasm is bounded by the inner nuclear membrane.

Each organelle has a unique composition of proteins (both membrane-bounded and soluble) and lipids, as well as other molecules that are in keeping with its function. Some proteins and lipids have covalently attached oligosaccharides. As cells grow and divide, new components must be synthesized in order for organelles to grow, divide, and eventually partition between the two daughter cells. Organelle components are also synthesized during differentiation, development, and in response to environmental stimuli, such as stress. However, synthesis of these components does not necessarily occur in the organelle where they function. Rather, the production of different macromolecules occurs at sites specialized for their synthesis. For example, most proteins are synthesized by ribosomes in the cytosol, which has evolved as the optimal environment for ribosome function and protein production.

The question then arises: How do organellar components get to the sites where they function? This has been a central question in cell biology since the early 1970s. As summarized in Figure 8.1, there are at least eight major types of organelles, each consisting of hundreds or thousands of distinct proteins and lipids, all of which must travel to the organelles in which they function. Most proteins are synthesized in the cytosol, so how are they delivered to the correct organelles, or, in the case of secreted

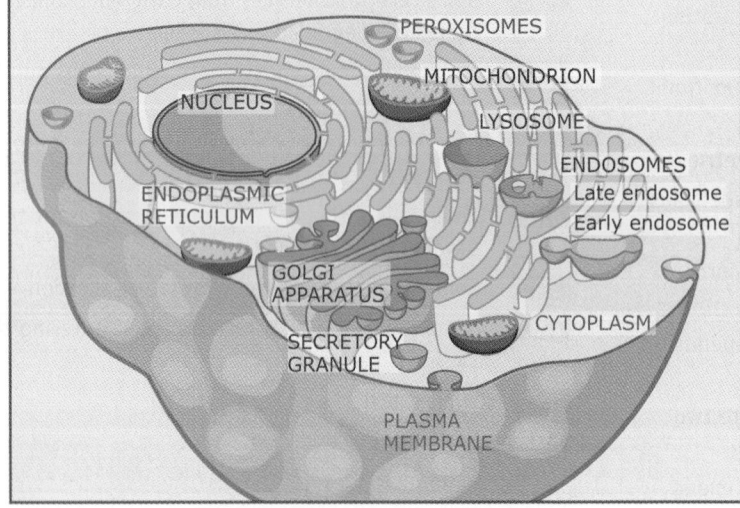

FIGURE 8.1 The membrane-bounded compartments in a typical animal cell.

proteins, to the outside of the cell? In many cases, the answer lies in a group of protein-based signals, typically called **sorting signals** or targeting signals. These signals are short sequences of amino acids present on proteins destined for locations other than the cytosol (for details see *8.5 The concepts of signal-mediated and bulk flow protein transport*). Each destination is associated with one or more distinct classes of signals. (For more on delivery of lipids see *7.22 Lipids must be moved from the endoplasmic reticulum to the membranes of other organelles*.)

Sorting signals are recognized by specific cellular machinery at one or more steps along the route taken to the protein's final destination. As shown in **FIGURE 8.2**, two of the main transport routes are the **exocytic pathway** (or secretory pathway) and the **endocytic pathway**, which transport material (**cargo**) out of and into the cell, respectively (for details see *8.2 Overview of the exocytic pathway* and *8.3 Overview of the endocytic pathway*). All newly synthesized proteins destined for secretion from the cell or for organelles in the exocytic and endocytic pathways share a common entry point at the ER membrane. Signals for protein translocation across the ER membrane are called signal sequences (see *7 Membrane targeting of proteins*). In this chapter, we will discuss the sorting signals that get proteins to their final destinations.

Once in the ER, proteins are no longer free to move through the cytoplasm; their only means of gaining access to other membrane-bounded organelles is through **vesicle-mediated transport**. **Transport vesicles** consist mainly of proteins and lipids and are said to form by "budding" from a membrane (donor compartment), as illustrated in **FIGURE 8.3** (for details see *8.4 Concepts in vesicle-mediated protein transport*). After budding, transport vesicles fuse with the next compartment (acceptor compartment) in the pathway.

Transport vesicles carry material from the ER either directly or indirectly to all the other compartments on the exocytic and endocytic pathways, as illustrated in **FIGURE 8.4**. Endocytosis relies on the formation of vesicles at the plasma membrane. These vesicles transport internalized material to **endosomes**, from which other vesicles form to move material to other compartments. Thus, the composition of transport vesicles differs depending on their origin and destination.

Vesicle-mediated transport poses a fundamental problem for the organelles that exchange vesicles. In order to function, organelles need to maintain their composition, but how can they do so when vesicles are shuttling material between them? The scale of the problem is evident from calculations of the extent of transport. The equivalent of all the plasma membrane's proteins and lipids can move through the organelles of the endocytic pathway in less than one hour. This rate is impressive when compared with the time required to synthesize a new organelle, typically one day.

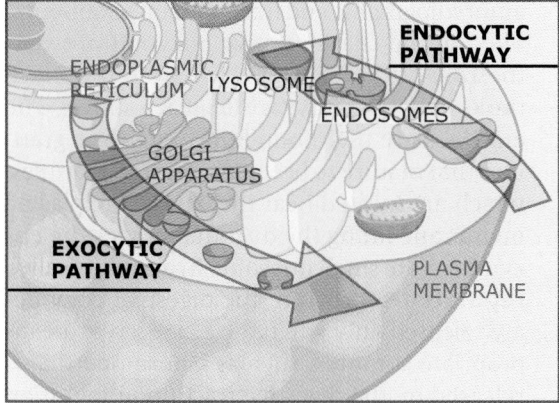

FIGURE 8.2 The exocytic and endocytic pathways. The exocytic pathway comprises the ER (including the nuclear envelope) and the Golgi apparatus (represented here as a single stack). The endocytic pathway comprises early and late endosomes and lysosomes.

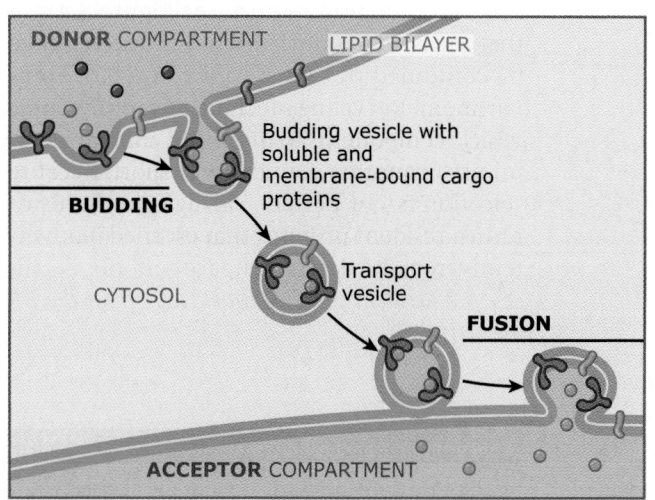

FIGURE 8.3 In vesicle-mediated transport, a membrane-bounded vesicle buds from one compartment and fuses with another.

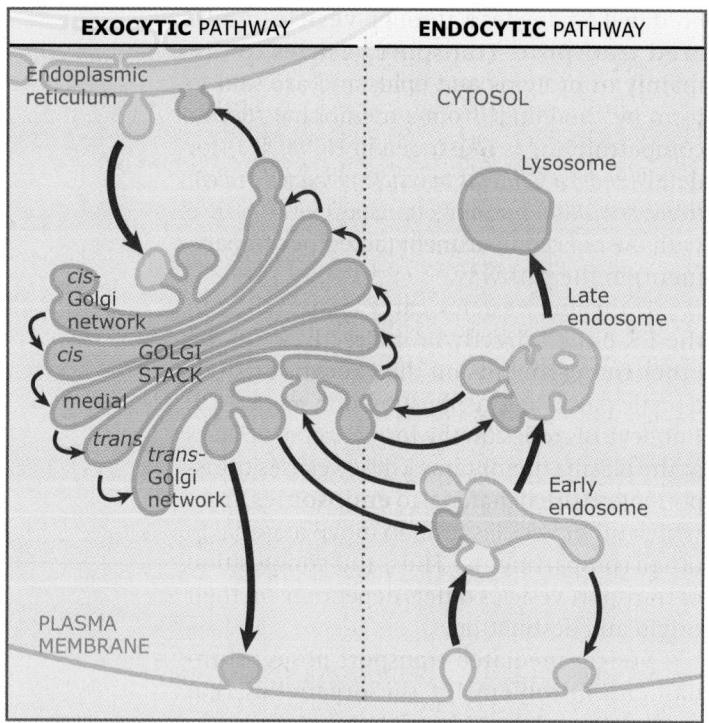

EXOCYTIC PATHWAY	ENDOCYTIC PATHWAY

Endoplasmic reticulum

CYTOSOL

Lysosome

cis-Golgi network

cis

GOLGI STACK

medial

trans

trans-Golgi network

Late endosome

Early endosome

PLASMA MEMBRANE

FIGURE 8.4 The known pathways for movement of proteins in a typical animal cell. Almost all of the flow pathways are bidirectional.

The answer to the problem lies in the mechanisms that impose selectivity on the transport process. Budding vesicles can selectively incorporate the proteins to be transported, leaving resident proteins of the organelle behind. The vesicles then dock and fuse selectively with the next, correct, compartment on the pathway. To maintain homeostasis between organelles, vesicular transport must always be bidirectional in nature (see Figure 8.4), so that the donor compartment is not consumed by continued transport to the acceptor compartment. Recycling mechanisms return some vesicle components to the donor compartment for use in another round of transport. Because selection is not perfect, salvage mechanisms return resident proteins that escaped inadvertently from the donor compartment (for details see *8.7 Resident proteins that escape from the ER are retrieved*).

Concept and Reasoning Checks

1. How are proteins transported to their correct intracellular location?
2. How is homeostasis maintained in intracellular protein trafficking?

Overview of the exocytic pathway

Key concepts

- All eukaryotes have the same complement of core exocytic compartments: the ER, the compartments of the Golgi apparatus, and post-Golgi transport vesicles.
- The amount and organization of exocytic organelles varies from organism to organism and cell type to cell type.
- Each organelle in the exocytic pathway has a specialized function.
- The ER is the site for the synthesis and proper folding of proteins.
- In the Golgi apparatus, proteins are modified, sorted, and carried by the post-Golgi transport vesicles to the correct destination.
- Cargo transport to the plasma membrane occurs directly by a constitutive process or indirectly by a regulated process that involves temporary storage in secretory granules until the cell receives an appropriate stimulus.

The exocytic pathway is the route taken by molecules whose destination is either the plasma membrane or the extracellular space. This pathway was originally called the secretory pathway because it was first defined in specialized cells, called pancreatic acinar cells, which secrete enzymes destined for the digestive tract. These enzymes represent a large fraction of the proteins synthesized. Palade and coworkers took advantage of this fact to trace the route taken by these proteins. They did pulse-chase experiments in which pancreatic tissue sections were incubated with radiolabeled amino acids, and the locations of radiolabeled proteins were followed over time using autoradiographic and fractionation techniques. They found that digestive enzymes are synthesized at the ER and move from the ER to the Golgi apparatus, as **FIGURE 8.5** shows. The enzymes are then found in condensing granules that mature to become zymogen granules, which are located near the part of the plasma membrane lining the duct that carries the enzymes to the small intestine. In a process called **regulated secretion**, the digestive enzymes are released only when the cell receives an appropriate stimulus, such as a hormone that is released by the digestive tract upon ingestion of food (for more on regulated secretion see *8.19 Some cells store proteins for later secretion*). In contrast to regulated secretion, **constitutive secretion** is the process by which proteins are continuously secreted. Constitutive secretion predominates in cells that have little

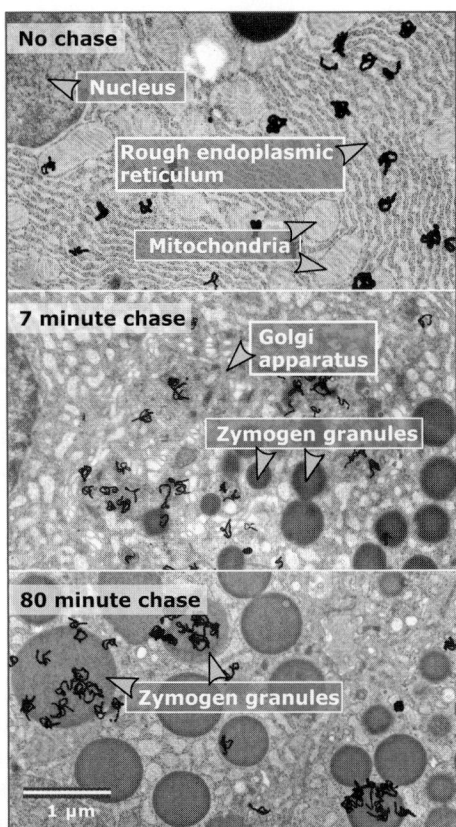

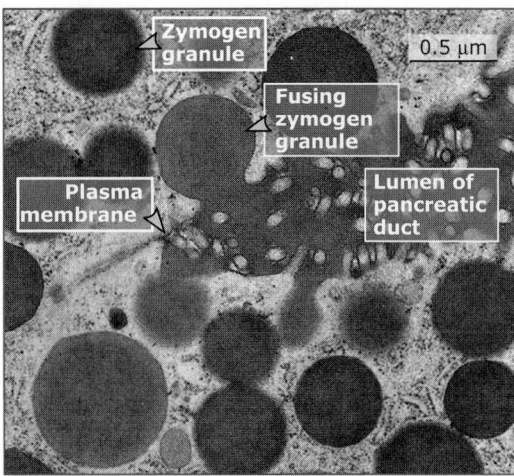

FIGURE 8.6 Transmission electron micrograph showing a zymogen granule fusing with the plasma membrane. The contents of the zymogen granule are released into the lumen of the pancreatic duct. Photo courtesy of Lelio Orci, University of Geneva, Switzerland.

FIGURE 8.5 The secretory pathway was defined by studies of the biogenesis of zymogen granules. Pancreatic slices were briefly pulsed with ^{3}H-leucine and the label chased for 0, 7, and 80 minutes before fixation and preparation for electron microscopic autoradiography. The autoradiographic grains representing newly synthesized secretory proteins were located first over the ER (top), then over the Golgi region (middle) and over the secretory/zymogen granules (bottom). Top photo © Jamieson and Palade, 1967. Originally published in **The Journal of Cell Biology**, 34: 597–615. Used with permission of Rockefeller University Press. Photo courtesy of James D. Jamieson, Yale University School of Medicine; middle photo courtesy of James D. Jamieson, Yale University School of Medicine; bottom photo © Jamieson and Palade, 1988. Originally published in **The Journal of Cell Biology**, 48: 503–522. Used with permission of Rockefeller University Press. Photo courtesy of James D. Jamieson, Yale University School of Medicine.

or no regulated secretion. Liver cells, for example, constitutively secrete proteins such as serum albumin into the blood plasma. In both regulated and constitutive secretion, vesicles containing the proteins to be secreted fuse with the plasma membrane in a process called **exocytosis**, as **FIGURE 8.6** shows.

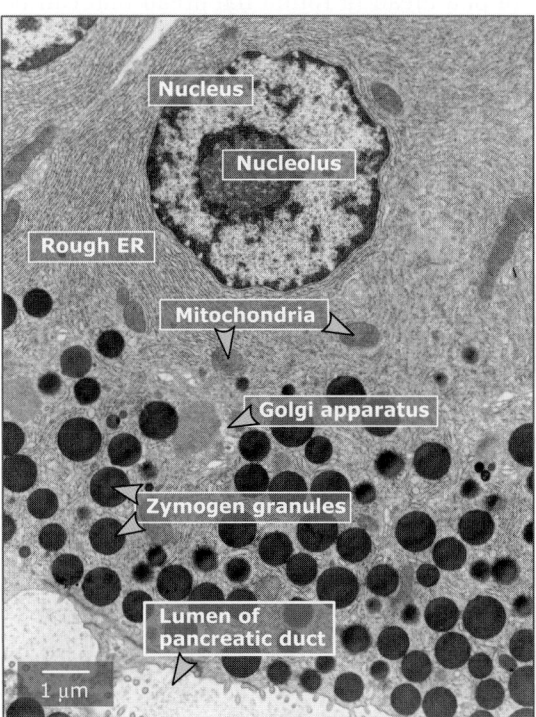

FIGURE 8.7 Transmission electron micrograph of the organelles in a pancreatic acinar cell. The exocytic pathway consists of the rough ER, Golgi apparatus and zymogen granules. The rough ER and zymogen granules are particularly abundant in these cells. Photo courtesy of Lelio Orci, University of Geneva, Switzerland.

The organization of the exocytic pathway in pancreatic acinar cells is shown in **FIGURE 8.7**. The ER is a single compartment with distinct domains, the most important for secretion being the rough ER (RER). The RER is "rough" because of bound ribosomes that are synthesizing secretory proteins, which are transferred across the membrane bilayer into the lumen of the ER (see 7 *Membrane targeting of proteins*). Flattened cisternal elements of the ER are packed together in the basal part of the cell. The large amount of ER elements in these cells provides sufficient ribosome binding and protein translocation sites to support the synthesis and translocation of the 10 million digestive enzymes that are made every minute. The ER membrane is the most abundant membrane in most eukaryotic cells. Even in cells not specialized for secretion, the ER typically comprises about 50% of a cell's membranes.

The newly synthesized secretory proteins are transported from the ER to the Golgi apparatus, which is a multicompartment organelle (see Figure 8.4). Its central feature is the **Golgi stack**, which consists of closely apposed and flattened cisternae, reminiscent of a stack of pita bread (a round flat bread that can be opened to form a pocket for filling). The Golgi stack is polarized and consists of *cis*, medial, and *trans* cisternae. The *cis* cisterna is nearest

the entry face of the Golgi apparatus, whereas the *trans* cisterna is nearest the exit face. The dilated cisternal rims are the sites from which transport vesicles bud and with which they fuse.

The Golgi stack contains a variety of enzymes that carry out posttranslational modifications to most of the transiting cargo. The best-characterized enzymes are those that modify the O- and N-linked oligosaccharides on proteins. These oligosaccharides have a number of functions, including the targeting of newly synthesized lysosomal enzymes to lysosomes (for details see *7.14 Sugars are added to many translocating proteins* and *8.17 Sorting of lysosomal proteins occurs in the* trans-*Golgi network*).

As shown in **FIGURE 8.8** for the N-linked pathway, the steps in the modification of the **high mannose oligosaccharides** added in the ER are accomplished in a stepwise fashion resulting in the sequential conversion of "immature," high mannose-containing oligosaccharides to complex, highly sialylated structures. To a first approximation, the enzymes responsible for these steps are arranged in sequence across the stack. Enzymes that act early in the process are more likely to be found at the face of the Golgi apparatus that first receives proteins exported from the ER. Enzymes that act later (e.g., those that add the final sialic acid residues) are found at the face from which the finished secretory proteins exit the Golgi apparatus.

The Golgi stack is bounded on each side by reticulotubular networks. The *cis*-**Golgi network (CGN)** on the entry face of the Golgi apparatus receives the output of proteins from the ER exit sites. It functions as a quality control device by allowing the retrieval of resident proteins that escape from the ER (see *8.7 Resident proteins that escape from the ER are retrieved*). The *trans*-**Golgi network (TGN)** on the exit face distributes proteins to different destinations. In the case of pancreatic acinar cells, the majority of proteins in the TGN are secretory proteins, which are incorporated into condensing vacuoles that mature to yield zymogen granules. In addition, the TGN sorts lysosomal proteins for transport to lysosomes (via the endosomal pathway) away from plasma membrane proteins and constitutively secreted proteins en route to the cell surface. In cells not specialized for regulated secretion, only the lysosomal and plasma membrane pathways exist.

Although the Golgi apparatus has a characteristic *cis-trans* polarity reflecting the

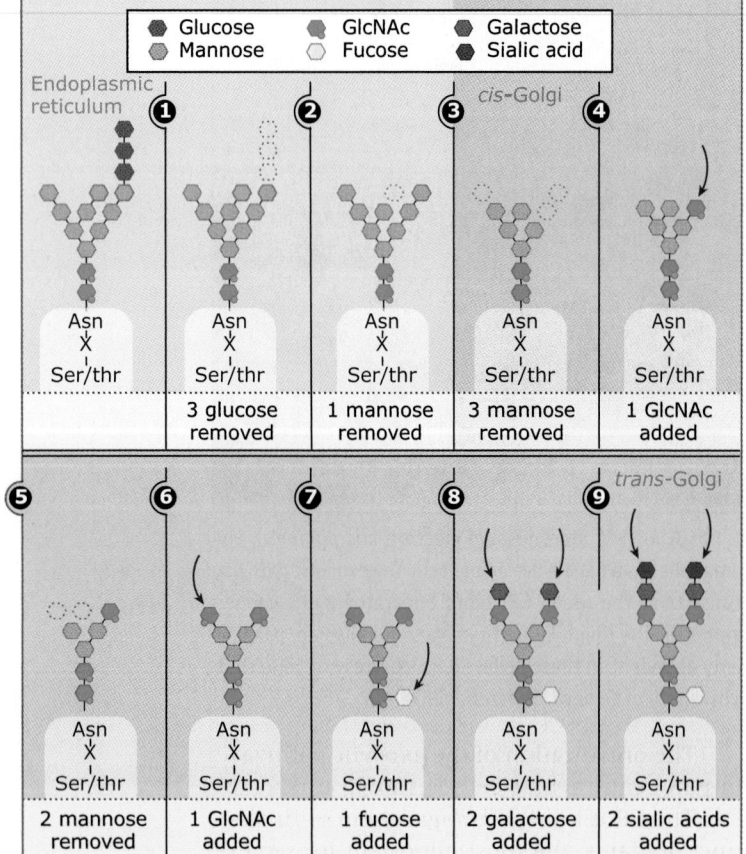

FIGURE 8.8 The steps in formation of N-linked oligosaccharides. Vesicle-mediated transport of the glycoprotein between compartments is not shown.

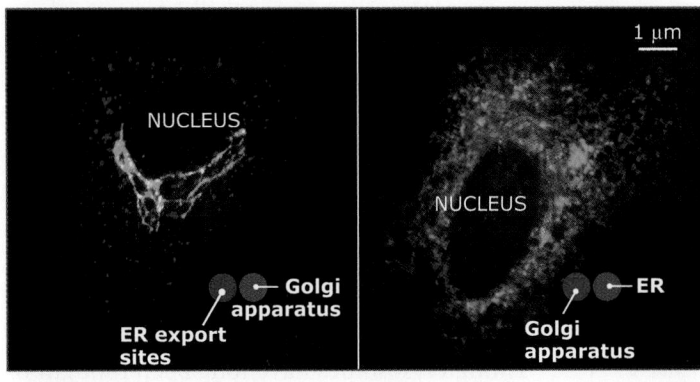

FIGURE 8.9 Immunofluorescence localization of proteins in the ER, Golgi apparatus, and ER export sites. Two different cells are shown. Photos courtesy of Laurence Pelletier, Yale University.

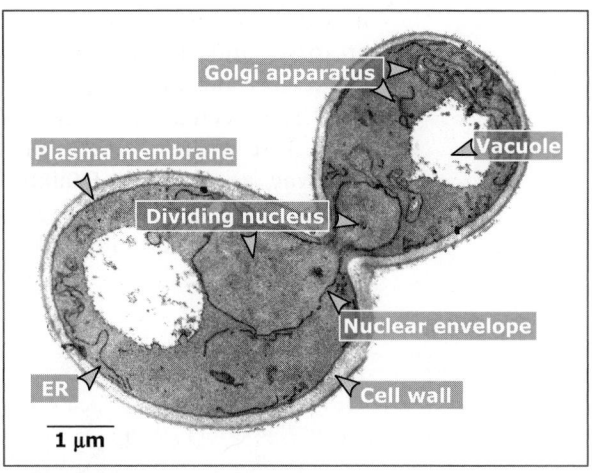

FIGURE 8.10 Transmission electron micrograph of a budding yeast cell. Photo courtesy of Francis A. Barr, University of Liverpool.

directionality of transport, the ER does not. Newly synthesized secretory proteins are directed toward ER exit sites, regions of the ER specialized for forming transport vesicles. In mammalian cells, these ER exit sites are located throughout the RER and are not often associated with the Golgi apparatus. In cells lacking a regulated secretory pathway, the ER exit sites are often located at a considerable distance from the Golgi apparatus. As shown in **FIGURE 8.9**, the ER pervades the entire cell cytoplasm and the hundred or so export sites are located randomly in it. In contrast, the Golgi apparatus is typically located near the nucleus. Therefore, some vesicles carrying secretory proteins from the ER must traverse distances of several microns before reaching the Golgi apparatus.

Just as pancreatic acinar cells were ideal cells to use in the microscopy and biochemical fractionation experiments that established the route taken by secretory proteins through the cell, so the budding yeast, *Saccharomyces cerevisiae*, has been a useful genetic system in which to identify the macromolecular machinery used in protein trafficking. As shown in **FIGURE 8.10**, budding yeast have a complement of exocytic organelles with functions similar to mammalian cells. However, their organization is somewhat different. The ER is rather sparse and mostly located just underneath the plasma membrane and around the nucleus. There are numerous small ER exit sites, which are used for biogenesis of transport carriers. The Golgi apparatus exists mostly as unstacked cisternae, with no obvious concentration near the nucleus. There are no obvious secretory granules in yeast, and regulated exocytosis is

rare. Despite these differences, the main protein machinery used for transport in yeast is homologous to that found in animal cells and is conserved among eukaryotes.

Concept and Reasoning Checks

1. Describe the polarized organization of the Golgi apparatus.
2. What are the vesicular pathways out of the *trans*-Golgi Network?

8.3 Overview of the endocytic pathway

Key concepts

- Extracellular material can be taken into cells by several different mechanisms.
- The low pH and degradative enzymes in endosomes and lysosomes are important in processing some endocytosed material.

Endocytosis is the process by which eukaryotic cells take up material from the extracellular environment by the formation of vesicles at the plasma membrane. In many ways the opposite of exocytosis, endocytosis has several functions as follows:

- internalization of nutrients
- regulating the cell surface expression of proteins, such as hormone receptors and glucose transporters, so that cells can control the uptake of ligands
- the uptake and digestion of extracellular debris
- recovery of membrane inserted into the plasma membrane during secretion.

In addition, pathogens, including bacteria, protozoa, and viruses, exploit the process of endocytosis for their entry into cells.

The vesicles that form at the plasma membrane fuse with organelles in the endocytic pathway, as shown in **FIGURE 8.11**. Two important features of these organelles are that the lumen is acidic and that they contain pro-

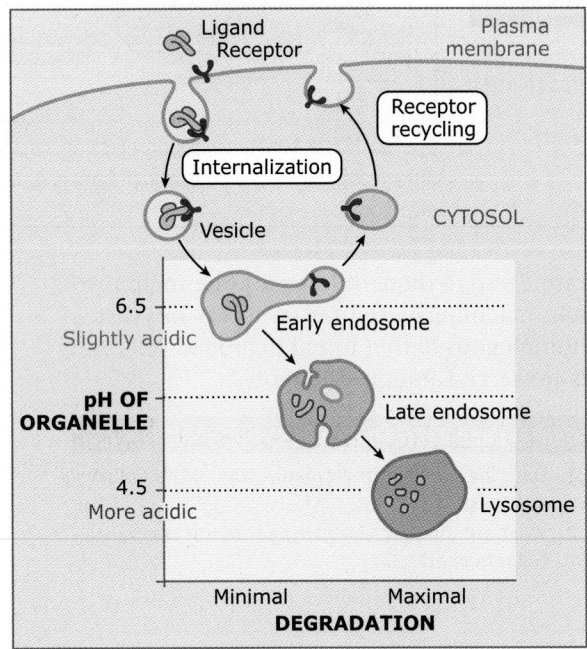

FIGURE 8.11 The organelles of the endocytic pathway constitute a gradient of acidity and degradative capacity. Endocytosed macromolecules are recycled to the plasma membrane or degraded.

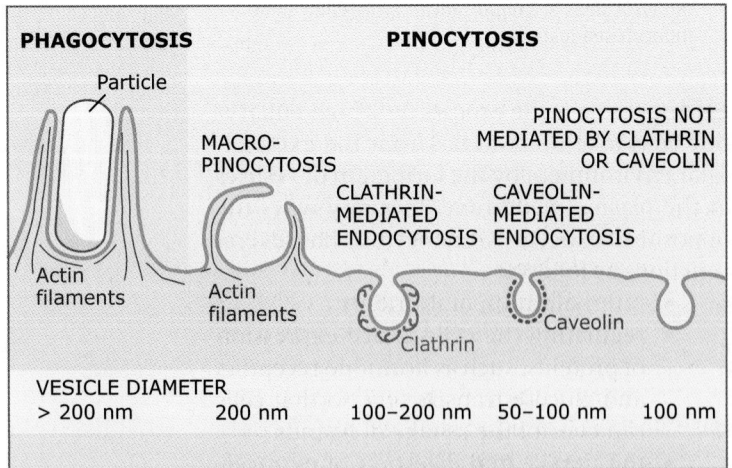

FIGURE 8.12 There are several mechanisms of endocytosis. Adapted from S. D. Conner and S. L. Schmid, *Nature* 422 (2003): 34–44.

teolytic and other degradative enzymes that function optimally at acidic pH. Together, these organelles form a continuum of acidity and concentration of degradative enzymes. They can be roughly divided into **early endosomes**, **late endosomes**, and **lysosomes**, which range from least to most degradative capacity. Lysosomes are a repository of degradative enzymes that together can degrade virtually any biologic polymer (proteins, lipid, carbohydrate, RNA, DNA) delivered to them by endocytosis. Though lysosomes have long been thought of as terminal organelles on the endocytic pathway, there is evidence that they can sometimes fuse with the plasma membrane.

An adenosine triphosphate (ATP)-driven proton pump called the vacuolar ATPase (v-ATPase) transfers H^+ ions from the cytosol to the lumen of endocytic organelles, thus decreasing their internal pH relative to the cytosol, which has a pH of 7.4. Early endosomes have a slightly acidic internal pH (6.5 to 6.8), whereas the internal pH of late endosomes and lysosomes can be as low as 4.5. Several factors contribute to the regulation of pH in the various endosomal compartments, including the density and activity of the v-ATPase itself as well as the ion conductance properties and other ion transport ATPases characteristic of different endosomal membranes. Importantly, the pH in each compartment is tailored to its functions (for details see *8.15 Some receptors recycle from early endosomes whereas others are degraded in lysosomes*). (For more on v-ATPases see *6.21 H^+-ATPases transport protons out of the cytosol*.)

Historically, endocytosis has been divided into **phagocytosis** ("cell eating") and **pinocytosis** ("cell drinking"), reflecting the size of the material being internalized, as shown in **FIGURE 8.12**. A professional phagocyte, such as a macrophage, can engulf material in transport vesicles as large as 10 μm in diameter. The analogy with eating is further supported by the internal environment of phagosomes, which is highly acidic and conducive to the activity of the enzymes that degrade the internalized proteins, lipids, and carbohydrates.

Nearly all cell types can mediate phagocytosis, but the process is most commonly associated with specialized cells of the immune system, such as macrophages and dendritic cells, which ingest pathogens to help mediate host defense mechanisms. Macrophages also participate in clearing senescent or apoptotic cells by phagocytosis, even in the absence of

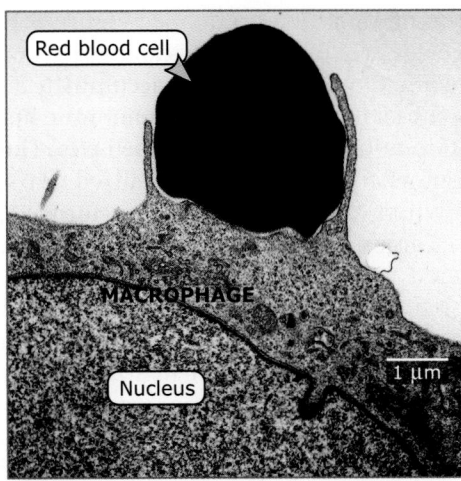

FIGURE 8.13 The electron micrograph shows a macrophage engulfing a red blood cell. Photo courtesy of John M. Robinson, Ohio State University.

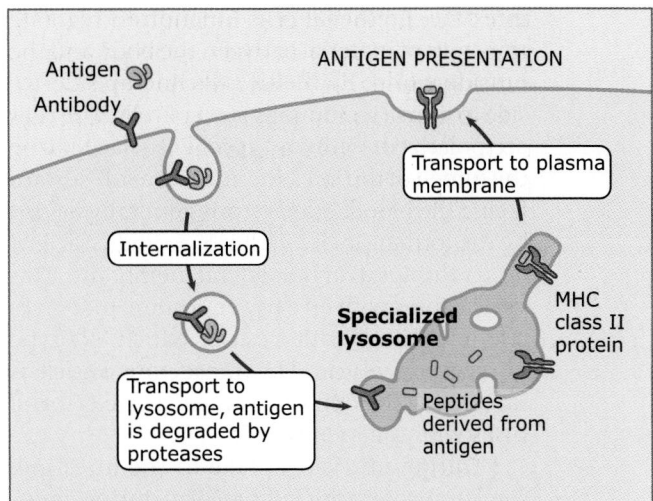

FIGURE 8.14 In cells of the immune system, specialized lysosomes degrade endocytosed protein antigens to peptides that bind to MHC class II molecules. The peptide-MHC class II complexes are transported to the plasma membrane for presentation to effector cells.

infection, as shown in **FIGURE 8.13**. Because professional phagocytes express specific receptors that trigger phagocytosis, they are more efficient at particle uptake than other cells. For example, macrophages and some other phagocytic cells express receptors for antibody molecules.

Pinocytosis is a general term that covers more than one mechanism of uptake, and typically involves the formation of small endocytic vesicles that are 0.1 to 0.3 μm in diameter. The best-studied type of pinocytosis is **receptor-mediated endocytosis**. Early studies on the uptake of low-density lipoprotein (LDL) by the LDL receptor provided much of the conceptual framework for our understanding of the endocytic pathway. At the cell surface, a wide variety of receptors binds ligands such as nutrients, growth factors, hormones, antibodies, or antigens. Receptor-ligand complexes are internalized by accumulating at specialized regions of the plasma membrane called "coated pits" (for details see *8.13 Endocytosis is often mediated by clathrin-coated vesicles*). Coated pits pinch off to form vesicles that fuse with early endosomes, where the acidity separates some ligands from their receptors, sending the receptors back to the cell surface and the ligands to the late endosomes and eventually the lysosomes. Some receptor-ligand complexes are not dissociated in early endosomes but instead move, as do released ligands, to the lysosome (see *8.15 Some receptors recycle from early endosomes whereas others are degraded in lysosomes*).

There are several variations on the basic endocytic pathway. Some cells have specialized lysosomes that carry out only partial degradation of internalized material, a feature essential for the stimulation of immune responses to internalized pathogens. Dendritic cells, which are white blood cells found in the circulation and in all tissues of the body, are a good example. They are responsible for initiating nearly all immune responses due to their unique ability to stimulate the B and T lymphocytes that recognize and mediate killing of pathogens. Dendritic cells capture circulating antigens or invading microorganisms and deliver them to a lysosomal compartment specialized by virtue of having an exceptionally low capacity for bulk proteolysis. These conditions favor the generation of short peptides (10–15 amino acids long) bound to major histocompatability complex (MHC) class II proteins. Another specialization then enables these peptide-MHC class II complexes to escape lysosomes by the formation of long tubules that carry their contents to the plasma membrane. This is shown in **FIGURE 8.14**. The peptide-class II complexes move to the cell surface where they stimulate effector cells.

Another variation of the endocytic pathway is to bypass lysosomes altogether, in a process called **transcytosis**. This process occurs in epithelial cells, the specialized cells that line all body cavities, such as the inner surface of the intestine. Epithelial cells function to regulate transport of material between the body and the outside world. Epithelial cells line up side-to-side to form a continuous layer of cells. The cells are polarized, with an "**apical**" surface facing the lumen of the intestine and a "**basal**" surface facing the blood. Transcytosis generally begins by the formation of a clathrin-coated vesicle at either the apical or basolateral membrane. This vesicle fuses with an early endosome that gives rise to a transcytotic vesicle, which is a type of recycling vesicle. The transcytotic vesicle is targeted to the opposite surface of the cell and fuses with the membrane.

During nutrient uptake in the intestine, some transport vesicles that form during endocytosis from the apical surface move directly to the basal surface, thereby delivering internalized components without risking their digestion in lysosomes, as illustrated in **FIGURE 8.15**. Another example of transcytosis is the transfer of humoral immunity from the mother to the newborn infant. Immunoglobulins in the mother's milk are captured in the infant gut by apical cell surface receptors that deliver them to the other side of the epithelium, into the blood plasma.

In addition to uptake by coated pits, other types of endocytic vesicles can form at the plasma membrane (see Figure 8.12). Caveolae are small invaginations coated with caveolin. Caveolae appear to accumulate a select subset of receptors and membrane lipids that do not accumulate at coated pits. These receptors and lipids are internalized into the cell when caveolae pinch off to form vesicles. Larger, more heterogeneous structures termed macropinosomes also form at the plasma membrane, particularly in response to certain growth factors. Macropinosomes are large vacuoles similar in size to phagosomes and can form bearing large droplets of extracellular fluid. Material internalized by caveolae and macropinosomes can reach the same endosomes and lysosomes as material internalized via coated pits. Lastly, some material enters cells via small pinocytic vesicles that form in the absence of known coats.

Interestingly, the acidity of the endocytic pathway is exploited by several enveloped animal viruses, which need to gain access to the cytoplasm in order to replicate. Viruses such as vesicular stomatitis virus and Semliki Forest virus gain entry by exploiting the acidic lumen of endosomes, which activates the spike glycoproteins on the viral surface, triggering fusion of the viral membrane with that of the endosome. This introduces the viral genome into the cell cytoplasm, initiating the process of infection.

Cells of the budding yeast *S. cerevisiae* carry out pinocytosis but not phagocytosis. This is not surprising since the yeast cell is surrounded by a thick cell wall. It is difficult to view most endocytic organelles in yeast morphologically because of their low abundance. However, the yeast vacuole, which is the equivalent of the lysosome, is readily seen by electron microscopy (see Figure 8.10). Genetic studies of endocytosis in yeast have lagged behind those of exocytosis until recently, when they have become particularly important in understanding the role played by ubiquitination in the degradation process (for details see *8.16 Early endosomes become late endosomes and lysosomes by maturation*).

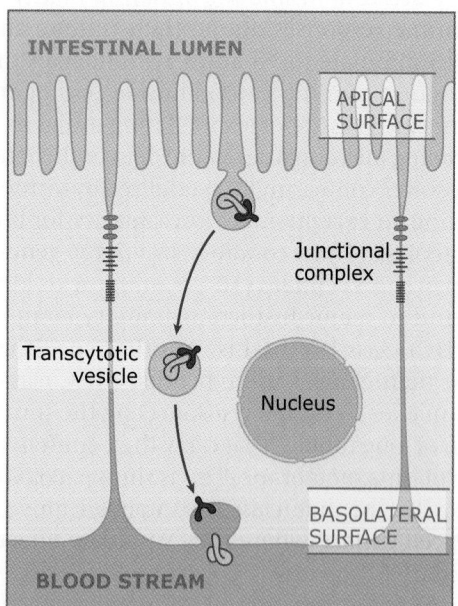

FIGURE 8.15 During transcytosis in polarized cells, material that has been endocytosed from one membrane domain is transported through the cell and exocytosed at a different membrane domain.

Concept and Reasoning Checks

1. What are five roles of endocytosis?
2. What are some variants on the endocytic pathway?
3. What is the pH gradient of the endosomal lumen from internalization to lysosome?

8.4 Concepts in vesicle-mediated protein transport

Key concepts

- Transport vesicles move proteins and other macromolecules from one membrane-bounded compartment to the next along the exocytic and endocytic pathways.
- Coats formed from cytoplasmic protein complexes help to generate transport vesicles and to select proteins that need to be transported.
- Proteins destined for transport to one compartment are sorted away from resident proteins and proteins that are destined for other compartments.
- Transport vesicles use tethers and soluble N-ethylmaleimide-sensitive factor attachment protein receptors (SNAREs) to dock and fuse specifically with the next compartment on the pathway.
- Retrograde (backward) movement of transport vesicles carrying recycled or salvaged proteins compensates for anterograde (forward) movement of vesicles.

The fact that eukaryotic cells consist of multiple membrane-bounded compartments, each of which has its own specific composition and function, has been discussed above. These compartments often exchange components, particularly during the processes of endocytosis and exocytosis. Because proteins and lipids are generally free to diffuse laterally in the lipid bilayer, any direct physical interaction between dissimilar membrane systems is an invitation to randomness. The concept that transport vesicles mediate this exchange (see Figure 8.3), and that in order for a compartment to maintain its composition, incorporation of cargo into vesicles must be selective has also been described above. In this section, the mechanism by which vesicle-mediated transport occurs is introduced.

Vesicle-mediated transport can be divided into a number of discrete steps, as illustrated in **FIGURE 8.16**. These steps include cargo selection, vesicle budding, fission, uncoating, tethering, docking, fusion, and recycling of fusion proteins.

Proteins to be transported must first be selected. Membrane proteins, such as receptors, contain distinct sorting signals in their cytoplasmic tails. Soluble proteins are selected by an appropriate receptor but some might also diffuse into the forming vesicle by bulk flow (see *8.5 The concepts of signal-mediated and bulk flow protein transport*). Some of the protein machinery needed at later steps is also selected for

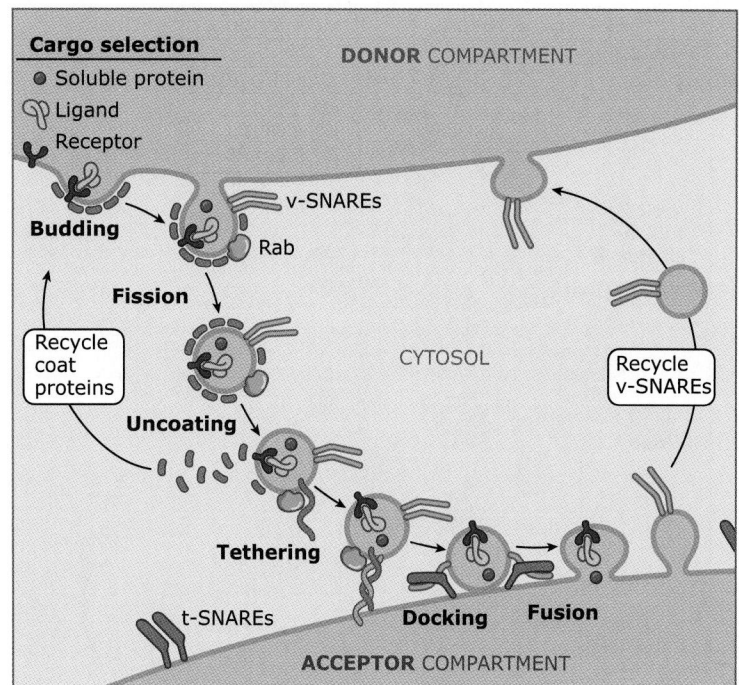

FIGURE 8.16 Vesicle-mediated transport consists of multiple steps that allow for the selective inclusion of cargo proteins into transport vesicles and the selective fusion of the transport vesicle with the target membrane.

transport. Some resident proteins, both membrane and soluble, have sequences that restrict them to a particular compartment and help prevent them from being selected or diffusing into forming transport vesicles.

Cargo selection occurs when cytoplasmic proteins, called **coat proteins**, bind to the sorting signals. Coat proteins bind either directly to sorting signals or indirectly by adaptor complexes that link the cargo proteins to the coat complexes. The coat proteins define the type of transport vesicle. Different coat proteins are used for different pathways. For example, coat protein I (COPI) and COPII coats are used in the exocytic pathway, whereas clathrin-coated vesicles are one type of vesicle used in the endocytic pathway, as summarized in **FIGURE 8.17**. The composition of the vesicles that mediate transport between compartments reflects the underlying mechanism of selective transfer. Not only does the cargo differ, the protein machinery involved in selecting the cargo, fission of the bud, vesicle docking, and fusion differ depending on the donor and target compartments. (For details see *8.6 Coat protein II–coated vesicles mediate transport from the ER to the Golgi apparatus; 8.8 Coat protein I–coated vesicles mediate*

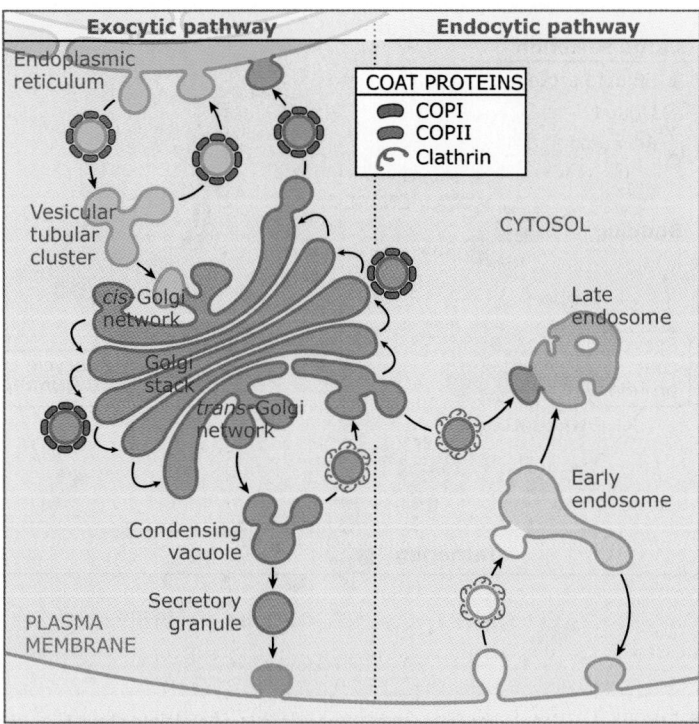

COAT PROTEINS
▬ COPI
▬ COPII
◠ Clathrin

Endoplasmic reticulum

Vesicular tubular cluster

CYTOSOL

cis-Golgi network

Golgi stack

trans-Golgi network

Condensing vacuole

Secretory granule

PLASMA MEMBRANE

Late endosome

Early endosome

FIGURE 8.17 The three main types of coats involved in vesicle-mediated transport are COPI, COPII, and clathrin coats.

retrograde transport from the Golgi apparatus to the ER; and *8.13 Endocytosis is often mediated by clathrin-coated vesicles*.) The coat proteins of Golgi to cell surface specific vesicles have not been identified and it is a distinct possibility that this transport step is mediated by vesicles that form by a process independent of coat proteins.

To form a vesicle, the organelle membrane must first be deformed to yield a "bud," as shown in Figure 8.16. The coat and adaptor complexes assist in this process, perhaps helped by associating with modified lipids. After a bud is formed, "fission" proteins help release the coated vesicle containing the selected material. The mechanisms by which the membrane deforms and fission occurs are largely unknown, but more is known about formation and fission of clathrin-coated vesicles than other types of coated vesicles.

After the vesicles have formed, the coat proteins are removed (in a process called "uncoating") and reused in subsequent rounds of vesicle budding. Uncoating appears to be required for vesicles to interact with the target membrane.

Transport vesicles become attached in two steps: the first is tethering, the second, docking.

Tethering is thought to provide the vesicle with an opportunity to sample the membrane, to determine whether it is the correct target. Docking is the process that brings the membranes close enough together to fuse. Tethers include small guanosine triphosphate-ases (GTPases) of the Rab family and tethering proteins that form protein complexes. Rabs and tethering complexes are specific for each pair of transport vesicle and destination organelle and thus help to determine the vesicle's destination (for details see *8.11 Rab guanosine triphosphate-ases and tethers are two types of proteins that regulate vesicle targeting*).

After tethering, SNARE proteins on the vesicle form complexes with their counterparts on the target organelle. This interaction "docks" the vesicle and plays an essential role leading to membrane fusion, which is necessary for a vesicle to complete its task of delivering membrane components and soluble internal molecules to its destination. Only by membrane fusion can two membranes physically coalesce and allow internal contents to come in contact. Different combinations of SNAREs mediate each of the many membrane fusion steps on the exocytic and endocytic pathways. The ATPase N-ethylmaleimide-sensitive factor (NSF) and the linking protein soluble NSF attachment protein (SNAP) subsequently disassemble the SNARE complexes formed during membrane fusion. The SNAREs from the donor membrane are then incorporated into budding transport vesicles for return to that membrane. (For details see *8.12 Soluble N-ethylmaleimide-sensitive factor attachment protein receptor proteins likely mediate fusion of vesicles with target membranes*.)

The elucidation of the molecular mechanisms of vesicle-mediated transport owes much to the development of innovative assays and analytical methods that were designed to provide new insights into each aspect of membrane traffic. Biochemical assays that reconstituted transport or fusion between organelles allowed the identification and purification of individual protein components required for transport (e.g., SNAREs, NSF, SNAP). The isolation of yeast (*S. cerevisiae*) mutants defective in protein secretion confirmed the physiologic significance of these proteins and led to the discovery of many other proteins. New methods involving genome wide screens in higher eukaryotes continue to reveal new components involved in protein secretion.

8.5 The concepts of signal-mediated and bulk flow protein transport

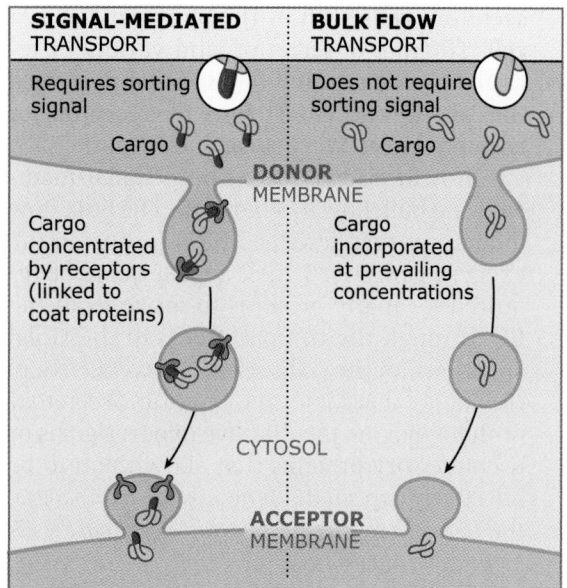

FIGURE 8.18 Cargo molecules are either concentrated in budding vesicles by receptors that bind to sorting signals on the cargo (signal-mediated transport) or incorporated into vesicles at the prevailing concentration (bulk flow).

There are two schools of thought concerning the mechanism by which proteins move through the endocytic and exocytic pathways. These are signal-mediated transport and bulk flow transport, illustrated in **FIGURE 8.18**. In signal-mediated transport, each cargo protein contains one or more short sequences that specify each destination along the route. The presence or absence of these "sorting sequences" (also called sorting signals) would dictate whether a protein stays in a given organelle as a resident component or moves in a vesicle to the next organelle, where it could stay or leave, and so on. If each step requires a different signal, then each protein must have multiple sequences reflecting the number of sequential transport steps it must navigate in order to reach its final destination (e.g., the plasma membrane in the exocytic pathway or lysosomes in the endocytic pathway).

For soluble proteins within the lumen of the vesicles and compartments in these pathways, the problem is more acute than for membrane proteins. Soluble proteins are, topologically, "outside" the cell. In order for their transport to be entirely signal-mediated, soluble proteins would be maneuvered by binding to membrane receptors that in turn bind to coat components that selectively accumulate cargo into budding vesicles. If each class of soluble protein needed a different receptor for each step, dozens or hundreds of receptors might

be required to accommodate all the proteins to be transported at a given time.

The other school of thought provides an alternative that reduces the complexity of the problem. In the bulk flow model, "default pathways" are defined as routes that do not require proteins to have sorting signals in order to move between membrane-bounded compartments. The exocytic and endocytic pathways would be considered as default pathways. In the exocytic pathway, for example, secreted proteins or plasma membrane proteins would not need sorting signals to enter budding vesicles. These proteins would be present at the same concentration in a vesicle budding from the ER as in the lumen of the ER. The bulk flow mechanism does not involve a sorting signal that would be used to concentrate the protein in the budding vesicle. Signals would only be required to stop a protein in a compartment within the exocytic pathway (e.g., retention in the Golgi apparatus) or to divert it onto another pathway (e.g., transport to lysosomes).

The concept of bulk flow can explain how the bacterial protein β-lactamase, introduced into *Xenopus* oocytes, can be transported from the ER to the cell surface. Bacteria have no internal membranes equivalent to those in eukaryotes, so their proteins presumably would lack an ER export signal. Bulk flow may also help explain the high rate of export by professional secretory cells such as the pancreatic acinar cell. The rate of secretion and variety of secreted proteins would be expected to exceed the capacity of the cell to provide receptors for each protein at every stage of transport.

Protein transport uses both signal-mediated and bulk flow mechanisms. The bulk flow model was first described for transport in the exocytic pathway. Signals appear not to be required for many proteins to move from the ER, through the compartments of the Golgi apparatus, to the cell surface. However, there is evidence that some membrane and secretory proteins leaving the ER have export signals or use accessory proteins that allow them to be selectively exported (for details see *8.6 Coat protein II–coated vesicles mediate transport from the ER to the Golgi apparatus*).

For proteins in the endocytic pathway, the available evidence suggests that signal-mediated transport is the preferred mechanism for some transport steps. For example, sorting signals in the cytoplasmic tails of membrane proteins mediate internalization at the plasma membrane (for more details see *8.13 Endocytosis is often mediated by clathrin-coated vesicles*). However, internalized soluble macromolecules move from endosomes to lysosomes without binding to receptors (see *8.16 Early endosomes become late endosomes and lysosomes by maturation*).

Signals increase the efficiency of internalization when the concentration of molecules is low. Small molecules, added to the medium surrounding nonphagocytic cells in culture, are internalized in the fluid phase (i.e., not bound to receptors) less efficiently than when they are bound to membrane receptors. Binding to receptors with internalization signals can increase the efficiency of uptake considerably (up to 1000-fold) by concentrating the molecules within the vesicles. However, when the concentration of molecules is high, significant amounts can be internalized even in the absence of receptor binding. These considerations also apply to export from the ER. Nonselective bulk flow of soluble cargo is also significant when the transport vesicles have a large diameter, for example, during macropinocytosis at the plasma membrane.

Concept and Reasoning Check

1. Describe two schools of thought on the mechanism of intracellular transport of proteins through the secretory pathway and their defining differences.

8.6 Coat protein II–coated vesicles mediate transport from the ER to the Golgi apparatus

Key concepts

- COPII vesicles are the only known class of transport vesicles originating from the ER.
- Assembly of the COPII coat proteins at export sites in the ER requires a GTPase and structural proteins.
- Export signals for membrane proteins in the ER are usually in the cytoplasmic tail.
- After fission, COPII vesicles may cluster, fuse, and then move along microtubule tracks to the cis-side of the Golgi apparatus.

The ER is specialized for the synthesis and proper folding of secretory proteins and other proteins destined for the compartments in the exocytic and endocytic pathways. Transport from the ER occurs only when folding is complete and, in the case of multisubunit protein complexes, only when they are fully assembled (see *7 Membrane targeting of proteins*). Proteins to be exported are gathered at ER exit sites, which are responsible for the assembly of the transport vesicles that carry the newly synthesized proteins to the Golgi apparatus (see Figure 8.9).

The transport vesicles are called **COPII** vesicles, and they provide the sole route of vesicular exit from the ER. The COPII components provide the means of deforming the membrane into a bud and may also interact specifically with some transmembrane cargo proteins to select them for export. The principles that govern the formation of COPII vesicles apply to other coated vesicles in the cell (see *8.4 Concepts in vesicle-mediated protein transport*).

The process of COPII vesicle budding is illustrated in **FIGURE 8.19**. Vesicle formation is initiated by the sequential recruitment of the soluble COPII components, Sar1p, Sec23/Sec24, and Sec13/31, from the cytosol. Sar1p is a small GTPase that exists in the cytosol in its inactive GDP-bound form. Sar1p binds to ER membranes by interacting with the guanine nucleotide exchange factor Sec12p (also called Sar-GEF), which is an intrinsic membrane protein. Sec12p activates Sar1p by exchanging the GDP for GTP.

Binding of Sar1p-GTP to the membrane triggers the sequential recruitment of the two

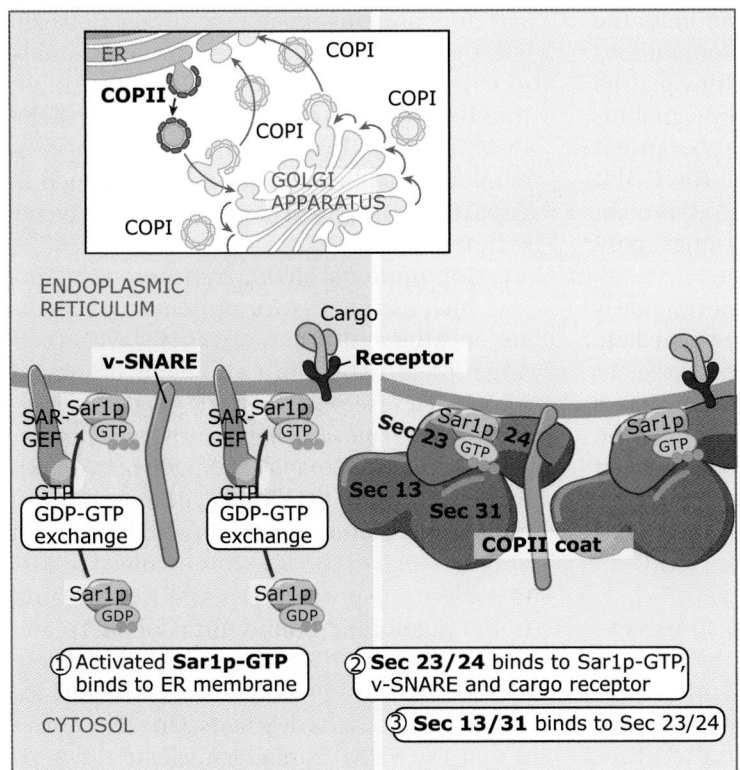

FIGURE 8.19 A model for the sequential recruitment of coat proteins and cargo molecules (v-SNARE and cargo receptor) during the formation of a COPII vesicle.

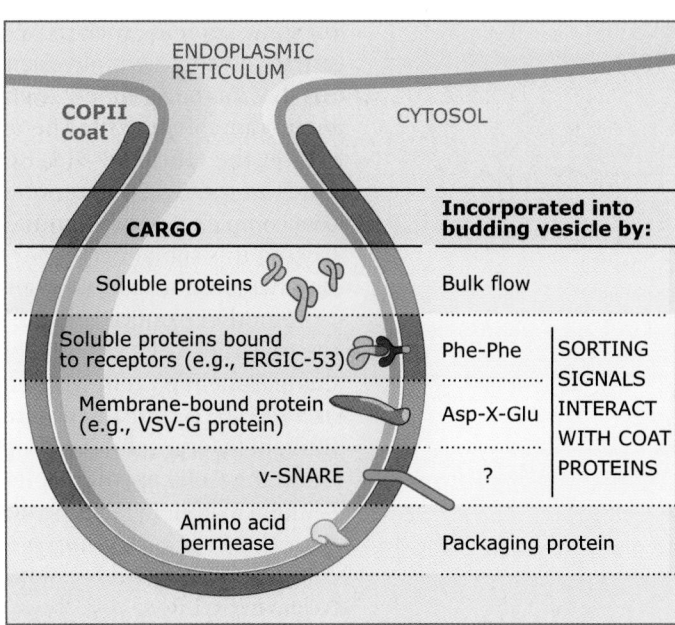

FIGURE 8.20 Cargo proteins are incorporated into nascent COPII vesicles by different mechanisms.

structural coat complexes, the Sec23/Sec24 and Sec13/Sec31 heterodimers. The crystal structure of the Sar1p-GTP-Sec23/24 complex reveals a bow-tie shape that would neatly fit the curved surface of a COPII-sized vesicle. Polymerizing these complexes with the Sec13/Sec31 complex might, therefore, help explain the deformation and budding of the vesicle. Sec24 binds the cargo that is incorporated into the COPI vesicle. There are multiple overlapping binding sites that help to explain the large number of different cargo molecules that can be incorporated. Sec23 binds directly to Sar1p and stimulates hydrolysis of the bound GTP during or after the formation of the vesicle. Sec23, therefore, acts as a GTPase-activating protein (GAP), uncoating the vesicle, and releasing the coat subunits for further cycles of budding.

Most cargo proteins are incorporated into budding COPII vesicles either by bulk flow or the use of sorting signals, as illustrated in **FIGURE 8.20**. The bulk flow mechanism seems to apply mostly to the soluble secretory proteins that are synthesized in large amounts in professional

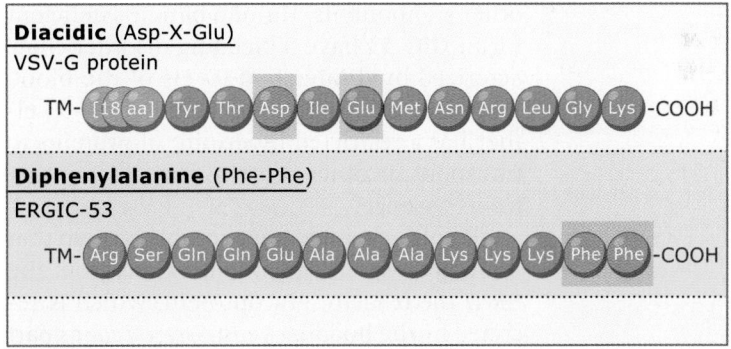

FIGURE 8.21 ER export signals include those that contain diacidic or diphenylalanine residues.

secretory cells (see *8.5 The concepts of signal-mediated and bulk flow protein transport* and *8.19 Some cells store proteins for later secretion*).

The sorting signals for ER export are usually short amino acid sequences in the cytoplasmic tails of membrane proteins, as shown in **FIGURE 8.21**. Of all sorting signals known, ER export signals are the least well characterized. One example is the diacidic signal found on some proteins such as VSV-G protein, the surface glycoprotein of vesicular stomatitis virus that has been widely used to study plasma membrane biogenesis. The diacidic signal increases the efficiency of ER export because it binds to the Sec24 component of the COPII coat complex. Export of mutant VSV-G protein lacking this signal occurs ~2 to 3 times more slowly than normal VSV-G protein.

Another example of an ER export signal is the one in ERGIC-53 (*ER*, *G*olgi, *I*ntermediate *C*ompartment protein with a molecular weight of 53 kDa), a transmembrane protein that is thought to ferry soluble glycoproteins from the ER to the Golgi apparatus. Its cytoplasmic tail contains a diphenylalanine signal that interacts with the Sec23/24 component of the COPII complex. This signal is essential for ERGIC-53 to leave the ER.

Soluble cargo proteins can also use export signals but do so indirectly by binding to receptors that are incorporated into COPII vesicles (see Figure 8.19). The soluble proteins that ERGIC-53 ferries from the ER to the Golgi apparatus contain such a signal. The lumenal domain of ERGIC-53 binds to these proteins via their high mannose oligosaccharides, which are present on most glycoproteins leaving the ER. However, because ERGIC-53 does not bind all proteins with high mannose oligosaccharides, the signal must have other components. Human patients deficient for ERGIC-53 have a bleeding disorder characterized by diminished levels of the blood coagulation Factors V and VIII. ERGIC-53 either has a restricted repertoire of proteins to transport, or other proteins can compensate for its absence.

Another example of a soluble protein that undergoes signal-dependent export from the ER is the α-factor glycoprotein, which is secreted by the budding yeast *S. cerevisiae* as part of the mating process. A membrane-spanning protein receptor has recently been identified that concentrates α-factor in COPII vesicles. Both the ERGIC-53 and the α-factor receptor (Erv29p) are packed into COPII vesicles and exit the ER with cargo.

Another way to incorporate cargo proteins into budding COPII vesicles is to use a packing protein that does not itself end up in the vesicle. An example is a protein called TANGO1, which localizes specifically to the ER exit sites and facilitates the loading of bulky collagen VII into COPII vesicles in mammalian cells. TANGO1 is likely to facilitate loading of a class of soluble secretory proteins and not specifically Collagen VII. Collagen VII is involved in the assembly of extra cellular matrix and TANGO1 might thus be involved in loading components of the ECM into COPII vesicles. It is likely that cells employ a small number of molecules such as TANGO1 to help load large number of diverse secretory cargo.

Components of the transport machinery, such as the SNARE proteins that mediate docking and fusion of vesicles with their target membrane, must also be incorporated into vesicles (see *8.12 Soluble* N-*ethylmaleimide-sensitive factor attachment protein receptor proteins likely mediate fusion of vesicles with target membranes*). How are SNAREs, which are integral membrane proteins, incorporated into COPII vesicles? Two of the SNAREs involved in ER-to-Golgi transport (Bet1p and Bos1p) bind to the membrane-bound forms of Sar1p and Sec23/24. Sec13/31 is recruited after these SNAREs, which argues that the SNAREs are incorporated as the vesicle buds. One hypothesis is that the SNAREs might nucleate the budding process and so guarantee the presence of targeting signals in all the COPII vesicles that form. However, there is no direct evidence for this attractive possibility.

After separation, by fission, from the ER, COPII vesicles appear to cluster and possibly fuse with each other at ER exit sites, forming "vesicular tubular clusters" (VTCs; see Figure 8.17). When COPII vesicles form at an export site distant from the Golgi apparatus, the VTCs may move along microtubule tracks to reach their destination. Both spatial apposition (proximity of the ER and Golgi apparatus) and microtubule-dependent transport may increase the efficiency of COPII vesicles finding their destination.

The ER of budding yeast does not have discrete export sites, and COPII vesicles can bud from all parts of the ER. In contrast to mammalian cells, the Golgi cisternae are dispersed throughout the cytoplasm, mostly as individual rather than stacked cisternal structures. Thus, in budding yeast, COPII vesicles may simply take the shortest distance between the site of budding in the ER and the first available cisterna.

Concept and Reasoning Check

1. What is the role of COPII vesicles?

8.7 Resident proteins that escape from the ER are retrieved

Key concepts

- Abundant, soluble proteins of the ER contain sequences (such as Lys-Asp-Glu-Leu [KDEL] or a related sequence) that allow them to be retrieved from later compartments by the KDEL receptor.
- Resident membrane proteins and cycling proteins are retrieved to the ER by a dibasic signal in the cytoplasmic tail.
- The ER retrieval signal for type I transmembrane proteins is a dilysine signal, whereas type II transmembrane proteins have a diarginine signal.

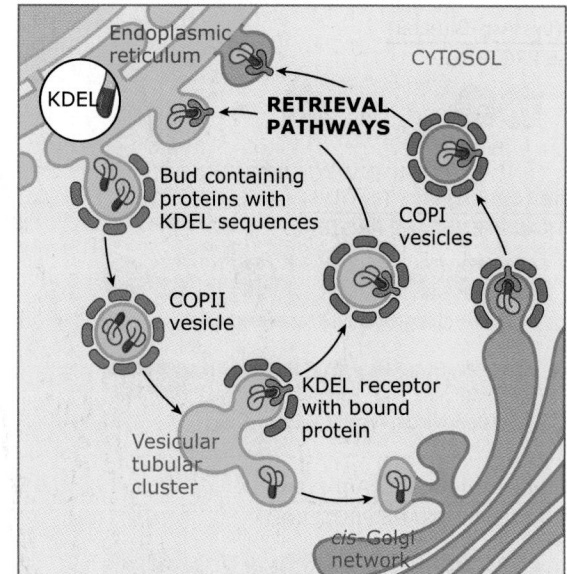

FIGURE 8.22 Proteins present at high concentrations in the ER lumen occasionally escape to the vesicular tubular clusters and *cis*-Golgi compartment. These proteins have a KDEL sequence that is recognized by the KDEL receptor, which returns them to the ER.

The lumen of the ER contains a high concentration of proteins, estimated to be >40 mg/ml. These include newly synthesized secretory proteins and, to a greater extent, resident ER proteins, in particular the chaperone proteins that help secretory proteins fold into functional conformations. Given the abundance of proteins and the likelihood that at least some ER export occurs by bulk flow, the cell is continuously "at risk" of exporting proteins that function in the ER. Indeed, such export does occur, but it rarely, if ever, results in the secretion of ER resident proteins to the extracellular environment. This is because cells have an effective "retrieval" mechanism that captures escapees from compartments further along the exocytic pathway and returns them to the ER.

The best-characterized retrieval mechanism involves a tetrapeptide sorting signal present at the carboxyl-terminus of most or all soluble resident proteins of the ER lumen. In mammalian cells, the **retrieval signal** has the sequence KDEL, which was first discovered for chaperone proteins. Variants of this signal are found in all other eukaryotes. In budding yeast, the ER retrieval signal is His-Asp-Glu-Leu (HDEL).

As shown in **FIGURE 8.22**, escaped proteins are recognized by a receptor that binds to the retrieval signal. In mammalian cells, the KDEL receptor is found mostly in the post-ER compartments of VTCs and the CGN, but a few KDEL receptors are found in more distal compartments of the Golgi apparatus. Binding of the KDEL receptor to the protein's KDEL sequence triggers the incorporation of the receptor-protein complex into a class of coated vesicles called COPI vesicles, which return the complex back to the ER (for details see *8.8 Coat protein I–coated vesicles mediate retrograde transport*

from the Golgi apparatus to the ER). There, the retrieved protein dissociates from the receptor, a process that may be triggered by the higher concentration of calcium ions in the lumen of the ER than in the VTCs and CGN. *In vitro*, dissociation can be triggered by a pH change, but it is not known if the ER and CGN differ in pH. The empty KDEL receptor exits the ER in COPII vesicles, returning to the VTCs and CGN for further rounds of retrieval.

Resident membrane proteins of the ER can also escape, and they, too, have retrieval signals. These are called dibasic signals. Resident ER proteins that are type I transmembrane proteins (N-terminus in the lumen) have a dilysine signal (Lys-X-Lys-X-X, where X is any amino acid) at the end of their cytoplasmic tails, as shown in **FIGURE 8.23**. The dilysine signal binds directly to the α subunit of the COPI coat (see *8.8 Coat protein I–coated vesicles mediate retrograde transport from the Golgi apparatus to the ER*). Resident ER proteins that are type II membrane proteins (C-terminus in the lumen) use a different retrieval signal that has two adjacent arginines.

Membrane proteins that normally cycle between the ER and the Golgi apparatus are also retrieved using a dibasic signal. An example is ERGIC-53, which transports some soluble proteins from the ER to the Golgi apparatus (see

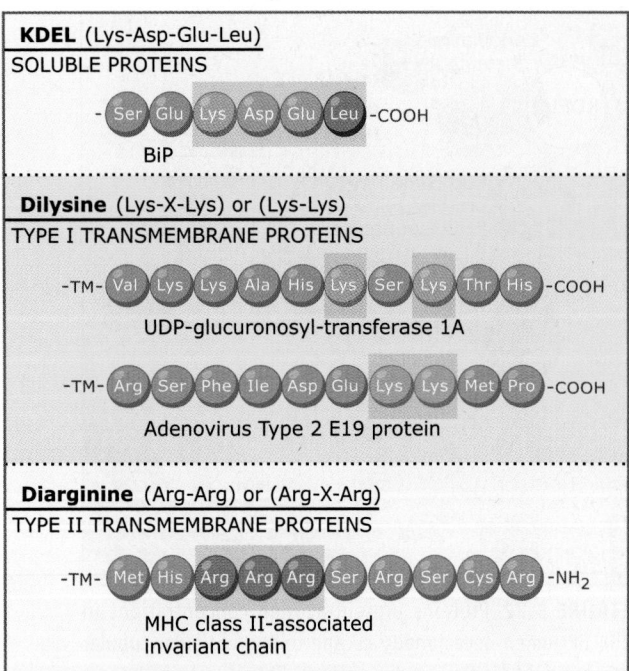

KDEL (Lys-Asp-Glu-Leu)
SOLUBLE PROTEINS

- Ser Glu Lys Asp Glu Leu -COOH
BiP

Dilysine (Lys-X-Lys) or (Lys-Lys)
TYPE I TRANSMEMBRANE PROTEINS

-TM- Val Lys Lys Ala His Lys Ser Lys Thr His -COOH
UDP-glucuronosyl-transferase 1A

-TM- Arg Ser Phe Ile Asp Glu Lys Lys Met Pro -COOH
Adenovirus Type 2 E19 protein

Diarginine (Arg-Arg) or (Arg-X-Arg)
TYPE II TRANSMEMBRANE PROTEINS

-TM- Met His Arg Arg Arg Ser Arg Ser Cys Arg -NH2
MHC class II-associated
invariant chain

FIGURE 8.23 Retrieval signals mediate the return of resident soluble and transmembrane proteins to the ER.

8.6 Coat protein II–coated vesicles mediate transport from the ER to the Golgi apparatus). ERGIC-53 has both an ER export signal and a dilysine retrieval signal in its cytoplasmic tail to exit and return to the ER, respectively.

Concept and Reasoning Check

1. What is the fate of resident proteins (e.g., of the ER) that "escape" into the secretory pathway?

8.8 Coat protein I–coated vesicles mediate retrograde transport from the Golgi apparatus to the ER

Key concepts

- COPI coat assembly is triggered by a membrane-bound GTPase called ARF, which recruits coatomer complexes, and disassembly follows GTP hydrolysis.
- COPI coats bind directly or indirectly to cargo proteins that are returned to the ER from the Golgi apparatus.

Just as COPII-coated vesicles are involved in the forward movement of newly synthesized and folded proteins from the ER to the Golgi apparatus, so **COPI**-coated vesicles are involved in **retrograde transport** from the

Golgi apparatus to the ER (see Figure 8.17). COPI vesicles play an important role in the retrieval of escaped ER proteins (see *8.7 Resident proteins that escape from the ER are retrieved*). In addition, they are thought to recycle essential vesicle components such as SNARE proteins from the Golgi apparatus back to the ER and from the *trans* side of the Golgi apparatus to the *cis* side. There is also evidence that COPI vesicles are involved in forward transport within the Golgi stack, though the precise role is still controversial (see *8.9 There are two popular models for forward transport through the Golgi apparatus*).

The best evidence of a role for COPI vesicles in retrograde transport comes from the use of genetics to manipulate the efficacy of mating in yeast. In order to initiate the mating process, the mating factor must bind to its receptor on the plasma membrane. An α-factor receptor that is engineered to contain a dilysine retrieval signal in its cytoplasmic tail is not expressed on the cell surface. Mating cannot occur in cells expressing this mutant receptor. Suppressor mutations that restore mating were identified. These mutations are in genes encoding components of the COPI coat, thereby implicating COPI vesicles in retrieval of dilysine-tagged proteins to the ER.

The formation of COPI vesicles requires the assembly of a GTPase and coat proteins on the donor membrane, as shown in **FIGURE 8.24**. The assembly process is analogous to that for COPII vesicles but uses different proteins. For COPI vesicles, the first step is recruitment of the GTPase adenosine diphosphate (**ADP**)–**ribosylation factor (ARF)** to the membrane. ARF is less well characterized than its ER counterpart, Sar1p, which plays an analogous role in the assembly of COPII coats (see *8.6 Coat protein II–coated vesicles mediate transport from the ER to the Golgi apparatus*). ARF was first identified as a cofactor needed for cell killing by cholera toxin.

ARF bound to GDP is a soluble protein in the cytoplasm. As with other small GTPases, ARF is activated by binding to a GTP-GDP exchange factor (GEF). The ARF-GEF is present on Golgi membranes and acts specifically on ARF to induce the dissociation of GDP and the binding of GTP. GTP loading causes a conformational change that exposes a fatty acid residue (a myristic acid) at the N-terminus of ARF that helps anchor the ARF-GTP in Golgi membranes. ARF-GEF is the specific target of Brefeldin A, a pharmacologic agent that blocks the assembly of COPI coats and causes a

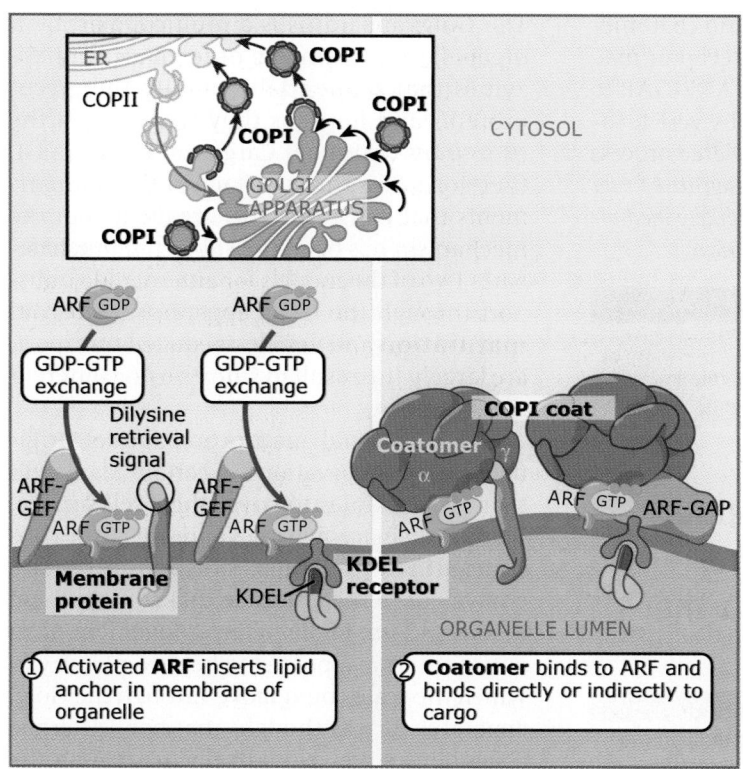

Labels in figure:

ER
COPII
COPI
COPI
COPI
COPI
COPI
GOLGI APPARATUS
CYTOSOL

ARF GDP   ARF GDP

GDP-GTP exchange   GDP-GTP exchange

Dilysine retrieval signal

COPI coat

Coatomer α   γ

ARF-GEF   ARF-GEF

ARF GTP   ARF GTP   ARF GTP   ARF GTP   ARF-GAP

Membrane protein   KDEL   KDEL receptor

ORGANELLE LUMEN

① Activated **ARF** inserts lipid anchor in membrane of organelle

② **Coatomer** binds to ARF and binds directly or indirectly to cargo

massive redistribution of Golgi membranes into the ER.

The membrane-bound ARF-GTP can then recruit COPI coat complexes (known as **coatomers**) to the membrane. Each coatomer complex is an assembly of seven proteins called α-, β-, β′-, γ-, δ-, ε-, and ζ-COP. Recruitment of coatomer to the membrane may be aided by the presence of negatively charged membrane lipids that bind to coatomer complexes. Coatomer binding helps to deform the membrane during vesicle budding. Thus, the role of coatomer in budding COPI vesicles is analogous to that of Sec23/24 and Sec13/31 for COPII vesicles.

COPI coats can bind directly or indirectly to proteins that are recycled back to the ER. For type I membrane proteins, the γ-subunit of coatomer binds directly to a dilysine retrieval signal in the cytoplasmic tail (see Figure 8.24). COPI coats also bind indirectly to the KDEL receptor, probably through an ARF-GTPase activating protein (ARF-GAP), to return KDEL-containing soluble proteins that have escaped the ER. This ARF-GAP also stimulates the hydrolysis of GTP on ARF, leading to eventual uncoating.

As shown in Figure 8.22, the retrieval process starts as soon as the vesicular tubular clusters (VTCs) are formed by the fusion of COPII vesicles, which bud from the ER, and continues as the VTCs move along microtubules to the CGN (see *8.6 Coat protein II–coated vesicles mediate transport from the ER to the Golgi apparatus*). Retrieval also occurs from the CGN and, to a lesser extent, from more distal parts of the Golgi apparatus.

Retrieval from distal Golgi compartments has been exploited by certain toxins, such as cholera toxin, a product of the bacterium *Vibrio cholera*. Cholera toxin is endocytosed, and it enters the cytoplasm from the ER, most likely using the retrotranslocation machinery in the ER membrane (see *7.19 Terminally misfolded proteins in the endoplasmic reticulum are returned to the cytosol for degradation*). One of the cholera toxin subunits has a C-terminal KDEL sequence. The endocytosed toxin is transported from endosomes to the TGN, where it binds the distal KDEL receptor and undergoes retrograde transport to the ER.

Before COPI vesicles can fuse with the ER, they must be uncoated so that the lipid bilayer of the transport vesicle can be closely apposed to its fusion partner. Fusion will not occur if uncoating is blocked. The mechanism of uncoating has not been fully elucidated, but it is thought that dissociation of coatomer from the vesicle is triggered by an ARF-GAP. The idea is that ARF-GAP stimulates GTP hydrolysis by ARF, resulting in release of ARF-GDP and coatomer from the membrane. The KDEL receptor may participate in the recruitment of ARF-GAP to the vesicle. Uncoating is followed by vesicle fusion with the ER, in a SNARE-mediated process (see *8.12 Soluble* N-*ethylmaleimide-sensitive factor attachment protein receptor proteins likely mediate fusion of vesicles with target membranes*).

Concept and Reasoning Checks

1. What is the role of COPI vesicles?
2. What class of proteins play a crucial role in assembly of COPI and COPII coatomers?

8.9 There are two popular models for forward transport through the Golgi apparatus

Key concepts

- Transport of large protein structures through the Golgi apparatus occurs by cisternal maturation.
- Individual proteins and small protein structures are transported through the Golgi apparatus either by cisternal maturation or vesicle-mediated transport.

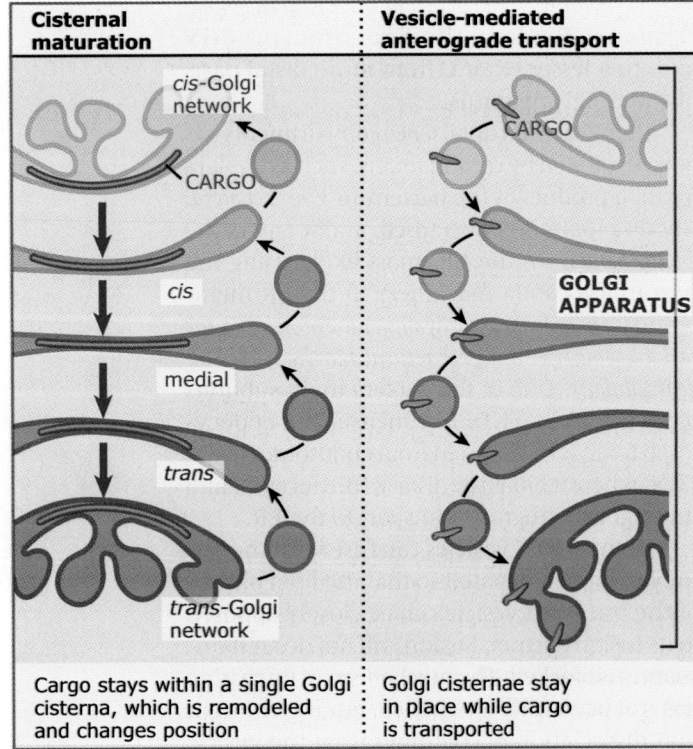

Cisternal maturation	Vesicle-mediated anterograde transport

cis-Golgi network

CARGO

CARGO

cis

medial

GOLGI APPARATUS

trans

trans-Golgi network

Cargo stays within a single Golgi cisterna, which is remodeled and changes position	Golgi cisternae stay in place while cargo is transported

FIGURE 8.25 Two models for transport through the Golgi apparatus are cisternal maturation and vesicle-mediated transport.

The Golgi apparatus is a multicompartment organelle containing an ordered array of enzymes that sequentially modify the glycoproteins and lipids as they transit from the *cis* to *trans* cisternae. Cargo molecules must, therefore, pass through each of the compartments that constitute this organelle, but the mechanism has been the subject of much debate. Two of the models for anterograde transport through the Golgi apparatus, **cisternal maturation** and vesicle-mediated transport, are largely the result of studies using different sized cargoes.

The cisternal maturation model originated from observations of certain plant cells that secrete scales to cover their cell surface. These proteinaceous scales appeared to be constructed in the Golgi apparatus and are of a size equivalent to the cisternae that contain them (about 1–2 μm in diameter). Movement of an entire cisterna appeared to be the only way in which the scale could move through the Golgi apparatus. It was thought that new cisternae assembling scales would be generated on the *cis*, or entry, side of the Golgi and mature cisternae containing scales would leave at the *trans*, or exit, side.

These observations gave rise to the idea of cisternal maturation, in which the cisterna carrying the scale would move through the stack, as illustrated in **FIGURE 8.25**. Maturation would not only involve physical movement of the cisterna but also remodeling of the cisternal membrane, to change the set of enzymes acting on the maturing scale to the next set in the sequence. Evidence in support of cisternal maturation mode of transport comes from live imaging of yeast. There is also now morphologic evidence that large proteins such as collagen fibrils may traverse the Golgi apparatus in mammalian cells by cisternal maturation, that is, without entering a typical transport vesicle of the COPI kind (see *19.2 Collagen provides structural support to tissues*).

The vesicle-mediated transport model arose from observations on the anterograde transport of the VSV-G protein in *in vitro* reconstitution experiments using Golgi membranes. These experiments demonstrated that COPI vesicles contain VSV-G protein under conditions that correlate with its transport between the *cis*- and *trans*-Golgi cisternae. Specifically, VSV-G protein was expressed in cells genetically deficient in a glycosyltransferase that catalyzes its terminal glycosylation. Golgi membranes from these cells were incubated in vitro with Golgi membranes from wild-type cells that contain the glycosyltransferase. Terminal glycosylation was observed and COPI vesicles appeared to serve as intermediates that mediate transport between the Golgi cisternae. Direct fusion of the Golgi membranes was not observed. These experiments gave rise to the idea, illustrated in Figure 8.25, that the Golgi apparatus comprises a stable stack of cisternae with cargo moving forward from cisterna to cisterna in COPI-coated vesicles.

A requirement of the vesicle transport mechanism, as derived from the *in vitro* reconstitution experiments, is that COPI vesicles mediate forward transport of secretory cargo across the Golgi apparatus in the *cis*-to-*trans* direction. However, the discovery that COPI vesicles mediate retrograde transport from the Golgi apparatus to the ER (see *8.8 Coat protein I-coated vesicles mediate retrograde transport from the Golgi apparatus to the ER*) suggests a mechanism for how cisternal remodeling might work in the cisternal maturation model. The COPI vesicles could retrieve resident components of the cis cisternae as they traverse the Golgi stack and "mature" to become *trans* cisternae. Indeed, Golgi enzymes have been found within COPI vesicles by biochemical techniques and by immunoelectron microscopy. Although, secretory cargo has been identified in COPI vesicles, the role of these vesicles in forward transport across the Golgi stack remains highly controversial.

Put simply, in vesicle-mediated transport, the Golgi membranes remain in place and the cargo is transported in vesicles. In contrast, in cisternal maturation, the cargo remains in place (within a cisterna) and the cisternal membrane is remodeled around it. It is important to realize that these two models are not mutually exclusive and that both cisternal maturation and vesicle-mediated transport may operate at the same time. Cargo too large to enter COPI vesicles could move through the Golgi apparatus by cisternal maturation, leaving smaller cargo molecules to transit the stack in COPI vesicles. This does not mean that cisternal maturation only moves large structures, but that it may be the best (if not only) means of doing so.

Concept and Reasoning Check

1. What is the critical difference between cisternal maturation and vesicle-mediated transport models of forward transport through the Golgi apparatus?

8.10 Retention of proteins in the Golgi apparatus depends on the membrane-spanning domain

Key concept

- The membrane-spanning domain and its flanking sequences are sufficient to retain proteins in the Golgi apparatus.

We have thus far considered sorting signals that program selective transport of cargo from one compartment to another in the exocytic pathway, either in the forward or reverse direction. However, there is another important mechanism that ensures the proper localization of individual proteins at defined intracellular sites, namely selective retention. This mechanism allows a given protein to be anchored in place upon reaching its final destination and prevents its further transport by either bulk flow or signal-dependent events. The Golgi apparatus is one of the places in the cell where selective retention operates.

There are many enzymes in the Golgi apparatus that modify the oligosaccharide chains, as illustrated in Figure 8.8 for the N-linked oligosaccharides. Different enzymes can exhibit distinct distributions in the *cis*, *trans*, or medial aspects of the stack of cisternae. In general, Golgi enzymes are type II transmembrane proteins with a short N-terminus exposed to the cytoplasm and the C-terminal, catalytic domain in the Golgi lumen.

The membrane-spanning domain is the signal anchor that initially directs the nascent Golgi enzyme to the ER (see 7 *Membrane targeting of proteins*). The amino acids flanking this domain ensure the correct topology of the enzyme in the membrane. The membrane-spanning domain and its flanking amino acids are also largely responsible for retention of the enzyme in the Golgi apparatus. When transplanted to reporter proteins that are not normally localized to the Golgi apparatus, the membrane-spanning domain, and flanking sequences suffice to locate the reporters in the Golgi apparatus. These sequences are considered targeting/**retention signals**. Although the mechanism by which proteins are retained in the Golgi is not clear, recent findings suggests that a protein Vps74p maintains the steady state distribution of type II Golgi specific glycosyltransferases. Vps74p binds the cytoplasmic tails of Golgi specific glycosyltransferases and regulates their incorporation into COPI vesicles, thereby controlling the kinetics of their export from Golgi membranes. Plants and protozoa lack orthologs of Vps74 and it is therefore likely that multiple mechanisms are involved to localize and retain Golgi-resident glycosyltransferases.

8.11 Rab guanosine triphosphate-ases and tethers are two types of proteins that regulate vesicle targeting

Key concepts

- Just as monomeric GTPases of the Sar/ARF family are involved in generating the coat that forms transport vesicles, so the Rab GTPases of another family, are involved in targeting these vesicles to their destination membranes.
- Different Rab family members are found at each step of vesicle-mediated transport.
- Proteins that are recruited or activated by Rabs, termed downstream effectors, include tethering proteins such as long fibrous proteins and large multiprotein complexes.
- Tethering proteins link vesicles to membrane compartments and compartments to each other.

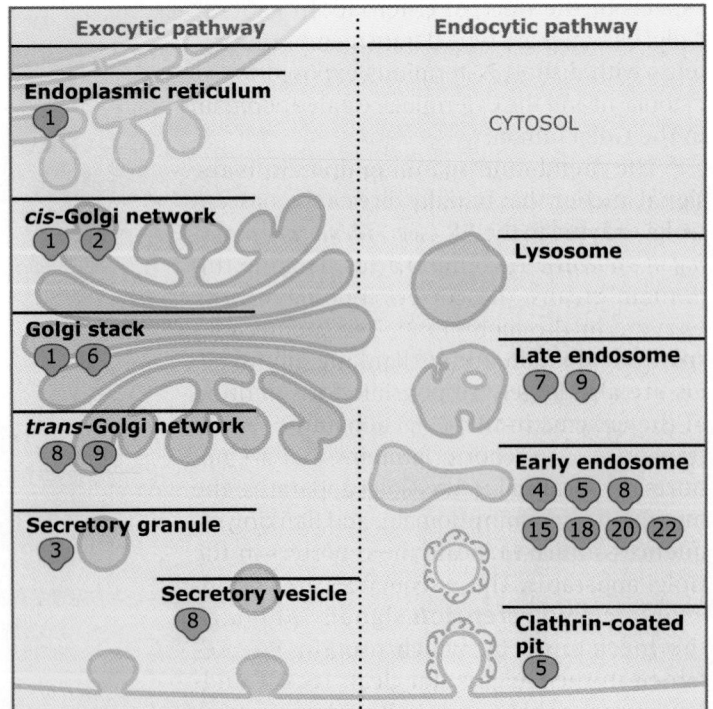

FIGURE 8.26 Localization of some mammalian Rab proteins to organelles.

Once cargo proteins have been selected and incorporated into a transport vesicle, the vesicle must fuse specifically with the next compartment on the pathway (see Figure 8.16). Given the abundance of intracellular membranes to choose from, this is no small task. Cells have at least two interdependent sets of proteins, the Ras-like GTPases called **Rabs** and the **tethers**, to accomplish the specificity of vesicle targeting. Both types of proteins must contribute to the accuracy of targeting before membrane fusion is allowed to occur (for more on fusion see *8.12 Soluble* N-*ethylmaleimide-sensitive factor attachment protein receptor proteins likely mediate fusion of vesicles with target membranes*).

The Ras-like GTPases were the first set of targeting proteins to be identified. They were originally identified in selecting for mutants that affect the secretory pathway in yeast. For example, Ypt1p is required for transport from the ER to the Golgi apparatus, whereas Sec4p is required for transport from the Golgi apparatus to the plasma membrane. In the absence of functional Sec4p, transport vesicles originating from the Golgi apparatus cannot fuse with the plasma membrane. The mammalian homologs were identified by screening rat brain cDNA libraries and hence are known as Rab ("*rat brain*") proteins.

There are now about a dozen known members of the Rab/Ypt family in budding yeast and about 60 in mammals. Given the number of individual transport steps that occur on the endocytic and secretory pathways, there are probably a sufficient number of Rab proteins such that at least one could be responsible for each step. In general, Rab proteins are localized to the organelles where they function, as shown in **FIGURE 8.26**.

As with other monomeric GTPases, Rab/Ypt proteins cycle between an inactive GDP-bound form and an active GTP-bound form. The GDP-bound Rab is cytosolic, whereas the GTP-bound form is membrane associated. As with Ras, the Rab/Ypt family proteins each contain a GTPase catalytic domain and a C-terminal lipid anchor, consisting of two 20-carbon-long prenyl (geranylgeranyl) chains, which helps maintain its association with membranes. Sequences at the C-terminus specify the compartment(s) to which a Rab protein will bind.

The association of Rab proteins with membranes is regulated. **FIGURE 8.27** illustrates one of the models that have been put forward for how Rab/Ypt proteins are localized. Rab in the

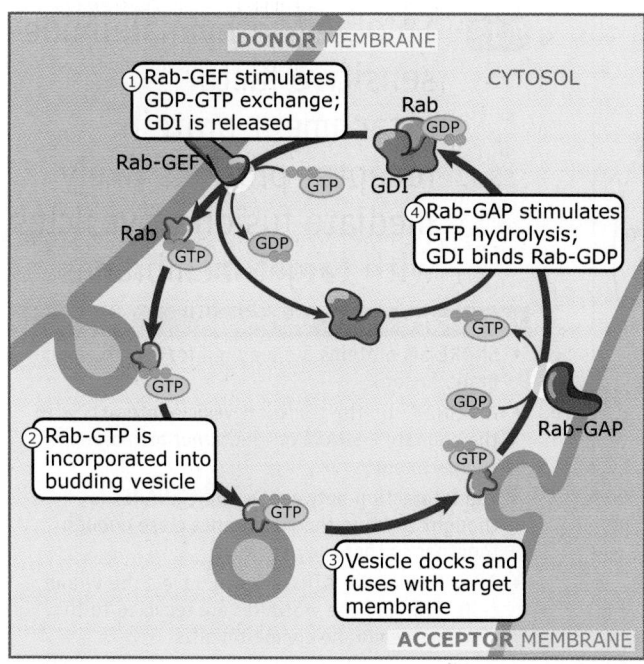

FIGURE 8.27 A model for the cycling of Rab proteins between the cytosol and membranes. (Other proteins involved in protein trafficking are not shown.)

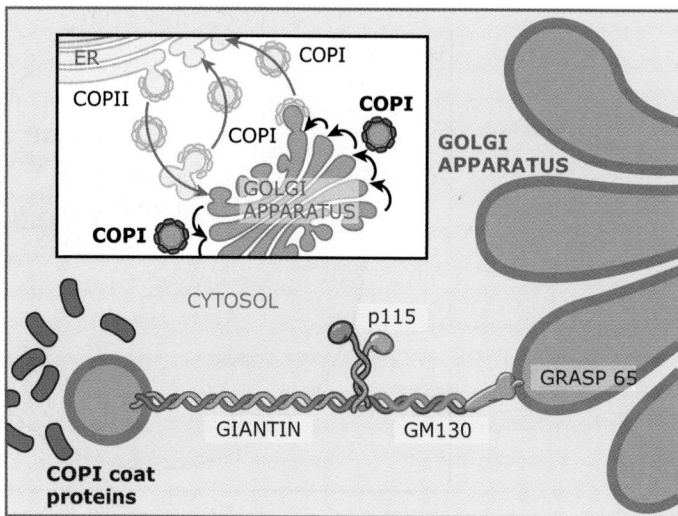

FIGURE 8.28 COPI vesicles are tethered to Golgi membranes by coiled-coil proteins. In this model, the movement of COPI vesicles is restricted by these tethers, which help to select the cisternal membranes between which the vesicles move. Rab1 binds to GM130, p115, and Giantin, but the precise organization is unclear, so it is not shown.

GDP-bound form is present in the cytosol as a complex with guanine nucleotide dissociation inhibitor (GDI). By sequestering the prenyl group that aids membrane association, GDI solubilizes what would otherwise be a membrane-bound Rab.

At the appropriate donor membrane, Rab-GDP is released from GDI, and GDP is exchanged for GTP by a membrane-bound GEF (see Figure 8.28). However, it remains unclear precisely what Rab proteins recognize on these membranes. Conceivably, the membrane-bound GEFs could serve as Rab "receptors." In any event, activated Rab-GTP is incorporated into the budding vesicle.

The Rab-GTP (and the tethers) targets the vesicle to the acceptor membrane. Correct targeting is followed by GTP hydrolysis, which is stimulated by a GTPase-activating protein specific for Rab (Rab-GAP; see Figure 8.28). The resulting Rab-GDP is then extracted by GDI and recycled back to the starting membrane compartment.

Genetic and biochemical analyses have identified a large number of **Rab effectors**, which are proteins that bind preferentially to the GTP-bound form of Rabs and carry out their downstream functions. One class of Rab effectors are the fibrous, coiled-coil proteins thought to serve as tethers that provide the initial link between a transport vesicle and its target membrane. The tethers that link transport vesicles to Golgi membranes are described here. (Tethers that link transport vesicles to endosomal membranes are described in *8.15 Some receptors recycle from early endosomes whereas others are degraded in lysosomes*).

Rab1 is the mammalian homolog of yeast Ypt1p and functions in both ER-to-Golgi and intra-Golgi transport. Rab1-GTP binds to p115, a coiled-coil protein that is a Rab effector. One of the functions of p115 is to tether COPI vesicles to Golgi membranes. Two other coiled-coil proteins, Giantin and GM130, also act at this tethering step. The interactions of Giantin and GM130 with Rab1 were first identified biochemically by allowing recombinant versions of the proteins to interact *in vitro*.

As illustrated in **FIGURE 8.28**, Giantin is anchored to COPI vesicle membranes by a membrane-spanning domain whereas GM130 is anchored to Golgi membranes by a myristoylated protein, GRASP65. Rab1 binds to Giantin, GM130, and p115, but the precise role that it plays in tethering is unclear. It is apparent that the Rab1-mediated assembly of the complex

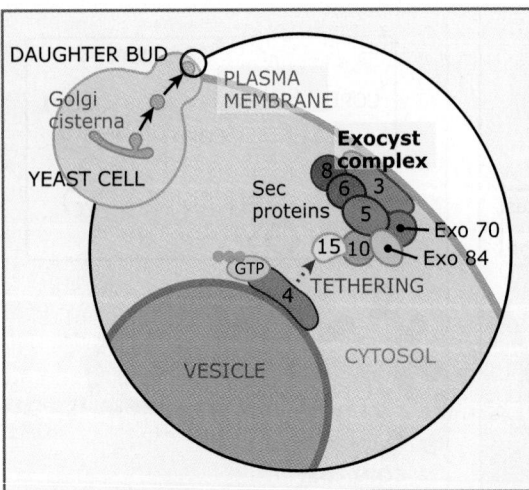

FIGURE 8.29 The exocyst is a complex of cytosolic proteins that mediates tethering of secretory vesicles at the plasma membrane.

literally tethers COPI vesicles to Golgi membranes, although it remains unclear whether this tethering is important for forward or retrograde transport events.

In addition to the fibrous tethers, there is a family of large, multiprotein complexes that are also involved in the early stages of vesicle targeting. These include the TRAPP complex at the entry face of the Golgi apparatus, the COG complex involved in intra-Golgi transport, and the exocyst complex at the plasma membrane.

In yeast, the **exocyst** is a complex of seven proteins that aids the interaction of post-Golgi vesicles with the plasma membrane at the tip of the growing daughter bud, as shown in **FIGURE 8.29**. In mammalian epithelial cells, the exocyst homolog is located at the junction of the apical and basolateral domains (for more on protein transport in polarized cells see *8.18 Polarized epithelial cells transport proteins to apical and basolateral membranes*). Blocking exocyst activity, either by mutation or by the injection of inhibitory antibodies, blocks or slows the release of secreted proteins. It is highly likely that this effect is the result of a failure of secretory transport vesicles to target to fusion sites at the plasma membrane.

Concept and Reasoning Check

1. What classes of proteins are involved in regulating vesicular transport, specifically in targeting vesicles to the correct compartment of destination?

8.12 Soluble *N*-ethylmaleimide-sensitive factor attachment protein receptor proteins likely mediate fusion of vesicles with target membranes

Key concepts

- SNARE-SM proteins are required for specific membrane fusion.
- A v-SNARE on the transport vesicle interacts with the cognate t-SNARE on the target membrane compartment.
- The interaction between v- and t-SNAREs is thought to bring the membranes close enough together so that they can fuse.
- After fusion, the ATPase NSF unravels the v- and t-SNAREs, and the v-SNAREs are recycled to the starting membrane compartment.

The Rabs and tethers initially bring a transport vesicle together with a membrane compartment with which it might fuse (see *8.11 Rab guanosine triphosphate-ases and tethers are two types of proteins that regulate vesicle targeting*). Although the two membranes thus become apposed, this is a targeting event that does not by itself lead to fusion. Membrane fusion is more closely associated with **SNAREs**, a family of organelle-specific membrane proteins. Fusion is regulated at two different levels. Tethering the vesicle to the destination compartment involves Rabs and tethering complexes that provide at least the initial specificity for vesicle-target membrane interactions. Docking is mediated by SNARE proteins, which provide further specificity and bring the membranes close enough together to fuse. Why two systems are needed is not known.

SNAREs are integral membrane proteins with a membrane (or occasionally a lipid) anchor at the C-terminus and an N-terminal domain projecting into the cytoplasm. SNARE proteins can be classified as v-SNAREs or t-SNAREs according to their presence in vesicles or in target membranes, respectively. The v-SNAREs are incorporated into budding transport vesicles (e.g., see *8.6 Coat protein II–coated vesicles mediate transport from the ER to the Golgi apparatus*).

SNARE proteins were initially identified as components of synaptic vesicles, which are essentially secretory transport vesicles found at the synapses of neurons. SNAREs were purified on the basis of their ability to form complexes with each other and with NSF and SNAP. The NSF and SNAP proteins had been previously identified as playing an important if poorly defined role in membrane fusion events *in vitro*. The SNARE proteins were identified by mass spectroscopy as founding members of a large family of proteins present in all eukaryotic cells. Interestingly, each family member has characteristically restricted patterns of distribution and, like Rab proteins, are associated with specific intracellular organelles, as shown in **FIGURE 8.30**. Tetanus and Botulinum neurotoxins are zinc proteases that cleave SNAREs and inhibit neurotransmitter release in neurons. Yeast mutants that map to genes encoding SNARE proteins are inhibited in specific steps of membrane transport in the endocytic and exocytic pathways. The blocks occur at sites corresponding to the organelle with which a given SNARE is associated. Based on these findings, the **SNARE hypothesis** was proposed to explain how SNAREs might confer specificity to the fusion of transport vesicles with their target membranes. It states that a v-SNARE in the vesicle interacts with its cognate t-SNARE in the target compartment and that this specific interaction leads to membrane fusion. Cognate SNARE pairs have now been identified for all vesicle-mediated transport steps in both mammals and yeast. These findings combined with the data from experiments in yeast provided genetic evidence that SNAREs are involved in all cellular membrane fusion events. There is some *in vitro* evidence for specificity in SNARE interactions: the only SNAREs that efficiently engage in complex formation are those that correspond to known intracellular transport events. Genome-wide searches may now have revealed all of the SNAREs expressed in yeast, *Drosophila*, and humans: 8–20 SNAREs, depending on the organism.

The crystal structures of SNARE complexes show that the coiled-coil domain is a four-helical bundle. The v-SNARE contributes one of the helices whereas the t-SNAREs contribute three. The neuronal t-SNAREs, shown in **FIGURE 8.31**, are unusual in that syntaxin contributes one helix and SNAP-25 contributes two helices. In all other cases, these latter two helices are provided by two SNAREs. A single v-SNARE, therefore, interacts with 3 t-SNAREs in order for fusion to occur.

The structural analysis revealed an unexpected relationship between SNARE complexes and the fusion proteins of enveloped viruses.

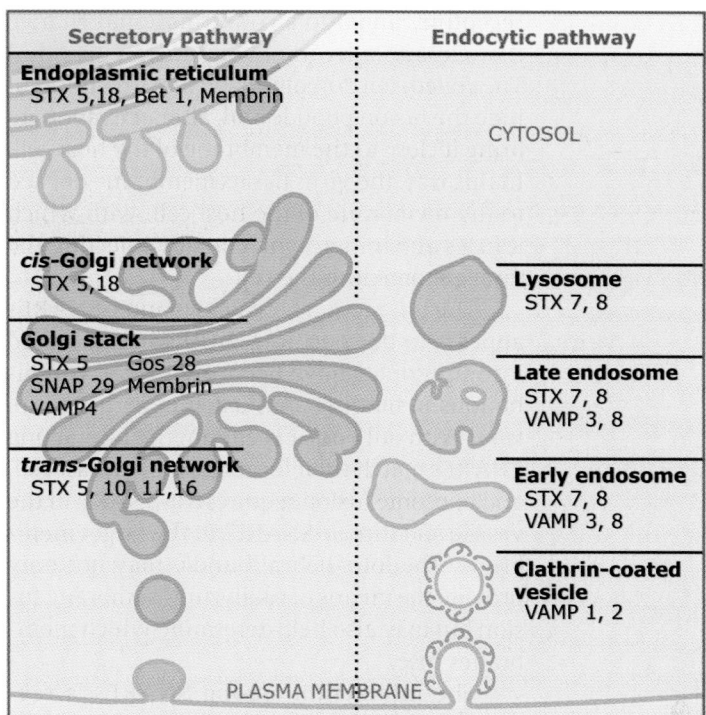

FIGURE 8.30 Localization of some SNARE proteins to organelles.

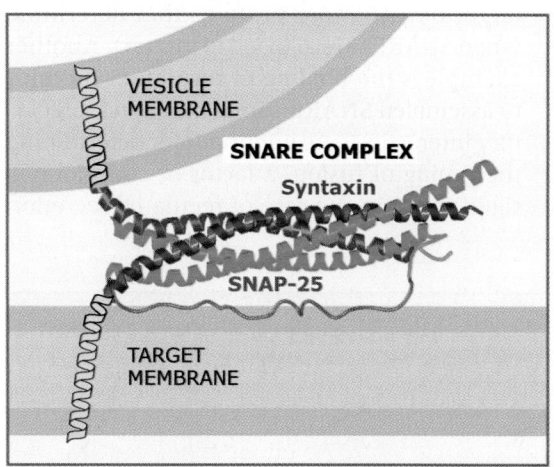

FIGURE 8.31 A model for the orientation of SNARE proteins during vesicle docking. The model is based on the X-ray structure of the cytosolic domains of SNARE proteins, which form a four-helical bundle. The t-SNAREs are syntaxin and SNAP-25; the v-SNARE is synaptobrevin. Their insertion into the membranes is inferred. The brown strand is the inferred link between the two helices of SNAP-25. Adapted from R. B. Sutton, et al., *Nature* 395 (1998): 347–353.

The best studied of these fusion proteins are the hemagglutinin of influenza virus and gp120 of HIV-1. In both cases, individual subunits form long coiled-coil structures with each other that, when triggered by low pH or binding to cellular receptors, undergo a conformational change that exposes a hydrophobic "fusion peptide." The coiled-coil domains have two functions: to hide the fusion peptide and, after activation, to bring it close to the membrane of the host cell. In this way, the virus has its membrane docked to the membrane of the host cell, with which it fuses and initiates infection by injecting the viral genome into the cell.

The interaction between v- and t-SNAREs appears to be specific. SNAREs incorporated into liposomes have been used in all combinations in binding and fusion experiments *in vitro*. With only one exception, the interacting SNAREs match their locations in the yeast cell, and liposome fusion requires a v-SNARE in the vesicle and three t-SNAREs in the target membrane. The four-helical bundle may not only provide the means of catalyzing membrane fusion but may also help determine which membranes fuse.

The interaction among SNAREs is regulated. For example, the t-SNAREs have N-terminal sequences that can dramatically inhibit the rate of formation of SNARE complexes in some cells. These sequences are the target of regulatory proteins that determine when SNAREs are allowed to interact. Another example is the binding of regulatory proteins to assembled SNARE complexes. Such regulatory interactions are important for determining the timing of fusion, a factor of considerable significance in the case of regulated secretion

as occurs in neurons or in the pancreas (for details see *8.19 Some cells store proteins for later secretion*).

How might SNARE proteins bring membranes together for fusion? The mechanism of SNARE action depends critically on their N-terminal cytoplasmic domains, which have the propensity to form coiled-coils with cognate SNAREs. Most importantly, the coils are arranged in parallel, which means that the interaction of cognate SNAREs would bring the membrane-anchored C-terminal domains close together, as illustrated in **FIGURE 8.32**. A "rosette" of such interacting coils between the vesicle and target membranes could suffice to bring them close enough together to fuse. *In vitro*, liposomes containing only one type of v-SNARE can fuse with liposomes containing its cognate t-SNARE. This result suggests that SNARE proteins are sufficient for fusion *in vitro*. Interestingly, however, new insights from in vivo experiments have revealed the essential requirement of SM (Sec1/Munc18-like) proteins in membrane fusion. It has been suggested that as the two SNAREs (v- and t-) are zippered up toward the membrane, the SM proteins cooperate in fusion, by clasping the assembling *trans*-SNARE complex. Thus, SNARE-SM provide the core (proteins) components of the fusion machinery. The activity of these proteins is regulated by a large number of diverse proteins that include complexins, RabGTPases, tethering proteins, and phosphoinositides.

After fusion has occurred, the coiled-coil SNARE complexes are unraveled and the v-SNAREs are recycled to the starting membrane for further rounds of vesicle budding, as illustrated in Figure 8.32. Unraveling is a highly energetic process because the SNARE complexes are extremely stable and are even resistant to low concentrations of strong ionic detergents such as sodium dodecyl sulfate. NSF is a cytosolic ATPase that catalyzes unraveling. NSF proteins form a barrel-shaped hexameric complex with a central channel. NSF binds indirectly to SNAREs via SNAP proteins. *In vitro* experiments using purified proteins or proteins embedded in biological membranes showed that hydrolysis of ATP by NSF separates the helices in the four-helix bundle. The mechanism of unraveling is unclear. One speculative possibility is that the SNAREs bind to the walls of the central cavity in NSF, and ATP hydrolysis opens the hole, thereby pulling the SNAREs apart.

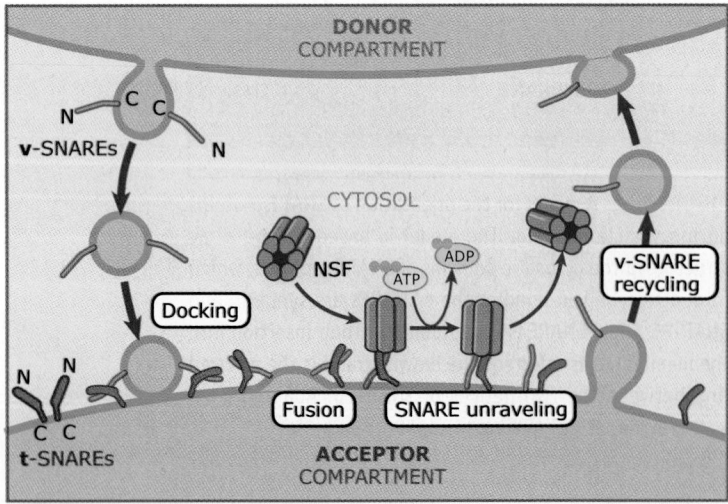

FIGURE 8.32 v-SNAREs cycle between donor and acceptor membranes. v-SNAREs incorporated into transport vesicles interact with t-SNAREs in the target membrane. This interaction brings the membranes close together, leading to fusion. After fusion, the SNARE pairs are unraveled by NSF/SNAPs in a reaction requiring ATP hydrolysis, and v-SNAREs are returned to the donor membrane.

After separation of the v- and t-SNAREs, the v-SNAREs are recycled to the donor membrane. That recycling occurs is indicated by the findings that v- and t-SNAREs do not accumulate at a single membrane and are not rapidly degraded. There are two possibilities for how v-SNAREs are transported. One possibility is that they become cargo in recycling vesicles, using a retrieval signal in the cytoplasmic domain. There is some evidence for retrieval signals in SNARE proteins, but it is largely unknown how this system works. Alternatively, some v-SNAREs may function in both directions, directing forward-moving vesicles to the target membrane and backward-moving vesicles to the starting membrane. In any case, SNAREs, like some plasma membrane receptors, are continuously recycled between the compartments in which they operate.

Concept and Reasoning Check

1. What is the likely role and mechanism of action of SNARE proteins?

8.13 Endocytosis is often mediated by clathrin-coated vesicles

Key concepts

- The stepwise assembly of clathrin triskelions may help provide the mechanical means to deform membranes into coated pits.
- Various adaptor complexes provide the means of selecting cargo for transport by binding to both sorting signals and clathrin triskelions.
- GTPases of the dynamin family help release the coated vesicle from the membrane.
- Uncoating ATPases remove the clathrin coat before docking and fusion.

Endocytosis provides the primary route by which extracellular macromolecules and plasma membrane proteins are internalized into cells (see *8.3 Overview of the endocytic pathway*). Endocytosis in its various forms represents another example of membrane traffic, with transport vesicles generated at sites where the plasma membrane invaginates to form a bud. Endocytosis is opposite in vector, but is otherwise similar, to the general scheme leading to vesicle formation from organelle membranes in the secretory pathway (see *8.4 Concepts in vesicle-mediated protein transport*).

The best-characterized mechanism of endocytosis involves transport vesicles whose coats consist of the protein **clathrin**, together with one or more types of **adaptor complexes** that link clathrin to the membrane. These adaptors also function in selecting cargo for inclusion in nascent endocytic vesicles. Clathrin and adaptor complexes are recruited from the cytosol to budding vesicles.

The cargo molecules in clathrin-coated vesicles that originate from the plasma membrane are often receptors for extracellular ligands intended for rapid uptake by a cell. In fact, the selective uptake of receptors by clathrin-coated vesicles was the first example of cargo selection during membrane traffic. Clathrin-coated vesicles also mediate transport at the *trans*-Golgi network; these vesicles have different adaptors and carry different cargo proteins than the vesicles that mediate endocytosis (see *8.17 Sorting of lysosomal proteins occurs in the* trans-*Golgi network*).

Coated pits and **coated vesicles** were first observed in the early 1960s by electron microscopy analysis of mosquito oocytes. Oocytes take up and store yolk granules for future embryogenesis, and the uptake rate is so high that much of the oocyte surface contains clathrin-coated pits, as shown in **FIGURE 8.33**.

In the early 1970s, the coat structure was revealed by freeze-substitution techniques, in nerve synapses during recycling of synaptic vesicles from the presynaptic membrane (for more on synaptic vesicles see *8.19 Some cells store proteins for later secretion*). The coat comprises an interlocking array of proteins, as shown in

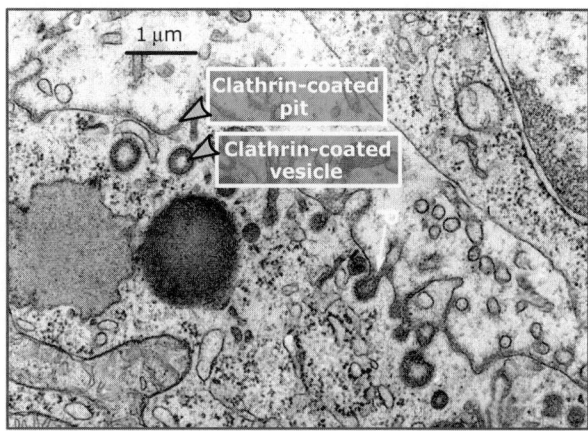

FIGURE 8.33 Transmission electron micrograph of clathrin-coated pits and vesicles at the oocyte surface. Photo © Roth and Porter, 1964. Originally published in **The Journal of Cell Biology**, 20: 313–332. Used with permission of Rockefeller University Press.

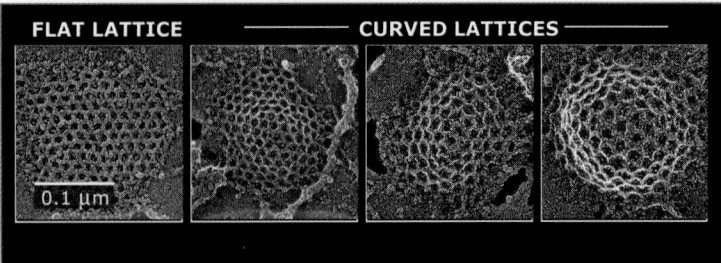

FIGURE 8.34 Electron micrographs of clathrin-coated structures found at the plasma membrane. The clathrin lattices vary in their degree of curvature. Each micrograph shows a different lattice. Photos courtesy of John Heuser, Washington University School of Medicine.

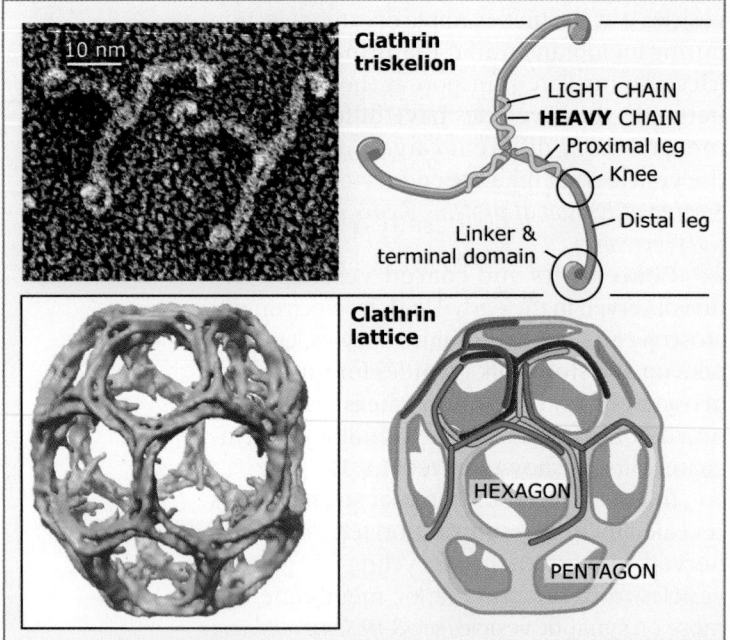

FIGURE 8.35 Clathrin triskelions assemble into closed cages. The triskelion structure can form pentagonal as well as hexagonal lattices so that closed cages can be generated. This figure shows the structure of a clathrin cage containing AP-2 adaptors. In this representation the adaptor density in the centre of the cage has been removed for clarity and a high contour level used to highlight the path of individual triskelions. Under these conditions the clathrin terminal domains are not visible. Top photo courtesy of John Heuser and Barbara Pearse, Washington University School of Medicine; bottom photo courtesy of Corinne J. Smith, The University of Warwick, UK.

FIGURE 8.34. These regular arrays inspired the name "clathrin" (from clathrate, used to describe a lattice).

Clathrin-coated vesicles were first isolated from brain in 1969, and the clathrin coats were later shown to consist primarily of two proteins, the clathrin heavy chain (180 kDa) and clathrin light chain (30 kDa), which are present in coats in equimolar amounts. Electron microscopy showed that clathrin proteins form a distinctive three-legged complex called a **triskelion**, as shown in **FIGURE 8.35**. Each "leg"

of the triskelion comprises one heavy chain and one light chain of clathrin. The proximal leg is found at the vertex of the triskelion and is the site at which the light chains bind. The distal leg extends from the vertex and consists entirely of heavy chain.

Although other proteins may help to assemble clathrin coats on membranes *in vivo*, clathrin exhibits the remarkable capacity to self-assemble into lattices *in vitro*. Triskelions isolated from cells and incubated under appropriate conditions of salt and pH will spontaneously form empty clathrin cages that look similar to those found surrounding membranes *in vivo*. Clathrin cages reconstituted from purified heavy and light chains show that the triskelion forms a hexagonal and pentagonal coat structure, as shown in the cryoelectron micrograph in Figure 8.34. These cages closely resemble the coats surrounding vesicles *in vivo*, which argues that clathrin heavy and light chains are both necessary and sufficient for lattice formation. Each vertex is the node of a single triskelion and each side of a pentagon or hexagon consists of four legs from different triskelions.

The geometry is such that hexagonal clathrin lattices are flat, but the introduction of pentagons produces the necessary curvature to form a cage (see Figure 8.35). However, the sequence of events leading to a fully curved coated pit in vivo is unknown. One model is that some of the hexagons within a flat coat are somehow modified to form pentagons, so as to introduce curvature. Another model is that proteins other than clathrin introduce curvature to the membrane, and this curvature is "locked" in place by the assembly of pentagons.

The assembly of clathrin cages does not require energy. Therefore, this assembly is thought to be associated with a favorable free energy change that may help to drive membrane invagination and vesicle formation. However, adaptor complexes also play a key role in deforming the plasma membrane prior to coated vesicle formation.

Clathrin coats do not interact directly with the membrane or membrane proteins. Instead, various adaptor complexes are intermediaries that link clathrin subunits and sorting signals in the cytoplasmic tails of cargo proteins, as shown in **FIGURE 8.36**. Different adaptors select transmembrane proteins at different transport steps (see *8.14 Adaptor complexes link clathrin and transmembrane cargo proteins*).

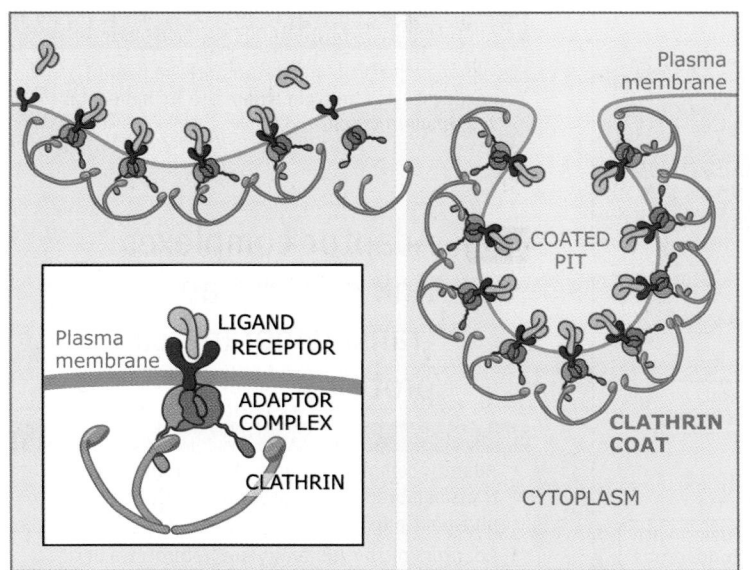

FIGURE 8.36 Clathrin binds to adaptors that recognize a sorting signal in the cytoplasmic tail of proteins to be transported.

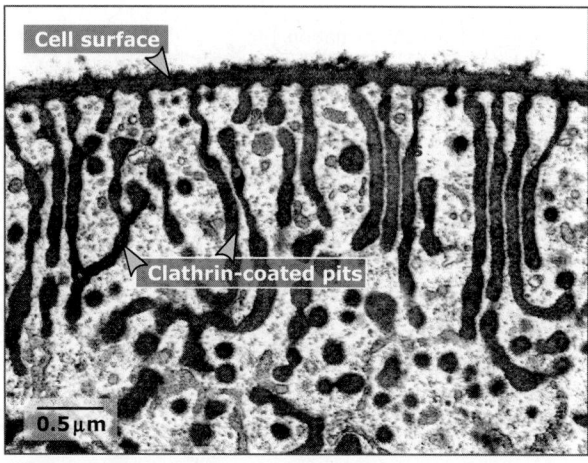

FIGURE 8.37 Electron micrograph of clathrin-coated pits in dynamin mutant flies at the nonpermissive temperature of 30°C. Note the tubes extending from the cell surface and ending as clathrin-coated buds. These elongated coated pits cannot pinch off because mutant dynamin cannot operate at this temperature. Photo © Kosaka and Ikeda, 1983. Originally published in **The Journal of Cell Biology**, 97: 499–507. Used with permission of Rockefeller University Press. Photo courtesy of Toshio Kosaka, Kyushu University.

Budding clathrin-coated vesicles must eventually pinch off from the plasma membrane to complete endocytosis. Fission is an energy-requiring process that is mediated by **dynamin**, a GTPase first identified in the *Drosophila shibire* mutant. This mutant is unable to fly at the nonpermissive temperature. In mutant flies at the nonpermissive temperature, the coated pits at the synapse of neurons have long necks, as shown in **FIGURE 8.37**, and do not pinch off from the plasma membrane. These structures are not present at the permissive temperature. This observation suggested that the phenotype of the shibire mutant was due to the inability of the synaptic vesicle components to recycle from the plasma membrane, thereby effectively stopping neurotransmission. The *D. shibire* mutation was traced to a defect in dynamin. Dynamin assembles into "collars" at the neck of a newly forming bud and polymerizes into helical polymers. A change in the conformation of helix upon GTP hydrolysis was suggested to break the bud neck. But it was not clear whether this was through constriction, expansion, or twisting of the neck. More recent findings indicate that dynamin helix brings the lipids in the neck region to close proximity and the disassembly of dynamin upon GTP hydrolysis promotes fission. In sum, dynamin dependent constriction of the bud neck followed by its dissociation is required for membrane fission

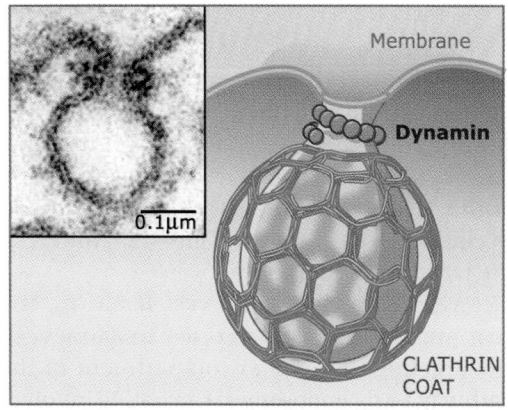

FIGURE 8.38 Dynamin can form spiral structures around the necks of budding vesicles, which leads to membrane fission. Photo courtesy of Pietro de Camilli, Yale University School of Medicine.

FIGURE 8.38. While the details are not clear there is no doubt that dynamin is a key component of the process that releases clathrin-coated vesicles from the plasma membranes and perhaps also at the *trans*-Golgi network (see *8.17 Sorting of lysosomal proteins occurs in the* trans-*Golgi network*).

After its release, the clathrin-coated vesicle may interact with elements of the actin cytoskeleton that assist in the movement of the vesicle into the cytosol. Newly formed endocytic

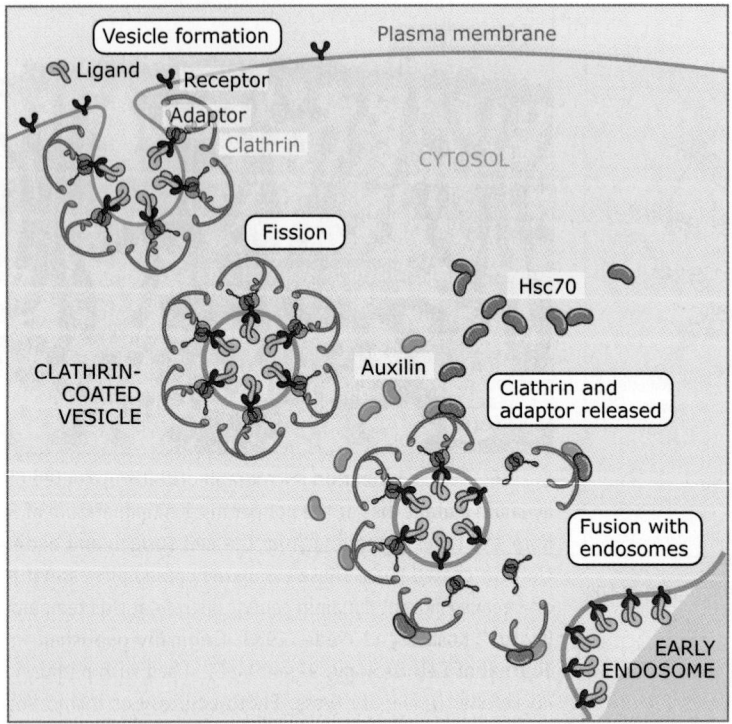

FIGURE 8.39 Hsc70 and auxilin are two of the proteins involved in the uncoating of clathrin-coated vesicles.

Concept and Reasoning Check

1. What is the best-understood mechanism of endocytosis and from where are its molecular components recruited?

8.14 Adaptor complexes link clathrin and transmembrane cargo proteins

Key concepts

- Adaptor complexes bind to the cytoplasmic tails of transmembrane cargo proteins, to clathrin, and to phospholipids.
- Adaptors of the AP family are heterotetrameric complexes of two adaptin subunits and two smaller proteins.
- The AP adaptors bind to sorting signals in the cytoplasmic tails of cargo proteins. The best-characterized of these signals contain tyrosine or dileucine residues.
- Adaptor complexes allow for the selective and rapid internalization of receptors and their ligands.

vesicles can serve as assembly sites for actin filaments that provide motility of this type. Although the necessity for such movements has yet to be established, it is clear that many of the "accessory proteins" that are recruited to clathrin-coated pits, notably dynamin, can nucleate actin assembly.

After clathrin-coated vesicles form, the coat must be removed in order to allow vesicle targeting to the next compartment of the pathway, early endosomes. Uncoating involves at least two proteins, the "uncoating ATPase," which is a cytosolic enzyme of the Hsp70 family of heat shock proteins, and auxilin, a protein that interacts with the uncoating ATPase, as shown in **FIGURE 8.39**. The precise mechanism of uncoating has yet to be resolved but is likely to involve the destabilization and dissociation of clathrin coats following Hsp70 binding. This would be consistent with the ability of this family of ATPases to bind and destabilize higher order structures in their function as chaperones in protein folding and unfolding reactions during protein synthesis and degradation. The hydrolysis of acidic phospholipids may also contribute to coat disassembly, particularly the release of the adaptor complexes (see *8.14 Adaptor complexes link clathrin and transmembrane cargo proteins*).

Adaptor complexes link the cytoplasmic tails of transmembrane cargo proteins and clathrin. The best-known adaptors are those of the "AP" family, which consists of four major species. These adaptors exhibit characteristic intracellular distributions, as shown in **FIGURE 8.40**. AP-1 and AP-2 were identified as major components of isolated clathrin-coated vesicles. AP-1 localizes to the TGN and endosomes, whereas AP-2 is limited to the plasma membrane. AP-1 functions in the transport of soluble lysosomal components and some membrane proteins from the TGN to endosomes and lysosomes (see *8.17 Sorting of lysosomal proteins occurs in the* trans-Golgi *network*). AP-2 functions in endocytosis.

AP-3 was identified by searching databases for cloned nucleic acid sequences homologous to known adaptor sequences. Like AP-1, AP-3 is associated with the TGN and may be involved in the biogenesis of some specialized lysosomes, such as melanosomes, which store the pigments in melanocytes. Mice with coat color defects often have mutations in AP-3 subunits. The function of a fourth complex, AP-4, is not known.

The different adaptor complexes bind to distinct organelles, at least in part, by interacting with specific phospholipids. AP-2 binds to

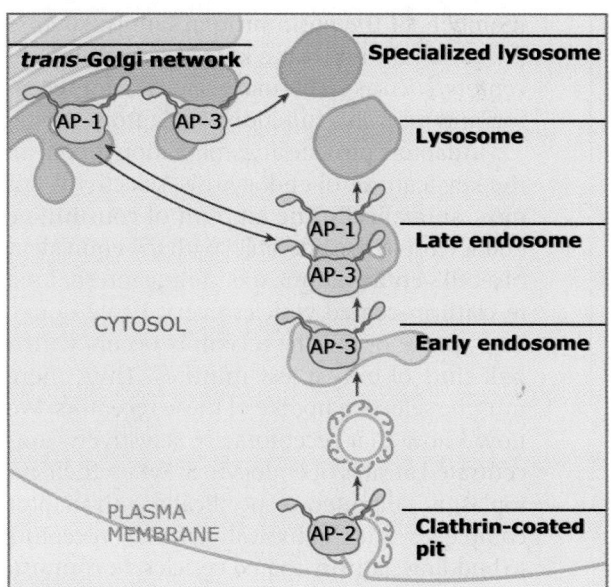

FIGURE 8.40 Different adaptors operate at different membrane traffic steps.

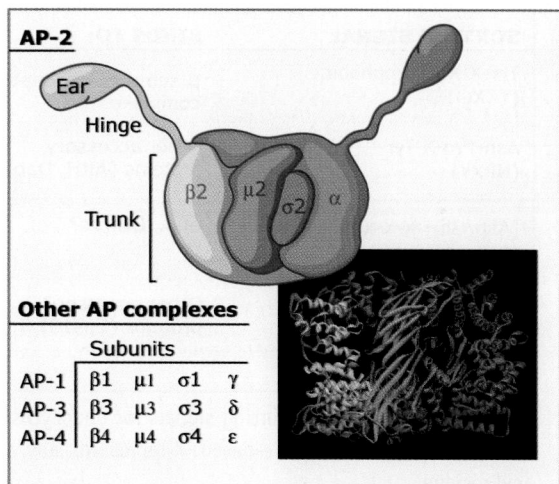

FIGURE 8.41 Adaptor complexes are involved in transport mediated by clathrin-coated vesicles. Adaptors are hetero-tetrameric structures constituting a "trunk" that binds to the signals in cytoplasmic tails of membrane receptors and hinge regions linked to "ears" that bind to clathrin. The X-ray structure is of the trunk region only. Structure reprinted from *Cell*, vol. 109, B. M. Collins, et al., Molecular Architecture and Functional Model..., pp. 523–535, Copyright (2002) with permission from Elsevier [http://www.sciencedirect.com/science/journal/009286740]. Photo courtesy of Brett M. Collins, Institute for Molecular Bioscience, Austria.

phosphatidylinositol-4,5-bis-phosphate (PI4,5P or PIP2), a lipid that is generated preferentially at the cytoplasmic leaflet of the plasma membrane. AP-1, in contrast, binds to PI4P, found primarily at intracellular membranes such as those of the Golgi complex and endosomes. Interactions with the proper lipid species may also induce a conformation change that enables adaptors to better interact with their cargos, further ensuring specificity.

Adaptors are complexes of four different subunits. The two large subunits in each complex are called **adaptins**. All members of the AP family share the same heterotetrameric structures illustrated in **FIGURE 8.41**. The over-all shape was revealed by electron microscopic and X-ray crystallographic analysis of purified complexes. The structure is that of a large brick or "trunk" domain from which two appendages ("ears") extend, attached to the trunk by long "hinge" domains. All four subunits are contained within the trunk, but the hinge and appendage domains are extensions of two of the chains (in AP-1, these are the β- and γ-chains; in AP-2, the α- and β-chains; in AP-3, the β- and δ-chains).

The different regions of adaptor complexes have distinct functions. The hinge and appendage domains contain binding sites for a specific region of clathrin called the β-propeller domain. These domains also allow the adaptor complex to interact with a variety of other accessory proteins that regulate the formation of clathrin-coated vesicles.

The trunk of AP-2 binds to the cytoplasmic tails of plasma membrane receptors and so links the clathrin coat to the plasma membrane. Phosphorylation of AP-2 exposes its binding site for the receptor tails. A specific kinase (called AAK1) phosphorylates AP-2 during coat assembly. Phosphorylation exposes the μ subunit of AP-2, which is the portion of the adaptor that recognizes the coated pit internalization signal on endocytic receptors. This phosphorylation also exposes binding sites for phosphoinositides. These negatively charged lipids in the plasma membrane provide additional anchors for the AP-2 complex.

Adaptors bind specifically to sorting signals in the cytoplasmic tails of cargo proteins, as shown in **FIGURE 8.42**. The critical function that adaptors play in selecting cargo for inclusion into forming coated vesicles is best understood for AP-2. The sorting signals that bind to the μ chain of AP-2 are found only in proteins that are incorporated into clathrin-coated pits and are called endocytic sorting signals. AP-2 appears to bind to two different types of sorting signals, tyrosine-based and dileucine-based.

The tyrosine-based signals were the first membrane sorting signals to be identified, by analysis of mutant LDL receptors that are

SORTING SIGNAL	BINDS TO:
Tyr-X-X-hydrophobic (YXXØ)	μ subunits of adaptor complexes
Asn-Pro-X-Tyr (NPXY)	Other accessory proteins (ARH, Dab2)
[Asp/Glu]-X-X-X-Leu-[Leu/Ile] (Dileucine)	GGA, Others?
Monoubiquitin	Other accessory proteins (Eps15)

FIGURE 8.42 The known sorting signals for endocytosis are either tyrosine- or dileucine-based. X represents any amino acid residue.

unable to enter clathrin-coated pits. The mutations are in a short tyrosine-containing sequence that is both necessary and sufficient for uptake. The lack of internalization of mutant receptors showed that this sequence is necessary for endocytosis. A chimeric protein, consisting of the sorting sequence and sequences from a plasma membrane protein that is not normally endocytosed, is internalized at the same rate as the normal LDL receptor. This experiment showed that the sequence is sufficient for endocytosis.

Further mapping experiments, in which mutant LDL receptors were generated and tested in internalization assays, emphasized the critical importance of the tyrosine residue and showed the importance of adjacent amino acid residues. These experiments yielded two consensus sequences for internalization: Tyr-X-X-f and Asn-Phe-X-Tyr where X is any amino acid, and f is an amino acid with a bulky hydrophobic side chain. However, only the former signal has been shown by yeast two-hybrid experiments to bind directly to the AP-2 μ chain.

Dileucine signals mediate the rapid endocytosis of a number of receptors. The AP-2 δ-chain, rather than the μ chain, may recognize these signals. However, this interaction is not as well characterized and may involve a second subunit. In addition, dileucine signals mediate sorting of lysosomal proteins in the *trans*-Golgi network (see *8.17 Sorting of lysosomal proteins occurs in the* trans-*Golgi network*).

Adaptor complexes other than AP family members link some endocytic receptors to clathrin in an analogous fashion to AP complexes. For example, β-arrestin binds to a signal on the cytoplasmic tail of β-adrenergic receptors and to clathrin and mediates receptor internalization in the absence of AP-2. Similarly,

members of the epsin protein family, such as Eps15, bind to tyrosine kinase-containing receptors, such as epidermal growth factor receptor, and may also function as adaptors.

Adaptors provide an explanation for one of the key features of endocytosis: selectivity. For most animal cells, the amount of constitutive endocytosis is considerable, with the equivalent of a cell's entire plasma membrane internalized in clathrin-coated vesicles every 1 to 2 hours. Yet, uptake of specific receptors occurs with a half-time of only a few minutes. Thus, there must be selective uptake of these receptors. We now know that receptors are selectively concentrated at sites of endocytosis when their cytoplasmic tails interact specifically with adaptor complexes, which physically link the receptors to budding clathrin-coated vesicles. In contrast, plasma membrane proteins that do not interact with adaptors are not internalized selectively. These proteins may be internalized at a much slower rate, or they may be sterically excluded from coated pits by selectively included receptors that are present at a higher concentration.

In addition, soluble extracellular proteins are incorporated into clathrin-coated vesicles at a low level, since it is impossible to form a vesicle without a lumenal content. Since the concentration of most ligands in the extracellular medium is so low, only by binding to receptors can appreciable quantities of ligand be internalized.

8.15 Some receptors recycle from early endosomes whereas others are degraded in lysosomes

Key concepts

- Early endosomes are mildly acidic and lack degradative enzymes, so internalized ligands can be dissociated without degradation of their receptors.
- Many receptors are recycled to the cell surface in transport vesicles that bud from the tubular extensions of early endosomes.
- Dissociated ligands are transferred from early endosomes to the more acidic and hydrolase-rich late endosomes and lysosomes for degradation.
- Receptors that are not recycled are segregated into vesicles within multivesicular bodies and move to late endosomes and lysosomes for degradation.
- Recycling endosomes are found adjacent to the nucleus and contain a pool of recycling receptors that can be transported rapidly to the cell surface when needed.

Receptors, ligands, and extracellular fluid internalized by clathrin-coated vesicles are delivered to early endosomes, a population of vacuoles with tubular extensions. Early endosomes are located throughout the peripheral cytoplasm. Delivery of incoming clathrin-coated vesicles to early endosomes occurs shortly after vesicle fission and removal of the clathrin coat, which happens rapidly (<1 min) after coated vesicle formation (for more on fission and uncoating see *8.13 Endocytosis is often mediated by clathrin-coated vesicles*). The tethering, docking, and fusion of uncoated vesicles with early endosomes is strikingly analogous to that described for the exocytic pathway in that it requires Rabs, tethers, and SNAREs (see *8.11 Rab guanosine triphosphatases and tethers are two types of proteins that regulate vesicle targeting*). However, different members of these protein families are used in the endocytic pathway. The relevant Rab protein is Rab5, the likely t-SNARE is syntaxin 13, and the tether is a long coiled-coil protein termed early endosomal antigen-1 (EEA1), as shown in **FIGURE 8.43** EEA1 binds to Rab5 (together with several accessory proteins), syntaxin 13, and phosphatidylinositol-3-phosphate (PI3P). Since these components are on different membranes, EEA1 links the membranes and facilitates the fusion event.

There are two main itineraries for internalized receptors and their ligands, as shown in **FIGURE 8.44**. In one pathway, internalized receptors discharge their ligands in early endosomes, which have a lumenal pH of 6.4 to 6.8. This mildly acidic pH triggers the dissociation of many types of ligands from their receptors, which are recycled to the cell surface from the tubules that extend from the vacuolar part of the early endosome. The dissociated ligands do not easily enter these tubules and accumulate within the early endosome lumen. The ligands move on to late endosomes and lysosomes and are degraded. Because early endosomes have a much lower concentration of hydrolytic enzymes and a higher pH than lysosomes, they provide a relatively safe haven for receptors to traverse repeatedly with less risk of damage or destruction. This pathway is used by nutrient receptors, for example, in the other pathway, internalized receptors and ligands move together to late endosomes and lysosomes and are degraded. This pathway is used for downregulation of growth factor receptors, for example.

An example of a recycling receptor is the transferrin receptor, which carries iron into the

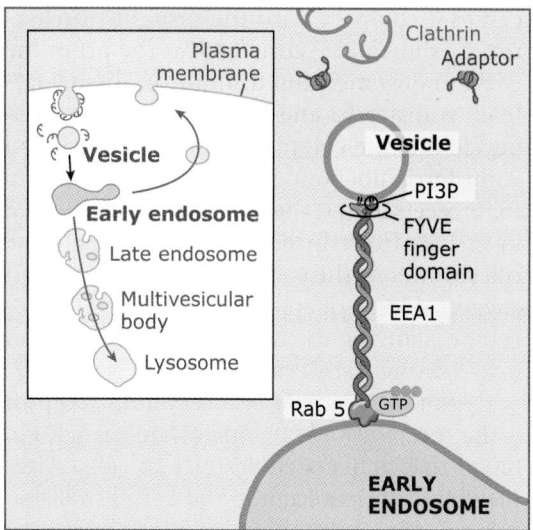

FIGURE 8.43 Clathrin-coated vesicles are uncoated and tethered to early endosomes via EEA1, which binds to phosphatidylinositol-3-phosphate and Rab5.

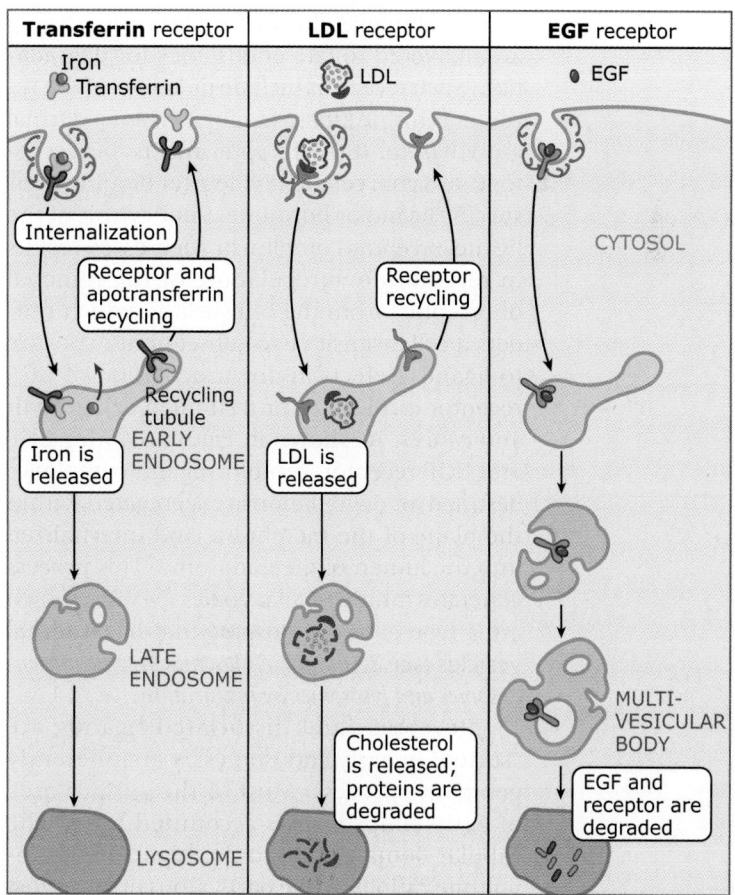

FIGURE 8.44 The fate of internalized receptors. The transferrin receptor and the LDL receptor cycle between the plasma membrane and endosomes, whereas the EGF receptor is degraded in lysosomes.

cell as a complex with the protein transferrin. As shown in Figure 8.45, at the pH of the early endosome lumen, the iron dissociates from transferrin and is subsequently transported into the cytoplasm. The apotransferrin (transferrin not bound to iron) remains bound to its receptor and the complex is transported in vesicles from the endosomal tubules to the cell surface. At the cell surface, the neutral pH triggers release of apotransferrin from the receptor, allowing the receptor to bind another iron-bearing transferrin molecule.

Another example of a recycling receptor is the receptor for LDL, one of the particles in blood that delivers cholesterol to cells. After LDL binds to its receptor and is internalized, it is released into the endosomal lumen and degraded, as shown in Figure 8.45. The receptor migrates into the tubular extensions of the endosomes. Transport vesicles bud from these tubules to recycle the receptors to the plasma membrane for further rounds of uptake.

In contrast to transferrin receptors and LDL receptors, which recycle to the plasma membrane from early endosomes, other receptor-ligand complexes do not recycle efficiently and are delivered to late endosomes for degradation. Such receptors include members of the receptor tyrosine kinase family such as epidermal growth factor (EGF) receptor and insulin receptor that signal cells to divide after binding their specific ligand or hormone. Endocytosis of the ligand-receptor complex in such cases results in receptor downregulation, or the removal of receptors from the cell surface, which renders a cell insensitive to subsequent exposure to ligand. Defects in downregulation of EGF receptor can lead to unrestrained cell growth and cancer. As shown in Figure 8.45 for EGF and EGF receptor, receptor-ligand complexes destined for degradation are segregated within the plane of the membrane and internalized into the lumen of the endosome. This process generates multivesicular bodies (MVBs), which are a type of late endosome that has internal vesicles (see *8.16 Early endosomes become late endosomes and lysosomes by maturation*).

Receptors and dissociated ligands are "sorted" in early endosomes by an inherently geometric process. Most of the surface area of early endosomes is accounted for by the tubular domains due to the high surface-to-volume ratio in these parts. Conversely, most of the internal volume of the early endosome is found within its more spherical vacuolar domain. Thus, receptors tend to be associated with the tubular extensions simply because that is where the bulk of the membrane resides. However, there is evidence that after ligand dissociation, receptors may also be selectively concentrated in the tubular regions. These tubules pinch off from early endosomes, giving rise to transport vesicles that return the ligand-free receptors back to the plasma membrane for reuse.

The bulk of recycling traffic proceeds directly and rapidly from early endosomes with a half-time of just a few minutes. However, ~25% of the receptors are transferred to a population of **recycling endosomes** (see Figure 8.46) where they can reside as an intracellular pool for up to 1 hour. In a typical cell, the membrane surface area of all early endosomes and recycling endosomes is only about 25% of the surface area of the plasma membrane. Since up to 100% of the plasma membrane is internalized every hour, the effective surface area of early endosomes and recycling endosomes must turn over several times per hour.

Dissociated ligands and macromolecules that are nonspecifically internalized from the extracellular fluid are largely excluded from the recycling tubules of early endosomes due to their very small internal volume. Some of this material does leak out of the cell via the transport vesicles that arise from these tubules, however. Nevertheless, the bulk of these fluid-dissolved components accumulate in the vacuolar portions of early endosomes as they mature to become late endosomes.

Concept and Reasoning Check

1. What is the difference in fate of transferrin and LDL receptors versus EGF and insulin receptors?

8.16 Early endosomes become late endosomes and lysosomes by maturation

Key concepts

- Movement of material from early endosomes to late endosomes and lysosomes occurs by "maturation."
- A series of endosomal sorting complex required for transport (ESCRT) protein complexes sorts proteins into vesicles that bud into the lumen of endosomes, forming multivesicular bodies that facilitate the process of proteolytic degradation.

The progression from early endosomes to late endosomes and finally to lysosomes results from a maturation process (shown in **FIGURE 8.45**) that is at least superficially similar to the concept of cisternal maturation in the Golgi apparatus (see *8.9 There are two popular models for forward transport through the Golgi apparatus*). Transport vesicles appear not to have a major role in the movement of proteins between endosomes. Instead, the loss of receptors by recycling to the plasma membrane converts early endosomes to late endosomes, which contain dissociated ligands such as LDL (see *8.15 Some receptors recycle from early endosomes whereas others are degraded in lysosomes*). Transport vesicles containing newly synthesized lysosomal enzymes and membrane components originate from the *trans*-Golgi network and fuse with late endosomes (see *8.17 Sorting of lysosomal proteins occurs in the* trans-*Golgi network*). This process converts late endosomes into lysosomes and leads to the digestion of the dissociated ligands.

Receptors are selectively and progressively removed from early endosomes due to the formation of the recycling tubules. Everything else remains in the vesicular portions of the early endosome, which binds to microtubule tracks and translocates to the center of the cell. They eventually accumulate in the perinuclear cytoplasm adjacent to the microtubule-organizing center. One reason for this movement is to limit further fusion with incoming clathrin-coated vesicles from the plasma membrane. Fusion with TGN-derived vesicles containing lysosomal components may continue, however. Thus, the organelles in the endocytic pathway become progressively more "lysosome-like" in terms of both lumenal content such as enzymes and H+ ions and membrane components such as v-ATPase. The conversion is gradual, generating late endosomes as an intermediate between early endosomes and lysosomes.

A critical feature of early endosome maturation is formation of **MVBs**. The later endocytic compartments (late endosomes and lysosomes) are often characterized by small vesicular inclusions that form from invaginations of the membrane, as shown in **FIGURE 8.46**. Thus, it is often the case that late endosomes and most lysosomes are referred to collectively as MVBs. The functional significance of these structures is clearest in the case of receptors of the receptor tyrosine kinase family, such as insulin receptor and EGF receptors, which are degraded after ligand uptake (see *8.15 Some receptors recycle from early endosomes whereas others are degraded in lysosomes*).

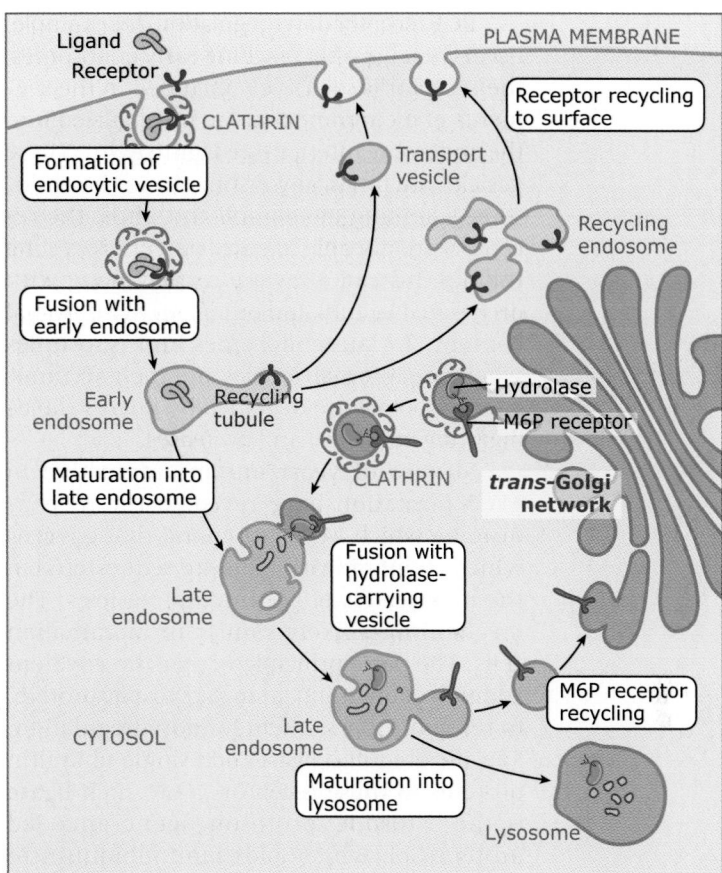

FIGURE 8.45 As endosomes mature from early endosomes to late endosomes, transport vesicles bring components to the endosomes and recycle receptors from the endosomes.

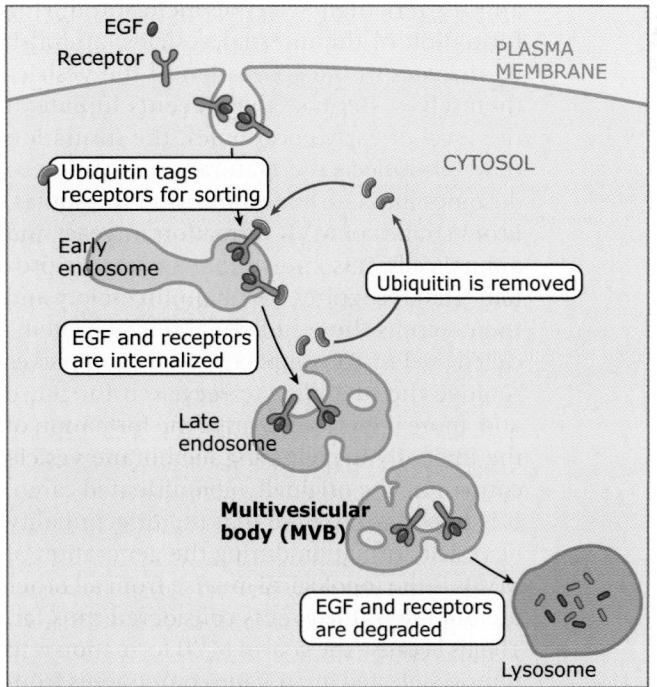

FIGURE 8.46 Ubiquitin-mediated degradation occurs in multivesicular bodies, which are late endosomal compartments of the endocytic pathway.

EGF receptor downregulation, for example, occurs because after reaching early endosomes, the receptor is selectively included in the segments of membrane that are internalized into the endosome interior (see Figure 8.46). These vesicles are physically distinct from the endosome's limiting membrane and, thus, the receptors are not able to enter nascent recycling tubules. Instead, they are carried along with dissociated ligands and other early endosomal contents to late endosomes and lysosomes where the internal vesicles and their accumulated receptors are degraded by lipases (lipid-digesting enzymes) and proteases.

Studies of yeast mutants defective in MVB formation have revealed the mechanism by which MVBs form and that governs which membrane proteins are sequestered in the membranes of the internal vesicles. The mechanism is likely similar in mammalian cells. The basic principle is that the covalent addition of ubiquitin, a small cytosolic protein, to receptors marks them for downregulation. Specific ubiquitin ligases add single ubiquitin proteins to these receptors. One such ligase is the cytosolic, proto-oncogene-encoded protein Cbl, which adds monoubiquitin to the EGF receptor. A defect in Cbl results in a lack of EGF receptor downregulation and unregulated cell growth. A complex of proteins detects these monoubiquitinated proteins and triggers their selective inclusion during formation of the internal vesicles, and also participates in the generation of the vesicles themselves. Because these events initiate at the level of early endosomes, the formation of MVBs reflects the maturation of early endosomes into late endosomes and lysosomes. From studies of MVB formation in yeast and animal cells, it is known that a cytosolic protein (Hrs) recognizes the ubiquitin moiety and then recruits three additional sets of proteins called the ESCRT complexes. These complexes remove the ubiquitin to recycle it for reuse and, more importantly, drive the formation of the inwardly invaginating membrane vesicle containing the originally ubiquitinated cargo. It is important to note that the directionality of vesicle formation during the generation of MVBs is the topological inverse from all other membrane traffic events considered thus far. That is because the goal of MVB formation is to remove selected membrane components from normal pathways of transport to ensure their digestion in lysosomes.

8.17 Sorting of lysosomal proteins occurs in the *trans*-Golgi network

Key concepts

- All newly synthesized membrane and secretory proteins share the same pathway up until the TGN, where they are sorted according to their destinations into different transport vesicles.
- Clathrin-coated vesicles transport lysosomal proteins from the *trans*-Golgi network to maturing endosomes.
- In the Golgi apparatus, mannose 6-phosphate is covalently linked to soluble enzymes destined for lysosomes. The mannose 6-phosphate receptor delivers these enzymes from the *trans*-Golgi network to the endocytic pathway.
- Lysosomal membrane proteins are transported from the *trans*-Golgi network to maturing endosomes but use different signals than the soluble lysosomal enzymes.

Proteins destined for lysosomes are synthesized at the ER, along with all the other proteins destined for compartments in the exocytic and endocytic pathways. The soluble degradative enzymes and transmembrane proteins of lysosomes are transported along the pathway from the ER through the Golgi apparatus and to the TGN (see Figure 8.4), where they are sorted away from proteins bound for the cell surface and other destinations.

In mammalian cells, sorting of the soluble lysosomal enzymes in the TGN requires a signal that is distinct from the cytoplasmic sorting signals discussed so far. Instead, the lysosomal sorting signal, called the mannose 6-phosphate (M6P) signal, is lumenal and is generated by modification of the oligosaccharides that are covalently attached to lysosomal enzymes. It was discovered through analysis of certain lysosomal storage diseases, such as I-cell disease (also called mucolipidosis type II). The lysosomes of patients with this disease lack the usual enzymes, which instead are secreted into the extracellular space. These lysosomes, therefore, accumulate undigested material, creating characteristic intracellular inclusions (the "I" in I-cell disease). Similar inclusions are found in other lysosomal storage diseases, such as Tay-Sachs disease, in which cells lack one of the enzymes (hexoseaminidase A), as shown in **FIGURE 8.47**.

The M6P signal is the result of phosphorylation of terminal N-linked mannose residues.

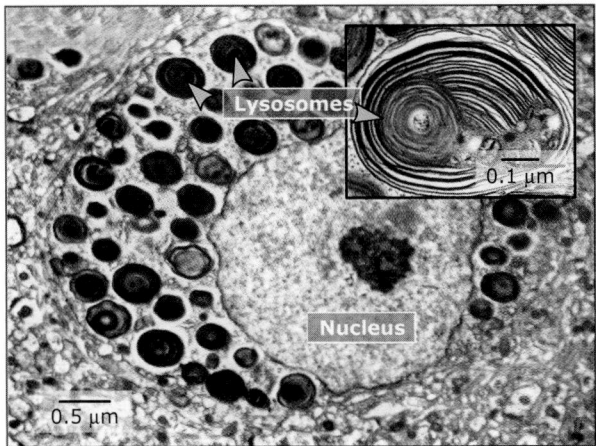

FIGURE 8.47 A neuron from the cerebrum of a Tay-Sachs patient. Note the large number of abnormal lysosomes, one of which is shown at higher magnification. Reproduced with permission from the *Journal of Neuropathology and Experimental Neurology* 22 (1963): 18–55.

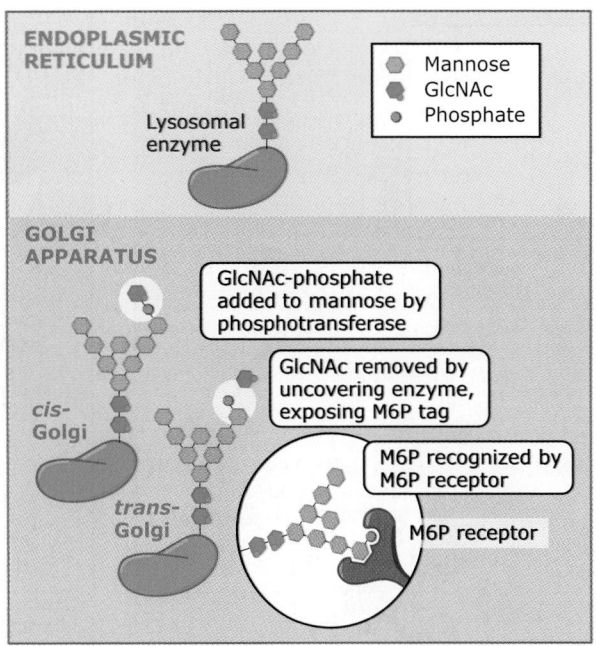

FIGURE 8.48 Assembly of the mannose 6-phosphate sorting signal. The high-mannose oligosaccharides attached to lysosomal enzymes are modified in the early Golgi cisterna by the addition of N-acetylglucosamine-phosphate. The N-acetylglucosamine residue is removed in the late Golgi cisterna, exposing the mannose 6-phosphate moiety that is recognized by M6P receptors.

As shown in **FIGURE 8.48**, N-acetylglucosamine-phosphate (GlcNAc-P) is added to the C-6-hydroxyl group of selected terminal mannose residues by a phosphotransferase. (In I-cell disease, most lysosomal enzymes do not receive the M6P signal due to a lack of this phosphotransferase.) The addition of GlcNAc-P most likely occurs in the early Golgi compartments. An "uncovering" enzyme in the TGN then removes the GlcNAc residue, thereby exposing the M6P moiety. Thus, the sorting signal is exposed only where sorting occurs.

Of course, many proteins in the exocytic pathway have high-mannose oligosaccharides and are not lysosomal enzymes, so the generation of the M6P sorting signal must require other information that only occurs in the lysosomal enzyme sequence. This information is not found in a linear amino acid sequence, as for many other sorting signals, but is in a "patch" formed by noncontiguous parts of the lysosomal enzyme. The phosphotransferase must first recognize this "patch" before modifying the bound oligosaccharides.

The M6P sorting signal is recognized by M6P receptors (of which there are two types), which are primarily located in the TGN. The cytoplasmic tails of both M6P receptors contain sorting signals for targeting to endosomes. One sorting signal is based on the tyrosine signal and may be recognized by AP-1 clathrin adaptor complexes at the TGN (for more on

adaptors see *8.14 Adaptor complexes link clathrin and transmembrane cargo proteins*).

Another sorting signal in the M6P receptor cytoplasmic tail is a dileucine signal in an acidic cluster. This signal is distinct from the dileucine-based signals that mediate internalization from the cell surface and interacts with a GGA protein, as shown in **FIGURE 8.49**. GGAs are a family of proteins so named because they localize to the Golgi apparatus, contain sequences homologous to the appendage domain of γ-adaptin, and bind to ADP-ribosylation factor. GGAs are thought to be packaging molecules that help load the M6P receptors, bound to lysosomal enzymes, into clathrin-coated vesicles budding from the TGN. It appears that GGAs hand over the M6P receptor and clathrin to the AP-1 complex on the nascent vesicle.

As shown in **FIGURE 8.50**, the TGN-derived clathrin-coated vesicles are probably targeted to late endosomes, where the M6P receptor and its bound enzyme dissociate, similar to receptor-ligand complexes that are endocytosed. Delivery of enzymes to early endosomes can also occur by a minor route from the TGN via

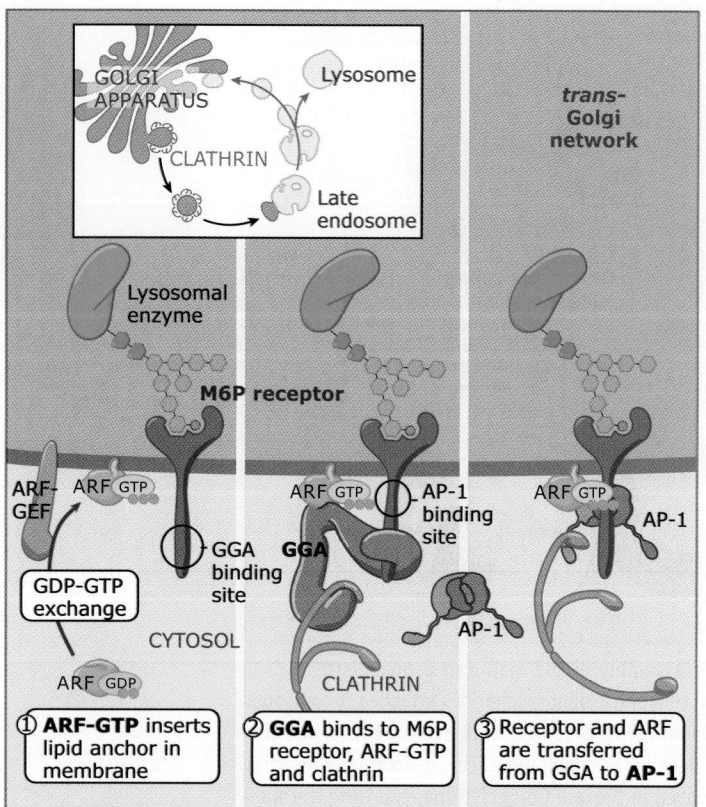

FIGURE 8.49 During formation of clathrin-coated vesicles at the *trans*-Golgi network, GGAs may transfer mannose 6-phosphate receptors and clathrin to AP-1 adaptor complexes.

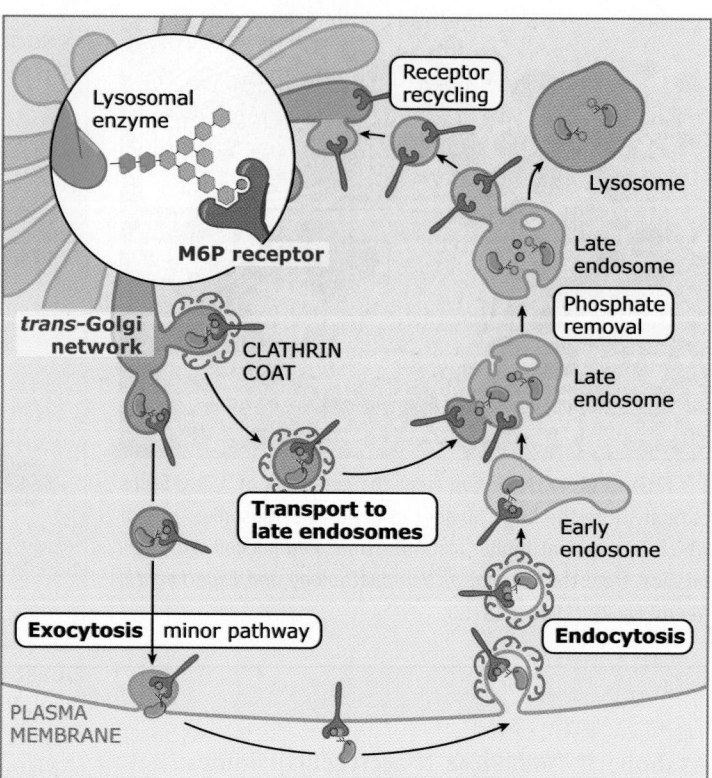

FIGURE 8.50 The mannose 6-phosphate sorting signal is used for the transport of lysosomal enzymes from the *trans*-Golgi network (TGN) to the endocytic pathway. The main pathway is directly from the TGN to late endosomes. There is also a minor pathway via the cell surface.

the cell surface. Endosomes are more acidic than the TGN, and the acidic pH causes dissociation of the enzymes from the M6P receptors, allowing the enzymes to accumulate free in the lumen of the endosome. The M6P receptors then recycle from endosomes to the TGN for further rounds of enzyme transport. Rab9 is involved in the targeting step for these vesicles.

The delivery of soluble lysosomal enzymes to endosomes via the exocytic pathway is part of the enzymes' normal maturation process. Until lysosomal enzymes reach their final destinations in the endocytic pathway, they are prevented from degrading proteins. There are at least three mechanisms to ensure that the enzymes are active only in the correct compartments as given below:

- Lysosomal enzymes are inactive at the pHs of the ER and Golgi apparatus and are active only at the acidic pH of endosomes and lysosomes. Thus, the acidic pH in endosomes not only facilitates their dissociation from

M6P receptors, but also is essential for enzyme activation.
- Some lysosomal enzymes are synthesized at the ER as proenzymes, which have short N-terminal peptide sequences that inhibit activity until they are cleaved off in endosomes. Many of these cleavages are "autocatalytic," meaning that a given protease activates itself upon reaching an organelle of sufficiently low pH.
- Some of the activated enzymes are phosphatases that cleave the M6P sorting signal. Removal of this signal helps limit the escape of active lysosomal enzymes to the TGN by preventing binding to recycling M6P receptors.

All of these maturation events are initiated at the time of enzyme entry into the endocytic pathway at the level of late endosomes. Subsequent movement of the enzymes to lysosomes increases the rapidity and efficiency of these events. There is a gradient of decreasing pH from the early endosomes (pH 6.5–6.8) through to the lysosomes (pH 4.5–5.0). Since most lysosomal enzymes exhibit optimal enzymatic activities at pH <5, they become progressively more active the further they penetrate into the endocytic pathway.

Resident lysosomal membrane proteins, such as lysosomal glycoproteins (lgps) and lysosomal-associated membrane proteins (lamps), are targeted from the TGN by an M6P-independent mechanism. These proteins have, in their cytoplasmic domains, tyrosine-containing signals that specify their interaction with one or more adaptor complexes (AP-1 or AP-3) or GGA proteins. They are delivered from the TGN to endosomes presumably via clathrin-coated vesicles and accumulate in lysosomes. Unlike M6P receptors, they do not recycle to the TGN. Although most other membrane proteins would be degraded in lysosomes, lgp and lamp proteins are resistant to lysosomal proteolysis. The lgp/lamp proteins are heavily glycosylated, and it is thought that either the carbohydrates are a protective coating or the dense packing of these glycoproteins in the lysosomal membrane confers this resistance.

Yeast cells and mammalian cells sort lysosomal proteins quite differently. The lysosome equivalent in yeast is called the vacuole. It also contains an array of soluble hydrolytic enzymes that are delivered by membrane receptors. However, transport from the TGN to the vacuole occurs in the absence of the M6P recognition marker and without the involvement of M6P receptors, which have no direct homologs in yeast. The signals that mediate targeting of soluble enzymes to the yeast vacuole are not known, but the enzymes appear to bind to a membrane receptor. Genetic evidence suggests that this receptor may interact with GGA proteins, indicating an overall conservation of sorting strategies. Vacuole membrane proteins, like lysosomal membrane proteins in animal cells, appear to rely on clathrin adaptors (such as AP-3) for transport from the Golgi apparatus. However, AP-3 in yeast may interact with dileucine rather than tyrosine-based targeting signals.

8.18 Polarized epithelial cells transport proteins to apical and basolateral membranes

Key concepts

- The plasma membrane of a polarized cell has separate domains with distinct sets of proteins, thereby necessitating a further sorting step.
- Sorting of cell surface proteins in polarized cells can occur at the TGN, endosomes, or one of the plasma membrane domains, depending on the cell type.
- Sorting in polarized cells is mediated by specialized adaptor complexes and perhaps lipid rafts and lectins.
- Transport to the basolateral membrane is mediated by Protein Kinase D (PKD) and Diacylglycerol.

In multicellular organisms, the plasma membrane of most cell types consists of multiple biochemically, structurally, and functionally distinct domains within a continuous lipid bilayer. Such cells are referred to as being "polarized." The most illustrative examples are epithelial cells and neurons. Epithelial cells line all body cavities (e.g., the intestine, kidney, airways) and, thus, possess two distinct surfaces. An apical domain faces the lumen of the organ, and a basolateral domain faces the blood or adjacent cells (see Figure 8.15).

In epithelia specialized for nutrient absorption (such as in the intestine), the apical plasma membrane domain is characterized by small outfoldings called microvilli, which increase the cell's surface area for absorption. The apical plasma membrane of such epithelial cells is enriched in membrane proteins that facilitate the uptake of nutrients such as amino acids, sugars, and other molecules. It also is enriched in a unique set of membrane glycolipids.

In contrast, the basolateral domain contains most of the membrane proteins (e.g., LDL receptor and EGF receptor) and lipids found in the plasma membrane of nonpolarized cells. In addition, transporters in the basolateral domain carry nutrients out of the cell into the blood plasma.

A junctional complex—consisting of a tight junction, adherens junction, and desmosome—demarcates the apical and basolateral domains (see *19.16 Tight junctions form selectively permeable barriers between cells*). One of the functions of these complexes is to prevent lateral diffusion of apical membrane components into the basolateral domain and vice versa. Therefore, lipids and proteins must be targeted to the correct domain. Traffic in the secretory and endocytic pathways is polarized to ensure that apical and basolateral membrane components are targeted to the appropriate destinations. Transport to the basolateral domain depends on targeting signals distinct from signals for apical transport.

Depending on whether they are intended for delivery to the apical or basolateral surfaces, newly synthesized membrane proteins are sorted into distinct classes of budding vesicles at the *trans*-Golgi network, as shown in **FIGURE 8.51**. Basolateral transport is well understood for many types of polarized epithelial cells. Most basolateral proteins have tyrosine- or dileucine-based signals in their cytoplasmic domains that are recognized by adaptor complexes, in a process that is similar for proteins that are endocytosed or targeted to lysosomes (see *8.14 Adaptor complexes link clathrin and transmembrane cargo proteins* and *8.17 Sorting of lysosomal proteins occurs in the* trans-*Golgi network*). The tyrosine-based signals and some non-tyrosine-based signals are recognized at the TGN by AP-1B, an epithelial cell-specific isoform of the clathrin AP-1 adaptor complex. AP-1B

couples with clathrin to yield a basolaterally directed transport vesicle. AP-1B differs from the ubiquitously distributed AP-1A complex only in the μ chain. The μ1A and μ1B subunits are nearly 80% identical but they bind different types of signals. The structural basis for this difference has not yet been identified.

Proteins destined for the apical plasma membrane do not have specific cytoplasmic domain signals. Rather, they contain critical N- or O-linked sugar residues in their lumenal domain or have a specific type of membrane anchoring domain that facilitates their inclusion in apically targeted transport vesicles. One example of such a membrane anchor is the GPI anchor that is added to certain proteins in the ER (see *7.13 The lipid glycosylphosphatidylinositol is added to some translocated proteins*). Such apically targeted vesicles often contain a unique lipid domain, called a lipid raft, which accumulates apical proteins, thus providing for their selective inclusion in apical vesicles (see Figure 8.52). Presumably, this mechanism also allows for a degree of lipid partitioning so that the complex glycolipids found at the apical, but not the basolateral, surface of epithelial cells are also selectively included in these vesicles.

A third type of polarized targeting exists for some membrane proteins that may have no targeting information and are transported out of the TGN in both apical and basolateral vesicles. These proteins can still become polarized, however, after delivery to either plasma membrane. This occurs by a process of domain-specific retention, in which the membrane protein interacts with a cytoskeletal scaffold that has previously been polarized. The scaffolds contain proteins that assemble at membrane domains in response to signals such as cell–cell attachment. Often, such scaffolds assemble at junctional complexes, thus allowing membrane proteins to accumulate asymmetrically. (For more on junctional complexes see *19.16 Tight junctions form selectively permeable barriers between cells*.) Membrane proteins that interact with scaffolds are stabilized at the appropriate membrane domain. However, membrane proteins delivered to a domain lacking its cognate scaffold are internalized by endocytosis and either degraded in lysosomes or recycled and thereby given another chance to arrive at the appropriate scaffold-containing domain.

Sorting of apical and basolateral proteins also occurs in endosomes. Since endocytosis occurs from both the apical and basolateral surfaces of epithelial cells, internalized proteins

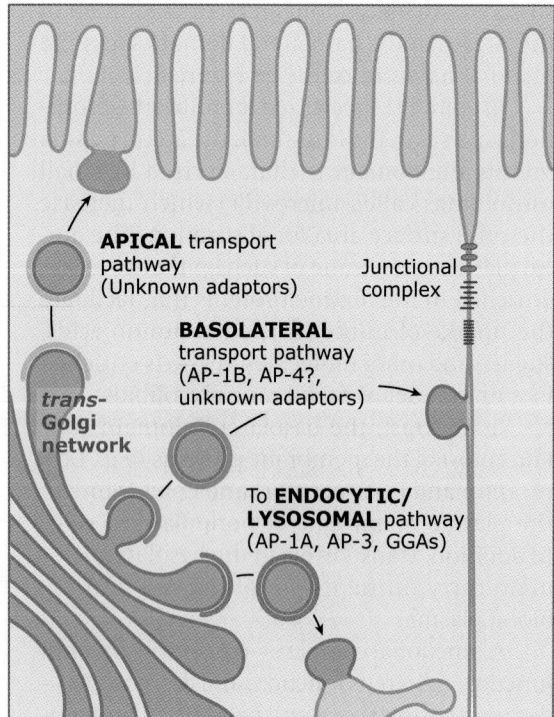

FIGURE 8.51 Three types of sorting events occur in the *trans*-Golgi network of polarized epithelial cells. Proteins are sorted to the endocytic pathway and to the apical and basolateral membranes. Different adaptor complexes are used for each pathway.

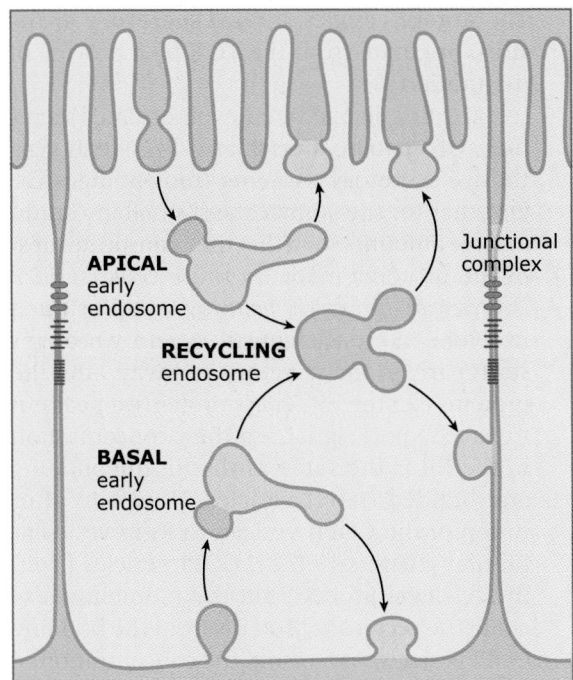

FIGURE 8.52 Molecules internalized from the apical or basolateral membranes can be recycled to their original membrane domain directly from early endosomes or sorted to a different membrane domain via recycling endosomes.

must be returned during recycling to their domain of origin. Endosomes, probably early endosomes, can act like the TGN in this sense, sorting apical and basolateral proteins into distinct recycling vesicles that are targeted to either the apical or basolateral membrane, as shown in **FIGURE 8.52**. In fact, polarized sorting in endosomes uses the same signals as polarized sorting in the TGN. In order for epithelial cells to maintain plasma membrane polarity in the face of continued endocytosis, polarized recycling from endosomes is essential. In fact, it also appears that even biosynthetic sorting occurs in endosomes, specifically, recycling endosomes. Thus, the pathway from the TGN to the basolateral plasma membrane may use recycling endosomes as an intermediate, perhaps explaining why the same sets of signals are used on both pathways.

How is cargo that is destined to plasma membrane packed into transport carriers and how are such transport carriers generated? There are many different exit routes from TGN to the cell surface. Genetic screens in yeast or in vitro reconstitution have not therefore been particularly successful to identify components involved in the biogenesis of TGN to cell surface specific transport carriers. A sea sponge metabolite called Ilimaquinone (IQ) was found

to convert Golgi cisternae into small vesicles in mammalian cells. In vitro reconstitution of IQ mediated Golgi vesiculation revealed the process to require trimeric G protein and a serine/threonine kinase called protein kinase D (PKD). Of interest is the finding that PKD is required for the formation of transport carriers at the TGN that contain basolateral plasma membrane specific cargo. When the kinase activity of PKD is compromised, cargo filled transport carriers form, but they fail to separate from the TGN. The carriers thus grow into large tubules that remain attached to the TGN. Thus, PKD is required for events leading to separation, by fission, of TGN to basolateral plasma membrane specific transport carriers. The binding of PKD to the TGN requires diacylglycerol (DAG) and the membrane-bound PKD activates the lipid kinase activity of an enzyme called phosphatidylinositol-4 kinaseIIIß (PI4KIIIß), PI4KIIIß converts phosphatidylinositol (PI) into PI4P. Both DAG and PI4P are required for Golgi to plasma membrane transport both in yeast and the mammalian cells. Thus the use of a chemical inhibitor IQ has revealed the components of the fission machinery for the generation of TGN to cell (basolateral) surface transport carriers. The identification of PKD, DAG and PI4P should help identify other components involved in this process.

Concept and Reasoning Check

1. Name other destinations to which vesicles depart from the TGN other than to the lysosome?

8.19 Some cells store proteins for later secretion

Key concepts

- Some cargo molecules are stored in secretory granules, which fuse with the plasma membrane and release their contents only upon stimulation.
- Storage of proteins for regulated secretion often involves a condensation process in which cargo self-associates, condensing to form a concentrated packet for eventual delivery to the outside of the cell.
- Condensation of proteins for regulated secretion often begins in the ER, continues in the Golgi apparatus, and is completed in condensing vacuoles that finally yield secretory granules.
- Condensation is accompanied by selective membrane retrieval at all stages of exocytosis.
- Fusion of synaptic vesicles with the plasma membrane involves SNARE proteins but is regulated by calcium-sensitive proteins such as synaptotagmin.

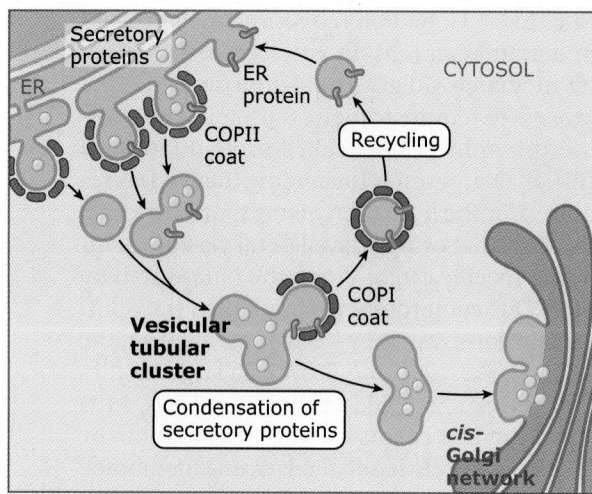

FIGURE 8.53 Proteins that are released from cells by regulated secretion become concentrated in vesicular tubular clusters as other proteins are removed in COPI-coated vesicles.

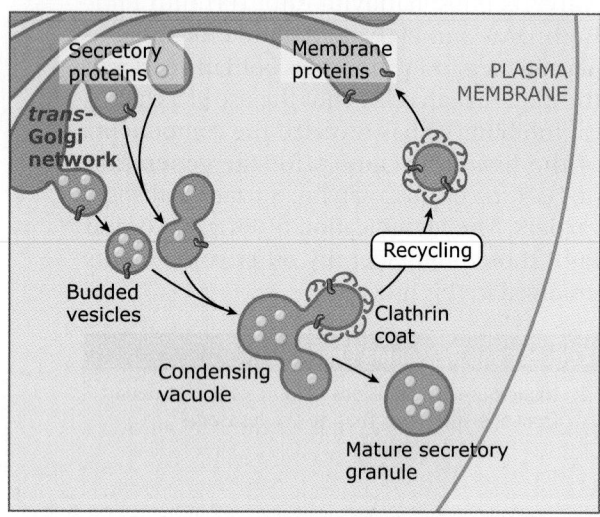

FIGURE 8.54 Proteins destined for regulated secretion are packaged into vesicles that bud from the *trans*-Golgi network and fuse with each other. Condensation of the secretory contents yields excess membrane that is recycled by clathrin-coated vesicles. The mature granules only fuse with the plasma membrane once an appropriate stimulus has been received by the cell.

Many eukaryotic cells can store secretory proteins in intracellular vesicles so that the proteins are released into the extracellular space only when the cell is stimulated by an appropriate signal (a secretagogue). This process is referred to as regulated secretion, which provides a means of rapidly delivering a large amount of material without the delay imposed by *de novo* synthesis. A wide variety of proteins, ranging from neurotransmitters to hormones and digestive enzymes, can be stored in this way.

The storage vesicles, termed **secretory granules**, originate from the TGN by a process of maturation.

The first detailed studies of regulated secretion used pancreatic acinar cells, which synthesize secretory proteins (mostly digestive enzymes for the stomach and small intestine) in large amounts such that they constitute most of the proteins made by these cells (see *8.2 Overview of the exocytic pathway*). As illustrated in **FIGURE 8.53**, packaging of secretory proteins starts early in the secretory pathway, after the proteins exit the ER. These proteins appear not to have export signals, so their concentration in the ER is the same as that in the budding and budded COPII vesicles. However, after fusion of the COPII vesicles to form vesicular tubular clusters (VTCs), COPI vesicles begin the retrieval process, but they somehow exclude the secretory proteins from the budding COPI vesicles. As a result, the concentration of the secretory proteins in the VTCs increases. Exclusion may be mediated by the colligative properties of the secretory proteins that cause them to associate with each other and exclude other types of protein from the complex. These protein complexes might be too large to enter the retrograde COPI vesicles.

The condensation process continues through the Golgi apparatus and into the condensing vacuoles that bud from the TGN. The secretory granules that form are remarkably homogeneous in size (about 0.5 µm in diameter), and they occupy a distinct region within the cell near the apical membrane, with which they eventually fuse once the cells have been stimulated with a secretagogue (see Figure 8.7). In the case of the exocrine pancreas, one such secretagogue is a peptide called CCK, which is released by the stomach upon ingestion of food. By stimulating the release of granules from the pancreas, CCK thus allows for the release of digestive enzymes into the gastrointestinal tract.

The maturation of secretory granules has also been studied in PC12 cells, a neuroendocrine cell line that is derived from adrenal gland and secretes hormones in response to nerve growth factor. As shown in **FIGURE 8.54**, condensing vacuoles bud from the TGN, then mature by further condensation and removal of membrane components not found in the final granule. These components are removed by the formation of clathrin-coated vesicles from the granule surface. These vesicles can salvage TGN proteins, such as furin and the M6P receptor,

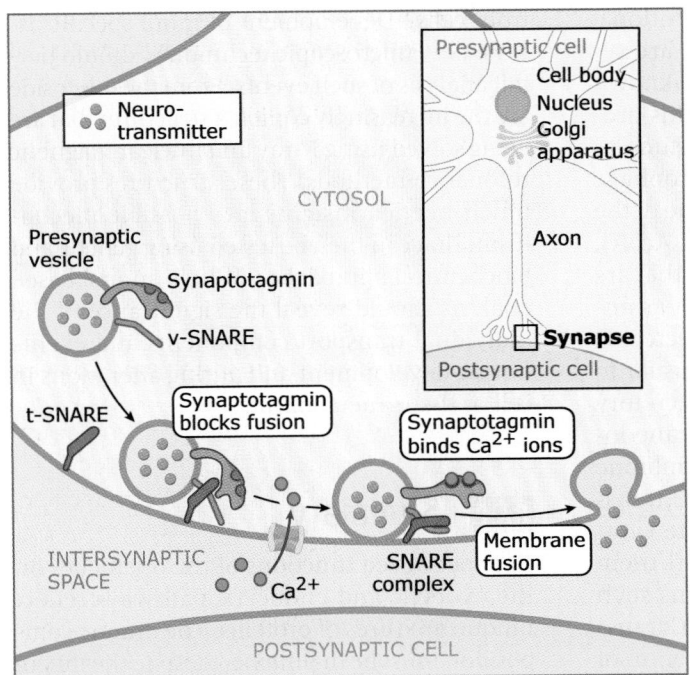

FIGURE 8.55 Synaptotagmin regulates the release of contents from synaptic vesicles. Synaptotagmin is thought to bind the complex formed by v- and t-SNAREs, thereby preventing membrane fusion. Calcium ions entering through the calcium channel bind to synaptotagmin, which releases the SNARE complex, allowing membrane fusion.

that were inadvertently incorporated into the budding granule membranes.

The regulated fusion of secretory granules with the plasma membrane uses the same underlying machinery as constitutive secretory processes. The major difference is that fusion is blocked at the stage where the SNARE complex is assembled. Fusion requires a signal mediated by an influx of calcium ions from the extracellular milieu. As shown in **FIGURE 8.55**, a calcium-binding protein, synaptotagmin, is a key player in this regulated process.

Synaptotagmin was first identified as a membrane component of synaptic vesicles, a specialized type of secretory granule that mediates neurotransmitter release at nerve terminals. Neurons are polarized cells with two distinct domains, a cell body and an axon, which are not separated by any obvious junction (see Figure 8.55). Axons are lengthy extensions (some are meters long) that carry electrical impulses from one neuron to the next. They end in synapses that form a unique type of junction with muscle cells or with the cell bodies (or extensions of the cell body called dendrites) of other neurons. The axonal and synaptic plasma membranes have biochemical compositions highly specialized to meet this function of neuronal transmission. Synaptotagmins participate in a regulated secretory process in neuronal and nonneuronal cells.

Synaptotagmin has a membrane anchor at one end, and the rest of the protein projects into the cytoplasm. This cytoplasmic domain contains two calcium-binding sites, forms a complex with all the SNARE proteins, and is thought to hold them in an inactive configuration, as shown in Figure 8.55. A nerve impulse that depolarizes the plasma membrane potential triggers the influx of calcium ions through calcium channels in the presynaptic membrane. The resulting increase in local calcium concentration is thought to be the signal that changes the conformation of synaptotagmin. This conformation change would allow synaptotagmin to disengage from the SNARE complexes, permitting them to mediate membrane fusion.

Concept and Reasoning Checks

1. How are neurotransmitters and hormones secreted from cells?
2. How do secretory storage granules form at the TGN and mature prior to fusion with the plasma membrane?

8.20 Some proteins are secreted without entering the ER–Golgi pathway

Key concepts

- Unconventional secretory pathways exist for specialized proteins.
- Some cytosolic proteins are secreted from cells.

A class of proteins that lack the conventional signal sequence for entering the ER are secreted from all eukaryotic cells. It is not known how many proteins are secreted by this process but this class includes proteins such as fibroblast growth factor (FGF), macrophage migration inhibitory factor (MIF), some of the interleukins, Acyl CoA binding protein (AcbA), and galectins. How are these proteins that are involved in important processes such as angiogenesis and immune surveillance secreted? Two proteins have been identified thus far to be involved in this unconventional secretory pathway. These are; a plasma membrane localized ABC transporter, and Golgi membrane associated protein called GRASP. ABC transporters, bind secretory proteins on the cytoplasmic side of the cells and transport them across the plasma membrane. Proteins such as α-factor in yeast are secreted by ABC transporter in the plasma membrane. Secretion of AcbA in dictyostelium (slime molds) during sporulation requires GRASP. In general the proteins secreted by this procedure are small in size and their release is developmentally regulated and mediated by signals that are cell type.

Concept and Reasoning Check

1. How are proteins that lack signal sequences transported out of cells?

8.21 What's next?

Approximately 30% of the genes in the human genome encode proteins that enter the secretory pathway. A large number of these proteins are secreted from the cells. Many, perhaps most, of the proteins that are part of the membrane trafficking machinery have been identified, and the functions they carry out have, in some instances, been elucidated. There are, however, many transport events, which remain poorly understood, and proteins for which the functions can only be guessed at, and there is a pressing need for higher resolution assays. Some of the obvious challenges are to understand how cargo is loaded into COPII carriers? How is the size of carrier regulated to accommodate bulky molecules such as Collagen fibres? What are the components of the membrane fission machinery? What is the biochemical composition of TGN to cell surface carriers? How are proteins that do not enter the conventional secretory pathway released

from cells? Development of more specific assays, new microscopic techniques should permit analysis of such events. From the other side are the increasingly complex structures that are being solved using X-ray and nuclear magnetic resonance methods. These structures provide much-needed insights into possible mechanisms that can then be tested using genetic and biochemical approaches. Whole animal based analysis should reveal the significance of the individual transport components and events during development and during alterations in cell or tissue metabolism.

8.22 Summary

The specialized functions of the organelles on the exocytic and endocytic pathways reflect unique mixtures of proteins. The unique composition must be maintained despite the flux of proteins that pass through the organelles after biosynthesis and as they function. The shuttling of vesicles between membranes solves the problem of compartmental identity in the face of continuous exchange. Proteins are selectively incorporated into vesicles that carry them to the next compartment and then fuse only with this compartment.

Proteins are selected using sorting signals that bind, either directly or indirectly, to the coat proteins that pinch off the vesicle. This mechanism is particularly prevalent on the endocytic pathway. In the exocytic pathway receptors and guides have been identified that help in cargo loading into COPII-coated vesicles. But many cargo molecules, especially those synthesized in large amounts might move by bulk flow through the pathway up to the *trans*-Golgi network. DAG and PI4P are required for the biogenesis of TGN to cell surface transport carriers. One of the key effectors of DAG in mammalian cells is PKD, which plays a key role in transport to the basolateral surface. Many soluble and abundant ER proteins have a C-terminal KDEL signal that is used to retrieve the proteins from early Golgi compartments.

Targeting the vesicles to the correct compartment requires Rab GTPases, membrane tethers, and SNARE proteins. Cognate SNAREs interact, bringing membranes together that eventually fuse. This final act of fusion can be controlled for purposes of regulated secretion of cargo ranging from digestive enzymes to neurotransmitters.

References

Bashkirov, P. V., Akimov, S. A., Evseev, A. I., Schmid, S. L., Zimmerberg, J., and Frolov, V. A. 2008. GTPase cycle of dynamin is coupled to membrane squeeze and release, leading to spontaneous fission. *Cell* v. 135 p. 1276–1286.

Bossard, C., Bresson, D., Polishchuk, R. S., and Malhotra, V. 2007. Dimeric PKD regulates membrane fission to form transport carriers at the TGN. *J. Cell Biol.* v. 179 p. 1123–1131.

Cai, H., Reinisch, K., and Ferro-Novick, S. 2007. Coats, tethers, Rabs, and SNAREs work together to mediate the intracellular destination of a transport vesicle. *Dev. Cell* v. 12 p. 671–682.

Flieger, O., Engling, A., Bucala, R., Lue, H., Nickel, W., and Bernhagen, J. 2003. Regulated secretion of macrophage migration inhibitory factor is mediated by a non-classical pathway involving an ABC transporter. *FEBS Lett.* v. 551 p. 78–86.

Hausser, A., Storz, P., Märtens, S., Link, G., Toker, A., and Pfizenmaier, K. 2005. Protein kinase D regulates vesicular transport by phosphorylating and activating phosphatidylinositol-4 kinase IIIbeta at the Golgi complex. *Nat. Cell Biol.* v. 7 p. 880–886.

Hayes, G. L., Brown, F. C., Haas, A. K., Nottingham, R. M., Barr, F. A., and Pfeffer, S. R. 2009. Multiple Rab GTPase binding sites in GCC185 suggest a model for vesicle tethering at the trans-Golgi. *Mol. Biol. Cell* v. 20 p. 209–217.

Malerød, L., and Stenmark, H. 2009. ESCRTing membrane deformation. *Cell* v. 136 p. 15–17.

Nickel, W., and Rabouille, C. 2009. Mechanisms of regulated unconventional protein secretion. *Nat. Rev. Mol. Cell Biol.* v. 10 p. 148–155.

Palade, G. 1975. Intracellular aspects of the process of protein synthesis. *Science* v. 189 p. 347–358.

Parton, R. G., and Simons, K. 2007. The multiple faces of Caveolae. *Nat. Rev. Mol. Cell Biol.* v. 8 p. 185–194.

Pfeffer, S. R. 2007. Unsolved mysteries in membrane traffic. *Annu. Rev. Biochem.* v. 76 p. 629–645.

Pucadyil, T. J., and Schmid, S. L. 2008. Real-time visualization of dynamin-catalyzed fission and vesicle release. *Cell* v. 135 p. 1263–1275.

Roux, A., and Antonny, B. 2008. The long and short of membrane fission. *Cell* v. 135 p. 1163–1165.

Saksena, S., Wahlman, J., Teis, D., Johnson, A. E., and Emr, S. D. 2009. Functional reconstitution of ESCRT-III assembly and disassembly. *Cell* v. 136 p. 97–109.

Saito, K., Chen, M., Bard, F., Chen, S., Zhou, H., Woodley, D., Polischuk, R., Schekman, R., and Malhotra, V. 2009. TANGO1 facilitates cargo loading at endoplasmic reticulum exit sites. *Cell* v. 136(5) p. 891–902.

Schekman, R., and Novick, P. 2004. 23 genes, 23 years later. *Cell* v. 116 p. 13–15.

Schmitz, R., Liu, J., Li, S., Setty, T. G., Wood, C. S., Burd, C. G., and Ferguson, K. M. 2008. Golgi localization of glycosyltransferases requires a Vps74p oligomers. *Dev. Cell* v. 14 p. 523–534.

Seelenmeyer, C., Stegmayer, C., and Nickel, W. 2008. Unconventional secretion of fibroblast growth factor 2 and galectin-1 does not require shedding of plasma membrane-derived vesicles. *FEBS Lett.* v. 582 p. 1362–1368.

Sudhof, T. C., and Rothamn, J. E. 2009. Membrane fusion: Grappling with SNARE and SM proteins. *Science* v. 323 p. 474–477.

Tu, L., Tai, W. C., Chen, L., and Banfield, D. K. 2008. Signal-mediated dynamic retention of glycosyltransferases in the Golgi. *Science* v. 321 p. 404–407.

The nucleus

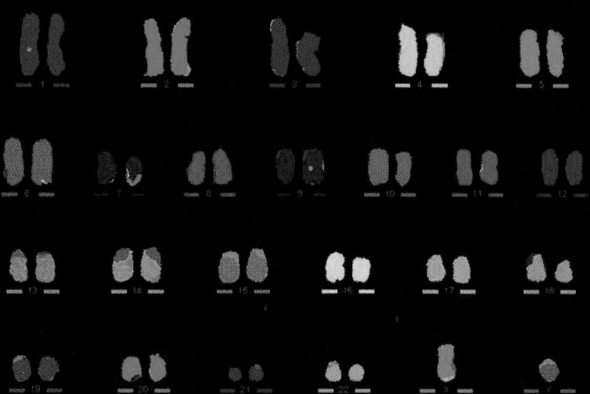

Photo courtesy of Dr. Mohammed Yusuf, Wellcome Trust
Centre for Human Genetics, University of Oxford.

PART

3

The nucleus

Nuclear structure and transport

Charles N. Cole
Departments of Biochemistry and Genetics, Dartmouth Medical School, Hanover, NH

Pamela A. Silver
Department of Systems Biology, Harvard Medical School, Boston, MA

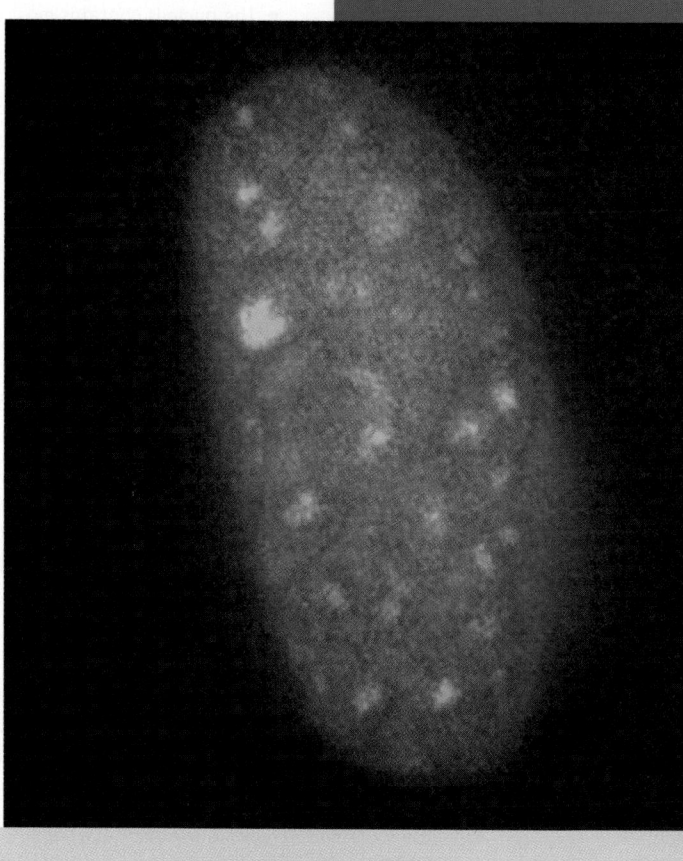

THE DISTRIBUTION OF DNA (red) and mRNA (green) in the nucleus of a mammalian cell is shown in this fluorescence micrograph. The mRNA is located in domains between those occupied by interphase chromatin. Reproduced from X. D. Shav-Tal, et al., *Science* 304 (2004): 1797–1800 [http://www.sciencemag.org]. Reprinted with permission from AAAS. Photo courtesy of Shailesh M. Shenoy and Robert Singer, Albert Einstein College of Medicine of Yeshiva University.

CHAPTER OUTLINE

9.1 Introduction

Key concepts

- The nucleus contains most of the cell's DNA, allowing for sophisticated regulation of gene expression.
- The nuclear envelope is a double membrane that surrounds the nucleus.
- The nucleus contains subcompartments that are not membrane-bounded.
- The nuclear envelope contains pores used for importing proteins into the nucleus and exporting RNAs and proteins from the nucleus.

When you look at a eukaryotic cell in a light microscope, the **nucleus** is the largest visible compartment, as shown in **FIGURE 9.1**. In fact, *eukaryotic* means "true nucleus," and having a nucleus is a defining feature of eukaryotic cells. The nucleus contains almost all of a eukaryotic cell's genetic material and acts as a center for controlling cellular activities. (A small amount of DNA is in mitochondria and for plants, also in chloroplasts.)

Antony van Leeuwenhoek (1632–1723) may have been the first to observe nuclei. He noted the presence of a centrally placed "clear area" when he examined blood cells of amphibians and birds. However, the actual discovery of the nucleus is credited to the abbot Felice Fontana (1730–1805), whose drawings in 1781 showed the nucleus as an ovoid structure in epidermal cells of eel skin. The Scottish botanist Robert Brown (1773–1858) noted that the plant cells he examined all contained "a single circular areola, generally somewhat more opaque than the membrane of the cell." He was the first to call these structures nuclei, a term derived from the Latin for *kernel*.

As the electron micrograph in **FIGURE 9.2** shows, a double membrane called the **nuclear envelope** encloses the nucleus. The two membranes are separated by a lumen that is contiguous with the endoplasmic reticulum (ER). The **nuclear pore complexes (NPCs)** that span the nuclear envelope are the channels through which macromolecules pass between the nucleus and cytoplasm. In contrast to proteins that pass across the ER or mitochondrial membranes (see *7 Membrane targeting of proteins*), proteins that pass through the NPCs are fully folded.

The nucleus contains subcompartments that are not enclosed by membranes. These subcompartments have specialized functions. The only nuclear subcompartment clearly seen by light microscopy is the **nucleolus** (see Figure 9.1), where ribosomal RNAs are synthesized

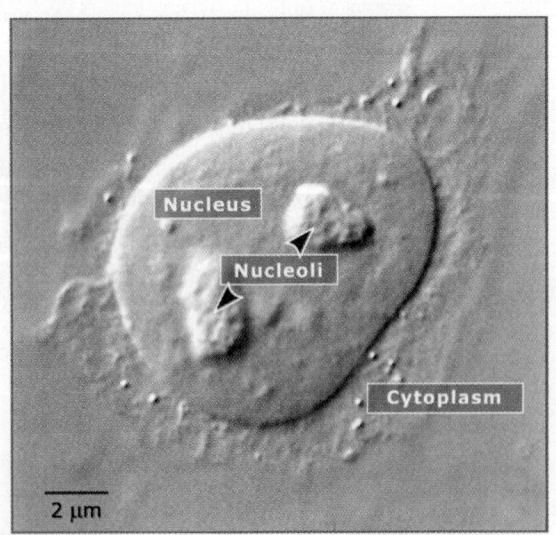

FIGURE 9.1 A cell from the human cervical carcinoma cell line HeLa has a nucleus that is easily seen using light microscopy. Photo courtesy of Zheng'an Wu and Joseph Gall, Carnegie Institution.

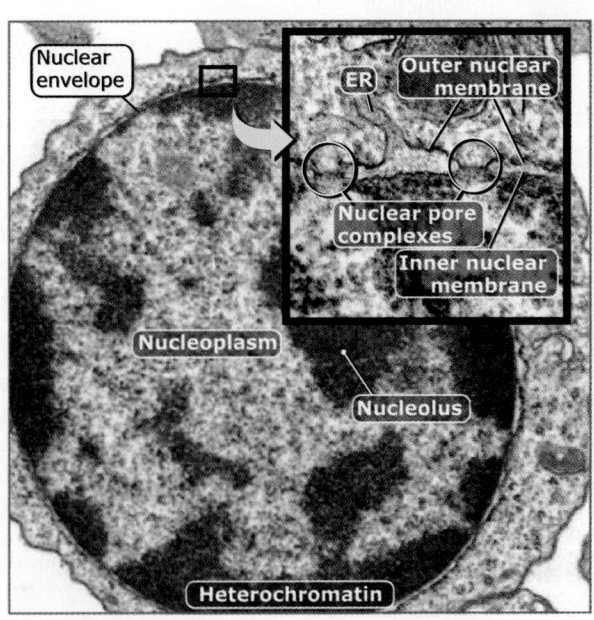

FIGURE 9.2 Many features of the nucleus of a lymphocyte are easily seen by electron microscopy. Photo courtesy of Terry Allen, University of Manchester.

and ribosomal subunits are assembled. Other subcompartments are revealed by immunofluorescence microscopy. These include speckles, which store RNA splicing factors, and replication factories. The non-nucleolar regions of the nucleus are often referred to as the **nucleoplasm**.

The DNA in the nucleus is found in different configurations (see *10.2 Chromatin is divided into euchromatin and heterochromatin*). In electron micrographs, some of the DNA appears darkly stained because it is relatively highly folded (see Figure 9.2). This DNA is called **heterochromatin** and is not actively transcribed. Much of the heterochromatin is found near the nuclear envelope. The remainder of the DNA is less densely compacted and is called **euchromatin**. Genes that are actively expressed are found in this fraction. In most cells, much more of the DNA is found in euchromatin than in heterochromatin.

What is the advantage to a eukaryotic cell of having a nucleus? The nucleus protects the DNA of the cell and allows for sophisticated gene regulation. Eukaryotic cells contain more DNA than prokaryotic cells, in some cases 10,000 times more. This DNA is packaged into chromosomes, each of which is a single DNA molecule (see *10 Chromatin and chromosomes* and *9.3 Chromosomes occupy distinct territories*). One double-stranded break in the DNA of one chromosome can be a lethal event for the cell. During interphase, the DNA is relatively loosely packaged so that the enzymes that replicate DNA and those that produce RNA can have access to the DNA. When it is loosely packaged, the DNA is more vulnerable to damage. The dynamic cytoskeleton generates shear forces that could break the DNA, were it not protected within the nucleus during interphase. In contrast, during mitosis, chromosomes are very compact, with the DNA folded into a highly ordered structure. Although the nuclear membrane breaks down during mitosis, exposing the DNA to the cytoplasmic environment, the condensed chromosomes are resistant to breakage by shear forces generated by the cytoskeleton.

Having a nucleus allows a cell to have much more sophisticated regulation of gene expression than is possible in prokaryotic cells. In prokaryotes, translation and transcription are coupled: translation of mRNAs begins before their synthesis is complete. As a result of the partitioning of the eukaryotic cell into cytoplasmic and nuclear compartments, many macromolecules must be transported between the nucleus and the cytoplasm. For example, mRNAs are transcribed and processed in the nucleus and then exported to the cytoplasm, where the machinery for protein synthesis is located. A comparison of the prokaryotic and eukaryotic processes appears in **FIGURE 9.3**. Many proteins are needed in the nucleus for replication, transcription, and other nuclear processes, so they must be imported from the cytoplasm. Ribosomal subunits are assembled in the nucleus from multiple RNAs, which are synthesized in the nucleus, and from more than a hundred proteins imported from the cytoplasm. The assembled subunits are then exported to the cytoplasm. All of these macromolecules move into and out of the nucleus through the NPCs. Importantly, the movement of molecules into and out of the nucleus can be regulated.

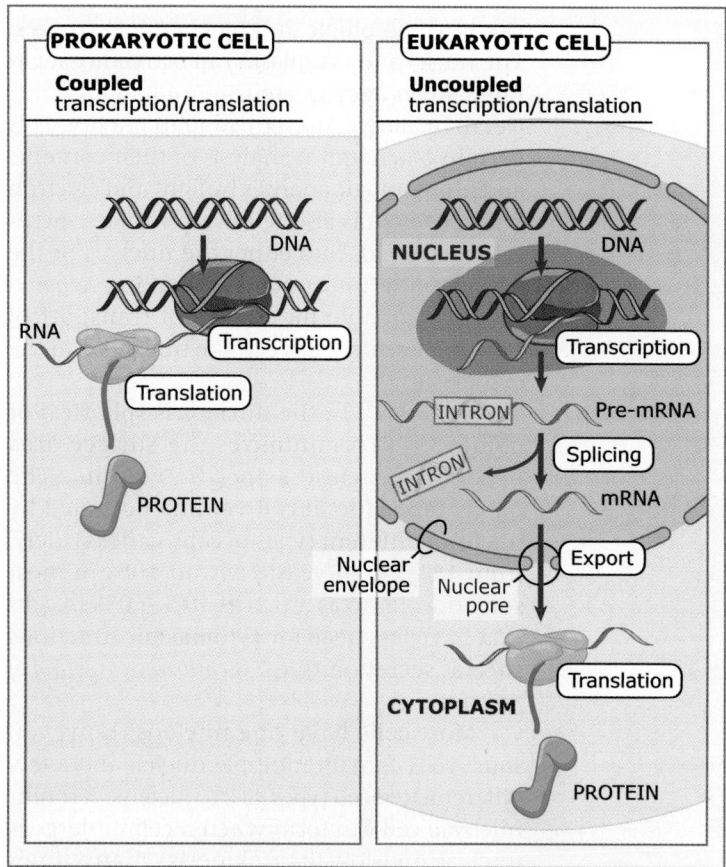

FIGURE 9.3 In prokaryotes, transcription and translation are coupled (left). In eukaryotes, transcription and translation occur in separate compartments (right).

9.2 Nuclei vary in appearance according to cell type and organism

Key concepts

- Nuclei range in size from about one micron (1 μm) to more than 10 μm in diameter.
- Most cells have a single nucleus, but some cells contain multiple nuclei, and a few cell types lack nuclei.
- The percentage of the genome that is heterochromatin varies among cells and increases as cells become more differentiated.

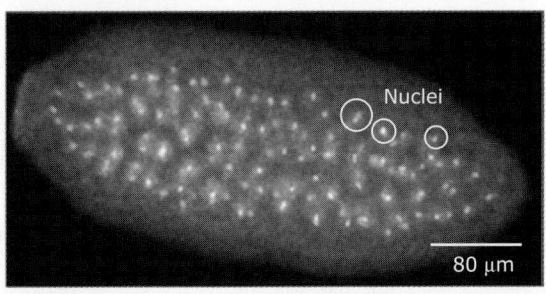

FIGURE 9.4 A *Drosophila* embryo at the multinucleate stage. DNA is stained with DAPI. Photo courtesy of Sharon E. Bickel, Dartmouth College.

The size of the nucleus is related to the amount of DNA it contains. The smallest nuclei are approximately 1 μm in diameter and are found in single-celled eukaryotes such as baker's yeast, *Saccharomyces cerevisiae*. The nuclei of the cells of many multicellular organisms are 5–10 μm in diameter. The nuclei of the oocytes of the frog *Xenopus laevis* are much larger and have a diameter of up to 1 mm. The availability of large numbers of *Xenopus* oocytes and the large size of its nuclei have led to their widespread use in cell biologic and biochemical studies. The nuclei and cytoplasm can be readily separated when oocytes are ruptured and the nuclei then allowed to settle due to gravity. This permits biochemical analysis of their contents and morphologic analysis by light and electron microscopy. It is also relatively easy to microinject materials into either the nucleus or the cytoplasm of intact oocytes, making *Xenopus* oocytes valuable for studies of transport of macromolecules between the nucleus and the cytoplasm.

In most cells, the nucleus is spherical or oblong, which minimizes the surface area needed to enclose a specific volume. The percentage of total cell volume occupied by nuclei of different types of cells varies widely, from 1%–2% in yeast cells, to 10% in most somatic cells, to as much as 40%-60% in cells that have less need for cytoplasmic functions such as secretion (see 7 *Membrane targeting of proteins*).

Most cells have one nucleus; however, some cells contain multiple nuclei, and a few differentiated cell types lack a nucleus. A multinucleate cell can form when a cell undergoes nuclear division (karyokinesis) many times without undergoing cell division (cytokinesis). For example, the early embryos of *Drosophila melanogaster* and related insects contain hundreds of nuclei within a common cytoplasm, as shown in **FIGURE 9.4**. This allows RNAs and proteins produced in a subset of the nuclei to move through the common cytoplasm to set up gradients, which later play a central role in proper development of the anterior/posterior axis of the fly. Other multinucleate cells form when cells fuse to form a syncytium. For example, mature muscle cells (myocytes) form by fusion of precursor cells (myoblasts). A few differentiated cell types, such as mammalian mature red blood cells, blood platelets, and some cells in the lens of the vertebrate eye, lack nuclei.

The shape and appearance of the nucleus are two of the factors that allow different types of cells to be distinguished from one another. For example, leukemias are blood diseases in which white blood cells are produced in much greater numbers than normal. There are many types of white blood cells, which go through several stages of differentiation. At each stage, the nucleus of each cell type has a characteristic appearance. One of the many tests performed in diagnosing the type of leukemia is to determine the morphology of the nuclei of the cells that are present in excess.

Another nuclear feature that helps to identify some cell types is the fraction of the genome that is heterochromatin. For example, as immature erythroblasts differentiate into mature erythrocytes, more and more of the DNA becomes heterochromatin, as shown in **FIGURE 9.5**. The increase in heterochromatin correlates with the permanent silencing of most genes, as almost all of the mRNA produced in mature erythrocytes is globin mRNA. In some species, including humans, the erythrocyte nucleus is ultimately ejected, leaving a membrane "bag" of hemoglobin and other proteins needed to carry oxygen and carbon dioxide through the bloodstream. Significantly, it is only after the nucleus is lost that red blood cells acquire their novel shape—a

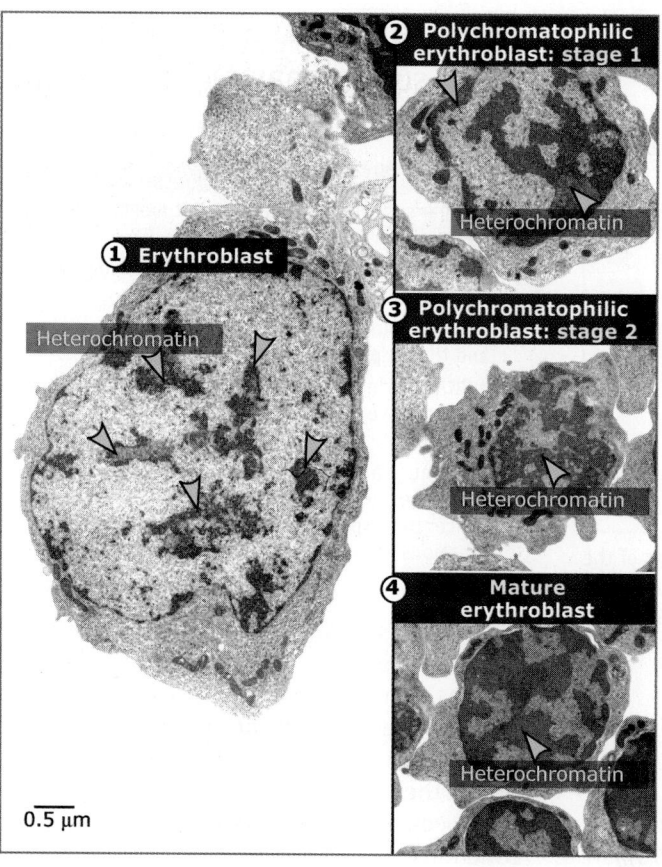

① Erythroblast

Heterochromatin

② Polychromatophilic erythroblast: stage 1

Heterochromatin

③ Polychromatophilic erythroblast: stage 2

Heterochromatin

④ Mature erythroblast

Heterochromatin

0.5 μm

FIGURE 9.5 As erythroid (red blood) cells differentiate, fewer genes are expressed, the fraction of the genome that is heterochromatin increases, and the cells become smaller. Photos courtesy of Terry Allen, University of Manchester.

biconcave disk—which is critical for their easy passage through tiny capillaries.

9.3 Chromosomes occupy distinct territories

Key concepts

- Although the nucleus lacks internal membranes, nuclei are highly organized and contain many subcompartments.
- Each chromosome occupies a distinct region or territory, which prevents chromosomes from becoming entangled with one another.
- The nucleus contains both chromosome domains and interchromosomal regions.

A remarkable feature of the cytoplasm is its compartmentalization into membrane-bounded organelles that provide optimal environments for particular cellular functions. In contrast, the nucleus lacks internal membranes, yet it is highly organized. There are many subcompartments that provide biochemically distinct environments optimized for the important tasks that occur in the nucleus. In this section, we consider the overall organization of chromosomes. In *9.4 The nucleus contains subcompartments that are not membrane-bounded* and *9.5 Some processes occur at distinct nuclear sites and may reflect an underlying structure*, we discuss the localization of nuclear processes such as ribosome subunit assembly and DNA replication.

The set of chromosomes in the nucleus is highly organized. The organization is revealed when fixed cells are treated so that different chromosomes bind dyes of different colors. The fluorescence micrograph in **FIGURE 9.6** reveals that chromosomes do not intertwine. Rather, they are ordered spatially, with each chromosome located in a separate area called a chromosome region, domain, or territory. If chromosomes were intertwined or tangled, they would need to be disentangled before chromosome segregation during mitosis in order to avoid breakage. Isolating each chromosome in its own territory avoids this problem. We do not know how chromosome territories are maintained, but in many types of cells, the ends of chromosomes, called telomeres, are anchored to the nuclear envelope. This likely helps to prevent tangling of chromosomes.

Chromatin does not fill the entire nucleus. Rather, there are regions where chromatin is located (**chromosome domains**), and adjacent chromatin-free regions called **interchromo-**

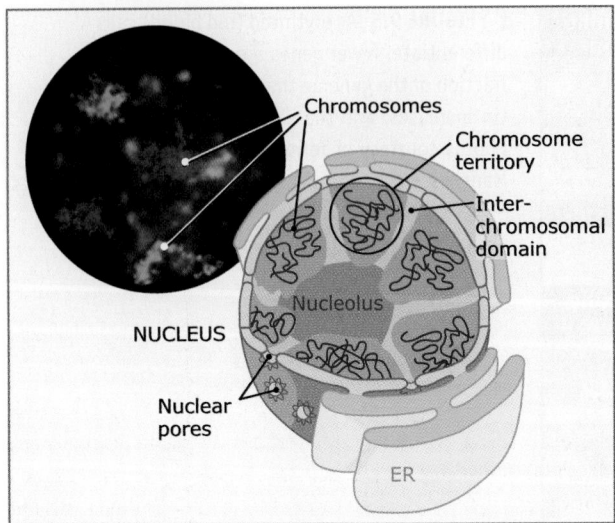

FIGURE 9.6 Individual chromosomes occupy distinct areas of the nucleus called chromosome territories. Photo reproduced with permission from *J. Cell Sci.*, vol. 114 (16): 2891–2893. [http://jcs.biologists.org/cgi/content/full/114/16/2891]. Photo courtesy of Thomas Reid, National Institutes of Health.

somal domains, depicted in Figure 9.6. The interchromosomal domains contain poly(A)+ RNAs that are undergoing the final steps of processing and are diffusing to the nuclear periphery for export (see Figure 9.53; *9.22 hnRNPs move from sites of processing to nuclear pore complexes*).

There is a further level of organization of chromosomes in terms of which genes lie immediately adjacent to the interchromosomal domains and which do not. Sensitive *in situ* hybridization and sophisticated imaging techniques allow us to determine where within the nucleus a specific gene and its transcripts are located. These studies indicate that highly transcribed genes tend to be located at the periphery of chromatin domains, adjacent to interchromosomal domains. Because different genes are active in different types of cells, the genes adjacent to interchromosomal regions vary according to cell type. The organization of chromosomes within their territories is not fixed and varies in response to changes in the pattern of gene expression (see *9.2 Nuclei vary in appearance according to cell type and organism*). The location of highly transcribed genes adjacent to interchromosomal domains facilitates diffusion of the most abundant mRNAs through these domains to the NPCs, thereby facilitating their export to the cytoplasm. In addition, some actively transcribed genes are located near NPCs, which may also enhance efficient export of their encoded mRNAs to the cytoplasm (for more on NPCs see *9.9 Nuclear pore complexes are symmetrical channels*).

9.4 The nucleus contains subcompartments that are not membrane-bounded

Key concepts

- Nuclear subcompartments are not membrane-bounded.
- rRNA is synthesized and ribosomal subunits are assembled in the nucleolus.
- The nucleolus contains DNA that encodes rRNAs and that is present on multiple chromosomes.
- mRNA splicing factors are stored in nuclear speckles and move to sites of transcription where they function.
- Other nuclear bodies have been identified using antibodies; some of these bodies are thought to concentrate specific nuclear proteins, but the functions of most nuclear bodies are unknown.

Key events that take place in the nucleus include the multiple steps of gene expression such as transcription and processing of RNAs. These processes occur at discrete locations in the nucleus. We know the most about the function of the nucleolus. Other subcompartments have been described, but for the most part their functions are unknown.

The most prominent subcompartment of the nucleus is the nucleolus (see Figure 9.1 and Figure 9.2). Most normal cells have a single nucleolus, but multiple nucleoli are sometimes found in cells, as shown in Figure 9.1. The nucleolus is where ribosomal RNAs are synthesized and processed and ribosomal subunits are assembled. The size of the nucleolus varies, depending on the amount of ribosome biogenesis in a given cell.

The nucleolus contains all of the components needed for assembly of ribosomal subunits, thereby providing a site for efficient assembly. These components include the rRNA genes from multiple chromosomes, rRNAs, enzymes for the synthesis and processing of rRNAs, and ribosomal proteins imported from the cytoplasm. There are multiple morphologically distinct regions within the nucleolus, visible in **FIGURE 9.7**, which reflect the fact that transcription of rRNA genes, rRNA processing, and subunit assembly take place in different regions of the nucleolus.

Although the nucleolus lacks a membrane, most proteins and RNAs present in it are found only there and not elsewhere in the nucleus. The nucleolus is detectable only when production of ribosomal subunits is occurring. It disappears when rRNA transcription is prevented

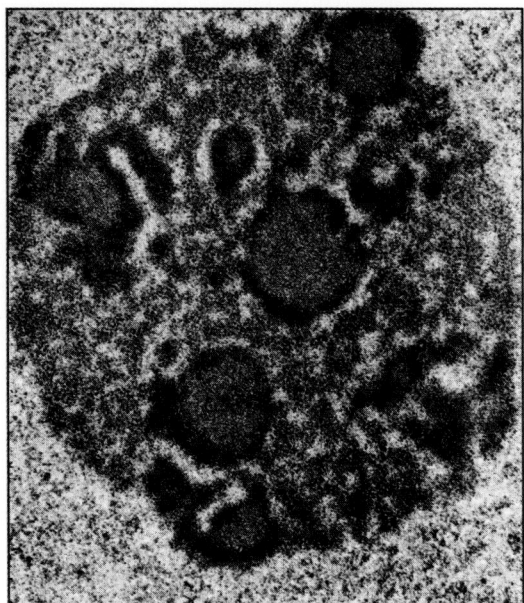

FIGURE 9.7 Subcompartments within the nucleolus, as visualized by electron microscopy. Reprinted courtesy of Dr. Don W. Fawcett. *The Cell*, 1981. Micrograph courtesy of David M. Phillips, The Population Council.

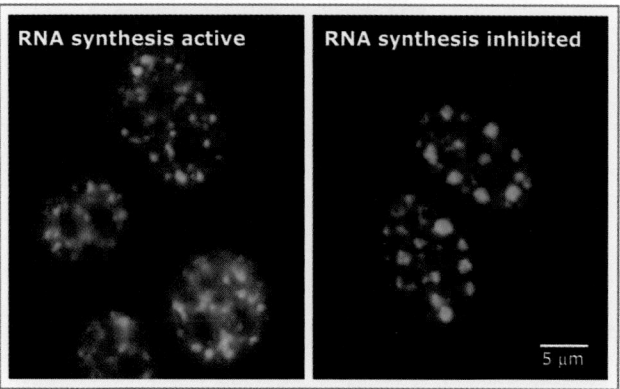

FIGURE 9.8 Splicing factors are concentrated in nuclear speckles. The more diffuse sites where these factors can be detected are sites of pre-mRNA processing (left). Actinomycin D causes a block in transcription (right). Speckles were localized by indirect immunofluorescence, using an antibody to the β" splicing factor component of U2 snRNP. Photos courtesy of David Spector, Cold Spring Harbor Laboratory.

experimentally and reappears when transcription is allowed to resume. Thus, the nucleolus is thought to arise from the coming together of rRNA genes, the transcription factors that bind to the promoters of rRNA genes, and RNA polymerase I molecules recruited by the transcription factors. The newly synthesized rRNAs would in turn attract ribosomal proteins and, in an ordered sequence, the many processing factors needed to assemble ribosomal subunits. We do not yet know the order of recruitment. By a mechanism not yet understood, the nucleolus is disassembled during mitosis and reassembled when mitosis has been completed.

The function of the nucleolus is not limited to biogenesis of ribosomal subunits. tRNA genes are clustered in nucleoli where they are transcribed and where tRNA processing begins. Many proteins that do not appear to participate in formation of ribosomal subunits are found in the nucleolus. In many cases, we do not know why they have a nucleolar location. In other cases, proteins that function in the nucleus might be sequestered in the nucleolus and could then be released rapidly into the nucleoplasm to function. Some cell cycle-regulated proteins fall into this category. Proteomic analyses of nucleoli from cultured human cells indicate that they contain more than 400 different polypeptides and that 30% of these are novel or uncharacterized.

Several other smaller, discrete subcompartments of the nucleus have been detected, primarily by using antibodies from patients with autoimmune diseases or antibodies prepared to react with specific nuclear proteins. These antibodies allow us to detect these discrete subcompartments by fluorescence and, in some cases, immunoelectron microscopy, as they are not visible by light microscopy alone. Sometimes called nuclear bodies, these subcompartments are present in many types of eukaryotic cells and are not membrane-bounded. They include speckles, Cajal bodies, Gemini bodies, and promyelocytic leukemia (PML) bodies. One function of nuclear bodies may be to increase the efficiency of biologic processes by concentrating multiple macromolecules that are required for a process.

RNA splicing factors are spatially organized in the nucleus in **speckles**. FIGURE 9.8 shows that splicing factors are concentrated at approximately twenty to fifty speckles per cell, but these factors are also located diffusely at many other sites called interchromatin granules. Because speckles do not contain pre-mRNA, they are believed to store splicing factors rather than be splicing factories. Splicing is thought to occur in the more diffusely stained areas, where polyadenylated RNA and splicing factors are both detected. An experiment that supports this hypothesis is shown in Figure 9.8. If transcription is blocked using an inhibitor of RNA polymerase II, the more diffuse distribution of splicing factors disappears as these factors relocate to speckles. When transcription

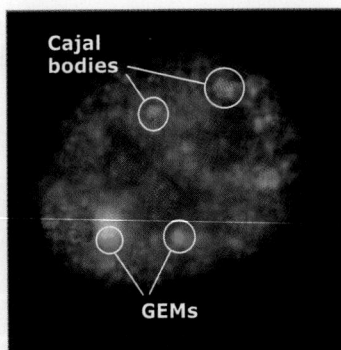

FIGURE 9.9 Cajal bodies and GEMs can be detected by using specific antibodies and indirect immunofluorescence. Reproduced with permission from *J. Cell Sci.*, vol. 114 (16): 2891–2893. [http://jcs.biologists.org/cgi/content/full/114/16/2891]. Photo courtesy of A. Greg Matera, University of North Carolina at Chapel Hill.

by RNA polymerase II is allowed to resume, the more diffusely distributed staining pattern reappears.

Some splicing factors are also in a different structure, the **Cajal body** or coiled body, shown in **FIGURE 9.9**. There are usually only one or a few Cajal bodies in the nucleus, often close to the nucleolus. They contain a protein called coilin, which is not present in speckles. We know that Cajal bodies are not involved in the enzymatic process of splicing because they do not contain pre-mRNA, but they do contain small nuclear RNAs (snRNAs) and small nucleolar RNAs (snoRNAs) and are thought to be sites where these RNAs are modified post-transcriptionally and assembled into ribonucleoprotein complexes. (For more on snRNAs see *9.24 U snRNAs are exported, modified, assembled into complexes, and imported.*) Another nuclear body, the **Gemini body** or GEM, is also visible in Figure 9.9. Gemini bodies are not found in all cells and some of their components are also found in Cajal bodies, suggesting that they may not perform distinct functions.

PML bodies are another nuclear subcompartment. Called PML bodies because they contain a protein related to one first found in patients with PML, they were first identified using antibodies from these patients. The PML protein recruits many other proteins, leading to formation of PML bodies, but the precise function of PML bodies is unknown. Among the proteins that localize to PML bodies are transcription factors and chromatin-modifying enzymes. These proteins are probably stored

in PML bodies since PML bodies do not appear to be sites of transcription or chromatin modification. Cells lacking PML bodies are normal in most ways so their functions cannot be essential.

Many of these smaller nuclear bodies appear to be absent from the small nuclei of unicellular eukaryotes such as yeast. However, yeast cells contain nuclear bodies similar to Cajal bodies. It may be that other nuclear bodies are also present, but because they are certain to be considerably smaller than in cells of metazoan animals, they would be more difficult to detect.

9.5 Some processes occur at distinct nuclear sites and may reflect an underlying structure

Key concepts
- The nucleus contains replication sites where DNA is synthesized.
- The nucleus may contain a nucleoskeleton that could help to organize nuclear functions.

We have discussed in previous sections some of the nuclear domains and subcompartments, which have unique compositions and functions (see *9.3 Chromosomes occupy distinct territories* and *9.4 The nucleus contains subcompartments that are not membrane-bounded*). Other processes, such as DNA replication, are also organized within the nucleus. It is thought that the macromolecular machinery for DNA replication and RNA splicing may be connected to an underlying nuclear structure.

During the early part of S phase, when DNA is synthesized, cells contain many sites of DNA replication. As S phase proceeds, these sites coalesce and only a few dozen much larger sites are detected. These larger sites are called replication factories. **FIGURE 9.10** shows the distribution of these replication factories at different stages of S phase. Because there are so many more origins of replication active at any one time than there are replication factories, each of these factories must contain dozens or hundreds of origins. Similar studies suggest that transcription may also occur in a limited number of sites, called transcription factories.

The localization of nuclear processes to discrete sites suggests that there may be an underlying structure in the nucleus. Nuclei do not contain a highly ordered skeleton

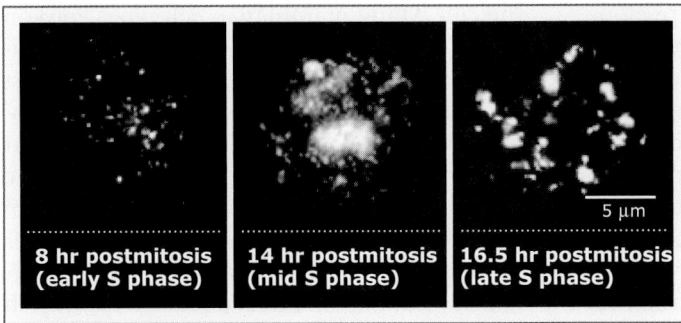

FIGURE 9.10 DNA replication occurs at a limited number of sites called replication factories. DNA was labeled with bromodeoxyuridine (BrdU) and detected by using fluorophore-conjugated antibodies to BrdU. The images show cells at different times after mitosis. Reproduced with permission from *J. Cell Sci.*, vol. 107 (8): 2191–2202. [http://jcs.biologists.org/cgi/content/abstract/107/8/2191]. Photos courtesy of Peter R. Cook, University of Oxford.

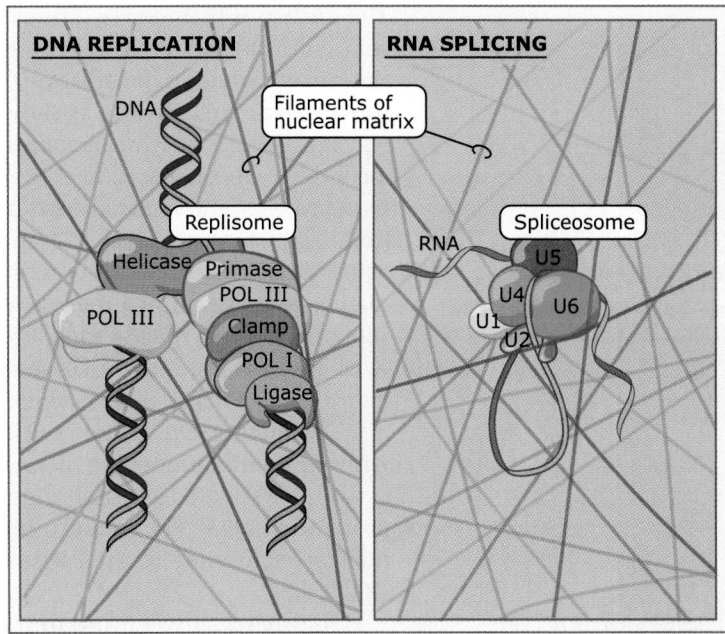

FIGURE 9.11 The enzymatic machines that replicate DNA and splice RNA may be anchored to a nuclear matrix.

resembling the cytoskeleton. However, some studies suggest that a sort of filamentous network, called the **nuclear matrix**, may exist in the nucleus. In contrast to the easily visualized cytoskeleton, this network is seen only if nuclei are treated with detergents, DNase, and high salt concentration. These treatments extract much material, including almost all DNA and all membranes, and leave only insoluble proteins and some of the RNA. The network contains short fibers approximately the size of intermediate filaments, actin (not in its filamentous form), and many other proteins. These components are not well organized into any larger structure. RNA appears to play a key organizing role, since this network is no longer seen if cells are treated with RNase in addition to detergents, DNase, and high salt.

Because the nuclear matrix is relatively insoluble, it is difficult to study it as a whole. Some cell biologists believe that the nuclear matrix is an artifact because it can only be seen after harsh extraction. However, because many important and complex processes occur in the nucleus and must be performed accurately, an underlying organization of some sort is likely (see *10.4 Eukaryotic DNA has loops and domains attached to a scaffold* and *10.5 Specific sequences attach DNA to an interphase matrix or a metaphase scaffold*).

One possible function for an underlying nuclear structure is the organization of the machinery for replication, transcription, and RNA processing, which are performed by replisomes, RNA polymerase II holoenzyme complexes, and spliceosomes, respectively. Although these large multisubunit complexes have much smaller masses than chromosomes, they have much greater diameters than their nucleic acid substrates. Structural studies indicate that these complexes contain clefts or passageways for nucleic acid chains. Many studies suggest that these complexes are attached to an underlying structure, as illustrated in **FIGURE 9.11**. This means that when replication, transcription, and splicing occur, the protein machineries may be fixed and the nucleic acids may move through the complexes.

Concept and Reasoning Checks

1. Describe the organization of chromosomes within the nucleus and how this was determined.
2. What subcompartments are found in the nucleus and what is the function of these sub**compartments**?

9.6 The nucleus is bounded by the nuclear envelope

Key concepts

- The nucleus is surrounded by a nuclear envelope consisting of two complete membranes.
- The outer nuclear membrane is continuous with the membranes of the ER, and the lumen of the nuclear envelope is continuous with the lumen of the ER.
- The nuclear envelope contains numerous NPCs, the only channels for transport of molecules and macromolecules between the nucleus and the cytoplasm.

The nucleus is bounded by a nuclear envelope that consists of two concentric membranes, the **outer nuclear membrane,** and the **inner nuclear membrane** as shown in Figure 9.2. Each of the nuclear membranes contains a complete phospholipid bilayers, some common proteins, and some proteins unique to the inner or outer nuclear membrane. Except in some single-celled eukaryotes, a network of filaments, woven into a meshwork structure, supports the inner nuclear membrane. This network is called the nuclear lamina (see *9.7 The nuclear lamina underlies the nuclear envelope*).

The outer nuclear membrane is continuous with the membranes of the ER and, like much of the ER, is covered with ribosomes engaged in protein synthesis. The connection of the outer nuclear membrane to the ER can be seen in **FIGURE 9.12**.

The space between the outer and inner nuclear membranes is the lumen of the nuclear envelope. Just as the outer membrane of the nuclear envelope is continuous with the ER membrane, the lumen of the nuclear envelope is continuous with the lumen of the ER, as shown in Figure 9.12. Each of the two membranes is 7-8 nanometers (nm) thick, and the **nuclear envelope lumen** is 20–40 nm wide.

The most conspicuous features of the nuclear envelope at the level of electron microscopy are the NPCs that serve as the channels for the movement of most molecules between the cytoplasm and the nucleus (see *9.9 Nuclear pore complexes are symmetrical channels*). The number of NPCs per cell varies widely and is correlated with the level of nuclear transport required by the cell. Many mammalian cell types contain about 3000 to 4000 NPCs. Some cells, however, have a much greater density of NPCs, most likely because the cells are very active transcriptionally and translationally, requiring transport of many macromolecules into and out of the nucleus. The much smaller nuclear envelope of yeast cells contains 150–250 NPCs. The oocytes of some amphibians, including Xenopus, contain several million NPCs (**FIGURE 9.13**), which has made them a favorites system to study NPC structure, composition, and function.

How is the double membrane of the nucleus thought to have arisen? It shares this property with two other organelles of eukaryotic cells—mitochondria and chloroplasts. The endosymbiont hypothesis proposes that these organelles arose during evolution when cells endocytosed other cells. The ingested cell would then be surrounded by two membranes: its own and that of the engulfing cell. One type of ingested cell may have been able to provide an activity, such as photosynthesis, that was lacking in the engulfing cell. The best evidence for the endosymbiotic origin of chloroplasts and mitochondria is that the ribosomes of both organelles are more closely related to those of modern prokaryotes than to the cytoplasmic ribosomes of eukaryotic cells. We are less certain about the origin of the nucleus. However, its having a double membrane, like mitochondria and chloroplasts, has led to the hypothesis that

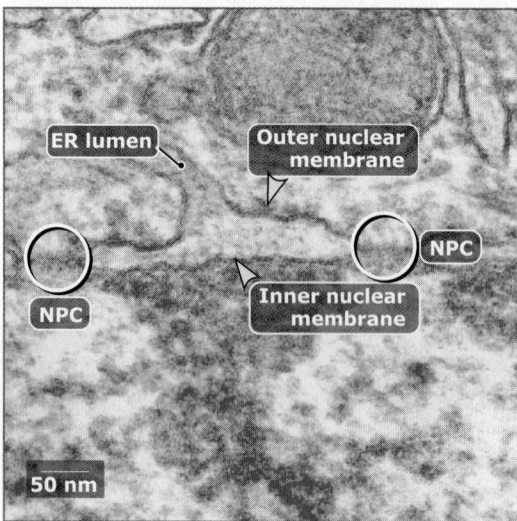

FIGURE 9.12 The nuclear envelope is continuous with the endoplasmic reticulum. Photo courtesy of Terry Allen, University of Manchester.

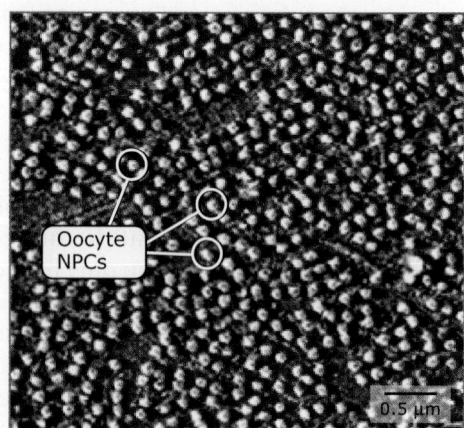

FIGURE 9.13 The surface of the nuclear envelope of a *Xenopus laevis* oocyte is covered with nuclear pore complexes. Reprinted from *J. Mol. Biol.*, vol. 287, D. Stoffler, et al., Calcium-mediated structural changes of native nuclear..., pp. 741–752, Copyright (1999) with permission from Elsevier [http://www.sciencedirect.com/science/journal/00222836]. Photo courtesy Ueli Aebi, University of Basel.

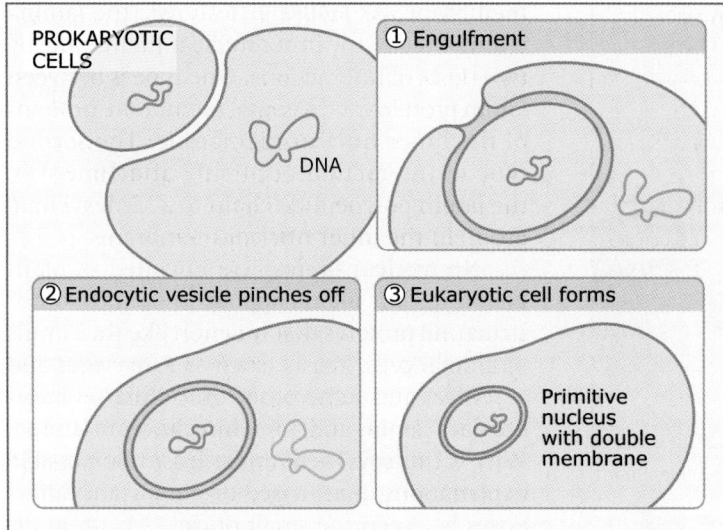

PROKARYOTIC CELLS

DNA

① Engulfment

② Endocytic vesicle pinches off

③ Eukaryotic cell forms

Primitive nucleus with double membrane

FIGURE 9.14 The nucleus may have arisen by endosymbiosis, a process in which one prokaryotic cell engulfs another cell, which then becomes a primitive nucleus.

an ingested prokaryote evolved to hold almost all the DNA of the cell, and became the nucleus, as illustrated in **FIGURE 9.14**.

9.7 The nuclear lamina underlies the nuclear envelope

Key concepts

- The nuclear lamina is constructed of intermediate filament proteins called lamins.
- The nuclear lamina is located beneath the inner nuclear membrane and is physically connected to it by lamina-associated integral membrane proteins.
- The nuclear lamina plays a role in nuclear envelope assembly and may provide physical support for the nuclear envelope.
- Proteins connect the nuclear lamina to chromatin; this may allow the nuclear lamina to organize DNA replication and transcription.
- Yeast and some other unicellular eukaryotes lack a nuclear lamina.

A common feature of the nuclei of metazoan organisms is the presence of the **nuclear lamina**, an intermediate filament meshwork lying just beneath the inner nuclear membrane (see *13 Intermediate filaments*). As **FIGURE 9.15** shows, the lamina is easily visualized by indirect immunofluorescence, using an antibody that recognizes one of the proteins in the lamina. Electron microscopy shows that the lamina has a filamentous, meshwork appearance. The figure also shows the typically disorganized lamin filaments lying directly beneath the inner nuclear membrane.

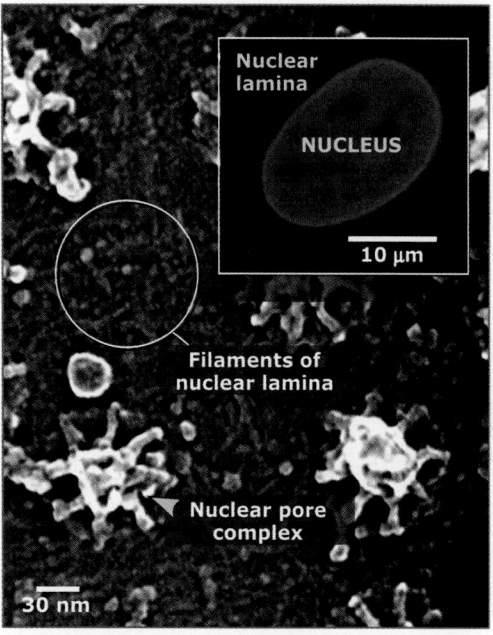

Nuclear lamina

NUCLEUS

10 µm

Filaments of nuclear lamina

Nuclear pore complex

30 nm

FIGURE 9.15 Antibodies to lamin proteins are used to visualize the nuclear lamina by fluorescence microscopy (inset). An electron micrograph reveals the nuclear baskets of NPCs and the filaments of the nuclear lamina. Main photo courtesy of Terry Allen, University of Manchester; inset photo courtesy of Anne and Bob Goldman, Department of Cell and Molecular Biology, The Feinberg School of Medicine, Northwestern University, Chicago, IL.

Nuclear lamina proteins are related to the keratin proteins of cytoplasmic intermediate filaments. Both lamin proteins and keratins are called intermediate filament proteins because the size of the filaments they form (10–20 nm in diameter) is intermediate between that of

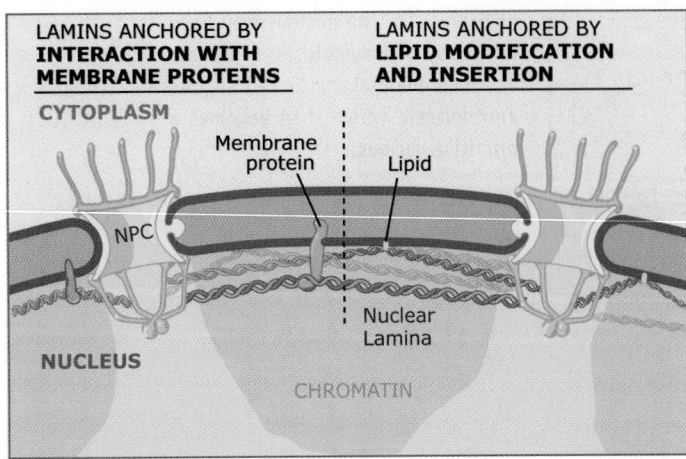

| LAMINS ANCHORED BY **INTERACTION WITH MEMBRANE PROTEINS** | LAMINS ANCHORED BY **LIPID MODIFICATION AND INSERTION** |

CYTOPLASM

Membrane protein · Lipid

NPC

Nuclear Lamina

NUCLEUS

CHROMATIN

FIGURE 9.16 The nuclear lamina is anchored to the inner nuclear membrane by two types of interactions.

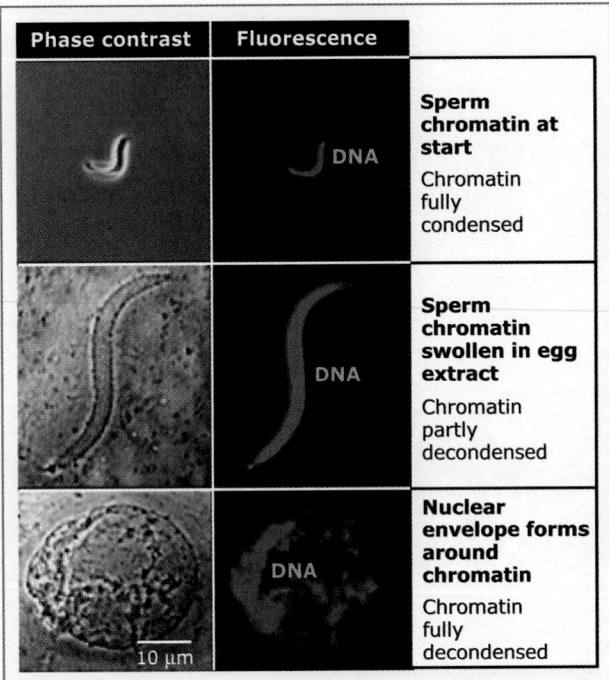

Phase contrast	Fluorescence	
	DNA	**Sperm chromatin at start** Chromatin fully condensed
	DNA	**Sperm chromatin swollen in egg extract** Chromatin partly decondensed
	DNA	**Nuclear envelope forms around chromatin** Chromatin fully decondensed

10 μm

FIGURE 9.17 When sperm cells from *Xenopus laevis* are demembranated (by removing the plasma membrane) and incubated in a *Xenopus* egg extract, the chromatin decondenses and a functional nuclear envelope forms around the chromatin. Fluorescence images show where DNA is located. Photos courtesy of Douglass Forbes, University of California, San Diego.

actin microfilaments (7 nm in diameter) and microtubules (25 nm in diameter). The nuclear lamina is interrupted by NPCs, which are anchored to the nuclear lamina, as also seen in Figure 9.15.

In addition to the lamins, the nuclear lamina also contains a set of integral membrane proteins called lamina-associated proteins (LAPs), some of which mediate interactions between the lamina and the inner nuclear membrane. As **FIGURE 9.16** shows, the lamina is anchored to the inner nuclear membrane by two types of interactions. One type is between lamin proteins and integral membrane proteins of the inner nuclear membrane. The second type of interaction occurs by attachment of the lamin polypeptide chain to a farnesyl lipid group in the inner nuclear membrane.

No nuclear lamins are encoded in plant genomes, but plants appear to contain other structural proteins that function like the lamins of animal cells. Yeasts (such as *S. cerevisiae* and *S. pombe*) and some other unicellular eukaryotes lack lamins and, therefore, have no lamina. Why is this so? There are at least two possible explanations, both based on important differences between the small nuclei (~1 μm in diameter) of yeast cells and much larger nuclei (averaging 10 μm in diameter) of cells of multicellular organisms. The first is that yeast cells undergo a closed mitosis, in which the nuclear envelope remains intact at all times. In contrast, in cells of multicellular eukaryotes, the nuclear envelope disassembles early in mitosis. After chromosome segregation, a new nuclear envelope forms around each set of chromosomes. The lamina is thought to play a central role in rebuilding nuclear organization at this time. The second explanation is that the lamina may provide essential structural support for the much larger nuclear envelope found in metazoan cells. (Further details on the role of lamins in mitosis are provided in *15.4. A cycle of cyclin-dependent kinases activities regulates cell proliferation.*)

In addition to its roles in nuclear reassembly and structural support, the nuclear lamina interacts with chromatin and may be needed for DNA replication to occur. Evidence for a role for the nuclear lamina in DNA replication comes from experiments showing that when sperm chromatin is added to *Xenopus laevis* egg extracts, a nuclear envelope forms around the sperm chromatin, as shown in **FIGURE 9.17**. These nuclei enlarge and the chromosomes, which were highly condensed in sperm nuclei, decondense. (This decondensation mimics what occurs during fertilization when the sperm chromatin encounters certain factors in oocytes.) The DNA in these nuclei is then replicated. Nuclear lamins and LAPs are abundant in these extracts. If the lamins are removed from these extracts by incubation of the extracts with immobilized antibodies to lamins, nuclear envelopes still assemble around the sperm chromatin. However, these nuclei are small and fragile, and DNA replication does not occur. These results indicate that the lamina might be important for organizing the chromatin so that it can be replicated.

Increased cellular fragility is associated with mutations affecting intermediate filament proteins. Mutations affecting lamins and LAPs are associated with several genetic diseases, primarily affecting muscle. These diseases are called laminopathies. An altered nuclear lamina appears to make the nucleus more fragile and more susceptible to injury. Muscle cells may be affected more than other cell types because the normal contraction of muscle cells exposes their nuclei to greater mechanical stress than occurs in other tissues.

9.8 Large molecules are actively transported between the nucleus and cytoplasm

Key concepts
- Uncharged molecules smaller than 100 daltons can pass through the membranes of the nuclear envelope.
- Molecules and macromolecules larger than 100 daltons cross the nuclear envelope by moving through NPCs.
- Particles up to 9 nm in diameter (corresponding to globular proteins up to 40 kDa) can pass through NPCs by passive diffusion.
- Larger macromolecules are actively transported through NPCs and must contain specific information in order to be transported.

Uncharged molecules smaller than 100 daltons, including water, can diffuse freely through phospholipid bilayers, but all other molecules and macromolecules that are transported across the nuclear envelope move through NPCs. The process of moving through the NPC is called translocation. FIGURE 9.18 illustrates the different classes of molecules that move into and out of the nucleus through NPCs.

The movement of molecules (>100 daltons) and macromolecules between the nucleus and the cytoplasm can be studied by labeling molecules either with radioactivity or with fluorescent dyes. These molecules are injected into the cytoplasm or nucleus of very large cells, such as amphibian oocytes. The localization of radiolabeled compounds is monitored by subcellular fractionation, whereas the locations of dye-labeled molecules in the cells are determined by fluorescence microscopy. Studies

using these techniques showed that relatively small molecules, such as glucose-6-phosphate or fluorescein, cross the nuclear envelope extremely rapidly, within several seconds. At equilibrium, the concentration of these types of small molecules is the same on the two sides of the nuclear envelope. We believe that movement of these small molecules occurs by simple diffusion because it occurs as readily at 4°C as at physiologic temperatures. Protein-assisted transport functions poorly or not at all at 4°C, whereas diffusion occurs at nearly equal rates at 4°C and at physiologic temperatures.

The maximum size of particles able to freely diffuse across the nuclear envelope was determined by injecting gold particles of precise sizes on one side of the nuclear envelope and assaying their ability to move to the other side. As summarized in FIGURE 9.19, these studies

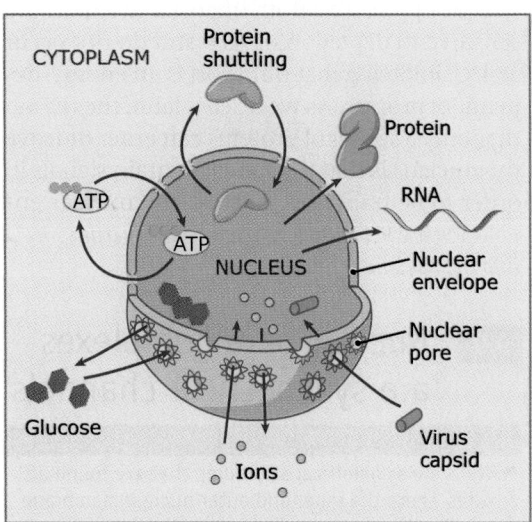

FIGURE 9.18 Many different classes of molecules and macromolecules are transported through NPCs. Not shown are small, uncharged molecules (<100 daltons) that can diffuse through the membranes of the nuclear envelope.

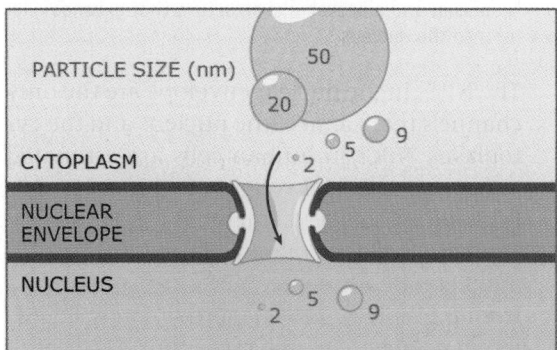

FIGURE 9.19 When polyvinyl-pyrrollidine-coated gold particles of various sizes are injected into cells, those smaller than 9 nm can pass through NPCs by passive diffusion.

have converged on an estimate that particles 9 nm diameter or smaller can move into and out of the nucleus by passive diffusion through the nuclear pore. This diameter corresponds to a globular protein of approximately 40 kDa. Because rates of diffusion are proportional to size, larger molecules diffuse into or out of the nucleus more slowly than smaller ones. The rate of diffusion is the same in both directions.

Proteins with dimensions larger than about 9 nm cannot diffuse freely through nuclear pores; these proteins are actively and selectively transported. This conclusion comes from studies in which proteins of various sizes were injected into the cytoplasm. These studies showed that only some proteins are imported into the nucleus, and their rate of entry is not proportional to their size. Export of proteins from the nucleus also has these properties. These studies showed that nuclear import and export of proteins are selective processes. Both import and export are sensitive to depletion of ATP and do not occur at 4°C, implying that transport is an energy-dependent process. As we discuss later, the reason that only a subset of proteins can enter or leave the nucleus is that they must contain signals in order to be transported across the nuclear envelope (see *9.11 Proteins are selectively transported into the nucleus through nuclear pores*).

9.9 Nuclear pore complexes are symmetrical channels

Key concepts

- NPCs are symmetrical structures that are found at sites where the inner and outer nuclear membrane are fused.
- Each NPC in human cells has a mass of ~120 × 10⁶ daltons, which is 40 times that of a ribosome, and is constructed from multiple copies of ~30 proteins.
- NPCs contain fibrils that extend into the cytoplasm, and a basket-like structure that extends into the nucleus.

The NPCs in the nuclear envelope are the only channels that connect the nucleus and the cytoplasm. NPCs in human cells are estimated to have a molecular weight of ~120 × 10⁶ daltons and an outside diameter of about 120 nm. Overall, an NPC has about 40 times the mass of a eukaryotic ribosome. NPCs are constructed from multiple copies of approximately 30 different polypeptides, the nucleoporins (see *9.10 Nuclear pore complexes are constructed from nucleoporins*). In contrast, each ribosome contains a single copy of four different types of RNA and approximately 80 different polypeptides.

NPCs are barrel-like structures that span the nuclear envelope and extend somewhat beyond the planes of both its membranes, creating a structure in the form of an annulus or a ring. Most NPCs have 8-fold rotational symmetry, as shown in **FIGURE 9.20**. As **FIGURE 9.21**

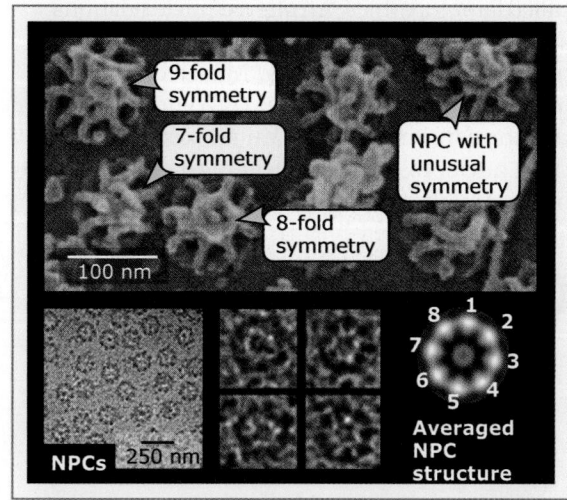

FIGURE 9.20 NPCs have 8-fold symmetry perpendicular to the nuclear envelope. Pores with 7- or 9-fold symmetry are seen occasionally. The 8-fold symmetry is easily visible in the enlarged images of individual NPCs (bottom panels). Averaging the electron micrographs from hundreds of NPCs yields an average electron density map (lower right). Top photo courtesy of Terry Allen, University of Manchester; bottom photos reprinted from *J. Mol. Biol.*, vol. 328, D. Stoffler, et al., Cryo-electron Tomography Provides Novel Insights..., pp. 119–130, Copyright (2003) with permission from Elsevier [http://www.sciencedirect.com/science/journal/00222836]. Photos courtesy Ueli Aebi, University of Basel.

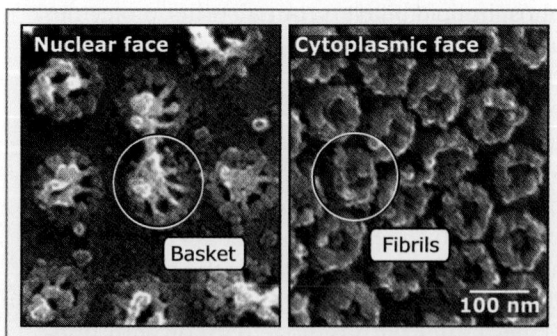

FIGURE 9.21 The terminal structures of the NPC differ. As seen by electron microscopy, there are baskets on the nuclear face (left) and fibrils on the cytoplasmic face (right). Reprinted from *J. Struct. Biol.*, vol. 140, B. Fahrenkrog, et al., Domain-specific antibodies reveal multiple-site topology..., pp. 254–267, Copyright (2002) with permission from Elsevier [http://www.sciencedirect.com/science/journal/10478477]. Photos courtesy Ueli Aebi, University of Basel.

FIGURE 9.22 The cytoplasmic fibrils and nuclear basket of nuclear pore as seen by transmission electron microscopy. Photo © Fahrenkrog, et al., 1988. Originally published in **The Journal of Cell Biology**, 143: 577–588. Used with permission of Rockefeller University Press. Photo courtesy of Ueli Aebi, University of Basel.

and **FIGURE 9.22** show, the cytoplasmic and nuclear faces of the NPC look quite different. The parts of the NPC that extend into the cytoplasm and nucleoplasm are called terminal structures. The terminal structures emanating from the cytoplasmic face of the NPC are eight relatively short fibrils that extend ~100 nm into the cytoplasm. On the nuclear face are similar fibrils joined to a ring, as shown in Figure 9.20. This terminal structure is referred to as the nuclear basket, or "fish trap." In some cells from multicellular organisms, additional fibrils extend from the basket structure deep into the interior of the nucleus. The terminal structures on both the cytoplasmic and nuclear sides are the sites where molecules to be transported first interact with the NPC and, after translocation through the NPC channel, have their final interactions with the NPC (see *9.11 Proteins are selectively transported into the nucleus through nuclear pores*).

Structural models of NPCs have been derived from the analysis of hundreds of high-resolution electron micrographs of individual NPCs. Mathematical methods were used to superimpose and average these images, arriving at an average electron density map, or average structure for the NPC core (this method does not resolve the terminal structures). **FIGURE 9.23** shows models for the yeast and *Xenopus* NPC cores, derived from analyses of electron micrographs. The NPCs of *S. cerevisiae* and other unicellular eukaryotes have a mass of approximately 50 × 10⁶ daltons—approximately half

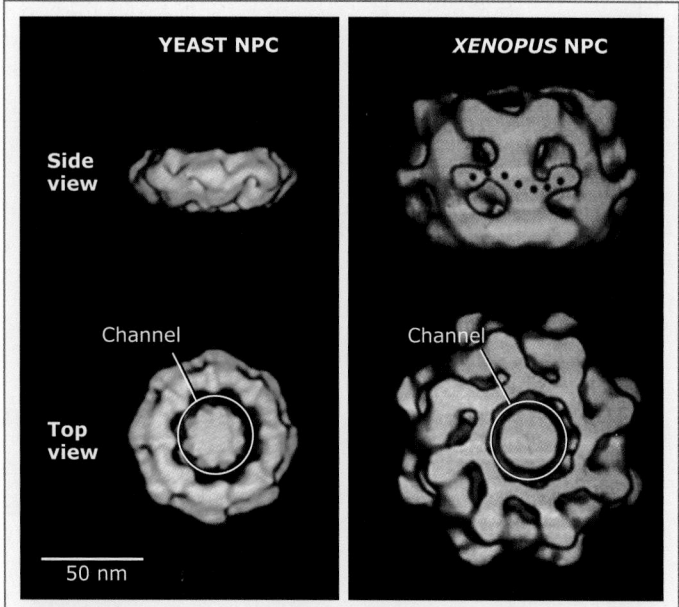

FIGURE 9.23 Computational methods are used to display the average electron density patterns of NPCs as three-dimensional models. These models are presented from the plane of the nuclear envelope (side views) and from the top of the nuclear envelope. Reprinted from *Curr. Opin. Cell Biol.*, vol. 11, D. Stoffler, B. Fahrenkrog, and U. Aebi., The nuclear pore complex: from molecular architecture..., pp. 391–401, Copyright (1999) with permission from Elsevier [http://www.sciencedirect.com/science/journal/09550674]. Photos courtesy Ueli Aebi, University of Basel.

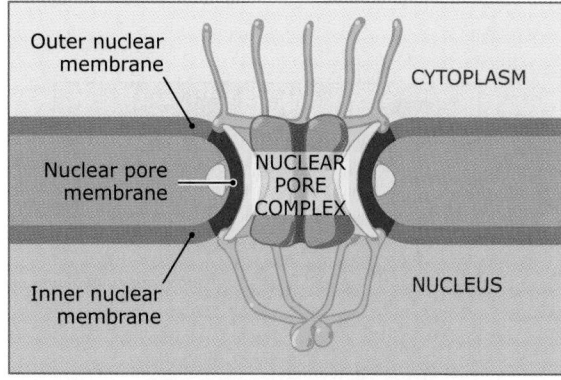

FIGURE 9.24 The inner and outer membranes of the nuclear envelope are fused at the nuclear pore complex.

the mass of NPCs from multicellular organisms. In spite of the difference in size, the overall structure is conserved. Metazoan and yeast NPCs do not differ in the size of the channels at the center of NPCs or in their transport properties. The best images currently available of NPCs are derived by cryo-electron microscopy.

As **FIGURE 9.24** depicts, wherever NPCs are located, the inner and outer nuclear membranes are fused, and the NPC can be thought of as a membrane tunnel lined with proteins. We do not know how fusion occurs, but it is an integral

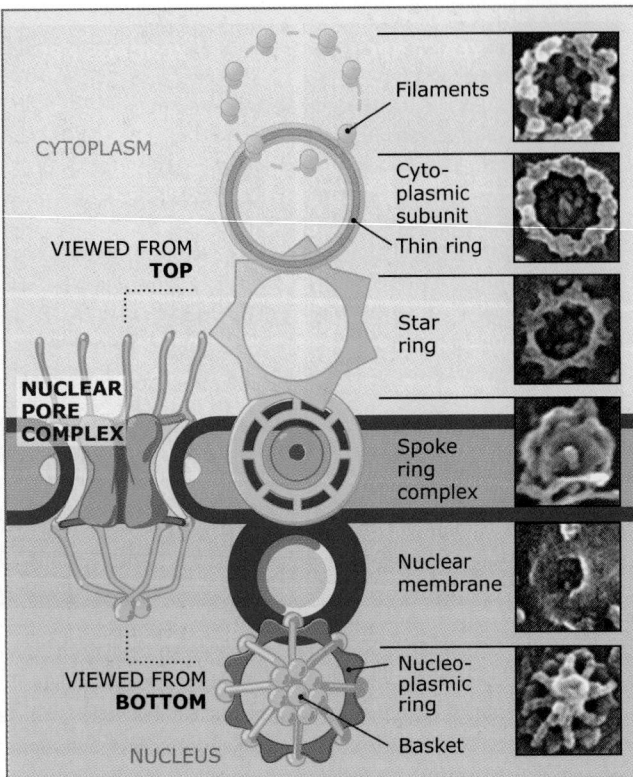

FIGURE 9.25 The NPC appears to be constructed from modular components. Electron micrographs of these components are shown at different stages of NPC reassembly following mitosis. Photos courtesy of Terry Allen, University of Manchester.

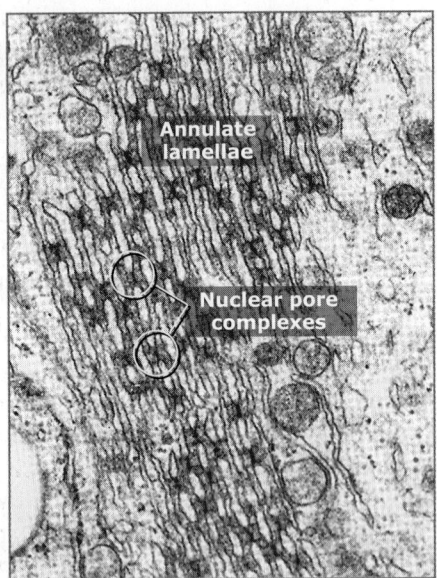

FIGURE 9.26 Annulate lamellae in *Xenopus* oocytes, as seen by transmission electron microscopy. Reprinted courtesy of Dr. Don W. Fawcett. *The Cell*, 1981. Micrograph courtesy of Héctor E. Chemes, Dept. Endocrinology, Children's Hospital, Buenos Aires, Argentina.

part of the process of assembling an NPC within the nuclear envelope. NPCs are anchored in the nuclear envelope by integral membrane proteins that are part of the core structure. These proteins extend into the lumen of the nuclear envelope. NPCs penetrate the nuclear lamina and are also anchored to it.

The best model of the NPC shows that it consists of multiple rings and spokes joined in an intricate manner, as **FIGURE 9.25** shows. Although models of NPC structure and organization are improving, we still know relatively little about the assembly process for NPCs.

In some cells, NPCs are found not only in the nuclear envelope, but also in structures called **annulate lamellae**, which are cytoplasmically located stacks of double membranes containing NPCs. Often, the NPCs in multiple layers of the annulate lamellae are aligned, as shown in **FIGURE 9.26**. Annulate lamellae are most commonly seen in oocytes of both invertebrates and vertebrates but have been observed in other types of cells as well. Their origin and function are unknown. It is difficult to isolate mammalian NPCs from the nuclear envelope because they are usually attached to the lamina, which is insoluble and difficult to work with. Because they lack an underlying lamina, annulate lamellae have proved to be a valuable source of NPCs for biochemical and cytologic studies. The NPCs in annulate lamellae appear to have the same structure and composition as NPCs in the nuclear envelope.

9.10 Nuclear pore complexes are constructed from nucleoporins

Key concepts

- The proteins of NPCs are called nucleoporins.
- Many nucleoporins contain repeats of short sequences such as Gly-Leu-Phe-Gly, X-Phe-X-Phe-Gly, and X-X-Phe-Gly, which are thought to interact with transport factors during transport.
- Some nucleoporins are transmembrane proteins that are thought to anchor NPCs in the nuclear envelope.
- All of the nucleoporins of yeast NPCs have been identified.
- NPCs are disassembled and reassembled during mitosis.
- Some nucleoporins are dynamic: they rapidly associate with and dissociate from NPCs.

The proteins that make up NPCs are called **nucleoporins**. We know the most about nucleoporins from baker's yeast, *S. cerevisiae*. Two major approaches contributed to identifying the complete set of yeast nucleoporins. One was a genetic approach to isolate mutants defective for nuclear transport. The other was biochemical, involving purification of NPCs from isolated nuclear envelopes. The solubilization of yeast NPCs was facilitated by yeast's lack of a nuclear lamina (see *9.7 The nuclear lamina underlies the nuclear envelope*). The proteins were separated by gel electrophoresis, as shown in **FIGURE 9.27**, and mass spectrometry was used to identify them. Together, the genetic and biochemical approaches have identified approximately thirty proteins.

Nucleoporins from metazoans have also been identified. Although metazoan NPCs have approximately twice the mass of yeast NPCs, the number of different nucleoporins is approximately the same and there are specific nucleoporins in metazoan NPCs that are orthologous to most or perhaps all yeast nucleoporins. Because genetic approaches are not easy to use with vertebrates, their nuclear pore proteins have been identified using biochemical and immunologic approaches, although in more recent studies, RNAi has been used to knock down expression of specific nucleoporins.

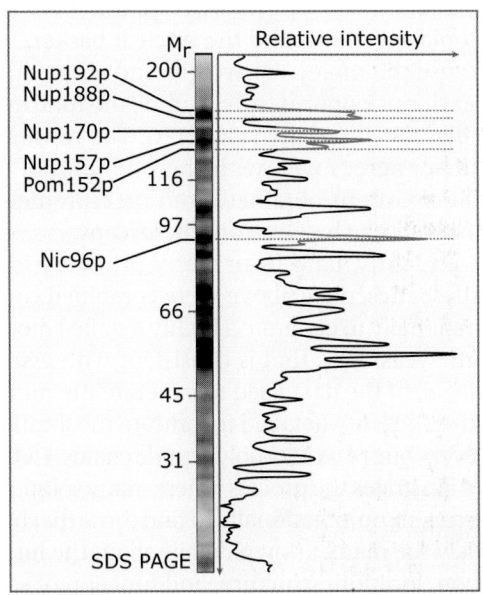

FIGURE 9.27 NPCs can be purified from isolated mammalian nuclear envelopes and the nucleoporins present can be separated by polyacrylamide gel electrophoresis. The stained gel (left) and the densitometric tracing of the gel (right) are shown. Some abundant nucleoporins are labeled. Photo © Aitchison, et al., 1995. Originally published in **The Journal of Cell Biology**, 131: 1133–1148. Used with permission of Rockefeller University Press. Photo courtesy of Michael P. Rout, Rockefeller University.

One characteristic of many nucleoporins is the presence of multiple repeats of short sequences. These repeats are thought to be sites where cargo molecules bind to the NPC during transport. About one-third of nucleoporins contain these repeats, whose sequences are Gly-Leu-Phe-Gly, X-Phe-X-Phe-Gly, or X-X-Phe-Gly, where X is any amino acid. Because they contain a phenylalanine-glycine (Phe-Gly) pair, these are often referred to as (phenylalanine-glycine) **FG repeats**. Spacers of 3 to 15 amino acids separate the repeats. In general, nucleoporins that contain FG repeats have 10 to 30 repeats, but some nucleoporins have only one or two. In nucleoporins with many FG repeats, the repeats are generally found together in one portion of the protein. These FG repeat regions are believed to be natively unstructured and to fill the central channel of the NPC, where they serve as docking sites for karyopherins (see *9.13 Cytoplasmic nuclear localization sequence receptors mediate nuclear protein import*). Other FG repeat-containing nucleoporins are components of the cytoplasmic filaments and the nuclear basket. Remarkably, a large number of the FG repeat regions can be deleted from yeast nucleoporins without affecting cell viability. In particular, the repeat region domains of the cytoplasmic filament and nuclear basket nucleoporins can all be deleted, whereas only a small number of the FG repeats within the central channel can be removed without compromising NPC function and cell viability.

Another feature of many nucleoporins is the presence of a type of alpha helical domain that is able to form a protein structure motif called the coiled coil by interacting with structurally similar regions of other nucleoporins. Coiled-coil interactions are thought to be very important in the overall structure of NPCs.

We are beginning to understand how nucleoporins are organized within NPCs. Biochemical fractionation has been used to isolate subcomplexes of NPCs. Mass spectrometry can then be used to determine the identity of nucleoporins in the subcomplex, thereby defining nearest neighbors for each nucleoporin. The location of a particular nucleoporin within NPCs can be determined by using immunoelectron microscopy. In this approach, antibodies coupled to gold particles are used to detect specific nucleoporins. These studies tell us which nucleoporins are located at the cytoplasmic fibrils, the nuclear basket, and the central framework. Most nucleoporins are present in both the nuclear and cytoplasmic halves of the pore, with only a few found on only one side, and these are present in the terminal structures, as shown in **FIGURE 9.28**. Immunoelectron microscopy has sufficient resolution that we can determine approximately how far the region of a nucleoporin recognized by an antibody is from the plane of the nuclear envelope and from the center of the NPC channel. This type of analysis indicates that some nucleoporins have highly mobile domains that appear to move through the NPC channel and can be detected on either side of the NPC.

A small number of nucleoporins in metazoan and yeast NPCs are transmembrane proteins. Three observations support this conclusion:

- Immunoelectron microscopy shows that these nucleoporins are located very close to the pore membrane.
- These nucleoporins have stretches of hydrophobic amino acids long enough to span the lipid bilayer.

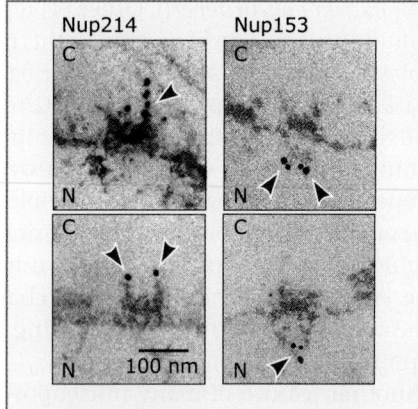

FIGURE 9.28 Individual nucleoporins can be located within NPCs by immunoelectron microscopy. Photos © Fahrenkrog, et al., 1988. Originally published in **The Journal of Cell Biology**, 143: 577–588. Used with permission of Rockefeller University Press. Photos courtesy of Ueli Aebi, University of Basel.

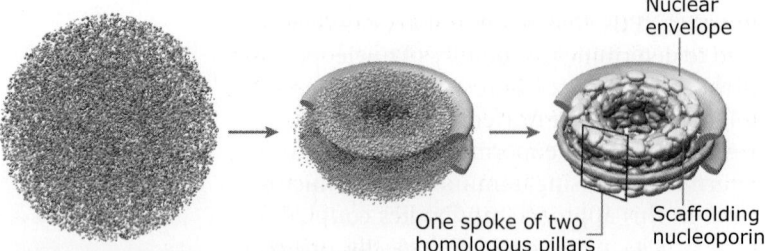

FIGURE 9.29 Illustration of the computational reconstruction of the architecture of the yeast nuclear pore complex based on analysis of all the available biochemical, biophysical, and immunoelectron microscopic data including location of each its 456 component proteins. Photo courtesy of Frank Alber, University of California, San Francisco.

- These nucleoporins can be identified in isolated nuclear envelopes that have been treated with a high salt concentration, which leaves only those proteins that are integral membrane proteins.

These transmembrane nucleoporins are thought to help anchor NPCs to the nuclear envelope.

At sites where pores are located, the nuclear membranes are tightly curved. Highly curved membranes are also found in coated vesicles. (see *12 Actin*). Recent evidence from structural biology shows that many nuclear pore proteins contain one of two additional structural motifs, the alpha solenoid and the beta propeller, that are also present in the proteins that form the coats of coated vesicles. In both coated vesicles and nuclear pores, these proteins are thought to play an important role in membrane curvature. This suggests that some components of coated vesicles and nuclear pore complexes share a common ancestry.

Some nucleoporins are much more abundant than others are. Because the NPC has 8-fold symmetry, no nucleoporin is present in fewer than 8 copies per NPC, and many are present in 16 or 32 copies per NPC. The least abundant nucleoporins are components of the cytoplasmic fibrils and the nuclear basket. We can use the molecular weight and abundance of each nucleoporin to determine the total molecular weight of the yeast NPC. The resulting number agrees well with the estimates of 50 million daltons obtained from measurements of overall size by electron microscopy.

Recently, data from many of the various analyses described above were combined computationally to generate a highly detailed model of the yeast NPC that is consistent with essentially all of the data used to generate the model and sufficiently detailed to indicate the location of every one of its 456-polypeptide chains. **FIGURE 9.29** illustrates the process where many solutions are tested computationally to find those that best satisfy the many known details about the interaction, location, structure, and function of specific nucleoporins and nucleoporin complexes. Although each of the data sets the contributes to the model has some degree of uncertainty, when the datasets are combined a model emerges that is likely to surpass all previous models in reflecting accurately how the NPC is structured.

It has been shown experimentally that some nucleoporins are stably associated with the NPC whereas others are dynamic. These studies used the technique of inverse fluorescence recovery after photobleaching. In this approach, mammalian cells engineered to express a GFP-tagged nucleoporin are used. A small circular area of the nuclear envelope is photobleached, and the kinetics with which fluorescence returns to NPCs in the bleached area is measured. These studies indicate that approximately half of the nucleoporins are stably associated and these are thought to constitute the framework components of the NPC. Other nucleoporin had a residence time of several hours, while a small number are highly dynamic and remained NPC-associated for a few minutes or less.

In some organisms, the nuclear envelope disassembles at mitosis, and NPCs are disassembled into subcomplexes. The NPCs reassemble when the nuclear envelope forms late in mitosis. The disassembly and reassembly processes each take less than one hour *in vivo*. Different stages in the reassembly process can be visualized by electron microscopy and appear to proceed through multiple, reproducible steps. We know little about the mechanisms used for reassembly or whether proteins not found within assembled pores help to assemble the NPC. New NPCs are also formed during interphase from newly synthesized nucleoporins. We do not yet know whether the same mechanism is used for *de novo* assembly as for reassembly. Because the NPCs of both budding and fission yeast remain intact during mitosis, all NPC assembly in yeast occurs by de novo construction of new NPCs within the nuclear envelopes of these organisms. Recent studies of the filamentous fungus, Aspergillus nidulans, indicate that its NPCs are partially disassembled during mitosis, even though the nuclear envelope remains largely intact. This may represent an intermediate stage between the complete NPC disassembly seen in metazoans and the complete stability of NPCs seen in yeasts.

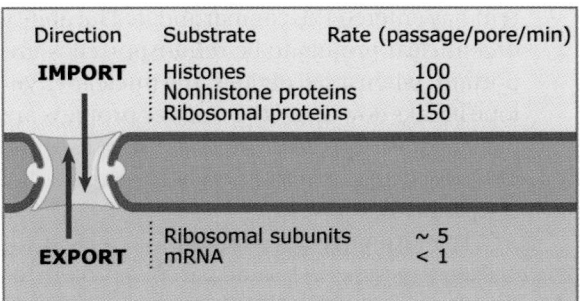

FIGURE 9.30 The rates of import and export for a few of the macromolecules that are transported through nuclear pores. Rates reflect primarily the relative amounts of each protein that are present and need to be transported.

All macromolecules enter and exit the nucleus through the large channels of nuclear pores. FIGURE 9.30 shows the rate of movement for some cargo proteins.

The process of nuclear import is thought to begin by association (docking) of cargo proteins with the NPC fibrils that extend into the cytoplasm. This was shown by using electron microscopy to follow the steps in the movement of proteins through NPCs. After electron-dense gold particles are coated with a nuclear protein and injected into the cytoplasm of an oocyte, they accumulate in the nucleus. Some can be seen lined up on the cytoplasmic fibrils or within the channel of the NPC.

Docking of cargo proteins at the NPC is energy independent, but transport is energy dependent. When cells are incubated at low temperature or treated to block energy-dependent processes, the protein-coated gold particles still associate with cytoplasmic components of the NPC but are not found within the channel of the NPC or in the nucleus.

Passage of proteins through the NPC is highly selective (see *9.8 Large molecules are actively transported between the nucleus and cytoplasm*). In many cases, the transported protein contains the information to target it to the nucleus. A protein without a targeting signal can also be imported by binding to a protein that contains one. That nuclear proteins contain import signals was best demonstrated by isolating nuclear proteins from *Xenopus* oocytes and injecting the proteins into the cytoplasm of oocytes. The injected proteins rapidly entered the nucleus, thereby showing that they contain information for transport into the nucleus.

This experiment also showed that unlike most signal sequences for translocation into the endoplasmic reticulum, targeting signals for the nucleus are not cleaved off when the pro-

9.11 Proteins are selectively transported into the nucleus through nuclear pores

Key concepts

- Mature nuclear proteins contain sequence information required for their nuclear localization.
- Proteins selectively enter and exit the nucleus through nuclear pores.
- Information for nuclear import lies in a small portion of the transported protein.

tein has entered the compartment. The ability of a nuclear protein to be retransported is important: when a cell divides, the nuclear envelope breaks down and the nuclear proteins are dispersed throughout the cell. Proteins must be reimported into the newly formed nuclei after mitosis.

The sequence information for nuclear localization resides in a small part of a protein, as first demonstrated by the classic experiments of Laskey. The experiments were carried out with nucleoplasmin, an abundant nuclear protein. When injected into the cytoplasm of a *Xenopus* oocyte, nucleoplasmin rapidly concentrates inside the nucleus (see FIGURE 9.31). Because the nucleus already contains a large amount of nucleoplasmin, the newly injected protein accumulates against a concentration difference. Nucleoplasmin is a relatively small protein of 33 kDa molecular weight. However, it always forms a pentamer, making its total molecular weight closer to 150 kDa; hence, it is too large simply to diffuse into the nucleus.

To test the idea that information for nuclear import might lie in only a part of the protein, the nucleoplasmin pentamer was cut by a protease into two parts—a 10-kDa "tail" derived from the C-terminus of each monomer, and a 100-kDa pentameric "core" containing the remainder of each monomer. As Figure 9.31 shows, the different parts of nucleoplasmin were injected into *Xenopus* oocytes to test whether they contain information for nuclear transport. When injected into the cytoplasm,

cores could not enter the nucleus. Therefore, cores do not contain information for nuclear entry. However, when injected directly into the nucleus, cores remained inside and did not leak back out to the cytoplasm. The tail fragments alone moved from the cytoplasm into the nucleus and were retained there. Partial treatment with protease yielded cores retaining one or more tails. The presence of a single tail was sufficient to direct entry into the nucleus.

The results of these experiments led to the conclusion that the nuclear pore is a selective channel that allows only proteins with the proper information in their amino acid sequence to enter. The information for nuclear transport lies in only a portion of the protein—in the case of nucleoplasmin, it is in the 10-kDa C-terminal tail. In addition, simple diffusion and retention within the nucleus are insufficient to target a protein from the cytoplasm to the nuclear interior. This is illustrated by the fact that the pentameric core stays in the nucleus if it is placed there but cannot concentrate there on its own.

9.12 Nuclear localization sequences target proteins to the nucleus

Key concepts

- A nuclear localization sequence (NLS) is often a short stretch of basic amino acids.
- NLSs are defined as both necessary and sufficient for nuclear import.

The signal within a protein that targets it to the nucleus is a short stretch of amino acids termed the **NLS**. The most common NLSs contain the basic amino acids lysine and arginine. The best-characterized NLS is from the viral nuclear protein SV40 large T antigen. When SV40 viruses infect cells, the viral large T antigen is synthesized in the cytoplasm and moves into the nucleus to promote viral growth. The large T antigen uses the host cell machinery to enter the nucleus.

The NLS of SV40 large T antigen has the sequence Pro-Lys-Lys-Lys-Arg-Lys-Val between amino acids 126 and 132, as shown in FIGURE 9.32. The identification of this NLS is based on two types of experiments. First, a mutation of lysine 128 to threonine in T antigen results in a protein that can no longer enter the nucleus because the NLS has been inactivated.

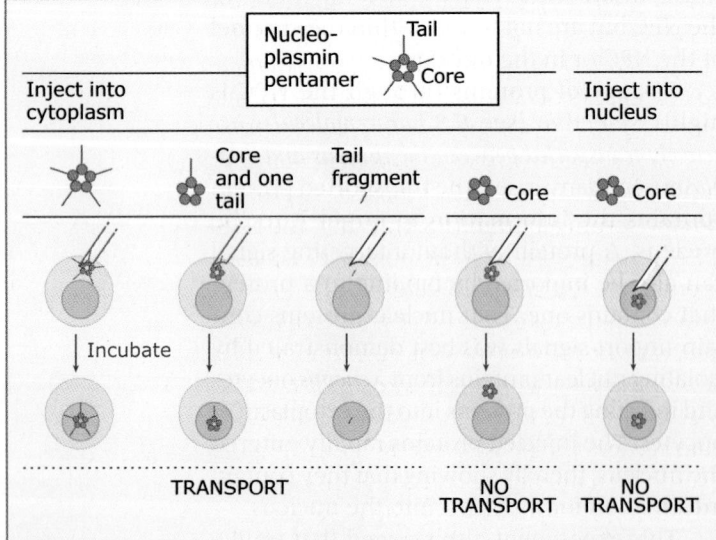

FIGURE 9.31 Injection of nucleoplasmin into the cytosol or the nucleus of *Xenopus* oocytes shows the importance of the C-terminal tail fragment for nuclear import.

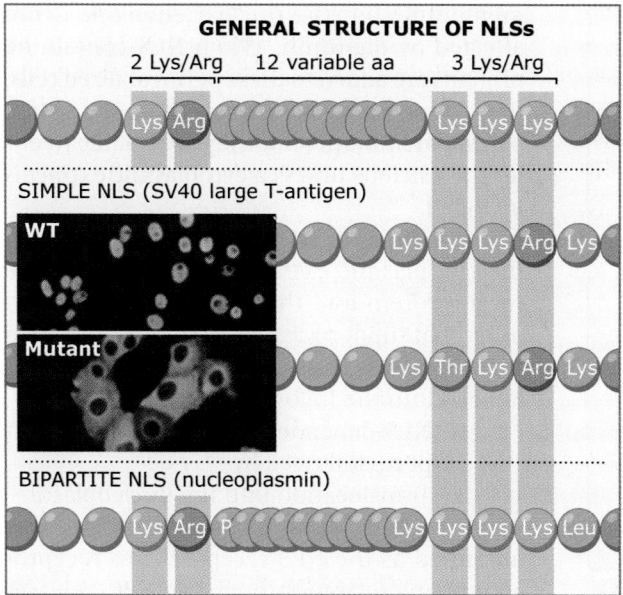

GENERAL STRUCTURE OF NLSs

SIMPLE NLS (SV40 large T-antigen)

WT

Mutant

BIPARTITE NLS (nucleoplasmin)

FIGURE 9.32 The NLS from SV40 large T antigen contains only one stretch of basic amino acids, and a single mutation from lysine to threonine renders it unable to localize T antigen to the nucleus. The NLS from nucleoplasmin is an example of a bipartite NLS in which two basic regions are separated by 12 variable amino acids. Inset photo reprinted from *Cell*, vol. 39, D. Kalderon, et al., A short amino acid sequence..., pp. 499–509, Copyright (1984) with permission from Elsevier [http://www.sciencedirect.com/science/journal/00928674]. Photo courtesy of Daniel Kalderon, Columbia University.

Mutations of amino acids outside of the NLS have no effect on nuclear import. These results show that the NLS is necessary for nuclear localization. Second, a peptide corresponding to the minimal NLS can be attached to a normally nonnuclear protein, causing the resulting chimeric protein to be imported into the nucleus. A chimeric protein with a mutated NLS is not imported. These results show that the NLS is sufficient for nuclear import. For a sequence of amino acids to be called an NLS, it must meet the criteria of being *necessary* and *sufficient* and for nuclear import.

Many nuclear proteins contain short, basic NLSs similar to that of T antigen. Another common type of NLS contains two shorter, basic stretches of amino acids separated by about 12 variable amino acids; this is called a bipartite NLS and is present in nucleoplasmin (see Figure 9.32). Both types are referred to as "classical" NLSs and are recognized by the same transport proteins during protein import. However, there are additional types of NLSs that are less well defined. Thus, like signal sequences for secretion, NLSs share certain general characteristics but do not have elaborate sequence requirements.

Concept and Reasoning Checks

1. Summarize the evidence that proteins that enter the nucleus carry information, a nuclear localization signal, that species their movement into the nucleus.
2. How was it determined that particles smaller than a certain size can enter the nucleus even if they lack an NLS and that they do so by diffusion?

9.13 Cytoplasmic nuclear localization sequence receptors mediate nuclear protein import

Key concepts

- Receptors for nuclear import are cytoplasmic proteins that bind to the NLS of cargo proteins.
- Nuclear import receptors are part of a large family of proteins often called karyopherins.

Import of proteins into the nucleus is a receptor-mediated process. An experiment showing that protein import is saturable suggested the involvement of receptors that recognize the nuclear localization signal. This experiment measured the kinetics of nuclear import in *Xenopus* oocytes, as illustrated in **FIGURE 9.33**. Short polypeptides containing NLSs were chemically cross-linked to a normally nonnuclear protein, bovine serum albumin (BSA). When the NLS-BSA was injected into the oocyte cytoplasm, it entered the nucleus, whereas BSA alone remained in the cytoplasm. The extent of import became saturated as the concentration of NLS-BSA was increased. This is consistent with the idea that there is a limiting factor (or factors) that recognizes the NLS and mediates import. This in turn implies the existence of a receptor that recognizes NLS-containing proteins.

We now know that NLS receptors are soluble, cytoplasmic proteins. Some NLS receptors were initially identified using an *in vitro* assay that recapitulates what happens in the

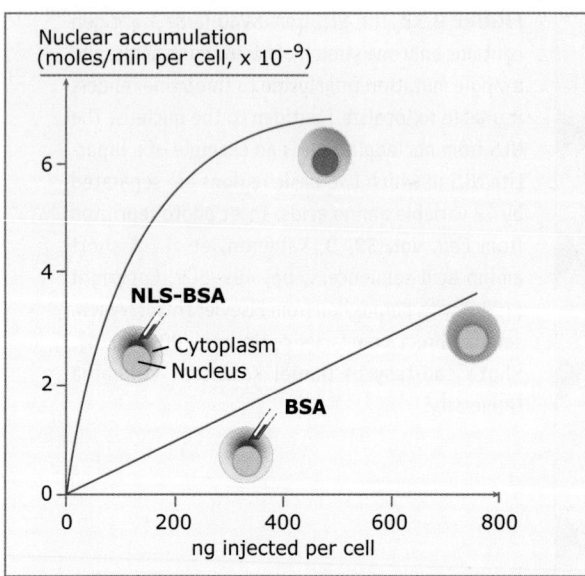

FIGURE 9.33 Nuclear import of NLS-bovine serum albumin injected into the cytoplasm of an oocyte is saturable.

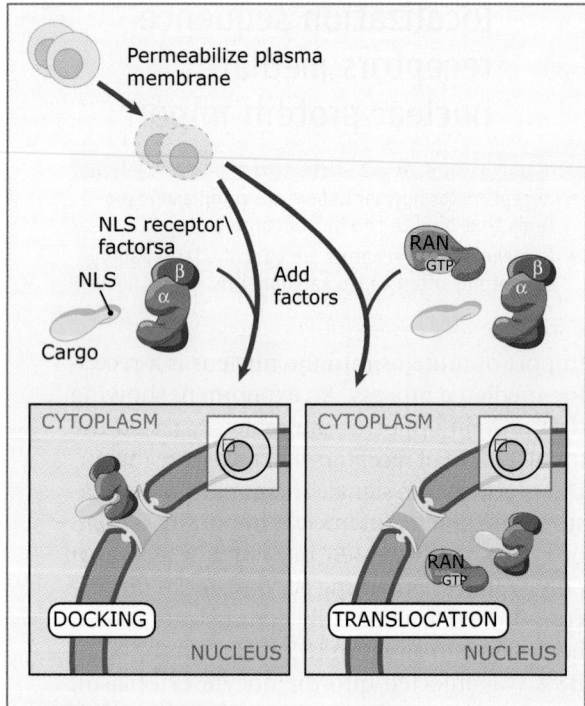

FIGURE 9.34 Semi-permeabilized cells were used to purify cytoplasmic proteins that are necessary for proper transport of proteins into the nucleus.

cell and allows purification of cytosolic factors that are necessary to support proper import. In this assay, shown in **FIGURE 9.34**, the plasma membranes of cells are permeabilized using the detergent digitonin. This lets cytoplasmic material escape through holes in the plasma membrane, but the nuclear envelope is not affected by digitonin. When NLS-containing proteins are added to these permeabilized cells, they enter the cells through the holes in the plasma membrane, but they are unable to enter the nucleus unless a cytoplasmic extract is included.

Two different activities were isolated from the cytoplasmic extract—one that binds the cargo protein and the nuclear pore and another that supports translocation, as illustrated in Figure 9.34. Thus, nuclear protein import is divided into the following two steps:

1. NLS-dependent docking at the nuclear pore, followed by
2. Translocation into the nucleoplasm.

The activity that promotes docking was identified as the NLS receptor. This receptor accompanies cargo through the NPC, releases the cargo in the nucleus, and then returns to the cytoplasm for another round of transport. The activity that promotes translocation through the NPC was identified as the Ran GTPase.

Two types of NLS receptors have been identified:

1. Those that can bind to the pore directly and
2. Those that require an adaptor to mediate docking at the pore.

The simplest NLS receptors consist of a single protein that binds to both the cargo's NLS and to the nuclear pore, as shown in **FIGURE 9.35**. These NLS receptors are most closely related to importin β. In contrast, the heterodimeric NLS receptors consist of one subunit that binds to the cargo and another that binds to nucleoporins. The first NLS receptor that was identified is composed of importin α and importin β. Importin α is about 55–60 kDa in size and binds directly to the NLS. Importin α is a specialized adaptor for classical NLSs. Proteins with classical NLSs, which are by far the most abundant, use importin α as an adaptor, allowing a large set of cargo proteins to be transported with importin β. Importin β is about 90 kDa in size and together with importin α docks at the nuclear pore. Importin α is but one of multiple proteins that recognize NLSs and interact with importin β.

The three-dimensional structure of importin α with an SV40 NLS shows two binding pockets for the NLS. This explains the ability of both simple and bipartite NLSs (see Figure 9.32) to function with the same receptor. Three critical amino acid contacts are made with the third lysine in the NLS, making this

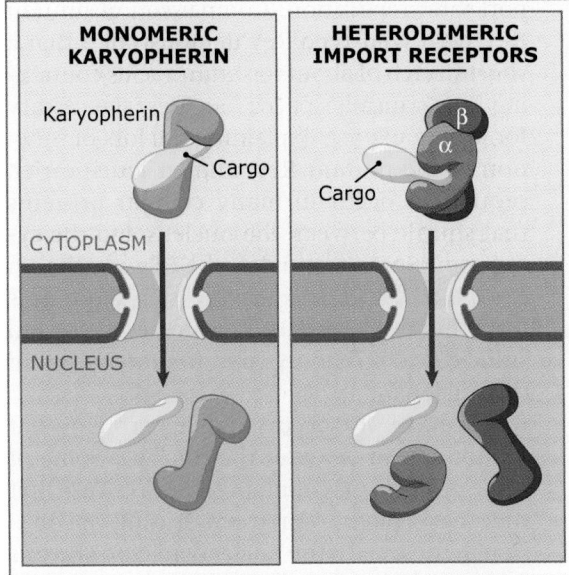

MONOMERIC KARYOPHERIN	HETERODIMERIC IMPORT RECEPTORS
Karyopherin	β α
Cargo	Cargo
CYTOPLASM	
NUCLEUS	

FIGURE 9.35 Many cargo proteins are imported by binding directly to karyopherins that are related to importin β (left). Other cargo proteins use the importin α β heterodimer (right).

lysine the most important of the residues for proper NLS binding. This helps explain how there can be slight variations in the NLS primary sequence, yet a single mutation of the third lysine can abolish function entirely (see *9.12 Nuclear localization sequences target proteins to the nucleus*).

After binding to cytoplasmic elements of the NPC, karyopherin-cargo complexes move through the pore channel. Although considerable progress has been made, the mechanism of passage through the NPC channel is not known. We believe that the NPC excludes macromolecules that are unable to interact with FG repeats. Further study will be required to determine how many sequential contacts a karyopherin/cargo complex makes nucleoporins during transport and by what mechanism it is moved along. Models for movement through the pore are discussed in *9.16 Multiple models have been proposed for the mechanism of nuclear transport*.

Some cargo proteins use more than one nuclear transport receptor for their import. The redundancy may help to ensure efficient import and/or offer the cell another level of regulation. Moreover, not all proteins are imported into the nucleus by karyopherins. Some proteins, such as the signaling molecule β-catenin, dispense with any kind of transporter and simply interact directly with nucleoporins for their import. Why some proteins have developed the ability to be transported independent of karyopherins is unknown but could reflect the need to regulate their transport separately from other nuclear transport.

Some viral proteins have adapted their own nuclear transport factors to interface with the nucleoporins and cellular transporters. For example, the viral Vpr protein, which is essential for immunodeficiency virus (HIV) infection of nondividing cells, binds importin α and nucleoporins to promote the import of the viral genetic material into the nucleus. Thus, Vpr mimics the action of importin β in nuclear transport.

Although originally thought to function solely in nuclear transport, we now know that importin β functions in the regulation of diverse cellular activities. Importin β can interact with both cargoes and microtubules and may use this interaction to facilitate movement of nuclear proteins along microtubules to the nuclear periphery. Importin β, along with other nuclear transport factors, plays a role in assembly of the mitotic spindle, reassembly of nuclei after mitosis, and assembly of NPCs.

9.14 Export of proteins from the nucleus is also receptor-mediated

Key concepts

- Short stretches of amino acids rich in leucine act as the most common nuclear export sequences.
- A nuclear export receptor binds proteins that contain nuclear export sequences (NESs) in the nucleus and transports them to the cytoplasm.

Some proteins are made in the cytoplasm, transported into the nucleus, and then exported back out of the nucleus. Many such shuttling proteins are transcription factors, whose activity is controlled by their localization. Some proteins even cycle continuously between the nucleus and the cytoplasm in a process referred to as nucleocytoplasmic shuttling. These proteins contain not only NLSs to target them into the nucleus but also signals for export from the nucleus. The latter signals are termed **NESs**. The general principles for nuclear protein export are similar to those for import. However, there are important differences in regulation that we will discuss later (see *9.17 Nuclear transport can be regulated*).

The NESs were discovered in the course of studies of the growth of human HIV in infected cells. HIV has an RNA genome, so some of the viral RNAs made in the nucleus are exported to the cytoplasm for translation and for packaging into new virions. A small HIV protein called Rev enters the nucleus of infected cells, where it binds specifically to viral RNA. The RNA-Rev complex moves out of the nucleus to the cytoplasm, where the RNA is released, and Rev reenters the nucleus for another round of RNA export. The Rev protein contains import information in the form of a classical NLS, and distinct information for export (an NES), as indicated in **FIGURE 9.36**.

The NESs were characterized by studying mutant versions of Rev that could not sup-

port viral replication. Comparison of mutant Rev with wild-type Rev demonstrated that a short stretch of about ten amino acids containing four critically spaced leucines is necessary for proper export, explaining the loss of function of the mutant Rev. Similar leucine-rich sequences occur in many cellular proteins that shuttle between the nucleus and the cytoplasm. Some examples of NESs are shown in **FIGURE 9.37**. These leucine-rich sequences form the "classical" NES and are necessary and sufficient to promote export of proteins out of the nucleus.

The NES also occurs in many proteins that do not appear to enter the nucleus. Some of these proteins may contain NESs to ensure that they are kept out of nucleus. During cell division in most cells, the nuclear envelope breaks down and, at the end of mitosis, forms again around each set of chromosomes. This process could result in nonnuclear proteins being trapped in the nucleus. The NESs of these proteins may promote their rapid export following reassembly of the nuclear envelope.

If an NES-containing protein enters the nucleus, it is recognized by a nuclear export receptor referred to as an **exportin**, a member of the karyopherin protein family. Crm1 is the major exportin in all cells studied to date. The cargo-exportin complex moves from the nucleus through the nuclear pore channel into the cytoplasm, where the cargo is released. The exportin then reenters the nucleus for another round of export, as shown for the Rev protein in Figure 9.36.

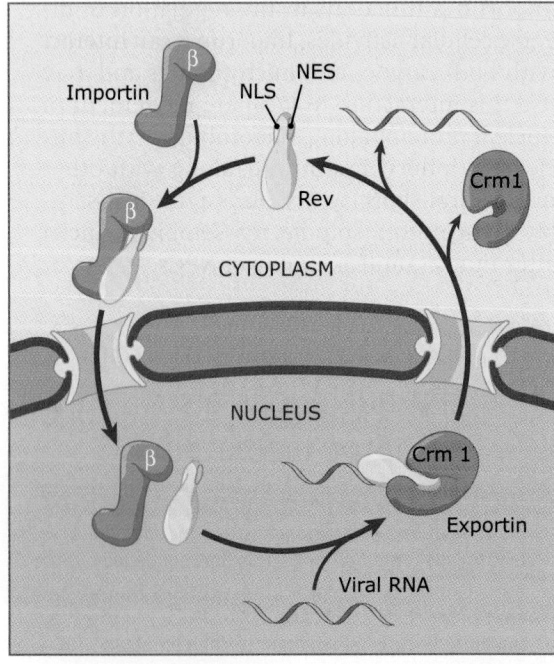

FIGURE 9.36 Some proteins, such as HIV Rev, shuttle between the nucleus and the cytoplasm. For simplicity, other proteins involved in transport are not shown.

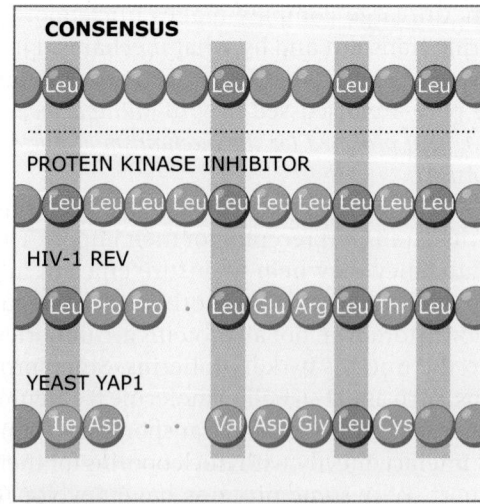

FIGURE 9.37 Some of the best-characterized nuclear export signals have a high leucine content. Other hydrophobic amino acids such as isoleucine and valine may appear in place of leucine.

Leptomycin is a natural product that was isolated from fungi by virtue of its ability to inhibit the action of Rev and, hence, the growth of HIV. Leptomycin blocks the nuclear export of Rev by covalently binding to Crm1 and disrupting its ability to bind NESs. Because it blocks all classical NES-dependent nuclear export, leptomycin is toxic to cells, which precludes its usefulness as a drug. However, it is a valuable compound for researchers to determine if the classical NES pathway mediates export of a given protein.

In addition to Crm1, a few other karyopherins also promote export. Consider the case of importin α, which enters the nucleus together with its cargo and importin β and then must be exported in order to function in further rounds of import. A dedicated exportin called CAS binds importin α and carries it back to the cytoplasm. Some RNAs are exported from nuclei as cargo for various exportins (see *9.19 Ribosomal subunits are assembled in the nucleolus and exported by exportin 1; 9.20 tRNAs are exported by a dedicated exportin; 9.24 U snRNAs are exported, modified, assembled into complexes, and imported; and 9.25 Precursors to microRNAs are exported from the nucleus and processed in the cytoplasm*).

9.15 The Ran GTPase controls the direction of nuclear transport

Key concepts

- Ran is a small GTPase that is common to all eukaryotes and is found in both the nucleus and the cytoplasm.
- The Ran-GAP promotes hydrolysis of GTP by Ran, whereas the Ran-GEF promotes exchange of GDP for GTP on Ran.
- The Ran-GAP is cytoplasmic, whereas the Ran-GEF is located in the nucleus.
- Ran controls nuclear transport by binding karyopherins and affecting their ability to bind their cargoes.

At any given time, there is a multitude of proteins moving into and out of the nucleus in a cell. Transport is mediated by the interaction of the cargo protein's NLS or NES with a member of the karyopherin transporter family. The transporters shuttle back and forth, binding cargo in one compartment and releasing cargo on the other side of the NPC. What controls the direction of transport? The answer lies in the action of a small monomeric GTP-binding protein called Ran, which binds to the transporters and moves between the nucleus and the cytoplasm. Ran itself does not have an NLS or NES.

Like many GTP-binding proteins, Ran on its own hydrolyzes GTP to GDP at a very slow rate. Ran has regulators to stimulate its enzymatic activity. There are two main proteins, Ran-GAP and Ran-GEF, which regulate the enzymatic activity of Ran. Ran-GAP is a *GTPase-activating* protein that stimulates Ran to hydrolyze bound GTP, resulting in the GDP-bound form of Ran. Ran-GEF, which is a guanine nucleotide exchange factor, reloads Ran with fresh GTP. Ran-GEF stimulates the removal of bound GDP and replaces it with GTP. Thus, depending on the regulator with which it interacts, Ran is found in either the GTP- or GDP-bound form. Ran-GEF (also called Rcc1) and Ran are also involved in assembly of the mitotic spindle.

The key to understanding how directionality of nuclear transport is conferred lies in the localization of Ran-GAP and Ran-GEF. As **FIGURE 9.38** shows, Ran-GAP is located in the cytoplasm, and some Ran-GAP is associated with the filaments on the cytoplasmic face of the NPC. This means that in the cytoplasm, Ran is expected to be in its GDP-bound form because Ran-GAP will have stimulated hydrolysis of the bound GTP. In contrast, Ran-GEF is located in the nucleus. This means that in the

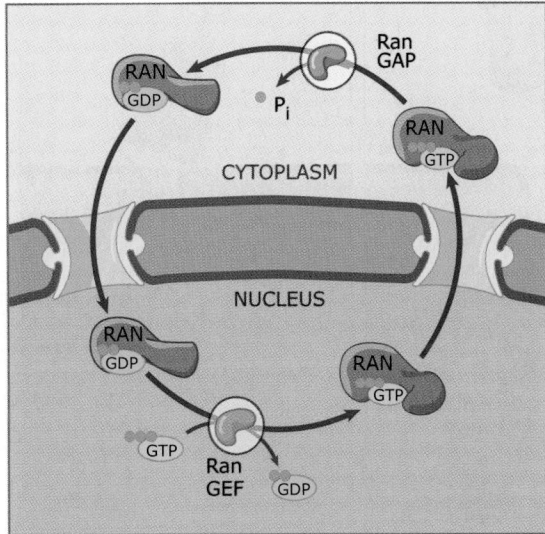

FIGURE 9.38 Ran exists in a GDP-bound form in the cytoplasm and in the GTP-bound form in the nucleus because of the distribution of the enzymes that affect its ability to bind and hydrolyze GTP. For simplicity, other proteins involved in transport are not shown.

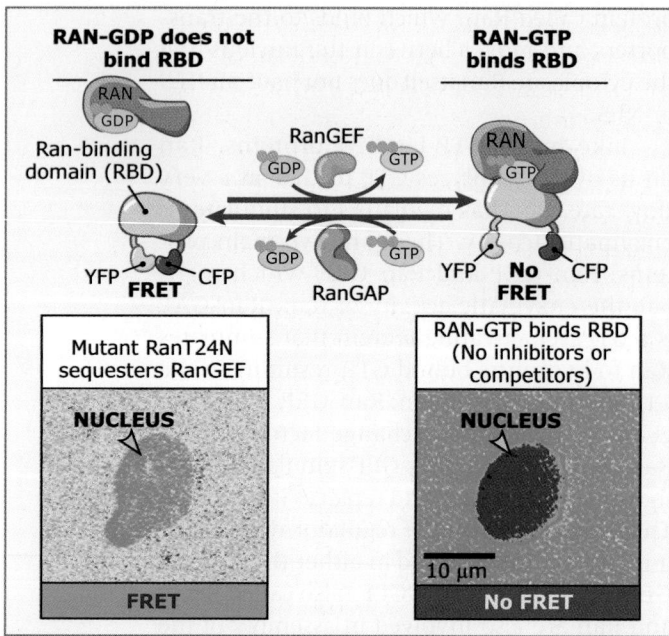

FIGURE 9.39 Use of a Ran-binding domain able to bind only to Ran-GTP permits analysis of the relative concentrations of Ran-GTP and Ran-GDP in different cellular compartments. Bottom photos reproduced from P. Kalab, K. Weis, and R. Heald, *Science* 295 (2002): 2452–2456 [http://www.sciencemag.org]. Reprinted with permission from AAAS. Photos courtesy of Rebecca Heald, University of California, Berkeley.

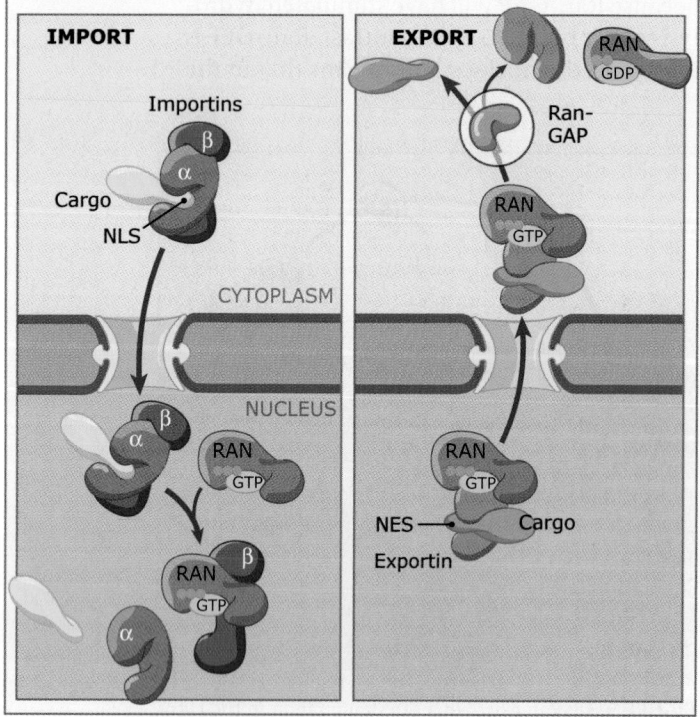

FIGURE 9.40 Importins bind cargos in the cytoplasm and release them in the nucleus after binding Ran-GTP. Conversely, export complexes form in the nucleus together with Ran-GTP. The exported cargo is released in the cytoplasm when the GTP is hydrolyzed.

nucleus, Ran is expected to be in its GTP-bound form because following nuclear import of Ran-GDP, Ran-GEF can replace the GDP bound to Ran with GTP. This asymmetric distribution of the effectors of Ran suggests that the concentration of Ran-GDP is higher in the cytoplasm and the concentration of Ran-GTP is higher in the nucleus. This was proved in an elegant experiment, shown in **FIGURE 9.39**, in which a protein that binds to Ran-GTP but not Ran-GDP was fused to two fluorescent proteins, yellow fluorescent protein (YFP) and cyan fluorescent protein (CFP), and expressed in cells. When this Ran-GTP binding protein is not associated with Ran-GTP, the fluorescent YFP and CFP domains of the fusion protein are in very close proximity and undergo energy transfer. This interaction, called fluorescence resonance energy transfer (FRET), can be detected using a fluorescence microscope. Thus, FRET will occur when the fusion protein is in the presence of Ran-GDP. In contrast, when Ran-GTP is present, the fusion protein will bind to it. This binding prevents the YFP and CFP domains from interacting and eliminates the FRET signal. The micrograph on the bottom right panel of the figure shows that the blue FRET signal is detected in the nucleus, indicating that Ran-GTP is present at high concentration there but not in the cytoplasm. In the control experiment shown in the bottom left panel, a cell expressing a mutant form of Ran (RanT24N), which binds Ran-GEF and prevents it from replacing GDP bound to wild-type Ran with GTP, has a reduced level of Ran-GTP in the nucleus. This is visible as a much weaker, green, FRET signal in the nucleus. These results prove that Ran is predominantly bound to GTP in the nucleus and bound to GDP in the cytoplasm. It is estimated that the concentration of Ran-GTP in the nucleus is approximately 200 times greater than in the cytoplasm.

Ran controls the direction of nuclear transport by binding to karyopherins and affecting their ability to bind their cargo. The effect of Ran is different depending on whether it binds an importer or an exporter. Consider the nuclear import receptor importin. In the cytoplasm, importin binds the NLS-containing cargo protein, as illustrated in **FIGURE 9.40**. Once in the nucleus, the importin-cargo complex binds to Ran-GTP. This causes rearrangement of the protein complex, leading to dissociation of the cargo. Thus, importin-cargo complexes are stable in the cytoplasm but dissociate after they bind Ran-GTP in the nucleus. Note that although the concentration of Ran-GDP in the cytoplasm is high, Ran-GDP is not needed for importin to bind its cargo or for the import process.

In contrast, Ran-GTP is needed in order for exportins to bind NES-containing cargo proteins. The presence of Ran-GEF maintains high levels of Ran-GTP in the nucleus. Ran-GTP, exportin, and cargo bind cooperatively to form a trimeric complex, as shown in Figure 9.40. After it has passed through the pore into the cytoplasm, the complex encounters Ran-GAP, which hydrolyzes the GTP. This triggers dissociation of the trimeric complex, thereby releasing the cargo protein.

The proper distribution of Ran in its GTP- and GDP-bound forms in the nucleus and the cytoplasm is critical for nuclear transport to occur. In most cells, Ran is concentrated in the nucleus. However, at any given time, some Ran-GTP moves out of the nucleus together with exportin. After GTP hydrolysis and cargo release, Ran-GDP is imported back into the nucleus by its own transport receptor called Ntf2, as shown in **FIGURE 9.41**. Once inside the nucleus, Ran-GEF replaces the GDP bound to Ran with GTP, resulting in dissociation of Ntf2 from Ran. Ntf2 then exits the nucleus for additional rounds of Ran-GDP import.

In addition to Ran-GAP and Ran-GEF, there are additional proteins that bind Ran and affect its activity. A Ran-binding protein (RanBP1) resides on the cytoplasmic surface of the NPC, where it stimulates the GTPase ac-

tivity of Ran by Ran-GAP to enhance the efficiency of cargo release. A second Ran-binding protein (RanBP3) resides in the nucleus, where it is important for facilitating the interaction of an exportin with the NPC. In order to ensure a high concentration of Ran at the NPC, certain nucleoporins also bind Ran on both the nuclear and cytoplasmic surfaces. One of these, Nup358, is also called RanBP2.

Ran also performs important roles during mitosis and, through its interactions with importin b, regulates delivery of key proteins to the mitotic spindle.

Concept and Reasoning Checks

1. How do the mechanisms of protein import and protein export differ from one another and how are they similar?
2. Explain how the asymmetric distribution of Ran-GAP and Ran-GEF controls the direction of nuclear transport and makes it irreversible.

9.16 Multiple models have been proposed for the mechanism of nuclear transport

Key concepts

- Interactions between karyopherins and nucleoporins are critical for translocation across the nuclear pore.
- Directionality may be conferred in part by distinct interactions of karyopherins with certain nucleoporins.

We now know that binding of cargo to a nuclear transport receptor and docking of the receptor at the pore complex are key events in the nuclear targeting of proteins. However, the mechanism by which cargo protein-karyopherin complexes move through the nuclear pore channel is still unclear. The traversed distance can be up to 200 nm, yet there is no known requirement for a molecular motor related to kinesins or myosins or for ATP hydrolysis.

Early experiments showed a requirement for ATP hydrolysis in transport through the NPC. However, later experiments showed that ATP is required to maintain sufficient levels of GTP to support transport. The only part of the nuclear transport process known to require GTP hydrolysis is in the function of Ran in controlling the direction of transport.

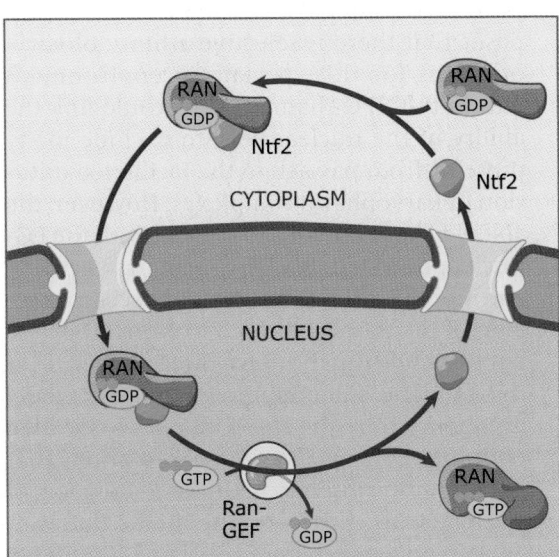

FIGURE 9.41 A small protein called Ntf2 mediates import of Ran into the nucleus.

How karyopherin-cargo complexes move through the nuclear pore channel is unknown. In considering possible mechanisms for translocation, we need to factor in the following that we do know:

- Karyopherins interact with most nucleoporins via the FG repeats and these FG repeats line the central channel of the pore.
- Binding studies have shown that some karyopherins have higher affinity for certain nucleoporins.
- Ran mediates not only the interaction between cargoes and karyopherins but also between karyopherins and nucleoporins.
- Karyopherins can move in both directions across the nuclear pore, although most probably carry cargo in only one direction.

One simple model suggests that movement through the channel occurs by facilitated diffusion. In this model, the transport receptor would make transient low-affinity interactions with nucleoporins with no directionality built into the binding reactions. The direction of movement would be conferred by Ran-dependent termination steps on either side of the pore as illustrated for import in **FIGURE 9.42**. A more complicated version of this model suggests that the direction of transport is due to a gradation of increasing affinities between the transport receptor and nucleoporins as the cargo/receptor complex moves through the NPC channel. By this model, the complex would encounter nucleoporins along a gradient of increasing affinity for the complex.

Another model builds on the observation that many of the hydrophobic FG-containing nucleoporins are located within the pore channel and the FG repeats are thought to fill the channel. There are about 3500 FG repeats in an NPC, not all of which are located in the NPC channel because some are found in cytoplasmic filament and nuclear basket nucleoporins. The unstructured FG domains are postulated to create a unique type of hydrophobic environment within the pore channel. It has long been known that the FG repeats interact with karyopherins, with Ran, and with other proteins involved in nuclear transport. More recently, evidence was obtained that there are important interactions among the FG repeats of the type found in the NPC core, but not among those that are parts of the terminal structures of the NPC. It is presumed that all or most of the FG repeats in the NPC core can interact with one another. Thus, karyopherins carrying cargo would need to disrupt some of these interactions in order to interact with FG repeats and work their way through the NPC.

In the selective phase model, shown in **FIGURE 9.43**, the attraction of the FG-rich regions of the nucleoporins for one another creates a central barrier through which most proteins cannot pass. This model further proposes that there is selective affinity of karyopherins for this specialized environment within the NPC channel. This would link the ability of the nuclear pore to exclude many proteins from passage to the facilitated diffusion of karyopherin complexes. However, the ability of karyopherins to interact with the FG-nucleoporins would allow them to selectively partition into and through this specialized region of the pore, and cargo to which they are bound would thus be transported. Evidence for this model is still preliminary, but the model is supported by the rapid kinetics associated with import, as well as the observations that karyopherins interact with the FG repeats of nucleoporins. However, the model does not explain how transport of very large complexes such as ribosomes and mRNA-protein complexes would occur.

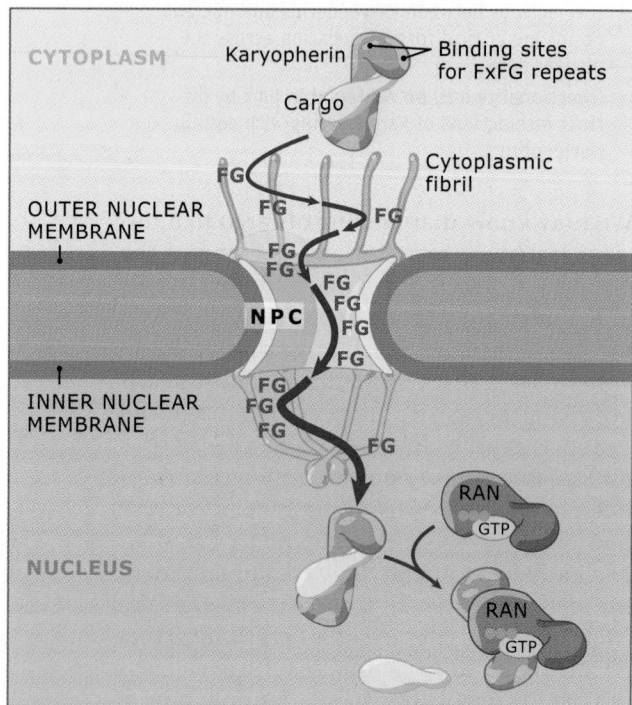

FIGURE 9.42 The contacts between karyopherins and nucleoporins via the phenylalanine-glycine repeats (FG repeats) are the key to understanding how translocation through the nuclear pore occurs.

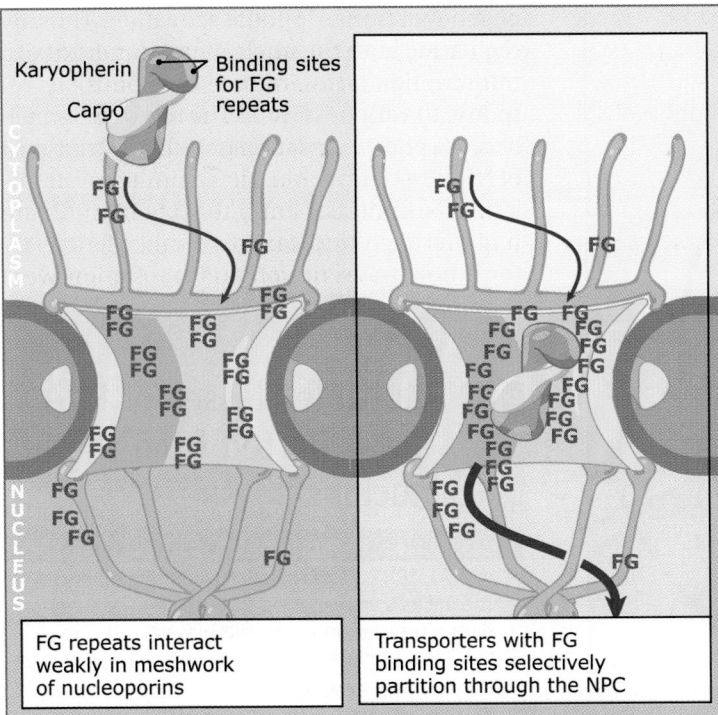

FIGURE 9.43 The selective phase model of translocation proposes that interactions among FG repeats prevent most proteins from translocating through the NPC. Proteins that contain binding sites for FG repeats can disrupt those interactions and partition through the NPC.

New biophysical approaches using fluorescence microscopy are beginning to allow detection of single molecules, and these methods are being applied to nuclear transport. These studies suggest that movement of karyopherin/cargo complexes through the NPC channel is very rapid (an average of 10 msec), most of the movement of the complex within the NPC channel is random, and the rate-limiting step may be escape from the central channel into the nucleus. These studies also suggest that an NPC can transport at least ten substrate molecules (with their receptors) at the same time and that an NPC can transport approximately 1000 molecules per second. The transport of very large complexes such as mRNPs appears to occur more slowly, on the order of seconds.

9.17 Nuclear transport can be regulated

Key concepts

- Both protein import and export are regulated.
- Cells use nuclear transport to regulate many functions, including transit through the cell cycle and response to external stimuli.
- The movement of the transcription factor NF-κB illustrates how nuclear transport is regulated.

The presence of the nuclear envelope allows for a level of regulation of gene expression and of the cell cycle in eukaryotes that is not possible in prokaryotes. Both protein import and export are regulated. There are many examples of the importance of regulated nuclear transport. Cells regulate the movement of transcription factors between the nucleus and cytoplasm to control transcription in response to stress and growth control signals. For example, circadian rhythms are controlled by regulated nuclear transport of the Period and Timeless transcription factors. In addition, movement of protein kinases and their regulators between the nucleus and the cytoplasm is important for progression through the cell cycle and response to external stimuli.

Regulated entry and exit of proteins from the nucleus can occur at several levels. First, the ability of the cargo to interact with its transport receptor can be regulated by direct modification of the cargo by, for example, phosphorylation. Second, the cargo can be anchored in one compartment and, thus, unable to move with its transporter until it is untethered. Third, the ability of the NPC itself to transport proteins is subject to regulation.

The movement of the transcription factor NF-κB illustrates many key aspects of regulated transport, as **FIGURE 9.44** shows. In response

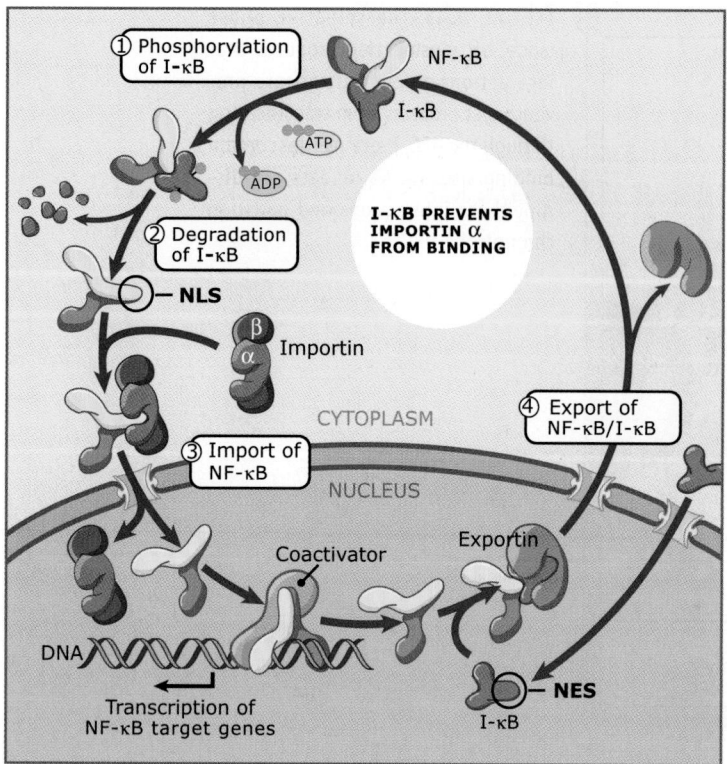

FIGURE 9.44 NF-κB moves into and out of the nucleus in a highly regulated manner that depends on covalent modifications of the cytoplasmic anchoring protein I-κB and on newly translated I-κB, respectively. For simplicity, Ran is not shown.

dependent transcriptional activation. Thus, by regulating both the nuclear entry and exit of a transcription factor, the cell can control its response to various stimuli. The response is rapid because phosphorylation of I-κB and transport of NF-κB occur within a few minutes of stimulation. In contrast, thirty to sixty minutes are required to produce an active transcription factor if new transcription and translation were required.

9.18 Multiple classes of RNA are exported from the nucleus

Key concepts

- mRNAs, tRNAs, and ribosomal subunits produced in the nucleus are exported through NPCs to function during translation in the cytoplasm.
- The same NPCs used for protein transport are also used for RNA export.
- Export of RNA is receptor-mediated and energy-dependent.
- Different soluble transport factors are required for transport of each class of RNA.

In eukaryotic cells, almost all of the RNAs produced in the nucleus are required in the cytoplasm for translation. Therefore, mRNAs, tRNAs, and rRNA-containing ribosomal subunits must be exported from the nucleus, as illustrated in **FIGURE 9.45**. Most RNAs do not contain export signals but must bind to proteins that have export signals in order to exit the nucleus. In fact, RNA molecules within cells are probably always in complexes with proteins that help to protect the RNA from degradation and mediate the interaction of RNA molecules with other cellular components.

The same NPCs are used for RNA export as for protein transport. This was shown by electron microscopy. Colloidal gold particles coated with RNA were injected into nuclei of *Xenopus* oocytes. When thin slices of the oocytes were subsequently examined by electron microscopy, gold particles could be seen within the channel of NPCs. Most NPCs contained one or more RNA-coated gold particles when a saturating amount of particles were injected into the nucleus. In a separate analysis, it was shown that most pores contain gold particles if enough NLS-coated gold particles were injected into the cytoplasm. These results suggested that RNA export and protein import use the same NPCs. The experiment that proved

to many stimuli, NF-κB moves from the cytoplasm to the nucleus and activates transcription of many genes important for the immune response. To ensure that NF-κB is activated only by the appropriate stimuli, it is bound to an inhibitory protein called I-κB (inhibitor of κB) in the cytoplasm. This interaction prevents the NLS of NF-κB from interacting with importin a. When cells are stimulated to activate NF-κB, I-κB becomes phosphorylated and rapidly degraded. This reveals NF-κB's NLS, allowing it to interact with its import receptor and enter the nucleus, where it activates transcription of its target genes.

Conversely, export of NF-κB occurs at an appropriate time to terminate the transcriptional response. I-κB contains both NLS and NES elements and shuttles between the nucleus and the cytoplasm. Newly synthesized I-κB enters the nucleus, where it binds NF-κB and promotes its nuclear export. In addition to hiding the NLS of NF-κB, the interaction between NF-κB and I-κB hides the NLS in I-κB. Because the NLSs of both proteins are sequestered, the complex of NF-κB and I-κB remains in the cytoplasm, preventing further NF-κB-

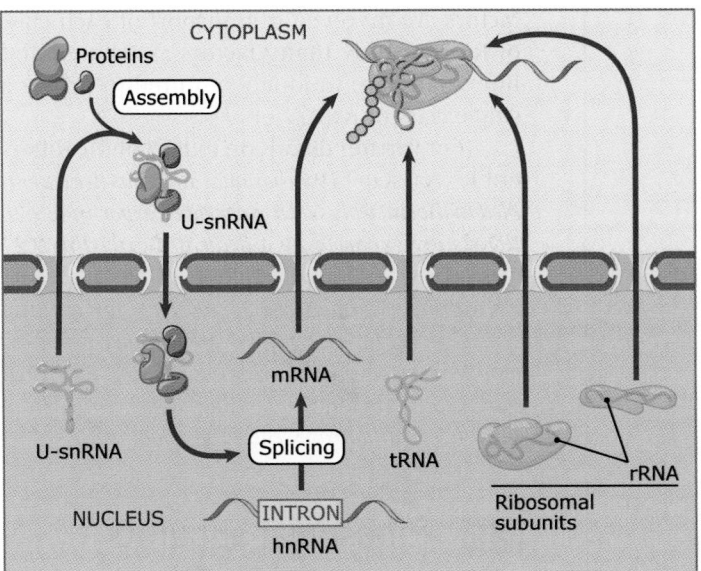

FIGURE 9.45 mRNA, tRNA, and ribosomal subunits are exported from the nucleus and function in protein synthesis in the cytoplasm. U-snRNAs are exported, processed, and assembled into RNA-protein complexes, and imported into the nucleus where they participate in RNA processing.

this showed that when protein-coated particles of one size and RNA-coated particles of another are injected into the cytoplasm and nucleus, respectively, many pores contain gold particles of both sizes, as shown in **FIGURE 9.46**.

Almost all types of RNAs are exported from the nucleus. This has been shown by studies in which radiolabeled RNA is injected into the nucleus of *Xenopus* oocytes. After incubation to allow transport to occur, cells are fractionated and the nuclear and cytoplasmic fractions analyzed to see whether the RNA has been exported to the cytoplasm. These studies show that tRNAs, mRNAs, rRNAs, and U-snRNAs are exported, as illustrated in **FIGURE 9.47**. In contrast, if the RNAs are injected into the cytoplasm, they remain there. This indicates that RNA transport is unidirectional. If nuclear injection studies are conducted at 0°C, no RNA export occurs. This indicates that RNA export, like protein transport, is an energy-dependent process.

Is the export of different types of RNAs mediated by the same type of transport receptors? This can be analyzed by determining whether one type of RNA can compete with another for export. In these experiments, radiolabeled RNA of one class is injected into nuclei along with unlabeled RNA of either the same or another class. After incubation, nuclear and cytoplasmic fractions are isolated and analyzed by electrophoresis to determine whether the injected RNA had been transported.

In the experiment diagrammed in **FIGURE 9.48**, radiolabeled tRNA was injected into oocyte nuclei, and the effect of increasing amounts of unlabeled tRNA or U1 snRNA was examined.

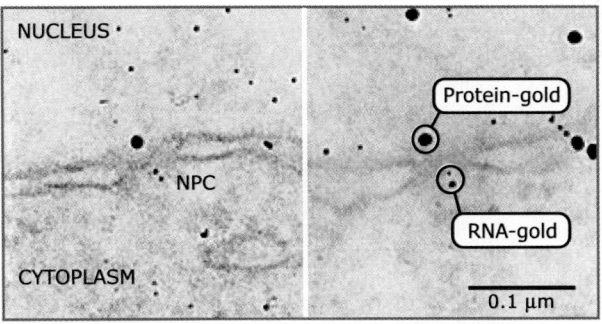

FIGURE 9.46 Large gold particles coated with nucleoplasmin, which has an NLS, were injected into the cytoplasm of a *Xenopus* oocyte and small gold particles coated with RNA were injected into the nucleus. Each panel shows one nuclear pore with protein-gold that has moved from the cytoplasmic to the nuclear side of the NPC and RNA-gold that has moved from the nuclear to the cytoplasmic side of the same NPC. Photos © Aitchison, et al., 1995. Originally published in **The Journal of Cell Biology**, 131: 1133–1148. Used with permission of Rockefeller University Press. Photos courtesy of Carl Feldherr and Steve Dwortezky, Rockefeller University.

Unlabeled tRNA competed with the labeled tRNA for export, but unlabeled U1 RNA (or RNAs of other classes) could not. The fact that unlabeled tRNA competed with labeled tRNA shows that tRNA export, like protein import, is a saturable process. This means that some protein or other cellular component with which tRNA interacts during export is present in a limiting amount. The finding that other types of RNAs do not compete for export with the labeled tRNA indicates that the factors limiting their rate of export are not required for tRNA export. Similar results were obtained when labeled RNA of any other class and unlabeled RNA of different classes were coinjected.

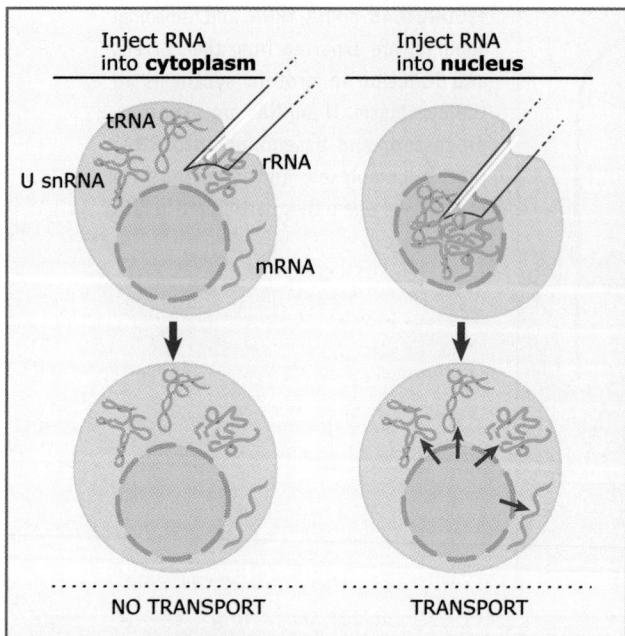

FIGURE 9.47 Transport of most RNAs is unidirectional from the nucleus to the cytoplasm.

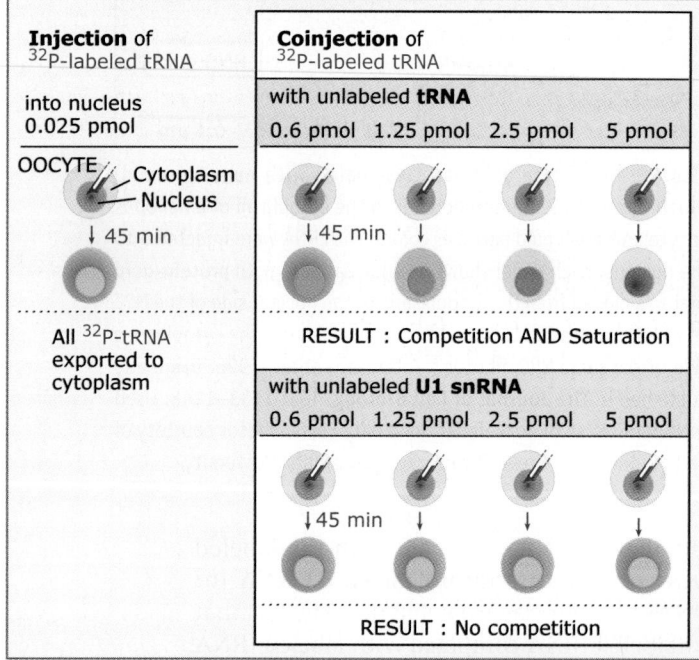

FIGURE 9.48 RNA export is saturable. The export of different classes of RNA is mediated by different transport proteins, whose abundance limits the rate of export.

The general conclusion from these studies is that export of each class of RNA requires at least one class-specific, rate-limiting factor. We know that all classes of exported RNA use the same NPCs, so the number of NPCs is not limiting. These studies do not tell us how many

factors are involved in transport of each class of RNA or how many factors are shared and how many are uniquely required to export a single class of RNAs.

(For further details on export of the different RNAs, see *9.19 Ribosomal subunits are assembled in the nucleolus and exported by exportin 1; 9.20 tRNAs are exported by a dedicated exportin; 9.21 Messenger RNAs are exported from the nucleus as RNA-protein complexes; 9.23 mRNA export requires several novel factors; 9.24 U snRNAs are exported, modified, assembled into complexes, and imported; and 9.25 Precursors to microRNAs are exported from the nucleus and processed in the cytoplasm.*)

9.19 Ribosomal subunits are assembled in the nucleolus and exported by exportin 1

Key concepts

- Ribosomal subunits are assembled in the nucleolus where rRNA is made.
- Ribosomal proteins are imported from the cytoplasm for assembly into the ribosomal subunits.
- Export of the ribosomal subunits is carrier-mediated and requires Ran.

Ribosomes are large complexes consisting of two subunits with a total of approximately 80 proteins and 4 ribosomal RNAs (rRNAs). The subunits are assembled separately in the nucleolus and exported to the cytoplasm for final assembly. The ribosomal subunits are among the largest complexes transported through NPCs, at the upper limit of what the nuclear pore channel can accommodate. On account of these subunits' large size, other macromolecules may be precluded from transport when a ribosomal subunit is passing through the channel. This would be in contrast with the localization of RNA and protein to the same NPC, which was described earlier (see *9.18 Multiple classes of RNA are exported from the nucleus*).

The assembly and export of the two ribosomal subunits are very complex processes and depend heavily on nuclear transport. Up to 50% of all nuclear transport may be involved in ribosome biogenesis. The ribosomal components are among the most abundant proteins and RNAs in the cell. Ribosomal proteins are synthesized in the cytoplasm, imported into the nucleus, and enter the nucleolus. There, they interact with rRNA precursors, which are transcribed within the nucleolus, and assembly factors. Following proper assembly, each subunit is exported through the NPC. Subunits appear to be retained in the nucleolus until they are assembled properly for export but what signals their release from the nucleolus into the nucleoplasm is not known. In the cytoplasm, maturation of the subunits is completed and subunits can then interact with translation initiation factors, tRNAs, and mRNAs to form mature translating ribosomes.

The nuclear import of all ribosomal proteins that have been studied so far employs members of the karyopherin protein family and the Ran GTPase. However, at least some ribosomal proteins use more than one dedicated importer. For example, in yeast, two different members of the karyopherin family mediate the import of the same ribosomal proteins. Because proper ribosome biogenesis is essential for cells to live, redundancy would ensure sufficient components for assembly.

As for other exported macromolecules, the export of ribosomal subunits is saturable, implying that it is receptor mediated. This was shown by using microinjection of ribosomal subunits into the nucleus of *Xenopus* oocytes. Moreover, ribosomal subunits do not compete with other RNA protein complexes for export, suggesting that distinct transport receptors are used for ribosomal subunits. Even bacterial ribosomes can be exported, suggesting that the eukaryotic export machinery can recognize these ribosomes. This may be coincidental, but it could indicate that the export machinery evolved early in eukaryotic evolution to accommodate the ancestral prokaryotic ribosome.

What is the export receptor for ribosomal subunits? Based on studies using yeast mutants, we know that export of the large (60S) subunits requires multiple receptors. One of these receptors is Crm1, which interacts with the 60S subunit through an adaptor protein called Nmd3, as illustrated in **FIGURE 9.49**. Export of the 60S subunit also requires two other export receptors, the Mtr2/Mex67 heterodimer that plays a central role in mRNA export (see below) and another protein. The export of the small 40S subunit also requires Crm1. Although it is likely that it interacts with the 40S subunit through an adapter, Nmd3 is not required for small subunit export, and an adapter has not yet been identified. It is also possible that other receptors besides Crm1 are needed for 40S subunit export. Other studies in yeast indicate that another protein, Rrp12,

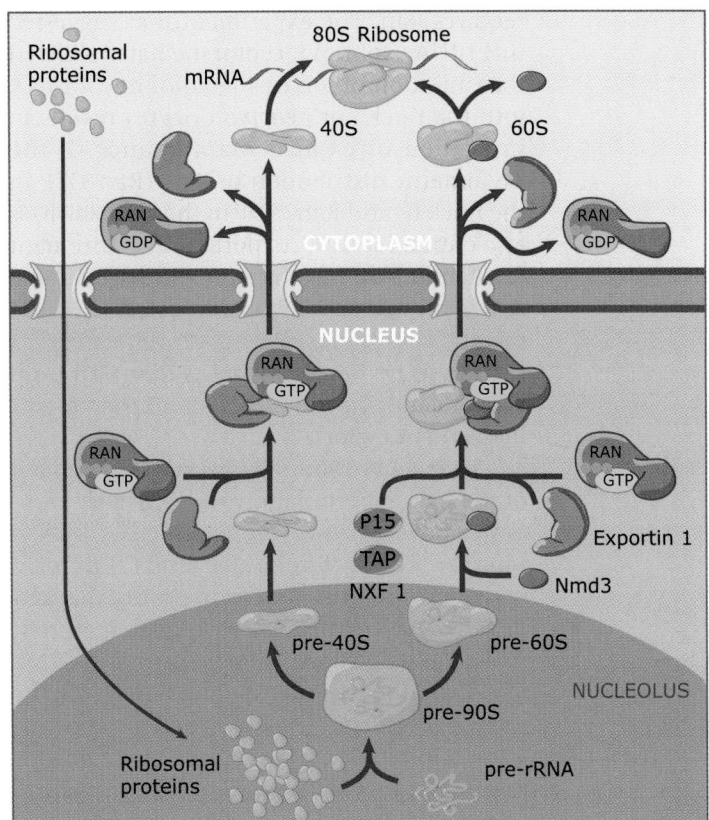

FIGURE 9.49 Ribosomal subunits assemble in the nucleolus from rRNA and imported ribosomal proteins. The subunits are then exported by the exportin Crm1 in a Ran-dependent manner. Nmd3 promotes the binding of Crm1 to the large subunit.

is involved in export of both 60S and 40S ribosomal subunits. This protein has structural similarities to the karyopherin family of receptors and, as is the case for karyopherins, binds to FG repeats of nucleoporins, to Ran and to ribosomal subunits.

9.20 tRNAs are exported by a dedicated exportin

Key concepts

- Exportin-t is the transport receptor for tRNAs.
- tRNA export requires Ran.
- tRNA export may be affected by modifications of the tRNAs.
- tRNAs may be reimported into the nucleus.

tRNAs are transcribed in the nucleus, where they are processed, and then exported to the cytoplasm, where they become aminoacylated and participate in translation. Only fully processed tRNAs are exported.

Export of tRNA from the nucleus is mediated by a dedicated transport receptor and requires Ran. The experiment that suggested that tRNA export is receptor mediated showed that injection of increasing amounts of tRNA into the nuclei of *Xenopus* oocytes results in saturation of export. Maintenance of the asymmetric distribution of Ran (Ran-GTP in the nucleus and Ran-GDP in the cytoplasm) is also critical for tRNA export. The requirement for Ran in tRNA export was shown by injection of the normally cytoplasmic Ran-GAP into the nucleus. Because Ran-GAP stimulates GTP hydrolysis by Ran, it prevents Ran-GTP from accumulating in the nucleus and thereby inhibits tRNA export.

The tRNA exporter is exportin-t, a member of the karyopherin family. Unlike most karyopherins, exportin-t binds directly to RNA. Moreover, it preferentially binds fully processed tRNAs, thus helping to ensure that unprocessed tRNAs are not exported prematurely. As **FIGURE 9.50** shows, exportin-t binds tRNA directly in the presence of Ran-GTP inside the nucleus to form a trimeric tRNA-exportin-Ran-GTP complex. This complex is analogous to those formed between other exportins and NES-containing proteins. The trimeric complex passes through the nuclear pore to the cytoplasm. After Ran-GAP stimulates GTP hydrolysis by Ran, the complex dissociates, releasing tRNA.

tRNAs are prepared for their role in translation by aminoacylation, which usually takes place in the cytoplasm. However, aminoacylation of tRNAs can also occur in the nucleus. Some studies indicate that aminoacylated tRNA is more efficiently exported and may be the preferred substrate for exportin-t. Because aminoacylation depends on prior complete and accurate processing of the tRNA precursor, this may act as a further proofreading mechanism to ensure that only functional tRNAs appear in the cytoplasm for translation.

Some tRNAs are matured by splicing using a splicing mechanism distinct from that used to splice pre-mRNAs. In yeast, some of the tRNA splicing enzymes are located in the cytoplasm and mutants defective for tRNA splicing accumulate unspliced tRNAs in the cytoplasm. Furthermore, mature forms of these tRNAs have been found in the nucleus; this suggests that tRNAs may be imported into the nucleus after splicing and then reexported. This process may permit more sophisticated proofreading to ensure that only correctly processed tRNAs take part in protein synthesis. There is also evidence that reimport is used as a mechanism to decrease protein synthesis under certain conditions, such as nutrient deprivation, by moving tRNAs into the nuclear compartment where they can no longer interact with ribosomes and other cytoplasmically located components of the translation machinery.

In yeast there is one gene, *LOS1* that encodes exportin-t. Interestingly, *LOS1* is not essential for yeast cells to survive, indicating that there is an additional export pathway for tRNAs. tRNAs are sufficiently small (~30,000 daltons) that they might be able simply to diffuse out of the nucleus. If this were to occur, it would lead to a uniform concentration of tRNA throughout the cell. However, we know that most tRNA is cytoplasmic. So is there some other receptor that mediates export of tRNA? If a yeast cell has mutations that slow tRNA synthesis, it cannot survive unless exportin-t is intact. This observation has led to the suggestion that enzymes that couple amino acids to tRNAs—the aminoacyl tRNA synthetases—may be able to function as export receptors for tRNA, as shown in Figure 9.50. This model would require shuttling of these tRNA

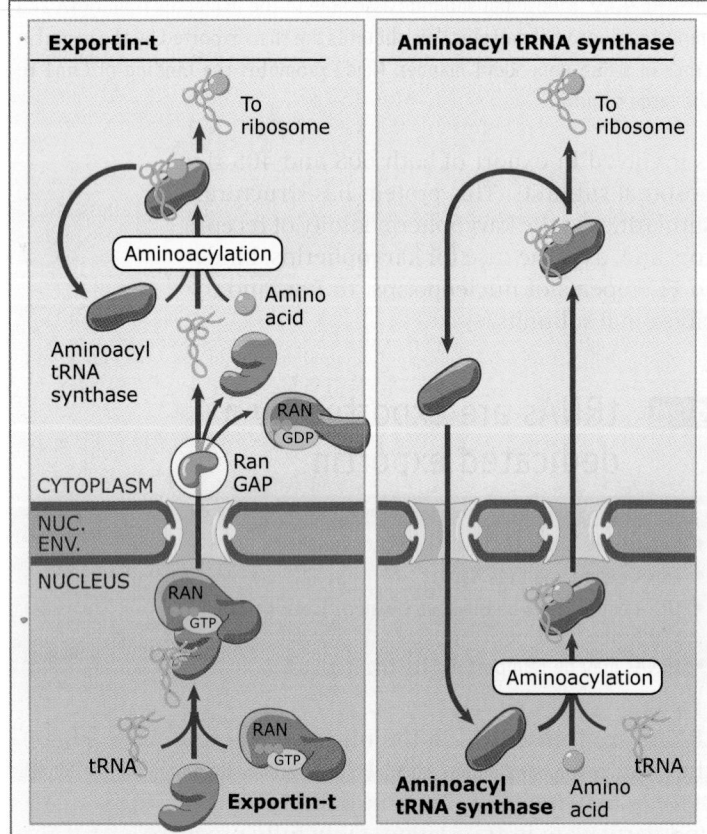

FIGURE 9.50 Movement of tRNA out of the nucleus uses a Ran-mediated pathway in which the tRNA binds to exportin-t (left). Aminoacylation of tRNAs also plays a role in export (right).

synthetases. They would bind and aminoacylate the tRNA inside the nucleus, then escort the aminoacylated tRNA out to the cytoplasm, release the tRNA for translation, and reenter the nucleus.

9.21 Messenger RNAs are exported from the nucleus as RNA-protein complexes

Key concepts

- Proteins that associate with mRNAs during transcription help to define sites of pre-mRNA processing and are also thought to package mRNAs for export.
- Most proteins that associate with mRNA in the nucleus are removed after export and returned to the nucleus. A few are removed immediately prior to export.
- Signals for mRNA export may be present in proteins bound to the mRNA.
- The export of mRNA can be regulated, but the mechanism for this is unknown.

As we have discussed previously, most RNAs synthesized in the nucleus are exported to function in the cytoplasm (see *9.18 Multiple classes of RNA are exported from the nucleus*). All nuclear events of mRNA biogenesis, from transcription through export, are coupled. RNAs are not exported as naked RNAs, but in complexes with proteins. In the case of mRNA, multiple proteins associate with a precursor to mRNA (pre-mRNA) as it is being synthesized, as illustrated in **FIGURE 9.51**. The resulting RNA-protein complexes are called **heterogeneous nuclear ribonucleoprotein particles (hnRNPs)**, and the proteins (other than cap-binding proteins and poly[A]-binding protein) are called hnRNP proteins.

There are at least 20 different hnRNP proteins in human cells. Some help to structure the pre-mRNA so that it can be processed properly. This is important because only completely and correctly processed mRNAs are exported (see *9.23 mRNA export requires several novel factors*). The cell has multiple mechanisms to detect improperly or incompletely processed mRNA. This prevents export of defective mRNAs that could direct formation of defective proteins, whose presence could be injurious to the cell.

It is thought that the function of many of the proteins that bind to mRNAs in the nucleus is to help package them for export.

Most eukaryotic cells contain between 3,000 and 15,000 different species of mRNA. These mRNAs vary greatly in size, sequence, and in the secondary structures into which they can fold. By forming RNA-protein complexes prior to export, these diverse mRNAs are thought to acquire common structural features predicted to be important for efficient mRNA export. Some of the bound proteins may contain specific export signals.

The association of these proteins with mRNAs is temporary. Some are removed before the hnRNP passes through the NPC. Others accompany the RNA out of the nucleus. After translocation through the NPC, most of the remaining proteins are removed, allowing them to return to the nucleus to participate in export of additional mRNAs.

Because hnRNPs are large complexes, their conformation may be altered so that they can pass through the NPC. Unique properties of the insect *Chironomus tentans* have allowed us to visualize stages in the export of mRNAs from the nucleus. In *Chironomus*, a set of extremely large mRNAs is produced at a specific stage of development when transcription of several genes is strongly induced by developmental signals. These highly active sites of transcription are called **Balbiani rings** (see *10.13 Polytene chromosomes expand at sites of gene expression*). The mRNAs produced are very large and, like other

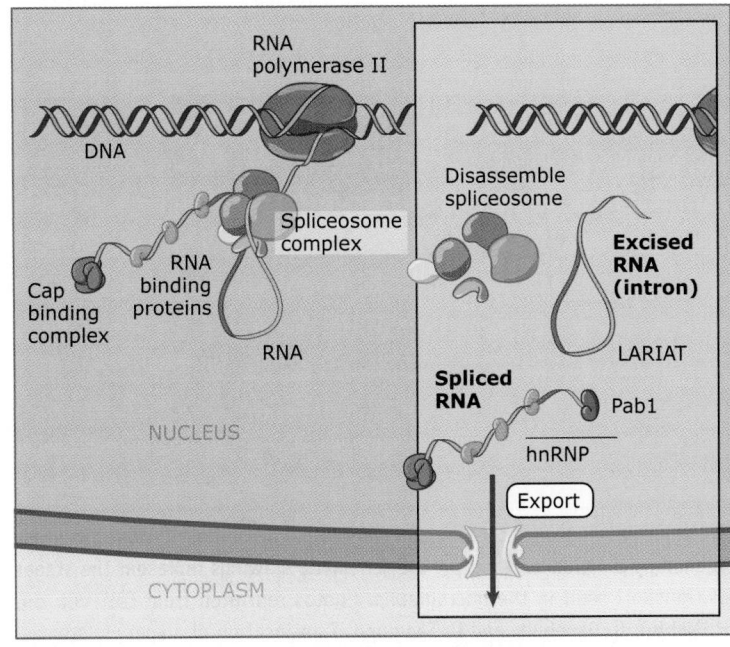

FIGURE 9.51 hnRNP proteins associate with mRNA during its synthesis. After splicing, the spliceosome and excised lariat remain in the nucleus and the mRNA with bound hnRNP proteins is exported.

mRNAs, associate with hnRNP proteins. These hnRNPs are called **Balbiani ring granules**. As the electron micrograph in **FIGURE 9.52** shows, they appear as small rings in the nucleus.

The diameter of a Balbiani ring granule is much larger than that of the channel of the NPC, which is ~27 nm in diameter. We can see Balbiani ring granules caught at different stages in the transport process by fixation performed prior to electron microscopy. As Figure 9.52 shows, the Balbiani ring granules become linear as they pass through NPCs. The linear hnRNP is long enough to span an NPC and extend into both the cytoplasm and nucleus. Because there are structural differences along the length of the Balbiani ring granule, we know that they always pass through NPCs in the same orientation, with the 5′ end of the mRNA in the lead. We do not know if other mRNAs are also exported with the 5′ end first, but we believe that the conformation of all mRNPs can be altered if necessary.

The various mRNAs produced in a cell share very few features that might function as export signals. However, they associate with many of the same proteins, so one possible mechanism for signaling export is that one or more hnRNP proteins might contain nuclear export signals. According to this model, proteins with NESs would bind to the mRNA and a transport receptor would recognize the NES in the protein. At least some hnRNP proteins have NESs, but others do not. Export of some mRNAs is mediated by signals found in mRNA-binding proteins that are not hnRNP proteins.

Although cellular mRNAs are generally not exported until all RNA processing has been completed, some viral RNAs are exported without removal of introns. Human immunodeficiency virus-1 (HIV-1) RNA is the best-understood example of this situation. Like all retroviruses, HIV-1 produces a set of overlapping mRNAs. Some mRNAs are spliced. However, full-length, unspliced viral mRNA must be exported because this is the form of viral RNA used to produce new infectious viral particles and it is also functions as an mRNA to produce some HIV proteins.

HIV-1 has developed a unique strategy to ensure export of unspliced viral mRNA. The viral Rev protein is produced by translation of one of the spliced viral mRNAs. HIV-1 Rev binds to RNA sequences of a type called the Rev response element (RRE), which are present in unspliced HIV-1 mRNAs. Rev contains a leucine-rich NES (see *9.14 Export of proteins from the nucleus is also receptor-mediated*). This NES is used to mediate export of Rev, both as a free protein and when it is bound to RRE-containing mRNAs. The RRE is sufficient to confer Rev binding and allow export to the cytoplasm. Export of RNA mediated by Rev uses the same export pathway and factors used for protein export, including exportin 1 (see Figure 9.36).

Just as protein transport is regulated in response to external signals and intracellular requirements (see *9.17 Nuclear transport can be regulated*), mRNA export can be regulated. The best example of regulated mRNA export occurs as part of the response of cells to stresses such as heat. This is called the heat shock response, although a similar response occurs following other kinds of stress as well (e.g., osmotic shock, or exposure to toxic metals or high concentrations of ethanol).

As part of the heat shock response, most polyadenylated RNAs are not exported and instead accumulate in the nucleus. The cell up-regulates the expression of heat shock genes, which encode proteins that protect the cell from possible damage. The heat shock mRNAs must be exported. Thus, after heat shock, the export of some mRNAs is prevented but the efficient export of heat shock mRNAs is permitted. The mechanism by which this occurs is not yet known. We do know that regulation of mRNA export by heat shock does not require new

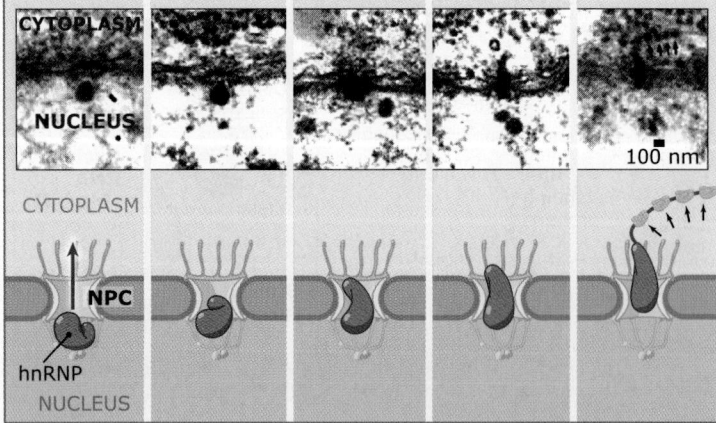

FIGURE 9.52 Export of the large Balbiani ring granule hnRNP from *C. tentans* as seen by electron microscopy. The schematic drawings represent the stages of transport seen in the micrographs. Photos reprinted from *Cell*, vol. 69, H. Mehlin, B. Daneholt, and U. Skoglund, Translocation of a specific premessenger..., pp. 605–613, Copyright (1992) with permission from Elsevier [http://www.sciencedirect.com/science/journal/00928674]. Photos courtesy of B. Daneholt, Karolinska Institutet.

protein synthesis and can be detected within a few minutes after heat shock. These observations suggest that a signal transduction pathway is induced either to modify the nuclear transport machinery so that it selectively exports stress response mRNAs or to alter mRNA biogenesis so that only stress response mRNAs are packaged correctly and identified for export.

Concept and Reasoning Checks

1. Describe the functions that TAP and Dbp5 play in mRNA export.
2. What are the major differences between export of proteins containing NESs and export of mRNAs?

9.22 hnRNPs move from sites of processing to nuclear pore complexes

Key concepts

- mRNAs are released from chromosome territories into interchromosomal domains following completion of pre-mRNA processing.
- mRNAs move to the nuclear periphery by diffusion through interchromosomal spaces.

Transcription and processing of mRNAs takes place in the nuclear interior, requiring that hnRNPs move from transcription sites to NPCs at the nuclear periphery. Sites of transcription are also the sites of most mRNA processing. 5' capping occurs as soon as the RNA is long enough to emerge from the RNA polymerase holoenzyme; splicing begins during transcription, and 3' ends are generated by cleavage of the RNA while it is still a growing chain. When splicing or 3' processing is defective, mRNAs are not exported.

We believe that hnRNPs diffuse from sites of transcription and processing through the interchromosomal spaces to NPCs and are then exported from the nucleus. We know that RNA is present in all of the nuclear spaces not filled by chromosomes, as shown in **FIGURE 9.53**. That hnRNPs diffuse through these spaces to the NPC is supported by studies using salivary gland cells of *Drosophila melanogaster*. The nuclei in these cells are very large because they contain many extra copies of the *Drosophila* genome. After processing, hnRNPs in these nuclei can be seen moving to the nuclear periphery at an equal rate in all directions. The rate of movement, approximately one micron per second (1 μm/sec), is what would be expected

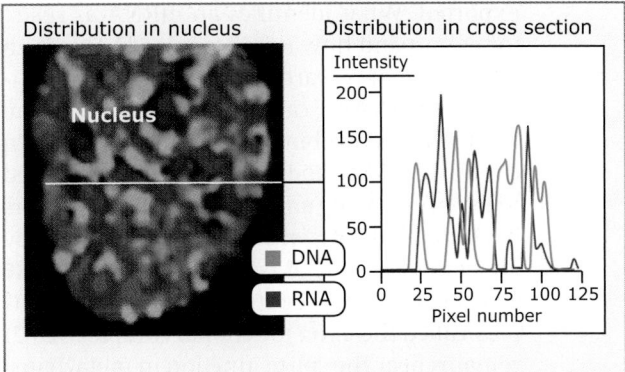

FIGURE 9.53 Polyadenylated RNA (red) and DNA (green) are stained with fluorescent dyes. By determining where the red and green signals are located along a straight line drawn through the nucleus (right), we can see that RNA and DNA do not overlap; rather, mRNA occupies a network of areas adjacent to and intertwined with the network of areas occupied by DNA. Photos reprinted from *Curr. Biol.*, vol. 9, J. C. Politz, et al., Movement of nuclear poly(A) RNA..., pp. 285–291, Copyright (1999) with permission from Elsevier [http://www.sciencedirect.com/science/journal/09609822]. Photos courtesy of Joan C. Ritland Politz, Fred Hutchinson Cancer Research Center.

for movement by passive diffusion. Molecular motors move cargoes at a wide range of rates, from a few tenths of a micron per second to ten microns per second. However, these motors generally move cargoes along filaments and in a highly directional manner.

9.23 mRNA export requires several novel factors

Key concepts

- Many factors required uniquely for mRNA export have been identified.
- Factors able to bind to both the mRNP and nuclear pore complex help to mediate mRNA export.
- One factor, Dbp5, is an ATPase and may use energy from ATP hydrolysis to remove mRNP proteins during transport.

mRNA export is considerably more complex than transport of proteins across the nuclear membrane. First, many more protein factors are required for mRNA export than for protein import and export. None of these factors is related to Ran or to the karyopherins involved in protein transport. Second, the mRNA export process is coordinated with transcription and mRNA processing so that only fully processed mRNAs are exported. In fact, some of the proteins required for mRNA export likely participate in processing events that must be completed correctly for the mRNA to be exported. What identifies an mRNA as ready for export and how is it recognized? This is an active field of research today.

The mRNA's 5' cap structure is not essential for mRNA export but enhances the process. In contrast, splicing and 3' processing are coupled to export. Splicing and polyadenylation factors begin to associate with mRNAs even as they are being transcribed (see Figure 9.51). **FIGURE 9.54** shows that after splicing, a protein complex called the exon junctional complex (EJC) remains near the splice junction in metazoans. The presence of this complex may be a signal that the mRNA has been spliced. However, it cannot be sufficient to signal exportability since mRNAs acquire an EJC after one intron has been removed but are not exported until all have been excised.

Among the proteins in the EJC is an mRNA export factor called Aly, which is a member of the REF family of RNA-binding proteins. Aly interacts with UAP56, a component of the spliceosome, and with NXF1/TAP, which binds to FG repeats in some nucleoporins (see *9.10 Nuclear pore complexes are constructed from nucleoporins*). NXF1/TAP, by binding to the **messenger ribonucleoprotein particle (mRNP)**, which is the complex of mature mRNA and mRNA-binding proteins, and to nucleoporins at the same time, may play a role in export similar to that of karyopherins such as exportin 1 (see Figure 9.54). Thus, NXF1/TAP can be thought of as a receptor for mRNPs. A small subset of mRNAs in mammalian cells may require exportin 1 for their export, but karyopherins do not appear to be essential for the export of most mRNAs.

In yeast, most genes do not have introns. However, the yeast homolog of Aly, Yra1, is important for mRNA export and is thought to play an analogous role. During transcription, Sub2, the yeast protein homolog of UAP56, is recruited to the mRNA, along with Yra1. Sub2 and Yra1 interact first with a protein complex, called THO, which is involved in transcription elongation and could transfer them to the nascent mRNA. Analogous to the model shown in Figure 9.54 for metazoan cells, the NXF1/TAP homolog Mex67 associates with Yra1. NXF1/TAP is a member of a family of proteins, the NXF proteins that are all thought to function during export of RNAs.

NXF1/TAP was originally identified in studies of the Mason-Pfizer monkey virus (MPMV), a retrovirus (see Medical Applications box on page 431). Although all retroviruses require export of both spliced and unspliced viral mRNA, most do not encode an export-directing protein like HIV-1 Rev (see *9.14 Export of proteins from the nucleus is also receptor-mediated*). Export of unspliced MPMV RNA requires a short sequence, called the constitutive transport element (CTE), which binds to NXF1/TAP. Although a sequence like the MPMV CTE is not a feature of cellular mRNAs, NXF1/TAP (Mex67 in yeast) is important for the export of almost all cellular mRNAs.

mRNAs are monitored for correct and complete processing before export from the nucleus. At least in yeast, mRNAs are retained near their sites of transcription if they are not fully processed. The exosome is a complex of ribonucleases that is located near transcription sites and degrades mRNAs that are processed incorrectly. Release of mRNAs from the retention sites appears to require accurate 3' processing and polyadenylation.

The composition of the mRNA-protein complex in the nucleus differs from the composition after passage through the NPC. Removal of some of the mRNA-associated proteins is a key event during transport. In contrast to protein transport, where the protein is generally

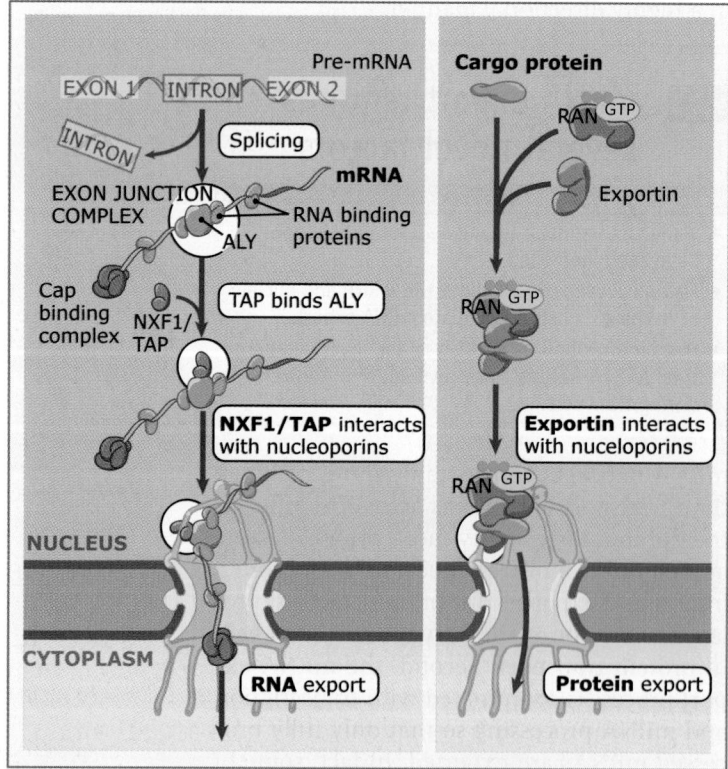

FIGURE 9.54 TAP binds to both mRNPs and to NPCs and functions as an export receptor for mRNA. The function of TAP is analogous to that of exportin for protein cargo.

Viruses are obligate intracellular parasites, growing only within cells, and they depend upon host cell translational machinery and the ability of the host to generate energy, stored as ATP. Some viruses also rely on host machinery for other processes required for viral replication, including DNA replication, transcription, and RNA processing. In contrast, other viruses are less dependent on host cell mechanisms because they encode viral proteins that perform functions like DNA and RNA replication and transcription. Viruses with the largest genomes encode most of the proteins needed for their propagation, relying on the host primarily for energy and the protein synthetic machinery. Depending on their strategies, viruses may inhibit host processes upon which they are not dependent, thereby optimizing the availability of host cell resources to enhance the production of more viruses. Viruses have been identified that infect cells from all branches of the tree of life, including animals, plants, bacteria, archaea, and fungi.

In the early days of molecular biology, bacteriophage, the viruses of bacteria, played fundamental roles in elucidating fundamental mechanisms of gene expression including DNA replication, RNA processing, and protein synthesis. Similarly, animal viruses were, and continue to be, of great importance in studies to understand the molecular biology of animal cells. Some key proteins involved in nuclear transport were first identified because they interact with viral proteins. In several cases, viruses enhance their own genetic program by inhibiting or modulating nuclear transport using a variety of mechanisms.

Retroviruses have RNA as their genetic material but produce a DNA-form of their genome, which becomes integrated into the host genome. Viral RNAs are produced by transcription from the integrated viral DNA, using the host's transcriptional machinery. Although mRNAs are generally not exported to the cytoplasm until all introns have been excised by splicing, retroviral propagation requires that unspliced viral RNA be exported, as this is the form of RNA that is assembled into new virus particles. In addition, some retroviral mRNAs produced through alternative splicing contain some introns and would normally be retained in the nucleus and degraded if not fully processed. Intron-containing mRNAs and viral genomic RNA encoded by the MPMV are readily exported because they contain a sequence, the cytoplasmic transport element (CTE) that binds the mRNA export receptor, TAP/NXF1. This receptor is usually recruited to the mRNA as a consequence of normal nuclear mRNA processing, but the presence of the CTE allows their direct recruitment to viral RNAs, and export, even though introns are still present. TAP/NXF1 was first identified because it bound a protein encoded by a simian herpesvirus. Studies using the CTE of MPMV and TAP/NXF1 were important for the initial discovery of the mRNA export pathway.

Human immunodeficiency virus type 1 (HIV-1), another retrovirus of the lentivirus family, uses a strategy similar to that of MPMV. A sequence present in HIV-1 viral genomic RNA and incompletely spliced viral mRNAs is able to bind a virus-encoded protein, called Rev. Rev forms a multimer by binding this sequence, the RRE. Each Rev monomer contains a nuclear export signal, which is able to interact with Crm1, the exportin responsible for most protein export. By this mechanism, the cell exports HIV-1 RNAs through the protein export pathway, and the retention of introns by viral RNAs is not detected by the surveillance mechanisms that operate to prevent export of incompletely spliced RNAs. Studies showing that export of HIV-1 RNAs involved interaction of Rev with Crm1 played an important role in the discovery of exportin 1 and the demonstration that protein export and mRNA export operate using different receptors and mechanisms.

Some RNA viruses that replicate in the cytoplasm can even replicate in enucleated cells, indicating that they are not dependent on nuclear processes. In some cases, RNA viruses produce proteins that inhibit nuclear transport. Poliovirus produces all of its proteins through proteolytic processing of a very large viral polyprotein. The polyprotein contains protease activities, which function to generate the viral replicase (an RNA polymerase) and capsid proteins that surround new viral RNA molecules in new virus particles. Many cellular pathways not required by poliovirus are inhibited because the viral protease digests key proteins. These include multiple components of the nuclear pore complex (Nups 62, 98, and 153). As a consequence, multiple protein transport pathways are inhibited, proteins required in the nucleus accumulate in the cytoplasm, and this leads to inhibition of host cell mRNA production, export and protein synthesis. Poliovirus infection also leads to the cleavage of a key translation factors and poly (A) binding protein, neither of which are required by poliovirus for translation of its own mRNA.

Another RNA virus that replicates in the cytoplasm and inhibits nuclear transport is vesicular stomatitis virus or VSV. An abundant viral protein found in the virus particle and also produced during infection, the matrix or M protein, inhibits export of host mRNAs by forming a complex with an important RNA export factor, Rae1, and Nup98. Over-production of Rae1 and Nup98 reverses this inhibition, indicating that VSV M protein inhibits host mRNA export by preventing Rae1 and Nup98 from performing their normal functions that are important for mRNA export. Additional examples continue to be discovered where viruses target steps of nuclear transport as part of their strategy to inhibit the host and maximize their own propagation.

It may be possible to discover or design antiviral drugs by screening for compounds that prevent viral proteins from targeting host processes, including nuclear transport. For example, a screen for low molecular weight inhibitors of Rev-mediated nuclear export led to the identification of leptomycin B as a potent inhibitor of HIV-1 infection. Further studies showed that leptomycin B binds covalently to the NES-binding pocket of Crm1, and therefore blocks all Crm1-mediated export. Unfortunately, leptomycin B cannot be used to treat HIV-1 infections because its target, Crm1-mediated export, is an essential process and inhibiting it causes lethality. In spite of this limitation, leptomycin B has been a useful tool for probing nuclear transport mechanisms. It should be possible to identify or design drugs that target viral proteins rather than components of the host process the viral proteins target.

the same on both sides of the nuclear envelope, and directionality is controlled by the Ran GTPase systems (see *9.15 The Ran GTPase controls the direction of nuclear transport*), directionality for mRNA export may be controlled by altering the composition of the mRNP so that it is recognized as cargo only in the nucleus. In addition, removal of the mRNP proteins may be necessary to allow the intimate interactions that occur between the ribosome and the mRNA during translation. It is thought that as the mRNP emerges from the NPC, the mRNP associates with ribosomes, which may facilitate export. However, since inhibitors of protein synthesis do not block mRNA export, ribosome function cannot be essential for export.

How are mRNP proteins removed? We do not know but there are at least three possibilities. Some hnRNP proteins are recognized by a karyopherin as proteins to be transported into the nucleus. The binding of the karyopherin to the protein causes a change in protein conformation that releases the protein from the mRNA. Other proteins may simply dissociate from the mRNP after transport and be transported back into the nucleus before they can rebind. However, many mRNA-binding proteins bind too tightly to mRNA for dissociation to allow efficient protein removal.

Most like, enzymes remove proteins bound to mRNA. One enzyme candidate is Dbp5, an essential mRNA export protein in yeast and mammals. Dbp5 shuttles between the nucleus and cytoplasm and binds to the cytoplasmic fibrils of the NPC. It is a member of the family of DEAD-box proteins, so called because they each contain the sequence Glu(D)-Asp(E)-Ala(A)-Glu(D) (or one very similar to it) and several other highly conserved motifs. DEAD-box proteins hydrolyze ATP and one or more are thought to function in each step of mRNA metabolism, from synthesis through turnover. *In vitro*, some DEAD-box proteins can use ATP energy to denature short double-stranded RNA substrates, and some have been shown to remove otherwise stably bound proteins from RNA. Recently, it was shown that Dbp5 can remove an mRNA-binding protein, Nab2, from polyA *in vitro*, and this supports the idea that Dbp5 mediates dissociation of proteins from mRNA *in* vivo as well.

A model for mRNP protein removal by Dbp5 is illustrated in **FIGURE 9.55**. In this model, Dbp5 binds to Nup159 (or to Nup214 in human cells), one of the nucleoporins that form the cytoplasmic fibrils of the NPC. Dbp5 then inter-

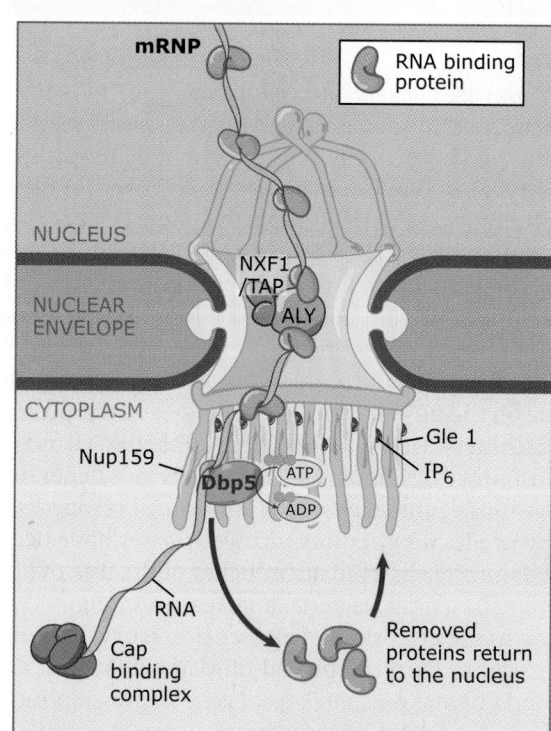

FIGURE 9.55 Dbp5 associates with the terminal filaments of the NPC where it may use ATP energy to remove mRNP proteins during mRNA export.

acts with another nucleoporin, Gle1, which has been shown to activate the ATPase activity of Dbp5 in vitro, a process that also involves inositol hexakisphosphate (IP_6), a phosphorylated inositide, acting as a cofactor that binds to Gle1. Surprisingly, it is the ADP bound form of Dbp5 produced as a consequence of ATP hydrolysis by Dbp5 that is active in removing protein from RNA in vitro. Because Dbp5 enters the nucleus and may associate with the mRNA during its synthesis, the Dbp5 that mediates mRNA export may travel through the NPC as part of the mRNP, and associate with the cytoplasmic filament nucleoporins upon exit from the NPC channel. Whether one or multiple molecules of Dbp5 mediate export of one mRNP is not known.

9.24 U snRNAs are exported, modified, assembled into complexes, and imported

Key concept
- U snRNAs produced in the nucleus are exported, modified, packaged into U snRNP RNA-protein complexes, imported into the nucleus, and processed further in order to function in RNA processing.

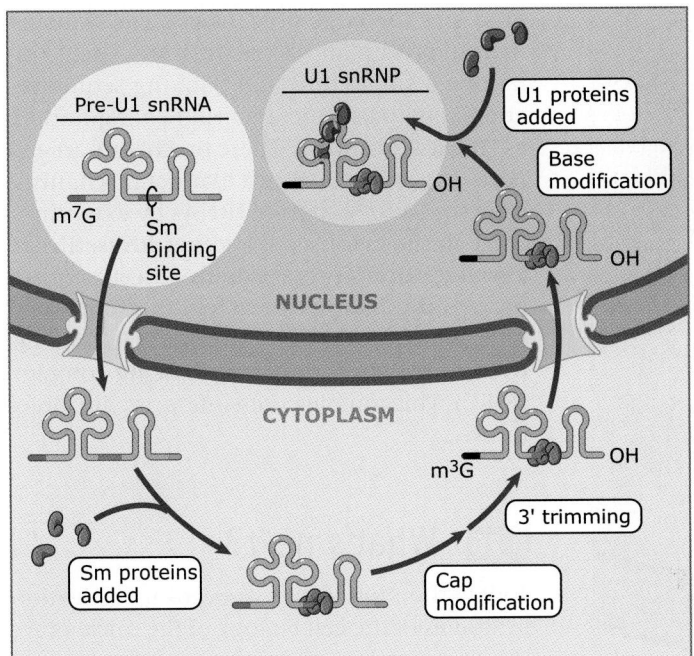

FIGURE 9.56 The production of snRNPs involves export of the pre-U1 snRNA, modification of the snRNA in the cytoplasm, and import of the snRNP into the nucleus for further modifications.

Small nuclear ribonucleoprotein particles (snRNPs) are RNA protein complexes that play a central role in pre-mRNA splicing and other types of nuclear RNA processing. The RNAs found in snRNPs, the U snRNAs, are produced in the nucleus. However, in metazoan cells the formation of functional snRNPs requires U snRNA export, modification in the cytoplasm, association with proteins in the cytoplasm, and import of the U snRNP complex, as **FIGURE 9.56** illustrates. The imported U snRNP complex then undergoes final assembly into the snRNP.

Most U snRNAs are produced by RNA polymerase II transcription. Like mRNAs, which are also RNA polymerase II products, U snRNAs have monomethylated 5′ caps, but they differ from those of mRNAs in lacking a poly(A) tail and their production uses a unique signal for 3′ processing that results in an RNA that is not polyadenylated. The U snRNA cap is a key signal for its export. When the RNA has entered the cytoplasm, its cap is methylated to become a trimethyl-G cap, and the RNA is assembled into an RNA-protein complex by interacting with a set of proteins called the Sm proteins. The trimethyl cap contributes to the splicing process in which U-snRNPs function, and the proteins help to create the important overall three-dimensional structure of each snRNP. These complexes, called U snRNPs, are imported into the nucleus by a transport receptor consisting of an adaptor (snurportin) and

importin β. The cap-binding and Sm proteins serve as dual signals for import.

We believe that U snRNAs are not exported from the nucleus in budding yeast. In yeast, U snRNPs are assembled in the nucleus from RNAs and imported proteins. We do not know why the pathway for making functional snRNPs in mammalian cells involves export and import, but it is possible that these multiple transport steps help insure that only properly processed and assembled snRNPs are allowed to participate in RNA processing. Such a mechanism to ensure proper U-snRNP biogenesis is similar to the multiple transport steps of tRNAs.

9.25 Precursors to microRNAs are exported from the nucleus and processed in the cytoplasm

Key concept

- MicroRNAs are produced by transcription in the nucleus, partial processing to generate a hairpin precursor, export of the precursor by exportin-V, and final processing in the cytoplasm.

A class of small RNAs called microRNAs (miRs) plays an important role in regulation of gene expression. miRs are 21–22 nucleotides in length and are found in multicellular organisms, both plant and animal. In humans, there

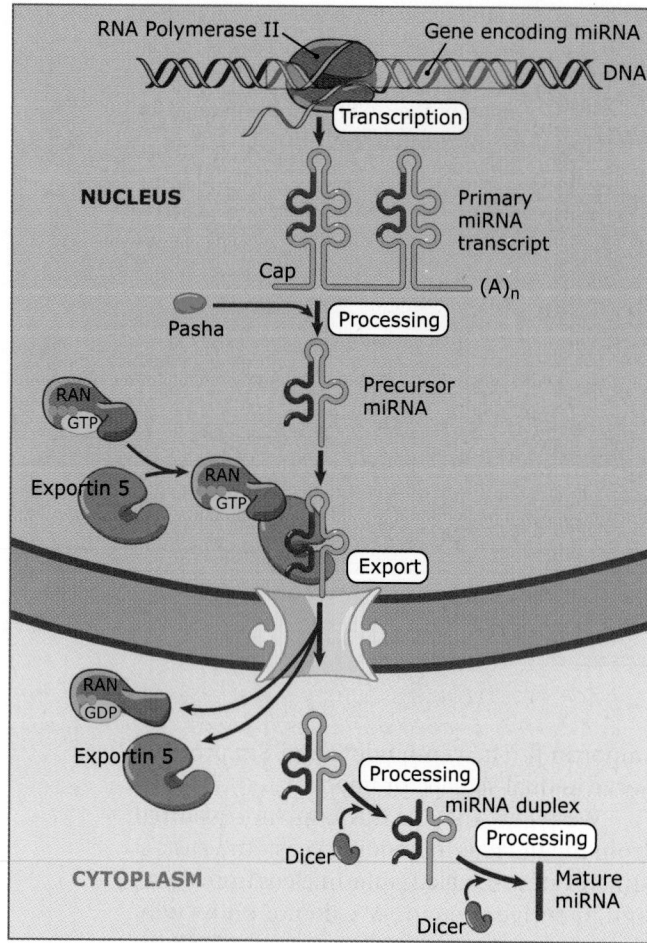

FIGURE 9.57 MicroRNAs are partially processed in the nucleus and exported to the cytoplasm, where they undergo further processing to become mature miRNAs.

processes the large precursor to yield smaller hairpin precursors, called pri-miRs, each containing a single miR and flanking sequences and likely to assume a hairpin configuration, as **FIGURE 9.57** shows. These precursors are exported by exportin-5, in a manner mechanistically similar to export of tRNAs by exportin-t. Once in the cytoplasm, another RNaseIII-like enzyme, Dicer, interacts with other factors to process the hairpin and yield the functional miR, which associates with multiple proteins to form the RNA-induced silencing complex (RISC). This complex then interacts with target mRNAs.

9.26 What's next?

Today, the most active areas of investigation in studies of the cell biology of the nucleus are the organization of the nucleus, the mechanism of NPC assembly, and the mechanisms by which macromolecules pass through the channel of the NPC. Controversy exists about the underlying structure of the nucleus. Does the nucleus contain a nucleoskeleton that plays a role similar to that performed by the cytoskeleton in organizing the cytoplasm and mediating intracellular transport? Extensive extraction of nuclei reveals an insoluble network of semi-organized short filaments, but what is the arrangement of these filaments in living nuclei? Are replication factories attached to an underlying nuclear structure? If so, then the DNA on which these factories operate is in motion, and the factories are fixed. Does transcription also take place in factories? Clearly, the nuclear steps in mRNA biogenesis are connected, but at the level of nuclear organization, how are various complexes (RNA polymerase, capping enzymes, splicing machinery, polyadenylation machinery, nuclease complexes to destroy improperly processed mRNAs) organized to facilitate their cooperative and often simultaneous action?

An increasing number of nuclear subcompartments or bodies have been identified immuno-histochemically, and more are likely to be discovered in the future. What is the structural organization of these bodies; what proteins are found in each; what determines which proteins can enter these bodies and under what conditions? Although some of these compartments appear to be associated with storage of factors required for RNA processing, we do not understand the movement of macromolecules

are more than 750 species of miRs. miRs have roles in the regulation of diverse pathways, including development, differentiation, programmed cell death (apoptosis), organogenesis, and cell proliferation. They function by binding to target mRNAs in the cytoplasm, in some cases blocking their translation and in others promoting their turnover.

miRs are products of RNA polymerase II transcription, and in animals they are generated by processing of a large precursor by RNase III-like enzyme complexes. Some genes encoding miRs are found in regions between protein-coding genes. Others are located within the intronic regions of protein-coding genes. Often, miR precursors contain multiple miRs and could assume a complex multiple hairpin structure.

miRs are generated from precursors by multiple processing steps. In the nucleus, the enzyme Drosha along with the Pasha protein

and macromolecular assemblies into and out of these locations. Investigators are beginning to purify these subnuclear compartments and are employing a proteomics approach to determine their composition, as has been done for the nucleolus and for nuclear speckles. This should provide clues to their functions and permit us to develop testable hypotheses about their roles.

We also do not understand how proteins are targeted to the inner and outer nuclear membranes. Translocation into the ER appears to be part of this mechanism of targeting, with some proteins destined for the inner nuclear membrane moving from the outer nuclear membrane through the curved membrane in the vicinity of the NPC (sometimes called the pore membrane) and ending up in the inner nuclear membrane. Do some proteins of the inner nuclear membrane cross through the lumen of the nuclear envelope? What signals that a protein should end up in the inner nuclear membrane?

NPCs are forty times the size of ribosomes, yet because of their symmetry, NPCs have far fewer different polypeptides than do ribosomes. NPCs are disassembled during mitosis when the nuclear envelope breaks down. Disassembly is incomplete and nucleoporins are thought to remain in subcomplexes. Progress is being made to define the mechanism for reassembling pores when the nuclear envelope forms again but there are major gaps in our knowledge.

The number of NPCs in a cell doubles during interphase. Are new ones constructed by the same mechanism that is used to reassemble pores at the end of mitosis? The relatively low abundance of nucleoporins makes this a challenging problem. In yeast, the nuclear envelope does not break down during mitosis, so NPCs must be assembled in the context of an existing double-membrane. What is the mechanism of the fusion event that brings the inner and outer membranes together to form a protein/membrane tunnel and what protein components, both of the NPC and others, are required for the fusion event? How do the biochemical and biophysical properties of the nuclear envelope contribute to NPC biogenesis and how are they regulated?

What is the overall structure of the NPC? Can an X-ray crystal structure of the NPC be determined? Several factors make this extremely unlikely. First, crystallization requires large amounts of highly purified material, and NPCs are difficult to purify, even from annulate

lamellae, which lack a nuclear lamina. Second, association with the nuclear envelope may be required in order to maintain the overall three-dimensional structure of the NPC. Although important advances have allowed scientists to solve the structures of some membrane-bound proteins, solving the structure of the NPC is much more difficult than determining the structure of soluble proteins or much smaller membrane-associated proteins. Since each NPC is usually surrounded by the nuclear membrane, it may not be possible to crystallize NPCs if membrane association is essential for it to retain its native structure. In addition, NPCs are very large; however, it has been possible to solve the structures of viruses of approximately the same size. Important advances have been made using high-resolution electron microscopy and mass spectrometry to define details of NPC structure, and a highly detailed model displaying the location within the NPC of all of its polypeptides has been generated. It is now important to develop rigorous tests of this model. To gain further understanding of NPC structure, we will likely need a combination of approaches, including further modeling determination of the structures of subcomplexes using X-ray crystallography, *in vitro* assembly of subcomplexes, and *in vivo* approaches to NPC organization such as fluorescence resonance energy transfer. An improved understanding of the structure of the NPC should provide further insight into the mechanisms that permit it to function as a gatekeeper for transport between nucleus and cytoplasm.

Although very rapid progress has occurred over the past decade in studies of nuclear transport, how macromolecules actually move through NPCs remains unknown. The findings that the FG-repeat regions of nucleoporins are natively unfolded, interact with one another, and largely fill the central channel has led to the development of models for the mechanism of movement of cargo/receptor complexes through this unusual biochemical milieu. These models now need to be tested rigorously (see *9.16 Multiple models have been proposed for the mechanism of nuclear transport* for descriptions of two of the models). One model postulates that the phenylalanine-glycine repeats interact to form an entropic barrier while another model suggests that the FG repeats form a sort of hydrogel. In both cases, karyopherins, which are known to interact with FG repeats, are thought to be uniquely able to penetrate the pore channel environment, allowing them to transport their cargoes through the NPC. Here, powerful approaches based on biophysics to examine transport of single molecules are beginning to provide some of the missing information. Today most studies on the mechanism of translocation through the NPC focus on karyopherins and their cargoes, and there is little understanding of how much larger macromolecules and complexes, including mRNPs and ribosomal subunits, traverse the pore channel.

Although some cargoes have been identified for most karyopherins, we have not determined the import signals that many karyopherins recognize. Are these signals usually primary sequences, or are structures sometimes recognized, as is the case for recognition of tRNAs by exportin-t. Proteomics coupled with bioinformatic approaches may allow us to define the complete set of cargoes for each karyopherin. In most situations where transport is regulated, it is modifications to or availability of the cargo that determines whether transport occurs. When, if ever, is transport regulated at the level of transport receptors or other components of the transport machinery? There are more karyopherins in multicellular organisms, including, for example, several distinct but closely related karyopherin α's. *S. cerevisiae* has a single karyopherin α. There is some evidence the some karyopherin alphas are tissue-specific but much more study is needed in order to understand the roles of the multiple karyopherin alphas. In the fission yeast *S. pombe*, there are two importin α's and only one is essential, suggesting that the two play distinct roles in this unicellular organism, but what those roles are is not known.

Transport of mRNAs and ribosomal subunits is more complex than transport of proteins and tRNAs. Many unique mRNA and ribosomal subunit export factors have been identified, but the functions of some are unknown. Some proteins that accompany mRNAs during export associate with RNA polymerase during transcription and associate with the mRNA as it is synthesized. In multicellular organisms, more than 20 proteins the can bind to mRNAs in the nucleus have been identified, although it is likely that not all associate with each type of mRNA. Not every mRNP is likely to contain an identical combination of bound proteins, nor is the overall structure of any one type of mRNP identical. These proteins are thought to define sites of processing and to package the mRNA for export. What overall structure do these proteins confer on mRNPs? What is the binding specificity of these mRNA-binding proteins? How is the binding of these proteins to RNA affected by which proteins are adjacent to them? Are there additional proteins that bind to the RNA-binding proteins before export? What functions do they play?

How dynamic are mRNPs during synthesis and export? Transcription, 5' capping, splicing, and 3' processing of mRNAs have all been studied *in vitro* separately and in detail. However, these processes are clearly integrated with one another and with mRNA export *in vivo*. Mechanistically, how are the processes coordinated? What actually marks an mRNA as accurately and completely processed? How are nuclear RNAs that should be degraded distinguished from those that should be exported? One possibility is that association with the splicing machinery physically restricts movement of the mRNP until processing by spliceosomes has been completed. A sort of kinetic proofreading could operate here, with nuclear exonucleases degrading any mRNA that is not released from spliceosomes within a limited amount of time. For export of ribosomal subunits, how are ribosomes recognized as ready to leave the nucleolus, and what is their mechanism of movement from the nucleolus to NPCs? It is known that at least three receptors, one of which is exportin 1, play roles in export of the large ribosomal subunit. Is the use of three receptors associated with one ribosomal subunit sufficient for movement of this very large complex through the NPC, or do more factors remain to be identified? In what ways is the interaction of the ribosomal subunit-export complex with the FG-repeat nucleoporins different from that of karyopherin-cargo complexes, where the cargoes are much smaller than ribosomal subunits?

Nuclear transport plays an important role in nuclear entry of many viruses that replicate in the nucleus. Viruses too large to pass through the NPC intact appear to be partially disassembled, but smaller viruses enter the nucleus intact. What information targets viruses to the nucleus? In some cases, only the viral nucleic acid is imported into the nucleus but little is known about the mechanism of import of viral DNA or what proteins from the cell might play important roles. Are some types of virus particles recognized by soluble receptors or other factors? Some viruses, such as adenoviruses, which are too large to enter the nucleus intact, appear to associate with the cytoplasmic face of the NPC, facilitating the entry of the viral DNA into the nucleus. It is important that the viral and NPC determinants of these interactions be defined. Is this a general strategy followed by most large viruses or for some, are the viral genome and associated proteins released into the cytoplasm followed by migration to and through the NPC? Might it be possible to develop antiviral compounds that target the nuclear entry stage of the virus life cycle while leaving cellular transport fully functional?

9.27 Summary

The nucleus is a defining feature of eukaryotic cells and contains all of the cell's chromosomes. Although it lacks membrane-bounded subcompartments, the nucleus contains distinct domains where specific functions are performed. It is bounded by a double membrane that is perforated by NPCs, the only channels for molecular trafficking into and out of the nucleus. The organization of the nucleus is dynamic. Replication and possibly mRNA transcription occur in a limited number of sites called factories. RNA processing factors cycle between storage sites and sites of transcription.

Small molecules and macromolecules smaller than ~40 kDa can diffuse through NPCs, but transport of larger macromolecules requires specific signals. Most nuclear transport is mediated by a set of related proteins called karyopherins, which recognize the specific signals and also interact with NPCs. Many proteins are imported into the nucleus, and those that shuttle are also exported.

To ensure that proteins are transported only in the proper direction, cells use a GTPase, Ran. Ran-GTP is nuclear and Ran-GDP primarily cytoplasmic. Ran-GTP interacts with import receptors to cause dissociation of transported proteins in the nucleus. It interacts cooperatively with export receptors and shuttling proteins to mediate formation of export complexes. Receptors return to their compartment of origin after delivering cargo, some as cargo for another receptor, and others on their own.

The regulation of nuclear transport is an important way in which cells control gene expression and other cellular properties. Many transcription factors are imported into the nucleus only when specific cell signaling occurs or another stimulus is present. Many mechanisms underlie regulated transport, and include requiring signaling-mediated modification (phosphorylation or dephosphorylation) of transcription factors or other proteins before they can be recognized by transport receptors. Sometimes proteins are held in the cytoplasm as part of a complex. In some cases, phosphorylation of other components of the complex leads to release of the transcription factor, which can then be transported into the nucleus.

Almost all of the RNAs produced in the nucleus function in the cytoplasm and must be exported. In general, export does not occur until all steps of nuclear RNA processing have been completed. tRNAs and microRNAs are exported by receptors that are members of the karyopherin family and are closely related to those that transport proteins. Ribosomal subunits are assembled in the nucleolus and exported through the action of several additional factors, includingCrm1. mRNAs are exported as RNA-protein complexes. Transcription, premRNA processing, and mRNA export occur in concert. Factors important for processing and export associate with RNA polymerase II during transcription, and some remain with the mRNA until export has occurred. mRNA export also requires multiple specific mRNA export factors. There is extensive proofreading within the nucleus to ensure that only accurately and completely processed RNAs are exported to the cytoplasm.

http://biology.jbpub.com/lewin/cells

To explore these topics in more detail, visit this book's Interactive Student Study Guide.

References

Alber, F., Dokudovskaya, S., Veenhoff, L. M., Zhang, W., Kipper, J., Devos, D., Suprapto, A., Karni-Schmidt, O., Williams, R., Chait, B. T., Sali, A., and Rout, M. P. 2007. The molecular architecture of the nuclear pore complex. *Nature* v. 450(7170) p. 695–701.

Boisvert, F. M., van Koningsbruggen, S., Navascues, J., and Lamond, A. I. 2007. The multifunctional nucleolus. *Nat. Rev. Mol. Cell Biol.* v. 8(7) p. 574–585.

Brown, C. R., and Silver, P. A. 2007. Transcriptional regulation at the nuclear pore complex. *Curr. Opin. Genet. Dev.* v. 17(2) p. 100–106.

Cole, C. N., and Scarcelli, J. J. 2006. Transport of messenger RNA from the nucleus to the cytoplasm. *Curr. Opin. Cell Biol.* v. 18(3) p. 299–306.

Cook, A., Bono, F., Jinek, M., and Conti, E. 2007. Structural biology of nucleocytoplasmic transport. *Annu. Rev. Biochem.* v. 76 p. 647–671.

D'Angelo, M. A., Raices, M., Panowski, S. H., and Hetzer, M. W. 2009. Age-dependent deterioration of nuclear pore complexes causes a loss of nuclear integrity in postmitotic cells. *Cell* v. 136(2) p. 284–295.

Darzacq, X., Yao, J., Larson, D. R., Causse, S. Z., Bosanac, L., de Turris, V., Ruda, V. M., Lionnet, T., Zenklusen, D., Guglielmi, T. B., Tjian, R., and Singer, R. H. 2009. Imaging transcription in living cells. *Annu Rev Biophys.* v. 38 p. 173–196.

Dernburg, A. F., and Misteli, T. 2007. Nuclear architecture—an island no more. *Dev. Cell* v. 12(3) p. 329–334.

Farny, N. G., Hurt, J. A., and Silver, P. A. 2008. Definition of global and transcript-specific mRNA export pathways in metazoans. *Genes Dev.* v. 22(1) p. 66–78.

Hopper, A. K., and Shaheen, H. H. 2008. A decade of surprises for tRNA nuclear-cytoplasmic dynamics. *Trends Cell. Biol.* v. 18(3) p. 98–104.

Jimeno, S., Luna, R., Garcia-Rubio, M., and Aguilera, A. 2006. Tho1, a novel hnRNP, and Sub2 provide alternative pathways for mRNP biogenesis in yeast THO mutants. *Mol. Cell Biol.* v. 26(12) p. 4387–4398.

Lim, R. Y., Aebi, U., and Fahrenkrog, B. 2008. Towards reconciling structure and function in the nuclear pore complex. *Histochem. Cell Biol.* v. 129(2) p. 105–116.

Stewart, M. 2007. Molecular mechanism of the nuclear protein import cycle. *Nat. Rev. Mol. Cell Biol.* v. 8(3) p. 195–208.

Terry, L. J., Shows, E. B., and Wente, S. R. 2007. Crossing the nuclear envelope: Hierarchical regulation of nucleocytoplasmic transport. *Science* v. 318(5855) p. 1412–1416.

Tran, E. J., Zhou, Y., Corbett, A. H., and Wente, S. R. 2007. The DEAD-box protein Dbp5 controls mRNA export by triggering specific RNA:protein remodeling events. *Mol. Cell* v. 28(5) p. 850–859.

Trinkle-Mulcahy, L., and Lamond, A. I. 2007. Toward a high-resolution view of nuclear dynamics. *Science* v. 318(5855) p. 1402–1407.

Worman, H. J., Fong, L. G., Muchir, A., and Young, S. G. 2009. Laminopathies and the long strange trip from basic cell biology to therapy. *J Clin Invest.* v. 119(7) p. 1825–1836.

Zenklusen, D., Larson, D. R., and Singer, R. H. 2008. Single-RNA counting reveals alternative modes of gene expression in yeast. *Nat. Struct. Mol. Biol.* v. 15(12) p. 1263–1271.

Chromatin and chromosomes

10

Benjamin Lewin
Founding Publisher/Editor, Cell Press and Virtual Text

Jocelyn E. Krebs
Associate Professor of Biological Sciences,
University of Alaska, Anchorage, AK

A HUMAN CHROMOSOME. Photo © Biophoto Associates/Photo Researchers, Inc.

CHAPTER OUTLINE

10.1 Introduction

All cellular genetic material exists as a compact mass in a relatively confined volume. In bacteria, the genetic material is seen in the form of a **nucleoid** that forms a discrete clump within the cell. In eukaryotic cells, it is seen as the mass of **chromatin** within the nucleus at interphase. The packaging of chromatin is flexible; it changes during the eukaryotic cell cycle. Interphase chromatin becomes even more tightly packaged at the time of division (mitosis or meiosis), when individual **chromosomes** become visible as discrete entities.

A chromosome is a device for segregating genetic material at cell division. The crucial structural feature by which this is accomplished is the **centromere**, visible as a constriction in the length of the chromosome under the light microscope. At a greater level of detail, the centromere can be seen to include the **kinetochore**, a structure by which it is attached to microtubules. A eukaryotic chromosome usually consists of a very long linear segment of DNA, and another crucial feature is the **telomere**, which stabilizes the ends and is extended by special mechanisms that bypass the difficulties of replicating the ends of linear DNA.

The density of DNA is high. In a bacterial nucleoid it is ~10 mg/ml, in a eukaryotic nucleus it is ~100 mg/ml, and in the head of the phage T4 virus it is >500 mg/ml. Such a concentration in solution would be equivalent to a gel of great viscosity and has implications (not fully understood) for the ability of proteins to find their binding sites on DNA. The various activities of DNA, such as replication and transcription, must be accomplished within this confined space. The organization of the material must accommodate transitions between inactive and active states. **FIGURE 10.1** shows the range of genome sizes and makes the point that they are divided into chromosomes varying greatly in DNA content.

The length of the DNA as an extended molecule would vastly exceed the dimensions of the region that contains it. Its condensed state results from its binding to basic proteins. The positive charges of these proteins neutralize the negative charges of the nucleic acid. The structure of the nucleoprotein complex is determined by the interactions of proteins that condense the DNA into a tightly coiled structure. So in contrast with the customary picture of DNA as an extended double helix, structural deformation of DNA to bend or fold it into a more compact form is the rule rather than exception.

Most of the chromatin has a relatively dispersed appearance; this material is called **euchromatin**, and it contains the active genes. Some regions of chromatin are more densely packed; this material is called **heterochromatin** and is usually not transcriptionally active.

What is the general structure of chromatin, and what is the difference between active and inactive sequences? The high overall packing ratio of the genetic material immediately suggests that DNA cannot be directly packaged into the final structure of chromatin. There must be *hierarchies* of organization. A major question concerns the *specificity* of packaging. Is the DNA folded into a *particular* pattern, or is it different in each individual copy of the genome? How does the pattern of packaging change when a segment of DNA is replicated or transcribed?

The fundamental subunit of chromatin has the same type of design in all eukaryotes. The **nucleosome** contains ~200 bp of DNA, organized by an octamer of small, basic proteins into a beadlike structure. The protein components are the **histones**. They form an interior core; the DNA lies on the surface of the particle. Nucleosomes are an invariant component of euchromatin and heterochromatin in the interphase nucleus and of mitotic chromosomes. The nucleosome provides the first level of organization. It packages 67 nm of DNA into a body of diameter 11 nm. Its components and structure are well characterized. A linear string of nucleosomes forms the "10-nm fiber."

The second level of organization is the coiling of the series of nucleosomes into a helical array to constitute the fiber of diameter ~30 nm that is found in both interphase chromatin and mitotic chromosomes (**FIGURE 10.2**). This condenses the nucleosomes by a factor of 6 to 73 per unit length.

The final packing ratio is determined by the third level of organization, the packaging of the 30-nm fiber itself. Euchromatin is about 50×

Organism	Genome (Mb)	Haploid chromosomes	Range of chromosome length (Mb DNA)	Total genes
E. coli	4.6	1	4.6	4,401
S. cerevisiae	12.1	16	(0.2) - 1.5	6,702
D. melanogaster	165	4	(1.3) - 28	14,399
Rice	389	12	24 - 45	37,544
Mouse	2,500	20	60 - 195	26,996
Man	2,900	23	49 - 245	24,194

FIGURE 10.1 The number of chromosomes in the haploid genome and the chromosome size vary extensively.

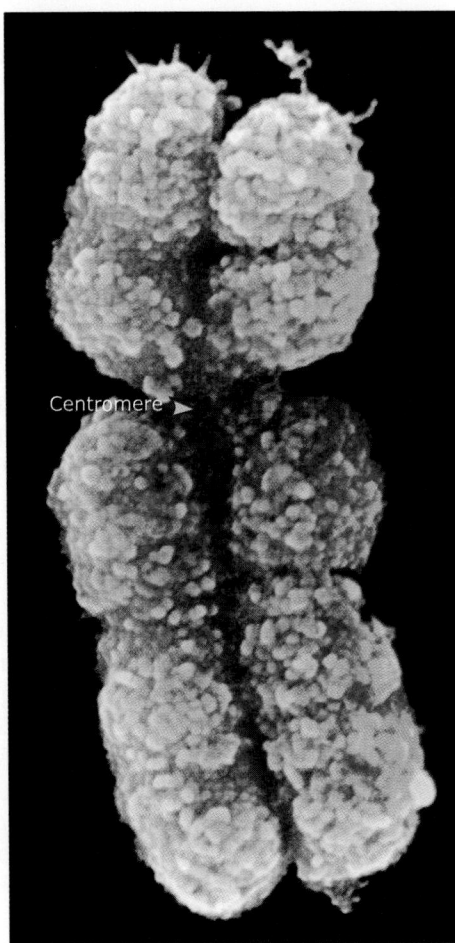

FIGURE 10.2 The sister chromatids of a mitotic pair each consist of a fiber (~30 nm in diameter) compactly folded into the chromosome. Photo © Biophoto Associates/Photo Researchers, Inc.

more condensed relative to the 30-nm fiber. Euchromatin is cyclically interchangeable with packing into mitotic chromosomes, which are about 5 to 10× more compact. Heterochromatin generally has the same packing density as mitotic chromosomes.

The mass of chromatin contains up to twice as much protein as DNA. Approximately half of the protein mass is accounted for by the nucleosomes. The mass of RNA is <10% of the mass of DNA. Much of the RNA consists of nascent transcripts still associated with the template DNA.

Changes in chromatin structure are accomplished by association with additional proteins or by modifications of existing chromosomal proteins. Both replication and transcription require unwinding of DNA and, thus, must involve an unfolding of the structure that allows the relevant enzymes to manipulate the DNA. This is likely to involve changes in all levels of organization.

The **nonhistones** include all the proteins of chromatin except the histones. They are more variable between tissues and species, and they comprise a smaller proportion of the mass than the histones. They also comprise a much larger number of proteins, so that any individual protein is present in amounts much smaller than any histone.

10.2 Chromatin is divided into euchromatin and heterochromatin

Key concepts

- Individual chromosomes can be seen only during mitosis.
- During interphase, the general mass of chromatin is in the form of euchromatin, which is less tightly packed than mitotic chromosomes.
- Regions of heterochromatin remain densely packed throughout interphase.

Each chromosome contains a single, very long duplex of DNA that is folded into a fiber that runs continuously throughout the chromosome. In accounting for interphase chromatin and mitotic chromosome structure, we have to explain the packaging of a single, exceedingly long molecule of DNA into a form in which it can be transcribed and replicated and can become cyclically more and less compressed.

Individual eukaryotic chromosomes are visible as such only during the act of cell division, when each can be seen as a compact unit. Figure 10.2 is an electron micrograph of a sister chromatid pair, captured at metaphase. (The sister chromatids are daughter chromosomes produced by the previous replication event, still joined together at this stage of mitosis.) Each consists of a fiber with a diameter of ~30 nm and a nubbly appearance. The DNA is 5 to 10× more condensed in chromosomes than in interphase chromatin.

During most of the life cycle of the eukaryotic cell, however, its genetic material occupies an area of the nucleus in which individual chromosomes cannot be distinguished. The 30-nm fiber from which chromatin is constructed is similar or identical with that of the mitotic chromosomes.

Chromatin can be divided into two types of material, which can be visualized by staining with DNA-specific dyes, as seen in the nuclear section of **FIGURE 10.3** and as given below:

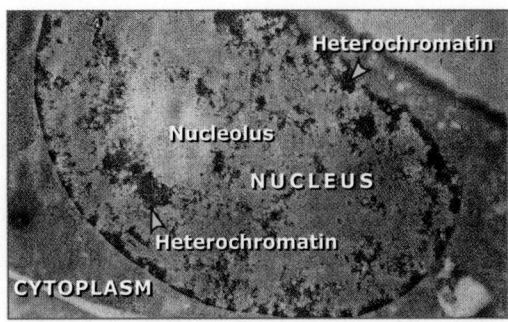

FIGURE 10.3 A thin section through a nucleus stained with a Feulgen-like material shows heterochromatin as compact regions clustered near the nucleolus and nuclear envelope. Photo courtesy of Edmund Puvion, Centre National de la Recherche Scientifique.

1. In most regions, the fibers are much less densely packed than in the mitotic chromosome. This material is called euchromatin. It has a relatively dispersed appearance in the nucleus, and occupies most of the nuclear region in Figure 10.3.
2. Some regions of chromatin are very densely packed with fibers, displaying a condition comparable to that of the chromosome at mitosis. This material is called heterochromatin. It is typically found at centromeres but occurs at other locations also. It passes through the cell cycle with relatively little change in its degree of condensation. It forms a series of discrete clumps in Figure 10.3, but often the various heterochromatic regions aggregate into a densely staining chromocenter. (This description applies to regions that are always heterochromatic, called constitutive heterochromatin; in addition, there is another sort of heterochromatin, called facultative heterochromatin, in which regions of euchromatin are converted to a heterochromatic state.)

The same fibers run continuously between euchromatin and heterochromatin, which implies that these states represent different degrees of condensation of the genetic material. In the same way, euchromatic regions exist in different states of condensation during interphase and during mitosis. So the genetic material is organized in a manner that permits alternative states to be maintained side by side in chromatin and allows cyclical changes to occur in the packaging of euchromatin between interphase and division. We discuss the molecular basis for these states later in this chapter.

The structural condition of the genetic material is correlated with its activity. The common features of constitutive heterochromatin are as follows:

- It is permanently condensed.
- It often consists of multiple repeats of a few sequences of DNA that are not transcribed or are transcribed at very low levels.
- The density of genes in this region is very much reduced compared with euchromatin and genes that are translocated into or near it are often inactivated. The one dramatic exception to this is the ribosomal DNA in the nucleolus, which has the general compacted appearance and behavior of heterochromatin (such as late replication), yet is engaged in very active transcription.
- Probably resulting from the condensed state, it replicates later than euchromatin and has a reduced frequency of genetic recombination.

We have some molecular markers for changes in the properties of the DNA and protein components (see *10.29 Heterochromatin depends on interactions with histones*). They include reduced acetylation of histone proteins, increased methylation of one histone protein, and hypermethylation of cytosine bases in DNA (see Figure 10.72). These molecular changes cause the condensation of the chromatin, which is responsible for its inactivity.

Although active genes are contained within euchromatin, only a small minority of the sequences in euchromatin are transcribed at any time. So location in euchromatin is *necessary* for gene expression (except for rRNA genes) but is not *sufficient* for it.

Concept and Reasoning Check

1. Why are genes not transcribed in heterochromatin? What do you predict happens to transcription during mitosis?

10.3 # Chromosomes have banding patterns

Key concepts

- Certain staining techniques cause the chromosomes to have the appearance of a series of striations called G-bands.
- The bands are lower in G · C content than the interbands.
- Genes are concentrated in the G · C-rich interbands.

Because of the diffuse state of chromatin, we cannot directly determine the specificity of its organization. But we can ask whether the structure of the mitotic chromosome is ordered. Do particular sequences always lie at particular sites, or is the folding of the fiber into the overall structure a more random event?

At the level of the chromosome, each member of the complement has a different and reproducible ultrastructure. When subjected to certain treatments and then stained with the chemical dye Giemsa, chromosomes generate a series of **G-bands**. **FIGURE 10.4** presents an example of the human set.

Until the development of this technique, chromosomes could be distinguished only by their overall size and the relative location of the centromere. G-banding allows each chromosome to be identified by its characteristic banding pattern. This pattern allows translocations from one chromosome to another to be identified by comparison with the original diploid set. **FIGURE 10.5** shows a diagram of the bands of the human X chromosome. The bands are large structures, each ~10^7 bp of DNA, which could include many hundreds of genes. This figure also shows the nomenclature used to identify genetic positions on individual chromosomes. A given location is indicated by its position on the long (q) or short (p) arm, then by the region of arm, the band, and subband(s).

The banding technique is of enormous practical use, but the mechanism of banding remains a mystery. All that is certain is that

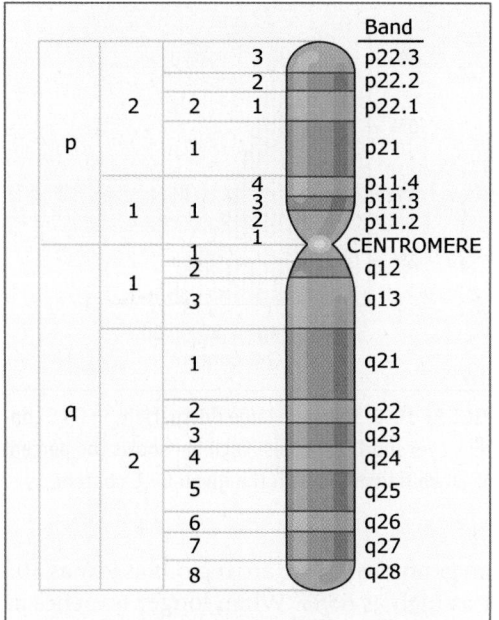

FIGURE 10.5 The human X chromosome can be divided into distinct regions by its banding pattern. The short arm is *p* and the long arm is *q*; each arm is divided into larger regions that are further subdivided. This map shows a low-resolution structure; at higher resolution, some bands are further subdivided into smaller bands and interbands; for example, *p21* is divided into *p21.1*, *p21.2*, and *p21.3*.

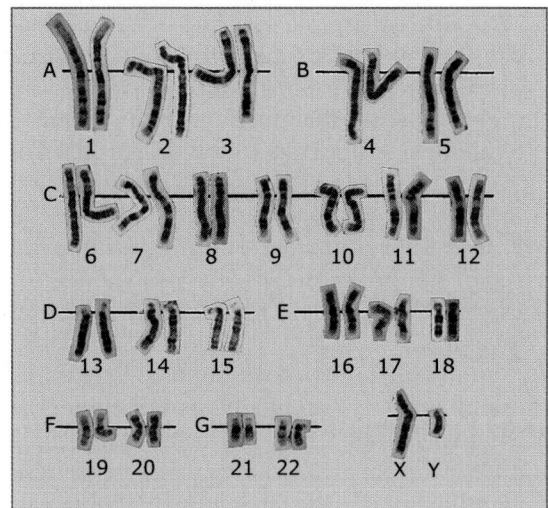

FIGURE 10.4 G-banding generates a characteristic lateral series of bands in each member of the chromosome set. Photo courtesy of Lisa Shaffer, Signature Genomic Laboratories, Spokane.

the dye stains untreated chromosomes more or less uniformly. So the generation of bands depends on a variety of treatments that change the response of the chromosome (presumably by extracting the component that binds the stain from the nonbanded regions). But similar bands can be generated by a variety of treatments.

The only known feature that distinguishes bands from interbands is that the bands have a lower G · C content than the interbands. If there are ~10 bands on a large chromosome with a total content of ~100 Mb, this means that the chromosome is divided into regions of ~5 Mb in length that alternate between low G · C (band) and high G · C (interband) content. There is a tendency for genes (as identified by hybridization with mRNAs) to be located in the interband regions. All of this argues for some long-range sequence-dependent organization.

The human genome sequence confirms the basic observation. **FIGURE 10.6** shows that there are distinct fluctuations in G · C content when the genome is divided into small tranches. The average of 41% G · C is common to mamma-

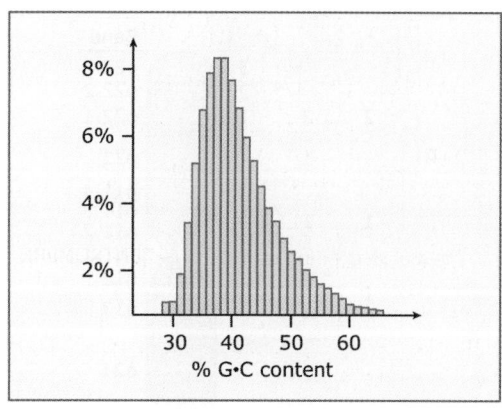

FIGURE 10.6 There are large fluctuations in G·C content over short distances. Each bar shows the percent of 20 kb fragments with the given G·C content.

lian genomes. There are regions as low as 30% or as high as 65%. When longer tranches are examined, there is less variation. The average length of regions with >43% G·C is 200 to 250 kb. This makes it clear that the band/interband structure does not represent homogeneous segments that alternate in G·C content, although the bands do contain a higher content of low G·C segments. Genes are concentrated in regions of higher G·C content. We have yet to understand how the G·C content affects chromosome structure.

Concept and Reasoning Check

1. How can G-banding be used to detect chromosomal deletions, inversions, and translocations?

10.4 Eukaryotic DNA has loops and domains attached to a scaffold

Key concepts

- DNA of interphase chromatin is negatively supercoiled into independent domains of ~85 kb.
- Metaphase chromosomes have a protein scaffold to which the loops of supercoiled DNA are attached.

The characteristic banded structure of every chromosome results from the folding of a deoxyribonucleoprotein fiber. The fiber is organized in a series of loops by a proteinaceous matrix. This looped organization can be revealed by the properties of material released by a gentle lysis of the cells.

When nuclei are lysed on top of a sucrose gradient, the eukaryotic genome can be iso-

lated as a single, compact body. As isolated from *D. melanogaster*, it can be visualized as a compactly folded fiber (10 nm in diameter), consisting of DNA bound to proteins. A revealing feature is that the DNA of the isolated chromatin behaves as a closed duplex structure, as judged by its response to ethidium bromide. This small molecule intercalates between base pairs to generate positive superhelical turns in "closed" circular DNA molecules, that is, molecules in which both strands have covalent integrity. (In "open" circular molecules, which contain a nick in one strand, or with linear molecules, the DNA can rotate freely in response to the intercalation, thus relieving the tension.)

Some nicks occur in the DNA during its isolation; they can also be generated by limited treatment with DNAse. But this does not abolish the ability of ethidium bromide to introduce positive supercoils. This capacity of the genome to retain its response to ethidium bromide even after accidental or DNAse-dependent nicking means that it must have many independent chromosomal domains, and that the supercoiling in each domain is not affected by events in the other domains. Each domain consists of a loop of DNA, the ends of which are secured in some (unknown) way that does not allow rotational events to propagate from one domain to another.

In a natural closed DNA that is negatively supercoiled, the intercalation of ethidium bromide first removes the negative supercoils and then introduces positive supercoils. The amount of ethidium bromide needed to achieve zero supercoiling is a measure of the original density of negative supercoils. In a typical eukaryotic genome, supercoiling measured by the response to ethidium bromide corresponds to about one negative supercoil per 200 bp. These supercoils can be removed by nicking with DNase, although the DNA remains in the form of the 10-nm fiber. This suggests that the supercoiling is caused by the arrangement of the fiber in space and represents the existing torsion.

Full relaxation of the supercoils requires one nick per 85 kb, identifying the average length of "closed" DNA. This region could comprise a loop or domain similar in nature to those identified in the bacterial genome. Loops can be seen directly when the majority of proteins are extracted from mitotic chromosomes. The resulting complex consists of the DNA associated with ~8% of the original protein content.

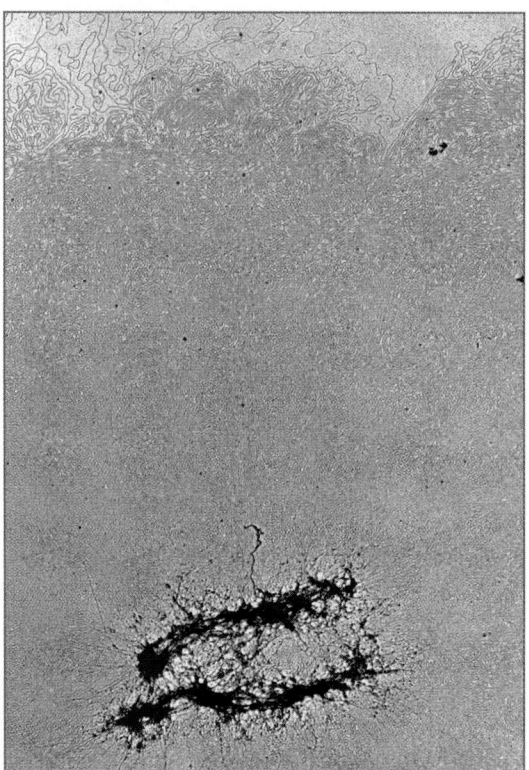

FIGURE 10.7 Histone-depleted chromosomes consist of a protein scaffold to which loops of DNA are anchored. Photo courtesy of Ulrich K. Laemmli, University of Geneva, Switzerland.

As seen in **FIGURE 10.7**, the protein-depleted chromosomes take the form of a central **scaffold** surrounded by a halo of DNA.

The metaphase scaffold consists of a dense network of fibers. Threads of DNA emanate from the scaffold, apparently as loops of average length 10 to 30 μm (30–90 kb). The DNA can be digested without affecting the integrity of the scaffold, which consists of a set of specific proteins. This suggests a form of organization in which loops of DNA of ~60 kb are anchored in a central proteinaceous scaffold.

The appearance of the scaffold resembles a mitotic pair of sister chromatids. The sister scaffolds usually are tightly connected, but sometimes are separate, joined only by a few fibers. Could this be the structure responsible for maintaining the shape of the mitotic chromosomes? Could it be generated by bringing together the protein components that usually secure the bases of loops in interphase chromatin?

Concept and Reasoning Check

1. Even though linear DNA can't be supercoiled, linear eukaryotic chromosomes have ~1 negative supercoil/200 bp. Explain.

10.5 Specific sequences attach DNA to an interphase matrix or a metaphase scaffold

Key concepts

- DNA is attached to the nuclear matrix at specific sequences called matrix attachment regions (MARs) or scaffold attachment regions (SARs).
- The MARs are A • T-rich but do not have any specific consensus sequence.

Is DNA attached to the scaffold via specific sequences? DNA sites attached to proteinaceous structures in interphase nuclei are called **MARs;** they are sometimes also called SARs, usually when referring to their attachment to the metaphase scaffold as discussed in *10.4 Eukaryotic DNA has loops and domains attached to a scaffold*. The nature of the structure in interphase cells to which they are connected is not clear. Chromatin often appears to be attached to a matrix, and there have been many suggestions that this attachment is necessary for transcription or replication. When nuclei are depleted of proteins, the DNA extrudes as loops from a residual proteinaceous structure. However, attempts to relate the proteins found in this preparation to structural elements of intact cells have not been successful.

Are particular DNA regions associated with this matrix? *In vivo* and *in vitro* approaches are summarized in **FIGURE 10.8**. Both start by isolating the matrix as a crude nuclear preparation containing chromatin and nuclear proteins. Different treatments can then be used to characterize DNA in the matrix or to identify DNA able to attach to it.

To analyze the existing MARs, the chromosomal loops can be decondensed by extracting the proteins. Removal of the DNA loops by treatment with restriction endonucleases leaves only the (presumptive) *in vivo* MAR sequences attached to the matrix.

The complementary approach is to remove all of the DNA from the matrix by treatment with DNAse. Next, isolated fragments of DNA can be tested for their ability to bind to the matrix *in vitro*.

The same sequences should be associated with the matrix *in vivo* or *in vitro*. Once a potential MAR has been identified, the size of the minimal region needed for association *in vitro* can be determined by deletions. We can also

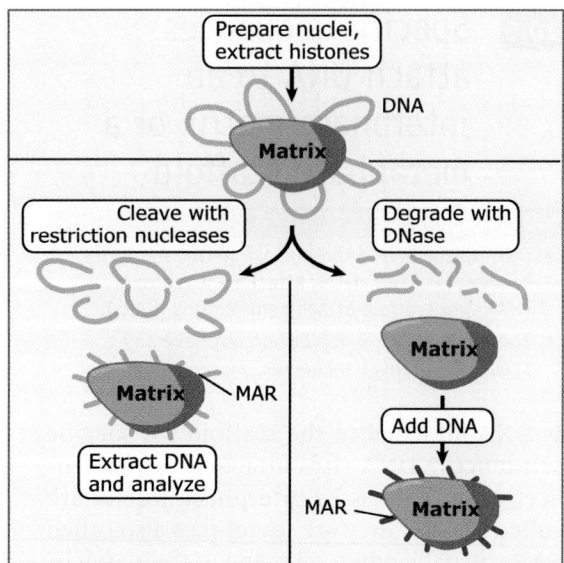

FIGURE 10.8 Matrix-associated regions may be identified by characterizing the DNA retained by the matrix isolated *in vivo* or by identifying the fragments that can bind to the matrix from which all DNA has been removed.

then identify proteins that bind to the MAR sequences.

A surprising feature is the lack of conservation of sequence in MAR fragments. They are usually ~70% A · T-rich, but otherwise lack any consensus sequences. However, other interesting sequences often are in the DNA stretch containing the MAR *cis*-acting sites that regulate transcription are common. A recognition site for topoisomerase II is usually present in the MAR. It is therefore possible that a MAR serves more than one function, not only providing a site for attachment to the matrix but also containing other sites at which topological changes in DNA are effected.

What is the relationship between the chromosome scaffold of dividing cells and the matrix of interphase cells? Are the same DNA sequences attached to both structures? In several cases, the same DNA fragments that are found with the nuclear matrix *in vivo* can be retrieved from the metaphase scaffold, and fragments that contain MAR sequences can bind to a metaphase scaffold. It therefore seems likely that DNA contains a single type of attachment site, which in interphase cells is connected to the nuclear matrix and in mitotic cells is connected to the chromosome scaffold. This single type of site is often referred to as an S/MAR. Research has shown that not all S/MARs are attached to the matrix at all times *in vivo*, and that in fact some S/MARs may dynamically associate and disassociate from the matrix depending on the transcription of genes in the area.

The nuclear matrix and chromosome scaffold consist of different proteins, although there are some common components. Topoisomerase II is a prominent component of the chromosome scaffold and is a constituent of the nuclear matrix, suggesting that the control of topology is important in both cases.

10.6 The centromere is essential for segregation

Key concepts

- A eukaryotic chromosome is held on the mitotic spindle by the attachment of microtubules to the kinetochore that forms in its centromeric region.
- Centromeres often have heterochromatin that is rich in satellite DNA sequences.

During mitosis, the sister chromatids move to opposite poles of the cell. Their movement depends on the attachment of the chromosome to microtubules, which are connected at their other end to the poles. (The microtubules constitute a cellular filamentous system, reorganized at mitosis so that they connect the chromosomes to the poles of the cell.) The sites in the two regions where microtubule ends are organized—in the vicinity of the centrioles at the poles and at the chromosomes—are called microtubule organizing centers (**MTOCs**).

FIGURE 10.9 illustrates the separation of sister chromatids as mitosis proceeds from metaphase to telophase. The region of the chromosome that is responsible for its segregation at mitosis and meiosis is called the centromere. The centromeric region on each sister chromatid is pulled by microtubules to the opposite pole, dragging its attached chromosome behind. The chromosome provides a device for attaching a large number of genes to the apparatus for division. It contains the site at which the sister chromatids are held together prior to the separation of the individual chromosomes. This shows as a constricted region connecting all four chromosome arms, as in the photograph of Figure 10.2, which shows the sister chromatids at the metaphase stage of mitosis.

The centromere is essential for segregation, as shown by the behavior of chromosomes that have been broken. A single break generates one piece that retains the centromere and another, an **acentric fragment**, which lacks

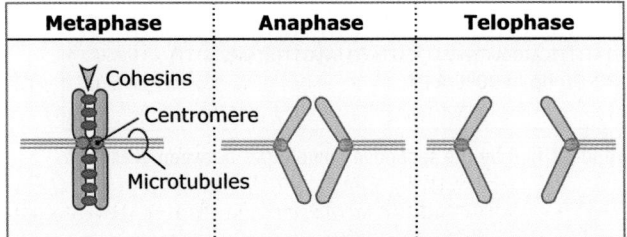

FIGURE 10.9 Chromosomes are pulled to the poles via microtubules that attach at the centromeres. The sister chromatids are held together until anaphase by glue proteins (cohesins). The centromere is shown here in the middle of the chromosome (metacentric) but can be located anywhere along its length, including close to the end (acrocentric) and at the end (telocentric).

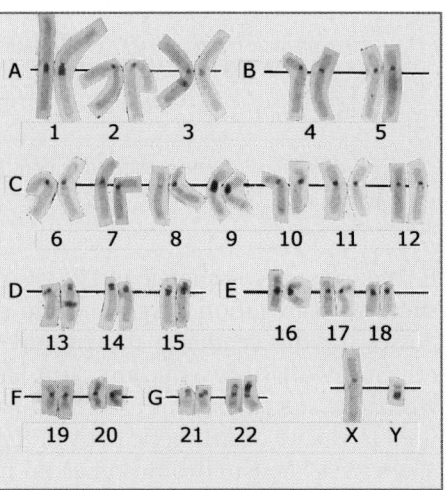

FIGURE 10.10 C-banding generates intense staining at the centromeres of all chromosomes. Photo courtesy of Lisa Shaffer, Signature Genomic Laboratories, Spokane.

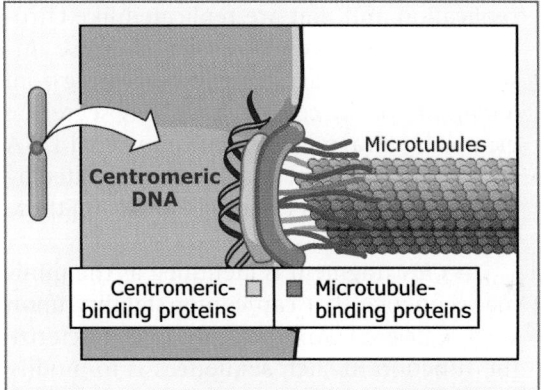

FIGURE 10.11 The centromere is identified by a DNA sequence that binds specific proteins. These proteins do not themselves bind to microtubules but establish the site at which the microtubule-binding proteins in turn bind.

it. The acentric fragment does not become attached to the mitotic spindle, and as a result, it fails to be included in either of the daughter nuclei.

The regions flanking the centromere are often rich in satellite DNA sequences and display a considerable amount of heterochromatin. Because the entire chromosome is condensed, centromeric heterochromatin is not immediately evident in mitotic chromosomes. However, it can be visualized by a technique that generates **C-bands**. In the example of FIGURE 10.10, all the centromeres show as darkly staining regions. Although it is common, heterochromatin cannot be identified around *every* known centromere, which suggests that it is unlikely to be essential for the division mechanism.

The region of the chromosome at which the centromere forms is defined by DNA sequences (although the sequences have been defined in only a very small number of cases). The centromeric DNA binds specific proteins that are responsible for establishing the structure that attaches the chromosome to the microtubules. This structure is called the kinetochore. It is a darkly staining fibrous object of diameter or length ~400 nm. The kinetochore provides the MTOC on a chromosome. FIGURE 10.11 shows the hierarchy of organization that connects centromeric DNA to the microtubules. Proteins bound to the centromeric DNA bind other proteins that bind to microtubules (see *10.8 The centromere binds a protein complex*).

When the centromeres on sister chromatids begin to be pulled to opposite poles, "glue" proteins called cohesins hold the sister chromatids together. Initially the sister chromatids separate at their centromeres, and then they are released completely from one another during anaphase when the cohesins are degraded.

Concept and Reasoning Check

1. What would happen to a chromosome that, as a result of a translocation, contains two centromeres?

10.7 Centromeres have short DNA sequences in *S. cerevisiae*

Key concepts

- *CEN* elements are identified in *S. cerevisiae* by the ability to allow a plasmid to segregate accurately at mitosis.
- *CEN* elements consist of short conserved sequences CDE-I and CDE-III that flank the T-rich region CDE-II.

```
TCACATGATGATATTTGATTTTATTATATTTTTAAAAAAAGTAAAAAATAAAAGTAGTTTATTTTTAAAAAATAAAATTTAAAATATTTCACAAAATGATTTCCGAA
AGTGTACTACTATAAACTAAAATAATATAAAAATTTTTTTCATTTTTTATTTTTCATCAAATAAAAATTTTTTATTTTAAATTTTATAAAGTGTTTTACTAAAGGCTT
```
CDE-I   *CDE-II* 80-90 bp, >90% A+T   *CDE-III*

FIGURE 10.12 Three conserved regions can be identified by the sequence homologies between yeast *CEN* elements.

If a centromeric sequence of DNA is responsible for segregation, any molecule of DNA possessing this sequence should move properly at cell division, while any DNA lacking it will fail to segregate. This prediction has been used to isolate centromeric DNA in the yeast, *S. cerevisiae*. Yeast chromosomes do not display visible kinetochores comparable to those of higher eukaryotes but otherwise divide at mitosis and segregate at meiosis by the same mechanisms.

Plasmids can be engineered in yeast by making circular DNAs that have origins of replication and that are replicated like chromosomal sequences. However, they are unstable at mitosis and meiosis, disappearing from a majority of the cells because they segregate erratically. Fragments of chromosomal DNA containing centromeres have been isolated by their ability to confer mitotic stability on these plasmids.

A *CEN* fragment is identified as the minimal sequence that can confer stability upon such a plasmid. Another way to characterize the function of such sequences is to modify them *in vitro* and then reintroduce them into the yeast cell, where they replace the corresponding centromere on the chromosome. This allows the sequences required for *CEN* function to be defined directly in the context of the chromosome.

A *CEN* fragment derived from one chromosome can replace the centromere of another chromosome with no apparent consequence. This result suggests that centromeres are interchangeable. They are used simply to attach the chromosome to the spindle and play no role in distinguishing one chromosome from another.

The sequences required for centromeric function fall within a stretch of ~120 bp. The centromeric region is packaged into a nuclease-resistant structure, and it binds a single microtubule. We may therefore look to the *S. cerevisiae* centromeric region to identify proteins that bind centromeric DNA and proteins that connect the chromosome to the spindle.

Three types of sequence element may be distinguished in the *CEN* region, as summarized in **FIGURE 10.12**. They are as follows:

1. *CDE-I* is a sequence of 9 bp that is conserved with minor variations at the left boundary of all centromeres.
2. *CDE-II* is a >90% A · T-rich sequence of 80 to 90 bp found in all centromeres; its function could depend on its length rather than exact sequence. Its constitution is reminiscent of some short tandemly repeated (satellite) DNAs.
3. *CDE-III* is an 11-bp sequence highly conserved at the right boundary of all centromeres. Sequences on either side of the element are less well conserved and may also be needed for centromeric function.

Mutations in *CDE-I* or *CDE-II* reduce but do not inactivate centromere function, but point mutations in the central CCG of *CDE-III* completely inactivate the centromere.

Concept and Reasoning Check

1. Why might the length of the *CDE-II* region, but not its precise sequence, have importance for *CEN* function?

10.8 The centromere binds a protein complex

Key concepts

- A specialized protein complex that is an alternative to the usual chromatin structure is formed at *CDE-II*.
- The CBF3 protein complex that binds to *CDE-III* is essential for centromeric function.
- The proteins that connect these two complexes may provide the connection to microtubules.

Can we identify proteins that are necessary for the function of *CEN* sequences? There are several genes in which mutations affect chromosome segregation and whose proteins are localized at centromeres. The contributions of these proteins to the centromeric structure are summarized in **FIGURE 10.13**.

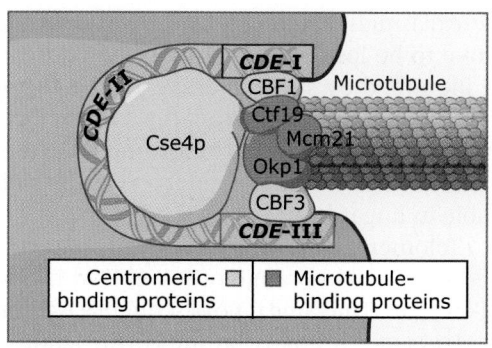

FIGURE 10.13 The DNA at *CDE-II* is wound around a protein aggregate including Cse4p, *CDE-III* is bound to CBF3, and *CDE-I* is bound to CBF1. These proteins are connected by the group of Ctf19, Mcm21, and Okp1.

A specialized chromatin structure is built by binding the CDE-II region to a protein called Cse4, which is a variant one of the histone proteins that constitute the basic subunits of chromatin (see *10.19 Histone variants produce alternative nucleosomes*). A protein called Scm3 is required for proper association of Cse4 with *CEN*. Inclusion of histone variants related to Cse4, such as CENP-A in higher eukaryotes, are a universal aspect of centromere construction in all species. A protein called Mif2 may also be part of this complex or connected to it; the higher eukaryotic homolog of Mif2, CENP-C, also localizes to centromeres, which suggests that this interaction may be a universal aspect of centromere construction. The basic interaction consists of bending the DNA of the *CDE-II* region around a protein aggregate; the reaction is probably assisted by the occurrence of intrinsic bending in the *CDE-II* sequence.

CDE-I is bound by the a homodimer of the Cbf1 protein; this interaction is not essential for centromere function, but in its absence the fidelity of chromosome segregation is reduced ~10×. A 240 kDa complex of four proteins, called CBF3, binds to *CDE-III*. This interaction is essential for centromeric function.

The proteins bound at *CDE-I* and *CDE-III* are connected to each other and to the protein structure bound at *CDE-II* by another group of proteins (Ctf19, Mcm21, and Okp1). The connection to the microtubule may be made by this complex.

The overall model suggests that the complex is localized at the centromere by a protein structure that resembles the normal building block of chromatin (the nucleosome). The bending of DNA at this structure allows proteins bound to the flanking elements to become part of a single complex. Some components of the complex (possibly not those that bind directly to DNA) link the centromere to the microtubule. The construction of kinetochores probably follows a similar pattern and uses related components in a wide variety of organisms.

10.9 Centromeres may contain repetitive DNA

Key concepts

- Centromeres in higher eukaryotic chromosomes contain large amounts of repetitive DNA.
- The function of the repetitious DNA is not known.

The length of DNA required for centromeric function is often quite long. (The short, discrete elements of *S. cerevisiae* may be an exception to the general rule.) In those cases where we can equate specific DNA sequences with the centromeric region, they usually include repetitive sequences. These consist of rather short sequences of DNA repeated many times in tandem, which have no genetic coding function.

S. cerevisiae is the only case so far in which centromeric DNA can be identified by its ability to confer stability on plasmids. However, a related approach has been used with the yeast *S. pombe*. *S. pombe* has only three chromosomes, and the region containing each centromere has been identified by deleting most of the sequences of each chromosome to create a stable minichromosome. This approach locates the centromeres within regions of 40 to 100 kb that consist largely or entirely of repetitive DNA. It is not clear how much of each of these rather long regions is required for chromosome segregation at mitosis and meiosis.

Attempts to localize centromeric functions in *Drosophila* chromosomes suggest that they are dispersed in a large region, consisting of 200 to 600 kb. The large size of this type of centromere suggests that it is likely to contain several separate specialized functions, including sequences required for kinetochore assembly, sister chromatid pairing, etc.

The size of the centromere in *Arabidopsis* is comparable. Each of the five chromosomes has a centromeric region in which recombination is very largely suppressed. This region occupies >500 kb. Clearly, it includes the centromere, but we have no direct information as to how much of it is required. There are expressed genes within these regions, which casts some doubt on whether the entire region is part of the centromere. At the center of the region is a series of 180-bp repeats; this is the type of structure generally associated with centromeres. It is too early to say how these structures relate to centromeric function.

The primary motif composing the heterochromatin of primate centromeres is the α satellite DNA, which consists of tandem arrays of a 170 bp repeating unit. There is significant variation between individual repeats, although those at any centromere tend to be better related to one another than to members of the family in other locations. It is clear that the sequences required for centromeric function reside within the blocks of α satellite DNA, but it is not clear whether the α satellite sequences themselves provide this function, or whether other sequences are embedded within the α satellite arrays.

10.10 Telomeres are replicated by a special mechanism

Key concepts
- The telomere is required for the stability of the chromosome end.
- A telomere consists of a simple repeat where a C+A-rich strand has the sequence $C_{>1}(A/T)_{1-4}$.
- The protein TRF2 catalyzes a reaction in which the 3′ repeating unit of the G+T-rich strand forms a loop by displacing its homologue in an upstream region of the telomere.

Another essential feature in all chromosomes is the telomere, which "seals" the end. We know that the telomere must be a special structure, because chromosome ends generated by breakage are "sticky" and tend to react with other chromosomes, whereas natural ends are stable.

We can apply the following two criteria in identifying a telomeric sequence:

1. It must lie at the end of a chromosome.
2. It must confer stability on a linear molecule.

The problem of finding a system that offers an assay for function again has been brought to the molecular level by using yeast. All the plasmids that survive in yeast (by virtue of possessing *ARS* and *CEN* elements) are circular DNA molecules. Linear plasmids are unstable (because they are degraded). We can identify telomeres as sequences that confer stability on

these plasmids. Fragments from yeast DNA that prove to be located at chromosome ends can be identified by such an assay. And a region from the end of a known natural linear DNA molecule—the extrachromosomal rDNA of *Tetrahymena*—is able to render a yeast plasmid stable in linear form.

Telomeric sequences have been characterized from a wide range of lower and higher eukaryotes. The same type of sequence is found in plants and humans, so the construction of the telomere seems to follow a universal principle. Each telomere consists of a long series of short, tandemly repeated sequences. There may be 100 to 1000 repeats, depending on the organism.

All telomeric sequences can be written in the general form $C_n(A/T)_m$, where $n > 1$ and $m = 1$-4. **FIGURE 10.14** shows one example (this is the telomeric sequence of the ciliate *Tetrahymena*). One unusual property of the telomeric sequence is the extension of the G-T-rich strand, usually for 14 to 16 bases as a single strand. The G-tail is probably generated by a specific limited degradation of the C-A-rich strand.

The telomere is replicated by a special mechanism. Telomerase is a ribonucleoprotein enzyme that carries a template RNA with the same sequence as the C-A-rich strand. The template RNA pairs with its complement in the telomere, which forms a primer that is extended by the enzyme's reverse transcriptase activity. The processivity of the enzyme and the number of repeats that are added are controlled by ancillary proteins.

Because DNA replication cannot start at the very end of a linear molecule, when a chromosome is replicated the number of repeats at the telomere is reduced. This can be demonstrated directly by eliminating telomerase activity. **FIGURE 10.15** shows that if telomerase is mutated in a dividing cell, the telomeres become gradually shorter with each cell division.

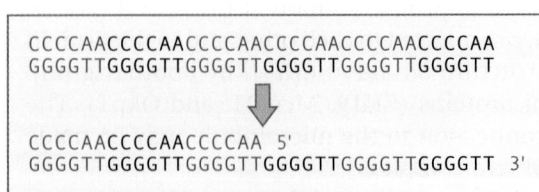

FIGURE 10.14 A typical telomere has a simple repeating structure with a G-T-rich strand that extends beyond the C-A-rich strand. The G-tail is generated by a limited degradation of the C-A-rich strand.

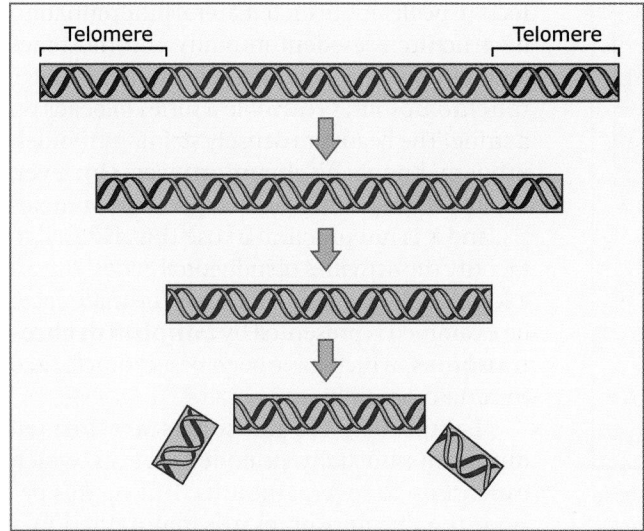

FIGURE 10.15 Mutation in telomerase causes telomeres to shorten in each cell division. Eventual loss of the telomere causes chromosome breaks and rearrangements.

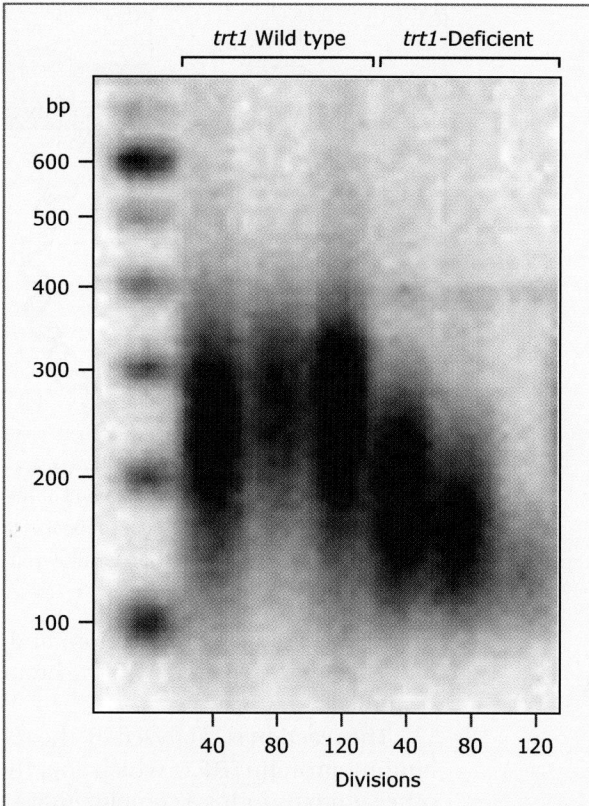

FIGURE 10.16 Telomere length is maintained at ~350 bp in wild-type yeast, but a mutant in the *trt1* gene coding for the RNA component of telomerase rapidly shortens its telomeres to zero length. Reproduced from T. M. Nakamura, et al., *Science* 277 (1997): 955–959 [http://www.sciencemag.org]. Reprinted with permission from AAAS. Photo courtesy of Thomas R. Cech, Howard Hughes Medical Institute.

An example of the effects of such a mutation in yeast is shown in FIGURE 10.16, where the telomere length shortens over ~120 generations from 400 bp to zero.

The ability of telomerase to add repeats to a telomere by *de novo* synthesis counteracts the loss of repeats resulting from failure to replicate up to the end of the chromosome. Extension and shortening are in dynamic equilibrium. If telomeres are continually being lengthened (and shortened), their exact sequence may be irrelevant. All that is required is for the end to be recognized as a suitable substrate for addition. Telomerase activity is found in all dividing cells and is generally turned off during cell differentiation in multicellular organisms. Differentiated cells divide slowly (if at all), and the telomere lengths in these cells is a measure of how many divisions they have undergone since differentiation.

In addition to providing the solution to the problem of replicating the ends of linear DNA, telomeres also stabilize the ends of the chromosome. Isolated telomeric fragments do not behave as though they contain single-stranded DNA; instead, they show aberrant electrophoretic mobility and other properties.

FIGURE 10.17 shows that a loop of DNA forms at the telomere. The absence of any free end may be the crucial feature that stabilizes the end of the chromosome. The average length of the loop in animal cells is 5 to 10 kb.

FIGURE 10.18 shows that the loop is formed when the 3′ single-stranded end of the telom-

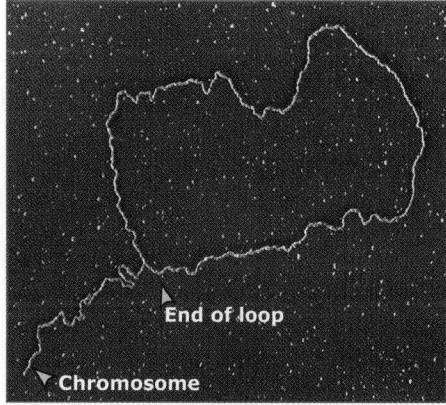

FIGURE 10.17 A loop forms at the end of chromosomal DNA. Photo courtesy of Jack Griffith, University of North Carolina at Chapel Hill.

ere (TTAGGG)$_n$ displaces the same sequence in an upstream region of the telomere. This converts the duplex region into a structure in which a series of TTAGGG repeats are displaced

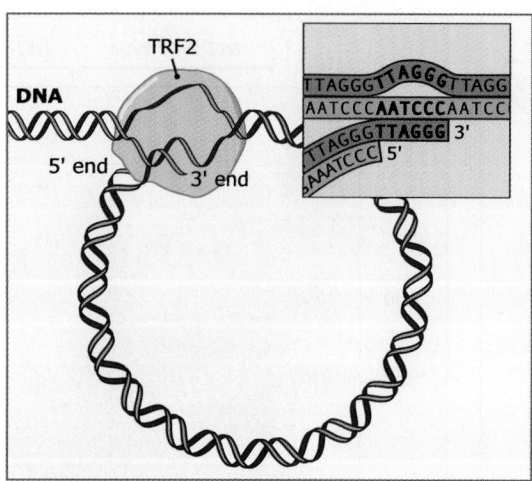

FIGURE 10.18 The 3' single-stranded end of the telomere $(TTAGGG)_n$ displaces the homologous repeats from duplex DNA to form a t-loop. The reaction is catalyzed by TRF2.

to form a single-stranded region, and the tail of the telomere is paired with the homologous strand.

The reaction is catalyzed by the telomere-binding protein TRF2, which together with other proteins forms a complex that stabilizes the chromosome ends. Its importance in protecting the ends is indicated by the fact that deletion of TRF2 causes chromosome rearrangements to occur.

Concept and Reasoning Check

1. What would be the effect on the telomere of a mutation in the telomerase template RNA that changes its sequence?

10.11 Lampbrush chromosomes are extended

Key concept

- Sites of gene expression on lampbrush chromosomes show loops that are extended from the chromosomal axis.

It would be extremely useful to visualize gene expression in its natural state, to see what structural changes are associated with transcription. The compression of DNA in chromatin, coupled with the difficulty of identifying particular genes within it, makes it impossible to visualize the transcription of individual active genes.

Gene expression can be visualized directly in certain unusual situations, in which the chromosomes are found in a highly extended form that allows individual loci (or groups of loci) to be distinguished. Lateral differentiation of structure is evident in many chromosomes when they first appear for meiosis. At this stage, the chromosomes resemble a series of beads on a string. The beads are densely staining granules, properly known as **chromomeres**. However, usually there is little gene expression at meiosis, and it is not practical to use this material to identify the activities of individual genes. But an exceptional situation that allows the material to be examined is presented by **lampbrush chromosomes**, which have been best characterized in certain amphibians.

Lampbrush chromosomes are formed during an unusually extended meiosis, which can last up to several months. During this period, the chromosomes are maintained in a stretched-out form in which they can be visualized in the light microscope. Later during meiosis, the chromosomes revert to their usual compact size. So the extended state essentially proffers an unfolded version of the normal condition of the chromosome.

The lampbrush chromosomes are meiotic bivalents, each consisting of two pairs of sister chromatids. **FIGURE 10.19** shows an example in which the sister chromatid pairs have mostly separated so that they are held together only by chiasmata. Each sister chromatid pair forms a series of ellipsoidal chromomeres, ~1 to 2 μm in diameter, which are connected by a very fine thread. This thread contains the two sister duplexes of DNA and runs continuously along the chromosome, through the chromomeres.

The lengths of the individual lampbrush chromosomes in the newt *Notophthalmus viridescens* range from 400 to 800 μm, compared with the range of 15 to 20 μm seen later in meiosis. So the lampbrush chromosomes are ~30 times less tightly packed. The total length of the entire lampbrush chromosome set is 5 to 6 mm, organized into ~5000 chromomeres.

The lampbrush chromosomes take their name from the lateral loops that extrude from

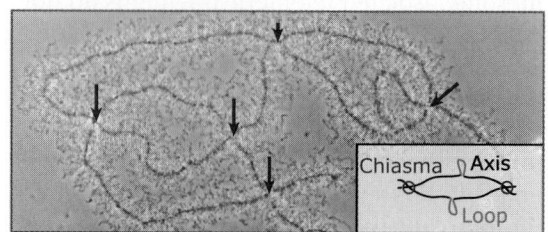

FIGURE 10.19 A lampbrush chromosome is a meiotic bivalent in which the two pairs of sister chromatids are held together at chiasmata (indicated by arrows). Photo courtesy of Joseph G. Gall, Carnegie Institution.

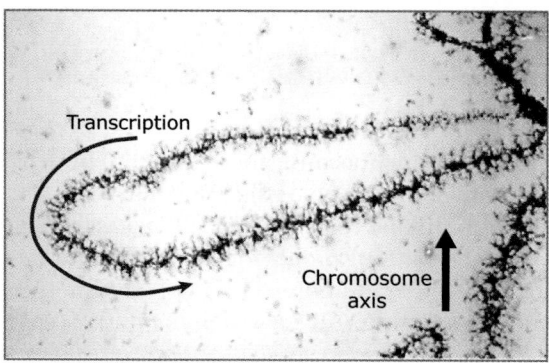

FIGURE 10.20 A lampbrush chromosome loop is surrounded by a matrix of ribonucleoprotein. Photo courtesy of Oscar Miller.

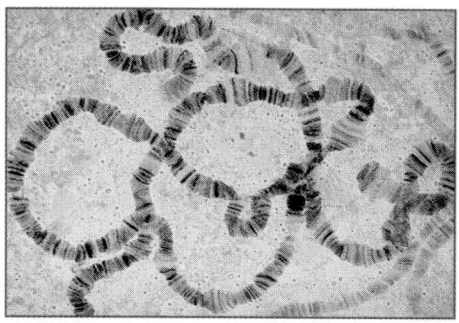

FIGURE 10.21 The polytene chromosomes of *D. melanogaster* form an alternating series of bands and interbands. Photo courtesy of José Bonner, Indiana University.

the chromomeres at certain positions. (These resemble a lampbrush, an extinct object.) The loops extend in pairs, one from each sister chromatid. The loops are continuous with the axial thread, which suggests that they represent chromosomal material extruded from its more compact organization in the chromomere.

The loops are surrounded by a matrix of ribonucleoproteins. These contain nascent RNA chains. Often a transcription unit can be defined by the increase in the length of the RNP moving around the loop. An example is shown in **FIGURE 10.20**.

So the loop is an extruded segment of DNA that is being actively transcribed. In some cases, loops corresponding to particular genes have been identified. Then the structure of the transcribed gene, and the nature of the product, can be scrutinized *in situ*.

10.12 Polytene chromosomes form bands

The interphase nuclei of some tissues of the larvae of Dipteran ("two-winged") flies contain chromosomes that are greatly enlarged relative to their usual condition. They possess both increased diameter and greater length. **FIGURE 10.21** shows an example of a chromosome set from the salivary gland of *D. melanogaster*. These are called polytene chromosomes.

Each member of the polytene set consists of a visible series of **bands** (more properly, but rarely, described as chromomeres). The bands range in size from the largest with a breadth of ~0.5 μm to the smallest of ~0.05 μm. (The smallest can be distinguished only under an electron microscope.) The bands contain most of the mass of DNA and stain intensely with appropriate reagents. The regions between them stain more lightly and are called **interbands**. There are ~5000 bands in the *D. melanogaster* set.

The centromeres of all four chromosomes of *D. melanogaster* aggregate to form a chromocenter that consists largely of heterochromatin (in the male it includes the entire Y chromosome). Allowing for this, ~75% of the haploid DNA set is organized into alternating bands and interbands. The DNA in extended form would stretch for ~40,000 μm. The length of the chromosome set is ~2000 μm, about 100× longer than the length of the haploid set at mitosis. This demonstrates vividly the extension of the genetic material relative to the usual states of interphase chromatin or mitotic chromosomes.

What is the structure of these giant chromosomes? Each is produced by the successive replications of a synapsed diploid pair. The replicas do not separate but remain attached to each other in their extended state. At the start of the process, each synapsed pair has a DNA content of 2C (where C represents the DNA content of the individual chromosome). Then this doubles up to nine times, at its maximum giving a content of 1024C. The number of doublings is different in the various tissues of the *D. melanogaster* larva. This process is known as *endoreduplication*.

Each chromosome can be visualized as a large number of parallel fibers running longitudinally, tightly condensed in the bands, less condensed in the interbands. Probably each fiber

Classical staining techniques such as G-banding (see Figure 10.4) have been used for decades to identify major chromosomal aberrations, such as large deletions, insertions, or translocations. However, these methods can only detect large changes in chromosome organization, and many smaller aberrations are undetectable. Many cancers (particularly leukemias) are characterized by specific chromosome translocations that can play causative roles in cancer development or progression. Researchers have long been interested in developing methods for sensitive mapping of chromosome aberrations across the genome.

FISH is a classic method for detecting individual genes or sets of repeated sequences in intact chromosomes (see Figure 10.22). An important breakthrough came when researchers realized that FISH did not need to be restricted to individual genes or segments of chromosomes, but could instead be applied to whole chromosomes at once. This variation of FISH, called **chromosome painting**, uses a fluorescent dye bound to DNA probes that bind all along a particular chromosome. **FIGURE 10.B1** shows an example of the use of chromosome painting to visualize a translocation between chromosomes 3 and 11 in a patient with chronic lymphocytic leukemia (CLL).

The major shortcoming of FISH and chromosome painting is that they cannot be used to study all chromosomes at the same time, because there are not enough fluorescent dyes with sufficient color differences to mark all 23 chromosomes in a unique color. This problem was solved in 1996 by labeling the painting probes for each chromosome with a different assortment of fluorescent dyes, a technique called **spectral karyotyping (SKY)**. When the fluorescent probes hybridize to a chromosome, each kind of chromosome is labeled with a different assortment of fluorescent dye com-

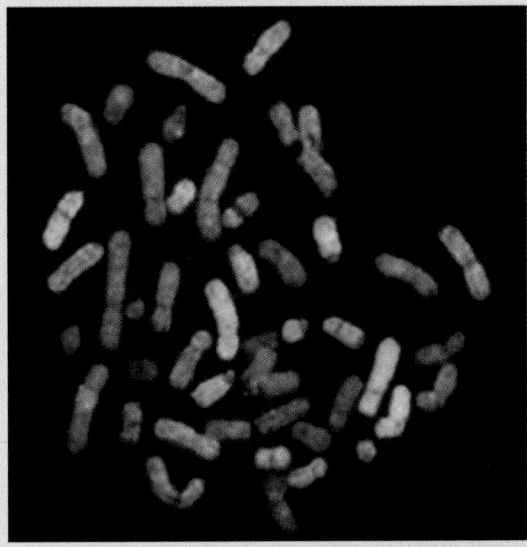

(A)

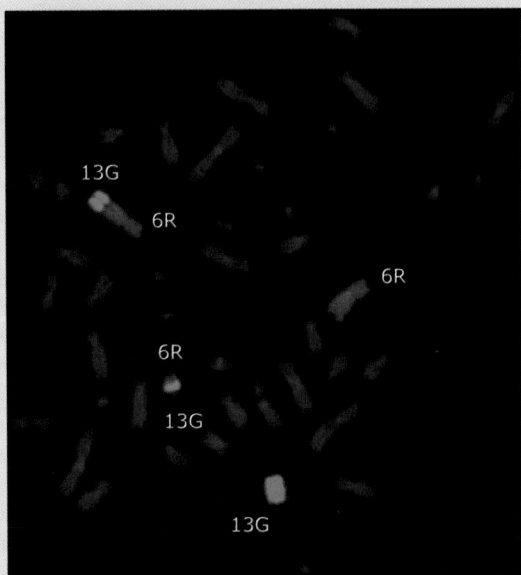

13G

6R

6R

6R

13G

13G

FIGURE 10.B1 Whole chromosome painting detects translocations between chromosome 6 (red) and chromosome 13 (green) in a patient with chronic lymphocytic leukemia. This study revealed loss of key sequences in the region of the translocation that would not be detectable by conventional banding techniques. Photo courtesy of Claudia Haferlach, Munich Leukemia Laboratory, Germany.

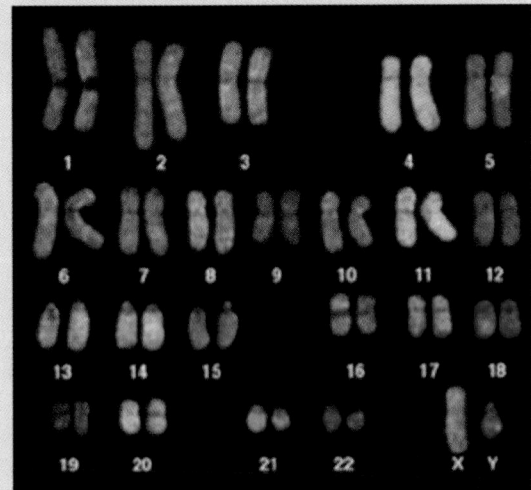

(B)

FIGURE 10.B2 SKY, an application of chromosome painting. (A) A chromosome spread hybridized with SKY fluorescent probes. (B) The same labeled chromosomes as in (A), sorted by size to show the karyotype. Photos courtesy of Johannes Wienberg, Ludwig-Maximilians-University, and Thomas Ried, National Institutes of Health.

binations. Stained chromosomes are then viewed through a series of filters, or an interferometer determines the full spectrum of light emitted by the stained chromosome. Then, a computer provides a composite picture that shows different chromosome pairs as if they were stained in different colors, as shown in **FIGURE 10.B2**.

SKY has allowed for systematic evaluation of chromosome aberrations in many types of cancers, revealing novel rearrangements, previously unrecognized translocations, and expansions or deletions of key chromosomal loci. Some of the chromosomal aberrations are highly complex, making SKY a powerful tool for unraveling these complicated rearrangements. An example is shown in **FIGURE 10.B3**, which shows the highly aberrant chromosomes from a pancreatic cancer cell line. As more and more data accumulates, SKY and other results can now be submitted to a public database (http://www.ncbi.nlm.nih.gov/sky/skyweb.cgi), which allows researchers to share results and identify recurrent patterns of aberrations that can lead to new understanding of how specific aberrations can contribute to carcinogenesis, and how ongoing chromosomal instability affects the progression of cancer.

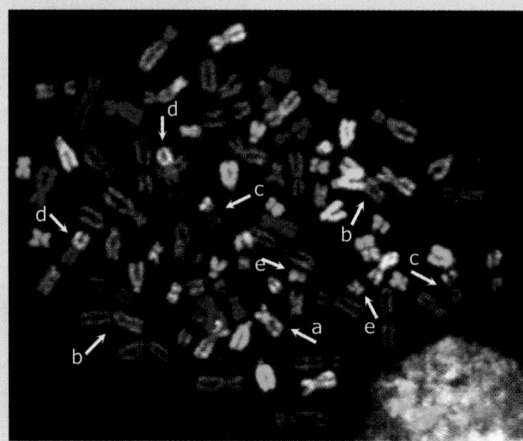

FIGURE 10.B3 SKY analysis of chromosome aberrations in a pancreatic cancer cell line. Arrows indicate numerous rearrangements, which are mostly unbalanced translocations. Reproduced from C. A. Griffin, et al., *Cytogenet. Genome Res.* 118 (2007): 148–156. Copyright 2007, S. Karger AG, Basel. Photo courtesy of Constance A. Griffin, Division of Molecular Pathology, Johns Hopkins Hospital.

represents a single (C) haploid chromosome. This gives rise to the name polytene. The degree of polyteny is the number of haploid chromosomes contained in the giant chromosome.

The banding pattern is characteristic for each strain of *Drosophila*. The constant number and linear arrangement of the bands was first noted in the 1930s, when it was realized that they form a *cytological map* of the chromosomes. Rearrangements—such as deletions, inversions, or duplications—result in alterations of the order of bands.

The linear array of bands can be equated with the linear array of genes. So genetic rearrangements, as seen in a linkage map, can be correlated with structural rearrangements of the cytological map. Ultimately, a particular mutation can be located in a particular band. Since the total number of genes in *D. melanogaster* exceeds the number of bands, there are probably multiple genes in most or all bands.

The positions of particular genes on the cytological map can be determined directly by the technique of **fluorescent *in situ* hybridization (FISH)**. The protocol is summarized in **FIGURE 10.22**. A fluorescent probe representing a gene (most often a labeled cDNA clone derived from the mRNA) is hybridized with the

denatured DNA of the polytene chromosomes *in situ*. The position or positions of the corresponding genes are then detected at a particular band or bands by visualizing the fluorescent probe using a microscope that excites the fluorescent label at a suitable wavelength. With this type of technique at hand, it is possible to determine directly the band within which a particular sequence lies. The Medical Applications box on pages 454–455 describes variations on

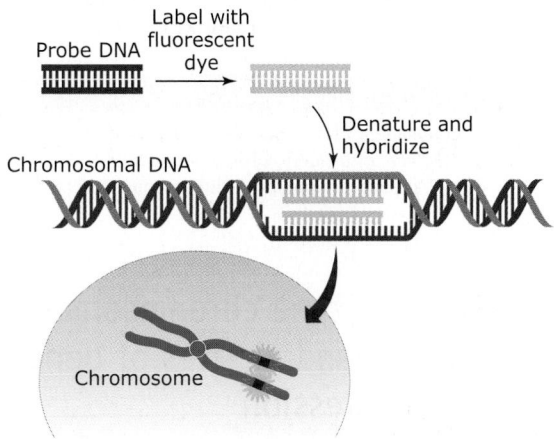

FIGURE 10.22 Fluorescence *in situ* hybridization (FISH). Adapted from an illustration by Daryl Leja, National Human Genome Research Institute (www.genome.gov).

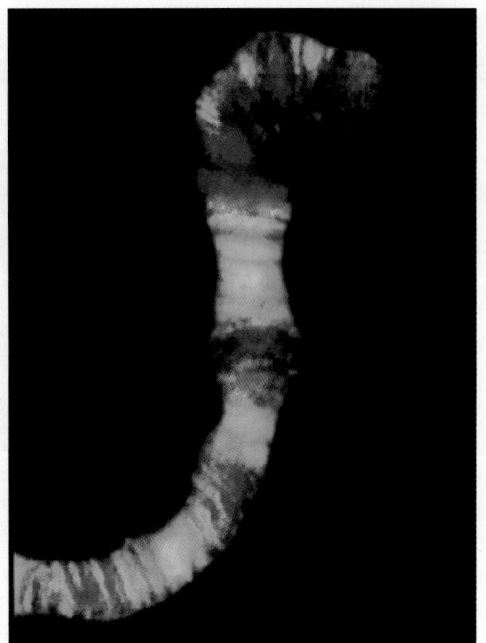

FIGURE 10.23 Detection of proteins bound to DNA *in situ*. In this example, two proteins involved in transcription are detected using antibodies that are conjugated to fluorescent dyes (red and green), while the DNA is counterstained (blue). Reproduced from *Mol. Cell. Biol.*, 2003, vol. 23, pp. 3305–3319, DOI and reproduced with permission from American Society for Microbiology. Photo courtesy of Jerry L. Workman, Howard Hughes Medical Institute.

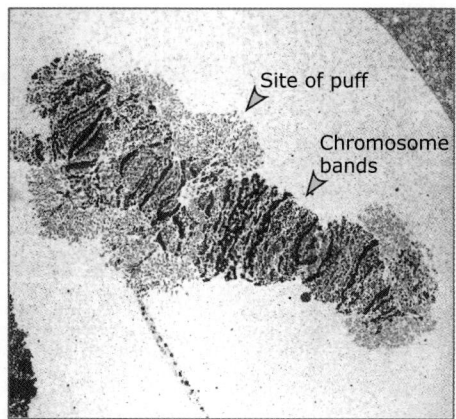

FIGURE 10.24 Chromosome IV of the insect *C. tentans* has three Balbiani rings in the salivary gland. Reprinted from *Cell*, vol. 4, B. Daneholt, Transcription in polytene chromosomes, pp. 1–9, Copyright (1975) with permission from Elsevier [http://www.sciencedirect.com/science/journal/00928674]. Photo courtesy of Bertil Daneholt, Karolinska Institutet.

FISH that allow fluorescent detection of whole chromosomes and how these methods are applied in the study of cancer. Another powerful in situ method (immunohistochemistry) allows detection of proteins bound to DNA, rather than the DNA sequences themselves. This method uses antibodies (which can also be fluorescently labelled) to detect the proteins. An example is shown in **FIGURE 10.23**.

Concept and Reasoning Check

1. If you made a FISH probe that hybridized to *Drosophila* centromeric DNA, where would the FISH signal appear in a polytene chromosome spread?

10.13 Polytene chromosomes expand at sites of gene expression

Key concept

- Bands that are sites of gene expression on polytene chromosomes expand to give "puffs".

One of the intriguing features of the polytene chromosomes is that active sites can be visualized. Some of the bands pass transiently through an expanded state in which they appear like a **puff** on the chromosome, when chromosomal material is extruded from the axis. An example of some very large puffs (called Balbiani rings) is shown in **FIGURE 10.24**.

What is the nature of the puff? It consists of a region in which the chromosome fibers unwind from their usual state of packing in the band. The fibers remain continuous with those in the chromosome axis. Puffs usually emanate from single bands, although when they are very large, as typified by the Balbiani rings, the swelling may be so extensive as to obscure the underlying array of bands.

The pattern of puffs is related to gene expression. During larval development, puffs appear and regress in a definite, tissue-specific pattern. A characteristic pattern of puffs is found in each tissue at any given time. Puffs are induced by the hormone ecdysone that controls *Drosophila* development. Some puffs are induced directly by the hormone; others are induced indirectly by the products of earlier puffs.

The puffs are sites where RNA is being synthesized. The accepted view of puffing has been that expansion of the band is a consequence of the need to relax its structure in order to synthesize RNA. Puffing has, therefore, been viewed as a consequence of transcription.

A puff can be generated by a single active gene. The sites of puffing differ from ordinary bands in accumulating additional proteins, which include RNA polymerase II and other proteins associated with transcription.

The features displayed by lampbrush and polytene chromosomes suggest a general conclusion. In order to be transcribed, the genetic material is dispersed from its usual more tightly packed state. The question to keep in mind is whether this dispersion at the gross level of the chromosome mimics the events that occur at the molecular level within the mass of ordinary interphase euchromatin.

Do the bands of a polytene chromosome have a functional significance, that is, does each band correspond to some type of genetic unit? You might think that the answer would be immediately evident from the sequence of the fly genome, since by mapping interbands to the sequence it should be possible to determine whether a band has any fixed type of identity. However, so far, no pattern has been found that identifies a functional significance for the bands.

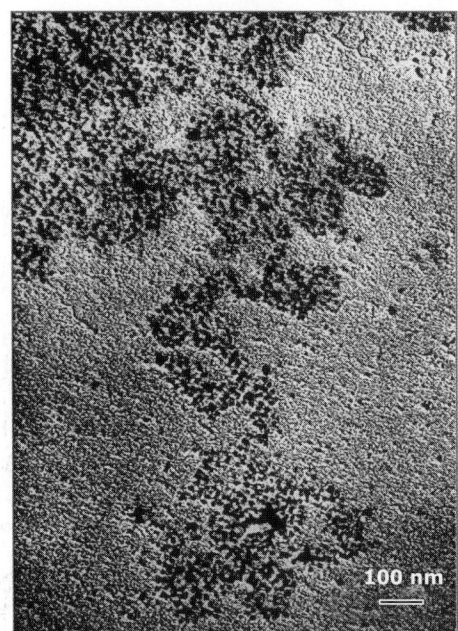

FIGURE 10.25 Chromatin spilling out of lysed nuclei consists of a compactly organized series of particles. Reprinted from *Cell,* vol. 4, P. Oudet, M. Gross-Bellard, and P. Chambon, Electron microscopic and biochemical evidence…, pp. 281–300, Copyright (1975) with permission from Elsevier [http://www.sciencedirect.com/science/journal/00928674]. Photo courtesy Pierre Chamon, College of France.

Concept and Reasoning Check

1. How do the bands on polytene chromosomes differ from the G-bands described in *10.13 Polytene chromosomes expand at sites of gene expression*?

10.14 The nucleosome is the subunit of all chromatin

Key concepts

- Micrococcal nuclease releases individual nucleosomes from chromatin as ~10-nm particles.
- A nucleosome contains ~200 bp of DNA, two copies of each core histone (H2A, H2B, H3, and H4), and one copy of H1.
- DNA is wrapped around the outside surface of the protein octamer.

Chromatin and chromosomes are constructed from a deoxyribonucleoprotein fiber that has several hierarchies of organization. Its most complex, fully folded state is seen in the banded structure of the mitotic chromosome. Its basic subunit is the same in all eukaryotes: a nucleosome consisting of ~200 bp of DNA and histone proteins. Nonhistone proteins are responsible for folding the thread of nucleosomes into the structures of higher-order fibers.

When interphase nuclei are suspended in a solution of low ionic strength, they swell and rupture to release fibers of chromatin. FIGURE 10.25 shows a lysed nucleus in which fibers are streaming out. In some regions, the fibers consist of tightly packed material, but in regions that have become stretched, they can be seen to consist of discrete particles. These are the nucleosomes. In especially extended regions, individual nucleosomes are connected by a fine thread, a free duplex of DNA. A continuous duplex thread of DNA runs through the series of particles.

Individual nucleosomes can be obtained by treating chromatin with the endonuclease **micrococcal nuclease**. It cuts the DNA thread at the junction between nucleosomes. First, it releases groups of particles; finally, it releases single nucleosomes. Individual nucleosomes can be seen in FIGURE 10.26 as compact particles of ~10 to 11 nm in diameter.

The nucleosome contains ~200 bp of DNA associated with a histone octamer that consists of two copies each of H2A, H2B, H3, and H4. These are known as the **core histones**. Their association is illustrated diagrammatically

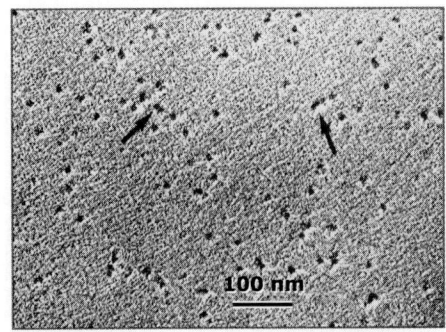

FIGURE 10.26 Individual nucleosomes are released by digestion of chromatin with micrococcal nuclease. Reprinted from *Cell*, vol. 4, P. Oudet, M. Gross-Bellard, and P. Chambon, Electron microscopic and biochemical evidence..., pp. 281–300, Copyright (1975) with permission from Elsevier [http://www.sciencedirect.com/science/journal/00928674]. Photo courtesy Pierre Chamon, College of France.

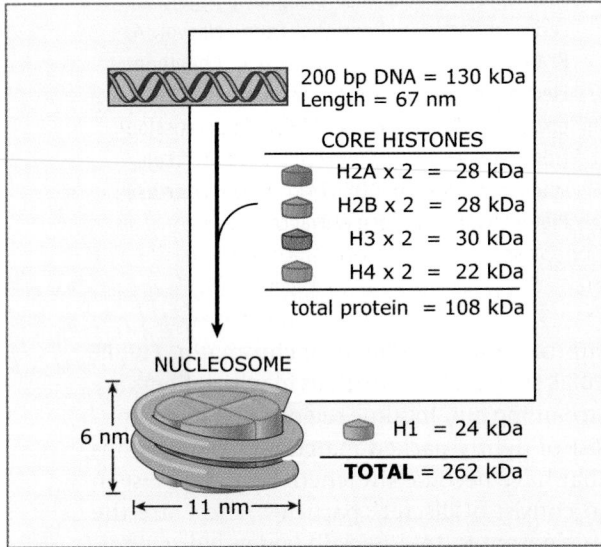

FIGURE 10.27 The nucleosome consists of approximately equal masses of DNA and histones (including H1). The predicted mass of the nucleosome is 262 kDa.

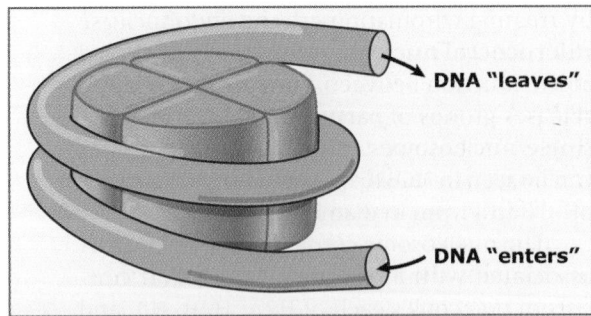

FIGURE 10.28 The nucleosome is roughly cylindrical with DNA organized into 1 ¾ turns around the surface.

in **FIGURE 10.27**. This model explains the stoichiometry of the core histones in chromatin: H2A, H2B, H3, and H4 are present in equimolar amounts, with 2 molecules of each per ~200 bp of DNA.

Histones H3 and H4 are among the most conserved proteins known. This suggests that their functions are identical in all eukaryotes. The types of H2A and H2B can be recognized in all eukaryotes but show appreciable species-specific variation in sequence.

Histone H1 comprises a set of closely related proteins that show appreciable variation between tissues and between species. The role of H1 is different from the core histones. It is present in half the amount of a core histone and can be extracted more readily from chromatin (typically with a moderate salt [0.5 M] solution). H1 can be removed without affecting the structure of the nucleosome, which suggests that its location is external to the particle. H1 and related proteins are known as **linker histones**.

The shape of the nucleosome corresponds to a flat disk or cylinder, of diameter 11 nm and height 6 nm. The length of the DNA is roughly twice the ~34 nm circumference of the particle. The DNA follows a symmetrical path around the octamer. **FIGURE 10.28** shows the DNA path diagrammatically as a helical coil that makes approximately 1¾ turns around the cylindrical octamer. Note that the DNA "enters" and "leaves" the nucleosome at points somewhat close together on one side of the nucleosome. Histone H1 may be located in this region (see *10.16 Nucleosomes have a common structure*).

Considering this model in terms of a cross section through the nucleosome, in **FIGURE 10.29** we see that the two circumferences made by the DNA lie close to one another. The height

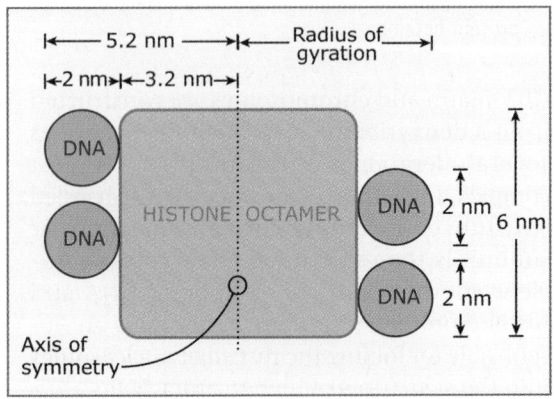

FIGURE 10.29 The two turns of DNA on the nucleosome lie close together.

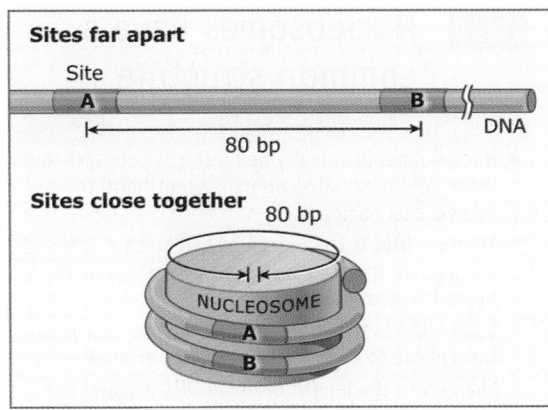

FIGURE 10.30 Sequences on the DNA that lie on different turns around the nucleosome may be close together.

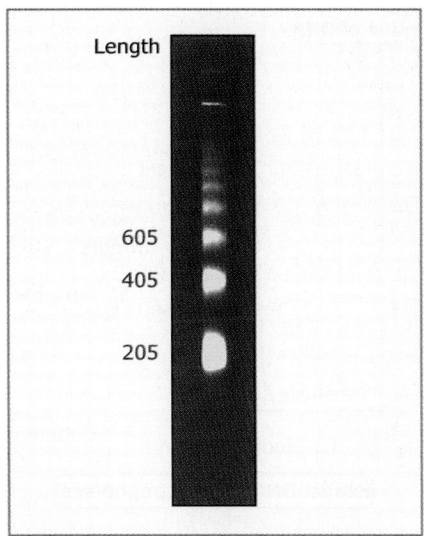

FIGURE 10.31 Micrococcal nuclease digests chromatin in nuclei into a multimeric series of DNA bands that can be separated by gel electrophoresis. Photo courtesy of Markus Noll, Universität Zürich.

of the cylinder is 6 nm, of which 4 nm is occupied by the two turns of DNA (each of diameter 2 nm).

The pattern of the two turns has a possible functional consequence. Since one turn around the nucleosome takes ~80 bp of DNA, two points separated by 80 bp in the free double helix may actually be close on the nucleosome surface, as illustrated in **FIGURE 10.30**.

10.15 DNA is coiled in arrays of nucleosomes

Key concepts

- Greater than 95% of the DNA is recovered in nucleosomes or multimers when micrococcal nuclease cleaves DNA of chromatin.
- The length of DNA per nucleosome varies for individual tissues (or species) in a range from 154 to 260 bp.

When chromatin is digested with the enzyme micrococcal nuclease, the DNA is cleaved into integral multiples of a unit length. Fractionation by gel electrophoresis reveals the "ladder" presented in **FIGURE 10.31**. Such ladders typically extend for ~10 steps, and the unit length, determined by the increments between successive steps, is ~200 bp.

FIGURE 10.32 shows that the ladder is generated by groups of nucleosomes. When nucleosomes are fractionated on a sucrose gradient, they give a series of discrete peaks that correspond to monomers, dimers, trimers, etc. When the DNA is extracted from the individual fractions and electrophoresed, each fraction yields a band of DNA whose size corresponds with a step on the micrococcal nuclease ladder.

The monomeric nucleosome contains DNA of the unit length, the nucleosome dimer contains DNA of twice the unit length, and so on.

So each step on the ladder represents the DNA derived from a discrete number of nucleosomes. We, therefore, take the existence of the 200-bp ladder in any chromatin to indicate that the DNA is organized into nucleosomes. The micrococcal ladder is generated when only ~2% of the DNA in the nucleus is rendered acid-soluble (degraded to small fragments) by the enzyme. So a small proportion of the DNA is specifically attacked; it must represent especially susceptible regions.

When chromatin is spilled out of nuclei, we often see a series of nucleosomes connected by a thread of free DNA (the beads on a string). However, the need for tight packaging of DNA *in vivo* suggests that probably there is usually little (if any) free DNA.

This view is confirmed by the fact that >95% of the DNA of chromatin can be recovered in the form of the 200-bp ladder. Almost all DNA must, therefore, be organized in nucleosomes. In their natural state, nucleosomes are likely to be closely packed, with DNA passing directly from one to the next. Free DNA is extremely rare *in vivo*, though some is probably generated by the loss of some histone octamers during isolation.

The length of DNA present in the nucleosome varies somewhat from the "typical"

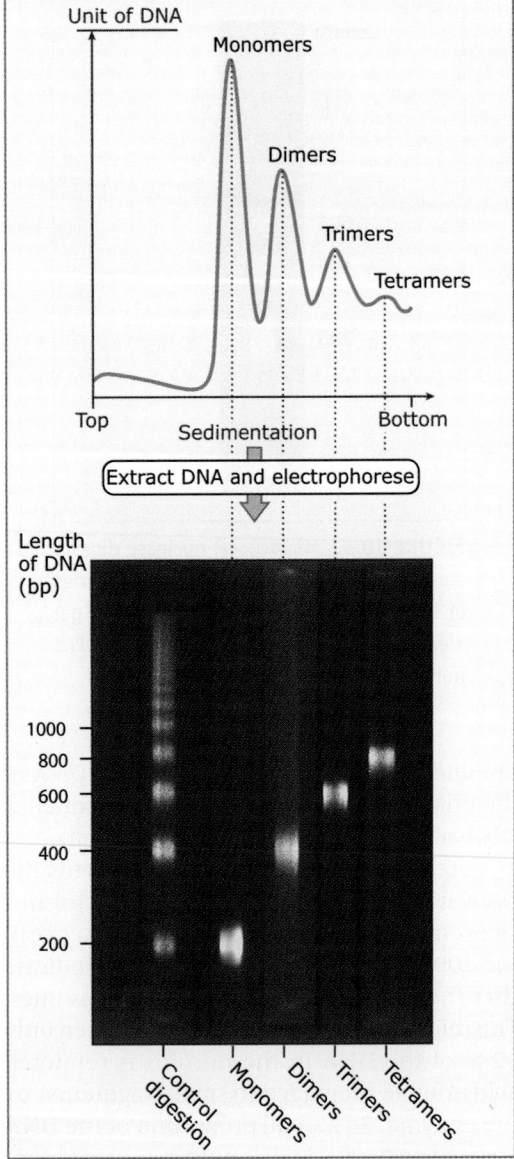

FIGURE 10.32 Each multimer of nucleosomes contains the appropriate number of unit lengths of DNA. Photo courtesy of John Finch, MRC Laboratory of Molecular Biology.

value of 200 bp. The chromatin of any particular cell type has a characteristic average value (±5 bp). The average most often is between 180 bp and 200 bp, but there are extremes as low as 154 bp (in a fungus) or as high as 260 bp (in a sea urchin sperm). The average value may be different in individual tissues of the adult organism. And there can be differences between different parts of the genome in a single cell type. Variations from the genome average include tandemly repeated sequences, such as clusters of 5S RNA genes.

10.16 Nucleosomes have a common structure

Key concepts

- Nucleosomal DNA is divided into the core DNA and linker DNA depending on its susceptibility to micrococcal nuclease.
- The core DNA is the length of 146 bp that is found on the core particles produced by prolonged digestion with micrococcal nuclease.
- Linker DNA is the region of 8 to 114 bp that is susceptible to early cleavage by the enzyme.
- Changes in the length of linker DNA account for the variation in total length of nucleosomal DNA.
- H1 is associated with linker DNA and may lie at the point where DNA enters or leaves the nucleosome.

A common structure underlies the varying amount of DNA that is contained in nucleosomes from different sources. The association of DNA with the histone octamer forms a core particle containing 146 bp of DNA, irrespective of the total length of DNA in the nucleosome. The variation in total length of DNA per nucleosome is superimposed on this basic core structure.

The amount of DNA in the core particle can be defined by the effects of micrococcal nuclease on the nucleosome monomer. The initial reaction of the enzyme is to cut between nucleosomes, but if it is allowed to continue after monomers have been generated, then it proceeds to digest some of the DNA of the individual nucleosome. This occurs by a reaction in which DNA is "trimmed" from the ends of the nucleosome.

The length of the DNA is reduced in discrete steps, as shown in **FIGURE 10.33**. With rat liver nuclei, the nucleosome monomers initially have 205 bp of DNA. Then some monomers are found in which the length of DNA has been reduced to ~165 bp. Finally this is reduced to the length of the DNA of the core particle, 146 bp. (The core is reasonably stable, but continued digestion generates a "limit digest", in which the longest fragments are the 146 bp DNA of the core, while the shortest are as small as 20 bp.)

This analysis suggests that the nucleosomal DNA can be divided into the following two regions:

1. Core DNA has an invariant length of 146 bp and is relatively resistant to digestion by nucleases.
2. Linker DNA comprises the rest of the repeating unit. Its length varies from as little as 8 bp to as much as 114 bp per nucleosome.

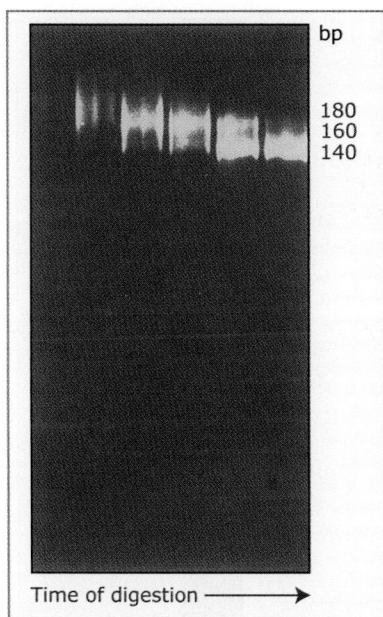

FIGURE 10.33 Micrococcal nuclease reduces the length of nucleosome monomers in discrete steps. Photo courtesy of Roger Kornberg, Stanford University.

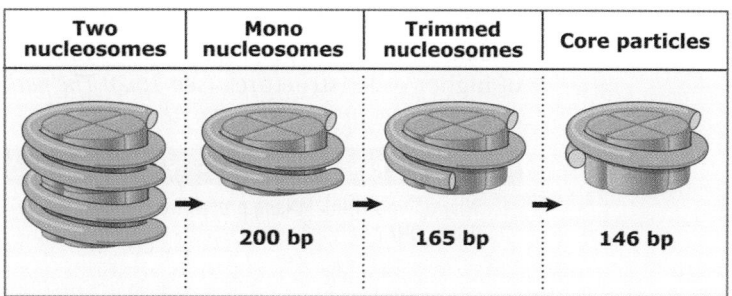

FIGURE 10.34 Micrococcal nuclease initially cleaves between nucleosomes. Mononucleosomes typically have ~200 bp DNA. End-trimming reduces the length of DNA first to ~165 bp and then generates core particles with 146 bp.

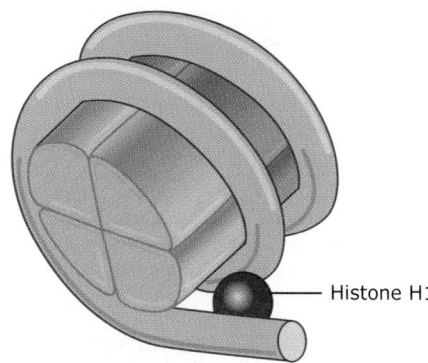

FIGURE 10.35 Model for histone H1 interaction with the nucleosome. H1 interacts with linker DNA near the entry or exit site, and may also interact with the central turn of DNA near the dyad axis of the nucleosome.

The sharp size of the band of DNA generated by the initial cleavage with micrococcal nuclease suggests that the region immediately available to the enzyme is restricted. It represents only part of each linker. (If the entire linker DNA were susceptible, the band would range from 146 bp to >200 bp). But once a cut has been made in the linker DNA, the rest of this region becomes susceptible, and it can be removed relatively rapidly by further enzyme action. The connection between nucleosomes is represented in FIGURE 10.34.

Core particles have properties similar to those of the nucleosomes themselves, although they are smaller. Their shapes and sizes are similar to nucleosomes, which suggest that the essential geometry of the particle is established by the interactions between DNA and the protein octamer in the core particle. Because core particles are more readily obtained as a homogeneous population, or can be easily assembled *in vitro*, they are often used for structural studies in preference to nucleosome preparations (see *10.18 Organization of the histone octamer*).

The existence of linker DNA depends on factors extraneous to the four core histones. Reconstitution experiments *in vitro* show that histones have an intrinsic ability to organize DNA into core particles but do not form nucleosomes with the proper unit length. The degree of supercoiling of the DNA is an important

factor. Histone H1 and/or nonhistone proteins influence the length of linker DNA associated with the histone octamer in a natural series of nucleosomes. And "assembly proteins" that are not part of the nucleosome structure are involved *in vivo* in constructing nucleosomes from histones and DNA, and these factors help control the proper spacing of nucleosomes (see *10.21 Reproduction of chromatin requires assembly of nucleosomes*).

As suggested by their names, linker histones such as H1 are thought to interact with linker DNA. H1 is lost during the degradation of nucleosome monomers. It can be retained on monomers that still have 165 bp of DNA but is always lost with the final reduction to the 146 bp core particle. This suggests that H1 could be located in the region of the linker DNA immediately adjacent to the core DNA. While the precise positioning of linker histones remains somewhat controversial, recent models suggest H1 may interact with either the entry or exit DNA in addition to the central turn of DNA on the nucleosome, as shown in FIGURE 10.35. In this position, H1 has the potential to influence the angle of DNA entry or exit, which may contribute to the formation of higher-order structures (see *10.20 The path of nucleosomes in the chromatin fiber*).

Concept and Reasoning Check

1. Why do you think MNase preferentially digests linker DNA?

10.17 DNA structure varies on the nucleosomal surface

Key concepts

- 1.65 turns of DNA are wound around the histone octamer.
- The structure of the DNA is altered so that it has an increased number of base pairs/turn in the middle, but a decreased number at the ends.
- Approximately 0.6 negative turns of DNA are absorbed by the change in bp/turn from 10.5 in solution to an average of 10.2 on the nucleosomal surface, explaining the linking number paradox.

The exposure of DNA on the surface of the nucleosome explains why it is accessible to cleavage by certain nucleases. The reaction with nucleases that attack single strands has been especially informative. These enzymes (such as DNase I) make single-strand nicks in DNA; they cleave a bond in one strand, but the other strand remains intact at this point. So no effect is visible in the double-stranded DNA. But upon denaturation, short fragments are released instead of full-length single strands. If the DNA has been labeled at its ends, the end

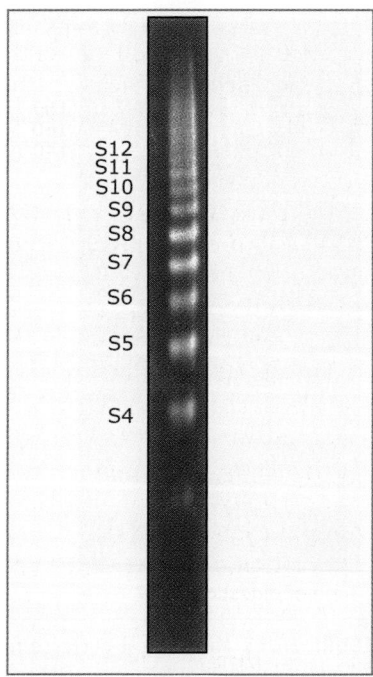

FIGURE 10.37 Sites for nicking lie at regular intervals along core DNA, as seen in a DNase I digest of nuclei. Photo courtesy of Leonard Lutter, Henry Ford Hospital.

fragments can be identified by autoradiography as summarized in **FIGURE 10.36**.

When free DNA in solution is treated with enzymes like DNase I, it is nicked (relatively) at random. The DNA on nucleosomes also can be nicked by the enzymes, *but only at regular intervals*. When the points of cutting are determined by using radioactively end-labeled DNA and then DNA is denatured and electrophoresed, a ladder of the sort displayed in **FIGURE 10.37** is obtained.

The interval between successive steps on the ladder is 10 to 11 bases. The ladder extends for the full distance of core DNA. The cleavage sites are numbered as S1 through S13 (where S1 is ~10 bases from the labeled 5′ end, S2 is ~20 bases from it, and so on).

Not all sites are cut with equal frequency: some are cut efficiently, while others are cut scarcely at all. The enzymes DNase I and DNase II generate the same ladder, although with some differences in the intensities of the bands. This shows that the pattern of cutting represents a unique series of targets in DNA, determined by its organization, with only some slight preference for particular sites imposed by the individual enzyme. The same cutting pattern is obtained by cleaving with a hydroxyl

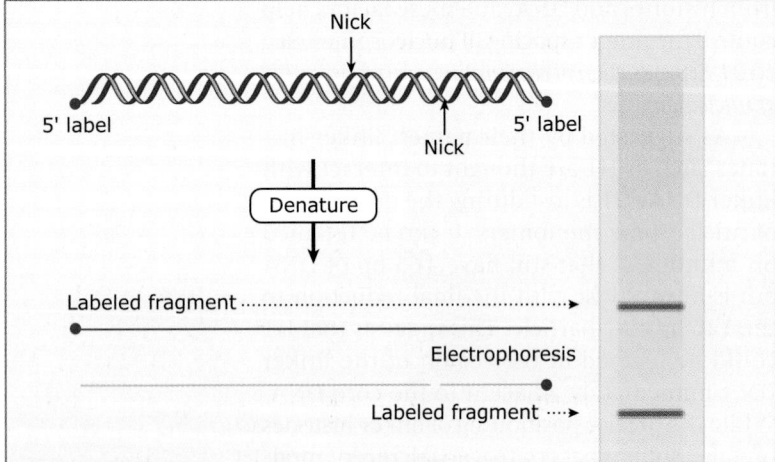

FIGURE 10.36 Nicks in double-stranded DNA are revealed by fragments when the DNA is denatured to give single strands. If the DNA is labeled at (say) 5′ ends, only the 5′ fragments are visible by autoradiography. The size of the fragment identifies the distance of the nick from the labeled end.

radical, which argues that the pattern reflects the structure of the DNA itself, rather than any sequence preference. The lack of reaction at particular target sites results from the structure of the nucleosome, in which certain positions on DNA are rendered inaccessible.

Since there are two strands of DNA in the core particle, in an end-labeling experiment both 5′ (or both 3′) ends are labeled, one on each strand. So the cutting pattern includes fragments derived from both strands. This is visible in Figure 10.36, where each labeled fragment is derived from a different strand. The corollary is that, in an experiment, each labeled band may actually represent two fragments, generated by cutting the *same* distance from *either* of the labeled ends.

What is the significance of discrete preferences at particular sites? One view is that the path of DNA on the particle is symmetrical (about a horizontal axis through the nucleosome drawn in Figure 10.28). So if (e.g.) no 80-base fragment is generated by DNase I, this must mean that the position at 80 bases from the 5′ end of *either* strand is not susceptible to the enzyme.

When DNA is immobilized on a flat surface, sites are cut with a regular separation. **FIGURE 10.38** suggests that this reflects the recurrence of the exposed site with the helical periodicity of B-form DNA (the classical form of duplex DNA discovered by Watson and Crick). The cutting periodicity (the spacing between cleavage points) coincides with—indeed, is a reflection of—the structural periodicity (the number of base pairs per turn of the double helix). So the distance between the sites corresponds to the number of base pairs per turn. Measurements of this type yield the average value for double-helical B-type DNA of 10.5 bp/turn.

A similar analysis of DNA on the surface of the nucleosome reveals striking variation in the structural periodicity at different points. At the ends of the DNA, the average distance between pairs of DNase I digestion sites is about 10.0 bases each, significantly less than the usual 10.5 bp/turn. In the center of the particle, the separation between cleavage sites averages 10.7 bases. This variation in cutting periodicity along the core DNA means that there is variation in the structural periodicity of core DNA. The DNA has more bp/turn than its solution value in the middle but has fewer bp/turn at the ends. The average periodicity over the entire nucleosome is only 10.17 bp/turn, which is significantly less than the 10.5 bp/turn of DNA in solution.

The crystal structure of the core particle suggests that DNA is organized as a flat superhelix, with 1.65 turns wound around the histone octamer (see *10.18 Organization of the histone octamer*). A high-resolution structure of the nucleosome core shows in detail how the structure of DNA is distorted. Most of the supercoiling occurs in the central 129 bp, which are coiled into 1.59 left-handed superhelical turns with a diameter of 80 Å (only 43× the diameter of the DNA duplex itself). The terminal sequences on either end make only a very small contribution to the overall curvature.

The central 129 bp are in the form of B-DNA, but with a substantial curvature that is needed to form the superhelix. The major groove is smoothly bent, but the minor groove has abrupt kinks, as shown in **FIGURE 10.39**. These conformational changes may explain why the central part of nucleosomal DNA is not usually a target for binding by regulatory proteins, which typically bind to the terminal parts of the core DNA or to the linker sequences.

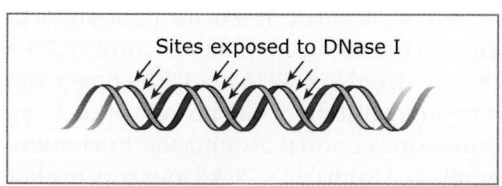

FIGURE 10.38 The most exposed positions on DNA recur with a periodicity that reflects the structure of the double helix. (For clarity, sites are shown for only one strand.)

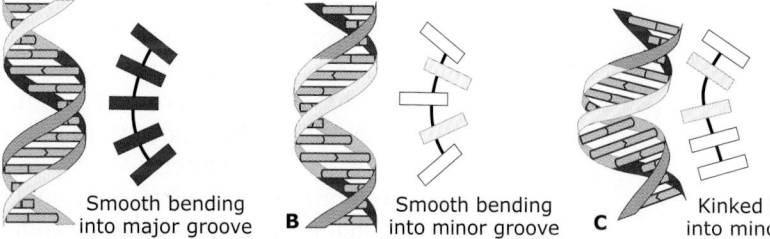

A Smooth bending into major groove **B** Smooth bending into minor groove **C** Kinked bending into minor groove

FIGURE 10.39 DNA bending in nucleosomal DNA. Structures (left) and schematic representations (right) show uniformity of curvature along the major groove (red) and both smooth and kinked bending into the minor groove (yellow). Also indicated are the DNA axes for the experimental (gold) and ideal (white) superhelices. Adapted from T. J. Richmond and C. A. Davey, *Nature* 423 (2003): 145–150.

DNA follows a path on the nucleosomal surface that generates ~1 negative supercoiled turn when the restraining protein is removed. But the path that DNA follows on the nucleosome corresponds to –1.67 superhelical turns (see Figure 10.28). This discrepancy is sometimes called the *linking number paradox*.

The discrepancy is explained by the difference between the 10.17 average bp/turn of nucleosomal DNA and the 10.5 bp/turn of free DNA. In a nucleosome of 200 bp, there are $200/10.17 = 19.67$ turns. When DNA is released from the nucleosome, it now has $200/10.5 = 19.0$ turns. The path of the less tightly wound DNA on the nucleosome absorbs –0.67 turns, and this explains the discrepancy between the physical path of –1.67 and the measurement of –1.0 superhelical turns. In effect, some of the torsional strain in nucleosomal DNA goes into increasing the number of bp/turn; only the rest is left to be measured as a supercoil.

10.18 Organization of the histone octamer

So far, we have considered the construction of the nucleosome from the perspective of how the DNA is organized on the surface. From the perspective of protein, we need to know how the histones interact with each other and with DNA. Do histones react properly only in the presence of DNA, or do they possess an independent ability to form octamers? Most of the evidence about histone-histone interactions is provided by their abilities to form stable complexes and by crosslinking experiments with the nucleosome.

The core histones form two types of complexes. H3 and H4 form a tetramer ($H3_2 \cdot H4_2$). H2A and H2B primarily form a dimer ($H2A \cdot$ H2B). Intact histone octamers can be obtained either by extraction from chromatin or by assembling histones *in vitro* under conditions of high salt and high protein concentrations. The octamer can dissociate to generate a hexamer of histones that has lost an $H2A \cdot H2B$ dimer. Then the other $H2A \cdot H2B$ dimer is lost separately, leaving the $H3_2 \cdot H4_2$ tetramer. This argues for a form of organization in which the nucleosome has a central "kernel" consisting of the $H3_2 \cdot H4_2$ tetramer. The tetramer can organize DNA *in vitro* into particles that display some of the properties of the core particle.

Structural studies show that the overall shape of the isolated histone octamer is similar to that of the core particle. This suggests that the histone-histone interactions establish the general structure. This can be seen in the space-filling model of the 3.1 Å-resolution crystal structure of the histone octamer, shown in **FIGURE 10.40**. Tracing the paths of the individual polypeptide backbones in the crystal structure suggests that the histones are not organized as individual globular proteins, but that each is interdigitated with its partner: H3 with H4 and H2A with H2B. This figure distinguishes the $H3_2 \cdot H4_2$ tetramer (white) from the $H2A \cdot H2B$ dimers (blue) but does not show individual histones.

The top view represents the same perspective that was illustrated schematically in Figure 10.40. The $H3_2 \cdot H4_2$ tetramer accounts for the diameter of the octamer. It forms the shape of a horseshoe. The $H2A \cdot H2B$ pairs fit in as two dimers, but only one can be seen in this view. The side view represents the same perspective that was illustrated in Figure 10.28. Here the responsibilities of the $H3_2 \cdot H4_2$ tetramer and of the separate $H2A \cdot H2B$ dimers can be distinguished. The protein forms a sort of spool, with a superhelical path that could correspond to the binding site for DNA, which would be wound in ~1 ¾ turns in a nucleosome. The model displays twofold symmetry about an axis that would run perpendicular through the side view.

A more detailed view of the positions of the histones (based on a crystal structure at 2.8 Å) is summarized in **FIGURE 10.41**. The upper view shows the position of one histone of each type relative to one turn around the nucleosome (numbered from 0 to +7). All four core histones show a similar type of structure in which three a-helices are connected by two loops: this is called the **histone fold**. These regions interact to form crescent-shaped heterodimers; each

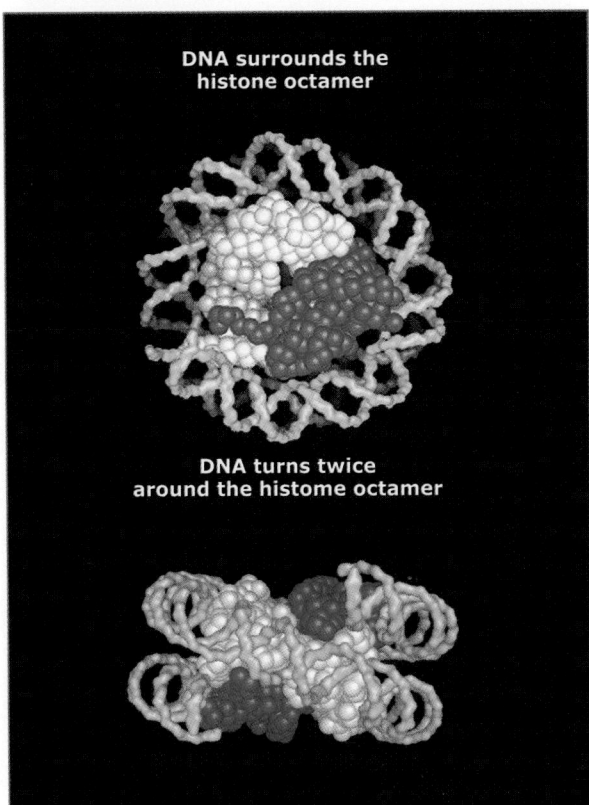

DNA surrounds the histone octamer

DNA turns twice around the histome octamer

FIGURE 10.40 The crystal structure of the histone core octamer is represented in a space-filling model with the $H3_2 \cdot H4_2$ tetramer shown in white and the H2A · H2B dimers shown in blue. Only one of the H2A · H2B dimers is visible in the top view, because the other is hidden underneath. The path of the DNA is shown in green. Photos courtesy of E. N. Moudrianakis, John Hopkins University.

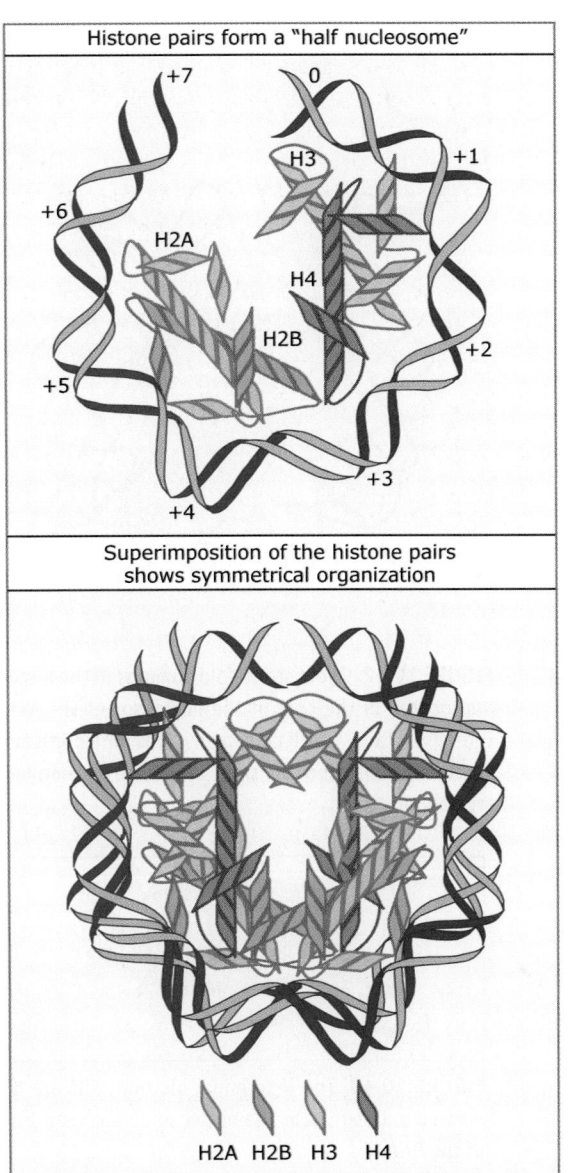

Histone pairs form a "half nucleosome"

+7 0 H3 +1 +6 H2A H4 +2 +5 H2B +3 +4

Superimposition of the histone pairs shows symmetrical organization

H2A H2B H3 H4

FIGURE 10.41 Histone positions in a top view show H3 · H4 and H2A · H2B pairs in a half nucleosome; the symmetrical organization can be seen in the superimposition of both halves.

heterodimer binds 2.5 turns of the DNA double helix (H2A · H2B binds at +3.5-+6; H3 · H4 binds at +0.5-+3 for the circumference that is illustrated). Binding is mostly to the phosphodiester backbones (consistent with the need to package any DNA irrespective of sequence). The $H3_2 \cdot H4_2$ tetramer is formed by interactions between the two H3 subunits, as can be seen in the lower part of the figure.

Each of the core histones has a globular histone core domain that contributes to the central protein mass of the nucleosome. Each histone also has a flexible N-terminal tail (H2A and H2B have C-terminal tails as well), which contains sites for modification that are important in chromatin function. The tails, which account for about one quarter of the protein mass, are too flexible to be visualized by X-ray crystallography. Therefore, their positions in the nucleosome are not well defined, and they are generally depicted schematically

as seen in **FIGURE 10.42**. However, the points at which the tails exit the nucleosome core are known, and the tails of both H3 and H2B can be seen to pass between the turns of the DNA superhelix and extend out of the nucleosome, as seen in **FIGURE 10.43**. The tails of H4 and H2A extend from both faces of the nucleosome. When histone tails are crosslinked to DNA by UV irradiation, more products are obtained with nucleosomes compared to core particles, which could mean that the tails contact the linker DNA. The tail of H4 appears to contact an H2A · H2B dimer in an adjacent

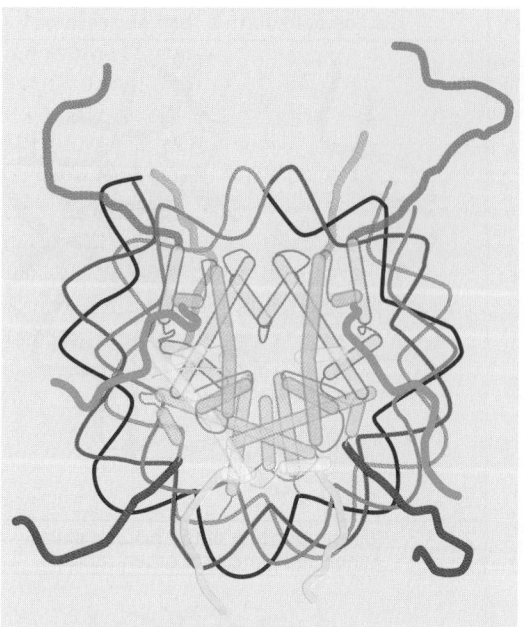

FIGURE 10.42 The histone-fold domains of the histones are located in the core of the nucleosome. The N- and C-terminal tails, which carry many sites for modification, are flexible and their positions cannot be determined by crystallography.

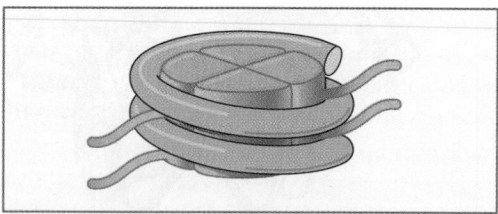

FIGURE 10.43 The N-terminal histone tails are disordered and exit from the nucleosome between turns of the DNA.

nucleosome, and the tails are required for formation of higher-order structure (see *10.20 The path of nucleosomes in the chromatin fiber*).

Concept and Reasoning Check

1. Histones contain many basic amino acids (and therefore carry a positive charge). Why?

10.19 Histone variants produce alternative nucleosomes

Key concepts

- All core histones except H4 are members of families of related variants.
- Histone variants can be closely related or highly divergent from canonical histones.
- Different variants serve different functions in the cell.

While all nucleosomes share a related core structure, some nucleosomes exhibit subtle or dramatic differences resulting from the incorporation of histone variants. Histone variants comprise a large group of histones that are related to the histones we have already discussed, but have differences in sequence from the "canonical" histones. These sequence differences can be small (as few as four amino acid differences) or extensive (such as alternative tail sequences).

Variants have been identified for all core histones except histone H4. The best-characterized histone variants are summarized in **FIGURE 10.44**. Most variants have significant differences between them, particularly in the N- and C-terminal tails. At one extreme, macroH2A is nearly three times larger than conventional H2A, and contains a large C-terminal tail that is not related to any other histone. At the other end of the spectrum, canonical H3 (also known as H3.1) differs from the H3.3 variant at only four amino acid positions, three in the histone core and one in the N-terminal tail.

Histone variants have been implicated in a number of different functions, and their incorporation changes the nature of the chromatin containing the variant. We have already discussed one type of histone variant, the centromeric H3 (or CenH3) histone, known as Cse4 in yeast and CENP-A in higher eukaryotes. CenH3 histones are incorporated into specialized nucleosomes present at centromeres in all eukaryotes (see *10.8 The centromere binds a protein complex*). In yeast, it has been shown that these centromeric nucleosomes consist of Cse4, H4, and a nonhistone protein Scm3, which replaces H2A/H2B dimers. In *Drosophila*, the centromeric chromatin appears to consist of "hemisomes" containing one copy each of CenH3, H4, H2A, and H2B.

The other major H3 variant is histone H3.3. H3.3 is expressed throughout the cell cycle, in contrast to most histones that are expressed exclusively during S phase, when new chromatin assembly is required during DNA replication. As a result, H3.3 is available for assembly at any time in the cell cycle, and is incorporated at sites of active transcription, where nucleosomes become disrupted. Because of this, H3.3 is often referred to as a "replacement" histone, in contrast to the "replicative" histone H3.1 (discussed further in *10.21 Reproduction of chromatin requires assembly of nucleosomes*).

The H2A variants are the largest and most diverse family of core histone variants, and

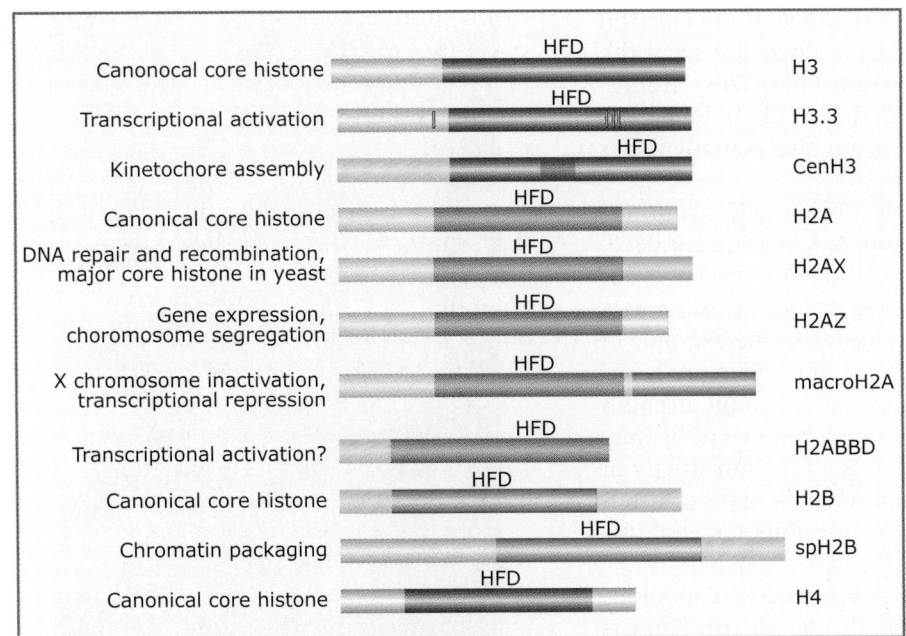

Canonocal core histone		H3
Transcriptional activation		H3.3
Kinetochore assembly		CenH3
Canonical core histone		H2A
DNA repair and recombination, major core histone in yeast		H2AX
Gene expression, choromosome segregation		H2AZ
X chromosome inactivation, transcriptional repression		macroH2A
Transcriptional activation?		H2ABBD
Canonical core histone		H2B
Chromatin packaging		spH2B
Canonical core histone		H4

FIGURE 10.44 The major core histones contain a conserved histone-fold domain. In the histone H3.3 variant, the residues that differ from the major histone H3 (also known as H3.1) are highlighted in yellow. The centromeric histone CenH3 has a unique N-terminus, which does not resemble other core histones. Most H2A variants contain alternative C-termini, except H2ABbd, which contains a distinct N-terminus. The sperm-specific SpH2B has a long N-terminus. Proposed functions of the variants are listed. Adapted from K. Sarma and D. Reinberg, *Nat. Rev. Mol. Cell Biol.* 6 (2005): 139–149.

have been implicated in a variety of distinct functions. One of the best studied is the variant H2AX. H2AX is normally present in only 10% to 15% of the nucleosomes in multicellular eukaryotes, though again (like H3.3) this subtype is the major H2A present in yeast. This variant has a C-terminal tail that is distinct from the canonical H2A, which is characterized by a SQEL/Y motif at the end. This motif is the target of phosphorylation by ATM/ATR kinases, activated by DNA damage, and this histone variant is involved in DNA repair, particularly repair of double strand breaks. H2AX phosphorylated at the SQEL/Y motif is referred to as "γ-H2AX," and is required to stabilize binding of various repair factors at DNA breaks and to maintain checkpoint arrest.

Other H2A variants have different roles. The H2AZ variant, which has ~60% sequence identity with canonical H2A, has been shown to be important in several processes, such as gene activation, heterochromatin-euchromatin boundary formation, and cell cycle progression. The vertebrate-specific macroH2A is named for its extremely long C-terminal tail, which contains a leucine-zipper dimerization motif that may mediate chromatin compaction by facilitating inter-nucleosome interactions. Mammalian macroH2A is enriched in the inactive X chromosome in females, which is assembled into a silent, heterochromatic state (see *10.30 X chromosomes undergo global changes*). In contrast, the mammalian H2ABbd variant is *excluded* from the inactive X, and forms a less

stable nucleosome than canonical H2A; perhaps this histone is designed to be more easily displaced in transcriptionally active regions of euchromatin.

Still other variants are expressed in limited tissues, such as SpH2B, present in sperm and required for chromatin compaction. The presence and distribution of histone variants shows that individual chromatin regions, entire chromosomes, or even specific tissues can have unique "flavors" of chromatin specialized for different function. In addition, the histone variants, like the canonical histones, are subject to numerous covalent modifications (see *10.27 Histone acetylation is associated with transcriptional activity*), adding levels of complexity to the roles chromatin plays in nuclear processes.

Concept and Reasoning Check

1. Why would a less stable nucleosome (such as one containing H2ABbd) be more likely to be present in euchromatin than in heterochromatin?

10.20 The path of nucleosomes in the chromatin fiber

Key concepts

- 10-nm chromatin fibers are unfolded from 30-nm fibers and consist of a string of nucleosomes.
- 30-nm fibers have 6 nucleosomes/turn, organized into a double solenoid.
- Histone H1 promotes the formation of the 30-nm fiber.

When chromatin is examined in the electron microscope, two types of fibers are seen: the 10-nm fiber and 30-nm fiber. They are described by the approximate diameter of the thread (that of the 30-nm fiber actually varies from ~25 to 30 nm).

The **10-nm fiber** is essentially a continuous string of nucleosomes. Sometimes, indeed, it runs continuously into a more stretched-out region in which nucleosomes are seen as a string of beads, as indicated in the example of **FIGURE 10.45**. The 10-nm fiber structure is obtained under conditions of low ionic strength and does not require the presence of histone H1. This means that it is a function strictly of the nucleosomes themselves. It may be visualized essentially as a continuous series of nucleosomes, as in **FIGURE 10.46**.

When chromatin is visualized in conditions of greater ionic strength, the **30-nm fiber** is obtained. An example is given in **FIGURE 10.47**. The fiber can be seen to have an underlying coiled structure. It has ~6 nucleosomes for every turn, which corresponds to a packing ratio of 40 (i.e., each μm along the axis of the

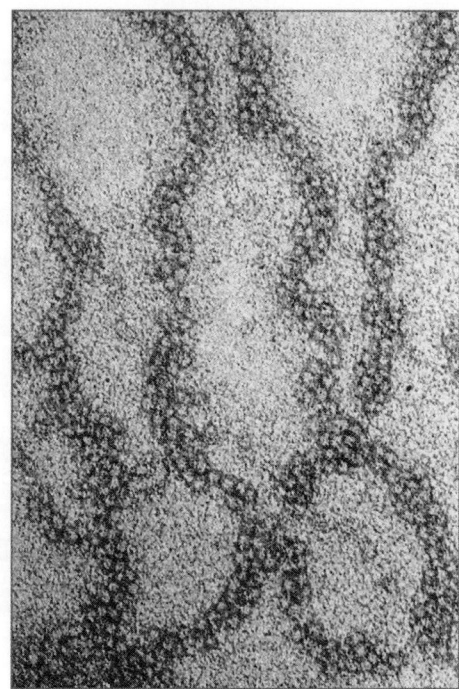

FIGURE 10.47 The 30-nm fiber has a coiled structure. Photo courtesy of Barbara A. Hamkalo, University of California, Irvine.

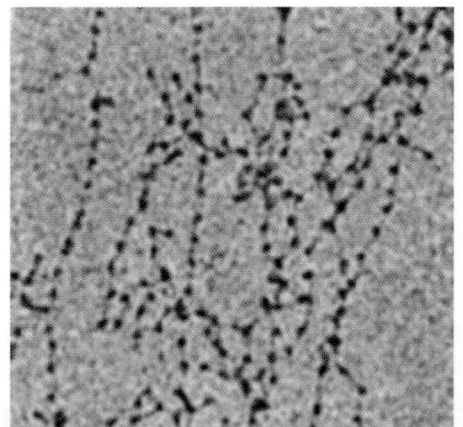

FIGURE 10.45 The 10-nm fiber in the partially unwound state can be seen to consist of a string of nucleosomes. Photo courtesy of Barbara A. Hamkalo, University of California, Irvine.

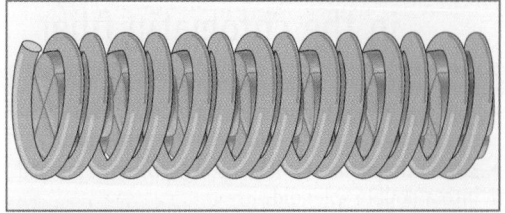

FIGURE 10.46 The 10-nm fiber is a continuous string of nucleosomes.

fiber contains 40 μm of DNA). The formation of this fiber from the 10-nm fiber requires the histone tails, which are involved in internucleosomal contacts, and is facilitated by high ionic strength and the presence of a linker histone such as H1. This fiber is thought to be the basic constituent of both interphase chromatin and mitotic chromosomes, though it has been difficult to observe this directly *in vivo*.

The most likely arrangement for packing nucleosomes into the fiber is a solenoid, in which the nucleosomes turn in a helical array, coiled around a central cavity. The two main forms of a solenoid are a single-start, which forms from a single linear array, and a two-start, which in effect consists of a double row of nucleosomes. **FIGURE 10.48** shows a two-start model suggested by crosslinking and X-ray crystallography data identifying a double stack of nucleosomes in the 30-nm fiber.

Although the presence of H1 may be necessary for the formation of the 30-nm fiber *in vivo*, information about its location is conflicting. Its relative ease of extraction from chromatin seems to argue that it is present on the outside of the superhelical fiber axis. But diffraction data, and the fact that it is harder to find in 30-nm fibers than in 10-nm fibers that retain it, would argue for an interior location.

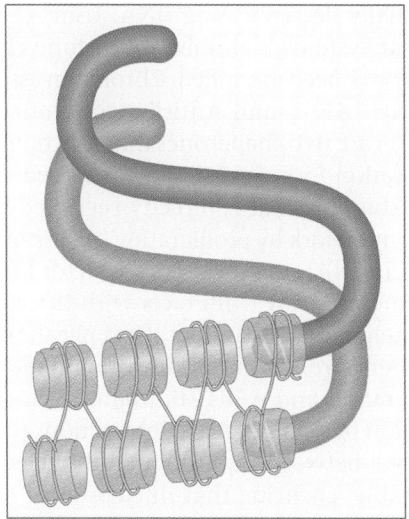

FIGURE 10.48 The 30-nm fiber is a helical ribbon consisting of two parallel rows of nucleosomes coiled into a solenoid.

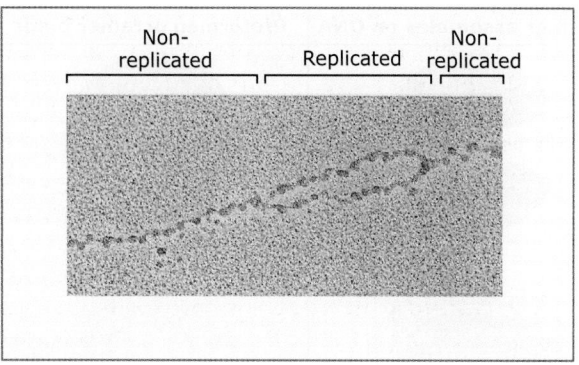

FIGURE 10.49 Replicated DNA is immediately incorporated into nucleosomes. Photo courtesy of Steven L. McKnight, UT Southwestern Medical Center.

10.21 Reproduction of chromatin requires assembly of nucleosomes

Replication separates the strands of DNA and therefore must inevitably disrupt the structure of the nucleosome. The structure of the replication fork is distinctive. It is more resistant to micrococcal nuclease and is digested into bands that differ in size from nucleosomal DNA. The region that shows this altered structure is confined to the immediate vicinity of the replication fork. This suggests that a large protein complex is engaged in replicating the DNA, but the nucleosomes reform more or less immediately behind as it moves along. This point is illustrated by the electron micrograph of FIGURE 10.49, which shows a recently replicated stretch of DNA, already packaged into nucleosomes on both daughter duplex segments.

Both biochemical analysis and visualization of the replication fork, therefore, suggest that the disruption of nucleosome structure is limited to a short region immediately around the fork. Progress of the fork disrupts nucleosomes, but they form very rapidly on the daughter duplexes as the fork moves forward. In fact, the assembly of nucleosomes is directly linked to the replisome that is replicating DNA.

How do histones associate with DNA to generate nucleosomes? Do the histones *preform* a protein octamer around which the DNA is subsequently wrapped? Or does the histone octamer assemble on DNA from free histones? FIGURE 10.50 shows that two pathways can be used *in vitro* to assemble nucleosomes, depending on the conditions that are employed. In one pathway, a preformed octamer binds to DNA. In the other pathway, a tetramer of H3$_2$ · H4$_2$ binds first, and then two H2A · H2B dimers are added. Both these pathways are related to reactions that occur *in vivo*. The first reflects the capacity of chromatin to be remodeled by moving histone octamers along DNA (see *10.26 Chromatin remodeling is an active process*). The second represents the pathway that is used in replication.

Accessory proteins assist histones to associate with DNA. Candidates for this role can be identified by using extracts that assemble histones and exogenous DNA into nucleosomes. Accessory proteins may act as "molecular chap-

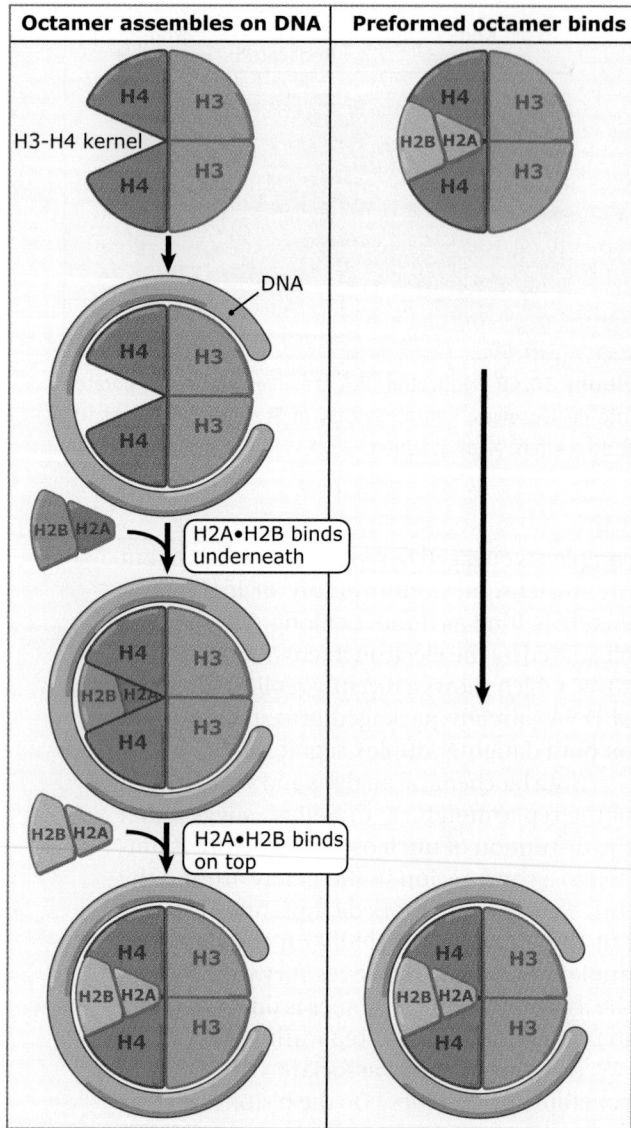

Octamer assembles on DNA	Preformed octamer binds

H3-H4 kernel

DNA

H2B•H2A

H2A•H2B binds underneath

H2B H2A

H2A•H2B binds on top

FIGURE 10.50 *In vitro*, DNA can either interact directly with an intact (crosslinked) histone octamer or can assemble with the $H3_2 \cdot H4_2$ tetramer, after which two $H2A \cdot H2B$ dimers are added.

erones" that bind to the histones in order to release either individual histones or complexes ($H3_2 \cdot H4_2$ or $H2A \cdot H2B$) to the DNA in a controlled manner. This could be necessary because the histones, as basic proteins, have a general high affinity for DNA. Such interactions allow histones to form nucleosomes without becoming trapped in other kinetic intermediates (i.e., other complexes resulting from indiscrete binding of histones to DNA).

A system that mimics nucleosome assembly during replication has been developed by using extracts of human cells that replicate SV40 DNA and assemble the products into chromatin. The assembly reaction occurs pref-

erentially on replicating DNA. Using this and similar systems, a number of histone chaperones have been identified. Chromatin assembly factor (CAF)-1 and Antisilencing function 1 (ASF1) are two chaperones that function at the replication fork. CAF-1 is a conserved 3-subunit complex that is directly recruited to the replication fork by proliferating cell nuclear antigen (PCNA), the processivity factor for DNA polymerase. ASF1 interacts with the replicative helicase that unwinds the replication fork. Furthermore, CAF-1 and ASF1 interact with each other, and ASF1 stimulates histone deposition by CAF-1. These interactions provide the link between replication and nucleosome assembly, ensuring that nucleosomes are assembled as soon as DNA has been replicated.

CAF-1 acts stoichiometrically and functions by binding to newly synthesized H3 and H4. This suggests that new nucleosomes form by assembling first the $H3_2 \cdot H4_2$ tetramer and then adding the $H2A \cdot H2B$ dimers. However, when chromatin is reproduced, a stretch of DNA *already associated with nucleosomes* is replicated, giving rise to two daughter duplexes, and the preexisting nucleosomes are displaced during this process. An important question is whether the histone octamers are dissociated into free histones for reuse, or remain assembled. The integrity of the octamer can be tested by growing cells in heavy amino acids, switching to light amino acids just before replication, and then using protein crosslinking to see whether octamers have only one type of amino acid or have a mixture. The results suggest that there is mixing of the histones made before replication with those synthesized during replication, which argues for at least some dissociation and reassociation of the components of the octamer.

The pattern of disassembly and reassembly has been difficult to characterize in detail, but our working model is illustrated in **FIGURE 10.51**. The replication fork displaces histone octamers, which then dissociate into $H3_2 \cdot H4_2$ tetramers and $H2A \cdot H2B$ dimers. These "old" tetramers and dimers enter a pool that also includes "new" tetramers and dimers, assembled from newly synthesized histones. Nucleosomes assemble ~600 bp behind the replication fork. Assembly is initiated when $H3_2 \cdot H4_2$ tetramers bind to each of the daughter duplexes, assisted by CAF-1 or ASF1. Some "old" tetramers may be transferred directly to the newly synthesized regions with the assistance of chaperones; ASF1 appears to play an important role in this

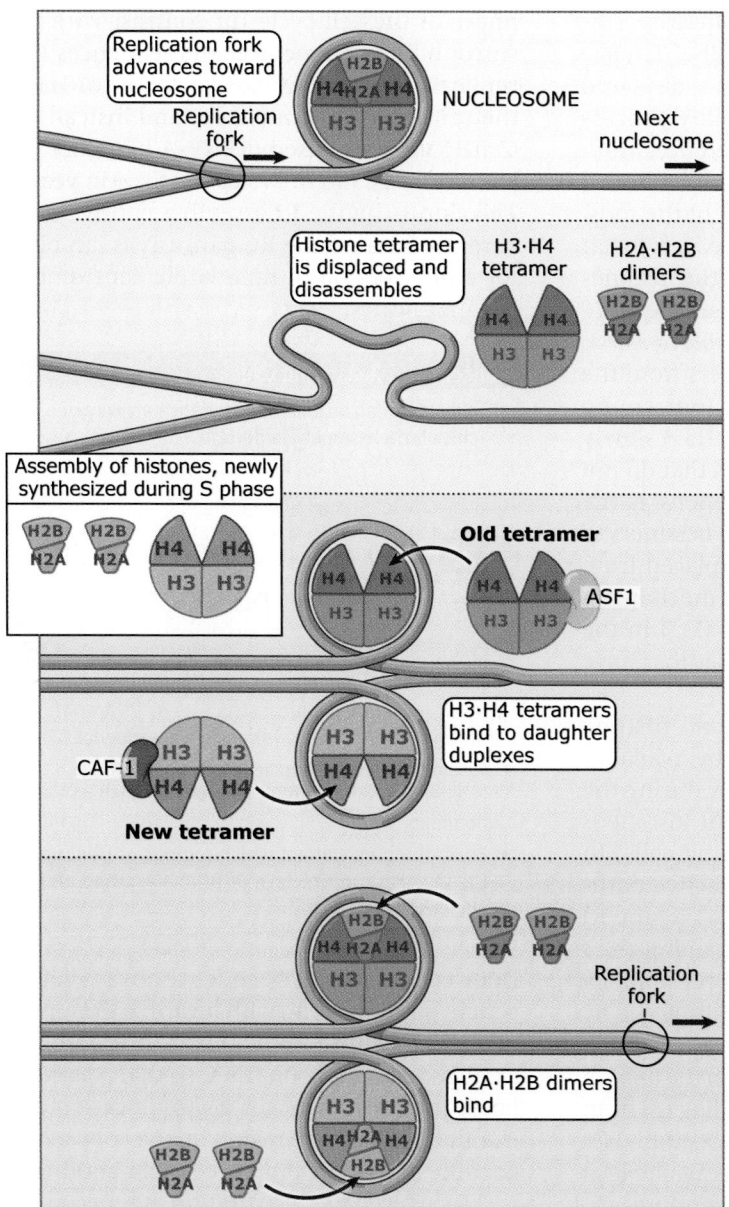

FIGURE 10.51 Replication fork passage displaces histone octamers from DNA. They disassemble into H3 · H4 tetramers and H2A · H2B dimers. Newly synthesized histones are assembled into H3 · H4 tetramers and H2A · H2B dimers. The old and new tetramers and dimers are assembled with the aid of CAF-1 and ASF1 at random into new nucleosomes immediately behind the replication fork. H2A-H2B chaperones have not been identified.

transfer of parental nucleosomes from ahead of the replication fork to the newly synthesized region behind the fork, although ASF1 can bind and assemble newly synthesized histones as well. Then two H2A · H2B dimers bind to each $H3_2 · H4_2$ tetramer to complete the histone octamer. The assembly of tetramers and dimers is random with respect to "old" and "new" subunits, explaining why old and new histones are mixed in the replicated octamers. It appears that nucleosomes are disrupted and reassembled in a similar way during transcription, using a different set of chaperones (see *10.24 Histone octamers are displaced and reassembled during transcription*).

During S phase (the period of DNA replication) in a eukaryotic cell, the duplication of chromatin requires synthesis of sufficient histone proteins to package an entire genome—basically, the same quantity of histones must be synthesized that are already contained in nucleosomes. The synthesis of histone mRNAs is controlled as part of the cell cycle and increases enormously in S phase. The pathway for assembling chromatin from this equal mix of old and new histones during S phase is called the replication-coupled (RC) pathway.

Another pathway, called the replication-independent (RI) pathway, exists for assembling nucleosomes during other phases of the cell cycle, when DNA is not being synthesized. This may become necessary as the result of damage to DNA or because nucleosomes are displaced during transcription. The assembly process must necessarily have some differences from the RC pathway, because it cannot be linked to the replication apparatus. One of the most interesting features of the RI pathway is that it uses different variants of some of the histones from those used during replication (see *10.19 Histone variants produce alternative nucleosomes*).

The histone H3.3 variant differs from the highly conserved H3 histone at four amino acid positions (see Figure 10.44). H3.3 slowly replaces H3 in differentiating cells that do not have replication cycles. This happens as the result of assembly of new histone octamers to replace those that have been displaced from DNA for whatever reason. The mechanism that is used to ensure the use of H3.3 in the RI pathway is different in two cases that have been investigated.

In the protozoan *Tetrahymena*, histone usage is determined exclusively by availability. Histone H3 is synthesized only during the cell cycle; the variant replacement histone is synthesized only in nonreplicating cells. In *Drosophila*, however, there is an active pathway that ensures the usage of H3.3 by the RI pathway. New nucleosomes containing H3.3 assemble at sites of transcription, presumably replacing nucleosomes that were displaced by RNA polymerase. The assembly process discriminates between H3 and H3.3 on the basis of their sequences, specifically excluding H3 from being utilized. By contrast, RC assembly uses both types of H3 (although H3.3 is available at much lower levels than H3 and, therefore, enters only a small proportion of nucleosomes).

CAF-1 is probably not involved in RI assembly. (And there are organisms such as yeast and *Arabidopsis* where its gene is not essential, implying that alternative assembly processes may be used in RC assembly.) A protein that may be involved in RI assembly is called HIRA. Depletion of HIRA from *in vitro* systems for nucleosome assembly inhibits the formation of nucleosomes on nonreplicated DNA but not on replicating DNA, indicating that the pathways do indeed use different assembly mechanisms.

RI assembly is also used for assembly of centromeric nucleosomes that incorporate the CenH3 variant (discussed in *10.19 Histone variants produce alternative nucleosomes*). Centromeric DNA replicates early during the replication phase of the cell cycle (in contrast with the surrounding heterochromatic sequences that replicate later). The incorporation of H3 at the centromeres is inhibited, and instead the CenH3 variant is used (CENP-A in higher eukaryotic cells, Cid in *Drosophila* Cse4 in yeast). This occurs by the RI assembly pathway, apparently because the RC pathway is inhibited for a brief period of time while centromeric DNA replicates.

Concept and Reasoning Check

1. Why is a replication-independent pathway of chromatin assembly important?

10.22 Do nucleosomes lie at specific positions?

Key concepts

- Nucleosomes may form at specific positions as the result either of the local structure of DNA or of proteins that interact with specific sequences.
- A common cause of nucleosome positioning is when proteins binding to DNA establish a boundary.
- Positioning may affect which regions of DNA are in the linker and which face of DNA is exposed on the nucleosome surface.

Does a particular DNA sequence always lie in a certain position *in vivo* with regard to the topography of the nucleosome? Or are nucleosomes arranged randomly on DNA, so that a particular sequence may occur at any location, for example, in the core region in one copy of the genome and in the linker region in another?

To investigate this question, it is necessary to use a defined sequence of DNA; more precisely, we need to determine the position relative to the nucleosome of a defined point in the DNA. **FIGURE 10.52** illustrates the principle of a procedure used to achieve this.

Suppose that the DNA sequence is organized into nucleosomes in only one particular configuration, so that each site on the DNA always is located at a particular position on the nucleosome. This type of organization is called **nucleosome positioning** (or sometimes nucleosome phasing). In a series of positioned nucleosomes, the linker regions of DNA comprise unique sites.

Consider the consequences for just a single nucleosome. Cleavage with micrococcal nuclease generates a monomeric fragment that constitutes a *specific sequence*. If the DNA is isolated

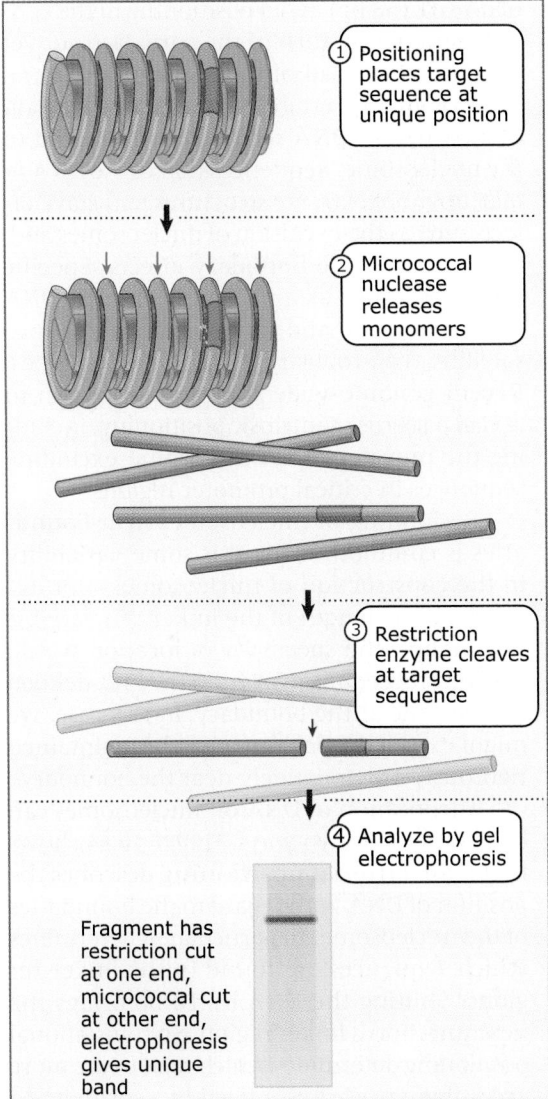

1. Positioning places target sequence at unique position

2. Micrococcal nuclease releases monomers

3. Restriction enzyme cleaves at target sequence

4. Analyze by gel electrophoresis

Fragment has restriction cut at one end, micrococcal cut at other end, electrophoresis gives unique band

FIGURE 10.52 Nucleosome positioning places restriction sites at unique positions relative to the linker sites cleaved by micrococcal nuclease.

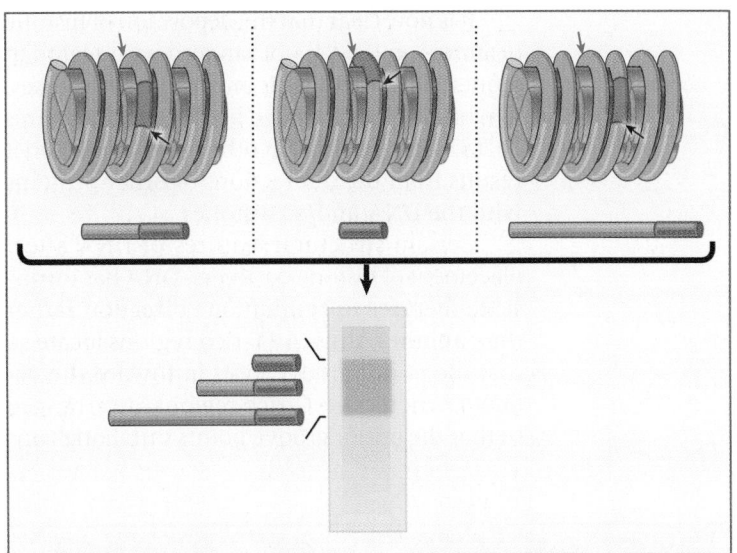

FIGURE 10.53 In the absence of nucleosome positioning, a restriction site lies at all possible locations in different copies of the genome. Fragments of all possible sizes are produced when a restriction enzyme cuts at a target site (red) and micrococcal nuclease cuts at the junctions between nucleosomes (green).

and cleaved with a restriction enzyme that has only one target site in this fragment, it should be cut at a unique point. This produces two fragments, each of unique size.

The products of the micrococcal/restriction double digest are separated by gel electrophoresis. A probe representing the sequence on one side of the restriction site is used to identify the corresponding fragment in the double digest. This technique is called **indirect end labeling**.

Reversing the argument, the identification of a single sharp band demonstrates that the position of the restriction site is uniquely defined with respect to the end of the nucleosomal DNA (as defined by the micrococcal

nuclease cut). So the nucleosome has a unique sequence of DNA.

What happens if the nucleosomes do *not* lie at a single position? Now the linkers consist of *different* DNA sequences in each copy of the genome. So the restriction site lies at a different position each time; in fact, it lies at all possible locations relative to the ends of the monomeric nucleosomal DNA. **FIGURE 10.53** shows that the double cleavage then generates a broad smear, ranging from the smallest detectable fragment (~20 bases) to the length of the monomeric DNA.

Nucleosome positioning might be accomplished in either of the following two ways:

1. It is intrinsic: every nucleosome is deposited specifically at a particular DNA sequence. This modifies our view of the nucleosome as a subunit able to form between any sequence of DNA and a histone octamer.

2. It is extrinsic: the first nucleosome in a region is preferentially assembled at a particular site. A preferential starting point for nucleosome positioning results from the presence of a region from which nucleosomes are excluded. The excluded region provides a boundary that restricts the positions available to the adjacent nucleosome. Then a series of nucleosomes may be assembled sequentially, with a defined repeat length.

It is now clear that the deposition of histone octamers on DNA is not random with regard to sequence. The pattern is intrinsic in some cases, in which it is determined by structural features in DNA. It is extrinsic in other cases, in which it results from the interactions of other proteins with the DNA and/or histones.

Certain structural features of DNA affect placement of histone octamers. DNA has intrinsic tendencies to bend in one direction rather than another; thus, A · T-rich regions locate so that the minor groove faces in toward the octamer, whereas G · C-rich regions are arranged so that the minor groove points out. Long runs

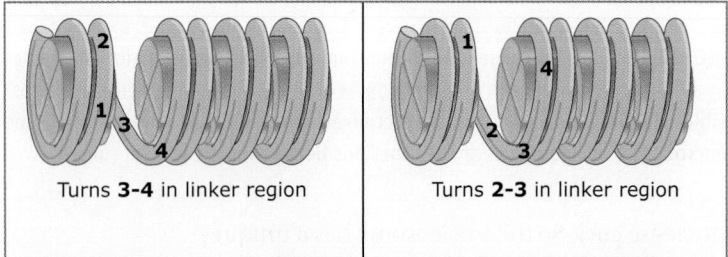

FIGURE 10.54 Translational positioning describes the linear position of DNA relative to the histone octamer. Displacement of the DNA by 10 bp changes the sequences that are in the more exposed linker regions but does not alter which face of DNA is protected by the histone surface and which is exposed to the exterior.

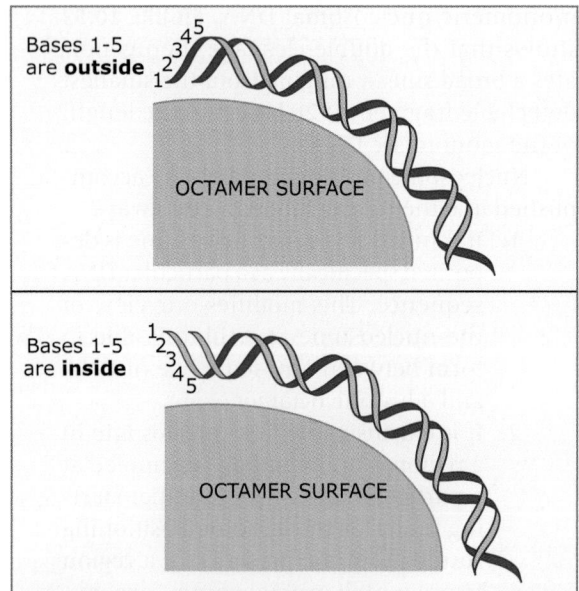

FIGURE 10.55 Rotational positioning describes the exposure of DNA on the surface of the nucleosome. Any movement that differs from the helical repeat (~10.2 bp/turn) displaces DNA with reference to the histone surface. Nucleotides on the inside are more protected against nucleases than nucleotides on the outside.

of dA · dT (>8 bp) avoid positioning in the central superhelical turn of the core. It is not yet possible to sum all of the relevant structural effects and, thus, entirely to predict the location of a particular DNA sequence with regard to the nucleosome. Sequences that cause DNA to take up more extreme structures may have effects such as the exclusion of nucleosomes and, thus, could cause boundary effects. Specific sequences, such as a portion of the 5S rDNA in some species, and certain simple sequence satellites, can robustly position nucleosomes. Recent genome-wide studies have begun to reveal patterns of intrinsic positioning, including the prevalence of nucleosome-excluding sequences in critical promoter regions.

Positioning of nucleosomes near boundaries is common. If there is some variability in the construction of nucleosomes—for example, if the length of the linker can vary by, say, 10 bp—the specificity of location would decline proceeding away from the first, defined nucleosome at the boundary. In this case, we might expect the positioning to be maintained rigorously only relatively near the boundary.

The location of DNA on nucleosomes can be described in two ways. FIGURE 10.54 shows that **translational positioning** describes the position of DNA with regard to the boundaries of the nucleosome. In particular, it determines which sequences are found in the linker regions. Shifting the DNA by 10 bp brings the next turn into a linker region. So translational positioning determines which regions are more accessible (at least as judged by sensitivity to micrococcal nuclease).

Because DNA lies on the outside of the histone octamer, one face of any particular sequence is obscured by the histones, but the other face is accessible. Depending upon its positioning with regard to the nucleosome, a site in DNA that must be recognized by a regulator protein could be inaccessible or available. The exact position of the histone octamer with respect to DNA sequence may, therefore, be important. FIGURE 10.55 shows the effect of **rotational positioning** of the double helix with regard to the octamer surface. If the DNA is moved by a partial number of turns (imagine the DNA as rotating relative to the protein surface), there is a change in the exposure of sequence to the outside.

Both translational and rotational positioning can be important in controlling access to DNA. The best-characterized cases of positioning involve the specific placement of nucleosomes at promoters. Translational positioning and/or the exclusion of nucleosomes from a particular sequence may be necessary to allow a transcription complex to form. Some regulatory factors can bind to DNA only if a nucleosome is excluded to make the DNA freely accessible, and this creates a boundary for translational positioning. In other cases, regulatory factors can bind to DNA on the surface of the nucleosome, but rotational positioning is important to ensure that the face of DNA with the appropriate contact points is exposed.

It is to be noted that promoters (and some other structures) often have short regions that exclude nucleosomes. These regions typically form a boundary next to which nucleosome positions are restricted. A survey of an extensive region in the *Saccharomyces cerevisiae* genome (mapping 2278 nucleosomes over 482 kb of DNA) showed that in fact 60% of the nucleosomes have specific positions as the result of boundary effects, most often from promoters.

Concept and Reasoning Check

1. How can a boundary effect be created by either extrinsic or intrinsic mechanisms?

10.23 Domains define regions that contain active genes

Key concept

- A domain containing a transcribed gene is defined by increased sensitivity to degradation by DNase I.

A region of the genome that contains an active gene may have an altered structure. The change in structure precedes, and is different from, the disruption of nucleosome structure that may be caused by the actual passage of RNA polymerase.

One indication of the change in structure of transcribed chromatin is provided by its increased susceptibility to degradation by DNase I. DNase I sensitivity defines a chromosomal **domain**, a region of altered structure including at least one active transcription unit and sometimes extending farther. (Note that use of the term *domain* does not imply any necessary connection with the structural domains identified by the loops of chromatin or chromosomes.)

When chromatin is digested with DNase I, it is eventually degraded into acid-soluble material (very small fragments of DNA). The progress of the overall reaction can be followed in terms of the proportion of DNA that is rendered acid soluble. *When only 10% of the total DNA has become acid soluble, more than 50% of the DNA of an active gene has been lost.* This suggests that active genes are preferentially degraded.

The fate of individual genes can be followed by quantitating the amount of DNA that survives to react with a specific probe. The protocol is outlined in **FIGURE 10.56**. The principle is that the loss of a particular band indicates that the corresponding region of DNA has been degraded by the enzyme.

FIGURE 10.57 shows what happens to β-globin genes and an ovalbumin gene in chromatin extracted from chicken red blood cells (in which globin genes are expressed and the ovalbumin gene is inactive). The restriction fragments representing the β-globin genes are rapidly lost, while those representing the

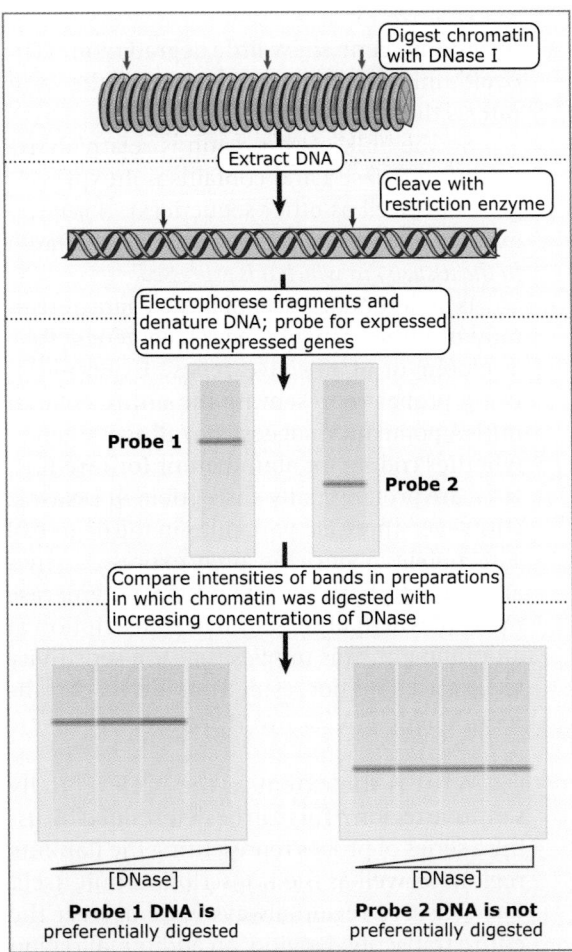

FIGURE 10.56 Sensitivity to DNase I can be measured by determining the rate of disappearance of the material hybridizing with a particular probe.

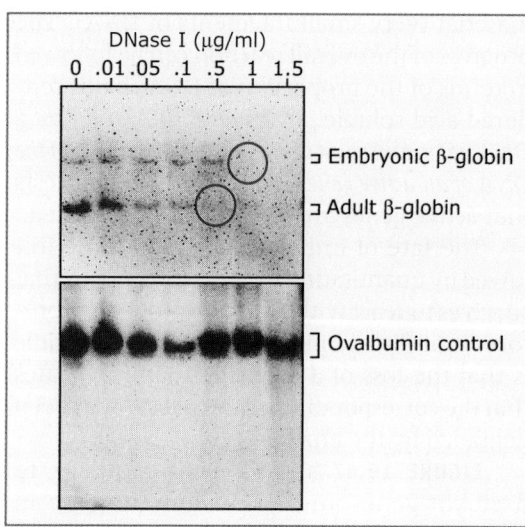

DNase I (µg/ml)

0 .01 .05 .1 .5 1 1.5

⊐ Embryonic β-globin

⊐ Adult β-globin

] Ovalbumin control

FIGURE 10.57 In adult erythroid cells, the adult β-globin gene is highly sensitive to DNase I digestion, the embryonic β-globin gene is partially sensitive (probably due to spreading effects), but ovalbumin is not sensitive. Photos courtesy of Harold Weintraub and Mark Groudine, Fred Hutchinson Cancer Research Center.

ovalbumin gene show little degradation. (The ovalbumin gene in fact is digested at the same rate as the bulk of DNA.)

So the bulk of chromatin is relatively resistant to DNase I and contains nonexpressed genes (as well as other sequences). *A gene becomes relatively susceptible to degradation specifically in the tissue(s) in which it is expressed.*

Is preferential susceptibility a characteristic only of rather actively expressed genes, such as globin, or of all active genes? Experiments using probes representing the entire cellular mRNA population suggest that all active genes, whether coding for abundant or for rare mRNAs, are preferentially susceptible to DNase I. (However, there are variations in the degree of susceptibility.) Since the rarely expressed genes are likely to have very few RNA polymerase molecules actually engaged in transcription at any moment, this implies that the sensitivity to DNase I does not result from the act of transcription but is a feature of *genes that are able to be transcribed.*

What is the extent of the preferentially sensitive region? This can be determined by using a series of probes representing the flanking regions as well as the transcription unit itself. The sensitive region always extends over the entire transcribed region; an additional region of several kilobases on either side may show an intermediate level of sensitivity (probably as the result of spreading effects).

The critical concept implicit in the description of the domain is that a region of high sensitivity to DNase I extends over a considerable distance. Often we think of regulation as residing in events that occur at a discrete site in DNA—for example, in the ability to initiate transcription at the promoter. Even if this is true, such regulation must determine, or must be accompanied by, a more wide-ranging change in structure. This is a difference between eukaryotes and prokaryotes.

10.24 Histone octamers are displaced and reassembled during transcription

Key concepts

- Most transcribed genes retain a nucleosomal structure.
- Some heavily transcribed genes appear to be exceptional cases that are devoid of nucleosomes.
- RNA polymerase displaces histone octamers during transcription in a model system, but octamers reassociate with DNA as soon as the polymerase has passed.
- Nucleosomes are reorganized when transcription passes through a gene.
- Ancillary factors are required both for RNA polymerase to displace octamers during transcription and for the histones to reassemble into nucleosomes after transcription.

Transcription involves the unwinding of DNA and may require the fiber to unfold in restricted regions of chromatin. A simple-minded view suggests that some "elbow-room" must be needed for the process.

Heavily transcribed chromatin adopts structures too extended to still be contained in nucleosomes. For example, in the intensively transcribed genes coding for rRNA, shown in **FIGURE 10.58**, the extreme packing of RNA polymerases makes it hard to see the DNA. We cannot directly measure the lengths of the rRNA transcripts because the RNA is compacted by proteins, but we know (from the sequence of the rRNA) how long the transcript must be. The length of the transcribed DNA segment, measured by the length of the axis of the "Christmas tree", is ~85% of the length of the rRNA. This means that the DNA is almost completely extended.

On the other hand, transcription complexes of SV40 minichromosomes can be ex-

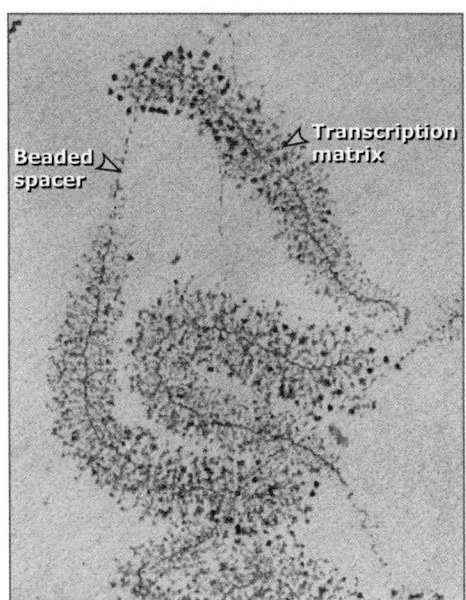

FIGURE 10.58 The extended axis of an rDNA transcription unit alternates with the only slightly less extended nontranscribed spacer. Photo courtesy of Yean Chooi and Charles Laird.

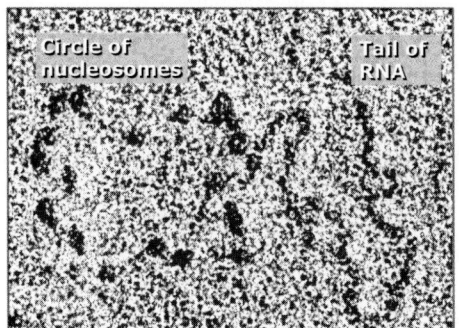

FIGURE 10.59 An SV40 minichromosome can be transcribed. Reprinted from *J. Mol. Biol.*, vol. 131, P. Gariglio, et al., The template of the isolated native simian virus 40…, pp. 75–105, Copyright (1979) with permission from Elsevier [http://www.sciencedirect.com/science/journal/00222836]. Photo courtesy of Pierre Chambon, College of France.

tracted from infected cells. They contain the usual complement of histones and display a beaded structure. Chains of RNA can be seen to extend from the minichromosome, as in the example of **FIGURE 10.59**. This argues that transcription can proceed while the SV40 DNA is organized into nucleosomes. Of course, the SV40 minichromosome is transcribed less intensively than the rRNA genes.

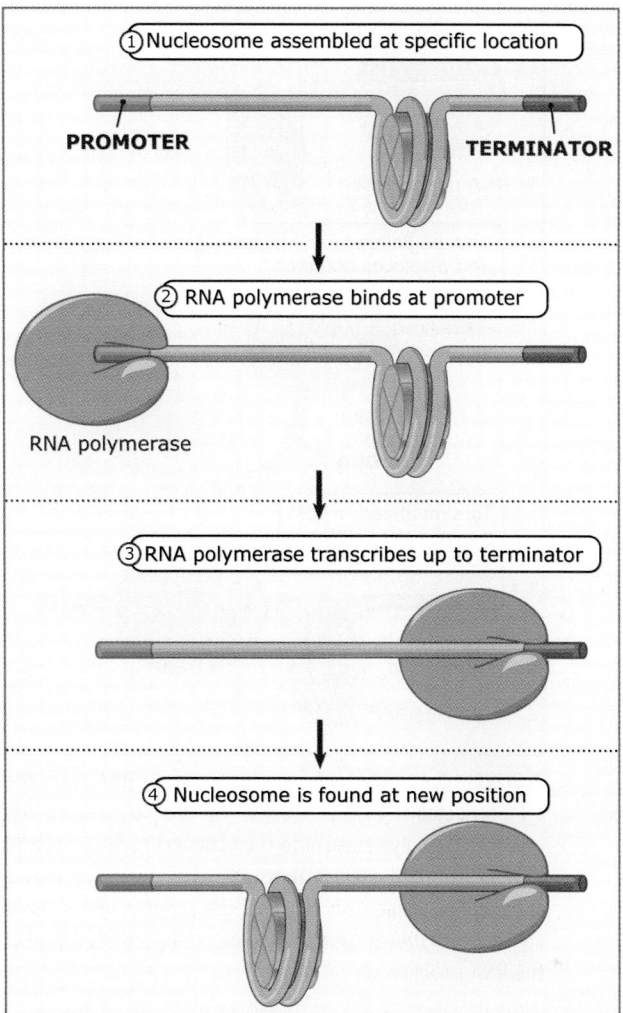

FIGURE 10.60 A protocol to test the effect of transcription on nucleosomes shows that the histone octamer is displaced from DNA and rebinds at a new position.

Experiments to test whether an RNA polymerase can transcribe directly through a nucleosome suggest that the histone octamer is displaced by the act of transcription. **FIGURE 10.60** shows what happens when the phage T7 RNA polymerase transcribes a short piece of DNA containing a single octamer core *in vitro*. The core remains associated with the DNA but is found in a different location. The core is most likely to rebind to the same DNA molecule from which it was displaced.

FIGURE 10.61 shows a model for polymerase progression based on these studies. DNA is displaced as the polymerase enters the nucleosome, but the polymerase reaches a point at which the DNA loops back and reattaches, forming a closed region. As polymerase advances further, unwinding the DNA, it creates positive supercoils in this loop; the effect

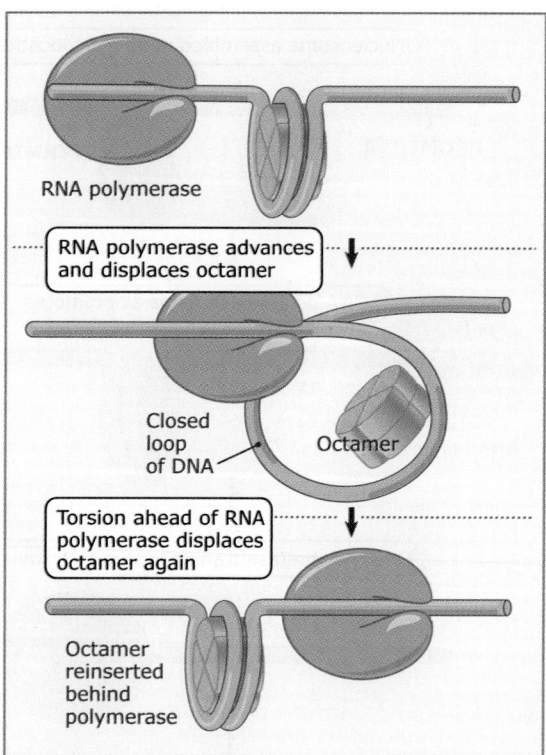

FIGURE 10.61 RNA polymerase displaces DNA from the histone octamer as it advances. The DNA loops back and attaches (to polymerase or to the octamer) to form a closed loop. As the polymerase proceeds, it generates positive supercoiling ahead. This displaces the octamer, which keeps contact with DNA and/or polymerase, and is inserted behind the RNA polymerase.

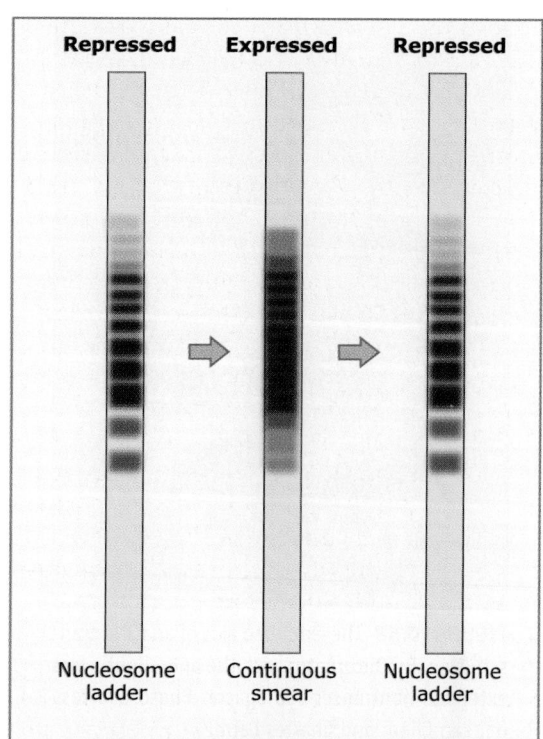

FIGURE 10.62 The *URA3* gene has positioned nucleosomes before transcription. When transcription is induced, nucleosome positions are randomized. When transcription is repressed, the nucleosomes resume their particular positions.

could be dramatic, because the closed loop is only ~80 bp, so each base pair through which the polymerase advances makes a significant addition to the supercoiling. In fact, the polymerase progresses easily for the first 30 bp into the nucleosome. Then it proceeds more slowly, as though encountering increasing difficulty in progressing. Pauses occur every 10 bp, suggesting that the structure of the loop imposes a constraint related to rotation around each turn of DNA. When the polymerase reaches the midpoint of the nucleosome (the next bases to be added are essentially at the axis of dyad symmetry), pausing ceases, and the polymerase advances rapidly. This suggests that the midpoint of the nucleosome marks the point at which the octamer is displaced (possibly, because positive supercoiling has reached some critical level that expels the octamer from DNA). This releases tension ahead of the polymerase and allows it to proceed. The octamer then binds to the DNA behind the

polymerase and no longer presents an obstacle to progress. Possibly, the octamer changes position without ever completely losing contact with the DNA.

These studies show that a small RNA polymerase can displace a single nucleosome, which reforms behind it, during transcription. Of course, the situation is more complex in a eukaryotic nucleus. RNA polymerase is very much larger, and the impediment to progress is a string of connected nucleosomes. Overcoming this obstacle requires additional factors that act on chromatin. There is also evidence in eukaryotes that H2A · H2B dimers are displaced more readily during transcription than $H3_2 \cdot H4_2$ tetramers, suggesting that tetramers and dimers may be reassembled sequentially as they are after passage of a replication fork (see *10.21 Reproduction of chromatin requires assembly of nucleosomes*).

The organization of nucleosomes can be dramatically altered by transcription. **FIGURE 10.62** shows what happens to the yeast *URA3* gene when it is transcribed under control of an inducible promoter. Positioning is examined by using micrococcal nuclease to examine cleavage sites relative to a restriction site at the 5′ end of the gene. Initially the gene displays a pattern of nucleosomes that are organized from the promoter for a significant distance across the gene; positioning is lost in the 3′ regions. When the gene is expressed, a general smear replaces the positioned pattern of nucleosomes. So, nucleosomes are present at the same density but are no longer organized in phase. This suggests that transcription destroys the nucleosomal positioning. When repression is reestablished, positioning appears within 10 min (although it is not complete).

The unifying model is to suppose that RNA polymerase displaces histone octamers (either as a whole or as dimers and tetramers) as it progresses. If the DNA behind the polymerase is available, the nucleosome is reassembled there. If the DNA is not available, for example, because another polymerase continues immediately behind the first, then the octamer may be permanently displaced, and the DNA may persist in an extended form.

The displacement and reassembly of nucleosomes does not occur solely as a result of the passage of RNA polymerase, but is facilitated by factors that help regulate this process. The first of these to be characterized is a heterodimeric factor called FACT that behaves like a transcription elongation factor. (FACT, an abbreviation for *facilitates chromatin transcription*, is not part of RNA polymerase, but associates with it specifically during the elongation phase of transcription.) FACT consists of two subunits that are well conserved in all eukaryotes. It is associated with the chromatin of active genes.

When FACT is added to isolated nucleosomes, it causes them to lose H2A · H2B dimers. During transcription *in vitro*, it converts nucleosomes to "hexasomes" that have lost H2A · H2B dimers. This suggests that FACT is part of a mechanism for displacing octamers during transcription. FACT may also be involved in the reassembly of nucleosomes after transcription, because it assists formation of nucleosomes from core histones, acting like a histone chaperone.

This suggests the model shown in **FIGURE 10.63**, in which FACT detaches H2A · H2B from a nucleosome in front of RNA polymerase and then helps to add it to a nucleosome that is reassembling behind the enzyme. Other factors must be required to complete the process.

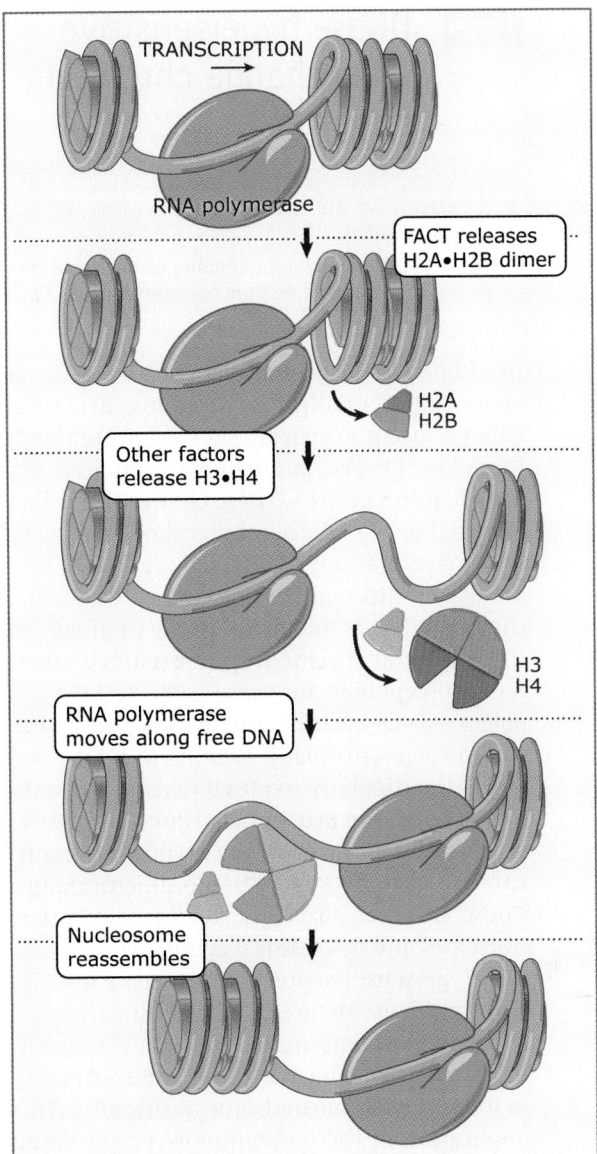

FIGURE 10.63 Histone octamers are disassembled ahead of transcription to remove nucleosomes. They re-form following transcription. Release of H2A · H2B dimers probably initiates the disassembly process.

Several other factors have been identified that play key roles in either nucleosome displacement or reassembly during transcription. These include the Spt6 protein, a factor involved in "resetting" chromatin structure after transcription. Spt6, like FACT, colocalizes with actively transcribed regions and can act as a histone chaperone to promote nucleosome assembly.

Concept and Reasoning Check

1. Why might it be important to reassemble nucleosomes behind an elongating RNA polymerase?

10.25 DNase hypersensitive sites change chromatin structure

Key concepts

- Hypersensitive sites are found at the promoters of expressed genes.
- They are generated by the binding of transcription factors that displace histone octamers.

In addition to the general changes that occur in active or potentially active regions, structural changes occur at specific sites associated with initiation of transcription or with certain structural features in DNA. These changes were first detected by the effects of digestion with very low concentrations of the enzyme DNase I.

When chromatin is digested with DNase I, the first effect is the introduction of breaks in the duplex at specific **hypersensitive sites**. Since susceptibility to DNase I reflects the availability of DNA in chromatin, we take these sites to represent chromatin regions in which the DNA is particularly exposed because it is not organized in the usual nucleosomal structure. A typical hypersensitive site is 100× more sensitive to enzyme attack than bulk chromatin. These sites are also hypersensitive to other nucleases and to chemical agents.

Hypersensitive sites are created by the (tissue-specific) structure of chromatin. Their locations can be determined by the technique of indirect end-labeling that we introduced earlier in the context of nucleosome positioning. This application of the technique is recapitulated in FIGURE 10.64. In this case, cleavage at the hypersensitive site by DNase I is used to generate one end of the fragment, and its distance is measured from the other end that is generated by cleavage with a restriction enzyme.

Hypersensitive sites are created by the local structure of chromatin, which may be tissue specific. Most hypersensitive sites are related to gene expression. Every active gene has a site, or sometimes more than one site, in the region of the promoter. Most hypersensitive sites are found only in chromatin of cells in which the associated gene is being expressed; they do not occur when the gene is inactive. The 5' hypersensitive site(s) appear before transcription begins; and the DNA sequences contained within the hypersensitive sites are required for gene expression, as seen by mutational analysis.

A particularly well-characterized nuclease-sensitive region lies on the SV40 minichromosome. A short segment near the origin of replication, just upstream of the promoter for the late transcription unit, is cleaved preferentially by DNase I, micrococcal nuclease, and other nucleases (including restriction enzymes).

The state of the SV40 minichromosome can be visualized by electron microscopy. In up to 20% of the samples, a "gap" is visible in the nucleosomal organization, as evident in FIGURE 10.65. The gap is a region of ~120 nm

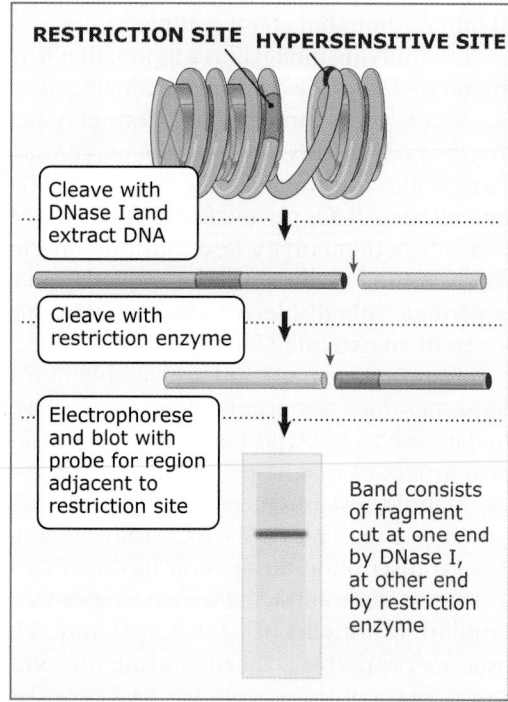

FIGURE 10.64 Indirect end-labeling identifies the distance of a DNase hypersensitive site from a restriction cleavage site. The existence of a particular cutting site for DNase I generates a discrete fragment, whose size indicates the distance of the DNase I hypersensitive site from the restriction site.

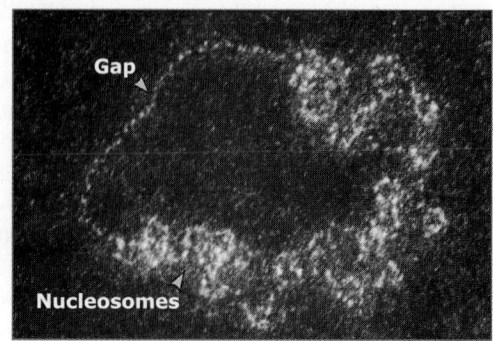

FIGURE 10.65 The SV40 minichromosome has a nucleosome gap. Photo courtesy of Moshe Yaniv, Pasteur Institute.

in length (about 350 bp), surrounded on either side by nucleosomes. The visible gap corresponds with the nuclease-sensitive region. This shows directly that increased sensitivity to nucleases is associated with the exclusion of nucleosomes.

A hypersensitive site is not necessarily uniformly sensitive to nucleases. FIGURE 10.66 shows the maps of two hypersensitive sites.

Within the SV40 gap of ~300 bp, there are two hypersensitive DNase I sites and a "protected" region. The protected region presumably reflects the association of (nonhistone) protein(s) with the DNA. The gap is associated with the DNA sequence elements that are necessary for promoter function.

The hypersensitive site at the β-globin promoter is preferentially digested by several enzymes, including DNase I, DNase II, and micrococcal nuclease. The enzymes have preferred cleavage sites that lie at slightly different points in the same general region. So a region extending from about −70 bp to −270 bp is preferentially accessible to nucleases when the gene is transcribable. Note that these hypersensitive sites frequently appear before a gene is actively transcribed, but it instead "poised" for transcription.

What is the structure of the hypersensitive site? Its preferential accessibility to nucleases indicates that it is not protected by histone octamers, but this does not necessarily imply that it is free of protein. A region of free DNA might be vulnerable to damage; and in any case, how would it be able to exclude nucleosomes? We assume that the hypersensitive site results from the binding of specific regulatory proteins that exclude nucleosomes. Indeed, the binding of such proteins is probably the basis for the existence of the protected region within the hypersensitive site.

The proteins that generate hypersensitive sites are likely to be regulatory factors of various types, since hypersensitive sites are found associated with promoters and other elements that regulate transcription, origins of replication, centromeres, and sites with other structural significance. In some cases, they are associated with more extensive organization of chromatin structure. A hypersensitive site may provide a boundary for a series of positioned nucleosomes. Hypersensitive sites associated with transcription may be generated by transcription factors when they bind to the promoter as part of the process that makes it accessible to RNA polymerase.

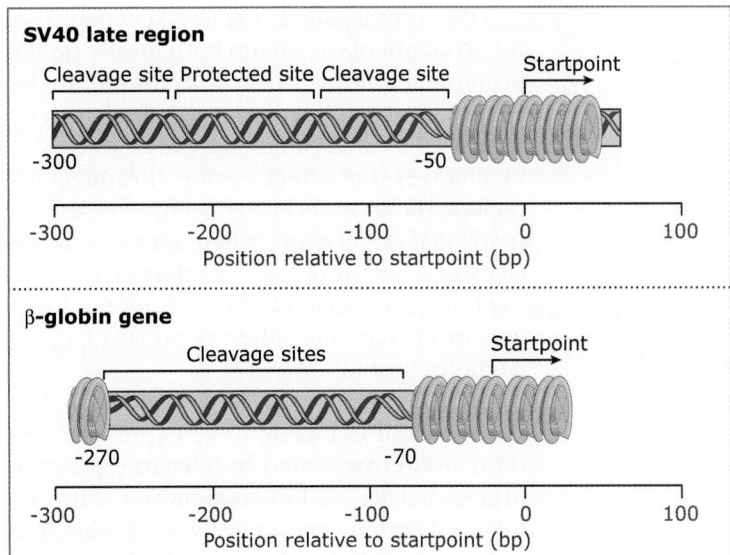

FIGURE 10.66 The SV40 gap includes hypersensitive sites, sensitive regions, and a protected region of DNA. The hypersensitive site of a chicken β-globin gene comprises a region that is susceptible to several nucleases.

Concept and Reasoning Check

1. What is the difference between a DNase I hypersensitive site and a DNase I-sensitive domain?

10.26 Chromatin remodeling is an active process

Key concepts

- Chromatin structure is changed by remodeling complexes that use energy provided by hydrolysis of ATP.
- All remodeling complexes contain a related ATPase subunit, and are grouped into subfamilies containing more closely related ATPases.
- A remodeling complex does not itself have specificity for any particular target site but must be recruited by a component of the transcription apparatus.
- Remodeling complexes are recruited to promoters by sequence-specific activators.
- Remodeling complexes can alter, slide, or displace nucleosomes.

The cellular genome is organized as nucleosomes, but initiation of transcription generally is prevented if the promoter region is packaged into nucleosomes. In this sense, histones function as generalized repressors of transcription. Activation of a gene requires changes in the state of chromatin: the essential issue is how the transcription apparatus gains access to the promoter DNA.

Whether a gene is expressed depends on the structure of chromatin both locally (at the promoter) and in the surrounding domain. Chromatin structure correspondingly can be regulated by individual activation events or by changes that affect a wide chromosomal region. The most localized events concern an individual target gene, where changes in nucleosomal structure and organization occur in the immediate vicinity of the promoter. More general changes may affect regions as large as a whole chromosome.

Changes that affect large regions control the potential of a gene to be expressed. The term **silencing** is used to refer to repression of gene activity in a local chromosomal region. Silenced regions are typically assembled into heterochromatin, which results from additional proteins binding to chromatin and either directly or indirectly preventing transcription factors and RNA polymerase from activating promoters in the region.

Changes at an individual promoter control whether transcription is initiated for a particular gene. These changes may be either activating or repressing.

Local chromatin structure is an integral part of controlling gene expression. Genes may exist in either of two structural conditions. Genes are found in an "active" state only in the cells in which they are expressed. The change of structure precedes the act of transcription and indicates that the gene is "transcribable." This suggests that acquisition of the "active" structure must be the first step in gene expression. Active genes are found in domains of euchromatin with a preferential susceptibility to nucleases (see *10.23 Domains define regions that contain active genes*). Hypersensitive sites are created at promoters before a gene is activated (see *10.25 DNase hypersensitive sites change chromatin structure*).

There is an intimate and continuing connection between initiation of transcription and chromatin structure. Some activators of gene transcription directly modify histones; in particular, acetylation of histones is associated with gene activation (see *10.27 Histone acetylation is associated with transcriptional activity*). Conversely, some repressors of transcription function by deacetylating histones. So a reversible change in histone structure in the vicinity of the promoter is involved in the control of gene expression. This is one mechanism by which a gene is maintained in an active or inactive state.

The general process of inducing changes in chromatin structure is called **chromatin remodeling**. This consists of mechanisms for moving or displacing histones that depend on the input of energy. Many protein-protein and protein-DNA contacts need to be disrupted to release histones from chromatin. There is no free ride: the energy must be provided to disrupt these contacts. **FIGURE 10.67** illustrates the principle of dynamic remodeling by a factor that hydrolyzes ATP. When the histone octamer is released from DNA, other proteins (in this case transcription factors and RNA polymerase) can bind.

FIGURE 10.68 summarizes the types of remodeling changes in chromatin that can be characterized *in vitro*:

- Histone octamers may slide along DNA, changing the relationship between the nucleic acid and protein. This alters the position of a particular sequence on the nucleosomal surface.
- The spacing between histone octamers may be changed, again with the result that the positions of individual sequences are altered relative to protein.
- The most extensive change is that an octamer(s) may be displaced entirely from DNA to generate a nucleosome-

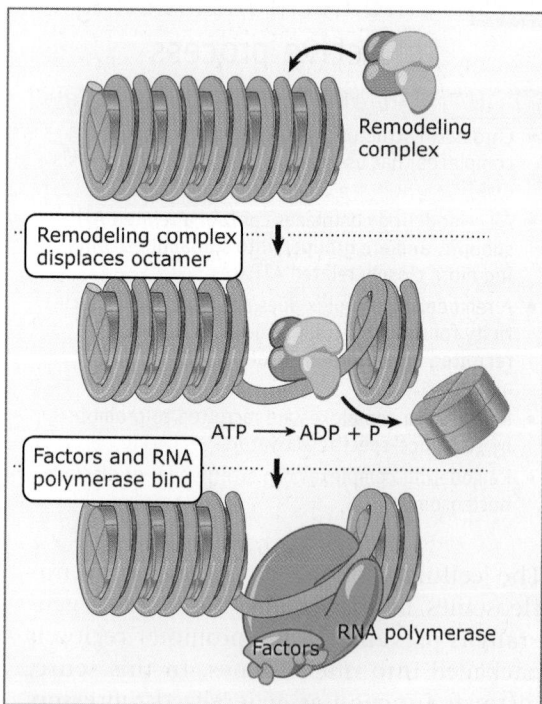

FIGURE 10.67 The dynamic model for transcription of chromatin relies upon factors that can use energy provided by hydrolysis of ATP to displace nucleosomes from specific DNA sequences.

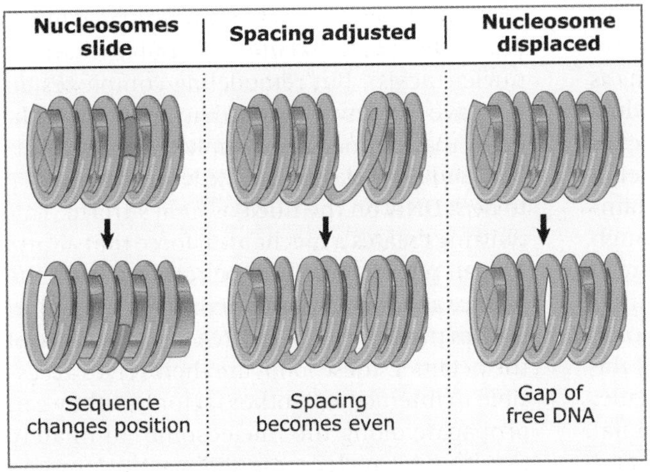

FIGURE 10.68 / FIGURE 10.69

FIGURE 10.68 Remodeling complexes can cause nucleosomes to slide along DNA, can displace nucleosomes from DNA, or can reorganize the spacing between nucleosomes.

Type of complex	SWI/SNF	ISW1	IN080	CHD
Yeast	SWI/SNF RSC	ISW1 ISW2	IN080 SWRI	CHD1
Fly	dSWI/SNF (Brahma)	NURF CHRAC ACF		CHD1 CHD3,4
Human	hSWI/SNF	RSF hACF/WCFR hCHRAC WICH NURF NORC	IN080	NURD CHD1-4
Frog		CHRAC ACF WICH		NURD

FIGURE 10.69 Remodeling complexes can be classified by their ATPase subunits.

free gap. Alternatively, one or both H2A-H2B dimers can be displaced.

A major role of chromatin remodeling is to change the organization of nucleosomes at the promoter of a gene that is to be transcribed. This is required to allow the transcription apparatus to gain access to the promoter. However, remodeling is also required to enable other manipulations of chromatin, including repair reactions to damaged DNA.

Remodeling often takes the form of displacing one or more histone octamers. This can be detected by a change in the micrococcal nuclease ladder where protection against cleavage has been lost. This can result in the creation of a site that is hypersensitive to cleavage with DNase I (see *10.25 DNase hypersensitive sites change chromatin structure*). Sometimes there are less dramatic changes, for example, involving a change in rotational positioning of a single nucleosome; this may be detected by loss of the DNase I 10-base ladder. So changes in chromatin structure may extend from altering the positions of nucleosomes to removing them altogether.

Chromatin remodeling is undertaken by **ATP-dependent chromatin remodeling complexes** that use ATP hydrolysis to provide the energy for remodeling. The heart of the remodeling complex is its ATPase subunit. The ATPase subunits of all remodeling complexes are related members of a large superfamily of proteins, which is divided into subfamilies of more closely related members. Remodeling complexes are classified according to the subfamily of ATPase that they contain as their cata-

lytic subunit. There are many subfamilies, but the four major subfamilies (SWI/SNF, ISWI, CHD, and INO80/SWR1) are shown in **FIGURE 10.69**. The chromatin-remodeling superfamily is large and diverse, and most species have multiple complexes in different subfamilies. Yeast has two SWI/SNF-related complexes and three ISWI complexes. Eight different ISWI complexes have been identified thus far in mammals. Remodeling complexes range from small heterodimeric complexes (the ATPase subunit plus a single partner) to massive complexes of 10 or more subunits. Each type of complex may undertake a different range of remodeling activities.

SWI/SNF was the first remodeling complex to be identified. Its name reflects the fact that many of its subunits are coded by genes originally identified by *swi* or *snf* mutations in *S. cerevisiae*. These mutations also show genetic interactions with mutations in genes that code for components of chromatin, in particular *SIN1*, which codes for a nonhistone protein, and *SIN2*, which codes for histone H3. The *SWI* and *SNF* genes are required for expression of a variety of individual loci (~120 or 2% of *S. cerevisiae* genes are affected). Expression of these loci may require the SWI/SNF complex to remodel chromatin at their promoters.

SWI/SNF acts catalytically *in vitro*, and there are only ~150 complexes per yeast cell. All of the genes encoding the SWI/SNF subunits are nonessential, which implies that yeast must also have other ways of remodeling chromatin. The RSC complex is more abundant and is essential. It acts at ~700 target loci.

Different subfamilies of remodeling complexes have distinct modes of remodeling, reflecting differences in their ATPase subunits as well as effects of other proteins in individual remodeling complexes. SWI/SNF complexes can remodel chromatin *in vitro* without overall loss of histones or can displace histone octamers. Both types of reaction may pass through the same intermediate in which the structure of the target nucleosome is altered, leading either to reformation of a (remodeled) nucleosome on the original DNA or to displacement of the histone octamer to a different DNA molecule. The SWI/SNF complex alters nucleosomal sensitivity to DNase I at the target site, and induces changes in protein-DNA contacts that persist after it has been released from the nucleosomes. The Swi2 (also known as Snf2) subunit is the ATPase that provides the energy for remodeling by SWI/SNF.

There are many contacts between DNA and a histone octamer—14 are identified in the crystal structure. All of these contacts must be broken for an octamer to be released or for it to move to a new position. How is this achieved?

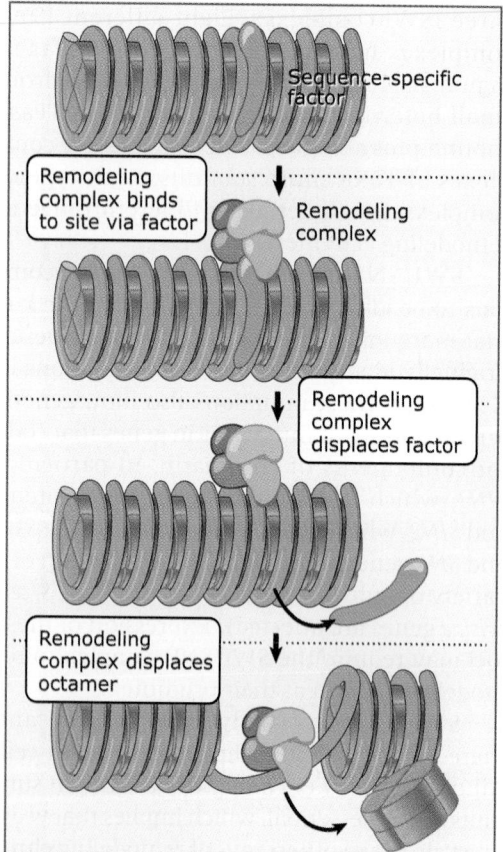

FIGURE 10.70 A remodeling complex binds to chromatin via an activator (or repressor).

The ATPase subunits are distantly related to helicases (enzymes that unwind double-stranded nucleic acids), but remodeling complexes do not have any unwinding activity. Present thinking is that remodeling complexes in the SWI/NSF and ISWI classes use the hydrolysis of ATP to *twist* DNA on the nucleosomal surface. This twisting creates a mechanical force that allows a small region of DNA to be released from the surface and then repositioned. This mechanism creates transient loops of DNA on the surface of the octamer; these loops are themselves accessible to interact with other factors, or they can propagate along the nucleosome, ultimately resulting in nucleosome sliding. Nucleosome sliding appears to be the major activity of the ISWI family of remodelers. ISWI complexes require the presence of both linker DNA and the N-terminal tail of histone H4, but we do not know exactly how the tail functions in this regard.

How are remodeling complexes targeted to specific sites on chromatin? They do not themselves contain subunits that bind specific DNA sequences. This suggests the model shown in **FIGURE 10.70** in which they are recruited by activators or (sometimes) by repressors. This is accomplished by a "hit and run" mechanism, in which the activator or repressor may be released after the remodeling complex has bound.

Different remodeling complexes have different roles in the cell. SWI/SNF complexes are generally involved in transcriptional activation, whereas some ISWI complexes act as repressors, using their remodeling activity to slide nucleosomes *onto* promoter regions to prevent transcription. Members of the CHD (chromodomain helicase DNA-binding) family have also been implicated in repression, particularly the Mi-2/NuRD complexes, which contain both chromatin remodeling and histone deacetylase activities (see *10.27 Histone acetylation is associated with transcriptional activity*). Remodelers in the SWR1/INO80 class have a unique activity: in addition to their normal remodeling capabilities, some members of this class also have *histone exchange* capability, in which individual histones (usually H2A/H2B dimers) can be replaced in a nucleosome, typically with a histone variant (see *10.19 Histone variants produce alternative nucleosomes*).

Concept and Reasoning Check

1. How can remodeling complexes act as either activators or repressors?

10.27 Histone acetylation is associated with transcriptional activity

Key concepts

- Histones are modified by acetylation, methylation, phosphorylation, and other modifications.
- Histone acetylation occurs transiently at replication.
- Histone acetylation is associated with activation of gene expression.
- Deacetylated chromatin may have a more condensed structure.
- Transcription activators are associated with histone acetyltransferase activities in large complexes.
- The remodeling complex may recruit the acetylating complex.
- Histone acetyltransferases vary in their target specificity.
- Deacetylation is associated with repression of gene activity.

In addition to the ATP-dependent chromatin remodeling discussed in *10.26 Chromatin remodeling is an active process*, chromatin structure is also modulated by making covalent changes on the histones, in particular by modifying the N- and C-terminal tails of the histones. The histone tails consist of 15 to 30 amino acids at the N-termini of all four core histones and the C-termini of H2A and H2B. The N-terminal tails of H2B and H3 pass between the turns of DNA (see Figure 10.43 in *10.18 Organization of the histone octamer*). They can be modified at several sites, by methylation, acetylation, phosphorylation, or a number of other modifications. Histone modification creates binding sites

for the attachment of nonhistone proteins that then change the properties of chromatin.

The range of nucleosomes that is targeted for modification can vary. Modification can be a local event, for example, restricted to nucleosomes at the promoter. Or it can be a general event, extending, for example, to an entire chromosome. FIGURE 10.71 shows that there is a general correlation in which acetylation is associated with active chromatin while methylation (at specific sites) is associated with inactive chromatin. However, this is not a simple rule, and the particular sites that are modified, as well as combinations of specific modifications, are very important. For example, there are a several cases in which histones methylated at a certain position are found in active chromatin; this is actually always the case in *S. cerevisiae*, which have no methylation associated with inactive chromatin.

The specificity of the modifications is indicated by the fact that many of the modifying enzymes have individual target sites in specific histones. FIGURE 10.72 summarizes the known effects of some of the modifications. Most modified sites are subject to only a single type of modification. In some cases, modification of one site may activate or inhibit modification of another site. The idea that combinations of signals may be used to define chromatin types has sometimes been called the *histone code*.

Of all of the many histone modifications, acetylation may be the best studies. All the core histones can be acetylated on lysine residues in the tails (and occasionally within the globular core). Acetylation occurs in the following two different circumstances:

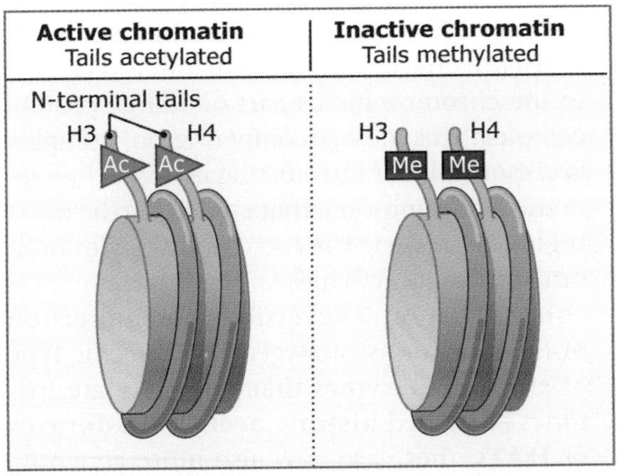

FIGURE 10.71 Acetylation of H3 and H4 is associated with active chromatin, while methylation is associated with inactive chromatin.

Histone	Site	Modification	Function
H3	Lys-4	Methylation	Activation
	Lys-9	Methylation	Chromatin condensation; required for DNA methylation
		Acetylation	Activation
	Ser-10	Phosphorylation	Activation
	Lys-14	Acetylation	Prevents methylation at Lys-9, activation
	Lys-79	Methylation	Telomeric silencing, DNA repair
H4	Arg-3	Methylation	
	Lys-5	Acetylation	Assembly
	Lys-12	Acetylation	Assembly
	Lys-16	Acetylation	Fly X activation

FIGURE 10.72 Most modified sites in histones have a single, specific type of modification, but some sites can have more than one type of modification. Individual functions can be associated with some of the modifications.

1. During DNA replication
2. When genes are activated.

When chromosomes are replicated, during the S phase of the cell cycle, histones are transiently acetylated on specific lysines, such as lysines 5 and 12 (K5 and K12) of histone H4 (see Figure 10.72). **FIGURE 10.73** shows that this acetylation occurs before the histones are incorporated into nucleosomes. We know that histones H3 and H4 are acetylated at the stage when they are associated with one another in the $H3_2 \cdot H4_2$ tetramer. The tetramer is then incorporated into nucleosomes. Quite soon after, the acetyl groups are removed.

The importance of the acetylation is indicated by the fact that preventing acetylation of both histones H3 and H4 during replication causes loss of viability in yeast. The two histones are redundant as substrates, since yeast can manage perfectly well so long as they can acetylate either one of these histones during S phase. There are two possible roles for the acetylation: it could be needed for the histones to be recognized by factors that incorporate them into nucleosomes, or it could be required for the assembly and/or structure of the new nucleosome.

The factors that are known to be involved in chromatin assembly do not distinguish between acetylated and nonacetylated histones, suggesting that the modification is more likely to be required for subsequent interactions. It has been thought for a long time that acetylation might be needed to help control protein-protein interactions that occur as histones are incorporated into nucleosomes. Some evidence for such a role is that the yeast SAS histone acetylase complex binds to chromatin assembly complexes at the replication fork, where it acetylates K16 of histone H4. This may be part of the system that establishes the histone acetylation patterns after replication.

Outside of S phase, acetylation of histones in chromatin is generally correlated with the state of gene expression. The correlation was first noticed because histone acetylation is increased in a domain containing active genes, and acetylated chromatin is more sensitive to nucleases. **FIGURE 10.74** shows that this involves the acetylation of histone tails in nucleosomes. We now know that this occurs largely because of acetylation of the nucleosomes in the vicinity of the promoter when a gene is activated.

In addition to events at individual promoters, widescale changes in acetylation occur on sex chromosomes. This is part of the mechanism by which the activities of genes on the X chromosome are altered to compensate for the presence of two X chromosomes in one sex but only one X chromosome (in addition to the Y chromosome) in the other sex (see *10.30 X chromosomes undergo global changes*). The inactive X chromosome in female mammals has under-acetylated H4. The super-active X chromosome in *Drosophila* males has increased acetylation of H4. This suggests that the presence of acetyl groups may be a prerequisite for a less condensed, active structure. In male *Drosophila*, the X chromosome is acetylated specifically at K16 of histone H4. The component that is responsible is an enzyme called MOF that is recruited to the chromosome as part of a large protein complex. This "dosage compensation" complex is responsible for introducing general changes in the X chromosome that enable it to be more highly expressed. The increased acetylation is only one of its activities.

Acetylation is reversible. Each direction of the reaction is catalyzed by a specific type of enzyme. Enzymes that can acetylate histones are called **histone acetyltransferases** or **HATs**; these are now also more generally referred to as lysine (K) acetyltransferases or KATs, as some of these enzymes have nonhistone substrates as well. The acetyl groups are

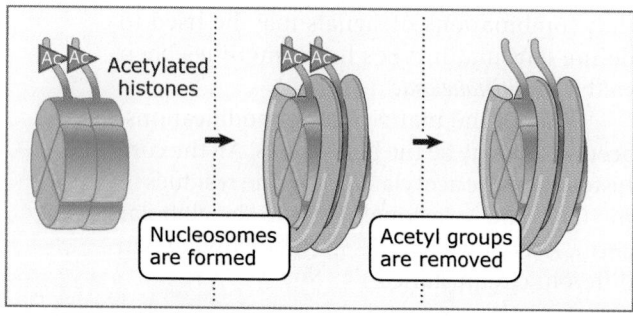

FIGURE 10.73 Acetylation at replication occurs on histones before they are incorporated into nucleosomes.

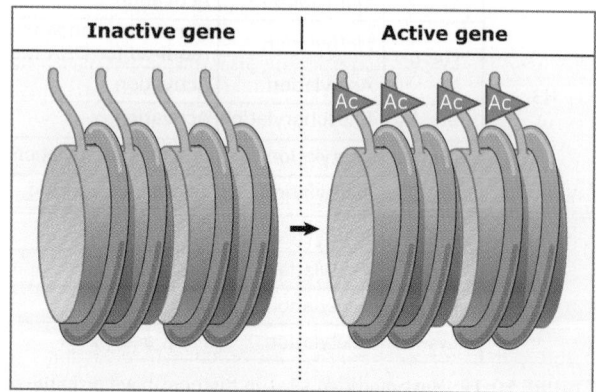

FIGURE 10.74 Acetylation associated with gene activation occurs by directly modifying histones in nucleosomes.

removed by **histone deacetylases** or **HDACs**. There are two groups of HAT enzymes: group A enzymes act on histones in chromatin and are involved with the control of transcription; group B enzymes act on newly synthesized histones in the cytosol and are involved with nucleosome assembly.

Like the chromatin-remodeling enzymes discussed previously, group A HATs are typically members of a large complexes that must be targeted to their sites of action in chromatin. FIGURE 10.75 shows a simplified model for their behavior. Typically, a site-specific binding protein recognizes its DNA target and directly interacts with the HAT complex. This determines the target for the HAT. HAT complexes sometimes also contain other subunits (and other enzymatic activities) that affect chromatin structure or act directly on transcription.

Just as activation of transcription is associated with acetyltransferases, so is inactivation associated with deacetylases. This is true both for individual genes and for heterochromatin. Repression at individual promoters may be accomplished by complexes that have deacetylase activities acting on localized regions in the vicinity of the promoter; these HDACs are typically recruited to their sites of action by sequence-specific repressors, just as HATs are recruited by activators. Absence of histone acetylation in heterochromatin is true of both constitutive heterochromatin (typically involving regions of centromeres or telomeres) and facultative heterochromatin (regions that are inactivated in one cell although they may be active in another). Typically, the N-terminal tails of histones H3 and H4 are not acetylated in heterochromatic regions.

There can also be direct interactions between remodeling complexes and histone-modifying complexes (or their modifications). Binding by the SWI/SNF remodeling complex may lead in turn to binding by an acetylase complex. Acetylation of histones may then in fact stabilize the association with the SWI/SNF complex, leading to a mutual reinforcement of the changes in the components at the promoter.

We can connect all of the events at the promoter into the series summarized in FIGURE 10.76. The initiating event is binding of a

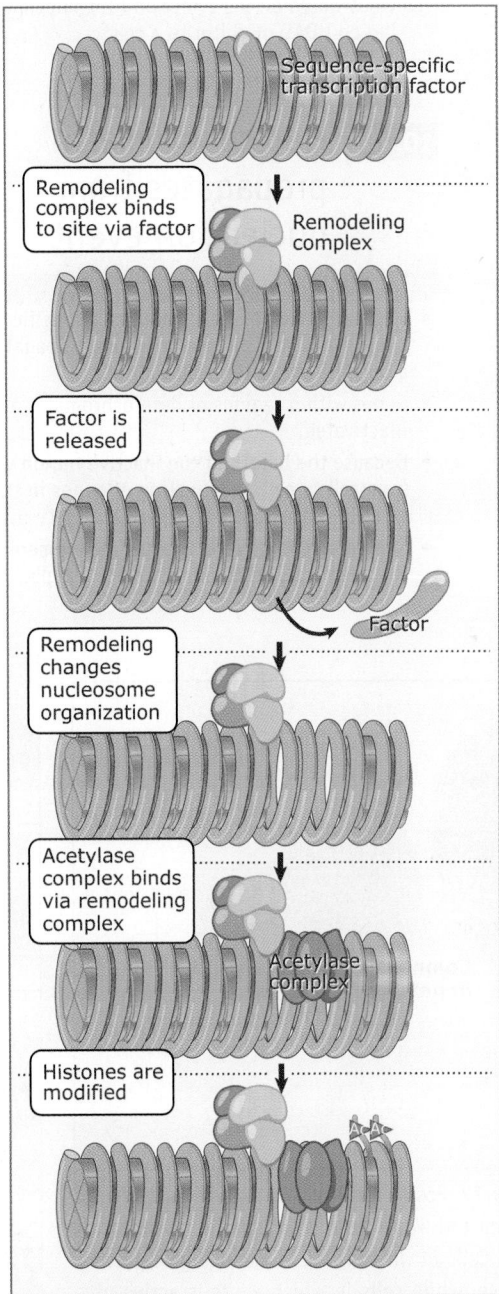

FIGURE 10.76 Promoter activation involves binding of a sequence-specific activator, recruitment and action of a remodeling complex, and recruitment and action of an acetylating complex.

Recruiting factor binds to DNA	(De)Acetylase acts on histone tails	Effectors act on chromatin or DNA
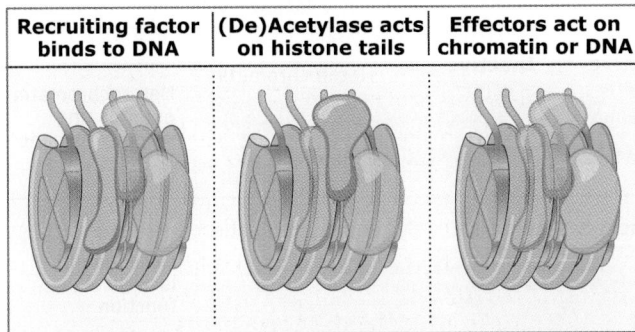		

FIGURE 10.75 Complexes that modify chromatin structure or activity are recruited to their sites of action. HAT or HDAC enzymes acetylate or deacetylate histones, and effector subunits may have other actions on chromatin or DNA.

sequence-specific component (which is able to find its target DNA sequence in the context of chromatin). This recruits a remodeling complex. Changes occur in nucleosome structure. An acetylating complex binds, and the acetylation of target histones provides a covalent mark that the locus has been activated. This is only one example; at some promoters histone modification precedes ATP-dependent remodeling.

Concept and Reasoning Check

1. What would be the effect on transcription of adding an HDAC inhibitor to a cell?

10.28 Heterochromatin propagates from a nucleation event

Key concepts

- Heterochromatin is nucleated at a specific sequence and the inactive structure propagates along the chromatin fiber.
- Genes within regions of heterochromatin are inactivated.
- Because the length of the inactive region varies from cell to cell, inactivation of genes in this vicinity causes position effect variegation.
- Similar spreading effects occur at telomeres and at the silent cassettes in yeast mating type.

An interphase nucleus contains both euchromatin and heterochromatin. The condensation state of heterochromatin is close to that of mitotic chromosomes. Heterochromatin remains condensed in interphase, is transcriptionally repressed, replicates late in S phase, and may be localized to the nuclear periphery. Centromeric heterochromatin typically consists of satellite DNAs. However, the formation of heterochromatin is not rigorously defined by sequence. When a gene is transferred, either by a chromosomal translocation or by transfection and integration, into a position adjacent to heterochromatin, it may become inactive as the result of its new location, implying that it has become heterochromatic.

Such inactivation is the result of an **epigenetic** effect. It may differ between individual cells in an animal and results in the phenomenon of **position effect variegation (PEV)**, in which genetically identical cells have different phenotypes. This has been well characterized in *Drosophila*. **FIGURE 10.77** shows an example of position effect variegation in the fly eye, in which some regions lack color while others are red, because the *white* gene is inactivated by adjacent heterochromatin in some cells, while it remains active in other cells.

The explanation for this effect is shown in **FIGURE 10.78**. Inactivation spreads from hetero-

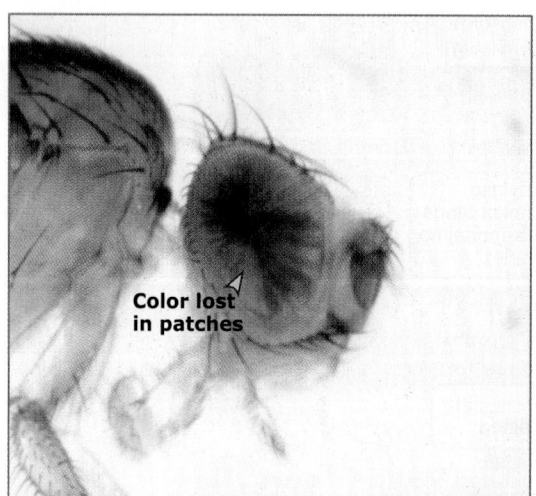

FIGURE 10.77 Position effect variegation in eye color results when the *white* gene is integrated near heterochromatin. Cells in which *white* is inactive give patches of white eye, while cells in which *white* is active give red patches. The severity of the effect is determined by the closeness of the integrated gene to heterochromatin. Photo courtesy of Steven Henikoff, Fred Hutchinson Cancer Research Center.

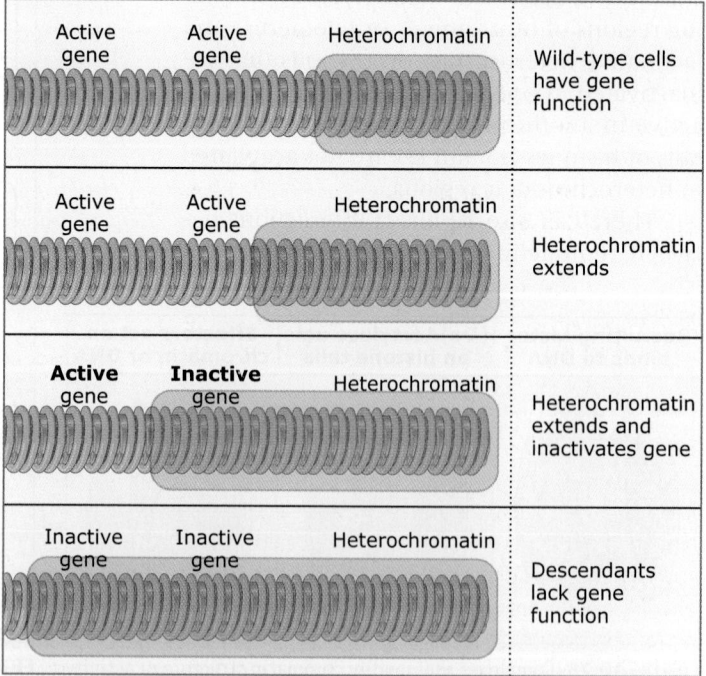

FIGURE 10.78 Extension of heterochromatin inactivates genes. The probability that a gene will be inactivated depends on its distance from the heterochromatic region.

chromatin into the adjacent region for a variable distance. In some cells, it goes far enough to inactivate a nearby gene, but in others, it does not. This happens at a certain point in embryonic development, and after that point, the state of the gene is inherited by all the progeny cells. Cells descended from an ancestor in which the gene was inactivated form patches corresponding to the phenotype of loss-of-function (in the case of *white*, absence of color).

The closer a gene lies to heterochromatin, the higher the probability that it will be inactivated. This suggests that the formation of heterochromatin may be a two-stage process: a *nucleation* event occurs at a specific sequence; and then the inactive structure *propagates* along the chromatin fiber. The distance for which the inactive structure extends is not precisely determined and may be stochastic, being influenced by parameters such as the quantities of limiting protein components. One factor that may affect the spreading process is the activation of promoters in the region; an active promoter may inhibit spreading.

Genes that are closer to heterochromatin are more likely to be inactivated and will, therefore, be inactive in a greater proportion of cells. On this model, the boundaries of a heterochromatic region might be terminated by exhausting the supply of one of the proteins that is required.

The effect of **telomeric silencing** in yeast is analogous to position effect variegation in *Drosophila*; genes translocated to a telomeric location show the same sort of variable loss of activity. This results from a spreading effect that propagates from the telomeres, described in *10.29 Heterochromatin depends on interactions with histones*.

A second form of silencing occurs in yeast. Yeast mating type is determined by the activity of a single active locus (*MAT*), but the genome contains two other copies of the mating type sequences (*HML* and *HMR*), which are maintained in an inactive form. The silent loci *HML* and *HMR* nucleate heterochromatin via binding of several proteins, which then lead to propagation of heterochromatin similar to that at telomeres. Heterochromatin in yeast exhibits features typical of heterochromatin in other species, such as transcriptional inactivity and self-perpetuating protein structures superimposed on nucleosomes (which are generally deacetylated). The only notable difference between yeast heterochromatin and that of most

other species is that histone methylation in yeast is not associated with silencing, whereas specific sites of histone methylation are a key feature of heterochromatin formation in most multicellular eukaryotes.

Concept and Reasoning Check

1. A strongly transcribed gene can sometimes heterochromatin spreading. Why might this be?

10.29 Heterochromatin depends on interactions with histones

Key concepts

- HP1 is the key protein in forming mammalian heterochromatin and acts by binding to methylated histone H3.
- Histone methylation and DNA methylation are linked in heterochromatin.
- Rap1 initiates formation of heterochromatin in yeast by binding to specific target sequences in DNA.
- The targets of Rap1 include telomeric repeats and silencers at *HML* and *HMR*.
- Rap1 recruits Sir4, which recruits Sir3 and the HDAC Sir2.
- Sir3/Sir4 interact with the N-terminal tails of H3 and H4.

Inactivation of chromatin occurs by the addition of proteins to the nucleosomal fiber. The inactivation may be due to a variety of effects, including condensation of chromatin to make it inaccessible to the apparatus needed for gene expression, addition of proteins that directly block access to regulatory sites, or the presence of proteins that directly inhibit transcription.

Two systems that have been characterized at the molecular level involve HP1 in mammals and the SIR complex in yeast. Although there are no detailed similarities between the proteins involved in each system, the general mechanism of reaction is similar: the points of contact in chromatin are the N-terminal tails of the histones.

HP1 (heterochromatin protein 1) was originally identified as a protein that is localized to heterochromatin by staining polytene chromosomes with an antibody directed against the protein. The original protein identified as HP1 is now called HP1α, since two related proteins, HP1β and HP1γ, have since been found.

HP1 contains a region called the *chromodomain* near the N-terminus and another domain that is related to it, called the *chromo-shadow* domain, at the C-terminus (see Figure 10.80). The importance of the chromodomain is indicated by the fact that it is the location of many of the mutations in HP1. The chromodomain is a common protein motif of 60 amino acids. It is found in proteins involved with either activating or repressing chromatin, suggesting that it represents a motif that participates in protein-protein interactions with targets in chromatin. Chromodomain(s) target proteins to heterochromatin by recognizing methylated lysines in histone tails.

Mutation of a deacetylase that acts on the H3 N-terminus prevents the methylation at K9. H3 that is methylated at K9 binds the protein HP1 via the chromodomain. This suggests the model for initiating formation of heterochromatin shown in **FIGURE 10.79**. First, the deacetylase acts to remove the acetylation in the H3 tail. Then a specific methylase acts on K9 of histone H3 to create the methylated signal to which HP1 will bind. **FIGURE 10.80** expands the reaction to show that the interaction occurs between the chromodomain and the methylated lysine. This is a trigger for forming inactive chromatin. **FIGURE 10.81** shows that the inactive region may then be extended by the

ability of further HP1 molecules to interact with one another.

Modification of DNA is also linked to silencing and heterochromatin formation in higher eukaryotes (and in the fission yeast *S. pombe*, but not in *S. cerevisiae*). Methylation of cytosine at CpG doublets is associated with gene inactivity. Methylation of DNA and methylation of histones is connected in a mutually reinforcing circuit. When HP1 binds to methylated K4 of histone H3, it can recruit DNA methyltransferases to modify CpG doublets. In turn, DNA methylation can in turn result in histone methylation. Some histone methyltransferase complexes (as well as some HDAC complexes) contain binding domains that recognize the methylated CpG doublet, so the DNA methylation reinforces the circuit by providing a target for the histone deacetylases and methyltransferases to bind.

While the yeast *S. cerevisiae* does not exhibit DNA methylation or heterochromatin-specific histone methylation, in other respects yeast heterochromatin is very similar to that of other eukaryotes. Heterochromatin formation at telomeres and silent mating type loci in yeast relies on an overlapping set of genes, known as *silent information regulators* (*SIR* genes).

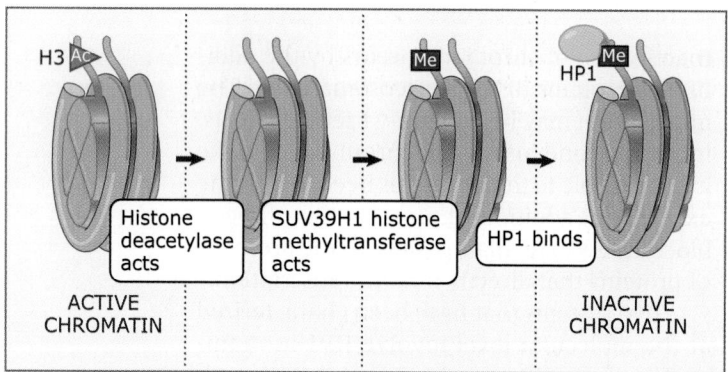

FIGURE 10.79 SUV39H1 is a histone methyltransferase that acts on [9]Lys of histone H3. HP1 binds to the methylated histone.

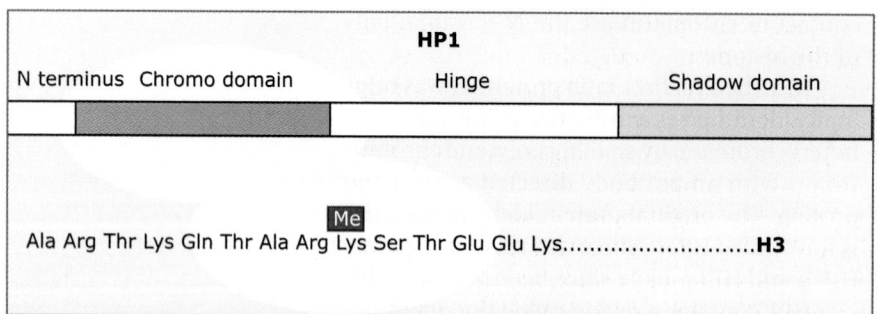

FIGURE 10.80 Methylation of histone H3 creates a binding site for HP1.

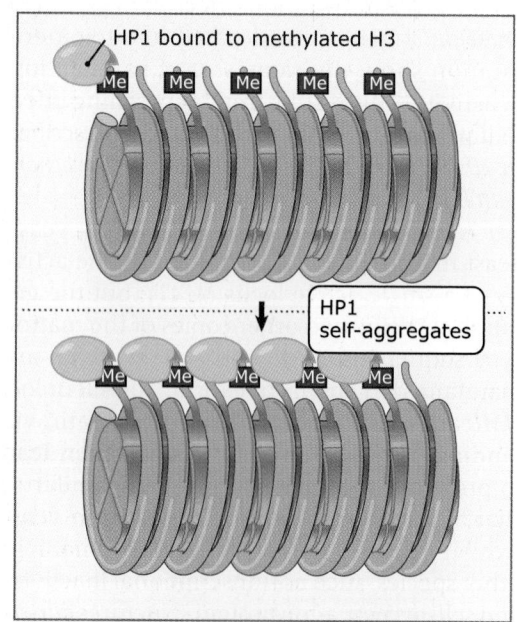

FIGURE 10.81 Binding of HP1 to methylated histone H3 forms a trigger for silencing because further molecules of HP1 aggregate on the nucleosome chain.

Mutations in *SIR2, SIR3,* or *SIR4* cause the two silent loci (*HML* and *HMR*) to become activated, and also relieve the inactivation of genes that have been integrated near telomeric heterochromatin. The products of these loci therefore function to maintain the inactive state of both types of heterochromatin.

FIGURE 10.82 proposes a model for actions of these proteins. Only one of them is a sequence-specific DNA-binding protein. This is Rap1, which binds to the $C_{1-3}A$ repeats at the telomeres and binds to the *cis*-acting silencer elements that are needed for repression of *HML* and *HMR*. The proteins Sir3 and Sir4 interact with Rap1 and with one another (they may function as a heteromultimer). Sir3/Sir4 interact with the N-terminal tails of the histones H3 and H4, with a preference for unacetylated tails. Another SIR protein, Sir2, is a deacetylase, and its activity is necessary to maintain binding of the Sir3/4 complex to chromatin.

Rap1 has the crucial role of identifying the DNA sequences at which heterochromatin forms. It recruits Sir4, which in turn recruits both its binding partner Sir3 and the HDAC Sir2. Sir3/Sir4 then interact directly with histones H3 and H4. Once Sir3/Sir4 have bound to histones H3/H4, the complex (including Sir2) may polymerize further and spread along the chromatin fiber. This inactivates the region, either because coating with Sir3/Sir4 itself has an inhibitory effect, or because Sir2-dependent deacetylation represses transcription. We do not know what limits the spreading of the complex. The C-terminus of Sir3 has a similarity to nuclear lamin proteins (constituents of the nuclear matrix) and may be responsible for tethering heterochromatin to the nuclear periphery.

A similar series of events forms the silenced regions at *HMR* and *HML*. Three sequence-specific factors are involved in triggering formation of the complex: Rap1, Abf1 (a transcription factor), and ORC (the origin replication complex). In this case, Sir1 binds to a sequence-specific factor and recruits Sir2, Sir3, and Sir4 to form the repressive structure.

How does a silencing complex repress chromatin activity? It could condense chromatin so that regulator proteins cannot find their targets. The simplest case would be to suppose that the presence of a silencing complex is mutually incompatible with the presence of transcription factors and RNA polymerase. The cause could be that silencing complexes block remodeling (and, thus, indirectly prevent fac-

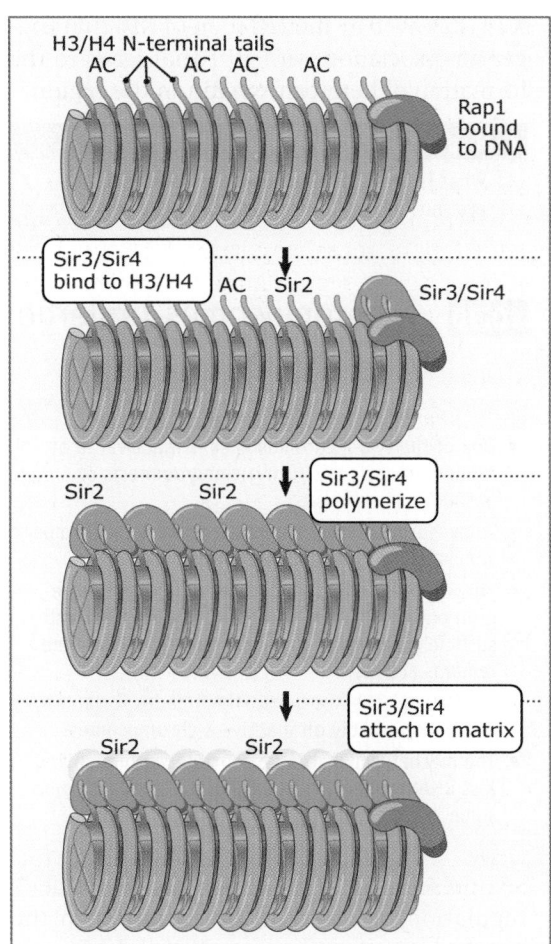

FIGURE 10.82 Formation of heterochromatin is initiated when RAP1 binds to DNA. SIR2 deacetylates histones to facilitate SIR3/SIR4 binding. SIR3/SIR4 bind to RAP1 and also to histones H3/H4. The complex polymerizes along chromatin and may connect telomeres to the nuclear matrix.

tors from binding) or that they directly obscure the binding sites on DNA for the transcription factors. However, the situation may not be this simple, because transcription factors and RNA polymerase can be found at promoters in silenced chromatin. This could mean that the silencing complex prevents the factors from working rather than from binding as such. In fact, there may be competition between gene activators and the repressing effects of chromatin, so that activation of a promoter inhibits spread of the silencing complex.

The specialized chromatin structure that forms at the centromere (see *10.8 The centromere binds a protein complex*) may be associated with the formation of heterochromatin in the region. In human cells, the centromere-specific protein CENP-B is required to initiate modifications of histone H3 (deacetylation of K9 and

K14, followed by methylation of K9) that trigger an association with HP1 that leads to the formation of heterochromatin in the region.

Concept and Reasoning Check

1. What is the role of sequence-specific binding proteins in heterochromatin formation?

10.30 X chromosomes undergo global changes

Key concepts

- One of the two X chromosomes is inactivated at random in each cell during embryogenesis of eutherian mammals.
- In exceptional cases where there are >2 X chromosomes, all but one are inactivated.
- The *Xic* (X-inactivation center) is a *cis*-acting region on the X chromosome that is necessary and sufficient to ensure that only one X chromosome remains active.
- *Xic* includes the *Xist* gene, which codes for an RNA that is found only on inactive X chromosomes.
- The mechanism that is responsible for preventing Xist RNA from accumulating on the active chromosome is unknown.

Sex presents an interesting problem for gene regulation, because of the variation in the number of X chromosomes. If X-linked genes were expressed equally well in each sex, females would have twice as much of each product as males. The importance of avoiding this situation is shown by the existence of **dosage compensation,** which equalizes the level of expression of X-linked genes in the two sexes. Mechanisms used in different species are summarized in **FIGURE 10.83** as given below:

- In mammals, one of the two female X chromosomes is inactivated. The result is that females have only one active X chromosome, which is the same situation found in males. The active X chro-

mosome of females and the single X chromosome of males are expressed at the same level.

- In *Drosophila*, the expression of the single male X chromosome is doubled relative to the expression of each female X chromosome.
- In *C. elegans*, the expression of each female X chromosome is halved relative to the expression of the single male X chromosome.

The common feature in all these mechanisms of dosage compensation is that *the entire chromosome is the target for regulation*. A global change occurs that quantitatively affects all of the promoters on the chromosome. We know most about the inactivation of the X chromosome in mammalian females, where the entire chromosome becomes heterochromatic.

The twin properties of heterochromatin are its condensed state and associated inactivity. It can be divided into the following two types:

1. Constitutive heterochromatin contains specific sequences that have no coding function. Typically these include satellite DNAs and are often found at the centromeres. These regions are invariably heterochromatic because of their intrinsic nature.

2. Facultative heterochromatin takes the form of chromosome regions or entire chromosomes that are inactive in one cell lineage, although they can be expressed in other lineages. The example par excellence is the mammalian X chromosome. The inactive X chromosome is perpetuated in a heterochromatic state, while the active X chromosome is part of the euchromatin. Thus, identical DNA sequences are involved in both states. Once the inactive state has been established, it is inherited by descendant cells. This is an example of epigenetic inheritance, because it does not depend on the DNA sequence.

Our basic view of the situation of the female mammalian X chromosomes was formed by the **single X hypothesis** in 1961. Female mice that are heterozygous for X-linked coat color mutations have a variegated phenotype in which some areas of the coat are wild-type but others are mutant. **FIGURE 10.84** shows that this can be explained *if one of the two X chromosomes is inactivated at random in each cell of a*

Mammals	Flies	Worms
Inactivate one ♀ X	Double expression ♂ X	Halve expression ♀ 2X

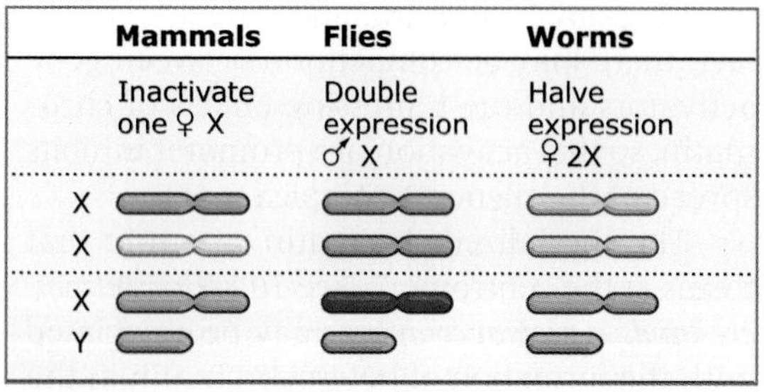

FIGURE 10.83 Different means of dosage compensation are used to equalize X chromosome expression in male and female.

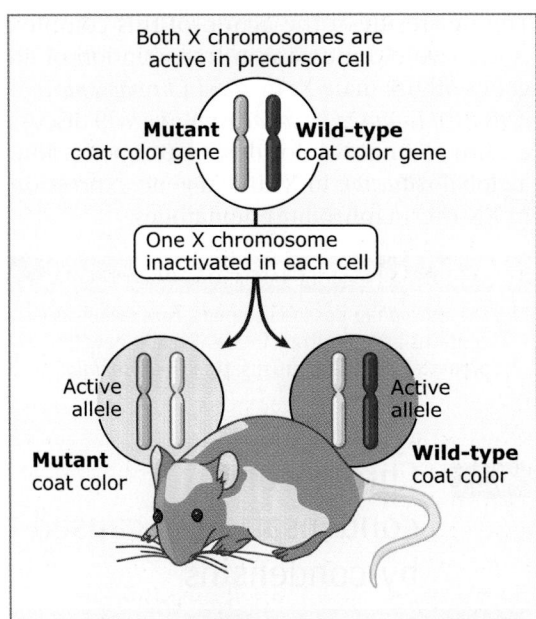

Both X chromosomes are
active in precursor cell

Mutant
coat color gene

Wild-type
coat color gene

One X chromosome
inactivated in each cell

Active
allele

Mutant
coat color

Active
allele

Wild-type
coat color

FIGURE 10.84 X-linked variegation is caused by the random inactivation of one X chromosome in each precursor cell. Cells in which the +allele is on the active chromosome have wild phenotype, but cells in which the −allele is on the active chromosome have mutant phenotype.

small precursor population. Cells in which the X chromosome carrying the wild-type gene is inactivated give rise to progeny that express only the mutant allele on the active chromosome. Cells derived from a precursor where the other chromosome was inactivated have an active wild-type gene. In the case of coat color, cells descended from a particular precursor stay together and, thus, form a patch of the same color, creating the pattern of visible variegation. In other cases, individual cells in a population will express one or the other of X-linked alleles; for example, in heterozygotes for the X-linked locus G6PD, any particular red blood cell will express only one of the two allelic forms. Interestingly, random inactivation of one X chromosome occurs in eutherian mammals, while in marsupials, the choice is directed: it is always the X chromosome inherited from the father that is inactivated.

Inactivation of the X chromosome in females is governed by the **n−1 rule**; however many X chromosomes are present, all but one will be inactivated. In normal females, there are of course two X chromosomes, but in rare cases where nondisjunction has generated a 3X or greater genotype, only one X chromosome remains active. This suggests a general model in which a specific event is limited to one X chromosome and protects it from an inactivation mechanism that applies to all the others.

A single locus on the X chromosome is sufficient for inactivation. When a translocation occurs between the X chromosome and an autosome, this locus is present on only one of the reciprocal products, and only that product can be inactivated. By comparing different translocations, it is possible to map this locus, which is called the *Xic* (X-inactivation center). A cloned region of 450 kb contains all the properties of the *Xic*. When this sequence is inserted as a transgene on to an autosome, the autosome becomes subject to inactivation (in a cell culture system).

Xic is a *cis*-acting locus that contains the information necessary to count X chromosomes and inactivate all copies but one. Inactivation spreads from *Xic* along the entire X chromosome. When *Xic* is present on an X chromosome-autosome translocation, inactivation spreads into the autosomal regions (although the effect is not always complete).

Xic is a complex genetic locus that expresses several long noncoding RNAs. The most important of these is a gene called *Xist* that is stably expressed only on the *inactive* X chromosome. The behavior of this gene is effectively the opposite from all other loci on the chromosome, which are turned off. Deletion of *Xist* prevents an X chromosome from being inactivated. However, it does not interfere with the counting mechanism (because other X chromosomes can be inactivated). So we can distinguish two features of *Xic*: an unidentified element(s) required for counting and the *Xist* gene required for inactivation.

FIGURE 10.85 illustrates the role of *Xist* RNA in X-inactivation. *Xist* codes for an RNA that lacks open reading frames. The *Xist* RNA "coats" the X chromosome, from which it is synthesized, suggesting that it has a structural role. Prior to X-inactivation, it is synthesized by both female X chromosomes. Following inactivation, the RNA is found only on the inactive X chromosome. The transcription rate remains the same before and after inactivation, so the transition depends on posttranscriptional events. An antisense RNA generated from *Xic*, *Tsix*, plays a role in regulating the stability of *Xist*. *Tsix* is active on the future active X, but downregulated on the future inactive X; this regulation allows persistence of *Xist* on the inactive X and thus leads to silencing.

Prior to X-inactivation, *Xist* RNA decays with a half-life of ~2 hr. X-inactivation is medi-

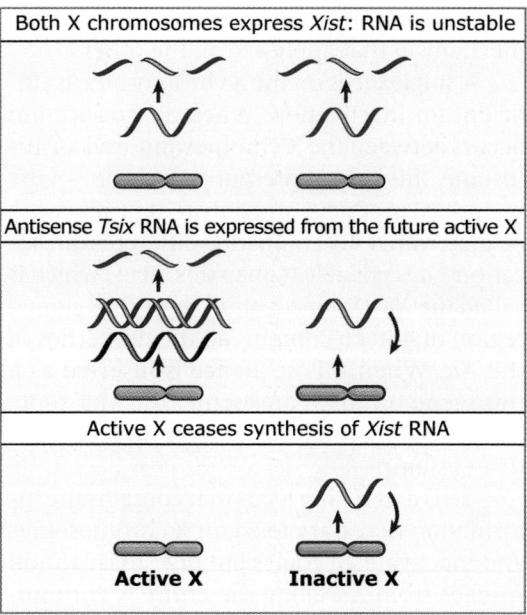

Both X chromosomes express *Xist*: RNA is unstable

Antisense *Tsix* RNA is expressed from the future active X

Active X ceases synthesis of *Xist* RNA

Active X **Inactive X**

FIGURE 10.85 X-inactivation involves stabilization of *Xist* RNA, which coats the inactive chromosome. *Tsix* prevents *Xist* expression on the future active X.

ated by stabilizing the *Xist* RNA on the inactive X chromosome. The *Xist* RNA shows a punctate distribution along the X chromosome, suggesting that association with proteins to form particulate structures may be the means of stabilization. We do not know yet what other factors may be involved in this reaction and how the *Xist* RNA is limited to spreading in *cis* along the chromosome. Accumulation of *Xist* on the future inactive X results in exclusion of transcription machinery (such as RNA polymerase II), and leads to a series of chromosome-wide histone modifications including H4 deacetylation, and specific methylation of both H3 and H4. Late in the process, an inactive X-specific histone variant, macroH2A, is incorporated into the chromatin (see Figure 10.44), and promoter DNA is methylated (see *10.29 Heterochromatin depends on interactions with histones*). At this point, the heterochromatic state of the inactive X is stable, and *Xist* is not required to maintain the silent state. Global changes also occur in other types of dosage compensation. In *Drosophila*, a large ribonucleoprotein complex, MSL, is found only in males, where it localizes on the X chromosome. This complex contains two noncoding RNAs, which appear to be needed for localization to the male X (perhaps analogous to the localization of *Xist* to the inactive mammalian X) and a histone acetyltransferase that acetylates histone H4 on K16 throughout the male X.

The net result of the action of this complex is the twofold increase in transcription of all genes on the male X. In *10.31 Chromosome condensation is caused by condensins,* we will discuss a third mechanism for dosage compensation, a global *reduction* in X-linked gene expression in XX (hermaphrodite) nematodes.

Concept and Reasoning Check

1. The *Drosophila* dosage compensation complex contains an acetyltransferase. Explain how this promotes dosage compensation in *Drosophila*.

10.31 Chromosome condensation is caused by condensins

Key concepts

- SMC proteins are ATPases that include the condensins and the cohesins.
- A heterodimer of SMC proteins associates with other subunits.
- Cohesins are responsible for holding sister chromatids together.
- Condensins cause chromatin to be more tightly coiled by introducing positive supercoils into DNA.
- Condensins are responsible for condensing chromosomes at mitosis.
- Chromosome-specific condensins are responsible for condensing inactive X chromosomes in *C. elegans*.

The structures of entire chromosomes are influenced by interactions with proteins of the **SMC** (structural maintenance of chromosome) family. They are ATPases that fall into two functional groups. **Condensins** are involved with the control of overall structure and are responsible for the condensation into compact chromosomes at mitosis. **Cohesins** are concerned with connections between sister chromatids that must be released at mitosis. Both consist of dimers formed by SMC proteins. Condensins form complexes that have a core of the heterodimer SMC2-SMC4 associated with other (non-SMC) proteins. Cohesins have a similar organization based on the heterodimeric core of SMC1-SMC3, and also interact with non-SMC partners.

FIGURE 10.86 shows that an SMC protein has a coiled-coil structure in its center, interrupted by a globular flexible hinge region. Both the amino and carboxyl termini have ATP- and DNA-binding motifs. SMC monomers fold at the hinge region, forming an antiparallel inter-

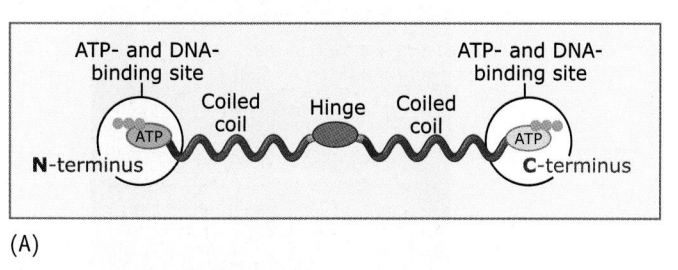

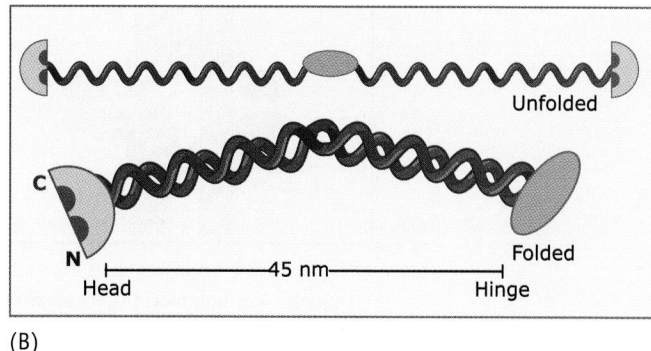

(A)   (B)

FIGURE 10.86 (A) An SMC protein has a "Walker module" with an ATP-binding motif and DNA-binding site at each end, connected by coiled coils that are linked by a hinge region. (B) SMC monomers fold at the hinge region and interact along the length of the coiled coils. The N- and C-termini interact to form a head domain. Reprinted, with permission, from the Annual Reviews of *Cell and Developmental Biology*, Volume 24 © 2008 by Annual Reviews www.annualreviews.org. Additional permission courtesy of Itay Onn, Howard Hughes Medical Institute.

action between the two halves of each coiled coil. This allows the amino and carboxyl termini to interact to form a "head" domain.

Folded SMC proteins form dimers via several different interactions. The most stable association occurs between hydrophobic domains in the hinge regions. **FIGURE 10.87** shows that these hing-hinge interactions result in V-shaped structures. Electron microscopy shows that in solution, cohesins tend to form V's with the arms separated by large angle, whereas condensins form more linear structures, with only a small gap between the arms. In addition, the heads of the two monomers can interact, closing the V, and the coiled coils of the individual monomers may also interact with each other. Various non-SMC proteins interact with SMC dimers and can influence the final structure of the dimer.

The function of cohesins is to hold sister chromatids together, but it is not yet clear how this is achieved. There are several different models for cohesin function. **FIGURE 10.88** shows one model in which a cohesin could take the form of extended dimers, interacting hinge-to-hinge, that crosslink two DNA molecules. Head-head interactions would create tetrameric structures, adding to the stability of cohesion. An alternative "ring" model is shown in **FIGURE 10.89**. In this model, dimers interact at both their head and hinge regions to form a circular structure. Instead of binding directly to DNA, a structure of this type could hold DNA molecules together by encircling them.

While cohesins act to hold separate sister chromatids together, condensins are responsible for chromatin condensation. **FIGURE 10.90** shows that a condensin could take the form

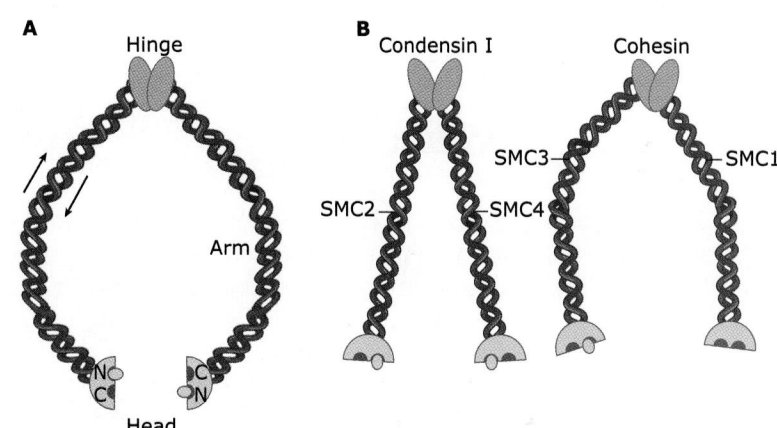

FIGURE 10.87 The basic architecture of condensin and cohesin complexes. Condensin and cohesin consist of V-shaped dimers of two SMC proteins interacting through their hinge domains. The two monomers in a condensin dimer tend to exhibit a very small separation between the two arms of the V, while cohesins have a much larger angle of separation between the arms. Adapted from T. Hirano, *Nat. Rev. Mol. Cell Biol.* 7 (2006): 311–322.

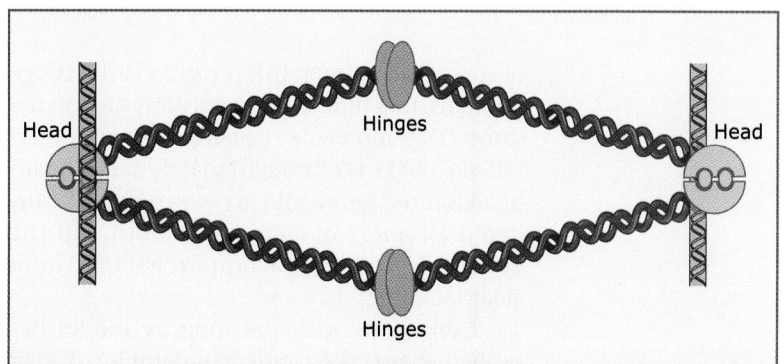

FIGURE 10.88 One model for DNA linking by cohesins. Cohesins may form an extended structure in which each monomer binds DNA and connects via the hinge region, allowing two different DNA molecules to be linked. Head domain interactions can result in binding by two cohesin dimers. Reproduced from Annual Reviews by Itay Onn, et al., Copyright 2008 by Annual Reviews, Inc. Reproduced with permission of Annual Reviews, Inc. in the format of Textbook via Copyright Clearance Center.

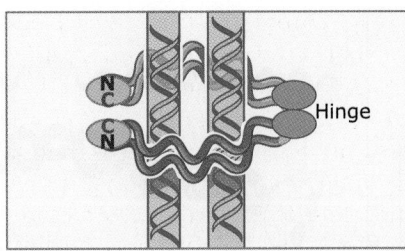

FIGURE 10.89 Cohesins may dimerize by intramolecular connections, then forming multimers that are connected at the heads and at the hinge. Such a structure could hold two molecules of DNA together by surrounding them.

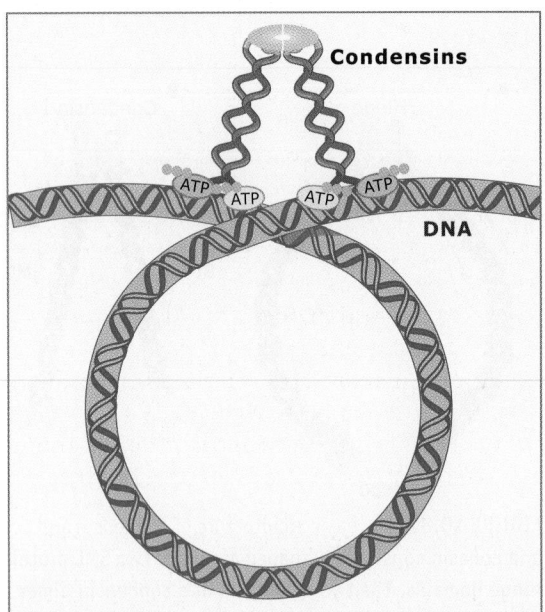

FIGURE 10.90 Condensins may form a compact structure by bending at the hinge, causing DNA to become compacted.

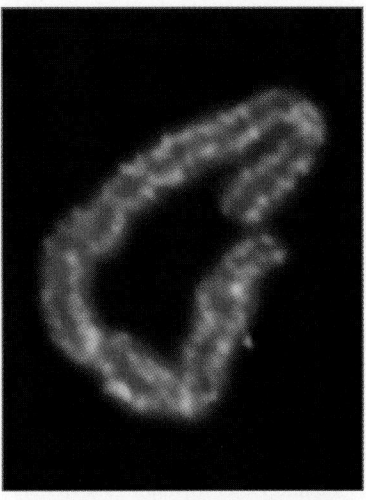

FIGURE 10.91 Condensins are located along the entire length of a mitotic chromosome. DNA is red; condensins are yellow. Photo courtesy of Ana Losada and Tatsuya Hirano, Cold Spring Harbor Laboratory.

of a V-shaped dimer, interacting via the hinge domains, that pulls together distant sites on the same DNA molecule, causing it to condense (FIGURE 10.91). It is thought that dynamic head-head interactions could act to promote the ordered assembly of condensed loops, but the details of condensin action are still far from clear.

Consistent with the looping model described above, the condensin complex has an ability to introduce positive supercoils into DNA in an action that uses hydrolysis of ATP and depends on the presence of topoisomerase I. This ability is controlled by the phosphorylation of non-SMC subunits, which occurs at mitosis. It is now know how this connects with other modifications of chromatin, such as the phosphorylation of histones, which is also linked to chromosome condensation.

We discussed in *10.30 X chromosomes undergo global changes*, the dramatic chromosomal changes that occur during X-inactivation in female mammals and in X chromosome upregulation in male flies (see *10.30 X chromosomes undergo global changes*). In the nematode *C. elegans*, a third approach is used: 2-fold *reduction* of X-chromosome transcription in XX hermaphrodites relative to XO males. A dosage compensation complex (DCC) is maternally provided to both XX and XO embryos, but it then associates with both X chromosomes only in XX animals, while remaining diffusely distributed in XO nuclei. The DCC contains a core of SMC proteins, and is similar to the condensin complexes that are associated with mitotic chromosomes in *C. elegans* and other species. This suggests that it has a structural role in causing the chromosome to take up a more condensed, inactive state. Multiple sites on the X chromosome may be needed for the complex to be fully distributed along it, and short DNA sequence motifs have been identified that appear to be key for localization of DCC. The complex binds to these sites and then spreads along the chromosome to cover it more thoroughly.

Dosage compensation in mammals and *Drosophila* both entail chromosome-wide changes in histone acetylation, and involve noncoding RNAs that play central roles in targeting X chromosomes for global change. In *C. elegans*, chromosome condensation by condensin homologs is used to accomplish dosage compensation. It remains to be seen whether there are also global changes in the histone acetylation or other modifications in XX *C. elegans* that reflect the two-fold reduction in transcription of the X chromosomes.

Concept and Reasoning Check

1. SMC homologs have not been implicated in mammalian or *Drosophila* dosage compensation. If SMCs did have a role in dosage compensation, would you expect it to be more likely in mammals or *Drosophila*? Why?

10.32 What's next?

Since the discovery of the nucleosome revolutionized the study of chromatin, there have been two concurrent lines of research as given below:

1. The analysis of structure, both to describe the nucleosome itself and to describe how the nucleosome is organized into higher-order structures, and
2. The analysis of function in terms of the nucleosome, to relate the events that occur in activating and during transcription to the structure of chromatin.

The most immediately pressing questions along these lines are to define the structure of the 30-nm chromatin filament in terms of the nucleosomes, linker histones and other constituent proteins; and to resolve how the transcription-activating and -inactivating apparati act via modification of nucleosomes to achieve their effects.

In the 1970s, histones were seen as general repressors of transcription whose effects had to be neutralized in order for genes to be expressed. There was, however, little insight into how this might be accomplished beyond the belief that they somehow had to be removed from chromatin in order to enable activators to get at DNA. Consider how much more we know now, and we can see that we are on the verge of achieving a definition of chromatin function in terms of structure. There is terrific progress in defining the roles of the acetylating, deacetylating, and other enzymes that modify histones, and we are beginning to see how they change chromatin structure in a localized way to enable activation of a promoter. We are close to defining promoter activation in terms of these structural interactions. And viewing histones as repressors, we are now beginning to understand how their interactions with other proteins create heterochromatic structures that propagate epigenetically and may inactivate local or even quite wide regions of chromatin. We can expect to see all these events described in terms of increased resolution of the structures of individual components, that is, in terms of the molecular changes in individual histones and other proteins of chromatin.

Understanding of hierarchical organization at higher levels, that is, of the chromosome itself, is still difficult. In spite of the enormous progress made in genome analysis, we do not understand the significance of the structural features of bands and interbands, of the concentrations of G-C-rich regions, and so on. We cannot yet relate these features to interactions in terms of DNA and proteins; even when the structure of the 30-nm fiber is resolved, there will still be much to learn about the higher orders of structure. Other distinctive features of the chromosome, most crucially the centromere and telomere, are yielding to description in terms of molecular components, although we have yet to understand the role of DNA sequences in the higher eukaryotic centromere.

As in other areas of biology, the ability to relate increased structural resolution to functional changes is providing powerful insights at deeper levels of understanding.

10.33 Summary

The genetic material of all organisms and viruses takes the form of tightly packaged nucleoprotein. In eukaryotes, transcriptionally active sequences reside within the euchromatin that constitutes the majority of interphase chromatin. The regions of heterochromatin are packaged ~5-10× more compactly and are transcriptionally inert. All chromatin becomes densely packaged during cell division, when the individual chromosomes can be distinguished. The existence of a reproducible ultrastructure in chromosomes is indicated by the production of G-bands by treatment with Giemsa stain. The bands are very large regions, ~10^7 bp that can be used to map chromosomal translocations or other large changes in structure.

In eukaryotes, interphase chromatin and metaphase chromosomes both appear to be organized into large loops. Each loop may be an independently supercoiled domain. The bases of the loops are connected to a metaphase scaffold or to the nuclear matrix by specific DNA sites.

The centromeric region contains the kinetochore, which is responsible for attaching a chromosome to the mitotic spindle. The centromere often is surrounded by heterochromatin. Centromeric sequences have been identified only in yeast *S. cerevisiae*, where they consist of short conserved elements—*CDE-I* and *CDE-III*, which bind Cbf1 and the CBF3 complex, respectively—and a long A · T-rich region called *CDE-II* that binds the histone variant Cse4 to form a specialized structure in chromatin. Another group of proteins that binds to this assembly provides the connection to microtubules.

Telomeres make the ends of chromosomes stable. Almost all known telomeres consist of multiple repeats in which one strand has the general sequence $C_n(A/T)_m$, where $n > 1$ and $m = 1$-4. The other strand, $G_n(T/A)_m$, has a single protruding end that provides a template for addition of individual bases in defined order. The enzyme telomerase is a ribonucleoprotein, whose RNA component provides the template for synthesizing the G-rich strand. This overcomes the problem of the inability to replicate at the very end of a duplex. The telomere stabilizes the chromosome end because the overhanging single strand $G_n(T/A)_m$ displaces its homologue in earlier repeating units in the telomere to form a loop, so there are no free ends.

Lampbrush chromosomes of amphibians and polytene chromosomes of insects have unusually extended structures with packing ratios <100. Polytene chromosomes of *D. melanogaster* are divided into ~5000 bands, varying in size by an order of magnitude, with an average of ~25 kb. Transcriptionally active regions can be visualized in even more unfolded ("puffed") structures, in which material is extruded from the axis of the chromosome. This may resemble the changes that occur on a smaller scale when a sequence in euchromatin is transcribed.

All eukaryotic chromatin consists of nucleosomes. A nucleosome contains a characteristic length of DNA, usually ~200 bp, wrapped around an octamer containing two copies each of histones H2A, H2B, H3, and H4. A single H1 protein can be associated with each nucleosome. Virtually all genomic DNA is organized into nucleosomes. Treatment with micrococcal nuclease shows that the DNA packaged into each nucleosome can be divided operationally into two regions. The linker region is digested rapidly by the nuclease; the core region of 146 bp is resistant to digestion. Histones H3 and H4 are the most highly conserved and an $H3_2 · H4_2$ tetramer accounts for the diameter of the particle. The H2A and H2B histones are organized as two H2A · H2B dimers. Octamers are assembled by the successive addition of two H2A · H2B dimers to the $H3_2 · H4_2$ kernel.

Nucleosomes are organized into a fiber of 30-nm diameter, which has six nucleosomes per turn and a packing ratio of 40. Removal of H1 allows this fiber to unfold into a 10-nm fiber that consists of a linear string of nucleosomes. The 30-nm fiber probably consists of the 10-nm fiber wound into a two-start solenoid. The 30-nm fiber is the basic constituent of both euchromatin and heterochromatin; nonhistone proteins are responsible for further organization of the fiber into chromatin or chromosome ultrastructure.

There are two pathways for nucleosome assembly. In the RC pathway, the PCNA processivity subunit of the replisome recruits CAF-1, which is a nucleosome assembly factor. CAF-1 assists the deposition of $H3_2 · H4_2$ tetramers onto the daughter duplexes resulting from replication. The tetramers may be produced either by disruption of existing nucleosomes by the replication fork or as the result of assembly from newly synthesized histones. Similar sources provide the H2A · H2B dimers that then assemble with the $H3_2 · H4_2$ tetramer to complete the nucleosome. Because the $H3_2 · H4_2$ tetramer and the H2A · H2B dimers assemble at random, the new nucleosomes may include both preexisting and newly synthesized histones. HIRA assembles nucleosomes outside of S phase, and ASF1 acts both during and outside replication to assemble chromatin.

RNA polymerase displaces histone octamers during transcription. Nucleosomes reform on DNA after the polymerase has passed, unless transcription is very intensive (such as in rDNA) when they may be displaced completely. The RI pathway for nucleosome assembly is responsible for replacing histone octamers that have been displaced by transcription. It uses the histone variant H3.3 instead of H3. A similar pathway, with another alternative to H3, is used for assembling nucleosomes at centromeric DNA sequences following replication.

Two types of changes in sensitivity to nucleases are associated with gene activity. Chromatin capable of being transcribed has a generally increased sensitivity to DNase I, reflecting a change in structure over an extensive region that can be defined as a domain containing active or potentially active genes. Hypersensitive sites in DNA occur at discrete locations and are identified by greatly increased sensitivity to DNase I. A hypersensitive site consists of a sequence of ~200 bp from which nucleosomes are excluded by the presence of other proteins. A hypersensitive site forms a boundary that may cause adjacent nucleosomes to be restricted in position. Nucleosome positioning may be important in controlling access of regulatory proteins to DNA.

Genes whose control regions are organized in nucleosomes usually are not expressed. In the absence of specific regulatory proteins, promoters and other regulatory regions are organized by histone octamers into a state in which they cannot be activated. This may explain the need for nucleosomes to be precisely positioned in the vicinity of a promoter, so that essential regulatory sites are appropriately exposed. Some transcription factors have the capacity to recognize DNA on the nucleosomal surface, and a particular positioning of DNA may be required for initiation of transcription.

Active chromatin and inactive chromatin are not in equilibrium. Sudden, disruptive events are needed to convert one to the other. Chromatin remodeling complexes have the ability to slide or displace histone octamers by a mechanism that involves hydrolysis of ATP. Remodeling complexes range from small to extremely large and are classified according to the type of the ATPase subunit. Two common types are SWI/SNF, ISWI, CHD, and INO80. A typical form of this chromatin remodeling is to displace one or more histone octamers from specific sequences of DNA, creating a boundary that results in the precise or preferential positioning of adjacent nucleosomes. Chromatin remodeling may also involve changes in the positions of nucleosomes, sometimes involving sliding of histone octamers along DNA.

Extensive covalent modifications occur on histone tails, all of which are reversible. Acetylation of histones occurs at both replication and transcription and could be necessary to form a less compact chromatin structure. Some coactivators that connect transcription factors to the basal apparatus have histone acetyltransferase activity. Conversely, repressors may be associated with deacetylases. The modifying enzymes are usually specific for particular amino acids in particular histones. The most common sites for modification are located in the N-terminal tails of histones H3 and H4, which extrude from nucleosomes between the turns of DNA. The activating (or repressing) complexes are usually large and often contain several activities that undertake different modifications of chromatin.

The formation of heterochromatin occurs by proteins that bind to specific chromosomal regions (such as telomeres) and that interact with histones. The formation of an inactive structure may propagate along the chromatin thread from an initiation center. Similar events occur in silencing of the inactive yeast mating type loci.

Formation of heterochromatin may be initiated at certain sites and then propagated for a distance that is not precisely determined. When a heterochromatic state has been established, it is inherited through subsequent cell divisions. This gives rise to a pattern of epigenetic inheritance, in which two identical sequences of DNA may be associated with different protein structures and, therefore, have different abilities to be expressed. This explains the occurrence of position effect variegation in *Drosophila*.

Modification of histone tails is a trigger for chromatin reorganization. Acetylation is generally associated with gene activation. Histone acetyltransferases are found in activating complexes, and histone deacetylases are found in inactivating complexes. Histone methylation at specific sites is associated with gene inactivation, as is DNA methylation. Some histone modifications may be exclusive or synergistic with others.

Inactive chromatin at yeast telomeres and silent mating type loci appears to have a common structure and involves the interaction of certain proteins with the N-terminal tails of histones H3 and H4. Formation of the inactive complex may be initiated by binding of one protein to a specific sequence of DNA; the other components may then polymerize in a cooperative manner along the chromosome.

Inactivation of one X chromosome in female (eutherian) mammals occurs at random. The *Xic* locus is necessary and sufficient to count the number of X chromosomes. The n−1 rule ensures that all but one X chromosome are inactivated. *Xic* contains the gene *Xist*, which codes for an RNA that is expressed only on the inactive X chromosome. Stabilization of *Xist* RNA is the mechanism by which the inactive X chromosome is distinguished.

SMC proteins can control the global structure of chromosomes. Cohesins hold sister chromatids together, while condensins are responsible for chromosome condensation during mitosis (and meiosis). Specialized homologs of condensins are used to reduce X chromosome gene expression for dosage compensation in *C. elegans*.

http://biology.jbpub.com/lewin/cells

To explore these topics in more detail, visit this book's Interactive Student Study Guide.

References

Ahmad, K., and Henikoff, S. 2002. The histone variant H3.3 marks active chromatin by replication-independent nucleosome assembly. *Mol. Cell* v. 9 p. 1191–1200.

Black, B. E., and Bassett, E. A. 2008. The histone variant CENP-A and centromere specification. *Curr. Opin. Cell Biol.* v. 20 p. 91–100.

Hirano, T. 2006. At the heart of the chromosome: SMC proteins in action. *Nat. Rev. Mol. Cell Biol.* v. 7 p. 311–322.

Li, B., Carey, M., and Workman, J. L. 2007. The role of chromatin during transcription. *Cell* v. 128 p. 707–719.

Luger, K., et al. 1997. Crystal structure of the nucleosome core particle at 28 Å resolution. *Nature* v. 389 p. 251–260.

Payer, B., and Lee, J. T. 2008. X chromosome dosage compensation: How mammals keep the balance. *Ann. Rev. Genet.* v. 42 p. 733–772.

Vignali, M., Hassan, A. H., Neely, K. E., and Workman, J. L. 2000. ATP-dependent chromatin-remodeling complexes. *Mol. Cell Biol.* v. 20 p. 1899–1910.

Workman, J. L. 2006. Nucleosome displacement during transcription. *Genes Dev.* v. 20 p. 2507–2512.

Yuan, G. C., et al. 2005. Genome scale identification of nucleosome positions in *S. cerevisiae*. *Science* v. 309 p. 626–630.

Zhang, Y., and Reinberg, D. 2001. Transcription regulation by histone methylation: Interplay between different covalent modifications of the core histone tails. *Genes Dev.* v. 15 p. 2343–2360.

The cytoskeleton

4

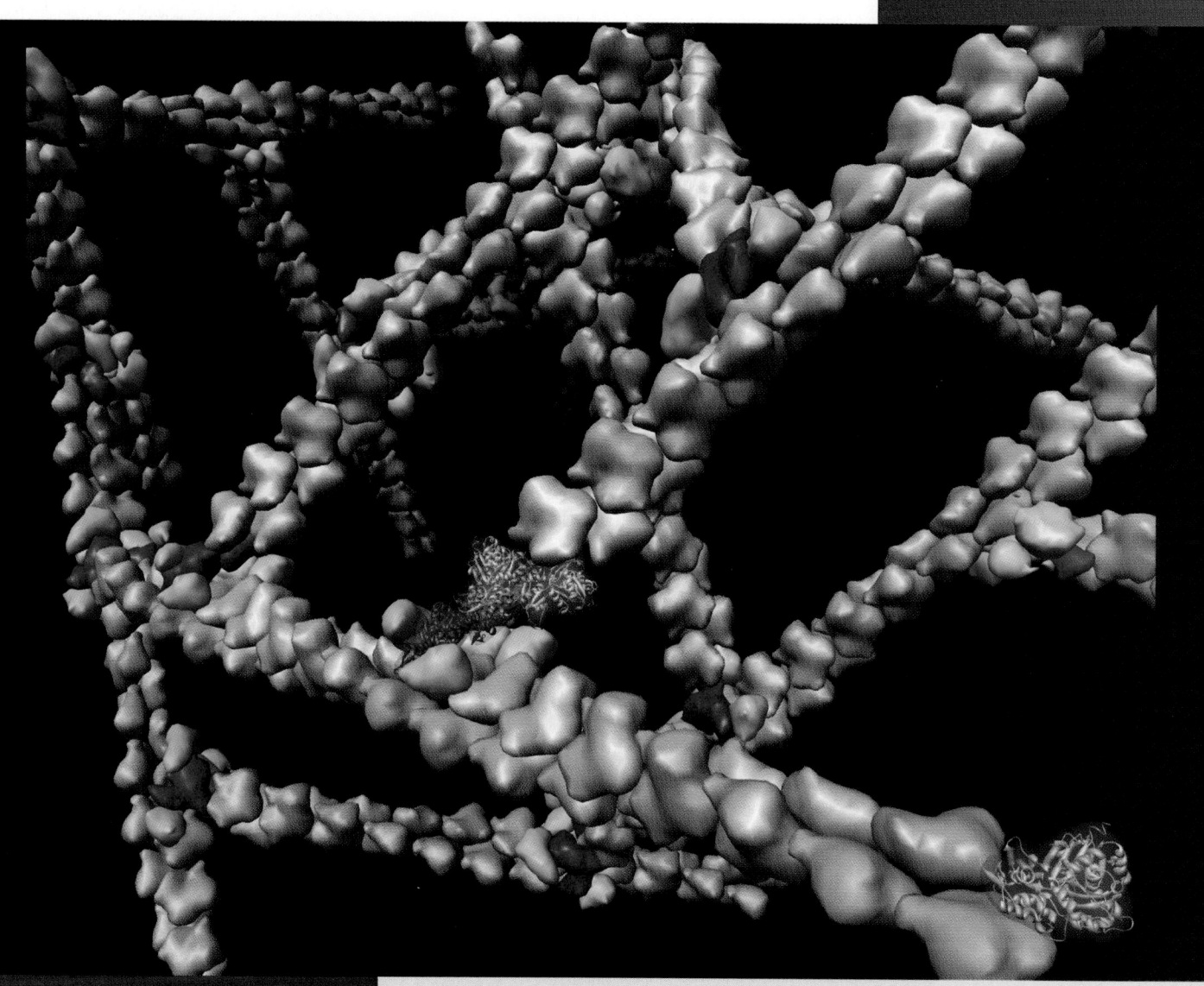

Photo courtesy of University of California in San Francisco. Image created with UCSF Chimera (www.rbvi.ucsf.edu/chimera/).

Microtubules

11

Lynne Cassimeris

Department of Biological Sciences, Lehigh
University, Bethlehem, PA

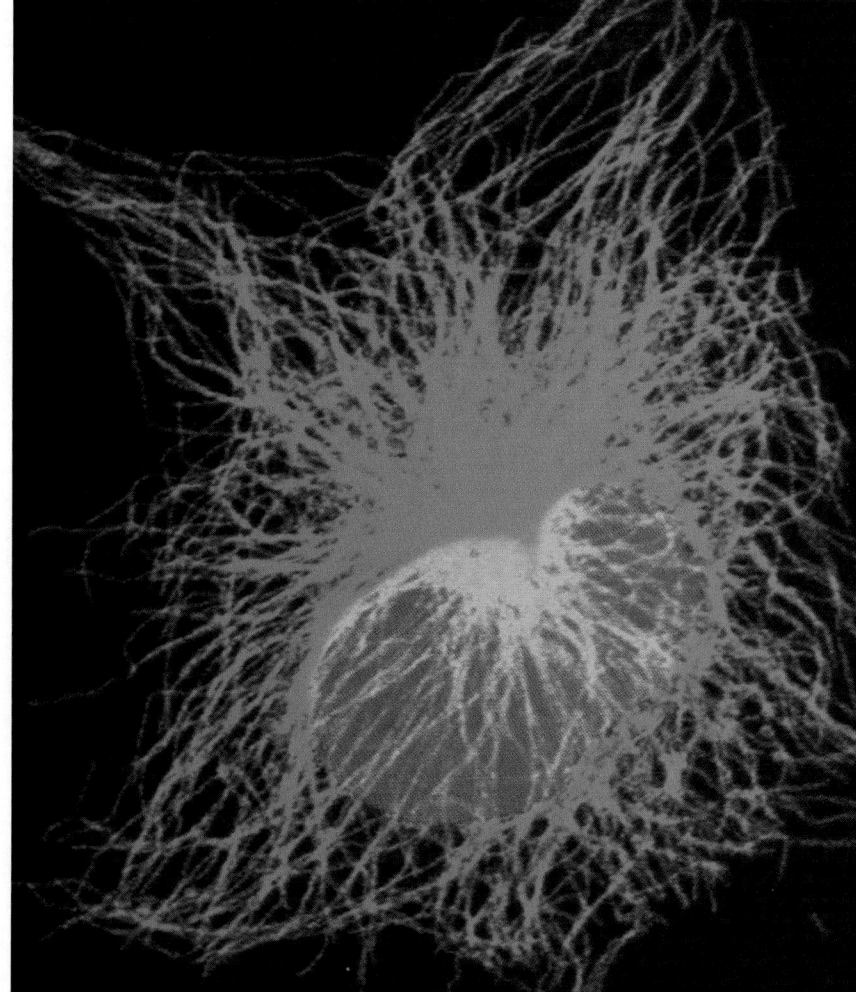

HUMAN EPITHELIAL CELL showing microtubules (green) and DNA (red). Photo courtesy of Lynne Cassimeris, Lehigh University.

CHAPTER OUTLINE

11.1 Introduction

Key concepts

- The cytoskeleton is made up of protein polymers. Each polymer contains many thousands of identical subunits that are strung together to make a filament.
- The cytoskeleton generates cell movements and provides mechanical support for the cell.
- Cells have three types of cytoskeletal polymers: actin filaments, intermediate filaments, and microtubules.
- All cytoskeletal polymers are dynamic; they continually gain and lose subunits.
- Microtubules are polymers of tubulin subunits.
- Microtubules almost always function in concert with molecular motors that generate force and move vesicles and other complexes along the microtubule surface.
- Cilia and flagella are specialized organelles composed of microtubules and motor proteins that propel a cell through fluid or move fluid over the surface of a cell.
- Drugs that disrupt microtubules have medicinal and agricultural uses.

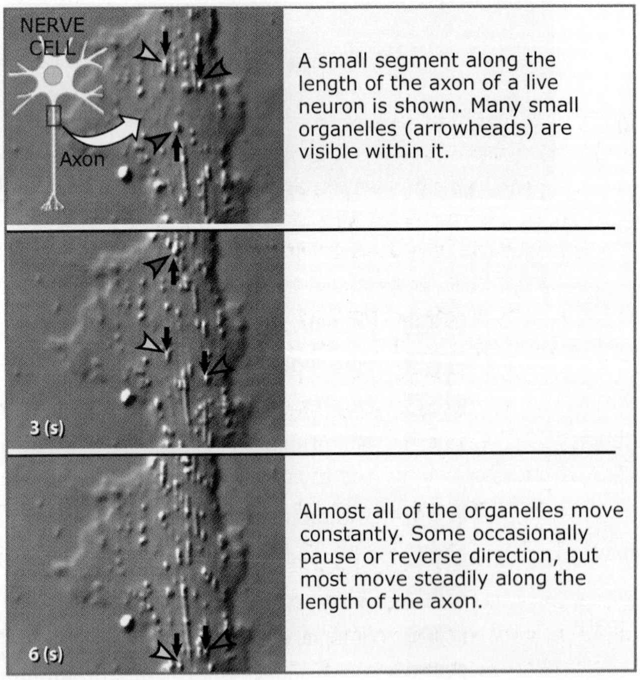

The cytoplasm of eukaryotic cells is in constant motion as organelles are continually transported from place to place. Such motion is particularly clear in the cytoplasm of a large, elongated cell such as the neuron shown in **FIGURE 11.1**. Motion occurs in all other types of cells as well. Movement of organelles serves many purposes. Secretory vesicles leave the Golgi apparatus near the center of the cell and are transported to the plasma membrane, where they empty their contents into the extracellular space. At the same time, vesicles internalized at the plasma membrane are transported to endosomes. Mitochondria constantly move about, and the ER stretches and reorganizes. In mitotic cells, chromosomes first align and then move to opposite sides of the cell. Movement of organelles and chromosomes to the right place, at the right time, is accomplished by the **cytoskeleton**, a collection of proteins that form the roadways of the cell's transportation system and the motors that run on them.

The cytoskeleton also has several other important functions: it powers the movements of motile cells and provides the organization and structural support that define the shapes of all cells. Many types of cells move themselves, either within the body (animal cells) or through the environment (single-celled organisms and some gametes). Cells like the white blood cells that track and destroy bacterial invaders crawl across a surface. Others, like sperm cells, swim through fluid to reach their destination. The cytoskeleton powers and guides both these forms of cell locomotion. In addition to its role in motility, the cytoskeleton organizes the internal structure of the cell and defines up from down, left from right and front from back. By determining the general organization of the cytoplasm, the cytoskeleton determines the overall shape of the cell, making it possible to have rectangular epithelial cells or neurons with long, thin axons and dendrites that can stretch for up to a meter in humans.

The cytoskeleton is composed of three major types of structural proteins: **microtubules**, microfilaments (see *12 Actin*), and intermediate filaments (see *13 Intermediate filaments*). These three types of proteins, shown in **FIGURE 11.2**, share several very general features. Each functions not as a single protein molecule, but as a polymer composed of many identical subunit proteins. Like stringing beads together to make a necklace, the cytoskeletal

NERVE CELL

Axon

A small segment along the length of the axon of a live neuron is shown. Many small organelles (arrowheads) are visible within it.

3 (s)

Almost all of the organelles move constantly. Some occasionally pause or reverse direction, but most move steadily along the length of the axon.

6 (s)

FIGURE 11.1 Three images, from a video, show an axon from a living neuron. An outline of the entire cell is shown in the sketch in the top panel. Three vesicles, marked by red, yellow, and blue arrowheads are followed over a 6 second interval. Two vesicles move toward the tip of the axon and one moves toward the cell body. Photos courtesy of Paul Forscher, Yale University.

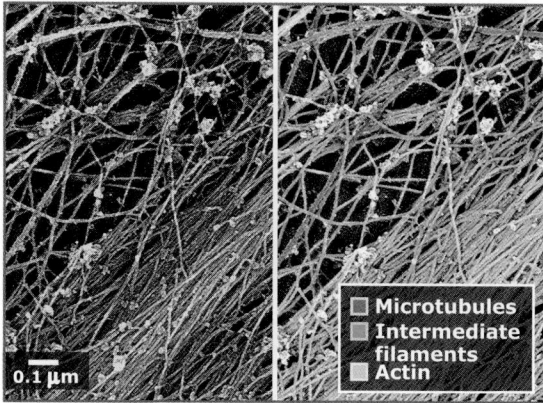

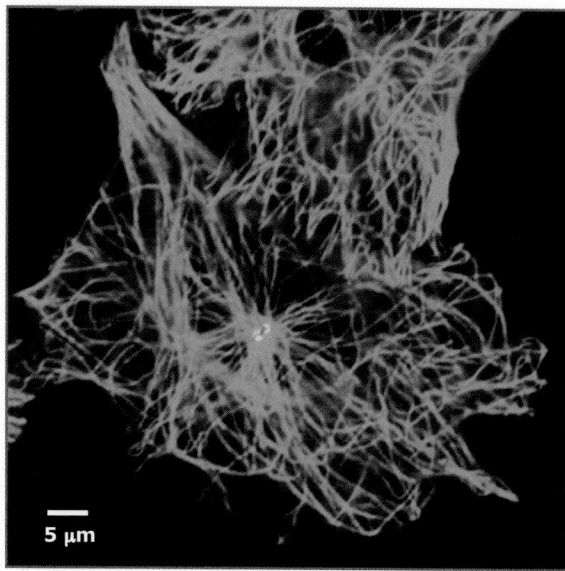

FIGURE 11.2 A small region of a fibroblast cell viewed by electron microscopy (left panel). Numerous filaments are visible. In the right panel, the three types of cytoskeletal polymers present in eukaryotic cells have been colored so that they can be easily distinguished from one another. Reprinted from *J. Struct. Biol.*, vol. 115, M. T. Svitkina, A. B. Verkhovsky, and G. G. Borisy, Improved Procedures for Electron Microsopic Visualization..., pp. 290–303, Copyright (1995) with permission from Elsevier [http://www.sciencedirect.com/science/journal/10478477]. Photos courtesy of Tatyana Svitkina, University of Pennsylvania.

FIGURE 11.3 The microtubules in a fibroblast cell, viewed after labeling them with a fluorescent dye so that they appear green. The microtubules are organized around a point (red) near the center of the cell and extend throughout the cytoplasm. Most of the microtubules are long enough to run from one part of the cell to another. Photo courtesy of Lynne Cassimeris, Lehigh University.

polymers are built in the cytoplasm by stringing together thousands of subunit proteins. A general feature of all cytoskeletal polymers is that they constantly gain and lose subunits, rather than form static structures. The dynamic turnover of the cytoskeletal polymers allows the cytoskeleton to reorganize itself, building new roadways for transport or new struts for support as needs change within a cell.

While the three cytoskeletal polymers share general features, each also has unique properties, making each polymer best suited to carry out specific tasks within the cell. The three polymer systems will be considered separately, although they often work together.

Here, we focus on microtubules. The basic subunit that forms a microtubule is the protein tubulin. Tubulin molecules assemble with one another to form a hollow tube about 25 nm in diameter, giving microtubules their name. A single microtubule can contain tens or hundreds of thousands of tubulin molecules and stretch for many microns, allowing it to extend over half the length of most eukaryotic cells. Interphase cells often contain hundreds of long microtubules that run throughout the

cytoplasm and connect one area of a cell to another, as shown in **FIGURE 11.3**.

Microtubules almost always function in concert with the molecular motors that move on them (see *11.12 How motor proteins work*). These motor proteins attach to various cargo, including organelles and vesicles, and pull them along the surface of the microtubule, much like trucks move cargo on a highway. Microtubules and motor proteins also work together to separate replicated chromosomes at mitosis (see *14 Mitosis*), and they form the core of motile structures used by cells to swim or move fluid over their surface (see *11.16 Cilia and flagella are motile structures*). Microtubules and motor proteins are even used by viruses, such as HIV and adenovirus, allowing the viruses to rapidly reach the nucleus and replicate.

Small organic molecules that disrupt microtubule polymerization have medicinal and agricultural uses. Drugs that make microtubules more or less stable block mitosis and are used to treat cancers. One such drug is paclitaxel (Taxol™) shown in **FIGURE 11.4**, which is used to treat ovarian and breast cancers. Taxol™ binds to microtubules and makes them more stable by preventing tubulin subunits from dis-

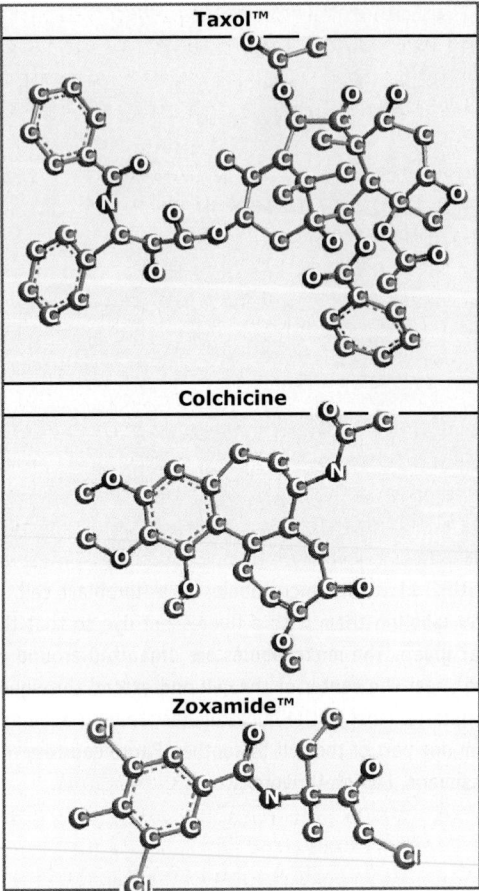

Taxol™

Colchicine

Zoxamide™

FIGURE 11.4 The structures of three small organic molecules that disrupt microtubule assembly or disassembly. Paclitaxel (Taxol™) and colchicine are natural products produced by specific plants (the Pacific yew tree and the meadow saffron, respectively). Zoxamide™ is a man-made molecule discovered by screening a large collection of small molecules for those with the ability to interfere with microtubules.

sociating. Colchicine (see Figure 11.4), another tubulin poison, has the opposite effect, causing a cell's microtubules to disappear. It is used to treat gout because disruption of the microtubule cytoskeleton blocks migration of the white blood cells responsible for inflammation in this disease. Small molecules targeted to tubulin also have important applications in agriculture. For example, Zoxamide™ (see Figure 11.4), a fungicide, binds specifically to fungal tubulins and prevents fungal growth. Zoxamide™ is used to control late blight in potatoes, the fungal disease responsible for the Irish potato famine of 1850. The search for new tubulin-binding drugs with medical or agricultural uses is an active area of research today.

11.2 General functions of microtubules

Key concepts

- Cells use microtubules to provide structural support because microtubules are the strongest of the cytoskeletal polymers. Microtubules resist compression.
- Cells also rely on the dynamic assembly and disassembly of microtubules to allow them to quickly reorganize the microtubule cytoskeleton.
- Cells can make their microtubules more or less dynamic, allowing them to take advantage of the adaptability of microtubules (when dynamic) or the strength of microtubules (when stable).
- Different cells can have unique organizations of microtubules to suit specific needs.

A simple experiment, to disrupt the cell's microtubules, illustrates the functions of this component of the cytoskeleton. The tubulin subunits strung together to make a microtubule can be depolymerized into individual subunits by treating cells with drugs like colchicine (see *11.1 Introduction*). These drugs block the formation of new microtubules, causing an imbalance in the continual formation and dissolution of the microtubule cytoskeleton. Microtubules that depolymerize cannot be replaced, soon leading to the loss of all microtubules from the cytoplasm. In general, most cells lose their shape and form round balls when their microtubules are depolymerized. The internal organization of the cell is also disrupted. The Golgi complex, which is normally present as a single structure located near the nucleus, fragments into pieces and disperses throughout the cell. The ER, normally a network that extends throughout the cytoplasm, collapses around the nucleus because it is connected to the nuclear envelope. All of these changes are reversed when the microtubule depolymerizing drug is subsequently removed: microtubules re-form into their original pattern, the cell regains its shape, and the ER and Golgi return to their normal positions. This simple experiment illustrates the broad functions of microtubules in cell organization, structure, and movement.

The functions of microtubules in cells depend upon two seemingly contradictory properties: microtubules can both act as stiff structural elements and they can easily fall apart. The tubular structure and relatively large diameter of microtubules make them relatively stiff and allow them to resist compression. Microtubules can be thought of as similar to a garden hose, which can bend over long distances but cannot be compressed. Unlike a hose, however, microtubules are exceptionally dynamic. If left to themselves, microtubules constantly either elongate or shorten by the addition or loss of subunits. The shortening of a microtubule is particularly dramatic, frequently covering most of its length and often making it fall apart and disappear altogether. Microtubules are inherently prone to falling apart once they have assembled, and cells often use other proteins to stabilize them and prevent their disassembly. Although a structural element designed to fall apart might seem odd, such instability has the great advantage of allowing the microtubules within a cell to be disassembled and reorganized within minutes when necessary. A common example of this, shown in **FIGURE 11.5**, is the dramatic rearrangement of microtubules that takes place at the beginning of mitosis, a reorganization that occurs within just a few minutes. Another example of microtubule reorganization occurs in some developing oocytes and illustrates how extensive it can be. A single oocyte of the frog *Xenopus laevis*, shown in **FIGURE 11.6**, is about 1 mm in diameter and contains roughly a half million microtubules of 600 μm average length. If all the subunits

in these microtubules were present as a single long microtubule, it would be about 300 m in length, close to the length of three football fields. Despite this large quantity of microtubules, the entire microtubule cytoskeleton is depolymerized and reorganized within 30 minutes when the oocyte is stimulated to mature into an egg.

In some cells, the dynamic nature of the microtubule cytoskeleton does more than just allow rapid remodeling from one type of array to another. A fibroblast, for example, must move within the body and be able to change direction. Its microtubules are organized in a starlike pattern, radiating outward in all directions from a single point near the nucleus, as shown in **FIGURE 11.7**. These microtubules are short-lived, most lasting only a fraction of the time it takes the cell to move any significant distance. A fibroblast can continue to move if its microtubules are all depolymerized. Intriguingly, however, without its microtubules it can no longer turn and navigate, suggesting that the dynamic nature of its microtubules is required to allow it to change direction.

A neuron is much different from a fibroblast in both shape (see Figure 11.7) and behavior and uses its microtubules for their strength.

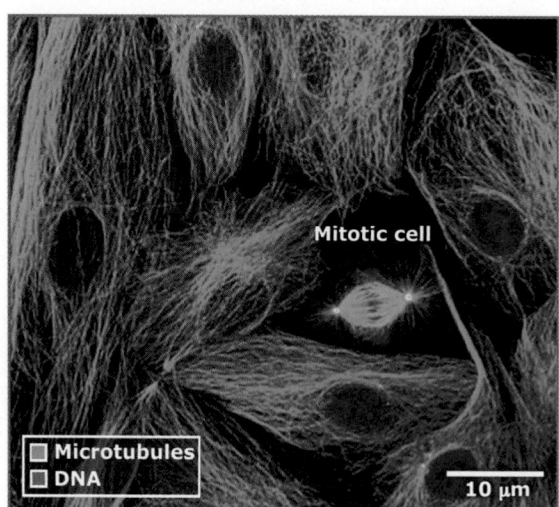

FIGURE 11.5 A field of about a dozen cells is shown, with their microtubules and chromosomes visualized by fluorescence microscopy. One mitotic cell—with its microtubules assembled into a very clear mitotic spindle—is surrounded by interphase cells. The reorganization of microtubules that takes place as a cell enters mitosis is dramatic but requires only a few minutes. Photo courtesy of Lynne Cassimeris, Lehigh University.

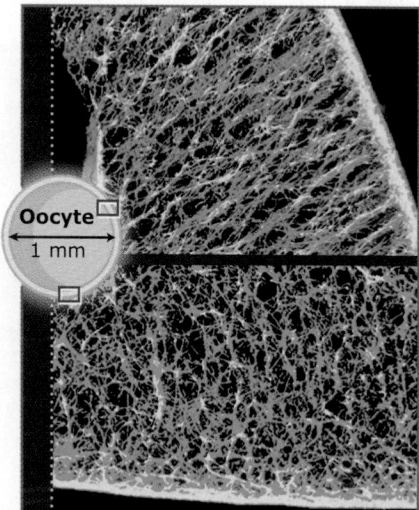

FIGURE 11.6 Mature oocytes of the frog *Xenopus laevis* are huge cells that are densely packed with microtubules. The two photographs show the microtubules at two points around the edge of an oocyte. Despite the enormous number of microtubules present in the oocyte and their great length, they can all be taken apart completely in only a few minutes. Photo courtesy of Dr. David L. Gard, University of Utah.

A neuron is immobile and has a small cell body from which long-lived processes (axons and dendrites) extend great distances. The interiors of the processes are packed with parallel microtubules. These microtubules carry large numbers of vesicles and other material to and from the synapse. Unlike the microtubules of the fibroblast, those in the process of a neuron are stable and are essential for the structure of the cell; if depolymerized, the processes slowly collapse. Neurons, then, take advantage of the strength of microtubules to use them as a stable structural element.

Although the mature neuron uses its microtubules for their strength, a growing neuron also uses the dynamic properties of microtubules. When neurons are first growing and forming synapses with other neurons, each extends from its cell body thin projections that will become the axon and dendrites. At the tip of each projection is a very active, motile region called a growth cone, which is clearly visible in **FIGURE 11.8**. Growth cones extend the projections by moving over long distances, the projections forming behind them as they travel. To help them move, growth cones have dynamic microtubules that function very much like those in a locomoting fibroblast. Thus, neurons can regulate when and where their microtubules are dynamic and when and where they are stable. The ability to regulate the dynamic turnover of microtubules in time and space is a general feature of all cells.

Microtubules are organized to suit the unique needs of each cell. An example, shown in **FIGURE 11.9**, is for two cells with similar shapes: a single-celled fission yeast, *Schizosaccharomyces pombe*, and a nucleated red blood cell from a nonmammalian vertebrate such as a chicken or a frog. In each case, the cell

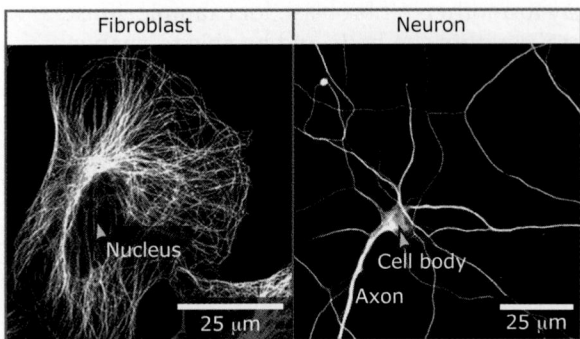

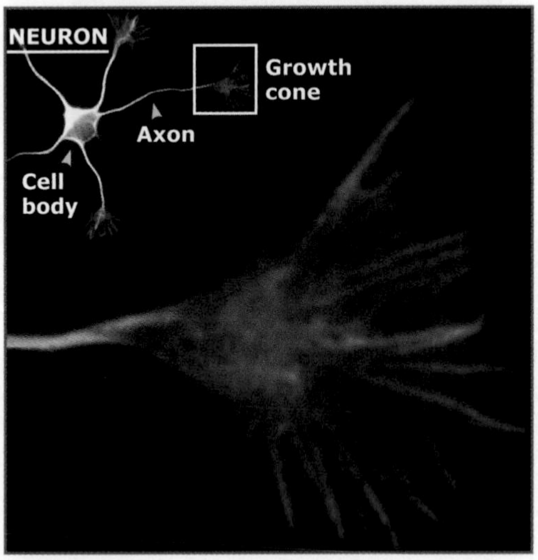

FIGURE 11.7 The extremely different shapes of these two types of cells require that their microtubules be organized differently. In the human fibroblast, individual microtubules are visible and run throughout the cytoplasm from a point near the nucleus. In the neuron, the microtubules are packed together within the long, thin projections that extend from the cell body. Left photo courtesy of Lynne Cassimeris, Lehigh University; right photo courtesy of Ginger Withers, Whitman College.

FIGURE 11.8 In the upper left is a picture of an entire neuron with several axons projecting from its cell body. Each axon has a growth cone (blue) at its tip. A single growth cone with its axon leading away to the right is shown enlarged in the main photograph. Microtubules are in red and actin filaments in blue. Photo courtesy of Drs. Leif Dehmelt, Max-Planck Institute, Dortmund, and Shelly Halpain, University of California, San Diego.

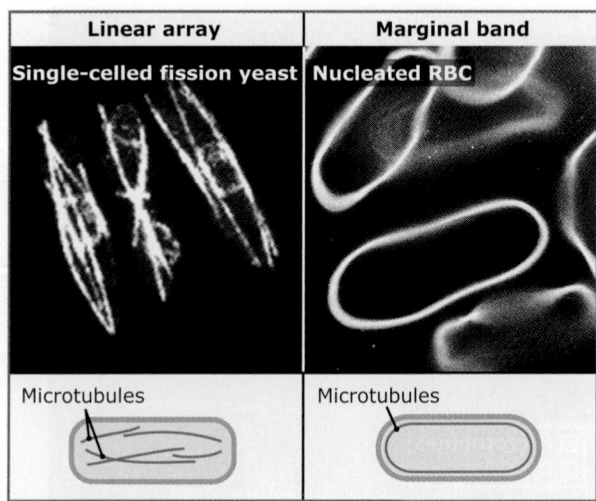

FIGURE 11.9 The yeast *S. pombe* (left) has a relatively small number of bundled microtubules that center the nucleus within the cell and transport factors that regulate growth to its ends. In an amphibian's red blood cells (right), a circular band of microtubules underlies the plasma membrane and helps a cell resist deformation as it squeezes through the capillaries. Left photo courtesy of Phong T. Tran, University of Pennsylvania; right photo courtesy of Lynne Cassimeris, Lehigh University.

is shaped like a sausage, but the microtubule organizations are very different. In *S. pombe*, bundles of microtubules extend toward the cell tips, where new material is deposited for polarized cell growth. The bundles of microtubules also position the nucleus in the center of the cell. *S. pombe* does not need to use its microtubules to resist compression because it has a cell wall to provide this function. A very different microtubule organization is found in the red blood cells, which, like all animal cells, lack any form of cell wall. These cells have bundles of microtubules associated with the plasma membrane in a structure called a marginal band. The marginal band microtubules provide strength to the cell membrane, much like ankryin and spectrin do in mammalian red blood cells.

These examples illustrate the general functions of the microtubule cytoskeleton and raise many questions. How are microtubules put together and taken apart so rapidly? How do cells regulate the dynamic assembly and disassembly of microtubules? What determines the organization of the microtubules within a cell? How does the microtubule cytoskeleton generate movement? This chapter will answer these questions.

11.3 Microtubules are polar polymers of α- and β-tubulin

Key concepts

- Microtubules are hollow polymers of tubulin heterodimers.
- Thirteen linear chains of subunits, called protofilaments, associate laterally to form the microtubule.
- Lateral bonds between protofilaments stabilize the microtubule and limit subunit addition and subtraction to microtubule ends.
- Microtubules are polarized polymers. The plus end is crowned by β-tubulin and assembles faster. The minus end is crowned by α-tubulin and assembles slower.

Microtubules participate in a wide range of cell functions, yet the underlying microtubule structure, and the subunits from which it is built, are virtually identical from yeast to humans. The next several sections will describe the structure of microtubules and how they assemble (and disassemble) from purified subunits in the test tube. Although how microtubules behave in a test tube may seem a bit remote from what happens in cells, it is critical

to understand because it establishes what the basic properties and behaviors of microtubules are. These in turn determine what a cell must do in order to organize its microtubules and how it can put them to use.

The building block of a microtubule is the protein tubulin. Tubulin is a heterodimer, made up of two closely related proteins, α- and β-tubulin. The two proteins share about 40% sequence identity and are never found alone. Instead, one molecule of α-tubulin and one of β-tubulin are always found associated together, making up the 100 kDa tubulin heterodimer. For simplicity, the heterodimer is usually just called "tubulin" in recognition of the fact that it always functions as a single unit. Since both α- and β-tubulin are nearly spherical in shape, the heterodimer resembles a peanut, as illustrated in **FIGURE 11.10**. Its structure is known at atomic resolution: α- and β-tubulin have very similar structures and are arranged "in line" in the dimer such that the front of one is bound to the back of the other.

Each molecule of α- and β-tubulin binds a molecule of GTP. The structure of the tubulin heterodimer shows that the GTP bound to α-tubulin is located near the interface with β-tubulin (see Figure 11.10). This GTP is never hydrolyzed and does not exchange with nucleotides in solution. In contrast, the GTP-bound to β-tubulin is exposed at one end of the heterodimer, where it can exchange

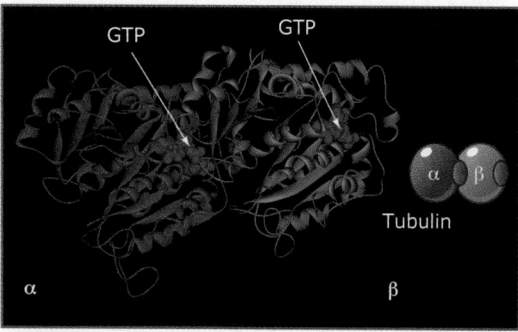

FIGURE 11.10 The three-dimensional structure of the tubulin heterodimer, the basic building block of a microtubule. In purple and green colors are the polypeptide backbones of the two protein subunits, and in blue are the two molecules of GTP that are bound to each heterodimer. Note the similarity in the structures of the two subunits, and that they are arranged head to tail in the complex. At the right is a schematic drawing that shows how the dimer will be represented in the figures in the chapter. Structure from Protein Data Bank 1TUB. E. Nogales, S. G. Wolf, and K. H. Downing, *Nature* 391 (1998): 199–203.

with nucleotide in solution (see Figure 11.10). β-tubulin's GTP is hydrolyzed to GDP during microtubule assembly. Hydrolysis of GTP to GDP is thought to result in a change in con-

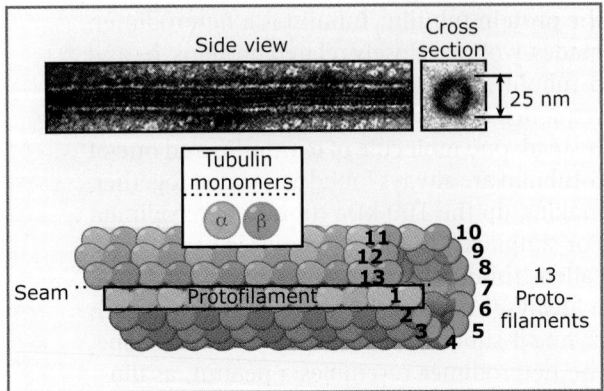

FIGURE 11.11 The structure of a small segment of a microtubule. The individual tubulin heterodimers are aligned end to end in straight protofilaments, and the protofilaments are arranged side by side to form a hollow tube. All of the heterodimers have the same orientation, with the β subunit toward one end of the microtubule and the alpha subunit toward the other. At the top are electron micrographs of microtubules assembled from pure tubulin. Left photo courtesy of Lynne Cassimeris, Lehigh University; right micrograph provided by Harold Erickson, Duke University School of Medicine.

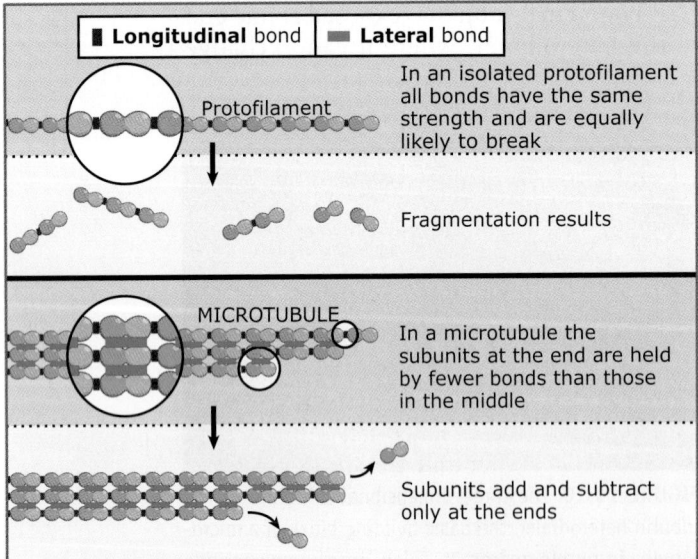

FIGURE 11.12 Tubulin heterodimers form both longitudinal bonds (along the length of the protofilament) and lateral bonds (between subunits in adjacent protofilaments). Single protofilaments are prone to fragmenting because all bonds are of equal strength and are equally likely to break. In a microtubule, the existence of lateral bonds makes it unlikely that the polymer will break in the middle. Subunits add or subtract only at the ends, where each subunit is held by fewer bonds than in the middle of the microtubule. For clarity, only three protofilaments are shown.

formation of the tubulin heterodimers, which plays a significant role in the dynamic turnover of microtubules.

Microtubules are protein polymers composed of thousands of tubulin subunits organized into a hollow tube. Typically, a microtubule is made up of 13 linear chains of subunits that run parallel to one another along its length; each linear chain is termed a **protofilament**. These protofilaments, shown in **FIGURE 11.11**, associate laterally to form the microtubule. While it is possible to form a microtubule from anywhere between 11 and 15 protofilaments, the large majority of microtubules found in cells have 13 protofilaments. With 13 protofilaments a microtubule has a diameter of 25 nm, about five times the thickness of its wall.

Tubulin heterodimers bind head to tail along the length of each protofilament (see Figure 11.11). Within most adjacent protofilaments, α-tubulins are next to other α-tubulins and β-tubulins next to other β-tubulins. Because adjacent protofilaments are slightly out of register with one another, there is a single discontinuity, or seam, where the α-tubulins in one protofilament are adjacent to the β-tubulins in the next. It is likely that this seam plays an important role in microtubule assembly.

Within a microtubule, each tubulin heterodimer forms extensive noncovalent bonds with its neighbors. As diagrammed in **FIGURE 11.12**, these noncovalent bonds form longitudinally as well as laterally between dimers in a protofilament, linking adjacent protofilaments. A single longitudinal bond is stronger than a single lateral bond, but it is the presence of numerous lateral bonds that makes the polymer strong. It is easy to illustrate why this is so. Subunits in an isolated protofilament have only longitudinal bonds. The bonds are all of equal strength, so they all have the same probability of breaking. This makes a single protofilament prone to fragmentation (see Figure 11.12). In a microtubule, subunits in the middle of the polymer also form lateral bonds to the subunits in the adjacent protofilaments. For a microtubule to break, all 13 of its protofilaments would have to simultaneously lose their longitudinal bonds at the same point. This is highly unlikely, making it very rare that a microtubule breaks.

The presence of lateral bonds between subunits in a microtubule also makes it unlikely that tubulin dimers in the center of the microtubule will come unbound from the polymer, since release of a dimer from the middle of the microtubule requires simultaneous breakage of several bonds. Instead, subunits add and subtract only at filament ends, where fewer bonds connect each subunit to the rest of the microtubule (see Figure 11.12).

Purified tubulin heterodimers will spontaneously form the lateral and longitudinal bonds necessary to form a microtubule. Therefore, microtubule assembly is a **self-assembly** process, in which all the information needed to build the final structure is contained within its subunits. Other examples of self-assembly processes include actin filament polymerization and the assembly of some viral capsids.

The arrangement of subunits within a microtubule makes its two ends different (see Figure 11.11). Within each protofilament all the tubulin heterodimers have the same orientation, and within a microtubule, all of the protofilaments run in the same direction. Thus, a cap of β-tubulins is exposed at one end of the microtubule (called the plus end), while at the opposite end a ring of α-tubulins is exposed (called the minus end). This organization gives a microtubule two very important properties. First, its two ends are structurally different and can behave differently. As we shall see, cells often make use of this, for example, by regulating assembly at the two ends independently. Second, each microtubule has a polarity—an inherent directionality—and can be thought of as pointing in one direction or another. This polarity is present throughout the microtubule, not just at its ends. Just by looking at the surface of a microtubule at any point—even far from either end—it is possible to tell which direction leads to the plus end and which leads to the minus end. The polarity of microtubules allows them to act as directional tracks for molecular motor proteins and is essential for the roles played by both the microtubules and the motors in organizing the interior of a cell.

Within cells, microtubules are highly organized with respect to their polarity, as illustrated in **FIGURE 11.13**. In a fibroblast or other type of cell with a radial array of microtubules, for example, all of the microtubules are arranged with their minus ends near the center of the cell and their plus ends near the periphery. In epithelial cells the microtubules are arranged parallel to one another, running from top to bottom of the cell, all with their plus ends at the bottom of the cell and their minus ends at the top. Similarly, all of the microtubules in an axon point in the same direction. In each case

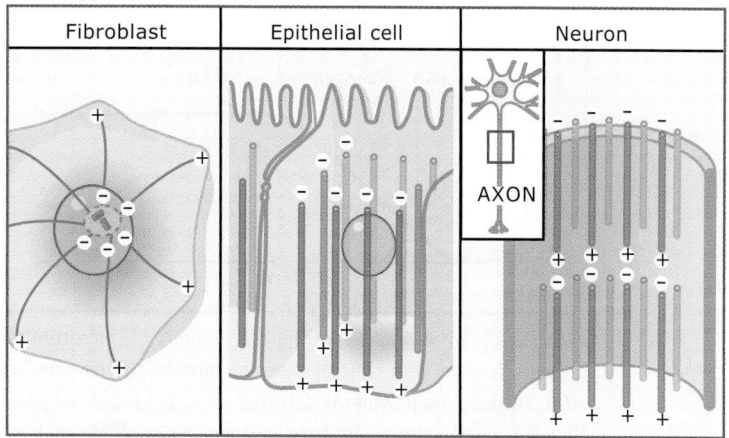

FIGURE 11.13 In each of the three types of cells, all of the microtubules within the cell have the same polarity. This allows the polarity of a microtubule to be used as an accurate indicator of direction in the cytoplasm. In the epithelial cell, for example, the direction + to – is "up." Cell shape and the ability of cells to specialize regions of their surface or interior depend on microtubule polarity.

the polarity of the microtubules in the array is essential for the organization and function of the cell.

11.4 Purified tubulin subunits assemble into microtubules

Key concepts

- Microtubule polymerization begins with the formation of a small number of nuclei (small polymers).
- Microtubules polymerize by addition of tubulin subunits to both ends of the polymer.
- A critical concentration of tubulin subunits always remains in solution. The concentration of tubulin must be above the critical concentration for assembly to occur.

Microtubules form by the polymerization of tubulin subunits. This process can be studied *in vitro* using purified tubulin. To form a microtubule, tubulin and GTP are first combined in an appropriate buffer and the solution is then warmed to 37°C to initiate polymerization (for mammalian tubulin). The formation of microtubules is easily detected by light scattering, because the assembled polymers scatter light, whereas individual tubulin molecules do not. The amount of light scattered is proportional to the amount of microtubule polymer.

A plot of the amount of microtubule polymer formed over time is shown in **FIGURE 11.14**. Initially, there is a lag phase during which no polymer is detected. Polymer then begins to

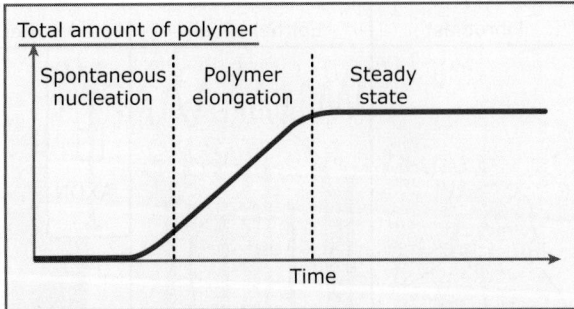

FIGURE 11.14 A graph showing the amount of microtubule polymer formed with time as purified tubulin polymerizes *in vitro*. Initially, no polymer is detected. The amount of polymer then increases linearly with time before leveling off. At each of the three stages different molecular events are occurring.

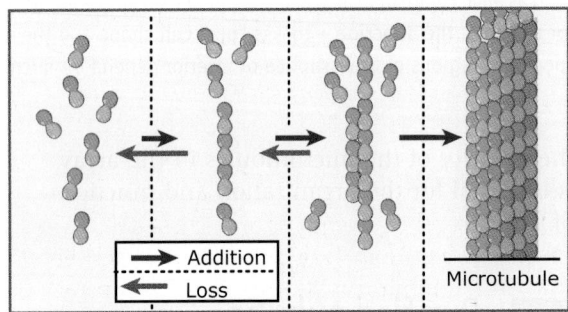

FIGURE 11.15 A simplified view of microtubule nucleation. A small number of dimers can associate, but the complex usually falls apart. In rare instances, however, additional dimers add on and dimers associate side by side. Once a complex of 6-12 dimers is formed, it is likely to continue growing. A small sheet of protofilaments results that will eventually close into a short microtubule, often called a "seed."

form and the amount rises linearly until it reaches a plateau. The initial stage, when no polymer is detected, is called the spontaneous nucleation phase. During this initial lag stage, small nuclei—assemblies of only a few tubulin subunits—begin to form, as shown in **FIGURE 11.15**. These nuclei are unstable because they are more likely to fall apart than they are to gain additional subunits. Some do increase in size, however. Once a sufficient number of subunits associate (6–12), a nucleus is stable because it is more likely to grow than to depolymerize. Because the formation of large enough nuclei is difficult, nucleation is the rate-limiting step in microtubule polymerization. Cells avoid being limited by the slow rate of spontaneous nucleation by having specialized protein complexes that accelerate the nucleation step and determine where it occurs.

During the linear rise in amount of polymer, tubulin subunits add to the ends of the

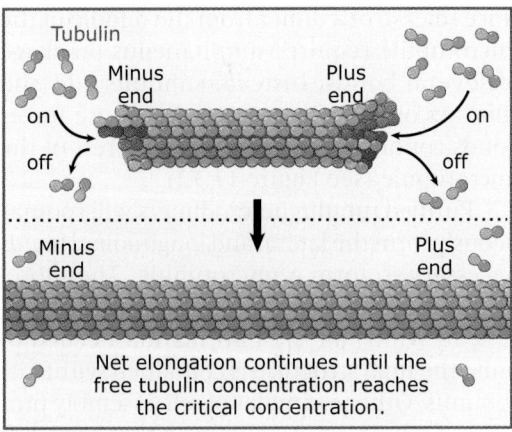

Net elongation continues until the free tubulin concentration reaches the critical concentration.

FIGURE 11.16 After a microtubule seed has formed, the microtubule elongates by the addition of subunits at its two ends. One end grows faster than the other. This growth leads to the linear increase in the total amount of polymer that follows the spontaneous nucleation phase.

microtubule seeds (nuclei) that formed during spontaneous nucleation. Each microtubule elongates by the addition of subunits onto both its plus and minus ends. Subunits also dissociate from the ends but do so less frequently than they are added. The net result is the addition of subunits, as illustrated in **FIGURE 11.16**. The net elongation rate is dependent on the rates of subunit addition and loss and is described by Equation 11.1:

$$dP/dt = k_{on}[\text{tubulin}] - k_{off} \qquad (11.1)$$

where dP/dt is the amount of polymer formed per unit time, [tubulin] is the concentration of tubulin molecules in solution, k_{on} is the on rate constant (units of $M^{-1}sec^{-1}$), and k_{off} is the off rate (units of sec^{-1}). The rate constants for the plus and minus ends differ, so Equation 11.1 must be written for each end of the polymer. As shown mathematically in Equation 11.1, microtubules polymerize faster at higher tubulin concentrations.

Eventually, the amount of microtubule polymer reaches a maximum as the system reaches a steady state in which subunit loss and addition are balanced. When the maximum amount of polymer is formed, some tubulin subunits will remain in solution. The concentration of subunits remaining in solution is called the **critical concentration (C_c)**, and this soluble tubulin concentration will be the same no matter what the starting tubulin concentration was. It is easy to see why this is so if we describe the critical concentration mathematically. At steady state, there is no net assembly of polymer, so dP/dt in Equation 11.1 is equal to zero. As shown in Equation 11.2, we can then solve for [tubulin] when dP/dt is zero:

$$[tubulin] = C_c = k_{off}/k_{on} \qquad (11.2)$$

For purified tubulin, the critical concentration is about 7 μM. Since this is the concentration of subunits that will always remain in solution at steady state, the concentration of tubulin dimers must be higher than the critical concentration to get assembly of polymer. Similarly, if microtubule polymers at steady state are diluted so that the total concentration of tubulin is reduced, they will begin to disassemble and the amount of polymer will decrease. The decrease will stop when the concentration of soluble tubulin subunits again equals the critical concentration. Thus, the critical concentration can also be thought of as the minimum concentration required for microtubule assembly.

An important question is what is happening to the individual microtubules once steady state is achieved. One possibility—the simplest—is that, like many chemical reactions, microtubule polymerization comes to a true equilibrium, with both ends of every microtubule in equilibrium with the tubulin molecules in solution. Were this the case, there could be only very minor exchange of subunits in and out of a microtubule at steady state, for reasons that will be discussed shortly. Experimental evidence, however, demonstrates just the opposite: at steady state, subunit addition and loss into and out of microtubules are extensive, much greater than that predicted by an equilibrium exchange. This indicates that something much more intriguing than a simple equilibrium is happening. The mechanisms that account for this enhanced exchange and polymer turnover are fundamental to understanding the behavior of microtubules and are discussed in the next section.

Concept and Reasoning Checks

Draw curves representing the amount of microtubule polymer over time under the following conditions:

1. You have a sample of microtubules and tubulin dimers at steady state and now dilute the sample such that the tubulin dimer concentration is below the critical concentration.

2. Again, you have a sample of microtubules and tubulin dimers at steady state and you dilute your sample such that the tubulin dimer concentration is now equal to the critical concentration.

11.5 Microtubule assembly and disassembly proceed by a unique process termed dynamic instability

Key concepts

- Microtubules constantly switch between phases of growth and shortening; this process is termed dynamic instability.
- The transition from growing to shortening states is called a catastrophe.
- The transition from shortening to growing states is called a rescue.
- A population of microtubules grows and shortens asynchronously; at any instant in time, most are growing and a few are shortening.
- The structures of growing and shortening microtubule ends are different: growing ends have extensions of protofilaments, whereas shortening ends have curling protofilaments that bend back, away from the microtubule lattice.

In vitro experiments to understand microtubule polymerization dynamics and kinetics initially used methods such as light scattering that could only measure the total amount of polymer present at any time. In the mid-1980s, methods that followed tubulin polymerization by immunofluorescence or electron microscopy were introduced. Unlike light scattering, these techniques allow visualization of the individual microtubules in a polymerization reaction. The microtubules can then be counted, and their lengths measured. The result was completely unexpected and provided an intriguing answer to why exchange of tubulin subunits into and out of the polymer was so much greater than predicted on the basis of equilibrium exchange. Were microtubules equilibrium polymers, they would have shown only tiny changes in length once at steady state, and the total number of microtubules could not have changed significantly with time. Using the new methods, however, it was clear that at steady state (the plateau in Figure 11.14) both the lengths and the number of microtubules changed. Over approximately 40 minutes, some microtubules grew ten times longer, while the total number of microtubules decreased. Additional experiments examining microtubules polymerized from centrosomes (stable nucleation sites; see also *11.7 Cells use microtubule-organizing centers to nucleate microtubule assembly*) also gave unexpected results. Microtubules were first allowed to polymerize from centrosomes, and the sample was then diluted to a concentration just below the criti-

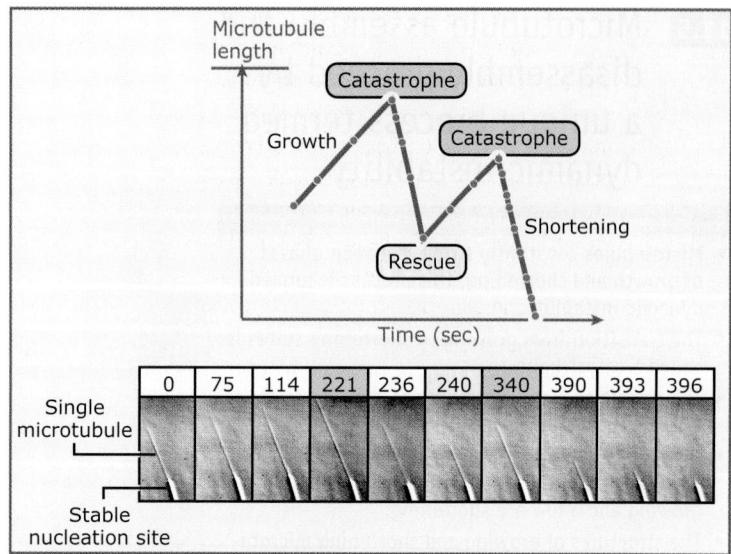

FIGURE 11.17 A single microtubule extends from a stable nucleation site *in vitro*. Its minus end is attached to the nucleating structure and its plus end is free. The length of the microtubule is recorded over time by video microscopy and is graphed at the top. The microtubule first grows, then abruptly begins to shorten. After losing most of its length it begins to grow again. A short while later it undergoes a second catastrophe, and this time depolymerizes completely. Note that the microtubule depolymerizes several times as fast as it polymerizes. The time in seconds is above each frame from the video. Photos courtesy of Lynne Cassimeris, Lehigh University.

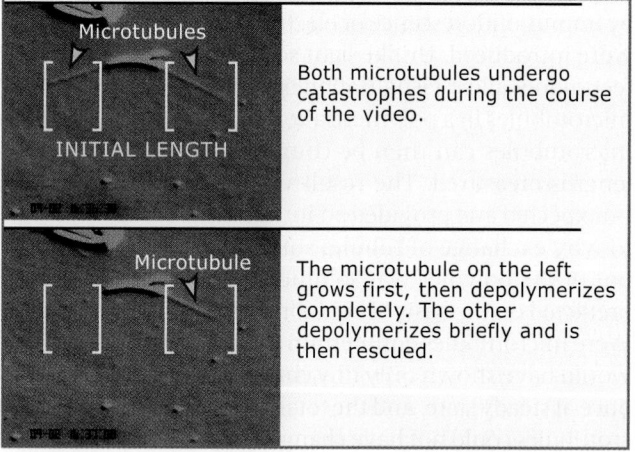

FIGURE 11.18 A small piece of a flagellar axoneme was used as a stable nucleation site for assembly of purified tubulin *in vitro*. Two frames of a video are shown. In the upper panel, a single microtubule has polymerized from each end of the axoneme. The microtubule on the left has its plus end free; the one on the right has its minus end free. In the time between the top and bottom video frames, the microtubule on the left has depolymerized completely; the microtubule on the right has depolymerized a short distance and then grows again. Photos courtesy of Lynne Cassimeris, Lehigh University.

until the critical concentration was reestablished. Instead, while some microtubules did depolymerize, the rest continued to polymerize, a result that is impossible with an equilibrium polymer. These results and others led to a model of microtubule polymerization called **dynamic instability**, in which microtubules exist in persistent phases of either growth or shortening, with abrupt transitions between them. These abrupt transitions are termed **catastrophe** for the switch from growing to shortening, and **rescue** for the switch from shortening to growth.

The dynamic instability model was confirmed using light microscopy to observe individual microtubules as they grow and shorten. **FIGURE 11.17** shows an experiment of this type. These experiments demonstrated that an individual microtubule grows steadily for a distance of several microns, then undergoes a catastrophe and shortens quickly. The microtubule may be rescued and resume growing, or it may depolymerize completely. **FIGURE 11.18** shows two images from a video recording of microtubules undergoing transitions between growing and shrinking phases. The two microtubules shown are nucleated from a stable structure (a short fragment of a flagellar axoneme, an organelle composed of stable microtubules), one growing at its plus end and the other at its minus end. This technique makes it possible to follow the polymerization dynamics at both the plus and minus ends of microtubules. Such experiments showed that the plus ends of microtubules grow faster than the minus ends. The plus ends of microtubules also undergo catastrophe more frequently, making them the more dynamic end of the microtubule.

The defining features of dynamic instability are the persistence of the elongation and shortening phases and the abruptness of transitions between them. Because catastrophes and rescues occur at random intervals, the behavior of the individual microtubules in a group of microtubules will be heterogenous and asynchronous. Most microtubules will be slowly growing, but at the same time a few will be shortening rapidly. Individual microtubules do not reach a steady-state length; instead, each is always becoming either longer or shorter.

The structure of microtubule ends is different for growing and shortening microtubules. During polymerization, microtubule ends often have sheetlike extensions in which some protofilaments have grown longer than others, as shown in **FIGURE 11.19**. Subunits add only to the ends of protofilaments. As each protofilament

cal concentration. The predicted result for an equilibrium polymer would be that all the microtubules would immediately begin to depolymerize and would continue to depolymerize

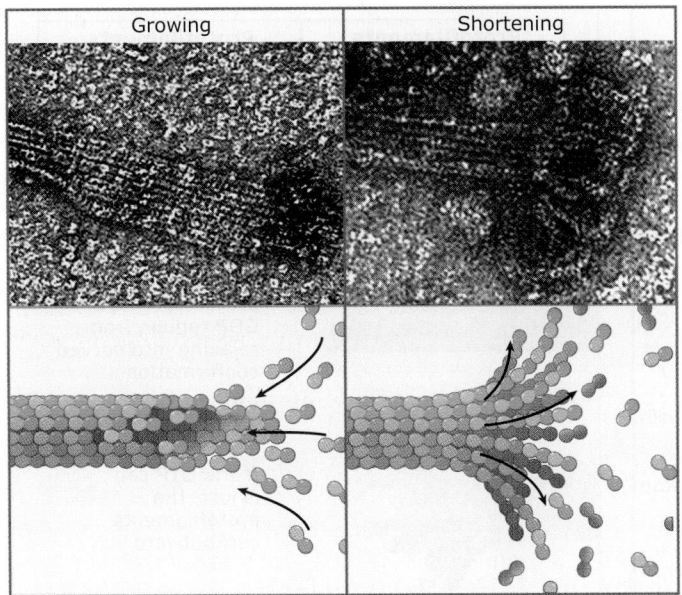

Growing	Shortening

FIGURE 11.19 At the top are electron micrographs of the plus ends of growing and shortening microtubules. Longer protofilaments extend from one side of the end of the growing microtubule. These protofilaments lie flat against the surface onto which the microtubule is attached for electron microscopy. The result is the appearance of a flat sheet of protofilaments at the end. At a shortening end, the protofilaments peel away from the wall of the microtubule and curl backward. Photos courtesy of Lynne Cassimeris, Lehigh University.

elongates, subunits form lateral bonds with their neighbors, eventually closing the sheet into the tubular shape of a microtubule. When microtubules depolymerize, individual protofilaments peel away from the polymer lattice. Subunits within the individual curling protofilaments are held together only by longitudinal bonds. As discussed above (see *11.3 Microtubules are polar polymers of α- and β-tubulin*), the bonds between subunits in a single protofilament are equally likely to break, leading to rapid disassembly of the peeling protofilaments by both dissociation of subunits from their ends and by breakage at other points.

11.6 A cap of GTP-tubulin subunits regulates the transitions of dynamic instability

Key concepts

- Growing microtubules have a cap of GTP-tubulins at their tip because the GTP associated with β-tubulin is hydrolyzed to GDP shortly after a subunit adds to a microtubule.
- The bulk of the microtubule is made up of GDP-tubulins.
- Hydrolysis of GTP is coupled to a structural change in the tubulin dimers.
- GTP-tubulins form straight protofilaments, which maintain contacts with subunits in adjacent protofilaments and allow these protofilaments to continue growing.
- GDP-tubulins curve away from the microtubule, breaking lateral bonds with adjacent subunits, causing protofilaments to peel apart.

Dynamic instability is possible because tubulin binds and hydrolyzes GTP. Microtubule assembly and disassembly by dynamic instability is a nonequilibrium process and requires energy input. The energy input is provided by GTP hydrolysis. As tubulin heterodimers are incorporated into a microtubule, β-tubulin is stimulated to hydrolyze its bound GTP to GDP. Hydrolysis does not occur immediately but lags behind polymerization slightly. Thus, the core of a growing microtubule is composed of GDP-tubulins, while the ends are capped by GTP-tubulins.

The cap of GTP-tubulins at microtubule ends regulates dynamic instability, determining whether a microtubule will grow or shorten, as shown in **FIGURE 11.20**. Shortening microtubules have lost this cap, exposing GDP-tubulin subunits at the end of the polymer. Whether GTP or GDP is associated with the subunits at the end of a microtubule determines the rate at which they dissociate. GDP-tubulin dissociates from an end approximately 50 times faster than GTP-tubulin. Exposure of GDP-tubulins at the end of a microtubule, therefore, results in rapid depolymerization. Catastrophes are thus the result of a growing microtubule losing its GTP cap, while rescue requires that GTP-tubulins recap the end of a shortening microtubule (see Figure 11.20).

The presence of GTP- or GDP-tubulins at microtubule ends regulates tubulin association and dissociation rates by changing the structure of the microtubule end. Structural studies indicate that GTP-tubulin forms straight protofilaments. GDP-tubulin, however, is most

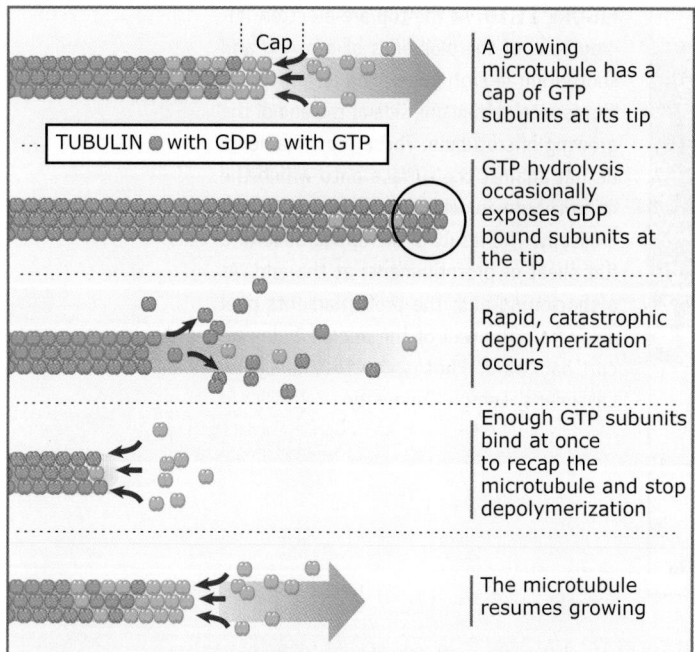

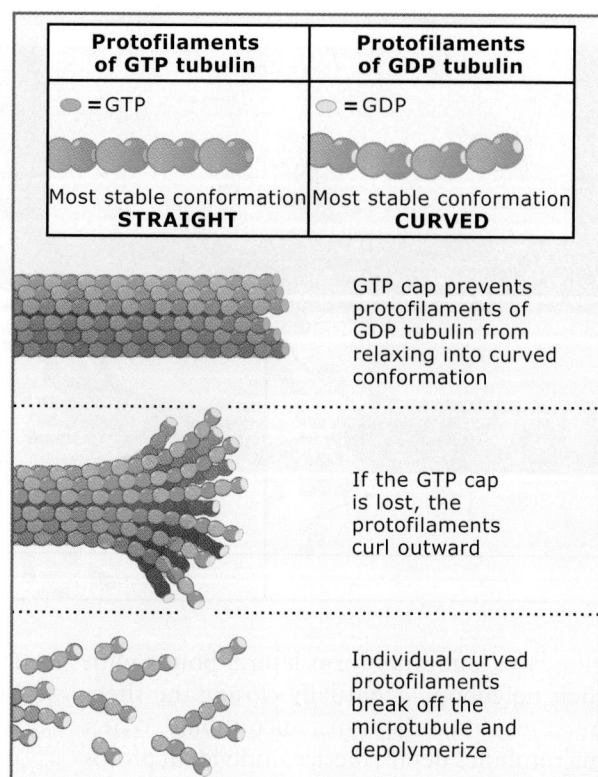

FIGURE 11.20 A schematic view of events at the end of a dynamic microtubule. The GTP bound to β-tubulin is hydrolyzed shortly after a tubulin dimer adds to a microtubule, creating a small cap of GTP-bound subunits at the microtubule's end as it grows. As long as the cap is present the microtubule will grow. However, if GDP-bound subunits ever become exposed at its end, the microtubule will begin to depolymerize very rapidly. The microtubule can be rescued if GTP subunits bind as it is depolymerizing. Depolymerization is fast enough, and rescue rare enough, so that a large fraction of a microtubule can depolymerize before it is rescued.

FIGURE 11.21 A GTP cap at the end of a microtubule forces the GDP-bearing protofilaments that make up the rest of the microtubule to lie straight. As soon as the GTP cap is lost, the protofilaments can relax into their more stable curved conformation. As they curl the protofilaments separate from one another, making them prone to breakage. At the top, the nucleotide associated with every β subunit is shown. For simplicity, in the bottom three panels only the nucleotide associated with the subunit at the end of each protofilament is shown.

stable as curved protofilaments, very similar to the curling protofilaments seen at the ends of depolymerizing microtubules, as illustrated in **FIGURE 11.21**. The GTP or GDP status of tubulins at the microtubule tip, therefore, determines the structure of the tip, and the structure of the tip determines whether a microtubule will grow or shorten. A cap of GTP-tubulins holds the protofilaments straight and prevents GDP-tubulins within the core of the microtubule from relaxing to their preferred curved conformation. If the GDP-tubulin core is ever exposed at the end of a microtubule, however, the GDP-tubulins are able to break the lateral bonds to their neighbors, and the protofilaments curl outward away from the core of the microtubule (see Figure 11.21). These curling protofilaments break in random positions and

the fragments depolymerize into individual subunits (see Figure 11.21).

How is a GTP cap generated and how can a microtubule lose it? We don't yet know exactly how large a GTP cap is, but do know that it is quite small (less than 200 subunits), perhaps as small as a single layer of subunits in depth. The existence of a cap requires that tubulin hydrolyze GTP shortly after entering the polymer, but only after another tubulin molecule has joined the same protofilament. One possible mechanism that would allow this depends on the structure of tubulin and where it binds GTP. At microtubule plus ends, the GTP of β-tubulin is exposed at the end of the microtubule. **FIGURE 11.22** shows that when a new tubulin heterodimer adds to the end, its α-tubulin makes contact with this GTP;

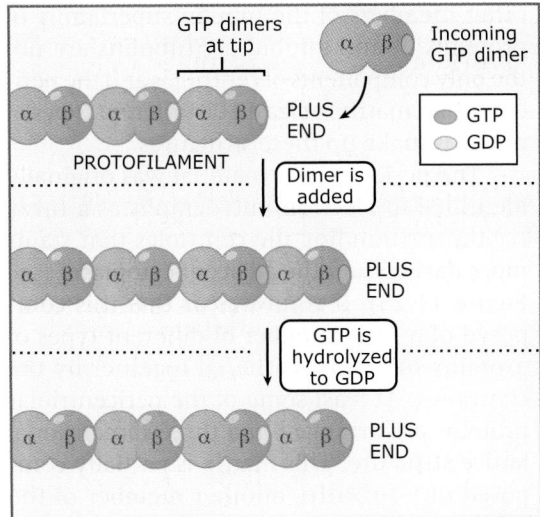

FIGURE 11.22 One possible mechanism that would create a GTP cap at the end of a microtubule. Because the GTP bound to β-tubulin is positioned at one end of the tubulin heterodimer, the GTP on the last dimer is exposed at one end of a protofilament. The α-tubulin of an incoming dimer contacts the GTP and may cause its hydrolysis. This mechanism would maintain a GTP cap of only a single layer of subunits.

this is thought to cause its hydrolysis. In this scenario, the GTP cap always remains one subunit deep, since each new subunit added stimulates hydrolysis by the subunit below it in the microtubule. If this is correct, how can a GTP cap ever be lost? One way would be through dissociation of GTP subunits from the polymer end. Another possibility would be by spontaneous GTP hydrolysis by the subunits at the end, which would occur slowly relative to the rate of hydrolysis when stimulated by another tubulin molecule. It is not yet known how many GDP-capped protofilaments must be exposed before the entire end structure begins peeling apart and shortening.

How can a shortening microtubule, with its protofilaments peeling outward, ever switch back to growth (i.e., be rescued)? GTP-tubulin subunits are likely adding to the ends of protofilaments even as they peel away during shortening. But addition of a GTP-tubulin to a peeling protofilament cannot stabilize the end because it cannot form lateral bonds with its neighbors. A rescue likely requires that some of the peeling protofilaments break off near the microtubule wall, allowing several incoming GTP-tubulins to form lateral bonds to each other and stabilize the end.

The molecular models describing catastrophe and rescue may suggest that these are unlikely events. In fact, they are. Catastrophes and rescues are rare events compared to the addition or loss of subunits from microtubule ends. For a microtubule depolymerizing in a fibroblast cell, approximately 11,000 subunits are lost before a rescue occurs. The infrequency of catastrophes and rescues allows rounds of polymerization and depolymerization to often cover significant fractions of the length of a microtubule and, sometimes, depolymerize it completely. As we shall see, the ability of microtubules to grow or shrink persistently over long distances is essential for the roles they play in cells.

Concept and Reasoning Check

1. You have microtubules growing from a stable nucleation site, as shown in Figure 11.18. What do you suppose would happen to a microtubule if you were able to cut off its end? Why?

11.7 Cells use microtubule-organizing centers to nucleate microtubule assembly

Key concepts

- In cells, microtubule-organizing centers (MTOCs) nucleate microtubules.
- The position of the MTOC determines the organization of microtubules within the cell.
- The centrosome is the most common MTOC in animal cells.
- Centrosomes are made up of a pair of centrioles surrounded by a pericentriolar matrix.
- The pericentriolar matrix contains γ-tubulin; it is γ-tubulin, in complex with several other proteins, that nucleates microtubules.
- Motile animal cells contain a second MTOC, the basal body.

In previous sections we saw how purified tubulin assembles into microtubules: the microtubules are first nucleated and then grow and shorten by adding and subtracting subunits. A similar sequence of events occurs inside cells, but cells start microtubule assembly using a specific organelle, the **MTOC** that functions to nucleate microtubules. Because spontaneous nucleation is very slow, almost all of a cell's microtubules are nucleated at MTOCs. As their name implies, MTOCs also organize the microtubules in a cell because they often remain associated with the minus ends of the microtubules they nucleate and, thus, dictate their position and orientation.

The most common MTOC in animal cells is the **centrosome**, as shown in **FIGURE 11.23**. The centrosome is composed of a pair of **centrioles** surrounded by the **pericentriolar material**. The centrioles are small organelles shaped like barrels and are organized at right angles to one another in the center of the centrosome. Centrioles are constructed from unusual microtubule structures called triplet microtubules, nine of which are arranged symmetrically to form the walls of the barrel. Each triplet microtubule contains one complete microtubule (the A tubule) and two partial microtubules (the B and C tubules). In addition to α- and β-tubulin, centrioles also contain two other members of the tubulin superfamily of proteins, δ- and ϵ-tubulins. Tubulins are not the only components of centrioles and the pericentriolar matrix; at least 100 different types of proteins make up these structures.

The pericentriolar material was originally identified in electron micrographs as a fuzzy region surrounding the centrioles that stains more darkly than the adjacent cytoplasm (see Figure 11.23). It is now clear that it is composed of a large number of different types of proteins somehow gathered together by the centrioles. At least some of the pericentriolar proteins are arranged in a three-dimensional lattice structure. This matrix is partially composed of γ-**tubulin**, another member of the tubulin superfamily. γ-tubulin is found in a complex with several other proteins called the γ-**tubulin ring complex (γTuRC)**.

The γTuRCs of the pericentriolar material bind to α- and β-tubulins and are the component of the centrosome that nucleates microtubules. The mechanism that the complex uses to nucleate microtubules is not yet clear, but its structure is suggestive. The γ-tubulin molecules within each γTuRC are arranged as one turn of a very shallow helix. This gives them the shape of a lock washer as illustrated in **FIGURE 11.24**. This arrange-

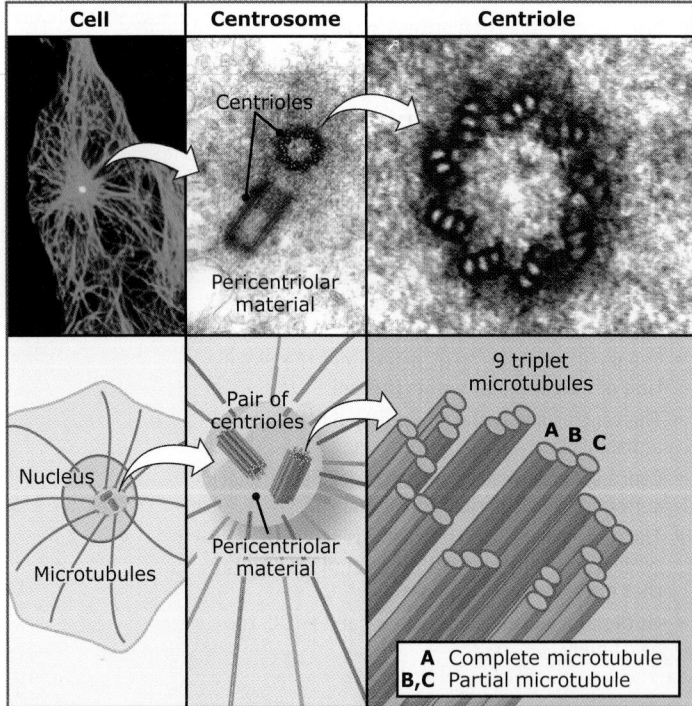

FIGURE 11.23 At the top left is a fluorescence micrograph of a whole cell with its microtubules labeled in green and its centrosome in yellow. Microtubules radiate from the centrosome. In the upper middle and right are electron micrographs of a whole centrosome and one of its two centrioles. The micrograph of the centrosome shows how the centrioles are arranged at right angles to one another. The pericentriolar material appears in the micrograph as a granular material immediately surrounding the two centrioles. Note how much clearer the cytoplasm is at the top and bottom of the picture. Photos courtesy of Lynne Cassimeris, Lehigh University.

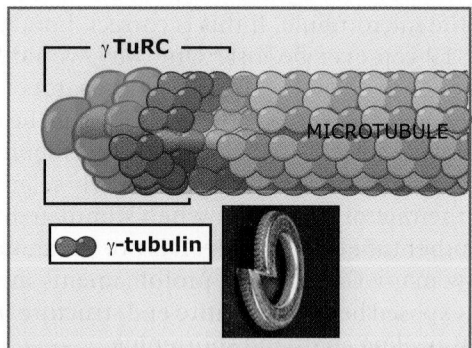

FIGURE 11.24 γ-tubulin (purple) and a number of associated proteins (green) form a large complex that serves as a template for the assembly of microtubules at their minus ends. Within the complex, γ-tubulin subunits are arranged as one turn of a helix, forming a structure that resembles a lock washer (bottom) and has the same diameter as a microtubule. The helical pitch of the ring formed by the γ-tubulins is the same as that of the subunits in a microtubule, suggesting that the complex nucleates a microtubule by positioning the first turn of subunits. Photo courtesy of Lynne Cassimeris, Lehigh University.

ment resembles one turn of the helix that results if tubulin subunits are traced side to side around the surface of a microtubule, suggesting that the γTuRC serves as a template for the formation of the end of a microtubule. However γTuRC nucleates microtubules; it is clear that the γTuRC do this from their minus ends.

The nucleation of microtubules by γTuRCs dictates the orientation of microtubules in many cells. Because γTuRCs associate only with microtubule minus ends, the plus ends of the microtubules nucleated by a centrosome all face away from it, as shown in **FIGURE 11.25**. When the centrosome is in the center of the cell, microtubules are found in a starlike array with all the plus ends located at the cell's edges.

Centrosomes reproduce themselves during each cell cycle in preparation for mitosis. The centrioles duplicate first, at the same time that the DNA is replicated, as shown in Figure 14.20 (see *15 Cell cycle regulation*). During duplication, a new centriole forms at right angles to each of the two original centrioles. As the centrosome splits into two, each new centrosome receives one original centriole and one of the newly formed ones. The two new centrosomes are separated at mitosis (see *14 Mitosis*) so that each daughter cell receives a single centrosome containing a pair of centrioles. Why new centrioles form only next to existing centrioles, why only one new centriole forms for each of the original centrioles, and why centrioles are positioned precisely at right angles to one another are all unclear. It is also not yet clear how centrioles contribute to the formation of the pericentriolar matrix.

The centrosome is a dynamic structure that changes in size during the cell cycle. Once duplicated, the centrosomes grow bigger as cells prepare for mitosis. At the beginning of mitosis, centrosomes increase their rate of microtubule nucleation approximately 5-fold. This large increase in the microtubule "birth rate" is likely to be important during mitosis because a very high density of microtubules is needed to build a mitotic spindle.

Motile animal cells (e.g., sperm cells) contain a second, more specialized MTOC, the basal body. Basal bodies serve as templates for assembly of the axoneme, a structurally complex bundle of microtubules that forms the core of cilia and flagella and is responsible for their movement (see *11.16 Cilia and flagella are motile structures*). Basal bodies are structurally

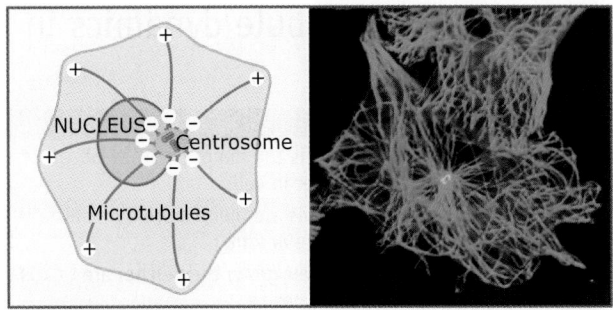

FIGURE 11.25 MTOCs orient microtubules in a cell. Photo courtesy of Lynne Cassimeris, Lehigh University.

very similar to centrioles, containing the same barrel-shaped arrangement of nine interconnected triplet microtubules. The similarity in structure reflects some overlap in function: in some cells, basal bodies can be converted to centrioles. Unlike centrioles, however, basal bodies need not act as pairs and nucleate microtubules directly rather than from a surrounding matrix. During the formation of a cilium or flagellum, microtubules grow directly from the triplet microtubules within the basal body. Basal bodies remain attached at the minus ends of the microtubules they have formed and are present at the base of cilia and flagella as they function.

Not all cells use centrosomes to nucleate microtubules, but all eukaryotic cells have one or more MTOCs of some sort to nucleate and organize microtubules. In fungi, the equivalent of the centrosome is a structure called the spindle pole body, which is embedded in the nuclear envelope. Plant cells lack a well-defined structure that acts as an MTOC but have a number of microtubule nucleating sites distributed throughout the cell cortex. Many types of differentiated animal cells, including neurons, epithelial cells, and muscle cells, have microtubule arrays that are not attached to the centrosome, suggesting that other, smaller types of MTOCs can be positioned within the cell to create specialized arrangements of microtubules. Epithelial cells, for example, have a number of microtubule nucleation sites located near the apical end of the cell. Microtubule plus ends grow out from the apical MTOCs toward the basal end of the cell. All the MTOCs of plants, animals, or fungi contain γ-tubulin, suggesting that all MTOCs use a similar mechanism to nucleate microtubules.

11.8 Microtubule dynamics in cells

Key concepts

- Dynamic instability is the major pathway of microtubule turnover in cells.
- Microtubule plus ends are much more dynamic in cells than they are *in vitro*.
- Free minus ends never grow; they either are stabilized or depolymerize.
- Cells contain a subpopulation of nondynamic, stable microtubules.

Microtubules in cells assemble and disassemble by dynamic instability. Since dynamic instability describes the simultaneous growth of some microtubules and shortening of others, it is necessary to observe individual microtubules in living cells in order to detect this mechanism of turnover. To visualize microtubule assembly in live cells, their tubulin is first made fluorescent by either expressing tubulin fused to a fluorescent protein or by injecting them with purified tubulin that has been covalently tagged with a fluorescent dye. Cells are then observed using a light microscope, and fluorescence images are collected every few seconds. **FIGURE 11.26** was made using these techniques and shows a static image of the microtubules in a living cell. At points around the cell's edges many individual microtubules can be seen repeatedly growing and shrinking, the two types of microtubules often occurring immediately adjacent to one another. Two such microtubules near the edge of a cell—one growing, the other shrinking—are shown in **FIGURE 11.27**.

Images like these can be used to measure microtubule lengths over time and allow the behavior of microtubules in cells and *in vitro* to be compared. They reveal that the dynamic instability of microtubules assembled *in vitro* from purified tubulin and that of microtubules in cells differ in several respects. In living cells, the plus ends of microtubules are much more dynamic than they are in microtubules polymerized from purified tubulin. Microtubule plus ends grow about 5 to 10 times faster in cells than *in vitro*. Microtubules in cells also switch between growth and shortening more frequently, as shown in **FIGURE 11.28**. Pauses,

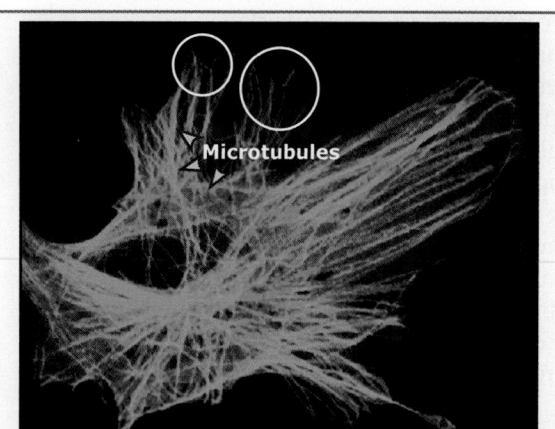

A live cell expressing α-tubulin fused to the green fluorescent protein. The individual microtubules are visible. Growing and shrinking microtubules are visible at many points around the cell's periphery but are clearest in the indicated regions.

FIGURE 11.26 Image from a video of a cell expressing fluorescent tubulin is shown. Microtubules are shown in green. In the video sequence many microtubules grow and shorten by dynamic instability. Two regions of the cell, marked by circles, have a number of dynamic microtubules. A series of images from one of these regions is shown in Figure 11.27. Photo courtesy of Michelle Piehl, Lehigh University.

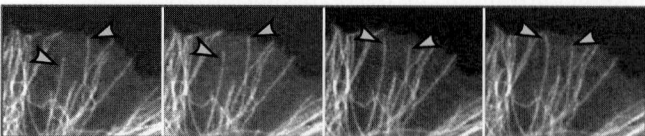

FIGURE 11.27 A small region at the very edge of a live cell expressing flourescent tubulin. The four frames are successive images taken over time. Two individual microtubules are indicated by the arrowheads. The microtubule on the left grows steadily even as the one next to it shrinks. Photos courtesy of Lynne Cassimeris, Lehigh University.

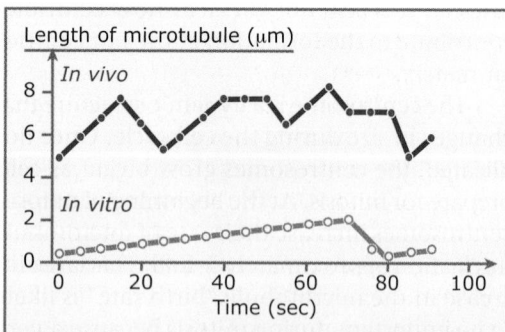

FIGURE 11.28 Each graph shows the length of a single microtubule over time. A microtubule assembled from pure tubulin *in vitro* (blue) grows steadily for more than a minute before depolymerizing almost completely and then beginning to grow again. A typical microtubule within a cell (red) is several times as long and undergoes many more transitions in the same amount of time. Unlike the *in vitro* microtubule, it loses only a fraction of its total length each time it depolymerizes, and sometimes pauses between growing and shrinking.

during which microtubules do not show detectable growth or shortening, are rare *in vitro* but are frequently observed for microtubules in living cells. These differences between microtubule dynamics *in vitro* and *in vivo* indicate that cells modify dynamic instability in order to speed it up or slow it down. As we will see later, this is done by proteins that bind to microtubules.

The ability of cells to regulate microtubule assembly dynamics was first demonstrated by comparing microtubule dynamics during interphase and mitosis. Initially, fluorescence recovery after photobleaching (FRAP)—a technique that measures how quickly new microtubules form in a particular region of the cell (see *11.20 Supplement: Fluorescence recovery after photobleaching*)—was used to follow microtubule turnover in living cells. An image of microtubule turnover detected by FRAP is shown in **FIGURE 11.29**. Photobleaching experiments demonstrated that interphase microtubules depolymerize and are replaced with newly polymerized microtubules with a half-time of approximately 5 to 10 minutes, whereas mitotic microtubules are replaced with a half-time of 0.5 to 1 minute. Direct observation of individual microtubules in mitotic cells—by the techniques described above—demonstrated that the increase in turnover is due to changes in the transition frequencies (e.g., more catastrophes, fewer rescues) and a large reduction in pauses. An example of a change in transition frequencies is shown in the graph in **FIGURE 11.30**.

The changes in microtubule dynamics that occur as a cell enters mitosis occur throughout its cytoplasm. Microtubule dynamics can also be regulated within specific regions of a cell. For example, microtubules near the center of the cell have a low probability of undergoing a catastrophe and show persistent growth toward the cell periphery. Switching between growth and shortening is much more frequent toward the edges of the cell in the regions near the plasma membrane. If the dynamics were not regulated differently in the cell's interior and at its periphery, few microtubules would reach all the way to its edges.

Not all microtubules in a cell show the same dynamics. Many interphase cells have two distinct populations of microtubules that differ in their rates of turnover. One population is dynamic and turns over rapidly (within minutes). The second consists of microtubules that are much more stable and often last for an

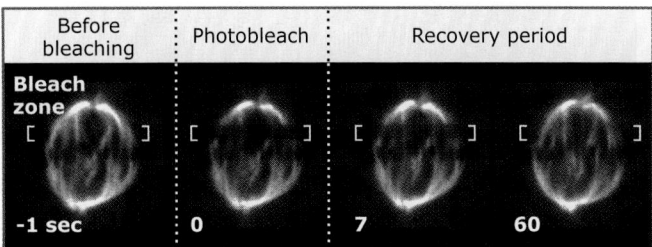

FIGURE 11.29 The rapid turnover of spindle microtubules demonstrated by fluorescence recovery after photobleaching (FRAP). Pictured are the microtubules of the mitotic spindle in a cell expressing fluorescent tubulin. The spindle's poles are at the top and bottom. Between the first two frames, the fluorescent tag on tubulin is locally destroyed (bleached) in a stripe across the spindle (indicated by the brackets). Within 60 seconds the flourescence in the bleached area has regained its original intensity, indicating that new microtubules with fluorescent subunits have grown through it. This is the result of constant assembly and disassembly of microtubules in the spindle. Photos courtesy of Lynne Cassimeris, Lehigh University.

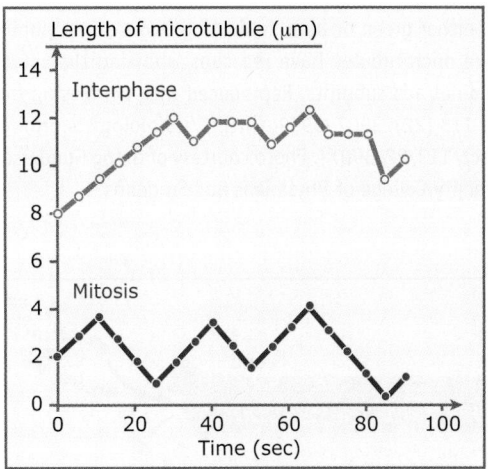

FIGURE 11.30 The two graphs show the length of a typical interphase (blue) or mitotic (red) microtubule over time. The interphase microtubule is much longer and undergoes relatively minor changes in length, sometimes neither growing nor shrinking. The mitotic microtubule never remains at one length and loses most of its length every time it depolymerizes. Because of cell cycle-dependent changes in the parameters of dynamic instability, mitotic microtubules tend to be shorter and more dynamic than interphase microtubules.

hour or more. These stable microtubules do not grow or shorten at their plus ends, suggesting that they are somehow capped there. The differences in these populations are apparent in **FIGURE 11.31**.

Stable microtubules arise from dynamic microtubules, although how this occurs is not

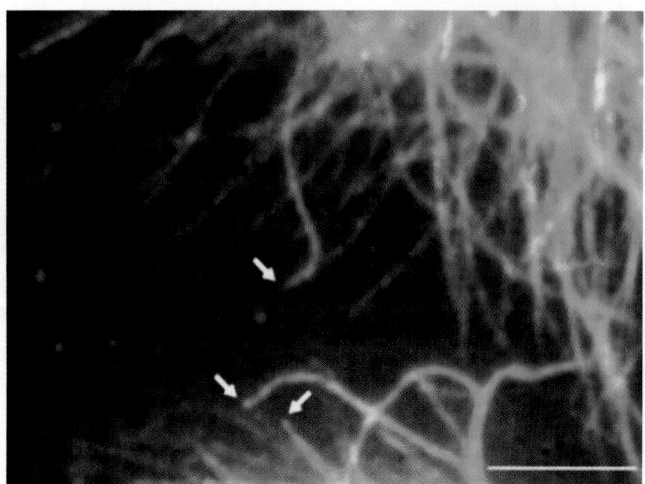

FIGURE 11.31 A fluorescence micrograph of the microtubules in a small region at the edge of a cell. A covalently modified form of tubulin found only in stable microtubules is in green; unmodified tubulin appears blue. Just before the cell was prepared for the photograph it was injected with a form of labeled tubulin that appears red in order to mark the ends of its growing microtubules. The microtubules in the cell are clearly either green or blue, indicating two distinct populations. Only blue microtubules have red caps, showing that stable microtubules do not add subunits. Reproduced with permission from *J. Cell Sci.*, vol. 113 (22): 3907–3919. [http://jcs.biologists.org/cgi/content/abstract/113/22/3907]. Photo courtesy of Gregg Gunderson, Columbia University College of Physicians and Surgeons.

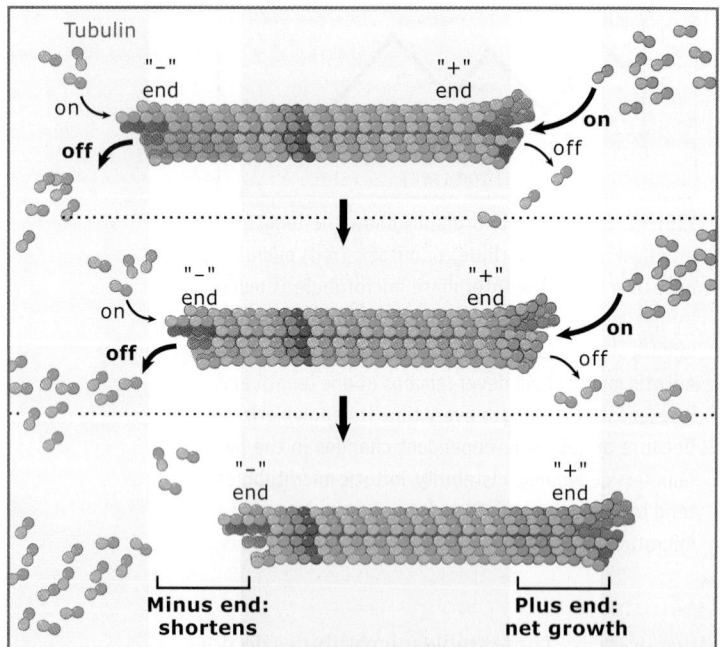

FIGURE 11.32 Under some conditions, microtubules undergo a treadmilling process where tubulin subunits preferentially add to the microtubule plus end and preferentially come off at the minus end. Having addition at one end and loss at the other means that tubulin subunits within the microtubule "treadmill" from the plus end to the minus end, as shown by the subunits marked in red. Treadmilling is a major pathway of microtubule turnover in plant cells.

yet clear. Cells contain several types of enzymes that covalently modify γ-tubulin in different ways, and these may play a role (see *11.21 Supplement: Tubulin synthesis and modification*). Stable microtubules contain much more modified tubulin than other microtubules, suggesting that they are preferred substrates for these enzymes. The functions of stable microtubules are not yet known, but it is clear that the number of stable microtubules varies in different cell types. In nondifferentiated cells, approximately 70% of the microtubules are dynamic and 30% are stable. Stable microtubules are much more abundant in nonmitotic, differentiated cells such as muscle, epithelial, or neuronal cells.

Microtubules are nucleated at the centrosome, but not all microtubules remain tethered there. Microtubules can be released from the centrosome, although the rate at which this occurs varies between cell types and cell cycle stages. Release of microtubules from the centrosome results in microtubules with both plus and minus ends free in the cytoplasm. Microtubules that are not anchored to the centrosome can also arise by breakage of existing centrosomal microtubules. Free microtubules only persist in cells if the minus end is stabilized, although the stabilizing mechanism has not been identified. In fibroblasts, free microtubules rapidly disassemble, and only those microtubules anchored to the centrosome persist. In epithelial cells and neurons, microtubule minus ends are stable and the free microtubules can persist in the cytoplasm. These free microtubules can be transported and organized by molecular motors (see *11.11 Introduction to microtubule-based motor proteins*), allowing cells to organize microtubules in patterns other than the radial array seen in fibroblasts (see Figure 11.7).

In some cells, microtubules that are not anchored to a centrosome undergo a form of turnover called treadmilling. For a treadmilling microtubule, the dynamic instability of the microtubule plus end is biased toward net growth whereas the minus end shortens. By this process, a tubulin subunit enters at the plus end and exits at the minus end, effectively moving the length of the microtubule, as **FIGURE 11.32** shows. Treadmilling is prominent in plant cells, which lack centrosomes (see *21.15 Cortical microtubules are highly dynamic and can change their orientation*). The treadmilling of microtubules observed in cells requires accessory proteins and is not observed in solutions of purified tubulin. (For details on treadmilling of actin filaments, see *12 Actin*.)

11.9 Why do cells have dynamic microtubules?

Evolution has clearly favored an unstable microtubule cytoskeleton because dynamic microtubules are found in every eukaryotic organism examined so far, suggesting that dynamic microtubules have been a feature of eukaryotic cells for at least 700 million years. Why might selection favor a dynamic polymer that consumes energy (GTP hydrolysis) over a static one that would be made just once? It is likely that cells need dynamic microtubules because they are easily adapted for new purposes. Dynamic microtubules can search intracellular space, they can reorganize, and they can even generate force. These properties make the dynamic microtubule cytoskeleton adaptable to a wide range of cellular functions. Figure 11.5 shows the dynamic remodeling of the microtubule cytoskeleton as cells move from interphase to mitosis; this is just one example of the adaptability of microtubules. (None of this would be possible if microtubules did not hydrolyze GTP. How important GTP hydrolysis is to making microtubules adaptable is emphasized by considering what their properties would be if they did not hydrolyze nucleotide; see *11.19 Supplement: What if tubulin did not hydrolyze GTP?*)

The ability of dynamic microtubules to search the interior of a cell is illustrated during the formation of the mitotic spindle. The formation of a spindle requires that microtubules originating from centrosomes find and connect their plus ends to kinetochores, the regions on each chromosome where microtubules are attached to the mitotic spindle. On the scale of a cell, kinetochores are tiny and the centrosome is a significant distance away, as shown in **FIGURE 11.33**. If cells were scaled up so that each kinetochore was the size of a 1-inch bull's eye in the center of a dartboard, the centrosome would be positioned at the throwing

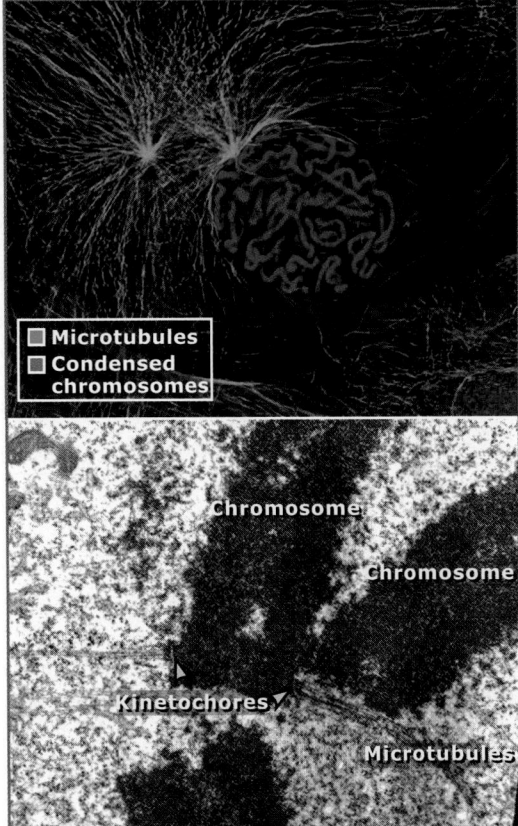

FIGURE 11.33 The upper photograph shows the microtubules and condensed chromosomes in a cell as it enters mitosis. At the bottom is an electron micrograph showing a small fraction of a condensed chromosome around its two kinetochores. Several microtubules can be seen leading into each kinetochore. These two photographs emphasize how randomly arranged the chromosomes are at the start of mitosis and how small the kinetochore is relative to an entire chromosome. The upper photograph illustrates the density of dynamic microtubules nucleated by each centrosome. The density and dynamics of the microtubules allow each of the kineotochores to be found reliably despite their small size and uncertain position. Top photo © Conly L. Rieder, Wadsworth Center; bottom photo courtesy of Lynne Cassimeris, Lehigh University.

line for a game of darts. If the centrosome had to aim its microtubules, both extraordinary aim as well as some preexisting "knowledge" of the location of the kinetochore would be required. Instead, dynamic instability allows centrosomes and kinetochores to be connected reliably without either. Centrosomes nucleate microtubules in all directions, essentially probing the entire volume of the cytoplasm with the tips of a large number of growing microtubules. Microtubules that do not hit a kinetochore rapidly fall apart, freeing their tubulin subunits to

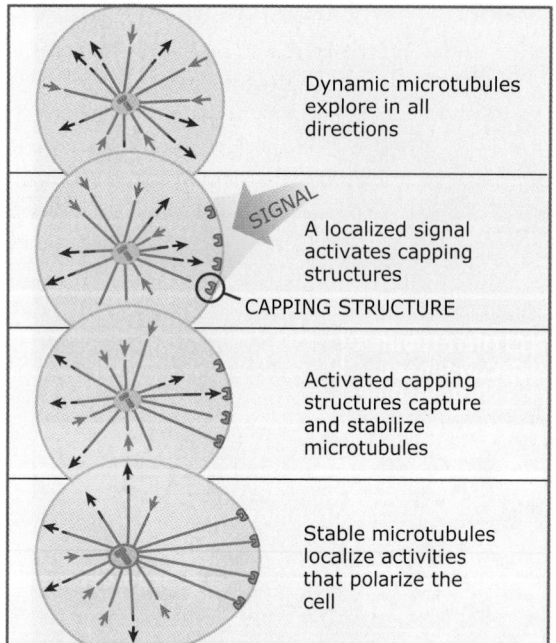

Dynamic microtubules explore in all directions

SIGNAL

A localized signal activates capping structures

CAPPING STRUCTURE

Activated capping structures capture and stabilize microtubules

Stable microtubules localize activities that polarize the cell

FIGURE 11.34 Localized changes in microtubule stability contribute to cell polarity and changes in cell shape. Here, microtubules are initially arranged radially in a spherical cell. The microtubules in effect explore the cell through their constant turnover by dynamic instability. A localized signal stabilizes a subset of microtubules. The stable microtubules then contribute to cell polarity by initiating the specialization of that region, for example, by causing the insertion of membrane there. This "selective stabilization" mechanism is one way that cells could modify the random assembly and disassembly of microtubules to generate a polarized cell.

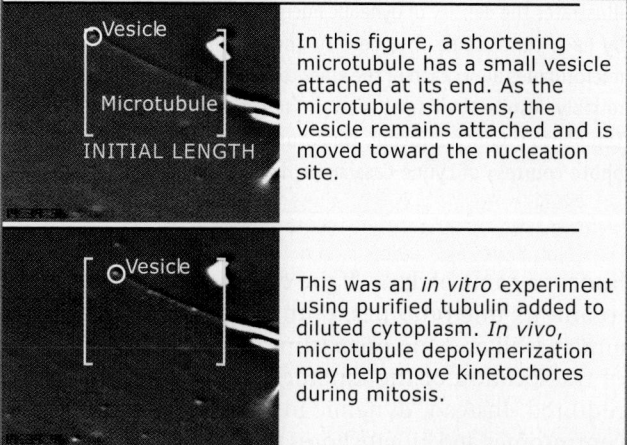

Vesicle

Microtubule

INITIAL LENGTH

In this figure, a shortening microtubule has a small vesicle attached at its end. As the microtubule shortens, the vesicle remains attached and is moved toward the nucleation site.

Vesicle

This was an *in vitro* experiment using purified tubulin added to diluted cytoplasm. *In vivo*, microtubule depolymerization may help move kinetochores during mitosis.

FIGURE 11.35 Two images from a video sequence are shown. In an *in vitro* experiment, a vesicle has bound to the end of a microtubule. As the microtubule shortens toward the nucleation site, the vesicle remains bound to the microtubule's tip and is transported toward the nucleation site. Photos courtesy of Lynne Cassimeris, Lehigh University.

assemble again. The rare microtubules that hit a kinetochore are stabilized, establishing the connections between the poles and the chromosomes. Although only a small fraction of the microtubules nucleated encounter a kinetochore, the continuous, rapid assembly and disassembly of microtubules made possible by dynamic instability allow all of the kinetochores to be found and connected to the centrosome within just a few minutes. Each kinetochore is connected with up to 40 microtubules in this short time frame. This "search-and-capture" mechanism of spindle formation has the advantage that it requires no preset positioning between the centrosomes and the kinetochores for a spindle to be constructed. This flexibility is put to use in every cell division, since the positions of the chromosomes at the start of mitosis are different in every cell and in each division.

FIGURE 11.34 shows how the growing, shrinking, and selective stabilization of microtubules is also useful in other contexts, particularly in allowing cells to respond to changes in their environment. Cells are often required to polarize toward signals that they detect at their plasma membrane, such as contact with another cell. The site at which a signal is received is unpredictable beforehand and may be only a small fraction of the cell's surface. The constant growth and shortening of dynamic microtubules throughout the cytoplasm makes it certain that some of them will encounter the signal, however. If the signal stabilizes these microtubules, they will serve as better highways for transport of vesicles, which will favor vesicle traffic toward that region of the cell. By inserting new membrane in one area, the cell becomes polarized and begins to change shape into a more elongated form. One example of local microtubule stabilization and cell polarization occurs as cells begin to heal an artificial wound caused by clearing cells from part of a petri dish. Healing requires that the cells at the edge of the "wound" move into it and begin to divide. Before the cells move, they first polarize in that direction. Polarization involves the reorientation of the cells' microtubules, caused by local stabilization of those microtubules that face the wound.

Dynamic microtubules may also have been selected during evolution because they can generate force and cause movement. Such movement is possible if an object is able to hold onto the tip of a growing or shrinking microtubule, allowing it to be either pushed or pulled

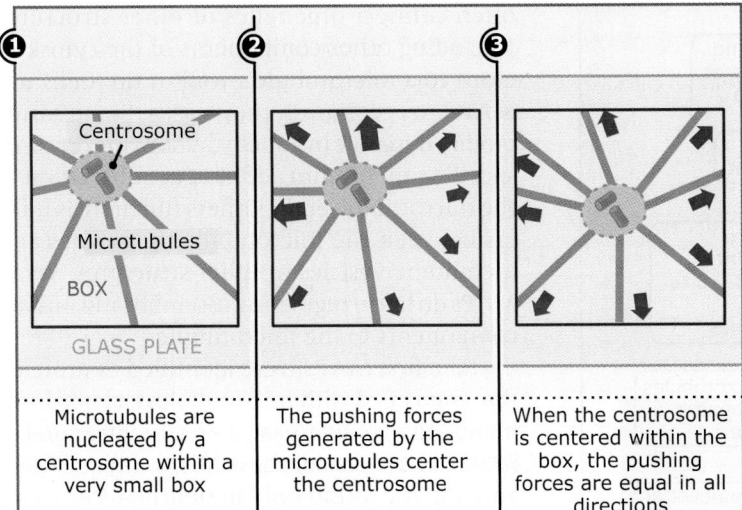

Microtubules are nucleated by a centrosome within a very small box	The pushing forces generated by the microtubules center the centrosome	When the centrosome is centered within the box, the pushing forces are equal in all directions

⬛⬛ **FIGURE 11.36** In this experiment, a centrosome is used to form a radial array of dynamic microtubules in a small chamber about the size of a cell. Over the course of several minutes, the radial array centers itself via pushing forces generated by microtubules polymerizing against the walls of the chamber. The relative magnitude of the forces are indicated by the size of the arrows. The centering or balancing ability of a three-dimensional microtubule array that this experiment demonstrates may sometimes be used to position structures within cells.

as the microtubule changes length. A number of proteins and organelles, including chromosomes and some types of vesicles, have this ability and can be transported on the tips of microtubules. An example of a vesicle being moved on the end of a shortening microtubule is shown in **FIGURE 11.35**. Movement coupled to microtubule shortening uses energy stored in the microtubule from the GTP hydrolysis that occurred as it polymerized. Most movements in cells are generated by molecular motors and not by the growing and shortening of microtubules, but it is interesting to speculate that dynamic microtubules may have evolved first because they allowed rapid reorganization of the cytoskeleton in response to signals, the ability to search for targets within the cytoplasm, and the ability to generate force. (For more on molecular motors see *11.11 Introduction to microtubule-based motor proteins*.)

The ability of dynamic microtubules to generate force can also be used by the microtubule cytoskeleton to position itself. This is true even in the absence of a cell. An experiment with pure centrosomes and tubulin illustrates this property and is shown in **FIGURE 11.36**. If a centrosome is placed in a very small box (formed by photolithography on a glass surface, the same technology that is used to make computer chips) and then stimulated to nucleate microtubules, the centrosome will move to the center of the box, regardless of where it was initially. This occurs because the polymerizing microtubules push against the box's walls. Since the walls are immobile, the centrosome and the microtubules move instead. Pushing forces are equal in all directions when the cen-

trosome is centered within the box. This process is similar to what happens in a cell, where many microtubules grow from the interior out to the plasma membrane. Much like the microtubules in the artificial box, microtubules in the cell polymerize against the plasma membrane in all directions, positioning the centrosome near the center of the cell.

11.10 Cells use several classes of proteins to regulate the stability of their microtubules

Key concepts

- Microtubule-associated proteins (MAPs) regulate microtubule assembly by stabilizing or destabilizing microtubules.
- MAPs determine how likely it is that a microtubule will grow or shorten.
- MAPs can bind to different locations on the microtubule. Some MAPs bind along the sides of the microtubule; others bind only at the tips of microtubules. Still others bind only to the tubulin dimers and prevent them from polymerizing.
- Changing the balance between active stabilizers and destabilizers regulates microtubule turnover.
- The activity of MAPs is regulated by phosphorylation.
- MAPs can also link membranes or protein complexes to microtubules.

Cells use their microtubule cytoskeleton for a wide range of functions. Some of these functions require stable microtubules, whereas others require dynamic microtubules. Cells also

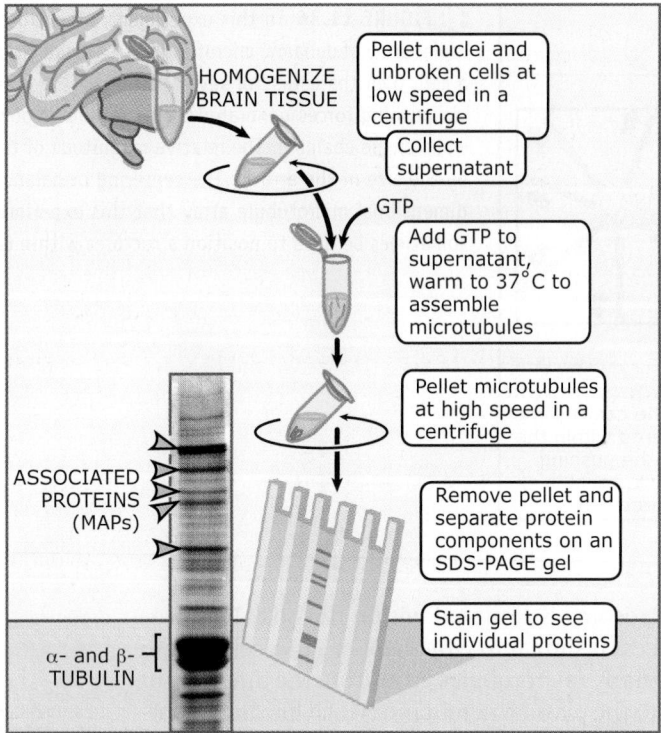

FIGURE 11.37 How the first MAPs were identified. Brain was used because of the large quantity of microtubules in neurons. Brain was homogenized under conditions in which the microtubules would depolymerize. Centrifugation then removed any large material and left only soluble proteins in the supernatant, including a large concentration of tubulin that resulted from all the microtubules in the brain tissue. When the tubulin was polymerized and the microtubules collected with a second centrifugation, many proteins in addition to α- and β-tubulin were also present. Photo courtesy of Lynne Cassimeris, Lehigh University.

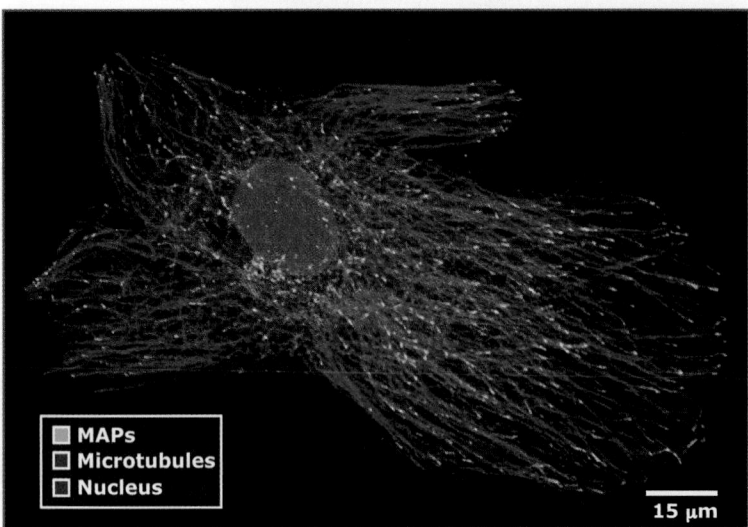

FIGURE 11.38 A fluorescence micrograph with tubulin in red and EB1, a MAP that binds only to the plus ends of growing microtubules, in green. The nucleus is in blue. EB1 appears only in short, elongated segments at the ends of microtubules. Photo courtesy of Lynne Cassimeris, Lehigh University.

often connect organelles or other structures (including other components of the cytoskeleton) to a microtubule's wall or tip. Cells use MAPs, to accomplish these tasks. Some MAPs modify dynamic instability by slowing down or speeding up tubulin addition or subtraction at the microtubule ends. Others function as linkers between the microtubule tip or sides and membrane vesicles or other structures. Some MAPs do both, regulating assembly and linking components to the microtubule.

The first MAPs were identified as proteins that copurified with microtubules isolated from mammalian brain tissue, a process illustrated in **FIGURE 11.37**. Two of these MAPs, called MAP2 and tau, are found only in neuronal cells, and are responsible for creating the very long-lived microtubules essential to the maintenance of axons and dendrites. They do this by binding along the sides of the microtubules. Because they bind to several tubulin subunits at once, they affect how frequently the microtubules undergo the transitions of dynamic instability. Both proteins suppress catastrophes and greatly increase the likelihood of rescue, making it less likely that a microtubule will fall apart. The result is long microtubules that last much longer than would be possible with tubulin alone. Proteins of this sort usually bind all along the length of a microtubule, coating its surface, and can be viewed as braces nailed to the wall of the microtubule that make it much harder to get the subunits apart.

Several proteins bind to microtubules only at their plus ends. These MAPs are referred to as "+TIPs." As shown in **FIGURE 11.38**, fluorescently labeled +TIPs appear as short segments at the plus ends of microtubules. +TIPs bind to a microtubule only when its plus end is growing and appear to ride along on the growing tip. **FIGURE 11.39** shows a still image of a cell expressing a fluorescently labeled +TIP. A video shows the protein appearing as many small comets moving through the cytoplasm, each marking the tip of an individual growing microtubule. For some +TIPs, it is thought that an individual +TIP attaches to new tubulin subunits as they add to the growing end of a microtubule. Each +TIP remains bound for only a short time before falling off the microtubule. While it is bound, the microtubule continues growing and new +TIPs are continuously added. The continuous addition of +TIP proteins at the end of the microtubule and their loss slightly behind it ensure that the concentration of bound +TIP proteins is always

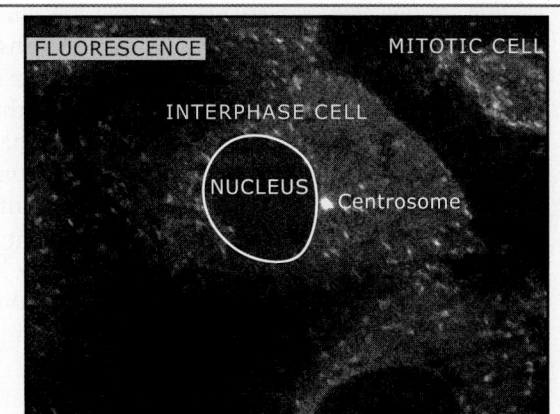

The cell viewed by fluorescence microscopy, showing the location of EB1 +TIP proteins bound to the ends of growing microtubules. Most of the fluorescence appears as many small, slighty elongated spots throughout the cytoplasm. The protein is also present at the centrosome.

FIGURE 11.39 A single image from a video sequence showing a fluorescent +TIP protein, EB1, bound to microtubule ends in an epithelial cell. Because the +TIP proteins bind to the tips of growing microtubule ends, they appear as fluorescent comets that "move" through the cytoplasm. The "movement" actually represents microtubule growth. Photo of EB1-GFP in epithelial cells provided by U. S. Tulu and P. Wadsworth.

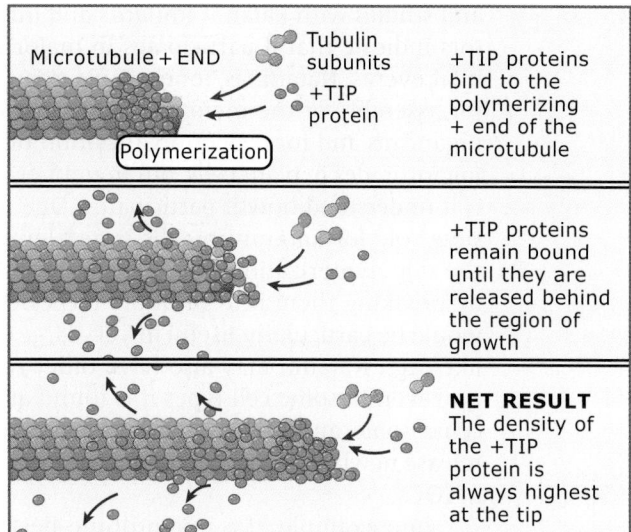

FIGURE 11.40 The binding and release mechanism that allows +TIPs to localize to the growing end of a microtubule.

highest at the microtubule's end, as shown in **FIGURE 11.40**. Some +TIPs are concentrated at the microtubule plus end through a different mechanism—they are transported to microtubule plus ends by the motor protein kinesin (see Section 11.11). Several different families of proteins show binding to microtubule ends and can regulate microtubule dynamics in different ways. Some +TIPs stabilize microtubules and promote their growth, while others make microtubule ends more prone to catastrophe. Additional proteins that destabilize the microtubule end are discussed next.

Cells can also speed up microtubule turnover when they need very dynamic microtubules, as they do during mitosis. Cells speed up microtubule turnover with MAPs that make microtubules less stable. These destabilizers make it more likely that microtubules will have a catastrophe and shorten. They also make rescues less likely, so that microtubules lose many subunits before they start to grow again. There are three general ways that destabilizers work: they disrupt the GTP cap to stimulate catastrophes; they cut microtubules into pieces to make more ends that can shorten; or they bind free tubulin subunits to decrease the amount of tubulin available for polymerization. These three processes are illustrated in **FIGURE 11.41**.

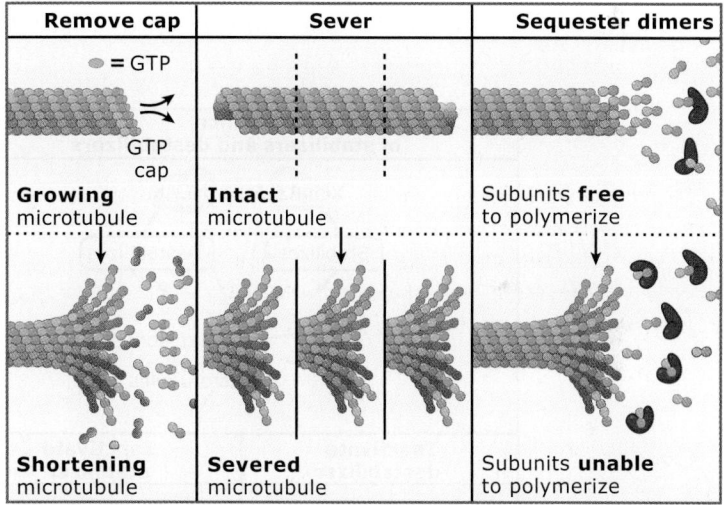

FIGURE 11.41 Three ways to destablize microtubules. Removing a microtubule's GTP cap or severing a microtubule in its interior immediately exposes GDP-bound subunits at an end and causes depolymerization to begin. Severing also increases the number of ends that can depolymerize at once. Sequestering free tubulin subunits slows polymerization and makes it more likely that GTP hydrolysis will occur in the subunits at the end of each protofilament.

Microtubules are cut by katanin, a protein named for the Japanese word for sword. Katanin cuts microtubules by binding to their walls and disrupting contacts between tubulin subunits. Several molecules of katanin must bind to a microtubule and hydrolyze ATP in order to break it. Although its cutting mechanism is well understood, it is not yet clear how katanin's microtubule severing ability is put to use in cells. Katanin is present in all cells, and studies with katanin mutants and inhibitors indicate that it participates in major cellular events. Katanin is, for example, required for assembly of the meiotic spindle in some organisms and for proper organization of the microtubules in plant cells, but in neither case is it understood how it participates. One possible role for katanin in cells is accelerating the depolymerization of long microtubules by breaking them into multiple pieces. This would be particularly useful in very large cells like eggs. Katanin may also have other roles, however. In some cell types it is found at the centrosome, and it is hypothesized that it could release newly polymerized microtubules from MTOCs.

One example of a microtubule-destabilizing protein that functions by disrupting the GTP cap is mitotic centromere associated kinesin (MCAK). MCAK is a member of the kinesin superfamily of molecular motors and plays a significant role in controlling microtubule dynamics during mitosis (see *11.11*

Introduction to microtubule-based motor proteins). Unlike most motors, MCAK does not transport cargo. Instead, MCAK is a +TIP protein; it binds at microtubule ends and destabilizes the microtubule tip structure by favoring formation of protofilaments that curve away from the microtubule wall. Curved protofilaments lose contact with their next-door neighbors, disrupting the GTP cap, and the microtubule begins to shorten. MCAK is then released from the depolymerizing tubulin subunits and is free to bind to the microtubule again.

How dynamic the microtubules in a cell are at any given time is determined by the balance between stabilizers and destabilizers. Shifting the balance by activating or inactivating various MAPs will lead to either greater microtubule stability or greater microtubule turnover. The idea that microtubule dynamics are regulated by a balance between stabilizing and destabilizing MAPs was first proposed for two MAPs found in frog eggs—XMAP215 (a microtubule stabilizer) and MCAK. Removing XMAP215 from frog cells allows MCAK to dominate; the result is very short microtubules with a high rate of catastrophe, as illustrated in **FIGURE 11.42**. Conversely, removal of MCAK tips the balance in favor of more stable microtubules that grow to longer lengths because catastrophes rarely occur. During mitosis, the proper balance of XMAP215 and MCAK is needed or a proper spindle does not form.

How is MAP activity regulated? In general, changes in microtubule stability occur too rapidly to involve changes in expression of the genes encoding MAPs. Instead, the activity of many MAPs is changed by phosphorylation. For example, phosphorylation of the tau protein reduces its affinity for microtubules and, therefore, reduces its ability to stabilize them. Phosphorylation of MAPs does not necessarily happen throughout the cell. Instead, local activation of a kinase in a small area of the cell can change MAP activity only in that area. By switching microtubule stabilizers and destabilizers between their "on" or "off" states, cells can tip the balance toward more stable or more dynamic microtubules. By locally changing MAP activity, a cell can respond to signals by making microtubules grow longer or disappear within a defined region of the cell. Local regulation of MAP activity occurs during cell locomotion and may happen during the assembly of the mitotic spindle.

Phosphorylation of tau protein is also associated with Alzheimer's disease, although it

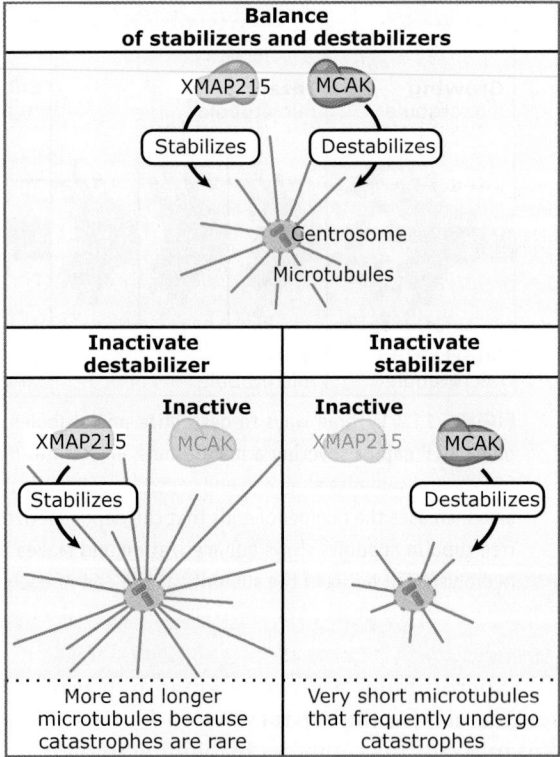

FIGURE 11.42 Microtubule-stabilizing and microtubule-destabilizing MAPs act in pairs to determine the size of a microtubule array. Inactivating the destabilizer makes catastrophes less likely and results in a larger array with more microtubules. Inactivating the stabilizer has the opposite effect. Using a balance of opposing components to determine the size of a microtubule array allows its size to be changed quickly.

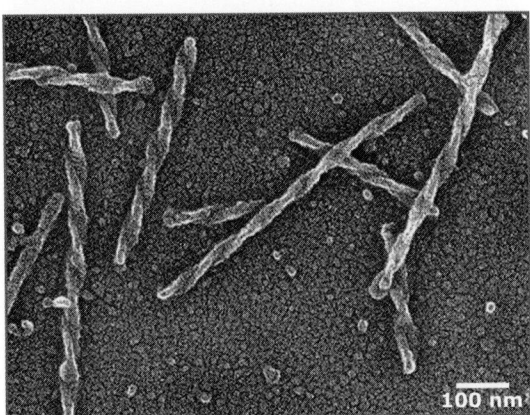

FIGURE 11.43 Tau protein, a MAP present in neuronal tissue, is hyperphosphorylated and aggregates into paired helical filaments during the neural degeneration associated with Alzheimer's disease. Here, purified filaments are viewed by electron microscopy. Photo courtesy of Denah Appelt and Brian Balin, Philadelphia College of Osteopathic Medicine.

is not clear that altered microtubule dynamics play a role in the disease. When tau protein is hyperphosphorylated, it aggregates into the neurofibrillary tangles seen in the brains of patients with the disease, as shown in **FIGURE 11.43**. It is not yet clear whether the tau defects observed in Alzheimer's disease are the cause or are an effect of the neurodegenerative process. Although it is not yet known if tau defects play a causal role in Alzheimer's disease, several other human neuronal diseases result from specific mutations in the tau gene, leading to formation of filamentous tangles of tau proteins and eventual neurodegeneration.

11.11 Introduction to microtubule-based motor proteins

Key concepts

- Almost every cell function that depends on microtubules requires microtubule-based motors.
- Molecular motors are enzymes that generate force and "walk" along microtubules toward the plus or minus ends.
- The motor "head" domain binds microtubules and generates force.
- The "tail" domain typically binds membrane or other cargo.
- Most kinesins "walk" toward the plus ends of microtubules.
- Dyneins "walk" toward the minus ends of microtubules.

One of the main functions of microtubules is to serve as tracks for the movement of material from place to place within a cell. The trucks that move cargo over these intracellular highways are called molecular motors. Molecular motors are microtubule-binding proteins that use repeated cycles of ATP hydrolysis to power continuous movement along the side of a microtubule. These motors deliver secretory vesicles to the plasma membrane, transport internalized vesicles to endosomes, and move mitochondria and the ER throughout the cell. One spectacular and very visible example of the work of molecular motors is the coordinated movement of pigment granules—small vesicles packed with pigment molecules—in cells within the scales of some fish and the skin of some amphibian species. In response to hormones or signals from the nervous system, microtubule-dependent molecular motors alternately collect these vesicles in the center of the cell or disperse them throughout its cytoplasm, allowing the animal to change color and avoid predators. Examples of this are shown in **FIGURE 11.44**.

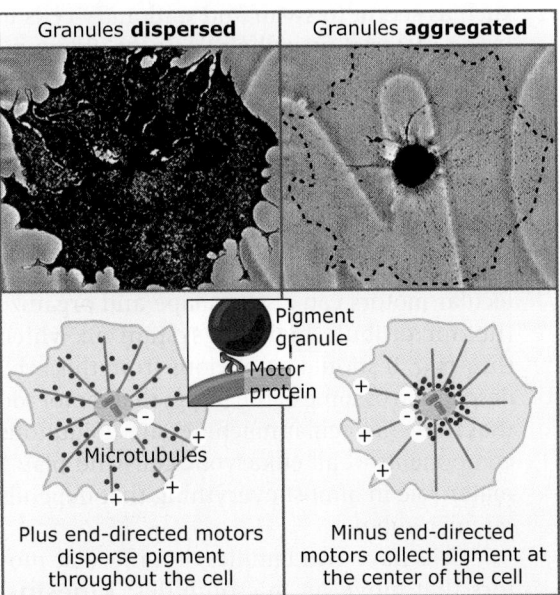

FIGURE 11.44 The photographs show a single cell containing thousands of pigment granules, small particles densely packed with dark pigment. A few individual granules are visible as very small dots in the clear area of the picture on the right. Each granule has both plus end- and minus end-directed motors attached to it. In response to the hormone melatonin, minus end-directed motors move the granules inward along the microtubules, aggregating the pigment in the center of the cell. In the absence of the hormone, other motors move the granules back out along the microtubules so that the pigment is again dispersed throughout the cell. Photos courtesy of Vladimir Rodionov, University of Connecticut Health Center.

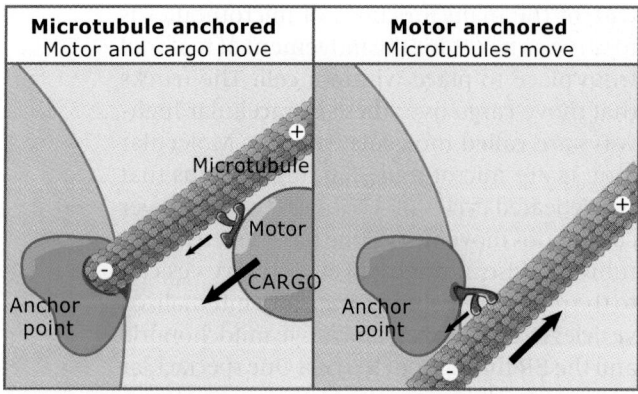

Microtubule anchored Motor and cargo move	**Motor anchored** Microtubules move

FIGURE 11.45 Whether it is the motor or the microtubule that moves when they interact depends upon which is anchored. Both cases are common in cells. Anchoring the microtubule allows vesicular traffic, whereas anchoring the motor results in rearrangement of the cytoskeleton.

In addition to moving many of the cell's internal membranes, motors move chromosomes during mitosis and position the spindle within the cell. Motors also power the beating of cilia and flagella, allowing specialized cells such as sperm to swim and stationary cells to move material over their surfaces. Some viruses hijack the cell's motors and use them to transport themselves to the nucleus of the cell; HIV is one example of a virus that uses the cell's trucking system.

Transport is not the only function of molecular motors, however. Unlike the trucks that transport cargo on our highway systems, molecular motors can also reshape and organize the microtubule highway system on which they run. It should be obvious from this brief description of some of the things that motors do that these molecular machines are ubiquitous components of all eukaryotic cells and play a major role in almost everything that depends on microtubules.

Cells have two families of molecular motors that move on microtubules: **kinesins**, which usually move toward the plus ends of microtubules, and **dyneins**, which move toward microtubules' minus ends. The organization of the microtubule array, together with the direction moved by a particular motor, provides the navigational information needed to direct cargo to the proper destination in the cell. For the radial array of microtubules found in a typical fibroblast cell (see Figure 11.7), motors that move toward the minus ends of microtubules will transport cargo to the center of the cell (e.g., to the nucleus or Golgi apparatus), while plus end-directed motors will move cargo to its margins (e.g., to the plasma membrane).

Movement in one direction on a polar polymer is an essential feature of all molecular motors, whether it is a motor that moves on microtubules or actin filaments. In this way, the polarity of the polymer contributes to the direction of movement and navigation by the motor. Intermediate filaments lack polarity (see *13 Intermediate filaments*), and no motors have been identified that use intermediate filaments as a roadway for movement.

The cargo pulled by kinesin or dynein can include the microtubules themselves, and motors frequently play a role in organizing and reorganizing the microtubules within a cell. As shown in **FIGURE 11.45**, if the microtubule is anchored (as it would be if attached to the centrosome), the motors move and can transport cargo along the microtubule. If the situation is reversed and the motor protein is anchored (e.g., to the cell cortex), the motor moves the microtubule instead, helping reorganize the microtubule array (see Figure 11.47). When it is the microtubule itself that is moved, its polarity is still important in providing navigational cues; in this case the microtubule's polarity determines the direction of its own movement.

All molecular motors, including the actin-based motor myosin (see *12 Actin*), have a characteristic shape that allows them to perform their task. This shape is clear when individual motor molecules are examined by electron microscopy, as shown in **FIGURE 11.46**. In each case, the motor consists of a pair of identical large globular domains attached to one end of a long rod-shaped domain, giving the motor a long (40–100 nm), thin shape overall. Many motors also have a second pair of smaller globular domains at the other end. The large globular domains contain both the polymer-binding (microtubule or actin) and ATP-binding sites of the motor and are referred to as the "head" or "motor" domains. They are the only parts of the motor that are needed to generate force; the other domains allow the force that they generate to be used for a specific purpose within the cell. Dynein motors are unique in having an extra "stalk" that extends from the globular head. For dyneins, it is the tip of the stalk that binds microtubules. The opposite end of a motor from the two heads is called the "tail" domain; it is here that cargoes, such as vesicles, are bound.

Typically, a single motor molecule contains several polypeptides of different sizes. Most of it

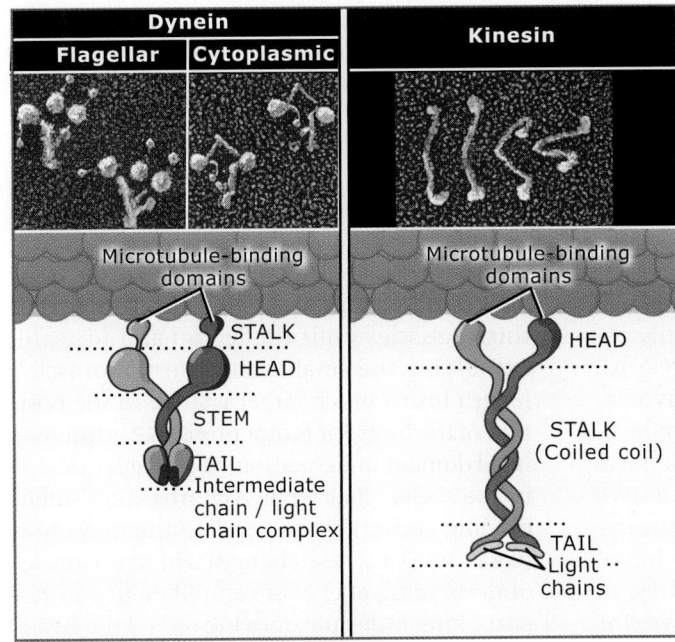

FIGURE 11.46 Structures of microtubule-based motor proteins are shown by rotary shadowing and electron microscopy (top panels). Schematic representations of motors bound to microtubules are shown in the lower panels. Each motor is made up of a two or more large polypeptides (heavy chains) and several smaller polypeptides (intermediate and light chains). Photos courtesy of John Heuser, Washington University, School of Medicine.

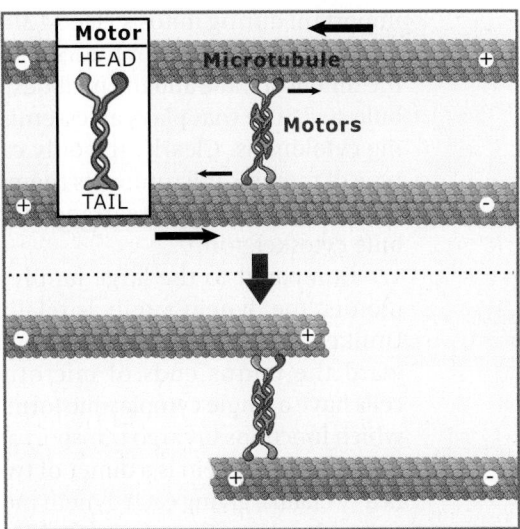

FIGURE 11.47 Some kinesins associate through their tail domains, forming bipolar motors with two motor head domains at each end. Such motors can simultaneously bind and move along two microtubules of opposite polarity. The net effect is to slide the microtubules past one another. Small arrows indicate the direction the motors move; large arrows indicate the direction the microtubules move as a result.

is composed of a dimer of the largest polypeptide, called the "heavy chain." The two heavy chains are connected by coiled-coil interactions over much of their length to form the central, rod-shaped region of the motor. Regions at each end of the coiled-coil form the head and tail domains. In each type of motor one or two different smaller polypeptides—called "light chains"—associate with each of the heavy chains and often play regulatory roles in motor function.

The kinesin family of microtubule-based motors is quite large; humans have about 45 different kinesin motors. This number alone suggests the variety of roles that microtubule-dependent motor proteins play in cells and how specialized some of them are. More than half of the kinesins are responsible for moving cargo to specific destinations in the cell, while the remainder function during mitosis. The kinesin family is defined by a high degree of sequence homology within the motor domain. Outside the motor domain, kinesin family members show much greater diversity in sequence, often being completely different from one another. These variable regions likely enable each kinesin to bind its specific cargo.

The kinesin superfamily can be broadly divided into three groups based upon the position of the motor domain within the heavy chain. The first kinesin identified has its motor domain near the N-terminus. This "conventional" kinesin moves vesicles toward the plus end of a microtubule. Other members of the kinesin superfamily have the motor domain near the C-terminus of the heavy chain. Having the motor domain at the C-terminus is correlated with kinesins that move toward the minus ends of microtubules. A few kinesins have the motor domain near the middle of the heavy chain (e.g., MCAK). Rather than powering movement, kinesins with an internal motor domain regulate microtubule dynamics by using ATP hydrolysis to weaken the structure at the end of a microtubule (see *11.10 Cells use several classes of proteins to regulate the stability of their microtubules*).

The tails of some kinesin motors can associate together to form a four-headed, bipolar motor. As shown in **FIGURE 11.47**, having motor domains facing in opposite directions allows these motors to bind two microtubules at once and slide them past each other. Sliding of microtubules past each other is particularly important during mitosis (see *14 Mitosis*). Such activities are essential for the formation of both the mitotic spindle and the midbody, a microtubule structure that plays an essential role during cytokinesis. Clearly, the only cargo of this type of motor is microtubules themselves, and their role is strictly to rearrange the microtubule cytoskeleton.

Compared to the large family of kinesin motors, the dynein family is relatively small. Unlike the kinesins, dyneins move only toward the minus ends of microtubules. All cells have a single cytoplasmic form of dynein, which functions in cargo transport and mitosis. Cytoplasmic dynein is a dimer of two identical heavy chains, giving each dynein molecule two motor domains. The other members of the dynein family, the axonemal dyneins, are found specifically in cilia and flagella. In contrast to cytoplasmic dynein, the axonemal dyneins are heterodimers or heterotrimers of different heavy chain subunits, and have either two or three motor domains per molecule. This, as well as how axonemal dyneins power flagellar or ciliary motion, will be discussed in *11.16 Cilia and flagella are motile structures*.

11.12 How motor proteins work

Key concepts

- Motor proteins use ATP hydrolysis to power movement.
- The nucleotide (ATP, ADP, or no nucleotide) bound to a motor's head domain determines how tightly the head binds to the microtubule.
- ATP hydrolysis also changes the shape of the head. This shape change is amplified to generate a larger movement of the motor molecule.
- Cycles of ATP hydrolysis and nucleotide release couple microtubule attachments with changes in the shape of the motor's head domain. By this mechanism the motor steps along the microtubule, taking one step for each ATP hydrolyzed.

Molecular motors use ATP as the fuel to power movement, but how do motor proteins convert the chemical energy stored in ATP into mechanical work? In this section, we will examine how motor proteins move on microtubules. We would know little about this if we did not have ways to watch motors move. We will not go into how microtubule-dependent motors are assayed here, but if you would like to find out, see the *Methods and Techniques* box on pages 534–535.

The most basic requirement for a motor is that it must undergo a very large confor-mational change between when it has ATP bound and when it has ADP bound. This is accomplished by changes in the motor domain and neighboring parts of the molecule that are analogous to the way we move our arms and legs. For both molecular motors and our limbs, a small change in shape in one place is amplified to produce a much larger change in shape or position in another place. As someone walks, for example, a small contraction of the thigh muscles pulls the leg up and forward, amplifying the small change in the muscle's length into a much larger change in the position of the foot. For motor proteins, the motor's head domain undergoes small changes in shape in the region that binds ATP (the nucleotide-binding pocket) as the result of the hydrolysis of ATP to ADP. These changes within the nucle-otide-binding pocket are amplified in another part of the molecule, moving one of the heads forward.

Molecular motors and walkers also share the requirement that they must be able to let go of the surface to which they are bound; otherwise, neither will be able to move forward. Just as a walker must pick her foot up off the ground in order to move her leg forward, a motor protein must release from the microtubule in order to move. To let go of a microtubule, a motor protein must decrease its binding affinity for it. How strongly a motor binds to a microtubule is determined by whether ATP, ADP, or no nucleotide is bound to the motor's nucleotide-binding pocket. For kinesin, binding to a microtubule is tightest when ATP is bound. By changing the strength of kinesin's hold on a microtubule, ATP hydrolysis and nucleotide release regulate the attachment of the motor to the microtubule. Because ATP hydrolysis also causes a shape change in the motor's head domain, the cycle of nucleotide binding, hydrolysis, and release, coordinates changes in the motor's shape with its binding to the microtubule. This allows motors to take one "step"—a cycle of binding to the microtubule, conformational change, and release—along the microtubule for each ATP hydrolyzed.

For a two-headed motor, one can imagine two different ways that steps could be used to produce movement along a microtubule. The two motors could move in a head-over-head motion, as shown in **FIGURE 11.48**, in which the rear head passes the forward head with each step forward. This type of motion is analogous to the way we walk, with each forward step moving one foot past the other. The alternative

Head-over-head (walking)	Inchworm (sliding)

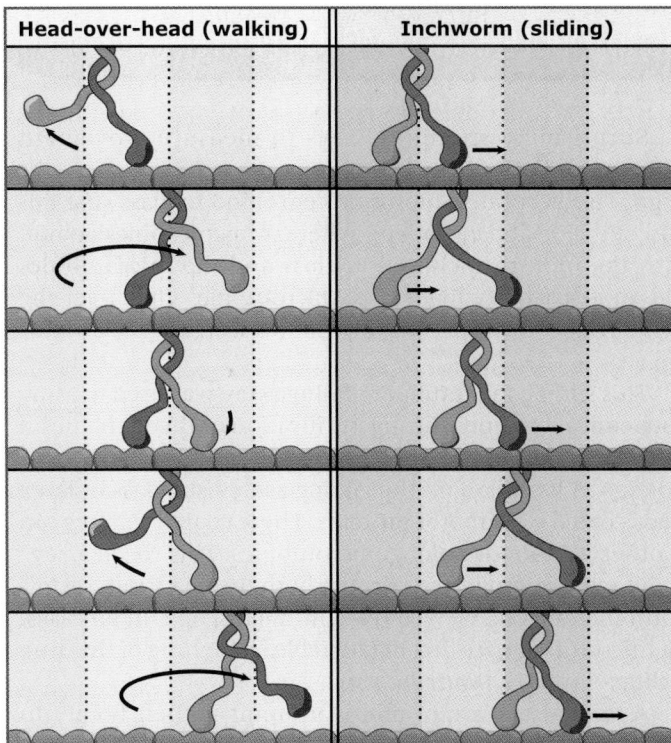

FIGURE 11.48 Two possible ways that a two-headed motor could move along a microtubule. Coordination of the activities of the two heads could result in an "inchworm"-type motility (right), in which the red head steps forward, the orange head catches up, and the cycle repeats itself. The orange head would never be in front of the red head. Although this mechanism is possible, a motor that uses it has yet to be found. Instead, all known two-headed motors move by a mechanism analogous to walking, the two heads stepping past one another and taking turns leading (left).

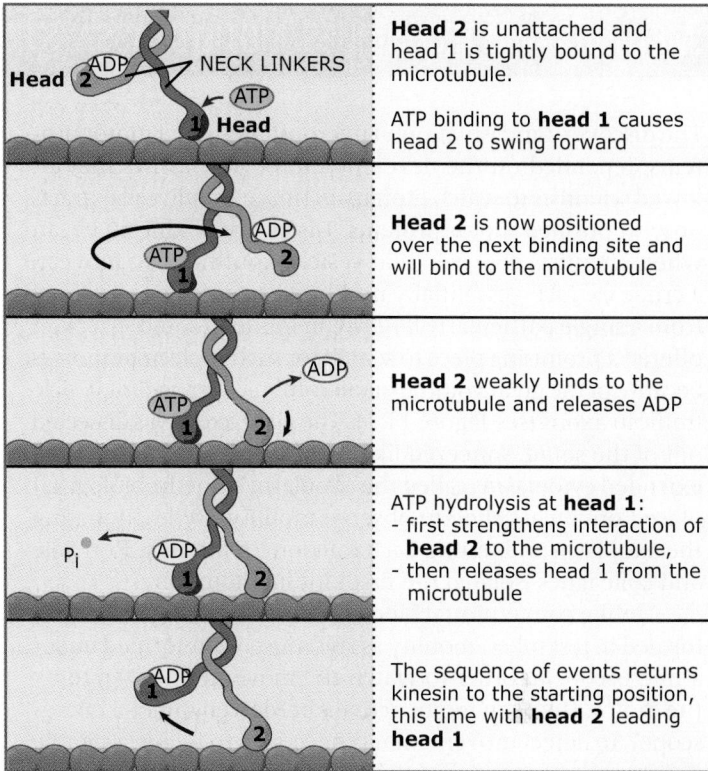

Head 2 is unattached and head 1 is tightly bound to the microtubule.

ATP binding to **head 1** causes head 2 to swing forward

Head 2 is now positioned over the next binding site and will bind to the microtubule

Head 2 weakly binds to the microtubule and releases ADP

ATP hydrolysis at **head 1**:
- first strengthens interaction of **head 2** to the microtubule,
- then releases head 1 from the microtubule

The sequence of events returns kinesin to the starting position, this time with **head 2** leading **head 1**

FIGURE 11.49 The sequence of events that allows kinesin to walk along a microtubule. Several times a change occurs in one head in response to an event in the other. Note that kinesin always has at least one of its two heads tightly bound to the microtubule.

mechanism would have the motor domains move in an "inchworm" motion, in which the rear head moves up to the forward head, the forward head advances, and the cycle repeats (see Figure 11.48). All two-headed motors studied to date use a head-over-head motion as they traverse a microtubule. In other words, kinesin and other motors can be viewed as "walking" along a microtubule.

Essential to kinesin's ability to walk is a small domain called the neck linker, an elongated sequence of 15 amino acids that connects the head domain of kinesin with its coiled-coil domain. The neck linker is the part of the kinesin molecule that amplifies small changes of structure in the nucleotide-binding pocket into the large changes necessary to enable kinesin to take steps of significant length. The large changes take the form of swings of the neck linkers back and forth, allowing one to view the neck linkers as the "legs" of the kinesin mol-

ecule. As we will now see, walking is achieved by using the ATPase cycles of the two heads to control when the neck linkers swing.

As kinesin moves along a microtubule, the two heads work in tandem. An event in one head often occurs as a result of a change in the other. To understand the cycle that enables a motor to walk along a microtubule, we begin just after kinesin has first landed on the microtubule, as shown in **FIGURE 11.49**. One head is tightly bound to the microtubule and has no nucleotide in its active site. Its neck linker trails behind it. The second head has ADP in its active site and is positioned behind the first, waving in the breeze alongside the microtubule. Kinesin is now ready for its first step and the coordination between the two heads comes into play. ATP binding at the forward head (head 1) causes its neck linker to swing forward toward the plus end of the microtubule. The consequence of this motion associated with head 1 is

The discovery and isolation of microtubule-based motor proteins depended on the development of new assays that allowed scientists to watch motility *in vitro*, either in cell extracts or with purified motor proteins. Discovery of kinesin began when scientists observed that vesicles continued to move in a crude extract—essentially undiluted cytoplasm—prepared from a single particularly large axon found in squid. An axon offered a promising place to search for such molecular motors because of the great volume of microtubule-dependent vesicle traffic in axons (see Figure 11.1). The giant axon was dissected out of the squid, squeezed like a tube of toothpaste, and the extruded cytoplasm (called the axoplasm by neurobiologists) placed on a glass slide to observe motility. By fractionating the axoplasm and testing each fraction for motility, Ron Vale and colleagues isolated the first kinesin motor.

Unlike conventional biochemical assays, which are performed in test tubes, motility assays must be performed under a microscope in order to watch the movement powered by the motor. How can motor activity be viewed under a microscope? To detect movement it is necessary to follow a sample over time, meaning that the proteins must be in their native form. This excludes electron microscopy, which requires extensive preparation of each sample with chemicals that leave the proteins inactive. Light microscopy does not have the resolution of electron microscopy, but it does have the advantage that it usually requires little if any modification of the proteins in a sample. It can, therefore, be used to follow events that require active proteins and last several minutes or more. Microtubules are too small to be seen with a conventional light microscope, however—a basic requirement of any assay for microtubule-dependent motility or the motors that drive it. This problem is solved by making microtubules visible in either of two ways. One way to visualize individual microtubules is to use differential interference contrast (DIC) microscopy (a special form of light microscopy) and enhance the images with video and computer methods. An example of a microtubule viewed by this method is shown in **FIGURE 11.B1**. Alternatively, microtubules can be polymerized from fluorescently tagged tubulin dimers and the fluorescent microtubules viewed using a fluorescence microscope.

Surprisingly, motility assays to measure movement generated by motors do not require something to serve as cargo. This is because motor proteins bind to glass (the microscope slide and coverslip), whereas microtubules do not. With the motors anchored to glass and the microtubules free, motor activity makes the microtubules glide over the glass surface. Examples of this can be seen in **FIGURE 11.B2** and **FIGURE 11.B3**.

This simple microtubule-gliding assay was used to assay biochemical fractions during purification of the first kinesin and is used routinely in the study of microtubule-based motor proteins. A variation on the gliding assay uses glass or latex beads coated with motor proteins. These coated spheres can be observed to move along microtubules. This "bead assay" is a bit more complicated because both the bead and the microtubule must be viewed by light microscopy. In all cases, the microtubules are first made stable by binding of the drug paclitaxel (see *11.1 Introduction*).

A critical feature of motor proteins is their ability to move in only one direction along a microtubule. To figure

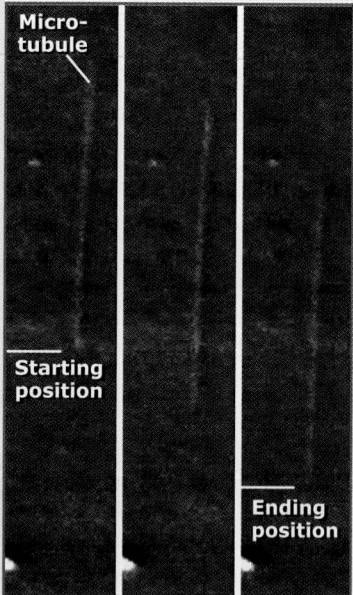

FIGURE 11.B2 A series of images taken from a video sequence show a microtubule gliding over a glass surface in an *in vitro* experiment. The microtubules are detected by differential interference microscopy and image processing. Motor proteins bound to the glass coverslip generate force to transport the microtubule. Photos courtesy of Lynne Cassimeris, Lehigh University.

FIGURE 11.B1 A microtubule viewed by differential interference contrast (DIC) microscopy and computer image enhancement. The image is grainy, but the position and length of the microtubule are clear. This form of microscopy can be used to measure how long microtubules are, watch them move, or watch particles move along them. Photo courtesy of Lynne Cassimeris, Lehigh University.

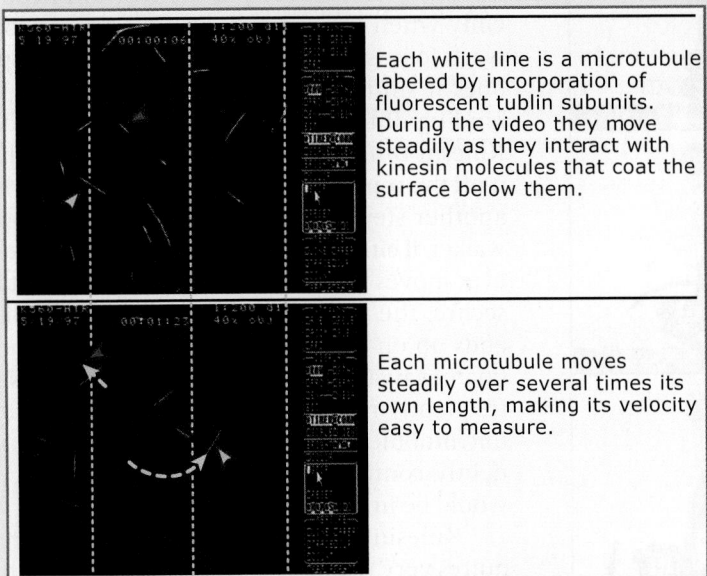

Each white line is a microtubule labeled by incorporation of fluorescent tublin subunits. During the video they move steadily as they interact with kinesin molecules that coat the surface below them.

Each microtubule moves steadily over several times its own length, making its velocity easy to measure.

FIGURE 11.B3 Two images taken from a video sequence show several microtubules gliding over a glass surface. The microtubules were polymerized from purified tubulin mixed with a small percentage of tubulin modified with a fluorescent group. Kinesin sticks to the glass simply because a positively charged region in its tail domain interacts with the negative charge on surface of the glass. Two microtubules are noted by the red and yellow arrowheads; the paths of their movements are noted by the dashed arrows. Animation courtesy of Ron Milligan, The Scripps Research Institute; Ronald Vale, Howard Hughes Medical Center; and, Graham Johnson, fiVth.com.

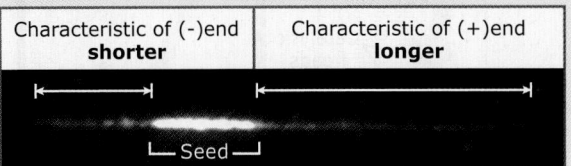

Characteristic of (-)end **shorter**	Characteristic of (+)end **longer**

— Seed —

FIGURE 11.B4 A single microtubule grown from a short seed, with a much higher concentration of fluorescent tubulin subunits at its plus end. The difference in length of the extensions at its plus and minus ends is clear. Use of microtubules labeled in this way provides an easy way of determining which way a motor moves along them. Photo courtesy of Arshad Desai, Ludwig Institute for Cancer Research.

out whether a motor moves toward the microtubule's plus or minus end, we need a way to know the polarity of the microtubule. One way to be certain of the polarity of the microtubules in an assay is to assemble a radial array of microtubules from a purified centrosome and then attach it to the microscope slide. Beads coated with the motor protein are then added; beads that move to the center of the aster move to the minus ends while beads that move away from the center move to the plus ends. Another way to mark the polarity of microtubules takes advantage of the faster polymerization at the plus end. To do this, short microtubules with a very high percentage of fluorescent subunits are first formed. These bright fluorescent microtubules are then used as seeds to nucleate longer microtubules in a solution in which a much lower percentage of the tubulin is fluorescent. Tubulin molecules add to both of a seed's ends but much more readily to its plus end. The result is a collection of microtubules each with a single short and very bright segment in the middle and dimmer but still clearly visible segments extending from its ends as shown in **FIGURE 11.B4**. The extension from the plus end is much longer than the one from the minus end, allowing the polarity of the microtubule to be determined at a glance. These fluorescent, polarity-marked microtubules can then be used in gliding assays to determine the directionality of a motor. When interpreting such assays, it is important to remember that the motor protein is anchored. A lawn of a plus-end directed motor will make a microtubule glide with its minus end leading.

Almost all motility experiments are some variant of one of these assays. The question being asked and the sophistication and resolution of the detection system may vary, but the basic principle of the assay is the same.

to move head 2 from the trailing to the leading position. There, it is positioned over the next binding site in the microtubule. It binds weakly and releases its ADP. ATP hydrolysis at head 1 then strengthens the interaction between head 2 and the microtubule, resulting in an intermediate with both heads strongly bound to the microtubule. Once head 2—that is now the leading head—is tightly bound, head 1—now trailing—releases the phosphate group generated when it hydrolyzed ATP. Release of phosphate causes head 1 to let go of the microtubule and results in a conformational change in head 2 that reopens its active site. This whole cycle of events returns kinesin to the starting condition, with the essential differences that head 2 is now in front and the kinesin molecule is 8 nm closer to the plus end of the microtubule. A second cycle will begin when head 2 binds ATP, and the two heads will then alternate roles for hundreds or thousands of rounds, each round producing a step and moving the motor along

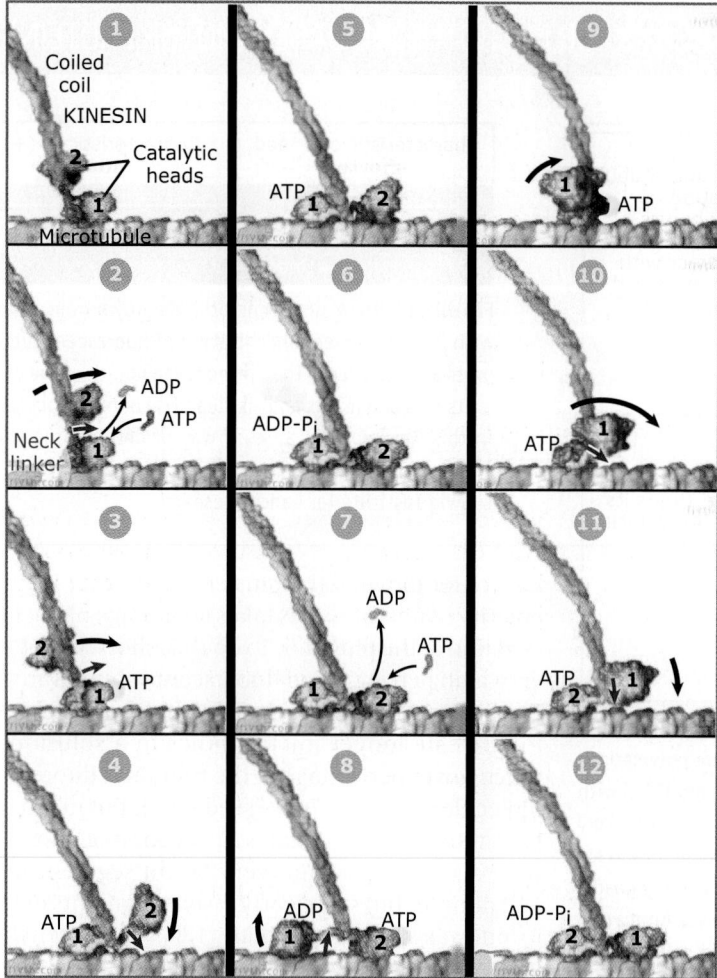

FIGURE 11.50 A series of frames from a video animation show the coordinated movement of kinesin's two heads along a microtubule. The two heads of kinesin (labeled 1 and 2) are shown in blue, the coiled-coil region shown in grey and the neck linker regions are shown in yellow when pointed forward and red when pointed backwards. For simplicity only one protofilament of the microtubule is shown. The α and β subunits of tubulin are shown in white and green, respectively, and the plus end is to the right. Reproduced from R. D. Vale and R. A. Milligan, *Science* 288 (2000): 88–95 [http://www.sciencemag.org]. Reprinted with permission from AAAS. Video courtesy of Ron Milligan, The Scripps Research Institute; Ronald Vale, Howard Hughes Medical Institute; and, Graham Johnson, fiVth.com.

one leg. It then swings the other leg forward (the change in position of the neck linker) and gingerly tests the rope with its foot (the initially weak binding of the forward head). Only when kinesin is convinced that its forward foot is properly positioned on the rope does it shift its weight onto that foot (tight binding of the newly landed forward head). Kinesin can then release its rear foot and lift it off the rope, putting itself in position to take another step. By analogy with the tightrope walker, if either kinesin or the tightrope walker ever moves its rear foot before its front foot is secure, the walk is over. The tightrope artist ends up on the ground, and kinesin ends up floating away from the microtubule. If kinesin frequently released its trailing head from the microtubule before its leading head was bound tightly, continuous motion over long distances would be impossible.

Kinesin's walking mechanism clearly requires very reliable coordination between the two heads. How is that coordination achieved? How does one head "know" what the other is doing? It appears to be the neck linkers that act as the paths of communication between the two heads. When both of kinesin's heads are bound tightly to the microtubule, the neck linkers are stretched and are under mechanical strain. It is apparently this strain that allows the two heads to communicate with one another and coordinate their activities (i.e., their ATPase cycles). The strain indicates to the trailing head that the leading head has bound tightly. The trailing head can, therefore, be safely released from the microtubule. Strain probably accomplishes the coordination of the heads by determining how fast different steps in the ATPase cycle occur. For example, were strain to accelerate phosphate release from a negligible rate to a significant one, release of the trailing head from the microtubule could not occur until the leading head was tightly bound.

Dynein motility is also based on amplification of a conformational change, but the structural change occurs over a much larger distance. Both kinesin and dynein walk along a microtubule in 8-nm steps, which is equal to the length of a single tubulin heterodimer. Kinesin walks "carefully," stepping from one heterodimer to the next along a single protofilament. By comparison, dynein "wanders" as it walks, stepping randomly between protofilaments as it traverses toward the microtubule's minus end.

the microtubule toward its plus end. The series of still frames, in **FIGURE 11.50**, of a kinesin molecule taking several successive steps show how the coordination of events in the two heads and their exchange of roles after every step allow it to walk along a microtubule.

The sequence of events that kinesin goes through to perform each step can be thought of as analogous to a tightrope artist taking a step on a rope high above the ground. Kinesin is initially stable and standing balanced on

Kinesin's mechanism of movement gives it the ability to walk continuously along a microtubule (i.e., its movement is highly "processive"). For example, in *in vitro* experiments, a single two-headed kinesin motor attached to a glass bead (which serves as a convenient cargo) can take hundreds or thousands of steps along a microtubule and move the bead for a considerable distance without falling off the microtubule and drifting away. The ability of a single kinesin to move cargo long distances along a microtubule is possible because each kinesin head spends approximately half its time bound to the microtubule, and the activities of the two heads are coordinated so that at least one head is always bound. Having one of the two heads always bound to the microtubule is a property found in motors that work individually or in small numbers, such as those that move vesicles. Because of the way they work, these motors can reliably move cargoes long distances within cells.

Motors that do not always keep one head bound to the microtubule, allowing the motor and its cargo to quickly lose contact with the microtubule surface, also have uses. Motors that act in large arrays, such as those in flagella (see *11.16 Cilia and flagella are motile structures*), spend much less time bound to the microtubule than those that move vesicles. In an array of dyneins (in a flagellum), some heads will be bound and generating force. Those that have completed their steps quickly let go of the microtubule so that their binding does not impede active motors from generating force on the same microtubule.

Many organelles move bidirectionally in cells, moving one direction along a microtubule for some distance and then turning around and traveling an appreciable distance in the opposite direction. These organelles bind both dynein and a member of the kinesin family, raising the question of how it is possible to achieve prolonged movement in either direction. Two possible models, shown in **FIGURE 11.51**, have been suggested to explain the bidirectional movement of organelles. Opposite-polarity motors may compete in a tug-of-war, in which both types of motors are always active and the one able to generate the stronger pulling force (because it is present in greater number) wins the war. Alternatively, motor activities may be coordinated such that one set of motors is turned off while the other is active. It appears that the second mechanism is present in cells, but it is not yet understood how

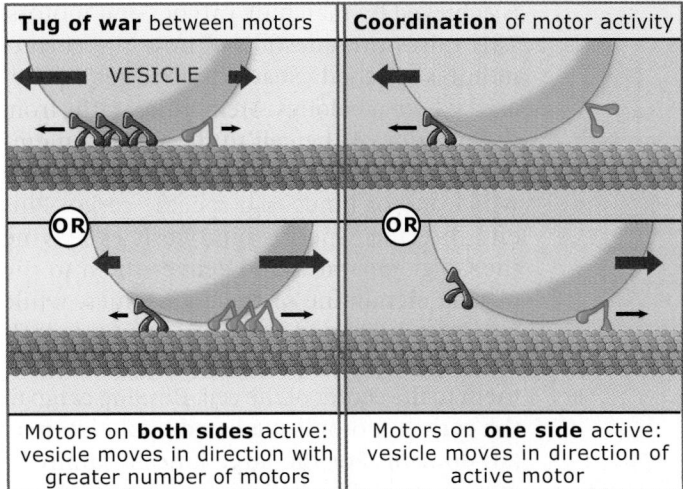

Tug of war between motors	Coordination of motor activity
Motors on **both sides** active: vesicle moves in direction with greater number of motors	Motors on **one side** active: vesicle moves in direction of active motor

FIGURE 11.51 Possible ways to generate bidirectional movement of a cargo along a microtubule. In each case, both plus end- and minus end-directed motors are bound to the vesicle surface. On the left they are active at the same time, and the one able to generate the larger pulling force (probably because it is present in a larger number) would determine the direction of vesicle movement. On the right the motors are coordinated so that only motors that pull in one direction are active at a time. Current evidence suggests that cells use the mechanism on the right, but how the motors are coordinated is not yet understood.

the activities of the motors are coordinated on the vesicle surface.

Concepts and Reasoning Checks

1. You have discovered a new molecular motor that walks on microtubules. How would you determine which way this motor transported cargo on a microtubule?
2. Kinesin is described as a processive motor—what does processive mean? How are the motions of kinesin's two heads coordinated to generate processive movement?

11.13 How cargoes are loaded onto the right motor

Key concepts

- Binding of motors to specific cargoes is mediated by the motor tail domain.
- Adaptor proteins associate with motors to regulate motor activity and to link motors to cargo.
- Coordination of plus end- and minus end-directed motor activities is used to generate bidirectional movement of organelles.

Cells need to move many different cargoes to specific locations in the cytoplasm. Specificity is achieved by matching cargoes and motors. This raises the question of how the correct motor is attached to each cargo to get it delivered where it belongs. Membrane traffic from the interior of the cell to the plasma membrane and back illustrates the need to specify which motor binds to a specific cargo. Plus end-directed kinesins bind vesicles leaving the Golgi apparatus and deliver them to the plasma membrane or the endosomes, while minus end-directed motors pick up internalized vesicles at the cell periphery and deliver them to the center of the cell. Binding cargo to the correct motor is mediated by the motor's tail domain. For the large kinesin family of motors, the tail domains of the family members are very different and distinguish each as a unique motor. The motor domains are much more similar and do not contribute to cargo specificity. In this way, the head domain is the engine common to all trucks, and the tail domains are unique trailers loaded with a select group of cargoes.

In general, the motor's tail domain does not bind directly to cargo. An adaptor protein typically binds to a membrane protein on one end and the motor's tail at the other end, indirectly linking the motor to the vesicle. For example, vesicles leaving the *trans*-Golgi network for the endosome contain the mannose-6-phosphate receptor in their membranes. The cytoplasmic domain of this receptor binds

the adaptor complex AP-1, and AP-1 binds to the tail of a kinesin. AP-1 is a familiar adaptor because it also links clathrin to regions of the *trans*-Golgi network where vesicles bud. In this way, AP-1 links vesicle budding to loading of the motor, ensuring that newly budded vesicles are properly equipped for transport. (For more on AP-1 see *8.14 Adaptor complexes link clathrin and transmembrane cargo proteins*.)

Adaptor proteins also link cytoplasmic dynein to membranes. The best-characterized adaptor is the dynactin complex. Dynactin is a complex of seven polypeptides and a short filament composed of Arp1, a protein closely related to actin. A recent model, shown in **FIGURE 11.52**, proposes that the Arp1 filament links dynein to membrane vesicles by binding to spectrin at the cytoplasmic surface of the membrane. This connection would be similar to the interaction between spectrin and actin filaments in the network they form on many membranes and would explain why dynactin includes an actin-like filament. In addition to attaching dynein to membranes, dynactin also helps dynein stay associated with a microtubule so that movement is possible.

Motor proteins are responsible for transporting more than just membrane vesicles. Additional cargoes include some mRNAs and virus particles, although the latter are clearly not among a cell's normal cargoes. Transport of mRNAs allows cells to restrict the synthesis of some proteins to specific locations and to ensure that mRNAs reach distant locations in very large cells. In some neurons, for example, specific mRNAs are sorted to the axons or dendrites, where proteins specific to these highly specialized cellular domains are then made. Because of the great length of axons and dendrites, molecular motors are required to transport the mRNAs. Movement of RNAs by diffusion alone would be both too slow and lack the specificity required to allow cells to construct and maintain such long, specialized structures. Similar to membrane vesicles, adaptor proteins are used to link mRNAs to a motor's tail. Viruses such as HIV, Herpes simplex virus, and adenovirus enter cells as a core particle of nucleic acid surrounded by a protein capsid. To replicate, the virus particles must reach the nucleus where they use the host cell's DNA replication machinery. Viruses hasten their replication process by binding dynein once they have entered the cell, allowing them to move directly from the plasma membrane to the nucleus.

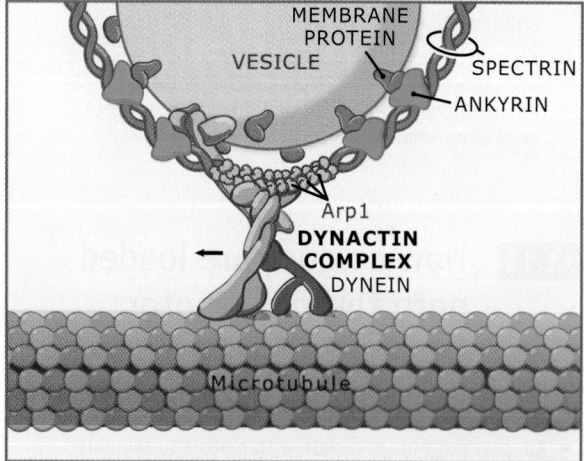

FIGURE 11.52 A model of how the dynactin complex (purple) links cytoplasmic dynein to membrane vesicles. The Arp1 filament of the dynactin complex is thought to associate with spectrin on the membrane in a manner similar to the way that actin filaments and spectrin interact on other membranes. Other components of the dynactin complex associate with both dynein and the microtubule.

11.14 Microtubule dynamics and motors combine to generate the asymmetric organization of cells

Key concepts

- Dynamic microtubules and motors work together to generate cell asymmetries.
- Microtubules work together with the actin cytoskeleton during processes such as cell locomotion and mitotic spindle positioning.

The positions of organelles within a cell and the overall shape of the cell as a whole often have a definite, intentional asymmetry. The ability of cells to make one end different from the other by arranging and orienting their organelles and specializing regions of cytoplasm is a fundamental and exceptionally important property. Although an isolated individual cell could live without this ability, such cells would be completely incapable of moving or forming any of the many highly shaped and specialized cell types that are necessary to construct and maintain an organism. Fibroblasts and other motile cells, for example, must extend at one end and retract at the other in order to allow them to move through the body and respond to injury or infection. Somehow, the different activities at the two ends of each cell must be set up and then coordinated.

The asymmetric organization of a cell usually depends on the arrangement of its microtubules, their dynamic turnover, and the movement of microtubule-dependent motor proteins along them. Actin and intermediate filaments also participate in organizing the internal architecture of a cell, and the three filament systems interact and control one another's behavior. In this section, we describe several examples of how the components of the microtubule cytoskeleton function to generate cell asymmetries, including cases that illustrate how the microtubule and actin cytoskeletons work together.

When the brain is wired during development, each nerve cell sends out long extensions called axons that make contacts with target neurons to form synapses and establish the appropriate circuits for neuronal communication. At the tip of each axon as it is elongating is a growth cone, a highly motile region rich in actin and microtubules that crawls over the surface (see Figure 11.8). Movement by the growth cone elongates the axon. As it moves, the growth cone explores the area around it and responds to navigational cues by turning. Through a series of turns stimulated by cues, the growth cone is led to its target, where it ceases to move and a synapse is formed.

The microtubule cytoskeleton plays a critical role in the steering process that directs movement of the growth cone. A growth cone is a large, flattened structure that spreads out over the surface across which it moves. In the absence of an external navigational cue, dynamic microtubules are nucleated at the rear of the growth cone, and these microtubules grow and shrink throughout the growth cone in a fanlike array. When a growth cone encounters a cue, a signal is generated only in the small area of the growth cone's plasma membrane that is in contact with the cue. As shown in **FIGURE 11.53**, a spectacular response follows. Microtubules within the growth cone respond by growing toward the source of the signal. Although the MAPs responsible are not known, it is likely that local activation of a stabilizing MAP by the signal keeps microtubules in the growth phase for longer periods of time and causes them to accumulate at the site where the signal originates. **FIGURE 11.54** shows the microtubules in the two growth cones in the video several minutes after the growth cones first come into contact. After the microtubules have reoriented and elongated, plus end-directed kinesins then move vesicles out to the cell periphery along them, causing the vesicles to fuse with the plasma membrane and expand the cell in a small region. A local asymmetry in the structure of the growth cone is created as a result. The actin cytoskeleton then provides the force to drive the membrane forward in the direction marked by the microtubules.

Another process where microtubules and motors contribute to cell asymmetry is in positioning the mitotic spindle in epithelial cells. The generation of new cells to expand or repair an epithelium requires that a dividing cell within the epithelial sheet produce two daugh-

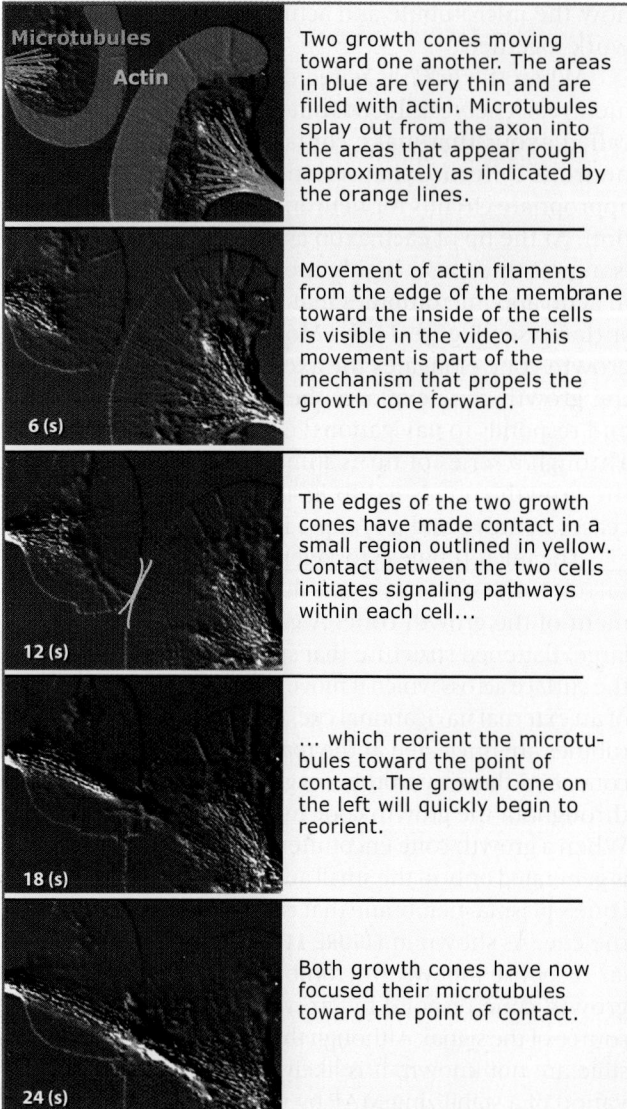

| | Two growth cones moving toward one another. The areas in blue are very thin and are filled with actin. Microtubules splay out from the axon into the areas that appear rough approximately as indicated by the orange lines. |

Microtubules **Actin**

| 6 (s) | Movement of actin filaments from the edge of the membrane toward the inside of the cells is visible in the video. This movement is part of the mechanism that propels the growth cone forward. |

| 12 (s) | The edges of the two growth cones have made contact in a small region outlined in yellow. Contact between the two cells initiates signaling pathways within each cell... |

| 18 (s) | ... which reorient the microtubules toward the point of contact. The growth cone on the left will quickly begin to reorient. |

| 24 (s) | Both growth cones have now focused their microtubules toward the point of contact. |

FIGURE 11.53 A series of images from a video sequence shows the growth cones of two neurons cultured *in vitro*. The response within the growth cones when they collide is very similar to the response that occurs *in vivo* when a growth cone encounters a navigational cue on another cell. Each growth cone rapidly extends microtubules toward the point of contact. Photos © Lin and Forscher, 1993. Originally published in **The Journal of Cell Biology**, 121: 1369–1383. Used with permission of Rockefeller University Press. Photos courtesy of Paul Forscher, Yale University.

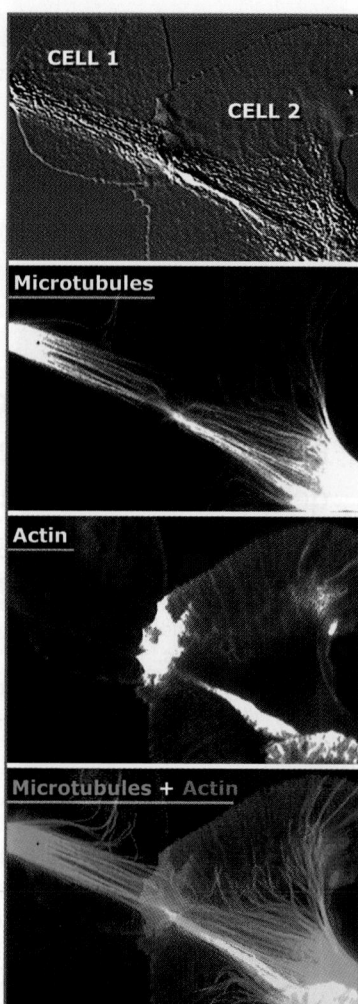

FIGURE 11.54 Actin and microtubules in the two growth cones after they have had several minutes to respond. Note how focused the microtubules are on the point of contact, and how massive polymerization of actin is also centered there. Photos © Lin and Forscher, 1993. Originally published in **The Journal of Cell Biology**, 121: 1369–1383. Used with permission of Rockefeller University Press. Photos courtesy of Paul Forscher, Yale University.

ter cells with the same elongated shape and the same orientation as those already present, as shown in **FIGURE 11.55**. Cells always divide perpendicular to the spindle, so for division to occur along the long axis of an epithelial cell, the spindle must be oriented from side to side within the cell before the chromosomes separate. The spindle forms with a random orientation, however. In order to achieve the required orientation, the spindle is rotated by its astral microtubules (the microtubules at each end of the spindle that extend away from the chromosomes). Astral microtubules are highly dynamic and are able to constantly search the cell's periphery. Cytoplasmic dynein is anchored to sites at the plasma membrane in a beltlike array that runs around the middle of the cell. Astral microtubules that encounter the belt become associated with the dynein/dynactin complex, which then generates a pulling force that in turn swings the spindle into the correct orientation.

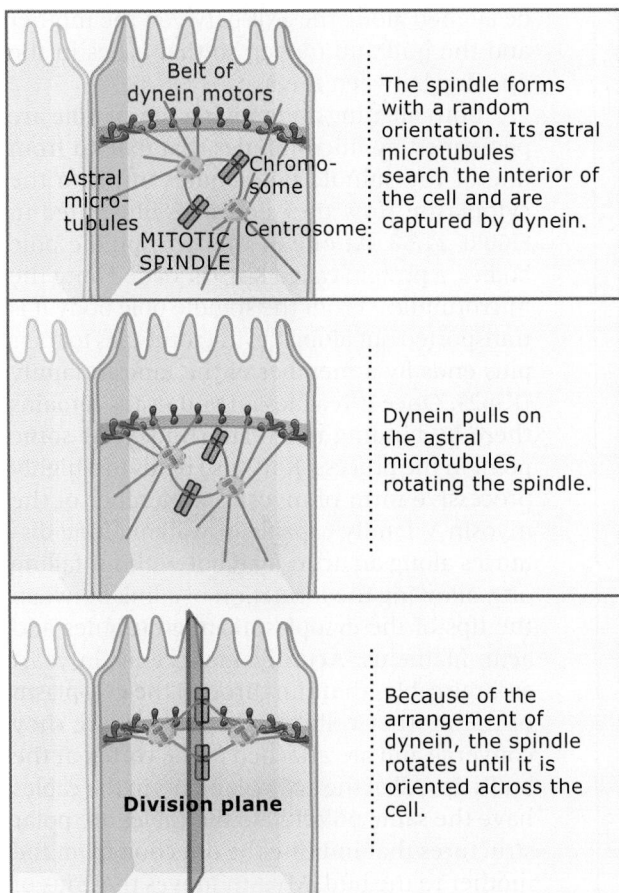

The spindle forms with a random orientation. Its astral microtubules search the interior of the cell and are captured by dynein.

Dynein pulls on the astral microtubules, rotating the spindle.

Because of the arrangement of dynein, the spindle rotates until it is oriented across the cell.

FIGURE 11.55 This cell is one of many within an epithelial sheet that extends to the left and right. The arrangement of dynein as a ring around the cell ensures that the spindle comes to rest oriented from side to side, regardless of how it was oriented when it formed. This orientation ensures that the cell will divide from top to bottom, creating two new cells for the sheet. The dynein is located at tight junctions, the points where adjacent cells in the sheet contact one another.

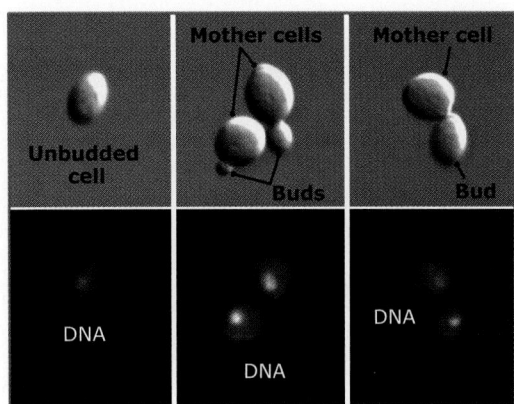

FIGURE 11.56 The three top panels show cells at successive stages of the cell cycle (from left to right). As the cell cycle progresses, the buds gradually increase in size until they are as large as the mother cells. The bottom panels show the DNA in the cells. The middle- and lower-right panels show that DNA appears in the bud only after it is as large as the mother cell. Photos courtesy of Robert Skibbens, Lehigh University.

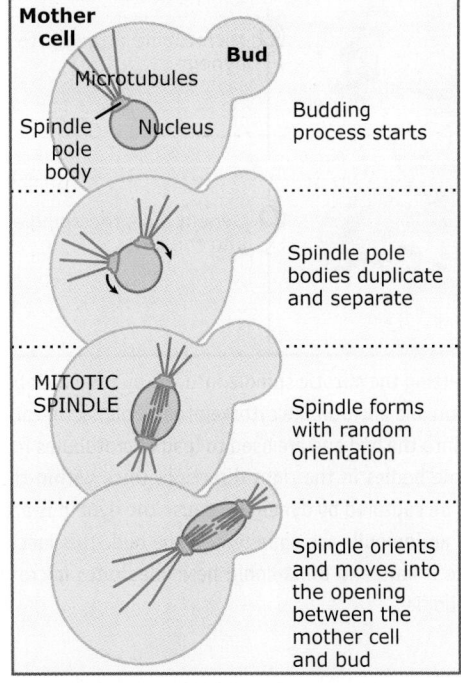

FIGURE 11.57 Budding yeast cells divide without breaking down their nucleus. The spindle pole body, a structure embedded in the nuclear envelope, nucleates microtubules both in the cytoplasm and inside the nucleus, where they form the mitotic spindle. The spindle forms in the mother cell and must be positioned in the opening between the mother cell and the bud. Properly positioning the spindle requires that it be both moved toward the bud and oriented along the axis between the bud and the mother cell.

The best understood example of how cell asymmetry is generated comes from studies of cell division in the yeast *Saccharomyces cerevisiae*, the yeast used to bake bread and brew beer. **FIGURE 11.56** shows that these yeast divide by a budding process in which a small area on the surface of a cell (the "mother" cell) grows outward, forming a bud that expands over the course of the cell cycle to eventually become the daughter cell. At about the same time the budding process starts, the lone microtubule-organizing center in the mother cell—a structure called the spindle pole body which is embedded in the nuclear membrane— duplicates, and the two spindle pole bodies that result separate and move to opposite ends of the nucleus as shown in **FIGURE 11.57**. Yeast

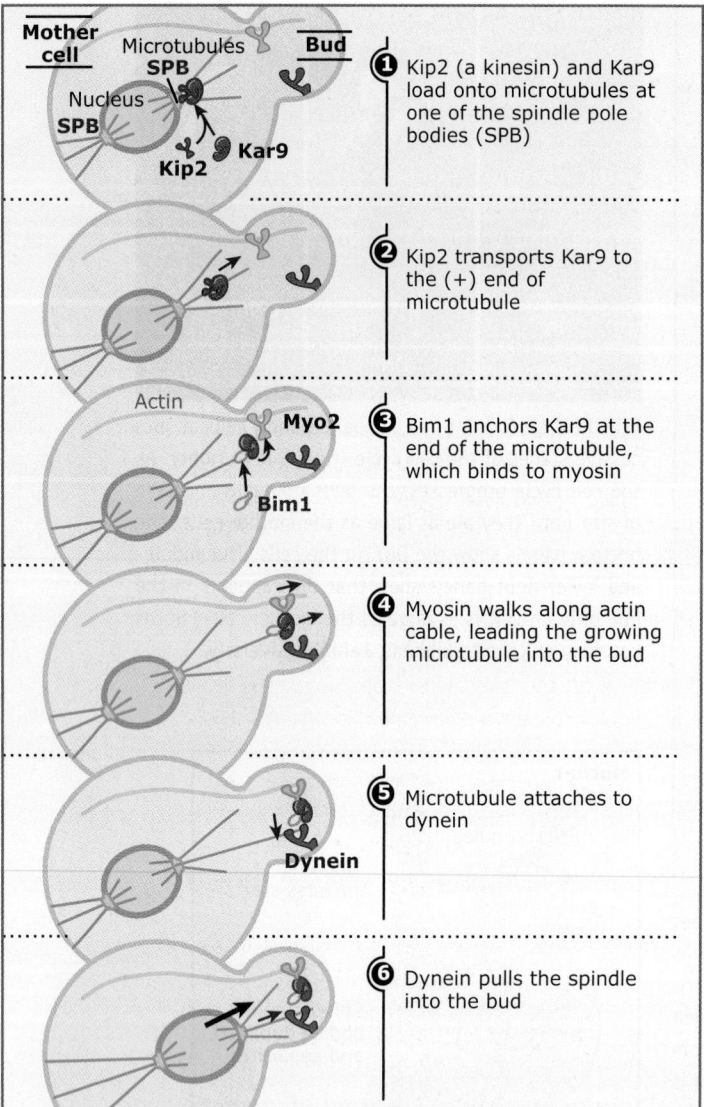

Mother cell · Microtubules · SPB · **Bud** · Nucleus · SPB · **Kar9** · **Kip2**

① Kip2 (a kinesin) and Kar9 load onto microtubules at one of the spindle pole bodies (SPB)

② Kip2 transports Kar9 to the (+) end of microtubule

Actin · **Myo2** · **Bim1**

③ Bim1 anchors Kar9 at the end of the microtubule, which binds to myosin

④ Myosin walks along actin cable, leading the growing microtubule into the bud

Dynein

⑤ Microtubule attaches to dynein

⑥ Dynein pulls the spindle into the bud

FIGURE 11.58 Getting the mitotic spindle into the bud is a collaborative effort between the actin and microtubule cytoskeletons. Polar actin cables run from the mother cell into the bud and are used to lead microtubules from one of the two spindle pole bodies in the right direction. Once within the bud, the microtubules can be captured by dynein. Because the dynein is attached to the membrane, its action pulls the spindle into the bud. This mechanism would not be possible if Kar9 and Bim1 could be loaded onto microtubules at both spindle pole bodies.

be aligned along the axis between the mother and the bud and moved so that it lies in the opening between them.

Both aligning and moving the spindle are performed by microtubules that extend from one of the spindle pole bodies out into the cytoplasm. How they do this is illustrated in **FIGURE 11.58**. At one of the two spindle pole bodies, a protein called Kar9 is loaded onto the microtubules. From the spindle pole body it is transported out along the microtubules to their plus ends by a member of the kinesin family (Kip2). Once it reaches an end, Kar9 remains there by binding to a +TIP (Bim1). At some point in the process, Kar9 also binds to a highly processive form of myosin (a member of the myosin V family capable of walking long distances along an actin filament without falling off), allowing the formation of a link between the tips of the cytoplasmic microtubules and actin filaments. Actin filaments exist in yeast as large cables that run through the cytoplasm of the mother cell into the bud, where they converge and are attached to the cortex at the bud's tip. All of the actin filaments in the cables have the same polarity, so the cables are polar structures that indicate the direction from the mother to the bud. Myosin moves the ends of the microtubules along the actin cables, leading the microtubules into the bud. Once there, the microtubules become attached to minus end-directed motors (dynein and a form of kinesin) anchored within the cortex. These then pull on the microtubules and hold onto their ends as they depolymerize. Thus, a combination of active pulling and force caused by controlled depolymerization of the microtubules moves the spindle through the cytoplasm of the mother cell and into the bud.

Microtubule dynamics plays a major role in orienting and positioning the spindle in both yeast cells and epithelial cells. The two situations have in common that microtubule ends must find a spot on the cortex of the cell, but the means by which dynamics are employed to do it are very different. The microtubules in yeast cells are much more stable and do not find their target by a search-and-capture mechanism that requires that the microtubules constantly grow and shrink. Rather, the cytoplasmic microtubules in yeast are led to their target. Microtubule dynamics comes into play as the tip of a microtubule is being towed along an actin filament by a myosin molecule. As the tip of the microtubule is being moved it must be able to add subunits without its connection

divide without breaking down their nuclear envelope, the spindle forming within the interior of the nucleus. The spindle is formed by microtubules nucleated by the two spindle pole bodies, which also nucleate microtubules on the outside of the nucleus. The spindle forms with a random orientation, uninfluenced by the location of the bud.

All of this occurs within the mother cell. In order for both the mother and daughter cells to inherit a set of chromosomes, the spindle must

with the actin filament being broken. This may be possible because of the Bim1 protein, the mammalian version of which appears to allow cargoes to hang onto the end of a growing microtubule.

The mechanism that positions the yeast spindle would clearly not work if the cytoplasmic microtubules from both spindle pole bodies formed links with actin. If so, the spindle would end up stuck in the mother cell with its axis perpendicular to that between the mother and the bud. The means by which this is prevented is apparently based upon the mechanism of spindle pole body duplication. Like centrioles, an old and a new spindle pole body can be distinguished after each spindle pole body duplication. And like the centrosomes that form around the mother and daughter centrioles, the compositions of the old and new spindle pole bodies differ. In yeast, this is used to ensure the presence of proteins that inactivate Kar9 only on the new spindle pole body. As a result, Kar9 can only be loaded onto microtubules that extend out from the original, older spindle pole body. Thus, only microtubules from that spindle pole body can be led into the bud.

11.15 Interactions between microtubules and actin filaments

Key concepts

- Microtubules and actin filaments function together during cell locomotion and cell division.
- In general, microtubules direct where and when actin assembles or generates contractile forces. Microtubules influence the actin cytoskeleton through direct binding or indirect signaling.
- The two cytoskeletal systems can be bound together by linker proteins that attach microtubules to actin filaments.
- The dynamic growth and shortening of microtubules can activate a subset of G proteins; these activated G proteins control actin assembly and cell contraction.

Many dynamic cell functions require cooperation between the different cytoskeletal filaments. For example, microtubules work with actin filaments to move a cell over a substrate or to divide a cell into two (see *11.14 Microtubule dynamics and motors combine to generate the asymmetric organization of cells*). Intermediate filaments also interact with both microtubules and actin filaments to maintain cell and tissue

integrity. In this section, we examine several aspects of microtubule-actin interactions during motility and division.

Several observations suggest that microtubules and actin filaments interact with each other within the cell. Researchers have known for about 30 years that when they depolymerize microtubules by adding a drug such as colchicine, the cell contracts. Contraction is driven by the actin cytoskeleton and the motor protein myosin, showing that microtubules normally resist or inhibit contraction. Cells whose microtubules are depolymerized also lose their polarized shape. For cells crawling over a surface, actin filaments are normally abundant at the front of the cell where their polymerization drives movement. When microtubules are depolymerized in these cells, the actin filaments are no longer properly localized to the front of the cell. These experimental observations suggest a general theme: microtubules function as directors, determining where actin should assemble and where it should contract. In this way, actin is used to provide force, whereas microtubules are used to organize or control where these forces act. By functioning together, actin and microtubules generate forces at the right place and time for specific cell functions.

How do the actin and microtubule cytoskeletons interact with each other at the molecular level? One way, shown in FIGURE 11.59, is through a linker, or a set of linkers, that binds an actin filament to a microtubule. A number of MAPs play this role by binding actin filaments as well as microtubules, forming static linkages

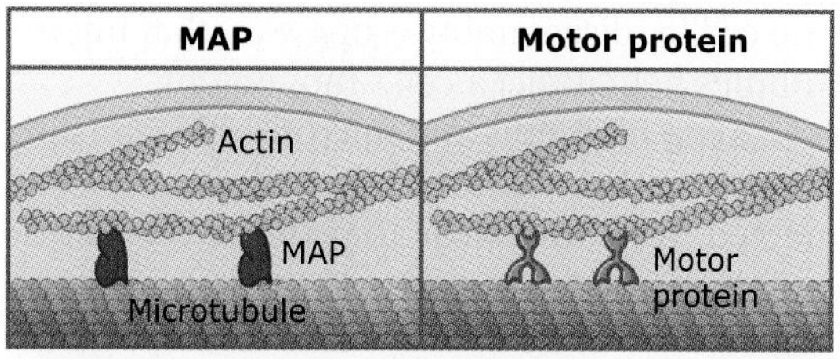

FIGURE 11.59 Several proteins or protein complexes bind to both microtubules and actin filaments and link them together. On the left is a protein that binds directly to both types of filaments and simply holds them together. Linkage can also occur through a motor, as on the right. In this case, the motor domains bind to one of the two types of filament while the tail—or other proteins bound to the tail—binds to the other. This type of interaction causes the microtubules and actin filaments to move relative to one another.

between the two. The neuronal MAP, MAP2c, is one such protein that can bind to both actin and microtubules. In growing neurons, binding of actin to microtubules is likely important as the neurons begin to form and send out long projections. Physical links between actin and microtubules may also occur through motor proteins. In this case, the linkage is dynamic, allowing one polymer to pull on the other. Such links can tether microtubules to the cell cortex, as we saw during spindle orientation in epithelial cells or spindle positioning in yeast (see *11.14 Microtubule dynamics and motors combine to generate the asymmetric organization of cells*). In both of those examples, a microtubule-based motor is anchored to the actin cytoskeleton and pulls on microtubules in order to move the spindle to the right place for cell division.

Linking microtubules to actin filaments may also guide growing microtubules to specific sites in the cell. In moving cells, some dynamic microtubules grow toward focal contacts, sites of cell adhesion to the extracellular matrix (see *12 Actin*). These dynamic microtubules are guided to the small focal contacts by the actin filament bundles attached to the contact sites. It is thought that +TIPs bound to microtubule ends may link microtubules to actin bundles, directing microtubule growth to the focal contact. Microtubules are targeted to adhesions at the rear of the cell and may deliver a signal that causes these adhesions to break down, selectively releasing the back of the cell from the substrate. Once the back of the cell is free it can contract, which moves the body of the cell forward. Many repetitions of this process coordinated with extension of the front of the cell allow the cell to move. Thus, specifically targeting adhesions at the rear of the cell for disassembly is one way that microtubules help direct a cell's movement.

Actin filaments and microtubules can also work together without being physically connected. It is now clear that the two types of polymer often relay signals to each other to regulate when and where the other polymer grows. The ability to signal and communicate with each other is critical; such communication allows microtubules and actin filaments to coordinate their activities and regulate when and where each polymer is built up, taken apart, or used to generate force. Although microtubules and actin filaments can relay signals to each other, they are also controlled by signaling pathways that respond to other inputs from inside or outside the cell. These signaling path-

ways act on many downstream targets in addition to the microtubule and actin cytoskeleton. We understand best the upstream signals that regulate actin filament assembly and organization. Most of the organization of actin filaments in cells is controlled by a small number of proteins called G proteins. When activated, these G proteins cause the formation of filopodia (spikelike actin projections that stick out at the front of a cell), lamellipodia (thin sheets of cytoplasm filled with actin filaments that also stick out at the front of the cell), or contractile actin bundles, such as the stress fibers that link up with focal adhesions (discussed above) and allow the cell to pull on the substrate. In general, an active G protein stimulates (often indirectly) an actin-binding protein, which then regulates the actin cytoskeleton. Remarkably, microtubule assembly or disassembly can control these G proteins by turning them on or off. In this way, dynamic microtubules direct actin assembly or contraction without physically binding to the actin filaments.

Signaling between microtubules and actin filaments is critical for cells to crawl over a substrate. Cell movement requires constant actin polymerization at the front of the moving cell to drive the cell forward, and contraction at the rear to move the body of the cell forward. At the front of the cell, actin polymerization pushes out a lamellipodium, and this polymerization is stimulated by the G protein Rac1. What activates Rac1 at the front of the cell, and why does the cell keep moving in the same direction? We now know that growing microtubules can activate Rac1, although we have no idea how they can do this. Microtubule activation of Rac1 is noteworthy because it demonstrates that the dynamic state of a microtubule can activate a signaling cascade in a specific region of the cell.

The communication between microtubules and Rac1 is not just one way; as shown in **FIGURE 11.60**, once activated, Rac1 may also relays a signal to microtubules to keep them in a growing state. Active Rac1 is thought to indirectly turns off a microtubule-destabilizing protein (oncoprotein 18), stimulating more microtubule growth. In this way, the communication between Rac1 and microtubules generates a local positive feedback loop where growing microtubules activate Rac1 and active Rac1 stimulates microtubule growth. This feedback loop keeps microtubules growing toward the front of the cell and stimulating actin polymerization there. As actin polymerization extends

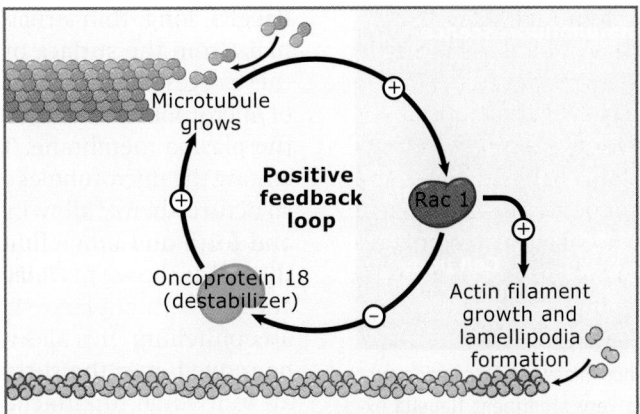

FIGURE 11.60 The polymerization state (growing or shrinking) of microtubules can indirectly change the dynamics and organization of actin filaments. A small G protein usually acts as an intermediate. In the example in this figure, growing microtubules activate Rac1, a small G protein that stimulates actin to polymerize with the organization needed to form a lamellipodium. Activated Rac1 also (indirectly) turns off the microtubule-destabilizing protein Oncoprotein 18, thus creating a positive feedback loop that helps maintain microtubule growth and the formation of filopodia. Shrinking microtubules activate a different small G protein that promotes a different type of actin structure.

the front of the cell, Rac1 also stimulates microtubules to grow into the newly expanded region. Thus, as a consequence of the feedback between microtubules and Rac1, the cell can maintain its polarity and move continuously in the same direction.

Depolymerizing microtubules also start a signaling cascade. When microtubules depolymerize, they activate another G protein, RhoA. Active RhoA stimulates stress fiber and focal adhesion assembly, and indirectly activates the actin-based motor, myosin. These changes to the actin cytoskeleton cause the cell to contract. It is interesting that active RhoA can also initiate a signaling cascade that stabilizes a subset of microtubules and makes them nondynamic. Whether active RhoA limits its own activity by stabilizing a subset of microtubules is not yet known.

We still have much to learn about how the actin and microtubule cytoskeletons communicate and signal to each other. By studying these interactions, and the signaling proteins that function as intermediaries between the two cytoskeletal polymers, we will learn much about how cell locomotion and cell division are regulated and how to control these cell processes in disease states.

11.16 Cilia and flagella are motile structures

Key concepts

- Cilia and flagella contain a highly ordered core structure called an axoneme.
- The axoneme is composed of nine outer doublet microtubules surrounding a pair of central microtubules.
- Radial spokes, a complex of several polypeptides, link each outer doublet to the center of the axoneme.
- Dyneins are bound to each outer doublet and extend their motor domains toward the adjacent outer doublet.
- Dynein slides the outer doublets past each other; the structural links between outer doublets converts the sliding motion into a bending of the axoneme.
- Kinesins participate in flagellar assembly by transporting axonemal proteins to the distal tip of flagella.
- Nonmotile primary cilia participate in sensory processes.

In addition to moving cargos within cells, microtubules also play a role in moving cells relative to their environment. As can be seen in **FIGURE 11.61**, this is accomplished by cilia and

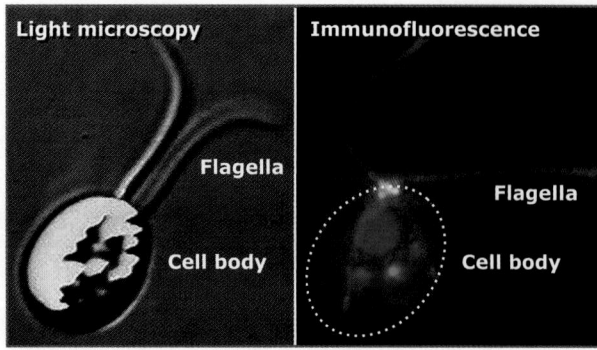

FIGURE 11.61 Light and fluorescence images of *Chlamydomonas reinhardtii*, a unicellular alga. Two very prominent flagella extend from the top of each cell. Microtubules are in red in the fluorescence image, showing that flagella are microtubule-based structures. *Chlamydomonas* cells swim by moving their flagella in a regular beating pattern. Left photo courtesy of Lynne Cassimeris, Lehigh University; right photo courtesy of Naomi Morrissette and Susan Dutcher, Washington University School of Medicine.

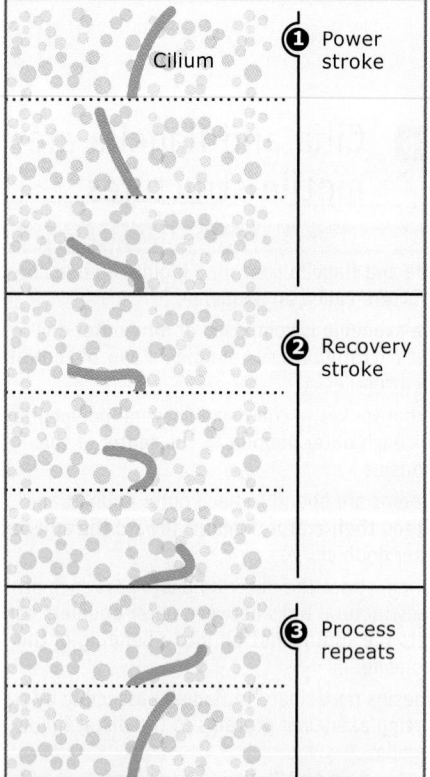

FIGURE 11.62 The beat of a cilium is divided into two parts. During the power stroke, the cilium is fully extended and moves liquid past the surface of the cell. During the recovery stroke that follows, the cilium bends from one end to the other, returning it to the starting position for another power stroke.

flagella, long, thin structures that project like hairs from the surface of many cells. Each of these organelles is composed of a long bundle of microtubules surrounded by an extension of the plasma membrane. Through interactions among the microtubules within the bundle the structures bend, allowing them to beat back and forth and move fluid past the surface of the cell as shown in **FIGURE 11.62**. For a stationary cell within a large group of cells, such as an epithelium, this allows fluid and objects to be moved over the surface of the tissue. For an individual, unattached cell, the cell itself is propelled through the fluid (i.e., it swims). Cilia and flagella are found on a number of unicellular organisms, such as *Paramecium* and *Chlamydomonas* (a green alga), as well as on the sperm cells of most eukaryotes. In mammals, cilia cover the apical domains of some types of epithelial cells and beat in synchrony, creating waves of ciliary motion that move across the surface of the tissue. Within the trachea this motion is used to clear mucus and debris from the respiratory tract; in the oviduct it transports eggs from the ovary to the uterus; and in the brain it circulates the cerebrospinal fluid.

Cilia and flagella share the same general structure and move by similar mechanisms but differ in several respects. Most significant are differences in length, the number of each per cell, and the beat pattern that they generate. Cilia are shorter (10–15 µm) and often number 100 or more per cell. Each cilium generates force by bending near its base (see Figure 11.62). The outer part of the cilium remains stiff and the bend at the base moves it in a motion resembling the powerstroke of an oar pulled through the water. This is followed by a recovery stroke, during which the cilium's bend is propagated from base to tip, readying the cilium for the next power stroke. **FIGURE 11.63** shows the movement of an actual cilium as it beats. (The online video has been slowed considerably in order to catch the stages of the motion. In reality, cilia beat so fast [many times per second] that they are little more than a blur.)

Flagella are usually longer (10–200 µm) than cilia and typically number only one or a few per cell. Flagella also generate force by bending; an S-shaped wave is propagated from the base to the tip of the flagellum, as illustrated in **FIGURE 11.64**. The beat patterns of both cilia and flagella share an underlying mechanism based on generation of the bend in the structure. Differences in the ways the bend is propagated along the length of the organelle generate the different waveforms of cilia and flagella. Because the two types of organelle are variations on the same theme, we will focus on their common properties and we will use the term *flagella* to describe the structure and motility of either organelle unless specifically referring to the ciliary waveform.

Flagella will continue to beat if removed from the cell, demonstrating that the motion is generated solely within the organelle. Isolated flagella will also continue to beat after removal of the plasma membrane, provided ATP is present. The results of these experimental manipulations demonstrate that force is generated by the protein core of the flagellum in concert with ATP hydrolysis.

The core of a flagellum is a highly ordered structure composed of at least 250 different types of polypeptides. This structure is termed an **axoneme**. The structure of the axoneme is well conserved between the flagella of organisms as diverse as *Chlamydomonas*, a single-celled protozoan, and humans.

The major structural features of the axoneme are shown in **FIGURE 11.65**. Most prominent, particularly in the cross section, is a precisely organized bundle of microtubules that run continuously for the structure's entire length. Arranged in a circle are nine unusual "doublet microtubules," each composed of one conventional 13-protofilament microtubule (called the A tubule) with a second, incomplete microtubule of 10–11 protofilaments (called the B tubule) attached to its wall (see Figure 11.65). At the center of the ring of doublet microtubules are two conventional microtubules with 13 protofilaments (the "central pair"). This characteristic arrangement of microtubules within the axoneme is described as "9 + 2." All of the microtubules have the same polarity, oriented with their plus ends at the tip of the flagellum and their minus ends at its base. A variety of proteins binds and stabilizes the microtubules.

The microtubules within the axoneme are extensively interconnected by several types of links (see Figure 11.65). The proteins that form these connections are essential for organizing the microtubules into a single coherent unit, allowing it to move, and coordinating its movement to generate a waveform. Adjacent doublet microtubules are connected around the circumference of the axoneme by a protein called nexin. The doublet microtubules are also connected to the central pair of microtubules by

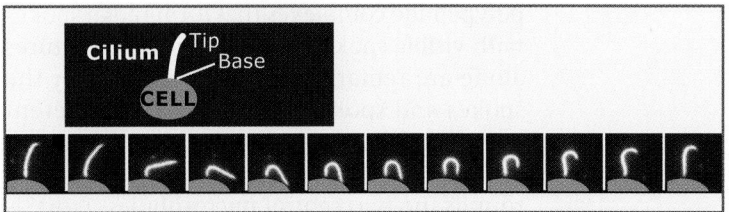

FIGURE 11.63 The beating of a cilium, shown using dark field microscopy. Note how sharply the cilium bends around its base during the power stroke and how smoothly it unfolds during the recovery stroke. Images of the cilium are taken from a video. Photos courtesy of D. R. Mitchell, SUNY Upstate Medical University.

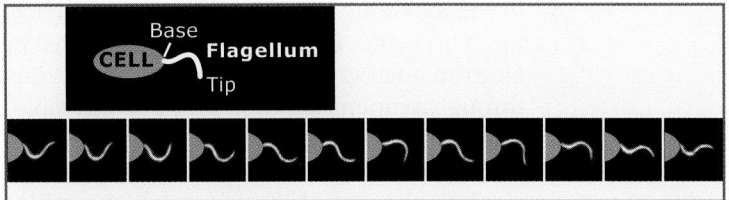

FIGURE 11.64 The beating of a flagellum, shown using dark field microscopy. Images of the flagellum are taken from a video. Photos courtesy of D. R. Mitchell, SUNY Upstate Medical University.

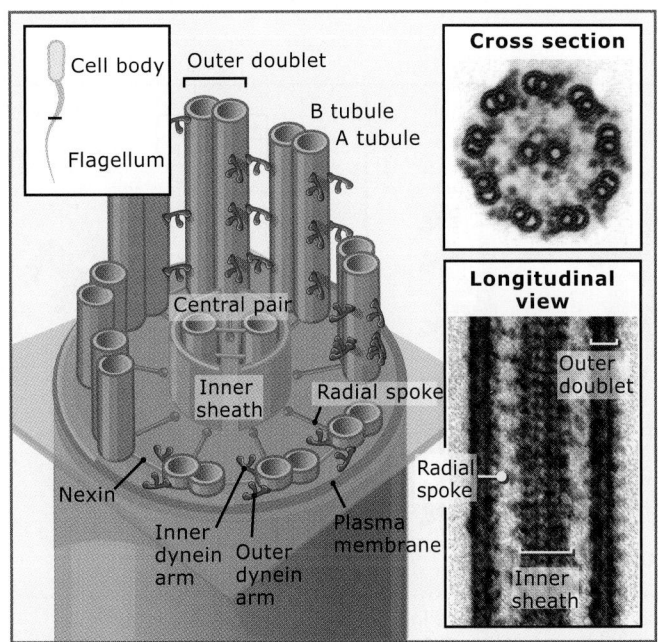

FIGURE 11.65 The structure of an axoneme, showing the highly ordered arrangement of microtubules within it. The microtubules are extensively interconnected by several different types of links. The different types of connections work together to create the beat pattern of a flagellum. On the right are electron micrographs. In the cross section the inner and outer dynein arms that link the outer doublets are visible. One very clear radial spoke and its head (on the lower left) is also visible. Photos courtesy of Dr. Gerald Rupp, Institute for Science and Health.

polypeptide complexes that form radial spokes with visible spokeheads. These two structures alone are remarkably complex: together the spokes and spokeheads contain 17 different polypeptides. The spokeheads are arranged around the inner sheath, a structure that surrounds the two central microtubules. Force is generated within the axoneme by axonemal (also called "ciliary" or "flagellar") dyneins that connect adjacent doublet microtubules; the tail domain binds to the A tubule of one doublet and the head domain to the B tubule of the next. The different connections formed by the nexins, the radial spokes, and the dyneins all occur at regular intervals along the length of the axoneme but have different periodicities. This makes it difficult to detect all three in electron micrographs that show cross sections through axonemes. When they are all visible, however, the structure resembles a wheel with thick spokes and a prominent hub.

Like the rest of the axoneme, the structure and arrangement of dyneins within it is complex. Axonemes contain more than one form of dynein, each larger and composed of more different polypeptides than cytoplasmic dynein. The different forms contain one, two, or three motor domains and are positioned at different places within the axoneme. Adjacent doublet microtubules are connected by two sets of dynein molecules, called the inner and outer arms (see Figure 11.65). The outer arms contain only dyneins with two or three heads, whereas dyneins in the inner arms have one or two.

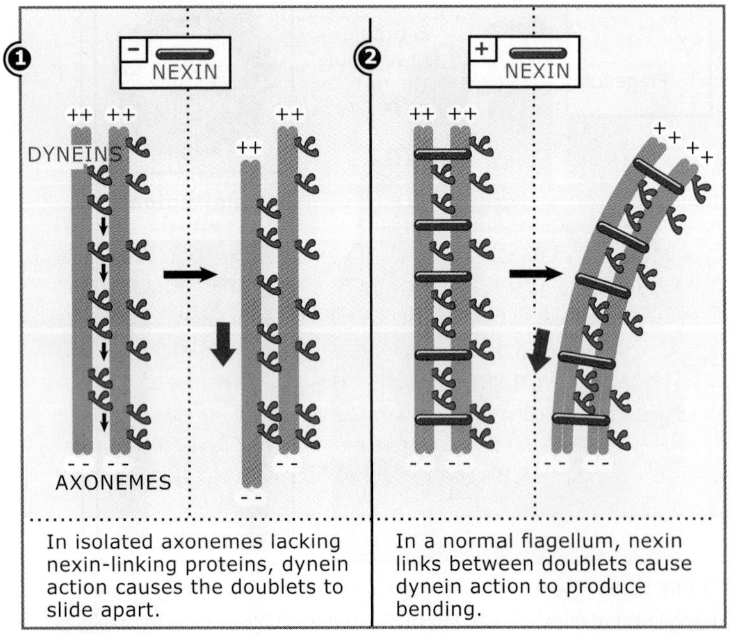

How do all these connections allow flagella to move and generate a beat? The most basic question to be asked is how dyneins work within the structure, since they are the motors and motion must start with them. To understand the contribution of dynein in flagellar motility, flagella can be isolated from cells and the membrane removed from around the axoneme. Such demembranated axonemes can be treated briefly with a protease to break the nexin linkages between the outer doublet microtubules. FIGURE 11.66 shows that if ATP is then added, those microtubules slide apart. Sliding is caused by the dyneins attached by their tails to one doublet microtubule generating force in a plus-to-minus end direction on the adjacent one. In an intact axoneme, dynein cannot slide the outer doublets apart because they are connected together by the nexin linkages. Instead, the force generated by dyneins is converted into a bending motion.

Cilia and flagella generate a beating motion by propagating a bend in the axoneme. A bend begins at the base of the cilium or flagellum and propagates toward the distal tip. The bend occurs because at any instant dyneins are active only within a small region of the axoneme. Dyneins are activated sequentially both along the length and around the circumference of the axoneme in order to propagate the bend. Dyneins are regulated by the central pair microtubules and the radial spokes; mutant flagella lacking these structures are paralyzed and unable to beat. In some organisms, the central pair microtubules rotate rapidly and as they spin may transmit signals to the radial spokes, which in turn activate dynein activity. Several kinases and phosphatases are localized to the central pair and radial spokes, and it is thought that rotation of the central pair activates a local signal transduction network to activate nearby dyneins. Through rapid, local activation and inactivation of specific dynein isoforms, axonemes generate either flagellar or ciliary beating and regulate the power and frequency of the beat.

At the base of a flagellum is a structure called a **basal body**. Basal bodies have the same structure as centrioles (see *11.7 Cells use*

FIGURE 11.66 These two frames from an animation show two outer doublets (gold) connected by dynein. The first part of the animation illustrates what happens in an experiment in which the doublets have been purified out of a flagellum and the nexin links selectively removed. The second part shows what happens in an intact flagellum. The presence of nexin links between the doublets causes the flagellum to bend when dynein exerts force.

microtubule-organizing centers to nucleate microtubule assembly). Each basal body is a cylinder composed of 9 triplet microtubules, each with a complete 13-protofilament A tubule and 11 protofilament B and C tubules. The A and B tubules of the basal body serve as templates for assembly of the 9 outer doublet microtubules of the axoneme. The basal body remains associated with the base of the axoneme it generates and serves to anchor the axoneme to the body of the cell.

The assembly of flagella can be understood by severing the existing flagella off a cell and watching new flagella grow. Flagella regenerate in less than an hour and are functional (i.e., they beat) during the regeneration process. Growth of new flagella occurs at the plus ends of axonemal microtubules, located at the distal tip of each flagellum. Assembly of a flagellum requires that the necessary axonemal components be transported to the tip and assembled into the axoneme as it grows. Transport occurs in large protein complexes that have been observed to move toward the tip along the outer surface of the axoneme, just beneath the plasma membrane. This movement has been termed **intraflagellar transport (IFT)** and is powered by kinesin. Protein complexes are also moved from the tip of the flagellum to the base (toward the microtubule minus ends), but the function of transport in this direction is not known. IFT toward the cell body is powered by cytoplasmic dynein.

Although most cilia are motile structures, a related nonmotile form of cilia that plays a much different role in cells exists. Primary cilia are nonmotile organelles found on nearly all vertebrate cells, with the exception of blood cells. Unlike the case with motile cilia, cells typically have only a single primary cilium. A particularly striking example is shown in **FIGURE 11.67**. The axoneme of a primary cilium lacks a central pair of microtubules and, therefore, is often referred to as having a "9 + 0" structure. Most primary cilia look much like a regular cilium from the exterior, simply extending from the surface of the cell like a single short hair. In some highly differentiated cell types, however, the distal tip of the primary cilium is highly expanded and elaborated into a specialized domain that can be as large as the cell body. This is the case, for example, in the rod and cone cells that contain the photoreceptors where light is absorbed in our retinas. In these cells, the tip of the cilium is expanded into a large domain, called the outer segment, which contains stacks of mem-

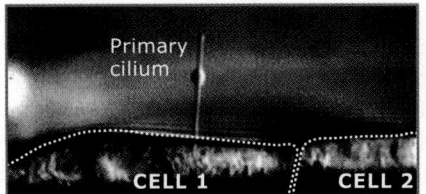

FIGURE 11.67 A very prominent primary cilium extending from the surface of a cell. The cell appears roughly in cross section, with its membrane and that of the neighboring cell indicated by dotted lines. The bulge in the primary cilium may be caused by cargo that is being moved between its axoneme and the surrounding membrane. Photo courtesy of Dr. Sam Bowser, Wadsworth Center.

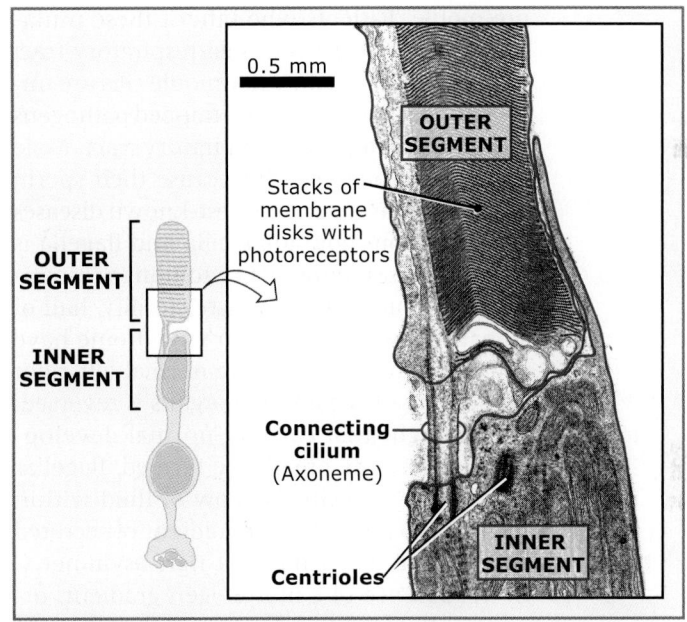

FIGURE 11.68 On the left is a drawing of an entire rod cell, showing its inner and outer segments and the thin connection between them. On the right is an electron micrograph of the region where the two are connected (indicated by the black box on the drawing). For a short distance near the point where it leaves the inner segment, the connecting cilium looks like a normal cilium. Its tip, however, is elaborated into the entire outer segment. Reprinted from *Histology of the Human Eye*, M. J. Hogan, J. A. Alvarado and J. E. Weddell, p. 425. Copyright 1971, with permission from Elsevier.

brane disks filled with the photoreceptor protein rhodopsin. An example of this is shown in **FIGURE 11.68**. The base of the primary cilium connects the outer segment to the rest of the cell, the axoneme extending only slightly beyond the point where the outer segment begins. IFT-type transport moves membrane vesicles containing rhodopsin from the cell body to the outer segment and is likely to be necessary for its construction and maintenance.

Use of the outer segment of a rod cell as a light-sensing device may be an extreme example of a property that is widespread among primary cilia. A possibly general function for primary cilia as sensory devices is just beginning to be appreciated; other cell types that have more modest, unelaborated primary cilia than rod cells also specifically localize various types of receptors there. Localizing receptors to primary cilia may turn them into a kind of antenna that can detect changes in the extracellular environment and relay this information to the cell body.

A number of rare human diseases result from mutations that leave cilia and flagella nonmotile. Patients who inherit these mutations usually suffer repeated respiratory tract infections because their nonmotile cilia are unable to transport mucus and trapped pathogens and irritants out of the respiratory tract. Male patients are often sterile because their sperm are nonmotile. One of the best-known diseases resulting from nonmotile cilia and flagella is Kartagener's syndrome. In addition to respiratory tract infections and male sterility, half of all patients with Kartagener's syndrome have situs invertus, in which the normal left-right asymmetry of the internal organs is reversed. It is thought that early in normal development, before the organs are formed, flagellar beating drives a circular flow of fluid within the embryo and creates a gradient of secreted morphogens that defines left-right asymmetry. In the absence of a morphogen gradient, organ positioning is random along the left-right axis. Mice with mutations in flagellar dyneins or the motors responsible for IFT also have situs invertus, indicating that mutations effecting either flagellar motility or flagellar assembly can result in developmental defects.

Concepts and Reasoning Check

1. Describe the core structure of a flagellum and how force is generated by this organelle.

11.17 What's next?

The dynamic instability of microtubule assembly/disassembly was first described in 1984, the same year that the first kinesin was discovered. These two discoveries began an explosion in our understanding of how the microtubule cytoskeleton is put together and how motor proteins generate motility. The rapid pace of discovery shows no signs of slowing and several recent discoveries are opening up new frontiers for study, expanding our knowledge of how microtubules function in diverse organisms and in a wide range of cellular functions.

Experiments over the last decade have identified a large number of proteins that associate with microtubules, regulating assembly, producing force, or anchoring microtubules to other cell components. Despite the number of microtubule-interacting proteins that we are already aware of, there are likely more to be found. Recently, proteomic approaches have been applied to the centrosome and the yeast kinetochore, identifying all the protein components of these organelles. Many of these newly identified proteins are also likely to interact with the microtubule cytoskeleton. A logical next step is to identify functions for these proteins. Once we understand how individual proteins interact with microtubules, it will then be critical to figure out how all the protein "parts" work in concert during various cell functions. In particular, understanding the interplay between MAPs and motors will help us understand how cells divide chromosomes during mitosis, deliver vesicles to the correct location, or change the shape of a cell.

Many MAPs, and perhaps even some motors, are likely to function only within specific regions of the cell. Are they active in one place because they are only located there? If so, how did they get there? Or, are MAPs and motors regulated so that they are turned on and off in local areas of the cell? What localizes the on/off signal and sets the boundary that keeps an internal signal confined to a small area of the cell?

Cytoskeletal proteins were identified in eukaryotic cells and were thought to be unique to eukaryotes. We now know that prokaryotes also have their own cytoskeletal polymers: FtsZ (a tubulin homolog) and MreB (an actin homolog). FtsZ subunits look very much like a tubulin monomer of α- or β-tubulin. These FtsZ monomers assemble into filaments that look like a microtubule's protofilament, and they even bind and hydrolyze GTP and assemble and disassemble continuously within the bacterial cell. FtsZ polymers are associated with the bacterial membrane, where they help a dividing cell cleave into two by constricting the membrane at the center of the dividing cell. How

FtsZ functions (and how bacteria divide) is an open question. Future comparisons between FtsZ and tubulins will provide insight into how each protein functions and what features make tubulin polymerize into a microtubule and FtsZ into a single protofilament-like filament. These comparative studies may also help us understand when a cytoskeleton first appeared in an ancestral cell and how that cytoskeleton evolved in eukaryotes and prokaryotes.

Another area that has not yet been well explored is how physical forces influence microtubule assembly dynamics. Physical forces can be imposed by motors pulling on microtubules or by polymerization against barriers, such as the plasma membrane. When a microtubule reaches the plasma membrane it often bends and follows the contour of the membrane. Other times the microtubule buckles, rather than bends, and it can even break if buckled too far. Membranes or other physical barriers may also inhibit tubulin subunit addition, blocking further assembly and resulting in a catastrophe. What determines whether a polymerizing microtubule bends, buckles, breaks, or starts shortening when it hits a physical barrier such as the plasma membrane? Does a motor pulling on a microtubule alter its dynamics? How do MAPs and physical forces work together to regulate microtubule assembly and organization?

We still have much to learn about how the microtubule and actin cytoskeletons function together within the cell. Interactions between these two cytoskeletal polymers have important functions in cell locomotion and cell polarization. The microtubule and actin cytoskeletons can physically interact, being linked together by additional proteins. The two cytoskeletal polymers also communicate indirectly, relaying signals to each other through activation of kinases or other signaling molecules, and possibly also through physical forces. The indirect communication between actin filaments and microtubules can generate a positive feedback loop, where polymerization of one filament signals the other to polymerize as well. How such feedback loops are generated and how they function in processes such as directed cell locomotion are presently under study.

Our understanding of how motors work individually, in groups, or when different motors attach to the same cargo will continue to expand. Understanding how motors work in groups and how the activities of motors, which pull cargo in opposite directions, are coordinated will be critical to understanding how cargoes, including chromosomes in mitosis, are moved to correct destinations. Recent experiments have shown that dynein can walk backwards—moving toward the plus end of a microtubule. Understanding how and when dynein walks backwards will add significantly to how we view motor functions.

We are now learning that mutations in genes encoding microtubule-based motors and MAPs contribute to defects in neuronal morphogenesis and cell growth control. For example, a mutation in one member of the dynactin complex has been linked to a rare familial motor neuron disease. How mutated forms of MAPs and motors disrupt normal intracellular transport and cytoskeletal organization are important questions for the future. The identification of more human diseases linked to mutations in microtubule cytoskeletal proteins is likely to occur in the coming years and add to our understanding of how changes to the microtubule cytoskeleton influence the cell's physiology and general health.

While defects in the microtubule cytoskeleton contribute to disease, the microtubule cytoskeleton can also be a target for development of new drugs to combat other types of disease. Screens for small molecules able to disrupt mitosis identified monastrol, a small molecule that blocks the motor activity of a specific kinesin whose function is limited to the mitotic spindle. Small molecules such as monastrol may be useful drugs to treat cancers by blocking mitosis. Monastrol and other small molecules that bind proteins other than tubulin may have benefits of reduced toxicity and fewer side effects when used to treat disease, compared to the tubulin-binding drugs in use today (see *11.1 Introduction*). Other drugs are in development that bind only to tubulins from a limited number of organisms. Although tubulins are highly conserved among eukaryotes, it has been possible to isolate small molecules specifically targeted to fungal tubulins or to those in some disease-causing parasites, including the parasite responsible for malaria.

11.18 Summary

Microtubules are dynamic, polarized polymers that function in cell organization, polarity, and motility. Centrosomes—the MTOCs of animal cells—nucleate microtubules and anchor them at their minus ends. In this way, the position of the centrosome dictates the overall pattern of microtubules present in a cell. Typically, the plus ends of microtubules are located near the plasma membrane and the minus ends are located near the cell center. The rapid assembly and disassembly of microtubules by dynamic instability makes them adaptable to new situations and allows them to easily reorganize into new patterns.

A major function of microtubules is to serve as a polarized track for the molecular motors kinesin and dynein. These motors attach to cargoes, including membrane vesicles, organelles, and chromosomes, and pull their cargoes toward the plus or minus ends of microtubules. The polarized organization of microtubules provides the navigational information necessary to direct cargo to the proper destination in the cell.

Specialized cells use microtubules as the major structural protein of cilia and flagella. Nine doublet microtubules and two center single microtubules form the core of the axoneme. Additional proteins link the doublet and center microtubules. Axonemal dyneins power ciliary or flagellar motility by forcing the outer doublets to slide relative to each other. In the highly crosslinked axoneme, sliding forces generate a bend in the axoneme. Rapid switching on and off of the axonemal dyneins allows propagation of the bend to the cilium or flagellum tip.

11.19 Supplement: What if tubulin did not hydrolyze GTP?

Key concepts

- If microtubules were equilibrium polymers they would depolymerize very slowly and would not easily reorganize.
- Tubulin dimers hydrolyze GTP when they assemble, making the microtubule a nonequilibrium polymer that can depolymerize rapidly.

Microtubules are not equilibrium polymers because tubulin subunits hydrolyze GTP after they assemble into polymers. GTP hydrolysis is not required for microtubules to assemble but instead functions to make it easy to take them apart.

Consider how hard it would be from a cell's perspective to take microtubules apart

if they were equilibrium polymers. If so, gain and loss of subunits at the ends of the microtubules would be a simple equilibrium. Recall that Equation 11.1 described the rate of microtubule polymer formation:

$$dP/dt = k_{on}[\text{tubulin}] - k_{off}$$

At equilibrium, the rate of polymer formation is zero, and, therefore, as Equation 11.2 indicates

$$[\text{tubulin}] = C_c = k_{off}/k_{on}$$

A microtubule could experience a net gain or loss of subunits (i.e., an increase or decrease in length) only if the concentration of free subunits changed. Particularly revealing is to consider what would be required to make the microtubules depolymerize, since that is what is needed in order to allow the interior of a cell to reorganize. The maximum disassembly rate would occur when the tubulin subunit concentration is reduced to zero (see Equation 11.1 above). We can calculate how long it would take for a microtubule to depolymerize using an off rate of 15 dimers per second (the off rate of GTP-tubulin) and the estimate that 1624 tubulin dimers make up one micron of a microtubule. Using these values, a microtubule that is 100 μm long (as they are in some interphase cells) would take three hours to depolymerize. Yet, cells completely depolymerize their entire interphase microtubule array within minutes in preparation for mitosis. If the off rate were faster it would be possible to have faster disassembly, but it would also be harder to have any microtubules in the first place. This is because a faster off rate would also require a higher critical concentration (see Equation 11.2 above). To make a hypothetical equilibrium microtubule depolymerize at the rates observed in cells, the off rate in Equation 11.1 would have to be 540 dimers/second. At this high off rate, the critical concentration (Equation 11.2) would be significantly increased. Using the off rates given above, the critical concentration would increase approximately 36-fold, to 250 μM—ten times the actual concentration of tubulin within cells and approaching the intracellular concentration of ATP (~1 mM). The cell would have little in it other than tubulin!

So, if microtubules were equilibrium polymers, a cell could have microtubules but would have great difficulty disassembling them in order to rearrange them. Because disassembling them would require altering the equilibrium between microtubules and free tubulin subunits, the only way the cell could disassemble its microtubules would be by destroying most of its tubulin! One consequence of this would be that cells would have great difficulty adapting their shape to changes in their environment. Cells would be sluggish and unable to adapt quickly.

These problems are avoided by having tubulin hydrolyze GTP after it has assembled, making microtubules nonequilibrium polymers. Unlike an equilibrium polymer, a nonequilibrium microtubule can both assemble and disassemble rapidly and can do both at the same tubulin concentration. This is because the gain and loss of subunits from the end of a microtubule are no longer one reaction and its exact reverse, as in a simple binding equilibrium. Instead, binding is by one species (GTP-tubulin) and loss is by another (GDP-tubulin). The release of energy by GTP hydrolysis between the two reactions means that the on and off rate constants can be independent of one another—in other words, each can be whatever is convenient for the cell. Evolution has selected the off rate constant to be very high so that a cell can take its microtubules apart quickly and without having to change the tubulin concentration, allowing fast rearrangements and a great deal of adaptability for the cell.

11.20 Supplement: Fluorescence recovery after photobleaching

Key concepts

- The fluorescent tag on proteins or lipids can be locally destroyed using very bright light from a laser.
- Recovery of fluorescence into the photobleached area occurs as unbleached proteins or lipids move into the bleached area, changing places with the photobleached protein or lipid.
- Recovery of photobleached regions on fluorescently tagged microtubules requires disassembly of the photobleached microtubule and new polymerization incorporating unbleached, fluorescent tubulin dimers.

FRAP is a method used to measure how fast a particular molecule or structure exchanges for others of its kind in a small region of a cell. With individual molecules (usually lipids or proteins that are not part of large structures), FRAP indicates how fast the molecules diffuse and what fraction of them are mobile. With

proteins that are components of large, immobile structures (such as cytoskeletal filaments), it indicates how frequently the structures disassemble and reassemble.

To perform a FRAP experiment, a fluorescently tagged version of the protein or lipid of interest is first introduced into the cell. The cell is viewed by fluorescence microscopy, and the fluorescent tag is then destroyed in the region of interest by exposing the area to a laser beam. (Destroying the tag with the intense light of the laser is called bleaching, hence the term *photobleaching*.) Only the tag is destroyed; the attached protein or lipid is still functional (and whatever structure it may be part of is still intact) but is now invisible by fluorescence microscopy. If unbleached fluorescently tagged molecules are able to diffuse or assemble in the bleached zone, then fluorescence recovers within the bleached zone.

For microtubules, fluorescently tagged tubulins are introduced into a cell and enough time is allowed for them to incorporate evenly throughout all the cell's microtubules. The fluorescent tag is either a small fluorescent chemical covalently attached to purified tubulin or a fusion protein of γ-tubulin and green fluorescent protein. **FIGURE 11.69** shows that

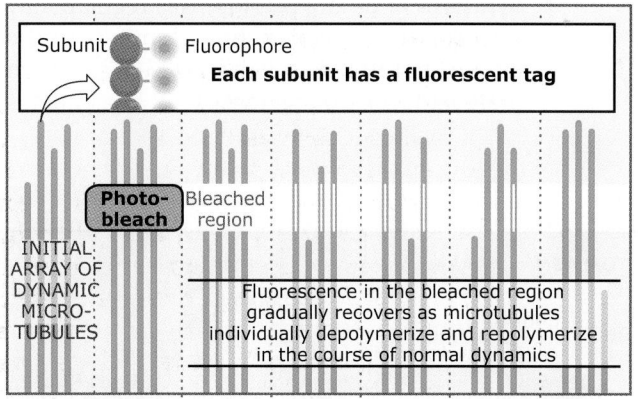

FIGURE 11.69 The green lines represent individual microtubules composed of subunits with fluorescent tags attached. To start an experiment, the tags in a small region are bleached by a very bright light. If the microtubules are not dynamic, the bleached region will remain indefinitely. If they are dynamic, as in the figure, fluorescence will gradually return to the region as bleached microtubules depolymerize and are replaced by new microtubules that polymerize and incorporate unbleached subunits. Because the number of bleached subunits in the cell is small compared to the total pool of fluorescently tagged subunits, the newly polymerized microtubules appear uniformly fluorescent along their lengths. The rate at which the fluorescence in a region recovers is thus a measure of how dynamic the microtubules there are.

after fluorescence is destroyed in a region, recovery of fluorescence to the photobleached area requires depolymerization of the photobleached microtubules and assembly of new microtubules that incorporate unbleached tubulins. The rate of fluorescence recovery is thus proportional to the rate of microtubule turnover.

11.21 Supplement: Tubulin synthesis and modification

Key concepts

- Synthesis of new tubulin is regulated by the concentration of dimers in the cytoplasm.
- α- and β-tubulins require cytosolic chaperonins and additional cofactors to fold properly and assemble into a heterodimer.
- Tubulins are subject to a number of posttranslational modifications.
- Some modifications only occur on tubulins in polymers. These modifications are associated with a stable subpopulation of microtubules.
- In some organisms, the presence of posttranslationally modified tubulins within a microtubule enhances the binding of motor proteins and provides an additional mechanism to regulate vesicle traffic in the cell.

Synthesis of α- and β-tubulins is regulated by a feedback mechanism that responds to the amount of tubulin available in a cell, as shown in **FIGURE 11.70**. During translation, a nascent tu-

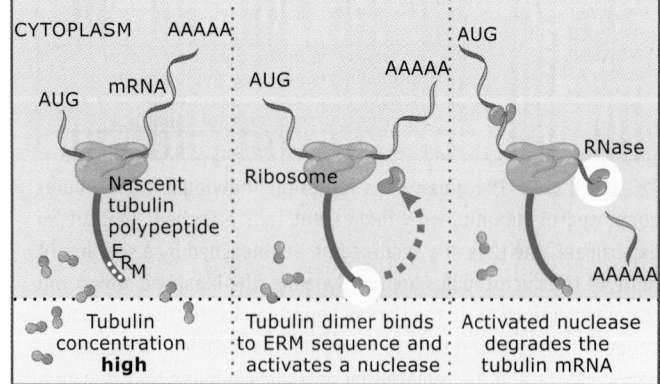

FIGURE 11.70 Interaction between a tubulin dimer and a short sequence at the end of tubulin polypeptides leads to the degradation of tubulin mRNA and decreased synthesis of tubulin. This feedback mechanism maintains the cytoplasmic tubulin concentration within narrow limits. In this figure, only tubulin and a nuclease are shown, but other components may participate as well.

bulin polypeptide emerging from the ribosome can be bound by an already assembled tubulin dimer, activating an RNase that specifically degrades the mRNA. Thus, in a cell with an elevated concentration of tubulin dimers, tubulin mRNAs will be less stable and less new tubulin will be synthesized. Conversely, a cell whose tubulin concentration has somehow dropped below normal will have more stable tubulin mRNAs and will synthesize tubulins to replenish the tubulin pool. By having tubulin protein levels determine the stability of tubulin mRNAs, a cell is able to maintain its pool of tubulin dimers within a concentration range that allows its microtubules to function properly.

α- and β-tubulin cannot fold or assemble properly by themselves. Several additional proteins are required, several of which are specific for tubulin. After translation, tubulin monomers are first bound and folded by CCT, a cytosolic chaperonin, in an ATP-dependent process. Several additional proteins that form a large complex (cofactors A-E) are then required to assemble the folded α and β subunits into heterodimers, as outlined in **FIGURE 11.71**. At some point in the assembly process each of the subunits binds a molecule of GTP. Release of the assembled dimer from the cofactor complex requires that the β subunit hydrolyze the GTP it has bound, but the GTP associated with the α subunit remains unhydrolyzed. Once the properly folded tubulin dimer is released, the β subunit rapidly exchanges GDP for GTP and the dimer is finally competent to assemble into microtubules.

Tubulin subunits are targets for a remarkable variety of posttranslational covalent modifications, including phosphorylation (on either α- or β-tubulin), detyrosination (removal of the terminal tyrosine residue from α-tubulin), acetylation (acetylation of a specific lysine residue within α-tubulin), and polyglutamylation and polyglycylation (covalent addition of chains of glutamate and glycine, respectively, to either α- or β-tubulin). It has long been recognized that certain modifications, such as acetylation and detyrosination, are always found on subsets of microtubules that are significantly more stable than most. High concentrations of modified tubulins are found in brain tissue and within the axonemal microtubules of cilia and flagella. Other cell types contain varying amounts of modified tubulins. In most cases, it is not yet clear how modification contributes to the function of a microtubule.

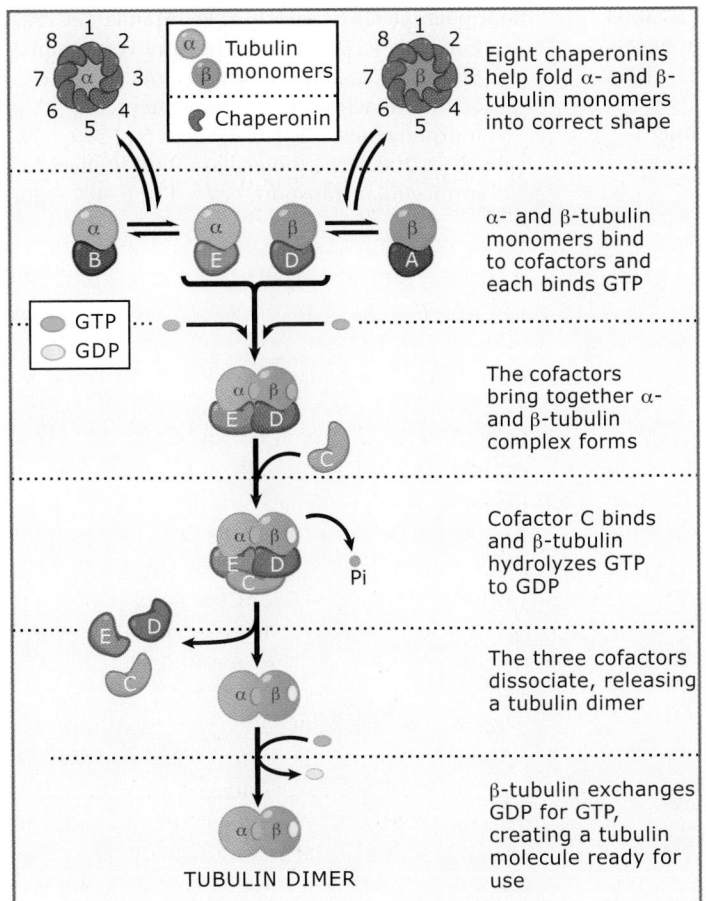

FIGURE 11.71 The steps required to assemble a tubulin heterodimer. The two tubulin subunits are folded separately and then brought together with the assistance of cofactors, several of which are specific to tubulin. The A, B, D, and E cofactors may be required because the folded conformations of the tubulin subunits are not stable until they are part of a tubulin dimer. Alternatively, the structures of the α and β subunits may need to be slightly contorted in order to get them to fit together. Cofactor C and energy input through hydrolysis of GTP by β-tubulin are necessary to dissociate the cofactors after assembly is complete.

Eight chaperonins help fold α- and β-tubulin monomers into correct shape

α- and β-tubulin monomers bind to cofactors and each binds GTP

The cofactors bring together α- and β-tubulin complex forms

Cofactor C binds and β-tubulin hydrolyzes GTP to GDP

The three cofactors dissociate, releasing a tubulin dimer

β-tubulin exchanges GDP for GTP, creating a tubulin molecule ready for use

The presence of modified tubulin subunits provides a convenient marker of microtubule stability, but the posttranslational modifications themselves are not responsible for making the microtubules more stable. At this time, it is not known how a subset of microtubules becomes much more stable, but it is likely that these microtubules are stabilized by the binding of specific MAPs to their walls or ends. Once the microtubule is stabilized, the tubulin subunits within it are modified in one or more of the ways previously described. In some cells, stable microtubules are preferentially bound by motor proteins, which enhances vesicle movement specifically on this subset of microtubules.

http://biology.jbpub.com/lewin/cells

To explore these topics in more detail, visit this book's Interactive Student Study Guide.

References

Addinall, S. G., and Holland, B. 2002. The tubulin ancestor, FtsZ, draughtsman, designer and driving force for bacterial cytokinesis. *J. Mol. Biol.* v. 318 p. 219–236.

Akhmanova, A., and Steinmetz, M. O. 2008. Tracking the ends: a dynamic protein network controls the fate of microtubule tips. *Nat. Rev. Mol. Cell Biol.* v. 9 p. 309–322.

Desai, A., and Mitchison, T. J. 1997. Microtubule polymerization dynamics. *Annu. Rev. Cell Dev. Biol.* v. 13 p. 83–117.

Hirokawa, N., and Noda, Y. 2008. Intracellular transport and kinesin superfamily proteins, KIFs: structure, function and dynamics. *Physiol. Rev.* v. 88 p. 1089–1118.

Hirokawa, N., and Takemura, R. 2003. Biochemical and molecular characterization of diseases linked to motor proteins. *Trends Biochem. Sci.* v. 28 p. 558–565.

Howard, J., and Hyman, A. A. 2003. Dynamics and mechanics of the microtubule plus end. *Nature* v. 422 p. 753–758.

Ibañez-Tallon, I., Heintz, N., and Omran, H. 2003. To beat or not to beat: Roles of cilia in development and disease. *Hum. Mol. Genet.* v. 12 Spec No 1 p. R27–R35.

Nogales, E. 2001. Structural insight into microtubule function. *Annu. Rev. Biophys. Biomol. Struct.* v. 30 p. 397–420.

Rodriguez, O. C., Schaefer, A. W., Mandato, C. A., Forscher, P., Bement, W. M., and Waterman-Storer, C. M. 2003. Conserved microtubule-actin interactions in cell movement and morphogenesis. *Nat. Cell Biol.* v. 5 p. 599–609.

Vale, R. D. 2003. The molecular motor toolbox for intracellular transport. *Cell* v. 112 p. 467–480.

Actin

12

Enrique M. De La Cruz
Molecular Biophysics and Biochemistry
Department, Yale University,
New Haven, CT

E. Michael Ostap
University of Pennsylvania School of
Medicine, Pennsylvania Muscle Institute,
Department of Physiology,
Philadelphia, PA

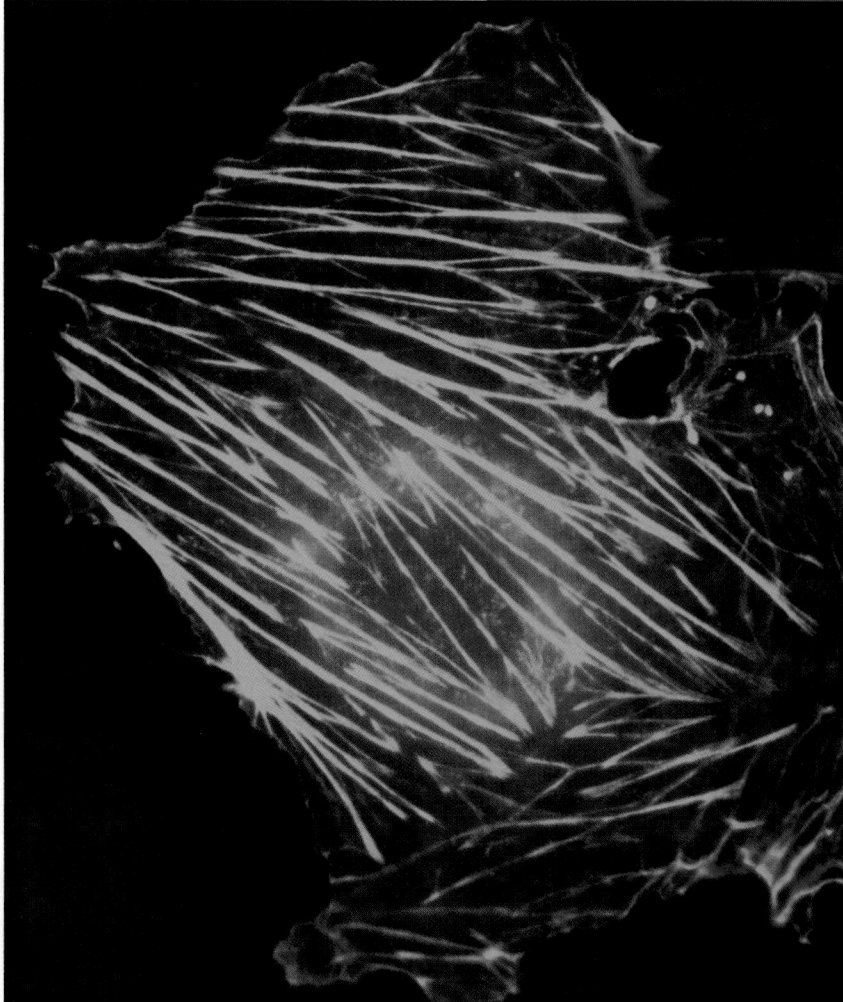

AN EPITHELIAL CELL WITH ACTIN filaments labeled in green by fluorescent phalloidin. The brightly labeled linear structures are myosin-containing contractile elements called stress fibers. (Note: The polymerization of actin can provide forces that drive the extension of cellular processes and movement of some organelles.) Photo courtesy of Nanyun Tang and E. Michael Ostap, University of Pennsylvania.

CHAPTER OUTLINE

12.1 Introduction

Key concepts

- Cell motility is a fundamental and essential process for all eukaryotic cells.
- Actin filaments form many different cellular structures.
- Proteins associated with the actin cytoskeleton produce forces required for cell motility.
- The actin cytoskeleton is dynamic and reorganizes in response to intracellular and extracellular signals.

A cell's internal **cytoskeleton** of microtubules, **actin** filaments (also called microfilaments), and intermediate filaments provides mechanical support and helps the cell maintain a defined shape. In addition, the cytoskeleton is the machinery used by a cell to crawl, change shape, and reposition its internal structures in a regulated way. Movements mediated by the cytoskeleton are called **cell motility**. Many cellular activities require the concerted contributions from two or three of the cytoskeletal filament networks (see *11.15 Interactions between microtubules and actin filaments*). However, actin filaments play a dominant role in many cellular functions, such as cytokinesis, phagocytosis, and muscle contraction.

The protein actin polymerizes into long, fiber-like filaments that are ~8 nm in diameter. These filaments can be cross-linked by other proteins to form a wide variety of cellular structures. **FIGURE 12.1** shows some of the structures in which the actin cytoskeleton is the primary component. These include the microvilli of the intestinal brush border, stereocilia of sensory epithelia, **stress fibers** of adherent cells, neu-

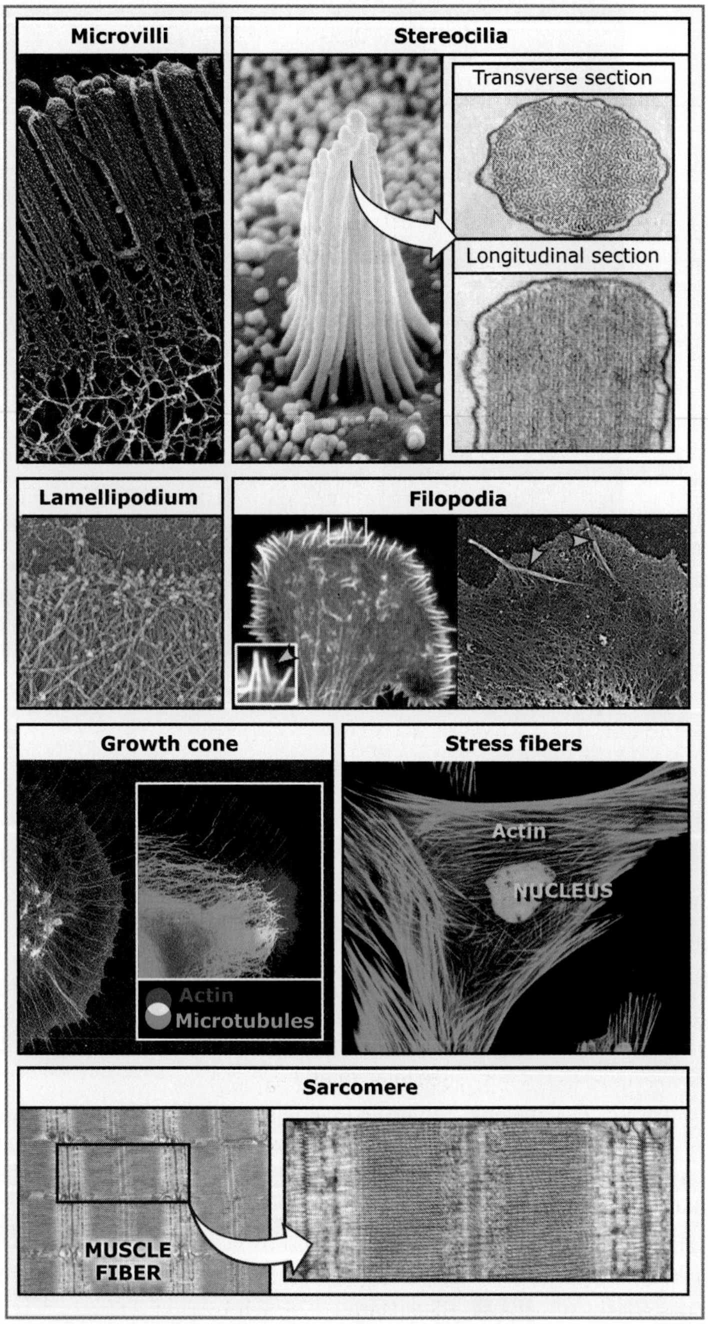

FIGURE 12.1 Actin filaments form the microvilli of the intestinal brush border, stereocilia of the inner ear, lamellipodia, filopodia, neuronal growth cones, stress fibers, and sarcomere. Electron micrograph of microvilli © Hirokawa, et al., 1982. Originally published in *The Journal of Cell Biology*, 94: 425–443. Used with permission of Rockefeller University Press. Photos courtesy of John E. Heuser, Washington University in St. Louis; scanning micrograph of stereocilia reproduced from A. J. Hudspeth and R. Jacobs, *Proc. Natl. Acad. Sci. USA* 76 (1979): 1506–1509. Photo courtesy of James A. Hudspeth, The Rockefeller University; electron micrographs of stereocilia © Tilney, DeRosier, and Mulroy, 1980. Originally published in *The Journal of Cell Biology*, 86: 244–259. Used with permission of Rockefeller University Press. Photos courtesy of Michael J. Mulroy, Medical College of Georgia; lamellipodium photo courtesy of Tatyana M. Svitkina, University of Pennsylvania; filopodia photos reprinted from *Cell*, vol. 118, M. R. Mejiliano, et al., Lamellipodial versus filopodial mode of the actin..., pp. 363–373, Copyright (2004) with permission from Elsevier [http://www.sciencedirect.com/science/journal/00928674]. Photo courtesy of Tatyana M. Svitkina, University of Pennsylvania; neuronal growth cone photos © Schaefer, Kabir, and Forscher, 2002. Originally published in *The Journal of Cell Biology*, 158: 139–152. Used with permission of Rockefeller University Press. Photos courtesy of Paul Forscher, Yale University; photo of cell with stress fibers courtesy of Michael W. Davidson and Florida State University Research Foundation; electron micrographs of sarcomere courtesy of Clara Franzini-Armstrong, University of Pennsylvania, School of Medicine.

ronal growth cones, protrusions (**lamellipodia** and **filopodia**) at the edge of migrating cells, and thin filaments of muscle fibers. Nearly all actin-containing structures are dynamic and are able to reorganize in response to intracellular and extracellular signals.

The actin cytoskeleton produces force and generates cell movement in two ways: by the polymerization of actin monomers into filaments and by the interaction of actin with the myosin family of molecular motors, which bind along the sides of actin filaments and use the energy from ATP hydrolysis to generate force and movement relative to the actin filament such as muscle contraction and vesicle transport. For example, the force generated by actin polymerization is used to push against the plasma membrane during cell locomotion, whereas actin and myosin work together in many processes, such as muscle contraction.

In this chapter, we discuss the mechanism of actomyosin-based movements in eukaryotic cells. Much of our understanding of actin-mediated movement in cells has come from biochemical experiments with purified proteins and components. Complicated cell movements can be modeled from what has been learned from these experiments. Therefore, we first introduce the molecular machinery of the actin cytoskeleton (the structure of actin, the assembly and disassembly of actin filaments, and the proteins that regulate actin dynamics in cells). We then discuss actin and its associated proteins in the context of cells. The actin cytoskeleton plays crucial roles in nearly every cellular process; here we focus on movements that are driven by the assembly of actin filaments and on actomyosin-driven contractions and transport. We include references to other chapters that discuss the contributions and dynamics of the actin cytoskeleton.

12.2 Actin is a ubiquitously expressed cytoskeletal protein

Key concepts

- Actin is a ubiquitous and essential protein found in all eukaryotic cells.
- Actin exists as a monomer called G-actin and as a filamentous polymer called F-actin.

Actin is a ubiquitous protein that is present in all eukaryotic cells. It is highly conserved; protein sequences are generally 90% identical between species. In addition, prokaryotes express highly conserved structural homologs of actin (see *20.6 The bacterial cell wall contains a cross-linked meshwork of peptidoglycan*). Actin constitutes up to 20% of total protein in muscle cells and reaches cytosolic concentrations greater than 100 μM in many nonmuscle cells.

Many organisms have multiple actin genes that encode different isoforms of actin (e.g., humans have 6 actin genes). In vertebrates, there are three isoforms of actin: α, β, and γ. These isoforms have very similar amino acid sequences but differ in function. The α isoform is expressed predominantly in muscle cells and is part of contractile structures. The β and γ isoforms are expressed primarily in nonmuscle cells.

Actin exists as a monomer (called **G-actin**, for globular) and as a linear polymer of monomers (called **F-actin**, for filamentous). The monomers and filaments are in a reversible chemical equilibrium, as an individual actin monomer can add to or leave the end of a filament. Within cells, actin polymerization is tightly regulated by a large number of proteins that interact preferentially with actin monomers or filaments. These regulatory actin-binding proteins control the transition between the monomeric and filamentous states (see *12.7 Actin-binding proteins regulate actin polymerization and organization*).

There are several actin-related proteins (Arps), which are similar in overall structure to actin. An example is Arp1, which is part of a protein complex that links the microtubule motor protein dynein to membranes (see *11.13 How cargoes are loaded onto the right motor*). Later in this chapter, we will discuss the function of a complex of proteins that includes Arp2 and Arp3 in regulating actin polymerization (see *12.9 Nucleating proteins control cellular actin polymerization*).

12.3 Actin monomers bind ATP and ADP

Key concepts

- The actin monomer is a 43 kDa molecule that has four subdomains.
- A nucleotide and a divalent cation bind reversibly in the cleft of the actin monomer.

The actin monomer is a 43 kDa protein with a bound nucleotide and divalent cation. Its overall shape can be described as two lobes

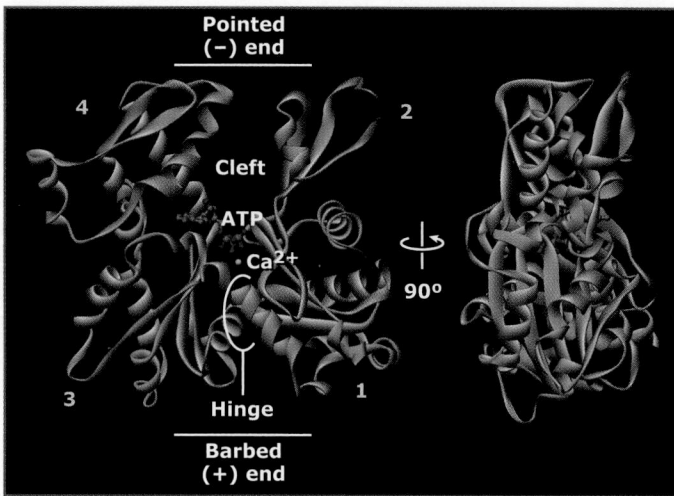

FIGURE 12.2 X-ray crystal structure of the actin monomer, depicted as a ribbon diagram of the protein backbone. The subdomains are numbered 1-4. Structures from Protein Data Bank 1ATN. W. Kabsch, et al., *Nature* 347 (1990): 21–22.

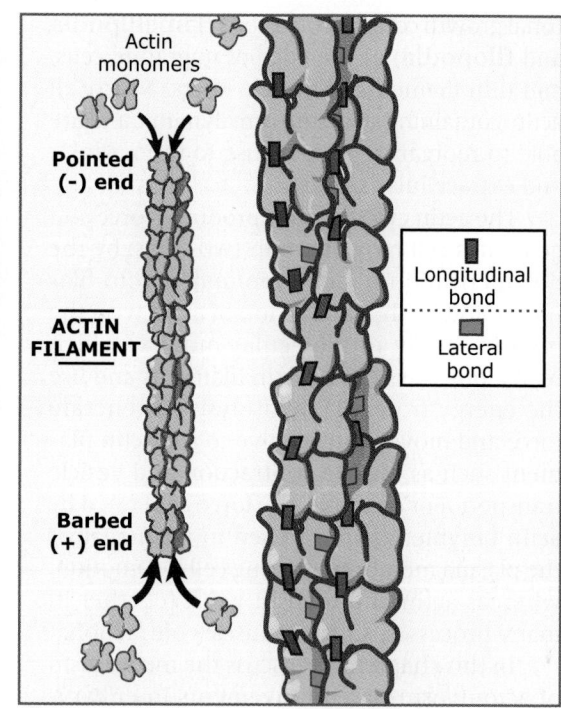

FIGURE 12.3 Actin monomers assemble head-to-tail into double-stranded, polar filaments.

separated by a large cleft, as the X-ray crystal structure in **FIGURE 12.2** shows. Each lobe is composed of two subdomains. The four subdomains are referred to as subdomains 1–4. Two strands, extending between subdomains 1 and 3, connect the two lobes. Together, these strands serve as a hinge, allowing the two lobes to move relative to each other.

A nucleotide (ATP or ADP) and a divalent cation (calcium or magnesium) bind at the center of the actin monomer, deep in the cleft between subdomains 2 and 4. Nucleotide binding is thought to close the cleft between the two lobes. Nucleotide binding to actin monomers is reversible, and bound nucleotide exchanges with free nucleotide in solution. Because actin binds ATP more tightly than ADP, ATP is present at a higher concentration than ADP, and the cytosolic concentration of Mg^{2+} is much higher than that of Ca^{2+}, MgATP occupies the nucleotide-binding site of actin monomers in cells.

12.4 Actin filaments are structurally polarized polymers

Key concepts

- In the presence of physiologic concentrations of monovalent and divalent cations, actin monomers polymerize into filaments.
- The actin filament is structurally polarized and the two ends are not identical.

At cellular concentrations of monovalent and divalent cations, actin monomers self-associate and polymerize into filaments that are ~8 nm in diameter. Polymerization is reversible, and monomers are constantly adding to and dissociating from the ends of filaments. This assembly and disassembly provide the driving force for cell motility based on actin polymerization.

FIGURE 12.3 and **FIGURE 12.4** show that monomers assemble into filaments in a "head-to-tail" fashion such that all subunits within a filament have the same orientation. An actin filament resembles a double-stranded string of beads that has a right-handed helical twist, as Figure 12.3 and **FIGURE 12.5** show. These features have been revealed by electron microscopy. Because the monomers are structurally polar (see Figure 12.2), the filament is also polar, that is, the two ends of the filament are different and are called the **barbed end** and the **pointed end**. This polarity is important for directional transport within cells and for establishing polarized cell shapes.

The end of the actin filament at which subdomains 1 and 3 are exposed is the barbed end, while the end with subdomains 2 and 4 exposed is the pointed end (see Figure 12.3). The terms *barbed* and *pointed* are based on the appearance of actin filaments that have bound myosin. Because the actin filament is helical

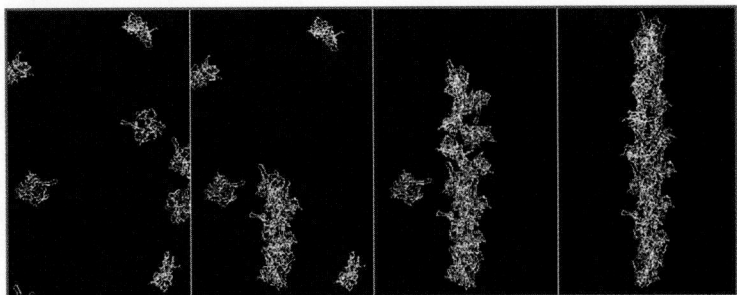

FIGURE 12.4 Frames from an animation showing the sequential association of actin monomers into an actin filament. Photos courtesy of Kenneth C. Holmes, Max Planck Institut für Medizinische Forschung.

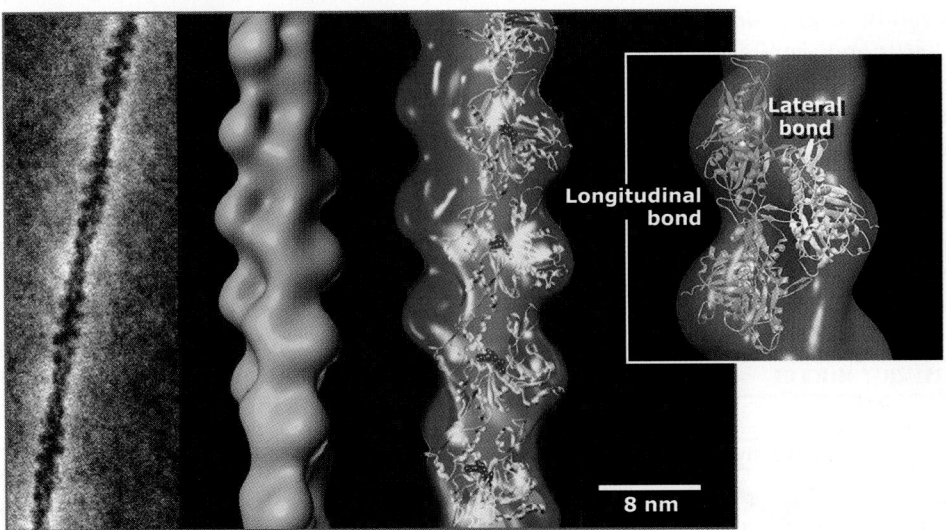

FIGURE 12.5 (Left) Electron micrograph of an actin filament. (Middle) Model of an actin filament derived from three-dimensional reconstruction of electron micrographs. (Right) Three-dimensional reconstruction model superimposed with atomic structures of actin monomers to show the proposed placement of subunits in one strand of an actin filament. Photos courtesy of Ueli Aebi, University of Basel.

and polar, the myosin proteins resemble arrowheads when visualized with an electron microscope (see Figure 12.9). The barbed and pointed ends of actin filaments are also called plus (+) and minus (−) ends, respectively, consistent with the nomenclature used for plus and minus ends of microtubules (see *11.3 Microtubules are polar polymers of α- and β-tubulin*).

Within an actin filament, an individual actin subunit contacts four neighboring subunits: one on each side of it on the same strand (longitudinal contacts) and two on the opposite strand (lateral contacts) (see Figure 12.3). These contacts help to make actin filaments very strong. Filaments are highly resistant to shearing by thermal forces and can grow to lengths of thousands of subunits. Filaments are

mechanically stiff and cannot be bent sharply without breaking. Although short filaments (<5 μm) do not show much curvature, long filaments (>5 μm) are often curved.

12.5 Actin polymerization is a multistep and dynamic process

Key concepts

- *De novo* actin polymerization is a multistep process that includes nucleation and elongation steps.
- The rates of monomer incorporation at the two ends of an actin filament are not equal.
- The barbed end of an actin filament is the fast growing end.

Cells migrate by extending their leading edge and retracting their trailing edge. The protrusions at the front of a migrating cell can be in the form of either a lamellipodium or a filopodium. A lamellipodium is a thin, membrane-enclosed, sheetlike extension containing a branched, brushlike actin filament network (see Figure 12.1). In contrast, a filopodium is a fingerlike, membrane-enclosed projection containing a bundle of parallel actin filaments (see Figure 12.1). The protrusion of both filopodia and lamellipodia requires actin polymerization, a process that we discuss in detail here.

Polymerization of actin monomers to form an actin filament is a multistep process. This process can be studied *in vitro*, starting with purified actin monomers in a solution of low ionic strength; polymerization can be initiated by increasing the salt concentration to a physiologic level. The time course of spontaneous polymerization *in vitro* can be divided into three distinct stages—**nucleation, elongation**, and **steady-state**—which are illustrated in **FIGURE 12.6**.

The nucleation phase reflects the formation of an actin oligomer that exhibits properties similar to that of a long filament. This oligomer is called the nucleus for actin polymerization. The nucleus is a trimer, which is the smallest unit that has the lateral and longitudinal contacts of a filament (see Figure 12.5). Consequently, nucleation occurs in two steps: formation of a dimer from two monomers, followed by the addition of a third subunit to yield a trimer. Actin dimers and trimers are unstable (K_d ~100 μM^{-1} mM) and exist at very low concentrations. Thus, the formation of nuclei results in a lag phase in the time courses of spontaneous polymerization. The lag phase can be eliminated if nuclei from which monomers can grow are available (e.g., if preformed filaments are present). In cells, actin-binding proteins initiate filament assembly (see *12.9 Nucleating proteins control cellular actin polymerization*).

The elongation phase represents the rapid, longitudinal growth of filaments (see Figure 12.6). Actin filaments grow by elongating from their ends, rather than their sides. As a result, filaments vary in length but all have identical widths. The elongation phase gradually slows as subunits polymerize and the free monomer concentration in solution diminishes.

In the steady-state phase of the polymerization reaction, there is no net filament growth. However, there is a slow and constant exchange of actin subunits at the filament ends with the pool of monomers. The concentration of unpolymerized actin monomers at this steady-state is referred to as the **critical concentration** (C_c), as shown in **FIGURE 12.7**. The critical concentration can also be defined as the concentration of actin subunits needed to form filaments (C_c for actin assembly). When the total concentration of monomeric actin is greater than the critical concentration, filaments will form. In contrast, when the total concentration of monomeric actin is below the critical concentration, only monomers will exist. The critical concentration is constant under a defined set of solution conditions but depends on the presence of regulatory proteins that bind actin monomers and filaments (see *12.8 Actin monomer-binding proteins influence polymerization* and *12.9 Nucleating proteins control cellular actin polymerization*).

At monomer concentrations above the critical concentration, filaments elongate at rates that depend linearly on the concentration of monomers in solution, as **FIGURE 12.8** shows. The rate at which an individual filament elongates is the rate constant for the addition

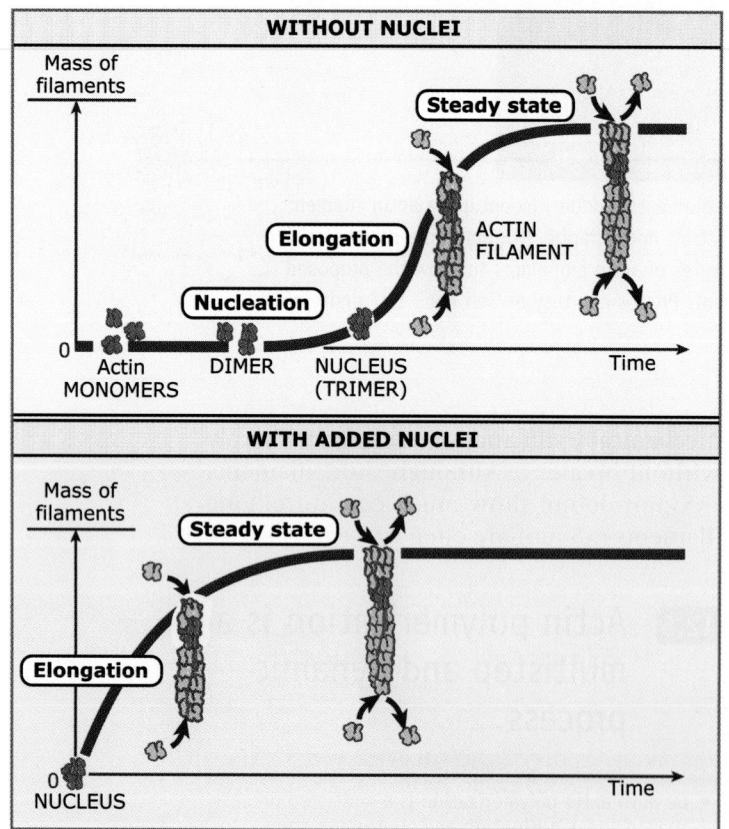

FIGURE 12.6 The process of actin assembly, starting from a solution of monomers, can be divided into the stages of nucleation, elongation, and steady state. The nucleation stage can be bypassed by the addition of a small amount of preformed filaments.

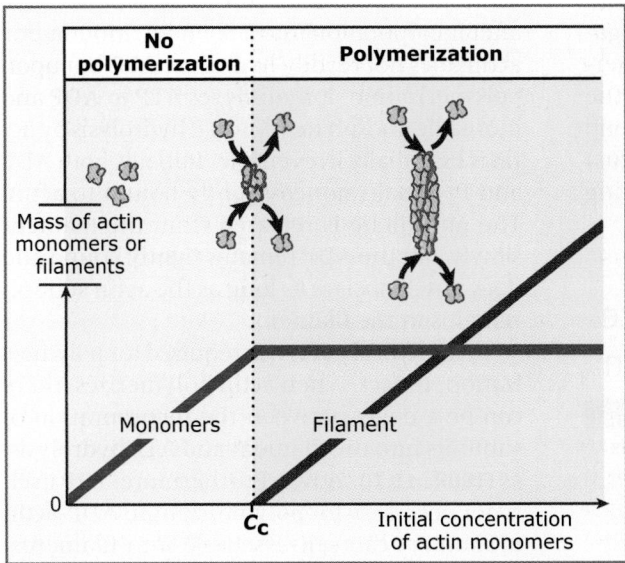

FIGURE 12.7 Assembly of actin filaments, as measured *in vitro*, occurs only when the starting concentration of actin monomers is above the critical concentration, C_c.

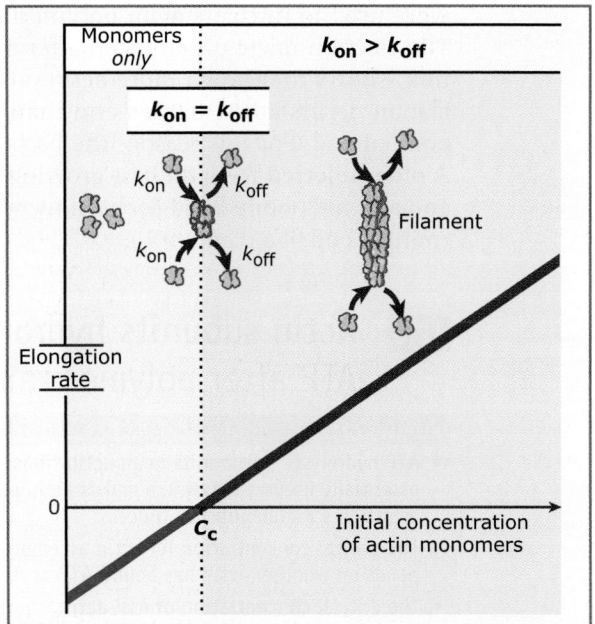

FIGURE 12.8 The elongation rate varies linearly with the starting concentration of actin monomers, as determined by *in vitro* experiments.

of monomers (k_{on}; units of $\mu M^{-1}\ sec^{-1}$) multiplied by the concentration of monomers. The depolymerization rate (k_{off}; units of sec^{-1}) is independent of the concentration of monomeric actin. When the actin monomer concentration is equal to the critical concentration, the elongation rate and the depolymerization rate are equal, and there is no net filament growth:

$$(k_{on})\,(C_c) = k_{off}$$

Rearranging gives:

$$C_c = \frac{k_{off}}{k_{on}}$$

For a reaction at equilibrium, the dissociation constant K_d equals k_{off}/k_{on}. Therefore, the critical concentration is equal to the dissociation equilibrium constant for binding of an actin monomer to the end of a filament:

$$C_c = k_d$$

The rates of monomer incorporation at the two ends of a filament are not equal. The values of the elongation rate constants k_{on} and k_{off} (and therefore their ratio, the critical concentration) depend on the nucleotide that is bound to actin (this is explained in detail in the next section). In the presence of ATP, actin monomers associate with the barbed end of a filament at a rate about ten times faster than at the pointed end, as demonstrated by the experiment shown in **FIGURE 12.9**. In this experiment, short actin filaments decorated with myosin

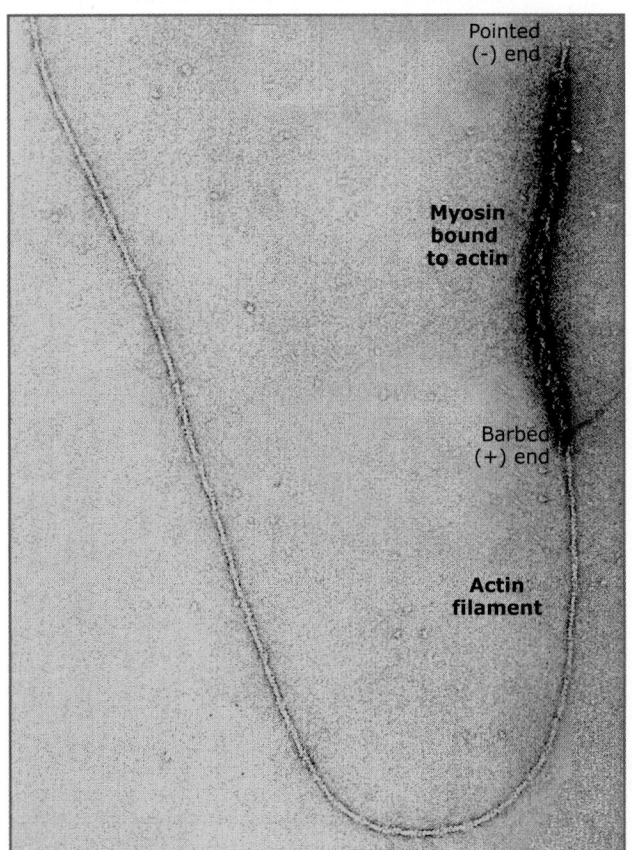

FIGURE 12.9 Electron micrograph from an experiment showing that actin polymerization occurs primarily from the barbed end of a filament. A myosin-decorated filament was used to nucleate filament elongation. Photo courtesy of Marschall Runge, John Hopkins School of Medicine and Thomas Pollard, Yale University.

were used to nucleate actin polymerization. The electron micrograph, taken after elongation, shows that much more actin polymerization occurs at the barbed end than at the pointed end. For this reason, the barbed end is often referred to as the fast growing (plus) end and the pointed end as the slow growing (minus) end of a filament.

12.6 Actin subunits hydrolyze ATP after polymerization

Key concepts

- ATP hydrolysis by subunits in an actin filament is essentially irreversible, which makes actin polymerization a nonequilibrium process.
- The critical concentration for actin assembly depends on whether actin has bound ATP or ADP.
- The critical concentration of ATP-actin is lower than that of ADP-actin.
- In the presence of ATP, the two ends of the actin filament have different critical concentrations.

As discussed in the previous section, spontaneous actin assembly *in vitro* can be accurately described by the familiar chemical principles of mass action and reversible reaction kinetics. However, additional levels of complexity are introduced by the enzymatic activity of the actin subunit, which hydrolyzes its bound ATP after incorporation into a filament. Monomeric actin does not readily hydrolyze ATP, but upon polymerization, it hydrolyzes ATP to ADP and inorganic phosphate (P_i). ATP hydrolysis by actin is essentially irreversible. Initially, both ADP and P_i remain noncovalently bound to actin. The phosphate is released from the filament slowly, but the ADP remains tightly bound and does not dissociate as long as the actin subunit remains in the filament.

ATP hydrolysis is not required for polymerization. In fact, when actin polymerizes, there can be a delay between the incorporation of subunits into the filament and ATP hydrolysis, as **FIGURE 12.10** shows. Furthermore, ATP itself is not necessary for polymerization: ADP-actin monomers can self-assemble into filaments. However, ATP hydrolysis is crucial for the regulation and function of actin in the cell (see *12.8 Actin monomer-binding proteins influence polymerization* and *12.9 Nucleating proteins control cellular actin polymerization*).

As illustrated in **FIGURE 12.11**, the critical concentration for actin assembly depends on which form of nucleotide (ATP, ADP-P_i, or ADP) is bound. The critical concentration values have been determined from the actin monomer concentration needed to form filaments and calculated from k_{off} and k_{on} values, which were determined using *in vitro* systems. At the barbed end of an actin filament, the critical concentration of ATP-actin (~0.1 µM) is lower than that of ADP-actin (1.7 µM). This means that purified ATP-actin monomers polymerize into actin filaments until the concentration of monomers reaches ~0.1 µM. Similarly, ADP-actin polymerizes until the concentration of monomers reaches 1.9 µM. The critical concentration of ADP-P_i actin at the barbed end (0.1 µM) is comparable to that of ATP-actin. Since $C_c = K_d$, these values indicate that the change in binding affinity of a terminal subunit occurs with the release of P_i. In other words, the energy from ATP hydrolysis is stored in the filament until the P_i is released, after which the terminal subunits (now with bound ADP) bind more weakly to a filament.

For actin filaments composed of ADP-actin, the critical concentration of ADP-actin monomers is roughly the same at both ends (~2 µM) since the subunits throughout the filament and the stabilizing contacts between them are identical. However, in the presence of ATP, the critical concentration at the barbed end (0.1 µM) is lower than at the pointed end

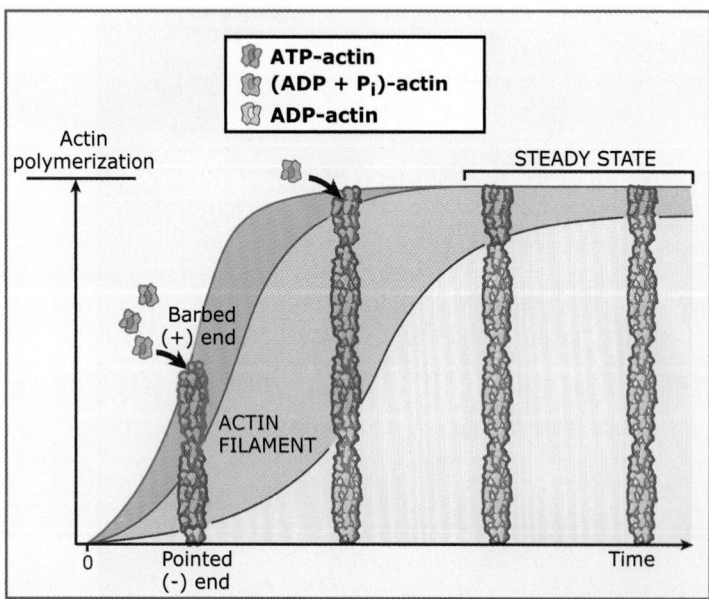

FIGURE 12.10 As ATP-actin monomers associate at the barbed end during the initial fast phase of filament elongation, actin subunits contain ATP or, after hydrolysis, ADP + P_i. As elongation continues, and the free monomer pool is depleted, elongation becomes slower than ATP hydrolysis. Thus at steady-state, actin filaments contain ADP-actin, except for those at the barbed end, which contain ADP + P_i. Adapted from M. F. Carlier, *J. Biol. Chem.* 266 (1991): 1–4.

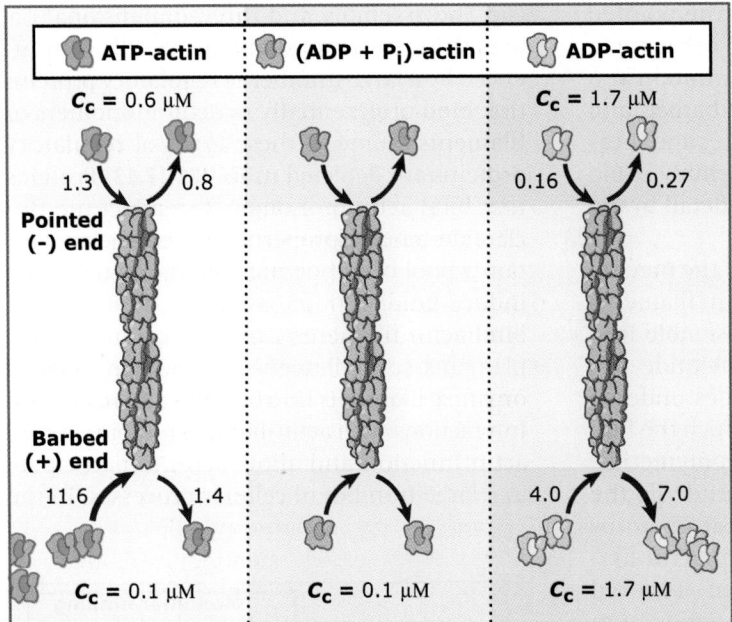

FIGURE 12.11 Under steady state conditions *in vitro*, the critical concentration (C_c) at the barbed end of a filament composed of ATP-actin or ADP + P_i-actin is lower than at the pointed end, whereas the C_c values at the barbed and pointed ends of an ADP-actin filament are similar. Also shown are the rate constants for association and dissociation of actin monomers, k_{on} and k_{off}, respectively, which are used to calculate C_c. Adapted from an illustration by Enrique De La Cruz, Yale University.

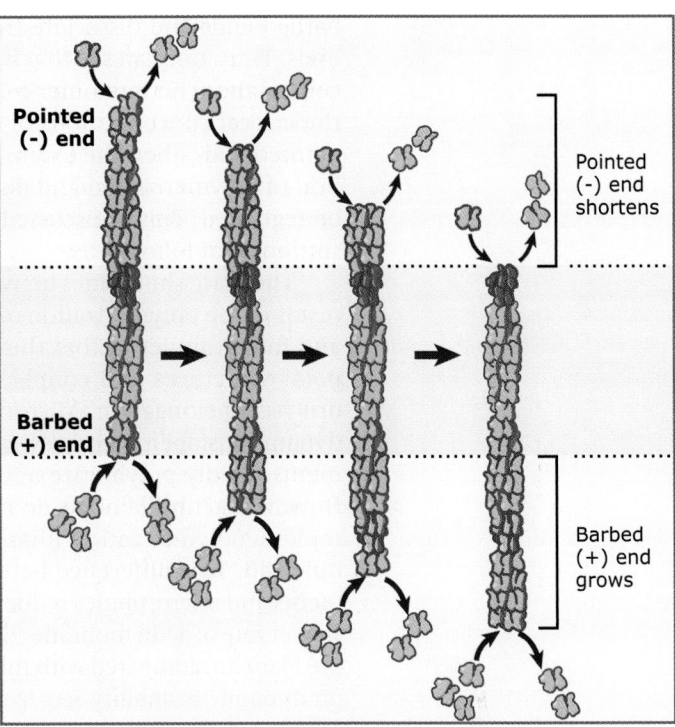

FIGURE 12.12 At steady state, actin filaments undergo treadmilling, in which ATP-actin monomers more rapidly associate at the barbed end and ADP-actin subunits more rapidly dissociate at the pointed end without changing the length of the filament.

(0.7 μM). This is because the actin subunits at the barbed end contain ATP (or ADP-P_i), and those at the pointed end have bound ADP (see Figure 12.10). This chemical polarization of the filament is due to the rapid addition of ATP-monomers at the barbed end, which results in formation of an "ATP cap." At the pointed end, monomer addition is slow, so ATP is hydrolyzed and phosphate is released faster than addition of ATP-actin monomers (unless polymerization is very rapid).

When the pool of actin monomers consists of ATP-actin (which predominates in cells), the critical concentration for the entire filament (i.e., the steady-state actin monomer concentration) is intermediate between the critical concentrations of the two ends: it is higher than at the barbed end and lower than at the pointed end (see Figure 12.11). Consequently, at steady state, actin monomers (as ATP-actin) associate with the barbed end and, after ATP hydrolysis and release of P_i, dissociate from the pointed end as ADP-actin. This polymerization and depolymerization generates a constant flux of subunits throughout the filament without changing the overall concentrations of polymer or mono-

mer. This flux is referred to as **treadmilling**, which is illustrated in **FIGURE 12.12**. Unregulated treadmilling does not occur in the cell but rather is controlled by actin-binding proteins.

Under experimental conditions in which filaments assemble from actin monomers containing ADP, treadmilling does not occur because the critical concentration is the same at both ends (see Figure 12.11). Thus, actin subunits depolymerizing from the pointed end must exchange bound ADP for ATP in order to add on to the barbed end. The energy from ATP hydrolysis is used, not for the polymerization itself, but rather to drive the assembly and disassembly of actin and vectorially cycle monomers through a filament.

Treadmilling occurs only when the actin monomer concentration is between the critical concentrations of the two ends. At monomer concentrations above the critical concentrations of the barbed and pointed ends, the filaments elongate at both ends. Filaments depolymerize from both ends when the monomer concentration is below the critical concentration. At monomer concentrations between the critical concentrations of the two ends, monomers add to the barbed ends and dissociate from the pointed ends. Thus, one can see that if the cell were to control the actin monomer concentration and the critical concentrations of the barbed and pointed ends, then the extent, rate, and location of polymerization and disassembly could be regulated. This is discussed in detail in the sections that follow.

There are similarities between the mechanisms of the polymerization of actin filaments and microtubules in that they assemble into polar structures and couple nucleotide hydrolysis to elongation. Microtubules undergo dynamic instability, a process in which the filaments rapidly polymerize and depolymerize. However, actin filaments do not undergo the rapid depolymerization phase that microtubules do. This difference between actin filaments and microtubules is due to an ~100-fold slower rate of actin monomer dissociation from the filament, compared with tubulin. (For more on dynamic instability see *11.5 Microtubule assembly and disassembly proceed by a unique process termed dynamic instability*.)

12.7 Actin-binding proteins regulate actin polymerization and organization

Key concepts

- For the actin cytoskeleton to drive motility, the cell must be able to regulate actin polymerization and depolymerization.
- Actin-binding proteins associate with monomers or filaments and influence the organization of actin filaments in cells.

The actin cytoskeleton is remodeled in response to environmental cues that stimulate cell division, differentiation, or locomotion. To control the remodeling process, cells must be able to assemble and disassemble actin filaments rapidly. For the actin cytoskeleton to drive motility and changes in cell shape, regulatory mechanisms must exist to:

- inhibit the spontaneous polymerization of the pool of monomeric actin,
- quickly nucleate new actin filaments,
- control the length of actin filaments, and
- elongate preexisting actin filaments.

Actin-binding proteins, many of which are regulated by intracellular signaling, mediate these mechanisms.

The assembly and three-dimensional organization of actin filaments in cells is influenced by a large number of regulatory proteins that bind preferentially to actin monomers or filaments. Some of these types of regulatory proteins are depicted in **FIGURE 12.13**. Proteins that bind actin monomers can affect the nucleotide-binding properties of actin, help maintain a pool of unpolymerized monomers, and induce unidirectional assembly. Proteins that bind actin filaments can stabilize nuclei and filaments, sever filaments, cap filament ends, or organize filaments into bundles and networks. Interactions with actin-binding proteins dictate actin function and allow actin to participate in a large number of cellular processes. In the

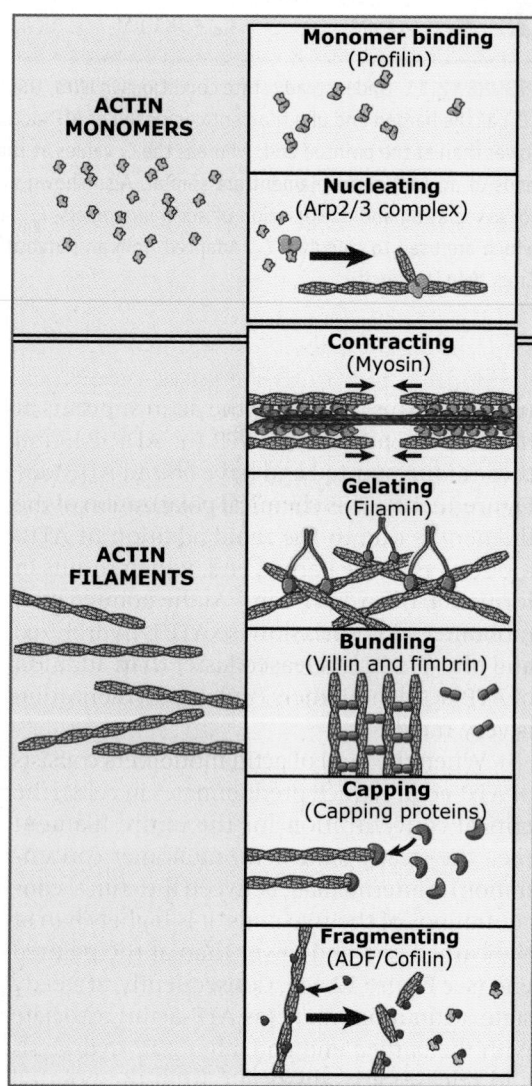

FIGURE 12.13 Actin-binding proteins regulate the dynamics of actin polymerization and the organization of filaments in cells. Some proteins function with actin monomers and others with filaments.

next few sections, we discuss the different types of actin-binding proteins, followed by a section that outlines the mechanisms by which these proteins work together (see *12.13 Actin and actin-binding proteins work together to drive cell migration*).

12.8 Actin monomer-binding proteins influence polymerization

Key concepts

- The two major actin monomer-binding proteins in many eukaryotic cells are thymosin β_4 and profilin.
- In metazoan cells, thymosin β_4 sequesters actin monomers and maintains a cytosolic pool of ATP-actin that can be utilized for rapid filament elongation.
- Profilin-actin monomer complexes contribute to filament elongation at barbed ends but not at pointed ends.

Cells maintain a pool of actin monomers that can be used for rapid elongation when extracellular cues initiate the timing and location for filament growth. The rate of actin filament elongation depends on the concentration of actin monomers available for polymerization (see Figure 12.7). To generate rapid elongation on the relevant cellular time scale, the concentration of actin monomers available for polymerization must be much greater than the critical concentration. Proteins that bind actin monomers help to regulate the rate of filament growth.

The two most abundant **actin monomer-binding proteins** in metazoan cells are thymosin β_4 and profilin. Thymosin β_4 is found only in higher eukaryotes and is most abundant in highly motile and phagocytic cells. Profilin is found in most eukaryotic cells, including those of plants, animals, and yeast. The concentration of these proteins is comparable to the total cellular actin concentration. Although they both bind actin monomers, thymosin β_4 and profilin have very different effects on the regulation of actin polymerization, and both appear to be critical for physiologic processes.

Thymosin β_4 is a small peptide ($M_r < 5$ kDa). It is a member of a highly conserved family of actin monomer-binding proteins that form 1:1 complexes with monomeric actin. Thymosin β_4 inhibits the spontaneous polymerization of actin monomers and prevents the addition of monomers to existing filaments. Although

thymosin β_4 is a small peptide, it is thought to bind a large area of the actin monomer surface, sterically blocking many regions necessary for forming important stabilizing filament contacts. The intracellular concentration of thymosin β_4 can reach several hundred micromolar in highly motile cells such as neutrophils, enabling the cell to maintain a large cytoplasmic pool of actin monomers. Thymosin β_4 binds with higher affinity to ATP-actin monomers ($K_d \sim 2$ µM) than to ADP-actin monomers ($K_d \sim 50$ µM), so the pool of unpolymerized actin in cells consists primarily of ATP-actin.

Profilin is also a small ($M_r \sim 15$ kDa) actin-binding protein that forms 1:1 complexes with actin monomers. However, in contrast to thymosin β_4, profilin-actin monomer complexes can associate with the barbed ends of actin filaments. Once the profilin-actin complex binds to the barbed end, profilin dissociates and can then bind another actin monomer. **FIGURE 12.14** shows that because profilin binds actin monomers at subdomains 1 and 3, its presence blocks the binding of a monomer to the pointed end of the filament. Thus, profilin-actin promotes elongation exclusively at the filament's barbed end.

Another key property of profilin is that it catalyzes the exchange of bound nucleotide from actin monomers. Because the intracellular

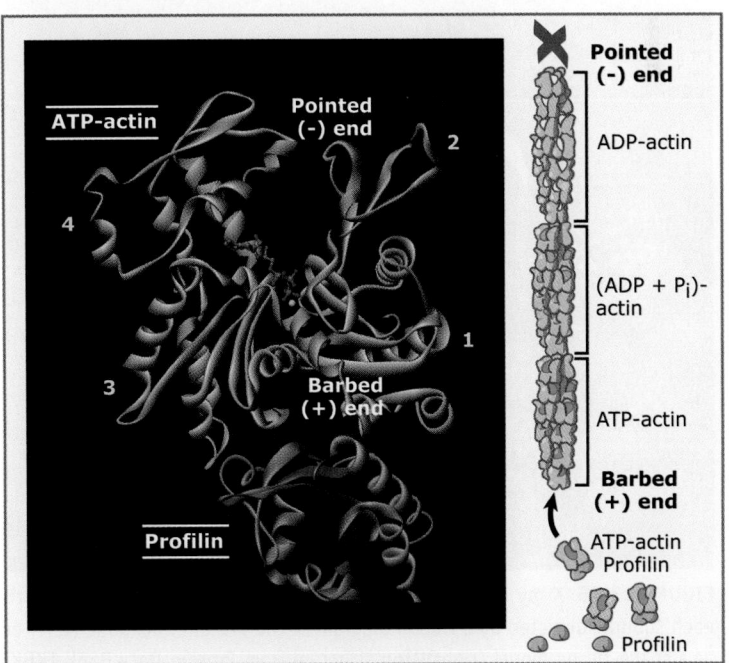

FIGURE 12.14 Profilin associates with the barbed end of actin monomers and helps regulate actin polymerization by allowing incorporation of the monomer only at the barbed end. Structure from Protein Data Bank 2ITF. J. C. Grigg, et al., *Mol. Microbiol.* 63 (2007): 139–149.

ATP concentration is much higher than that of ADP, profilin allows the actin monomers to rapidly equilibrate with cytoplasmic ATP and converts ADP-actin monomers into ATP-actin monomers, helping maintain a large pool of ATP-actin monomers for rapid filament growth. (For a discussion of the role of thymosin β_4 and profilin during actin remodeling in cells see *12.13 Actin and actin-binding proteins work together to drive cell migration*.)

12.9 Nucleating proteins control cellular actin polymerization

Key concepts

- Nucleating proteins allow the cell to control the time and place of *de novo* filament formation.
- The Arp2/3 complex and formins nucleate filaments *in vivo*.
- Arp2/3 nucleation generates a branched filament network, whereas formin proteins nucleate unbranched filaments.
- Arp2/3 is activated at cell membranes by suppressor of *cAMP* receptor (Scar), *Wiskott-Aldrich syndrome protein* (WASP), and *WASP-verprolin homologs* (WAVE) proteins.

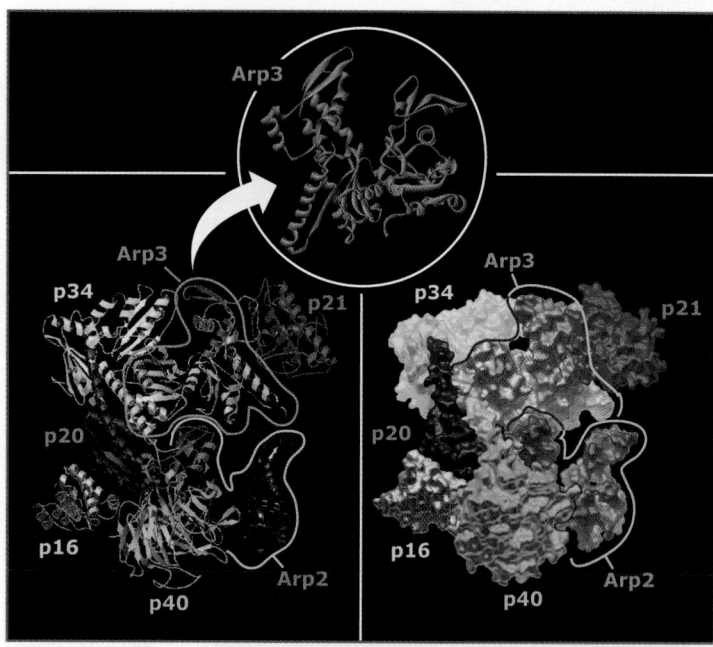

FIGURE 12.15 X-ray crystal structure of the Arp2/3 protein complex, with each subunit depicted by a ribbon diagram of the protein backbone (left) or as a space-filling model (right). Top structure from Protein Data Bank 1KBK. E. H. Rydberg, et al., *Biochemistry* 41 (2002): 4492–4502. Bottom photos reproduced from R. C. Robinson, et al., *Science* 294 (2001): 1679–1684 [http://www.sciencemag.org]. Reprinted with permission from AAAS. Photos courtesy of Thomas D. Pollard, Yale University.

The nucleation step is the slowest step in the polymerization of new actin filaments *in vitro* (see Figure 12.6) and is an important control point in the regulation of cellular actin polymerization. To rapidly polymerize new actin filaments in a spatially regulated manner, the nucleation step must be accelerated. Proteins that facilitate *de novo* filament formation are called **nucleating proteins**. The Arp2/3 complex and formins are two types of actin filament nucleators that play prominent roles in regulating cell motility and are well characterized. These proteins nucleate filaments by different mechanisms and yield different types of filament networks throughout the cell.

The Arp2/3 complex is a macromolecular protein complex consisting of Arp2, Arp3, and five additional proteins, as the X-ray crystal structure in **FIGURE 12.15** shows. Arp2 and Arp3 are actin-related proteins. Although Arp2 and Arp3 are similar in structure to actin (see Figure 12.2), they lack the ability to polymerize into filaments. It is thought that Arp2 and Arp3 associate in the complex in a way that mimics a stable actin dimer with an exposed barbed end, thus enabling a stable nucleus to form the moment an actin monomer binds the complex. As the new filament elongates at the barbed end, the Arp2/3 complex remains associated with the pointed end.

The nucleation activity of the Arp2/3 complex is activated by interactions with regulatory proteins and with the sides of preexisting actin filaments. Nucleation by the Arp2/3 complex can occur at the plasma membrane (see Figure 12.22). Newly nucleated "daughter" filaments can grow in the barbed end direction as 70° branches on the sides of preexisting "mother" filaments. *In vitro* experiments show that this side-binding nucleation activity of the Arp2/3 complex results in a dendritic meshwork of actin filaments, similar to what is seen at the rapidly growing edges of motile cells (see lamellipodium in Figure 12.1). The Arp2/3 complex binds more tightly to filament subunits with bound ATP or ADP-P_i ("young" filaments) than to filament subunits with bound ADP ("old" filaments) (see *12.6 Actin subunits hydrolyze ATP after polymerization*). Therefore, Arp2/3 preferentially binds and branches from young filaments. As P_i is released and the filaments age, Arp2/3 dissociates from the actin filaments, resulting in the dissociation of the branches and disassembly of the actin filament network.

Several proteins activate the Arp2/3 complex at membranes in response to signal

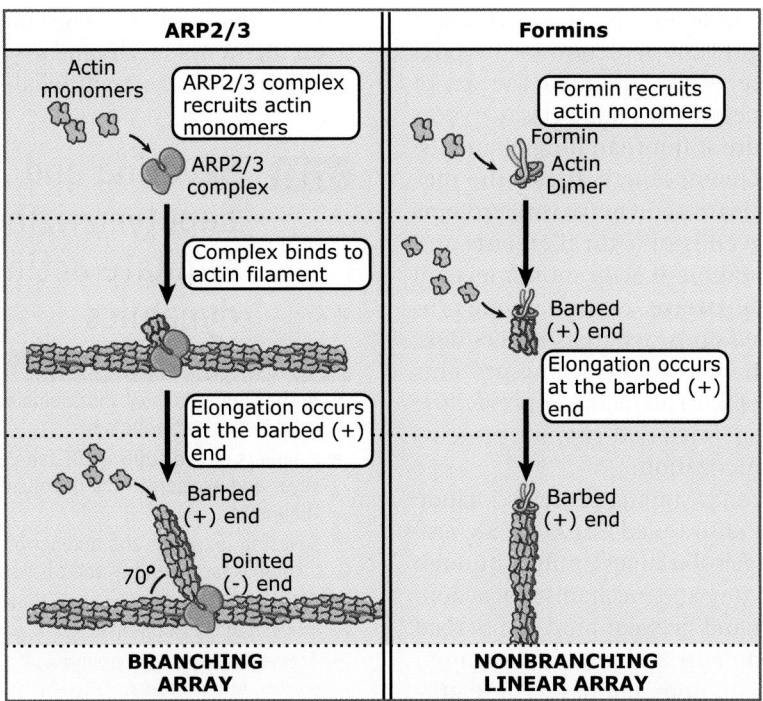

ARP2/3	Formins
Actin monomers ARP2/3 complex recruits actin monomers ARP2/3 complex	Formin recruits actin monomers Formin Actin Dimer
Complex binds to actin filament	Barbed (+) end Elongation occurs at the barbed (+) end
Elongation occurs at the barbed (+) end Barbed (+) end Pointed (–) end 70°	Barbed (+) end
BRANCHING ARRAY	**NONBRANCHING LINEAR ARRAY**

FIGURE 12.16 The Arp2/3 complex nucleates the formation of branching actin structures, whereas formins nucleate nonbranching, linear actin structures.

transduction pathways. Arp2/3 activation at the membrane is crucial for the role of actin in membrane protrusion (see *12.13 Actin and actin-binding proteins work together to drive cell migration*). These proteins are activated by small G proteins and include Scar, WASP, and WAVE. These proteins function with an existing actin filament to activate the Arp2/3 complex.

Formins are a family of structurally conserved proteins named for the mouse gene, *limb deformity*. Members of the formin family contain two unique homology domains called formin homology-1 and -2 (FH1 and FH2). The FH1 domain binds profilin, and the FH2 domain nucleates actin polymerization. Formins are remarkable in that they stay associated with the barbed end of the filament during barbed-end elongation, protect the barbed end from capping proteins during elongation, and increase the elongation rate by direct association with profilin.

This mechanism of nucleation and elongation by formins is unlike the Arp2/3 complex, which remains at the pointed end of the filament during elongation. Because formins nucleate new actin filaments without binding to the sides of preexisting filaments, they regulate the assembly of unbranched filaments

rather than the branched networks created by the Arp2/3 complex, as **FIGURE 12.16** shows.

Concept and Reasoning Check

1. Compare and contrast actin filament nucleation by Arp2/3 complexes and formins.

12.10 Capping proteins regulate the length of actin filaments

Key concepts

- Capping proteins inhibit actin filament elongation.
- Capping proteins function at either the barbed or pointed ends of actin filaments.
- Capping protein and gelsolin inhibit elongation at barbed ends and are inhibited by phospholipids of the plasma membrane.
- Tropomodulin is a protein that caps the pointed end of actin filaments.

In vitro, unregulated actin filaments grow rapidly and continuously until the free monomer concentration reaches the critical concentration (see Figure 12.6 and Figure 12.7). *In vivo*, cellular mechanisms exist to control the number of free barbed ends. This regulation is necessary to prevent depletion of the pool of actin monomers and to regulate the size of specific actin structures. In addition, since short actin filaments are stiffer than long filaments, regulation of filament length affects the mechanical properties of actin networks. Proteins that bind to the ends of actin filaments and prevent incorporation of actin monomers are called **capping proteins**. Some capping proteins bind barbed ends, whereas others bind pointed ends. Barbed end-capping proteins limit the length of actin filaments by preventing elongation, and pointed end-capping proteins prevent depolymerization.

The barbed end-capping proteins include capping protein (also called CapZ), EPS8, and members of the gelsolin superfamily. Although different in structure and mechanism of action, capping protein and gelsolin bind the barbed ends of actin filaments and inhibit elongation by preventing monomer addition. These proteins have high affinity for the barbed ends of actin filaments and inhibit the addition of actin subunits, even at the high cellular concentrations of actin monomers. (For more on CapZ see *12.21 Myosin-II functions in muscle contraction*.)

The capping activities of both gelsolin and capping protein may be regulated by phospholipids at the plasma membrane. The lipid phosphatidylinositol 4,5-bisphosphate (PIP$_2$) competes with the barbed ends of actin filaments for binding capping protein, thereby possibly providing a mechanism for inhibiting capping at the plasma membrane. PIP$_2$ is located in the inner leaflet of the plasma membrane and is important in intracellular signaling. The level of PIP$_2$ is modulated in response to signaling through some G-protein coupled cell-surface receptors. Regulated capping allows for the control of filament elongation at the cell membrane, a key event for cell migration and membrane protrusion (see *12.13 Actin and actin-binding proteins work together to drive cell migration*). The level of capping protein activity may influence the type of protrusion (lamellipodia or filopodia) formed by a cell. (For more on PIP$_2$ see *18 Principles of cell signaling*.)

The tropomodulin family of proteins are widely expressed capping proteins that bind the pointed ends of actin filaments with high affinity in the presence of the actin regulatory protein, tropomyosin. Tropomodulins help determine the lengths of actin filaments in striated muscles (see *12.21 Myosin-II functions in*

muscle contraction). Tropomodulins also regulate actin filament lengths and assembly dynamics in erythrocytes and epithelial cells.

12.11 Severing and depolymerizing proteins regulate actin filament dynamics

Key concepts

- Actin filaments must disassemble to maintain a soluble pool of monomers.
- Members of the cofilin/ADF family of proteins sever and accelerate the depolymerization of actin filaments.
- Severing increases the number of filament ends available for assembly and disassembly.
- Cofilin/ADF binds cooperatively and changes the twist of actin filaments.
- Actin filaments with bound ADP are targets for cofilin/ADF proteins.

For cells to replenish rapidly the soluble pool of actin monomers, polymerized actin filaments must be disassembled. Although *de novo* protein synthesis is also necessary for maintaining the pool of actin monomers, it would occur too slowly for rapid remodeling of the actin cytoskeleton. Members of the cofilin/actin depolymerizing factor (ADF) family of actin regulatory proteins bind actin filaments and accelerate their disassembly in two ways: by increasing the rate of subunit dissociation from filament ends and by fragmenting (severing) filaments. The latter also increases the total number of filament ends available for elongation. The binding of cofilin/ADF proteins to actin filaments is cooperative, which allows low concentrations of these proteins to be very effective at targeting filaments for disassembly.

Electron microscopy of native and cofilin-decorated actin filaments indicates that cofilin increases the average helical twist of filaments, as **FIGURE 12.17** shows. The change in helical twist presumably promotes cooperative binding of cofilin, which exerts physical strain and helps dissociate subunits and sever filaments by disrupting stabilizing longitudinal and lateral contacts among actin subunits. The mechanism by which cofilin/ADF proteins mediate severing is yet to be fully elucidated.

Binding of cofilin/ADF to actin filaments is highly dependent on the nucleotide bound to the actin subunits. Filaments with bound ADP are selectively recognized and severed by cofi-

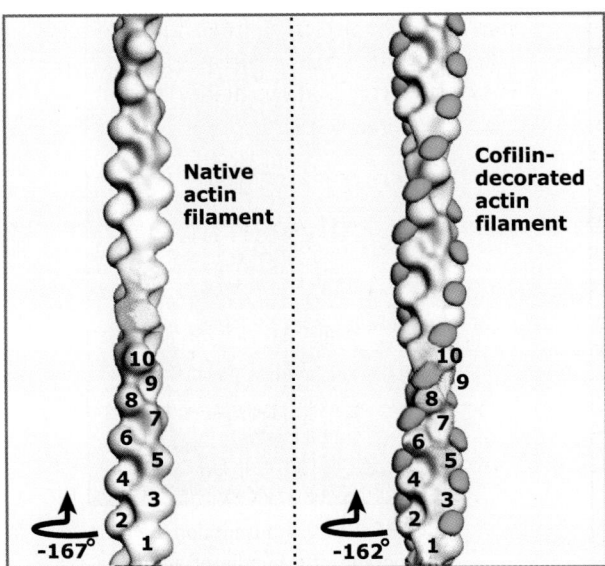

FIGURE 12.17 Binding of multiple cofilin molecules to actin filaments increases the degree of helical twist, which may destabilize the subunit interactions and break the filament. Reprinted from *Trends Cell Biol.*, vol. 9, J. R. Bamburg, A. McGough, and S. Ono, Putting a new twist on actin..., pp. 364–370, Copyright (1999) with permission from Elsevier [http://www.sciencedirect.com/science/journal/09628924]. Photos courtesy of James Bamburg, Colorado State University.

GROUP	PROTEIN	MOLECULAR WEIGHT (kDa)		LOCATION
I	**Fascin**	55		• Acrosomal process • Filopodia • Lamellipodia • Microvilli • Stress fibers
	Scruin	102		• Acrosomal process
II	**Villin**	92		• Intestinal and kidney brush border microvilli
III Calponin homology-domain superfamily	**Fimbrin**	68		• Adhesion plaques • Microvilli • Stereocilia • Yeast actin cables
	Dystrophin	427		• Muscle cortical networks
	ABP120 (Dimer)	92		• Pseudopodia
	α - Actinin (Dimer)	102		• Adhesion plaques • Filopodia • Lamellipodia • Stress fibers
	Filamin (Dimer)	280		• Filopodia • Pseudopodia
	Spectrin (Tetramer)	α 280 β 246-275		• Cortical networks

FIGURE 12.18 Cross-linking proteins, which organize actin filaments into arrays, have modular structures with actin-binding domains (ABDs) separated by spacer regions of varying lengths. These proteins can be divided into three groups based on the nature of their ABDs. Group I proteins have unique ABDs, whereas the group II protein villin has an ABD of ~7 kDa. The group III proteins have calponin homology domains (~26 kDa each), which are a type of ABD present in many different actin-binding proteins.

lin/ADF proteins. The affinity of cofilin/ADF for filaments with bound ADP-P_i is weak, so these filaments are resistant to severing. Preferential binding to ADP-actin ensures that cofilin/ADF proteins do not disassemble newly formed filaments composed of ATP- and ADP-P_i-actin, but rather, target ADP-actin subunits. Thus, during a cell's response to environmental cues, the older filaments are preferentially disassembled as the actin cytoskeleton is remodeled.

12.12 Cross-linking proteins organize actin filaments into bundles and orthogonal networks

Key concepts

• Cross-linking proteins connect actin filaments to form bundles and orthogonal networks.
• Actin bundles and networks are mechanically very strong.
• Actin cross-linking proteins have two binding sites for actin filaments.
• Actin bundles help form stereocilia and filopodia.
• Orthogonal actin networks form sheets (lamellae) and gels.

For actin filaments to function in giving cells their shape and strength, they must be organized into complex higher order structures, or arrays. The two main types of structures found in cells are **actin bundles** and **actin networks**. Actin filaments in bundles are arranged in parallel, whereas filaments in networks crisscross and are organized into orthogonal, netlike meshworks (see lamellipodium in Figure 12.1). Both structures provide mechanical strength to the cell and can push the membrane forward as they grow. Actin arrays can span relatively large areas within the cytoplasm.

Bundles and networks are formed by the interaction of actin filaments with **cross-linking proteins**. Different cell types can express different cross-linking proteins. Cross-linking proteins interact with two actin filaments at once and must therefore possess two binding sites for actin filaments, as seen in **FIGURE 12.18**. In several cases, actin-binding and network

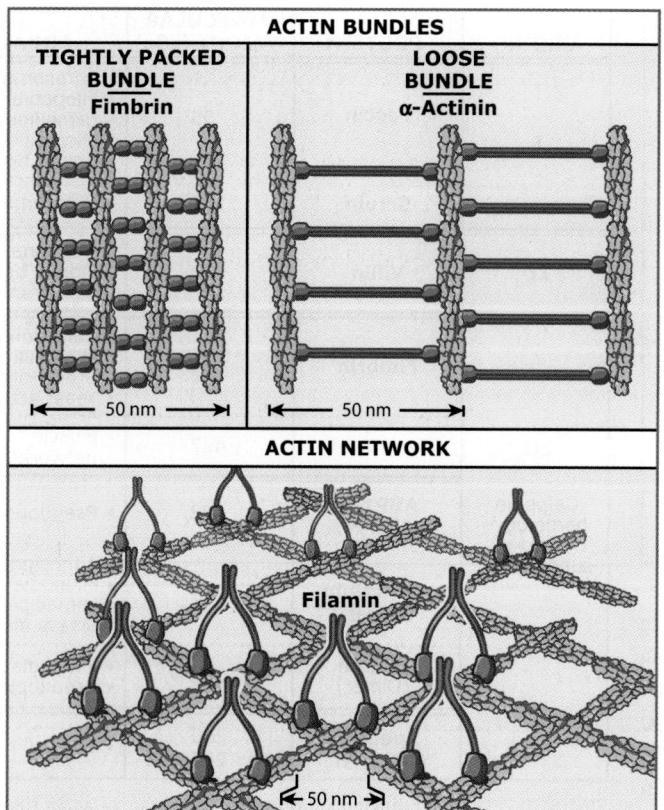

ACTIN BUNDLES

TIGHTLY PACKED BUNDLE
Fimbrin

|← 50 nm →|

LOOSE BUNDLE
α-Actinin

|← 50 nm →|

ACTIN NETWORK

Filamin

|← 50 nm →|

FIGURE 12.19 Cross-linking proteins interact with actin filaments to form two types of arrays: bundles of parallel filaments and networks.

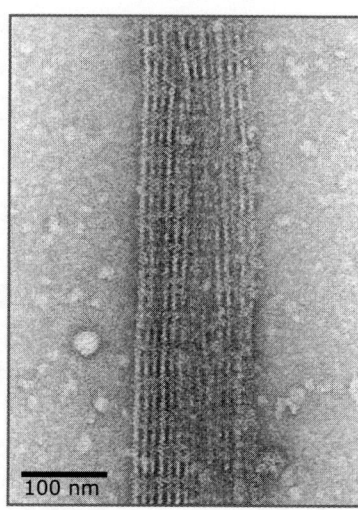

100 nm

FIGURE 12.20 Transmission electron micrograph of a bundle of actin filaments crosslinked by fascin. Photo © Cant, et al., 1994. Originally published in **The Journal of Cell Biology**, 125: 369–380. Used with permission of Rockefeller University Press. Photo courtesy of Lynn Cooley, Yale University School of Medicine.

formation are inhibited by phosphorylation of some cross-linking proteins.

Figure 12.18 shows that proteins that cross-link actin filaments can be divided into three groups. Some actin filament–cross-linking proteins are members of the actin-binding domain (ABD) family of proteins. These are the group III proteins and include fimbrin, α-actinin, spectrin, filamin, and ABP120, which have a conserved 27 kDa ABD consisting of two α-helical calponin homology domains. Most ABD-family proteins form homo- or heterodimers, so that each molecule possesses two ABDs and can bind and crosslink two different filaments. The type of actin array formed depends on the geometric organization and spacing of the domains, as illustrated in **FIGURE 12.19**. For example, the two ABDs in a fimbrin monomer are close together, which allows fimbrin to form tightly packed bundles of actin such as those in microvilli. In contrast, spacer regions of greater length separate the ABDs of α-actinin, filamin, spectrin, and dystrophin dimers so these proteins form loose bundles and networks of actin.

Classes of actin filament–cross-linking proteins distinct from the ABD-family include fascin and villin, which are group I and group II proteins, respectively (see Figure 12.18). Fascin organizes filaments into tightly packed bundles, as **FIGURE 12.20** shows; these bundles shape filopodia such as those found in the growth cones of migrating neurons (see Figure 12.1). Villin is enriched in the microvilli of some types of epithelial cells (see Figure 12.1). In villin, as in fimbrin, the domains that bind to actin are close together, and both proteins form the cross-links between actin filaments in microvilli.

12.13 Actin and actin-binding proteins work together to drive cell migration

Key concepts

- Interactions among actin and proteins that bind actin monomers and filaments regulate the growth and organization of protrusive structures in cells.
- The addition of actin monomers to the barbed ends of actin filaments located at the cell's plasma membrane pushes the membrane outward.

Crawling of a cell along a surface requires the formation and extension of actin-rich protrusive structures, adhesion of the protrusions to the surface (also see *19 The extracellular matrix and cell adhesion*), and retraction of the cell

How does the addition of an actin monomer to a filament generate force to drive a membrane protrusion? Two "Brownian ratchet" models have been proposed to describe how actin polymerization can generate force and displace a particle or boundary. Both models require some component (the actin filament or particle being moved) to undergo random thermal (Brownian) motion, and require the actin monomer concentration to be greater than the critical concentration so that elongation is favored. However, the models differ in which component (the actin filament or the particle) diffuses.

In one model, the actin filament is immobile and sufficiently stiff so that it does not bend, and a particle (such as a vesicle) or a boundary (such as the membrane at the leading edge of motile cells) fluctuates in position due to Brownian motion, as **FIGURE 12.B1** shows. When the filament's barbed end is in contact with the membrane, filament elongation is prevented because free actin monomers cannot add to the barbed end. When diffusion of the particle or boundary occurs, a space between it and the filament's barbed end will emerge. When this space is large enough, a monomer will add to the barbed end, thus preventing the particle from diffusing back (to the left in Figure 12.B1). By preventing backward diffusion and filling in the gaps between barbed ends and the particle as they form, polymerization rectifies particle diffusion so that it moves in a vectorial manner (to the right in Figure 12.B1). The velocity of particle movement depends on how fast the particle can diffuse and the probability of adding a new actin monomer (which is dependent on the actin concentration and the probability that space is available for elongation); the faster the particle diffuses, the faster a barbed end elongates, and the faster the particle moves to the right.

A second model considers the ability of both the filament and the particle to undergo Brownian motion. Because the degree of motion is proportional to size (large molecules diffuse more slowly than small ones), particles much larger than an actin filament can be considered relatively immobile. Entire actin filaments do not diffuse very rapidly but thermal energy does cause them to bend. Therefore, filament bending can open up a space between the particle and the filament, allowing a monomer to add to the barbed end. The bent filament is under elastic strain and wants to straighten. This elastic strain exerts a force on the particle that, if large enough, will move the particle as the filament straightens.

Force generation arises from the polymerization reaction, and it does not require the energy of ATP hydrolysis. ATP hydrolysis is required to recycle the actin subunits so that the process can happen again (see *12.13 Actin and actin-binding proteins work together to drive cell migration*). Without hydrolysis and monomer recycling, this process can happen only once.

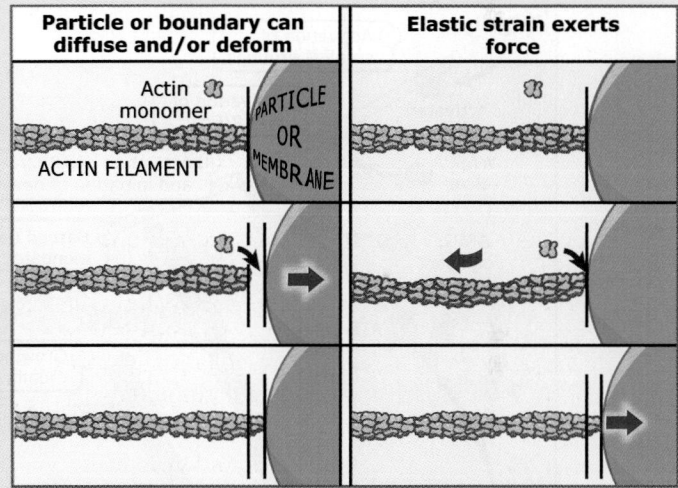

FIGURE 12.B1 Two models for how polymerization of actin filaments that are anchored at the barbed end can generate sufficient force to move a membrane or vesicle.

body. In this section, we describe how protrusion of the leading edge of a lamellipodium of a migrating cell, shown in **FIGURE 12.21**, is driven by the regulated polymerization of actin, which results in remodeling of the actin cytoskeleton. The mechanisms of lamellipodia formation and extension are well characterized. The nucleation of actin filaments by the Arp2/3 complex is needed for the formation of a lamellipodium. The subsequent addition of actin monomers to the barbed end of an actin filament located at the cell edge effectively pushes the plasma membrane forward. (For a discussion of how this might occur see the *Methods and Techniques* box on page 573.) Remarkably, motility can be reconstituted *in vitro* with purified actin, activated Arp2/3, cofilin, and capping protein.

In many cells, the lamellipodium maintains a relatively constant size as the cell migrates forward because the rate of monomer addition at the leading edge is equal to the rate

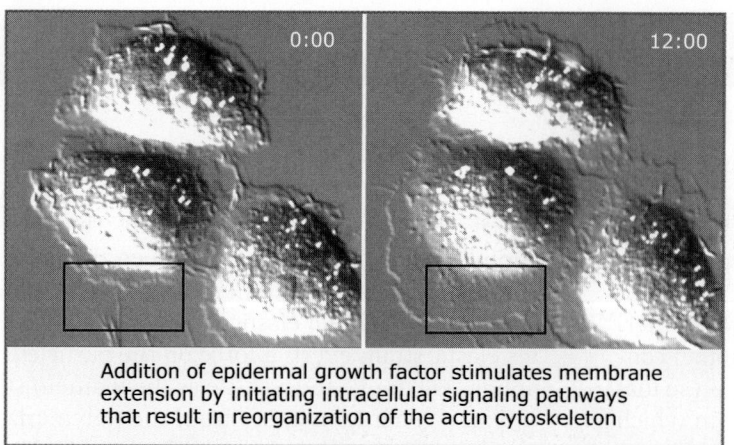

Addition of epidermal growth factor stimulates membrane extension by initiating intracellular signaling pathways that result in reorganization of the actin cytoskeleton

FIGURE 12.21 Epidermal growth factor (EGF) stimulates the migration and division of cells that express EGF receptors. These frames from a video show epithelial cells responding to EGF by membrane extension that is due to actin polymerization. Photos courtesy of E. Michael Ostap, University of Pennsylvania.

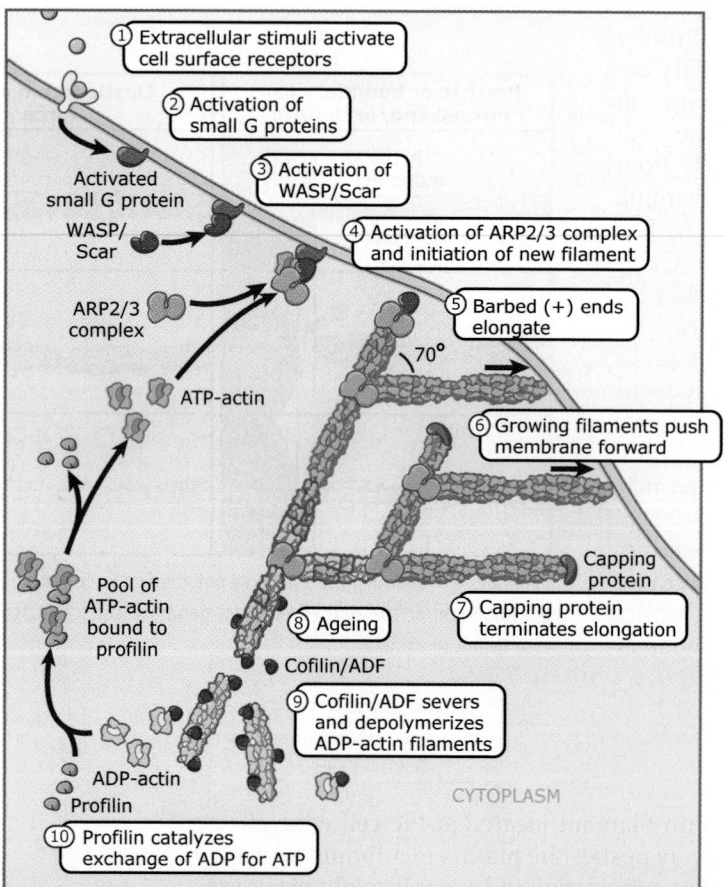

FIGURE 12.22 Cell locomotion occurs in response to extracellular signals. The dendritic nucleation model proposes how Arp2/3 nucleates the assembly of a branched actin filament network and how filament assembly drives the outward movement of the leading edge of a cell. Reproduced from Annual Reviews by Thomas Pollard, et al., Copyright 2000 by Annual Reviews, Inc. Reproduced with permission of Annual Reviews, Inc. in the format of Textbook via Copyright Clearance Center. Modified from a figure by Thomas D. Pollard, Yale University.

of filament disassembly at the rear of the lamellipodium. Interactions among actin, nucleators, capping proteins, sequestering proteins, and severing proteins regulate this growth and define the geometry of the protrusion.

The series of events that lead to remodeling of the actin cytoskeleton begin with extracellular signals that initiate signal transduction pathways, which activate WASP/Scar proteins, as illustrated in **FIGURE 12.22**. These proteins bind and activate Arp2/3 complexes, which nucleate new filaments that branch from the mother filament. The geometry of activation ensures that the barbed ends of the new filaments are oriented toward the plasma membrane.

Filaments elongate rapidly and push the membrane forward until the barbed ends are capped by capping protein. The Arp2/3 complex and newly polymerized, short, capped actin filaments form a meshwork with a morphology that is mechanically suited for persistent protrusion of the membrane over a surface (see lamellipodium in Figure 12.1).

The large soluble pool of ATP-actin complexed with thymosin β_4 supplies the monomers used for elongation. The ATP-actin monomers equilibrate between thymosin β_4 and profilin. Actin, in the profilin-actin complex, associates with the free barbed end of actin filaments. Once the profilin-actin complex binds to the barbed end, profilin dissociates, thereby allowing the addition of another monomer to the filament. The ATP bound to the polymerized actin subunits is hydrolyzed and phosphate is released slowly, resulting in ADP-actin. Phosphate release increases the rate of dissociation of the Arp2/3 complex from the actin filament; it also allows cofilin/ADF to bind ADP-actin and to sever and depolymerize the actin filament. Profilin then catalyzes the exchange of ADP for ATP on the actin monomers that have been released from the filament. The ATP-actin monomers can then bind thymosin β_4 and replenish the pool of actin monomers.

12.14 Small G proteins regulate actin polymerization

Key concepts

- Members of the Rho family of small G proteins regulate actin polymerization and dynamics.
- Activation of Rac, Cdc42, and Rho proteins induces formation of lamellipodia, filopodia, and contractile filaments, respectively.

As discussed in the previous section, the series of events that lead to remodeling of the actin cytoskeleton begins with extracellular signals that initiate signal transduction pathways, which in turn activate WASP/Scar proteins. Many extracellular factors, such as growth factors, chemoattractants, and chemorepellants that promote cell migration and changes in cell shape, bind to transmembrane receptors. Signaling through these receptors activates members of the Rho family of monomeric G proteins, with different cell-surface receptors activating different Rho-family proteins. Rho proteins are GTPases and are members of the Ras superfamily, which are activated upon the exchange of GDP for GTP (see *18.23 Small, monomeric GTP-binding proteins are multiuse switches*). At least twenty different genes encoding Rho-family proteins have been identified in humans. Based on amino acid sequence and cellular functions, these proteins can be classified into subfamilies called Rho, Rac, and Cdc42. Many of these proteins play critical roles in coordinating changes in the actin cytoskeleton in response to extracellular signals.

Proteins of the Rac and Cdc42 subfamilies promote actin reorganization by activating WASP/Scar and WAVE proteins, which in turn are activators of the Arp2/3 complex. **FIGURE 12.23** shows the result of activation of Rac, Cdc42, and Rho, each of which has different downstream targets and, therefore, induces the formation of different actin structures. Rac-like proteins induce the formation of lamellipodia and membrane ruffles, which are lamellipodia-like structures that extend and retract back to the dorsal cell surface. Rac-like proteins do not interact with WAVE proteins directly but release the WAVE proteins from inactive complexes so that they can then interact with the Arp2/3 complex. Cdc42-like proteins bind WASP directly; this binding changes WASP from an inactive form to an active form that stimulates the Arp2/3 complex. Cdc42 activation mediates the formation of actin filament bundles and filopodia extensions.

The Rho subfamily proteins stimulate the formation of contractile filaments that contain actin and myosin-II (such as those in stress fibers; see Figure 12.1). These contractile filaments are thought to sustain mechanical stresses in cells and contribute to forces required for cell shape and adhesion (for details on stress fibers and focal adhesions see *19.14 Integrin receptors participate in cell signaling*). Signaling by Rho proteins results in phosphorylation of myosin light chains, which activates nonmuscle myosin (see *12.20 Myosins are regulated by multiple mechanisms*). Rho proteins also stabilize myosin-containing filaments by activating a kinase (LIM kinase) that phosphorylates cofilin and inhibits actin filament severing and depolymerization. In addition, Rho-family proteins contribute to activation of formin-induced nucleation of actin filaments in yeast.

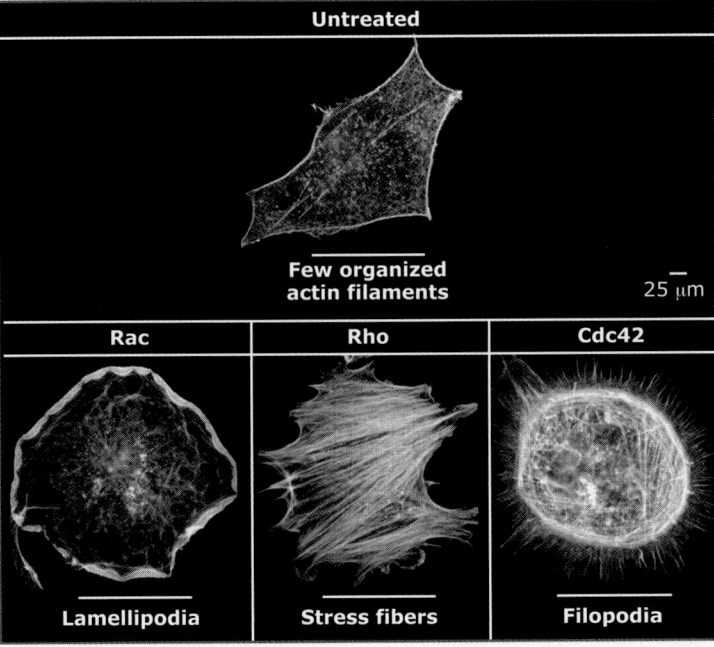

FIGURE 12.23 Activation of Rac, Rho, or Cdc42 in cells induces the formation of different types of actin structures, as revealed by fluorescence micrographs of cells stained for actin. Reproduced from A. Hall, *Science* 279 (1998): 509–514 [http://www.sciencemag.org]. Reprinted with permission from AAAS. Photos courtesy of Alan Hall, Memorial Sloan Kettering Cancer Center, New York.

12.15 Myosins are actin-based molecular motors with essential roles in many cellular processes

Myosins are molecular motor proteins that use the energy from ATP binding and hydrolysis to generate force and power motility along actin filaments. Myosins are best known for their role in muscle contraction (described in *12.21 Myosin-II functions in muscle contraction*). However, myosin expression is not limited to muscle; rather, myosins constitute a large family with members that are widely expressed in nearly all eukaryotic cell types. Different myosin family members have structural and biochemical adaptations that have evolved for specific cellular roles.

The myosin superfamily of actin-based molecular motors consists of at least eighteen classes (or families), with many classes having multiple isoforms. The myosin classes are distinguished by phylogenetic analysis of the protein sequences. All characterized myosins, with the exception of myosin-VI, move toward the barbed ends of actin filaments. The types and numbers of myosins expressed in a given cell or organism vary considerably. For example, the yeast *Saccharomyces cerevisiae* expresses five myosins from three different classes, whereas humans have thirty-nine genes from eleven families.

Members of the myosin superfamily have essential and remarkably diverse cellular functions. They have in common three domains (the head, or motor, domain, the regulatory domain, and the tail domain) that have evolved such that they are tuned for various mechanical and regulatory roles in the cell. The motor and regulatory domains power motility, and the tail domain functions in myosin assembly or attachment to other cellular components for transport (see *12.16 Myosins have three structural domains*). Much progress has been made in determining the cellular functions of myosins and in determining how the motor and tail domains mediate these functions. **FIGURE 12.24** shows the functions of the different myosin families, **FIGURE 12.25** summarizes the features of the myosin domains, and **FIGURE 12.26** shows the expression pattern of human myosins. The characterized myosins can be classified into four broad functional groups:

Myosins power muscle and cellular contractions. The isoforms of the myosin-II family generate the powerful contractile forces of skeletal, cardiac, and smooth muscles (see *12.21 Myosin-II functions in muscle contraction*). Members of the myosin-II family are also responsible for constriction of the contractile ring during cytokinesis, for cell migration, and for other whole-cell contractile events.

Myosins power membrane and vesicle transport. Microtubule-based motors carry out the long distance transport of membrane vesicles through the cytosol (for details see *11 Microtubules*). However, several myosins play essential roles in the short-range transport and the regulation of the distribution of vesicles and organelles. For example, the processive motor myosin-V transports the pigment-containing organelles that give color to skin and hair, whereas several other myosins, such as isoforms of myosin-I, -VI, -IX, and -X, function in the formation and transport of endocytic and phagocytic vesicles.

Myosins play key roles in regulating cell shape and polarity. Myosins are required for the formation and dynamics of actin-rich cell-surface specializations such as filopodia, stereocilia, and pseudopodia. For example, some myosin-I isoforms link the lipid membrane with the actin cytoskeleton and power the retraction of actin-rich membrane protrusions; myosin-II isoforms provide the contractile forces in stress fibers and cortical actin cables that give cells their shape; and myosin-VII provides a contractile link between the actin cytoskeleton and extracellular attachments.

Myosins participate in signal transduction and sensory perception pathways. Myosins participate

Myosin Family	Organisms	Cellular functions
I	Fungi; Protozoa; All metazoans	• Links actin cytoskeleton to lipid membranes • Membrane trafficking • Mechano-signal transduction • Plasma membrane dynamics

II
(Conventional myosins)
	Fungi; Protozoa; All metazoans	• Powers cardiac, skeletal, and smooth muscle contraction
		• Nonmuscle isoforms have numerous functions, including:
		  - cytokinesis;
		  - cell crawling;
		  - determination of cell shape;
		  - regulation of cell polarity

III	Vertebrates; *Limulus*; *Drosophila*	• Sensory cell function, including transport of signaling molecules in photoreceptors • Mutations of a human isoform result in deafness
IV	Acanthamoeba (a protist)	• Unknown
V	Most eukaryotic cells	• Powers short-range vesicle transport and dispersion • Role in mRNA localization
VI	Metazoans	• Role in endocytosis and membrane trafficking • May have a role in the structural organization of the actin cytoskeleton
VII	Metazoans; *Dictyostelium*	• Cell adhesion, phagocytosis, and structural organization of actin cytoskeleton • Mutations in the human gene result in deafness
VIII	Plants	• May function in endocytosis and in the organization of actin at the cell periphery
IX	Vertebrates; *C. elegans*	• Regulation of signaling activity
X	Vertebrates	• Proposed membrane-associated cargo transporter • Links phosphoinositide signaling to phagocytosis • Involved in nuclear positioning
XI	Plants; *Dictyostelium*	• Proposed cargo transporter, similar to myosin-V
XII	*C. elegans*	• Unknown
XIII	*Acetabularia* (a green alga)	• May function in organelle transport and algal tip growth
XIV	*Toxoplasma*; *Plasmodium*	• Powers gliding motility and host-cell invasion
XV	Vertebrates; *Drosophila*	• May play a role in the formation or maintenance of actin-rich structures, such as the sensory hair cells of the inner ear • Mutations result in deafness in humans and mice
XVI	Vertebrates	• May target phosphatases to specific regions in the cells of developing brains
XVII	Fungi	• May function in polarized cell wall synthesis and fungal morphogenesis
XVIII	Vertebrates; *Drosophila*	• Unknown

FIGURE 12.24 Overview of the 18 myosin families, the organisms in which they are expressed, and their functions. The functions of some myosins are not known.

in signaling pathways via association with signaling proteins. For example, myosin-I regulates the activity of some mechanically sensitive ion channels, myosin-III interacts with signaling molecules in the photoreceptors of the eye, myosin-IX is proposed to be a regulator of Rho, and myosin-XVI may target phosphatases to specific cellular regions. Myosins are also important in sensory perception. Naturally occurring mutations in the genes encoding myosins-VI, -VII, and -XV result in hearing loss due to the disruption of actin-containing structures in the sensory hair cells of the ear.

Myosin Family	Structural features	Biochemical and motor properties
I	• Single-headed molecule • All characterized isoforms have a tail domain that binds acidic phospholipids • Some isoforms contain Src homology 3 (SH3) domains and nucleotide-insensitive actin-binding sites	• Nonprocessive low duty ratio motor
II (Conventional myosins)	• The coiled-coil domain polymerizes to form bipolar filaments	• Characterized muscle isoforms are low duty ratio motors • At least one nonmuscle isoform has a relatively high duty ratio and appears to be appropriate for slow and sustained contractions
III	• Contains a kinase domain at the N-terminus of the motor domain	• Appears to be a low duty ratio motor
IV	• The tail domain includes an SH3 domain and other protein-protein interaction domains	• Unknown
V	• The coiled-coil domain dimerizes to form a two-headed molecule • Proteins have been identified that bind to the tail domain and allow organelle-specific targeting	• High duty ratio motor • Two-headed myosin-V is able to take several steps along an actin filament before detaching
VI	• The motor domain contains an insertion that allows myosin-VI to move toward the pointed end of actin filaments	• High duty ratio motor that may exist as a monomer and a dimer • Dimers are able to take several steps along an actin filament before detaching
VII	• May dimerize to form a two-headed molecule • The tail domain of at least one isoform binds the cytoskeletal protein talin	• High duty ratio motor that is able to take several steps along an actin filament before detaching
VIII	• Contains a coiled-coil domain for dimerization • The motor domain has an N-terminal extension of unknown function	• Unknown
IX	• Proposed to be a RhoGAP • The motor domain has a large insert at the actin-binding site and a second insert structurally homologous to a Ras binding domain	• Proposed to be a single-headed processive motor • May move toward the pointed end of actin filaments
X	• The tail domain contains pleckstrin homology and protein-protein interaction domains	• Low duty ratio motor
XI	• The coiled-coil domain dimerizes to form a two-headed molecule	• Processive high duty ratio motor
XII	• The motor domain contains unique insertions of unknown function • The tail domain contains protein-protein interaction domain	• Unknown
XIII	• The motor domain contains an N-terminal extension of unknown function	• Unknown
XIV	• The short tail domain has a cryptic regulatory domain	• Nonprocessive low duty ratio motor
XV	• The tail domain contains MyTH4 and FERM domains	• Unknown
XVI	• The motor contains an N-terminal extension composed of phosphatase-binding ankyrin repeats • The tail domain contains several stretches of polyproline residues	• Unknown
XVII	• The tail domain contains a chitin synthase domain	• Unknown
XVIII	• One isoform has an N-terminal extension consisting of a PDZ domain that acts as a protein interaction scaffold • The tail domain contains a coiled-coil region and a C-terminal globular tail	• Unknown

FIGURE 12.25 Overview of the structural features and kinetic properties of myosin families.

Myosin Family	Number of genes	Expression
I	8	Widely expressed
II (Conventional myosins)	14	Divided into muscle and nonmuscle genes; nonmuscle genes are widely expressed
III	2	Retina (eye), testis, kidney, intestine, and sacculus (ear)
V	3	Widely expressed
VI	1	Widely expressed
VII	2	Differentially expressed in several tissues, including cochlea (ear), retina (eye), lung, testis, and kidney
IX	2	Widely expressed
X	1	Widely expressed
XV	2	Differentially expressed in cochlea, pituitary gland, stomach, kidney, intestine, and colon
XVI	2	Brain, kidney, and liver
XVIII	2	Differentially expressed in hematopoietic (blood forming) cells, muscle, and intestine

FIGURE 12.26 Overview of the myosin family proteins that are expressed in humans and their expression patterns.

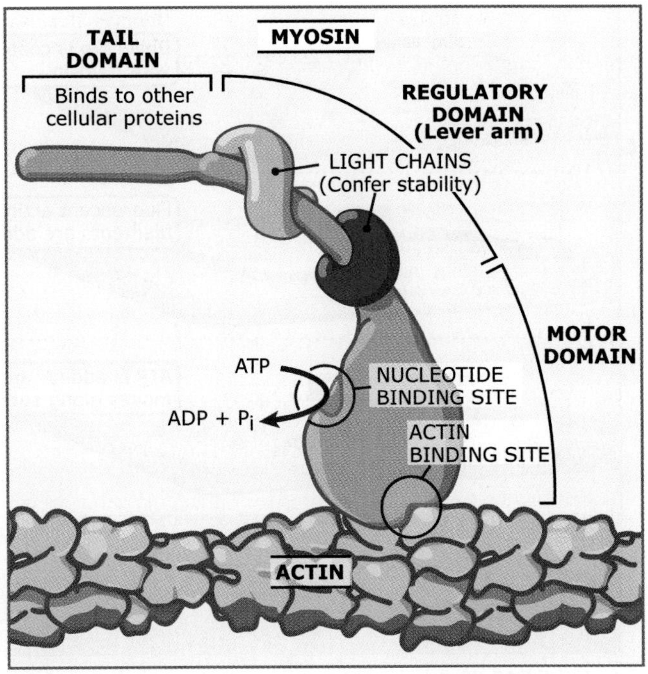

FIGURE 12.27 Myosin proteins have three structural domains (motor, regulatory, and tail domains) with different functions.

In the following sections we will discuss myosin structure, the basic force-producing properties of all characterized myosins, and how these properties relate to their cell biological roles. (For a discussion of motor proteins that interact with microtubules see *11.11 Introduction to microtubule-based motor proteins*; for a discussion of the role of myosin during cell division in yeast see *11.14 Microtubule dynamics and motors combine to generate the asymmetric organization of cells*.)

12.16 Myosins have three structural domains

Key concepts

- Myosin family members have three structural domains termed the head (or motor), regulatory, and tail domains.
- The motor domain contains the ATP- and actin-binding sites and is responsible for converting the energy from ATP hydrolysis into mechanical work.
- In most myosins, the regulatory domain acts as a force transducing lever arm.
- The tail domain of myosin interacts with cargo proteins or lipid and determines its biologic function.

Myosin family members can generally be described as having three structural domains. These are called the motor, regulatory, and tail domains and are illustrated in FIGURE 12.27. The ~80 kDa motor domain contains the ATP- and actin-binding sites and is responsible for converting chemical energy from ATP hydrolysis to mechanical work. This catalytic motor domain is highly conserved among all members of the myosin superfamily and is what defines a protein as a myosin.

The regulatory domain is a region that binds proteins known as light chains (because they are of a lower molecular weight than the myosin "heavy chain"). Most light chains are calmodulin or calmodulin-like proteins (see *18.15 Ca²⁺ signaling serves diverse purposes in all eukaryotic cells*). The identity of the associated light chains depends on the myosin isoform and the developmental stage of the organism. The number of bound light chains depends on the myosin isoform. Light chains usually remain tightly bound to the myosin and are considered subunits of the myosin molecule. Modifications of the light chains by phosphorylation or calcium binding regulate the motility and ATPase activity of some myosins (see *12.20 Myosins are regulated by multiple mechanisms*).

The motor and regulatory domains are necessary and sufficient for force generation, as demonstrated by the *in vitro* motility assay shown in FIGURE 12.28.

The X-ray crystal structure of a myosin motor and regulatory domain, shown in

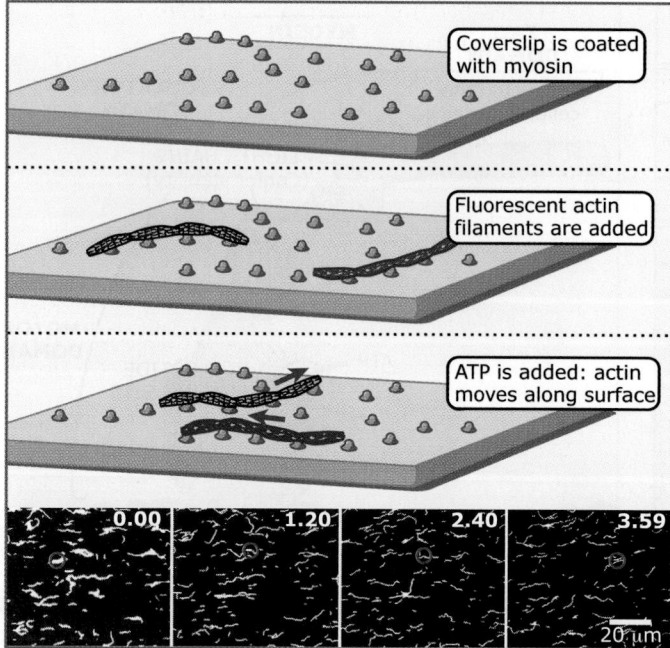

FIGURE 12.28 An experiment showing that myosin fragments containing only the motor and regulatory domains propel actin filaments in an *in vitro* motility assay. The photos are frames from a video showing movement of actin over time. A single actin filament is circled. Photos courtesy of Tianming Lin and E. Michael Ostap, University of Pennsylvania.

FIGURE 12.29, reveals important clues to the mechanism of force generation. Myosin has an elongated structure, with a motor domain consisting of a core β sheet surrounded by α helices. This structure is very similar to that of the microtubule-based motor kinesin (see Figure 11.51), despite the absence of sequence conservation between these motors (see *11.11 Introduction to microtubule-based motor proteins*). The ATP-binding site, like that in other ATPases and G proteins, binds to the phosphates of ATP and the associated magnesium ion. Nucleotide binding to the ATP-binding site alters the conformation of both the actin-binding site and the regulatory domain. The actin-binding site spans a large cleft at the end of the molecule, ~4 nm from the ATP binding site. Nucleotide binding to myosin affects the opening and closing of this cleft, which in turn affects myosin binding to actin. When ATP binds to myosin, the cleft opens slightly, thereby lowering the affinity of myosin for actin.

The regulatory domain extends from the motor domain as a long α helix (see Figure 12.29). The regulatory domain of myosin plays a crucial mechanical role in myosin function, acting as a "lever arm" for force generation. Conformational changes in the nucleotide-binding site, which are determined by whether ATP, ADP-P_i, or ADP is bound, are transmitted to the regulatory domain from the motor. These conformational changes cause rotation of the lever arm, resulting in a force-generating **powerstroke**, shown in **FIGURE 12.30**. (The

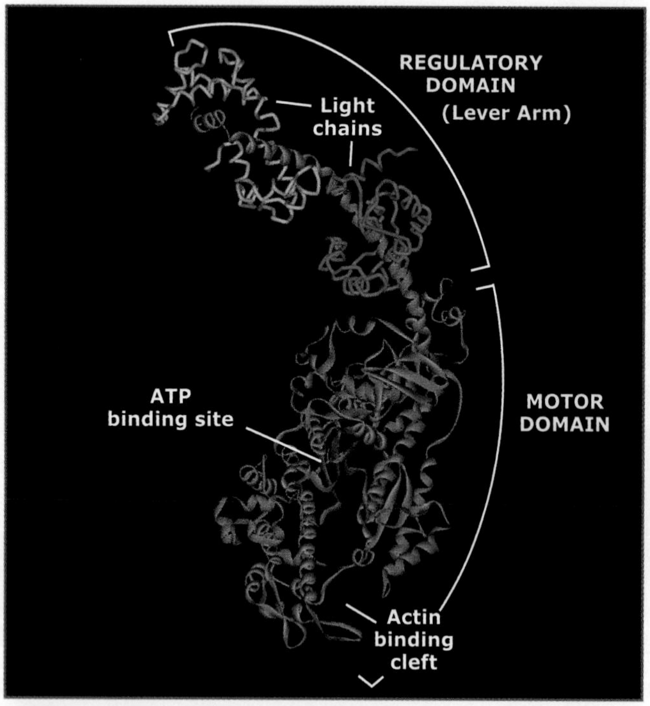

FIGURE 12.29 X-ray crystal structure of a myosin fragment, containing the motor and regulatory domains, depicted as a ribbon diagram of the protein backbone. Structure from Protein Data Bank 2MYS. I. Rayment, et al., *Science* 261 (1993): 50–58.

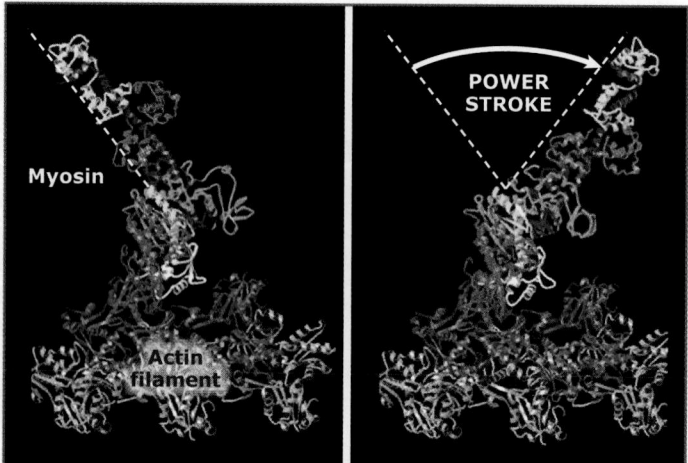

FIGURE 12.30 The regulatory domain (lever arm) of myosin undergoes a large conformational change that generates force for movement along an actin filament. ATP hydrolysis and P_i release induce a small conformational change at the actin-binding cleft that results in this large conformational change. Reproduced from K. Holmes, et al., *Phil. Trans. R. Soc.*, vol. 359, figure 7, © 2004, The Royal Society. Photos courtesy of Kenneth C. Holmes, Max Planck Institute für Medizinische Forschung.

powerstroke is discussed in more detail in *12.17 ATP hydrolysis by myosin is a multistep reaction.*) The light chains structurally stabilize the regulatory domain, allowing it to act as a stiff lever arm.

The tail domains of myosin are responsible for binding other cellular proteins and lipids and/or for self-association of the myosin molecules. The sequences and structures of myosin tail domains are highly divergent within the myosin superfamily. In many myosins, the tail domain contains well-defined subdomains known to be responsible for protein-protein interactions, as **FIGURE 12.31** shows. The tails

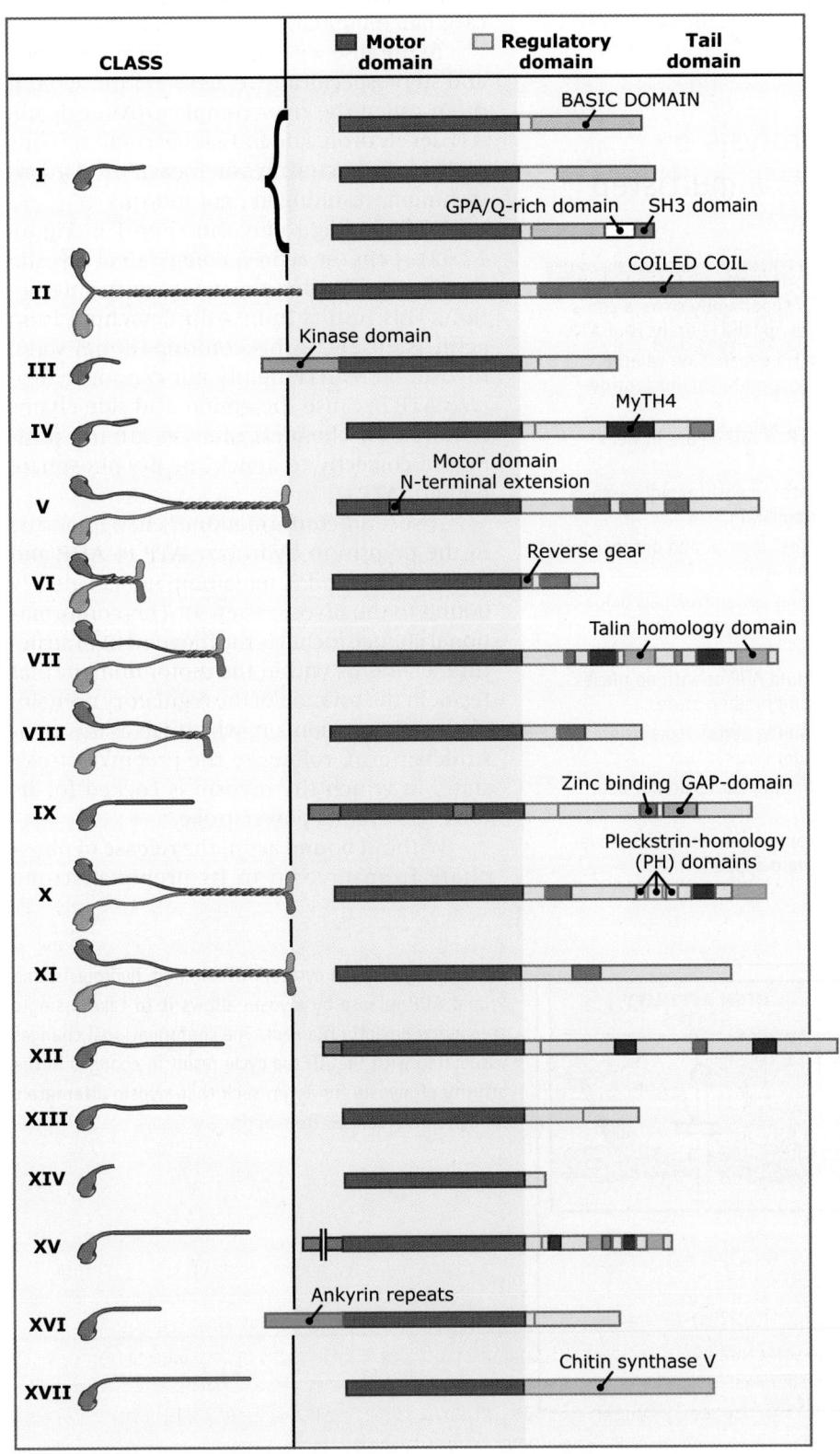

FIGURE 12.31 The tail domains of myosins confer specific functions. Some tail domains have regions that oligomerize to form myosin filaments, whereas others bind to cargo proteins or function as enzymes.

specify the nature of the transported cargo (proteins or lipids) and allow some myosins to dimerize or oligomerize into filaments, so that they possess two or more catalytic motor domains. The thick filaments of striated muscle are an example of such myosin filaments (described in *12.21 Myosin-II functions in muscle contraction*). Sequence variation in the tail domains is an adaptation to specific cellular roles for each myosin type (see Figure 12.24).

12.17 ATP hydrolysis by myosin is a multistep reaction

Key concepts

- Members of the myosin superfamily share a conserved reaction pathway for the hydrolysis of ATP.
- Myosin's affinity for actin depends on whether ATP, ADP-P_i, or ADP is bound to the nucleotide-binding site of myosin.
- Myosins with bound ATP or ADP-P_i are in weak binding states.
- In its weak binding states, myosin rapidly associates and dissociates from actin.
- ATP hydrolysis "activates" myosin and occurs while myosin is detached from actin.
- Myosin's force-generating powerstroke accompanies phosphate release after myosin-ADP-P_i rebinds actin.
- Myosins with either bound ADP or with no nucleotide bound are in strong binding states.
- Myosin in its strong binding states remains attached to actin for longer times.
- Myosins in the weak binding states do not bear force.
- Myosins in the strong binding states resist movement if external forces are applied.

All myosins that have been characterized so far share a similar overall reaction pathway for the hydrolysis of ATP to ADP and P_i. The reaction scheme outlined in **FIGURE 12.32** shows the mechanism of the myosin ATPase pathway coupled to lever arm rotation. This coupling of ATP hydrolysis with conformational changes allows myosins to move themselves and their cargo incrementally along actin filaments.

In the absence of ATP, myosin binds tightly and stereospecifically to actin to form what is often called the rigor complex. (After death, ATP levels drop, and muscles become stiff due to this strong actin-myosin [or actomyosin] attachment, resulting in rigor mortis.)

ATP binding to myosin (step 1 in Figure 12.32) opens the actin-binding cleft of myosin, thereby weakening the actomyosin interaction. This results in myosin detaching from actin (step 2). In this conformational state, myosin binds ATP tightly but cannot hydrolyze ATP because the amino acid side chains required for chemical catalysis are not positioned correctly to attack the β-γ phosphate bond of ATP.

A second conformational change occurs in the myosin to hydrolyze ATP to ADP and P_i, with ADP and P_i remaining noncovalently bound to the myosin (step 3). This conformational change includes the movement of structural elements within the motor domain that result in the rotation of the regulatory domain. The regulatory domain, which acts as the myosin lever arm, rotates to the prepowerstroke state, in which the myosin is cocked for its force-generating powerstroke.

Without bound actin, the release of phosphate from myosin in its prepowerstroke

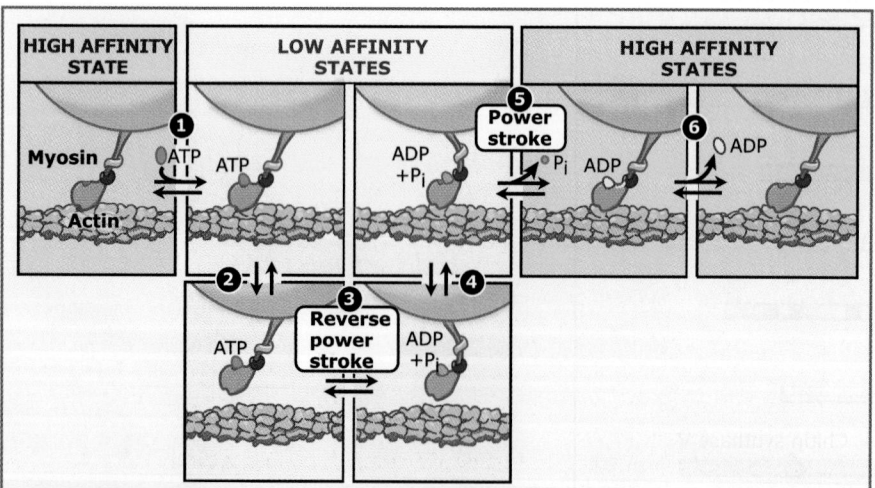

FIGURE 12.32 Each cycle of ATP binding, hydrolysis, and P_i and ADP release by myosin allows it to take a single step along an actin filament. The conformational changes associated with the ATPase cycle result in changes in the affinity of myosin for actin, such that myosin alternately binds to and releases from actin.

state is very slow. This ensures that myosin does not waste ATP when it is not interacting with actin. Upon binding to actin (step 4), the rate of phosphate release from myosin is greatly accelerated. Myosin's force-generating powerstroke, that is, rotation of the regulatory domain (step 5), accompanies the phosphate release step. Upon completion of the power-stroke, ADP is released (step 6), resulting in actin-attached myosin that lacks nucleotide. The cycle repeats upon ATP binding to myosin. Depending on the type of myosin, each round of ATP binding, ATP hydrolysis, and phosphate release results in movement of between 5 and 25 nm along an actin filament (see *12.19 Myosins take nanometer steps and generate piconewton forces*).

The affinity of myosin for actin changes during the ATP hydrolysis cycle. The biochemical intermediates in this cycle can be defined according to their affinities for actin filaments. Myosins that have bound ATP or ADP-P$_i$ are in the "weak binding" or "preforce generating" states (steps 2 and 4) that attach and detach from the actin filament on a timescale >100 times faster than the ATP hydrolysis cycle. These weak binding states cannot bear force. If the actin filament is pulled relative to myosins in these states, the myosins slide past the filament because the myosins attach and detach from actin very rapidly. It is in the weak binding states that myosin rotates to its prepowerstroke state, thus cocking itself for the force-generating step. If myosin underwent this reverse powerstroke step while attached to an actin filament, it would move backward.

Myosin-ADP and nucleotide-free myosin bind tightly to a specific site on the actin subunits in a filament. These strong binding states of myosin are the "force-bearing" intermediates. It is in the strong binding states that myosin undergoes its powerstroke. If an actin filament were pulled, myosin that is strongly bound to it will resist movement and effectively act as an actin filament anchor. (For a discussion of the mechanism by which microtubule-based motors function see *11.12 How motor proteins work*.)

Concept and Reasoning Check

1. Compare the general features of myosins and kinesins in how each motor type generates force from ATP hydrolysis.

12.18 Myosin motors have kinetic properties suited for their cellular roles

Key concepts

- The ATPase cycle mechanism is conserved among all myosins.
- The ATPase cycle kinetics of different myosins are tuned for specific biologic functions.
- Myosins with high-duty ratios spend a large fraction of their cycle time attached to actin.
- Low duty ratio myosins spend most of their time detached from actin.
- Some high-duty ratio myosins are processive and "walk" along actin filaments for long distances.

All myosins undergo the mechanochemical pathway described in the previous section. However, the motor domains of the different myosin isoforms have biochemical and mechanical properties that have evolved to be "tuned" for specific cellular roles, as summarized in Figure 12.24 and Figure 12.25. For example, myosins that work individually or in small groups to move organelles long distances over actin filaments have different properties than myosins that work in organized filaments to generate large forces and support rapid rates of contraction (e.g., in muscle). The specific properties of the different myosins are conferred by differences in the rate constants that define the transitions between the steps in the ATPase cycle, rather than by differences in the overall ATPase reaction pathway. Different rate constants result in different rates at which myosins go through the ATPase cycle and in different myosin **duty ratios**. The duty

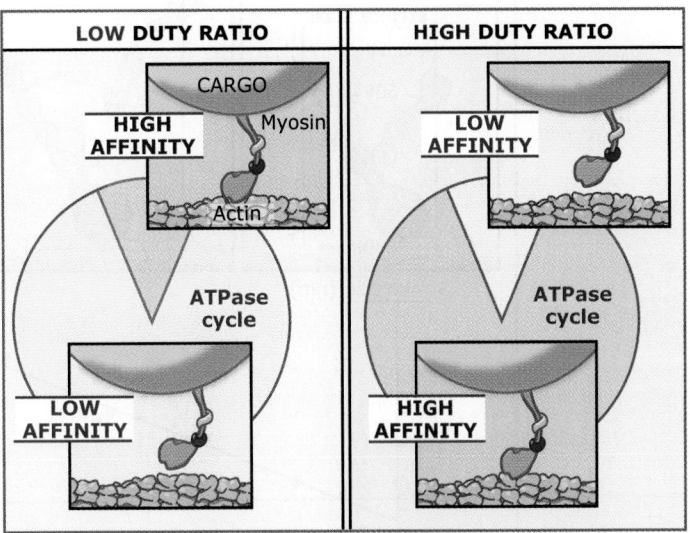

FIGURE 12.33 The duty ratio is the fraction of time during the ATPase cycle that a myosin motor is bound to actin.

ratio can be defined as the fraction of the total ATPase cycle time that an individual motor is attached to actin, usually in the strong binding state, as **FIGURE 12.33** shows.

Myosins that work in large assemblies of many motors usually have low duty ratios. In a myosin filament that contains hundreds of motor domains (e.g., in the thick filaments of muscle), each motor needs to attach to actin, perform a powerstroke, and detach from actin rapidly without interfering with other myosins in the same assembly. If one of the myosin motors remains bound to actin too long, it would impede other myosins in the macromolecular assembly from performing their powerstrokes, and inhibit rapid sliding such as occurs during muscle contraction (see *12.21 Myosin-II functions in muscle contraction*).

Myosins with high-duty ratios are able to take multiple steps along an actin filament before detaching. Consider a dimeric myosin (such as myosin-V) transporting an organelle along an actin filament. If both of its motor domains were to detach from actin, thermal forces could cause the myosin molecule to diffuse away from its track. Therefore, in order for transport to occur, at least one motor must remain attached to actin at all times. These processive two-headed myosins are thought to coordinate the ATPase activities of the two heads to allow hand-over-hand walking activity that is well synchronized.

12.19 Myosins take nanometer steps and generate piconewton forces

Key concepts

- A single myosin motor generates enough force (several piconewtons) to transport biologic molecules and vesicles.
- The stroke size of a myosin is proportional to the length of its "lever arm."

Although myosin is an enzyme that catalyzes the hydrolysis of ATP into ADP and P_i, the product of biologic consequence is not ADP and P_i, but rather, force generation. Under physiologic solution conditions, the amount of work available from the hydrolysis of a molecule of ATP is approximately 80 pN-nm, where pN is the abbreviation for a piconewton (10^{-12} newtons) and nm is a nanometer (10^{-9} meters). For reference, one pN is approximately the weight of a single red blood cell, and 1 nm is about the radius of a strand of DNA. Myosins are able to convert ~50% of the total energy from ATP hydrolysis into useful work (the rest of the energy is dispersed as heat), so myosin actually generates ~40 pN-nm per ATP. This may not seem like a lot of work, yet this energy is 10-fold greater than thermal energy at body temperature, and it is enough energy to allow a myosin molecule to translocate an organelle

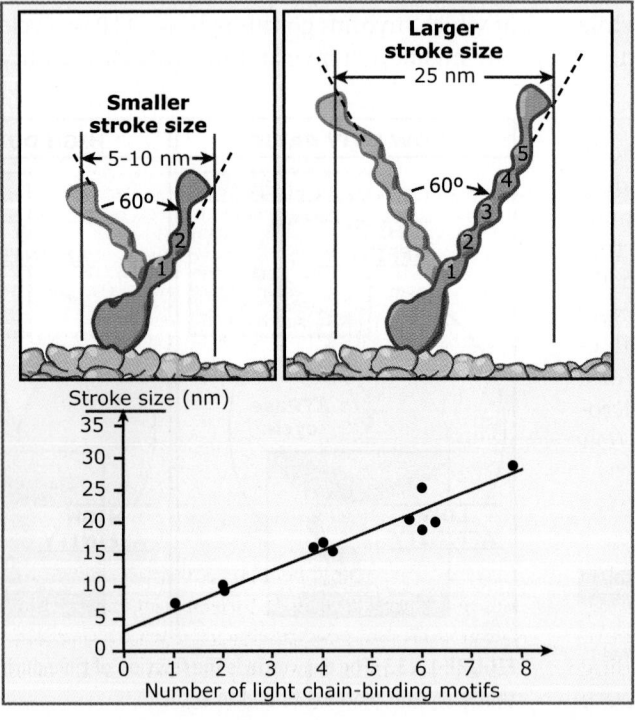

FIGURE 12.34 The distance traversed by myosin per ATP hydrolyzed depends on the length of the lever arm (regulatory domain). Adapted from D. Washaw, *J. Muscle Res. Cell Motil.* 25 (2004): 467–474.

through the cellular cytoplasm. However, many millions of myosins must work together in muscle tissues to do work on the scale with which we are more familiar; for example, lifting a full grocery bag from the floor to the table requires ~50 N-m of energy.

The amount of force generated by single myosin molecules has been determined using optical trap force spectroscopy. This technique allows measurement of the force exerted by a single myosin molecule as it moves along an actin filament. Different myosin isoforms generate approximately 5-10 pN of force per ATP hydrolyzed (unitary force). Although all myosins generate about the same amount of force, the amount of time that different myosins remain strongly bound to the actin filament after undergoing its powerstroke varies significantly, as predicted by biochemical kinetic measurements.

In addition, the unitary stroke size, that is, the distance that a myosin moves per ATP hydrolyzed, varies significantly. Single molecule measurements have also allowed for the measurement of the unitary stroke size and have provided support for the lever arm mechanism of myosin motility. The myosin stroke size also depends on the angle that the regulatory domain rotates through, and **FIGURE 12.34** shows that myosins with long regulatory domains have larger stroke sizes than myosins with short regulatory domains. The sliding velocity (v) of a myosin is equal to the distance translocated (d) per unit time (t), and $v = d/t$. Therefore, a myosin that has a large stroke will have a faster sliding velocity than a myosin having the same ATPase rate but a smaller stroke size.

12.20 Myosins are regulated by multiple mechanisms

Key concepts

- The force-generating activity and cellular localization of myosins are regulated.
- Myosin function is regulated by phosphorylation and by interactions with actin- and myosin-binding proteins.

The force-generating activity and cellular localization of myosins are regulated by their phosphorylation and by their interactions with actin- and myosin-binding proteins. The most extensively characterized regulators of myosin activity are the actin filament-binding proteins, tropomyosin and troponin, which are expressed in striated muscle cells (for details see *12.21 Myosin-II functions in muscle contraction*). In this section, we will discuss the regulation of myosins via their regulatory and tail domains.

The activities of some myosins are regulated by phosphorylation of the myosin light chains that bind to the regulatory domain. The best characterized of these are cytoplasmic and smooth muscle myosin-II. Each heavy chain of these myosins binds to one light chain called the essential light chain and another called the regulatory light chain. **FIGURE 12.35** shows that phosphorylation of the regulatory light chains by myosin light chain kinase (MLCK) activates myosin, and dephosphorylation by myosin light chain phosphatase

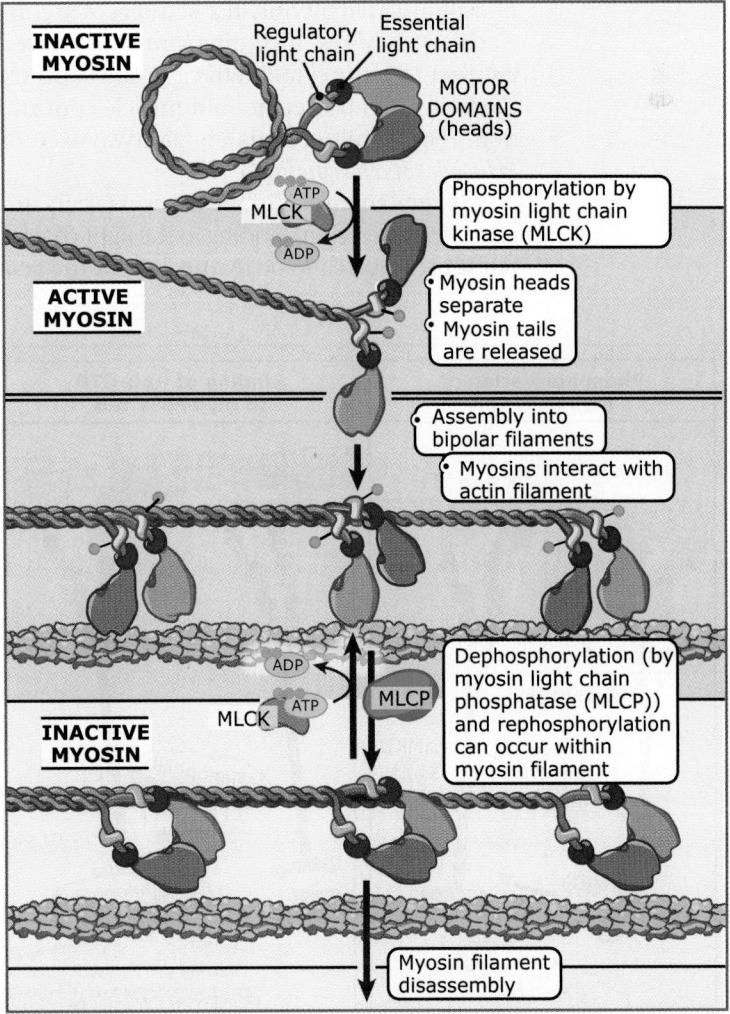

FIGURE 12.35 Phosphorylation of nonmuscle myosin-II by myosin light chain kinase activates myosin-II by promoting myosin filament formation and allowing myosin to interact with actin. When dephosphorylated by myosin light chain phosphatase, the two motor domains of the myosin molecule interact in a manner that inhibits their activity.

(MLCP) inhibits myosin. When the regulatory light chain is unphosphorylated, the two motor domains of the myosin molecule interact in a manner that inhibits their ability to hydrolyze ATP and bind actin. Additionally, it has been shown *in vitro* that the unphosphorylated myosin molecule can adopt a "folded" conformation, which may inhibit myosin filament formation. However, it is not known if this folded conformation exists within the cell. Phosphorylation of the regulatory light chain disrupts inhibitory interactions within the myosin, induces myosin filament formation, and activates the myosin ATPase activity.

Signal transduction pathways involving calcium and the Rho family of small G proteins regulate the activities of MLCK and MLCP, thus affecting the total amounts of phosphorylated and dephosphorylated myosin-II. Phosphorylated myosin-II assembles into contractile structures, including stress fibers (see Figure 12.1), the cytokinetic ring, and contractile fibers that power smooth muscle contraction. (For details of signaling pathways see *18 Principles of cell signaling*.)

Several members of the myosin family are regulated by calcium binding to the light chains, but there is much to be learned about the cellular significance of this mode of regulation. However, it is clear that calcium modulates the ATPase and motility activities of myosin in response to conformational changes in the light chains.

The tail domains of some myosins bind to cargo, such as intracellular membranes and proteins, which are transported by myosin along actin filaments. The interactions of myosin with this transported cargo can be regulated by phosphorylation of the myosin tail or by association with myosin-binding proteins. The tail of myosin-V, a two-headed myosin that transports cargo, is regulated by both of these mechanisms, as shown in **FIGURE 12.36**. Phosphorylation of the myosin-V tail weakens binding to cellular cargo, causing dissociation of the complex. Dephosphorylation of myosin results in cargo binding and transport along actin filaments. Myosin-V also interacts with its cellular cargo via association with the Rab family of small G proteins. Myosin-V binds to GTP-bound Rab with high affinity but dissociates from GDP-bound Rab. When myosin-V is not bound to cargo, it may adopt a folded conformation that inactivates its ATPase activity.

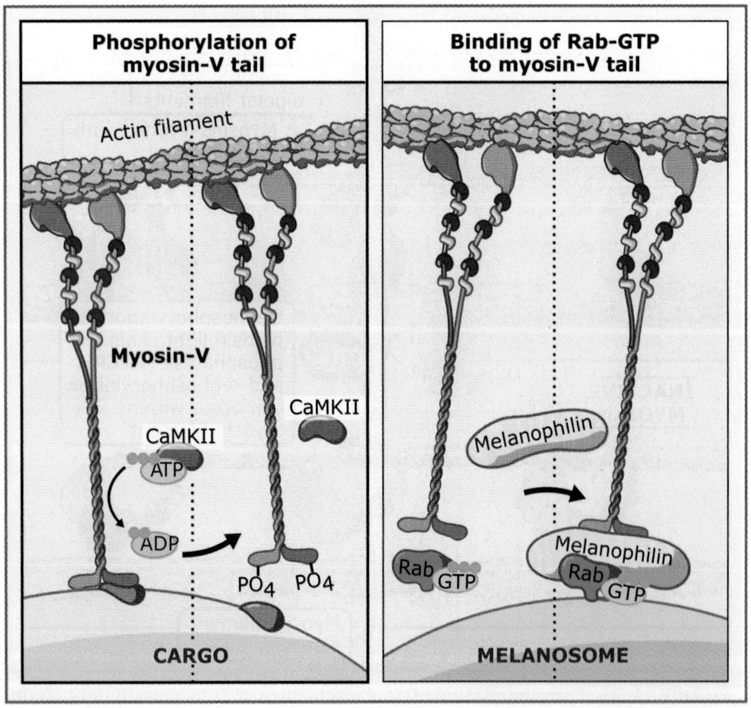

12.21 Myosin-II functions in muscle contraction

Key concepts

- Myosin-II is the motor that powers muscle contraction.
- Actin and myosin-II are the major components of the sarcomere, the fundamental contractile unit of striated muscle.

Muscles are contractile tissues that power movements of a body and movements within a body. Muscles can be divided into two major classes based on the appearance of their contractile fibers: striated and smooth. Striated muscle fibers are so named because of their striped appearance when viewed under high magnification, and include skeletal and cardiac muscles. Skeletal muscles are responsible for skeletal movements, and cardiac muscles are responsible for contraction of the heart. Smooth muscle fibers are not striated, are spindle shaped, and are found in the walls of organs such as the bladder, blood vessels, and gastrointestinal tract.

The motor protein that powers smooth and striated muscle contraction is a member of the myosin-II family (see Figure 12.24). Myosin-II

FIGURE 12.36 Myosin-V associates with and transports membrane-bounded vesicles along actin filaments. For some vesicles, phosphorylation regulates myosin-V binding, whereas for others, such as melanosomes, Rab proteins regulate binding.

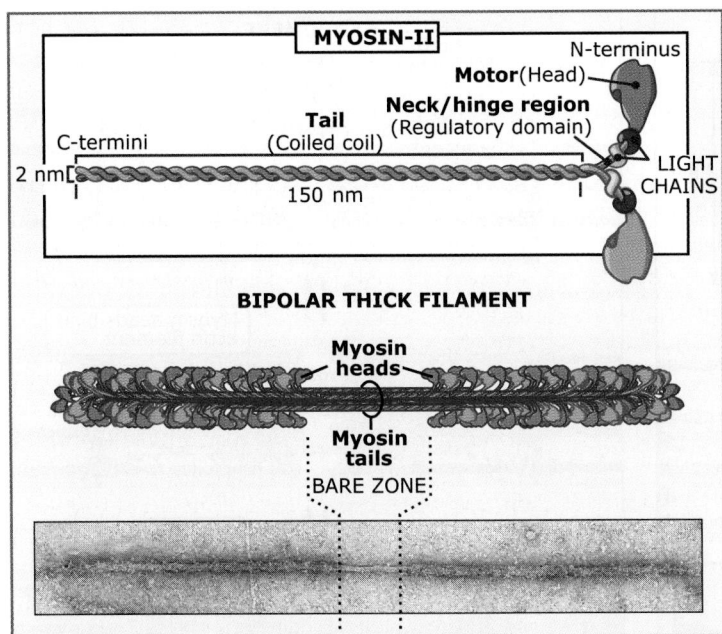

FIGURE 12.37 Myosin-II is a hexameric complex consisting of two heavy chains and two pairs of different light chains. These complexes assemble into bipolar thick filaments. Photo courtesy of Andrea Weisberg and Saul Winegrad, University of Pennsylvania.

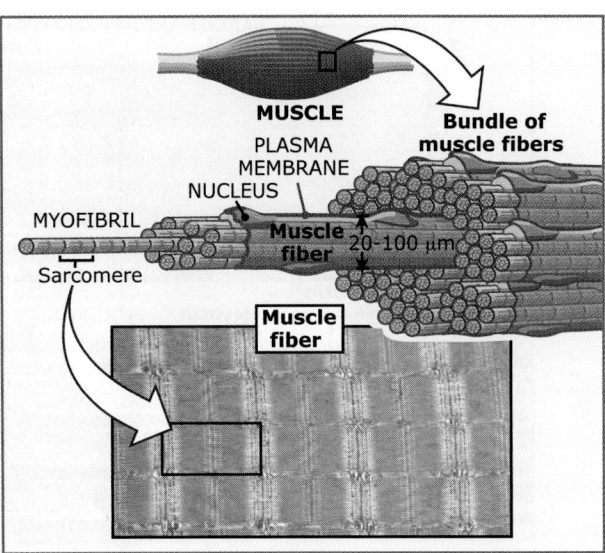

FIGURE 12.38 Skeletal muscle consists of muscle fiber cells, which contain myofibrils that run the length of the cell. Myofibrils are the contractile apparatuses, and they have repeating units called sarcomeres. Photo courtesy of Clara Franzini-Armstrong, University of Pennsylvania, School of Medicine.

is one of the most abundant proteins in vertebrates, is easy to isolate biochemically, and consequently is one of the best-characterized myosins. FIGURE 12.37 shows that a single myosin-II molecule is composed of six polypeptide chains: two heavy chains, and two sets of two light chains. The distal region of the muscle myosin-II tail associates with other myosin-II molecules to form filaments that consist of ~300 myosin molecules. Myosin filaments are bipolar, with the motor domains of all myosins directed away from a central bare zone. These myosin filaments are called **bipolar thick filaments**. In the remainder of this section, we discuss the organization and role of myosin-II in contraction of striated muscle, which are well understood.

Striated muscle tissue consists of bundles of muscle fibers. Muscle fibers are large multinucleate cells, having lengths that range from a few millimeters to many centimeters, and diameters between 20–100 μm. FIGURE 12.38 shows that each muscle fiber contains >1000 myofibrils, which are rod-shaped contractile organelles. Myofibrils are composed of repeating units called **sarcomeres** that are arranged end-to-end and give the muscle its striated appearance.

Sarcomeres are the fundamental units of striated muscle contraction. These structures contain actin and myosin-II, and they change their length during muscle contraction and relaxation. The ends of sarcomeres are defined by structures called Z-discs, which are aligned throughout the myofibril via the intermediate filament, desmin. The Z-discs at the outer edges of muscle cells are linked to multi-protein complexes called costameres, which couple force-generating sarcomeres with the muscle membrane and extracellular matrix. Proteins in this complex include those found in focal adhesions (see Figure 12.39).

Sarcomeres contain thick filaments, which are composed mainly of bipolar myosin-II filaments, and thin filaments, which contain actin filaments and actin regulatory proteins, as FIGURE 12.39 shows. The barbed ends of actin filaments are anchored on either side of the sarcomere to the Z-disc, such that all of the actin filaments on one side of the Z-disc have the same polarity. The actin filaments are anchored to the Z-disc and are capped by binding to capping protein (CapZ), which prevents depolymerization of the actin filaments. The pointed ends of the actin filaments are oriented toward the center of the sarcomere and are capped by tropomodulin. The protein nebulin also interacts with the actin filaments; it may regulate the assembly and the length of the thin filaments.

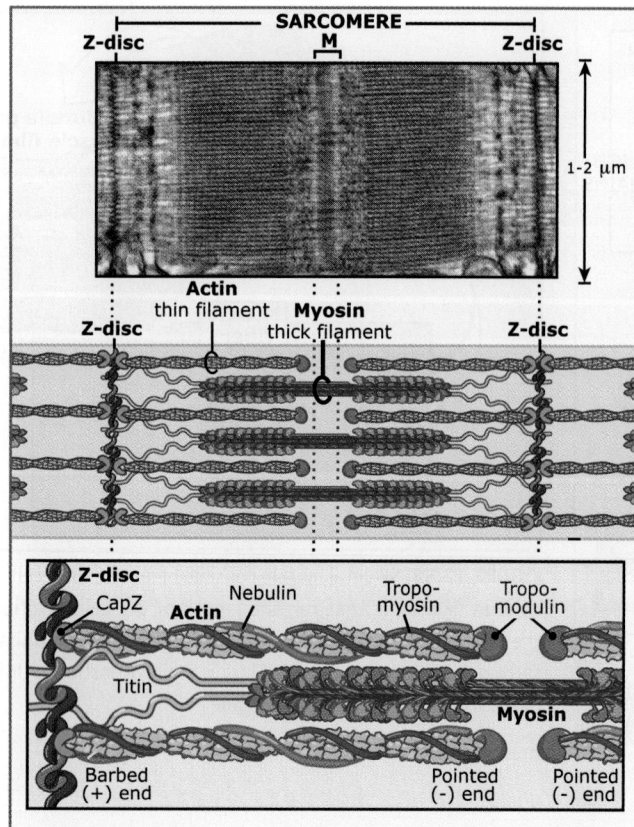

FIGURE 12.39 The ends of each sarcomere are defined by the Z-disc, to which actin filaments are attached via CapZ (capping protein). Myosin thick filaments connect to the Z-disc via the protein titin and interdigitate between actin filaments. Nebulin extends from Z-disc to tropomodulin. Photo courtesy of Clara Franzini-Armstrong, University of Pennsylvania, School of Medicine.

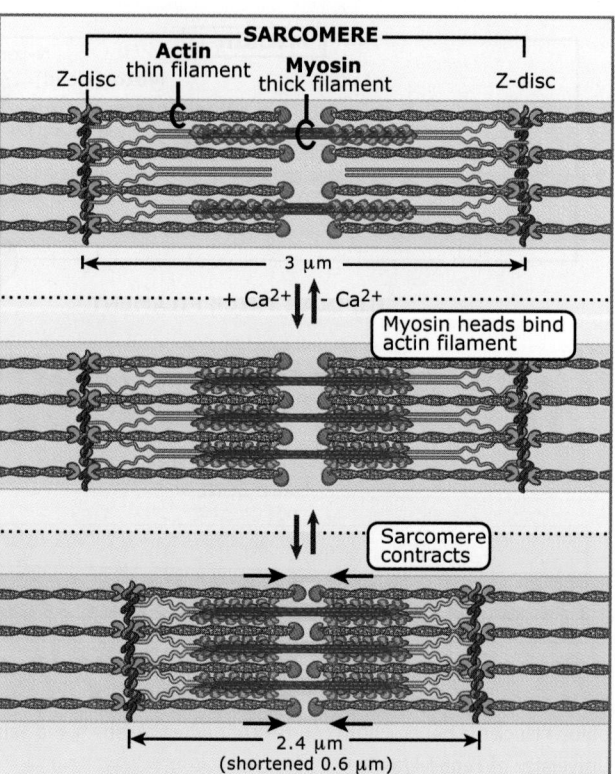

FIGURE 12.40 Muscle contraction occurs when the myosin thick filaments attach to and pull on the actin filaments so that the Z-discs move toward each other, thereby decreasing the length of the sarcomere.

The thick filaments are centered between the Z-discs on the M-line. The M-line is a structure that consists of flexible links between the bipolar thick filaments; these links hold the thick filaments in register in a hexagonal array. In addition, titin, a giant filamentous protein, forms elastic connections between the Z-discs and the myosin filament. Titin keeps the thick filaments centered in the sarcomere and acts as a spring that resists overstretching of the sarcomere.

The thick and thin filaments interdigitate in a precise three-dimensional lattice. The sarcomere is bipolar, so that the orientation of the myosin motor relative to actin is the same in both halves. During contraction, the myosin motor domains reach from the thick filament to interact with the actin in the thin filaments, as **FIGURE 12.40** shows. Contraction shortens the sarcomere by the sliding of the thick and thin filaments past each other, pulling the Z-discs

toward the center of the sarcomere. The thick and thin filaments maintain constant lengths as the myosin heads move toward the barbed ends of the actin filaments. In vertebrates, sarcomeres in relaxed muscle are ~3 μm wide and shorten to ~2.4 μm during contraction.

Thousands of sarcomeres within a muscle fiber shorten in series, causing the entire muscle to shorten. Two factors determine the total length that a muscle fiber shortens: the length by which each sarcomere shortens and the number of sarcomeres in series, as **FIGURE 12.41** shows. The percentage by which muscle fibers are shortened is the same, regardless of the length of the fiber.

The amount of force that a sarcomere generates is proportional to the number of actin-myosin interactions in a half-sarcomere, and the amount of force a muscle fiber is able to produce is proportional to the number of sarcomeres in parallel. Thus, weightlifters increase their strength by increasing the cross-sectional area, not the length, of their muscles.

Striated muscle contraction is regulated by the troponin-tropomyosin complex that is

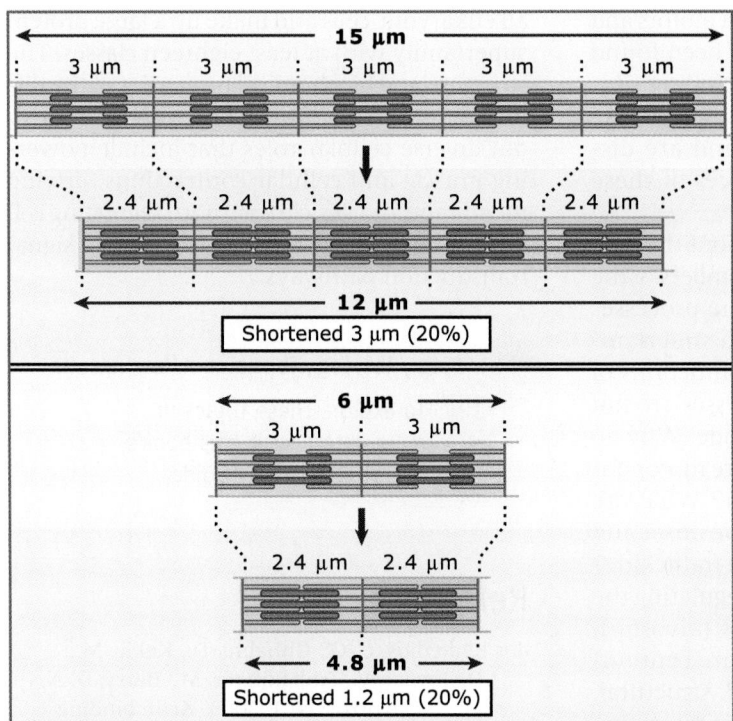

FIGURE 12.41 The longer the myofibril (that is, the more sarcomeres), the more the myofibril decreases in length upon contraction. However, the percent decrease is independent of sarcomere number.

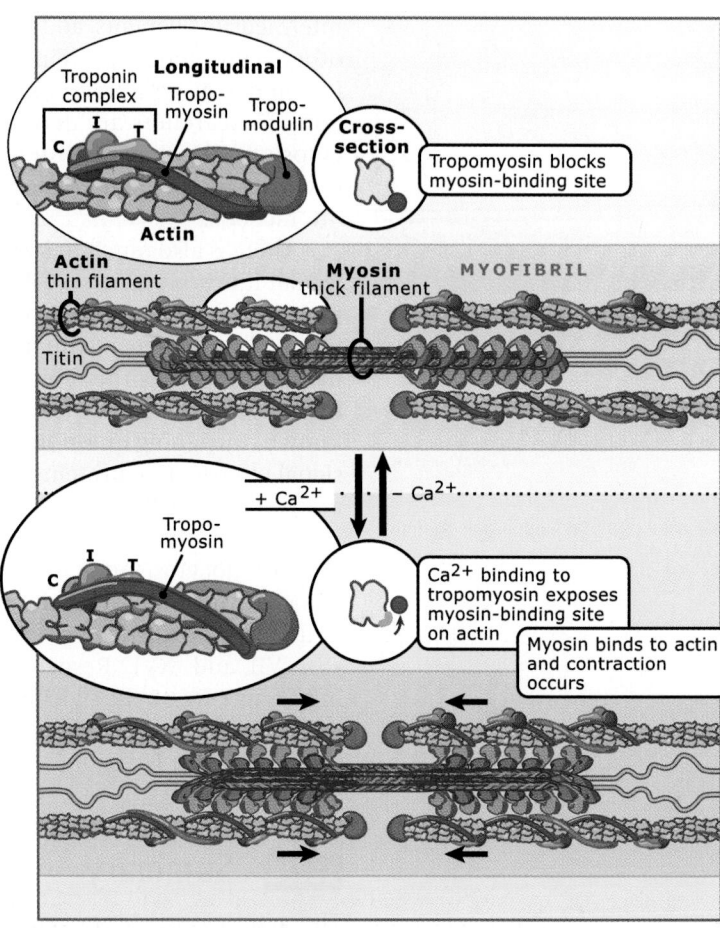

FIGURE 12.42 The cytosolic Ca²⁺ level regulates the contraction of striated muscle by controlling the position of the troponin/tropomyosin complex in relation to myosin and actin.

bound to actin in the thin filaments, as FIGURE 12.42 shows. Tropomyosin molecules are 40 nm long, coiled-coil polypeptides that lie end-to-end along the actin helix. Troponin is a complex of three different proteins: troponin-C, troponin-I, and troponin-T. One troponin complex binds to each tropomyosin so that they are spaced at 40 nm intervals along the thin filament.

At low calcium concentrations, the tropomyosin is in a position that sterically blocks the myosin-binding site on actin, so the muscle is relaxed and ATP hydrolysis by myosin occurs at a very low rate. Relaxed sarcomeres can be stretched passively with little resistance from actin-myosin interactions.

Nerve impulses trigger calcium release from the sarcoplasmic reticulum, a calcium storage organelle in muscle, to the cytosol. At the resulting higher calcium concentration, calcium binds to troponin-C, inducing a conformational change in troponin. This causes tropomyosin to move away from the myosin-binding site on actin, enabling myosin to interact with actin and produce force through its mechanochemical cycle (see Figure 12.32). (For details on the sarcoplasmic reticulum and calcium release see *7.24 The endoplasmic recticulum is*

morphologically and functionally subdivided and *6.13 Cardiac and skeletal muscles are activated by excitation-contraction coupling.*)

12.22 What's next?

Remarkable progress has been made in determining the mechanism and control of actin polymerization. However, there is still much to learn about how different signal transduction pathways specify the organization of the actin cytoskeleton into functionally distinct structures. For example, which nucleating, capping, and cross-linking proteins are required to make filopodia? Which proteins are required to organize a contractile ring?

The three cytoskeletal filament networks (actin filament, microtubule, and intermediate filament) interact extensively. For example, some translocating vesicles and organelles can move along actin filaments and microtubules, microtubule motors bind and transport intermediate filaments, and myosin motors and other actin-binding proteins have been found in complexes with microtubule-binding proteins. Researchers are determining how the cytoskeletal networks interact and are discovering the cellular consequences of these interactions.

There is also much to learn about the molecular functions of most of the members of the myosin superfamily. We know the processes in which members of the myosin superfamily participate, yet the molecular functions of the motor proteins in these processes are not known. Intriguing questions include: Why do signal transduction proteins require motor domains (myosin-II, -IX, and -XVI)? What are the roles of myosins, such as myosins-I and -VI, in endocytosis and membrane trafficking? What are the roles of myosins in regulating the structure of actin-rich protrusions (myosin-I, -VI, -VII, and -XV)? Researchers are continuing to investigate the biochemical, structural, and cellular properties of myosins to determine their molecular function.

12.23 Summary

The actin cytoskeleton is a mechanical support that allows the cell to dynamically define its shape, migrate, and reposition its internal structures and organelles. These dynamic properties are mediated via the assembly and disassembly of actin filaments, which are structurally polarized polymers that are composed of actin monomers. The assembly of actin filaments from monomers, and the disassembly of filaments to monomers, is tightly controlled within the cell by a multitude of actin-binding proteins.

Actin-binding proteins regulate polymerization of new actin filaments, prevent polymerization of actin monomers, control filament length, and crosslink actin filaments. The interaction of signal transduction pathways with actin-binding proteins provides a mechanism for controlling the dynamics and structure of the cytoskeleton.

Myosins are actin-binding proteins that use the energy released by ATP hydrolysis to perform mechanical work. Myosins are found in
all eukaryotic cells and make up a large protein superfamily with at least eighteen classes. The structure and biochemical properties of the different myosin isoforms have evolved to carry out diverse cellular roles that include powering muscle and cellular contractions, driving membrane and vesicle transport, regulating cell shape and polarity, and participating in signal transduction pathways.

http://biology.jbpub.com/lewin/cells

To explore these topics in more detail, visit this book's Interactive Student Study Guide.

References

dos Remedios, C. G., Chhabra, D., Kekic, M., Dedova, I. V., Tsubakihara, M., Berry, D. A., and Nosworthy, N. J. 2003. Actin binding proteins: Regulation of cytoskeletal microfilaments. *Physiol. Rev.* v. 83 p. 433–473.

Hirokawa, N., and Takemura, R. 2003. Biochemical and molecular characterization of diseases linked to motor proteins. *Trends Biochem. Sci.* v. 28 p. 558–565.

Holmes, K. C., and Geeves, M. A. 2000. The structural basis of muscle contraction. *Philos. Trans. R. Soc. Lond. B Biol. Sci.* v. 355 p. 419–431.

Holmes, K. C., Popp, D., Gebhard, W., and Kabsch, W. 1990. Atomic model of the actin filament. *Nature* v. 347 p. 44–49.

Howard, J. 2001. *Mechanics of Motor Proteins and the Cytoskeleton.* Sunderland, MA: Sinauer Associates.

Krendel, M., and Mooseker, M. S. 2005. Myosins: Tails (and heads) of functional diversity. *Physiology (Bethesda)* v. 20 p. 239–251.

Oda, T., Iwasa, M., Aihara, T., Maéda, Y., and Narita, A. 2009. The nature of the globular-to fibrous-actin transition. *Nature* v. 457 p. 441–445.

Pollard, T. D., and Borisy, G. G. 2003. Cellular motility driven by assembly and disassembly of actin filaments. *Cell* v. 112 p. 453–465.

Vale, R. D., and Milligan, R. A. 2000. The way things move: Looking under the hood of molecular motor proteins. *Science* v. 288 p. 88–95.

Wegner, A. 1976. Head to tail polymerization of actin. *J. Mol. Biol.* v. 108 p. 139–150.

Intermediate filaments

13

Birgit Lane
Institute of Medical Biology, Singapore

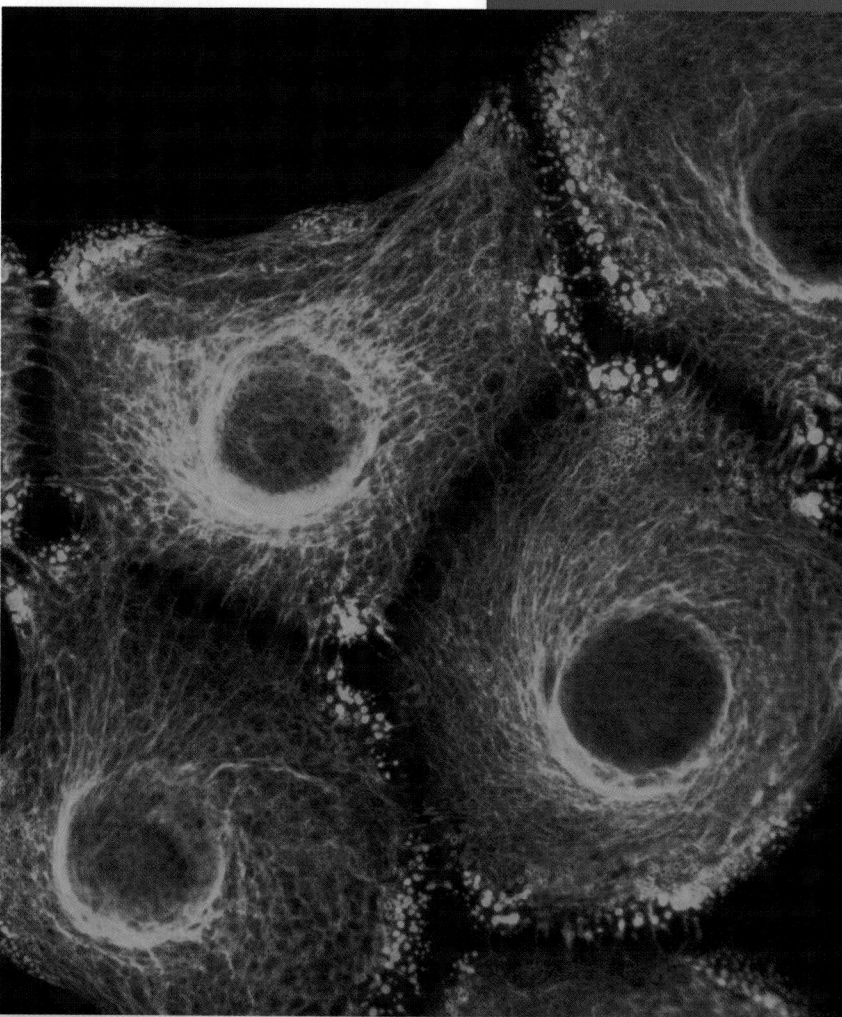

KERATIN MUTATIONS that cause skin fragility *in vivo* lead to filament breakdown on mechanical stress in culture. Reproduced with permission from *J. Cell Sci.*, vol. 117 (22): 5233–5243. [http://jcs.biologists.org/cgi/content/abstract/117/22/5233]. Photo courtesy of Birgit Lane, Centre for Molecular Medicine, Singapore.

CHAPTER OUTLINE

13.1 Introduction

Key concepts

- Intermediate filaments, or nanofilaments, are major components of the cytoplasmic and nuclear cytoskeletons.
- Intermediate filaments are 8-12 nm thick and form resilient networks in the cytoplasm and nucleus.
- Intermediate filaments are essential to maintain correct tissue structure and function.
- Intermediate filament proteins are heterogeneous and are encoded by a large and complex gene superfamily.
- Defective intermediate filaments are the cause of many human diseases.

Animal cells have no outer reinforcing cell wall, yet they are highly mobile and many—like us— live in a dry, abrasive environment, without the buoyancy of water, exposed to much mechanical stress. There are many specializations in animal cells that facilitate their survival in this hostile environment. One of these is the evolution of **intermediate filaments**: a system of sinuous, strain-resistant microscopic ropes than course through the cell to help maintain cell shape and resist mechanical rupture. Intermediate filaments (or nanofilaments as they are sometimes called) are major structural components of the cytoskeleton in almost all animal cells. They make up the third cytoskeleton filament system—after microtubules and actin filaments—but the one about which we still know the least. Within the cell, some form cytoplasmic meshworks (e.g., vimentin) and some are only in the nucleus (e.g., lamins), as seen in **FIGURE 13.1**, but their subcellular distribution is quite distinct from that of actin or tubulin. Histologists recorded them by light microscopy (as "neurofibrils" in neurons and "tonofila-

ments" in epidermal cells) long before individual filaments were described in the 1960s by electron microscopy of muscle tissue. In the 1970s, "intermediate" filaments were observed as intermediate in diameter between "thick filaments" of myosin II and "thin filaments" of actin in muscle cells. With an average diameter of 8-12 nm, they are thicker than actin filaments (~8 nm) and thinner than microtubules (~25 nm). The three systems can be seen together in a cell by electron microscopy in **FIGURE 13.2**.

Animals have evolved multiple parallel and alternative systems of intermediate filaments so that cells can differentiate into appropriate physical states without loosing their supporting filaments. Seventy genes have been identified in the human genome; with alternative splicing for 5 of these genes, the total number of intermediate filament proteins is ~78. This represents considerably more variety and heterogeneity than either actin or tubulin proteins. The gene family has 6 subclasses (types I–VI), each showing dramatic tissue-specificity of expression. It is this last feature, especially with its implicit potential for cancer biomarkers, which has attracted attention to these proteins since the 1970s. This specificity can readily be detected with monoclonal antibodies, and yet the proteins share a very similar overall molecular structure, and assemble into filaments with almost identical appearance by electron microscopy.

Within the cell, nanofilaments form bundles and networks that exhibit considerable tensile

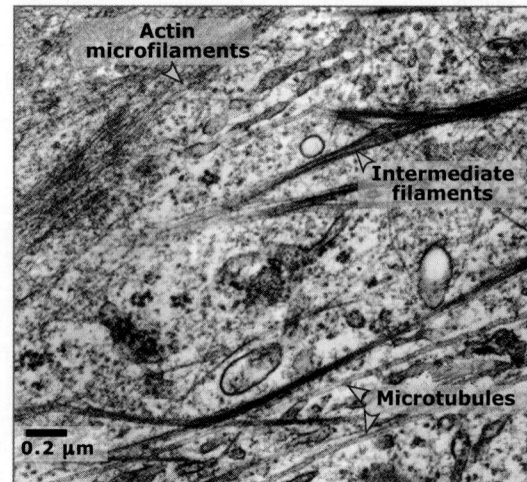

FIGURE 13.2 The major components of the cytoskeleton as seen by transmission electron microscopy. A thin section of a kidney epithelial cell shows actin microfilaments, intermediate filaments of K8/K18, and microtubules. Reproduced from I. M. Leigh, et al. *The Keratinocyte Handbook*. New York, 1994. Reprinted with permission of Cambridge University Press. Photo courtesy of Birgit Lane, College of Life Sciences, University of Dundee.

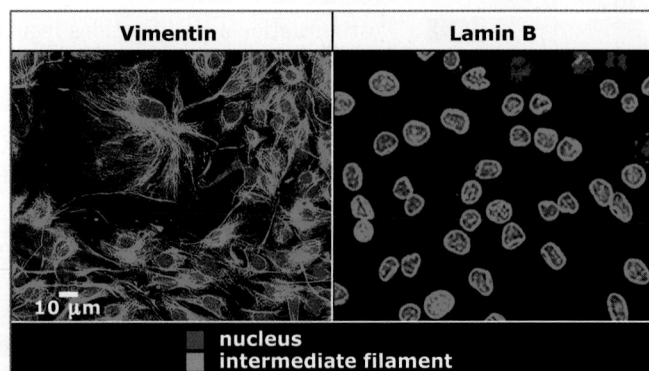

FIGURE 13.1 Immunofluorescence micrographs of vimentin and lamin B in cultured fibroblasts show the distribution of different types of intermediate filaments. Vimentin is restricted to the cytoplasm, whereas lamins are restricted to the nucleus. Photos courtesy of John Common and Birgit Lane, Institute of Medical Biology, Singapore.

strength; they have no recognizable organizing centre catalyzing directional assembly, and free ends of intermediate filaments are rarely if ever seen. Intermediate filaments are implicated in diverse cell functions, including chemical and mechanical stress resilience, signal transduction and cytoplasm compartmentalization, all of which are probably real functions of intermediate filaments. Yet in contrast to microtubules and actin filaments, these nanofilaments are not required for short-term cell survival in tissue culture. Their true function only kicks in at the level of multicellular tissues, where they are essential for effective and sustained tissue and organ function. Most intermediate filaments are expressed in epithelial tissues that form sheets at interfaces or barrier tissues of the body—all locations where stress is rife.

The discoveries of disease-causing mutations in intermediate filament genes have regenerated great interest in these proteins. There are now known to be more than 80 distinct clinical entities caused by defective intermediate filament genes, ranging from fragile blistering skin (keratins) to premature aging (lamins). This strongly suggests that having the correct physical resilience is essential for tissue function *in vivo* and that a significant part of this resilience is due, directly or indirectly, to the cells' intermediate filaments. Add to this the fact that expression of intermediate filament genes is highly tissue-specific, and it becomes very likely that all of these proteins give subtly different properties to a tissue cell. The different cellular engineering requirements—for rigidity, plasticity, or speed of assembly or disassembly of their reinforcement system—may be the reason that so many different intermediate filament genes have arisen throughout evolution.

13.2 Similarities in structure define the intermediate filament family

Key concepts

- Intermediate filament proteins show common domain structures.
- Type I–type VI sequence homology groups are based on DNA and amino acid features.
- Type I–type VI groups show restricted tissue expression.
- Sequence features in subdomains can identify candidate functional sites.
- Head and tail domain divergence may underlie tissue specificity.

By electron microscopy, all mammalian intermediate filaments examined have a similar appearance of long sinuous filaments of approximately 10 nm in diameter, running alone or in bundles through the cell. Most intermediate filament proteins are between 40 kDa and 70 kDa in molecular weight. From their primary amino acid or DNA sequences, we can see that all intermediate filament proteins share a similar domain structure, as shown in **FIGURE 13.3**. There is a long α-helical rod domain, divided into four helical sections (1A, 1B, 2A, 2B) by three or four nonhelical "linker" regions (L1, L12, L2; S). This whole α-helical rod domain is about 45nm long, of 310 amino acids in cytoplasmic proteins and 350 amino acids in lamins (which always have a longer rod domain).

Two well-conserved sequence motifs mark the boundaries of the rod domain (see Figure 13.3). The last 12 C-terminal residues or so of the rod domain constitutes the helix termination motif, usually Glu-Ile-Ala-Thr-Tyr-Arg-(X)-Leu-Leu-Glu-Gly-Glu (where X is any amino acid), which is a hallmark sequence

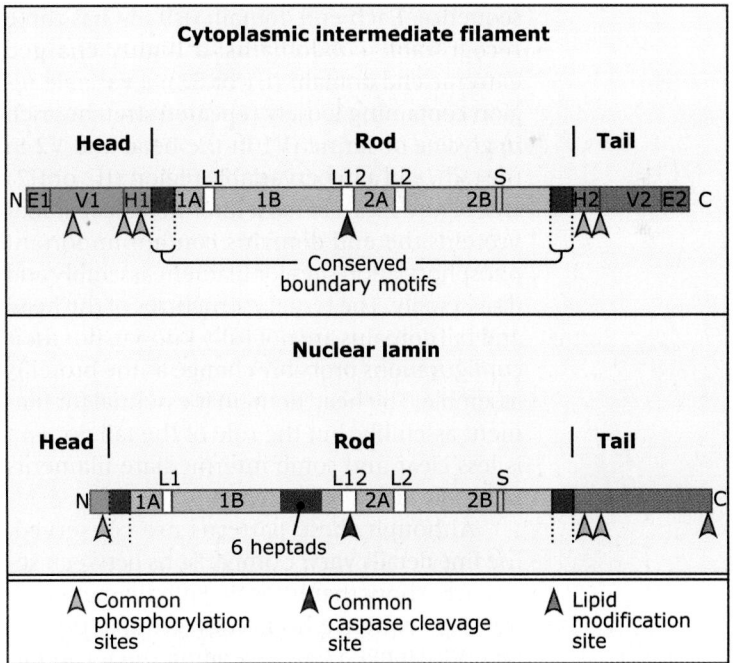

FIGURE 13.3 The general structure of human intermediate filament proteins, as predicted from primary amino acid sequence data. All intermediate filament proteins have a central rod domain and flanking head and tail domains. This general structure best represents type II, type III, and type VI proteins, as type I proteins lack H1/H2 regions and type V proteins lack the E/V domains. For the lamin protein, the position of the additional amino acids lengthening the rod domain in vertebrate lamins and invertebrate filaments is shown. Arrowheads denote common phosphorylation sites (some intermediate filament proteins have additional phosphorylation sites), the common caspase cleavage site, and the common lamin lipid modification site.

that is recognizable in all intermediate filament proteins, and has even been used to trawl for unknown intermediate filament proteins in databases. The helix initiation motif at the N-terminus of the rod is slightly more variable. The conservation of these two helix boundary motifs argues for their functional importance, and experiments have shown that they serve as docking sites for end-to-end interaction in filament assembly. Mutations in these domains are highly disruptive and cause severe disease (see *13.5 Mutations in keratins cause epithelial cell fragility*). Two other well-conserved sequence features of the rod are an irregularity in helix 2B known as the "stutter" (S in Figure 13.3), and patches of alternating positively and negatively charged clusters of amino acids along the surface of the α-helix, repeating at about every 9.5 residues, which are important for lateral interactions of rod domains during filament assembly (see *13.3 Intermediate filament subunits assemble with high affinity into strain-resistant structures*).

The N-terminal head domain and C-terminal tail domain are much more variable than the rod domains, both in length and in amino acid sequence. Each end domain usually has three recognizable subdomains: a highly charged extreme end domain (E1 or E2), a variable region containing loosely repeated stretches rich in glycine or serine (V1 in the head and V2 in the tail), and a hypervariable region (H1 or H2) (see Figure 13.3). In most intermediate filament proteins the end domains contain important phosphorylation sites for filament assembly and disassembly. The tertiary structures of the head and tail domains are not fully known, but their configurations probably change as the proteins assemble. The head domain is essential for filament assembly, but the role of the tail domain is less clear and some intermediate filaments naturally lack a tail domain.

Although these patterns are conserved, the fine details vary. Comparisons between sequences show that intermediate filaments fall into six sequence homology groups, type I–type VI (**FIGURE 13.4**), extending the terminology used when the first intermediate filament protein sequence fragments, of wool keratins, were identified. (For comparison, members of the type I and type II groups known today are listed in Figure 13.8 and Figure 13.9.) Sequence homology between genes in the same group is mostly around 60%, but in some cases (e.g., some keratins) it is over 95%. Between genes in two different sequence groups, the homol-

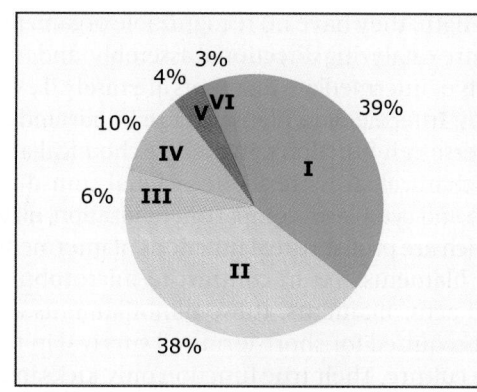

FIGURE 13.4 The intermediate filament genes can be divided into six groups based on sequence similarities. The type I and type II genes constitute the majority of intermediate filament genes.

ogy drops to as low as 20%. The rod domain sequence homology is generally high within the same group, especially in the helix boundary motifs, but the end domains are much more variable. There are often greater similarities in the head and tail domains between filaments sharing the same expression range than between sequences in the same homology group, suggesting that evolution for tissue-specific functions has preferentially targeted these head and tail end domains. End domain structures can also be quite different: type I keratins have short or nonexistent H1 and H2 domains, while type IV proteins lack well-defined E1/V1 and E2/V2 subdomains. One type I keratin, K19, has no tail domain whilst tail domains of some type IV neurofilament proteins can be enormous (NF-H at over 600 amino acid residues, synemin α at twice that and nestin at 1308 residues).

The divisions into sequence homology groups also fit well with tissue-specific expression of the proteins and with their proposed evolutionary history. Most intermediate filament genes (54 of 70) are in type I and II, encoding the many different keratins. These are expressed in epithelia (sheet tissues of the body), and are listed in Figure 13.8 and Figure 13.9. The type III group includes four to six highly related proteins (vimentin, desmin, GFAP, peripherin, and probably now also syncoilin and its isoforms), each of which is cell-type specific. The type IV group contains six to eight neurofilament-related proteins (NF-L, NF-M, NF-H, α-internexin, nestin, and synemin isoforms). Type V contains the ubiquitous nuclear lamin genes, encoding lamins A/C1/C2, B1, and B2/B3, evolutionarily the oldest group of intermediate filament genes. Finally, the type VI group accommodates two divergent eye lens filament protein genes that do not fit into the other groups (filensin/CP115 and CP49/phakinin). Further tissue specificity is seen between individual members of each group, and some intermediate filament proteins are expressed upon inflammatory stress or wound injury (see Figure 13.8, Figure 13.9, and Figure 13.15; for descriptions of the different types of epithelia see *13.4 Two-thirds of all intermediate filament proteins are keratins*).

Intermediate filament proteins are probably the most robust biomarkers of tissue and cell differentiation available today. Antibodies to individual intermediate filament proteins are widely used to assess cell and tissue differentiation in applications from cell biology to pathology, because tissue-specific expression of intermediate filaments can be traced through development and differentiation, and usually persists even into metastatic tumors. Abnormal differentiation, which can often be detected as altered intermediate filament expression, is an early clue to serious pathological changes. Tissue specificity is a striking characteristic of intermediate filament proteins, which above all other properties was responsible for initiating research on these major structural proteins, before the extensive links with genetic diseases were established.

13.3 Intermediate filament subunits assemble with high affinity into strain-resistant structures

Key concepts

- Intermediate filament assembly is rapid and requires no additional factors.
- Dimer formation uses the long "leucine zipper" of the rod domain.
- Keratin pairs assemble with high affinity.
- Assembly from antiparallel tetramers determines the apolar nature of cytoplasmic intermediate filaments.
- Intermediate filament networks are stronger than actin filaments or microtubules and exhibit strain hardening under stress.

Intermediate filaments assemble spontaneously from a solution of soluble subunits without the need for any accessory protein or catalytic action of other components. They are apolar and multistranded polymers, which has made the analysis of their assembly challenging. Most

available information, as discussed here, comes from *in vitro* studies and only relates to the rod domains.

Assembly begins with dimer formation, either homodimers or heterodimers, depending on the protein. Whilst some neurofilaments (NF-H and NF-M) and all keratins can only assemble as heteropolymers (starting from heterodimers), the type III proteins can assemble as homopolymers or (more commonly) as heteropolymers, with type III or type IV proteins. Selective coexpression of intermediate filament proteins in tissues suggests a very specific assembly pairing *in vivo*, yet *in vitro* assembly appears to be more flexible and a wider assortment of partnering still leads to filament assembly.

The coiled coil dimers form by interactions of the two α-helical rod domains, shown in **FIGURE 13.5**. (The α-helix is the most common protein secondary structure seen in nature and was first described in an intermediate filament protein.) The α-helices of the nanofilament rod domains are exceptionally long, with periodic hydrophobic residues along their length, and on their own, would misfold back on themselves and become targets for proteolytic destruction (for more on protein folding see *7.18 The assembly of proteins into complexes is monitored*). Instead, by twisting around another similar α-helix, and sequestering the hydrophobic residues in the seam between the two long helices, a more stable extended structure is generated. This

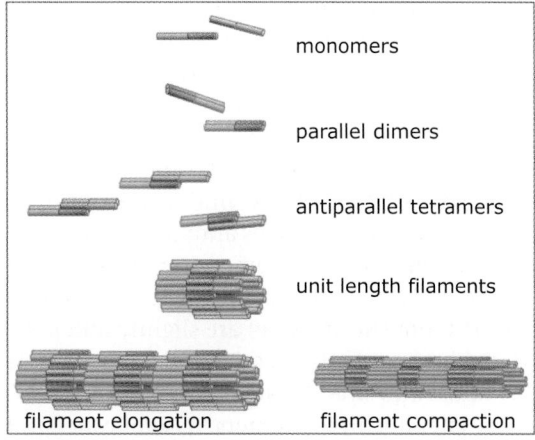

FIGURE 13.5 Consensus model for assembly of intermediate filaments from monomeric proteins. Assembly into dimers is immediate and necessary to prevent protein degradation. Tetramers are the most likely minimal state of proteins in vivo. For simplicity, nonhelical domains are not shown. The two halves of the rod domain are colored green (N-terminal half) and red (C-terminal half). Adapted from R. Kirmse, et al., *J. Biol. Chem.* 282 (2007): 18563–18572.

interaction can be predicted from the "heptad repeat" pattern typically seen in the primary sequence of coiled-coil proteins. In every seven residues (denoted *a-b-c-d-e-f-g* then repeating), the sequence will show hydrophobic residues (usually leucine, isoleucine, valine, alanine, or methionine) at positions *a* and *d*. This feature is sometimes referred to as a "leucine zipper" motif.

Keratin (type I/II) proteins have been useful in deducing many aspects of early filament assembly, because researchers can take advantage of the absolute requirement for two proteins, forming a type I/type II heterodimer, to initiate the process. A single keratin transfected into a cell is unstable and gets degraded, so no protein appears until both a type I and a type II keratin are present in the cytoplasm—whereupon assembly follows rapidly. Thus, transfecting in modified second keratin constructs after a first (copolymerizing) partner keratin can be used to determine which molecular features facilitate dimerization. Such experiments demonstrated that monomers were indeed unstable, proved that keratin homodimers cannot make filaments, and proved that the type I and type II polypeptides in a dimer lie in parallel and in register. Keratin heterodimerization *in vitro* can also be used to analyze affinities between different monomers. The associations between type I and type II biologic partner keratins show extremely high affinities, exceeding that of good antibody-antigen interactions—an important point to consider when interpreting analyses of biochemical mixtures of keratin proteins. Thus, spontaneous assembly of keratin intermediate filaments is extremely rapid and difficult to slow down.

Dimers then rapidly assemble to antiparallel, half-staggered tetramers (possibly the smallest stable unit). This antiparallel arrangement of dimers into tetramers immediately precludes polarity in the overall filament. Nuclear lamins (not shown here) are slightly different, assembling as end-to-end parallel dimers that associate into strands: lamin dimers do not associate laterally into tetramers. Assembly beyond these subunits into polymers is likely to be very fast inside cells, but in systems where it can be slowed down (e.g., by experimentally dropping the ambient temperature), assembly appears to proceed through the following three stages:

1. rapid lateral association of eight tetramers into short (~60 nm long) and stumpy (20 nm wide) "unit length filaments" (ULFs), followed by

2. end-to-end association of ULFs into a loose, thick immature filament, and then

3. a further conformational compaction of the immature filament to give a long, smooth-profiled 10 nm mature filament, with an average thickness of 32 polypeptide chains.

Although the speed and efficiency of different stages varies between proteins, this assembly model is consistent with known sequence features of intermediate filament proteins. Lateral association of tetramers to form ULFs can be driven by charge interactions along the lengths of the rod domains, and tetramer end-to-end association into filaments takes place through the conserved helix boundary motifs. The predicted overlap of the helix boundary peptides (see Figure 13.3) as subunits assemble into a filament would explain why mutations in the helix boundary motifs are likely to be so detrimental and why mutation of these sites in K5/K14 leads to such severe skin fragility in *epidermolysis bullosa simplex* (EBS) (see *13.5 Mutations in keratins cause epithelial cell fragility*).

Analysis of near-neighbor interactions from mature filaments isolated from tissues reveals many more lateral interactions than those described in a simple tetramer as in Figure 13.15. Some of these may arise during compaction, with closer subunit packing, longitudinal shifting of strands, and cross-strand interactions taking place as the filament adopts its preferred mature configuration. Other alignments may be driven by end domain interactions, not addressed by the model in Figure 13.5. Structural data suggest that head domains may interact with rod domains in the same or neighboring molecules and that tail domains may be more compact (as they are in lamins).

FIGURE 13.6 shows a model of vimentin that has been generated by amalgamating the crystal structure of part of the vimentin rod domain with structures predicted from homology modeling. In this model, we can see how the long α-helical subdomains coil around each other in each of the two dimers, drawn here in an antiparallel half-stagger as they are thought to lie in the tetramer.

Very little is known about intermediate filament assembly *in vivo*. There is not much unpolymerized intermediate filament protein detectable in cells (estimated as up to 5% of simple keratins per cell, in contrast to 25–50% of actin and tubulin in a cell). The instability of the long α-helical rod domain requires very

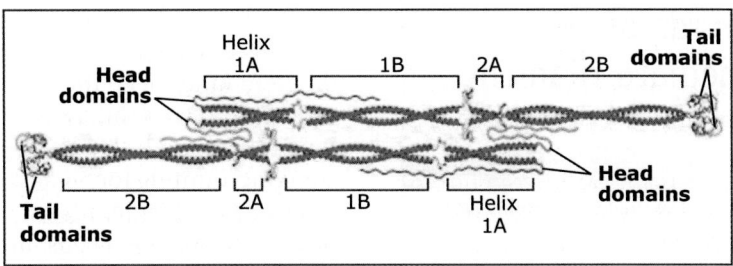

FIGURE 13.6 Atomic model of the homopolymeric type III protein vimentin, based on crystallographic modeling of parts of the protein complemented by homology modeling of the rest. When the protein assembles into filaments, the head domains probably associate with rod domains in the regions indicated. Reproduced from Annual Reviews by Harald Herrmann, et al., Copyright 2004 by Annual Reviews, Inc. Reproduced with permission of Annual Reviews, Inc. in the format of Textbook via Copyright Clearance Center. Photo courtesy of Harald Herrmann, German Cancer Research Center.

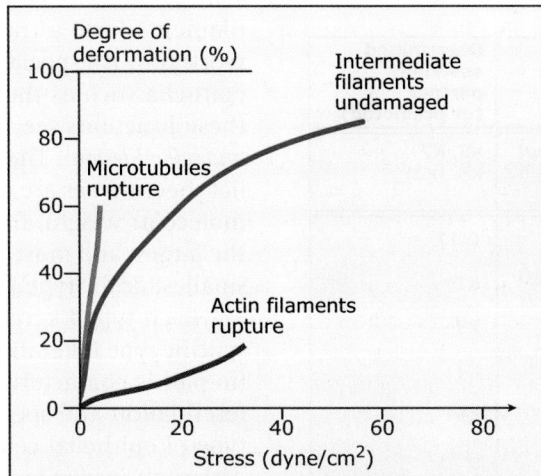

FIGURE 13.7 Intermediate filaments are much more strain resistant than either microtubules or actin filaments, which rupture at lower stress levels. Strain was measured as the extent of deformation (stretching) caused by applying shear stress to filaments purified from cells. Adapted from P. A. Janmey, et al., *J. Cell Biol.* 113 (1991): 155–160.

rapid assembly after synthesis or the involvement of chaperone proteins. Some chaperone proteins, such as the small heat shock proteins Hsp27 and αB crystallins, do interact with intermediate filament proteins, especially upon cell stress. Because intermediate filament proteins depend on each other for correct folding, they may act as their own chaperones.

Finally, the filaments formed by all these processes are undoubtedly very tough. *In vitro* biophysical analyses has shown that intermediate filaments, or nanofilaments, are much more resistant to strain than microtubules or actin filaments, as illustrated in **FIGURE 13.7**. Unlike microtubules and actin filaments, intermediate filaments show increasing resistance to distor-

tion as more strain is applied (a property called strain hardening). And direct measurements of stretched keratin intermediate filaments have revealed high tensile strength, long-range elasticity, and plastic irreversible deformation in extending filaments. Emerging data suggests that different types of intermediate filaments have different biophysical properties when mechanically stressed *in vitro*, which may be an important property driving the evolution of so many apparently similar intermediate filament genes.

Concept and Reasoning Check

1. How do intermediate filaments assemble? And how can you stop this happening?

13.4 Two-thirds of all intermediate filament proteins are keratins

Key concepts

- Most of the intermediate filament proteins are keratins, expressed in epithelia.
- Keratins are obligate heteropolymers of type I and type II proteins.
- Keratin expression is an indicator of epithelial differentiation and proliferative status.
- Simple keratins K8 and K18 are the least specialized keratins.
- Barrier keratins have the most complex and varied expression of all intermediate filaments.
- Structural keratins of hard appendages are different and may be late evolving.

Protein (old name)	Tissue differentiation program (*example in normal tissue*)	Determined assembly partner (or predicted)
K18	Simple epithelia: all (*intestinal lining*)	K8, K7
K20	Simple epithelia: some gastrointestinal (*small intestine*)	K8, (K7)
K9	Stratified epithelia: suprabasal; cornifying (*palm, sole*)	(K1)
K10	Stratified epithelia: suprabasal; cornifying (*epidermis*)	K1
K12	Stratified epithelia: noncornifying (*cornea*)	K3
K13	Stratified epithelia: suprabasal; noncornifying (*oral*)	K4
K14	Stratified and complex epithelia: basal cells; all (*epidermis*)	K5
K15	Stratified epithelia: some basal cells (*epidermis*)	(K5)
K16	Stratified epithelia: suprabasal; stress, fast turnover (*oral*)	K6a
K17	Stratified epithelia: stress, fast turnover (*deep hair follicle*)	K6b
K19	Stratified epithelia: basal cells; some simple epithelia (*mammary gland*)	K8
K23	Epithelia (*unmapped*)	
K24	Epithelia (*unmapped*)	
K25 (K25irs1) K26 (K25irs2) K27 (K25irs3) K28 (K25irs4)	Structural epithelial cells: appendage-forming epithelia (*inner hair follicle*)	
K31 (Ha1) K32 (Ha2) K33a (Ha3-I) K33b (Ha3-II) K34 (Ha4) K35 (Ha5) K36 (Ha6) K37 (Ha7) K38 (Ha8) K39 K40	Structural epithelial cells: hard structures and appendages (*hair, nail, tongue*)	Type II trichocyte keratins

FIGURE 13.8 The human intermediate filament proteins in the type I sequence homology group (type I keratins). Proteins are grouped into simple keratins (simple epithelial cells), barrier keratins (in stratified squamous epithelia and complex epithelia), and the two types of structural keratins associated with epidermal appendages.

Most of the intermediate filament genes in humans encode keratins, expressed in all the different kinds of epithelial tissues of the body (see below). There are 28 type I keratins and 26 type II keratins, as **FIGURE 13.8** and **FIGURE 13.9** show, accounting for 54 of the 70 genes in the intermediate filament family. Keratins, sometimes called cytokeratins, are coexpressed as type I/type II pairs in epithelial tissues. Keratins are such a fundamental characteristic of epithelia that the presence of keratins can be used to define an epithelial tissue; sheet tissues that do not express keratins (e.g., the endothelium of blood vessels) are not classified as epithelia. Keratin filaments are associated with cell-to-cell anchorage junctions (desmosomes) and with cell-to-matrix junctions (hemidesmosomes); together with these junctions, keratin filaments form a trans-tissue structural network that is especially apparent in stratified epithelia such as the epidermis (for more on these junctions see *19 The extracellular matrix and cell adhesion*). The major keratins of epithelial sheet tissues are catalogued by charge and molecular weight, from K1 (type II group) as the largest and most basic protein, to K19, the smallest acidic type I protein.

Each type I keratin is coexpressed with a specific type II keratin partner, and each keratin pair is characteristic and indicative of differentiation and specialization of a particular type of epithelial cell, as **FIGURE 13.10** shows. Although *in vitro* any keratin of one type will form filaments with a wide range of keratins of the complementary type, *in vivo* they show much more selective coassembly, in specific preferred and predetermined pairs. Expression of these keratins pairs is tightly associated with specific differentiation pathways of epithelia, or even with a stage in that pathway, and the presence of one member of a pair is nearly always diagnostic of the presence of the other (**FIGURE 13.11**). Functionally, the keratin pairs can also be divided up into at least three groups: simple, barrier, and structural keratins.

Keratins are expressed in epithelial cells. An epithelium is a common type of tissue organization in which cells are packed closely together in one or more layers. Epithelia mark the limits of organs or form the boundaries of secretion and absorption channels. These are often one-layered **simple epithelia**, in which the cells are in direct contact with the underlying specialized carpet of extracellular matrix (the basal lamina) and have a free surface facing the lumen of a duct or intestine, as illus-

Protein (old name)	Tissue differentiation program (*example in normal tissue*)	Determined assembly partner (or predicted)
K7	Simple epithelia: many (*mammary gland*)	K18 (K19)
K8	Simple epithelia: all (*intestinal lining*)	K18 (K19, K20)
K1	Stratified epithelia: suprabasal; cornifying (*epidermis*)	K10 (K9)
K2 (K2e)	Stratified epithelia: suprabasal; cornifying; late (*epidermis*)	(K10)
K3	Stratified epithelia: noncornifying (*cornea*)	K12
K4	Stratified epithelia: suprabasal; noncornifying (*oral*)	K13
K5	Stratified and complex epithelia: basal cells; all (*epidermis*)	K14 (K15)
K6a	Stratified epithelia: suprabasal; stress, fast turnover (*oral*)	K16
K6b	Stratified epithelia: stress, fast turnover (*deep hair follicle*)	K17
K6c (K6e/h)	Epithelia (*unmapped*)	
K75 (K6hf)	Stratified epithelia (*hair follicle sheath*)	(K16, K17)
K76 (K2p)	Stratified epithelia: suprabasal; cornifying (*oral palate*)	(K10)
K77 (K1b)	Epithelia (*sweat gland*)	
K78 (K5b)	Epithelia (*unmapped*)	
K79 (K6l)	Epithelia (*unmapped*)	
K80 (Kb20)	Epithelia (*unmapped*)	
K71 (K6irs1) K72 (K6irs2) K73 (K6irs3) K74 (K6irs4)	Structural epithelial cells: appendage-forming epithelia (*inner hair follicle*)	
K81 (Hb1) K82 (Hb2) K83 (Hb3) K84 (Hb4) K85 (Hb5) K86 (Hb6)	Structural epithelial cells: hard structures and appendages (*hair, nail, tongue*)	Type I trichocyte keratins

FIGURE 13.9 The human intermediate filament proteins in the type II sequence homology group (type II keratins). Proteins are grouped into simple keratins (simple epithelial cells), barrier keratins (in stratified squamous epithelia and complex epithelia), and the two types of structural keratins associated with epidermal appendages.

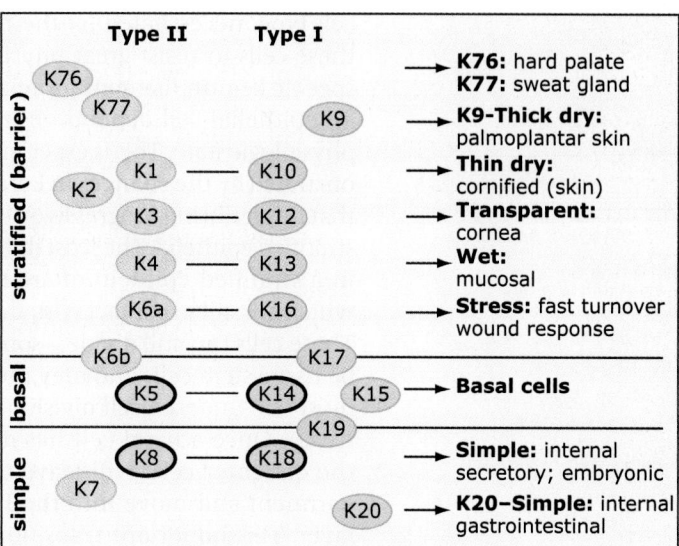

FIGURE 13.10 Keratins are expressed as tissue-specific pairs of type I/type II proteins. Each keratin pair is characteristic of a particular type of epithelial differentiation. Bold rings indicate the primary keratins.

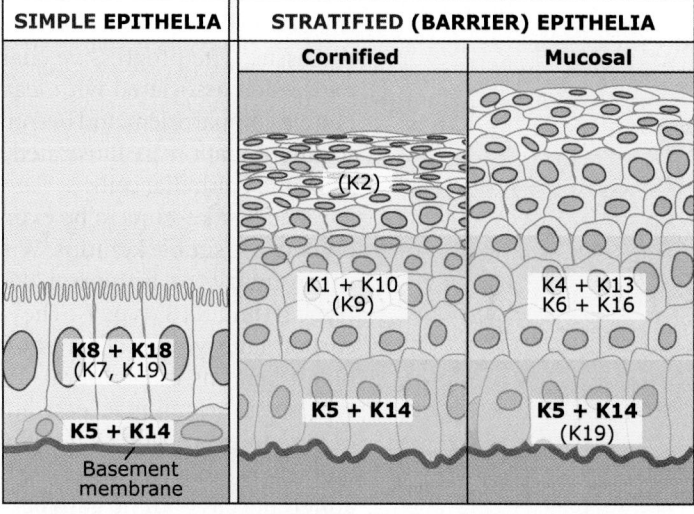

FIGURE 13.11 The sequential expression of major epithelial keratins in simple (for example, glandular) and barrier (for example, cornified or mucosal) epithelia. Primary keratins are shown in bold. Variable or minor keratins are not shown here. Keratin expression is linked to the cell's position in the tissue and thus also to its proliferative state: loss of contact with the basement membrane initiates exit from the cell cycle and progression of differentiation.

trated in Figure 13.11. At the other extreme, multilayered or **stratified epithelia** form the major physical barrier tissues of the body, from the epidermis that covers the external body surface to the specialized epithelial surfaces that line body orifices and the outer regions of contiguous ducts. The cells of stratified epithelia are usually called keratinocytes. Stratified epithelial tissues are also the embryologic origin of specialized complex epithelia of accessory structures such as glands, hair, and nail, which may include domains of simple epithelial structure. When fully developed, the outer cells of the stratified barrier epithelia are flattened or squamous. Stratified squamous epithelia are usually 6 to 10 cell layers thick and exhibit constant turnover to protect the body from physical, chemical, and carcinogenic damage.

Epithelial cells are held together at the plasma membrane by cell–cell desmosome junctions and cell-substrate hemidesmosomes, all connected across the cytoplasm of each cell by dense networks of keratin filament bundles (see *19 The extracellular matrix and cell adhesion*). Keratinocytes of stratified epithelia express a larger amount per cell and a wider range of intermediate filament proteins than any other cell type, necessitated by the requirement of these cells to resist great physical forces. The specific keratin filament proteins expressed in an epithelial cell depend on its location and physiologic state. This is especially clearly demonstrated by the changes in keratin expression that occur during progressive differentiation of stratified epithelia. The least differentiated cells in a stratified epithelium are the basal cells, which are still in contact with the basal lamina. These cells can still divide—some of them will be tissue stem cells and may rarely divide, but most will undergo cell division to bulk up the tissue. Once a basal cell has divided, one of the daughter cells will leave the basal compartment and move into the first suprabasal layer. This important transition removes the cell from the direct influence of the basement membrane and its supply of growth signals, and the cell becomes committed to terminal differentiation. It then begins its journey out to the surface of the epithelium. This is a process of terminal differentiation, during which the cell will eventually die and be lost from the tissue. The progressive changes in keratin expression associated with leaving the proliferative compartment and becoming committed to differentiation are illustrated in Figure 13.11 for some epithelial cells.

The first keratins to be expressed in development are simple keratins. Within this group, the primary, or universal, keratins are K8 (type II) and K18 (type I). They are seen in the earliest embryonic cells and are probably the oldest keratins in evolution. Their expression represents the minimal, least differentiated state of keratin expression that defines a functional epithelium, that is, a sheet of polarized, tightly adherent cells with no gaps between them. K8 and K18 are the most widely conserved keratins among vertebrates and are present throughout development from the oocyte to adult tissues. All embryonic cells express K8 and K18 until gastrulation, when some ectodermal cells differentiate to form the mesoderm layer, shutting off K8/K18 synthesis and initiating expression of the type III protein vimentin.

K8 and K18 continue to be expressed in embryonic epithelial cells until these cells become committed to specific morphogenetic pathways, at which point they switch to expression of more tissue-specific intermediate filament proteins. In the adult, K8 and K18 are characteristic of simple epithelial cells that have a secretory and/or absorptive function, such as those of glands, liver, respiratory epithelium,

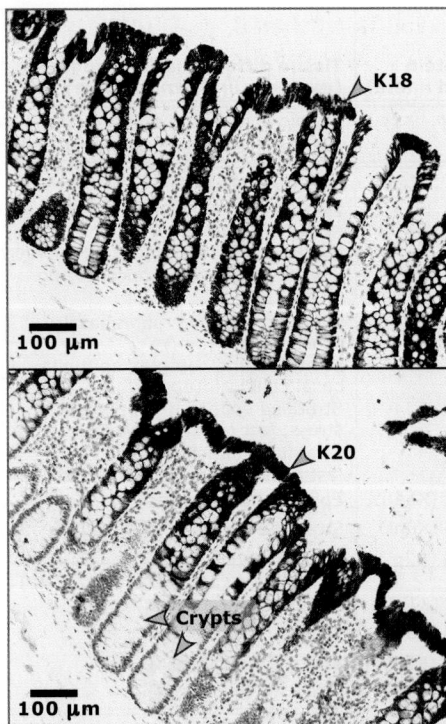

FIGURE 13.12 Detection of simple keratins in the epithelial lining of the large intestine by immunohistochemistry. The locations of two type I keratins, K18 and K20, are revealed by immunoperoxidase staining with two specific monoclonal antibodies. Keratins are stained brown, with blue haematoxylin counterstain. The primary keratin K18 is present in all simple epithelia, whereas K20 is specific for certain gastrointestinal differentiation lineages and seen here to be restricted to the later parts of the progressively differentiating epithelium as cells move upward from the bottom of the crypt. Photos courtesy of Declan Lunny and Birgit Lane, College of Life Sciences, University of Dundee.

and gastrointestinal tract—see the example in **FIGURE 13.12**. They are expressed in a wide range of carcinomas, and antibodies to K8 and K18 are widely used in diagnostic pathology. In addition to the primary keratins, there are at least two other simple keratins, K7 (K8-like, mostly in gland ducts) and K20 (K18-like, in parts of the gastrointestinal tract).

The barrier keratins are characteristic of stratified epithelia. In this group, the primary or essential keratins are K14 (type I) and K5 (type II), which are present in basal layer keratinocytes of stratified squamous epithelia such as the epidermis of skin. This single layer of K5/K14–expressing basal cells is the least-differentiated tissue compartment and retains the capacity to proliferate. In complex glandular epithelia, basal cells expressing K5/K14 are also found alongside simple epithelial cells expressing K8/K18 (see Figure 13.11). In some tissues, basal proliferative cells can express a few other keratins, such as K19, K15, and K6/K17.

As cells leave the basal layer, they cease production of K5/K14 and switch to synthesis of a secondary, or differentiation-specific, keratin pair (see Figure 13.11). In the epidermis, this will be the type II keratin K1 and the type I keratin K10, which is shown in **FIGURE 13.13**. Secondary keratin expression in suprabasal cell layers depends on the tissue type (see Figure 13.10 and Figure 13.11). One set of keratins, the K6 isoforms with K16 and K17, behave as stress response proteins in the epidermis and are rapidly induced upon wound injury or inflammation. Other tissues constitutively express these "stress" keratins such that their inductive stimulus must be intrinsic to these tissues.

Thus, the proliferative compartment (basal cell layer) in multilayered barrier epithelia has a different keratin synthesis program from that of the differentiating compartment (suprabasal layers). Exit from the cell cycle in these barrier tissues is closely correlated with cessation of primary keratin synthesis and initiation of secondary differentiation-specific keratin synthesis. The suprabasal keratins are specialized to provide great tissue resilience, and expression of some secondary keratins may impede cell separation in mitosis and so may be incompatible with proliferation.

The third group of "structural" keratins consists of the large number of intermediate filament proteins expressed only in and around specialized hard structures and appendages such as hair and nail (listed in Figure 13.8 and Figure 13.9). These are the hair cell or trichocyte keratins that form these structures, as well as the specialized keratins made by the epithelial cells producing the structures. These two groups of structural keratins have distinct sequence types and are thought to be recent evolutionary acquisitions of vertebrates.

Sequential expression of structural keratins has been well defined in the hair follicle. The first group of structural keratins are selectively expressed in the layers of concentric epithelial tubes of the hair follicle inner root sheath. This tube structure becomes quite firm and molds the forming hair shaft. The second group, of hard (trichocyte or hair) keratins, are expressed in the hair shaft itself, nail, some hard cells

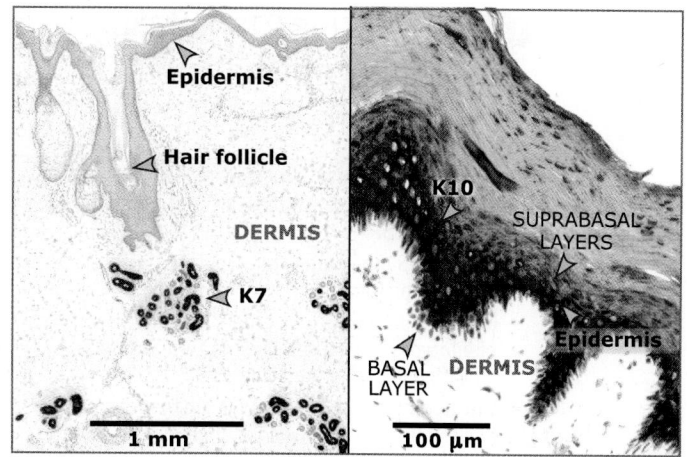

FIGURE 13.13 Tissue specificity of keratin expression in skin. Left: Tissue section stained with an antibody to K7, a type II simple keratin, highlights only the secretory cells of the sweat gland (dark brown, using indirect immunoperoxidase staining). The stratified squamous barrier epithelium of the epidermis is unrecognized by the antibody, only stained pale blue by the haematoxylin counterstain. Right: A section of thick epidermis stained with an antibody to K10, a type I secondary or tissue-specific keratin expressed in postmitotic suprabasal cells of cornified barrier tissues. The basal cell layer is unstained by the antibody, and the nuclei are counterstained blue. Photos courtesy of Declan Lunny and Birgit Lane, College of Life Sciences, University of Dundee.

of surface papillae on the tongue (very apparent on a cat's tongue), and small amounts in the thymus. These trichocyte keratins have a high content of cysteine and proline residues in their head and tail domains, allowing for disulfide cross-linking to nonfilament hair proteins, called keratin-associated proteins, in the cytoplasm of hair cells as they differentiate and harden. This cross-linking leads to a very robust structure that forms the hard tissue of the appendage.

Thus the great diversity of keratin intermediate filament proteins simply reflects the diversity of specializations in the epithelial sheet tissues of the body—or vice versa, if the varying properties of the different keratin filament networks define the function of different epithelia. The ability of keratins to interact with desmosomes and hemidesmosomes places them in a context to provide mechanical continuity and sustain mechanical tension across these sheets of cells. This could function as a tension-transducing or tension-sensing system—an interesting but as yet unproven concept, and another way in which intermediate filaments may contribute to tissue function and homeostasis.

13.5 Mutations in keratins cause epithelial cell fragility

Key concepts

- Mutations in K5 or K14 cause the skin blistering disorder EBS.
- Severe EBS mutations are associated with aggregates of misfolded keratin protein.
- Many tissue fragility disorders with diverse clinical phenotypes are caused by structurally similar mutations in other keratin genes.
- Cell fragility disorders provide clear evidence of a tissue-reinforcing function for keratin intermediate filaments.

There is no argument so compelling of the importance of a thing, as the catastrophic consequences of its absence or destruction. In the early 1990s, a major breakthrough in our understanding of the function of intermediate filaments occurred when it was discovered that mutations in either keratin K5 or K14, the two keratins expressed in basal layer cells in epidermis, lead to an inherited skin-blistering disorder, namely EBS. People with EBS have very fragile skin that cannot withstand everyday stress, such as scratching or rubbing, or even walking. When their skin is mechanically stressed, the basal cells fracture; fluid accumulates between the cell fragments on the basement membrane and the overlying layers of intact epithelial cells, leading to fluid-filled intra-epidermal blisters, as shown in **FIGURE 13.14**. These intra-epidermal blisters grow and spread if not punctured but usually heal without scarring. The finding that this phenotype could be caused by a single (i.e., dominant) mutation in one of the two co-assembling keratins showed at a stroke—and for the first time—that keratin intermediate filaments played an essential role in the epidermis, in equipping the keratinocytes to resist physical trauma.

The clinical severity of EBS varies widely, depending on the position of the mutation in the keratin molecule. The most severe cases, in which spreading blistering occurs in response to mild scratching or rubbing of the skin, are associated with mutations in the helix boundary peptides of keratin proteins, which as discussed above are important in filament assembly (see Figure 13.3 and Figure 13.5). In these patients the mutant protein induces diagnostic misfolded keratin protein aggregates in the basal cells. In general, milder phenotypes, in which blistering occurs only on the hands and feet (where the stress is greatest), arise from mutations elsewhere in the rod domain or in the nonhelical domains, where variations in amino acid sequence are presumably less critical. Most of the disease-associated keratin mutations are dominant, meaning that only one of the two copies of the gene has to be faulty for the phenotype to be expressed: like weak links in a chain, only a scattering of defective subunits can reduce the function of a polymeric filament. Thus, EBS can be passed on through inheritance of only one defective gene copy (from either parent). There are, however, several recorded cases of recessive EBS, in which both copies of the gene are defective. These are usually K14-null mutations, and patients with null mutations in both copies of their K14 gene have blistering but, remarkably, survive with no K14 at all. No K5 null cases have ever been found, however, strongly suggesting that this is lethal, probably because there is no back-up type II keratin that can adequately stand in for K5 (whilst there are some type I minor keratins that could be induced in these cells to replace the function of K14).

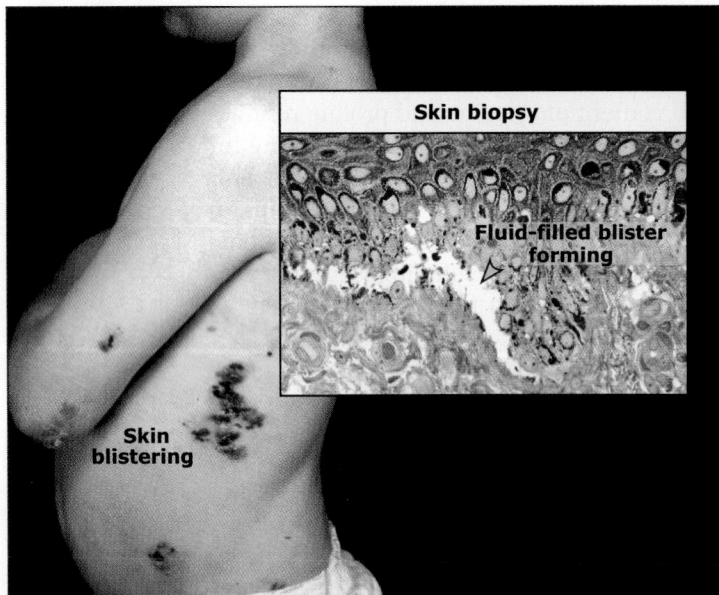

FIGURE 13.14 The condition illustrated here is *epidermolysis bullosa simplex* (EBS), the first disorder linked to intermediate filament gene mutations, caused by mutations in keratins K5 or K14. The main photo shows the characteristic skin blistering caused by scratching, rubbing, or tight clothing. The inset shows a section through a diagnostic skin biopsy taken after rubbing the skin surface with a pencil eraser: the basal layer of epidermal keratinocytes has a fluid-filled blister due to rupturing of the cells. Photos courtesy of Robin A. J. Eady, St. John's Institute of Dermatology, St. Thomas' Hospital.

Discovering the link between K5 and K14 mutations and EBS was a significant advance in understanding the function of intermediate filaments, as it clearly showed that keratin filaments are important for physically strengthening epithelial cells in tissues. When the integrity of the intermediate filament network is compromised, cells are fragile and susceptible to breakdown on mild physical trauma. The implications for the rest of the intermediate filament field were immediately obvious, and this started a search for other causative nanofilament mutations in cell fragility conditions that is still going on today. Based on the known expression patterns of keratins, analysis of many other skin fragility conditions following a "candidate gene" approach revealed that mutations in at least 19 keratin genes are associated with at least 40 clinically distinct disorders. The clinical appearance of these disorders can vary greatly, depending on the cell type expressing the keratin that is compromised by the mutation. For example, mutations in K1 or K10 lead to fragility of all suprabasal keratinocyte layers (called bullous congenital ichthyosiform erythroderma), whereas more superficial, flatter blisters arise from mutations in the secondary, late-expressed keratin K2 (ichthyosis bullosa of Siemens). Mutations in several keratins, including K9 (site-specific for palm and sole) and K16, cause severely thickened skin of palms and soles. Fragility of suprabasal cells in oral and genital epithelia arises from mutations in K4 or K13 (white sponge naevus) and from mutations in K16 or K6a. Other keratin disorders have phenotypes ranging from greatly thickened nails (mutations in K6a, K6b, K16, K17 in forms of pachyonychia congenita) to corneal surface blisters (mutations in K3 or K12, in Meesman corneal dystrophy).

In contrast to the wide range of mutations in the primary keratins K5 and K14, disease-causing mutations in genes for differentiation-specific keratins are mostly found within the helix boundary motifs—that is, the highly conserved regions where mutations should be very disruptive in all keratins. In disorders of secondary keratins, mutations equivalent to those that cause mild EBS are rare, although they must, of course, occur in the population, suggesting that the effect of mutations in the secondary keratins is attenuated. The most likely explanation is that residual K5/K14 filaments persisting from the basal cells may reinforce the suprabasal keratin cytoskeleton. Morphologically related skin fragility disorders can also be caused by mutations in keratin-associated proteins, such as plectin or the proteins of desmosome and hemidesmosome junctions (see *13.10 Interacting proteins facilitate secondary functions of intermediate filaments*).

The simple epithelial keratins K8 and K18 are highly conserved in all vertebrates and are essential for normal development, but mutations of these keratins are less clearly implicated in human disease. In mice, ablation of K8 is lethal due to placental insufficiency, and K8 and K18 protect against a wide variety of cell stresses; the same is probably true in humans. Severely disruptive mutations (in the helix boundary motifs) in K8 and K18 have not been observed in humans and are likely to be lethal; "mild" mutations do occur and may be risk factors for multifactorial disorders affecting the liver, pancreas, and gastrointestinal epithelium. These mildly defective keratins may be the only ones that are compatible with embryonic survival and may not be pathogenic by themselves.

Concept and Reasoning Check

1. Discuss the different strands of cross-talk "from bench to bedside" that have bridged intermediate filament research and human diseases.

13.6 Intermediate filament proteins of nerve, muscle, and connective tissue often show overlapping expression

Key concepts

- Some type III and type IV intermediate filament proteins have overlapping expression ranges.
- Many type III and type IV proteins can coassemble with each other.
- Coexpression of multiple types of intermediate filament proteins may obscure pathology.
- Desmin is an essential muscle protein.
- Vimentin is often expressed in solitary cells.
- Mutations in type III or type IV genes are usually associated with muscular or neurologic degenerative disorders.

The members of the nonkeratin homology groups of intermediate filament proteins, types III to VI, are listed in **FIGURE 13.15**. In this section, we discuss the intermediate filament proteins of the type III and type IV sequence homology groups. The type III proteins are desmin, vimentin, glial fibrillary acidic protein (GFAP, shown in **FIGURE 13.16**), peripherin, and probably syncoilin, which are differentially expressed in connective tissue cells, muscle cells,

	Protein	Tissue differentiation program	Assembly partners (where known)
Type III	Vimentin	Widespread	self
	GFAP	Astroglial cells	self
	Desmin	All muscle cell types	self
	Peripherin	Peripheral nervous system; some CNS; injured axons	self, NF-L
	Syncoilin	Muscle cells	type III, IV
Type IV	NF-H	Neurons	NF-L
	NF-M	Neurons	NF-L
	NF-L	Neurons	self
	Nestin	Widespread: neuroepithelial stem cells, glial cells, muscle	type III
	α-internexin	Neurons	self
	Synemin α, Desmuslin/synemin β	Muscle cells	type III
Type V	Lamin A	Nuclei: many cell types, differentiated	Lamin A, C
	Lamin C1,C2	Nuclei: many cell types, differentiated	Lamin A, C
	Lamin B1	Nuclei: many cell types, less differentiated	Lamin B
	Lamin B2, B3	Nuclei: from early development	Lamin B
Type VI	Filensin/ CP115	Eye lens	CP49
	CP49/ phakinin	Eye lens	Filensin

FIGURE 13.15 The human intermediate filament proteins in the types III–VI sequence homology groups.

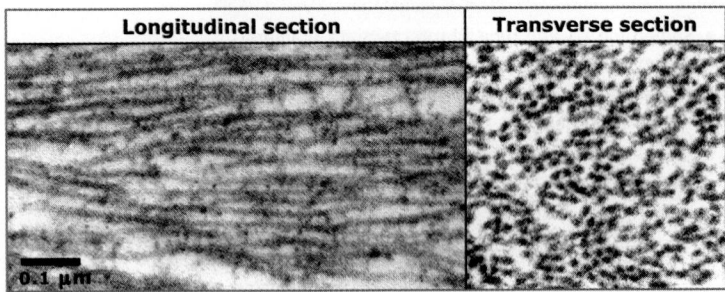

Longitudinal section	Transverse section

0.1 μm

FIGURE 13.16 Electron micrographs of *glial fibrillary acidic protein* (GFAP) filaments in astrocytes of the spinal cord. Reproduced from C. Eliasson, et al., *J. Biol. Chem.* 274 (1999): 23996–24006. © 1999 The American Society for Biochemistry and Molecular Biology. Photos courtesy of Dr. Milos Pekny and Dr. Claes-Henric Berthold, Sahlgrenska Academy, University of Gothenburg.

nerve cells, and certain other differentiated cells. The type IV proteins are the neurofilament proteins NF-L, NF-M, NF-H, nestin and synemin and isoforms. Some proteins have overlapping expression ranges, and some type III and type IV proteins can copolymerize with each other. However, none of them can coassemble with either keratins (types I/II) or lamins (type V). As with other intermediate filaments, it is not yet known exactly how the sequence variations among these proteins relate to their specific function in the different cell types, although mutations in these proteins are clearly associated with tissue failure.

The best understood of the type III intermediate filament proteins is desmin. Desmin is essential for the function of cells in all types of muscle (striated, cardiac, and smooth). Like keratins in epithelia, desmin provides critical physical resilience to tissue cells, as demonstrated by the catastrophic effect of desmin mutations which lead to cardiomyopathies and muscular dystrophies. There are over 70 different mutations reported in desmin, associated with 8 different diseases to date. Desmin filaments run between anchorage points of the contractile units (e.g., of sarcomeres in striated muscle) that are the foci of considerable mechanical stress in muscle cells, consistent with a role in resisting mechanical stress in muscle cells. Desmin coassembles with other type III proteins; in some cells, desmin also interacts with type IV proteins. Syncoilin is also found in muscle cells, particularly in association with the membrane-associated dystroglycan-associated protein complexes. The three isoforms of syncoilin have a very short or nonexistent tail domain and cannot assemble as a homopolymer. (For more on muscle contraction see *12.21 Myosin-II functions in muscle contraction*.)

In contrast to desmin, vimentin expression is typical of cells that function as single cells, or more loosely associated groups or sheets of cells, rather than in dense tissue masses such as muscle and most epithelia. During development, vimentin is expressed after the embryonic keratins (K8/K18), and it persists in the adult in many mesenchymal and connective tissue cells, from fibroblasts (see Figure 13.1) to hematopoietic cells, melanocytes and blood vessel wall endothelial cells, as well as in some epithelia. Because of the widespread use of fibroblasts in tissue culture, vimentin is often taken as a model of intermediate filament behavior, even though vimentin actually represents a minority type with significant differences from keratins.

Astrocytes and glial cells are the nonneuronal cells found in the nervous system and are required for growth, differentiation, and regeneration of neurons. All astroglial cells express GFAP, which is usually expressed with either vimentin or the type IV protein nestin. This coexpression protects cells from the full consequences of mutation or ablation of one of the genes. Compound double knockout experiments in animal models have shown that these type III intermediate filaments are essential for normal astrocyte behavior during wound healing in the central nervous system and for resistance to osmotic stress. Normal astrocyte function requires outgrowth of cell processes, which is defective in the absence of intermediate filaments.

Peripherin is predominantly expressed in the peripheral nervous system. During axonal outgrowth, the axons initially express peripherin and vimentin, which are superseded by expression of the neurofilament triplet proteins (NF-L, NF-M, and NF-H). However, peripherin is reexpressed rapidly upon nerve injury. The amount of peripherin appears to be important for normal function: in mouse models, too much peripherin is lethal and results in a neurodegenerative phenotype, whereas absence of peripherin results in loss of some small sensory axons.

The type IV sequence homology group is now considered to include the neurofilament triplet proteins of low, mid, and high molecular weights (NF-L, NF-M, and NF-H), plus α-internexin, nestin, and synemin (see Figure 13.15), with some researchers favoring a further subdivision of this group. Most type IV proteins are preferentially heteropolymeric in tissues and form polymers much more efficiently in combination with a type III or another type IV protein. The neurofilament triplet proteins are almost always expressed together in mature neurons, whereas syncoilin and the two alternatively spliced forms of synemin (α and β) are predominantly expressed in muscle cells and interact with type III proteins. Like desmin and syncoilin, synemin is localized to stress points in muscle cells as seen by immunohistochemistry, suggesting that these intermediate filament proteins also have a function in stress resistance.

Neurofilaments clearly demonstrate the phenomenon of overlapping expression patterns in development. The extreme length of the axonal processes in neurons (up to 1 m in the case of the human sciatic nerve) emphasizes the importance of maintaining adequate expression of neurofilaments to reinforce the cytoplasm. During development, neurofilament expression follows a complex, staggered, overlapping sequence of one protein on top of another, shifting in a stepwise manner such that neurons continuously express intermediate filaments. Nestin, together with vimentin, is expressed first. (Like GFAP in astroglia, nestin in neurons is reexpressed upon wounding.) As neuronal processes extend, α-internexin expression supplants nestin and vimentin and is itself followed by NF-L and, lastly, the higher molecular weight neurofilament proteins. The neurofilament triplet proteins are important for stabilizing the extended axons and neurites.

Several of the type IV proteins have long tail domains, which may function to organize the cytoplasmic architecture. The tail domain of NF-H has a series of repeats of the motif Lys-Ser-Pro, which is a target for serine phosphorylation. When these repeats are phosphorylated, the tail becomes highly charged and extends at right angles to the body of the filament; this may help space out the axonal cytoplasm. In mice, the number and diameter of axons correlate with the level of neurofilament expression. Greater axonal diameter is associated with faster conductance of the nerve—an especially important feature for the evolution of the larger body size in vertebrates.

Human diseases are associated with mutations in nearly all of the type III and IV proteins; most of these are neurologic or muscular phenotypes. Mutations of type III genes are linked to a wide range of pathological conditions whose mechanisms have in some cases been clarified by studies of animal models. In humans, mutations in GFAP are associated with the fatal neurodegenerative disorder Alexander disease; animal experiments have shown that without GFAP, the astroglial response to injury is abnormal, and astrocytes fail to extend cytoplasmic processes. Many pathogenic mutations have been documented in desmin and are scattered throughout the protein structure. Human desmin mutations are associated with cardiovascular defects, in particular with heart failure from dilated cardiomyopathy, and with some forms of muscular dystrophy. In mice lacking desmin, the major blood vessels are too soft-walled to maintain efficient blood pressure; this places strain on the already weakened heart, which is then prone to developing dilated cardiomyopathy. In many desmin myopathies, muscle cells have intracellular desmin aggregates that resemble the keratin aggregates seen in basal keratinocytes from patients with severe EBS. The absence of known disease associated with vimentin mutations probably reflects a degree of redundancy from its coexpression with other filament proteins.

Mutations in type IV neurofilament proteins, nearly all in the head or tail domains, are associated with the neurodegenerative diseases amyotrophic lateral sclerosis, Charcot-Marie-Tooth disease types 1 and 2E, and Parkinson's disease. The length of the neuronal cytoplasm renders it vulnerable to dysfunction by a variety of routes, not least physical fragility. Neurofilament proteins are synthesized in the cell body and then transported down the axons by microtubule motor proteins. Neurofibrillary tangles, a common indicator of neurodegeneration, are accumulations of neurofilaments in the cell body, but it is not known whether these are causal: they may be secondary consequences of other factors that disturb axonal transport, such as failure of microtubule function. It is much more difficult to conclude a cause-and-effect connection with neurofilament mutations than with keratins. "Keratinopathies" are usually visible at or soon after birth, whereas most neurodegenerative disorders are of late onset, which confounds genetic analysis.

13.7 Lamin intermediate filaments reinforce the nuclear envelope

Key concepts

- Lamins are intranuclear, forming the lamina that internally lines the nuclear envelope.
- Lamin genes undergo alternative splicing.
- Membrane anchorage sites are generated by post-translational modifications of lamins.
- Lamin filaments are depolymerized by phosphorylation with Cdk1 during mitosis.

The type V group of intermediate filaments contains the lamins, which differ significantly from vertebrate cytoplasmic intermediate filaments in a number of ways—they are intranuclear, have an unusually long rod domain, and arise by alternative splicing and complex posttranslational modifications (see Figure 13.3).

Three mammalian lamin genes (*LMNA*, *LMNB1*, and *LMNB2*) encode at least six proteins. The *LMNA* gene gives rise to three or four alternatively spliced mRNAs for lamin A, AΔ10, C1, and C2 proteins (collectively called A-type lamins). *LMNB1* encodes lamin B1, and the lamin B2 mRNA can be spliced to produce B2 or B3. B-type lamins are expressed in all cell types from the early embryo onward, whereas A-type lamins are restricted to more differentiated cell types. For example, in the epidermis of the skin, basal keratinocytes only express lamin B2, which is supplemented with A-type lamins as the cells differentiate. The C2 and B3 lamins, and their counterparts in other vertebrates, are restricted to male germ cells (sperm).

Lamins are strictly intranuclear (see Figure 13.1), and get into the nucleus with a nuclear localization signal in the tail domain. Even if the lamin proteins did not go into the nucleus but stayed in the cytoplasm, they could not copolymerize with cytoplasmic filament proteins. Lamin filament proteins have significant structural differences from cytoplasmic intermediate filament proteins, and lamin filaments assemble in a different way than cytoplasmic filaments (see *13.3 Intermediate filament subunits assemble with high affinity into strain-resistant structures*). The helix 1 subdomain of lamin proteins contains six heptads (42 residues total) that are missing from the vertebrate cytoplasmic intermediate filaments (see Figure 13.3). This rod subdomain size difference alone would prevent lamins from copolymerizing with other intermediate filaments. The longer helix 1 is also found in invertebrate cytoplasmic filament proteins, which is strong evidence that it is an "old" feature in evolutionary terms. It is believed, therefore, that lamins are related to the oldest evolutionary form of intermediate filaments. A mechanism to protect the fragile threads of DNA from breakage was probably an essential early step in the evolution of complex organisms.

In order to dock at the inner nuclear membrane, the lamin proteins must undergo complex posttranslational modifications to generate membrane attachment sites, which once they are located at the inner nuclear envelope site are then cleaved off. B-type lamins interact with the nuclear membrane through a posttranslational lipid modification at a well-conserved site at the end of the tail domain (see Figure 13.3). In A-type lamins, this site is removed by RNA splicing or by posttranslational proteolytic cleavage. Thus, in most cells only B lamins can interact directly with the nuclear membrane. However, in germ cells (sperm nuclei), the C2 lamin has an additional membrane anchorage mechanism in its head domain, allowing it to bind to the nuclear membrane. (For more on the nucleus see *9 Nuclear structure and transport*; for more on cell division see *14 Mitosis* and *15 Cell cycle regulation*.)

Once in place in the nucleus, the lamins form a stable meshwork of filaments along the inside surface of the nuclear membrane, forming the structural framework for this highly active filtering interface between the nucleus and cytoplasm. Lamins in the nucleus interact with a large and growing number of specialized proteins at the inner nuclear membrane, including lamin-associated proteins (LAPs), lamin B receptor, emerin and SUN proteins, plus several others. The Sun proteins connect with the giant nesprins in the lumen between the inner and outer nuclear membrane, and the nesprins cross into the cytoplasm and link up with other cytoskeleton filament proteins, to give a continuity from inside to outside the nucleus.

During mitosis, the nuclear envelope of mammalian cells is broken down, and for this to happen the nuclear lamina must be disassembled, as **FIGURE 13.17** shows. During prophase, lamins are phosphorylated in the head and tail domains (see Figure 13.3) by the mitotic kinase Cdk1, leading to lamin disassembly. A-type and B-type lamins behave differently in mitosis: B-type lamins remain associated with nuclear membrane vesicle fragments throughout mitosis, whereas A-type lamins are dispersed throughout the cytoplasm. At the end of mitosis when the nuclear envelope reforms, the lamins progressively associate with the condensed chromatin.

Like other intermediate filaments, mutations in lamins are pathogenic, but in the case of these type V genes, the diversity of disease phenotypes is bewildering. Mutations in A-type lamins are associated with diverse genetic disorders that variably affect muscle, nerve, and adipose tissues (mostly). These range from familial partial lipodystrophy (loss of fat from some body areas plus diabetes) to forms of progressive peripheral neurodegeneration (Charcot–Marie–Tooth disease type 2B1) to premature aging (Hutchinson–Gilford progeria). It has been very hard to understand how disruption of one gene can give rise to so many different diseases. Mapping of mutations suggests some functional clustering, but this remains to be confirmed. In cultured cells, disruptive mutation or ablation of lamins results in

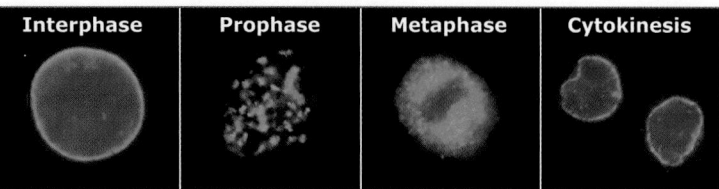

FIGURE 13.17 Immunofluorescence labeling of lamin B in fibroblast cells at progressive stages of the cell cycle. Lamin B staining localized to the nuclear envelope appears veil-like in interphase but fragments as the lamins become phosphorylated in prophase. Lamin B stays dispersed through metaphase and reassociates with chromatin in telophase to produce a new nuclear envelope around each daughter nucleus at cytokinesis. Photos courtesy of John Common and Birgit Lane, Institute of Medical Biology, Singapore.

a less robust nuclear envelope that is unable to maintain a regular shape, shows spontaneous herniation, and is less resistant to mechanical stress. And as more lamin-interacting proteins are identified at the inner nuclear membrane, genetic disorders are emerging with phenotypes mimicking the laminopathies but with genetic basis in defects of another nuclear envelope protein (such as emerin mutations causing a not-uncommon form of muscular dystrophy). Mutations in human B-type lamins have not been identified and are probably lethal since B lamins are expressed very early in embryogenesis and in all cells. However, mice expressing a truncated lamin B1 have a pleiotropic range of developmental defects that are reminiscent of many of the human disorders associated with mutations in A-type lamins and are lethal soon after birth. This mouse model may shed some light on the mechanism underlying the human diseases.

The study of pathologic consequences of failure of protein function is uniquely informative in understanding a protein's function in biology, and this has certainly been the case with intermediate filaments. Although we cannot claim to understand the disease mechanisms fully, in the case of keratins, desmin, GFAP, and neurofilaments, there is an intuitive link between the clinical phenotype of the disease and an underlying cell fragility. However, the increasing identification of highly diverse "laminopathies" raises a question of whether all these disorders are really only caused, primarily or secondarily, by insufficient mechanical resilience of cells and tissues. Many physiologic stresses could of course translate into mechanical stress at the cellular level, such as chemical stress leading to osmotic swelling. Alternatively, the laminopathies may reflect accelerated, selective loss of damaged cells from

13.8 Even the divergent lens filament proteins are conserved in evolution

Key concepts

- The eye lens contains two highly unusual intermediate filament proteins, CP49 and filensin, which constitute the type VI sequence homology group.
- These unusual intermediate filament proteins are conserved in evolution of vertebrates.

The cells of the vertebrate eye lens show an extreme form of differentiation with highly restrictive criteria for the tissue to function correctly. The cells must have significant stiffness with some elasticity, to allow accommodation of the lens to different focal distances, they must remain as translucent as possible to allow the undistorted passage of light, and they must persist in this pristine state throughout the lifetime of the organism.

The cells of the eye lens contain two unusual intermediate filament proteins that coassemble to form "beaded filaments," so called because of their lumpy outline when visualized by electron microscopy (other intermediate filaments are smooth in outline). These proteins are CP49 (or phakinin) and filensin, which are categorized as type VI proteins. Both proteins share structural features of various other sequence homology types, but neither is consistently aligned to any of the other groups. One of the most striking differences lies in the "hallmark" amino acid sequence motif of intermediate filaments, the helix termination motif. In CP49, the rod domain terminates with the sequence Tyr-His-Gly-Ile-Leu-Asp-Gly-Glu, instead of the highly conserved Tyr-Arg-Lys-Leu-Leu-Glu-Gly-Glu seen in other intermediate filament proteins.

Although the CP49 and filensin genes are so divergent from the other intermediate filament genes in the body, their unusual characteristics may have some selective advantage in the eye lens, as clearly homologous lens proteins are conserved throughout vertebrates. A CP49/phakinin sequence is recognizable in the genome of the puffer fish *Fugu ribripes*, a vertebrate that is very distant in evolution from mammals. The sequence conservation is evident even in the unusual sequence of the CP49/phakinin helix termination motif.

Lens cell proteins require unusual properties for a number of reasons. First, they allow cells of the eye lens to develop with great optical clarity. In addition, because the lens is long-lived, the polymer structures in lens cells must have exceptional biochemical stability to avoid proteolytic degradation and the accompanying changes in protein conformation that could lead to changes in the optical or physical properties of the lens and loss of function. It is presumed that the unusual sequence and morphologic characteristics of CP49 and filensin in some way adapt them to meet these challenges. Animal models in which lens filament protein expression is altered have confirmed a consequential deterioration in the function of the eye lens, and dominant mutations in CP49 have been identified as causing early onset familial cataracts in humans.

13.9 Posttranslational modifications regulate and remodel intermediate filament networks

Key concepts

- Intermediate filament networks are stable but filaments are dynamic.
- Phosphorylation is the main mechanism for intermediate filament remodeling.
- Phosphorylation and O-glycosylation of head and tail domains modulates assembly.
- Proteolytic degradation regulates protein quantity and facilitates apoptosis.

Intermediate filaments were long considered to be inactive and inert due to the difficulty in disrupting them experimentally. Improved techniques for live cell imaging have revealed a different picture. Although the overall network configuration may be stable over a long period of time, with only slight fluctuations detectable, at the level of protein subunits there is a great deal of flux in and out of the filaments. Unlike microtubules, there is no evidence for any organizing center that preferentially triggers nanofilament assembly in cells. The proteins assemble readily and without a need for cofactors *in vitro*, and the apolar nature of the filaments means that assembly can take place from either end. Free ends of long intermediate filaments are rarely if ever seen in cells, but tiny fragments and moving particles can be observed by fluorescence tagging of living cells in culture, as shown in **FIGURE 13.18**. Techniques such as fluorescence recovery after photobleaching confirm that intermediate filament protein constantly moves in and out of the filament network. Such fragments are especially apparent at the cell periphery, which may be a preferred zone for assembly or disassembly.

This background flux, as well as large-scale remodeling for specific events like mitosis or initiation of cell migration is likely to be mediated by phosphorylation—the major mechanism for nanofilament remodeling. To reshape a filament network or redistribute the proteins in the cell, localized and controlled disassembly and reassembly is required. Cell division is one event during which this occurs, and phosphorylation of intermediate filament proteins is important during mitosis. To break down the nuclear lamina and allow chromosome separation, phosphorylation of lamin proteins (A-type and B-type) by the mitotic kinase Cdk1 drives nuclear envelope disassembly. At mitosis the cytoplasmic intermediate filaments must also temporarily become looser, to allow partitioning of the cell into two daughters. The type I, type II, and type III proteins are phosphorylated at prophase and cytokinesis; vimentin, GFAP, nestin, and K18 each contain a consensus site for phosphorylation by Cdk1, and other kinases may also play a role. In some epithelial cells in culture, disruption of the cytoplasmic filament system can be clearly seen during the lead up to mitosis as a redistribution of keratins from filaments into aggregates, which then reform into filaments after cytokinesis. Within tissues, the cytoplasmic intermediate filaments may not completely break down to form aggregates but may instead just loosen or be disassembled locally at the cleavage furrow.

Phosphorylation is also the principle mechanism by which intermediate filaments are remodeled outside of cell division. The regulation of assembly and disassembly of intermediate filaments does not depend on cycles of nucleotide hydrolysis (as for actin and tubulin) but rather on phosphorylation and dephosphorylation. Multiple serine and threonine phosphorylation sites have been identified in intermediate filament proteins from all sequence homology groups, and their effector kinases are known in several cases, but complete mapping has been done for only a few intermediate filament proteins. Phosphorylation target sites in keratins, nearly all Ser/Thr, have been studied in K8, K19 and K18, and are predominantly situated in the nonhelical head domain and to a lesser

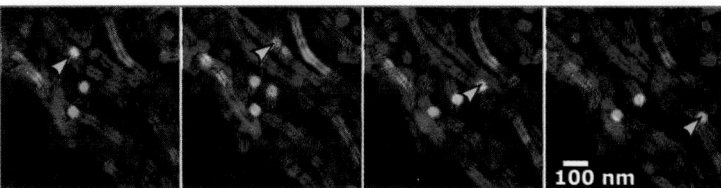

FIGURE 13.18 Particles of intermediate filament protein move around the cytoplasm along different types of cytoskeleton filaments, probably carried by motor proteins. In a time-lapse sequence, a keratin particle labeled with green fluorescent protein (arrowheads) is tracked moving across the field, following a microtubule (in red; imaged as edges) that goes in and out of the plane of focus. Photos courtesy of Mirjana Liovic and Birgit Lane, College of Life Sciences, University of Dundee.

extent in the tail domain (see Figure 13.3). Type III proteins have most active sites in the head domain, and the long tail domains of type IV and V proteins have many important regulatory phosphorylation sites. Neurofilament tail domains are highly phosphorylated; this is thought to determine effective axonal architecture because the resulting negatively charged tail domains repel one another and extend at right angles from the axis of the filament. Phosphorylation also regulates the interaction of intermediate filament proteins with some signaling molecules. Simple keratins are phosphorylated in response to stress by stress-activated protein kinases. Nestin is phosphorylated to modulate its expression during myoblast differentiation. Phosphorylation would result in an increase in negative charge and is usually expected to decrease in filament assembly. However the complex nature of some of the predicted target sites, such as the multiple runs of serines in the head domain of type II and III proteins, suggests that multiple sequential phosphorylation is taking place and the outcome cannot be easily predicted. Average occupancy appears to be low and dwell time of the phosphates is very variable between proteins, again suggesting a highly dynamic situation which is difficult to analyze experimentally. A high capacity for phosphate absorption by intermediate filaments may also be an end in itself, allowing these proteins—well known to protect cells against many forms of stress—to act as phosphate "sinks" to buffer detrimental or destructive phosphorylation storms.

Other posttranslational modifications detected on intermediate filament proteins include farnesylation and myristoylation of lamins for membrane anchorage, transglutamination of certain keratins, disulfide bond formation, and glycosylation which may work in concert

Intermediate filaments, by comparison with the better known actin and tubulin systems, form networks in the cell that resisted analysis for a long time as they appear to be inert and nonresponsive to experimental drug treatment. The identification of the link between intermediate filaments and human genetic diseases was a critical one that changed the way people thought about these cytoskeleton structures. As is the case with many paradigm shifts in science, when the discoveries came they arose within a year from several directions. Three independent labs, using three different approaches—traditional genetics, animal models and immunochemistry—all came separately to the realization of the disease link between keratins K5 and K14 and the skin blistering disease EBS. None of these studies would have been possible without close links between the basic scientists and the clinical community.

In one study, Ervin Epstein, a well-connected San Francisco dermatologist with a wide patient base, was researching the genetic bases of various skin disorders. He and his colleagues were able to map the disease association to the type II keratin gene cluster on chromosome 12 from a traditional genetic mapping analysis of patient families. From there they were able to identify a dominant mutation in helix 2B of K14 in a patient with EBS.

Studying keratin biology in Chicago, Elaine Fuchs's lab became the first into print with the EBS-keratin association after their attempts to make mouse models with defective keratins led to a dramatic skin loosening phenotype: on seeing this phenotype a clinical colleague alerted them to the resemblance with EBS. This helped the lab to map a dramatic phenotype in a mouse model to a documented but rare human disease, and thence by sequencing to identify K14 mutations in human patients and uncovering the major dominant hotspot for severe EBS mutations in K14, in the helix initiation motif of rod domain 1A.

In the third study, Birgit Lane's lab in the UK had been investigating disruption of keratin filaments in culture, looking at the keratin aggregates formed during mitosis and after microinjecting antibodies. A dermatology collaborator, Robin Eady, showed them electron micrographs of EBS with basal cell aggregates remarkably similar to the experimentally disrupted keratin. After demonstrating by immuno-electron microscopy that the aggregates were indeed basal cell keratins, they homed in on the EBS defect by immunoblotting, noting selective loss of reactivity of patient skin keratins to a monoclonal antibody with a known epitope target. This alerted them to a K5 defect and to the domain in which the mutation was located. Sequencing then uncovered the second hotspot, in the helix termination motif, and this time in K5.

Leaving aside the tensions over publication once it became clear what everyone was doing, the three convergent publications within 6 months made an unequivocal case for the causative role of keratin mutations in EBS, and clearly showed that keratin intermediate filaments must have a significant function in maintaining the physical resilience of epidermal keratinocytes *in situ*. This was the first clear demonstration of intermediate filament function, confirming the prevailing hypotheses that these filament proteins were involved in mechanical stability of the tissues. The results implicated both conserved ends of the keratins, and both of the copolymerizing partner keratins of the basal epidermal cells that break down in EBS, in the disease mechanism.

The EBS findings naturally opened a floodgate of interest in genetic skin diseases and intermediate filaments, as many laboratories now went straight for a wide variety of tissue fragility disorders and found that, indeed, very many of them were caused by keratin mutations. The clinical phenotypes of some of these diseases were so diverse, and some diseases so rare, that only an experienced dermatologist in a wide clinical network would be able to find them. In the first few years many of these mutations were published from the UK—not because the technology was particularly advanced but because of the wide clinical network of the National Health Service, with its good communications and nationally integrated patient records, that allowed scientists to work with a coherent clinical network.

Taking advantage of the extensive knowledge accumulated since the 1970s on tissue-specific expression of intermediate filament proteins, the candidate disease approach paid off in spades—first in keratins, and then increasingly with other intermediate filament genes. Today there are over 1500 separate mutations published (see http://www.interfil.org), involving 36 genes of all intermediate filament types. At least 86 distinct clinically distinct pathologic conditions link intermediate filaments to human diseases—rare diseases individually, but together adding up to a significant health burden with common challenges to society and most likely, ultimately, with common routes to therapy.

The role of patient advocacy groups cannot be underestimated here either. These groups, such as the highly successful organization DebRA (Dystrophic Epidermolysis Bullosa Research Association) rooted in the UK, are highly motivated to support research into their rare diseases and can provide strong advocacy for the research. Scientists can help the patients too through these networks. Although we may still be a few years from any practical treatment for EBS and similar diseases, feedback from the scientific community can be hugely supportive to the patient families, in a world where very few people understand their rare disorder.

with phosphorylation. These modifications to intermediate filament proteins are mostly tissue specific. Transglutamination of keratins occurs in epidermal keratinocytes and in the hair follicle. In the epidermis, this modification contributes to the formation of the cornified envelope of the terminally differentiating keratinocytes that makes them highly strain resistant. Transglutamination also contributes to the formation of hard cellular structures in the concentric layers of the forming hair. Cysteine residues are not common in intermediate filament proteins, but disulfide bond formation occurs in some keratins during maturation of hard keratin structures and terminal differentiation of keratinocytes. Glycosylation (O-linked N-acetyl glucosamine) has been detected on K18, K13, NF-M, and NF-L. In K18, it appears to be associated with the soluble or unpolymerized state of the protein and to allow a reservoir of monomers to accumulate.

Proteolysis of intermediate filaments is important for clearance of apoptotic cells and helps regulate the quantitative balance between two coexpressed keratins, as demonstrated by transfection experiments. Keratin filaments can only be formed from type I–type II heterodimers, and if excess keratin of one type is synthesized in a cell, the excess unpolymerized keratin is cleared by proteolysis. Ubiquitylation of intermediate filament proteins marks them for proteolysis and has been detected in the protein accumulations characteristic of a variety of diseases from neurodegenerative disorders to liver cirrhosis. Ubiquitylation can also be inhibited by phosphorylation and like phosphorylation, ubiquitylation can be associated with stress.

Intermediate filament proteins are also targeted by caspases as part of the rapid clearance of cells by apoptosis, and caspase cleavage sites have been identified in vimentin, lamins, and K18 (see Figure 13.3). Caspase action also appears to be modulated by phosphorylation events. The fate of intermediate filaments in apoptosis has been most studied for lamins and keratins in epithelial cells, which are the cells most often involved in major morphogenetic events of development and cancer (where failure of apoptosis contributes to disease progression). Both lamins and type I keratins have a conserved caspase 6 target site in the linker at the midpoint of the α-helical rod domain. Type II keratins lack this site, but as they are unable to polymerize without type I keratins, it is only necessary to destroy one of the two types of keratin in order to disrupt the whole network in a cell. There is also a caspase target site in the C-terminal part of K18, which is cleaved by caspases 3 and 7 before any other evidence of apoptosis is apparent (such as DNA fragmentation or loss of membrane polarity) and creates a useful and unique antibody-detectable epitope for monitoring early stages of apoptotic activity. (For more on caspases see *16 Apoptosis*; for more on cancer see *17 Cancer—Principles and overview*.)

Concept and Reasoning Check

1. How, where, when and why are intermediate filament systems remodeled?

13.10 Interacting proteins facilitate secondary functions of intermediate filaments

Key concepts

- Intermediate filament proteins do not need associated proteins for their assembly.
- Intermediate filaments connect to protein complexes for effecting secondary functions.
- Specific intermediate filament-associated proteins include cell–cell and cell–matrix junction proteins and terminal differentiation matrix proteins of keratinocytes.
- Transiently associated proteins include the plakin family of diverse, multifunctional cytoskeletal linkers.

Cytoskeleton-associated proteins can serve many purposes. Actin-associated protein often catalyze or restrict filament assembly, but intermediate filament proteins do not require association with other proteins in order to polymerize and form filaments, at least *in vitro*. Filament bundling activities are also intrinsic to certain intermediate filaments, for example, keratins, which have a tendency to aggregate laterally and form bundles. However, intermediate filaments do interact with other proteins for the cell to function effectively, such as specific multiprotein complexes anchoring at the cell periphery. These interacting proteins are diverse, often multifunctional, and only rarely specific for a particular type of intermediate filament proteins. Intermediate filaments interact transiently with plakin proteins and microtubule motor proteins. Specific and selective interactions have been best defined between lamins and nuclear envelope components, between keratins or desmin and junction proteins, and between keratins and terminal differentiation matrix molecules. In the latter case interactions are essentially irreversible.

The most widespread intermediate filament-associated proteins are the cytoskeleton linker proteins known as plakins. Plakin proteins, also called cytolinker proteins and in a larger grouping spectraplakins, form a loose family of large multifunctional proteins that interact with intermediate filaments, actin filaments, and microtubules. Those binding to intermediate filaments include plectin, desmoplakin, BPAG1 (BP230), envoplakin, and periplakin. Most of these proteins have multiple isoforms generated by alternate splicing. They also contain protein-binding domains for multiple cytoskeleton proteins, in many cases binding actin filaments and/or microtubules at the same time as intermediate filament proteins, and most are expressed in the epidermis as well as other tissues. Desmoplakins are major structural components of the desmosomes of keratinocytes in stratified squamous epithelia, and BPAG1 and plectin occupy equivalent locations in hemidesmosomes. Plectin forms links between intermediate filaments and microtubules, as the electron micrograph in **FIGURE 13.19** shows, and contributes to the integration of different cytoskeleton systems in the cell. Members of the plakin family have also been associated with tissue fragility disorders, such as muscular dystrophy with skin blistering caused by mutations in plectin.

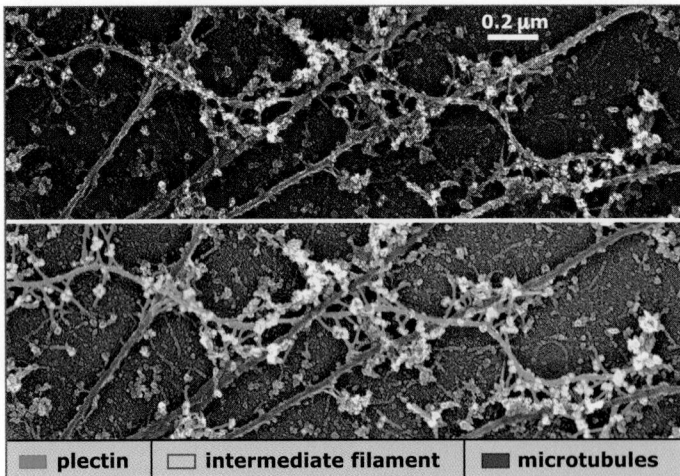

FIGURE 13.19 Plakins bind to intermediate filaments and integrate them with actin filaments and microtubules. These scanning electron micrographs show microtubules and intermediate filaments interacting with the large plakin protein, plectin. The bottom image is a false colored version of the top micrograph. Plectin has been conjugated to gold particles (yellow). Photos © Svitkina, Verkhovsky, and Borisy, 1996. Originally published in **The Journal of Cell Biology**, 135: 991–1007. Used with permission of Rockefeller University Press. Photos courtesy of Tatyana M. Svitkina, University of Pennsylvania.

Another group of proteins that transiently associate with intermediate filament proteins are the dynein and kinesin motor proteins, which move "cargo" along microtubules; intermediate filament proteins may be one type of cargo. Particles containing green fluorescent protein (GFP)–labeled keratin have been observed by live cell imaging to move along microtubules (see Figure 13.18), as well as actin filaments.

In addition, some proteins, such as filaggrin, interact with keratin filaments during terminal differentiation, supposedly by facilitating compaction and flattening of the late-stage keratinocyte to form dense and physically resilient complexes in the uppermost protective layers of the epidermis of the skin. Mutations in filaggrin cause poor compaction, dry skin and eczema. In the formation of special epithelial appendage structures, keratins interact with proteins specific to that differentiation pathway, as in the extensive interaction between trichocyte keratins and keratin-associated proteins (KAPs) of the forming hair shaft. KAPs are small proteins that have either high sulfur content or many glycine and tyrosine residues. These proteins are rich in cysteine residues and make extensive cross-links with keratins to form a dense matrix in cells such as those of hair and nails to produce the overall hard tissue structure.

Intermediate filaments polymerize rapidly and readily and often in great numbers, but for intermediate filaments to function optimally they need to connect to effectors and protein complexes of these secondary functions, to link their robust and resilient filament systems into the specific protein machinery carrying out the differentiated function of the cell. Ultimately, it's not what you are, but where you are.

13.11 Intermediate filament genes are represented through metazoan evolution

Key concepts

- Intermediate filament genes are present in all metazoan genomes that have been analyzed.
- The intermediate filament gene family evolved by duplication and translocation, followed by further duplication events.
- Humans have 70 genes encoding intermediate filament proteins.
- Human keratin genes are clustered, but nonkeratin intermediate filament genes are dispersed.

Cytoplasmic intermediate filaments appear to have evolved as a way of reinforcing cells and tissues in motile multicellular organisms that have not gone down the evolutionary route of external cellular support (such as a hard exoskeleton). Within the Metazoa there are two lineages of intermediate filament genes: one is lamin-like (L) with the longer helix 1B domain, and the other has the short helix (S) like the human cytoplasmic intermediate filaments. These two forms diverged early after the split between invertebrate protostomes, and the deuterostomes, which include vertebrates (animals with backbones) and other chordates with a notochord structure (the forerunner of a proper backbone). The protostomes have only the L form, whereas both forms are present in all chordates, with the L form in the modern lamins and the S form in modern cytoplasmic intermediate filaments. This strongly supports the probability of a lamin-like ancestor of the intermediate filament gene family.

The intermediate filament proteins of invertebrates and vertebrates have many other similarities, indicating features that were present before the divergence of protostomes and deuterostomes. For example, some intermediate filament proteins of the worm *Caenorhabditis elegans* have assembly properties similar to keratins, indicating an early evolution of obligate heteropolymeric interactions. In contrast, the fruit fly *Drosophila melanogaster* has two lamin genes but no cytoplasmic filaments: in animals with an exoskeleton, evolution of an intermediate filament has not been so necessary. *Drosophila* appears instead to have adapted microtubules to fulfill some of the reinforcing functions that its cells still require.

The expansion of intermediate filament gene families can be traced in parallel with the increasing evolutionary complexity of the chordates. For example, one of the most primitive chordates, the tunicate *Ciona intestinalis*, has 5 intermediate filament genes, whereas *Branchiostoma*, an organism closer to vertebrates and invertebrates, has 13, the puffer fish *Fugu ribripes* has over 40, and *Homo sapiens* has 70.

The location of introns in genes can give clues to their evolutionary history. Most intermediate filament genes have 5 to 7 introns in similar positions spread through the coding region of the rod domain, as **FIGURE 13.20** shows. At least 5 of these intron positions are conserved between the human genome and a distant vertebrate genome (the puffer fish, *Fugu ribripes*). The different intron pattern in

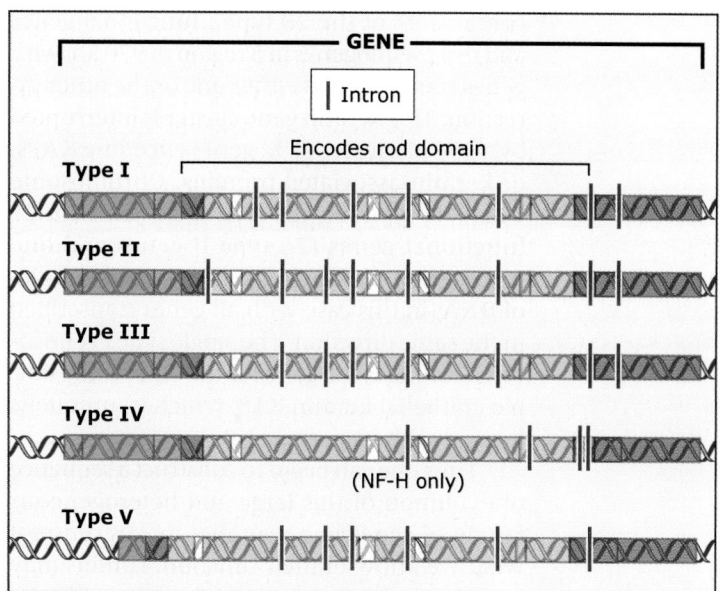

FIGURE 13.20 Relative positions of conserved introns (shown as vertical red lines) in types I–V human intermediate filament genes (excluding some variable tail introns).

the type IV group suggests a different evolutionary origin of neurofilaments from that of the other intermediate filaments, possibly a reverse-transcriptase–based mechanism, for example, as several introns appear to be missing.

Pseudogenes lacking introns are interpreted as processed pseudogenes introduced by the action of retroviruses early in development (retroviruses infect cells with their RNA genomes, which can be reverse transcribed to generate DNA that is integrated into the host genome). Several intermediate filament genes have one or two pseudogenes, but the primary simple keratins K8 and K18 have an exceptionally large number of pseudogenes: 35 for K8 and 62 for K18, plus many fragments, scattered throughout the genome. This is typical for genes expressed at the start of development. Because the *KRT8* and *KRT18* genes are transcribed early in embryogenesis when there are few cells, any retroviral transpositions derived from them can still be incorporated and retained in the germ line and thus fixed in the genome.

The proximity of related genes in the genome can give indications of how these genes arose. The two families of human keratin genes are tightly grouped in two compact loci on chromosomes 12 and 17, suggesting that they originated through duplication events. The type I keratin cluster on chromosome 17q21.2

contains 27 of the 28 type I functional genes and five pseudogenes in a region of 970 kb, with genes transcribed in either one or the other direction. This keratin gene cluster is interrupted by a 350 kb cluster of 32 genes encoding KAPs, or keratin-associated proteins. Chromosome 12q13.13 contains the type II cluster, with 27 functional genes (26 type II genes and one type I gene) plus eight pseudogenes in 780 kb of DNA, in this case with all genes transcribed in the same direction. The single type I gene located on 12q is the gene for the embryonic simple epithelial keratin K18, which is, uniquely, located with its partner K8.

Thus one can begin to construct a sequence of evolution of this large and heterogeneous family of genes in a way that makes sense of what we know of their function. Lamins may have been the first "priority" in evolution, as a reinforced cage to preserve the integrity of long threads of DNA would be favored early on by natural selection. A lamin gene must have shed both its nuclear targeting and anchoring sequence features to evolve into a cytoplasmic intermediate filament gene. Intermediate filament sequences are well conserved in mammals and obviously diverged before mammalian radiation and speciation. During chordate evolution, the keratins diversified by duplication and translocation events as epithelial complexity became advantageous. K8 (type II) and K18 (type I) are probably the oldest keratins and appear to have diverged first. Embryonic K8

is more closely related to the type III proteins, and a duplication plus loss of H1/H2 domains would have resulted in a keratin no longer able to assemble on its own—an ancestral K18, or precursor type I protein. Duplication and translocation then possibly separated K18 from the proliferation of type I keratins, and more extensive parallel duplication events amplified both loci, now on chromosomes 12 and 17 in humans. The trichocyte or hair keratins appear to be the latest product of this keratin evolution, leading to fur, hair, and claws in mammals, and "recent" changes between humans and the great apes shows that this process is still ongoing. This proposed evolutionary sequence is illustrated in **FIGURE 13.21**. Because they have a unique gene structure, the type IV proteins are difficult to place in this scheme and are not shown.

Finally, in the plant or fungal genomes examined so far, there do not appear to be any sequences with recognizable characteristics of intermediate filaments. Therefore, although lamins appear to be so old and so essential in animal cell lineages, adequate nuclear organization is clearly achievable without lamins, providing a tough cell wall is in place. An intermediate filament-like protein has been identified in bacteria—the so-called crescentin protein in the curved *Caulobacter crescentus* species, and an analogous gene in *Helicobacter pylori*, the common gastric pathogen linked to stomach ulcer susceptibility. In both cases, these proteins are likely responsible for maintaining cell shape, and as such they show some functional homology with animal intermediate filament proteins. However, they are not recognizable as intermediate filaments from their amino acid sequence. Such convergent evolution provides further support for the hypothesis that the intermediate filament gene family has evolved to be so large and diverse in the Metazoa because these proteins are essential for maintaining cell shape in tissues.

Concept and Reasoning Check

1. Discuss possible reasons why mammals have evolved so many different intermediate filament genes.

13.12 What's next?

After a burst of classificatory information gathering in the 1970s, the acquisition of significant understanding of intermediate filament func-

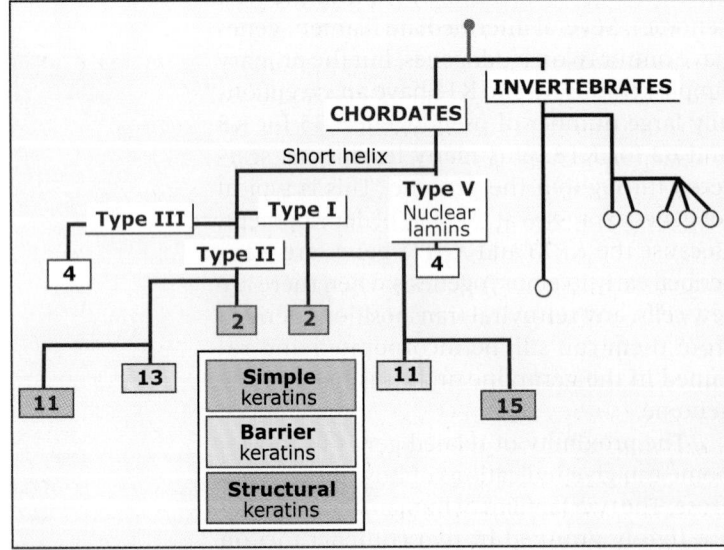

FIGURE 13.21 Possible evolutionary relatedness between human types I–III and V intermediate filament genes. Blue lines (invertebrates) are only indicative; boxed numbers indicate the number of genes in each group.

tion was slow until the technological advances of the 1990s. DNA sequencing became faster, microscopy became more ambitious, and live cell imaging began to take off. A leap forward in clarity came from recognition of the devastating effect of defective intermediate filaments in pathologic situations. The use of live cell imaging, increasingly in thick specimens, is now taking us into another age of cell biology, in which protein dynamics of the cytoskeleton will be better understood in a functional context.

What next? The questions are being defined by these new technologies. We need to look at intermediate filament kinetics in living cells, and we need to know more about how the kinetics are regulated in living tissues. Only then can we begin to understand how intermediate filaments, which evolved for mechanical resilience, can alter cell behavior and gene expression when the external environment demands it, be it during development or in disease. Little of a cell's resilience can be critically evaluated when it is grown in a petri dish. Microscopy and culture techniques for thick specimens must be improved so we can observe cells within living organs at high resolution and avoid fixation artifacts.

The function of simple epithelial keratins is a matter of current debate. Expression of K8/K18 clearly protects tissues and cells against a variety of stresses, from chemical toxic stress to apoptosis, and human mutations in K8/K18 are risk factors for several disorders. Signal transduction pathways that involve intermediate filaments are now beginning to be elucidated, but whether the stresses can all ultimately be described as mechanical stress (as is clear for the keratinocyte keratins in barrier tissues) is an area of debate. Intermediate filaments also protect cells against osmotic stress, and it may be that toxic stress signals through intermediate filaments by mechanically distorting osmotic stress, if cell metabolism is stalled and membrane ion pumps are inactivated.

There will also need to be further studies on expression of the newly discovered intermediate filament proteins, and how significant these will be depends to some extent on why these last few genes have been overlooked until now. Some of these proteins have been difficult to analyze *in vitro* because of their extreme insolubility. Some may be minor keratins, either of low functional importance (perhaps newly evolving genes?) or of critical importance at transitional points during development (like nestin), or in very specific tissue situations (like synemin, or lens proteins). Others may have been overlooked because they are expressed in tissues not usually chosen for analysis, or their biochemical characteristics are similar to those of another, known protein.

Over 80 human diseases are already known to be caused by intermediate filament mutations, but there are almost certainly more remaining to be recognized. Some may be clinically cryptic, due to the protective effects of overlapping expression patterns of multiple intermediate filament proteins, or due to inaccessible tissue location so early stage disease is not visible, as probably happens in inflammatory bowel disease. The full spectrum of intermediate filament diseases may only become apparent when we consider the consequences of failure of cell resilience in different tissues, and so can recognize these downstream consequences as they present to clinicians and pathologists. As we come to understand just how widespread intermediate filament failure pathology is, the number of emerging intermediate filament diseases will put more pressure on the development of new therapeutic strategies for these genetic disorders—again requiring some creative thinking to get around the cost implications for individually tailored gene therapy.

13.13 Summary

Intermediate filaments are a major component of the cell cytoskeleton and are essential for maintenance of correct tissue structure and function. They form robust networks of 8 to 12 nm thick filaments (intermediate in diameter between actin filaments and microtubules) in the cytoplasm and nucleus. The proteins are encoded by a large gene family and all share a similar core structure consisting of a central, long α-helical rod domain divided into sections. This is flanked by a head and a tail domain, which are subject to many posttranslational modifications and regulate assembly by their phosphorylation. Assembly *in vitro* is rapid and requires no additional factors, although the kinetics vary from protein to protein. Dimers form antiparallel, tetrameric assembly subunits (therein establishing the nonpolar nature of the final filament) and *in vitro*, these rapidly associate laterally and longitudinally to form a 10-nm filament. Mature intermediate filament networks are highly strain resistant and exhibit strain hardening.

Different kinds of intermediate filaments have highly differentiation-specific tissue expression patterns, and antibodies to them are useful tools for monitoring differentiation in cell biology and pathology. Most of the intermediate filament proteins are keratins (type I and type II) and are expressed in epithelia (sheet tissues). Simple keratins (K8/K18) are the least specialized and probably the oldest keratins evolutionarily, whereas structural trichocyte (hair) keratins are probably the most recently evolved. The keratinocyte group of keratins found in barrier tissues is the most varied, well-developed, and resilient of all intermediate filament protein groups. Type III and type IV groups show overlapping expression ranges and are often heteropolymeric in tissues; they are expressed in connective, nerve, muscle, and haematopoietic tissues. Vimentin represents minimal, nonepithelial intermediate filament expression and is characteristic of solitary cells. Type V proteins, the ubiquitous and ancient lamins, reinforce the nuclear envelope and interact with it via post-translational modifications that produce membrane anchorage sites. Lamins, and probably all other intermediate filament proteins, are disassembled by phosphorylation during mitosis to allow chromosome separation and cytokinesis. RNAs generated from some intermediate filament genes, most notably the lamins, are alternatively spliced.

Some 86 different clinically identified diseases are now associated with mutations in intermediate filament genes. Individually these disorders are quite rare, but taken together, the burden of disease associated with intermediate filaments is quite significant. Disease studies have proved that intermediate filaments provide physical resilience to cells, as mutations leave the tissue cells fragile and easily destroyed in disorders such as the well-studied EBS. Many diseases associated with mutations in cytoplasmic intermediate filaments are characterized by protein aggregates. The extent to which these aggregates contribute to the disease phenotype, in terms of the direct contribution of compromised filament resilience as opposed to secondary pathological signs of disease, is a matter of current debate.

Remodeling of intermediate filaments occurs during mitosis and in cell migration and is principally driven by phosphorylation and dephosphorylation. Proteolytic degradation modulates protein quantity and facilitates apoptosis. Associated proteins are optional for intermediate filament function and are usually differentiation specific; proteins such as filaggrin may modify the filaments for tissue function, whereas others, such as plakins, serve to restrict the filament deployment in a cell.

Intermediate filament genes are present in all metazoan genomes analyzed and probably originated from a lamin-like ancestral gene. Today's intermediate filament gene family has evolved in the genome by duplication, translocation, and further duplication events, as seen most clearly in the dominant intermediate filament group of modern mammals, the keratins. Human keratin genes are clustered on chromosomes 12 (the type II genes and also the type I K18) and chromosome 17 (the other type I genes), whereas the 16 nonkeratin genes are dispersed throughout the genome. Intermediate filament genes are compact, intron positions are generally conserved within sequence homology groups, and most of the genes encode only a single protein.

Finally, concepts of intermediate filament function are gradually becoming clearer, generally centering around mechanical resilience, and through this or in parallel, to resistance to a variety of stresses. They provide potential reaction platforms for efficient solid-phase reactions and can sequester and tether many reactive molecules. Spanning the cytoplasm with their long resilient fibers, their links to functional complexes at the cell periphery positions them well for signaling conduits—evidence for which is growing. And it is possible that before long we may have novel therapies for genetic disorders arising from intermediate filament mutations.

http://biology.jbpub.com/lewin/cells

To explore these topics in more detail, visit this book's Interactive Student Study Guide.

References

Broers, J. L., Ramaekers, F. C., Bonne, G., Yaou, R. B., and Hutchison, C. J. 2006. Nuclear lamins: Laminopathies and their role in premature ageing. *Physiol. Rev.* v. 86 p. 967–1008.

Burke, B., and Stewart, C. L. 2006. The laminopathies: the functional architecture of the nucleus and its contribution to disease. *Ann. Rev. Genom. Hum. Genet.* v. 7 p. 369–405.

Coulombe, P. A., and Omary, M. B. 2002. "Hard" and "soft" principles defining the structure, function and regulation of keratin intermediate filaments. *Curr. Opin. Cell Biol.* v. 14 p. 110–122.

Fuchs, E., and Cleveland, D. W. 1998. A structural scaffolding of intermediate filaments in health and disease. *Science* v. 279 p. 514–519.

Herrmann, H., Hesse, M., Reichenzeller, M., Aebi, U., and Magin, T. M. 2003. Functional complexity of intermediate filament cytoskeletons: from structure to assembly to gene ablation. *Int. Rev. Cytol.* v. 223 p. 83–175.

Lane, E. B., and Alexander, C. M. 1990. Use of keratin antibodies in tumor diagnosis. *Semin. Cancer Biol.* v. 1 p. 165–179.

Omary, M. B., Coulombe, P. A., and McLean, W. H. 2004. Intermediate filament proteins and their associated diseases. *N. Engl. J. Med.* v. 351 p. 2087–2100.

Omary, M. B., Ku, N.-O., Tao, G. Z., Toivola, D. M., and Liao, J. 2006. "Heads and tails" of intermediate filament phosphorylation: multiple sites and functional insights. *Trends Biochem. Sci.* v. 31 p. 383–394.

Owens, D. W., and Lane, E. B. 2004. Keratin mutations and intestinal pathology. *J. Pathol.* v. 204 p. 377–385.

Rezniczek, G. A., Janda, L., and Wiche, G. 2004. Plectin. *Methods Cell Biol.* v. 78 p. 721–755.

Cell division, apoptosis, and cancer

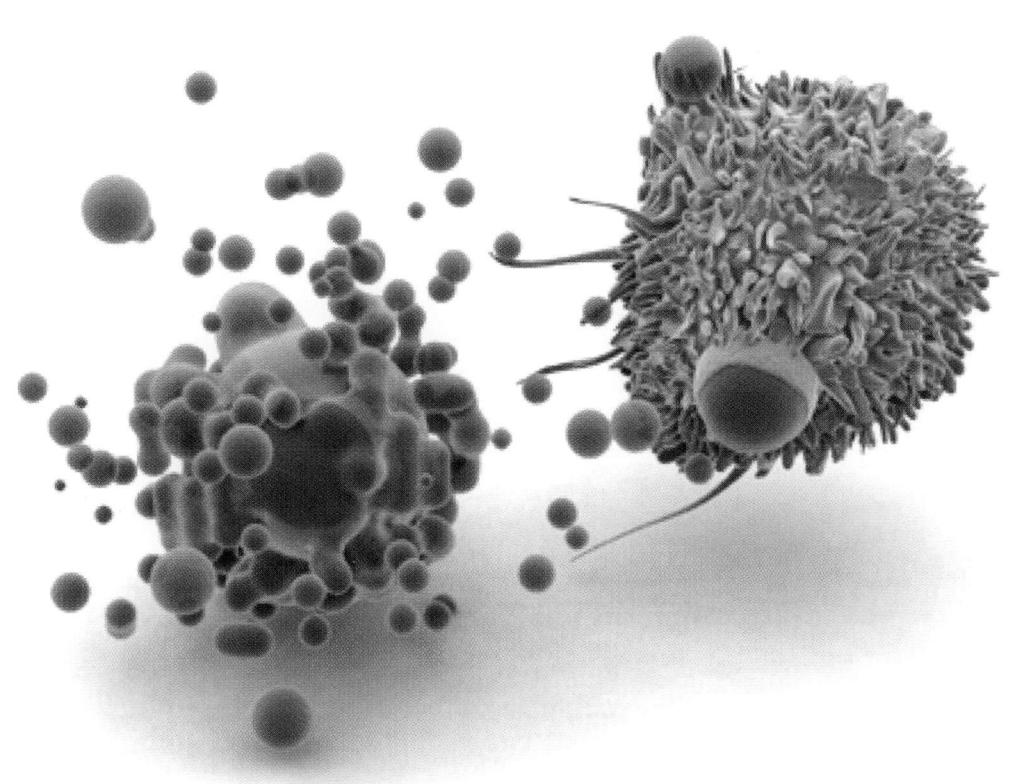

Mitosis

14

Conly L. Rieder

Wadsworth Center, NYS Department of
Health, Albany, NY

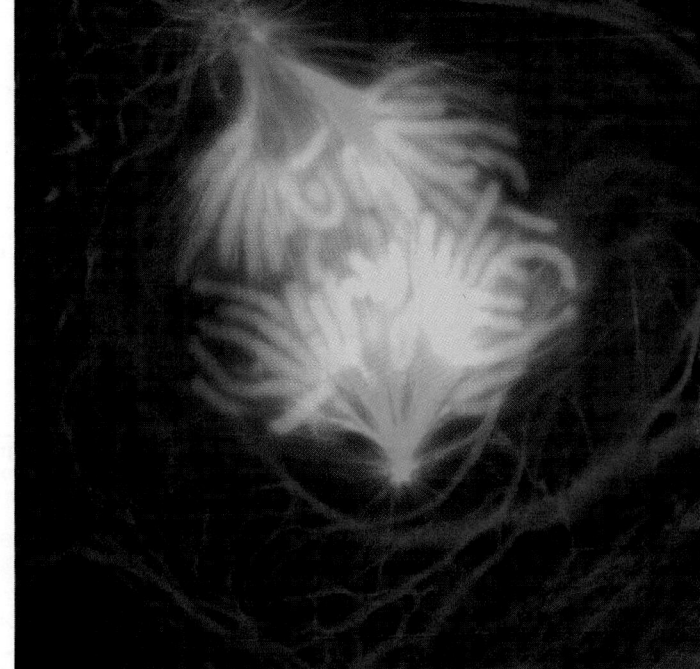

THIS FLUORESCENCE MICROGRAPH shows the anaphase stage of cell division
in a salamander lung cell. During anaphase, the chromosomes are equally segregated into two well-separated daughter nuclei. The cell is stained for DNA (blue),
microtubules (green), and intermediate filaments (red). (Note: Acentrosomal
spindle formation involves nucleation of microtubules by the chromosomes and
the functions of several different types of microtubule-dependent motor proteins.)
Photo © Conly L. Rieder, Wadsworth Center.

CHAPTER OUTLINE

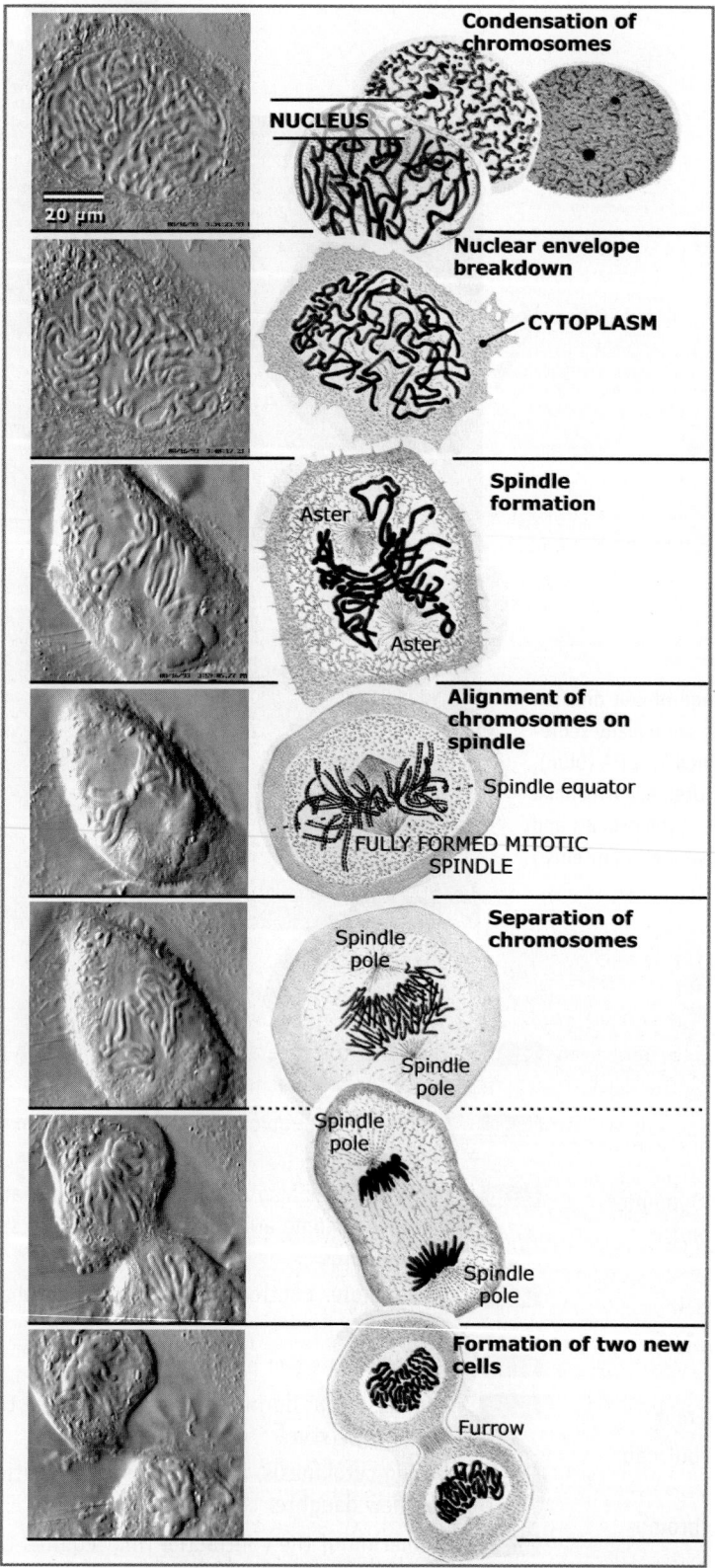

Condensation of chromosomes

NUCLEUS

Nuclear envelope breakdown

CYTOPLASM

Spindle formation

Aster

Aster

Alignment of chromosomes on spindle

Spindle equator

FULLY FORMED MITOTIC SPINDLE

Separation of chromosomes

Spindle pole

Spindle pole

Spindle pole

Spindle pole

Formation of two new cells

Furrow

20 µm

FIGURE 14.1 In the top panel, only the nucleus is shown. The remaining panels show the entire cell. After spindle formation, the two spindle poles are located in the center of the clear areas in the cytoplasm at the upper left and lower right of the cell. Photos © Conly L. Rieder, Wadsworth Center. Illustrations from W. Flemming, *Archiv für Mikroskopische Anatomie*. Berlin: J. Springer (1871).

14.1 Introduction

Key concepts

- All cells are produced by the division of other cells through a process called mitosis.
- Mitosis occurs after a cell has replicated its chromosomes. Mitosis separates the chromosomes into two equal groups and then divides the cell between them to form two new cells.
- Errors in mitosis are catastrophic, and mechanisms have evolved to ensure its accuracy.

Perhaps the most fundamental activity of cells is reproduction: life depends on the ability of cells to divide. In single-celled organisms, to divide is to reproduce. In complex multicellular organisms, division is required not only to produce the cells essential for development and growth but also to replace cells as they die.

The term *cell* was coined in 1665 by Robert Hooke, who used it to describe the hollow cubicles seen in thin slices of cork when viewed under a microscope. It took 175 years of further microscopic observations before Schleiden and Schwann recognized with their *Cell Theory* that cells are the fundamental building blocks of life. As this major landmark of nineteenth-century science gained general acceptance, a logical next question was: How are new cells formed? Although some people believed that new cells arose spontaneously, in 1855 the German physician Virchow made the definitive argument *omnis cellula e cellula*—that every cell is the offspring of a preexisting parent cell.

With the invention and widespread use of the compound light microscope in the late nineteenth century, progress in describing the events that take place as a cell divides accelerated rapidly. In 1879, the German anatomist Walther Flemming used the term **mitosis** to characterize the formation of what resembled paired threads (Greek: *mitos* = threads) inside the nucleus of dividing salamander cells, and he described a series of changes that they underwent, shown in **FIGURE 14.1**. These threads, which formed from a substance within the nucleus that Flemming called **chromatin**, came to be known as the **chromosomes** (Greek: *chroma* = color; *soma* = body). Flemming noted that during the early stages of mitosis, every chromosome consisted of two identical threads, or **chromatids**, that are stuck to one another along their length, as shown in **FIGURE 14.2**. In higher organisms, every chromosome contains a small but conspicuous region at which

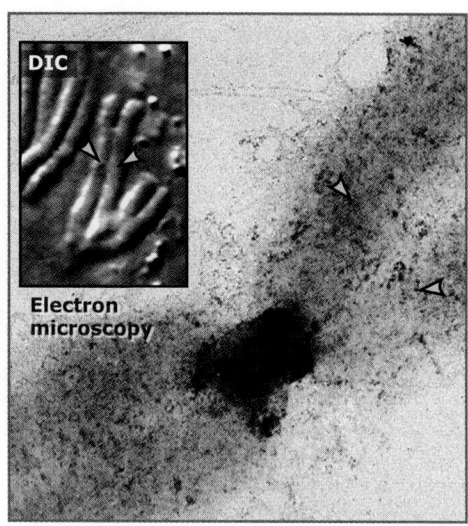

FIGURE 14.2 The inset shows an entire metaphase chromosome in a living newt cell. The larger photograph shows the region around the primary constriction of another metaphase chromosome. In both photographs the arrowheads indicate the paired sister chromatids. Main photo courtesy of Jerome B. Rattner, University of Calgary, Canada; inset photo © Conly L. Rieder, Wadsworth Center.

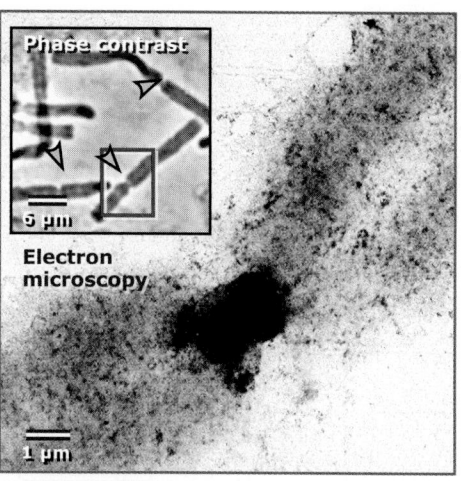

FIGURE 14.3 The inset shows several entire metaphase chromosomes in a live newt cell. Each narrows at a unique point called the primary constriction. The electron micrograph shows a highly magnified view of the primary constriction of a single chromosome. Main photo courtesy of Jerome B. Rattner, University of Calgary, Canada; inset photo © Conly L. Rieder, Wadsworth Center.

it narrows, known as the primary constriction or **centromere**, as shown in **FIGURE 14.3**. Every cell within an organism has the same number of chromosomes, which is the same for all the members of a species. The number of chromosomes per cell differs among species, however—some having many times more chromosomes than others.

As early as 1880, Flemming argued that all cells reproduce through the "metamorphosis of the nuclear mass into threads." By 1883 observations on the fertilization of sea urchin eggs proved that the egg and sperm contribute an equal number of chromosomes to the embryo. Two years later it was shown that all the nuclei in an organism are generated by repeated divisions of the single nucleus formed within the embryo from a fusion between the egg and sperm nuclei. Thus, by 1885 it was evident that every cell contains chromosomes from both parents. This conclusion connected the *Cell Theory* (1838) of Schleiden and Schwann with Darwin's *Theory of Evolution* (1859). The nature of this connection was later established with the discovery that the chromosomes contain a cell's genes, the units that transmit properties between generations.

With the exception of sperm and eggs, all of the cells in the body are **diploid** (di = 2), in that two copies of each chromosome are present: one inherited from the mother via the egg and the other from the father via the sperm. (Human cells contain 23 pairs of chromosomes, giving humans a total of 46 chromosomes.) *The purpose of mitosis is to preserve the diploid number of chromosomes over repeated generations of cells.* Since sperm and eggs are **haploid**, in that they contain only half the number of chromosomes found in tissue cells, they cannot be produced by mitosis. Rather, these specialized cells (called gametes) are formed by a process known as **meiosis**, as shown in **FIGURE 14.4**. During meiosis four haploid cells, each containing just one copy of every chromosome, are produced from one precursor cell. This reduction in chromosome number results from dividing the cell twice after its chromosomes have been replicated, rather than just once as in mitosis. Unlike mitosis, the purpose of meiosis is to maintain the diploid number of chromosomes *over repeated generations of the organism.* In practice, mitosis and meiosis involve many of the same mechanisms—the major difference being how the chromosomes are organized at the start of the process.

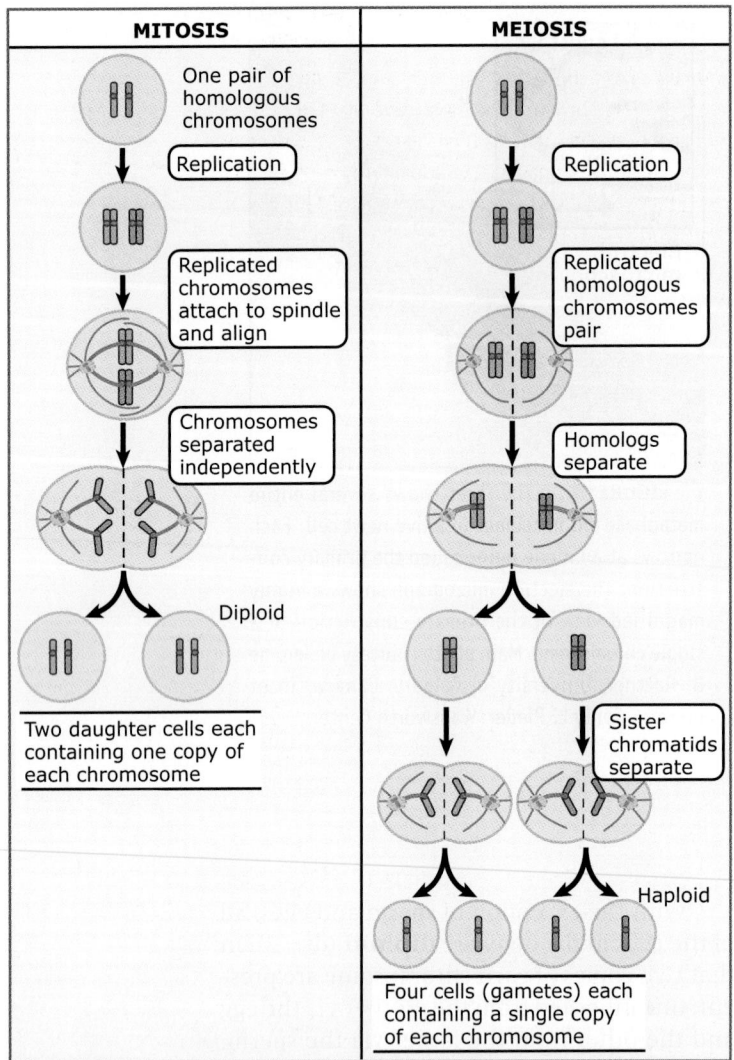

MITOSIS	MEIOSIS

One pair of homologous chromosomes

Replication

Replicated chromosomes attach to spindle and align

Chromosomes separated independently

Diploid

Two daughter cells each containing one copy of each chromosome

Replication

Replicated homologous chromosomes pair

Homologs separate

Sister chromatids separate

Haploid

Four cells (gametes) each containing a single copy of each chromosome

FIGURE 14.4 Meiosis involves two cell divisions in sequence. The first separates homologous chromosomes, the second the individual chromatids of each chromosome. In mitosis only the chromatids are separated.

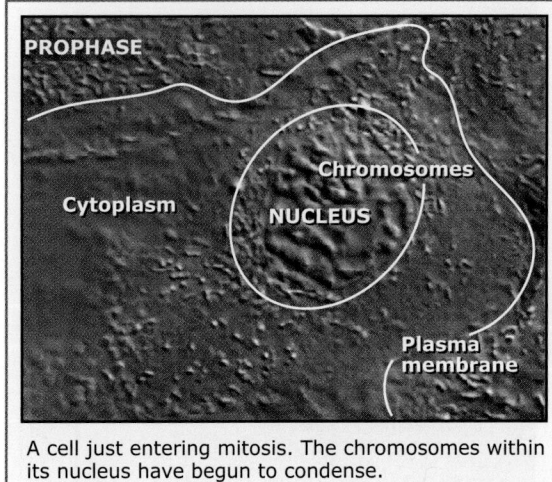

PROPHASE

Cytoplasm

Chromosomes

NUCLEUS

Plasma membrane

A cell just entering mitosis. The chromosomes within its nucleus have begun to condense.

FIGURE 14.5 The first frame of a video that follows the chromosomes through the initial stages of mitosis. Photo © Conly L. Rieder and Alexey Khodjakov, Wadsworth Center.

This chapter will focus on how mitosis works in higher animals, specifically vertebrates. Although the details can vary depending on the organism, the fundamental aspects of mitosis are similar in all cells. The stages can be seen in Figure 14.1. In higher animals, the first visible sign of an impending division is the appearance of the replicated chromosomes within the nucleus. Once this condensation of chromosomes is well underway, the envelope surrounding the nucleus disintegrates, dispersing the chromosomes into the cytoplasm. Next, the chromosomes become attached to a structure called the **spindle**, named because it is shaped like two cones joined at their wide ends. This spindle, or mitotic apparatus, is re-sponsible for generating the forces for moving the chromosomes and also for directing where in the cell they will move. Once attached, the chromosomes gradually become aligned across the middle of the spindle, which is referred to as its equator. A video, whose first frame is shown in **FIGURE 14.5**, shows the entire sequence of events from chromosome condensation to alignment.

After all the chromosomes are aligned, each splits lengthwise (i.e., the chromatids separate), and the two independent groups of chromosomes that result move away from each other toward the opposite ends of the spindle, called the **spindle poles**. Finally, the chromosomes within each of the two separating groups decondense and a new envelope is formed around each. The multiple small nuclei formed at each end of the cell fuse together, giving rise to two independent daughter nuclei. The definition of mitosis has been expanded over the years to also include **cytokinesis**, the series of events by which the cell cytoplasm is partitioned after the nucleus has divided (see Figure 14.1).

Even though chromosome segregation occurs with a high level of accuracy, errors do occasionally occur. Mistakes during mitosis or meiosis can arise at several stages of the processes and can lead directly to cells that contain too few or too many chromosomes. *This condition is known as* **aneuploidy**, *and its consequences vary depending on the organism and the time when*

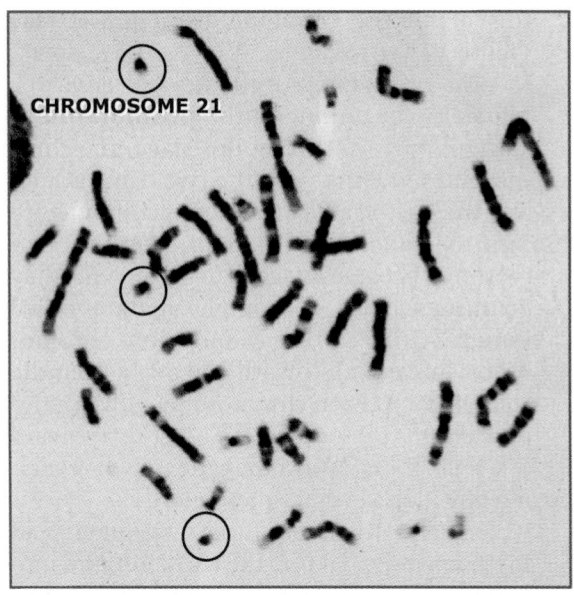

CHROMOSOME 21

FIGURE 14.6 The mitotic chromosomes from a single cell of a human with Down's syndrome. The different chromosomes can be distinguished by the position of their primary constriction, their size, and the pattern of dark and light bands in each. There are three copies of the small chromosome 21, but only two of each of the other chromosomes. Photo courtesy of Ann Willey, Wadsworth Center.

the mistake occurs. When it arises during the production of gametes (meiosis), it will lead to an embryo with a birth defect syndrome in which all of its cells have at least one extra or missing chromosome. An example of aneuploidy in humans is Down syndrome, in which all of the cells of an individual contain an extra copy of chromosome 21, as shown in **FIGURE 14.6**. In most cases, however, aneuploidy in the embryo leads to death before development is complete. By contrast, when aneuploidy arises during development, a **mosaic** organism is formed, in which different tissues consist of cells containing different numbers of chromosomes. Finally, there is good evidence that the formation of aneuploid cells in adult organisms plays a role in initiating some cancers.

Because the equal distribution of the chromosomes is essential to an organism's viability, mitosis includes processes that are devoted solely to enhancing its accuracy. In all organisms, the accuracy of chromosome segregation is increased by **checkpoint controls**. Checkpoints are biochemical pathways that stop or delay division until a specific event is completed or corrected. The need for great accuracy is also reflected by the existence of multiple pathways for accomplishing the same goal, be it forming the spindle or moving the chromosomes. Although mitosis always proceeds through the sequence of events just described, several different possible routes are present to complete the more critical processes. This duplication of mechanisms, which has only recently been recognized, adds an extra layer of complexity to the mitotic process but gives it a flexibility that allows it to withstand conditions that would otherwise result in errors.

14.2 Mitosis is divided into stages

Key concepts

- Mitosis proceeds through a series of stages that are characterized by the location and behavior of the chromosomes.
- Some of the conversions between stages correspond to cell cycle events and are irreversible transitions.

Mitosis occurs through the initiation and completion of two separate and distinct processes. In the first, which is sometimes termed **karyokinesis** (Greek: *karyo* = nut; *kinesis* = division), the replicated chromosomes are separated into two distinct daughter nuclei. During the second process, called cytokinesis, the cytoplasm is divided between these two nuclei to form two independent daughter cells. Historically, the division of the nucleus is broken down into several stages that are defined by the structure and position of the chromosomes. Breaking down a complex series of events like mitosis into stages is useful because some transitions mark irreversible changes that have occurred within the cell. Most of these changes depend on the activation or inactivation of particular enzymes and sometimes also require the timely destruction of specific proteins that play strategic roles during division. Viewing mitosis as a series of stages is also useful because the chromosomes and the spindle both alter their behavior between stages, suggesting that each stage is characterized by specific mechanisms at the molecular level. As we discuss the stages in detail, refer to Figure 14.1.

The first visible sign of an impending division is the appearance of condensing chromosomes within the nucleus. This starts the **prophase** stage of mitosis. In cold-blooded animals that contain large chromosomes (e.g., salamanders, grasshoppers), this stage takes several hours; in warm-blooded creatures with small chromosomes (e.g., mice, humans), it may last less than 15 minutes. At some point in prophase, biochemical changes occur within the cell that commit it to mitosis. Before this **point of no return** is reached, chromosome condensation can be reversed by physical or chemical insults that damage the cell.

Prophase is also commonly marked by the appearance of the **centrosomes**. In many prophase cells, two of these organelles become visible in the cytoplasm as small dots surrounded by a clear area. As we will see, these centrosomes play an important role in spindle formation: not only will they define the two poles of the spindle; they will also nucleate many of the **microtubules** used in its construction.

Cells are driven into mitosis by the addition of phosphate groups to some proteins and their removal from others. Enzymes known as kinases and phosphatases accomplish this phosphorylation and dephosphorylation. During mitosis, the most important kinase is the **cyclin B/CDK1** complex. This enzyme is considered to be the master mitotic regulator because, when it is injected into cells, mitosis is induced. (The discovery of this kinase and the mechanisms by which it is regulated was the subject of the 2001 Nobel Prize in physiology and medicine.) Near the end of prophase, cyclin B/CDK1 accumulates within the nucleus in an inactive form. Shortly thereafter, another enzyme, the cdc25 phosphatase, enters the nucleus, where it activates cyclin B/CDK1. Once activated, cyclin B/CDK1 phosphorylates many nuclear proteins, including those that provide structural support for the membrane surrounding the nucleus. As a result, these proteins lose their association with the nuclear membrane, causing the nucleus to swell until its surrounding membrane envelope disintegrates (see Figure 14.1).

The breakdown of the nuclear envelope marks the beginning of the **prometaphase** stage of mitosis. During this stage, the chromosomes interact with the two centrosomes and their associated arrays of microtubules to form the spindle (see Figure 14.1). As the chromosomes become attached to the spindle, they go through a series of complex motions called **congression**. During congression, chromosomes move both toward and away from the spindle poles. Each chromosome moves independently of the others, moving first toward one pole, then toward the other, often reversing direction several times before the process is complete. Ultimately, these movements lead to the *congregation* of all the chromosomes into a plane, or "plate," at the spindle equator, halfway between the two poles. In most cells, prometaphase is the longest stage of mitosis, since it lasts until all of the chromosomes are positioned at the equator. This may take just a few minutes in embryos or up to several hours in highly flattened tissue cells.

Once all of the chromosomes are near the spindle equator, the cell is considered to be in **metaphase** (see Figure 14.1). Metaphase can last for different lengths of time, depending on the cell type. Surprisingly, the complex events that have brought the cell to this point are reversible. When the spindle in a metaphase or prometaphase cell is destroyed by treating it with drugs (e.g., colcemid or nocodazole) or other agents (e.g., cold or high pressure) that depolymerize microtubules, it reforms and the chromosomes repeat the congression process as soon as the treatment is stopped. Dissolution of the spindle in metaphase cells prevents the cell cycle from advancing and is often used to experimentally produce cells that are considered as being "locked in metaphase." These cells are actually in prometaphase since their condensed chromosomes are scattered throughout the cytoplasm. (For more on microtubules, see *11 Microtubules*.)

Metaphase ends when the two sister chromatids of each chromosome separate, beginning the **anaphase** stage of mitosis (see Figure 14.1). Although each chromosome was replicated before mitosis, its two constituent chromatids normally only become visible as distinct units shortly before metaphase ends (see Figure 14.2). In movies, the separation of chromatids seems to occur suddenly and simultaneously for all chromosomes; in reality,

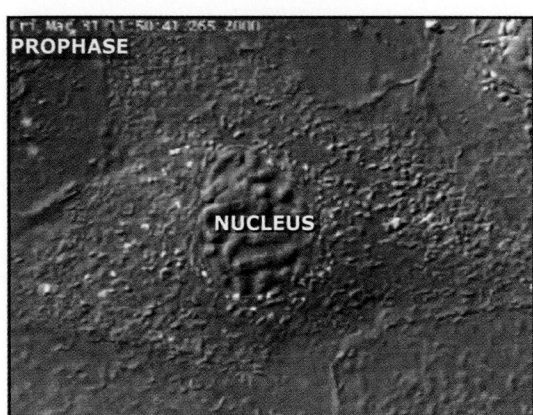

PROPHASE

NUCLEUS

FIGURE 14.7 The first frame of a video that shows mitosis from beginning to end. Photo © Conly L. Rieder and Alexey Khodjakov, Wadsworth Center.

it usually takes several minutes and occurs at a slightly different time for each. *The separation of chromatids at the beginning of anaphase marks another point-of-no-return in mitosis: it coincides with the destruction of "glue" proteins that hold the chromatids together and with the deactivation of the master regulatory kinase that drives cells into mitosis* (for details, see *15 Cell cycle regulation*). After the sister chromatids have separated, they move away from one another toward different poles of the spindle. This movement occurs by a combination of two different mechanisms. During **anaphase A**, the distance between each chromatid and the pole to which it is attached decreases. At the same time, the two spindle poles themselves move farther apart, pulling their attached groups of chromosomes with them in a process known as spindle elongation, or **anaphase B** (see Figure 14.1). As the two groups of chromosomes move apart, the spindle begins to disassemble, and new structures known in animal cells as stem bodies (see Figure 14.55) form between them.

The final **telophase** (Greek: *telo* = end) stage of mitosis begins as the chromosomes start to reform nuclei near the poles (see Figure 14.1). In cases in which neighboring anaphase chromosomes are not touching as telophase begins (as in large cells), each chromosome forms its own small nucleus. These then fuse to form a single, larger nucleus. During telophase, the events that will divide the cell in two also begin. Initially, a furrow forms around the surface of the cell in the same plane in which the chromosomes were aligned at metaphase. In this position the furrow is located midway between the two new nuclei and encircles the stem bodies (see Figure 14.1). Once formed,

the furrow gradually constricts, dividing the cell into two roughly equal lobes in the process of cytokinesis. As the furrow constricts, the stem bodies are gathered together into a tight bundle called the midbody, the last structure that connects the two cells (see Figure 14.55 and Figure 14.56). The events of telophase require the inactivation of cyclin B/CDK1, and they signal that the cell is leaving the mitotic state.

Discussing mitosis as a series of stages and looking at still photographs of live or fixed cells may make it appear as a somewhat static, discontinuous process. However, in reality it is continuous and highly dynamic, something that can only be fully appreciated by viewing movies of dividing cells such as the one whose first frame is shown in **FIGURE 14.7**.

14.3 Mitosis requires the formation of a new apparatus called the spindle

Key concepts

- The chromosomes are separated by the mitotic spindle.
- The spindle is a symmetrical, bipolar structure composed of microtubules that extend between two poles. At each pole is a centrosome.
- Chromosomes attach to the spindle via interactions between their kinetochores and the microtubules of the spindle.

The spindle is a dynamic and complex structure that suddenly appears as cell division begins and then quickly disassembles as the process is completed (see Figure 14.54). *The spindle is required for mitosis and serves two distinct functions: (1) it is responsible for separating the replicated chromosomes into daughter nuclei during division of the nucleus (karyokinesis), and (2) it directs the process of dividing the cytoplasm (cytokinesis).* When the spindle is prevented from forming (e.g., by treatment of the cell with various drugs), the chromosomes condense but do not undergo any of the movements of a normal mitosis and progression through division stops. In many ways, the spindle is a kind of biological machine that converts chemical energy into the mechanical work needed to move the chromosomes and divide the cell. Its function is reflected in its structure. The symmetrical structure of the spindle—with two poles—is essential for a successful mitosis.

Indeed, it defines the inherent "two-ness" of cell division, in which one cell and its replicated DNA are divided equally into two separate daughter cells.

The spindle can be viewed by several means. Microtubules, the spindle's primary structural component, are too small to be seen with the light microscope (i.e., they cannot be resolved). As a result, although the condensed chromosomes can often be seen within the cells of higher animals by traditional forms of light microscopy, the spindle cannot. However, in many cells, the shape of the spindle can be inferred because it excludes most of the visible organelles from its volume. As FIGURE 14.8 shows, this causes the space occupied by the spindle to appear clear relative to the surrounding cytoplasm. Although scientists initially suspected that the spindle consisted of fibers, this idea was not proved until the early 1950s. At that time, refinements in polarization light microscopy allowed the spindle to be seen in living cells. A typical photograph of a spindle taken with this method is shown in Figure 14.8, where the spindle appears dark black because of the interaction between its microtubules and the polarized light. Since the 1970s, powerful fluorescence-tagging techniques have been developed that allow components of the spindle to be viewed in three dimensions, even in living cells (see Figure 14.22). With these techniques the positions of one or more specific proteins within the spindle can be determined and each followed over the course of mitosis.

One of the proteins followed is almost always tubulin because it allows the microtubules to be visualized.

When viewed with the electron microscope, the mature animal cell spindle contains three primary structural components, as shown in FIGURE 14.9. Each of the two polar areas is defined by a centrosome, as shown in FIGURE 14.10. This beautiful organelle consists of a pair of small, densely staining structures known as **centrioles** surrounded by a diffuse cloud of more lightly staining material. Positioned between the centrosomes are the chromosomes, which in most organisms are the largest structures in the spindle (see Figure 14.9). Chromosomes are composed of compacted, tightly coiled and heavily staining chromatin fibers 25 nm in diameter, and each has two small structures called **kinetochores** (Greek: *kineto* = movable; *chora* = space) (see Figure 14.3) that are attached to opposite sides of its centromere. A dense array of roughly parallel microtubules runs between the two poles of the spindle. This can be seen particularly

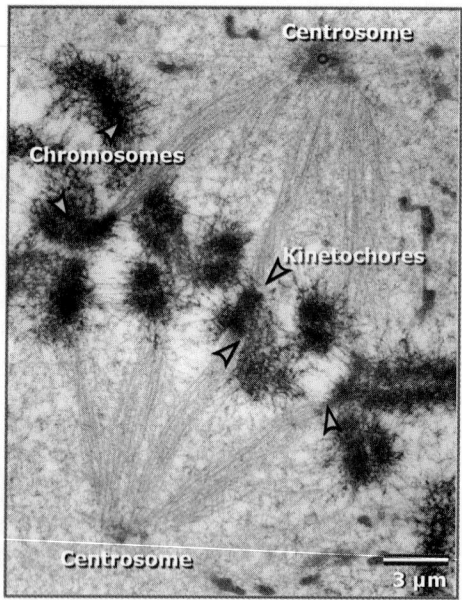

‥ **FIGURE 14.9** An electron micrograph showing the basic structural elements of a mitotic spindle. Large bundles of microtubules connect each centrosome to the kinetochores on the chromosomes. The labeled kinetochores in the center of the picture illustrate how the two kinetochores on a chromosome face opposite poles of the spindle. Reproduced from C. Rieder and A. Khodjakov, *Science* 300 (2003): 91–96. Photo courtesy of Conly Rieder and Alexey Khodjakov, Wadworth Center.

‥ **FIGURE 14.8** A metaphase spindle in a living newt cell viewed by phase contrast and polarized light microscopy. A similar cell with its spindle in the same orientation is shown in part after it was stained by immunofluorescence methods for microtubules (green), chromosomes (blue), and keratin filaments (red). Note that the spindle is invisible in the phase image but detected by polarized light. Spindle microtubules are most clearly seen after immunofluorescence staining. Left and middle photos reproduced from C. Rieder and A. Khodjakov, *Science* 300 (2003): 91–96. Photos courtesy of Conly Rieder and Alexey Khodjakov, Wadworth Center; right photo courtesy of Conly L. Rieder and Alexey Khodjakov, Wadworth Center.

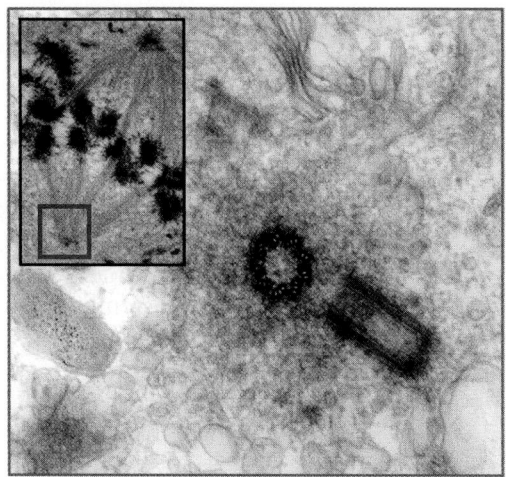

FIGURE 14.10 The large photograph shows an electron micrograph of a centrosome. The two centrioles are at right angles to one another so that one appears as a circle and the other as a rectangle. Around the first is a cloud of material that appears granular. (Compare the region immediately adjacent to the centriole with the more distant parts of the cytoplasm, which stain more lightly and where many membrane vesicles are visible.) Reproduced from C. Rieder and A. Khodjakov, *Science* 300 (2003): 91–96. Main photo © Conly L. Rieder, Wadsworth Center; inset photo reproduced from C. Rieder and A. Khodjakov, *Science* 300 (2003): 91-96. Photo courtesy of Conly Rieder and Alexey Khodjakov, Wadworth Center.

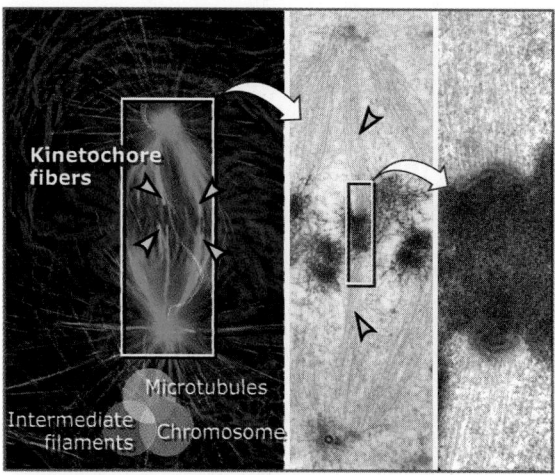

FIGURE 14.11 Kinetochore fibers on sister kinetochores seen by immunofluorescence (left) and electron microscopy (center and right). Photos © Conly L. Rieder, Wadsworth Center.

clearly in Figure 14.9. These spindle microtubules have two ends, one of which is usually located within or near a pole. The other is either free within the spindle or is associated with a kinetochore. Microtubules emanate from each of the two poles, making the spindle a symmetrical structure that is formed by two opposing and overlapping arrays of microtubules. Each of these arrays is termed a **half-spindle**. In most vertebrates half-spindles contain 600–750 microtubules, of which 30% to 40% end on kinetochores.

In addition to the microtubules in each half-spindle, other microtubules radiate outward from each pole (see Figure 14.61). These microtubules extend in all directions, forming a radial array called an aster that is centered on each pole. Like the spindle microtubules, all **astral microtubules** are oriented with one end at the pole while the other is located at a distant point within the cytoplasm. The asters play several roles during mitosis. In addition to positioning the spindle within the cell, which defines the plane of cytokinesis, they are also involved in separating the poles (centrosomes) during spindle formation and anaphase B.

The two kinetochores on each chromosome also play critical roles during mitosis. Their importance for chromosome motion was recognized very early because chromosome fragments lacking kinetochores cannot undergo directed movements. Critical to their role is how they are positioned relative to one another. Because they are located on opposite sides of the centromere, they face different spindle poles, allowing a replicated chromosome to become attached to both poles. This positional relationship between the two kinetochores is essential for ensuring that the two chromatids become segregated into different nuclei. During spindle formation each kinetochore binds to the ends of multiple microtubules emanating from one of the poles, forming a bundle of microtubules called a **kinetochore fiber** that extends between it and the pole, as shown in FIGURE 14.11. Kinetochore fibers and kinetochores do not act simply as ropes and hooks that allow the chromatids to be pulled to the poles. Rather, through various interactions they play a vital active role not only in defining the direction the chromosome will move but also in generating the force for moving the chromosome.

The major questions that must be answered in order to understand mitosis at a molecular level are: How does the spindle form and how is its bipolarity ensured? How are the forces that move chromosomes produced and regulated? How is the fidelity of chromosome segregation ensured? How is the cytoplasm partitioned into two daughter cells after the chromosomes have been segregated?

14.4 Spindle formation and function depend on the dynamic behavior of microtubules and their associated motor proteins

Key concepts

- The spindle is a complex assembly of microtubules and microtubule-dependent motor proteins. The microtubules are highly organized with respect to their polarity.
- Spindle microtubules are very dynamic. Some exhibit dynamic instability, while others experience subunit flux.
- Interactions between microtubules and motors generate forces that are required to assemble the spindle.

The formation and proper function of the spindle depend on both the dynamic properties of its microtubules and the function of microtubule-dependent motor proteins. Although microtubules form the basic structural elements of the spindle, motor proteins are involved in organizing the microtubules into a spindle and in moving the chromosomes. Some motors play a direct role in assembling the spindle and linking its components into a coherent unit, while others are responsible for attaching the chromosomes to the spindle and generating forces for their motions. Even though the spindle has traditionally been regarded as a microtubule structure, it is more accurate to consider it as a collection of microtubules, motors, and other proteins.

Although motors play an essential role in generating forces within the spindle, the microtubules are far more than a static framework over which motors move. Throughout mitosis, the microtubules are extremely dynamic, and this quality is essential both for assembling the spindle and for separating the chromosomes.

Within the spindle, microtubules are organized with respect to their polarity. As discussed in *11 Microtubules*, the two ends of a microtubule differ both chemically and structurally, imparting a structural "polarity" to the microtubule; a microtubule can be thought of as pointing in one direction or the other. The microtubules of each half-spindle, and those in its associated aster, are all arranged with the same polarity: their minus ends are near the pole while their plus ends are located at a distance from it, as shown in **FIGURE 14.12**. Where the two polarized arrays cross, their microtubules overlap, creating a region in the center of the spindle where adjacent microtubules are of opposite polarity. The uniform orientations of the microtubules in each opposing half-spindle are necessary for microtubule-dependent motors to participate in division. If the polarities of the microtubules within each half-spindle were random, different molecules of each type of motor would simply oppose one another, making any net movement chaotic, if not impossible.

The dynamic properties of microtubules play an important role during all stages of mitosis. Work on cultured vertebrate cells and extracts made from the eggs of the frog *Xenopus laevis* reveals that the microtubules in each aster are undergoing dynamic instability and are shorter and much more dynamic than the microtubules in interphase cells. Some of this difference can be attributed to an increase during mitosis in the frequency of catastrophes, when microtubule plus ends switch from a growing or polymerizing state to a shrinking or depolymerizing state. It is also partly due to a decrease in the frequency of rescues, when depolymerizing or shrinking microtubules switch back to a polymerizing or growing state. *This increase in dynamics occurs as cells enter mitosis because microtubule-associated proteins that normally dampen catastrophe are inhibited, while others that promote microtubule growth are activated.* The balance between

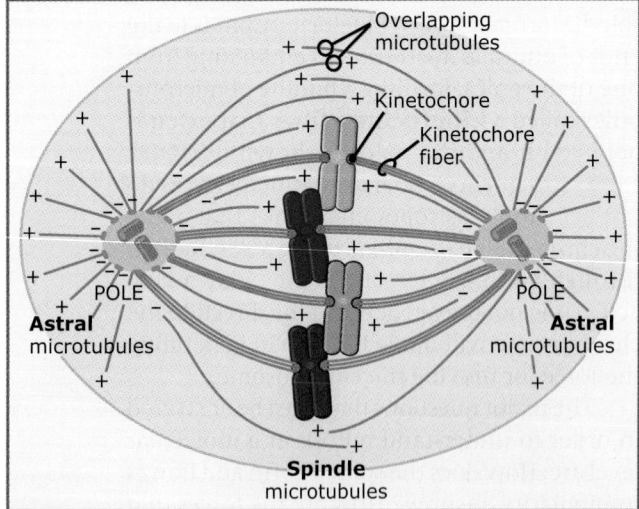

FIGURE 14.12 The microtubules within a spindle are organized with respect to their polarity. All have their minus ends near one of the two centrosomes and their plus ends at a distance from it. In the center of the spindle microtubules from the two centrosomes overlap, placing microtubules of opposite polarity adjacent to one another.

these two opposing activities is controlled by the master mitotic regulatory kinase, cyclin B/CDK1, which becomes active near the time of nuclear envelope breakdown (see *14.2 Mitosis is divided into stages*). As will be discussed next, the increase in microtubule dynamics that occurs as cells enter mitosis plays a major role in assembling the spindle.

As the spindle forms, a second type of microtubule dynamics arises. At this time, microtubules begin to exhibit a behavior known as **subunit flux**. In this curious form of behavior, tubulin subunits are incorporated at the plus end of a microtubule and then move through the microtubule to its minus end, where they are released. Flux occurs in all the microtubules of the spindle but is particularly prevalent in the microtubules within kinetochore fibers, as shown in **FIGURE 14.13** and **FIGURE 14.14**. The origin of flux is unclear, but it may be due to the interaction of spindle microtubule plus and minus ends with other components (such as motor proteins) that are involved in organizing the spindle. Even as spindle microtubules undergo flux, astral microtubules continue to undergo dynamic instability. In some types of cells the entire force for poleward chromosome movement is produced by flux, while in others flux contributes a more minor force-producing component. Flux may also play a role in chromosome alignment at the spindle equator as well as in regulating spindle length and symmetry.

Many different types of motor proteins interact within the spindle microtubule framework. Both the minus end-directed motor cytoplasmic dynein and motors of the kinesin superfamily (most of which move toward a microtubule's plus end) participate in mitosis. The spindle is sufficiently complex, and motors are so inherently involved in its formation and function, that *in higher organisms there are more than 15 kinesin family members that function only during the division process.*

Motor proteins are found throughout the spindle. This includes at the kinetochores, along the arms of the chromosomes, at the poles, and along the microtubules between the poles and the chromosomes. Many types of motors are found in only one location, but some are found in several. Cytoplasmic dynein, for example, is found at both the kinetochores and the poles, as well as in the cell's cortex, where it interacts with astral microtubules. The kinesin motor protein CENP-E, on the other hand, is concentrated in the kinetochore, while

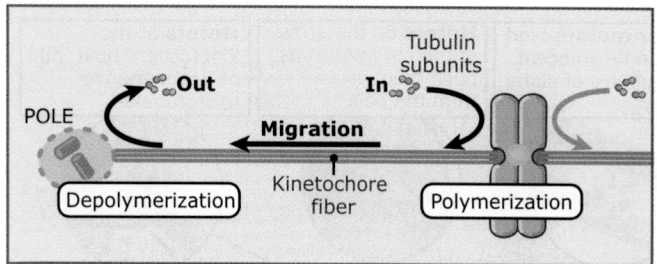

FIGURE 14.13 Tubulin subunits are continuously incorporated into microtubules at the kinetochores and move progressively toward the poles, where they are released. Tubulin subunits thus constantly flow from kinetochore to pole within the microtubules of a kinetochore fiber. During metaphase, the length of a kinetochore microtubule remains constant as long as subunit assembly at its plus end matches disassembly at its minus end. If subunit assembly at a kinetochore decreases with no change in disassembly at the pole, the kinetochore will move toward the pole. Flux is thus one possible means of moving a chromosome.

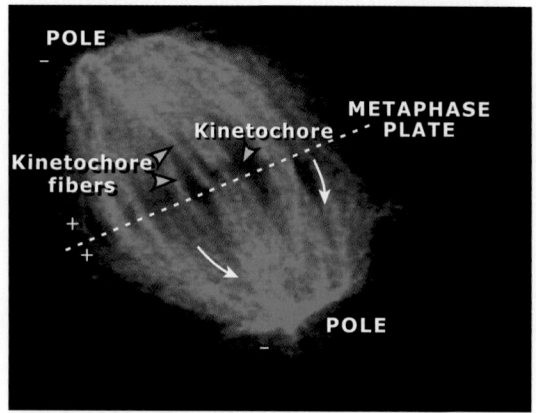

FIGURE 14.14 The first frame of a video shows the mitotic spindle of a cell, with a small percentage of its tubulin fluorescently labeled (green). The kinetochores are in orange. The video shows the poleward flow of green dots along kinetochore fibers throughout the spindle. Photo courtesy of Paul Maddox, University of Montreal.

chromokinesins are found only on the arms of the chromosomes.

During mitosis, motor proteins perform several basic functions, as shown in **FIGURE 14.15**. Some, such as cytoplasmic dynein, bind to objects—including kinetochores and the plasma membrane—and move them in one direction along a microtubule (although in the case of the plasma membrane, it is the microtubule that actually moves). Others have multiple motor domains organized so that the motor can bind to two microtubules at once and crosslink them together. Depending on the structure of these motors, the adjacent microtubules may

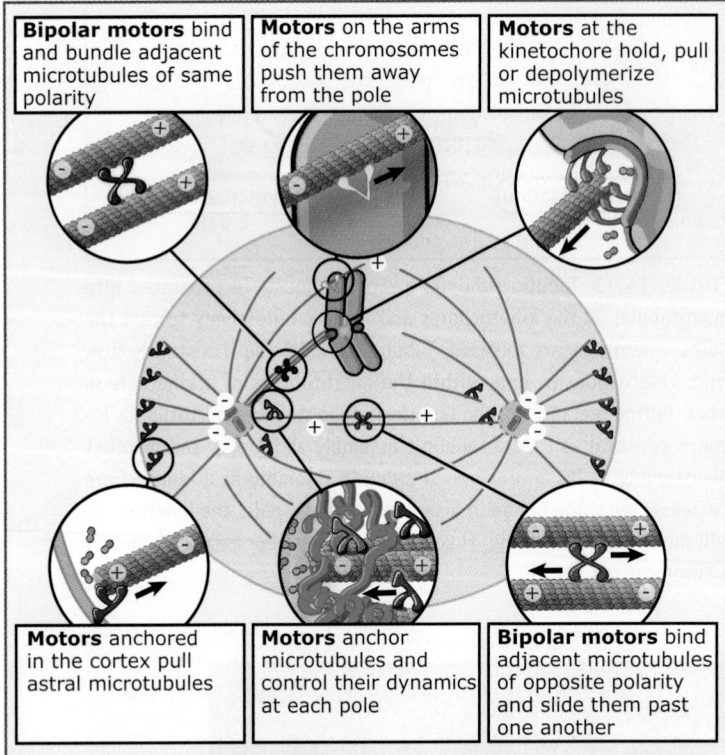

Bipolar motors bind and bundle adjacent microtubules of same polarity	**Motors** on the arms of the chromosomes push them away from the pole	**Motors** at the kinetochore hold, pull or depolymerize microtubules

Motors anchored in the cortex pull astral microtubules	**Motors** anchor microtubules and control their dynamics at each pole	**Bipolar motors** bind adjacent microtubules of opposite polarity and slide them past one another

FIGURE 14.15 The spindle is packed with motor molecules that work on microtubules. Specific interactions between these motors and the microtubules form the spindle and are required for the movements it makes and the forces that it generates. Arrows indicate the direction in which a motor moves.

have the same or opposite polarity. If a motor binds to microtubules of opposite polarity, it will try to move (slide) them past one another until they no longer overlap. One example of this type of motor is the kinesin family member Eg5, which can bind to different microtubules at its two ends. Alternatively, if a motor is organized so that it binds to two microtubules with the same polarity, the result will be a collection of microtubules of the same polarity connected at one of their ends so that they form a radial array (like an aster). Other kinesin family proteins do not move on microtubules but instead stimulate the disassembly of their plus ends. A good example is *m*itotic *c*entromere *a*ssociated *k*inesin (MCAK), which is found at the centromere of each chromosome. From motors with these basic properties, and the strategic positioning of different types of motors relative to one another, the spindle is constructed and the forces that move chromosomes are generated.

It is not always obvious how motors participate in spindle function. In some cases, for example, different motors are positioned such that they appear to oppose one another. Regardless of the details of the system, how-

ever, it is clear *that the formation and function of the spindle require multiple forces that must be balanced, and that those forces are generated by the activity of microtubule motors that work on a scaffold of dynamic spindle microtubules.*

14.5 Centrosomes are microtubule organizing centers

Key concepts

- Centrosomes define the poles of the spindle and play a role in spindle formation.
- Centrosomes nucleate microtubules and often remain bound to their minus ends afterward.

A number of changes occur in the cell near the time it becomes committed to mitosis. In animal cells, one of the most spectacular things is that the interphase array of long cytoplasmic microtubules disappears and is replaced by two radial arrays of shorter microtubules, often referred to as asters. This conversion is shown in Figure 14.21. Each of these radial arrays surrounds a centrosome. The spindle in animal cells is formed from these two asters as the centrosomes separate. Since the two centrosomes will define the poles of the spindle, it is extremely important that two and only two are present during mitosis. As the two asters interact to begin the formation of the spindle, its structure is stabilized by the chromosomes, with a major role being played by their kinetochores.

When it is born, each cell contains a single centrosome that defined one of the spindle poles during the previous division. In nondividing cells, this minute organelle normally resides near the center of the cell, where it is closely associated with the nucleus. During interphase, the centrosome acts as a **microtubule organizing center**, generating and organizing an array of cytoplasmic microtubules that extends throughout the cell, as shown in **FIGURE 14.16**. These arrays are involved in organizing the cytoplasm and in moving material and organelles within the cell.

Centrosomes form microtubule arrays by acting as sites of microtubule nucleation. Within a centrosome, a microtubule begins to grow from a ring-shaped complex that contains a type of tubulin called γ-tubulin (see *11.7 Cells use microtubule-organizing centers to nucleate microtubule assembly*). After a microtubule has been nucleated, its minus end usually remains

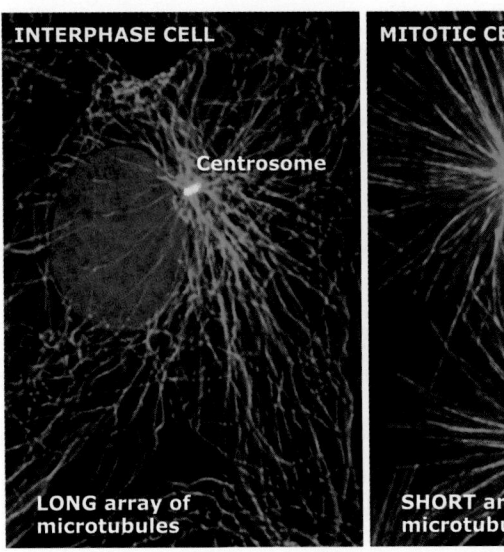

INTERPHASE CELL

Centrosome

LONG array of microtubules

MITOTIC CELL

Centrosome

Centrosome

SHORT array of microtubules

FIGURE 14.16 In interphase cells, the centrosome (the yellow dots near the nucleus) nucleates an extensive array of long microtubules that extend throughout the cytoplasm. In mitotic cells, the capacity of a centrosome to nucleate microtubules increases and each of the replicated centrosomes nucleates a dense radial (astral) array of short, straight microtubules. Microtubules are in green, DNA in blue. Photos © Conly L. Rieder, Wadsworth Center.

anchored to the centrosome. The microtubule then elongates or shortens by the addition or removal of tubulin molecules primarily at its plus end, which is positioned at a distance from the centrosome. Microtubules remain anchored to the centrosome for variable periods and, in some cell types, are actively released by enzymes located within its structure. The anchoring mechanism involves several structural proteins, as well as minus end-directed motors, including cytoplasmic dynein and a member of the kinesin family (HSET). (For more on microtubule motors see *11.11 Introduction to microtubule-based motor proteins*.)

14.6 Centrosomes reproduce about the time the DNA is replicated

Key concepts

- Centrosomes are composed of two centrioles surrounded by the pericentriolar material.
- The formation of a new centrosome requires duplication of the centrioles.
- Centriole duplication is controlled by the cell cycle and is coordinated with DNA replication.
- Centrioles duplicate by the formation and growth of a new centriole immediately adjacent to each existing one.

During mitosis, every centrosome present in the cell has the potential to form a spindle pole. Spindles are normally bipolar only because cells usually enter mitosis with just two centrosomes. If a cell enters mitosis with more than two centrosomes, a spindle with too many

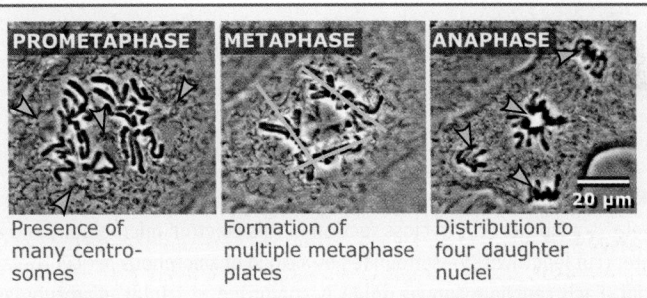

PROMETAPHASE

METAPHASE

ANAPHASE

20 µm

Presence of many centrosomes

Formation of multiple metaphase plates

Distribution to four daughter nuclei

FIGURE 14.17 Mitosis in a rat kangaroo cell with four centrosomes (yellow arrowheads in the left panel). Three metaphase plates (yellow lines in the center panel) and four groups of chromosomes (arrowheads in the right panel) result. Cytokinesis will produce four aneuploid cells. Reproduced with permission from *J. Cell Sci.*, vol. 110 (4): 421–429. [http://jcs.biologists.org/cgi/content/abstract/110/4/421]. Photos © Conly Rieder, Wadsworth Center.

poles will form and the cell stands a good chance of producing aneuploid progeny, as shown in **FIGURE 14.17**. To prevent this, mechanisms exist to ensure that the centrosome is replicated only once during the cell cycle. When these control mechanisms break down, too many centrosomes are formed, which may lead to genetic defects that result in cancer cells and tumors. In order to understand how the bipolar nature of a normal spindle is ensured, it is necessary to examine the structure of the centrosome and how the cell controls its replication.

Under the light microscope the centrosome appears in most living cells as one or two dots. To reveal more detail and fully appreciate the structural complexity of this organelle, the electron microscope (EM) is used. With this tool, the core of the centrosome is seen to

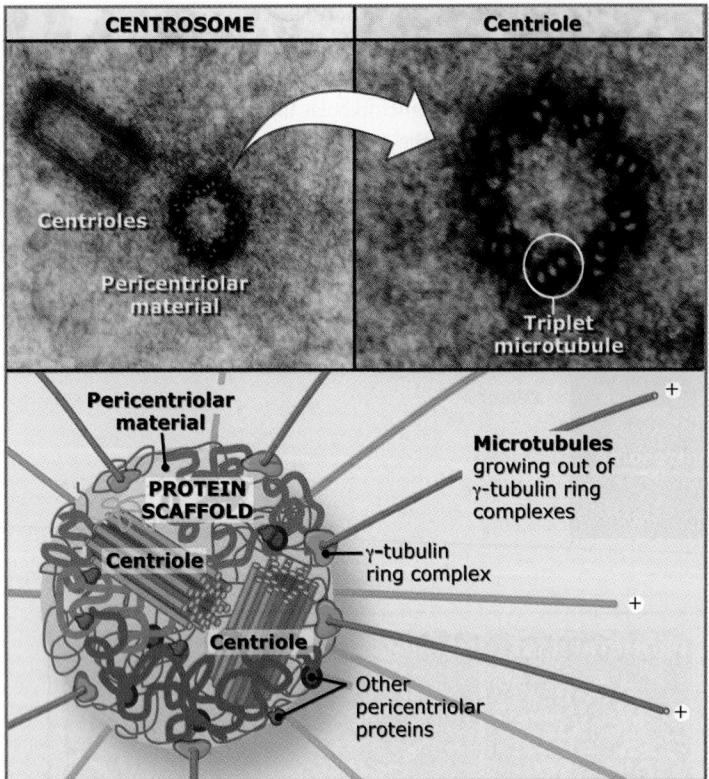

CENTROSOME	Centriole

Centrioles

Pericentriolar material

Triplet microtubule

Pericentriolar material

PROTEIN SCAFFOLD

Centriole

Centriole

Microtubules growing out of γ-tubulin ring complexes

γ-tubulin ring complex

Other pericentriolar proteins

FIGURE 14.18 The centrosome in a mitotic cell (upper left) consists of two centrioles—a mother (cut in cross section in this electron micrograph) and a daughter (cut lengthwise)—surrounded by a cloud of amorphous pericentriolar material. Each centriole (upper right) is composed of triplet microtubules arranged as the wall of a cylinder. The drawing shows how the centrioles are positioned at right angles to one another, with many different proteins collected around them to form the pericentriolar material. Photos © Conly L. Rieder, Wadsworth Center.

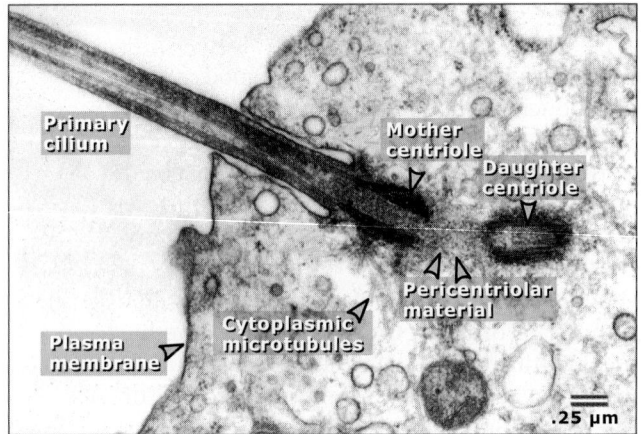

Primary cilium

Mother centriole

Daughter centriole

Pericentriolar material

Cytoplasmic microtubules

Plasma membrane

.25 μm

FIGURE 14.19 An electron micrograph of the base of a primary cilium. The structure projects from the mother centriole, which is located immediately beneath the plasma membrane. The mother and daughter centrioles are connected by the pericentriolar material, visible as dense, granular material extending between the two. Only a small part of the entire cell and the length of the primary cilium are shown. Photo © Conly L. Rieder, Wadsworth Center.

contain a pair of structures called centrioles, as shown in FIGURE 14.18. Each centriole consists of a pinwheel of nine triplet microtubule blades distributed evenly around the perimeter of a cylinder ~0.3 μm in diameter. As early as 1888, Boveri and others concluded that the centriole, which was just barely visible with their light microscopes, was a permanent and independent organelle formed only by the division of a preexisting centriole. Indeed, with few exceptions, new centrioles are only formed in association with the wall of an existing centriole. In some cells, this close physical relationship between the two centrioles persists throughout the cell cycle. However, in many cells it is lost during interphase, and the two centrioles wander independently throughout the cell.

Each centriole is associated with a diffuse cloud of material that appears in the EM as an opaque substance. It is clearly visible around the centriole seen "end-on" in Figure 14.10. This **pericentriolar material** consists of a large number of proteins attached to a scaffold. As a rule, the older (mother) centriole contains more of this material than its daughter, at least until the next round of centriole replication is completed. Among the proteins included in the pericentriolar material are several types of microtubule-dependent molecular motors and the γ-tubulin rings used in microtubule nucleation (see Figure 14.18). The centrioles themselves also contain a number of specific structural and enzymatic proteins, some of which are also present in the pericentriolar material.

During the interphase portion of the cell cycle, the centrosome serves several concurrent functions. In most cells, during G1, the mother centriole initiates the formation of a long, thin, membrane-enclosed structure called a **primary cilium**, which projects from the surface of the cell, as shown in FIGURE 14.19. Although often ignored, primary cilia are so common that it is easier to list the cells that lack them than those that possess them. In some epithelia these structures can protrude more than 20 μm from the dorsal cell surface. Since they are not found in many transformed cells, primary cilia are not essential for cell survival, a fact that initially led scientists to speculate that they are simply vestigial appendages like the appendix. However, the outer segments of the rod and cone cells of the eye—the highly specialized structures where photons are absorbed—are formed from derivatives of primary cilia, and primary cilia are required for proper development and tissue function.

In animal cells, the number of centriole pairs defines the number of centrosomes. *Cells, therefore, control the number of centrosomes they contain by regulating the replication of centrioles.* Research is just beginning to uncover the mechanisms that control the precise doubling of centrioles and how this reproduction is coordinated with the nuclear activity of the cell cycle (e.g., DNA replication). It is now evident that the timing of centriole replication is governed by changes in the cytoplasm, indicating that a soluble factor regulates their duplication. In addition, conditions that allow centrioles to replicate are found only during the S phase of the cell cycle, when the cell's DNA is also being replicated. The primary regulator of centrosome replication appears to be the CDK2 kinase and its cyclin A and E activators. These regulators become active near the beginning of S phase and are also responsible for driving the cell into DNA synthesis (for details see *15 Cell cycle regulation*). The fact that the same regulator initiates the replication of both DNA and centrioles ensures that these two activities are coordinated, so that a cell enters mitosis having replicated both its centrosome and its chromosomes. Although it is evident from work on the nematode worm, *C. elegans,* how the duplication of centrioles is initiated, it is not yet clear how their replication is limited to the formation of a single new centriole from each preexisting one. One idea is that the size of the pericentriolar cloud surrounding the mother centriole limits the number of daughters that it can simultaneously form. Another nonmutually exclusive possibility is that the amount of a specific "starter" protein (SAS-6 in the worm), that accumulates at the wall of the mother centriole to initiate procentriolar formation, is limited.

Once started, centriole duplication proceeds by the gradual formation of a new centriole adjacent to each of the two centrioles with which the cell entered the S phase. Of those, one, termed the daughter, is younger because it was formed in the previous cell cycle. The other, termed the mother, was formed in an earlier cell cycle and carries the primary cilium (see Figure 14.19). The first sign of centriole replication is the appearance of two short procentrioles, each of which extends at a right angle from the wall of one of the existing centrioles, as shown in **FIGURE 14.20**. This process does not depend on a physical relationship between the two original centrioles because it can occur even when they are separated. Once procentrioles have formed, they

slowly elongate until they reach the length of a mature centriole near the time of mitosis. Although the mother and daughter centrioles form and grow procentrioles identically, most of the pericentriolar material remains associated with the more mature mother centriole. The daughter centriole ultimately gathers new pericentriolar material during the duplication process, in part from the microtubule array that it organizes. By late interphase the cell contains two centrosomes, each containing a pair of closely associated centrioles and their pericentriolar material. In some cells, these two centrosomes remain physically tethered together, and function as a single unit until the cell enters mitosis. In other cells, this connection is broken, and the two centrosomes move apart before there is any visual evidence that the cell is entering mitosis. The timing of when the two centrosomes separate with respect to nuclear envelope breakdown is highly variable, even in genetically identical cells that are adjacent to one another.

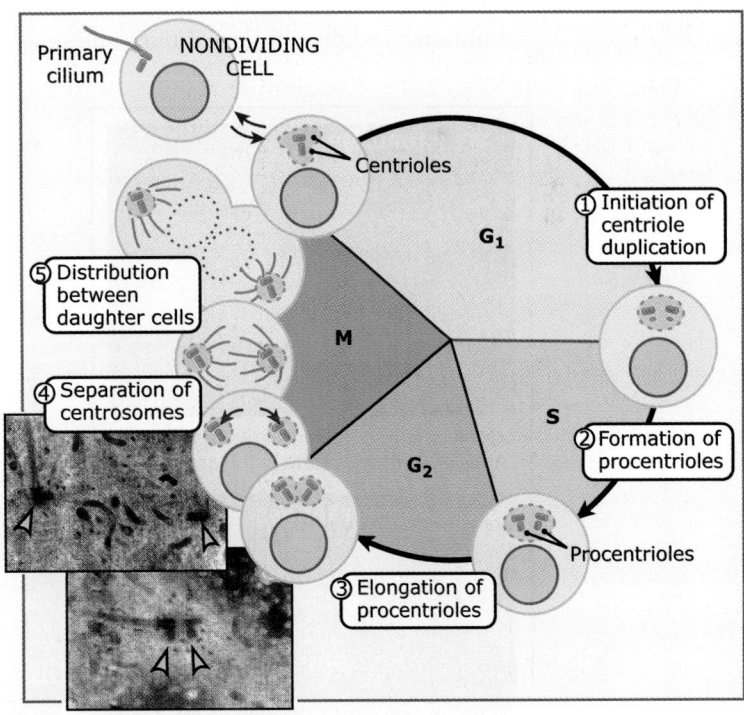

FIGURE 14.20 The centriole cycle in a mammalian cell. The two centrioles in the mother cell's centrosome are duplicated and the two centrosomes that result are then separated into the daughter cells. The insets show electron micrographs of replicated centrosomes before and after they separate early in mitosis. Photos reprinted from *J. Ultrastruct. Res.*, vol. 68, C. Rieder, C. G. Jensen, and L. C. W. Jensen, The resorption of primary cilia during mitosis..., pp. 173–185, Copyright (1979), with permission from Elsevier. [http://www.sciencedirect.com/science/journal/00225320]. Photos courtesy of Conly Rieder, Wadsworth Center.

14.7 Spindles begin to form as separating asters interact

Key concepts

• As mitosis begins, changes in both the centrosomes and the cytoplasm cause a radial array of short, highly dynamic microtubules to form around each centrosome.

• Interactions between the asters formed by the two centrosomes initiate the formation of the mitotic spindle.

• Separation of the centrosomes depends on microtubule-dependent motor proteins.

• The pathway of spindle formation depends on whether the centrosomes separate before or after the nuclear envelope breaks down.

As cells progress from interphase into mitosis, the distribution of microtubules goes through a rapid and striking change. The array of long cytoplasmic microtubules typical of interphase cells disassembles, and each of the two centrosomes nucleates a dense radial array of shorter microtubules, as shown in **FIGURE 14.21** (see also Figure 14.16). As illustrated in **FIGURE 14.22**, these two asters will ultimately contribute many of the microtubules for spindle formation. The changes in number and distribution of microtubules at the onset of mitosis are mediated both by changes within the centrosomes and by others that occur throughout the cell.

Near the time when the cell becomes committed to mitosis, the two centrosomes change so that they are capable of nucleating many more microtubules than during interphase. As this happens, proteins associated with the centrosome become heavily phosphorylated, the centrosome's γ-tubulin content increases, and the pericentriolar material surrounding the centrioles expands. It is not clear how this "maturation" process occurs; however, it probably involves specific kinases activated as a cell progresses from G2 into M phase, including the master mitotic regulatory kinase (for details see *15 Cell cycle regulation*).

At about the same time, due to changing conditions in the cytoplasm, microtubules become less stable, leading to the replacement of the long interphase array by two shorter astral arrays growing from the centrosomes, as discussed earlier. The result is that by the start of prometaphase, the total length of microtubules in the cell has decreased, and the rate at which new microtubules are formed and old microtubules disassemble (i.e., turn over) has increased. This means that as the nuclear envelope begins to break down, the region surrounding the nucleus is constantly being probed by a large number of very dynamic microtubules growing from each aster. As we will see, this dynamic behavior facilitates connecting the asters with the chromosomes.

As the two centrosomes separate, interactions between the astral microtubule arrays growing from them begin to form the spindle. Spindle formation is a remarkably flexible process that can occur by either of two pathways, depending on whether the centrosomes separate before or after

LATE PROPHASE

LONG microtubules that extend throughout much of the cytoplasm

EARLY PROMETAPHASE

Dense array of **SHORT** microtubules

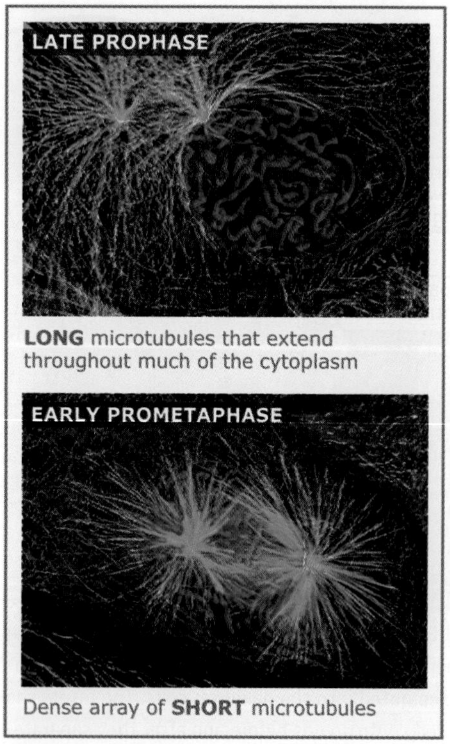

FIGURE 14.21 The conversion from one array to the other takes only a few minutes. Microtubules are in green, chromosomes in blue, and intermediate filaments in red. Photos © Conly L. Rieder and Alexey Khodjakov, Wadsworth Center.

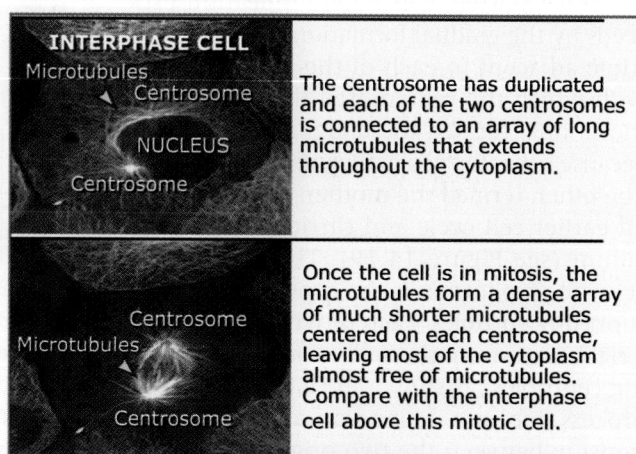

INTERPHASE CELL
Microtubules
Centrosome
NUCLEUS
Centrosome

The centrosome has duplicated and each of the two centrosomes is connected to an array of long microtubules that extends throughout the cytoplasm.

Centrosome
Microtubules
Centrosome

Once the cell is in mitosis, the microtubules form a dense array of much shorter microtubules centered on each centrosome, leaving most of the cytoplasm almost free of microtubules. Compare with the interphase cell above this mitotic cell.

FIGURE 14.22 Frames from a video that shows the conversion of a microtubule array into a mitotic array. Photos courtesy of Patricia Wadsworth, University of Massachusetts, Amherst.

the nuclear envelope breaks down, as shown in **FIGURE 14.23**. Both pathways involve the interaction of microtubules with motor proteins.

In cases where the nuclear envelope dissolves before the centrosomes have begun to separate, the liberated chromosomes are distributed throughout the cytoplasm and are exposed to only a single large aster. As **FIGURE 14.24** shows, a "monopolar" or half-spindle is formed; it lasts until the two centrosomes finally separate and convert it into a bipolar spindle. The separation of centrosomes after nuclear envelope breakdown involves two forces: one that pushes them apart, caused by the activity of the kinesin family protein Eg5 as it interacts with adjacent microtubules of opposite polarity between the two centrosomes, and a second that pulls, caused by the activity of cytoplasmic dynein anchored at the periphery of the cell (i.e., in the cortex), as shown in **FIGURE 14.25** (for more on kinesin and dynein, see *11.11 Introduction to microtubule-based motor proteins*). If unopposed, these pushing and pulling forces move the centrosomes apart until the two astral arrays no longer overlap. However, separation of the centrosomes is restricted by other motors that bind to the overlapping microtubule arrays, as well as by the formation of kinetochore fibers

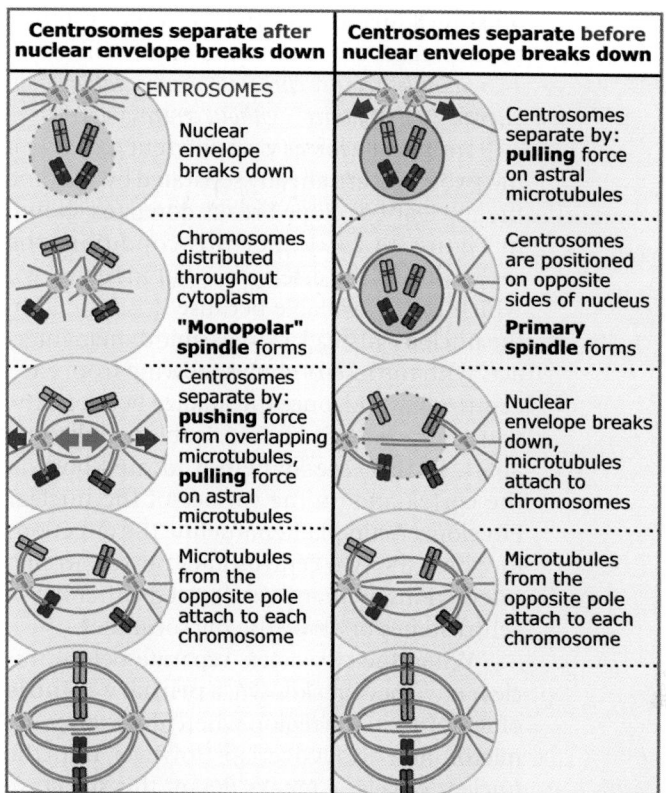

FIGURE 14.23 The two pathways differ in when the centrosomes separate relative to nuclear envelope breakdown. The ability of a spindle to form regardless of when the centrosomes separate emphasizes how remarkably flexible the process of spindle formation is.

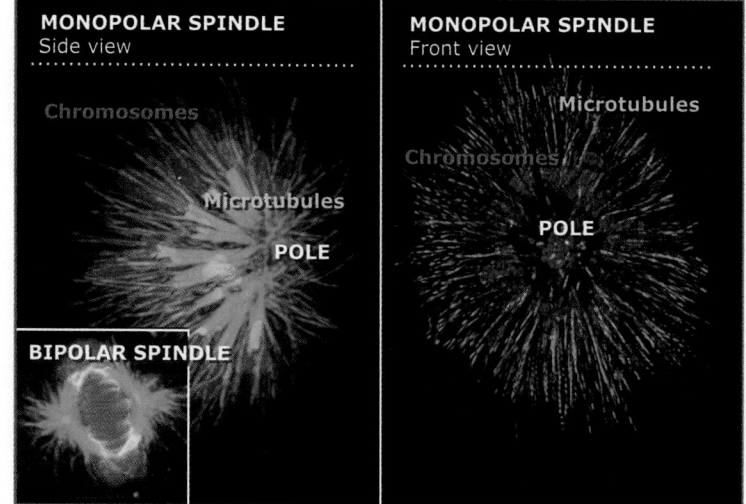

FIGURE 14.24 On the left is a side view of the monopolar spindle formed in a rat kangaroo cell when the centrosomes were prevented from separating. The chromosomes (orange) attach to the single polar region. Note the thick kinetochore fibers. A normal bipolar spindle is shown in the inset for comparison. An end-on view of a similar monopolar spindle in a human cell is shown on the right. The centrosomes are in blue in the center. Left photo reprinted by permission from Macmillan Publishers Ltd: *Nature*, J. C. Canman, et al., Determining the position of the cell division plain, 2003, vol, 424 (6952): 1074–1078. Photo courtesy of Julie C. Canman and E. D. Salmon, University of North Carolina at Chapel Hill; right photo courtesy of Alexei Mikhailov, Wadsworth Center; inset photo courtesy of Lynne Cassimeris, Lehigh University.

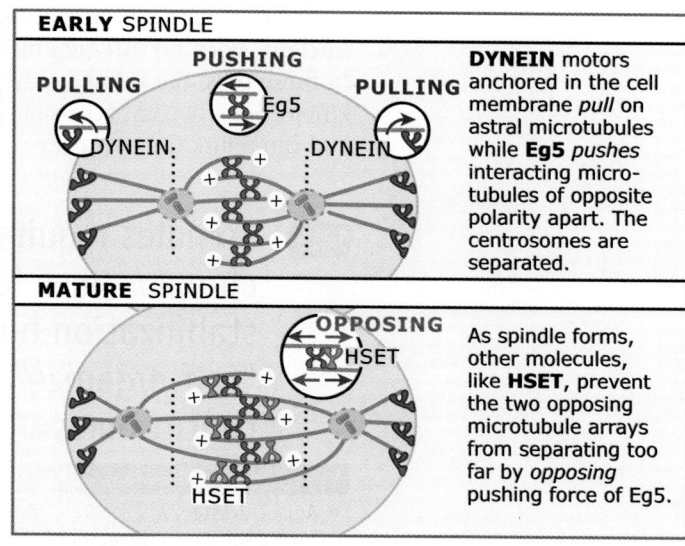

FIGURE 14.25 The process begins as soon as the nuclear envelope breaks down. Initially the two centrosomes are immediately adjacent to one another. Both Eg5 and HSET are members of the kinesin family of microtubule-dependent motors, but Eg5 moves toward the plus (+) end of a microtubule and HSET toward the minus (−) end. The spindle's length is determined by a balance among the forces of the three motors.

on sister kinetochores (which tether the two centrosomes to one another, as we will see in *14.8 Spindles require chromosomes for stabilization but can "self-organize" without centrosomes*).

The spindle forms via a different pathway if the two asters are already separated by the time the nuclear envelope breaks down, as shown in Figure 14.23. Under this condition, the separation of the asters does not involve Eg5, which is not available because it is located in the nucleus. Instead, cytoplasmic dynein interacts with the microtubules growing from each centrosome and, once the linkage between the two centrosomes is severed, is able to pull them apart. In this case, dynein is located both in the cortex and on the surface of the nuclear envelope. Actin filaments define the directions in which the two centrosomes move through interactions, with myosin located either in the centrosomes or along the microtubules.

When the two asters separate before nuclear envelope breakdown, a **primary spindle** often forms in the region where their opposing microtubule arrays overlap. However, until the nuclear envelope breaks down, this structure is not stable, and the microtubule arrays of the separating centrosomes can be pulled far enough apart so that they no longer overlap. The reason for this is that the stability of the spindle requires proteins that are sequestered in the nucleus during interphase and are only released into the cytoplasm after nuclear envelope breakdown. As a result, in many late prophase cells the two centrosomes and their associated asters are positioned on opposite sides of the nucleus, with no interaction between them. In these cells, the spindle only forms after the kinetochores become accessible to microtubules and can relink the two asters.

14.8 Spindles require chromosomes for stabilization but can "self-organize" without centrosomes

Key concepts

- Adjacent asters will separate completely and fail to form a spindle in the absence of chromosomes.
- Chromosomes stabilize both the basic geometry of the spindle and the microtubules in it by binding astral microtubules at their kinetochores.
- Spindles can form in the absence of centrosomes, although they form more slowly and lack astral microtubules.

As it forms, the spindle is stabilized by the chromosomes and their kinetochores and by molecular motors that bind to the microtubules and attract additional components into the spindle.

Particularly important for establishing the basic shape of a spindle are the motor proteins that connect neighboring microtubules of opposite polarities. Such microtubules are present where two asters overlap, and crosslinking them creates a spindle-shaped structure similar in length to a normal spindle even if there are no chromosomes nearby. *However, spindles that lack chromosomes are not stable and progressively lose their microtubules.*

How does the association of chromosomes with the nascent spindle prevent the loss of microtubules? The answer is not fully understood, but it appears to involve several different overlapping mechanisms. Each of the kinetochores on a chromosome recruits astral microtubules into a specialized bundle called a kinetochore fiber that connects the kinetochore to a pole (discussed in *14.9 The centromere is a specialized region on the chromosome that contains the kinetochores*). Inclusion in a kinetochore fiber increases the stability of those microtubules relative to other microtubules in the spindle. This happens to a significant fraction of the microtubules in each aster; by metaphase, ~30% to 40% of the 1200–1500 microtubules in a typical spindle are stabilized by attachment to a kinetochore. Because there are two kinetochores on each chromosome, the formation of kinetochore fibers tethers the two spindle poles together, causing the opposing astral arrays of microtubules to interact even more.

While kinetochore fibers are being formed, each aster is also recruiting and concentrating a wide variety of proteins that help stabilize its microtubules. Some of these are structural proteins that are organized into a loose **spindle matrix** that surrounds the microtubules (see Figure 14.39). For example, when the *nu*clear *m*itotic *a*pparatus (NuMA) protein is released from the nucleus and phosphorylated by CDK1, it becomes concentrated within the spindle. By binding various motor molecules in the spindle, NuMA anchors and stabilizes microtubules. The stabilizing influence of matrix components is significant, but is not nearly as great as that of kinetochores. In a mature spindle, the lifetime of a microtubule associated with a kinetochore is 10 times longer than for other spindle microtubules.

Surprisingly, bipolar spindles can form in the absence of centrosomes. This occurs through a remarkable process of self-organization in which randomly nucleated microtubules are assembled into a bipolar structure by the chromosomes and microtubule-dependent motor proteins. This "acentrosomal" route for spindle assembly is used by all higher plants and is also found during meiosis and the early developmental stages of some animals. One possibility is that it is an evolutionary ancestor of centrosome-mediated spindle formation, and that it is normally masked by the presence of centrosomes. This idea is supported by the fact that even animal tissue cells, which normally contain centrosomes, can construct a bipolar spindle in their absence. This is an excellent example of how cells have evolved multiple mechanisms by which to accomplish the same task.

During acentrosomal spindle assembly, short microtubules form near each chromosome via a pathway that involves proteins on the chromosome surface, including the guanine nucleotide exchange factor, RCC1, which is particularly enriched in the centromere region. These microtubules are initially randomly oriented but are then organized into parallel arrays by the action of microtubule-dependent motor proteins, as shown in FIGURE 14.26. A central role is played by motors that bind two oppositely oriented microtubules at once and simultaneously move toward the plus end of each. These proteins first align the microtubules around a chromosome simply by binding along their length and crosslinking them. By then moving toward the plus end of each microtubule, these motors sort them into two groups of parallel, polarized microtubules that meet at their plus ends. The motor proteins may remain bound there, helping maintain the structure. Motor proteins on the arms of the chromosomes—called **chromokinesins**—also participate. Part of their role may be to tether the microtubules in the vicinity of the chromosomes so that they remain at the center of the spindle as it forms. Once sorted, the microtubules within each of the two arrays become bunched together at their minus ends by minus end-directed motors (like cytoplasmic dynein and HSET), thus giving the entire array the shape of a spindle. This self-assembly process takes place independently around each chromosome. The multiple spindles thus formed then fuse into a single large spindle. Structural proteins and other matrix components—like NuMA—that are transported toward the minus ends of microtubules (i.e., toward the forming pole) glue the ensemble together and stabilize the polar regions.

If centrosomes are not needed to form a bipolar spindle, why are they found at the spindle poles during mitosis in most animal cells? One reason is that centrosomes provide a kinetic advantage and allow the spindle to form more quickly. This is important because, during the development of an organism, spindles must often form synchronously and very rapidly. Another reason is that the poles of spindles formed by centrosomes contain asters, which are not present in the polar regions of spindles formed by the acentrosomal route. These asters collect and transport all the chromosomes into a common region after nuclear envelope breakdown, which minimizes chromosome loss in large cells like those from amphibians. They also define the position of the furrow during cytokinesis by positioning the spindle within the cell. This aspect of aster function is critical for proper development. Lastly, in addition to its role as a microtubule organizing center, the

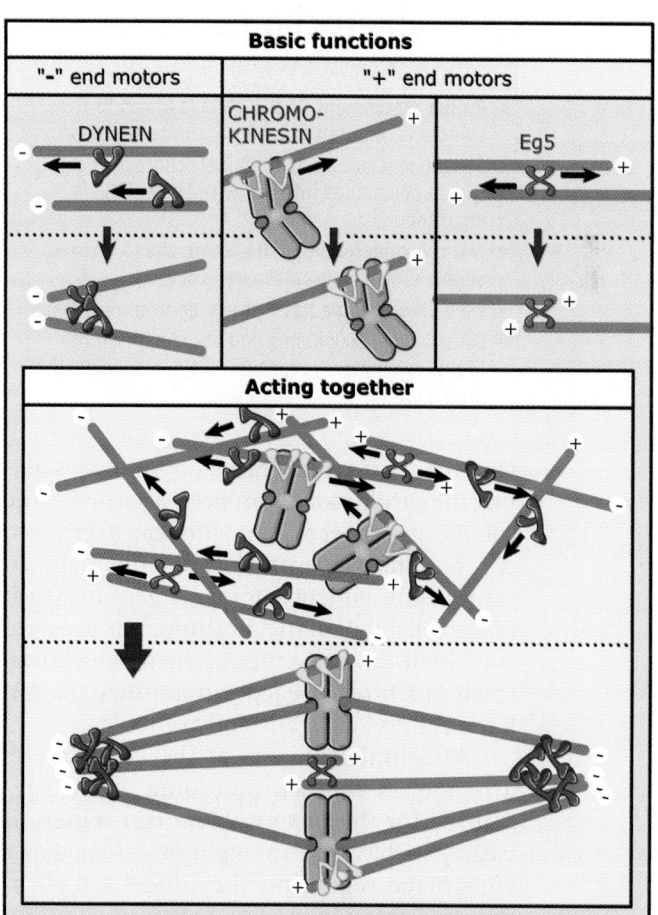

FIGURE 14.26 Microtubules are nucleated with random orientations around the chromosomes. Once microtubules have formed, three types of motors work together to organize them into a bipolar array around the chromosomes.

centrosome plays other roles in the cell (e.g., in primary cilia formation). Associating a centrosome with each spindle pole is a convenient and reliable mechanism to ensure that each new cell will inherit a copy of this important organelle.

Concept and Reasoning Checks

In the absence of centrosomes, vertebrate somatic cells form functional bipolar spindles by the chromosome-mediated self assembly mechanism.

1. Why, when two centrosomes are present but inhibited from separating, do vertebrate cells form monopolar spindles?
2. Why are multipolar spindles seldom if ever formed in plants?

14.9 The centromere is a specialized region on the chromosome that contains the kinetochores

Key concepts

- Proper attachment of the chromosomes to the spindle is required for their accurate segregation.
- Attachment occurs at the kinetochores, where the chromosomes interact with the spindle's microtubules.
- The centromere is the site where the two kinetochores on each chromosome form.
- Each chromosome has a single centromeric region.
- Centromeres lack genes and are composed of highly specialized, repetitive DNA sequences that bind a unique set of proteins.

One the most critical events of mitosis is getting the chromosomes properly attached to the spindle. Improper attachments lead to errors in chromosome segregation, with potentially disastrous consequences for the organism. Much of the reliability of the attachment process results from the properties of the kinetochores, small structures on each chromosome specialized for interacting with microtubules.

Attachment occurs at the centromere, a region on each chromosome that is specialized for the purpose. The centromere is clearly visible with the light microscope as a constricted region on the condensed chromosome (see Figure 14.3). The constriction gives mitotic chromosomes their characteristic shape. Depending on the chromosome, the centromere may be located near the middle of a chromosome (metacentric), near its end (acrocentric), or somewhere in between (submetacentric). Its position within a chromosome does not change.

The centromere region differs chemically from the rest of the chromosome. It contains few if any genes, and is composed almost entirely of a large number of highly repetitive DNA sequences called α satellite repeats, or satellite DNA. A unique group of proteins called *cen*tromere *p*roteins, or CENPs, associate with these sequences. The CENPs include CENP-A, which is a modified form of histone H3; CENP-B and CENP-G, which are involved in packaging the satellite DNA; and CENP-C, the function of which is unknown. Most of the CENPs play structural roles, binding to the repeat sequences and organizing them into a highly compacted form of chromatin called heterochromatin that is found nowhere else in the chromosome.

In addition to the CENPs the centromere region also contains another class of proteins known as **chromosomal passengers**. Passenger proteins are unusual in that they change their location during the course of mitosis. Unlike the CENPs, which remain at the centromere throughout mitosis, the passenger proteins are present at the centromere during prophase and metaphase but relocate onto microtubules between the two groups of separating chromosomes as anaphase begins. All of the passenger proteins that have so far been identified appear to form a single complex within the centromere that includes the Aurora B kinase, an enzyme whose inactivation has a dramatic effect on the fidelity of chromosome segregation. This complex is important for correcting errors in kinetochore attachment early in mitosis, a role it must play while located at the centromere. Much later, after the complex has moved onto the microtubules in the spindle midzone, it participates in cytokinesis.

In addition to the CENPs and passenger proteins, the centromere region also contains the kinesin family protein MCAK (mitotic centromere-associated kinesin). Unlike most members of the kinesin family, which move along the sides of microtubules, this protein stimulates microtubule plus ends to disassemble (for more on kinesin see *11.12 How motor proteins work*). Recent evidence suggests that this motor protein, along with its Aurora kinase activator, is involved in correcting errors in kinetochore attachment.

Perhaps the most important role of the centromere is to organize the kinetochores on the chromosome. Kinetochores are highly defined structures and the surface of each centromere is specialized for their assembly. They are extremely small relative to the size of the entire chromosome and their shape and internal structure can only be seen when they are viewed with an electron microscope (as in Figure 14.9). Each centromere forms two kinetochores, positioned on opposite sides of the chromosome. The kinetochores are the site where the chromosomes are actually attached to the spindle, and their precise positioning relative to one another on each chromosome is fundamentally important to the success of mitosis.

14.10 Kinetochores form at the onset of prometaphase and contain microtubule motor proteins

Key concepts

- Kinetochores change structure as mitosis begins, forming a flat plate or mat on the surface of the centromere.
- Unattached kinetochores have fibers extending out from them (the corona) that contain many proteins that interact with microtubules.
- The corona helps kinetochores capture microtubules.

Associated with each centromere are two "sister" kinetochores arranged back-to-back so that they are on opposite sides of the centromere and face in opposite directions. This back-to-back positioning of sister kinetochores helps to ensure both that each attaches to only one pole, and that they attach to different poles. A chromosome's sister chromatids will only move to opposite poles of the spindle if the chromosome has acquired this proper bipolar attachment. The structure and composition of kinetochores is complex and changes both over the course of the cell cycle and during the various stages of mitosis.

The composition of the kinetochore remained a mystery until the early 1980s. At that time it was discovered that some patients with an autoimmune disease called CREST syndrome (a variant of systemic sclerosis) have antibodies to kinetochore proteins in their blood. Immunofluorescence studies with these antibodies showed pairs of adjacent dots in the nuclei of interphase cells that had replicated their chromosomes, with the number of pairs of dots in each nucleus equal to the number of chromosomes. These antibodies identify precursor structures present in interphase cells that will mature into mitotic kinetochores. These "prekinetochores" contain several of the CENPs and appear in the electron microscope as spheres of tightly packed fibrillar material embedded in the heterochromatin of the centromere. As the cell enters mitosis and the chromosomes begin to condense, additional components associate with the prekinetochores.

Still more components are added to the prekinetochore when the nuclear envelope breaks down, and it undergoes a physical change as a result. *The spherical mass of fibers characteristic of the prophase kinetochore is replaced by a very thin (50–75 nm thick) circular or sometimes rectangular fibrous plate or "mat,"* as shown in **FIGURE 14.27**, on the surface of the centromere. The diameter of this new form of the kinetochore is usually about 0.2–0.5 μm, although it varies considerably, even on different chromosomes within the same cell. (For comparison, a microtubule is about 0.025 μm in diameter and a mitotic chromosome can be up to about 40 μm from end to end.) Several proteins that are important for proper assembly of the plate, including CENP-A and CENP-C, are found on the surface of the centromere to which the mat is attached. The mat itself consists of numerous

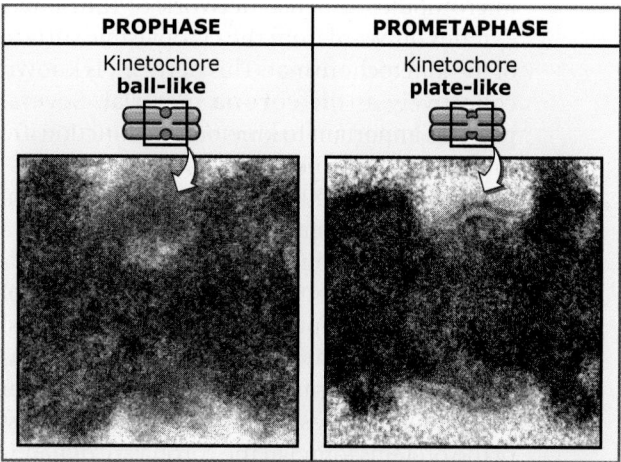

FIGURE 14.27 Kinetochores in a rat kangaroo cell during prophase (left) and prometaphase (right). The prometaphase kinetochore has no microtubules attached to it because microtubule assembly was inhibited in the cell before the picture was taken. The kinetochore changes from a ball of material in prophase to a plate-like structure in prometaphase. Photos © Conly L. Rieder, Wadsworth Center.

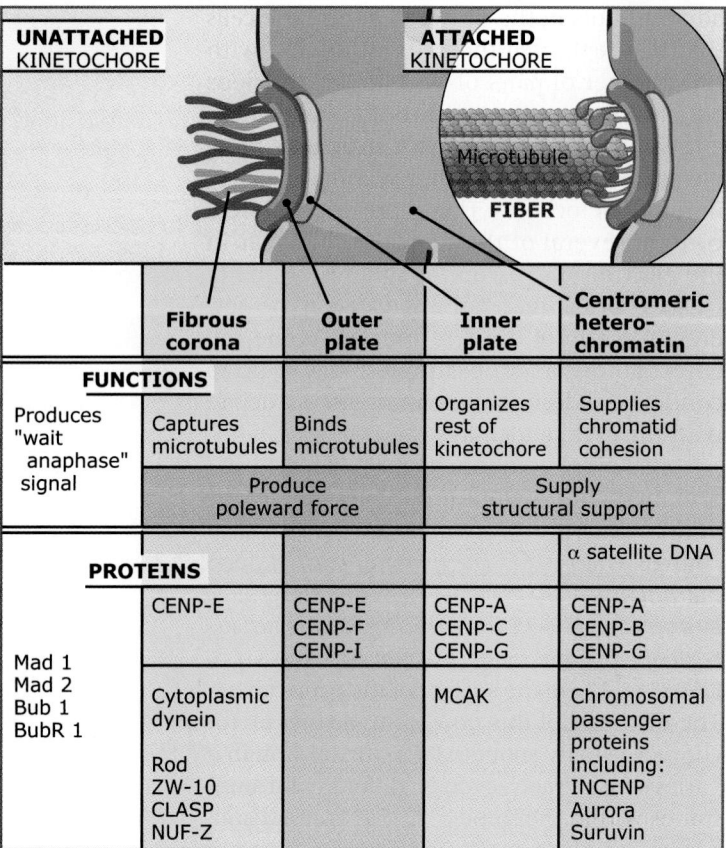

	Fibrous corona	Outer plate	Inner plate	Centromeric hetero-chromatin
FUNCTIONS				
Produces "wait anaphase" signal	Captures microtubules	Binds microtubules	Organizes rest of kinetochore	Supplies chromatid cohesion
	Produce poleward force		Supply structural support	
PROTEINS				α satellite DNA
	CENP-E	CENP-E CENP-F CENP-I	CENP-A CENP-C CENP-G	CENP-A CENP-B CENP-G
Mad 1 Mad 2 Bub 1 BubR 1	Cytoplasmic dynein Rod ZW-10 CLASP NUF-Z		MCAK	Chromosomal passenger proteins including: INCENP Aurora Suruvin

FIGURE 14.28 The functions of the different parts of the vertebrate kinetochore. The number of different proteins found in each part indicates how complex the structure is.

protein complexes including one that contains the Spc24,Spc25, Nuf2 and Ndc80 proteins.

When kinetochores are not attached to microtubules, a dense network of thin fibers extends outward from the cytoplasmic surface of the kinetochore mat. This network is known collectively as the **corona** material. Several proteins important to kinetochore function are found in the corona. These include cytoplasmic dynein (a minus-end microtubule motor), CENP-E (a member of the kinesin family and a plus-end microtubule motor), and several additional proteins that facilitate the attachment of microtubules to kinetochores, including at least one +TIP. In addition, the corona also contains several components of a cell cycle checkpoint that monitors the assembly of the spindle. Most of the proteins found in the corona are dynamically associated with it, constantly dissociating and later rebinding. This continuous turnover makes the corona a steady state structure, its overall form and composition remaining the same but its individual components continuously changing.

The function of the corona is, in part, to help the kinetochore capture microtubules. The presence of the corona in early mitosis greatly increases the surface area of the kinetochore during the period when the spindle is forming and the chromosomes must become attached to it. The concentration within the corona of motors and other proteins that bind microtubules accelerates the attachment process by creating a large surface around each kinetochore capable of catching and trapping microtubules, much as flypaper traps flies.

As a kinetochore acquires microtubules and becomes attached to the spindle, many of the components of the corona begin to disappear and/or become redistributed. At the same time, the amount of kinetochore-associated molecular motor proteins also decreases. These proteins reappear, however, if the microtubules associated with the kinetochore are removed by disassembling them with drugs.

The composition of the various parts of the kinetochore and the role each plays are shown in **FIGURE 14.28**. The number of different proteins in each of the parts emphasizes how complex a structure, the kinetochore is. Note that throughout it there are proteins that interact with microtubules.

14.11 Kinetochores capture and stabilize their associated microtubules

Key concepts
- Kinetochores and microtubules become connected by a search-and-capture mechanism made possible by the dynamic instability of astral microtubules. The search-and-capture mechanism gives spindle assembly great flexibility.
- Capturing an astral microtubule that is anchored in the centrosome causes a kinetochore to move poleward toward the centrosome. This expedites the capture of additional microtubules and starts the formation of a kinetochore fiber.
- One sister kinetochore usually captures microtubules and develops a kinetochore fiber before the other does.
- The microtubules for constructing a kinetochore fiber also come from microtubules nucleated in the vicinity of the kinetochore.
- The ability of kinetochores to stabilize associated microtubules is essential for the formation of a kinetochore fiber.
- Kinetochores under tension are much more effective at stabilizing microtubules than kinetochores that are not under tension.

The attachment of a chromosome to the spindle requires that each of its kinetochores bind at least one microtubule that is anchored at its minus end in one of the two spindle poles. Normally this occurs as a growing astral microtubule makes contact with a kinetochore surface. However, this is a very demanding spatial problem for the cell to solve reliably. Chromosomes are very large and diffuse very slowly, so the kinetochore cannot move to aid the process. It is, therefore, a stationary target that must be found by microtubules. Considered on the scale of the cell kinetochores are very small, and all 92 of them (in the case of human cells) must be found and associated with microtubules if the chromosomes are to be segregated properly. The problem is compounded by the fact that, as mitosis begins, the kinetochores are in unpredictable locations. After nuclear envelope breakdown the chromosomes are distributed throughout the cytoplasm, with their positions and orientations different from cell to cell and division to division. A spindle must be formed correctly regardless of how the chromosomes are arranged. Clearly the mechanism by which the microtubules find and attach to the kinetochores must be extremely flexible as well as exceptionally reliable.

These problems are solved by the dynamic behavior of the spindle's microtubules. Shortly before mitosis begins, the two centrosomes become modified so that they are capable of nucleating many more microtubules than in interphase. At roughly the same time, the microtubules become much more dynamic. Catastrophes occur more frequently and shrinking microtubules are rarely rescued and often depolymerize completely. These two changes create a situation in which large numbers of microtubules are constantly polymerizing in random directions from each of the two centrosomes, then depolymerizing and disappearing completely if they fail to be stabilized. Microtubules that are lost are replaced by others growing in other directions. The dynamic, searching behavior of the microtubules shortly before spindle assembly begins is shown in the video whose first frame is shown in **FIGURE 14.29**. The result of these dynamics is that after nuclear envelope breakdown the entire interior of the cell is continuously being probed by growing microtubule ends. Under these conditions, it is simply a matter of time before every kinetochore is encountered by an astral microtubule. This search-and-capture

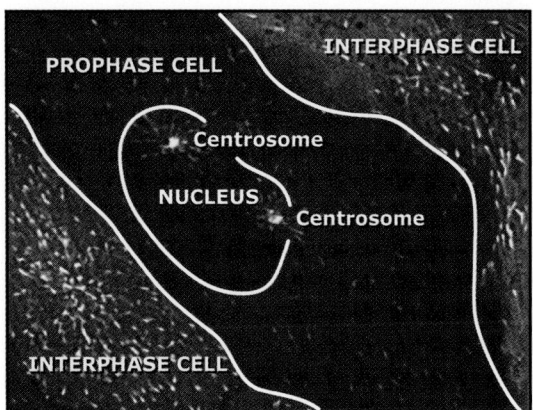

FIGURE 14.29 The first frame of a video that shows live bird cells expressing a fluorescent version of a protein, EB1, which binds to the tips of growing microtubules. Each white dot is the tip of a growing microtubule. The nucleus can be seen as a slightly darker region between and slightly to the left of the two centromeres. Photo courtesy of Patricia Wadsworth, University of Massachusetts, Amherst.

mechanism ensures that all the kinetochores become attached to microtubules, and it allows a spindle to assemble regardless of how the chromosomes are positioned and oriented at the start of mitosis.

When a growing astral microtubule encounters a kinetochore, it becomes trapped by molecular motors associated with the kinetochore's corona. In some cases, the kinetochore attaches to the wall of the microtubule, while in others, it becomes attached directly to its plus end. Under both conditions, *the kinetochore immediately begins to move rapidly poleward along the microtubule*, as illustrated in **FIGURE 14.30** and the video whose first frame is shown in **FIGURE 14.31**. As a result of this motion, the chromosome is dragged toward the pole with the kinetochore leading the way. At some point during this process, kinetochores that initially attached to the side of a microtubule become attached to the plus ends of other astral microtubules. The poleward motion of a kinetochore as it attaches to the spindle orients it so that it now faces the pole. As the oriented kinetochore continues to move toward the pole, it captures more microtubules, beginning the formation of a kinetochore fiber. Because the kinetochore is facing the pole when the new microtubules are captured, most of them bind the kinetochore plate at their tips and terminate there. The gradual formation of

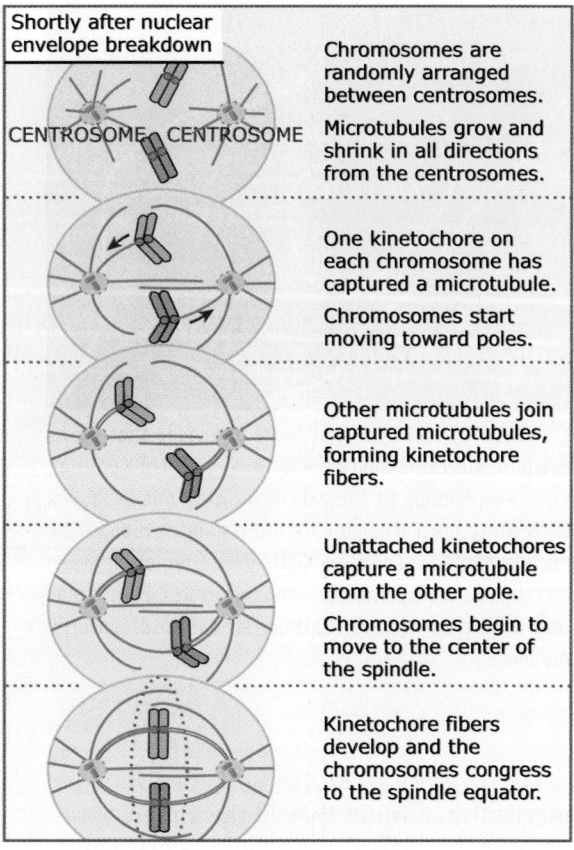

FIGURE 14.30 Dynamic microtubules search for kinetochores throughout the cell by growing and shrinking in random directions from the centrosomes. Microtubules that encounter a kinetochore are captured and stabilized. This search-and-capture mechanism of spindle assembly allows a spindle to be formed regardless of the shape of the cell or the positions of the chromosomes at the start.

The figure panels are labeled:

Shortly after nuclear envelope breakdown

CENTROSOME   CENTROSOME

Chromosomes are randomly arranged between centrosomes.

Microtubules grow and shrink in all directions from the centrosomes.

One kinetochore on each chromosome has captured a microtubule.

Chromosomes start moving toward poles.

Other microtubules join captured microtubules, forming kinetochore fibers.

Unattached kinetochores capture a microtubule from the other pole.

Chromosomes begin to move to the center of the spindle.

Kinetochore fibers develop and the chromosomes congress to the spindle equator.

kinetochore fibers early in mitosis is shown in **FIGURE 14.32**.

Owing to the random nature of the search-and-capture attachment mechanism, sister kinetochores rarely attach to the forming spindle simultaneously. After the first has attached, the chromosome is considered to be **mono-oriented** (see Figure 14.30, **FIGURE 14.33**, **FIGURE 14.34**, and **FIGURE 14.35**). The other kinetochore remains unattached until it begins to acquire a kinetochore fiber of its own that anchors it to the opposing spindle pole. Once this occurs, the chromosome is considered to be **bi-oriented** (see Figure 14.30). *Bi-orientation is the only orientation that ensures that the two chromatids of a replicated chromosome will be distributed to opposite poles during anaphase.* Once a chromosome is bi-oriented, it begins to move toward the middle of the spindle, as shown in **FIGURE 14.36**. As it does so, the two kinetochores function differently: one must move toward the pole to which it is attached, requiring that the kinetochore fiber shorten, while the other must move away from its pole on an elongating fiber. Because bi-orientation is essential for the fidelity of chromosome distribution, cells have evolved a cell cycle checkpoint that detects when sister kinetochores are not stably attached to the spindle (discussed below).

Kinetochores change the properties of the microtubules they bind, and their influence is essential for converting the initial attachment of a chromosome to a single microtubule, into

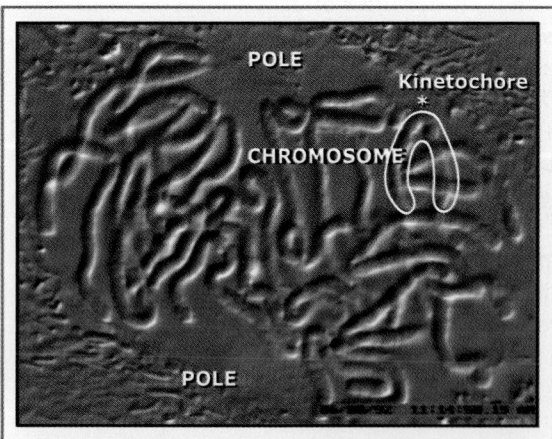

A prometaphase cell early in the process of attaching its chromosomes to the spindle. The kinetochore on a chromosome that is not yet associated with a pole is indicated (*).

FIGURE 14.31 The first frame of a video that depicts the attachment of a kinetochore to a microtubule and the subsequent movement of the chromosome to a pole. Photo © Conly L. Rieder, Wadsworth Center.

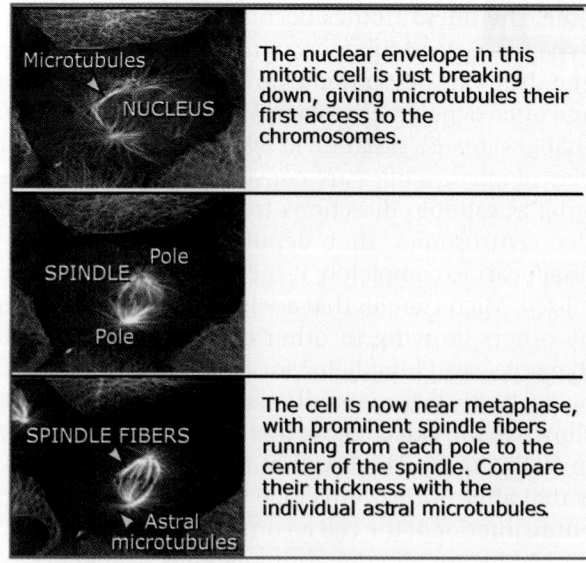

The nuclear envelope in this mitotic cell is just breaking down, giving microtubules their first access to the chromosomes.

The cell is now near metaphase, with prominent spindle fibers running from each pole to the center of the spindle. Compare their thickness with the individual astral microtubules.

FIGURE 14.32 Frames from a video that shows how microtubules attach to chromosomes and form spindle fibers. Photos courtesy of Patricia Wadsworth, University of Massachusetts, Amherst.

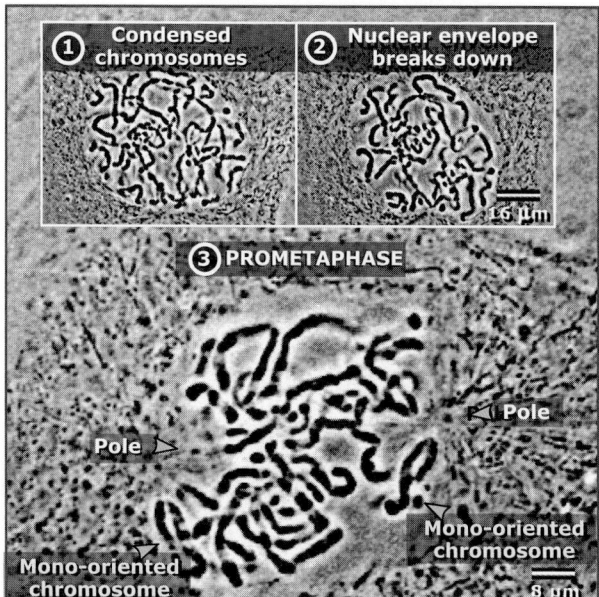

① **Condensed chromosomes** ② **Nuclear envelope breaks down**

16 μm

③ **PROMETAPHASE**

Pole

Pole

Mono-oriented chromosome

Mono-oriented chromosome

8 μm

FIGURE 14.33 A sequence of light micrographs of a cell as its nuclear envelope breaks down and its chromosomes first become accessible to microtubules from the poles. Shortly after nuclear envelope breakdown several of the cell's chromosomes are clearly positioned near one of the poles and oriented with one kinetochore pointing toward it. The rest of the chromosomes are either still unattached or have already attached to both poles and moved to the middle of the spindle. Photos © Conly L. Rieder, Wadsworth Center.

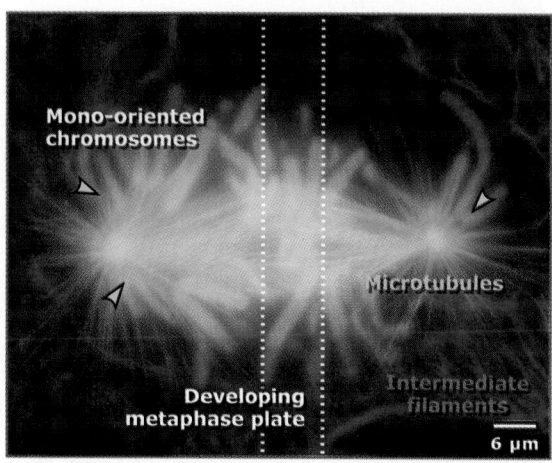

Mono-oriented chromosomes

Microtubules

Developing metaphase plate

Intermediate filaments

6 μm

FIGURE 14.34 An immunofluorescence micrograph of a prometaphase cell. Many of its chromosomes have already attached to both poles and moved to the center of the spindle, but several—including those indicated with arrowheads—are still associated with only one pole. Note their clear "V" shape and proximity to the pole. Photo © Conly L. Rieder, Wadsworth Center.

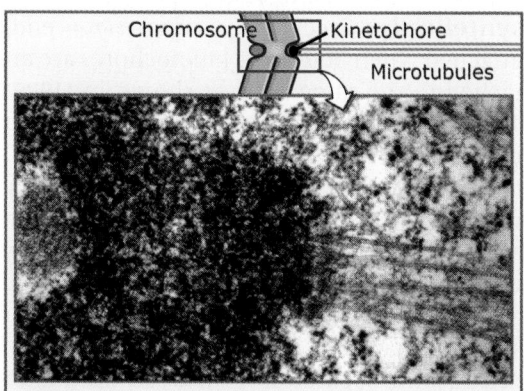

Chromosome — Kinetochore

Microtubules

FIGURE 14.35 An electron micrograph of the centromere and kinetochores of a mono-oriented chromosome like the ones shown in Figure 14.33 and Figure 14.34. A bundle of parallel microtubules terminates in the kinetochore on the right; the other kinetochore has none. Photo © Conly L. Rieder, Wadsworth Center.

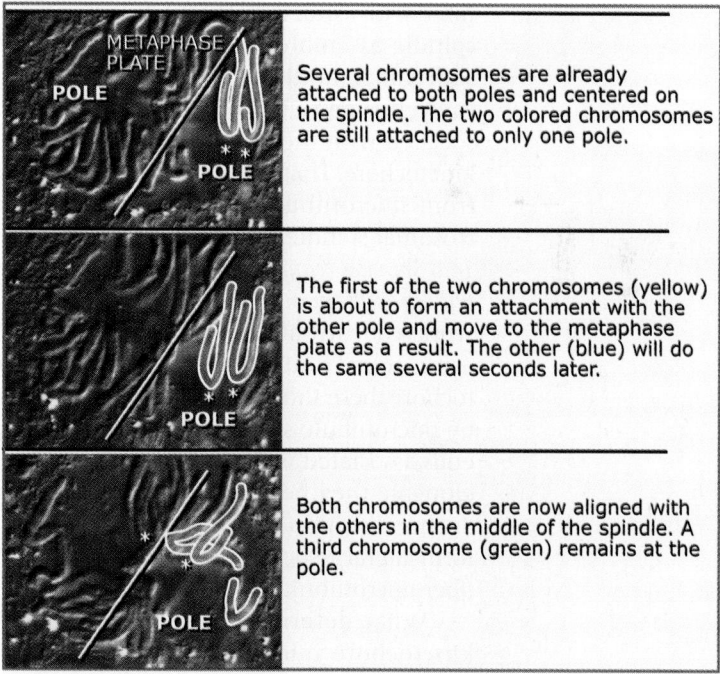

METAPHASE PLATE

POLE

POLE

Several chromosomes are already attached to both poles and centered on the spindle. The two colored chromosomes are still attached to only one pole.

POLE

The first of the two chromosomes (yellow) is about to form an attachment with the other pole and move to the metaphase plate as a result. The other (blue) will do the same several seconds later.

POLE

Both chromosomes are now aligned with the others in the middle of the spindle. A third chromosome (green) remains at the pole.

FIGURE 14.36 Frames from a video that shows the capture of microtubules by two mono-oriented chromosomes and the subsequent movement of the chromosomes to the center of the spindle. Photos © Conly L. Rieder, Wadsworth Center.

the type of connection present in a fully formed mitotic spindle. The most important effect of interacting with a kinetochore is to make a microtubule much longer lived. Microtubules associated with a kinetochore have half-lives of about five minutes, while those in the rest of the spindle typically last less than one minute. The increased stability causes microtubules to accumulate at kinetochores, leading to the formation of kinetochore fibers. Even in kinetochore fibers, however, the microtubules are dynamic, with individual microtubules occasionally detaching from the kinetochore and being lost and new microtubules being incorporated.

The number of microtubules that ultimately bind to a kinetochore is influenced by the size of the kinetochore and the rate at which the microtubules attached to it turn over. The larger the kinetochore, the more microtubules it can engage at once. The kinetochores of higher animal cells generally have the capacity to bind between 20 and 40 microtubules, but the kinetochore fibers may have fewer because of the constant disassociation and reassociation of microtubules taking place.

In vertebrates the microtubules within a mature kinetochore fiber are not just derived from those astral microtubules captured by the kinetochore. If this were the case, many hours would be required for each of the 92 kinetochore in a human cell to become saturated with astral microtubules. Instead, during spindle assembly, kinetochores also capture short microtubules nucleated in the vicinity of the centromere by the action of proteins associated with the chromosome, especially the kinetochore. That is, the fiber is also constructed from microtubules generated by the acentrosomal spindle assembly route described in *14.8 Spindles require chromosomes for stabilization but can "self-organize" without centrosomes*, which works concurrently in the background with the centrosomal route. Once captured by the kinetochore these short microtubules then elongate by microtubule subunit addition at their plus ends associated with the kinetochore. As they elongate they become incorporated into and anchored within the kinetochore fiber as they form lateral association with other kinetochore fiber microtubules and the spindle matrix.

What determines the rate with which kinetochore microtubules detach from the kinetochore? There is good evidence that it is due, in part, to the degree of tension between the kinetochore and its associated centromere. For example, if a fine needle is used to pull the centromere of a bi-oriented chromosome away from one of the poles as the spindle is forming, the number of kinetochore microtubules directed toward that pole increases. Clearly, *tension* somehow promotes the stability (and, thus, the accumulation) of microtubules at a kinetochore. One important implication of this

is that it provides a means for selectively stabilizing the proper attachment of a chromosome to the spindle: sister kinetochores will be under maximum tension—and their kinetochore microtubules will be the most stable—when they are attached to and being pulled toward the opposite poles, that is, *when they are properly bi-oriented for a successful mitosis.*

14.12 Mistakes in kinetochore attachment are corrected

Key concepts

- Improper attachments often occur transiently as the kinetochores attach to the spindle.
- Improper attachments are unstable because they do not allow kinetochores to stabilize attached microtubules.
- On a bipolar spindle, only the correct bipolar attachment of a chromosome produces a stable kinetochore attachment.

The sensitivity of the microtubule/kinetochore junction to tension plays an important role in correcting attachment errors that occur during spindle formation. The search and capture mechanism can produce two types of improper attachments, both of which occur during the course of a normal mitosis. A chromosome is **syntelically** (Greek: *syn* = same; *telos* = end) attached when both of its kinetochores are attached to the same pole, as shown in **FIGURE 14.37**. This condition usually occurs just after nuclear envelope breakdown, when the chromosomes face in random directions and may be much closer to one centrosome than the other, allowing sister kinetochores to simultaneously capture microtubules emanating from the same aster. It is also commonly seen on monopolar spindles that form on a transient basis when centrosome separation is delayed until after nuclear envelope breakdown.

A single kinetochore may also become simultaneously attached to both poles (merotelic attachment; Greek: *mero* = part) (see Figure 14.37). Chromosomes containing one (or even two) merotelically oriented kinetochores move normally to the spindle equator and can pose a serious problem for the cell. If a merotelic attachment persists, the two chromatids separate as the cell enters anaphase, but the improperly attached chromatid remains stuck near the spindle equator. It then stays there until its attachment to one of the poles is broken,

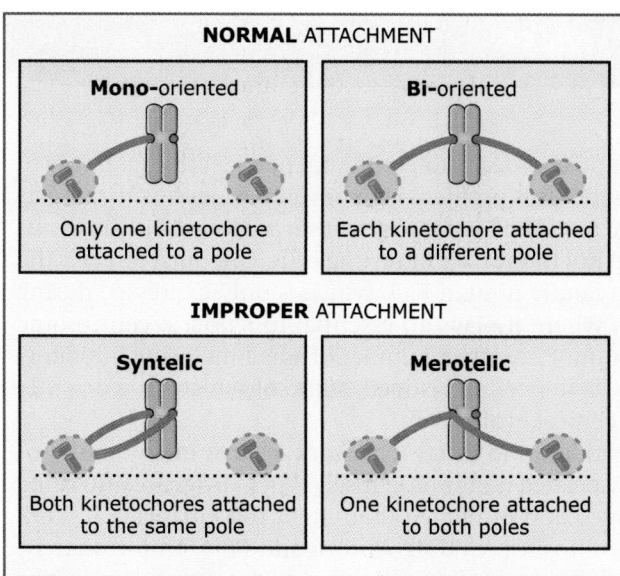

FIGURE 14.37 Bi-orientation is the only type of attachment that ensures that the chromosomes will be equally distributed to the two daughter cells; all chromosomes must be bi-oriented before anaphase begins. Mono-orientation is a normal intermediate on the way to bi-orientation. Syntelic and merotelic attachments often occur early in prometaphase but are corrected to bi-orientation before metaphase.

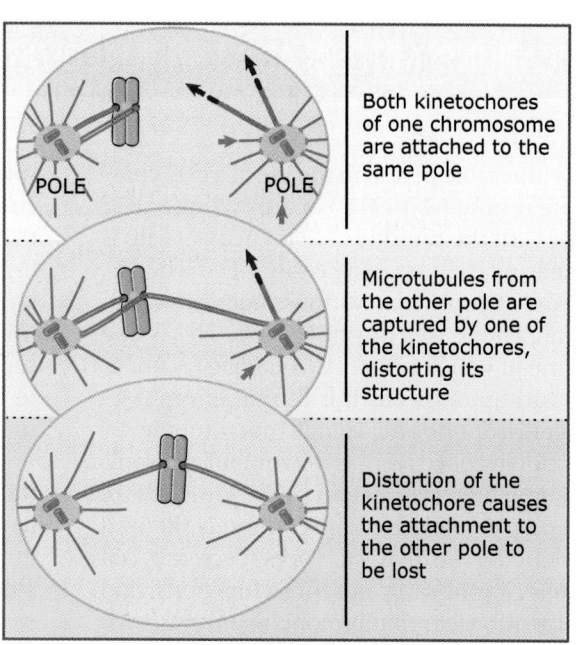

FIGURE 14.38 The ability of an improperly attached chromosome to reorient reliably depends on both the dynamics of microtubules during mitosis and the sensitivity of kinetochores to distortion. Only when the configuration at the bottom is achieved can kinetochore fibers develop and the chromosome be stably attached to the spindle.

giving it a 50% chance of segregating to the same pole as its sister.

Both merotelic and syntelic attachments are normally corrected soon after they arise. In both types of improper attachment, the kinetochore fiber microtubules attach to the kinetochore at a sharp angle, rather than perpendicularly. This distorts the kinetochore, which destabilizes the connection between it and the microtubules within the fiber. As a result, the microtubules detach faster and are replaced less frequently than in a proper attachment. Under this condition, the number of microtubules in a kinetochore fiber will sooner or later dwindle to zero, and one of the attachments will be broken. Depending on which type of improper attachment is involved, loss of an attachment either remedies the problem by creating a bi-oriented chromosome, or results in a chromosome that is attached to only one pole via a single kinetochore. This mono-oriented chromosome can then become bi-oriented by the usual mechanism.

The sensitivity of kinetochores to tension also plays a significant role in correcting improper attachments. Improper attachments do not allow the kinetochores to face the poles,

preventing the tension of a proper attachment from developing. Syntelic attachments (both kinetochores attached to the same pole) in particular are inherently unstable because there can be little tension exerted on the two kinetochores, certainly much less than if the chromosome were properly oriented and its centromere stretched between the two poles. This type of error is often corrected by the spontaneous loss of one of the connections because of lack of tension. Alternatively, the problem may be resolved when one of the two kinetochores acquires an additional attachment to the far pole, as shown in **FIGURE 14.38**. This causes the kinetochore to experience a sudden force in that direction, which distorts its structure and destabilizes the original attachment.

Although the distortion of kinetochore structure and improper tension likely facilitate the correction of syntelic and merotelic attachments, a more potent active mechanism also exists for error correction. As described in *14.9 The centromere is a specialized region on the chromosome that contains the kinetochores*, the centromere region is enriched for a passenger protein complex, consisting of INCENP, survivin, borealin and the Aurora kinase, which

Mitosis is the most vulnerable period in the life of a cell: compared to cells that are resting (G$_o$ or G2$_o$) or are cycling and in interphase (G1, S or G2), the reproductive capacity of cells is more easily abolished and they are more easily killed when irradiated, heat shocked or exposed to various chemicals when the chromosomes are condensed during mitosis. This is due to several factors, one of which is that DNA damage is not as readily repairable during mitosis as it is during interphase. Also, chemicals that disrupt normal spindle microtubule assembly, which are known collectively as microtubule poisons, prolong mitosis by impeding satisfaction of the kinetochore attachment checkpoint. Since gene transcription is silenced during mitosis, prolonging the division process decreases cell viability because cells cannot carry out those functions vital for maintenance. It is also clear that prolonging mitosis with microtubule poisons triggers apoptotic and nonapoptotic death pathways in some cancer cell types, and also to a lesser extent in normal cells, that either kill the cell in mitosis or in the subsequent G1. Finally, a cell cycle checkpoint is present during G1 in cells that contain a functional p53 pathway. In the continuous presence of microtubule poisons this pathway arrests those cells that successfully slip through mitosis in G1 until they die of senescence.

Cancer is characterized by an uncontrolled proliferation of cells: that is, during the transforming process cells lose those control mechanisms that normally prevent them from progressing continuously through unlimited mitotic cell cycles. Since cell proliferation is directly dependent on mitosis, and since mitosis is the most vulnerable period of a cell's life, treatments that block or arrest cells in mitosis are potentially deadly, especially when combined with subsequent radiation treatments. Indeed, over the years a number of spindle poisons, especially those that significantly prolong mitosis in nM concentrations, have proven clinically effective against a variety of cancers. These include, for example, the vinca alkaloids (notably vincristine) as well as the taxanes (paclitaxel). The former, is a potent poison of the microtubule assembly process, and has been used to treat Hodgkin's and non-Hodgkin's lymphomas, acute lymphocytic leukemia and nephroblastoma since its approval by the FDA in 1963. Unlike vincristine, paclitaxel stabilizes spindle microtubules, and since its clinical introduction in 1992 it has proven useful against a broad spectrum of cancers (lung to breast, including Kaposi's sarcoma).

A major problem with drugs that target microtubules is that these structures are also present during interphase in most cells, including nondividing differentiated cells. As a result while drugs like vincristine and paclitaxel may kill cancer cells during or shortly after mitosis, they are also toxic to normal nondividing cells in the body, especially neurons, which have long cytoplasmic processes (axons and dendrites) that rely heavily on the transport of proteins along microtubules. As a result, prolonged treatment of patients with

spindle poisons leads to peripheral neuropathy. One way to circumvent this problem is to develop drugs against targets that are required for mitosis but that are found only in dividing and not interphase or resting cells. One such target is the kinesin family protein Eg5, which is only expressed during mitosis where it plays an essential function in centrosome separation. Recently a number of small molecule inhibitors to Eg5 have been developed, some of which are currently under clinical evaluation.

In addition to drugs that work against microtubules or their associated motor proteins like Eg5, drugs are also being developed and evaluated that target the kinases that work during mitosis, including Aurora and Polo. Unfortunately, although these antimitotic agents may circumvent the peripheral neuropathy associated with microtubule poisons, they too have their limitations: the major problem with mitosis as a target for cancer chemotherapy is that, at any given time, many normal cells in the body are also dividing. This is especially true for the gut and blood tissues, which in a healthy individual are renewed every few days. As a result, any drug that disrupts mitosis in cancer cells also produces side effects, some of which can be quite severe. One such side effect is neutropenia which, by depleting the body of white blood cells (neutrophils), leaves it open to bacterial infections.

Work in the future will include defining how and why delaying satisfaction of the kinetochore attachment checkpoint, as occurs when mitosis is prolonged by the drugs noted above, triggers cell death, and why this differs between normal and cancer cells. In this regard, in the clinic the concentrations of microtubule poisons used are minimal (often just several nM), and while progression through mitosis is delayed at these concentrations most cells ultimately satisfy the kinetochore attachment checkpoint and divide into 2-3 daughters. Although under these conditions it is known that, compared to normal cells, cancer cells exhibit a higher rate of death during or shortly after mitosis, the reasons why remain unknown. It is also unclear what happens to normal cells that complete mitosis in the presence of very low concentrations of spindle poisons: many of the progeny from these divisions will be aneuploid and may themselves contain genetic changes that foster transformation. Since for a given cell type different antimitotic drugs accumulate to different levels within the cell, it will also be important to determine the mechanisms by which membrane drug pumps work, how they discriminate between drugs, and how their regulation differs between cancer and normal cells: are cancer cells more sensitive to paclitaxel than normal cells because they accumulate the drug to higher internal concentrations? Finally, it will be important to develop methods for delivering antimitotic drugs selectively to cancer cells, either by utilizing antigens unique to the surface of cancer cells or by somehow restricting the area of drug application.

regulates the activity of the microtubule plus end destabilizing motor protein, MCAK. When the passenger protein complex is disrupted by genetic means, or when Aurora kinase activity is inhibited with drugs, syntelic and merotelic attachments are not corrected. The current view is that during spindle assembly the passenger complex activates MCAK which, in turn, destabilizes the plus ends of those microtubules that pass over or contact the centromere. Since both the merotelic and syntelic conditions are generated by microtubules that ultimately pass by the centromere (see Figure 14.37), destabilizing these microtubules rapidly corrects the error. Here it is noteworthy that merotelic attachments that form during a tripolar mitosis when a kinetochore becomes attached to two spindle poles that are positioned to its front are stable. This is because the two sets of microtubules that connect the merotelically attached kinetochore to the two spindle poles do not contact or cross the centromere. Also, even though the kinetochore is attached to two poles, the pulling force it exerts on its sister, which is attached to a single pole, generates tension across the centromere, as occurs during a normal bipolar attachment.

The intricacies of the attachment process demonstrate the complexity of kinetochores and the central role they play in mitosis. In fact, the separation of replicated chromosomes equivalently into two groups results directly from behaviors and functions associated with the kinetochores. In 1961, Mazia—one of the pioneers in thinking about the mechanisms of mitosis—emphasized their importance by stating that the kinetochore is "the only essential part of the chromosome so far as mitosis is concerned." He likened the rest of the chromosome to the "corpse at a funeral," in that it "provides the reason for the proceedings but does not take an active part in them." Although Mazia was right that the real goal of mitosis is to segregate sister kinetochores (the rest of the chromosome can be viewed largely as a "bag of genes" coming along for the ride), at the time he could not envision the extent to which these macromolecular assemblies are involved in directing this process.

In conclusion, kinetochore fibers attach the chromosome to the spindle and define the direction the chromosome will move. The equal segregation of chromosomes during mitosis can be related directly to the following facts: (1) each replicated chromosome contains two sister kinetochores; (2) these two kinetochores

lie on opposite sides of the chromosome's primary constriction and face in opposite directions; (3) kinetochore fiber microtubules bind to the kinetochore only on its outer surface, which faces away from the chromosome; and (4) mechanisms exist to correct improper kinetochore-to-pole connections that would otherwise lead to aneuploidy.

Concept and Reasoning Checks

By preventing centrosome separation with an inhibitor of the Eg5 motor protein, you create a monopolar spindle in a human cell that is enriched for syntelic attachments.

1. Why is the spindle enriched for syntelic attachments?
2. How are these errors corrected as the spindle becomes bipolar after washing out the Eg5 inhibitor?
3. What would happen if you wash out the inhibitor in the presence of an Aurora inhibitor?
4. Why do abnormal kinetochore attachments not occur during spindle assembly in plant cells?

14.13 Kinetochore fibers must both shorten and elongate to allow chromosomes to move

Key concepts

- Poleward forces are exerted on attached kinetochores during all stages of mitosis.
- Kinetochore fibers are anchored near the poles.
- Anchorage may depend on the spindle matrix, composed of the NuMA protein and a number of molecular motors.
- Kinetochore fibers change length by addition or loss of tubulin subunits at their ends.
- Both kinetochores and poles can remain attached to the ends of kinetochore fibers as the fibers change length.

The question "How do chromosomes move?" has been asked since the process of mitosis was discovered. In 1880, Flemming summarized the problem when he stated, "We do not know, in the movements or changes of position of the threads of a nuclear figure, whether the immediate causes lie within the threads themselves, outside of them, or both." (By "threads" Flemming was referring to the condensed chromosomes, and by "nuclear figure" he meant the mitotic spindle.)

We now know that a force directed toward the spindle pole begins to act on a kinetochore *as soon as it attaches to the spindle.* We also know *that this force acts during all stages of mitosis,* and *that the mechanism that generates it is the same throughout.* Thus, the poleward motion of a chromosome as it congresses during prometaphase occurs by the same mechanism that moves it toward a pole during anaphase.

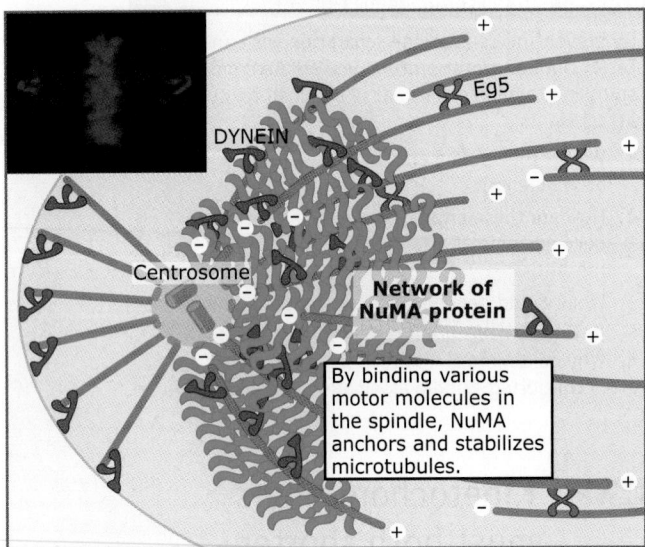

FIGURE 14.39 Within the spindle the NuMA protein forms a highly branched and crosslinked network—the matrix—that surrounds the microtubules and helps anchor and organize them at the poles. Motors associated with the matrix are likely to be responsible for microtubule dynamics that occur at the poles, including pulling kinetochore fibers poleward. The inset shows an immunofluorescence photograph of a metaphase cell with NuMA in red and the chromosomes in blue. Photo courtesy of Duane Compton, Dartmouth Medical School.

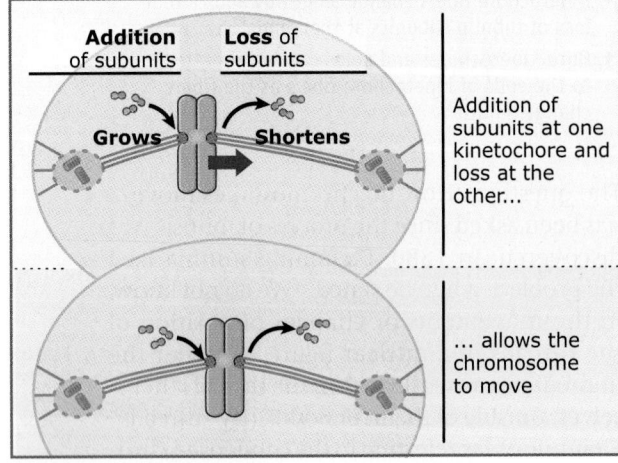

FIGURE 14.40 Movement of a bi-oriented chromosome within the spindle requires the simultaneous and coordinated growth and shortening of its two kinetochore fibers. The fibers change length by the gain or loss of tubulin subunits at the kinetochores.

In order for the chromosomes to move in response to forces exerted at their kinetochores, the kinetochore fibers must somehow be anchored. Without some form of anchorage, the chromosomes would remain in one place and reel in the microtubules rather than moving toward the poles along them. Micromanipulation experiments, in which individual chromosomes are tugged on by a very fine glass needle, reveal that it is difficult to pull a chromosome away from a pole but is relatively easy to move it from side to side. This implies that kinetochore fiber microtubules are most firmly anchored at their minus ends, near the poles of the spindle.

All of the microtubules in the spindle, including those in kinetochore fibers, are surrounded by the spindle matrix, as shown in **FIGURE 14.39**. The proteins of the spindle matrix likely play an important role in anchoring the kinetochore fibers. One of the major constituents of the matrix is the NuMA protein, which is critical for maintaining the integrity of the spindle. The concentration of NuMA within the spindle is related to microtubule density. As seen in Figure 14.39, microtubule density decreases gradually from the polar regions to the spindle equator, so that NuMA is particularly concentrated near the poles. The matrix also contains the microtubule-dependent motors Eg5 and HSET, both of which are kinesin family proteins. HSET is unusual because, like cytoplasmic dynein, it moves toward the minus ends of spindle microtubules. As a result, HSET also accumulates in the spindle poles. The current model for how kinetochore fiber microtubules are anchored, envisions that these motor proteins are attached to NuMA, which surrounds the microtubules near their minus ends. While bound to NuMA, the motors also interact with the walls of microtubules, creating a drag that resists their movement and serves as an effective anchor.

When a kinetochore moves away from its associated pole, as each does intermittently during congression, its fiber must elongate. Similarly, when it moves toward the pole, its fiber must shorten, as shown in **FIGURE 14.40**. *The elongation of kinetochore fiber microtubules occurs by the addition of tubulin subunits at the kinetochore. Shortening occurs by the loss of tubulin subunits at both the kinetochore and the poles.* In both cases, in order to remain continuously attached to its fiber, the kinetochore must somehow hold onto the ends of the microtubules as they gain or lose subunits. *The polar attachment at the other end of the fiber must also be complex in order to allow for the loss of subunits observed during shortening.* The mechanisms for these remarkable behaviors at the ends of the fibers are just beginning to be explored. The current view is that the attachment of microtubule plus ends to a kinetochore requires the NDC80 protein complex (consisting of 4 proteins including Nuf2 and Hec1), while microtubule subunit addition and deletion at this end require the CLASP protein, as well as microtubule plus end binding proteins that promote microtubule shortening. Microtubule subunit deletion at the minus ends of kinetochore fiber microtubules, located in the polar region, is thought to be mediated by another kinesin family motor protein (KIF 2a) that functions as a minus microtubule end depolymerase.

14.14 The force to move a chromosome toward a pole is produced by two mechanisms

Key concepts

- A kinetochore pulls the chromosome toward the pole but can move only as fast as the microtubules in the kinetochore fiber can shorten.
- Dynein at the kinetochore pulls a chromosome poleward on the ends of depolymerizing microtubules.
- Force generated along the sides of the kinetochore fiber also moves the entire fiber poleward, pulling the kinetochore and chromosome behind it.

Once a kinetochore fiber is fully formed, a kinetochore pulls its associated chromosome toward the pole at about 1-2 μm/min. At this rate it takes between 5 and 10 minutes to cover 10 μm, or about half the length of a spindle. Theoretically, the force required to move an object the size of a large chromosome at this rate for 10 μm is only about 10^{-8} dynes. Surprisingly, the amount of energy that would be necessary to generate this force could be obtained by the hydrolysis of just 20 ATP molecules!

What actually occurs during mitosis, however, is much different. The force on a moving chromosome can be measured by determining how far it bends a fine glass needle placed in its path. Such experiments yield a surprising result: the maximum pole-directed force that the spindle can exert on a chromosome is ~10,000

times stronger than that theoretically needed to move it at the speeds typically observed during mitosis. Thus, *the rate of poleward motion is not determined by the amount of force acting on the kinetochore, but rather by some other influence that keeps the velocity constant regardless of the force applied.* As an analogy, no matter how many horsepower the engine of a car can produce, the gear ratio limits how fast it can move; most cars cannot do 60 mph in first gear. In the case of the spindle, the limiting factor is the rate at which the microtubules in a kinetochore fiber can depolymerize: a kinetochore can move poleward only as fast as its microtubules can shorten.

Poleward chromosome motion is powered by two mechanisms. Both are present throughout mitosis, and both can act simultaneously, as shown in **FIGURE 14.41**. One involves the kinetochore; the other, its associated fiber. The contribution each makes varies depending on the type of cell. In vertebrate cells, the force for poleward chromosome motion is produced both from the activity of a microtubule-dependent motor located at the kinetochore and from movement of the kinetochore fiber as a whole toward the pole.

In the "Pac-man" mechanism, motors in the kinetochore produce the force, moving the

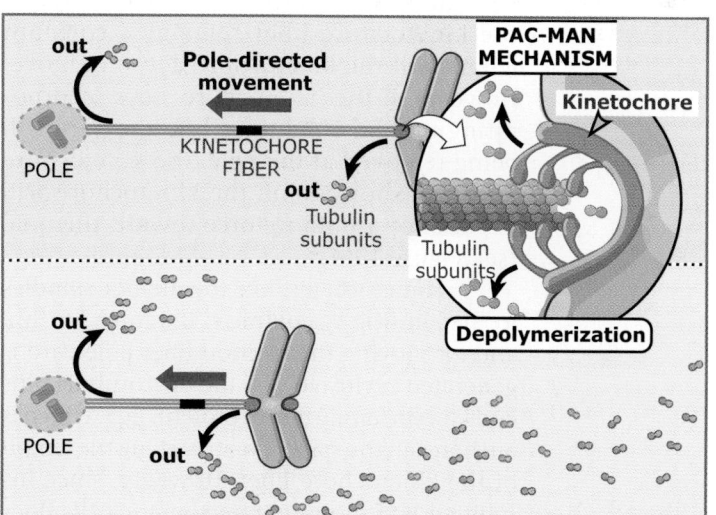

FIGURE 14.41 Motors at the kinetochore move it poleward along the microtubules within the kinetochore fiber. As the kinetochore moves, the microtubules depolymerize behind the motors. At the same time, other forces move the entire kinetochore fiber toward the pole, where subunits are also lost from the microtubules. Were a mark (shown here in black) made on the kinetochore fiber, both the distance between it and the kinetochore and the distance between it and the pole would decrease as the chromosome moved.

kinetochore poleward at the end of the kinetochore fiber as the microtubules in it shorten by depolymerization. Thus, as the kinetochore moves poleward, it can be viewed as "chewing up" the end of the fiber, giving the process its name. In the other "traction fiber" mechanism, force is produced along the length of the kinetochore fiber. As a result the whole kinetochore fiber slides toward the pole, pulling the kinetochore behind it.

The Pac-man mechanism for chromosome motion is powered by cytoplasmic dynein, a minus end-directed microtubule motor. Dynein located at the kinetochore actively pulls it poleward on the disassembling plus ends of its microtubules. During this motion, compression against the kinetochore, and/or catastrophe-promoting factors contained within its structure, induce its associated microtubule plus ends to disassemble. Dynein also plays a role during the initial attachment of kinetochores to the spindle, when they can move poleward along the side of a single microtubule at a velocity of more than 40 μm/min.

The traction fiber mechanism is based on the movement (flux) of tubulin subunits that occurs from one end of a kinetochore fiber to the other. Tubulin subunits are incorporated into each microtubule at the kinetochore and then migrate through it to the pole, where they are released (see Figure 14.13). As long as the incorporation of subunits at the kinetochore is equal to their loss at the pole, the kinetochore fiber remains a constant length and the kinetochore does not move. However, if the kinetochore stops incorporating tubulin subunits while they are still being removed at the pole, the kinetochore fiber will shorten and the kinetochore will experience a pulling force toward the pole (see Figure 14.41).

In some systems, like the meiotic spindles formed in frog egg extracts, the force for flux and for moving the traction fiber poleward is generated exclusively by the microtubule plus end kinesin family motor protein, Eg5, which is anchored in the spindle matrix along the length of the kinetochore fiber. However, since inhibiting Eg5 in vertebrate somatic cells does not affect the rate of flux, the force for flux in this system must occur via a different route. At the moment, the mechanism favored envisions that this force is generated locally in the spindle pole by a "pulling-in' mechanism mediated by a microtubule depolymerase complex located on and anchored at the kinetochore fiber microtubule minus ends.

14.15 Congression involves pulling forces that act on the kinetochores

Key concepts

- The balance of several forces aligns the chromosomes at metaphase.
- A plausible model suggests that poleward forces proportional to the length of each kinetochore fiber positions and/or maintains the chromosomes in the center of the spindle.
- This mechanism may align the chromosomes in some types of cells.
- In many types of cells other forces must participate, including forces generated by the kinetochore and others that push chromosomes away from poles.

In 1945, Ostergren—another of the pioneers in thinking about the mechanisms of mitosis—offered an elegant explanation for congression. He proposed that "the equilibrium position of [a chromosome] strongly indicates that centromeres [kinetochores] are attracted to spindle poles by forces increasing in strength with an increasing distance between centromere and pole, and that each centromere is attracted only by that pole toward which it is turned." In other words, *the magnitude of the poleward pulling force that acts on each kinetochore would be proportional to the length of its associated kinetochore fiber*. In his view, a chromosome moves to the spindle equator because that is where its opposing kinetochore fibers are of equal length, and, thus, where the net force on the chromosome is zero, as shown in **FIGURE 14.42**.

Ostergren's "traction fiber" model draws support from the movement (flux) of tubulin subunits through kinetochore fibers. In some systems, flux is known to be mediated by molecular motors anchored in the spindle matrix that surround and extend along the length of the kinetochore fiber microtubules (see *14.14 The force to move a chromosome toward a pole is produced by two mechanisms*). Under this condition, the motors exert a pole-directed pull on tubulin subunits along the entire length of the microtubules, and the resulting movement of the microtubule subunits toward the pole is what we observe as flux. In these cells it would

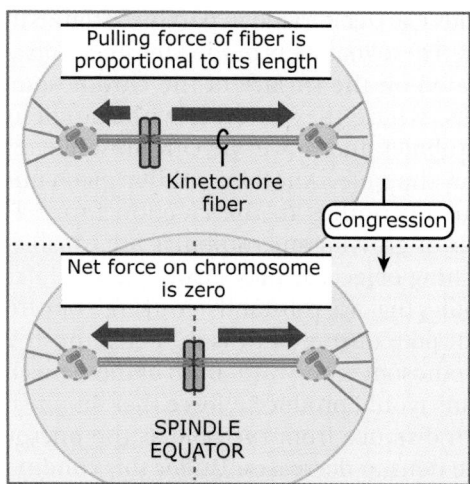

FIGURE 14.42 Ostergren's hypothesis for the mechanism of congression. A bi-oriented chromosome moves from left to right because the pulling force on each sister kinetochore is proportional to the length of its kinetochore fiber, with a longer fiber producing a greater pulling force. Here the length and direction of the red arrows indicate the magnitude and direction of the force. The chromosome moves until the two forces are balanced, which will occur in the middle of the spindle.

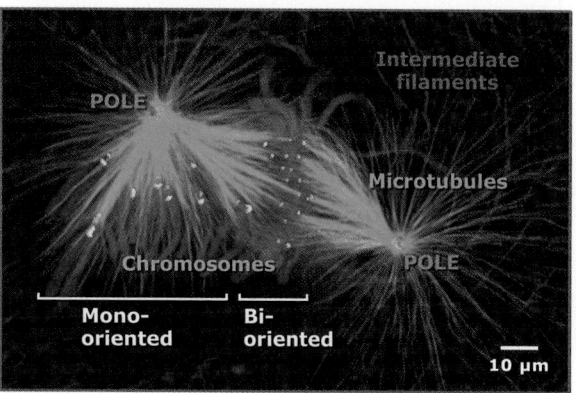

FIGURE 14.43 In this prometaphase cell, about half the chromosomes are already bi-oriented and aligned in the middle of the spindle. The others (at least 7) are still mono-oriented. Although clearly associated with only the pole on the left, all the mono-oriented chromosomes remain some distance from it. Kinetochores appear as yellow dots. Note how the pairs of dots on the bi-oriented chromosomes are all aligned parallel to the axis of the spindle and those on the mono-oriented chromosomes are oriented radially from the pole. Photo courtesy of Alexey Khodjakov, Wadsworth Center.

be expected that the longer the microtubules, the greater the pole-directed pulling force that will act on the kinetochore, just as Ostergren hypothesized.

The traction fiber model remains the most feasible model for how chromosomes congress in cells like some insect spermatocytes and frog eggs, where all kinetochore fibers in a half-spindle focus onto a common discrete spindle pole and where microtubule subunit flux is the only mechanism for moving the chromosomes toward a pole. In other systems like plants, where a common spindle with broad "diffuse" poles is formed from the coalescence of many small spindles built around each chromosome, flux may be responsible for maintaining the position of chromosomes on the equator once a common spindle is formed. However, in such cells congression results from two characteristic features inherent to acentrosomal spindle assembly. The first is that the kinetochore fibers formed on all sister kinetochore pairs by this route will be the same length, due to the global nature of regulatory influences on microtubule and kinetochore dynamics. The second is that the actual alignment of chromosomes onto a common equator or plate occurs from the fusion of multiple smaller spindles of a similar length, followed by their alignment along a common parallel axis. In these systems congression does not occur from the movement of a chromosome to the middle of the spindle. Rather it occurs instead from the alignment of many spindles of the same length into one common spindle that is the same length as the original mini-spindles.

Since inhibiting microtubule flux in vertebrate somatic cells does not inhibit chromosome congression, other forces must work in these cells to position the chromosomes on the spindle equator. In these cells the forces that pull on kinetochores are not the only ones acting on the chromosomes. This is evident from the behavior of chromosomes attached to just one pole. Were the position of a chromosome determined solely by pulling forces at its kinetochores, as predicted by Ostergren's model, such mono-oriented chromosomes would be expected to move all the way to the pole. Instead, as **FIGURE 14.43** shows, they often become stably positioned many micrometers away from it, suggesting the existence of a second force that pushes chromosomes away from poles.

14.16 Congression is also regulated by forces that act along the chromosome arms and the activity of sister kinetochores

Key concepts

- Forces that act on the arms of chromosomes push them away from a pole.
- These forces arise from interactions between a chromosome's arms and spindle microtubules.
- The unattached kinetochore on a mono-oriented chromosome can participate in congression
- Kinetochores can switch between active and passive states.
- Switching of sister kinetochores between the two states is coordinated.

In addition to the forces that act at the kinetochores, the arms of a mitotic chromosome also experience a force. This can be demonstrated experimentally by using a very finely focused laser beam to cut the arms of a mono-oriented chromosome away from its kinetochore, as shown in **FIGURE 14.44**, Once freed, the arms are rapidly expelled from the polar region. This means that *during mitosis, chromosomes also experience forces that work to push them away from each pole.* These ejection forces, called "polar winds," are generated in part by chromokinesins. These members of the kinesin family are located on the surface of the chromosome's arms, where they interact with spindle microtubules and move the chromosome away from the pole. Another, subtler mechanism also contributes to the ejection force. The ends of growing microtubules are capable of pushing objects as they grow, and the microtubules that are constantly growing away from each pole during mitosis may help push the chromosomes outward. Both of these mechanisms would produce a force that diminishes with distance from the pole as the microtubule density decreases. Under this condition, a chromosome attached to only one pole will come to rest at the position where the ejection forces of the polar winds are balanced by the pole-directed forces acting on the kinetochore, accounting for the inability of such chromosomes to move all the way to the pole, as seen in **FIGURE 14.45**.

In addition to the polar winds, a mono-oriented chromosome can experience another force that acts to move it away from the pole it is attached to. On these chromosomes the corona material associated with the unattached kinetochore contains CENP-E, a type of kinesin motor protein with microtubule plus end motility. When this CENP-E containing corona material contacts the surface of a kinetochore fiber on an adjacent bi-oriented chromosome it

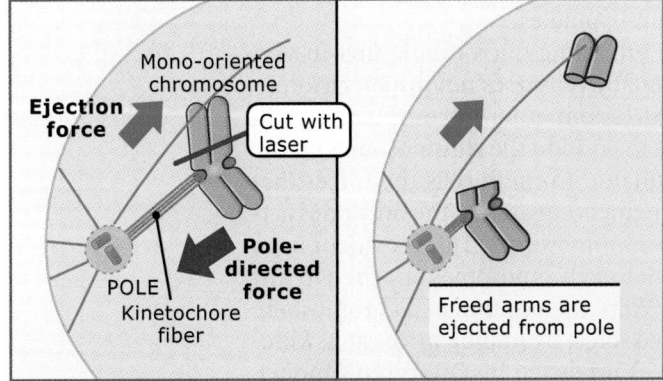

FIGURE 14.44 An experiment that demonstrates that chromosome arms experience a force that pushes them away from a pole. A mono-oriented chromosome is stably positioned at a distance from the pole to which it is attached. If the chromosome is cut, the piece without a kinetochore moves rapidly away from the pole while the rest of the chromosome moves to a new stable position closer to the pole. Pole-directed forces at the kinetochore are thus opposed by ejection forces that act on the chromosome's arms.

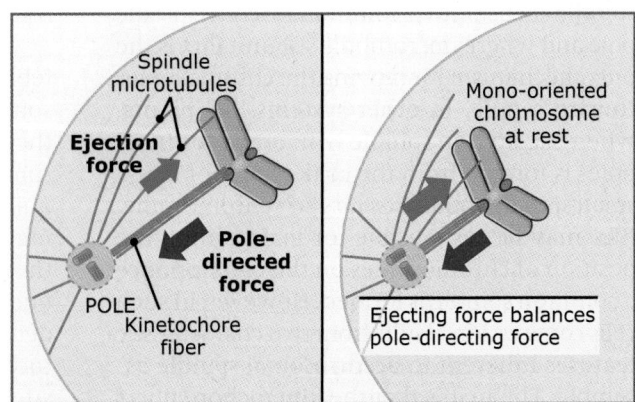

FIGURE 14.45 Schematic depicting the forces that work on a mono-oriented chromosome. Kinetochore-based forces (red arrow) pull the chromosome toward the pole while ejection forces (green arrow) push the chromosome away from the pole. The chromosome becomes stably positioned where these two forces are balanced (right). If one of the two forces is greater than the other, the chromosome will move toward or away from the pole (left).

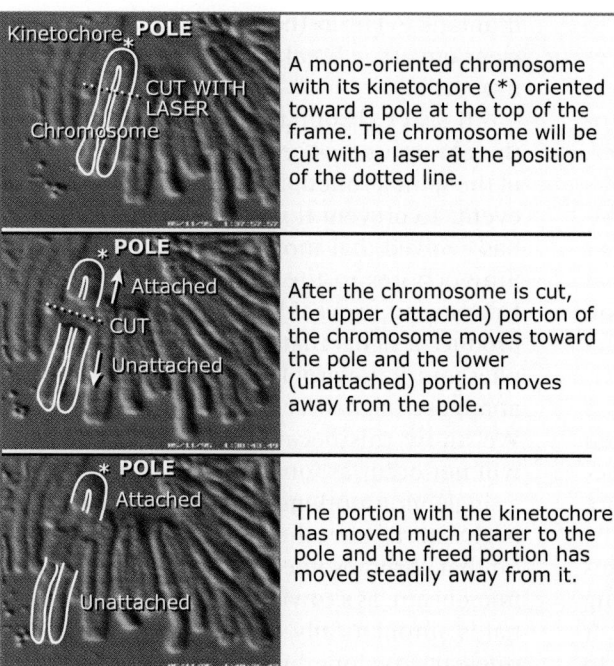

Kinetochore **POLE** | A mono-oriented chromosome with its kinetochore (*) oriented toward a pole at the top of the frame. The chromosome will be cut with a laser at the position of the dotted line.

After the chromosome is cut, the upper (attached) portion of the chromosome moves toward the pole and the lower (unattached) portion moves away from the pole.

The portion with the kinetochore has moved much nearer to the pole and the freed portion has moved steadily away from it.

: **FIGURE 14.46** Frames from a video showing that mitotic chromosomes experience a force that pushes them away from a pole. Photos © Conly L. Rieder, Wadsworth Center.

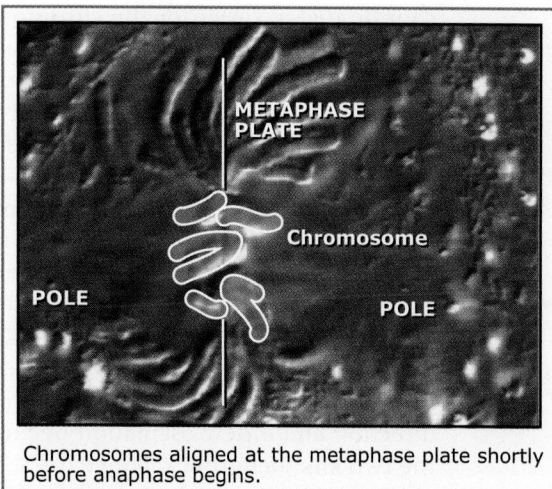

Chromosomes aligned at the metaphase plate shortly before anaphase begins.

: **FIGURE 14.47** The first frame of a video showing that during metaphase, chromosomes do not remain stationary at the center of the cell but instead move back and forth constantly. Photo © Conly L. Rieder, Wadsworth Center.

can slide along the fiber toward the plus ends of its microtubules at the spindle equator, as seen in **FIGURE 14.46**. Not only does this CENP-E mediated sliding contribute to congression, but it also helps the mono-oriented chromosome become bi-oriented by positioning its unattached kinetochore closer to the spindle equator and thus distal pole.

Ostergren's original traction fiber model also predicts that once a kinetochore is attached to the spindle, it constantly experiences a pole-directed force. Were this the case, once attached to both poles, chromosomes would move smoothly and continuously to the middle of the spindle and then remain motionless there until anaphase begins. This appears to be true for many types of cells, including those from plants and insects, in which the chromosomes do not move after they become positioned on the metaphase plate. However, in vertebrate cells congressed chromosomes constantly move small distances back and forth across the middle of the spindle, as seen in the movie whose first frame is shown in **FIGURE 14.47**.

These movements are caused by the kinetochores and reveal something important about how they work. In order for oscilla-

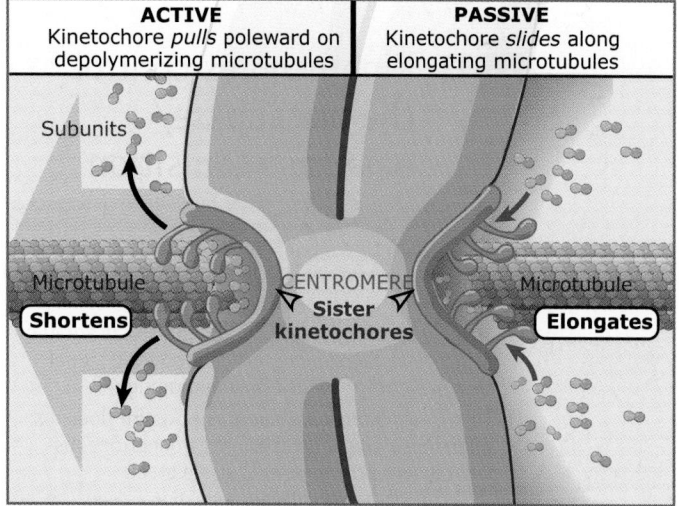

FIGURE 14.48 Once attached to the spindle, a kinetochore can exist in one of two activity states. One (left) allows it to move toward a pole on a shortening kinetochore fiber. The other (right) allows it to remain stationary or move away from its pole on an elongating fiber. The activities of the two sister kinetochores are somehow coordinated so that both kinetochores are rarely in the active state at the same time.

tions to occur, each kinetochore must have two different states that it can switch between. As illustrated in **FIGURE 14.48**, the switching of sister kinetochores between states must be coordinated in order for the chromosome to reverse direction. As a chromosome moves during an oscillation, the kinetochore that is moving toward its associated pole is producing or experiencing a pole-directed force. As the kinetochore moves, tubulin subunits must be lost from the microtubules in its kinetochore fiber. At the same time, the other kinetochore must be in a "neutral" state that allows it to be dragged away from its pole on the end of its kinetochore fiber, which is actively incorporating subunits in order to allow it to elongate. The chromosome reverses direction and a new oscillation begins when the two kinetochores exchange roles. The periodic switches in activity by a kinetochore are known as kinetochore directional instability, and the position of a chromosome on the spindle is heavily influenced by how the switching between sister kinetochores is coordinated. It is not known what causes a kinetochore to switch between the neutral and active states, or what coordinates this activity between the sister kinetochores. Switching must be sensitive to the position of the chromosome relative to the spindle equator, however; otherwise the chromosomes could not remain reliably centered.

14.17 Kinetochores control the metaphase/anaphase transition

Key concepts

- Progression through mitosis is controlled by a cell cycle checkpoint, which functions to minimize the production of aneuploid cells.
- This checkpoint detects kinetochores that are not attached to the spindle, which produce a signal that prevents anaphase from beginning.
- This kinetochore attachment checkpoint does not distinguish proper attachments from erroneous kinetochore attachments that are stable.
- When all the kinetochores in a cell are stably attached to the spindle, the checkpoint is satisfied which, in turn, allows anaphase promoting complexes (APCs) to be activated.
- Activation of the APCs leads to the destruction of proteins that hold sister chromatids together.

The separation of chromatids at the start of anaphase provides visual evidence that a cell has undergone the metaphase-to-anaphase transition. With the exception of nuclear envelope breakdown, this is the most visually

dramatic event in the cell cycle. As with nuclear envelope breakdown, this event is also irreversible.

If sister chromatids separated before all the chromosomes are attached to both poles of the spindle, aneuploidy would be a frequent event. To prevent this, a cell cycle checkpoint has evolved that monitors the attachment of kinetochores to the spindle during mitosis. Unattached or weakly attached kinetochores produce a signal that delays the onset of anaphase until the problem is corrected—a "wait anaphase" signal (see *15 Cell cycle regulation*). We know this because anaphase normally will not occur as long as a cell contains even a single mono-oriented chromosome but will begin shortly after the unattached kinetochore on that chromosome is destroyed with a laser microbeam, as shown in **FIGURE 14.49**. The signal is automatically produced as soon as the nuclear envelope breaks down because the kinetochores of the newly exposed chromosomes are all unattached. Therefore, this signal

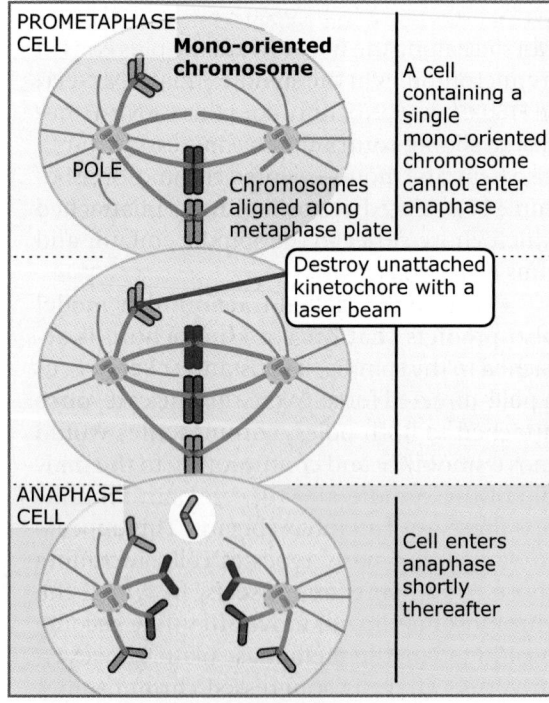

FIGURE 14.49 An experiment demonstrating that an unattached kinetochore produces a signal that prevents anaphase from beginning. The cell is stuck in prometaphase at the top. After an extremely fine laser beam (red) is used to inactivate the components of its one unattached kinetochore, anaphase soon begins.

is continuously present until the last kinetochore attaches to the spindle.

Although this checkpoint is commonly referred to as the spindle assembly checkpoint, it is more accurately termed the kinetochore attachment checkpoint. This is because the checkpoint does not monitor spindle assembly, but rather unattached kinetochores, and all kinetochores can become stably attached on grossly abnormal or incompletely assembled spindles. Also, the checkpoint does not prevent anaphase until all sister kinetochores are properly attached to the spindle, as commonly assumed, because it cannot differentiate between a kinetochore that is stably attached to one pole or to multiple poles simultaneously (i.e., merotelically oriented). The inability of the checkpoint to detect this type of malorientation means that anaphase sometimes occurs before the problem can be corrected. Although mistakes in the segregation of chromosomes are normally rare, they do occur approximately once every 10,000 divisions. Current evidence suggests that merotelic attachment is a leading cause of aneuploidy in tissue cells.

The kinetochore attachment checkpoint works by detecting kinetochores that have not yet acquired an attachment to the spindle, or which have lost their attachment because it was unstable. When a kinetochore is not attached to the spindle, it produces the wait anaphase signal and this signal is turned off once it becomes attached. The checkpoint is sensitive to whether a kinetochore is under tension because tension across the centromere enhances the stability of microtubules associated with a kinetochore, thereby decreasing the chance that it will loose its attachment. The attachment of sister kinetochores to the spindle is most stable when they are connected to opposite poles, because at this time they experience maximum tension. As a result the checkpoint is able to indirectly assess whether all the chromosomes are bi-oriented on the spindle. Remember, however, that a kinetochore that is stably attached to two spindle poles, as often occurs during a tripolar mitosis, is just as invisible to the checkpoint as one that is stably attached to only one pole.

When spindle formation is inhibited by microtubule poisons, the kinetochore attachment checkpoint remains active for a prolonged period. However, even in the absence of microtubules the checkpoint cannot keep the cell in mitosis indefinitely, and after many hours the cell exits mitosis and enters the next G1 as a 4N tetraploid cell. This process, termed **mitotic slippage**, occurs because the checkpoint is not 100% efficient at preventing a slow but continuous APC-mediated background destruction of cyclin B. The persistence of the checkpoint signal in the presence of drugs that prevent spindle microtubule assembly was used as a screen to isolate yeast mutants that are defective in genes needed for the checkpoint. The genes that were discovered code for three Mad (*m*itosis *a*rrest *d*eficient) and three Bub (*b*udding *u*ninhibited by *b*enzamidazole) proteins. These proteins have counterparts (homologs) in vertebrate cells, including those of humans. When the activity of any Bub or Mad protein is inhibited, the checkpoint is overridden, and anaphase occurs shortly thereafter. During mitosis, several of these proteins, including Mad2, constantly cycles on and off unattached kinetochores but are not found on attached kinetochores.

The critical event that initiates anaphase, which in turn allows sister chromatids to separate, is the activation of a large macromolecular assembly known as the **anaphase-promoting complex**. The job of the anaphase-promoting complex is to target selected proteins for destruction, and it does so by attaching a string of ubiquitin molecules to them. This tag then allows the protein to be recognized and degraded by the cell's proteolytic machinery (proteosomes). The anaphase-promoting complex cannot work by itself, however. Rather, a cofactor is required to specify which proteins are to be targeted for destruction and at what time. Anaphase onset requires that the anaphase-promoting complex be activated by the Cdc20 cofactor. *The spindle assembly checkpoint works by inhibiting the activity of Cdc20, which, in turn, prevents the anaphase-promoting complex from recognizing and targeting for destruction several proteins including cyclin B and **securin**.* Cyclin B destruction begins as soon as the kinetochore attachment checkpoint is satisfied, and its progressive destruction ultimately drives the cell out of mitosis. The destruction of securin starts only after cyclin B levels have fallen below a critical threshold. We know this because when a cell is filled with a type of cyclin B that cannot be destroyed chromatid disjunction is blocked even though the kinetochore attachment checkpoint is satisfied. It is the destruction of securin that ultimately allows the glue proteins that actually hold the replicated chromatids together to be dissolved.

Once the checkpoint is satisfied, the destruction of cyclin B begins. During this time there is a short (5 min) window of time, before the cell becomes committed to anaphase, in which the kinetochore attachment checkpoint can be turned back on and the process of cyclin B degradation stopped. However, once the activity of the cyclin B/CDK1 kinase has dropped to the point that securin is destroyed the cell becomes committed to anaphase.

Exactly how the kinetochore and the components of the checkpoint pathway influence Cdc20 is not yet clear. The question is how the signal is transmitted from the kinetochore into the spindle, where the anaphase-promoting complexes are found. One view is that complexes composed of Mad2, BubR1, and Cdc20 are formed at the kinetochore and then released. Another possibility is that unattached kinetochores bind and activate one or more of the checkpoint proteins and then release the activated proteins into the spindle where they form complexes with Cdc20 that prevent it from activating the APC. Although it is not yet clear where complexes between Cdc20 and the components of the checkpoint pathway are formed, it is clear that they are formed continuously as long as an unattached kinetochore exists, but are short lived so that inhibition of the APC is quickly relieved once the last kinetochore becomes stably attached. How the wait-anaphase signal emitted by the last unattached kinetochore is amplified so that it is heard throughout the spindle remains to be discovered.

Concept and Reasoning Checks

1. Why is it not unexpected that yeast Mad and Bub proteins would have counterparts in humans?
2. Would sister chromatids separate once the spindle assembly checkpoint is satisfied if the cell contains a functional but nondegradable form of cyclin B?
3. Why do cells lacking spindle microtubules ultimately escape mitosis?
4. (a) What would be the outcome if a replicating gut epithelial cell suddenly developed a spontaneous mutation in one of its Mad2 genes? (b) In both genes?

14.18 Anaphase has two phases

Key concepts

- Destroying the connections between sister chromatids allows them to begin moving toward opposite poles.
- Movement occurs because pulling forces that act on sister kinetochores throughout mitosis no longer oppose one another.
- Elongation of the mitotic spindle during anaphase increases the distance between the separating chromosomes.
- Spindle elongation is caused by both pushing forces that act on midzone microtubules and pulling forces that act on astral microtubules.

After the last kinetochore has attached to the spindle, there is a lag period before the chromatids separate. During this period, securin and other proteins are degraded. Once chromatid separation begins, the process is completed throughout the cell within several minutes. Normally, sister chromatids separate first in the centromere region, because that is where the opposing pole-directed forces act on the chromosome. After this region separates, the sister chromatids are then "peeled" apart as they move toward their respective poles, as shown in **FIGURE 14.50**.

The process by which the two sets of newly separated chromosomes move closer to the poles is called anaphase A. This is to distinguish it from anaphase B, the process by which the poles themselves move apart. Anaphase's A and B are illustrated in **FIGURE 14.51**. In vertebrate cells the two processes are not different temporal stages of anaphase but are instead two independent mechanisms that work simultaneously to separate the chromosomes.

Although the chromosomes abruptly begin moving toward the poles at the start of anaphase, the force-producing mechanism that moves them is not suddenly turned on at the metaphase-anaphase transition. This is evident from an experiment in which a laser beam is used to destroy one kinetochore of a bi-oriented chromosome during prometaphase, well before the chromatids normally separate. Freed of its connection to one of the poles, the chromosome immediately moves toward the other pole in exactly the same way as an anaphase chromosome. Thus, the same mechanism(s) that moves a chromosome toward a pole during spindle formation and congression also moves it toward a pole during anaphase A.

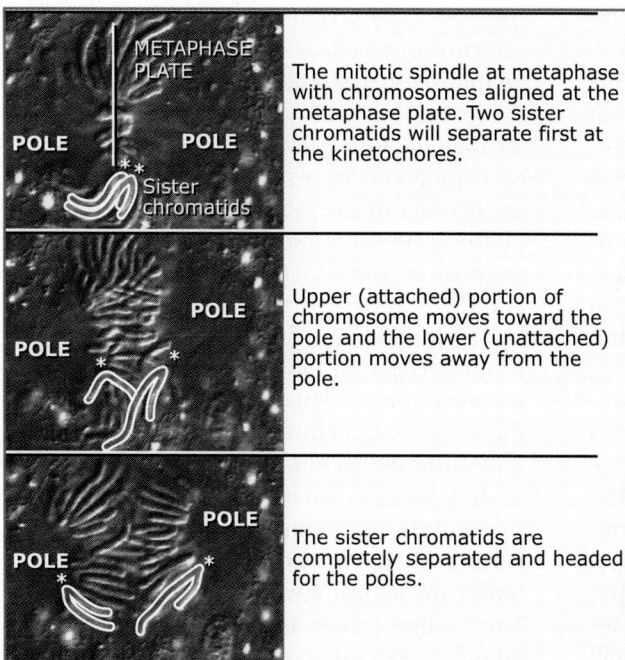

FIGURE 14.50 Frames from a video that follows the chromosomes as they are pulled apart and away from the metaphase plate. Photos © Conly L. Rieder, Wadsworth Center.

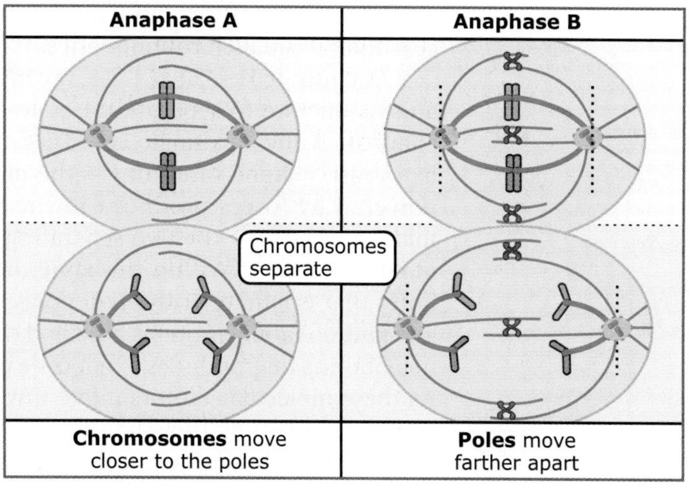

The mitotic spindle at metaphase with chromosomes aligned at the metaphase plate. Two sister chromatids will separate first at the kinetochores.

Upper (attached) portion of chromosome moves toward the pole and the lower (unattached) portion moves away from the pole.

The sister chromatids are completely separated and headed for the poles.

FIGURE 14.51 As the chromosomes move to the poles (anaphase A), the poles themselves move farther apart (anaphase B), increasing the separation between the two groups of chromosomes. The poles are moved by pulling forces on the astral microtubules and motors in the center of the spindle that slide overlapping microtubules past one another. Together, anaphases A and B ensure that the two new nuclei are far enough apart so that the cell can be reliably divided between them.

Poleward forces on the kinetochores are continuously present throughout mitosis. The only difference during anaphase is that the forces acting on sister kinetochores no longer oppose one another and can now act independently. The result is that the chromatids abruptly begin moving poleward as soon as they are separated. As in movements during earlier stages of mitosis, in vertebrates this poleward motion is powered by both an activity at the kinetochore and microtubule flux.

As the two separating groups of chromatids move toward their respective poles during anaphase A, the poles themselves move farther apart, as shown in **FIGURE 14.52**. This process of spindle elongation is anaphase B. *Anaphase B increases the separation between the two groups of chromatids,* ensuring that the furrow that will later divide the cytoplasm and create two new cells occurs between the two reforming nuclei. In some organisms, anaphase B begins only after the completion of anaphase A. However, in vertebrates and most other cells, the distance between the spindle poles begins to increase as soon as the chromatids disjoin, so that the two phases occur at the same time. In general, the extent that the spindle elongates varies widely, even among adjacent cells in the same culture. Some of the variation is due to shape; the spin-

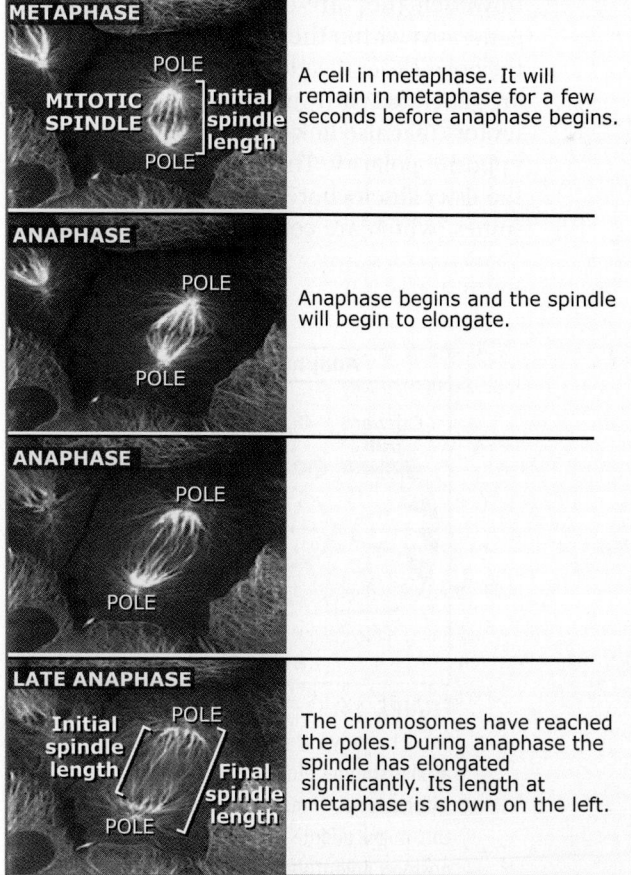

FIGURE 14.52 Frames from a video that depicts the lengthening of the spindle during anaphase. Photos courtesy of Patricia Wadsworth, University of Massachusetts, Amherst.

dle tends to elongate more in long, rectangular cells than in smaller, rounder ones.

As illustrated in **FIGURE 14.53**, several mechanisms operate to separate the poles during anaphase B. In many single-celled organisms—like yeast, diatoms, and fungi—the process is powered by forces produced in the spindle midzone, between the two separating groups of chromosomes. Within this zone, microtubules that originate at the two poles overlap and kinesin family proteins bind and crosslink neighboring microtubules of opposite polarity. As these molecular motors move toward microtubule plus ends, they slide adjacent microtubules past one another, pushing each in the direction of the pole to which it is attached. The spindle elongates as a result. While this is happening the interzonal microtubules also grow at their plus ends so that the zone of overlap is maintained. The growth of these microtubules determines how much the entire spindle will elongate.

As with the forces that drive anaphase A, these pushing forces are also present during the earlier stages of mitosis. Before anaphase, however, they are opposed by other forces generated within the spindle that work to pull the poles closer together. These opposing forces are generated partly by microtubule minus end motors that also link adjacent microtubules of opposite polarity. They are also produced by the sister kinetochores on bi-oriented chromosomes, which are constantly working to pull the poles toward the metaphase plate. When the chromatids finally separate at the onset of anaphase, this balance of forces is disrupted as the forces that act to pull the poles together are weakened. Also at this time global biochemical changes occur within the cell that diminish the activity of the minus end motors that link microtubules of opposite polarity. As a result, the pushing forces between the poles dominate and the poles are pushed apart.

How much this "pushing" mechanism contributes to spindle elongation in vertebrate cells is controversial. This is because in these cells the minus ends of the spindle's nonkinetochore microtubules become detached from the poles as they separate during anaphase B. Thus, by the middle of anaphase the spindle poles in vertebrate cells are not being pushed apart but rather are being pulled apart (see Figure 14.53). The pulling forces are generated by interactions between the spindle's astral microtubules, which remain connected to the poles during anaphase, and cytoplasmic dynein anchored in the cell periphery (i.e., within the cell cortex). Dynein molecules anchored in the cortex "reel in" astral microtubules, pulling the two poles apart.

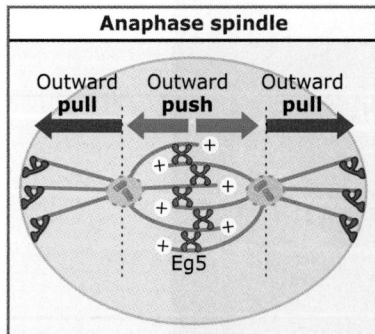

FIGURE 14.53 Two mechanisms are used to move the spindle poles apart as the spindle elongates in anaphase B. Bifunctional kinesin-like molecular motors in the middle of the spindle (orange) push on microtubules of opposing polarity. At the same time, cytoplasmic dynein (purple) anchored at the cell cortex pulls on astral microtubules.

Concept and Reasoning Checks

Using photo-activation methods you mark a small region on the kinetochore fiber microtubules in an early anaphase human cell midway between the chromosomes and a spindle pole.

1. As anaphase A progresses what happens to the mark on the kinetochore fibers relative to the chromosomes?
2. Relative to the pole?

14.19 Changes occur during telophase that lead the cell out of the mitotic state

Key concepts

- The same cell cycle controls that initiate anaphase also initiate events that lead to cytokinesis and prepare the cell to return to interphase.
- Inactivation of CDK1 by destruction of cyclin B reverses the changes that drove the cell into mitosis.
- Destruction of cyclin B begins when the spindle assembly checkpoint is satisfied, but a lag prevents telophase from beginning before the chromosomes have separated.

After the chromosomes have been separated the cell must start cytokinesis and ultimately leave mitosis. The coordination of chromosome separation with these events is due to the same checkpoint that controls the transition from metaphase to anaphase. In addition to delaying the separation of the chromatids, the checkpoint delays the activation of biochemical pathways that cause the cell to undergo cytokinesis and drive it out of mitosis. Once those pathways are activated, however, they cannot be stopped. As a result, once a cell enters anaphase it is committed to leaving mitosis.

As with chromatid disjunction, the events that cause a cell to leave mitosis are initiated with the targeted destruction of cyclin B by the APC. Once the APCs become active, CDK1 activity begins to slowly decline as the cyclin B level progressively falls. At some point in this decline, securin destruction begins and the chromatids separate. At yet lower cyclin B levels, the last stage of mitosis, telophase, begins. During telophase the chromosomes within each anaphase group swell and decondense, while a new nuclear envelope is formed around their periphery (see Figure 14.1). At the same time, the spindle disassembles and the microtubules in the asters grow longer, as shown in **FIGURE 14.54**, and a new microtubule-based structure called the midbody forms in the area between the two separated groups of chromosomes, as illustrated in **FIGURE 14.55** and **FIGURE 14.56**. As these events are occurring within the cell other events at its surface prepare it for cytokinesis.

Basically, cells are driven out of mitosis by a reversal of the events that drove them into it. As discussed earlier, cells are driven into mitosis by the sudden activation of the cyclin B/CDK1 kinase. This enzyme then phosphorylates target proteins, which changes their activities and/or

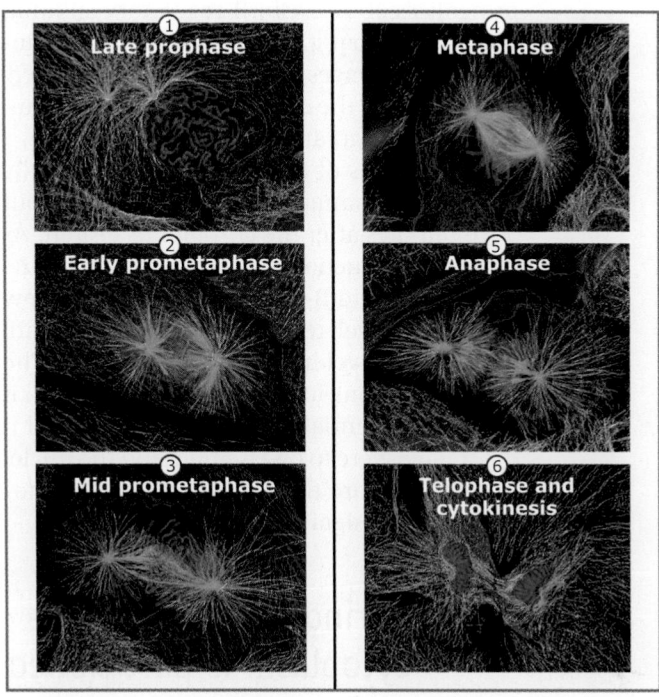

FIGURE 14.54 The immunofluorescence micrographs show microtubules (green), keratin filaments (red), and the chromosomes (blue) at successive stages of mitosis. Photos © Conly L. Rieder and Alexey Khodjakov, Wadsworth Center.

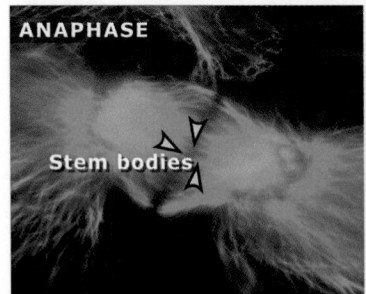 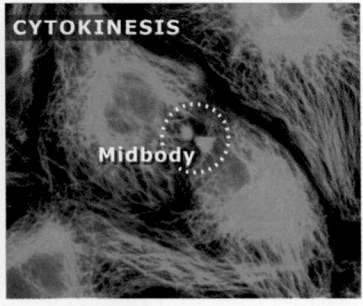

FIGURE 14.55 Microtubules are in green and DNA in blue. In the anaphase cell many small bundles of microtubules—stem bodies (several are indicated by arrowheads)—extend between the two recently separated groups of chromosomes. By the end of cytokinesis, the stem bodies have coalesced to form the midbody, a single structure between the two nuclei. The center of the midbody and each stem body is dark because it has a specialized structure that the green dye used to stain the microtubules cannot penetrate. The small nonstaining region in the middle of each stem body produces an apparent dark line across the middle of the anaphase cell. Photos © Conly L. Rieder, Wadsworth Center.

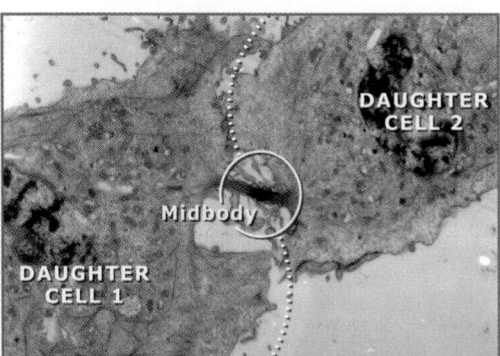

FIGURE 14.56 An electron micrograph showing a midbody. Cytokinesis is almost complete and only a thin bridge of cytoplasm occupied by the midbody connects the two daughter cells. The dotted line indicates the approximate boundary between the two cells. Photo © Conly L. Rieder, Wadsworth Center.

locations. The nuclear envelope and the cell's endomembrane system, including the Golgi complex and the endoplasmic reticulum, are induced to break up into small vesicles. The properties of the centrosomes and their associated microtubule arrays are also changed in ways that promote the formation of the spindle. When the kinase is progressively inactivated by proteolysis of its cyclin B regulatory subunit, the proteins that it had phosphorylated to induce these changes become dephosphorylated. As this occurs, the events that produce the mitotic state are gradually reversed.

The events of telophase do not normally begin until anaphase A is completed, about 5–10 minutes after the chromatids disjoin. As with the destruction of securin, which occurs only after cyclin B levels have fallen to below a threshold level, telophase does not start until the cyclin B level has fallen even further. The practical reasons for this delay are obvious: it would be potentially disastrous for the cell if the spindle were to disassemble and the nuclei to reform before the two groups of chromosomes were a significant distance apart.

14.20 During cytokinesis, the cytoplasm is partitioned to form two new daughter cells

Key concepts

- The two newly formed nuclei that are the products of karyokinesis are separated into individual cells by the process of cytokinesis.
- As with the other events of telophase, cytokinesis is initiated by declining levels of cyclin B/CDK1 kinase activity.
- Cytokinesis involves two new cytoskeletal structures: the midbody and the contractile ring.
- The mitotic spindle, the midbody, and the contractile ring are all highly coordinated with one another.
- Cytokinesis has three stages: definition of the plane of cleavage, ingression of the cleavage furrow, and separation of the two new cells.

After a cell separates its chromosomes, it must divide. Animal cells accomplish division by constricting between the two newly separated sets of chromosomes. Near the end of anaphase A, the surface of the cell begins to constrict in the same plane that the chromosomes occupied at metaphase. Over the next 10–15 minutes the cell is pinched in two in the process known as cytokinesis, as seen in the video whose first frame is shown in FIGURE 14.57. [Note: Although the term "cleavage" is often used as a synonym for cytokinesis, it actually describes the process by which the cytoplasm is partitioned at the end of the first few mitotic divisions in a fertilized egg. There are important but subtle differences between cytokinesis and cleavage. Cytokinesis always produces two separate and independent cells. Although this is also true for eggs that undergo "holoblastic" cleavage during embryogenesis, it is not true for those that undergo "meroblastic cleavage". In the latter the cytoplasm between the newly formed nuclei is partitioned into two regions that remain connected at their base by yolk.] Although the onset of cytokinesis coincides with other changes that occur during telophase, the process is considered to be the last stage of mitosis because it is not completed until after all the other changes have occurred. It is triggered by the progressive decline of cyclin B/CDK1 kinase activity which occurs throughout anaphase and telophase. We know this because microinjecting a form of cyclin B that cannot be degraded into anaphase cells prevents the events of telophase, including cytokinesis.

Like the separation of the chromosomes, cytokinesis not only requires the presence of the mitotic spindle but also involves the formation of two new structures, the **midbody** and

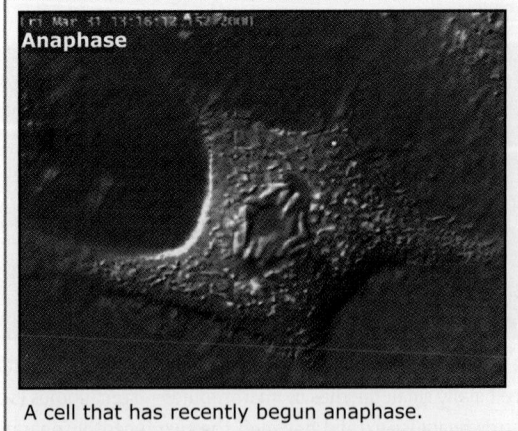

A cell that has recently begun anaphase.

FIGURE 14.57 The first frame of a video that shows that cytokinesis begins when the chromosomes have finished separating. The process divides the cell in two by a deep furrow that forms between them. Photo © Conly L. Rieder and Alexey Khodjakov, Wadsworth Center.

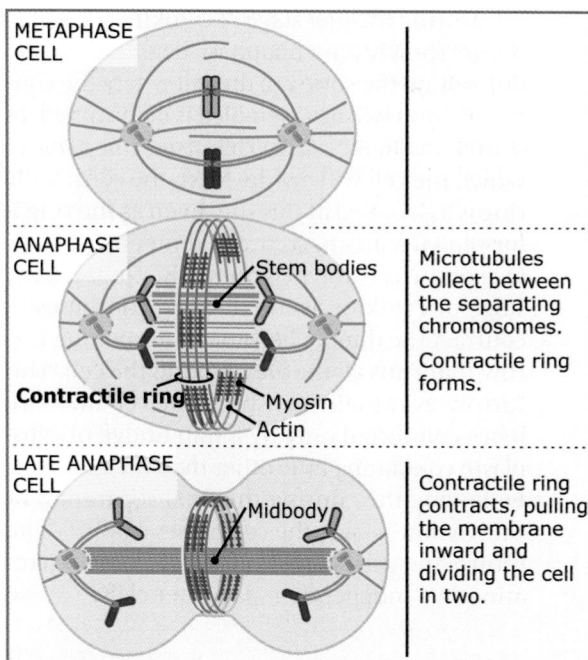

METAPHASE CELL	
ANAPHASE CELL — Stem bodies / **Contractile ring** — Myosin / Actin	Microtubules collect between the separating chromosomes. Contractile ring forms.
LATE ANAPHASE CELL — Midbody	Contractile ring contracts, pulling the membrane inward and dividing the cell in two.

FIGURE 14.58 The midbody is a large microtubule structure that forms from bundles of microtubules that collect between the separating chromosomes during anaphase. At the same time the contractile ring, a tight band of actin and myosin filaments, forms around the cell immediately beneath its plasma membrane. The midbody directs the placement of the ring, which will divide the cell into two.

the **contractile ring**, as shown in **FIGURE 14.58**. The midbody is formed during anaphase as microtubules from the spindle are reorganized into a large bundle of parallel microtubules that extend between the two groups of separating chromosomes. It forms gradually as numerous independent small bundles coalesce, as illustrated in **FIGURE 14.59**. The contractile ring is composed of actin filaments bundled together to form a tight band immediately beneath the plasma membrane. Bipolar myosin filaments similar to those in muscle connect them. Both the midbody and the contractile ring contain many proteins in addition to those that form these principal structural features.

The contractile ring is responsible for powering the constriction process that divides a cell in two. As its name implies, the ring is capable of contraction, driven by interactions between the actin and myosin within it. Because it is attached to the plasma membrane, the ring works in a manner similar to that of a purse string: contraction decreases its diameter, gradually decreasing the size of the opening between the

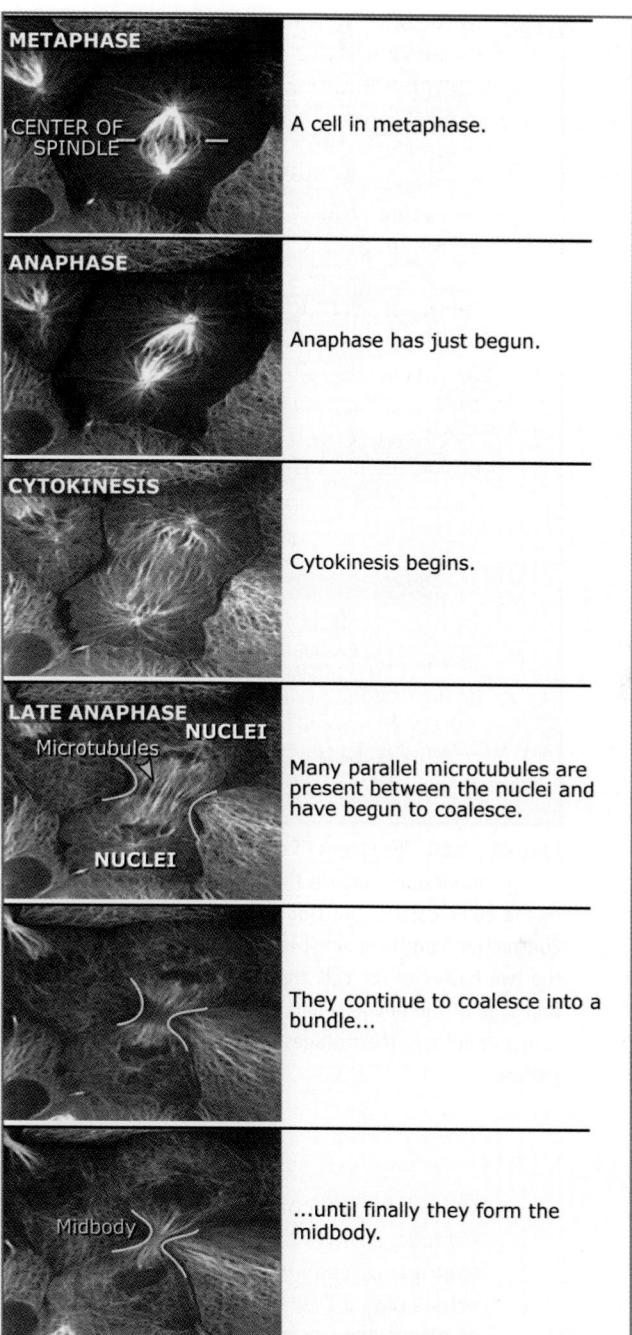

METAPHASE — CENTER OF SPINDLE	A cell in metaphase.
ANAPHASE	Anaphase has just begun.
CYTOKINESIS	Cytokinesis begins.
LATE ANAPHASE — Microtubules — **NUCLEI** — **NUCLEI**	Many parallel microtubules are present between the nuclei and have begun to coalesce.
	They continue to coalesce into a bundle...
Midbody	...until finally they form the midbody.

FIGURE 14.59 Frames from a video that shows the midbody forming during anaphase. Photos courtesy of Patricia Wadsworth, University of Massachusetts, Amherst.

two halves of the cell. Because contraction at the wrong place or time would be catastrophic, the formation and function of the ring depend on interactions with the other two cytoskeletal elements present during cytokinesis. The position of the spindle determines where the midbody forms, which in turn defines the

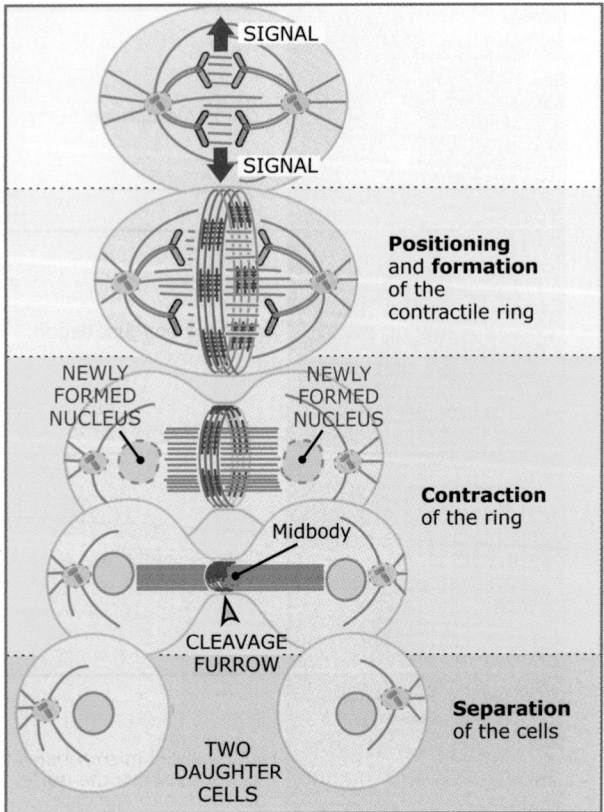

During the first stage of cytokinesis, which occurs shortly after anaphase begins, the location within the cortex of the cell where the contractile ring will be assembled is determined, as shown in **FIGURE 14.60**. This *defines the plane* in which the cell will divide. Next, the contractile ring is assembled at this site. Even as the ring is forming it starts to contract, dimpling the surface of the cell and beginning the "ingression" stage of cytokinesis. As the ring continues to contract the dimple becomes a deepening furrow that runs all the way around the cell. This furrow eventually constricts the cell into two lobes connected only by a thin bridge of cytoplasm containing little other than the midbody. Following this, during the final separation or "abscission" stage, the cell passes a point of no return and the cytoplasmic bridge breaks, creating two independent daughter cells.

FIGURE 14.60 The stages of cytokinesis. Signals from the separating chromosomes induce the formation of the contractile ring in the cell's cortex. The ring immediately begins contracting. Contraction continues until it is tight around the midbody and the two halves of the cell are connected by only a thin bridge. Breakage of the bridge separates the two new cells. As with the stages of mitosis, these stages form a continuous process without pauses.

14.21 Formation of the contractile ring requires the spindle and stem bodies

Key concepts

- The location of the mitotic spindle determines where the contractile ring forms.
- The mitotic spindle is positioned by interactions between its astral microtubules and the cortex of the cell.
- Bundles of parallel microtubules called stem bodies form between the two separating groups of chromosomes in anaphase.
- As anaphase progresses the stem bodies coalesce into one large bundle called the midbody.
- Stem bodies signal to the cortex to cause the formation of the contractile ring.

Where the contractile ring forms is determined by an interaction between the mitotic spindle and the cortex of the cell. Early in anaphase the entire cortex is competent to support the formation of a contractile ring and furrow, but only a small area of it later does. The significance of the spindle in determining which area is selected is demonstrated by experiments in which an anaphase cell is manipulated to move its spindle to a new location. If the spindle is moved within an anaphase cell that has already begun to form a contractile ring, a second contractile ring will form wherever the spindle comes to rest. The first ring gradually disappears, and as long as the spindle is not moved again, the cell will divide at the position of the

site where the contractile ring assembles. *This sequence ensures that the ring forms between the separated chromosomes.* The chromosomes themselves play a role in the timing of contraction; as anaphase begins they contribute factors to the ring that are apparently required for it to function. The failure of any of these events prevents cytokinesis and results in a cell with two nuclei. Such binucleate cells exist in some human tissues (e.g., lung, liver), but they rarely divide again.

Like the separation of the chromosomes, the process of cytokinesis can be subdivided into several sequential stages, each of which is defined by one or more recognizable events. Among them there is only one point of no return, which occurs at the very end of the process. Thus, cytokinesis *is fully reversible until the daughter cells are separate entities.*

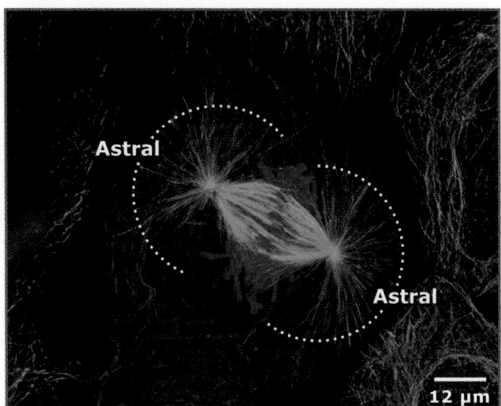

FIGURE 14.61 Astral microtubules radiating from each pole of a metaphase spindle. Microtubules are in green, chromosomes in blue, and intermediate filaments in red. The edge of the intermediate filament array defines the boundary of the cell. Note that many of the astral microtubules extend all the way to the edge of the cell. Photo © Conly L. Rieder and Alexey Khodjakov, Wadsworth Center.

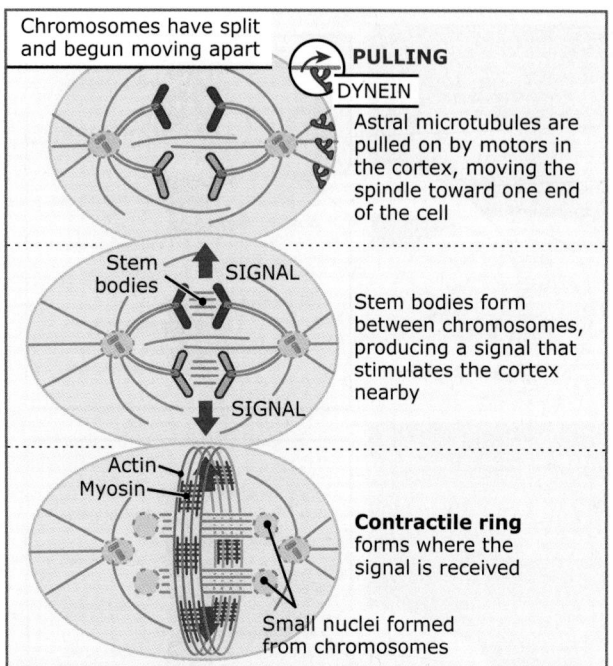

FIGURE 14.62 The position of the contractile ring is determined by a sequence of events. The position of the spindle in the cell is first determined by interactions between its astral microtubules and dynein in the cell cortex. Stem bodies that form as the spindle is being positioned then produce a signal that initiates the assembly of the contractile ring in the nearby cortex.

second. This result is achieved regardless of the distance the spindle is moved from its original position but does not occur if the spindle is moved after a certain point in anaphase. The entire cortex is thus capable of responding to the spindle early in anaphase but later becomes unresponsive.

The spindle itself is positioned within the cell by its astral microtubules. These radiate from both of its poles and are long enough to make extensive contact with much of the cell's cortex, as shown in **FIGURE 14.61**. Spindle positioning in animal cells occurs as the result of interactions between astral microtubules and dynein anchored within the cortex, as **FIGURE 14.62** illustrates. Because dynein moves toward the minus ends of microtubules—which are located at the poles of the spindle—the result is a pulling force on each of the microtubules. Where dynein is located or activated within the cortex of a cell, the shape of the cell and the relative number of astral microtubules that emanate from the two poles of the spindle determine how the spindle is oriented. It is possible that the spindle's position is determined by a mechanical equilibrium achieved when the pulling forces on the microtubules of its two asters are equal. While there are exceptions, the spindle tends to position itself with its long axis parallel to that of a cell.

In some developing systems, tissues and stem cells, cytokinesis results in the formation of two daughter cells that differ greatly in size. These asymmetric divisions are produced from the sudden movement of the spindle during mitosis: in mid-anaphase the spindle quickly shifts its position and moves closer to one side of the cell. As a result of this shift, the cleavage furrow forms off the cell's center. What regulates when the spindle will move and how it moves to a particular position are not yet clear.

How does the spindle determine where the contractile ring forms? Until recently, it was thought that the site of cytokinesis was defined by the spindle's asters. A large number of experiments had led to the conclusion that, as a rule, cytokinesis occurs between two adjacent asters, whether or not they are connected by a mitotic spindle. Factors essential for the contractile ring were proposed to accumulate at the cell membrane where microtubules from the two asters overlap. However, it is now clear that the position of the contractile ring depends not on the asters but instead on another type of microtubule structure that forms between the separating groups of chromosomes early in anaphase.

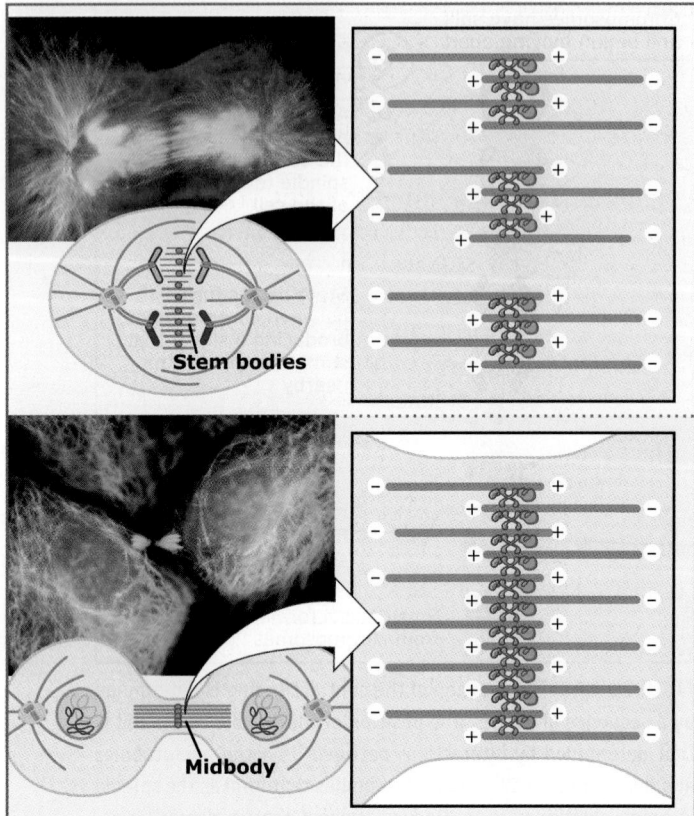

FIGURE 14.63 The upper photograph shows a cell in mid- to late anaphase, its chromosomes (blue) just arriving at the poles. Between the two groups of chromosomes are a broad array of stem bodies (green). The center of each stem body does not stain green because the dense collection of proteins there prevents the stain from binding to the microtubules. The lower photograph shows two sister cells that remain connected only by a midbody. Photos © Conly L. Rieder, Wadsworth Center.

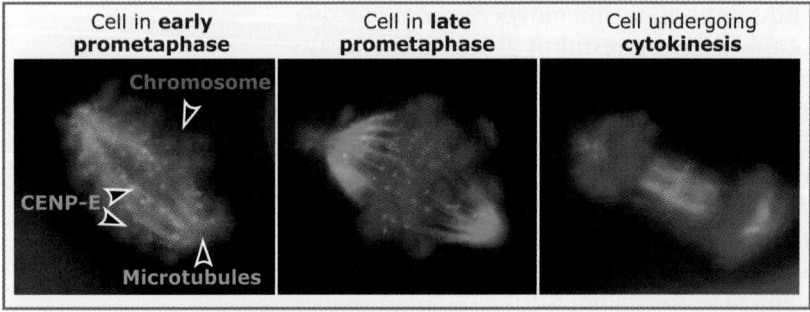

FIGURE 14.64 A series of immunofluorescence micrographs showing how the distribution of CENP-E changes over the course of mitosis. Before and during metaphase it is a component of kinetochores and appears as discrete dots. At the start of anaphase, however, it leaves the kinetochores and accumulates across the equator of the spindle in the narrow zone where the midzone microtubules overlap. CENP-E staining appears purple when it overlaps with DNA and orange when it overlaps with microtubules. Photos courtesy of Bruce F. McEwen, Wadsworth Center.

As the chromosomes separate, many small bundles of microtubules aligned with the spindle form in the region where the chromosomes were aligned at metaphase, as

shown in **FIGURE 14.63**. Called stem bodies or midzone microtubule bundles, these bundles may either be assembled from microtubules shed from the two centrosomes as they separate during anaphase B, or formed *de novo* by the polymerization of new microtubules. Each bundle is composed of microtubules of both polarities, with the plus ends of microtubules of opposite polarity overlapping in a small region in the middle of the bundle. These regions of overlap are located midway between the poles of the spindle, so that the bundles are all centered on its midline. The overlap region of each bundle is specialized and contains a unique set of proteins, including microtubule-dependent motor proteins. A kinesin family motor protein that binds adjacent microtubules of opposite polarity (MKLP1, for *m*itosis *k*inesin-*l*ike *p*rotein 1) is found there and plays a role in the formation of stem bodies by moving to the plus ends of two microtubules simultaneously. At the same time CENP-E and the chromosomal passenger complexes, which consist in part of INCENP protein and Aurora B kinase, leave the centromere and relocate to the stem bodies, as shown in **FIGURE 14.64**.

As cytokinesis progresses, the individual stem bodies coalesce in register to form a single large bundle of microtubules, the midbody, that is positioned between the two groups of separated chromosomes (see Figure 14.63 and Figure 14.59). Like each of the individual stem bodies, the central region of the midbody where microtubules of opposite polarity overlap is a specialized structure containing a large number of different types of proteins.

Stem bodies are required for the formation of the contractile ring and play the important role of determining where it forms. Assembly of the contractile ring involves the organization of a large number of proteins, including its main structural components, actin and myosin. These proteins are initially recruited to a site underneath the plasma membrane where they then assemble into the ring. Both the accumulation and the assembly of the proteins require the presence of stem bodies nearby. In experiments in which stem bodies are prevented from forming, actin and myosin do not accumulate at any site and a contractile ring does not form. Cells can also be experimentally manipulated so that stem bodies are close to the cortex outside the context of a mitotic spindle. When this is done, actin and myosin are gathered and a contractile ring forms at any point in the cortex that is near a stem body.

Stem bodies, then, appear to be responsible for producing a signal that stimulates the formation of the contractile ring. The molecular mechanism is not yet known, but it is likely that the signal originates within the specialized region at the middle of each stem body where the plus ends of the microtubules are concentrated. Among the proteins localized, there are the Aurora B and polo-like kinases which, by activating other midbody components, recruit and activate the small GTPase Rho under the plasma membrane. In many other cellular contexts Rho regulates the formation of structures that contain actin, and it is likely that the continuous production and release of active Rho by stem bodies leads to the reorganization of actin and myosin into a contractile ring.

14.22 The contractile ring cleaves the cell in two

Key concepts

- Contraction of the contractile ring causes it to constrict and produces a furrow around the surface of a dividing cell.
- The contractile ring is composed largely of actin and myosin. Its constriction is driven by their interaction.
- Constriction by the contractile ring requires signals from the stem bodies or the midbody.
- A significant amount of membrane fusion is required during cytokinesis.

Once its components have been recruited and assembled at the correct position, the contractile ring begins the task of dividing the cytoplasm in two. Almost as soon as it forms, it begins to contract. As it contracts, its diameter decreases steadily until only a small opening is present between the two halves of the cell, each containing one of the two newly formed nuclei (see Figure 14.60 and Figure 14.57). Because the ring is attached to the overlying plasma membrane, its contraction draws the membrane inward between the two nuclei, creating a deep indentation around the surface of the cell, the cleavage furrow. In many unicellular organisms and in cells within the tissues of animals, the cytokinetic furrow is broad and its sides gently sloping, giving a dividing cell the appearance of a dumbbell. In other cells, particularly the large eggs of animals such as sea urchins and frogs, it is a very sharp and deep cleft. In some cases, a furrow forms on only one side of a cell and divides it by slicing from one side to the other rather than by constriction. These cases reveal that although the contractile apparatus is usually present as a ring, other forms—such as a crescent extending only partway around a cell—can contract as well.

The force for the contraction of the ring is provided by actin and myosin. Like the sarcomeres of muscle, the contractile ring is composed of overlapping filaments of myosin II and actin, and the force for contraction is generated as the two interact and move past one another. Many other proteins are also present in the ring in smaller quantities and serve to organize the actin and myosin or to control their ability to interact and cause contraction. The contractile ring is not simply a small, circular version of a muscle, however. The actin and myosin filaments in the ring are not as precisely arranged as in muscle and are not organized into sarcomeres. The ring is also a much more dynamic structure, quickly disappearing if actin polymerization is blocked with drugs. The dynamic nature of the ring is probably essential for its function, since as it contracts its thickness remains constant. Components must be steadily lost as it decreases in diameter, so much so that as cytokinesis nears completion the great majority of the material that was initially in the ring has been released.

Contraction of the ring is controlled by the microtubules that extend between the separated groups of chromosomes. Ingression of the cleavage furrow requires the continuous presence of stem bodies and may be driven by their coalescence to form the midbody. This is suggested by experiments that demonstrate that the stem bodies collect into a single large bundle even when the contractile ring is prevented from forming by treatment of cells with a drug that prevents actin polymerization.

The signal from the stem bodies to the ring that drives its contraction originates within the special region of microtubule overlap at the middle of each bundle. Ingression requires the continuous activity of several components found there, including a complex of passenger proteins containing the Aurora B kinase. These proteins relocated from the centromere to the overlap region of the stem bodies at the start of anaphase. How they stimulate ingression—whether they are part of the signaling system or are somehow required for the formation of stem bodies—are not yet known. Why these proteins move from one place to another at a critical transition, and what it implies about

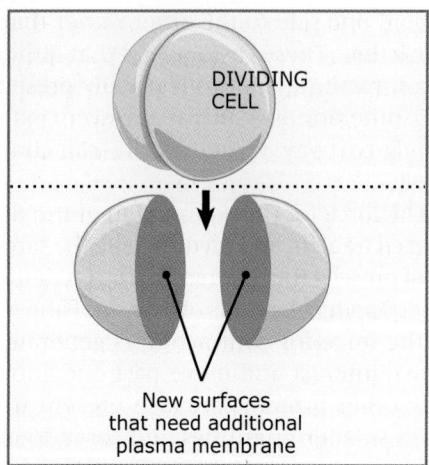

FIGURE 14.65 The combined surface area of the two daughter cells that result from cell division is greater than the surface area of the mother cell. As a result, a significant amount of new plasma membrane must be added during cytokinesis.

how the stem bodies are formed and positioned, are intriguing questions that remain to be answered.

Contraction of the ring continues and the furrow advances until the two halves of a dividing cell are connected by only a thin bridge that contains the midbody (see Figure 14.57 and Figure 14.56). The primary role of this structure is to govern the terminal stage of cytokinesis, when the two daughter cells are finally completely separated. The intercellular bridge can persist for many hours, and cytokinesis is not complete until it finally breaks. This is especially evident when the bridge forms around a piece of chromatin, which sometimes occurs when the chromosomes fail to separate completely during anaphase. When this happens, the bridge fails to break even after many hours, and the furrow finally relaxes to produce a binucleated cell. In normal divisions there appear to be two ways that the bridge can be broken. In some cases, it is severed by forces generated as the two daughter cells begin to migrate apart. In others, there exists a specific mechanism in which vesicles delivered to the bridge fuse to close the gap between the two halves of the cell. Different types of cells must use these two mechanisms to different extents. Cells that are not motile, for example, could not rely upon the first mechanism.

In addition to all the cytoskeletal events that take place during cytokinesis, a great deal of membrane fusion must also occur. Some fu-sion events must occur simply to completely separate the two daughter cells. However, membrane fusion events must also occur for a less obvious reason. If a sphere or cube is cleaved in half (which is basically what is happening in cytokinesis), the total volume enclosed by the two halves produced is the same as that of the original object. Their total surface area, however, is much greater since the cleavage produces new surfaces, as **FIGURE 14.65** illustrates. This means that a significant amount of new plasma membrane must be added as cytokinesis occurs. This process has been studied in detail in animal cells. There, the expansion of surface area occurs by the fusion of internal vesicles with the plasma membrane immediately behind the leading edge of the furrow. This process is particularly clear in the eggs of some amphibians, such as frogs. These eggs are extremely large and must undergo many successive divisions as rapidly as possible. Because this strategy does not allow time for the synthesis of large amounts of new material during each cell cycle, the cytoplasm of each egg contains a large stockpile of vesicles waiting to be added to the plasma membrane each time the cell divides.

14.23 The segregation of nonnuclear organelles during cytokinesis is based on chance

Key concept

- Many of the cell's internal membranes break down during mitosis and are distributed between the two daughter cells as small vesicles. These vesicles reform the organelle after mitosis is finished.

The division of a cell entails more than just separation of the replicated centrosomes and chromosomes into two daughter cells. Each new daughter cell must also inherit enough of the original cytoplasm and organelles to ensure its viability. For many organelles *this distribution is accomplished by first breaking them down into a large number of small subunits and then dispersing these subunits randomly throughout the volume of the mother cell.* In general, disassembly and dispersion occur in response to the activation of the master mitotic regulatory kinase, cyclin B/CDK1.

Almost all of a cell's structural components and membrane-bound organelles are affected. During the mitotic state, the components of the endomembrane system, including the Golgi apparatus, the rough and smooth endoplasmic reticula, and the nuclear envelope, all fragment. The fragments generated then become randomly distributed throughout the cell, often by associating with the highly dynamic microtubules growing from the two asters. With the exception of the microtubules (which become remodeled to form the spindle) the components of the cytoskeletal system also undergo significant disassembly. The subunits released from the disassembly of actin and intermediate filaments also diffuse throughout the cell. In most cases, the breakdown of the cytoskeleton induces the cell to round up into a sphere. This radical morphological change during the early stages of mitosis undoubtedly helps disperse the components throughout the volume of the cell.

About the only cytoplasmic organelles that do not disassemble or fragment as mitosis begins are mitochondria. Mitochondria appear unaffected at the structural level during mitosis and probably do not break down because they are already present in multiple small copies. Individual mitochondria function throughout the cell cycle as independent units, and there are hundreds or thousands of them per cell. In essence, the function of ATP generation is already extensively fragmented and dispersed. In contrast, the Golgi apparatus is present in only a single copy in animal cells, requiring that it be broken down in order to distribute its function between the two daughter cells.

As a result of all the disassembly events that take place early in mitosis, by the time a cell enters anaphase most of its components are present in multiple copies that are distributed randomly throughout its volume. After cytokinesis occurs and the cytoplasm is divided into two roughly equivalent volumes, the two cells each end up with a centrosome, a single complement of chromosomes, and roughly similar quantities of precursors for their organellar and cytoskeletal systems. Then, as cyclin B/CDK1 is inactivated and the cell exits the mitotic state, the processes that led to the breakdown of its internal systems are reversed and they reform.

14.24 What's next?

The quest to understand how mitosis works is not simply an academic challenge. The genetic changes associated with the origins of many diseases can often be traced directly to mistakes that occurred during the division process. With modern genetic and molecular tools, it is now relatively easy to discover new molecules that are involved in chromosome segregation and cytokinesis. The key will be to determine the function of these molecules, how they work together to accomplish a particular event, and then how the multiple events of mitosis are integrated to effect the most fundamental processes of life.

We still have much to learn about how the individual proteins that participate in mitosis work during the process. The development of increasingly sophisticated imaging systems, as well as methods for fluorescently tagging specific proteins in the living cell, will provide increasingly accurate views of how the components of the spindle interact. Coupled with broadly applicable methods that allow the inactivation or removal of a specific protein from cells—such as RNAi—these techniques should produce a much clearer picture of how each player affects entry into and/or progression through mitosis. No doubt, one area of emphasis will be to define the molecular mechanism behind centrosome replication, how this replication is coordinated with DNA synthesis, and how the cell controls the number of copies of this important organelle. It is now clear that early in the development of many cancerous cells, centrosomes are overproduced. Extra centrosomes lead to the formation of multipolar spindles and the production of aneuploid cells. In the future it will be important to determine how this overproduction of centrosomes occurs and whether it has a causative role in the early stages of tumor formation.

Another exciting area for future investigation will be defining how the spindle assembly checkpoint works at the molecular level. How can the signal produced by just one unattached kinetochore inhibit anaphase onset throughout the cell? Malfunctions in this checkpoint also lead directly to aneuploidy and its disastrous consequences, including tumor formation. Many cancer cells contain a compromised spindle assembly checkpoint. Thus, understanding how the wait-anaphase signal inhibits the anaphase-promoting complex will assist in the development of drugs to kill selected dividing cells.

We also have an incomplete understanding of the checkpoint controls that regulate entry into mitosis, or the G2/M transition. It is clear that normal cells contain a DNA damage checkpoint, based on the ATM/ATR kinase that prevents G2 cells from entering mitosis if their DNA is damaged. However, it is increasingly clear that the G2/M transition is also controlled by other cell cycle checkpoints independent of those that detect DNA damage. For example, the p38-mediated stress-activated protein kinase pathway rapidly but transiently halts progression through late G2 in response to various stresses, from a sudden cold shock to drugs that effect, for example, microtubules, topoisomerase II activity, histone acetylation and protein synthesis. When the stress is long term or if the damage is nonrepairable, this pathway gives the cell time to activate other more permanent checkpoints like those based on p53 which work via their effects on transcription. Understanding these pathways and how they stop the cell from entering mitosis will help lead to great advances in the prevention and cure of many devastating diseases.

Finally, as noted in the box, several spindle poisons that perturb microtubule assembly like vincristine and paclitaxel have proven to be effective in treating cancers. Although it is evident from clinical studies that these drugs are effective, how and why they work remains largely an important but unsolved mystery. The notion that they work simply by arresting cancer cells in mitosis, where they then die, is likely incorrect because the therapeutic concentrations of the drug are below those needed to permanently arrest normal and most cancer cells in mitosis. It is more likely that these drugs work by delaying progression through mitosis which then somehow triggers pathways that lead to death in G1. Unraveling the changes that occur in cancer cells that make them more sensitive than normal cells to microtubule poisons is also an important area for future work.

14.25 Summary

The process of cell division is an intensively investigated area of biomedical research because errors in mitosis lead to aneuploidy and the genetic instability behind cancer. Mitosis occurs through two processes, nuclear division and cytokinesis. Nuclear division, or karyokinesis, equally distributes the replicated sister chromatids into two daughter nuclei. Near the end of nuclear division, cytokinesis partitions the cell and its cytoplasm between the two daughter nuclei.

The mitotic apparatus, or spindle, mediates nuclear division and cytokinesis by moving the chromosomes and defining the plane that bisects the cell. This bipolar structure is composed primarily of microtubules, microtubule associated proteins (including motors), and structural proteins. The microtubules within the mature spindle are of two types: those that firmly connect the sister kinetochores on each chromosome to the opposing spindle poles, and those that have free ends within the spindle. The dynamic nature of microtubules is critical to spindle formation and function.

Two distinct mechanisms underlie the motion of kinetochores, and thus their associated chromosomes, toward a spindle pole. One force is generated by shortening of the kinetochore-associated microtubules, which occurs through subunit deletion in the spindle pole region. The other involves microtubule motors associated with the kinetochore. In some cells, like those of vertebrates, these mechanisms work concurrently.

A complex cell cycle checkpoint control delays the metaphase/anaphase transition until all of the kinetochores are stably attached to the spindle. When the checkpoint is turned off, a series of biochemical changes leads to the disjunction of sister chromatids, the dissolution of the spindle proper, and the formation of stem bodies between the separating groups of chromosomes. The stem bodies then initiate the process of cytokinesis.

http://biology.jbpub.com/lewin/cells

To explore these topics in more detail, visit this book's Interactive Student Study Guide.

References

Barr, F. A., and Gruneberg, U. 2007. Cytokinesis: placing and making the final cut. *Cell* v. 131 p. 847–860.

Flemming, W. 1965. Contributions to the knowledge of the cell and its vital processes. (Historical paper translated from German to English) *J. Cell Biol.* v. 25, p. 3–69.

Lloyd, C., and Chan, J. 2006. Not so divided: the common basis of plant and animal cell division. *Nat. Rev. Mol. Cell Biol.* v. 7 p. 147–152.

Maiato, H., DeLuca, J., Salmon, E. D., and Earnshaw, W. C. 2004. The dynamic kinetochore-microtubule interface. *J. Cell Sci.* v. 117 p. 5461–5477.

Musacchio, A., and Salmon, E. D. 2007. The spindle-assembly checkpoint in space and time. *Nat. Rev. Mol. Cell Biol.* v. 8 p. 379–393.

Nigg, E. A. 2007. Centrosome duplication: of rules and licenses. *Tr. Cell Biol.* v. 17 p. 215–221.

Pines, J., and Rieder, C. L. 2001. Re–staging mitosis: a contemporary view of mitotic progression. *Nat. Cell Biol.* v. 3 p. E3–E6.

Rieder, C. L., and Khodjakov, A. 2003. Mitosis through the microscope: advances in seeing inside live dividing cells. *Science* v. 300 p. 91–96.

Schrader, F. 1953. *Mitosis: the movements of chromosomes in cell division.* 2nd Edn. N.Y.: Columbia Univ. Press. p. 1–170.

Walczak, C. E., and Heald, R. 2008. Mechanisms of mitotic spindle assembly and function. *Int. Rev. Cytol.* v. 265 p. 111–158.

Cell cycle regulation

Cell cycle regulation

15

Kathleen L. Gould

Kathleen L. Gould
Vanderbilt University School of Medicine,
Nashville, TN

Susan L. Forsburg
University of Southern California,
Los Angeles, CA

FISSION YEAST CELLS at various stages of the cell cycle expressing mCherry-tagged Cdc15 (in red), GFP-tagged alpha-tubulin (in green), and RFP-tagged Sid4 (also in red). Cdc15 and Sid4 are essential components of the actomyosin contractile ring and spindle pole body, respectively. Interphase cells contain cytoplasmic longitudinal arrays of microtubules and Cdc15 at cell tips. Dividing cells exhibit Cdc15 localization in the medially placed actomyosin ring and either a mitotic spindle or postanaphase arrays of microtubules. Photo courtesy of Anne Feoktistova and Kathleen L. Gould, Vanderbilt University Medical Center.

CHAPTER OUTLINE

15.1 Introduction

Key concepts

- A cell contains all the information necessary for making a copy of itself during a cell division cycle.
- The eukaryotic cell division cycle (cell cycle) is composed of an ordered set of events that results in the generation of two copies of a preexisting cell.
- The cell cycle is partitioned into distinct phases during which different events take place.
- Replication of a cell's chromosomes and chromosome segregation are two important events in the cell cycle.

Central to all of biology is the "cell theory" put forth by Theodor Schwann (1810–1882) and Matthias Schleiden (1804–1881), which states that "cells arise from preexisting cells." New cells arise not by some form of spontaneous generation—as was widely believed up to that time—but by a process of division in which one cell divides to give rise to two. In unicellular organisms each division produces a completely new and independent organism. In large multicellular organisms thousands or more cell divisions beginning with a single cell are required to construct an animal. Many more divisions are required over the lifetime of the animal to replace cells lost in the course of its life.

Cells reproduce by carrying out an ordered sequence of events called the **cell cycle**, shown in **FIGURE 15.1**. The nuclear DNA is replicated in **S phase**. The two copies of the genome are separated from one another in **M phase**, or mitosis, when the mitotic spindle forms and undergoes an elaborate series of motions that end with the chromosomes separated into two equal sets on opposite sides of the cell. Cytokinesis—usually considered part of M phase—then results in the formation of two independent cells by the division of the cytoplasm. The topic of mitosis is covered in detail in *14 Mitosis*.

In most cells, "Gap" phases separate the S and M phases: **G1** between the M phase that ends one cell cycle and the S phase of the next, and **G2** between S phase and the M phase that follows it within the same cycle. Gap phases serve several purposes, including giving the cell time to grow (i.e., synthesize all the components required to double its mass) and providing time to ensure that each of the major cell cycle events is completed properly before the next is begun. In some specialized cell divisions, Gap phases are skipped and the cells simply alternate back and forth between S and M phases. This occurs, for example, in many embryonic cell cycles, when the cells are very large and do not have to increase in mass between divisions.

Cell division cycles do not occur continuously. Entry into the cell cycle is usually controlled by conditions in a cell's surroundings, which can generate either stimulatory or inhibitory signals. Such external control is essential for several reasons. For example, cells must not be allowed to enter the cell cycle when the resources are not available to complete the cycle. This is not usually a problem for the cells of multicellular animals, but for a unicellular organism it would likely be catastrophic to enter the cell cycle with insufficient nutrients available. It would be equally disastrous were the cells of an animal allowed to divide continuously without regard for what other cells around them were doing. Organisms are communities of cooperating cells, and the cooperation includes strict controls on when cells divide. The consequences of breakdown in these controls in even a small number of cells can be seen in cancer, a disease of uncontrolled cell division.

Within the cell cycle it is obvious *why* the basic events—replication, segregation, and

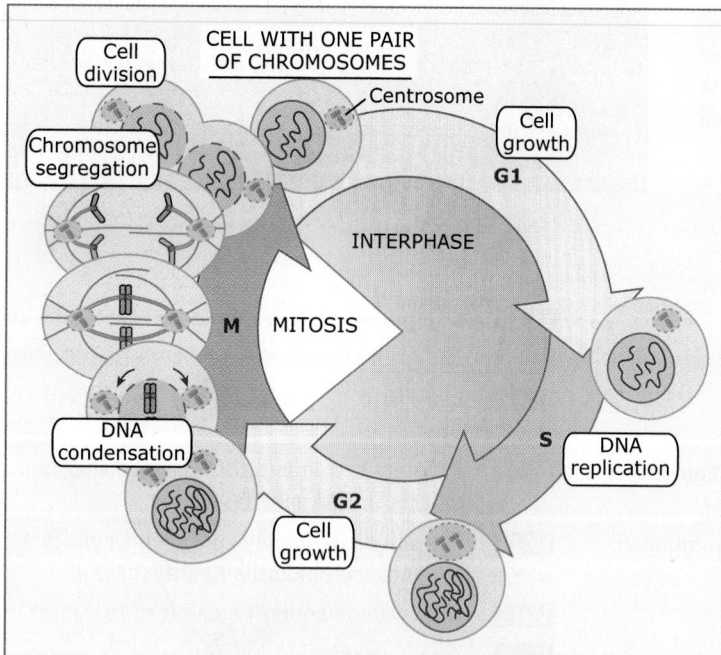

FIGURE 15.1 This cycle lasts about 24 hours. The DNA replicative phase (S) and mitosis (M) are separated by GAP phases (G1 and G2). The chromosomes, indicated by blue and red lines, are contained within the nuclear envelope until mitosis. Centrosomes serve as organizing centers, or poles, for the mitotic spindle during mitosis and are duplicated during S phase. Each centrosome contains two rod-shaped centrioles in its center.

Mitosis has been recognized as a distinct stage of the cell cycle for more than a hundred years, based on microscopic observations of chromosome condensation and segregation. Until the early 1950s, however, when chromosomes were replicated during a cell division cycle remained a mystery. Because mitosis comprises such a small fraction of every cell division cycle, it was conceivable that the remainder of the cycle, interphase, was devoted to chromosome replication or that chromosome duplication and segregation both occurred during mitosis.

By 1950, it had been established that chromosomes are composed of DNA, although the structure of DNA was still to be discovered. The key to determining when chromosomes were duplicated within a division cycle was to devise a method to detect DNA in the process of being made within cells while simultaneously detecting mitoses. Though several dyes were available to stain DNA, these had not proven useful for quantitative analysis of DNA synthesis. It was the advent of ^{32}P-labeling that allowed Alma Howard, a chromosome biologist, in collaboration with Stephen Pelc, a pioneer in the use of autoradiography (a method to image the spatial localization of radioactive compounds using a photographic emulsion), to label DNA in the process of its synthesis and furthermore, to visualize individual nuclei in which DNA was replicated.

In their famous work on DNA replication conducted from 1951 to 1953, Howard and Pelc studied cell proliferation in roots of the broad bean, *Vicia faba*. In this tissue, rapidly dividing meristematic cells are found within ~2 mm of the root tip and differentiated, nondividing cells can be distinguished above them on the basis of their elongated shape. By microscopic observation of cells in fixed root samples over an experimental time course, the researchers first determined the average length of a cell cycle in the meristem (~30 h) and then estimated the length of mitosis (4 h) by scoring the percentage of cells in mitosis, that is the mitotic index. In subsequent experiments, Howard and Pelc added radioactive inorganic phosphate to the water in which the roots were growing in order to label synthesizing DNA and collected samples periodically from 2 to 48 hours. Root samples were extracted with HCl at 60°C to eliminate unincorporated ^{32}P and other labeled macromolecules such as proteins, and then tissue squashes were prepared and placed in a photographic emulsion (typically for 28 days!). Silver grains developed over the nuclei that had synthesized ^{32}P-labeled DNA and by phase contrast microscopy, mitotic nuclei in the meristem were identified. Control experiments in which tissue squashes were treated with DNaseI allowed the researchers to be confident that the silver grains resulted solely from labeled DNA. By analyzing samples at the earliest time point, Howard and Pelc determined that only a percentage of cells were engaged in synthesizing DNA at the time of labeling and that none of the labeled nuclei were in mitosis, ruling out that DNA replication took all of interphase or occurred during mitosis. They coined the discrete period of the cell cycle in which DNA is replicated as S (synthesis) phase, a term still used, and estimated its length (6 h). As concluded by their contemporary J. M. Thoday, "Howard and Pelc have discovered a phase of the mitotic cycle that is of great significance" (Thoday, J. M. 1954. Radiation-induced chromosome breakage, desoxyribosenucleic acid synthesis and the mitotic cycle in root-meristem cells of Vicia faba. *New Phytol.* v. 53 p. 511–516). By plotting the number of labeled nuclei in mitosis at each of their experimental time points, these researchers were also the first to recognize that there was a time gap between S phase and mitosis, estimated to be 8 hours, and also between cell division and the next S phase (~12 h). They called these "gap" periods G_2 and G_1 respectively and hence, the concept that a cell cycle consists of four discrete phases was born.

The Howard and Pelc strategy of determining the *frequency* of *labeled mitoses* (FLM) was used in the following decades to analyze cell cycle parameters in many tissues and cell lines. These subsequent experiments confirmed their original findings and led to the general conclusion that the lengths of S, G_2 and M phases are fairly constant within any given cell type but G_1-phase length can vary. The difficulty of the FLM strategy was diminished significantly in 1957 when ^{3}H-thymidine supplanted ^{32}P as a DNA label. Now, DNA labels that can be detected based on fluorescence, or antibodies to proteins produced only in S phase, are typically used to identify or mark cells in the process of DNA replication. But, it remains routine in cell cycle research to identify and measure the length of S phase under varying conditions.

It is noteworthy how important the groundbreaking studies of Howard and Pelc on plant roots were to the development of our current understanding of the cell cycle, and that a number of other salient observations were made by these investigators. For example, they noted that differentiated cells had synthesized their DNA in their last cell cycle and therefore had differentiated out of G_1 phase. Also, while studying radiation effects on the cell cycle, Howard and Pelc reported a dose-dependent delay in mitotic entry in irradiated bean roots, a first hint of a G_2 DNA damage checkpoint.

division—are performed in the order that they are. For example, there would be no copies of the genome to segregate if a cell's chromosomes were not already replicated. But how does the cell *make* the events occur in that order?

Cell cycle transitions are regulated by a set of proteins that form a central cell cycle control

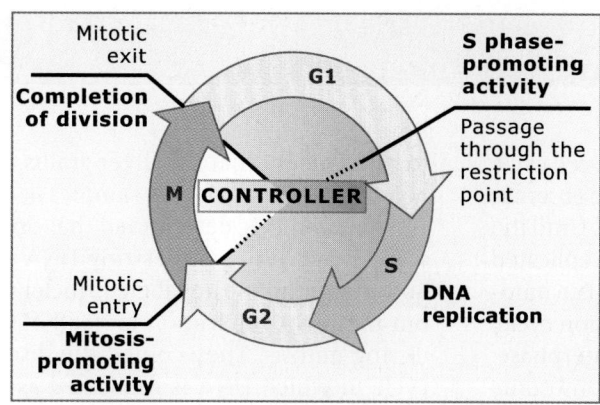

FIGURE 15.2 Major transitions in the cell cycle are regulated by a cell cycle control system.

system. This system regulates the machineries that replicate and segregate the DNA and that divide the cell, telling each when to perform its functions. **Checkpoints**, which impinge on the central controller, exist for two critical purposes. One is to ensure that a cell cycle event does not start before the previous one is completed successfully. The other is to impose dependence of the initiation of the cell cycle on the cell's surroundings, as was discussed earlier. **FIGURE 15.2** shows the major points in the cell cycle regulated by the cell cycle control system and checkpoints.

The central cell cycle control system of eukaryotes and the checkpoints that affect it are discussed here in *15 Cell cycle regulation*. What are their components, and what is the nature of their interactions that allows them to be activated and inactivated in sequence? How are they regulated from outside the cell? How is the decision to start the cell cycle made? How does the cell cycle control system activate each cell cycle event? Finally, what are the consequences when the system malfunctions?

15.2 Several experimental systems are used for cell cycle research

Key concepts

- Studies in a wide variety of organisms have contributed to our knowledge of cell cycle regulation.
- A combination of genetic and biochemical analyses of the cell cycle synergized to provide key insights into the molecular nature of cell cycle regulators.
- The use of synchronized cells is important for analyzing cell cycle events.

Before we proceed with a discussion of how the cell cycle is controlled, we must introduce the experimental systems that enabled researchers to make their discoveries. These have ranged widely from single-cell organisms to amphibian eggs to human tissue culture cells. Each system has distinct advantages for cell cycle research and, by and large, what is learned in one organism is applicable to most because the underlying mechanisms of cell cycle regulation are conserved throughout evolution.

Some of the experimental organisms that attracted researchers have cell divisions that are inherently synchronous. For example, oocytes of sea urchins (*Arbacia punctulata*), frogs (*Xenopus laevis*), and clams (*Spisula salidissima*) can be induced to undergo **meiotic maturation** synchronously by treatment with appropriate hormones. Hormone stimulation drives the oocyte from an interphase-arrested state (immature oocyte) to a metaphase-arrested state in which it awaits fertilization. The stages of oocyte maturation and fertilization are depicted in **FIGURE 15.3**. Following fertilization, the early embryonic cell divisions of these oocytes are also synchronous. Synchronous division allows one to study the behavior of populations of cells rather than that of an individual cell. Another advantage of using mollusk or amphibian oocytes or eggs (which are fertilized oocytes) for cell cycle studies is that sizable populations of them can be collected to provide large quantities of material for biochemical analyses. Their large size is also amenable to injection of biologic molecules such as proteins and drugs, and the effect of such treatments on cell cycle progression can be studied. Cytoplasmic extracts can be prepared from these oocytes or eggs and stored for future use. These extracts retain their ability to assemble nuclear envelopes around DNA, replicate exogenously added DNA, and form mitotic spindles, all in the correct cell cycle order. Thus, cell cycle events can be recapitulated in a test tube. The extracts can also cycle more than once. These properties of amphibian and mollusk oocytes are widely exploited to purify and identify proteins that are involved in cell cycle processes, to manipulate their levels, and to test various hypotheses regarding cell cycle control *in vitro*.

Budding yeast (*Saccharomyces cerevisiae*) and fission yeast (*Schizosaccharomyces pombe*) are used extensively for cell cycle studies. These single-cell eukaryotes carry out all the basic steps of the cell cycle, and they offer many

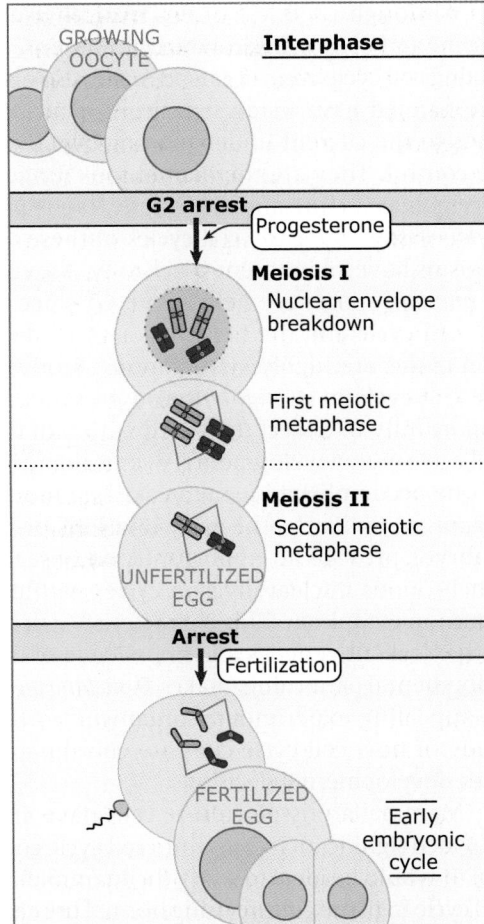

Interphase

GROWING OOCYTE

G2 arrest

Progesterone

Meiosis I
Nuclear envelope breakdown

First meiotic metaphase

Meiosis II
Second meiotic metaphase

UNFERTILIZED EGG

Arrest

Fertilization

FERTILIZED EGG

Early embryonic cycle

FIGURE 15.3 A growing oocyte is arrested in interphase. Progesterone signals the oocyte to transit through meiosis I and begin meiosis II. It becomes arrested in metaphase of meiosis II awaiting fertilization. Fertilization alleviates this second cell cycle arrest and the fertilized egg then proceeds through the embryonic cell cycles.

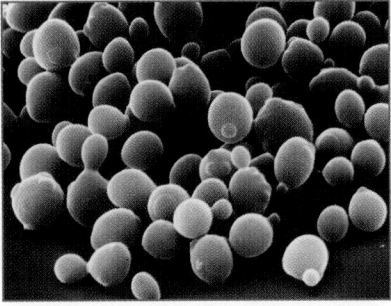

FIGURE 15.4 Scanning electron micrograph of budding yeast cells. Unbudded cells are in G1, whereas large-budded cells are in G2 or M. Photo courtesy of the estate of Ira Herskowitz, University of California, San Francisco, with permission of Eric Schabtach, University of Oregon.

experimental advantages over multicellular eukaryotes. They are easy to grow and manipulate under lab conditions, fast growing with a division cycle time of 1 to 4 hours, amenable to simple synchronization procedures such as size selection, and genetically tractable. One of the biggest differences between the cell cycles of these organisms and multicellular eukaryotes is that the nuclear envelope of yeasts does not break down during mitosis; these yeasts undergo a "closed" mitosis. However, the cell cycle control system and checkpoints are all present.

A budding yeast cell, as the name implies, replicates itself by forming a bud that grows during interphase and separates from the preexisting "mother" cell during mitosis to form a new daughter cell. The size of the bud is an indicator of the stage of the cell cycle that the cell is in. For example, unbudded cells are in G1, whereas large-budded cells are in G2 or M. A group of budding yeast cells is shown in **FIGURE 15.4**.

Both budding and fission yeast can be grown as haploid cells and **conditional** loss-of-function mutants defective in any process can be isolated as long as a strategy is developed to identify them. Hartwell and colleagues isolated mutants on the basis that after shift from a **permissive growth condition** of 25°C to a nonpermissive temperature of 37°C, each mutant colony arrested growth and died with a uniform bud size. Arrest with a uniform bud size indicates that the mutant cells are defective in progression through a specific stage of the cell cycle. For example, a mutant that arrests with a large bud is defective in a mitotic process such as chromosome segregation. The **temperature-sensitive mutants** isolated by Hartwell and colleagues were called *cell division cycle* (*cdc*). The concept of a temperature-sensitive *cdc* mutation is illustrated in **FIGURE 15.5**. The *cdc* screen was highly productive in identifying genes regulating diverse cell cycle events. Another useful feature of the temperature-sensitive mutations is that the uniform arrest phenotype and the reversibility of certain mutations (they continue on with their cell cycle once they switch from restrictive to permissive conditions) provide an excellent way to synchronize entire populations of these cells.

Unlike budding yeast cells, fission yeast cells are cylindrical, grow by tip elongation, and divide using a medially placed septum. In a wild-type population growing in rich medium, cell length at division is fairly constant

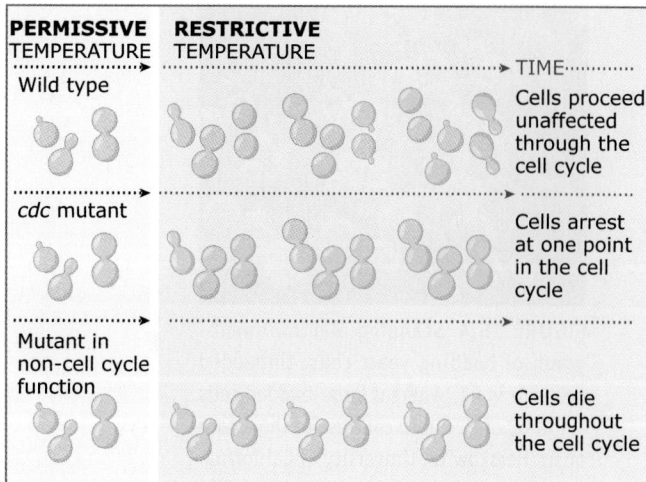

FIGURE 15.5 When grown at a permissive temperature of 25°C, the budding yeast, *S. cerevisiae*, is a mixture of unbudded, small-budded, and large-budded cells. If a temperature-sensitive mutation exists in a gene that does not affect the cell cycle, cells will stop growing and die throughout the cell cycle. However, if they carry a temperature-sensitive mutation in, for example, a gene regulating mitosis, they arrest uniformly as large-budded cells when shifted to the restrictive temperature of 37°C.

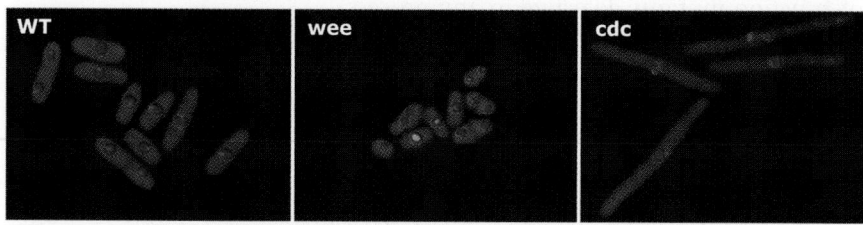

FIGURE 15.6 In fission yeast, *cdc* mutants fail to divide and grow longer than wild-type yeast. *wee* mutants enter mitosis earlier than wild-type (WT) yeast, resulting in shorter cells. These phenotypes indicate a mutation in a cell cycle regulator. The images of the cells have DNA in blue and the spindle pole bodies in red. Photos courtesy of K. Adam Bohnert and Kathleen L. Gould, Vanderbilt University Medical Center.

and, therefore, *S. pombe* cell length is a good indication of cell cycle stage. The predictable length of *S. pombe* cells provided an ideal morphologic marker for the visual identification of *cdc* mutants by Nurse and colleagues. These mutants did not divide but became much longer than wild-type cells. The predictable length of *S. pombe* cells at division also aided in the isolation of "wee" mutants that divided at a reduced cell length. The difference in length of wild-type and mutant fission yeast cells is illustrated in **FIGURE 15.6**. The isolation of both *cdc* and *wee* mutations within a single genetic locus (*cdc2*+) led to the proposition that there was an essential rate-limiting factor for the initiation of mitosis. We will discuss Cdc2 in greater detail later in *15 Cell cycle regulation*.

Although used less often, two other experimental organisms in which mutations affecting cell cycle control can be readily isolated and studied have made significant contributions to the current understanding of cell cycle control. They are the filamentous fungus, *Aspergillus nidulans*, and the fruit fly, *Drosophila melanogaster*. The early cell cycles of these organisms have an inherent synchrony. Asexual *A. nidulans* spores are held in the G1 stage of the cell cycle and the first two or three division cycles are highly synchronous. Study of the cell cycle in *A. nidulans* also presents an opportunity to dissect the coordination of the cell cycle with developmental events that cannot be accomplished using yeast as a model organism. Similarly, the early divisions of fly embryos provide an opportunity to observe synchronous nuclear division cycles within a common cytoplasm. This characteristic, along with excellent genetics and knowledge of developmental patterning, makes *D. melanogaster* an appealing experimental organism for the study of how cell cycle cues are coordinated with developmental decisions.

Mammalian tissue culture cells have also provided significant insights into cell cycle control. It would be ideal to study the mammalian cell cycle in tissue culture using normal primary cells, that is, those obtained directly from an organism that are not subjected to any genetic alterations. However, normal primary cells do not proliferate indefinitely in culture. Rather, they stop dividing after 25 to 40 cell divisions and enter senescence (discussed later in *15 Cell cycle regulation*). For this reason, **immortalized** cell lines derived from normal or tumor cells are used widely for cell cycle analyses. As the name suggests, immortalized cells have undergone genetic alterations that allow them to divide indefinitely in culture when supplied with appropriate media and growth-promoting agents. Although it is always important to bear in mind that their growth properties do not necessarily reflect those of cells within an organism, immortalized cell lines such as HeLa have proved to be extremely valuable for cell cycle analyses. They can be synchronized readily with various drugs that inhibit DNA replication or mitosis and/or size selection procedures, and they are also amenable to biochemical and cell biological studies.

Indeed, elegant mammalian cell fusion experiments laid the foundation for the concepts of checkpoint controls and the coordination of cell cycle events. These experiments involved fusing cells from different stages of the cell cycle (G1, S, G2, and M) to each other. Relevant to understanding cell cycle controls, it was found that fusion of an S-phase cell with a G1 cell accelerated the G1-phase nucleus into DNA replication. This result led to the proposal that S-phase promoting factors are transmittable through the cytoplasm of the fused G1/S cell.

Fusion of mitotic cells with G1, S, or G2 cells also provided important insight into dominant factors influencing cell cycle progression. In all cases, chromosomes from the interphase cells prematurely condensed and the cells entered a pseudo-mitotic state. These observations suggested the presence of a mitotic inducer that was dominant over other states, in that it could induce chromosome condensation from any stage of the cell cycle. Such an activity turned out to be present in all eukaryotes and was purified and characterized from many types of cells.

It was also discovered from these experiments that when G2 cells were fused with S-phase cells, the G2-phase nucleus did not inhibit ongoing replication in the nucleus from the S-phase cell, but the G2-phase nucleus also did not begin to rereplicate its own DNA. This observation developed into the hypothesis that DNA is "licensed" to replicate once and only once per cell cycle (see Figure 15.22). These fusion experiments are illustrated in **FIGURE 15.7**.

In addition to establishing these important principles of autonomous cell cycle regulation, tissue culture cell lines also enable the study of how external stimuli regulate the cell cycle and have facilitated the study of how tumor-suppressor genes and oncogenes regulate cell growth and division in mammals.

Lastly, deletion of genes in mice can be used to determine the function of cell cycle genes in mammals. This approach can reveal whether cell cycle proteins are required for all cell cycles or only for the development and function of particular tissues. It can also reveal whether the loss of a particular cell cycle protein in a mammal can be compensated for by another protein and whether mammals lacking a cell cycle regulator are more prone to the development of diseases such as cancer.

Each experimental system has its advantages and disadvantages. For example, the amenability of frog, clam, and sea urchin oocytes to biochemical studies has given them an advantage over other experimental systems in reconstructing cellular processes *in vitro*. On

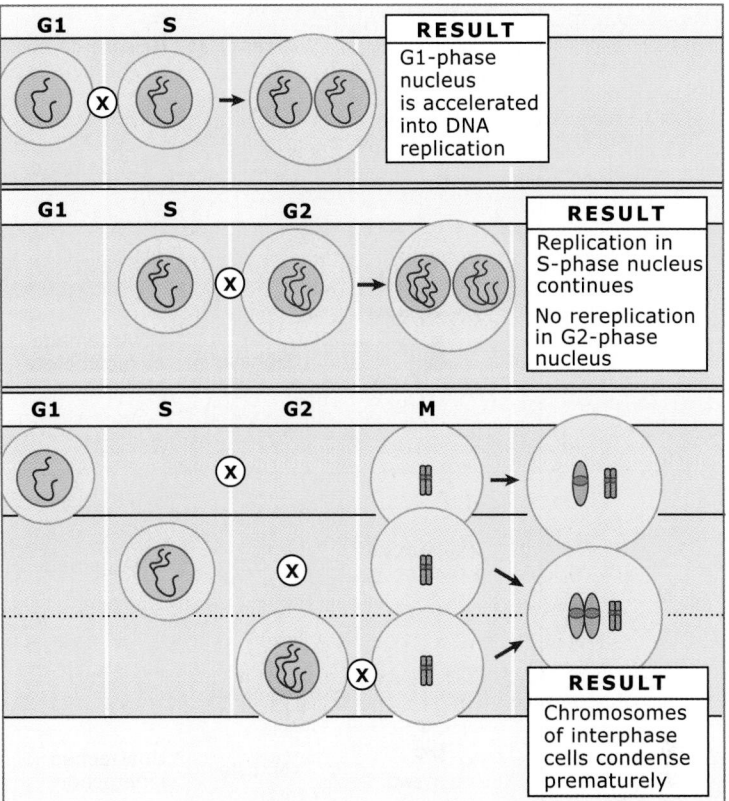

FIGURE 15.7 (Top row) When a G1 mammalian cell is fused with a cell in S phase, the G1 nucleus accelerates into DNA replication. (Middle row) When an S-phase cell is fused with a G2-phase cell, the cell in G2 does not rereplicate its DNA, nor does it inhibit ongoing DNA replication in the S-phase cell. It also does not enter mitosis. Rather, it waits for the S-phase nucleus to complete DNA replication before entering mitosis. (Bottom row) When an interphase mammalian cell (G1, S, or G2) is fused with a mitotic cell, the interphase cell immediately undergoes chromosome condensation and enters a pseudomitotic state. This indicated that mitotic cells contain a dominant mitotic-promoting activity.

the other hand, the powerful genetics of yeast, fungi, and fruit fly paved the way for the identification of key regulators of cell cycle progression. Together these experimental systems have provided a wealth of information about mechanisms of cell cycle control.

15.3 Events of the cell cycle are coordinated

Key concepts

- Processes in the cell cycle occur in an irreversible order.
- The major events of the cell cycle occur once and only once in each round of cell division.
- Checkpoints ensure the error-free completion of one cell cycle event before the next process begins.

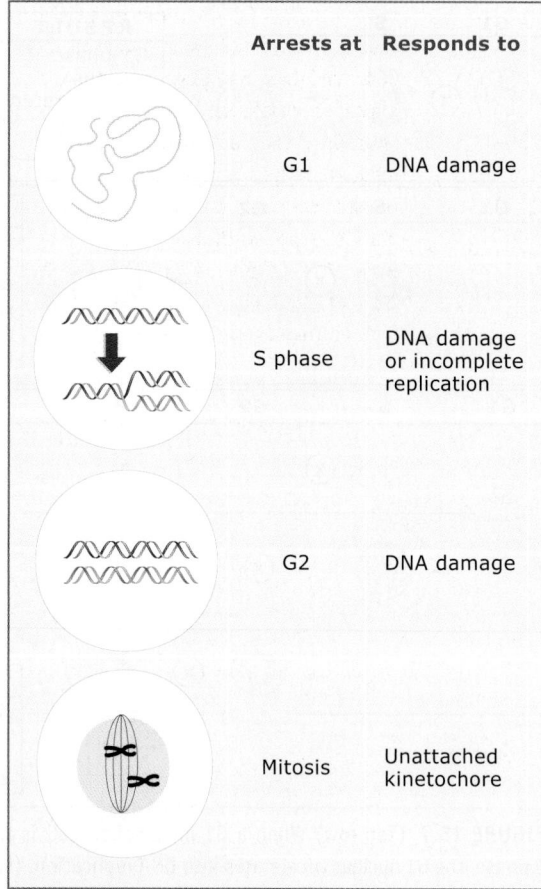

	Arrests at	Responds to
	G1	DNA damage
	S phase	DNA damage or incomplete replication
	G2	DNA damage
	Mitosis	Unattached kinetochore

FIGURE 15.8 Each step in the cell cycle is monitored to prevent the cycle from proceeding when there is trouble.

For successful completion of a cell cycle, several events must occur in the proper order like dominoes falling in a line, and events that occur in parallel must be linked to each other. How are these requirements incorporated into cell cycle control? The concept of checkpoints that monitor the proper completion of distinct cell cycle processes and control the transition from one cell cycle stage to another are discussed in *15.3 Events of the cell cycle are coordinated.*

The central cell cycle control system receives constant feedback throughout the cycle. As long as an event is still in progress, a signal is sent to the central control system that prevents it from moving on to initiate the next event. Only after a process is completed successfully does the inhibitory signal cease, allowing the central control system to proceed to the next phase.

Several such checkpoints exist. One checkpoint is activated by incompletely replicated or damaged DNA and acts to prevent the cell from entering mitosis until the problem is corrected. Another acts during mitosis and monitors the

attachment of the chromosomes to the mitotic spindle, preventing the separation of any sister chromatid pair until every chromosome is properly attached to both poles of the spindle. Other checkpoints monitor conditions, such as the position of the mitotic spindle, to ensure that the two separate groups of chromosomes produced by mitosis end up in separate cells after cytokinesis. In each case, if the condition is not met, the cell cycle is halted in order to provide time for the defect to be corrected before the cell commits itself to the next stage. Together, all of the checkpoints, summarized in **FIGURE 15.8**, ensure that each daughter cell receives a complete and intact copy of the genome.

15.4 A cycle of cyclin-dependent kinases activities regulates cell proliferation

Key concepts

- Cyclin-dependent kinases (CDKs) are master regulators of the cell cycle.
- CDKs are active only when complexed with a cyclin protein.
- Cyclins are proteins that oscillate in abundance during the cell cycle.
- A CDK partners with different cyclins during different phases of the cell cycle to phosphorylate distinct sets of target proteins.

As described in *15.1 Introduction*, the cell cycle is driven by a central control system. Key components of this control system are a small family of protein kinases called the CDKs. A CDK is a complex of two polypeptides. One, the CDK, binds ATP and contains the active site. However, it possesses kinase activity only when bound to the other, called a **cyclin**. **FIGURE 15.9** shows that cyclin binding causes a conformational change in the CDK that permits substrates to access the catalytic center. Each CDK-cyclin complex is active only during a specific stage of the cell cycle and is inactive for the reminder of the cell cycle. Some are active during G1, some while the cell is replicating its DNA, and others during mitosis. During the period that each kinase is active, it phosphorylates a large number of proteins, either activating them to go on to execute one of the major events of the cell cycle or inhibiting their activity to prevent repetition of a previous cell cycle event. For example, the

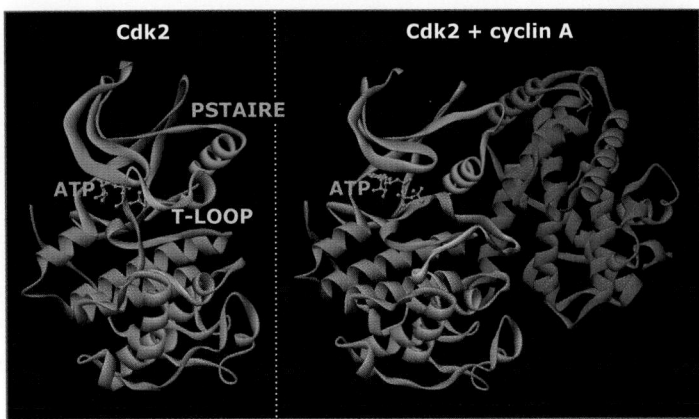

FIGURE 15.9 Cyclin binding to CDK causes a conformational change in the CDK. The T-loop changes position, allowing substrate access. In addition, reorientation of some amino acid side chains induces changes required for phosphate transfer. Left structure from Protein Data Bank 1B38. N. R. Brown, et al., *J. Biol. Chem.* 274 (1999): 8746–8756. Right structure from Protein Data Bank 1FIN. P. D. Jeffrey, et al., *Nature* 376 (1995): 313–320.

CDK that initiates mitosis phosphorylates the lamin proteins so that the nuclear lamina can break down, as well as phosphorylating a large collection of other proteins that regulate the assembly of the mitotic spindle.

CDKs were first identified through the characterization of *cdc* mutants in yeast. DNA sequencing of the *S. pombe cdc2+* gene revealed that it was homologous to the *CDC28* gene of the budding yeast, *S. cerevisiae*. In fact, the budding yeast gene was shown to functionally substitute for the fission yeast gene. In other words, when *CDC28* was introduced into a fission yeast *cdc2ts* mutant, the mutant strain could grow at the nonpermissive temperature, demonstrating that these two proteins were **orthologs**, that is, that they had a shared function. These proteins were then found to function as protein kinases. When a human relative, *CDC2*, was isolated by **complementation** of the *S. pombe cdc2* mutant, it became apparent that the kinase is an evolutionarily conserved regulator of mitosis. This finding spurred further discoveries of conserved proteins involved in cell cycle control and common strategies used for the regulation of eukaryotic cell division. Rather than use the organism-specific name of the kinase gene (*cdc2*, *CDC28*, or *CDC2*), we will use the generic name "*Cdk1*" for the remainder of *15 Cell cycle regulation*.

During the course of their studies on sea urchin and clam embryonic divisions in the early 1980s, Tim Hunt and colleagues and Joan Ruderman's laboratory, noticed the periodic change in the abundance of particular proteins as a function of cell cycle progression. These proteins were later named cyclins. The detection of such periodicity in protein abundance is possible only when examining single cells or cells that are synchronized to proceed together through the cell cycle.

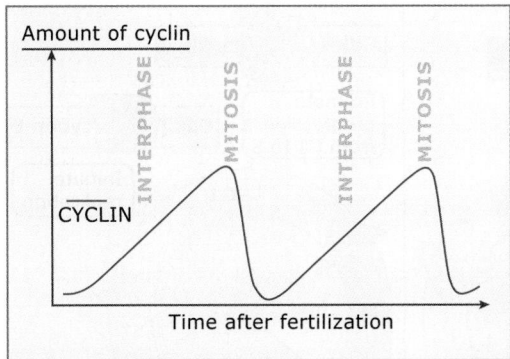

FIGURE 15.10 Scientists discovered the first cyclins when they noted that high cyclin levels were associated with the onset of mitosis in embryos. Cyclin levels dropped sharply after mitosis.

The change in protein levels is dramatic for most cyclins; the proteins accumulate steadily due to protein synthesis during interphase before abruptly disappearing just as the chromosomes separated to begin anaphase, as shown in **FIGURE 15.10**. The appearance and disappearance of the two proteins, occurring in perfect synchrony with the cell cycle, happens in each division. The periodicity of cyclin availability means that a CDK can be complexed with its cyclin and, hence, be active only during a short window of each cell cycle.

Further progress in understanding the role of CDKs and cyclins came from elegant studies of cell cycle control in *Xenopus laevis*. Maturation promoting factor (MPF) was identified as a biochemical activity in the cytosol of unfertilized frog eggs that, when injected into frog oocytes, was able to promote their **meiotic maturation**. Subsequent purification and biochemical characterization of MPF revealed that it consisted of Cdk1 and a cyclin.

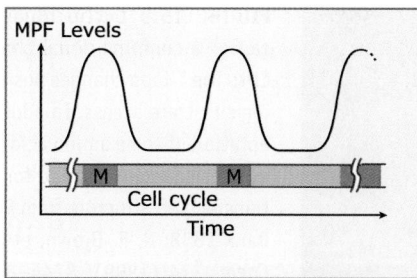

FIGURE 15.11 Variations in the level of MPF during the mitotic cycle suggested that it might be a regulator of mitosis.

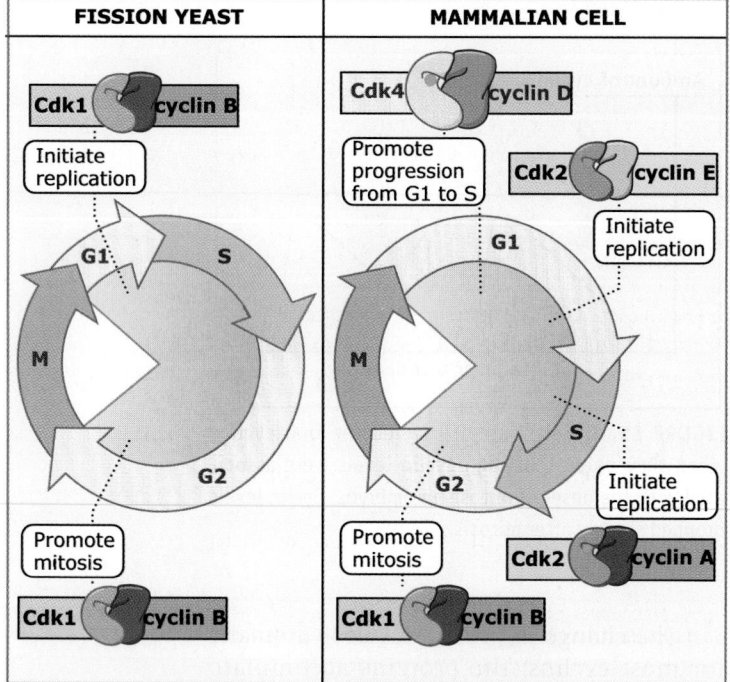

FIGURE 15.12 A single CDK-cyclin complex (Cdk1-cyclin B) can drive different cell cycle transitions in fission yeast (left), whereas different CDK-cyclin complexes accomplish these tasks in mammalian cells (right).

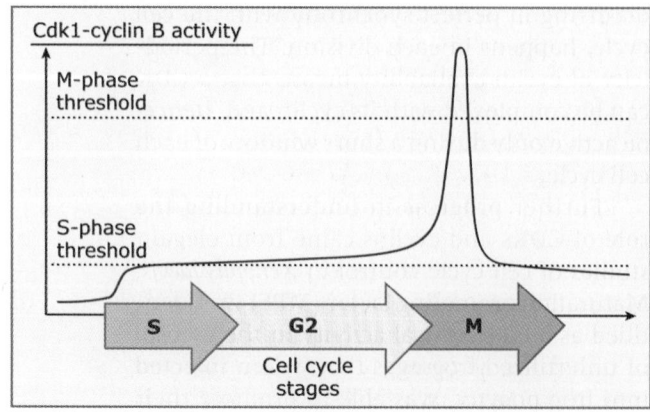

FIGURE 15.13 The graph illustrates that different thresholds of a single Cdk1 activity stimulates different events during a fission yeast cell cycle.

As diagrammed in **FIGURE 15.11**, MPF activity is now known to rise and fall precipitously with each meiosis or mitosis. The precipitous decline is due to cyclin proteolysis (see *15.11 Exit from mitosis requires more than cyclin proteolysis*), and the increase is due in part to rising levels of cyclin. The elucidation of MPF identity bolstered the revelation that similar proteins controlled the cell cycle of all eukaryotes.

In some organisms, a single cyclin and a single CDK can be enough to organize the cell cycle. For example, *S. pombe* has only one CDK and only one essential cyclin, and the same CDK-cyclin complex can drive both the G1/S and G2/M transitions, as seen in the left panel of **FIGURE 15.12**. This observation differs from the view that CDK-cyclin complexes are active at only one transition. In this case, it is believed that different levels of CDK-cyclin activity are required for the two transitions: a lower level promoting S phase and a higher level promoting mitosis, as diagrammed in **FIGURE 15.13**. Since the amount of cyclin increases over time, this mechanism would allow S phase and mitosis to be triggered in the correct order by the same CDK-cyclin complex.

In most organisms, multiple cyclins act over the course of each cell cycle, with different cyclins appearing and disappearing at different stages. Based upon the phase of the cell cycle during which they are present, they form three classes: G1 cyclins are responsible for moving the cell cycle through G1 and toward S phase; S-phase cyclins are required to initiate and maintain DNA replication; and M phase or mitotic cyclins initiate mitosis. Many cells have more than one member of each class; even a eukaryote as simple as the yeast *S. cerevisiae* has nine different cyclins. All nine cyclins interact with the same catalytic subunit. Because the catalytic subunit is the same in each *S. cerevisiae* CDK complex, the cyclin subunits must not only activate the kinase subunit, but also be responsible for determining the target proteins that it phosphorylates. In metazoans, the different classes of cyclin are typically designated by letter. For example, Cyclin D is associated with G1, cyclins A and E with S phase, and cyclins A and B with mitosis.

In addition to having multiple types of cyclins, many multicellular eukaryotes have multiple members of each type (designated cyclin B1, B2, etc). This leads to a large number of cyclins. The human genome, for example, encodes at least 12 cyclins implicated in cell cycle regulation. Why so many? There is now evidence from many organisms that much of the variation allows the cyclins to be regulated differently. For example, some cyclins are found only in restricted areas within the cell—such as within the nucleus or at the **centrosome**—whereas others are present in some tissues of an animal but not in others. Cyclins may also be temporally regulated, that is, different cyclins present during the same phase may appear or disappear at different times. The purpose of this fine regulation is not yet clear, but one possibility is that it is necessary to specialize the cell cycle slightly according to cell type. Thus, large animals with many kinds of cells might need many different cyclins.

Although different members of the cyclin family appear at different times and places, all cyclins share common features at the molecular level. All contain a stretch of approximately 150 amino acids that is similar in sequence among all family members and is termed the **cyclin box**. This sequence is the region of the molecule through which cyclins bind to a catalytic CDK subunit. Outside of the cyclin box, the primary sequences of cyclins diverge considerably, although many cyclins contain a hydrophobic patch in this region that is involved in substrate recognition. Despite differences, there appears to be a large degree of functional redundancy among cyclins with one being able to substitute for another, although not always particularly well. Such redundancy was first observed in the yeasts and is now evident in mice. For example, mice develop and live normally in the absence of cyclin E. It is likely that other cyclins compensate for cyclin E loss in the initiation of S phase, although they might ordinarily not perform the same function as cyclin E.

Metazoan cells also contain several cyclin-dependent kinases. Whereas CDK1 is important for mitosis, CDK2, CDK4, and CDK6 have roles earlier in the cell cycle. Most CDKs associate with one or two different cyclins, each CDK partnering with a distinct set. As a result, a large number of different CDK-cyclin complexes are present at one time or another. The different combinations of CDK subunits and cyclins apparently allow very fine regulation over when and where each acts—both in an individual cell and among the cells in a multicellular organism—as well as enabling each to have a distinct set of substrates. The stages in which various mammalian CDK-cyclins act are illustrated in the right panel of Figure 15.12.

To summarize, cell cycle transitions are regulated by CDK-cyclin complexes. Cyclins are proteins that oscillate in abundance during the cell cycle and they impart cell cycle specificity to CDK activities.

15.5 Cyclin-dependent kinases-cyclin complexes are regulated in several ways

Key concept
- CDK-cyclin complexes are regulated by phosphorylation, inhibitory proteins, proteolysis, and by subcellular localization.

Although the binding of cyclins is a primary means of regulating CDK activities, other control mechanisms are also used. We will discuss here in *15.5 Cyclin-dependent kinases-cyclin complexes are regulated in several ways* the various ways that CDK-cyclin complexes are regulated and how such regulation contributes to the cell cycle-specific activities of these complexes.

For the most part, these regulatory mechanisms control the activity of a CDK after it has bound a cyclin. The additional layers of regulation help make the transitions between the phases of the cell cycle sharp and irreversible. They are used to pause or advance the cell cycle in response to conditions inside the cell or in its environment.

One way to control the activity of CDK-cyclin complexes is phosphorylation. CDKs can be phosphorylated on two distinct surfaces, with phosphorylation at one of them activating the kinase and phosphorylation at the other inhibiting it. Activating phosphorylation occurs on a threonine residue within the T-loop of the kinase, a conserved domain in all protein kinases, and causes a conformational change in the CDK that is required for it to become active. This positive phosphorylation event is catalyzed by a CDK-activating kinase. The inhibitory phosphorylation occurs at another region of the CDK either on a conserved tyrosine residue, or on an adjacent threonine residue, as shown in **FIGURE 15.14**.

Addition of the inhibitory phosphates is mediated by members of the Wee1 family of kinases and their removal is performed by the Cdc25 family of phosphatases. At any given time, the activity of a CDK-cyclin complex is controlled by the phosphorylation state of these sites, and different signals relevant to whether the cell cycle should go forward are integrated

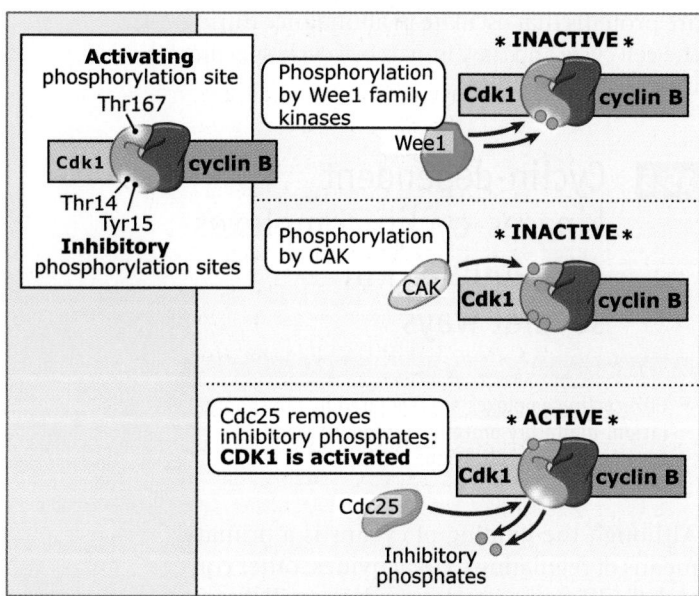

FIGURE 15.14 Phosphorylation of Thr167 of Cdk1 by CAK activates the Cdk1-cyclin B complex. However, phosphorylation of Thr14 and Tyr15 by Wee1 family kinases keeps the Cdk1 complex inactive in G2. At the G2-to-M transition, this phosphorylation by Wee1 is reversed by the action of Cdc25 family phosphatases, resulting in active Cdk1 complex.

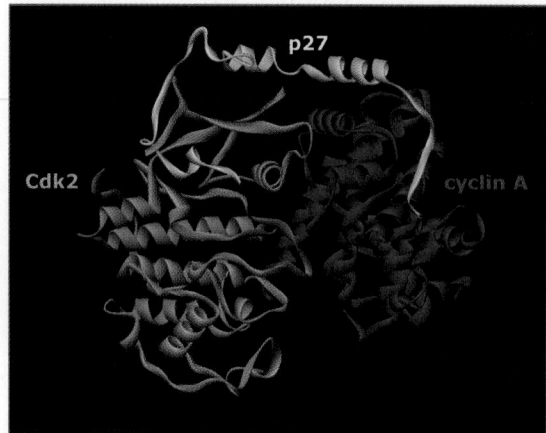

FIGURE 15.15 p27 inhibits the CDK-cyclin complex. Structure from Protein Data Bank 1JSU. A. A. Russo, et al., *Nature* 382 (1996): 295–296.

to control Wee1 and/or Cdc25 activities. As we shall see, for example, both environmental conditions and checkpoints have their effects on the cell cycle by regulating one or another of these phosphorylation events.

Two families of small inhibitory proteins, called cyclin kinase inhibitors (CKIs), also regulate CDKs. The p16 family of inhibitors interacts with the kinase subunits and prevents cyclin association. In contrast, members of the Cip/Kip family of CKIs, such as p27, bind and

inhibit CDK-cyclin complexes, as shown in **FIGURE 15.15**. Most of the CKIs identified so far act during G1 and/or S phase and block cell cycle progression until conditions allow them to be overcome. p16 CKIs present in G1, for example, block the cell cycle until enough G1 cyclins are synthesized to displace them from the G1 CDKs.

Cyclin-dependent kinases are also controlled by segregating them in different regions of the cell from their activators, inhibitors, or substrates. One of the best-understood examples is the regulation of cyclin B1 localization during interphase. In order for a cell to enter mitosis, cyclin B1 must associate with Cdk1 and phosphorylate substrates within the nucleus. Cyclin B1 constantly shuttles in and out of the nucleus, but because it contains both a cytoplasmic retention signal and a nuclear export signal, it accumulates in the cytoplasm during interphase (for details on nuclear export, see *9.14 Export of proteins from the nucleus is also receptor-mediated*). Just prior to the onset of mitosis, the nuclear export signals are inactivated by phosphorylation, allowing the bulk of cyclin B-Cdk1 to accumulate in the nucleus. It is possible that this type of regulation serves also to bring CDK complexes into contact with their substrates.

As highlighted earlier, the periodic availability of cyclins is a key mechanism of regulating the catalytic activity of CDK subunits. Cyclins accumulate at certain periods of the cell cycle to activate their CDK partners. At cell cycle transition points, cyclins become highly unstable and cyclin destruction irreversibly compels the cell cycle forward. The abrupt instability of cyclins is due to activation of ubiquitin ligases that target cyclins for proteasome-mediated degradation. Briefly, **ubiquitin** (Ub) is covalently attached to substrate proteins such as cyclins via an enzyme cascade consisting of Ub-activating (E1), conjugating (E2), and ligating (E3) enzymes. Once multiple ubiquitins are attached to a protein, the protein is recognized and degraded by a protein complex called the proteasome. Indeed, in a great example of feed-forward control, Cdks program their own destruction by activating cyclin ubiquitin ligases that lead to cyclin elimination (for one example, see *15.6 Cells may withdraw from the cell cycle*). The cyclin destruction machinery comes in different varieties, specialized for the various cyclin types. **FIGURE 15.16** summarizes major mechanisms used to regulate CDK activity.

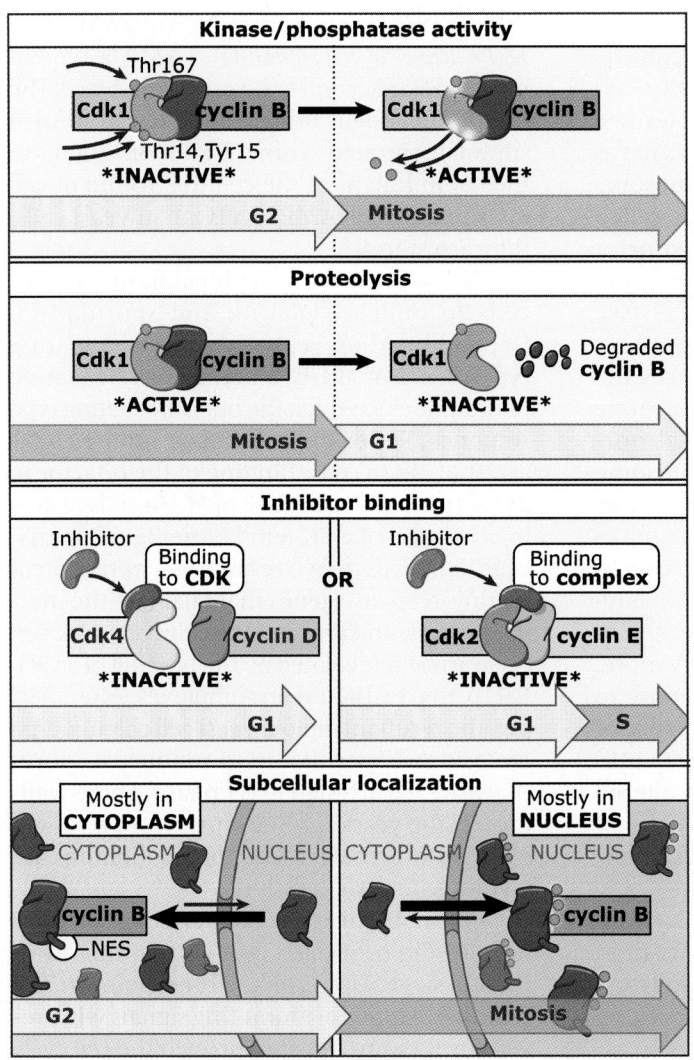

Kinase/phosphatase activity

Thr167

Cdk1 | cyclin B → Cdk1 | cyclin B

Thr14,Tyr15

INACTIVE → *ACTIVE*

G2 → Mitosis

Proteolysis

Cdk1 | cyclin B → Cdk1 | Degraded cyclin B

ACTIVE → *INACTIVE*

Mitosis | G1

Inhibitor binding

Inhibitor → Binding to **CDK**   OR   Inhibitor → Binding to **complex**

Cdk4 | cyclin D   Cdk2 | cyclin E

INACTIVE   *INACTIVE*

G1   G1 → S

Subcellular localization

Mostly in **CYTOPLASM**   Mostly in **NUCLEUS**

CYTOPLASM | NUCLEUS   CYTOPLASM | NUCLEUS

cyclin B   cyclin B

—NES

G2   Mitosis

FIGURE 15.16 (Top) CDK activity is regulated by both activating and inhibitory phosphorylation. (Second panel) Proteolysis of cyclin B during mitotic exit contributes to inactivation of mitotic Cdk1. (Third panel) Binding of CDK inhibitors (CKI) either to the CDK subunit (left) or to the CDK-cyclin complex (right) inhibits CDK activity. (Bottom) During G2, the NES-mediated nuclear export of cyclin B dominates, resulting in the separation of cyclin B and Cdk1. At mitotic commitment, phosphorylation of key residues in cyclin B attenuates its export resulting in its nuclear accumulation and accessibility to substrates.

In the rest of *15 Cell cycle regulation*, we will see how the central cell cycle machinery is controlled to perform the various transitions of the cell cycle. Cell cycle regulation is complex, and we will see that the complexity is required for some of its fundamental properties: cell cycle transitions must be complete, irreversible, and properly timed.

15.6 Cells may withdraw from the cell cycle

Key concepts

- Cell divisions are controlled by external stimuli and nutrient availability, and are not continuous.
- Cells can be in a nonproliferating state called quiescence, or G0.
- Cells reenter the cell cycle primarily in G1.
- Cells can permanently leave the cell cycle, becoming senescent.

We have now introduced the core cell cycle machinery that is necessary for cell cycle progression. This machinery, however, is not always "turned on." Normal cells proliferate only when conditions are appropriate to do so. For unicellular organisms, the availability of sufficient nutrients creates a permissive environment for proliferation. In multicellular organisms, additional environmental cues signal whether or not proliferation is appropriate. Such external signals can be either stimulatory or inhibitory in nature. We will discuss here in *15.6 Cells may withdraw from the cell cycle* how information from a cell's environment is integrated into a decision regarding whether or not to turn the cell cycle "engine" on and to start dividing.

When signals indicating that a cell should proliferate become limiting, metazoan cells enter a nondividing state called **quiescence**, or **G0**. This can be a temporary and reversible withdrawal from the cell cycle. However, as cells age, or differentiate *in vivo* or in culture, they typically lose proliferative capacity and enter a permanent state of quiescence termed **senescence**. Replicative senescence occurs as the repetitive DNA sequences at chromosome ends, which organize protective caps called telomeres, becomes progressively shorter with each round of cell division, eventually becoming too short to form telomeres (see *10 Chromatin and chromosomes*). Progressive shortening of telomeric DNA sequences occurs because the telomeric DNA replicative enzyme, telomerase, is not produced normally in somatic cells. When chromosome ends eventually become uncapped due to loss of telomeres, they are recognized by the cell as broken chromosomes and a cell cycle checkpoint is turned on (*15.12 Checkpoint controls coordinate different cell cycle events*). This checkpoint then prevents the cells from cycling any more. Cellular senescence can also be a response to an excess of proliferative signals in the case of oncogene-induced senescence. Senescent cells remain metabolically active but display altered phenotypes, including distinctive regions of chromatin called senescence-associated heterochromatic foci.

In yeast, nutrient starvation inhibits cell proliferation. Yeast cells respond to starvation conditions (low nitrogen and carbon sources) by reducing the activity of adenylyl cyclase,

the enzyme that converts ATP to cAMP (see *18 Principles of cell signaling*). cAMP, a type of intracellular second messenger, activates the cAMP-dependent protein kinase that in turn stimulates protein synthesis and contributes to the commitment of the cell to a round of cell division. When cAMP levels fall in yeast cells, they arrest in G1.

Arrest in G1 is also a preparation for yeast cells to undergo mating and sporulation. Haploid budding yeast exists in two mating types, *MAT***a** or *MAT*α, and they secrete mating pheromones to which the opposite mating type responds. For example, α factor induces a G1 arrest of *MAT***a** cells. Binding of the α factor to its receptor on the surface of *MAT***a** cells results in activation of a protein kinase signaling cascade that leads to two responses: expression of mating-response genes that facilitate the mating process, and arrest of the cell cycle. The cell cycle arrest is mediated by the binding of a CKI, Far1p, to G1 CDK-cyclin complexes. After mating, the resultant diploid yeast cell can undergo meiosis and sporulation to produce 4 spores that are each arrested in G1 phase. Spore walls protect the yeast chromosomes from a variety of environmental stresses until conditions are favorable for proliferation.

When stimulated to reenter a division cycle from a quiescent state, cells do so primarily in the G1 stage. Up until a certain point in G1, removal of growth-promoting signals will prevent further progression through the cell cycle and return cells to a quiescent state. However, cells reach a point late in G1 at which withdrawal of these signals no longer affects cell cycle progression; the cell is committed to a round of cell division regardless of extracellular cues. This point in G1, at which cells commit to beginning (and completing) a cycle of replication, is called **START** in yeast and the **restriction point** in multicellular eukaryotes. The concept of a commitment point in the cell cycle is depicted in **FIGURE 15.17**.

In summary, extracellular conditions normally govern whether or not cells divide. When conditions are not favorable for proliferation, cells enter a quiescent state called G0. When favorable conditions are encountered, cells reenter the cycle in the G1 stage. Reentry into the cell cycle and passage through G1 requires activation of G1 CDKs, which is accomplished by producing G1-phase cyclins and inactivating CKIs that inhibit CDK-cyclin complexes. Cellular senescence is the permanent withdrawal of cells from the division cycle.

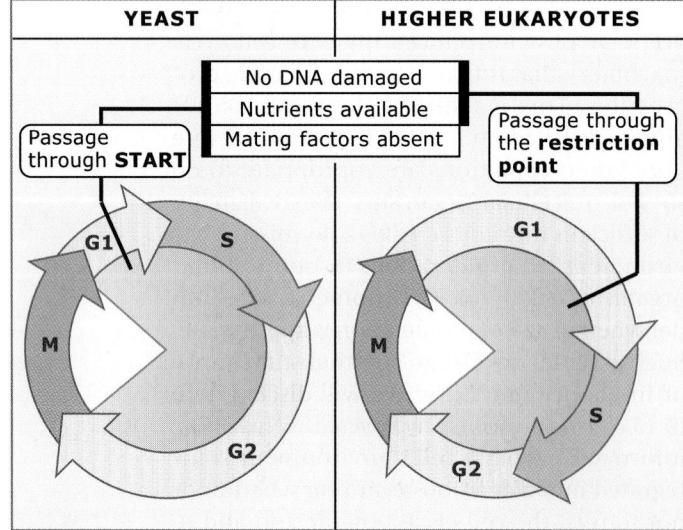

FIGURE 15.17 Passage through the START point (in yeast) or the restriction point (in higher eukaryotes) signals commitment to the cell cycle and is regulated by external factors such as the availability of nutrients for growth or absence of mating hormones (yeast). It is also regulated by intrinsic factors such as the integrity of the DNA.

15.7 Entry into cell cycle is tightly regulated

Key concepts

- Cells detect the presence of chemical signals in their environment using a variety of cell surface receptors.
- Extracellular signals elicit intracellular biochemical responses that can impinge on the activity of the cell cycle control engine
- Some extracellular signals induce cells to self-destruct by apoptosis.

As discussed earlier, withdrawal from the cell cycle can be temporary. Many cells can reenter the cell cycle from quiescence when nutrients and signals necessary for cell division are provided again. These signals include appropriate growth factors and hormones. Peptides and hormones that stimulate proliferation are called **growth factors**. These factors are present in serum, which is why serum is used for culturing cells *in vitro*. One of the first peptide growth factors to be identified from serum was platelet-derived growth factor (PDGF). PDGF is released by platelets upon blood clotting and contributes to the rapid cell proliferation required for wound healing. PDGF and other growth factors bind to specific receptors on the surface of target cells. These receptors are coupled to intracellular signaling pathways that can induce cellular proliferation.

How do growth factors stimulate the CDK activities that are required for commitment to S phase and mitosis? Following growth factor binding and receptor activation, many biochemical events are set into motion and intracellular second messengers are generated (see *18.32 Mitogen-activated protein kinases are central to many signaling pathways*). Ultimately, these lead to changes in gene expression as diagrammed in **FIGURE 15.18**. A set of genes that is expressed very quickly after serum addition is termed the **early response genes**. Many of these encode transcription factors, such as Fos, Jun, and Myc, which activate the transcription of other genes called **delayed early response genes**. One delayed early response gene is a G1 cyclin, cyclin D. Although the activation of CDKs in response to growth factors requires other events, stimulating cyclin D expression is one important mechanism driving cells through the G1/S transition.

Another mechanism involved in reactivating the cell cycle control system is the ubiquitin-mediated destruction of inhibitory CKIs that

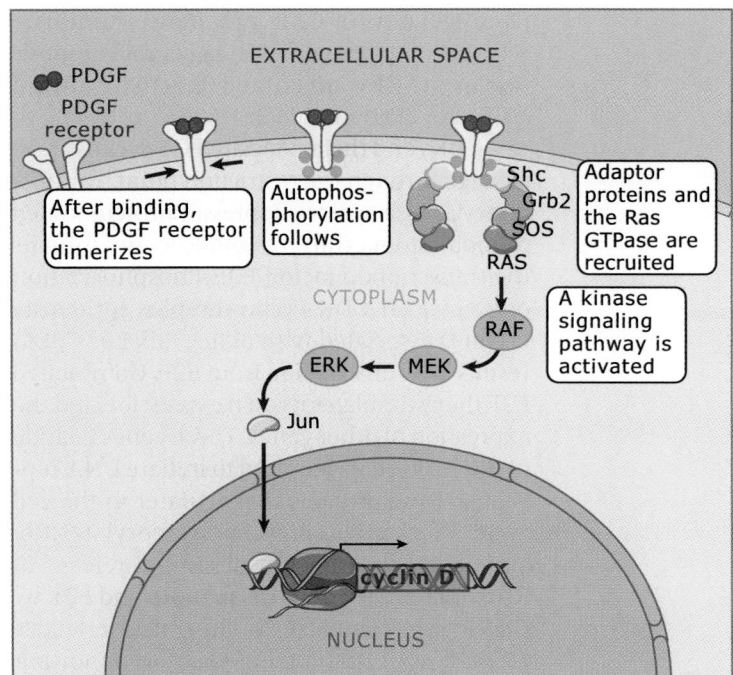

FIGURE 15.18 Binding of ligand to a receptor on the extracellular surface results in dimerization and activation of the receptor. In this example, a Tyr-kinase receptor, the PDGF receptor, undergoes intramolecular phosphorylation that recruits adaptor proteins called Shc, Grb2, and SOS. These recruit and activate a GTPase called RAS. These proteins activate a kinase signaling cascade that includes the kinases RAF, MEK, and ERK that ultimately results in a transcriptional response. Genes that regulate cell proliferation and passage through the restriction point are induced during this response contributing to commitment through a division cycle.

have accumulated during the G0 cell cycle arrest. The conserved E3 ubiquitin ligase required for CKI proteolysis is called **SCF** and is distinct from the E3 ubiquitin ligase involved in mitotic cyclin ubiquitination.

The SCF is composed of four core subunits: *S*kp1, *C*dc53, an F-box protein, and a RING finger protein, Rbx1. Multiple F-box proteins are present in every cell type, and they bind to different substrates, thereby providing substrate specificity to SCF complexes. The RING finger protein interacts with the E2 ubiquitin conjugating enzyme.

Prior phosphorylation of a substrate is a prerequisite for its recognition by an F-box protein and the SCF. In the case of CKIs, multisite phosphorylation by CDK-cyclins results in recognition by SCF and subsequent proteolysis. With its CKI destroyed, the G1 CDK is now active. In this manner, a small amount of CDK-cyclin activity is amplified and ultimately CDK activities reach the threshold required to induce passage through a stage of the cell cycle. Therefore, certain SCF complexes act as negative regulators of CKIs and, hence, stimulate return to the cell cycle. The general composition of an SCF complex and its activity toward a CKI are diagrammed in FIGURE 15.19.

CDKs and their associated G1 cyclins assist passage through the restriction point by phosphorylating a tumor-suppressor protein, called retinoblastoma (Rb) that binds to and inhibits the transcription factor, E2F. Phosphorylation of Rb by a G1 CDK-cyclin complex (primarily cyclin D associated with either Cdk4 or Cdk6) results in its dissociation from E2F. Uninhibited E2F then stimulates its own expression and the expression of other genes. These genes include cyclin E, proteins required to initiate DNA replication, and proteins that act later in the cell cycle. Cdk2-cyclin E also phosphorylates Rb, amplifying the effect of the initial release of Rb from E2F. The regulation of Rb and E2F by CDK-cyclin complexes is illustrated in FIGURE 15.20. Loss of Rb function plays an important part in the development of tumors (for details see *17 Cancer—Principles and overview*).

Some extracellular factors induce yet a different cell fate, that of **apoptosis** (for details, see *16 Apoptosis*). While apoptosis is a normal part of the development of multicellular organisms, it can also be induced by particular extracellular signals, such as tumor necrosis factor-α (TNF-α), that like growth factors, bind specific cell surface receptors and engage a cell death signaling pathway. Apoptosis can also occur when cells receive conflicting extracellular or intracellular cues, for example, to proliferate in the absence of sufficient nutrients.

In summary, extracellular proliferative signals engage the cell cycle control machinery. A major mechanism is to signal production of cyclins, which promote G1-CDK activities. Additionally, the SCF clears the cell of G1 CDK inhibitors by directing them for proteasome-mediated degradation. If proliferative cues are provided to cells in the absence of sufficient nutrients to ensure a successful round of cell division, apoptosis is triggered.

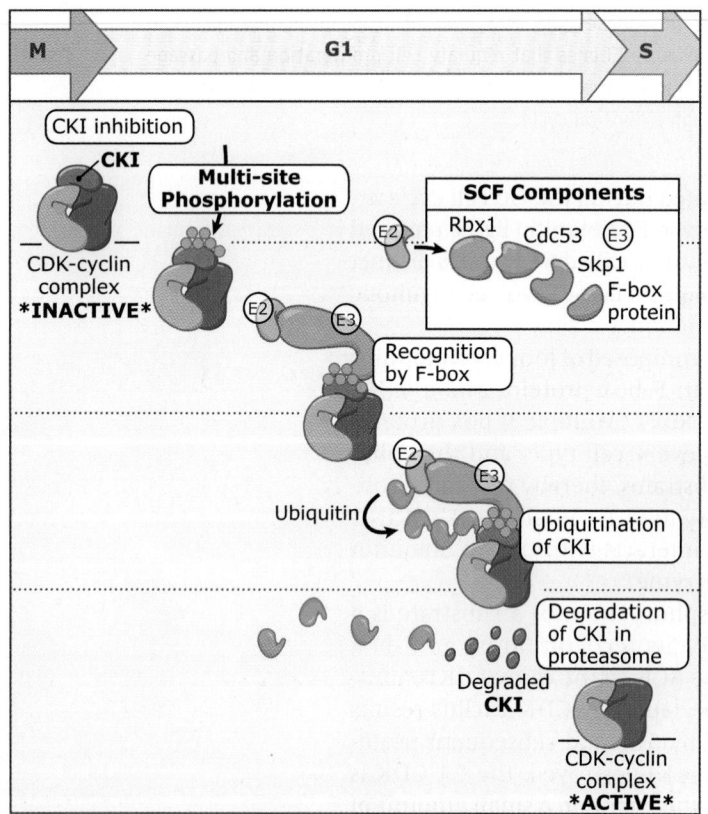

FIGURE 15.19 An SCF complex consists of four core subunits: Rbx1 (ring finger protein), a Cullin (for example, Cdc53), Skp1, and an F-box protein. The F-box subunits recognize specific phosphorylated substrates and link them to the SCF complex by also binding Skp1 through the F-box domain.

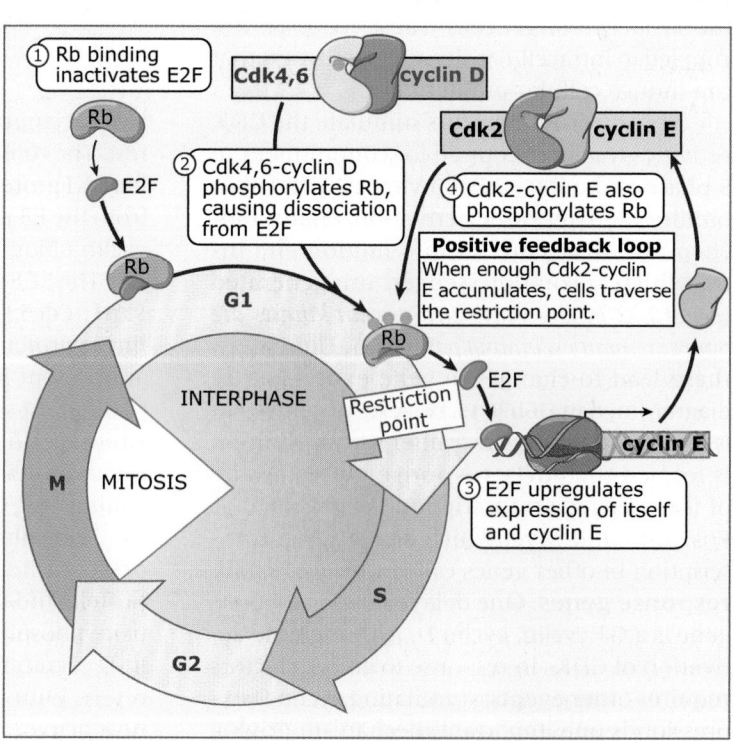

FIGURE 15.20 The transcription factor E2F is inactivated by Rb binding. When Rb is phosphorylated, it cannot bind to E2F, thus allowing E2F to upregulate expression of numerous genes including itself and cyclin E. Cdk2-cyclin E can then phosphorylate more Rb, resulting in an amplification loop.

15.8 DNA replication requires the ordered assembly of protein complexes

Key concepts

- Replication occurs after cells progress through the restriction point or START.
- Replication is regulated in a stepwise fashion and is coordinated with the completion of mitosis.
- Replication occurs at origins that may be defined by sequence, by position, or by spacing mechanisms.
- Initiation occurs only at origins that are licensed to replicate.
- Once fired, origins cannot be reused until the next cell cycle.

Once cells have taken survey of their environment and decide to embark on a division cycle, they traverse the G1/S transition and initiate DNA replication. How do cells assemble and activate factors required for DNA replication? What control mechanisms operate to ensure that cells replicate their DNA once, and only once, per division cycle?

Although complete answers to these questions remain unknown, the identification and analyses of yeast DNA sequences that are able to replicate independently of the context of a chromosome has provided a great deal of information about the process of DNA replication. These sequences, called *autonomously replicating sequences* (ARS), are part of an **origin** of replication in the chromosome. An origin of replication is a sequence of DNA where replication begins. Origins in budding yeast are determined by small consensus sequences, whereas origins in most other species are not. In fission yeast, origins are specified by large regions of DNA that are rich in A/T bases but not by particular sequences. In other eukaryotes, origins may be determined at random by mechanisms that distribute nonsequence specific DNA binding proteins across the genome.

A chromosome must have enough origins to ensure timely duplication of the genome. In bacteria, a single circular chromosome requires just one origin, but in eukaryotes with large genomes dispersed on multiple linear chromosomes, multiple origins are required. In budding yeast, with a genome of about 13 Mb, there are approximately 400 origins distributed on the 16 chromosomes. This imposes several regulatory problems. Replication origin usage must be coordinated with the cell cycle so that origins are fired only during S phase. The cell needs to be sure replication is complete before proceeding into mitosis. In addition, each origin must fire just once, to ensure that the DNA is replicated only once per cell cycle.

Replication origins direct the binding of factors required for the activation of replication and initiation of DNA synthesis. Initiation of DNA replication can occur only at those origins that have these necessary factors bound and are, therefore, considered licensed to replicate. However, DNA replication during each round of replication is restricted to a subset of potential origins within the chromosomes. Moreover, initiation events are temporally separated as different licensed origins are activated at different times. For example, some origins are activated early during S phase, whereas others are activated later. It is not clear what determines the temporal order, but it appears as if the position of a specific origin along a chromosome plays a role in specifying the timing of replication.

In order to initiate DNA replication, a prereplication complex (pre-RC) must be assembled on the origin. Genetic studies in yeast, complemented by biochemical studies using extracts from *Xenopus* eggs, have provided a picture of pre-RC assembly. It begins with the binding of a six-protein complex, the *o*rigin *r*ecognition *c*omplex (ORC) to the DNA. ORC marks potential origins but is not sufficient for activation. It serves as a platform for binding of two other conserved proteins: Cdc6, which belongs to the family of AAA+ ATPases, and Cdt1. (Many proteins with ATPase domains use the energy liberated by ATP hydrolysis to perform work.) The minichromosome maintenance complex (MCM), a ring made up of six closely related proteins that are also members of the large AAA+ ATPase family, is recruited next. The MCMs are abundant, and MCM complexes may spread beyond the origin. Once MCMs are loaded, ORC and Cdc6 are dispensable, and the pre-RC is poised for activation. The ordered assembly of the pre-RC at origins is diagrammed in **FIGURE 15.21**.

Assembly of the pre-RC is restricted to the window between the end of M phase and early S phase by several mechanisms. First, the amount of Cdc6 protein is controlled so that it is available only during this time. The MCM proteins cannot bind to the replication origin in the absence of Cdc6. Second, in metazoan cells, the Cdt1 protein is negatively regulated by a protein called geminin, which prevents its activity outside of the G1 window. Finally,

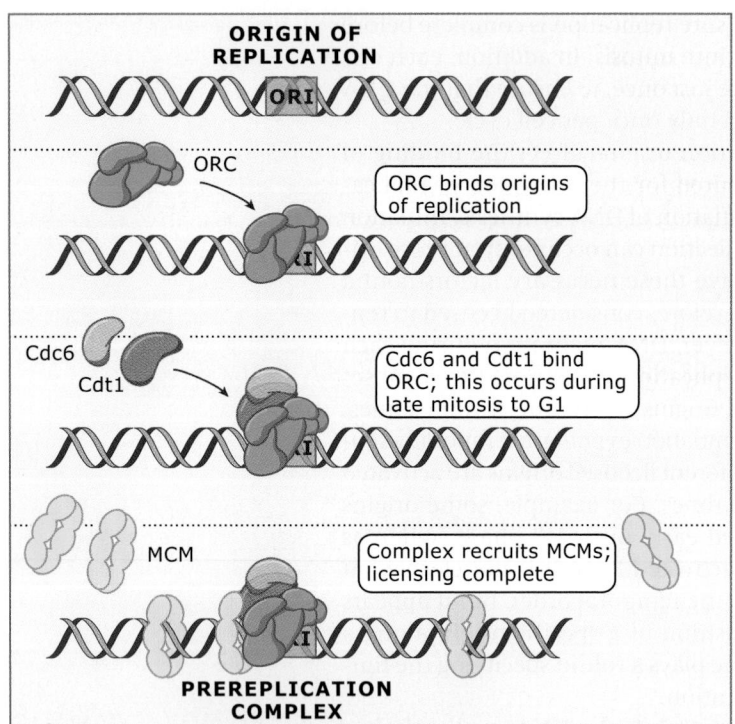

FIGURE 15.21 ORC binds the replication origin on a chromosome throughout the cell cycle. During a short window of the cell cycle spanning late mitosis to G1, the licensing proteins Cdc6 and Cdt1 also bind to the origin. This, in turn, recruits the hexameric MCM helicase complex (MCM2-7). This completes the licensing process and pre-RC assembly.

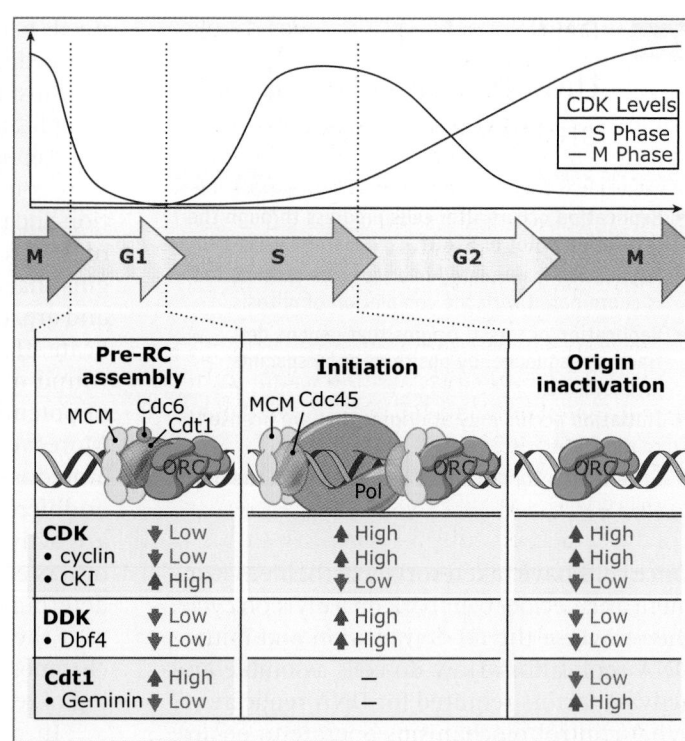

FIGURE 15.22 The pre-RC is assembled during late M phase and G1 when CDK and DDK activities are low. As CDK and DDK activities rise, DNA synthesis is initiated. The pre-RC is disassembled following initiation, thus inactivating the origin, and can be reassembled only when CDK activity falls again at the end of mitosis.

pre-RC assembly itself is restricted by mitotic CDK-cyclin activity. The CDK-cyclin complex targets subunits of ORC, Cdc6, and MCMs. In response to CDK phosphorylation, Cdc6 is inactivated and CDK phosphorylation of MCMs during S phase correlates with removal of MCMs from the DNA. The pre-RC, therefore, can only be established when CDK-cyclin activity is low, between the times of high mitotic CDK-cyclin activity in M phase and S phase, when the activity increases again to contribute to origin firing.

How is the origin transformed from a prereplicative to a replicative state? This transition requires the assembly of many additional proteins and is under the control of two kinases, a CDK-cyclin and Cdc7-Dbf4 (also known as Dbf4-dependent kinase, or DDK). Thus, CDK-cyclin activity couples replication to cell cycle progression both negatively (preventing pre-RC assembly and thus inappropriate origin reuse), and positively (promoting origin activation). An unanswered question is what are the substrates of these kinases that promote initiation of replication?

If CDK-cyclin activities provide global cell cycle coupling, the DDK acts at individual origins to initiate DNA synthesis. The most prominent substrates known so far are the MCM proteins themselves. Interestingly, a single-point mutation in Mcm5 can bypass the requirement for the DDK, which suggests that DDK phosphorylation results in changes in MCM structure to allow initiation. However, structural evidence suggests that this change is very subtle. The control of replication by CDK and DDK activities is illustrated in FIGURE 15.22.

The rate-limiting step for initiation at individual origins appears to be the binding of the Cdc45 protein, which requires both CDK-cyclin and DDK activity. Cdc45 binding, accompanied by loading of an additional complex called GINS, results in unwinding of DNA at the origin by activation of the MCM complex as a helicase. Thus, the MCMs are converted from an assembly factor in the pre-RC to a helicase that will be part of the elongation complex. The initial unwinding at the origin exposes single-strand DNA that is bound by the ssDNA binding protein RPA, which in turn allows loading of the primase/DNA polymerase alpha complex that initiates DNA synthesis. The MCM complex and Cdc45 move with the expanding replication fork, which forms a large **replisome** that now includes primarily DNA polymerase δ, rather than polymerase α. The initiation of DNA replication is illustrated in **FIGURE 15.23**. The cell relies on checkpoint and repair mechanisms to maintain the replisome and protect the DNA at the fork.

Once the MCMs move away from the replication origin, the origin is "used" and cannot be reactivated until the end of the next M phase when the pre-RC is once again established at the origin. If the moving replication fork is disrupted and the MCMs are dislodged, they cannot regain access until origins are relicensed in the next cell cycle. Thus, preservation of replication fork integrity is a crucial requirement for genome stability (see *15.13 DNA replication and DNA damage checkpoints monitor defects in DNA metabolism*).

Additionally, the moving replication fork is connected with establishment of **cohesion** that connects the newly synthesized sister chromatids together until mitosis. Thus, completion of S phase is linked to establishment of structures required for faithful chromosome segregation in mitosis, helping tie different phases of the cell cycle together.

Origin function is also regulated by chromosomal packaging. It is useful to remember that DNA in the cell is packaged by nucleosomes into chromatin, which imposes structural constraints (see *10 Chromatin and chromosomes*). This structure can affect origin timing; for example, not all origins replicate at the same time during S phase. In some cases, the relative timing of origin firing appears to be determined by the position of the origin along the chromosome and not by the origin itself. Origins in the vicinity of transcriptionally active euchromatin are initiated early, whereas those in the vicinity of transcriptionally inert heterochromatin are generally initiated late in S phase. For example, origins near telomeric regions of chromosomes (which are generally transcriptionally inert) are replicated late during S phase. This general rule is supported by elegant experiments performed in budding yeast in which an origin that normally replicated late could be induced to replicate early by moving it into a euchromatic region and vice versa. However, origin timing may also be affected by intrinsic elements, and the role of chromatin structure and nuclear position-

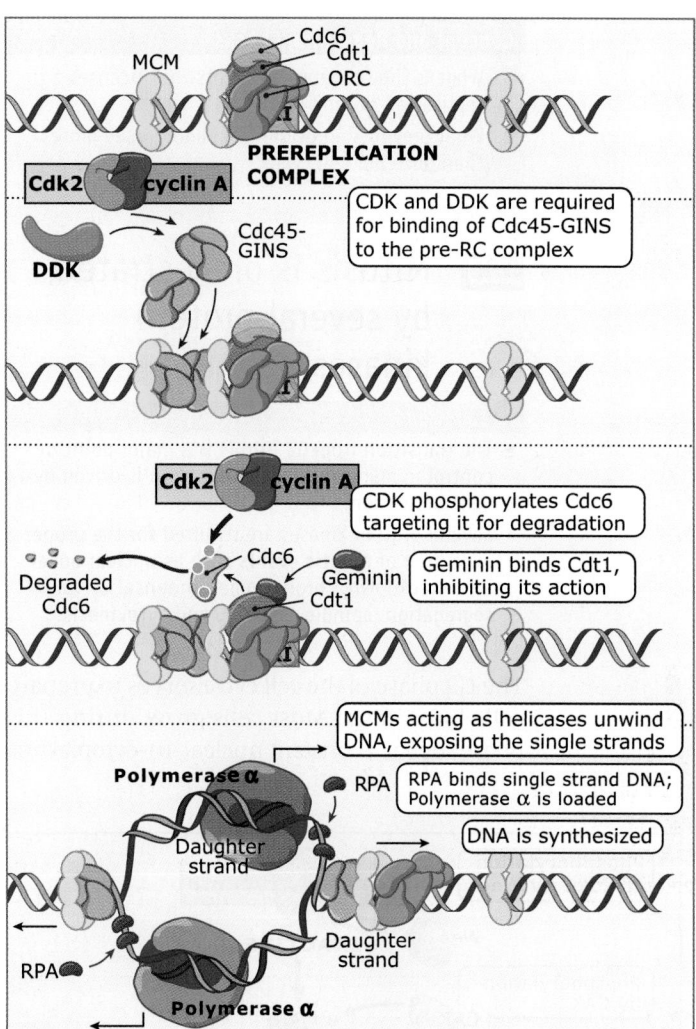

FIGURE 15.23 Initiation of DNA synthesis is regulated by CDK and DDK.

ing in replication dynamics is incompletely understood.

Additionally, for replication to proceed, chromatin remodeling enzymes are required to remove the nucleosomes ahead of the replication fork, and to reestablish chromatin immediately behind the fork. The synthesis of DNA is therefore tightly coordinated with the synthesis of new histone proteins to ensure that the packaging occurs efficiently.

Thus, genome duplication requires integration of a variety of signals, linking global cell cycle regulators that monitor the status of the cell (the CDKs), with chromatin-specific binding proteins that regulate individual *cis*-acting sites and general chromatin remodeling proteins that facilitate access. The coordination of CDK and DDK activity shows how different types of kinases can provide a convergent signal for cell cycle progression.

15.9 Mitosis is orchestrated by several protein kinases

The G2 phase of the cell cycle serves to prepare cells for mitosis. Most cells grow during this phase so that a constant nuclear-to-cytoplasmic

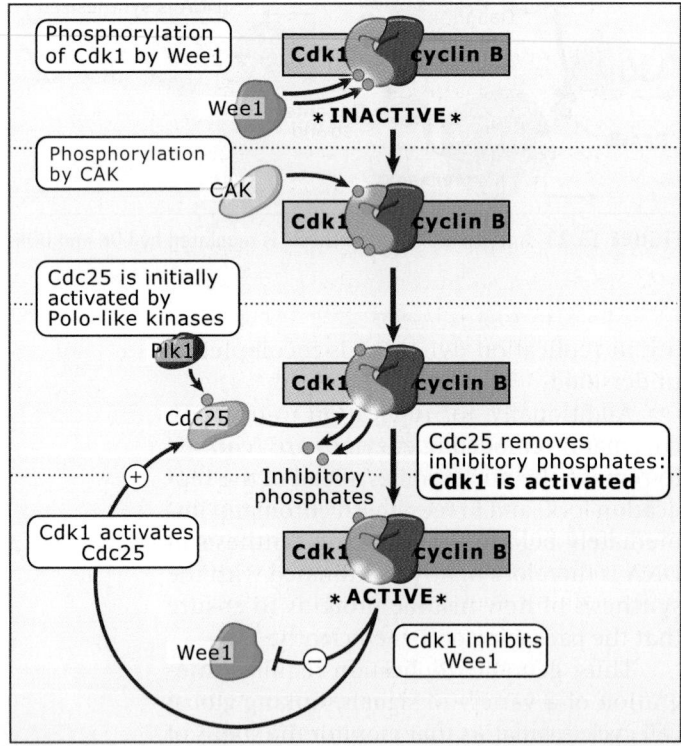

FIGURE 15.24 Phosphorylation of Cdk1 primes it for activation but also keeps it in an inactive state. Polo-like kinases (Plks) activate the Cdc25 phosphatase, which then activates a small amount of Cdk1 by removing its inhibitory phosphate. Once activated, Cdk1 phosphorylates Cdc25 and augments its activity. In addition, once some Cdk1 is activated, it can phosphorylate and inactivate Wee1. This autoamplification loop results in precipitous activation of Cdk1.

ratio is maintained upon cell division. In addition, errors in DNA replication are detected and corrected during G2 before cells begin the processes of chromosome condensation and segregation. When all the conditions are met, how do cells initiate mitosis? We will discuss here in *15.9 Mitosis is orchestrated by several protein kinases* the protein kinases that contribute to mitotic progression.

Cdk1-cyclin B is the major kinase that promotes the G2-M transition. As mitotic cyclins accumulate, bind Cdk1, and the complex accumulates, it is kept inactive by inhibitory phosphorylation mediated by the Wee1 family of kinases. When cells lack Wee1 activity, Cdk1 is not inhibited and, as a consequence, cells enter mitosis at a reduced cell size and are "wee." Mik1 is a homolog of Wee1, important for phosphorylating and inhibiting Cdk1 when S phase is prolonged. One mammalian Wee1 homolog, Myt1, is localized exclusively at the endoplasmic reticulum (ER) whereas other members of the family are localized in the nucleus. These kinases phosphorylate either a tyrosine (Tyr15) or a threonine (Thr14) residue, inhibiting Cdk1 kinase function.

The rate-limiting step for activation of Cdk1 and mitotic entry is the removal of these inhibitory phosphates by the Cdc25 phosphatase. Unlike *wee1*, the *cdc25* gene in fission yeast is essential. If the inhibitory phosphate(s) is not removed, mitosis cannot occur. Mammals have three different Cdc25 isoforms to ensure regulation of CDK activities.

Of the steps in Cdk1 activation, Cdc25 regulation is the most thoroughly investigated. In higher eukaryotes, Cdc25 activation involves the Polo-like kinases (PLKs), a family of conserved kinases distinct from the CDK family. After PLK activation of Cdc25 and consequent increase in Cdk1 activity, further activation of Cdc25 may be mediated by Cdk1 itself. Such a positive feedback loop of Cdk1 activation results in a dramatic increase in its kinase activity, leading to mitotic entry. The activity of Wee1 is also negatively regulated by Cdk1 phosphorylation to ensure a precipitous activation of Cdk1 for mitosis. A schematic of these steps of Cdk1 activation is depicted in **FIGURE 15.24**.

Two different CDK-cyclin complexes govern mitotic entry and progression in multicellular eukaryotes. For example, in mammalian cells, Cdk1-cyclin A regulates some aspects of mitosis, including chromosome condensation and alignment on the mitotic spindle, whereas variants of cyclin B with different localizations and cell cycle periodicities presumably regulate other mitotic events by phosphorylating different substrates. A number of Cdk1 substrates are known, involved in almost every aspect of mitosis and cytokinesis.

While Cdk1 is considered the master mitotic regulator, other protein kinase families also play critical roles in various aspects of mitosis and we will introduce three of them here. First, the PLKs, mentioned earlier, are one such family. First identified by a mutant in *Drosophila* that failed in several aspects of mitosis, PLKs were subsequently found to control entry into mitosis, centrosome maturation, spindle formation, chromosome segregation, and cytokinesis in eukaryotes. Whereas the genomes of *Drosophila* and the yeasts encode a single PLK enzyme, vertebrates produce up to four distinct family members with Plk1 appearing most similar functionally to its yeast and *Drosophila* equivalents. All members of the family contain an N-terminal kinase domain and a C-terminal domain consisting of one or more conserved motifs called **Polo boxes**. The domain architecture and activation mode of PLKs are shown in **FIGURE 15.25**.

Polo boxes serve as intracellular targeting domains for PLKs and are responsible for directing them to multiple docking sites at centrosomes, **kinetochores** (proteinaceous structures that serve to link the chromosomes to the microtubule ends: refer *14 Mitosis*), the mitotic spindle, and the cytokinetic ring. Polo boxes bind to consensus phosphorylation sequences for Cdk1 and other proline-directed kinases, called phospho-Ser/Thr-Pro modules. While it is likely that *in vivo* docking sites for PLKs involve additional information present within its binding partners, it is clear that PLKs generally dock onto proteins that were previously phosphorylated by Cdk1 or other kinases. In fact, PLKs can prime their own binding events. This mechanism can provide convergence points for the activities of different mitotic kinases and can also provide a means of coregulating Plk1 with other protein kinases. In terms of coregulation, PLKs are activated by phosphorylation within their T loops, similar to CDKs, and this stimulatory event on Plk1 is directed by Aurora kinases (discussed below), providing another means of costimulating mitotic phosphorylation events. PLKs are likely to have numerous substrates at their various sites of action. Those studied in detail so far are all consistent with its established roles in cen-

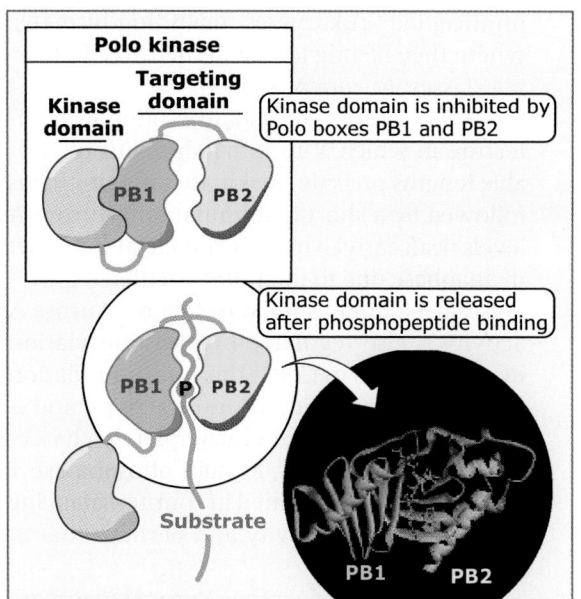

FIGURE 15.25 The Polo family of kinases possesses an N-terminal kinase domain and a C-terminal region consisting of two conserved regions called Polo boxes (PB1 and PB2). Polo boxes act in targeting the kinase to its intracellular partners. Inset structure from Protein Data Bank 1UMW. A. E. Elia, et al., *Cell* 115 (2003): 3–4.

trosome duplication and maturation, spindle formation and cytokinesis.

A second family of kinases acting in mitosis, the NimA-like kinases (NEKs), was discovered in *A. nidulans*. Their discovery and importance serve as prime illustrations of why the use of multiple organisms to study cell cycle control is so important, with each providing unique contributions that turn out to be broadly informative. In *A. nidulans*, *nimA* mutants arrest in G2 even though Cdk1 activity is high, suggesting that Cdk1 activation is not sufficient to drive all events of mitosis on its own. Indeed, the protein kinase activity of NIMA parallels that of Cdk1 during mitosis. Based on sequence homology, NEKs have now been identified in many eukaryotes and found to be involved in several mitotic events including chromosome condensation and centrosome separation.

A third family of kinases that has garnered much attention for its involvement in mitotic events is the Aurora family. Similar to Cdk1 and PLKs, Aurora kinases participate in a plethora of activities including chromosome condensation and segregation, kinetochore function, centrosome maturation, spindle formation, and cytokinesis. Aurora kinases were first identified in budding yeast, in which there is a single member, and were later identified in multicellular eukaryotes, including humans, where they belong to one of the three following classes: Aurora A, B, or C.

All Aurora enzymes share a common architecture in which N-terminal domains of variable lengths precede the kinase domain that is followed by a short C-terminal tail. Aurora A levels peak early in mitosis and fall at the onset of anaphase due to ubiquitin-mediated proteolysis in a manner similar to cyclins. Aurora A activity is also regulated by phosphorylation in its T-loop. As in Cdk1, this phosphorylation event is essential for Aurora A activity and is tightly controlled by an incompletely characterized set of proteins. Protein phosphatase 1 (PP1) activity is implicated in counterbalancing Aurora A kinase activity and perhaps that of Aurora B as well.

Although it is clear that Aurora kinases are critical for multiple steps of mitosis, much remains to be learned about how these kinases are regulated and the mechanistic details of their action. However, an interesting connection links Aurora A and activation of PLK. Together with a transiently produced cofactor called Bora, Aurora A is able to phosphorylate PLK and trigger its activation. PLK then phosphorylates Bora and induces Bora degradation. In this manner, Aurora A, PLK and CDK1 activities are tightly coupled during mitosis.

An important development is the recognition that misregulation of Aurora kinases likely contributes to tumorigenesis. Many tumor types contain elevated levels of Aurora kinases and overexpression of Aurora A can transform rodent cells. Overexpression of Aurora kinases results in centrosome amplification and chromosome segregation defects and this is likely due, at least in part, to its activation of PLKs. Moreover, Aurora A is a cancer susceptibility gene (see *17 Cancer—Principles and overview*).

Consistent with the multiple roles played by protein kinases during mitosis, they are found at many intracellular sites. One important site is the centrosome. In addition to their role as microtubule organizing centers, centrosomes have emerged as signaling centers where regulators of the cell cycle and their substrates accumulate. In this manner, signals impinging on the cell cycle can be rapidly integrated to generate an appropriate proliferative response. Cdk1-cyclin, PLK, Aurora, and NIMA family members all localize to centrosomes.

Aurora B exemplifies the dynamic localization pattern exhibited by the major mitotic protein kinases. Together with three other proteins, namely INCENP, Survivin, and Borealin, Aurora B moves from kinetochores in metaphase to the central spindle during anaphase and then to the midbody region of dividing cells. Based on their characteristic movement from the chromosomes to the spindle midzone, these proteins are called **chromosome passenger proteins**. In the absence of any member of the complex, chromosomes fail to condense properly, align at the metaphase plate, and make connections to both centrosomes.

A key function of Aurora B and the other chromosome passenger proteins at kinetochores is to dismantle kinetochore-spindle microtubule connections that do not lead to tension across the kinetochores—that is, attachments that do not result in bipolar attachment (refer to *14 Mitosis*). Additionally, cells lacking the function of any chromosome passenger protein fail at cytokinesis because Aurora B kinase activity is required to organize the central spindle and the cleavage furrow. Though PLKs and Cdk1 do not exhibit the same localization pattern as Aurora B, these proteins also localize dynamically during the course of mitosis to interact with their targets.

In summary, initiation and progression through mitosis requires the function of the CDK, PLK, NEK, and Aurora families of protein kinases. Because controlling these activities is critical for maintaining genomic integrity, each kinase is tightly regulated to prevent premature activation and ensure timely inactivation. Emerging evidence points to a dependency of PLK on other kinase activities, particularly Cdk1 and Aurora A.

15.10 Sister chromatids are held together until anaphase

Key concepts

- A mitotic chromosome consists of two sister chromatids held together by cohesive forces.
- Ubiquitin-mediated proteolysis drives the release of sister chromatid cohesion and marks the onset of anaphase.
- The anaphase promoting complex or cyclosome (APC/C) is the E3 ubiquitin ligase that directs proteolysis of key mitotic proteins including securin and cyclin.

During its duplication during S phase, each chromatid is tethered to its sister. This cohesion is important for the equal partitioning of chromatids to daughter cells during mitosis for if sisters were to separate prior to anaphase, there would be no way to keep track of them and ensure that each daughter cell inherits one copy of each chromosome. The "glue" that zippers sister chromatids together is the **cohesin complex**. The cohesin complex is related to the condensin complex discussed in *6 Transport of ions and small molecules across membranes*. It is composed of two SMC proteins with ATPase activity that are distinct from, but related to, the SMC proteins of the condensin complex. The cohesin complex contains three additional proteins, Scc1, Scc3, and Pds5. It is proposed that the two SMC proteins, along with the Scc1 subunit, encircle sister chromatids to glue them together. Though the mechanism of cohesion is still under investigation, it does require that a component of the cohesin complex become acetylated during S phase.

At the onset of anaphase, sister chromatid cohesion must be dissolved and at least two mechanisms are employed. First, cohesin is largely removed from chromosome arms in prophase, leaving centromere-bound cohesin in place. This first wave of cohesin removal is orchestrated by Plk1 phosphorylation of Scc1. At centromeres, however, the Scc1 cohesin subunit is protected from Plk1 phosphorylation and to remove it, it must be proteolyzed. If the ring model of cohesion is correct, the cleavage of Scc1 could open up the cohesin ring, permitting the physical separation of the sister chromatids. Whatever mechanism of cohesion is correct, it is clear that Scc1 cleavage liberates the sister chromatids to allow anaphase to begin.

Separase is the site-specific protease that cleaves Scc1 to allow sister chromatid separation. For much of the cell cycle, separase is inhibited by the binding of another protein called **securin**. Securin is ubiquitinated and targeted for proteosome-mediated degradation at the metaphase-anaphase transition. Securin proteolysis liberates separase and enables it to cleave Scc1 that is bound to DNA. Thus, the event that induces chromosome segregation is securin proteolysis. The process of chromatid separation is illustrated in **FIGURE 15.26**.

Securin is targeted for destruction by the E3 ubiquitin ligase known as the **anaphase-promoting complex** (APC). The APC recognizes proteins containing short "destruction sequences," including destruction boxes (D boxes) and KEN boxes. When appended to an otherwise stable protein, these sequences cause APC-mediated ubiquitination and proteolysis.

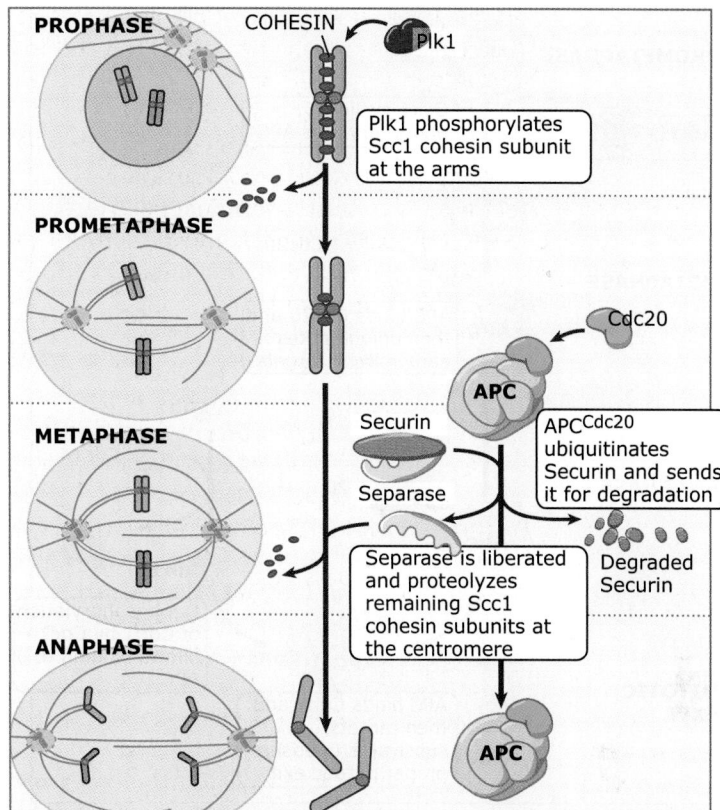

FIGURE 15.26 Until metaphase, the protease called separase is bound by securin and kept inactive. At metaphase, the APC targets securin for proteolysis, thus releasing separase from inhibition. Active separase then cleaves cohesin enabling sister chromatid separation.

Unlike the SCF (see *15.7 Entry into cell cycle is tightly regulated*), prior phosphorylation of the substrate is not required for its recognition by the APC. In fact, substrate phosphorylation can block recognition of some destruction boxes. In budding yeast, the essential function of the APC is to mediate the elimination of cyclins and securin, although many other targets exist.

The APC prevents the accumulation of cyclins and securing, but only during mitosis and G1. How is APC activity regulated? At least three mechanisms are known. First, the binding of adaptor proteins called Cdc20 and Cdh1 (also referred to as APC^{Cdc20} or APC^{Cdh1}) to the APC provides much of the temporal and substrate specificity to the APC. These proteins are only transiently available to the core APC during mitosis and G1 phases. Second, the APC is regulated via specific phosphorylation of some of its multiple subunits and its adaptor proteins. Both Cdk1 and Plk1 are implicated in the phosphorylation and activation of the APC. Third, a signaling pathway that monitors chromosome attachment, called the spindle

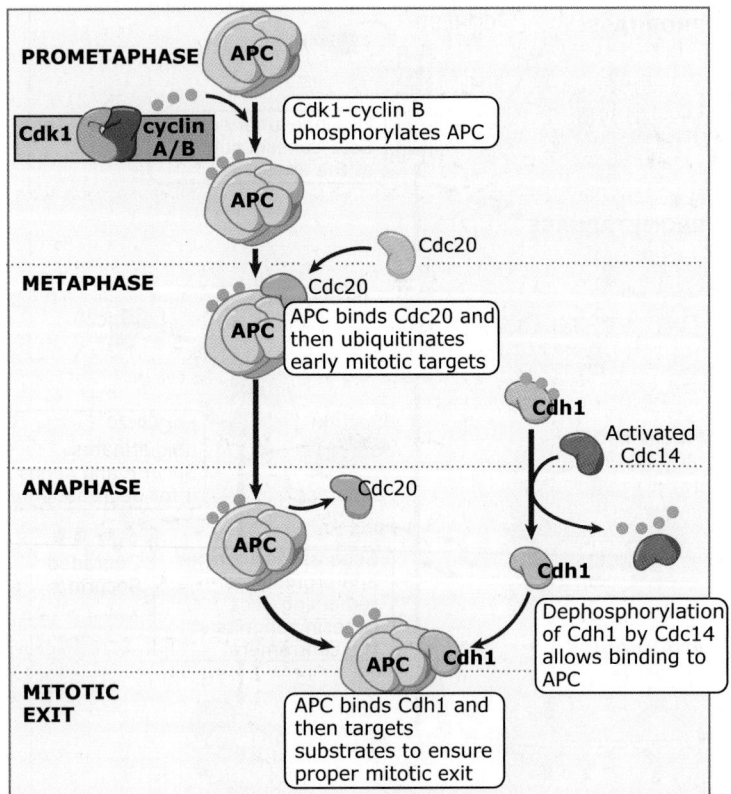

FIGURE 15.27 During interphase, the APC ubiquitin ligase is inactive. During mitosis, Cdk1 phosphorylates the APC and the APC binds its activator, Cdc20. APC^{Cdc20} then targets substrates such as securin, ensuring irreversible progression through mitosis. Later in mitosis, APC binds to the Cdh1 activator and targets other substrates that ensure proper mitotic exit.

In the figure:

PROMETAPHASE — APC — Cdk1 cyclin A/B — Cdk1-cyclin B phosphorylates APC — APC

METAPHASE — Cdc20 — Cdc20 APC — APC binds Cdc20 and then ubiquitinates early mitotic targets

ANAPHASE — Cdc20 — APC — Cdh1 — Activated Cdc14

MITOTIC EXIT — Cdh1 — APC Cdh1 — APC binds Cdh1 and then targets substrates to ensure proper mitotic exit — Dephosphorylation of Cdh1 by Cdc14 allows binding to APC

assembly checkpoint (SAC), controls APC activity (for a discussion on the SAC, see *15.14 The spindle assembly checkpoint monitors defects in chromosome-microtubule attachment*). These multiple levels of regulation ensure that the APC becomes active only during mitosis to trigger proteolysis of securin and cyclin, liberation of separase, cleavage of cohesin and, ultimately, separation of sister chromatids. The APC remains active during G1 to keep Cdk1 activity low. Low Cdk1 activity permits formation of the pre-RC that is required for the next round of DNA replication. Also, in postmitotic cells, the APC is implicated in enforcing the quiescent state by preventing accumulation of cell cycle promoting factors. The activation of the APC is diagrammed in **FIGURE 15.27**.

Once sister chromatids are released from one another by dissolving cohesion, they are transported to the opposite poles of the cell by the mitotic spindle (for a review see *14.18 Anaphase has two phases*). Cytokinesis ensues once the chromosomes are well separated and Cdk1 activity has fallen.

In summary, sister chromatids are held together by the cohesin complex until their separation in anaphase. The dissolution of cohesion requires cohesin phosphorylation and the cleavage of the cohesin subunit Scc1 by the protease, Separase. Activation of Separase requires APC-mediated destruction of its inhibitor, securin, by ubiquitin-mediated proteolysis.

15.11 Exit from mitosis requires more than cyclin proteolysis

Key concepts

- Exit from mitosis requires Cdk1 inactivation.
- Mitotic exit also involves the reversal of Cdk1 phosphorylation.
- Inactivation of Cdk1 and reversal of Cdk1 phosphorylation are coordinated with disassembly of the mitotic spindle and cytokinesis.

For completion of one cell cycle, cells must exit mitosis and divide. We will now discuss how these tasks are accomplished.

High Cdk1 activity defines the mitotic state. In order for cells to return to interphase and enter a new cell cycle, Cdk1 must be inactivated and its phosphorylation events reversed. This process of return to interphase is called mitotic exit. Cdk1 inactivation is achieved primarily through ubiquitin-dependent proteolysis of cyclin B, as discussed earlier. However, other mechanisms are also employed.

One additional way in which mitotic exit is promoted is by reversal of mitotic phosphorylation. One key protein involved in this process is the Cdc14 phosphatase, which specifically antagonizes Cdk1 activity. Like so many other key cell cycle regulators, *CDC14* was identified through characterization of a cell cycle mutant. The *cdc14-3* mutant arrests in telophase with high levels of Cdk1 activity. *CDC14* encodes a dual-specificity phosphatase (having the capacity to dephosphorylate serine, threonine, and tyrosine residues *in vitro*), and it desphosphorylates Cdk1 substrates. None of its known targets is phosphorylated on tyrosine residues despite its structural similarity to tyrosine phosphatases. Among its targets in budding yeast are the transcription factor, Swi5, the APC activator, Cdh1, and the Cdk1 inhibitor (CKI), Sic1.

Dephosphorylation of Swi5 permits its nuclear accumulation, whereas Cdk1 phosphorylation inhibits its nuclear localization. Once in the nucleus, Swi5 acts to increase expression of Sic1. In addition, dephosphorylation of Sic1 by Cdc14 prevents its degradation by the SCF. Thus, Cdc14 and Swi5 act in a concerted fashion to inhibit Cdk1 by increasing the amount of one of its inhibitors. In this way, even if cyclin is not completely destroyed by the APC, the Cdk1-cyclin complex is inactivated. In addition, Cdc14 dephosphorylation of the substrate-specific activator of APC, Cdh1, allows it to bind the APC and target mitotic cyclins for proteolysis. In this manner, the APC continues to be activated as Cdk1 levels fall and Cdc14 activity rises. In fission yeast, the Cdc14 phosphatase (termed Clp1/Flp1) promotes Cdk1 inactivation and mitotic exit in a distinct manner from its budding yeast ortholog. The major target of Clp1/Flp1 is the Cdc25 phosphatase. When Cdc25 is dephosphorylated at the end of mitosis, it is inactivated and targeted for APC-mediated degradation. Thus, Clp1/Flp1 silences the Cdk1 autoamplification loop.

To achieve a coordinated exit from mitosis, the appropriate balance of Cdk1 and the opposing phosphatase activity must be optimal at different stages of mitosis. In budding yeast, this is achieved by sequestering Cdc14 in the nucleolus such that it cannot combat Cdk1 phosphorylation events at other locations until anaphase, when it is fully released and activated. However, in other organisms such as mammals and fission yeast, Cdc14 enzymes are released from their interphase locations at the onset of mitosis. Though it is not clear in mammals what prevents the Cdc14 phosphatases from reversing Cdk1 phosphorylation during early mitosis, this is now understood in fission yeast. Clp1/Flp1 activity is attenuated at the beginning of mitosis because Clp1/Flp1 is a Cdk1 target. Cdk1 phosphorylates Clp1/Flp1 and inhibits its phosphatase activity. However, Clp1/Flp1 can auto-dephosphorylate its Cdk1-mediated phosphorylation sites when Cdk1 activity begins to drop. As it does so, it becomes maximally active. In this way, there is a transition during anaphase from a state of high Cdk1 activity to a state of high Clp1/Flp1 activity. Studies of Cdc14 homologs in other eukaryotic cells indicate that the ability of members of the Cdc14 family of phosphatases to reverse Cdk1 phosphorylation is conserved. However, while it participates in reversing Cdk1-mediated phosphorylation in all organisms and antagonizing Cdk1, it is not an essential activity in most of the organisms

in which it has been studied such as fission yeast, *D. melanogaster*, and *C. elegans*. Therefore, other phosphatases must play important roles in reversing Cdk1 phosphorylation events, as well as the phosphorylation events catalyzed by other mitotic kinases such as Plk, NEK, and Aurora.

Protein phosphatase 2A (PP2A) is implicated in cell cycle regulation among its many functions, in part because of the plethora of cell cycle defects displayed by mutants of this enzyme in yeast. Several distinct regulatory subunits help this very abundant protein target a wide range of substrates. For example, PP2A is targeted to centromeres to protect centromeric cohesin until anaphase. However, the abundance of regulatory subunits and the widespread effects of PP2A on cellular processes limit progress in gaining a detailed understanding of its contributions to mitotic exit.

Similarly, it is challenging to tease apart specific mitotic roles for PP1, which has even more regulatory subunits than PP2A. Nevertheless, genetic and biochemical studies have demonstrated its requirement in mitotic regulation. PP1 activity is involved in reversing Aurora kinase-mediated phosphorylation of histone H3 and Aurora kinase activation. Indeed, genetic and physical interactions between PP1 and Aurora kinase in budding yeast have suggested that the balance in activities of the two enzymes is likely to regulate some mitotic processes. However, there are likely to be non-Aurora kinase-mediated phosphorylation events reversed by PP1 as cells proceed out of mitosis.

Coordination of sister chromatid separation with Cdk1 inactivation is a critical cell cycle timing issue. For example, Cdk1 inactivation prior to sister chromatid separation would lead to erroneous chromosome segregation events by permitting premature spindle disassembly and cytokinesis. Studies in budding yeast demonstrate that securin and cyclin B are targeted by different forms of the APC (APC[Cdc20] and APC[Cdh1], respectively) at different times, and suggest that stable securin inhibits cyclin proteolysis. Thus, the destruction of securin by APC[Cdc20] in this organism links sister chromatid separation and the relief of inhibition on cyclin proteolysis.

In addition, since the bulk of the cyclin B is stable at the metaphase-anaphase transition, Cdk1 is active and Cdh1 is maintained in a phosphorylated state that prevents it from interacting with the APC. Persistent phosphorylation of Cdh1 until an initial decline in Cdk1 activity and Cdc14 activation ensures that the bulk of cyclin proteolysis—and hence Cdk1 inactivation—occurs only after securin degradation and chromosome separation. The temporal control and the substrate specificity of APC-dependent proteolysis in budding yeast are illustrated in **FIGURE 15.28**. Whether all of the details of this scheme apply in all organisms is unknown, but it is a good illustration of the complexity involved in regulating these key cell cycle events properly. It is known however that in animal cells, APC substrates are also degraded at different points during mitotic progression. Exactly how sequential ubiquitination of target proteins is accomplished is unclear.

As described earlier (see *15.10 Sister chromatids are held together until anaphase*), destruction of securin liberates the active form of separase. Separase cleaves cohesin, but at least in budding yeast, it has a second role in triggering the release of Cdc14 phosphatase from its anchoring compartment, the nucleolus, in early anaphase. The release of Cdc14 from the nucleolus leads to its activation in budding yeast. Once out of the nucleolus, Cdc14 is then maintained in its active and broadly distributed localization by a signaling cascade called the *mitotic exit network* (MEN).

An analogous pathway operates in fission yeast to control the timing of cytokinesis and septum formation and is called the *septation initiation network* (SIN). The MEN and SIN are protein kinase signaling cascades regulated by the action of a small GTPase. Assembled at the spindle poles, the overall goal of the MEN and SIN pathways is to coordinate cytokinesis with chromosome segregation. Given the need for this coordination, it seems likely that a similar pathway will be found to operate in multicellular eukaryotes. Indeed, homologs of several of the pathway's components are known in plants and animal cells. The SIN and the MEN pathways are diagrammed in **FIGURE 15.29**. One of the important unanswered questions

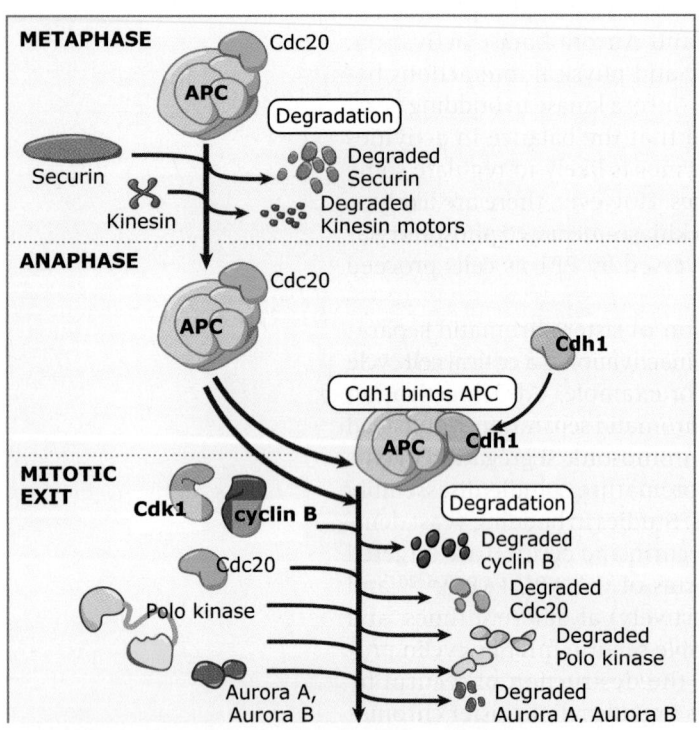

FIGURE 15.28 The APC^{Cdc20} mediates the proteolysis of securin, a subset of kinesin-related motor proteins, and a small fraction of cyclin B. During most of the cell cycle, including early mitosis, a second APC activator, Cdh1, is phosphorylated by Cdk1 and is unable to bind the APC. A drop in Cdk1 activity due to the initiation of cyclin B proteolysis by APC^{Cdc20} at anaphase results in Cdh1 dephosphorylation. Dephosphorylated Cdh1 can then bind to the APC, and the APC^{Cdh1} mediates cyclin proteolysis and Cdk1 inactivation.

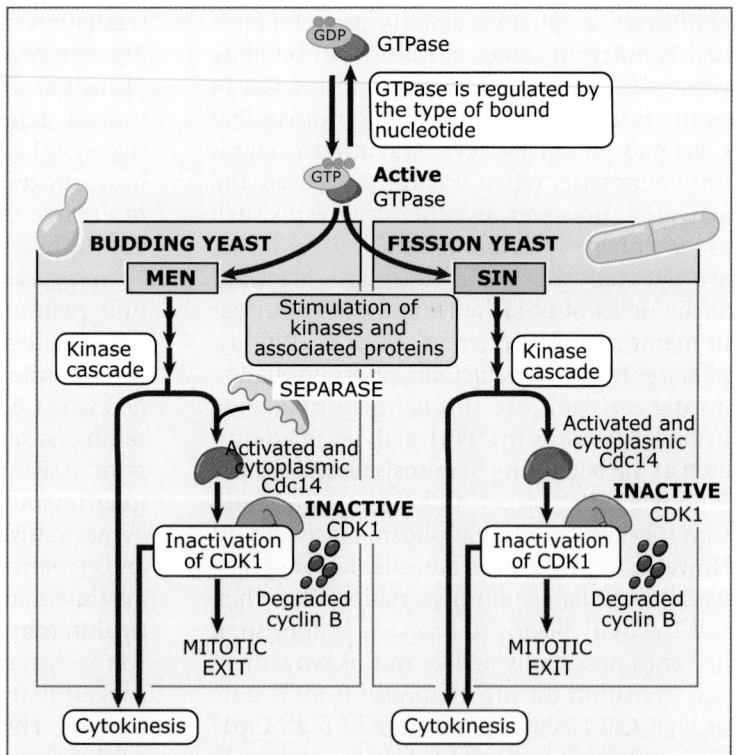

FIGURE 15.29 The MEN (budding yeast) and SIN (fission yeast) pathways are analogous signaling cascades regulated by a GTPase that switches between active (GTP-bound) and inactive (GDP-bound) forms. The active form of the GTPase stimulates downstream protein kinases (three in the MEN and four in the SIN). MEN and SIN signaling promotes activation of the Cdc14 family of phosphatase and cytokinesis. It is not clear if the phosphatase plays a role in cytokinesis in both yeasts.

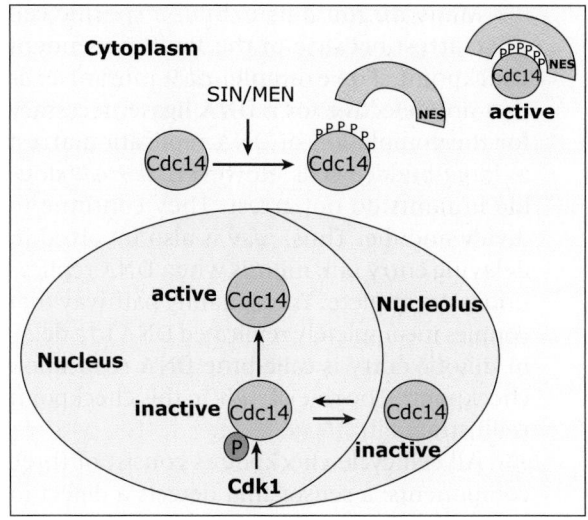

FIGURE 15.30 Cdc14 phosphatases are sequestered at the nucleolus and centrosomes/spindle pole bodies during interphase. During mitosis, they are released from these locations and regulated by several protein kinases. The figure combines regulatory pathways from different organisms although it is not yet clear whether all pathways are utilized in all organisms. Cdk1 inhibits Cdc14, Plk1 is thought to activate Cdc14, and in yeast, the SIN and MEN prevent their return to the nucleolus during anaphase.

is what are the specific targets of the MEN and SIN pathways that serve to trigger cytokinesis? Interestingly, both pathways target the Cdc14 phosphatase to promote its cytoplasmic localization during cytokinesis. This is important in yeast because the nuclear envelope does not break down during mitosis and Cdc14 phosphatases must be cytoplasmic to promote cytokinesis. The dynamic localization and activation cycle of Cdc14 enzymes is diagrammed in FIGURE 15.30.

In summary, it is critical for cells to inactivate Cdk1 and reverse phosphorylation events in a stepwise manner so that sister chromatid separation, spindle disassembly, and cytokinesis occur in a coordinated manner.

15.12 Checkpoint controls coordinate different cell cycle events

Key concepts

- Cell cycle events are coordinated with one another.
- The coordination of cell cycle events is achieved by the action of specific biochemical pathways called checkpoints that delay cell cycle progression if a previous cell cycle event is not completed.
- Checkpoints may be essential only when cells are stressed or damaged but may also act during a normal cell cycle to ensure proper coordination of events.

During each mitotic division cycle, cells duplicate and segregate their DNA, and then the cell divides. How is this order of cell cycle events maintained? Are there times when these rules are broken, and with what consequence?

As mentioned earlier, specific surveillance mechanisms, or checkpoints, operate to entrain cell cycle events with one another. They do so by affecting the activity of major cell cycle regulatory factors such as CDKs and the APC.

The simple but elegant cell fusion experiments of Rao and Johnson (see *15.2 Several experimental systems are used for cell cycle research*) pointed not only to dominant regulators of cell cycle progression, as discussed earlier, but also to the existence of checkpoint controls. It was found that when a G2 cell was fused with an S-phase cell, the G2-phase nucleus "waited" for the S-phase nucleus to finish DNA replication before undergoing nuclear envelope breakdown and entering mitosis. This suggested that a mechanism exists to prevent mitosis until DNA replication is complete.

The concept that a checkpoint pathway imposes the order of cell cycle events and delays cell cycle progression when a problem arises was fully articulated with the characterization of the *rad9* mutant in budding yeast. Several budding yeast mutants, termed "*rad*" mutants, were isolated based on enhanced sensitivity to ionizing radiation, which causes DNA damage. It was expected that mutants sensitive to ionizing radiation would have defects in DNA repair. However, careful microscopic examination of *rad* mutants revealed differences in their terminal arrest phenotypes. Most *rad* mutants never divided following DNA damage. In contrast, *rad9* mutant cells divided a few times before dying, forming microcolonies.

The behavior of the majority of *rad* mutants indicated that damage to the DNA signaled a cell cycle delay that, if the damage were not too great, would provide time for repair of the lesion. Because the majority of *rad* mutants had defective DNA repair enzymes, however, damaged DNA could not be repaired, and, therefore, these mutants became terminally cell cycle arrested. The continued proliferation of *rad9* mutant cells under the same circumstances suggested that, in this mutant, the damage was not sensed by the cell cycle machinery. Thus, the presence of a Rad9-dependent checkpoint that monitors the integrity of DNA was established; this checkpoint detects DNA damage and signals a halt to cell cycle progression. The checkpoint is not required in a normal yeast cell cycle and becomes essential only when DNA integrity is compromised.

Many *cdc* mutants exhibit a specific cell cycle arrest because of the Rad9-dependent checkpoint. For example, *cdc9* mutant cells that are defective for a DNA ligase necessary for the completion of DNA replication arrest as large-budded cells. However, *cdc9rad9* double mutants do not arrest. They continue to divide and die. Thus, *rad9* is also involved in delaying entry into mitosis when DNA replication is incomplete. The signaling pathway that couples incompletely replicated DNA to a delay in mitotic entry is called the DNA replication checkpoint. The role of *rad9* in this checkpoint is illustrated in **FIGURE 15.31**.

All cell cycle checkpoints consist of three components: a sensor that detects a defect in a cell cycle process, a signaling module that relays information upon detection of an error, and a target of the signal that is part of the cell cycle engine. Inhibition of the target halts cell cycle progression. The basic framework of a checkpoint that responds to damaged DNA or incomplete DNA replication is diagrammed in **FIGURE 15.32**.

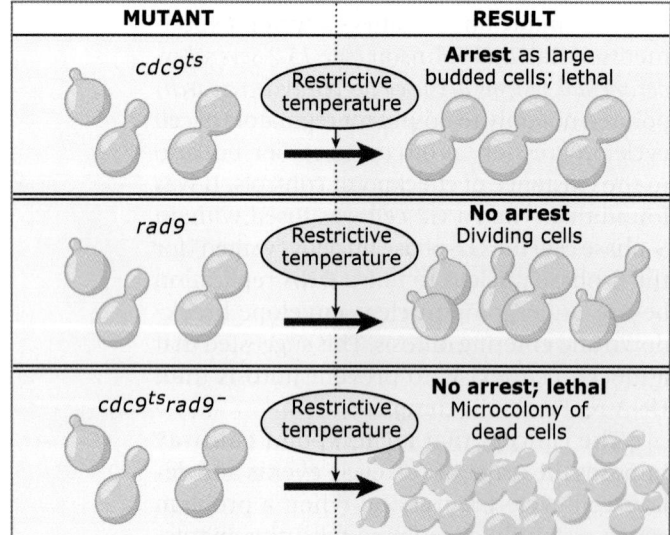

FIGURE 15.31 Inactivation of Cdc9p DNA ligase function, accomplished by shifting a *cdc9ts* mutant to the restrictive temperature, results in incomplete DNA replication. The mutant cells arrest uniformly as large-budded cells. Cells lacking Rad9p function display no discernable phenotype when shifted to the same temperature. However, inactivation of *cdc9* in combination with *rad9* results in a different lethal phenotype. Cells fail to cell cycle arrest and proceed to form microcolonies of dead cells. This establishes that the cell cycle arrest of a *cdc9ts* mutant is due to a checkpoint that requires Rad9.

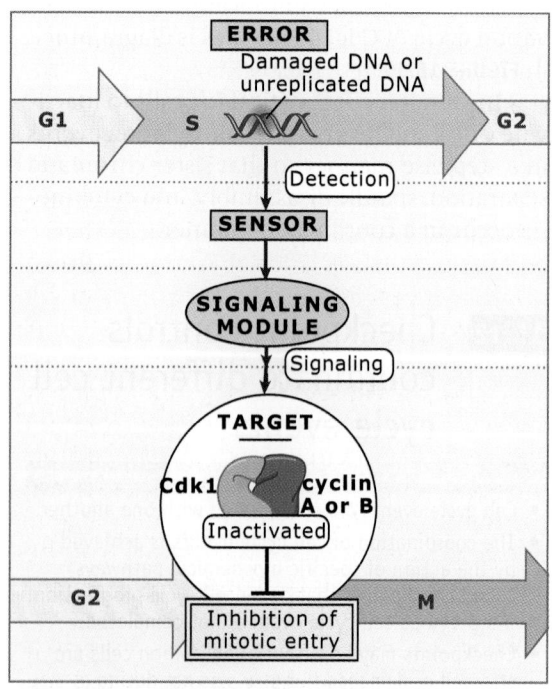

FIGURE 15.32 Upon detection of an error (such as damaged DNA), cells activate a checkpoint. The checkpoint is a signaling pathway that inhibits cell cycle progression. In this example, the checkpoint pathway acts to prevent mitotic entry by inhibiting Cdk1 activation.

15.13 DNA replication and DNA damage checkpoints monitor defects in DNA metabolism

Key concepts

- Incomplete and/or defective DNA replication activates a cell cycle checkpoint.
- Damaged DNA activates a different checkpoint that shares some components with the replication checkpoint.
- The DNA damage checkpoint halts the cell cycle at different stages depending on the stage during which the damage occurred.

CLASS OF PROTEIN	YEAST		MAMMALS
	S. cerevisiae	*S. pombe*	
Sensors	Rad24 Ddc1 Rad17 Mec3 Tel1 Mec1	Rad17 Rad9 Rad1 Hus1 Tel1 Rad3	Rad17 Rad9 Rad1 Hus1 ATM ATR
Mediators/Transducers	Rad9 Mrc1 Chk1 Rad53	Crb2 Mrc1 Chk1 Cds1	BRCA1 CLASPIN Chk1 Chk2
Targets	Pds1/APC	Cdc25	Cdc25 family

FIGURE 15.33 A number of evolutionarily conserved proteins have been found to play a role in each stage of the checkpoint response.

In *15.12 Checkpoint controls coordinate different cell cycle events*, we discussed checkpoints in general terms. Now we will discuss the individual checkpoints that protect the cell. Damage to DNA is clearly a significant threat to the cell and the status of the genome is monitored continuously. There is also a specific checkpoint that operates at S phase to monitor the DNA replication process. A large number of proteins participate in the cell's response to agents that damage DNA or slow DNA replication, and many of them are listed in **FIGURE 15.33**.

At present, the molecular nature of damage sensing is the least understood aspect of the DNA damage and DNA replication checkpoints. It is clear, however, that alterations of DNA structure—such as regions of single-stranded DNA (ssDNA) or double-stranded breaks (DSBs)—are recognized by different sensing molecules to activate the DNA damage checkpoint. Formation of these aberrant structures results in the recruitment and activation of the conserved related protein kinases ATR (with its adaptor ATRIP) and ATM.

In higher eukaryotes, a region of ssDNA elicits an ATR-dependent response, whereas DSBs elicit a response that is ATM-dependent; in contrast, in the yeasts, both responses are predominantly ATR-mediated. Studies conducted in budding yeast have shown that the presence of a single DSB and its subsequent processing into overhanging ssDNA or the generation of ssDNA in *cdc13* mutants, leads to a checkpoint response that is dependent on the Mec1 kinase, an ATR homolog. On the other hand, γ-irradiation of mammalian cells, which generates DSBs, results in an initial increase in ATM kinase activity, followed by an increase in ATR activity, presumably due to the processing of DSBs to ssDNA. In all cases, the ATM and ATR kinases are recruited directly to the sites of damage to initiate the response. Clearly, these kinases play pivotal roles in sensing damage and transducing the checkpoint response, and there is significant crosstalk between the pathways stimulated by the two. Even in the case of stalled replication forks, it is likely that ssDNA in these regions is sensed by the DNA replication checkpoint resulting in the activation of the ATM/ATR family of kinases.

In addition to the ATM/ATR family, the evolutionarily conserved Rad17-RFC and 9-1-1 complexes independently sense damaged DNA. The Rad17 family of proteins interacts with components of Replication Factor C to create a checkpoint-specific complex with DNA binding activity. The Rad17-RFC complex then loads the 9-1-1 complex (composed of three conserved proteins called Rad9, Rad1, and Hus1 in humans) onto DNA at sites of damage. The 9-1-1 complex forms a ring-like structure similar to the PCNA clamp that is involved in DNA replication. Despite recognizing and binding to regions of damaged DNA independently, all of these proteins are required for DNA damage checkpoints. How they communicate molecularly at the site of DNA damage is still the subject of intense investigation.

Checkpoint mediators are a second group of proteins, important for aspects of DNA damage recognition, signal transduction, and DNA repair that are generally considered to act after damage is sensed. They are also listed in Figure 15.33. In this group, the line is blurred between functions in the checkpoint and in DNA repair because several are important for both activities. Mediators may also be involved in a sustained response to DNA damage by nucleating stable multiprotein signaling complexes. These proteins link the ATM and ATR signaling to downstream effectors. These include repair proteins, chromatin remodelers, and of particular interest, additional kinases that transduce the checkpoint signal to affect cell cycle progression. The major transducers in the DNA damage and DNA replication checkpoints are the Chk1 and Chk2 kinases. These directly phosphorylate components of the cell cycle engine to block their activities.

Identical defects in DNA structure can induce delays or arrests at different cell cycle stages depending on the stage at which they are detected. For example, if damaged DNA is detected in G1, cell cycle progression is halted at the G1/S transition. However, if the same lesion is detected during G2, mitotic entry is inhibited. In fact, although they have overlapping sets of components, there are three distinct checkpoints that delay cell cycle progression in response to DNA damage: the G1/S checkpoint, the S-phase checkpoint, and the G2/M checkpoint. These are conceptually similar, resulting in arrest of cell cycle progression (including but not limited to regulation of CDK activity) and initiation of repair.

The G1/S checkpoint, which is studied extensively in mammalian cells, delays entry into S phase by inhibiting the activities of G1 CDK-cyclin complexes. Inhibition can even occur after cells pass the restriction point (for details on G1 CDK, see *15.6 Cells may withdraw from the cell cycle*). G1 arrest is brought about in at least two ways. One is to inhibit the activation of G1 CDK-cyclin complexes. Phosphorylation and activation of the checkpoint effector kinases Chk1 and/or Chk2 result in phosphorylation of Cdc25A, resulting either in its nuclear exclusion or its degradation by ubiquitin-mediated proteolysis. The end result is the maintenance of the Cdk2-cyclin E complex in its inactive tyrosine-phosphorylated state.

A second way of delaying the G1-S transition in response to DNA damage is upregulation of CKI expression. Following checkpoint activation, ATM/ATR is activated to phosphorylate p53. p53 is a transcription factor and an important tumor suppressor (for details on p53, see *15.15 Cell cycle deregulation can lead to cancer*). ATM/ATR-mediated phosphorylation of p53 prevents its turnover, permitting p53 to accumulate in the nucleus. This leads to increased transcription of p53 target genes, including a G1 CDK-cyclin inhibitor called p21. Therefore, two separate pathways contribute to inhibition of the CDK-cyclin complexes necessary for DNA replication. The G1/S checkpoint response to damaged DNA is diagrammed in **FIGURE 15.34**.

The S-phase checkpoint functions primarily to delay the firing of late origins when damaged DNA is detected and to prevent replication

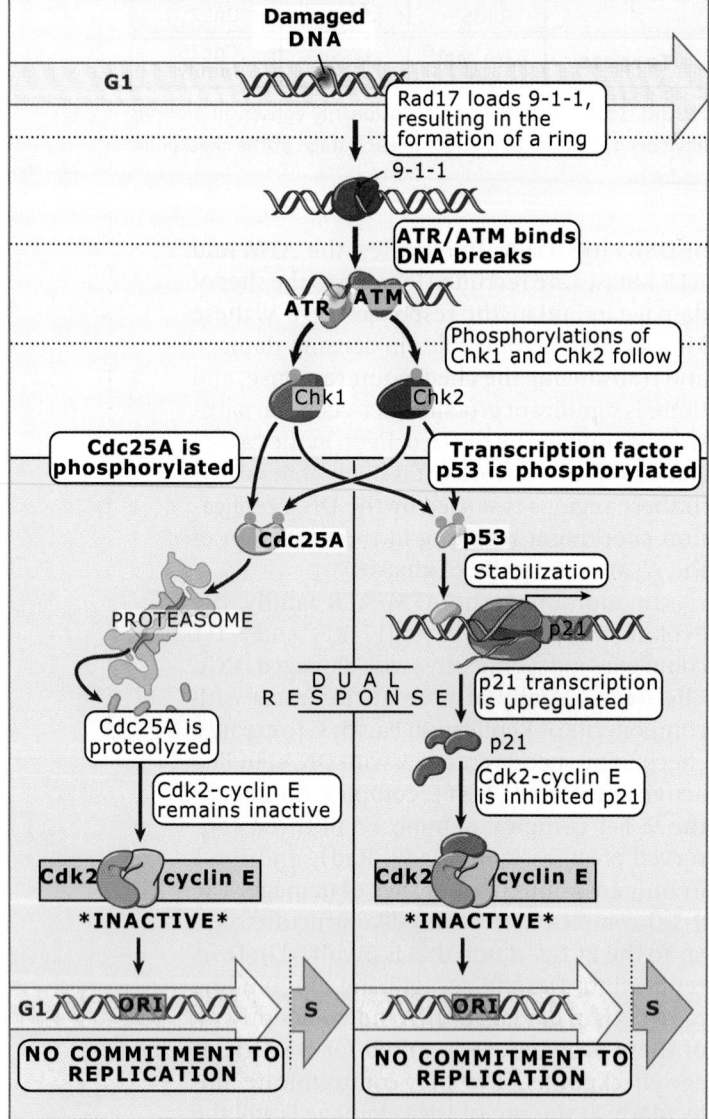

FIGURE 15.34 Upon detection of damaged DNA in G1, cells activate an ATM- and ATR-dependent checkpoint pathway that has at least two effects. In one response, Cdc25A phosphatase is phosphorylated. This leads to its proteolysis and Cdk2-cyclin E is therefore maintained in an inactive Tyr-phosphorylated state. In a parallel response, the p53 transcription factor is phosphorylated, and thereby stabilized. p53 stabilization leads to upregulation of p21 transcription. This results in the accumulation of the p21 CKI, which then binds to and inhibits Cdk2-cyclin E. These dual responses contribute to inhibition of the G1/S transition.

fork disassembly. As in the G1 checkpoint, the ATM/ATR kinases and the effector kinases Chk1 and Chk2 are key participants that mediate inhibition of Cdc25A and CDK-cyclin complexes. In addition, however, active ATM/ATR kinases phosphorylate replication proteins to halt the initiation of DNA replication. For example, a protein complex containing ATR and ATR Interacting Protein (ATR-ATRIP) inhibits the DDK kinase and therefore prevents loading of the essential replication protein, Cdc45. In this manner, S phase initiation is repressed. Another important target of the checkpoint kinases is a conserved exonuclease complex that processes double-strand breaks termed the MRN (Mre11/Rad50/Nbs1).

In the absence of a functional S-phase checkpoint, DNA synthesis continues in the presence of DNA damage. This process is called radio-resistant DNA synthesis (RDS). The extent of RDS is increased if both the cell cycle branch (ATM/ATR-Chk1/Chk2-Cdc25A) and the branch that halts replication are nonfunctional, suggesting that the pathways triggered by DNA damage in S phase act in parallel. S-phase checkpoint responses are illustrated in FIGURE 15.35.

An additional S-phase associated checkpoint mechanism, called the replication checkpoint, responds specifically to agents that arrest replication fork progression even if they do not specifically cause damage. Studies in multiple organisms have demonstrated that stalled replication forks and not incompletely replicated DNA activate the replication checkpoint. This may occur by depletion of the dNTP pool by drugs such as hydroxyurea, replication stalling lesions or other mechanisms that stall replication after it has begun. However, the checkpoint response cannot be activated by these treatments in cells that fail to assemble a replication fork. This specialized response reflects the critical role for the replication fork in maintaining genome stability. If forks are disrupted or fall apart, there is no way to reassemble them because loading of the MCM helicase is strictly dependent upon the licensed replication origin, and affected by the "once and only once" rule described in *15.8 DNA replication requires the ordered assembly of protein complexes*. Moreover, collapse of the replication fork renders DNA susceptible to DNA DSBs; thus the S phase damage checkpoint becomes a backup for cell cycle arrest if the replication checkpoint is missing.

Once triggered, there are four outcomes of the DNA replication checkpoint:

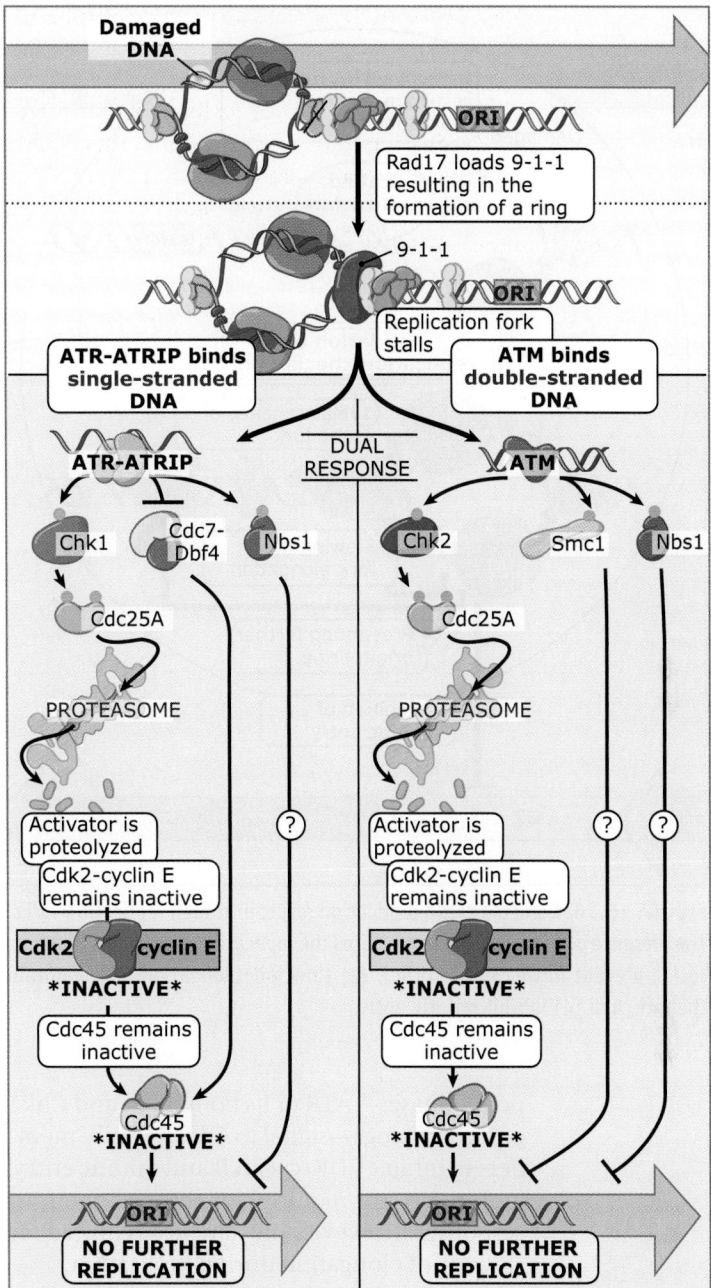

FIGURE 15.35 Incomplete or erroneous DNA replication triggers an inhibitory signal, the S-phase checkpoint. This signal prevents entry into mitosis until the errors are corrected.

1. prevention of further origin firing,
2. slowing of replication elongation,
3. maintenance of stalled replication forks, and
4. inhibition of mitotic entry.

These functions of the replication checkpoint are summarized in FIGURE 15.36.

Many of the components are shared between the replication and the DNA damage checkpoints. For example, the central check-

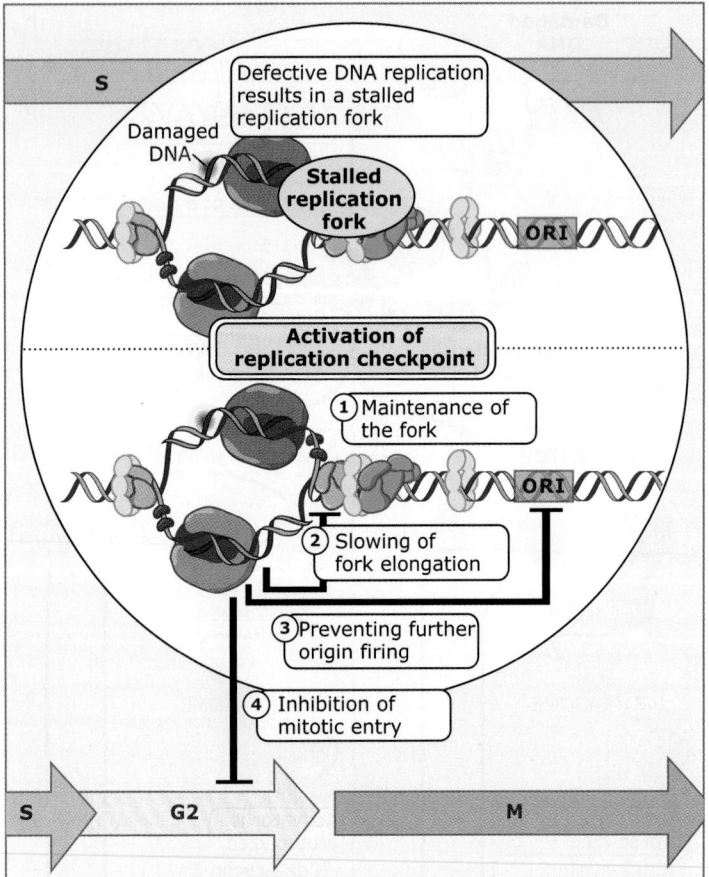

FIGURE 15.36 Defects in DNA replication result in stalled replication forks. The presence of these structures activates the replication checkpoint that acts to (1) prevent further origin firing, (2) slow fork elongation, (3) maintain the fork, and (4) inhibit mitotic entry.

point kinases—ATR or its homologs, and Chk1 or its homologs—signal to Cdc25 family members to inhibit CDK-cyclin B and mitotic entry, and in fact are required for three of the four responses. However, they are not required for slowing of elongation during replication.

A specialized trio of proteins first identified in yeast is critical for replication fork stabilization. The first protein, Mrc1 (human Claspin) contributes to replication fork progression, possibly by linking leading and lagging strand synthesis. It has an independent role in activating the replication checkpoint kinases in response to replication fork stalling; this checkpoint function may be distinguished from its elongation function by genetic separation-of-function mutations. Two additional proteins, Tof1 and Csm3 (human Timeless and Tipin), work together in a "fork protection complex." These are not essential for viability but in their absence, replication forks become fragile under conditions where they pause or stall. The fork

protection complex proteins associate with the MCM helicase and may provide a physical link to the polymerases that synthesize the DNA. They ensure the fork is robust and resistant to collapse.

The G2-M checkpoint is analogous to that of the G1 and S-phase checkpoints (refer to Figure 15.34 and Figure 15.35) in that the activation of the ATM/ATR-Chk1/Chk2 pathways leads to inhibition of a CDK-cyclin complex, in this case CDK1-cyclin B. Again, Chk1 and Chk2 kinases target a Cdc25 family member. In the case of the G2-M checkpoint, both Cdc25C and Cdc25A are inhibited by sequestration and/or degradation.

Live cell imaging methods and green fluorescent protein fusions have provided additional insight into when and for how long each component acts in checkpoint pathways. A large number of checkpoint proteins accumulate in foci at sites of damaged DNA or stalled replication forks. With the presumption that these foci correspond to sites of checkpoint signaling, researchers are examining the order in which checkpoint components arrive, the length of their stay, their relative turnover, and the dependence of these parameters on other checkpoint proteins. The visualization of these foci is also a convenient experimental approach to verify that DNA damage is induced in cells and to gauge its severity.

Phosphorylation of the C-terminal tail of Histone H2AX by ATM/ATR to form γ-H2AX appears to provide the cue for the development of DNA damage foci. Though recruitment of some checkpoint proteins is independent of γ-H2AX, the sustained presence of these checkpoint proteins in foci and the checkpoint response depends on it. Hence, γ-H2AX might organize a signaling center of checkpoint and repair proteins to effect and monitor DNA repair.

As might be imagined, the DNA damage checkpoints are critical for maintaining genome stability. Malfunction of checkpoint proteins can lead to the accumulation of mutations and ultimately cancer. Indeed, a genetic disorder in humans named ataxia-telangiectasia (A-T) is associated with cerebellar degeneration, immunodeficiency, radio sensitivity, genome instability, and cancer predisposition. Cloning of the gene mutated in A-T patients revealed it to be ATM and established a role for this protein in the maintenance of genome stability. ATR is not a tumor suppressor, but rather an essential protein that is required for every round of DNA replication. However, partial inactivation of ATR function is associated with the rare human autosomal recessive disorder, Seckel syndrome that is marked by dwarfism and microencephaly.

15.14 ## The spindle assembly checkpoint monitors defects in chromosome-microtubule attachment

Key concepts

- The mitotic spindle attaches to individual kinetochores of chromosomes during mitosis.
- Proper attachment of microtubules to kinetochores is a prerequisite for accurate chromosome segregation.
- Defects in spindle-MT attachment are sensed by the "spindle assembly checkpoint," which subsequently halts the metaphase-anaphase transition to prevent errors in sister chromatid separation.

During the early stages of mitosis, spindle microtubules (MTs) attach to kinetochores, which are protein complexes assembled on the centromeric DNA of chromosomes. For chromosomes to be equally segregated to daughter cells, it is imperative that the kinetochores of sister chromatids ultimately be attached to MTs emanating from opposite poles of the spindle before anaphase begins. Such a pattern of attachment is called amphitelic or bipolar attachment.

However, other patterns of kinetochore-MT attachment develop during metaphase since attachment of MTs to kinetochores is a random process. The different kinetochore-MT attachment possibilities are defined and illustrated in **FIGURE 15.37**.

Therefore, cells must have methods to distinguish between correct (bipolar) and incorrect (monopolar) attachments and the ability to correct improper ones when they develop. Indeed, several proteins participate in converting incorrect patterns of kinetochore-MT attachment to an amphitelic pattern during metaphase. One key protein for destabilizing incorrect attachments is the protein kinase, Aurora B. As discussed earlier, Aurora B is a component of the chromosomal passenger complex and localizes to the centromere during mitosis (for details, see *15.9 Mitosis is orchestrated by several protein kinases*). How Aurora B kinase functions to promote bipolar spindle attachment is presently the subject of intense research.

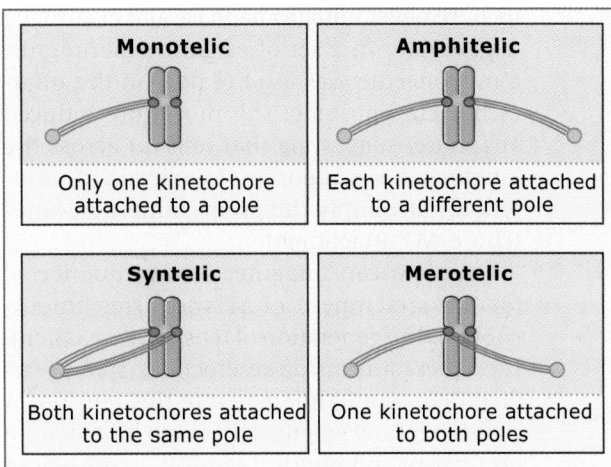

FIGURE 15.37 In a monotelic configuration, only one of the kinetochores (shown in purple) is attached to spindle fibers that are anchored at one pole. In an amphitelic configuration, the kinetochores of sister chromatids are bound to spindle microtubules emanating from opposite poles, resulting in proper bipolar orientation of the chromosomes. In a syntelic configuration, both kinetochores of sister chromatids attach to spindle microtubules emanating from the same pole. In a merotelic configuration, one kinetochore is attached to spindle microtubules emanating from both poles. All the incorrect configurations are detected by the spindle assembly checkpoint, and as a result, mitotic progression is halted.

What happens during prometaphase when some chromosomes are not properly attached to spindles? In these cases, a "wait-anaphase" signal is generated by the "spindle assembly checkpoint" (SAC). The purpose of this checkpoint is to delay sister chromatid separation, that is, the onset of anaphase, until each and every kinetochore is properly attached to spindle microtubules. If a "wait-anaphase" signal were not generated, sister chromatids could separate prior to reaching bipolar attachment and this would result in unequal chromosome segregation. The SAC prevents anaphase onset by inhibiting the APC and, therefore, the cleavage of securin (for details on APC, see *15.10 Sister chromatids are held together until anaphase*).

How is the SAC activated? Kinetochores that are unattached, or improperly attached, to spindle MTs initiate SAC activation. Also, some studies show that a lack of physical tension across each kinetochore triggers the SAC. Before a molecular understanding of the SAC was possible, the effects of SAC activation were observed in grasshopper spermatocytes. These cells have three sex chromosomes, and in some cells, one of these three chromosomes remains unpaired and connected to only one spindle pole. The cells containing an unattached kinetochore never initiate anaphase and eventually degenerate. In a set of elegant experiments, a microneedle was used to pull on the unattached chromosome. This procedure-induced anaphase, suggesting that tension across the kinetochores is important to relieve SAC inhibition, not simply the establishment of kinetchore-MT attachment.

As one can imagine, a consequence of bipolar attachment of MTs to sister kinetochores is the generation of tension across them. Incorrect patterns of kinetochore-spindle attachment would not generate this tension. A model has now emerged that both kinetochore occupancy and physical tension across paired kinetochores are important parameters that are monitored by the SAC.

The SAC and its components were first identified in genetic screens designed to isolate mutants sensitive to microtubule-destabilizing drugs in budding yeast. Homologs of its six yeast components were later also found in many eukaryotes, including mammals. Interestingly, SAC components are dispensable for vegetative growth in yeast, but in the examples so far examined, their murine counterparts are essential for embryonic viability. Cell lines derived from mouse embryos lacking

certain SAC components display elevated levels of chromosome missegregation events. The necessity of the SAC in metazoan cells is likely due in part to the larger number of MTs bound per kinetochore and the longer times required to establish correct kinetochore-MT attachments, during which the APC must be kept inactive. The role of the SAC is illustrated in **FIGURE 15.38**.

How does the SAC work to delay anaphase? Although much remains to be learned about the sensing and signaling branches of the SAC, it is clear that the signal for SAC activation emanates from kinetochores. All SAC components are detected at kinetochores either during each normal cell cycle, or upon checkpoint activation, suggesting that this is their primary site of action. Moreover, kinetochore targeting of SAC components is essential for the establishment and maintenance of the checkpoint signal. Certain defects in kinetochore structure that result from mutation of kinetochore proteins prevent SAC components from assembling on the kinetochore. These mutants have an inactive checkpoint. Other structural defects in the kinetochore that compromise MT attachment have the opposite effect; they induce SAC activation.

We know that SAC signaling involves a phosphorylation cascade because some SAC components are protein kinases and several SAC components are phosphorylated. However, the SAC does not signal through a strictly linear pathway and the role of each component is not fully understood. Different components are important for monitoring different aspects of correct bipolar attachment and mice heterozygous for the deletion of different SAC components display different propensities to develop tumors. With multiple targets of some protein kinase components known, it is also possible that some components function outside the SAC. Mad2 is perhaps the best understood of the SAC components. It is recruited to kinetochores by another SAC component, Mad1. Binding to Mad1 at kinetochores induces a conformation change in one subunit of a Mad2 dimer. This allows that subunit to bind Cdc20. By recruiting dimers and releasing monomers to bind Cdc20, Mad1 amplifies and broadcasts the SAC. However, overproduction of Mad1 inactivates the SAC by sequestering all available Mad2 and preventing its interaction with Cdc20.

Cdc20 is the downstream target of SAC activity. As introduced previously, Cdc20 is one of the substrate-specific activators of the APC

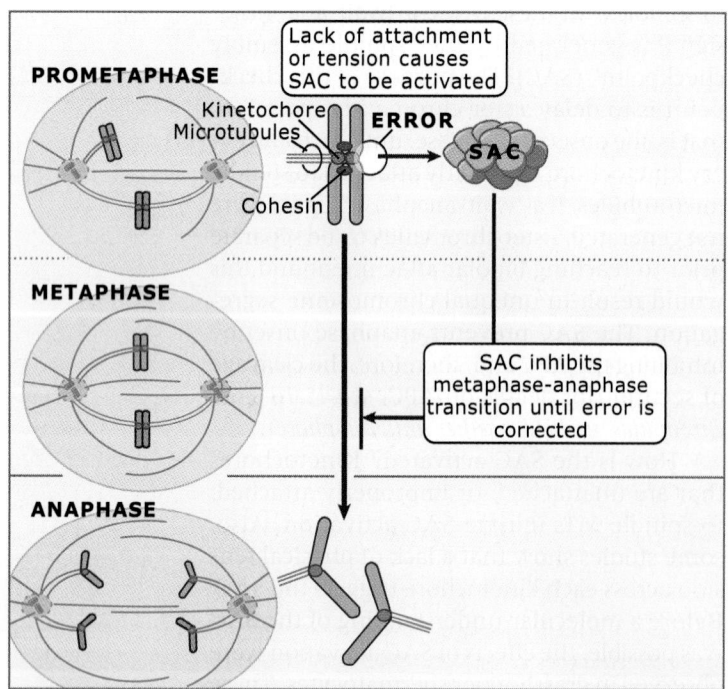

FIGURE 15.38 Lack of microtubule attachment to one kinetochore or proper tension across sister kinetochores activates the spindle assembly checkpoint (SAC). The SAC inhibits the metaphase-anaphase transition until the error is corrected.

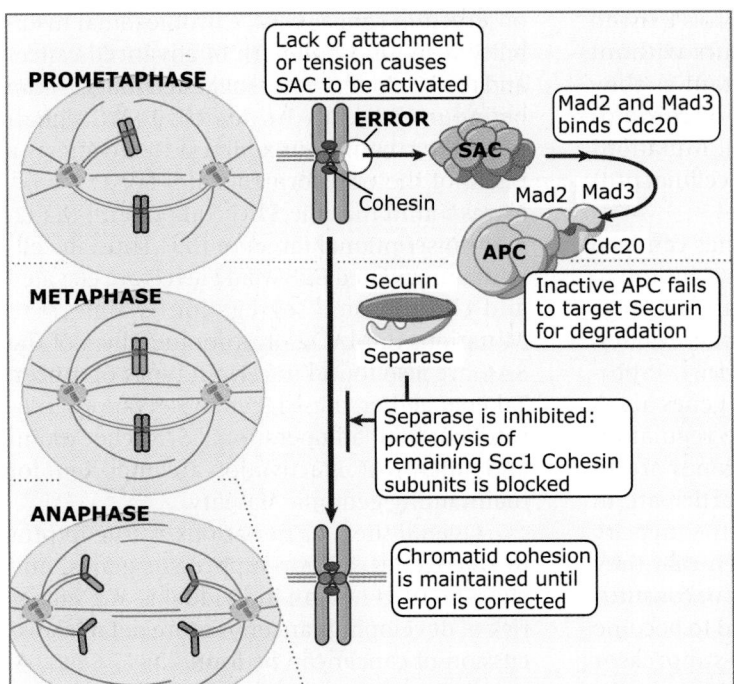

PROMETAPHASE

Lack of attachment
or tension causes
SAC to be activated

ERROR

SAC

Cohesin

METAPHASE

Securin

Separase

ANAPHASE

Mad2 and Mad3
binds Cdc20

Mad2   Mad3

APC   Cdc20

Inactive APC fails
to target Securin
for degradation

Separase is inhibited:
proteolysis of
remaining Scc1 Cohesin
subunits is blocked

Chromatid cohesion
is maintained until
error is corrected

FIGURE 15.39 When the spindle assembly checkpoint (SAC) is activated, binding of Mad2, one SAC component, or a complex of SAC components (MCC) to Cdc20, an activator of the APC, inhibits APC^{Cdc20} activity. APC^{Cdc20} is required to target securin for proteolysis and to allow sister chromatid separation.

and Mad2 was the first SAC component shown to interact with it. Later, three checkpoint proteins—Mad2, BubR1, and Bub3—were identified in a mitotic checkpoint complex (MCC) that helps form the SAC. Although various models are proposed to explain how Mad2 alone, or with the MCC, signals to delay anaphase, it is not yet settled how this occurs or whether it can happen in multiple ways. One suggests Cdc20 is lost from the core APC while another proposes that the MCC binds the core APC preventing substrate access, and in still another model, the MCC stimulates Cdc20 ubiquitination and degradation. While the mechanism might not be resolved, the result of SAC activity is clear. Upon checkpoint activation, APC^{Cdc20} is inactivated, resulting in the stabilization of securin and the persistence of sister chromatid cohesion. A model of SAC inhibition of APC activity is summarized in FIGURE 15.39.

The primary function of the SAC is to prevent chromosome segregation in the absence of proper bipolar attachment of sister kinetochores. As one might predict, in the absence of proper SAC function, the resultant defects in kinetochore attachment could lead to erroneous chromosome segregation and cell death or to **aneuploidy**, in which daughter cells receive either less or more than an equal share of genetic material. Indeed, many forms of cancer are typified by aneuploidy, and mutations in SAC components are associated with some forms of colorectal cancer. Strong evidence that

the SAC is involved in preventing chromosome damage in mammals comes from experiments in mice. Mice lacking one copy of either Mad1 or Mad2 are tumor-prone and cells from these animals show high levels of chromosome missegregation events. Mice lacking one copy of other SAC components show genetic interactions in various cancer models suggesting that they too play a protective role against cancer formation.

In summary, the spindle assembly checkpoint ensures that chromosome segregation occurs only after all the chromosomes are properly attached to the mitotic spindle. Since chromosome segregation is triggered by APC activation, the checkpoint pathway prevents APC activation until proper chromosome attachment is achieved. The checkpoint signal is initiated at the kinetochore and leads to the direct binding of some checkpoint components to the APC to inhibit its function.

15.15 Cell cycle deregulation can lead to cancer

Key concepts

- Proto-oncogenes encode proteins that drive cells into the cell cycle.
- Tumor-suppressor genes encode proteins that restrain cell cycle events.
- Mutations in proto-oncogenes, tumor suppressor genes, or checkpoint genes may lead to cancer.

As highlighted earlier, many controls exist to make sure that cell division occurs without errors. These controls are important as they ensure the integrity and fidelity of chromosome duplication and segregation. Mutations in genes that control the fidelity of cellular proliferation can result in cancer.

Mutations in two types of genes can lead to unrestrained cell proliferation. These two types are **proto-oncogenes** and **tumor-suppressor** genes (for details on these genes, see *17 Cancer—Principles and overview*). Proto-oncogenes and tumor-suppressor genes act in opposition to one another in the regulation of cell proliferation. Proto-oncogenes are so named because they normally participate in promoting proliferation and have the capacity to be upregulated or mutated such that they signal proliferation inappropriately or constitutively. Indeed, they can be mutated to become oncogenes. In contrast, tumor-suppressor genes normally act to restrain cellular proliferation and ensure genome stability. Hence, a mutation(s) that abrogates the function of both copies of a tumor-suppressor gene can lead to loss of growth control and be involved in tumor formation. Environmental factors play a significant role in cancer formation. Exposure of human cells to mutagens/carcinogens such as UV light and cigarette smoke can cause mutations in a proto-oncogene or a tumor-suppressor gene.

p53 is a key tumor-suppressor gene that normally functions to restrain cell cycle progression. It is also a checkpoint gene. The mechanism and regulation of p53 protein function is the subject of intense investigation because most human tumors contain inactivating mutations in p53. In the absence of p53 function, cells escape cell cycle regulatory control and can proliferate in the presence of DNA damage. This renders them vulnerable to the acquisition of additional mutations in subsequent cell divisions. The accumulation of several mutations in important growth controlling genes can lead to tumor formation.

Mutations in other checkpoint proteins may lead to cell cycle progression in the presence of DNA damage. One example is the inactivation of the checkpoint protein, Hus1, in either mouse or human cell lines, which sensitizes the cells to agents that damage DNA and allows progression through the cell cycle in the presence of damaged DNA.

Mutations of genes involved in DNA segregation also contribute to the genetic alterations observed in cancer cells. Chromosomal instability (CIN) is a hallmark of advanced cancer and recent evidence has suggested that CIN can be induced directly by defective SAC signaling rather than, or in addition to, arising as a result of the transformed state. For example, it was found that the SAC component, Mad2, is a transcriptional target of E2F. Thus, in cells with unregulated E2F, Mad2 levels are elevated and chromosome segregation is abnormal. Mutations in MAD2 or other members of the SAC are also found in certain types of cancer. Taken together, these recent observations indicate that both an operational SAC checkpoint and timely SAC inactivation are important for maintaining genomic stability.

Overall then, as mutations accrue in proto-oncogenes, tumor-suppressor genes, and checkpoint genes, an individual is at a higher risk of developing cancer. A more detailed discussion of cancer can be found in *17 Cancer—Principles and overview*.

Concept and Reasoning Checks

1. Explain why cell cycle checkpoints are important, identify and define their three components, and describe the consequences of their action.
2. What are the two major posttranslational mechanisms that drive cell cycle progression?

15.16 What's next?

With the completion of genome sequencing projects, approaches to understanding changes in gene and protein function during the cell cycle at a global level have begun. Microarray techniques have helped researchers identify genome-wide changes in the transcriptional program during the cell cycle, upon treatment of cells with drugs that inhibit cell cycle progression, and in proliferative disorders. The most comprehensive analyses of gene expression profiles have been done using *S. cerevisiae* and *S. pombe*, and many genes that are transcriptionally regulated during the cell cycle are now known. Similarly, microarray analyses have been used to identify genes that are transcriptionally up- or downregulated upon activation of a checkpoint. This strategy is also applied to other organisms to search for common regulatory paradigms involved in cell cycle regulation and checkpoint activation, and to define differences between normal and transformed animal cells. Microarray strategies are now used for cancer profiling and for defining therapeutic approaches to patient treatment.

Another approach used to define proteins involved in cell cycle regulation is global analyses of protein localization. In *S. cerevisiae* and *S. pombe* for example, the localization of nearly every protein encoded by the genome was determined by fusing each open reading frame to sequences encoding green fluorescent protein. On the horizon are large-scale proteomic analyses to determine the protein composition of all multimeric complexes that contribute to a specific cellular function. In combination with localization studies, this will provide a wealth of knowledge regarding undefined players in each cell cycle transition.

Just as genetic screens in yeast defined key players in cell cycle control, large-scale RNAi screens in mammalian cells have identified new proteins involved in the process. Proteomic approaches will be certain to shed additional light on the function of uncharacterized proteins.

Though unbiased screens will provide new insight, targeting particular pathways will also be important. Given the importance of checkpoints in coordinating cell cycle events, much effort will be directed at understanding their mechanism of action thoroughly and in molecular detail. The DNA-damage, S-phase, and spindle-assembly checkpoints require the function of a number of protein kinases. Proteomic screens have identified many targets of these kinases but it will take longer to parse out their relevance in each case and to explain how the signaling pathways transmit their information in molecular detail.

Similarly, despite the wealth of information on the regulation of mitotic kinases, much remains to be learned about their targets. A combination of genetic studies from yeast and biochemical analyses from frog and sea urchin extracts has shed important light onto mitotic Cdk1 function. However, the picture is still not complete. How does Cdk1 regulate so many essential mitotic processes such as chromosome condensation, spindle formation, and sister chromatid separation? Identifying and analyzing the mitotic substrates of Cdk1 will help us further understand the molecular nature of these processes. It is equally important to determine how exit from mitosis is linked to down-regulation of Cdk1 function. How is the decline in Cdk1 activity related to spindle breakdown and cytokinesis? Are these processes brought about by the reversal of Cdk1 phosphorylation events? If so, how does dephosphorylation regulate these processes?

Continued development of mass spectrometric techniques has made a significant impact on the ease of detecting protein modifications such as phosphorylation. This will be certain to expand to tackle yet other types of posttranslational modifications. Chemical genetics has emerged as a powerful technology to help identify substrates of protein kinases. This method uses specific inhibitors of a modified kinase to determine the role of the kinase in specific phosphorylation events. Such analog-sensitive alleles (mutant alleles of kinases that do not use ATP but an analog of it) have proved useful for selective inhibition of that specific kinase *in vivo* and for identifying *in vivo* targets of the kinase. These methodologies will doubtless help researchers tackle the following questions: What proteins are phosphorylated during specific stages of the cell cycle? Is the phosphorylation of a specific protein altered upon abolishing the function of a specific kinase? Answers to these questions will help put together a comprehensive picture of how protein kinases regulate essential processes during a cell cycle.

While protein kinases are critical regulators of cell cycle events, directed proteolysis drives the cell cycle forward irreversibly. The SCF and the APC are two well-established cell cycle-specific ubiquitin ligases critical for normal cell cycle progression. Defects in ubiquitin-dependent proteolysis are also associated with tumor progression. Therefore, building a comprehensive list of proteins targeted by ubiquitin-dependent proteolysis during a cell division cycle is critical for our understanding of cell cycle and tumor progression. Attempts to accomplish this goal have already begun. For example, using an elegant *in vitro* system, a subset of proteins subjected to mitosis-specific proteolysis was identified in frog extracts. Extending such analyses to generate a more comprehensive list, along with rigorous hunting for substrates of the SCF and the APC, should better define the role of ubiquitin-dependent proteolysis in cell cycle events. Moreover, a better understanding of cell cycle-specific proteolysis could provide additional targets for therapeutics.

A major obstacle in cancer therapy is the targeting of therapeutic agents that will selectively kill cancer cells. Continued research into the intricacies of cell cycle regulation should provide clues into how the cell cycle could be manipulated to prevent inappropriate cell divisions. The identification of proteins participating in various aspects of cell cycle progression will provide a list of targets for the development of anticancer therapeutics. Also, the extended list of genes involved in cell cycle regulation may help in genetic fingerprinting to identify predisposal to a specific type of cancer and the opportunity for pharmacologic intervention.

15.17 Summary

The cell cycle consists of an ordered set of events that lead to the duplication of a cell's contents and its division into two parts. The events of a cell cycle are regulated in both time and space. Surveillance mechanisms known as checkpoints ensure the order and fidelity of cell cycle processes. In the event of errors in the duplication process, checkpoints delay cell cycle progression so that corrections can be made.

Replication of the genetic information occurs during S phase and equal segregation of the duplicated information occurs during the mitotic phase. These two phases are separated by Gap phases, G1 and G2. This order is imposed by checkpoints and is set by the activity of the major cell cycle regulatory kinases, CDKs. Some events can occur only when Cdk1 activity is low (assembling the prereplication complex) and some only when it is high (mitosis). An environment permissible for DNA replication is established only when Cdk1 activity is low and mitotic processes are initiated once Cdk1 activity is high. Entry into the mitotic phase is accomplished upon full activation of Cdk1 and a number of other conserved protein kinases. These kinases together regulate the function of the chromosome segregation machinery, the mitotic spindle. After bipolar attachment of the sister chromatids to the mitotic spindle, sister chromatids are separated at the onset of anaphase and are further carried to the opposite poles of the cell due to elongation of the spindle. The separation of the sister chromatids during anaphase requires the function of a conserved ubiquitin ligase, the APC. Cdk1 activity starts to decline with anaphase, thus permitting the establishment of the ensuing S phase in the next cell cycle.

Extracellular stimuli dictate whether a normal cell proceeds into a division cycle or adopts a quiescent nondividing state. Such signals include the availability of nutrients, cell-to-cell contacts, and the presence of growth factors. The biochemical signals generated from extracellular stimuli can be either stimulatory or inhibitory in nature. In the majority of cases, these signals influence the G1 to S transition point that is the major control point in most cell cycles. The commitment to S phase, the G2 to mitosis transition point, and the onset of anaphase are all targets of checkpoint controls that delay cell cycle progression until certain criteria are met to ensure fidelity of cell cycle events. Defective checkpoint mechanisms allow errors in the cell duplication process to persist into the next generation and can lead to unregulated proliferation and the development of cancer. Two different types of mutations contribute to cancer formation: inactivating mutations in tumor-suppressor genes and activating mutations in proto-oncogenes. Single mutations seldom act alone to cause cancer but require the presence of other mutations that compromise the genetic stability of a particular cell and potentiate the risk of cancer formation.

http://biology.jbpub.com/lewin/cells

To explore these topics in more detail, visit this book's Interactive Student Study Guide.

References

Archambault, V., and Glover, D. M. 2009. Polo-like kinases: conservation and divergence in their functions and regulation. *Nat. Rev. Mol. Cell Biol.* v. 10 p. 265–275.

Berthet, C., and Kaldis, P. 2007. Cell-specific responses to loss of cyclin-dependent kinases. *Oncogene* v. 26 p. 4469–4477.

Besson, A., Dowdy, S. F., and Roberts, J. M. 2008. CDK inhibitors: cell cycle regulators and beyond. *Dev. Cell* v. 14 p. 159–169.

Campisi, J., and d'Adda di Fagagna, F. 2007. Cellular senescence: when bad things happen to good cells. *Nat. Rev. Mol. Cell Biol.* v. 8 p. 729–740.

Deshaies, R. J., and Joazeiro, C. A. 2009. RING domain E3 ubiquitin ligases. *Annu. Rev. Biochem.* v. 78 p. 399–434.

Elledge, S. J. 1996. Cell cycle checkpoints: preventing an identity crisis. *Science* v. 274 p. 1664–1672.

Hallstrom, T. C., and Nevins, J. R. 2009. Balancing the decision of cell proliferation and cell fate. *Cell Cycle* v. 8 p. 532–535.

Harrison, J. C., and Haber, J. E. 2006. Surviving the breakup: the DNA damage checkpoint. *Annu. Rev. Genet.* v. 40 p. 209–235.

Hartwell, L. H., and Weinert, T. A. 1989. Checkpoints: controls that ensure the order of cell cycle events. *Science* v. 246 p. 629–634.

Morgan, D. O. 1997. Cyclin-dependent kinases: engines, clocks, and microprocessors. *Annu. Rev. Cell Dev. Biol.* v. 13 p. 261–291.

Musacchio, A., and Salmon, E. D. 2007. The spindle-assembly checkpoint in space and time. *Nat. Rev. Mol. Cell Biol.* v. 8 p. 379–393.

Peters, J. M., Tedeschi, A., and Schmitz, J. 2008. The cohesin complex and its roles in chromosome biology. *Genes Dev.* v. 22 p. 3089–3114.

Ruchaud, S., Carmena, M., and Earnshaw, W. C. 2007. Chromosomal passengers: conducting cell division. *Nat. Rev. Mol. Cell Biol.* v. 8 p. 798–812.

Sclafani, R. A., and Holzen, T. M. 2007. Cell cycle regulation of DNA replication. *Annu. Rev. Genet.* v. 41 p. 237–280.

Trinkle-Mulcahy, L., and Lamond, A. I. 2006. Mitotic phosphatases: no longer silent partners. *Curr. Opin. Cell Biol.* v. 18 p. 623–631.

Apoptosis

Apoptosis

16

16

Apoptosis

16

Douglas R. Green

Department of Immunology, St. Jude
Children's Research Hospital, Memphis, TN

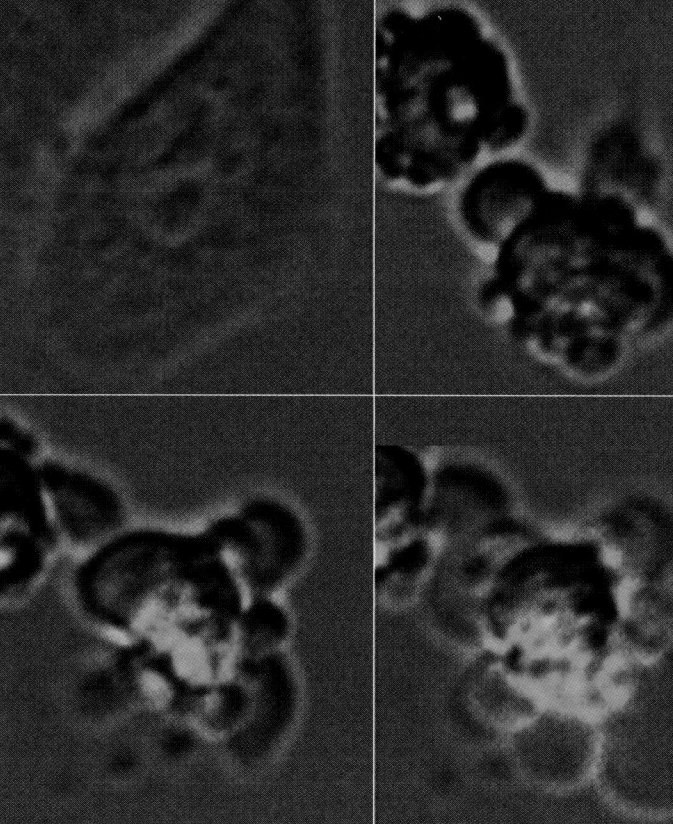

HeLa CELLS UNDERGOING APOPTOSIS. The cells in this series of photos were treated with a cytotoxic agent to induce apoptosis and images were taken every hour (left to right, top to bottom). Images are overlaid phase contrast and fluorescent micrographs. Cell cultures contain annexin V-FITC (green) to detect phosphatidylserine externalization and propidium iodide to detect loss of plasma membrane integrity. Photos courtesy of Nigel J. Waterhouse, Mater Medical Research Institute.

CHAPTER OUTLINE

16.1 Introduction

Key concepts

- Programmed cell death is a developmental process that usually proceeds by apoptosis.
- Apoptosis is also the mode of cell death occurring in a variety of other settings and has roles in normal homeostasis, inhibition of cancer, and disease processes.
- Most animal cells possess the molecules comprising the pathways that can cause death by apoptosis, and these pathways are activated by appropriate stimuli.

During the development of animals, *some cells die*. Unwanted cells are eliminated during embryogenesis, metamorphosis, and tissue turnover. This process is called **programmed cell death**, and, in general, it occurs by the mechanism of cell death called **apoptosis**. For example, during the development of the vertebrate limb, cell death can "sculpt" fingers and toes by the removal of the interdigital web, as shown in **FIGURE 16.1**. Although most programmed cell deaths occur by apoptosis, *these are not the same thing*. Programmed cell death refers specifically to cell death occurring at a defined point in development, whereas apoptosis is defined by the morphologic features of the cell death. During apoptosis, the dying cell becomes more compact, blebbing of the plasma membrane occurs, the nucleus and cell often fragment, and the chromatin becomes condensed, as seen in **FIGURE 16.2**. Ultimately, the dead cells may become fragmented into membrane-bounded pieces and engulfed by surrounding cells.

Death by apoptosis is not restricted to cells programmed to die during development. In addition to environmental cues, apoptosis can be triggered by a variety of stimuli (Figure 16.3), including withdrawal of essential growth factors, treatment with glucocorticoids, DNA damage by γ-irradiation or chemotherapeutic drugs, drugs that affect the cytoskeleton, endoplasmic reticulum dysfunction, heat shock, cold shock, and many other stresses on the cell.

Cells that are aberrant in still other ways are also eliminated by apoptosis. Adherent cells that become detached from their substrates also die by apoptosis, and this process is called **anoikis** ("homelessness"). In the immune system, cytotoxic lymphocytes attack

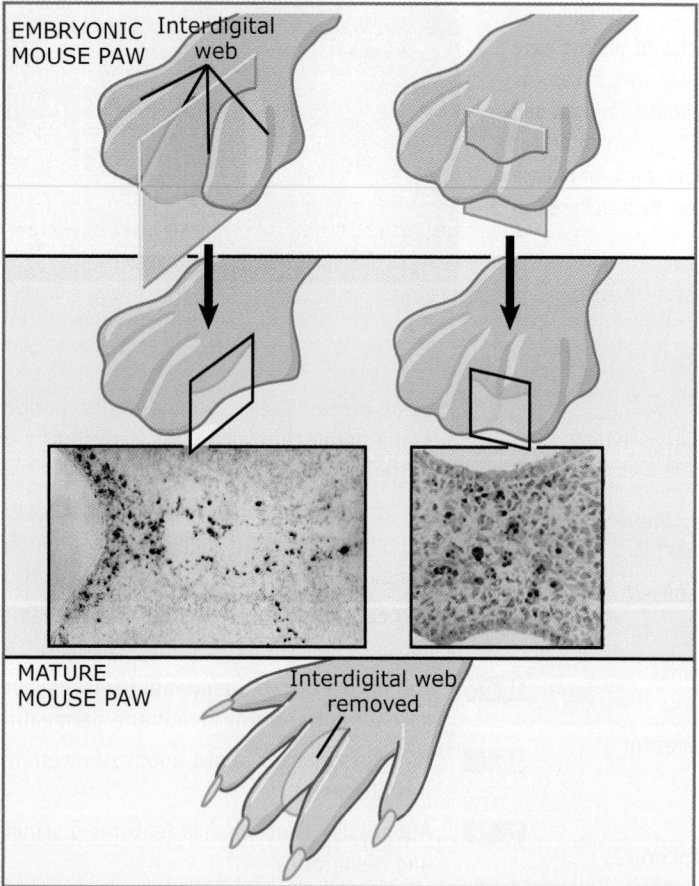

FIGURE 16.1 Sections of the interdigital web show cell death (dark-staining nuclei). This cell death has the characteristics of apoptosis. Photos reprinted from *Curr. Biol.*, vol. 9, M. Chautan, et al., Interdigital cell death can occur through..., pp. 967–970, Copyright (1999) with permission from Elsevier [http://www.sciencedirect.com/science/journal/09609822]. Photos courtesy of Pierre Golstein, Centre d'Immunologie.

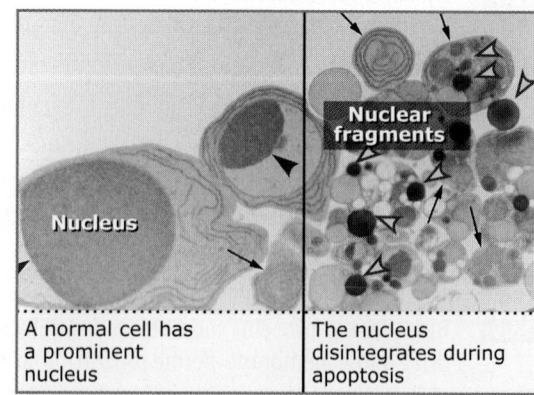

FIGURE 16.2 Cell structure changes during apoptosis. The left panel shows a normal cell. The right panel shows an apoptosing cell; gold arrows indicate condensed nuclear fragments. Photos courtesy of Shigekazu Nagata, Kyoto University Medical School.

target cells and initiate the apoptotic pathway in those cells. Apoptosis is also an important mechanism for removing tumorigenic cells. For example, the ability of the tumor suppressor p53 to trigger apoptosis is a key defense against cancer (see *17.11 Mutation of DNA repair and maintenance genes can increase the overall mutation rate*). Apoptosis is important, therefore, not only in tissue development but also in immune defense and in elimination of cancerous cells. However, apoptosis must be carefully regulated. Inappropriate activation of apoptosis can cause the destruction seen in some neurodegenerative diseases and in the aftermath of **ischemic** injury to the brain, heart, and other organs.

It is important to note that not all cell death is apoptotic. For example, cells that die as a direct result of ischemia do so by **necrosis**. In general, necrosis occurs when a cell is damaged to such an extent that it simply cannot survive. The difference between apoptotic death and necrotic death is evident in the dying cell's morphology. As mentioned above, apoptotic cells are compact, characterized by membrane blebbing, chromosome condensation, and fragmentation into membrane-bounded pieces that are engulfed by surrounding cells, as illustrated in **FIGURE 16.3**. A frequent biochemical feature of apoptosis is the fragmentation of the DNA into a characteristic "ladder," as seen in **FIGURE 16.4**. In contrast to apoptotic cells, necrotic cells usually swell and do not display chromatin condensation. Another important distinction between necrosis and apoptosis, especially in the vertebrates, is that necrosis invokes an inflammatory response, whereas death by apoptosis does not. For this reason, apoptosis is often referred to as being "silent."

Apoptosis is a property of animal cells. Most animal cells possess the molecules involved in the pathways that can cause death by apoptosis, and these pathways are activated by appropriate stimuli. The existence of apoptosis in other organisms remains controversial. For the most part, the molecular pathways responsible for apoptosis are delineated in vertebrates, insects (primarily *Drosophila*), and nematodes (*Caenorhabditis elegans*). Many of the key players are well conserved, although their precise roles and functions may differ significantly.

In this chapter, we will describe the main effectors and inhibitors of apoptosis. Then we will discuss the pathways by which the ef-

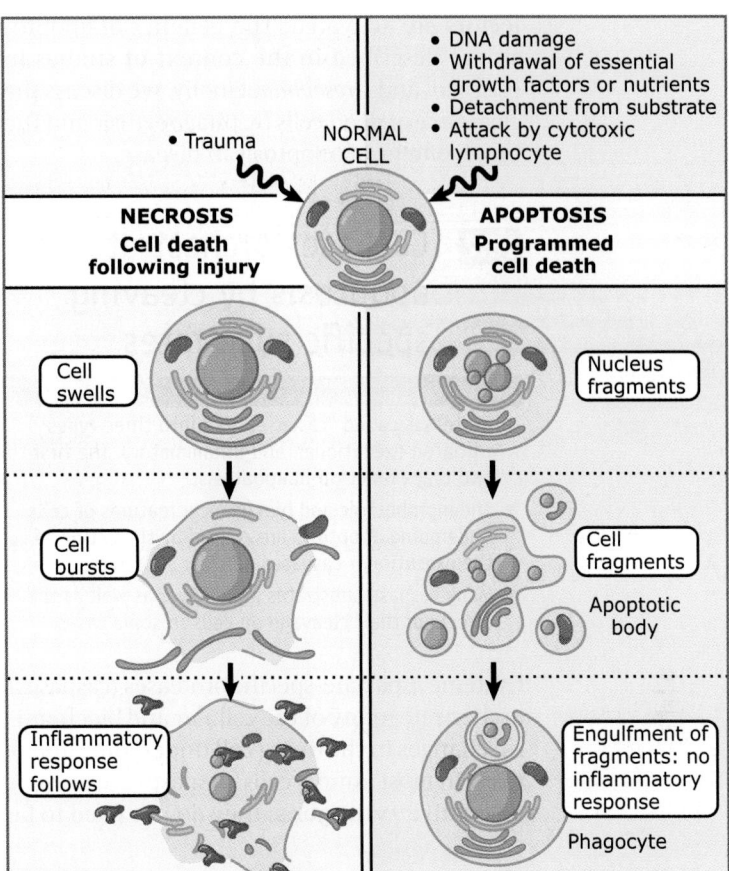

FIGURE 16.3 Cellular damage can result in necrosis, which has a different appearance than apoptosis, as organelles swell and the plasma membrane ruptures, without chromatin condensation.

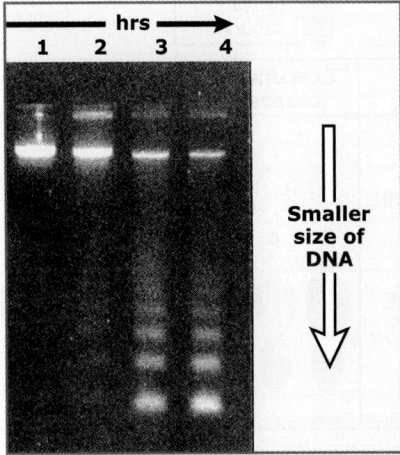

FIGURE 16.4 Fragmentation of DNA occurs in cells undergoing apoptosis. Photo courtesy of Shigekazu Nagata, Kyoto University Medical School.

fectors are activated. The genetics of apoptosis are described in the context of studies in *C. elegans* and *Drosophila*. Finally, we discuss the clearance of dead cells by phagocytosis and the involvement of apoptosis in disease.

16.2 Caspases orchestrate apoptosis by cleaving specific substrates

Key concepts

- Proteases called "caspases" fall into three types: initiator, executioner, and inflammatory. The first two types function in apoptosis.
- The morphologic and biochemical features of cells undergoing apoptosis are caused by the action of the executioner caspases on their substrates.
- Many caspase substrates are known, as well as the effects of their cleavage on cells in some cases.

Cysteine aspartate-specific proteases (caspases) orchestrate many of the cellular and biochemical changes in the dying cell undergoing apoptosis. In most animal cells, caspases are present as inactive **zymogens**; they do not need to be newly synthesized upon activation of apoptosis. There are three main types of caspases with different functions in cells: executioner, initiator, and inflammatory caspases, shown in **FIGURE 16.5**. The executioner caspases (predominantly caspases-3 and -7 in vertebrates) are responsible for cleaving many different proteins to effect apoptosis. These caspases generally cleave substrates at Asp-Xaa-Xaa-Asp/Gly, Asp-Xaa-Xaa-Asp/Ser, or Asp-Xaa-Xaa-Asp/Ala sites. (The "/" indicates the cleavage and Xaa stands for any amino acid.) It is estimated that there are ~500 substrates for caspases in mammalian cells, although the effects of cleavage of most of them are unknown. In several cases, these cleavage events have defined roles in apoptosis, and produce the changes associated with this form of cell death. Others may be irrelevant "bystander" cleavage events.

DNA fragmentation during apoptosis is a result of a caspase-mediated cleavage event. A DNase in cells, *c*aspase-dependent *D*Nase (CAD), exists as a complex with an inhibitor, iCAD. In fact, CAD must be folded by iCAD into this complex for the DNase to have potential activity, which manifests only when iCAD is cleaved by an executioner caspase. The now active CAD digests DNA at accessible sites between the nucleosomes, resulting in DNA degradation into the characteristic "ladder" seen in apoptosis. Cells lacking either CAD or iCAD do not show this feature as they undergo apoptosis.

The myosin-light chain kinase ROCK-1 is a substrate for executioner caspases, but in this case cleavage activates the kinase. Gelsolin, another enzyme, is similarly cleaved and activated by executioner caspases. Both of these affect changes in the cytoskeleton, and as a result, cells undergo the characteristic "blebbing" associated with apoptosis (for more on cytoskeletal regulation see *12.10 Capping proteins regulate the length of actin filaments*).

Although each caspase-cleavage event may participate in features of apoptosis, no one substrate, or even a collection of substrates, can be shown to be responsible for cell death. How, ultimately, caspases kill a cell may be a result of many different substrate cleavage events, and to date, cell death resulting from caspase activation is not effectively blocked by manipulating any known set of specific substrates. Either there are many that contribute to death (which seems likely), or we have

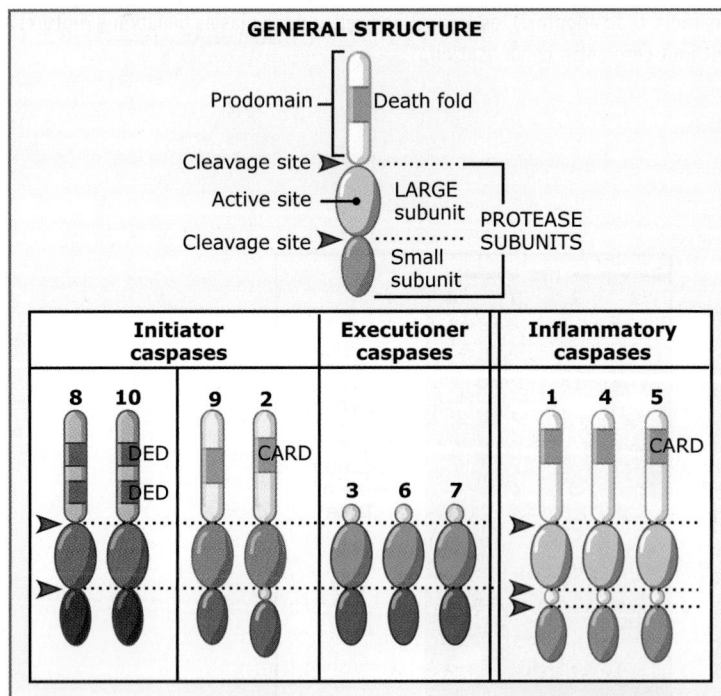

FIGURE 16.5 Different types of vertebrate caspases are shown schematically. Note the prodomains and protein-protein interaction regions of the initiator and inflammatory caspases. Death effector domain (DED); caspase-recruitment domain (CARD).

simply not identified those that are most important. However, pharmacologic inhibitors of caspases, including valine-alanine-aspartate-fluoromethylketone (VAD-fmk), inhibit cellular and biochemical changes and, in some cases, can even prevent the death of the cell (but not all; see *16.15 Mitochondrial outer membrane permeabilization can cause caspase-independent cell death*).

16.3 Executioner caspases are activated by cleavage, whereas initiator caspases are activated by dimerization

Key concepts

- Cleavage of executioner caspases at specific sites is necessary and sufficient for their activation.
- This cleavage is usually mediated by the initiator caspases.
- Initiator caspases are activated by adaptor molecules that contain protein-protein interaction domains called death folds.

The zymogen form of the mammalian executioner caspases, caspases-3 and -7, as well as another, caspase-6, preexist in the cell as inactive dimers. Activation of the executioner caspase occurs through cleavage of the zymogens at specific aspartate residues between what will be the large subunit (containing the cysteine-histidine dimer that constitutes the active site) and the small subunit (containing the specificity-determining region) of the mature, active caspase. Cleavage of the zymogen brings the catalytic cysteine-histidine dimer, required for protease activity, into proximity with the region that transiently binds the substrate. The arginine labeled "R" in **FIGURE 16.6** is responsible for binding to the aspartate in the substrate for the proteolytic cleavage event. Cleavage of an executioner caspase thereby permits the formation of the active sites in the mature, active caspase.

One protease that can produce this activating cleavage of executioner caspases is granzyme B, present in the granules of cytotoxic lymphocytes (cytotoxic T cells and natural killer cells). When cytotoxic lymphocytes target a cell for death (e.g., a virally infected cell), granzyme B is released into the cytoplasm of the target. Granzyme B can directly activate executioner

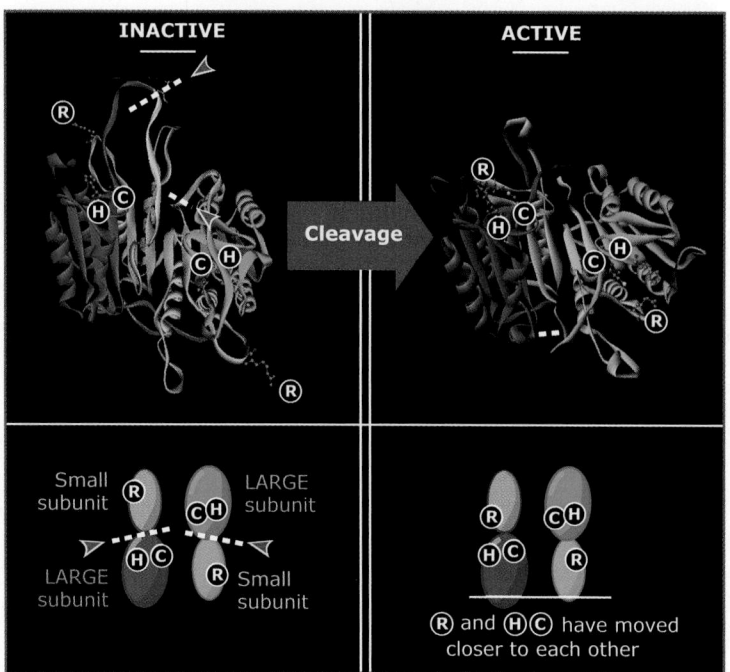

FIGURE 16.6 The upper illustrations are the structure of caspase-7 before (left) and after (right) cleavage resulting in activation. C (cysteine) and H (histidine) show the catalytic dyad, and R (arginine) is the specificity-determining site where aspartate in the substrate binds. Note that cleavage results in a movement of the R into proximity with the catalytic dyad. The lower drawings are cartoons of executioner caspase activation, showing how the cleavage event allows formation of active protease sites. Left structure from Protein Data Bank 1GQF. S. J. Riedl, et al., *Proc. Natl. Acad. Sci. USA* 98 (2001): 14790–14795. Right structure from Protein Data Bank 1I51. J. Chai, et al., *Cell* 104 (2001): 769–780.

caspases in the target to cause apoptosis. This, however, is probably not the only way in which granzyme B kills cells (see *16.14 The death receptor pathway of apoptosis can engage mitochondrial outer membrane permeabilization through the cleavage of the BH3-only protein Bid*).

Similarly, caspase-6 is activated during apoptosis through cleavage by active caspases-3 and -7. This still leaves the problem of how caspases-3 and -7 are usually activated. In most cases of apoptosis, the cleavage of these executioner caspases is mediated by the action of another set of caspases, the "initiator" caspases. It is the activation of the executioner caspases by the initiator caspases that defines and coordinates the different pathways of apoptosis.

Unlike the executioner caspases, the initiator caspases (caspases-2, -8, -9, and -10 in mammals) preexist in cells as inactive monomers. Cleavage of these monomers does not cause the formation of active sites in these proteases, but instead, activation of the initiator caspases occurs when two monomers are forced into dimers, which then interact to create the active sites, as shown in **FIGURE 16.7**. Subsequent cleavage at aspartates within the molecules can stabilize the resulting dimer, once it has formed. This mechanism of activation by dimerization is called **induced proximity**.

Initiator caspases contain large prodomains (see Figure 16.5 and Figure 16.7) that bear protein-protein interaction motifs, including the caspase-recruitment domain (CARD), death effector domain (DED), and pyrin domain (PYR). (The latter is not found in the prodomains of mammalian initiator caspases but is found in one of the initiator caspases found in fish, as well as in other proteins.) All of these are structurally related and are collectively called "death folds," as shown in **FIGURE 16.8**. The prodomain of a specific initiator caspase interacts with a specific adaptor molecule to define an apoptotic pathway. Such interactions tend to occur in a like-like manner (e.g., CARD-CARD, DED-DED). Two such pathways are known in detail: the death receptor pathway and the mitochondrial pathway (see Figure 16.7 and *16.6 The death receptor pathway of apoptosis transmits external signals* and *16.8 The mitochondrial pathway of apoptosis*).

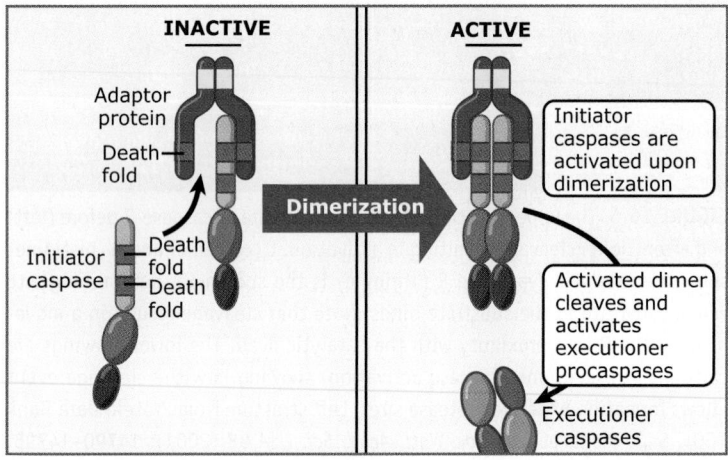

FIGURE 16.7 Initiator caspases are activated when adaptor molecules bind to them and force them into dimers ("induced proximity"), allowing the formation of active sites in the protease. The active initiator caspase can now cleave and activate executioner caspases.

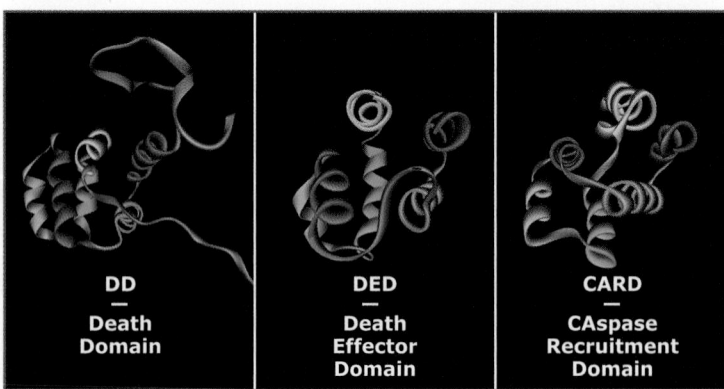

FIGURE 16.8 The structures of death effector domains (DEDs), caspase-recruitment domains (CARDs), and death domains (DD) as well as pyrin domains (not shown) all resemble one another. Collectively, these structures are called "death folds." Death folds function in proteins as protein-protein interaction regions and tend to interact with related regions on other molecules. Left structure from Protein Data Bank 1DDF. B. Huang, et al., *Nature* 384 (1996): 638–641. Middle structure from Protein Data Bank 1E41. H. Berglund, et al., *J. Mol. Biol.* 302 (2000): 171–188. Right structure from Protein Data Bank 1CY5. D. E. Vaughn, et al., *J. Mol. Biol.* 293 (1999): 439–447.

16.4 Some inhibitors of apoptosis proteins block caspases

Key concepts

- The inhibitors of apoptosis proteins (IAPs) comprise a family of proteins with different functions; some of these proteins bind to and inhibit caspases and induce their degradation by the proteasome.
- Since executioner caspases are activated by cleavage, and since these caspases can cleave and activate each other, any proteolytic activity of the caspases will be rapidly amplified in cells, resulting in their death by apoptosis. It is, therefore, important that there be mechanisms present to limit potential "accidental" activation of caspases in cells that are not signaled to die.

The first IAP was identified in an insect virus (baculovirus), where it functions to prevent apoptosis in infected cells (this idea, that viruses encode proteins that block apoptosis is explored in more detail later in this chapter). Subsequently, IAPs were identified by sequence homology in other organisms, including mammals. Many of the IAPs, despite their name and domain organization, have functions that are distinct from the regulation of apoptosis. Nevertheless, some IAPs—including X-linked IAP (XIAP), which is found in mammals—act as potent inhibitors of caspases, especially the initiator caspase-9 and the executioner caspases-3 and -7. **FIGURE 16.9** shows the structure of XIAP bound to caspase-3. (For more on IAPs, see *16.18 Apoptosis in insects has features distinct from mammals and nematodes*.)

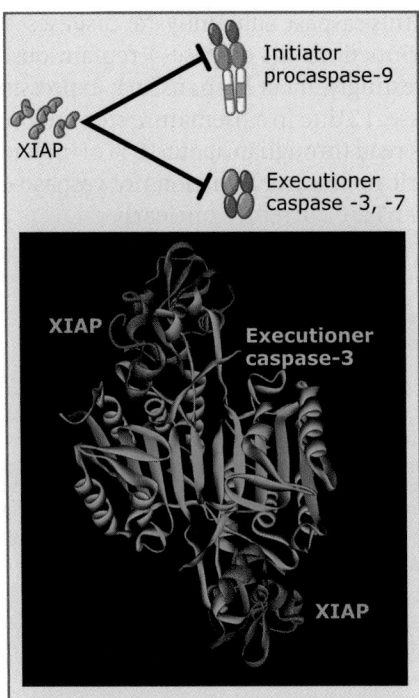

FIGURE 16.9 XIAP binds to and inhibits caspases-9, -3, and -7. The structure of the interaction between caspase-3 and XIAP is shown. Structure from Protein Data Bank 1I30. M. A. Seefeld, et al., *Bioorg. Med. Chem. Lett.* 11 (2001): 2241–2244.

IAPs share domains called baculovirus IAP repeats (BIR) and usually also carry a characteristic RING finger domain. Those IAPs with RING domains, including XIAP, also function as ubiquitin E3-ligases, and XIAP can effectively target both itself and its target caspases for ubiquitination and degradation by the proteasome. Therefore, the activation of a caspase need not result in apoptosis if that caspase is inhibited and ultimately degraded by the actions of IAPs.

Despite the potent caspase-inhibiting activity of XIAP, however, mice lacking this IAP develop normally and do not display any obvious abnormalities associated with apoptosis. Therefore, the importance of the XIAP in controlling apoptosis (and the role of the regulation of this molecule) remains unclear.

Concept and Reasoning Check

1. Proteases can do considerable damage to a cell if they are active in the wrong location and/or time period. Compare the mechanisms used to control the activity of proteases in lysosomes and proteasomes with those used to control caspase activity. What properties make the caspase mechanisms especially effective in controlling programmed cell death?

16.5 Some caspases have functions in inflammation

Key concept

• In addition to the initiator and executioner caspases, another set of proteases in this family acts to process cytokines rather than regulate apoptosis.

Some caspases appear to lack significant roles in most forms of apoptosis but instead function predominantly in inflammation. In fact, the first caspase was identified, not as an apoptosis regulator, but as a protein required for the processing and secretion of a cytokine, interleukin-1β. This protease, caspase-1 (originally called interleukin-1β-converting enzyme, or ICE), is also required for the processing and secretion of interleukin-18.

FIGURE 16.10 shows that the activation of caspase-1 involves the formation of a complex that includes another caspase (caspase-5 in humans and caspase-11 in rodents) and two adaptor molecules, NALP and ASC. Mice lacking caspase-1, caspase-11, or ASC fail to secrete interleukins-1 or -18, but show no obvious defects in development or apoptosis. In contrast, humans with activating mutations in NALP-1

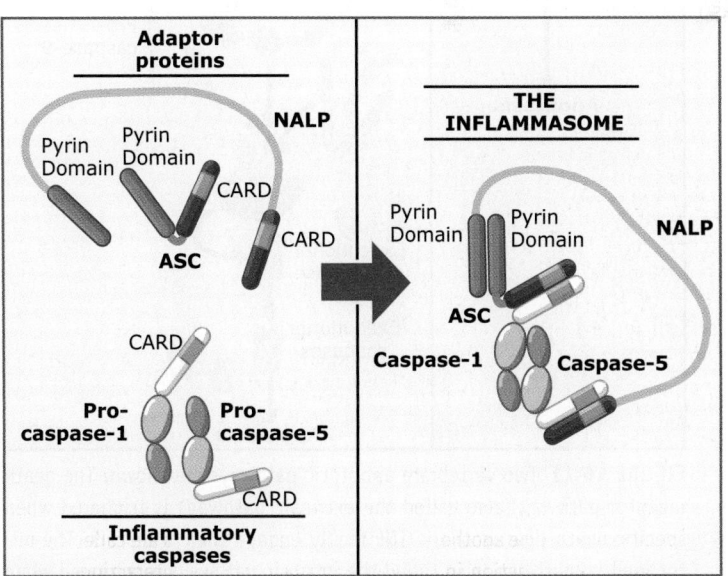

FIGURE 16.10 Caspase-1, which is required for the processing and secretion of interleukins-1β and -18, is activated by a complex that includes another caspase (caspase-5) and two adaptor molecules (ASC and NALP). Each of these proteins binds with others via protein-protein interaction regions. Several different inflammasomes have been described, but all resemble this model.

have inflammatory syndromes relating to elevated cytokine secretion.

Two other caspases that map close to caspases-1 and -5 and show sequence homology with this caspase subfamily are caspases-4 and -12. Functions for caspase-4 remain obscure. Interestingly, most humans lack expression of caspase-12 due to a premature stop codon, but this is read through in about 10% of individuals of African descent. Functions for caspase-12 in humans are currently unclear.

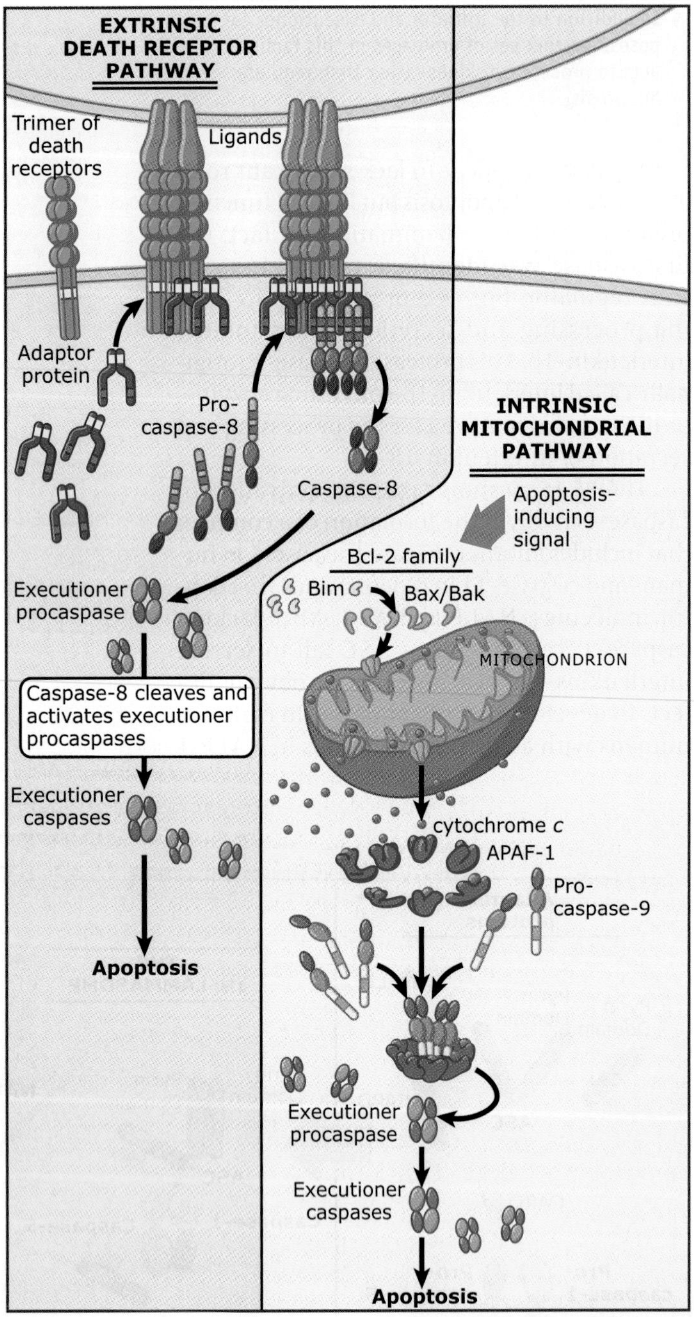

16.6 The death receptor pathway of apoptosis transmits external signals

Key concepts

- Two well-characterized pathways of apoptosis are the death receptor (extrinsic) pathway and the mitochondrial (intrinsic) pathway.
- Caspase activation and apoptosis are induced by the binding of specialized ligands in the TNF family to their receptors (death receptors).

Caspase activation is triggered in vertebrate cells by different pathways of apoptosis. Two well-defined pathways are the death receptor pathway (also called the extrinsic pathway) and the mitochondrial pathway (also called the intrinsic pathway), which are shown in **FIGURE 16.11**. Although differing in several important ways, these two pathways have similarities in that each involves the activation of an initiator caspase by induced proximity, leading to activation of the executioner caspases. In addition, there is crosstalk between the pathways, as the death receptor pathway can engage the mitochondrial pathway.

The death receptors are a subset of the tumor necrosis factor receptor (TNFR) family in vertebrates, and include TNFR1, Fas (also called CD95 or APO-1), and the TRAIL receptors (TRAIL-R1 and -R2 in humans, also called DR4 and DR5). **FIGURE 16.12** shows the different types of death receptors. These trimeric receptors are bound by specific ligands (TNF, Fas-ligand, or TRAIL, respectively) and can rapidly trigger apoptosis in cells. These ligands are produced by a variety of different cell types, including cells of the immune system, often in response to inflammatory stimuli.

Death receptors share a death domain in the intracellular portion of the molecule (see Figure 16.8 and Figure 16.12). The death domains are another example of death folds, like CARD, DED and PYR domains, and these interact with death domains in adaptor molecules (for introduction to death folds see *16.3 Executioner*

FIGURE 16.11 Two vertebrate apoptotic pathways are shown. The death receptor pathway (also called the extrinsic pathway) is triggered when specific death ligands of the TNF family engage their receptors. The mitochondrial pathway (also called the intrinsic pathway) is triggered when the mitochondrial outer membrane becomes permeable as a result of Bcl-2 family protein interactions, and proteins of the intermembrane space are released. These include cytochrome *c*, which triggers caspase activation by interacting with cytosolic proteins. These pathways are considered in detail in the following sections.

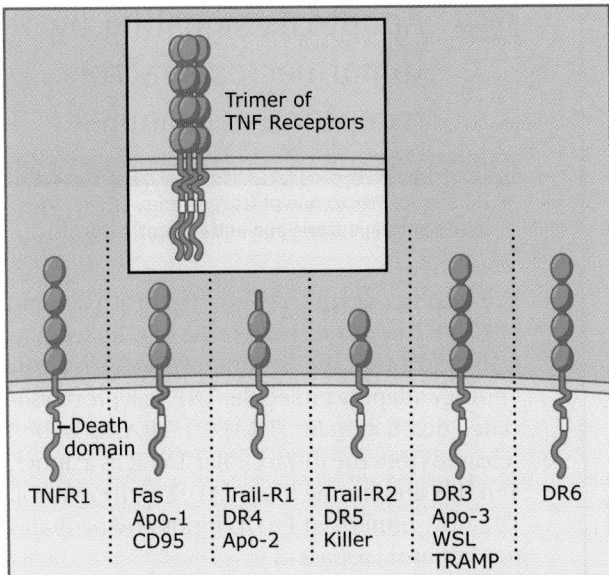

FIGURE 16.12 The death receptors are members of the TNF-receptor family that carry death domains in their intracellular regions. These exist as trimers on the cell surface of many cell types in vertebrates.

caspases are activated by cleavage, whereas initiator caspases are activated by dimerization).

The death receptors exist as trimers on the cell surface, and it is likely that the ligands for these receptors bind as clusters to crosslink two or more of these trimers. This clustering exposes the death domains for interaction with intracellular proteins. The death domains of Fas/CD95 and the TRAIL receptors, following ligation, bind to the adaptor protein FADD (which stands for Fas-associated death domain), via the death domain in FADD. This brings together FADD molecules within the cell and also exposes another region of FADD, containing a DED.

The DED of FADD now binds to the DEDs in the prodomain of caspase-8 monomers, resulting in dimer formation and activation of the initiator caspase by induced proximity. The complex of the ligated death receptor and FADD (via death domain interactions), and of FADD with caspase-8 (via DED interactions), occurs rapidly following death receptor ligation and is called a death-inducing signaling complex (DISC). The activated caspase-8 can now cleave substrates in the cell, including the executioner caspases-3 and -7, and apoptosis proceeds. **FIGURE 16.13** summarizes the death receptor pathway.

Many functions for death receptor-mediated apoptosis are known, especially in

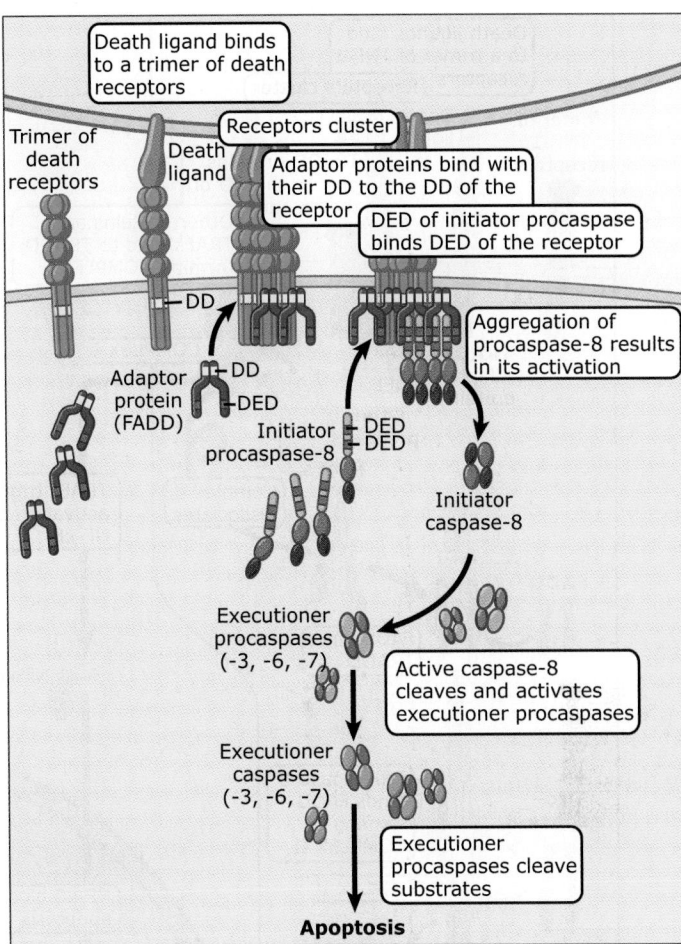

FIGURE 16.13 Ligation of death receptors causes the recruitment of the adaptor protein FADD to the intracellular region of the death receptor, via death domain (DD)-(DD) interactions. Caspase-8 is then recruited to FADD via death effector domain (DED)-(DED) interactions. The dimerization of caspase-8 activates it through induced proximity. The active caspase-8 can cleave and activate executioner caspases to cause apoptosis. The complex of death receptor, FADD, and caspase-8 is called a death-inducing signaling complex (DISC).

immune effector mechanisms and immune regulation. Additional roles are suggested in a variety of cell types, including neurons. TRAIL is currently being investigated as a possible anticancer therapeutic, by virtue of its ability to trigger apoptosis in some tumor cells.

In keeping with the proposed role of Fas in the immune system, humans or mice with mutations in either Fas or its ligand display a disease involving massive enlargement of lymphoid organs due to the accumulation of an unusual T cell population. These individuals can also have B lymphocyte abnormalities including production of autoimmune antibodies and development of B cell lymphomas.

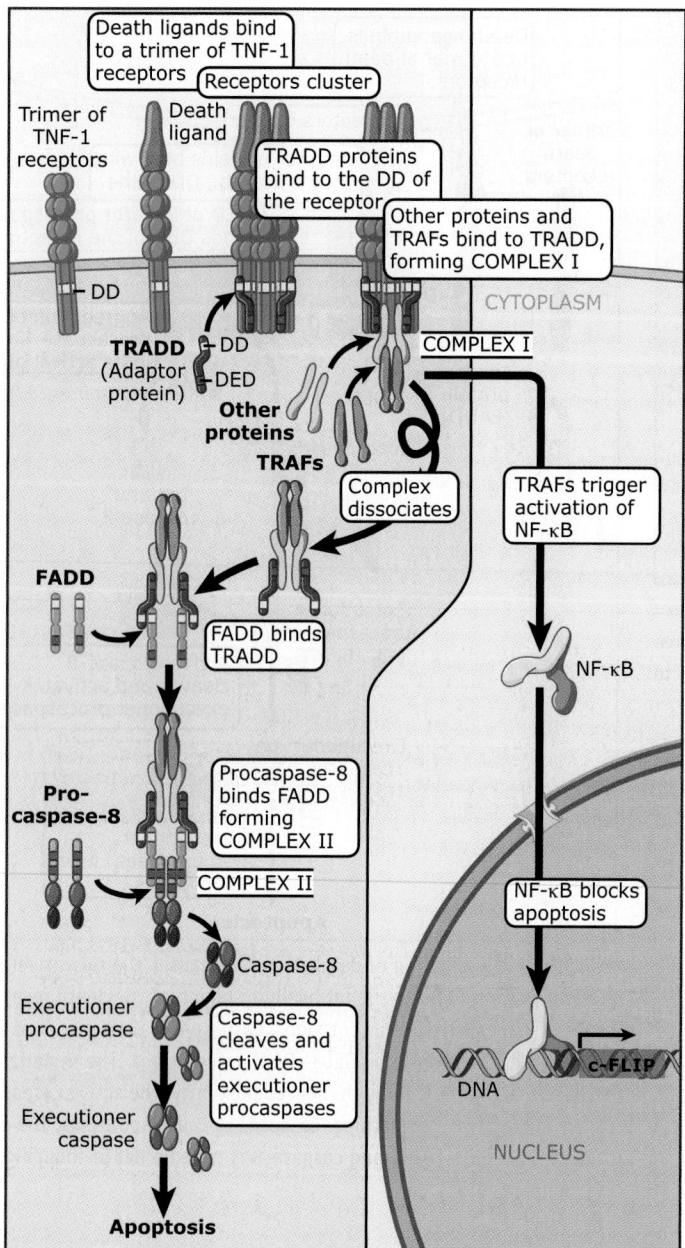

FIGURE 16.14 Binding of TNF to TNF receptor 1 (TNFR1) can trigger nuclear factor-κB activation, blocking apoptosis, or caspase activation leading to apoptosis. The ligation leads to the formation of a complex (complex I) composed of the intracellular region of the receptor, an adaptor protein (TRADD), and other proteins. This complex dissociates, and in the cytosol binds to FADD, which in turn can bind and activate caspase-8 (complex II). The active caspase-8 cleaves and activates executioner caspases to cause apoptosis. However, the activation of NF-κB can also induce antiapoptotic molecules that prevent caspase activation and death.

Key concept

- Binding of TNF to one of its receptors, TNFR1, induces both apoptotic and antiapoptotic signals.

Unlike other death receptors, the death domain of TNFR1 does not bind to FADD. Instead, as **FIGURE 16.14** shows, ligation of TNFR1 recruits another adaptor molecule, TNF receptor-associated death domain (TRADD). TRADD is then released from the intracellular TNFR as a dimer, and this dimer now finds FADD in the cytosol. The now dimerized FADD binds and activates the initiator caspase-8.

Ligation of TNFR1 also causes the binding of other molecules to the intracellular region of the receptor, including TNFR-associated factors (TRAFs). The TRAFs now trigger the activation of nuclear factor-κB (NF-κB), which functions as a transcription factor. Several proteins expressed by the action of NF-κB act to prevent the formation of the DISC and the activation of caspase-8.

The ligation of TNFR1 by TNF is therefore capable of producing signals that both block and promote apoptosis. If NF-κB is activated, TNF fails to trigger apoptosis, but instead participates in the inflammatory response. Alternatively, if NF-κB is blocked, or if RNA or protein synthesis is inhibited, TNF triggers apoptosis.

One important protein expressed after NF-κB activation is called c-FLIP. This protein is interesting in that it is closely related to caspase-8 but lacks elements required for protease activity (including the active site cysteine). **FIGURE 16.15** shows that when c-FLIP is strongly expressed, it can compete with caspase-8 for FADD and thereby prevent caspase activation. Not surprisingly then, c-FLIP is also capable of blocking apoptosis induced by other death receptors as well.

One such case of c-FLIP-mediated suppression of apoptosis appears to occur in activated lymphocytes. These T lymphocytes express Fas, but for the first few days after activation, they are resistant to Fas-mediated apoptosis. After this time, they become susceptible. This shift from resistance to susceptibility is inversely proportional to the expression of c-FLIP in these cells, but perhaps more importantly, inhibition of the expression of c-FLIP can sensi-

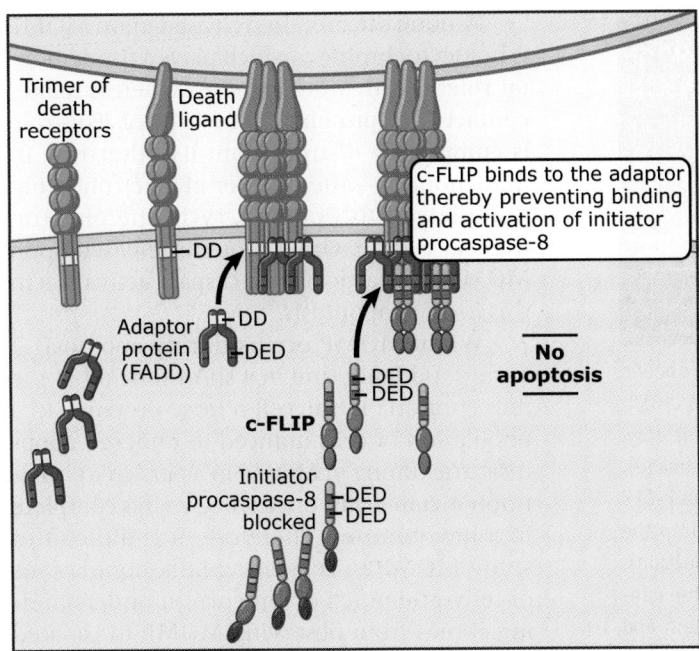

FIGURE 16.15 c-FLIP resembles caspase-8 but lacks a catalytic site. It binds to FADD via a DED-DED interaction and can restrict access of caspase-8 to the DISC. Expression of c-FLIP in response to NF-κB activation is one way that death receptor-induced apoptosis is inhibited by this transcription factor.

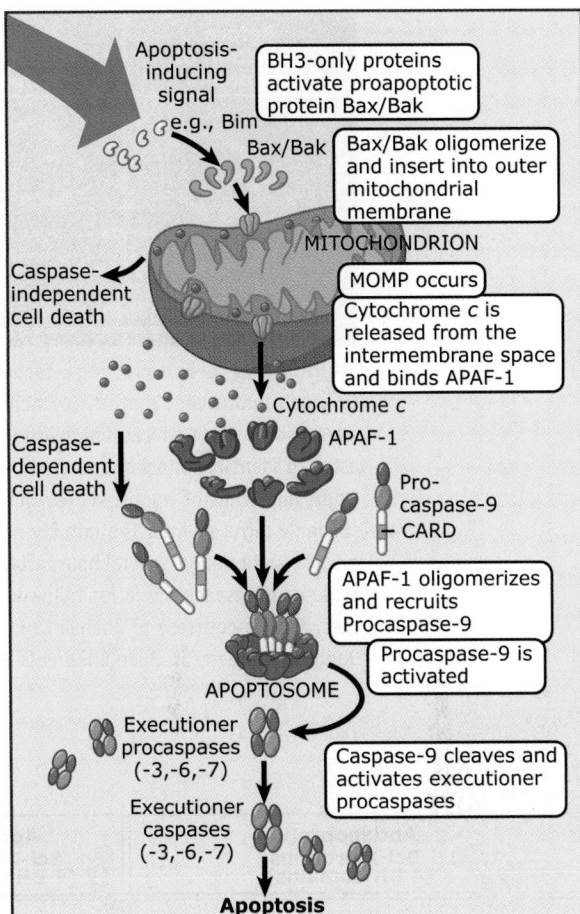

FIGURE 16.16 Signals for the induction of apoptosis trigger changes in the Bcl-2 family proteins, which function to inhibit (antiapoptotic proteins) or promote (proapoptotic proteins) apoptosis. As a result, the proapoptotic multidomain proteins of the Bcl-2 family may be activated and, if so, cause permeabilization of the outer membranes of all mitochondria in the cell. Mitochondrial outer membrane permeabilization (MOMP) allows proteins of the intermembrane space to diffuse into the cytosol, including cytochrome *c*, which activates APAF-1. This leads to recruitment and activation of the initiator caspase-9, which cleaves and activates executioner caspases to cause apoptosis.

tize activated T cells to death induced by Fas ligation.

Although death receptor signaling is well understood (or perhaps because of it), it came as a surprise that knockouts for FADD, c-FLIP, or caspase-8 showed a striking phenotype distinct from that of any death receptor-deficient animals. Each of these mutant mice die at the same early stage of embryonic life, and this death is clearly not due to defects in apoptosis. Instead, it appears that these molecules are required for signaling events that participate in cell survival at a critical time in embryogenesis. Exactly what these signals might be remains obscure.

16.8 The mitochondrial pathway of apoptosis

Key concepts

- Most apoptosis in mammalian cells proceeds via a pathway in which the mitochondrial outer membranes are disrupted, releasing the contents of the mitochondrial intermembrane space into the cytosol.
- Mitochondrial outer membrane permeabilization (MOMP) is a key feature of this pathway.

In vertebrates, most forms of apoptosis are triggered not by death receptors, but by engaging the mitochondrial pathway of apoptosis. When this pathway is engaged, the outer membranes of the mitochondria become disrupted, so that soluble proteins in the mitochondrial intermembrane space (between the inner and outer mitochondrial membranes) diffuse into the cytosol. This MOMP is tightly regulated and is a defining event in this pathway. FIGURE 16.16 shows the mitochondrial pathway of apoptosis, which we will describe in detail in the next four sections.

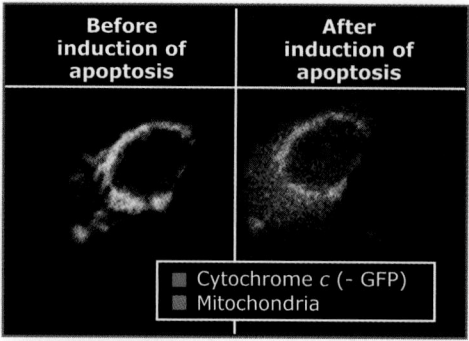

FIGURE 16.17 Cells expressing cytochrome *c* bound to green fluorescent protein (cytochrome *c*-GFP) were also stained with tetramethylrhodamine ethyl ester to identify mitochondria (red) (left image). Upon induction of apoptosis, cytochrome *c*-GFP suddenly diffused from the mitochondria into the cytosol (right image, several hours after induction). Evidence of caspase activation followed within minutes. Photos courtesy of Joshua C. Goldstein and Douglas R. Green, St. Jude Children's Hospital.

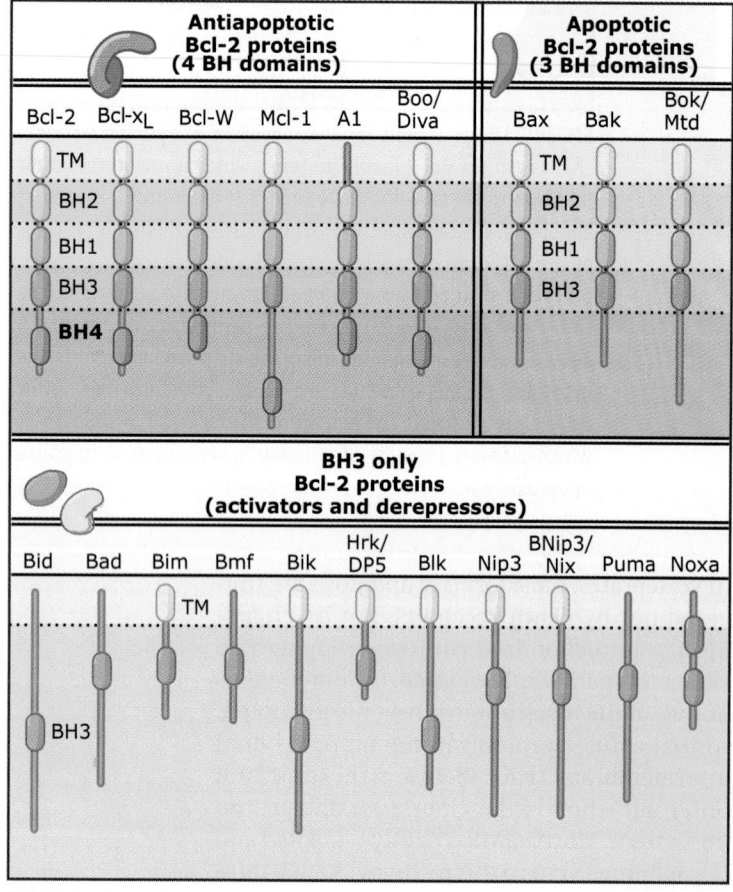

FIGURE 16.18 The Bcl-2 family proteins share up to four Bcl-2 homology domains (BH) and can be antiapoptotic or proapoptotic. The proapoptotic proteins include multidomain proteins and the BH3-only proteins.

Among the proteins released upon MOMP is holocytochrome *c*, which plays a fundamental role in activating caspases when it makes contact with proteins in the cytosol. This role is completely distinct from its other role in mitochondria—the transfer of electrons from Complex III to Complex IV in the electron transport chain. Other proteins released upon MOMP also participate in caspase activation in this form of apoptosis.

When MOMP occurs during apoptosis it occurs suddenly, and in a short time all of the mitochondria in the cell release proteins. In a population of cells induced to undergo apoptosis, the timing of MOMP in a particular cell is unpredictable, but once it begins it is complete in a few minutes. Therefore, it is difficult to study MOMP in cells using bulk populations of cells, and much of our current understanding comes from observing MOMP in isolated mitochondria or in single cells. **FIGURE 16.17** shows cytochrome *c* release (green) from mitochondria (red).

16.9 Bcl-2 family proteins mediate and regulate mitochondrial outer membrane permeabilization and apoptosis

Key concepts

- The Bcl-2 family proteins are central to the mitochondrial pathway of apoptosis.
- There are 3 classes of Bcl-2 proteins that induce, directly cause, or inhibit MOMP.

The process of MOMP is tightly controlled during apoptosis, and whether this event occurs is at the heart of the decision between life and death of the cell. This decision is the function of the Bcl-2 family proteins.

The proteins of the Bcl-2 family were known to be important in the regulation of apoptosis (both positive and negative), but their functions were obscure until the mitochondrial pathway was elucidated. We now know that a major way these proteins exert their control is by regulating MOMP.

Bcl-2 family proteins share up to four Bcl-2 homology (BH) domains, as shown in **FIGURE 16.18**. **FIGURE 16.19** shows that they have striking structural similarities with each other. They also share structural features with some bacterial pore-forming proteins, including the diph-

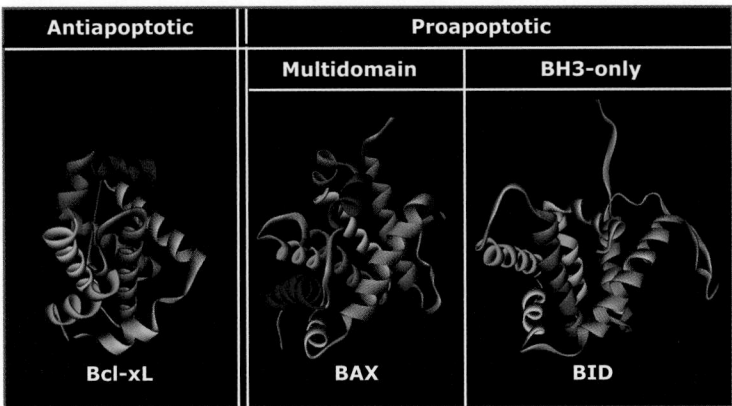

Antiapoptotic	Proapoptotic	
	Multidomain	BH3-only
Bcl-xL	BAX	BID

FIGURE 16.19 The structures of antiapoptotic Bcl-2 proteins (e.g., Bcl-xL) are similar to those of proapoptotic Bcl-2 family proteins (Bax, Bid). Left structure from Protein Data Bank 1PQ0. X. Liu, et al., *Immunity* 19 (2003): 341–352. Middle structure from Protein Data Bank 1F16. M. Suzuki, R. J. Youle, and N. Tjandra, *Cell* 103 (2000): 645–654. Right structure from Protein Data Bank 1DDB. J. M. McDonnell, et al., *Cell* 96 (1999): 625–634.

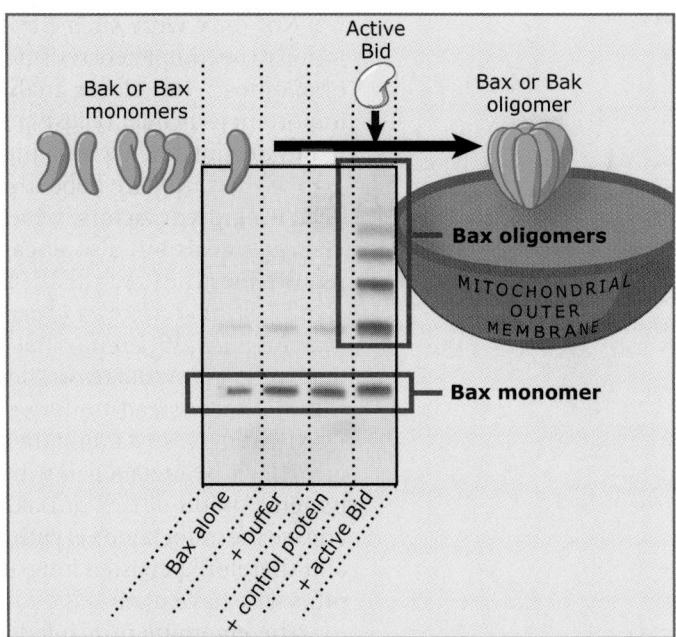

FIGURE 16.20 BAX or BAK, when activated by BH3-only proteins (or even peptides corresponding to the BH3 region) insert into the mitochondrial outer membrane and oligomerize. This appears to result in permeabilization of the membrane through an unknown mechanism. Oligomerization is shown in the experiment on the left; BAX protein was cultured with mitochondrial lipid vesicles and mixed with active BH3-only protein, BID (or controls). Cross-linked proteins were resolved by electrophoresis and immunoblot for BAX. Active BID induced oligomerization of BAX. Photo courtesy of Jerry Chipuk and Douglas R. Green, St. Jude Children's Hospital.

theria toxin B chain. Bcl-2 family proteins may therefore have functions relating to pores.

Three general subfamilies of Bcl-2 proteins are recognized. These include the antiapoptotic proteins, the proapoptotic "multidomain" proteins (also called BH1, -2, and -3 proteins, as they carry BH1, -2, and -3), and the proapoptotic BH3-only proteins (which carry only BH3). The antiapoptotic Bcl-2 proteins possess the BH4 domain and prevent MOMP, whereas the proapoptotic Bcl-2 proteins lack the BH4 domain and promote MOMP. However, the inhibition of MOMP is not simply a function of the BH4 domain.

16.10 The multidomain Bcl-2 proteins Bax and Bak are required for mitochondrial outer membrane permeabilization

Key concepts

- Bax and Bak are essential for the permeabilization of the mitochondrial outer membrane and are required for the mitochondrial pathway of apoptosis.
- Bax and Bak probably directly cause the membrane disruption associated with MOMP.

Bax and Bak are two of the multidomain proteins in the Bcl-2 family. These two proteins are responsible for MOMP and probably form the pores through which the proteins of the mitochondrial intermembrane space diffuse. FIGURE 16.20 shows that Bax or Bak are induced to oligomerize in the mitochondrial membrane in the presence of a BH3-only protein, such as Bid.

Studies in mice in which the multidomain proteins were eliminated by gene knockout experiments helped define the role of these proteins in MOMP. Mice deficient in the multidomain protein Bak were found to be profoundly (and disappointingly) normal, and mice lacking Bax, although showing some developmental defects, were also fairly normal in their development. An important advance occurred when double knockout (DKO) mice were generated that lacked both of these proteins. The result was extensive embryonic and perinatal mortality with a variety of defects in apoptosis. Perhaps most importantly, cells from these mice failed to undergo MOMP or apoptosis in response to a variety of apoptotic stimuli.

Not only were such Bax-Bak DKO cells found to be completely resistant to many forms of apoptosis, but closer analysis revealed an important principle. Wild-type cells in culture, if they are not derived from tumors (and often, even if they are), are dependent for their survival on growth factors, which not only prevent apoptosis but also engage mechanisms required for nutrient uptake and metabolism (see *17.2 Cancer cells have a number of phenotypic characteristics*). When Bax-Bak DKO cells were deprived of growth factors, however, they did not die but instead underwent **autophagy** ("self-eating"), sustaining themselves for several weeks by metabolizing intracellular components. Without Bax or Bak, they could not engage the mitochondrial pathway of apoptosis and therefore persisted long after they would otherwise have died.

The emerging principle is that in the absence of apoptosis, cell life (even if it requires autophagy) is the default condition. If the mitochondrial pathway of apoptosis is intact, cells deprived of survival factors die. Since the

growth and survival factors for one cell type in a multicellular organism are usually made by another, the "society of cells" that we call an animal is largely maintained by apoptosis (and in vertebrates, at least, by the mitochondrial pathway of apoptosis).

16.11 The activation of Bax and Bak are controlled by other Bcl-2 family proteins

Key concepts

- The antiapoptotic members of the Bcl-2 family block the permeabilization of the mitochondrial outer membrane by Bax and Bak.
- The BH3-only proteins of the Bcl-2 family either directly activate Bax and Bak or interfere with the antiapoptotic Bcl-2 protein functions.

Bax and Bak, in general, do not function to cause MOMP unless activated by another protein (or, perhaps, through other mechanisms). The BH3-only proteins, Bid and Bim, are particularly good at activating Bax and Bak by inducing oligomerization, although other activators almost certainly exist (probably also including non-Bcl-2 family members).

The activation and function of Bax and Bak are inhibited by the antiapoptotic members of the Bcl-2 family, including Bcl-2, Bcl-xL, Mcl-1, and others. These bind to the activator proteins and prevent their function and also bind to active forms of Bax and Bak, as shown in the top half of **FIGURE 16.21**. In this way, the antiapoptotic Bcl-2 family members prevent MOMP and prevent apoptosis.

Other members of the BH3-only protein subfamily, in turn, promote apoptosis, not by directly activating Bax and Bak, but by blocking the functions of antiapoptotic Bcl-2 family members (i.e., inhibiting the inhibitors). These are derepressor (or "sensitizer") BH3-only proteins that sensitize cells to apoptosis induction.

The combined effects of the activator and derepressor BH3-only proteins, acting through Bax and Bak, and antagonized by the antiapoptotic Bcl-2 proteins, are to cause MOMP and apoptosis. The BH3-only proteins are regulated at the levels of transcription, protein stability, interactions with other proteins, and modifications that affect their functions. Because of the variety of ways in which these are regu-

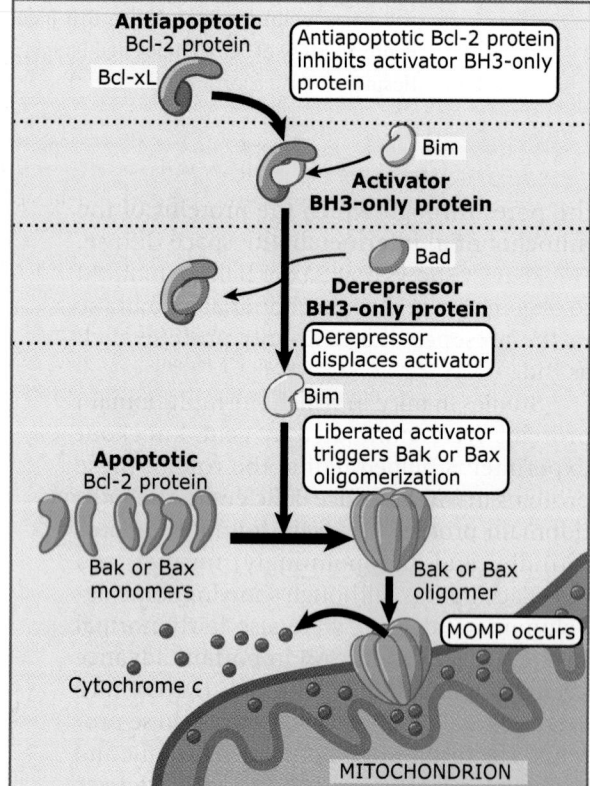

FIGURE 16.21 The BH3-only proteins of the Bcl-2 family function either to directly activate Bax and Bak (activators) or to interfere with the inhibitory functions of the antiapoptotic Bcl-2 family proteins.

lated, these can be thought of as the "stress sensors," linking specific stimuli to MOMP and apoptosis.

16.12 Cytochrome c, released upon mitochondrial outer membrane permeabilization, induces caspase activation

Key concept

- Holocytochrome c triggers the activation of cytosolic apoptotic protease activating factor-1 (APAF-1), which binds and activates caspase-9.

As a result of MOMP, soluble proteins in the mitochondrial intermembrane space diffuse into the cytosol, where they interact with cytosolic proteins. How does this protein release lead to caspase activation? In the cytosol, holocytochrome c (cytochrome c with the heme group attached) binds to APAF-1, inducing it to undergo a conformational change that allows a molecule of dATP to bind to the APAF-1. This causes a further change in APAF-1, so that it now exposes an oligomerization domain, causing it to form a complex of seven APAF-1 molecules, called an apoptosome, the center of which has exposed CARD domains. These bind to the CARD domain of an initiator caspase, caspase-9. The binding of caspase-9 molecules to the CARD domains of the apoptosome forces the caspases to dimerize, thus activating them by induced proximity. **FIGURE 16.22** shows the formation of the apoptosome and its binding of caspase-9. Once activated, caspase-9 cleaves and activates the executioner caspases-3 and -7, and apoptosis proceeds.

Cytochrome c is encoded in the nucleus, and the apocytochrome c molecule is transported into the mitochondrial intermembrane space, where the heme group is enzymatically attached. Only holocytochrome c, not apocytochrome c, can activate APAF-1 to bind to dATP and form an apoptosome, thereby ensuring that activation of apoptosis by this pathway occurs only upon MOMP.

Mice lacking APAF-1, caspase-9, or caspase-3 all have similar developmental defects, relating to a failure to activate the mitochondrial pathway of apoptosis in a timely manner. This phenotype is also shared by a mutant mouse in which cytochrome c is altered so that it can still maintain mitochondrial respiration

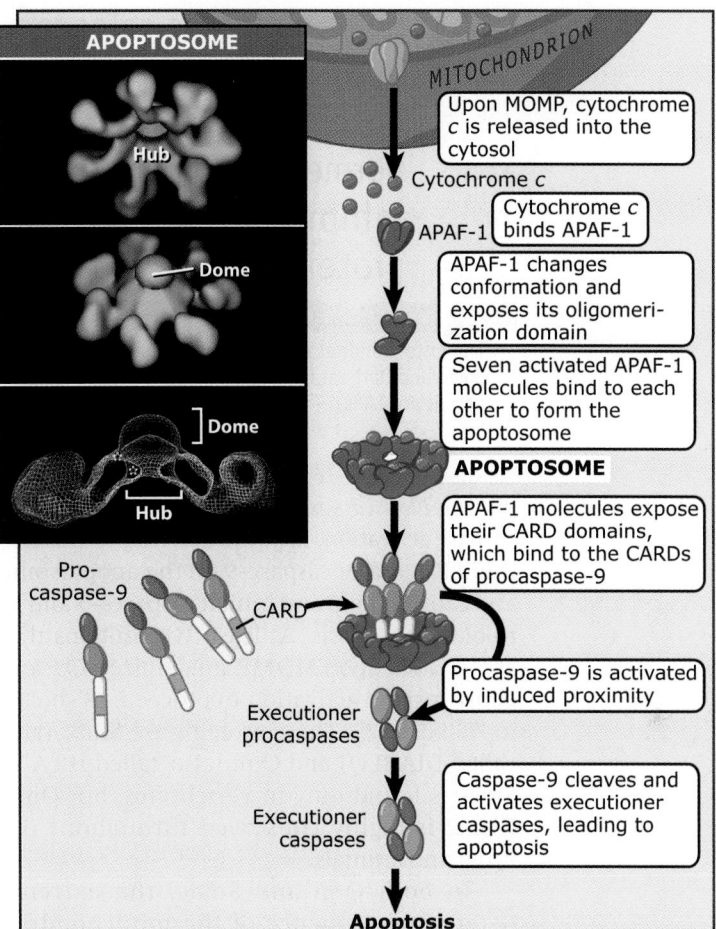

FIGURE 16.22 Cytochrome c activates APAF-1 to form an apoptosome. When cytochrome c is released from the mitochondria into the cytosol, it contacts APAF-1, which then undergoes a conformational change to expose an oligomerization domain. The activated APAF-1 molecules bind to one another to form a complex (the apoptosome). Each APAF-1 molecule in the apoptosome exposes a CARD domain which binds to the CARD domain in caspase-9, thereby activating the caspase. The activated caspase-9 now cleaves and activates executioner caspases. In the inset, the structure of the apoptosome as determined by electron microscopy techniques is shown. Photos reprinted from *Mol. Cell*, vol. 9, D. Acehan, et al., Three-dimensional structure of the apoptosome..., pp. 423–432, Copyright (2002) with permission from Elsevier [http://www.sciencedirect.com/science/journal/10972765]. Photos courtesy of Christopher W. Akey, Boston University School of Medicine.

but is less active in promoting apoptosome formation. These mice show defects in craniofacial development and display a marked enlargement of the forebrain. Interestingly, though, programmed cell death still proceeds in these animals, albeit more slowly and without the characteristics of apoptosis. This "caspase-independent" cell death is discussed in more detail in *16.15 Mitochondrial outer membrane permeabilization can cause caspase-independent cell death*.

16.13 Some proteins released upon mitochondrial outer membrane permeabilization block inhibitors of apoptosis proteins

Key concept

- The mitochondrial intermembrane space proteins Smac and Omi antagonize the caspase-inhibitory activity of IAPs.

As discussed in *16.4 Some inhibitors of apoptosis proteins block caspases*, some IAPs interfere with caspase activation. In particular, XIAP binds and inhibits initiator caspase-9 on the apoptosome and can also bind and inhibit caspases-3 and -7 to block apoptosis. At least two proteins that are released upon MOMP antagonize XIAP and allow caspase activation to proceed, as shown in **FIGURE 16.23**. These proteins are Smac (also called DIABLO) and Omi (also called HtrA2). Smac is found only in vertebrates, but Omi/HtrA2 is highly conserved throughout the fungi and animals.

In both Omi and Smac, the extreme N-terminal sequence of the mitochondrial protein is responsible for inhibiting XIAP. This sequence is called the AVPI motif after the sequence found in mature Smac. The AVPI motif is generated only after proteolytic removal of the mitochondrial localizing sequence that occurs after the nuclear-encoded proteins translocate to the mitochondrial intermembrane space. Therefore, it is only when Smac and Omi are released from mitochondria upon MOMP that these proteins can interfere with the function of XIAP.

The knockouts of XIAP and Smac have little or no phenotypes, whereas that of Omi shows neurologic defects, most likely due to loss of an important mitochondrial activity unrelated to its IAP-inhibitory function. This interpretation is suspected to be correct because the IAP-inhibitory motif in Omi is missing in the protein in some vertebrates, such as bovids. Similarly, cells deficient in both Smac and Omi do not seem to have defects in apoptosis. The biologic importance of caspase inhibition by IAPs and their repression by the IAP antagonists such as Smac and Omi remains an unresolved question.

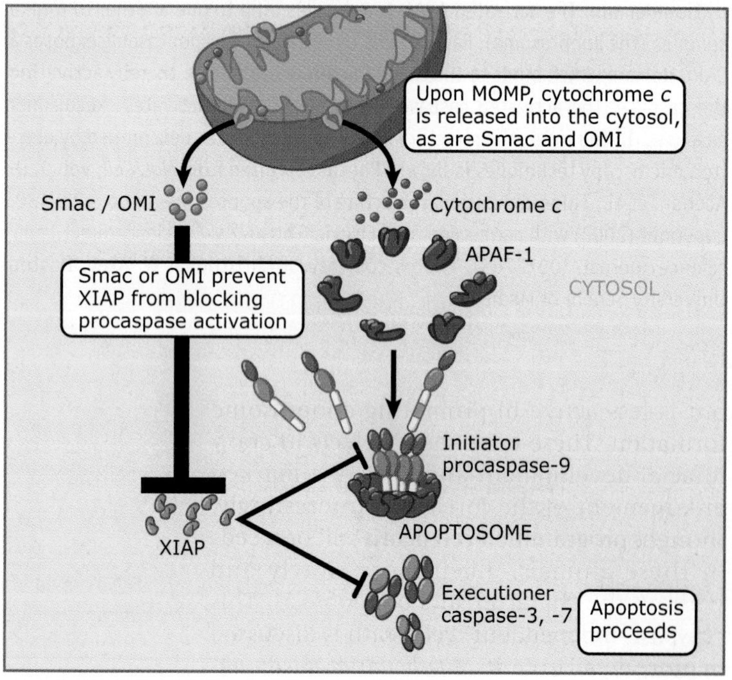

Concept and Reasoning Check

1. Though the evolutionary origins of apoptosis are still poorly understood, many of the proteins in the mitochondrial apoptosis pathway share structural similarity with those involved in the cellular response to infection. This leads some cell biologists to suggest that intrinsic apoptosis arose as a modification of an immune response in unicellular organisms. What aspects of the intrinsic pathway support this hypothesis?

16.14 The death receptor pathway of apoptosis can engage mitochondrial outer membrane permeabilization through the cleavage of the BH3-only protein Bid

Key concepts

- Caspase-8, activated upon ligation of death receptors, cleaves the BH3-only protein Bid, resulting in its activation.
- Bid then triggers Bax and Bak to cause MOMP, thereby engaging the mitochondrial pathway of apoptosis.
- Bid acts as a link between the two apoptotic pathways.

FIGURE 16.23 Proteins released upon MOMP interfere with IAP-mediated inhibition of caspases.

Because XIAP can inhibit the executioner caspases-3 and -7, it can function to inhibit apoptosis triggered by death receptors and the initiator caspase-8 (see *16.6 The death receptor pathway of apoptosis transmits external signals*). However, caspase-8, once activated by the death receptor pathway, can engage the mitochondrial pathway of apoptosis, such that MOMP occurs. It does this by cleavage and activation of the BH3-only protein Bid. The subsequent release of IAP antagonists upon MOMP can be important in allowing apoptosis to proceed.

Bid, like Bim discussed earlier, can activate Bax and Bak to cause MOMP. However, Bid is inactive unless proteolytically cleaved. Several proteases can cleave and activate Bid to cause MOMP and apoptosis via the mitochondrial pathway. These include cathepsins, calpains, and caspases. Caspase-8 is extremely efficient in cleaving and activating Bid; in fact, Bid is a much better substrate for caspase-8 than are the executioner caspases.

Therefore, when caspase-8 cleaves Bid to induce MOMP, one effect is to release the IAP antagonists, Smac and Omi, which block XIAP and promote the activation of the executioner caspases by caspase-8. At the same time, apoptosis is further amplified by cytochrome *c*-mediated activation of the apoptosome and caspase-9. **FIGURE 16.24** shows this process.

The importance of MOMP in promoting the death receptor pathway varies among cells. In Type II cells, inhibition of MOMP by the antiapoptotic Bcl-2 proteins, such as Bcl-2 and Bcl-xL, blocks death receptor-triggered apoptosis. In Type I cells, Bcl-2 has no effect on the death receptor pathway. Based on such considerations, pharmacologic agents that mimic the IAP antagonists are currently being tested for their ability to enhance cancer therapy by death ligands such as TRAIL.

Such effects may also be important in apoptosis induced by cytotoxic lymphocytes. Although granzyme B effectively activates executioner caspases, this activity can be inhibited by XIAP. However, granzyme B, like caspase-8, very effectively cleaves and activates Bid (albeit at a different site) to trigger MOMP. Evidence suggests that Bid-mediated MOMP is the preferred mechanism of apoptosis induction by granzyme B and cytotoxic lymphocytes in at least some target cells. The action of cytotoxic lymphocytes, however, generally involves more than the release of granzyme B, because these cells use additional mechanisms to trigger cell death as well.

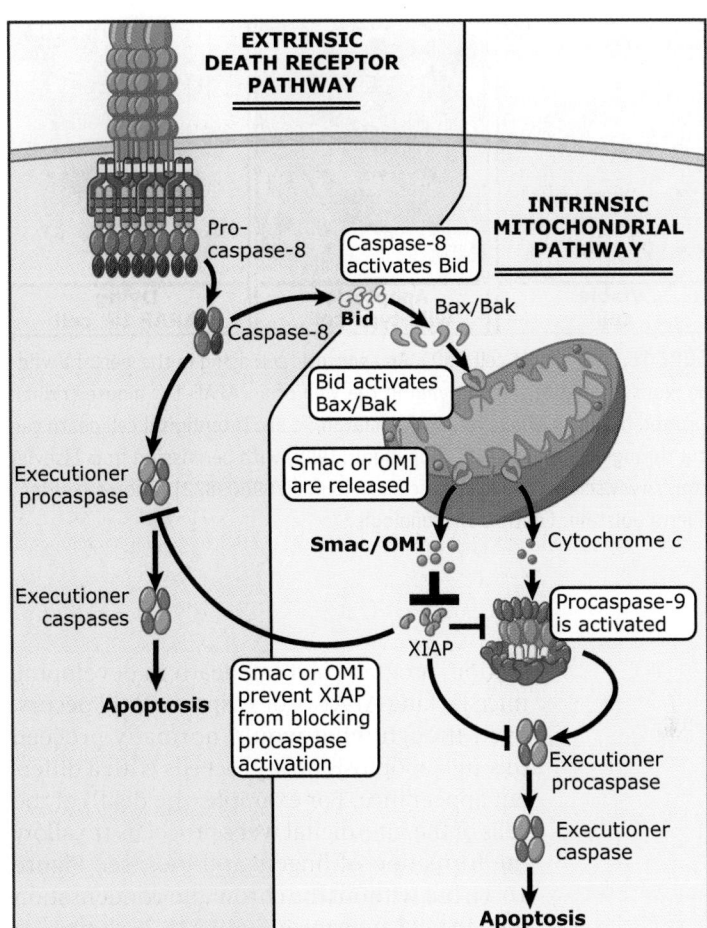

FIGURE 16.24 Caspase-8, activated as a result of the death receptor pathway, cleaves the BH3-only protein, Bid, thereby activating it. Activated Bid acts as an activator of Bax and Bak to promote MOMP and cytochrome *c* release. As a result, caspase-9 is activated. In addition, Smac and Omi, released upon MOMP, effectively de-repress any IAP-mediated inhibition of executioner caspase activation.

16.15 Mitochondrial outer membrane permeabilization can cause caspase-independent cell death

Key concept

- Once MOMP occurs, cells generally die even if caspase activation is blocked or disrupted. The precise mechanisms of this cell death are not fully known.

Once the mitochondrial pathway of apoptosis is engaged and MOMP has occurred, cell death can proceed even if caspase activation is blocked or disrupted. As noted in the previous

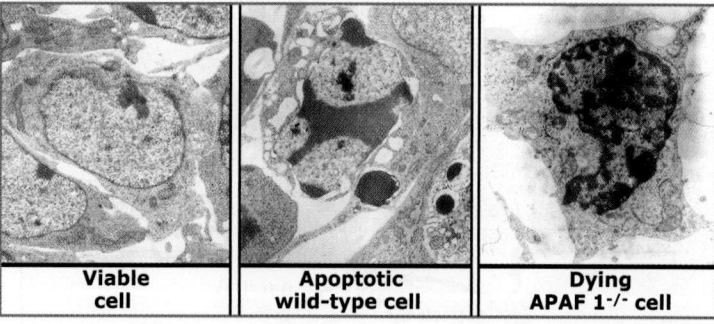

FIGURE 16.25 A viable cell (left). An apoptotic cell dying in the web of a wild-type mouse (middle). A cell dying in the web of an APAF-1⁻/⁻ mouse (right). Reprinted from *Curr. Biol.*, vol. 9, M. Chautan, et al., Interdigital cell death can occur through..., pp. 967–970, Copyright (1999) with permission from Elsevier [http://www.sciencedirect.com/science/journal/09609822]. Photos courtesy of Pierre Golstein, Centre d'Immunologie.

section, programmed cell death in developing mice lacking APAF-1 or caspase-9 still occurs, even though what would normally proceed through apoptosis now proceeds with a different appearance. For example, the death of the cells of the interdigital webs proceeds (to allow the formation of fingers and toes; see Figure 16.1), but without the chromatin condensation seen in wild-type mice. **FIGURE 16.25** shows the appearance of these cells.

Similarly, caspase inhibitors that block apoptotic cell death can fail to rescue cells that have undergone MOMP from this "caspase-independent" cell death. In contrast, Bcl-2, which blocks MOMP, can preserve cell survival under the same conditions.

Two general mechanisms for this caspase-independent cell death are proposed. One is that following MOMP, mitochondria slowly lose function. (A rapid loss of mitochondrial function following MOMP is caused by activated caspases.) This slow loss of mitochondrial function may condemn the cell to die even if caspases are not activated.

Alternatively (or perhaps at the same time), proteins that are released from the mitochondrial intermembrane space may kill the cell in a manner not involving caspases. These include a DNase, endonuclease G, Omi, and apoptosis-inducing factor (AIF), all of which are able to cause death if over-expressed in cells. (The death-inducing activity of Omi is independent of its IAP-inhibitory activity.) However, because each of these proteins also appears to perform other critical functions in mitochondria required for proper cellular

physiology, it is difficult to prove which, if any, are actually involved in cell death following MOMP. None unambiguously cause cell death following MOMP in any physiologic situation.

Some cells undergo MOMP without undergoing caspase-independent cell death. In postmitotic neurons, for example, growth factor deprivation induces MOMP, but if caspase activation is blocked, the cells can subsequently be rescued by re-addition of the growth factors. At present, it is not known if cells that are capable of proliferation can survive MOMP.

16.16 The mitochondrial permeability transition can cause mitochondrial outer membrane permeabilization

Key concept

- In some forms of cell death, the mitochondria are disrupted by a change in the mitochondrial inner membrane, leading to swelling and rupture of the organelle.

If isolated mitochondria are exposed to a variety of agents, including high calcium concentrations, reactive oxygen species, or certain drugs, the mitochondrial inner membrane can undergo a change that allows small solutes to pass through a poorly defined channel in the membrane. As a result, the electrochemical gradient across the membrane dissipates and the matrix swells. This swelling can be sufficient to rupture the outer membrane to release intermembrane space proteins. This mitochondrial permeability transition (mPT) plays roles in cell death caused by extreme conditions, such as occur in ischemic injury (e.g., stroke and heart attack) resulting in necrosis. It remains unclear if, and under what conditions, it may function in apoptosis.

Although many of the molecular components of the mPT are controversial, it is well established that a matrix enzyme, cyclophilin D, facilitates this event (although it is not absolutely essential). Mice lacking cyclophilin D show defects in mPT and are resistant to ischemic injury, but show normal development and apoptosis. Drugs, such as cyclosporine, that inhibit cyclophilin D can prevent or delay mPT and are being tested for efficacy in some forms of cell injury.

16.17 Many discoveries about apoptosis were made in nematodes

Key concept

- Apoptosis in nematodes follows a simple pathway with similarities to the mitochondrial pathway of apoptosis in the vertebrates.

During development in *C. elegans*, 131 of the 1090 adult somatic cells undergo programmed cell death: cells die predictably at a defined time and place in each animal, and this developmental cell death proceeds through apoptosis. The molecular elements of the *C. elegans* apoptotic pathway are absolutely required for this programmed cell death. The genetic dissection of apoptosis in *C. elegans* helped to lay the foundations for understanding apoptosis in the vertebrates, especially the mitochondrial pathway, although there are some striking differences.

The apoptotic pathway of *C. elegans* is illustrated in **FIGURE 16.26**. The single caspase responsible for cell death during nematode development is the product of the *C. elegans* death gene *ced-3* (CED-3). Like caspase-9 in vertebrates, CED-3 contains a CARD domain and is activated by dimerization. The adaptor molecule responsible for activating CED-3 is homologous to APAF-1 and binds the caspase via its own CARD domain. Unlike APAF-1, this protein, called CED-4, does not appear to require activation by cytochrome *c* (in fact, CED-4 lacks the equivalent domain in APAF-1 that is thought to interact with cytochrome *c*). It is likely that CED-4 requires dATP for its oligomerization, which is in turn needed for CED-3 activation.

In cells not destined to die, CED-4 is bound by another protein, CED-9. Remarkably, CED-9 is a Bcl-2 homolog, but unlike the antiapoptotic Bcl-2 proteins in vertebrates, CED-9 does not regulate MOMP. Instead, it appears to control apoptosis by its direct binding to CED-4. In contrast, none of the antiapoptotic Bcl-2 family members in vertebrates bind or directly inhibit APAF-1 (and CED-9 does not inhibit apoptosis if expressed in vertebrate cells).

The binding of CED-9 to CED-4 is disrupted by another protein, the product of the *egl-1* gene (egl stands for "egg laying deficient"). EGL-1 is a BH3-only protein. When it is transcriptionally expressed, EGL-1 protein displaces CED-4

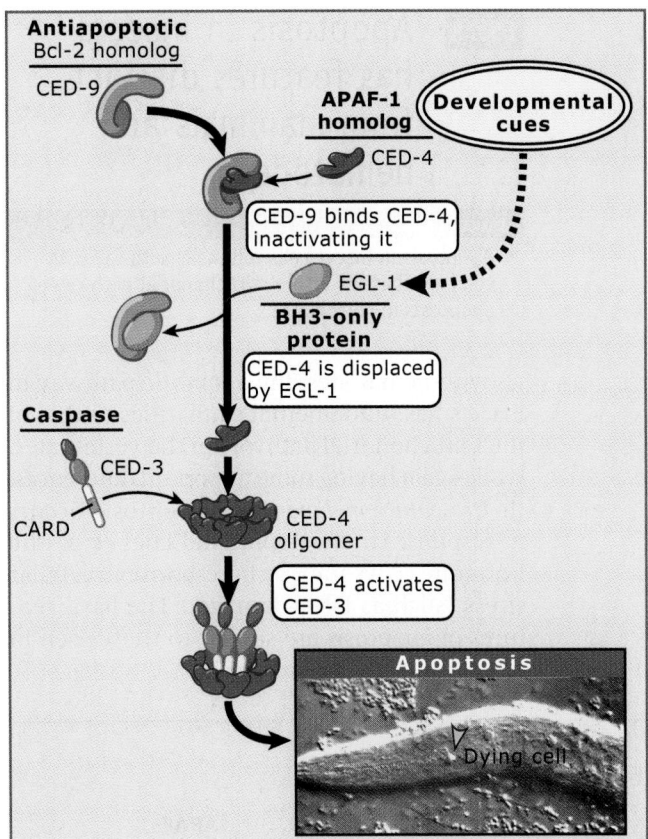

FIGURE 16.26 The photograph shows a dying cell in a nematode embryo. The scheme shows the relationships between the proteins that control programmed cell death in the worm. Photo reprinted from *Cell*, vol. 44, H. M. Ellis and H. R. Horvitz, Genetic control of programmed cell death..., pp. 817–829, Copyright (1986) with permission from Elsevier [http://www.sciencedirect.com/science/journal/00928674]. Photo courtesy of H. Robert Horvitz, Massachusetts Institute of Technology.

from CED-9, and CED-4 now activates CED-3 to kill the cell.

Loss of function mutations in *egl-1*, *ced-4*, or *ced-3* result in cell survival, and these cells can differentiate into functional cells. Gain-of-function mutations in *ced-9* have the same effect. Loss of *ced-9* function results in early embryonic lethality that is suppressed by loss of function mutations in *ced-4* or *ced-3* (but not *egl-1*).

Additional genes regulate what happens to the dying cell, including phagocytosis and degradation. These are discussed in more detail in *16.19 The clearance of apoptotic cells requires cellular interaction*.

16.18 Apoptosis in insects has features distinct from mammals and nematodes

Key concept

• Apoptosis in insect cells follows a pathway with some similarities to the mitochondrial pathway of apoptosis in vertebrates.

As with nematodes, the apoptotic pathway in insects has fundamental similarities to that of the mitochondrial pathway in the vertebrates, while again having some important differences. In *Drosophila melanogaster*, apoptosis occurs in response to developmental cues (e.g., the hormone ecdysone), or in response to cellular stress (such as DNA damage). The basic features of apoptosis are similar to those seen in vertebrates and nematodes.

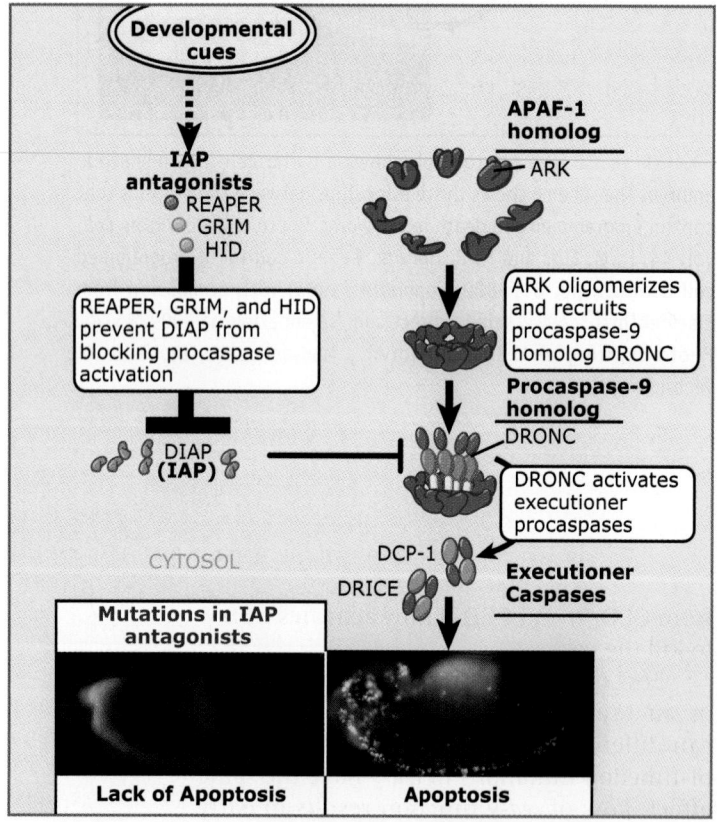

FIGURE 16.27 The photo on the right shows programmed cell death, which proceeds by apoptosis, in a wild-type fly. Dying cells (bright green) were detected with acridine orange staining and death was confirmed by a number of other techniques. The photo on the left shows that a fly embryo bearing mutations in the region that includes the genes encoding Reaper, Hid, and Grim does not have cells undergoing programmed cell death. The schematic illustrates the major apoptotic pathway in *Drosophila*. Photos courtesy of John Abrams, UT Southwestern Medical Center at Dallas.

The apoptotic pathway in *Drosophila* is shown in FIGURE 16.27. The executioner caspases that orchestrate apoptosis in insects are DRICE and DCP-1. These may be activated by cleavage, though this is not definitively known. The major initiator caspase responsible for activating these executioner caspases is called DRONC, and like caspase-9 (and CED-3) has a CARD domain. Another potential initiator caspase is DREDD, and like caspase-8, the prodomain of DREDD has DEDs. However, its role in any form of apoptosis in flies remains obscure.

DRONC is activated by the insect APAF-1 homolog, ARK (APAF-1-related killer). Although ARK has all of the regions found in APAF-1, there is currently no biochemical evidence that ARK is activated by cytochrome *c*. Similarly, neither MOMP nor the release of cytochrome *c* is observed in insect cells undergoing apoptosis. It is possible that ARK is either constitutively active or activated by some mechanism that does not depend on MOMP. Like APAF-1 and CED-4, ARK probably forms oligomeric structures that bind to DRONC via a CARD-CARD interaction to activate the initiator caspase. Flies lacking ARK or DRONC, or cells in which their expression is suppressed, do not display apoptosis.

The activation of DRONC is not sufficient for induction of apoptosis, however. IAPs, and in particular DIAP1, block DRONC and the executioner caspases and sustain cell survival. This is an important key to the *Drosophila* apoptotic pathway.

When apoptosis is triggered, one or more proteins are expressed that have the ability to antagonize IAP function. These proteins— Reaper, Head involution defective (Hid), Grim, and Sickle—all carry a motif related to the AVPI motif of Smac and Omi but otherwise have no similarities to any vertebrate molecule. Also, unlike Smac and Omi, the *Drosophila* proteins are cytosolic and are not found in the mitochondria. These apoptosis inducers block the IAPs, DRONC becomes active, and apoptosis proceeds.

Interestingly, Reaper, Hid, Grim, and Sickle are part of a cluster of genes. A deletion, *Df(3L) H99*, that eliminates the genes for Reaper, Hid, and Grim, was identified as being defective in apoptosis induction, leading to the discovery of the genes.

Apoptosis in insect cells appears to be controlled mostly by the IAPs, in contrast to vertebrates, for which the Bcl-2 family proteins predominantly regulate apoptosis. Two Bcl-2 family proteins were identified in *Drosophila*, called Debcl (also called dBorg-1 and dBok) and Buffy (also called dBorg-2). Their roles in regulating apoptosis remain unresolved.

16.19 The clearance of apoptotic cells requires cellular interaction

Key concept
- The removal of apoptotic cells from the body occurs by an active process.

The clearance, or "burial," of the dying cell is the ultimate function of apoptosis, as the cell is removed with almost no indication that it had ever been there. This occurs by **phagocytosis** ("cell eating"), either by professional phagocytes (macrophages, dendritic cells) or "amateurs" (e.g., epithelial cells). The importance of efficient clearance of apoptotic cells is underscored by the characteristics of apoptotic cell death that probably contribute to this process, as the cell is packaged for removal. DNA is chopped into small pieces for facilitated digestion, cells fragment into "bite-sized" morsels, and the cell generates signals to solicit its engulfment. The latter process can be summarized as "find me" signals, "eat me" signals, and the disruption of "don't eat me" signals, shown in **FIGURE 16.28**.

Studies in *C. elegans* have identified genes that are important in the clearance of apoptotic cells, and these fall into two general **complementation groups**. The first set of genes includes *ced-1*, *ced-6*, and *ced-7*, which appear to be involved in early recognition and signaling events. The genes in the second set, *ced-2*, *ced-5*, *ced-10*, and *ced-12* are probably involved in the machinery of engulfment rather than recognition. All of these genes act at the level of the phagocyte, affecting the ability of the phagocytic cell to recognize or engulf the dying cell, although one (*ced-7*) also appears to function in the dying cell as well.

CED-1 may be a receptor that responds to changes in the apoptotic cell membrane. It is related to a class of receptors in mammalian cells called scavenger receptors. Two of these in mammals, SREC (scavenger receptor expressed by endothelial cells) and CD91, make some contribution to apoptotic cell engulfment by mammalian phagocytes in culture. The only gene identified in *Drosophila* that contributes to apoptotic cell clearance, called *croquemort*, is also a member of the scavenger receptor family.

Studies in mammalian cells have shown that a critical "eat me" signal (although probably not the only one) in the dying cell is the externalization of phosphatidylserine, which occurs when the plasma membrane scrambles as a consequence of executioner caspase function. **FIGURE 16.29** shows that phosphatidylser-

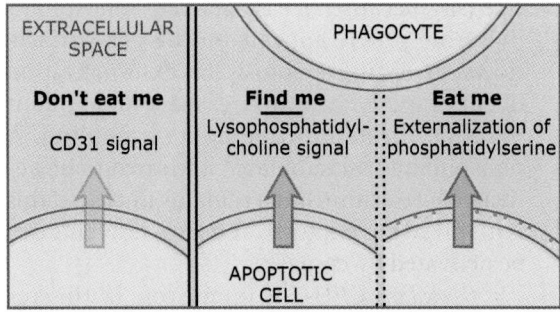

FIGURE 16.28 As cells undergo apoptosis, they produce "find me" signals, such as lysophosphatidylcholine, they downregulate "don't eat me" signals, such as CD31, and they express "eat me" signals, such as the externalization of phosphatidylserine. As a result of all of these, the dying cell is rapidly eaten by "professional" phagocytes such as macrophages, or other cells (such as epithelial cells).

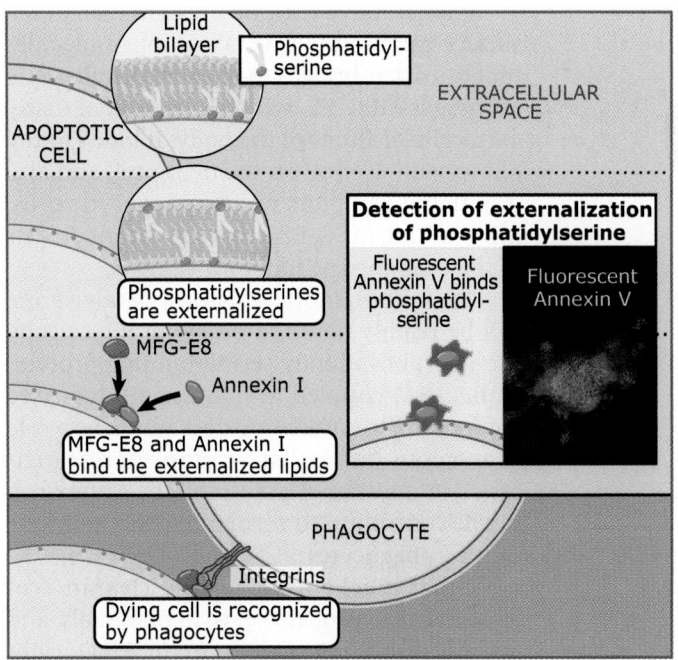

FIGURE 16.29 During apoptosis, the plasma membrane scrambles its lipids, such that phosphatidylserine, which is normally on the inner leaflet, appears on the outer leaflet. There it is bound by proteins such as Annexin I and MFG-E8, which in turn are bound by phagocytic cells, which then engulf the apoptotic cells. The externalization of phosphatidylserine can be detected by the use of fluorescent Annexin V. Photo courtesy of Joshua C. Goldstein and Douglas R. Green, St. Jude Children's Hospital.

ine is normally localized to the inner leaflet of the plasma membrane by an ATP-dependent "flippase" that flips the phospholipid from the outer to inner leaflet of the membrane. (The term flip refers specifically to this outside-inside translocation of lipids. It is often misused in the apoptosis literature to refer to exposure of phosphatidylserine as a "flip." The latter is, technically, a "flop.") If the translocase is inactive (say, because ATP is depleted), phosphatidylserine slowly appears on the cell surface; however, during apoptosis this externalization occurs rapidly, and in a caspase-dependent manner, through active lipid scrambling. A phospholipid "scramblase" is known to be activated by calcium, but it remains unclear if this is involved in apoptosis, and if so, how this can be activated by caspases.

C. elegans CED-7 is interesting in this regard, as it functions not only in the phagocytic cell but also in the dying cell. The mammalian homolog of CED-7 is ABCA1, a protein that is involved in lipid reorganization in the plasma membrane, and it was suggested that this protein is important in externalization of phosphatidylserine in mammalian cells. However, recent studies have suggested that it is not involved in this event in *C. elegans*.

Phosphatidylserine on the mammalian cell surface can bind to several soluble molecules that can act as bridges to phagocytic cells. One of these is MFG-E8, which is present in many extracellular fluids of the body. It binds to the phosphatidylserine on the dying cell and also to integrin receptors on the phagocytic cells. Mice lacking MFG-E8 show significant defects in the clearance of apoptotic cells.

The externalization of phosphatidylserine can be readily detected on apoptotic cells by use of a phosphatidylserine-binding protein, Annexin V, coupled to fluorescent dyes (see Figure 16.29). Annexin V does not have a role in apoptosis (other than as a tool for researchers), but another related protein, Annexin I, might contribute to recognition of apoptotic cells by phagocytes.

Other mechanisms for the clearance of apoptotic cells probably exist in mammals, and in some cases these are important. Phagocytes from mice lacking a receptor with tyrosine kinase activity, MER, show defects in apoptotic cell clearance in culture, and these animals accumulate apoptotic cells. How MER functions in this regard remains unclear.

Healthy cells can express "eat me" signals; for example, elevated calcium can trigger phospholipid scrambling to temporarily externalize phosphatidylserine. These are not removed, however, because healthy cells also seem to bear "don't eat me" signals that probably become inactive as cells undergo apoptosis. One such "don't eat me" signal is CD31, which is present on both the phagocytes and on most healthy cells. The interaction of CD31 with itself may inhibit the clearance of healthy cells displaying "eat me" signals.

As cells undergo apoptosis, they can actively recruit "professional" phagocytic cells, such as macrophages, by producing a "find me" signal. One of these signals is lysophosphatidylcholine, which can be produced by activated phospholipase A. Executioner caspases can activate phospholipase A by cleaving the enzyme. In this way, apoptotic cells can increase the likelihood that they will be found and efficiently cleared from the body.

16.20 Apoptosis plays a role in diseases such as viral infection and cancer

Key concept

- Viral infection and cancer are conditions in which apoptotic pathways may be blocked.

Viruses are obligate intracellular parasites, and, therefore the death of the host cell is a strategy for eliminating viral infections. For this reason, viruses have a vested interest in keeping cells alive. Many viruses do this by blocking apoptosis, and the ways in which they do so make sense when we consider the cell death pathways that are important in this context.

As we have seen, apoptosis in insects is regulated at the level of caspase activation by the IAPs. The insect baculoviruses block apoptosis in precisely this manner, by expressing an IAP that inhibits DRONC and the executioner caspases. In addition, these viruses also express another protein, p35, which binds and inhibits active caspases. By blocking caspase activation, these viruses keep the cell alive until sufficient virus is produced and the cell is lysed.

Viruses in vertebrate cells, in general, do not use this strategy, perhaps because caspase inhibition does not block MOMP or caspase-independent cell death (and might promote immune responses to the virus). Instead, some vertebrate viruses produce antiapoptotic proteins in the Bcl-2 family, such as the adenovirus

E1B19K protein, and the Epstein-Barr virus BHRF protein. These and other viral Bcl-2 proteins function, like Bcl-2, to prevent MOMP and apoptosis.

Viruses in vertebrate cells often also operate to prevent apoptosis induced by immune effector cells. Pox viruses, for example, produce protease inhibitors, called serpins, that are capable of blocking granzyme B, as well as caspase-8 (but do not effectively block caspase-9 or the executioner caspases). By blocking granzyme B, the virus evades the apoptosis-inducing effects of cytotoxic lymphocytes that survey the body for virally infected cells. Similarly, inhibition of caspase-8 blocks apoptosis induced by death ligands binding to their receptors; these ligands are often produced by cytotoxic lymphocytes and other cells in response to viral infection. Another way in which viruses block the death receptor pathway is through expression of molecules related to c-FLIP, such as the v-FLIP protein produced by herpes viruses.

Another situation in which disease is associated with a subversion of the mechanisms of apoptosis is cancer. As mentioned at the start of the chapter, the potent tumor suppressor, p53, which is mutated in ~50% of human cancers, exerts its suppressive effects in part by inducing apoptosis in the transformed cell. Similarly, the antiapoptotic protein Bcl-2 was originally discovered as a breakpoint of chromosomal translocations in follicular B cell lymphomas. Cancer involves changes in the control of apoptosis.

But the relationships between apoptosis and cancer are more fundamental than these simple associations suggest. Cells in vertebrates are sensitized for apoptosis by the signals that cause them to enter the cell cycle, and the decision to live or die comes from survival signals imparted to the cell by the surrounding tissue (e.g., by growth factors). When these factors are limited, apoptosis occurs in the cycling cells, and this limits tissue expansion. For example, FIGURE 16.30 shows that the c-Myc protein promotes not only cell proliferation but also cell death by apoptosis. Therefore, simply expressing c-Myc in a tissue does not necessarily promote expansion (or cancer) unless other survival signals are present. It is likely that this basic relationship between the cell cycle and apoptosis is fundamental to how we can exist as large, long-lived, multicellular organisms without cancer being ubiquitous. It is not that the cell cycle causes apoptosis; instead, the same molecules that promote the entry into

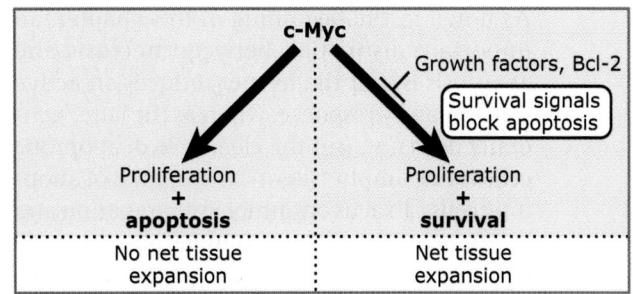

FIGURE 16.30 c-Myc and related signals not only drive cells into cycle but also sensitize cells to die via apoptosis. As a result, in the absence of additional signals, no net tissue expansion occurs. When apoptosis is blocked, for example, by growth factors, cell number can increase. The ability of c-Myc to sensitize cells for apoptosis is a fundamental property of this protein and not a side effect of engaging the cell cycle. Mechanisms such as this are probably central for normal homeostasis and the prevention of cancer.

the cell cycle also trigger apoptosis that can be suppressed by antiapoptotic signals.

Although c-Myc promotes both proliferation and cell death, Bcl-2 or Bcl-xL can work with c-Myc to promote oncogenesis by blocking apoptosis. Indeed, the antiapoptotic members of the Bcl-2 family are frequently upregulated in human cancers. This cooperation between these two types of oncogenes (proapoptotic and proliferative, plus antiapoptotic) is likely to be a common principle in cancer. In general, oncogenes are like Myc (promoting both proliferation and apoptosis), like Bcl-2 (blocking apoptosis without promoting proliferation), or like both (promoting proliferation while blocking apoptosis, as a consequence of multiple signals). The potent Ras protein, in its active form, can engage distinct signaling pathways that lead to these two outcomes. Thus, Ras can both promote or inhibit apoptosis in different settings.

Concept and Reasoning Check

1. If cells can induce apoptosis when they become cancerous, what apoptosis-associated proteins could be good targets for anti-cancer treatments? What are some of the complications you would expect from these treatments?

16.21 Apoptotic cells are gone but not forgotten

Key concept

- The uptake and clearance of apoptotic cells has lasting effects on the immune system.

As noted in the beginning of this chapter, an important distinction between necrosis and apoptosis is that the former induces an active inflammatory response, whereas the latter generally does not. But the clearance of apoptotic cells is not simply "silent"; the uptake of apoptotic cells can actively inhibit inflammation and immune responses. For example, apoptotic cell uptake can inhibit the generation of inducible nitric oxide synthase, an enzyme that produces nitric oxide, which is an important signaling molecule that participates in inflammation.

In addition, the uptake of apoptotic cells by dendritic cells, the key antigen-presenting cells for T lymphocytes, can result in associated antigens being presented in a manner that induces a state of immune tolerance. The immune system is essentially "instructed" not to generate lasting immune memory to the antigens associated with cells dying by apoptosis. Some current theories of immune function suggest that this is an important mechanism for self-tolerance by the immune system. The use of apoptotic cells to "trick" the immune system into tolerating antigens is being explored for practical purposes.

A failure to effectively clear apoptotic cells may be at the root of systemic lupus erythematosis (SLE), an autoimmune disease characterized by autoantibodies to a number of intracellular antigens, including nuclear proteins and double-stranded DNA. A diagnostic feature of this disease is the presence of LE bodies in the circulation. LE bodies are apoptotic cells, either free or engulfed in blood phagocytes. Mice lacking MFG-E8 or MER and, therefore, displaying defective clearance of apoptotic cells (see above) develop a disease that resembles SLE in many ways.

16.22 What's next?

Predicting "what's next" is, at best, problematic. As one sage put it, "Difficult to see; always in motion is the future." While prognostication is hardly a scientific discipline, we may be able to speculate on those questions under current investigation that are likely to illuminate those aspects of cell death that we still do not understand. But bear in mind that one of the things that makes biologic science so interesting is its unpredictability. With those caveats in mind, we can look at the future of cell death research.

While two central pathways of apoptosis are characterized in detail, there are almost certainly more. Caspase-2 is an initiator caspase that is bound and activated by the adaptor molecule RAIDD (another molecule PIDD, is implicated in the activation of caspase-2). Nevertheless, we know very little about how or when this putative pathway is engaged, although hints are emerging. Caspase-2 may be involved in apoptosis induced by heat shock, by metabolic changes in the cell, and in some cases by DNA damage. Similarly, almost nothing is known about the activation of caspase-4 and roles it might play in apoptosis or inflammation. In contrast, while we know a great deal about the activation of caspase-8 and its role in apoptosis, we do not know how this protease and its partner molecules FADD and c-FLIP contribute to development (as mentioned above, the knockouts of these genes are all lethal at an early stage of embryogenesis).

Not all active cell deaths in the body are apoptotic. Some forms of necrosis may be biologically controlled (and not simply due to accidental damage). In addition, recent efforts have focused on another form of cell death that is accompanied by the process of autophagy. Autophagy is seen in all eukaryotes and is best understood to be a mechanism for short-term generation of energy for survival when nutrients are lacking. At present, we do not know if "autophagic cell death" is caused by autophagy, or something else.

The next few years will also see a deeper dissection of the connections between the central apoptotic pathways and other important events in the cell. Signaling from the cell cycle, cytoskeleton, repair pathways, stress pathways, calcium and other ions, metabolic pathways, and others undoubtedly impact on elements of the apoptotic pathways we have discussed, but little is known about such signals and their targets in the apoptosis machinery.

Similarly, the signals that maintain survival in the cell are important and in need of better understanding. Signaling pathways that block apoptosis are crucial for normal tissue development, but if they go awry they can cause unwanted tissue expansion and cancer. While many of these survival mechanisms are identified, the precise ways in which they function to block the apoptotic pathways will remain an area of intense research.

We still have a lot to learn about apoptosis, but we also understand a great deal, as this chapter attests. Despite this, our knowledge of apoptosis is still not being effectively applied to combat disease. Caspase inhibitors are currently in clinical trials for prevention of apoptosis in some settings of tissue damage. Ultimately, inhibitors of Bax and Bak may emerge as agents to prevent unwanted cell death at an early (and recoverable) stage of the process. Conversely, inhibitors of the antiapoptotic Bcl-2 family proteins are currently being tested as anticancer agents, an approach that shows promise. Similarly, pharmacologic inhibitors of IAPs are being used to promote caspase activation in tumors, and these too appear to be effective (although at this point, we do not know if they actually work as suspected.)

One prediction regarding the future of cell death research can be made with confidence. The field will continue to be interesting.

16.23 Summary

Apoptosis is a form of cell death in animals that plays roles in normal development and homeostasis, as well as in diseases including viral infection and cancer. During apoptosis, cells are "packaged" for rapid removal by phagocytic cells, which involves the action of a set of proteases called caspases. The activation of the caspases occurs through the engagement of one or more apoptotic pathways. In mammals, the two best-defined apoptotic pathways are the extrinsic death receptor pathway and the intrinsic mitochondrial pathway. In the death receptor pathway, the ligation of a specialized death receptor—a member of a subset of the TNF-receptor family—causes initiator caspase-8 to be dimerized and activated, and this in turn cleaves and activates executioner caspases-3 and -7 to orchestrate apoptotic cell death. In the mitochondrial pathway, proteins of the Bcl-2 family interact to ultimately cause a disruption of the mitochondrial outer membrane, releasing proteins into the cytosol. Cytochrome c, so released, engages APAF-1, which oligomerizes and in turn binds and activates initiator caspase-9. This in turn cleaves and activates executioner caspases-3 and -7 to orchestrate apoptosis.

http://biology.jbpub.com/lewin/cells

To explore these topics in more detail, visit this book's Interactive Student Study Guide.

References

Boatright, K. M., and Salvesen, G. S. 2003. Mechanisms of caspase activation. *Curr. Opin. Cell Biol.* v. 15 p. 725–731.

Bodmer, J. L., Schneider, P., and Tschopp, J. 2002. The molecular architecture of the TNF superfamily. *Trends Biochem. Sci.* v. 27 p. 19–26.

Clem, R. J. 2005. The role of apoptosis in defense against baculovirus infection in insects. *Curr. Top. Microbiol. Immunol.* v. 289 p. 113–129.

Danial, N. N., and Korsmeyer, S. J. 2004. Cell death: Critical control points. *Cell* v. 116 p. 205–219.

Hay, B. A., Huh, J. R., and Guo, M. 2004. The genetics of cell death: Approaches, insights and opportunities in *Drosophila*. *Nat. Rev. Genet.* v. 5 p. 911–922.

Hill, M. M., Adrain, C., and Martin, S. J. 2003. Portrait of a killer: The mitochondrial apoptosome emerges from the shadows. *Mol. Interv.* v. 3 p. 19–26.

Kinchen, J. M., and Hengartner, M. O. 2005. Tales of cannibalism, suicide, and murder: Programmed cell death in *C. elegans*. *Curr. Top. Dev. Biol.* v. 65 p. 1–45.

Lauber, K., Blumenthal, S. G., Waibel, M., and Wesselborg, S. 2004. Clearance of apoptotic cells: Getting rid of the corpses. *Mol. Cell* v. 14 p. 277–287.

Martinon, F., and Tschopp, J. 2004. Inflammatory caspases: Linking an intracellular innate immune system to autoinflammatory diseases. *Cell* v. 117 p. 561–574.

Vaughn, D. E., Rodriguez, J., Lazebnik, Y., and Joshua-Tor, L. 1999. Crystal structure of Apaf-1 caspase recruitment domain: An alpha-helical Greek key fold for apoptotic signaling. *J. Mol. Biol.* v. 293 p. 439–447.

Cancer—Principles and overview

17

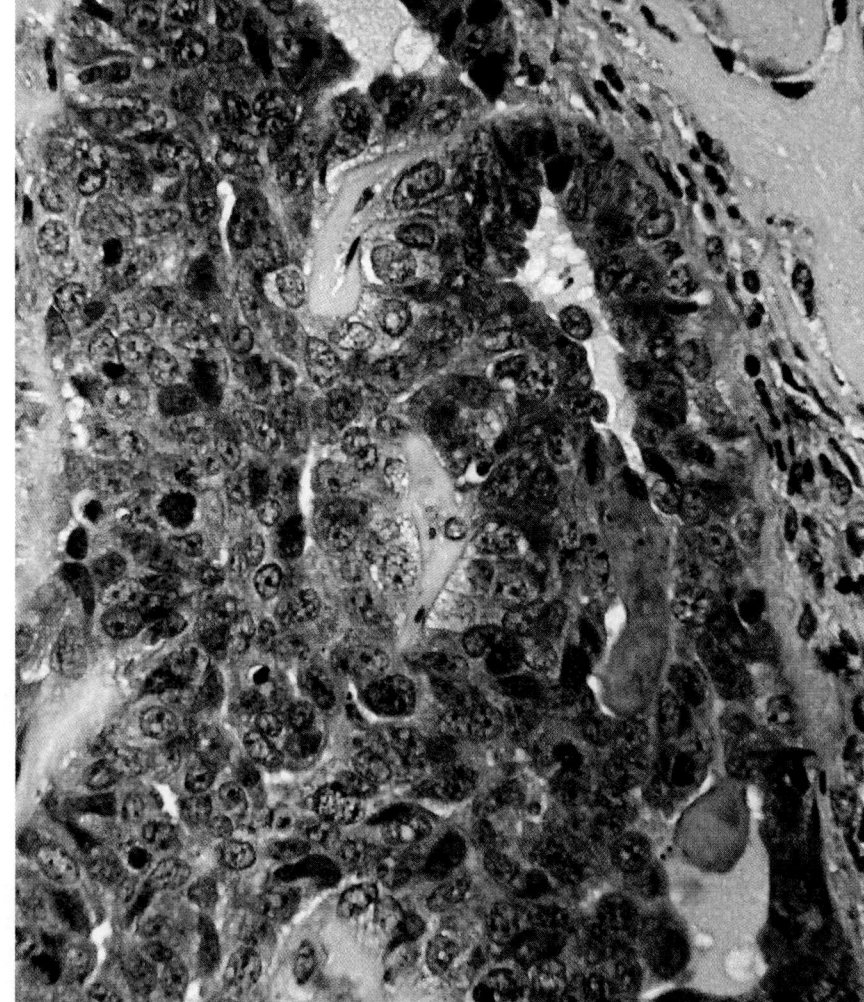

Robert A. Weinberg

Whitehead Institute for Biomedical
Research and Massachusetts Institute of
Technology, Cambridge, MA

EXPERIMENTAL TRANSFORMATION of human mammary epithelial cells through the introduction of a set of oncogenes and the gene encoding telomerase has yielded cells that create tumors very reminiscent of the invasive ductal carcinomas frequently encountered in the oncology clinic. The tumor seen here shows several ducts and, in addition, a significant number of mesenchymal cells of host origin that have been recruited into the tumor-associated stroma (upper right). Photo courtesy of Tan A. Ince, Harvard Medical School.

CHAPTER OUTLINE

17.1 Tumors are masses of cells derived from a single cell

Key concepts

- Cancers progress from a single mutant cell to a tumor and then to metastasis.
- Tumors are clonal.
- Tumors are classified by cell type.

Cancer is a disease of cells that proliferate at inappropriate times and locations in the body. When cells acquire mutations that abolish regulation of cell division, the cells multiply to form masses that we call tumors. Benign tumors are noninvasive and therefore do not affect distant tissues, whereas malignant tumors proceed along a destructive pathway that involves local invasion. Tumor growth is accompanied by the growth of new blood vessels that nourish the tumor, a process called angiogenesis. Finally, malignant cells can detach from their original location and establish tumors in new locations in the body, doing so in a process termed metastasis. This sequence of steps is shown in **FIGURE 17.1**.

The aberrantly growing cells within a tumor mass represent a **monoclonal** outgrowth, in that all the cells within a cancer descend from a single common progenitor cell, sometimes termed a "cell-of-origin." This progenitor cell sustained the initial damage that launched it and its descendants on a path toward the aberrant growth associated with malignancy. The initial damage, in the form of a genetic mutation, may occur decades before the tumor becomes detectable; subsequently, the descendants of this founding cell continue to exhibit many of the aberrations that were acquired initially by the founder cell and acquire additional ones during this time. These additional aberrations are acquired during a succession of steps that is termed "tumor progression." Viewed in this way, cancer becomes a disease of cells, and many if not all of the traits of tumors encountered in the oncology clinic may be deduced from the traits of its individual constituent cells.

Many different cell types within the human body can form tumors, each with its own characteristic histologic (i.e., microscopic) appearance and biological behavior. Accordingly, cancers appear in more than 100 distinct tissue sites and exhibit a wide spectrum of distinctive traits. Even within a given tumor type, there is great variability in behaviors and biochemistry. Nevertheless, while the particular characteristics of the specific cell type may affect some behaviors of the tumor (e.g., melanomas are more invasive than basal cell carcinomas), all cancer cells share certain fundamental properties, such as uncontrolled cell growth and proliferation.

Tumors are classified into four major groupings, defined by the cell type from which these various **neoplasms** derive. In humans, the most frequently observed tumors are the **carcinomas**, which derive from the transformation of epithelial cells lining various organ cavities and surfaces. Among the most common carcinomas are those of the lung, colon, breast, prostate, stomach, pancreas, and skin. **Sarcomas**, which are much less frequent, arise from the mesenchymal tissues formed from fibroblasts and closely related cell types; tumors of the bone and muscle are in this class. The hematopoietic (blood-forming) organs represent frequent targets for oncogenic events, and malignant transformation of associated cells leads to **leukemias**, **lymphomas**, **myelomas**, and related neoplasias. The tumors of **neuroectodermal** cells form a fourth class. These tumors include **neuroblastomas, glioblastomas, neuromas, neurofibrosarcomas**, and **melanomas**. **FIGURE 17.2** shows the estimated incidence of different types of cancers in the United States for the year 2005.

Cancer research is rapidly evolving from a description of diverse, ostensibly unrelated phenomena into a logical science in which the behavior of cells and tumors can be rationalized in terms of a small number of underlying

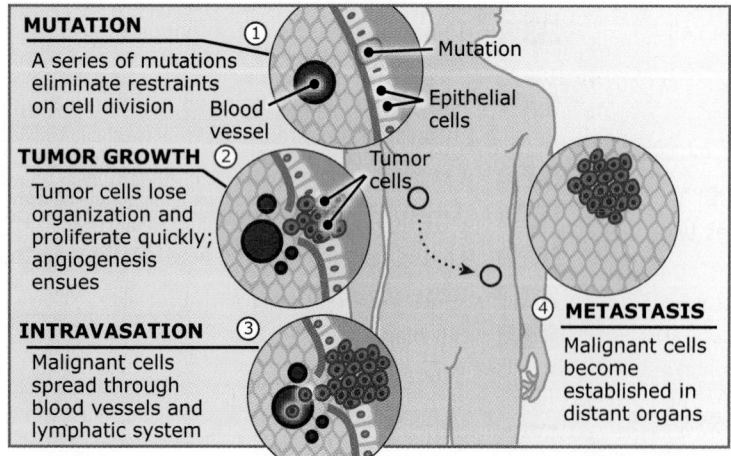

MUTATION ① — Mutation
A series of mutations eliminate restraints on cell division   Blood vessel   Epithelial cells

TUMOR GROWTH ②   Tumor cells
Tumor cells lose organization and proliferate quickly; angiogenesis ensues

INTRAVASATION ③
Malignant cells spread through blood vessels and lymphatic system

④ **METASTASIS**
Malignant cells become established in distant organs

FIGURE 17.1 In general, cancer progresses from a local tumor to widespread metastasis. Here we show the basic steps that occur after an epithelial cell has been mutated.

SITE	NEW CASES	DEATHS
All sites	1,479,350	562,340
Oral cavity & pharynx	35,720	7,600
Digestive system	275,720	135,830
Respiratory system	236,990	163,790
Soft tissue (including heart)	10,660	3,820
Bones and joints	2,570	1,470
Skin (excluding basal & squamous)	74,610	11,590
Breast	194,280	40,610
Genital system	282,690	56,160
Urinary system	131,010	28,100
Eye & orbit	2,350	230
Brain & other nervous system	22,070	12,920
Endocrine system	39,330	2,470
Lymphoma	74,490	20,790
Multiple myeloma	20,580	10,580
Leukemia	44,790	21,870
Other & unspecified primary sites*	31,490	44,510

* More deaths than cases suggests lack of specificity in recording underlying causes of death on death certificates.

FIGURE 17.2 The frequency with which certain tissue types become cancerous, the course of disease development, and the efficacy of treatment is quite variable. Reproduced from "Estimated New Cancer Cases and Deaths by Sex, US, 2009," from the ACS online publication *Cancer Facts and Figures 2009*. Used with permission of the American Cancer Society.

principles. This chapter describes these principles and how they help us to rationalize the apparently complex and varied behaviors of human tumors.

17.2 Cancer cells have a number of phenotypic characteristics

Key concepts

- Cancer cells are characterized by several distinct properties.
- Unlike normal cells, cancer cells do not stop dividing when they contact a neighboring cell when such cells are propagated in a Petri dish.
- Cancer cells have a greatly reduced requirement for growth factors to sustain growth and proliferation.
- Unlike normal cells, cancer cells in culture do not require attachment to a physical substrate in order to grow—the trait of anchorage independence.
- Unlike normal cells in culture, which halt division after a certain number of growth-and-division cycles, cancer cells are immortal and, therefore, do not stop dividing after a predetermined number of generations.
- Cancer cells often have chromosomal aberrations, including changes in chromosome number and structure.

Over the years, researchers have cataloged the traits that are specifically associated with cancer cells. These traits are summarized in **FIGURE 17.3**. Many of the traits can be most readily studied by propagating the cells in culture (*in*

Characteristics	Normal cells	Cancer cells
Contact inhibition of growth	Present	Absent
Growth factor requirements	High	Low
Anchorage-dependence	Present	Absent
Cell cycle checkpoints	Intact	Absent
Karyotypic profile	Normal	Abnormal
Proliferative life span	Finite	Indefinite

FIGURE 17.3 Cancerous cells are distinguished from normal cells by a number of traits.

vitro) rather than studying them within living host tissues (*in vivo*). Conveniently, many types of cancer cells are readily cultured in Petri dishes. This is in contrast to many types of normal cells, which are more difficult to propagate *in vitro*. (Note that "normal" cells growing in culture undergo some changes that promote growth in culture, as we will discuss in *17.7 The genesis of tumors is a complex process*.)

Normal cells that are introduced into Petri dishes with nutritive media will sit on the bottom of the dishes and multiply until they cover the entire bottom surface, at which point they stop growing. This growth behavior is termed **contact inhibition** to indicate that when cells contact one another, they shut down their own proliferation. Consequently, such normal cells form a cell sheet one cell thick—a monolayer. When the layer of cells covers the entirety of the bottom of a Petri dish, it is called a **confluent monolayer**.

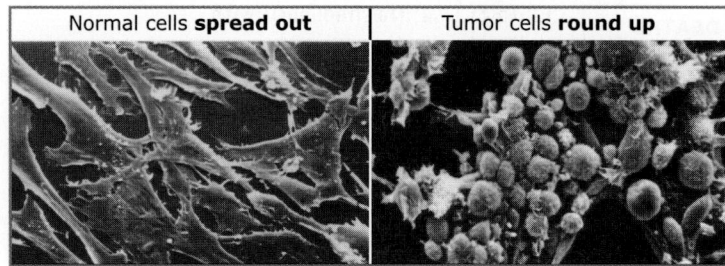

| Normal cells **spread out** | Tumor cells **round up** |

FIGURE 17.4 Scanning electron microscopy shows that normal cells spread out and form long processes, whereas tumor cells round up into tightly balled masses. Photos courtesy of G. Steven Martin, University of California, Berkeley.

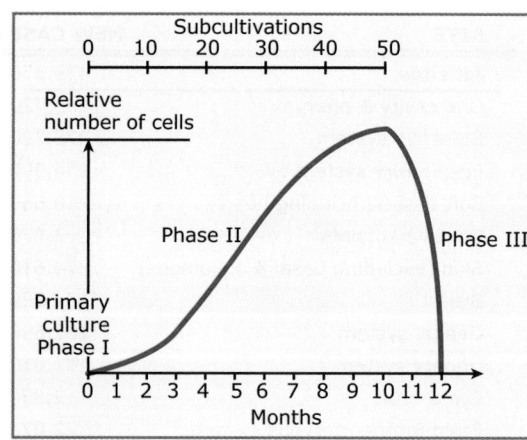

FIGURE 17.5 In a normal cell population, doubling will cease after 30 to 40 rounds. The population then enters a phase called senescence. Adapted from L. Hayflick and P. S. Moorhead, *Exp. Cell Res.* 25 (1961): 588–621.

Cancer cells behave in a dramatically different way. **FIGURE 17.4** shows that when placed in a Petri dish, these cells continue to proliferate and divide long after they touch one another. Failure to stop dividing causes them to pile up on top of one another. If a cancer cell is growing amid a large number of normal cells in a culture dish, eventually it and its descendants form a thick clump, termed a **focus**, which is surrounded by a thin monolayer sheet of the normal cells. These foci are easily scored, even by the naked eye. Thus, cancer biologists can easily quantify the number of transformed, cancerous cell clones present within a population of normal cells.

Loss of contact inhibition is not the only distinctive growth trait exhibited by cancer cells. When these cells are placed in suspension and mixed with a jelly-like agar matrix before being introduced into a Petri dish, they form spherical colonies. Normal cells, in contrast, are unable to grow in these suspension cultures; instead, they must be firmly attached to the bottom of the dish before they will begin to multiply. Normal cells are, therefore, called **anchorage dependent** in contrast to cancer cells whose growth is **anchorage independent**.

One of the easiest cell types to grow *in vitro* is a connective tissue cell called a fibroblast. When fibroblasts are introduced into culture and passaged from one dish to another each time they reach confluence, they will grow and divide robustly for a number of passages and then cease proliferating. This cessation of growth implies that somehow cells are limited to double a fixed number of divisions, and that once a lineage of cells has exhausted its allotted number of growth-and-division cycles, it stops growing, even if conditions are otherwise optimal for further proliferation. This is indicated by observations showing a halt in proliferation after 30, 40, or 50 cell generations,

depending on the cell type and donor organism. An example of this behavior is shown in **FIGURE 17.5**. Cancer cells present a stark contrast to the limited growth potential of healthy cells. Once adapted to culture, cancer cells can proliferate without limit and, therefore, are considered to be **immortalized**.

Cancer cells in culture have yet other peculiarities that set them apart from their normal counterparts. When normal cells are propagated in culture, they require a medium supplemented with more than just the essential nutrients (amino acids, glucose, and vitamins). The medium must also contain serum prepared from cow blood. This serum contains a number of **growth factors (GFs)**, which are proteins that are normally released by one cell in order to stimulate the growth of another.

One prominent growth factor present in serum, platelet-derived GF (PDGF), is released into serum by blood platelets during the clotting process, which occurs after tissues are damaged. Normally, when PDGF is released by the platelets in a wound site, it stimulates the growth of nearby fibroblasts, which then are induced to participate in tissue reconstruction. In a similar fashion, the presence of PDGF is required in tissue culture medium to stimulate the growth of fibroblasts in a Petri dish; without added PDGF, these cells might remain alive for weeks but would never grow and divide. Other types of growth factors, such as epidermal growth factors (EGFs), are required for growth of various types of epithelial cells in culture. **FIGURE 17.6** shows a generalized growth factor signaling pathway.

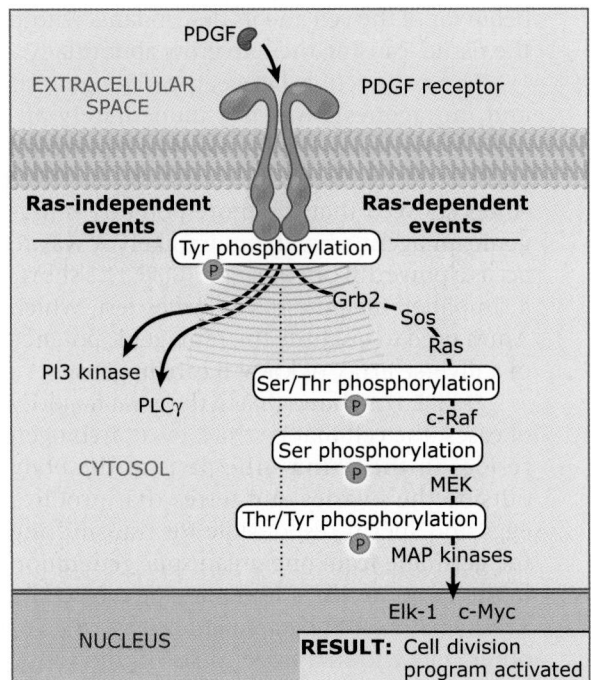

FIGURE 17.6 In a typical signaling pathway, a ligand (PDGF shown here) binds its receptor, which activates a cascade of reactions inside the cell leading to changes in cell behavior. Protein phosphorylation and lipid and GTP hydrolysis are the main signaling methods within the cell.

In fact, virtually all types of healthy cells require stimulation by one or more growth factors present in their culture medium before they will multiply. This has led to an important insight about the biology of most kinds of cells: they do not proliferate unless induced to do so by exogenous stimuli, specifically the signals carried by GFs present in their surroundings. Cancer cells present a striking contrast to this generalization: They require far smaller amounts of GFs in their medium than their normal counterparts. Cancer cells are able to stimulate their own proliferation, and their growth and division is, therefore, largely independent of exogenous, growth-promoting signals.

Importantly, cancer cells rarely become fully independent of exogenous growth-stimulating factors. More often than not, they continue to depend on certain growth factors that their normal antecedents required for proliferation. Thus, the majority of human breast cancers depend on estrogen to drive their proliferation, whereas prostate cancers depend on androgen. (Although these two agents are normally considered to be "hormones," they function essentially as GFs to drive cell proliferation of these cancer cells.) In addition, insulin-like growth factor 1 (IGF-1) is required by many types of human cancer cells in order to sustain their survival.

Unlike PDGF and EGF, which stimulate growth, some types of extracellular factors actually inhibit cell proliferation. The best studied of these growth-inhibitory factors is transforming growth factor-β (TGF-β), which is potent in its ability to stop the proliferation of various types of epithelial cells. It operates, in effect, like a negative GF. (Its ability to induce the anchorage-independent growth of certain cells led to its original naming as a tumor "growth" factor.) Once again, cancer cells respond differently than their normal counterparts: many types of cancer cells have acquired the ability to resist inhibition by TGF-β and continue to multiply in the presence of rather high concentrations of this negative GF.

So, cancer cells really show two very distinct changes in their relationship to the signals present in their extracellular environment. They are relatively independent of exogenous growth-stimulatory signals, and at the same time, they acquire a resistance to exogenous growth-inhibitory signals. These two changes suggest a general theme about cancer cell biology: the cancer cell is no longer well connected with the signals (notably positively and negatively acting GFs) present in its immediate environment.

Another important feature of cancer cells is that the chromosomal complement in their nuclei is often aberrant, with some chromosomes missing, others present in excessive number, and strange chromosomes formed from the fusions of segments of normal chromosomes. This chromosomal disorder—termed **aneuploidy**—stands in marked con-

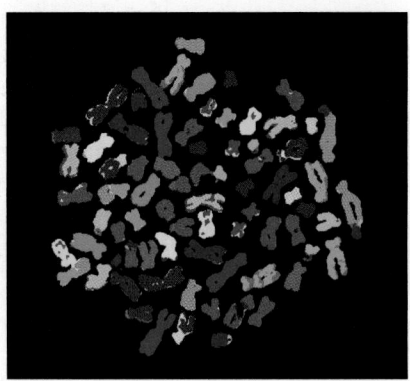

FIGURE 17.7 In this micrograph, chromosomes from a colorectal cancer cell are visualized with chromosome-specific hybridization probes labeled with distinct fluorochromes, allowing detection of defects in both chromosome structure and number. Reprinted by permission from Macmillan Publishers Ltd: *Nat. Rev. Cancer*, P. V. Jallepalli and C. Lengauer, Chromosome segregation and cancer..., 2001, vol, 1 (2): 109–117. Photo courtesy of Prasad V. Jallepalli, Memorial Sloan-Kettering Cancer Center.

trast with the normal chromosomal state—**euploidy**. FIGURE 17.7 shows a fluorescent *in situ* hybridization image of the aneuploidy of a tumor cell.

In sum, cancer cells exhibit a substantial number of aberrations. Indeed, the preceding list is far from complete. However, it is important to remember that cancer cells and normal cells also share many features, which often makes it very difficult to kill cancer cells without injuring healthy cells as well.

17.3 Cancer cells arise after DNA damage

Key concepts

- Agents that cause cancer may do so by damaging DNA.
- Mutations in certain genes cause a cell to grow abnormally.
- Ames devised a test to determine the carcinogenicity of chemical agents.
- Cancers usually arise in somatic cells.

Cancer-causing agents, such as chemicals or X-rays, are termed **carcinogens**. Many carcinogens exert their effects by acting as **mutagens**. Carcinogens enter the body and strike a cell in a target organ, mutating critical genes in the cell. Such mutated genes subsequently influence the behavior of the cell and its descendants within the tissue, causing them to grow abnormally.

The connection between carcinogenesis and mutagenesis was not immediately apparent to researchers but received a powerful boost from the work of Bruce Ames in 1975. Ames showed that the more potently mutagenic an agent was, the more likely it was to act as a powerful carcinogen. FIGURE 17.8 shows a simplified diagram of the Ames test, which Ames used to quantify the mutagenic potency of a diverse array of known carcinogens.

FIGURE 17.9 shows that in the great majority of cases, the cell that is struck by a carcinogen resides in the **soma**—the parts of the body outside the ovaries and testes that produce eggs and sperm responsible for transmitting the germline from one organismic generation to the next. Hence, a mutant gene created by a mutagenic carcinogen might be passed on to other cells within a target tissue of the soma; however, this gene will not be transmitted to the progeny of the organism since it will not be present in sperm or egg. Such a mutation is called a **somatic mutation**, distinguishing it from the **germline** mutations that are passed in heritable fashion from an organism to its

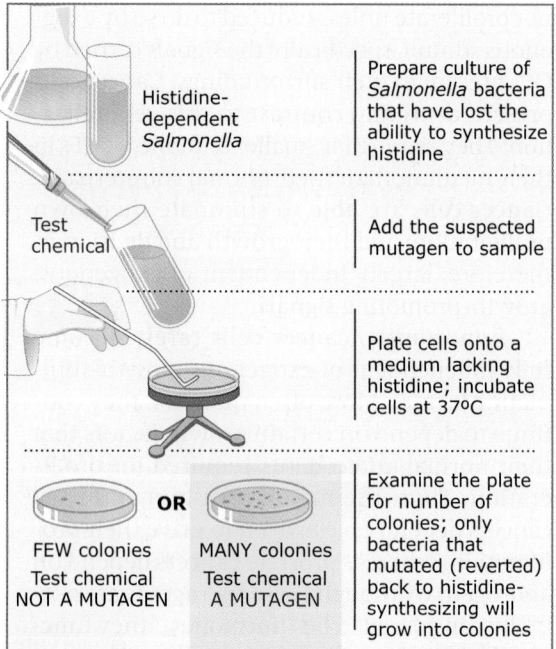

FIGURE 17.8 Mutagens can be detected by the frequency with which they are able to mutate the histidine gene in *Salmonella* from a defective version back to the normal version. Not shown here is addition of rat liver extract, which is added to carryout the biochemical reactions in the liver that can change the properties of an agent, by activating its mutagenic powers.

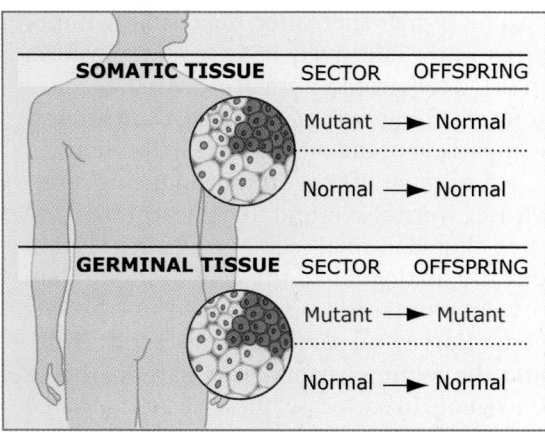

FIGURE 17.9 Carcinogens usually strike somatic cells; thus, cancers are limited to the afflicted individual rather than passed on to offspring.

offspring. Thus, most cancers arise as a consequence of somatic mutations afflicting the cells located in various tissues throughout the body. However, as we will see later, mutant genes passed through the germline can lead to an inborn susceptibility to cancer.

17.4 Cancer cells are created when certain genes are mutated

Key concepts

- Oncogenes promote cell growth and division.
- Tumor suppressors inhibit cell growth and division.
- Cellular genomes harbor multiple protooncogenes.
- Tumor viruses carry oncogenes.
- Genetic alterations can convert protooncogenes into potent oncogenes.

When a cell is exposed to a mutagen, genes that serve as either positive or negative regulators of cell growth and proliferation may be damaged. The two types of regulatory genes normally work in opposing fashions to ensure a well-balanced control system. The positive regulators that act to promote normal cell growth and division are termed **protooncogenes**; following mutation, these genes become activated into **oncogenes**. The negative regulators that normally serve to constrain proliferation are termed **tumor suppressors**; these genes become involved in carcinogenesis when they are inactivated by mutations, depriving the cells of their growth-suppressing powers.

We will take a short trip through history to describe how work with retroviruses led to the discovery of oncogenes. Oncogenes were discovered earlier than tumor suppressors. The origins of this research can be traced to the work of Peyton Rous, who reported in 1910 on a virus that he had extracted from a connective tissue tumor—a sarcoma—in the wing of a chicken brought to his lab by a Long Island chicken farmer. When Rous ground up the sarcoma and filtered the resulting extract, he discovered a substance that passed through the filter and that, upon subsequent injection into a second chicken, induced a sarcoma in the connective tissues of that bird. Rous repeated this experiment, introducing filtered extract from the tumor of the second bird into a third bird, which soon also showed a sarcoma at the site of injection. Because this agent was filterable and could multiply, it was by definition a *virus* rather than a bacterium (whose far larger size would have caused it to be trapped in a filter).

As was realized much later, the virus found by Rous, which came to be called **Rous sarcoma virus (RSV)**, clearly behaved very differently from most other viruses, which enter into a host cell, multiply, and then kill the cell, releasing progeny virus particles that infect other cells nearby. Instead, RSV spares the life of the cell it infects. This infected cell rapidly takes on many of the traits that are associated with cancer cells, including the ability to grow in suspension, to change shape, and to form tumors (to become **tumorigenic**); in other words, it becomes **transformed**.

Moreover, when the initially infected, transformed cell grows and divides, its progeny continue to show the traits of cancer. The genome of RSV persists in the descendant cells, in which it continues to drive malignant growth. In effect, cancerous growth becomes a heritable trait, passed from one cell to its offspring and requiring the continued presence of the RSV genome.

Analysis of RSV in the early 1970s showed that it is a **retrovirus** with a very small single-stranded RNA genome. A single gene, termed *src*, was found to be responsible for all of the cancer-inducing properties of RSV and was, for this reason, considered to be an oncogene. The far-reaching powers of this gene were quite unexpected to some researchers because it indicated that such a gene can act **pleiotropically**; that is, it is capable of simultaneously inducing a large number of altered phenotypes in cells. This spectrum of alterations enables RSV-transformed cells to multiply within the tissues of a chicken, eventually resulting in a large sarcoma.

In 1975, researchers found to their surprise that healthy cells also carry a normal version of the *src* gene, which plays an ostensibly essential role in normal cell and organismic development. This normal version of the virus was called a protooncogene, since it could serve as the precursor of the oncogene carried by RSV.

It became clear that the *src* gene became part of the RSV genome after an ancestral retrovirus that lacked its own *src* gene infected a chicken cell, made a copy of the cellular src gene (sometimes called c-*src*), and incorporated this copy into its genome. Having acquired the *src* gene, the hybrid virus converted it into an on-

cogene (v-*src*). Thereafter, the resulting tumor virus—now effectively RSV—could transform infected cells from a normal growth state into a tumorigenic growth state. **FIGURE 17.10** shows a comparison of the v-Src and c-Src proteins.

A number of other tumor-inducing retroviruses were also found to have acquired and then altered normal cellular genes in a fashion directly analogous to the case of RSV. Avian myelocytomatosis virus picked up the *myc* gene; Harvey rat sarcoma virus carries H-*ras*; and the feline sarcoma virus harbors the *fes* oncogene. In each case, these virus-associated oncogenes derived from a preexisting normal cellular protooncogene. Consequently, the

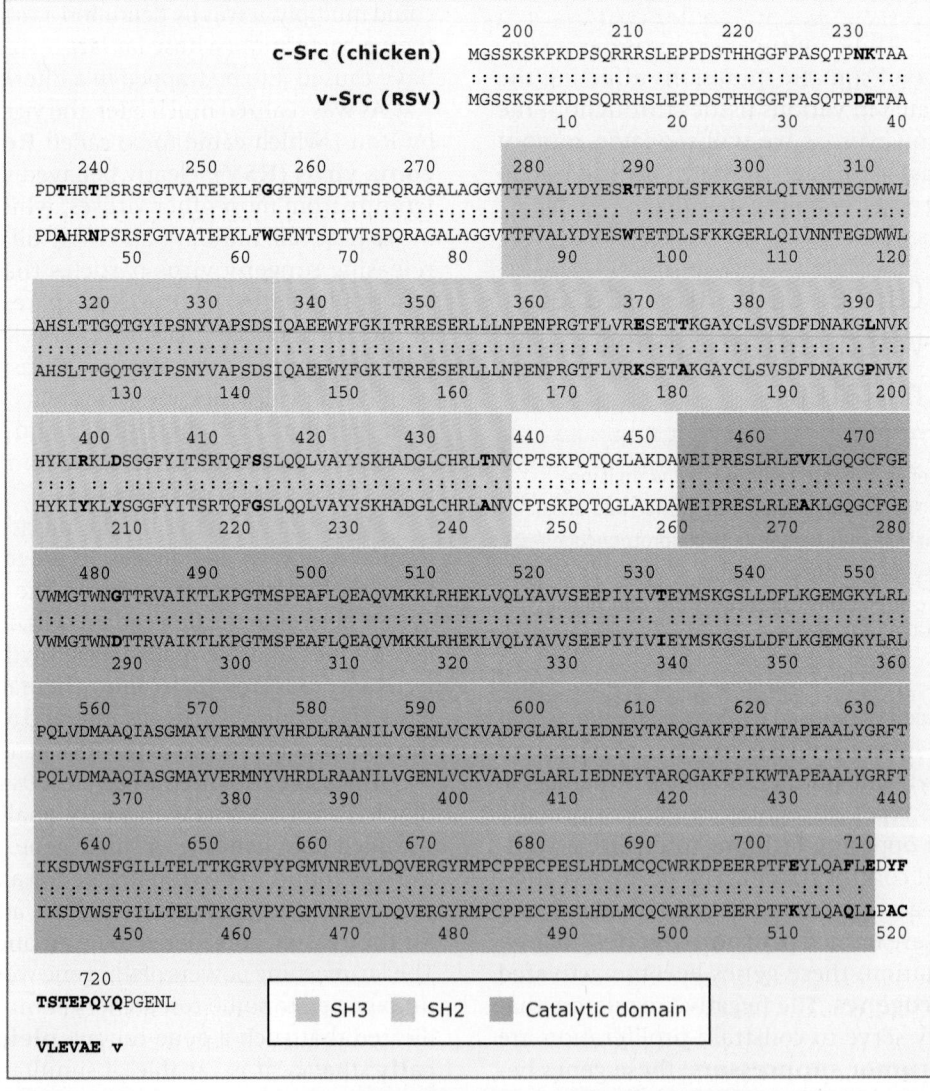

FIGURE 17.10 Sequence alignment of the viral Src protein with the cellular Src protein shows that the two proteins vary by several amino acids. Sequence data for v-scr from S. Broome and W. Gilbert, *Cell* 40 (1985): 537–546. Sequence data for c-src from T. Takeya and H. Hanafusa, *Cell* 32 (1983): 881–890. Sequences were aligned with the FASTA program, courtesy of W. R. Pearson, *Meth. Enzymol.* 183 (1990): 63–98.

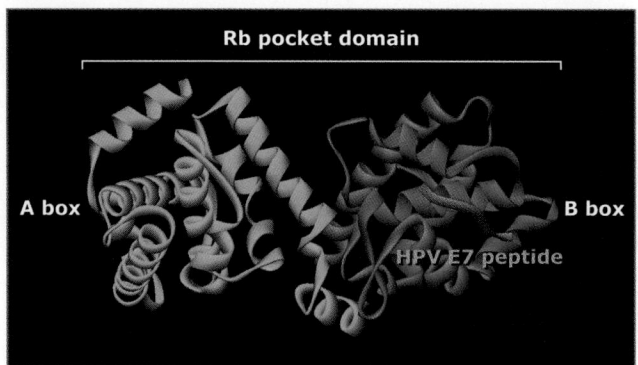

FIGURE 17.11 The human papilloma virus (HPV) interferes with cell cycle control by inserting the E7 peptide (orange) into the pocket domain (blue) of the Rb protein, which causes it to release the E2F transcription factor. Structure from Protein Data Bank 1GUX. J. O. Lee, A. A. Russo, and N. P. Pavletich, *Nature* 391 (1998): 859–865.

genomes of animals harbor a relatively large number of these protooncogenes whose existence can be revealed by their presence in the genomes of one or another transforming retrovirus.

Another unrelated class of tumor viruses causes transformation by a very different mechanism. This class of viruses is characterized by the double-stranded DNA genomes that they possess and includes papillomaviruses, which cause warts and malignant skin cancers in rabbits and cervical carcinomas in humans; SV40 and polyoma virus, which cause a variety of tumors in rodents; and viruses distantly related to herpesvirus, such as Epstein-Barr virus (EBV), which cause lymphomas in African populations and nasopharyngeal carcinomas in Southeast Asia. DNA tumor viruses produce cancer-inducing proteins (**oncoproteins**) that are unrelated to normal cellular growth-regulating proteins. Instead, the DNA tumor virus proteins bind and perturb normal cellular proteins. For example, SV40, adenovirus, and papillomaviruses all produce proteins that are able to bind and inactivate cellular tumor suppressor proteins, creating a state that mimics the condition seen when cells lose tumor suppressor gene function through mutational inactivation of these genes in their genomes. **FIGURE 17.11** shows how viral oncoproteins bind cellular proteins.

Three ideas emerged from work on tumor viruses that were to have a profound effect on our understanding of the molecular origins of cancer. First, cellular genomes harbor multiple protooncogenes. Second, genetic alterations, at least those wrought by retroviruses, can convert each of these into potent oncogenes. Third, once converted into an activated oncogene, each of these genes can act pleiotropically to elicit a number of the abnormal traits of cancer cells. In addition, research with DNA tumor

viruses helped elucidate the molecular mechanisms of tumor suppressor gene action.

17.5 Cellular genomes harbor a number of protooncogenes

Key concepts

- Gain-of-function mutations can activate protooncogenes.
- Overexpression of protooncogenes can cause tumors.
- Translocations can create hybrid proteins that are oncogenic.

Most human cancers are not caused by infections launched by either retroviruses or DNA tumor viruses. This raised the question of how human cancer cells acquire oncogenes and lose tumor suppressor gene function. In 1979, it was shown that DNA taken from mouse tumor cells that were transformed with a chemical carcinogen could be transferred into normal cells (by the procedure termed **transfection**), whereupon the recipient cells would undergo transformation. This indicated that the chemically transformed cells carried oncogenes that behaved much like the oncogenes that were associated with RNA and DNA tumor viruses. However, these chemically transformed cells were not associated with viral infections. This suggested a notion that was proved within several years: the transforming oncogenes present in the chemically transformed cells were mutant versions of normal cellular protooncogenes that was created through the mutagenic actions of the chemical carcinogens.

In fact, the initially studied oncogene that was activated by chemical carcinogens was closely related to the Ki-*ras* oncogene carried by Kirsten rat sarcoma virus, a mouse retrovirus. Similarly, an oncogene detected by transfection of human bladder carcinoma DNA was discovered to be virtually identical to the Ha-*ras* oncogene borne by Harvey rat sarcoma virus. Soon thereafter, the *myc* oncogene that was originally uncovered through its presence in the genome of avian myelocytomatosis virus was also found in altered form in a variety of human hematopoietic tumors.

These discoveries led to a clear and by now obvious lesson: protooncogenes could be activated into oncogenes through alternative mechanisms. In some animals, their activation was caused by their incorporation into retrovirus genomes. In nonviral malignancies, such as most of those occurring in humans, the same genes could be activated by mutational processes that altered normal genes residing within cellular chromosomes.

The mutations in human tumors that are responsible for the activation of protooncogenes into oncogenes are quite varied. In the case of the *ras* oncogenes, a simple point mutation in the reading frame is invariably responsible for converting a protooncogene into an oncogene. This mutation operates as a **gain-of-function** mutation, in that it causes the encoded Ras protein to remain locked for an extended period of time in a growth-stimulating state rather than being in such a state only transiently. In the case of the *myc* oncogene, at least two mechanisms can activate it in human tumors; the several mechanisms that lead to its formation affect the extent of its expression and thereby result in elevated levels of the encoded Myc protein. Thus, in some cases, the copy number of the *myc* gene is increased by the process of **gene amplification**. In others, the *myc* gene becomes fused to another gene, such as an immunoglobulin (antibody) gene, by a **translocation** event; as a consequence, expression of the *myc* gene, which is normally quite low and well-controlled, is now driven by a foreign transcriptional promoter that forces its expression at a high, constant, and, hence, uncontrolled levels.

Chromosomal translocations can also create oncogenes by an entirely different mechanism. A good example is the Philadelphia chromosome, which is found in many patients with chronic myelogenous leukemia (CML). This chromosome is the result of a reciprocal translocation—an exchange of chromosomal material—between Chromosome 9 and Chromosome 22. The translocation results in the fusion of the *Bcr* and *Abl* genes and the associated formation of a hybrid reading frame. In particular, the hybrid protein contains the N-terminus of the Bcr protein plus most of the Abl protein at its C-terminus. The mechanisms through which this hybrid protein becomes oncogenic are not entirely clear; however, it is known that the Abl protein encodes a signaling molecule that operates much like the Src protein encoded by Rous sarcoma virus. **FIGURE 17.12** shows the different ways in which the gain-of-function mutations in protooncogenes can be generated.

As we have seen, both qualitative changes (as in the case of *ras* and *Bcr-Abl*) and quantitative changes (as in the case of *myc*) in genes can contribute to cell transformation.

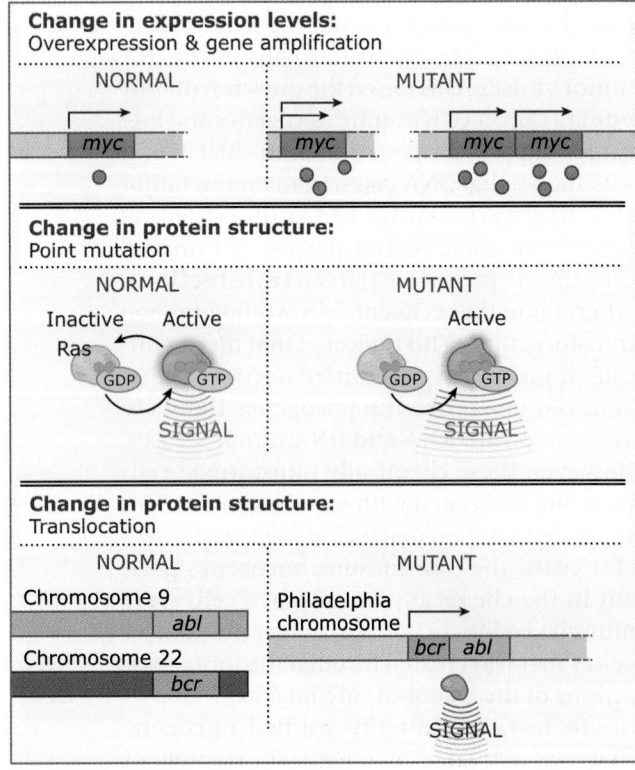

FIGURE 17.12 Oncogenes can be activated by both quantitative and qualitative changes.

17.6 Elimination of tumor suppressor activity requires two mutations

Key concepts

- Both copies of a tumor suppressor gene must usually be inactivated to see a phenotype.
- Mechanisms that result in loss of heterozygosity are often responsible for the loss of the remaining normal copy of the tumor suppressor gene.
- Cancer susceptibility can be caused by the inheritance of a mutant copy of a tumor suppressor gene.

While activating mutations serve to convert protooncogenes into oncogenes, it is clear that other types of mutations must lead to the *inactivation* of tumor suppressor genes (TSGs). The two processes are not symmetrical, however, because the activation of oncogenes usually requires a single genetic change, whereas the elimination of tumor suppressors requires two.

The reasons for this are indicated by the laws of Mendelian genetics. The activating mutation that converts a protooncogene into an oncogene invariably creates a dominant allele that is able to influence cell phenotype, even in the continued presence of a coexisting wild-type allele residing on the homologous chromosome. In contrast, the inactivating mutation that eliminates the functioning of one copy of a TSG usually creates a recessive null allele whose presence is not felt by the cell because of the continued activities of the surviving wild-type allele. Only when this second gene copy is inactivated or eliminated will a cell be deprived of both redundantly acting copies of the TSG, and only then will it be liberated from the inhibitory actions of this gene.

All types of mutations, including those that lead to TSG inactivation, occur with a relatively low probability per cell generation (often ~10^{-6}). This would seem to make the elimination of both copies of a TSG a very rare event. The problem here is a mathematical one: the probability of two such mutational events occurring is the square of the probability of each, that is, $10^{-6} \times 10^{-6}$. This product represents such an improbable event (occurring in 1 in 10^{12} cells) that it is highly unlikely to occur in the history of any population of premalignant cells.

Nonetheless, evolving premalignant cell populations do succeed in eliminating both copies of the critical TSGs that otherwise would hold back their proliferation. More often than not, they succeed in doing so through a mechanism that does not depend on the improbable 10^{-6} elimination of the second TSG gene copy. Instead, the second, still-wild-type gene copy is lost in a process termed **loss of heterozygosity (LOH)**. Through mechanisms involving recombination in somatic cells, a chromosomal arm may be discarded and replaced by a duplicated copy of its homolog, the latter deriving from the homologous chromosome, as shown in **FIGURE 17.13**. This process, which occurs with frequencies of at least 10^{-3} per cell generation, results in the conversion of genes that previously were present in a heterozygous genetic configuration into a homozygous state. In the case of a chromosomal arm carrying a TSG, this may result in the loss of the wild-type gene copy and its replacement by a duplicated copy of the already-mutated copy. The result is a cell lacking any functional, wild-type copies of the TSG.

In many cancers, the loss of TSG function may occur through the silencing of expression of these genes. This silencing is achieved through the **methylation** of cytidines residing in CpG sequences in the vicinity of the pro-

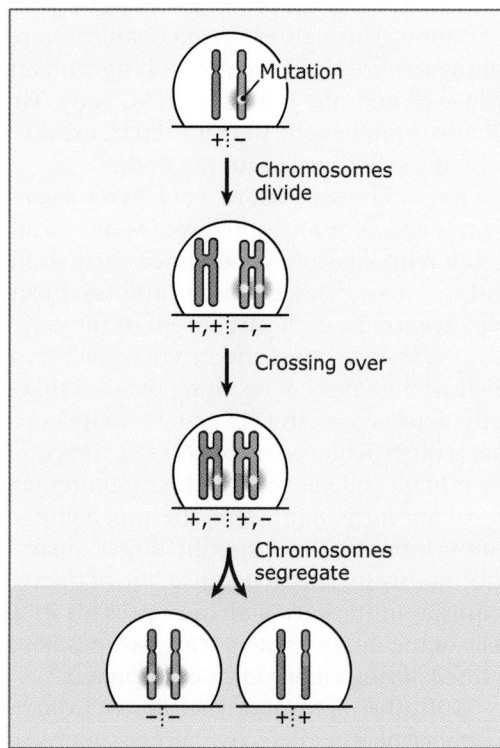

FIGURE 17.13 A heterozygous mutation may become homozygous during a somatic cell division if there is recombination between the homologs and the two mutant chromosomes segregate together.

moters of the affected TSGs. Such promoter methylation is frequently associated with many of the well-studied TSGs and is as effective in eliminating TSG function as are the mutations in the nucleotide sequences composing such genes. For example, in some tumors, the promoter region of the retinoblastoma (*Rb*) gene is methylated, but its DNA sequence is unaltered. Methylation of the promoter of the *Rb* gene results in shutdown of transcription of the gene. On occasion, one copy of the *Rb* gene is repressed through promoter methylation, and the surviving wild-type copy is then discarded with duplication of the already-methylated gene copy (for more on *Rb* see *17.10 Tumor suppressors block inappropriate entry into the cell cycle*).

A somatic cell may lose both functional copies of a critical TSG by other mechanisms as well. These are associated with hereditary cancer syndromes in which greatly increased susceptibility is passed in a heritable fashion from parent to offspring. Many types of these familial cancer syndromes derive from the passage through the germline of a defective copy of a TSG. In the much-studied example of **familial retinoblastoma**, a defective allele of the *Rb* gene is inherited from a parent or is generated during the processes of egg or sperm formation. Through either mechanism, a resulting fertilized egg (zygote) ends up with one wild-type and one mutant *Rb* TSG copy. This heterozygous genetic state is then transmitted to all the cells throughout the body.

For unknown reasons, such heterozygosity at the *Rb* gene locus predisposes specifically to the retinoblastoma eye tumor early in life and to osteosarcomas (bone tumors) during adolescence. In each case, a cell in the target tissue is known to lose the surviving wild-type allele of the *Rb* gene, resulting in a cell that is fully deprived of any *Rb* gene activity; such loss is often achieved through LOH. This doubly mutant cell has met the first requirement for tumor formation. Here, the probability of tumor formation in a specific target organ is enormously increased because one of the two requisite mutations is already present in all cells of the target tissue as this tissue is being formed during embryonic development.

Still other familial cancers derive from inheritance of other defective tumor suppressor genes. Many cases of Li-Fraumeni syndrome, which involves increased cancer susceptibility in a number of epithelial and mesenchymal organs, derive from inheritance of a mutant allele of the *p53* tumor suppressor gene (For more on *p53* see *17.11 Mutation of DNA repair and maintenance genes can increase the overall mutation rate*). Two types of neurofibromatosis arise through transmission of mutant alleles of the *NF-1* or the *NF-2* TSG from parent to offspring. Familial adenomatous polyposis, which is manifested in the appearance of dozens to hundreds of polyps in the colon, some of which may progress to carcinomas, derives from inheritance of a defective *APC* gene. Once the surviving *APC* gene copy is eliminated from a colonic epithelial cell, descendants of the resulting *APC*⁻/⁻ cell then proceed through the same succession of genetic alterations that occur during the development of nonfamilial (sporadic) colon carcinomas.

17.7 The genesis of tumors is a complex process

Key concepts

- Cancer is a multistep process that requires four to six different mutations to reach the tumor state.
- Tumorigenesis progresses by clonal expansion, where increasingly abnormal clones of cells outgrow their less mutant neighbors.

Cancer formation is a complex, multistep process that usually takes decades to reach completion. In some cancerous tissues, there are fully normal cells, frankly (clinically evident) malignant growths, and a number of regions within the tissue that exhibit intermediate degrees of abnormality along this spectrum. For example, in the colon this spectrum begins with the normal colonic epithelium that lines the gut; proceeds with mild **hyperplasia** (excess numbers of cells); progresses further to **dysplasia** (cells forming marginally abnormal epithelium including relatively undeveloped **polyps**); and further still to advanced, large polyps (localized, noninvasive growths that often protrude into the colonic cavity). Finally there are frank carcinomas that have invaded the underlying muscle layer; those that have crossed *through* the layer; and yet others that have seeded colonies (metastases) in distant sites such as the liver.

When genetic mutations were cataloged in colonic epithelial cells at various stages of abnormality, mutations at four distinct genetic loci were implicated in driving tumor formation. The more advanced a growth was (in terms of its progression toward frank malignancy), the larger was the number of mutant versions of these four genes that could be found in its genome. **FIGURE 17.14** shows the correlation between progression and mutations. This evidence, although only correlative, suggested that each sustained mutation drives a cell incrementally closer to full-fledged malignancy.

Several other lines of evidence also suggest that multiple steps are required for cancer progression. First, the frequency with which most adult tumors appear in the human population can be predicted by a mathematical function that depends upon the fourth to sixth power of elapsed lifetime. Such a relationship indicates that there are a comparable number of rate-limiting steps in tumor formation, each of which occurs with relatively low probability per unit time, and all of which must occur in order for a tumor to become clinically apparent. Stated in more concrete terms, cancer might be the end result of a sequence of four to six mutations, each of which is relatively improbable per cell generation, and all of which need to strike the genome of a cell before it grows in a truly malignant fashion. **FIGURE 17.15** shows the graph of incidence of cancer at different ages.

Second, unlike the immortalized, cultured cells that we discussed before, normal cells cannot be transformed into tumor cells through the introduction of a single oncogene. For example, at least two distinct oncogenes are required to transform fully normal rodent cells into a tumorigenic state. Two concomitantly introduced oncogenes, such as *ras* and *myc* or *ras* and E1A, can collaborate with one another to transform the cell, whereas neither can do it on its own. Thus, when the starting point in the transformation process is a fully normal cell, at least two and perhaps more genetic changes are necessary to make a cancer cell. **FIGURE 17.16** illustrates such cooperation between oncogenes. Human cells require even more mutant genes in order to become tumorigenic—by some accounts as many as five.

A third line of evidence of the multistep nature of cancer has come from studies of **transgenic** mice. These mice transmitted in their germline a transgene that encoded either a *ras* or a *myc* oncogene. Expression of these genes was controlled by a transcriptional promoter that was active in epithelial cells of the mammary gland. As anticipated, this transgene induced high rates of breast cancer in these mice. However, even though the *ras* or *myc* gene was strongly expressed in virtually all mammary epithelial cells from early in life, mammary carcinomas only became apparent

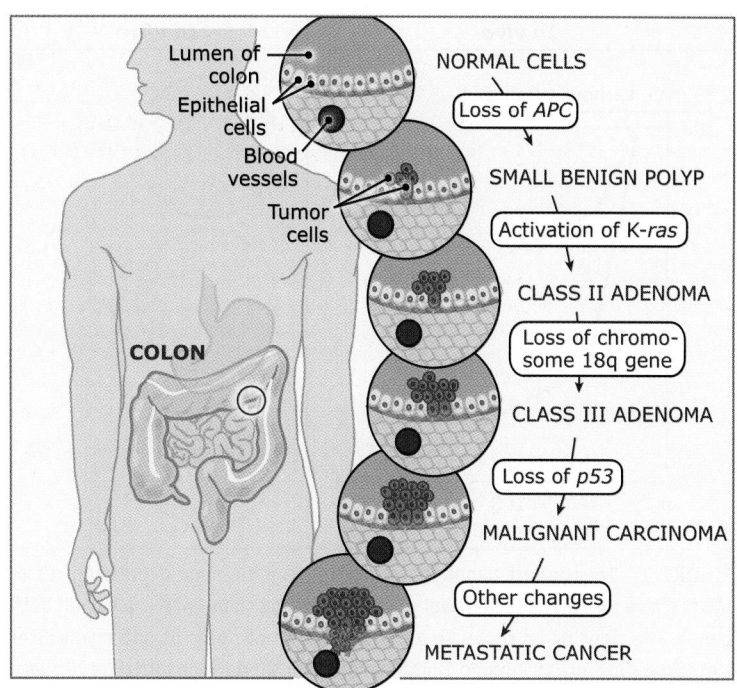

FIGURE 17.14 In colorectal cancer, mutations in the APC gene initiate the process while additional mutations are associated with various stages of progression.

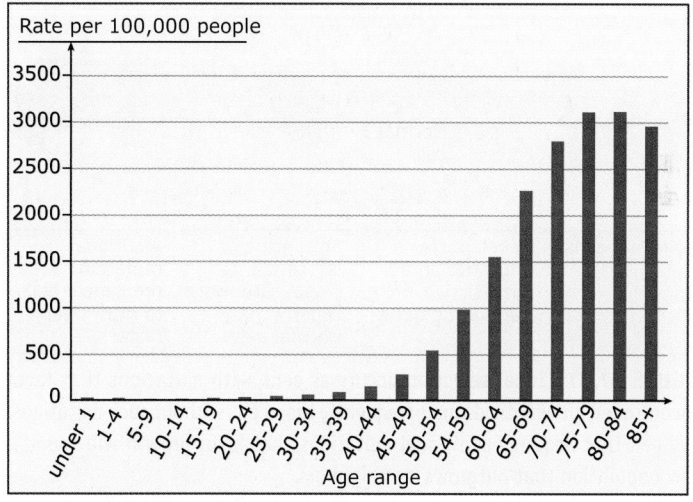

FIGURE 17.15 One indication that cancer is a multistep process is the number of years it takes for the disease to develop. Data courtesy the National Cancer Institute.

three or four months after birth. It was clear that, in addition to the transgenic oncogene that was active in these cells, at least one more and perhaps two somatic mutations needed to occur in the mammary epithelial cells before they would begin to multiply into a rapidly expanding tumor mass. These somatic mutations were presumed to affect specific target protooncogenes or tumor suppressor genes in

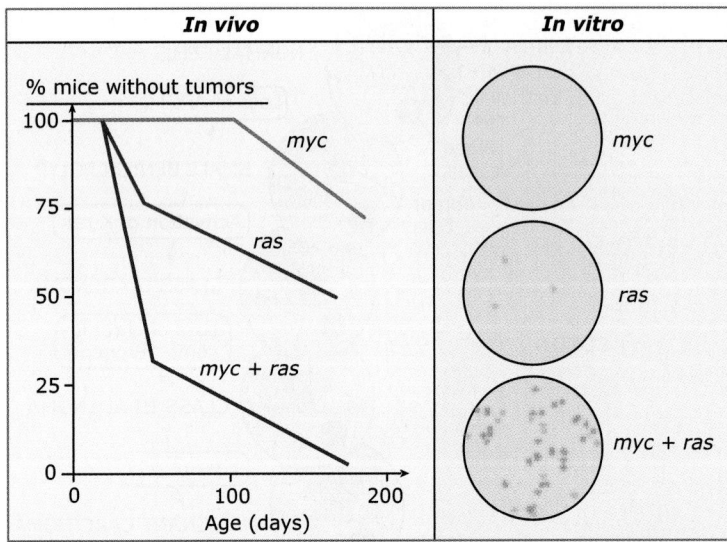

FIGURE 17.16 The left panel shows that transgenic mice expressing *ras* or hyperactive *myc* alone have fewer tumors over time than mice expressing both genes. The right panel shows that cotransfection of both *ras* and *myc* causes transformation of rat embryo fibroblasts; either alone is insufficient. *In vivo* adapted from E. Sinn, et al., *Cell* 49 (1987): 465–475. *In vitro* adapted from H. Land, L. F. Parada, and R. A Weinberg, *Nature* 304 (1983): 596–602.

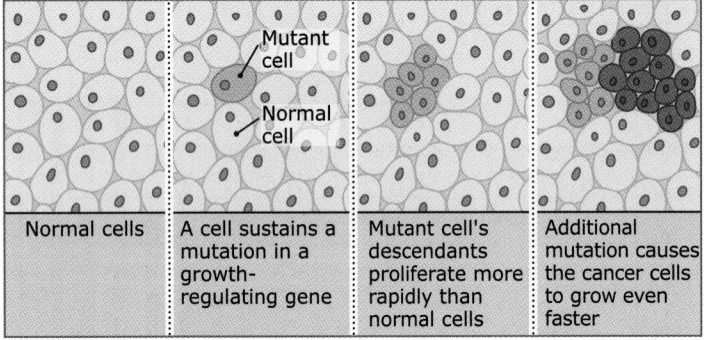

FIGURE 17.17 Clonal selection occurs as cells with mutations that favor uncontrolled growth divide more frequently than cells without such mutations. Any cell that acquires additional mutations that favor growth will found a new population that outgrows its neighbors.

the genomes of these cells. Significantly, when mice were bred that simultaneously expressed both transgenic *ras* and *myc* oncogenes in their mammary epithelial cells, the tumor formation was greatly accelerated and the lag time for the appearance of a tumor was concomitantly compressed, as seen in Figure 17.16. However, even with these two oncogenic transgenes, tumor formation occurred with a lag, indicating that yet another event, possibly a randomly occurring somatic mutation, was required in order for tumors to arise.

These various lines of evidence can be woven together into a simple, coherent scheme, in which tumor development (often called **progression**) is a process formally similar to Darwinian evolution. A cell sustains a mutation in one of its growth-regulating genes, which now enables it and its descendants to proliferate more rapidly than their normal neighbors do. With time, these cells form a clone of cells in the population that may number many millions of cells and dominates a small area within the tissue. Then another somatic mutation strikes one member of this clone, enabling this particular cell to grow even faster or to survive more effectively. Soon these doubly mutant cells will have outpaced all their neighbors, and after a period will monopolize an even larger area of the tissue, as shown in **FIGURE 17.17**. Each round of mutation followed by **clonal expansion**—sometimes termed *clonal succession*—yields a population of cells that is more abnormal and more adapted to proliferate. The increasing abnormality of the individual cells is manifested in the progressively more aberrant tissue architectures that they create. As in Darwinian evolution, randomly occurring mutations create a genetically heterogenous population, and the occasional rare cell that happens to have acquired a growth-advantaging mutation becomes the ancestor of a dominant successor population.

These observations of multistep tumorigenesis, when taken together, seem to conflict with the known abilities of tumor viruses to trigger cell transformation with a single virus-borne oncogene. In truth, these two sets of observations are difficult to reconcile with one another. The number of cell phenotypes that are altered during multistep tumor progression far exceeds the number of genes—perhaps half a dozen in humans—that must be mutated in order to create a tumor cell. Hence, each of these mutant genes is acting pleiotropically to elicit multiple cellular changes. Still, none of these genes acts in such a far-reaching manner to induce full transformation on its own. Although transformation of chicken fibroblasts by RSV yielded the seminal observations revealing the existence of cellular protooncogenes, in other respects, RSV proved misleading by suggesting to researchers that transformation could be a simple, single-hit event.

Concept and Reasoning Check

1. Explain why pleiotropy alone is insufficient to explain how oncogenes cause cancer.

17.8 Cell growth and proliferation are activated by growth factors

Key concepts

- Cell signaling requires extracellular factors, receptors, and other proteins that transmit the signal to the nucleus.
- Extracellular signals may be growth promoting or growth inhibiting.
- Many genes encoding cell signaling molecules are protooncogenes and tumor suppressor genes.

To understand how mutations in oncogenes and tumor suppressors affect cell growth and proliferation, we must first discuss how cells normally regulate these processes. Cells respond to a variety of extracellular signals, notably those conveyed by growth factors. These factors are usually released by one cell in a tissue, traverse the space between cells, and impinge upon a target cell that responds by up- or downregulating its proliferation. This ability of the target cell to respond in both ways implies the existence of a complex cellular machinery that receives the extracellular signals, processes them, and then executes the decision to alter the proliferation rate of the cell. We will use the Ras and the TGF-β signaling pathways as examples of how signaling pathways function and how they can go awry.

Cells display an array of growth factor receptors on their surfaces that sense and bind growth factors, each receptor binding specifically to its own cognate **ligand** (see *18.24 Protein phosphorylation/dephosphorylation is a major regulatory mechanism in the cell; 18.30 Diverse signaling mechanisms are regulated by protein tyrosine kinases;* and *18.34 Diverse receptors recruit protein-tyrosine kinases to the plasma membrane*). For example, PDGF binds specifically to the cell surface PDGF receptor while EGF binds to the EGF receptor. Once the receptor has bound its ligand, it **transduces** a signal through the plasma membrane into the cell interior. This signal, in turn, is passed down a complex cascade of signal-transducing proteins that operates much like a molecular bucket brigade, in which each member receives a signal from its upstream partner, processes and amplifies the signal, and then passes it on to one or several downstream partners, as shown in **FIGURE 17.18**. Protooncogene-encoded proteins may be found

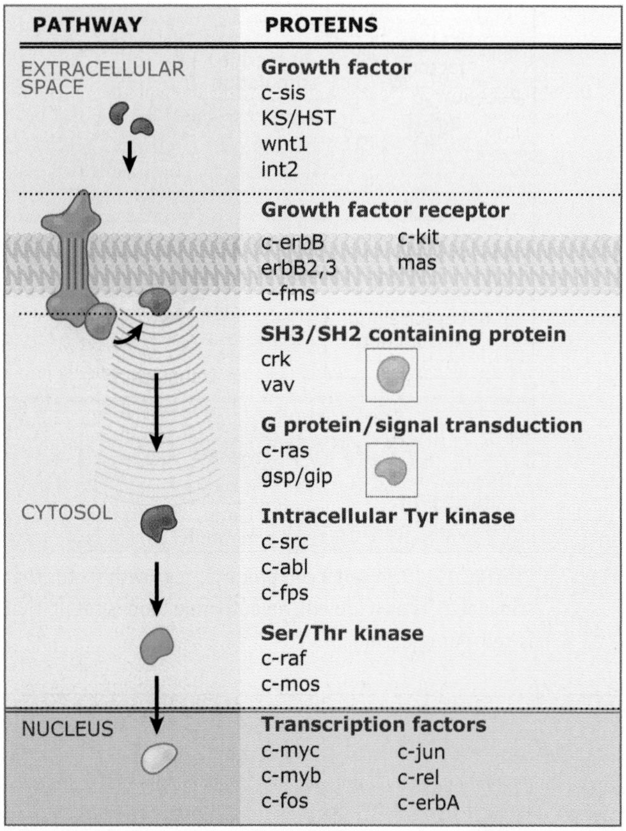

PATHWAY	PROTEINS
EXTRACELLULAR SPACE	**Growth factor** c-sis KS/HST wnt1 int2
	Growth factor receptor c-erbB c-kit erbB2,3 mas c-fms
	SH3/SH2 containing protein crk vav
	G protein/signal transduction c-ras gsp/gip
CYTOSOL	**Intracellular Tyr kinase** c-src c-abl c-fps
	Ser/Thr kinase c-raf c-mos
NUCLEUS	**Transcription factors** c-myc c-jun c-myb c-rel c-fos c-erbA

FIGURE 17.18 Many proteins that play a role in transducing growth-promoting signals can become oncoproteins.

as components of these growth-promoting signaling cascades.

One class of protooncogenes encodes growth factor proteins. If a gene of this class is inappropriately expressed in a cell that also displays the receptor for this particular growth factor, then a positive feedback loop is established. Thus, a cell releases large amounts of an oncogene-encoded mitogenic growth factor into the space around the cell. This growth factor then attaches to and activates receptors on the surface of the same cell, which in turn stimulate the cell to grow unrelentingly, as shown in **FIGURE 17.19**. Here independence from externally derived growth factors results because the cell is manufacturing its own mitogenic signals. Such self-stimulation of growth is termed **autocrine** signaling. This distinguishes it from **paracrine** signaling, which involves signaling between neighboring cells, and **endocrine** signaling, which depends upon factors that are released in one part of the body and travel great distances through the circulation before they impinge upon a responding target cell. As an

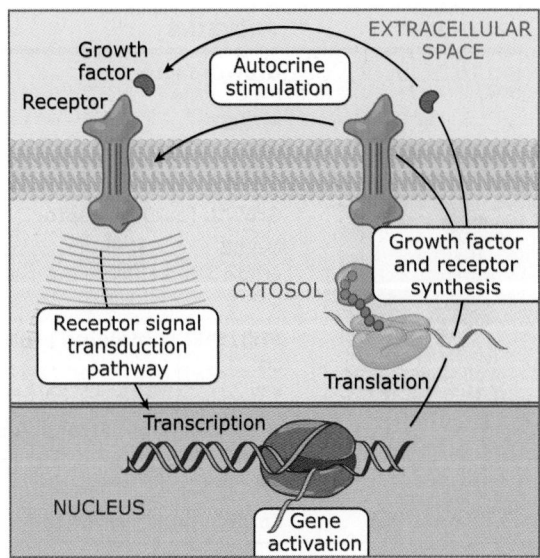

FIGURE 17.19 When a cell produces a growth factor that stimulates its own growth, an autocrine loop is set up that is self-perpetuating.

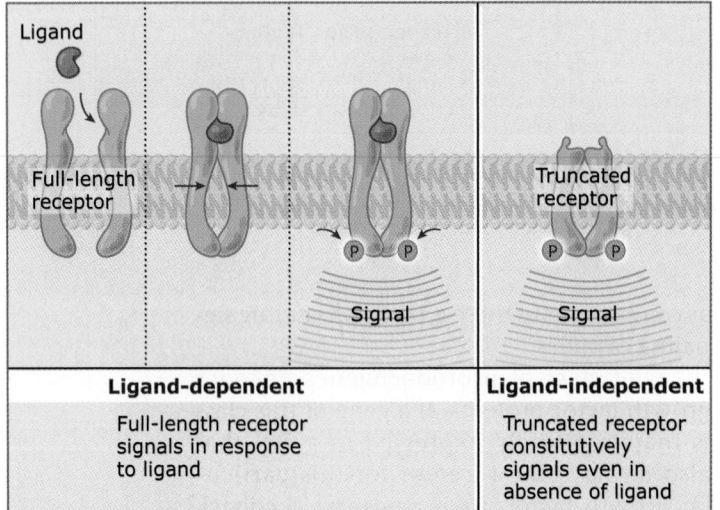

Ligand-dependent	Ligand-independent
Full-length receptor signals in response to ligand	Truncated receptor constitutively signals even in absence of ligand

FIGURE 17.20 In some cases, removal of the ligand-binding domain of a receptor causes constitutive signaling because ligand binding is no longer required for receptor dimerization.

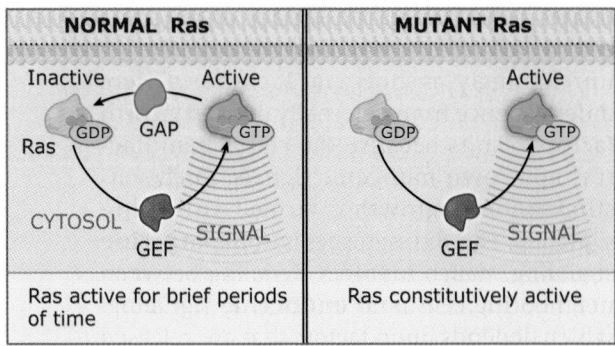

NORMAL Ras	MUTANT Ras
Ras active for brief periods of time	Ras constitutively active

FIGURE 17.21 Some mutant Ras proteins are stuck in a GTP-bound conformation and constitutively activate downstream signal transduction pathways.

example, the *sis* oncogene encodes a form of PDGF that, once secreted, triggers the growth of the secreting cell.

A second class of protooncogenes encodes growth factor receptors. When malfunctioning, these receptors send growth-stimulating signals into the cell, even when not bound to their ligand. Once again, the cell becomes growth factor–independent, since its proliferation occurs in the absence of the mitogenic factors normally required to trigger this process. In fact, the protooncogenes encoding GF receptors may be altered in two quite different ways. **FIGURE 17.20** shows that some oncogenes encode truncated GF receptors. Many human tumors, for instance, express truncated EGF receptors are found that lack their extracellular domains of these proteins; such truncated receptors fire continuously in a ligand-independent fashion. In other tumors, such as breast, brain and stomach tumors, the EGF receptor may be overexpressed to levels far higher than those encountered in normal cells. Once again, ligand-independent receptor firing occurs.

The intracellular proteins that lie downstream of growth factor receptors and process signals emitted by these receptors are also targets for oncogenic activation. The protein encoded by the *ras* protooncogene presents a good example of this kind of target (see also *18.23 Small, monomeric GTP-binding proteins are multiuse switches*). The normal Ras protein sits in a resting state within the cytoplasm, awaiting signals that originate from a cell-surface receptor. When the latter is activated by its ligand, it emits a signal that is passed through several intermediary proteins to the Ras protein, which responds by converting itself into an activated, signal-emitting configuration. This active signaling state of the Ras protein persists for only a short period of time—second to minutes—after which time the Ras protein shuts itself off, stopping further signal release and converting itself back into its inactive (**quiescent**) state.

In cells carrying a mutant *ras* oncogene, the structurally abnormal Ras oncoprotein is, as before, able to become activated into its signal-emitting configuration, but now has lost the ability to shut itself off. Therefore, it remains in this active, signal-releasing state for many minutes, even hours, flooding the cell with an unrelenting stream of growth-stimulating signals. This contrasts with the brief pulse of mitogenic signals released by the normal Ras protein. **FIGURE 17.21** shows how activated Ras differs from normal Ras. Such malfunction-

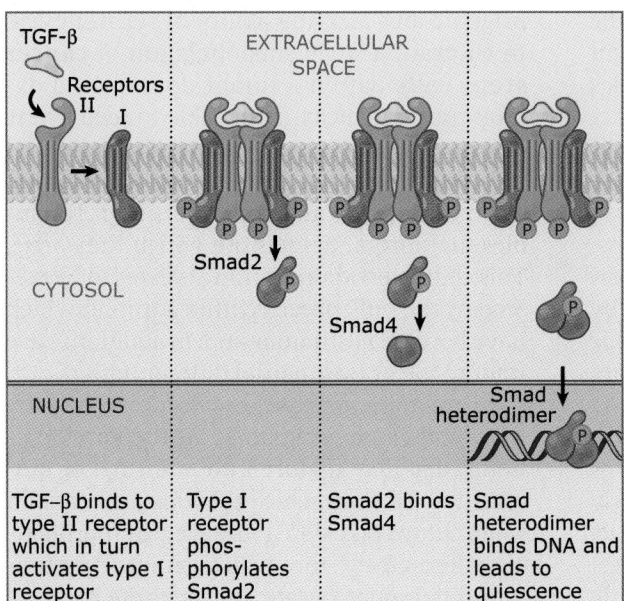

FIGURE 17.22 TGF-β is one of the growth-suppressing signals that a cell may receive. After reception a signal cascade activates the Smad proteins, which lead to transcription of proteins that inhibit growth.

TGF–β binds to type II receptor which in turn activates type I receptor	Type I receptor phosphorylates Smad2	Smad2 binds Smad4	Smad heterodimer binds DNA and leads to quiescence

ing Ras proteins are found in about a quarter of human tumors arising in a wide variety of tissues.

Cells may also respond to external signals by inhibiting their growth and proliferation. As mentioned earlier, TGF-β is a signaling molecule that has such a negative growth effect on the receiving cells, as shown in FIGURE 17.22. As a result of TGF-β activity, the cell retreats temporarily into an inactive, quiescent state from which it may emerge on some future occasion. Cancer cells become resistant (refractory) to such growth-inhibitory signals in the same ways that they can become independent of externally derived growth factors. Some cancer cells have lost expression of TGF-β receptors. Others no longer synthesize a critical downstream signaling partner, which belongs to a class termed Smad proteins. These Smad proteins are essential for conveying signals from the TGF-β receptor to the cell nucleus. Sometimes, defective Smad proteins are made that are no longer capable of properly conveying signals emanating from the TGF-β receptor. Some cells lack components of the nuclear machinery that allow cells to halt their proliferation in response to TGF-β-initiated signals. Here, we see situations that are the mirror images of those involved in receiving and processing mitogenic signals. Rather than hyperactivating signaling by growth-stimulatory proteins, in these cases tumor suppressor proteins responsible for inhibiting cell proliferation are inactivated.

Concept and Reasoning Check

1. If the protooncogene encoding a growth factor *and* the protooncogene encoding a signaling protein downstream of the receptor for this growth factor were both activated by mutation, would this mutant cell be *twice* as likely to form a tumor, compared to a cell that contained only one of these mutations? Explain.

17.9 Cells are subject to growth inhibition and may exit from the cell cycle

Key concepts

- Cells that have differentiated have reached their final specialized form.
- Differentiated cells are usually postmitotic; thus, differentiation reduces the pool of dividing cells.
- Cells can commit suicide by apoptosis.
- Apoptosis eliminates healthy cells during development and at other times in an organism's lifetime.
- Apoptosis eliminates damaged cells that can pose a threat to the organism.
- Mutations that compromise a cell's ability to carry out apoptosis can result in malignancy.

In *17.8 Cell growth and proliferation are activated by growth factors*, we saw how TGF-β signaling can cause a cell to become quiescent. An alternative strategy for limiting proliferation is associated with the process of **differentiation**, during which primitive, embryonic cells throughout the body, termed **stem cells**, acquire specific traits that are typical of the bulk of the cells in their host tissue. Indeed, differentiation enables these cells to carry out the tissue-specific functions assigned to the tissues in which these cells are located. Most types of differentiated cells are **postmitotic**; that is, they have lost the ability to ever divide again. Stem cells, in contrast, have an essentially unlimited ability to renew themselves through growth and division. Hence, by inducing its stem cells to enter into postmitotic differentiation programs, a tissue can decrease its pool of replicating cells. **FIGURE 17.23** shows that, in contrast to stem cells, most types of differentiated cells do not replicate.

Many types of tumor cells behave similarly to stem cells whose entrance into postmitotic, differentiated states is blocked. Therefore, genes and proteins that induce postmitotic differentiation operate like tumor suppressors, having growth-inhibitory effects like the various components of the TGF-β pathway. In both cases, the replicative potential of tissues is constrained; in both cases, deletion or inactivation of the responsible genes and proteins enhances the ability of incipient cancer cells to proliferate, yielding large numbers of abnormal descendants.

Recent research has revised and elaborated upon this scheme. More specifically, it seems likely that the organization of normal tissues, involving self-renewing, undifferentiated stem cells and more differentiated, postmitotic cells, is recapitulated within tumors as well. For example, the great bulk of cancer cells within breast carcinomas are unable to seed new tumors when transplanted into appropriate mouse hosts; in this sense they are not tum-

origenic and lack the ability of self-renewal. In contrast, a small subpopulation of **cancer stem cells** within a tumor does indeed exhibit tumorigenicity (being able to seed a new tumor) and possess an essentially unlimited self-renewing capability. This explains why the structural organization of most tumors resembles the normal tissue from which they arise. This characteristic appearance seems to be created by the bulk of cells within a tumor, which have become postmitotic and, in addition, have undergone at least partial differentiation.

Until now, we have discussed two ways by which cells limit or circumscribe their proliferation: They may mature into a resting, or quiescent, state from which they may emerge on some future occasion and once again proliferate. Alternatively, they may enter into a postmitotic differentiated state. In fact, there is a third mechanism that limits cell proliferation. This one depends on the process of cell suicide called **apoptosis**. The machinery that activates and executes the apoptotic program is hard-wired into virtually every cell type in the body. Its purpose is to facilitate the elimination of cells within the tissue, doing so quickly and efficiently. Apoptosis is used for multiple purposes during normal development and during adult life. Thus, this cell suicide program eliminates unneeded healthy cells during development, such as the skin webs that are initially formed between the developing digits. In addition, the apoptotic machinery inside a cell may become activated following substantial, possibly irreparable damage to the cell's genome, by metabolic imbalances such as those caused by lack of oxygen (*anoxia*); by imbalances in the fluxes of growth-promoting and growth-inhibitory signals within the cell; and even by the cell's inability to tether properly to solid substrates, specifically those formed by the proteins of the extracellular matrix. **FIGURE 17.24** shows an apoptotic cell.

The body's tissues also use apoptosis to eliminate cells that have begun to behave abnormally and, therefore, carry the potential of eventually seeding a tumor. In the end, the actions of misbehaving cells (such as an incipient cancer cell) may exact a far higher price on the tissue and organism than is paid by simply eliminating these cells from tissues. Consequently, tissues are continually discarding cells that show even minimal deviation from normal behavior, using these cells' built-in apoptotic machinery to do so.

The apoptotic machinery is also attuned to the signaling imbalances that occur when,

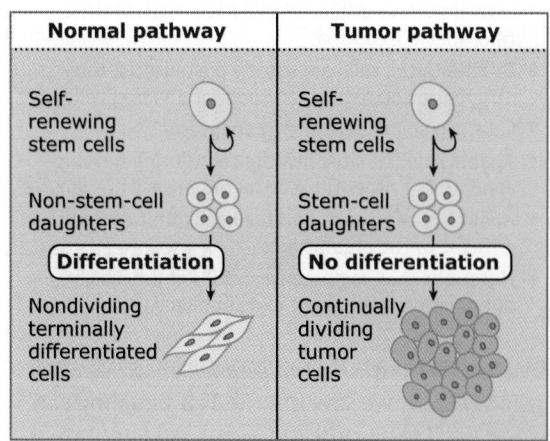

FIGURE 17.23 Differentiation is one way to limit the pool of dividing cells. Failure of cells to differentiate can lead to tumors.

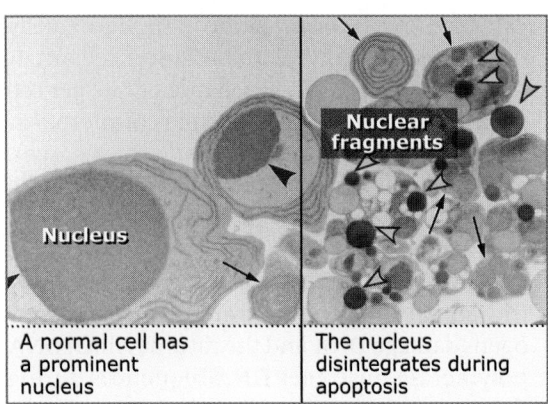

FIGURE 17.24 Cell structure changes during apoptosis. The left panel shows a normal cell. The right panel shows an apoptosing cell; gold arrows indicate condensed nuclear fragments. Photos courtesy of Shigekazu Nagata, Kyoto University Medical School.

for example, an oncogene becomes activated within a cell. It is likely that most cells that happen to acquire a mutant oncogene never succeed in spawning a large population of descendants because of the actions of an ever-vigilant apoptotic machinery, which eliminates the mutant cells before they have the chance to multiply. This indicates that an important class of tumor-suppressing proteins operates to activate and execute apoptosis; in many cell types, these proteins must be inactivated or deleted from the cells' normal repertoire of proteins before these cells can succeed in forming a large population of malignant descendants. We conclude that another important group of tumor suppressor genes acts to favor apoptosis, and that these genes must suffer some type of inactivating damage to allow certain types of tumors to arise. Indeed, it is plausible that all types of human cancer cells have inactivated components of their apoptotic machinery in order to survive and proliferate.

17.10 Tumor suppressors block inappropriate entry into the cell cycle

Key concepts

- Cells decide whether or not to divide at the restriction point.
- pRb is a tumor suppressor that can prevent passage through the restriction point.
- pRb can be inactivated by mutations, sequestration by oncoproteins, or by hyperactivity of the Ras pathway.

The diverse signals impinging on a cell, both positive and negative, must be integrated and processed, and ultimately used to make the final decisions about growth, quiescence, and differentiation. As mentioned previously, two alternatives to cell division are to exit from the cell cycle *reversibly* into a quiescent, nongrowing state or to exit *irreversibly* into a postmitotic, differentiated state. The decisions made by the cell—growth versus quiescence—occur during one particular phase of the cell cycle, termed G1. At a specific decision point toward the end of G1, just before the G1-to-S transition, the cell decides whether it should divide or, alternatively, retreat into the nongrowing phase termed G_0. This decision point is termed the **restriction (R) point** (see **FIGURE 17.25**). Once the cell has passed through the R point, the cell commits itself to complete the remainder of the cell cycle (i.e., the last part of the G1 phase, the S, G2, and M phases) in an essentially automatic fashion unless the cell encounters some genetic or metabolic disaster. Figure 11.13 shows the cell cycle and the restriction point.

The decision to pass through the R point is positively influenced by the mitogenic signals that the cell has received since the beginning of G1. Conversely, this decision is discouraged when the cell receives a substantial boost of antimitogenic signals such as those conveyed by TGF-β. It seems that the decision to execute the transition through the R point is made in a faulty fashion in virtually all types of human cancer cells.

One protein that plays a major role in the R point decision is the product of the retinoblastoma TSG, **pRb**. pRb exists in two states: **hypophosphorylated** (low levels of phosphorylation) and **hyperphosphorylated** (high levels of phosphorylation). When

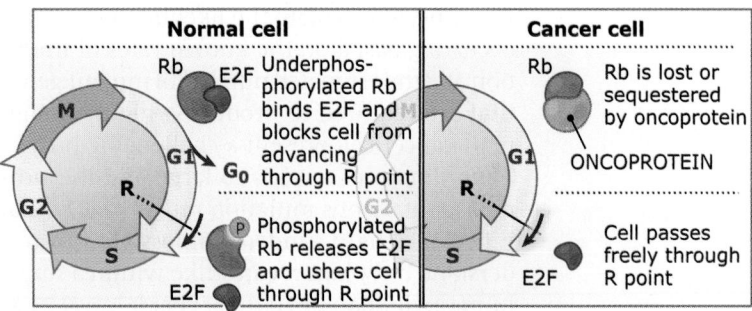

FIGURE 17.25 *Rb* is a major tumor suppressor gene that blocks cell cycle progression by binding to the E2F transcription factors, thereby preventing E2Fs from stimulating transcription of genes required for entrance into S phase.

hypophosphorylated, pRb blocks advance through the R point; when hyperphosphorylated, pRb opens the R point gate and ushers the cell through into late G1, as shown in Figure 17.25. In a variety of human tumors, including the previously mentioned retinoblastomas and osteosarcomas, pRb is lost because of inactivating mutations affecting both copies of the *Rb* gene. In many cervical carcinomas, pRb is rendered functionally inactive, being sequestered by an oncoprotein made by a human papilloma virus, the **etiologic** (provoking) **agent** in virtually all of these tumors. In the absence of functional pRb, cells pass through the R point without responding to the physiologic signals that normally are able to regulate the passage.

In addition to controlling cell division (by regulating advance through the G1 phase of the cell cycle), pRb may collaborate with a number of transcription factors to allow expression of tissue-specific differentiation programs. For example, in the case of muscle differentiation, pRb may associate with transcription factors that are responsible for the induction of muscle-specific genes. Consequently, pRb may couple cessation of proliferation with differentiation.

17.11 Mutation of DNA repair and maintenance genes can increase the overall mutation rate

Key concepts

- DNA repair proteins keep the spontaneous mutation rate low.
- Defects in DNA repair genes increase the basal rate of mutation in the cell.
- Mutations in checkpoint proteins compromise chromosome integrity.

Attempts to calculate the likelihood of cancer occurring based on the known rates of mutation and the estimated number of mutant genes that are required to produce a highly malignant cancer cell present a challenging puzzle. The number of steps is so large and the normal spontaneous mutation rate is so low that it becomes hard—indeed impossible—to understand how cancer can strike within a single human life span of 70 to 80 years.

One solution to this paradox derives from a critical parameter that is included in these calculations: the normal rate of spontaneously occurring mutations. Perhaps cancer cells accumulate mutations at a much higher rate per cell generation than do their normal counterparts. Such increased mutability could, on its own, expedite the rate of tumor progression and make the appearance of cancer mathematically plausible.

Spontaneous mutation rates are driven by the rates at which physical and chemical carcinogens damage DNA and the rates at which DNA polymerase miscopies DNA sequences during the process of DNA replication. Operating in the opposite direction, it is clear that the great majority of the erroneous sequences resulting from miscopied or damaged DNA are quickly erased and restored to wild-type sequence configuration. Cells have a surveillance system that continuously monitors the integrity of the genome, removes errant nucleotides, and replaces them with the proper nucleotides, resulting in the reconstruction of the original DNA sequence. In addition, this DNA repair system appears to be highly effective in detecting and removing bases that are damaged by oxidative processes and by other chemical reactions that affect DNA and its bases. Working in concert, it is clear that the DNA repair apparatus succeeds in holding down the rate of mutation to less than 10^{-6} per gene per cell generation.

In addition, several "checkpoint controls" ensure that critical steps in the cell cycle must be successfully executed before a cell is allowed to proceed with subsequent step. For example, DNA synthesis is halted if the DNA being replicated is damaged, mitosis is not allowed to begin if DNA replication is not completed, and the anaphase of mitosis is halted if the chromosomes are not properly attached to the mitotic spindle.

Like other cellular machines discussed here, these custodians of the genome are susceptible to corruption. The mutation of genes that encode components of this DNA repair apparatus results in defective repair and the accumulation of mutant alleles at an abnormally high rate. To be sure, these mutations strike all parts of the genome with comparable frequency, but among the genes affected will be many growth-controlling genes. As a direct consequence, many and perhaps all of the steps in tumor progression will occur more rapidly. Instead of requiring five or ten years to occur, each of these steps may succeed in happening in a much shorter period. The result will be that the multistep process of tumor progression will reach its endpoint far more rapidly, and the likelihood of clinical cancer will be increased commensurately.

One protein that is critical in arresting cell cycle progression following DNA damage is the **p53** tumor suppressor protein. In response to many types of cell-physiologic stress, p53 may halt cell proliferation temporarily; alternatively, p53 may serve to activate the apoptotic program. The disease of ataxia telangiectasia, a complex trait that results in other pathologies in addition to cancers, is caused by inheritance of a defective gene encoding a protein kinase responsible for minimizing DNA damage by phosphorylating, and thereby activating, p53 after damage occurs. Once activated, p53 may induce expression of a protein (termed p21) that blocks further advance through the cell cycle; often, once DNA repair is completed, p53 no longer induces p21 expression and the cell will be permitted to advance into the S or M phases of the cell cycle. In the absence of p53 function, a cell may therefore enter into S phase and replicate still-damaged DNA. Massive irreparable DNA damage normally leads to p53-induced apoptosis; once p53 is lost, however, cells with extensively damaged genomes may survive, since these cells are liberated from the threat of p53-induced apoptosis.

We also know about a number of familial cancer syndromes in which the inherited culprit gene encodes one or another component of the complex apparatus responsible for maintenance of genomic integrity. The first identified cancer syndrome of this class to be identified is xeroderma pigmentosum (XP). Individuals suffering from XP have extraordinary sensitivity to the ultraviolet (UV) radiation in sunlight because they have inherited defective copies of genes that encode one of the ten or more proteins that are responsible for detecting UV-induced DNA damage, excising the damaged bases, and replacing them with the bases that preexisted prior to the sun exposure. Sunlight provokes hundreds of skin lesions in these individuals, some of which ultimately progress to form skin carcinomas.

The *BRCA1* and *BRCA2* genes, well known because inheritance of defective alleles predisposes to breast and ovarian carcinomas, encode proteins that are responsible for repairing double-strand DNA breaks. The disease of hereditary nonpolyposis colon cancer (HNPCC)

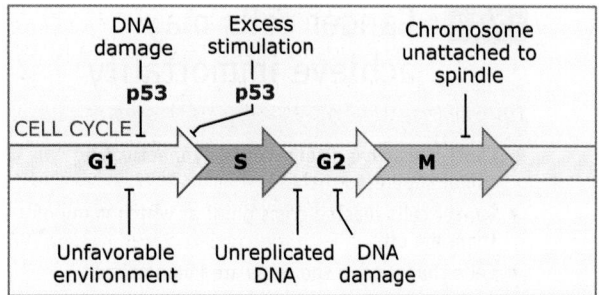

FIGURE 17.26 The cell has a number of systems that monitor the external environment and the state of the chromosomes throughout the cell cycle. These systems can slow or halt cell division until necessary repairs are made.

is caused by inheritance of defective versions of any one of four genes delegated to orchestrate mismatch DNA repair, which is concerned largely with copyediting and correcting recently replicated DNA. **FIGURE 17.26** shows the various genome surveillance systems that protect the chromosomal DNA and various mutations in the responsible genes that lead to cancer.

Finally, the great majority of human cancer cells are aneuploid, having lost their normal diploid karyotype at some point *en route* to becoming highly malignant. Some argue that this aneuploidy represents another route through which cancer cells, by continually reshuffling their chromosomal cards, can generate ratios of chromosomes and defective chromosomal arms that are highly favorable for malignant growth. Indeed, there is evidence from some types of tumors that virtually every malignant growth exhibits genetic instability manifested either at the level of DNA sequence or karyotype.

These genetic instabilities may represent a defect in cancer cells that is as important and as intrinsic to malignant growth as the defects associated with the actions of the specific growth-regulating genes.

Concept and Reasoning Check

1. Where is the "Magic Bullet"? Several decades ago, the concept of a "Magic Bullet" drug that would kill cancer cells by specifically targeting only the differences between cancer cells and their normal counterparts, was a driving motivation for a great deal of cancer research. In the intervening years, the details described in this chapter have been discovered, giving us many possible "therapeutic targets," yet very few current treatments can be considered "Magic Bullets." Why can we not find more "Magic Bullets"?

17.12 Cancer cells may achieve immortality

Key concepts

- Cancer cells avoid senescence by inactivating tumor suppressor genes.
- Cancer cells reach a crisis point at which many of them die off.
- Cells that survive the crisis are immortalized.
- Telomeres become shorter each generation unless telomerase is activated.
- When telomeres become too short to protect the chromosomes, the chromosomes fuse, which provokes crisis.
- Most cancer cells activate telomerase transcription, thereby escaping death.

Oncogenes and tumor suppressor genes explain much about the growth deregulation of cancer cells except for one critical aspect. As mentioned earlier in *17.2 Cancer cells have a number of phenotypic characteristics*, cancer cells have an unlimited proliferative potential, whereas normal cells are able to multiply in culture for only a limited number of doublings before they cease growing. At this point, these cells enter into a nongrowing state termed **senescence**. In human cells, senescence can be circumvented through the inactivation of the *p53* and *Rb* tumor suppressor genes. Once this inactivation occurs, cells will proliferate for another period of time before entering into a period called **crisis**, at which point they die in large numbers. On rare occasion, a clone of cells emerges from crisis and begins to grow

robustly; these cells now have gained unlimited replicative potential—that is, they are *immortalized*. Presumably, many of these steps also precede the immortalization of cancer cells that occurs during the tumor progression occurring within human tissues.

The ability of a cell lineage to tally the number of replicative doublings through which it has passed is hard to reconcile with the known functions of oncogenes and tumor suppressor genes. These genes are involved in receiving, processing, and rapidly responding to the signals that cells continually receive from their extracellular surroundings. The generation-counting ability of normal cells, however, implies the workings of some cellular clock that operates within the cell, independent of extracellular environment.

This cell-autonomous counting device resides in the **telomeres** at the ends of chromosomes. These structures are formed from a hexanucleotide sequence that is repeated several thousand times in each telomere. In the absence of functional telomeres, chromosomes will undergo end-to-end fusion, leading to chaotic chromosomal structures and often the death of the cell. Functioning together with bound telomere-specific proteins, the telomeric DNA serves to protect the ends of the chromosomes from end-to-end fusion with other chromosomes.

Importantly, the DNA replication apparatus of the cell is unable to faithfully copy the absolute ends of the telomeric DNA. As a consequence, each time the cell passes through an S phase, the telomeric DNA grows progressively shorter by 100–150 nucleotides, as shown in **FIGURE 17.27**. Ultimately, the cumulative telomere erosion that has occurred following a long succession of growth-and-division cycles leaves a descendant cell with telomeres that are so short that they can no longer function effectively to protect the ends of chromosomal DNA. The resulting end-to-end fusions of chromosomes are the catastrophic molecular events that trigger entrance of cells into crisis.

Cancer cell populations must break through this highly effective barrier to continue growing. They do so by de-repressing the expression of the **telomerase** enzyme. Normally this enzyme is only detectable at significant levels in early embryonic cells and in germ cells in the testes. In these cells, the telomerase enzyme operates to maintain and elongate telomeric DNA, and its expression is repressed in most tissue lineages as they differ-

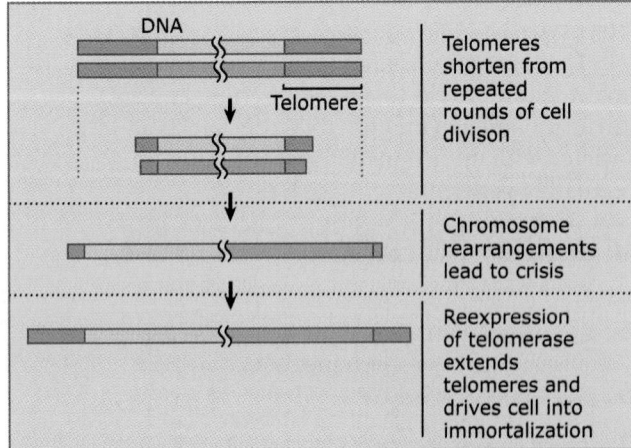

FIGURE 17.27 The shortened telomeres that occur after many cell divisions destabilize the chromosome. Chromosomes with shortened telomeres are prone to catastrophic rearrangements. Reexpression of telomerase in cancer cells restores the length of chromosomes and permits the cells to continue to divide.

entiate. However, ~90% of human cancer cells exhibit significant levels of telomerase, having de-repressed expression of this gene at some point during tumor progression. It is clear that in these cells, the continued function of the telomerase enzyme is essential to their ongoing proliferation. Thus, by deploying this enzyme to continuously elongate their telomeres, these cancer cells subvert the generational clock that depends on progressive telomere shortening. Once deprived of telomerase, cancer cells reenter crisis and die in large numbers. This makes it clear that the proliferative power of cancer cells is as dependent on telomerase as it is on the hyperactivity of oncogenes and the inactivity of tumor suppressor genes. (In fact, the residual 10% of human cancer cells circumvent the work generational clock by extending their telomeres through an alternative, telomerase-independent mechanism.)

17.13 Access to vital supplies is provided by angiogenesis

Key concepts

- Tumor growth is limited by access to nutrients and waste removal mechanisms.
- Tumors can stimulate blood vessel growth (angiogenesis), which enables them to expand.

The ability of the cells in a small incipient tumor mass to proliferate is compromised by a number of factors, the most important of which is the access by these cells to an adequate blood supply. It is estimated that the initially formed tumor mass—the **primary tumor**—can grow only to a size of 0.2 mm before it begins to encounter limitations in its access to nutrients and oxygen; at the same time, the cancer cells in such a mass begin to have difficulty evacuating metabolic wastes and carbon dioxide. Cells that experience extreme hypoxia may be induced to enter apoptosis or die by alternative means, often generating large areas of dead cells—**necrosis**—within a tumor mass.

These problems are avoided if the tumor mass succeeds in acquiring a network of blood vessels. Once in place, these vessels can address the problem of oxygen and nutrient supply and the elimination of the waste products of tumor cell metabolism. Unlike normally developing tissues, the pathologic tissue represented by the tumor mass does not naturally develop a network of arteries and veins. Instead, tumor cells must develop *de novo* an ability to attract the ingrowth of these vessels from adjacent normal tissues.

This process of developing new blood vessels is termed neo-angiogenesis, or simply **angiogenesis**. It occurs when tumor cells begin to secrete specialized growth factors termed **angiogenic factors** that, once released by the tumor cells, impinge on the endothelial cells that form the capillaries in adjacent normal tissues. In response to angiogenic factors, these endothelial cells begin to proliferate and to thread their way into the tumor mass, beginning the task of forming a vasculature that will ultimately support the further growth of the tumor mass, as shown in **FIGURE 17.28**.

It is clear that in small incipient tumor masses, the cancer cells multiply at a constant rate for many years. In spite of this, the tumor mass as a whole does not increase in size during this period. Lacking a blood supply, cells in this mass will die at the same rate as they are formed, killed by the lack of oxygen (*anoxia*), and poisoned by their own metabolic waste. Indeed, this state may be the ultimate fate of the majority of incipient tumor masses throughout our tissues. During this period of stasis, the process of diffusion succeeds in supporting the needs of the small tumor mass by providing it access to oxygen and nutrients from nearby well-vascularized tissues.

After years of such futile cycles of cell division, some cells within the small mass suddenly acquire the ability to stimulate angiogenesis.

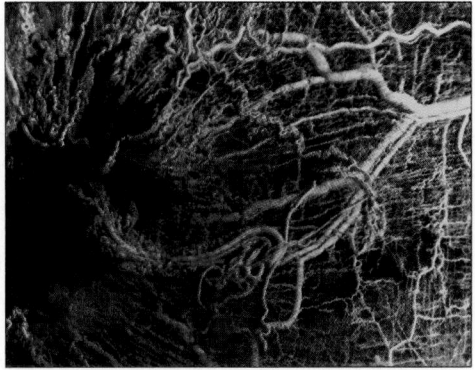

FIGURE 17.28 Blood vessels grow toward a sarcoma tumor in rat muscle. The sarcoma appears as a dark area on the left of the photo. Reproduced from J. Marx, *Science* 301 (2003): 452–454. Photo courtesy of Judah Folkman, Louis Heiser, and Robert Auckland, Karp Family Research Laboratories, Children's Hospital Boston.

When this occurs, the tumor mass as a whole begins a program of rapid expansion. This abrupt shift in behavior that occurs during multistep tumor progression is termed the "angiogenic switch." It indicates the workings of yet another barrier that the body's normal tissues place in the path of tumor cells intent on unlimited expansion. In other tumors, the acquisition of blood vessels does not result from a single, discrete event (like the angiogenic switch) during the course of multistep tumor progression. Instead, as tumor progression proceeds, tumor cells seem to acquire angiogenic ability progressively. (Indeed, in most tumors, it may be that once the angiogenic switch is flipped on, some type of biologic rheostat progressively increases the angiogenic powers of the tumor, resulting in ever-increasing angiogenesis and capillary density.)

The processes that govern angiogenesis involve complex cooperative interactions between the cancer cells and nearby normal cells that are recruited into the tumor mass and then co-opted by the cancer cells. Such normal cells, which provide vital physiologic support to the cancer cells, are largely of mesenchymal origin and constitute the tumor-associated stroma. It is thought that in addition to the angiogenic factors released directly by the cancer cells, other essential factors are released by nearby fibroblasts and **macrophages** within the tumor stroma. Working together, these signals succeed in inducing the formation of a blood vessel network that supplies all the needs of the tumor mass and gives it access to an essentially unlimited supply of nutrients and oxygen. Lymphatic vessels may also be induced to infiltrate into the tumor mass to aid in the evacuation of wastes and interstitial fluids from the tumor, but their role appears to be relatively minor when compared to that of blood vessels. In fact, within most large tumors, lymphatic vessels appear to be absent, resulting in inefficient drainage of the interstitial fluids that accumulate in the extracellular spaces throughout the tumor; this results, in turn, in the elevated interstitial hydrostatic pressure that is typically found in most large tumors.

Tumors can be graded by the density of capillaries that they have managed to recruit. Those with a densely woven capillary network are said to be highly vascularized, and this characteristic often bodes ill for the patient carrying such a mass, indicating the behavior of an aggressively growing population of cancer cells. Even more ominous, the presence of

these vessels reveals that the tumor mass has, for the first time, gained access to the highways in the body that will enable it to migrate to distant sites and wreak havoc there.

17.14 Cancer cells may invade new locations in the body

Key concepts
- Some cells from a primary tumor can gain entrance to blood and lymphatic vessels (intravasation).
- The process of intravasation often requires breaking through barriers of neighboring tissue.
- Cells that survive the trip through the blood vessels may colonize other organs.
- Metastasis, or colonization of other tissues, usually results in death of the individual.

The cells in a tumor mass may continue to proliferate for years, even decades, at the site where the tumor was initiated. Such primary tumors can grow to a size where they become life threatening through their abilities to displace vital tissue and compromise critical physiologic functions of the organism. Still, primary tumors are responsible for only about 10% of cancer mortality. The vast majority of cancer deaths are caused by the migration of cancer cells from primary tumors to other sites in the body.

As the tumor grows larger, some cells at the edge of the tumor begin to break down the physical barriers that constrain expansion. In the case of an epithelial tumor (a carcinoma), for example, this barrier is formed by the basement membrane, also known as the basal lamina that normally separates the epithelial cells from the underlying stroma, the latter being composed largely of fibroblasts and other mesenchymal cells (for more about the basement membrane see *19.11 The basal lamina is a specialized extracellular matrix*). Having broken through the basement membrane, some cancer cells may gain access to blood and lymphatic vessels—the process of **intravasation**—and use these as avenues to migrate to distant sites.

The wandering cells encounter a hostile environment once they leave the primary tumor mass. Many die while in circulation or when they become lodged in a capillary or small lymphatic vessel at some distant organ site in the body. Some succeed in escaping from the vessel—**extravasation**—and gain a foothold in the surrounding tissue. Here too, the vast majority of attempts fail. On rare occasions, however, such a cell will indeed succeed in finding a hospitable organ in which it can lodge and in which it can found a new, large colony of tumor cells—a **metastasis**.

The site of the resulting metastatic growth may be predetermined by anatomy (e.g., the paths of blood vessels leading from the primary tumor site to a target tissue) or by the growth factor environment in this new tissue. Colon carcinoma cells tend to form metastases in the liver, gaining access by the portal vein that drains blood from the colon directly into the liver. Breast cancer cells will often form metastases in the bone, the brain, and the lung. Prostate cancer cells will also form bone metastases. The precise reasons for these peculiar predilections are not well understood. But the consequences are clear: these metastases create much of the suffering that is associated with cancer and, in the great majority of cancer cases, these growths are responsible for the lethal endpoints of tumor progression.

The formation of a metastasis signals successful completion of a complex succession of steps. How do cancer cells with an epithelial tumor acquire the ability to do so? Increasingly it appears that the ability to execute most of the steps of cancer cell dissemination are acquired when the cancer cell activates a normally latent cell-biologic program, termed the **epithelial-mesenchymal transition (EMT)**; this program is normally activated during multiple steps of embryonic morphogenesis and is programmed by a set of embryonic transcription factors with names like Snail, Slug, Twist, and Zeb1. Each of these transcription factors is able, on its own, to act pleiotropically to activate an EMT. Once activated, this EMT program imparts to an epithelial cancer cell critical cell phenotypes, such as motility, invasiveness, and an ability to resist apoptosis. In fact, an EMT program appears to empower a carcinoma cell within a primary tumor to execute all of the steps of the invasion-metastasis cascade except the last one, which involves the growth of a micrometastasis into a macroscopic growth. This last step depends on the ability of cells within a micrometastasis to adapt to the foreign tissue microenvironment in which they have landed, an attribute that would not seem to be within the purview of EMT-inducing transcription factors. This suggests that EMT-inducing transcription factors and the programs that they organize explain most of the steps of metastasis. However, the precise mechanisms of adaptation of the cells from one tissue to another remain obscure.

17.15 What's next?

Cancer research will move in several directions over the next decade. To begin, there will be a further mapping of the interconnections between signal-transducing proteins in the cell, which will provide us with greater insight into how these signaling circuits operate. At the same time, this information will provide new power for identifying attractive targets for therapeutic intervention.

Indeed, the development of anticancer drugs will be changed dramatically by a confluence of several changes—the aforementioned insights into cancer cell circuitry, the faster ways of generating the three-dimensional structures of target proteins, the development of new ways of designing drugs *in silico* that will circumvent some of the laborious and expensive high-throughput screening of hundreds of thousands of agents, and better preclinical models to test drug candidates prior to introducing them into patients in Phase I clinical trials. At the same time, the development of sophisticated functional genomics screens will make it possible to stratify tumors into various subtypes that have differing responsiveness to this new generation of tailored drugs.

Accompanying these more practical advances will be increasing insights into the basic theory of cancer and how it arises. For example, it seems highly likely that we will be able to formulate some rules that govern the transformation of all types of human cells and allow us to rationalize why different sets of mutant alleles coexist and collaborate in the genomes of different types of cancer cells. We will understand more thoroughly the process of tumor pathogenesis. Traditionally, this depended on conceptual models in which mutational events were the driving forces in tumor progression. But it now seems highly likely that nongenetic mechanisms, notably inflammation, are likely to be responsible for the presence of tumors in some tissues and not others. Accordingly, we will gain extensive insights into how these nongenetic mechanisms, which are classified as **tumor-promoting** processes, conspire with the mutant alleles that accumulate in the genomes of evolving preneoplastic cell populations to create full-blown cancers.

In addition, this should allow us further insights into the causative mechanisms of human cancer, which at present are only poorly understood, with the exception of tobacco-induced tumors. These insights will, in turn, yield the greatest benefits, by allowing us modifications of diet and lifestyle that will ultimately yield far greater reductions in cancer mortality than all the drugs that biochemists, synthetic organic chemists, and pharmacologists can conjure up as novel agents for treating already-diagnosed tumors.

17.16 Summary

The process of tumor progression represents a drama in many acts. At least half a dozen steps may be involved in the progressive evolution of normal cells into fully malignant cells; many of these steps are likely to involve a change in cell genotype and phenotype. The logic behind this complexity is obvious. There are a large number of barriers in the path of the incipient tumor mass, each one designed to obstruct its advance to the next stage of tumor progression. Accordingly, each of the steps in multistep tumor progression represents the successful breaching of yet another barrier that impedes the advance of cells to the aggressive malignant growth state.

This complex series of antitumor defenses appears to be highly effective. It is likely that the vast majority of initiated tumors throughout the human body never succeed within a human life span in progressing through all the steps that lead to clinically detectable, life-threatening malignancies. It seems that all humans of advanced age carry hundreds, maybe thousands, of clones of premalignant cells in various organs throughout the body whose progression toward cancer is prevented by the concentric lines of defense that are hardwired into all cells and tissues. Included among these are the mechanisms that trigger senescence and apoptosis in response to oncogene activation; the mechanisms that trigger senescence and crisis following repeated rounds of growth and division; the mechanisms that block ready access by the incipient tumor mass to blood vessels; and the mechanical barriers that serve to physically constrain all but the most aggressive and invasive tumor cells.

Still, cancer does occur and these various steps do reach completion and become life threatening in 20% of people in Western societies. Clearly the likelihood of their occurrence can be increased substantially by carcinogenic influences that impinge upon the body throughout life—the diet and lifestyle factors. We now know that factors such as tobacco use, high-fat and high-meat diets, exposure to radiation, obesity, and lack of childbearing greatly increase cancer risk in various organ sites. It is known or presumed that each of these factors acts directly or indirectly to induce the genetic alterations that create the mutant genes implicated in the various steps of tumor progression.

Tobacco use is by far the most important of these causative factors. For example, as many as 30% of cancer-related deaths registered in recent years in the United States could be traced directly to the use of tobacco by the afflicted individuals. It is estimated that cessation of tobacco use and improvements in diet and lifestyle (e.g., increased exercise) could lead to as much as a 50% reduction in the incidence and, thus, mortality of cancer in the United States. Even with all these identified causative mechanisms, there are major puzzles that remain unsolved about the connections between genetic alterations and tumor progression. For example, it is quite unclear how disseminated cancer cells acquire the ability to become established and thrive in a foreign tissue environment.

An overall goal of basic cancer research is to finally be able to enumerate a complete list of all of the defects that are required to convert a normal human cell into a highly malignant derivative. This in turn will enable the cancer researcher to write a biography of the cancer cell in which each step in its formation can be documented in detail and with great precision. In the end, numerous molecular targets for therapeutic intervention will be found as the details of these complex pathways are elucidated.

Within recent years the discovery of tumor-initiating cancer stem cells (CSCs) has changed the landscape of cancer research. Because they have the power to generate or regenerate entire tumor masses, it has become evident that elimination of these cells is critical to the development of curative cancer therapies. Ominously, CSCs appear to be more resistant to existing therapies than the bulk of the more differentiated cells within tumors. This suggests that in the future, truly effective antitumor therapies will need to be able to eliminate CSCs as well as the non-CSC cells that form the bulk of tumor masses.

References

Aaronson, S. A. 1991. Growth factors and cancer. *Science* v. 254 p. 1146–1153.

Bergers, G., and Benjamin, L. E. 2003. Tumorigenesis and the angiogenic switch. *Nat. Rev. Cancer* v. 3 p. 401–410.

Cech, T. R. 2004. Beginning to understand the end of the chromosome. *Cell* v. 116 p. 273–279.

Evan, G., and Littlewood, T. 1998. A matter of life and cell death. *Science* v. 281 p. 1317–1322.

Foulds, L. 1954. *The Experimental Study of Tumor Progression*, Vols. I–III. London: Academic Press.

Hayflick, L., and Moorhead, P. S. 1961. The serial cultivation of human diploid cell strains. *Exp. Cell Res.* v. 25 p. 585–621.

Kastan, M. B., and Bartek, J. 2004. Cell-cycle checkpoints and cancer. *Nature* v. 432 p. 316–323.

Kinzler, K. W., and Vogelstein, B. 1996. Lessons from hereditary colorectal cancer. *Cell* v. 87 p. 159–170.

Knudson, A. G. 2002. Cancer genetics. *Am. J. Med. Genet.* v. 111 p. 96–102.

Land, H., Parada, L. F., and Weinberg, R. A. 1983. Tumorigenic conversion of primary embryo-fibroblasts requires at least two cooperating oncogenes. *Nature* v. 304 p. 596–602.

Lee, J. O., Russo, A. A., and Pavletich, N. P. 1998. Structure of the retinoblastoma tumour-suppressor pocket domain bound to a peptide from HPVE7. *Nature* v. 391 p. 859–865.

Massagué, J. 2004. G1 cell-cycle control and cancer. *Nature* v. 432 p. 298–306.

Martin, G. S. 2001. The hunting of the Src. *Nat. Rev. Mol. Cell Biol.* v. 2 p. 467–475.

Nowell, P. C. 1976. The clonal evolution of tumor cell populations. *Science* v. 194 p. 23–28.

Sherr, C. J. 2004. Principles of tumor suppression. *Cell* v. 116 p. 235–246.

Shi, Y., and Massague, J. 2003. Mechanisms of TGF-beta signaling from cell membrane to the nucleus. *Cell* v. 113 p. 685–700.

Shih, C., Shilo, B. Z., Goldfarb, M. P., Dannen-berg, A., and Weinberg, R. A. 1979. Passage of phenotypes of chemically transformed cells via transfection of DNA and chromatin. *Proc. Natl. Acad. Sci. USA* v. 76 p. 5714–5718.

Shih, C., and Weinberg, R. A. 1982. Isolation of a transforming sequence from a human bladder carcinoma cell line. *Cell* v. 29 p. 161–169.

Sinn, E., Muller, W., Pattengale, P., Tepler, I., Wallace, R., and Leder, P. 1987. Coexpression of MMTV/v-Ha-ras and MMTV/c-myc genes in transgenic mice: synergistic action of onco-genes in vivo. *Cell* v. 49 p. 465–475.

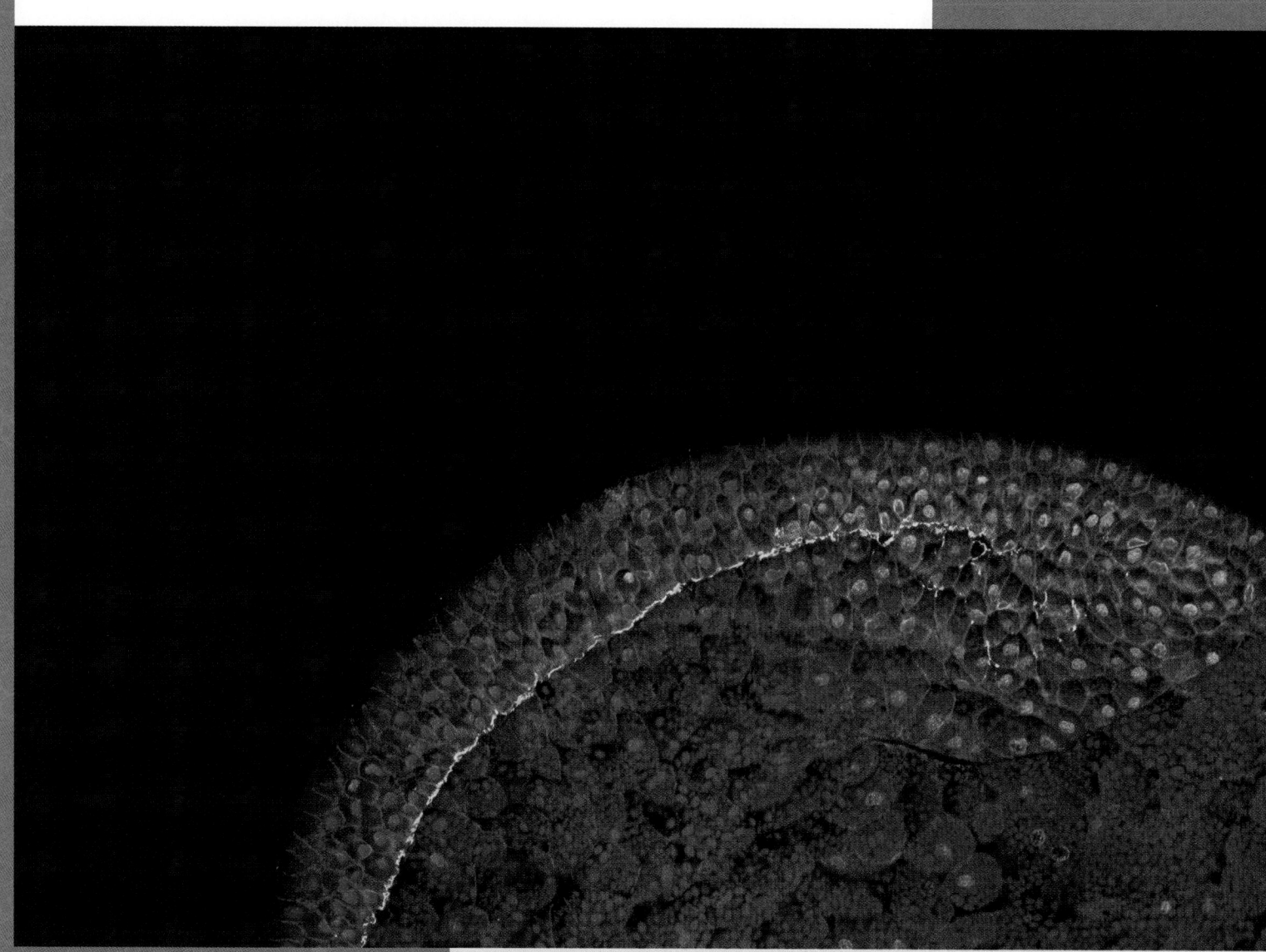

PART 6

Cell communication

Principles of cell signaling

18

Elliott M. Ross and Melanie H. Cobb
The University of Texas Southwestern Medical Center
at Dallas

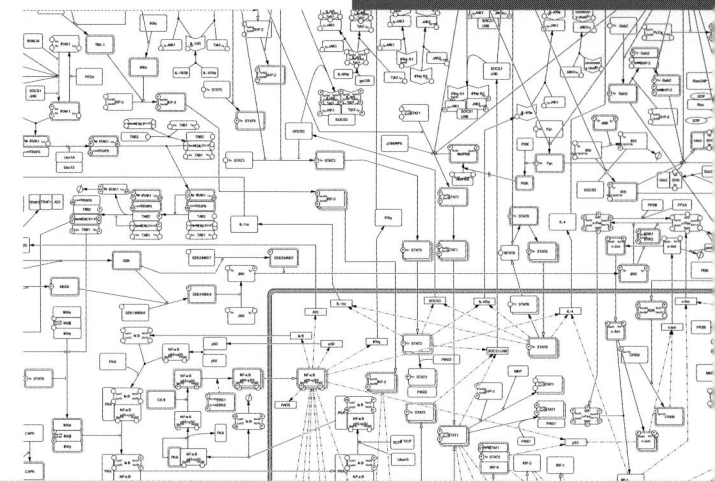

ABOUT 10% OF THE MAP of the known signaling interactions and reactions in the mouse macrophage. Preparing such a map in a computable format is the first step in analyzing a large signaling network. Map courtesy of Kanae Oda, Yukiko Matsuoka, and Hiroaki Kitano, The Systems Biology Institute.

CHAPTER OUTLINE

18.1 Introduction

All cells, from prokaryotes through plants and animals, sense and react to stimuli in their environments with stereotyped responses that allow them to survive, adapt, and function in ways appropriate to the needs of the organism. These responses are not simply direct physical or metabolic consequences of changes in the local environment. Rather, cells express arrays of sensing proteins, or **receptors** that recognize specific extracellular stimuli. In response to these stimuli, receptors regulate the activities of diverse intracellular regulatory proteins that in turn initiate appropriate responses by the cell. The process of sensing external stimuli and conveying the inherent information to intracellular targets is referred to as cellular **signal transduction**.

Cells respond to all sorts of stimuli. Microbes respond to nutrients, toxins, heat, light, and chemical signals secreted by other microbes. Cells in multicellular organisms express receptors specific for hormones, neurotransmitters, **autocrine** and **paracrine** agents (hormone-like compounds from the secreting cell or cells nearby), odors, molecules that regulate growth or differentiation, and proteins on the outside of adjacent cells.

A mammalian cell typically expresses about fifty distinct receptors that sense different inputs, and, overall, mammals express several thousand receptors.

Despite the diversity of cellular lifestyles and the enormous number of substances sensed by different cells, the general classes of proteins and mechanisms involved in signal transduction are conserved throughout living cells, as shown in **FIGURE 18.1**.

- **G protein-coupled receptors**, composed of seven membrane-spanning helices, promote activation of heterotrimeric GTP-binding proteins called **G proteins**, which associate with the inner face of the plasma membrane and convey signals to multiple intracellular proteins.
- **Receptor protein kinases** are often dimers of single membrane-spanning proteins that **phosphorylate** their intracellular substrates and, thus, change the shape and function of the target proteins. These protein kinases frequently contain protein interaction domains that organize complexes of signaling proteins on the inner surface of the plasma membrane.
- **Phosphoprotein phosphatases** reverse the effect of protein kinases by removing the phosphoryl groups added by protein kinases.
- Other single membrane-spanning enzymes, such as **guanylyl cyclase**, have an overall architecture similar to the receptor protein kinases but different enzymatic activities. Guanylyl cyclase catalyzes the conversion of GTP to 3':5'-cyclic GMP, which is used to propagate the signal.
- **Ion channel receptors**, although diverse in detailed structure, are usually oligomers of subunits that each contain several membrane-spanning segments. The subunits change their conformations and relative orientations to permit ion flux through a central pore.
- **Two-component systems** may either be membrane spanning or cytosolic. The number of their subunits is also variable, but each two-component system contains a histidine kinase domain or subunit that is regulated by a signaling molecule and a response regulator

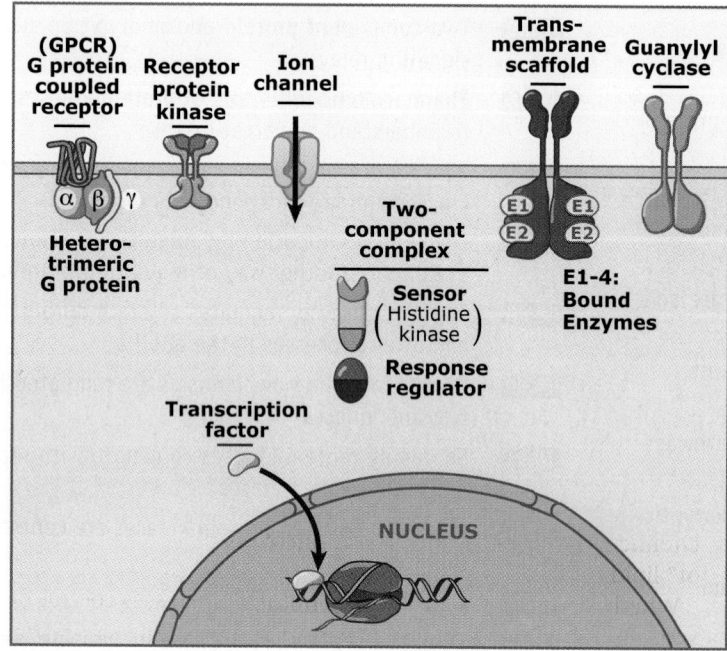

FIGURE 18.1 Receptors form a rather small number of families that share common mechanisms of action and overall similar structures.

that contains a phosphorylatable aspartate (Asp) residue.

- Some receptors are transmembrane **scaffolds** that change either the conformation or oligomerization of their intracellular scaffold domains in response to extracellular signaling molecules, or **ligands**, and, thus, recruit interacting regulatory proteins to a common site on the membrane.
- **Nuclear receptors** are **transcription factors**, often heterodimers that may reside in the cytoplasm until activated by agonists or may be permanently located in the nucleus.

The biochemical processes of signal transduction are strikingly similar among cells. Bacteria, fungi, plants, and animals use similar proteins and multiprotein modules to detect and process signals. For example, evolutionarily conserved heterotrimeric G proteins and G protein-coupled receptors are found in plants, fungi, and animals. Similarly, 3':5' cyclic AMP (cAMP) is an intracellular signaling molecule in bacteria, fungi, and animals; and Ca^{2+} serves a similar role in all eukaryotes. Protein kinases and phosphoprotein phosphatases are used to regulate enzymes in all cells.

Although the basic biochemical components and processes of signal transduction are conserved and reused, they are often used in wildly divergent patterns and for many different physiologic purposes. For example, cAMP is synthesized by distantly related enzymes in bacteria, fungi, and animals, and acts on different proteins in each organism; it is a **pheromone** in some slime molds.

Cells often use the same series of signaling proteins to regulate a given process, such as transcription, ion transport, locomotion, and metabolism. Such signaling pathways are assembled into signaling networks to allow the cell to coordinate its responses to multiple inputs with its ongoing functions. It is now possible to discern conserved reaction sequences in and between pathways in signaling networks that are analogous to devices within the circuits of analog computers: amplifiers, logic gates, feedback and feed-forward controls, and memory.

This chapter discusses the principles and strategies of cellular signaling first and then discusses the conserved biochemical components and reactions of signaling pathways and how these principles are applied.

18.2 Cellular signaling is primarily chemical

Key concepts
- Cells can detect both chemical and physical signals.
- Physical signals are generally converted to chemical signals at the level of the receptor.

Most signals sensed by cells are chemical, and, when physical signals are sensed, they are generally detected as chemical changes at the level of the receptor. For example, the visual photoreceptor rhodopsin is composed of the protein opsin, which binds to a second component, the colored vitamin A derivative *cis*-retinal (the **chromophore**). When *cis*-retinal absorbs a photon, it **photoisomerizes** to *trans*-retinal, which is an activating ligand of the opsin protein. (For more on rhodopsin signaling see *18.20 G protein-signaling modules are widely used and highly adaptable*). Similarly, plants sense red and blue light using the photosensory proteins phytochrome and cryptochrome, which detect photons that are absorbed by their tetrapyrrole or flavin chromophores. Cryptochrome homologs are also expressed in animals, where they mediate adjustment of the diurnal cycle.

A few receptors do respond directly to physical inputs. Pressure-sensing channels, which exist in one form or another in all organisms, mediate responses to changes in pressure or shear by changing their ionic conductance. In mammals, hearing is mediated indirectly by a mechanically operated channel in the hair cell of the inner ear. The extracellular domain of a protein called cadherin is pulled in response to acoustic vibration, generating the force that opens the channel.

Cells sense mechanical strain through a number of cell surface proteins, including integrins. Integrins provide signals to cells based on their attachment to other cells and to molecular complexes in the external milieu.

One major group of physically responsive receptors is made up of channels that sense electric fields. Another interesting group are heat/pain-sensing ion channels; several of these heat-sensitive ion channels also respond to chemical compounds, such as capsaicin, the "hot" lipid irritant in hot peppers.

Whether a signal is physical or chemical, the receptor initiates the reactions that change the behavior of the cell. We will discuss how these effects are generated in the rest of the chapter.

18.3 Receptors sense diverse stimuli but initiate a limited repertoire of cellular signals

Key concepts

- Receptors contain a ligand-binding domain and an effector domain.
- Receptor modularity allows a wide variety of signals to use a limited number of regulatory mechanisms.
- Cells may express different receptors for the same ligand.
- The same ligand may have different effects on the cell depending on the effector domain of its receptor.

Receptors mediate responses to amazingly diverse extracellular messenger molecules; hence, the cell must express a large number of receptor varieties, each able to bind its extracellular ligand. In addition, each receptor must be able to initiate a cellular response. Receptors, thus, contain two functional **domains**: a **ligand-binding domain** and an **effector domain**, which may or may not correspond to definable structural domains within the protein.

The separation of ligand-binding and effector functions allows receptors for diverse ligands to produce a limited number of evolutionarily conserved intracellular signals through the action of a few effector domains. In fact, there are only a limited number of receptor families, which are related by their conserved structures and signaling functions (see Figure 18.1).

There are several useful correlates to the two-domain nature of receptors. For example, a cell can control its responsiveness to an extracellular signal by regulating the synthesis or degradation of a receptor or by regulating the receptor's activity (see *18.10 Cellular signaling is remarkably adaptive*).

In addition, the nature of a response is generally determined by the receptor and its effector domain rather than any physicochemical property of the ligand. **FIGURE 18.2** illustrates the concept that a ligand may bind to more than one kind of receptor and elicit more than one type of response, or several different ligands may all act identically by binding to functionally similar receptors. For example, the neurotransmitter acetylcholine binds to two classes of receptors. Members of one class are ion channels; members of the other regulate G proteins. Similarly, steroid hormones bind both to nuclear receptors, which bind chromatin and regulate transcription, and to other receptors in the plasma membrane.

Conversely, when multiple ligands bind to receptors of the same biochemical class, they generate similar intracellular responses. For example, it is not uncommon for a cell to express several distinct receptors that stimulate production of the intracellular signaling molecule cAMP. The effect of the receptor on the cell will also be determined significantly by the biology of the cell and its state at any given time.

Ligand-binding and effector domains may evolve independently in response to varied selective pressures. For example, mammalian and invertebrate rhodopsins transduce their signal through different effector G proteins (G_t and G_q, respectively). Another example is calmodulin, a small calcium-binding regulatory protein in animals, which in plants appears as a distinct domain in larger proteins.

The receptor's two-domain nature allows the cell to regulate the binding of ligand and the effect of ligand independently. **Covalent modification** or **allosteric** regulation can alter ligand-binding affinity, the ability of the ligand-bound receptor to generate its signal or both. We will discuss these concepts further in *18.13 Cellular signaling uses both allostery and covalent modification*.

Receptors can be classified according to either the ligands they bind or the way in which they signal. Signal output, which is characteristic of the effector domain, usually correlates best with overall structure and sequence conservation. (Receptor families grouped by their

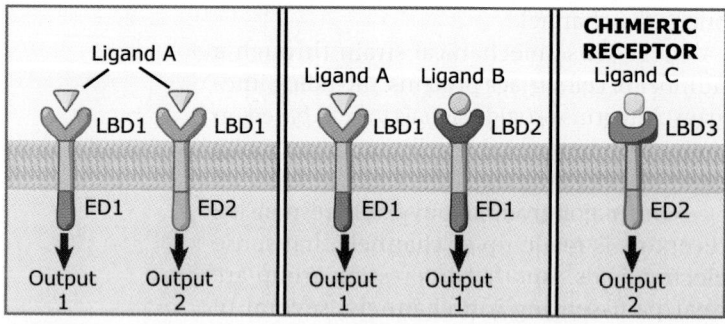

FIGURE 18.2 Receptors can be thought of as composed of two functional domains, a ligand-binding domain (LBD) and an effector domain (ED). The two-domain property implies that two receptors that respond to different ligands (middle) could initiate the same function by activating similar effector domains, or that a cell could express two receptor isoforms (left) that respond to the same ligand with distinct cellular effects mediated by different effector domains. It also implies that one can create an artificial chimeric receptor with novel properties.

functions are the organizational basis of the second half of this chapter.) However, classifying receptors pharmacologically, according to their specificity for ligands, is particularly useful for understanding the organization of endocrine and neuronal systems and for categorizing the multiple physiologic responses to drugs.

Expression of a receptor that is not normally expressed in a cell is often sufficient to confer responsiveness to that receptor's ligand. This responsiveness often occurs because the cell expresses the other components necessary for propagating the intracellular signal from the receptor. The precise nature of the response will reflect the biology of the cell. Experimentally, responsiveness to a compound can be induced by introducing the cDNA that encodes the receptor. For example, mammalian receptors may be expressed in yeast, such that the yeast responds visibly to receptor ligands, thus providing a way to screen for new chemicals (drugs) that activate the receptor.

Finally, it is possible to create chimeric receptors by fusing the ligand-binding domain from one receptor with the effector domain from a different receptor (see Figure 18.2). Such chimeras can mediate novel responses to the ligand. With genetic modification of the ligand-binding domain, receptors can be re-engineered to respond to novel ligands. Thus, scientists can manipulate cell functions with nonbiologic compounds.

18.4 Receptors are catalysts and amplifiers

Key concepts
- Receptors act by increasing the rates of key regulatory reactions.
- Receptors act as molecular amplifiers.

Receptors act to accelerate intracellular functions and are, thus, functionally analogous to enzymes or other catalysts. Some receptors, including the protein kinases, protein phosphatases, and guanylate cyclases, are themselves enzymes and thus classical biochemical catalysts. More generally, however, receptors use the relatively small energy of ligand binding to accelerate reactions that are driven by alternative energy sources. For example, receptors that are ion channels catalyze the movement of ions across membranes, a process driven by the electrochemical potential developed by distinct ion pumps. G-protein-coupled receptors and other guanine nucleotide exchange factors catalyze the exchange of GDP for GTP on the G protein, an energetically favored process dictated by the cell's nucleotide energy balance. Transcription factors accelerate the formation of the transcriptional initiation complex, but transcription itself is energetically driven by multiple steps of ATP and dNTP hydrolysis.

As catalysts, receptors enhance the rates of reactions. Most signaling involves **kinetic** rather than **thermodynamic** regulation; that is, signaling events change reaction rates rather than their equilibria (see *18.5 Ligand binding changes receptor conformation*). Thus, signaling is similar to metabolic regulation, in which specific reactions are chosen according to their rates, with thermodynamic driving forces playing only a supportive role.

In all signaling reactions, receptors use their catalytic activities to function as molecular amplifiers. Directly or indirectly, a receptor generates a chemical signal that is huge, both energetically and with respect to the number of molecules recruited by a single receptor. Molecular amplification is a hallmark of receptors and many other steps in cellular signaling pathways.

18.5 Ligand binding changes receptor conformation

Key concepts
- Receptors can exist in active or inactive conformations.
- Ligand binding drives the receptor toward the active conformation.

A central mechanistic question in receptor function is how the binding of a signaling molecule to the ligand-binding domain increases the activity of the effector domain. The key to this question is that receptors can exist in multiple molecular conformations, some active for signaling and others inactive. Ligands shift the conformational equilibrium among these conformations. The structural changes that occur during the receptor's inactive-active isomerization and how ligand binding drives these changes are exciting areas of biophysical research. However, the basic concept can be described simply in terms of coupling the conformational isomerizations of the ligand-binding and effector domains.

How do ligands activate (or not activate) a receptor? Most of the basic regulatory activities of receptors can be described by a simple scheme that considers the receptors as having two interconvertible conformations, inactive (R) and active (R*). R and R* are in equilibrium, which is described by the equilibrium constant J.

$$R \xrightleftharpoons{J} R*$$

Because unliganded receptors are usually minimally active, J<<1 and an unliganded receptor spends most of its time in the R state. When a signaling molecule (L) binds, it drives the receptor toward the active conformation, R*, in which the effector domain is functional. The ligand-bound receptor thus spends most of its time in the active R* state.

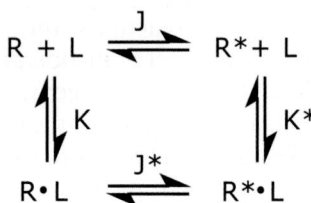

The mechanism whereby a ligand can activate receptor is a consequence of its relative affinities for the receptor's active and inactive conformations. A ligand can bind to the receptor in either of its conformations, described here by association constants K for the R state and K* for the R* state. Any ligand that binds with higher affinity for the R* conformation than for R will be an activator. If K* is greater than K, the ligand is an agonist. According to

the Second Law of Thermodynamics, a system of coupled equilibria displays path independence: the net free energy difference between two states is independent of which intermediary reactions take place. For the receptor, any path from R to R*L therefore has the same free energy change, and the products of the equilibrium constants along each path are equal. For the example above, path independence means the following:

J•K* = K•J*

Therefore, J*/J = K*/K.

Thus, if binding to the R* configuration is preferred (i.e., K*/K>>1), then ligand binding will shift the conformation to the R* state to an equivalent extent (i.e., J*/J>>1). The relative activation by a saturating concentration of ligand, J*/J, will exactly equal the ligand's relative selectivity for the active receptor conformation, K*/K. This argument is generally valid for the regulation of a protein's activity by any regulatory ligand.

This model explains many properties of receptors and their ligands both simply and quantitatively.

- First, J must be greater than zero for the equilibrium to exist. Thus, even unliganded receptor has some activity. Overexpressed receptors frequently display their intrinsic low activity.

- Because physiologic receptors are nearly inactive in the absence of ligand, J must be much less than 1 and is probably less than 0.01; most receptors are less than 1% active without agonist.

- Ligands can vary in their selectivities between R and R*. Their abilities to activate will also vary. Some ligands, referred to as agonists, can drive formation of appreciable R*. Others, known as **partial agonists**, will promote submaximal activation. Chemical manipulation of a ligand's structure will often alter its activity as an agonist. These relationships are depicted graphically in **FIGURE 18.3**.

- A ligand that binds equally well to both the R and R* states will not cause activation. However, such a ligand may still occupy the binding site and thereby competitively inhibit binding of an activating ligand. Such competitive inhibitors, referred to as **antagonists**, are frequently used as drugs to block unwanted activation of a receptor in various disease states.

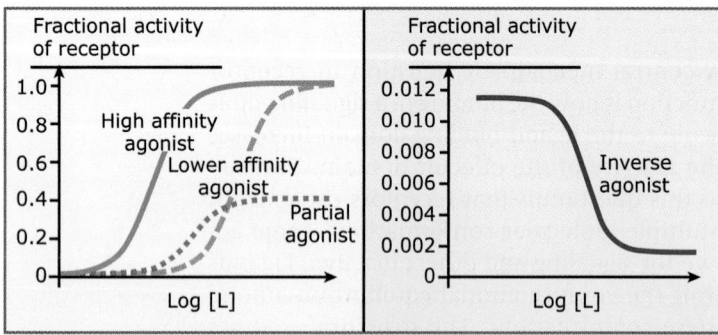

FIGURE 18.3 The simple two-state model shown here can describe a wide variety of behaviors displayed by receptors and their various regulatory ligands. The left panel shows fractional activity of a receptor exposed to two agonists with different affinities and one partial agonist. The right panel shows the effect of an inverse agonist. If the low fractional activity of unliganded receptor is detected as significant biological activity, then its inhibition by the inverse agonist would be easily detectable.

- A ligand that binds preferentially to R relative to R* will further shift the conformational equilibrium to the inactive state and cause net inhibition. Such ligands are called **inverse agonists**. Because J is already low, effects of inverse agonists may only be noticeable if a receptor is overexpressed or if the receptor is mutated to increase its intrinsic activity (i.e., if the mutation increases J).
- The extent to which an agonist stimulates a receptor is unrelated to its affinity. Both agonists and antagonists may bind with either high or low affinity. Affinity does determine the receptor's sensitivity—that is, how low a concentration of ligand can the receptor detect. Affinities of receptors for natural regulatory ligands vary enormously, with physiologic K_d values ranging from $<10^{-12}$ M for some hormones to about 10^{-3} M for some bacterial chemoattractants. Another aspect of sensitivity is how abruptly or gradually the receptor is activated as the concentration of agonist increases. The above model predicts that a receptor is activated significantly at agonist concentrations between 0.1 and 10 times its K_d. A variety of cellular mechanisms can convert such a conventional response range of about 100-fold to either a more gradual response or a very steep, switchlike response.
- This model only describes equilibria. It makes no predictions about the rates of ligand binding or release, or of the conformational isomerization that leads to activation.

This model shows how three important aspects of receptor action are independently determined. As mentioned above, affinity for ligand, which determines the concentration range over which the ligand functions, is independent of the ligand's net effectiveness at driving receptor activation. The rate of response is also largely independent of these other two properties. Each aspect of receptor function can thus be independently regulated in response to other incoming signals or by the metabolic or developmental state of the cell. Such control of signal input is central to whole-cell coordination of signal transduction. Examples and mechanisms will recur throughout this chapter.

Concept and Reasoning Check

1. A partial agonist drug P elicits about 30% of the response characteristic of "full" agonists. What will be the effect on cellular response of combining increasing concentrations of P with a fixed concentration of a full agonist that is added at 4 times its half-maximally effective concentration? It will be easiest to answer this question by sketching a plot of fractional response against the concentration of P.

18.6 Signals are sorted and integrated in signaling pathways and networks

Key concepts

- Signaling pathways usually have multiple steps and can diverge and/or converge.
- Divergence allows multiple responses to a single signal.
- Convergence allows signal integration and coordination.

Receptors rarely act directly on the intracellular processes that they ultimately regulate. Rather, receptors typically initiate a sequence of regulatory events that involve intermediary proteins and small molecules. The use of multistep signaling pathways allows cells to amplify signals, adjust signaling kinetics, insert control points, integrate multiple signals, and route signals to distinct effectors.

Branched pathways give cells the ability to integrate multiple incoming signals and to direct information to the correct control points. As **FIGURE 18.4** illustrates, branching can be either convergent, with multiple signals regulating common end points, or divergent, with a single pathway branching to control more than one process. In multicellular organisms, divergent branching allows a single hormone receptor to initiate distinct cell-appropriate patterns of responses in different cells and tissues. Divergent signaling also allows a receptor to regulate qualitatively different cellular responses with quantitatively distinct intensities, each dependent on signal amplification in the intermediary pathway.

Convergent branching—when several receptors activate the same pathway to elicit the same regulatory responses—is also common. Convergent branching allows multiple incoming signals, both stimulatory and inhibitory, to be integrated and coordinately regulated at a common site downstream of the receptors.

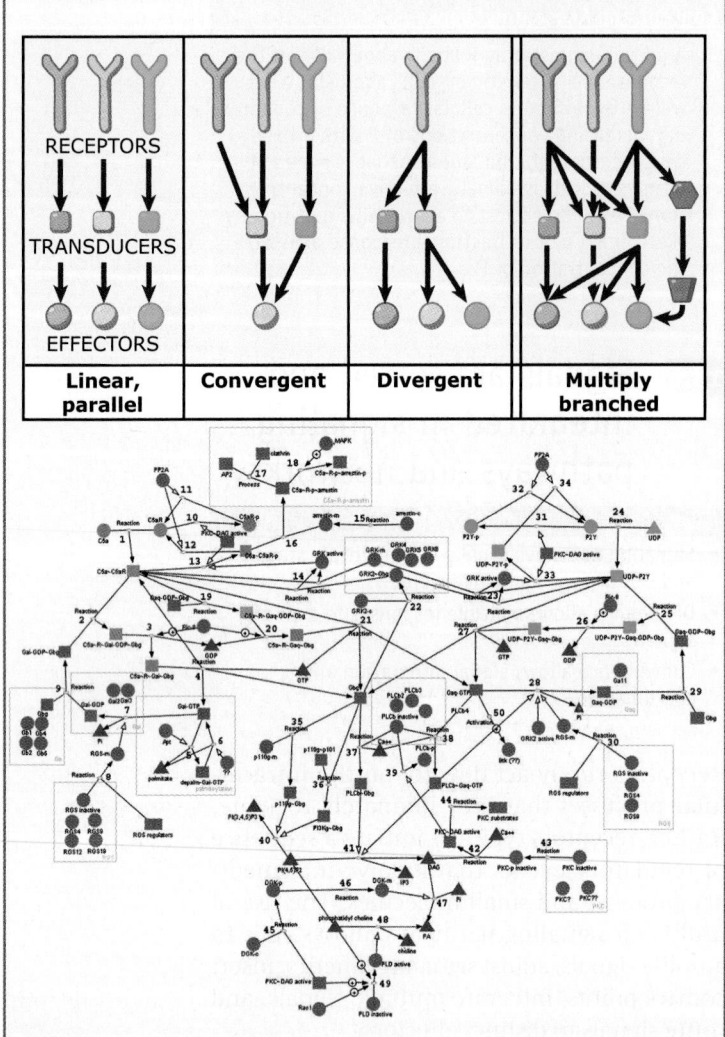

FIGURE 18.4 Signaling pathways use convergent and divergent branching to coordinate information flow. The diagrams at top show how even a simple, three-level signaling network can sort information. Convergence or divergence can take place at multiple points along a signaling pathway. As an example of complexity, the lower portion of the figure shows a small segment (~10%) of the G protein-mediated signaling network in a mouse macrophage cell line. It omits several interpathway regulatory mechanisms and completely ignores inputs from non-G protein-coupled receptors. Pathway map courtesy of Lily Jiang, University of Texas Southwestern Medical Center.

Receptors for several different hormones frequently initiate similar or overlapping patterns of signaling in a single target cell.

Overlapping converging and diverging signaling pathways create signaling networks within cells that coordinate responses to multiple inputs (see Figure 18.4). Typically, such pathways are complex in the number and diversity of their components and in the topology of their circuit maps. Signaling networks are also spatially complex. They may include components in various subcellular locations, with receptors and associated proteins in the plasma membrane, but with downstream proteins in the cytoplasm or intracellular organelles. Such complexity is necessary to allow the cells to integrate and sort incoming signals and to regulate multiple intracellular functions simultaneously.

The complexity and adaptability of signaling networks, like the one shown in the lower half of Figure 18.4, make their dynamics at the whole-cell level difficult or impossible to grasp intuitively. Signaling networks resemble large analog computers, and investigators are increasingly depending on computational tools to understand cellular information flow and its regulation. Many signaling interactions that include only two or three proteins exert functions analogous to traditional computational logic circuits (see *18.7 Cellular signaling pathways can be thought of as biochemical logic circuits*). The theory and experience with such circuits in electronics facilitate understanding biologic signaling functions as well.

The enormous complexity of cellular signaling networks can be simplified by considering them to be composed of interacting signaling modules, that is, groups of proteins that process signals in well-understood ways. A cellular signaling module is analogous to an integrated circuit in an electronic instrument that performs a known function, but whose exact components could be changed for similar use in another device. The concept of modular construction facilitates both qualitative and quantitative understanding of signaling networks. We will refer to many standard signaling modules later in the chapter. Examples include monomeric and heterotrimeric G protein modules, mitogen-activated protein kinase (MAPK) cascades, tyrosine (Tyr) kinase receptors and their binding proteins, and Ca^{2+} release/uptake modules. In each case, despite the numerous phylogenetic, developmental, and physiologic variations, understanding the basic function of that class of module conveys understanding its incarnations. Last, the evolutionary importance of modules is significant; once the architecture of a module is established it can be reused.

Multiplexed, high-throughput measurements on living cells are combined with powerful kinetic modeling strategies to allow an increasingly accurate quantitative depiction of information flow within signaling modules or entire networks. Such models, with sound and experimentally based parameter sets, can describe signaling processes in systems too complex for intuitive or *ad hoc* analysis. They are also vital as tests of understanding because they can predict experimental results in ways that can be used to test the validity of the model. Well-grounded models can then be used (cautiously) to suggest the mechanisms of systems for which data sets remain unattainable. At even greater levels of complexity, the theories and tools of computer science are increasingly giving useful systems-level analyses of signal flow in cells. Using computational tools to analyze large arrays of quantitative data allows us to understand cellular information flow and its regulation. Developing quantitative models of signaling networks is a frontier in signaling biology. These models both help describe network function and pinpoint experiments to clarify mechanism.

18.7 Cellular signaling pathways can be thought of as biochemical logic circuits

Key concepts

- Signaling networks are composed of groups of biochemical reactions that function as mathematical logic functions to integrate information.
- Combinations of such logic functions combine as signaling networks to process information at more complex levels.

As introduced in the preceding section, processes that signaling pathways use to integrate and direct information to cellular targets are strikingly analogous to the mathematical logic functions that are used to design the individual circuits of electronic computers. Indeed, there are biologic equivalents of essentially all of the functional components that computer scientists and engineers consider in the design of computers and electronic control devices. To understand signaling pathways, it is, therefore, useful to consider groups of reactions within a pathway as constituting logic circuits of the sort used in electronic computing, as illustrated in **FIGURE 18.5**. The simplest example is when two stimulatory pathways converge. If sufficient input from either is adequate to elicit the response, the convergence would constitute an "OR" function. If neither input is sufficient by itself but the combination of the two elicits the

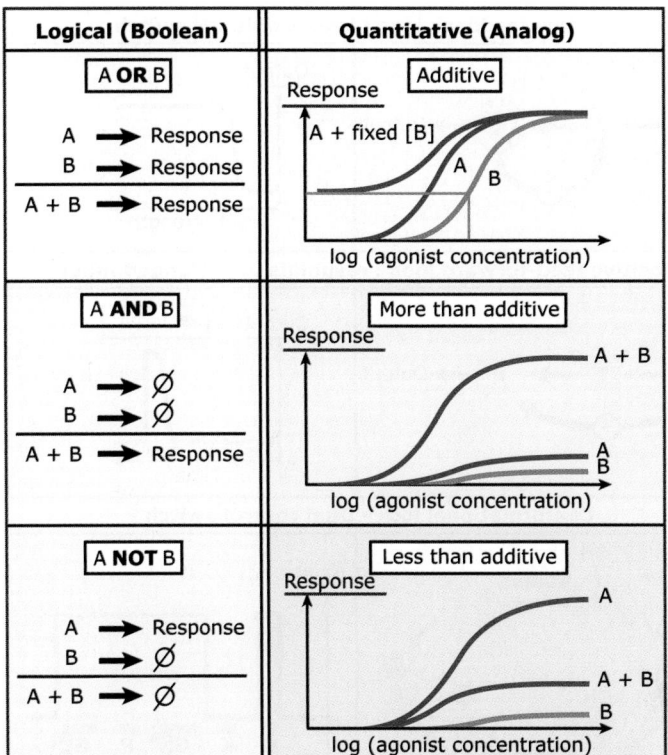

FIGURE 18.5 Signaling networks use simple logic functions to process information. Boolean OR, AND, and NOT functions (left) correspond to the quantitative interactions between converging signals that are shown on the right.

response, then the converging pathways would create "AND" functions. AND circuits are also referred to as **coincidence detectors**—a response is elicited only when two stimulating pathways are activated simultaneously.

AND functions can result from the combination of two similar but quantitatively inadequate inputs. Alternatively, two mechanistically different inputs might both be required to elicit a response. An example of the latter would be a target protein that is allosterically activated only when phosphorylated, or that is activated by phosphorylation but is only functional when recruited to a specific subcellular location.

The opposite of an AND circuit is a NOT function, where one pathway blocks the stimulatory effect of another. Simple logic gates are observed at many locations in cellular signaling pathways.

We can also think about convergent signaling in quantitative rather than Boolean terms by considering the additivity of inputs to a distinct process (see Figure 18.5, right). The OR function referred to above can be considered to

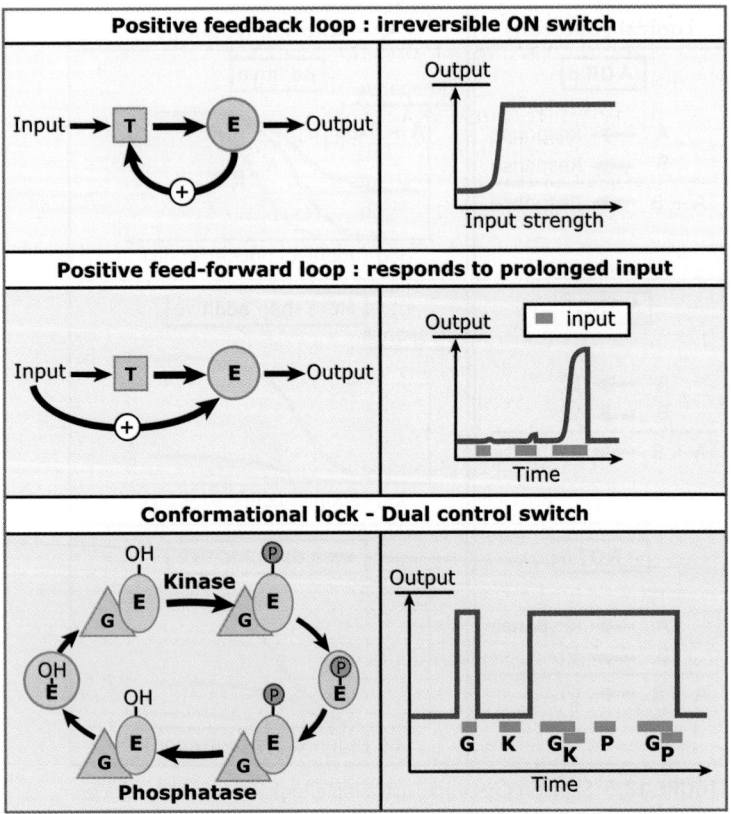

FIGURE 18.6 Relatively complex signal processing can be executed by simple multiprotein modules. The figure depicts three types of signaling modules (left) and their behavior in response to agonist (right). (top) In a positive feedback module, a transducer protein (T) stimulates an effector (E) to produce a cellular output, but the effector also stimulates the activity of the transducer. The result can be an all-or-none switch, where input up to a threshold has little effect, but then becomes committed when feedback from the effector is sufficient to maintain transducer activity even in the absence of continued input from the receptor. (center) In a positive feed-forward module, the effector requires input both from the transducer *and* from upstream in the pathway. When stimulation is brief (short horizontal bar under trace at right), significant amounts of active transducer do not accumulate, and output is minimal. When stimulation is prolonged (longer bar), signal output is substantial. (bottom) In some dual-control switching modules, the binding of one regulator (G) can both activate the effector and expose another regulatory site, shown here as a Ser substrate site (–OH) for a protein kinase. The effector can only be phosphorylated or dephosphorylated when G is bound. Therefore, as shown at the right, addition of G alone will activate but activation of the kinase (K) alone will not. If kinase is active while G is bound, phosphorylation is resistant to phosphatase activity unless G is again present to reexpose the phosphoserine residue (shown on the graph at the right as a bold P).

be the additive positive inputs of two pathways. Such additivity could represent the ability of several receptors to stimulate a pool of a particular G protein or the ability of two protein kinases to phosphorylate a single substrate. Additivity may be positive, as in the examples above, or negative, such as when two inhibi-

tory inputs combine. Inhibition and stimulation may also combine additively to yield an algebraically balanced output. Alternatively, multiple inputs can combine with either more or less than an additive effect. The NOT function, discussed above, is analogous to describing a blockade of stimulation. The AND function describes synergism, where one input potentiates another but alone has little effect.

Even simple signaling networks can display complex patterns of information processing. One good example is the creation of "memory": making the effect of a transient signal more or less permanent. Signaling pathways have multiple ways of setting memories, and of forgetting. One mechanism, common in protein kinase pathways, is the positive feedback loop, illustrated in the top panel of **FIGURE 18.6**. In a positive feedback loop, the input stimulates a transducer (T), which in turn stimulates the effector protein (E) to create the output. If the effector can also activate the transducer, sufficient initial signal can be fed back to the transducer that it can maintain the effector's full signal output even when input is removed. Such systems typically display a threshold behavior, as shown on the right.

A positive feed-forward loop can generate memory of another type (see Figure 18.6, middle panel), indicating the duration of input. In such circuits, the effector requires simultaneous input from both the receptor and from the intermediary transducer. If the pathway from receptor through transducer is relatively slow, or if it requires the accumulation of a substantial amount of transducer, only a prolonged input will trigger a response, as shown in the time-base output diagram at the right.

A third way to establish memory is to allow one input to control the reversibility of a second regulatory event (see Figure 18.6, bottom panel). WASP, a protein that initiates the polymerization of actin to drive cellular motion and shape change, is activated both by phosphorylation and by the binding of Cdc42, a small GTP-binding protein (G). However, the phosphorylation site on WASP is only exposed when WASP is bound to Cdc42. Phosphorylation thus requires both activated Cdc42 and activated protein kinase. If Cdc42 dissociates, the phosphorylated state of WASP persists until another signaling molecule, whose identity remains uncertain, binds again to expose the site to a protein phosphatase. As shown in the time-base graph, exposure to Cdc42 will activate, but exposure to kinase alone will not. If Cdc42 is present, then the kinase can activate WASP. Phospho-WASP is relatively insensitive to protein phosphatase (P) alone, but can be dephosphorylated if Cdc42 or another G protein binds to expose the site to phosphatase.

18.8 Scaffolds increase signaling efficiency and enhance spatial organization of signaling

Key concepts

- Scaffolds organize groups of signaling proteins and may create pathway specificity by sequestering components that have multiple partners.
- Scaffolds increase the local concentration of signaling proteins.
- Scaffolds localize signaling pathways to sites of action.

The proteins in a signaling pathway are frequently colocalized within cells such that their mutual interactions are favored and their interactions with other proteins are minimized. Many signaling pathways are organized on scaffolds. **Scaffolds** bind several components of a signaling pathway in multiprotein complexes to enhance signaling efficiency. Scaffolds promote interactions of proteins that have a low affinity for each other, accelerate activation (and often inactivation) of the associated components, and localize the signaling proteins to appropriate sites of action. Colocalization may be tonic or regulated, and stimulus-dependent scaffolding often determines signaling outputs.

The binding sites on a scaffolding protein are often localized in distinct modular protein-binding domains, giving the impression that the protein is designed simply to hold the components of the pathway together. Many scaffolding proteins do lack intrinsic enzymatic activity, but some signaling enzymes also act as scaffolds.

Binding to a scaffold facilitates signaling by increasing the local concentrations of the components, so that diffusion or transport of molecules to their sites of action is not necessary. In the photoreceptor cells of *Drosophila*, scaffolding of signaling components is critical for rapid signal transmission. These cells contain the InaD scaffolding protein, which has five modular binding domains, known as PDZ domains. Each of its PDZ domains binds to a

C-terminal motif of a target protein, thereby facilitating interactions among the associated proteins. **FIGURE 18.7** shows a model for how InaD organizes the signaling proteins. The mutational loss of InaD produces a nearly blind fly, and deletion of a single PDZ domain can yield a fly with a distinct visual defect characteristic of the protein that binds to the missing domain.

A second example is Ste5p, a scaffold for the pheromone-induced mating response pathway in *S. cerevisiae*. **FIGURE 18.8** illustrates how Ste5p binds and organizes components of a MAPK cascade, including a MAP3K (Ste11p), a MAP2K (Ste7p) and a MAPK (Fus3p). (The MAPK cascade will be discussed in *18.32 Mitogen-activated protein kinases are central to many signaling pathways*). The function of Ste5p is partially retained even if the positions of its binding sites for the kinases are shuffled in the linear sequence of the protein, indicating that a major role is to bring the enzymes into proximity, rather than to precisely orient them. Ste5p also binds to the βγ subunits of the heterotrimeric G protein that mediates the actions of mating pheromones, linking the membrane signal to the intracellular transducers. Yeast that lack Ste5p cannot mate, demonstrating that Ste5p is required for this biologic function (but not all functions) carried out by the pathway.

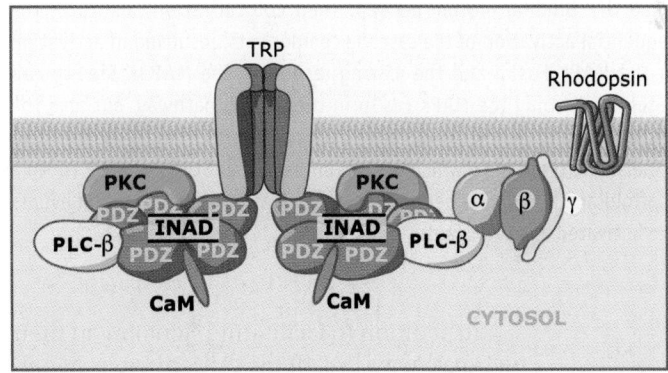

FIGURE 18.7 The scaffold InaD organizes proteins that transmit visual signals in the fly photoreceptor cell. InaD is localized to the photoreceptor membrane and coordinates light sensing and visual transduction. In invertebrate eyes, the visual signaling pathway goes from rhodopsin through G_q to a phospholipase C-β, and Ca^{2+} release triggered by PLC action initiates depolarization. This system is specialized for speed, and requires that the relevant proteins are nearby. InaD contains five PDZ domains, each of which binds to the C-terminus of a signal transducing protein. The TRP channel, which mediates Ca^{2+} entry, PLC-β, and a protein kinase C isoform that is involved in rapid desensitization all bind constitutively to InaD. Rhodopsin and a myosin (NinaC) also bind, and G_q binds indirectly.

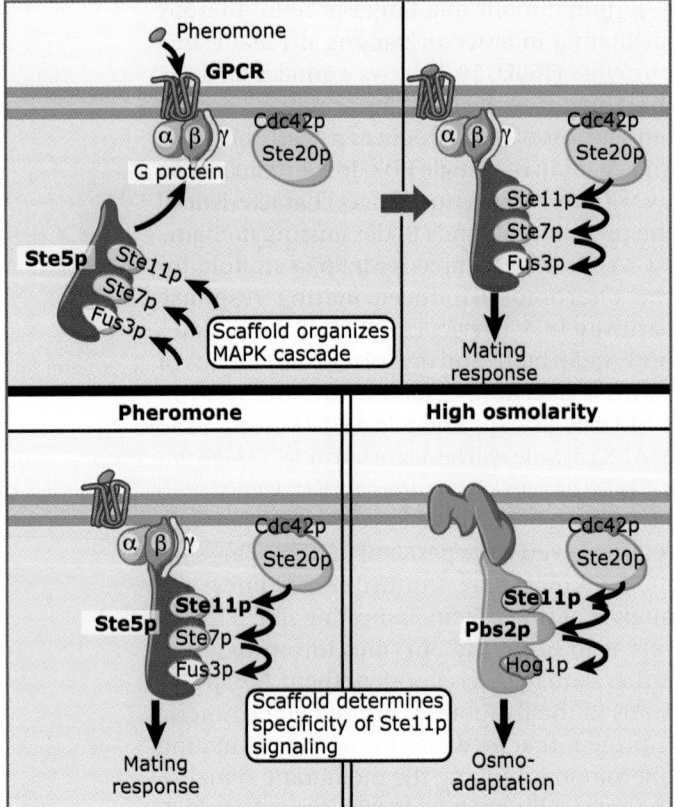

FIGURE 18.8 The scaffold Ste5p organizes the components of the MAPK cascade that mediates the pheromone-induced mating response in *Saccharomyces cerevisiae*. In the top left panel, Ste5p brings the components of the MAPK cascade to the membrane in response to pheromone. In the top right panel, binding to the heterotrimeric G protein brings loaded Ste5p in proximity to the protein kinase Ste20p bound to the activated small GTP binding protein Cdc42p. Their colocalization facilitates the sequential activation of the cascade components, resulting in activation of the MAPK Fus3p and the mating response. The MAP3K Ste11p can regulate not only the MAPK Fus3p in the mating pathway, but also the MAPK Hog1p in the high osmolarity pathway, as shown in the bottom two panels. The scaffold to which Ste11p binds, either Ste5p or Pbs2 (both a scaffold and a MAP2K), determines which MAPK and downstream events are activated as the output.

In addition to facilitating signaling in their own pathways, scaffolds can enhance signaling specificity by limiting interactions with other signaling proteins. Scaffolds thus insulate components of a signaling pathway both from activation by inappropriate signals and from producing incorrect outputs. For example, the mating and osmosensing pathways in yeast share several components, including the MAP3K Ste11p, but each pathway maintains specificity because it employs different scaffolds that restrict signal transmission.

In contrast, the presence of excess scaffold can inhibit signaling because the individual signaling components will more frequently bind to distinct scaffold proteins rather than forming a functional complex. Such dilution among scaffolds causes separation rather than concentration of the components, preventing their productive interaction.

18.9 Independent, modular domains specify protein-protein interactions

Key concepts

- Protein interactions may be mediated by small, conserved domains.
- Modular interaction domains are essential for signal transmission.
- Adaptors consist exclusively of binding domains or motifs.

Modular protein interaction domains or motifs occur in many signaling proteins and confer the ability to bind structural motifs in other molecules, including proteins, lipids, and nucleic acids. Some of these domains are listed in **FIGURE 18.9**. In contrast to scaffolds, which bind specific proteins with considerable selectivity, modular interaction domains generally recognize not a single molecule but a group of targets that share related structural features.

Modular interaction domains important for signal transduction were first discovered in the protein-tyrosine kinase proto-oncogene Src, which contains a protein-tyrosine kinase domain and two domains named Src homology (SH) 2 and 3 domains. The modular **SH2** and **SH3** domains were originally identified by comparison of Src to two other tyrosine kinases. One or both of these domains appear in numerous proteins and both are critically involved in protein-protein interactions.

SH3 domains, which consist of ~50 residues, bind to specific short proline-rich sequences. Many cytoskeletal proteins and proteins found in focal adhesion complexes contain SH3 domains and proline-rich sequences, suggesting that this targeting motif may send proteins with these domains to these sites of action within cells. In contrast to phosphotyrosine-SH2 binding, the proline-rich binding sites for SH3 domains are present in resting and activated cells. However, SH3-proline interactions may be negatively regulated by phosphorylation within the proline-rich motif.

SH2 domains, which consist of ~100 residues, bind to Tyr-phosphorylated proteins, such

Domain	Characteristics	Cellular involvement
14-3-3	Binds protein phosphoserine or phosphothreonine	Protein sequestration
Bromo	Binds acetylated lysine residues	Chromatin-associated proteins
CARD	Dimerization	Caspase activation
C1	Binds phorbol esters or diacyl-glycerol	Recruitment to membranes
C2	Binds phospholipids	Signal transduction, vesicular trafficking
EF hand	Binds calcium	Calcium-dependent processes
F-Box	Binds Skp1 in a ubiquitin-ligase complex	Ubiquitination
FHA	Binds protein phosphothreonine or phosphoserine	Various; DNA damage
FYVE	Binds to PI(3)P	Membrane trafficking, TGF-β signaling
HECT	Binds E2 ubiquitin-conjugating enzymes to transfer ubiquitin to the substrate or to ubiquitin chains	Ubiquitination
LIM	Zinc-binding cysteine-rich motif that forms two tandemly repeated zinc fingers	Wide variety of processes
PDZ	Binds to the C-terminal 4-5 residues of proteins that have a hydrophobic residue at the terminus; may bind to PIP2	Scaffolding diverse protein complexes often at the membrane
PH	Binds to specific phosphoinositides, esp. PI-4,5-P_2, PI-3,4-P_2 or PI-3,4,5-P_3.	Recruitment to membranes and motility
RING	Binds zinc and may be found in E3 ubiquitin ligases	Ubiquitination, transcription
SAM	Homo- and hetero-oligomerization	Wide variety of processes
SH2	Binds to protein phosphotyrosine (pY)	Tyrosine protein kinase signaling
SH3	Binds to PXXP motifs	Various processes
TPR	Degenerate sequence of ~34 amino acids with residues WL/GYAFAP; forms a scaffold	Wide variety of processes
WW	Binds proline-rich sequences	Alternative to SH3; vesicular trafficking

FIGURE 18.9 The table describes a subset of known modular protein interaction domains found in many proteins. Interactions mediated by these domains are essential to controlling cell function. Few if any of these domains exist in prokaryotes. Adapted from the Pawson Lab, Protein Interaction Domains, Mount Sinai Hospital (http://pawsonlab.mshri.on.ca/).

as cytoplasmic tyrosine kinases and receptor tyrosine kinases. Thus, Tyr phosphorylation regulates the appearance of SH2 binding sites and, thereby, regulates a set of protein-protein interactions in a stimulus-dependent manner.

A clever strategy was used to identify the binding specificity of SH2 domains. An isolated recombinant SH2 domain was incubated with cell lysates and then recovered from the lysates using a purification tag. The proteins associated with the SH2 domain were some of the same proteins that were recognized by antiphosphotyrosine antibodies. By this and other methods, it was discovered that SH2 domains recognize sequences surrounding Tyr phosphorylation sites and require phosphorylation of the included Tyr for high affinity binding.

Information on specific amino acid sequences that recognize and bind to modular binding domains is being accumulated as these individual interactions are identified. In addition, screening programs using cDNA and/or peptide libraries to assess binding capabilities yield such motifs. Consensus target sequences for individual domains are identified based on the sequence specificity of their binding to arrayed sequences. These consensus sequences can then be used to predict whether the domain will bind a site in a candidate protein.

Adaptor proteins, which lack enzymatic activity, link signaling molecules and target them in a manner that is responsive to extracellular signals. Adaptor proteins are generally made up of two or more modular interaction domains or the complementary recognition motifs. Unlike scaffolds, adaptors are usually multifunctional because their modular interaction domains and motifs are not as highly specific. Adaptors bind to two or more other signaling proteins via their protein-protein interaction domains to colocalize them or to facilitate additional interactions.

Grb2 is a prototypical adaptor protein that was identified as a protein that bound to the C-terminal region of the EGF receptor. Grb2 has one SH2 and two SH3 domains. It binds constitutively to specific proline-rich segments of proteins through its SH3 domain, although this binding can be negatively regulated. One target of Grb2 is SOS, a guanine nucleotide exchange factor that activates the small GTP-binding protein Ras in response to EGF signaling. Through its SH2 domain, Grb2 binds Tyr-phosphorylated proteins, including the receptors themselves in a stimulus-dependent manner. Thus, Tyr phosphorylation of these receptors in response to ligand will enable the binding of Grb2 to the receptors, which, in turn, will recruit SOS to the membrane-localized receptor. Once at the membrane, SOS can activate its target, Ras.

18.10 Cellular signaling is remarkably adaptive

Key concepts

- Sensitivity of signaling pathways is regulated to allow responses to change over a wide range of signal strengths.
- Feedback mechanisms execute this function in all signaling pathways.
- Most pathways contain multiple adaptive feedback loops to cope with signals of various strengths and durations.

A universal property of cellular signaling pathways is adaptation to the incoming signal. Cells continuously adjust their sensitivity to signals to maintain their ability to detect changes in input. Typically, when a cell is exposed to a new input, it initiates a process of desensitization that dampens the cellular response to a new plateau lower than the initial peak response, as illustrated in **FIGURE 18.10**. When the stimulus

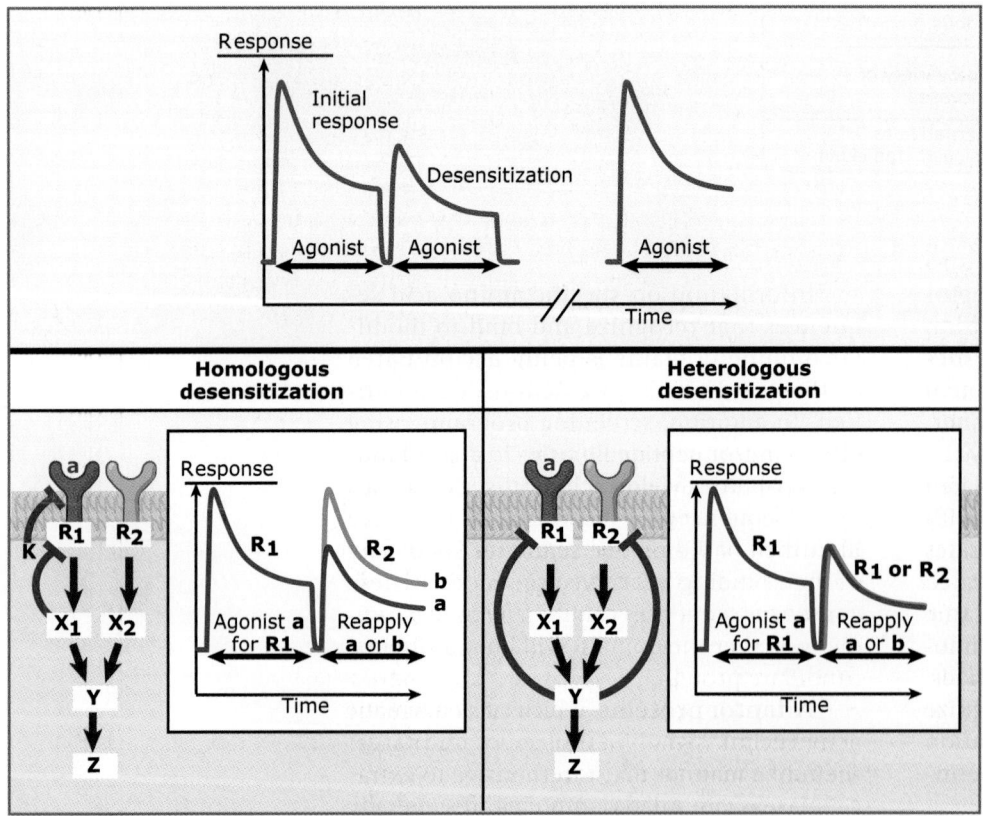

FIGURE 18.10 Top: Upon exposure to a stimulus, signaling pathways adjust their sensitivities to adapt to the new level of input. Thus, the response decays after initial stimulation. A second similar stimulus will elicit a smaller response unless adequate time is allowed for recovery. Bottom: Some adaptation mechanisms feed back only on the receptor that is stimulated and do not alter parallel pathways. Such mechanisms are referred to as homologous. At left, agonist **a** for receptor R1 can initiate either of two feedback events that desensitize R1 alone. In other cases, a stimulus will also cause parallel or related systems to desensitize. At the right, agonist **a** initiates desensitization of both R1 and R2. The response to agonist **b**, which binds to R2, is also desensitized. Such heterologous desensitization is common.

is removed, the desensitized state can persist, with sensitivity slowly returning to normal. Similarly, the removal of a tonic stimulus can hypersensitize signaling systems.

Adaptation in signaling is one of the best examples of biologic homeostasis. The adaptability of cellular signaling can be quite impressive. Cells commonly regulate their sensitivity to physiologic stimuli over more than a 100-fold range, and the mammalian visual response can adapt to incoming light over a 10^7-fold range. This remarkable ability allows a photoreceptor cell to detect a single photon, and allows a person to read in both very dim light and intense sunlight. Adaptability is observed in bacteria, plants, fungi, and animals. Many of its properties are conserved throughout biology, although the most complex adaptive mechanisms are found in animals. The general mechanism for adaptation is the negative feedback loop, which biochemically samples the signal and controls the adaptive process.

Adaptation varies with both the intensity and the duration of the incoming signal. Stronger or more persistent inputs tend to drive greater adaptive change and, often, adaptation that persists for a longer time. Cells can modulate adaptation in this way because adaptation is exerted by a succession of independent mechanisms, each with its own sensitivity and kinetic parameters.

G protein pathways offer excellent examples of adaptation. **FIGURE 18.11** shows that the earliest step in adaptation is receptor phosphorylation, which is catalyzed by G protein-coupled receptor kinases (GRKs) that selectively recognize the receptor's ligand-activated conformation. Phosphorylation inhibits the receptor's ability to stimulate G protein activation and also promotes binding of arrestin, a protein that further inhibits G protein activation. Moreover, arrestin binding primes receptors for endocytosis, which removes them from the cell surface. Endocytosis can also be the first step in receptor proteolysis. Along with these direct effects, many receptor genes display feedback inhibition of transcription, such that signaling by a receptor decreases its own expression.

Stimulation thus causes multiple adaptive processes that range from immediate (phosphorylation, arrestin binding) through delayed (transcriptional regulation), and include both reversible and irreversible events. This array of adaptive events is used by many G protein-coupled receptors, and may control output from

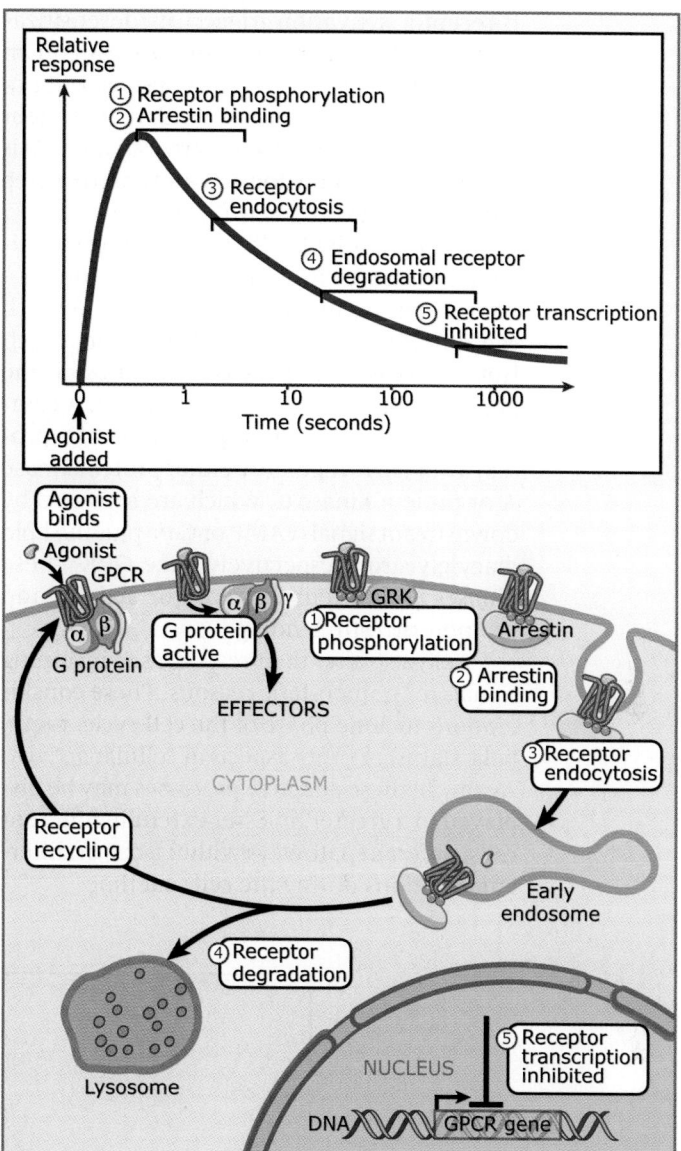

FIGURE 18.11 Multiple adaptation processes are invoked during a stimulus, and multiple nested mechanisms for adaptation are the rule. They are usually invoked sequentially according to the duration and intensity of the stimulus. For GPCRs, at least five desensitizing mechanisms are known, with others acting on the G protein and effectors.

a single receptor in many cells. The speed, extent, and reversibility of adaptation are selected by a cell's developmental program.

Cells can change their patterns of adaptation both qualitatively and quantitatively by altering the points in a pathway where feedback is initiated and exerted. In a linear pathway, changing these points will alter the kinetics or extent of adaptation (see Figure 18.10). In branched pathways, changing these points can determine whether adaptation is unique to one input or is exerted for many similar inputs.

If receptor activation triggers its desensitization directly, or if an event downstream on an unbranched pathway triggers desensitization, then only signals that initiate with that receptor will be altered. Receptor-selective adaptation is referred to as **homologous adaptation** (see Figure 18.10).

Alternatively, feedback control can initiate downstream from multiple receptors in a convergent pathway and thus regulate both the initiating receptor and the others. Such **heterologous adaptation** regulates all the possible inputs to a given control point. A common example is the phosphorylation of G protein-coupled receptors by either protein kinase A or protein kinase C, which are activated by downstream signals cAMP or Ca^{2+} plus the lipid diacylglycerol, respectively. Like GRK, these kinases both attenuate receptor activity and promote arrestin binding.

Cells also alter their responses to incoming signals for homeostatic reasons. These considerations include phase of the cell cycle, metabolic status, or other aspects of cellular activity. Again, all these adaptive processes may be displayed to a greater or lesser extent in different cells, different pathways within a cell or different situations during the cell's lifetime.

18.11 Signaling proteins are frequently expressed as multiple species

Key concepts
- Distinct species (isoforms) of similar signaling proteins expand the regulatory mechanisms possible in signaling pathways.
- Isoforms may differ in function, susceptibility to regulation or expression.
- Cells may express one or several isoforms to fulfill their signaling needs.

Cells increase the richness, adaptability, and regulation of their signaling pathways by expressing multiple species of individual signaling proteins that display distinct biochemical properties. These species may be encoded by multiple genes or by multiple mRNAs derived from a single gene by alternative splicing or mRNA editing. The numerical complexity implicit in these choices is impressive. Consider the neurotransmitter serotonin: In mammals, there are thirteen serotonin receptors, each of which stimulates a distinct spectrum of G proteins of the G_i, G_s, and G_q families. (A fourteenth serotonin receptor is an ion channel.) **FIGURE 18.12**

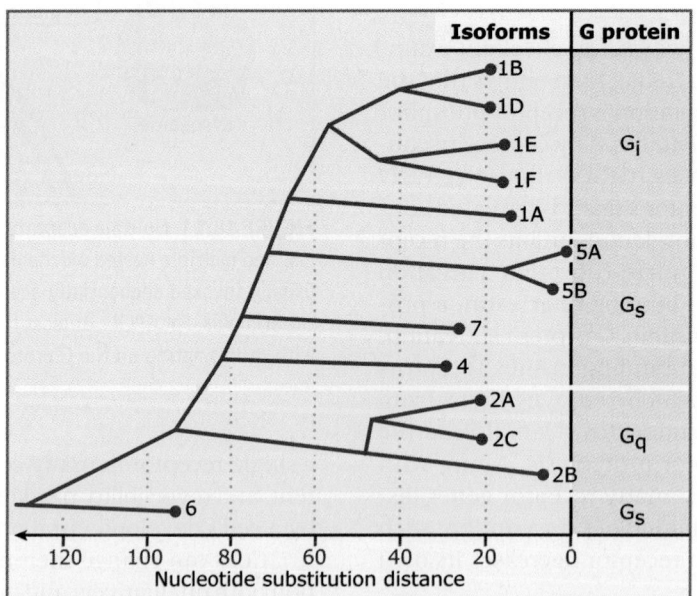

FIGURE 18.12 Receptors for serotonin have evolved in mammals as a family of 13 genes that regulate three of the four major classes of G proteins. While all respond to the natural ligand serotonin, the binding sites have evolved sufficient differences that drugs have been developed that specifically target one or more isoforms. The type 3 serotonin receptors, not shown here, are ligand-gated ion channels and are not obviously related to the others.

shows the relationship of serotonin receptors to these G protein families.

There is also tremendous diversity among the G proteins and adenylyl cyclases. There are three genes for $G\alpha_i$ and one each for the closely related $G\alpha_z$ and $G\alpha_o$. Furthermore, the $G\alpha_o$ mRNA is multiply spliced. There are four G_q members. In addition, there are five genes for $G\beta$ and twelve for $G\gamma$, and most of the possible $G\beta\gamma$ dimers that include the four common $G\beta$s are expressed naturally. There are ten genes for adenylyl cyclases, which are direct targets of G_s and either direct or indirect targets of the other G proteins. While all nine membrane-bound adenylyl cyclase isoforms are stimulated by $G\alpha_s$, they display diverse stimulatory and inhibitory responses to $G\beta\gamma$, $G\alpha_i$, Ca^{2+}, calmodulin, and several protein kinases, as illustrated in **FIGURE 18.13**. Thus, stimulation by serotonin can lead to diverse responses depending upon the various forms of the proteins that are engaged at a particular time and location.

Sometimes isoforms of a signaling protein are subject to quite different kinds of inputs. For example, all of the members of the phospholipase C family (PLC) hydrolyze phos-phatidylinositol-4,5-bisphosphate to form two second messengers, diacylglycerol and inositol-1,4,5 trisphosphate (see *18.16 Lipids and lipid-derived compounds are signaling molecules*). The distinct isoforms may be regulated by diverse combinations of $G\alpha_q$, $G\beta\gamma$, phosphorylation, monomeric G proteins, or Ca^{2+}.

Because a cell has multiple options when expressing a form of a signaling protein, it can use expression of particular isoforms to alter how it performs otherwise identical signaling functions. Different cells express one or more isoforms to allow appropriate responses, and expression can vary according to other inputs or the cell's metabolic status. In addition, signaling pathways are remarkably resistant to mutational or other injuries because loss of a single species or isoform of a signaling protein can often be compensated for by increased expression or activity of another species. Similarly, engineered overexpression can result in the reduced expression of endogenous proteins. The existence of multiple receptor species can, thus, substantially add to adaptability and the consequent resistance of signaling networks to damage.

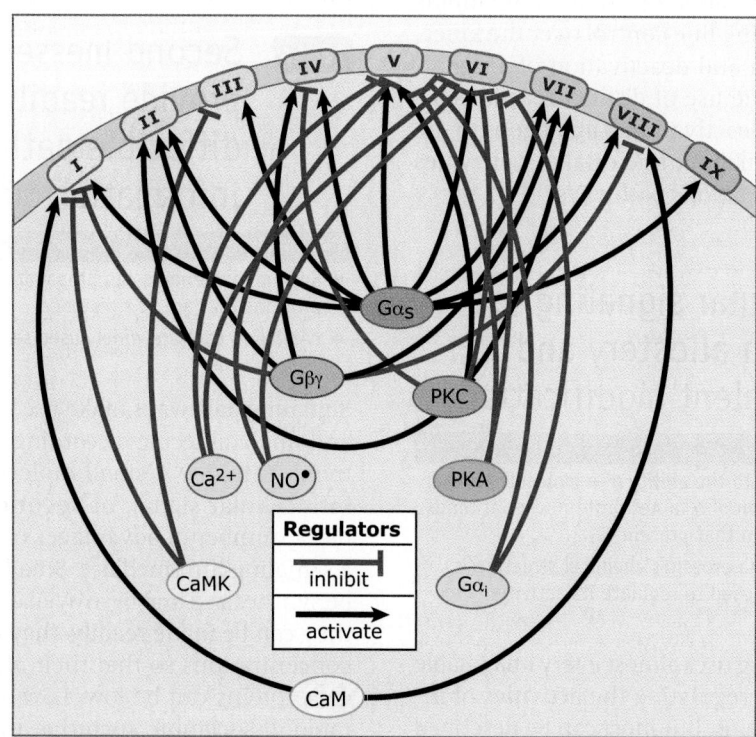

FIGURE 18.13 All of the mammalian membrane-bound adenylyl cyclases are structurally homologous and catalyze the same reaction, and all are stimulated by $G\alpha_s$. Their responses to other inputs (protein kinases CaMK, PKA and PKC; Ca^{2+}; calmodulin (CaM); NO•) are specific to each isoform, allowing a rich combinatoric input to cellular cAMP signaling. Adapted from Paul Sternweis, Alliance for Cellular Signaling.

18.12 Activating and deactivating reactions are separate and independently controlled

Key concepts

- Activating and deactivating reactions are usually executed by different regulatory proteins.
- Separating activation and inactivation allows for fine-tuned regulation of amplitude and timing.

In signaling networks, individual proteins are frequently activated and deactivated by distinct reactions, a feature that facilitates separate regulation. Common examples include using protein kinases and phosphoprotein phosphatases to catalyze protein phosphorylation and dephosphorylation; using adenylyl cyclase to create cAMP while using phosphodiesterases to hydrolyze it or anion transporters to pump it out of the cell; or using GTP/GDP exchange factors (GEFs) to activate G proteins and GTPase-activating proteins (GAPs) to deactivate them. Depending on stoichiometry and detailed mechanism, these strategies can convey either additive or nonadditive inputs while maintaining fine control over the kinetics of activation and deactivation of a signaling pathway. The use of distinct reactions for activation and deactivation is analogous to the use of distinct anabolic and catabolic enzymes in reversible metabolic pathways.

18.13 Cellular signaling uses both allostery and covalent modification

Key concepts

- Allostery refers to the ability of a molecule to alter the conformation of a target protein when it binds noncovalently to that protein.
- Modification of a protein's chemical structure is also frequently used to regulate its activity.

Cellular signaling uses almost every imaginable mechanism for regulating the activities of intracellular proteins, but most can be described as either allosteric or covalent. Individual signaling proteins typically respond to multiple allosteric and covalent inputs.

Allostery refers to the ability of a molecule to alter the conformation of a target protein when it binds noncovalently to that protein.

Because a protein's activity reflects its conformation, the binding of any molecule that alters conformation can change the target protein's activity. Any molecule can have allosteric effects: protons or Ca^{2+}, small organic molecules, or other proteins. Allosteric regulation can be both inhibitory and stimulatory.

Covalent modification of a protein's chemical structure is also frequently used to regulate its activity. The change in the protein's chemical structure alters its conformation and, thus, its activity. Most regulatory covalent modification is reversible. The classic and most common regulatory covalent event is phosphorylation, in which a phosphoryl group is transferred from ATP to the protein, most often to the hydroxyl group of serine (Ser), threonine (Thr), or tyrosine (Tyr). Enzymes that phosphorylate proteins are known as protein kinases. Their actions are opposed by phosphoprotein phosphatases, which catalyze the hydrolysis of the phosphoryl group to yield free phosphate and restore the unmodified hydroxyl residue. Other forms of covalent modification are also common and will be addressed throughout the chapter.

18.14 Second messengers provide readily diffusible pathways for information transfer

Key concepts

- Second messengers can propagate signals between proteins that are at a distance.
- cAMP and Ca^{2+} are widely used second messengers.

Signaling pathways make use of both proteins and small molecules according to their distinctive attributes. A small molecule used as an intracellular signal, or **second messenger**, has a number of advantages over a protein as a signaling intermediary. Small molecules can be synthesized and destroyed quickly. Because they can be made readily, they can act at high concentrations so that their affinities for target proteins can be low. Low affinity permits rapid dissociation, such that their signals can be terminated promptly when free second messenger molecules are destroyed or sequestered. Because second messengers are small, they also can diffuse quickly within the cell, although many cells have developed mechanisms to spatially restrict such diffusion. Second messengers are, thus, superior to proteins in mediating fast responses, particularly at a distance. Second messengers are also useful when signals have to be addressed to large numbers of target proteins simultaneously. These advantages often overcome their lack of catalytic activity and their inability to bind multiple molecules simultaneously.

FIGURE 18.14 lists intracellular second messengers developed through evolution. This number is surprisingly low. Several are nucleotides synthesized from major metabolic nucleotide precursors. They include cAMP, cyclic GMP, ppGpppp, and cyclic ADP-ribose. Other soluble second messengers include a sugar phosphate, inositol-1,4,5-trisphosphate (IP$_3$), a divalent metal ion Ca^{2+}, and a free radical gas nitric oxide (NO$^{\bullet}$). Lipid second messengers include diacylglycerol and phosphatidylinositol-3,4,5-trisphosphate, phosphatidylinositol-4,5-diphosphate, sphingosine-1-phosphate and phosphatidic acid.

The first signaling compound to be described as a second messenger was cAMP. The name arose because cAMP is synthesized in animal cells as a second, intracellular signal in response to numerous extracellular hormones, the first messengers in the pathway. cAMP is used by prokaryotes, fungi, and animals to convey information to a variety of regulatory proteins. (It has not been found in higher plants.)

Adenylyl cyclases, the enzymes that synthesize cAMP from ATP, are regulated in various ways depending on the organism in which they occur. In animals, adenylyl cyclase is an integral protein of the plasma membrane whose multiple isoforms are stimulated by diverse agents (see Figure 18.13). In animal cells, adenylyl cyclase is generally stimulated by G$_s$, which was originally discovered as an adenylyl cyclase regulator. Some fungal adenylyl cyclases are also stimulated by G proteins. Bacterial cyclases are far more diverse in their regulation.

cAMP is removed from cells in two ways. It may be extruded from cells by ATP-driven anion pumps but is more often hydrolyzed to 5'-AMP by members of the cyclic nucleotide phosphodiesterase family, a large group of proteins that are themselves under multiple regulatory controls.

The prototypical downstream regulator for cAMP in animals is the cAMP-dependent protein kinase, but a bacterial cAMP-regulated transcription factor was discovered shortly thereafter, and other effectors are now known (see Figure 18.14). The cAMP system remains the prototypical eukaryotic signaling pathway in that its components exemplify almost all of the recognized varieties of signaling molecules and their interactions: hormone, receptor, G

Second messenger	Targets	Synthesis/ Release	Pre-cursor	Removal
3':5'-cyclic AMP (cAMP)	Protein kinase A	Adenylyl cyclase	ATP	Phospho-diesterase
	Bacterial transcription factors			Organic anion transporter
	Cation channel			
	Cyclic nucleotide phospho-diesterase			
	Rap GDP/GTP exchange factor (Epac)			
Magic spot (ppGpp, ppGppp)	RNA polymerase	Rel1A	GTP	SpoT-catalyzed hydrolysis
	ObgE transcription arrest detector	SpoT		
Inositol-1,3,5-trisphosphate (IP$_3$)	IP3-gated Ca^{2+} channel	Phospho-lipase C	PIP$_2$	Phosphatase
Diacylglycerol (DAG)	Protein kinase C	Phospho-lipase C	PIP$_2$	Diacylglycerol kinase
	Trp cation channel			Diacylglycerol lipase
Phosphatidyl-inositol-4,5-bisphosphate (PIP$_2$)	Ion channel	PIP 5-kinase	PI-4-P	Phospho-lipase C
	Transporters			Phosphatase
3':5'-Cyclic GMP (cGMP)	Protein kinase G	Guanylyl cyclase	GTP	Phospho-diesterase
	Cation channel			
	Cyclic nucleotide phospho-diesterase			
Cyclic ADP-ribose	Ca^{2+} channel	ADP-ribose cyclase	NAD	Hydrolysis
Cyclic diguanosine-monophosphate	Various two component system proteins	Diguanylate cyclase	GTP	Cyclic di-GMP phospho-diesterase
Nitric oxide (NO$^{\bullet}$)	Guanylyl cyclase	NO$^{\bullet}$ synthase	arginine	Reduction
Ca^{2+}	Numerous calmodulin	Release from storage organelles or plasma membrane channels	Stored Ca^{2+}	Reuptake and extrusion pumps
Phosphatidyl-inositol-3,4,5-trisphosphate	Akt (protein kinase B)	PI 3-kinase	PIP$_2$	Phosphatase
	Other PH domains/proteins			

FIGURE 18.14 Major second messengers, some of the proteins that they regulate, their sources and their disposition.

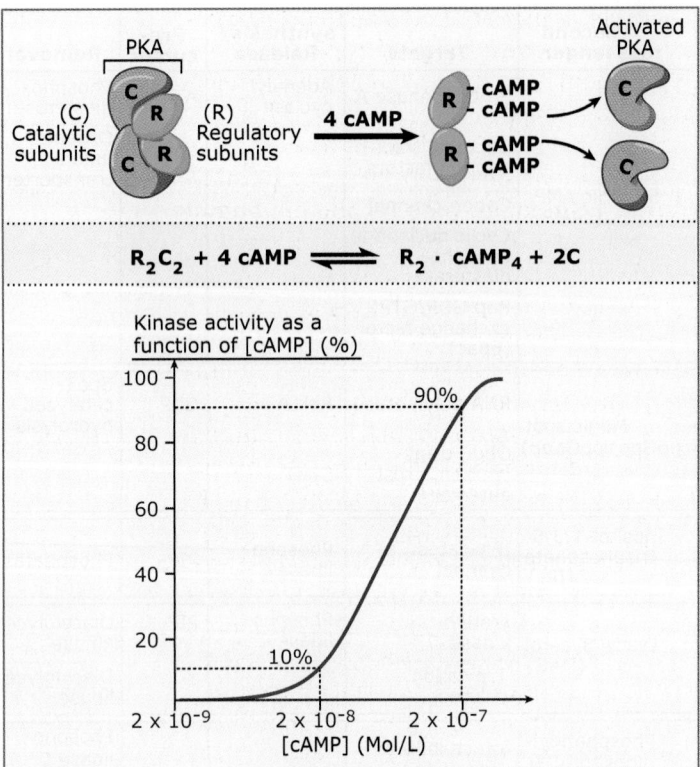

FIGURE 18.15 PKA is a heterotetramer composed of two catalytic (C) and two regulatory (R) subunits. Binding of four molecules of cAMP to the regulatory subunits induces dissociation of two molecules of C, the active form of PKA, from the cAMP-bound regulatory subunit dimer. In the bottom panel, the cooperative binding of four molecules of cAMP generates a steep activation profile. Activity increases from approximately 10% to 90% as the cAMP concentration increases only 10-fold. An apparent threshold is introduced because there is little change in activity at low concentrations of cAMP.

protein, adenylyl cyclase, protein kinase, phosphodiesterase, and extrusion pump.

The second messenger-stimulated protein kinase (PKA) is a tetramer composed of two catalytic (C) subunits and two regulatory (R) subunits, as illustrated in **FIGURE 18.15**. The R subunit binds to the catalytic subunit in the substrate-binding region, maintaining C in an inhibited state. Each R subunit binds two molecules of cAMP, four cAMP molecules per PKA holoenzyme. When these sites are filled, the R subunit dimer dissociates rapidly, leaving two free catalytic subunits with high activity. The difference in affinity of R for C in the presence and absence of cAMP is ~10,000-fold. The strongly cooperative binding of cAMP generates a very steep activation curve with an apparent threshold below which no significant activation of PKA occurs, as illustrated in Figure 18.15. PKA activity, thus, increases dramatically over a narrow range of cAMP concentrations. PKA is also regulated by phosphorylation

of its activation loop. Phosphorylation occurs cotranslationally, and the activation loop phosphorylation is required for assembly of the R_2C_2 tetramer.

The PKAs are mostly cytosolic, but are also targeted to specific locations by binding organelle-associated scaffolds (A-kinase anchoring proteins, or AKAPs). These AKAPs facilitate phosphorylation of membrane proteins including G protein-coupled receptors (GPCRs), transporters, and ion channels. AKAPs can also target PKA to other cellular locations including mitochondria, the cytoskeleton, and the centrosome. AKAPs often harbor binding sites for other regulatory molecules such as phosphoprotein phosphatases and additional protein kinases, which allows for coordination of multiple signaling pathways and integration of their outputs.

PKA generally phosphorylates substrates with a primary consensus motif of Arg-Arg-Xaa-Ser-Hydrophobic, placing it in a large group of kinases that recognize basic residues preceding the phosphorylation site. PKA regulates proteins throughout the cell ranging from ion channels to transcription factors, and its conserved substrate preference frequently permits prediction of substrates by sequence analysis. The cAMP response element binding protein CREB is phosphorylated by PKA on Ser 133 and mediates the effect of cAMP on transcription of numerous genes.

18.15 Ca²⁺ signaling serves diverse purposes in all eukaryotic cells

Key concepts

- Ca²⁺ serves as a second messenger and regulatory molecule in essentially all cells.
- Ca²⁺ acts directly on many target proteins and also regulates the activity of a regulatory protein calmodulin.
- The cytosolic concentration of Ca²⁺ is controlled by organellar sequestration and release.

Ca²⁺ is used as a second messenger in all cells, and is, thus, an even more widespread second messenger than cAMP. Many proteins bind Ca²⁺ with consequent allosteric changes in their enzymatic activities, subcellular localization, or interaction with other proteins or with lipids. Direct targets of Ca²⁺ regulation include almost all classes of signaling proteins described in this chapter, numerous metabolic enzymes, ion channels and pumps, and contractile proteins. Most noteworthy may be muscle actomyosin fibers, which are triggered to contract in response to cytosolic Ca^{2+} (see *12.21 Myosin-II functions in muscle contraction*).

Although free Ca^{2+} is found at concentrations near 1 mM in most extracellular fluids, intracellular Ca^{2+} concentrations are maintained near 100 nM by the combined actions of pumps and transporters that either extrude free Ca^{2+} or sequester it in the endoplasmic reticulum or mitochondria. Ca^{2+} signaling is initiated when Ca^{2+}-selective channels in the endoplasmic reticulum or plasma membrane are opened to allow Ca^{2+} to enter the cytoplasm. The most important entrance channels include electrically gated channels in animal plasma membranes; a Ca^{2+} channel in the endoplasmic reticulum that is opened by another second messenger, inositol 1,4,5-trisphosphate (see below); and an electrically gated channel in the endoplasmic (**sarcoplasmic**) reticulum of muscle that opens in response to depolarization of nearby plasma membrane, a process known as excitation-contraction coupling (see

6.9 Plasma membrane Ca^{2+} channels activate intracellular and intercellular signaling processes).

In addition to the proteins that are regulated by binding Ca^{2+} directly, many other proteins respond to Ca^{2+} by binding a widespread Ca^{2+} sensor, the small, ~17 kDa protein calmodulin. Calmodulin requires the binding of four molecules of Ca^{2+} to become fully active, and binding is highly cooperative, generating a sigmoid activation profile illustrated in **FIGURE 18.16**. Calmodulin generally binds its targets in a Ca^{2+}-dependent manner, but Ca^{2+}-free calmodulin may remain bound but inactive in some cases. For example, calmodulin is a constitutive subunit of phosphorylase kinase that is activated upon Ca^{2+} binding. Higher plants again make major modifications to this paradigm. Calmodulin is not expressed as a distinct protein but, instead, is found as a domain in Ca^{2+}-regulated proteins. In yet another variation, the adenylyl cyclase secreted by the pathogenic bacterium *Bordetella pertussis* is inactive outside cells but is activated by Ca^{2+}-free calmodulin in animal cells, where its rapid production of cAMP is highly toxic.

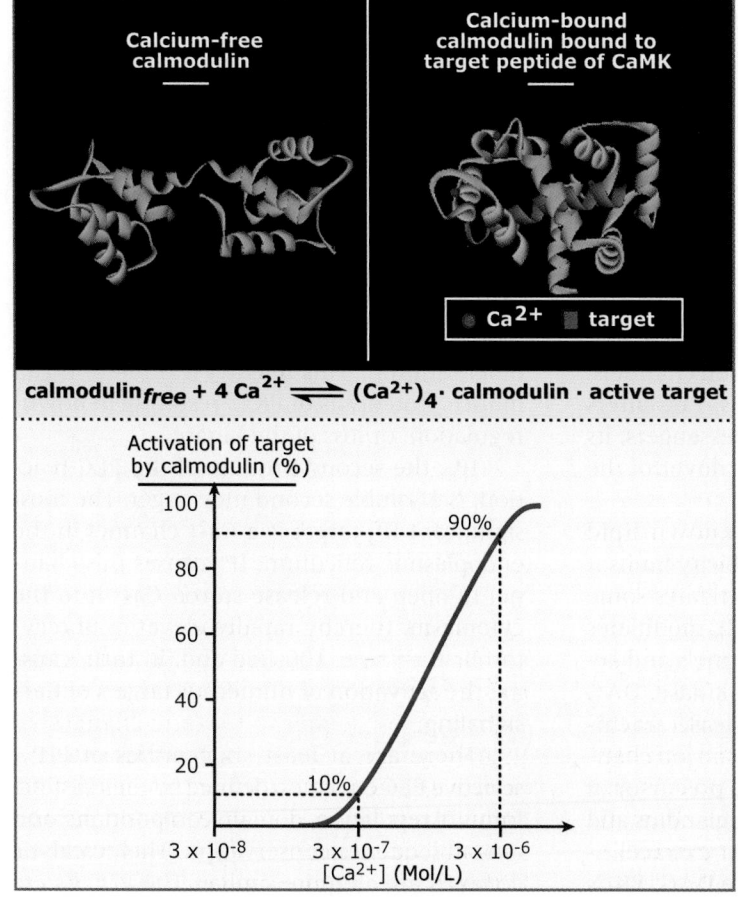

FIGURE 18.16 Ribbon diagrams representing the crystal structures of calmodulin free of Ca^{2+} and bound to four Ca^{2+} ions reveal the huge conformational change that calmodulin undergoes upon Ca^{2+} binding. Ca^{2+}-calmodulin causes activity changes in target proteins. The bottom panel shows the activation of a target by calmodulin as a function of the intracellular free Ca^{2+} concentration. The requirement for binding four Ca^{2+} ions to induce the conformational transition results in cooperative activation of targets. Activity increases from 10% to 90% as the Ca^{2+} concentration increases only 10-fold. Left structure from Protein Data Bank 1CFD. H. Kuboniwa, et al., *Nat. Struct. Biol.* 2 (1995): 768–776. Right structure from Protein Data Bank 1MXE. J. A. Clapperton, et al., *Biochemistry* 41 (2002): 14669–14679.

18.16 Lipids and lipid-derived compounds are signaling molecules

Key concepts

- Multiple lipid-derived second messengers are produced in membranes.
- Phospholipase Cs release soluble and lipid second messengers in response to diverse inputs.
- Channels and transporters are modulated by different lipids in addition to inputs from other sources.
- PI 3-kinase synthesizes PIP_3 to modulate cell shape and motility.
- Phospholipases D and A_2 create other lipid second messengers.

Signals that originate at the plasma membrane may have soluble regulatory targets in the cytoplasm or intracellular organelles, but integral plasma membrane proteins are also subject to acute controls. For these targets, lipid second messengers may be primary inputs. Lipids derived from membrane phospholipids or other lipid species play numerous roles in cell signaling. Because their analysis is more difficult than for soluble messengers, many probably remain to be discovered and understood. **FIGURE 18.17** shows the structure of some of these lipids.

Phospholipase Cs (PLCs) are the prototypical lipid signaling enzymes. PLC isoforms catalyze the hydrolysis of phospholipids between the 3-*sn*-hydroxyl and the phosphate group to yield a diacylglycerol and phosphate ester. In animals and fungi, PLCs specific for the substrate phosphatidylinositol-4,5-bisphosphate (PIP_2) hydrolyze PIP_2 to form two second messengers: 1, 2-*sn*-diacylglycerol (DAG) and inositol-1,4,5-trisphosphate (IP_3). The PLC substrate PIP_2 is itself an important regulatory ligand that modulates the activity of several ion channels, transporters, and enzymes. Thus, PLC alters concentration of three second messengers; its net effect depends on the net turnover of the substrate and products.

DAG is probably the best-known lipid second messenger; its hydrophobicity limits it to action to membranes. DAG activates some isoforms of protein kinase C (PKC), modulates the activity of several cation channels and activates at least one other protein kinase. DAG can be further hydrolyzed to release arachidonic acid, which can regulate some ion channels. Arachidonic acid is also the precursor of oxidation products, such as prostaglandins and thromboxanes, which are potent extracellular signaling agents. In addition to DAG, PKCs require interaction with Ca^{2+} and an acidic phospholipid, such as phosphatidylserine, to become activated. Thus, activation of PKC requires the coincidence of multiple inputs both to generate DAG and to increase intracellular Ca^{2+}. There are more than a dozen PKCs, classified together according to highly conserved sequences in the catalytic domain. Three subgroups of PKCs, also identifiable by sequence, share different patterns of regulation. Their regulation provides examples of many ways in which other mammalian protein kinases are regulated.

The first of these groups, canonical PKCs, are generally soluble or very loosely associated with membranes prior to the appearance of DAG. DAG causes their association with membranes and permits activation upon binding of other regulators. The second group of PKCs requires similar lipids but not Ca^{2+}, and the third group requires other lipids but neither DAG nor Ca^{2+} for activation.

The N-terminal region of PKCs contains a pseudosubstrate domain, a sequence that resembles that of a typical substrate except that the target Ser is replaced with Ala. The pseudosubstrate region binds to the active site to inhibit the kinase. Activators cause the pseudosubstrate domain to flip out of the active site. PKCs are also activated by proteolysis, as are many protein kinases with discrete autoinhibitory domains. Proteases clip a flexible hinge region, which results in loss of the regulatory domain and consequent activation of the kinase.

PKC is the major receptor for phorbol esters, a class of powerful tumor promoters. Phorbol esters mimic DAG and cause a more massive and prolonged activation than physiologic stimuli. This massive stimulation can induce proteolysis of PKC, resulting in downregulation, or loss of the kinase.

IP_3, the second product of the PLC reaction, is a soluble second messenger. The most significant IP_3 target is a Ca^{2+} channel in the endoplasmic reticulum. IP_3 causes this channel to open and release stored Ca^{2+} into the cytoplasm, thereby rapidly elevating the cytosolic Ca^{2+} over 100-fold and, in turn, causing the activation of numerous targets of Ca^{2+} signaling.

There are at least six families of PIP_2-selective PLC enzymes, defined by their distinct forms of regulation, domain compositions, and overall sequence conservation. Their catalytic domains are all quite similar. The PLC-βs are

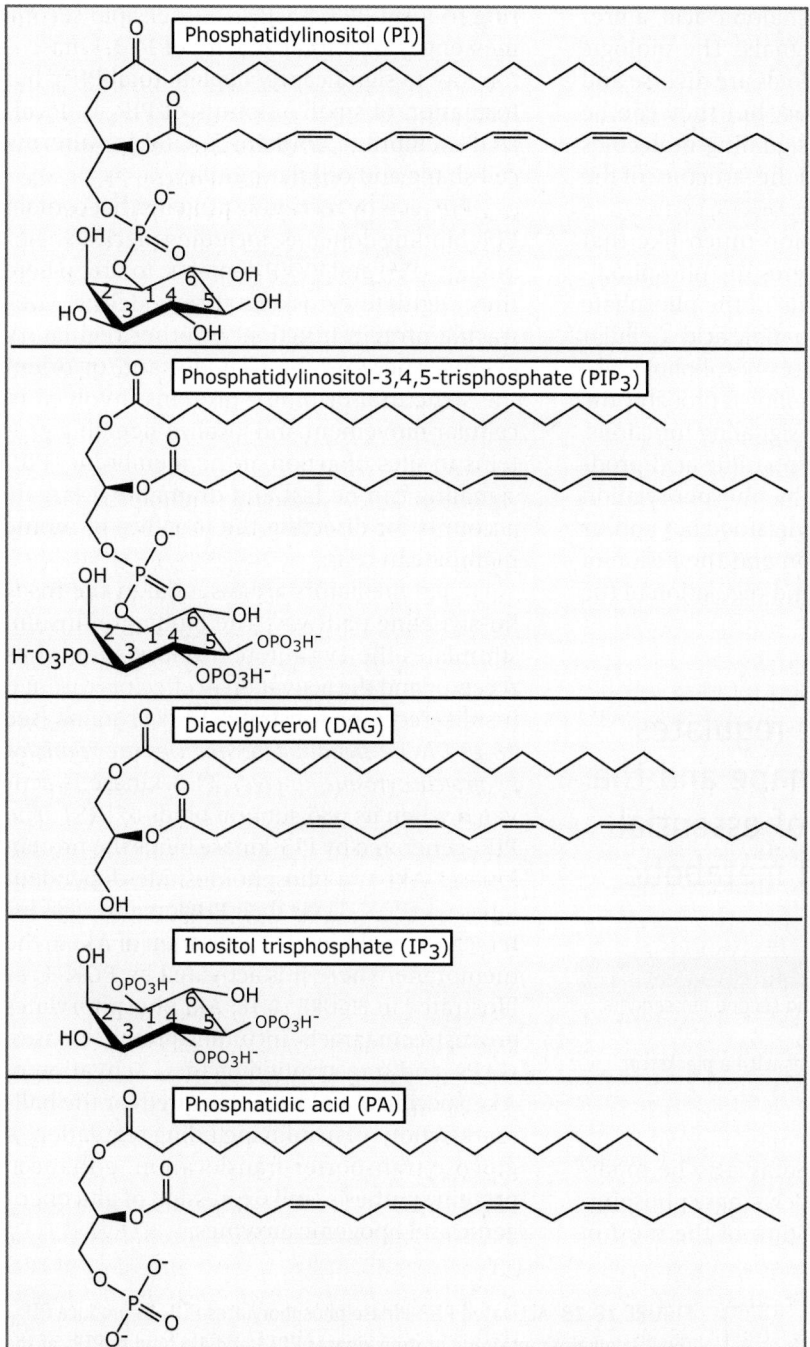

FIGURE 18.17 Structures of some lipid second messengers and the common precursor phosphatidylinositol. The acyl side chain structures shown here are the most common for mammalian PI lipids. Much of the PA in cells is derived from phosphatidylcholine, and its acyl chains may differ from those shown.

stimulated primarily by $G\alpha_q$ and $G\beta\gamma$ (to individually varying extents). Several are also modulated by phosphorylation. PLC-γ isoforms are stimulated by phosphorylation on Tyr residues, frequently by receptor tyrosine kinases. The PLC-ε isoforms are regulated by small, monomeric G proteins of the Rho family and $G\beta\gamma$. The regulation of the PLC-δs is still incompletely understood. Two other classes similar to the PLC-δs, PLC-η and -ζ, have also been defined recently. (There is no PLC-α.) In addition to their distinct modes of regulation, all of the PLCs are stimulated by Ca^{2+}, and Ca^{2+} often acts synergistically with other stimulatory inputs. This synergy underlies the intensification and prolongation of Ca^{2+} signaling observed in many cells.

Phospholipases A_2 and D (PLA_2 and PLD) also hydrolyze glycerol phospholipids in cell membranes to form important signaling compounds. PLA_2 hydrolyzes the fatty acid at the *sn*-2 position of multiple phospholipids to produce the cognate lysophospholipid and the free fatty acid, which is generally unsaturated. The free fatty acid is often arachidonic acid, a precursor of extracellular signals. The biologic roles of free lysophospholipids are diverse and not completely understood, but they can be released as extracellular signaling molecules and are linked to effects on the structure of the membrane bilayer.

PLD catalyzes a reaction much like that of PLC but instead hydrolyzes the phosphodiester on the substituent side of the phosphate group to form 3-sn-phosphatidic acid. Cellular PLDs act on multiple glycerol phospholipid substrates, but phosphatidylcholine is probably the substrate most relevant to signaling functions. The functions of the phosphatidic acid product, which is also formed by phosphorylation of DAG, remain poorly understood but appear to include roles in secretion and the fusion of intracellular membranes and regulation of the cytoskeleton.

18.17 PI 3-kinase regulates both cell shape and the activation of essential growth and metabolic functions

Key concepts

- Phosphorylation of some lipid second messengers changes their activity.
- PIP$_3$ is recognized by proteins with a pleckstrin homology domain.

Lipid second messengers may also be modified by phosphorylation. PI 3-kinase phosphorylates PIP$_2$ on the 3-position of the inositol ring to form PI 3,4,5-P$_3$, another lipid second messenger. The total activity of PI 3-kinase is too low to significantly deplete total PIP$_2$, but formation of small amounts of PIP$_3$ in localized membrane domains is vital for altering cell shape and cellular motility.

PIP$_3$ acts by recruiting proteins that contain PIP$_3$ binding domains, including pleckstrin homology (PH) and FYVE domains, to sites where they regulate cytoskeletal remodeling, contractile protein function, or other regulatory events. These proteins anchor and/or orient the structural or motor proteins involved in cellular movement and localize signaling proteins to sites of action at the membrane. PIP$_3$ signaling can be fast and dramatic; it largely accounts for directing the mobility of motile mammalian cells.

Lipid mediators are essential in the insulin-signaling pathway. The binding of insulin stimulates the Tyr autophosphorylation of its receptor and the activation of effectors through insulin receptor substrate (IRS) proteins (see *18.30 Diverse signaling mechanisms are regulated by protein tyrosine kinases*). PI 3-kinase is activated when its p85 subunit binds to IRS1. The PIP$_3$ generated by PI3-kinase binds the protein kinases Akt and phosphoinositide-dependent kinase-1 (PDK-1) via their PH domains. This interaction results in the localization of Akt to the membrane where it is activated by PDK-1, as illustrated in **FIGURE 18.18**. Akt phosphorylates downstream targets, including protein kinases, GAPs, and transcription factors. Activation of Akt, specifically Akt-2, is required for the hallmark actions of insulin including regulation of glucose transporter translocation, enhanced protein synthesis, and expression of gluconeogenic and lipogenic enzymes.

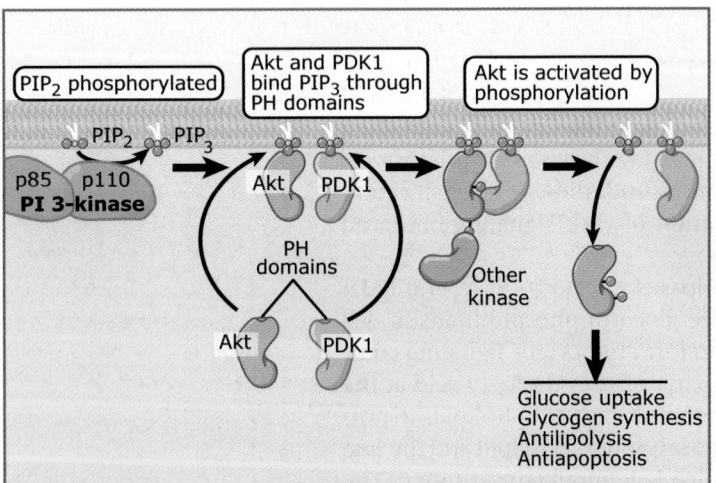

FIGURE 18.18 Activated PI 3-kinase phosphorylates PIP$_2$ to produce PIP$_3$. The PH domain-containing protein kinases PDK1 and Akt bind to PIP$_3$ at the plasma membrane. Their colocalization facilitates the phosphorylation of Akt by PDK1. A second phosphorylation within a hydrophobic motif results in Akt activation by one of several candidate protein kinases. The Akt-2 isoform is required to elicit hallmark actions of insulin.

18.18 Signaling through ion channel receptors is very fast

Key concepts

- Ion channels allow the passage of ions through a pore, resulting in rapid (microsecond) changes in membrane potential.
- Channels are selective for particular ions or for cations or anions.
- Channels regulate intracellular concentrations of regulatory ions, such as Ca^{2+}.

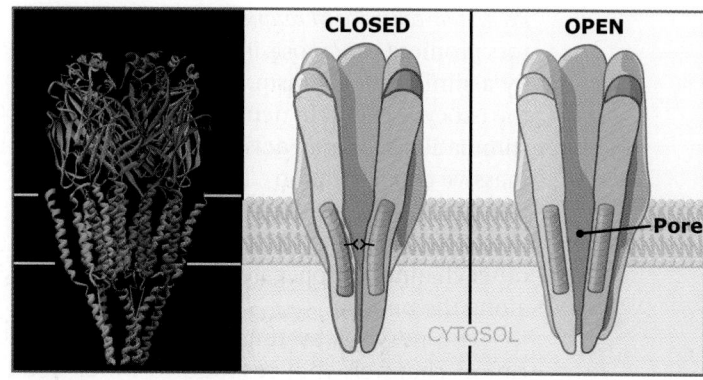

FIGURE 18.19 The nicotinic cholinergic receptor is a cation-selective channel that is composed of five homologous but usually nonidentical subunits that oligomerize to form a primarily α-helical membrane-spanning core. The channel itself is created within this core, and its opening and closing are executed by cooperative changes in subunit arrangement. Structure from Protein Data Bank 2BG9. N. Unwin, *J. Mol. Biol.* 346 (2005): 967–989.

Ligand-gated ion channels are multisubunit, membrane-spanning proteins that create and regulate a water-filled pore through the membrane, as illustrated in the x-ray crystal structure of the nicotinic acetylcholine receptor in **FIGURE 18.19**. When stimulated by extracellular agonists, the subunits rearrange their conformations and orientations to open the pore and, thus, connect the aqueous spaces on either side of the membrane. The pore has a diameter that allows ions to diffuse freely from one side of the membrane to the other, driven by the electrical and chemical gradients that are established by ion pumps and transporters. (For more about channel, pump and transporter mechanics, see *6 Transport of ions and small molecules across membranes*.) Channels maintain selectivity among ions by regulating the pore diameter precisely and by lining the walls of the pore with appropriate hydrophilic residues. Receptor ion channels can, thus, provide a diffusion path for only cations or anions, or select among different ions.

Ligand-gated ion channels provide the fastest signal transduction mechanism found in biology. Upon binding an agonist ligand, channels open within microseconds. At synapses, where neurotransmitters need to diffuse less than 0.1 micron, a signal in the postsynaptic cell can be generated in 100 microseconds. In contrast, receptor-stimulated G proteins require about 100 milliseconds to exchange GDP for GTP, and the action of receptor protein kinases is even slower. Ligand-gated ion channels are important receptors in many cells in addition to neurons and muscle, and other ion channels play equally vital roles in signaling pathways triggered by other classes of ligands.

Ion channel signaling differs from that of the other receptors mentioned in this chapter in that there is no immediate protein target nor, in most cases, is there a specific second messenger involved. In most cases, channel-mediated ion flow acts to increase or decrease the cell's membrane potential and, thus, modulates all transport processes for metabolites or ions that are electrically driven.

Animal cells maintain an inside-negative membrane potential by pumping out Na^+ ions and pumping in K^+ ions (for more on membrane potential see *6.4 Electrochemical gradients across the cell membrane generate the membrane potential*). The opening of a channel selective for Na^+ will thus depolarize cells, and the opening of a channel for K^+ will hyperpolarize cells. Similarly, because Cl^- is primarily extracellular, opening Cl^- channels will also cause hyperpolarization. These electrical effects convey information to effector proteins that are energetically coupled to the membrane potential, or to specific ion gradients, or that bind a specific ion (such as Ca^{2+}) whose concentration changes upon channel opening.

The nicotinic acetylcholine receptor is the prototypical receptor ion channel and was the first receptor that was shown to be a channel. It is a relatively unselective cation channel that causes depolarization of the target cell by allowing Na^+ influx. It is best known as the excitatory receptor at the neuromuscular synapse, where it triggers contraction, but alternative isoforms are also active in neurons and many other cells. In muscles, nicotinic depolarization acts via a voltage-sensitive Ca^{2+} channel to allow Ca^{2+} release from the sarcoplasmic reticulum into the cytosol. Calcium acts as a second (or third) messenger to initiate contraction (see *6.13 Cardiac and skeletal muscles are activated by* excitation-contraction coupling). Nicotinic receptors promote exocytosis in some secretory cells by a similar mechanism, where Ca^{2+} triggers the exocytic event. In neurons, where nicotinic stimulation causes an **action potential** (rapid, massive depolarization), the initial depolarization is sensed by voltage-sensitive Na^+ channels. Their opening (along with the action of other channels) propagates the action potential along the neuron.

The nervous system is rich in receptor cation channels that respond to other neurotransmitters, the most common of which is the amino acid glutamate (Glu). The three different families of glutamate receptors share the property of cation conductance, but each family has its own spectrum of drug responses. All operate as neuronal activators, with one interesting twist: The NMDA family of receptors, named for their response to a selective drug, is permeant to Ca^{2+} in addition to Na^+. A significant component of its activity is to permit the inward flow of Ca^{2+}, which acts as a second messenger that regulates a wide variety of targets. Persistent stimulation of NMDA channels by glutamate released during injury, or by drugs, can cause toxic amounts of Ca^{2+} to enter, resulting in neuronal death.

A second functional group of receptor channels is selective for anions and, by allowing inward flux of Cl^-, hyperpolarizes the target cell. Anion-selective receptors include those for γ-aminobutyric acid (GABA) and glycine (Gly). In neurons, hyperpolarization can inhibit the initiation of an action potential and/or neurotransmitter release.

Perhaps the most diverse family of ligand-gated channels is the TRP family, of which about 30 are found in mammals. The TRP channels are Ca^{2+}-selective channels that are formed by tetramers of identical subunits that surround the central channel. Each subunit is composed of a homologous bundle of six membrane-spanning helices. The N and C termini are least similar among TRP isoforms and contain isoform-specific collections of regulatory and protein interaction domains, including protein kinase domains (whose substrates are currently unknown).

All TRP channels allow transmembrane flux of Ca^{2+} to permit its action as a second messenger, but different TRP isoforms serve numerous physiologic functions. The prototypical TRP, found in invertebrate photoreceptors, gates Ca^{2+} flow into the cytoplasm to initiate visual signaling. Others admit Ca^{2+} from out-side the cell, and still others allow Ca^{2+} to enter the endoplasmic reticulum virtually directly from the extracellular space because they form a bridge between the plasma membrane and channels in the endoplasmic reticulum at points where the membranes abut each other.

Regulation of TRP channels is perhaps even more diverse. Various TRP channels respond to heat, cold, painful stimuli, pressure, and high or low osmolarity. Many TRPs are regulated either positively or negatively by lipids, such as eicosanoids, diacylglycerol, and PIP_2. For example, capsaicin, the hot compound in chilis, is an agonist for some vanilloid receptors (TRPVs). Still other TRP channels are mechanosensors that allow cilia to sense fluid flow. The most famous of these is the sensory channel of the hair cell of the inner ear. This channel opens when the apical cilia on the hair cell are bent in response to sound-driven fluid flow.

18.19 Nuclear receptors regulate transcription

Key concepts

- Nuclear receptors modulate transcription by binding to distinct short sequences in chromosomal DNA known as response elements.
- Receptor binding to other receptors, inhibitors, or coactivators leads to complex transcriptional control circuits.
- Signaling through nuclear receptors is relatively slow, consistent with their roles in adaptive responses.

Nuclear receptors are unique among cellular receptors in that their ligands pass unaided through the plasma membrane. These receptors, when complexed with their ligands, enter the nucleus and regulate gene transcription. Ligands for nuclear receptors include sex steroids (estrogen and testosterone) and other steroid hormones, thyroid hormone, vitamins A and D, retinoids and other fatty acids, oxysterols, and bile acids.

Nuclear receptors are structurally conserved. They consist of a C-terminal ligand-binding domain, an N-terminal interaction region that recognizes components of the transcriptional machinery and acts as a transactivation domain, a centrally located zinc finger domain that binds DNA, and, often, another transactivation domain nearer the C-terminus. In the absence of ligand, these receptors are bound to corepressor proteins that suppress their activity. Upon hormone binding, corepres-

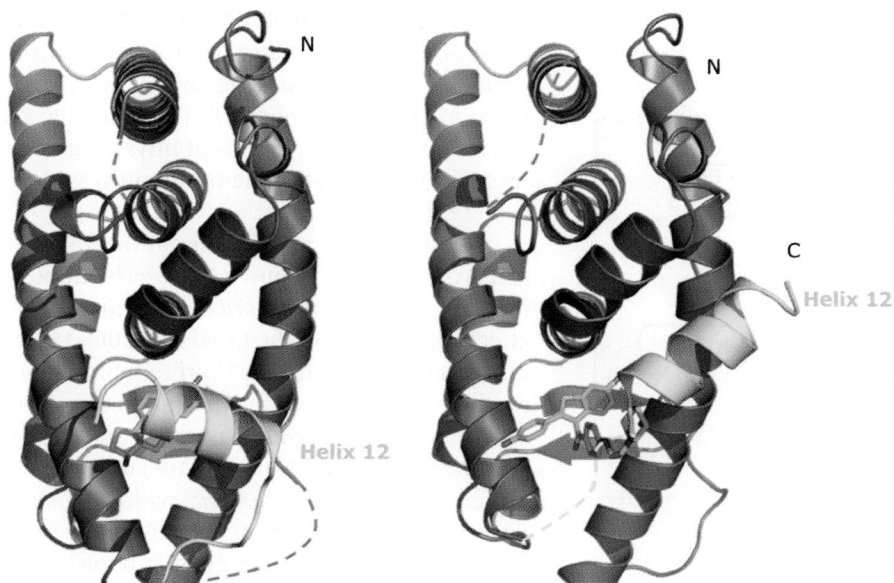

N

N

C

Helix 12

Helix 12

FIGURE 18.20 The estrogen receptor adopts different conformations when bound to agonists and antagonists. The ligand-binding domain of the estrogen receptor is bound to the agonist estradiol on the left and to the antagonist raloxifene on the right. Note the marked difference in position of helix 12, shown in yellow in the active structure and green in the inhibited structure. Structures from Protein Data Bank 1ERE. A. M. Brzozowski, et al., *Nature* 389 (1997): 753–758. Figures created by Yu-Chi Juang, UT Southwestern Medical Center.

sors dissociate and the receptors are assembled in multiprotein complexes with coactivators that modulate receptor action and facilitate transcriptional regulation. As illustrated in **FIGURE 18.20**, agonists and antagonists bind to distinct receptor conformations (see *18.5 Ligand binding changes receptor conformation*). Receptor agonists favor the binding of receptors to coactivators and DNA, and antagonists favor conformations that block coactivator-receptor binding.

Nuclear receptors bind with high specificity to hormone **response elements** in the 5′ untranscribed region of regulated genes. Response elements are typically short direct or inverted repeat sequences, and a gene may contain response elements for several different receptors in addition to binding sites for other transcriptional regulatory proteins.

The sex steroid estrogen can bind to two different nuclear receptors, the estrogen receptors ERα and ERβ. Coactivator and corepressor proteins differentially regulate ERα and ERβ in transcriptional complexes that are expressed in specific cell types. Other ligands that bind to these receptors include valuable therapeutic agents. For example, 4 hydroxy-tamoxifen is an estrogen receptor antagonist used in the therapy of estrogen-receptor-positive breast cancer to inhibit growth of residual cancer cells. However, unlike its antagonistic effects on the estrogen receptor in breast, 4 hydroxy-tamoxifen displays weak partial agonist activity in uterus. In the estrogen receptor system, partial agonists are known as selective estrogen

receptor modulators (SERMs). Properties that contribute to partial agonist activity include the relative expression of the two estrogen receptors, ERα and ERβ as well as the expression of repressors and coactivators that interact with each receptor type. Thus, the behavior of nuclear receptor ligands must be considered in the tissue, cellular, and signaling context.

18.20 G protein-signaling modules are widely used and highly adaptable

Key concepts

- The basic G protein module is a receptor, a G protein and an effector protein.
- Cells express several varieties of each class of proteins.
- Effectors are heterogeneous and initiate diverse cellular functions.

Activation of GPCRs and their associated heterotrimeric G proteins is one of the most widespread mechanisms of communicating extracellular signals to the intracellular environment. G protein-signaling modules are found in all eukaryotes. Depending on the species, mammals express 500–1000 GPCRs that respond to hormones, neurotransmitters, pheromones, metabolites, local signaling substances, and other regulatory molecules. Essentially all chemical classes are represented among the GPCR ligands. In addition, a roughly equal

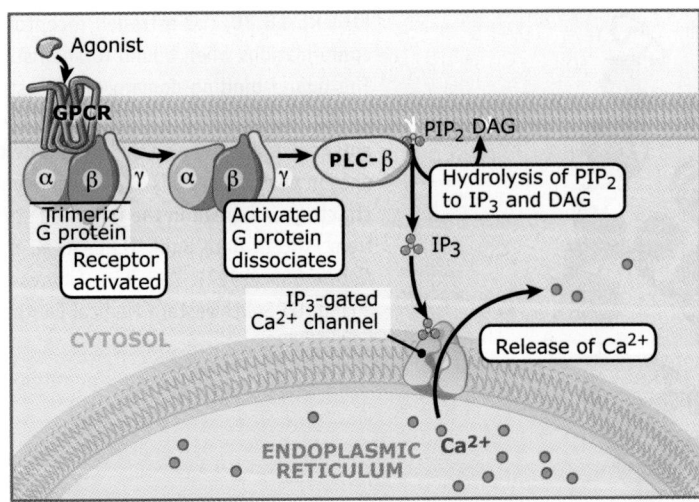

FIGURE 18.21 G protein-mediated signal transduction follows a path of agonist to receptor to heterotrimeric G protein to effector to the effector's output. Both Gα and Gβγ subunits regulate distinct effectors. In the example shown here, G_q regulates a phospholipase C-β to produce two second messengers, diacylglycerol (DAG) and inositol-trisphosphate (IP_3). IP_3 triggers Ca^{2+} release from the endoplasmic reticulum.

number of olfactory GPCRs are expressed in olfactory neurons and work in combination to screen compounds in the animal's environment via the sense of smell. Because GPCRs are involved in many kinds of physiologic responses, they are also one of the most widely used targets for drugs.

A minimal G protein-signaling module consists of three proteins: a G protein-coupled receptor, the heterotrimeric G protein, and an effector protein, as illustrated in FIGURE 18.21. The receptor activates the G protein on the inner face of the plasma membrane in response to an extracellular ligand. The G protein then activates (or occasionally inhibits) an effector protein that propagates a signal within the cell. Thus, signal conduction in the simplest G protein module is linear. However, as depicted in FIGURE 18.22, a typical animal cell may express a dozen GPCRs, more than six G proteins, and a dozen effectors. Each GPCR regulates one or more G proteins, and each G protein regulates several effectors. Moreover, distinct efficiencies and rates govern each interaction. Thus, a cell's G protein network is actually a signal-integrating computer whose output is a

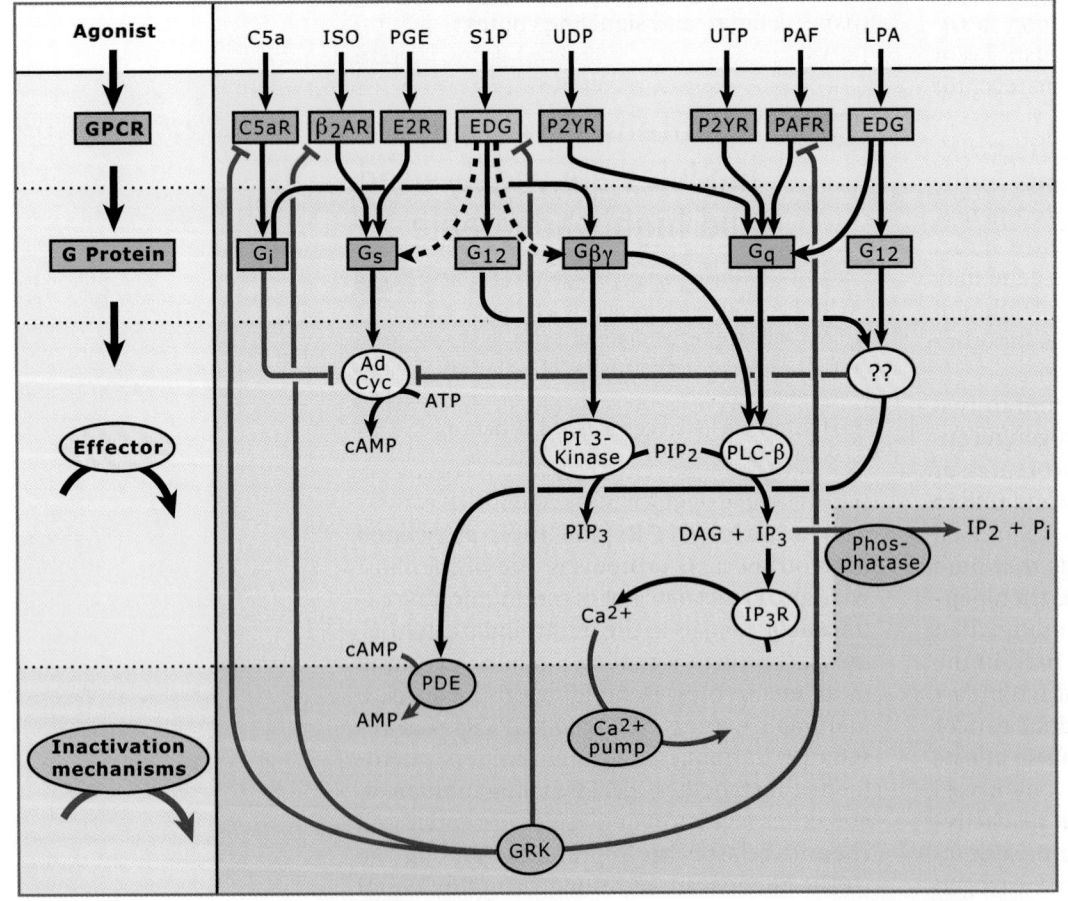

FIGURE 18.22 A portion of the G protein-mediated signaling network in macrophages highlights some of the complexity of interactions possible in such systems. Several receptors and G protein subunits are omitted. Where a named G protein is shown, its signaling output is probably mediated by its Gα subunit. Activation of any G protein also activates its Gβγ subunit, although Gβγ-mediated signaling is usually most prominent from G_i trimers. In addition, several G proteins modulate the activities of others through poorly understood pathways. Only a small sampling of effectors is shown, and the only adaptive mechanism shown is GRK-catalyzed phosphorylation of receptors. Data from Paul Sternweis, Alliance for Cellular Signaling.

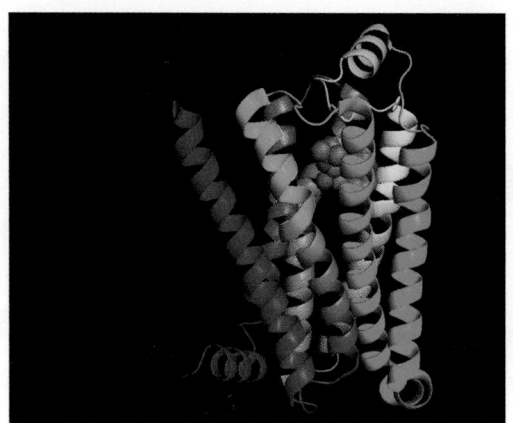

FIGURE 18.23 This crystal structure is of the beta1-adrenergic receptor, a prototypical GPCR that is a major receptor for adrenaline and which activates G$_s$. Each membrane-spanning helix is a different color, from blue (N-terminal) to red (C-terminal amphipathic helix on the cytosolic face). Some structures on the cytoplasmic and extracellular faces are not resolved by crystallography and are not shown. Cyanopindolol, a synthetic beta-adrenergic antagonist, was present in the crystal and its structure is shown as atomic spheres within the bundle of helices. GPCR sequence similarity separates the mammalian GPCRs into at least four structural families that are so diverse that there is sometimes little obvious sequence similarity among the classes. Within a family, similarity is greatest in the membrane-spanning helices, less in the interhelical loops, and least in the N- and C-terminal domains and in the cytoplasmic loop that connects spans five and six. Regardless, the generalizations about functional domains in receptors seem to hold true within different families. GPCRs frequently form dimers, occasionally heterodimers, and dimerization can be crucial for function. Structure from Protein Data Bank 2VT4. A. Warne, et al., *Nature* 454 (2008): 486. Photo courtesy of Elliott Ross, University of Texas Southwestern Medical Center.

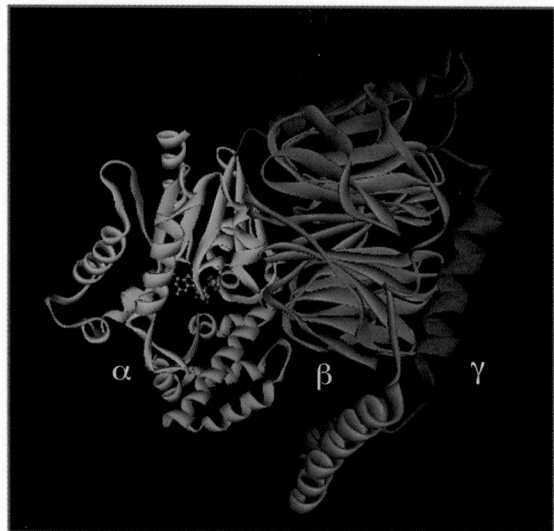

FIGURE 18.24 The structure of the nonactivated G$_i$ heterotrimer, the G protein that is responsible for inhibition of adenylyl cyclase and for most Gβγ-mediated signaling, is shown with each subunit colored as shown. GDP is shown bound to the Gα$_i$ subunit. Structure from Protein Data Bank 1GP2. M. A. Wall, et al., *Cell* 83 (1995): 1047–1058.

spectrum of cellular signals that is complex in both amplitude and kinetics. Because of their conserved parts list, G protein modules are well suited to initiating a wide variety of intracellular signals in response to diverse molecular inputs and can do so over a wide range of time scales (milliseconds to minutes).

GPCRs are integral plasma membrane proteins composed of a bundle of seven hydrophobic membrane-spanning helices with an extracellular N terminus and cytosolic C-terminus, as depicted in FIGURE 18.23. Based on structural homology and copious biochemical and genetic data, it is likely that all GPCRs share the same basic mechanism of conformational activation and deactivation in response to activating ligands (see *18.5 Ligand binding changes receptor conformation*). Binding of agonist ligand on the extracellular face of the receptor drives realignment of the helices to alter the structure of a binding site for the heterotrimeric G protein on the cytoplasmic face, and this altered conformation of the G protein-binding surface promotes G protein activation.

The heterotrimeric G proteins to which GPCRs are coupled are composed of a nucleotide-binding Gα subunit and a Gβγ subunit dimer, as illustrated in FIGURE 18.24. The structure of the trimer and each subunit is known for several states of activation and in complex with several interacting proteins. A Gαβγ heterotrimer is named according to its α subunit, which largely defines the G protein's selectivity among receptors. Each subunit also regulates a distinct group of effector proteins.

Gα subunits are globular, two-domain proteins of 38–44 kDa. The GTP-binding domain belongs to the GTP-binding protein superfamily that includes the small, monomeric G proteins (such as Ras, Rho, Arf, Rab; see *18.23 Small, monomeric GTP-binding proteins are multiuse switches*) as well as the GTP-binding translational initiation and elongation factors. A second domain modulates GTP binding and hydrolysis. Gα subunits are only slightly hydrophobic, but

G protein	EFFECTOR PROTEIN	
	Stimulated	Inhibited
G_s G_{olf}	Adenylyl cyclase	
G_i (3) G_o G_z	K$^+$ channel, PI 3-kinase	Adenylyl cyclase
G_{gus}	Other cation channel	
G_t (2)	Cyclic GMP phospho-diesterase	
G_q (4)	Phospholipase-Cβ	
G_{12} G_{13}	Rho GEF	

FIGURE 18.25 G protein-regulated effectors do not share structural similarities. They may be ion channels or membrane-spanning enzymes in the plasma membrane, peripheral proteins on the inner face of the membrane, or fundamentally soluble proteins that can bind to Gα subunits. The chart shows the major groups of G proteins, sorted according to sequence similarity, and some of the effectors that they are known to regulate.

they are predominantly membrane-associated because of constitutive N-terminal fatty acylation and because they bind to the membrane-attached G$\beta\gamma$ subunits. Mammals have 16 Gα genes that are grouped in subfamilies according to similar sequence and function. These subfamilies are listed in FIGURE 18.25.

Gβ and Gγ subunits associate irreversibly soon after translation to form stable G$\beta\gamma$ dimers, which then associate reversibly with a Gα. Gβ subunits are 35 kDa proteins composed of seven β-strand repeats that form a cylindrical structure known as a β propeller. There are five Gβ genes in mammals. Four encode strikingly similar proteins that naturally dimerize with the twelve Gγ subunits (see Figure 18.24). The fifth, Gβ5, is less closely related to the others and interacts primarily with a Gγ-like domain in other proteins rather than with Gγ subunits themselves.

Gγ subunits are smaller (~7 kDa) and far more diverse in sequence than are the Gβ's. The last three amino acid residues of Gγ subunits are proteolyzed to leave a conserved C-terminal cysteine that is irreversibly S-prenylated and carboxymethylated, helping to anchor G$\beta\gamma$ to the membrane. Gβ and Gγ subunits can associate in most possible combinations. Because almost all cells express multiple Gβ and Gγ subunits, it is difficult to assign specific roles to individual G$\beta\gamma$ combinations. The best-recognized protein interactions of G$\beta\gamma$ subunits

occur at sites on Gβ, although distinct functions of individual Gγ subunits are also indicated by some experiments.

18.21 Heterotrimeric G proteins regulate a wide variety of effectors

Key concepts

- G proteins convey signals by regulating the activities of multiple intracellular signaling proteins known as effectors.
- Effectors are structurally and functionally diverse.
- A common G protein binding domain has not been identified among effector proteins.
- Effector proteins integrate signals from multiple G protein pathways.

G protein-regulated effectors include enzymes that create or destroy intracellular second messengers (adenylyl cyclase, cyclic GMP phospho-diesterase, PLC-β, PI 3-kinase), protein kinases, ion channels (K$^+$, Ca^{2+}), and possibly membrane transport proteins (see Figure 18.25). Effectors may be integral membrane proteins or intrinsically soluble proteins that bind G proteins at the membrane surface. No conserved G protein-binding domain or sequence motif has been identified among effector proteins, and most effectors are related to proteins that have similar functions but that are not regulated by G proteins. Sensitivity to G protein regulation, thus, evolved independently in multiple families of regulatory proteins.

Because they can respond to a variety of Gα and G$\beta\gamma$ subunits, effector proteins can integrate signals from multiple G protein pathways. The different Gα or G$\beta\gamma$ subunits may have opposite or synergistic effects on a given effector. For example, some of the membrane-bound adenylyl cyclases in mammals are stimulated by Gα_s and inhibited by Gα_i (see Figure 18.13). Many effectors are further regulated by other allosteric ligands (e.g., lipids, calmodulin) and by phosphorylation, contributing even more to integration of information.

Effectors are usually represented as multiple isoforms, and each isoform may be regulated differently, adding to the complexity of G protein networks. For example, some isoforms of adenylyl cyclase are stimulated by G$\beta\gamma$, whereas others are inhibited. All phospholipase C-βs are stimulated both by Gα_q family members and by G$\beta\gamma$, but the potency and maximal effect of these two inputs vary dramatically among the four PLC-β isoforms.

18.22 Heterotrimeric G proteins are controlled by a regulatory GTPase cycle

Key concepts

- Heterotrimeric G proteins are activated when the Gα subunit binds GTP.
- GTP hydrolysis to GDP inactivates the G protein.
- GTP hydrolysis is slow, but is accelerated by proteins called GAPs.
- Receptors promote activation by allowing GDP dissociation and GTP association; spontaneous exchange is very slow.
- Regulators of G protein-signaling (RGS) proteins and phospholipase C-βs are GAPs for G proteins.

The key event in heterotrimeric G protein-signaling is the binding of GTP to the Gα subunit. GTP binding activates the Gα subunit, which allows both it and the Gβγ subunit to bind and regulate effectors. The Gα subunit remains active as long as GTP is bound, but Gα also has GTPase activity and hydrolyzes bound GTP to GDP. Gα-GDP is inactive. G proteins thus traverse a GTPase cycle of GTP binding/activation and hydrolysis/deactivation, as depicted in **FIGURE 18.26**. Therefore, the control of G protein-signaling is intrinsically kinetic. The relative signal strength, or amplitude, is proportional to the fraction of G protein that is in the active, GTP-bound form. This fraction equals the balance of the rates of GTP binding and GTP hydrolysis, the activating and deactivating arms of the GTPase cycle. Both limbs are highly regulated over a range of rates greater than 1000-fold.

Receptors promote G protein activation by opening the nucleotide-binding site on the G protein, thus accelerating both GDP dissociation and GTP association. This process is referred to as GDP/GTP exchange catalysis. Exchange proceeds in the direction of activation because the affinity of G proteins for GTP is much higher than that for GDP and because the cytosolic concentration of GTP is about 20-fold higher than that of GDP. Spontaneous GDP/GTP exchange is very slow for most G proteins (many minutes), which maintains basal signal output at a low level. In contrast, receptor-catalyzed

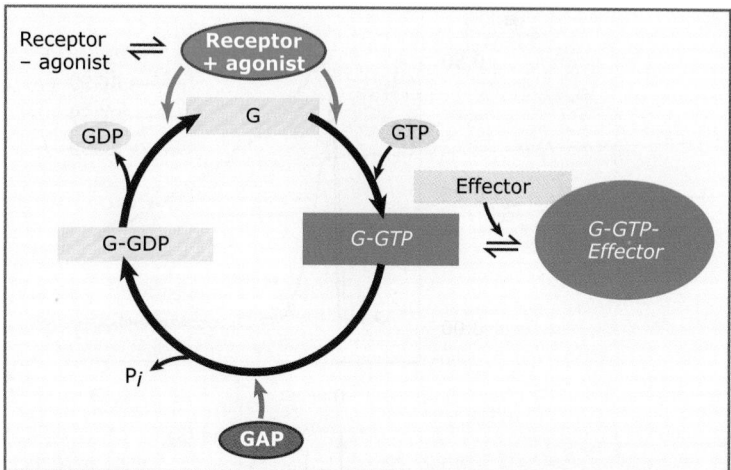

FIGURE 18.26 G proteins are activated when GTP binds to the Gα subunit, such that both Gα-GTP and Gβγ can bind and regulate the activities of appropriate effector proteins. Gα subunits also have intrinsic GTPase activities, and the primary deactivating reaction is hydrolysis of bound GTP to GDP (rather than GTP dissociation). Thus, the steady-state signal output from a receptor-G protein module is the fraction of the G protein in the GTP-bound state, which reflects the balance of the activation and deactivation rates. Both GTP binding and GTP hydrolysis are intrinsically slow and highly regulated. GDP binds tightly to Gα, such that GDP dissociation is rate-limiting for binding of a new molecule of GTP and consequent reactivation. Both GDP release and GTP binding are catalyzed by GPCRs. Hydrolysis of bound GTP is accelerated by GTPase-activating proteins (GAPs). Receptors and GAPs coordinately control both the steady-state level of signal output and the rates of activation and deactivation of the module.

exchange can take place in a few tens of milliseconds, which allows rapid responses in cells such as visual photoreceptors, other neurons, or muscle.

Because receptors are not directly required for a G protein's signaling activity, a receptor can dissociate after GDP/GTP exchange and catalyze the activation of additional G protein molecules. In this way, a single receptor may maintain the activation of multiple G proteins, providing molecular amplification of the incoming signal. Other receptors may remain bound to their G protein targets, which means that they do not act as amplifiers. However, more tightly bound receptors can initiate signaling more quickly and promote G protein reactivation when hydrolysis of bound GTP is rapid.

In the absence of stimulus, Gα subunits hydrolyze bound GTP slowly. The average activation lifetime of the Gα-GTP complex is about 10–150 seconds, depending on the G protein. This rate is far slower than rates of deactivation often observed in cells when an

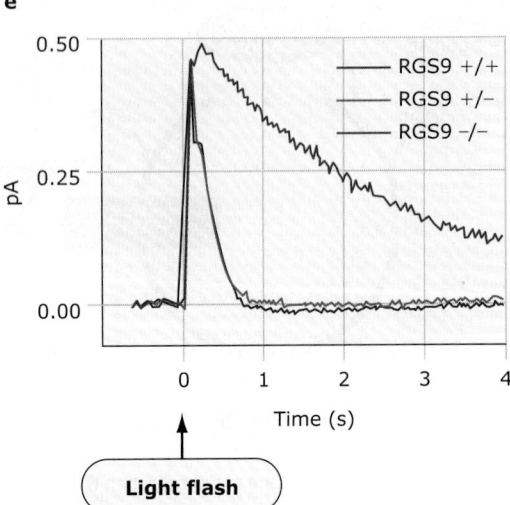

e

Light flash

FIGURE 18.27 G protein GAPs can accelerate signal termination upon removal of agonist, and often do not act as inhibitors during the response to receptor. The figure shows the electrical response of a mouse photoreceptor (rod) cell to a single photon of light. In mice that lack RGS9, the GAP for the photoreceptor G protein G_t, the signal is prolonged for many seconds because hydrolysis of GTP bound to G_t is slow. In wild-type or heterozygous mice, hydrolysis takes place in about 15 milliseconds, and the decay of the signal is much faster. Note that the maximal output is similar in wild-type and mutant mice, indicating that the GAP does not act as an inhibitor in rod cells. In humans, genetic loss of RGS9 leads to severe loss of vision that is particularly marked in bright light. Adapted from C. -K. Chen, et al., *Nature* 403 (2000): 557–560.

agonist is removed. For example, visual signaling terminates in about 10 milliseconds after stimulation by a photon, and many other G protein systems are almost as fast. GTP hydrolysis is accelerated by GTPase-activating proteins (GAPs), which directly bind $G\alpha$ subunits. In some cases, acceleration exceeds 2000-fold. Such speed is necessary in systems like vision or neurotransmission, which must respond to quickly changing stimuli. Because G protein-signaling is a balance of activation and deactivation, GAPs deplete the pool of GTP-activated G protein and can thereby also act to inhibit G protein-signaling. GAPs can thus inhibit signaling, quench output upon signal termination, or both. What behavior predominates depends on the GAP's intrinsic activity and its regulation.

There are two families of GAPs for heterotrimeric G proteins. The RGS proteins (regulators of G protein-signaling) are a family of about 30 proteins, most or all of which have GAP activity and regulate G protein-signaling

rates and amplitudes. The role of RGS proteins in terminating the G protein signal can be seen in **FIGURE 18.27**. Some proteins with RGS domains also act as G protein-regulated effectors. These include activators of the Rho family of monomeric GTP-binding proteins (see below) and GPCR kinases, which are feedback regulators of GPCR function. The second group of G protein GAPs are phospholipase C-βs. These enzymes are effectors that are stimulated by both $G\alpha_q$ and by $G\beta\gamma$, but they also act as G_q GAPs, probably to control output kinetics.

While the GTPase cycle described in Figure 18.26 is general, it is highly simplified. Interactions among receptor, $G\alpha$, $G\beta\gamma$, GAP, and effector are frequently simultaneous and often demonstrate complex cooperative interactions. For example, $G\beta\gamma$ inhibits the release of GDP (to minimize spontaneous activation), promotes the exchange catalyst activity of the receptor, inhibits GAP activity, and helps initiate receptor phosphorylation that leads to desensitization. The other components can be nearly this multifunctional. In addition, inputs from other proteins can alter the dynamics of the GTPase cycle at several points. The core G protein module is, thus, functionally versatile as a signal processor in addition to being versatile in the scope of its targets.

Concept and Reasoning Check

1. A major mechanism of cholera pathogenesis in the gut is mediated by cholera toxin, a protein toxin that is secreted by Vibrio cholerae and endocytosed into intestinal epithelial cells. Cholera toxin is an enzyme that catalyzes the ADP-ribosylation of $G\alpha_s$, the G protein that activates adenylyl cyclase. ADP-ribosyl-$G\alpha_s$ functions normally in most respects, but it hydrolyzes bound GTP very slowly compared with unmodified $G\alpha_s$. What is the primary effect of cholera toxin on cellular signaling? Why?

18.23 Small, monomeric GTP-binding proteins are multiuse switches

Key concepts

- Small GTP-binding proteins are active when bound to GTP and inactive when bound to GDP.
- GDP/GTP exchange catalysts known as GEFs (guanine nucleotide exchange factors) promote activation.
- GAPs accelerate hydrolysis and deactivation.
- GDP dissociation inhibitors (GDIs) slow spontaneous nucleotide exchange.

Monomeric GTP-binding proteins, which are encoded by about 150 genes in animals, modulate a wide variety of cellular processes including signal transduction, organellar trafficking, intra-organellar transport, cytoskeletal assembly, and morphogenesis. The small GTP-binding proteins that most clearly function in signal transduction are the Ras and Ras-related proteins (Ral, Rap) and the Rho/Rac/Cdc42 proteins, about 10–15 in all. They are usually about 20–25 kDa in size and are homologous to the GTP-binding domains of $G\alpha$ subunits.

The regulatory activities of the small GTP-binding proteins are controlled by a GTP binding and hydrolysis cycle like that of the heterotrimeric G proteins, with similar regulatory inputs. They are activated by GTP, and hydrolysis of bound GTP to GDP terminates activation. GDP/GTP exchange catalysts, known as GEFs (guanine nucleotide exchange factors, functionally analogous to GPCRs) promote activation, and GAPs accelerate hydrolysis and consequent de-activation. In addition, GDIs slow spontaneous nucleotide exchange and activation to dampen basal activity, an activity shared by $G\beta\gamma$ subunits for the heterotrimeric G proteins.

While the underlying biochemical regulatory events are essentially identical for monomeric and heterotrimeric G proteins, monomeric G proteins use the basic GTPase cycle in additional ways. Signal output by heterotrimeric G proteins and many monomeric G proteins is usually thought to reflect a balance of their active (GTP-bound) and inactive (GDP-bound) states in a rapidly turning-over GTPase cycle. GEFs favor formation of more active G protein, and GAPs favor the inactive state. In contrast, probably an equal number of the monomeric G proteins behave as acute on-off switches. Upon binding GTP, they initiate a process (regulation, recruitment of other proteins). They then maintain this activity, sometimes for many seconds or minutes, until they are acted upon by a GAP. For example, the monomeric G protein Ran regulates nucleocytoplasmic trafficking of protein and RNA in both directions, cooperating with carrier proteins known as karyopherins (see *9.15 The Ran GTPase controls the direction of nuclear transport*). In the nucleus, high Ran GEF activity promotes GTP binding. Nuclear Ran-GTP then binds import karyopherins to drive dissociation of newly arrived cargo and promote return of the karyopherin to the cytoplasm. It also binds export karyopherins to permit binding of outgoing cargo. Outside the nucleus, high Ran GAP activity promotes GTP hydrolysis. Cytoplasmic Ran-GDP dissociates from the export karyopherins to allow dissociation of outgoing cargo and from the import karyopherins to allow them to bind cargo for import. Thus, for monomeric G proteins such as Ran, each phase of the GTPase cycle determines a specific, coupled step in a parallel regulatory cycle.

A second major difference between the monomeric and the heterotrimeric G proteins is the structures of the GEFs, GAPs, and GDIs. Both GEFs and GAPs for monomeric GTP-binding proteins are structurally heterogeneous (although some clearly related families are evident). In addition, mechanisms for regulating these GEFs and GAPs are equally diverse. They include phosphorylation by protein kinases; allosteric regulation by heterotrimeric and/or monomeric G proteins, by second messengers and by other regulatory proteins; subcellular sequestration or recruitment to scaffolds; and assorted other mechanisms.

The Ras proteins were the first small GTP-binding proteins to be discovered. They were identified as oncogene products because they can cause malignant growth if they are either overexpressed or persistently activated by mutation; they are among the most commonly mutated genes in human tumors. Several viral *ras* genes figure prominently as oncogenes.

Mammalian cells contain three *ras* genes (H, N, and K). They may share inputs and outputs to varying extents, and they can compensate for each other in some genetic screens. Inputs to the Ras proteins are diverse and speak to the importance of Ras proteins as a crucial node in signaling.

Ras GEFs and GAPS are regulated by both receptor and nonreceptor Tyr kinases through direct phosphorylation and by recruitment of the regulators to the plasma membrane. Other cytoplasmic serine/threonine kinases also converge on Ras activation. Rap1, another member of the Ras family, may also fit directly into this network because it is suspected of competing with Ras proteins for protein kinase targets; *in vivo* it can suppress the oncogenic activity of Ras. Rap1 is regulated independently, however, and acts on independent signaling pathways as well. One of its GAPs is stimulated by the G_i class of G proteins, for example, and its several GEFs are stimulated by Ca^{2+}, diacylglycerol, cAMP and, perhaps, other monomeric G proteins.

Ras proteins generally regulate cell growth, proliferation, and differentiation by modulating the activities of multiple effector proteins. The best-known Ras effector is the protein

Function	Effector	Target
Protein kinase cascade	Raf	MAPK
Lipid kinase	PI 3-kinase	Akt
Exchange factor	RalGDS	Exocyst

FIGURE 18.28 Ras-GTP binds to many proteins. Three well-established effectors include Raf, PI 3-kinase, and RalGDS. Activation of these effectors activates a MAPK pathway, increases PI 3-kinase activity, and promotes assembly of a protein complex involved in exocytosis of secretory vesicles.

kinase Raf, which initiates a MAPK cascade. **FIGURE 18.28** shows well-established Ras effectors. Other effectors include PI-3 kinase (see *18.30 Diverse signaling mechanisms are regulated by protein tyrosine kinases*) and a GEF for Ral (RalGDS), which may both be required for the oncogenic potential of Ras.

Rho, Rac, and Cdc42 are related monomeric GTP-binding proteins that are involved in generating signals that affect cell morphology. Each class of proteins regulates its own array of effectors and is controlled by largely separate groups of GEFs, GAPs, and GDIs. Effectors regulated by this family include phospholipases C and D, multiple protein and lipid kinases, proteins that nucleate or reorganize actin filaments, and components of the neutrophil oxygen activating system, among others (see *12.14 Small G proteins regulate actin polymerization*).

18.24 Protein phosphorylation/dephosphorylation is a major regulatory mechanism in the cell

Key concepts
- Most protein kinases are members of the same protein family.
- Protein kinases phosphorylate Ser and Thr, or Tyr, or in a few cases all three.
- Protein kinases may recognize the primary sequence surrounding the phosphorylation site.
- Protein kinases may preferentially recognize phosphorylation sites within folded domains.

Protein phosphorylation is the most common form of regulatory posttranslational modification. It occurs in all organisms, and it is estimated that at least one-third of proteins in animals are at some time phosphorylated. Phosphorylation can stimulate or inhibit the catalytic activity of an enzyme, the affinity with which a protein binds other molecules,

its subcellular localization, its ability to be further covalently modified, or its stability. Single phosphorylations may cause 500-fold or greater changes in activity, and proteins are often phosphorylated on multiple residues in complex and interacting patterns.

Most protein phosphorylation in eukaryotes, and essentially all in animals, is catalyzed by protein kinases; dephosphorylation is catalyzed by phosphoprotein phosphatases. Both classes of enzymes are controlled by diverse mechanisms. In addition, proteins are often phosphorylated by multiple protein kinases, resulting in the generation of a range of activity states. This complexity allows inputs from different signaling pathways to be integrated into the resulting activity of the target.

In bacteria, plants, and fungi, an additional protein phosphorylating system known as two-component signaling is vital. The protein kinases involved in two-component signaling are unrelated to the eukaryotic protein kinase superfamily and phosphorylate aspartate residues rather than serine, threonine, or tyrosine.

Protein kinases transfer a phosphoryl group from ATP to Ser, Thr, and Tyr residues of protein substrates to form chemically stable phosphate esters, as shown in **FIGURE 18.29**. In animals, the distribution of phosphate among these three amino acid residues is uneven: ~90%–95% is on Ser, 5%–8% on Thr, and less than 1% on Tyr residues. The human genome contains ~500 related genes that encode protein kinases, and many protein kinase mRNAs undergo alternative splicing. This makes the protein kinase gene superfamily one of the largest functional gene groups. The number and diversity of these enzymes emphasize the great and varied uses of protein kinases to regulate cellular functions. Although some protein kinases have a limited tissue and/or developmental distribution, many are ubiquitously expressed.

Protein kinases are grouped according to their residue specificity. Protein kinases that phosphorylate Ser will usually also recognize Thr, hence the name protein Ser/Thr kinase. Multicellular organisms have protein Tyr kinases, which only recognize Tyr. Dual specificity protein kinases can phosphorylate Ser, Thr, and Tyr in the appropriately restricted conformational context and are generally the most selective of the protein kinases.

The analysis of the **kinomes** of several organisms has led to a more elaborate grouping derived from sequence relationships, shown in **FIGURE 18.30**, that also reflects to some extent

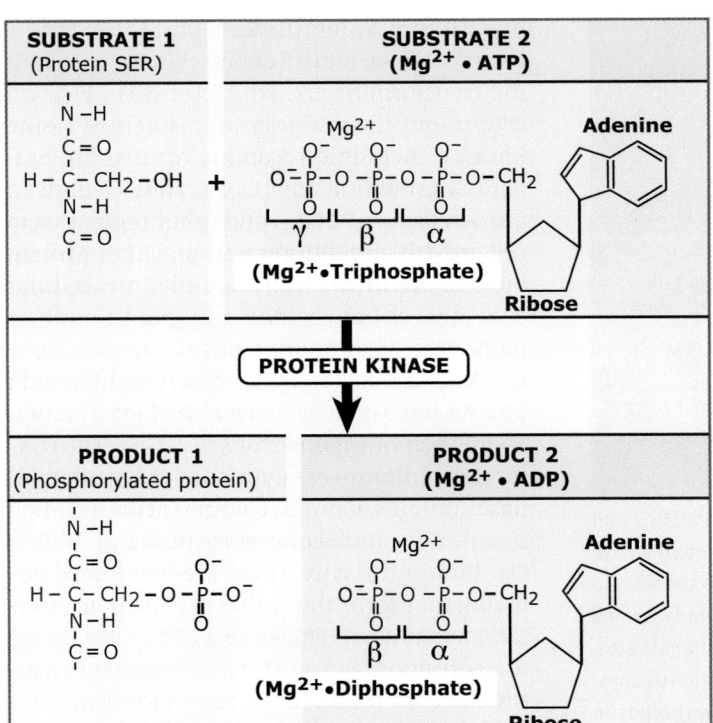

FIGURE 18.29 Protein kinases transfer the γ-phosphoryl group from ATP to serine, threonine, or tyrosine residues in protein substrates.

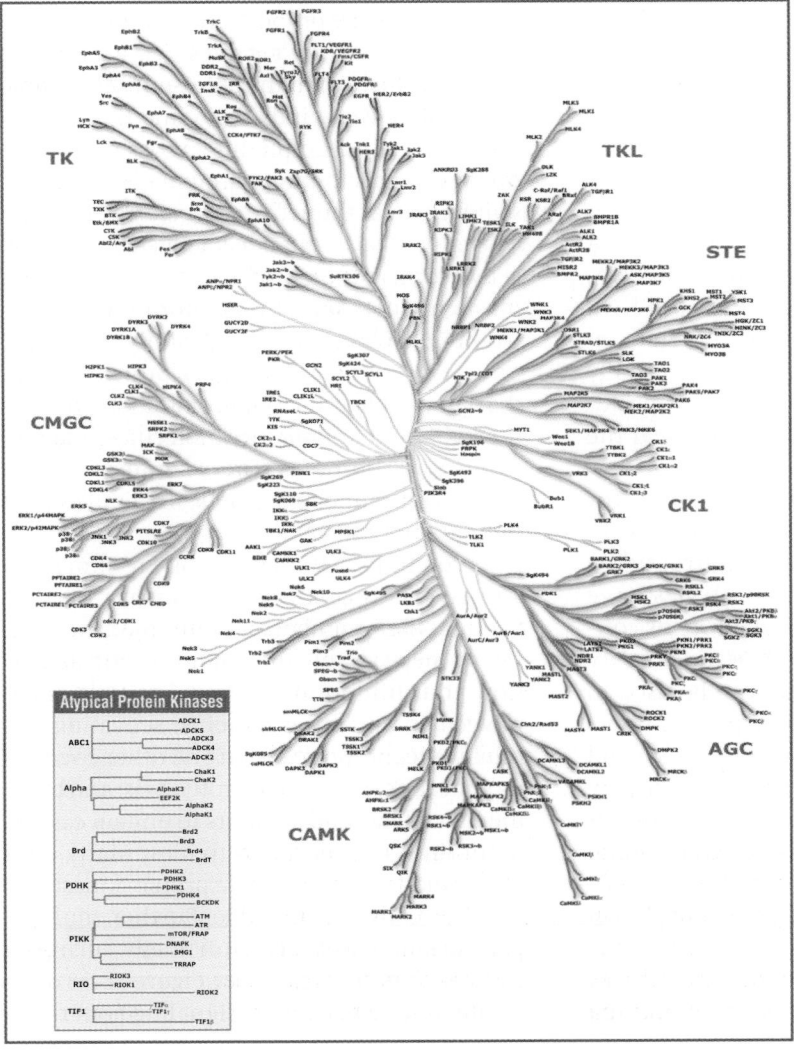

FIGURE 18.30 The protein kinases in the human genome can be grouped according to sequence relationships that reveal seven major branches. The tyrosine kinases are contained within one major branch. The others are Ser/Thr-specific or dual specificity, and are named for the best described members: AGC from PKA, PKG, and PKC; CAMK from the calcium, calmodulin-dependent kinases; CMGC from CDKs, MAPKs, GSK3, Clks; CK1 from casein kinase-1; STE from Ste20, Ste11, and Ste7, the MAP4K, MAP3K, and MAP2K in the yeast mating pathway; and TKL, the Tyr kinase-like enzymes. Kinome illustration reproduced courtesy of Cell Signaling Technology, Inc. (www.cellsignal.com).

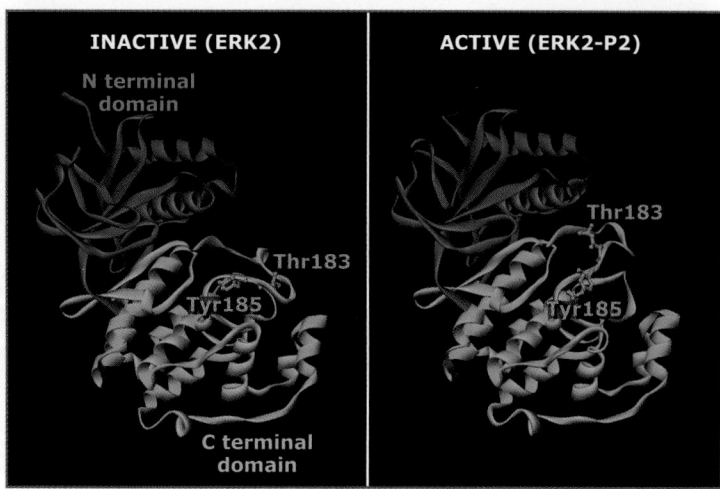

INACTIVE (ERK2)	ACTIVE (ERK2-P2)
N terminal domain	
Thr183	Thr183
Tyr185	Tyr185
C terminal domain	

FIGURE 18.31 The structures of unphosphorylated, inactive MAPK ERK2 and phosphorylated, active ERK2 are compared. ERK2 has a typical protein kinase structure. The smaller N-terminal domain is composed primarily of β strands and the larger C-terminal domain is primarily α-helical. The active site is formed at the interface of the two domains. The activation loop emerges from the active site and is refolded following phosphorylation of the Tyr and Thr residues, inducing the repositioning of active site residues. ATP (not shown) binds in the interior of the active site; productive binding of protein substrates to the surface of the C-terminal domain is also facilitated by the reorganization of the activation loop. Left structure from Protein Data Bank 1ERK. F. Zhang, et al., *Nature* 367 (1994): 704–711. Right structure from Protein Data Bank 2ERK. B. J. Canagarajah, et al., *Cell* 90 (1997): 859–869.

on regulatory mechanisms and substrate specificity. For example, the AGC group is named for its founding members, cAMP-dependent PKA, cyclic GMP-dependent protein kinase (PKG), Ca²⁺, and phospholipid-dependent protein kinase (PKC). These protein kinases are usually regulated by second messengers and prefer substrates that contain basic residues near the phosphorylation site.

In addition to substrate specificity for amino acid residues, most protein kinases are also selective for local sequence surrounding the substrate site. Screening strategies have resulted in methods to predict if proteins contain consensus substrate sites for a wide variety of protein kinases. Antibodies can be used to identify and roughly quantitate protein phosphorylation at specific sites in proteins. Beyond local recognition, protein kinases often display marked substrate selectivity among similar proteins based on overall three-dimensional structure, or among proteins that are differentially covalently modified by phosphorylation or ubiquitination.

In animal cells, some protein kinases are hormone receptors that span the plasma membrane. Some protein kinase receptors are protein serine/threonine kinases, such as the transforming growth factor-β (TGF-β) receptor, but the majority are protein-tyrosine kinases, including receptors for insulin, epidermal growth factor (EGF), platelet-derived growth factor (PDGF), and other regulators of cell growth and differentiation. Other protein kinases are intrinsically soluble intracellular enzymes, although they may bind to one or more organellar membranes.

X-ray crystallographic structures of protein kinases have revealed a wealth of information about their mechanism of activation. The conserved minimum catalytic core of a protein kinase contains about 270 amino acids, yielding a minimum molecular mass of about 30,000 Da. Within this core, there are two folded domains that form the active site at their interface, as shown in FIGURE 18.31. One or both of the conserved lysine (Lys) or aspartate (Asp) residues are required for phosphoryl transfer and they can be mutated to disrupt kinase activity. Protein kinase activity is frequently controlled by the proper juxtaposition of these and other active site residues. For example, α helix C, which contains an essential glutamic acid residue that binds the α and β phosphoryl groups of ATP, often undergoes conformational changes to modulate kinase activity. A sequence near the active site, referred to as the activation loop, is often rearranged to generate active forms of the protein kinases and is the most common site of regulatory phosphorylation in the protein kinase family. There are unique inserts on the surface of protein kinases that generate specificity in localization, interaction with other regulatory molecules, and recognition of substrates. These landmarks allow both classification and genetic manipulation of protein kinases.

Protein kinases have evolved numerous and diverse regulatory mechanisms to complement their number and multiple functions. These mechanisms include allosteric activation and inhibition by lipids, soluble small molecules, and other proteins; activating and inhibitory phosphorylation and other covalent modifications, including proteolysis; and binding to scaffolds and adaptors to enhance activity or limit nonspecific activities. Many such inputs may regulate a single protein kinase in a complex combinatoric code. Further, multiple protein kinases that act sequentially, such as in a protein kinase cascade (see Figure 18.38), can create uniquely complex signaling patterns.

18.25 Two-component protein phosphorylation systems are signaling relays

Key concepts

- Two-component signaling systems are composed of sensor and response regulator components.
- Upon receiving a stimulus, sensor components undergo autophosphorylation on a histidine (His) residue.
- Transfer of the phosphate to an aspartyl residue on the response regulator serves to activate the regulator.

Prokaryotes, plants, and fungi share an alternative mechanism for regulatory phosphorylation and dephosphorylation known as two-component signaling. **FIGURE 18.32** shows a typical two-component system. In this system, the receptor, referred to as a sensor, responds to a stimulus by catalyzing its own phosphorylation on a His residue. Sensors include chemoattractant receptors in bacteria, a regulator of osmolarity in fungi, light-sensitive proteins, the receptor for the plant-ripening hormone ethylene, and other receptors for diverse environmental, hormonal, and metabolic signals. The mammalian mitochondrial dehydrogenase kinases are related in sequence to the bacterial histidine kinases, although the mammalian enzymes phosphorylate serine or threonine residues, not histidine. The phosphorylated sensor next transfers its covalently bound phosphate to an aspartyl residue on a second protein known as a response regulator. Response regulators initiate cellular responses, usually by binding to other cytoplasmic proteins and allosterically regulating their activities.

Although all two-component systems follow this same general pattern, their structures, and precise reaction pathways vary enormously. Some two-component systems are composed of only one protein (sensor and response regulator in a single polypeptide chain). Others are composed of a sensor protein and two aspartyl-phosphorylated proteins, in which the first or the second may display response regulatory activity. Finally, two-component systems usually lack conventional protein phosphatases. Hydrolysis of the aspartyl-phosphate bond may be spontaneous or regulated by the response regulator itself.

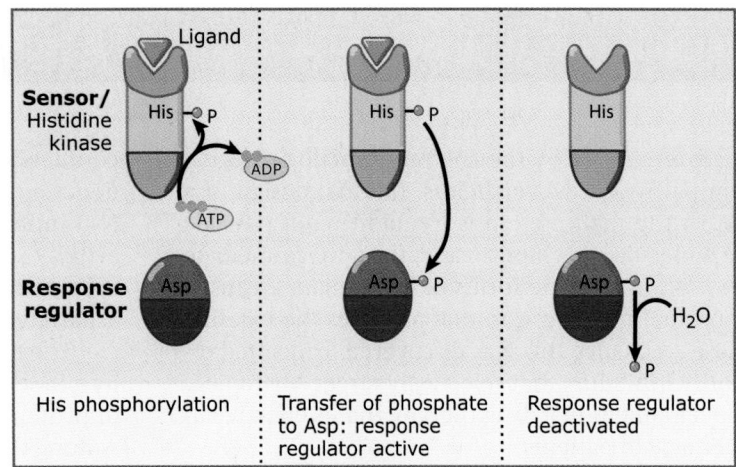

FIGURE 18.32 The basic two-component system is composed of a signal-activated histidine kinase, referred to as a sensor, and an effector protein, the response regulator, that is activated when it is phosphorylated on an aspartate residue by the sensor. The activity of the response regulator is terminated when the aspartyl-phosphate is hydrolyzed.

18.26 Pharmacologic inhibitors of protein kinases may be used to understand and treat disease

Key concepts

- Protein kinase inhibitors are useful both for signaling research and as drugs.
- Protein kinase inhibitors usually bind in the ATP binding site.

Many inhibitors developed for basic research purposes help investigators explore the functions of protein kinases. The importance of these enzymes in disease processes has also made them targets of drug screening projects yielding inhibitors for many protein kinases. The majority of pharmacologic inhibitors of protein kinases compete with ATP binding. Because of the huge number of ATP-binding proteins in a cell, there are inevitable concerns about inhibitor specificity not only with respect to the other protein kinases but also to the other proteins that bind nucleotides. This problem is partially mitigated by variable success through chemical library screening, structure-based modification of lead compounds, and inhibitor testing against panels of protein kinases.

Many inhibitors with actions on PKA or PKCs, for example, have effects on several other members of the AGC family. Although pharma-

Many diseases are caused by the aberrant function of signal transduction pathways. Nevertheless, rational design of a drug to correct the malfunction is frequently not possible because the molecular mechanism and, thus, the immediate target are not known. Some forms of leukemia are caused by a chromosomal translocation that activates the tyrosine protein kinase Abl. Once this was discovered, concerted efforts to identify inhibitors that selectively target Abl rapidly led to the development of a drug therapy that is effective in a large percentage of patients.

Chronic myelogenous leukemia (CML) is a form of cancer that results from an expansion of white blood cells of myeloid lineage. Most cases of CML result from a reciprocal translocation between chromosomes 9 and 22 in a stem cell. This translocation generates a shortened form of chromosome 22, called the Philadelphia chromosome, which is diagnostic for CML in adults. The translocation occurs at the breakpoint cluster region (bcr) of chromosome 22 and the Abl gene on chromosome 9, and creates a Bcr-Abl fusion. The sequence contributed from the Bcr gene constitutively activates the Abl tyrosine kinase activity. In the chronic phase of CML, which may continue for several years, the stem cell with the translocation continues to proliferate. Eventually immature myeloid cells, which normally differentiate to neutrophils, monocytes, macrophages, and mast cells, among others, begin to appear in the circulation. The expansion of these Bcr-Abl-containing cells that do not differentiate eventually causes progression to advanced phases of the disease. Expression of Bcr-Abl in mice recapitulates leukemia, supporting the conclusion that activated Abl tyrosine kinase activity is sufficient to cause the disease. Because the Philadelphia chromosome can be detected by visualizing chromosomes using light microscopy, activation of Abl can be deduced simply and without DNA sequencing.

Inhibitors that block Abl tyrosine kinase activity were identified in a panel of compounds that inhibited the PDGF receptor tyrosine kinase. The most potent of those was imatinib, also known as Gleevec or STI-571. Imatinib, an aminopyrimidine derivative with high affinity for Abl (see **FIGURE 18.B1**), has proven remarkably effective in treating CML patients in the chronic phase of the disease.

Imatinib, like many protein kinase inhibitors, binds in the nucleotide-binding pocket. Because this pocket is one of the essential conserved features among protein kinases, specificity depends on interactions with a few variable residues surrounding the pocket. Characteristics of an effective inhibitor are well illustrated in the crystal structure of the protein kinase domain of Abl bound to an imatinib-related compound shown in the Figure. There are extensive interactions of the inhibitor with residues in the Abl kinase domain. A portion of the inhibitor binds in the site normally occupied by the adenine ring of ATP. In addition, the inhibitor makes contacts with other, less conserved regions of the kinase domain. It extends under α-helix C and interacts with the N-terminal portion of the activation loop stabilizing the low activity state. In this structure of Abl, its activation loop is in an autoinhibitory conformation that pushes an essential aspartate residue out of the active site.

The mechanisms used by imatinib to generate high affinity and selectivity for Abl highlight important general features of protein kinases that make them attractive drug targets. Among the most important is a flexible N-terminal structure that can allow drugs to bind in sites other than or in addition to the nucleotide-binding pocket. Such sites may occur naturally or may be induced by a drug, for example, by pushing an essential structural element away from its necessary orientation with other active site residues. The relative sequence diversity outside of the nucleotide-binding pocket increases the potential selectivity for compounds that bind at other sites. Reorganization of the activation loop is essential to create

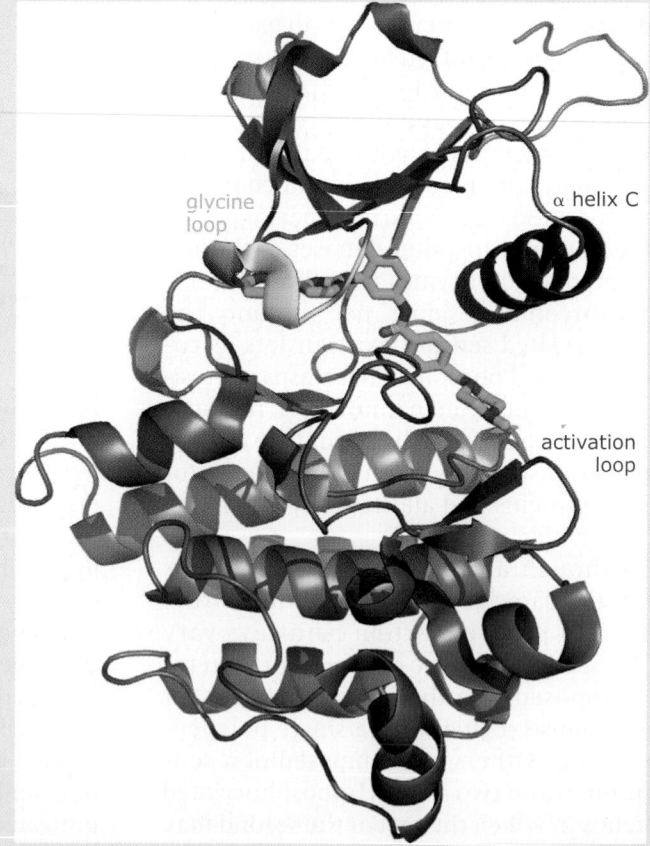

FIGURE 18.B1 The crystal structure of the protein kinase domain of Abl bound to an imatinib-related inhibitor. Structure from Protein Data Bank 1IEP. N. B. Bornmann, et al., *Cancer Res.* 62 (2002): 4236–4243. Figure created by Yu-Chi Juang, UT Southwestern Medical Center.

active conformations of most protein kinases. Inhibitors that lock protein kinases in autoinhibited conformations through interaction with the activation loop may gain selectivity by taking advantage of the sequence diversity among activation loops. Another important feature of several protein kinases is that they bind ATP poorly in their inactive conformations. If a protein kinase conformation binds ATP poorly, an inhibitor that binds to that inactive conformation will compete better with intracellular ATP, which is present at millimolar concentrations. Imatinib is estimated to work 100-fold better on the inactive conformation of Abl.

In contrast to the successful treatment of patients in the chronic phase of CML, patients in advanced phases of the disease frequently develop resistance to imatinib. In many cases, resistance is due to point mutations in the Abl protein kinase domain that decrease drug affinity by interfering with the ability of the inhibitor to bind in the ATP-binding pocket. Second generation inhibitors are effective against many imatinib-resistant mutants of Bcr-Abl, except mutations of threonine 315. Threonine 315 and the homologous residue (frequently methionine) in other protein kinases are often termed the gatekeeper residue due to their impact on the size of what may bind in the ATP site. The side chain of this residue determines the depth of the ATP-binding pocket. A residue at this position with a larger side chain such as isoleucine, the residue often found in Bcr-Abl in drug-resistant CML, prevents inhibitors like imatinib from penetrating deeply into this pocket. In common with Abl, other tyrosine kinases sensitive to imatinib, including the PDGF receptor, have a small gatekeeper residue. Third generation Abl inhibitors are currently being developed that bypass the gatekeeper mutation in drug-resistant Bcr-Abl.

Before imatinib, CML was inevitably a fatal illness. With the discovery of Bcr-Abl and its activated protein kinase activity, the directed drug discovery process rapidly generated inhibitors that successfully treat this disease.

---

cologic inhibitors with effects on PKA abound, the most selective are derived from the naturally occurring small inhibitory protein known as PKI or the Walsh inhibitor. *In vitro* and cell-based screens have identified much more selective inhibitors for MAP2Ks in the ERK1/2 pathway. These inhibitors have fewer known protein kinase cross reactivities, probably due to the fact that they do not bind in the ATP site. Among inhibitors that have progressed in the clinic, compounds developed against the EGF receptor, Abl and certain other protein-tyrosine kinases have had considerable success (see the *Medical Applications* box on pages 38–39).

18.27 Phosphoprotein phosphatases reverse the actions of kinases and are independently regulated

Key concepts

- Phosphoprotein phosphatases reverse the actions of protein kinases.
- Phosphoprotein phosphatases may dephosphorylate phosphoserine/threonine, phosphotyrosine, or all three.
- Phosphoprotein phosphatase specificity is often achieved through the formation of specific protein complexes.

Protein phosphorylation is reversed by phosphoprotein phosphatases. Phosphoprotein phosphatases can be considered in two broad groups based on their specificity and sequence relationships: protein-serine/threonine phosphatases and protein-tyrosine phosphatases.

Most protein-serine/threonine phosphatases are regulated by association with other proteins. Targeted localization is the major determinant of substrate specificity. Phosphoprotein phosphatase 1 (PP1) associates with a variety of regulatory subunits that specifically direct it to relevant organelles. One subunit, for example, specifies association with glycogen particles. The interaction with this subunit is itself regulated by phosphorylation. Small protein inhibitors can suppress PP1 activity.

Phosphoprotein phosphatase 2A (PP2A) is composed of a catalytic subunit, a scaffolding subunit, and one of a large number of regulatory subunits. The regulatory subunit modulates activity and localization of the phosphatase. Some viruses alter the behavior of the cells they infect by interfering with phosphatase activity. For example, cells transformed with the SV40 virus express a viral protein known as small t antigen. Small t displaces the regulatory subunit from PP2A and alters the activity and the subcellular localization of the phosphatase. In addition, natural toxins such as okadaic acid, calyculin, and microcystin inhibit PP2A and PP1 to varying extents both *in vitro* and in intact cells.

Another major protein-serine/threonine phosphatase, called calcineurin (also known as phosphoprotein phosphatase 2B), is regulated by Ca^{2+}-calmodulin (see *18.15 Ca^{2+} signaling serves diverse purposes in all eukaryotic cells*), and plays essential roles in cardiac development and T-cell activation, among other events. The major mechanism of action of the immunosuppressants cyclosporin and FK506 is inhibition of calcineurin.

The protein-tyrosine phosphatases (PTPs) are cysteine-dependent enzymes that utilize a conserved Cys-Xaa-Arg motif to hydrolyze phosphoester bonds in their substrates. The PTPs are encoded by over 100 genes in humans and are classified in four subfamilies: the phosphotyrosine-specific phosphatases, the Cdc25 phosphatases, the dual specificity phosphatases (DSPs), and the low molecular weight phosphatases.

Thirty-eight of the PTPs are highly selective for phosphotyrosine residues within substrates. Some of the phosphotyrosine-selective phosphatases are transmembrane proteins, whereas others are membrane associated. The most obvious function of the PTPs is to reverse the functions of tyrosine kinases; however, some have primary functions in transducing tyrosine kinase signals. For example, the protein-tyrosine phosphatase SHP2 (also known as SHPTP2), binds to certain tyrosine kinase receptors through its SH2 domain and is itself tyrosine phosphorylated, thereby creating a binding site for the SH2 domain-containing adaptor protein, Grb2, which leads to activation of Ras (see *18.32 Mitogen-activated protein kinases are central to many signaling pathways*).

The Cdc25 phosphatases recognize cyclin-dependent kinase (CDK) family members as substrates and play a critical role in increasing CDK activity at key junctures of the cell cycle (see Figure 18.39 and *15.4 A cycle of cyclin-dependent kinases activities regulates cell proliferation*). Similar to the dual specificity kinases, the dual specificity phosphatases are specific for a restricted number of substrates. A number of DSPs dephosphorylate MAPKs; these DSPs are called MAP kinase phosphatases, or MKPs. Several of these are implicated in MAPK nuclear entry and exit. Some MKPs are encoded by early response genes, whose products are active near the initiation of the cell cycle (see *15.7 Entry into cell cycle is tightly regulated*).

Some PTP family members hydrolyze non-protein substrates. The tumor suppressor PTEN hydrolyzes **phosphoinositides**, which are phosphorylated derivatives of the glycerolipid phosphatidylinositol that serve as second messengers (see *18.16 Lipids and lipid-derived compounds are signaling molecules*). Removal of the phosphate group inactivates the second messenger. It remains unclear whether members of this group work exclusively on phophoinositides or also on protein-tyrosine phosphate.

18.28 Covalent modification by ubiquitin and ubiquitin-like proteins is another way of regulating protein function

Key concepts

- Ubiquitin and related small proteins may be covalently attached to other proteins as a targeting signal.
- Ubiquitin is recognized by diverse ubiquitin binding proteins.
- Ubiquitination can cooperate with other covalent modifications.
- Ubiquitination may regulate signaling.

An important mechanism for control of protein function is through covalent modification with small proteins of the **ubiquitin** family. Ubiquitin is one of a family of proteins referred to as ubiquitin-like (Ubl) proteins. Ubiquitin itself is highly conserved among species, suggesting the functional importance of all of its 76 residues. In addition to the long-established role of ubiquitin in initiating protein degradation, ubiquitin modification also has a variety of functions in signal transduction.

Ubl proteins are conjugated to the substrate protein by a peptide-like amide bond between an amino group on the substrate, usually from a Lys side chain, and the C-terminal Gly residue of the processed Ubl protein. E1, E2, and E3 proteins are required to catalyze conjugation to Ubl proteins. Several Ubl proteins may be attached to one substrate, often by serial formation of a polyubiquitin chain. Mono- and polyubiquitination both change the protein's behavior to induce downstream signals. Monoubiquitination is a significant regulatory modification in vesicular trafficking and DNA repair. For example, the monoubiquitinated form of the FANCD2 protein becomes associated with the repair protein BRCA1 at sites of DNA repair. Modification by the Ubl protein SUMO has roles in nuclear transport, transcription, and cell cycle progression, among others.

Polyubiquitin chains are formed when the Lys residues of ubiquitin itself, particularly K48 and K63, are ubiquitinated. Addition of polyubiquitin with a K48 linkage generally directs proteins to the proteasome for degradation, whereas conjugation to polyubiquitin chains with a K63 linkage promotes signal transmission, not proteolysis. Protein-bound ubiquitin is recognized by a variety of ubiquitin-binding domains, including UIM (ubiquitin-interacting motif), UBA (ubiquitin association), and certain zinc finger domains. Such domains have the capacity to act as receptors for ubiquitin within modified proteins.

Activation of the transcription factor NF-κB occurs by a mechanism dependent on modification both by the addition of Ubl proteins and phosphorylation. This fascinating example of regulation by ubiquitin is depicted in **FIGURE 18.33**. Prior to stimulation, NF-κB is retained in the cytoplasm in an inactive form by binding to its inhibitor, IκB. Phosphorylation of IκB by the IκB kinase (IKK) complex promotes its recognition by a multisubunit E3 ligase, which directs its ubiquitination and subsequent proteasomal degradation. Destruction of IκB allows NF-κB to move to the nucleus to mediate changes in transcription.

IκB can be stabilized in response to certain signals through covalent attachment of the Ubl SUMO. Sumoylation occurs on the same Lys residues that must be conjugated to ubiquitin to achieve IκB degradation. Thus, SUMO attachment stabilizes IκB and attenuates NF-κB action. This is one of numerous examples of crosstalk between Ubl conjugates.

A key regulatory event in NF-κB signaling is activation of the IKK complex. IKK is itself regulated by ubiquitination and phosphorylation. The cytokine interleukin-1β causes association of adaptor proteins with its receptor to create a receptor activation complex. The interleukin-1β receptor activation complex recruits another adaptor complex that contains TRAF6. A phosphorylation event releases a TRAF6 complex from the receptor activation complex into the cytoplasm.

TRAF6 contains a RING domain, and is an E3 ubiquitin ligase that catalyzes formation of K63 polyubiquitin chains on the protein kinase TAK1. Polyubiquitinated TAK1 can then recruit

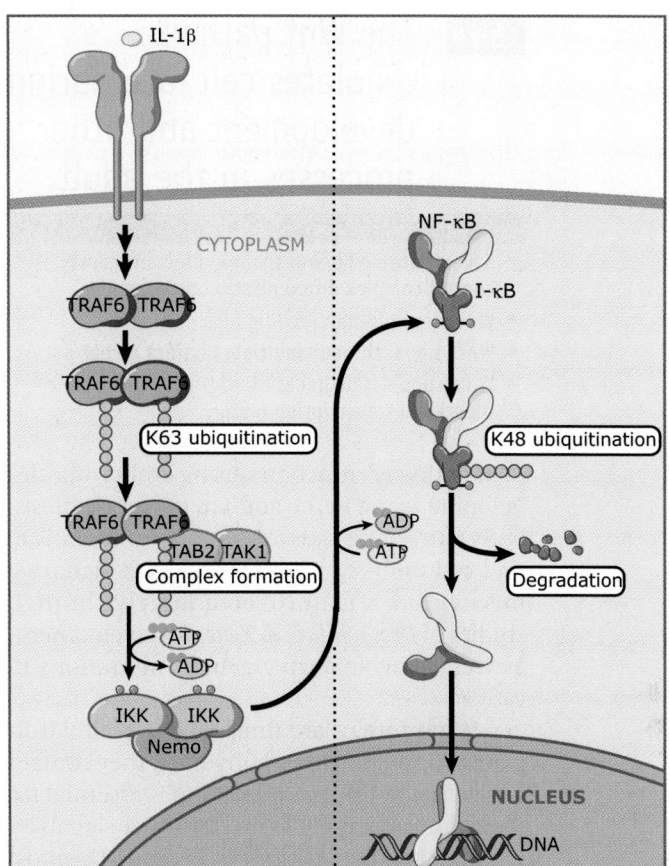

FIGURE 18.33 Activation of NF-κB involves steps dependent on the interaction of proteins attached to ubiquitin through ubiquitin-binding proteins, competition by sumoylation, phosphorylation, and ubiquitin-mediated protein degradation.

TAB2 and TAB3, which are adaptor proteins with conserved zinc finger domains. These particular zinc finger domains bind to polyubiquitinated TAK1 and enhance its activity. TAK1, thus activated, phosphorylates and activates IKK, which then phosphorylates IκB, targeting it for degradation. Thus, ubiquitin-binding domains, such as the TAB2 and TAB3 zinc fingers, may selectively recognize K63 polyubiquitin chains to promote signal transmission.

Naturally occurring small molecules may control ubiquitin ligase activity directly. Auxin (indole 3-acetic acid) is a plant hormone that regulates development by promoting the transcription of a large number of genes. Rather than stimulating transcription factors, however, auxin accelerates the degradation of several specific transcriptional repressors. The auxin receptor is in fact a ubiquitin ligase complex that targets the auxin-regulated transcriptional repressors for proteolysis. F-box proteins account for all of the auxin binding activity in plant extracts.

18.29 The Wnt pathway regulates cell fate during development and other processes in the adult

Key concepts
- Seven transmembrane-spanning receptors may control complex differentiation programs.
- Wnts are lipid-modified ligands.
- Wnts signal through multiple distinct receptors.
- Wnts suppress degradation of β-catenin, a multifunctional transcription factor.

Wnt pathways function during embryonic development and in the adult in morphogenesis, body patterning, axis formation, proliferation, and cell motility. The classical Wnt signaling mechanism was uncovered largely through studies of *Drosophila* and *Xenopus* development, as well as by analyzing genetic alterations in cancer.

Wnt proteins are unusual extracellular ligands. In addition to carbohydrate, they contain covalently bound palmitate that is essential for their biologic activity. Wnts transduce signals by binding to multiple distinct receptors. The most significant are members of the Frizzled family of seven-transmembrane-span receptors.

Wnts regulate the stability of β-catenin, which either is rapidly degraded or, in response to Wnt, is stabilized to enter the nucleus and induce transcription by interacting with T-cell factor. Genes induced include c-*jun, cyclin D1*, and many others.

The coordinated activities of the protein kinases glycogen synthase kinase 3 (GSK3) and casein kinase 1(CK1), the scaffolding proteins axin and adenomatous polyposis coli (APC), and the protein disheveled (DSH) are key to β-catenin stability. In the absence of Wnt, phosphorylation of β-catenin by CK1 and GSK3 promotes its ubiquitination and subsequent destruction by the proteasome. Axin and APC are required for phosphorylation of β-catenin by GSK3.

In contrast to most seven transmembrane-span receptors, the Frizzled family has not yet been shown to have major functions mediated by a heterotrimeric G protein, and G proteins may not be central to this pathway. Instead a proximal step in signaling by Frizzled involves binding to DSH, which inactivates the β-catenin destruction mechanism.

Mutations that cause changes in the amounts of components of the classical Wnt pathway are common in a wide variety of cancers. Both Wnts and β-catenin may be viewed as proto-oncogenes. APC is a tumor suppressor and is mutated in the majority of human colorectal cancers, for example, either too little or too much axin can also disrupt Wnt signaling, and axin, like APC, is a tumor suppressor.

Wnts utilize additional signaling mechanisms. The receptor proteins Lrp5/6 (which are related to the low-density lipoprotein receptor) are Wnt receptors and also bind axin. Wnts bind to tyrosine kinase receptors to influence axon guidance and to other proteins that inhibit their function. Through DSH, Wnts can regulate the JNK MAPK pathway and Rho family G proteins to control planar cell polarity. Certain Wnts increase intracellular calcium to activate calcium-dependent signaling pathways.

18.30 Diverse signaling mechanisms are regulated by protein tyrosine kinases

Key concepts
- Many receptor protein-tyrosine kinases are activated by growth factors.
- Mutations in receptor tyrosine kinases can be oncogenic.
- Ligand binding may promote receptor oligomerization and autophosphorylation.
- Signaling proteins bind to the phosphotyrosine residues of the activated receptor.

Large groups of protein tyrosine kinases are receptors that span the plasma membrane and bind extracellular ligands, as shown in **FIGURE 18.34**. The receptors are generally activated by factors whose normal physiologic functions are to promote growth, proliferation, development, or maintenance of differentiated properties. This group includes receptors for insulin, epidermal growth factor (EGF), and PDGF. These receptors both control the activities of many other protein kinases of all families and directly regulate other classes of signaling proteins.

Because receptor tyrosine kinases have physiologic roles as growth regulators, mutations that activate them are often oncogenic. For example, the oncogene *erbB* results from the mutational loss of the extracellular ligand-binding domain of a kinase closely related to the EGF receptor. This mutation causes constitutive activation of the protein kinase domain.

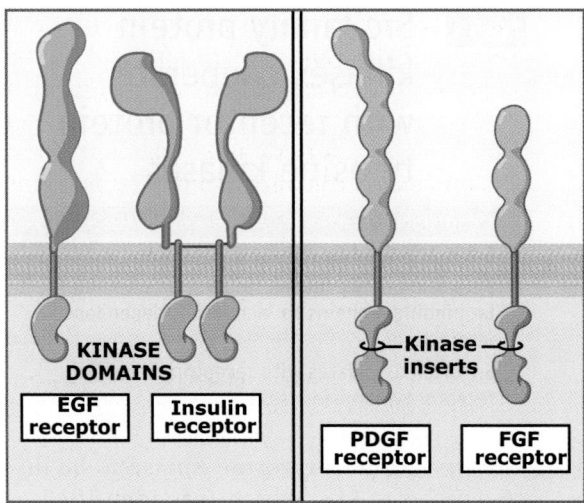

FIGURE 18.34 The monomeric tyrosine kinase receptors consist of a globular extracellular domain that binds ligand, a single transmembrane-span, and a globular intracellular region containing the protein kinase domain. The intracellular regions contain additional sequences preceding, following, and, in the case of the PDGF and FGF receptor groups, inserted into the protein kinase domain. These regions contain sites of tyrosine phosphorylation-dependent interactions. The insulin receptor is encoded by a single gene. The precursor is proteolyzed into α and β subunits, which are disulfide bonded to each other. Disulfide bonds also link two α subunits, yielding an obligate heterotetramer.

Point mutations that affect the transmembrane domain can also cause oncogenic activation, as is found in the EGF receptor-related *neu/HER2* oncogene (see *17.8 Cell growth and proliferation are activated by growth factors*).

Receptor tyrosine kinases are diverse both in their extracellular ligand-binding domains and, with the exception of a conserved tyrosine protein kinase domain, their intracellular regulatory regions. These receptors usually have one membrane span per monomer but some, such as the insulin receptor, which is a disulfide-bonded heterotetramer, have two. Ligand binding to receptor tyrosine kinases favors receptor oligomerization and enhances kinase activity leading to increased Tyr phosphorylation of the intracellular domain of the receptor and of associated molecules. These tyrosine-phosphorylated motifs create docking sites for additional signal transducers and adaptors.

A comparison of the PDGF and insulin receptors reveals common themes and a range of behaviors of receptor tyrosine kinases. The two PDGF receptors are monomeric receptor tyrosine kinases. The insulin receptor exists in two alternatively spliced forms each of which is a heterotetramer of two α and two β subunits. In each case, the receptor isoforms utilize some unique signaling mechanisms.

PDGF and insulin each stimulate the kinase activity of their receptors, causing oligomerization and autophosphorylation. Seven or more sites are phosphorylated on the PDGF receptor, and each phosphotyrosine residue generates a binding site for one or more SH2 domain-containing proteins, as illustrated in **FIGURE 18.35**. The PDGF receptor binds PI 3-kinase, p190 Ras

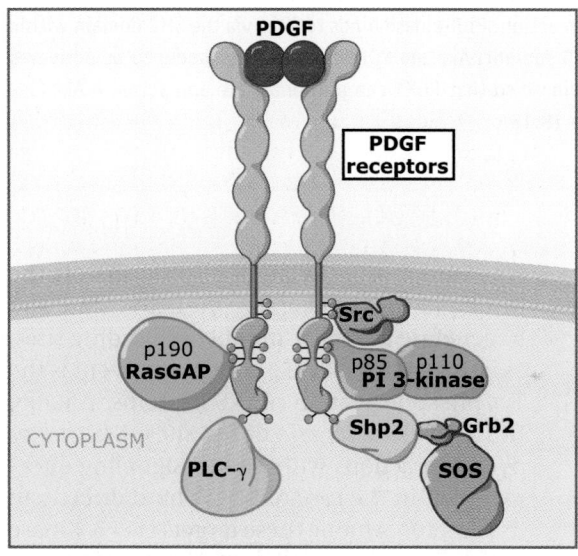

FIGURE 18.35 PDGF binds to its receptor and induces receptor autophosphorylation. The autophosphorylated receptor binds target proteins that contain SH2 domains.

GAP, PLC-γ, Src (which may catalyze additional Tyr phosphorylation of the receptor), and the SHP2 tyrosine phosphatase which itself binds the adaptor Grb2 (see *18.32 Mitogen-activated protein kinases are central to many signaling pathways*). With the exception of Src, all of these proteins are also receptor substrates. Thus, substrates are recruited to the receptor as a consequence of specific interactions of substrate SH2 domains with receptor phosphotyrosine producing changes in activities and distributions of numerous intracellular signal transducers. This array of signaling events leads to increased proliferation of connective tissue during development and in wound healing.

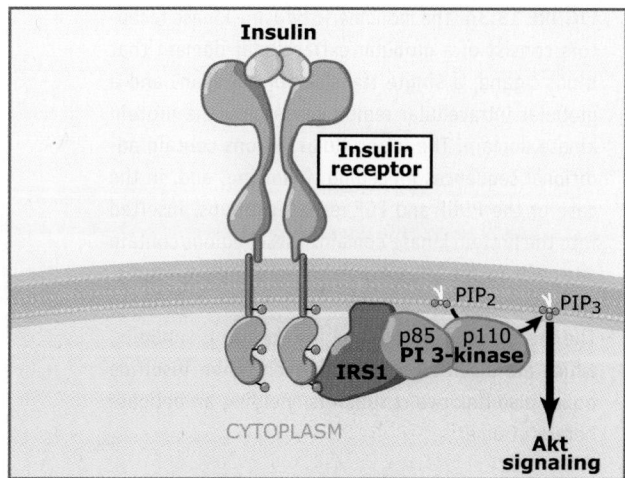

FIGURE 18.36 Insulin binding to its receptor causes activation of the receptor tyrosine protein kinase and autophosphorylation. The receptor kinase also phosphorylates IRS1, a large adaptor with many potential phosphorylation sites. IRS1 is an essential intermediate in insulin action. PI 3-kinase binds to IRS1 via the SH2 domain within its p85 subunit. Akt and PDK1 bind to PIP$_3$ produced by activated PI 3-kinase so that PDK1 can phosphorylate and activate Akt (see Figure 18.18).

Autophosphorylation also occurs on the insulin receptor to stabilize the active state and to generate a smaller number of binding sites, as illustrated in **FIGURE 18.36**. A key event is the Tyr phosphorylation of IRS proteins, notably IRS1, on as many as a dozen sites. IRS1 takes over interactions with several signaling effectors that, in the case of PDGF, bind directly to its receptor. Among these targets is PI 3-kinase which leads to activation of Akt-2 and several essential metabolic actions of insulin (see *18.16 Lipids and lipid-derived compounds are signaling molecules*). IRS proteins are also phosphorylated by serine/threonine protein kinases to modulate their signaling capability.

Tyr phosphorylation often enhances the enzymatic activity of the associated proteins. Other proteins gain enhanced function primarily as a consequence of greater proximity to targets achieved by binding through their SH2 domains to phosphotyrosine sites on either the receptors or the IRS adaptors. The precise actions of the many tyrosine kinase receptors are determined by the overlapping sets of signal transducers with which they interact, as well as by detailed differences in amounts of signal transducers, adaptor accessory proteins, and receptor expression patterns (see Figure 18.43).

18.31 Src family protein kinases cooperate with receptor protein tyrosine kinases

Key concepts

- Src is activated by release of intrasteric inhibition.
- Activation of Src involves liberation of modular binding domains for activation-dependent interactions.
- Src often associates with receptors, including receptor tyrosine kinases.

The first protein-tyrosine kinase to be discovered was Src, which was identified as the transforming entity in the Rous sarcoma virus. Src is the prototype of a number of related enzymes, the Src family kinases. It participates in signaling pathways regulated by numerous cell surface receptors, including those that lack their own kinase domain (see in *18.34 Diverse receptors recruit protein-tyrosine kinases to the plasma membrane*). Src is bound to the plasma membrane via an N-terminal myristoyl group. In the inactive state, Src is phosphorylated on Tyr527, C-terminal to its catalytic domain, by the C-terminal Src kinase.

The structure and regulation of Src is depicted in **FIGURE 18.37**. Phosphorylation of Tyr527 causes it to bind to its own SH2 domain. The SH2 and SH3 domains suppress the kinase activity through interactions on the surface of the protein. The SH3 domain binds to an SH3-binding site distant from the active site. Activation of Src by dephosphorylation of Tyr527 causes its SH2 to dissociate; this causes a conformational change in the SH3 domain to dissociate it from the binding site. Viral isolates of Src are often truncated prior to Tyr527, which increases their activity.

Conformational changes in the kinase domain resulting from dissociation of the SH3 promote Src autophosphorylation on Tyr416 in its activation loop and further increase protein kinase activity. An important consequence of the interaction of Src with its own SH2 and SH3 domains is that these domains cannot bind anything else when in the autoinhibited state; therefore, other interactions are promoted when the SH2 and SH3 domains are released from their associations with the Src kinase domain. The heterologous interactions of the SH2 and SH3 domains contribute to Src localization and signaling.

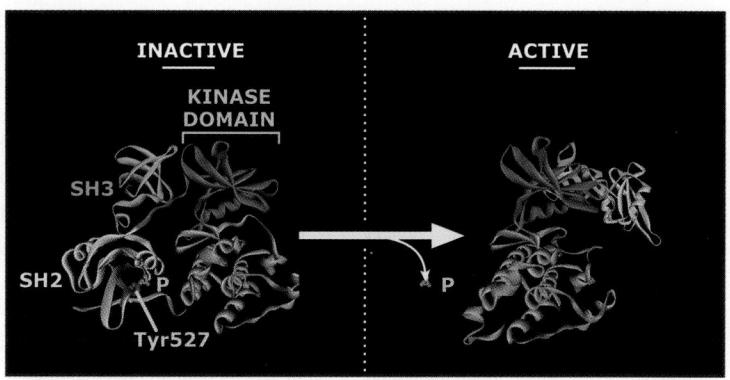

FIGURE 18.37 The structures of inactive and active Src are compared. The inactive protein is autoinhibited by binding to its own SH2 and SH3 domains. The SH2 domain binds to phosphorylated Tyr527. The SH3 domain binds to a noncanonical SH3-binding motif on the opposite side of the kinase domain active site. In contrast to the steric inhibition of PKA caused by its R subunit, inhibition of Src by its SH2 and SH3 domains is allosteric. In the active structure, the SH2 and SH3 domains are not bound to the kinase domain and are available for heterologous interactions. Left structure from Protein Data Bank 1FMK. W. Xu, S. C. Harrison, and M. J. Eck, *Nature* 385 (1997): 595–602. Right structure from Protein Data Bank 1Y57. S. W. Cowan-Jacob, et al., *Structure* 13 (2005): 861–871.

18.32 Mitogen-activated protein kinases are central to many signaling pathways

Key concepts

- MAPKs are activated by Tyr and Thr phosphorylation.
- The requirement for two phosphorylations creates a signaling threshold.
- The ERK1/2 MAPK pathway is usually regulated through Ras.

MAPKs are present in all eukaryotes. They are among the most common multifunctional protein kinases, mediating cellular regulatory events in response to many ligands and other stimuli. MAPKs are activated by protein kinase cascades consisting of at least three protein kinases acting sequentially, as illustrated in **FIGURE 18.38**. Activation of a MAPK is catalyzed by a MAPK kinase (MAP2K), which is itself activated by phosphorylation by a MAP3K. MAP3Ks are activated by a variety of mechanisms including phosphorylation by MAP4Ks, oligomerization, and binding to activators such as small G proteins.

MAP2Ks are activated by phosphorylation on two Ser/Thr residues; MAP2Ks then activate MAPKs by dual phosphorylation on Tyr and Thr residues (see Figure 18.30). Both Tyr and Thr phosphorylations are required for maxi-

mum MAPK enzymatic activity. Each MAP2K phosphorylates a limited set of MAPKs and few or no other substrates. The great specificity of MAP2Ks is one means of insulating MAPKs from activation by inappropriate signals.

Studies on the MAPK ERK2 led to an understanding of the events induced by phosphorylation that are important for increased activity. Conformational changes include refolding of the activation loop to improve substrate positioning and realignment of catalytic residues; this is most obvious in the repositioning of α helix C (see *18.24 Protein phosphorylation/dephosphorylation is a major regulatory mechanism in the cell*).

Amplification occurs moving down the cascade from the MAP3K to the MAP2K step because the MAP2Ks are much more abundant than the MAP3Ks. The MAP2K to MAPK step may also amplify the signal if the MAPK is present in excess of the MAP2K. In addition, the phosphorylation of a MAPK by a MAP2K on a Tyr and a Thr residue creates cooperative activation of the MAPK; this is another mechanism, in addition to those described for PKA and calmodulin, to introduce a threshold and apparently cooperative behavior into the pathway over a narrow range of input signal. This multistep cascade also provides multiple sites for modulatory inputs from other pathways.

Stabilized interactions between components are also important for regulation and

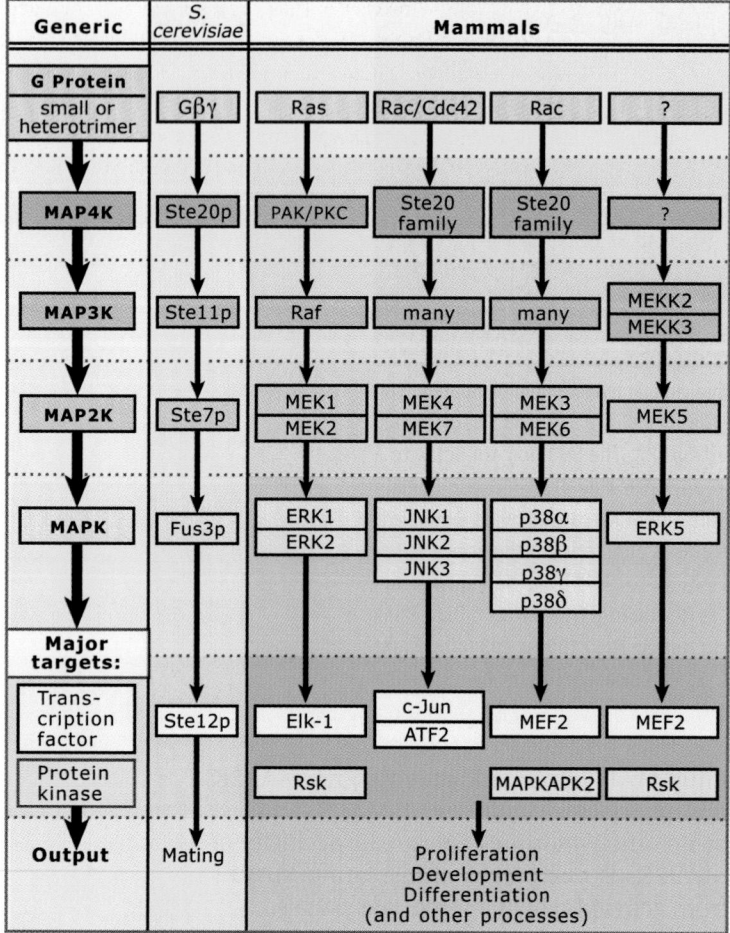

Generic	S. cerevisiae	Mammals			
G Protein small or heterotrimer	Gβγ	Ras	Rac/Cdc42	Rac	?
MAP4K	Ste20p	PAK/PKC	Ste20 family	Ste20 family	?
MAP3K	Ste11p	Raf	many	many	MEKK2 MEKK3
MAP2K	Ste7p	MEK1 MEK2	MEK4 MEK7	MEK3 MEK6	MEK5
MAPK	Fus3p	ERK1 ERK2	JNK1 JNK2 JNK3	p38α p38β p38γ p38δ	ERK5
Major targets:					
Transcription factor	Ste12p	Elk-1	c-Jun ATF2	MEF2	MEF2
Protein kinase		Rsk		MAPKAPK2	Rsk
Output	Mating	Proliferation Development Differentiation (and other processes)			

FIGURE 18.38 MAPK pathways can be regulated by a diverse group of upstream regulatory mechanisms that often include adaptors, small G proteins, and MAP4Ks. These molecules impinge on the activities of MAP3Ks. MAP3Ks regulate one or more MAP2Ks depending on localization and scaffolding. The MAP2Ks display great selectivity for a single MAPK type. MAPKs have overlapping and unique substrates and participate in signaling cascades leading to many cellular responses.

activity of the cascade. MAP2Ks and MAPK phosphatases usually contain a basic/hydrophobic docking motif that interacts with acidic residues and binds in a hydrophobic groove on the MAPK catalytic domain. Many MAPK substrates also contain docking motifs that bind to MAPKs outside of the enzyme active site. Additional components including scaffolds are necessary for the efficient activation of MAPK cascades in cells and usually have additional functions. Several scaffolds bind to two or more components for each of the three major MAPK cascades, the ERK1/2, JNK1-3, and p38 α, β, γ, and δ cascades. Scaffolds and other binding proteins may escort MAPKs to sites of action.

The ERK1/2 pathway is regulated by most cell surface receptors, including receptors that employ tyrosine kinases, GPCRs, and others.

The PDGF receptor, like most receptor systems, activates the ERK1/2 cascade through Ras. PDGF stimulates autophosphorylation of its receptor and the subsequent association of effectors with its cytoplasmic domain (see *18.30 Diverse signaling mechanisms are regulated by protein tyrosine kinases*). In response to PDGF, ERK1/2 promotes cell proliferation and differentiation by phosphorylation of membrane enzymes, proteins involved in determining cell shape and motility, and regulatory factors that control transcription.

Concept and Reasoning Check

1. Pathogenic *Yersinia* infects eukaryotic cells and inactivates the host cell immune response. Several *Yersinia* effector proteins contribute to virulence. One of these is YopJ. YopJ has serine/threonine acetylase activity and specifically catalyzes acetylation of the serine/threonine residues in several MAP2Ks that are phosphorylated to activate them. Predict possible effects of YopJ on MAPK activities.

2. Anthrax lethal factor is an effector protein of *Bacillus anthracis*. Lethal factor is a protease that cleaves proteins within basic/hydrophobic motifs. Among the most significant targets of lethal factor are several MAP2Ks. The cleavage site is in a basic/hydrophobic motif that is required for docking to MAPKs. Cleavage does not inhibit the catalytic activity of the MAP2Ks. What effect will infection with anthrax have on MAPK activity?

3. Compare and contrast the potential effects on signaling of infection with *Yersinia* and anthrax.

18.33 Cyclin-dependent protein kinases control the cell cycle

Key concepts

- The cell cycle is regulated by CDKs.
- Activation of CDKs involves protein-binding, dephosphorylation, and phosphorylation.

Cell division is regulated positively and negatively by factors that stimulate proliferation and inputs that monitor cell state. The sum of these factors is integrated in the regulation of CDKs. CDKs are protein serine/threonine kinases that are major regulators of cell cycle progression. Most CDKs are regulated both by kinases and phosphatases and by association with other proteins called **cyclins**. Cyclins are synthesized and degraded during every cell cycle. Because most CDKs are dependent upon cyclin binding for activation, the timing

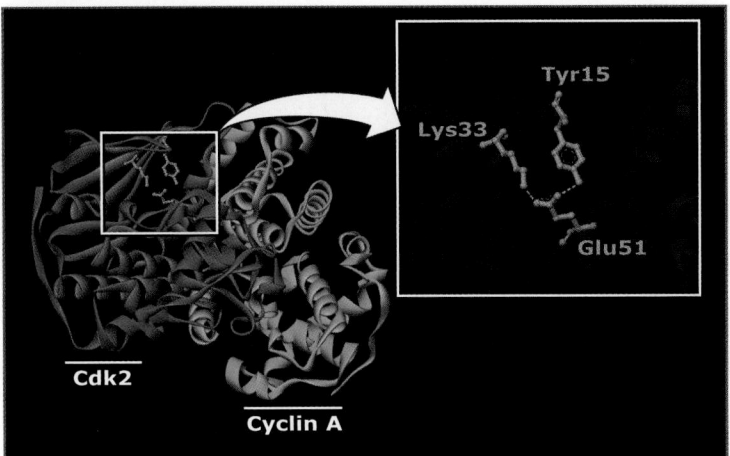

FIGURE 18.39 The view of the crystal structure of CDK2 bound to cyclin A shows residues in the ATP-binding site. The enlargement on the right shows the interaction between Lys33 and Glu51, catalytic residues that interact with ATP to promote phosphoryl transfer. Tyr15 is phosphorylated in inactive forms of CDK2. A phosphoryl group on Tyr15 inhibits CDK activity by interfering with ATP binding. Structure from Protein Data Bank 1JST. A. A. Russo, P. D. Jeffrey, and N. P. Pavletich, *Nat. Struct. Biol.* 3 (1996): 696–700.

of the synthesis and degradation of individual cyclins determines when a CDK will function. An exception is Cdk5, the most notable non-cycling member of the CDK family, which is highly expressed in terminally differentiated neurons. Cdk5 binds the noncyclin protein p35 as its activating subunit.

Cdc2 is a major cell cycle kinase in both mammals and yeast. The first step in regulation of Cdc2 is the association with cyclin. A second step required for activation of Cdc2 is phosphorylation of a Thr residue in its activation loop by another CDK type kinase. In spite of its association with cyclin, this form of Cdc2 is not yet active due to inhibitory phosphorylation of Tyr and Thr residues in the ATP-binding pocket. Release of inhibition by dephosphorylation of the residues in the ATP pocket is catalyzed by the Cdc25 family of phosphoprotein phosphatases, resulting in activation of Cdc2. The proximity of the Tyr residue to catalytic residues is shown in **FIGURE 18.39**. The complexity of activation of CDKs makes possible the imposition of cell cycle checkpoints. For more on CDKs and cyclins see *15.5 Cyclin-dependent kinases-cyclin complexes are regulated in several ways.*

18.34 Diverse receptors recruit protein-tyrosine kinases to the plasma membrane

Key concepts

- Receptors that bind protein tyrosine kinases use combinations of effectors similar to those used by receptor tyrosine kinases.
- These receptors often bind directly to transcription factors.

Many receptors act through protein-tyrosine kinases, but their cell surface receptors lack kinase activity. Instead, these receptors act by recruiting and activating protein-tyrosine kinases at the plasma membrane. In this group of receptors are integrins, which are key molecules involved in cell adhesion, growth hormone receptors, and receptors that mediate inflammatory and immune responses. While their structures vary enormously, their mechanisms of action are related.

Integrins are receptors whose major function is to attach cells to the extracellular matrix. They also mediate some interactions with proteins on other cells, as depicted in **FIGURE 18.40**. Ligands for integrins include a number of extracellular matrix proteins, such as fibronectin, as well as cell surface proteins that cooperate in cell-cell interactions. Integrin ligation provides cells with information about their environment that influences cell behavior. Ligation of integrins initiates signals that control cell programs, including cell cycle entry, proliferation, survival, differentiation, changes in cell shape, and motility, as well as fine-tuning responses to other ligands. For more details on integrins see *19.13 Most integrins are receptors for extracellular matrix proteins* and *19.14 Integrin receptors participate in cell signaling.*

Talin and α-actinin are among cytoskeletal proteins that bind certain integrin subunits and link integrins to complex cytoskeletal structures known as **focal adhesions**. Focal adhesions connect the cytoskeleton to signal transduction cascades that coordinate states of cellular attachment with regulation of cellular responses. Focal adhesion complexes contain the focal adhesion kinase FAK, which is activated by integrin ligation. Autophosphorylation of FAK recruits signaling proteins containing

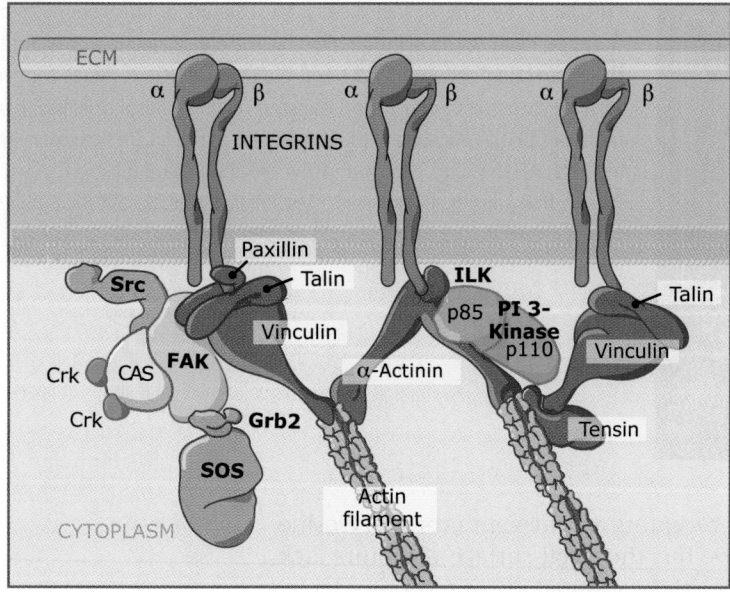

FIGURE 18.40 Integrins bind to an array of cytoplasmic proteins to regulate the cytoskeleton and intracellular signaling pathways. The associated cytoskeletal elements include actin filaments and focal adhesion proteins α-actinin, vinculin, paxillin, and talin. Signaling molecules include the focal adhesion kinase FAK; the adaptors Cas, Crk and Grb2; Src and CSK (see *18.31 Src family protein kinases cooperate with receptor protein tyrosine kinases*), PI 3-kinase (see *18.16 Lipids and lipid-derived compounds are signaling molecules*); and the Ras exchange factor SOS. Stimulation of GTP binding of Ras by SOS leads to activation of the MAPK pathway (see *18.32 Mitogen-activated protein kinases are central to many signaling pathways*).

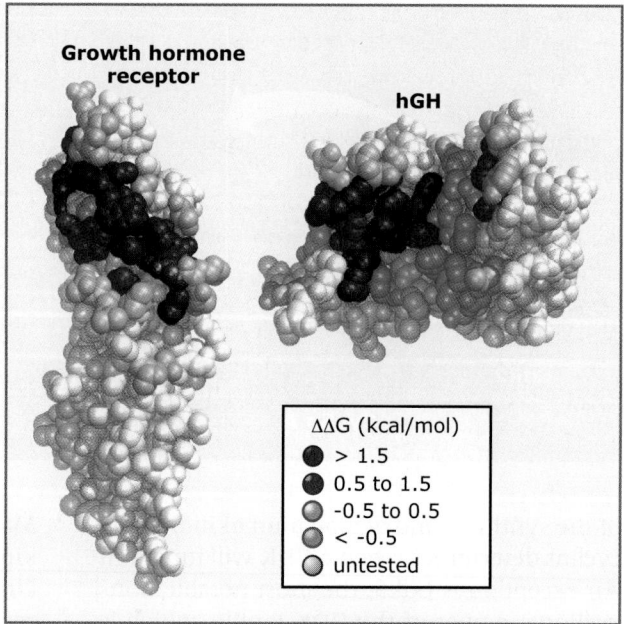

FIGURE 18.41 Proteins often interact over a large surface area. Growth hormone binding to its receptor is an example of the energy of binding coming primarily from a small number of the contacts between the two proteins, creating an interaction hot spot. The complex of growth hormone bound to the growth hormone receptor-binding domain determined by crystallography has been peeled apart in this figure to show the binding energy associated with residues in the binding interface from each protein determined by mutagenesis and binding studies. Fewer than half of the residues in the interface contribute the majority of binding energy. Reproduced from T. Clackson and J. A. Wells, *Science* 267 (1995): 383–386 [http://www.sciencemag.org]. Reprinted with permission from AAAS. Photos courtesy of Tim Clackson, ARIAD Pharmaceuticals, Inc.

SH2 domains, especially the p85 subunit of PI 3-kinase and Src family protein kinases. The signaling molecules associated with the integrin-bound cytoskeletal proteins, whether focal adhesions or other structural complexes, mediate the diverse actions of integrins. The association of cytoskeletal proteins with integrin receptors also causes functional changes to the receptors.

Signals that act over a distance, such as hormones, can also employ nonreceptor tyrosine kinases to transmit their message inside a cell. Growth hormone (GH) is a protein hormone secreted by the anterior pituitary gland that regulates bone growth, fat metabolism, and other cellular growth phenomena. Absence of growth hormone results in short stature, whereas hypersecretion causes acromegaly, a form of gigantism. The GH receptor

is a member of the cytokine receptor family, which includes receptors for prolactin, erythropoietin, leptin, and interleukins. All these receptors display similar biochemical functions, such as association with members of the JAK/TYK family of protein-tyrosine kinases, but select for different but overlapping sets of cytoplasmic signaling proteins. Signal transduction by the GH receptor provides a model for receptors that lack enzymatic function and act as agonist-promoted scaffolds for intracellular signaling proteins.

FIGURE 18.41 shows the structure of growth hormone bound to the extracellular domain of its receptor. The majority of binding energy comes from only a small number of residues in the binding interface. Inside the cell, signaling by the GH receptor depends significantly on its association with the cytoplasmic tyrosine protein kinase, Janus kinase 2 (JAK2). **FIGURE 18.42** shows that JAK2 binds to a proline-rich region of the receptor. Ligand binding induces receptor dimerization, which then promotes activation of JAK2 through intermolecular autophosphorylation.

GH signaling is thus mediated primarily by initiating Tyr phosphorylation. In addition to JAK2 autophosphorylation, the receptor itself becomes Tyr phosphorylated. As is true for receptor tyrosine kinases, Tyr phosphorylation of the growth hormone receptor creates binding sites for signaling proteins that contain phosphotyrosine-binding domains. Primary targets are transcription factors known as *s*ignal *t*ransducers and *a*ctivators of *t*ranscription, or STATs. STATs contain SH2 domains and bind Tyr-phosphorylated motifs on the growth hormone receptor. While receptor bound, STATs are Tyr phosphorylated by JAK2 and then released to travel to the nucleus to mediate changes in transcription.

The growth hormone receptor and the associated JAK2 also activate other signaling pathways. For example, the adaptor Shc is Tyr phosphorylated by JAK2. Engagement of Shc leads to activation of Ras and the ERK1/2 MAPK pathway. Adaptors specialized for insulin-signaling pathways, IRS 1, 2, and 3, are also growth hormone targets, perhaps reflecting the ability of growth hormone to induce certain insulin-like metabolic actions.

Feedback circuits are also engaged during GH signaling. The growth hormone receptor complex binds the adaptor SH2-B, which has a stimulatory effect on growth hormone signaling. On the other hand, suppressors of cytokine signaling (SOCS proteins) are among the genes whose transcription is induced by growth hormone. As the name indicates, SOCS proteins inhibit cytokine signaling, in some if not all cases, by inhibiting the activity of JAK2. SOCS proteins contain an SH2 domain that facilitates their binding either to phosphorylated JAK2 or cytokine receptors. The mechanism of signaling inhibition may differ among SOCS proteins because some require the GH receptor to interfere with JAK2 signaling. SOCS-1, on the other hand, binds directly to the JAK2 activation loop and does not require a receptor to inhibit JAK2 activity. This mechanism may be particularly important in GH signaling because, in contrast to the ligand-induced downregulation mechanisms controlling many receptors, the GH receptor is degraded in a ligand-independent manner.

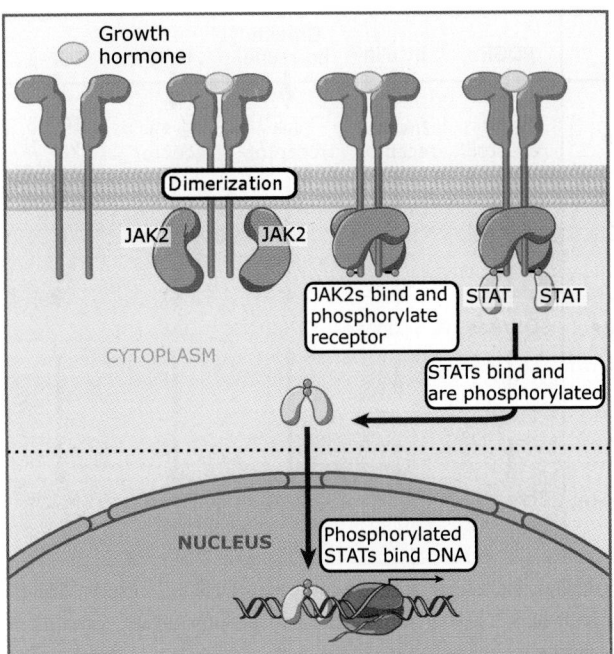

FIGURE 18.42 The growth hormone receptor binds to JAK2. Many GH signals are mediated by Tyr phosphorylation of the receptor by JAK2, which creates binding sites for signaling molecules with SH2 domains, notably STATs. STATs then enter the nucleus to cause changes in gene transcription.

Receptors for cytokines also act by recruiting tyrosine kinases. The cytokines—signaling proteins that modulate inflammation and cell growth and differentiation—include interleukins, leukemia inhibitory factor, oncostatin M, cardiotrophin-1, cardiotrophin-like cytokine, and ciliary neurotrophic factor (CNTF). Each cytokine binds a unique receptor, but each receptor binds a transmembrane protein called gp130. Mechanisms of signaling by gp130 involve interactions with tyrosine kinases of the JAK/TYK types and transcription factors in the STAT family. This mechanism is similar to those employed by the growth hormone receptor.

Unlike many cytokine receptors in this class, the CNTF receptor does not itself span the membrane. Instead, it is **glycosylphosphatidylinositol (GPI)**-linked to the outer face of the plasma membrane. The GPI linkage is a covalent bond, and the receptor can be released into the extracellular fluid by a specific phospholipase. The freed receptor may interact with membranes of other cells to induce signals.

The use of a common signal transducing subunit, gp130, suggests that unique mechanisms exist to create ligand-specific responses; under some circumstances competition among ligand-binding subunits for interaction with

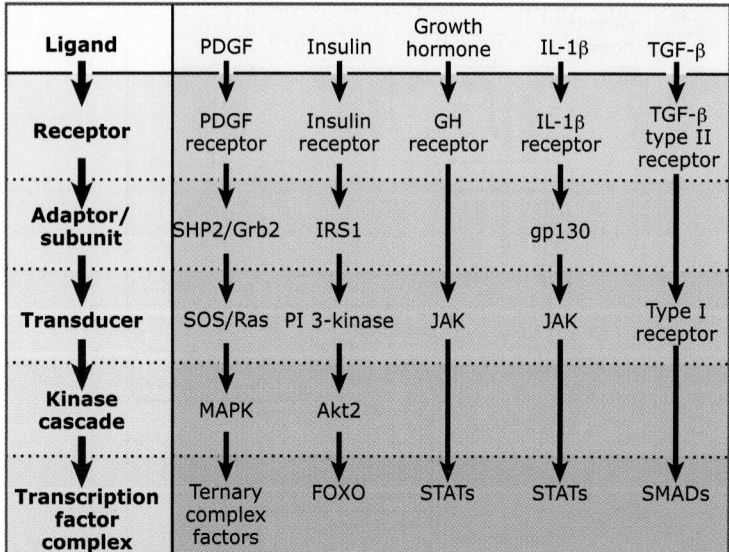

Ligand	PDGF	Insulin	Growth hormone	IL-1β	TGF-β
Receptor	PDGF receptor	Insulin receptor	GH receptor	IL-1β receptor	TGF-β type II receptor
Adaptor/ subunit	SHP2/Grb2	IRS1		gp130	
Transducer	SOS/Ras	PI 3-kinase	JAK	JAK	Type I receptor
Kinase cascade	MAPK	Akt2			
Transcription factor complex	Ternary complex factors	FOXO	STATs	STATs	SMADs

FIGURE 18.43 Major signaling cascades controlled by PDGF, insulin, TGF-β, IL-1β, and growth hormone are compared. Each receptor either contains or interacts with a protein kinase that associates with or recruits a transducer. The transducer regulates downstream effectors either directly or through an intermediate protein kinase cascade. The effectors shown are transcriptional regulators. Phosphorylation by the transducer or kinase cascade activates all of the effectors except the FOXO proteins, which may be excluded from the nucleus by phosphorylation. The table only shows snapshots of much more complex signaling networks controlled by these ligands. Many of these and other intermediates serve multiple ligands. For example, IRS proteins also contribute to growth hormone and IL-1β signaling, and MAPK pathways are regulated by all of these ligands.

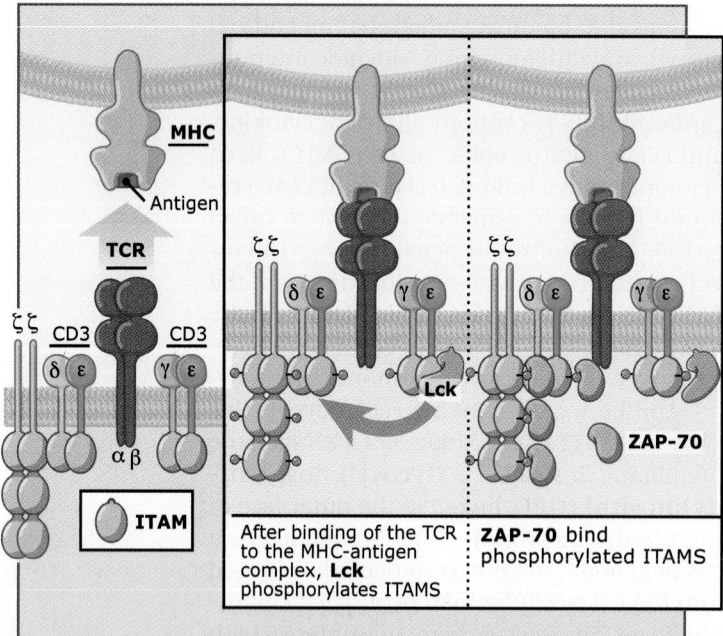

FIGURE 18.44 The T-cell receptor (TCR) is a multisubunit receptor. It is phosphorylated on activation motifs or ITAMs by Lck, or a related Src family protein kinase. The phosphorylated residues create binding sites for another tyrosine protein kinase ZAP-70. ZAP-70 then recruits other signaling molecules to the complex including phospholipase Cγ, PI 3-kinase, and a Ras exchange factor to activate downstream signaling pathways.

the gp130 signal transducer may influence signaling outcomes. FIGURE 18.43 illustrates some parallels in signaling pathways initiated by receptors with associated or intrinsic protein kinases.

The multiprotein T-cell receptor (TCR) is among the most complex of the receptors that act by recruiting yrosine protein kinases, as shown in FIGURE 18.44. It is found uniquely on T lymphocytes and is responsible for the ability of these cells to recognize and respond to specific antigens. The TCR is composed of eight subunits that can be described as an assembly of four dimers, αβ, γε, δε, and ζζ. The α and β subunits, which are different for each cell, determine the specificity of antigen recognition; the remaining subunits are invariant in TCRs. The γ, δ, and ε subunits, referred to as the CD3 complex, are similar in sequence to one another. The ζ chain, unlike the other subunits, appears on certain other cell types and can also be a component of other receptors, such as the Fc receptor, which binds a portion of certain immunoglobulins.

The immunoreceptor tyrosine-based activation motif (ITAM), which features closely spaced pairs of Tyr residues, is key to signaling by the TCR. The CD3 subunits each contain one ITAM and the ζ chain contains three ITAMs, for a total of ten motifs in each TCR. Engagement of the TCR causes the Src family kinases Lck and Fyn to phosphorylate the pairs of Tyr residues in the ITAMs. Next, another protein kinase called ZAP-70 binds these phosphorylated ITAMs through its tandem SH2 domains, which promotes Tyr phosphorylation of ZAP-70 by Src. Phosphorylated Tyr sites on ZAP-70 bind to other adaptors and signaling molecules, and Tyr phosphorylation by ZAP-70 activates additional signal transducers. The sum of these events leads to the downstream responses of T cells to antigen engagement, which include cell cycle progression and the elaboration of cytokines that modulate activities of other immune cells.

18.35 What's next?

New signaling proteins and new regulatory interactions seem to show up every day. The challenge now is to understand how cells organize these proteins and their individual interactions to create adaptable information-processing networks. How do cells use simple chemical reactions to sort and integrate multiple simultaneous inputs and then direct this information to diverse effector machinery? How do they interpret the inputs in the context of their growth and metabolic activities? In principle, three areas of research have to contribute to allow us to understand integrative cellular signaling.

First, we need real-time, noninterfering biosensors to measure intracellular signaling reactions. Most current sensors use combinations of fluorescent moieties and signal-binding protein domains to provide fast optical readouts. For many pathways, several reactions can be monitored within cells over subsecond time scales. We need more, better, and faster sensors and sensors that can report with single-cell and subcellular resolution. Genetically encoded sensors will be complemented by synthetic molecules.

Our ability to manipulate signaling networks is also improving dramatically but still falls short. We can manipulate signaling networks by overexpression, knockout, and knockdown of genes, but signaling pathways are wonderfully adaptive and frequently circumvent our best efforts to control them. We still need chemical regulators that can act quickly and selelctively in cells. Structure-based design of such regulatory molecules will be vital. Regulation of signaling by photochemistry is even faster, and fiber optics allows such manipulation of signaling molecules even in intact animals.

Last, our ability to analyze the behavior of signaling networks depends on our ability to measure and interpret signaling quantitatively. It is ironic but true that really complex systems cannot be described without explicit quantitative models for how they work. Computational modeling and simulation of signaling networks requires both better theoretical understanding of network dynamics and better algorithmic implementation.

The goal is to understand how cells think.

18.36 Summary

Signal transduction encompasses mechanisms used by all cells to sense and react to stimuli in their environment. Cells express receptors that recognize specific extracellular stimuli, including nutrients, hormones, neurotransmitters, and other cells. Upon receptor-binding, signals are converted to well-defined intracellular chemical or physical reactions that change

the activities and the organization of protein complexes within cells. The changes directed by the stimuli lead to altered cell behavior. The behavior of the cell is determined then by its intracellular state and the integrated information from extracellular stimuli so that the appropriate responses are achieved.

The basic biochemical components and processes of signal transduction are conserved throughout biology. Families of proteins are used in a variety of ways for many different physiologic purposes. Cells often use the same series of signaling proteins to regulate multiple processes, such as transcription, ion transport, locomotion, and metabolism.

Signaling pathways are assembled into signaling networks to allow the cell to coordinate its responses to multiple inputs with its ongoing functions. It is now possible to discern conserved reaction sequences in and between pathways in signaling networks that are analogous to devices within the circuits of analog computers: amplifiers, logic gates, feedback and feed-forward controls, and memory.

http://biology.jbpub.com/lewin/cells

To explore these topics in more detail, visit this book's Interactive Student Study Guide.

References

Alon, U. 2007. *An Introduction to Systems Biology. Design Principles of Biological Networks*. Boca Raton, FL: Chapman and Hall/CRC Press.

Downward, J. 2004. PI 3-kinase, Akt and cell survival. *Semin. Cell Dev. Biol.* v. 15 p. 177–182.

Heasman, S. J., and Ridley, A. J. 2008. Mammalian Rho GTPases: New insights into their functions from in vivo studies. *Nat. Rev. Mol. Cell Biol.* v. 9 p. 690–701.

Manning, G., Whyte, D. B., Martinez, R., Hunter, T., and Sudarsanam, S. 2002. The protein kinase complement of the human genome. *Science* v. 298 p. 1912–1934.

Milo, R., Shen-Orr, S., Itzkovitz, S., Kashtan, N., Chklovskii, D., and Alon, U. 2002. Network motifs: Simple building blocks of complex networks. *Science* v. 298 p. 824–827.

Pawson, T., and Nash, P. 2003. Assembly of cell regulatory systems through protein interaction domains. *Science* v. 300 p. 445–452.

Raman, M., Chen, W., and Cobb, M. H. 2007. Differential regulation and properties of MAPKs. *Oncogene* v. 26(22) p. 3100–3112.

Ross, E. M., and Wilkie, T. M. 2000. GTPase-activating proteins for heterotrimeric G proteins: Regulators of G protein signaling (RGS) and RGS-like proteins. *Annu. Rev. Biochem.* v. 69 p. 795–827.

Sebolt-Leopold, J. S., and English, J. M. 2006. Mechanisms of drug inhibition of signalling molecules. *Nature* v. 441 p. 457–462.

Smith, C. L., and O'Malley, B. W. 2004. Coregulator function: A key to understanding tissue specificity of selective receptor modulators. *Endocr. Rev.* v. 25 p. 45–71.

Unwin, N. 2005. Refined structure of the nicotinic acetylcholine receptor at 4Å resolution. *J. Mol. Biol.* v. 346–967.

Yang, C., and Kazanietz, M. G. 2003. Divergence and complexities in DAG signaling: Looking beyond PKC. *Trends Pharmacol. Sci.* v. 24 p. 602–608.

The extracellular matrix and cell adhesion

19

George Plopper
Rensselaer Polytechnic Institute, Troy, NY

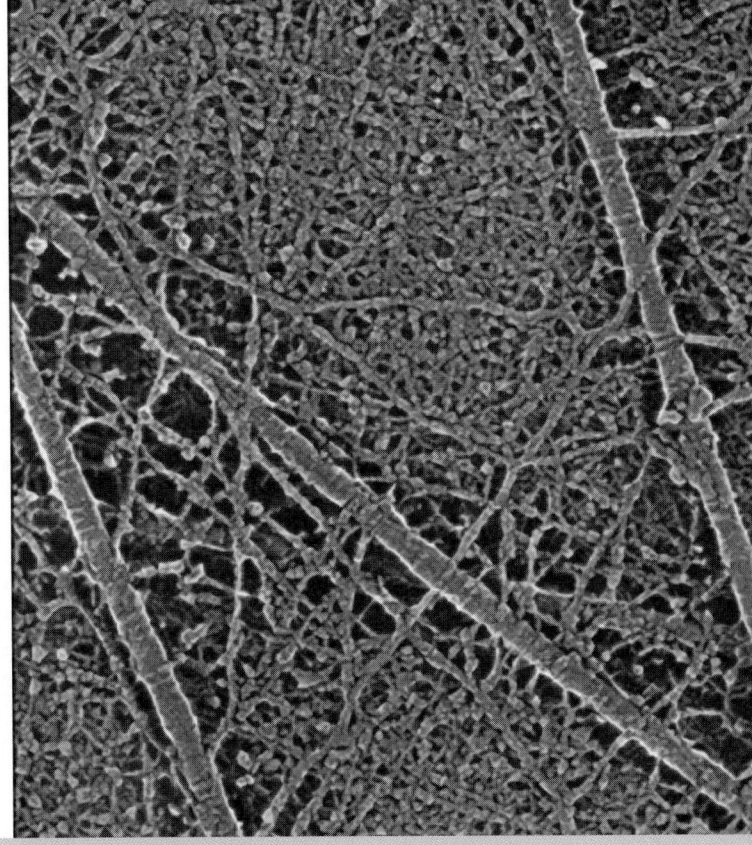

SCANNING ELECTRON MICROGRAPH OF BASAL LAMINA. Photo courtesy of John Heuser, Washington University School of Medicine.

CHAPTER OUTLINE

19.1 Introduction

Key concepts

- The evolution of multicellularity required functional, cooperative interactions between groups of cells.
- The extracellular matrix (ECM) is a dense network of molecules that lies between cells in a multicellular organism and is made by the cells within the network.
- Cells express receptors for ECM molecules.
- Cell-cell junctions are specialized protein complexes that allow neighboring cells to adhere to and communicate with one another.
- Nonjunctional adhesion takes place via receptors that do not form large junctional complexes.
- All cell-ECM and cell-cell receptors contribute to signaling pathways linking cells.
- The molecules in the ECM and their receptors, plus proteins in cell-cell junctions control the three-dimensional organization of cells in tissues as well as the growth, movement, shape, and differentiation of these cells.

One of the most important events in the evolution of life on earth was the advent of multicellularity. Current theories suggest that the evolution of multicellular life occurred in at least six stages: (1) formation of a cell group, (2) increase in cooperation between different cell groups, (3) evolution of "conflict mediator" mechanisms to protect against "cheater (selfish) cells" in the groups, (4) increases in group size, (5) specialization of cells within each group, and (6) spatial organization of specialized cell types in the groups. The evolution from unicellular to multicellular organisms required even more time than the appearance of unicellular life on the planet. While one billion years passed between the formation of earth and the appearance of the first unicellular organisms, more than a billion years of unicellular evolution transpired before the ancestors of today's multicellular organisms appeared. This observation alone strongly suggests that successful navigation of all six steps to multicellularity is a terribly complicated process.

Fortunately, rudiments of this evolutionary journey still exist to help us visualize what the earliest multicellular organisms looked like. One excellent example is the tremendous variation in the group of green algae known as Chlorophyta. When one examines organisms in this group representing different stages of multicellularity, as shown in **FIGURE 19.1**, three emergent properties of multicellular lifeforms are recognizable. The first is the formation of stable bonds between neighboring cells. To function effectively as a group, each cell in a cluster must be able to maintain contact with, and/or close proximity to, other cells in the group for a relatively long time. The second property is that, as the number of cells in an organism increases, the material that fills the extracellular space plays an increasingly important role in governing the location and behavior of the cells. For example, the small cluster of cells formed by *Gonium pectoral*, shown in Figure 19.1B, adopts a fairly simple "grape cluster" shape with very little space in between cells, while cells in the more complex *Volvox* species shown in Figure 19.1E and Figure 19.1F are organized into complex spherical shapes defined by large spaces between cells. That cells can organize in this fashion suggests they possess a means for communicating with one another, either directly via cell-cell contact or indirectly via signaling molecules that span the extracellular spaces. This intercellular communication is the third emergent property of multicellularity.

The goal of this chapter is to explain the molecular basis for these three properties of multicellular organisms. We will first examine the most common molecules found in the extracellular spaces, which are collectively called

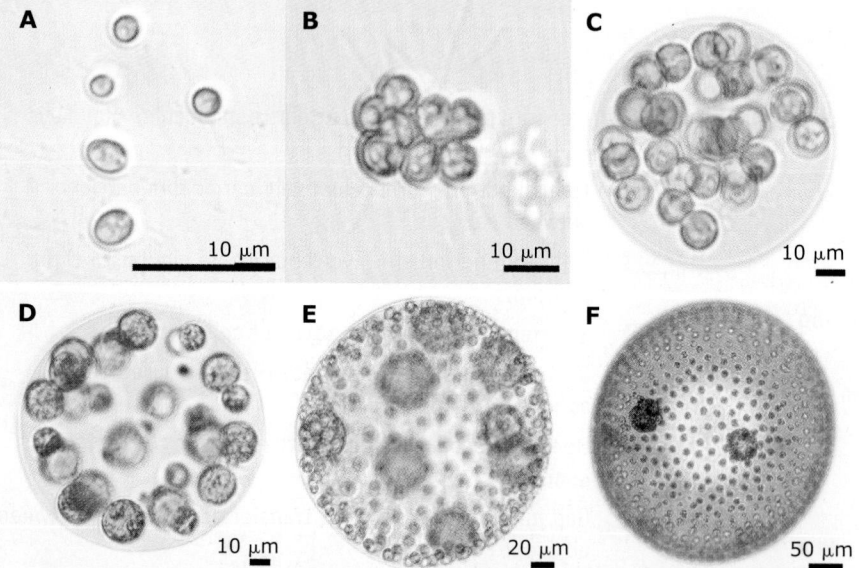

FIGURE 19.1 Examples of species representing different stages of multicellular evolution. A, *C. reinhardtii*, a unicellular, photosynthetic, eukaryotic organism. B, *Gonium pectorale*, a multicellular organism composed of a small group of 8–32 similar cells. C, *Eudorina elegans*, a larger colony of 16–64 undifferentiated cells. D, *Pleodorina californica*, a colony of somatic and reproductive cells. E, *Volvox carteri*, a larger differentiated spherical organism. F, *Volvox aureus*. In D–F, small cells are somatic cells, large cells are reproductive cells. Photos courtesy of the laboratory of Richard E. Michod, University of Arizona.

the ECM, shown in **FIGURE 19.2**. The ECM is a complex network of proteins, sugars, minerals, and fluids that simultaneously provide stable adhesion sites for cells and transmit signals between them. The ECM is organized into fibers, layers, and sheetlike structures. In some tissues, the ECM is organized into a complex sheet named the basal lamina, which is in direct contact with cell layers. The protein portion of the ECM is classified into two types: the structural glycoproteins, such as collagens and elastin, and the proteoglycans. Together, these proteins impart tremendous strength and flexibility to tissues while also serving as a selective filter to control the flow of particulate (undissolved) material between cells. The proteoglycans also attract water, thereby maintaining a hydrated environment around cells. When cells migrate, the ECM functions as a scaffold over which cells crawl. Cells secrete these ECM molecules, thereby building their own external support network, and can reshape the ECM as necessary by degrading and replacing the ECM around them. Controlling the assembly and degradation of ECM molecules is currently a subject of great interest, because it plays a major role in development and in many pathological states such as wounds and cancer.

We will then turn our attention to the cell surface receptor proteins that permit both cell-ECM and cell-cell binding (see **FIGURE 19.3**). Most of these receptor proteins form clusters on the cell surface that are specialized to perform a subset of the tasks required to establish and maintain functional cell groups. Receptor clusters that permit adhesion to the ECM, called **cell-ECM junctions**, link the ECM on the external face of the plasma membrane to the cytoskeleton on the cytosolic surface. One class of these junctions, known as focal contacts, contain transmembrane integrin receptors that bind to structural glycoproteins in the ECM and actin filaments (see *12 Actin*) in the cytosol. A second class, called hemidesmosomes, connects the structural glycoproteins with intermediate filaments (see *13 Intermediate filaments*).

Receptor clusters that permit stable attachment of one cell to another are classified as **cell-cell junctions**. There are several different types of cell-cell junctions, each of which serves a specific role in cell-cell contact and communication. The proteins in gap junctions allow adjacent cells to communicate directly by forming channels that permit exchange of small cytoplasmic molecules. The proteins that form tight junctions serve as selective barri-

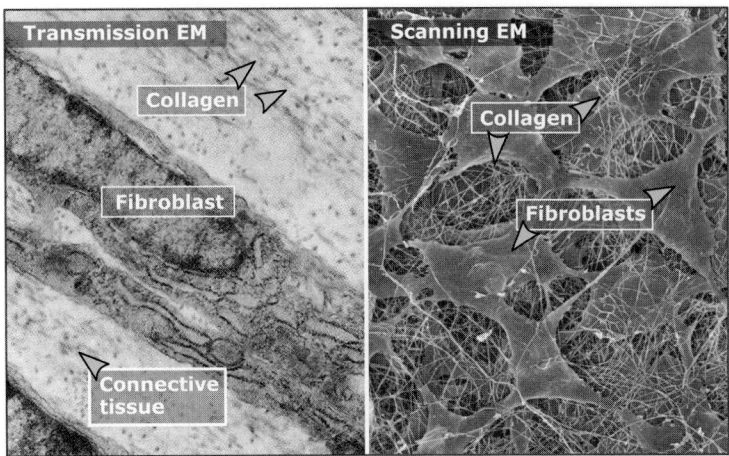

FIGURE 19.2 Electron microscopy reveals that the space between cells is filled with fibrous material. These images show collagen fibrils in the extracellular space between fibroblasts in connective tissue (left) and in the cornea (right). Left photo reproduced from W. Bloom and D. W. Fawcett. *A Textbook in Histology*, 1986. Used with permission of Don W. Fawcett, Harvard Medical School; right photo courtesy of Junzo Desaki, Ehime University School of Medicine.

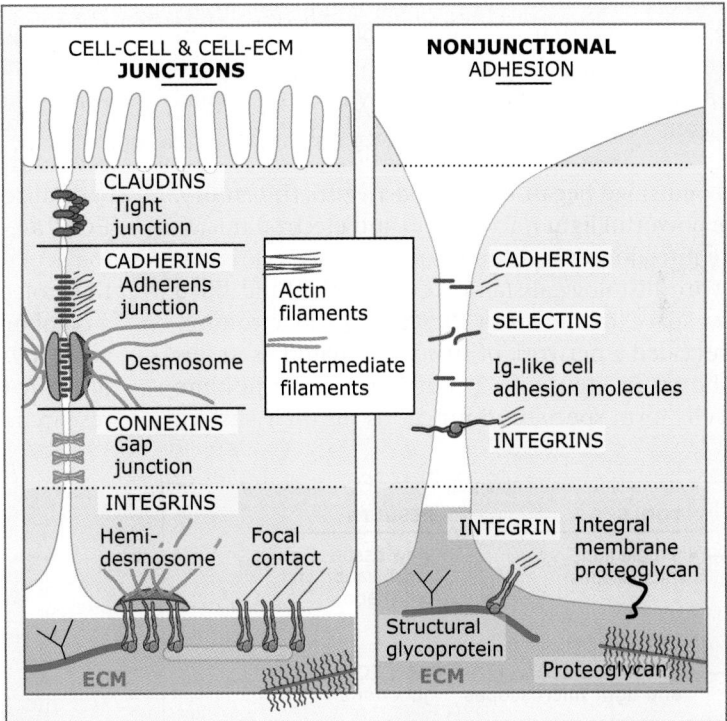

FIGURE 19.3 Schematic representation of epithelial cell-cell junctions (left), nonepithelial cell-cell adhesive complexes (right), and cell-extracellular matrix complexes (bottom). Major classes of extracellular matrix components are also shown.

ers to regulate passage of molecules across cell layers and as barriers to diffusion of proteins in the plasma membrane. Adherens junctions and desmosomes impart mechanical strength by linking the cytoskeletons of adjacent cells

The study of the ECM and cell junctions, like many fields in cell biology, has occurred in four historical stages, as illustrated in **FIGURE 19.B1**. The first of these began in the mid-seventeenth century with the invention of microscopes powerful enough to resolve single cells. As more sophisticated methods were developed for visualizing subcellular structures, biologists came to appreciate the complexity of the cell surface and cell interior. Coincident with the development of the cell theory, biologists began to appreciate the central role played by cells in the development of complex organisms. Biologists also began to understand, at a structural level, the tremendous variation in shapes, sizes, and organization of cells within tissues. In the mid-nineteenth century, a new subdiscipline of biology was coined, called histology, which was concerned with the minute structure (often called the ultrastructure) of the tissues of multicellular organisms.

But something was missing from this picture of tissue architecture: What about the space between cells? Emphasis was placed on the structures that were most apparent in the microscope, and those that were not visible received scant attention. In most tissues, the space between cells appeared rather pale and amorphous when viewed with conventional light microscopes. Early histology texts do not mention this space at all.

The second stage began in the mid-twentieth century, when more powerful light microscopes and electron microscopes were introduced. Light microscopes, used on tissues incubated with histological stains, revealed the fluid-filled extracellular space, as shown in **FIGURE 19.B2**. Electron microscopes revealed a network of structural materials in this space, as shown in Figure 19.2. Furthermore, it became clear that cells form specialized junctions on their surface to interact with these materials as well as with one another (see Figure 19.45). Tissues were finally accepted as being comprised of cells, fluids, and these extracellular materials. A name was given to this group of structural materials: the ECM. But microscopes could not reveal the components of the ECM.

The third stage began in the 1970s when a wealth of new technologies was developed to fractionate, isolate, and characterize the components of cells. As these new techniques in biochemistry, genetics, molecular biology, and microscopy were applied to cell biology problems, the pace of discoveries accelerated rapidly. For example, the development of rapid DNA sequencing methods has allowed researchers to decode the entire genome of several species, so it is now conceivable that we will soon be able to identify every gene in these organisms.

Having used these techniques to identify hundreds of proteins that constitute the ECM and cell junctions, we are now confronted with the next big question: What are the functions of these proteins? It is now widely accepted that the ECM plays a critical role not only in determining the three-dimensional arrangement of cells in tissues but also in controlling the growth, movement, differentiation, and cooperation of each cell in those tissues. Furthermore, the specialized junctions that join cells to each other and to the ECM are key regulators of these functions. The principal focus of research in this area is now directed at determining the molecular mechanisms underlying these functions, using a combination of genetics, molecular biology, and light microscopy. One area that has greatly improved is the imaging of living cells. This is the fourth stage of ECM/cell junction research.

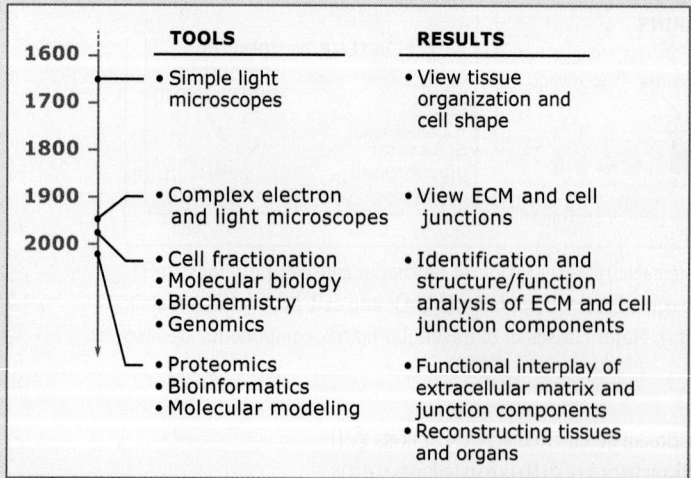

FIGURE 19.B1 Advances in extracellular matrix and cell junction research have accelerated in recent history as new research methods and technologies have become available.

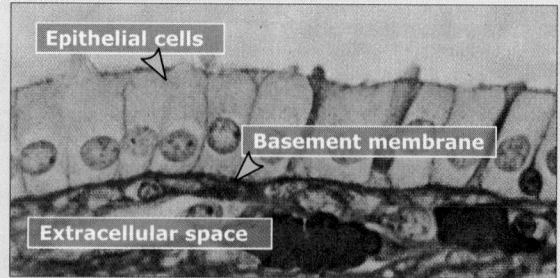

FIGURE 19.B2 Histological stains allow microscopists to visualize multiple cellular features in tissues. This image shows an epithelial tissue stained with histological stains to reveal cellular shape and arrangement within the epithelial sheet. Photo reproduced from W. Bloom and D. W. Fawcett. *A Textbook in Histology*, 1986. Used with permission of Don W. Fawcett, Harvard Medical School.

so that a layer of cells can function as a single unit. Finally, some receptor proteins mediate adhesion without clustering into large complexes. Examples of these **nonjunctional proteins** include some integrins, cadherins, selectins, and cell adhesion molecules related to immunoglobulins.

The receptor proteins that link cells to the ECM and to other cells can also serve as signal transducers, translating changes in cell surface binding into biochemical signals that are propagated throughout the rest of the cell. In fact, cell-cell and cell-ECM receptors activate and control many of the signaling networks and pathways and these are discussed in another chapter (see *18 Principles of cell signaling*). In this chapter, we will focus on the mechanisms linking these receptors to some of the most common signaling pathways found in cells. This is how the third property of multicellular organisms evolved.

The combination of different ECM molecules and different cell surface receptors for the ECM and other cells gives rise to the great diversity of structure and function in multicellular organisms, and permits specialization of subgroups within a single organism to form tissues. For example, cartilage, bone, and other connective tissues are able to withstand tremendous mechanical stress, while others, such as the lining of the lung, are fragile yet highly flexible. The balance between strength, flexibility, and three-dimensional complexity is carefully regulated, so that the components of each tissue work as a specialized team. Thus, the organization and composition of tissues conforms to the function of the organ in which the tissue is found.

19.2 Collagen provides structural support to tissues

Key concepts

- The principal function of collagens is to provide structural support to tissues.
- Collagens are a family of over 20 different ECM proteins that, together, are the most abundant proteins in the animal kingdom.
- All collagens are organized into triple helical, coiled-coil "collagen subunits" composed of three separate collagen polypeptides.
- Collagen subunits are secreted from cells and then assembled into larger fibrils and fibers in the extracellular space.
- Mutations of collagen genes can lead to a wide range of diseases, from mild wrinkling to brittle bones to fatal blistering of the skin.

We begin our discussion of the ECM with **collagens**, which have been present in multicellular organisms for at least 500 million years and may be the first ECM proteins to evolve in animal cells. The collagen family consists of at least 27 proteins that are collectively the most abundant proteins in the animal kingdom. Nearly all animal cells synthesize and secrete at least one form of collagen.

Collagens provide structural support to tissues and come in a variety of shapes organized into different structures. All proteins of the collagen family have a common property: they are bundled together as thin (approximately 1.5 nm diameter), triple-helical coiled coils composed of three collagen protein subunits, held together by both noncovalent and covalent bonds.

The coiled coils form three kinds of collagen structures—fibrillar, sheetlike, and fibril-associated, as illustrated in **FIGURE 19.4** and as given below:

1. In fibrillar collagens, the coiled coils are organized into fibrils, or "ropes," that provide great strength along a single axis. (This is analogous to the bundling of wires to form strong steel cables.) Parallel bundles of these fibrils, such as in those in tendons, impart tremendous strength capable of resisting the strain imposed by muscles on bones.
2. Sheetlike collagens are coiled coils organized into networks that, while less able to resist muscular force, are better able to withstand stretching in multiple directions. These networks are found in skin, for example.

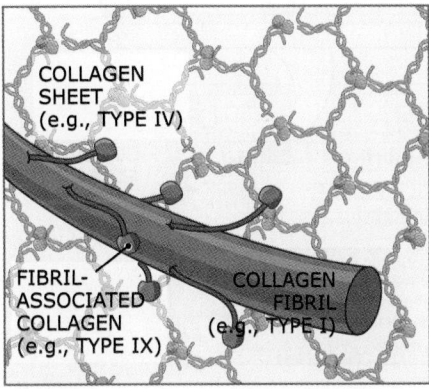

FIGURE 19.4 Collagen subunits are assembled into triple-helical coiled coils that can be organized into fibrils or sheets joined together by other extracellular matrix proteins, including fibril-associated collagens.

3. A third type of collagen, known as "fibril-associated," forms coiled coils used to bind fibrillar collagens together.

Regardless of the organization, collagens form the main structural scaffold in the ECM. Other ECM proteins, such as fibronectin and vitronectin, bind to collagens and are woven

Class	Example	Location
FIBRIL-FORMING (FIBRILLAR)	$[\alpha1(I)]_2\alpha2(I)$	Bone, cornea, internal organs, ligaments, skin, tendons
FIBRIL-ASSOCIATED	$\alpha1(IX)\alpha2(IX)\alpha3(IX)$	Cartilage
NETWORK-FORMING	$[\alpha1(IV)]_2\alpha2(IV)$	Basal lamina
TRANS-MEMBRANE	$[\alpha1(XVII)]_3$	Hemidesmosomes

FIGURE 19.5 Collagens are organized into four major classes, which vary according to their molecular formula, polymerized form, and tissue distribution. Some classes encompass several types of collagens.

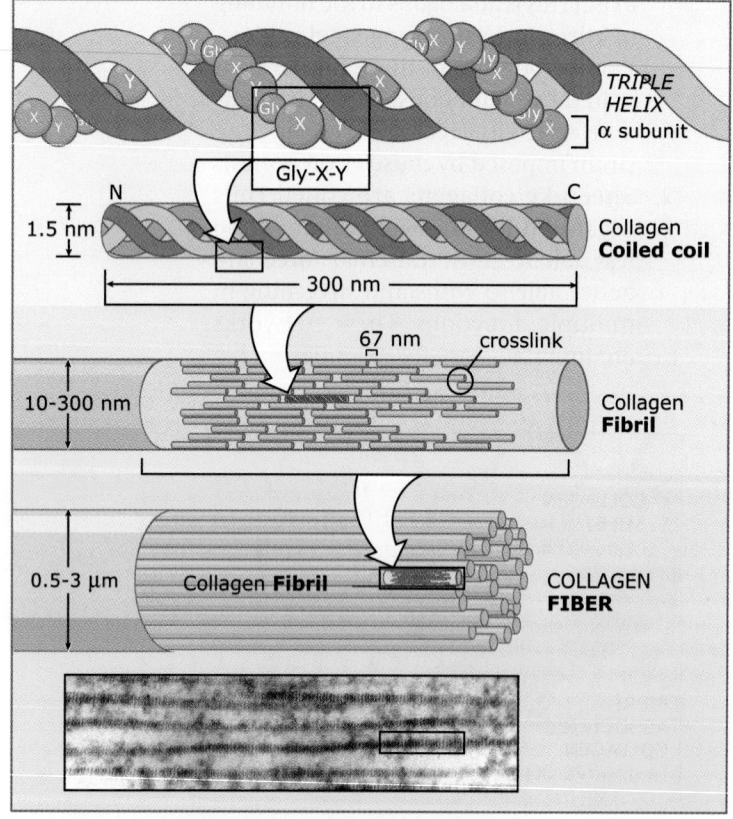

FIGURE 19.6 Schematic diagram of collagen triple-helical coiled coil (top), organization of coiled coils within a fibril (middle), and fibrils in a collagen fiber (bottom). The 67-nm gap between adjacent coiled coils results in a striated appearance in the fibrils that compose the fiber. Photo courtesy of Robert L. Trelstad, Robert Wood Johnson Medical School.

into the patterns established by the collagen scaffolds (see *19.3 Fibronectins connect cells to collagenous matrices* and *19.6 Vitronectin facilitates targeted cell adhesion during blood clotting*). One member of the collagen family is a transmembrane protein that is part of a cell junction (see *19.20 Hemidesmosomes attach epithelial cells to the basal lamina*).

There are at least 27 different types of collagens, most of which can be grouped into the four classes listed in FIGURE 19.5. Each triple-helical structure is given a Roman numeral type designation (I, II, III, etc.). Each collagen subunit is called an α subunit, and the type of subunit is designated by a number (α1, α2, α3, etc.), which is followed by the Roman numeral of the type in which it is found. For example, the principal fibrillar collagen found in rat tails (and other tissues), named Type I-collagen, consists of two copies of the α1(I)-subunit and one copy of the α2(I)-subunit.

The structure of collagen fibers is shown in FIGURE 19.6. Three polypeptide subunits are wrapped together in parallel to form a 300-nm long coiled coil. Collagens contain a characteristic repeating sequence of amino acids consisting of glycine-X-Y, where X and Y can be any amino acid but are usually proline and hydroxyproline, respectively. This sequence enables tight packing of the three subunits and facilitates coiled-coil formation. These 300-nm-long units are held together by covalent bonds formed between the N-terminus of one unit and the C-terminus of an adjacent unit. The coiled coils are arranged in parallel with small (64–67 nm) gaps between them. These gaps give the fibrils their characteristic striped, or striated, appearance when viewed with an electron microscope.

Fully assembled collagen structures, be they fibrils or sheets, are much larger than the cells that synthesize them; some fibrils can be several millimeters in length. Collagen subunits are synthesized and secreted as coiled coils, and the final steps in assembly occur outside of the cell. Collagen synthesis and processing occurs along the secretory pathway, as illustrated in FIGURE 19.7. During synthesis, the collagen proteins are directed to the rough endoplasmic reticulum (ER) by the signal recognition particle and its associated protein machinery (for details on protein translocation into the ER see *7 Membrane targeting of proteins*). The collagen subunits are synthesized as extra-long polypeptides called **procollagens** that contain propeptides, called N-peptides and C-peptides,

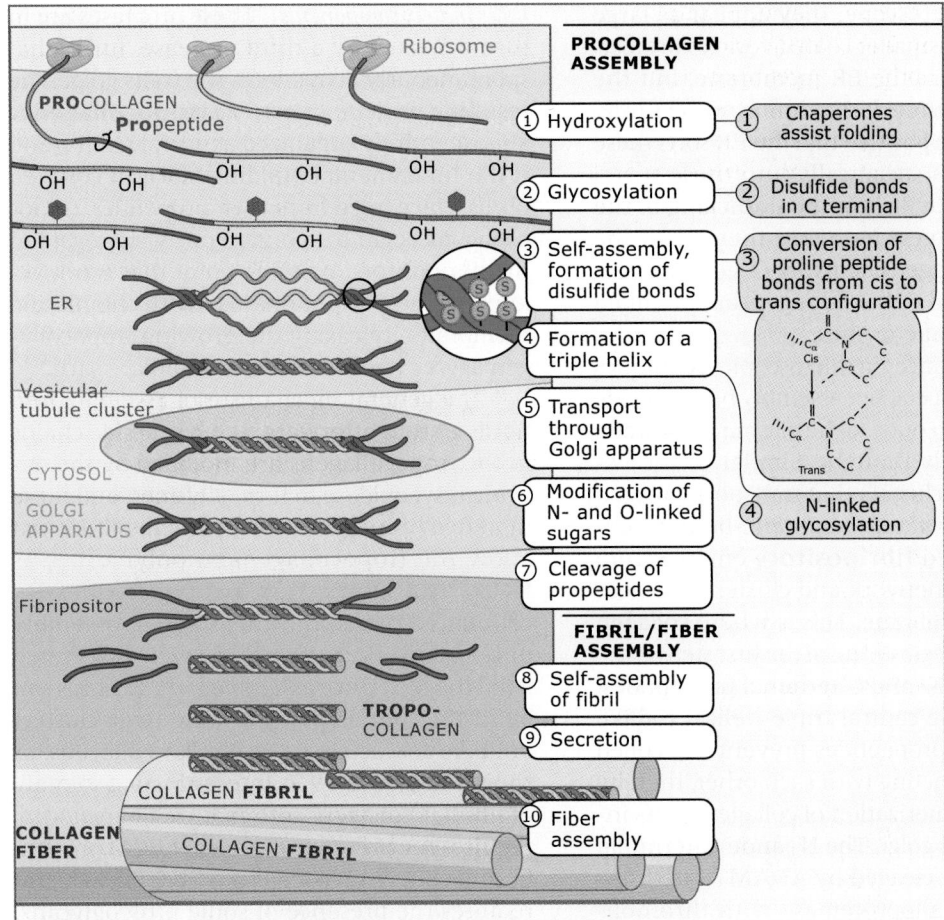

FIGURE 19.7 Posttranslational modification and assembly of procollagen subunits into triple-helical coiled coils occurs during intracellular trafficking through the secretory pathway, but fibril assembly occurs in the extracellular space following secretion of the coiled coils. For simplicity, hydroxyl and sugar groups are not shown in the triple-stranded structures.

which are extended "tails" of amino acids at the amino- and carboxyl-termini, respectively.

Once the procollagens are inserted into the lumen of the ER, a number of chaperones and enzymes assist in their folding into triple-helical coiled coils (see *7.15 Chaperones assist folding of newly translocated proteins*). The folding begins at the C-termini of three procollagen polypeptides and progresses towards the N-termini. Six types of posttranslational modification are required for successful assembly of the coiled coils. First, the C-peptides are stabilized by the formation of interchain disulfide bonds. Second, the enzyme **peptidyl proline cis-trans isomerase** converts the peptide bonds formed by proline amino acids in the polypeptides from the cis configuration to the trans configuration, thereby preserving the linear configuration of the polypeptides necessary for their assembly into coiled coils (see Figure 19.7). Third, the enzyme **prolyl-4-hydroxylase** adds hydroxyl groups to the gamma carbon of proline side chains, and this stabilizes the triple coils by creating additional

hydrogen bonds with neighboring water molecules. Fourth, a family of **lysyl hydroxylases** adds hydroxyl groups to lysine side chains. Fifth, some of these hydroxylysyl amino acids are further modified by the addition of glucose or galactose to the hydroxyl group. And finally, some collagens are N-glycosylated (see *7.14 Sugars are added to many translocating proteins*). All of these steps must be completed for the procollagen chains to exit the ER; procollagen molecules that fail to properly assemble into triple coils are degraded.

Transport of collagen triple helices from the ER to the golgi often presents a challenge for cells. Most trafficking between the ER and golgi occurs in small (~60–80 nm diameter) coated vesicles (see *8.6 Coat protein II-coated vesicles mediate transport from the ER to the Golgi apparatus*), but the triple-helical portion of most collagens is ~300 nm long, so collagens don't fit into these vesicles. The solution to this problem is to assemble a different membrane-bound compartment, called a **vesicular tubular cluster**, to carry these long molecules. When viewed with an electron microscope, they appear as large aggregations of smaller coated vesicles projecting directly from the ER membrane, but the mechanism of their formation is not yet fully understood. It is possible that the ER sorts these large molecules specifically into these structures so as to not disrupt the trafficking of the smaller coated vesicles. Once they reach the golgi, some collagen chains are O-glycosylated, and both N- and O-linked sugars are modified by golgi-resident enzymes.

Recent advances in microscopic techniques reveal that collagen fiber assembly begins inside cells, rather than outside the plasma membrane as was originally thought. Similar to the vesicular tubular clusters that transport collagen to the golgi, elongated membrane-bound compartments called **fibripositors** emerge from the trans-golgi network and cluster at or near the plasma membrane. This is where collagen fibrillogenesis begins. One of the first steps is the removal of the N- and C-terminal propeptides, leaving only the central triple-helical rodlike structure. The propeptides prevent the coiled coils from interacting with each other, thereby inhibiting polymerization of collagen fibers inside the ER and golgi. The N- ande C-terminal propeptides are cleaved by ADAMTS (a disintegrin and metalloproteinase with thrombospondin type-1 motifs) and tolloid proteases, respectively (see *19.12 Proteases degrade extracel-*

lular matrix components). These proteases are in turn activated by a third protease, furin, that spontaneously activates in the trans golgi. The resulting protein, known as **tropocollagen** is almost entirely organized as a triple helix, and is the fundamental building block of collagen fibrils. Once freed from their propetides, tropocollagens begin to spontaneously aggregate in the fibripositors. At some point that is not yet known, the fibripoitors fuse with the plasma membrane, releasing the growing tropocollagen aggregates into the extracellular space.

The general mechanism of assembly into fibrils is straightforward: the lysine side chains in the tropocollagens are modified by the enzyme lysyl oxidase to form allysines, and these modified lysines form covalent crosslinks that allow the tropocollagens to polymerize, as **FIGURE 19.8** shows. Lysyl oxidase is an extracellular enzyme, and this stage of assembly only occurs after the procollagen is secreted from the cell. Once assembled, these fibrils can be further bundled to form the large clusters of fibers characteristic of fibrillar collagens, as shown in Figure 19.7. Interestingly, assembly of fibrillar collagens into mature fibers *in vitro* requires the presence of another ECM protein, fibronectin. In vivo, collagen fiber growth also requires the presence of some proteoglycans. How fibronectin and proteoglycans assist in fiber growth, and whether the same fibronectin dependency takes place in collagen growth in vivo have yet to be determined.

Given their central importance in providing structural support to tissues, it is not surprising that failure to form collagen fibers leads to devastating consequences in vivo. Even a simple deficiency in vitamin C, an essential cofactor for proline hydroxylase, prevents proper collagen assembly and leads to scurvy. Mutations in the collagen genes or in the enzymes that modify procollagen can give rise to a wide range of genetic disorders that affect nearly all tissues. For example, Type I collagen is the primary structural protein in bones; mutations in Type I collagen genes give rise to osteogenesis imperfecta, the so-called brittle bone disease. On the other hand, mutations in Type IV collagen genes result in poorly assembled basal laminae in most epithelial tissues and lead to blistering skin diseases such as epidermolysis bullosa (see *19.11 The basal lamina is a specialized extracellular matrix*).

Cells bind to collagens through specific receptors called integrins. These receptors provide a means of reversibly binding to and releasing collagens as cells crawl through the

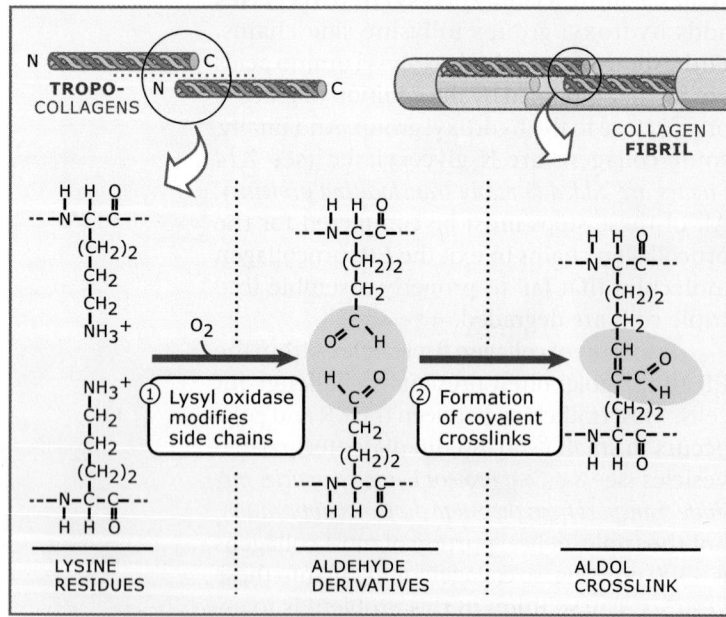

FIGURE 19.8 Lysyl oxidase catalyzes the covalent linkage of two lysine side chains by forming allysines (aldehyde derivatives of lysine), which then form an aldol crosslink.

ECM. Integrin receptors also activate signaling pathways, so that binding to collagens (and other ECM proteins) alters the biochemical activity inside cells and, thus, helps control cell growth and differentiation. (Integrins are discussed in more detail in *19.13 Most integrins are receptors for extracellular matrix proteins; 19.14 Integrin receptors participate in cell signaling*; and *19.15 Integrins and extracellular matrix molecules play key roles in development.*)

19.3 Fibronectins connect cells to collagenous matrices

Key concepts

- The principal function of the ECM protein fibronectin is to connect cells to matrices that contain fibrillar collagen.
- At least 27 different forms of fibronectin are known, all of which arise from alternative splicing of a single fibronectin gene.
- The soluble forms of fibronectin are found in tissue fluids, while the insoluble forms are organized into fibers in the ECM.
- Fibronectin fibers consist of crosslinked polymers of fibronectin homodimers.
- Fibronectin proteins contain six structural regions, each of which has a series of repeating units.
- Fibrin, heparan sulfate proteoglycan, and collagen bind to distinct regions in fibronectin and integrate fibronectin fibers into the ECM network.
- Some cells express integrin receptors that bind to the Arg-Gly-Asp (RGD) sequence of fibronectin.

Fibronectins (from the Latin, *fibra*, fiber, and *nectere*, to connect) are expressed in nearly all animal connective tissues. Fibronectins are synthesized by several cell types, including fibroblasts, hepatocytes (liver cells), endothelial cells, and some support cells in the nervous system. In humans, at least 27 different fibronectin proteins can arise from alternative splicing at four sites within the primary transcript of a single fibronectin gene These variants are cell type-specific. The fibronectins are classified into two groups: soluble (or plasma) fibronectins, found in a variety of tissue fluids (such as plasma, cerebrospinal fluid, amniotic fluid), and insoluble (or cellular) fibronectins, which form fibers in the ECM of virtually all tissues.

Fibronectins attach cells to extracellular matrices in tissues, regulate the shape and cytoskeletal organization of these cells, assist in blood clot formation, and help control the behavior of many cells during development and wound

healing. At sites of injury, fibronectin binds to platelets during blood clotting, and later, during wound healing, it supports the migration of new cells as they crawl to cover the wound area. Many tumor cells also express fibronectins, which may provide a substrate upon which the cells crawl during metastasis. Fibronectin is essential for proper development: fibronectin-null mice die early in embryogenesis.

Cells bind to fibronectin through specific receptors called integrins. Like other integrin receptors, the fibronectin receptors are responsible for activating intracellular signaling pathways that control cell growth, movement, and differentiation. (Integrins are discussed in more detail in *19.13 Most integrins are receptors for extracellular matrix proteins; 19.14 Integrin receptors participate in cell signaling*; and *19.15 Integrins and extracellular matrix molecules play key roles in development.*)

To fulfill their various functions, fibronectins bind to many other molecules in the ECM. Binding assays, using fragments of fibronectin generated by limited proteolysis, have revealed the functional organization of the fibronectin protein. Fibronectin is organized into a set of domains called **fibronectin repeats**. The exact order of these repeats varies due to alternative splicing. The repeating sequences are classified into three groups, named Type I, II, and III, and are numbered consecutively beginning at the amino-terminus of the protein. The functions of these repeats are shown in **FIGURE 19.9** and given below:

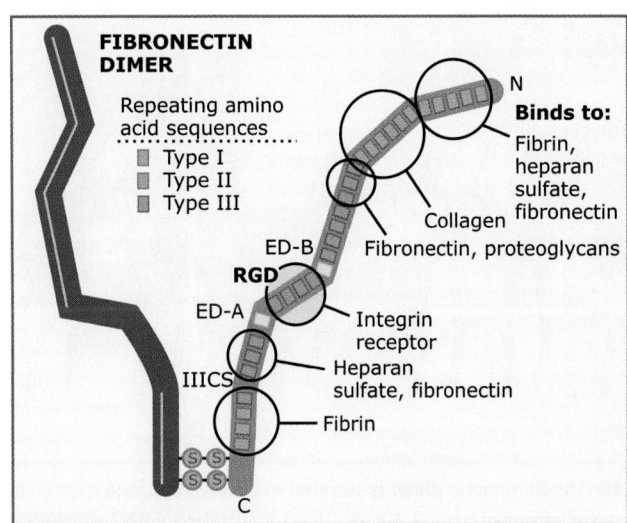

FIGURE 19.9 Two fibronectin polypeptides are covalently linked via disulfide bonds near the carboxyl terminus. Each polypeptide is organized into six domains that consist of small repeating sequences. The major protein binding regions are indicated.

- Near the amino-terminus, a glutamine residue is a substrate for factor XIIIa, an enzyme that crosslinks fibronectin to fibrin, fibrinogen, or other fibronectins as part of the blood clotting process.
- Type I repeats 1–5 (often abbreviated as I_{1-5}) bind to fibronectin, heparan sulfate proteoglycan and fibrin, a protein involved in blood clotting (see *19.8 Proteoglycans provide hydration to tissues*).
- Type I repeats 6–9 and Type II repeats 1–2 bind to collagens.
- The extradomain-B (ED-B) module is found predominantly in fetal tissues, in healing wounds, and in tumors, suggesting that it may be important for tissue remodeling in areas of substantial cell growth. It is not present in plasma fibronectins.
- Type III repeats 1–2 bind to I_{1-5} on other fibronectin molecules.
- Type III repeats 8–11 bind to cell surface integrin receptors by means of a three-amino acid sequence (Arg-Gly-Asp, or RGD) in Type III repeat 10. This region plays an important role in supporting cell adhesion, growth, and migration, and is critical for fibronectin fiber formation.

- Extradomain-A (ED-A), like ED-B, is not present in the plasma fibronectins. Its likely function is to enhance cellular binding to fibronectin, though this has not been demonstrated conclusively.
- Type III repeats 12–14 binds heparan sulfateproteoglycan and I_{1-5} on other fibronectin molecules.
- The Type III connecting segment (IIICS) can be spliced to give rise to modules of different sizes, and hence, multiple forms of fibronectin. At least five different IIICS splice variants are known in humans, some of which regulate apoptosis in some cell types. This module binds to two of the integrin receptors via a Leu-Asp-Val sequence.
- One Type II repeat and three Type I repeats constitute a second binding site for fibrin, which is involved in blood clotting.
- Near the carboxyl terminus of the protein, two cysteine amino acids form disulfide bonds with another fibronectin polypeptide.

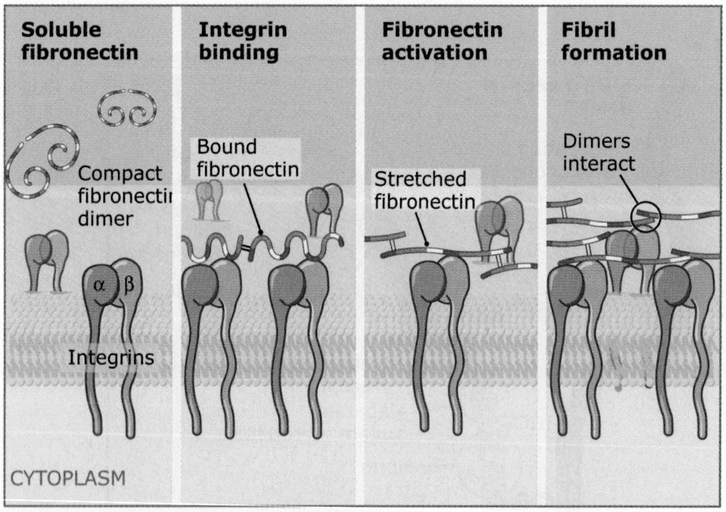

The mature fibronectin protein secreted by cells is always a soluble dimer, held together by two disulfide bridges near the carboxy termini of each fibronectin molecule, and usually contains two copies of the same splice variant of fibronectin (shown in Figure 19.9). Moreover, dimerization of fibronectin is essential for proper formation of the insoluble fibronectin fibers. Assembly of soluble fibronectins into insoluble fibronectin networks requires direct contact with cells. Though the mechanism of fibronectin fiber formation is not yet fully understood, most models suggest that fibronectin dimers first bind to the cell surface via integrin receptors, as shown in **FIGURE 19.10**. Attachment to integrins triggers assembly of an actin-based cytoskeletal complex on the cytosolic side of the integrin. Pulling of the actin filaments by myosins (see *12 Actin*) causes the integrin receptors to pull on the fibronectin, stretching the fibronectin dimers into a more exposed, nearly linear shape. Additional fibronectin dimers are attached to the outstretched dimers, via interactions between the I_{1-5}, III_{2-3}, and III_{12-14} domains, thus forming a dense network. Also, the force applied by the interin-actin network elongates each fibronectin molecule, most likely by unwinding the Type III repeats. The result ap-

FIGURE 19.10 The fibronectin dimer is secreted in a folded conformation that discourages association with other dimers. Upon association with cell surface integrin receptors, the fibronectin dimers are stretched out to reveal binding regions that attract other fibronectin dimers. The resulting accumulation of fibronectin dimers is organized into a fibril associated with the cell surface.

FIGURE 19.11 Primary mouse dermal fibroblasts preincubated with fluorescently labeled-fibronectin (green) for 48 hr and counter stained with rhodamine-labeled phallodin to detect actin microfilaments (red). Photo courtesy of Melinda Larsen and William Daley, University at Albany, SUNY, and Mary Ann Stepp, George Washington University.

pears under the microscope as a collection of fibers that line up with the actin fibers inside the cells, as **FIGURE 19.11** shows. Because fibronectin also binds to other ECM molecules, these fibers can weave the matrix into a strong, supportive structure.

19.4 Elastic fibers impart flexibility to tissues

Key concepts

- The principal function of elastin is to impart elasticity to tissues.
- Elastin monomers (known as tropoelastin subunits) are organized into fibers that are so strong and stable they can last a lifetime.
- The strength of elastic fibers arises from covalent crosslinks formed between lysine side chains in adjacent elastin monomers.
- The elasticity of elastic fibers arises from the hydrophobic regions, which are stretched out by tensile forces and spontaneously reaggregate when the force is released.
- Assembly of tropoelastin into fibers occurs in the extracellular space and is controlled by a seven-step process.
- Mutations in elastin give rise to a variety of disorders, ranging from mild skin wrinkling to death in early childhood.

Elastin, as its name implies, is the ECM protein primarily responsible for imparting elasticity

to tissues. Elastin allows tissues to stretch and return to their original size without expending any additional energy. It is particularly abundant in tissues such as blood vessels, skin, and lungs, where this flexibility is critical for proper organ function. For example, the flexibility of blood vessels is important for maintaining proper blood pressure, and flexibility in lungs allows proper filling and evacuation of the lungs with each breath.

Elastin is synthesized and secreted by fibroblasts, vascular smooth muscle cells, endothelial cells, and cartilage-forming cells (chondrocytes). These cells also secrete collagens, which resist stretching (see *19.2 Collagen provides structural support to tissues*). By varying the proportion of elastin to collagen in the ECM, cells can regulate the flexibility and strength of organs.

Elastin is organized into elastic fibers, which consist of a core region enriched in elastin proteins surrounded by a microfiber sheath, as illustrated in **FIGURE 19.12**. The sheath is comprised of at least 10 different proteins. The most abundant sheath proteins are large (~350 kDa) glycoproteins known as fibrillins, which fold into linear microfibrils containing "bead-like" globular domains. These proteins contain an Arg-Gly-Asp sequence like that

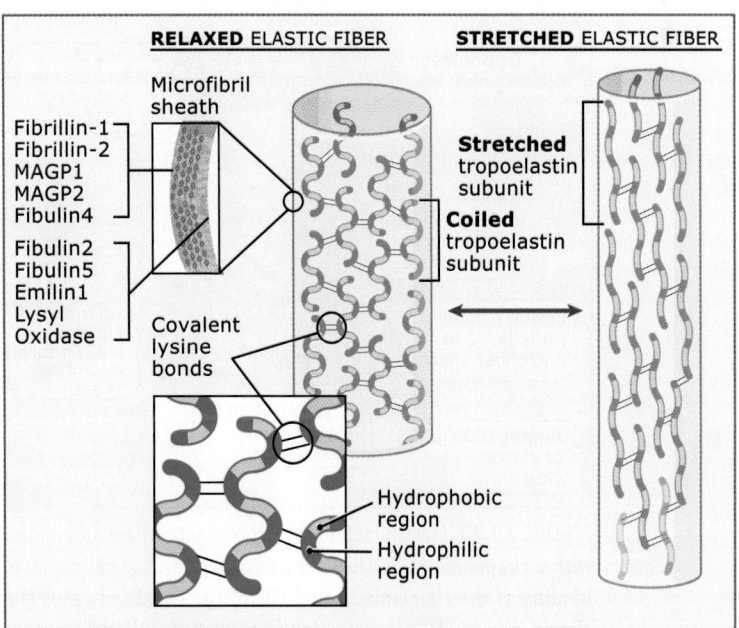

FIGURE 19.12 Schematic representation of relaxed and stretched elastic fibers. Note the dramatic difference in elastin subunit structure in each condition. The exact structure of these subunits is still not known.

in fibronectin, allowing them to bind to cell surface integrin receptors. Microfilament-associated glycoproteins (MAGPs) bind to the beaded regions in fibrillin microfibers. Fibulin proteins lie on either side of the microfibrils, serving to bridge the microfibrils with the elastin core on the inside and with ECM proteins in basement membranes on the outside (see *19.11 The basal lamina is a specialized extracellular matrix*). Another protein, called elastin microfibril interface located protein-1 (EMILIN-1) also bridges the elastin core with the surrounding sheath. Lysyl oxidase, an enzyme that forms covalent links between lysine side chains in elastin molecules, is also found at the core/sheath interface. These fibers are so strong and stable that until recently they were thought to last a lifetime. More recent studies demonstrated that elastin is constantly undergoing a slow turnover (i.e., degradation and replacement) in healthy tissues. The elastin in these

fibers is also the most insoluble protein in the vertebrate body.

How can elastin be tremendously strong and stable, yet be highly flexible? The answer lies in its structural organization, as shown in Figure 19.12. The elastin gene contains 36 exons, which encode two very different types of sequences: some are hydrophilic and contain a high density of lysine amino acids, while others are rich in hydrophobic amino acids, especially glycine, proline, alanine, and valine. The hydrophobic sequences are interspersed amongst the hydrophilic regions, giving rise to a large protein with two different properties. Alternative splicing of the elastin gene gives rise to multiple forms, each customized to the specific needs of a tissue. Much of the strength in elastic fibers arises from the covalent crosslinks formed by lysyl oxidase between lysine side chains in the hydrophilic portions of adjacent elastin proteins, just as is found in collagens (see Figure 19.8). In contrast, the hydrophobic regions impart elasticity by clustering into coils under low stretch conditions, and uncoiling when stretching force is applied, as shown in Figure 19.12. These regions spontaneously coil up again when the stress is removed. Even after many years of study, researchers have not agreed on the exact conformation of elastin proteins within the elastic fibers.

Assembling such an insoluble protein poses special problems for a cell. If these proteins spontaneously aggregate before they are secreted, they could interfere with secretion of other proteins by "clogging" the secretory pathway or rupture organelles or the plasma membrane. Cells synthesize and secrete the elastin proteins as monomers but assemble them into fibers only in the extracellular space, after they are secreted and pose no threat to the cell interior.

A current model of elastin fibrilogenesis suggests it occurs in seven steps, as shown in **FIGURE 19.13**:

1. Fibrilins and MAGPs are secreted into the extracellular space, and form the microfibrillar lattice that nucleates elastin fiber assembly. The lattices are strengthened by chemical crosslinks formed between glutamine and other free amine groups (e.g., on lysine side chains) by transglutaminase enzymes.

2. Elastin monomers (known as **tropoelastin**) are synthesized in the

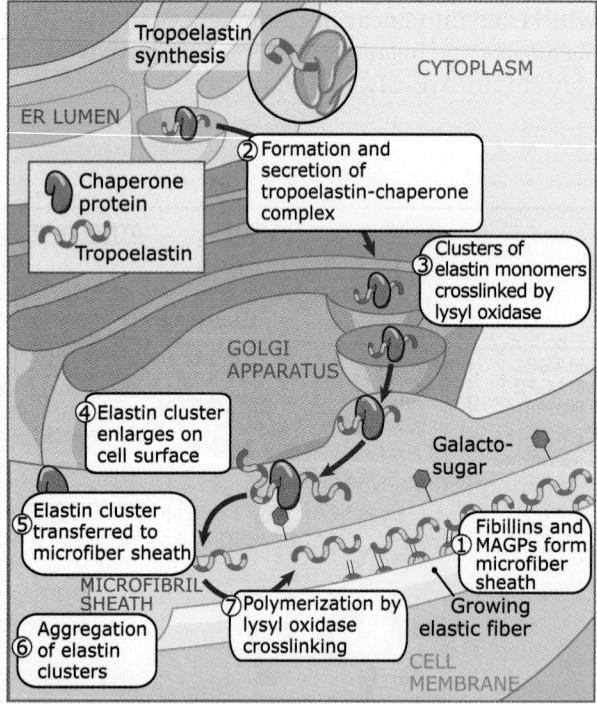

FIGURE 19.13 Elastin monomers (tropoelastins) are associated with a chaperone while they are transported to the cell surface. Binding of the chaperone to the microfiber sheath releases the elastin monomer. Polymerization is catalyzed by lysyl oxidase-mediated crosslinking of tropoelastin. Adapted from J. E. Wagenseil and R. P. Mecham, *Birth Defects Res. C Embryo Today* 81 (2008): 229–240.

rough ER, and carried to the plasma membrane via the endomembrane system (see 8.X). They bind to a chaperone/receptor protein complex in the ER. This chaperone complex remains attached to the tropoelastin throughout the secretory pathway and prevents aggregation of tropoelastin monomers in the cell.

3. Upon secretion, the tropoelastin is held at the cell surface by elastin-binding proteins. This includes the chaperone complex and at least three different integrin receptors. It is crosslinked there by a lysyl oxidase (very similar to collagen crosslinking, see Figure 19.8) to generate small disorganized clusters. Fibulin 4 and/or fibulin 5, as well as heparan sulfate proteoglycans and the cell surface receptors, play a role in determining the degree of crosslinking and the size of the aggregates.

4. Over time as additional tropoelastins are secreted and added, the cell surface cluster grows in size, and is further crosslinked by lysyl oxidase.

5. At some still-undefined point, the aggregate is moved from the external face of the plasma membrane and binds to microfibrils. The microfibrils are attached to the cell surface by integrin receptors that bind the Arg-Gly-Asp (RGD) sequence in fibrillin proteins in the microfiber sheath.

6. Aggregates on the microfibers condense to form large complexes.

7. Aggregates are further crosslinked by lysyl oxidase to form the final structure, which is then covalently linked to the sheath by transglutaminase enzymes. The term **mature elastin** is reserved for the elastin proteins that are modified by lysyl oxidase and assembled into this polymer.

During the period when tropoelastin is attached to the cell surface, it is capable of activating cell surface signaling receptors, including integrins (see *19.14 Integrin receptors participate in cell signaling*), G protein-coupled receptors, and guanylyl cyclases (see *18.21 Heterotrimeric G proteins regulate a wide variety of effectors*). This signaling triggers a number of responses, from cell proliferation to chemotaxis. After they are fully assembled, elastin fibers

play an important role in cellular recognition of transforming growth factor-β (TGFβ), a signaling molecule that controls the synthesis and secretion of ECM molecules. Due to the continuous, slow turnover of elastin, elastin fragments circulate throughout the body and can trigger responses in a large number of cell types, even those that do not synthesize elastin themselves (e.g., lymphocytes), and these fragments can often elicit the same responses as partially assembled elastin.

Elastin is closely integrated into tissues—at least 30 different proteins either bind to elastin fibers or form a portion of the microfibrillar sheath. Alterations in the assembly or function of elastin and elastic fibers can therefore have dramatic consequences. Cutis laxa, a disease involving loss of elastic fibers in the skin and connective tissues, ranges in severity from slightly disrupted fibers and mild skin wrinkling to nearly undetectable amounts of elastin fibers. Patients with little or no elastin cannot maintain tissue integrity and die in early childhood. Patients with Williams syndrome produce truncated forms of elastin that lack some crosslinking domains and are poorly organized into fibers. These patients develop severe narrowing of their large arteries, possibly due to abnormal growth of smooth muscle cells around the arteries to compensate for the loss of the elastic fibers normally found in the artery walls.

19.5 Laminins provide an adhesive substrate for cells

Key concepts

- Laminins are a family of ECM proteins found in virtually all tissues of vertebrate and invertebrate animals.
- The principal functions of laminins are to provide an adhesive substrate for cells and to resist tensile forces in tissues.
- Laminins are heterotrimers comprising three different subunits wrapped together in a coiled-coil configuration.
- Laminin heterotrimers do not form fibers; rather, they bind to linker proteins that enable them to form complex webs in the ECM.
- A large number of proteins bind to laminins, including more than 20 different cell surface receptors.

The **laminins** are a diverse family of large (>100 kDa) ECM proteins found in the basal laminae (hence the name) and other sites of ECM deposition in many tissue types (for more detail on the basal lamina see *19.11 The basal lamina is a specialized extracellular matrix*). Laminins are expressed in invertebrate and vertebrate animals, and the degree of homology between laminin family members is quite low, suggesting that laminins have a long evolutionary history. Laminin experts believe that the ancestral gene that gave rise to current laminins closely resembles the single laminin gene in the invertebrate *Hydra vulgaris*.

Like collagens, laminins consist of three polypeptide subunits wrapped together to form a triple-helical coiled coil. In laminins, the sequence responsible for establishing this coiled coil is seven amino acids long and is found in multiple repeats in each of the three subunits. The coiled coil maximizes the number of noncovalent bonds formed between the subunits and confers structural stability on the completed trimer. Once the coiled coil is formed, the subunits are covalently linked via disulfide bonds. Only a portion of each subunit is organized into the coiled coil; each subunit also extends "arms" out from the coil, giving rise to a cross-shaped structure, as illustrated in **FIGURE 19.14**. Like collagens and elastins, laminins are routinely cleaved by proteases to generate mature, functional forms of these proteins.

Laminin proteins are heterotrimers: the three subunits contained in a single laminin protein are products of different genes, and are classified into three groups, α, β, and γ. So far, five α, three β, and three γ subunits are known, and some of these can be differentially spliced to yield additional variants. Together, these subunits could theoretically combine to form over 100 different heterotrimer combinations, yet only 16 combinations have been identified so far. Still, this allows for a wide range of laminin networks to be constructed in a single organism. The most recent method for naming the laminins moves away from simple numbering (laminin-1, laminin-2, etc.) to a more descriptive system that identifies the constituent chains in each (e.g., a laminin isoform previously listed as laminin-5, is now laminin-332, to indicate that it consists of α3-, β3-, and γ2-chains).

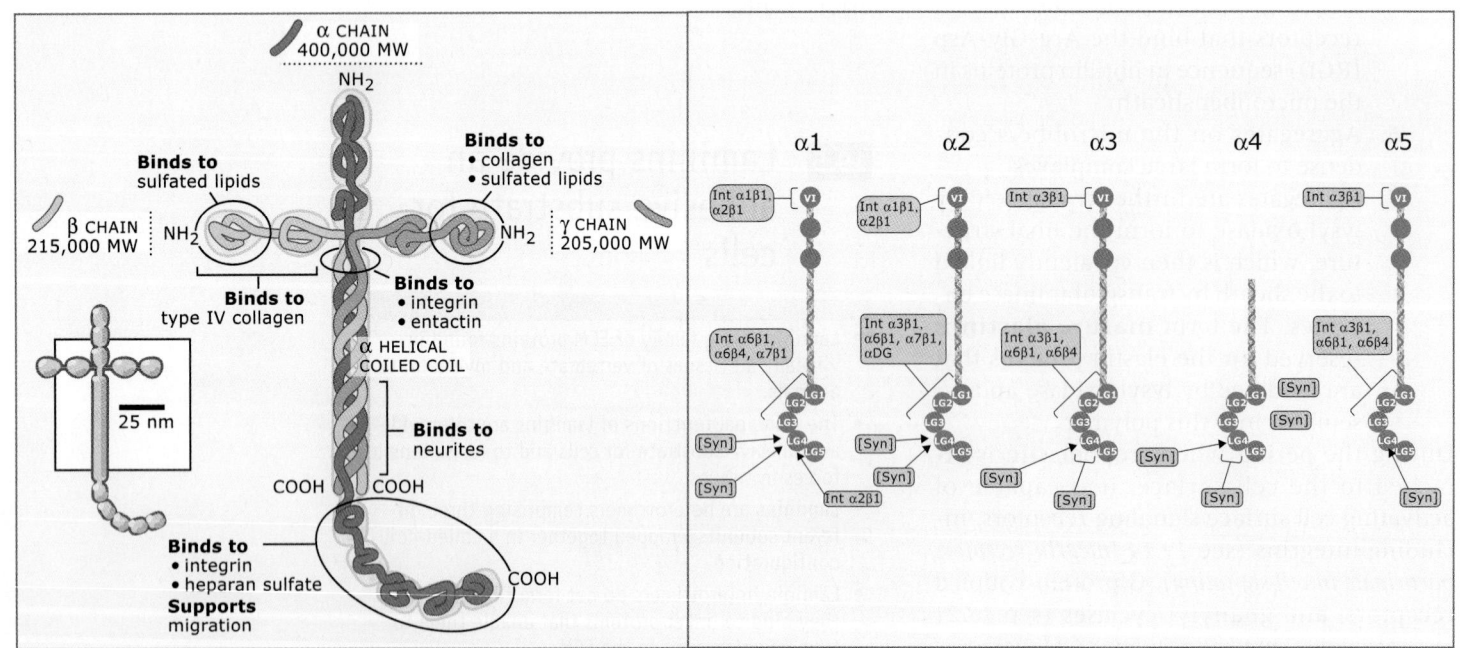

FIGURE 19.14 The three chains of the laminin molecules are wrapped into a central core. The amino-terminal portion of each chain extends from the central core to form a cross-shaped structure. The carboxyl tail of the α chain extends beyond the central core to form up to five globular domains. Important binding regions are indicated. (Syn) = syndecan and Int = integrin.

Unlike the other major classes of glycoproteins found in the ECM, laminins do not form fibrils. Rather, they are organized into weblike networks that are able to resist tensile (stretching) forces from many directions at once. The short arms of the laminin heterotrimer, which represent the amino-terminal portion of each subunit, contain domains that associate with other components in the ECM to form this large web. One excellent example, shown in **FIGURE 19.15**, is the web formed by laminin-111 in the basal lamina, where it interacts with other ECM components such as entactin (also called nidogen), perlecan, and collagen IV. Some laminin subunits lack these domains, and the mechanism for polymerization of these laminins is not yet known. Recent evidence suggests that some of these laminins cooperate to assemble into a matrix (e.g., binding of a proteolytically cleaved form of laminin-332 promotes deposition and matrix assembly of laminin-511).

Structurally, laminins resemble both collagens (both have a triple helical, central rod domain) and fibronectins (both have mul-

tiple binding sites for cell surface receptors and other components of the ECM), and these similarities are reflected in the functions of laminins as well. Early morphological and biochemical studies showed that laminin-111 is widely expressed in basal laminae and that it supports the attachment and spreading of many epithelial cell types through integrin receptors. Cell spreading on laminins requires that the laminin network be strong enough to resist the tension caused by cytoskeletal remodeling. Many proteins bind to laminin-111 and some of these (most notably entactin) play a critical role in assembling laminin-111 into networks. Immunohistochemical studies showed that laminin-111 is expressed as early as the 8-cell stage in mouse embryos, suggesting that it plays an important role in development.

Further studies identified the portions of laminin that support cell adhesion and migration and the roles of laminins in development. As with the fibronectins, limited digestion of the laminins with mild proteases has been used to break the intact molecule into smaller

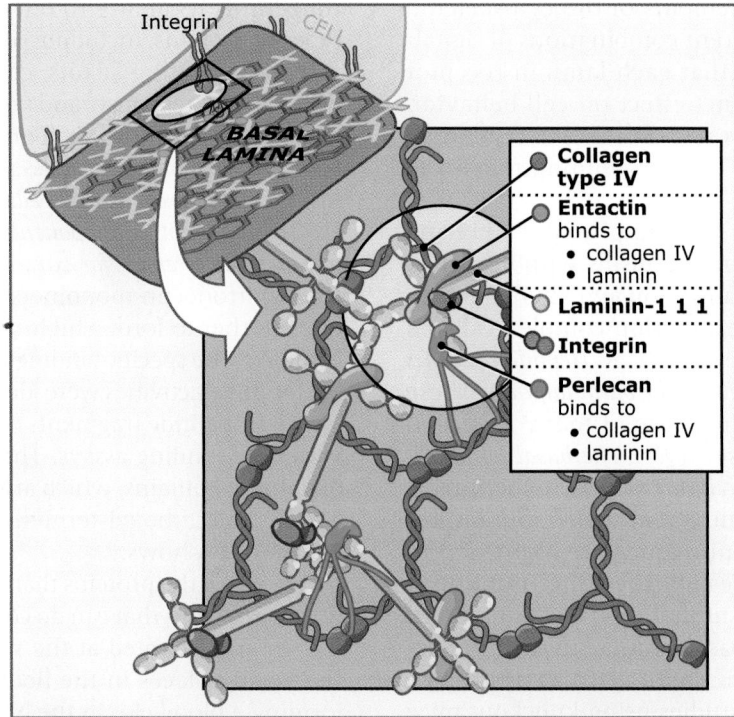

FIGURE 19.15 Laminin associates with at least three other extracellular matrix proteins to form a network within the basal lamina. Laminins also bind to integrin receptors projecting outward from cells attached to the basal lamina.

functional units that are used to assay binding to other proteins. The result of these studies so far is a functional map of laminin domains that is less detailed than for proteins such as fibronectin (see *19.3 Fibronectins connect cells to collagenous matrices*). This is because laminin forms a continuous coiled coil so that individual regions can be disrupted even by limited proteolysis.

We know a large number of the functions and binding partners associated with regions of laminins. For example, more than 20 different receptors for laminin-111 have been identified. In addition, as shown in Figure 19.14, multiple sites on laminins, such as the globular domains at the carboxyl terminus of the α chains, play a role in regulating cell migration. Some of these domains are "cryptic," meaning they are only exposed after proteolytic cleavage. In fact, selective cleavage of laminins is an important mechanism for controlling cellular adhesion and migration in development and cancer (see *17 Cancer—Principles and overview*). The major laminin receptors include three of the integrins, a 67 kDa nonintegrin receptor, and the syndecan family of heparin sulfate proteoglycans (see *19.10 Heparan sulfate proteoglycans are cell surface coreceptors*). These receptors in turn associate with different elements of the cytoskeleton and bind to different combinations of signaling proteins, so that each laminin receptor can have a distinct effect on cell behavior. The mechanisms responsible for regulating cell response to laminin binding remain to be elucidated.

Mutations in subunits of laminin-111 and -332 are found in organisms with disrupted basal lamina organization and altered formation of a specific adhesion complex called a hemidesmosome (see *19.20 Hemidesmosomes attach epithelial cells to the basal lamina*). These mutations are the cause of many inherited skin disorders (see *19.11 The basal lamina is a specialized extracellular matrix*). Experiments using reverse genetics, in which mutant, recombinant laminin genes are expressed in organisms, has established that mutations in laminin-211 subunits cause disruption of the muscle cell basal laminae in the most severe cases of inherited muscular dystrophy. Developmental studies using knockout mice have established a role for laminins in the targeting of neurons to their specific cellular partners, and the formation of the basal lamina in the kidney.

19.6 Vitronectin facilitates targeted cell adhesion during blood clotting

Key concepts

- Vitronectin is an ECM protein that circulates in blood plasma in its soluble form.
- Vitronectin can bind to many different types of proteins, such as collagens, integrins, clotting factors, cell lysis factors, and extracellular proteases.
- Vitronectin facilitates blood clot formation in damaged tissues.
- In order to target deposition of clotting factors in tissues, vitronectin must convert from the soluble form to the insoluble form, which binds clotting factors.

Vitronectin is a relatively small (75 kDa), multifunctional ECM glycoprotein that is found in blood plasma as well as in wounds and other areas undergoing tissue remodeling. It is also heavily glycosylated: approximately one-third of its mass is contributed by N-linked sugars. Unlike most other ECM proteins, which are synthesized by cells in many different tissues, vitronectin is synthesized primarily by the liver, which secretes it directly into the bloodstream.

One of the remarkable properties of vitronectin is its ability to bind to many other types of proteins, including collagens, integrin receptors, clotting factors, cell lysis factors of the immune response, and proteases involved in degradation of ECM. (For more details on some of these proteins see *19.2 Collagen provides structural support to tissues; 19.12 Proteases degrade extracellular matrix components;* and *19.13 Most integrins are receptors for extracellular matrix proteins*.) Vitronectin monomers can also bind to one another to form a high-molecular weight complex. The specific binding regions responsible for these activities were identified by testing synthetic peptide fragments of vitronectin in a variety of binding assays. The result is a map of binding domains, which are clustered at the amino- and carboxyl-termini of the protein, as **FIGURE 19.16** shows.

Many of the proteins that bind to vitronectin are enzymes that can do considerable harm if they are activated at the wrong time or in the wrong places in the body. For example, forming a blood clot in the brain can result in stroke. It is, therefore, very important to control when and where vitronectin binds to its partners. The most fundamental means of accomplishing this is elegantly simple: vitronectin

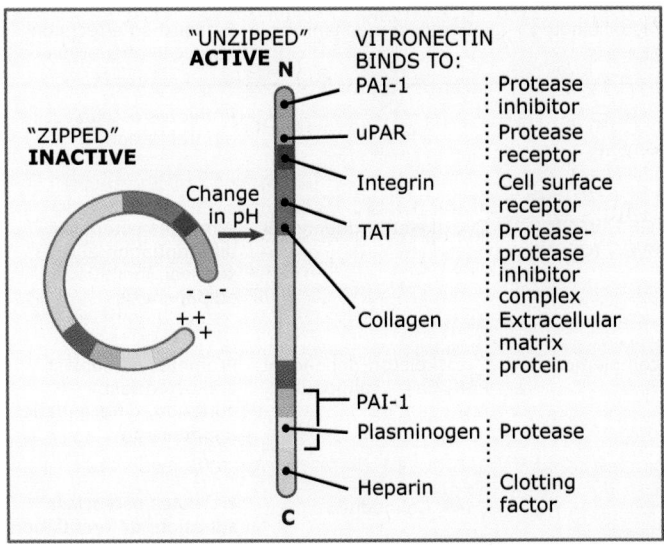

monomers bind to themselves, thereby "zipping up" until they are needed. The folding is thought to occur as a result of ionic attraction between negatively charged amino acids near the N-terminus and positively charged amino acids at the C-terminus. The map of binding sites in Figure 19.16 reveals how contact between the two ends of vitronectin could prevent these sites from recognizing their binding partners.

The folded vitronectin monomers circulate in the bloodstream and do not readily unfold. While it is generally accepted that vitronectin must at adopt an intermediate, partially unfolded state to initiate binding to its partners, it is not clear how this unfolding is accomplished. Studies with purified vitronectin have demonstrated that its structure is sensitive to changes in pH and ion concentration *in vitro*, suggesting that changes in the blood plasma may initiate unfolding. It is also possible that partial unfolding occurs spontaneously within a small subset of the circulating vitronectin monomers.

Once vitronectin is unfolded to expose its binding domains, numerous outcomes are possible, including clustering of vitronectin into large aggregates within the bloodstream and attachment of cells to these clusters. In the case of blood clot formation, vitronectin serves as a scaffold for initiating proper placement of clots. The partially unfolded vitronectin molecules reveal binding sites for the integrin receptors on platelets (see Figure 19.16), thereby recruiting platelets to injured blood vessels. The bound platelets become activated and release other factors that promote clot formation.

19.7 Matricellular proteins regulate cell-extracellular matrix interactions

Key concept

- Matricellular proteins are ECM molecules that bind to cell surface receptors and regulate their signaling activities.

The term "matricellular proteins" is used to describe a number of structurally distinct proteins that are found in the ECM but that do not play any significant structural role. Instead, these proteins bind to cell surface receptors for ECM molecules and regulate their signaling activities. While they are found distributed throughout most mammalian tissues, expression of matricellular proteins is most pronounced during embryonic development, wound healing, and in tissues with a high proportion of ECM, such as cartilage and bone.

FIGURE 19.17 lists some of the more common matricellular proteins and their functions in bone cells. Like vitronectin, matricellular proteins exist in two states. The soluble forms of these molecules circulate throughout the body and exert a mild antiadhesive effect on cells. That is, they prevent strong cellular adhesion to other ECM molecules most likely by competitively binding to receptors for collagens, fibronectins, laminins, and proteoglycans. The insoluble forms of matricellular proteins form part of the ECM surrounding cells, where they bind to and sequester soluble growth factors and proteases. Like cell surface heparin sulfate

MP	ECM interactions	Cell Surface interactions	Soluble factor binding	Signaling mechanisms	Established effects on bone cell phenotype
Tenascin C	Fibronectin	Integrins heparan sulfated proteoglycans			Promotes osteoblast differentiation
SPARC	Collagens	A cell surface receptor for SPARC has not been identified	PDGF, TGF·I, VEGF, MMP2, bFGF, IGF	Cytoskeleton-dependent	Promotes osteoblast differentiation and survival, inhibits adipogenesis
OSTEOPONTIN OPN	Fibronectin, collagens	Integrins, CD44	MMP3, complement, EGF	Cytoskeleton-dependent, PBK, Ca²⁺, calmodulin	Promotes osteoblast and osteoclast adhesion, differentiation and function
BONE SIALO PROTEIN BSP	Collagen	Integrins	MMP2	Ca²⁺, Calmodulin	Promotes osteoclast adhesion, differentiation and function Promotes angiogenesis
THROMBOSPONDIN-1 TSP1	Fibrinogen, collagens, heparin sulfated proteoglycans, fibronectin, laminin	Integrins, HSPG, calreticulin, chondroitin sulfate, syndecan1	Clotting cascade factors, TGF-βI, elastase, PDGF, bFGF, MMP2, IGF-1, IGF-BP	Focal adhesion kinase, G protein, MAP kinases, JNK, Caspase	Promotes osteoclast function
THROMBOSPONDIN-2 TSP2	Collagens, proteoglycans	Integrins, HSPG, chondroitin sulfate	MMP2	Caspase	Inhibits MSC proliferation, promotes osteoblast differentiation, inhibits adipogenesis

FIGURE 19.17 Binding interactions and functions of six common matricellular proteins in bone tissue (see *18 Principles of cell signaling*). Abbreviations: BSP, bone sialoprotein; MMP, matrix metalloproteinase; OPN, osteopontin; SPARC, Secreted Protein Acidic and Rich in Cysteine; TSP, thrombospondin. Reprinted from *Bone*, vol. 38, A. I. Alford and K. D. Handenson, Matricellular proteins: Extracellular modulators…, pp. 749–757, Copyright (2006) with permission from Elsevier [http://www.sciencedirect.com/science/journal/87563282].

proteoglycans, they can immobilize signaling proteins at the cell surface, thereby serving as coreceptors (see *19.10 Heparan sulfate proteoglycans are cell surface coreceptors*).

Concept and Reasoning Check

Complete the table below by briefly comparing each of the glycoproteins discussed in this chapter with respect to how they perform these five functions.

Function	Glycoprotein type					
	Collagens	Fibronectins	Elastins	Laminins	Vitronectin	Matricellular proteins
1. Impart mechanical strength to tissue						
2. Provide adhesion sites for cells						
3. Provide elasticity to tissues						
4. Direct cell movement during embryonic development						
5. Trigger cell signaling pathways controlling cell growth, migration, differentiation, etc.						

19.8 Proteoglycans provide hydration to tissues

Key concepts

- Proteoglycans consist of a central protein "core" to which long, linear chains of disaccharides, called glycosaminoglycans (GAGs), are attached.
- GAG chains on proteoglycans are negatively charged, which gives the proteoglycans a rodlike, bristly shape due to charge repulsion.
- The GAG bristles act as filters to limit the diffusion of viruses and bacteria in tissues.
- Proteoglycans attract water to form gels that keep cells hydrated and cushion tissues against hydrostatic pressure.
- Proteoglycans can bind to a variety of ECM components, including growth factors, structural proteins, and cell surface receptors.
- Expression of proteoglycans is cell type specific and developmentally regulated.

Proteoglycans are the counterpart to the structural glycoproteins (such as collagen and elastin) in the ECM. While structural glycoproteins provide tensile strength, proteoglycans ensure that the ECM is a hydrated gel. This is important for tissues to resist compressive forces.

Like other glycoproteins that are abundantly expressed on the cell surface, a proteoglycan is composed of a single polypeptide core (hence the term, *proteo*) to which sugars (glycans) are attached. Over 40 different proteoglycan core proteins are known, and each contains modular structural domains that can

bind to other components of the ECM, such as carbohydrates, lipids, structural proteins, integrin receptors, and other proteoglycans. **FIGURE 19.18** shows examples of the types of proteoglycans. Most proteoglycans, such as decorin and aggrecan, are secreted from the cell, but two types are membrane bound. Members of the syndecan family of glycoproteins contain a transmembrane domain, and the glypicans are anchored to the membrane via a glycosylphosphatidylinositol linkage.

Proteoglycans are distinguished from glycoproteins by the type and arrangement of the sugars attached to them. The sugars attached

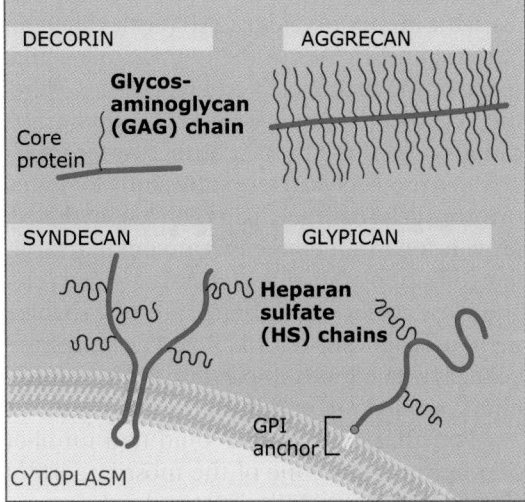

FIGURE 19.18 Summary of proteoglycan structures. All proteoglycans have GAG chains attached to a core protein. Proteoglycans are either secreted or attached to cell membranes by a membrane-spanning core domain or by a glycosylphosphatidylinositol anchor attached to the core protein.

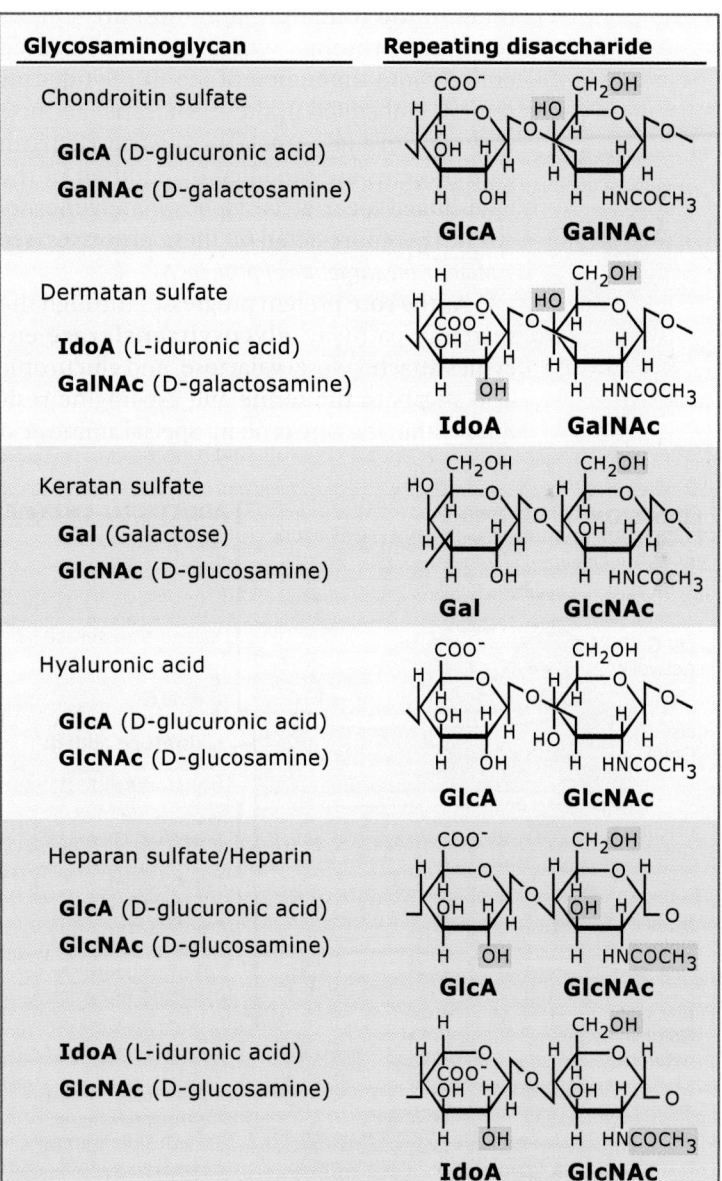

FIGURE 19.19 GAGs are classified according to the type of repeating disaccharide they contain. Sulfate groups are added at the highlighted positions in the disaccharides. Adapted from K. Prydz and K. T. Dalen, *J. Cell Sci.* 113 (2000): 193–205.

LIVERPOOL JOHN MOORES UNIVERSITY
LEARNING SERVICES

to proteoglycans are termed **glycosaminoglycans (GAGs)**, which are arranged as long, linear chains of repeating disaccharides. These chains may contain hundreds of linked sugars and can reach molecular weights of up to 1,000 kDa. As **FIGURE 19.19** shows, GAGs are organized into five classes based on the disaccharides they contain and all but one (hyaluronic acid) can be attached to proteins to form proteoglycans. All GAGs contain acidic sugars and/or sulfated sugars, which gives them a highly negative charge.

The steps involved in building proteoglycans are shown in **FIGURE 19.20**. The core protein is synthesized at the rough ER. All core proteins contain signal sequences that target them to the rough ER, and most are soluble secretory proteins, which are translocated entirely into the lumen of the ER. Syndecans remain embedded in the membrane, because they contain a stop-transfer sequence. Glypican core proteins are modified by addition of the lipid-linked sugar, glycosylphosphatidylinositol (GPI). (For more detail on these processes, see *7 Membrane targeting of proteins*.)

As the core protein progresses through the secretory pathway, **glycosyltransferase** enzymes attach xylose, galactose, and glucuronic acid sugars to the serine and asparagine residues within the core protein. Special amino acid sequences within the core protein determine the type and location of the sugars that are attached. These sugars serve as the attachment sites for additional sugars, such as N-acetyl glucosamine, that make up the GAG chains. The GAGs may be modified by still more enzymes that rearrange the structure of the sugars (epimerases) or add sulfate groups to the sugars (sulfotransferases). Some proteoglycans also have N- and O-linked oligosaccharides typical of glycoproteins (for more on N-linked sugars see *7.14 Sugars are added to many translocating proteins*). The newly synthesized proteoglycans are sorted at the *trans*-Golgi network into the regulated secretory pathway and are stored in secretory granules until they are released by exocytosis. Various signals, including direct application of pressure, stimulate secretion of proteoglycans. (For more on regulated secretion see *8 Protein trafficking between membranes*.)

A proteoglycan can have from one to over 100 large GAGs attached to it. Because many of the sugars are negatively charged, GAGs repel each other. On proteoglycans containing many GAGs, this forces the core protein into a linear, rodlike shape, with GAGs projecting outward. The result is that the mature proteoglycan resembles a bristly rod, much like a hairbrush, as seen in Figure 19.18.

This distinctive shape provides proteoglycans with special properties that help define the nature of the ECM. First, their relatively rigid structure helps them act as structural scaffolds that define the overall shape of the tissues in which they are found. Second, proteoglycans assist the immune system: GAG bristles filter out bacteria and viruses in the extracellular fluid, reducing the chance of infection in tissues. Third, the negative charge of the GAG chains attracts cations, and these in turn attract water molecules, so proteoglycans are sufficiently hydrated to form a gel. These gels help keep cells hydrated, provide for an aqueous environment that facilitates diffusion of small molecules between cells, and allow tissues to absorb large pressure changes without significant deformation. These pressure changes occur, for example, during blunt force, injury, or vigorous exercise.

Fourth, proteoglycans bind to a number of other proteins. One of the most important classes of these proteins is the growth factors. Cells secrete growth factors into the bloodstream and tissue fluid, and they circulate throughout the body. As **FIGURE 19.21** shows, proteoglycans bind to and trap these growth factors, thereby increasing their concentration

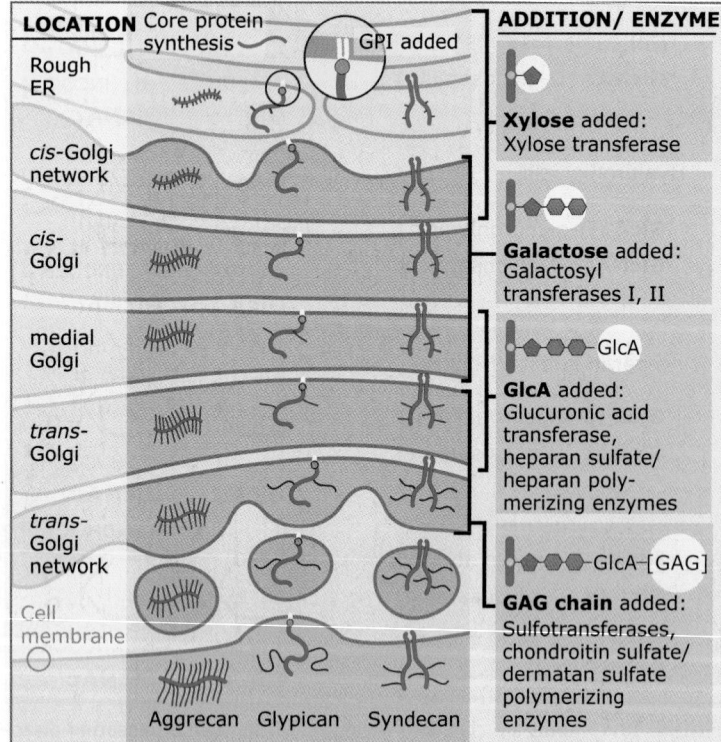

FIGURE 19.20 Proteoglycans are assembled during their transport through the secretory pathway. The locations of important enzymes are indicated.

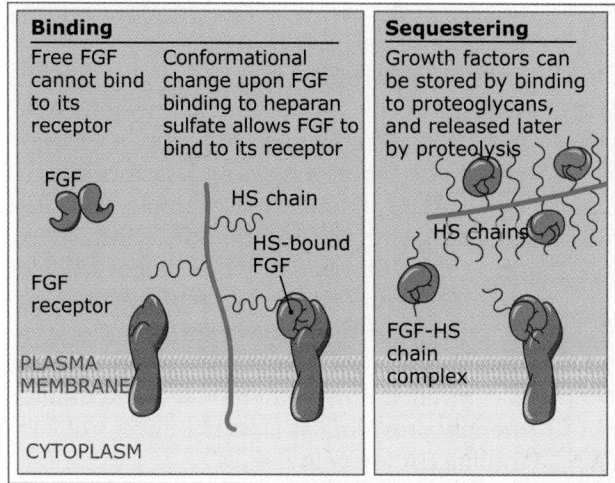

FIGURE 19.21 Proteoglycans help sequester growth factors near the cell surface and control their binding to cell surface receptors. This is a model of how heparan sulfate assists in the binding of fibroblast growth factor (FGF) to cell receptors.

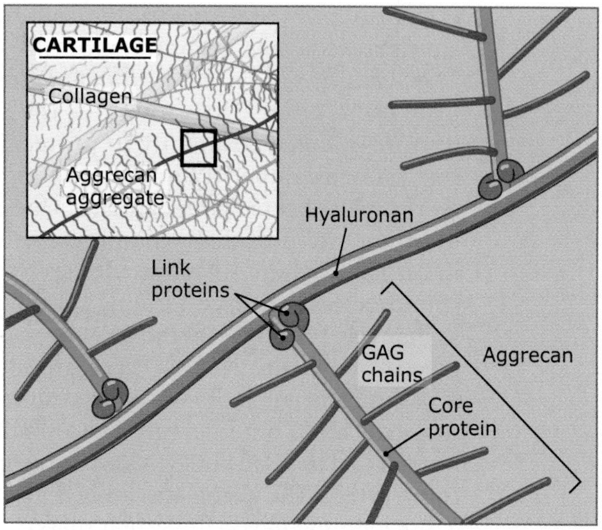

FIGURE 19.22 Proteoglycans such as aggrecan complex with collagen II fibers in cartilage. The aggrecan complexes bind to hyaluronan molecules and attract water, which absorbs compressive forces and acts as a lubricant.

within the ECM. This binding also localizes growth factors to specific regions in tissues, and protects them from degradation by extracellular proteases. In some cases, this trapping is necessary for cells to bind the growth factors. Proteoglycans can, therefore, function as coreceptors for growth factors and thus, indirectly control the growth of cells within tissues. The growth factors may even be stored in this state, to be released later when the proteoglycans break down (see *19.12 Proteases degrade extracellular matrix components*).

Proteoglycans also bind to and direct the assembly of other ECM proteins. For example, the proteoglycans aggrecan and decorin bind to collagen (see *19.2 Collagen provides structural support to tissues*). Aggrecan forms large aggregates with fibers of collagen Type II in cartilage, as shown in **FIGURE 19.22**. To form the aggregates, aggrecan molecules bind to hyaluronan (HA) via linker proteins (see *19.9 Hyaluronan is a glycosaminoglycan enriched in connective tissues*). Decorin acts as a spacer between collagen fibers and controls the fiber diameter as well as the rate of their assembly. Mice that have had their decorin gene "knocked out" develop irregularly shaped collagen fibrils and have especially fragile skin as a result.

Expression of proteoglycans is developmentally regulated and cell type specific. For example, in the developing chick embryo, aggrecan is expressed primarily in cartilage tissues and is maximally expressed at day 5 when chondrocytes, the cells that synthesize

cartilage, differentiate. However, it is also expressed at lower levels in the developing brain and spinal cord, and this expression peaks at day 13. Expression of proteoglycans such as aggrecan is regulated by the same growth factors that bind proteoglycans, suggesting that proteoglycans may play a role in regulating their own expression.

19.9 Hyaluronan is a glycosaminoglycan enriched in connective tissues

Key concepts

- HA is a glycosaminoglycan that forms enormous complexes with proteoglycans in the ECM. These complexes are especially abundant in cartilage, where HA is associated with the proteoglycan aggrecan, via a linker protein.
- HA is highly negatively charged and thus binds to cations and water in the extracellular space; this increases the stiffness of the ECM and provides a water cushion between cells that absorbs compressive forces.
- HA consists of repeating disaccharides linked into long chains.
- Unlike other glycosaminoglycans, HAs chains are synthesized on the cytosolic surface of the plasma membrane and translocated out of the cell.
- Cells bind to HA via a family of receptors known as hyaladherins, which initiate signaling pathways that control cell migration and assembly of the cytoskeleton.

HA, also known as hyaluronic acid or hyaluronate, is a GAG. (For more on GAGs see *19.8 Proteoglycans provide hydration to tissues*.) Unlike the other GAGs found in the ECM, it is not coupled covalently to proteoglycan core proteins; rather, it forms enormous complexes with secreted proteoglycans. One of the most important of these complexes is found in cartilage, where HA molecules secreted by chondrocytes (cartilage-forming cells) bind to as many as 100 copies of the proteoglycan aggrecan (see Figure 19.22). The aggrecan core proteins associate indirectly with a single HA molecule at 40 nm intervals, via a small link protein that binds to both HA and the aggrecan core protein. The aggregate can be more than 4 mm long and have a molecular mass exceeding 2×10^8 Da. In this way, HA acts to create large, hydrated spaces in the ECM of cartilage. These spaces are especially important in tissues with a low density of blood vessels because they facilitate diffusion of nutrients and wastes through the extracellular spaces.

The structure of HA is quite simple. Like all GAGs, it is a linear polymer of a disaccharide, specifically glucuronic acid linked to N-acetylglucosamine by a β(1–3) bond. HA molecules contain an average of 10,000 (and up to 50,000) of these disaccharides linked by a β(1–4) bond, as shown in Figure 19.19. Because these disaccharides are negatively charged, they bind up cations and water. Like proteoglycans, HA increases the stiffness of the ECM and serves as a lubricant in connective tissues such as joints. The hydrated HA molecules also form a water cushion between cells that enables tissues to absorb compressive forces.

HA molecules are much larger than other GAGs. Because they are so large, cells must expend an enormous amount of energy to make them. It is estimated that 50,000 ATP equivalents, 20,000 NAD cofactors, and 10,000 acetylCoA groups are required to make one average-sized HA chain. Consequently, synthesis of HAs is tightly controlled in most cells.

Synthesis of HA is catalyzed by transmembrane HA synthase enzymes in the plasma membrane. These enzymes are somewhat unusual, in that they assemble the HA polymer on the cytosolic face of the plasma membrane and then translocate the assembled polymer across the membrane into the extracellular space. This is entirely different from synthesis of other GAGs, which are synthesized in the Golgi complex and are covalently attached to proteoglycan core proteins as they pass through the secretory pathway (see *19.8 Proteoglycans provide hydration to tissues*).

An important means of regulating HA synthesis is by varying the expression of the HA synthase enzymes. Expression of these enzymes is induced by growth factors in a cell type-specific manner. For example, fibroblast growth factor and interleukin-1 induce expression in fibroblasts, while glucocorticoids suppress expression in these cells; epidermal growth factor stimulates expression in keratinocytes but not fibroblasts. Secretion of HA is controlled independently of HA synthesis, thereby providing at least two means of controlling HA levels in tissues.

In addition to its role in tissue hydration, HA binds to specific cell surface receptors, which results in stimulation of a multitude of intracellular signaling pathways that control processes such as cell growth, survival, differentiation, and migration. Virtually all human cells express at least one of these receptors. The principal HA receptor is CD44, which belongs to a family of related proteins, known as **hyladherins**, all of which bind to HA. Other members of this family include proteoglycans (e.g., versican, aggrecan, brevican) and the link protein that couples HA and aggrecan in cartilage. Multiple forms of CD44 are generated by alternative splicing of transcripts from a single CD44 gene, though the functional differences between these isoforms are not clear. CD44 exists as homodimers expressed on many cell types or as a heterodimer with the ErbB tyrosine kinase that is expressed on epithelial cells.

The cytoplasmic tail of CD44 has several functions. It is required for correct binding to HA and for sorting of the receptor to the cell surface. It is also required for effective intracellular signaling, as diagrammed in **FIGURE 19.23**. Mapping of the functional regions in the cytoplasmic tail of CD44 was accomplished by expressing mutated forms of CD44 in cultured cells and testing for activation of signaling pathways following adhesion to HA. From these studies, we know that CD44 homodimers and CD44/ErbB heterodimers activate nonreceptor tyrosine kinases such as Src, as well as members of the Ras family of small G proteins. These kinases activate downstream signaling proteins such as protein kinase C, MAP kinase, and nuclear transcription factors.

In addition, CD44-mediated signals can alter the assembly of the actin cytoskeleton at the cell surface, as Figure 19.23 shows, by activating actin binding proteins such as fodrin and the

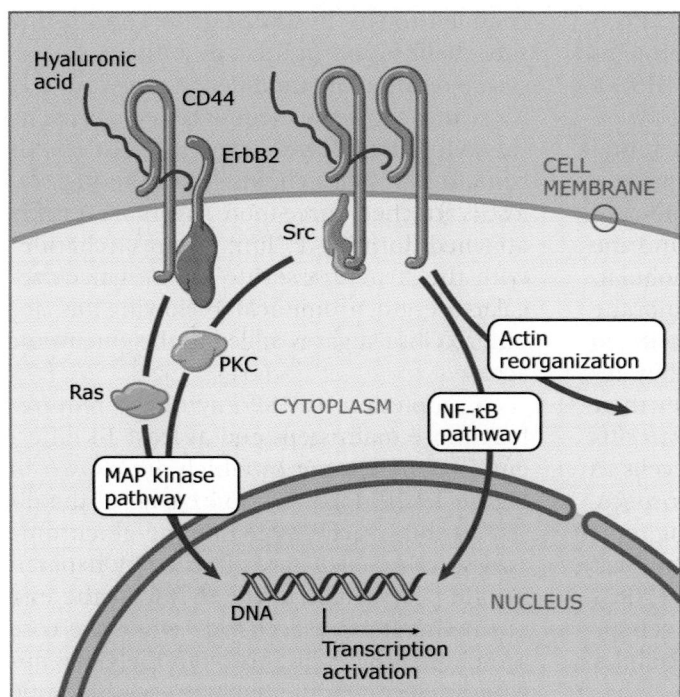

FIGURE 19.23 CD44 forms homodimers or heterodimers with ErbB2 receptors. These complexes bind to a number of signaling molecules that control cytoskeletal organization and gene expression.

small G protein Rac1 (see *12 Actin*). One of the consequences of this actin reorganization is that CD44-mediated binding to HA promotes cell migration. In tumors, increased CD44 expression and HA secretion correlate with increased tumor aggressiveness and poor prognosis.

It is generally thought that HA plays two roles in promoting cell migration. First, by binding to ECM molecules, it disrupts cell-cell and cell-matrix interactions. Mice that fail to express HA have much smaller spaces between cells and consequently do not develop properly. Because HA has such a large hydrated volume, increased secretion of HA in a tumor may disrupt the integrity of the ECM and create large spaces through which tumor cells may crawl. Second, HA binding to CD44 receptors may activate intracellular signaling pathways that lead directly to cytoskeletal rearrangements and increased cell migration (see *18 Principles of cell signaling*). Evidence for this comes from experiments in which HA is added to cells growing in culture. Cells that express CD44 migrate almost immediately after coming into contact with HA, and drugs that disrupt intracellular signaling molecules known to associate with CD44 inhibit this migration.

An intracellular pool of HA has been detected, but its function remains to be determined. Some HA remains in the cytosol after synthesis, and some newly secreted HA can be internalized by endocytosis. The amount of intracellular HA changes with the cell cycle and some cytoplasmic proteins that are involved in controlling cell proliferation also bind to HA. These observations suggest that HA may also serve as an internal signaling molecule that regulates cell division.

19.10 Heparan sulfate proteoglycans are cell surface coreceptors

Key concepts

- Heparan sulfate proteoglycans are a subset of proteoglycans that contain chains of the glycosaminoglycan heparan sulfate.
- Most heparan sulfate is found on two families of membrane-bound proteoglycans, the syndecans, and the glypicans.
- Heparan sulfates are composed of distinct combinations of more than 30 different sugar subunits, allowing for great variety in heparan sulfate proteoglycan structure and function.
- Cell surface heparan sulfate proteoglycans are expressed on many types of cells and bind to over 70 different proteins.
- Cell surface heparan sulfate proteoglycans act as coreceptors for soluble proteins such as growth factors and insoluble proteins such as ECM proteins, and they assist in the internalization of some proteins.
- Genetic studies in fruit flies show that heparan sulfate proteoglycans function in growth factor signaling and development.

Heparan sulfate proteoglycans (HSPGs) are defined as those proteoglycan core proteins that are attached to heparan sulfate (HS), a GAG (for more on GAGs see *19.8 Proteoglycans provide hydration to tissues*). HS is mostly found on two families of membrane-bound proteoglycans, the syndecans (which are proline-rich, extended transmembrane proteins) and the glypicans (which are cysteine-rich, globular, and weakly linked to the plasma membrane via their glycosylphosphatidylinositol-linked sugars) (see Figure 19.19). Because they remain attached to the cell surface after their assembly, these proteoglycans play a critical role in regulating the adhesion of cells to other components in the extracellular space, including structural proteins, signaling molecules, and other cells. In this section, we will first explore the structural diversity of HSPGs and then describe the biochemical and genetic evidence linking HSPGs to a variety of cellular functions.

As Figure 19.20 shows, HSPG synthesis begins in the medial Golgi when a xylose sugar is added to the hydroxyl group of a serine side chain found in the core protein. Not all serine residues are modified in this way, only those in a consensus sequence recognized by the xylose transferase enzyme. Most HSPGs contain 3–7 sugar chains. After the first xylose is attached, three more sugars are quickly attached, forming a "linker tetrasaccharide" with the structure serine-xylose-galactose-galactose-glucuronic acid. Following this, another xylose sugar is added to the glucuronic acid.

Completion of HSPG synthesis requires four more main steps and at least 14 different enzymes, some of which are shown in Figure 19.20. First, 50–150 copies of the disaccharide N-acetylglucosamine-glucuronic acid (GlcNAc-GlcA) are attached by heparan sulfate polymerase to the xylose at the end of the linker tetrasaccharide while the core protein is progressing through the Golgi apparatus. Second, other enzymes modify some of the GlcNAc sugars (at specific amino acid sequences) by replacing their N-acetyl groups with sulfate groups. Third, some GlcA sugars in the chain are epimerized to form iduronic acid. Finally, just before the proteoglycan leaves the Golgi apparatus, additional sulfates may be added to the iduronic acid and the remaining unmodified GlcNAc sugars. (The most heavily sulfated forms of HS are called heparin, a naturally occurring anticoagulant that is also used clinically.)

The result of all this sugar modification is a tremendous amount of structural variability in HSPGs. Due to the five structural modifications, 32 different disaccharide "building blocks" are possible, which gives rise to greater structural complexity in HS chains than is found in proteins, which are made up of 20 different amino acids. Because so many different forms of HS can be made on a single HSPG molecule, cells can express multiple different forms of HSPG at the same time, with each form folding into a slightly different shape and, thus, having different binding properties for extracellular proteins.

Not surprisingly, then, HSPGs bind specifically to over 70 extracellular proteins, some of which are listed in **FIGURE 19.24**. In many cases, the binding of a particular ligand depends on the exact sequence of sugars in the HS chains. The function of the HSPG binding can be organized into three classes, illustrated in **FIGURE 19.25**:

Category	Binding proteins
Morphogens	Wnt proteins
Coagulation	Factor Xa, Thrombin
ECM components	Fibrin, Fibronectin, Interstitial collagens, Laminins, Vitronectin
Growth factors (GFs)	Epidermal GFs, Fibroblast GF, Insulin-like GF, Platelet-derived GF
Tissue remodeling factors	Tissue plasminogen activator, Plasminogen activator inhibitor
Proteinases	Cathepsin G, Neutrophil elastase
Growth factor binding proteins (GF BPs)	Insulin-like GF BPs, Transforming GF BP
Anti-angiogenic factors	Angiostatin, Endostatin
Cell adhesion molecules	L-selectin, Neural cell adhesion molecule (NCAM)
Chemokines	C-C, CXC
Cytokines	Interleukin-2, -3, -4, -5, -7, -12; Interferon γ, Tumor necrosis factor-α
Energy metabolism	Apolipoproteins B and E, Lipoprotein lipase, Triglyceride lipases

FIGURE 19.24 HSPGs bind to many extracellular proteins and control a broad range of biological functions. A partial list of binding proteins is shown here. Adapted from M. Bernfield, et al., *Annu. Rev. Biochem.* 68 (1999): 729–777.

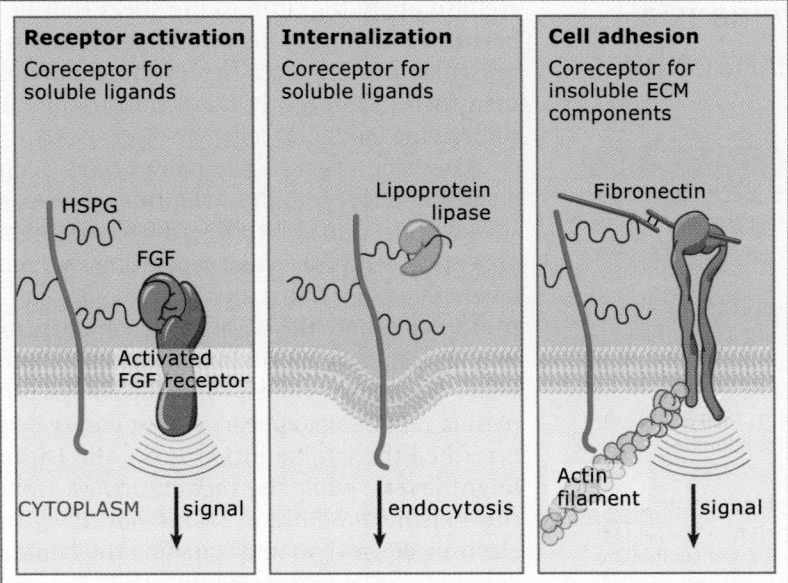

FIGURE 19.25 HSPGs act as coreceptors for growth factors, enzymes, and extracellular matrix proteins.

- HSPGs can act as coreceptors for soluble proteins such as growth factors by stabilizing the binding between the growth factor and its signaling receptor. This effectively increases the local concentration of a growth factor at the cell surface, thereby enhancing responsiveness to a given dose. For example, this interaction occurs between syndecans, fibroblast growth factor (FGF), and the FGF receptor.
- HSPGs can enhance internalization of some soluble proteins, such as those found in low-density lipoprotein particles.
- HSPGs can act as coreceptors for insoluble proteins such as ECM structural proteins or cell adhesion receptors. This helps establish a link between the extracellular domain of the receptor and the actin cytoskeleton, thereby preserving structural integrity between cells. (Due to their weak association with the plasma membrane, glypicans do not participate in this third function.)

Biochemical evidence for the interaction between HSPGs and their ligands comes largely from *in vitro* binding assays and coimmunoprecipitation data. These studies are complicated, however, by the fact that it is difficult to purify large quantities of a single type of HSPG, and their extensive posttranslational modification makes it difficult to functionally overexpress them in cells. Genetic analysis is a much more powerful tool for elucidating the function of HSPGs in development and disease.

One of the best model organisms for genetic analysis is the fruit fly, *Drosophila melanogaster*. To investigate the role of HSPGs in *Drosophila* development, researchers generated strains of flies with mutations either in the HSPG core protein or in the sugar processing enzymes required for HS synthesis. These flies exhibit the same phenotype as flies with mutations in genes for growth factors or their receptors, including a loss of activity in key enzymes linked to the receptors. Expression of additional copies of the wild-type growth factor receptors rescues flies bearing the HSPG mutations, strongly suggesting that the two phenotypes are linked.

Similar studies in mice demonstrate a multitude of functions for HSPGs that are difficult to detect in simpler model systems. For example, "knockout mice" that fail to express syndecan-1, a major cell surface HSPG, have compromised immune systems and a severely weakened wound healing response. Knockout of the HSPG perlecan in mice disrupts proper cartilage formation during development, and leads to severe skeletal deformities and early death.

19.11 The basal lamina is a specialized extracellular matrix

Key concepts

- The basal lamina is a thin sheet of ECM found at the basal surface of epithelial sheets and at neuromuscular junctions and is composed of at least two distinct layers.
- The basement membrane consists of the basal lamina connected to a network of collagen fibers.
- The basal lamina functions as a supportive network to maintain epithelial tissues, a diffusion barrier, a collection site for soluble proteins such as growth factors, and a guidance signal for migrating neurons.
- The components of the basal lamina vary in different tissue types, but most share four principal ECM components: sheets of collagen IV and laminin are held together by heparan sulfate proteoglycans and the linker protein nidogen.

The term **basal lamina** refers to a thin sheet (or lamina) of ECM that lies immediately adjacent to, and is in contact with, many cell types. The basal lamina is recognized as a distinct form of ECM because it contains proteins, such as collagen IV and nidogen, found only in this structure and because it adopts a distinct, sheet-like arrangement. Originally this term applied only to the sheet of ECM in contact with the basal surface of epithelial cells (hence the term *basal*) where it was first seen with an electron microscope. Now that the major constituents of the basal lamina are identified, we also apply the term to the sheet that lies between muscle and nerve cells at the neuromuscular junction, because this sheet contains many of the same proteins as the basal laminae underlying epithelial cells.

Over the years, many names have been given to this layer of ECM. When viewed with a scanning electron microscope, the basal lamina appears as a distinct sheet separating two cell layers. When viewed with a transmission electron microscope, the basal lamina appears as two layers, each approximately 40 to 60 nm wide. The region closest to the epithelial cell plasma membrane appears almost empty and is termed the lamina lucida (from the Latin, bright layer), while the region furthest from the plasma membrane stains darkly with electron-dense dyes and is named the lamina densa (dense layer) (see Figure 19.57). Beyond the lamina densa lies a network of collagen fibers that is sometimes called the reticular lamina; under a light microscope, the basal lamina and reticular lamina appear as a single boundary, often called the basement membrane, as shown in **FIGURE 19.26**. Often, the terms *basal lamina* and *basement membrane* are used interchangeably.

The basal lamina performs the following four principal functions:

1. It serves as the structural foundation underneath epithelial cell layers. Cells attach to laminin and collagen fibers in the basal lamina through specialized structures known as hemidesmosomes, which also connect to the intermediate filament network (see *19.20 Hemidesmosomes attach epithelial cells to the basal lamina*). In this way, the basal lamina connects the intermediate filament networks of several cells, which strengthens the tissue. This is especially prevalent in the skin, which is a very tough organ.

2. It is a selectively permeable barrier between epithelial compartments. The proteoglycans in the basal lamina trap particulate matter (dead cells, bacteria, etc.), thereby containing infections, and assisting the immune system.

3. Proteoglycans in the basal lamina bind, immobilize, and concentrate soluble ligands (such as growth factors) from tissue fluid. This enhances cellular access to growth factors and, in some cases, facilitates binding by growth

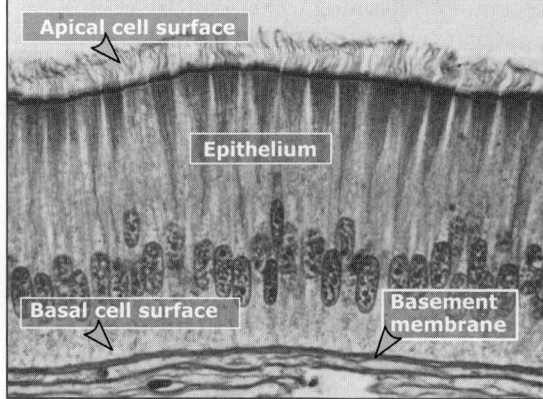

FIGURE 19.26 The basement membrane appears as a thin layer of protein immediately under epithelial cells. Photo reproduced from W. Bloom and D. W. Fawcett. *A Textbook in Histology*, 1986. Used with permission of Don W. Fawcett, Harvard Medical School.

factor receptors (see *19.8 Proteoglycans provide hydration to tissues*).

4. Laminin proteins in the basal lamina serve as a guidance signal to the growth cones of developing neurons. This is one of the ways in which the long projections extending from neurons find their cellular targets.

Given this broad range of functions, it is not surprising that the molecular components of the basal lamina can vary in differing tissues, or even over time in the same tissue. Isolating these components has proved difficult because the basal lamina represents such a small percentage of the ECM found in most tissues. Fortunately for researchers, the discovery of a mouse chondrosarcoma tumor that secretes large amounts of "basement membrane" proteins allowed detailed analysis of basal lamina components. Nearly 20 different proteins have now been identified in the basal lamina.

There are four principal components found in nearly all basal laminae. These are type IV collagen, laminin, heparan sulfate proteoglycans, and entactin (also known as nidogen). (For more on collagens, laminins, and heparan sulfate proteoglycans see *19.2 Collagen provides structural support to tissues*; *19.5 Laminins provide an adhesive substrate for cells*; and *19.10 Heparan sulfate proteoglycans are cell surface coreceptors*.) A current model suggests how these components are woven into a sheetlike configuration that defines the basal lamina.

In this model, shown in Figure 19.15, Type IV collagen and laminin polymerize to form networks. These networks are stacked upon one another to form layers and are held together by bridging proteins such as the heparan sulfate proteoglycan perlecan and entactin, which are able to bind to both networks. Other components, such as the laminin-332 and Type VII collagen filaments that bind to hemidesmosome proteins, are interwoven between the layers. How these additional proteins associate with the principal components is unknown, though there is some evidence that cell contact, via integrin receptors, is responsible for proper assembly of an intact basal lamina, while the presence of entactin is not required. Once assembled, the basal lamina forms a tightly woven, complex web of proteins that provides enough structural stability to support epithelial tissues yet is porous enough to act as a selective filter of the extracellular fluid.

19.12 Proteases degrade extracellular matrix components

Key concepts

- Cells must routinely degrade and replace their ECM as a normal part of development and wound healing.
- ECM proteins are degraded by specific proteases, which cells secrete in an inactive form.
- These proteases are only activated in the tissues where they are needed. Activation usually occurs by proteolytic cleavage of a propeptide on the protease.
- The matrix metalloproteinase (MMP) family is one of the most abundant classes of these proteases and can degrade all of the major classes of ECM proteins.
- MMPs can activate one another by cleaving off their propeptides. This results in a cascade-like effect of protease activation that can lead to rapid degradation of ECM proteins.
- ADAMs (a disintegrin and metalloproteinases) are a second class of proteases that degrade the ECM; these proteases also bind to integrin ECM receptors and, thus, help regulate ECM assembly as well as its degradation.
- Cells secrete inhibitors of these proteases to protect themselves from unnecessary degradation.
- Mutations in the matrix MMP-2 gene give rise to numerous skeletal abnormalities in humans, reflecting the importance of ECM remodeling during development.

So far in this chapter, we have discussed the critical roles of ECM molecules in regulating cellular behavior in multicellular organisms. But organisms also make proteases that destroy the ECM. Why would an organism want to remove the very molecules that are holding it together? The simplest answer is that like cells, the ECM has to be plastic, that is, able to respond to changes in environmental conditions. For example, the ECM that surrounds a developing neuron may not be the type required to maintain that neuron once it has fully differentiated and is replaced when the neuron matures. Or the ECM that forms in the webbing between our fingers and toes during early development is no longer required at later stages, so it is removed altogether. In addition, injuries and infections can do considerable damage to tissues, and during the wound healing process, a damaged part is sometimes degraded and a new one built in its place. Finally, the small peptides produced by the action of these proteases on ECM proteins promote cell migration and, thus, stimulate wound repair. These pep-

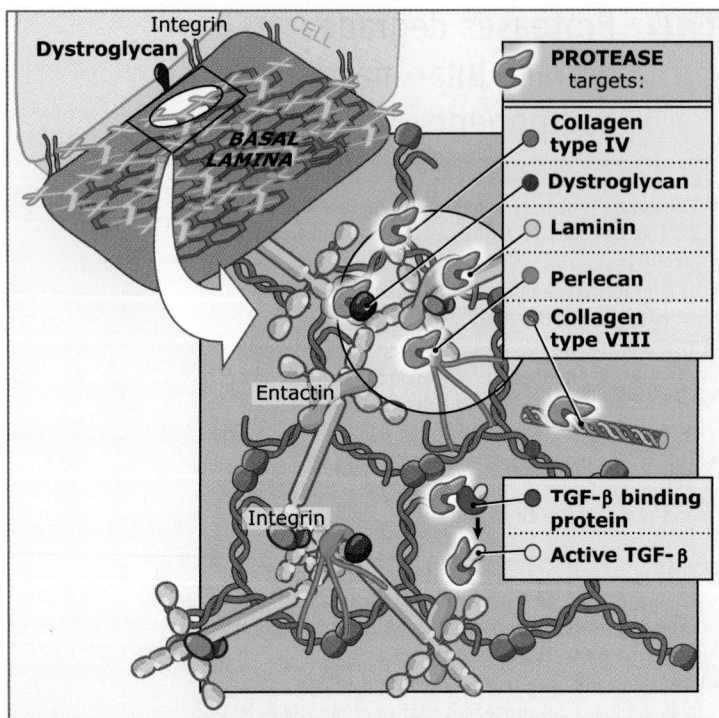

FIGURE 19.27 There are several possible targets in the ECM for extracellular proteases. In some cases, digestion of ECM proteins liberates fragments that are functionally active.

Enzymes		Examples of substrates
COLLAGENASES	Collagenase-1 (MMP-1)	• Collagen I, II, III, VII, VIII, X
	Collagenase-2 (MMP-8)	• Collagen I, II, III
	Collagenase-3 (MMP-13)	• Aggrecan, collagen I, II, III, IV, IX, X, XIV; fibronectin, gelatin, laminin
GELATINASES	Gelatinase A (72 kDa) (MMP-2)	• Collagen I, IV, V, VII, X; gelatin, fibronectin
	Gelatinase B (92 kDa) (MMP-9)	• Collagen IV, V, VII, XI, XIV; elastin, gelatin
MEMBRANE-TYPE MMPs	MT1-MMP (MMP-14)	• Aggrecan, collagen I, II, III; gelatin, entactin, fibrin, laminin, perlecan, vitronectin
	MT2-MMP (MMP-15)	• Aggrecan, entactin, fibronectin, laminin, perlecan
	MT3-MMP (MMP-16)	• Cartilage, collagen III, fibronectin, gelatin, laminin
	MT4-MMP (MMP-17)	• Gelatin
	MT5-MMP (MMP-24)	• Not determined
	MT6-MMP (MMP-25)	• Not determined
STROMELYSINS	Stromelysin-1 (MMP-3)	• Aggrecan, collagen IV, V, IX, X; elastin, entactin, fibronectin, gelatin, laminin
	Stromelysin-2 (MMP-10)	• (Same as for MMP3)
STROMELYSIN-LIKE MMPs	Stromelysin-3 (MMP-11)	• Serpins
	Matrilysin (MMP-7)	• Collagen IV, elastin, entactin fibronectin, laminin
	Metalloelastase (MMP-12)	• Collagen IV, elastin, gelatin, fibronectin, laminin, vitronectin
OTHER MMPs	MMP-19	• Gelatin
	Enamelysin (MMP-20)	• Amelogenin
	MMP-23	• Synthetic MMP substrate
	MMP-26	• Gelatin, synthetic MMP substrate

FIGURE 19.28 The six classes of matrix metalloproteinases in humans. Adapted from L. Ravanti and V. -M. Kähäri, *Int. J. Molec. Med.* 6 (2000): 391–407.

tides can also stimulate tumor cell migration (see *17 Cancer—Principles and overview*). Lastly, digestion of the ECM sometimes liberates other useful compounds, such as growth factors and hormones, which become "trapped" in the web formed by large ECM proteins. The range of possible targets in the extracellular space for proteolysis is shown in **FIGURE 19.27**.

As you may expect, there are nearly as many proteases produced by cells as there are extracellular proteins to degrade. Dozens of proteases are secreted into the extracellular space by a host of cell types, and most of these are free to circulate in the bloodstream. The potential danger, of course, is that these proteases may bind to and degrade perfectly healthy, normal tissues. So, many of them are secreted in an inactive form and are only activated in the tissues where they are needed. This way, random binding to healthy tissue is not a problem. Think of the protease as a pocket knife that is normally folded into the handle to hide the blade; the blade is exposed only when it is needed, and the knife is used until it wears out or is jammed by another object (in the case of the protease, this would be an inhibitor molecule).

There are three main families of ECM proteases, the MMPs, the ADAM family, and the closely related ADAMTS, which contain both metalloproteinase and disintegrin domains. While all three families degrade ECM proteins, the ADAM proteases may also support integrin-based cell adhesion. ADAMs are transmembrane proteases expressed on the cell surface, where they degrade ECM proteins as well as other cell surface receptors. ADAMTSs are secreted into the extracellular space, where they bind to glycosaminoglycans and degrade proteoglycans.

At least 24 MMPs are identified in humans, and based on their structure and substrate specificity, they are classified into six groups, listed in **FIGURE 19.28**.

All MMPs share some common features, as **FIGURE 19.29** shows, which are listed below:

- A signal peptide that directs the proteases to be secreted from the cells that make them.
- A highly conserved zinc ion-binding site in the catalytic domain.
- An N-terminal propeptide that acts as the "handle" in our pocket knife analogy: this portion of the protein folds back so that a conserved cysteine amino acid can form a covalent bond with a

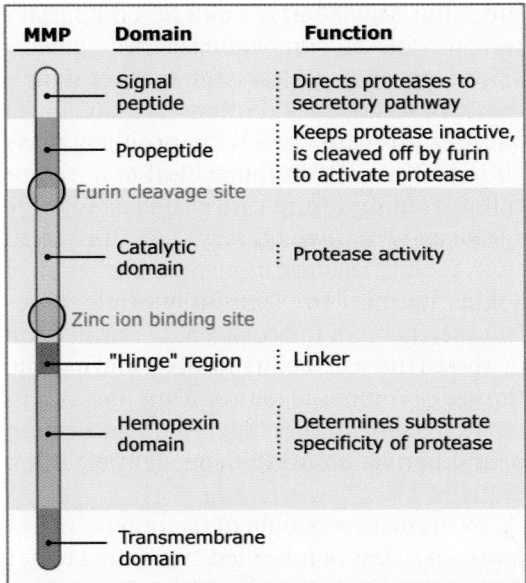

MMP	Domain	Function
	Signal peptide	Directs proteases to secretory pathway
	Propeptide	Keeps protease inactive, is cleaved off by furin to activate protease
	Furin cleavage site	
	Catalytic domain	Protease activity
	Zinc ion binding site	
	"Hinge" region	Linker
	Hemopexin domain	Determines substrate specificity of protease
	Transmembrane domain	

FIGURE 19.29 All six classes of matrix metalloproteinases share common structural features. A "generic" MMP is shown to indicate the relative positions of these features within MMP family members. Some MMPs lack one or more of these features, but all MMPs have the propeptide and catalytic domain.

zinc ion in the catalytic site, thereby inhibiting activity of the protease (i.e., known as the "cysteine switch" mechanism for regulating activation of MMPs). This propeptide is cleaved off by furin or a related enzyme when the protease is activated.

- A hemopexin domain, which determines the substrate specificity of the protease.
- A proline-rich "hinge" region that links the catalytic domain to the hemopexin domain.

The membrane-type MMPs also contain a transmembrane domain at the carboxyl terminus that anchors the protein in the plasma membrane.

Together, the MMP family of proteases is capable of digesting every ECM glycoprotein discussed in this chapter, as well as several proteoglycans. In many cases, the MMPs work together in groups. What makes them especially interesting is that they can activate other MMPs by cleaving off their propeptides. MMP-3 can activate MMP-7, which in turn can activate MMP-2, for example. This allows for rapid degradation of multiple substrates. For example, the combination of MMP-2, -3, and -7 can digest over a dozen different ECM proteins. Usually, the cascade of proteolysis begins with capture of an activated protease at the site of degradation. For instance, the protease plasmin is activated in blood clots and activates multiple MMPs near the clot to initiate the process of tissue remodeling during wound healing.

Many MMPs cleave more than just ECM proteins and each other. Additional MMP targets include cell-cell adhesion molecules (e.g., E-cadherin; see *19.22 Calcium-dependent cadherins mediate adhesion between cells*), signaling proteins (e.g., EGF, IGF, TGFβ; see *18.24 Protein phosphorylation/dephosphorylation is a major regulatory mechanism in the cell*) and numerous mediators of the inflammatory response (e.g., fibrin and monocyte chemoattractant proteins).

The ADAM family consists of almost 30 different proteases identified in a range of species, and 15 contain the conserved, zinc-binding catalytic site found in MMPs and many other proteases. ADAM proteases also contain a propeptide (or prodomain) that, as in MMPs, functions as an inhibitor of the catalytic site. ADAM proteases can degrade a host of extracellular proteins, including those found in the ECM. One important role for these proteases during development and disease progression is the cleavage of Notch receptors on the external side of the cell surface. This step is required for further cleavage of the receptor in the cytosol by γ-secretase and subsequent translocation of the cleavage product, intracellular domain of Notch (NICD), into the nucleus. Upon reaching the nucleus, NICD controls the expression of numerous genes and thereby influences the development of numerous organ systems in multicellular animals (see *18 Principles of cell signaling*). ADAMs can also shut the Notch signaling pathway down by cleaving the Notch ligands (in mammals, these are Deltalike 1, 3, and 4, and Jagged 1 and 2). They also contain a domain that is found in a class of proteins known as disintegrins, which bind to the major integrin receptor found on platelets and inhibit platelet aggregation during blood clot formation. The disintegrin domain in ADAMs is likely to function as an integrin substrate, thereby allowing cell-cell adhesion. Thus, while the functions of ADAMs are quite diverse, they do serve as regulators of ECM structural integrity.

ADAMTS proteases exhibit more substrate specificity. One of the most clinically significant substrates is aggrecan, an extremely important proteoglycan in cartilage (see *19.8 Proteoglycans provide hydration to tissues*). Overexpression and/or enhanced activity of these so-called aggrecanases is closely linked to the onset and progression of osteoarthritis caused by the degradation of articular cartilage. In particular, knockout

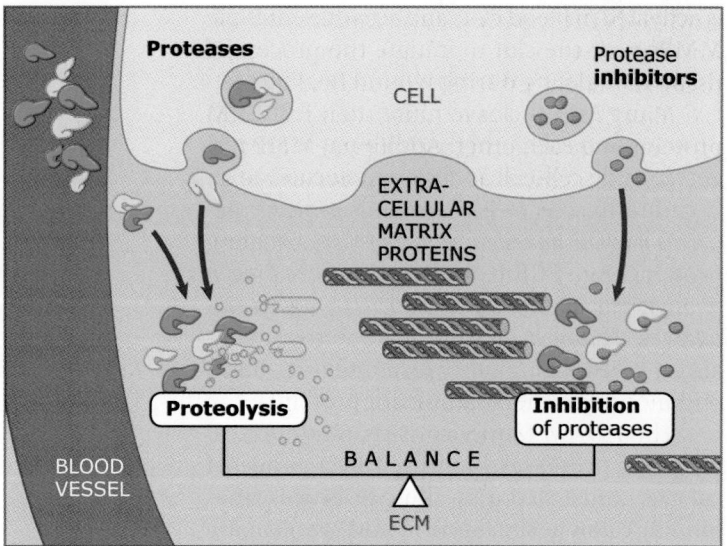

Proteases

CELL

Protease inhibitors

EXTRA-CELLULAR MATRIX PROTEINS

Proteolysis

Inhibition of proteases

BLOOD VESSEL

BALANCE

ECM

FIGURE 19.30 Cells secrete proteases to digest extracellular matrix proteins and also secrete protease inhibitors to inhibit this degradation. This allows fine-tuned control of extracellular matrix degradation and remodeling within tissues.

mice lacking functional ADAMTS-4 and -5 genes failed to develop osteoarthritis, even when a surgical method was used to intentionally destabilize a knee joint in these animals.

MMPs, ADAMs, and ADAMTSs can be inhibited by a class of soluble proteins known as tissue inhibitors of metalloproteases (TIMPs). Cells secrete TIMPs to protect themselves (and their surrounding matrix) from degradation by activated proteases. In some cases, cells secrete ECM proteins as well as the proteases that degrade them and their inhibitors, thereby establishing a highly sensitive balance between matrix assembly and degradation, as illustrated in **FIGURE 19.30**. When this system becomes unbalanced, as in tumors, tissues can be rapidly degraded and tumor cells can more readily escape into the circulation. For this reason, TIMPs are being investigated as potential anticancer drugs.

Because cells use integrin receptors to bind to a majority of ECM proteins, digestion of these proteins may have significant consequences on integrin-mediated functions such as adhesion, migration, and signaling (see *19.13 Most integrins are receptors for extracellular matrix proteins*). In some cases, digestion of the ECM molecules results in the liberation of functionally active proteolytic fragments that are hidden in the intact molecules. This can then trigger a cascade of matrix digestion events. For example, receptor-mediated binding of an ECM molecule can trigger a signaling cascade resulting in the secretion of a protease that partially digests the ECM molecule, revealing a cryptic binding site in the fragments for a different cell surface receptor. Subsequent binding of the fragments via this receptor triggers yet another signaling cascade, and may result in secretion of yet another protease. This scenario likely plays an important role in the evolution of tumor cell migration during cancer metastasis (see *17 Cancer—Principles and overview*). One recent study found that integrin-mediated adhesion of a skin cancer cell to a laminin substrate stimulated secretion of the collagenase MMP-9 and increased the cells' ability to digest and migrate through a composite matrix. Thus, one receptor-ECM combination triggered the activation of another via an MMP-dependent digestion of the ECM.

A dramatic example of the importance of MMPs is a class of inherited "vanishing bone" diseases known as Winchester syndrome, Torg syndrome, Nodulosis-Arthropathy-Osteolysis syndrome, and Al-Aqeel Sewairi syndrome. Afflicted individuals exhibit numerous skeletal problems associated with bone loss, including resorption of wrist and ankle bones, crippling arthritic changes, severe osteoporosis, and distinctive facial abnormalities. The source of these inherited disorders is traced to autosomal recessive inheritance of mutated forms of the MMP-2 gene that results in no active MMP-2 production.

19.13 Most integrins are receptors for extracellular matrix proteins

Key concepts

- Virtually all animal cells express integrins, which are the most abundant and widely expressed class of ECM protein receptors.
- Some integrins associate with other transmembrane proteins.
- Integrins are composed of two distinct subunits, known as α and β chains. The extracellular portions of both chains bind to ECM proteins, while the cytoplasmic portions bind to cytoskeletal and signaling proteins.
- In vertebrates, there are many α and β integrin subunits, which combine to form at least 24 different αβ heterodimeric receptors.
- Most cells express more than one type of integrin receptor, and the types of receptor expressed by a cell can change over time or in response to different environmental conditions.
- Integrin receptors bind to specific amino acid sequences in a variety of ECM proteins; all of the known sequences contain at least one acidic amino acid.

Cells bind to ECM proteins through specific receptors. The **integrin** family of proteins is the best known group of these receptors. Integrins bind to ECM proteins and, in some cases, to membrane proteins expressed on the surface of other cells. Virtually all animal cells express integrin receptors. Integrins appear to be the principal cell surface proteins responsible for holding tissues together (see *19.15 Integrins and extracellular matrix molecules play key roles in development*). Integrins connect the ECM to intracellular signaling proteins and the cytoskeleton (see *19.14 Integrin receptors participate in cell signaling*).

To understand how integrins function, we need to know their structural organization. Integrin receptors consist of two different polypeptides, called α and β subunits, which cross the membrane once and associate non-covalently to form the heterodimeric receptor. Based on several types of experimental data, including X-ray analysis of crystals of five different integrin receptors, a detailed and complex model of integrin structure has emerged, as **FIGURE 19.31** shows. Each subunit contains several domains that contribute to the function of the intact receptor. For the α-chain, these domains include a structure called a β-propeller, which is found at the N-terminus of the subunit, in the extracellular portion of the protein. The propeller contains seven 60-amino acid repeats that form the "blades" of the propeller, plus three or four motifs called EF hands, which bind to divalent cations such as Ca^{2+}. Half of the known α subunits have an additional domain termed the I domain, which interacts with an Mg^{2+}/Mn^{2+} metal

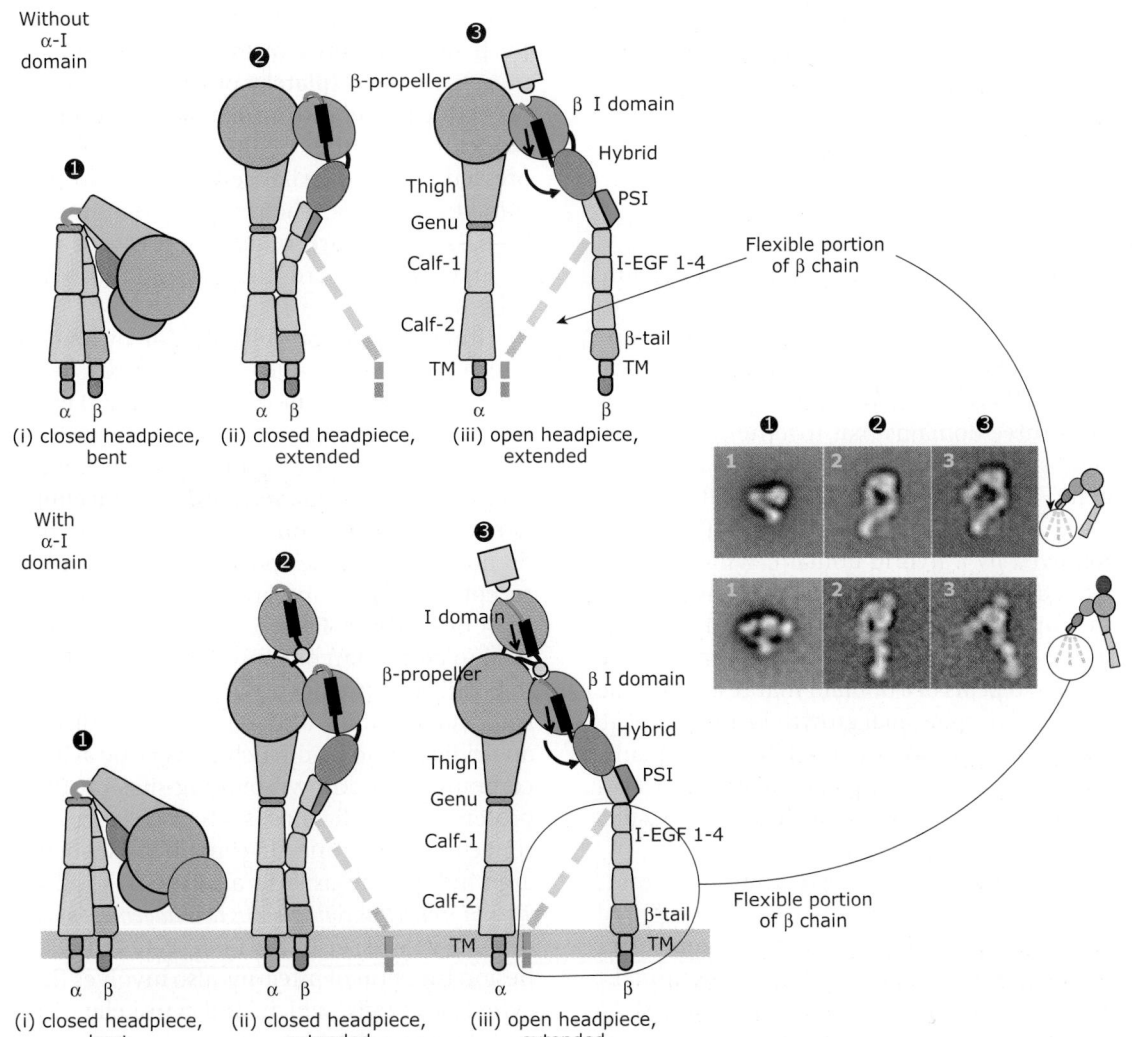

FIGURE 19.31 Model of integrin structure. Top row, three conformations of integrins lacking the alpha-I domain. Bottom row, three conformations of integrins containing alpha-I domain. Note that the juxtamembrane portion of the beta chain in each is flexible, and can adopt several different orientations in the "open" conformations of the integrins. Right, electron micrographs of integrins corresponding to each drawing. Reprinted, with permission, from the *Annual Reviews of Immunology*, Volume 25 © 2007 by Annual Reviews www.annualreviews.org. Courtesy of Bing-Hao Luo, Loyola University of Chicago.

Class		Ligands	Location/Function
β₁	α₁	Collagens, laminin	Extracellular matrix
	α₂	Collagens, laminin	
	α₃	Fibronectins, laminin, thrombospondin	
	α₄	Fibronectin, vascular cell adhesion molecule-1	Cell-cell adhesion
	α₅	Collagen, fibronectin, fibrinogen	Extracellular matrix / Blood clotting
	α₆	Laminin	Extracellular matrix
	α₇	Laminin	
	α₈	Cytotactin/tenascin-C, fibronectin	
	α₉	Cytotactin/tenascin-C	
	α₁₀	Collagens	
	α₁₁	Collagens	
β₂	α_D	Intercellular adhesion molecule-3, vascular adhesion molecule-1	Cell-cell adhesion
	α_L	Intercellular adhesion molecules 1-5	
	α_M	C3b	Host defense
		Fibrinogen, factor X, intercellular adhesion molecule-1	Blood clotting / Cell-cell adhesion
	α_X	Fibrinogen,	Blood clotting
		C3b	Host defense
β₃	α_Ib	Collagens	Extracellular matrix
	α_IIb	Collagens, fibronectin, thrombospondin, vitronectin, fibrinogen, von Willebrand factor, plasminogen, prothrombin	Blood clotting
	α_V	Collagen, fibronectin, laminin, osteopontin, thrombospondin, vitronectin,	Extracellular matrix
		disintegrin, fibrinogen, prothrombin, von Willebrand factor,	Blood clotting
		matrix metalloproteinase-2	Protease

FIGURE 19.32 Integrins are organized into subgroups that share β subunits.

ion-dependent adhesion site (MIDAS) found in the β subunit. For these integrins, the αI domain interactions with MIDAS determine the ligand specificity for each receptor. Closer to the plasma membrane, the α subunit contains three domains that together make up a "leg" structure, and these are called the thigh, genu, calf1, and calf2 domains. All β chains have a conserved N-terminal PSI domain, followed by a hybrid domain, which is connected to a globular βI domain that contacts the β-propeller domain of the α chain. Closer to the membrane, the β chain contains three flexible repeats of a domain that resembles the structure of epidermal growth factor (the EGF domain), followed by a β-tail domain. Finally, both the α- and β-chains contain a single transmembrane domain, and a short cytoplasmic domain at the C-terminus.

There are currently 18 α- and 8 β-subunits known in vertebrates. (Most are numbered consecutively, while some are given letter names that reflect how they were identified.) These 26 subunits interact to form at least 24 different αβ-receptor combinations. In ad-

dition, variants of some subunits arise from alternative splicing, which yields further alternatives in subunit composition. Most cells express more than one type of integrin receptor, and the types of receptors expressed can vary during development or in response to specific signals.

Why are there so many integrins? Genetic knockout of some integrin subunits is lethal in developing organisms, while knockout of other integrins appears to have a mild effect, suggesting that some of these receptors may be able to compensate for one another. This ability to compensate is known as functional redundancy. (For further details see *19.15 Integrins and extracellular matrix molecules play key roles in development.*)

Integrins are classified into three subfamilies based on the β subunits, as listed in **FIGURE 19.32**. The β₁-integrins bind mostly to ECM proteins and are by far the most widely expressed group of integrins. The β₂-integrins are expressed only by leukocytes and some bind to other cell surface proteins. Some of the β₃-integrins are expressed on platelets and megakaryocytes (platelet precursor cells), and play critical roles during platelet adhesion and blot clotting. Other β₃-integrins are also expressed on endothelial cells, fibroblasts, and some tumor cells. The receptors that include β₄–β₈-subunits are relatively few and very diverse, so they are not classified into any subgroups.

Integrins support cell adhesion by binding directly to an ECM protein using the extracellular domains of both the α- and β-chains. Ligand specificity is determined almost exclusively by the αI-domain for those integrins that contain it, and by a combination of the extracellular domains of the α- and β-chains for those that do not. With the exception of the fibronectin receptor α₅ β₁, all integrins can bind to more than one ligand. Each ECM protein also can bind to more than one integrin. Although it is impossible to predict an integrin binding site based on the amino acid sequence of the ligand, an acidic amino acid (such as aspartic acid) is common to all known binding sites on ECM proteins. Many ligands, such as collagen, vitronectin, and fibronectin contain the sequence arginine-glycine-aspartic acid (RGD) (e.g., see *19.3 Fibronectins connect cells to collagenous matrices*). As we will see in the next section, the adhesion function of integrins also involves their cytoplasmic tails, which bind cytoskeletal and signaling proteins.

19.14 Integrin receptors participate in cell signaling

Key concepts

- Integrins are signaling receptors that control both cell binding to ECM proteins and intracellular responses following adhesion.
- Integrins have no enzymatic activity of their own. Instead, they interact with adaptor proteins that link them to signaling proteins.
- Two processes, affinity modulation (varying the binding strength of individual receptors) and avidity modulation (varying the clustering of receptors) regulate the strength of integrin binding to ECM proteins.
- Changes in integrin receptor conformation, resulting from changes at the cytoplasmic tails of the receptor subunits or in the concentration of extracellular cations, are central to both types of modulation.
- In inside-out signaling, changes in receptor conformation result from intracellular signals that originate elsewhere in the cell (e.g., at another receptor).
- In outside-in signaling, signals initiated at a receptor (e.g., upon ligand binding) are propagated to other parts of the cell.
- Cells form five different integrin-based adhesion complexes with different structure and signaling functions.
- Many of the integrin signaling pathways overlap with growth factor receptor pathways.

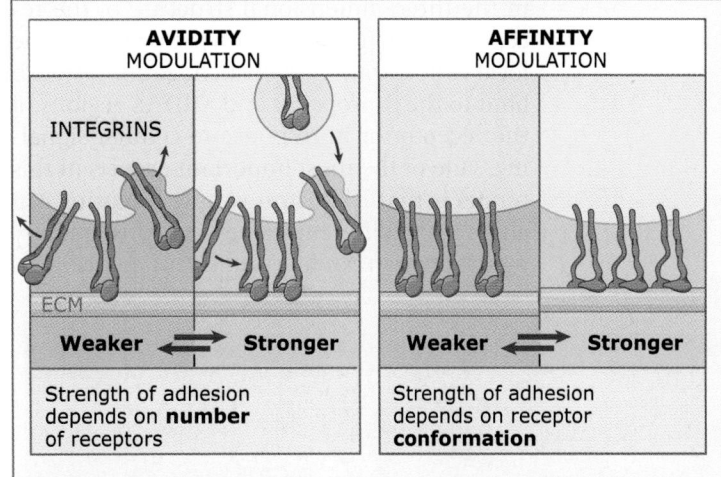

FIGURE 19.33 Avidity modulation refers to changes in the density of receptors on the cell surface, while affinity modulation refers to changes in binding strength by individual receptor proteins.

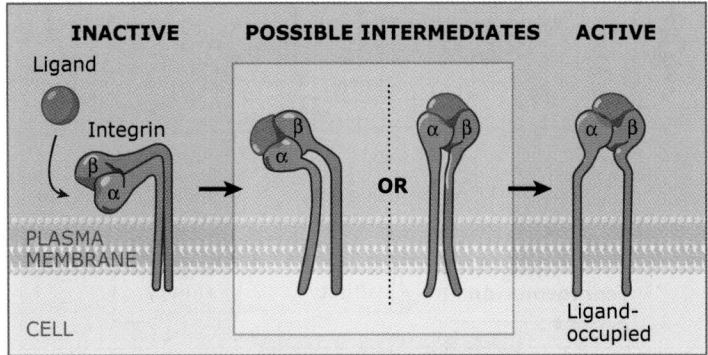

FIGURE 19.34 In this model for integrin activation, the extracellular portion of an inactive receptor is folded in toward the plasma membrane. As the integrin receptor straightens, it binds ligand and becomes active. The fully straightened form has the highest affinity for ligand.

The binding of integrin receptors to ECM proteins can be altered in response to environmental signals. This is useful, for example, to control the adhesion of platelets circulating in the blood stream or to permit the cyclical binding and release of the ECM during cell migration (see *17 Cancer—Principles and overview*). Cells use two complementary mechanisms for controlling the degree of integrin binding, as illustrated in **FIGURE 19.33** and they are listed below:

1. **Affinity modulation**, in which a change in receptor conformation results in altered affinity for its ligands.
2. **Avidity modulation**, in which the number of contacts formed between integrins and ECM proteins changes.

Signaling pathways inside the cell can control both processes, and changes in integrin conformation are central to both types of modulation.

Based on X-ray analysis of crystallized integrin receptors, a model for integrin activation has emerged, as shown in **FIGURE 19.34**.

In the inactive form, the extracellular portion of both the α- and β-chains is bent backward toward the plasma membrane, and the receptor does not bind to its ligand. The genu and hybrid domains act as hinges for the α- and β-subunits, respectively. Activation of an integrin requires that this extracellular portion of the receptor be straightened out; it is thought that the straighter the receptor, the higher its affinity state for ligand. At some point in this straightening process, the receptor binds to its ligand with low affinity, and it is possible that the ligand binding will even assist in the further straightening of the receptor. One conformational change that correlates well with integrin activation is the separation of the α- and β-cytoplasmic tails. Further changes in the three-dimensional structure of the receptor can be triggered either by altering the local concentration of divalent cations, which bind to the β-propeller and MIDAS regions of the receptor or in response to cellular signaling. One of the most important aspects of this model is that it illustrates how integrins can adopt many different shapes and consequently many different binding states. This model helps

explain how cells can fine-tune their binding to ECM proteins.

The concentration of divalent cations in the extracellular fluid can change, as part of maintaining the salt balance in the body. Since integrins bind to Ca^{2+} and Mg^{2+} ions, this change can alter their conformation and, therefore, affect their binding to extracellular proteins (see *19.13 Most integrins are receptors for extracellular matrix proteins*). Typically, in *in vitro* assays, removal of cations results in folding of the receptor and loss of adhesion, while addition of cations restores adhesion by straightening the receptor.

Modulation of integrin function also occurs in response to signaling initiated by other receptors, such as growth factor receptors. This is known as "inside-out signaling," which is illustrated in **FIGURE 19.35**. The signal is propagated in the cytoplasm and terminates at the extracellular portions of the integrin subunits. The biochemical basis for integrin activation during inside-out signaling is based on the binding of cytsolic proteins such as talin or tensin to the cytoplasmic tail of the β-subunit, which separates it from the α subunit and forces straightening of the extracellular portion of the receptor.

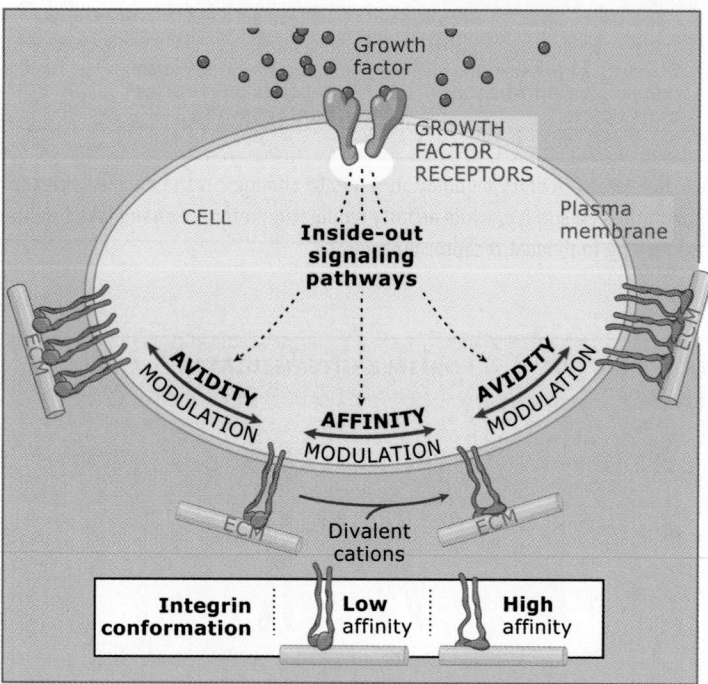

FIGURE 19.35 Signals that initiate in the cytoplasm (e.g., from growth factor signaling) are communicated to the cytoplasmic portion of integrin receptors. These signals result in avidity or affinity modulation of integrins at the cell surface. Thus, information is said to pass inside-out, from the cytoplasm to the extracellular space.

This transmembrane shape change is an important part of signal transduction, because it affects how integrins interact with other cytosolic proteins. A model for how the cytoplasmic tails of integrin receptors undergo conformational changes in response to changes in extracellular conditions or to inside-out signaling is shown in **FIGURE 19.36**. This model suggests that the spreading apart of the cytoplasmic tails during integrin activation reveals binding sites for intracellular signaling and adaptor molecules. In the inactive state, the integrin is in a closed conformation, and signaling molecules do not bind.

In avidity modulation, the clustering of integrin receptors on the cell surface increases upon activation, as diagrammed in Figure 19.33. Individual integrins widely distributed over the cell surface, form comparatively weak bonds with their substrate. However, if the same number of integrins diffuse in the plane of the membrane to form clusters on the cell surface, the dense concentration of bonds they form is more likely to withstand tensile forces. Thus, widely distributed receptors are less likely to support adhesion than clustered receptors, even when the number and affinity of the receptors does not vary.

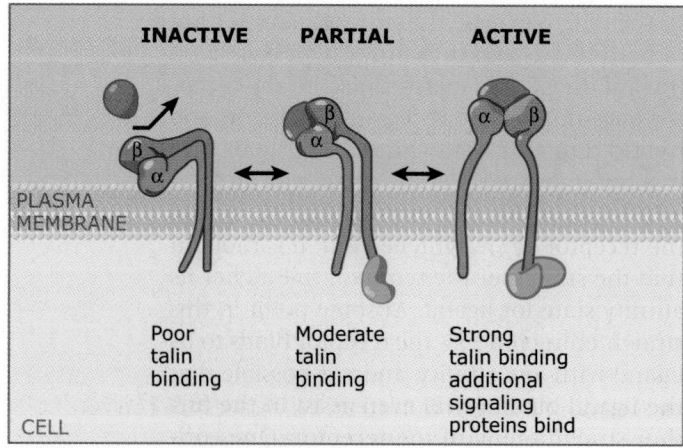

FIGURE 19.36 Two models for how integrins might be activated to bind an extracellular ligand and to expose binding sites for cytosolic proteins. The activation signal itself may be extracellular or intracellular.

When integrin receptors cluster on the cell surface, the cytoplasmic tails of the α- and β- subunits serve as docking sites for the assembly of various proteins. Based on studies conducted with purified ECM proteins in cell culture, plus in vivo studies, integrin clusters are classified into five types, as illustrated in **FIGURE 19.37**. Those that contain β_1-, β_2-, and β_3-integrins form four different linkages with the actin cytoskeleton (see *12 Actin*), and hemidesmosomes, which contain the $\alpha_6\beta_4$-integrin, link to the intermediate filament network (see *19.20 Hemidesmosomes attach epithelial cells to the basal lamina*). **Focal contacts** are the first integrin clusters to form at the leading edge of migrating cells, and are induced by stimulation of actin filament growth near the plasma membrane. If a focal contact

generates a stable link between the ECM and actin filaments, sufficient to resist mechanical force (supplied by myosin pulling on the actin filaments: see *12.15 Myosins are actin-basedmolecular motors with essential roles in many cellular processes*), the focal contact increases in size to form a **focal adhesion**. Focal adhesions were initially thought to exist only in cells grown in culture on flat surfaces, but recently they have also been found in vascular endothelial cells in vivo, specifically in areas of high fluid sheer stress. Cells cultured in 3D ECM gels form elongated integrin clusters called **3D matrix adhesions**. These most likely resemble the type of integrin clusters formed by cells in vivo.

To date, over 50 proteins have been identified in these complexes, and they are classi-

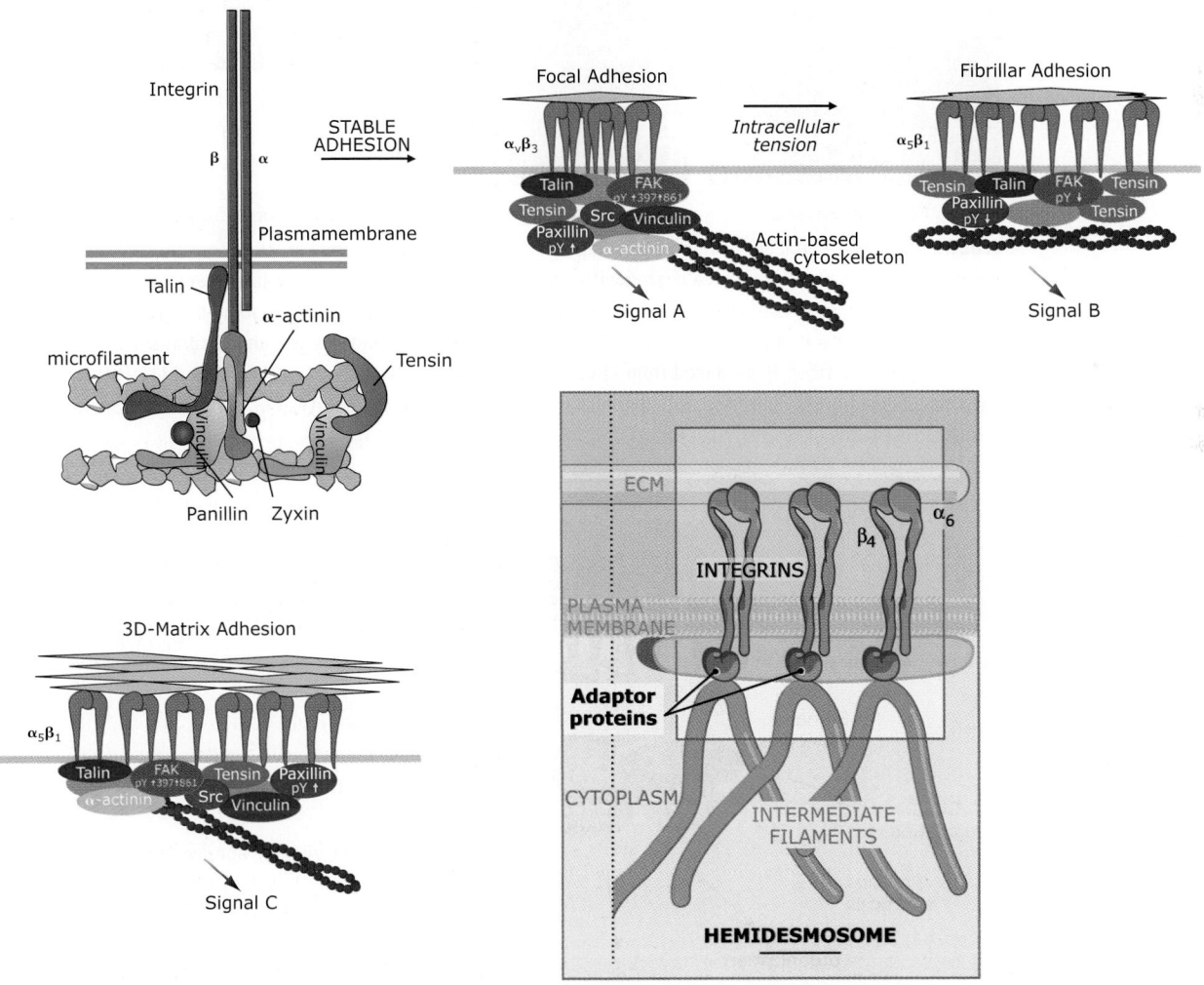

FIGURE 19.37 Five types of integrin clusters. Focal contacts are short-lived complexes that form at the leading edge of cells. Focal adhesions are mature focal contacts capable of resisting mechanical force. Fibrillar adhesions are elongated structures found in cells exposed to sustained intracellular tension. 3D matrix adhesions are formed by cells in 3D ECM gels and in vivo. Hemidesmosomes are formed on the basal surface of polarized epithelial cells and link to the intermediate filament network. Note that focal adhesions, fibrillar adhesions, and 3D matrix adhesions all participate in inside-out signaling.

fied in four groups: transmembrane receptors (e.g., growth factor receptors, syndecans), and three types of cytosolic proteins: structural proteins, "adaptors," and enzymes. The exact composition of each cluster varies depending on the type(s) of integrins in the cluster, the type of ECM bound by the integrins, the degree of tensile strain imposed on the cluster, the location of the cluster in the cell, and the type of cell in which the cluster forms, as illustrated in **FIGURE 19.38**. Collectively, these proteins function to control a vast range of cellular functions, as shown in **FIGURE 19.39**.

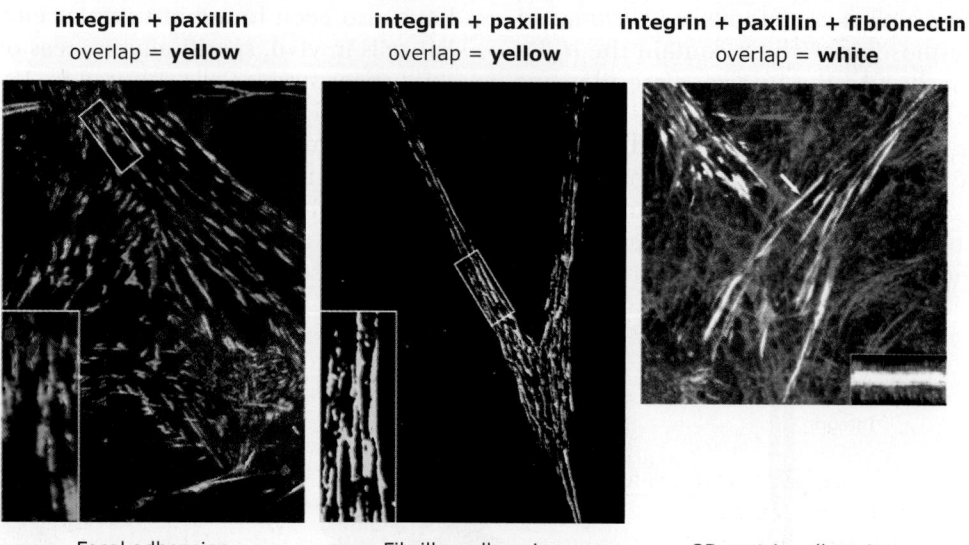

| **integrin + paxillin** overlap = **yellow** | **integrin + paxillin** overlap = **yellow** | **integrin + paxillin + fibronectin** overlap = **white** |

Focal adhesion   Fibrillar adhesion   3D matrix adhesion

FIGURE 19.38 Differences in shape and composition in integrin clusters. Left, mouse fibroblasts were cultured *in vitro* on a rigid surface coated with fibronectin. Middle, mouse fibroblasts were cultured in a 3D matrix derived from cultured cells. Right, mouse fibroblasts were cultured in a 3D matrix derived from mouse tissues. Arrow in right image shows 3D matrix adhesion containing integrin and paxillin and aligned with a fibronectin fiber. Reproduced from E. Cukierman, et al., *Science* 294 (2001): 1708–1712 [http://www.sciencemag.org]. Reprinted with permission from AAAS. Photos courtesy of Kenneth M. Yamada, National Institutes of Health/NIDCR.

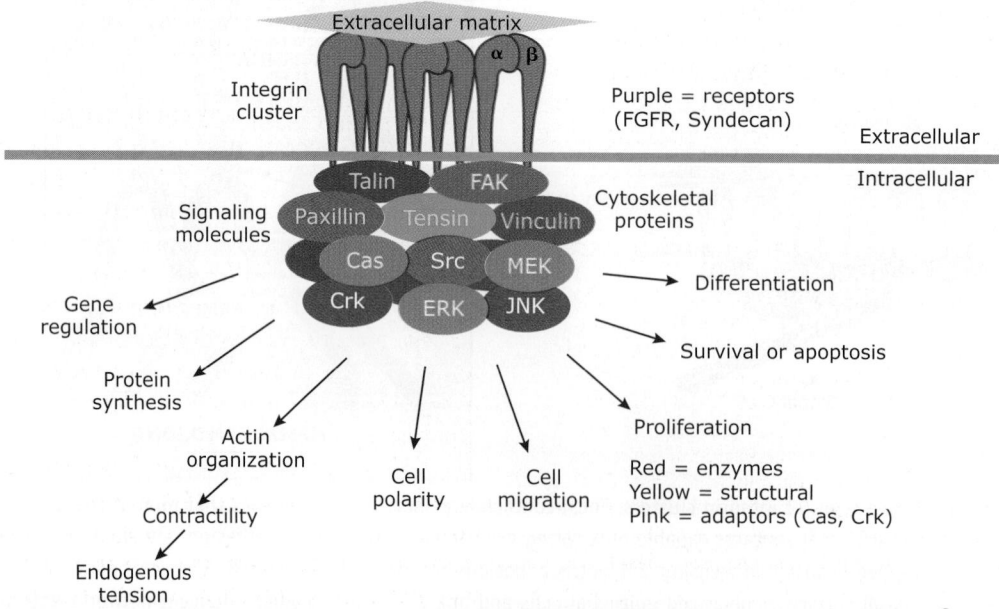

FIGURE 19.39 Summary of integrin cluster components and the cellular activities they control.

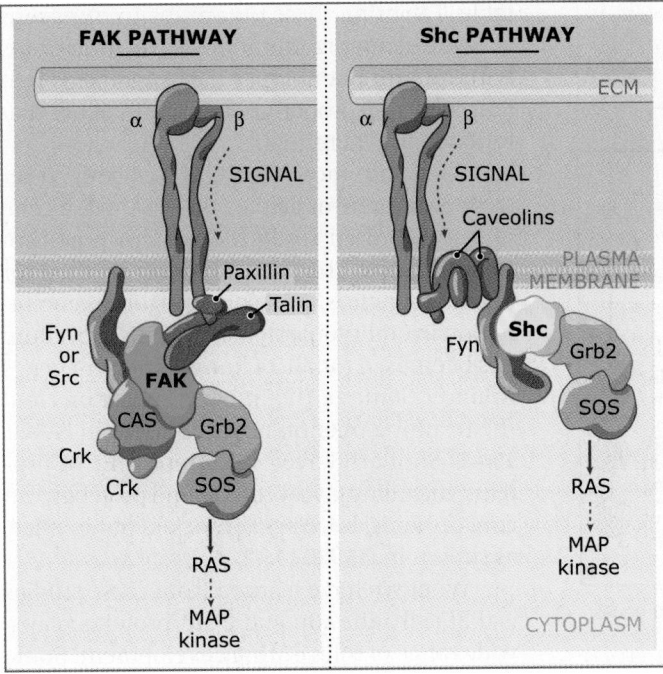

FIGURE 19.40 The cytoplasmic tails of integrin receptors associate with many signaling and adaptor proteins. Shown are two models of signaling complexes. In one pathway, the principal signaling protein is FAK, which binds to integrins directly or through the proteins talin and paxillin. In the other pathway, the principal signaling protein is Shc, which associates with integrins through an adaptor protein, caveolin. Both signaling complexes control MAP kinase signaling.

Many of the signaling proteins associated with integrin clusters, shown in **FIGURE 19.40**, are also found in signaling complexes that form around growth factor receptors. Indeed, many integrin signaling pathways overlap extensively with those assembled by growth factor receptors and some growth factor receptors are recruited into integrin clusters, so it is often hard to discriminate between the contributions of the two classes of receptors. For example, focal adhesion kinase (FAK), which is found in integrin clusters, serves as an organizer of several signaling pathways and plays a major role in controlling cell growth and migration. As a result of overlapping signaling pathways, integrin binding to ECM molecules also helps control cell growth and suppresses programmed cell death.

Concept and Reasoning Check

1. One good example of tissue remodeling in adults is the proliferation of milk ducts at the onset of lactation in pregnant women, followed by reduction of these ducts following weaning. Describe a scenario whereby integrins, structural glycoproteins in the ECM, proteoglycans, and proteases cooperate to complete this remodeling process. The goal here is not to describe the actual mechanisms involved, but rather to practice integrating these different molecules into a realistic, functional system. You may wish to review the concepts of cellular signaling in chapter 18 to assist in answering this question.

19.15 Integrins and extracellular matrix molecules play key roles in development

Key concepts

- Gene knockout by homologous recombination has been applied to over 40 different ECM proteins and all 24 integrin genes in mice. Some genetic knockouts are lethal, while others have mild phenotypes.
- Targeted disruption of the β1 integrin gene has revealed that it plays a critical role in the organization of the skin and in red blood cell development.

Much of what is known about cell adhesion proteins and their extracellular ligands comes from *in vitro* experiments using cultured cells, isolated proteins, and conditions that differ significantly from an organism's normal physiological state. This reductionist approach often unmasks a function of a protein that could not be observed in more complex, realistic systems. Yet the ultimate goal is to determine how these proteins function in normal, intact organisms.

Two common approaches have been used to understand what ECM and cell junctions do in intact organisms. One approach is to study model organisms (such as fruit flies and worms) whose developmental patterns are well described and that are genetically trac-

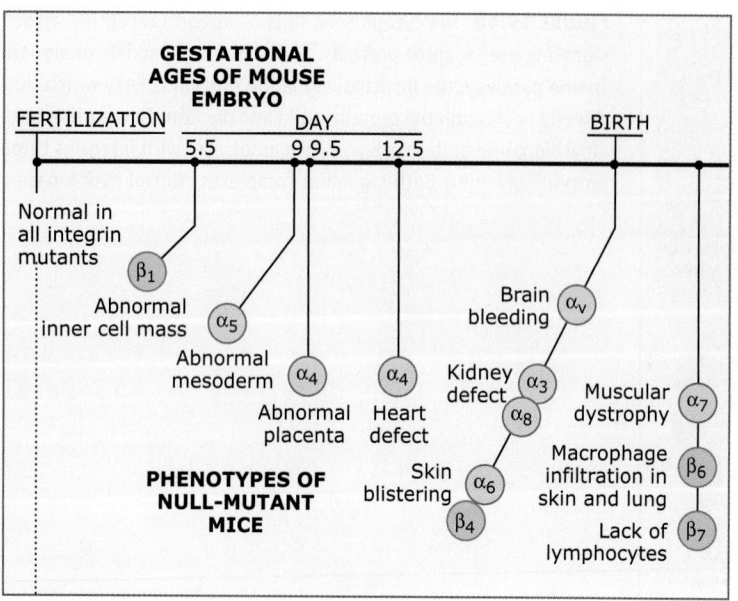

FIGURE 19.41 Some of the phenotypes of mice with mutated integrin genes.

ECM protein	Phenotype of null mutant mice	
Brevican	• Neural defects	Viable
Collagen 1a1	• Vascular defects	
Collagen 2a1	• Cartilage abnormalities; intervertebral disc defects	Lethal
Collagen 3a1	• Vascular and skin defects	
Collagen 4a3	• Kidney failure	
Collagen 5a2	• Skin fragility; skeletal abnormalities	
Collagen 6a1	• Mild muscular dystrophy	Viable
Collagen 7a1	• Skin blistering	Lethal
Collagen 9a1	• Cartilage defects	
Collagen 10a1	• Mild skeletal phenotype	Viable
Collagen 11a2	• Hearing loss	
Decorin	• Skin fragility; abnormal collagen fibrils	Viable
Elastin	• Vascular defects	Lethal
Fibrinogen	• Hemostatic defects; uterine bleeding in pregnant mice	Viable
Fibronectin	• Mesodermal and cardiovascular defects; lack of somites	Lethal
Laminin α2	• Muscular dystrophy	
Laminin α3	• Skin blistering	
Laminin α5	• Skull abnormalities; fused toes; failure to develop placenta, kidneys	Lethal
Laminin β2	• Neuromuscular, renal and neural defects	
Laminin γ1	• No basement membrane formation, failure of endoderm differentiation	
Link Protein	• Cartilage abnormalities	Lethal
Perlecan	• Defects in heart and brain; cartilage abnormalities	Lethal
Vitronectin	• No obvious phenotype	Viable

FIGURE 19.42 A sample of phenotypes from knockouts of extracellular matrix proteins in mice. Adapted from E. Gustafsson and R. Fässler, *Exp. Cell Res.* 261 (2000): 52–68.

table. Exposing these organisms to mutagens generates random mutations. From organisms with mutant phenotypes (i.e., organisms that fail to develop properly), the gene or genes that mutated are identified.

A second, more direct, approach is to selectively mutate or delete ("knock out") a gene of interest, then study the mutant organism as it develops. The most common method for deleting a gene is to use homologous recombination in embryonic stem cells. This procedure has been used for all 24 known integrin genes in mice. Some of the integrin mutations and the resulting phenotypes are shown in FIGURE 19.41. Similarly, over 40 different ECM proteins, including glycoproteins and proteoglycan core proteins, have been knocked out in mice, as shown in FIGURE 19.42.

What we have learned from these studies is that cell adhesion and ECM proteins play a wide range of roles during development. Some genetic knockouts (e.g., β_1-integrin, laminin γ1 chain, perlecan proteoglycan) are lethal, while others (e.g., α_1-integrin, Type X collagen, decorin proteoglycan) have relatively mild phenotypes. In the case of no or mild phenotype, there may be functional redundancy, in which other integrins may compensate for the missing one.

Because the traditional gene knockout studies target the gene of interest in all cells that develop from an embryo, a lethal mutation makes it impossible to assess the function of a given protein in an adult organism. For example, laminin γ1-null mice fail to develop beyond five days after fertilization. The solution to this problem is to develop knockouts that allow the normal copy of the gene to be expressed in all tissues except those of interest. These targeted knockouts are generated using induced genetic recombination mediated by the Cre/lox system, rather than spontaneous recombination as is used to make traditional knockout mice.

This strategy has been used to study the function of the β_1 integrin receptor exclusively in skin. Mice lacking β_1 integrin only in keratinocytes develop to adulthood but suffer numerous problems, including severe hair loss, destruction of hair follicles, abnormal basal lamina assembly, reduction in hemidesmosome assembly, and severe skin blistering. This phenotype reflects the functions of integrins and ECM proteins. Skin cells lacking β_1 integrins cannot form many types of integrin receptors (see Figure 19.34), and this in turn prevents these cells from adhering to the basal lamina and other components of the ECM. Cells that cannot adhere to ECM often die, which explains why these mice exhibit so many problems. For example, the epithelial cells that form hair die, resulting in hair loss. (For further details on the role of integrins in cell growth signaling see *19.14 Integrin receptors participate in cell signaling*.) The study of these mice showed that one of the crucial functions of the β_1 integrin gene is to organize entire tissues and the ECM proteins in them, even those that do not bind to β_1 integrins. A conclusion this broad could not have been reached in any *in vitro* system developed so far.

The induced knockout system has been developed a step further so that genes from healthy adult organisms can be inactivated. This approach was used to remove the β_1 integrin gene from red blood cell precursors in fetal and adult mice, and showed that adhesion and migration of these cells into the bone marrow is inhibited. This experiment established that β_1 integrin is required for homing of red blood cell precursors. Thus, the generation of different types of β_1 integrin null mutants has revealed a role of β integrin in specific tissues at different times of development.

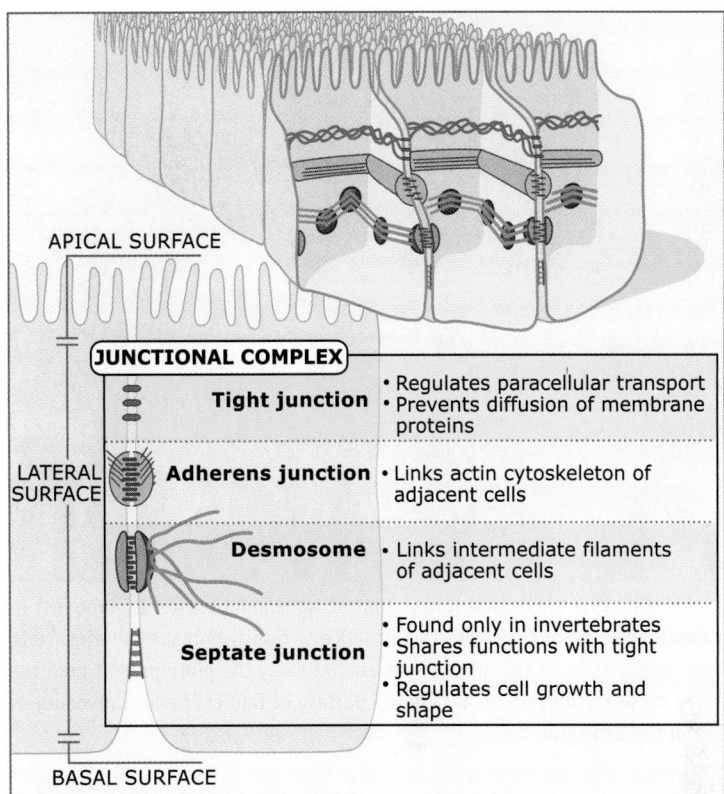

FIGURE 19.43 The junctional complex is composed of at least three distinct cell-cell junctions. It permits epithelial cells to provide structural support and to function as a selective barrier to transport. The septate junction is found only in invertebrates; often, it is present instead of the tight junction.

19.16 Tight junctions form selectively permeable barriers between cells

Key concepts

- Tight junctions are part of the junctional complex that forms between adjacent epithelial cells or endothelial cells.
- Tight junctions regulate transport of particles between epithelial cells.
- Tight junctions also preserve epithelial cell polarity by serving as a "fence" that prevents diffusion of plasma membrane proteins between the apical and basal regions.

Cell-cell junctions play a critical role in establishing and maintaining multicellularity. Along the lateral surfaces of adjacent cells in epithelial and endothelial cell layers, three separate cell-cell junctions function as a group called the **junctional complex**. In vertebrates, these junctions are the **tight junction**, **adherens junction**, and **desmosome**; in invertebrates, the septate junction often acts in place of the tight junction (see *19.17 Septate junctions in invertebrates are similar to tight junctions*). The

relative positions of these junctions are shown in **FIGURE 19.43**. Together, these junctions help to segregate a multicellular organism into discrete, specialized regions and to regulate the transport of molecules between them. These junctions also help protect the cells from physical and chemical damage. We will discuss each type of junction within this complex in turn, beginning with the tight junction (for adherens junction see *19.18 Adherens junctions link adjacent cells*; for desmosomes see *19.19 Desmosomes are intermediate filament-based cell adhesion complexes*).

When viewed in a thin section of cells with a transmission electron microscope, tight junctions appear as a series of small contacts (sometimes referred to as "kisses") between the opposed lateral membranes of neighboring cells, as **FIGURE 19.44** shows. Proteins on the cytoplasmic face of the membrane adjacent to these contacts are seen as electron-dense "clouds." Freeze fracture of cells provides a different view by revealing the protein distribution in the two lipid monolayers separated

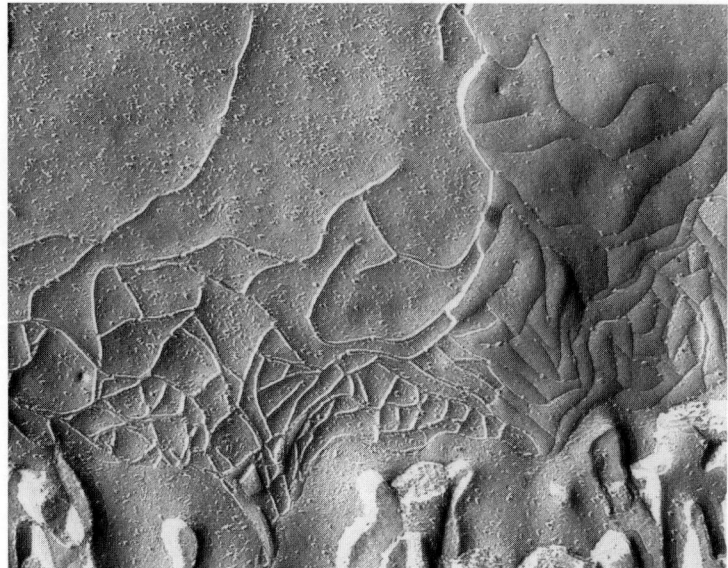

FIGURE 19.44 Electron micrograph of tight junction fibrils (grooves) in a membrane. The pits indicate the presence of transmembrane proteins. Note that some pits lie in the grooves—these are likely the proteins that hold the fibrils close to the membrane. Photo courtesy of Daniel Friend, University of California, San Francisco.

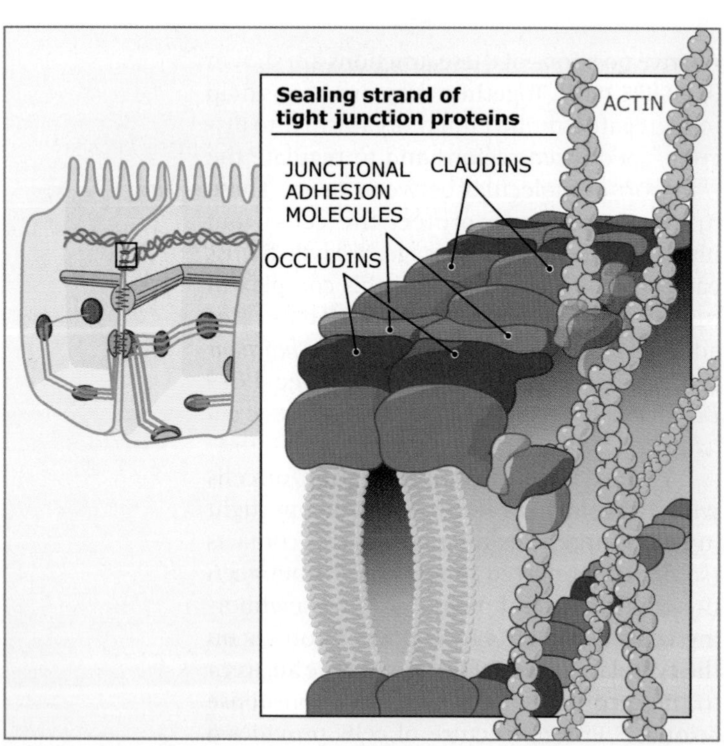

through the middle of the plasma membrane. The tight junctions appear as a weblike network of thin fibrils (or strands) where the proteins remain embedded in the membrane or as a network of grooves if the proteins have come off during the fracture process.

The molecular composition of tight junctions is complex. Over 24 proteins have been identified in tight junctions, and they are classified into four groups: transmembrane proteins, polarity proteins, cytoskeletal proteins and signaling proteins. Thus far, three types of transmembrane proteins are found in the tight junction: claudins, occludins, and the junctional adhesion molecule (JAM), as shown in **FIGURE 19.45**. Claudins form the core protein in the tight junction fibrils by clustering their extracellular domains into loops that form selective channels in the fibrils. There are at least 24 different claudin proteins in mammals, and they are arranged in different combinations to form channels with different ion selectivity. Transfection of claudin genes, into cells that normally do not express them, results in formation of tight junctions. Occludins copolymerize laterally with claudins along tight junction fibrils, but their function is unknown.

The three transmembrane proteins attach stably to nine or more structural proteins, including actin. Occludins also bind to connexin proteins, suggesting that tight junctions and gap junctions are both structurally and functionally related (see *19.21 Gap junctions allow direct transfer of molecules between adjacent cells*). The transmembrane proteins also bind transiently to over a dozen signaling proteins. This suggests that the tight junction may have an additional role as an organizer for signaling at the cell surface, akin to the role played by focal adhesion complexes at the basal surface of cells (for more on focal adhesions see *19.14 Integrin receptors participate in cell signaling*).

Many of the other tight junction proteins, such ZO-1, have sequences that place them in the membrane-associated guanylate kinase (MAGUK) family of proteins, which contain three domains in a characteristic order. These domains allow MAGUK proteins to bind to many types of target proteins, including signaling proteins and elements of the actin cytoskeleton. Some tight junction proteins also contain a PDZ domain, which enables them to bind to one another. *In vitro* binding experiments using intact and truncated forms of these proteins suggest a variety of possible binding combinations at the tight junction.

FIGURE 19.45 Tight junctions are held together by occludin, claudin, and junctional adhesion molecules.

Tight junctions play two important roles. First, they are the molecular structures responsible for regulating paracellular transport (the transport of material in the space between cells) in an epithelial or endothelial cell layer. (Historically, they were thought of as barriers that occluded [prevented] this transport and, thus, were given the name *zonula occludens*.) In this role, tight junctions may be thought of as a "molecular sieve" through which extracellular molecules are filtered as they cross epithelial and endothelial boundaries. Not all of these sieves are the same, however, since each tissue serves to filter a unique set of diffusing molecules: for example, smoke particles need not be filtered in the kidney. In fact, the cut-off size for free diffusion across tight junctions ranges between ~4 and 40 Å, depending on the tissue in which they are found.

The physical barriers for transport of ions and other solutes are quite different: ions are transported instantaneously, but other solutes require minutes or even hours to cross the tight junction. How is this possible? A recent model proposes that the tight junction permeability barrier is composed of rows of charge-selective pores that form the web of fragile strands seen in Figure 19.44. Ions are transported through the pores, but other solutes must wait until the strands break before they can move through the junction. As the strands break and reseal, the solutes move stepwise through the barrier, as illustrated in **FIGURE 19.46**.

The second role played by tight junctions is to structurally and functionally separate the plasma membrane of polarized cells into two domains, as shown in Figure 19.44. The **apical** (from the Greek word *apex*, or top) surface is the portion of the plasma membrane that is oriented toward a cavity or space on one side of the epithelial sheet. The **basal** (or bottom) surface is the region on the opposite side, in contact with the ECM. The lateral surface makes up the "sides" between these two regions. Tight junctions completely encircle epithelial and endothelial cells along the lateral surface at the apical-lateral border, thereby separating the cell into two zones: the apical domain and the basolateral domain. These domains effectively divide the cell surface into "top" and "bottom" regions and play different roles in controlling molecular traffic via the transcellular route (see *8 Protein trafficking between membranes*). Although membrane proteins are able to diffuse in the plane of each domain, they are prevented from diffusing from one domain to

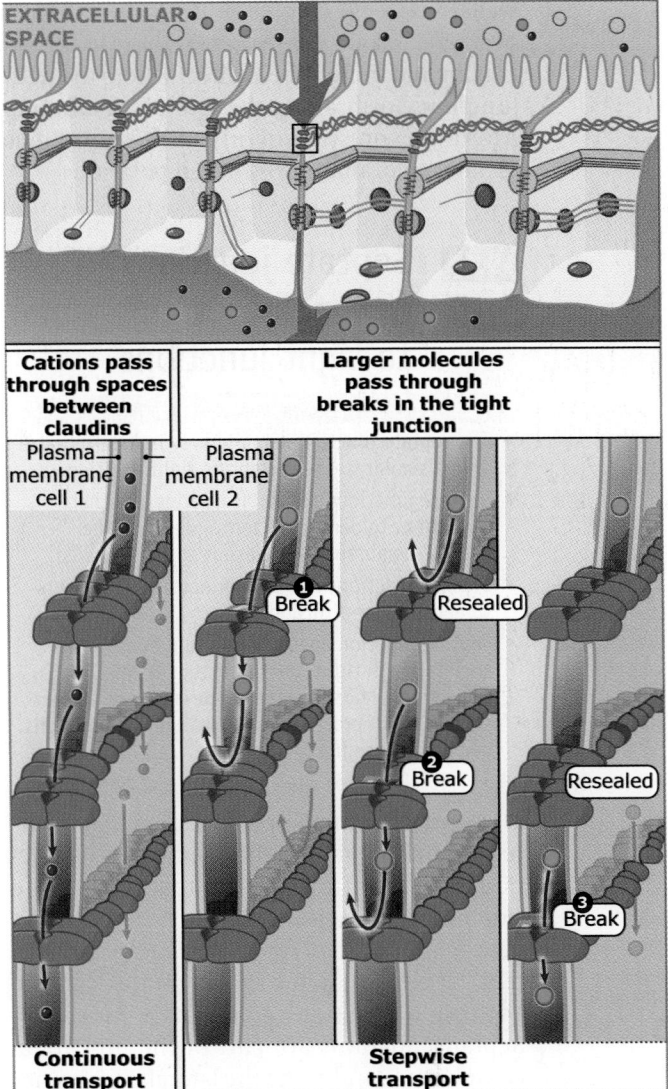

FIGURE 19.46 A model of fast and slow transport of solutes through tight junctions. Fast transport of some ions occurs through ion channels embedded within the fibrous strands of the junction. Slow transport of the solutes that cannot pass through channels occurs when the strands break, allowing these solutes to flow through the breaks. Because there are many layers of strands, this form of transport occurs in stages.

the other across the tight junction. In this role, tight junctions serve as a "fence" that maintains the unique molecular composition of the two membrane domains.

Three proteins, known as aPKC (atypical protein kinase C), Par (partitioning defective)-3, and Par-6, form a "core cluster" that plays at least two roles in establishing polarity in the plasma membrane. First, these molecules activate a signaling protein called cdc42, which stimulates actin polymerization near the plasma membrane (see *18.23 Small,* monomeric GTP-binding proteins are multiuse switches). Second, these proteins bind to the occludins and thereby help direct where the tight junctions form along the lateral membrane. Altering expression of any of these proteins results is a dramatic loss of cell polarity.

19.17 Septate junctions in invertebrates are similar to tight junctions

Key concepts

- The septate junction is found only in invertebrates and is similar to the vertebrate tight junction.
- Septate junctions appear as a series of either straight or folded walls (septa) between the plasma membranes of adjacent epithelial cells.
- Septate junctions function principally as barriers to paracellular diffusion.
- Septate junctions perform two functions not associated with tight junctions: they control cell growth and cell shape during development. A special set of proteins unique to septate junctions performs these functions.

The **septate junction** is found only in invertebrate animals and is thought to be the functional analog of the vertebrate tight junction (for more on tight junctions see *19.16 Tight junctions form selectively permeable barriers between cells*). However, some invertebrates contain both tight junctions and septate junctions. Septate junctions are part of the junctional complexes found along the lateral membranes of epithelial cells and, like tight junctions, play

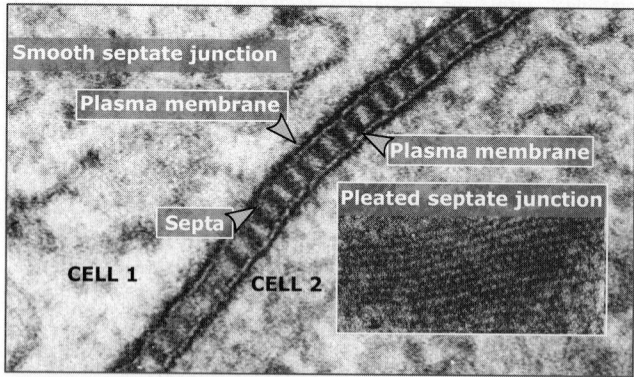

FIGURE 19.47 Smooth septate junctions appear as linear walls between adjacent cells. Pleated septate junctions (inset) appear as folded walls between adjacent membranes. Reprinted from *Tissue Cell*, vol. 13, C. R. Green, A clarification of the two types of invertebrate..., pp. 173–188, Copyright (1981) with permission from Elsevier [http://www.sciencedirect.com/science/journal/00408166]. Photo courtesy of Colin R. Green, University of Auckland.

a role in controlling paracellular transport across epithelial sheets. When viewed under an electron microscope, they appear as stacks of parallel walls (or septa, hence the name) bridging a 15- to 20-nm-wide gap between the plasma membranes of adjacent cells. As **FIGURE 19.47** shows, the septa can appear either relatively straight as in so-called smooth septate junctions or folded into a pleated pattern.

Septate junctions and tight junctions differ in at least three ways. First, septate junctions are composed of proteins that are not found in tight junctions. Second, septate junctions are found near the basal edge of the lateral membrane, whereas the tight junction is found near the apical edge, "above" the zonula adherens (see Figure 19.43). Third, septate junctions play two roles not associated with tight junctions.

Some cells contain both tight junctions and septate junctions in the same junctional complex, suggesting that they serve distinct functions. When cells contain both tight junctions and pleated septate junctions, the tight junctions act as the primary barrier to paracellular diffusion. This observation comes from electron microscopy studies using electron-dense tracer dyes to observe how small molecules pass between adjacent cells. Septate junctions may also serve as attachment sites for actin filaments, similar to adherens junctions. What, then, is the function of the septate junction?

Like tight junctions, septate junctions contribute to the "gate" and "fence" functions of the junctional complex. They act as "gates" by restricting the flow of extracellular particulate matter between adjacent cells and as "fences" by restricting the flow of phospholipids and membrane proteins between the apical and basal membrane domains.

But, unlike tight junctions, septate junctions also perform at least two other functions. One is to help control cell growth. For example, *Drosophila* or *C. elegans* that express mutant septate junction proteins often develop epithelial cell tumors. Another is to control cell shape: mutations in a different set of septate junction proteins results in extra wide tubules in *Drosophila*. How does the septate junction perform these functions? To answer this question, we need to discuss the proteins that comprise the septate junction. As **FIGURE 19.48** shows, several septate junction proteins have been identified in *Drosophila*.

The gene *scribble* encodes Scrib, one of the most important septate junction proteins. Unlike normal fly embryos that develop a smooth outer layer (cuticle), embryos homozygous for mutated *scribble* form bumpy, lined patterns on the cuticle that look as though someone has scribbled on the surface, as **FIGURE 19.49** shows. These deformations appear because septate junctions fail to form in the cuticle. The epithelial cells that form the cuticle are not properly aligned, and they pile up to form ridges. Other epithelial tissues are also much larger and more disorganized in Scrib mutants compared to normal embryos.

In addition to its function in septate junctions, the Scrib protein interacts with at least two other junctional complex proteins, Disks large (Dlg) and lethal giants larvae (lgl), to initiate the formation of epithelial cell polarity. When these proteins are mutated, epithelial cells appear to lose their septate junctions. Moreover, they form tight junctions in unexpected locations (at the basal side of the lateral membrane) and apical membrane proteins are not properly sorted. As a result, the cells fail to polarize properly, and the integrity of the epithelial sheet is lost. Interestingly, these mutant cells also divide more often than normal cells, and this gives rise to abnormally large clusters of epithelial cells.

A second set of septate junction genes, encoding at least eight different proteins including neurexin, control cell shape during formation of the epithelial tubes that constitute the tracheal system in *Drosophila*. Mutations in these genes result in epithelial cells with enlarged apical surfaces, which then form tubes of exceptionally large diameter. Curiously, the growth of these cells is unaffected, and the septate junctions in these mutants retain the barrier functions associated with tight junctions, demonstrating that these functions are distinct. Two current models suggest that the septate junction controls cell shape by controlling the behavior of proteins that form apical membrane domains or by controlling the formation of ECM on the apical side of these cells.

The precise relationship between septate junction formation and control of cell division and shape is still unknown. However, the answer may shed light on the mechanism of epithelial cell growth in many tissues, including diseases such as human cancers and kidney disease. (This is an example of how fruit fly research can contribute to human health.)

Another interesting feature of Scrib is that it belongs to a family of proteins that contain a **PDZ domain**. This domain is found on many proteins located in cell-cell junctions and is

Septate junction protein	Function
Coracle	• Cell spreading, and dorsal closure during development
Discs large 1	• Formation of septate junctions in epithelial cells and in neurons
Discs lost	• Polarization of embryonic epithelia during cellular blastoderm formation
Expanded	• Regulates the growth of imaginal discs by inhibiting cell proliferation
Fascilin 3	• Homophilic cell adhesion molecule
Lethal (2) giant larvae	• Formation of the cytoskeletal network • Tumor suppressor
Neurexin	• Formation of pleated septate junctions • Regulation of cell shape
Polychaetoid	• Dorsal closure during development
Scribble	• Control of epithelial growth and polarity
α-spectrin	• Membrane cytoskeletal protein required for septate junction formation

FIGURE 19.48 The functions of septate junction proteins in *Drosophila*.

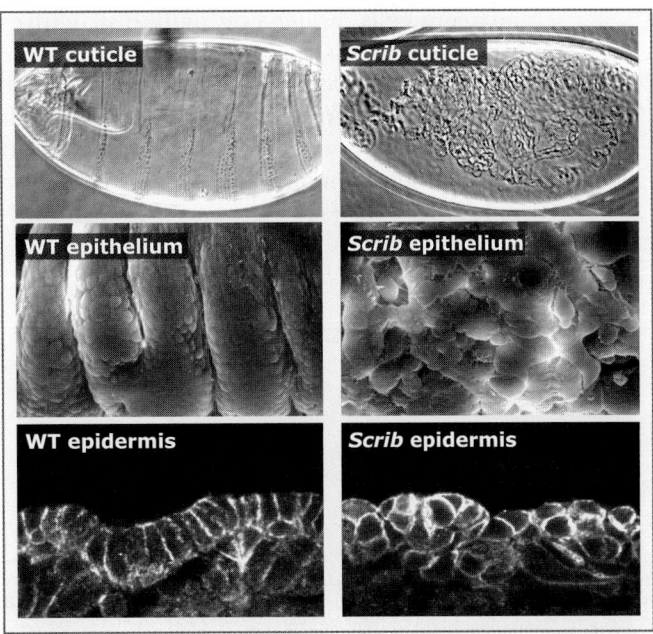

FIGURE 19.49 Comparison of normal (wild-type, WT) *Drosophila* embryos (left) and *scribble* mutant embryos (right). Top: Phase contrast images of embryos. Middle: scanning electron micrographs of 14-day embryos. Bottom: immunofluorescence images of spectrin distribution in stage 15 embryos. Spectrin is a marker for epithelial cells. Reprinted by permission from Macmillan Publishers Ltd: *Nature*, D. Bilder and N. Perrimon, Localization of apical epithelial..., 2000, vol. 403 (6770): 611–612. Photos courtesy of David Bilder, University of California, Berkeley.

thought to mediate binding between transmembrane and cytosolic proteins, similar to other domains found on signaling proteins. Efforts are under way to understand the molecular mechanism of how septate junction proteins such as Scrib play such diverse and important roles in epithelial cell function.

19.18 Adherens junctions link adjacent cells

Key concepts

- Adherens junctions are a family of related cell surface domains that link neighboring cells together.
- Adherens junctions contain transmembrane cadherin receptors.
- The best known adherens junction, the zonula adherens, is located within the junctional complex that forms between neighboring epithelial cells in some tissues.
- Within the zonula adherens, adaptor proteins called catenins link cadherins to actin filaments.

Adherens junctions are components of the junctional complex that hold epithelial and endothelial cells together. In the electron microscope, adherens junctions appear as dark, thick bands that lie near the plasma membranes of adjacent cells, bridged by rodlike structures that project into the intercellular space. The most well-known adherens junction is the **zonula adherens**, shown in **FIGURE 19.50**. It is found just beneath the tight junctions in the junctional complex formed between some epithelial cells (see Figure 19.43). (The mi-

croscopists who discovered it gave it the name belt desmosome because it looked like a large desmosome. We now know that it is quite different from the desmosome junction, and this name is no longer used.) As **FIGURE 19.51** shows, other examples of adherens junctions include the adhesive junctions in the synapses between neurons in the central nervous system, in intercalated disks between adjacent cardiac muscle cells, and in junctions formed between layers of the myelin sheath surrounding peripheral nerves.

Regardless of their location, adherens junctions share two properties. First, they contain transmembrane receptor proteins known as cadherins that bind to identical cadherins on neighboring cells, as Figure 19.51 shows. (For more details on cadherins see *19.22 Calcium-dependent cadherins mediate adhesion between cells*.) Binding of receptors on one cell to the same type of receptor on another cell is called **homophilic binding**. It is thought to play an important role in determining the cellular organization of tissues by helping cells find specific binding partners. This can be demonstrated for cadherins by expressing the genes for two different cadherin receptors in two populations of the same cell type, then mixing the two populations of cells together in the same petri dish,

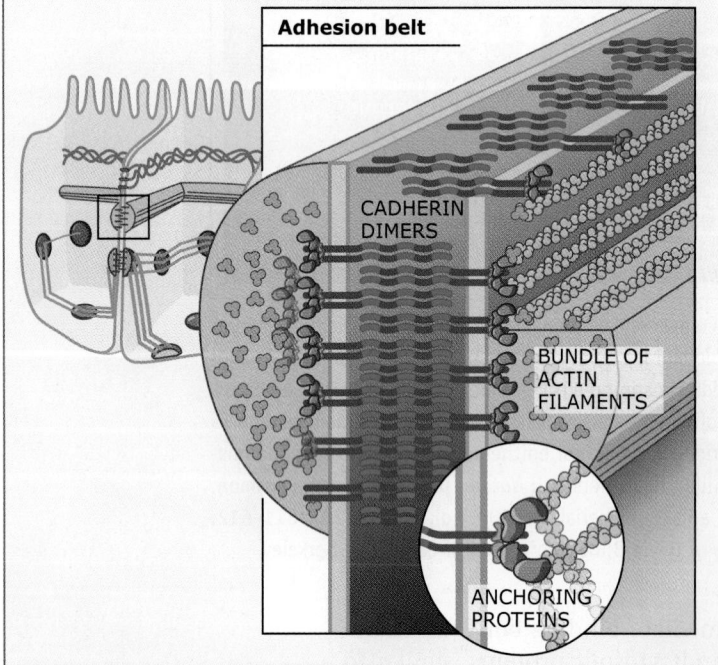

FIGURE 19.50 The zonula adherens is part of the junctional complex. Actin bundles are adjacent to the junction at the cytoplasmic face of the plasma membrane. Cadherins form the rods between cells and are linked to the actin cytoskeleton by anchoring proteins such as catenins.

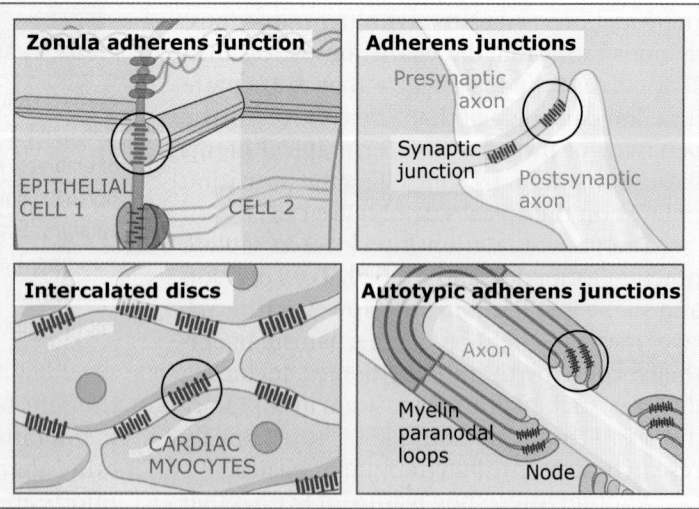

FIGURE 19.51 Each type of adherens junction functions to hold adjacent cells together tightly.

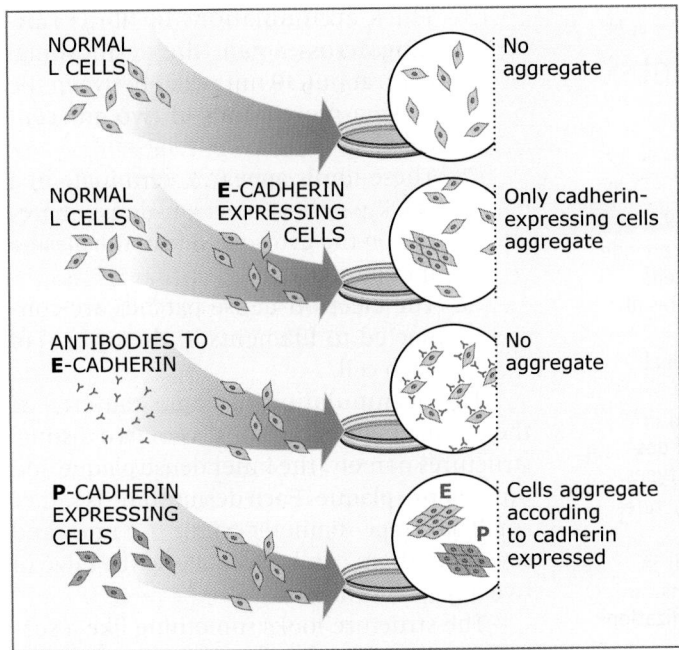

FIGURE 19.52 Cells expressing identical cadherin receptors will recognize and selectively bind to one another. This homophilic binding plays an important role in tissue formation during development.

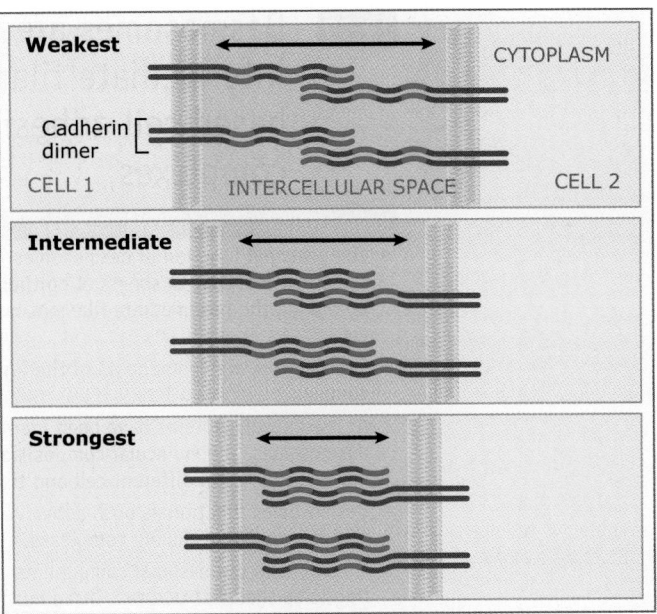

FIGURE 19.53 Cadherin proteins form dimers that bind to one another. Three different configurations of cadherin interaction are shown. Direct measurement of the strength of adhesion shows that the configuration with the most overlap is also the strongest.

as shown in **FIGURE 19.52**. After several hours, the cells sort themselves into groups expressing the same cadherin receptor. If antibodies that block the homophilic binding sites of cadherins are added to the culture, these clusters will not form.

The dimeric cadherin receptors used in adherens junctions contain five extracellular domains that determine exactly how homophilic binding occurs. As shown in **FIGURE 19.53**, three different overlapping arrangements of these domains are possible. The strongest binding occurs when the receptors overlap completely in an antiparallel arrangement, while weaker binding interactions form when the receptors partially overlap. By changing the number of cadherin receptors clustered on their surface, cells can change the strength of their binding to their neighbors (a process called avidity modulation). There is no evidence that cadherins undergo conformation changes that alter binding affinity (which is known as affinity modulation and is used by integrin receptors) to control binding strength in adherens junctions. (For more on avidity and affinity modulation see *19.14 Integrin receptors participate in cell signaling*.)

The second property shared by adherens junctions is that they form adhesions strong enough to allow tissues to change shape and/or resist sheer stress. For example, the zonula adherens uses anchor proteins, known as catenins, to link the cytoplasmic tails of cadherin receptors to bundles of actin, as diagrammed in Figure 19.50. These actin filaments are, in turn, attached to myosin proteins that cause the actin filaments to slide past one another. This is thought to result in contractions that can change the shape of the apical pole of epithelial cells. This may be important in development of the neural tube, for example, when epithelial cells invaginate to close the neural groove (see *19.22 Calcium-dependent cadherins mediate adhesion between cells*).

Beyond the function of cadherin-based adhesion, the specific functions of adherens junctions remain to be determined. Genetic analysis in fruit flies suggests that proteins other than cadherins and catenins are required to form morphologically distinct zonula adherens junctions. These additional proteins may also be involved in regulating the assembly of the cytoskeleton at sites quite far from the adherens junctions themselves. For example, they may be involved in establishing the polarity of epithelial cells and, thus, may indirectly affect the assembly of other cell junctions, including the tight junction. How this may be accomplished is currently under investigation.

19.19 Desmosomes are intermediate filament-based cell adhesion complexes

Key concepts

- The principal function of desmosomes is to provide structural integrity to sheets of epithelial cells by linking the intermediate filament networks of cells.
- Desmosomes are components of the junctional complex.
- At least seven proteins have been identified in desmosomes. The molecular composition of desmosomes varies in different cell and tissue types.
- Desmosomes function as both adhesive structures and as signal transducing complexes.
- Mutations in desmosomal components result in fragile epithelial structures. These mutations can be lethal, especially if they affect the organization of the skin.

The **desmosome** is a component of the junctional complex in epithelial cells (see Figure 19.43) and is also located in some nonepithelial cells such as myocardial, liver, spleen, and some neural cells. Three features of desmosomes are immediately apparent in electron micrographs, as **FIGURE 19.54** shows and as given below:

1. Thick accumulations of fibrils running across a gap (the desmosomal core, about 30 nm wide) between the plasma membranes of two adjacent cells.
2. These fibrils appear to terminate in a thick patch of electron-dense material on the cytosolic side of the plasma membrane.
3. The electron-dense patches are connected to filaments in the cytosol of each cell.

The accumulation of dense material at the plasma membrane consists of two distinct structures namely, the inner dense plaque and outer dense plaque. Each desmosome is rather small (average diameter about 0.2 μm), and several of them can be seen along the edge of two adjacent cells.

The structure looks something like a suspension bridge: cytosolic filaments in neighboring cells are linked together by bridging extracellular filaments connected to supporting "anchors" on the plasma membrane. For this reason, the structure was given the name desmosome, derived from the Greek words *desmos* (bond, fastening, chain) and *soma* (body). It seems obvious that the purpose of such a structure is to link two cells together.

What function might such a linkage serve in cells? Remember the two main functions of the junctional complex shown in Figure 19.43: controlling paracellular transport and resisting physical stresses imposed on the epithelium. Because desmosomes are especially abundant in cells exposed to physical stress such as skin and cardiac muscle, cell biologists thought that they contributed to the latter function. Consequently, the cytoplasmic filaments attached to the dense plaques were called tonofilaments to reflect the supposition that they were under strain (Greek, *tonos*). Later, it was determined that these filaments are intermediate filaments, a major class of cytoskeletal structures, though they are still sometimes called tonofilaments (for more on intermediate filaments see *13 Intermediate filaments*).

In addition to the intermediate filament fibers, at least seven other protein types have been identified in desmosomes, and these are organized into three families. Three of these proteins (the desmogleins, desmocollin 1, and desmocollin 2) are members of the cadherin superfamily of cell surface receptors (see *19.22 Calcium-dependent cadherins mediate adhesion between cells*). These proteins are the major trans-

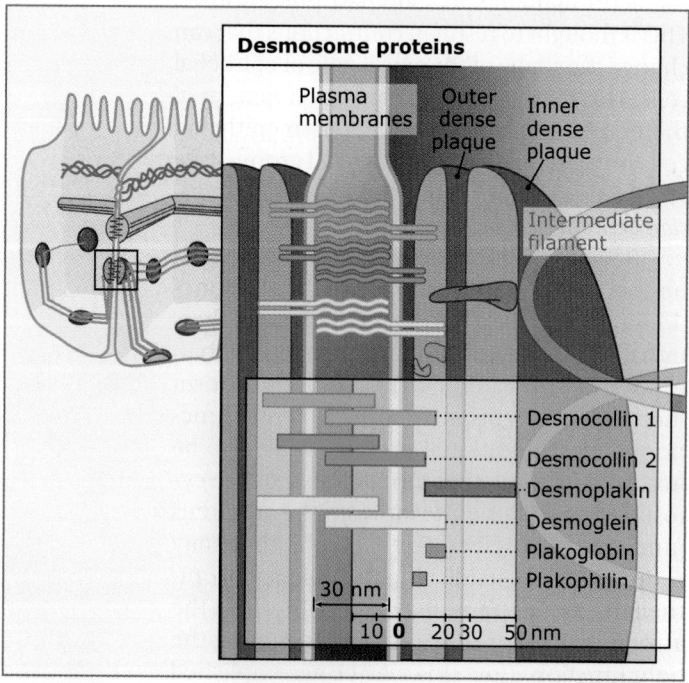

FIGURE 19.54 Desmosome proteins are distributed in the plasma membrane and a distinctive double plaque arrangement at the cell surface.

membrane proteins found in desmosomes, and they are major components of the outer dense plaque, as Figure 19.54 shows. They form the "bridging filaments" that stretch across the intercellular space and serve as binding sites for cytoplasmic proteins that are members of the armadillo (plakoglobin, the plakophilins) and plakin (desmoplakin) families. Desmoplakin, in turn, binds to the intermediate filament proteins in the inner dense plaque. The exact makeup of the desmosome, as well as the number of desmosomes formed, varies in different cell types, reflecting the wide variety of stresses that cells must endure.

A common description of the desmosome is that it serves as a "spot weld" between two neighboring cells. In addition to this structural role, desmosome proteins also play important signaling roles at the cell surface. For example, plakoglobin is a close relative of the protein β catenin, which binds to "classical" cadherins in adherens junctions. β-catenin is a structural protein in the adherens junction and transduces signals into the nucleus (for more on adherens junctions see *19.18 Adherens junctions link adjacent cells*). Plakoglobin and plakophilins likewise move into the nucleus upon activation of signaling receptors at the cell surface, and plakoglobin even binds growth factor receptors directly. As a result of this signaling activity, desmosomes can control the expression of multiple genes and significantly affect the function of proteins located elsewhere in the cell, including other junctions.

The most dramatic proof of desmosome function comes from cases where desmosome structure is compromised. In these cases, epithelial sheets are especially fragile, and the organs they cover are easily damaged. This is especially true of the skin, which is severely blistered. When viewed under a microscope, epithelial cells lacking desmosomes are badly disorganized, lack junctional complexes, and are detached into small clusters rather than forming a single continuous sheet.

Patients with damaged or missing desmosomes have a wide range of diseases, which are grouped into two general classes based on their origin. **Genodermatosis**, such as palmoplantar keratoderma or junctional epidermolysis bullosa, arises from a mutation in either desmosomal or hemidesmosomal proteins, respectively. (For more on hemidesmosomes see *19.20 Hemidesmosomes attach epithelial cells to the basal lamina*.) **Autoimmune bullous dermatosis**, such as pemphigus vulgaris or bullous

pemphigoid, arises when patients develop autoantibodies against the proteins in their desmosomes or hemidesmosomes, respectively. In both classes, the structure and function of cell junctions is severely compromised, and the diseases can be lethal.

A combination of molecular genetics and tissue engineering is used to detect and treat these diseases. At least in the case of genodermatoses, prenatal screens are available to detect mutations in the desmosome genes. These screens use the polymerase chain reaction to amplify the gene of interest (e.g., desmocollin-1) from a sample of the fetal tissue. The DNA is then analyzed by restriction fragment length polymorphism and/or Southern blotting techniques.

Current treatments for patients with these diseases are focused primarily on protecting the skin and avoiding risky behaviors that might induce blister formation, and the result is a rather poor quality of life. One experimental treatment currently being evaluated is the application of tissue engineered skin. By substituting the damaged skin with a fresh layer of living, normal skin cells embedded in an engineered ECM, researchers hope to develop more stable, trauma-resistant skin that forms normal desmosomes.

19.20 Hemidesmosomes attach epithelial cells to the basal lamina

Key concepts

- Hemidesmosomes, like desmosomes, provide structural stability to epithelial sheets.
- Hemidesmosomes are found on the basal surface of epithelial cells, where they link the ECM to the intermediate filament network via transmembrane receptors.
- Hemidesmosomes are structurally distinct from desmosomes and contain at least six unique proteins.
- Mutations in hemidesmosome genes give rise to diseases similar to those associated with desmosomal gene mutations.
- The signaling pathways responsible for regulating hemidesmosome assembly are not well understood.

The **hemidesmosome** is a cell surface junction found at the basal surface of the plasma membrane of epithelial cells. As seen in **FIGURE 19.55**, this structure is a complex interweaving of "plaques" and filaments and looks like half of a desmosome (see *19.19 Desmosomes are in-*

termediate filament-based cell adhesion complexes). Despite appearances, however, the hemidesmosome is not half of a desmosome. The primary function of hemidesmosomes is to anchor epithelial sheets to the basal lamina. This further distinguishes the hemidesmosome from the desmosome. While both structures hold epithelial sheets together, they are oriented

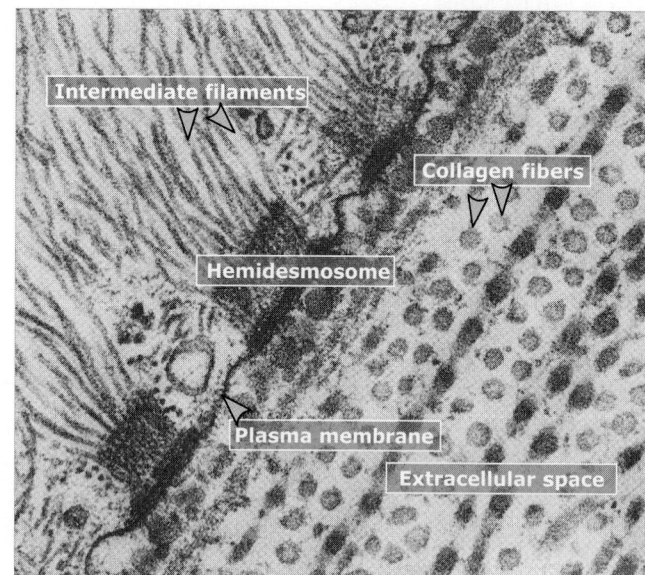

FIGURE 19.55 Hemidesmosomes are specialized structures that form at the junction of epithelial cells and the specialized extracellular matrix called the basal lamina. The hemidesmosome is characterized by the collection of filamentous material that terminates in a dense plaque at the cell surface. Photo © Kelly, 1966. Originally published in **The Journal of Cell Biology**, 28: 51–72. Used with permission of Rockefeller University Press. Photo courtesy of Dr. William Bloom and Dr. Don W. Fawcett. *A Textbook in Histology*, 1986.

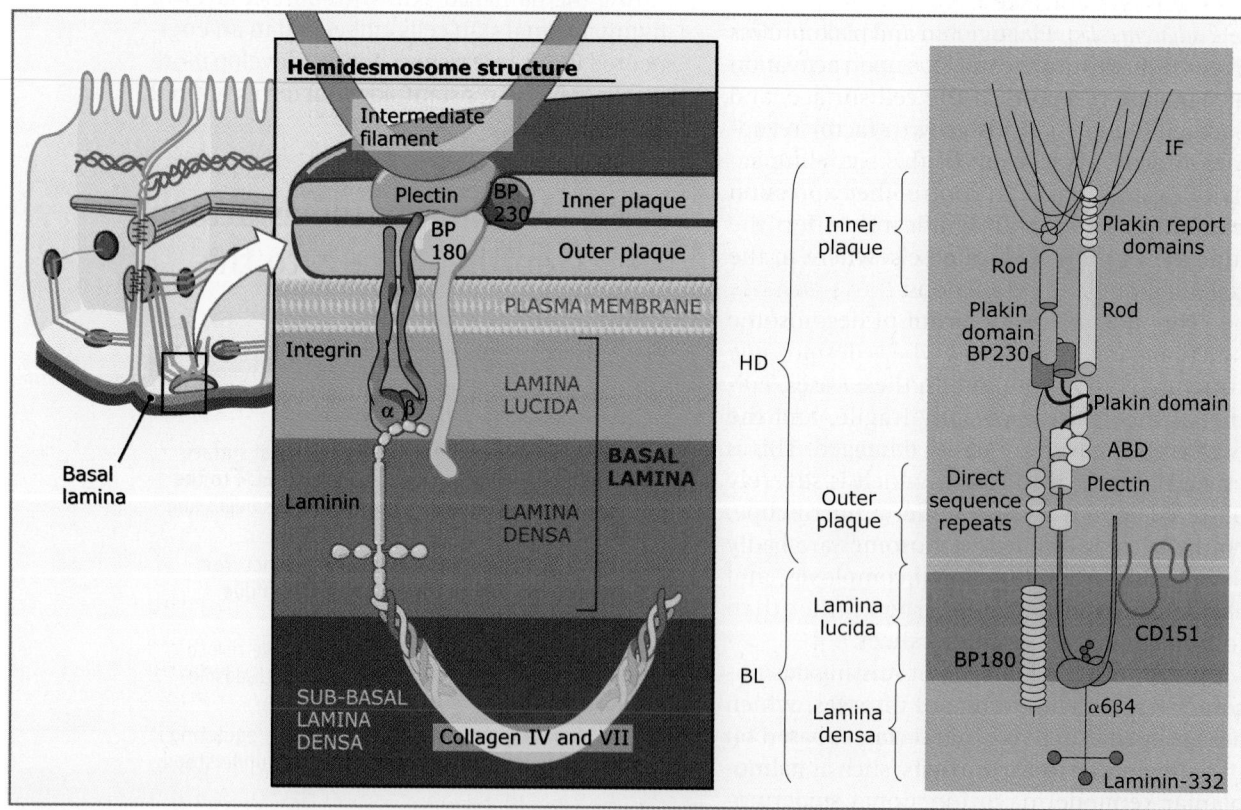

FIGURE 19.56 Hemisdesmosomes connect to the basement membrane, which consists of the basal lamina and a network of collagen fibers. The drawing on the right shows some of the molecular components found in the basal lamina (BL) and hemidesmosome (HD). In the basal lamina, BP180 (light green) and α6β4 integrin (purple) binds to laminin-332, which is connected to collagens IV and VII. In the cytosolic portion of the HD, BP180 and α6β4 integrin bind to BP230 (dark green) and plectin (light blue), respectively. The plakin domain, plakin report domain, and actin binding domain (ABD) mediate the interactions between these proteins. Adapted from C. Margadant, et al., *Curr. Opin. Cell Biol.* 20 (2008): 589–596.

at right angles to one another in the cell (see Figure 19.43) and, thus, resist different types of mechanical stresses. Together, both types of structures, linked by intermediate filaments, form a remarkably tough network. Lack of functional hemidesmosomes results in severe blistering of many epithelial tissues, including the skin, and these diseases can be fatal.

FIGURE 19.56 shows the composition of hemidesmosomes. At the cytoplasmic side of a "typical" (type I) hemidesmosome, we see a cluster of intermediate filaments (keratins 5 and 14) attached to an inner plaque (see Figure 19.55). This plaque is composed of the proteins BP230 and plectin, which bind to the intermediate filament proteins. The outer plaque contains two types of transmembrane proteins: an integrin receptor ($\alpha_6\beta_4$) and a member of the collagen family (known as collagen XVII or BP180). In the extracellular space, anchoring filaments (composed of BP180 and the ECM protein laminin-332) project outward from the plasma membrane, through the lamina lucida, and into the lamina densa. The lamina densa contains a variety of basal lamina proteins (see *19.11 The basal lamina is a specialized extracellular matrix*). Finally, anchoring fibrils of Type VII collagen connect the lamina densa to the subbasal lamina densa, which is composed of several ECM proteins. Type II hemidesmosomes, which lack the BP180 and BP230 proteins, are found in a subset of tissues, such as the intestine.

Despite their critical importance in maintaining tissue integrity, hemidesmosomes are not static structures. For example, cuts in the skin require that neighboring cells detach from the underlying basal lamina and migrate to the wound area. There the cells divide to repopulate the wound area, and then they reattach to the basal lamina and to each other. This change in phenotype requires that the cells be able to disassemble, and then later reassemble, cell-cell and cell-matrix junctions such as desmosomes and hemidesmosomes, respectively.

One model for the mechanism of disassembly and assembly of hemidesmosomes is that phosphorylation of the cytoplasmic tail of β_4 integrin causes it to curl inward, releasing its grip on plectin, and disassembling the hemidesmosome. Most of the hemidesmosome proteins remain in a complex that is then internalized by endocytosis. Later, when the cells come to rest and reattach to the basal lamina, these complexes are recycled back to the cell surface, the β_4 integrin chain is dephosphorylated, and hemidesmosomes reform.

19.21 Gap junctions allow direct transfer of molecules between adjacent cells

Key concepts

- Gap junctions are protein structures that facilitate direct transfer of small molecules between adjacent cells. They are found in most animal cells.
- Gap junctions consist of clusters of cylindrical gap junction channels, which project outward from the plasma membrane and span a 2–3 nm gap between adjacent cells.
- The gap junction channels consist of two halves, called connexons or hemichannels, each of which consists of six protein subunits called connexins.
- Over 20 different connexin genes are found in humans, and these combine to form a variety of connexon types.
- Gap junctions allow for free diffusion of molecules 1200 Da in size and exclude passage of molecules 2000 Da in size.
- Gap junction permeability is regulated by opening and closing of the gap junction channels, a process called "gating." Changes in intracellular pH, calcium ion flux, or direct phosphorylation of connexin subunits controls gating.
- Two additional families of nonconnexin gap junction proteins have been discovered, suggesting that gap junctions evolved more than once in the animal kingdom.

Gap junctions are specialized structures on the cell surface that facilitate the direct transfer of ions and small molecules between adjacent cells. They are found in most vertebrate and invertebrate cell types and are the only known means of cell-to-cell transport for animal cells. (Plant cells use plasmodesmata—see *21.9 Plasmodesmata are intercellular channels that connect plant cells*.) The gap junctions between cardiac muscle cells also facilitate transmission of electrical signals during muscle contractions.

Gap junctions were discovered in the 1960s when researchers learned that electric current passes directly between neighboring cells rather than through the extracellular fluid separating them. This suggested that cells exchange electrically charged ions and other small molecules via a channel that penetrates the plasma membranes and directly connects adjacent cells. Support for this idea came from electron micrographs of adherent cells that revealed the presence of closely apposed 2–3 nm "gaps" separating the adjacent plasma membranes.

As shown in **FIGURE 19.57**, these gaps are bridged by **gap junction channels**, which cluster into patches (or plaques) that project out of the plasma membrane, as shown by freeze fracture of the plasma membrane. Gap junctions can contain from a few dozen to many thousand gap junction channels and can extend to several micrometers in diameter on the cell surface. The gap junction channels are made up of two halves called hemichannels or **connexons**, which dock together in the intercellular gap. Each connexon is composed of six protein subunits, called **connexins**.

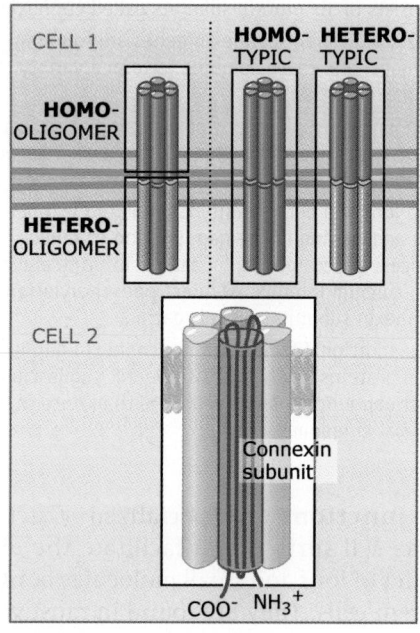

FIGURE 19.57 The principal structural unit of the gap junction is the connexon, which consists of six membrane-spanning connexin subunits. Each connexon is 17 nm long and 7 nm in diameter.

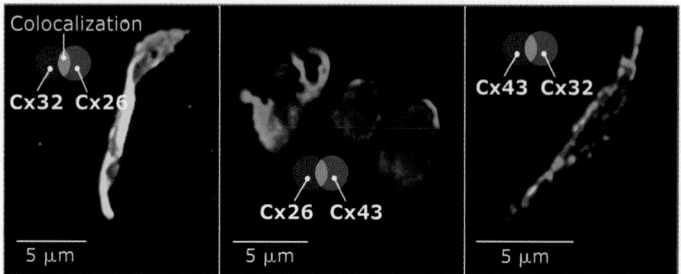

FIGURE 19.58 Double label immunofluorescence staining of connexin (Cx) subunits in gap junction plaques. Cells were transfected with the indicated pairs of connexin genes, then stained with antibodies to the connexins. Cx32 colocalizes with Cx26 but not with Cx43, for example. The cell bodies are not visible. Photos courtesy of Matthias Falk, Lehigh University.

A connexon is a 17-nm long, hydrophilic, cylindrical channel measuring 7 nm in diameter at its widest and about 3 nm in diameter at its narrowest point. The central pores of the channels can be visualized using negative stain techniques. Connexin subunits contain four membrane-spanning α helices linked by two extracellular loops. High-resolution structures suggest that the extracellular loops of opposing connexins bind to each other via antiparallel b sheets, thereby forming a β-barrel.

Gap junction channels can vary in composition, as shown in Figure 19.57. The human genome sequence suggests that at least 20 different connexin proteins exist in humans. Many cells express more than one connexin type, allowing for the formation of homo-oligomeric connexons (consisting of only one subunit type) and hetero-oligomeric connexons (containing multiple subunit types). In addition, connexons can dock with connexons of the same (homotypic channels) or different composition (heterotypic channels). A single gap junction plaque can contain connexons of different connexin composition. Within the plaque, the connexons are either homogenously mixed or are spatially segregated according to their connexin composition, as shown in **FIGURE 19.58**.

The specific domains that are involved in connexon/connexon docking, connexin/connexin recognition and oligomerization, and connexin-subunit compatibility (selectivity) have been identified. The experiments that identified these domains were binding assays that used either recombinant mutant connexins that lack specific regions of the protein or chimeric connexins composed of regions derived from different connexin-types. How the different domains function is currently being investigated.

In the original experiment to test the hypothesis that cells use channels to exchange small molecules, fluorescent molecules were injected into cells growing in culture. The diffusion of the molecules was followed over time by microscopy. These experiments showed that the molecules diffuse between neighboring cells much faster than would be expected if the molecules had to pass through the lipid bilayer of each plasma membrane. This result implicated the presence of a direct channel joining the cytosol of the neighboring cells. These channels were later identified as gap junctions. By using fluorescent molecules of different sizes, it was determined that gap junctions allow passage of molecules up to 1,200 Da in size (corresponding to a molecule approximately 2 nm in diameter) but exclude molecules larger than 2,000 Da. A recent version of the experiment showing exchange of fluorescent molecules between cells expressing connexins is shown in **FIGURE 19.59**.

Experiments such as these showed that ions could pass freely between the cytosolic compartments of cells linked by gap junctions. Many small molecules, including sugars, nucleotides, and second messenger molecules such as cAMP and cGMP may be exchanged as well. Communication through gap junctions can be critical when rapid, well-coordinated responses are required of a large number of cells. For example, rapid reflex reactions in the brain are mediated by neurons linked by gap junctions that allow nearly instantaneous exchange of ions, and the carefully controlled timing of contraction by heart muscle fibers is mediated by rapid exchange of ions as well.

Gap junction permeability appears to be controlled by channel opening and closing (a process referred to as channel **gating**). There is good evidence that gap junction channel gating is controlled by protein kinases that phosphorylate connexin subunits, changes in intracellular pH, and alterations in intracellular calcium ion concentration. For example, as calcium concentrations rise from 10^{-7} M to 10^{-5} M, gap junction permeability falls; concentrations above 10^{-5} M result in complete closure of the junctions. This may serve as a self-defense mechanism, since cells undergoing apoptosis typically experience a burst of cytosolic calcium, and closure of gap junctions may protect neighboring cells from initiating their apoptotic signaling mechanisms by accident. Connexins are also rapidly degraded in most cells: the half-life for a connexin protein is estimated to be ~15 hr.

Two additional families of gap junction proteins have been discovered. The innexins (invertebrate connexins) are found only in invertebrate animals and, despite their name, share no sequence homology with connexins. Nonetheless, they are capable of forming intercellular junctions that look and behave like vertebrate gap junctions. Taking a cue from the "nexin" nomenclature, the second family is called pannexins (from the Latin *pan*, meaning all). Pannexins are found in both vertebrates and invertebrates and are structurally distinct from both connexins and innexins. Pannexins are found almost entirely in neuronal cells, suggesting that they may have an important role in neuronal development and function, even

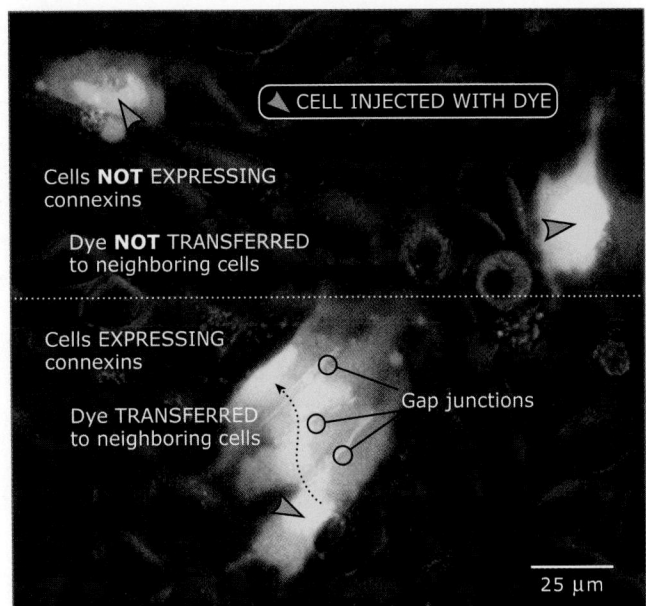

FIGURE 19.59 Fluorescent imaging of dye transfer through gap junctions. Cells were transiently transfected with DNA encoding for a connexin, so not all cells express connexin protein. Photo courtesy of Matthias Falk, Lehigh University.

in organisms with very primitive nervous systems. These observations also suggest that gap junctions arose at least twice during evolution of animals, through completely separate means, a process known as convergent evolution.

19.22 Calcium-dependent cadherins mediate adhesion between cells

Key concepts

- Cadherins constitute a family of cell surface transmembrane receptor proteins that are organized into eight groups.
- The best-known group of cadherins, called "classical cadherins," plays a role in establishing and maintaining cell-cell adhesion complexes such as the adherens junctions.
- Classical cadherins function as clusters of dimers, and the strength of adhesion is regulated by varying both the number of dimers expressed on the cell surface and the degree of clustering.
- Classical cadherins bind to cytoplasmic adaptor proteins, called catenins, which link cadherins to the actin cytoskeleton.
- Cadherin clusters regulate intracellular signaling by forming a cytoskeletal scaffold that organizes signaling proteins and their substrates into a three-dimensional complex.
- Classical cadherins are essential for tissue morphogenesis, primarily by controlling specificity of cell-cell adhesion as well as changes in cell shape and movement.

The **cadherin** superfamily consists of over 70 structurally related proteins, all of which share two properties: the extracellular regions of these proteins bind to calcium ions to fold properly (hence Ca, for calcium) and these proteins adhere to other proteins (hence, adherin). The cadherins are involved in cell-cell adhesion, cell migration, and signal transduction. The first group of cadherins discovered includes those found in the zonula adherens junctions formed between epithelial cells (see *19.18 Adherens junctions link adjacent cells*). These are now termed "classical cadherins" to distinguish them from their more distantly related family members.

All classical cadherins are transmembrane receptors with a single membrane-spanning domain, five extracellular domains at the amino end of the protein, and a conserved cytoplasmic C-terminal tail, as diagrammed in Figure 19.53. In vertebrates, the five classical cadherins are termed E-, P-, N-, R-, and VE-cadherins, based on the sites where they were first discovered: epithelium, placenta, nerve, retina, and vascular endothelium, respectively. We now know that expression of each type is more widespread, but the names have "stuck."

Classical cadherins function as clusters of dimers on the cell surface (see Figure 19.54). These dimers bind to identical dimers on neighboring cells (this is called **homophilic binding**). N- and R-cadherin pairs will also bind to each other (**heterophilic binding**). Cells can control their strength of adhesion by avidity modulation, which involves varying both the total number of receptors on the cell surface and the lateral diffusion of the receptors within the plasma membrane. Cadherins that are not clustered will not form strong adhesions with neighboring cells (see also *19.18 Adherens junctions link adjacent cells*). The mechanism of adhesion is biphasic: first, a low-affinity association is made almost immediately after opposing cadherins come into contact, then a second, more stable association is formed. The first association requires only the amino-terminus of the receptors to form, but the second requires the entire extracellular portion of the receptor.

There is direct evidence for the importance of cadherin clustering in cell-cell adhesion. The experiment that provided this evidence is based on the fact that the cadherin cytoplasmic tails are important for dimerization. As shown in **FIGURE 19.60**, a recombinant chimeric protein, containing the extracellular and transmembrane domains of the cadherin receptor and the cytoplasmic tail of an unrelated protein called FKBP12, was expressed in cells that normally do not bind to each other. Because the FKBP12 cytoplasmic tails do not bind to one another, the chimeric protein does not form dimers. As a result, the cells do not aggregate. However, addition of a chemical that links the FKBP12 cytoplasmic tails brings the cadherin portions of the chimeric protein close enough together to form clusters of dimers. In this case, the cells adhere to one another.

A 56-amino acid region at the C-terminus of the conserved cytoplasmic tail of classical cadherins binds to a cytosolic protein known as **β-catenin**. When attached to cadherins, β-catenin acts as an adaptor molecule by binding to **α-catenin**, which in turn binds to actin filaments, as shown in **FIGURE 19.61**. (α-catenin and β-catenin are unrelated in sequence.) β-catenin, therefore, is a link in the chain of proteins formed between the cadherin dimers at the cell surface and the actin cytoskeleton. As we saw above, breaking a link in this chain by disrupting dimer formation results in decreased cell adhesion. Another experiment in which this link is broken is shown in Figure 19.61.

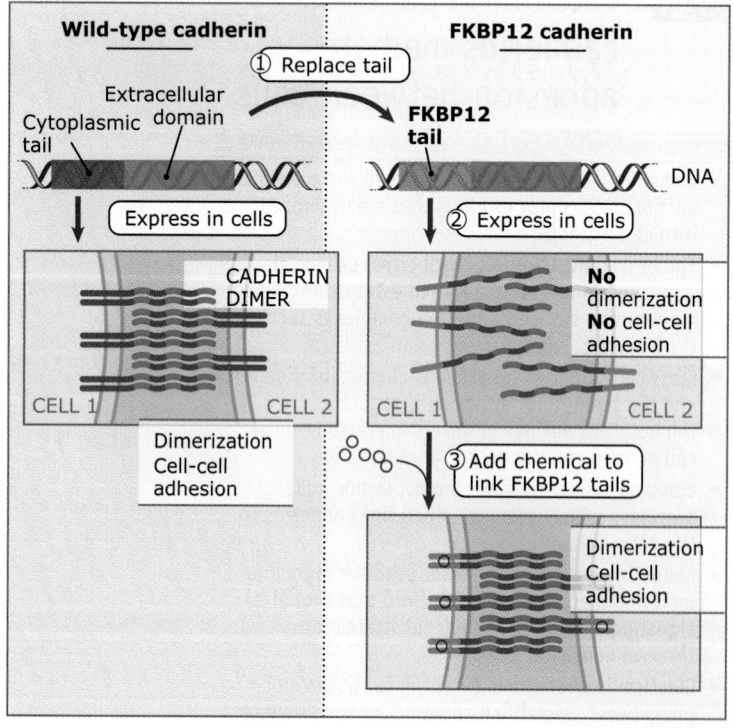

FIGURE 19.60 The cytoplasmic tail of cadherins is important for adhesion mediated by the extracellular domain.

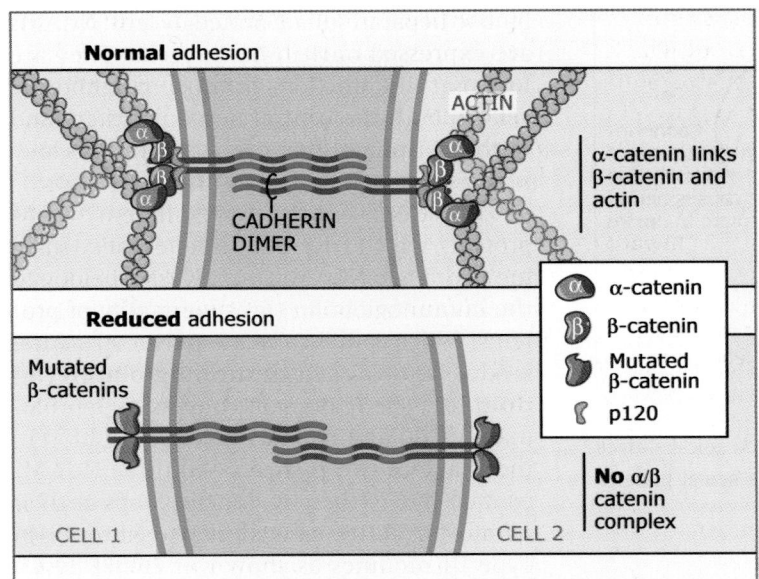

FIGURE 19.61 Cadherin cytoplasmic tails are linked to actin filaments via catenin proteins. Cadherin complexes also contain other linkers and signaling proteins that are not shown. The α-catenin binding domain of β catenin is required for normal cadherin-mediated cell adhesion.

Cells expressing a mutated β-catenin gene lacking the sequence encoding the α-catenin binding site do not adhere to one another. This is because the mutant β-catenin can bind to the cytoplasmic tails of cadherins but will not link with α-catenin.

When not complexed with cadherins, β-catenin plays a role in signal transduction, specifically in the pathway linking the growth factor Wnt to changes in gene expression. In Wnt-stimulated cells, β-catenin that is not bound to cadherins and α-catenin binds instead to a transcription factor. The β-catenin enters the nucleus and binds to transcription factors that control expression of genes required for cell growth. Cadherins are thought to indirectly play a role in controlling cell growth by sequestering β-catenin in adhesion complexes and limiting the activity of Wnt growth factors.

A second major protein that binds to cadherin cytoplasmic tails is the signaling adaptor protein p120^{CAS} shown in Figure 19.61. This protein is also a substrate for the Src family of tyrosine kinases, which are found in cadherin-based adherens junctions. Other signaling proteins are also found in cadherin clusters, suggesting that cadherins may be involved directly in signal transduction. In fact, β-catenin is phosphorylated by Src tyrosine kinases. However, there is no evidence that cadherins are signaling proteins per se. Rather, it is likely that because they can associate with the cytoskeleton and with signaling adaptors such as p120^{CAS}, cadherins and the junctions they form may act as scaffolds that organize

signaling molecules to regulate their activity. The formation of complexes of interacting signaling proteins is a common theme in signal transduction, and is seen, for example, in the focal adhesion associated with clusters of integrin receptors (see *19.14 Integrin receptors participate in cell signaling*). To use an analogy: cadherins may act as switchboards rather than as telephones.

Classical cadherins play a significant role during development by controlling the strength of cell-cell adhesion and by providing a mechanism for specific cell-cell recognition. For example, during development, E-cadherins are expressed when the blastocyst forms, and are thought to increase cell-cell adhesion when tight junctions form and epithelial cells subsequently polarize in the developing embryo. Not surprisingly, genetic knockout of E-cadherin genes is lethal early in development.

Functional mutations or knockout of other cadherin family members affect development of a wide variety of organs including brain, spinal cord, lung, and kidney. An important theme common to all of these developmental events is a process of cellular movement known as **invagination**. For example, the first nervous tissue arises in vertebrates when the cells composing the ectoderm form a ridge along the outer surface of the embryo that deepens into a cleft and then pinches off to form the neural tube. To form this tube, epithelial cells must constrict their apical domains and bend inward, forming a groove, then dissociate and move to new locations to close the tube, as illustrated

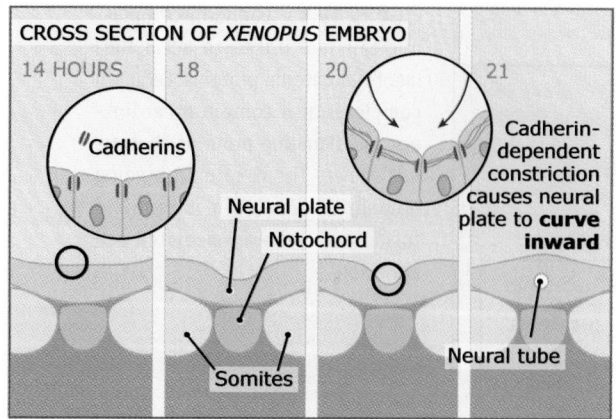

FIGURE 19.62 As the neural tube is formed, the apical surface of the neural plate cells constricts, causing the neural plate to curve inward.

in **FIGURE 19.62**. Similar movements occur in the formation of many ectodermally derived tissues, and all require variations in the types of cell-cell contacts. Deletion of cadherin genes results in a wide variety of developmental abnormalities, such as poor motor skills due to mistargeted neurons, which also result from errors in epithelial invaginations.

19.23 Calcium-independent neural cell adhesion molecules mediate adhesion between neural cells

Key concepts

- Neural cell adhesion molecules (NCAMs) are expressed only in neural cells and function primarily as homotypic cell-cell adhesion and signaling receptors.
- Nerve cells express three different types of NCAM proteins, which arise from alternative splicing of a single NCAM gene.
- Some NCAMs are covalently modified with long chains of polysialic acid (PSA), which reduces the strength of homotypic binding. This reduced adhesion may be important in developing neurons as they form and break contacts with other neurons.

While some cell adhesion proteins, such as cadherins and integrins, must bind to extracellular calcium ions to promote adhesion, this is not true for all of them. One major class of calcium-independent cell adhesion proteins is the NCAMs. NCAMs function primarily as cell-cell adhesion receptors, though they can also

bind to heparan sulfate proteoglycans. NCAMs are expressed only in neural cells. They are found at the junctions between neighboring cells in both the central nervous system and in the peripheral nervous system, especially in nerve fibers.

Nerve cells express three different NCAM proteins, which are formed by alternative splicing of a single NCAM gene. NCAMs belong to the immunoglobulin (Ig) superfamily of proteins, which contain a characteristic structural module known as the **immunoglobulin (Ig) domain**. The Ig domain consists of approximately 100 amino acids that form a loop in the shape of two β-sheets. All three NCAMs contain five of these Ig domain loops at their amino-terminus, as well as two fibronectin Type III modules as shown in **FIGURE 19.63**. NCAM-180 and NCAM-140 (the numbers refer to the size of the proteins in kilodaltons) contain a single transmembrane domain and differ in their C-terminal cytoplasmic tails. In contrast, NCAM-120 is anchored to the cell surface by a glycosylphosphatidyl inositol tail. All three forms of NCAM can be cleaved from the cell surface—NCAM-180 and NCAM-140 by proteolysis, NCAM-120 by phospholipases— and released as soluble molecules that diffuse through cerebrospinal fluid and plasma. The soluble NCAMs promote adhesion of neuronal cells and neurite outgrowth (the extension of axons from the nerve cell body).

Cell surface NCAMs can function as signaling receptors, by binding to a specific region on fibroblast growth factor receptor. This lateral association on the cell surface leads to an opening of cell surface calcium channels, presumably via a signaling pathway containing phospholipase C, diacyl glycerol, and arachidonic acid. At least one form of NCAM, NCAM-140, also associates with the nonreceptor tyrosine kinase p59[fyn], which links the receptor to activation of focal adhesion kinase. Thus, NCAM receptors activate signaling pathways shared by other cell surface receptors. The exact role of this signaling is not known.

Following their synthesis in the ER, NCAMs undergo many posttranslational modifications in the Golgi apparatus on their way to the cell surface. These modifications include phosphorylation, sulfation, and glycosylation. Perhaps the most significant of these modifications is the addition of long, linear chains of sialic acid sugars, PSA, to N-linked sugars on two asparagine residues in the fifth Ig domain, as shown in Figure 19.62.

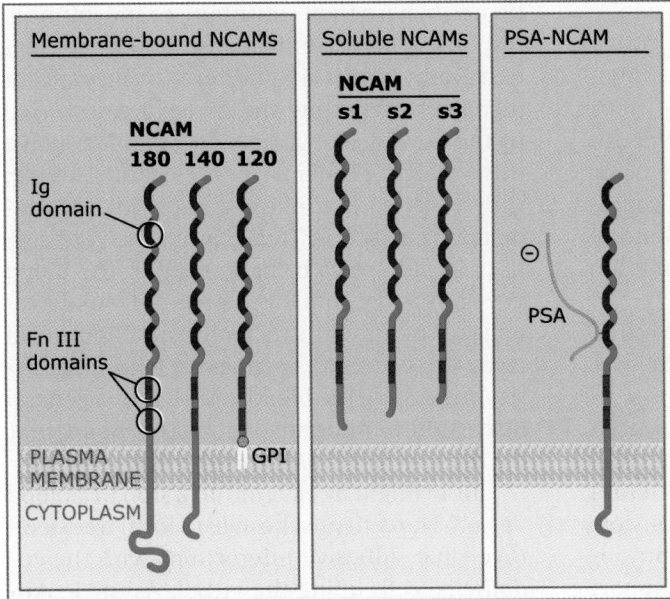

FIGURE 19.63 NCAMs are produced as both membrane-bound and soluble proteins of different sizes. The domain organizations of NCAMs are shown. The extracellular portions of NCAMs can be modified by addition of polysialic acid (PSA), which is attached to asparagine residues during transit through the secretory pathway (shown is a model of the 140 kDa transmembrane form of PSA-NCAM).

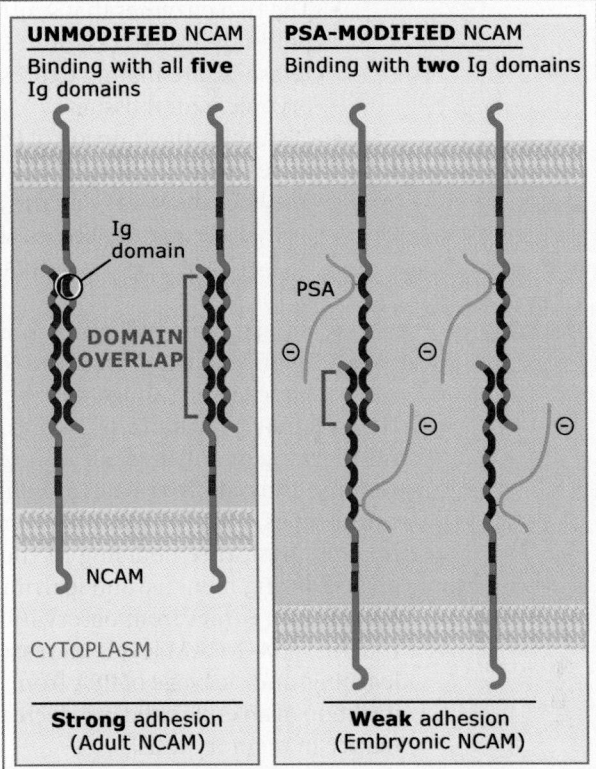

FIGURE 19.64 Unmodified NCAMs can bind to each other with five Ig domains, which results in strong cell-cell adhesion. PSA-modified NCAMs bind with only two Ig domains, which reduces the strength of cell-cell adhesion. The presence of PSA disrupts overlap of selectins, resulting in weak cell-cell adhesion during development. In adults, PSA is not added to selectins, so that firm cell-cell attachment is maintained.

Addition of the PSA chains significantly changes both the shape and function of NCAMs. Because sialic acid is negatively charged, PSA chains project outward from the NCAM protein, attract cations and bind to water molecules, similar to the glycosaminoglycan chains found on proteoglycans. The most striking effect of PSA addition to NCAMs is on the adhesive function of the receptors. Membrane-bound NCAMs bind primarily to identical NCAM receptors on neighboring cells. The exact mechanism of this homophilic binding is not known, but it does involve the Ig domains at the amino-terminus of each receptor. One current model for this interaction suggests that all five Ig domains on NCAM receptors overlap, resulting in a strong, stable adhesion between neighboring cells. However, PSA-NCAM receptors do not overlap completely, presumably because the large hydration volume and negative charge of the PSA chains repel the Ig domains of the complementary receptors on the neighboring cells, as shown in **FIGURE 19.64**. Consequently, cells expressing PSA-NCAMs bind less strongly to their neighbors than do cells expressing NCAMs lacking PSA.

What advantage is gained by having both strongly and weakly adhering forms of the same receptor? Remember that during development, cells must grow and move about the body to form tissues. While this process is taking place, cells may form and break contacts with each other many times. This is especially true of cells that must form multiple, highly specific cell contacts in the mature organism, such as neurons. Consequently, a low affinity but highly specific interaction such as occurs between PSA-NCAMs may be quite useful to developing neurons.

Directly proving that PSA is used to control neural cell development is extremely difficult. But several lines of evidence support this idea as given below:

- Immunohistochemistry, with antibodies that specifically recognize NCAM with or without PSA chains, have shown that about 30% of NCAMs expressed in embryonic mice contain PSA and that this figure drops to 10% in adults.

- The two enzymes that synthesize PSA, polysialyltransferase and sialyltransferase-X, are mainly expressed in embryonic neural tissues.
- Disrupting the function of PSA-NCAMs in developing animals, either by using enzymes that cleave off the PSA chains or by injecting antibodies that bind to PSA-NCAMs, results in malformation of the brain.
- Similar brain malformations are found in mice that have had the NCAM gene knocked out altogether.

NCAMs may play an important role in neural rearrangement in adult animals as well. Because they allow rearrangement of neural connections, PSA-NCAMs may be responsible for the physical remodeling of the brain observed during memory and learning. Support for this idea comes from observations that, in rodents, PSA-NCAM levels increase following learning, and cleavage of PSA from NCAM (by injection of an endosialidase) decreases performance in memory tests.

19.24 Selectins control adhesion of circulating immune cells

Key concepts

- Selectins are cell-cell adhesion receptors expressed exclusively on cells in the vascular system. Three forms of selectin, called L-, P-, and E-selectin, have been identified.
- Selectins function to arrest circulating leukocytes in blood vessels so that they can crawl out into the surrounding tissue.
- In a process called discontinuous cell-cell adhesion, selectins on leukocytes bind weakly and transiently to glycoproteins on the endothelial cells such that the leukocytes come to a "rolling stop" along the blood vessel wall.

Selectins are highly specialized cell surface receptors that are expressed exclusively on cells in the vascular system. Three different types of selectins have been identified so far, and they are named according to the cells that express them: L-selectin (leukocytes), P-selectin (platelets), and E-selectin (endothelial cells). Endothelial cells can express both E- and P-selectins on their surface after the cells have become activated by cytokines during inflammation.

The function of selectins is to facilitate the movement of leukocytes out of blood vessels (a process called extravasation) and into inflamed tissues, where they contribute to the immune response. This is a difficult task: leukocytes that extravasate must first adhere to the walls of blood vessels despite the sheer forces exerted by the flowing blood. How do leukocytes solve this problem? The answer is elegantly simple: they come to a "rolling stop," which is mediated by selectins. In this way, they are able to gradually reduce their speed in the blood vessel. As they come to a stop, the leukocytes engage the integrin receptors on endothelial cells. These receptors increase adhesion and assist in escaping the blood vessel. (For a general discussion of integrins see *19.13 Most integrins are receptors for extracellular matrix proteins*.)

How might a cell "roll" to a stop in a blood vessel? The cell must be able to form transient, reversible, adhesive interactions with the endothelial cells lining the vessel. As the leukocyte forms these associations, it drags (or rolls) along the vessel wall until it forms enough associations to come to a complete stop. This is known as **discontinuous cell-cell adhesion**. As shown in **FIGURE 19.65**, leukocytes express selectin ligands such that they will bind only to endothelial cells that express the E- and P-selectins on their cell surface and, thus, will not adhere to vessel walls in noninflamed tissues.

The key to this selective adhesion is the use of protein-sugar binding interactions, such as those used by proteoglycans and their receptors. Selectins are so named because the ligand-binding portion of these receptors resembles that found in lectins, a group of proteins that bind specifically to cell surface oligosaccharides (see Figure 19.64). The ligand-binding region of selectins lies at the N-terminus of the protein, connected to a series of short consensus repeats, followed by a single membrane-spanning domain and a short cytoplasmic domain at the C-terminus. Like cadherins and integrin receptors, selectins require extracellular calcium to fold properly and bind to their ligands.

Selectins bind to a complex and specific arrangement of sialic acid and fucose sugars, known as **sialyl Lewis(x) (sLex)**, which is attached to "carrier proteins" expressed on the surface of target cells. Selectins can distinguish between subtly different forms of sLex that are attached to different core proteins and, thus, can establish high binding specificity. The predominant selectin ligands include P-selectin glycoprotein ligand 1 (PSGL-1), glycosylation-dependent cell adhesion molecule-1 (GlyCAM-1), and mucosal addressin cell

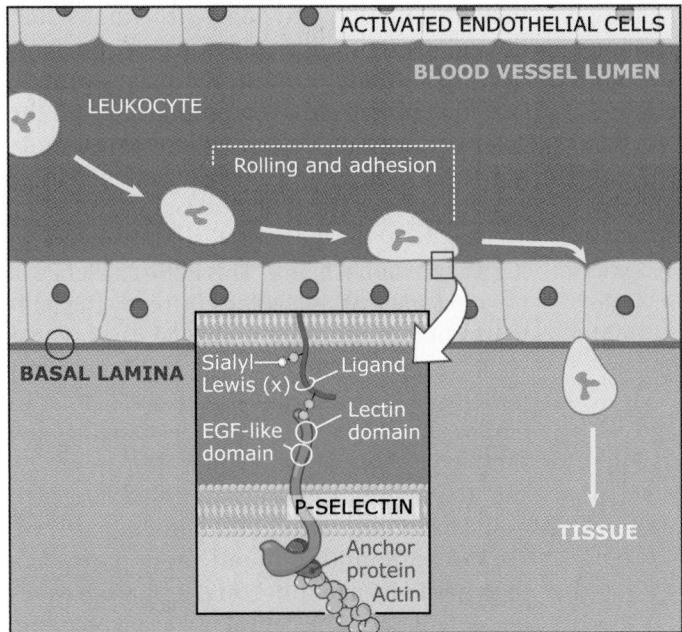

FIGURE 19.65 An illustration of the "rolling stop" function of selectins. Inset: a model of selectin structural organization and binding to a leukocyte ligand.

adhesion molecule-1 (MadCAM-1). There is some speculation that binding to these ligands activates intracellular signaling pathways that assist in the activation of the integrin receptors involved in later stages of extravasation, but this has not been proved.

Concept and Reasoning Check

1. One theme running throughout the last 10 sections of this chapter is that cells form specialized protein complexes on their plasma membrane to assist in their direct adhesion to other cells. Taken to the extreme, this suggests that (1) all cells in a multicellular organism should make at least one of these complexes, and (2) the more cell-cell adhesion complexes a cell creates the better it functions in a multicellular organism. Do you agree with these statements? Why/why not?

19.25 What's next?

As we mentioned at the beginning of this chapter, the study of the ECM and cell junctions has advanced considerably in the past 100 years (see Figure 19.B1). We now know the major components of these structures and have learned a considerable amount about their structure and function. Recent advances in molecular genetics have given us insight into the function of these molecules during development.

In our discussion of **what's next** in this field, let's look at a very practical example. Have a look at your hand, and consider how complex it is. Move it around, tap your fingers on a table, snap your fingers. Each of these seemingly simple movements requires careful cooperation between the several billion cells that make up the tissues in your hand. These cells are supported and interconnected by the ECM and cell junctions.

We now know that hundreds of different types of proteins compose the ECM and cell junctions. The era of discovering new proteins in these structures is drawing to a close. Having gathered all of this information, what shall we do with it? Once every protein is identified in an organism, the reductionist approach to cell biology will reach a milestone, theoretically allowing the *in vitro* assembly of the tissues and organs that contain these proteins. Here is where our understanding lags far behind our technology. To understand the function of proteins and the mechanisms that control them, we need to focus our attention on how these proteins work together in teams; in other words, we must learn how multicellularity is generated and maintained.

Fortunately, the field of systems biology, which emerged late in the 20th century, is focused on understanding these dynamic interactions. Such an approach can allow us to ask how contact with an ECM protein via one receptor can control the function of a cell junction and change the expression of ECM genes, for example. The first phase of systems biology is aimed at understanding cytsoslic networks, such as the transcription factors controlling gene expression and the enzymes controlling cellular metabolism. In the future, the scope of systems biology will broaden to include the issues discussed in this chapter.

More immediately, the field of tissue engineering is one popular area where our current understanding of cell-cell and cell-ECM adhesion is being applied to create realistic functional tissues in the lab. Examples include advances in the development of replacement organs such as skin, bone, cartilage, liver, cornea, blood vessels, and even spinal cord. One of the hallmarks of the next stage of ECM and cell junction research will be the expansion of the basic knowledge gained previously to develop fully functional organs that can replace those damaged by injury or disease.

19.26 Summary

The ECM consists of hundreds of different molecules that interact in complex and highly organized ways. The two major classes of macromolecules in the ECM are the structural glycoproteins (such as collagens, elastins, fibronectins, and laminins) and the proteoglycans (e.g., heparan sulfate) that impart structural stability and provide a hydrophilic environment in tissues. Each of these molecules consists of structural modules that mediate attachment of cell receptors, growth factors, and other ECM molecules. These molecules control cell behavior by determining the three-dimensional arrangement of cells in tissues, by activating intracellular signaling pathways, and by providing substrates upon which cells migrate. We also know that the constituents of this matrix change over time and that cells within it are responsible for both producing and degrading the matrix in response to specific signals.

Of the proteins that combine to form specialized cell surface complexes that mediate attachment to the ECM and between adjacent cells, well over 100 are known—the integrin clusters alone interact with over 50 of them. These complexes perform a variety of specialized functions. Tight junctions and septate junctions regulate paracellular transport across epithelial sheets; adherens junctions and focal adhesions link the cell surface to the actin cytoskeleton, thereby controlling cell movement; and desmosomes and hemidesmosomes link the cell surface to the intermediate filament network, thereby providing structural stability and distributing tensile forces across large networks. Many of these complexes contain signaling proteins that communicate with the cell interior to regulate a variety of cell functions such as growth. The organization of these signaling networks is highly complex.

But how do these molecules combine to form a functional being? This is the challenge of the next stage of cell biology. Already the hallmarks of this effort are visible: in cases where we have identified most of the molecular components of a structure (e.g., the basal lamina), emphasis is now shifting to understanding how these parts interact to form a functional tissue. Having reduced a cellular structure such as the basal lamina to its constituent parts, we are now attempting to put these parts back together to form a functional unit, such as replacement skin. The field of tissue engineering, which is devoted to applying this knowledge to manufacture biological structures *de novo*, will play a major role in advancing cell biology in the fourth stage of research.

http://biology.jbpub.com/lewin/cells

To explore these topics in more detail, visit this book's Interactive Student Study Guide.

References

Chen, C., and Sheppard, D. 2007. Identification and molecular characterization of multiple phenotypes in integrin knockout mice. *Meth Enzymol.* v. 426 p. 291–305.

Daley, W. P., Peters, S. B., and Larsen, M. 2008. Extracellular matrix dynamics in development and regenerative medicine. *J Cell Sci.* v. 121(Pt 3) p. 255–264.

Ditlevsen, D. K., Povlsen, G. K., Berezin, V., and Bock, E. 2008. NCAM-induced intracellular signaling revisited. *J Neurosci Res.* v. 86 p. 727–743.

Leckband, D. 2008. Beyond structure: mechanism and dynamics of intercellular adhesion. *Biochem Soc Trans.* v. 36 p. 213–220.

Melrose, J., Hayes, A. J., Whitelock J. M., and Little C. B. 2008. Perlecan, the "jack of all trades" proteoglycan of cartilaginous weight-bearing connective tissues. *Bioessays* v. 30 p. 457–469.

Michod, R. E. 2007. Evolution of individuality during the transition from unicellular to multicellular life. *Proc Natl Acad Sci USA* v. 104 Suppl 1 p. 8613–8618.

Paris, L., Tonutti, L., Vannini, C., and Bazzoni, G. 2008. Structural organization of the tight junctions. *Biochim Biophys Acta* v. 1778 p. 646–659.

Posey, K. L., Hankenson, K., Veerisetty, A. C., Bornstein, P., Lawler, J., and Hecht, J. T. 2008. Skeletal abnormalities in mice lacking extracellular matrix proteins, thrombospondin-1, thrombospondin-3, thrombospondin-5, and type IX collagen. *Am J Pathol.* v. 172 p. 1664–1674.

Vakonakis, I., and Campbell, I. D. 2007. Extracellular matrix: from atomic resolution to ultrastructure. *Curr Opin Cell Biol.* v. 19 p. 578–583.

Zarbock, A., and Ley, K. 2008. Mechanisms and consequences of neutrophil interaction with the endothelium. *Am J Pathol.* v. 172 p. 1–7.

PART

7

Prokaryotic and plant cells

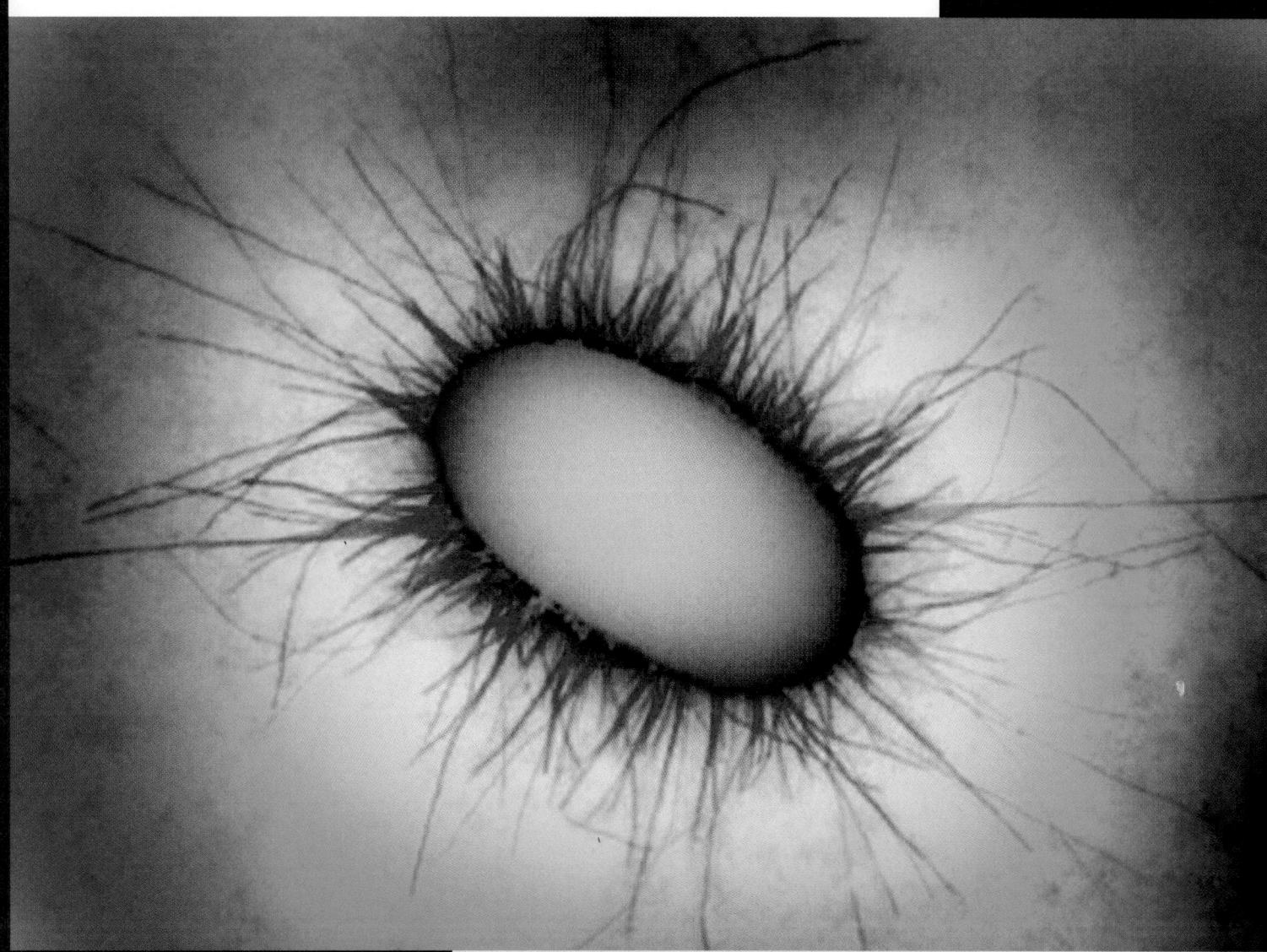

Prokaryotic cell biology

20

Matthew Chapman
University of Michigan, Ann Arbor, MI

Jeff Errington
Institute for Cell and Molecular Biosciences,
Newcastle University, UK

BACILLUS SUBTILIS CELLS as visualized by fluorescence microscopy. The cells express green fluorescent protein-coupled FtsZ, a protein that localizes to the site of cell division. Cells were stained for DNA (blue) and cell membrane (red). Photo courtesy of Ling Juan Wu and Jeffery Errington, Newcastle University.

CHAPTER OUTLINE

Did you know that on a cell basis that we are outnumbered in our own bodies? Depending on the particular estimate, there are 10 to 100 times the number of bacterial cells occupying our bodies than human cells. Not surprisingly, this complex microbial population contributes significantly to our physiology, development, health, and genomics.

In 2003, the science community celebrated the completion of the Human Genome Project, providing an accurate sequence of the human genome. Although a truly monumental effort whose human health benefits will be realized over the next few decades, this analysis is far from complete. This is because we only sequenced part of the genetic material that humans carry with them, leaving the microbial components of our bodies unsequenced, and in many cases, unidentified. In fact, it has been estimated that the micro genome contains 100× as many genes as is encoded by our cellular genome.

Between 500 and 1000 species of bacteria call our bodies home. *Bacteriodes* and *Firmucutes* are the two major genera of bacteria found in the human microbiota, predominately colonizing the human GI tract. However, bacteria also colonize our skin, teeth and just about every other niche in our bodies, save for the privileged circulatory system.

We are just beginning to understand the contribution of this microbiota on human development. Germ free mice that lack a normal microbiota have many documented defects in gut development. Intestinal epithelial cells in germ free mice have defective microvilli formation and do not properly glycosylate surface proteins. Furthermore, germ free mice have several immunological defects such as reduction in lymphoid tissues and smaller Peyer's patches. This large microbiota also protects against opportunistic pathogens that cannot efficiently compete for resources with the well-established resident bacteria.

Metagenomics and 16 Ribosomal RNA (rRNA) sequencing have allowed scientists to probe the diversity of organisms in the human microbiota. However, we are far from knowing how the diversity of the microbiota contributes to human health, or how we might be able to manipulate this population through probiotics. Elucidating these issues will remind us that as human beings we are part of and tied in separately to a remarkably diverse microbial ecosystem.

20.1 Introduction

Key concepts

- The relative simplicity of the prokaryotic cell architecture compared with eukaryotic cells belies an economical but highly sophisticated organization.
- A few prokaryotic species are well described in terms of cell biology, but these represent only a tiny sample of the enormous diversity represented by the group as a whole.
- Many central features of prokaryotic cell organization are well conserved throughout cellular life.
- Diversity and adaptability have been facilitated by a wide range of optional structures and processes that provide some prokaryotes with the ability to thrive in specialized and sometimes harsh environments.
- Bacteria colonize nearly every environmental niche on Earth, including our own bodies. On average bacteria in our bodies outnumber our own cells 10–100:1.
- Prokaryotic genomes are highly flexible and a number of mechanisms enable prokaryotes to adapt and evolve rapidly.

Prokaryotes are single-celled organisms that are defined by the lack of a nucleus, which in eukaryotic cells is the membrane-bounded compartment that contains DNA. Prokaryotic cells differ in other fundamental ways from eukaryotic cells. Prokaryotes are relatively simple in terms of genetic information and cell architecture, as **FIGURE 20.1** shows (for a schematic of a eukaryotic cell see Figure 8.1). They generally have a single circular chromosome that together with bound proteins constitutes the nucleoid. Most prokaryotes do not have internal membranes, but a few, such as photosynthetic bacteria, do. Prokaryotic cells are no longer viewed as "bags of enzymes"; it is now clear that prokaryotic cells are highly organized, with many proteins specifically targeted to subcellular domains. Furthermore, essential cellular processes such as DNA replication, transcription, translation, protein folding, and protein secretion are fundamentally the same in prokaryotes and eukaryotes. Prokaryotic cells even have actin and tubulin homologs that function as simple cytoskeletal elements.

Our current understanding of the organization of prokaryotic cells is a major topic of this chapter. The **cell envelope** comprises the layers surrounding the cytoplasm of prokaryotes. These layers include the **cytoplasmic membrane**, the **cell wall**, **capsule,** and in the case of Gram-negative bacteria, the outer membrane as **FIGURE 20.2** shows. In contrast, Gram-positive bacteria have a thicker cell wall and no outer membrane. We will discuss in detail the composition of the various cell envelope layers (see *20.5 Most prokaryotes produce a polysaccharide-rich layer called the capsule; 20.6 The bacterial cell wall contains a cross-linked meshwork of peptidoglycan; 20.7 The cell envelope of Gram-positive bacteria has unique features; 20.8 Gram-negative bacteria have an outer membrane and a periplasmic space;* and *20.9 The cytoplasmic membrane is a selective barrier for secretion*).

So far, detailed information on prokaryotic structure and function is mostly restricted to a small number of tractable model organisms. However, prokaryotes are an evolutionarily ancient and diverse group of organisms and constitute a wide variety of forms, as shown for bacteria in **FIGURE 20.3**. Understanding prokaryotic phylogeny has been problematic: because prokaryotes are primarily asexual organisms, the species concept used for higher organisms could not readily be applied to prokaryotes. In addition, prokaryotes have a range of mechanisms whereby genes can be transferred horizontally (i.e., genetic material is transferred between two organisms, neither of which is the offspring of the other). This horizontal gene transfer confounds attempts to classify prokaryotes based on any single criterion. Molecular methods, particularly rRNA sequencing and more recently, complete genome sequencing, have revolutionized prokaryotic phylogenetics and are beginning to allow progress toward a systematic description of the whole domain. Prokaryotes can be divided into two lineages, bacteria and archaea, which are discussed in the next section (*20.2 Molecular phylogeny techniques are used to understand microbial evolution*).

Work on the cell biology of prokaryotes has been mainly focused on a handful of organisms that are experimentally tractable and have some practical importance, either medically or industrially. The knowledge base is currently dominated by two bacterial species, *Escherichia coli* and *Bacillus subtilis*. These bacteria are Gram-negative and Gram-positive, respectively, and probably diverged about 2 billion years ago. However, *E. coli* and *B. subtilis* represent only the "tip of the iceberg," in terms of the myriad of prokaryotic forms. Although most of *20 Prokaryotic cell biology*, focuses on what we know about *E. coli, B. subtilis*, and closely related organisms, a diverse range of organisms will be mentioned where there are

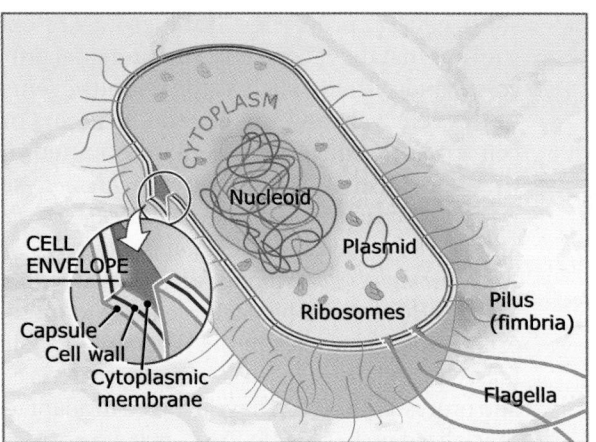

FIGURE 20.1 Prokaryotic cells are characterized by the lack of a membrane-enclosed nucleus.

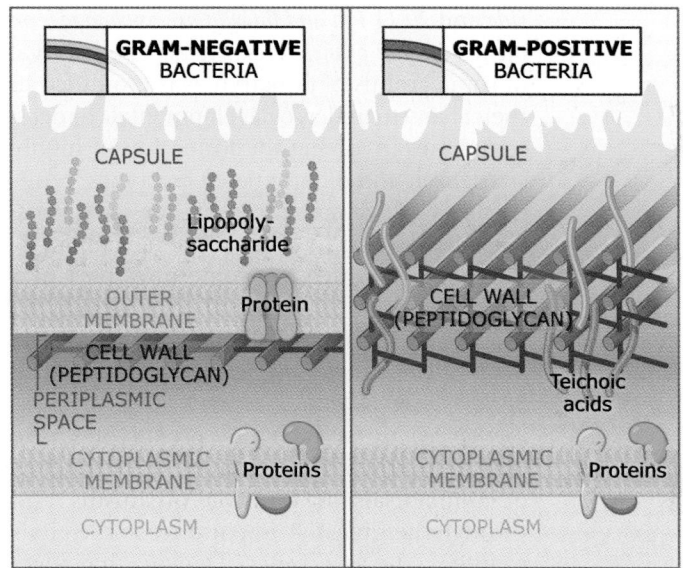

FIGURE 20.2 Bacterial cell envelopes can be categorized as either Gram-negative or Gram-positive, depending on the results of the Gram staining procedure. In contrast to Gram-positive bacteria, Gram-negative bacteria do not take up the violet dye and have an outer membrane and a thinner cell wall.

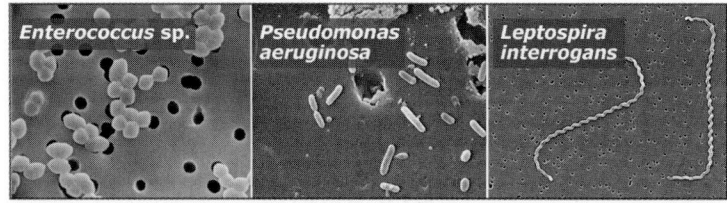

FIGURE 20.3 The three main shapes of bacteria are spherical, rod-shaped, and spiral. Photos courtesy of Janice Carr/NCID/HIP/CDC.

significant differences from the paradigms. Nevertheless, a huge amount of diversity undoubtedly awaits analysis.

The *E. coli* and *B. subtilis* paradigms have been very powerful tools in understanding some general properties of the groups of organisms that they represent. Fundamental processes and systems, such as the various events of the cell cycle and the general elements of cell structure, have been well determined in these organisms; *20.15 Prokaryotic chromosome segregation occurs in the absence of a mitotic spindle* and *20.16 Prokaryotic cell division involves formation of a complex cytokinetic ring* are devoted to this system. However, a big factor leading to interest in prokaryotes lies in the diversity of their lifestyles and, thus, in the additional structures and processes that they have evolved to adapt to specialized niches. Structures, such as capsules and flagella, and their roles in various adaptive functions are described in *20.5 Most prokaryotes produce a polysaccharide-rich layer called the capsule* and *20.11 Pili and flagella are appendages on the cell surface of most prokaryotes*. The survival of some groups of prokaryotes is grounded in their ability to undergo developmental changes, sometimes involving the formation of highly specialized and differentiated cell types, reminiscent of development in higher organisms. We will describe in detail these developmental changes, which are sometimes part of a prokaryote's developmental cycle or may be induced by stress (see *20.17 Prokaryotes respond to stress with complex developmental changes* and *20.18 Some prokaryotic life cycles include obligatory developmental changes*). Lastly, prokaryotes have critical interactions with man, as pathogens or commensals, as industrial organisms, and through their massive importance in the environment. The aspects of cell biology that affect these interactions are described in *20.19 Some prokaryotes and eukaryotes have endosymbiotic relationships*; *20.20 Prokaryotes can colonize and cause disease in higher organisms*; and *20.21 Biofilms are highly organized communities of microbes*.

20.2 Molecular phylogeny techniques are used to understand microbial evolution

Key concepts

- Only a fraction of the prokaryotic species on earth has been analyzed.
- Unique taxonomic techniques have been developed for classifying prokaryotes.
- Ribosomal RNA (rRNA) comparison has been used to build a three-domain tree of life that consists of bacteria, archaea, and eukarya.

The study of evolution is grounded in the ability to determine accurately how organisms are related to one another. Prokaryotic **taxonomy** presents special problems because methods that assess morphology, genetic features, or fossil records are not practical. Furthermore, we understand very little or nothing about the vast majority of microbes that inhabit the Earth. Instead, several unique methods of taxonomy have been used to classify prokaryotic organisms into groups. These include numeric taxonomy, which is based on comparison of many characteristics (such as the presence or absence of a capsule layer, differential Gram staining, oxygen requirements, properties of nucleic acids and proteins, ability to form spores, motility, and the presence or absence of certain enzymes) to assess similarities and differences between microbes. However, because numeric taxonomy requires significant knowledge of an organism, it is not feasible for microbes that have not been cultured and studied in detail.

Molecular phylogeny techniques have been pioneered to permit the classification of organisms about which very little is known. Molecular phylogeny assumes that evolutionary differences are reflected by variations in an organism's genetic material. There are several ways to classify organisms using DNA. Organisms can be grouped according to the ratios of bases contained in their chromosomal DNA. DNA contains four bases—adenine (A), thymine (T), cytosine (C), and guanine (G)—that pair with each other so that the ratio of adenine and thymine is equal, as is the ratio of guanine and cytosine. The percentage of guanines and cytosines (also known as G-C content) present in chromosomal DNA varies between 45% and 75%. Not surprisingly, the G-C content of related organisms is similar. However, similar percentages of bases do not in themselves mean that the organisms are closely related. For example, *B. subtilis* and humans have approximately the same G-C content, although they are obviously not closely related evolutionarily. However, in many cases, G-C content is a relatively easy way to make an initial assessment of evolutionary similarity between organisms.

Arguably the most accurate and powerful method of molecular phylogeny involves the comparison of gene sequences that are highly conserved. The degree of similarity between two organisms' small subunit ribosomal RNA (SSU rRNA) sequences indicates their evolutionary relatedness. The SSU rRNA genes are ideal for these analyses because of their ubiquitous nature and their evolutionary stability. Either the rRNA genes are isolated and directly sequenced, or the DNA is first amplified using the polymerase chain reaction and cloned. The latter technique is useful for characterization of organisms that are present in limiting amounts or for those that cannot be easily cultured. Once sequence information is obtained, it can be analyzed using computer programs that compare rRNA sequences to generate phylogenetic trees, such as the one shown in **FIGURE 20.4**.

Sequence comparisons have revealed some surprises about phylogenetic relationships. Using traditional phenotypic relationships (including numerical taxonomy), biologists had grouped the living world into five kingdoms, only one of which was prokaryotic. In contrast, molecular phylogeny has revealed that cellular life has evolved along three major lineages or domains, two of which consist of prokaryotes. Bacteria and archaea make up the two prokaryotic lineages, joining the single eukaryotic lineage called Eukarya (see Figure 20.4). Interestingly, archaeal rRNA is more similar to eukaryotic rRNA than to bacterial rRNA. A note about nomenclature: in the scientific literature, the bacteria lineage is often referred to as *eubacteria*, and the archaea lineage was called *archaebacteria*. We will use bacteria and archaea when referring to these specific prokaryotic lineages, although we will use the word *prokaryote* when referring to bacteria and archaea in general.

Pioneered by Carl Woese and colleagues, the phylogenetic tree based on rRNA sequences describes the evolutionary history of all organisms. At the root of this tree is the postulated universal ancestor, which would be the common ancestor to all life on earth. Many genes have been found that are shared among all three domains, suggesting that early in life's evolution, horizontal gene transfer was extensive. Thus, genes that coded for core cellular functions such as transcription and translation were presumably passed freely throughout the population of primitive organisms. This proposal helps to explain why all cells, regardless of their lineage, share many common genes. As each lineage continued to grow and evolve, certain biological traits would be lost and others would be gained, giving each lineage its own unique set of genetic material.

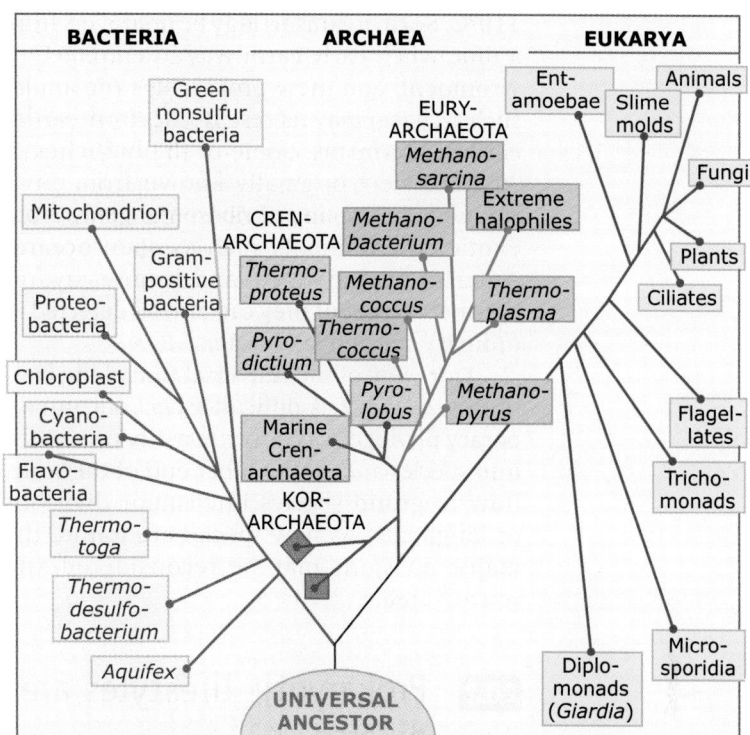

FIGURE 20.4 This universal phylogenetic tree shows the relationships among types of organisms based on the sequences of their small subunit ribosomal RNAs.

The domain bacteria is divided into at least 10 major divisions in the universal tree of life (see Figure 20.4). However, this number is probably low, as our view of the microbial world has been limited by the ability to culture strains in the laboratory, and only a small fraction of the microbial world can be grown using current laboratory methods. Some of the divisions within bacteria, as distinguished by phylogenetic comparisons, consist of species that lack strong phenotypic cohesiveness. For example, the proteobacteria kingdom contains species that have a mixture of physiological traits whose breadth nearly equals that of all known prokaryotic physiology.

The second prokaryotic domain is the archaea, which consists of three major divisions, the crenarchaeota, euryarchaeota, and korarchaeota. Physiologically, bacteria and archaea are most easily differentiated by the presence (in bacteria) or absence (in archaea) of a peptidoglycan-containing cell wall (for details on peptidoglycan see *20.6 The bacterial cell wall contains a cross-linked meshwork of peptidoglycan*). Members of the domain eukarya also lack peptidoglycan in their cell walls. Branching very close to the base of the archaeal tree is Methanopyrus, an extreme thermophile that is capable of growing in temperatures up to

110°C. Such organisms may be holdovers from a time when early earth was an extreme environment, and these prokaryotes (or similar such archaea) may represent relics from earth's earliest life-forms. Some of the branches of Archaea were originally known from environmental sampling of ribosomal genes from exotic environments such as open oceans, Antarctic marine waters, and deep-sea hydrothermal vents, but they can also be detected in ordinary soil and lake water.

Our view of the universal tree of life continues to evolve. A difficult aspect of contemporary phylogeny is how to classify organisms into species and, at the other end of the scale, how to group species into major divisions or kingdoms. As new species are found, the major divisions may be reconsidered and restructured.

20.3 Prokaryotic lifestyles are diverse

Key concepts

- The inability to culture many prokaryotic organisms in the laboratory has hindered our knowledge about the true diversity of prokaryotic lifestyles.
- DNA sampling has been used to better gauge the diversity of microbial life in different ecological niches.
- Prokaryotic species can be characterized by their ability to survive and replicate in environments that vary widely in temperature, pH, osmotic pressure, and oxygen availability.

TYPE OF PROKARYOTE	FAVORED GROWTH CONDITIONS
Psychrophile	~ 0 - 20° C
Mesophile	~ 25 - 40° C
Thermophile	~ 50 - 110° C
Acidophile	pH < 5.4
Neutrophile	pH ~ 5.4 - 8.0
Alkalophile	pH > 8.0
Obligate aerobe	Requires O_2
Microaerophile	Low O_2 and high CO_2
Aerotolerant	O_2 not required but tolerated
Obligate anaerobe	Cannot survive if O_2 is present
Facultative anaerobe	Aerobic respiration or fermentation
Barophile	High hydrostatic pressure

FIGURE 20.5 Prokaryotes can be classified based on their preferred growth conditions.

Our view of the prokaryotic world has been skewed by the ability (or inability) to culture organisms in the laboratory. It has been estimated that over 99% of the prokaryotes in the environment are uncultivatable using standard techniques. This is a problem since the vast majority of what we know about prokaryotic diversity comes from studies conducted in laboratory environments. Despite this, through DNA sampling it can be appreciated that prokaryotes are amazingly diverse organisms that live in many different locations. They can live on our bodies, in ponds, lakes, and rivers, in deep-sea vents at the bottom of the ocean, and in hot springs that reach temperatures of over 100°C. Equally remarkable is how many different nutrient resources prokaryotes can utilize to support their growth. Prokaryotic growth can be influenced by many factors including pH, temperature, oxygen availability, water availability, and osmotic pressure. Within these varied physical environments, the availability of nutrients such as carbon, nitrogen, sulfur, phosphorous, as well as many vitamins and trace elements, also influence growth. We will discuss the major restrictions on prokaryotic growth, and give examples of organisms that live on the edges of these boundaries. FIGURE 20.5 summarizes some of the major classifications based on growth conditions.

Prokaryotes can be classified into three groups according to their ability to grow at different temperatures: psychrophiles, mesophiles, and thermophiles. Psychrophiles are cold-loving organisms that grow best at temperatures between 15°C and 20°C, with some species living quite happily at 0°C. *Bacillus glodisporus* is an obligate psychrophile, which means that it cannot grow in temperatures above 20°C. Psychrophiles are unable to live in the human body because of its relatively high temperature but instead inhabit cold water and soil. As expected, enzymes made by psychrophiles are adapted to work best at low temperatures. The biotechnology industry has taken advantage of these characteristics. For example, proteases isolated from psychrophiles are employed as reagents for contact lens cleaning because of their robust activity at room temperature. "Ice-minus" psychrophiles have been used as a sort of biological antifreeze to prevent crops from freezing.

The majority of known prokaryotic species, called mesophiles, grow best between 25°C and 40°C. This group of bacteria has been extensively studied because it includes nearly every human pathogen. Some of these organisms can withstand high temperatures for short periods, making inadequate heating during canning or pasteurization potentially dangerous.

Thermophiles are heat-loving microbes that usually grow at temperatures between 50°C and 60°C, but some can tolerate temperatures as high as 110°C. Members of the *archaea* genus *Sulfolobus*, which are commonly found in geothermal vents such as those shown in FIGURE 20.6 (top panel), grow best at 80°C to 85°C. Such organisms are referred to as hyperthermophiles, and enzymes from these organisms have been used for many applications. Probably the most recognized is the DNA polymerase from *Thermus aquaticus* that is used in polymerase chain reactions (PCRs) carried out routinely in laboratories all over the world. The enzymes from these organisms are stable at high temperatures, which allows them to remain active during the repeated cycles of heating and cooling that are necessary to denature and renature double-stranded DNA templates during PCR.

Prokaryotes can also be classified by their ability to grow in acidic and alkaline environments. As we saw with temperature, many microbes can withstand variations in pH that allow them to occupy niches not suitable for other organisms. Acidophiles are acid-loving organisms that grow best at a pH below 5.4. These microbes often produce acid as a byproduct of fermentation. *Lactobacillus*, for example, ferments glucose to lactic acid, which decreases the local pH, and thus inhibits the growth of nonacidophilic organisms. Neutrophiles, including most of the pathogens that cause disease in humans, thrive in pH 5.5 to 8.0. Alkalophiles are alkali-loving (base-loving) organisms that grow in pH 8.0 and higher. Some soil prokaryotes are among the most alkaline tolerant organisms known, sometimes growing in pH 12. The human pathogen *Vibrio cholerae* grows best at a pH of about 9. For the most part, pH tolerance exhibited by microbes is the result of protection provided by the cell wall. Although microbes may prefer a certain environmental pH for optimal growth, in most cases, the pH inside the cell remains around 7. Deviations from near neutral pH can lead to protein denaturation and can interfere with proton gradients across the cytoplasmic membrane. The intracellular pH is maintained by ion transport proteins in the cytoplasmic membrane (see *20.9 The cytoplasmic membrane is a selective barrier for secretion*).

Yellowstone National Park: Fountain Paint Pots Area

Salt Lake on Kangaroo Island, South Australia

FIGURE 20.6 Archaea in two of their environments. Top photo courtesy of J. Schmidt/Yellowstone National Park; bottom photo courtesy of Mike Dyall-Smith, Max Planck Institute for Biochemistry.

Oxygen can be used as an efficient electron acceptor during respiration. However, not all organisms require oxygen to carry out respiration, and some prokaryotes forgo respiration altogether, producing their energy through fermentation. Because oxygen requirements vary widely among prokaryotes, its usage provides a convenient system of classification. Obligate aerobes are organisms that absolutely require oxygen to grow. Some *Pseudomonads*, which are a common cause of hospital-acquired infections, are examples of obligate aerobes. Microaerophiles, on the other hand, grow best when free oxygen levels are low and carbon dioxide levels are high. Some prokaryotes can grow in the presence of oxygen but do not utilize it in their metabolism. These aerotolerant organisms, such as *Lactobacillus*, exclusively use fermentation regardless of environmental levels of free oxygen. Anaerobes are prokaryotes that grow in the absence of free oxygen. Anaerobes such as *Bacteroides*, which are killed in the presence of free oxygen, are called obligate anaerobes. Facultative anaerobes carry out aerobic respiration in the presence of oxygen but shift to fermentation when no terminal electron acceptor is present. Prokaryotes that inhabit the oxygen-poor environments in our intestinal and urinary tracts are usually facultative anaerobes. These organisms, which include *E. coli*, have complex enzyme systems that confer this diversity in energy production.

All prokaryotes require some moisture to grow. However, microbes that make their homes in water have to cope with hydrostatic pressure and limited nutrients. A typical 50-meter-deep lake exerts hydrostatic pressure 32 times that of atmospheric pressure—enough to make anyone uncomfortable, including most microbes. Barophiles are prokaryotes found at ocean depths that exceed 7000 m. At these depths, the hydrostatic pressure is sufficient to turn all other life forms into biological sludge. Interestingly, barophiles are unable to survive at atmospheric pressure because their cell walls only function properly at high pressure.

Another force that affects prokaryotic growth is osmotic pressure. Osmotic pressure develops due to the difference in concentration of dissolved substances in the environment relative to that within the cell. If prokaryotes are placed in an environment containing a high level of dissolved substances, water escapes from the cytoplasm, and the cells undergo plasmolysis (shrinking of the cell). Conversely, microbes that find themselves in environments with few dissolved substances will take on water and swell. A major function of the cytoplasmic membrane is to cope with osmotic pressure by limiting the exchange of solutes between the environment and the cell cytoplasm.

20.4 Archaea are prokaryotes with similarities to eukaryotic cells

Key concepts

- Archaea tend to be adapted to life in extreme environments and to utilize "unusual" energy sources.
- Archaea have unique cell envelope components and lack peptidoglycan cell walls.
- Archaea resemble bacteria in their central metabolic processes and certain structures, such as flagella.
- Archaea resemble eukaryotes in terms of DNA replication, transcription, and translation, but gene regulation involves many bacteria-like regulatory proteins.

Prokaryotes can be divided into two groups, now generally known as the bacteria and the archaea (see *20.2 Molecular phylogeny techniques are used to understand microbial evolution*). The distinction between these two groups of organisms was originally based on their rRNA sequences but is also reflected in important physiological and biochemical properties. In particular, as summarized in **FIGURE 20.7**, the archaea have a surprising number of properties in common with the eukaryotes. Moreover, although the archaea are clearly prokaryotic, in lacking a nuclear membrane, they differ from the better-known bacteria in fundamental properties, particularly the lack of peptidoglycan in their cell walls and the presence of ether-linked lipids bonded to glycerol in their membranes. In general, much less is known about archaea than bacteria.

Most of the best-known archaeal species are characterized by metabolic versatility and the ability to inhabit extreme environments. Some can grow at very high temperatures (>80°C). Apparently, only a single protein, reverse gyrase, is specifically found in hyperthermophilic archaea, so the ability to overwind DNA may be the single critical factor needed to allow adaptation to life at high temperature. Many archaeal species are specialized for environments that have extremes of pH (acidophiles and alkalophiles) or of salt (halophiles). Among the metabolic specializations are those of the methanogens (e.g., *Methanococcus janaschii*), which utilize CO_2, methyl compounds, or acetate anaerobically; the thermophilic sulfur reducers (e.g., *Archaeglobus fulgidis*), which respire using oxidized sulfur species as electron acceptors; and the halophiles (e.g., *Halobacterium salinarum*), which are adapted for extremely saline environments. The Crenarchaeotes (e.g., *Sulfolobus solfataricus*) represent a well-separated, deeply rooted branch of the archaea. The best-characterized members of this group are sulfur-dependent thermophiles. For a simple phylogeny of the main groups of archaea, see Figure 20.4. (For more on prokaryotic diversity see *20.3 Prokaryotic lifestyles are diverse*.)

The genomes of many archaeal species have now been completely sequenced. Like bacterial chromosomes, archaeal chromosomes tend to be circular and relatively compact, in the range of about 1.5 to 3 Mbp. The genome sequences have allowed the total protein coding capacity of each organism to be deduced. From the set of predicted proteins, one can

	BACTERIA	ARCHAEA	EUKARYA
Cell structure			
Flagella	Flagellum filament		Microtubule-based
Nuclear membrane	Absent		Present
Cell division	FtsZ ring*		Actomyosin
Nucleic acids			
Chromosome(s)	Many plasmids, usually single, circular		Multiple, linear
mRNA processing			mRNA splicing, polyadenylation, capping
Gene organization	Operons		Monocistronic
DNA packaging	Histone-like proteins*	Nucleosomes*	
DNA replication initiation	Dna A / Ori C	Origin recognition complex/PCNA	
Core RNA polymerase	Simple	Complex	
Basal promoter recognized by	σ factor	TATA-binding protein	
Protein synthesis			
Ribosomes	70S		80S
Translation initiation	N-formyl-methionine		
	Shine-Dalgarno sequence		
		5′ AUG	

*Except Crenarcheota

FIGURE 20.7 Archaea share some features with bacteria and others with eukaryotes.

make predictions about archaeal structure and function. In general, archaea and bacteria are most similar in terms of cell structure and genome organization. Archaea and bacteria have many membrane and cell envelope components, such as ATP-binding cassette (ABC) transporters and capsular polysaccharides, in common. Archaea also resemble bacteria in terms of their central metabolic pathways and in certain adaptive functions such as flagella-based motility and chemotaxis (see *20.11 Pili and flagella are appendages on the cell surface of most prokaryotes*). In addition, archaeal genomes are rich in insertion sequences and extrachromosomal elements (e.g., plasmids) that are similar to those of bacteria.

Other archaeal systems appear to have aspects of both bacterial and eukaryotic organization. Protein secretion systems and protein chaperone systems appear to fall into this category. Most ribosomal subunits are universally conserved, but archaea have a few subunits that are present in eukaryotes but not bacteria. Archaeal translation initiation and elongation factors also resemble their eukaryotic counterparts rather than bacterial ones. However, some aspects of archaeal initiation are more bacteria-like. In bacteria (and in mitochondria and chloroplasts), start codons are preceded by a "ribosome-binding site" sequence (the Shine-Dalgarno sequence) that allows genes to be organized in operons, with more than one start site in a single mRNA. In the cytoplasm of eukaryotes, recognition of the translation initiation site is based on a tracking mechanism in which the AUG closest to the 5′ end of the mRNA is chosen. This mechanism seems to preclude translation of more than one gene per mRNA. In archaea, most transcripts are monocistronic and translation appears to be initiated by the eukaryotic mechanism. However, Archaea also have operons, where the "downstream" genes are preceded by a bacterial ribosome-binding site sequence.

In terms of DNA organization, most Archaea contain readily recognizable homologs of eukaryotic core histone proteins. This is consistent with the chromosome being organized by nucleosome-like entities that bacteria do not possess (see *20.13 The bacterial nucleoid and cytoplasm are highly ordered*). The Crenarchaeota are an exception, in having neither histones, nor bacterial "histone-like" proteins. DNA replication in archaea appears to be mediated by polymerases that are similar to those of eukaryotes. Similarly, factors involved in the initiation of DNA replication and processivity of the polymerase are of the eukaryotic rather than bacterial types.

The transcriptional machinery of archaea is relatively complex, and the core RNA polymerase and associated transcription initiation factors are like those of eukaryotes. Thus, the core RNA polymerase typically comprises 11 subunits, the majority of which are missing from the bacterial polymerase. Initiation of transcription appears to be controlled largely by factors homologous to those of eukaryotes, of which TATA-binding protein (TBP) is the central factor involved in basal promoter recognition. There are no sigma factors, which are the key players in transcription initiation and promoter recognition in bacteria. Presently, little is known about the regulation of transcription in archaea. Surprisingly, however, archaeal genome sequences appear to contain many transcriptional regulators that are similar to those in bacteria. This suggests that activation and repression of gene expression in archaea may function as in bacteria, even though the polymerase itself resembles the eukaryotic one.

Most archaea contain the key cell division protein FtsZ (see *20.16 Prokaryotic cell division involves formation of a complex cytokinetic ring*) but no other recognizable bacterial cell division proteins or eukaryotic division proteins. Furthermore, the Crenarchaeota differ from other archaea in the absence of FtsZ. This suggests that the mechanisms of cell division in archaea may be quite different from all other organisms.

To summarize, the archaea are hybrids in the sense of having a mixture of properties characteristic of eukaryotes and bacteria, yet they have some unique features. This mixture probably reflects their evolutionary position (see Figure 20.4). The basic cellular organization of the archaea is prokaryotic, but many of their fundamental cellular processes, particularly those involved in the flow of genetic information, are much more similar to those of eukaryotes than of bacteria. There is currently much interest in studying some of these latter processes because of the insights they might provide into the evolution of eukaryotes. Furthermore, because these systems are inherently simpler than those of eukaryotes, certain experimentally tractable archaea are likely to be increasingly important as model systems for fundamental work on processes such as DNA replication and transcription.

Concept and Reasoning Check

1. Outline the characteristics of archaea that resemble bacteria and those characteristics that resemble eukaryotes.

20.5 Most prokaryotes produce a polysaccharide-rich layer called the capsule

Key concepts

- The outer surface of many prokaryotes consists of a polysaccharide-rich layer called the capsule or slime layer.
- The proposed functions of the capsule or slime layer are to protect bacteria from desiccation, to bind to host cell receptors during colonization, and to help bacteria evade the host immune system.
- *E. coli* capsule formation occurs by one of at least four different pathways.
- In addition to or in place of the capsule, many prokaryotes have an S-layer, an outer proteinaceous coat with crystalline properties.

Most, if not all, prokaryotes produce a glycocalyx or capsule external to their cell wall. The term glycocalyx applies to both prokaryotic and eukaryotic cells and refers to a mixture of extracellular polysaccharides and protein. For prokaryotic cells, there is a somewhat arbitrary distinction between different types of glycocalyx. Polysaccharides that are covalently linked to the cell are referred to as the "capsule." Loosely formed polysaccharides that are released from the cell are referred to as the "slime layer" or "extracellular polysaccharides." The composition of the capsule varies among prokaryotes, but it generally contains polysaccharides, including polyalcohols and amino sugars, proteoglycans, and glycoproteins. The thickness and plasticity of this layer can vary depending on its chemical nature. Capsular polysaccharides are formed by joining monosaccharides into long chains. Since any monosaccharide can be joined to another, there is a large diversity of capsular polysaccharides. Capsular serotypes can be used to differentiate between closely related organisms. For example, over 80 different capsular polysaccharides or K antigens have been described for *E. coli*. Strains that express certain K antigens are known to be associated with specific infections.

Capsule layers have several proposed functions including protection from desiccation, phagocytosis, detergents, and bacteriophages. Protection from desiccation may be important for transmission of encapsulated organisms from one host to the next. Capsules can play an important role in promoting adherence to host tissues and environmental surfaces. Capsules also promote adherence to other prokaryotes, which leads to biofilm formation. For example, *P. aeruginosa* produces copious amounts of alginate that allows biofilm formation in the lung. This heavy alginate coat may also help protect the pathogen from antibiotics and host defenses. (For more on biofilms see *20.21 Biofilms are highly organized communities of microbes*.)

In addition to *P. aeruginosa*, other pathogens rely on capsule function while colonizing the host. For example, the first virulence factor recognized for *Bacillus anthracis*, the causative agent of anthrax, was its capsule, shown in **FIGURE 20.8**. The *B. anthracis* capsule is composed of polymeric D-glutamic acid, making it one of the only prokaryotic capsules composed primarily of peptides. Capsule material is produced *in vivo*, and *B. anthracis* strains that cannot synthesize capsule have greatly attenuated pathogenicity. It has been proposed that because the *B. anthracis* capsule is only weakly immunogenic, it protects the pathogen from the host immune response. The capsule also inhibits binding of complement proteins made by the host immune system, thereby allowing the pathogen to survive in the human circulatory system.

The capsule of *Mycobacterium tuberculosis* is also a major virulence factor. The *Mycobacterium* capsule mediates adhesion to macrophages, a critical step in host cell invasion. Since mac-

rophages are phagocytes that can engulf and destroy microbes, microbes have evolved various strategies to survive within macrophages. For *Mycobacterium*, the capsule is critical for this process. First, the *Mycobacterium* capsule promotes binding to the macrophage's CR3 receptor, which elicits a signaling cascade that results in a "safe" route of infection. The capsule is also important in excluding reactive oxygen intermediates that are present in the macrophage and that might otherwise kill the pathogen. This feature helps to explain how encapsulated *Mycobacteria* are able to avoid immune system detection and persist within the host, while nonencapsulated strains are less virulent. In fact, capsule-less *Mycobacterium* strains have been used in vaccines to confer immunity to tuberculosis infection.

Capsule biogenesis has been studied in several species, but it is probably best understood in *E. coli*. The *E. coli* capsules have been classified into four groups based on genetic and biogenesis criteria. The biogenesis of group 1 and group 2 capsules has been studied extensively. Transport of *E. coli* group 1 and group 2 capsules to the cell surface is thought to occur at positions where the cytoplasmic and outer membranes come into close contact with one another. Wza, a lipoprotein in the outer membrane, forms a high-molecular weight complex that plays a role in the translocation of group 1 capsules. Wza is predicted to be a β-barrel protein that together with Wzc forms a channel through which the capsule precursors are secreted. This secretion system is functionally and genetically related to the outer membrane usher protein, PapC (see *20.10 Prokaryotes have several secretion pathways*). Group 2 capsules are shuttled across the outer membrane via the KpsE and KpsD proteins. KpsE is anchored to the cytoplasmic membrane via its N-terminus, while a large portion of the protein resides in the periplasmic space. Although KpsE does not span the outer membrane, the C-terminus of KpsE associates with it. This secretion apparatus is dependent on close proximity of the cytoplasmic and outer membranes. Although its exact function is currently unresolved, KpsD is a periplasmic protein that may be required for recruiting KpsE to the outer membrane. The biosynthesis of group 3 and group 4 capsules is poorly understood.

As an alternative or adjunct to the capsule, the outer surface of many bacteria and archaea is bounded by a proteinaceous structure called a surface layer (S-layer). S-layers are formed

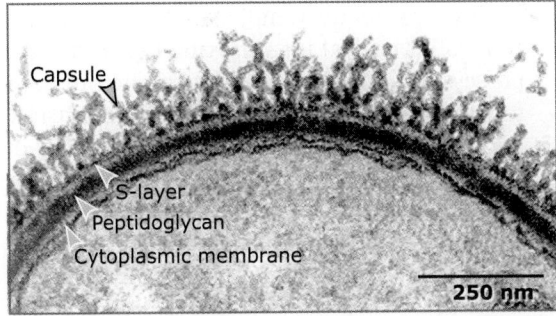

FIGURE 20.8 The capsule is the outer layer of most prokaryotic cell envelopes. The electron micrograph shows the cell envelope of a strain of *B. anthracis*. Reproduced from *J. Bacteriol.*, 1998, vol. 180, pp. 52–58, DOI and reproduced with permission from American Society for Microbiology. Photo courtesy of Agnés Fouet, Pasteur Institute.

from a single protein or glycoprotein species that self-assembles into a crystalline lattice. The protein sequences are not especially conserved but are often rich in acidic and hydrophobic amino acid residues. Some proteins, particularly in archaea and Gram-positive bacteria, are glycosylated, sometimes at multiple sites, with chains composed of 20 to 50 identical repeating units of a range of sugar subunits. FIGURE 20.9 shows a freeze-etched preparation of an S-layer with hexagonal symmetry, along with a cross-sectional view of the S-layer. The S-layer is usually between 5 and 25 nm thick. The outside surface is relatively smooth and the inner surface is corrugated where it interacts with the underlying cell envelope layers. Assembly of the S-layer appears to be coordinated with the cell envelope through interaction with cell surface molecules such as teichoic acids (in Gram-positive bacteria) or lipopolysaccharide (in Gram-negative bacteria) (for more on these molecules see *20.7 The cell envelope of Gram-positive bacteria has unique features* and *20.8 Gram-negative bacteria have an outer membrane and a periplasmic space*). How the assembly and positioning of the S-layer relate to the capsule (in cases where the two layers coexist) is not yet clear. (Figure 20.8 shows a bacterium with both a capsule and an S-layer.) In archaea that lack peptidoglycan or an equivalent molecule, the S-layer may be the principle component of the cell envelope. Although S-layer proteins can represent the most abundant protein syn-

thesized by the cell, there is still no clear understanding of their function. Many laboratory strains seem to have lost the ability to make an S-layer but nevertheless seem to be perfectly capable of growth. Conceivably, there is a strong selective advantage in not making this layer under conditions in which it is not needed.

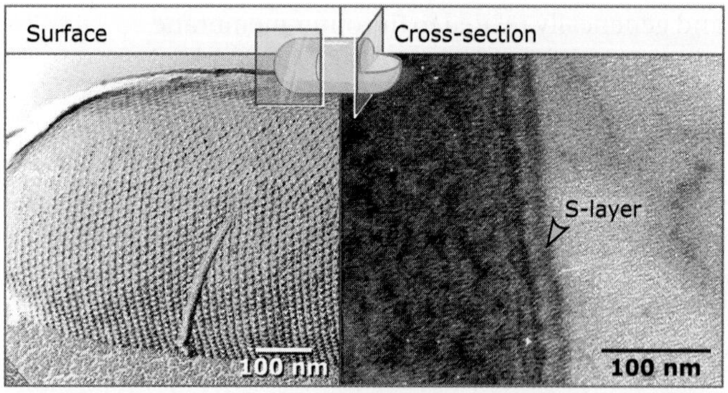

20.6 The bacterial cell wall contains a cross-linked meshwork of peptidoglycan

Key concepts

- Most bacteria have peptidoglycan, a tough external cell wall made of a polymeric meshwork of glycan strands cross-linked with short peptides.
- The disaccharide pentapeptide precursors of peptidoglycan are synthesized in the cytoplasm, exported, and assembled outside the cytoplasmic membrane.
- One model for cell wall synthesis is that a multiprotein complex carries out insertion of new wall material following a "make-before-break" strategy.
- Many autolytic enzymes remodel, modify, and repair the cell wall.
- For some bacteria, the peptidoglycan cell wall is important for maintaining cell shape.
- A bacterial actin homolog, MreB, forms helical filaments in the cell cytoplasm that direct the shape of the cell through control of peptidoglycan synthesis.

FIGURE 20.9 The outer layer of some prokaryotic cell envelopes is an S-layer instead of the capsule layer. Left: An electron micrograph showing the hexagonal symmetry of the outer surface of the S-layer. Right: An electron micrograph showing a cross-section of a cell envelope with an S-layer. Left photo reproduced from *J. Bacteriol.*, 2000, vol. 182, pp. 859–868, DOI and reproduced with permission from American Society for Microbiology. Photo courtesy of Margit Sára, University of Agricultural Sciences—Austria; right photo courtesy of Christina Schäffer, Department of NanoBiotechnology, University of Natural Resources and Applied Life Sciences, Vienna.

The major component of the cell envelope of most bacteria is called **peptidoglycan**. It is composed of a meshwork of glycan strands cross-linked by peptides. The polymer covers the whole surface of the organism, forming a tough protective shell. The peptidoglycan cell wall is critical for survival of bacteria that possess it because it counteracts the outward force exerted on the underlying cytoplasmic membrane by the high internal osmolarity of these cells. Therefore, rupture of the wall is catastrophic for the cell. The peptidoglycan is also important in maintaining the shape of most bacterial cells.

The best-known bacteria mainly fall into two groups that can be distinguished from each other on the basis of staining properties. The different responses of the Gram-positive and Gram-negative bacteria to the Gram staining protocol reflect fundamental differences in their cell wall organization (see Figure 20.2).

Gram-negative bacteria tend to have a thin layer of peptidoglycan, probably comprising a single layer of glycan strands. The architecture is, therefore, essentially two-dimensional, and the peptide cross-bridges lie only in the same two-dimensional plane as the glycan strands. Gram-positive bacteria, in contrast, have a much thicker layer of peptidoglycan, comprising multiple layers of glycan strands. In this case, the cross-bridges probably lie in different planes, connecting strands in the same plane and connecting an overlying layer to the one below. It is not clear why Gram-positive and Gram-negative bacteria organize peptidoglycan synthesis in such different ways. One possible advantage of the thicker cell wall is

that it is tougher and can, therefore, provide greater protection against physical damage and provide support against the higher internal osmotic pressure of the Gram-positive cell.

As illustrated in **FIGURE 20.10**, the glycan strands usually consist of a repeating disaccharide, N-acetylglucosamine-N-acetylmuramic acid (NAG-NAM). The strands average about 30 disaccharides in length, but there is a broad length distribution. Covalently linked to the NAM residues is a peptide that is synthesized by a nonribosomal route and incorporates several unusual amino acids, typically including D-glutamate, D-alanine, and diaminopimelate (DAP). (The amino acids used for protein synthesis are invariably of the L-form.) The DAP

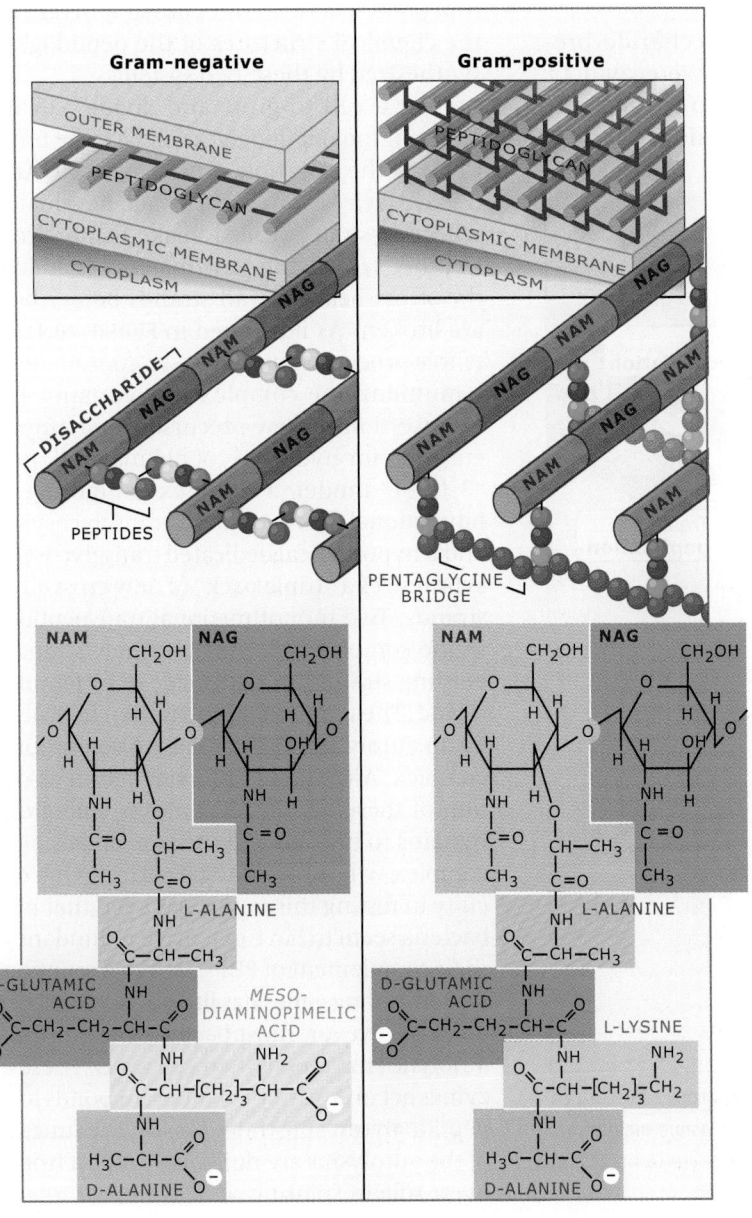

FIGURE 20.10 The peptidoglycan layers of Gram-negative and Gram-positive bacteria contain the same repeating disaccharide but can differ in their peptide links. Some Gram-positive bacteria contain L-lysine in place of *meso*-diaminopimelic acid. The disaccharide is N-acetylglycosamine-N-acetylmuramic acid (NAG-NAM).

provides an amino group that is used for the cross-linking reaction. Cross-linking results in release of the last D-alanine from the peptide chain (see **FIGURE 20.11**). In some bacteria, the peptide cross-links incorporate additional amino acid residues. In Gram-positive bacteria, one or more glycines in the cross-bridges can connect the strands of NAM-NAG.

The disaccharide precursor is synthesized in the cytosol, starting from uridine 5′-diphosphate (UDP)-NAM, as Figure 20.11 shows. The amino acid side chain is built up sequentially with one enzyme for each step. The resulting UDP-NAM-pentapeptide is attached to a specialized lipid (bactoprenol) in the cell membrane. Addition of NAG completes the disaccharide precursor, which is then flipped to the exterior of the membrane for assembly into the existing peptidoglycan.

Incorporation of the disaccharide precursor into polymers requires two enzymatic reactions: transglycosylation, to produce the glycan chains; and transpeptidation, to gener-ate the cross-links. The enzymes responsible for transpeptidation were discovered through binding of penicillin, which kills bacterial cells by inhibiting the transpeptidases, and hence are called penicillin-binding protein (PBPs). The high-molecular weight "class A" PBPs are bifunctional enzymes that have an additional, separate transglycosylase domain, as well as a transpeptidation domain.

In rod-shaped bacteria, the cylindrical parts of the cell and the hemispherical poles (formed during cell division) seem to require specialized machineries for peptidoglycan synthesis. In *E. coli*, these dedicated transpeptidases are PBP2 and PBP3, respectively. There is some evidence that these transpeptidases use slightly different substrates, with the former preferring a pentapeptide side chain and the latter a tetra- or tripeptide. There are no known differences in the chemical structures of the peptidoglycan synthesized by these two systems.

For a cell to grow and divide, covalent bonds in the peptidoglycan need to be broken to allow the insertion of new material. An attractive idea on how to safely add new material relies on a strategy called "make-before-break." In other words, the new material is added to the stress-bearing wall strands before bonds are broken. As illustrated in **FIGURE 20.12**, the whole process is thought to be coordinated by a multienzyme complex that contains PBPs for insertion of new precursors and autolytic enzymes for hydrolysis of old material. In the "3-for-1" model, a complex containing two bifunctional transpeptidase-transglycosylases, and a hypothetical dedicated transglycosylase, synthesizes a "triple pack" of new cross-linked strands. Two monofunctional transpeptidases in the complex link the outer new strands to existing strands on either side of a "template" strand. The template strand is then hydrolyzed by an autolytic enzyme that is also part of the complex. Although some evidence for association of these various enzymes is emerging, it remains to be seen whether the details of this complex will turn out to be correct. One difficulty in testing this model has been that many bacteria seem to have extensive redundancy in their complement of PBPs, so that disruption of PBP-encoding genes has little or no phenotypic effect. However, most bacteria have multiple autolytic enzymes in their cell walls. These enzymes act on a variety of different bonds in the peptidoglycan substrate. Mostly, the functions of the autolysins are not known, apart from an overt role in splitting of the cell wall to allow

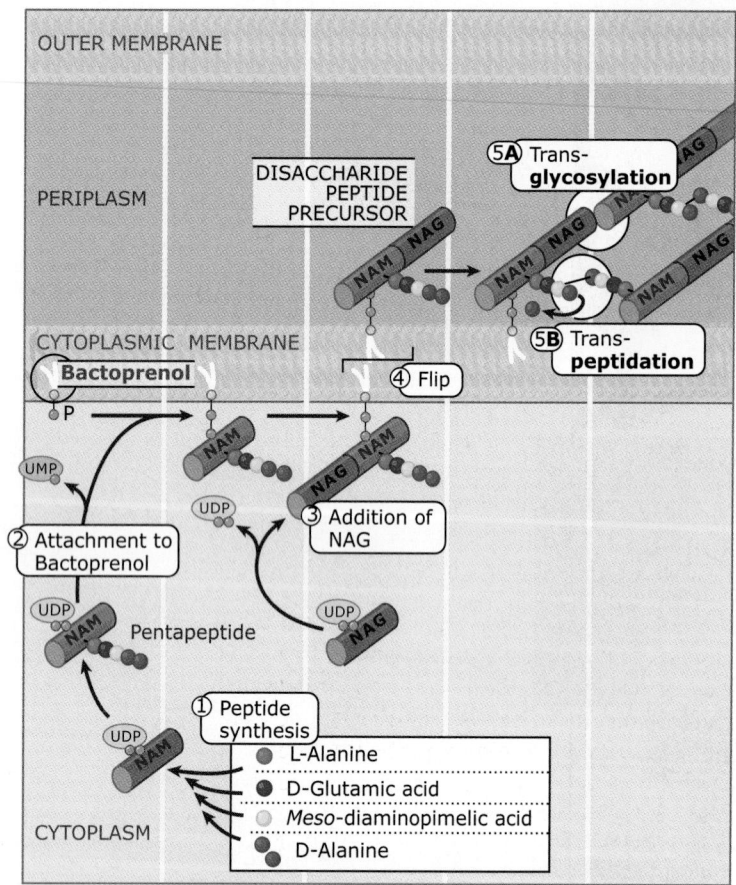

FIGURE 20.11 Peptidoglycan synthesis occurs in several steps, starting in the cytoplasm and ending on the exterior side of the cytoplasmic membrane. The bactoprenol is released from NAM during the transglycosylation reaction (not shown).

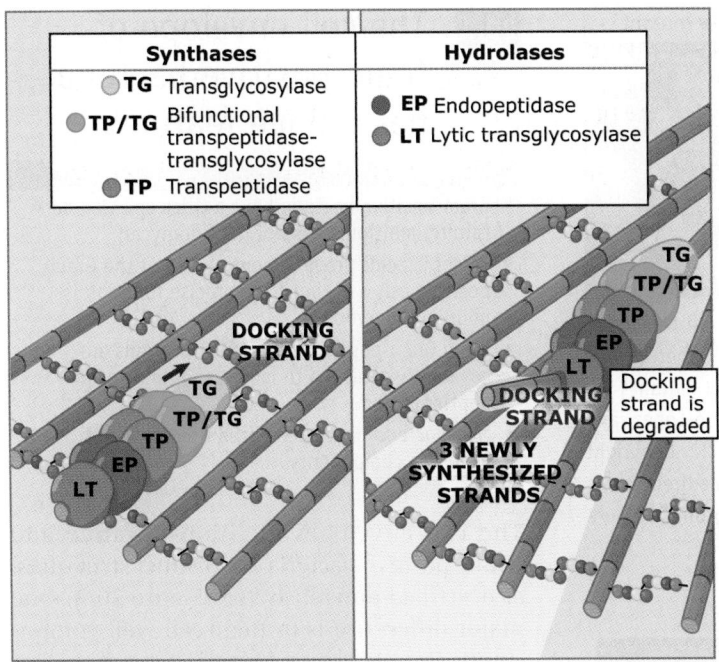

Synthases		Hydrolases
TG Transglycosylase		**EP** Endopeptidase
TP/TG Bifunctional transpeptidase-transglycosylase		**LT** Lytic transglycosylase
TP Transpeptidase		

FIGURE 20.12 The "3-for-1" model proposes a "make-before-break" strategy for insertion of newly synthesized peptidoglycan. In this model, a multienzyme complex forms on an existing strand (the docking strand), synthesizes three cross-linked strands, and inserts them into the existing layer just before degrading the docking strand.

separation of sister cells at the end of cell division (see *20.16 Prokaryotic cell division involves formation of a complex cytokinetic ring*). As for the PBPs, knockout of multiple autolysin-encoding genes usually has little or no phenotypic effect, other than a thickening of the cell wall and delay in cell separation.

Several lines of evidence support the view that the peptidoglycan cell wall has a critical role in maintenance of bacterial cell shape. First, as **FIGURE 20.13** shows, the isolated cell wall sacculus retains the shape of the cell from which it was obtained. Second, treatment of cells with enzymes such as lysozyme, which hydrolyzes the cell wall, results in loss of shape. Third, mutations that alter the shape of the cell frequently lie in genes that are involved in cell wall synthesis.

For decades, it was assumed that bacteria do not possess an actin cytoskeleton, which is responsible for cell shape in eukaryotes (see *12 Actin*). Recently, however, it has been shown that a bacterial protein, MreB, which has very weak sequence similarity to actin, is in fact a functional homolog of actin. Indeed, the three-dimensional fold determined by x-ray crystallography is congruent with that of actin. The mreB genes were first identified by mutations that affect cell shape in *E. coli* and *B. subtilis*. **FIGURE 20.14** shows the effects on cell shape of mutations in two different *B. subtilis mreB* homologs, *mreB* and *mbl* (*mreB-like*). The *mreB* gene is present in almost all bacteria

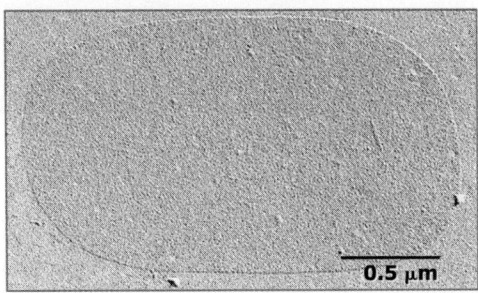

FIGURE 20.13 The isolated cell wall sacculus is intact but adopts a two-dimensional structure. The sacculus was prepared using agar filtration. Reproduced from *Microbiol. Mol. Biol. Rev.*, 1998, vol. 62, pp. 181–203, DOI and reproduced with permission from American Society for Microbiology. Photo courtesy of Joachim-Volker Höltje, Max-Planck-Institut für Entwicklungsbiologie, Germany.

that have a nonspherical shape but is absent in cocci (round bacteria). If the spherical symmetry of a round cell is taken as the default, then the MreB protein might be involved in a system that actively determines the more complex shapes of rod-shaped (*E. coli* and *B. subtilis*), curved (*Vibrio*), and helical (*Helicobacter*) bacteria. **FIGURE 20.15** shows *B. subtilis* cells expressing a construct containing a fusion of coding regions for Mbl and green fluorescent protein (GFP). The proteins form filamentous structures that follow helical paths just under

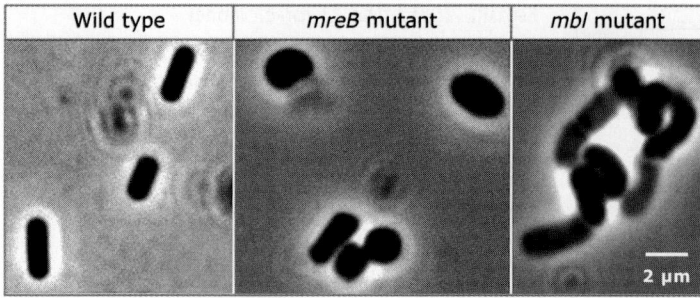

| Wild type | *mreB* mutant | *mbl* mutant |

2 μm

FIGURE 20.14 Effects of mutations in *mreB* or *mbl* genes on the shape of *Bacillus subtilis* cells. Wild-type cells are rod-shaped, whereas the *mreB* mutants have a more rounded morphology and the mbl mutants are bent. Reprinted from *Cell*, vol. 104, L. J. F. Jones, R. Caballido-López, and J. Errington, Control of cell shape in bacteria..., pp. 913–922, Copyright (2001) with permission from Elsevier [http://www.sciencedirect.com/science/journal/00928674]. Photo courtesy of Jeffrey Errington, Institute for Cell and Molecular Biosciences, Newcastle University.

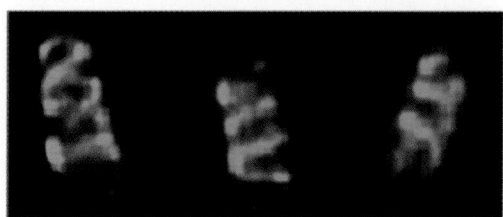

FIGURE 20.15 The bacterial Mbl protein, which is weakly homologous to eukaryotic actin, forms helical filamentous structures in cells. Shown are three *Bacillus subtilis* cells expressing Mbl-green fluorescent protein. Photo courtesy of Rut Carballido-López, University of Oxford.

the cell membrane. These structures do not in themselves possess sufficient structural rigidity to directly determine the shape of the cell. If the cylindrical shape of the cell is perturbed artificially, for example, by removing the cell wall and making protoplasts, the Mbl cables loose their helical configuration. They seem to work by directing synthesis of new cell wall material in such a way that the overall shape and dimensions of the cell are maintained during growth.

It is not yet known precisely how MreB filaments interact with the cell wall. However, *mreB* genes almost always lie immediately upstream of two other conserved bacterial genes, *mreC* and *mreD*. Both of these genes are also required for correct cell shape, and they encode transmembrane proteins that could, in principle, couple information associated with the helical MreB filaments inside the cell to the cell wall synthetic machinery outside.

20.7 The cell envelope of Gram-positive bacteria has unique features

Key concepts

- Gram-positive bacteria have a thick cell wall containing multiple layers of peptidoglycan.
- Teichoic acids are an essential part of the Gram-positive cell wall, but their precise function is poorly understood.
- Many Gram-positive cell surface proteins are covalently attached to membrane lipids or to peptidoglycan.
- *Mycobacteria* have specialized lipid-rich cell envelope components.

The cell envelopes of Gram-negative and Gram-positive bacteria are distinct structures, as described previously (see Figure 20.2) One major difference is in their cell wall composition: Gram-positive bacteria have a thicker layer of peptidoglycan than Gram-negative bacteria (for more on peptidoglycans see *20.6 The bacterial cell wall contains a cross-linked meshwork of peptidoglycan*). Another difference is that, in addition to peptidoglycan, the cell walls of Gram-positive bacteria have a second essential class of polymeric macromolecules. Generally, these polymers are polyanionic. The best-studied and most prominent molecules of this class are the teichoic acids. **FIGURE 20.16** shows that teichoic acids have a simple repeating structure comprising a polymer of phosphorylated sugar or glycerol moieties. The polymer is attached to the cell in one of two ways. Lipoteichoic acids are attached to a fatty acid in the cytoplasmic membrane, whereas wall teichoic acids are attached via a linkage group to the peptidoglycan. A wide range of different repeating units is found on different bacteria and even in closely related strains of the same species. The polymers have the common property of being polyanionic, which is due to the presence of phosphate in the repeating unit. However, phosphate is not essential and, in phosphate-limiting conditions, can be replaced by another repeat unit, such as glucuronate, to yield teichuronic acid. Similar to peptidoglycan, teichoic acids are synthesized on a UDP-linked precursor in the cytoplasm and then transported to the exterior for assembly.

Teichoic acids are essential for cell viability, as shown by the fact that deletion of genes involved in their synthesis is lethal. Their precise function is not yet clear, though it seems likely

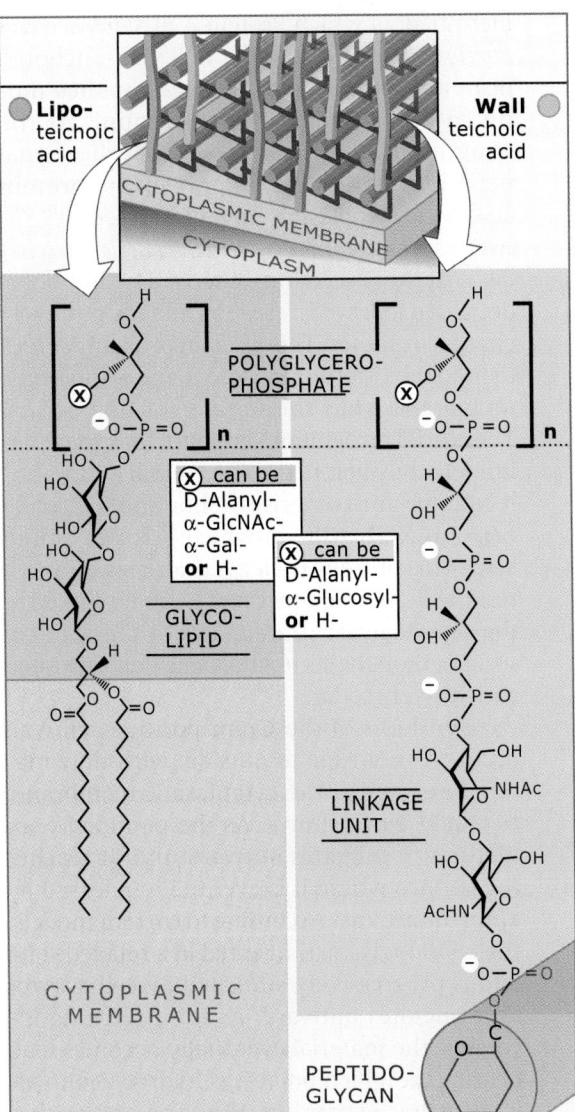

Figure labels:
Lipo-teichoic acid
Wall teichoic acid
CYTOPLASMIC MEMBRANE
CYTOPLASM
POLYGLYCERO-PHOSPHATE
Ⓧ can be D-Alanyl-α-GlcNAc-α-Gal- or H-
Ⓧ can be D-Alanyl-α-Glucosyl- or H-
GLYCO-LIPID
LINKAGE UNIT
CYTOPLASMIC MEMBRANE
PEPTIDO-GLYCAN

that it relates in some way to maintenance of the charge on the cell wall. Several functions have been proposed: that they are important to capture divalent cations, to control the activity of autolytic enzymes acting on the peptidoglycan, or to maintain the general permeability of the cell wall layer.

Gram-positive bacteria, which lack the outer membrane, have two mechanisms to anchor cell surface proteins and prevent them from being lost into the surrounding medium. First, as shown in **FIGURE 20.17**, a subset of proteins that are secreted via the general secretory (Sec) pathway are cleaved by a specialized, "type II" signal peptidase. (For more on the general secretory pathway see *20.9 The cytoplasmic membrane is a selective barrier for secretion.*) Before cleavage, the enzyme phosphatidylglycerol: prolipoprotein diacylglyceryl transferase

couples the protein to a phospholipid in the outer leaflet of the cytoplasmic membrane. Proteins destined for this "lipo modification" can be recognized by the presence of a sorting signal containing an invariant cysteine residue immediately after the signal peptidase cleavage site. Covalent linkage of the protein to the outer surface of the cytoplasmic membrane prevents it from diffusing away from the cell surface. A large number of proteins are processed in this way in Gram-positive bacteria. For example, *B. subtilis* is predicted to have well over 100 such proteins. (Although Gram-positive bacteria use this mechanism frequently for surface-associated proteins, Gram-negative bacteria also possess lipo-modified proteins.)

A second approach to retaining proteins at the cell surface in Gram-positive bacteria is to anchor the protein to the cell wall peptido-

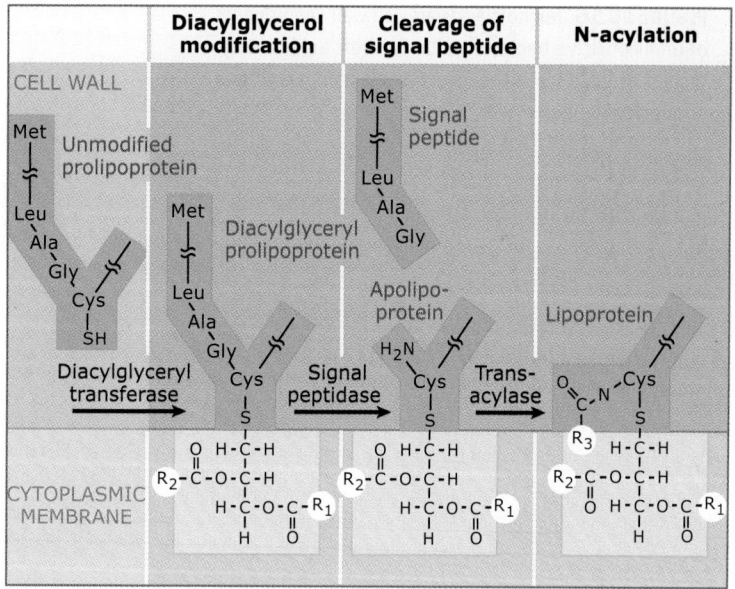

Diacylglycerol modification	Cleavage of signal peptide	N-acylation

CELL WALL

Met — Unmodified prolipoprotein
Leu
Ala
Gly
Cys
SH

Diacylglyceryl transferase →

Met — Diacylglyceryl prolipoprotein
Leu
Ala
Gly
Cys
S

Signal peptidase →

Met — Signal peptide
Leu
Ala
Gly

Apolipoprotein

H_2N — Cys
S

Trans-acylase →

Lipoprotein

$\underset{R_3}{O=C-N}$ — Cys
S

CYTOPLASMIC MEMBRANE

FIGURE 20.17 One of the ways that Gram-positive bacteria anchors proteins to the cell surface is to couple the proteins to phospholipids of the cytoplasmic membrane. R_1, R_2, and R_3 represent the long chain fatty acids.

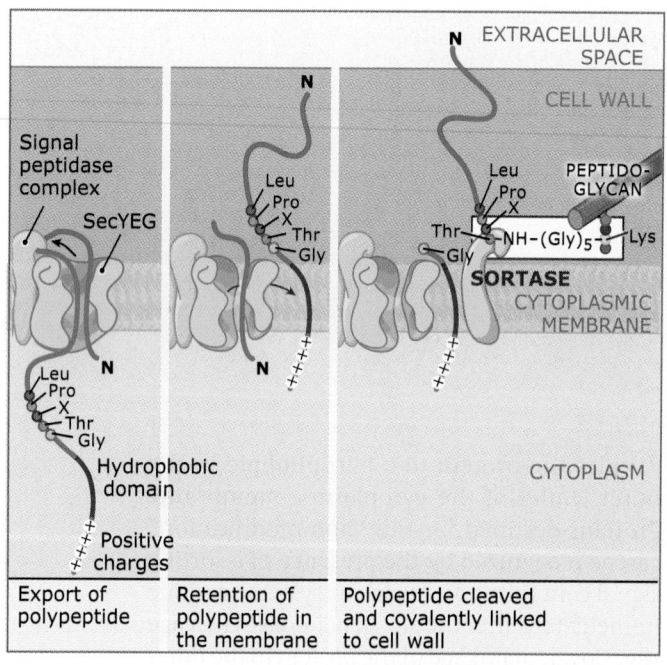

Export of polypeptide	Retention of polypeptide in the membrane	Polypeptide cleaved and covalently linked to cell wall

EXTRACELLULAR SPACE
CELL WALL
CYTOPLASMIC MEMBRANE
CYTOPLASM

Signal peptidase complex
SecYEG
Hydrophobic domain
Positive charges

Leu Pro X Thr Gly

Leu Pro X Thr Gly — NH–(Gly)$_5$ — Lys
SORTASE
PEPTIDO-GLYCAN

FIGURE 20.18 In Gram-positive bacteria some cell surface proteins are anchored to the peptidoglycan cell wall. It has been proposed that the enzyme sortase cleaves the Thr-Gly bond and covalently links the protein to the cell wall.

glycan. In general, it appears that many of the cell wall-anchored proteins are associated with pathogen-host interactions (however, this tendency may reflect a bias in genome sequencing toward pathogens). The key enzyme involved in this process is called sortase, as **FIGURE 20.18** shows. A large number of substrates have been

identified, of which Protein A of *Staphylococcus aureus* is the best studied. Cell wall-anchored proteins possess recognizable sequence motifs at both their N- and C-termini. At the N-terminus, there is a classic cleavable signal peptide, required for export of the protein across the cytoplasmic membrane (see *20.9 The cytoplasmic membrane is a selective barrier for secretion*). At the C-terminus, a second hydrophobic domain is followed by several, mainly positively charged residues. This domain probably serves to hold the C-terminus close to the cytoplasmic membrane while the sortase reaction occurs. The crucial recognition sequence for sortase lies immediately before the C-terminal hydrophobic stretch and comprises a short motif, usually Leu-Pro-X-Thr-Gly, in which X is any amino acid. The protein is cleaved after the fourth residue in this sequence and is amide-linked to the peptidoglycan, presumably by the action of sortase, though this has not been demonstrated conclusively so far.

Synthesis of the Gram-positive cell wall occurs from inside to outside, with new material inserted at the cytoplasmic membrane, as **FIGURE 20.19** shows. As the peptidoglycan matures, it migrates outward until it reaches the surface where it is eventually released by autolytic activity. According to current models, the peptidoglycan is inserted in a relaxed state and is progressively stretched by cell growth as it migrates outward. As the stretching increases, the material eventually becomes load bearing, at which point it is hydrolyzed to allow further expansion. How the maturation of the peptidoglycan is controlled by autolytic enzymes and coordinated with cell growth remains an important unsolved question. It is also not clear whether the "3-for-1" model for peptidoglycan synthesis, which was developed to explain the growth of Gram-negative cell walls, is relevant to the multilayered Gram-positive cell wall (for more on this model see *20.6 The bacterial cell wall contains a cross-linked meshwork of peptidoglycan*).

The detailed structure of bacterial cell walls is surprisingly poorly understood, even in the paradigms of *B. subtilis* and *E. coli*. This is despite the fact that the most important of our antibiotics (particularly β-lactams and glycopeptides) work by inhibiting steps in cell wall synthesis. In the main, this seems to be due to a focus on the biochemistry of interaction of inhibitors with purified enzymes. How the PBPs synthesize peptidoglycan in the context of the three-dimensional structure of the cell wall

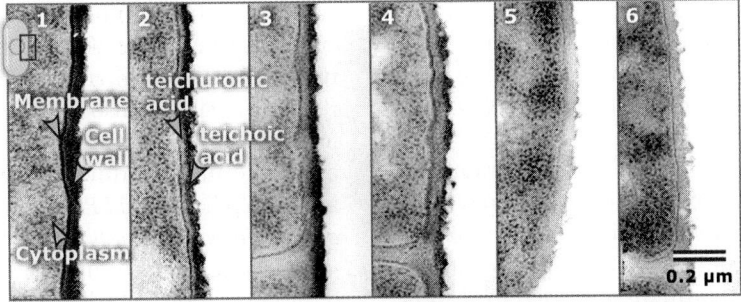

FIGURE 20.19 Experimental demonstration of the time course of deposition of new cell envelope material in *B. subtilis* cells during transition from magnesium to phosphate limitation. Old cell wall (containing teichoic acid) is gradually replaced uniformly along the entire cell wall by newly synthesized wall material (containing teichuronic acid). Reproduced from T. Merad, et al., *J. Gen. Microbiol.* 135 (1989): 645–644. Used with permission of the Society for General Microbiology. Photo courtesy of Colin R. Harwood, Newcastle University.

of a growing bacterium remains an important unanswered question. Also, important variations in cell wall structure are seen in diverse organisms. The extremes of specialization may be illustrated by the mycoplasmas, which lack a cell wall altogether, and the mycobacteria, which have complex fatty acids covalently linked to the peptidoglycan.

Concept and Reasoning Check

1. How do Gram-positive bacteria retain proteins at their cell surface?

20.8 Gram-negative bacteria have an outer membrane and a periplasmic space

Key concepts

- The periplasmic space is found between the cytoplasmic and outer membranes in Gram-negative bacteria.
- Proteins destined for secretion across the outer membrane often interact with molecular chaperones in the periplasmic space.
- The outer membrane is a lipid bilayer that prevents the free dispersal of most molecules.
- Lipopolysaccharide is a component of the outer leaflet of the outer membrane.
- During infection by Gram-negative bacteria, lipopolysaccharide activates inflammatory responses.

Gram-negative bacteria have an outer membrane, in contrast to Gram-positive bacteria (see Figure 20.2). The gap between their cytoplasmic and outer membranes is called the periplasmic space, or **periplasm**, as **FIGURE 20.20** shows. The

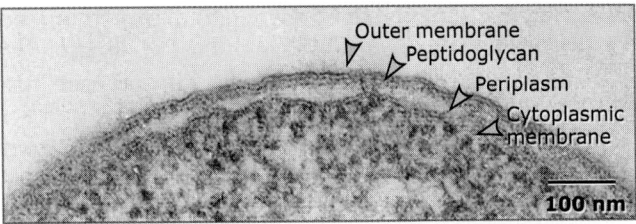

FIGURE 20.20 Thin section electron micrograph showing the cell envelope of the Gram-negative Archaea, *Methanococcus voltae*. Photo © Dr. Terry Beveridge/Visuals Unlimited.

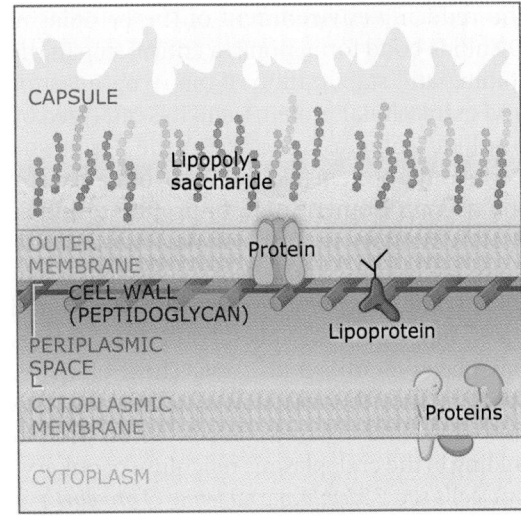

FIGURE 20.21 The cell envelope of Gram-negative bacteria consists of cytoplasmic (inner) and outer membranes separated by the periplasm and peptidoglycan cell wall. The two membranes differ greatly in composition.

components of the outer membrane differ from the cytoplasmic (inner) membrane, as summarized in **FIGURE 20.21**. In *20.8 Gram-negative bacteria have an outer membrane and a periplasmic space*, we highlight some of the components of the periplasm and outer membrane, and the processes that occur in the periplasm. (For more on the cytoplasmic membrane see *20.9 The cytoplasmic membrane is a selective barrier for secretion.*)

The periplasmic space is packed with many types of proteins, including digestive enzymes, transport proteins, and proteins involved in metabolism. In addition, many of the proteins found in the periplasm are destined for insertion into or secretion across the outer membrane. (For details on protein secretion systems see *20.10 Prokaryotes have several secretion pathways.*) Here we focus on the periplasmic space as a "folding center" for these proteins. The periplasm is somewhat analogous to the endoplasmic reticulum of eukaryotic cells in terms of the oxidizing environment and protein folding (for more on the endoplasmic reticulum see *7 Membrane targeting of proteins*). Many proteins that function as molecular **chaperones** are present in the periplasm. Chaperones can act in several ways: they can prevent proteins from adopting their final folded conformation, they can prevent proteolysis, or they can prevent unwanted interactions with other proteins that might lead to aggregation.

For example, many proteins contain free cysteines that do not form disulfide bonds in the reducing environment of the cytoplasm. Disulfide bond formation is a critical step in the folding and stabilization of many periplasmic and extracellular proteins, and it is catalyzed by disulfide isomerases located in the periplasm. Periplasmic protein folding is also influenced by the *cis-trans* isomerization of proline residues. Proline residues in unfolded proteins are predominately found in the *trans* conformation, but folded proteins contain a mixture of the *cis* and *trans* conformations. Several prolyl isomerases have been identified in *E. coli.* These enzymes speed up the otherwise slow *cis-trans* isomerization of proline residues. (For more on protein folding in the endoplasmic reticulum of eukaryotic cells see *7 Membrane targeting of proteins.*)

Another example of a chaperone is the protein DegP, which helps stabilize proteins that have become unfolded in the periplasm. Homologs to DegP exist in most prokaryotes, and some eukaryotes. DegP is a general chaperone, acting on many types of proteins. Other periplasmic chaperones, such as the PapD chaperone that is required for P-pili biogenesis, are specific for one type of substrate. PapD binds to pilus subunits as they are secreted into the periplasm and prevents them from interacting

with each other (nonproductively) or with other subunits (see Figure 20.28). Preventing premature subunit-subunit interactions is only part of the story. PapD is also required for the correct folding of pilus subunits, and without it subunits aggregate and are proteolytically degraded. (For more on pili see *20.11 Pili and flagella are appendages on the cell surface of most prokaryotes.*)

Despite the presence of chaperones, protein aggregation can still occur in the periplasm, and when it does, the Cpx system is activated. The Cpx system consists of at least three proteins, as **FIGURE 20.22** shows. The periplasmic protein CpxP is thought to bind to aggregated or misfolded proteins in the periplasm, which then leads to activation of the cytoplasmic membrane-localized CpxA protein. When activated, the cytoplasmic domain of CpxA becomes phosphorylated and this phosphate is then transferred to the DNA-binding protein CpxR. Phosphorylated CpxR induces the expression of many genes, including those that code for periplasmic proteases that can degrade aggregated or misfolded proteins. Other proteins induced by the Cpx pathway include those that facilitate protein folding, such as the disulfide oxidase DsbA, and the peptidyl-prolyl-isomerases PpiA and PpiD.

The outer membrane is an asymmetric membrane. The inner leaflet of the outer membrane is composed of phospholipids similar to those in the inner membrane, whereas

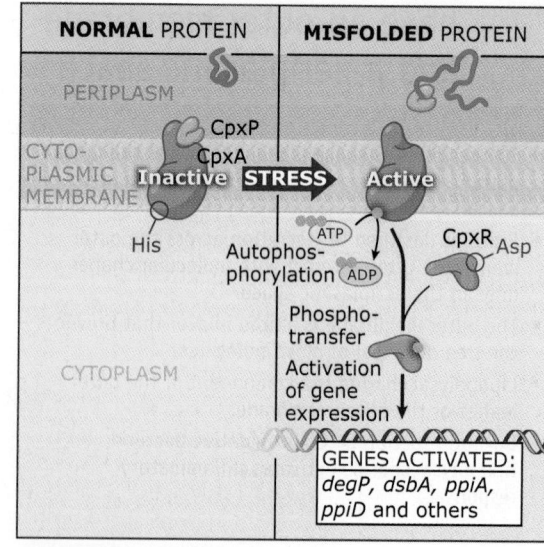

FIGURE 20.22 The Cpx pathway of Gram-negative bacteria is activated when cellular stress causes misfolding of envelope proteins. Activation of the Cpx pathway results in upregulated expression of genes involved in protein folding and protein degradation.

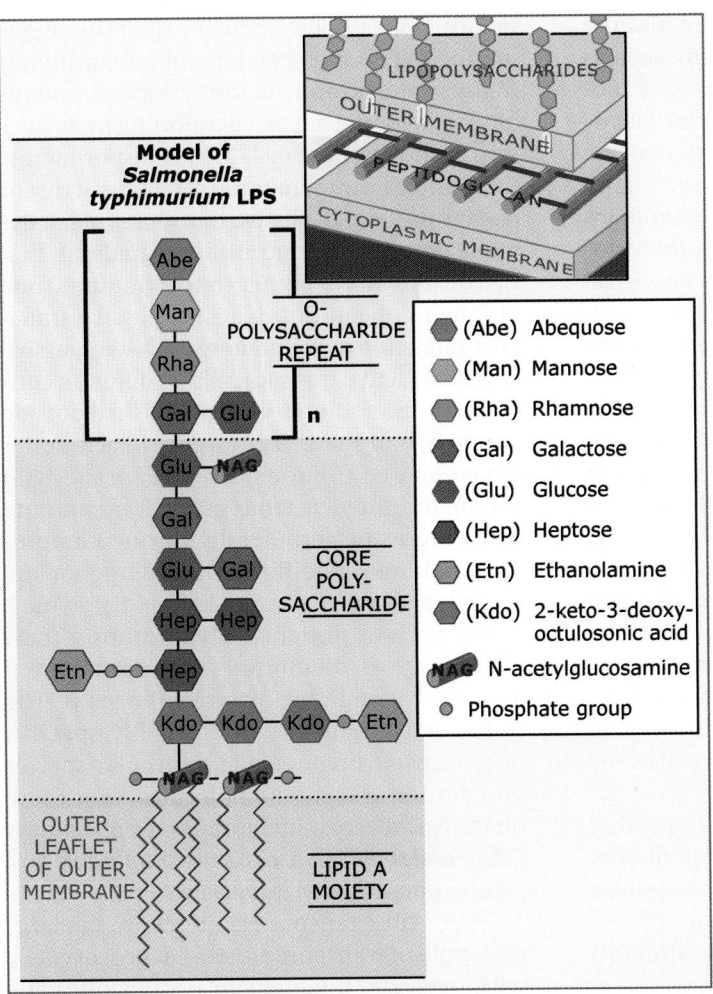

FIGURE 20.23 The proposed structure of lipopolysaccharide from *Salmonella typhimurium* is shown.

the outer leaflet consists of lipopolysaccharide (LPS; see Figure 20.21). Depending on the species, the exact composition of LPS can vary. However, in most cases, LPS consists of the O-polysaccharide, the core polysaccharide, and a lipid A moiety, as **FIGURE 20.23** shows. The lipid A moiety consists of fatty acids such as caproic, lauric, myristic, palmitic, and various steric acids. Lipid A is connected to the O-polysaccharide via ester amide linkages to the core region. The O-polysaccharide is usually made up of a repeating array of galactose, glucose, rhamnose, and mannose. These sugars are arranged in branched, four- or five-membered sequences, which are repeated to form the long O-polysaccharide. The core polysaccharide is composed of N-acetylglucosamine, glucosamine, phosphate, heptose, and ketodeoxyoctonate. This membrane structure is anchored to the peptidoglycan layer via lipoproteins that insert into and span the membrane.

An important biological property of LPS is that it can be toxic to animals. Thus, it is an important determinant of diseases caused by Gram-negative pathogens. Lipid A is the toxic moiety in LPS, and it is often referred to as endotoxin. Humans and other animals have receptors on their epithelial and white blood cells that recognize LPS and activate inflammatory responses. These responses include high fever; decreased numbers of lymphocytes, leukocytes, and platelets; and general inflammation.

20.9 The cytoplasmic membrane is a selective barrier for secretion

Key concepts

- Molecules can pass the cytoplasmic membrane by passive diffusion or active translocation.
- Specialized transmembrane transport proteins mediate the movement of most solutes across membranes.
- The cytoplasmic membrane maintains a proton motive force between the cytoplasm and the extracellular milieu.

All living cells have a cytoplasmic (or plasma) membrane that prevents soluble molecules from diffusing into and out of the cell. For prokaryotes, the ~8-nm-thick cytoplasmic membrane defines the barrier between the inside and the outside of the cell (see Figure 20.1). Cytoplasmic membranes contain lipids and proteins. The basic structure of the cytoplasmic membrane, like most other biological membranes, is that of a phospholipid bilayer. Phospholipids contain a phosphate attached to a three-carbon glycerol backbone. Hydrophobic fatty acid chains are attached to free carbons in the glycerol backbone, orienting themselves toward each other and away from the aqueous extracellular environment. In contrast, the hydrophilic phosphate groups are exposed to the aqueous environment. The cytoplasmic membrane prevents most biological molecules and ions from passively diffusing into or out of the cell (see Figure 6.1). One exception is water, which is able to diffuse slowly across the cytoplasmic membrane; its small size and overall lack of charge allow it to pass freely across the phospholipid bilayer.

Many types of proteins are associated with the cytoplasmic membrane. Membrane proteins often have stretches of hydrophobic amino acids that interact with the hydrophobic fatty acid chains in the membrane. Proteins whose hydrophobic stretches span the membrane are called integral membrane proteins. Many proteins found in the cytoplasmic membrane play some role in transporting molecules into and out of the cell. This transport across the cytoplasmic membrane can be an active or passive process. In passive transport, molecules move down their concentration gradient, that is, from an area of higher concentration to one of lower concentration. Thus, passive transport does not require energy. Unlike passive diffusion, active transport allows different concentrations of solutes to be established outside and inside of the cell. Transport machineries are composed of proteins that associate with the membrane as stable integral membrane constituents or peripherally through amino-terminal lipid modifications that anchor the protein to the membrane surface. Large molecules such as proteins, which cannot freely diffuse across membranes, are transported across the membrane by active transport. The transport systems are often highly specific, only transporting a single molecular species or a conserved class of molecules. (For more on transport across membranes see *6 Transport of ions and small molecules across membranes* and *7 Membrane targeting of proteins*.)

The ABC transporters are easily the largest family of transport proteins in prokaryotes, with over 200 members in *E. coli* alone. The ABC transporters can move substrates into cells, as **FIGURE 20.24** shows, or out of cells (see *20.10 Prokaryotes have several secretion pathways*). ABC transporters have numerous types of substrates, ranging from ions to entire proteins. In Gram-negative bacteria, the ABC transport apparatus that moves substrates into cells typically contains three components: a membrane-spanning transporter protein, a periplasmic substrate-binding protein, and a cytoplasmic ATP-hydrolyzing protein. The periplasmic-binding proteins have remarkably high affinity for their specific substrate, which allows the transport of substrates even when they are present in extremely low concentrations. ATP-binding proteins on the cytoplasmic side of the membrane provide energy for transport. These proteins associate stably with the membrane, usually via a lipid tail at the amino terminus of the protein.

The cytoplasmic membrane plays an important role in energy metabolism in prokaryotes. Electrons produced during respiration are coupled to electron receptors in the membrane. Protons are moved through membrane trans-

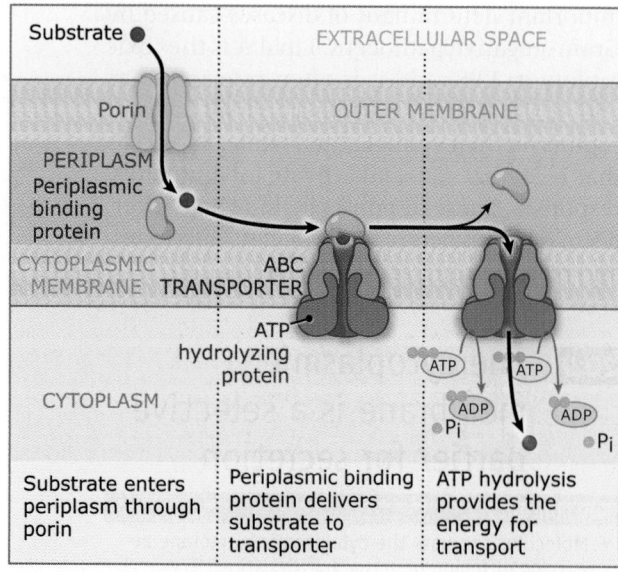

FIGURE 20.24 In Gram-negative bacteria, solutes enter the periplasm through porins, which are proteins that form size-selective pores in the outer membrane. Solutes are bound by their specific periplasmic-binding proteins, which in turn bind to ABC transporters in the cytoplasmic membrane. These transporters use the energy of ATP hydrolysis to move solutes into the cytoplasm.

port proteins to the outside the cell, causing the outside of the membrane to become slightly positively charged. The inside of the membrane is negatively charged, creating a proton gradient that energizes the membrane. This electrochemical gradient generates the **proton motive force** across the membrane. The energetically favorable movement of protons down this gradient, from outside to inside the cell, drives a variety of cellular reactions. In other words, the energy stored at the cytoplasmic membrane can be used by the cell for other processes. For example, enzymes use the proton motive force to generate ATP from ADP (see *6.20 The F_1F_0-ATP synthase couples H^+ movement to ATP synthesis or hydrolysis*). Several membrane-bound enzyme complexes help generate the proton motive force during oxidative phosphorylation. During oxidative phosphorylation, the terminal electron acceptor is oxygen. However, under anaerobic conditions most prokaryotes are able to use other electron acceptors including sulfur, nitrogen, iron, and manganese. The energy that is captured by enzymes in the cytoplasmic membrane fuels most of the activities of a growing cell, such as synthesis of molecules, protein and molecule transport, and motility. Understanding the molecular details of how these pathways provide energy for the microbe is an active area of ongoing research.

20.10 Prokaryotes have several secretion pathways

Key concepts

- Gram-negative and Gram-positive species use the Sec and Tat pathways for transporting proteins across the cytoplasmic membrane.
- Gram-negative bacteria also transport proteins across the outer membrane.
- Pathogens have specialized secretion systems for secreting virulence factors.

All living cells require correct targeting of macromolecules. For prokaryotes, nearly 20% of their proteins are transported across the cytoplasmic membrane. Protein transport across a biological membrane presents the same potential problem for all cells: how to move a relatively hydrophilic molecule (the protein) through a relatively hydrophobic area (the membrane). Not surprisingly, both prokaryotes and eukaryotes have evolved similar systems for protein transport across membranes.

The protein transport machinery generally consists of an integral membrane protein or protein complex that forms a channel through which substrates pass in an energy-dependent manner. The translocated proteins become membrane bound or are released into the extracellular space. The majority of proteins translocated across the cytoplasmic membrane by prokaryotes use two general secretion systems: the Sec pathway, also called the general secretory pathway (GSP), and the twin arginine translocation (Tat) pathway. In addition, at least five major classes of specialized secretion machinery, designated types I to V, are found in bacteria.

As **FIGURE 20.25** shows, the Sec pathway consists of several integral membrane proteins including SecY, SecE, SecG, and a peripherally bound ATPase, SecA, which provides the energy required for translocation. (For more on the Sec YEG system see *7 Membrane targeting of proteins*.) In Gram-negative bacteria, some Sec substrates require the cytoplasmic chaperone SecB in order to reach the secretion complex in an unfolded and secretion-competent state. No SecB homolog has been found in Gram-positive organisms. All proteins translocated by the Sec pathway contain an amino-terminal signal sequence that targets them to the cytoplasmic membrane. Signal sequences are generally hydrophobic but include a short, amino-terminal hydrophilic region that contains one or more positively charged residues. Signal sequences from Gram-positive and Gram-negative species are similar, although Gram-positive signal sequences tend to be longer and have more

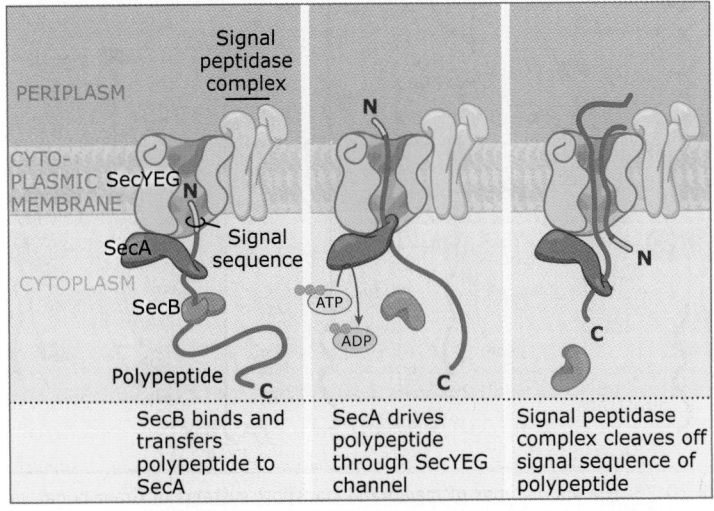

FIGURE 20.25 The proteins of the general secretory pathway, or Sec pathway, translocate newly synthesized proteins across the cytoplasmic membrane.

positive residues near the amino terminus. Proteins translocated by the Sec pathway can become membrane-localized, released into the periplasm, or secreted into the extracellular milieu.

Besides Sec, another general secretion system called Tat is conserved among bacteria. The Tat system, unlike Sec, uses only the proton gradient to transport folded, or even oligomeric proteins, to the cell surface. The Tat system does not use ATP hydrolysis to drive secretion and does not secrete unfolded proteins. In fact, the Tat system may reject substrates that are not properly folded. For example, the PhoA protein is normally secreted by the Sec system, but it can become a substrate for the Tat system if a Tat-specific signal sequence is added to its N-terminus. However, PhoA is only a substrate for Tat if it is engineered to fold correctly in the cytoplasm before secretion to the periplasm. Therefore, it has been hypothesized that all substrates must be folded before the Tat secretion complex recognizes them.

The proteins that comprise the Tat secretion complex include TatA, B, C, and E. TatABC make up a membrane complex, while TatE appears to be a functional homolog of TatA. TatA is thought to form the channel, and TatBC form a complex that binds to the substrate's signal peptide. The Tat-specific signal peptides are similar to the Sec signal peptides but have additional features. The most distinguishing feature is the conserved Ser-Arg-Arg-X-Phe-Leu-Lys motif located just after the N-terminal region. The two arginine residues are what give this system its name. Nearly every Tat substrate contains two arginine residues, although a few contain only a single arginine.

In addition to general secretion systems, prokaryotes have numerous other systems that direct protein secretion across the cytoplasmic membrane, or in the case of Gram-negative organisms, both the cytoplasmic and outer membranes. The five major classes of these secretion systems are shown in **FIGURE 20.26** for Gram-negative bacteria. Many of these secretion systems are multisubunit complexes.

Type I secretion machinery are ABC-type transporters and can be found throughout the domain bacteria. (There are also many ABC transporters in eukaryotic organisms.) A typical type I secretion apparatus consists of a two-domain cytoplasmic membrane protein. One domain interacts with the membrane, while the other contains one or two ATP-binding cassettes. Type I secretion systems in Gram-negative organisms have an outer membrane protein that transports substrates across the outer membrane. An accessory protein, present in most type I systems, associates with the cytoplasmic membrane and may provide specificity for substrate transport. Type I substrates are varied, but the best studied include pore-forming toxins and degradative enzymes secreted by pathogenic organisms.

In Gram-negative bacteria, type II secretion (also called "the main terminal branch" of the GSP) is responsible for directing many Sec-dependent substrates across the outer membrane. The main terminal branch features an integral outer membrane protein called a secretin. Secretins are stable multimers (10–14 subunits) that form a translocation channel through the outer membrane. The secretins often require a "pilot" protein for proper targeting and insertion into the outer membrane. Secretins have been visualized by electron microscopy, and each oligomeric secretin appears as a ring-shaped complex with a central pore of 10 to 15 nm in diameter. Presumably, this pore size is large enough to allow folded or partially folded proteins to pass.

Because it is involved in host-pathogen interactions, the type III secretion system has been extensively studied. In the type III system, substrates are transported from the bacterial

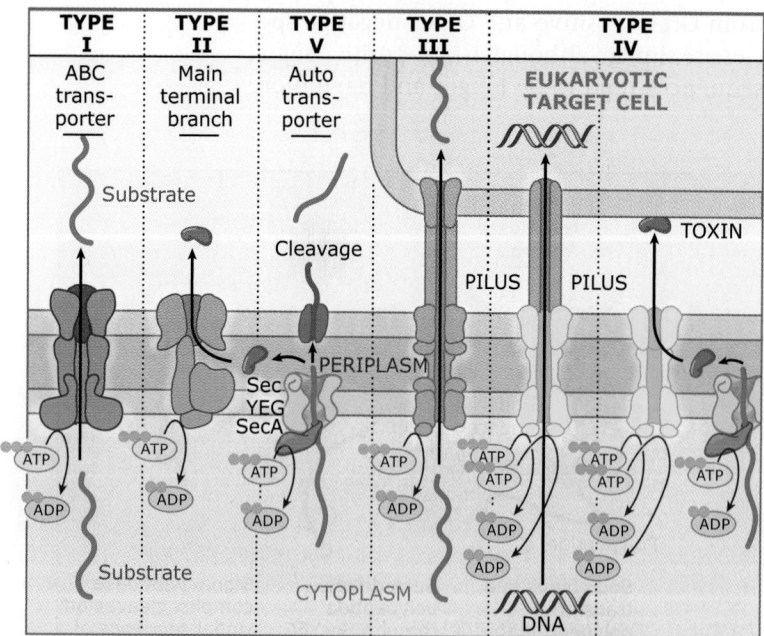

FIGURE 20.26 The major types of membrane transport systems of Gram-negative bacteria are shown. Some secretion systems are protein complexes that span the inner membrane, periplasm, and outer membrane.

cytoplasm directly into the host cytoplasm. Many virulence factors, which are molecules that allow pathogens to invade and colonize host organisms, are delivered in this manner. Proteins secreted by the type III machinery are thought not to have a periplasmic intermediate. Instead, the ~20 proteins involved in type III secretion form a complex that spans the bacterial cytoplasmic and outer membranes, which allows the secretion of "effector" proteins out of the cell. An analogous system has recently been described in Gram-positive bacteria. (For more on host-pathogen interactions see *20.20 Prokaryotes can colonize and cause disease in higher organisms.*)

Type IV protein secretion systems are homologous to DNA transfer systems. Protein substrates of the type IV system are first secreted across the cytoplasmic membrane by the Sec system. Like the type III secreted proteins, type IV substrates can be secreted directly into the cytoplasm of a eukaryotic host. Some type IV secretion systems, such as the *Agrobacterium tumefaciens* T-DNA transfer system, can deliver both proteins and DNA to host cells. Although type IV secretion systems have been identified in many Gram-negative bacteria, only a few secreted substrates have been discovered to date. One substrate is pertussis toxin, which is secreted by *Bordetella pertussis* in a type IV-dependent pathway. Unlike T-DNA transfer by *Agrobacterium*, which occurs directly from the bacterial cytoplasm to the host cytoplasm, pertussis toxin is secreted in two steps. The toxin transverses the cytoplasmic membrane via the Sec pathway and then interacts with the type IV machinery to cross the outer membrane.

In type V secretion, a single protein is able to direct its own secretion across the outer membrane. Called autotransporters, these proteins are first delivered to the periplasm by the Sec machinery. For example, pathogenic *Neisseria* species use the autotransporter system to secrete a protease that helps in evasion of the host immune response. Upon delivery to the periplasm, the C-terminal end of the protein inserts into the outer membrane and functions as a transporter for the N-terminal domain. After the N-terminal protease domain is secreted to the cell surface, it cleaves itself from the C-terminal domain. The protease domain is then released from the bacterial surface, while the C-terminal domain remains in the outer membrane. Autotransporters similar to the protease in *Neisseria* have been found in many other Gram-negative organisms.

20.11 Pili and flagella are appendages on the cell surface of most prokaryotes

Key concepts

- Pili are extracellular proteinaceous structures that mediate many diverse functions, including DNA exchange, adhesion, and biofilm formation by prokaryotes.
- Many adhesive pili are assembled by the chaperone/usher pathway, which features an outer membrane, usher proteins that form a pore through which subunits are secreted, and a periplasmic chaperone that helps to fold pilus subunits and guides them to the usher.
- Flagella are extracellular apparati that are propellers for motility.
- Prokaryotic flagella consist of multiple segments, each of which is formed by a unique assembly of protein subunits.

Two types of appendages, **pili** and **flagella**, extend from the surface of prokaryotic cells. Pili are filamentous protein oligomers expressed on the cell surface, as shown in **FIGURE 20.27**. There are different types of pili. For example, F-pili mediate cell-cell conjugation and DNA transfer. When these appendages were first described, they were designated "fimbriae" (plural, from the Latin for thread or fiber) and their presence was correlated with the ability of *E. coli* to agglutinate red blood cells. The term *pilus* (from the Latin for hair) was later introduced to describe the fibrous structures (F-pili) associated with the conjugative transfer of genetic material between organisms. Since then *pilus* has become a generic term used to describe all types of nonflagellar filamentous appendages,

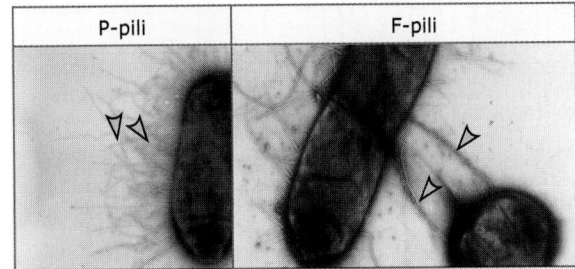

FIGURE 20.27 Two types of pili on prokaryotes are shown. P-pili are involved in adhesion and are shorter than F-pili. F-pili are involved in conjugation and DNA transfer between cells. Left photo courtesy of Matthew Chapman, University of Michigan; right photo courtesy of Ron Skurray, School of Biological Sciences, University of Sydney.

and it is used interchangeably with the term *fimbria*.

Pili-mediated interactions of prokaryotic cells with both eukaryotic and prokaryotic cells often represent essential steps leading to colonization of an epithelial surface, entry into host cells, exchange of DNA, and development of biofilms (see *20.21 Biofilms are highly organized communities of microbes*). Pili can act as receptors for bacteriophages. The primary function of most pili is to act as scaffolding for the presentation of specific adhesive moieties. Adhesive pilus subunits (adhesins) are often incorporated as minor components into the tips of pili, but major structural subunits can also function as adhesins. Adhesive pili are often important colonization factors. For example, type I pili are required for uropathogenic *E. coli* to adhere to bladder epithelial cells during a urinary tract infection. Type I pili are assembled by many Gram-negative species. The type I pili are composite structures consisting of a thick pilus rod attached to a thin fibrillar tip. At the distal end, the adhesin FimH confers binding to mannose moieties on host cells.

Pilus assembly is complex, involving structural proteins that will become part of the fiber and accessory proteins that facilitate the as-sembly of pilus subunits into fibers on the cell surface. All pilins (structural components of pili) destined for assembly on the surface of Gram-negative organisms must be translocated across the cytoplasmic membrane, through the periplasm, and across the outer membrane. To accomplish these steps, two specialized assembly proteins are used: a periplasmic chaperone and an outer membrane transport protein called usher. Chaperone/usher assembly pathways are involved in the biogenesis of over 30 different types of pilus structures.

Chaperone-subunit complexes form in the periplasm and interact with usher in the outer membrane, where the chaperone is released, exposing interactive surfaces on the subunits that facilitate their assembly into the pilus, as shown in **FIGURE 20.28**. Studies in the P- and type 1 pilus systems have demonstrated that the adhesin-chaperone complexes (PapDG or FimCH) bind with great affinity to the usher and that the adhesins are the first subunits assembled into the pilus. Additional subunits are incorporated into the pilus depending, in part, upon the kinetics with which they are partitioned to the usher in complex with the chaperone. Besides acting as an assembly platform for the growing pilus, the usher protein appears to have additional roles in pilus biogenesis. High-resolution electron microscopy revealed that the PapC usher is assembled into a 15-nm-diameter ring-shaped complex with a 2-nm-wide central pore. After dissociating from the chaperone at the usher, subunits are incorporated into a growing pilus structure that is predicted to be extruded as a one-subunit-thick linear fiber through the central pore of the usher complex.

A second extracellular fiber assembled by Gram-negative bacteria is curli. Curli are remarkably stable, aggregative fibers that are part of an extracellular matrix that contributes to biofilm and other community behaviors. Curli are also the first described bacterial functional amyloid fiber. Amyloids are readily associated with human diseases such as Alzheimer's, but biochemically and structurally similar fibers, such as curli, are found throughout cellular life.

Curli are assembled by a unique and highly regulated biogenesis pathway. At least six proteins, encoded by the divergently transcribed csgBA and csgDEFG operons (csg, curli specific genes), are dedicated to curli formation in *E. coli*. The major curli fiber subunit protein is CsgA. CsgA is nucleated into a fiber by the minor fiber subunit, CsgB. After nucleation, CsgB becomes

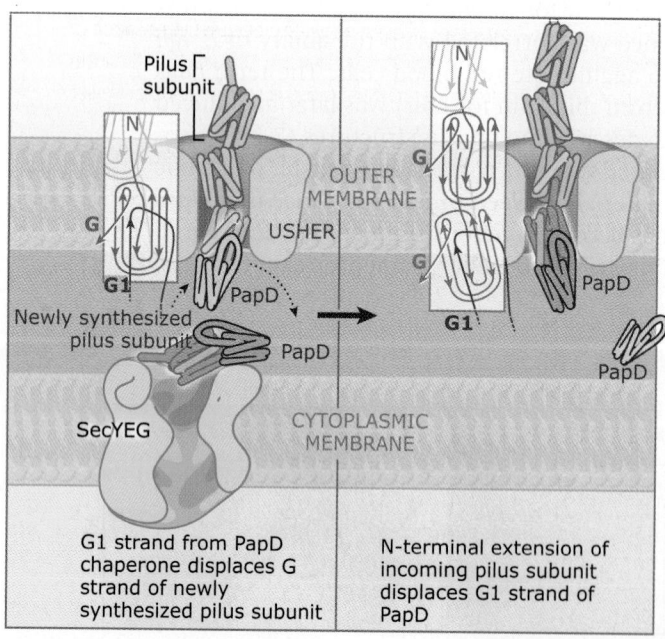

FIGURE 20.28 The PapD chaperone and membrane protein usher are required for the proper assembly of pili. Left: the G1 strand of PapD completes the folding of a newly synthesized pilus subunit by inserting between two strands of the subunit. Right: during pilus assembly, the N-terminal strand of the newly synthesized subunit displaces the G1 strand of the PapD chaperone associated with the previous subunit.

incorporated into the fiber. CsgA fibers are extraordinarily stable, and can remain intact after "boiling in SDS, a strong detergent," concentrated urea, and other harsh treatments. CsgA monomers are only liberated from the fiber after treatment with at least 75% formic acid (FA).

CsgA and CsgB do not have to be expressed from the same cell in order to assemble curli. In the absence of the CsgB nucleator, CsgA is secreted from the "donor" cell in a soluble, unassembled state. This unassembled CsgA can be polymerized into curli fibers if it contacts the CsgB nucleator expressed by an adjacent "acceptor" cell in a process called interbacterial complementation. Interbacterial complementation provides an indirect method for testing the activity of CsgA or CsgB in noncurli producing strains. In the simplest example, a csgB deletion strain will not make curli; however, this strain will still produce CsgA and act as a donor strain during interbacterial complementation.

The csgDEFG operon encodes CsgD, a transcriptional activator of curli synthesis, and three putative assembly factors. Transcriptional regulation of these operons is complex and responds to a number of environmental cues including osmolarity, temperature, acid shock, and protein aggregation in the periplasm. Secretion of both CsgA and CsgB is dependent on the outer membrane-localized CsgG protein. In the absence of CsgG, the CsgA and CsgB curli subunits are not secreted to the extracellular environment and they are not detectable by whole cell western blotting. CsgG forms a pore in the outer membrane that is proposed to allow the CsgA and CsgB proteins to cross the membrane.

Most microorganisms are motile, and often motility depends on long appendages called flagella, shown in **FIGURE 20.29**. Gram-positive and Gram-negative microbes assemble flagella at their cell surface. A single flagellum located at one pole is called monotrichous (or polar), while multiple flagella distributed around the cell surface are peritrichous. A group of flagella that arises at one pole is called lophotrichous (from the Latin for "tuft of hair"). Bacterial flagella differ from eukaryotic flagella, which are composed of microtubules and associated proteins and are enclosed by the plasma membrane (see *11 Microtubules*).

Flagella vary in length, but they are usually around 20 nm in diameter and cannot be seen with a light microscope unless they are first stained with reagents that increase their diameter. Flagella can be divided into three distinct domains: filament, hook, and basal

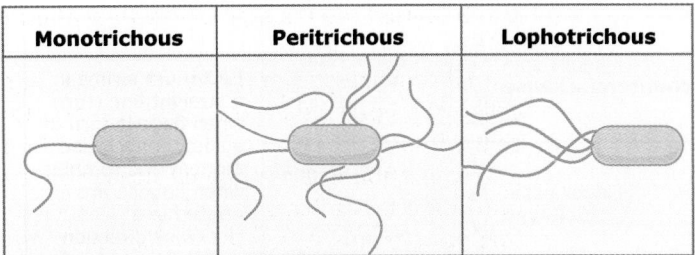

Monotrichous	Peritrichous	Lophotrichous

FIGURE 20.29 Different types of bacteria assemble flagella in distinct arrangements.

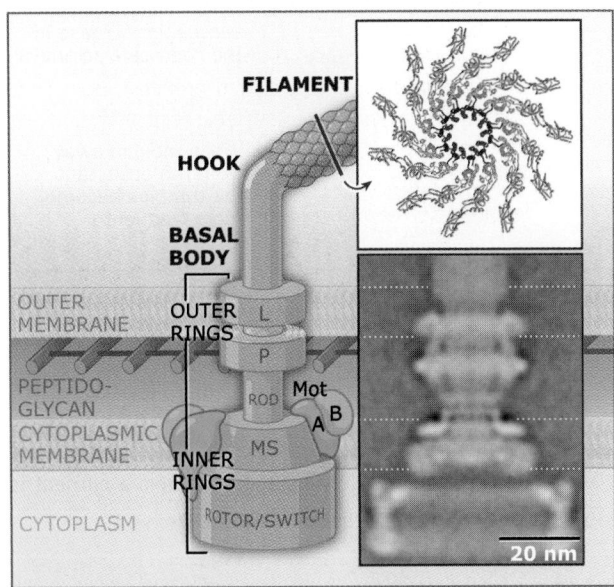

FIGURE 20.30 Prokaryotic flagella have distinct segments, each of which consists of multiple proteins. Top inset shows the flagellin structure obtained using electron cryomicroscopy, and bottom inset shows an electron micrograph of the flagella basal body and hook regions. The flagellum filament consists of flagellin subunits that assemble into a helix with 11 subunits per turn. Top photo reprinted by permission from Macmillan Publishers Ltd: *Nature*, K. Yonekura, S. Maki-Yonekura, K. Namba, Complete atomic model of the bacterial flagellar filament..., 2003, vol, 424 (6949): 643–650. Photo courtesy of Keiichi Namba, Osaka University; bottom photo courtesy of David DeRosier, Professore Emeritus, Brandeis University.

body, as **FIGURE 20.30** shows. The flagella filament is composed of a repeating array of flagellin proteins. Flagellins are highly conserved throughout the domain bacteria, suggesting that flagellar-mediated motility is a primitive trait. At the point where the flagella attaches to the cell is the basal body, which is a complex, multiprotein structure. The flagellum filament is tethered to the basal body by the hook region. The basal body spans the outer membrane, cell wall peptidoglycan, and cytoplasmic membrane of Gram-negative organisms. The L ring anchors the flagellum to the

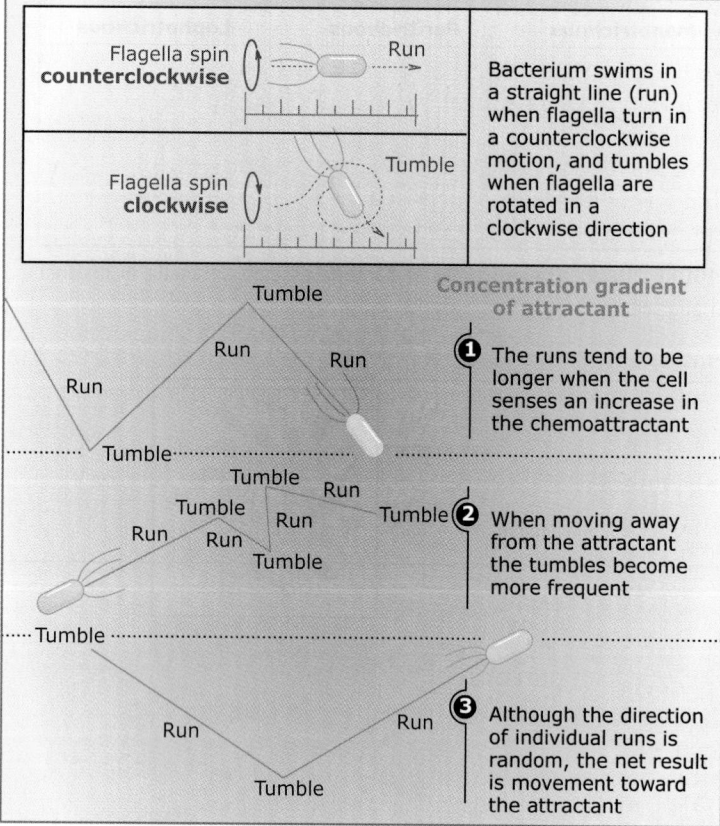

Flagella spin **counterclockwise** | Run

Bacterium swims in a straight line (run) when flagella turn in a counterclockwise motion, and tumbles when flagella are rotated in a clockwise direction

Flagella spin **clockwise** | Tumble

Concentration gradient of attractant

Tumble

Run | Run
Run

Tumble

Tumble | Run
Tumble
Tumble | Run | Tumble
Run | Run
Tumble

Tumble

Run | Run

Tumble

1 The runs tend to be longer when the cell senses an increase in the chemoattractant

2 When moving away from the attractant the tumbles become more frequent

3 Although the direction of individual runs is random, the net result is movement toward the attractant

FIGURE 20.31 Bacterial movement toward and away from a nutrient is characterized by longer runs and more tumbles, respectively.

outer membrane. There are two pairs of rings, the S-M and P rings, which help anchor the flagellum to the cytoplasmic membrane and to the cell wall, respectively. Multiple membrane proteins make up each ring. At the cytoplasmic membrane are two Mot proteins that act as the engines to drive flagellar motion. Another set of proteins found embedded in the cytoplasmic membrane act to reverse the flagellar motor. Since Gram-positive organisms lack an outer membrane, they only have the S-M rings.

There are several dozen genes involved in the expression and assembly of the flagellar filament. These genes are coordinately regulated in the order in which they are assembled. Thus, the first genes expressed are those involved in the assembly of the basal body and hook followed by the expression of the flagellin subunit. The expression of the flagellin subunit is suppressed until after the assembly of the hook apparatus, at which point a negative regulator of transcription is pumped out of the cell via the hook channel, thus relieving the suppression of flagellin expression. Flagellin subunits are exported through the growing flagellum and are added to the tip. This mechanism en-

sures that the filament will not be produced until after hook assembly. The hook apparatus is related to other protein secretion apparatuses (see *20.10 Prokaryotes have several secretion pathways*).

The presence of nutrients is sensed by the chemotaxis machinery, which then determines the direction of flagella rotation. In the absence of nutrients, the default direction of flagellum rotation is clockwise, resulting in tumbling. The movement of cells toward or away from a chemical is known as **chemotaxis**. Here we will consider the movement of a prokaryotic cell in the presence of a nutrient attractant. The rigid flagellum, using energy provided by proton motive force, is rotated like a propeller in order to facilitate this movement. Movement consists of a series of straight runs, followed by short random tumbles. Cells swim in a straight line when flagella turn in a counterclockwise motion and tumble when flagella are rotated in a clockwise direction, as **FIGURE 20.31** shows. Since tumbling randomly positions the organism, one might predict that the net movement would be zero. However, the periodicity of the runs is regulated according to nutrient availability: longer runs occur when the cell travels toward a nutrient source and the number of tumbles increases when it travels away from the nutrient gradient. Although the direction of individual runs is still random, the net result is movement toward the attractant.

The chemotaxis signal transduction pathway is remarkably conserved in prokaryotes. The only known species whose genome does not include chemotaxis genes is *Mycoplasma*. Chemotaxis proteins that are conserved in virtually all prokaryotes are CheR, CheA, CheY, CheW, and CheB. Using an intricate cascade of phosphorylation or methylation modifications, these proteins coordinate a complex and highly adaptable response to attractants and repellants. Here we describe the pathway in *E. coli*. Receptors in the cytoplasmic membrane bind attractants or repellants in the extracellular milieu. The kinase CheA, which resides in the cytoplasmic membrane, interacts with these receptors. CheA phosphorylates CheY, which can then bind to the flagella motor, causing it to switch directions and the cell to tumble. The phosphate group is removed from CheY by the phosphatase CheZ. At low attractant concentrations, CheA is autophosphorylated, the phosphate is transferred to CheY, and CheY migrates to the flagella motor and changes the direction of the motor causing a tumble.

There is another level of complexity to the chemotaxis system that allows for constant adaptation so that the cell can respond to small fluctuations as it swims toward a chemical gradient. This short-term memory is achieved through methylation of membrane receptors. CheR methylates membrane receptors, and CheB removes the methyl group. Receptor methylation leads to increased kinase activity by CheA, which results in desensitization of the system. In turn, CheB can also become phosphorylated by CheA; this increases CheB's methylesterase activity and provides a feedback loop for the signaling cascade.

20.12 Prokaryotic genomes contain chromosomes and mobile DNA elements

Key concepts

- Most prokaryotes have a single circular chromosome.
- Genetic flexibility and adaptability is enhanced by transmissible plasmids and by bacteriophages.
- Transposons and other mobile elements promote the rapid evolution of prokaryotic genomes.

Most prokaryotes have a single circular chromosome and are haploid. The organization of the chromosome into a discrete body, the nucleoid, is described in the next section (*20.13 The bacterial nucleoid and cytoplasm are highly ordered*). More than 200 complete genome sequences are now available, and their sizes range from 580 kbp (*Mycoplasma genitalium*) to 9 Mbp (*Streptomyces, Myxococcus*). Well-characterized bacteria such as *E. coli* and *B. subtilis* are in the middle of the size range (4 to 5 Mbp). The relatively small size of prokaryotic chromosomes compared with higher eukaryotes is due to their genomes being compact, with negligible amounts of noncoding DNA. In general, genes that are needed for the everyday growth and maintenance of prokaryotes are carried on the chromosome, and genetic flexibility is enabled by a variety of mobile elements.

A few bacterial species have linear chromosomes or multiple chromosomes or both of these features. For example, *Streptomyces* has a linear chromosome. The ends of the chromosome are connected via a protein bridge, explaining why the long-standing genetic map is circular. *Rhodobacter sphaeroides* has two large circular chromosomes (of 3.0 and 0.9 Mbp), each of which contains many essential housekeeping genes. The Lyme disease pathogen, *Borrelia burgdorferi*, has multiple linear chromosomes.

Stable extrachromosomal DNA elements that do not carry essential housekeeping genes and are, therefore, dispensable are called **plasmids**. Examples of some well-known bacterial plasmids, illustrating the range of genes that they carry, are provided in **FIGURE 20.32**. Plasmids are usually relatively small, ranging from 2 to 100 kbp, and circular. As with chromosomes, there are exceptions, with some huge plasmids reaching 1 Mbp or more, and a minor fraction has a linear structure. All plasmids carry genes that direct their own replication, usually harnessing various elements of the host cell DNA replication machinery. Well-known genes carried on plasmids include those conferring antibiotic resistance, pathogenicity, or degradation of exotic carbon sources.

Plasmids can spread between organisms by a variety of mechanisms. **Conjugation** is a mating-like process involving direct DNA transfer between a plasmid-containing donor cell and a recipient cell. The donor plasmid encodes the functions needed to bring donor and recipient cells together, to initiate replicative DNA transfer, and to transfer the DNA into the recipient cell. Plasmids can also spread by direct DNA uptake (**transformation**) or by a bacteriophage-mediated process (**transduc-**

FUNCTION	PLASMID	SIZE (kb)	ORIGINAL HOST ORGANISM
Antibiotic resistance (for example, ampicillin, chloramphenicol, kanamycin, and tetracycline)	RP4	60	*Pseudomonas aeruginosa*
Antibiotic production	SCP1	356	*Streptomyces coelicolor*
Conjugation	F	100	*Escherichia coli*
Heavy metal resistance	pI258	28	*Staphylococcus aureus*
Nodulation (symbiosis) and nitrogen fixation	pSym1	1,400	*Sinorhizobium meliloti*
Pathogenicity genes	pWR501	222	*Shigella flexneri*
Plant tumorigenesis	pTiC58	214	*Agrobacterium tumefaciens*
Prophage (bacterial virus)	P1	94	*Shigella*
Utilization of exotic carbon sources (for example, toluene)	TOL	117	*Pseudomonas putida*
UV (ultraviolet light) resistance	pKM101	35	*Salmonella*

FIGURE 20.32 Examples of bacterial plasmids and their functions.

tion). Irrespective of the existence of extra-chromosomal elements, bacterial genomes can also undergo change via the processes of homologous and site-specific recombination.

For many bacteria, **bacteriophages** (bacterial viruses) appear to be a significant source of genetic variation. Many sequenced genomes contain obvious integrated bacteriophages (**prophages**): for example, *E. coli* carries at least 9 and *B. subtilis* has 10. Some of these prophages are clearly defective, carrying obvious deletions and other mutations, and appear unlikely to be capable of being activated into forming infective bacteriophage particles. In some cases, prophages carry genes that could confer a selective advantage on the host cell; these include restriction and modification systems, UV resistance, and pathogenicity determinants, such as toxins.

Lastly, bacterial genomes also contain many mobile genetic elements that rely on **transposition** to spread. Insertion sequences comprise the minimal elements, which, in their simplest form, possess only a single transposase gene together with flanking sequences that the transposase protein recognizes in initiating the transposition reaction. Host cell enzymes that function in DNA replication and repair are then recruited to complete insertion at the target site. More complex transposons carry additional genes that can provide an adaptive advantage to the host cell. Antibiotic resistance genes are the best-known examples of transposable genes in bacteria but a wide diversity of other genes can be carried as transposons. Elements related to transposons can catalyze several other kinds of DNA rearrangements, such as inversions and deletions.

Integrons are a particularly important source of adaptive genome rearrangement. As **FIGURE 20.33** shows, integrons generally comprise an integrase gene, an adjacent target site for capture of gene cassettes, and a strong promoter that drives expression of the captured genes. Gene cassettes, often conferring antibiotic resistance, have sequences allowing them to be inserted at the target site, via the action of the integrase protein. Integrons can grow by sequential capture of different gene cassettes, facilitating rapid evolution of bacteria, such as ones that are resistant to multiple antibiotics.

20.13 The bacterial nucleoid and cytoplasm are highly ordered

Key concepts

- The bacterial nucleoid appears as a diffuse mass of DNA but is highly organized, and genes have non-random positions in the cell.
- Bacteria have no nucleosomes, but a variety of abundant nucleoid-associated proteins may help to organize the DNA.
- In bacteria, transcription takes place within the nucleoid mass; translation, the peripheral zone—analogous to the nucleus and cytoplasm of eukaryotic cells.
- RNA polymerase may make an important contribution to nucleoid organization.

Prokaryotic cells differ fundamentally from eukaryotes in lacking a nuclear membrane. In eukaryotes, the presence of this membrane results in distinct compartments in which transcription and translation are essentially separated (see Figure 5.3). In prokaryotes, these processes are not separated by a membrane, and mRNA can be translated while its transcription is ongoing. This simultaneous transcription and translation has important consequences for the regulation of some genes.

Bacterial chromosomal DNA has the appearance of an amorphous mass, the **nucleoid**, which fills much of the central part of the cytoplasm, as **FIGURE 20.34** shows. The nucleoid consists of the chromosomal DNA and its associated proteins. Bacteria do not have nucleosomes, which package the DNA of eukaryotic and archaeal chromosomes (see *10 Chromatin and chromosomes*). However, the DNA is compacted and organized by a number of abundant nucleoid-associated proteins (some are histone-like proteins), listed in **FIGURE 20.35**.

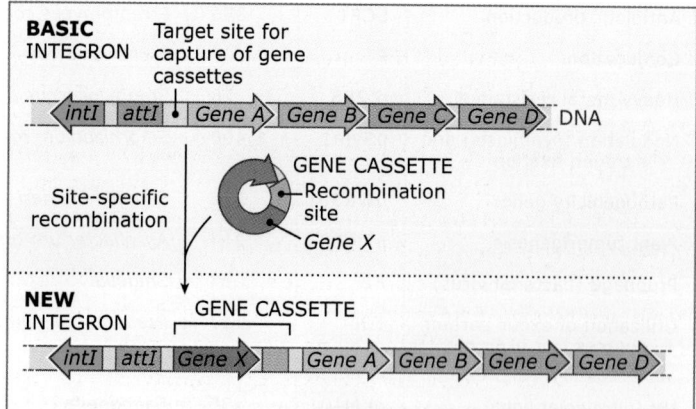

FIGURE 20.33 Structure of a typical integron that has accumulated several gene cassettes.

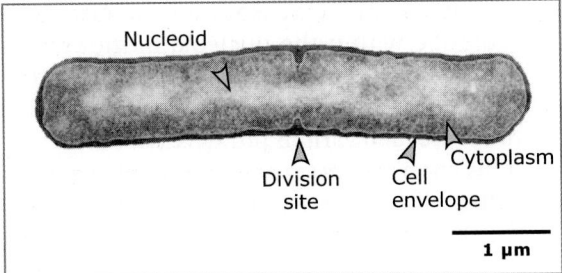

Nucleoid
Division site
Cell envelope
Cytoplasm

1 μm

FIGURE 20.34 The electron micrograph shows that the nucleoid is a diffuse mass within a bacterial cell. Photo reproduced from J. Errington, et al., *Phil. Trans. R. Soc. B.*, vol. 29, figure 2, © 2005, The Royal Society. Photo courtesy of Jeffrey Errington, Newcastle University.

PROTEIN(S)	GENE(S)	FUNCTION
HUα, HUβ	*hupA, hupB*	Abundant nucleoid-associated protein
H-NS	*hns*	*Histone-like nucleoid structuring protein*; abundant nucleoid-associated protein
IHF	*ihf*	*Integration host factor*; abundant nucleoid-associated protein
Gyrase	*gyrA, gyrB*	DNA supercoiling
Topo-isomerase IV	*parC, parE*	DNA supercoiling; possible role in decatenation (removing interlinks between sister chromosomes)
Topo-isomerase I	*topA*	DNA supercoiling
MukB	*mukB*	Chromosome condensation
MukE	*mukE*	Chromosome condensation
MukF	*mukF*	Chromosome condensation
RNA polymerase	*rpoA, rpoB, rpoC*	Transcribing RNA polymerase may be a major source of restraints on DNA structure

FIGURE 20.35 Proteins involved in organizing the nucleoid of *Escherichia coli*. For most other bacterial species, the MukB, MukE, and MukF proteins are replaced by structural maintenance of chromosome (SMC) proteins and associated factors that are related to eukaryotic cohesin and condensin proteins.

Among the most important of these proteins are the topoisomerases. These proteins control DNA supercoiling, which plays an important role in compaction of the DNA, and allow processes such as replication and transcription, which require unwinding, to proceed. Proteins of the structural maintenance of chromosomes (SMC) family also contribute to nucleoid organization (including supercoiling) as indicated by the phenotype of mutants, but the mechanisms remain unclear. In eukaryotes, proteins closely related to the SMC proteins are involved in both chromosome cohesion and condensation during meiosis and mitosis (see *10.31 Chromosome condensation is caused by condensins*). These various nucleoid-associated proteins cooperate to maintain the level of supercoiling and overall compaction of the nucleoid. However, details of how this overall nucleoid homeostasis is achieved remain to be worked out.

Although the nucleoid appears to have an amorphous structure, individual genes tend to have well-defined positions in it. Their positions in the nucleoid reflect their relative positions in the chromosome map. The first evidence for this emerged serendipitously from the properties of a *B. subtilis* mutant in the gene *spoIIIE*, as **FIGURE 20.36** shows. This mutant fails to segregate its chromosome properly during the asymmetric division that accompanies the early stages of spore formation in this organism. Instead, the division septum near the cell pole closes around one of the chromosomes. In these mutant cells, certain genes are almost always trapped in the small compartment next to the pole, whereas others are always excluded. This observation suggests that the chromosome always takes up a particular position and orientation prior to septation. (For details on spore formation see *20.17 Prokaryotes respond to stress with complex developmental changes*.)

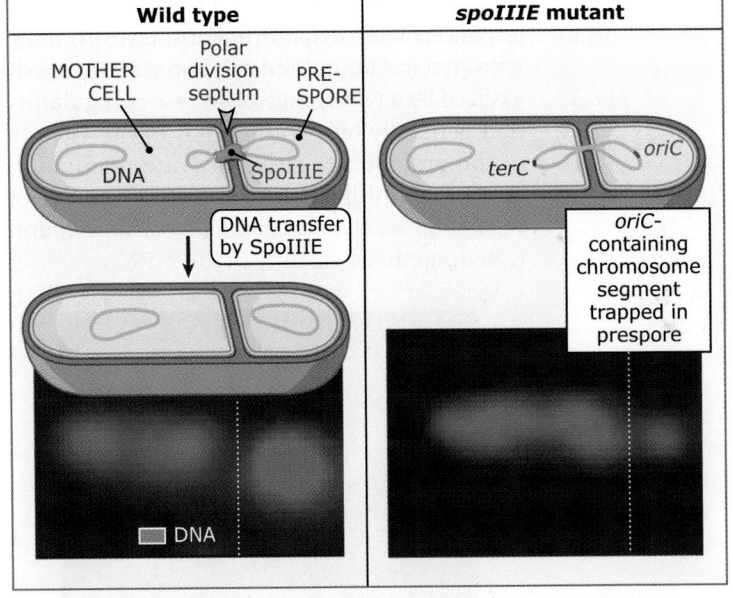

FIGURE 20.36 Segregation of chromosomes after polar septation at the onset of sporulation. When *B. subtilis* cells sporulate, they divide asymmetrically to produce a mother cell and a smaller prespore. Each cell acquires a single copy of the chromosome. Chromosome segregation into the prespore is a two-step process. First, the polar division septum closes around the chromosome, then SpoIIIE protein actively transports the remaining 2/3 of the chromosome into the prespore compartment. In *spoIIIE* mutants, only about 1/3 of the chromosome is segregated into the prespore. Analysis of the DNA trapped in the small compartment of *spoIIIE* mutant cells revealed that a specific segment of DNA is always trapped, indicating that the chromosome must be strictly orientated and ordered prior to cell division. The fluorescence micrographs show sporulating wild-type and *spoIIIE* mutant cells stained for DNA. Photos reproduced from J. Errington, et al., *Phil. Trans. R. Soc. B.*, vol. 360, figure 3, © 2005, The Royal Society. Photo courtesy of Jeffrey Errington, Newcastle University.

More direct evidence has come from studies using fluorescence *in situ* hybridization (FISH), which can be used to look directly at the positions of particular genes in cells. However, this technique requires fixation and other harsh treatments to allow the probe to hybridize with its target DNA. Yet another approach has been to use a fusion of green fluorescent protein to a DNA-binding protein (LacI) that can be targeted to an array of binding sites placed at different positions in the cell. On the basis of all these experiments, it is clear that genes are not free to diffuse around the bacterial cell but have positions that are highly constrained. In general, the *oriC* region of the chromosome lies at one end of the nucleoid and the *terC* region opposite. Genes between these locations on the genetic map are positioned more or less proportionately within the nucleoid.

All transcription in bacterial cells uses a single catalytic core RNA polymerase comprising two α, one β, and one β′ subunits. Promoter specificity is determined at a primary level by a range of sigma (σ) factors, which are also required for the initiation of transcription but dissociate from the core after initiation. Transcriptional regulation is exerted by a wide range of accessory regulators that generally bind to the DNA in the vicinity of the promoter to either activate (stimulate) or repress (inhibit) initiation. Other modes of regulation act at the termination of transcription (attenuation) or stability of the mRNA.

Most of the core RNA polymerase molecules lie within the nucleoid in the central part of the cell, so it is presumably here that the bulk of transcription tends to occur. In contrast, ribosomes and various proteins associated with translation are enriched in the peripheral areas of the cell. Thus, as **FIGURE 20.37** shows, even in the absence of a nuclear membrane, transcription and translation are spatially separated in bacterial cells, in a manner analogous to that of eukaryotes. However, various lines of evidence have shown that transcription and translation are sometimes closely coupled in bacteria. This observation may not be inconsistent with the observation that RNA polymerases and ribosomes do not colocalize. One possibility is that both processes may occur at the interface between core and peripheral zones of the cell. Little is yet known about how the overall organization of the nucleoid and how the core and peripheral zones of the cell are maintained.

20.14 Bacterial chromosomes are replicated in specialized replication factories

Key concepts

- Initiation of DNA replication is a key control point in the bacterial cell cycle.
- Replication takes place bidirectionally from a fixed site called oriC.
- Replication is organized in specialized "factories."
- Replication restart proteins facilitate the progress of forks from origin to terminus.
- Circular chromosomes usually have a termination trap, which ensures that replication forks converge in the replication terminus region.
- Circular chromosomes require special mechanisms to coordinate termination with decatenation, dimer resolution, segregation, and cell division.
- The SpoIIIE (FtsK) protein completes the chromosome segregation process by transporting any trapped segments of DNA out of the closing division septum.

Chromosome replication in bacteria initiates at a fixed site called *oriC* (*ori*gin of *c*hromosome replication) and is bidirectional, that is, replication forks simultaneously set off in both clockwise and counterclockwise directions. Overall, replication is highly processive and the replication forks meet at a site diametrically opposite *oriC*, called *terC* (*ter*minus of chromosome replication).

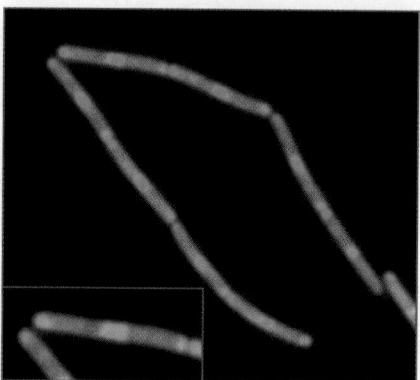

FIGURE 20.37 Despite the absence of a nuclear membrane, the transcription and translation machineries are localized to separate regions of bacterial cells. Shown are *B. subtilis* cells in the process of dividing. They express the ribosomal subunit RpsB fused to green fluorescent protein (GFP) and RNA polymerase subunit RpoC fused to GFP-UV, which fluoresce differently and are represented here as green and red, respectively. Photos courtesy of Peter Lewis, University of Newcastle, Australia.

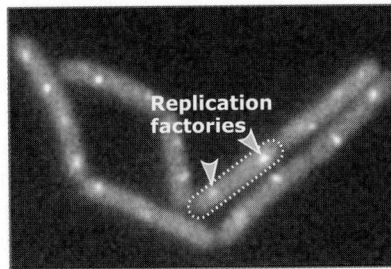

FIGURE 20.38 Fluorescence micrograph showing DNA replication factories in dividing cells. Cells express a fusion protein consisting of green fluorescent protein and the DnaE subunit of DNA polymerase holoenzyme. Under the growth conditions used, each cell has two replicating nucleoids, each with a central replication factory. Photo courtesy of Richard Daniel, University of Oxford.

Replication takes place in discrete sites dubbed "replication factories," where the subunits of the replication machinery accumulate, as **FIGURE 20.38** shows. Cell cycle analysis of replication factories in *B. subtilis* has shown that each round of replication takes place in a single factory. Thus, there is a correlation of approximately one focus of labeled replication protein for each ongoing round of DNA replication. This observation is contrary to the traditional view of replication forks setting off in opposite directions from the bidirectional replication origin *oriC*. Instead, it suggests that both clockwise and counterclockwise forks are localized close together. This localization opens up the possibility that the two forks might share a common pool of replication protein subunits and even the pool of nucleotide triphosphate precursors for replication.

In terms of the overall organization of replication in the cell, a model that takes into account this localization is shown in **FIGURE 20.39**. At the onset of chromosome replication, replication factory assembles at the origin, *oriC*. (The origin region defines where the replisome assembles. The replisome does not assemble prior to positioning of the origin). Newly replicated copies of *oriC* emerge on opposite sides of the replication complex, and both replication forks are localized together in the complex. As replication progresses, the unreplicated chromosome spools into the replication factory at midcell and newly replicated DNA is extruded toward opposite cell poles. The *oriC* regions are segregated toward opposite cell poles. The *terC*

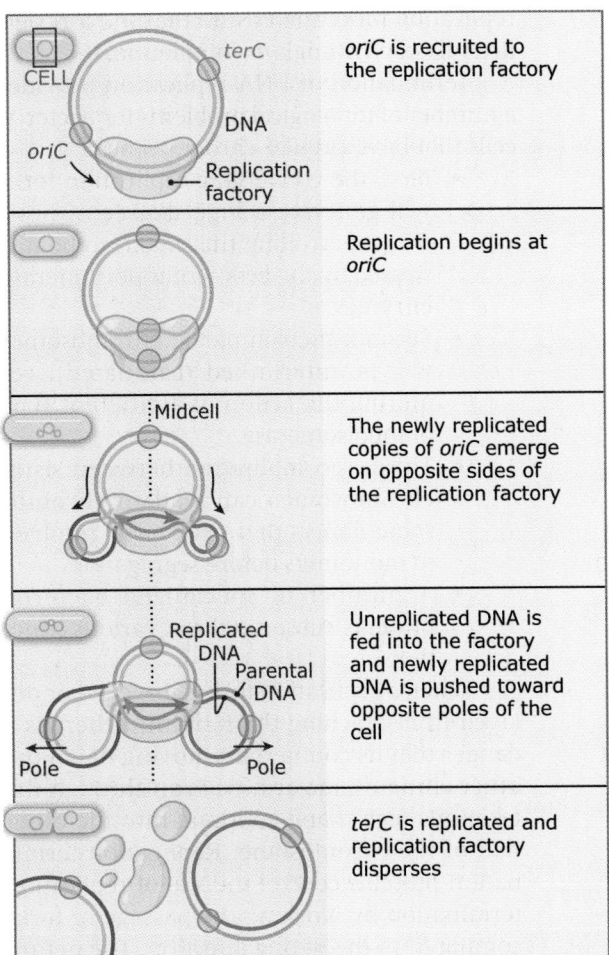

FIGURE 20.39 A model for replication of circular chromosomes in which the two replication forks are localized together at a replication factory. A blue circle represents one strand of DNA and a red circle represents its complement; dark red and dark blue lines represent newly replicated DNA.

region is located at midcell in the shrinking mass of unreplicated DNA. Eventually, the *terC* region is replicated, the round of chromosome replication is completed, and the replication factory disperses.

The replication factory model may help to account for the remarkable processivity of DNA replication *in vivo*. Replication forks rarely pass from origin to terminus without interruption. Both DNA damage and stalled RNA polymerases can result in the arrest of DNA replication. Bacterial cells have a range of mechanisms at their disposal that allow replication forks to be restarted, before or after repair of the damaged DNA strand. Retaining all of these proteins in a loose complex, together with the nascent replicated strands of DNA, could facilitate the reestablishment of collapsed replication forks and ensure that the forks efficiently run through to completion.

Termination of DNA replication provides a number of topological problems for bacterial cells that have circular chromosomes:

- First, the converging replication forks will generate a high degree of positive supercoiling that needs to be dissipated by the action of topoisomerase enzymes.
- Second, the completed chromosomes will be interlinked (catenated), requiring the action of a different type of topoisomerase.
- Third, recombination between sister chromosomes can lead to chromosome dimers that need to be resolved to monomers before segregation.
- A number of specialized bacterial proteins subserve these various functions.

When a replication fork is impeded or delayed from reaching the terminus, there is a danger that its counterpart moving down the other chromosome arm will run through the terminus region and continue into the other half of the chromosome. Replication termination proteins control the site of replication termination by blocking the passage of forks coming from the wrong direction. The net result is that the replication forks are forced to meet in a specific region of the chromosome, where the cell can manage the subsequent functions of dimer resolution and decatenation. Decatenation is probably achieved by the action of a specialized topoisomerase, Topo IV. Dimer resolution involves a specialized site-specific recombination system, comprising a heterodimeric catalytic resolvase (XerCD) and a target site, *dif*, located in the terminus domain. When a dimeric chromosome has been formed, the two *dif* sites are paired, and XerCD cuts and exchanges the two chromosome strands, generating two separate circles. Obviously, this system needs somehow to distinguish dimeric and monomeric chromosomes and act only on dimers (because recombination of monomers will generate a dimer!).

In *E. coli,* topological discrimination seems to be achieved by use of the FtsK protein, which plays several roles in the late stages of cell division. First, FtsK is needed for formation of the division septum (see *20.16 Prokaryotic cell division involves formation of a complex cytokinetic ring*). Second, FtsK has a DNA translocation function that can pump DNA out of the

division septum if it has been trapped by the closing septum. This DNA pumping activity may help to bring the two *dif* sites of a dimeric chromosome together, and it is likely that DNA transfer is directional in order to facilitate this association. Third, FtsK triggers the activity of the XerCD recombinase, if needed, to resolve the dimer and allow the sister, now monomeric, chromosomes to separate into their respective cells. Lastly, FtsK also interacts with Topoisomerase IV, which may allow it to coordinate all of the other processes with chromosome decatenation.

20.15 Prokaryotic chromosome segregation occurs in the absence of a mitotic spindle

Key concepts

- Prokaryotic cells have no mitotic spindle, but they segregate their chromosomes accurately.
- Measurements of oriC positions on the chromosome show that they are actively separated toward opposite poles of the cell early in the DNA replication cycle.
- The mechanisms of chromosome segregation are poorly understood probably because they are partially redundant.
- The ParA-ParB system is probably involved in chromosome segregation in many bacteria and low-copy-number plasmids.

Like all cells, prokaryotes need to ensure that before cell division their genetic information has been completely replicated and that pairs of replicated sister DNA molecules are accurately segregated, one to each daughter cell. Measurements of wild-type bacterial cells have shown that segregation is an extremely efficient process, with anucleate cells being generated at almost undetectable frequencies ($<10^{-4}$ per cell, per generation). Unlike eukaryotes, prokaryotes have no overt mitotic spindle to drag replicated chromosomes apart. Probably the smaller dimensions of a prokaryotic cell preclude the need for a spindle. However, even in eukaryotic cells, duplicated centromeres need to be spatially separated so as to achieve bipolar attachment to the spindle. It is possible that prokaryotic chromosome segregation could be analogous, or even homologous, to this poorly understood, early part of the eukaryotic segregation process. (For more on mitosis in eukaryotic cells see *14 Mitosis*.)

Early models of bacterial chromosome segregation involved attachment of newly replicated chromosomes via their *oriC* regions to the cell envelope on either side of a central zone of cell growth. Elongation of the cell would then result in sister *oriC* regions being slowly dragged apart in parallel with cell growth. However, it was subsequently shown that growth of the cell envelope does not occur in a zonal manner. Moreover, in some situations, chromosome segregation occurs over extreme distances, such as during polar division in sporulating cells of *B. subtilis* (see *20.17 Prokaryotes respond to stress with complex developmental changes*), or through an elongated tube, as in certain stalked bacteria.

One reason that progress in understanding bacterial chromosome segregation has been relatively slow is because it has proved to be difficult to isolate mutations that specifically affect chromosome segregation. Most of the candidate mutations that have been obtained turn out to have an indirect effect on chromosome segregation by interfering with either chromosome replication or with the global organization of the nucleoid. The SMC (MukB) system is a good example of this effect. Mutants in *mukB* were isolated on the basis of an elegant genetic screen for mutants that generate anucleate cells. However, it turns out that the primary function of MukB and associated proteins is probably in maintenance of nucleoid organization (chromosome condensation), though a direct role in segregation cannot be completely excluded (see Figure 20.35; see also *20.13 The bacterial nucleoid and cytoplasm are highly ordered*). Two approaches to the problem of chromosome segregation in bacteria that have proved to be fruitful in recent years are as given below:

1. Studies of the segregation of plasmids.
2. The application of microscopical methods to look directly at the localization and movement of specific sites in the chromosome during cell cycle progression.

There are varieties of mechanisms that allow plasmids to achieve stable maintenance. The replication of low-copy-number plasmids is such that the number of copies per cell is maintained close to that of the host chromosome (1–2 copies per cell). Plasmids have the same problems of decatenation and dimer resolution that the chromosome faces. In general, plasmids use host-encoded systems, such as the XerCD

recombinase, to overcome these problems (see *20.14 Bacterial chromosomes are replicated in specialized replication factories*). Many plasmids also encode an interesting "poison-antidote" system, which uses a different approach to solve the problem of unstable maintenance. With this system, daughter cells that fail to acquire a copy of the plasmid are killed. Ultimately, most low-copy-number plasmids have an active, dedicated segregation mechanism comprising two proteins with the general names of ParA and ParB. The ParA proteins have a weak ATPase activity, and they frequently have additional duties as transcriptional regulators. The ParB proteins are DNA-binding proteins that bind to specific *cis*-acting sites required for segregation. ParA can interact with ParB bound to its segregation site and is also required for segregation. Unfortunately, despite nearly two decades of work, the mechanism whereby ParA and ParB bring about stable segregation is unclear.

It turns out that most bacteria have homologs of ParA and ParB, which are required for chromosome segregation (though, interestingly, *E. coli* and close relatives do not!). In *B. subtilis*, the homologs Soj and SpoOJ, respectively, are involved in both chromosome segregation and sporulation. Although their chromosomes are frequently abnormally positioned, null mutants of *spoOJ* are viable, which reinforces the idea that chromosome segregation is a redundant process. The SpoOJ protein binds to a series of preferred sites in a large region of about 800 kbp in the vicinity of *oriC*. As **FIGURE 20.40** shows, the bound proteins are

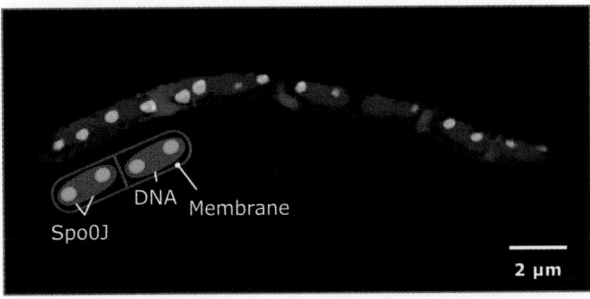

FIGURE 20.40 Visualization of SpoOJ proteins forming compact foci associated with the oriC regions of the chromosome in *Bacillus subtilis*. DNA is labeled in blue (stained with DAPI), membrane in red (stained with FM5–95), and SpoOJ in green (cells express a fusion protein consisting of SpoOJ and green fluorescent protein). The cartoon shows an interpretation of the organization of the cell, with duplicated oriC/SpoOJ foci arranged at the outer edges of each nucleoid. Photo courtesy of Alison Hunt, Royal Infirmary of Edinburgh.

condensed into compact "foci" that are visible by fluorescence microscopy, and this condensation requires Soj protein. As for the plasmid ParAB system, how Soj and Spo0J facilitate chromosome segregation is not yet known. However, observations of Spo0J foci revealed that *oriC* regions are actively segregated soon after a round of DNA replication. The foci rapidly take up positions at opposite ends of the replicating nucleoid. Similar conclusions have been drawn from observations of *oriC* regions visualized by other methods. However, at least one report suggests that movement of foci is gradual, which may imply passive segregation. At present there is no clear picture as to how the chromosomal ParAB proteins help to ensure faithful chromosome segregation, and indeed rapid initial segregation of *oriC* regions can take place in the absence of this system. One suggestion has been that extrusion of DNA from the fixed replisome could provide the driving force for segregation. However, this kind of mechanism would appear to be at odds with the notion of using sister chromosome repair as an important mechanism underpinning processivity of chromosome replication (see *20.14 Bacterial chromosomes are replicated in specialized replication factories*). In some species of bacteria, the MreB (actin) proteins may be involved in active segregation of chromosomes; this is an important area for future work.

Concept and Reasoning Check

1. How does chromosome segregation in prokaryotes differ from that in eukaryotes?

20.16 Prokaryotic cell division involves formation of a complex cytokinetic ring

Key concepts

- At the last stage of cell division, the cell envelope undergoes either constriction and scission, or septum synthesis followed by autolysis, to form two separate cells.
- A tubulin homolog, FtsZ, orchestrates the division process in bacteria, forming a ring structure at the division site.
- A set of about 8 other essential division proteins assemble at the division site with FtsZ.
- The cell division site is determined by two negative regulatory systems: nucleoid occlusion and the Min system.

Most prokaryotic cells divide precisely at mid-cell to produce two identical daughter cells. Division is closely coordinated with the completion of DNA replication and segregation. Division generally occurs after a period of growth in which the cell doubles in mass. After chromosome segregation, the cell undergoes **cytokinesis**, the process by which its contents are separated into two cells. Cytokinesis involves the annular (ringlike) ingrowth of all of the cell envelope layers. As **FIGURE 20.41** shows, cytokinesis can occur by at least two different pathways. In organisms such as Gram-negative *E. coli*, division occurs by concerted constriction of the existing cell envelope layers followed by scission. In other bacteria, such as Gram-positive *B. subtilis*, a newly synthesized annulus of wall material grows inward to form a septum. When septum formation is completed, the sister cells are physiologically separated by a pair of membranes but remain attached to each other. Cell separation is a distinct event that occurs by autolytic cleavage of the material in the septum. Depending on the growth conditions, dissolution of the septum may occur slowly enough that long chains of connected cells are produced.

A number of genes required for division were discovered by isolation and characterization of *fts* (filamentous temperature-sensitive) mutations. Cells with *fts* mutations grow as long nondividing filaments at the nonpermissive temperature. About 8 *fts* genes are present in most bacteria. A seminal discovery was the observation by Lutkenhaus that the FtsZ protein forms a ringlike structure just under the cell membrane at the site of impending division. The other division proteins are then recruited to this "Z ring" in the order illustrated in **FIGURE 20.42** for *E. coli*. The functions of most of these proteins are not known.

FtsZ, the key player in division, is a homolog of eukaryotic tubulin, the cytoskeletal protein used to make microtubules. Like tubulin, FtsZ is a GTPase and polymerizes in the presence of GTP to form linear protofilaments that make up a variety of bundles and sheets *in vitro*. The FtsZ ring appears to be a dynamic structure that is continuously remodeled *in vivo* (with a half-life of <10 sec!). This is reminiscent of the behavior of tubulin in eukaryotic cells (see *11 Microtubules*).

The FtsA protein interacts directly with FtsZ to assemble at the Z ring and presumably has a function in stabilizing the ring. FtsA is closely related to eukaryotic actin but possesses

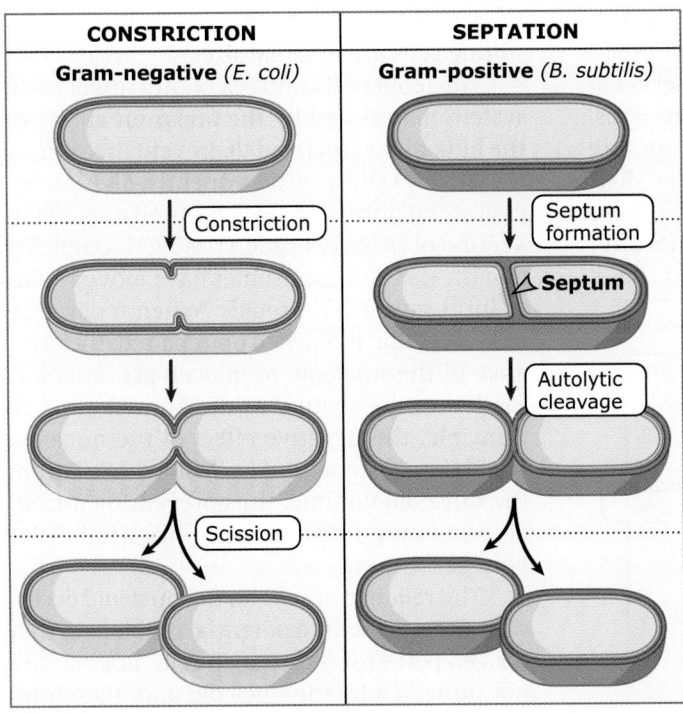

CONSTRICTION	SEPTATION
Gram-negative (*E. coli*)	**Gram-positive** (*B. subtilis*)

FIGURE 20.41 For prokaryotic cells, cytokinesis occurs by constriction or septation. For simplicity, the capsule layer of the cell envelope is not shown.

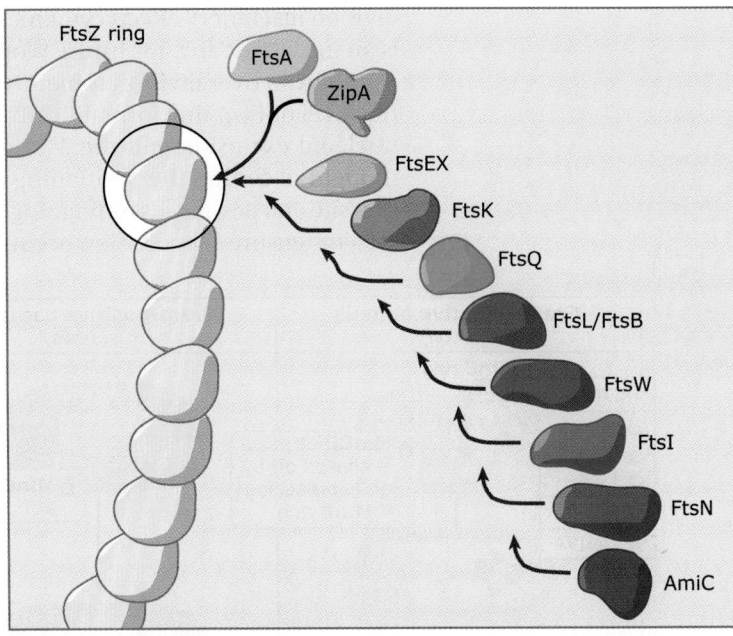

FIGURE 20.42 Cell division proteins and their assembly pathway in *E. coli*. FtsZ proteins form a ring to which the other proteins are recruited in the order shown, as determined using genetic techniques.

an additional domain of unknown function. FtsA forms dimers but does not appear to polymerize. Although FtsA is not essential for formation of the Z ring, a double mutant deficient in both FtsA and the ZipA protein no longer makes Z rings. Thus, FtsA and ZipA have partially redundant functions, and at least one of these proteins is needed to stabilize the Z ring. The ZipA protein has also been shown to interact directly with FtsZ, and it differs from both FtsZ and FtsA in being a transmembrane protein. ZipA may, therefore, serve to couple the Z ring to the cell membrane.

The remaining division proteins are all transmembrane proteins. FtsL and FtsQ do not have well-characterized functions. FtsW probably supplies precursors for FtsI, an enzyme involved in septal cell wall synthesis. FtsI is a specialized penicillin-binding protein that interacts with the general cell wall synthetic machinery to synthesize the new wall material during division (for more on cell wall synthesis see *20.6 The bacterial cell wall contains a cross-linked meshwork of peptidoglycan*). In *E. coli*, FtsK and FtsN are also required for cell division, but in *B. subtilis*, the FtsK homolog (SpoIIIE) is not required for division, and there is no FtsN homolog.

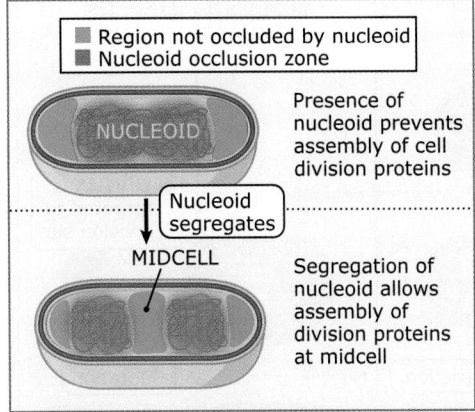

FIGURE 20.43 Negative spatial regulation of cell division by nucleoid occlusion.

There are some interesting differences between the assembly pathways for division proteins in the two well-characterized organisms, *E. coli* and *B. subtilis*. Thus, the pathway appears almost linear in *E. coli* (see Figure 20.42), whereas proteins are extensively interdependent for assembly at the Z ring in *B. subtilis*. These differences may reflect the distinctive cell envelope organization of these Gram-negative and Gram-positive organisms. So far, little is known about how the completely assembled division machinery effects cytokinesis, and this is an important area for future work.

Control over division is mainly exerted at the level of FtsZ ring formation. Two systems, nucleoid exclusion and the Min system, are thought to control the positioning of the division site and possibly its timing. Together, these systems ensure that division occurs only after DNA replication is complete and that the resulting cells are of equal size.

Nucleoid occlusion is a poorly understood system manifested by the apparent ability of the bulk of the nucleoid to prevent division, as **FIGURE 20.43** shows. Consequently, division occurs at the normal midcell position only when a round of DNA replication has been completed and the sister chromosomes have moved apart to form separate nucleoids. When replication or segregation is blocked or impaired, the presence of the nucleoid at midcell prevents formation of the septum at the normal time. In principle, the negative effect of the nucleoid could simply stem from exclusion of FtsZ from the nucleoid volume, thus preventing it from accumulating to its critical concentration for polymerization.

The reliance of cells on the nucleoid occlusion effect presents a potential problem in that the cell poles (of rod-shaped cells, at least) are not protected by the nucleoid and, therefore, are susceptible to aberrant polar division. To overcome this, many bacteria have proteins constituting the Min system, which serves to prevent division at the poles. The system derives its name from the minicells that are produced by *min* mutants when division occurs abnormally at the poles.

The key effector of the Min system is an inhibitor of cell division called MinC. This protein can inhibit formation of a Z ring, probably by directly reversing the polymerization of FtsZ. MinC activity is regulated by the MinD protein. MinD does this probably by controlling the localization of MinC in the cell, and there may be two distinct effects. First, MinD brings MinC to the periphery of the cell (near the cytoplasmic membrane), which is where FtsZ ring assembly takes place. Second, MinD restricts MinC action to the cell poles, so that it prevents polar division but allows midcell division.

Many rod-shaped bacteria use a MinCD system to control the site of cell division. The *E. coli* and *B. subtilis* MinCD systems are well characterized. Remarkably, the way in which MinD restricts MinC action to the cell poles occurs by completely different mechanisms in these organisms. *B. subtilis* uses a simple mechanism in which a polar anchor protein, DivIVA, attracts the MinCD complex to the cell poles and retains it there in a static manner during the cell cycle. As **FIGURE 20.44** shows, DivIVA and MinD are localized to the cell poles in a newborn cell, and the presence of the MinC inhibitor prevents formation of the FtsZ ring

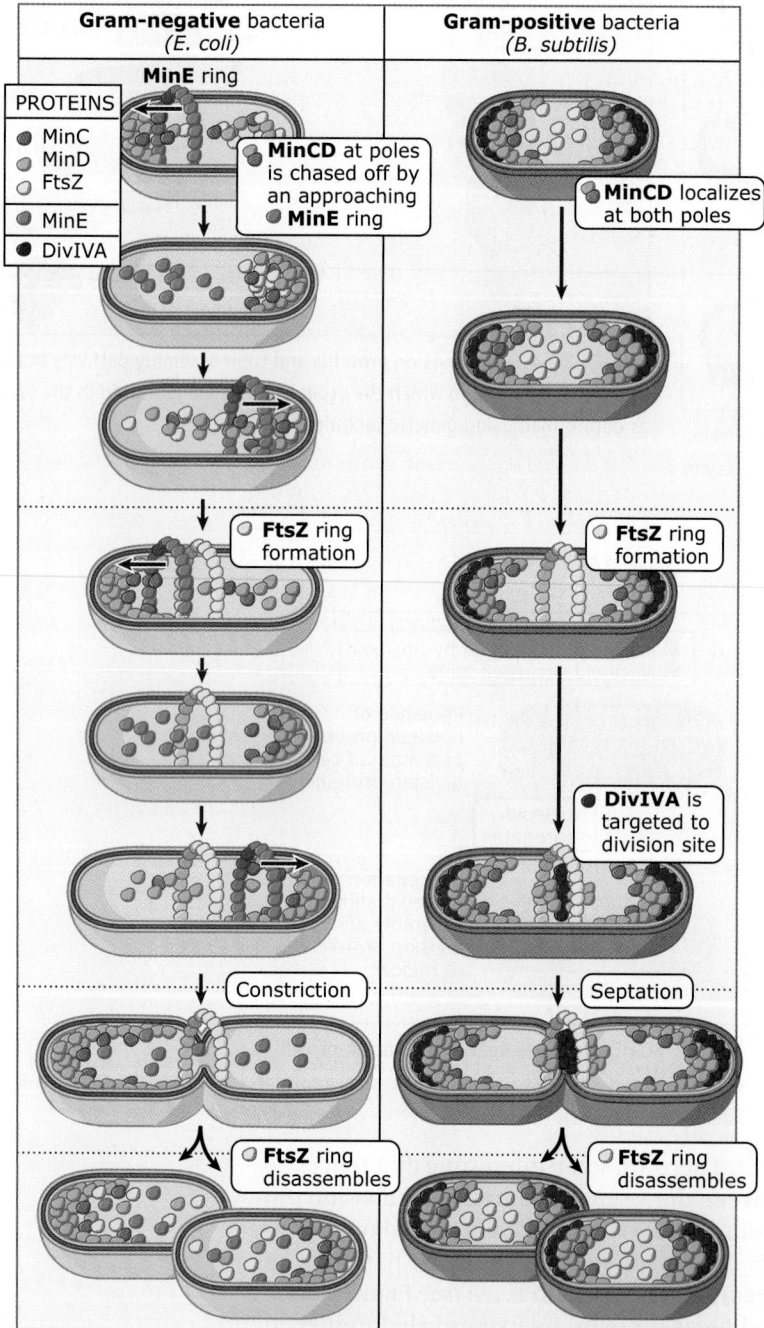

FIGURE 20.44 The localization of key proteins that determine the cell division site are shown at different stages of the cell cycle, beginning with a newborn cell and ending with cell division that produces two daughter cells. Cutaway views show half of each cell, but entire FtsZ and MinE rings are shown. For clarity, nucleoids are not shown.

at the cell poles. Presumably after completion of DNA replication, a new potential division site is created at midcell. The sequestration of the MinC inhibitor to the poles allows assembly of the FtsZ ring at midcell and recruitment of other cell division proteins. At this point, the division machinery presumably becomes resistant to the MinC inhibition. DivIVA and MinD proteins then are recruited to midcell. Therefore, when a new pair of cell poles is generated by division, DivIVA is built into the new poles, and it directly sets up a new zone of MinCD inhibition. When constriction is completed, the FtsZ ring disassembles, but DivIVA and MinD remain at the newly formed poles, preventing further divisions from taking place in these polar sites. Thus, targeting of DivIVA to the division site and then its retention at the cell pole are key features of this mechanism. Remarkably, the DivIVA protein localizes to division sites when it is expressed in a eukaryotic cell (fission yeast). This promiscuity suggests that DivIVA may recognize a topology, such as membrane curvature, rather than a specific protein target.

In contrast, *E. coli* has a dynamic MinCD system in which the complex assembles at one pole transiently. It then disassembles and reassembles at the opposite pole in an oscillating manner. The oscillation is driven by a ring of MinE protein that sweeps toward each pole in turn, driving MinCD off the pole and allowing it to reassemble at the opposite pole. The alternation of MinD localization from one pole to the other occurs at a frequency on the order of tens of seconds. As Figure 20.44 shows, MinD accumulates alternately at the membrane periphery on either side of a ring of MinE. The rapid relocation of MinD ensures that no FtsZ ring is assembled at either of the cell poles. The general presence of MinE around the midcell region precludes the MinD inhibitory activity at this site, allowing assembly of the FtsZ ring there. It is not yet obvious why *E. coli* has chosen to use such an energetically expensive mechanism to control MinCD and define the cell poles.

MinD belongs to an interesting family of proteins with a common nucleotide binding function that includes the ParA chromosome partitioning protein. The ParA-like Soj protein also shows dynamic movement. It seems likely that the common function of these proteins might be to use nucleotide binding and hydrolysis to control polymerization and depolymerization reactions, rather like control of the dynamic turnover instability of actin filaments and microtubules in eukaryotes (see *12 Actin* and *11 Microtubules*). These proteins might, therefore, comprise another general class of bacterial cytoskeletal proteins with a range of functions, particularly aspects of morphogenesis during the cell cycle.

A protein responsible for nucleoid occlusion in Gram-positive bacteria has been identified recently. Noc is a relatively non-sequence-specific DNA-binding protein that colocalizes with the bulk of the nucleoid. Noc is also an inhibitor of cell division. *noc* mutant cells grow normally unless chromosome replication is impaired. In this case, division occurs straight through the nucleoid in *noc⁻* cells, whereas in wild-type cells, cell division does not occur. Noc works together with the MinCD system to localize the FtsZ ring at midcell, as **FIGURE 20.45** shows. In wild-type cells, DivIVA nucleates polymerization of the MinD protein, which spreads from the cell pole toward midcell along the membrane. MinC associated with MinD prevents FtsZ accumulation or polymerization in the vicinity of the cell pole. It is proposed that the Noc protein associates with the nucleoid and inhibits FtsZ accumulation or activity in the vicinity of the nucleoid. In *noc⁻* cells, the Min system prevents FtsZ ring assembly except at midcell, and cells grow normally. In *min⁻* cells, however, Noc prevents FtsZ assembly only around the nucleoid, and

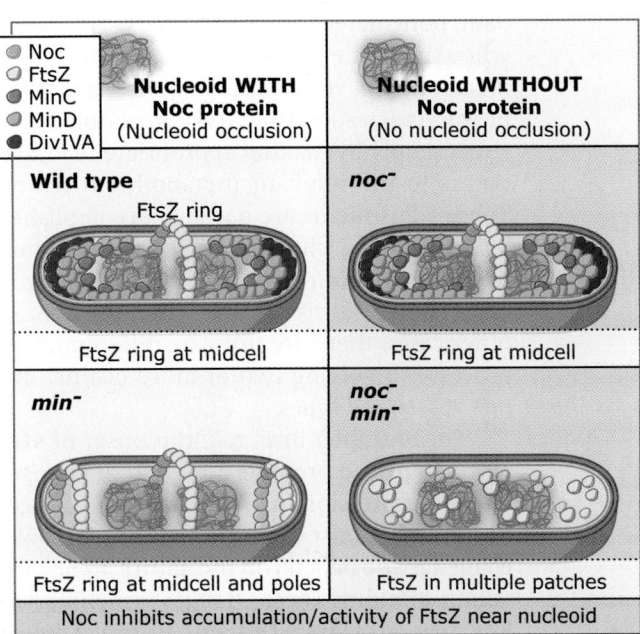

FIGURE 20.45 A model for topological control of the cell division machinery in the Gram-positive bacterium, *B. subtilis*.

FtsZ assembles at midcell and in the nucleoid-free spaces at the cell poles. In cells that lack both topological inhibitors (*min⁻noc⁻* double mutants), FtsZ assembly is unrestrained, and multiple patches form throughout the cells. None of these patches are productive, leading to loss of ability to divide. Noc is not present in Gram-negative bacteria, but a protein that governs nucleoid occlusion in a way similar to Noc has been found in *E. coli*.

20.17 Prokaryotes respond to stress with complex developmental changes

Key concepts

- Prokaryotes respond to stress, such as starvation, with a wide range of adaptive changes.
- The simplest adaptative responses to stress involve changes in gene expression and metabolism, and a general slowing of the cell cycle, preparing the cell for a period of starvation.
- In some cases, starvation induces formation of highly differentiated specialized cell types, such as the endospores of *B. subtilis*.
- During starvation, mycelial organisms such as actinomycetes have complex colony morphology and produce aerial hyphae, spores, and secondary metabolites.
- *Myxococcus xanthus* exemplifies multicellular cooperation and development of a bacterium.

Many prokaryotic populations undergo periods of rapid growth when nutrients are abundant, punctuated by periods of stasis or decline when the nutrients have been used up or toxic metabolites have accumulated. In some cases, prokaryotes respond to starvation or nutritional stress simply by making appropriate readjustments to the relevant metabolic processes. These adjustments are partially accomplished by widespread changes in gene expression to redirect resources into processes that are essential for survival rather than growth. In other cases, there are intricate differentiation processes involving two or more distinct differentiated cell types.

For example, in *E. coli*, the onset of **stationary phase** prompts a myriad of changes in gene expression. At one level, the changes occur as responses to a particular metabolic challenge and depend on the nature of the starvation stimulus. These changes sometimes allow a period of limited growth. However, when all possibilities for growth are exhausted, the cell turns to a more generalized response to

starvation, which is mainly orchestrated by a dedicated **sigma factor**, σ^S, which is an RNA polymerase subunit encoded by the *rpoS* gene. As summarized in **FIGURE 20.46**, σ^S is subject to a complex set of regulatory inputs that act at the transcriptional, translational, and post-translational levels. An increase in the amount of cellular σ^S protein can be obtained either by stimulating σ^S synthesis at the levels of rpoS transcription or rpoS mRNA translation or by inhibiting σ^S proteolysis. (Under nonstress conditions, proteolysis is extraordinarily rapid.) The most rapid and strongest reaction can be achieved by a combination of these processes, for example, upon hyperosmotic or pH shifts. This complex regulation allows a common response to be elicited by a variety of distinct physiological stresses.

A large number of genes are known to be under the control of σ^S. **FIGURE 20.47** lists some of the better known of these, with an indication of their likely function in adaptation to nongrowing conditions. The net result of the stationary phase response is a relatively dormant cell that has high levels of resistance to environmental insults that would rapidly kill growing cells.

In contrast to *E. coli*, the ultimate response of *B. subtilis* to starvation lies in **sporulation**,

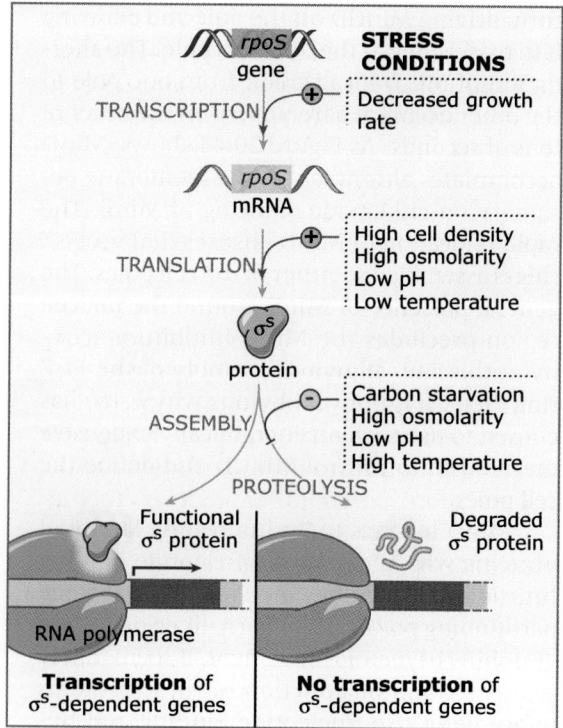

FIGURE 20.46 Various levels of σ^S regulation are differentially affected by various stress conditions.

the process in which a spore (a dormant cell) is formed, as illustrated in **FIGURE 20.48**. In the case of *B. subtilis,* the spore develops as an **endospore** within the bacterium. The onset of sporulation from **vegetative phase** growth is under complex control, with more than a score of regulatory genes interacting to measure many distinct intracellular and extracellular factors. The pivotal positive regulators of sporulation gene expression are the sigma factor sH and a response regulator, Spo0A. Spo0A is activated by phosphorylation, and the delivery of phosphate to Spo0A is a key control point, with at least two kinases, two intermediate phosphate carriers, and three or more phosphatases, all capable of intervening to promote or inhibit sporulation. In addition to positive regulation, a large number of transcriptional repressors have the ability to reduce the expression of key components of the machinery.

The information that this complex regulatory system monitors is of at least three kinds. First, the major signals seem to be nutritional, with the intracellular levels of GTP being one of the key indicators. A GTP-binding protein, CodY, is a key player in this regulatory system. Second, the cells can sense the population density through a quorum-sensing system, which monitors the external accumulation of a secreted peptide (for more on quorum-sensing see *20.21 Biofilms are highly organized communities of microbes*). Only when the population of cells is dense does the concentration of peptide reach a high enough level to allow sporulation. Third, cell cycle progression is critical for sporulation, so that initiation is allowed only at a particular point in the cycle. Otherwise, another round of medial division takes place. Additionally, various "checkpoint" mechanisms appear to prevent sporulation if chromosome replication or segregation is impaired.

The decision to initiate sporulation leads to a modified asymmetric cell division (see Figure 20.48). The switch to asymmetric division is apparently driven by an increase in FtsZ concentration, together with synthesis of a sporulation-specific protein SpoIIE, though precisely how these factors bring about the positional switch is not understood. This is a critical point in the developmental process because it generates the two distinct cell types that will then cooperate in the formation of the spore. The small prespore cell is destined to become the mature endospore. The much larger mother cell devotes all of its resources to maturation of the spore. This cooperative

GENE(S)	PROTEIN FUNCTION (KNOWN OR LIKELY)
csgA, csgB	Components of extracellular fibrous structures (curli) involved in adhesion and biofilm formation
dacC	Carboxypeptidase; catalyzes formation of crosslinks that stabilize peptidoglycan
dps	DNA-binding protein involved in regulation of gene expression and protection against DNA damage
emrA, emrB	Multiple drug resistance pump (membrane protein that transports some drugs out of cells)
ftsA, ftsQ, ftsZ	Cell division proteins; allow stationary phase cells to divide after growth arrest so that ultimately unigenomic, minimal-sized cells are formed
glgS	Involved in synthesis of glycogen (storage form of glucose)
katG	Catalase-peroxidase; catalyzes the breakdown of H_2O_2
otsA, otsB	Synthesis of trehalose, which is involved in osmoprotection and stationary-phase thermotolerance
proU	Membrane protein that transports proline- and glycine-betaine into cells, conferring osmoprotection

FIGURE 20.47 Some σ^S-dependent genes in *Escherichia coli*.

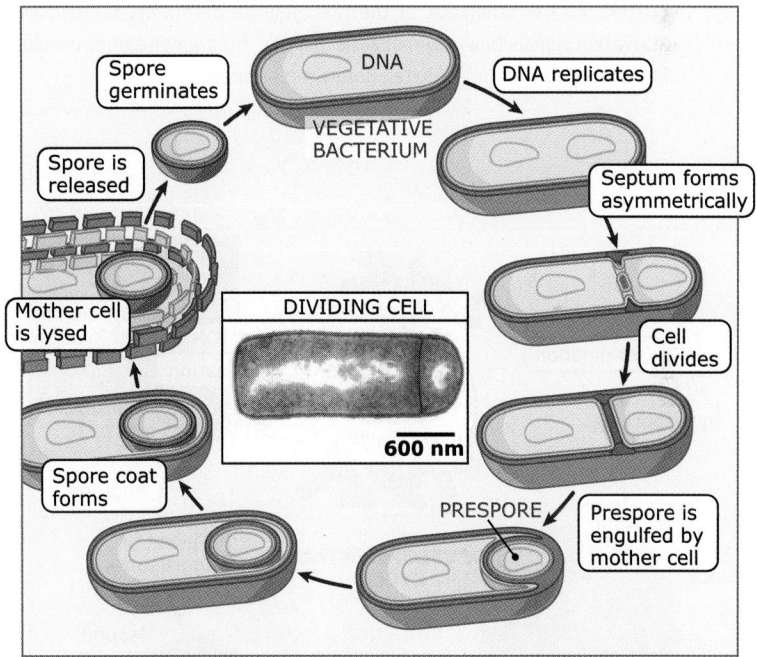

FIGURE 20.48 *B. subtilis* endospore formation involves formation of a prespore that is engulfed by the mother cell. Upon formation of the spore coat, the mother cell is lysed and the mature spore is released. The electron micrograph shows a *B. subtilis* cell dividing asymmetrically during sporulation. Photo reproduced from J. Errington, et al., *Phil. Trans. R. Soc. B.*, vol. 360, figure 2, © 2005, The Royal Society. Photo courtesy of Jeffrey Errington, Newcastle University.

approach provides the secret to the success of bacteria that form endospores because the endospore is much tougher and more resistant than the specialized cells made by an "everyperson-for-himself" approach. The resistance is generated by a combination of factors, including dehydration and mineralization of the

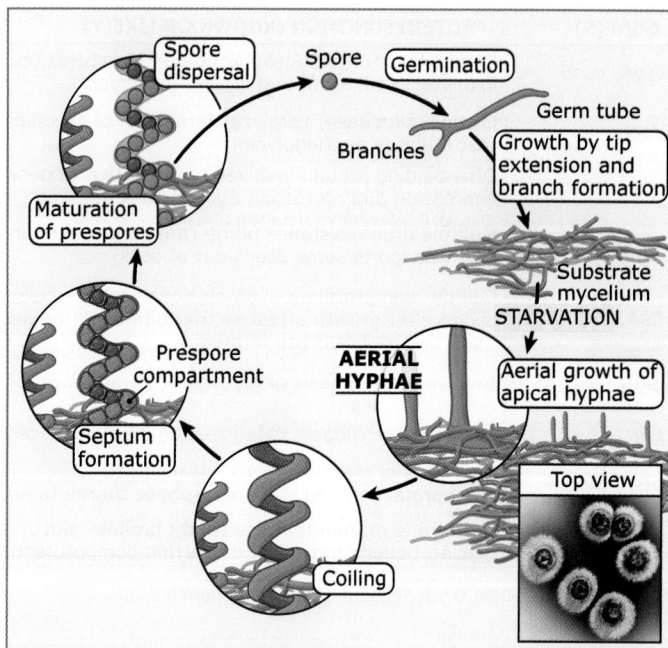

FIGURE 20.49 Schematic of the life cycle of *Streptomyces coelicolor*. Starvation signals the growth of aerial hyphae, from which spores develop. Photo courtesy of Keith Chater, John Innes Centre.

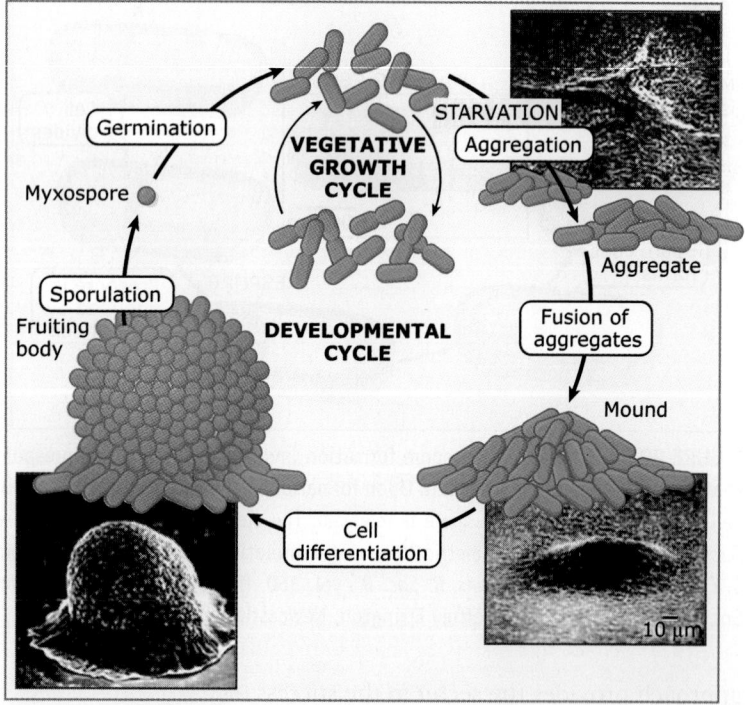

FIGURE 20.50 Photographs show *Myxococcus xanthus*, starvation signals cell aggregation and formation of a fruiting body, in which spores develop. Photos show cells at 7, 12, and 72 hours after starvation. The 10 mm scale bar applies to all three photos. Photos reproduced from *J. Bacteriol.*, 1982, vol. 151, pp. 458–461, DOI and reproduced with permission from American Society for Microbiology. Photo courtesy of Dale Kaiser, Stanford University.

spore's core and formation of protective outer layers called the cortex (a modified form of cell wall) and of a proteinaceous coat layer. The complex programs of gene expression that occur in the *B. subtilis* prespore and mother cell are now well understood.

Streptomyces coelicolor, another Gram-positive bacterium, has a yet more complex starvation response. This organism is an actinomycete and has a filamentous, branching growth pattern rather similar to that of filamentous fungi. In contrast to the case of free-living bacteria such as *E. coli* and *B. subtilis*, starvation of *S. coelicolor* is almost an inevitable consequence of growth into a meshwork (mycelium) of filaments called hyphae. As **FIGURE 20.49** shows, colony development begins with formation of a substrate mycelium that spreads over the surface of the growth medium (soil). Hyphae within the body of the colony face starvation at the same time that hyphae at the periphery are thriving. The starvation response involves elaboration of specialized aerial hyphae, which grow vertically up from the surface of the colony. Although the hyphae of the substrate mycelium have very few division septa, cell division is dramatically upregulated in the aerial hyphae to generate masses of uninucleate cells that differentiate into resistant spores. Again, a complex set of regulators (even more complex than those of *B. subtilis*) are involved in the decision to initiate development of aerial hyphae and spores. The actinomycete life cycle is important from an industrial perspective because it is at about the transition from vegetative mycelial growth to starvation that these organisms make a number of important secondary metabolites. In particular, they make a wide range of antibiotics, which are presumably important in suppressing competing bacteria in the soil.

One of the most dramatic bacterial developmental systems is exemplified by the myxobacteria. The best-studied example of this group of organisms is *M. xanthus*. As **FIGURE 20.50** and **FIGURE 20.51** show, impending starvation triggers the population of cells to migrate together to form focal aggregates. Large numbers of cells are attracted into these aggregates, which pile up to form a large mound of cells. A mound becomes a fruiting body when the rod-shaped cells round off and differentiate into myxospores, which are dormant and resistant to desiccation. Some myxobacteria, such as *Stigmatella*, form complex fruiting

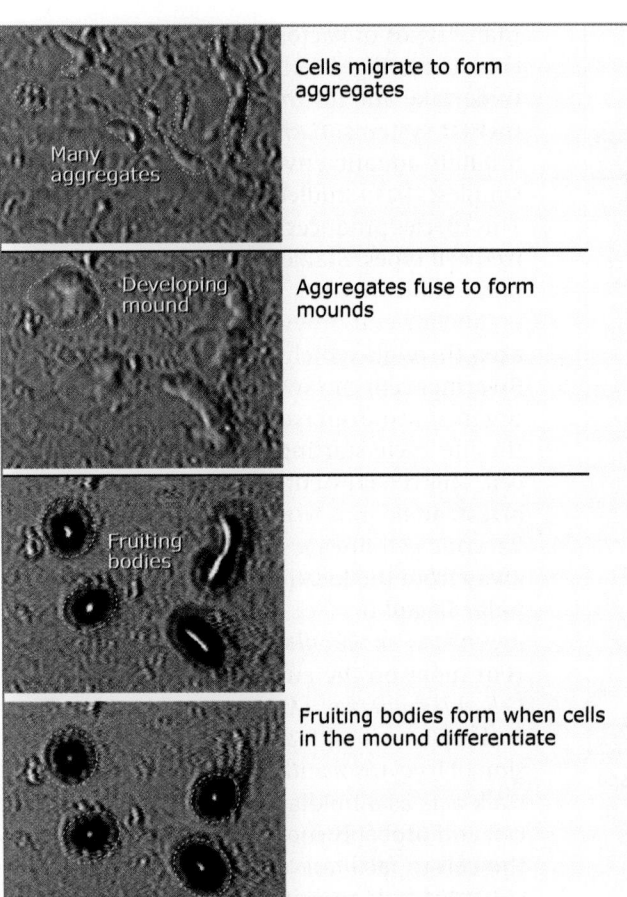

Cells migrate to form aggregates

Many aggregates

Developing mound

Aggregates fuse to form mounds

Fruiting bodies

Fruiting bodies form when cells in the mound differentiate

FIGURE 20.51 Frames from a video showing formation of *Myxococcus xanthus* fruiting bodies as viewed from above the culture dish. In starvation medium, cells form aggregates, which fuse and the cells inside differentiate to form spores. Movie by Roy Welch [http://cmgm.stanford.edu/devbio/kaiserlab/movies/develpment.mov].

bodies with multiple sporangioles borne on branched stalks.

As for *B. subtilis*, a large number of genes involved in the development of *M. xanthus*. have been identified. Some of the most interesting developmental problems that can be studied in this organism lie in the intercellular signaling processes. At various steps in development, intercellular signals are exchanged so as to facilitate aggregation and morphogenesis of the fruiting body. Genetic analysis of mutants affected in intercellular signaling identified signaling systems, of which the ones designated "A" and "C" are now quite well understood. The A signal comes into play early on and corresponds to a complex set of amino acids, generated by one or more cell-bound proteases. It is probably used to estimate the general nutritional status of the population of cells. Some of the molecular mechanisms controlling signal production and sensing are summarized in **FIGURE 20.52**. The C signal comes into play later and, as illustrated in **FIGURE 20.53**, is a quite different signal requiring intimate contact between the signaling cells. C factor is a 25 kDa cell surface-associated protein with similarity to a family of short-chain alcohol

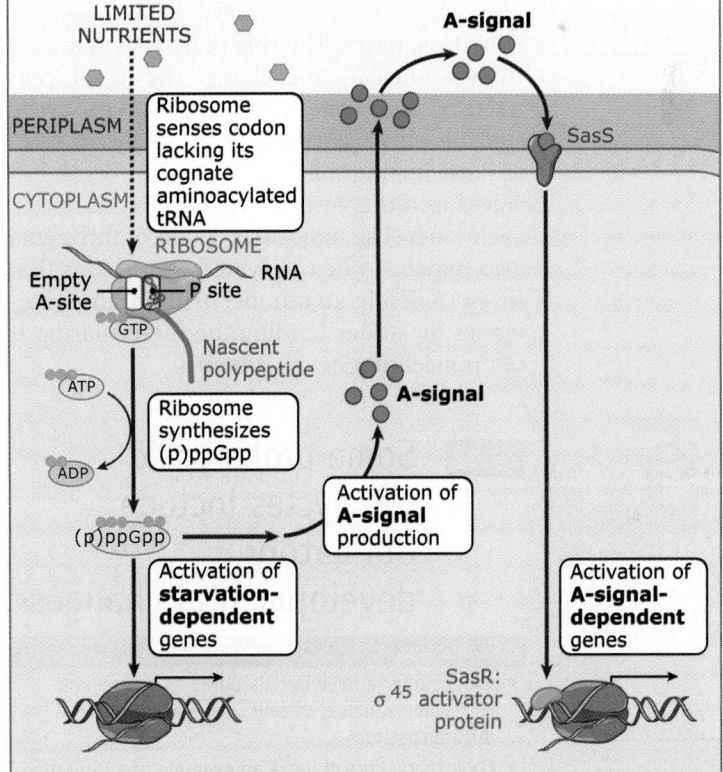

FIGURE 20.52 Overview of A-signaling during aggregation and fruiting body formation of *Myxococcus xanthus*.

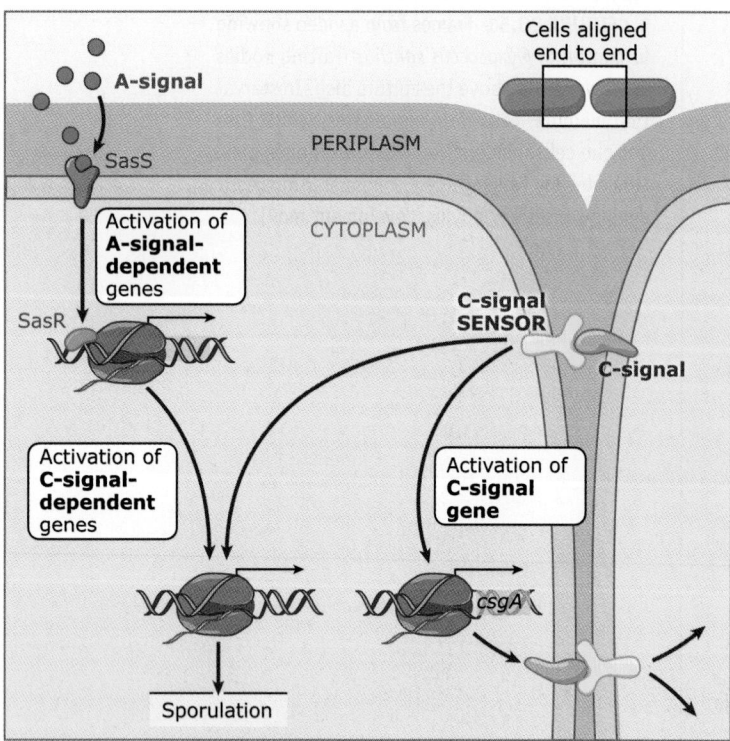

FIGURE 20.53 Overview of the *Myxococcus xanthus* C-signaling system for sporulation. Both cells respond to the signals for sporulation, but for clarity, the signals are shown in only one cell.

dehydrogenases. The role of its enzymatic activity is unclear. To exchange the signal, cells need to be motile, apparently because the signal is only passed between the tips of cells, so participating cells need to maneuver themselves into the correct orientation relative to each other. The molecular details of this signal also remain to be resolved, but it seems that this system offers a number of interesting challenges for understanding the molecular basis of multicellular development.

20.18 Some prokaryotic life cycles include obligatory developmental changes

Key concepts

- Many bacteria have been studied as simple and tractable examples of cellular development and differentiation.
- *Caulobacter crescentus* is an example of an organism that produces specialized cell types at every cell division.

Many types of bacteria have been studied for the interesting developmental processes they undertake and for their tractability as experimental systems. *C. crescentus*, which normally inhabits aquatic environments, is one of the simplest, best-studied, and tractable systems. This species produces two quite distinct specialized cell types. Stalked cells have a specialized polar stalk that has two functions: it acts as an anchor and provides an enhanced surface area through which nutrients are absorbed. Swarmer cells can swim and thereby allow the organism to disperse. **FIGURE 20.54** illustrates the life cycle starting with a motile swarmer cell. This cell is essentially in a state of cell cycle arrest, analogous to the G_0 state of certain eukaryotic cell lineages, during which it swims away from the location of its birth via a single polar flagellum (see *20.11 Pili and flagella are appendages on the cell surface of most prokaryotes*). (For more on the eukaryotic cell cycle see *15 Cell cycle regulation*.) It then sheds its flagellum and replaces it with a stalk, which is an extension of its cytosol and cell envelope layers. The stalk acts as a holdfast for the now sedentary cell and probably increases the surface area of the cell to facilitate uptake of nutrients. The swarmer-stalk transition also signals the onset of the cell division cycle, with initiation of a round of DNA replication. As the cell divides, the newer pole of the cell (i.e., the pole with no stalk) begins to differentiate, most strikingly by the targeted assembly of a new flagellar apparatus. Cell division gives rise to a swarmer cell and a stalked cell. The swarmer cell has an active polar flagellum and can swim off to continue the dispersal process. The stalked cell differs from the swarmer in being immediately ready to reenter the cell cycle by reinitiation of DNA replication.

In *C. crescentus*, differentiation, and cell cycle progression are intimately linked. For example, treatment with agents that block DNA replication prevent formation of the flagellum. Cell cycle progression is governed by a number of regulators, of which the key player is a response regulator protein, CtrA. Like Spo0A of *B. subtilis*, CtrA is regulated by phosphorylation, and various kinases and phosphoproteins play important roles in the regulation of *C. crescentus* development. Protein turnover is also important, with a number of key players in the cell cycle being degraded immediately after the period in which their products act. CtrA works by positively or negatively regulating the transcription of many genes. The changes in

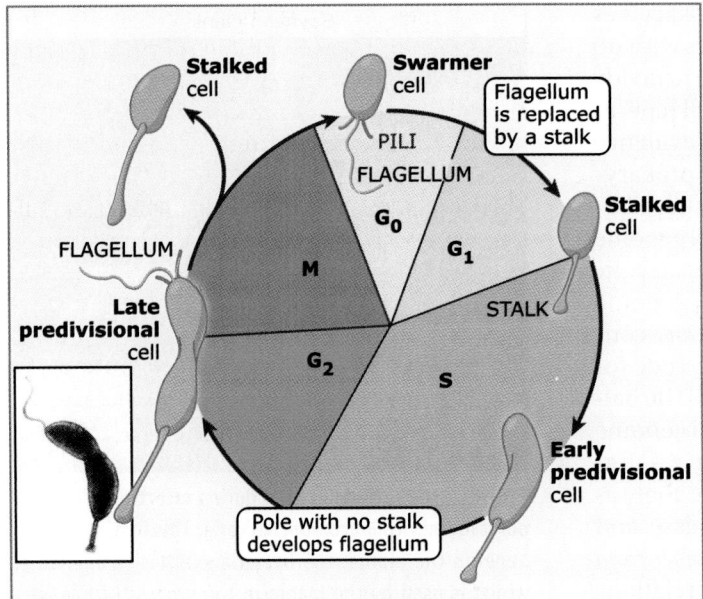

transcription during cell cycle progression and the effects of *ctrA* mutation have been cataloged in detail using microarray-based experiments. CtrA also acts directly to control the initiation of DNA replication, by binding to sites in the *oriC* region.

An important experimental advantage of *C. crescentus* as a model organism lies in the ability to obtain synchronous populations of cells, by isolating swarmer cells. This makes it possible to follow changes in protein accumulation and gene expression during cell cycle progression. These changes have now been comprehensively mapped and cataloged, which will greatly facilitate reaching a detailed understanding of the cell cycle of this organism and its developmental process.

20.19 Some prokaryotes and eukaryotes have endosymbiotic relationships

Key concepts

- Mitochondria and chloroplasts arose by the integration of free-living prokaryotes into the cytoplasm of eukaryotic cells where they became permanent symbiotic residents.
- Rhizobia species form nodules on legumes so that elemental nitrogen can be converted into the biologically active form of ammonia.
- The development and survival of pea aphids depends on an endosymbiotic event with *Buchnera* bacteria.

Endosymbiosis describes the relationship between two symbiotic organisms in which one of the organisms lives within the other. Usually, these interactions benefit one or both of the organisms. Eukaryotic evolution has been steered in many ways by endosymbiotic events with prokaryotes. Most notable are the endosymbiotic relationships that led to mitochondria and chloroplast acquisition in the eukaryotic cell. Mitochondria and chloroplasts serve as the sites of oxidative phosphorylation, a mechanism of ATP formation. Without mitochondria, the eukaryotic cell would be dependent on anaerobic glycolysis, which in comparison to aerobic respiration is terribly inefficient. There are 15 times as many ATP molecules produced by aerobic respiration compared to glycolysis. Not surprisingly, high-energy animal tissues such as the heart and skeletal muscle require a large number of mitochondria. In plant cells, chloroplasts convert light energy into ATP in a process called photosynthesis.

Lynn Margulis has provided an elegant and convincing argument that mitochondria and chloroplasts are remnants of microbes that were once free-living but at some point began residing symbiotically within eukaryotic cells. The evidence supporting the endosymbiotic theory comes from many fronts. These organelles can arise only from preexisting mitochondria or chloroplasts, because nuclear genes encode only a fraction of the proteins of which they are composed—the rest are encoded by genetic material within the organelle itself. Mitochondria and chloroplasts have their own genomes, which resemble that of prokaryotes in being single circular DNA molecules with no associated histones, and they are able to divide and replicate independently of host cell replication. The most convincing molecular evidence suggesting that mitochondria have a prokaryotic origin came from the discovery that ribosomal RNA genes in the mitochondrial genome are clearly of prokaryotic and not eukaryotic origin. The sequenced mitochondrial genome of the protozoan *Reclinomonas americana* contains ~70 genes, almost half of which code for translational machinery components. The balance of the *R. americana* mitochondrial genome codes for enzymes that are involved in respiratory energy production. Phylogenetic analysis of ribosomal proteins, cytochrome oxidase, and NADH dehydrogenases from the *R. americana* mitochondria genome reveals a close relationship between mitochondria and members of the α-proteobacteria (see Figure 20.4). Most members of the α-proteobacteria group live in close association with eukaryotic cells, either as symbionts or as parasites of plants and animals. The nearly identical energy systems present in α-proteobacteria (such as *Rickettsia prowazekii*) and aerobic mitochondria reflect a common ancestry, suggesting that the prokaryotic ancestor for mitochondria was a facultative aerobe in the α-proteobacteria family. However, the prokaryotic ancestor to mitochondria would have had a genome quite different from that of contemporary mitochondria because mitochondrial evolution has been volatile, with many genes being transferred to the nuclear genome or lost outright.

Not all endosymbiotic events are captured permanently by evolution as for mitochondria and chloroplasts. Some endosymbiotic relationships are formed on a "need" basis. Such is the case for nitrogen-fixing prokaryotes and their plant hosts. Nitrogen fixation (the process of converting atmospheric nitrogen to ammonia) can only be carried out by certain prokaryotes, yet all living organisms require ammonia to survive. Members of the prokaryotic genera *Rhizobium* are able to infect legume root hairs, forming collections of microbes called nodules, as **FIGURE 20.55** shows. Secreted signal molecules (called Nod factors) produced by *Rhizobium* initiate root infection and nodule formation. Nod factors are expressed when the *Rhizobium* detects plant flavonoids, which are specific to individual plant species. Therefore, nodule formation only occurs if the prokaryote can sense the particular plant flavonoid

Soybean plant

Nodules

FIGURE 20.55 Nodules form during infection of certain plant roots with *Rhizobium* bacteria. This endosymbiosis benefits the plants: the bacteria synthesize ammonia, which is used by the plants in the process of nitrogen fixation. Photo courtesy of Harold Evans.

and respond to it by making Nod factors. In turn, the Nod factors will only initiate endosymbiosis with the plant species that secreted the flavonoid. Nod factors are oligosaccharide molecules. The search is on to identify Nod factor receptors in plant root cells. Candidate Nod receptors have been identified by binding studies using cellular extracts. However, the proteins identified by this approach bind to Nod factors nonspecifically and are, therefore, not likely Nod receptors *in vivo*. Another approach to finding Nod receptors has been to first identify plant sugar-binding proteins, then to test their affinity and binding specificity to Nod factors. This approach has been used to identify a root-localized protein called lectin-nucleotide phosphohydrolase (LNP), which specifically binds to Nod factors. Antibodies against LNP block nodulation, suggesting that LNP is an important determinant of nodulation. Attempts to identify other Nod receptors are currently underway.

Many insects rely on endosymbionts for their development and survival. The best-studied insect-prokaryote endosymbiotic relationship is the one between the pea aphid and bacteria of the species *Buchnera*. *Buchnera* is a Proteobacterium and an obligate endosymbiont of its aphid host. *Buchnera* is maintained in bacteriocytes, which are specialized cells in the aphid body cavity, as **FIGURE 20.56** shows. Maternal transfer of *Buchnera* to eggs and embryos ensures that the bacterium colonizes every pea aphid. Why does the pea aphid

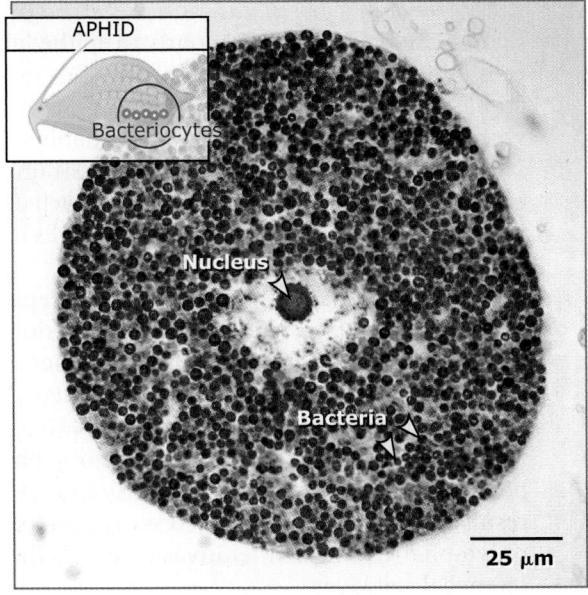

APHID

Bacteriocytes

Nucleus

Bacteria

25 μm

FIGURE 20.56 Bacteriocytes are cells in the pea aphid (an insect) that harbor the *Buchnera* bacteria. The photograph shows a bacteriocyte stained to show its nucleus and the bacteria that fill its cytoplasm. Photo courtesy of Takema Fukatsu, National Institute of Advanced Industrial Science and Technology (AIST), Japan.

tolerate rampant colonization by *Buchnera*? It turns out that prokaryotic invaders supplement the aphid's diet by providing several necessary nutrients. The diet of aphids is primarily plant sap, which is rich in carbohydrates but poor in amino acids. Experimentally, it has been difficult to demonstrate exactly what nutrients *Buchnera* provides for the host, but *Buchnera's* genome contains genes encoding the enzymes to produce amino acids that are essential for the host. Furthermore, *Buchnera*-infected aphids can survive on synthetic diets that lack essential amino acids. However, if aphids are treated with antibiotics that eliminate *Buchnera*, the aphids quickly become starved and die. Clearly, this implies that *Buchnera* provides essential amino acids for the host. The caveat is that *Buchnera*-free aphids grown on rich diets that include all the essential amino acids still develop poorly, suggesting that *Buchnera* endosymbiosis is providing the host with more than just essential amino acids.

20.20 Prokaryotes can colonize and cause disease in higher organisms

Key concepts

- Although many microbes make their homes in or on the human body, only a small fraction cause harm to us.
- Pathogens are often able to colonize, replicate, and survive within host tissues.
- Many pathogens produce toxic substances to facilitate host cell damage.

TYPE OF SYMBIOSIS	Characteristics	Damage to host	Examples
MUTUALISTIC	Mutual benefits	None	Bacteria that digest cellulose in the cow gut
COMMENSAL	One organism benefits; no effect on other organism	None	Bacteria in the human gut that live on nutrients ingested by host
PARASITIC	Microbe depends on host for survival and may damage the host	Some	
PATHOGENIC	Microbe causes damage to the host	Major	Bacteria such as *Bacillus anthracis* (anthrax) and *Yersinia pestis* (plague)

FIGURE 20.57 There are several types of symbiotic relationships. Depending on the interaction, the impact on the host ranges from beneficial to major damage.

Animal bodies are rich in nutrients and they provide relatively stable conditions of pH, osmotic pressure, and temperature. These features provide an optimal environment for the growth of a variety of prokaryotes. A close biological relationship between two different organisms, such as that between a microbe and its host, is called a symbiotic relationship. Depending on the degree of benefit or damage, **symbiosis** between host and microbe can be classified as **mutualistic**, **commensal**, **parasitic**, or **pathogenic**, as **FIGURE 20.57** shows. Endosymbiosis is discussed in the previous section (see *20.19 Some prokaryotes and eukaryotes have endosymbiotic relationships*). The vast majority of prokaryotes in symbiotic relationships are commensal residents that have little or no effect on the host. Some commensal dwellers provide helpful, and in some cases, essential services for the host. For example, enteric *E. coli* strains aid in food digestion.

Microorganisms that grow on or in another organism while providing no benefits to the host organism are loosely defined as parasites. A small number of these parasitic relationships prove harmful for the host. Microbes that cause harm to their hosts are called pathogens, and their ability to colonize and cause disease is the result of many factors present in both the host and the microbes. Pathogens can be classified as opportunistic or primary. Opportunistic pathogens cause disease only when the host is compromised in some way, such as severe burn victims, and AIDS and some cancer patients. Primary pathogens are able to cause disease in normal individuals, and sometimes their replication is completely host dependent. Regardless of the type of pathogen, to cause an infection, the microbe must enter the host, colonize, avoid the immune system's countermeasures, and replicate. In addition, primary pathogens must position themselves for transmission to another host.

Before a pathogen can cause damage to the host, it must gain access to host tissues and multiply. The primary sites of infection are usually those that are exposed to the external environment, such as the skin or mucous membranes in the respiratory, genitourinary, and intestinal epithelium. Microbes express several different surface molecules that bind to receptors present on host tissues. These adhesive molecules are composed of either polysaccharide or protein. For example, *Streptococcus mutans* adheres via its polysaccharide-rich capsule to tooth surfaces, where it causes tooth decay. Surface protein structures, such as pili or fimbriae, are used by a wide variety of microbes to attach to host tissues. The presence of these adhesive organelles often defines a pathogen's **virulence** capabilities, and without them the pathogens are unable to establish an infection and are usually swept away by physical forces. (For more on capsules, see *20.5 Most prokaryotes produce a polysaccharide-rich layer called the capsule*; for more on pili see *20.11 Pili and flagella are appendages on the cell surface of most prokaryotes*.)

Microbes can adhere with great tissue and species specificity. This specificity for certain tissues is generally referred to as **tropism**. For example, the causative agent of gonor-rhea, *Neisseria gonorrhoeae*, adheres strongly to urogenital epithelia and poorly to epithelial cells derived from other bodily tissues. An example of species specificity in binding occurs with pyelonephritic strains of *E. coli* that bind kidney epithelial cells. Pyelonephritic strains express one of three varieties of P-pili, each of which are specific for kidney epithelial cells in humans, dogs, or rats.

Some bacterial pathogens invade the epithelium to initiate pathogenicity. Invasion awards the microbe access to cellular nutrients that support their replication. Invasion across the epithelium to blood vessels can allow bacterial growth at sites distant from the original point of entry. Systemic infections are often the result of pathogens that gain access to the blood or lymphatic systems after invasion across the epithelial cell layer.

Since the initial inoculum is rarely high enough to cause damage, a pathogen must multiply within the host to cause disease. Host tissues are a potentially suitable place to support bacterial growth. However, several key nutrients are in short supply in host tissues, and pathogens must be able to best utilize the resources that are available, while resisting host immune defenses. Organisms that are able to use complex nutrient sources such as glycogen are at an advantage. Pathogens can compete with host cells for trace elements like iron. Animals have two proteins, transferrin and lactoferrin, that bind and transfer iron into host cells, so that very little free iron exists in host tissues. To combat this, bacterial pathogens often have efficient iron-chelating complexes called siderophores that help them gather iron from the environment.

Bacterial pathogens produce a variety of **virulence factors** that help to establish and maintain an infection. Gram-positive pathogens, including *Streptococci, Staphylococci*, and *Pneumococci*, produce an enzyme that breaks down host polysaccharides, which allows dissemination through tissues. *Clostridium* makes a collagenase that depolymerizes the host collagen networks that hold tissues together, allowing this organism to spread through the body. *Streptococcus pyogenes* promotes its dissemination through tissues by producing streptokinase, an enzyme that weakens host fibrin networks, which the host assembles to prevent pathogen spreading. Alternatively, some bacterial pathogens localize themselves (to enclose themselves in a protective case) by producing enzymes that promote fibrin clotting. One well-studied fibrin-clotting enzyme, produced by *Staphylococcus aureus*, is coagulase. Coagulase-producing organisms become coated in fibrin, and it is thought that this coating protects the microbe from host defense mechanisms.

In addition to bacterial colonization and replication in host tissues, bacterial toxins cause damage to host cells. Secreted toxins, called exotoxins, are able to cause tissue damage in distant parts of the body. Exotoxins can be categorized into three classes based on their mechanism of action. Cytolytic exotoxins, such as hemolysin, act on the host cytoplasmic membrane, resulting in cell lysis. Hemolytic strains can be easily identified when grown on blood agar plates. When hemolysin is released, it lyses the red blood cells, resulting in a zone of clearance that appears as a white area on the otherwise red plate. A second type of exotoxin is exemplified by that produced by *Corynebacterium diphtheriae*, the causative agent of diphtheria. When *Corynebacterium* detects low iron levels (indicative of the host environment), it secretes diphtheria toxin, which translocates into the host cell and poisons translation. A third class of toxins are the neurotoxins, which includes botulinum produced by *Clostridium botulinum*. Botulinum toxin is one of the most toxic substances known to man. Botulinum toxin blocks acetylcholine release from neuronal cells, causing irreversible muscle relaxation and paralysis.

Successful pathogens each use a unique repertoire of virulence determinants in order to cause disease. Pathogens that produce powerful toxins do not have to invade and replicate to high numbers in order to cause disease. For example, *Clostridium tetani* is non-invasive, yet because of the toxin it produces, *C. tetani* infections are often fatal. On the other hand, *Streptococcus pneumoniae* does not produce a toxin; instead, its copious replication in lung tissue causes disease. The large numbers of invaders in the lung tissue elicit a host immune response and lead to pneumonia. Understanding the mechanisms of prokaryotic virulence will allow the development of new therapies for combating infections.

In response to pathogens, the host mounts immune defenses of two general types, innate and adaptive. Innate defenses are "constitutive," meaning they are always present. Innate defenses offer general protection against invasion, colonization, and infection by pathogenic organisms. By protecting internal tissues from exposure to pathogens, the skin is an organ that serves as an effective innate defense. Mucous membranes provide protection against invasion and colonization by producing substances that limit the growth of these organisms. Adaptive defenses are turned on only after the host has been infected. Adaptive responses are characterized by cell- and antibody-mediated immune responses. These immune responses are specific and provide the host with the ability to recognize pathogens they have seen previously, allowing for efficient removal of these pathogens. Understanding the dynamics of the host-pathogen interaction is an active area of research.

20.21 Biofilms are highly organized communities of microbes

Key concepts

- It has been estimated that most of the Earth's prokaryotes live in organized communities called biofilms.
- Biofilm formation involves several steps including surface binding, growth and division, polysaccharide production and biofilm maturation, and dispersal.
- Organisms within a biofilm communicate by quorum-sensing systems.

Most of the earth's microbial population is present in closely associated, multispecies groups called **biofilms**. Most surfaces that have access to sufficient moisture and nutrients can support biofilm formation. Structurally, biofilms consist of mushroom-shaped microcolonies with a complex web of water channels that extend throughout the structure and provide its residents with a constant nutrient supply. The individual organisms within a biofilm are embedded in a protective polysaccharide matrix. Formation of **sessile** (immobilized) biofilm communities contributes to many human diseases, such as otitis media (ear infection) and cystic fibrosis (for more on cystic fibrosis see *6.26 Supplement: Mutations in an anion channel cause cystic fibrosis*). Biofilm formation in indwelling catheter devices, prosthetic heart valves, and other medical devices poses additional medical problems.

Sessile life has its advantages over a **planktonic** existence. First, the sticky polymeric matrix surrounding the biofilm helps to concentrate carbon, nitrogen, phosphate, and other nutrients within the biofilm. At the same time, microbes within the biofilm are largely protected from antibiotics, sheer stress, and host defenses. Biofilms are also able to disperse, thereby promoting the colonization of new surfaces.

For the most part, biofilm formation has been studied using laboratory conditions. There are notable differences between biofilms found in nature and those that have been used for the majority of biofilm studies. Notably, biofilms found in nature almost invariably consist of multiple species, while the majority of laboratory studies use single-species systems. Additionally, biofilms in nature will form on virtually any inorganic surface, while laboratory biofilm formation has used almost exclusively biofilms grown on plastic or glass. Nonetheless, laboratory studies have given us insight into the complexity of biofilm formation.

Biofilm formation by *E. coli* and *P. aeruginosa* has been genetically dissected by targeting mutants that can no longer form biofilms on plastic. These studies have helped to define the stepwise progression of biofilm formation, shown in **FIGURE 20.58**. Initial surface contact and binding is reversible and requires flagella. It remains unclear whether flagella play a direct role in adhesion, or if flagella-dependent motility is required for surface attachment. Once attached, a diffuse monolayer forms that progresses into a densely packed monolayer interspersed with microcolonies. For *P. aeruginosa*, microcolony formation depends on type IV pili. Type IV pili confer "twitching" motility to the microbe and promote cell-cell interactions. Therefore, type IV pili may contribute to biofilm maturation in two ways: by promoting cell-cell interactions within the growing biofilm and/or by facilitating movement of the microbe toward the growing microcolony. Microcolony growth leads to mature biofilm structures characterized by fluid channels that allow nutrient access to embedded organisms. High-molecular weight polysaccharides hold the biofilm together. In *E. coli* and *Salmonella* sp. proteinaceous fibers called curli provide a strong scaffold for cellulose polymers. This curli/cellulose extracellular matrix provides protection from environmental assaults such as dessication that would cripple planktonic microbes. A mature biofilm is a dynamic structure with many organisms detaching from the structure in a process called diffusion or dispersal. Diffusion can occur *en masse*, in which a group of organisms breaks from the biofilm and remains encased in a protective polysaccharide matrix. The organisms in the exiting cluster remain protected and can attach to nearby surfaces to continue growing. Alternatively, motile organisms can swim away from the biofilm planktonically, leaving the biofilm structure relatively unchanged. This method of diffusion may promote "exploratory" moves by the community in order to sample the environment for places of new colonization, without compromising the entire community's well-being.

Biofilms represent the pinnacle of prokaryotic community behavior. As with any community, communication between individual members is vital to the success of the population. Prokaryotes communicate by secreting diffusible signaling molecules that are sensed by nearby microbes. Signaling molecules called autoinducers allow for coordinated gene expression changes throughout the group. Because autoinducers are only sensed when they are present in relatively large amounts, it takes many organisms in a small area to secrete enough autoinducer to cause gene expression changes. Because of this requirement, autoinducer gene regulation is called **quorum sensing**. Only three types of autoinducers have been described, acylhomoserine lactone (acyl-HSL) systems in Gram-negative organisms, the autoinducer 2 system in Gram-negative and Gram-positive species, and the peptide system in Gram-positive species. Quorum sensing has been shown to play a role in regulating several bacterial processes. For Gram-positive bacteria, these include *B. subtilis* sporulation, conjugation by *Enterococcus faecalis*, virulence of *Staphylococcus aureus*, and competence of *Streptococcus pneumoniae*. Gram-negative species also have a long list of processes that are

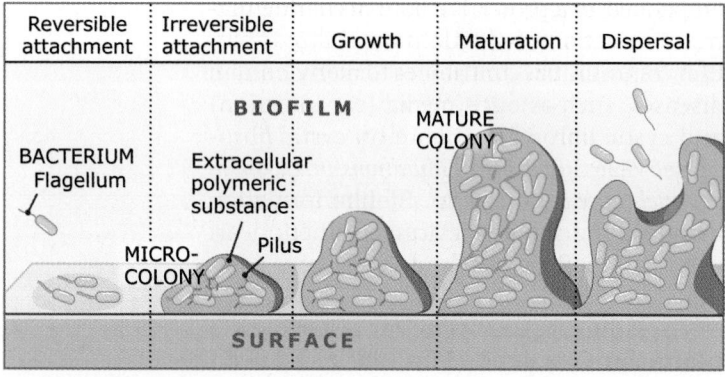

FIGURE 20.58 Biofilm development involves adhesion of bacteria to a surface and formation of colonies that are held together by extracellular polysaccharides.

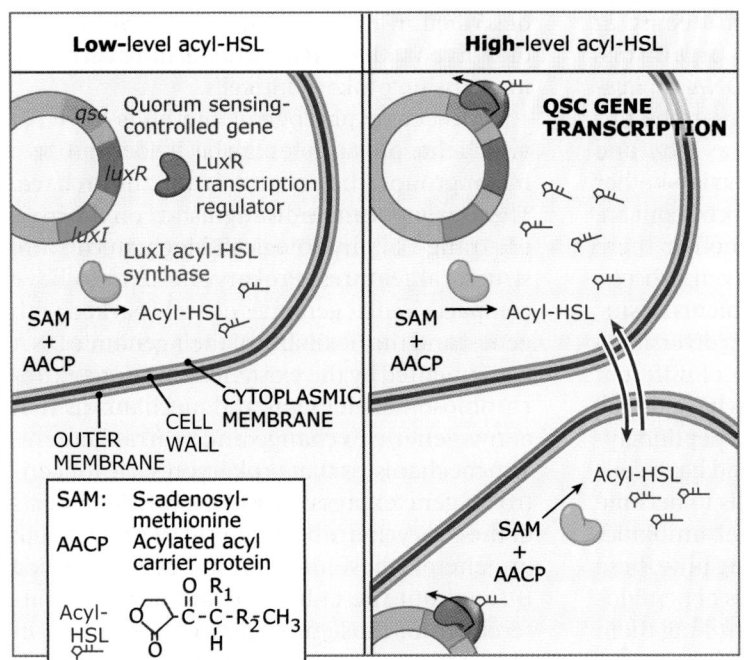

FIGURE 20.59 In the acylhomoserine lactone (acyl-HSL) quorum-sensing system, acyl-HSL is the signaling molecule synthesized by bacteria. As cells accumulate, the acyl-HSL reaches a threshold concentration, binds to a transcription regulator, and activates expression of genes involved in biofilm formation. For acyl-HSL, R_1 is H, OH, or O- and R_2 is $CH_2\text{-}CH_2\text{-}CH=CH\text{-}CH_2CH_2$ or $(CH_2)_{2-14}$.

regulated by quorum-sensing: light production by *Vibrio fischeri*, virulence of *P. aeruginosa*, and crown gall tumor formation by *Agrobacterium tumefaciens*, among others.

A general scheme for acyl-HSL quorum sensing is shown in **FIGURE 20.59**. In this system, an enzyme of the LuxI family of acyl-HSL synthases catalyzes the formation of an acyl-HSL. At low cell densities, quorum-sensing regulated genes are not activated. However, above a threshold level of acyl-HSL, transcription regulators of the LuxR family bind acyl-HSL and activate transcription of these genes.

HSL signaling plays a role in biofilm formation by organisms including *P. aeruginosa* and *Streptococcus mutans*. A *P. aeruginosa* strain with a mutant *lasI* gene (*luxI* homolog) that is deficient in autoinducer production is unable to form mature biofilms. Biofilm formation by these mutant cells is restored upon addition of exogenous homoserine lactone to the cells. *Streptococcus mutans* forms biofilms on dental surfaces and has a *luxS* homolog that is important for correct biofilm formation. However, quorum sensing does not always promote biofilm formation. For example, in *Vibrio cholera*, homoserine lactone production is required for efficient biofilm dissolution. These signaling pathways clearly dictate community behavior in bacteria.

Much of the work to understand the molecular details of biofilm formation is still in its infancy. Most notably is work aimed at dissecting the relationships between different bacterial species found within a common biofilm. Recent studies suggest that bacterial biofilms present as part of our natural gut microflora may play an important role in precluding colonization by pathogenic organisms. The complex interplay between organisms within a biofilm, and between those competing for resources in the same niche, will be an important focus of research going forward.

20.22 What's next?

Whole genome sequences are being completed at an amazing rate, but there is every sign that the rate will continue to escalate. The diversity of organisms sequenced extends to many corners of the bacterial lineage, with a natural emphasis on pathogenic organisms. It is estimated that only a fraction of all bacterial diversity is available in pure culture under laboratory conditions. Perhaps even unculturable organisms will eventually be sequenced. Then the complete coding capacity of the bacterial subkingdoms will become available. Sequencing of closely related organisms will allow the identification of subsets of genes that are characteristic of the group to be identified and distinguished from "accessory" genes that contribute to the diversity and adaptability of the group.

So far we have only looked deeply at gene function in a handful of organisms. Fundamental processes such as DNA replication and cell division are beginning to be understood in model organisms such as *E. coli* and *B. subtilis*. However, a major question is whether the mechanisms that have been worked out are common to all prokaryotes or whether there are multiple mechanisms for a given process. Finding different solutions to problems of survival might have been important in diversification of lifestyles and exploitation of different niches. For example, Mycoplasmas have no cell wall and so have no need for the peptidoglycan synthesis machinery. This could have contributed to the ability of these cells to become pathogenic or to avoid the action of antibiotics in the environment. Understanding how these organisms have adapted to the loss of peptidoglycan could be important in controlling them as pathogens.

Studies of whole microorganisms have mainly concentrated on pure cultures using classical principles of strain isolation. However, it is becoming increasingly clear that many bacteria live in communities in which competition and cooperation play important roles. Unraveling the players in these communities and their dynamic interactions will require the development of novel methods. Some of the tools are probably available now but will need to be adapted and modified to make them work in complex samples. Development of methods to investigate these communities is likely to open up many previously intractable problems, particularly in areas such as intercellular communication and microbial competition. Increased understanding of single organisms and communities will allow us to exert increasing control over prokaryotes in disease and provide ways to manipulate them for biotechnology, agriculture, and medicine.

20.23 Summary

Prokaryotic cells are distinguished from eukaryotic cells by the lack of a nuclear membrane and by a simpler cellular architecture. Their relative simplicity, along with ease of manipulation in the laboratory, has allowed the study of prokaryotes to contribute to our understanding of more complex organisms. The basic cellular processes of replication, transcription, and translation are evolutionarily conserved in all organisms and were first described in molecular detail using prokaryotes. Even today, many seminal discoveries are made using prokaryotic cells.

Molecular phylogenic methods have revealed that prokaryotes can be divided into two major groups: the eubacteria and the archaea. These groups can be distinguished on the basis of a range of physiological, biochemical, and structural features. Prokaryotes tend to have compact circular genomes densely packed with genes, and the flexibility of their genomic DNA is augmented by the existence of various extrachromosomal elements and mechanisms that allow genetic exchange and rearrangement. The mechanisms that prokaryotes use for control of gene expression and for various aspects of the cell cycle are beginning to be understood; in general, these mechanisms are conserved throughout the eubacteria but are quite different from those of eukaryotes. Archaea, in contrast, tend to share certain features with eubacteria and others with eukaryotes.

Prokaryotic cells were once thought to consist of only a surrounding membrane, a cytoplasm full of enzymes, and an unorganized mass of DNA. However, cell biological methods have recently demonstrated that prokaryotic cells are highly organized, in some ways similar to eukaryotic cells. For example, many proteins are targeted to specific locations in the cell, some proteins display organized dynamic changes in distribution, and cytoskeletal-like proteins maintain cell structure. The chromosome appears as a diffuse mass, but in reality it too is organized, so that genes also occupy predictable positions in the cell. Proteins associated with prokaryotic DNA help to organize the chromosome; in bacteria, these components form the nucleoid, whereas archaea have nucleosomes, similar to eukaryotic cells.

Important to many aspects of the existence of prokaryotic cells is their ability to secrete proteins and polysaccharides across their cell membranes and to assemble structures such as flagella and pili on their surfaces. Flagella are motility factors that provide the means to converge on food sources in a process called chemotaxis. Pili are fibrous structures that allow cells to interact with each other, with their environment, and sometimes with their host during colonization. Assembly of flagella and pili occurs via specialized protein secretion and assembly systems.

Prokaryotes are amazingly diverse in terms of their metabolic and catabolic activities. Many prokaryotes are free-living, and they occupy an incredible variety of ecological niches. Other prokaryotes enter into sophisticated symbiotic relationships, such as the communities in biofilms, or act as pathogens against hosts ranging from other prokaryotes to humans. We interact with prokaryotes every day—our surrounding environment, food, and bodies are home to countless species of prokaryotes—yet we have only begun to understand their diversity and impact.

http://biology.jbpub.com/lewin/cells

To explore these topics in more detail, visit this book's Interactive Student Study Guide.

References

Aldridge, P., and Hughes, K. T. 2002. Regulation of flagellar assembly. *Curr. Opin. Microbiol.* v. 5 p. 160–165.

Baron, C., and Zambryski, P. C. 1996. Plant transformation: A pilus in Agrobacterium T-DNA transfer. *Curr. Biol.* v. 6 p. 1567–1569.

Barry, C. E. 2001. Interpreting cell wall "virulence factors" of Mycobacterium tuberculosis. *Trends Microbiol.* v. 9 p. 237–241.

Bassler, B. L. 2002. Small talk. Cell-to-cell communication in bacteria. *Cell* v. 109 p. 421–424.

Brock, T. D., and Darland, G. K. 1970. Limits of microbial existence: temperature and pH. *Science* v. 169 p. 1316–1318.

Ding, Z., Atmakuri, K., and Christie, P. J. 2003. The outs and ins of bacterial type IV secretion substrates. *Trends Microbiol.* v. 11 p. 527–535.

Gordon, G. S., and Wright, A. 2000. DNA segregation in bacteria. *Annu. Rev. Microbiol.* v. 54 p. 681–708.

Woese, C. R. 2000. Interpreting the universal phylogenetic tree. *Proc. Natl. Acad. Sci. USA* v. 97 p. 8392–8396.

Woese, C. R. 2002. On the evolution of cells. *Proc. Natl. Acad. Sci. USA* v. 99 p. 8742–8747.

Plant cell biology

21

Clive Lloyd

Department of Cell and Developmental
Biology, John Innes Centre, Norwich, UK

AN ONION ROOT TIP CELL in prophase, with fluorescently labeled chromosomes in red and both the Golgi apparatus and
plasma membrane in green, as visualized by laser scanning con-
focal microscopy. Because it has a rigid cell wall, this plant
cell maintains its shape after being isolated from a tissue. The
dispersed Golgi system allows for secretion of wall components
over the large surface area of the cell. Photo courtesy of Chris
Hawes, Oxford Brookes University.

CHAPTER OUTLINE

21.1 Introduction

Key concepts

- Plant and animal cells grow in fundamentally different ways.
- The tough cell wall prevents cell movement and uptake of large molecules as food.
- Plant development depends upon how immobile cells manipulate the cell wall.

Single-celled organisms lived on Earth for many years before they developed into multicellular forms that diverged into plants and animals. By the time of this divergence, most of the features of the eukaryotic cell had already been established. As a result, we see common features in organisms as diverse as yeast, green plants, and vertebrates. For example, all eukaryotic cells store most of their genetic material in a nucleus and contain mitochondria for the production of energy. These organelles are basically the same whether they are in a root cell or a liver cell. These features, which are common to plants and animals, are covered in depth in other chapters. However, because plant cells grow in quite different ways than animal cells, we will see that other common components are used differently, and this chapter describes some plant-specific features. For instance, the Golgi system in plant cells is not centralized as it is in animals but is instead dispersed over the cell cortex to support the insertion of growth materials over large surface areas. The distribution of the cytoskeleton also reflects this mode of growth, with microtubules attached to the cytoplasmic face of the growing cell surface. Actin filaments are not used for attachment and cell motility but for stirring the cytoplasmic contents in immobile cells that grow to much larger sizes than animal cells can achieve. One major organelle found in animal cells, the centrosome, is absent in higher plants. On the other hand, plant cells possess organelles and structures that animal cells do not, and it is these unique features that provide us with a key for understanding the very different way in which plant cells construct an organism:

- *The vacuole*. A problem faced by all cells is that water passes across the plasma membrane by osmosis, diluting the cytoplasmic contents. Animal cells solved this problem by using membrane pumps to eject the accumulated ions that are responsible for the influx of water. Plant cells found a completely different solution. They contain special membrane-bounded organelles called vacuoles that take up excess water. Vacuoles swell as they absorb water, causing plant cells to expand to far greater sizes than animal cells. In many cases vacuoles occupy most of the volume of an expanded cell, squashing the cytoplasm into a thin layer adjacent to the plasma membrane and leaving only thin strands passing from one side of the cell to the other. Filling its vacuole with water is an economical way for a cell to increase its size; far less energy is required than producing the same volume of protein-rich cytoplasm. This is an important consideration since many plants grow in soils deficient in the inorganic nitrogen-containing compounds that are essential for making the amino acids required for protein synthesis.

- *The cell wall*. To prevent the vacuole from swelling until the cell bursts, plant cells encase themselves within a cell wall. This wall is exceptionally tough, composed of fibers as strong as steel—much stronger than anything possessed by animal cells. These fibers are wrapped around the cell to prevent it from bursting. However, being enclosed within the "box" formed by its cell wall, also prevents a cell from moving. Instead, the internal pressure (or "turgor") which builds up as a result of water influx is used to expand plant cells. The cell wall can yield to allow growth so that cell shape is ultimately controlled according to the way that the cell wall is remodeled. This is where the unique arrangement of organelles and cytoskeletal elements found in plant cells plays a part.

- *The chloroplast*. The presence of rigid walls around all its cells prevents a plant from having the type of flexibility required to move and largely limits it to one location. As a consequence it cannot chase or gather its food. This problem is overcome by the ability of green plants to make their own. This is performed by chloroplasts, another type of organelle unique to plant cells. Chloroplasts allow plants to live without moving by using light to convert CO_2 into carbohydrates that can either be used to provide energy or be converted into the complex polymers that form the cell wall.

These three interrelated aspects of plant cells contribute to a unit that is quite different from a "typical" animal cell. This is true both of the plant cell's internal organization and of the way in which it participates in the construction of an organism. The differences arise because the walls that surround plant cells bond them together, making the elaborate shape changes and movements of animal cells unavailable to plant cells. As a result they cannot undergo the type of complex rearrangements that occur among cells as an animal develops. Instead, plants must develop and grow with very little change in the positions of their cells relative to one another and their growth must be propelled by very different mechanisms than the ones that animal cells use to move. Plant cells are able to accomplish this through the use of basic mechanisms like expansion and control in the direction in which they divide.

Because the cells within a plant are always in intimate contact with one another, it is artificial to consider a "typical" plant cell in isolation. For this reason, we shall start by looking at the way in which plants grow so that we can see the properties of their cells in context.

21.2 How plants grow

Key concepts

- Plants extend into the environment using apical meristems.
- Plant development continues beyond the embryonic stage.
- Plant growth is sensitive to the environment.

The first and most obvious thing to appreciate about plants is that they do not explore space by walking, crawling, or swimming, but by growing into it.

As humans, when we grow, the number of cells increases more or less uniformly throughout our bodies. Our organs and limbs all grow proportionately, so that as adults, we are simply bigger versions of who we were as infants. Plants do things differently. Rather than growing evenly throughout, so that all

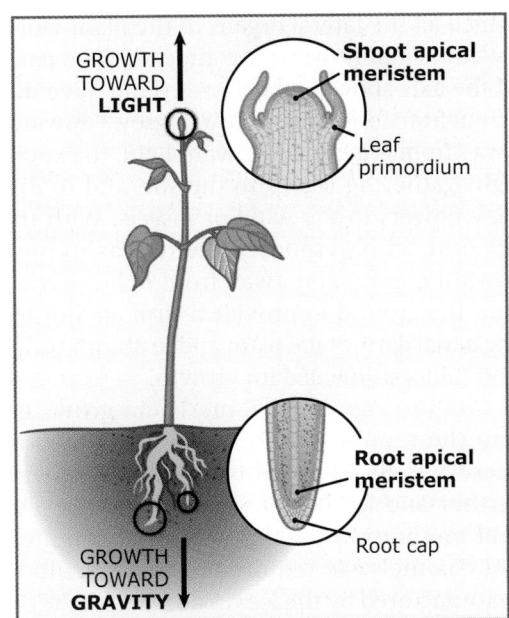

FIGURE 21.1 The cell division that drives the growth of a plant is limited to small, specialized regions (meristems) located at the tips of the roots and the shoot. Their structures are shown in the insets.

parts contribute equally to an increase in size, plants grow at only a few specialized points that remain "juvenile" throughout their lifetime.

These specialized points are called **meristems. FIGURE 21.1** shows their locations within a plant. Meristems found at the tips of the roots and shoots are called "primary" or "apical," and are particularly active sites of cell division, providing the new cells that are required for growth. Through the production of new cells, these meristems extend away from the older parts of the plant. In this way, roots grow down into the ground and shoots extend up into the atmosphere and toward light. While primary meristems provide material for increasing the height of a plant, cell division in "secondary" meristems (called "cambia"; singular, cambium), positioned along the sides of the mature roots and stems, produce cells that increase the organ's girth.

In order to form a plant, growth at primary meristems must be highly directed. Random growth would result in a jumble of disorganized tissue. Instead, growth is channeled along an axis that extends between root and shoot. This is the major growth axis, to which all the lateral organs of the plant (such as leaves and flowers) are attached. The parts of the axis above and below ground have different functions. The shoot axis grows upward, away from gravity and toward light, to expose light-gathering leaves to the sun and to display flowers to the wind or insects. With the opposite set of responses, the root axis grows toward gravity and away from light, driving into the ground to provide a firm anchor for the aerial parts of the plant and to absorb water and minerals needed for growth.

As the shoot apical meristem grows up and the root apical meristem grows down, these two main growing points are moved farther and farther apart. This creates several mechanical problems. Special conducting channels are required to transport food manufactured by the leaves down to the roots and to transport water and minerals from the roots up to the leaves. In addition, as the plant becomes bigger, the parts behind the growing points must be toughened to provide structural support for the ever-advancing tips. We will see how specialized thickening of cell walls strengthens new sections of a plant and allows them to support the load of further growth.

As well as being organized differently than animals, plant growth is also much more sensitive to the environment. The rate and/or direction of plant growth are strongly influenced by gravity, temperature, the length of day, and the direction of light. So, while the fixed body plan of an animal can be seen as early as the embryo, the body plan of plants is plastic; it continues to evolve in response to these external factors as the shape changes through branching and the production of flowers and leaves. This ability to adapt through the arrangement of organs is clearly dependent on the ability of plants to grow continuously. Another consequence of the ability to maintain juvenile growing points is that plants can grow far bigger and live much longer than any animal that has ever lived on Earth. The giant redwood trees of North America, for example, can weigh as much as 2000 tons and grow to heights of more than 100 meters (~330 feet), while reaching ages measured in thousands of years.

Concept and Reasoning Check

1. How does the growth pattern of plants differ from animals?

21.3 The meristem provides new growth modules in a repetitive manner

Key concepts

- Apical meristems divide to produce new cells at the growing points.
- Growth occurs by repeated addition of new growth modules.
- Cells divide, expand, then differentiate.
- Massive expansion of cells behind the tips drives the growing points onward.

A glance at a tree covered in buds is enough to know that the youngest parts are those extending farthest out into the environment and that they are supported by older parts closer to the ground. The pattern of plant growth is repetitive in that new modules are continually added onto the older parts. This mode of growth depends upon the cyclic activity of the apical meristems, where the production of new cells extends the newest parts of the plant away from the older ones and out into the environment. Before we look at the cellular processes involved in this, we will briefly take a closer look at a primary meristem and how its activity determines the form of the whole plant.

The surface of the shoot apical meristem is shaped like a dome, as shown in **FIGURE 21.2**. Cells in the center of this apical dome divide to add new tissue to the main axis. Cells on the flanks of the dome divide to form lateral bulges that grow out sideways and, depending on environmental conditions, become either leaf buds or flower buds projecting from the major axis. These lateral organs often arise at regular intervals, forming complex patterns that vary from plant to plant. For example, successive leaves may alternate between opposite sides of a stem, or as pairs at right angles to the previous pair, or they may spiral along its length. **FIGURE 21.3** shows lateral organs arising in a regular pattern behind an apical meristem, and **FIGURE 21.4** and **FIGURE 21.5** illustrate the types of patterns that can result in the plant. Each lateral bud, together with the portion of shoot that carries it, constitutes a growth module that can be produced time after time by the apical meristem to make the plant bigger. This repetitive basis of plant growth involves complex interactions between the rate of cell division and the external conditions, mediated by

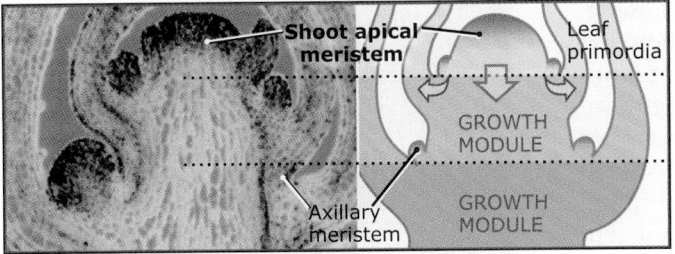

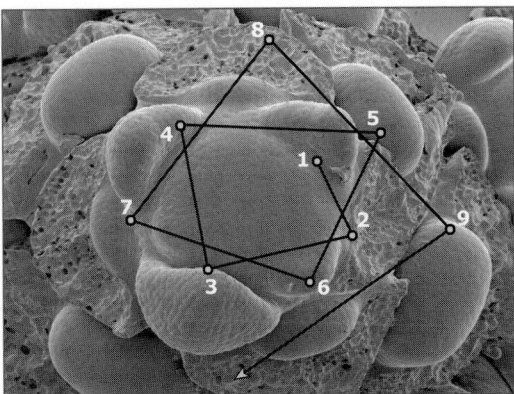

FIGURE 21.2 The photograph shows a longitudinal section of the growing tip of a snapdragon shoot. Sites of cell division have been stained black by *in situ* hybridization for cyclin mRNA, labeling the apical and axillary meristems. The diagram on the right illustrates how the apical meristem repeatedly produces growth modules as the plant grows. The arrows indicate that cell division in the center of the apical meristem produces the cells of the stem in each module, while division at its edges forms primordia that develop into leaves or flowers. Left photo reproduced from V. Gaudin, et al., *Plant Physiology* 122 (2000): 1137–1148. Copyright 2000 by American Society of Plant Biologists. Reproduced with permission of the American Society of Plant Biologists in the format of Textbook via Copyright Clearance Center. Photo courtesy of John H. Doonan, John Innes Centre.

FIGURE 21.3 A scanning electron micrograph of the tip of a growing snapdragon shoot. The apical meristem is in the center, and the leaves and leaf primordia in successive growth modules that it has produced are arranged in a spiral pattern around it. They are numbered from the newest to the oldest. Differentiation into a leaf occurs between numbers 6 and 7. Photo courtesy of Enrico Coen, John Innes Centre.

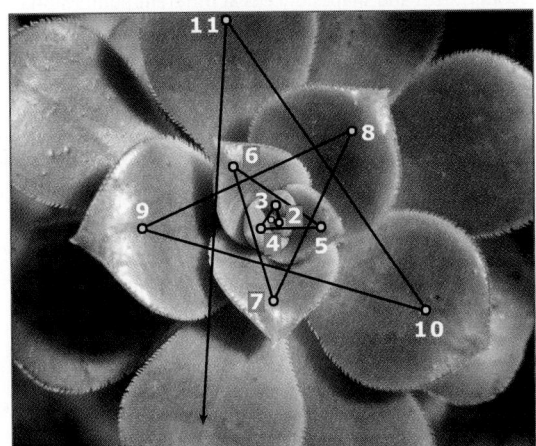

FIGURE 21.4 Successively produced leaves organized in a spiral behind the growing point of a plant. Photo by Hans Meinhardt, see http://www.eb.tuebingen.mpg.de/meinhardt.

FIGURE 21.5 The pattern of leaves along the stem of a plant. Photo by Hans Meinhardt, see http://www.eb.tuebingen.mpg.de/meinhardt.

plant growth hormones such as auxin and cytokinin.

Cells produced by division in the center of the apical dome are used to construct the main **root-shoot axis** of the plant in stages. One of the cells produced remains in the meristem to maintain its structure and mitotic activity, while the other provides material from which new parts of the plant will be constructed. The meristem is, therefore, a self-perpetuating

mobile mitotic site. As the dividing cells at the apex move ahead by division, the division products that are left behind gradually stop dividing. They then undergo a period of dramatic expansion to many times their original size and finally differentiate into the tissues that make the root or shoot. These three processes occur in overlapping zones that can be seen in sequence in a section made along the length of a meristem. At the tip is a zone of cell division where the bulk of mitosis occurs. Following it is a zone of cell expansion, and following that, nearest the mature parts of the plant, is the zone in which the cells have begun to differentiate.

The repeated division of cells at the apex produces ladder-like files of cells extending away behind it. The cells in the zone of cell division are small and either have very small vacuoles or lack them completely. By the time the cells enter the zone of cell expansion, however, vacuoles have formed within all of them. Inflation of the vacuoles then drives a remarkable expansion of the cells to many tens or hundreds of times the size of those in the division zone. If expansion were a generalized process, like blowing up a spherical balloon, growth would result in nothing more than a ball of cells, and plants would be unable to grow more than a few inches/centimeters off the ground. Expansion does not, however, occur evenly in all directions. Crucially, it is channeled along the growth axis, elongating the walls along the axis and, in the process, lengthening the axis itself. So not only do cells become bigger in the zone of expansion, they become very much longer. **FIGURE 21.6** illustrates the dramatic lengthening of cells that occurs behind a meristem. During this process the vacuoles expand to occupy up to 95% of each cell's volume, compressing the cytoplasm into a thin layer just beneath the plasma membrane. It is this massive directional expansion of cells just behind the apex that pushes the dividing cells at the tip farther out into the environment, carrying the lateral organs with them. Directional cell expansion is a key morphogenetic strategy in plants, and later in this chapter we will look at the way in which it is regulated by a special relationship between the cytoskeleton and the cell wall.

Expansion and differentiation of cells behind the meristem are accompanied by changes in the walls that surround the cells. Expansion requires flexibility of the wall: it must be strong enough to hold the shape of the cell but still capable of yielding to the forces that cause it to expand. To allow this, the still-growing "primary" cell walls in the zones of division and expansion are relatively thin and plastic. In the zone of differentiation, however, the cells stop expanding and the cell walls become thickened and chemically hardened. These rigidified "secondary" walls form architecturally stronger material that holds and fixes the mature shape of the plant and supports the ever-advancing growth zone.

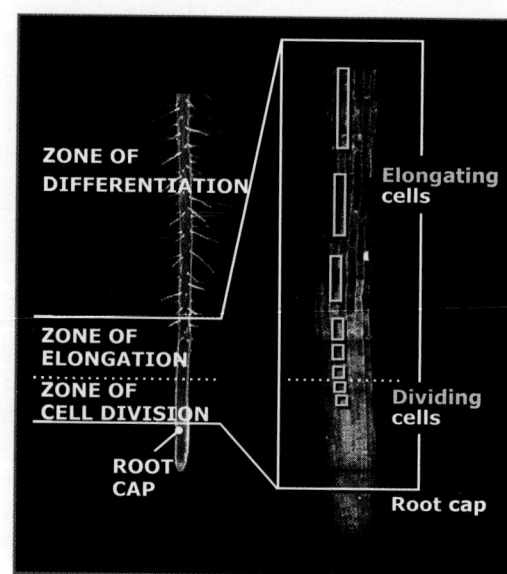

FIGURE 21.6 On the left is a light micrograph of an *Arabidopsis* root showing the three zones of cell behavior that occur in sequence behind a growing point. The root on the right is shown at higher magnification and has been stained to show microtubules, allowing the outlines of individual cells to be distinguished. The rough sizes and shapes of cells at different points along its length are indicated. Left photo courtesy of Keke Yi, Zhejiang University; right photo reproduced from K. Sugimoto, R. E. Williamson, and G. O. Wasteneys, *Plant Physiology* 124 (2000): 1493–1506. Copyright 2000 by American Society of Plant Biologists. Reproduced with permission of the American Society of Plant Biologists in the format of Textbook via Copyright Clearance Center. Photo courtesy of Geoffrey O. Wasteneys, University of British Columbia.

Although the cell division essential for the growth of a plant is limited to the meristems, many of the differentiated cells within a plant retain the ability to divide. If a leaf is browsed by an animal, for example, the cells around the wound can restart division activity in order to close the injury. The ability of mature plant cells to divide, however, goes well beyond the limited kind of wound-healing that we exhibit after we scratch our skin. Many nonmeristematic plant cells can resume mitosis and rerun the developmental program to form new organs or even whole new plants. These cells are therefore totipotent. Horticulturalists make use of this property to regenerate multiple identical copies (or clones) of a plant, bypassing the normal route of sexual reproduction. For example, by chopping a leaf from an African Violet into small pieces it is possible to reproduce many new, identical plants.

21.4 The plane in which a cell divides is important for tissue organization

Key concepts

- In the absence of cell movement, orientation of the division plane helps determine shape.
- Formative divisions generate new cell types; proliferative divisions add more cells.

The presence of a wall around each cell has fundamental consequences for how plants develop and how they determine the shapes of their organs. Unlike animal cells, which can organize the body plan by changing their shape or location, the cells in the meristem of a plant are immobile. Their walls fix their shape and bond them together, so that they cannot move apart. As a consequence, cell division represents the only opportunity for plant cells to position themselves relative to one another. The presence of the cell wall means that division cannot occur by the type of cortical constriction seen during cytokinesis in animal cells. Instead, a plant cell divides by constructing a new wall across its interior. The plane of this cross-wall is carefully controlled because once it is laid down across the cell and attaches to the parental wall it cannot be swiveled or moved to a new position. Orienting cell divisions within the meristem

is an important process used to arrange cells within a plant and organize them so that they can differentiate into functional tissues.

To see the effect that changes in the division plane have on the resulting tissue, consider the simple example of the division of the rectangular cell shown in **FIGURE 21.7**, **FIGURE 21.8**, and **FIGURE 21.9**. If the cell and its daughters all

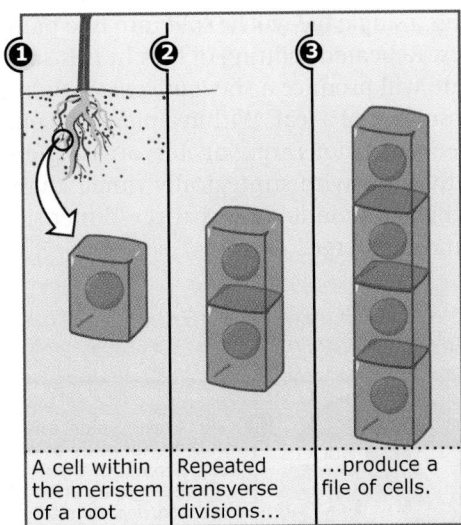

FIGURE 21.7 Sequence showing how repeated divisions with the same orientation can contribute to the structure of a root. Transverse divisions such as these are used to produce cells to allow a root or shoot to increase in length.

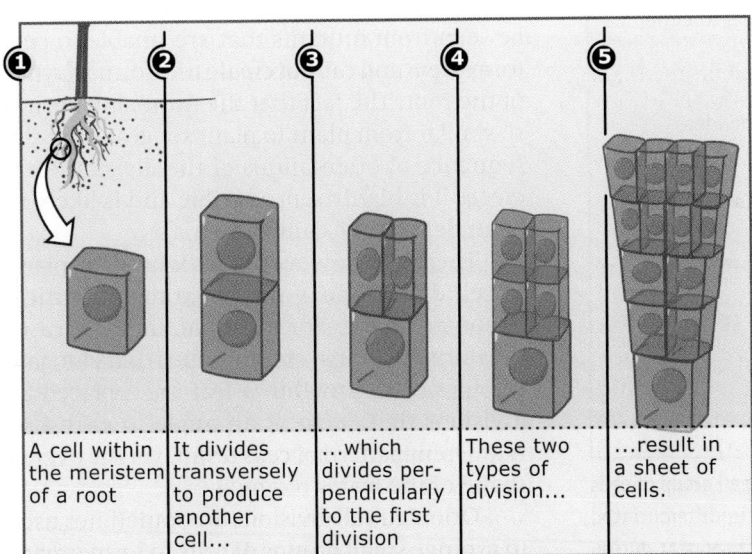

FIGURE 21.8 Sequence showing how coordinated divisions in two orientations can create new files of cells. These two types of division eventually form a sheet of cells.

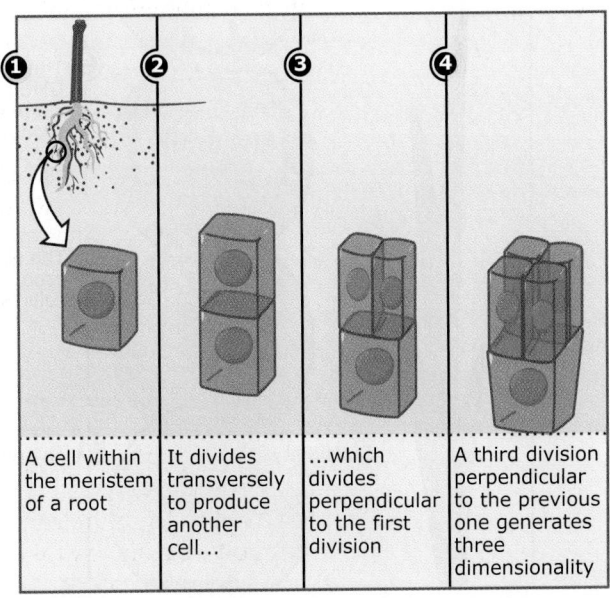

FIGURE 21.9 Sequence showing how coordinated divisions can create three-dimensional groups of cells. Divisions with the different orientations do not have to occur in the sequence shown. Complex shapes are generated within plants by different sequences and numbers of the three types of division, coordinated with cell expansion.

divide with the same transverse orientation, a ladder-like file of cells, joined end to end, will be produced. Directional expansion can stretch this file, but as long as the cells within it continue dividing with the same orientation, it will remain a single file of cells. This form of growth is seen in simple plants like filamentous algae. If at some point however, one of the cells divides perpendicularly to the previous divisions, a single file will be split into two parallel ones. Repeated splitting of files in this second plane will produce a sheet of cells, rather like the surface of a leaf. Within a plant, of course, three dimensions are available, and sequences of divisions with strategically timed changes in their orientation allow three-dimensional forms to emerge.

Generating a root reliably in three dimensions requires that different types of division be coordinated. Roots can be considered as bundles of cell files. As in the preceding example, the cells within a file can divide either perpendicular to its axis or along it. The two types of divisions are called "transverse" and "longitudinal," respectively. Each type contributes differently to the formation of a root, and the location and the number of times each type occurs determines its ultimate shape. Transverse (or "proliferative") divisions increase the length of the root by increasing the number of one type of cell within a file. If we pick an individual file well behind the meristem and trace it toward the apex of the root, we can see that the file itself originates in the division of a special founder cell in a different orientation. Such divisions that create new files of cells occur longitudinally and are called "formative" divisions. The two daughter cells and their descendants that result from a formative division can adopt different cell fates. So, regardless of the tissue, each file of cells within it is created long before differentiation by the longitudinal division of a founder cell near the apex of the root. By occurring circumferentially as well as radially, longitudinal divisions give rise to the three-dimensional form of the plant. Cross sections of roots of *Arabidopsis thaliana*, a small plant used extensively for genetic experiments, illustrate this three-dimensionality. They have a stereotypical pattern of cells arranged in concentric layers, each with a highly predictable number of cells. The significance of formative divisions in creating this pattern can be seen from mutants that are unable to perform them and cannot create the normal layout of the root. The fact that the normal pattern is so similar from plant to plant suggests that the sequence of orientations of the divisions that create it is highly reproducible and is likely to be under genetic control.

There are, however, cases where the plane of cell division may not be so highly critical for organ shape. For example, in some cases the normally transverse divisions that elongate the files of cells within a leaf are replaced by divisions that occur at an angle to each file. A different pattern of cells result, yet the overall shape of the leaf is retained.

Oriented cell divisions are sometimes used to arrange small groups of cells to form organs or to give particular cells a specific shape. Both occur during the formation of **stomata**, the

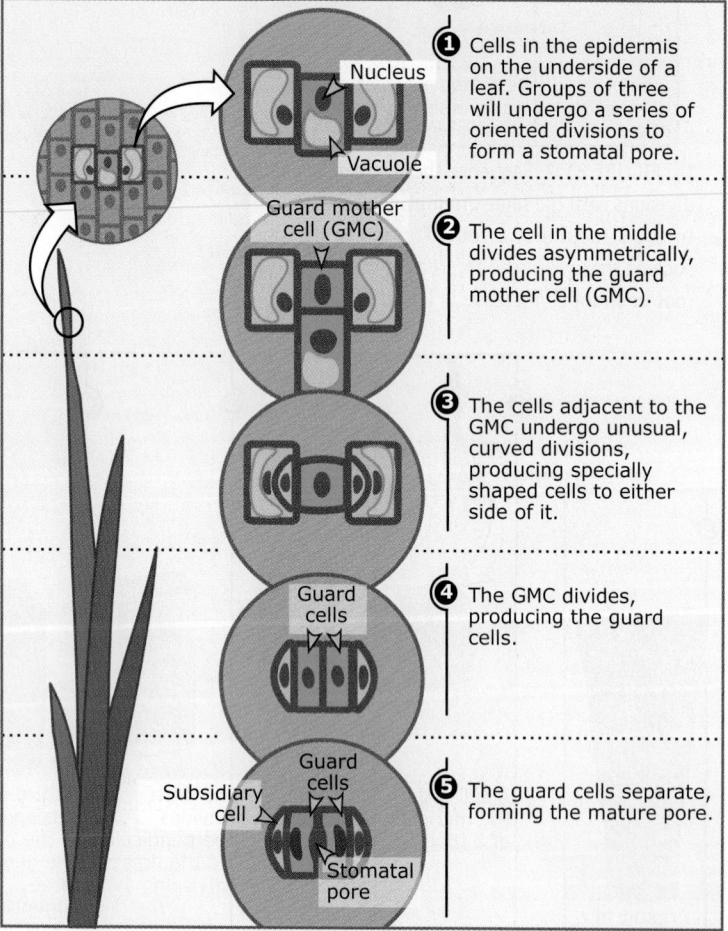

FIGURE 21.10 A sequence of oriented divisions within a small group of cells creates a specialized organ, the stomatal pore. Three typical undifferentiated cells are focused upon initially. The division products that disappear during the sequence remain within the leaf as normal epidermal cells and do not play a role in the function of the pore.

The labels in the figure:

Nucleus
Vacuole

① Cells in the epidermis on the underside of a leaf. Groups of three will undergo a series of oriented divisions to form a stomatal pore.

Guard mother cell (GMC)

② The cell in the middle divides asymmetrically, producing the guard mother cell (GMC).

③ The cells adjacent to the GMC undergo unusual, curved divisions, producing specially shaped cells to either side of it.

Guard cells

④ The GMC divides, producing the guard cells.

Guard cells
Subsidiary cell

⑤ The guard cells separate, forming the mature pore.

Stomatal pore

pores on the underside of a leaf that open and close to allow the flow of gas and water vapor in and out of the interior of a leaf. They are composed of a small number of cells: a variable number (depending on the species) of "subsidiary" cells plus two crescent-shaped "guard cells," which undergo turgor-driven movements to open and close the pore formed between them. This group of cells is organized and the pore formed by a series of oriented cell divisions that occur in a specific sequence, as shown in FIGURE 21.10. The sequence begins with the transverse division of a cell within an epidermal file; the division is asymmetric and the cell that will form the guard cells—called the guard mother cell—is the smaller of the two. Neighboring cells in the two adjacent files also undergo asymmetric curved divisions to form cells that cup the guard mother cell. It then divides symmetrically and longitudinally, parallel to the cell file. This forms two guard cells, which then detach from one another to form the stomatal pore. The various planes of division that give rise to the stomatal complex are, therefore, crucial to the function of this structure.

Concept and Reasoning Check

1. Why is the plane of cell division so important to the morphogenesis of plants?

21.5 Cytoplasmic structures predict the plane of cell division before mitosis begins

Key concepts

- The plane of cell division is predicted before mitosis by a ring of microtubules and actin filaments around the cortex.
- A sheet of cytoplasm (the phragmosome) also predicts the plane of division in vacuolated cells.

What determines the position where the new wall will divide a plant cell? Surprisingly, it is not the mitotic spindle, since the plane in which the chromosomes align at metaphase does not always predict the one in which the new cross-wall will form. This indicates that the mechanism that positions the plant cell wall is thus much different from the one that

determines the site of division in animal cells. There, interactions between the spindle and the cortex form and position the contractile ring, an actomyosin-based structure that divides the cell by constriction.

In plant cells, the position of the new wall is first indicated before mitosis even begins, by the formation of the **preprophase band**, a structure unique to plants. This structure is formed during the G2 phase of the cell cycle when microtubules immediately beneath the plasma membrane accumulate and align in a broad band that encircles the cell. These events are shown in FIGURE 21.11 and FIGURE 21.12. Actin filaments are also concentrated within the band and may be responsible for narrowing it until it becomes a well-defined, tight bundle of microtubules around the circumference of the cell. However, actin filaments appear to be excluded from a narrow ring in the middle of the larger preprophase band and is known as the actin-deficient zone. The preprophase band forms halfway along the cell when the division will produce two cells of equal size, but forms nearer one end of the cell when an asymmetric division will take place. In addition to the microtubules and actin filaments within the band itself, microtubules and actin filaments extend radially outward from the surface of the nucleus to connect it to the preprophase band. Thus, by the time the band has

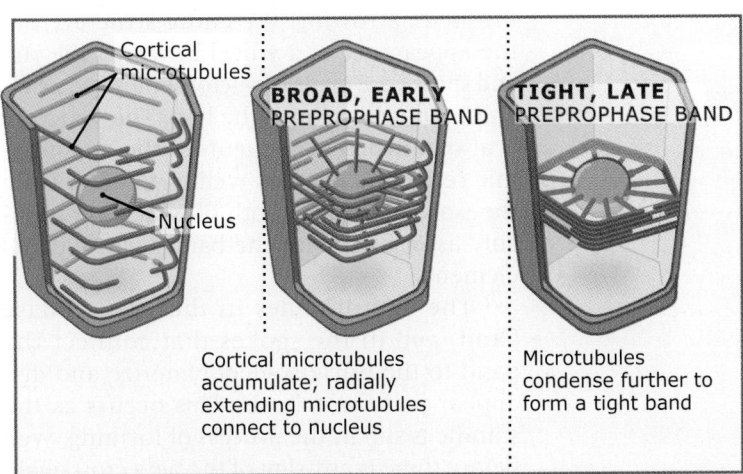

FIGURE 21.11 The microtubules in the interphase cell on the left are distributed evenly throughout its cortex. In the period leading up to mitosis they gradually disappear from the ends of the cell and accumulate around its center, eventually forming a tight band. The cell's nucleus is located within the same plane as the band and is connected to it by radially oriented microtubules. Mitosis begins shortly afterward.

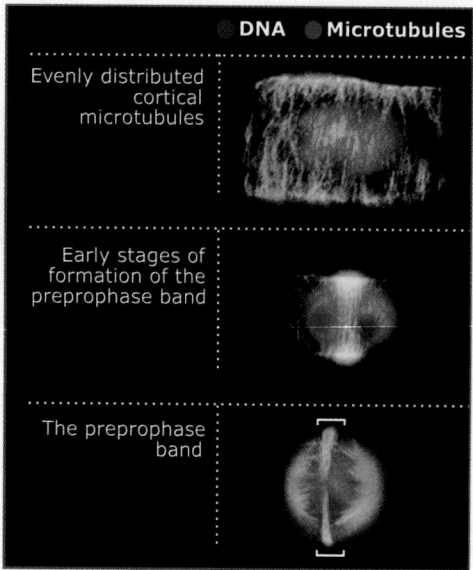

FIGURE 21.12 The formation of the preprophase band in an isolated cell. Microtubules are in green and the nucleus is in blue. In the lowest frame, the band has finished tightening and appears as a thin band running vertically over the middle of the nucleus. The more diffuse microtubule structures to either side of it are the two halves of the prophase spindle. Note that the nucleus remains intact throughout the formation of the preprophase band and that there are no other microtubules in the cortex of the cell while it is present. Photos courtesy of Dr. Sandra McCutcheon, The Roslin Institute, University of Edinburgh and R(D)SVS.

finished narrowing, the entire structure has the appearance of a wheel in which the rim and spokes are made of microtubules and actin, and the nucleus forms the hub. The structure is also enriched in elements of the endoplasmic reticulum (ER) as well as vesicles from the Golgi apparatus, both of which are probably associated with the band's cytoskeletal elements.

The microtubules in the preprophase band, and in the spokes that connect the band to the nucleus, depolymerize and disappear early in mitosis. This occurs as the spindle is still in the process of forming, well before there is any sign of the new cross-wall. Despite the disappearance of the band, when the new cross-wall forms after mitosis, it does so in precisely the same plane that the band defined earlier. This suggests that although the most visible elements of the band have disappeared, it has somehow marked the cortex of the cell. The protein TANGLED was

the first marker found to remain at the cortex, "remembering" the site where the preprophase band depolymerised. In principle, the band could leave behind molecules that prepare that region of the cortex for the attachment of the new cross-wall during cytokinesis. Alternatively, molecules left at the position of the band could somehow guide the growth of the new wall outward from the center of the cell. Actin filaments are also candidates for a role in guiding the new cross-wall outward, since actin filaments that connect the dividing nucleus to the cortex remain throughout mitosis.

The preprophase band is not the only structure to forecast the plane in which the cell will divide. Meristematic cells are usually small, with dense cytoplasm, and have no conspicuous vacuoles at this stage. However, larger, vacuolated cells can also divide (for instance, when they are induced to reenter the division cycle as a result of tissue wounding) and dividing such a cell poses the problem of how to construct a cross-wall across the liquid-filled vacuole. This problem is solved in the period leading up to mitosis by a dramatic rearrangement that creates a sheet of cytoplasm across the interior of the cell in the plane in which division will occur. Within this sheet, which sometimes occupies less than 5% of the cell's volume, the mitotic spindle forms and the cross-wall is constructed.

Division in a cell with a large vacuole requires that the nucleus first be moved. The nucleus of a highly vacuolated cell is normally confined to a thin layer of cortical cytoplasm during interphase and is usually repositioned to the center of the cell before mitosis begins. Initially, only a few thin strands of cytoplasm stretch from the nucleus across the vacuole and connect with the opposite side of the cell. In preparation for division, the number of strands increases, and the nucleus migrates from the cortex out into the center of the cell. **FIGURE 21.13** shows a cell that has reached this stage. Actin filaments are prominent within the strands and it seems likely that the contractile properties of actomyosin are used to pull the nucleus into position.

When the nucleus first arrives in the center of the cell, strands of cytoplasm radiate from it in all directions. **FIGURE 21.14** shows that as division approaches, their cortical ends move across the cortex and the strands collect and coalesce in a plane to form a continuous sheet of cytoplasm across the vacuole. This

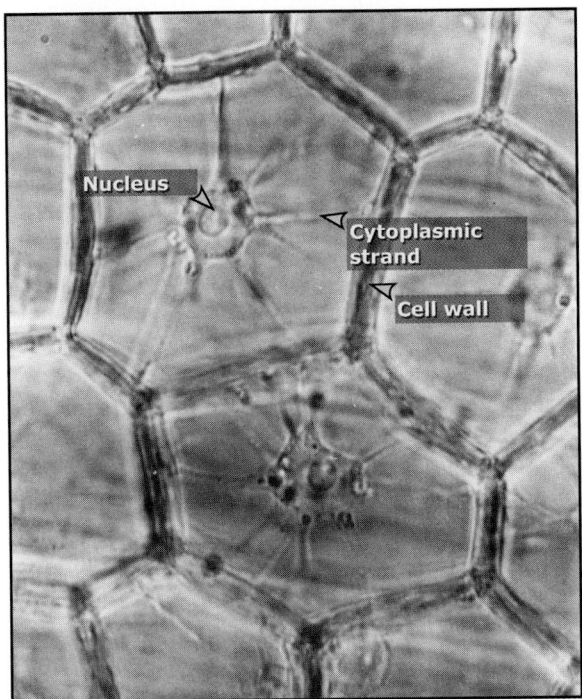

FIGURE 21.13 A light micrograph of an epidermal cell stimulated to divide. The nucleus has migrated to the center of the cell and division is about to begin. Most of the interior of the cell is occupied by a vacuole, and strands of cytoplasm that cross it and connect the nucleus with the periphery of the cell are visible. Photo courtesy of Clive Lloyd , John Innes Centre.

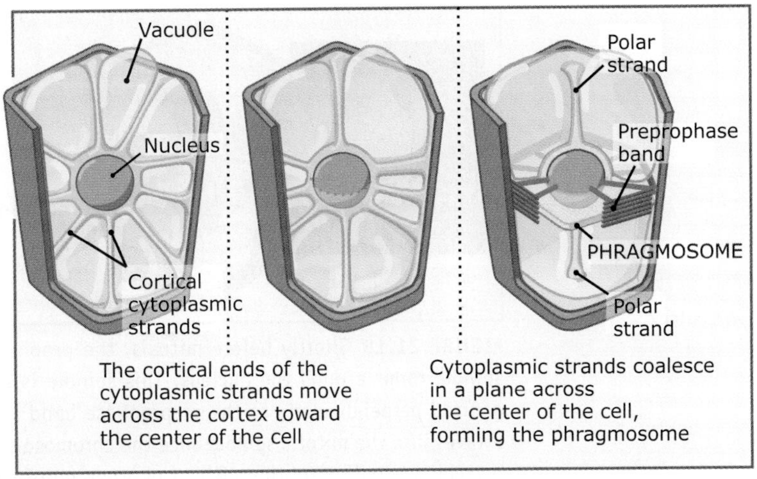

FIGURE 21.14 After the nucleus has been positioned in the center of a vacuolated cell preparing to divide, cytoplasmic strands radiating in all directions connect it with the cortex. To form the phragmosome, the ends of most of the strands migrate across the cortex and the strands coalesce into a continuous sheet of cytoplasm across the cell. A few strands (the "polar strands") are not included and remain connecting the nucleus to the cortex at the ends of the cell. The phragmosome forms in the same plane as the preprophase band.

cytoplasmic structure is supported by microtubules as well as actin filaments and is called the **phragmosome** (from the Greek *phragma* = barrier). The phragmosome forms at the same time as the preprophase band, and the two occupy the same plane. The formation of the two structures is evidently coordinated. The actin filaments and microtubules within the phragmosome that connect the nucleus with the cortex are likely to play a role and appear to be a more developed version of cytoskeletal elements associated with the nucleus of small, densely cytoplasmic, meristematic cells.

Not all of the cytoplasmic strands coalesce into the phragmosome. A few remain separate and extend from the surface of the nucleus at right angles to the plane in which the phragmosome has formed. These "polar" strands, connecting nucleus and cortex, are aligned with the pole-to-pole axis of the spindle. The spindle becomes misaligned when these strands are destroyed by centrifugation or actin-depolymerizing drugs, suggesting that they are involved in aligning the mitotic apparatus relative to the other structures that participate in cell division.

21.6 Plant mitosis occurs without centrosomes

Key concept
• The poles of plant mitotic spindles do not contain centrioles and can be much more diffuse than the poles of animal spindles.

Before we discuss how a cell wall is laid down in the plane predicted by premitotic structures, we must briefly cover the intervening process of mitosis. Although mitosis is essentially similar in plant and animal cells, plants differ in important respects.

At the beginning of a plant mitosis, the nucleus is enclosed within a so-called "prophase spindle" formed by microtubules. **FIGURE 21.15** shows this structure. Despite its suggestive shape, it is not a real mitotic spindle since it is not yet connected to the chromosomes. The prophase spindle gives way to a proper spindle

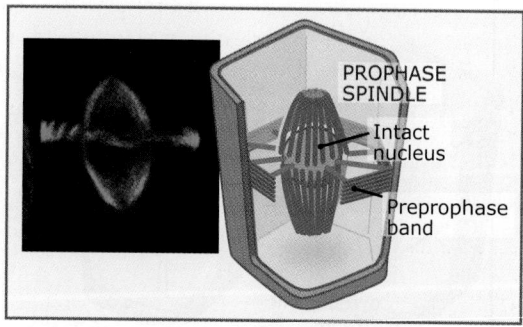

FIGURE 21.15 Shortly before mitosis, the prophase spindle forms around the nucleus. This spindle is organized perpendicularly to the preprophase band and evolves into the mitotic spindle once the chromosomes become accessible. On the left is a photograph of the prophase spindle (running from top to bottom) and the preprophase band (running horizontally) of a tobacco cell grown in suspension. Microtubules are in green and DNA in blue. Photo courtesy of Sandra McCutcheon and Clive Lloyd, John Innes Centre.

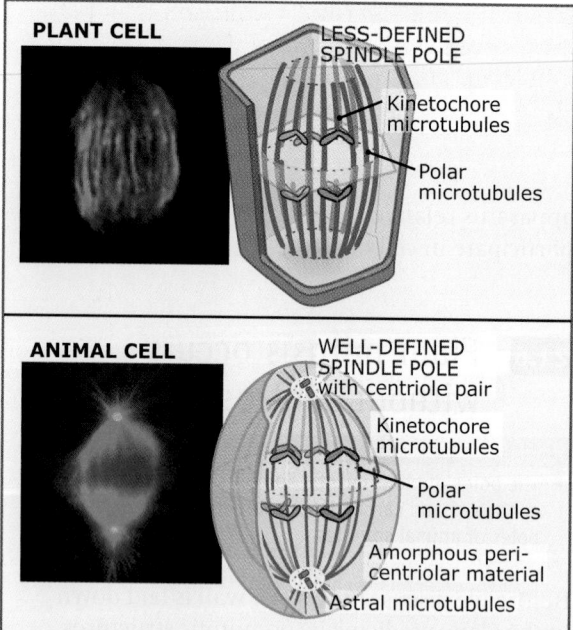

FIGURE 21.16 Plant and animal mitotic spindles differ mainly in the organization of their poles. Animal spindle poles are focused tightly on centrioles and have large numbers of astral microtubules, while the poles of plant cells are much more diffuse and have many fewer astral microtubules. The difference in shape that results can be seen in the photographs at the left. Microtubules are in green, DNA is in blue, and the areas immediately around the centrioles are seen as two bright yellow spots at the poles of the animal spindle. Top photo courtesy of Andrey Korolev, John Innes Centre; bottom photo courtesy of Christian Roghi, University of Cambridge.

as the nuclear envelope breaks down and the microtubules gain access to the chromosomes. At the same time, the microtubules of the preprophase band depolymerize. As in animal cells, once the mitotic spindle has formed there are no other cytoplasmic microtubules in the cell apart from those associated with the nucleus.

At their most basic level of organization, the mitotic spindles of plant cells are essentially similar to those of animal cells. As shown in **FIGURE 21.16**, in both cases two oppositely oriented sets of microtubules meet in the middle, where they attach to the paired chromosomes. The poles of the two types of spindles, however, are significantly different. In most animal cells, the spindle microtubules converge to a point at the poles, each of which is defined by possession of an organelle called the centrosome. Each centrosome is based upon a pair of centrioles surrounded by a cloud of amorphous pericentriolar material that acts as a microtubule nucleation site during the assembly of the spindle. The separation of microtubule arrays radiating away from duplicated centrosomes plays a central role in the formation and the basic bipolar organization of an animal cell's spindle. Plant cells do not possess centrioles, and so the structure of the poles is different. Possibly because there is no centriole to act as a focal point, the poles of plant spindles are often broader than those in animal cells. In some cases the "poles" of a plant spindle are so delocalized that they are barely any narrower than the rest of the spindle.

How do plant spindles form if they have no centrosomes to organize the poles? There are perhaps two general ways that their bipolarity could be set up. One possibility is that the formation of plant spindles could be initiated by the chromosomes themselves, as can be seen in some types of animal cells that lack centrosomes. A mechanism for this is suggested by in vitro experiments in which spindles were formed in the absence of centrosomes. **FIGURE 21.17** shows the sequence of events that would take place. Microtubules could first be nucleated with random orientations in the vicinity of the chromosomes. Microtubule-based motors, capable of binding and moving two microtubules of opposite polarity simultaneously, could then sort microtubules with opposite orientations to opposite sides of each chromosome. The nature of the sorting motors is such that the microtubules would be sorted with their plus (+) ends facing inward

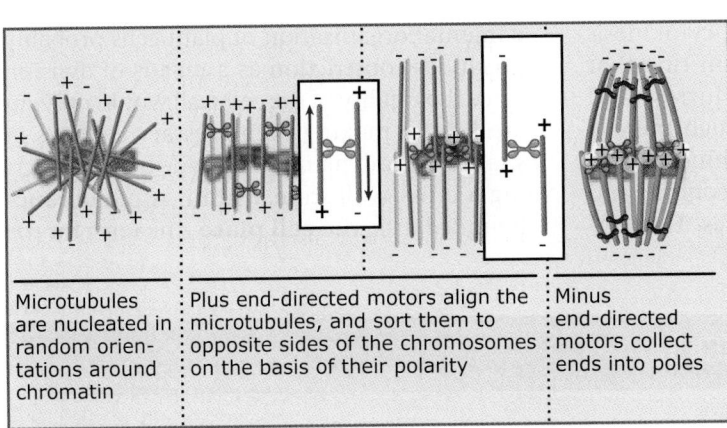

FIGURE 21.17 In vitro experiments suggest that a combination of micro-tubule nucleation by chromosomes and the actions of two different types of microtubule-dependent motors could create a bipolar mitotic spindle in the absence of centrosomes. Sorting of microtubules into two groups with their plus ends facing one another requires a motor that can bind two microtubules of opposite polarity and move toward their plus ends (green). The poles are formed by motors that bind to two microtubules of the same polarity and move toward their minus ends.

Microtubules are nucleated in random orien-tations around chromatin	Plus end-directed motors align the microtubules, and sort them to opposite sides of the chromosomes on the basis of their polarity	Minus end-directed motors collect ends into poles

toward the chromosomes, allowing some to be captured at kinetochores—the microtubule-binding sites on chromosomes. At the same time that sorting is taking place, other proteins could bundle the minus (–) ends of microtubules together. The net effect of these three types of activities—nucleating, sorting, and bundling—working in concert, is to form a bipolar spindle around each chromosome. Thus, with the participation of a limited number of basic activities, a bipolar spindle can be reliably constructed in the absence of centrioles and centrosomes. It is important to note that the bipolar organization of such a spindle would be based upon the inherent polarity of microtubules, which is what allows them to be sorted and bundled into two opposing populations.

Alternatively, plant spindles could form by a mechanism much more similar to that which occurs in animal cells. It is possible that one function of centrioles in animal cells is to gather the amorphous pericentriolar material into one place to form a centrosome, thereby sharpening the spindle poles. Despite the absence of centrioles in plants, it is possible that they contain a less visible form of the pericentriolar material. If this is sorted into two opposite poles, it could organize a bipolar spindle, although the spindle poles would be more diffuse without centrioles. This mechanism is suggested by the two "polar caps" composed of proteins such as γ tubulin that mark the sites of the spindle poles in some plant cells before the nuclear envelope breaks down (see preprophase cell in Figure 21.12).

Plant spindles also differ from animal spindles in that they have relatively few astral microtubules. These microtubules, which radiate from the back end of the spindle poles and extend throughout the cytoplasm, interact with the cortex in animal cells and are used to position and orient the spindle. The astral microtubules of animal cells also play an essential role in forming and positioning the contractile machinery that divides the cell during cytokinesis. Together these roles make the astral microtubules responsible for determining the plane in which a cell will divide. In contrast, the site at which a plant cell will divide is determined before the mitotic spindle even begins to form, during the cytoplasmic reorganization that results in the formation of the preprophase band. What roles astral microtubules may play in plant cells is not yet known, although they may be involved in aligning the spindle axis.

21.7 The cytokinetic apparatus builds a new wall in the plane anticipated by the preprophase band

Key concepts

- The cytokinetic apparatus—the phragmoplast— is a ring of cytoskeletal filaments that expand outward.
- Vesicles directed to the midline of this double ring fuse to form the new cross-wall.
- The plane in which the cell plate grows conforms to the preprophase band and not to the spindle midzone.

After nuclear division (mitosis), the cytoplasm divides (cytokinesis). In plant cells, this is when the new cross-wall is laid down in the plane predicted by the preprophase band.

Cytokinesis in plant and animal cells could hardly be more different. Unlike mitosis, which involves a spindle in both types of cells, the basic design of cytokinesis is different in the two types of cells. In animal cells, cytokinesis involves the formation of an actin ring that contracts, forming the cleavage furrow and pinching the cell into two. Although there is actin alongside microtubules within the pre-prophase band, the band does not contract and disappears well before a cell divides. The fun-damental organization of plant cells probably precludes constriction as a means of division since the rigidity of the cell wall would prevent a cell from making the necessary changes in shape. Instead, plant cells divide by construct-ing a cross-wall from within. A membrane-bounded disk, the **cell plate** (the term for the

HISTORICAL PERSPECTIVES: THE PLANE OF CELL DIVISION

The way that plant cells are packed together in tissues has fascinated scientists for hundreds of years. In the seventeenth century the great English polymath, Robert Hooke, placed a section of cork under one of the first microscopes. He saw a network of material and named each hollow pore a "cell", after the resemblance to the compartment in which a monk lived. We now know that the pores in this dead tissue were the spaces once occupied by cells but his drawing is important for illustrating how the cell walls are arranged. In some sec-tions, the cells looked like bricks in a wall where the bricks are not stacked on top of one another, but offset between rows to make three-way T-junctions. This give plant tissues a rather regular appearance and over the years they have been likened to other systems, such as honeycombs and soap foams, that seem to share some physical rule for the way that walls meet. Various supposedly cell-like analogues have been squashed together to see if this will reveal how plant cells combine. In 1727, Stephen Hales compressed peas together in a pot. By counting the faces of the squashed peas he sug-gested that they have 12 sides but it is now thought he might have been influenced by the perfection of carefully stacked cannon balls where 12 surround a central one. When irregu-larly sized balls of lead are pressed together, each ball tends to have about 14 sides and this agrees with the finding that undifferentiated plant tissues show an average of 14 sides. This is significant since the most efficient way of using the least amount of surface to envelope the maximum volume, without leaving gaps, is possessed by an ideal 14-sided fig-ure called the tetrakaidecahedron. When these are packed together, their corners, in section, are seen to be made by three walls meeting at a point. However, the big difference between living plant tissue and lead shot, or cannon balls, or perfect geometrical figures, is that the shape of cells in divid-ing tissue constantly evolves out of many decisions where to place the cross-wall. Real plant cells only approach the ideal in averages, with some cells having more than 14 sides and some having less (epidermal cells without outer neighbors have an average of 11). But regardless of the exact number of sides, plant cells strive to obey the "rule" that within a section only three walls meet at a point, making strong three-way T-junctions instead of weak four-way X-junctions. There has been a long search for the rule by which a new cross-wall makes a junction with the parental wall. In 1863, Hofmeister stated that the cross-wall is generally perpendicular to the direction of growth. Sachs' rule (1878) subtly restated this, saying that each new plane of division is perpendicular to the previous plane. And Errera's rule (1888) said that the new wall should be the minimum area for halving the cell's volume.

The first cytoplasmic structure shown to be involved in setting the plane of cell division was the phragmosome. Sinnott and Bloch (1940) found that when large vacuolated cells are induced back into the division cycle, the division plane is predicted before mitosis by the gradual coalescence of transvacuolar strands to form the cytoplasmic raft within which the new cross-wall will be laid across the vacuole. The first advance of the electron microscopic era was the obser-vation by Pickett-Heaps and Northcote (1966) that the pre-prophase band of microtubules (PPB) transiently marks the site at the cortex where the cross-wall will later attach to the parental wall. This cortical division site, marked by the PPB, coincides with the site where the phragmosome attaches, underlining the extensive cytoplasmic reorganization that plant cells undergo prior to mitosis. The great puzzle about the preprophase band concerns the nature of the "memory" it is presumed to leave behind when the band depolymerizes at prophase. After the advances brought about by light, then electron microscopy, we have had to wait for the molecular approaches of the twenty-first century to start providing the answer. In 2007, Laurie Smith and her colleagues described TANGLED as the first protein found to remain at the cortical division site after the PPB had depolymerized. Other mol-ecules are slowly being added to the list and we can expect to find out how the cortical division site is memorized and prepared for the insertion of the new cross-wall some hours later. Perhaps this will inform us whether microtubules and/or actin filaments guide the leading edge of the phragmoplast to the prepared landing site. But as this brief survey suggests, this kind of description has to be seen against the historical backdrop. Not only do we need to know the fine detail of how a cross-wall attaches to the parental wall, but how the cell plate avoids lining up with a cross-wall in the neighboring row of cells, preserving the three-way junctions first seen by Hooke over 300 years ago.

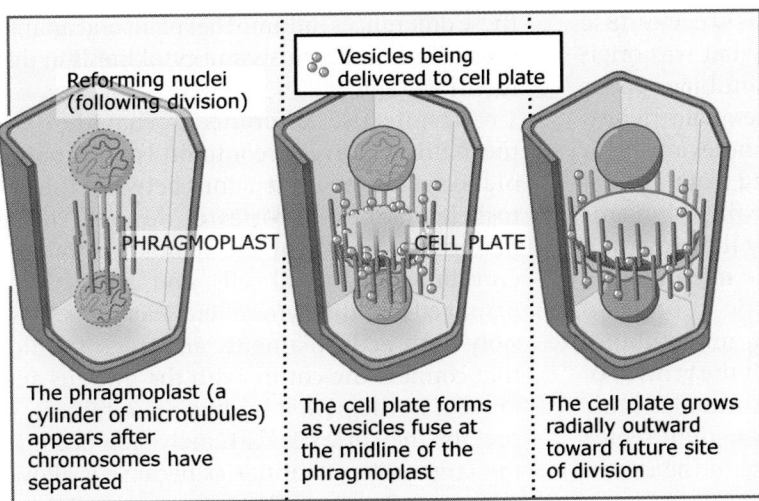

Reforming nuclei
(following division)

PHRAGMOPLAST

The phragmoplast (a
cylinder of microtubules)
appears after
chromosomes have
separated

Vesicles being
delivered to cell plate

CELL PLATE

The cell plate forms
as vesicles fuse at
the midline of the
phragmoplast

The cell plate grows
radially outward
toward future site
of division

FIGURE 21.18 The microtubules (orange) of the two halves of the phragmoplast overlap at its center. Vesicles directed by the microtubules accumulate there and fuse together to form the cell plate. The phragmoplast is initially a tight cylinder of microtubules directly between the two reforming nuclei, but expands as the cell plate does so that vesicles are always delivered at its edge.

immature cross-wall), forms in the center of the cell and grows outward until it meets the wall of the mother cell.

The events of cytokinesis are shown in **FIGURE 21.18**. They begin during late anaphase as a cylindrical bundle of microtubules appears between the newly separated chromosomes. This structure is called the **phragmoplast** and is responsible for constructing the wall that will create two new cells. In some cell types, the phragmoplast appears to be assembled from the remnants of the mitotic spindle, analogous to the midbody—the remnant of the mitotic spindle in animal cells. In other, specialized plant cells (such as the multinucleate endosperm that surrounds and nourishes the embryo), cytokinesis can form several days after the mitotic phase. In this case, when the phragmoplast forms, it occurs at the point of overlap between microtubules that radiate from sister and nonsister nuclei alike. Therefore, like the mitotic spindle, the phragmoplast (not to be confused with the phragmosome) is composed of two oppositely oriented sets of microtubules with their plus ends overlapping midway between the nuclei. Positioned among the microtubules are fingerlike tubules of ER. Actin filaments are also found among the microtubules and share their polarity. Vesicles move down the two sets of microtubules to their plus ends, where they fuse so that the membrane and the cell wall precursors they contain can begin to form the new cell wall and the membranes that will eventually separate the two cells. **FIGURE 21.19** shows a region of a dividing cell in which many vesicles have been delivered to the plane of division but have not yet begun to fuse together.

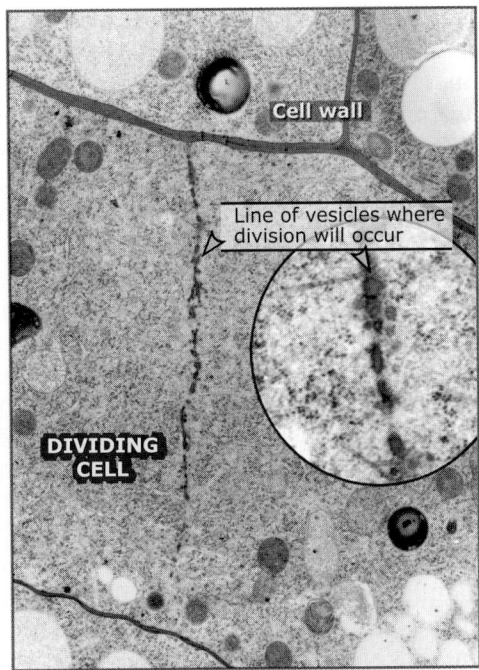

Cell wall

Line of vesicles where
division will occur

DIVIDING
CELL

FIGURE 21.19 An electron micrograph of a dividing cell in the process of constructing the cell plate. A large number of vesicles has accumulated in the plane in which division will occur, but most have not yet fused to form the cell plate. The inset shows a higher magnification view from within the line of vesicles. Photo courtesy of T. H. Giddings and L. A. Staehelin, University of Colorado.

The structure of the phragmoplast allows it to guide the formation of the new wall. Vesicles fuse only in the narrow region where the plus ends of microtubules from the two halves of the phragmoplast come together, giving the newly formed wall the shape of a flattened disk (hence its description as a "plate"). This disk grows radially outward as vesicles fuse around its circumference. So what was originally a central column of microtubules opens up as a ring that continues to expand, "like a ripple in a pond," with microtubules always at the leading edge of the growing plate. At first glance, it might seem that the ring of microtubules is widened passively by the growth of the new cell plate, and that the microtubules would not need to be dynamic in order to serve their function. However, microtubule-stabilizing drugs actually inhibit the growth of the plate, suggesting that microtubules must remain dynamic to allow its growth. It seems that microtubules depolymerize in the center of the plate where vesicle delivery is no longer needed, while new microtubules polymerize at the outer edge.

The key feature of the phragmoplast, therefore, is that it possesses mirror image symmetry with opposing sets of microtubules and actin filaments defining the midline where the new cross-wall will be laid down by the fusion of vesicles. In the next section, we will look at the contribution of membranes and cell wall molecules to this process.

Now that we have seen how the mitotic and cytokinetic processes work in plant cells, we can appreciate how much the overarching positional controls that connect them differ from those in animal cells. In animal cells the cleavage furrow is positioned by the mitotic spindle and is usually perpendicular to the spindle axis; if it is not, the result is catastrophic. The exact orientation of the spindle axis is controlled by the interaction of the spindle's astral microtubules with the cortex of the cell. The connections between events and structures in plant cells are different in several ways. First, the division plane in plant cells is predicted well before the spindle even forms, so the spindle plays no role in determining it. Second, it follows that it is a plant cell's nucleus rather than its spindle that is moved to the position where division will occur. Third, the connection between the orientations of the spindle and the new cross-wall is not nearly as tight as in animal cells. In mitotic plant cells, the axis of the spindle is sometimes not perpendicular to the plane defined across the cell by the preprophase band, and the chromosomes align at metaphase in a plane that is tilted with respect to it. This makes no difference: the new cross-wall starts to grow in a correspondingly tilted manner but then corrects itself and grows to the position that the preprophase band occupied. Together,

these differences indicate that plant and animal cells coordinate mitosis and cytokinesis in different ways.

Despite the differences, what the two mechanisms have in common is positioning of the spindle by interactions between the cytoskeleton and the cortex of the cell. This is performed directly by the spindle's astral microtubules in animal cells, and indirectly in plant cells by the cytoskeletal elements, possibly both actin filaments and microtubules that connect the cortex with the nucleus and move it in the period leading up to mitosis. A role for the cortex is extremely significant in the larger scheme of things because it allows influences from outside a cell to orient division and determine the relative sizes of the resulting cells. Without this, cell divisions could not be controlled spatially to organize the development of either a plant or an animal.

Concept and Reasoning Checks

1. Compare and contrast mitosis in plants and animals.
2. What is the function of the phragmoplast and how does it differ from the phragmosome?

21.8 Secretion during cytokinesis forms the cell plate

Key concepts

- The Golgi apparatus continues to make secretory vesicles throughout cytokinesis.
- These vesicles fuse to make a cell plate lined with new plasma membrane.

The membrane fusion events that are guided by the phragmoplast are an important part of cytokinesis. Fusion of vesicles provides material for both the cell wall that will separate the two new cells and the plasma membranes that will line both sides of the new wall. This section examines these fusion events and the way they form the cell plate in more detail.

The Golgi-derived vesicles that move along the microtubules of the phragmoplast are loaded with cell wall precursors. When the vesicles fuse, this material is released to contribute to the formation of the immature wall within the cell plate. The newly formed wall has a different composition than the existing, more mature cell walls. It is made largely of callose, a more flexible carbohydrate polymer than the relatively rigid cellulose microfibrils that are the primary component of older walls. Callose has a jelly-like consistency, giving the growing wall much greater flexibility than the cross-wall between the two new cells will possess once it is complete. Although other wall components arrive packaged inside vesicles, callose cannot be detected within them, despite the fact that it will be the primary component of the cell plate. It is thought that this polymer is synthesized directly at the plasma membrane of the cell plate. The fusion of vesicles with the new cross-wall, therefore, fulfills three roles by delivery of the following:

1. The noncallose contents destined for insertion into the wall.
2. The new plasma membrane required to line the two daughter cells.
3. The membrane-associated callose-synthesizing enzymes.

In animal cells, a single central Golgi apparatus breaks down at the beginning of mitosis and reforms after cytokinesis. Plant cells, in contrast, have multiple small Golgi stacks called "dictyosomes" that are dispersed throughout the cytoplasm during all phases of the cell cycle. These elements are essential for building the new cross-wall and continue to make secretory vesicles throughout division. Several mechanisms ensure that during cytokinesis they are targeted only to the cell plate. One way in which this is accomplished involves incorporating specialized fusion molecules into their membranes. The syntaxin KNOLLE, whose expression is tightly restricted to this part of the cell cycle, ensures that vesicles can fuse only to the new plasma membrane that lines the cell plate and not to the preexisting plasma membrane of the mother cell. Once the vesicles have been delivered to the narrow region where the two sets of microtubules in the phragmoplast face each other, a dynamin-like protein, phragmoplastin, is suggested to form spirals around the vesicles, squeezing them into elongated tubules and assisting in their fusion. In this way, a collection of membrane tubules and vesicles forms at the midline of the phragmoplast and extends outward at the edges by the arrival of new vesicles, as shown in **FIGURE 21.20**. The result is an expanding disk of plasma membrane elements midway between the two reforming nuclei. Toward the center of this disk, in the wake of the outward-moving phragmoplast microtubules, continuing fusion events among the membrane elements consolidate them into a network of tubules.

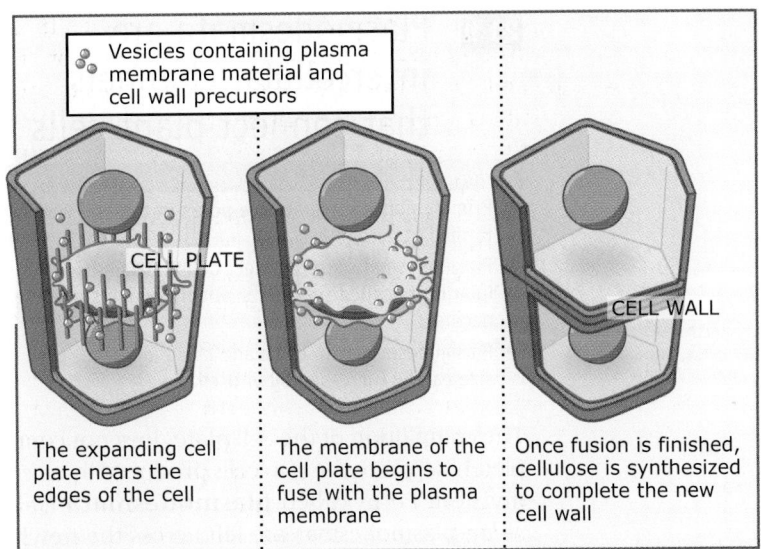

FIGURE 21.20 Fusion between the membranes of the growing cell plate and the plasma membrane of the mother cell creates two new cells. They are initially separated by a flexible cross-wall based upon the flexible polymer callose, which lacks the strength of cellulose. Cellulose synthesis strengthens the newly formed cross-wall and converts it into a mature cell wall.

Finally, the tubular network fuses further to form a more-or-less complete plate, composed of "holey" sheets of membrane sandwiching a layer of callose. There is, therefore, a gradient of membrane fusion events that is highest at the periphery of the cell plate and diminishes toward its center. Similarly, the oldest, most consolidated part of the cell plate is at its center, with newly formed regions at its outer edges.

The final stages of cytokinesis are shown in Figure 21.20. Conversion of the cell plate into a wall between two cells begins with the fusion of the fingerlike tubules of membrane at the edge of the expanding plate with the mother cell's plasma membrane. Because one side of the phragmoplast often contacts the mother cell's membrane first, fusion begins at one side and continues as other regions reach the cortex. While it is growing, the cell plate can curve or appear wavy due to the flexible consistency of callose. During these last stages of its growth, however, it straightens and flattens to make perpendicular contact with the original wall and to extend directly across the cell. This process is probably guided by cytoskeletal elements. The completed cross-wall continues to mature during the following cell cycle as cellulose is increasingly deposited within it, giving it the stiffer composition of a regular primary cell wall.

21.9 Plasmodesmata are intercellular channels that connect plant cells

Key concepts

- Primary plasmodesmata are pores in the cell wall formed at cytokinesis.
- Plasmodesmata interconnect cells into multicellular units called symplasts, within which signaling occurs.
- Plasmodesmata can open and close and their pore size can be increased by viruses.

The completion of the cell plate does not completely separate the two cells produced by a cell division. Pores called **plasmodesmata** (*singular*, plasmodesma) are left across the newly formed wall and allow material to move between the cytoplasms of the two sister cells.

Plasmodesmata form during cytokinesis when tubules of ER are trapped across the cell plate as it is constructed, as shown in **FIGURE 21.21**. The membranes of the cell plate and the trapped ER do not fuse with one another, leaving the segment of ER surrounded by a tube of plasma membrane that is continuous with the plasma membranes of the two cells. Between the tube of plasma membrane and the ER is a thin ring of cytoplasm that connects the cytoplasms of the cells. The channel formed by the three elements together is a plasmodesma. Plasmodesmata formed during cytokinesis are called "primary" plasmodesmata. A second mechanism of making plasmodesmata exists because they are also found connecting cells that are not products of the same cell division.

Formation of these "secondary" plasmodesmata begins with a localized thinning of the walls of the two cells and must somehow involve the insertion of ER into the resulting opening.

Plasmodesmata connect the cytoplasm of many cells into a single large domain called a **symplast**, as shown in **FIGURE 21.22**. Within this domain, plasmodesmata allow the transmission of electrical signals and the unrestricted movement of small water-soluble molecules. Some mRNAs and proteins can also move between cells. Not all large molecules can pass through plasmodesmata, however, and the regulated movement of specific mRNAs and proteins between cells may be used during development to determine cell fate. Movement of large molecules is likely to be controlled by regulating the size of the plasmodesmata.

Molecules may be brought to plasmodesmata by actin filaments. Plasmodesmata are concentrated in the end walls of cells, which are also a focus of the actin cytoskeleton. As we will see in *21.17 Actin filaments form a network for delivering materials around the cell*, actin filaments form a very active system for transporting materials over the large cytoplasmic distances of vacuolated plant cells. Prominent among the actin structures in a plant cell are cables that run from one end of a cell to the other and seem to be anchored at the end walls, possibly by a form of myosin unique to plants that is concentrated there. Together, the actin cables and the plasmodesmata may collaborate to facilitate the movement of molecules between cells. In the general flow of cytoplasmic contents along actin cables, material that had

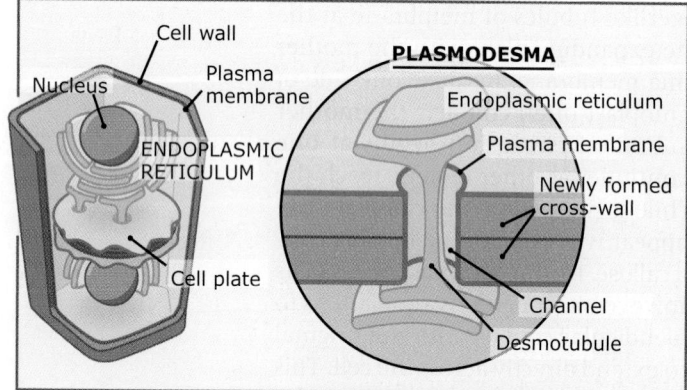

FIGURE 21.21 Plasmodesmata result when tubules of ER are trapped across the cell plate as it forms (left). A mature plasmodesma (right) has a tubule of ER (the desmotubule) running down the middle of a channel connecting the two cells. The plasma membranes of the two cells are continuous and line the channel. Molecules can move between the cytoplasms of the two cells in the space between the desmotubule and the walls of the channel.

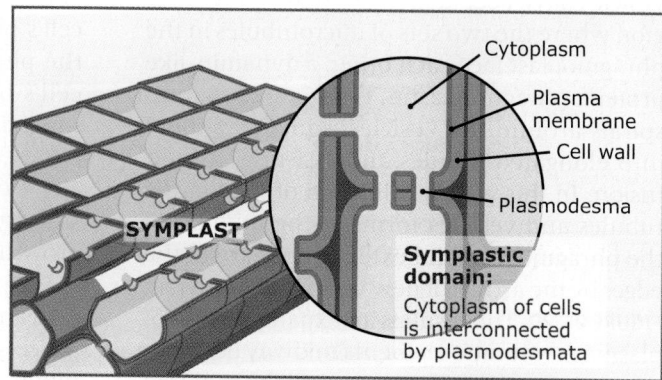

FIGURE 21.22 Plasmodesmata connect the cells in the foreground into a group with a shared cytoplasm (a symplast). Materials can move from one cell to another within the symplast but cannot move to any of the cells in the background. Such connections allow domains of cells within a plant to be given common properties.

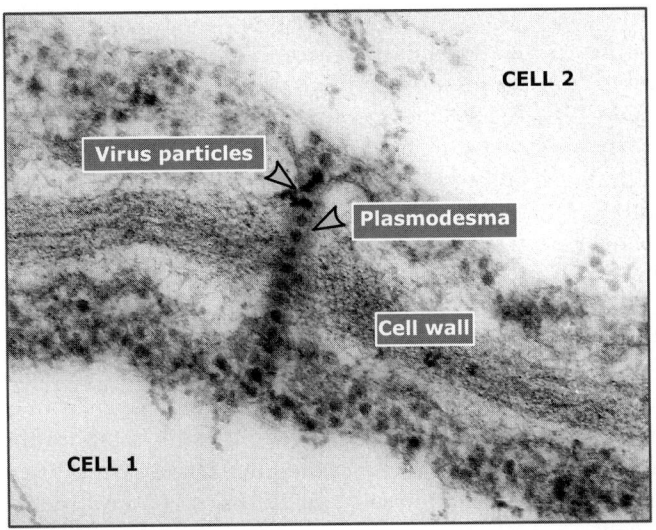

CELL 2

Virus particles

Plasmodesma

Cell wall

CELL 1

FIGURE 21.23 An electron micrograph shows cowpea mosaic virus particles moving through a plasmodesma that connects two leaf cells. Both cells have large vacuoles, which appear as the featureless areas in the lower left and upper right of the photograph. Each cell has a thin layer of cytoplasm running along the interior of its cell wall. The viral particles appear as densely stained black dots within the cytoplasm of the lower cell and line up within the plasmodesma, suggesting that the infection is spreading from the lower cell to the upper. Photo courtesy of Kim Findlay, John Innes Centre.

entered a cell from one end would be moved to the other, where the plasmodesmata could move it on to the next cell. Although there is no evidence that the large actin cables themselves pass from cell to cell via the plasmodesmata, there are suggestions that the pores contain finer actin filaments as well as myosin molecules. Their contractile activity is hypothesized to change the pore size, thereby regulating the flow of material between cells.

Many plant viruses take advantage of the connections between cells provided by plasmodesmata. Despite their large size, viruses can pass rapidly from cell to cell by way of plasmodesmata, as shown in **FIGURE 21.23**. This is possible because plant viruses express proteins—called "movement proteins"—that allow plasmodesmata to transmit much larger particles than they are capable of otherwise. How these proteins work is not yet clear. One, expressed by tobacco mosaic virus, is associated with the ER—which is modified during infection to become a virus factory—as well as with microtubules and actin filaments. The virus, therefore, takes over both the synthetic and the transport systems of its host cell, increasing the size of the plasmodesmatal pore to enable infection to spread to neighboring cells.

21.10 Cell expansion is driven by swelling of the vacuole

Key concepts

- Uptake of water into the vacuole provides a unique, pressure-driven mechanism of cell expansion.
- There is more than one type of vacuole.

So far, we have focused exclusively on the processes of mitosis and cytokinesis that occur in the apical zone of cell division. What happens as cells stop dividing? The most obvious change is the enormous increase in cell size in which vacuoles play a fundamental role. Cells in the zone of cell division are relatively small and contain at most a few small vacuoles, with some cells lacking them completely. After cells leave the zone of cell division, they enter the zone of cell expansion, where they enlarge by a mechanism in which water is taken up into a central vacuole. This expansion process involves three coordinated processes: (1) the swelling of the vacuole to provide the force for growth; (2) controlled loosening of the wall to allow the cell to expand; and (3) the channeling of the swelling force so that it can produce elongation in a particular direction. In this section, we will examine the vacuolar part of this partnership before examining the role of the cell wall.

The ability of the vacuole to drive expansion depends on one of the most basic physical properties of cells. Cells depend upon the existence of a barrier to separate their internal contents from the surrounding environment. This enables cells to concentrate metabolic reactions so that they can proceed under stable conditions unthreatened by changes in the external medium. Both plant and animal cells are surrounded by lipid-rich boundary membranes. Macromolecules and charged molecules do not pass freely across the membrane, allowing them to be concentrated within the cell. The membrane is, however, permeable to water and small, uncharged molecules, which cross it by **osmosis**. Osmosis is the process by which water passes across a semi-permeable membrane like the cell's, driven by differences in the concentrations of solutes on the two sides of the membrane. The net flow of water occurs from the less to the more concentrated solution. By concentrating salts from the environment, as well as other molecules that they make themselves, cells set up an osmotic imbalance between the cytoplasm and the external medium, causing water to flow in, to equalize the difference. Because of this influx of water, a cell is in danger of "death by flooding," in which its cytosolic reactions would be diluted and it would eventually burst. Animal cells prevent this by constantly expending energy to pump out ions like sodium, thereby lowering the internal solute concentration and counteracting the entry of water. Plant cells use a completely different strategy to deal with this problem and exploit the influx of water to drive growth.

In plants, water that has entered the cell is stored in the vacuole. By actively accumulating inorganic salts from the environment, the vacuole encourages the further uptake of water by osmosis and enlarges dramatically as a result. After a cell stops dividing and becomes fully expanded, the large central vacuole can account for up to 95% of its volume, as in the cell shown in FIGURE 21.24. The vacuole can, therefore, easily be the largest plant organelle. Surrounded by its own membrane, the **tonoplast**, the enlarged vacuole displaces the cytoplasm to a thin layer beneath the plasma membrane. Because vacuoles are filled largely with water and simple salts, vacuolation is an economical, energy-efficient way of increasing cell size.

Although they are essential for the large sizes achieved by plant cells, vacuoles also serve purposes in addition to simply occupying space. Vacuoles are often used to sequester small molecules that the cell either needs to store for use at a later time, or which it needs to segregate from its cytoplasm on a permanent basis. Some toxic chemicals accumulated from the environment are broken down or stored in vacuoles that serve a detoxifying role. In other cases, the plant itself makes toxic chemicals as a defense against microorganisms, and such toxins are stored in the vacuole. Some pigments that color parts of a plant are also deposited in vacuoles. Not all materials stored in vacuoles are sequestered there on a long-term basis, however. Vacuoles can also act as temporary reservoirs for useful materials such as inorganic ions and amino acids.

Other vacuoles are used as locations to perform reactions that cannot take place in the cytosol. Lytic and digestive vacuoles, for example, are analogous to animal lysosomes and yeast vacuoles and are responsible for the breakdown of metabolites into components that the cell can put to use. One situation in which storage and lytic vacuoles work together occurs in seeds. There, one type of vacuole performs the plant-specific function of accumulating and storing protein. When mature, such vacuoles are packed with one or a few special types of proteins at extremely high density. As the seed begins to develop, fusion of a protein storage vacuole with a lytic vacuole leads to the breakdown of the protein, releasing amino acids that are used for protein synthesis during the early stages of the plant's growth.

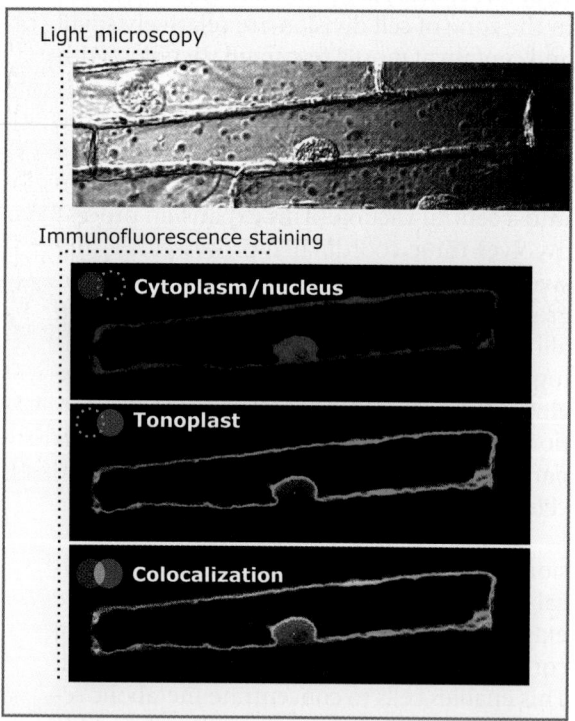

FIGURE 21.24 The uppermost photograph shows a fully elongated cell within a root or shoot. Its nucleus is visible as a semicircle in the middle of its lower border. Below are pictures of the same cell with fluorescent probes showing the region of its interior occupied by its cytoplasm and nucleus (in red) and the position of the tonoplast (the membrane that surrounds the vacuole; in green). The large unstained region that occupies most of the interior of the cell is the vacuole. Photos courtesy of Sebastien Thomine and Imagif (https://www.imagif.cnrs.fr/pole-1-Cellular_Biology.html).

21.11 The large forces of turgor pressure are resisted by the strength of cellulose microfibrils in the cell wall

Key concepts

- The plant cell wall is based largely on carbohydrate unlike the protein-rich extracellular matrix of animal cells.
- The nonrandom arrangement of stiff cellulose microfibrils controls the swelling force of turgor pressure.
- Proteins loosen the cell wall to allow cell expansion.
- The orientation of cellulose microfibrils can change from layer to layer.

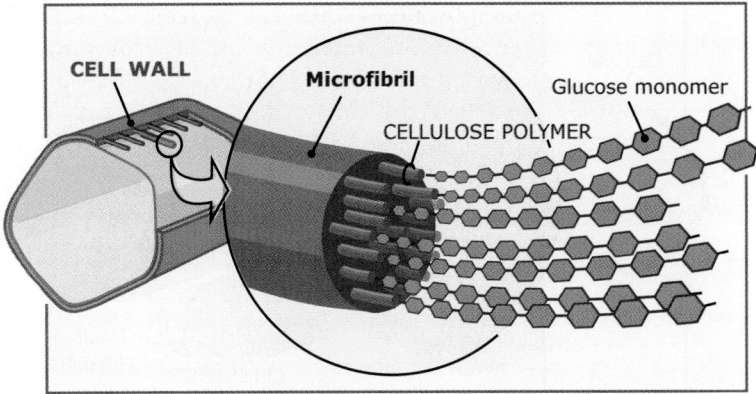

FIGURE 21.25 Individual cellulose polymers are tightly packed together to form a cellulose microfibril, the basic structural component of the cell wall. Within the microfibril the cellulose polymers all have the same orientation, allowing extensive hydrogen bonding among them. These connections make a microfibril rigid and give it great strength.

A plant cell can withstand the force of an expanding vacuole because of its cell wall. To give it this capability, the wall must possess exceptional strength. Despite its strength, however, the wall must also be able to allow the cell to expand. The details of the structure of the wall reveal the sources of these properties and show how the wall is capable of guiding expansion in a particular direction to produce an elongated cell.

The plant cell wall is thicker and stronger than an animal cell's extracellular matrix. The latter is made largely of protein, but proteins only make up about 10% of the primary plant cell wall. One such protein is extensin, which contributes to the structure of the wall. However, the primary plant cell wall is made mostly (about 90%) of carbohydrate derived from atmospheric CO_2 by photosynthesis and then converted into polymeric form.

The basic component of the cell wall is **cellulose**, a polymer of glucose residues linked end to end to form a linear chain. The large number and regular spacing of the hydroxyl groups on each chain allows many such chains to be hydrogen-bonded side by side to form a linear, semi-crystalline structure called a microfibril, as shown in **FIGURE 21.25**. It is these cellulose **microfibrils** that are the major structural component of plant cell walls. **FIGURE 21.26** shows how densely they are packed together within one. The highly hydrogen-bonded structure of microfibrils gives them both stiffness and enormous strength—their strength under tension exceeds that of steel—and it is their strength that provides the cell wall with its extraordinary toughness. The properties

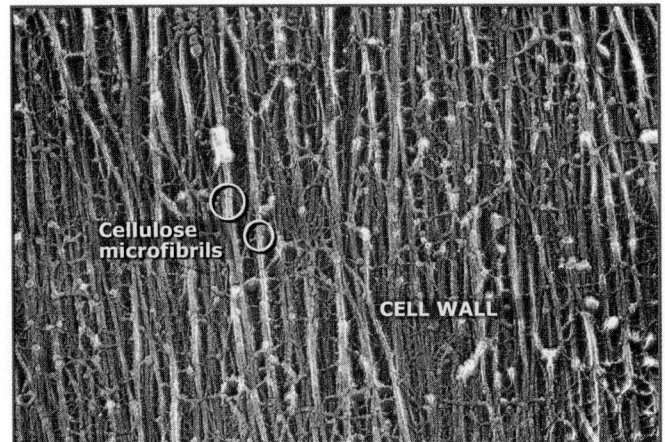

FIGURE 21.26 An electron micrograph of a cell wall. Cellulose microfibrils are visible as prominent cables of constant diameter running continuously from top to bottom of the photograph. Connections made from other wall polymers are visible between them. Photo courtesy of Brian Wells and Keith Roberts, John Innes Centre.

of cellulose are of use to humans as well as plants: wood, cotton, cardboard, and paper are all composed primarily of cellulose and depend on its stiffness and strength.

In addition to cellulose microfibrils, the cell wall contains a number of other components. **FIGURE 21.27** shows how the major components of a wall are organized. The cellulose microfibrils are crosslinked together by another form of carbohydrate polymer, the crosslinking glycans (also known as hemicelluloses). These polymers form a filamentous meshwork that is united by hydrogen bonds and extends throughout the wall. Surrounding this

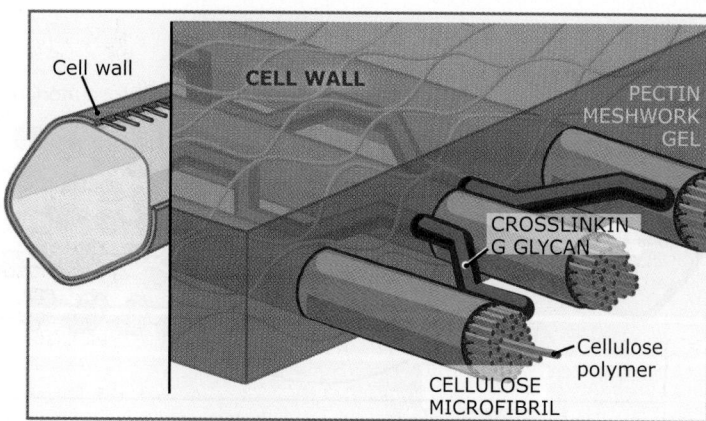

FIGURE 21.27 Within the cell wall extensive connections are present between cellulose microfibrils and meld the wall into a single large unit. Crosslinking glycans connect adjacent microfibrils along their lengths, and both components are embedded within a continuous meshwork of pectin molecules that forms a gel throughout the wall. The gel is indicated in green.

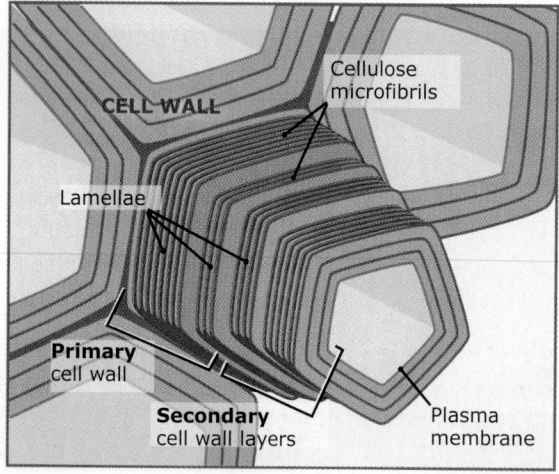

FIGURE 21.28 Cell walls consist of a number of layers, each laid down independently. Each new layer is formed inside the previous one. The oldest layer is on the outside (the primary wall) and was laid down while the cell was still expanding. After the cell has finished expanding, several new layers are added (secondary cell wall) to stiffen the wall.

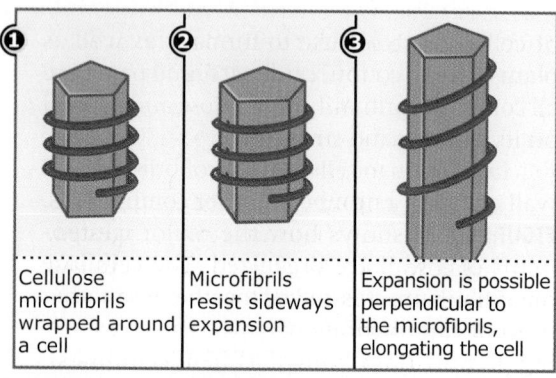

FIGURE 21.29 The red helix represents cellulose microfibrils wrapped around a cell. Their orientation determines the direction in which the cell can expand.

structural core of crosslinked fibrils are members of another class of complex carbohydrates, the pectins. Their branched structures cause them to be highly hydrated, and they form a gel that surrounds the cellulose. The tendency of pectins to gel is easy to visualize because they are the component that is responsible for the consistency of the edible jams made from many types of fruit. The gel that they form within the cell wall allows the diffusion of water and small molecules but prevents the movement of larger molecules. The pectins bind cations such as calcium, and this stiffens the gel and contributes to the strength of the wall.

The great strength given to the cell wall by cellulose microfibrils allows a plant cell to withstand the force created by an expanding vacuole without bursting. The vacuole pressing outward against the resisting wall places the contents of the cell under an internal pressure called **turgor** (also often called turgor pressure). Turgor pressure can be remarkably high. The turgor (0.6 MPa) inside a garden plant like the tulip can be several times higher than the air pressure inside a car tire (0.2 MPa), and in some plants turgor pressure has been measured at more than 3 MPa. Turgor distends the cell, puts the wall under tension, and makes the cell stiff. This stiffness is used to support many parts of the plant; the significance of turgor to the ability of non-woody plants to stand upright becomes evident when they wilt after not being watered.

Turgor pressure is an isotropic or nondirectional force. When placed in pure water, a cell deprived of its wall (called a "protoplast") will swell isotropically, like a spherical balloon, until it bursts. A cell can use turgor to elongate but to do so must construct its wall nonrandomly. **FIGURE 21.28** shows that the cell wall is made of several concentric layers—called lamellae— laid down in succession as a cell grows. Within any one layer, the cellulose microfibrils do not cross over one another, so that each lamella is not woven but is instead composed of more-or-less parallel microfibrils. As is illustrated in **FIGURE 21.29**, the orientation of these microfibrils determines the direction in which the cell is allowed to swell. When they are wrapped transversely around a cell, as in the figure, the cell is prevented from expanding sideways. Adjacent cellulose microfibrils can move apart, however, allowing the cell to elongate in the direction perpendicular to them. This is the major mechanism that plants use to direct their growth. Within a growing root, for example, the microfibrils are oriented perpendicular (transverse) to the root-shoot axis throughout the zone of cell expansion, causing the elongation of the cells along the axis.

21.12 The cell wall must be loosened and reorganized to allow growth

Key concepts

- Proteins loosen the cell wall to allow cell expansion.
- The orientation of cellulose microfibrils can change from layer to layer.

In order for plant cells to expand, the polymers in their walls must either move relative to one another or be broken. Although the cellulose microfibrils within a cell wall resist stretching, the connections that hold adjacent microfibrils together within a layer can be teased apart in a controlled fashion so that expansion can occur perpendicular to them. Two major candidates are thought to contribute to this wall loosening effect. The protein **expansin** has been demonstrated to allow sheets of pure cellulose (pieces of filter paper) to stretch when placed under tension with weights. Expansin is thought to do this in a nonenzymatic fashion by breaking hydrogen bonds that hold the closely packed cellulose microfibrils in paper together. However, cell walls are not made purely of cellulose; within a wall, cellulose is surrounded by materials like pectins and crosslinking glycans that separate the microfibrils. In the heterogeneous environment of the plant cell wall, active expansin is thought to "unzip" the hydrogen bonds between the crosslinking glycans and the cellulose microfibrils, allowing the microfibrils to move apart, as shown in **FIGURE 21.30**. The secretion of protons (H+ ions) by plant cells is thought to promote this activity, since expansins work best under acidic conditions. Because of this effect the secretion of protons could be used to control the expansion of cells. The **acid growth theory** suggests that the plant growth hormone auxin stimulates cell expansion in this manner.

Another cell wall protein is also thought to play a part in weakening the wall to allow growth. The enzyme xyloglucan endotransglycosylase has a "cut and paste" function. It works by severing crosslinking glycans and rejoining the cut ends with the ends of other cleaved glycan chains, thereby loosening the wall matrix and allowing expansion. This enzyme is likely to work in conjunction with expansins.

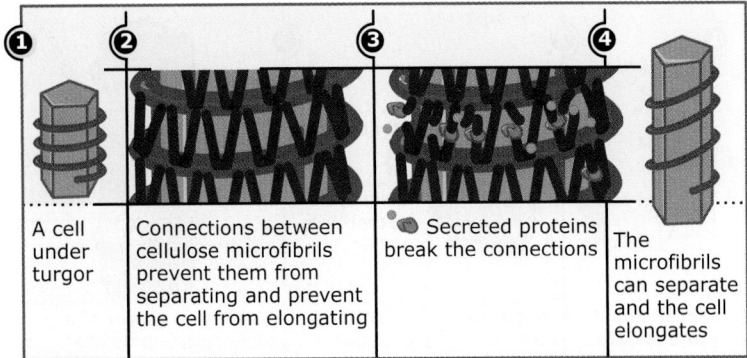

① A cell under turgor

② Connections between cellulose microfibrils prevent them from separating and prevent the cell from elongating

③ Secreted proteins break the connections

④ The microfibrils can separate and the cell elongates

FIGURE 21.30 The connections between adjacent microfibrils within a cell wall prevent them from moving apart and prevent the cell from expanding. Secreted proteins break the connections and allow a cell to expand.

What happens to the layers of cellulose as the cell grows? Like nested Russian dolls, the youngest wall lamella is on the inside, next to the plasma membrane, with several older generations surrounding it. But, unlike the dolls, the cell wall continues to grow. The first layers can be deposited around the cell when it is quite small, while the most recent layers can be put down after the cell has expanded several-fold. Something must happen to the arrangement of the cellulose microfibrils in the older layers as the cell continues to expand. One long-standing idea, called the multinet growth hypothesis, was that microfibrils were initially deposited around the cell, perpendicular to the direction of expansion (or in a "flat" helix, like a compressed bedspring). Then as newer layers of cellulose microfibrils took their place next to the plasma membrane, to match the increased cell size, the displaced older layers were realigned by the forces of growth. This process would be very much like stretching a spring, as shown in **FIGURE 21.31**. According to this model, cellulose microfibrils would show a gradient of orientations: those next to the membrane would be perpendicular to the direction of cell elongation, while those microfibrils in layers farther out would tend to become parallel to the direction of growth, with intermediate oblique angles in between.

However, cell wall growth appears to be more complex than this early model suggested. New cellulose microfibrils are not always laid down upon the plasma membrane perpendicular to the direction of cell expansion. They can be deposited in alternating layers of different orientations, forming a criss-cross pattern not easily explained by the multinet growth hypothesis. Plant growth hormones can trigger shifts in the

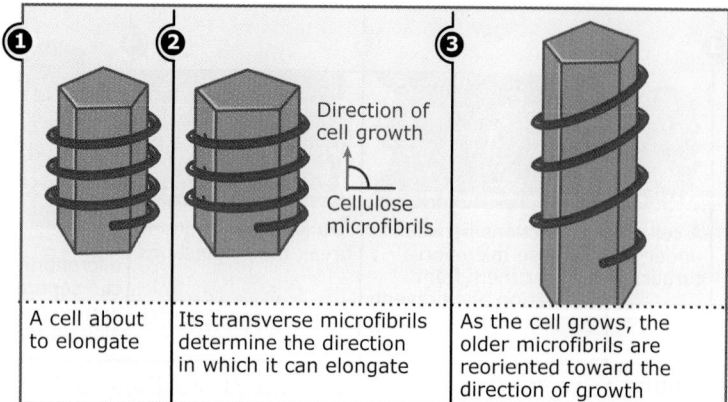

| A cell about to elongate | Its transverse microfibrils determine the direction in which it can elongate | As the cell grows, the older microfibrils are reoriented toward the direction of growth |

FIGURE 21.31 The multinet growth hypothesis suggested that as a cell expands, older microfibrils reorient toward the axis along which expansion is occurring. This reorientation is similar to the change in angle that occurs in a spring when its ends are pulled apart.

orientation of both microtubules and cellulose microfibrils. These shifts suggest that the orientation of cellulose microfibrils is actively controlled by the cell and is not the result of subsequent realignment as the cell expands. Breakage of the bonds binding adjacent cellulose microfibrils, rather than the reorientation of older layers of microfibrils, is likely to play a more important part in allowing walls to expand.

21.13 Cellulose is synthesized at the plasma membrane, not preassembled and secreted like other wall components

Key concepts

- Cellulose is polymerized by complexes embedded in the plasma membrane.
- The synthesizing complexes move along the face of the plasma membrane.

We have seen that a cell wall is of enormous importance to a plant cell and that the organization of the components within the wall plays a role in how a plant grows. How does a cell construct its wall? In particular, how does it assemble the major components of its wall, the cellulose microfibrils that give it strength?

It was initially thought that cellulose must be synthesized within the wall itself in order to patch layers that were broken as the cell grew. It is now known that cellulose is actually polymerized from multienzyme **cellulose synthesizing complexes** embedded in the plasma membrane. The sites of cellulose biosynthesis were first seen when a technique called freeze-fracture electron microscopy was applied to the membranes of plant cells synthesizing new wall. This technique splits the two leaflets of a membrane apart so that complexes of integral membrane proteins can be easily seen as particles when one of the leaflets is examined in the electron microscope. As shown in FIGURE 21.32, the plasma membranes of plant cells contained many very large and characteristically shaped transmembrane complexes containing six particles arranged as a hexagon. Each of these hexagonal "rosettes" is about the size of a ribosome and contains multiple copies of at least three different isoforms of CESA cellulose-synthesizing enzymes. FIGURE 21.33 shows how the individual copies of the enzyme are likely to be arranged within a rosette. The complex takes in activated glucose precursors on its cytoplasmic face and uses them to polymerize multiple polyglucan chains that immediately crystallize to form the cellulose microfibril that is continuously extruded onto the extracellular face of the membrane. As a rosette synthesizes cellulose, the complex moves within the plane of the membrane, trailing the newly synthesized microfibril along behind it, as shown in

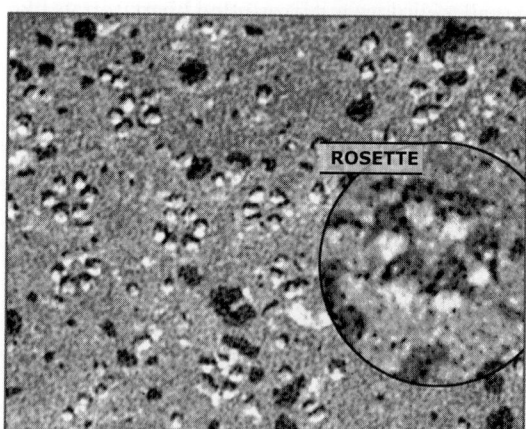

FIGURE 21.32 A freeze-fracture electron micrograph shows the integral membrane proteins within the plasma membrane of a plant cell. Several hexagonal "rosettes" are visible. A single rosette is shown at higher magnification in the inset. Reproduced from C. H. Haigler and R. L. Blanton, *Proc. Natl. Acad. Sci. USA* 93 (1996): 12082–12085. Copyright © 1996 National Academy of Sciences, U.S.A. Photo courtesy of Candace H. Haigler, North Carolina State University and appreciation to M. Grimson for technical assistance.

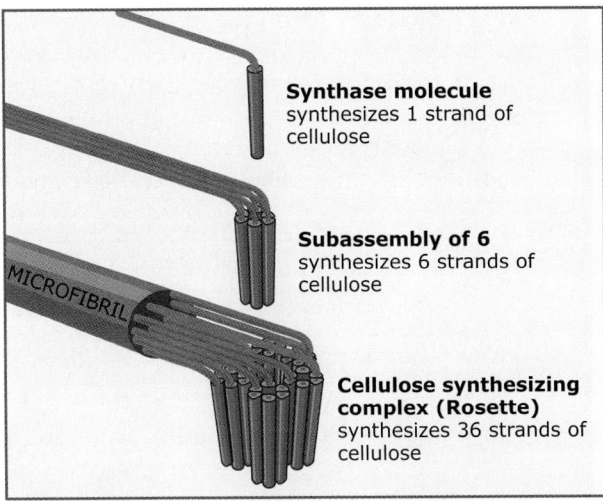

Synthase molecule
synthesizes 1 strand of cellulose

Subassembly of 6
synthesizes 6 strands of cellulose

MICROFIBRIL

Cellulose synthesizing complex (Rosette)
synthesizes 36 strands of cellulose

FIGURE 21.33 Each complete rosette is assembled from 36 copies of the cellulose synthase enzyme. Their arrangement within the structure allows the cellulose strands they produce to assemble together into a microfibril.

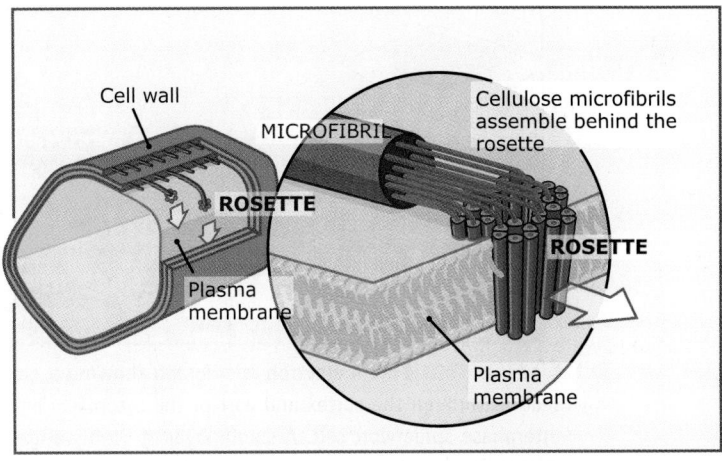

Cell wall

MICROFIBRIL

Cellulose microfibrils assemble behind the rosette

ROSETTE

Plasma membrane

ROSETTE

Plasma membrane

FIGURE 21.34 Microfibrils are assembled continuously behind cellulose-synthesizing complexes that move within the plane of the membrane. The close proximity of the cellulose polymers as they leave the rosette causes them to assemble almost immediately after they are synthesized.

FIGURE 21.34. One idea about how rosettes are propelled is that the act of extruding a linear polymer, which becomes anchored into the cell wall, provides the motive force for moving the rosette across the face of the fluid plasma membrane.

Each rosette contains enough cellulose-synthesizing enzymes to extrude the 30–50 cellulose chains that make up a microfibril. The best guess is that each of the six particles in a rosette makes 6 polyglucan chains, so that a total of 36 makes up the microfibril extruded from each rosette. These unbranched polyglucan chains pack together in parallel fashion to form a linear semi-crystalline structure—the cellulose microfibril. Since individual chains of cellulose are about 1–5 micrometers in length, while the microfibrils they comprise can be much longer than 100 micrometers (enough to encircle most cells), the ends of the individual cellulose chains must be staggered along the length of the microfibril. The use of membrane-bound enzymes for producing the major fiber of the plant cell's extracellular matrix is unique and contrasts with the secretion of the major fibrous protein of the animal cell's extracellular matrix, collagen (collagen is secreted in precursor form and assembled on the cell surface). With the exception of callose, which is also synthesized directly at the plasma membrane, but by other synthases, the production of cellulose is quite different from that of the rest of the cell wall. All other

components of the cell wall are preassembled inside the cell and exported within Golgi vesicles that fuse with the plasma membrane to release their contents.

21.14 Cortical microtubules organize components in the cell wall

Key concepts

- During interphase the microtubules in plant cells are primarily located immediately beneath the plasma membrane.
- Cortical microtubules are often coaligned with the newest cellulose microfibrils.
- Cortical microtubules provide tracks for the synthesis and assembly of cellulose microfibrils.

We have seen that the nondirectional force created by vacuolar swelling is channeled into directional cell elongation by transversely wound cellulose microfibrils. How are the microfibrils arranged upon the cell surface in this highly nonrandom way? The answer is thought to lie with the way that microtubules are organized in plant cells.

Unlike many animal cells, plant cells do not possess microtubules that radiate from a single point near the nucleus and extend throughout the cytoplasm. Instead, during interphase the microtubules of a plant cell are largely restricted to the inner face of the plasma membrane

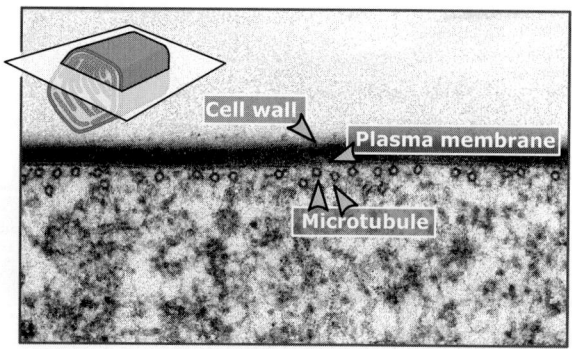

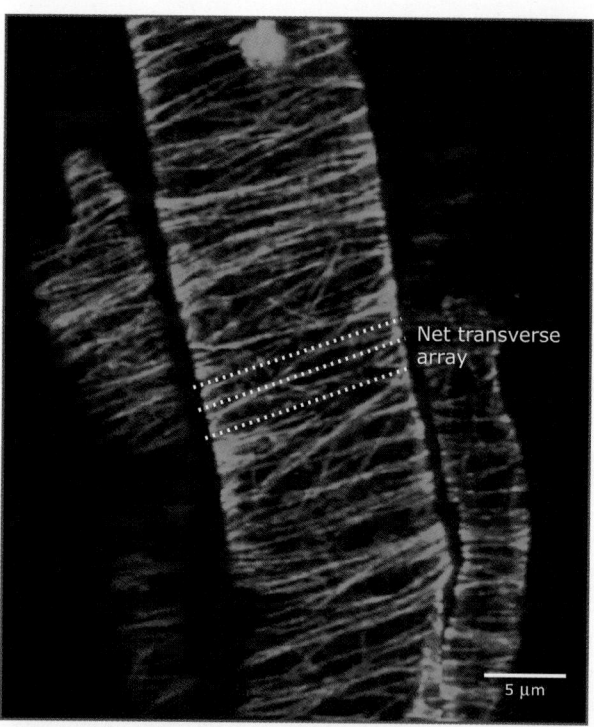

: **FIGURE 21.35** An electron micrograph showing a cross section through the cortex and part of the cytoplasm of an interphase spiderwort cell. A single layer of highly aligned microtubules lies immediately beneath the plasma membrane. No other microtubules are present in the cytoplasm. Reproduced from S. A. Lancelle, D. A. Callaham, and P. K. Hepler, *Protoplasma* 131 (1986): 153–165. Used with permission of Springer Business + Media. Photo courtesy of Peter K. Hepler, University of Massachusetts—Amherst.

FIGURE 21.36 The microtubules within an individual *Arabidopsis* cell expressing a fluorescent form of tubulin. Most of the microtubules run around the cell in a parallel array. Reproduced from D. H. Burk and Z. H. Ye, *The Plant Cell* 14 (2002): 2145–2160. Copyright 2002 by American Society of Plant Biologists. Reproduced with permission of the American Society of Plant Biologists in the format of Textbook via Copyright Clearance Center. Photo courtesy of Zheng-Hua Ye, University of Georgia.

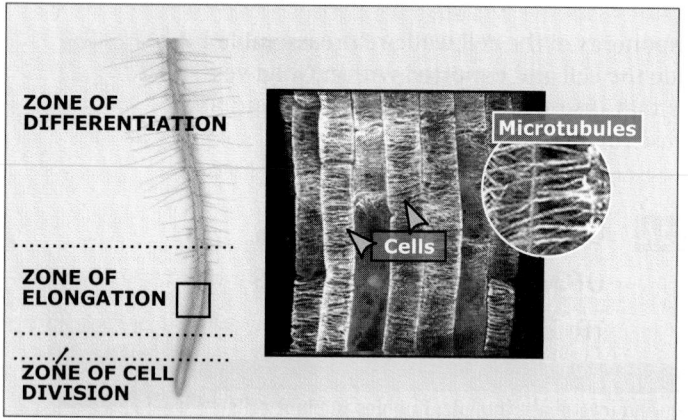

: **FIGURE 21.37** A growing *Arabidopsis* root is shown on the left. A group of cells from the zone of elongation of another *Arabidopsis* root is shown in the box on the right. The cells within the box have been stained to reveal the organization of their microtubules. Almost all of the microtubules within them are aligned perpendicular to the direction in which the cells and the root are elongating. The microtubules within a small area of one or two of the cells are shown in the circular inset. Left photo courtesy of John Schiefelbein, University of Michigan; right photos courtesy of Keiko Sugimoto, RIKEN Plant Science Center.

and run parallel to its surface, as shown in **FIGURE 21.35**. Multiple connections between each microtubule and the membrane are apparently responsible for keeping the microtubules immediately adjacent to it over most of their length. Microtubules are distributed over the entire plasma membrane and are highly organized into one large array in which all the individual microtubules can be roughly parallel

to one another, as shown in **FIGURE 21.36**. The orientation of this cortical array is not random and often appears to be related to what a cell is doing. In elongating cells, for example, the microtubules are usually arranged perpendicular to the direction of expansion so that they encircle the axis along which expansion is occurring, as shown in **FIGURE 21.37**. In addition to a large number of highly organized cortical microtubules, some cell types have microtubules extending outward from around the surface of the nucleus. Their function is unclear, however, and it is the cortical array of microtubules that plays a major role in organizing the cell wall.

The connection between microtubules and the organization of the cell wall is shown most clearly when a cell's microtubules are depolymerized. This is most easily caused by the use of small molecules that interact specifically with microtubules, such as the natural compound colchicine or one of several synthetic chemicals that are used as weed killers. Cells in a root treated with such drugs continue to expand but lose their ability to elongate and instead swell in all directions. As a result, the root tip becomes a disorganized mass of cells. At the cellular level, this occurs because the ability to order newly synthesized cellulose in the cell wall is lost when the cell's microtubules are depolymerized. In a cell treated with the drug, new cellulose is laid down randomly and the cell cannot generate the organized pattern of cellulose microfibrils that is needed to convert expansion into elongation.

How can microtubules influence the organization of cellulose? The answer appears to involve the coalignment of cytoplasmic microtubules with cellulose microfibrils on the other side of the membrane. Microtubules within the cortical array tend to parallel the newest microfibrils being laid down on the external face of the membrane, as shown in **FIGURE 21.38**. This observation suggests that microtubules act as a template for the organization of the wall by somehow guiding the synthesis and assembly of new cellulose. Two related models have been proposed for how this could occur and result in the parallel arrangement of microtubules and cellulose microfibrils as shown in **FIGURE 21.39**. In one, cellulose synthesizing complexes move in the plasma membrane in lanes created by underlying microtubules. Either the microtubules themselves or a series of links formed by proteins connecting them to the membrane would serve as boundaries that the complex cannot cross. In this model, the microtubules don't help the cellulose synthesizing complex to move; they simply act to keep a self-propelled cellulose synthesizing complex on course. The act of spinning out a stiff, linear cellulose microfibril would propel the complex within the fluid membrane. In a second model, the microtubules play a more active role and provide the motive force for moving the complex. By fluorescently labeling both the cellulose synthase enzyme and the microtubules, it has been shown that the rosettes do move along the plasma membrane and they do so along tracks that match the underlying microtubules. Sometimes, the synthases continued moving past the ends of microtubules, suggesting that microtubules are not required for the motility of the enzymes but are required to orient their movement.

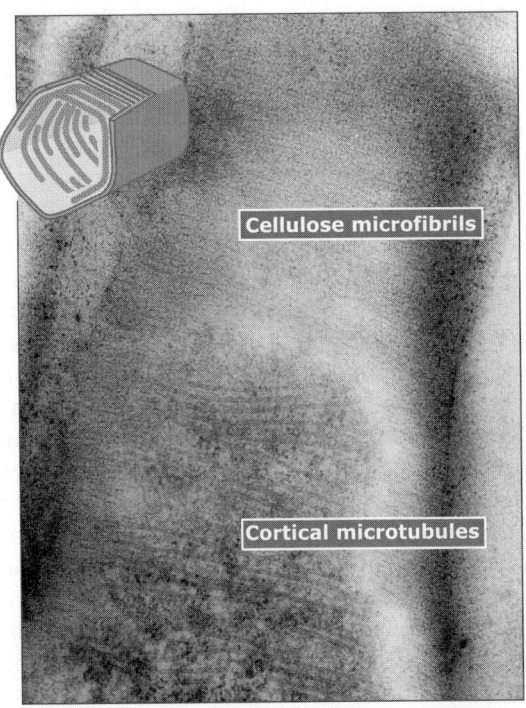

FIGURE 21.38 An electron micrograph showing a section that enters an *Arabidopsis* root cell at a very shallow angle. The long axis of the cell is from top to bottom of the photograph. The section gradually crosses the cell wall in the upper half of the picture before entering the cortical cytoplasm and exposing cortical microtubules. The microtubules within the cortex are aligned with one another and run parallel to the strands of cellulose within the wall. Photo courtesy of Brian Wells and Keith Roberts, John Innes Centre.

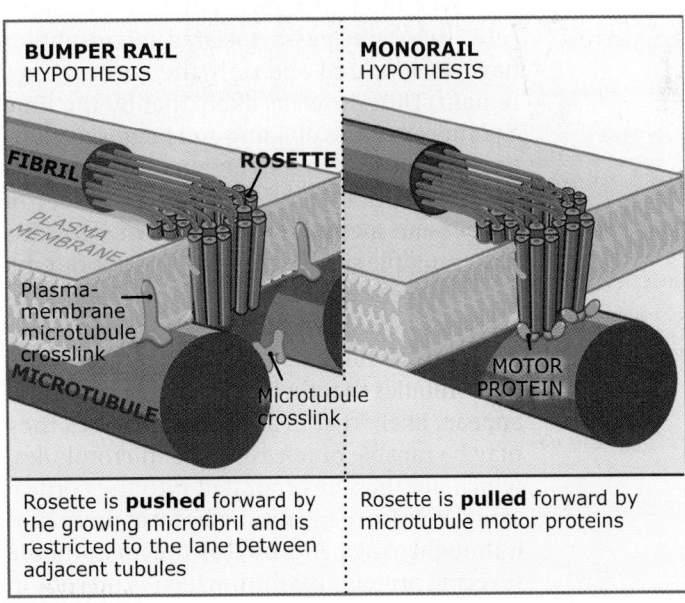

FIGURE 21.39 In the bumper rail model, cortical microtubules and their connections with the membrane form boundaries that a rosette cannot cross. Rosettes are thus restricted to move in the lanes between adjacent microtubules. Because of the close spacing of the microtubules in the cortex, they are closely aligned with newly formed microfibrils. In the monorail model a rosette is simply pulled along a microtubule by microtubule-dependent motor proteins.

21.15 Cortical microtubules are highly dynamic and can change their orientation

Key concepts

- Plant microtubules polymerize from multiple sites.
- Microtubules can move along the cortex after they have been nucleated.
- Microtubule-associated proteins organize microtubules into parallel groups.
- The microtubule array can reorient in response to hormones, gravity, and light.

How do the microtubules within the plant cell cortex become so highly organized? To answer this question, we need to know how the cortical microtubule array forms, starting with how the individual microtubules within it are generated. Plant cells do not have a single, conspicuous site of microtubule nucleation, like the animal centrosome, but are thought to have multiple nucleation sites dispersed around the cell. The dispersion of the microtubule nucleation process seems to be related to the large sizes achieved by vacuolated plant cells. Some nucleating material is apparently located around the nuclear surface since microtubules are sometimes seen to radiate from it. This often occurs just after cytokinesis. However, in many kinds of mature elongated cells, these nucleus-associated microtubules have disappeared and only the cortical ones remain. Thus, it seems likely that by the time a plant cell ceases dividing and begins to elongate, the nucleation sites for microtubules have become distributed over the inner surface of the plasma membrane. Consistent with this idea, multiple sites distributed throughout the cortex have been observed to nucleate microtubules. These sites may be able to move within the cortex while attached to the ends of the microtubules that they have nucleated. It also appears likely that in at least some cases they may be capable of releasing the microtubules, which are then incorporated into the cortical array. Release of microtubules from these sites is thought to be performed by the microtubule severing protein, katanin, which is present in plant cells and is known to affect the organization of cortical microtubules.

Cortical microtubules are not static tracks but are highly dynamic. Fluorescently tagged tubulin microinjected into plant cells incorporates rapidly throughout the entire array. This is due to dynamic instability—the alternating bouts of rapid shrinkage and steady regrowth of the fast-growing plus ends of microtubules (see *11 Microtubules*). The ability of microtubules to shrink and re-grow in new directions allows the microtubule array as a whole to explore different configurations. However, plant microtubules are not attached to a centrosome, and this additional freedom of their minus ends allows cortical microtubules to move upon the inner surface of the plasma membrane by treadmilling. In treadmilling, the loss of tubulin subunits from the minus end of a microtubule is matched by addition of subunits to the plus end, giving rise to a net movement of polymer. However, it is important to note that plant microtubules do not treadmill exactly like microtubules in vitro or in animal cells. In plants, the plus end grows faster than the minus end shrinks. This so-called modified or hybrid form of treadmilling allows the microtubules—not to slide—but to move along the membrane by addition of subunits at one end and subtraction at the opposite end.

Microtubule nucleation is, therefore, dispersed over the cortex and the microtubules then treadmill in various directions; how do these behaviors generate an organized array? When two "crawling" microtubules meet, the outcome seems to depend upon the angle of the collision. Microtubules colliding at steep angles can cross over one another. In some cases this leads to severing of the crossing microtubule, with the lagging end shrinking back from the cross-roads and the leading end surviving to make a fresh contact. Another possibility is that collisions at a steep angle more than 40 degrees might lead to depolymerization before crossing over. But when growing microtubules make contact at shallow angles less than 40 degrees, they adjust their paths to grow together on parallel paths. It is known that the microtubules of cortical arrays can be highly organized in parallel groups (see Figure 21.38). It is likely that this parallel organization is maintained by regular cross-bridges, such as those formed by the microtubule-associated protein MAP65, as **FIGURE 21.40** shows. The properties of treadmilling, angle-dependent contact behavior, and cross-bridging begin to explain how microtubules that initially grow in various directions can be channeled locally into growing in a common direction. However, it is not yet clear how these behaviors are coordinated so that microtubules come to share the same orientation throughout the cortex of a large cell and even between the cells within

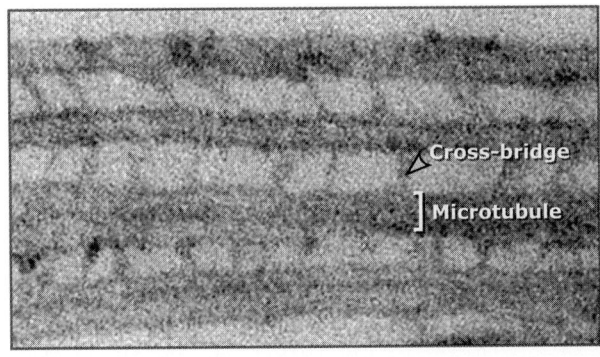

FIGURE 21.40 An electron micrograph showing several microtubules formed *in vitro* from pure tubulin and crosslinked together by the filamentous microtubule-associated protein, MAP65. The protein connects adjacent microtubules at multiple regularly spaced points along their lengths. Reproduced from J. Chan, et al., *Proc. Natl. Acad. Sci. USA* 96 (1999): 14931–14936. Copyright © 1999 National Academy of Sciences, U.S.A. Photo courtesy of Clive Lloyd. John Innes Centre.

a group. Plant growth hormones are likely to play a part in this.

Although microtubules are generally arranged perpendicular to the long axis (transverse) in elongating cells, they can adopt other patterns. In some cases they are oblique to the long axis, forming helices like the stripes on a barber's pole. They can also align parallel to the long axis and be accompanied by new layers of cellulose microfibrils in the same orientation. A switch to this pattern from the transverse pattern is often seen in cells that are ceasing to elongate. As shown in **FIGURE 21.41**, one situation in which this occurs is when cells stop expanding and progress out of the zone of cell expansion and into the zone of differentiation.

Many of the influences that affect the shape and growth of a plant act at the cellular level by controlling the orientation of microtubules, as shown in **FIGURE 21.42**. One example is the gas ethylene, which acts as a hormone in plants. The hormone is made within plants and is released in areas where they are wounded but can also be added experimentally as a gas. Plants treated with ethylene are short and stunted because cell elongation is inhibited. Ethylene causes the microtubules within the cells of the plant to reorient with respect to the root-shoot axis, changing an array in which the microtubules are perpendicular to the axis to one in which they are parallel with it. Concomitant with the rearrangement of the microtubules, cellulose is deposited such that cells expand perpendicularly to the axis ("sideways") rather than along it, producing a much shorter and thicker plant than normal. Another hormone, gibberellic acid, is responsible for stem elongation and stimulates a reversal of this orientation, reorienting the microtubules so that they are again perpendicular to the axis and cellulose can be deposited so that cells cease expanding sideways and revert to elongation. Several other hormones that affect

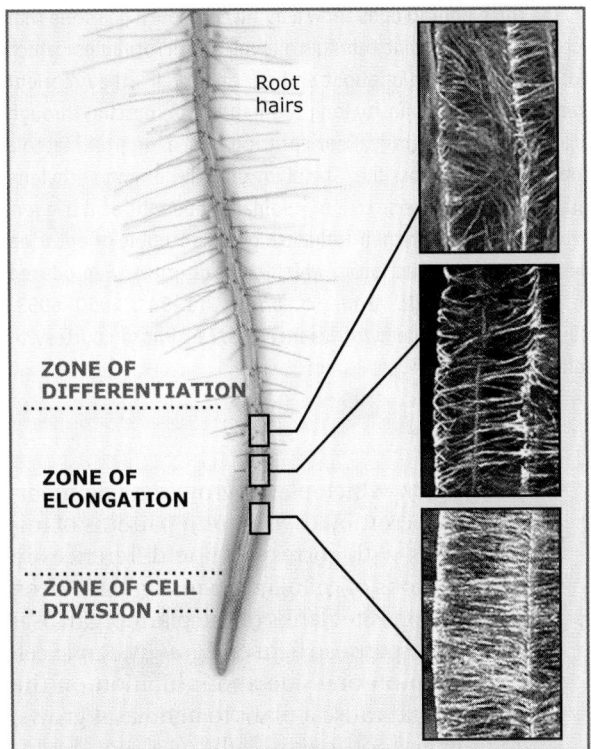

FIGURE 21.41 A growing *Arabidopsis* root is shown on the left. Areas within cells from a similar root stained to reveal its tubulin show that the microtubules remain transversely oriented while a cell is actively elongating but reorient toward a longitudinal arrangement as the cell nears its full length. Left photo courtesy of John Schiefelbein, University of Michigan; right photos reproduced from K. Sugimoto, R. E. Williamson, and G. O. Wasteneys, *Plant Physiology* 124 (2000): 1493–1506. Copyright 2000 by American Society of Plant Biologists. Reproduced with permission of the American Society of Plant Biologists in the format of Textbook via Copyright Clearance Center. Photo courtesy of Geoffrey O. Wasteneys, University of British Columbia.

the growth of plants are also known to change the alignment of cortical microtubules and affect the direction of cell expansion. Even the well known effects of light and gravity on the

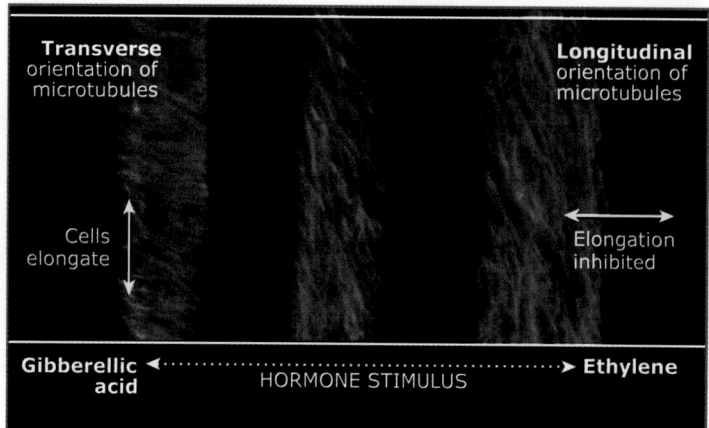

FIGURE 21.42 The three photographs show a living cell from a pea stem that has been injected with fluorescent tubulin to label its microtubules, which are seen in red. Over the course of about an hour and a half, they reorient from mostly transverse (left) to mostly longitudinal (right), passing through an intermediate orientation (center). Reorientation from a longitudinal to a transverse orientation is also possible. The plant growth hormone ethylene causes reorientation from transverse to longitudinal; gibberellic acid triggers the opposite transition. Through their influence on microtubule orientation these hormones control the direction in which the cell grows. Reproduced from M. Yuan, et al., *Proc. Natl. Acad. Sci. USA* 91 (1994): 6050–6053. Copyright © 1994 National Academy of Sciences, U.S.A. Photo courtesy of Clive Lloyd, John Innes Centre.

direction in which plants grow may be mediated by microtubules. Different patterns of microtubules with corresponding differences in the directions of cellular expansion have been seen on opposite flanks of the plant organ as it bends in response to light and gravity. Rapid cell elongation on one side and inhibition on the other would cause a plant to bend as it grows, guiding a shoot toward light or a root downward with gravity. Stationary plants can, therefore, be thought of as "steering" their direction of growth by controlling the sites of cell elongation at strategic points in their roots or stems, and microtubules could play a significant role.

21.16 A dispersed Golgi system delivers vesicles to the cell surface for growth

Key concepts

- The plasma membrane and cell wall materials needed for growth are provided by the ER/Golgi system.
- The Golgi apparatus is dispersed in plants.
- The actin system propels the dynamic Golgi apparatus over the ER network.

We have so far concentrated on how microtubules and cellulose microfibrils contribute to the growth of the cell wall as a cell expands. In this section, we consider how the plasma membrane grows along with the wall and how components of the wall other than cellulose are delivered to it.

Enlargement of the cell requires the production of a great deal of new membrane. Membranes in general are not very elastic and stretch very little, so new membrane must be constantly inserted as the cell grows. This occurs by **exocytosis**, which is the fusion of cytoplasmic membrane-bounded vesicles with the plasma membrane. Not only does exocytosis insert vesicle membrane into the plasma membrane, but at the same time it releases the contents of the vesicle to the exterior of the cell, where they can be incorporated into the growing cell wall. These vesicles are produced by the secretory pathway of the cell's endomembrane system. This is a highly dynamic system of membrane-bound compartments in which materials pass sequentially from one compartment to the next either via interconnected tubules or by vesicles that bud off one compartment and fuse with another. The ER is a major component of the endomembrane system, comprising a highly three-dimensional network of membranous tubules and sheets that extends from the nuclear envelope to just beneath the plasma membrane. Proteins synthesized on ribosomes associated with the ER enter the organelle and pass from it to the **Golgi apparatus**, where they are sorted to the plasma membrane for growth or to the vacuole for storage or breakdown.

The Golgi apparatus of a growing plant cell is responsible for synthesizing much of the material used to construct the new cell wall. In addition to adding sugar side chains (glycosylating) and sorting proteins imported from the ER—functions it shares with that of animal cells—the Golgi apparatus of a plant cell also performs the synthesis of all the large, complex cell wall polysaccharides other than cellulose and callose. Carbohydrate wall components are synthesized and preassembled inside the Golgi apparatus, packaged in vesicles, and exported. Although an animal cell's extracellular matrix contains proteins with sugar side chains, the large quantities of new wall components that are required during the growth and differentiation of a plant cell mean that the plant Golgi apparatus is more heavily devoted to carbohydrate synthesis than that of all but a few types of specialized animal cells.

Despite the fact that it does not synthesize cellulose, the Golgi apparatus seems to sometimes play a role in its localization. The hexagonal rosettes that contain the cellulose synthase enzymes are assembled in the Golgi apparatus and arrive at the plasma membrane in the membranes of Golgi vesicles. By localizing the delivery of such vesicles, plant cells selectively thicken their walls in one area but not another.

In addition to sending vesicles to the plasma membrane, the Golgi apparatus routes vesicles to the vacuole so that it can expand. The membrane of these vesicles supplies the vacuolar membrane—the tonoplast—while their contents can help specialize the interior of a vacuole for functions other than simply filling space. In seeds, for example, vacuoles are used to store large quantities of certain proteins that will later be hydrolyzed to provide amino acids during the initial stages of the development of the seed into a plant. Both the storage proteins and the enzymes that hydrolyze them are delivered to the vacuole in Golgi vesicles.

The general arrangement of ER and the Golgi apparatus is different in the cells of plants and animals. In animal cells, the ER extends from the nucleus throughout the cytoplasm in three dimensions and the Golgi apparatus is usually a single coherent unit that is centrally located well within the interior of the cell near the nucleus. In plant cells, the ER is also associated with the nucleus, but the sections in other parts of the cell form an essentially two-dimensional system flattened by the large, turgid central vacuole into a thin layer of cytoplasm beneath the plasma membrane. **FIGURE 21.43** shows the two-dimensional nature of the ER in a vacuolated plant cell, and **FIGURE 21.44** illustrates how closely apposed the ER and the plasma membrane can be in such a cell. The plant Golgi apparatus is also near the plasma membrane. But unlike animal cells where the Golgi apparatus is gathered around the nucleus, the plant cell's Golgi is scattered throughout the cortical layer of cytoplasm in the form of multiple bodies called dictyosomes, each one composed of stacks of cisternae. Several independent Golgi stacks within the cytoplasm of a plant cell are shown in **FIGURE 21.45**. This dispersion of both the ER and the Golgi apparatus near the plasma membrane allows plant cells to have multiple biosynthetic sites for wall components and to secrete them over the very large areas of expanding wall in vacuolated plant cells.

The distribution of the ER and the Golgi apparatus within a plant cell depends on the

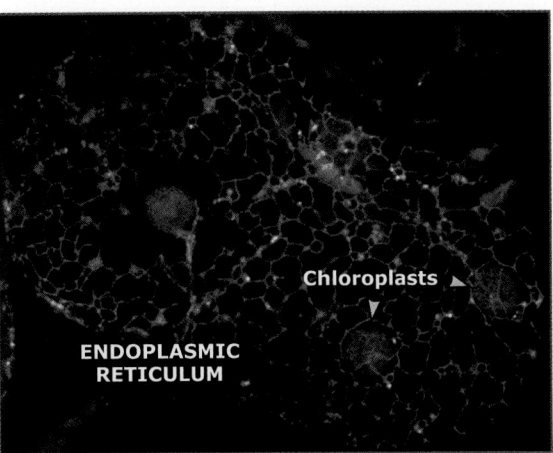

FIGURE 21.43 A single live epidermal cell expressing a form of green fluorescent protein (GFP) restricted to its ER. The plane of focus is just beneath the surface of the cell, and an extensive two-dimensional network of ER extending throughout its entire cortex is visible. Several chloroplasts are present near the cortex and are surrounded by the ER. The chloroplasts do not contain GFP; they are green because they autofluoresce at the wavelength used to stimulate GFP fluorescence. Reproduced from *The Plant Journal*, P. Boevink, et al., vol. 10 (5), pp. 935–941. Copyright © 1996 John Wiley & Sons, Ltd. Reproduced with permission of Blackwell Publishing Ltd. Photo courtesy of Karl J. Oparka, The University of Edinburgh.

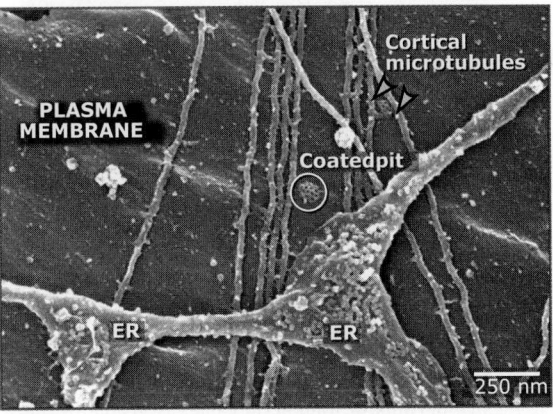

FIGURE 21.44 A scanning electron micrograph of a region of the cortex of a plant cell. The background is the interior surface of the plasma membrane. Several tubules of ER and two points at which they branch are visible. Photo courtesy of Tobias Baskin, UMASS—Amherst.

actin cytoskeleton. In contrast to animal cells, where these two organelles are organized by microtubules, their organization in plant cells results from association with actin filaments. **FIGURE 21.46** shows how the Golgi elements in a plant cell are almost exclusively associated with actin cables. Actin-dependent movement extends both the ER and the Golgi apparatus

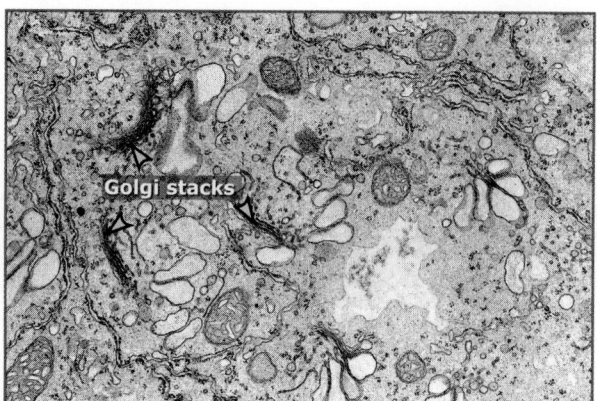

FIGURE 21.45 An electron micrograph of part of the cytoplasm of a maize root cap cell shows multiple independent Golgi stacks. Many other similar stacks are distributed throughout the rest of cytoplasm, many cells containing 30 or more. ER is visible as long, thin elements along the left edge of the picture and in the upper right. About six mitochondria appear as round or oval objects with highly convoluted interiors. Photo courtesy of Chris Hawes, Oxford Brookes University.

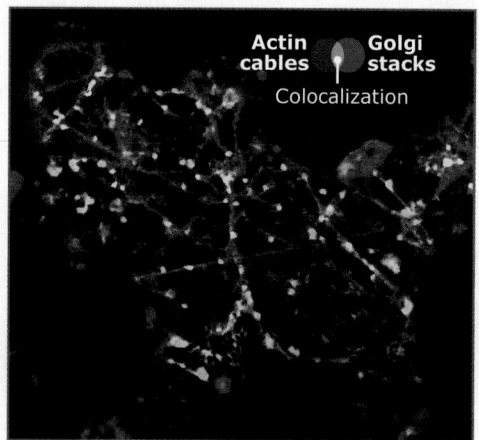

FIGURE 21.46 A single live leaf epidermal cell expressing a fluorescent protein in its Golgi apparatus (green) and simultaneously labeled for actin (red). Yellow indicates where both are present. An extensive network of long actin cables is visible in the cell. The Golgi apparatus is visible as dots, almost all of which colocalize with actin cables. Photo courtesy of Chris Hawes, Oxford Brookes University.

into new regions of a cell as it grows. However, movement is not limited to this type of gradual motion. The Golgi stacks within plant cells are highly motile and are in constant motion over the surface of the less mobile ER. This motion may increase the area of ER from which the Golgi stack can receive cargo and increase the rate at which material is transferred between the two organelles.

21.17 Actin filaments form a network for delivering materials around the cell

Key concepts

- Organelles and vesicles move around the cell by cytoplasmic streaming, powered by actin-myosin interaction.
- Plants have two unique classes of myosin.

As cells grow by vacuolation, they begin to face problems distributing material around their interiors. Plant cells can become extremely large, and as they grow their cytoplasm is spread very thinly over large expanding surfaces. A cell's large dimensions and great surface area require active means to ensure both that wall components can be incorporated wherever they are needed and that materials in general can be distributed throughout the cell. Plant cells solve these problems by **cytoplasmic streaming**, a special mechanism, based on actin filaments, that constantly propels materials around the interior of the cell.

Actin filaments form a variety of different networks in plant cells. They are found associated with all the microtubule arrays, to which they seem to be linked. We have already seen that actin filaments run parallel to the microtubules of the phragmoplast and are found among those of the preprophase band. Similarly, during cell expansion, fine actin filaments run alongside the cortical microtubules. Because these filaments are arranged perpendicularly to the direction of cell elongation, transport along them would probably not be an efficient means of delivering materials to the ends of elongated cells measuring hundreds of micrometers in length. This long-distance role seems to be fulfilled by thicker cables of actin, which run deeper in the cytoplasm and are not generally associated with microtubules. These cables, which support the thin strands of cytoplasm stretched across the vacuole, radiate from the nucleus and extend to the edge of the cell, often running all the way to the ends of elongated cells. Several such large actin cables are visible in the cell shown in **FIGURE 21.47**. Prominent cables often run to the ends of expanding cells, parallel to the direction of cell expansion. Using a simple light microscope, it is easy to see rapid bidirectional motion of many cytoplasmic contents along these cytoplasmic strands. This constant flow of particles around the cell constitutes the process of cytoplasmic streaming. Vesicles and organelles such as chloroplasts, the ER, the Golgi apparatus, and the nucleus are moved about the cell in this way. Movement of organelles along actin contrasts with the situation in animal cells, where relatively long-distance movements of organelles usually occur along microtubules. In general, the two types of cells organize and use their actin cytoskeletons differently. Instead of forming the actin structures that drive the motility of animal cells—such as sheet-like lamellipodia and finger-like filopodia—actin in plant cells is directed toward internal movements.

Actin-dependent motility is provided by members of the myosin family. In the case of organelle movement the myosins involved are located on the surfaces of the organelles, which are "walked" along the actin strands. Perhaps reflecting the different uses to which plant cells put their actin cytoskeletons, they contain classes of myosin not found in animal cells, and lack at least one class that is prominent in animals. Myosin VIII, for example, a class of "unconventional" myosin (meaning that it does not form filaments), occurs only in plants and is concentrated at newly formed cross-walls, which are rich in plasmodesmata. The association of actin filaments with these cell-cell junctions explains how the actin cables can be stretched from one end of a cell to the other. Plants, however, appear not to possess myosin II, the filament-forming myosin that can simultaneously pull on two actin filaments to cause contraction, as occurs in muscle. Animal cells usually use myosin II to move or change shape, so the absence of this type of myosin from plant cells is probably a reflection of their stationary lifestyle.

Not only do particles move over actin tracks, but the actin filaments themselves appear to be highly dynamic. The actin filaments in many actin structures in plant cells are constantly depolymerizing and repolymerizing, allowing the overall structure to change its shape when required. The dynamics of these actin structures are likely to be caused by many of the same proteins that affect actin dynamics in animal cells. Despite the different ways in which the plant and animal actomyosin systems are organized, they share many of the same actin-binding proteins. Evolutionarily conserved molecules like profilin and actin depolymerizing factor (ADF) have been shown in plants to affect the dynamics of filament assembly and disassembly. Villin, a major protein

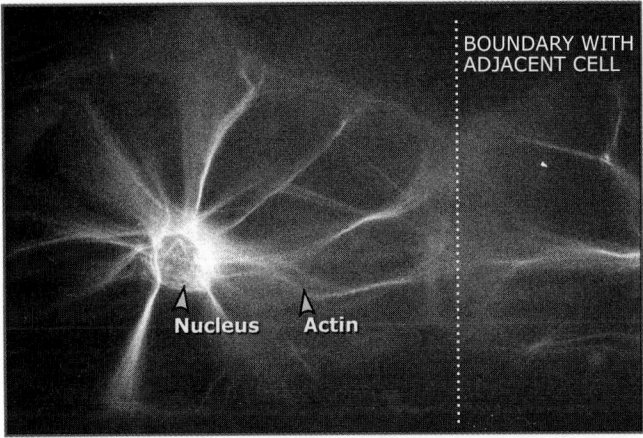

FIGURE 21.47 A fluorescence micrograph of a vacuolated cell with fluorescently labeled actin shows large actin cables extending from near the nucleus to the periphery of the cell. These cables run within thin strands of cytoplasm that cross the vacuole and connect the nucleus with the thin layer of cytoplasm adjacent to the plasma membrane. Photo courtesy of Clive Lloyd, John Innes Centre.

in the microvilli of the gut, is also present in plants, where it may be responsible for packing actin filaments into bundles of uniform polarity. By doing so, villin may be indirectly responsible for determining the direction in which particles move along these highly dynamic strands, since each type of myosin motor can move in only one direction along actin filaments. The action of conserved proteins like these emphasizes the fact that although plant cells are immobile their contents are anything but static.

Concept and Reasoning Check

1. Describe how actin filaments are put to different uses according to the different lifestyles of plants and animals.

21.18 Differentiation of xylem cells requires extensive specialization

Key concepts

- Files of xylem cells undergo programmed cell death to form water-conducting tubes.
- The tubes are prevented from inward collapse by transverse patterns of secondary wall thickening.
- Cortical microtubules bunch-up to form patterns that anticipate the pattern of secondary thickening.

So far, we have looked at the cell walls and the cytoplasmic organization of cells as they divide, then expand in the meristem. However, as cells stop expanding they start to differentiate into specialized tissues. Differentiation involves modification of the cell wall for both general and specialized purposes. Although the walls of immature cells in the zones of division and expansion are extremely strong, so that they can withstand the forces of turgor, they are not nearly thick or rigid enough to play the roles that they must in the mature

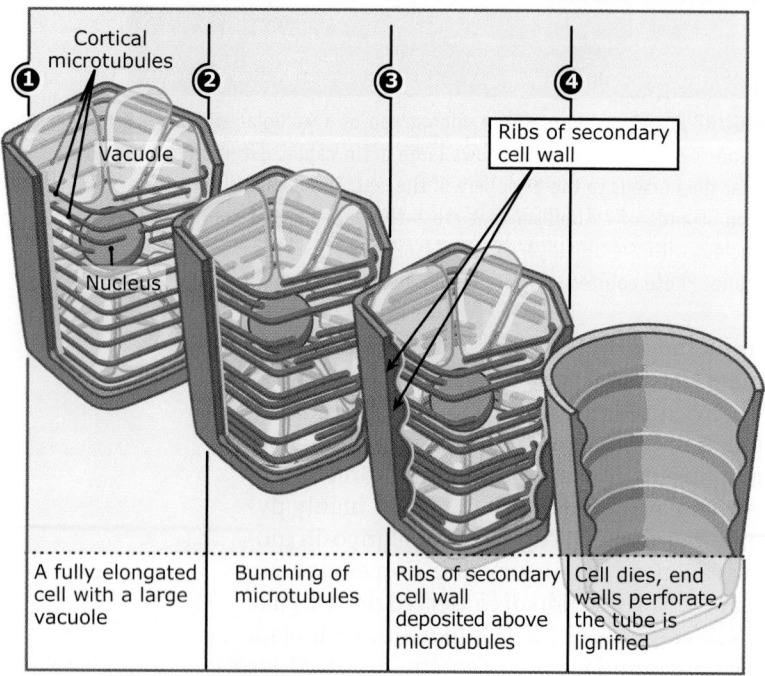

FIGURE 21.48 The sequence of cellular events that occurs during the differentiation of a xylem tube element.

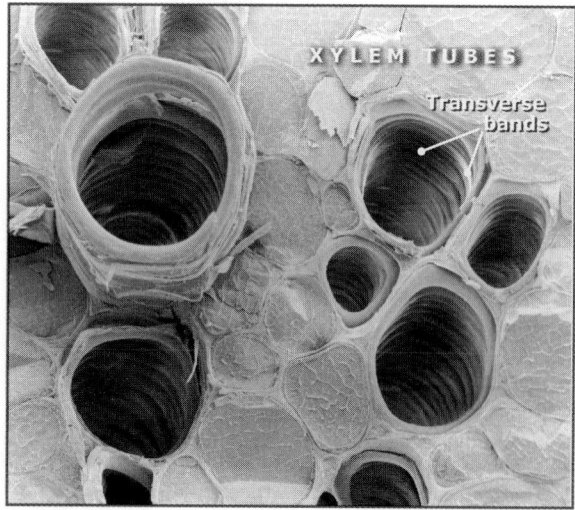

FIGURE 21.49 A scanning electron micrograph showing several xylem tubes in a *Zinnia* leaf. The tubes are seen in cross section, and thickened rings can be seen on the interiors of their walls. Photo courtesy of Kim Findlay, John Innes Centre.

parts of the plant. As cells differentiate, their walls are thickened by the addition of new, chemically modified layers in a process known as secondary thickening. The thickened walls of these cells provide the strength to prevent a plant from collapsing under its own weight as it grows.

In addition to a general thickening of all plant cell walls as they differentiate, some cells also modify their walls in other ways as they differentiate into highly specialized tissues. One example is the formation of the **xylem**, named after the classical Greek word for wood. Files or columns of elongated cells within roots and shoots differentiate to form the vascular tissues—the xylem and phloem. The xylem tissue consists of several cell types, but here we will consider just the tracheary elements. To the early microscopists these hollow tubes resembled the breathing tubes or trachea of animals. However, these dead cells, which are joined end-to-end to form tubes, do not contain air but conduct water from the root system throughout the plant, all the way to its aerial parts. It is the presence of this conducting tissue that distinguishes vascular plants (e.g., flowering plants) from nonvascular plants (e.g., green algae).

Water and dissolved minerals are absorbed over the large surface area formed by the root hairs and pass through to the core of the root and into the xylem via the plasmodesmata that connect cells within the root. Above ground, the heat of the sun vaporizes water, which is lost through the stomatal pores of the leaves. This loss of water by evaporation, or "transpiration," creates a suction force that "drags" water up the xylem. This transpiration-driven flow of water up through the plant is structurally important for maintaining turgor and preventing wilting.

In order to form these continuous tubes (which in giant trees can be many tens of meters in length) the individual xylem cells undergo several interesting features of cell differentiation. The process is illustrated in **FIGURE 21.48**. It begins in living cells that are leaving the zone of cell expansion. The cell walls of these elongated, no-longer-expanding cells undergo secondary thickening. Rather than being thickened evenly throughout, the walls of these cells are thickened in a variety of patterns (hooped, ladderlike, webbed, spiral, pitted). The common feature of the patterns is that they are essentially transverse bands or ribs that will reinforce the walls of the mature xylem. These wall thickenings are visible in **FIGURE 21.49**. The reinforced walls prevent tracheary elements from collapsing inward under the suction force of transpiration and, as hollow vessels, from compression by neighboring cells.

Xylem differentiation is difficult to study deep inside the plant but some cell cells can be induced to form tracheary elements in culture. How is the cytoplasm reorganized as the xylem tracheary element undergoes this remarkable process of differentiation? As shown in **FIGURE 21.50**, the cortical microtubules that were uniformly spread beneath the plasma membrane bunch together to form several predominantly transverse bands. Cellulose microfibrils are then laid down above these bundles of microtubules to form massive secondarily thickened ribs, of which cellulose is the major fibrous component. The pattern of microtubules exactly matches the pattern of wall thickenings, and when the pattern of microtubules is altered by microtubule-depolymerizing or stabilizing drugs, correspondingly abnormal wall patterns are formed, providing one of the clearest examples of the role of microtubules in determining wall patterning. Actin filaments are also found among the microtubules, suggesting that the two filament systems may interact in organizing the secondary wall. The cytoskeletal elements almost certainly channel the wall-synthesizing machinery to particular parts of the cell by focusing the delivery of exocytotic vesicles. Such vesicles contain cellulose-synthesizing complexes for insertion into the plasma membrane, and localized delivery of them would result in the restricted buildup of cell wall above the microtubule bands. Vesicles also contain secretory products such as the other polysaccharides and structural proteins of the secondary wall. Once the thickenings are complete, the entire wall is modified by a substance called lignin (derived from "lignum", the Latin name for wood). Lignin is a hydrophobic polymer made by the oxidative polymerization of up to three aromatic alcohols (monolignols) and is a major (20%–35%) component of wood. Because of its hydrophobicity, it serves to waterproof the inner surface of the tube. Lignification serves the additional purpose of strengthening cell walls, helping them to support the weight of the growth above them. After these modifications to the wall have been completed, the cell lyses and dies, and the end walls dissolve, leaving the thickened, hollow tube that is the mature specialized tracheary element. The development of these strong, transport vessels undoubtedly

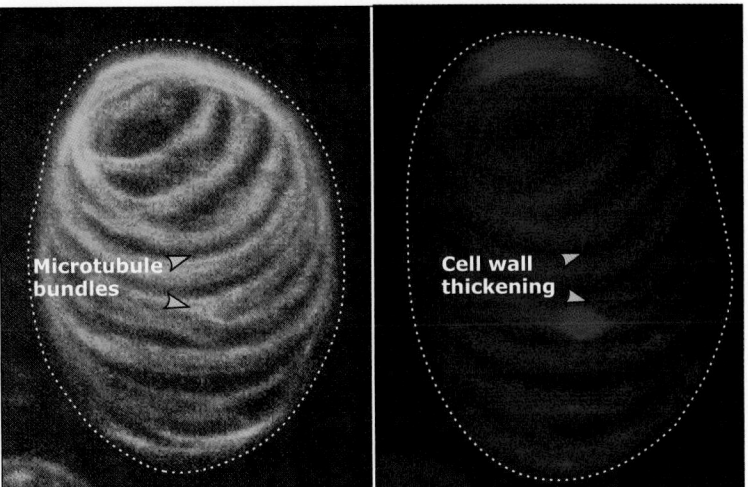

FIGURE 21.50 An *Arabidopsis* xylem cell (tracheary element) in culture. The microtubules have aggregated into thick bundles (green) that act as templates for identical bands of secondarily thickened wall (blue). Photos courtesy of Edouard Pesquet, John Innes Centre.

contributed to the ability of plants to conquer the earth, providing both resistance to gravity and long-distance transport of water to the aerial parts.

21.19 Tip growth allows plant cells to extend processes

Key concepts

- Highly localized secretion of cell wall materials allows plant cells to extend long processes.
- In tip-growing cells, actin filaments and microtubules generally run parallel to the direction of outgrowth.
- Bundles of actin filaments direct the movement of vesicles to the tip, where they fuse with the plasma membrane, driving extension.
- Microtubules seem to control the number and location of cell tips.
- Symbiotic bacteria turn tip growth in on itself to gain access into the plant.

We have seen that major aspects of a plant cell's metabolic machinery are dedicated to the synthesis of cell wall components that are added uniformly over the expanding side walls of elongating cells. This form of diffuse growth is the major mechanism by which plant cells are shaped by expansion, but it is not the only one. Certain specialized cells restrict expansion to a small area, allowing them to extend narrow processes that grow only at their tips. Such **tip growth** involves a much different organization of the cell from that which supports general elongation.

Tip growth is possible because of highly localized delivery of cell wall precursors. As the name implies, material for growth is inserted only at the dome-shaped tip of the cell, as shown in **FIGURE 21.51**. This type of growth is used to form root hairs and pollen tubes. In roots, the hair develops as a bulge at the apical end of a specialized epidermal cell (called a trichoblast) in the zone of differentiation hairs. The resulting perpendicular outgrowth forms a tubular hair that can be many times the length of the trichoblast itself, as shown in **FIGURE 21.52**. These fine white hairs greatly increase the surface area of the root for the absorption of water and ions.

The formation of a tip-growing pollen tube offers a plant cell a form of motility under critical circumstances. Fertilization during reproduction in plants occurs after a pollen grain from a male flower, borne either by an insect or the wind, sticks to the stigma, the structure in the female flower that is the receptor for pollen. In order to increase the chances of catching a pollen grain, the stigma is often large and extends some distance from the flower's ovules, where the eggs are found. As illustrated in **FIGURE 21.53**, a tip-growing tube germinates from the pollen grain and is guided from the stigma to the ovule where the tube delivers sperm cells that fuse with an egg to produce a diploid embryo.

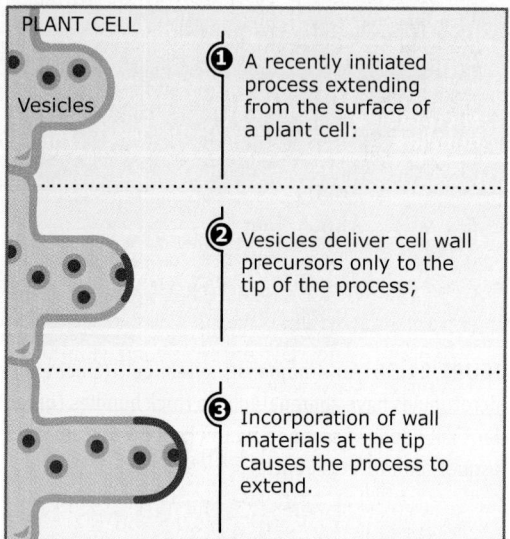

FIGURE 21.51 Tip growth is possible because of highly localized delivery and fusion of secretory vesicles containing cell wall precursors, indicated in red. Continuous fusion of vesicles with the plasma membrane at the tip of a process drives its elongation, producing a long extension of constant diameter off the side of a cell.

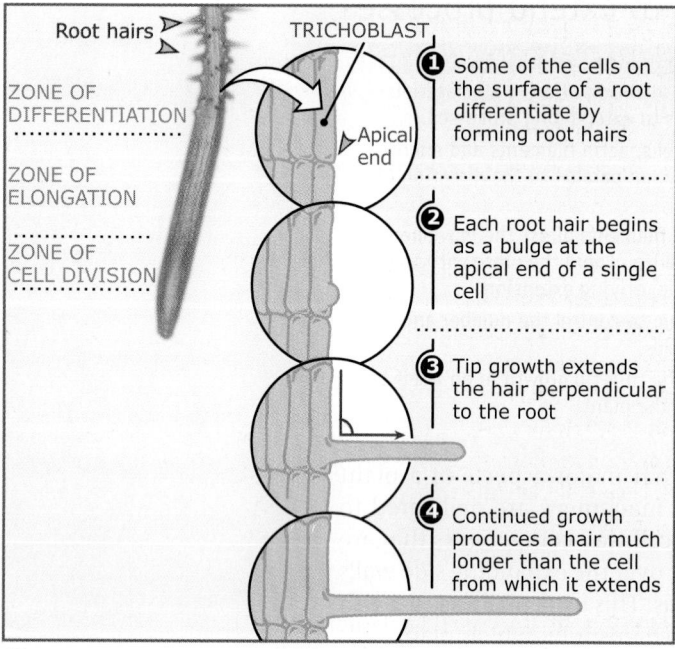

FIGURE 21.52 On the left is an *Arabidopsis* root. As part of the differentiation of the root, root hairs are formed and extend well beyond its surface. The sequence shows the origin of the individual hairs. Each is an extension of a single cell. Photo courtesy of John Schiefelbein, University of Michigan.

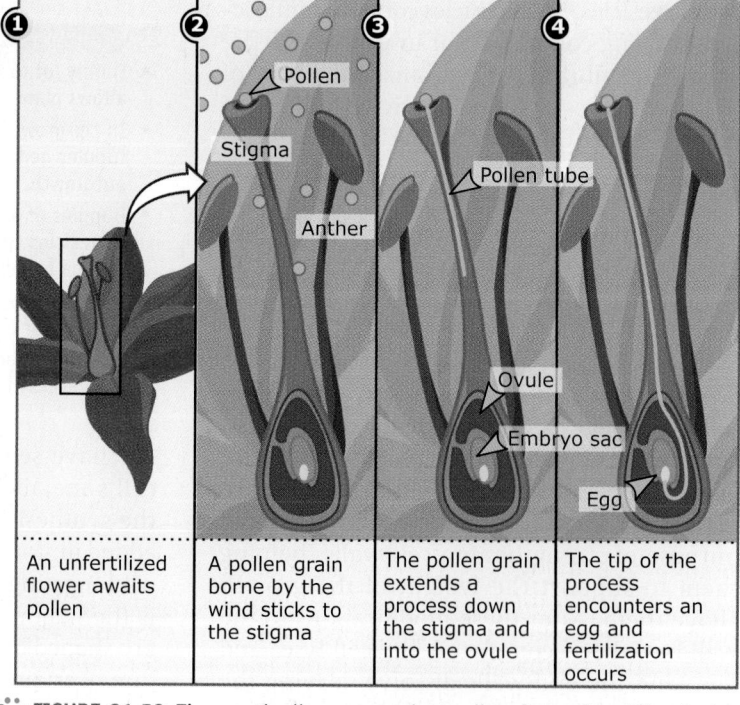

FIGURE 21.53 Tip growth allows some plant cells a form of motility. During fertilization, pollen grains captured by a flower use tip growth to extend a process to the egg. The length of the process can be many times the size of the pollen grain.

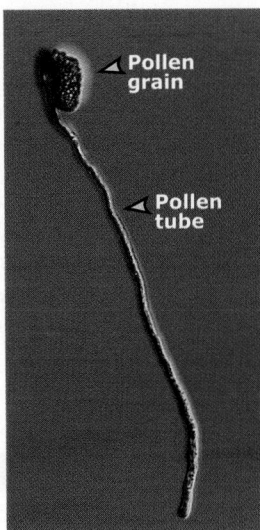

FIGURE 21.54 A pollen grain that has been induced to extend a pollen tube. The original pollen grain is the small granule at the top left of the picture. The length of the tube it extends can become many times the size of the grain itself. Photo courtesy of Norbert de Ruijter , Wageningen University.

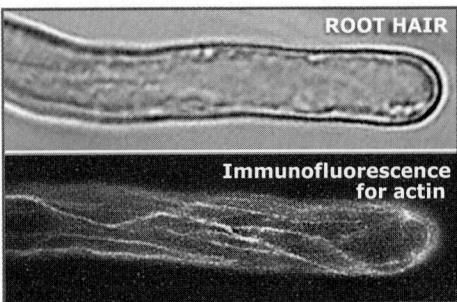

FIGURE 21.55 The upper photograph shows the tip of a growing root hair. The lower shows fluorescently labeled actin in a similar hair. Actin cables are oriented along its length but stop in a dense mass just behind the tip where exocytosis takes place. Photos courtesy of Norbert de Ruijter, Wageningen University.

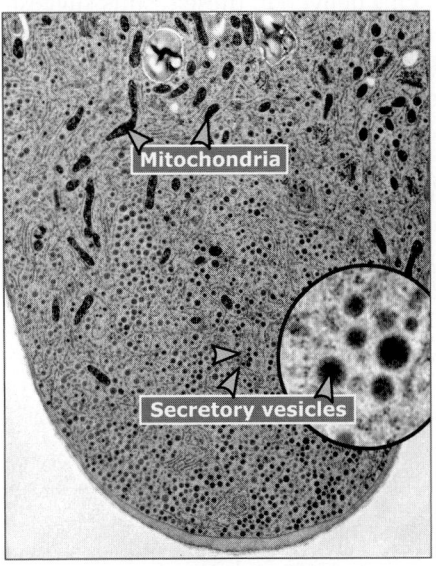

FIGURE 21.56 An electron micrograph of the interior of the tip of a lily pollen tube. Secretory vesicles transported by the actin cytoskeleton are highly concentrated at the tip, excluding most other organelles such as mitochondria. The vesicles continuously fuse with the plasma membrane at the tip, driving extension. Reproduced from S. A. Lancelle, D. A. Callaham, and P. K. Hepler, *Protoplasma* 131 (1986): 153–165. Used with permission of Springer Business + Media. Photo courtesy of Peter K. Hepler, University of Massachusetts—Amherst.

A pollen grain that has extended a pollen tube is shown in **FIGURE 21.54**.

The interior of a cell undergoing tip growth is highly polarized. While most of the tube is occupied by the vacuole, the growing tip and the region immediately behind it are filled with cytoplasm. Actin filaments and microtubules are present within the process and—in contrast to elongating cells in the main body of the plant—often run parallel to its direction of growth. The orientation of actin filaments parallel to the direction of growth of a tip is shown in **FIGURE 21.55**. In the case of both actin filaments and microtubules, their faster growing (plus) ends are directed towards the growing tip of the hair. This organization allows the cytoskeleton to be devoted to the task of delivering secretory vesicles to the advancing tip. Molecules of the motor protein myosin, which move along actin filaments, are bound to the surfaces of Golgi vesicles. This actin-myosin interaction drives cytoplasmic streaming, transporting the vesicles along the actin filaments to a subapical region of dense actin that stops short of the very tip. The transport of vesicles occurs along the flanks of the hair but once the vesicles are re-leased the actin filaments are funnelled into the centre of the hair in a process known as reverse fountain streaming. As shown in **FIGURE 21.56**, such a high concentration of vesicles accumulates within the tip that most other organelles are excluded. Calcium influx is high at the very tip and this gradient encourages fusion of the accumulating vesicles with the plasma membrane, causing localized tip growth. Because of their central role in vesicle delivery, actin filaments must grow continuously at the tip of the tube if they are to keep pace with extension. Evidence from pollen tubes indicates that the gradient of calcium ions originating at the tip regulates several actin-binding proteins to ensure a supply of free filament ends for the polymerization of actin.

The role of microtubules in tip growth is not as clear as that of actin. When drugs are used to depolymerize microtubules during the growth of root hairs, the hairs continue to grow but do so in a zigzag pattern and in some cases even form multiple tips. Since growth of the tube in any form indicates that vesicles are being delivered, these results suggest that microtubules play no role in the mechanics of vesicle movement and fusion but are somehow involved in the overall spatial coordination of tip growth.

Symbiotic nitrogen-fixing bacteria gain entry into legumes by exploiting the tip growth of root hairs. As illustrated in **FIGURE 21.57**, attachment of a bacterium to the tip of a root hair causes the hair to curl around the bacterial cell and encircle it like a shepherd's crook. Once snared in this manner the bacterium can in-

vade the hair. Tip growth is arrested and turned back in on itself (like the inverted finger of a rubber glove) to form an "infection thread" that travels back up the interior of the hair and penetrates the cell. In order to cause this inverted form of growth, the bacterium must somehow reorganize the events involved in vesicle delivery and fusion at the tip. The arrival of the infection thread in the main body of the plant cell triggers the first of a series of cell divisions that eventually form a nodule in which a colony of the bacteria live and supply the plant with reduced nitrogen.

Concept and Reasoning Check

1. How is the cytoskeleton adapted for tip growth and how does this differ from the way that the cytoskeleton is used in regular tissue cells?

21.20 Plants contain unique organelles called plastids

Key concepts

- Plastids are membrane-bounded organelles that are unique to plants.
- Several types of plastid exist, each with a different function.
- All plastids differentiate from proplastids.
- Plastids arose during evolution by an endosymbiotic event.

Plastids are membrane-bounded organelles that are unique to plant cells. There are various kinds that differ depending on the tissue in which they are found. In general, plastids perform specialized functions that have no equivalents in animal cells.

Regardless of their function, all plastids share a number of features. All are bounded by two membranes, which are closely apposed over the entire surface of the organelle. Within the interior of plastids (called the stroma) are free-floating disks of membrane formed by invagination and pinching off of sites on the inner of the two bounding membranes. Plastids differ from most other organelles in that they possess their own genomes, each plastid having multiple copies of a small circular genome containing about 100 genes. The genome encodes proteins required for the highly specialized functions of plastids (such as membrane proteins required for photosynthesis) and other proteins and RNAs for transcription and translation (such as ribosomal proteins, RNA poly-

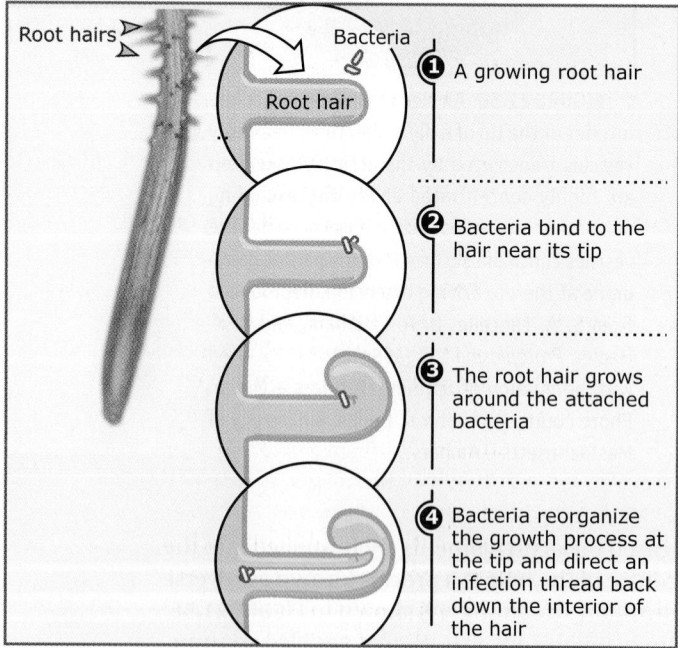

Root hairs

Bacteria

Root hair

❶ A growing root hair

❷ Bacteria bind to the hair near its tip

❸ The root hair grows around the attached bacteria

❹ Bacteria reorganize the growth process at the tip and direct an infection thread back down the interior of the hair

FIGURE 21.57 The sequence shows how symbiotic bacteria are able to reorganize the growth of a root hair to gain entry into the interior of a root. Photo courtesy of John Schiefelbein, University of Michigan.

merases, and transfer and ribosomal RNAs). Transcription and translation of genes in the plastid genome occur within plastids, but the great majority of their proteins are encoded by the nucleus. They are synthesized in the cytoplasm and must be imported into the organelle and properly localized among the various compartments that its membranes define. Like mitochondria, plastids are not connected via vesicular traffic with any of the organelles of the secretory pathway.

All types of plastids differentiate from a common precursor organelle, called a **proplastid**, found in actively dividing cells. Proplastids are small, round organelles with only rudimentary internal membranes and no form of specialization. Their primary function appears to be to generate differentiated forms of plastid when needed. Proplastids differentiate and acquire specialized functions after cells have left a meristem and begin to form a particular tissue. The type of plastid that develops depends on the cell type. In the presence of light, **chloroplasts** develop in leaves and other green parts of a plant that will be devoted to light gathering and photosynthesis. The chloroplasts within the photosynthetic cells of a leaf are shown in FIGURE 21.58. A different type of plastid, the **amyloplast**, develops for the purpose of synthesizing and storing starch in nonphotosynthetic tissues. These plastids, found in seeds and tubers (e.g., a potato), store starch as granules free in the stroma. In specialized cells within the root, amyloplasts serve as gravity-sensing

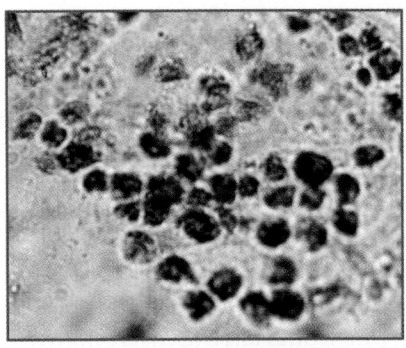

FIGURE 21.59 A single cell from the pericarp of a ripe tomato. The red spots are chromoplasts within the cell's cytoplasm that have accumulated a red pigment. Cells such as this one are distributed throughout the tomato and give it its bright red color. Reproduced from K. A. Pyke and C. A. Howells, "Plastid and stromule morphogenesis in tomato," *Ann. Bot.*, 2002, vol. 90 (5), pp. 559–566, by permission of Oxford University Press. Photo courtesy of Kevin A. Pyke, University of Nottingham.

devices, initiating turning or growth of the root downward by sedimenting within the cells.

Other types of plastid are largely responsible for the synthesis of small chemicals that plants use for a variety of purposes. **Chromoplasts**, as their name suggests, accumulate red, orange, or yellow pigmented molecules (carotenoids) that are responsible for giving many flowers and fruits their color. FIGURE 21.59 shows chromoplasts from a tomato. Another type of plastid, the **leucoplast**, synthesizes small, often volatile organic compounds rather than pigments. These compounds are often put to use as drugs or flavors, and many give particular plants their characteristic odors or tastes. The molecule responsible for the odor of peppermint, for example, is synthesized within leucoplasts. The cells that synthesize and secrete such compounds are specialized for the purpose and are collected into glands positioned to aid their release, such as within the peel of an orange.

Although plastids are highly specialized they do have some reactions in common. For reasons that are not clear, plants perform many of their basic metabolic reactions within plastids. The synthesis of fatty acids, of many amino acids, and of purines and pyrimidines all take place within plastids in plants, whereas they take place in the cytoplasm of animal cells.

Once plastids have differentiated, they show a remarkable ability to interconvert. Chromoplasts, amyloplasts, or chloroplasts

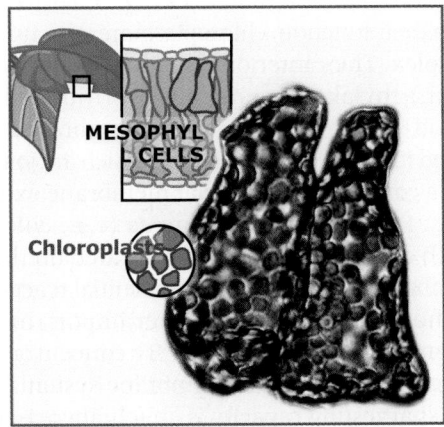

FIGURE 21.58 Two photosynthetic cells isolated from the interior of a tomato leaf. The green disks throughout each are chloroplasts. A schematic cross section of a leaf illustrating where these two cells would be located is shown in the upper left. Photo courtesy of Kevin A. Pyke, University of Nottingham.

can all interconvert depending on the environmental or developmental circumstances. For example, the chloroplasts present in unripe (green) tomatoes convert to chromoplasts as the fruit ripens, gradually giving it the red color of a ripe tomato.

Plastids are thought to have arisen early in evolution by an endosymbiotic event in which a photosynthetic prokaryote was engulfed by a primitive eukaryotic cell. The internalized prokaryote would have used its photosynthetic ability to perform chemical reactions that provided the host cell with a source of energy. In return the prokaryote could exploit the nutrients in the host's cytoplasm. After the association was well established, a gradual transfer of genes out of the bacterial/organellar genome occurred, eventually producing the current situation in which most plastid proteins are encoded by genes in the nucleus.

The bacterial origins of plastids are clear when they are examined at the molecular level. Plastids contain their own transcription and translation components, and the proteins involved are clearly of bacterial origin. Plastid ribosomes, for example, strongly resemble those of *E. coli*, and the RNA polymerases within plastids are also similar to bacterial ones. The similarities with bacteria extend to the plastid genome, in which the elements that control plastid gene expression, such as the promoters, are almost identical to those in bacteria.

Bacterial origins can also be seen in the way that plastids propagate. Unlike the organelles of the secretory pathway, plastids divide by constriction. To do this, the plastid employs components related to those used in bacterial division. For example, plant homologs of the bacterial FtsZ protein form a ring around the inside of the chloroplast, resembling the ring of FtsZ seen at the midpoint of dividing bacteria. Bacterial FtsZ is a distant relative of eukaryotic tubulin.

21.21 Chloroplasts manufacture food from atmospheric CO_2

Key concepts

- Photosynthesis occurs in specialized plastids called chloroplasts.
- Leaves maximize the amount of light for photosynthesis.
- Mesophyll cells are shaped for maximal gas exchange.

In the first section of this chapter, we saw that the presence of a cell wall limits the options a plant cell has for acquiring food. An immobile cell cannot pursue or search for food, and the cell wall prevents a cell from engulfing particles that it could digest within itself. Filamentous fungi (which also have cell walls) solve the problem by secreting enzymes that break down organic material in their surroundings into molecules that are small enough to be transported across the plasma membrane. Plants found a different solution: they acquired the ability to capture light and exploit it as a source of energy. This frees them of the need to find material in the environment that they can absorb and digest. Instead they use the energy of captured photons to polymerize the single carbon atom of atmospheric CO_2 into the carbon chains of sugars. These can be stored and transported internally to be used as a source of energy or to be metabolized into the carbon chains of other organic compounds, such as lipids. A major destination for the photosynthetically derived carbon is the various classes of carbon-based cell wall polysaccharides, particularly cellulose—the most abundant biopolymer on this planet. The ability to synthesize their own food, termed **autotrophy**, allows the stationary lifestyle of plants.

At the cellular level, photosynthesis occurs in chloroplasts—highly specialized membrane-bound organelles unique to plant cells. Cells within the leaves and needles of plants contain dozens of these organelles, with the proteins within them sometimes constituting more than half the total protein in a leaf. In order to perform their function, chloroplasts are structurally complex. Their interiors contain two compartments: thylakoid membranes surrounded by a fluid stroma. The thylakoid membranes are folded into stacks, called grana, which are joined into a continuum by tubular membrane extensions. Photosynthetic pigments (e.g., chlorophyll) and enzymes are concentrated on these membranes, which perform the initial reactions of photosynthesis, while other important reactions occur in the stroma. By concentrating proteins on the folded membrane system, the light-harvesting capacity is greatly increased.

The architecture of the tissues and cells that perform photosynthesis is often specialized in order to facilitate the process. Leaves are thin and flat, thereby spreading the chloroplast-containing cells over a wide area, only a few cells thick. This organization maximizes the exposure of the cells to light. Photosynthesis occurs in mesophyll cells in the interior of the leaf. Unlike almost all other plant cells, these cells largely separate from each other as they develop, acquiring irregular, multi-lobed shapes and creating large air spaces within the leaf. **FIGURE 21.60** shows the organization of such cells in the interior of a leaf. The spaces between the cells are connected with the external atmosphere via stomatal pores, as shown in **FIGURE 21.61**. As with other shape-determining events in plants, cortical microtubules help establish the morphology of mesophyll cells. As the cells begin to differentiate, their evenly distributed cortical microtubules bunch together to produce bands that encircle each cell. Expansion is prevented at the sites of the bands, but the cell can still bulge out between them, giving rise to an irregular, lobed shape. The lobing increases the surface area available for the exchange of the gases required for (CO_2), and produced by (O_2), photosynthesis.

Within cells the organization of chloroplasts is often adjusted depending on the

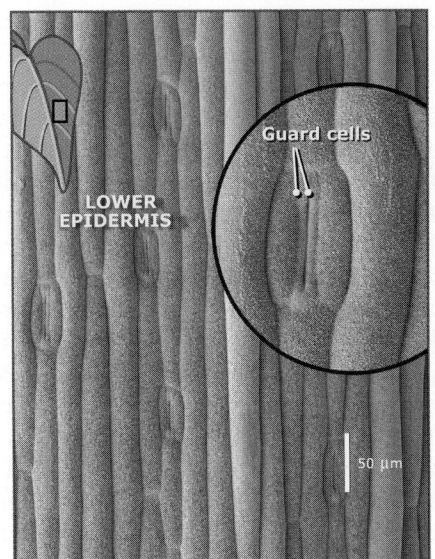

FIGURE 21.61 A scanning electron micrograph of the epidermis on the underside of an oat leaf shows numerous stomata evenly spaced over its surface. Each is formed by four specialized cells within the epidermis. The insert shows an individual stoma at higher magnification. The slit in the middle of each stoma is the pore through which gases are exchanged between the interior of the leaf and the surrounding atmosphere. Stomata are capable of opening or closing depending on the conditions. Their density on the surface allows gas exchange to be efficient. Photo courtesy of Kim Findlay, John Innes Centre.

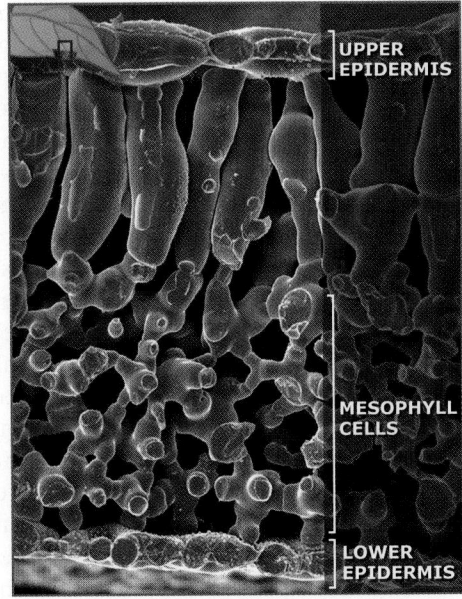

FIGURE 21.60 A scanning electron micrograph of a cross section of a bean leaf. Thin sheets of closely connected cells arranged horizontally form the upper and lower epidermis of the leaf. Its interior is filled with irregularly shaped mesophyll cells whose walls have detached from those of adjacent cells, minimizing contacts between them. Large spaces that extend throughout the leaf are created between the cells, allowing for efficient exchange of the gases during photosynthesis. Reproduced from C. E. Jeffree, et al., *Planta* 172 (1987): 20–37. Used with permission of Springer Business + Media. Photo courtesy of Chris Jeffree, Institute of Molecular Plant Sciences, The University of Edinburgh.

direction or the intensity of the light to which the cell is exposed. To catch more light, chloroplasts can be lined up along a cell surface facing the direction from which the light is coming. To catch less, they can be moved to a surface parallel to it. This movement requires actin filaments. How light influences the actin cytoskeleton is unknown, but it is thought that there are light-responsive receptors in the cytoplasm that can reorganize it.

21.22 What's next?

The publication of the genome sequence for the thale cress, *Arabidopsis thaliana*, is providing a great impetus for plant cell and developmental biology. It is now possible to identify genes that are homologous to those more intensively studied in yeast and animal cells. Plant scientists will initially continue to catalog the function of the coded proteins in plants. However, not all plant proteins share a high degree of homology with those of other eukaryotes and even those that do may be used in different ways.

One of the big challenges is to determine not just where an individual protein is located but how it functions as part of a complex. An important example for plant biology is the directional control of cellulose biosynthesis. Not only will this involve characterization of all the enzymes that constitute the cellulose synthesizing particles, and its association with the membrane lipids, but also exactly how the movement of the particles is influenced by the cytoskeleton. Cellulose synthesizing particles are now known to move along microtubule tracks but it is not clear if there is a direct connection between the two systems. Over and above this is the lingering question of what orients the cytoskeleton. How do microtubules assemble into regular transverse or helical arrays and how do plant hormones and external factors like light and gravity cause the array to reorient, thereby shifting the direction of growth? We may start to see how the shape of individual cells might be regulated, but to obtain a clearer picture of whole-plant physiology will require a far greater understanding of the signaling systems that integrate cell and organ.

Another gargantuan challenge in the postgenomic era is to understand how the cell wall is assembled and functions. The problem lies in the large proportion of complex polysaccharides that make up the wall, for monosaccharides can be linked and branched in chains in a virtually infinite number of combinations. We need to know not only how these are first synthesized and then interact to form the extracellular matrix, but how different cell walls—and even the walls on different faces of the same cell—come to have different polysaccharide components.

Control of the plane of cell division is where cell biology flows into developmental biology. There is the specific problem of knowing how the preprophase band anticipates the division plane. We now know of at least one "molecular tidemark" that remains after the band has depolymerized, promising a fuller description of the way that the division plane is memorized. But transcending the question of how the division plane is achieved in one cell is the bigger question of how the division planes are coordinated over fields of cells. Is the division plane aligned by morphogens, such as hormones, that can override individual cell boundaries? Might physical forces that cause strain over large regions of a tissue be involved?

One of the biggest gaps in our knowledge concerns the way in which plant hormones function. We need to know how the balance between division and cell expansion is regulated in meristems and how the presumed hormonal controls are readjusted when mature cells are triggered back into division (e.g., after wounding). Also, how are the effects of environmental factors, such as light and gravity, translated into changes in growth and form? The cytoskeletal response seems to occur somewhere between the initial perception of the external trigger and the morphological response, and perhaps this will be better understood when more is known about the signaling chains that link the primary effects of hormones to the cytoskeleton. This is likely to involve a variety of cell signaling pathways.

21.23 Summary

Plant and animal cells grow and behave in fundamentally different ways. The differences between them are the result of the strong and rigid wall that surrounds each plant cell and prevents it from moving or changing shape. Plant cells compensate for their immobility by making their own food, using unique organelles—chloroplasts—to convert CO_2 into compounds that can be metabolized or used to construct cell wall polymers. The rigidity of cell walls imposes a stationary lifestyle on plants, but gives them much of their strength and shape. How the cells within a plant organize and remodel their walls determines its shape by determining the direction in which it will grow.

Unlike animals, plants grow throughout their lifetimes. Small groups of cells called meristems divide continuously and generate the new cells required for growth. Primary meristems at the tips of the root and shoot increase the height of a plant, while secondary meristems around their sides allow it to increase in thickness. The meristem at the tip of the shoot produces modules repetitively, each forming atop the previous one and consisting of a leaf or flower bud plus the portion of the shoot that carries them. This type of growth is influenced by the environment, allowing the shape of a plant to change throughout its lifetime. Although the cell divisions that allow a plant to grow occur in specialized regions, many plant cells retain the capacity to divide when necessary and can generate new organs or even entire new plants.

Within a primary meristem, cells go through a sequence of division and expansion that generates columns—or files—of cells along the plant's growth axis. The rigidity of the cell wall prevents division from occurring by constriction, as occurs in animal cells. Instead, a new wall—the cell plate—is constructed across the interior of the cell from within. The plane in which the cell will divide is forecast beforehand by the preprophase band, a cortical band of microtubules and actin filaments that both forms and disappears before mitosis begins, leaving behind molecules that "memorize" the division plane. In cells with large vacuoles a sheet of cytoplasm—the phragmosome—forms across the cell in coordination with the preprophase band but remains throughout mitosis, allowing the mitotic spindle and the cell plate to form within it.

The one structure of a plant cell division that resembles its animal counterpart is the mitotic spindle. Plant spindles, however, often have much broader poles because of the absence of centrioles and centrosomes in plant cells. As the spindle completes anaphase, the phragmoplast—the plant cytokinetic apparatus—appears between the separating chromatids. The phragmoplast consists of two oppositely oriented bundles of microtubules whose ends overlap. Vesicles derived from the Golgi apparatus move along the microtubules and fuse in the small region where they overlap, forming the cell plate. The cell plate grows outward from the center of the cell, eventually reaching and fusing with the original cell wall. As the cell plate forms, tubules of ER are trapped across it, creating channels—plasmodesmata—that will connect the cytoplasms of the two sister cells and allow the passage of material between them.

As the older cells within an apical meristem stop dividing, they begin to expand, often to many dozens of times their original size. This massive expansion of cells just behind the dividing cells pushes the growing points of a plant out into the environment. Expansion is based on the osmotically driven uptake of water. The excess water is stored in a vacuole—a membrane-bounded organelle unique to plant cells—which can occupy up to 95% of the volume of the cell. The pressure of the swollen vacuole—called turgor—is resisted by the high tensile strength of cellulose microfibrils in the cell wall. Turgor pressure is converted into a directional force by the nonrandom orientation of cellulose. Cellulose microfibrils wrapped perpendicularly to the cell's long axis provide "hoop-reinforcement" that prevents a cell from increasing in width but allows it to expand along the axis, causing the cell to elongate. Cellulose is synthesized from multienzyme complexes embedded in the plasma membrane and immediately assembles into microfibrils. Within the wall the stiff, strong cellulose microfibrils are connected by cross-linking glycans, and these fibrous components are surrounded by a gel formed by pectins. In addition to these three classes of polysaccharides, the wall contains structural proteins and enzymes. The protein expansin allows the cell to remodel its wall by breaking the connections between adjacent microfibrils, allowing them to move apart.

The orientation of the cellulose microfibrils in the wall is often matched by cortical microtubules attached to the inner surface of the plasma membrane. Microtubules guide the cellulose-synthesizing particles that move across the face of the plasma membrane, forming microfibrils as they go. Microtubules in plant cells do not radiate from a single central point as in many animal cells but are formed from multiple nucleation sites scattered throughout the cortex (and around the nucleus). This dispersion of the interphase microtubule array is consistent with the synthesis of new cell wall over the expanding surface of a cell. The microtubules in the cortex run parallel to one another and are highly dynamic, allowing the entire array to rearrange. Rearrangement usually accompanies changes in the direction of cell expansion. Such reorganizations may be stimulated by light and gravity in a manner coordinated by plant growth hormones.

The secretory system is also decentralized in plant cells. Instead of a single Golgi apparatus, as in animals, plants have multiple Golgi bodies dispersed throughout the cell. This dispersion supports the requirement for new plasma membrane and wall components over large areas of the surface of a cell as it expands. The Golgi stacks are moved around the cell on actin cables. Actin filaments in plant cells run parallel to all of the different microtubule arrays but are also present in large bundles of actin alone that cover large distances within the cell. These support constant and vigorous movement of vesicles and organelles within the cell in a process called cytoplasmic streaming. This motion appears to be powered by myosin on the surface of the moving particles.

Cell differentiation occurs after expansion and is exemplified by the formation of xylem cells. Files of cells along the center of the growth axis undergo programmed cell death and their end walls dissolve, forming continuous tubes that conduct water and dissolved nutrients from the roots to the aerial parts of the plant. Before the cells die, their cortical microtubules bunch together and direct the formation of special thickenings of the cell wall that will strengthen each tube and prevent it from collapsing. Unlike the primary walls found in expanding cells, such secondary walls formed in differentiating cells have extra layers of cellulose and are strengthened and waterproofed by chemical modification with molecules such as lignin.

Although most plant cells grow by diffuse expansion, some cells, such as root hairs and pollen tubes, grow by directing secretion to a small expansion site in a process known as tip growth. This site elongates to form a tube that grows only at its tip. Microtubules and actin filaments both run along the length of the tube, and the actin filaments carry vesicles to the tip, driving its growth.

Plants contain unique organelles called plastids. During cell differentiation, unspecialized precursor organelles—proplastids—differentiate into a variety of plastids, depending on the cell type. Pigmented plastids, such as chromoplasts, provide the coloration of flowers; the unpigmented amyloplast stores starch. Plastids arose during evolution by the phagocytosis of a photosynthetic prokaryote by a primitive eukaryotic cell. The prokaryotic ancestry of plastids is reflected by the fact that both their proteins and their genetic elements closely resemble those of modern bacteria. The most common type of plastid is the chloroplast. Chloroplasts are highly specialized for their function, with very highly elaborated internal membranes—called grana—on which the initial reactions of photosynthesis take place.

http://biology.jbpub.com/lewin/cells

To explore these topics in more detail, visit this book's Interactive Student Study Guide.

References

Baskin, T. I. 2005. Anisotropic expansion of the plant cell wall. *Annu Rev Cell Dev Biol*. v. 21 p. 203–222.

Ehrhardt, D. W., and Shaw, S. L. 2006. Microtubule dynamics and organization in the plant cortical array. *Annu. Rev. Plant Biol*. v. 57 p. 859–875.

Hawes, C., Osterrieder, A., Hummel, E., and Sparkes, I. 2008. The plant ER-Golgi interface. *Traffic* v. 9 p. 1571–1580.

Hussey, P. J., Ketelaar, T., and Deeks, M. J. 2006. Control of the actin cytoskeleton in plant cell growth. *Annu Rev Plant Biol*. v. 57 p. 109–125.

Lloyd, C. W., and Chan, J. 2006. Not so divided: the common basis of plant and animal mitosis. *Nat Rev Mol Cell Biol*. v. 7 p. 147–152.

Mutwil, M., Debolt, S., and Persson, S. 2008. Cellulose synthesis: a complex complex. *Curr Opin Plant Biol*. v. 11 p. 252–257.

Paredez, A. R., Somerville, C. R., and Ehrhardt, D. W. 2006. Visualization of cellulose synthase demonstrates functional association with microtubules. *Science* v. 312 p. 1491–1495.

Sussex, I. M., and Kerk, N. M. 2001. The evolution of plant architecture. *Curr. Opin. Plant Biol*. v. 4 p. 33–37.

Van Damme, D., Vanstraelen, M., and Geelen, D. 2007. Cortical division zone establish-ment in plant cells. *Trends Plant Sci*. v. 12 p. 458–464.

Glossary

(6–4) photoproducts (6–4PPs) are DNA derivatives induced by UV radiation in which adjacent pyrimidines become covalently linked through the 6 and 4 positions.

The **–10 hexamer** is a 6 bp consensus sequence (TATAAT) located ~10 bp upstream of the transcription start site; it is found in *E. coli* promoters utilized by RNA polymerase holoenzyme containing the σ^{70} subunit.

The **–35 hexamer** is a 6 bp consensus sequence (TTGACA) located around ~35 bp upstream of the transcription start site; it is found in *E. coli* promoters utilized by RNA polymerase holoenzyme containing the σ^{70} subunit.

The **10 nm fiber** is a linear array of nucleosomes, generated by unfolding from the natural condition of chromatin.

The **3′ CCA** is a sequence of three nucleotides located at the 3′ end of transfer RNA. The hydroxyl group on the ribose sugar of the terminal adenosine residue of the CCA sequence serves as the acceptor for an amino group from an amino acid during the tRNA aminoacylation reaction.

The **3′ untranslated region** (**UTR**) is the section of untranslated messenger RNA (mRNA) located after the protein-coding region.

The **30 nm fiber** is a coiled coil of nucleosomes. It is the basic level of organization of nucleosomes in chromatin.

3D matrix adhesions contain integrin receptors as well as a collection of cytoskeletal and signal transduction proteins. They are only observed in cells growing in 3D matrices or in vivo, and are distinct from the focal adhesions typically formed by cells cultured on a flat (2D) surface.

The **5′ untranslated region** (**UTR**) is the section of untranslated messenger RNA (mRNA) located before the protein-coding region.

The **7-methylguanosine (7-MeG) cap** is added post-transcriptionally to a pre-mRNA and functions in mRNA export from the nucleus to the cytoplasm, translation initiation, and mRNA stability.

α-catenin is an adaptor protein that binds to β-catenin and actin filaments in cadherin receptor complexes.

β-catenin is an adaptor protein that binds to cadherins and α-catenin. It also participates in the Wnt signaling pathway.

γ-tubulin is a member of the tubulin protein superfamily. It is found at the centrosome, where it functions to nucleate microtubules as part of a complex with several other proteins. Many cells also contain γ-tubulin in other locations.

The **γ-tubulin ring complex (γTuRC)** is a complex of about ten proteins that includes γ-tubulin. It nucleates microtubules and is a component of centrosomes.

An **abasic site** is a site in DNA in which the base has been either spontaneously lost or removed by the action of a glycosylase. Also known as an AP site.

An **acceptor compartment** receives cargo from one or more donor compartments by the fusion of incoming membrane-bounded transport vesicles carrying the cargo.

An **acentric fragment** of a chromosome (generated by breakage) lacks a centromere and is lost at cell division.

Acetaldehyde can be created in a cell from pyruvate, the end product of glycolysis, through fermentation. It acts as an electron receptor for NADH and is subsequently converted into ethanol, which diffuses out of the cell.

Acetyl-CoA is a critical intermediate in fatty acid biosynthesis and degradation, in the citric acid cycle, and in glycolysis. It is a thiol ester of coenzyme A (CoA) and acetic acid.

The **acid growth theory** suggests that the plant growth hormone auxin stimulates the expansion of cells by causing them to secrete protons. Acidification of the outside of the cell would weaken connections between adjacent cellulose microfibrils, allowing them to move apart. This, in turn, would allow the cell to expand.

Actin is a protein that is expressed in eukaryotic cells and forms microfilaments, which are filaments involved in some types of cell motility.

Actin bundles are arrays formed by crosslinking of actin filaments into parallel or antiparallel assemblies. Actin bundles are present in filopodia, stereocilia, and other cellular structures.

Actin crosslinking proteins connect individual actin filaments to form actin bundles and networks. These proteins can be divided into three groups based on the structure of their actin-binding domain(s).

Actin filaments are two-stranded helical filaments containing subunits of actin. They are one of the three main components of the cytoskeleton and function in cell motility and contraction.

Actin monomer-binding proteins associate preferentially with actin monomers rather than with actin subunits in filaments. Monomer-binding proteins regulate the rates of filament elongation by controlling the concentration of free actin subunits available for polymerization. Thymosin β_4 and profilin are the major actin monomer-binding proteins in metazoan cells.

Actin networks are a type of array formed when actin filaments are crosslinked to form meshworks, such as those found in the cell cortex and lamellipodia.

An **action potential** is a wave of electrical signals that is propagated rapidly and transiently along the plasma membrane of excitable cells such as neurons and some muscle and endocrine cells. It occurs when the membrane potential increases to a threshold level that triggers the opening and closing of voltage-sensitive ion channels.

An **activation domain** is a protein domain that functions through protein-protein interactions to increase the activity of a component within the transcription machinery.

The **active pulling model** for posttranslational translocation into the endoplasmic reticulum proposes that ATP hydrolysis elicits a conformational change in BiP that pulls the substrate through the translocation channel into the ER lumen.

An **acyl group** is a functional group with the formula RCO, where R is an alkyl group (a hydrocarbon). A fatty acid, when bound to another molecule, is an acyl group.

Adaptins are the individual subunits of the cytosolic adaptor complexes that help mediate formation of clathrin-coated vesicles. There are several types of adaptin subunits.

Adaptor complexes bind to signals in the cytoplasmic tails of transmembrane cargo proteins and recruit clathrin molecules during the assembly of clathrin-coated pits. Different types of adaptor complexes function at different compartments. Each adaptor complex contains four different subunits.

Adaptor proteins link signaling molecules to mediate responses to extracellular signals. Adaptor proteins may be made up of two or more modular interaction domains or the complementary recognition motifs.

An **adherens junction** functions to couple adjacent cells together. It consists of membrane proteins that connect to actin filaments.

ADP-ribosylation factor (**ARF**) is a component of COPI coats. It was first identified as a cofactor in cholera toxin action.

Aerobes are organisms that use oxygen gas (O_2) for metabolism.

In **affinity modulation**, changing the binding strength of individual receptors varies the strength of cell adhesion.

Alkaloids are nitrogen-containing natural products, many of which are synthesized by plants.

Alkylating agents are chemicals that modify DNA or other targets with alkyl groups; most commonly methyl and ethyl modifications.

Alkyltransferases are repair enzymes that transfer alkyl groups from damaged DNA to cysteine residues within the enzymes.

Allosteric modification is a process by which a molecule binds to a secondary site on a protein and changes the activity of the protein's primary active site.

Allosteric regulation describes the ability of a protein to change its conformation (and therefore its activity) as the result of binding a small molecule or another protein at a site located elsewhere on the protein.

An **aminoacyl tRNA** is a tRNA that has become esterified to an amino acid through the 3′ CCA end.

Aminoacyl-tRNA synthetases are a family of enzymes responsible for covalently linking amino acids to the 2′- or 3′-OH position of tRNA.

AMP-dependent kinase is a regulatory kinase enzyme that is activated by AMP concentrations.

AMP-kinase *See* AMP-dependent kinase.

Amyloplasts are plastids that store starch.

Anabolism is the process by which the body synthesizes needed biochemicals.

Anaphase is the stage of mitosis in which the two chromatids of each replicated chromosome separate and move to opposite poles of the spindle.

During **anaphase A**, each chromatid moves closer to the pole to which it is connected.

During **anaphase B**, the two spindle poles move away from one another. Anaphase B is also known as the spindle elongation phase.

The **anaphase-promoting complex** (**APC**) is a large complex of proteins responsible for advancing the cell cycle from metaphase to anaphase. The complex targets specific proteins for destruction by attaching chains of ubiquitin molecules to them.

An **anaplerotic** reaction is a filling reaction, by which intermediates of a metabolic cycle are increased in order to balance their loss at other steps in the cyclic process.

Anchorage dependence describes the need of normal eukaryotic cells for a solid surface to attach to in order to grow in culture.

Anchorage-independent cells do not require attachment to a surface in order to proliferate.

Aneuploidy is the condition of having too many or too few chromosomes. Twice the normal number is known as tetraploidy, and half the normal number is known as haploidy.

Angiogenesis is the process by which new blood vessels are formed.

Angiogenic factors are secreted factors, usually proteins, that serve to stimulate the formation of new blood vessels.

The **angiogenic switch** is a point at which cells in a tumor suddenly acquire the ability to stimulate angiogenesis.

Annulate lamellae are stacks of nuclear membranes containing nuclear pore complexes (NPCs) and are located in the cytoplasm. They may be storage sites for excess NPCs.

Anoikis is apoptosis that occurs when adherent cells become detached from their substrates.

Anoxia is a lack of oxygen.

An **antagonist** is a nonactivating ligand that occupies a binding site on a receptor and thereby competitively inhibits binding of an activating ligand.

The **antenna complex** is composed of light-absorbing molecules (mostly chloroplasts) that funnel photons to the reaction center.

The **anticodon loop** is a region on a transfer RNA (tRNA) that contains a three-nucleotide sequence that is complementary to a codon in an mRNA. During translation, the tRNA recognizes the mRNA through base pairing of the anticodon with the codon.

An **antiporter** is a type of carrier protein that simultaneously moves two or more different types of solutes in opposite directions across a membrane. Antiporters are also called exchangers.

AP endonuclease is an enzyme in the BER pathway that hydrolyzes the phosphodiester bond 5′ to the AP site to generate a nick.

The **apical** surface of a polarized cell faces the outside world. For example, the apical surface of intestinal epithelial cells is the region next to the lumen. The opposite end is called the basal surface.

AP lyase is an enzyme that cleaves the DNA phosphodiester backbone between carbon and oxygen at AP sites.

Apoptosis is the capacity of a cell to respond to a stimulus by initiating a signaling pathway that leads to its death through the activation of a characteristic set of reactions.

An **AP site** (apurinic/apyrimidinic site) is a site in DNA in which the base has been either spontaneously lost or removed by the action of a glycosylase. It is also known as an abasic site.

An **aquaporin** is a transmembrane channel protein that transports water or other molecules across the plasma membrane.

Archaea are prokaryotes that differ from eubacteria in the chemistry of their cell wall, ribosomal RNA, lipids, and certain enzymes. This group of prokaryotes includes halophiles that live in salty environments, methanogens that live in oxygen-free environments, and thermoacidophiles that live in hot, acidic environments.

Argonaute (**Ago**) is a family of proteins that assemble into the RISC to effect RNA interference in both the miRNA and siRNA pathways.

ASF/SF2 is a protein factor that regulates mRNA splicing. The protein was discovered separately by two different research teams, who each gave it a different name, ultimately leading to the fusion of the two names.

Astral microtubules are microtubules that radiate outward from a centrosome, creating a radially symmetric array called an aster. The term "astral microtubules" is often used to refer specifically to the microtubules that radiate from each pole of a mitotic spindle but are not directed toward the chromosomes.

An **ATP-dependent chromatin remodeling complex** is a complex of one or more proteins associated with an ATPase of the SW12/SNF2 superfamily that uses the energy of ATP hydrolysis to alter or displace nucleosomes.

ATP synthetase is the enzyme involved in forming ATP by using the energy in a proton gradient.

An **AU-rich element** (**ARE**) is a eukaryotic mRNA sequence consisting largely of A and U ribonucleotides that acts as an RNA destabilizing element.

Autocrine signaling is signaling in which the cell that secretes a signal can also respond to that signal.

Autoimmune bullous dermatosis is a disease that arises when patients develop autoantibodies against the proteins in their desmosomes or hemidesmosomes.

An **autonomously replicating sequence** (**ARS**) is a name given to origins of replication in budding yeast.

Autophagy is a mechanism for nonspecifically degrading cellular components so that the degradation products can be used as nutrients for the cell. Autophagy usually only occurs in starved or otherwise stressed cells.

Autophosphorylation is the phosphorylation of a protein kinase catalyzed by the same protein kinase molecule. Autophosphorylation does not necessarily occur on the same polypeptide chain as the catalytic site; for example, in a dimer, each subunit may phosphorylate the other.

Autotrophy is the ability to make organic compounds from inorganic sources. During photosynthesis, plants convert atmospheric carbon dioxide into more complex molecules using the energy of sunlight.

In **avidity modulation**, changing the number of receptors engaged in a cell junction varies the strength of cell adhesion.

The **axial element** is a proteinaceous structure around which the chromosomes condense at the start of synapsis.

The **axoneme** is a bundle of microtubules and other proteins forming the core of a cilium or flagellum.

Bacteria (*singular*: bacterium) are unicellular organisms that lack membrane-bounded organelles such as a nucleus.

A **bacteriophage** is a virus that infects bacteria.

A **Balbiani ring** is an extremely large puff or zone of active transcription located at a specific band of a polytene chromosome.

A **Balbiani ring granule** is a specific, extremely large pre-mRNA and its associated RNA-binding proteins. The pre-mRNA is transcribed from a *Chironomus tentans* gene activated during development. The individual mRNPs are seen as ring-shaped macromolecular assemblies.

Bands of polytene chromosomes are visible as dense regions that contain the majority of DNA. They include active genes.

The **barbed end** of an actin filament is the fast-growing end. It was defined based on the appearance of myosin-coated actin filaments in electron micrographs.

The **basal** surface of polarized cells faces the extracellular fluids. It is opposite the apical surface and is in contact with the basal lamina.

The **basal body** is a short cylindrical array of microtubules and other proteins found at the base of a eukaryotic cilium or flagellum. It has a role analogous to the microtubule-organizing center in organizing the assembly of microtubules into the axoneme that forms the core of the cilium or flagellum.

The **basal lamina** is a thin sheet of extracellular matrix proteins that binds to the basal surface of epithelial and endothelial cell layers and separates them from other tissues. It is also found in neuromuscular junctions.

The **basal transcription apparatus** refers to RNA polymerase and the complete set of general transcription factors necessary for transcription initiation at a core promoter.

Basal transcription factors or **general transcription factors** are the set of transcription initiation factors that are required to facilitate transcription initiation by RNA polymerase at a core promoter. For eukaryotic RNA polymerase II, these factors include TFIIA, TFIIB, TFIID, TFIIE, TFIIF, and TFIIH.

Base excision repair (**BER**) is a multistep pathway that repairs damage to DNA bases caused by deamination, oxidation, and alkylation.

The **basic helix-loop-helix** (**bHLH**) **motif** participates in the dimerization of a class of transcription factors called HLH proteins. A bHLH protein has a region of basic amino acids (lysine and arginine) that makes sequence-specific contacts with the transcription factor-binding site on the DNA and is often found proximal to the HLH dimerization motif.

The **basic leucine zipper** (**bZIP**) **motif** is a structure located on a number of transcription factors that facilitates dimerization of two subunits through stretches of leucine-rich segments located on each of the two protein monomers. The interaction of the two segments resembles a zipper. A region of basic amino acids (lysine and arginine) that makes sequence-specific contacts with the transcription factor-binding site on the DNA is often found proximal to the zipper dimerization motif.

A **benign** tumor is one that is noninvasive and, therefore, does not spread to other areas of the body.

Beta oxidation is the breakdown of fatty acids during cellular metabolism through the successive removal from one end of two carbon units.

Bidirectional replication occurs when two growing points start at the same site and move in opposite directions until they meet at the opposite side of the circle.

A **biofilm** is a structure that consists of many prokaryotic cells surrounded by an extracellular polysaccharide matrix secreted by the cells. Biofilms adhere to many types of surfaces.

A mitotic chromosome is **bi-oriented** when its two kinetochores are attached to opposite poles of the mitotic spindle. All of a cell's chromosomes must be bi-oriented before anaphase begins.

Bipolar thick filaments are assemblies of myosin-II that function in muscle contraction. Thick filaments interdigitate with actin thin filaments in the sarcomeres of muscle cells. The myosin-II tails associate with one another, and the myosin heads extend from both ends of the filament and pull on the thin filaments during contraction.

The **bottom strand** in DNA sequence is by convention the noncoding or nonsense strand of a DNA molecule and is always written in the 3' to 5' direction. The bottom strand serves as the template strand during transcription.

Branch migration is the process of exchange of base-pairing partners at a Holliday junction formed during homologous recombination.

A **branch point** is specific adenosine residue located within a short conserved sequence just before the end of an intron in a pre-mRNA. During splicing, the lariat intermediate is formed by covalent joining of the 5' end of the intron to the 2' position of the ribose on the adenosine at the branch point.

The **BRE** or **TFIIB recognition element** is a loosely conserved short G-C-rich sequence in eukaryotic RNA polymerase II core promoters that binds the general transcription factor TFIIB. If a TATA box is present in the core promoter, the BRE is generally located in close proximity to the TATA box.

The **Brownian ratchet model** for posttranslational translocation into the endoplasmic reticulum proposes that movement of a translocating protein through the channel occurs by diffusion. Diffusion inward is unhindered, whereas diffusion outward is limited by BiP molecules bound to the part of the protein already inside the ER. Brownian ratchet models, where diffusion in one direction is limited, favoring diffusion in the opposite direction, have been proposed for other movements in cells including extension of the front of a moving cell or movement of chromosomes during mitosis.

A **cadherin** is a cell surface receptor that forms dimers and participates in cell-cell adhesion. Some cadherins are found in adherens junctions.

A **Cajal body** (or coiled body) is a subnuclear structure defined by the presence of the protein coilin; it contains additional proteins and RNA molecules and may be the site of assembly of snRNAs and snoRNAs that assemble with proteins in preparation for RNA processing.

The **cAMP response element binding** (**CREB**) protein is a gene-specific transcription factor that stimulates RNA polymerase II transcription initiation of a number of genes in mammalian cells.

Cancer stem cells are self-renewing cancer cells within a tumor mass that have the ability to seed new tumors upon implantation in an appropriate animal host.

Capping proteins are a class of proteins that bind either to the barbed or pointed ends of actin filaments and slow filament elongation. These proteins help control the length of actin filaments in cells.

The prokaryotic **capsule** is the polysaccharide-rich outer layer of the cell envelope.

Carbon assimilation is the conversion of inorganic carbon dioxide gas (CO_2) to simple organic carbon molecules that cells can use.

A **carcinogen** is a chemical that increases the frequency with which cells are converted to a cancerous condition.

A **carcinoma** is a malignant tumor that arises from epithelial cells (which form the skin and the cell layers lining ducts and cavities of internal organs).

Cargo describes any macromolecule (such as RNA, soluble or membrane protein, or lipid) that is transported from one compartment to another. Cargo may contain sequences or modifications that specify their destination. Some cargo molecules are carried in transport vesicles, but others (such as those that move between the nucleus and cytosol) are not.

The **Carnot cycle** is an idealized path for a perfect engine between heat and work portions, used to derive engine efficiency and later, the notion of entropy.

Carriers, or **carrier proteins**, move a solute directly from one side of a membrane to the other. In the process, the protein undergoes conformational changes. Carrier proteins can be divided into two groups: transporters and pumps.

A **caspase** is a cysteine aspartate-specific protease. Several types of caspases participate in apoptosis.

Catabolism is an energy-liberating process in which larger organic compounds are broken down into smaller ones.

Catabolite repressor protein (**CRP**), also known as cAMP activator protein (CAP), activates transcription initiation of the *E. coli lac* operon in response to a decrease in cellular glucose levels. This phenomenon is known as catabolite repression.

During microtubule dynamic instability, the transition step between growth and shortening is a **catastrophe**.

C-bands are generated by staining techniques that react with centromeres. The centromere appears as a darkly staining dot.

The term **cell-cell junction** refers to any of several specialized, protein-based structures that join neighboring cells together.

The **cell cycle** is an ordered set of stages that results in the accurate division of one cell into two.

The term **cell-ECM junctions** refers to regions on the plasma membrane where extracellular matrix molecules bind to their cellular receptors. This includes interactions between structural glycoproteins in the ECM (e.g., collagens, laminins) and their integrin receptors, as well as proteoglycans and their receptors.

Cell motility refers to movement mediated by the cytoskeleton and encompasses the movement of organelles and macromolecular complexes within cells as well as the locomotion of entire cells.

The **cell plate** is the precursor of the wall that will divide a plant cell after mitosis. It forms initially as a small disk between the two nuclei and grows radially, from inside to outside, until it reaches the plasma membrane. Its formation is directed by the phragmoplast.

Cellulose is a fiber made of multiple linear polymers of glucose. Cellulose is the major structural constituent of plant cell walls.

Cellulose-synthesizing complexes are large complexes within a plant cell's membrane that synthesize multiple strands of polymeric glucose that then crystallize, forming a cellulose fiber that incorporates into the cell wall.

A **cell wall** is a rigid layer that surrounds the plasma membrane of some cells (such as bacteria, yeast, and plant cells) and consists of extracellular molecules produced by the cells. Most animal cells do not have a cell wall. The cell wall of bacteria consists mainly of peptidoglycan, which forms a layer of the cell envelope that protects and supports the cell. The major component of plant cell walls is the polymer cellulose, which is organized into extremely strong fibers.

The **Central Dogma of Molecular Biology** is a historical term that refers to the original hypothesis that DNA contains the information to make RNAs (transcription) and RNAs contain the information to make proteins (translation). A corollary to the Central Dogma is that proteins cannot be used as templates for the synthesis of RNA.

The **central spindle** is a structure that is composed of the interdigitated plus ends of microtubules and associated proteins and that forms during mitosis.

A **centriole** is a small organelle found in animal cells. Each centriole is composed of nine specialized microtubules arranged to form a cylinder. Centrioles are found in the center of centrosomes and usually exist in pairs.

The **centromere** is a constricted region of a chromosome that includes the site of attachment (the kinetochore) to the mitotic or meiotic spindle. The centromere consists of unique DNA sequences and proteins not found anywhere else in the chromosome.

A **centrosome** is a small organelle that acts as a microtubule-organizing center. Centrosomes form the poles of the mitotic spindle in animal cells. Each centrosome consists of a pair of centrioles surrounded by a large number of proteins.

The **channel pore** is the part of a channel protein through which ions or solutes move. It is sometimes referred to as the ion conduction or ion permeation pathway.

Chaperones are a class of proteins (also called molecular chaperones) that bind to incompletely folded or assembled proteins in order to assist their folding or prevent them from aggregating. Chaperones do not remain associated with their substrate proteins afterward.

Charge density refers to the amount of electric charge per unit area or unit volume.

A **checkpoint** is a biochemical pathway that delays or arrests the cell cycle in response to a specific problem or condition.

Chemical potential is a measure of free energy for a component of a reaction or system.

Chemiosmosis is a mechanism that uses the energy of a hydrogen ion gradient across a membrane to drive a cellular process, such as ATP synthesis.

The **chemiosmotic hypothesis** explains energy formation by mitochondria and chloroplasts as an electrochemical gradient of protons.

Chemotaxis describes behavior in which the organism moves toward or away from certain chemicals in its environment.

Chlorophyll molecules are light-absorbing pigments within photosystems.

The **chloroplast genome** is a double-stranded circular DNA that is found in multiple copies in the chloroplasts of plants. Chloroplast genomes vary in size from 140-200 kb.

Chloroplasts are the plant organelles which house photosynthesis.

Chromatids are the copies of a chromosome produced by replication. The name is usually used to describe each of the copies in the period before they separate at the subsequent cell division.

Chromatin refers to the assembly of DNA and histones within chromosomes.

Chromatin remodeling describes the energy-dependent displacement or reorganization of nucleosomes that occurs in conjunction with activation of genes for transcription.

The **chromocenter** is an aggregate of heterochromatin from different chromosomes.

Chromokinesins are members of the kinesin family of molecular motors that are attached to the arms of mitotic chromosomes.

Chromomeres are densely staining granules visible in chromosomes under certain conditions, especially early in meiosis, when a chromosome may appear to consist of a series of chromomeres.

A **chromophore** is the light-absorbing part of a molecule. Chromophore frequently refers to a light-absorbing cofactor in a protein.

Chromoplasts are plastids that accumulate pigments.

Chromosomal passenger proteins are a group of proteins located at the centromere of each chromosome until the onset of anaphase, after which they relocate to microtubules that extend between the two groups of separating chromosomes. Chromosomal passenger proteins are important for proper attachment of kinetochores to the spindle and appear to play a role in cytokinesis.

Chromosomal translocation is the exchange of segments between nonhomologous chromosomes.

A **chromosome domain** is the region of the nucleus occupied by an individual chromosome.

Chromosome painting is a variation of FISH in which a fluorescent dye is bound to DNA pieces that bind all along a particular chromosome.

Chromosomes are structures in the nucleoid or cell nucleus that carry hereditary information in the form of genes.

A **cilium** (*plural*: cilia) is a whiplike structure that is involved in locomotion. It extends from the cell surface and consists of an internal array of microtubules surrounded by plasma membrane.

The *cis*, or entry, face of the Golgi apparatus is a reticulum of interconnected tubules called the *cis*-**Golgi network** (**CGN**).

Cisternal maturation is one of two popular models for the mechanism of cargo transport through the Golgi stack. It is also called cisternal migration or cisternal progression. In this model, a new Golgi cisterna forms at the *cis* face, then moves forward in the stack as the enzyme content of the cisterna changes from *cis* to *medial* to *trans*. Enzymes that belong in earlier cisternae are retrieved by retrograde transport vesicles.

The **clamp loader** is a multisubunit complex that loads sliding clamps at replication forks.

Clathrin proteins interact with adaptor complexes to form the coat on some of the vesicles that bud from the cytoplasmic face of the plasma membrane and the *trans*-Golgi network. Clathrin is composed of heavy and light chains that form triskelions, which then assemble into polyhedral-curved lattices during the formation of clathrin-coated pits and vesicles.

Cleavage factors I and II (**CFI and CFII**) are required for the cleavage of the pre-mRNA prior to polyadenylation.

Cleavage/polyadenylation specificity factor (**CPSF**) binds to the polyadenylation signal AAUAAA and regulates the synthesis of the poly(A) tail.

Clonal expansion is the proliferation of cells originating from a single progenitor cell.

Clonal succession represents the overgrowth by one cell clone of a previously dominant cell clone within a tissue or organ.

The **cloverleaf structure** is a highly conserved RNA folding pattern found in all transfer RNAs (tRNAs) that resembles a four-leaf clover.

The **c-Myc family** is a group of transcription factors that function as dimers. The family includes the proteins Myc, Max, and Mad, each of which can form homo- or heterodimers. Depending on the specific combination, these dimeric transcription factors function as either activators or repressors of genes involved in cell growth and proliferation.

CoA (**coenzyme A**) is a small, organic molecule of cellular respiration that functions in the release of carbon dioxide gas (CO_2) and the transfer of electrons and protons to another coenzyme.

A **coactivator** is a transcription factor that functions in conjunction with gene-specific transcription factors to increase the activity of a core promoter.

A **coated pit** is an invagination of membrane that is pinched off to form a coated vesicle.

Coated vesicles are formed by the pinching off of coated pits from a membrane. The membrane of a coated vesicle has on its cytosolic surface proteins such as clathrin, COPI, or COPII.

Coatomer is the complex of coat proteins on COPI-coated vesicles. It consists of seven proteins.

Coat proteins are multiprotein complexes that bind directly or indirectly to proteins that are incorporated into transport vesicles. Lateral association of these complexes helps deform the membrane into a bud that eventually becomes a vesicle.

The **coding** (**sense**) **strand** is the DNA strand that has the same sequence as the mRNA and is complementary to the noncoding (nonsense) strand, which is used as the template in transcription.

A **codon** is a particular sequence of three adjacent nucleotides in an mRNA that codes for a specific amino acid in translation.

A **cofactor** is an inorganic substance that acts with and is essential to the activity of an enzyme; examples include metal ions and some vitamins.

Cohesin proteins form a complex that holds sister chromatids together until anaphase. The cohesin complex includes two SMC proteins.

The **cohesin complex** binds sister chromatids together until mitosis.

Cohesion keeps two newly synthesized sister chromatids together until mitosis.

A **coincidence detector** is a signaling component that responds only when two different stimulating inputs are received simultaneously.

Collagen is the principal protein component of the extracellular matrix in most animal tissues.

Collision theory is an explanation for reaction rate based upon the frequency that reactants collide.

A **commensal** relationship is a symbiotic relationship between organisms in which one species benefits while the other is unaffected.

A **compartment** is a membrane-enclosed space.

Complementation refers to the ability of a gene to provide a product that converts a mutant phenotype to wild type.

A **complementation group** is a series of mutations unable to complement when tested in pairwise combinations in *trans*; it defines a genetic unit (the cistron).

An enzymatic **complex** is a noncovalently associated group of proteins which catalyze a multistep process without the intermediates entering other pathways (e.g., the pyruvate dehydrogenase complex).

Condensation is a reaction in which a small chemical moiety is lost. In biology, the most common example is peptide bond formation where each amino-acid addition is accompanied by loss of a single water molecule.

Condensin proteins are components of a complex that binds to chromosomes and causes their condensation during meiosis or mitosis. The condensing complex includes members of the SMC family of proteins.

A **conditional mutation** displays a wild-type phenotype under certain (permissive) conditions and a mutant phenotype under other (restrictive) conditions.

A **confluent monolayer** is a layer of cells one cell thick that has covered the entirety of the bottom surface of a culture dish.

Congression is the series of movements exhibited by a mitotic chromosome as it attaches to the spindle and moves to its equator.

Conjugation is a process by which two cells come in contact and exchange genetic material. In bacteria, DNA is transferred from a donor to a recipient cell. In protozoa, DNA passes from each cell to the other.

Connexins are a class of proteins that comprise the gap junction. Six connexins form a hemichannel.

Connexon is another name for a hemichannel in a gap junction.

A **conserved sequence** is a sequence of nucleotides in DNA (or RNA) that serves the same function in different locations within a genome or across the genomes of different species.

Constitutive heterochromatin describes the inert state of permanently nonexpressed sequences (usually satellite DNA).

Constitutive secretion is the process by which macromolecules are transported to the plasma membrane or secreted at a relatively constant rate. These include lipids and soluble and membrane proteins that exit to the plasma membrane from the *trans*-Golgi network but are not secreted by regulated exocytosis.

The **contractile ring** is a ring of actin and myosin filaments that forms just under the plasma membrane of animal cells during telophase. It is positioned halfway between the poles of the mitotic spindle and produces the forces for constricting the cell in half during cytokinesis.

Control points are metabolically irreversible reactions within a chemical pathway.

A **COPI** coat consists of coatomer and ADP-ribosylation factor. COPI-coated vesicles are transport vesicles that bud from the cytoplasmic face of the Golgi complex and mediate retrograde transport from the Golgi complex to the endoplasmic reticulum. COPI-coated vesicles may also mediate transport between Golgi cisternae.

A **COPII** coat is made up of protein complexes comprising a GTPase (Sar1p) and two heterodimers (Sec 23/24 and Sec 13/31). COPII-coated vesicles are transport vesicles that bud from the cytoplasmic face of the rough endoplasmic reticulum and mediate anterograde transport to the Golgi apparatus.

A **core histone** is one of the four types of histone (H2A, H2B, H3, H4) found in the core particle derived from the nucleosome. (This excludes histone H1.)

Core promoter elements are short sequences within a promoter region that are essential to direct transcription initiation of a gene.

The **corona** is a dense group of thin fibers that project outward from the surface of a kinetochore before it attaches to microtubules. The corona contains a number of microtubule-dependent motors and other proteins involved in attaching microtubules to the kinetochore.

Cotranslational translocation describes the movement of a protein across a membrane as the protein is being synthesized. The term is usually restricted to cases in which the ribosome binds to the channel. This form of translocation may be restricted to the endoplasmic reticulum.

Crisis is a state reached when primary cells placed into culture are unable to replicate their DNA because their telomeres have become too short. Most cells enter apoptosis, but a few variants emerge by a process of immortalization that usually involves an acquired ability to elongate and maintain telomeric DNA.

In a polymerization reaction, the **critical concentration** is the total concentration of subunits (for example, of actin or tubulin dimers) remaining in solution at steady state. The monomer concentration must be greater than the critical concentration in order for polymers to form.

cyclic AMP (**cAMP**) is a metabolite derived from ATP that functions as a coactivator by binding and activating the CRP protein, which in turn activates transcription initiation of the *lac* operon. cAMP also functions as a second messenger in eukaryotic cell signaling.

The **cyclin box** is a stretch of approximately 150 amino acids, similar in sequence among all cyclins, through which a cyclin binds to a catalytic CDK subunit. The CDK-cyclin complex controls the progression of events in the cell cycle.

Cyclin B/Cdk1 is a cytoplasmic enzyme that, when activated, drives the cell into mitosis. It consists of the Cdk1 kinase and its cyclin B regulatory subunit. The enzyme is inactivated by the destruction of cyclin B, an event that is required to allow a cell to leave mitosis.

A **cyclin-dependent kinase** (**Cdk**) is one of a family of kinases that are inactive unless bound to a cyclin molecule. Most cyclin-dependent kinases participate in some aspect of cell cycle control.

Cyclins are proteins that bind to and help activate cyclin-dependent kinases. Cyclin concentrations vary throughout the cell cycle and their periodic availability plays an important role in regulating cell cycle transitions.

Cyclobutane pyrimidine dimers (**CPDs**) are DNA derivatives induced by UV radiation in which adjacent pyrimidines become covalently linked via cyclobutane (4-membered) rings.

Cystic fibrosis (**CF**) is a potentially lethal human disease characterized by secretion of excess, higher viscosity, lung mucus and related defects in other exocrine organs. CF is caused by mutations in the gene CFTR (cystic fibrosis membrane conductance regulator), which codes for a transmembrane protein involved in ion transport.

A **cytokine** is a small polypeptide secreted by cells of the immune system that stimulates the proliferation of target cells.

A **cytokine response element** is a DNA sequence located proximal to promoter regions of genes that is stimulated by cytokine-mediated signal transduction.

Cytokinesis is the process that divides the cytoplasm of a cell in half at the end of mitosis, after the replicated chromosomes have been segregated. In animal cells, cytokinesis occurs after the chromosomes have been segregated by the spindle into two independent nuclei.

The **cytoplasm** is the content of a cell inside the plasma membrane; in eukaryotic cells, the cytoplasm includes the cytosol and organelles except the nucleus.

The **cytoplasmic membrane** of prokaryotes is the membrane (lipid bilayer and associated components) surrounding the cytoplasm. It is analogous to the plasma membrane of eukaryotes.

Organelles and particles within the cytoplasm of a plant cell are constantly moved and mixed by a process called **cytoplas-**

mic streaming. Cytoplasmic streaming is based upon actin filaments.

The **cytoskeleton** is the collection of structural filaments in the cytoplasm of eukaryotic cells. It confers the properties of cell shape, mechanical rigidity, and cell motility and is mainly composed of microtubules, actin filaments, and intermediate filaments.

The **cytosol** refers to the volume of the cytoplasm that surrounds the organelles of a cell. Cytosol also refers to the supernatant fraction obtained after centrifugation of homogenized cells at $100,000 \times g$ for one hour.

The **dark reactions** of photosynthesis utilize the energy molecules NADPH and ATP and synthesize sugars from CO_2.

DCGR8 protein acts in concert with Drosha in the processing of RNAi precursor molecules in the nucleus.

A **deacetylase** is an enzyme that removes acetyl groups from proteins.

Delayed early response genes are activated by transcription factors encoded by early response genes.

Membrane **depolarization** occurs when the membrane potential of a cell increases above its resting potential, such as during an action potential. The inside of a cell becomes less negative during depolarization.

Desmosomes function to couple adjacent cells together. They consist of membrane proteins that connect to intermediate filaments.

Development refers to the series of changes that occur as a fertilized egg becomes an adult organism.

Diacylglycerol (**DAG**) forms the lipid backbone of most membrane phospholipids. DAG is made up of two fatty acids covalently linked to a glycerol molecule.

Dicer is an endonuclease that processes double-stranded precursor RNA molecules into siRNAs and miRNAs.

Dielectric constant is a measure of the polarity of a chemical substance.

Differentiation is the process that occurs when a cell becomes more specialized.

A **dimerization domain** is a protein domain that functions through protein-protein interactions to facilitate the dimerization of transcription factors. Dimerization can result in the formation of a homodimer of identical transcription factors or a heterodimer of two different transcription factors.

The term **diploid** refers to cells that contain pairs of homologous chromosomes. A diploid set of chromosomes contains two copies of each autosome and two sex chromosomes. Most eukaryotic cells other than gametes are diploid.

In **direct repair**, DNA lesions are directly reversed to restore the original sequence. Photoreactivation is one example of direct repair.

Discontinuous cell-cell adhesion is the rapid attachment and release of leukocytes as they come to a "rolling stop" in blood vessels.

A **DNA-binding domain** is a protein domain that makes sequence-specific contacts with a segment of DNA. Gene-specific transcription factors usually contain a DNA-binding domain that interacts with the control elements in a gene's promoter or enhancer.

DNA glycosylases are enzymes that remove damaged bases from DNA in the first step of base excision repair, leaving an intact phosphodiester backbone (resulting in an AP site).

A **domain** is an independently folding unit within a large polypeptide chain that often has a specific biological activity. A protein may contain multiple domains separated by intervening flexible polypeptide regions.

A **domain** of a *chromosome* may refer either to a discrete structural entity defined as a region within which supercoiling is independent of other domains or to an extensive region including an expressed gene that has heightened sensitivity to degradation by the enzyme DNAase I.

A **domain** of a *protein* is a discrete continuous part of the amino acid sequence that folds into an identifiable tertiary structure. A structural domain often performs a discrete function.

A **donor compartment** buds membrane-bounded transport vesicles that carry cargo destined for acceptor compartments.

Dosage compensation describes mechanisms employed to compensate for the discrepancy between the presence of two X chromosomes in one sex but only one X chromosome in the other sex.

Double-strand break repair (**DSBR**) refers to a group of DNA repair pathways that recognize and repair DNA double-strand breaks.

Downstream is the term used to describe those sequences on the coding strand of a DNA that are in the same direction (towards the 3' end) in which RNA polymerase is moving during transcription.

A **downstream promoter element** (**DPE**) is a sequence motif present in many (but not all) eukaryotic RNA polymerase II core promoters that is located downstream from the transcription start site. The DPE can help to compensate for the lack of a TATA box in some promoters.

Drosha is an endonuclease that processes double-stranded primary miRNAs and siRNAs into short (~70 bp) precursors for Dicer processing.

A myosin's **duty ratio** is the amount of time during its ATPase cycle that the myosin head is attached to an actin filament. Low duty ratio myosins spend little time attached to actin; high duty ratio myosins spend most of their time attached to actin filaments and may function processively.

Microtubules continually switch between growth and shortening states. This process is termed **dynamic instability**.

Dynamin is a cytosolic protein that is a GTPase and is required for clathrin-mediated vesicle formation. Although the exact role of dynamin is debated, dynamin polymers are involved in the scission of clathrin-coated pits from membranes. A variant of dynamin functions in mitochondrial septation.

Dyneins are a family of molecular motor proteins that bind and translocate along microtubules in a plus-end to minus-end direction.

Dysplasia describes an early stage of abnormal cell growth in which both cellular and tissue architecture have become moderately abnormal.

Early endosomes are the first compartments of the endocytic pathway in which internalized molecules appear. They are slightly acidic (pH 6.5 to 6.8). In early endosomes, internalized molecules are sorted for recycling to the plasma membrane or for delivery to late endosomes and lysosomes.

Early response genes are a set of genes that are expressed very quickly after serum addition. Many of these encode transcription factors.

Elastin is an extracellular matrix protein capable of elongating and retracting in response to tensile strain.

A difference in the concentration of ions across a cell membrane produces an **electrochemical gradient**. The term indicates that there is a difference in the concentrations of both electrical charge and chemical species across a lipid bilayer.

The **electron transport chain** (**ETC**) is a series of proteins that transfer electrons in cellular respiration to generate ATP.

Elongation is the stage in a macromolecular synthesis reaction (replication, transcription, or translation) when the polynucleotide or polypeptide chain is extended by the addition of individual monomers (nucleotides or amino acids).

An **endergonic** reaction is one that absorbs energy, so that the change in free energy is positive. The reaction written in reverse is exergonic.

Endocrine signaling is signaling that occurs when the signal, such as a hormone, travels through the bloodstream to affect cells distant from the secreting cell.

The **endocytic pathway** is the route from the plasma membrane through early and late endosomes to lysosomes. It is involved in the internalization of material from the extracellular space.

Endocytosis is the process by which cells internalize small molecules and particles from their surroundings. There are several forms of endocytosis, all of which involve the formation of a membranous vesicle from the plasma membrane.

The **endoplasmic reticulum** (**ER**) is an organelle involved in the synthesis of lipids, membrane proteins, and secretory proteins. It is a single compartment extending from the outer layer of the nuclear envelope into the cytoplasm and has subdomains such as the rough ER and the smooth ER. It is also the intracellular storage compartment for calcium. In striated muscle cells in the heart and skeletal muscle, it is commonly referred to as the sarcoplasmic reticulum, which has a specialized structure and function.

An **endosome** is a membrane-bounded organelle that sorts molecules received via endocytosis or delivery from the trans-Golgi network and transfers them to other compartments, such as lysosomes.

An **endospore** is a spore (dormant cell) that develops within a bacterium and is released upon lysis of the mother cell.

Endosymbiosis is a symbiotic relationship in which an organism lives inside the cells of another organism. It occurs when one cell captures another, and the second cell (typically a bacterium) becomes integrated into the first cell.

The **endosymbiotic theory** is a hypothesis that chloroplasts and mitochondria evolved from prokaryotes that lived symbiotically in free-living single cells.

Energy is a transfer of a mixture of work and heat between systems.

An **enhancer** is a transcription factor-binding site that increases the efficiency of a core promoter. In contrast to other types of transcriptional regulatory sites, enhancers can work at large distance (up to thousands of bp), and function in either orientation and in any location (upstream or downstream) relative to the core promoter region. Enhancers generally function through DNA looping mediated by protein-protein interactions between transcription factors bound at the enhancer and the core promoter region.

Enthalpy is heat exchange at constant pressure; it is a state function.

Entropy is commonly viewed as a "measure of disorder" but is more properly viewed as a measure of the number of energy states into which a system may be distributed.

Envelopes surround some organelles (for example, nucleus or mitochondrion) and some cells (such as prokaryotes) and consist of concentric membranes, each membrane consisting of the usual lipid bilayer. The cell envelope of prokaryotes consists of the cytoplasmic membrane and layers outside of it (cell wall, outer membrane, capsule). The number of outer layers and their composition depends on the type of prokaryote.

Epigenetic changes influence the phenotype without altering the genotype. They consist of changes in the properties of a cell that are inherited but that do not represent a change in genetic information.

The **epithelial-mesenchymal transition (EMT)** is the transdifferentiation process whereby an epithelial cell sheds many epithelial characteristics and acquires mesenchymal ones in their stead; passage through an EMT often results in the acquisition of stem cell properties as well.

An **epithelium** (*plural*: epithelia) is a sheet of polarized cells that is one or more layers thick and lines a body cavity (for example, the intestine) or the outer surface (skin). Epithelial cells have a basal surface and an apical surface. The functions of epithelial cells include secretion, absorption, and transcellular transport.

The **equilibrium constant** is the ratio of forward to reverse rate constants for a reaction. Alternatively, it is the ratio of products multiplied together, divided by substrates multiplied together.

The degradation of proteins that have been in the endoplasmic reticulum (ER) is called **ER-associated degradation (ERAD)**. ERAD commonly refers to retrograde translocation of a protein out of the ER followed by its degradation in the cytosol. No proteases have been found within the lumen of the ER.

Essential nutrients are those which cannot be synthesized by an organism, usually applied to humans.

Ethanol, an alcohol, is the end product of fermentation when pyruvate is converted to acetaldehyde, but it diffuses out of the cell rather than accumulate to toxic concentrations.

An **etiologic agent** is a substance or process that functions to help cause a disease or pathologic condition.

Euchromatin comprises most of the genome in the interphase nucleus, is less tightly coiled than heterochromatin, and contains most of the active or potentially active single copy genes.

A **eukaryote** is an organism that is composed of one or more cells, each containing a nucleus. The term eukaryote means "true nucleus."

Euploid describes a cell or organism that has a completely normal set of chromosomes.

Excision repair is a repair system in which one strand of DNA is directly excised and then replaced by resynthesis using the complementary strand as a template.

Excitation-contraction coupling is the process by which muscle fibers contract in response to membrane depolarization. Contraction is regulated by changes in the cytosolic Ca^{2+} concentration.

An **exergonic** reaction is one that releases energy, so the change in free energy is negative. The reaction written in reverse is endergonic.

The **exocyst** is a complex of eight proteins that is found at sites on the plasma membrane where secretion occurs. It tethers secretory vesicles to the membrane as the first step in the process of membrane fusion.

The **exocytic pathway** is the transport route from the endoplasmic reticulum, through the Golgi apparatus, to the plasma membrane. It is also called the secretory pathway.

Exocytosis is the process by which cells secrete cargo molecules that are originally synthesized at the endoplasmic reticulum. Exocytosis occurs by the fusion of secretory vesicles with the plasma membrane.

An **exon** is an expressed sequence of an intron-containing gene that is represented in the mature mRNA product.

Exonucleases are nucleases that cleave nucleotides one at a time from one end of a polynucleotide chain.

An **exosome** is an exonuclease complex involved in mRNA processing and degradation.

Expansin is a protein that loosens the connections between the structural elements of a plant cell's wall by displacing the hydrogen bonds that hold them together. The wall is weakened as a result, allowing the cell to expand.

Exportins are transport receptors that bind their cargo and associate with Ran-GTP in the nucleus. The trimeric complex translocates across the nuclear envelope into the cytoplasm, where hydrolysis of GTP bound to Ran results in release of cargo.

The **extracellular matrix** (**ECM**) is a relatively rigid layer of insoluble glycoproteins that fill the spaces between cells in multicellular organisms. These glycoproteins connect to plasma membrane proteins.

Extravasation is the movement of cells out of blood vessels into the parenchyma of tissues.

F-actin refers to polymerized filamentous actin.

Facultative heterochromatin describes the inert state of sequences that also exist in active copies—for example, one mammalian X chromosome in females.

FAD is the oxidized flavin cofactor found in reactions in which one as well as two-electron transfer is required.

FADH$_2$ is the reduced coenzyme partner to FAD.

Familial retinoblastoma is a heritable tendency to develop retinoblastomas before the age of 7, resulting from inheritance of a defective copy of the *RB* gene.

(Fatty) acyl-CoA is the molecule formed when a fatty acid is joined to coenzyme A.

Ferritin is an iron storage protein found primarily in the liver and kidneys.

An **FG repeat** is an amino acid motif with four or five residues (often Gly-Leu-Phe-Gly or X-Phe-X-Phe-Gly) found in multiple copies separated by short spacer sequences in approximately one-third of nucleoporin species.

A **fibripositor** is a membrane-bound compartment where collagen fibers assemble immediately before their secretion into the extracellular space. They bud from the Golgi and resemble elongated exocytic vesicles.

Fibronectin is an extracellular matrix protein found in most animal tissues. It is primarily responsible for supporting cell adhesion and migration.

Fibronectin proteins are composed of a linear arrangement of repeating structural modules that are classified into groups and are referred to as **fibronectin repeats**.

Filopodia (*singular*: filopodium) are thin, spiky projections that form at the leading edge of cells as they crawl along a surface. The shape and rigidity of filopodia are conferred by the underlying bundles of actin filaments.

The **first law of thermodynamics** states that energy changes are conserved between systems.

A **flagellum** (*plural*: flagella) is a whiplike structure that provides locomotion. A bacterial flagellum is a filamentous structure that extends from the surface. A eukaryotic flagellum contains an array of microtubules and is sheathed in the plasma membrane.

Fluorescent hybridization (**FISH**) is a technique used to detect and localize specific DNA sequences on chromosomes, using fluorescent probes.

A **focal adhesion** is a large protein complex that develops from a focal complex and has assembled a strong link with actin stress fibers. The proteins in focal adhesions initiate intracellular signaling pathways in response to cell-matrix interactions.

A **focal complex** consists of integrin, cytoskeletal, and signaling proteins clustered at the plasma membrane at points of cell contact with the extracellular matrix.

Focal contacts are regions on the plasma membrane where integrin receptors cluster upon contract with an extracellular ligand. Focal contacts are also enriched in cytosolic proteins that link integrins with cytoskeletal and signaling proteins.

Transformed cells grow as a compact mass of rounded-up cells that pile up on one another. They appear as a distinct **focus** on a culture plate, in contrast to normal cells, which grow as a spread-out monolayer attached to the substratum.

A **forward** reaction signifies the direction as written from left to right.

Free energy is a combination of enthalpy and entropy changes for a reaction.

G0 is a noncycling state in which a cell has ceased to divide (also called quiescent).

G1 is the period of the eukaryotic cell cycle between the last mitosis and the start of DNA replication.

G2 phase is the period of the cell cycle separating the replication of a cell's chromosomes (S phase) from the following mitosis (M phase).

G-actin refers to globular actin (actin monomer).

A **gain-of-function** mutation usually refers to a mutation that causes an increase in the normal gene activity. It sometimes represents acquisition of certain abnormal properties. It is often, but not always, associated with the creation of a dominant phenotype.

A **gap junction** is a cell-cell junction that allows the direct transfer of small cytosolic components between adjacent cells.

The **gap junction channel** is the structural foundation of the gap junction. Each cell constructs half of the channel (a hemichannel), and two hemichannels are joined together in the extracellular space.

Gating refers to the regulated opening and closing of channel proteins (such as ion channels and gap junctions) that control the passage of molecules through the channels.

G-bands are generated on eukaryotic chromosomes by staining techniques and appear as a series of lateral striations. They are used for karyotyping (identifying chromosomes and chromosomal regions by the banding pattern).

A **Gemini body** (**GEM**) is a subnuclear structure, similar to and often adjacent to Cajal bodies and containing specific proteins and small RNAs.

A **gene** is a segment of DNA that encodes the information for the synthesis of a functional RNA transcript. The transcript can either be a noncoding RNA (ncRNA) or it can be an mRNA that is translated to make a protein.

Gene amplification is the presence of an increased number of gene copies in a cell, resulting from inappropriate re-replication of DNA.

Genodermatosis is a genetic disease that arises from mutations in either desmosomal or hemidesmosomal proteins.

The **genome** is the complete set of sequences in the genetic material of an organism. It includes the sequence of each chromosome plus any DNA in organelles.

Genomics is the systematic study of the structure and function of genomes.

A **germ cell** is a sex cell or gamete that results from meiosis.

The **germline** consists of the reproductive cells (egg and sperm) that serve as the vehicles for the transmission of genetic material from one organismic generation to the next.

Glioblastoma is a highly malignant type of astrocytoma (brain tumor).

Global genome repair (**GG-NER**) is a subtype of NER that repairs DNA damage anywhere in the genome.

Gluconeogenesis is the pathway for glucose formation from non-sugar precursors.

Glycolysis is a metabolic pathway in which glucose is broken down into two molecules of pyruvate with a net gain of two ATP molecules.

A **glycosaminoglycan** (**GAG**) is a sugar polymer consisting of repeating disaccharide subunits. GAGs are often attached to the core proteins of proteoglycans.

Glycosylphosphatidylinositol (**GPI**) is a phospholipid based on phosphatidylinositol. GPI is covalently attached to some proteins and serves to tether/anchor them to a membrane.

Glycosyltransferases are a class of enzyme that transfers sugars onto proteoglycan core proteins.

The **Golgi apparatus** is an organelle that receives newly synthesized proteins from the endoplasmic reticulum and processes them for subsequent delivery to other destinations. It is composed of several flattened membrane disks arranged in a stack. Plants contain multiple Golgi bodies, which are called dictyosomes.

The central feature of the Golgi apparatus comprises flattened, membrane-bounded cisternal structures that in some cells are arranged in the form of a stack. The **Golgi stack** carries out an array of posttranslational modifications to the transiting cargo.

G protein-coupled receptors are proteins composed of seven membrane-spanning helices. These receptors promote activation of associated heterotrimeric G proteins by catalyzing exchange of bound GDP for GTP.

G proteins are guanine nucleotide-binding regulatory proteins. Their active form is bound to GTP. Hydrolysis of GTP to GDP via intrinsic GTPase activity inactivates the G protein. Heterotrimeric G proteins are composed of α, β, and γ subunits and are membrane-bound. The α subunit binds GTP. Monomeric G proteins—structural homologs of α subunits—may be cytosolic or membrane-bound.

A **growth factor** is a ligand (usually a small polypeptide) that activates a receptor (generally in the plasma membrane) to stimulate growth of the target cell. Growth factors were originally isolated as the components of serum that enabled cells to grow in culture.

Guanylyl cyclase is an enzyme that catalyzes the conversion of GTP to the second messengers 3′,5′-cyclic GMP and pyrophosphate.

A **guide strand** refers to the strand of an miRNA or siRNA that is assembled into the RISC after processing of a double-stranded RNA precursor by Dicer. The guide strand contains the complementary RNA sequence that targets the RISC to a specific mRNA sequence.

A **half-spindle** is the region of a mitotic spindle to one side of its equator. Each of a spindle's two poles organizes a half-spindle.

The **haploid** set of chromosomes contains one copy of each autosome and one sex chromosome; the haploid number n is characteristic of gametes of diploid organisms.

Heat is a type of energy exchanged between systems that acts as a result of temperature differences. In general terms it is not a state variable, but with restrictions such as constant pressure, it can become one.

A **helicase** is an enzyme that uses energy provided by ATP hydrolysis to separate the strands of a double-stranded nucleic acid substrate.

The **helix-turn-helix** (**HTH**) **motif** is found in a number of transcription factors and describes an arrangement of two a-helices that form a site that binds to DNA, one fitting into the major groove of DNA and the other lying across it.

The **hemidesmosome** is a cell junction that links intermediate filaments in cells to the basal lamina via cell surface integrins.

Heparan sulfate proteoglycans are a subset of proteoglycans that are found predominantly attached to the cell surface. They function primarily as coreceptors for extracellular matrix proteins and other cell receptors.

Heterochromatin describes regions of the genome that are highly condensed, are not transcribed, and are late-replicating. Heterochromatin is divided into two types, which are called constitutive and facultative.

A **heterogeneous nuclear ribonucleoprotein particle** (**hnRNP**) is the ribonucleoprotein form of hnRNA (heterogeneous nuclear RNA), in which the hnRNA is complexed with proteins. Because pre-mRNAs are not exported until processing is complete, hnRNPs are found only in the nucleus.

Heterologous adaptation is a form of feedback control that initiates downstream of multiple receptors in a convergent pathway, regulating the activity of the initiating receptor as well as the others.

Heterophilic binding is the binding of different protein receptors to one another.

A **high mannose oligosaccharide** is an N-linked oligosaccharide that contains N-acetylglucosamine linked only to mannose residues. It is covalently added to transmembrane proteins in the rough endoplasmic reticulum and is trimmed and modified in the Golgi apparatus.

Histone acetyltransferase (HAT) enzymes modify histones by addition of acetyl groups; some transcriptional coactivators have HAT activity.

Histone deacetylases (HDACs) remove acetyl groups from histones; they may be associated with repressors of transcription.

The **histone fold** is a motif found in all four core histones in which three α helices are connected by two loops.

Histones are conserved DNA-binding proteins that form the basic subunit of chromatin in eukaryotes. Histones H2A, H2B, H3, and H4 form an octameric core around which DNA coils to form a nucleosome. Histone H1 is external to the nucleosome.

Holoenzyme usually refers to either (1) The DNA polymerase complex that is competent to initiate replication, or (2) The RNA polymerase complex that is competent to initiate transcription.

The **homeobox** is a conserved ~180 bp DNA sequence within homeotic genes that specifies the 60–amino acid residue helix-turn-helix DNA-binding domain.

The **homeodomain** is a carboxy-terminal domain of about 60-amino acid residues found in homeobox proteins that folds into a helix-turn-helix DNA binding motif.

Homeostasis refers to the ability of a system (for example, a cell, organ, or organism) to maintain relatively constant internal conditions even as its external environment changes.

Homologous adaptation is a form of feedback control that affects only the initiating receptor in a signaling pathway.

Homologous recombination (HR) involves a reciprocal exchange of homologous DNA sequences.

Homophilic binding is the binding of identical protein receptors to one another.

A **hormone-response element** is a DNA sequence that specifically binds a steroid hormone receptor transcription factor. Such factors include the receptors for estrogen, progesterone, and retinoic acid.

Hyaluronan is a glycosaminoglycan that interacts with proteoglycans but is not covalently coupled to proteoglycan core proteins. It is especially abundant in cartilage.

Hybridization is the pairing of complementary nucleic acid strands from different DNAs or RNAs to produce a stable duplex structure.

The **hydration shell** consists of the water molecules that are attracted and oriented by an ion's charge and surround the ion in solution.

Hydrophilic substances dissolve in or mix easily with water.

Hydrophobic substances do not dissolve in or mix easily with water.

Hyladherins are cell surface proteins that bind to hyaluronan.

Hyperphosphorylated means having relatively many covalently attached phosphate groups.

Hyperplasia is an abnormal increase in the number of cells in a tissue without any obvious abnormality in the morphology of individual cells.

A **hypersensitive site** is a short region of chromatin detected by its extreme sensitivity to cleavage by DNAase I and other nucleases; it comprises an area from which nucleosomes are excluded.

Hypophosphorylated means having relatively few covalently attached phosphate groups.

Immortalization describes the process by which a eukaryotic cell line attains the ability to divide an indefinite number of times in culture.

The **immunoglobulin (Ig) domain** is a characteristic domain found in antibodies and a variety of other proteins that are members of the immunoglobulin superfamily. It consists of two layers of β sheets connected by a disulfide bond.

Importins are transport receptors that bind cargo molecules in the cytoplasm and translocate into the nucleus, where they associate with RanGTP, resulting in release of the cargo.

Indirect end labeling is a technique for examining the organization of DNA by making a cut at a specific site and identifying all fragments containing the sequence adjacent to one side of the cut; it reveals the distance from the cut to the next break(s) in DNA.

Induced proximity is a mechanism of caspase activation whereby two monomers are brought close to one another, forming active dimers.

An **inducer** is a small molecule that triggers gene transcription by binding and inactivating a transcriptional repressor protein.

Initiation includes the stages of transcription up to synthesis of the first ribonucleotide monomer unit in RNA. This includes binding of RNA polymerase to the promoter and melting a short region of DNA into single strands.

The **initiator region (INR)** is a short, loosely conserved sequence found in eukaryotic RNA polymerase II core promoters that overlaps the transcription start site and is required for the location of correct initiation.

The **inner nuclear membrane** is the membrane on the nuclear side of the nuclear envelope, which consists of two concentric nuclear membranes. The inner nuclear membrane interacts with chromatin and with the nuclear lamina.

In situ **hybridization** (also called cytological hybridization) is performed by denaturing the DNA of cells squashed on a microscope slide so that reaction is possible with an added single-stranded RNA or DNA; the added preparation is radioactively labeled and its hybridization is followed by autoradiography.

An **integrated, complex dynamic system** refers to a group of cellular processes that share common functional and structural components and that can change their activities in response to signals from the internal or external environment.

Integration of membrane proteins refers to the process by which a protein's hydrophobic, transmembrane domain(s) is inserted into a lipid bilayer, such that the protein extends across the membrane on both sides.

Integrins are a family of αβ heterodimeric cell surface proteins that mediate cell adhesion. Some integrins attach to extracellular matrix proteins, whereas others interact with ICAMs expressed on the surface of a partner cell. Integrins are the most abundant class of extracellular matrix protein receptors. Some integrins bind to other cell receptors.

An **integron** is a genetic element that allows integration of gene cassettes. A basic integron consists of an integrase gene, a target site for integration, and a promoter that drives expression of the integrated gene.

Interbands are the relatively dispersed regions of polytene chromosomes that lie between the bands.

Intercellular gene control is the process by which gene expression in a cell is regulated by other proximal or distal cells through diffusible signaling molecules.

An **interchromosomal domain** or interchromosomal channel is a region of the nucleus that does not contain chromatin.

Intermediate filaments are one of three types of cytoskeletal filaments that run throughout the cytoplasm of a eukaryotic cell. Intermediate filaments are the strongest of the three types and appear devoted to helping give tissues mechanical strength.

The **intermembrane space** is the space between the inner and outer membranes of a mitochondrion or chloroplast.

Intraflagellar transport describes the movement of molecular complexes toward the distal tip or proximal base of the axoneme. Movement occurs just underneath the plasma membrane surrounding the axoneme and is necessary to deliver new axoneme components for flagellar growth.

Intravasation refers to the entrance of cells from a primary tumor into blood and lymphatic vessels.

Intrinsic terminators terminate transcription in the absence of any additional factors.

An **intron** is a segment of RNA that is removed through a set of ribonucleoprotein-catalyzed reactions subsequent to transcription and is not included in the mature functional RNA transcript.

Invagination refers either to the infolding of a sheet of cells to form a groove as part of development or to the formation of clathrin-coated pits at the plasma membrane during endocytosis.

An **inverse agonist** is a ligand that binds preferentially to the inactive form of a receptor and thereby causes net inhibition of signaling through that receptor.

Inward rectification means that an ion channel conducts inward current better than outward current.

An **ion channel** is a transmembrane protein that selectively allows the passage of one or more types of ions across the membrane. Ions pass through a central aqueous pore of the channel.

An **iron-response element** (IRE) is a sequence in the mRNAs encoding ferritin and the transferrin receptor that binds to the iron response element binding protein IRP.

Ischemia is a lack of oxygen flow to a tissue; it is usually caused by blockage of a blood vessel.

An **isoform** is a protein that shares close structural and functional homology with other similar proteins (other isoforms), and frequently shares the same name, with a specifying number or letter.

Isomerization refers to the step during transcription when RNA polymerase opens up (melts) a small region of duplex DNA at the promoter to expose the template strand as a guide for RNA synthesis.

The **JAK-STAT pathway** is a signal transduction pathway that controls the expression of a number of genes involved in cell growth and proliferation.

Janus Kinase (JAK) is a family of proteins that function as a subunit of the JAK-STAT transcription factor dimer and the principal target of many growth factor and cytokine cellular signaling pathways.

The **junctional complex** is a collection of cell-cell junctions found along the lateral surface of adjacent epithelial or endothelial cells. These junctions are responsible for controlling permeability between cells and imparting structural stability to cell layers.

Karmellae (*singular*: karmella) are stacks of endoplasmic reticulum that develop around the nucleus in response to the overexpression of certain membrane proteins.

Karyogamy is the fusion of nuclei during the mating of two yeast cells.

Karyokinesis is the process by which the replicated chromosomes become equally segregated into two daughter nuclei. Mitosis consists of karyokinesis followed by cytokinesis.

Karyopherins are a family of related proteins that bind cargo molecules to be transported in and out of the nucleus, thereby acting as nuclear transport receptors.

The **Kennedy pathway** is a sequence of reactions that generates phospholipids from water-soluble precursors. Because it includes the reactions that first incorporate lipids into a membrane, the Kennedy pathway is essential for expanding the membranes of cell.

A **kinase** is an enzyme that catalyzes the transfer of a phosphoryl group from one substrate (usually ATP) to another substrate.

Kinesins are a family of molecular motor proteins that bind and translocate along microtubules. Most kinesins move in a minus-end to plus-end direction along microtubules.

Kinetic control of a reaction, or among several reactions, indicates that the choice of reaction or its extent is determined by the relevant reaction rates rather than by the poise of equilibria.

The **kinetochore** is a proteinaceous structure that associates with the centromere and attaches a chromosome to the microtubules of the mitotic spindle. Each mitotic chromosome contains two "sister" kinetochores, one on each centromere.

A **kinetochore fiber** is a bundle of microtubules that attaches a kinetochore to a spindle pole. Kinetochore fibers are also known as chromosomal fibers or spindle fibers.

A **kinome** is the total complement of protein kinases encoded in an organism's genome.

The **Ku** proteins (Ku70 and Ku80) form a heterodimer and bind to the free ends of DNA breaks to initiate the pathway of NHEJ.

The *lac* **operator** is a DNA sequence to which the *lac* repressor binds and inhibits transcription initiation at the promoter for the *lac* operon.

The *lac* **repressor** is a negative gene regulator encoded by the *lacI* gene that binds to the *lac* operator and inhibits the transcription of the *lac* operon.

Lactate results when cells oxidize NADH by reducing pyruvate (the end product of glycolysis) through fermentation. Lactate diffuses out of the cell to avoid buildup.

The **lactose** (*lac*) **operon** is a group of genes in *E. coli* that are required for the metabolism of the disaccharide lactose when the bacterium is starved for glucose. The genes in the *lac* operon include β-galactosidase, lactose permease, and transacetylase.

The **lagging strand** is the strand of DNA that must grow overall in the 3′-to-5′ direction and is synthesized discontinuously in the form of short fragments (5′ to 3′) that are later connected covalently.

Lamellipodia (*singular*: lamellipodium) are protrusions that form at the leading edge of cells moving along a surface; they contain dynamic actin networks.

Laminins are extracellular matrix proteins that are especially abundant in the basal lamina of many tissue types. Laminins form weblike networks rather than fibers.

Lampbrush chromosomes are the extremely extended meiotic bivalents of certain amphibian oocytes.

Late endosomes are organelles that are part of the endocytic pathway. They are formed upon the maturation of early endosomes as proteins are recycled to the plasma membrane, are more acidic than early endosomes, and contain some hydrolytic enzymes.

Lateral elements are structures in the synaptonemal complex that forms when a pair of sister chromatids condenses onto an axial element.

The **leading strand** is the strand of DNA that is synthesized continuously in the 5′-to-3′ direction.

Lectins are proteins that selectively bind to particular sugars. They are often used to bind sugars that are part of a glycoprotein.

Leucoplasts are a type of plastid that synthesizes and accumulates a variety of types of small organic molecules.

Leukemia is a cancer of the white blood cells.

In general, a **ligand** is a molecule that binds noncovalently to a protein. The term ligand often refers to an extracellular molecule that binds to and activates a receptor, thereby effecting a change in the cytoplasm.

A **ligand-gated ion channel** is a transmembrane protein that selectively allows the passage of one type of ion across the membrane in response to ligand binding.

Ligase is an enzyme that forms covalent phosphodiester bonds between 5′ and 3′ ends of DNA in an ATP-dependent reaction.

The **light reactions** of photosynthesis convert sunlight at specific wavelengths into energy in the form of NADPH and ATP.

Linker histones are a family of histones (such as histone H1) that are not components of the nucleosome core; linker histones bind nucleosomes and/or linker DNA and promote 30 nm fiber formation.

Lipoic acid is a cofactor in the pyruvate dehydrogenase complex (and in similar complexes) which transiently binds a 2-carbon portion of pyruvate and releases it to CoA.

A **lipoprotein** consists of a covalently linked protein and lipid. The term may also refer to particles formed by association between lipids and proteins.

Loss of heterozygosity describes a change that converts a heterozygous state of a genetic locus to a homozygous state.

The **lumen** describes the interior of a compartment (such as the endoplasmic reticulum or the Golgi apparatus) that is bounded by a membrane.

Lymphoma is a cancer arising from lymphocytes.

A **lysosome** is an organelle that contains a high concentration of hydrolytic enzymes and has an acidic lumen (pH as low as 4.5). Its primary function is the degradation of endocytosed material.

Lysyl hydroxylases hydroxylate the amino group in the side chain of selected lysines in procollagen to form 5-hydroxylysine, which can later be glycosylated by addition of an N-linked carbohydrate.

A **macrophage** is a phagocytic cell that moves through tissues, ingesting debris and foreign materials such as bacteria; macrophages may be involved in antigen presentation to T lymphocytes.

The **major groove** of DNA is ~17 Å wide and contains sufficient information to determine the DNA sequence.

A **malignant** tumor is one that is locally invasive and may metastasize to other areas of the body.

The **Mass Action Ratio** is the ratio of the product concentrations multiplied together to the substrate concentrations multiplied together, at any state of the reaction.

A **matrix attachment region** (MAR) is a region of DNA that attaches to the nuclear matrix. It is also known as a scaffold attachment region (SAR).

Matrix metalloproteinases are a class of enzyme that digests extracellular matrix proteins.

Mature elastin is an elastin protein that has been modified by lysyl oxidase and assembled into elastin fibers.

Secretory and membrane proteins that have had their signal sequences removed are called **mature proteins**.

A **mature mRNA** is the final functional product of mRNA processing that includes 5' 7-MeG capping, splicing, and polyadenylation.

Mediator is a large multi-protein complex that associates with the basal transcription apparatus on RNA polymerase II core promoters to mediate the effects of gene-specific transcription factors.

Meiosis occurs by two successive divisions (meiosis I and II) that reduce the starting number of $4n$ chromosomes to $1n$ in each of four product (haploid) cells. These cells may mature to germ cells (sperm or eggs) in animals or to spores in plants.

Meiotic maturation of an animal oocyte occurs when intercellular signals trigger the oocyte to exit from an arrested state at prophase of meiosis I and advance through meiosis in preparation for fertilization.

Melanoma is a cancer of the pigment-producing cells of the skin.

A cellular **membrane** consists mainly of a lipid bilayer. The bilayer is asymmetrical, with hydrophobic heads facing out on either side and lipid tails pointing into the center. It has lateral fluidity, contains proteins, and is selectively permeable.

Membrane potential is the electrical potential across a membrane.

Membrane pumps are proteins that use energy (such as the energy of ATP hydrolysis) to transport solutes from one side of a membrane to another, against their electrochemical gradients.

Meristems are small, highly organized regions in a plant where the cell division that makes the plant grow occurs. Meristems are located at the tips of a plant's roots and shoots, where they produce cells that increase its height. Other meristems around the plant's circumference produce cells that increase its girth.

Merotelic attachment is the condition when one kinetochore of a mitotic chromosome is attached to both poles of the mitotic spindle. Merotelic attachments occur early in mitosis and are usually corrected before anaphase begins.

Messenger RNA (**mRNA**) is the RNA molecule specified by a gene that programs a ribosome to synthesize a specific polypeptide.

A **metabolically irreversible** reaction is one in which the mass action ratio is greatly displaced from the reaction equilibrium constant by at least two orders of magnitude less. As a result, such a reaction is not reversed under cellular metabolic conditions and is a candidate for a regulatory site.

Metabolomics is the systematic measurement and computerized analysis of metabolites in cells or organisms.

Metaphase is the stage of mitosis or meiosis when all the chromosomes are aligned on the equator of the spindle, halfway between its two poles.

Metastasis describes the ability of tumor cells to leave the primary tumor and migrate to other locations in the body, where a new colony of tumor cells may be established.

Methylation is the covalent attachment of a methyl group to a substrate—in the context of DNA, to the C moiety of a CpG sequence.

A **microarray** is an arrayed series of thousands of tiny DNA oligonucleotide samples imprinted on a small silicon chip or glass slide. mRNAs can extracted from cells and hybridized to microarrays to assess the amount and level of gene expression.

Micrococcal nuclease is an endonuclease that cleaves DNA; in chromatin, it preferentially cleaves DNA between nucleosomes.

In plants, multiple glucose chains, polymerized by each cellulose-synthesizing complex, are hydrogen bonded together into extremely strong fibers called **microfibrils**.

Microfilaments are dynamic polymers of actin that function in the cytoskeleton to confer cell shape and plasticity, as well to transport proteins, RNA and organelles to specific locations.

A **microRNA** (**miRNA**) is a member of a class of small RNAs (18-23 nucleotides) that regulate the expression of specific genes by inhibiting translation or stimulating degradation of specific mRNAs.

Microtubule-associated proteins (**MAPs**) are proteins that bind to microtubules and influence their stability and organization.

Microtubule-dependent molecular motors are proteins that move in one direction along microtubules by hydrolyzing ATP. They move cargoes along microtubules and often play a major role in organizing the microtubules themselves.

A **microtubule-organizing center** (**MTOC**) is a region from which microtubules emanate. In animal cells, the centrosome is the major microtubule-organizing center.

Microtubules are thin, hollow filaments composed of the proteins α- and β-tubulin. Microtubules help organize the interior of the cell and determine its shape, in part by acting as tracks for the movement of the kinesin or dynein motor proteins.

Microtubule subunit flux is a behavior observed in the microtubules of mitotic spindles. Tubulin subunits are constantly added at the plus end of a microtubule and removed from the minus end. At steady state, subunits move or flow continuously from one end of the microtubule to the other without changing its length.

The **midbody** consists of a large bundle of microtubules and associated proteins that forms between the two separated daughter nuclei late in mitosis. The midbody plays an important role during cytokinesis.

The **minor groove** of DNA is ~12 Å wide, and is typically bound by non-sequence-specific binding proteins.

Mismatch repair (**MMR**) corrects recently inserted bases that do not pair properly. The process preferentially corrects the sequence of the daughter strand by distinguishing the daughter strand and parental strand, sometimes on the basis of their states of methylation.

The **mitochondrial-associated membrane** is a specialized region of the endoplasmic reticulum membrane that is in very close proximity to the outer membranes of mitochondria, and is involved in certain lipid synthesis reactions.

The **mitochondrial genome** is a double-stranded circular DNA that is found in the mitochondria of eukaryotes. Mitochondrial genomes vary in size from 16-20 kb (mammals) up to 80-100 kb (higher plants).

The **mitochondrial matrix** is the space enclosed by the inner mitochondrial membrane.

Mitosis is part of the process of eukaryotic cell division by which cells reproduce themselves, resulting in daughter cells that contain the same amount of genetic material as the parent cell. During mitosis, the replicated chromosomes separate and the nucleus divides.

Mitotic slippage is the process by which a cell ultimately exits mitosis, after a prolonged mitotic arrest, when the mitotic checkpoint cannot be satisfied. It occurs from the slow but continuous ubiquitinylation and destruction of cyclin B even in the presence of an active checkpoint, and it usually results in the formation of a single 4N G_1 cell.

Mobile cofactors are coenzymes that transiently bind to an enzyme for its reaction and subsequently dissociate, making them available for other enzymes.

Monocistronic refers to a mature messenger RNA (mRNA) that contains only one open reading frame (ORF) and usually encodes only one protein.

A **monoclonal** population of cells is one in which all the cells are derived from a single common progenitor cell.

A **monolayer** describes the growth of eukaryotic cells in culture as a layer only one cell deep.

A mitotic or meiotic chromosome is **mono-oriented** when only one of its kinetochores is attached to the spindle. Mono-orientation is a normal intermediate stage in the attachment of chromosomes to the spindle.

A **mosaic** organism is one that contains cells of different genotypes.

M phase is the period of the eukaryotic cell cycle during which nuclear and cytoplasmic division occurs.

mRNA editing is the process by which specific nucleotides are added post-transcriptionally to specific locations in a pre-mRNA sequence before the mRNA is translated to synthesize a protein.

An **mRNP** (*mRNA ribonucleoprotein particle*) is a complex of mature (spliced) mRNA and mRNA-binding proteins. Cytoplasmic and nuclear mRNPs differ greatly in their protein composition.

The **multivesicular body** (**MVB**) represents a morphological form of the late endosome inside of which many vesicles have accumulated. The internal vesicles are derived by the involution and pinching off of the limiting membrane of the endosome.

Mutagens increase the rate of mutation by inducing changes in DNA structure and thus, base sequence.

A **mutator** is a mutation or a mutated gene that increases the basal level of mutation. Such genes often code for proteins that are involved in repairing damaged DNA.

A **mutualistic** relationship is a symbiotic relationship in which both species benefit.

A **myeloma** is a malignant tumor that develops in the antibody-producing cells of the bone marrow; the cells are used in the procedure to produce monoclonal antibodies.

Myosins are a family of molecular motor proteins that bind and carry cargo by translocation along microfilaments in the cytoplasm.

The **n-1 rule** states that only one X chromosome is active in female mammalian cells; any other(s) are inactivated.

A **nascent protein** has not yet completed its synthesis; the polypeptide chain is still attached to the ribosome via a tRNA.

A **near-equilibrium** reaction is one which has all of its reactants and products at concentrations such that its mass action ratio is close to the reaction equilibrium constant. As a result, such a reaction can be reversed under changing metabolic conditions.

Necrosis is a type of cell death after injury in which cells swell and break open, causing inflammation in the surrounding area.

Neoplasm. *See* tumor.

A **neural cell adhesion molecule** (**NCAM**) is one of a class of cell surface receptors found only in cell-cell junctions of neuronal cells.

Neuroblastoma is a childhood cancer of neuroendocrinal cells of the sympathetic nervous system.

The **neuroectoderm** is the portion of the early embryo that gives rise to the central and peripheral nervous systems.

Neurofibrosarcoma is a malignant tumor of the Schwann cells forming the sheathing around peripheral nerves.

Nicotinamide adenine dinucleotide phosphate (**NADPH**), the reduced form of the cofactor $NADP^+$, is an essential ingredient for the dark reactions of photosynthesis. In animals, NADPH is used for reductive synthesis.

Nitrogenase is the nitrogen-fixing enzyme that converts nitrogen gas (N_2) to ammonia (NH_3).

The **nitrogen cycle** converts atmospheric nitrogen (N_2), which can be used by very few organisms, into more useful forms such as NH_3, NO_2^-, and NO_3^-, and back into N_2. The process is performed largely by soil bacteria, but also by metals in the Haber process, and by lightning in the atmosphere.

N-linked glycosylation is the covalent linkage of a carbohydrate group to asparagine residues within a protein as it is translocated into the endoplasmic reticulum.

Noncoding RNA (**ncRNA**) refers to several classes of RNAs that do not code for proteins but function in a variety of different cellular processes.

Nonessential nutrients are those which can be synthesized by an organism, usually applied to humans.

A **nonhistone** protein is any structural protein (with the exception of histones) found in a chromosome.

Nonhomologous end joining (**NHEJ**) is a repair mechanism used to fix double-strand breaks where no homology is present.

Nonjunctional proteins are proteins that are not contained in cell-cell junctions (e.g., tight junctions, adherens junctions, etc.). Most cellular proteins are nonjunctional.

Nonsense-mediated decay (**NMD**) is a pathway that degrades an mRNA that has acquired a nonsense mutation prior to the last exon.

Nonstop mRNA decay is a pathway that rapidly degrades an mRNA that lacks an in-frame termination codon.

The **nuclear envelope** is made of two concentric membranes (the inner and outer nuclear membranes) and is the boundary between the nucleus and the cytoplasm. Beneath the nuclear envelope is an intermediate filament lattice (the nuclear lamina), and proteins embedded in the inner nuclear membrane contact the lamina. The nuclear envelope is penetrated by nuclear pores, which are also sites where the inner and outer nuclear membranes are fused. The outer membrane is continuous with the membrane of the rough endoplasmic reticulum.

The **nuclear envelope lumen** is the space between the inner and outer nuclear membranes and is continuous with the lumen of the endoplasmic reticulum.

A **nuclear export signal** (**NES**) is a protein domain (usually a short amino acid sequence) that interacts with an exportin, resulting in the transport of the protein from the nucleus to the cytoplasm.

The **nuclear lamina** is an organized proteinaceous layer on the nucleoplasmic side of the nuclear envelope. It consists of (up to) three related intermediate filament proteins called lamins.

A **nuclear localization signal** (**NLS**) is a protein domain (usually a short amino acid sequence) that interacts with an importin, allowing the protein to be transported into the nucleus.

The **nuclear matrix** is the name given to the observed network of fibers that remains when the nucleus has been treated to remove all soluble proteins, all lipids, and almost all DNA. It is only observed after such extraction. It may provide structure and organization for the nucleus, but its existence is controversial.

A **nuclear pore complex** (**NPC**) is a very large, proteinaceous structure that extends through the nuclear envelope, providing a channel for bidirectional transport of molecules and macromolecules between the nucleus and the cytosol.

Nucleating proteins help initiate and control the formation of polymers, such as actin filaments, in cells. The Arp2/3 complex and formins are the two main types of nucleating protein complexes.

Nucleation refers to the step in a polymerization reaction in which formation of a small multimer (nucleus) allows for the elongation of the polymer. For example, nucleation occurs in the assembly of microtubules and actin filaments.

The **nucleoid** is the structure in a prokaryotic cell that contains the genome. The DNA is bound to proteins and is not enclosed by a membrane.

The **nucleolus** (*plural*: nucleoli) is a discrete region of the nucleus where ribosomes are produced and some additional nuclear functions take place.

The **nucleoplasm** refers to the content of the nucleus, excluding the nucleolus.

Nucleoporin was the term originally used to describe those components of the nuclear pore complex that bind to the inhibitory lectins, but now is used to mean any polypeptide component of the nuclear pore complex.

Nucleoside diphosphokinase (**NDP kinase**) is an enzyme-catalyzing transfer of phosphoryl groups from ATP to a nucleoside diphosphate.

The **nucleosome** is the basic structural unit of chromatin in the interphase chromosome. The nucleosome core particle consists of ~140 bp of DNA and two copies each of the histones (H2A, H2B, H3 and H4).

Nucleosome positioning describes the placement of nucleosomes at defined sequences of DNA instead of at random locations with regard to sequence.

A **nucleotide** is a component of a nucleic acid consisting of a carbohydrate molecule, a phosphate group, and a nitrogenous base.

Nucleotide excision repair (**NER**) is a repair system that recognizes and repairs bulky lesions in DNA.

The **nucleus** is a cellular organelle that contains the genomic DNA. It is surrounded by a nuclear envelope consisting of a double layer of membranes and is the site of DNA replication, gene transcription, and most RNA processing.

Okazaki fragments are short stretches of 1000–2000 bases produced during discontinuous bacterial DNA replication, which are later joined into a covalently intact strand by DNA ligase.

Oligosaccharyltransferase (**OST**) is a multisubunit complex in the membrane of the endoplasmic reticulum. It transfers carbohydrate groups onto proteins as they are translocated into the ER.

Oncogenes are genes whose products have the ability to transform eukaryotic cells so that they grow in a manner analogous to tumor cells.

An **oncoprotein** is the product of a gene (oncogene) that when overexpressed or mutated can lead to unregulated cell growth and proliferation and cancer.

The **One Gene, One Protein Hypothesis** states that a gene us the unit of genetic information that directs the synthesis of a specific messenger RNA (mRNA). With the discovery of alternative splicing and noncoding RNAs (ncRNAs), this hypothesis is only of historical significance.

An **open reading frame** (**ORF**) is the complete set of codons read by the ribosome in an mRNA that begins with an initiation (start) codon and ends with a termination (stop) codon.

Open systems are those which can exchange energy and matter with their surroundings.

An **operon** is a cluster of genes within a bacterial genome that encodes proteins or RNAs belonging to the same functional pathway. In most cases, the genes within an operon are transcribed as a single mRNA with consecutive open reading frames, each of which codes for a separate protein or functional noncoding RNA (ncRNA).

Organelles are subcellular structures in the cytoplasm of eukaryotic cells. Most organelles have a limiting membrane. An organelle may consist of one or more distinct compartments.

The **origin** of replication is the specific site within the replicator at which replication is initiated.

The **origin recognition complex** (**ORC**) is a multisubunit complex that functions as a replication initiator protein.

Orthologs can be related genes in different species, or proteins with a shared function.

Osmosis is the process by which water passes across a semipermeable membrane, such as a cell's plasma membrane. The flow of water is driven by differences in the concentrations of solutes on the two sides of the membrane.

Osmotic pressure is a measure of the tendency of a solution to take up water when separated from pure water by a membrane.

The **outer nuclear membrane** is the membrane of the nuclear envelope that is on the cytoplasmic side and is continuous with the membrane of the rough endoplasmic reticulum.

Oxidative phosphorylation is a series of sequential steps in which energy is released from electrons as they pass from coenzymes to cytochromes, and ultimately, to oxygen gas (O_2). The energy is used to combine phosphate ions with ADP molecules to form ATP molecules.

Oxidative respiration is the process by which aerobes metabolize their food sources.

The **oxygen-evolving complex** is a cofactor that strips electrons from water, releasing protons and oxygen gas. It provides the driving force for electrons moving from PSII to PSI and sets up the proton gradient.

p53 is a tumor suppressor gene that is lost or inactivated in a variety of neoplasms, including breast, colon, and lung carcinomas, sarcomas, and leukemias.

p300/CBP is a complex between p300 and CREB binding protein (CBP) that functions as a coactivator for the expression of a number of genes involved in cell growth and differentiation.

The **p300/CBP-associated factor** (**PCAF**) is a lysine acetyltransferase that acetylates histones, loosening the binding of nucleosomes to DNA in promoter regions and thereby activating RNA polymerase transcription initiation of specific genes.

P680 *See* PS II.

P700 *See* PS I.

Paracrine signaling is signaling that occurs between neighboring cells.

A **parasitic** relationship is a symbiotic relationship in which one species, called the parasite, benefits at the expense of the other species, called the host.

A **partial agonist** is a ligand that binds preferentially to the active receptor but promotes submaximal activation because its selectivity for the active form is modest.

A **pathogen** is an organism that causes disease in another organism.

The **PDZ domain** is a structural domain found in many signaling proteins that allows them to bind to one another.

Peptidoglycan is the major component of the rigid cell wall of bacteria. It is a polymer of carbohydrate strands crosslinked by peptides and is important for resisting osmotic pressure and for maintaining cell shape.

Peptidyl proline cis-trans isomerase is an enzyme in the endoplasmic reticulum that facilitates the folding of procollagen by interconverting the *cis* and *trans* forms of the peptide bonds linking prolines to other amino acids in the procollagen polypeptide chains.

The **peptidyl transferase reaction** is the step during translation when the growing polypeptide chain linked to the tRNA (peptidyl tRNA) is transferred to the amino acid linked to the tRNA bound to the next codon in the open reading frame (ORF).

The **pericentriolar material** is a large collection of proteins assembled into a poorly defined structure that surrounds centrioles. The pericentriolar material is where microtubules are nucleated by the centrosome.

The **periplasm** (or periplasmic space) is the region between the inner and outer membranes in the cell envelope of Gram-negative bacteria.

Permissive growth conditions allow conditional lethal mutants to survive.

The interior space of a peroxisome is the **peroxisomal matrix**.

Phagocytosis is a type of endocytosis in which a cell internalizes a large particle such as a bacterium.

A **phosphatase** is an enzyme that removes phosphate groups from substrates by hydrolysis.

A **phosphoinositide** is a derivative of the glycerolipid phosphatidylinositol that is phosphorylated on the inositol ring.

Phosphorylation refers to the addition of a phosphoryl group to a molecule.

Photoisomerization is the change of a compound from one isomer to another in response to absorption of light. (An isomer is one of two or more forms of a compound with the same composition of atoms that can be arranged in a different manner.)

Photolyase is an enzyme that uses energy from visible light wavelengths to reverse UV-induced photodamage in DNA; this reaction is called photoreactivation.

Photoreactivation is a repair mechanism that uses a white light-dependent enzyme to split cyclobutane pyrimidine dimers formed by ultraviolet light.

Photosystems are membrane-bound complexes within the thylakoid membrane of chloroplasts which absorb light and transport electrons.

The **phragmoplast** is a microtubule-based structure that directs the formation of the new wall that divides a plant cell after mitosis.

Many plant cells contain vacuoles that occupy most of their volume. The **phragmosome** is a sheet of cytoplasm that forms across the vacuole of such a cell so that it can divide. The mitotic spindle and the new cell wall (which divides the cell) both form within the phragmosome.

Physical interference is the process by which a protein or other biological molecule blocks the assembly of another component in a macromolecular complex.

A **pilus** (*plural*: pili) is a fibrous appendage that extends from the surface of prokayotes and mediates interactions with other cells. Pili are composed of pilin proteins. During conjugation, pili are used to transfer DNA from one bacterium to another. Pili are also called fimbriae (*singular*: fimbria).

Pinocytosis is a type of endocytosis in which the cell takes up material dissolved in the extracellular fluid in variously sized endocytic vesicles.

Planktonic organisms are free to move about (in contrast to sessile organisms).

The **plasma membrane** is the continuous membrane defining the boundary of every cell.

A **plasmid** is a circular, extrachromosomal DNA. It is autonomous and can replicate itself.

Plasmodesmata (*singular*: plasmodesma) are narrow channels that connect the cytoplasm of adjacent plant cells and allow material to pass between them. Each plasmodesma is formed by a tube of plasma membrane extending between the cells and has a thin tubule of endoplasmic reticulum running down its center.

Plastids are a family of membrane-bounded organelles unique to plant cells. The different types of plastids all differentiate from a common precursor organelle called a proplastid but have different functions. Only one type is found in a cell. Chloroplasts are the most widely known form of plastid.

Pluripotent is used to describe precursor cells that have the ability to differentiate into a wide variety of cell types.

The **pointed end** of an actin filament is its slow-growing end.

A **point of no return** is a transition point in the cell cycle which, when passed, irreversibly commits the cell to the process that follows. For example, the cell becomes committed to mitosis when it passes a point of no return in late G2 of the cell cycle.

Pol a/primase is a four-subunit protein complex essential for eukaryotic replication, which contains both a primase and a DNA polymerase.

A **polarized cell** is a cell that has two or more distinct membrane domains with distinct functions, such as the apical and basolateral membranes of epithelial cells, or the axon and cell body of a neuron.

Polo boxes serve as intracellular targeting domains for PLKs and are responsible for directing them to multiple docking sites. Polo boxes bind to consensus phosphorylation sequences for Cdk1

and other proline-directed kinases, called phospho-Ser/Thr-Pro modules.

The **polyadenylation signal** is a hexamer sequence AAUAAA located in the 3' UTR of mRNAs that is required for the synthesis of the poly(A) tail.

The **poly (A) tail** is a segment of adenosine monophosphate residues that is added post-transcriptionally to a pre-mRNA and functions in the export of mRNA from the nucleus to the cytoplasm, as well as in translation and mRNA stability.

Polycistronic refers to a mature messenger RNA (mRNA) that contains multiple open reading frames (ORFs) and encodes more than one protein.

A **polyp** is a benign growth or tumor protruding from the mucous lining of an organ, such as the colon.

Polysialic acid (**PSA**) is a long, linear chain of sialic acid sugars found on neural cell adhesion molecules (NCAMs). Its presence results in reduced NCAM adhesion strength.

Polytene chromosomes are generated by successive replications of a chromosome set without separation of the replicas.

Positional information describes the localization of macromolecules at particular places in an embryo. The localization may itself be a form of information that is inherited.

Position effect variegation is silencing of gene expression that occurs as the result of proximity to heterochromatin.

Postmitotic cells are cells that have lost the ability to ever divide again.

Posttranslational translocation is the movement of a protein across a membrane after the synthesis of the protein is completed and it has been released from the ribosome.

The **powerstroke** is the force-generating movement of myosin. During the powerstroke, the lever arm (regulatory domain) of myosin undergoes a large rotation or conformational change.

pRb is the protein product of the retinoblastoma gene, which undergoes periodic phosphorylation and dephosphorylation as cells progress through their growth and division cycles.

The **preinitiation complex** (**PIC**) is group of proteins that assemble onto a eukaryotic promoter DNA and function to help RNA polymerase recognize and bind to the core promoter.

A **pre-miRNA** is generated in the nucleus from a pri-miRNA by the Drosha-DCGR8 complex and is exported to the cytoplasm by the Exportin 5 protein.

A **pre-mRNA** is the immature nuclear transcript that is processed by various modifications and splicing to yield a mature functional mRNA.

The **preprophase band** is a band of microtubules and actin filaments that forms within the cortex of a plant cell shortly before mitosis begins. The band is usually continuous and encircles the cell in the same plane in which it will later divide.

A protein to be imported into an organelle or secreted from bacteria is called a **preprotein** until its signal sequence has been removed.

A carrier protein that mediates **primary active transport** couples the energy of ATP hydrolysis with the transport of a solute against its electrochemical gradient. These carrier proteins help to maintain solute gradients across membranes.

The **primary cilium** is a single large cilium found on almost all cells. In some cells specialized for sensory purposes the primary cilium is highly modified to act as a sensory structure.

The **primary spindle** is a spindle-shaped array of microtubules formed early in mitosis when the microtubules from the two centrosomes interact before the nuclear envelope breaks down.

The **primary tumor** is the tumor mass that was initially formed in an individual with cancer.

Primases are RNA polymerases that synthesize short segments of RNA that are used as primers for DNA replication.

A **primer** is any nucleic acid that provides a free 3′ hydroxyl for elongation by a DNA polymerase.

A **pri-miRNA** is the primary miRNA transcript that is generated through transcription of an individual gene or cleaved from an intron.

Procollagen is a form of collagen containing sequences of amino acids, at the amino and carboxyl termini, that prevent collagen polymerization.

Programmed cell death is cell death occurring at a defined point in development.

A **prokaryote** is a unicellular organism that does not have a membrane-bounded nucleus.

Proliferating cell nuclear antigen (**PCNA**) is the eukaryotic protein complex that forms a sliding clamp to tether the DNA polymerase holoenzyme to the DNA.

Prolyl-4-hydroxylase assists in the folding of procollagen by converting accessible proline residues in the Y position of –Gly–X–Y– triplets to peptidyl 4-hydroxyproline.

Prometaphase is the stage of mitosis during which the spindle is formed. Prometaphase starts at nuclear envelope breakdown, when the chromosomes are first able to interact with microtubules, and ends when all of the chromosomes are attached to the spindle and aligned in the middle of the spindle.

A **promoter** is a DNA sequence at the beginning of a gene that binds to RNA polymerase and functions as a signal for RNA polymerase to initiate transcription.

Promoter escape occurs when RNA polymerase becomes free of the promoter and moves downstream to begin transcription elongation.

Proofreading is a mechanism for correcting errors in protein or nucleic acid synthesis that involves removing individual units after they have been added to the chain.

Prophage is a phage genome that is covalently integrated as a linear part of the bacterial chromosome or maintained as an autonomous plasmid-like circular DNA.

Prophase is the period of mitosis when the chromosomes condense and form visible structures within the nucleus. Prophase ends and prometaphase begins when the nuclear envelope breaks down.

A **proplastid** is a small membranous organelle that is found in plant cells and can give rise to any of several different forms of plastid.

Prosthetic groups are compounds that are tightly bound to and which form an integral structural/functional part of many proteins, particularly but not exclusively enzymes.

Protein disulfide isomerase (**PDI**) is a member of a family of related proteins that catalyze disulfide bond formation and rearrangement in the lumen of the endoplasmic reticulum.

Protein sorting is the direction of different types of proteins for transport into or between specific organelles.

Protein targeting is the process by which proteins are selectively recognized and brought to their sites of translocation across a membrane.

Protein translocation describes the movement of a protein across a membrane. This occurs across the membranes of organelles in eukaryotes or across the cytoplasmic membrane in bacteria. Each membrane across which proteins are translocated has a channel specialized for the purpose.

Proteoglycans are heavily glycosylated proteins of the extracellular matrix. They are composed of a single core protein covalently linked to one or (usually) more glycosaminoglycan sugar chains.

Proteoliposomes are lipid vesicles that contain proteins. The term usually refers to vesicles that have been reconstituted from proteins and lipids solubilized in detergent.

Proteomics is the study of protein sequences and structural characteristics in a systemized (and usually computerized) way.

Protofilaments are linear, polarized assemblies of protein subunits. Microtubules and actin filaments are formed by the lateral association of protofilaments. For example, α/β-tubulin heterodimers associate to form protofilaments, and thirteen protofilaments form the wall of a microtubule.

The **proton motive force** is the sum of electrical and chemical potentials of protons across a membrane.

A **proto-oncogene** is a normal gene that can be mutated to become an oncogene.

PSI is the first discovered photosystem. It actually occurs later in the sequence of electron flow than PSII.

PSII was the second photosystem discovered, but it is the first step in the light reactions, reacting directly with H_2O to produce O_2.

A **puff** is an expansion of a band of a polytene chromosome associated with the synthesis of RNA at some locus in the band.

A **purine** is a heterocycle with a five-membered imidizole ring fused to a six member pyrimidine ring that is the parent compound for both adenine and guanine.

The **purine nucleotide cycle** is a pathway for generation of Krebs Cycle intermediates.

A **pyrimidine** is a six-atom nitrogen-containing heterocycle that is the parent compound for both thymine and cytosine.

Pyruvate dehydrogenase complex is a complex of three enzymes and two regulatory enzymes that convert pyruvate to acetyl-CoA without intermediate metabolites connected to other pathways.

Quality control of proteins refers to the identification and either repair or elimination of aberrant proteins. It is most commonly used in reference to protein biogenesis in the endoplasmic reticulum.

Quantitative PCR (qPCR) is a polymerase chain reaction method in which a specific segment of DNA is amplified by a heat-stable DNA polymerase using sequence-specific oligodeoxyribonucleotides as primers. The amplification process in qPCR is monitored by fluorescence spectroscopy, which quantitatively detects dyes that become highly fluorescent when bound to the amplified DNA.

The **quantitative reverse-transcriptase polymerase chain reaction (qRT-PCR)** is a automated process by which mRNA extracted from cells and is converted to complementary DNA (cDNA) by the reverse transcriptase enzyme; the amount of a specific cDNA in the population is quantified by qPCR.

A **quiescent** cell is one that is not proliferating. *See also* G0.

Quinone ($coQH_2$) is a mobile cofactor used in electron chains, a lipid-soluble molecule within the interior of membranes. It is an important intermediate in the Krebs Cycle; when acetyl CoA is oxidized, quinone and NADH accept the electrons.

Quorum sensing refers to the ability of bacteria to monitor their population density through signaling systems.

Rab proteins are small Ras-like GTPases. Different Rabs are required for protein trafficking between different membrane compartments. Although their exact role is not clear, Rabs appear to regulate membrane targeting and fusion.

Rab effectors are proteins that bind to the GTP-bound form of Rabs and help to carry out their function(s).

The **rate** of a reaction is proportional to the substrate concentrations; the constant of proportionality is thus called the rate constant.

A **rate constant** signifies the proportionality between reaction rate and reactant concentrations.

The **reaction center** is a small region of a photosystem to which photons from the antenna system are donated and cause an elevation of the energy state (potential) of electrons.

Reaction coupling is a key feature of living organisms. The process uses reactants of higher energy (e.g., ATP) to drive what would otherwise be an energetically unfavorable reaction.

A **reaction equilibrium** occurs when the forward and reverse rates are equal.

A **receptor** is generally a plasma membrane protein that binds a ligand in a domain on the extracellular side and, as a result, changes the activity of the cytoplasmic domain. The same term is also used for nuclear receptors, which are transcription factors that are activated by binding ligands such as steroids or other small molecules.

Receptor-mediated endocytosis is the process by which soluble macromolecules are taken up after binding to cell surface receptors.

Receptor protein kinases are membrane-spanning proteins that phosphorylate their intracellular substrates and thus change the function of the target proteins.

Recruitment refers to the process by which a protein interacts with another protein to facilitate the assembly of a functional macromolecular complex.

Rectification is the property of ion channels in which the ion conductance changes in response to a change in membrane voltage.

Recycling endosomes serve as storage depots for endocytosed receptors and are the site of polarized membrane sorting in the endosomal system.

Redox potential is a molecule's affinity for electrons. A high redox potential indicates that a molecule is easily reduced.

Regulated secretion (also called regulated exocytosis) is the triggered release from the cell of cargo molecules stored in secretory granules.

Repair of damaged DNA can take place by repair synthesis, when a strand that has been damaged is excised and replaced by the synthesis of a new stretch. It can also take place by recombination reactions, when the duplex region containing the damaged strand is replaced by an undamaged region from another copy of the genome.

The **replication fork** is the site of active replication, where the two parental DNA strands have been separated by the action of helicase.

Replication licensing is the process that ensures origins of replication fire once and only once per cell cycle.

Replication protein A (RPA) is a single-stranded binding protein present in eukaryotes that is involved in DNA replication, DNA repair, and recombination.

Replication slippage is a phenomenon in repetitive sequences in which the daughter strand misaligns with the parental strand, resulting in insertions or deletions of repeats.

The **replisome** is the protein complex that assembles at the replication fork to synthesize DNA.

Membrane **repolarization** is the phase of an action potential when the membrane potential returns to its resting value.

During microtubule dynamic instability, the transition step between shortening and growing phases is a **rescue**. In genetics, to rescue a mutant means to restore the wild-type phenotype.

Resolution occurs when the Holliday junction structure is cleaved to generate two separate nicked duplexes.

A **response element** is a sequence in a promoter or enhancer that is recognized by a specific transcription factor.

The **restriction point** is the point during G1 at which a cell becomes committed to division. In yeast, this point is known as START.

A **retention signal** is the part (or parts) of a protein that locates it in the correct membrane and prevents it from leaving the compartment. An example is the transmembrane domain of Golgi resident proteins.

A **retrieval signal** is a type of sorting signal. Should a protein be transported inadvertently to another compartment, the retrieval signal is the part (or parts) of the protein that mediates its return to the compartment in which it normally functions.

Retrograde translocation, or reverse translocation, is the translocation of a protein from the lumen of the endoplasmic reticulum to the cytosol. It usually occurs to allow misfolded or damaged proteins to be degraded by the proteasome. It is also called retrotranslocation or dislocation.

Retrograde transport is the movement of material from the Golgi apparatus (or from the plasma membrane) to the endoplasmic reticulum.

A **retrovirus** is an RNA virus with the ability to copy the sequences of its RNA genome into a DNA version through the process of reverse transcription.

A **reverse** reaction signifies the direction as written from right to left.

A **Rho-utilization (rut) sequence** binds to bacterial transcription termination factor Rho and activates the Rho ATPase activity required for transcript release.

A **ribonucleoprotein (RNP)** refers to a complex between RNA and one or more proteins.

The **ribonucleoside triphosphates (NTPs)** are the four monomer precursors adenosine triphosphate (ATP), cytidine triphosphate, (CTP), guanosine triphosphate (GTP), and uridine triphosphate (UTP) that are used by RNA polymerase to synthesize RNA during transcription.

Ribosomal RNAs (rRNAs) are a class of noncoding RNAs (ncRNAs) that have structural and functional roles in ribosomes during protein synthesis.

The **ribosome** is a large complex of proteins and RNAs that directs the synthesis of proteins from messenger RNA (mRNA) in the process known as translation.

The **ribosome-binding site** is the sequence on an mRNA that binds to the small ribosomal subunit during the initiation phase of protein synthesis. In bacteria, this sequence is known as the Shine-Dalgarno sequence, named after the researchers who discovered it.

The **RNAi complex (RISC)** is an assembly of either a miRNA or siRNA with an Ago protein to form a complex that can inhibit protein synthesis or degrade a specific mRNA.

RNA interference (RNAi) refers to a set of pathways for the generation of short (18-23 nucleotides) RNAs that base pair with specific mRNA sequences and modulate expression of the mRNA by translation inhibition or degradation.

RNA polymerase is the enzyme that copies the information present in a DNA into RNA in the process known as transcription.

RNA polymerase core enzyme is the set of RNA polymerase subunits necessary to catalyze the transcription elongation reaction. However, core enzyme cannot recognize a promoter and initiate transcription.

RNA polymerase holoenzyme is the complete set of RNA polymerase subunits that is required for the enzyme to bind to a promoter and initiate transcription. In bacteria, the RNA polymerase holoenzyme consists of the core enzyme plus a σ factor.

An **RNA recognition motif (RRM)** is found in the RNA binding domains of a large group of RNA binding proteins involved in RNA processing.

An **RNP granule** is a large aggregate of ribonucleoprotein particles that is involved in storage, transport, degradation or processing of RNA.

The **root-shoot axis** is the axis extending between the root and the shoot of a plant. Most plant growth is oriented along this axis.

Rotational positioning describes the location of the histone octamer relative to turns of the double helix, which determines which face of DNA is exposed on the nucleosome surface.

Rough endoplasmic reticulum (RER) refers to the region of the ER to which ribosomes are bound. It is the site of synthesis of membrane proteins and secretory proteins.

Rous sarcoma virus (RSV) is an acutely transforming retrovirus, in which the first oncogene was identified.

RS domains are polypeptide regions found in SR proteins that are rich in the amino acids serine and threonine.

Rubisco is the abbreviation for the enzyme ribulose-1,5-bisphosphate carboxylase/oxygenase, which catalyzes the carbon fixation reaction.

A **sarcoma** is a tumor that arises from mesodermal cells (usually of connective tissue lineages).

Sarcomeres are the repeating units of striated myofibrils that are responsible for muscle contraction. They contain interdigitating bipolar thick filaments of myosin and actin thin filaments, as well as other proteins. The positions of two Z-discs define each sarcomere.

The **sarcoplasmic reticulum (SR)** is a form of smooth endoplasmic reticulum specialized for the storage and release of calcium in skeletal and cardiac muscle cells. Calcium release from the SR initiates muscle contraction.

A chromosome **scaffold** is a proteinaceous structure in the shape of a sister chromatid pair, generated when chromosomes are depleted of histones.

A **scaffolding protein** binds multiple protein components of a signaling pathway.

SCF is the conserved E3 ubiquitin ligase required for proteolysis of cyclin-dependent kinase inhibitors and other cell cycle regulators.

Carrier proteins that mediate **secondary active transport** use the free energy stored in transmembrane electrochemical gradients to drive solute transport.

Secondary metabolism is the name for metabolic pathways specific to a small number of plants, usually involved in forming products that are not essential to immediate survival but rather ones that may protect the plant or the species from predators.

Secondary operators are transcription factor binding sites in bacterial operons that can function to increase the activity of the primary operator in the regulation of gene transcription. Secondary operators are usually located distal to the primary operator and function through protein-mediated DNA looping.

Secondary structures are local structural elements within proteins. Most common examples are β-strands and α-helices.

A **second messenger** is a small molecule (such as cyclic AMP or Ca^{2+}) that is generated or released when a signal transduction pathway is activated.

A **secretory granule** is a membrane-bounded compartment that contains molecules to be released from cells by regulated exocytosis (that is, the molecules are concentrated and stored in secretory granules and are released only in response to a signal).

Secretory pathway. *See* exocytic pathway.

Securins are proteins that prevent the initiation of anaphase by binding to and inhibiting separase, a protease that cleaves a subunit of the cohesin complex that is responsible for holding sister chromatids together. Inhibition of separase by securin ends when securin is itself proteolyzed as a result of activation of the anaphase-promoting complex.

Selectins are a class of cell-cell adhesion receptors expressed only on vascular cells.

The **selectivity filter** is the part of an ion channel that confers specificity for one species of ion over another.

Self-assembly refers to the ability of a protein (or of a complex of proteins) to form its final structure without the intervention of any additional components (such as chaperones). The term can also refer to the spontaneous formation of any biological structure that occurs when molecules collide and bind to each other.

Semiconservative replication is a model of replication in which each double-stranded daughter DNA molecule will have a conserved DNA strand, which is derived from the parental DNA, and a newly synthesized strand.

Senescence is the process by which cells lose the ability to divide, often after prolonged propagation in culture *in vitro*.

Separase is protease that plays a direct role in initiating anaphase by cleaving and inactivating a component of the cohesin complex that holds sister chromatids together.

The **septate junction** is found in invertebrate animal cells, and it closely resembles the tight junction found in vertebrate animal cells. It is typically formed along the lateral membrane of adjacent epithelial cells.

Sessile organisms are attached to a substrate and are not free to move about (in contrast to planktonic organisms).

An **SH2** domain is a protein domain of ~100 amino acids that recognizes sequences surrounding tyrosine phosphorylation sites and that usually requires phosphorylation of the included tyrosine for high affinity binding.

SH3 domains are protein interaction domains that bind to specific short proline-rich sequences. SH3 domains consist of approximately 50 residues.

Sialyl Lewis(x) (**sLe^x**) is a complex arrangement of sialic acid sugars that is found on selectin ligands and is bound by selectin receptors.

Sickle-cell anemia is a genetic disease in which mutations in hemoglobin cause red blood cells to adopt a characteristic shape resembling a sickle.

Sigma (s) factor is a component of the bacterial RNA polymerase holoenzyme that is required for the enzyme to bind a promoter site and initiate transcription.

A **signal anchor sequence** is a signal sequence that remains attached to a protein and serves as a transmembrane domain.

Signal peptidase is an enzyme within the membrane of the endoplasmic reticulum. It specifically removes the signal sequences from proteins as they are translocated. Analogous activities are present in bacteria, archaea, and in each organelle in a eukaryotic cell into which proteins are targeted and translocated by means of removable targeting sequences. Signal peptidase is one component of a larger protein complex.

The **signal recognition particle** (**SRP**) is a ribonucleoprotein complex that recognizes signal sequences during translation and guides the ribosome to the translocation channel. SRPs from different organisms may have different compositions, but all contain related proteins and RNAs.

A **signal sequence** is a short region of a protein that directs it to the endoplasmic reticulum for cotranslational translocation. The term signal sequence is also used more generally to refer to a sorting signal.

Signal Transducer and Activator of Transcription (**STAT**) is a family of proteins that function as a subunit of the JAK-STAT transcription factor dimer and the principal target of many growth factor and cytokine cellular signaling pathways.

Signal transduction describes the process by which a stimulus or cellular state is sensed by and transmitted to pathways within the cell.

A **silencer** is a DNA sequence element that selectively attenuates the effects of an enhancer on other promoters, thereby allowing an enhancer to upregulate a single transcription unit.

Silencing describes the repression of gene expression in a localized region, usually as the result of a structural change in chromatin.

Simple epithelia have a single layer of cells.

Single-strand binding proteins (**SSBs**) are proteins that bind to and stabilize single-stranded DNA.

The **single X hypothesis** describes the inactivation of one X chromosome in female mammals.

siRNAs are single-stranded, noncoding RNAs generated from longer double-stranded precursors. An siRNA assembles into the RISC and binds to specific regions on mRNAs, leading to mRNA degradation.

The **sliding clamp** subassembly forms a ring around DNA, tethering the remainder of the polymerase holoenzyme to the DNA.

Small nuclear RNAs (**snRNAs**) are a class of noncoding RNAs (ncRNAs) that participate in mRNA splicing and other RNA processing reactions.

Small nuclear RNP complexes (**snRNPs or snurps**) are assemblies of proteins with the individual snRNAs that function during splicing or other RNA processing reactions.

Small nucleolar RNAs (**snoRNAs**) are a class of noncoding RNAs (ncRNAs) in the nucleolus that serve as guides for site-specific ribosomal RNA (rRNA) processing reactions.

The **smooth endoplasmic reticulum** (**SER**) is the part of the ER that is free of bound ribosomes. It is involved in lipid biosynthesis.

The **SNARE hypothesis** proposes that the specificity of a transport vesicle for its target membrane is mediated by the interaction of SNARE proteins. In this hypothesis, a SNARE on the vesicle (v-SNARE) binds specifically to its cognate SNARE on the target membrane (t-SNARE).

SNAREs are membrane proteins that project into the cytosol and help mediate membrane fusion. The specific interaction of SNAREs in one membrane with cognate SNAREs in another is thought to bring the membranes close enough together to fuse.

Somatic cells are all the cells of an organism except those of the germline. Their genes cannot be passed on to future generations.

A **somatic mutation** is a mutation occurring in a somatic cell, therefore affecting only its direct descendents. Since such a mutation does not affect cells in the germ line, it is not inherited by descendants of the organism.

A **sorting signal**, or targeting signal, is the part of a protein that allows it to be transported from one location in the cell to another. Sorting signals consist of either a short sequence of amino acids or a covalent modification.

Sp1 is gene-specific transcription factor that binds to GC-rich elements near RNA polymerase II promoter regions of a number of highly expressed mammalian genes.

The **spacer region** refers to a 17±1 bp segment located between the -10 and -35 hexamers in an *E. coli* promoter utilized by RNA polymerase holoenzyme containing the σ^{70} subunit. The spacer region generally lacks any consensus sequence elements.

The **specificity** of a protein interaction is its tendency to bind to one or a few of many potential partner molecules. In the case of an enzyme, the substrate specificity of its active site can be conferred by a lock-and-key mechanism or by induced fit.

Speckles are a type of interchromatin granule. They are structures that store, and may help to assemble, components of the spliceosomal machinery.

Spectral karyotyping (**SKY**) is a technique in which fluorescent probes, each containing a different fluorescent dye, are hybridized to chromosomes. The spectrum of light emitted by the stained chromosomes is analyzed using a computer, which generates a composite picture that shows different chromosome pairs as if they were stained in different colors.

S phase is the part of the eukaryotic cell cycle during which replication of DNA occurs.

The **spindle** is a highly dynamic structure that forms during mitosis and meiosis to segregate the chromosomes and divide the cell. Spindles have three main structural elements: microtubules, centrosomes (in animals), and chromosomes.

The **spindle matrix** is a collection of proteins thought to form a network interspersed among the microtubules of the mitotic spindle. The matrix contains the NuMA protein and uses motor proteins to anchor microtubules and provide integrity to the spindle. Matrix components are more concentrated near the poles of the spindle.

A **spindle pole** is a region in the cell toward which chromosomes move during mitosis. A normal bipolar mitotic spindle has one pole at each end. All poles contain a collection of microtubule minus ends. In animal cells, the poles are formed from centrosomes and the microtubules focus to a tight point at each of the two ends of the spindle. In plant cells, centrosomes are absent and the poles of the spindle are much broader.

A **splice junction** is the location on a pre-mRNA of the boundary between an exon and an intron. For each intron, there is a 5′ and a 3′ splice junction.

A **spliceosome** is a large assembly of ribonucleoproteins that catalyzes the removal of pre-mRNA introns and ligates the exons together to form a mature mRNA.

A **splicing enhancer** is a region near an exon-intron boundary that increases the efficiency of the removal of the intron.

Splicing signals are short sequences required for mRNA intron removal from pre-mRNAs during mRNA splicing.

A **splicing silencer** is a region near an exon-intron boundary that decreases the efficiency of the removal of the intron, causing an alternative mature transcript to be formed.

Sporulation is the generation of a spore by a bacterium (by morphological conversion) or by a yeast cell (as the product of meiosis).

The **SR family** is a group of proteins with RS domains that regulate mRNA splicing.

SRP receptor is a protein that is within the membrane of the endoplasmic reticulum, binds to signal recognition particle, and mediates the targeting of secretory and membrane proteins during cotranslational translocation.

Standard states are those set by convention to compare individual reactions under fixed conditions: 1 atm pressure, 1 M concentrations, and 298K.

Standard reduction potential *See* redox potential.

START is the point during G1 at which a yeast cell becomes committed to division. In other organisms, this point is called the restriction point.

A **start codon** is the initiation codon in an mRNA that signals the ribosome to begin translation of an mRNA. Generally, the codon AUG, which codes for methionine, functions as the initiation codon.

A **state**, in thermodynamics, is a characteristic condition with fixed variables such as volume, temperature, and pressure.

State functions are those which produce values independent of how transitions are made from one state to another.

State variables, in thermodynamics, are attributes of a state; collectively, they define the state.

Stationary phase describes the plateau of the growth curve that is reached after log growth. During this phase, cell number remains constant.

Steady state is a condition in which the concentrations of reactant(s) and product(s) no longer change with time.

Stem cells are relatively undifferentiated cells that can divide continuously to generate more stem cells as well as more differentiated (specialized) cells. In many organisms, different populations of stem cells give rise to different cell lineages.

Stoichiometry is the relationship among the numbers of reactant and/or product molecules in a chemical reaction.

Stomata (*singular*: stoma) are pores usually on the underside of a leaf that allow the exchange of gases in and out of its interior so that photosynthesis can occur. Each stoma is formed by the precise arrangement of a small number of cells and can be opened or closed by changes in turgor within them.

A **stop codon** is one of three base triplets (UAA, UAG, UGA) that signal the ribosome to stop translation and to dissociate from the mRNA.

Strand invasion is the process by which single-stranded DNA pairs to homologous sequences that can be in a single stranded form or a double stranded form.

Strand resection is a step in DSBR in which the 5′ ends of DNA at a breakpoint are removed by exonuclease action, resulting in single-stranded 3′ overhangs. These 3′ overhangs are subsequently used in strand invasion.

Stratified epithelia have more than one layer of cells.

Stress fibers are contractile bundles of actin filaments, myosin filaments, and other proteins. They are anchored to protein complexes (focal adhesions) at the plasma membrane and function in cell adhesion.

The interior space of the chloroplast is called the **stroma**. It is highly enriched in enzymes involved in photosynthesis. Stroma also refers to the supporting connective tissue of organs.

Structural maintenance of chromosomes (SMCs) describes a group of proteins that include the cohesins, which hold sister chromatids together, and the condensins, which are involved in chromosome condensation.

A **structural motif** refers to a secondary structure found in a variety of different proteins with related functions.

Substrate-level phosphorylation is the formation of ATP resulting from the transfer of phosphate from a substrate to ADP.

Suicide enzymes are proteins that lose enzymatic activity after acting only one time.

Supercoiling is the coiling of a closed duplex DNA in space so that it crosses over its own axis.

A **super-secondary structure** is a group of individual protein secondary structures that function in concert to form a functional motif.

Surroundings, in thermodynamics, comprise all portions of the universe except for the system.

SWI/SNF is a chromatin remodeling complex; it uses hydrolysis of ATP to change the organization of nucleosomes.

Symbiosis is an extremely close relationship between two species. Commensalism, mutualism, and parasitism are examples.

A **symplast** is a group of cells within a plant that are connected by plasmodesmata.

A **symporter** is a type of carrier protein that moves two different solutes across a membrane in the same direction. The two solutes can be transported simultaneously or sequentially. Symporters are also called cotransporters.

Synapsis is the association of two pairs of sister chromatids that occurs at the start of meiosis.

The **synaptonemal complex** refers to the morphological structure of synapsed chromosomes.

A **syncytium** is a cell with more than one nucleus.

Syntelic attachment is the condition in which sister kinetochores are attached to the same spindle pole. Such mistakes in attachment are common during the early stages of spindle formation but are corrected before anaphase begins.

A **system**, in thermodynamics, is the specific portion of the universe under study.

Taq DNA polymerase is a heat stable DNA polymerase isolated from the thermophilic bacterium *Thermus aquaticus* that is often employed in the polymerase chain reaction (PCR) because it can withstand the high heat used during the denaturation step of each amplification cycle.

The **TATA box** is a short A-T-rich sequence found in many (but not all) eukaryotic RNA polymerase II core promoters, which is located ~25–30 base pairs upstream of the transcription start site. The general transcription factor TFIID binds to the TATA box.

Taxonomy is the science of naming and classifying organisms.

Telomerase is a reverse transcriptase that contains its own RNA template, which it uses to create repeating units of one strand at the telomere by adding nucleotides to the 3′ overhang.

A **telomere** is the natural end of a chromosome, consisting of a DNA sequencing of a simple repeating unit with a protruding single stranded end.

Telomeric silencing describes the repression of gene activity that occurs near a telomere.

Telophase is the last stage of mitosis. It begins when a full set of chromosomes arrives at each of the two poles of the spindle, and ends when the nuclear membranes of the two daughter nuclei form.

A **temperature-sensitive mutation** creates a gene product (usually a protein) that functions at some low temperature but poorly or not at all at some high temperature. The converse is a cold-sensitive mutation.

The **template strand** is the noncoding or nonsense strand of DNA utilized by RNA polymerase as a guide to synthesize an RNA during transcription.

A **terminally differentiated** cell is a specialized cell that usually is no longer able to divide.

Termination is the stage of a macromolecular synthesis reaction (replication, transcription, or translation) where polymerization ceases and the synthetic apparatus disassembles.

Tethered tracking is a process by which a protein remains bound to a primary site on DNA or RNA while moving along the nucleic acid.

Tethers are proteins that provide the initial means of anchoring transport vesicles to their destination membranes. Two well-characterized types are long, fibrous proteins and large multiprotein complexes.

Thermodynamic control of a reaction occurs when the composition of the product is determined by the equilibrium thermodynamics of the system. The formation of a thermodynamically more stable product is favored.

Thiamine pyrophosphate is a cofactor in the pyruvate dehydrogenase reaction which assists in binding the pyruvate substrate at the carbonyl and releasing CO_2.

Thiol-mediated retention is a form of quality control that retains proteins in the endoplasmic reticulum if they have not formed the appropriate intermolecular disulfide bonds.

The **thylakoids** of plants are organelles in which the light reactions of photosynthesis, along with the proteins responsible for electron transfer and NADPH and ATP formation, are localized.

The multiprotein complex that mediates the translocation of proteins across the inner membrane of chloroplasts is called the **TIC complex** (for *t*ranslocon *i*nner *c*hloroplasts).

Tight junctions are cell-cell junctions that prevent most molecules from diffusing in the spaces between the cells.

The **TIM complex** resides in the inner membrane of mitochondria and is responsible for transporting proteins from the intermembrane space into the interior of the organelle.

Tip growth is used by some specialized plant cells to extend long, thin processes such as root hairs. Growth of a process is initiated when the deposition of cell wall materials and new membrane is limited to a very small area of the cell's surface, forming a bulge. Deposition continues only at its tip, resulting in the formation of a tube of constant diameter.

The multiprotein complex that mediates the translocation of proteins across the outer membrane of chloroplasts is called the **TOC complex** (for *t*ranslocon *o*uter *c*hloroplasts).

The **TOM complex** resides in the outer membrane of the mitochondrion and is responsible for importing proteins from the cytosol into the space between the inner and outer membranes.

The **tonoplast** is the membrane that surrounds a plant cell's vacuole.

Topoisomerases are enzymes that catalyze the conversion of one topoisomer into another.

The **top strand** in a DNA sequence is by convention the coding or sense strand and is always written in the 5' to 3' direction. Often only the top strand of a DNA sequence is presented, while the bottom strand, noncoding or nonsense strand is inferred from the Watson-Crick base pairing rules (A base pairs with T; G base pairs with C).

A **totipotent** cell is able to differentiate into any of the cells within an organism and can give rise to an entire organism.

The individual tubes that form the xylem of a plant are called **tracheary elements**. The name arose because the thickenings of their walls resemble the rings that reinforce the trachea of an animal. The tracheary elements die and their end walls dissolve to form a hollow tube that transports water and dissolved materials between roots and shoots.

Transaminases are the enzymes that catalyze transamination.

Transamination is the process by which an amine group is transferred to either create or metabolize an amino acid.

Transcription is the enzymatic process by which RNA polymerase uses a DNA template to direct the polymerization of RNA from ribonucleoside triphosphate (NTP) precursors.

Transcription-coupled repair (TC-NER) is a subtype of NER that preferentially repairs DNA damage in the transcribed strand of active genes.

A **transcription factor** is a protein that directly or indirectly regulates the ability of RNA polymerase to initiate transcription at specific promoter(s), but is not itself part of the enzyme.

The **transcription start site** is the base pair within the promoter of the DNA that serves to guide the polymerization of the first nucleotide or an RNA transcript by RNA polymerase.

The **transcription stop site** corresponds to the position of the last template-directed polymerization event during transcription termination.

A **transcription termination signal** stimulates the release of RNA polymerase and the RNA transcript from the DNA template at the end of the transcription cycle.

Transcytosis is the transport of molecules from one type of plasma membrane domain to another in a polarized cell, such as a neuron or epithelial cell.

Transduction refers to the transfer of a bacterial gene from one bacterium to another by a phage; a phage carrying host genes as well as its own is called a transducing phage. The word transduction also describes the acquisition and transfer of eukaryotic cellular sequences by retroviruses.

Transesterification is a reaction that breaks and makes chemical bonds in a coordinated transfer so that no energy is required. Lariat formation during mRNA splicing involves two transesterification reactions.

Transfection of eukaryotic cells is the process by which naked DNA or RNA is introduced into eukaryotic cells, often in the presence of calcium phosphate precipitates.

Transferrin is an iron transport protein found primarily in the serum fraction of blood.

A **transfer RNA (tRNA)** is a small RNA molecule that carries a specific amino acid to the ribosome for incorporation into the growing polypeptide chain.

Transformation of bacteria is the acquisition of new genetic material by incorporation of added DNA.

A **transgenic** organism is one whose germline genome has been changed, usually by insertion of a cloned DNA segment into its chromosomal DNA.

The *trans*, or exit, face of the Golgi apparatus is a reticulum of interconnected tubules called the ***trans*-Golgi network (TGN)**. It serves as a sorting station where molecules destined for different locations are incorporated in different vesicles.

Transitional endoplasmic reticulum is the smooth ER at the interface between the ER and Golgi apparatus. It is where protein cargo transits between the two organelles. The transitional ER is sometimes known as the ER-Golgi intermediate compartment, or ERGIC.

The signal sequences of proteins transported into chloroplasts are called **transit peptides**.

Translation is the process by which the ribosome decodes the information contained within a messenger RNA (mRNA) to synthesize a protein.

Translational positioning describes the location of a histone octamer at successive turns of the double helix, which determines which sequences are located in linker regions.

Movement of a protein across a lipid bilayer usually requires a **translocon**, an integral membrane protein that provides a channel for displacement of polypeptide segments across the membrane.

The **transmembrane domain** is the part of a protein that spans the membrane bilayer. It is hydrophobic and in many cases contains approximately 20 amino acids that form an α helix. It is also called the transmembrane region.

A **transmission electron microscope (TEM)** allows electrons to pass through a thin section of the object, resulting in a detailed view of the object's structure.

A **transporter** is a type of receptor that moves ions and small molecules across a membrane. It binds the molecules on one side of the membrane, and then releases them on the other side.

Transport vesicles carry material from one compartment to the next on the exocytic and endocytic pathways.

Transposition refers to the movement of a transposon to a new site in the genome.

A polymer undergoes **treadmilling** when the rate of subunit association at one end is comparable or equal to the dissociation rate at the other end, such that vectorial subunit flux occurs while the filament maintains a constant length.

Triacylglycerol is a more chemically exact name for triglyceride (or fat).

Triglycerides (fats or oils) contain three fatty acids esterified to a glycerol backbone.

The functional subunit of a clathrin coat is a **triskelion**. It is a three-legged structure; each leg is composed of a clathrin heavy chain and a clathrin light chain.

Tropism refers to the specificity of a virus or microbe for a particular host cell (or tissue or species) during infection.

Tropocollagen consists of collagen proteins that are organized as triple helices and lack procollagen tails; they are the fundamental building blocks of collagen fibers.

Tropoelastin is an elastin protein that has not been modified by lysyl oxidase or assembled into elastin fibers. It is a precursor form of elastin.

A **tumor** is an abnormal growth of tissue.

Tumor progression is the process by which a tumor evolves to progressively higher grades of malignancy.

Tumor suppressor proteins usually act by blocking cell proliferation or promoting cell death. Cancer may result when a tumor suppressor gene is inactivated by a loss-of-function mutation or promoter methylation.

Turgor is pressure that develops within a plant cell. It is the result of an osmotic imbalance between the inside and the outside of the cell, causing water to enter into the vacuole. Turgor distends a cell and is used to expand cells during plant growth.

A **two-component signaling system** contains a histidine kinase domain or subunit that is regulated by a signaling molecule and a response regulator that contains a phosphorylatable aspartate residue.

Ubiquitin is a highly conserved, 76-residue protein that is covalently attached to other proteins.

The **unfolded protein response (UPR)** is an intracellular signaling pathway that increases the expression of chaperones in the endoplasmic reticulum when unfolded proteins accumulate in the lumen of the ER.

A **uniporter** is a type of carrier protein that moves only one type of solute across a membrane.

The **universe**, in thermodynamics, is the combination of system and surroundings.

A **UP element** is DNA sequence that is sometimes found upstream of the *E. coli* core promoter. The UP element increases promoter efficiency but is not essential for transcription.

Upstream is the term used to describe those sequences on the coding strand of a DNA that are in the opposite direction (towards the 5' end) from which RNA polymerase is moving during transcription.

Vacuoles are membrane-bounded organelles that drive the growth of plant cells by absorbing water and expanding. Vacuoles can occupy more than 90% of the interior of a large plant cell.

The **vegetative phase** describes the period of normal growth and division of a bacterium. For a bacterium that can sporulate, this contrasts with the sporulation phase, when spores are being formed.

Vesicle-mediated transport is a major mechanism of transport between cellular compartments. It involves budding of a membrane-bounded vesicle from one compartment and fusion of the vesicle with another compartment.

Vesicles are small bodies that are bounded by membrane, derived by budding from one membrane, and often able to fuse with another membrane in a cell.

The **vesicular tubular cluster** is an organelle that serves as an endoplasmic reticulum–Golgi intermediate compartment during the trafficking of vesicles between these two organelles.

Virulence refers to the extent to which a microbe overcomes host defense mechanisms.

Virulence factor is a general term for molecules that are produced by pathogens and allow pathogens to invade host organisms, cause disease, or evade immune responses.

Vitronectin is a multifunctional extracellular matrix protein that circulates in the bloodstream. It plays a central role in regulating blood clot formation.

Wilms tumor protein is a gene-specific transcription factor that is over-expressed in a number of human cancers.

Work is a type of energy exchange between systems that acts as a result of a difference in potentials other than temperature, such as volume or pressure.

Xeroderma pigmentosum (**XP**) is an autosomal recessive disease in which the body's normal ability to repair damage caused by UV irradiation is defective.

The **xylem** is a vascular element in plants that moves water and nutrients absorbed by the roots to the parts of the plant above ground. The xylem is composed of a collection of long thin tubes that run the entire height of a plant. Water moves through each tube by capillary action, driven by evaporation from the leaves.

The **zinc finger** (**ZF**) **motif** is found within the DNA binding domains of a number of transcription factors and describes a structure in which a segment of the polypeptide chain forms a loop stabilized by a Zn^{2+} ion chelated by the side chains on the amino acids cysteine or histidine. The ZF motif and an adjacent segment of basic amino acids act in concert to form sequence-specific interactions with the transcription factor-binding site on the DNA.

A **zipcode** is a sequence in the 3' UTR of an mRNA required for localization of the mRNA to a specific region of the cytoplasm prior to translation.

A **zipcode-binding protein** binds to an mRNA zipcode and directs the cytoplasmic localization of the mRNA.

The **zonula adherens** is a beltlike cell-cell junction that surrounds epithelial and endothelial cells. It is found in the junctional complex.

The **Z ring** is an assembly of proteins that forms at the site of division, under the cytoplasmic membrane of prokaryotes, and is responsible for cytokinesis.

A **zymogen** is an inactive enzyme precursor.

Index

Note: An *f* following a page number indicates a figure.

HIV (human immunodeficiency virus), 416, 416*f*, 428
HML gene mutations, 491
HMR gene mutations, 491
hnRNPs (heterogeneous nuclear ribonucleoprotein particles), 427–429, 427*f*–429*f*
Holliday junctions, 97, 98*f*
    resolution of, 101, 102*f*, 104
Holoenzyme, 76, 77, 115
Homeobox, 126
Homeodomain motif, 126, 126*f*
Homeostasis
    defined, 15, 27
    osmotic pressure and, 15
Homologous adaptation, in signaling, 782*f*, 783–784
Homologous recombination (HR) pathway, 97
    for repair and meiotic recombination, 99–101, 99*f*
Homophilic binding, 864, 865*f*, 872
Hooke, Robert, 622
Hormone-response element, 134, 795
Hormones, 27, 743, 965
HP1 (heterochromatin protein 1), 489–490, 490*f*
HPV (human papilloma virus), 747, 747*f*
Hrs cytosolic protein, 380
Hsc70, in uncoating clathrin-coated vesicles, 374
Hsp47, 318
Hsp70, in uncoating clathrin-coated vesicles, 374*f*
HSPGs (heparan sulfate proteoglycans), 843–845, 844*f*, 845*f*
Human genome, 13
Human immunodeficiency virus (HIV), 416, 416*f*, 428
Human papilloma virus (HPV), 747, 747*f*
Humans, remodeling complexes, 483, 483*f*
Humoral immunity, transcytosis and, 354
Huntington's disease, 221
Hyaluronan (hyaluronic acid; hyaluronate), 841–843, 843*f*
Hyaluronan receptor (CD44), 842–843, 843*f*
Hybridizes, 132
Hydration, ion, transmembrane transport and, 235–236
Hydration shells, 236
Hydrogen ion, and lactose, symport of, 266
Hydrophilic head, of lipid bilayer, 5, 5*f*
Hydrophobic molecules, lipid bilayer permeability and, 14, 14*f*
Hydrophobic tail, of lipid bilayer, 5, 5*f*
Hydroxyl radicals, and DNA damage, 86
Hyperplasia, 750
Hypersensitive sites
    chromatin structural changes and, 480–481, 480*f*, 481*f*
    creation, 483
Hyphae, 924, 924*f*
Hypopolarization. *See* Repolarization

I

IAPs (inhibitors of apoptosis proteins), 718–719, 719*f*, 728, 728*f*
IFT (intraflagellar transport), 549
IKK complex, 809, 809*f*
Immortalization, of cancer cells, 760–761, 760*f*
Immortalized cells, 742
Immune system
    clearance of apoptotic cells and, 735–736

proteoglycans and, 840
Immune tolerance, 736
Immunoglobulin domain, 874
Immunoreceptor tyrosine-based activation motif (ITAM), 818, 818*f*
Impermeability, 5, 7, 11
Import, protein
    into chloroplasts, 337–338
    peroxisomal, 338–339
Importin-cargo complex, 418, 418*f*
Importin α, 414–415, 415*f*
Importin β, 414–415, 415*f*, 417
Inborn errors of metabolism, 220
Incretin, 60
Induced fit model, 207*f*, 208, 212, 213*f*
Induced proximity, 718, 718*f*
Inflammation, caspase-1 in, 719, 719*f*
Information transfer, cellular, 6*f*
Inhibitors of apoptosis proteins (IAPs), 718–719, 719*f*, 728, 728*f*
Innate defenses, 931
Inner nuclear membranes
    defined, 402, 402*f*
    nuclear pore complexes, 407–408, 407*f*
Innexins, 871
Inorganic phosphate, 43, 49
Inositol 1,4, 5-trisphosphate receptors (IP3Rs), 262, 263
Inositol synthesis, 332
INR (initiator regions), 121
Insects, endosymbionts, 928, 929*f*
Insertion, of membrane proteins, 304
in situ hybridization, 455, 455
Insulin receptor, 811–812, 812*f*, 818*f*
Insulin signaling pathway, 792, 818*f*
Integration, of membrane proteins, 304
Integrins (integrin receptors)
    activation, 853–854, 853*f*, 854*f*
    α6β4, 869
    binding
        by affinity modulation, 853, 853*f*
        by avidity modulation, 853, 853*f*
    cell adhesion and, 852
    classification, 851–852, 852*f*
    defined, 829
    functions, 851
    gene knockout studies, 858–859
    ligation, signaling and, 815, 816*f*
    protein complexes
        formation of, 854–855
    protein digestion and, 848
    signaling, 853–857, 852*f*–857*f*
        inside-out pathway, 854, 854*f*
    structure
        α and β subunits, 851–852, 851*f*
        domain organization, 851–852, 851*f*
Integrons, 912, 912*f*
Interbands, 443, 453
Intercalated discs, 864, 864*f*
Interchromosome domains, 398, 398*f*
Interleukins
    inflammatory caspases and, 719, 719*f*
    interleukin 1-β, signaling pathway, 818*f*

phosphorylation, 526–527, 527*f*
research, future directions, 548–549
Microtubule-based motor proteins. *See* Motor proteins
Microtubule-destabilizing protein, 526
Microtubule gliding assay, 554, 554*f*
Microtubule-interacting drugs, 506, 506*f*
Microtubule-organizing centers (MTOCs)
    basal bodies, 518–519
    centromere. *See* Centromere
    centrosomes, 632–633, 633*f*
    defined, 446
    microtubule assembly nucleation, 517–519, 517*f*, 518*f*
Microtubules, 503–555
    anchoring, motor proteins and, 530, 530*f*
    assembly. *See also* Tubulin, polymerization
        amount over time, 511–512, 511*f*
        dynamic instability and, 513–514, 514*f*
        dynamic instability of, 548
        physical forces and, 549
        steady state, 512–513
        tubulin subunits and, 511–513, 512*f*
    astral, 629
        formation of, 693
        in mitosis, 634–635634*f*
        in plant cells, 949
        spindle positioning, 665, 665*f*
    attachment to centromeres, 446–447, 447*f*
    in axoneme, 542, 543*f*
    cell division and, 25, 25*f*
    copurified with MAPs, 524, 524*f*
    cortical
        dynamic nature of, 964–966, 965*f*
        orientation in plant cell, 965, 965*f*, 966*f*
        plant cell wall organization and, 961–963, 962*f*, 963*f*
    cytoskeleton, 522
        defects in, 549
        movement of growth cone and, 535–536, 536*f*, 537*f*
    depolymerization, 506, 523, 524*f*, 538–539, 540, 652
    disassembly, dynamic instability of, 513–514, 514*f*, 548
    dissolution, 25
    dynamic instability, 519–522
        regulation of, 515–517, 515*f*, 516*f*
        in vitro, 519–520, 520*f*
        in vivo, 519, 519*f*
    dynamics, 520–524, 537–538
        in mitosis, 520–521, 521*f*, 630
        motor proteins and, 534–538, 535*f*–537*f*
        radial array of, 524–525, 524*f*
        selective stabilization of, 523–524, 523*f*
        turnover rates, 521, 521*f*
    ends, regulation of dynamic instability, 515–517, 515*f*, 517*f*
    as equilibrium polymers, 550–551, 551*f*
    in fibroblasts, 25, 25*f*
    FRAP (fluorescence recovery after photobleaching), 551–552
    free, 522
    functions, 506–508, 507*f*, 508*f*
    interactions
        with actin filaments, 538–540, 539*f*, 540*f*, 549, 558
        with motor proteins, 636, 636*f*
    length
        elongation of, 512, 512*f*, 513–514, 514*f*
        MAPs and, 526, 527*f*
        shortening of, 507, 513–514, 514*f*
        in vitro vs. in vivo, 519–520, 520*f*

minus end, 510–511, 511*f*, 522
    during mitosis, 633–634, 633*f*
    motor proteins and, 505
    in neurons, 25, 25*f*
    nucleation, 512, 512*f*, 518, 518*f*, 523, 626
    number, MAPs and, 526, 527*f*
    organization
        by cell need, 508, 508*f*
        cell type and, 507, 507*f*
        motor proteins and, 530, 531*f*
    in plant tip growth, 971–974
    plus end
        MAPs associated with, 524–525, 525*f*
        structure of, 510–511, 511*f*
    polarity, 510–511, 511*f*, 524, 630, 630*f*
    polymerization, *vs.* actin polymerization, 564–565
    positional effects, 26
    in prophase, 643, 643*f*
    rearrangement during mitosis, 507, 507*f*
    release from centromere, 521–522
    reorganization, motor proteins and, 530, 531*f*
    research, future directions, 548–549
    self-assembly process, 510
    spindle
        capture by kinetochore, 642–646, 643*f*, 644*f*, 645*f*
        detachment from kinetochore, 649
        number bound to kinetochore, 649
        rapid turnover of, 520–521, 520*f*
        stabilization by kinetochore, 650
    in spindle, 627, 628, 628*f*
    stability, regulation of, 524–527, 524*f*, 525*f*
    stable, 521, 521*f*
    strength, tubulin bonding and, 510, 510*f*
    structure, 25, 55, 505*f*, 509–511
        α-β tubulin dimer, 510, 510*f*
        protofilament, 509–510, 510*f*
        subunits. *See* Tubulin
    treadmilling, 522, 522*f*
    turnover, MAPs and, 525–526, 526*f*
Microvilli, actin filaments in, 558*f*
Midbody
    contractile ring constriction and, 666
    formation, 661, 661*f*, 662, 662*f*, 663*f*, 694
MinCD system, 920, 920*f*
MinC protein, 920–921, 920*f*
MinD protein, 920–921, 920*f*
Minichromosome maintenance complex (MCM), 689–690, 690*f*
miRs (microRNAs), 433–434, 243*f*
Mismatch repair system, 90–92, 91*f*, 92*f*
Mitochondria
    calcium ion concentration, 12*f*
    cytochrome c release, 724, 724*f*
    defined, 12, 279
    division, 17, 29, 29*f*
    endosymbiotic origin of, 17
    energy production in, 19, 19*f*
    evolutionary origin, 334, 334*f*
    functions, 21*f*
    importing, by posttranslational translocation, 342
    protein importing, 334–335
    protein "sink," 280
    signal sequences, 333, 341, 342
Mitochondrial-associated membrane (MAM), 326, 327

Cα plot representation, 184, 184f
defined, 170
determination of. *See* Cryo-electron microscopy, of biomolecules; nuclear magnetic resonance spectroscopy; X-ray crystallography
DNA-binding protein, electrostatic potential of, 184–185, 185f
folding, 178, 178f
"molecular" or "Connolly" surface representation, 184, 185f
primary, 185–190
quaternary, 199–204
ribbons representation, 184, 184f
secondary, 190–192
space-filling or CPK representation, 183, 184f
tertiary, 192–196
topology diagram, 185, 185f
synthesis, by ribosomes, 292
targeting by signal sequences, 20, 21f
terminally misfolded, return to cytosol for degradation, 319–321, 319f, 321
translocated
folding of, 315, 321f
glycosylation, 313–314
glycosylphosphatidylinositol additions, 312–313, 313f
translocation. *See* Translocation
transport, 22, 22f
unconventional secretory pathway, 388
unfolded, preventing accumulation in ER, 322–324, 128f, 323f
Protein sorting, 307
polarized epithelial cells
domain-specific retention and, 384
in endosomes, 384–385, 384f
in *trans*-Golgi network, 384, 384f
Protein targeting, 295
mitochondrial, 333
in organelles, 335, 335f
signal sequences, 334–335, 334f
Protein trafficking, 22–24
between membranes, 345–388
Protein translocation. *See* Translocation, protein
Protein transport
across mitochondrial membranes, 333–334, 334f
vesicle-mediated, 23, 23f, 23f
Protein tyrosine kinases, recruitment, 815–818, 816–818f
Protein tyrosine phosphatases (PTPs), 807
Protein tyrosine phosphatase SHP2, 807
Proteoglycans, 823, 836–837, 836–837f
in basal lamina, 846–847
binding, to growth factors, 840–841, 841f
distribution in cartilage, 841, 841f
glycosaminoglycans, 839–840, 839f
hydration, gel formation and, 842
negative charge, 840
secretion, 840
shape, 840
structure, 839–840, 839f
synthesis, 840
Proteolipsomes, 301
Proteomics, 59
Protofilament, 510
Proton motive force, 279, 905
Proton motive force, 53
ATP conversion, 52f
Proton transport, H⁺-ATPase and, 280–283, 281f

Protooncogenes, 708
encoding
growth factor proteins, 753
growth factor receptors, 753, 753f
mutations creating oncogenes, 745, 749
gain-of-function mutations and, 748
Protoplast, 958
Pseudohypoaldoseronism type I, 249
Pseudomonas aeruginosa, 893, 932
Psychrophiles, 9, 888, 888f
Puffs, 256, 256f
Pumps. *See* Transporters
Purine, 67
Purine nucleotide cycle, 57, 58, 58f
Putative promoters, 118
Pyrin domain (PYR), 718, 718f
Pyruvate, 19, 43
Pyruvate carboxylase, 48, 58
Pyruvate dehydrogenase complex, 44–45

Q

qRT-PCR, 132, 133–134
Quality control, for protein folding, 296
Quasi-equivalance, 203
Quaternary structure, of proteins, 199–204. *See also* structure of protein
Quiescence (G0), 685–686, 757
Quinon (coQH₂), 48
Quorum sensing, 932–933, 933f

R

R (restriction point), 686, 687f, 757, 757f
Rab/Ypt proteins, 366–367
Rab1, 367
Rab effectors
defined, 367
tethering proteins. *See* Tethers
Rab-GTP, 367
Rab GTPase regulation, of vesicle targeting, 356, 366–368, 367f
Rab proteins
cycling, 366–367, 367f
distribution, in organelles, 365, 365f
Rac1, 544
Rac protein, regulation of actin polymerization, 575
Rad17-RFC complex, 701
Radiation, and DNA damage, 87, 88f
Ran, direction of nuclear transport and, 417–419, 417f
Ran-GAP, 417, 217f, 426
Ran-GDP, 418419, 418f
Ran-GEF (Rcc1), 418–419, 418f, 419f
Ran-GTP, 418–419, 418f, 419f, 426
Ran GTPase
direction of nuclear transport, 423
nuclear transport control, 414, 417–419, 417–418f
RAP1, 491
ras genes, 801
Ras-GTP, 801f
ras oncogene, 752, 752f, 753

Trombone model, DNA replication, 78, 79*f*
Tropism, 930
Tropocollagen, 628, 682*f*
Tropoelastin, 831, 832, 832*f*
Tropomodulins, 570
Troponin C, 262
Troponins, 589
Troponin-tropomyosin complex, in muscle contraction, 588–589, 589*f*
TRP channels, 794
Tubulin
    α-β dimer, 511, 511*f*
    β-tubulin
        microtubule polarity and, 511
        synthesis, 554, 554*f*
    critical concentration, 512–513
    depolymerization, 515
    fluorescence recovery after photobleaching, 553–554
    GTP hydrolysis, 552
    γ-tubulin, 518, 632, 634
    heterodimer assembly, 554, 555*f*
    in microtubules, 24, 504
    polymerization, 552–553
        amount over time, 511–512, 512*f*
        dynamic instability of, 514, 514*f*
        in microtubule assembly, 511–512, 512*f*
    posttranslational covalent modifications, 554–555
    in protofilaments, 511, 512*f*
    self-assembly process, 511
    in stable microtubules, 521, 522*f*
    structure, 510*f*, 511
    synthesis, 554, 555*f*
    α-tubulin
        microtubule polarity and, 510, 511*f*
        synthesis, 554, 555*f*
γ-Tubulin ring complex (γTURC), 518, 518*f*
Tumorigenesis
    Aurora kinase misregulation and, 693–694
    multistep, 750–752, 751*f*, 752*f*
Tumorigenic cells, 745
Tumor necrosis factor receptor (TNFR), 729
Tumor-promoting processes, 763
Tumors
    benign, 740
    classification, 740
    defined, 740
    formation and progression, 740, 740*f*
    genesis of, 750–752, 750*f*, 751*f*
    growth, angiogenesis and, 740, 740*f*, 760, 760*f*
    malignant, 740. *See also* Cancer
Tumor suppressor genes, 708
    apoptosis and, 756
    inactivation, 749–750, 749*f*
    restriction points, 757, 757*f*
Tumor suppressors, 745
Turgor pressure, in plant cells, 957–959, 958*f*
Twin-Arginine-Translocation system (Tat system), 338, 905
Two-component systems, 770–771
Two-state model of receptor-ligand binding, 774, 774*f*
Tyr527, phosphorylation, 812
Tyrosine kinase inhibitors
    in chronic myelogenous leukemia (CML) treatment, 806–807

U

U2AF, 146
UAP56, in mRNA export, 430
UBA (ubiquitin association), 809
Ubiquitin
    covalent modification, 808–809, 809*f*
    defined, 684
    multivesicular body formation, 380
Ubiquitin association (UBA), 809
Ubiquitin-interacting motif (UIM), 809
Ubiquitin-like proteins (Ubl proteins), covalent modification, 80–809, 809*f*
Ubiquitinylation, 210
UDP-NAM-pentapeptide, 896, 896*f*
UGGT (UDP-glucose-glycoprotein glucosyl transferase), 317–318
UIM (ubiquitin-interacting motif), 809
Uncoating, in vesicle-mediated protein transport, 355, 356, 355*f*
Uncoating ATPase, 374
Uncoupling reactions, 53, 53*f*
Unfolded protein response (UPR)
    activation, 332
    description, 322–324
Unfolded protein response element (UPRE), 323
Unicellular organisms. *See also* Bacteria; Cells; Yeast; *specific unicellular organisms*
    cilia, 545–546
    environment, 26
        adaptations to, 4
        external, 5
    homeostasis, 15
    signal transduction, 27
Uniporters, 234, 235*f*. *See also* Glucose transporters
Universe, in thermodynamics, 37
UPR. *See* Unfolded protein response
UPRE (unfolded protein response element), 323
URA3 gene expression, nucleosome organization and, 478, 478*f*
Urea, 57*f*
Uric acid, 57*f*
UsnRNP complex, 432–433, 433*f*
Uvr excision, 94, 94*f*

V

Vacuolar-type proton pumps (V-ATPases), 280–281, 280*f*, 352
Vacuole membrane proteins, 383
Vacuoles
    enzymes, 383
    functions, 938, 956
    swelling, 955–956, 956*f*
VAD-fmk, 717
"Vanishing bone" diseases, 850
Van Leeuwenhoek, Antony, 394
Vasopressin (antidiuretic hormone), 256, 258
Vav oncogene product, molecular architecture of, 197, 197*f*
Vegetative phase, 923
Vertebrates
    apoptosis pathways, 720–721, 720*f*, 721*f*
    chromosome movement in, 652, 653*f*
    integrin subunits, 851, 852
    intermediate filament proteins, 593*f*, 613

Z

ZAP-70, 818, 818*f*
Z-disc, 587
Zinc finger (ZF) motif, 126, 127*f*, 134
Zinnia elegans, 970*f*
ZipA protein, 919, 919*f*
Zipcode, 161
Zipcode-binding protein (ZBPs), 161

Zip proteins, 100
Zonula adherens, 864, 864*f*
Zonula occludens, 861, 862
Zoxamide™, 506, 506*f*
Z ring, 918–920, 919*f*, 920*f*
Zygotes, totipotent, 30
Zymogens, 716, 717